Aviation Main
Technician Handbook—Airframe

MW01092919

2023

U.S. Department of Transportation
Federal Aviation Administration
Flight Standards Service

Preface

The Aviation Maintenance Technician Handbook—Airframe (FAA-H-8083-31B) is one of a series of three handbooks for persons preparing for certification as an airframe or powerplant mechanic. This handbook provides basic information on principles, fundamentals, and technical procedures in the subject matter areas relating to the airframe rating. It is designed to aid students enrolled in a formal course of instruction, as well as the individual who is studying on their own. Since the knowledge requirements for the airframe and powerplant ratings closely parallel each other in some subject areas, the chapters which discuss fire protection systems and electrical systems contain some material which is also duplicated in the Aviation Maintenance Technician Handbook—Powerplant (FAA-H-8083-32B).

This handbook contains information on airframe construction features, assembly and rigging, fabric covering, structural repairs, and aircraft welding. The handbook also contains an explanation of the units that make up the various airframe systems. Because there are so many different types of aircraft in use today, it is reasonable to expect that differences exist in airframe components and systems. To avoid undue repetition, the practice of using representative systems and units is carried out throughout the handbook. Subject matter treatment is from a generalized point of view and should be supplemented by reference to manufacturers' manuals or other textbooks if more detail is desired. This handbook is not intended to replace, substitute for, or supersede official regulations or manufacturers' instructions. Occasionally the word "must" or similar language is used where the desired action is deemed critical. The use of such language is not intended to add to, interpret, or relieve a duty imposed by Title 14 of the Code of Federal Regulations (14 CFR).

The subject of Human Factors is contained in the Aviation Maintenance Technician Handbook—General (FAA-H-8083-30) (as revised).

This handbook is available for download, in PDF format, from www.faa.gov.

This handbook is published by the United States Department of Transportation, Federal Aviation Administration, Airman Testing Standards Branch, AFS-630, P.O. Box 25082, Oklahoma City, OK 73125.

Comments regarding this publication should be emailed to AFS630comments@faa.gov.

Acknowledgments

The Aviation Maintenance Technician Handbook—Airframe (FAA-H-8083-31B) was produced by the Federal Aviation Administration (FAA). The FAA wishes to acknowledge the following contributors:

Mr. Chris Brady (www.b737.org.uk) for images used throughout this handbook

Captain Karl Eiríksson for image used in Chapter 1

Cessna Aircraft Company for image used in Chapter 1

Mr. Andy Dawson (www.mossie.org) for images used throughout Chapter 1

Mr. Bill Shemley for image used in Chapter 1

Mr. Bruce R. Swanson for image used in Chapter 1

Mr. Burkhard Domke (www.b-domke.de) for images used throughout Chapter 1 and 2

Mr. Chris Wonnacott (www.fromtheflightdeck.com) for image used in Chapter 1

Mr. Christian Tremblay (www.zodiac640.com) for image used in Chapter 1

Mr. John Bailey (www.knots2u.com) for image used in Chapter 1

Mr. Rich Guerra (www.rguerra.com) for image used in Chapter 1

Mr. Ronald Lane for image used in Chapter 1

Mr. Tom Allensworth (www.avsim.com) for image used in Chapter 1

Navion Pilots Association's Tech Note 001 (www.navionpilots.org) for image used in Chapter 1

U.S. Coast Guard for image used in Chapter 1

Mr. Tony Bingelis and the Experimental Aircraft Association (EAA) for images used throughout Chapter 2

Mr. Benoit Viellefon (www.johnjohn.co.uk/compare-tigermothflights/html/tigermoth_bio_aozh.html) for image used in Chapter 3

Mr. Paul Harding of Safari Seaplanes–Bahamas (www.safariseaplanes.com) for image used in Chapter 3

Polyfiber/Consolidated Aircraft Coatings for images used throughout Chapter 3

Stewart Systems for images used throughout Chapter 3

Superflite for images used throughout Chapter 3

Cherry Aerospace (www.cherryaerospace.com) for images used in Chapters 4 and 7

Raytheon Aircraft (Structural Inspection and Repair Manual) for information used in Chapter 4

Mr. Scott Allen of Kalamazoo Industries, Inc. (www.kalamazooind.com) for image used in Chapter 4

Miller Electric Mfg. Co. (www.millerwelds.com) for images used in Chapter 5

Mr. Aaron Novak, contributing engineer, for charts used in Chapter 5

Mr. Bob Hall (www.pro-fusiononline.com) for image used in Chapter 5

Mr. Kent White of TM Technologies, Inc. for image used in Chapter 5

Safety Supplies Canada (www.safetysuppliescanada.com) for image used in Chapter 5

Smith Equipment (www.smithequipment.com) for images used in Chapter 5

Alcoa (www.alcoa.com) for images used in Chapter 7

Mr. Chuck Scott (www.itwif.com) for images used throughout Chapter 8

Mr. John Lagerlof of Paasche Airbrush Co. (paascheairbrush.com) for image used in Chapter 8

Mr. Philip Love of Turbine Products, LLC (www.turbineproducts.com) for image used in Chapter 8

Consolidated Aircraft Coatings for image used in Chapter 8

Tianjin Yonglida Material Testing Machine Co., Ltd for image used in Chapter 8

Mr. Jim Irwin of Aircraft Spruce & Specialty Co. (www.aircraftspruce.com) for images used in Chapters 9, 10, 11, 13, 14, 15

Mr. Kevan Hashemi for image used in Chapter 9

Mr. Michael Leasure, Aviation Multimedia Library (www2.tech.purdue.edu/at/courses/aeml) for images used in Chapters 9, 13, 14

Aircraft Owners and Pilots Association (AOPA) (www.aopa.org) for image used in Chapter 10

Cobra Systems Inc. (www.cobrasys.com) for image used in Chapter 10

www.free-online-private-pilot-ground-school.com for image used in Chapters 10, 16

DAC International (www.dacint.com) for image used in Chapter 10

Dawson Aircraft Inc. (www.aircraftpartsandsalvage.com) for images used throughout Chapter 10

Mr. Kent Clingaman for image used in Chapter 10

TECNAM (www.tecnam.com) for image used in Chapter 10

TGH Aviation-FAA Instrument Repair Station (www.tghaviation.com) for image used in Chapter 10

The Vintage Aviator Ltd. (www.thevintageaviator.co.nz) for image used in Chapter 10

ACK Technologies Inc. (www.ackavionics.com) for image used in Chapter 11

ADS-B Technologies, LLC (www.ads-b.com) for images used in Chapter 11

Aviation Glossary (www.aviationglossary.com) for image used in Chapter 11

AT&T Archives and History Center for image used in Chapter 11

Electronics International Inc. (www.buy-ei.com) for image used in Chapter 11

Excelitas Technologies (www.excelitas.com) for image used in Chapter 11

Freestate Electronics, Inc. (www.fse-inc.com) for image used in Chapter 11

AirTrafficAtlanta.com for image used in Chapter 11

Western Historic Radio Museum, Virginia City, Nevada (www.radioblvd.com) for image used in Chapter 11

Avidyne Corporation (www.avidyne.com) for image used in Chapter 11

Kintronic Laboratories (www.kintronic.com) for image used in Chapter 11

Mr. Dan Wolfe (www.flyboysalvage.com) for image used in Chapter 11

Mr. Ken Shuck (www.cessna150.net) for image used in Chapter 11

Mr. Paul Tocknell (www.askacfi.com) for image used in Chapter 11

Mr. Stephen McGreevy (www.auroralchorus.com) for image used in Chapter 11

Mr. Todd Bennett (www.bennettavionics.com) for image used in Chapter 11

National Oceanic and Atmospheric Administration, U.S. Department of Commerce for image used in Chapter 11

RAMI (www.rami.com) for image used in Chapter 11

Rockwell Collins (www.rockwellcollins.com) for image used in Chapter 11, Figure 11-73

Sarasota Avionics International (www.sarasotaavionics.com) for images used in Chapter 11

Southeast Aerospace, Inc. (www.seaerospace.com) for image used in Chapter 11

Sporty's Pilot Shop (www.sportys.com) for image used in Chapter 11

Watts Antenna Company (www.wattsantenna.com) for image used in Chapter 11

Wings and Wheels (www.wingsandwheels.com) for image used in Chapter 11

Aeropin, Inc. (www.aeropin.com) for image used in Chapter 13

Airplane Mart Publishing (www.airplanemart.com) for image used in Chapter 13

Alberth Aviation (www.alberthaviation.com) for image used in Chapter 13

AVweb (www.avweb.com) for image used in Chapter 13

Belle Aire Aviation, Inc. (www.belleaireaviation.com) for image used in Chapter 13

Cold War Air Museum (www.coldwarairmuseum.org) for image used in Chapter 13

Comanche Gear (www.comanchegear.com) for image used in Chapter 13

CSOBeech (www.csobeech.com) for image used in Chapter 13

Desser Tire & Rubber Co., Inc. (www.desser.com) for image used in Chapter 13

DG Flugzeugbau GmbH (www.dg-flugzeugbau.de) for image used in Chapter 13

Expedition Exchange Inc. (www.expeditionexchange.com) for image used in Chapter 13

Fiddlers Green (www.fiddlersgreen.net) for image used in Chapter 13

Hitchcock Aviation (hitchcockaviation.com) for image used in Chapter 13

KUNZ GmbH aircraft equipment (www.kunz-aircraft.com) for images used in Chapter 13

Little Flyers (www.littleflyers.com) for images used in Chapter 13

Maple Leaf Aviation Ltd. (www.aircraftspeedmods.ca) for image used in Chapter 13

Mr. Budd Davisson (Airbum.com) for image used in Chapter 13

Mr. C. Jeff Dyrek (www.yellowairplane.com) for images used in Chapter 13

Mr. Jason Schappert (www.m0a.com) for image used in Chapter 13

Mr. John Baker (www.hangar9aeroworks.com) for image used in Chapter 13

Mr. Mike Schantz (www.trailer411.com) for image used in Chapter 13

Mr. Robert Hughes (www.escapadebuild.co.uk) for image used in Chapter 13

Mr. Ron Blachut for image used in Chapter 13

Owls Head Transportation Museum (www.owlshead.org) for image used in Chapter 13

PPI Aerospace (www.ppiaerospace.com) for image used in Chapter 13

Protective Packaging Corp. (www.protectivepackaging.net, 1-800-945-2247) for image used in Chapter 13

Ravenware Industries, LLC (www.ravenware.com) for image used in Chapter 13

Renold (www.renold.com) for image used in Chapter 13

Rotor F/X, LLC (www.rotorfx.com) for image used in Chapter 13

SkyGeek (www.skygeek.com) for image used in Chapter 13

Taigh Ramey (www.twinbeech.com) for image used in Chapter 13

Texas Air Salvage (www.texasairsalvage.com) for image used in Chapter 13

The Bogert Group (www.bogert-av.com) for image used in Chapter 13

W. B. Graham, Welded Tube Pros LLC (www.thefabricator.com) for image used in Chapter 13

Zinko Hydraulic Jack (www.zinkojack.com) for image used in Chapter 13

Aviation Institute of Maintenance (www.aimschool.com) for image used in Chapter 14

Aviation Laboratories (www.avlab.com) for image used in Chapter 14

AVSIM (www.avsim.com) for image used in Chapter 14

Eggenfellner (www.eggenfellneraircraft.com) for image used in Chapter 14

FlightSim.Com, Inc. (www.flightsim.com) for image used in Chapter 14

Fluid Components International LLC (www.fluidcomponents.com) for image used in Chapter 14

Fuel Quality Services, Inc. (www.fqsinc.com) for image used in Chapter 14

Hammonds Fuel Additives, Inc. (www.biobor.com) for image used in Chapter 14

Jeppesen (www.jeppesen.com) for image used in Chapter 14

MGL Avionics (www.mglavionics.com) for image used in Chapter 14

Mid-Atlantic Air Museum (www.maam.org) for image used in Chapter 14

MISCO Refractometer (www.misco.com) for image used in Chapter 14

Mr. Gary Brossett via the Aircraft Engine Historical Society (www.enginehistory.org) for image used in Chapter 14

Mr. Jeff McCombs (www.heyeng.com) for image used in Chapter 14

NASA for image used in Chapter 14

On-Track Aviation Limited (www.ontrackaviation.com) for image used in Chapter 14

Stewart Systems for image used in Chapter 14

Prist Aerospace Products (www.pristaerospace.com) for image used in Chapter 14

The Sundowners, Inc. (www.sdpleecounty.org) for image used in Chapter 14

Velcon Filters, LLC (www.velcon.com) for image used in Chapter 14

Aerox Aviation Oxygen Systems, Inc. (www.aerox.com) for image used in Chapter 16

Biggles Software (www.biggles-software.com) for image used in Chapter 16

C&D Associates, Inc. (www.aircrafttheater.com) for image used in Chapter 16

Cobham (Carleton Technologies Inc.) (www.cobham.com) for image used in Chapter 16

Cool Africa (www.coolafrica.co.za) for image used in Chapter 16

Cumulus Soaring, Inc. (www.cumulus-soaring.com) for image used in Chapter 16

Essex Cryogenics of Missouri, Inc. (www.essexind.com) for image used in Chapter 16

Flightline AC, Inc. (www.flightlineac.com) for image used in Chapter 16

IDQ Holdings (www.idqusa.com) for image used in Chapter 16

Manchester Tank & Equipment (www.mantank.com) for image used in Chapter 16

Mountain High E&S Co. (www.MHoxygen.com) for images used throughout Chapter 16

Mr. Bill Sherwood (www.billzilla.org) for image used in Chapter 16

Mr. Boris Comazzi (www.flightgear.ch) for image used in Chapter 16

Mr. Chris Rudge (www.warbirdsite.com) for image used in Chapter 16

Mr. Richard Pfiffner (www.craggyaero.com) for image used in Chapter 16

Mr. Stephen Sweet (www.stephensweet.com) for image used in Chapter 16

Precise Flight, Inc. (www.preciseflight.com) for image used in Chapter 16

SPX Service Solutions (www.spx.com) for image used in Chapter 16

SuperFlash Compressed Gas Equipment (www.oxyfuelsafety.com)

Mr. Tim Mara (www.wingsandwheels.com) for images used in Chapter 16

Mr. Bill Abbott for image used in Chapter 17

Additional appreciation is extended to Dr. Ronald Sterkenburg, Purdue University; Mr. Bryan Rahm, Dr. Thomas K. Eismain, Purdue University; Mr. George McNeill, Mr. Thomas Forenz, Mr. Peng Wang, and the National Oceanic and Atmospheric Administration (NOAA) for their technical support and input.

Table of Contents

Chapter 3
Aircraft Fabric Covering

Chapter 4
Aircraft Metal Structural Repair

Chapter 5
Aircraft Welding

Chapter 6
Aircraft Wood & Structural Repair

Chapter 7
Advanced Composite Materials

Chapter 12
Hydraulic & Pneumatic Power Systems

Chapter 14
Aircraft Fuel Systems

Chapter 17
Fire Protection Systems

Chapter 1
Aircraft Structures

A Brief History of Aircraft Structures

The history of aircraft structures underlies the history of aviation in general. Advances in materials and processes used to construct aircraft have led to their evolution from simple wood truss structures to the sleek aerodynamic flying machines of today. Combined with continuous powerplant development, the structures of "flying machines" have changed significantly.

The key discovery that "lift" could be created by passing air over the top of a curved surface set the development of fixed and rotary-wing aircraft in motion. George Cayley developed an efficient cambered airfoil in the early 1800s, as well as successful manned gliders later in that century. He established the principles of flight, including the existence of lift, weight, thrust, and drag. It was Cayley who first stacked wings and created a tri-wing glider that flew a man in 1853.

Earlier, Cayley studied the center of gravity of flying machines, as well as the effects of wing dihedral. Furthermore, he pioneered directional control of aircraft by including the earliest form of a rudder on his gliders. *[Figure 1-1]*

In the late 1800s, Otto Lilienthal built upon Cayley's discoveries. He manufactured and flew his own gliders on over 2,000 flights. His willow and cloth aircraft had wings designed from extensive study of the wings of birds. Lilienthal also made standard use of vertical and horizontal fins behind the wings and pilot station. Above all, Lilienthal proved that man could fly. *[Figure 1-2]*

Octave Chanute, a retired railroad and bridge engineer, was active in aviation during the 1890s. *[Figure 1-3]* His interest was so great that, among other things, he published a definitive work called "Progress in Flying Machines." This was the culmination of his effort to gather and study all of the information available on aviation. With the assistance of others, he built gliders similar to Lilienthal's and then his own. In addition to his publication, Chanute advanced aircraft structure development by building a glider with stacked wings incorporating the use of wires as wing supports.

The work of all of these men was known to the Wright Brothers when they built their successful, powered airplane in 1903. The first of its kind to carry a man aloft, the Wright Flyer had thin, cloth-covered wings attached to what was primarily truss structures made of wood. The wings contained forward and rear spars and were supported with both struts and wires. Stacked wings (two sets) were also part of the Wright Flyer. *[Figure 1-4]*

Powered heavier-than-air aviation grew from the Wright design. Inventors and fledgling aviators began building their own aircraft. Early on, many were similar to that constructed by the Wrights using wood and fabric with wires and struts to support the wing structure. In 1909, Frenchman Louis Bleriot produced an aircraft with notable design differences. He built a successful mono-wing aircraft. The wings were still supported by wires, but a mast extending above the

Figure 1-1. *George Cayley, the father of aeronautics (top) and a flying replica of his 1853 glider (bottom).*

Figure 1-2. *Master of gliding and wing study, Otto Lilienthal (top) and one of his more than 2,000 glider flights (bottom).*

Figure 1-3. *Octave Chanute gathered and published all of the aeronautical knowledge known to date in the late 1890s. Many early aviators benefited from this knowledge.*

Figure 1-4. *The Wright Flyer was the first successful powered aircraft. It was made primarily of wood and fabric.*

Figure 1-5. *The world's first mono-wing by Louis Bleriot.*

fuselage enabled the wings to be supported from above, as well as underneath. This made possible the extended wing length needed to lift an aircraft with a single set of wings. Bleriot used a Pratt truss-type fuselage frame. *[Figure 1-5]*

More powerful engines were developed, and airframe structures changed to take advantage of the benefits. As early as 1910, German Hugo Junkers was able to build an aircraft with metal truss construction and metal skin due to the availability of stronger powerplants to thrust the plane forward and into the sky. The use of metal instead of wood for the primary structure eliminated the need for external wing braces and wires. His J-1 also had a single set of wings (a monoplane) instead of a stacked set. *[Figure 1-6]*

Leading up to World War I (WWI), stronger engines also allowed designers to develop thicker wings with stronger spars. Wire wing bracing was no longer needed. Flatter, lower wing surfaces on high-camber wings created more lift. WWI expanded the need for large quantities of reliable aircraft. Used mostly for reconnaissance, stacked-wing tail draggers

Figure 1-6. *The Junker J-1 all metal construction in 1910.*

with wood and metal truss frames with mostly fabric skin dominated the wartime sky. *[Figure 1-7]* The Red Baron's Fokker DR-1 was typical.

In the 1920s, the use of metal in aircraft construction increased. Fuselages able to carry cargo and passengers were developed. The early flying boats with their hull-type construction from the shipbuilding industry provided the blueprints for semimonocoque construction of fuselages. *[Figure 1-8]* Truss-type designs faded. A tendency toward cleaner mono-wing designs prevailed.

Into the 1930s, all-metal aircraft accompanied new lighter and more powerful engines. Larger semimonocoque fuselages were complimented with stress-skin wing designs. Fewer truss and fabric aircraft were built. World War II (WWII) brought about a myriad of aircraft designs using all metal technology. Deep fuel-carrying wings were the norm, but the desire for higher flight speeds prompted the development of thin-winged aircraft in which fuel was carried in the fuselage. The first composite structure aircraft, the De Havilland Mosquito, used a balsa wood sandwich material in the construction of the fuselage. *[Figure 1-9]* The fiberglass radome was also developed during this period.

After WWII, the development of turbine engines led to higher altitude flight. The need for pressurized aircraft pervaded aviation. Semimonocoque construction needed to be made even stronger as a result. Refinements to the

Figure 1-7. *World War I aircraft were typically stacked-wing fabric-covered aircraft like this Breguet 14 (circa 1917).*

Figure 1-8. *The flying boat hull was an early semimonocoque design like this Curtiss HS-2L.*

all-metal semimonocoque fuselage structure were made to increase strength and combat metal fatigue caused by the pressurization-depressurization cycle. Rounded window and door openings were developed to avoid weak areas where cracks could form. Integrally machined copper alloy aluminum skin resisted cracking and allowed thicker skin and controlled tapering. Chemical milling of wing skin structures provided great strength and smooth high-performance surfaces. Variable contour wings became easier to construct. Increases in flight speed accompanying jet travel brought about the need for thinner wings. Wing loading also increased greatly. Multispar and box beam wing designs were developed in response.

In the 1960s, ever larger aircraft were developed to carry passengers. As engine technology improved, the jumbo jet was engineered and built. Still primarily aluminum with a

Figure 1-9. *The De Havilland Mosquito used a laminated wood construction with a balsa wood core in the fuselage.*

Figure 1-10. *The nearly all composite Cessna Citation Mustang very light jet (VLJ).*

semimonocoque fuselage, the sheer size of the airliners of the day initiated a search for lighter and stronger materials from which to build them. The use of honeycomb constructed panels in Boeing's airline series saved weight while not compromising strength. Initially, aluminum core with aluminum or fiberglass skin sandwich panels were used on wing panels, flight control surfaces, cabin floor boards, and other applications.

A steady increase in the use of honeycomb and foam core sandwich components and a wide variety of composite materials characterizes the state of aviation structures from the 1970s to the present. Advanced techniques and material combinations have resulted in a gradual shift from aluminum to carbon fiber and other strong, lightweight materials. These new materials are engineered to meet specific performance requirements for various components on the aircraft. Many airframe structures are made of more than 50 percent advanced composites, with some airframes approaching

100 percent. The term "very light jet" (VLJ) has come to describe a new generation of jet aircraft made almost entirely of advanced composite materials. *[Figure 1-10]* It is possible that noncomposite aluminum aircraft structures will become obsolete as did the methods and materials of construction used by Cayley, Lilienthal, and the Wright Brothers.

General

An aircraft is a device that is used for, or is intended to be used for, flight in the air. Major categories of aircraft are airplane, rotorcraft, glider, and lighter-than-air vehicles. *[Figure 1-11]* Each of these may be divided further by major distinguishing features of the aircraft, such as airships and balloons. Both are lighter-than-air aircraft but have differentiating features and are operated differently.

The concentration of this handbook is on the airframe of aircraft; specifically, the fuselage, booms, nacelles, cowlings,

Figure 1-11. *Examples of different categories of aircraft, clockwise from top left: lighter-than-air, glider, rotorcraft, and airplane.*

fairings, airfoil surfaces, and landing gear. Also included are the various accessories and controls that accompany these structures. Note that the rotors of a helicopter are considered part of the airframe since they are actually rotating wings. By contrast, propellers and rotating airfoils of an engine on an airplane are not considered part of the airframe.

The most common aircraft is the fixed-wing aircraft. As the name implies, the wings on this type of flying machine are attached to the fuselage and are not intended to move independently in a fashion that results in the creation of lift. One, two, or three sets of wings have all been successfully utilized. *[Figure 1-12]* Rotary-wing aircraft such as helicopters are also widespread. This handbook discusses features and maintenance aspects common to both fixed-wing and rotary-wing categories of aircraft. Also, in certain cases, explanations focus on information specific to only one or the other. Glider airframes are very similar to fixed-wing aircraft. Unless otherwise noted, maintenance practices described for fixed-wing aircraft also apply to gliders. The same is true for lighter-than-air aircraft, although thorough coverage of the unique airframe structures and maintenance practices for lighter-than-air flying machines is not included in this handbook.

The airframe of a fixed-wing aircraft consists of five principal units: the fuselage, wings, stabilizers, flight control surfaces, and landing gear. *[Figure 1-13]* Helicopter airframes consist of the fuselage, main rotor and related gearbox, tail rotor (on helicopters with a single main rotor), and the landing gear.

Airframe structural components are constructed from a wide variety of materials. The earliest aircraft were constructed primarily of wood. Aircraft structures were later made of steel tubing and are now most commonly made of aluminum. Many newly certified aircraft are built from molded composite materials, such as carbon fiber. Structural members of an aircraft's fuselage include stringers, longerons, ribs, bulkheads, and more. The main structural member in a wing is called the wing spar.

The skin of aircraft can also be made from a variety of materials, ranging from impregnated fabric to plywood, aluminum, or composites. Under the skin and attached to the structural fuselage are the many components that support airframe function. The entire airframe and its components are joined by rivets, bolts, screws, and other fasteners. Welding, adhesives, and special bonding techniques are also used.

Major Structural Stresses

Aircraft structural members are designed to carry a load or to resist stress. In designing an aircraft, every square inch of wing and fuselage, every rib, spar, and even each metal fitting

Figure 1-12. *A monoplane (top), biplane (middle), and tri-wing aircraft (bottom).*

must be considered in relation to the physical characteristics of the material of which it is made. Every part of the aircraft must be planned to carry the load to be imposed upon it.

A single member of the structure may be subjected to a combination of stresses. In most cases, the structural members are designed to carry end loads rather than side loads: that is, to be subjected to tension or compression rather than bending.

The determination of such loads is called stress analysis. Although planning the design is not the function of the aircraft technician, it is, nevertheless, important that the technician understand and appreciate the stresses involved in order to avoid changes in the original design through improper repairs.

The term "stress" is often used interchangeably with the word "strain." While related, they are not the same thing. External loads or forces cause stress. Stress is a material's

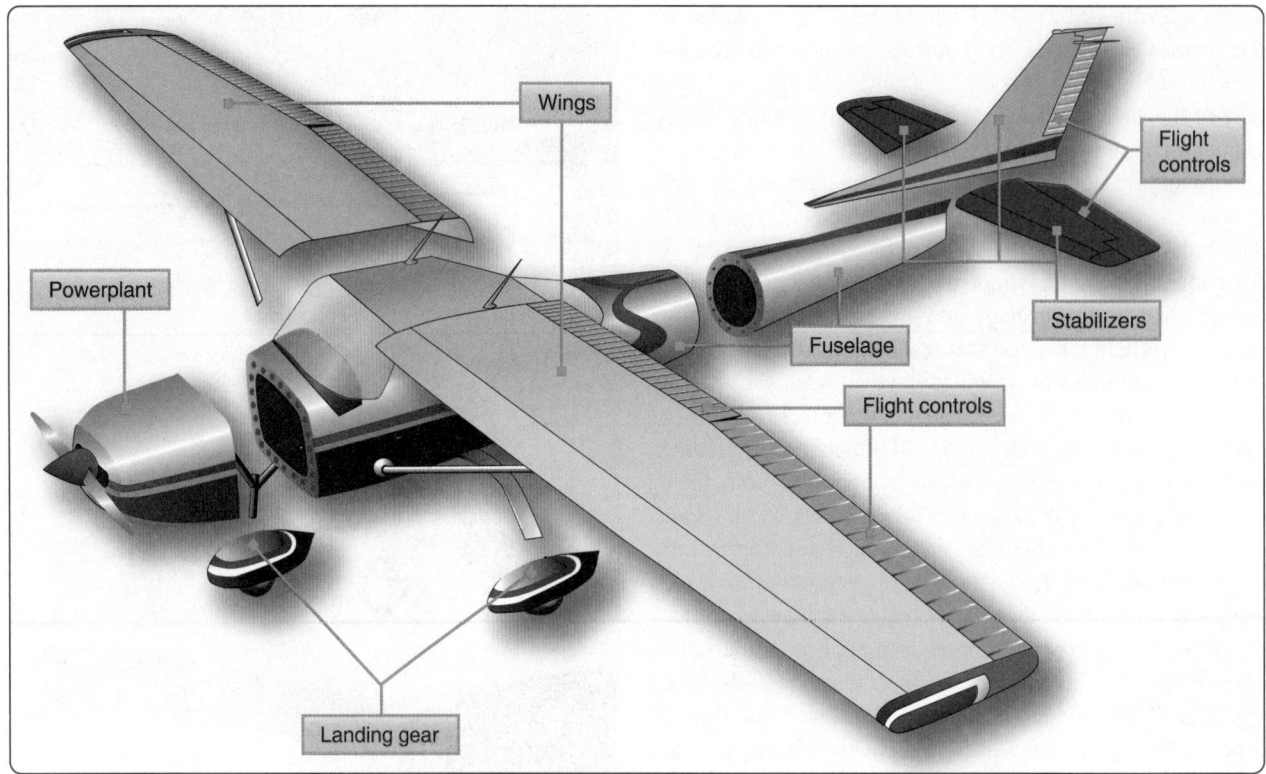

Figure 1-13. *Principal airframe units.*

internal resistance, or counterforce, that opposes deformation. The degree of deformation of a material is strain. When a material is subjected to a load or force, that material is deformed, regardless of how strong the material is or how light the load is.

There are five major stresses *[Figure 1-14]* to which all aircraft are subjected:

- Tension
- Compression
- Torsion
- Shear
- Bending

Tension is the stress that resists a force that tends to pull something apart. *[Figure 1-14A]* The engine pulls the aircraft forward, but air resistance tries to hold it back. The result is tension, which stretches the aircraft. The tensile strength of a material is measured in pounds per square inch (psi) and is calculated by dividing the load (in pounds) required to pull the material apart by its cross-sectional area (in square inches).

Compression is the stress that resists a crushing force. *[Figure 1-14B]* The compressive strength of a material is also measured in psi. Compression is the stress that tends to shorten or squeeze aircraft parts.

Torsion is the stress that produces twisting. *[Figure 1-14C]* While moving the aircraft forward, the engine also tends to twist it to one side, but other aircraft components hold it on course. Thus, torsion is created. The torsion strength of a material is its resistance to twisting or torque.

Shear is the stress that resists the force tending to cause one layer of a material to slide over an adjacent layer. *[Figure 1-14D]* Two riveted plates in tension subject the rivets to a shearing force. Usually, the shearing strength of a material is either equal to or less than its tensile or compressive strength. Aircraft parts, especially screws, bolts, and rivets, are often subject to a shearing force.

Bending stress is a combination of compression and tension. The rod in *Figure 1-14E* has been shortened (compressed) on the inside of the bend and stretched on the outside of the bend. A single member of the structure may be subjected to a combination of stresses. In most cases, the structural members are designed to carry end loads rather than side loads. They are designed to be subjected to tension or compression rather than bending.

Strength or resistance to the external loads imposed during operation may be the principal requirement in certain

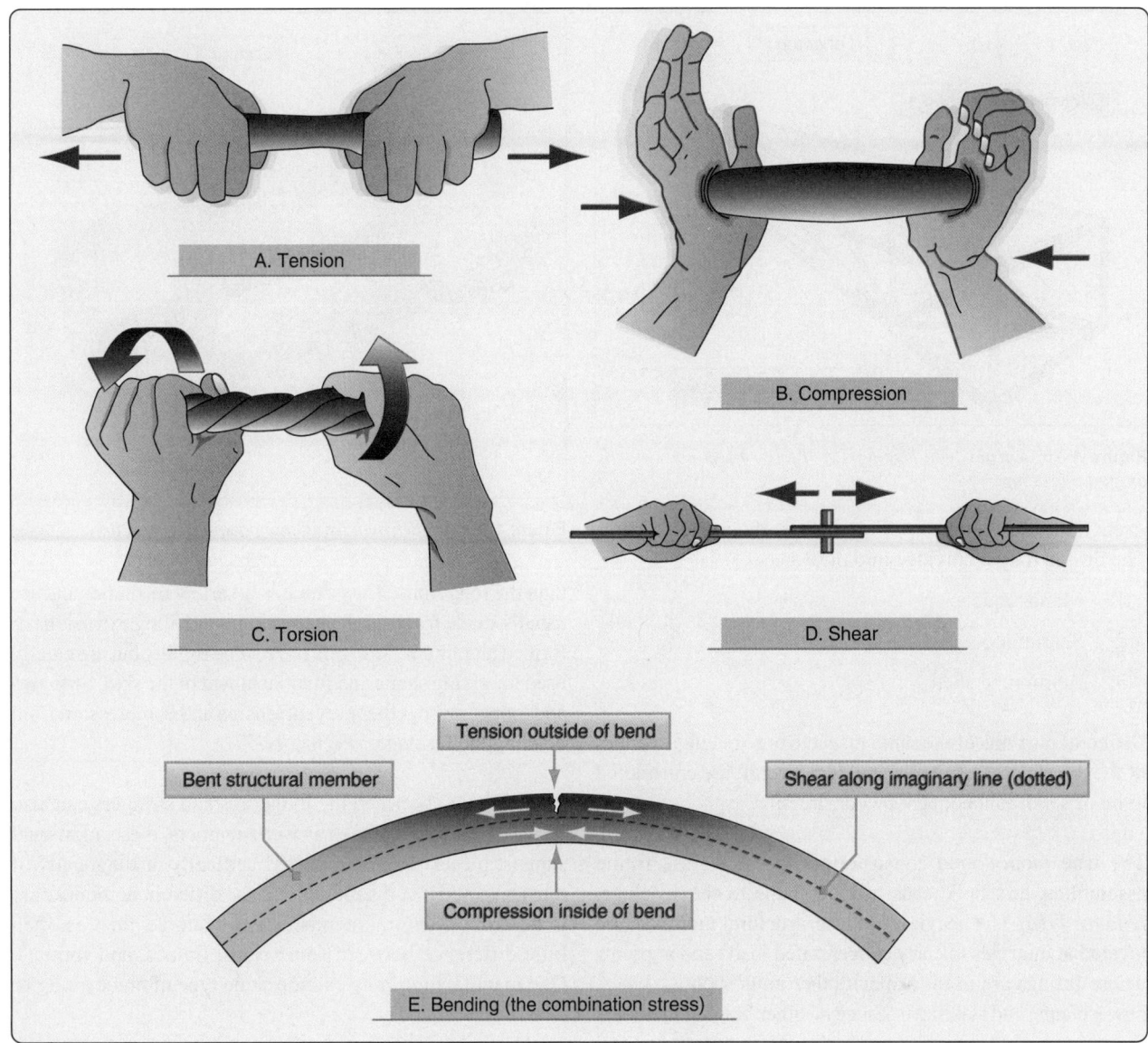

Figure 1-14. *The five stresses that may act on an aircraft and its parts.*

structures. However, there are numerous other characteristics in addition to designing to control the five major stresses that engineers must consider. For example, cowling, fairings, and similar parts may not be subject to significant loads requiring a high degree of strength. However, these parts must have streamlined shapes to meet aerodynamic requirements, such as reducing drag or directing airflow.

Fixed-Wing Aircraft

Fuselage

The fuselage is the main structure or body of the fixed-wing aircraft. It provides space for cargo, controls, accessories, passengers, and other equipment. In single-engine aircraft, the fuselage houses the powerplant. In multiengine aircraft, the engines may be either in the fuselage, attached to the fuselage, or suspended from the wing structure. There are two general types of fuselage construction: truss and monocoque.

Truss-Type

A truss is a rigid framework made up of members, such as beams, struts, and bars to resist deformation by applied loads. The truss-framed fuselage is generally covered with fabric. The truss-type fuselage frame is usually constructed of steel tubing welded together in such a manner that all members of the truss can carry both tension and compression loads. [Figure 1-15] In some aircraft, principally the light, single-engine models, truss fuselage frames may be constructed of aluminum alloy and may be riveted or bolted into one piece, with cross-bracing achieved by using solid rods or tubes.

Monocoque Type

The monocoque (single shell) fuselage relies largely on the

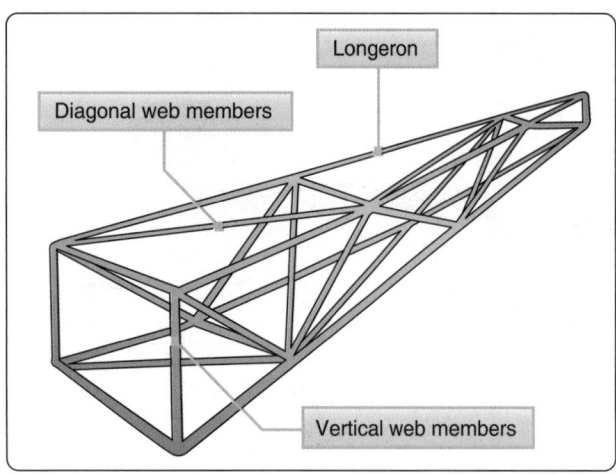

Figure 1-15. *A truss-type fuselage. A Warren truss uses mostly diagonal bracing.*

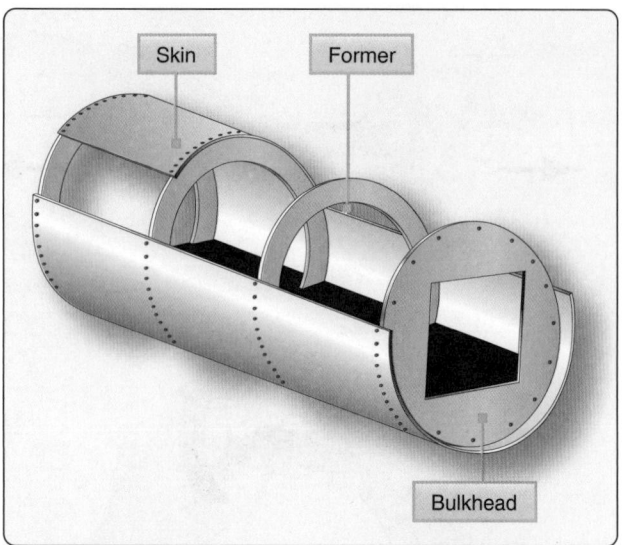

Figure 1-16. *An airframe using monocoque construction.*

strength of the skin or covering to carry the primary loads. The design may be divided into three classes:

1. Monocoque

2. Semimonocoque

3. Reinforced shell

Different portions of the same fuselage may belong to either of the three classes, but most modern aircraft are considered to be of semimonocoque type construction.

The true monocoque construction uses formers, frame assemblies, and bulkheads to give shape to the fuselage. *[Figure 1-16]* The heaviest of these structural members are located at intervals to carry concentrated loads and at points where fittings are used to attach other units such as wings, powerplants, and stabilizers. Since no other bracing members are present, the skin must carry the primary stresses and keep the fuselage rigid. Thus, the biggest problem involved in monocoque construction is maintaining enough strength while keeping the weight within allowable limits.

Semimonocoque Type

To overcome the strength/weight problem of monocoque construction, a modification called semimonocoque construction was developed. It also consists of frame assemblies, bulkheads, and formers as used in the monocoque design but, additionally, the skin is reinforced by longitudinal members called longerons. Longerons usually extend across several frame members and help the skin support primary bending loads. They are typically made of aluminum alloy either of a single piece or a built-up construction. The longerons are supplemented by other longitudinal members called stringers.

Stringers are typically more numerous and lighter in weight

than the longerons. They come in a variety of shapes and are usually made from single piece aluminum alloy extrusions or formed aluminum. Stringers have some rigidity but are chiefly used for giving shape and for attachment of the skin. Stringers and longerons together prevent tension and compression from bending the fuselage. *[Figure 1-17]*

Other bracing between the longerons and stringers can also be used. Often referred to as web members, these additional support pieces may be installed vertically or diagonally. It must be noted that manufacturers use different nomenclature to describe structural members. For example, there is often little difference between some rings, frames, and formers. One manufacturer may call the same type of brace a ring or

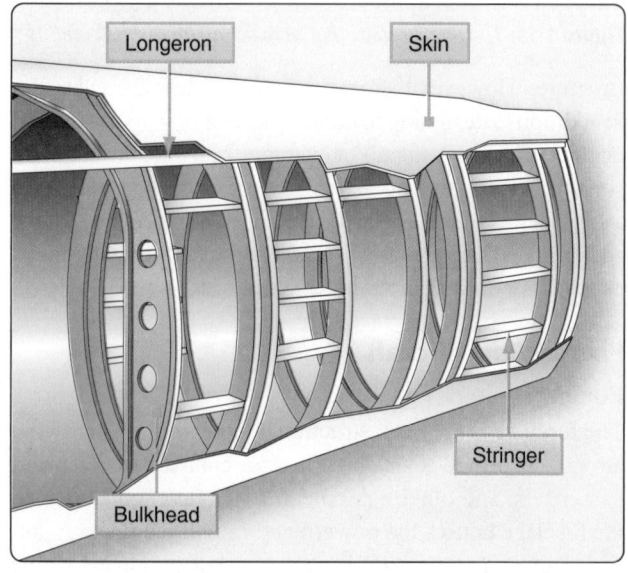

Figure 1-17. *The most common airframe construction is semimonocoque.*

a frame. Manufacturer instructions and specifications for a specific aircraft are the best guides.

The semimonocoque fuselage is constructed primarily of alloys of aluminum and magnesium, although steel and titanium are sometimes found in areas of high temperatures. Individually, not one of the aforementioned components is strong enough to carry the loads imposed during flight and landing. But, when combined, those components form a strong, rigid framework. This is accomplished with gussets, rivets, nuts and bolts, screws, and even friction stir welding. A gusset is a type of connection bracket that adds strength. *[Figure 1-18]*

To summarize, in semimonocoque fuselages, the strong, heavy longerons hold the bulkheads and formers, and these, in turn, hold the stringers, braces, web members, etc. All are designed to be attached together and to the skin to achieve the full-strength benefits of semimonocoque design. It is important to recognize that the metal skin or covering carries part of the load. The fuselage skin thickness can vary with the load carried and the stresses sustained at a particular location.

The advantages of the semimonocoque fuselage are many. The bulkheads, frames, stringers, and longerons facilitate the design and construction of a streamlined fuselage that is both rigid and strong. Spreading loads among these structures and the skin means no single piece is failure critical. This means that a semimonocoque fuselage, because of its stressed-skin construction, may withstand considerable damage and still be strong enough to hold together.

Fuselages are generally constructed in two or more sections. On small aircraft, they are generally made in two or three sections, while larger aircraft may be made up of as many as six sections or more before being assembled.

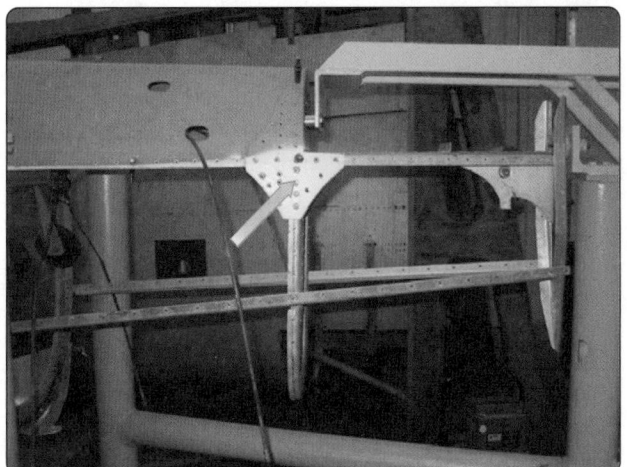

Figure 1-18. *Gussets are used to increase strength.*

Reinforced Shell Type

The reinforced shell has the skin reinforced by a complete framework of structural members.

Pressurization

Many aircraft are pressurized. This means that air is pumped into the cabin after takeoff and a difference in pressure between the air inside the cabin and the air outside the cabin is established. This differential is regulated and maintained. In this manner, enough oxygen is made available for passengers to breathe normally and move around the cabin without special equipment at high altitudes.

Pressurization causes significant stress on the fuselage structure and adds to the complexity of design. In addition to withstanding the difference in pressure between the air inside and outside the cabin, cycling from unpressurized to pressurized and back again on each flight causes metal fatigue. To deal with these impacts and the other stresses of flight, nearly all pressurized aircraft are semimonocoque in design. Pressurized fuselage structures undergo extensive periodic inspections to ensure that any damage is discovered and repaired. Repeated weakness or failure in an area of structure may require that section of the fuselage be modified or redesigned.

Wings

Wing Configurations

Wings are airfoils that, when moved rapidly through the air, create lift. They are built in many shapes and sizes. Wing design can vary to provide certain desirable flight characteristics. Control at various operating speeds, the amount of lift generated, balance, and stability all change as the shape of the wing is altered. Both the leading edge and the trailing edge of the wing may be straight or curved, or one edge may be straight and the other curved. One or both edges may be tapered so that the wing is narrower at the tip than at the root where it joins the fuselage. The wing tip may be square, rounded, or even pointed. *Figure 1-19* shows a number of typical wing leading and trailing edge shapes.

The wings of an aircraft can be attached to the fuselage at the top, mid-fuselage, or at the bottom. They may extend perpendicular to the horizontal plane of the fuselage or can angle up or down slightly. This angle is known as the wing dihedral. The dihedral angle affects the lateral stability of the aircraft. *Figure 1-20* shows some common wing attach points and dihedral angle.

Wing Structure

The wings of an aircraft are designed to lift it into the air. Their particular design for any given aircraft depends on a number of factors, such as size, weight, use of the aircraft,

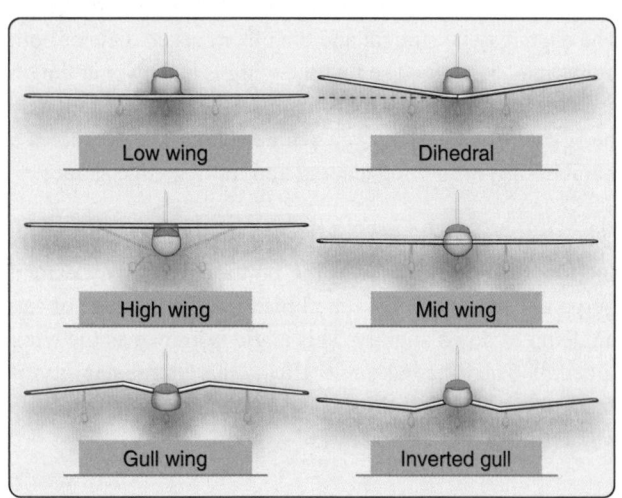

Figure 1-19. *Various wing design shapes yield different performance.*

| Tapered leading edge, straight trailing edge | Tapered leading and trailing edges | Delta wing |
| Sweptback wings | Straight leading and trailing edges | Straight leading edge, tapered trailing edge |

Figure 1-20. *Wing attach points and wing dihedrals.*

Low wing	Dihedral
High wing	Mid wing
Gull wing	Inverted gull

desired speed in flight and at landing, and desired rate of climb. The wings of aircraft are designated left and right, corresponding to the left and right sides of the operator when seated in the flight deck. *[Figure 1-21]*

Often wings are of full cantilever design. This means they are built so that no external bracing is needed. They are supported internally by structural members assisted by the skin of the aircraft. Other aircraft wings use external struts or wires to assist in supporting the wing and carrying the aerodynamic and landing loads. Wing support cables and struts are generally made from steel. Many struts and their attach fittings have fairings to reduce drag. Short, nearly vertical supports called jury struts are found on struts that attach to the wings a great distance from the fuselage. This serves to subdue strut movement and oscillation caused by the air flowing around the strut in flight. *Figure 1-22* shows samples of wings using external bracing, also known as semicantilever wings. Cantilever wings built with no external

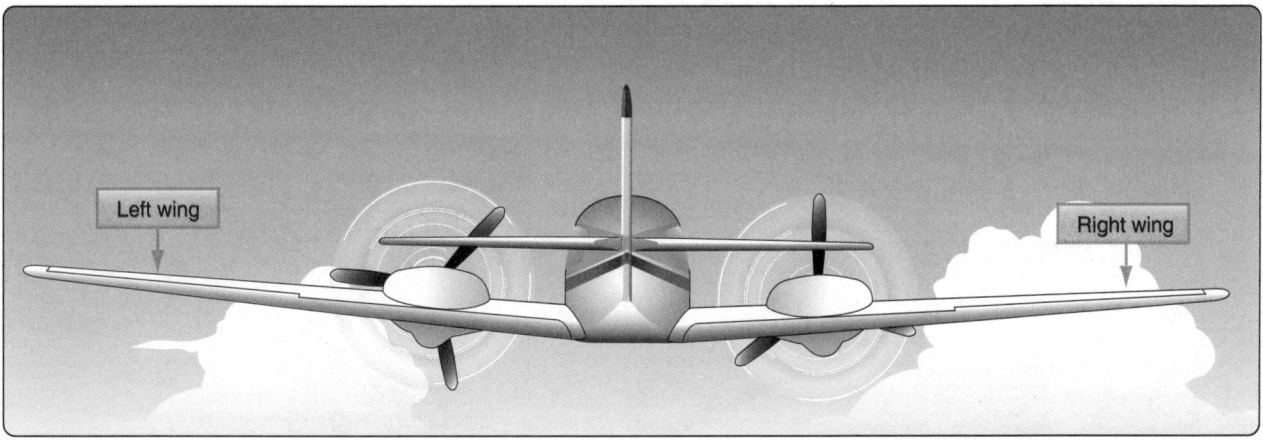

Figure 1-21. *"Left" and "right" on an aircraft are oriented to the perspective of a pilot sitting in the flight deck.*

bracing are also shown.

Aluminum is the most common material from which to construct wings, but they can be wood covered with fabric, and occasionally a magnesium alloy has been used. Moreover, modern aircraft are tending toward lighter and stronger materials throughout the airframe and in wing construction. Wings made entirely of carbon fiber or other composite materials exist, as well as wings made of a combination of materials for maximum strength to weight performance.

The internal structures of most wings are made up of spars and stringers running spanwise and ribs and formers or bulkheads running chordwise (leading edge to trailing edge). The spars are the principle structural members of a wing. They support all distributed loads, as well as concentrated weights such as the fuselage, landing gear, and engines. The skin, which is attached to the wing structure, carries part of the loads imposed during flight. It also transfers the stresses to the wing ribs. The ribs, in turn, transfer the loads to the wing spars. *[Figure 1-23]*

In general, wing construction is based on one of three fundamental designs:

1. Monospar
2. Multispar
3. Box beam

Modification of these basic designs may be adopted by various manufacturers.

The monospar wing incorporates only one main spanwise or longitudinal member in its construction. Ribs or bulkheads supply the necessary contour or shape to the airfoil. Although the strict monospar wing is not common, this type of design modified by the addition of false spars or light shear webs along the trailing edge for support of control surfaces is sometimes used.

The multispar wing incorporates more than one main longitudinal member in its construction. To give the wing contour, ribs or bulkheads are often included.

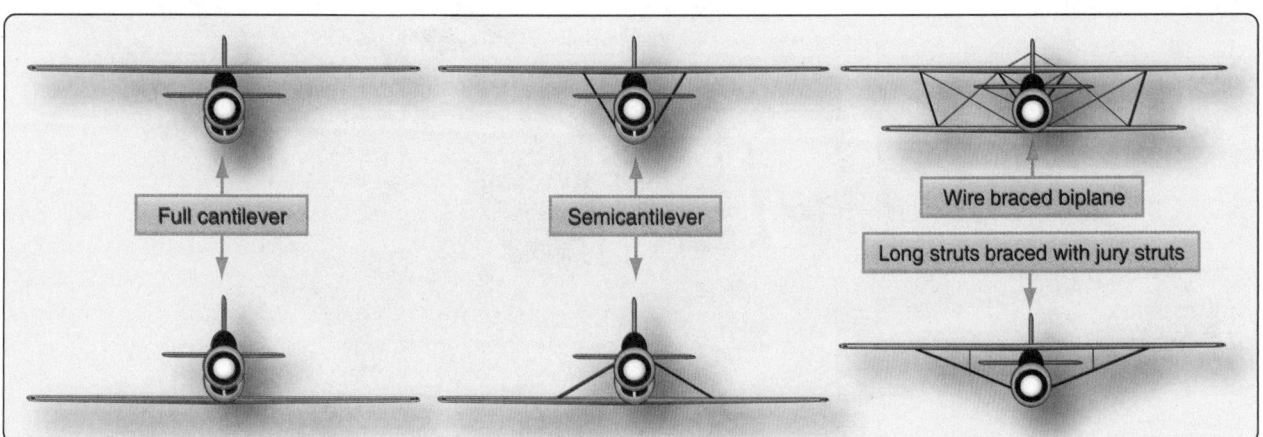

Figure 1-22. *Externally braced wings, also called semicantilever wings, have wires or struts to support the wing. Full cantilever wings have no external bracing and are supported internally.*

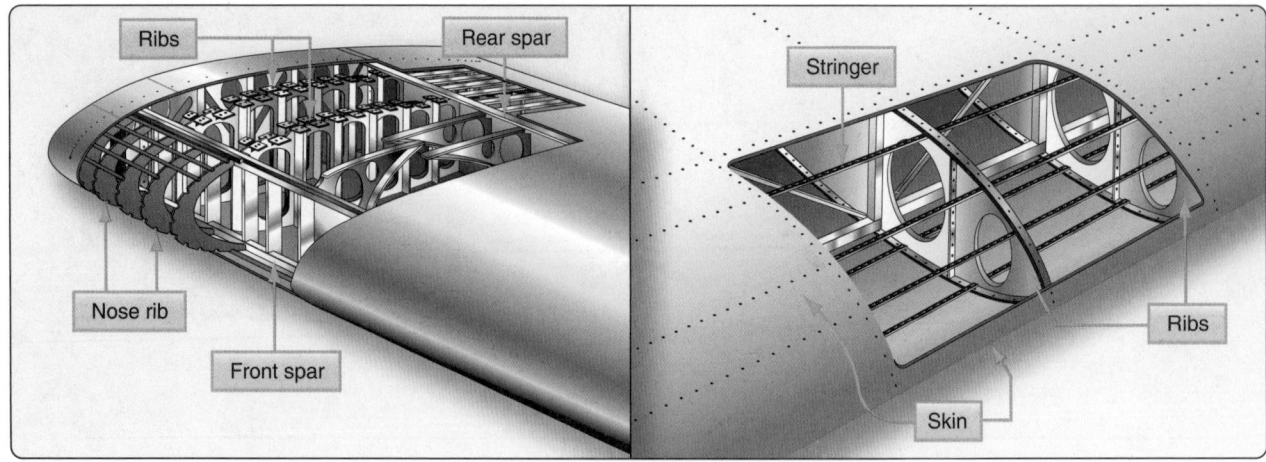

Figure 1-23. *Wing structure nomenclature.*

The box beam type of wing construction uses two main longitudinal members with connecting bulkheads to furnish additional strength and to give contour to the wing. *[Figure 1-24]* A corrugated sheet may be placed between the bulkheads and the smooth outer skin so that the wing can better carry tension and compression loads. In some cases, heavy longitudinal stiffeners are substituted for the upper surface of the wing and stiffeners on the lower surface corrugated sheets. A combination of corrugated sheets on the upper surface of the wing and stiffeners on the lower surface is sometimes used. Air transport category aircraft often utilize box beam wing construction.

Wing Spars

Spars are the principal structural members of the wing. They correspond to the longerons of the fuselage. They run parallel to the lateral axis of the aircraft, from the fuselage toward the tip of the wing, and are usually attached to the fuselage by wing fittings, plain beams, or a truss.

Spars may be made of metal, wood, or composite materials depending on the design criteria of a specific aircraft. Wooden spars are usually made from spruce. They can be generally classified into four different types by their cross-sectional configuration. As shown in *Figure 1-25*, they may be (A) solid, (B) box-shaped, (C) partly hollow, or (D) in the form of an I-beam. Lamination of solid wood spars is often used to increase strength. Laminated wood can also be found in box-shaped spars. The spar in *Figure 1-25E* has had material removed to reduce weight but retains the strength of a rectangular spar. As can be seen, most wing spars are basically rectangular in shape with the long dimension of the cross-section oriented up and down in the wing.

Currently, most manufactured aircraft have wing spars made of solid extruded aluminum or aluminum extrusions riveted together to form the spar. The increased use of

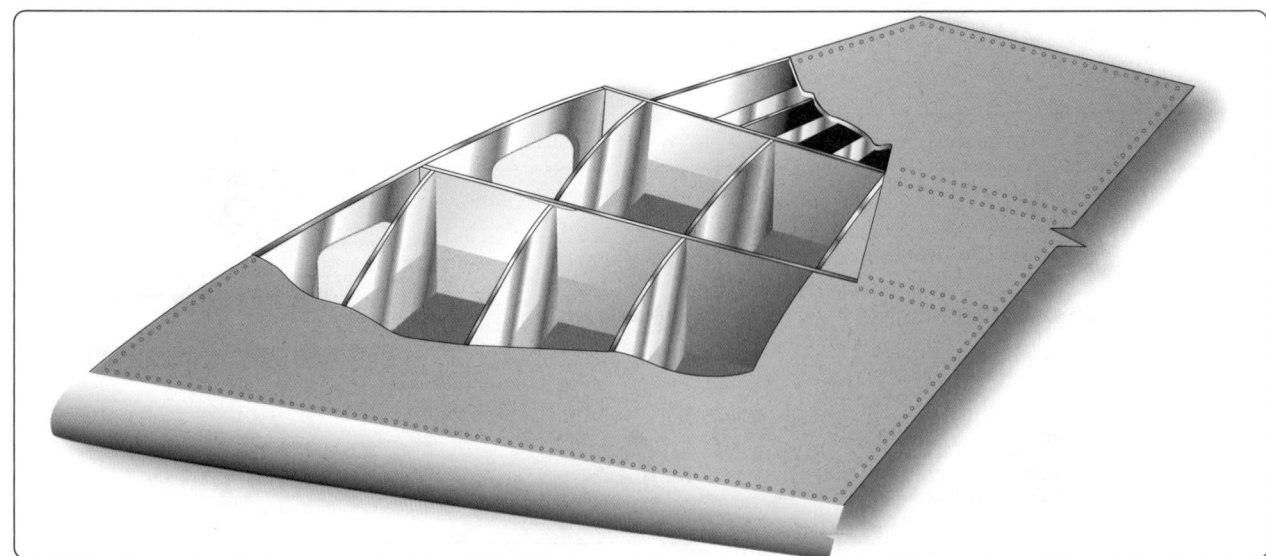

Figure 1-24. *Box beam construction.*

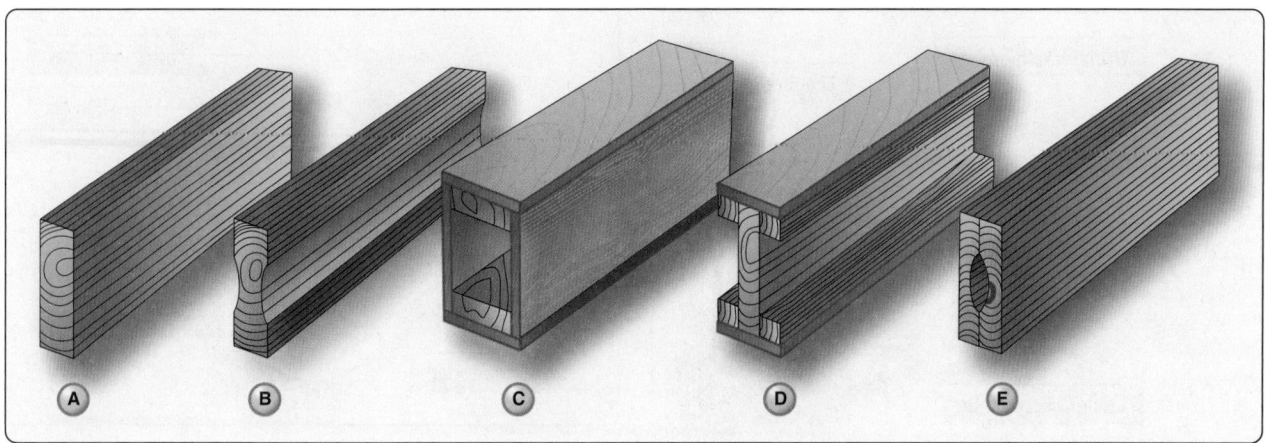

Figure 1-25. *Typical wooden wing spar cross-sections.*

composites and the combining of materials should make airmen vigilant for wings spars made from a variety of materials. *Figure 1-26* shows examples of metal wing spar cross-sections.

In an I–beam spar, the top and bottom of the I–beam are called the caps and the vertical section is called the web. The entire spar can be extruded from one piece of metal but often it is built up from multiple extrusions or formed angles. The web forms the principal depth portion of the spar and the cap strips (extrusions, formed angles, or milled sections) are attached to it. Together, these members carry the loads caused by wing bending, with the caps providing a foundation for attaching the skin. Although the spar shapes in *Figure 1-26* are typical, actual wing spar configurations assume many forms. For example, the web of a spar may be a plate or a truss as shown in *Figure 1-27.* It could be built up from lightweight materials with vertical stiffeners employed for strength. *[Figure 1-28]*

It could also have no stiffeners but might contain flanged holes for reducing weight while maintaining strength. Some metal and composite wing spars retain the I-beam concept but use a sine wave web. *[Figure 1-29]*

Additionally, fail-safe spar web design exists. Fail-safe means that should one member of a complex structure fail, some other part of the structure assumes the load of the failed member and permits continued operation. A spar with fail-safe construction is shown in *Figure 1-30.* This spar is made in two sections. The top section consists of a cap riveted to the upper web plate. The lower section is a single extrusion consisting of the lower cap and web plate. These two sections are spliced together to form the spar. If either section of this type of spar breaks, the other section can still carry the load. This is the fail-safe feature.

As a rule, a wing has two spars. One spar is usually located near the front of the wing, and the other about two-thirds of the distance toward the wing's trailing edge. Regardless of type, the spar is the most important part of the wing. When other structural members of the wing are placed under load, most of the resulting stress is passed on to the wing spar.

False spars are commonly used in wing design. They are longitudinal members like spars but do not extend the entire spanwise length of the wing. Often, they are used as hinge

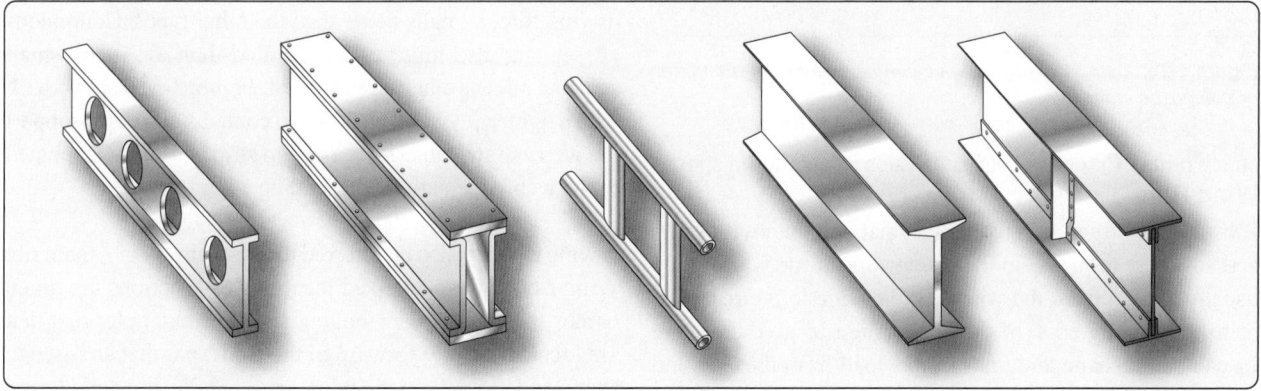

Figure 1-26. *Examples of metal wing spar shapes.*

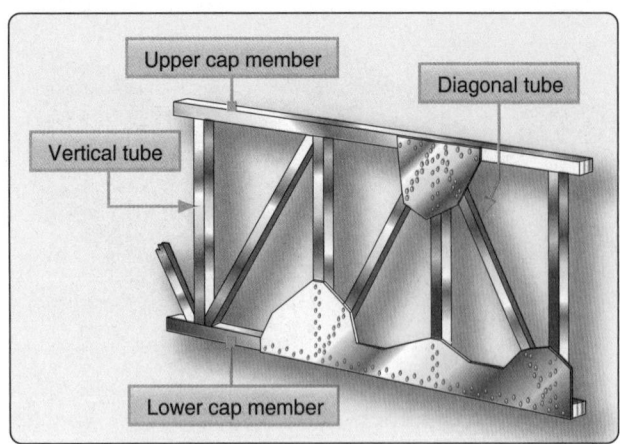

Figure 1-27. *A truss wing spar.*

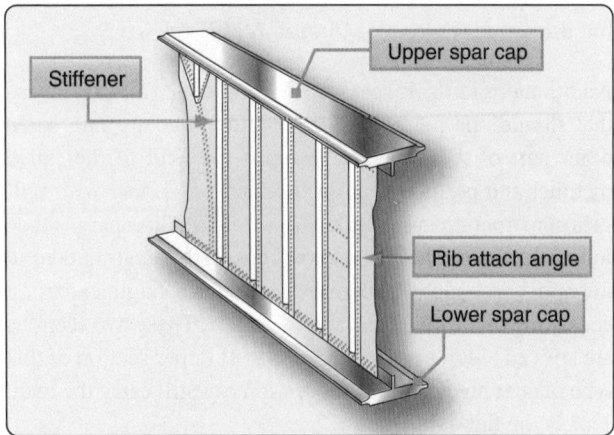

Figure 1-28. *A plate web wing spar with vertical stiffeners.*

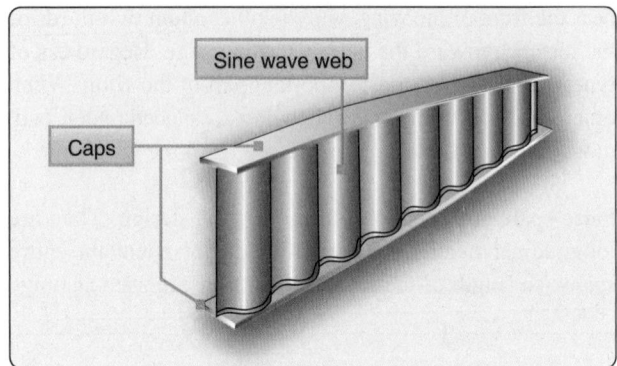

Figure 1-29. *A sine wave wing spar can be made from aluminum or composite materials.*

attach points for control surfaces, such as an aileron spar.

Wing Ribs

Ribs are the structural crosspieces that combine with spars and stringers to make up the framework of the wing. They usually extend from the wing leading edge to the rear spar or to the trailing edge of the wing. The ribs give the wing its cambered shape and transmit the load from the skin and stringers to the spars. Similar ribs are also used in ailerons,

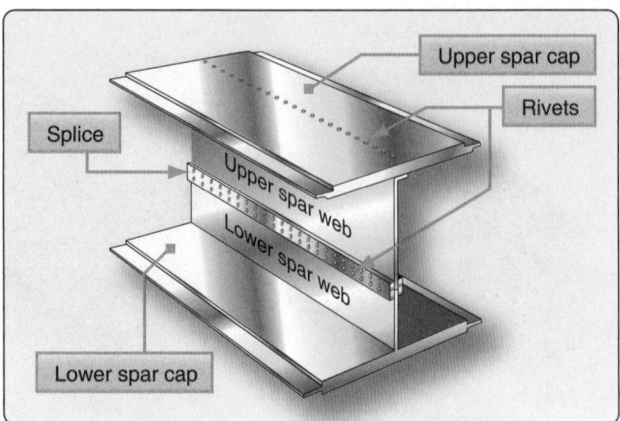

Figure 1-30. *A fail-safe spar with a riveted spar web.*

elevators, rudders, and stabilizers.

Wing ribs are usually manufactured from either wood or metal. Aircraft with wood wing spars may have wood or metal ribs while most aircraft with metal spars have metal ribs. Wood ribs are usually manufactured from spruce. The three most common types of wooden ribs are the plywood web, the lightened plywood web, and the truss types. Of these three, the truss type is the most efficient because it is strong and lightweight, but it is also the most complex to construct.

Figure 1-31 shows wood truss web ribs and a lightened plywood web rib. Wood ribs have a rib cap or cap strip fastened around the entire perimeter of the rib. It is usually made of the same material as the rib itself. The rib cap stiffens and strengthens the rib and provides an attaching surface for the wing covering. In *Figure 1-31A,* the cross-section of a wing rib with a truss-type web is illustrated. The dark rectangular sections are the front and rear wing spars. Note that to reinforce the truss, gussets are used. In *Figure 1-31B,* a truss web rib is shown with a continuous gusset. It provides greater support throughout the entire rib with very little additional weight. A continuous gusset stiffens the cap strip in the plane of the rib. This aids in preventing buckling and helps to obtain better rib/skin joints where nail-gluing is used. Such a rib can resist the driving force of nails better than the other types. Continuous gussets are also more easily handled than the many small separate gussets otherwise required. *Figure 1-31C* shows a rib with a lighten plywood web. It also contains gussets to support the web/cap strip interface. The cap strip is usually laminated to the web, especially at the leading edge.

A wing rib may also be referred to as a plain rib or a main rib. Wing ribs with specialized locations or functions are given names that reflect their uniqueness. For example, ribs that are located entirely forward of the front spar that are used to shape and strengthen the wing leading edge are called nose ribs or false ribs. False ribs are ribs that do not span the entire

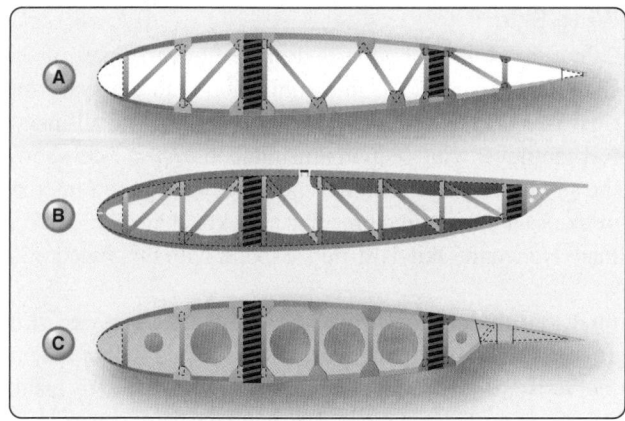

Figure 1-31. *Examples of wing ribs constructed of wood.*

wing chord, which is the distance from the leading edge to the trailing edge of the wing. Wing butt ribs may be found at the inboard edge of the wing where the wing attaches to the fuselage. Depending on its location and method of attachment, a butt rib may also be called a bulkhead rib or a compression rib if it is designed to receive compression loads that tend to force the wing spars together.

Since the ribs are laterally weak, they are strengthened in some wings by tapes that are woven above and below rib sections to prevent sidewise bending of the ribs. Drag and anti-drag wires may also be found in a wing. In *Figure 1-32,* they are

shown crisscrossed between the spars to form a truss to resist forces acting on the wing in the direction of the wing chord. These tension wires are also referred to as tie rods. The wire designed to resist the backward forces is called a drag wire; the anti-drag wire resists the forward forces in the chord direction. *Figure 1-32* illustrates the structural components of a basic wood wing.

At the inboard end of the wing spars is some form of wing attach fitting as illustrated in *Figure 1-32.* These provide a strong and secure method for attaching the wing to the fuselage. The interface between the wing and fuselage is often covered with a fairing to achieve smooth airflow in this area. The fairing(s) can be removed for access to the wing attach fittings. *[Figure 1-33]*

The wing tip is often a removable unit, bolted to the outboard end of the wing panel. One reason for this is the vulnerability of the wing tips to damage, especially during ground handling and taxiing. *Figure 1-34* shows a removable wing tip for a large aircraft wing. Others are different. The wing tip assembly is of aluminum alloy construction. The wing tip cap is secured to the tip with countersunk screws and is secured to the interspar structure at four points with ¼-inch diameter bolts. To prevent ice from forming on the leading edge of the wings of large aircraft, hot air from an engine is often channeled through the leading edge from wing root to wing tip. A louver on the top surface of the wing tip allows this warm air to be

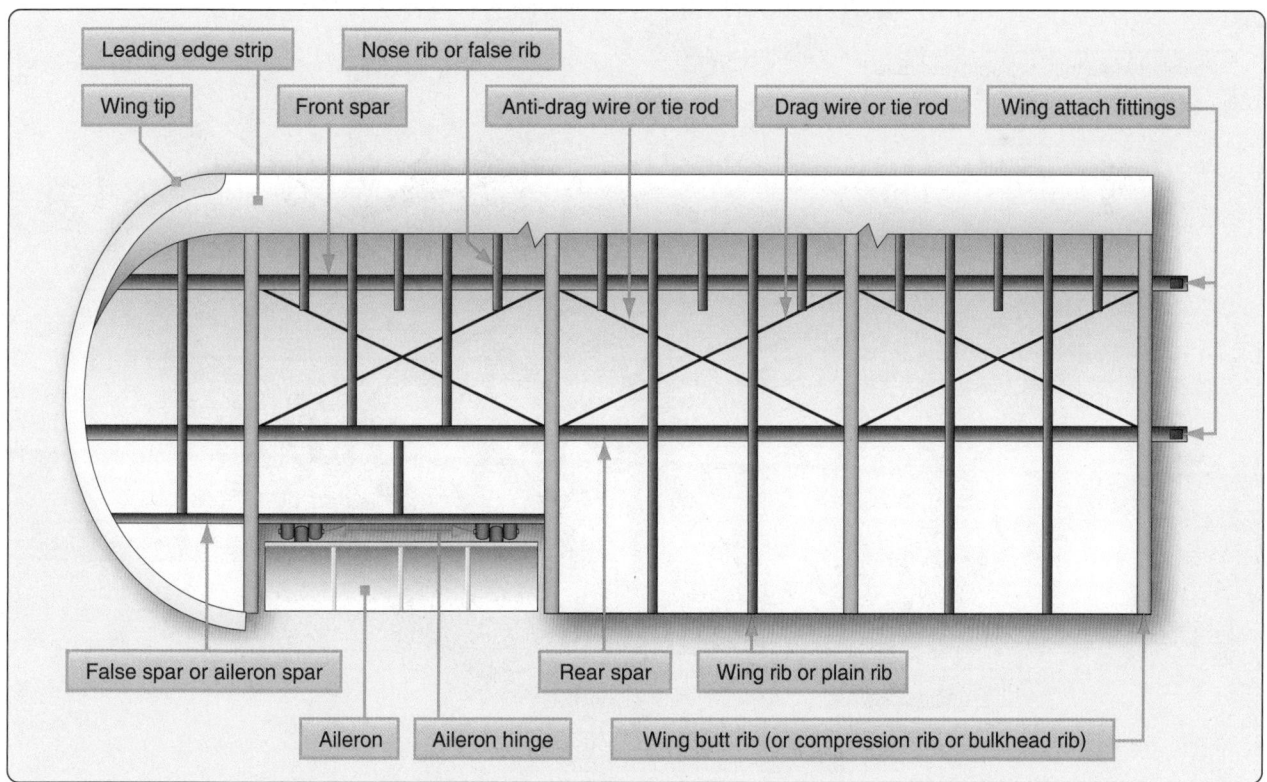

Figure 1-32. *Basic wood wing structure and components.*

Figure 1-33. *Wing root fairings smooth airflow and hide wing attach fittings.*

Wing Skin

Often, the skin on a wing is designed to carry part of the flight and ground loads in combination with the spars and ribs. This is known as a stressed-skin design. The all-metal, full cantilever wing section illustrated in *Figure 1-35* shows the structure of one such design. The lack of extra internal or external bracing requires that the skin share some of the load. Notice the skin is stiffened to aid with this function.

Fuel is often carried inside the wings of a stressed-skin aircraft. The joints in the wing can be sealed with a special fuel resistant sealant enabling fuel to be stored directly inside the structure. This is known as wet wing design. Alternately, a fuel-carrying bladder or tank can be fitted inside a wing. *Figure 1-36* shows a wing section with a box beam structural design such as one that might be found in a transport category aircraft. This structure increases strength while reducing weight. Proper sealing of the structure allows fuel to be stored in the box sections of the wing.

The wing skin on an aircraft may be made from a wide variety of materials such as fabric, wood, or aluminum. But a single thin sheet of material is not always employed. Chemically

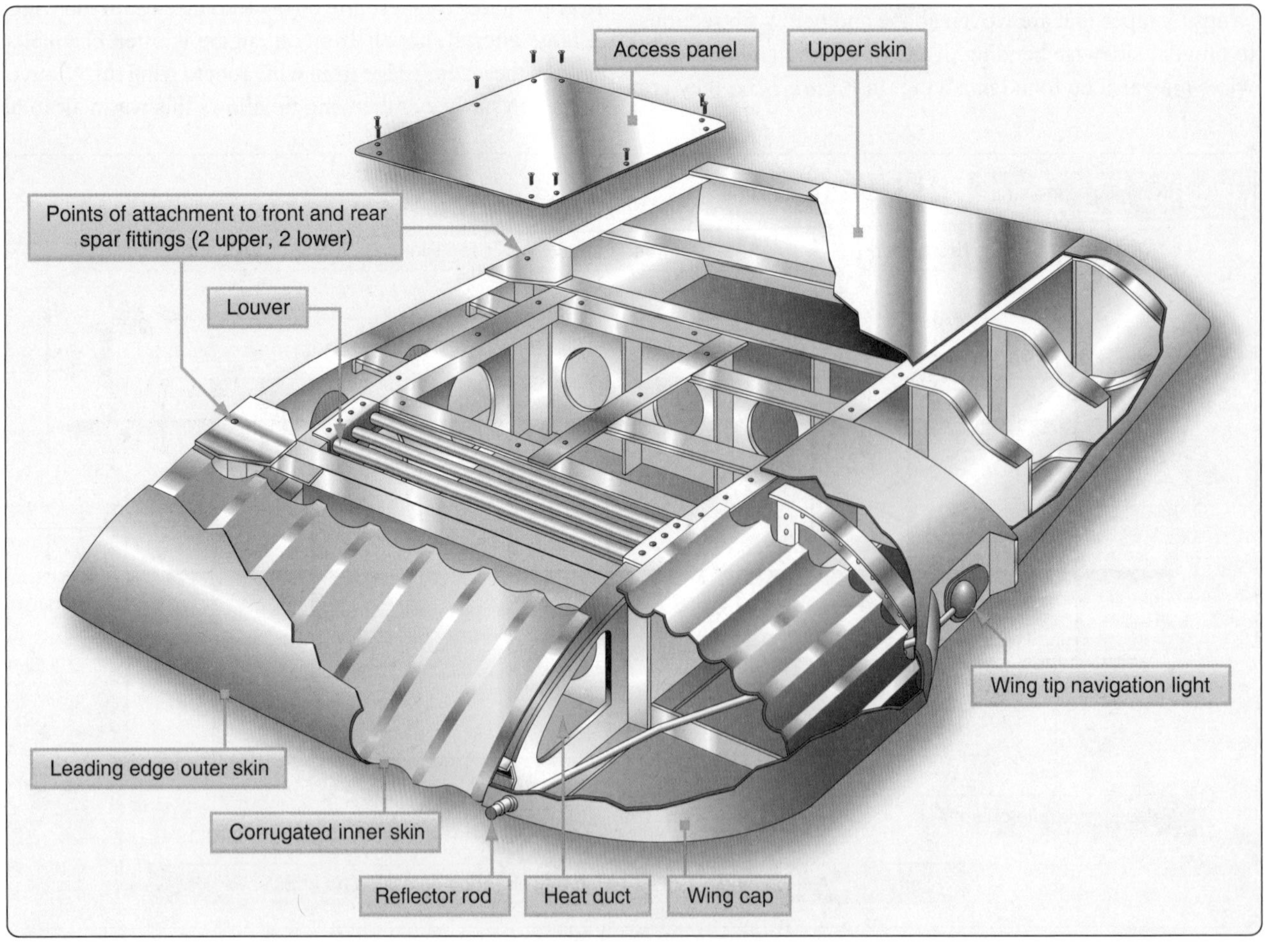

exhausted overboard. Wing position lights are located at the center of the tip and are not directly visible from the flight deck. As an indication that the wing tip light is operating, some wing tips are equipped with a Lucite rod to transmit the light to the leading edge.

Access panel

Upper skin

Points of attachment to front and rear spar fittings (2 upper, 2 lower)

Louver

Wing tip navigation light

Leading edge outer skin

Corrugated inner skin

Reflector rod

Heat duct

Wing cap

Figure 1-34. *A removable metal wing tip.*

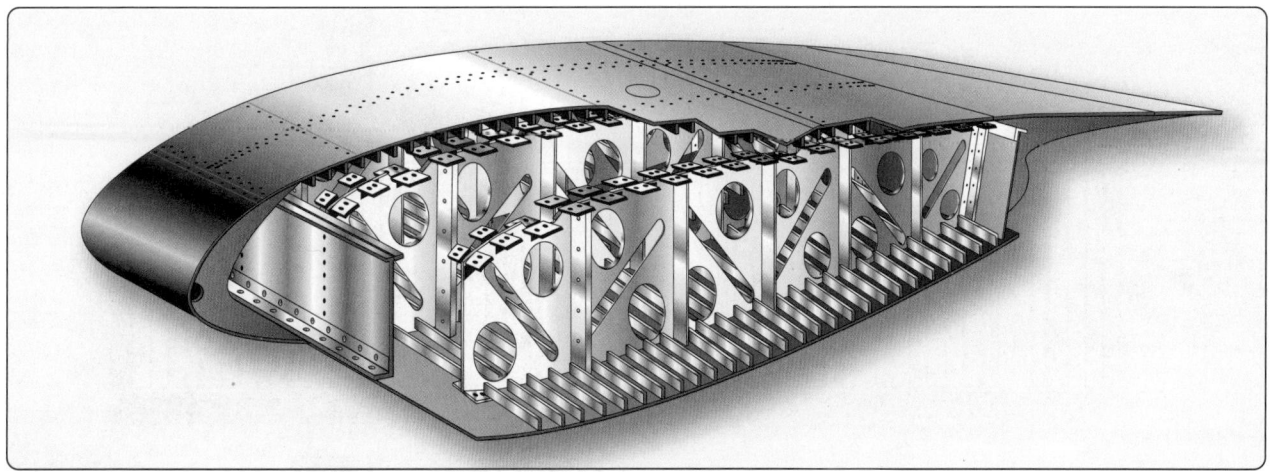

Figure 1-35. *The skin is an integral load carrying part of a stressed skin design.*

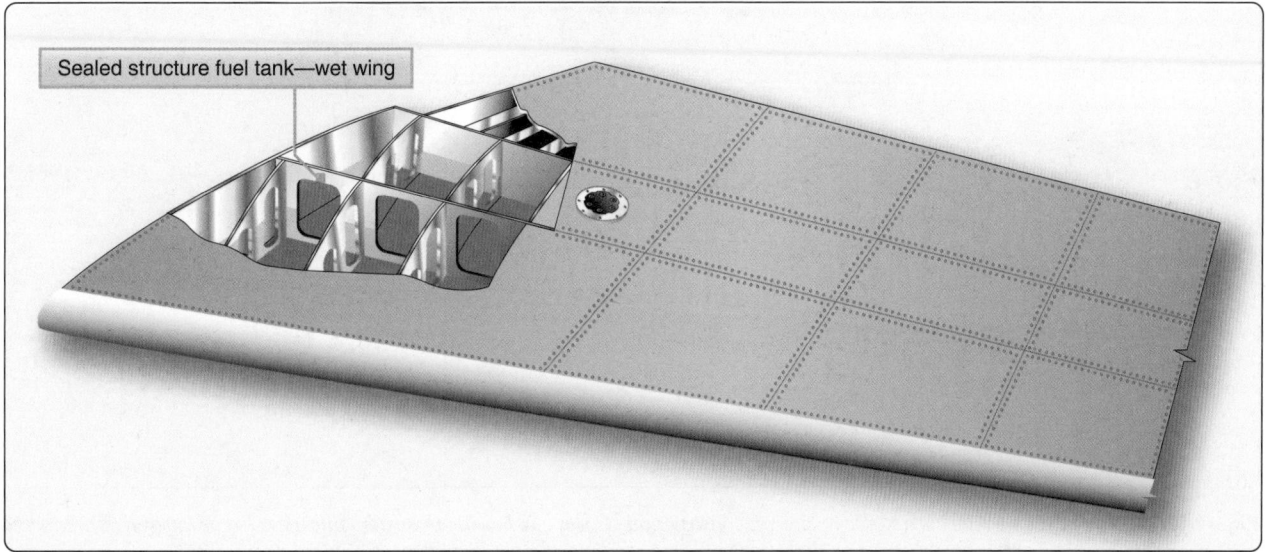

Sealed structure fuel tank—wet wing

Figure 1-36. *Fuel is often carried in the wings.*

milled aluminum skin can provide skin of varied thicknesses. On aircraft with stressed-skin wing design, honeycomb structured wing panels are often used as skin. A honeycomb structure is built up from a core material resembling a bee hive's honeycomb which is laminated or sandwiched between thin outer skin sheets. *Figure 1-37* illustrates honeycomb panes and their components. Panels formed like this are lightweight and very strong. They have a variety of uses on the aircraft, such as floor panels, bulkheads, and control surfaces, as well as wing skin panels. *Figure 1-38* shows the locations of honeycomb construction wing panels on a jet transport aircraft.

A honeycomb panel can be made from a wide variety of materials. Aluminum core honeycomb with an outer skin of aluminum is common. But honeycomb in which the core is an Arimid® fiber and the outer sheets are coated Phenolic® is common as well. In fact, a myriad of other material combinations such as those using fiberglass, plastic, Nomex®,

Kevlar®, and carbon fiber all exist. Each honeycomb structure possesses unique characteristics depending upon the materials, dimensions, and manufacturing techniques employed. *Figure 1-39* shows an entire wing leading edge formed from honeycomb structure.

Nacelles

Nacelles (sometimes called "pods") are streamlined enclosures used primarily to house the engine and its components. They usually present a round or elliptical profile to the wind thus reducing aerodynamic drag. On most single-engine aircraft, the engine and nacelle are at the forward end of the fuselage. On multiengine aircraft, engine nacelles are built into the wings or attached to the fuselage at the empennage (tail section). Occasionally, a multiengine aircraft is designed with a nacelle in line with the fuselage aft of the passenger compartment. Regardless of its location, a nacelle contains the engine and accessories, engine mounts, structural members, a firewall, and skin and cowling on the

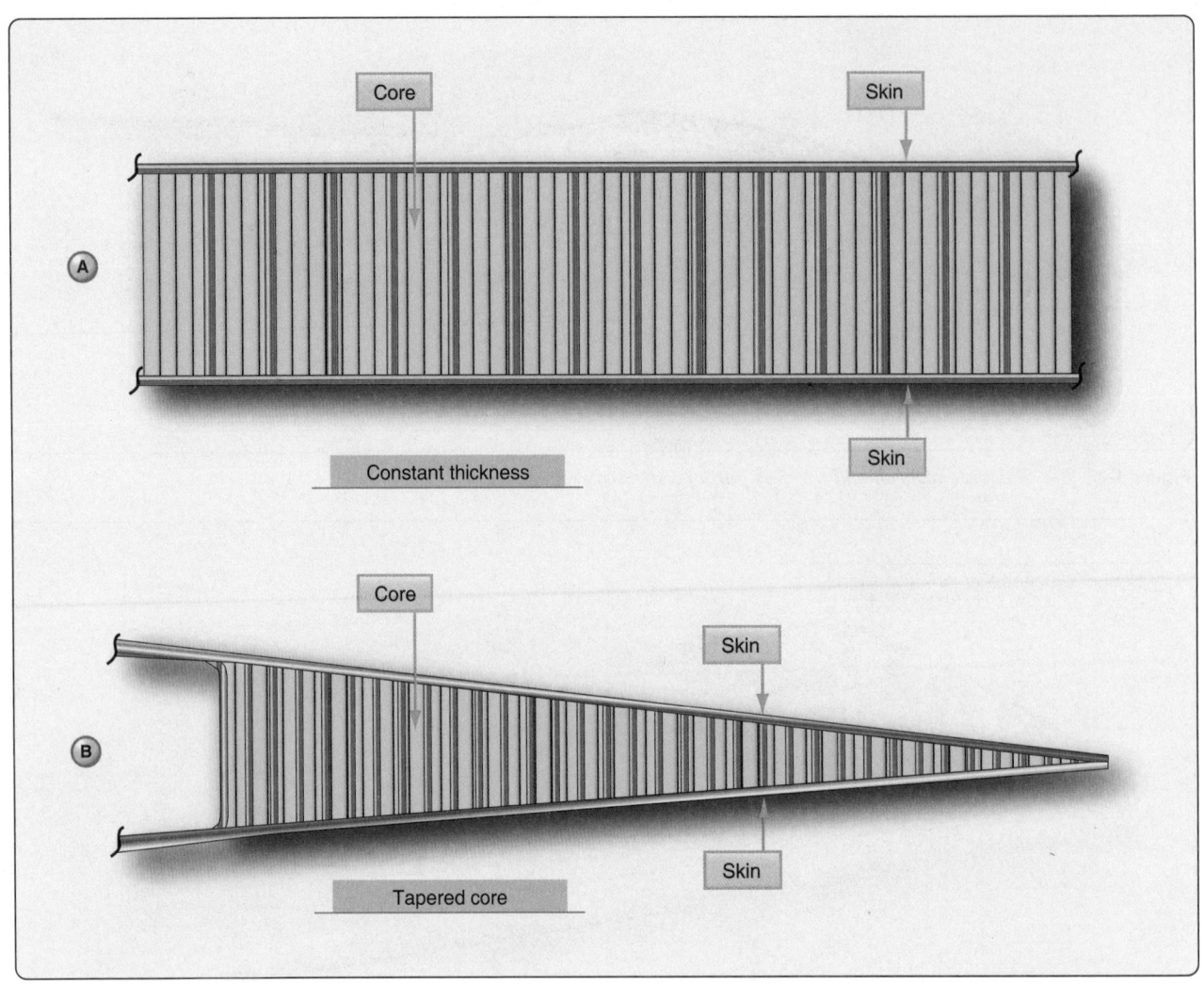

Figure 1-37. *The honeycomb panel is a staple in aircraft construction. Cores can be either constant thickness (A) or tapered (B). Tapered core honeycomb panels are frequently used as flight control surfaces and wing trailing edges.*

exterior to fare the nacelle to the wind.

Some aircraft have nacelles that are designed to house the landing gear when retracted. Retracting the gear to reduce wind resistance is standard procedure on high-performance/high-speed aircraft. The wheel well is the area where the landing gear is attached and stowed when retracted. Wheel wells can be located in the wings and/or fuselage when not part of the nacelle. *Figure 1-40* shows an engine nacelle incorporating the landing gear with the wheel well extending into the wing root.

The framework of a nacelle usually consists of structural members similar to those of the fuselage. Lengthwise members, such as longerons and stringers, combine with horizontal/vertical members, such as rings, formers, and bulkheads, to give the nacelle its shape and structural integrity. A firewall is incorporated to isolate the engine compartment from the rest of the aircraft. This is basically a

stainless steel or titanium bulkhead that contains a fire in the confines of the nacelle rather than letting it spread throughout the airframe. *[Figure 1-41]*

Engine mounts are also found in the nacelle. These are the structural assemblies to which the engine is fastened. They are usually constructed from chrome/molybdenum steel tubing in light aircraft and forged chrome/nickel/molybdenum assemblies in larger aircraft. *[Figure 1-42]*

The exterior of a nacelle is covered with a skin or fitted with a cowling which can be opened to access the engine and components inside. Both are usually made of sheet aluminum or magnesium alloy with stainless steel or titanium alloys being used in high-temperature areas, such as around the exhaust exit. Regardless of the material used, the skin is typically attached to the framework with rivets.

Cowling refers to the detachable panels covering those areas

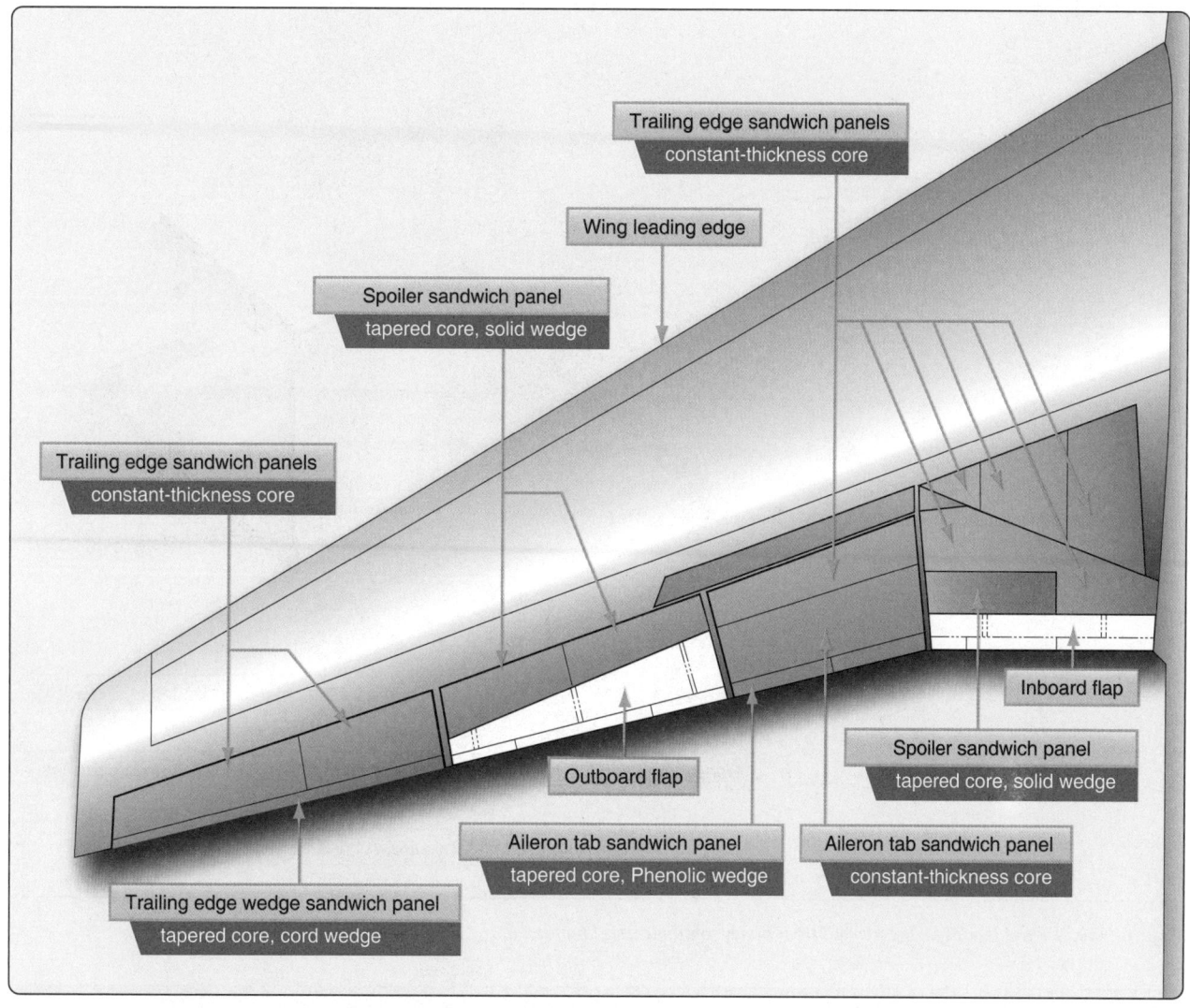

Figure 1-38. *Honeycomb wing construction on a large jet transport aircraft.*

into which access must be gained regularly, such as the engine and its accessories. It is designed to provide a smooth airflow over the nacelle and to protect the engine from damage. Cowl panels are generally made of aluminum alloy construction. However, stainless steel is often used as the inner skin aft of the power section and for cowl flaps and near cowl flap openings. It is also used for oil cooler ducts. Cowl flaps are moveable parts of the nacelle cowling that open and close to regulate engine temperature.

There are many engine cowl designs. *Figure 1-43* shows an exploded view of the pieces of cowling for a horizontally opposed engine on a light aircraft. It is attached to the nacelle by means of screws and/or quick release fasteners. Some large reciprocating engines are enclosed by "orange peel" cowlings which provide excellent access to components inside the nacelle. *[Figure 1-44]* These cowl panels are attached to the forward firewall by mounts which also serve as hinges for opening the cowl. The lower cowl mounts are

secured to the hinge brackets by quick release pins. The side and top panels are held open by rods and the lower panel is retained in the open position by a spring and a cable. All of the cowling panels are locked in the closed position by over-center steel latches which are secured in the closed position by spring-loaded safety catches.

An example of a turbojet engine nacelle can be seen in *Figure 1-45*. The cowl panels are a combination of fixed and easily removable panels which can be opened and closed during maintenance. A nose cowl is also a feature on a jet engine nacelle. It guides air into the engine.

Empennage

The empennage of an aircraft is also known as the tail section. Most empennage designs consist of a tail cone, fixed aerodynamic surfaces or stabilizers, and movable aerodynamic surfaces.

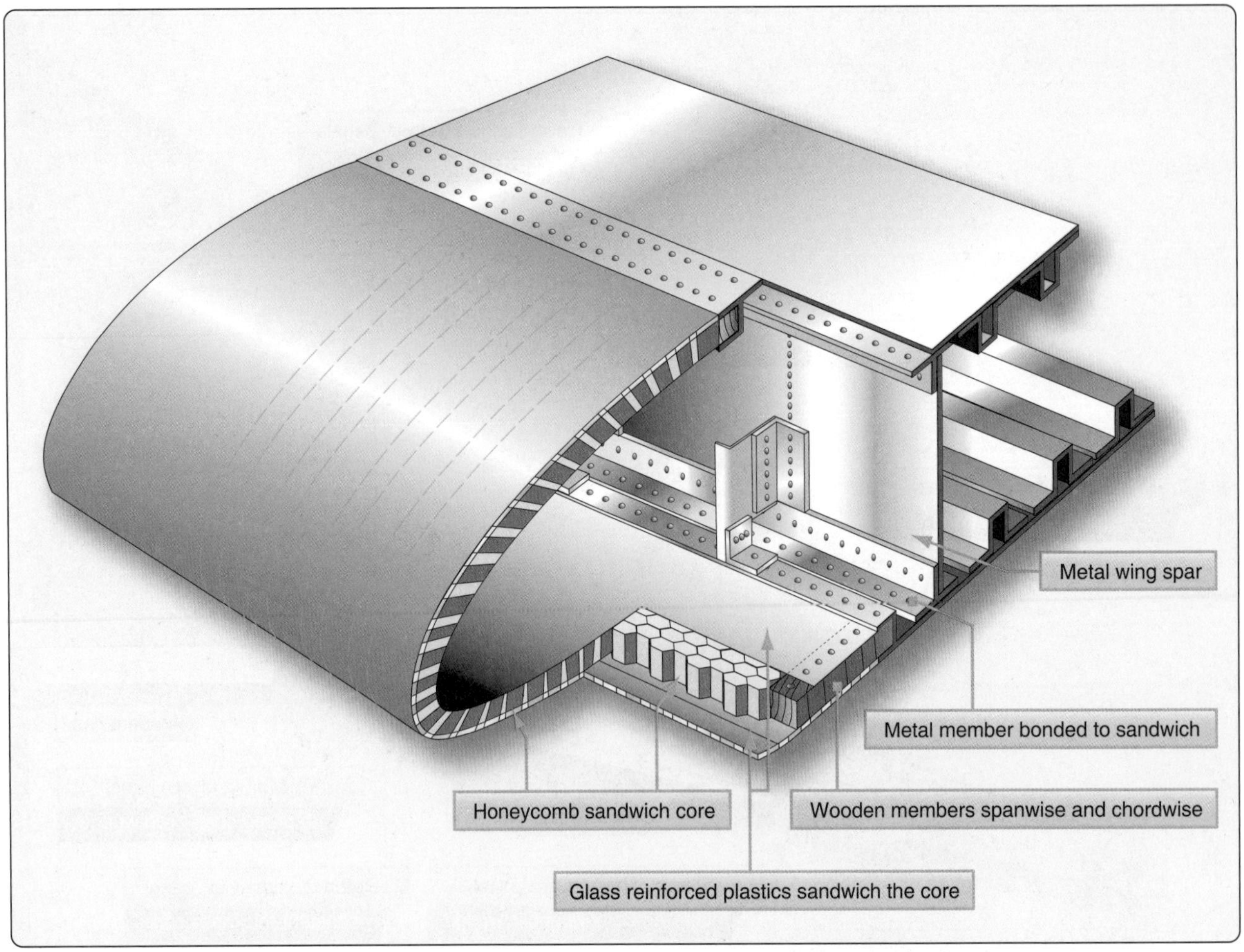

Figure 1-39. *A wing leading edge formed from honeycomb material bonded to the aluminum spar structure.*

Metal wing spar

Metal member bonded to sandwich

Honeycomb sandwich core

Wooden members spanwise and chordwise

Glass reinforced plastics sandwich the core

Figure 1-40. *Engine nacelle incorporating the landing gear with the wheel well extending into the wing root.*

Figure 1-41. *An engine nacelle firewall.*

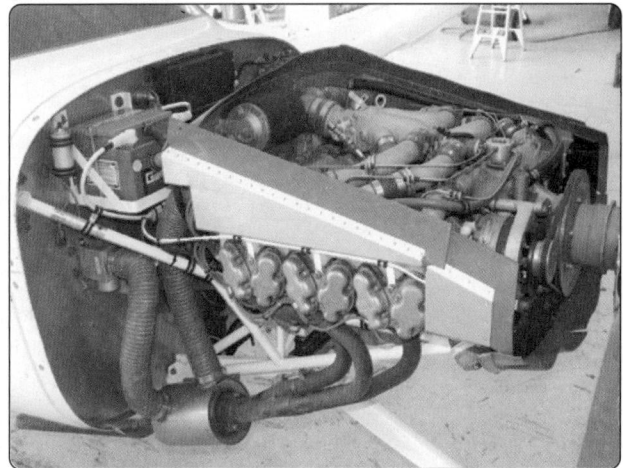

Figure 1-42. *Various aircraft engine mounts.*

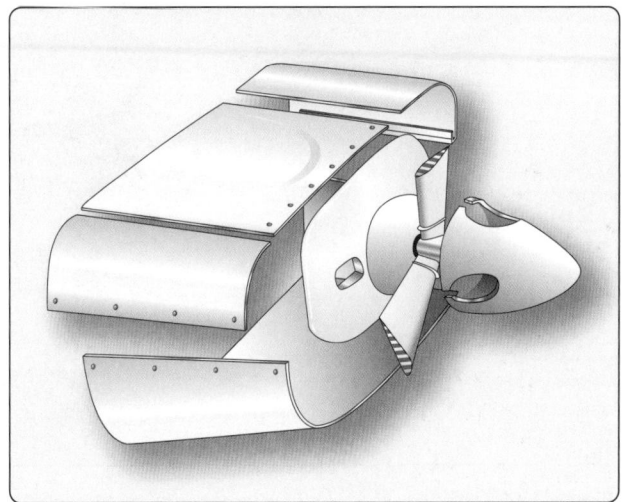

Figure 1-43. *Typical cowling for a horizontally opposed reciprocating engine.*

The tail cone serves to close and streamline the aft end of most fuselages. The cone is made up of structural members like those of the fuselage; however, cones are usually of lighter construction since they receive less stress than the fuselage. *[Figure 1-46]*

The other components of the typical empennage are of heavier construction than the tail cone. These members include fixed surfaces that help stabilize the aircraft and movable surfaces that help to direct an aircraft during flight. The fixed surfaces are the horizontal stabilizer and vertical stabilizer. The movable surfaces are usually a rudder located at the aft edge of the vertical stabilizer and an elevator located at the aft edge the horizontal stabilizer. *[Figure 1-47]*

The structure of the stabilizers is very similar to that which is used in wing construction. *Figure 1-48* shows a typical vertical stabilizer. Notice the use of spars, ribs, stringers, and skin like those found in a wing. They perform the same functions shaping and supporting the stabilizer and transferring stresses. Bending, torsion, and shear created by air loads in flight pass from one structural member to another. Each member absorbs some of the stress and passes the remainder on to the others. Ultimately, the spar transmits any overloads to the fuselage. A horizontal stabilizer is built

the same way.

The rudder and elevator are flight control surfaces that are also part of the empennage discussed in the next section of this chapter.

Flight Control Surfaces

The directional control of a fixed-wing aircraft takes place around the lateral, longitudinal, and vertical axes by means of flight control surfaces designed to create movement about these axes. These control devices are hinged or movable surfaces through which the attitude of an aircraft is controlled during takeoff, flight, and landing. They are usually divided into two major groups: 1) primary or main flight control surfaces and 2) secondary or auxiliary control surfaces.

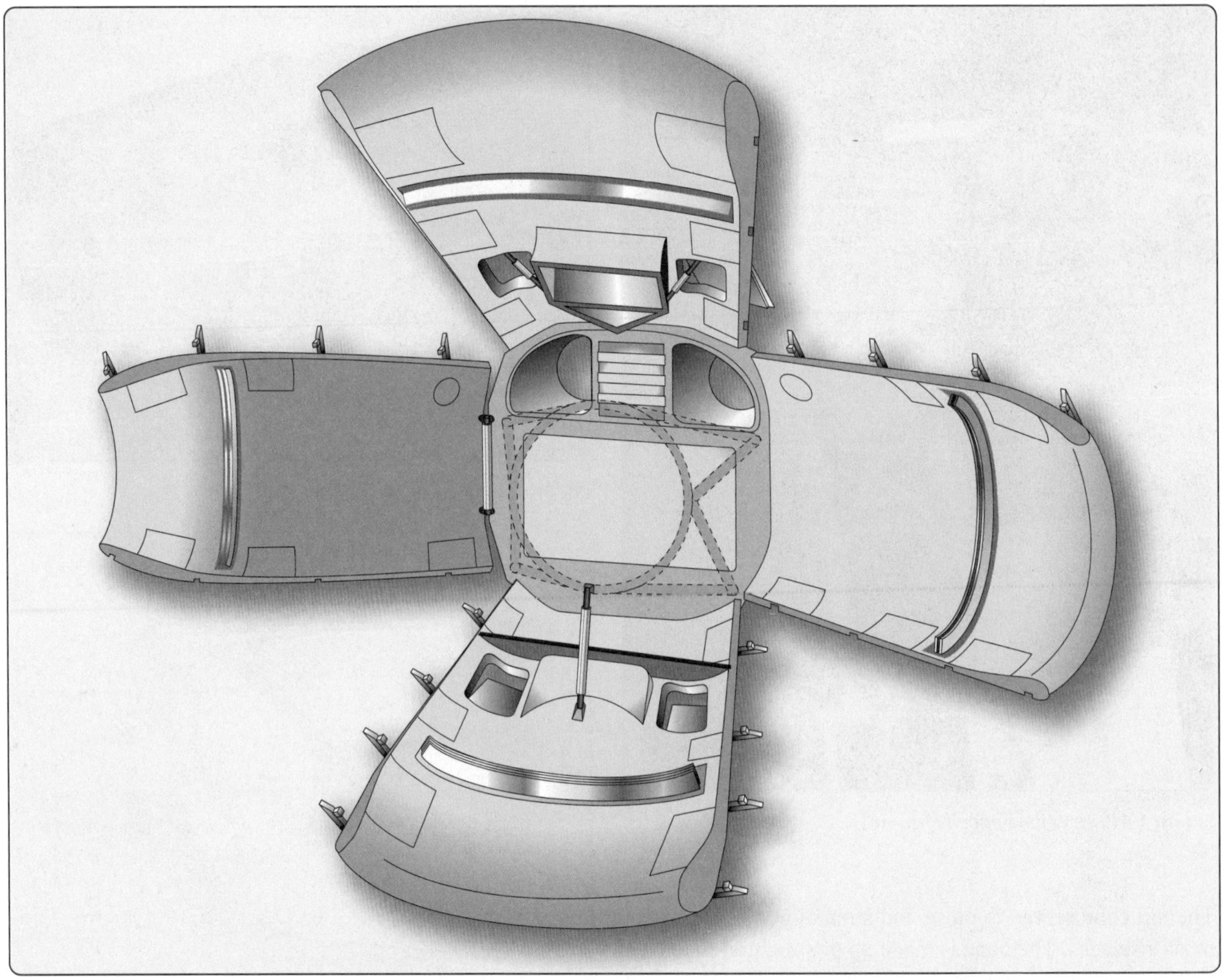

Figure 1-44. *Orange peel cowling for large radial reciprocating engine.*

Figure 1-45. *Cowling on a transport category turbine engine nacelle.*

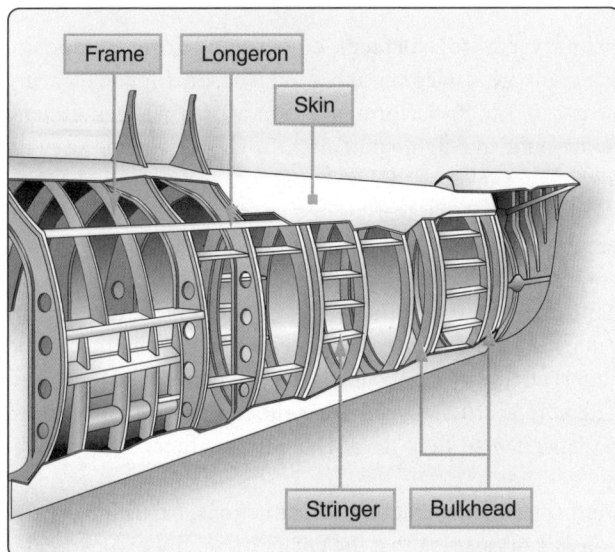

Figure 1-46. *The fuselage terminates at the tail cone with similar but more lightweight construction.*

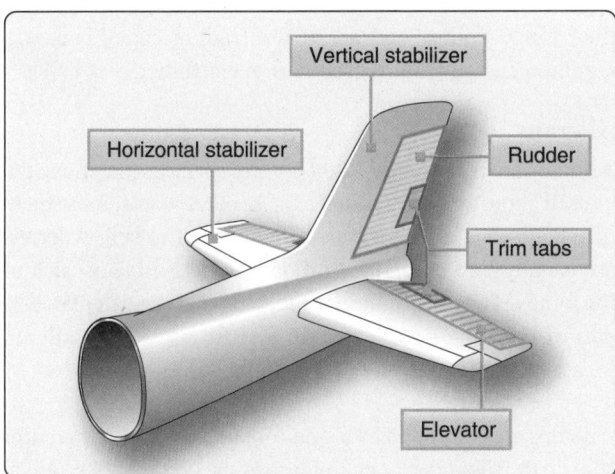

Figure 1-47. *Components of a typical empennage.*

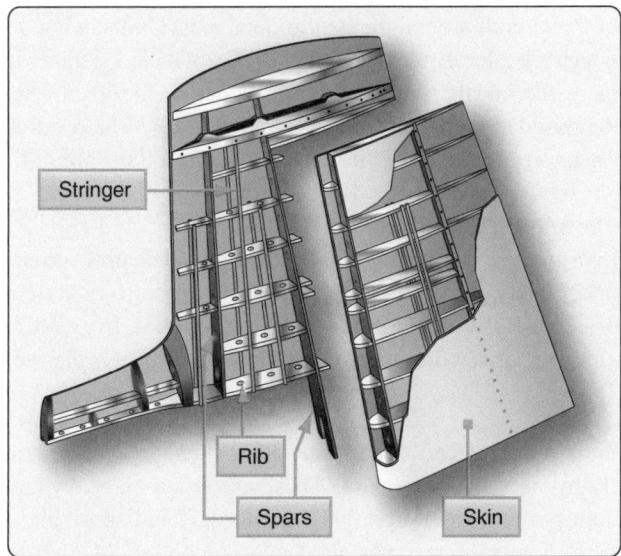

Figure 1-48. *Vertical stabilizer.*

Primary Flight Control Surfaces

The primary flight control surfaces on a fixed-wing aircraft include: ailerons, elevators, and the rudder. The ailerons are attached to the trailing edge of both wings and when moved, rotate the aircraft around the longitudinal axis. The elevator is attached to the trailing edge of the horizontal stabilizer. When it is moved, it alters aircraft pitch, which is the attitude about the horizontal or lateral axis. The rudder is hinged to the trailing edge of the vertical stabilizer. When the rudder changes position, the aircraft rotates about the vertical axis (yaw). *Figure 1-49* shows the primary flight controls of a light aircraft and the movement they create relative to the three axes of flight.

Primary control surfaces are usually similar in construction to one another and vary only in size, shape, and methods of attachment. On aluminum light aircraft, their structure is often similar to an all-metal wing. This is appropriate because the primary control surfaces are simply smaller aerodynamic devices. They are typically made from an aluminum alloy structure built around a single spar member or torque tube to which ribs are fitted and a skin is attached. The lightweight ribs are, in many cases, stamped out from flat aluminum sheet stock. Holes in the ribs lighten the assembly. An aluminum skin is attached with rivets. *Figure 1-50* illustrates this type of structure, which can be found on the primary control surfaces of light aircraft as well as on medium and heavy aircraft.

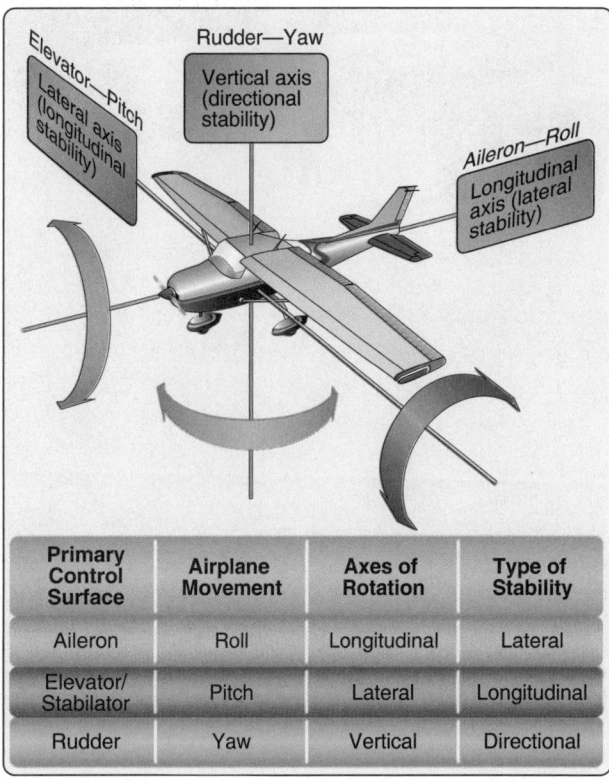

Primary Control Surface	Airplane Movement	Axes of Rotation	Type of Stability
Aileron	Roll	Longitudinal	Lateral
Elevator/Stabilator	Pitch	Lateral	Longitudinal
Rudder	Yaw	Vertical	Directional

Figure 1-49. *Flight control surfaces move the aircraft around the three axes of flight.*

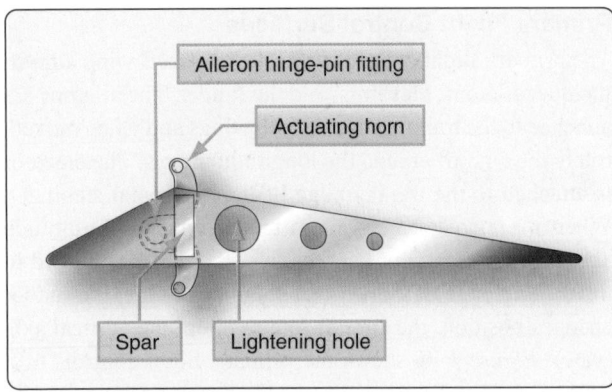

Figure 1-50. *Typical structure of an aluminum flight control surface.*

Figure 1-51. *Composite control surfaces and some of the many aircraft that utilize them.*

Primary control surfaces constructed from composite materials are also commonly used. These are found on many heavy and high-performance aircraft, as well as gliders, home-built, and light-sport aircraft. The weight and strength advantages over traditional construction can be significant. A wide variety of materials and construction techniques are employed. *Figure 1-51* shows examples of aircraft that use composite technology on primary flight control surfaces. Note that the control surfaces of fabric-covered aircraft often have fabric-covered surfaces just as aluminum-skinned (light) aircraft typically have all-aluminum control surfaces. There is a critical need for primary control surfaces to be balanced so they do not vibrate or flutter in the wind.

Performed to manufacturer's instructions, balancing usually consists of assuring that the center of gravity of a particular device is at or forward of the hinge point. Failure to properly balance a control surface could lead to catastrophic failure. *Figure 1-52* illustrates several aileron configurations with their hinge points well aft of the leading edge. This is a common design feature used to prevent flutter.

Ailerons

Ailerons are the primary flight control surfaces that move the aircraft about the longitudinal axis. In other words, movement of the ailerons in flight causes the aircraft to roll. Ailerons are usually located on the outboard trailing edge of each of the wings. They are built into the wing and are calculated as part of the wing's surface area. *Figure 1-53* shows aileron locations on various wing tip designs.

Ailerons are controlled by a side-to-side motion of the control stick in the flight deck or a rotation of the control yoke. When the aileron on one wing deflects down, the aileron on the opposite wing deflects upward. This amplifies the movement of the aircraft around the longitudinal axis. On the wing on which the aileron trailing edge moves downward, camber is increased, and lift is increased. Conversely, on the other wing, the raised aileron decreases lift. *[Figure 1-54]* The result is a sensitive response to the control input to roll the aircraft.

The pilot's request for aileron movement and roll are transmitted from the flight deck to the actual control surface in a variety of ways depending on the aircraft. A system of control cables and pulleys, push-pull tubes, hydraulics, electric, or a combination of these can be employed. *[Figure 1-55]*

Simple, light aircraft usually do not have hydraulic or electric fly-by-wire aileron control. These are found on heavy and high-performance aircraft. Large aircraft and some high-performance aircraft may also have a second set of ailerons located inboard on the trailing edge of the wings. These

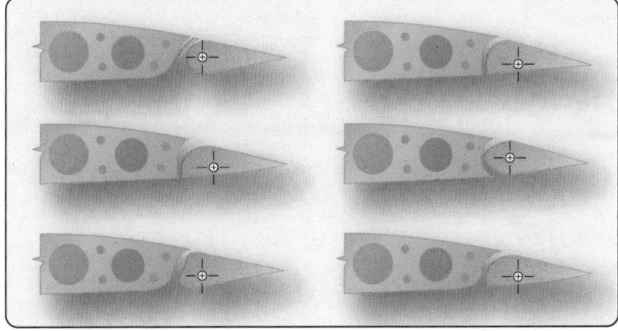

Figure 1-52. *Aileron hinge locations are very close to but aft of the center of gravity to prevent flutter.*

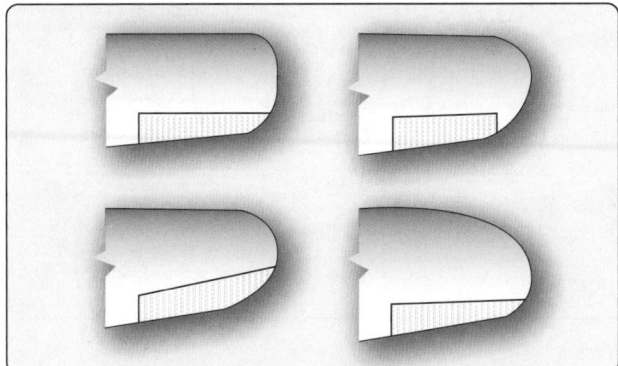

Figure 1-53. *Aileron location on various wings.*

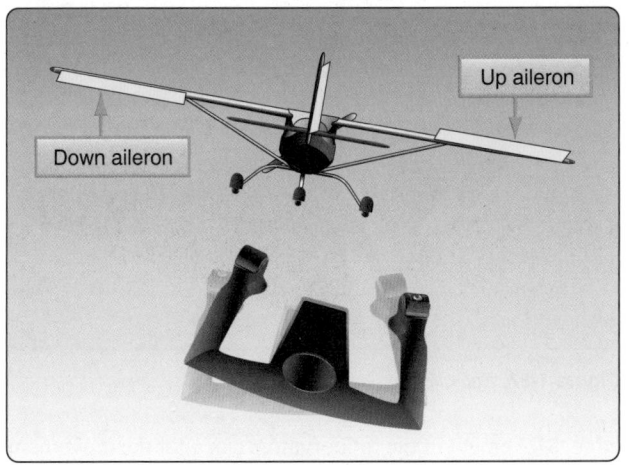

Figure 1-54. *Differential aileron control movement. When one aileron is moved down, the aileron on the opposite wing is deflected upward.*

are part of a complex system of primary and secondary control surfaces used to provide lateral control and stability in flight. At low speeds, the ailerons may be augmented by the use of flaps and spoilers. At high speeds, only inboard aileron deflection is required to roll the aircraft while the other control surfaces are locked out or remain stationary. *Figure 1-56* illustrates the location of the typical flight control surfaces found on a transport category aircraft.

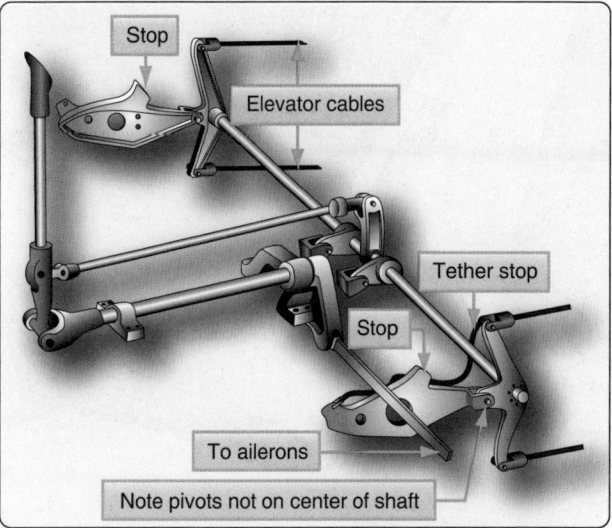

Figure 1-55. *Transferring control surface inputs from the flight deck.*

Elevator

The elevator is the primary flight control surface that moves the aircraft around the horizontal or lateral axis. This causes the nose of the aircraft to pitch up or down. The elevator is hinged to the trailing edge of the horizontal stabilizer and typically spans most or all of its width. It is controlled in the flight deck by pushing or pulling the control stick or yoke forward or aft.

Light aircraft use a system of control cables and pulleys or push-pull tubes to transfer flight deck inputs to the movement of the elevator. High-performance and large aircraft typically employ more complex systems. Hydraulic power is commonly used to move the elevator on these aircraft. On aircraft equipped with fly-by-wire controls, a combination of electrical and hydraulic power is used.

Rudder

The rudder is the primary control surface that causes an aircraft to yaw or move about the vertical axis. This provides directional control and thus points the nose of the aircraft in the direction desired. Most aircraft have a single rudder hinged to the trailing edge of the vertical stabilizer. It is controlled by a pair of foot-operated rudder pedals in the flight deck. When the right pedal is pushed forward, it deflects the rudder to the right which moves the nose of the aircraft to the right. The left pedal is rigged to simultaneously move aft. When the left pedal is pushed forward, the nose of the aircraft moves to the left.

As with the other primary flight controls, the transfer of the movement of the flight deck controls to the rudder varies with the complexity of the aircraft. Many aircraft incorporate the directional movement of the nose or tail wheel into the rudder control system for ground operation. This allows the operator

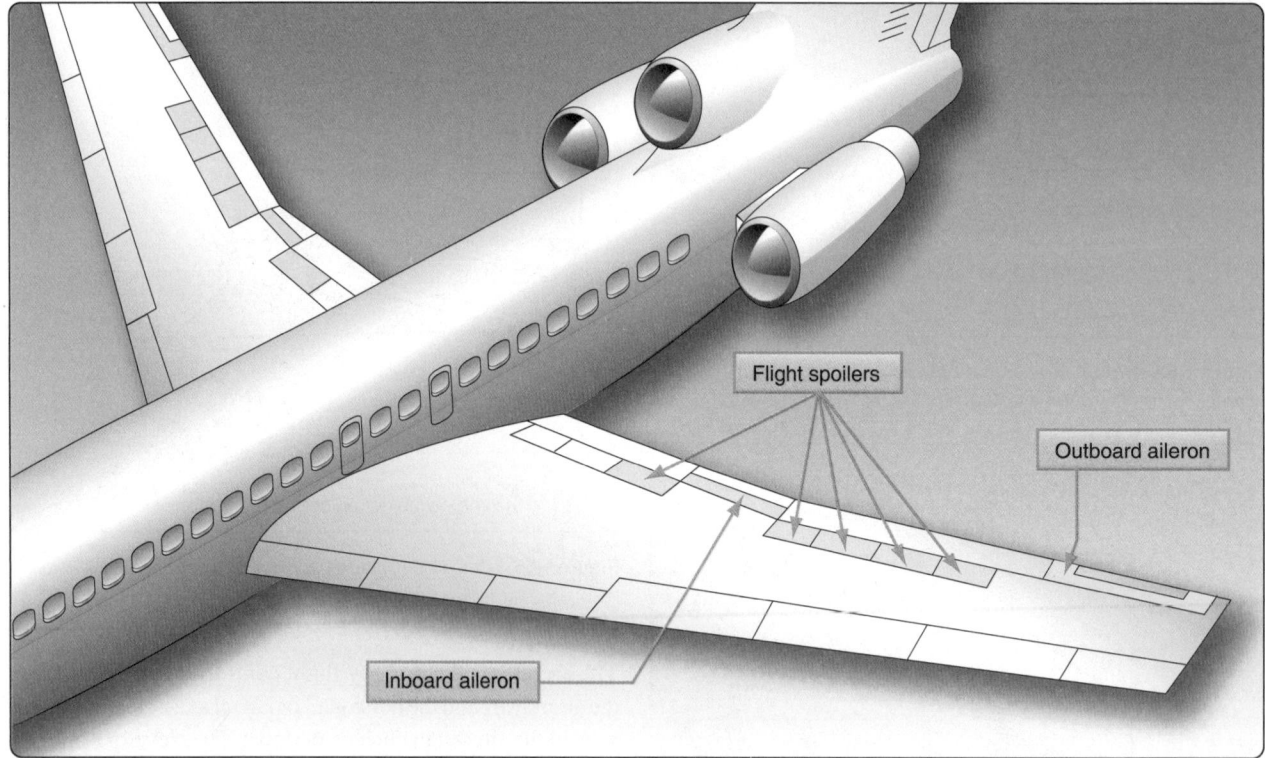

Figure 1-56. *Typical flight control surfaces on a transport category aircraft.*

to steer the aircraft with the rudder pedals during taxi when the airspeed is not high enough for the control surfaces to be effective. Some large aircraft have a split rudder arrangement. This is actually two rudders, one above the other. At low speeds, both rudders deflect in the same direction when the pedals are pushed. At higher speeds, one of the rudders becomes inoperative as the deflection of a single rudder is aerodynamically sufficient to maneuver the aircraft.

Dual Purpose Flight Control Surfaces

The ailerons, elevators, and rudder are considered conventional primary control surfaces. However, some aircraft are designed with a control surface that may serve a dual purpose. For example, elevons perform the combined functions of the ailerons and the elevator. *[Figure 1-57]*

A movable horizontal tail section, called a stabilator, is a control surface that combines the action of both the horizontal stabilizer and the elevator. *[Figure 1-58]* Basically, a stabilator is a horizontal stabilizer that can also be rotated about the horizontal axis to affect the pitch of the aircraft.

A ruddervator combines the action of the rudder and elevator. *[Figure 1-59]* This is possible on aircraft with V–tail empennages where the traditional horizontal and vertical stabilizers do not exist. Instead, two stabilizers angle upward and outward from the aft fuselage in a "V" configuration. Each contains a movable ruddervator built into the trailing

Figure 1-57. *Elevons.*

edge. Movement of the ruddervators can alter the movement of the aircraft around the horizontal and/or vertical axis. Additionally, some aircraft are equipped with flaperons. *[Figure 1-60]* Flaperons are ailerons which can also act as flaps. Flaps are secondary control surfaces on most wings, discussed in the next section of this chapter.

Secondary or Auxiliary Control Surfaces

There are several secondary or auxiliary flight control surfaces. Their names, locations, and functions of those for most large aircraft are listed in *Figure 1-61*.

Flaps

Flaps are found on most aircraft. They are usually inboard

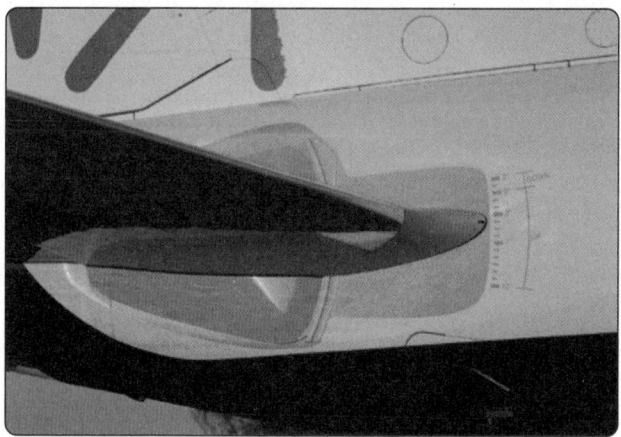

Figure 1-58. *A stabilizer and index marks on a transport category aircraft.*

Figure 1-60. *Flaperons.*

Figure 1-59. *Ruddervator.*

on the wings' trailing edges adjacent to the fuselage. Leading edge flaps are also common. They extend forward and down from the inboard wing leading edge. The flaps are lowered to increase the camber of the wings and provide greater lift and control at slow speeds. They enable landing at slower speeds and shorten the amount of runway required for takeoff and landing. The amount that the flaps extend and the angle they form with the wing can be selected from the flight deck. Typically, flaps can extend up to 45–50°. *Figure 1-62* shows various aircraft with flaps in the extended position.

Flaps are usually constructed of materials and with techniques used on the other airfoils and control surfaces of a particular aircraft. Aluminum skin and structure flaps are the norm on light aircraft. Heavy and high-performance aircraft flaps may also be aluminum, but the use of composite structures is also common.

There are various kinds of flaps. Plain flaps form the trailing edge of the wing when the flap is in the retracted position. *[Figure 1-63A]* The airflow over the wing continues over the upper and lower surfaces of the flap, making the trailing edge of the flap essentially the trailing edge of the wing. The plain flap is hinged so that the trailing edge can be lowered. This

increases wing camber and provides greater lift.

A split flap is normally housed under the trailing edge of the wing. *[Figure 1-63B]* It is usually just a braced flat metal plate hinged at several places along its leading edge. The upper surface of the wing extends to the trailing edge of the flap. When deployed, the split flap trailing edge lowers away from the trailing edge of the wing. Airflow over the top of the wing remains the same. Airflow under the wing now follows the camber created by the lowered split flap, increasing lift.

Fowler flaps not only lower the trailing edge of the wing when deployed but also slide aft, effectively increasing the area of the wing. *[Figure 1-63C]* This creates more lift via the increased surface area, as well as the wing camber. When stowed, the Fowler flap typically retracts up under the wing trailing edge similar to a split flap. The sliding motion of a Fowler flap can be accomplished with a worm drive and flap tracks.

An enhanced version of the Fowler flap is a set of flaps that actually contains more than one aerodynamic surface. *Figure 1-64* shows a triple-slotted flap. In this configuration, the flap consists of a fore flap, a mid flap, and an aft flap. When deployed, each flap section slides aft on tracks as it lowers. The flap sections also separate leaving an open slot between the wing and the fore flap, as well as between each of the flap sections. Air from the underside of the wing flows through these slots. The result is that the laminar flow on the upper surfaces is enhanced. The greater camber and effective wing area increase overall lift.

Heavy aircraft often have leading edge flaps that are used in conjunction with the trailing edge flaps. *[Figure 1-65]* They can be made of machined magnesium or can have an aluminum or composite structure. While they are not installed or operate independently, their use with trailing edge flaps can greatly increase wing camber and lift. When stowed, leading edge flaps retract into the leading edge of the wing.

Secondary/Auxiliary Flight Control Surfaces		
Name	**Location**	**Function**
Flaps	Inboard trailing edge of wings	Extends the camber of the wing for greater lift and slower flight. Allows control at low speeds for short field takeoffs and landings.
Trim tabs	Trailing edge of primary flight control surfaces	Reduces the force needed to move a primary control surface.
Balance tabs	Trailing edge of primary flight control surfaces	Reduces the force needed to move a primary control surface.
Anti-balance tabs	Trailing edge of primary flight control surfaces	Increases feel and effectiveness of primary control surface.
Servo tabs	Trailing edge of primary flight control surfaces	Assists or provides the force for moving a primary flight control.
Spoilers	Upper and/or trailing edge of wing	Decreases (spoils) lift. Can augment aileron function.
Slats	Mid to outboard leading edge of wing	Extends the camber of the wing for greater lift and slower flight. Allows control at low speeds for short field takeoffs and landings.
Slots	Outer leading edge of wing forward of ailerons	Directs air over upper surface of wing during high angle of attack. Lowers stall speed and provides control during slow flight.
Leading edge flap	Inboard leading edge of wing	Extends the camber of the wing for greater lift and slower flight. Allows control at low speeds for short field takeoffs and landings.
NOTE: An aircraft may possess none, one, or a combination of the above control surfaces.		

Figure 1-61. *Secondary or auxiliary control surfaces and respective locations for larger aircraft.*

Figure 1-62. *Various aircraft with flaps in the extended position.*

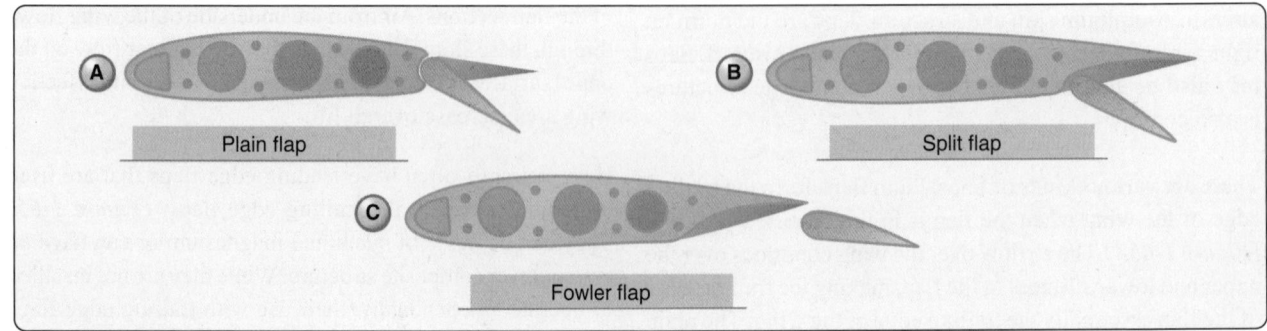

Figure 1-63. *Various types of flaps.*

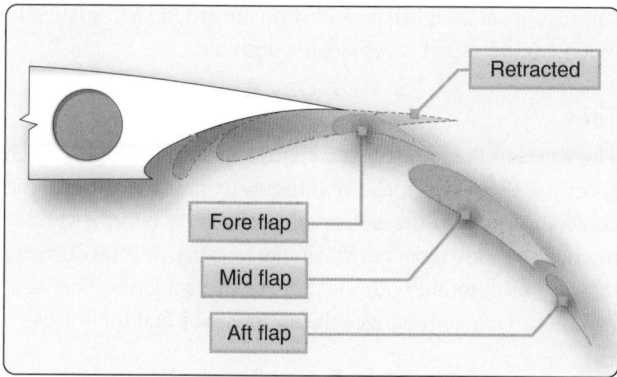

Figure 1-64. *Triple-slotted flap.*

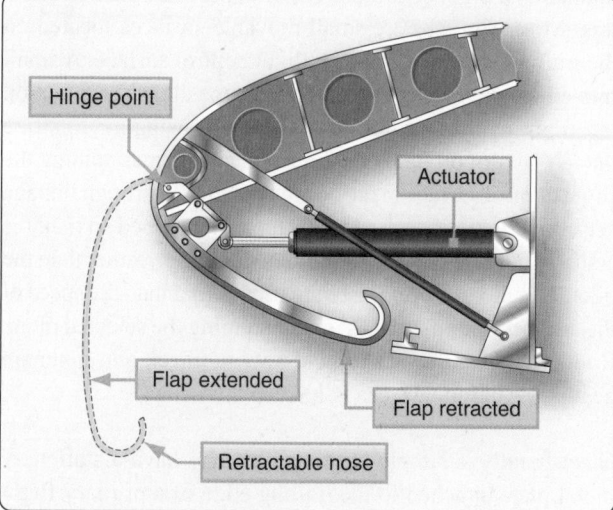

Figure 1-65. *Leading edge flaps.*

The differing designs of leading edge flaps essentially provide the same effect. Activation of the trailing edge flaps automatically deploys the leading edge flaps, which are driven out of the leading edge and downward, extending

the camber of the wing. *Figure 1-66* shows a Krueger flap, recognizable by its flat mid-section.

Slats

Another leading edge device which extends wing camber is a slat. Slats can be operated independently of the flaps with their own switch in the flight deck. Slats not only extend out of the leading edge of the wing increasing camber and lift, but most often, when fully deployed leave a slot between their trailing edges and the leading edge of the wing. *[Figure 1-67]* This increases the angle of attack at which the wing will maintain its laminar airflow, resulting in the ability to fly the aircraft slower with a reduced stall speed, and still maintain control.

Spoilers & Speed Brakes

A spoiler is a device found on the upper surface of many heavy and high-performance aircraft. It is stowed flush to the wing's upper surface. When deployed, it raises up into the airstream and disrupts the laminar airflow of the wing, thus reducing lift.

Spoilers are made with similar construction materials and techniques as the other flight control surfaces on the aircraft. Often, they are honeycomb-core flat panels. At low speeds, spoilers are rigged to operate when the ailerons operate to assist with the lateral movement and stability of the aircraft. On the wing where the aileron is moved up, the spoilers also raise thus amplifying the reduction of lift on that wing. *[Figure 1-68]* On the wing with downward aileron deflection, the spoilers remain stowed. As the speed of the aircraft increases, the ailerons become more effective and the spoiler interconnect disengages.

Spoilers are unique in that they may also be fully deployed on both wings to act as speed brakes. The reduced lift and increased drag can quickly reduce the speed of the aircraft in

Figure 1-66. *Side view (left) and front view (right) of a Krueger flap on a Boeing 737.*

Figure 1-67. *Air passing through the slot aft of the slat promotes boundary layer airflow on the upper surface at high angles of attack.*

flight. Dedicated speed brake panels similar to flight spoilers in construction can also be found on the upper surface of the wings of heavy and high-performance aircraft. They are designed specifically to increase drag and reduce the speed of the aircraft when deployed. These speed brake panels do not operate differentially with the ailerons at low speed. The speed brake control in the flight deck can deploy all spoiler and speed brake surfaces fully when operated. Often, these

Figure 1-68. *Spoilers deployed upon landing on a transport category aircraft.*

surfaces are also rigged to deploy on the ground automatically when engine thrust reversers are activated.

Tabs

The force of the air against a control surface during the high speed of flight can make it difficult to move and hold that control surface in the deflected position. A control surface might also be too sensitive for similar reasons. Several different tabs are used to aid with these types of problems. The table in *Figure 1-69* summarizes the various tabs and their uses.

While in flight, it is desirable for the pilot to be able to take their hands and feet off of the controls and have the aircraft maintain its flight condition. Trims tabs are designed to allow this. Most trim tabs are small movable surfaces located on the trailing edge of a primary flight control surface. A small movement of the tab in the direction opposite of the direction the flight control surface is deflected, causing air to strike the tab, in turn producing a force that aids in maintaining the flight control surface in the desired position. Through linkage set from the flight deck, the tab can be positioned so that it is actually holding the control surface in position rather than the pilot. Therefore, elevator tabs are used to maintain the speed of the aircraft since they assist in maintaining the selected pitch. Rudder tabs can be set to hold yaw in check and maintain heading. Aileron tabs can help keep the wings level.

Occasionally, a simple light aircraft may have a stationary metal plate attached to the trailing edge of a primary flight control, usually the rudder. This is also a trim tab as shown in *Figure 1-70*. It can be bent slightly on the ground to trim the aircraft in flight to a hands-off condition when flying straight and level. The correct amount of bend can be determined only by flying the aircraft after an adjustment. Note that a small amount of bending is usually sufficient.

The aerodynamic phenomenon of moving a trim tab in one direction to cause the control surface to experience a force moving in the opposite direction is exactly what occurs with the use of balance tabs. *[Figure 1-71]* Often, it is difficult to move a primary control surface due to its surface area and the speed of the air rushing over it. Deflecting a balance tab hinged at the trailing edge of the control surface in the opposite direction of the desired control surface movement causes a force to position the surface in the proper direction with reduced force to do so. Balance tabs are usually linked directly to the control surface linkage so that they move automatically when there is an input for control surface movement. They also can double as trim tabs, if adjustable in the flight deck.

A servo tab is similar to a balance tab in location and effect, but it is designed to operate the primary flight control surface, not just reduce the force needed to do so. It is usually used as

Flight Control Tabs			
Type	**Direction of Motion** (in relation to control surface)	**Activation**	**Effect**
Trim	Opposite	Set by pilot from cockpit. Uses independent linkage.	Statically balances the aircraft in flight. Allows "hands off" maintenance of flight condition.
Balance	Opposite	Moves when pilot moves control surface. Coupled to control surface linkage.	Aids pilot in overcoming the force needed to move the control surface.
Servo	Opposite	Directly linked to flight control input device. Can be primary or back-up means of control.	Aerodynamically positions control surfaces that require too much force to move manually.
Anti-balance or Anti-servo	Same	Directly linked to flight control input device.	Increases force needed by pilot to change flight control position. De-sensitizes flight controls.
Spring	Opposite	Located in line of direct linkage to servo tab. Spring assists when control forces become too high in high-speed flight.	Enables moving control surface when forces are high. Inactive during slow flight.

Figure 1-69. *Various tabs and their uses.*

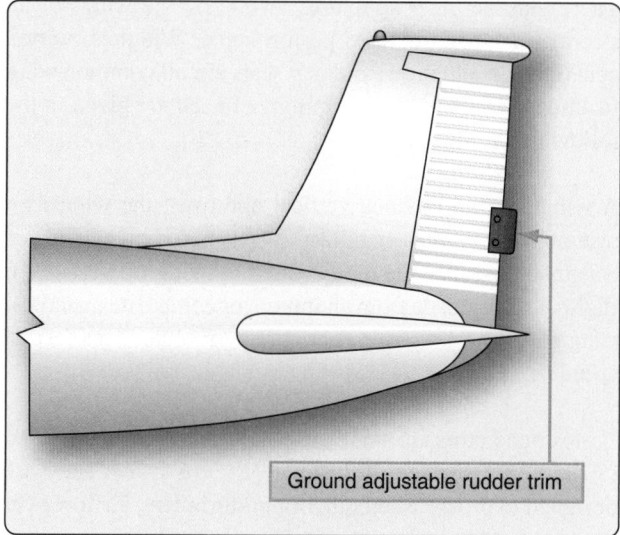

Ground adjustable rudder trim

Figure 1-70. *Example of a trim tab.*

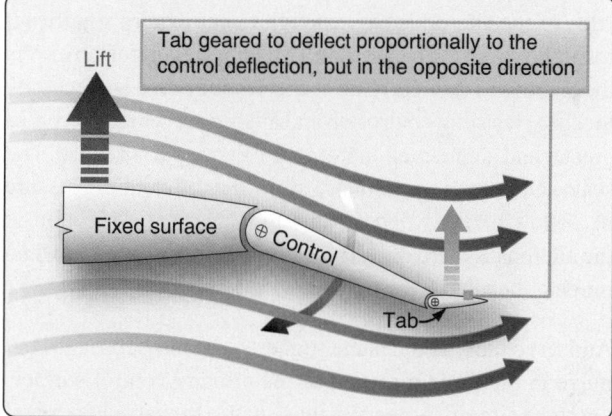

Tab geared to deflect proportionally to the control deflection, but in the opposite direction

Lift

Fixed surface

Control

Tab

Figure 1-71. *Balance tabs assist with forces needed to position control surfaces.*

a means to back up the primary control of the flight control surfaces. *[Figure 1-72]*

On heavy aircraft, large control surfaces require too much force to be moved manually and are usually deflected out of the neutral position by hydraulic actuators. These power control units are signaled via a system of hydraulic valves connected to the yoke and rudder pedals. On fly-by-wire aircraft, the hydraulic actuators that move the flight control surfaces are signaled by electric input. In the case of hydraulic system failure(s), manual linkage to a servo tab can be used to deflect it. This, in turn, provides an aerodynamic force that moves the primary control surface.

A control surface may require excessive force to move only in the final stages of travel. When this is the case, a spring tab can be used. This is essentially a servo tab that does not activate until an effort is made to move the control surface beyond a certain point. When reached, a spring in line of the control linkage aids in moving the control surface through the remainder of its travel. *[Figure 1-73]*

Figure 1-74 shows another way of assisting the movement of an aileron on a large aircraft. It is called an aileron balance panel. Not visible when approaching the aircraft, it is positioned in the linkage that hinges the aileron to the wing.

Balance panels have been constructed typically of aluminum skin-covered frame assemblies or aluminum honeycomb structures. The trailing edge of the wing just forward of the leading edge of the aileron is sealed to allow controlled airflow

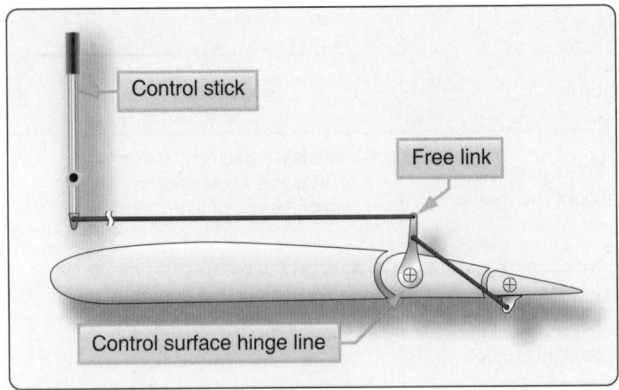

Figure 1-72. *Servo tabs can be used to position flight control surfaces in case of hydraulic failure.*

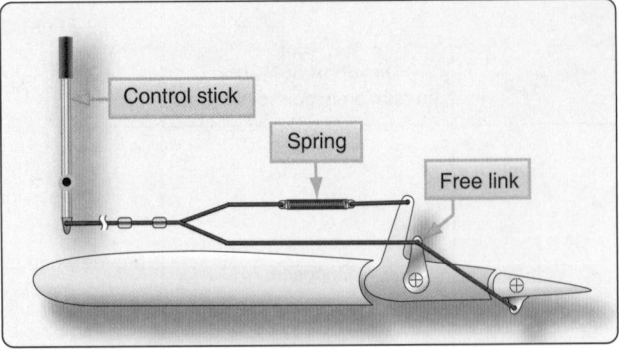

Figure 1-73. *Many tab linkages have a spring tab that kicks in as the forces needed to deflect a control increase with speed and the angle of desired deflection.*

in and out of the hinge area where the balance panel is located. *[Figure 1-75]* When the aileron is moved from the neutral position, differential pressure builds up on one side of the balance panel. This differential pressure acts on the balance panel in a direction that assists the aileron movement. For slight movements, deflecting the control tab at the trailing edge of the aileron is easy enough to not require significant assistance from the balance tab. (Moving the control tab moves the ailerons as desired.) But, as greater deflection is requested, the force resisting control tab and aileron movement becomes greater and augmentation from the balance tab is needed. The seals and mounting geometry allow the differential pressure of airflow on the balance panel to increase as deflection of the ailerons is increased. This makes the resistance felt when moving the aileron controls relatively constant.

Antiservo tabs, as the name suggests, are like servo tabs but move in the same direction as the primary control surface. On some aircraft, especially those with a movable horizontal stabilizer, the input to the control surface can be too sensitive. An antiservo tab tied through the control linkage creates an aerodynamic force that increases the effort needed to move the control surface. This makes flying the aircraft more

stable for the pilot. *Figure 1-76* shows an antiservo tab in the near neutral position. Deflected in the same direction as the desired stabilator movement, it increases the required control surface input.

Other Wing Features

There may be other structures visible on the wings of an aircraft that contribute to performance. Winglets, vortex generators, stall fences, and gap seals are all common wing features. Introductory descriptions of each are given in the following paragraphs.

A winglet is an obvious vertical upturn of the wing's tip resembling a vertical stabilizer. It is an aerodynamic device designed to reduce the drag created by wing tip vortices in flight. Usually made from aluminum or composite materials, winglets can be designed to optimize performance at a desired speed. *[Figure 1-77]*

Vortex generators are small airfoil sections usually attached to the upper surface of a wing. *[Figure 1-78]* They are designed to promote smooth, or non-turbulent, airflow over the wing and control surfaces. Usually made of aluminum

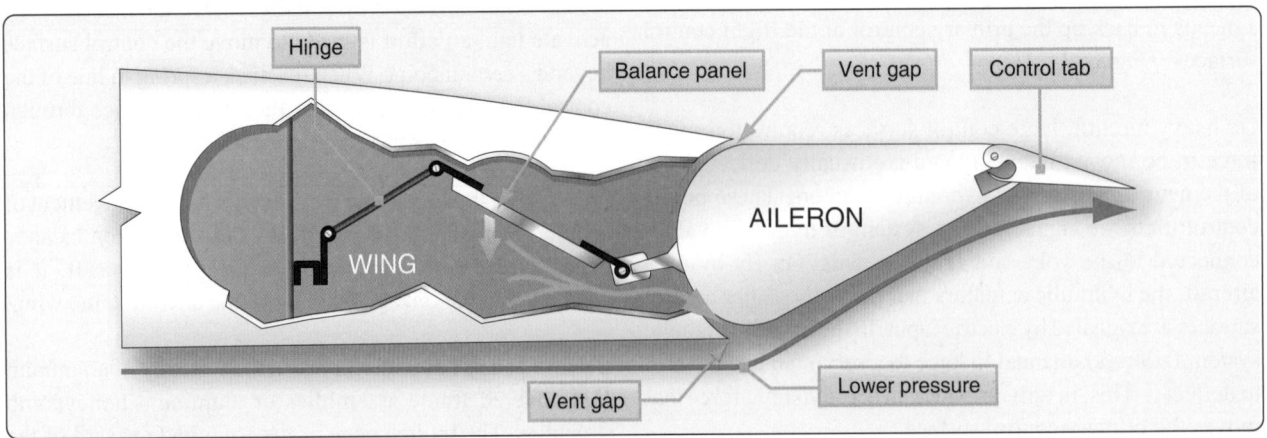

Figure 1-74. *An aileron balance panel and linkage uses varying air pressure to assist in control surface positioning.*

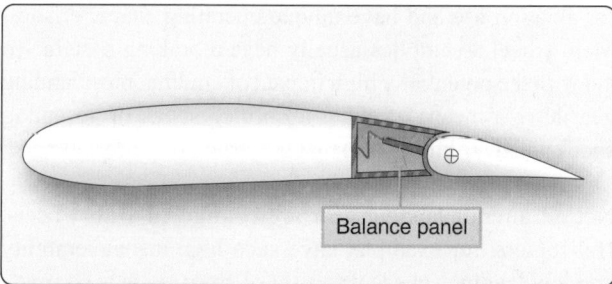

Figure 1-75. *The trailing edge of the wing just forward of the leading edge of the aileron is sealed to allow controlled airflow in and out of the hinge area where the balance panel is located.*

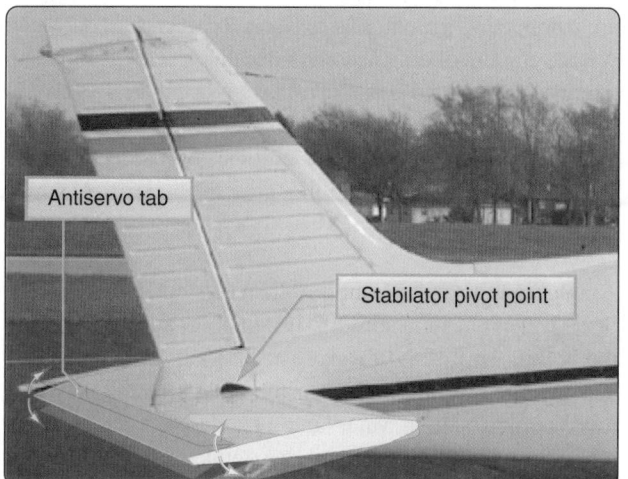

Figure 1-76. *An antiservo tab moves in the same direction as the control tab. Shown here on a stabilator, it desensitizes the pitch control.*

Figure 1-77. *A winglet reduces aerodynamic drag caused by air spilling off of the wing tip.*

Figure 1-78. *Vortex generators.*

Figure 1-79. *The Symphony SA-160 has two unique vortex generators on its wing to ensure aileron effectiveness through the stall.*

and installed in a spanwise line or lines, the vortices created by these devices swirl downward assisting maintenance of the boundary layer of air flowing over the wing. They can also be found on the fuselage and empennage. *Figure 1-79* shows the unique vortex generators on a Symphony SA-160 wing.

A chordwise barrier on the upper surface of the wing, called a stall fence, is used to halt the spanwise flow of air. During low speed flight, this can maintain proper chordwise airflow reducing the tendency for the wing to stall. Usually made of aluminum, the fence is a fixed structure most common on swept wings, which have a natural spanwise tending boundary air flow. *[Figure 1-80]*

Often, a gap can exist between the stationary trailing edge of a wing or stabilizer and the movable control surface(s). At high angles of attack, high pressure air from the lower wing surface can be disrupted at this gap. The result can be turbulent airflow, which increases drag. There is also a tendency for some lower wing boundary air to enter the gap and disrupt the upper wing surface airflow, which in turn reduces lift and control surface responsiveness. The use of gap seals is common to promote smooth airflow in these gap areas. Gap seals can be made of a wide variety of materials

ranging from aluminum and impregnated fabric to foam and plastic. *Figure 1-81* shows some gap seals installed on various aircraft.

Landing Gear

The landing gear supports the aircraft during landing and while it is on the ground. Simple aircraft that fly at low speeds generally have fixed gear. This means the gear is stationary and does not retract for flight. Faster, more complex aircraft

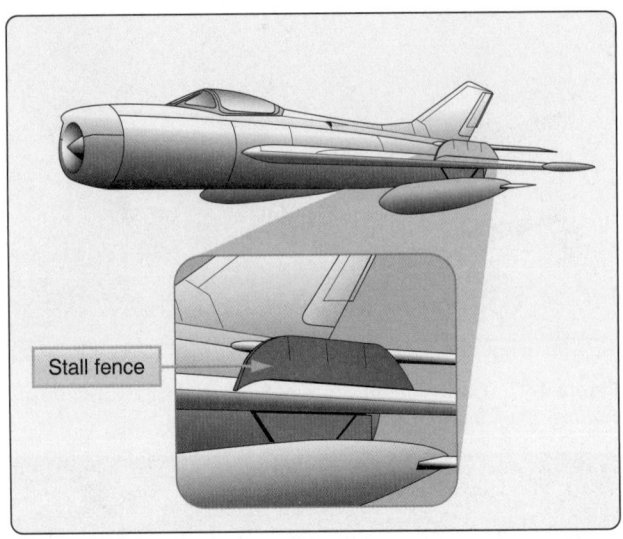

Figure 1-80. *A stall fence aids in maintaining chordwise airflow over the wing.*

for aviation use and have unique operating characteristics. Main wheel assemblies usually have a braking system. To aid with the potentially high impact of landing, most landing gear have a means of either absorbing shock or accepting shock and distributing it so that the structure is not damaged.

Not all aircraft landing gear are configured with wheels. Helicopters, for example, have such high maneuverability and low landing speeds that a set of fixed skids is common and quite functional with lower maintenance. The same is true for free balloons which fly slowly and land on wood skids affixed to the floor of the gondola. Other aircraft landing gear are equipped with pontoons or floats for operation on water. A large amount of drag accompanies this type of gear, but an aircraft that can land and take off on water can be very useful in certain environments. Even skis can be found under some aircraft for operation on snow and ice. *Figure 1-83* shows some of these alternative landing gear, the majority of which are the fixed gear type.

Amphibious aircraft are aircraft than can land either on land or on water. On some aircraft designed for such dual usage, the bottom half of the fuselage acts as a hull. Usually, it is accompanied by outriggers on the underside of the wings near the tips to aid in water landing and taxi. Main gear that retract into the fuselage are only extended when landing on the ground or a runway. This type of amphibious aircraft is sometimes called a flying boat. *[Figure 1-84]*

Many aircraft originally designed for land use can be fitted with floats with retractable wheels for amphibious use. *[Figure 1-85]* Typically, the gear retracts into the float when not needed. Sometimes a dorsal fin is added to the aft underside of the fuselage for longitudinal stability during water operations. It is even possible on some aircraft to direct this type of fin by tying its control into the aircraft's rudder pedals. Skis can also be fitted with wheels that retract to allow landing on solid ground or on snow and ice.

have retractable landing gear. After takeoff, the landing gear is retracted into the fuselage or wings and out of the airstream. This is important because extended gear create significant parasite drag which reduces performance. Parasite drag is caused by the friction of the air flowing over the gear. It increases with speed. On very light, slow aircraft, the extra weight that accompanies a retractable landing gear is more of a detriment than the drag caused by the fixed gear. Lightweight fairings and wheel pants can be used to keep drag to a minimum. *Figure 1-82* shows examples of fixed and retractable gear.

Landing gear must be strong enough to withstand the forces of landing when the aircraft is fully loaded. In addition to strength, a major design goal is to have the gear assembly be as light as possible. To accomplish this, landing gear are made from a wide range of materials including steel, aluminum, and magnesium. Wheels and tires are designed specifically

Figure 1-81. *Gap seals promote the smooth flow of air over gaps between fixed and movable surfaces.*

Figure 1-84. *An amphibious aircraft is sometimes called a flying boat because the fuselage doubles as a hull.*

Figure 1-82. *Landing gear can be fixed (top) or retractable (bottom).*

Tail Wheel Gear Configuration

There are two basic configurations of airplane landing gear: conventional gear or tail wheel gear and the tricycle gear. Tail wheel gear dominated early aviation and therefore has become known as conventional gear. In addition to its

two main wheels which are positioned under most of the weight of the aircraft, the conventional gear aircraft also has a smaller wheel located at the aft end of the fuselage. *[Figure 1-86]* Often this tail wheel is able to be steered by rigging cables attached to the rudder pedals. Other conventional gear have no tail wheel at all using just a steel skid plate under the aft fuselage instead. The small tail wheel or skid plate allows the fuselage to incline, thus giving clearance for the long propellers that prevailed in aviation through WWII. It also gives greater clearance between the propeller and loose debris when operating on an unpaved runway. But the inclined fuselage blocks the straight-ahead vision of the pilot during ground operations. Until up to speed where the elevator becomes effective to lift the tail wheel off the ground, the pilot must lean his head out the side of the flight deck to see directly ahead of the aircraft.

The use of tail wheel gear can pose another difficulty. When

Figure 1-83. *Aircraft landing gear without wheels.*

Figure 1-85. *Retractable wheels make this aircraft amphibious.*

landing, tail wheel aircraft can easily ground loop. A ground loop is when the tail of the aircraft swings around and comes forward of the nose of the aircraft. The reason this happens is due to the two main wheels being forward of the aircraft's center of gravity. The tail wheel is aft of the center of gravity. If the aircraft swerves upon landing, the tail wheel can swing out to the side of the intended path of travel. If far enough to the side, the tail can pull the center of gravity out from its desired location slightly aft of but between the main gear. Once the center of gravity is no longer trailing the mains, the tail of the aircraft freely pivots around the main wheels causing the ground loop.

Conventional gear is useful and is still found on certain models of aircraft manufactured today, particularly aerobatic aircraft, crop dusters, and aircraft designed for unpaved runway use. It is typically lighter than tricycle gear which requires a stout, fully shock absorbing nose wheel assembly. The tail wheel configuration excels when operating out of unpaved runways. With the two strong main gear forward providing stability and directional control during takeoff roll, the lightweight tail wheel does little more than keep the aft end of the fuselage from striking the ground. As mentioned, at a certain speed,

Figure 1-86. *An aircraft with tail wheel gear.*

the air flowing over the elevator is sufficient for it to raise the tail off the ground. As speed increases further, the two main wheels under the center of gravity are very stable.

Tricycle Gear Configuration

Tricycle gear is the most prevalent landing gear configuration in aviation. In addition to the main wheels, a shock absorbing nose wheel is at the forward end of the fuselage. Thus, the center of gravity is then forward of the main wheels. The tail of the aircraft is suspended off the ground and clear view straight ahead from the flight deck is given. Ground looping is nearly eliminated since the center of gravity follows the directional nose wheel and remains between the mains.

Light aircraft use tricycle gear, as well as heavy aircraft. Twin nose wheels on the single forward strut and massive multistrut/multiwheel main gear may be found supporting the world's largest aircraft, but the basic configuration is still tricycle. The nose wheel may be steered with the rudder pedals on small aircraft. Larger aircraft often have a nose wheel steering wheel located off to the side of the flight deck. *Figure 1-87* shows aircraft with tricycle gear. Chapter 13, Aircraft Landing Gear Systems, discusses landing gear in detail.

Maintaining the Aircraft

Maintenance of an aircraft is of the utmost importance for safe flight. Certificated technicians are committed to perform timely maintenance functions in accordance with the manufacturer's instructions and under Title 14 of the Code of Federal Regulations (14 CFR). At no time is an act of aircraft maintenance taken lightly or improvised. The consequences of such action could be fatal, and the technician could lose their certificate and face criminal charges.

Airframe, engine, and aircraft component manufacturers are responsible for documenting the maintenance procedures that guide managers and technicians on when and how to perform maintenance on their products. A small aircraft may only require a few manuals, including the aircraft maintenance manual. This volume usually contains the most frequently used information required to maintain the aircraft properly. The Type Certificate Data Sheet (TCDS) for an aircraft also contains critical information. Complex and large aircraft require several manuals to convey correct maintenance procedures adequately. In addition to the maintenance manual, manufacturers may produce such volumes as structural repair manuals, overhaul manuals, wiring diagram manuals, component manuals, and more.

Note that the use of the word "manual" is meant to include electronic as well as printed information. Also, proper maintenance extends to the use of designated tools and fixtures called out in the manufacturer's maintenance documents. In

Figure 1-87. *Tricycle landing gear is the most predominant landing gear configuration in aviation.*

the past, not using the proper tooling has caused damage to critical components, which subsequently failed and led to aircraft crashes and the loss of human life. The technician is responsible for using the correct information, procedures, and tools needed to perform appropriate maintenance or repairs.

Standard aircraft maintenance procedures do exist and can be used by the technician when performing maintenance or a repair. These are found in the Federal Aviation Administration (FAA) approved advisory circulars (AC) 43.13-2, *Acceptable Methods, Techniques, and Practices - Aircraft Alterations* and AC 43.13-1, *Acceptable Methods, Techniques, and Practices - Aircraft Inspection and Repair*. If not addressed by the manufacturer's literature, the technician may use the procedures outlined in these manuals to complete the work in an acceptable manner. These procedures are not specific to any aircraft or component and typically cover methods used during maintenance of all aircraft. Note that the manufacturer's instructions supersede the general procedures found in AC 43.13-2 and AC 43.13-1.

All maintenance related actions on an aircraft or component are required to be documented by the performing technician in the aircraft or component logbook. Light aircraft may have only one logbook for all work performed. Some aircraft may have a separate engine logbook for any work performed on the engine(s). Other aircraft have separate propeller logbooks. Large aircraft require volumes of maintenance documentation comprised of thousands of procedures performed by hundreds of technicians. Electronic dispatch and recordkeeping of

maintenance performed on large aircraft such as airliners is common. The importance of correct maintenance recordkeeping should not be overlooked.

Location Numbering Systems

Even on small, light aircraft, a method of precisely locating each structural component is required. Various numbering systems are used to facilitate the location of specific wing frames, fuselage bulkheads, or any other structural members on an aircraft. Most manufacturers use some system of station marking. For example, the nose of the aircraft may be designated "zero station," and all other stations are located at measured distances in inches behind the zero station. Thus, when a blueprint reads "fuselage frame station 137," that particular frame station can be located 137 inches behind the nose of the aircraft.

To locate structures to the right or left of the center line of an aircraft, a similar method is employed. Many manufacturers consider the center line of the aircraft to be a zero station from which measurements can be taken to the right or left to locate an airframe member. This is often used on the horizontal stabilizer and wings.

The applicable manufacturer's numbering system and abbreviated designations or symbols should always be reviewed before attempting to locate a structural member. They are not always the same. The following list includes location designations typical of those used by many manufacturers.

- Fuselage stations (Fus. Sta. or FS) are numbered in inches from a reference or zero point known as the reference datum. *[Figure 1-88]* The reference datum is an imaginary vertical plane at or near the nose of the aircraft from which all fore and aft distances are measured. The distance to a given point is measured in inches parallel to a center line extending through the aircraft from the nose through the center of the tail cone. Some manufacturers may call the fuselage station a body station, abbreviated BS.

- Buttock line or butt line (BL) is a vertical reference plane down the center of the aircraft from which measurements left or right can be made. *[Figure 1-89]*

- Water line (WL) is the measurement of height in inches perpendicular from a horizontal plane usually located at the ground, cabin floor, or some other easily referenced location. *[Figure 1-90]*

- Aileron station (AS) is measured outboard from, and parallel to, the inboard edge of the aileron, perpendicular to the rear beam of the wing.

- Flap station (KS) is measured perpendicular to the rear

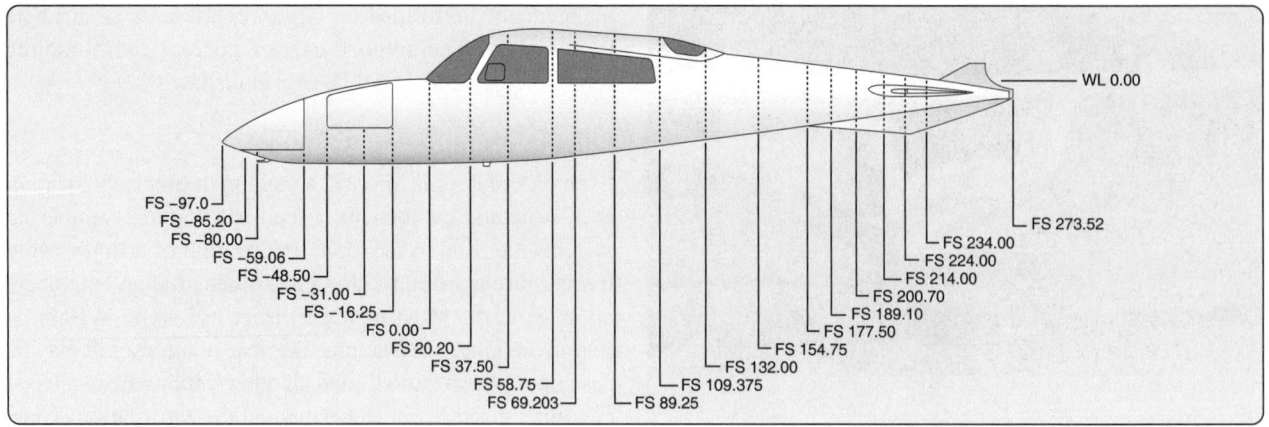

Figure 1-88. *The various fuselage stations relative to a single point of origin illustrated in inches or some other measurement (if of foreign development).*

beam of the wing and parallel to, and outboard from, the inboard edge of the flap.

- Nacelle station (NC or Nac. Sta.) is measured either forward of or behind the front spar of the wing and perpendicular to a designated water line.

In addition to the location stations listed above, other measurements are used, especially on large aircraft. Thus, there may be horizontal stabilizer stations (HSS), vertical stabilizer stations (VSS) or powerplant stations (PPS). *[Figure 1-91]* In every case, the manufacturer's terminology and station location system should be consulted before locating a point on a particular aircraft.

Another method is used to facilitate the location of aircraft components on air transport aircraft. This involves dividing the aircraft into zones. These large areas or major zones are further divided into sequentially numbered zones and subzones. The digits of the zone number are reserved and indexed to indicate the location and type of system of which the component is a part. *Figure 1-92* illustrates these zones and subzones on a transport category aircraft.

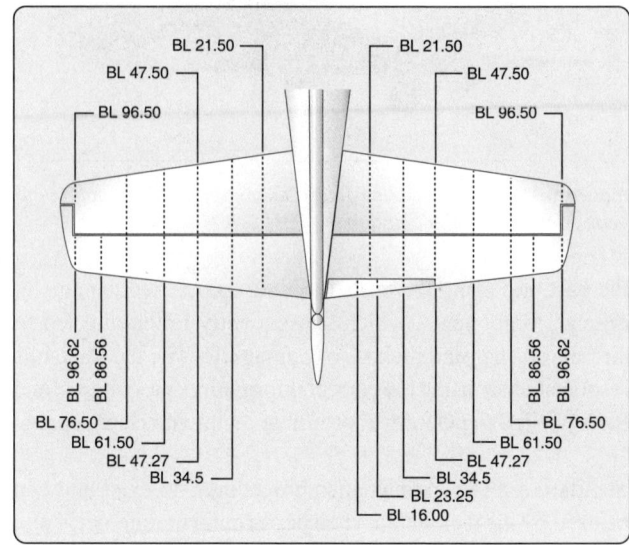

Figure 1-89. *Butt line diagram of a horizontal stabilizer.*

Access & Inspection Panels

Quick access to the accessories and other equipment carried in the fuselage is provided for by numerous access doors, inspection plates, landing wheel wells, and other openings.

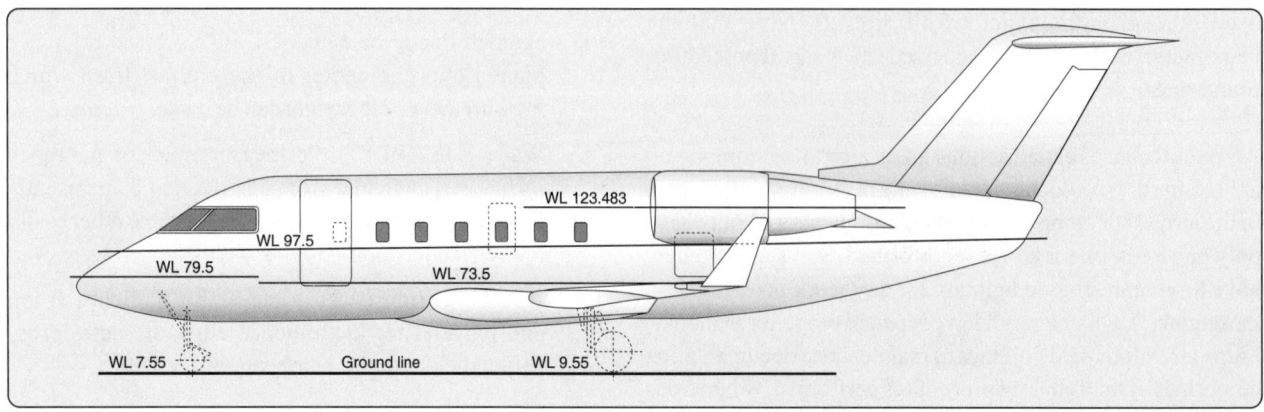

Figure 1-90. *Water line diagram.*

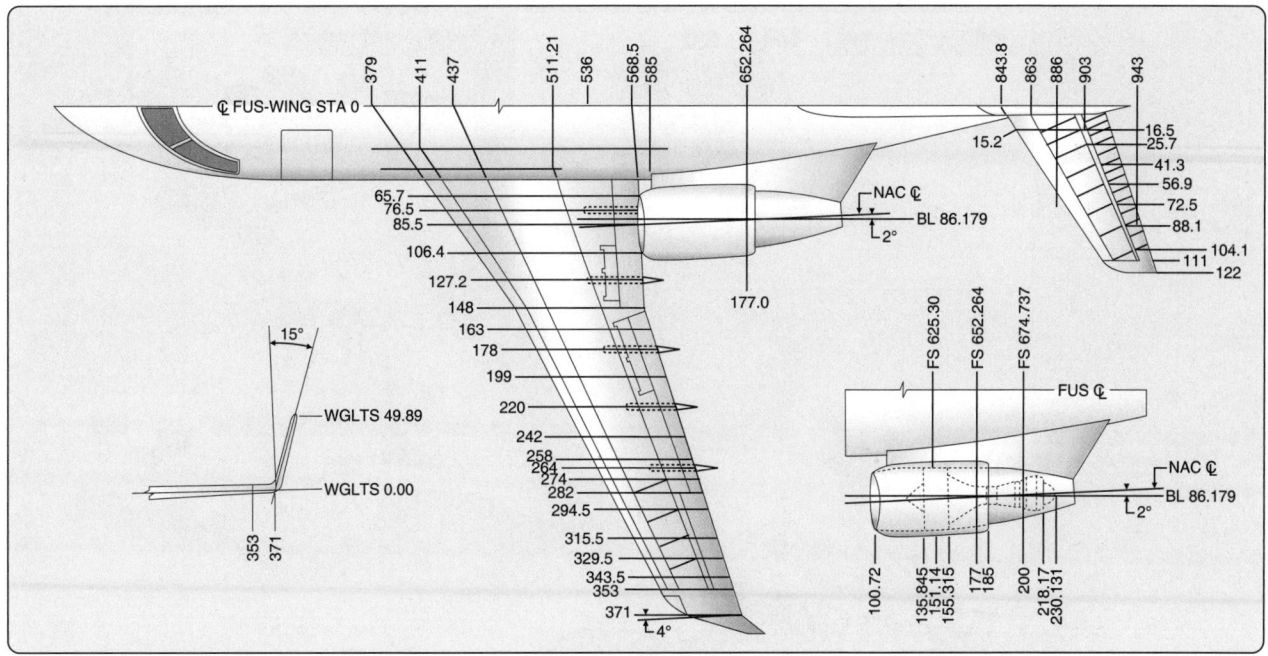

Figure 1-91. *Wing stations are often referenced off the butt line, which bisects the center of the fuselage longitudinally. Horizontal stabilizer stations referenced to the butt line and engine nacelle stations are also shown.*

Servicing diagrams showing the arrangement of equipment and location of access doors are supplied by the manufacturer in the aircraft maintenance manual.

Knowing where a particular structure or component is located on an aircraft needs to be combined with gaining access to that area to perform the required inspections or maintenance. To facilitate this, access and inspection panels are located on most surfaces of the aircraft. Small panels that are hinged or removable allow inspection and servicing. Large panels and doors allow components to be removed and installed, as well as human entry for maintenance purposes.

The underside of a wing, for example, sometimes contains dozens of small panels through which control cable components can be monitored and fittings greased. Various drains and jack points may also be on the underside of the wing. The upper surface of the wings typically have fewer access panels because a smooth surface promotes better laminar airflow, which causes lift. On large aircraft, walkways are sometimes designated on the wing upper surface to permit safe navigation by mechanics and inspectors to critical structures and components located along the wing's leading and trailing edges. Wheel wells and special component bays are places where numerous components and accessories are grouped together for easy maintenance access.

Panels and doors on aircraft are numbered for positive identification. On large aircraft, panels are usually numbered sequentially containing zone and subzone information in the panel number. Designation for a left or right side location on the aircraft is often indicated in the panel number. This could be with an "L" or "R," or panels on one side of the aircraft could be odd numbered and the other side even numbered. The manufacturer's maintenance manual explains the panel numbering system and often has numerous diagrams and tables showing the location of various components and under which panel they may be found. Each manufacturer is entitled to develop its own panel numbering system.

Helicopter Structures

The structures of the helicopter are designed to give the helicopter its unique flight characteristics. A simplified explanation of how a helicopter flies is that the rotors are rotating airfoils that provide lift similar to the way wings provide lift on a fixed-wing aircraft. Air flows faster over the curved upper surface of the rotors, causing a negative pressure and thus, lifting the aircraft. Changing the angle of attack of the rotating blades increases or decreases lift, respectively raising or lowering the helicopter. Tilting the rotor plane of rotation causes the aircraft to move horizontally. *Figure 1-93* shows the major components of a typical helicopter.

Airframe

The airframe, or fundamental structure, of a helicopter can be made of either metal or wood composite materials, or some combination of the two. Typically, a composite component consists of many layers of fiber-impregnated resins, bonded to form a smooth panel. Tubular and sheet metal substructures are usually made of aluminum, though stainless steel or

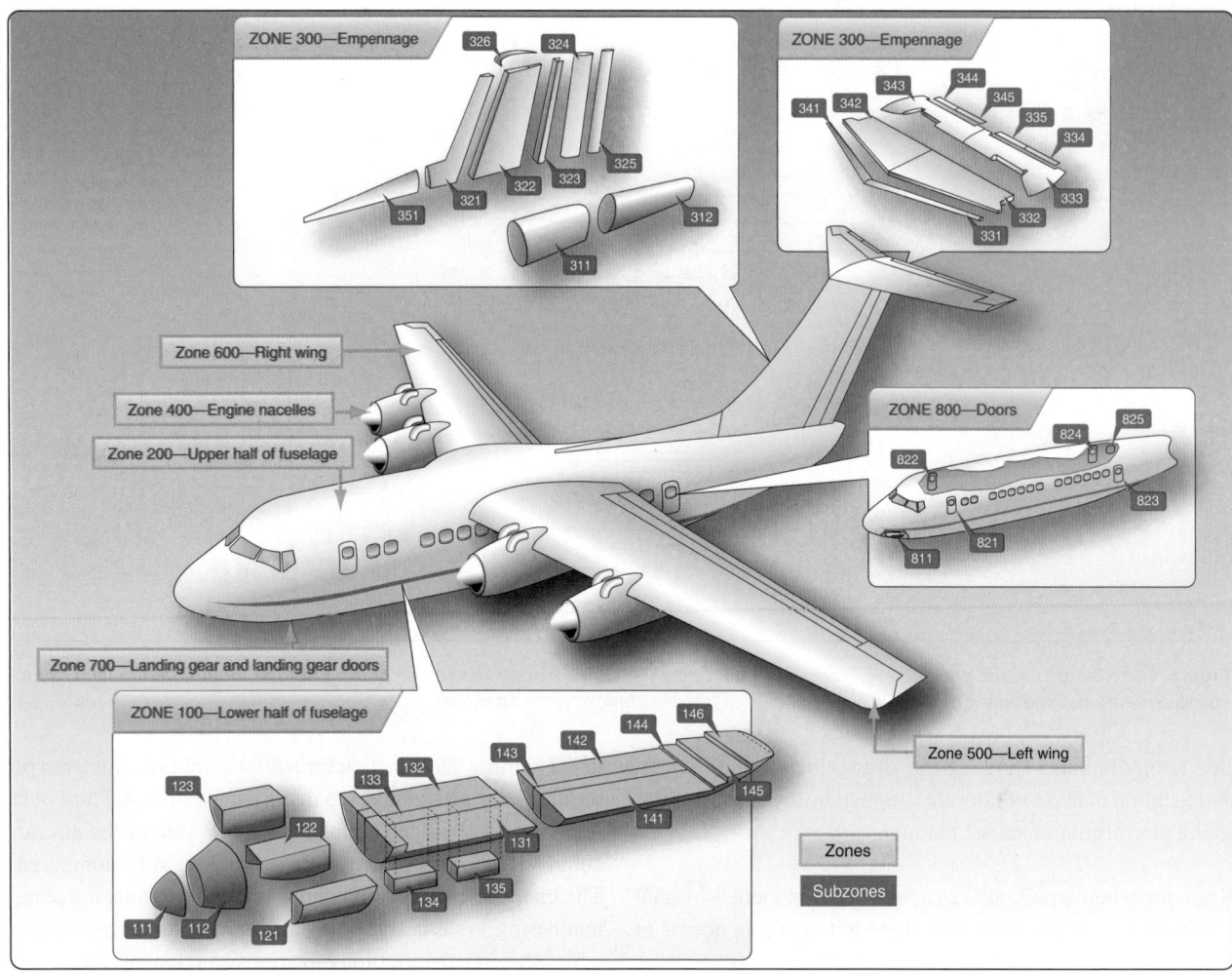

Figure 1-92. *Large aircraft are divided into zones and subzones for identifying the location of various components.*

titanium are sometimes used in areas subject to higher stress or heat. Airframe design encompasses engineering, aerodynamics, materials technology, and manufacturing methods to achieve favorable balances of performance, reliability, and cost.

Fuselage

As with fixed-wing aircraft, helicopter fuselages and tail booms are often truss-type or semimonocoque structures of stress-skin design. Steel and aluminum tubing, formed aluminum, and aluminum skin are commonly used. Modern helicopter fuselage design includes an increasing utilization of advanced composites as well. Firewalls and engine decks are usually stainless steel. Helicopter fuselages vary widely from those with a truss frame, two seats, no doors, and a monocoque shell flight compartment to those with fully enclosed airplane-style cabins as found on larger twin-engine helicopters. The multidirectional nature of helicopter flight makes wide-range visibility from the flight deck essential. Large, formed polycarbonate, glass, or plexiglass windscreens are common.

Landing Gear or Skids

As mentioned, a helicopter's landing gear can be simply a set of tubular metal skids. Many helicopters do have landing gear with wheels, some retractable.

Powerplant & Transmission

The two most common types of engine used in helicopters are the reciprocating engine and the turbine engine. Reciprocating engines, also called piston engines, are generally used in smaller helicopters. Most training helicopters use reciprocating engines because they are relatively simple and inexpensive to operate.

Turbine Engines

Turbine engines are more powerful and are used in a wide variety of helicopters. They produce a tremendous amount of power for their size but are generally more expensive to operate. The turbine engine used in helicopters operates differently than those used in airplane applications. In most applications, the exhaust outlets simply release expended

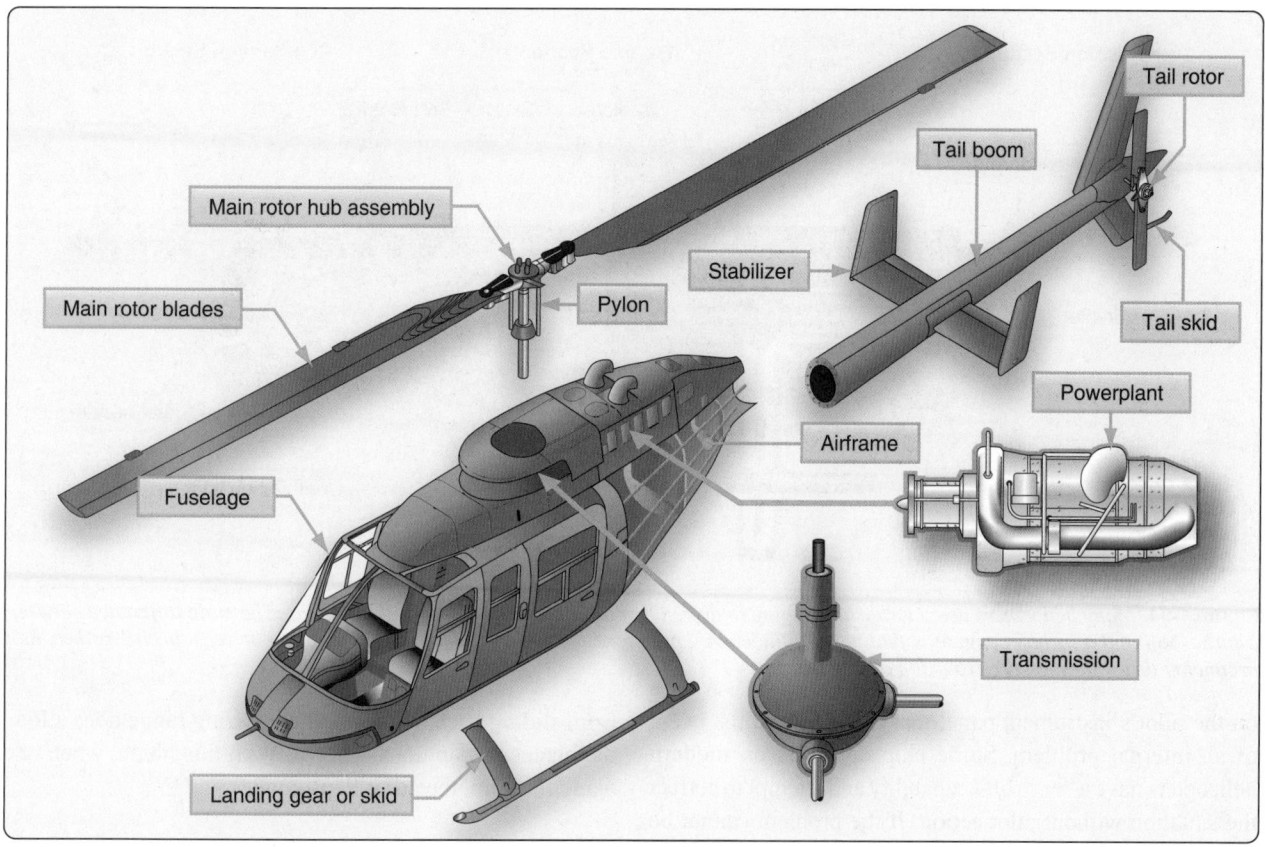

Figure 1-93. *The major components of a helicopter are the airframe, fuselage, landing gear, powerplant/transmission, main rotor system, and antitorque system.*

gases and do not contribute to the forward motion of the helicopter. Because the airflow is not a straight line pass through as in jet engines and is not used for propulsion, the cooling effect of the air is limited. Approximately 75 percent of the incoming airflow is used to cool the engine.

The gas turbine engine mounted on most helicopters is made up of a compressor, combustion chamber, turbine, and accessory gearbox assembly. The compressor draws filtered air into the plenum chamber and compresses it. Common type filters are centrifugal swirl tubes where debris is ejected outward and blown overboard prior to entering the compressor, or engine barrier filters (EBF), a paper element type filter, encased in a frame with a screen/grill over the inlet, and usually coated with an oil. This design significantly reduces the ingestion of foreign object debris (FOD). The compressed air is directed to the combustion section through discharge tubes where atomized fuel is injected into it. The air-fuel mixture is ignited and allowed to expand. This combustion gas is then forced through a series of turbine wheels causing them to turn. These turbine wheels provide power to both the engine compressor and the accessory gearbox. Depending on model and manufacturer, the rpm range can vary from a range low of 20,000 to a range high of 51,600.

Power is provided to the main rotor and tail rotor systems through the freewheeling unit which is attached to the accessory gearbox power output gear shaft. The combustion gas is finally expelled through an exhaust outlet. The temperature of gas is measured at different locations and is referenced differently by each manufacturer. Some common terms are: inter-turbine temperature (ITT), exhaust gas temperature (EGT), or turbine outlet temperature (TOT). TOT is used throughout this discussion for simplicity purposes. *[Figure 1-94]*

Transmission

The transmission system transfers power from the engine to the main rotor, tail rotor, and other accessories during normal flight conditions. The main components of the transmission system are the main rotor transmission, tail rotor drive system, clutch, and freewheeling unit. The freewheeling unit, or autorotative clutch, allows the main rotor transmission to drive the tail rotor drive shaft during autorotation. Helicopter transmissions are normally lubricated and cooled with their own oil supply. A sight gauge is provided to check the oil level. Some transmissions have chip detectors located in the sump. These detectors are wired to warning lights located

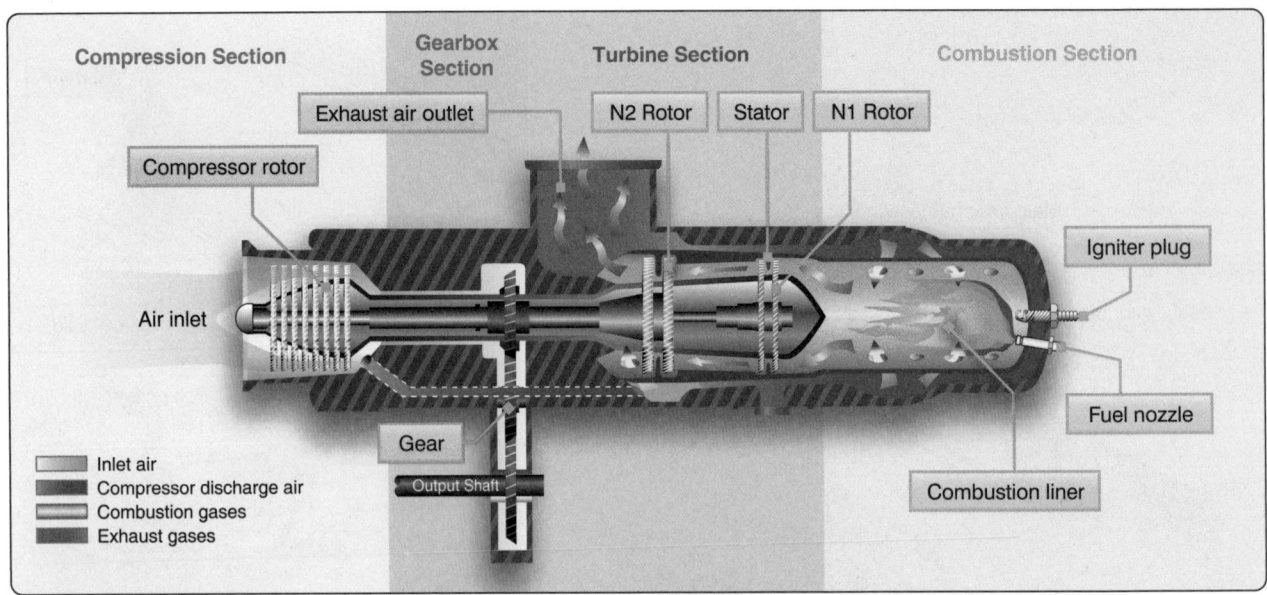

Figure 1-94. *Many helicopters use a turboshaft engine to drive the main transmission and rotor systems. The main difference between a turboshaft and a turbojet engine is that most of the energy produced by the expanding gases is used to drive a turbine rather than producing thrust through the expulsion of exhaust gases.*

on the pilot's instrument panel that illuminate in the event of an internal problem. Some chip detectors on modern helicopters have a "burn off" capability and attempt to correct the situation without pilot action. If the problem cannot be corrected on its own, the pilot must refer to the emergency procedures for that particular helicopter.

Main Rotor System

The rotor system is the rotating part of a helicopter which generates lift. The rotor consists of a mast, hub, and rotor blades. The mast is a cylindrical metal shaft that extends upwards from and is driven, and sometimes supported, by the transmission. At the top of the mast is the attachment point for the rotor blades called the hub. The rotor blades are then attached to the hub by any number of different methods. Main rotor systems are classified according to how the main rotor blades are attached and move relative to the main rotor hub. There are three basic classifications: rigid, semirigid, or fully articulated.

Rigid Rotor System

The simplest is the rigid rotor system. In this system, the rotor blades are rigidly attached to the main rotor hub and are not free to slide back and forth (drag) or move up and down (flap). *[Figure 1-95]* The forces tending to make the rotor blades do so are absorbed by the flexible properties of the blade. The pitch of the blades, however, can be adjusted by rotation about the spanwise axis via the feathering hinges.

Semirigid Rotor System

The semirigid rotor system in *Figure 1-96* makes use of a teetering hinge at the blade attach point. While held in check

from sliding back and forth, the teetering hinge does allow the blades to flap up and down. With this hinge, when one blade flaps up, the other flaps down.

Flapping is caused by a phenomenon known as dissymmetry of lift. As the plane of rotation of the rotor blades is tilted and the helicopter begins to move forward, an advancing blade and a retreating blade become established (on two-bladed systems). The relative windspeed is greater on an advancing blade than it is on a retreating blade. This causes greater lift to be developed on the advancing blade, causing it to rise up or flap. When blade rotation reaches the point where the blade becomes the retreating blade, the extra lift is lost and the blade flaps downward. *[Figure 1-97]*

Fully Articulated Rotor System

Fully articulated rotor blade systems provide hinges that allow the rotors to move fore and aft, as well as up and down. This lead-lag, drag, or hunting movement as it is called is in response to the Coriolis effect during rotational speed changes. When first starting to spin, the blades lag until centrifugal force is fully developed. Once rotating, a reduction in speed causes the blades to lead the main rotor hub until forces come into balance. Constant fluctuations in rotor blade speeds cause the blades to "hunt." They are free to do so in a fully articulating system due to being mounted on the vertical drag hinge.

One or more horizontal hinges provide for flapping on a fully articulated rotor system. Also, the feathering hinge allows blade pitch changes by permitting rotation about the spanwise axis. Various dampers and stops can be found on

1-42

Figure 1-95. *Four-blade hingeless (rigid) main rotor. The hub is a single piece of forged rigid titanium.*

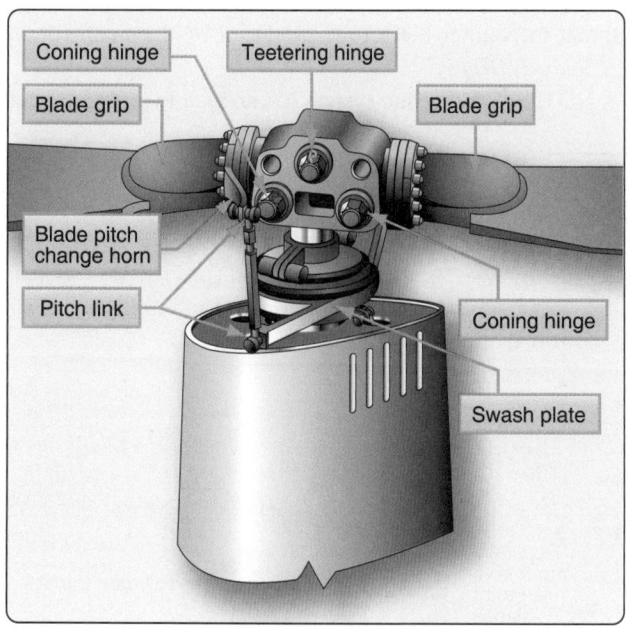

Figure 1-96. *The semirigid rotor system of the Robinson R22.*

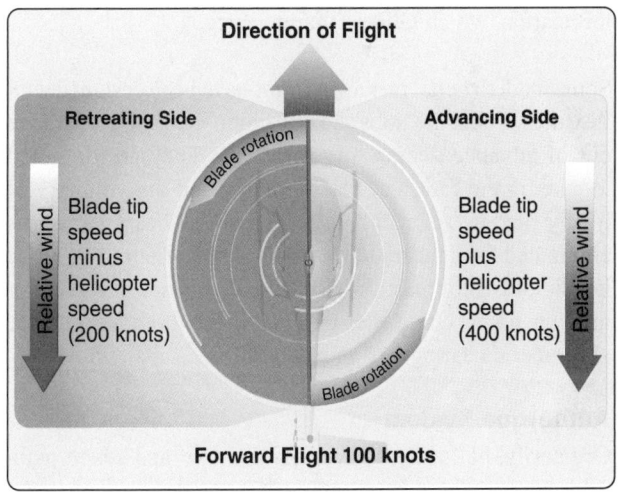

Figure 1-97. *The blade tip speed of this helicopter is approximately 300 knots. If the helicopter is moving forward at 100 knots, the relative windspeed on the advancing side is 400 knots. On the retreating side, it is only 200 knots. This difference in speed causes a dissymmetry of lift.*

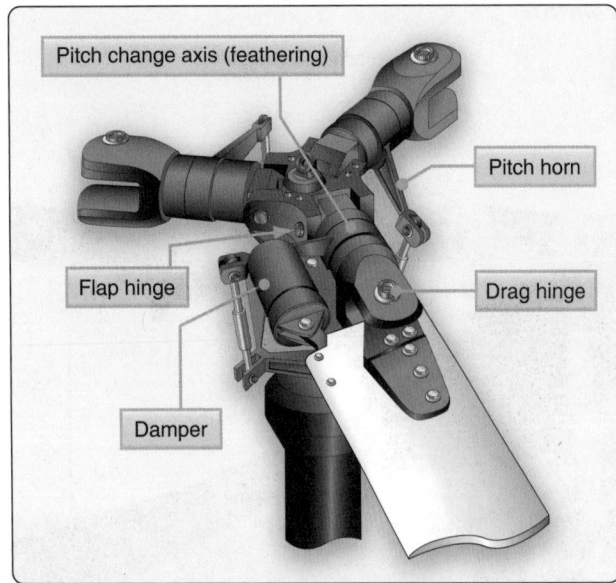

Figure 1-98. *Fully articulated rotor system.*

Figure 1-99. *Five-blade articulated main rotor with elastomeric bearings.*

different designs to reduce shock and limit travel in certain directions. *Figure 1-98* shows a fully articulated main rotor system with the features discussed.

Numerous designs and variations on the three types of main rotor systems exist. Engineers continually search for ways to reduce vibration and noise caused by the rotating parts of the helicopter. Toward that end, the use of elastomeric bearings in main rotor systems is increasing. These polymer bearings have the ability to deform and return to their original shape. As such, they can absorb vibration that would normally be transferred by steel bearings. They also do not require regular lubrication, which reduces maintenance.

Some modern helicopter main rotors have been designed with flextures. These are hubs and hub components that are made out of advanced composite materials. They are designed to take up the forces of blade hunting and dissymmetry of lift by flexing. As such, many hinges and bearings can be eliminated from the traditional main rotor system. The result is a simpler rotor mast with lower maintenance due to fewer moving parts. Often, designs using flextures incorporate elastomeric bearings. *[Figure 1-99]*

Antitorque System

Ordinarily, helicopters have between two and seven main rotor blades. These rotors are usually made of a composite structure. The large rotating mass of the main rotor blades of a helicopter produce torque. This torque increases with engine power and tries to spin the fuselage in the opposite direction. The tail boom and tail rotor, or antitorque rotor, counteract this torque effect. *[Figure 1-100]* Controlled with foot pedals, the countertorque of the tail rotor must be modulated as engine power levels are changed. This is done

by changing the pitch of the tail rotor blades. This, in turn, changes the amount of countertorque, and the aircraft can be rotated about its vertical axis, allowing the pilot to control the direction the helicopter is facing.

Similar to a vertical stabilizer on the empennage of an airplane, a fin or pylon is also a common feature on rotorcraft. Normally, it supports the tail rotor assembly, although some tail rotors are mounted on the tail cone of the boom. Additionally, a horizontal member called a stabilizer is often constructed at the tail cone or on the pylon.

A Fenestron® is a unique tail rotor design which is actually a multiblade ducted fan mounted in the vertical pylon. It works the same way as an ordinary tail rotor, providing sideways thrust to counter the torque produced by the main rotors. *[Figure 1-101]*

A NOTAR® antitorque system has no visible rotor mounted

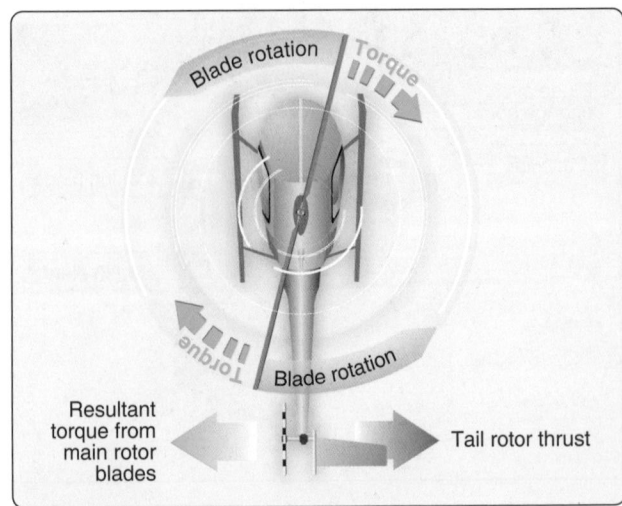

Figure 1-100. *A tail rotor is designed to produce thrust in a direction opposite to that of the torque produced by the rotation of the main rotor blades. It is sometimes called an antitorque rotor.*

Figure 1-101. *A Fenestron or "fan-in-tail" antitorque system. This design provides an improved margin of safety during ground operations.*

on the tail boom. Instead, an engine-driven adjustable fan is located inside the tail boom. NOTAR® is an acronym that stands for "no tail rotor." As the speed of the main rotor changes, the speed of the NOTAR® fan changes. Air is vented out of two long slots on the right side of the tail boom, entraining main rotor wash to hug the right side of the tail boom, in turn causing laminar flow and a low pressure (Coanda Effect). This low pressure causes a force counter to the torque produced by the main rotor. Additionally, the remainder of the air from the fan is sent through the tail boom to a vent on the aft left side of the boom where it is expelled. This action to the left causes an opposite reaction to the right, which is the direction needed to counter the main rotor torque. *[Figure 1-102]*

Controls

The controls of a helicopter differ slightly from those found in an aircraft. The collective, operated by the pilot with the left hand, is pulled up or pushed down to increase or decrease the angle of attack on all of the rotor blades simultaneously. This increases or decreases lift and moves the aircraft up or down. The engine throttle control is located on the hand grip at the end of the collective. The cyclic is the control "stick" located between the pilot's legs. It can be moved in any direction to tilt the plane of rotation of the rotor blades. This causes the helicopter to move in the direction that the cyclic is moved. As stated, the foot pedals control the pitch of the tail rotor blades thereby balancing main rotor torque. *Figures 1-103* and *1-104* illustrate the controls found in a typical helicopter.

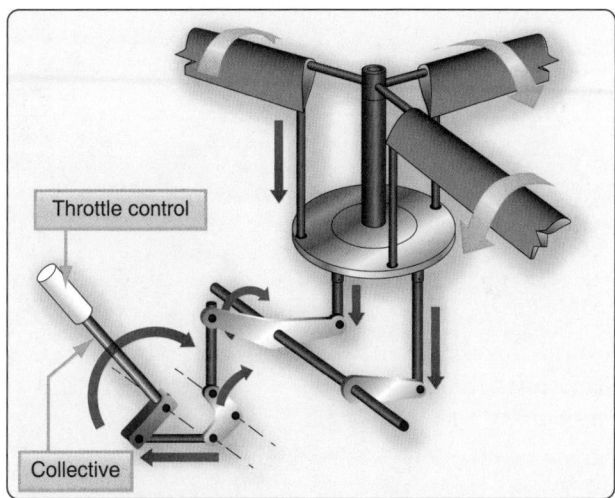

Figure 1-103. *The collective changes the pitch of all of the rotor blades simultaneously and by the same amount, thereby increasing or decreasing lift.*

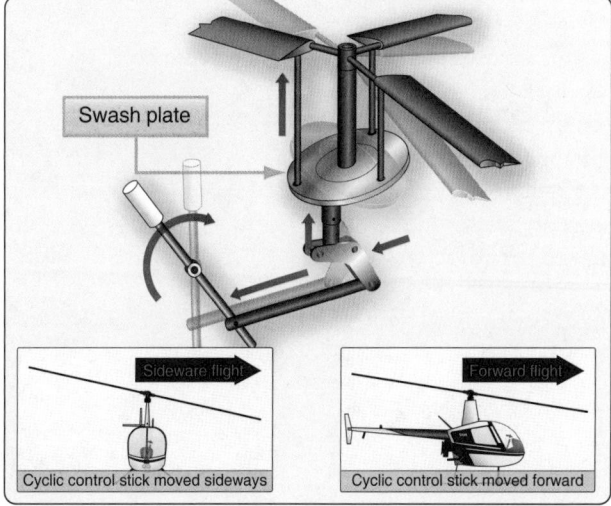

Figure 1-104. *The cyclic changes the angle of the swash plate which changes the plane of rotation of the rotor blades. This moves the aircraft horizontally in any direction depending on the positioning of the cyclic.*

Figure 1-102. *While in a hover, Coanda Effect supplies approximately two-thirds of the lift necessary to maintain directional control. The rest is created by directing the thrust from the controllable rotating nozzle.*

Chapter 2
Aerodynamics, Aircraft Assembly, & Rigging

Introduction

Three topics that are directly related to the manufacture, operation, and repair of aircraft are: aerodynamics, aircraft assembly, and rigging. Each of these subject areas, though studied separately, eventually connect to provide a scientific and physical understanding of how an aircraft is prepared for flight. A logical place to start with these three topics is the study of basic aerodynamics. By studying aerodynamics, a person becomes familiar with the fundamentals of aircraft flight.

Basic Aerodynamics

Aerodynamics is the study of the dynamics of gases, the interaction between a moving object and the atmosphere being of primary interest for this handbook. The movement of an object and its reaction to the air flow around it can be seen when watching water passing the bow of a ship. The major difference between water and air is that air is compressible and water is incompressible. The action of the airflow over a body is a large part of the study of aerodynamics. Some common aircraft terms, such as rudder, hull, water line, and keel beam, were borrowed from nautical terms.

Many textbooks have been written about the aerodynamics of aircraft flight. It is not necessary for an airframe and powerplant (A&P) mechanic to be as knowledgeable as an aeronautical engineer about aerodynamics. The mechanic must be able to understand the relationships between how an aircraft performs in flight and its reaction to the forces acting on its structural parts. Understanding why aircraft are designed with particular types of primary and secondary control systems and why the surfaces must be aerodynamically smooth becomes essential when maintaining today's complex aircraft.

The theory of flight should be described in terms of the laws of flight because what happens to an aircraft when it flies is not based upon assumptions, but upon a series of facts. Aerodynamics is a study of laws which have been proven to be the physical reasons why an airplane flies. The term aerodynamics is derived from the combination of two Greek words: "aero," meaning air, and "dyne," meaning force of power. Thus, when "aero" joins "dynamics" the result is "aerodynamics"—the study of objects in motion through the air and the forces that produce or change such motion.

Aerodynamically, an aircraft can be defined as an object traveling through space that is affected by the changes in atmospheric conditions. To state it another way, aerodynamics covers the relationships between the aircraft, relative wind, and atmosphere.

The Atmosphere

Before examining the fundamental laws of flight, several basic facts must be considered, namely that an aircraft operates in the air. Therefore, those properties of air that affect the control and performance of an aircraft must be understood.

The air in the earth's atmosphere is composed mostly of nitrogen and oxygen. Air is considered a fluid because it fits the definition of a substance that has the ability to flow or assume the shape of the container in which it is enclosed. If the container is heated, pressure increases; if cooled, the pressure decreases. The weight of air is heaviest at sea level where it has been compressed by all of the air above. This compression of air is called atmospheric pressure.

Pressure

Atmospheric pressure is usually defined as the force exerted against the earth's surface by the weight of the air above that surface. Weight is force applied to an area that results in pressure. Force (F) equals area (A) times pressure (P), or F = AP. Therefore, to find the amount of pressure, divide area into force (P = F/A). A column of air (one square inch) extending from sea level to the top of the atmosphere weighs approximately 14.7 pounds; therefore, atmospheric pressure is stated in pounds per square inch (psi). Thus, atmospheric pressure at sea level is 14.7 psi.

Atmospheric pressure is measured with an instrument called a barometer, composed of mercury in a tube that records atmospheric pressure in inches of mercury ("Hg). *[Figure 2-1]* The standard measurement in aviation altimeters

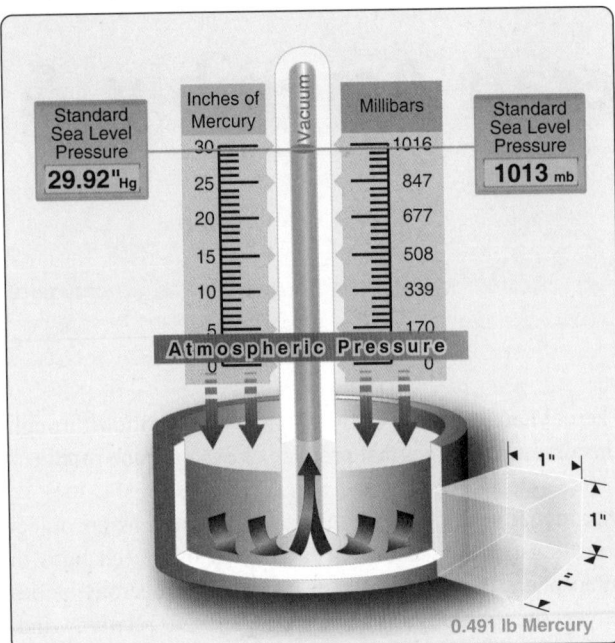

Figure 2-1. *Barometer used to measure atmospheric pressure.*

and U.S. weather reports has been "Hg. However, worldwide weather maps and some non-U.S. manufactured aircraft instruments indicate pressure in millibars (mb), a metric unit. At sea level, when the average atmospheric pressure is 14.7 psi, the barometric pressure is 29.92 "Hg, and the metric measurement is 1013.25 mb.

An important consideration is that atmospheric pressure varies with altitude. As an aircraft ascends, atmospheric pressure drops, oxygen content of the air decreases, and temperature drops. The changes in altitude affect an aircraft's performance in such areas as lift and engine horsepower. The effects of temperature, altitude, and density of air on aircraft performance are covered in the following paragraphs.

Density

Density is weight per unit of volume. Since air is a mixture of gases, it can be compressed. If the air in one container is under half as much pressure as an equal amount of air in an identical container, the air under the greater pressure weighs twice as much as that in the container under lower pressure. The air under greater pressure is twice as dense as that in the other container. For the equal weight of air, that which is under the greater pressure occupies only half the volume of that under half the pressure.

The density of gases is governed by the following rules:

1. Density varies in direct proportion with the pressure.

2. Density varies inversely with the temperature.

Thus, air at high altitudes is less dense than air at low altitudes, and a mass of hot air is less dense than a mass of cool air.

Changes in density affect the aerodynamic performance of aircraft with the same horsepower. An aircraft can fly faster at a high altitude where the density is low than at a low altitude where the density is greater. This is because air offers less resistance to the aircraft when it contains a smaller number of air particles per unit of volume.

Humidity

Humidity is the amount of water vapor in the air. The maximum amount of water vapor that air can hold varies with the temperature. The higher the temperature of the air, the more water vapor it can absorb.

1. Absolute humidity is the weight of water vapor in a unit volume of air.

2. Relative humidity is the ratio, in percent, of the moisture actually in the air to the moisture it would hold if it were saturated at the same temperature and pressure.

Assuming that the temperature and pressure remain the same, the density of the air varies inversely with the humidity. On damp days, the air density is less than on dry days. For this reason, an aircraft requires a longer runway for takeoff on damp days than it does on dry days.

By itself, water vapor weighs approximately five-eighths as much as an equal amount of perfectly dry air. Therefore, when air contains water vapor, it is not as heavy as dry air containing no moisture.

Aerodynamics & the Laws of Physics

The law of conservation of energy states that energy may neither be created nor destroyed.

Motion is the act or process of changing place or position. An object may be in motion with respect to one object and motionless with respect to another. For example, a person sitting quietly in an aircraft flying at 200 knots is at rest or motionless with respect to the aircraft; however, the person and the aircraft are in motion with respect to the air and to the earth.

Air has no force or power, except pressure, unless it is in motion. When it is moving, however, its force becomes apparent. A moving object in motionless air has a force exerted on it as a result of its own motion. It makes no difference in the effect then, whether an object is moving with respect to the air or the air is moving with respect to the object. The flow of air around an object caused by the movement of either the air or the object, or both, is called the relative wind.

Velocity & Acceleration

The terms "speed" and "velocity" are often used interchangeably, but they do not have the same meaning. Speed is the rate of motion in relation to time, and velocity is the rate of motion in a particular direction in relation to time.

An aircraft starts from New York City and flies 10 hours at an average speed of 260 miles per hour (mph). At the end of this time, the aircraft may be over the Atlantic Ocean, Pacific Ocean, Gulf of Mexico, or, if its flight were in a circular path, it may even be back over New York City. If this same aircraft flew at a velocity of 260 mph in a southwestward direction, it would arrive in Los Angeles in about 10 hours. Only the rate of motion is indicated in the first example and denotes the speed of the aircraft. In the last example, the particular direction is included with the rate of motion, thus, denoting the velocity of the aircraft.

Acceleration is defined as the rate of change of velocity. An aircraft increasing in velocity is an example of positive acceleration, while another aircraft reducing its velocity is an example of negative acceleration, or deceleration.

Newton's Laws of Motion

The fundamental laws governing the action of air about a wing are known as Newton's laws of motion.

Newton's first law is normally referred to as the law of inertia. It simply means that a body at rest does not move unless force is applied to it. If a body is moving at uniform speed in a straight line, force must be applied to increase or decrease the speed.

According to Newton's law, since air has mass, it is a body. When an aircraft is on the ground with its engines off, inertia keeps the aircraft at rest. An aircraft is moved from its state of rest by the thrust force created by a propeller, or by the expanding exhaust, or both. When an aircraft is flying at uniform speed in a straight line, inertia tends to keep the aircraft moving. Some external force is required to change the aircraft from its path of flight.

Newton's second law states that if a body moving with uniform speed is acted upon by an external force, the change of motion is proportional to the amount of the force, and motion takes place in the direction in which the force acts. This law may be stated mathematically as follows:

$$\text{Force} = \text{mass} \times \text{acceleration} \quad (F = ma)$$

If an aircraft is flying against a headwind, it is slowed down. If the wind is coming from either side of the aircraft's heading, the aircraft is pushed off course unless the pilot takes corrective action against the wind direction.

Newton's third law is the law of action and reaction. This law states that for every action (force) there is an equal and opposite reaction (force). This law can be illustrated by the example of firing a gun. The action is the forward movement of the bullet while the reaction is the backward recoil of the gun.

The three laws of motion that have been discussed apply to the theory of flight. In many cases, all three laws may be operating on an aircraft at the same time.

Bernoulli's Principle & Subsonic Flow

Bernoulli's principle states that when a fluid (air) flowing through a tube reaches a constriction, or narrowing, of the tube, the speed of the fluid flowing through that constriction is increased and its pressure is decreased. The cambered (curved) surface of an airfoil (wing) affects the airflow exactly as a constriction in a tube affects airflow. [Figure 2-2] Diagram A of Figure 2-2 illustrates the effect of air passing through a constriction in a tube. In Diagram B, air is flowing past a cambered surface, such as an airfoil, and the effect is similar to that of air passing through a restriction.

As the air flows over the upper surface of an airfoil, its velocity increases and its pressure decreases; an area of low pressure is formed. There is an area of greater pressure on the lower surface of the airfoil, and this greater pressure tends to move the wing upward. The difference in pressure between the upper and lower surfaces of the wing is called lift. Three-fourths of the total lift of an airfoil is the result of the decrease in pressure over the upper surface. The impact of air on the under surface of an airfoil produces the other one-fourth of the total lift.

Airfoil

An airfoil is a surface designed to obtain lift from the air through which it moves. Thus, it can be stated that any part of the aircraft that converts air resistance into lift is an airfoil. The profile of a conventional wing is an excellent example of an airfoil. [Figure 2-3] Notice that the top surface of the wing profile has greater curvature than the lower surface.

The difference in curvature of the upper and lower surfaces of the wing builds up the lift force. Air flowing over the top surface of the wing must reach the trailing edge of the wing in the same amount of time as the air flowing under the wing. To do this, the air passing over the top surface moves at a

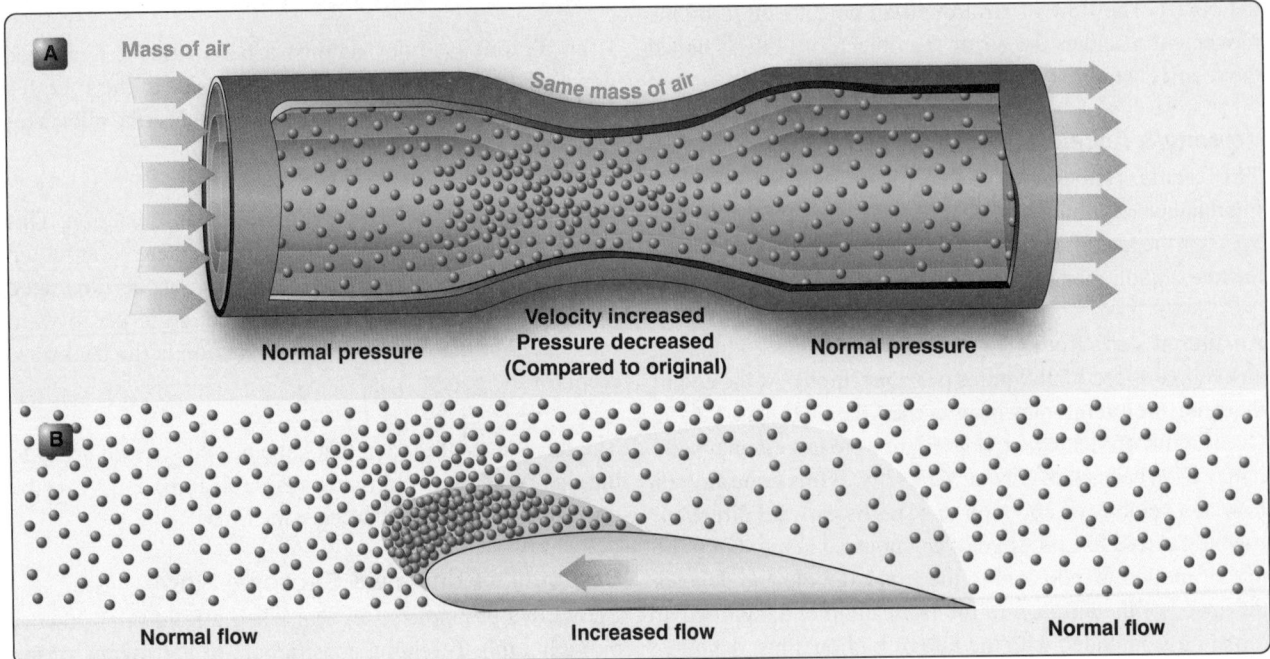

Figure 2-2. *Bernoulli's Principle.*

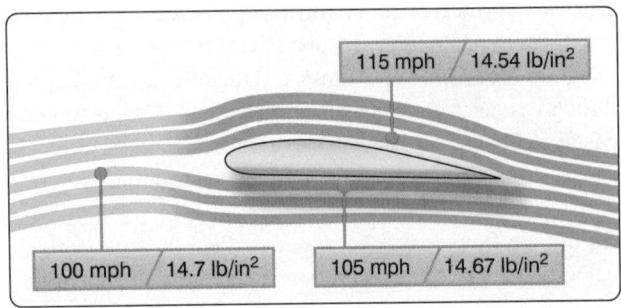

115 mph / 14.54 lb/in^2

100 mph / 14.7 lb/in^2

105 mph / 14.67 lb/in^2

Figure 2-3. *Airflow over a wing section.*

greater velocity than the air passing below the wing because of the greater distance it must travel along the top surface. This increased velocity, according to Bernoulli's Principle, means a corresponding decrease in pressure on the surface. Thus, a pressure differential is created between the upper and lower surfaces of the wing, forcing the wing upward in the direction of the lower pressure.

Within limits, lift can be increased by increasing the angle of attack (AOA), wing area, velocity, density of the air, or by changing the shape of the airfoil. When the force of lift on an aircraft's wing equals the force of gravity, the aircraft maintains level flight.

Shape of the Airfoil

Individual airfoil section properties differ from those properties of the wing or aircraft as a whole because of the effect of the wing planform. A wing may have various airfoil sections from root to tip, with taper, twist, and sweepback.

The resulting aerodynamic properties of the wing are determined by the action of each section along the span.

The shape of the airfoil determines the amount of turbulence or skin friction that it produces, consequently affecting the efficiency of the wing. Turbulence and skin friction are controlled mainly by the fineness ratio, which is defined as the ratio of the chord of the airfoil to the maximum thickness. If the wing has a high fineness ratio, it is a very thin wing. A thick wing has a low fineness ratio. A wing with a high fineness ratio produces a large amount of skin friction. A wing with a low fineness ratio produces a large amount of turbulence. The best wing is a compromise between these two extremes to hold both turbulence and skin friction to a minimum.

The efficiency of a wing is measured in terms of the lift to drag ratio (L/D). This ratio varies with the AOA but reaches a definite maximum value for a particular AOA. At this angle, the wing has reached its maximum efficiency. The shape of the airfoil is the factor that determines the AOA at which the wing is most efficient; it also determines the degree of efficiency. Research has shown that the most efficient airfoils for general use have the maximum thickness occurring about one-third of the way back from the leading edge of the wing.

High-lift wings and high-lift devices for wings have been developed by shaping the airfoils to produce the desired effect. The amount of lift produced by an airfoil increases with an increase in wing camber. Camber refers to the curvature of an airfoil above and below the chord line surface. Upper camber refers to the upper surface, lower camber to the lower surface, and mean camber to the mean line of the section. Camber is positive when departure from the chord line is outward and negative when it is inward. Thus, high-lift wings have a large positive camber on the upper surface and a slightly negative camber on the lower surface. Wing flaps cause an ordinary wing to approximate this same condition by increasing the upper camber and by creating a negative lower camber.

It is also known that the larger the wingspan, as compared to the chord, the greater the lift obtained. This comparison is called aspect ratio. The higher the aspect ratio, the greater the lift. In spite of the benefits from an increase in aspect ratio, it was found that definite limitations were defined by structural and drag considerations.

On the other hand, an airfoil that is perfectly streamlined and offers little wind resistance sometimes does not have enough lifting power to take the aircraft off the ground. Thus, modern aircraft have airfoils which strike a medium between extremes, the shape depending on the purposes of the aircraft for which it is designed.

Angle of Incidence
The acute angle the wing chord makes with the longitudinal axis of the aircraft is called the angle of incidence, or the angle of wing setting. *[Figure 2-4]* The angle of incidence in most cases is a fixed, built-in angle. When the leading edge of the wing is higher than the trailing edge, the angle of incidence is said to be positive. The angle of incidence is negative when the leading edge is lower than the trailing edge of the wing.

Angle of Attack (AOA)
Before beginning the discussion on AOA and its effect on airfoils, first consider the terms chord and center of pressure (CP) as illustrated in *Figure 2-5*.

The chord of an airfoil or wing section is an imaginary straight line that passes through the section from the leading edge to the trailing edge, as shown in *Figure 2-5*. The chord line provides one side of an angle that ultimately forms the AOA. The other side of the angle is formed by a line indicating the direction of the relative airstream. Thus, AOA

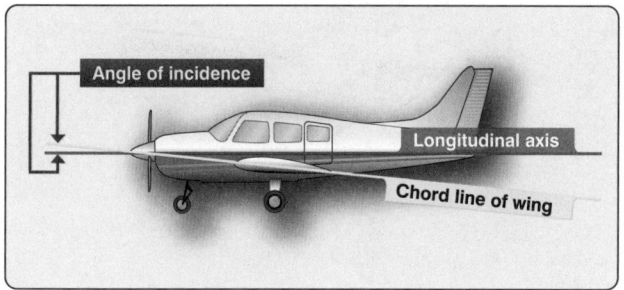

Figure 2-4. *Angle of incidence.*

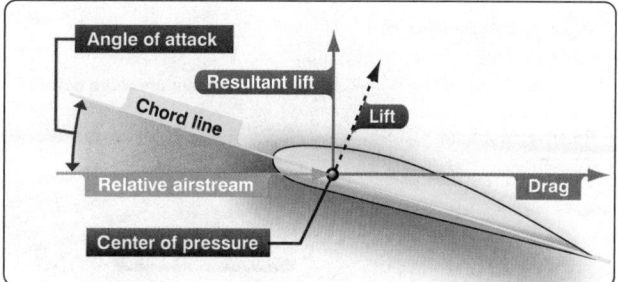

Figure 2-5. *Airflow over a wing section.*

is defined as the angle between the chord line of the wing and the direction of the relative wind. This is not to be confused with the angle of incidence, illustrated in *Figure 2-4*, which is the angle between the chord line of the wing and the longitudinal axis of the aircraft.

On each part of an airfoil or wing surface, a small force is present. This force is of a different magnitude and direction from any forces acting on other areas forward or rearward from this point. It is possible to add all of these small forces mathematically. That sum is called the "resultant force" (lift). This resultant force has magnitude, direction, and location, and can be represented as a vector, as shown in *Figure 2-5*. The point of intersection of the resultant force line with the chord line of the airfoil is called the center of pressure (CP). The CP moves along the airfoil chord as the AOA changes. Throughout most of the flight range, the CP moves forward with increasing AOA and rearward as the AOA decreases. The effect of increasing AOA on the CP is shown in *Figure 2-6*.

The AOA changes as the aircraft's attitude changes. Since the AOA has a great deal to do with determining lift, it is given primary consideration when designing airfoils. In a properly designed airfoil, the lift increases as the AOA is increased. When the AOA is increased gradually toward a positive AOA, the lift component increases rapidly up to a certain point and then suddenly begins to drop off. During this action the drag component increases slowly at first, then rapidly as lift begins to drop off.

When the AOA increases to the angle of maximum lift, the burble point is reached. This is known as the critical angle. When the critical angle is reached, the air ceases to flow smoothly over the top surface of the airfoil and begins to burble or eddy. This means that air breaks away from the upper camber line of the wing. What was formerly the area of decreased pressure is now filled by this burbling air. When this occurs, the amount of lift drops and drag becomes excessive. The force of gravity exerts itself, and the nose of the aircraft drops. This is a stall. Thus, the burble point is the stalling angle.

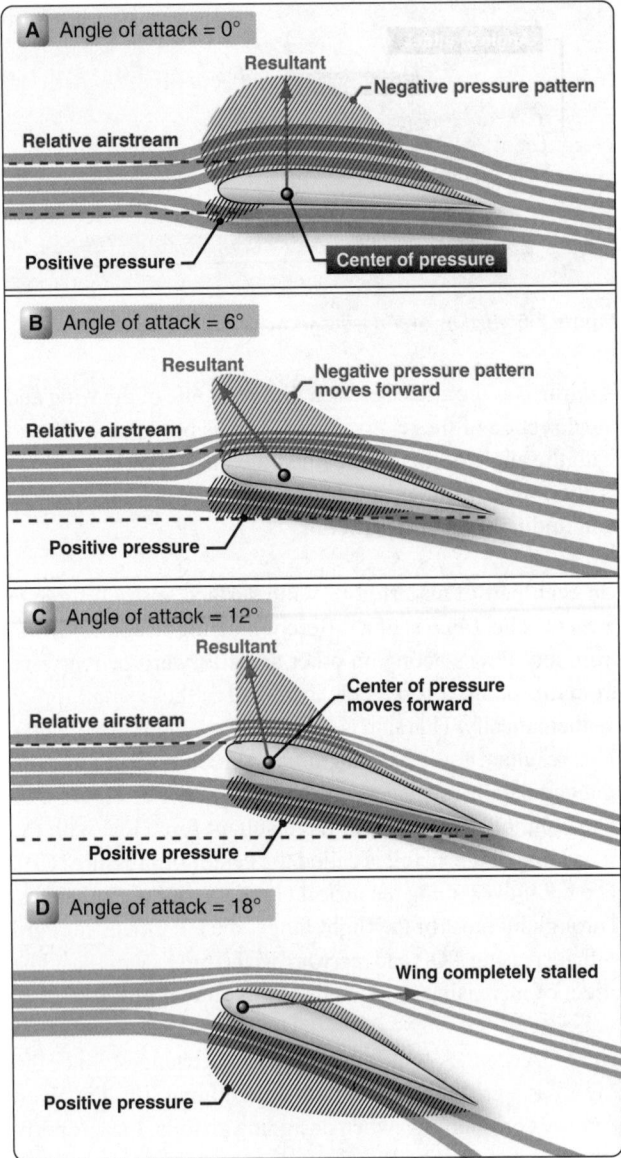

A | Angle of attack = 0°

Resultant
Negative pressure pattern
Relative airstream
Positive pressure
Center of pressure

B | Angle of attack = 6°

Resultant
Negative pressure pattern moves forward
Relative airstream
Positive pressure

C | Angle of attack = 12°

Resultant
Center of pressure moves forward
Relative airstream
Positive pressure

D | Angle of attack = 18°

Wing completely stalled
Positive pressure

Figure 2-6. *Effect on increasing angle of attack.*

As previously seen, the distribution of the pressure forces over the airfoil varies with the AOA. The application of the resultant force, or CP, varies correspondingly. As this angle increases, the CP moves forward; as the angle decreases, the CP moves back. The unstable travel of the CP is characteristic of almost all airfoils.

Boundary Layer

In the study of physics and fluid mechanics, a boundary layer is that layer of fluid in the immediate vicinity of a bounding surface. In relation to an aircraft, the boundary layer is the part of the airflow closest to the surface of the aircraft. In designing high-performance aircraft, considerable attention is paid to controlling the behavior of the boundary layer to minimize pressure drag and skin friction drag.

Thrust & Drag

An aircraft in flight is the center of a continuous battle of forces. Actually, this conflict is not as violent as it sounds, but it is the key to all maneuvers performed in the air. There is nothing mysterious about these forces; they are definite and known. The directions in which they act can be calculated, and the aircraft itself is designed to take advantage of each of them. In all types of flying, flight calculations are based on the magnitude and direction of four forces: weight, lift, drag, and thrust. *[Figure 2-7]*

An aircraft in flight is acted upon by four forces:

1. Gravity or weight—the force that pulls the aircraft toward the earth. Weight is the force of gravity acting downward upon everything that goes into the aircraft, such as the aircraft itself, crew, fuel, and cargo.

2. Lift—the force that pushes the aircraft upward. Lift acts vertically and counteracts the effects of weight.

3. Thrust—the force that moves the aircraft forward. Thrust is the forward force produced by the powerplant that overcomes the force of drag.

4. Drag—the force that exerts a braking action to hold the aircraft back. Drag is a backward deterrent force and is caused by the disruption of the airflow by the wings, fuselage, and protruding objects.

These four forces are in perfect balance only when the aircraft is in straight-and-level unaccelerated flight.

The forces of lift and drag are the direct result of the relationship between the relative wind and the aircraft. The force of lift always acts perpendicular to the relative wind, and the force of drag always acts parallel to and in the same direction as the relative wind. These forces are actually the components that produce a resultant lift force on the wing. *[Figure 2-8]*

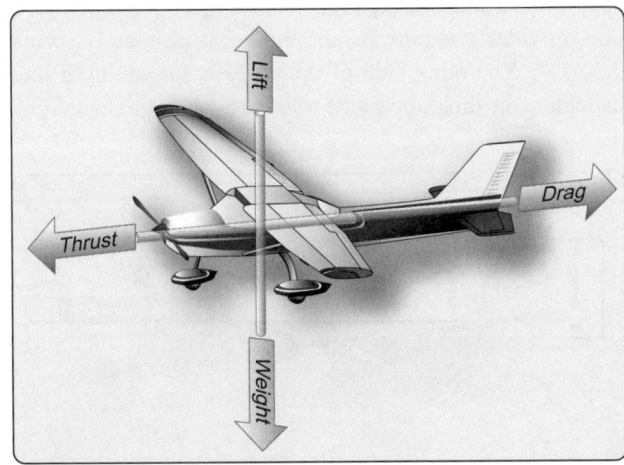

Figure 2-7. *Forces in action during flight.*

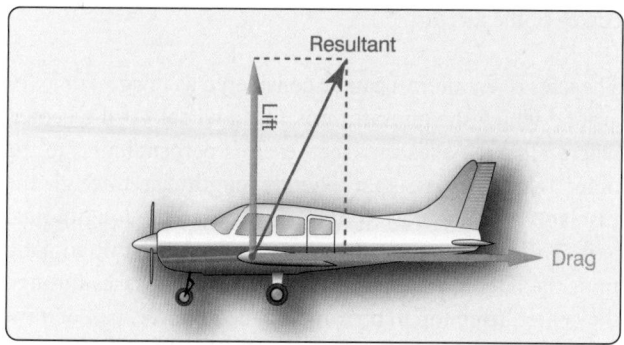

Figure 2-8. *Resultant of lift and drag.*

Weight has a definite relationship with lift, and thrust with drag. These relationships are quite simple, but very important in understanding the aerodynamics of flying. As stated previously, lift is the upward force on the wing perpendicular to the relative wind. Lift is required to counteract the aircraft's weight, caused by the force of gravity acting on the mass of the aircraft. This weight force acts downward through a point called the center of gravity (CG). The CG is the point at which all the weight of the aircraft is considered to be concentrated. When the lift force is in equilibrium with the weight force, the aircraft neither gains nor loses altitude. If lift becomes less than weight, the aircraft loses altitude. When the lift is greater than the weight, the aircraft gains altitude.

Wing area is measured in square feet and includes the part blanked out by the fuselage. Wing area is adequately described as the area of the shadow cast by the wing at high noon. Tests show that lift and drag forces acting on a wing are roughly proportional to the wing area. This means that if the wing area is doubled, all other variables remaining the same, the lift and drag created by the wing is doubled. If the area is tripled, lift and drag are tripled.

Drag must be overcome for the aircraft to move, and movement is essential to obtain lift. To overcome drag and move the aircraft forward, another force is essential. This force is thrust. Thrust is derived from jet propulsion or from a propeller and engine combination. Jet propulsion theory is based on Newton's third law of motion *(page 2-3)*. The turbine engine causes a mass of air to be moved backward at high velocity causing a reaction that moves the aircraft forward.

In a propeller/engine combination, the propeller is actually two or more revolving airfoils mounted on a horizontal shaft. The motion of the blades through the air produces lift similar to the lift on the wing, but acts in a horizontal direction, pulling the aircraft forward.

Before the aircraft begins to move, thrust must be exerted. The aircraft continues to move and gain speed until thrust and

drag are equal. In order to maintain a steady speed, thrust and drag must remain equal, just as lift and weight must be equal for steady, horizontal flight. Increasing the lift means that the aircraft moves upward, whereas decreasing the lift so that it is less than the weight causes the aircraft to lose altitude. A similar rule applies to the two forces of thrust and drag. If the revolutions per minute (rpm) of the engine is reduced, the thrust is lessened, and the aircraft slows down. As long as the thrust is less than the drag, the aircraft travels more and more slowly until its speed is insufficient to support it in the air.

Likewise, if the rpm of the engine is increased, thrust becomes greater than drag, and the speed of the aircraft increases. As long as the thrust continues to be greater than the drag, the aircraft continues to accelerate. When drag equals thrust, the aircraft flies at a steady speed.

The relative motion of the air over an object that produces lift also produces drag. Drag is the resistance of the air to objects moving through it. If an aircraft is flying on a level course, the lift force acts vertically to support it while the drag force acts horizontally to hold it back. The total amount of drag on an aircraft is made up of many drag forces, but this handbook considers three: parasite drag, profile drag, and induced drag.

Parasite drag is made up of a combination of many different drag forces. Any exposed object on an aircraft offers some resistance to the air, and the more objects in the airstream, the more parasite drag. While parasite drag can be reduced by reducing the number of exposed parts to as few as practical and streamlining their shape, skin friction is the type of parasite drag most difficult to reduce. No surface is perfectly smooth. Even machined surfaces have a ragged uneven appearance when inspected under magnification. These ragged surfaces deflect the air near the surface causing resistance to smooth airflow. Skin friction can be reduced by using glossy smooth finishes and eliminating protruding rivet heads, roughness, and other irregularities.

Profile drag may be considered the parasite drag of the airfoil. The various components of parasite drag are all of the same nature as profile drag.

The action of the airfoil that creates lift also causes induced drag. Remember, the pressure above the wing is less than atmospheric pressure, and the pressure below the wing is equal to or greater than atmospheric pressure. Since fluids always move from high pressure toward low pressure, there is a spanwise movement of air from the bottom of the wing outward from the fuselage and upward around the wing tip. This flow of air results in spillage over the wing tip, thereby setting up a whirlpool of air called a "vortex." *[Figure 2-9]*

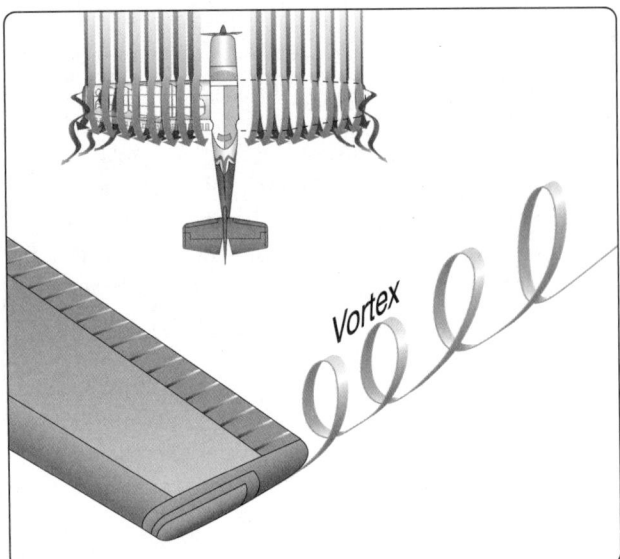

Figure 2-9. *Wingtip vortices.*

The air on the upper surface has a tendency to move in toward the fuselage and off the trailing edge. This air current forms a similar vortex at the inner portion of the trailing edge of the wing. These vortices increase drag because of the turbulence produced, and constitute induced drag.

Just as lift increases with an increase in AOA, induced drag also increases as the AOA becomes greater. This occurs because, as the AOA is increased, the pressure difference between the top and bottom of the wing becomes greater. This causes more violent vortices to be set up, resulting in more turbulence and more induced drag.

Center of Gravity (CG)

Gravity is the pulling force that tends to draw all bodies within the earth's gravitational field to the center of the earth. The CG may be considered the point at which all the weight of the aircraft is concentrated. If the aircraft was supported at its exact CG, it would balance in any position. CG is of major importance in an aircraft, for its position has a great bearing upon stability.

The CG is determined by the general design of the aircraft. The designers estimate how far the CP travels. They then fix the CG in front of the CP for the corresponding flight speed in order to provide an adequate restoring moment for flight equilibrium.

The Axes of an Aircraft

Whenever an aircraft changes its attitude in flight, it must turn about one or more of three axes. *Figure 2-10* shows the three axes, which are imaginary lines passing through the center of the aircraft.

The axes of an aircraft can be considered as imaginary axles around which the aircraft turns like a wheel. At the center, where all three axes intersect, each is perpendicular to the other two. The axis that extends lengthwise through the fuselage from the nose to the tail is called the longitudinal axis. The axis that extends crosswise from wing tip to wing tip is the lateral, or pitch, axis. The axis that passes through the center, from top to bottom, is called the vertical, or yaw, axis. Roll, pitch, and yaw are controlled by three control surfaces. Roll is produced by the ailerons, which are located at the trailing edges of the wings. Pitch is affected by the elevators, the rear portion of the horizontal tail assembly. Yaw is controlled by the rudder, the rear portion of the vertical tail assembly.

Stability & Control

An aircraft must have sufficient stability to maintain a uniform flightpath and recover from the various upsetting forces. Also, to achieve the best performance, the aircraft must have the proper response to the movement of the controls. Control is the pilot action of moving the flight controls, providing the aerodynamic force that induces the aircraft to follow a desired flightpath. When an aircraft is said to be controllable, it means that the aircraft responds easily and promptly to movement of the controls. Different control surfaces are used to control the aircraft about each of the three axes. Moving the control surfaces on an aircraft changes the airflow over the aircraft's surface. This, in turn, creates changes in the balance of forces acting to keep the aircraft flying straight and level.

Three terms that appear in any discussion of stability and control are: stability, maneuverability, and controllability. Stability is the characteristic of an aircraft that tends to cause it to fly (hands off) in a straight-and-level flightpath. Maneuverability is the characteristic of an aircraft to be directed along a desired flightpath and to withstand the stresses imposed. Controllability is the quality of the response of an aircraft to the pilot's commands while maneuvering the aircraft.

Static Stability

An aircraft is in a state of equilibrium when the sum of all the forces acting on the aircraft and all the moments is equal to zero. An aircraft in equilibrium experiences no accelerations, and the aircraft continues in a steady condition of flight. A gust of wind or a deflection of the controls disturbs the equilibrium, and the aircraft experiences acceleration due to the unbalance of moment or force.

The three types of static stability are defined by the character

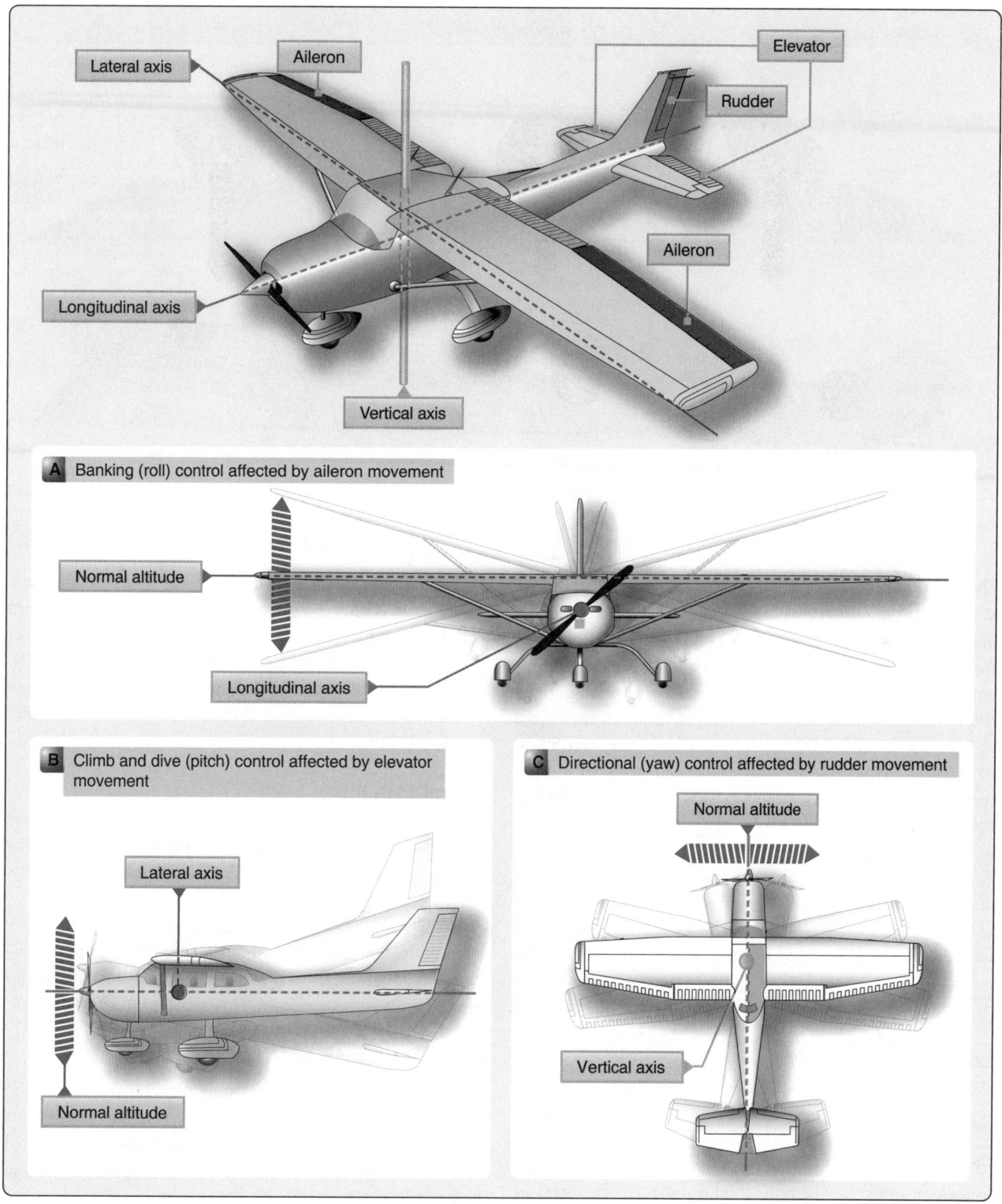

Figure 2-10. *Motion of an aircraft about its axes.*

of movement following some disturbance from equilibrium. Positive static stability exists when the disturbed object tends to return to equilibrium. Negative static stability, or static instability, exists when the disturbed object tends to continue in the direction of disturbance. Neutral static stability exists when the disturbed object has neither tendency, but remains in equilibrium in the direction of disturbance. These three types of stability are illustrated in *Figure 2-11*.

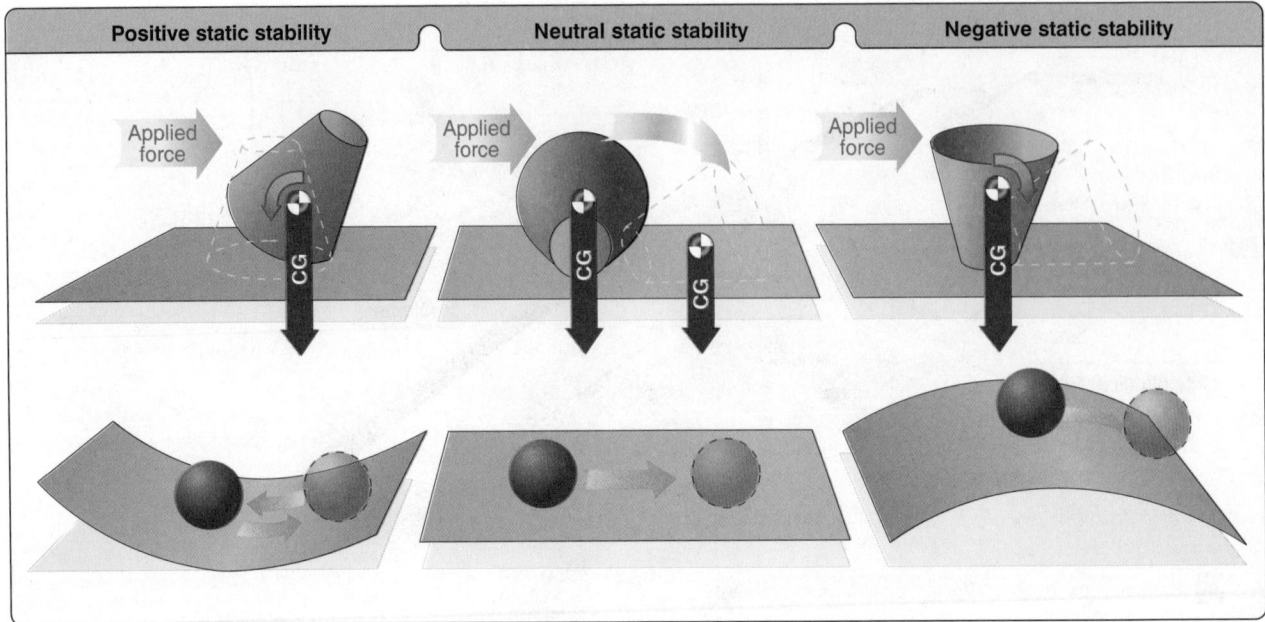

| Positive static stability | Neutral static stability | Negative static stability |

Figure 2-11. *Three types of stability.*

Dynamic Stability

While static stability deals with the tendency of a displaced body to return to equilibrium, dynamic stability deals with the resulting motion with time. If an object is disturbed from equilibrium, the time history of the resulting motion defines the dynamic stability of the object. In general, an object demonstrates positive dynamic stability if the amplitude of motion decreases with time. If the amplitude of motion increases with time, the object is said to possess dynamic instability.

Any aircraft must demonstrate the required degrees of static and dynamic stability. If an aircraft were designed with static instability and a rapid rate of dynamic instability, the aircraft would be very difficult, if not impossible, to fly. Usually, positive dynamic stability is required in an aircraft design to prevent objectionable continued oscillations of the aircraft.

Longitudinal Stability

When an aircraft has a tendency to keep a constant AOA with reference to the relative wind (i.e., it does not tend to put its nose down and dive or lift its nose and stall); it is said to have longitudinal stability. Longitudinal stability refers to motion in pitch. The horizontal stabilizer is the primary surface which controls longitudinal stability. The action of the stabilizer depends upon the speed and AOA of the aircraft.

Directional Stability

Stability about the vertical axis is referred to as directional stability. The aircraft should be designed so that when it is in straight-and-level flight it remains on its course heading even

though the pilot takes their hands and feet off the controls. If an aircraft recovers automatically from a skid, it has been well designed for directional balance. The vertical stabilizer is the primary surface that controls directional stability. Directional stability can be designed into an aircraft, where appropriate, by using a large dorsal fin, a long fuselage, and sweptback wings.

Lateral Stability

Motion about the aircraft's longitudinal (fore and aft) axis is a lateral, or rolling, motion. The tendency to return to the original attitude from such motion is called lateral stability.

Dutch Roll

A Dutch Roll is an aircraft motion consisting of an out-of-phase combination of yaw and roll. Dutch roll stability can be artificially increased by the installation of a yaw damper.

Primary Flight Controls

The primary controls are the ailerons, elevator, and the rudder, which provide the aerodynamic force to make the aircraft follow a desired flightpath. *[Figure 2-10]* The flight control surfaces are hinged or movable airfoils designed to change the attitude of the aircraft by changing the airflow over the aircraft's surface during flight. These surfaces are used for moving the aircraft about its three axes.

Typically, the ailerons and elevators are operated from the flight deck by means of a control stick, a wheel, and yoke assembly and on some of the newer design aircraft, a joystick. The rudder is normally operated by foot pedals on most aircraft. Lateral control is the banking movement or roll of an aircraft that is controlled by the ailerons. Longitudinal control is the climb and dive movement or pitch of an aircraft that is controlled by the elevator. Directional control is the left and right movement or yaw of an aircraft that is controlled by the rudder.

Trim Controls

Included in the trim controls are the trim tabs, servo tabs, balance tabs, and spring tabs. Trim tabs are small airfoils recessed into the trailing edges of the primary control surfaces. *[Figure 2-12]* Trim tabs can be used to correct any tendency of the aircraft to move toward an undesirable flight attitude. Their purpose is to enable the pilot to trim out any unbalanced condition which may exist during flight, without exerting any pressure on the primary controls.

Servo tabs, sometimes referred to as flight tabs, are used primarily on the large main control surfaces. They aid in moving the main control surface and holding it in the desired position. Only the servo tab moves in response to movement by the pilot of the primary flight controls.

Balance tabs are designed to move in the opposite direction of the primary flight control. Thus, aerodynamic forces acting on the tab assist in moving the primary control surface.

Spring tabs are similar in appearance to trim tabs, but serve an entirely different purpose. Spring tabs are used for the same purpose as hydraulic actuators—to aid the pilot in moving the primary control surface.

Figure 2-13 indicates how each trim tab is hinged to its parent primary control surface, but is operated by an independent control.

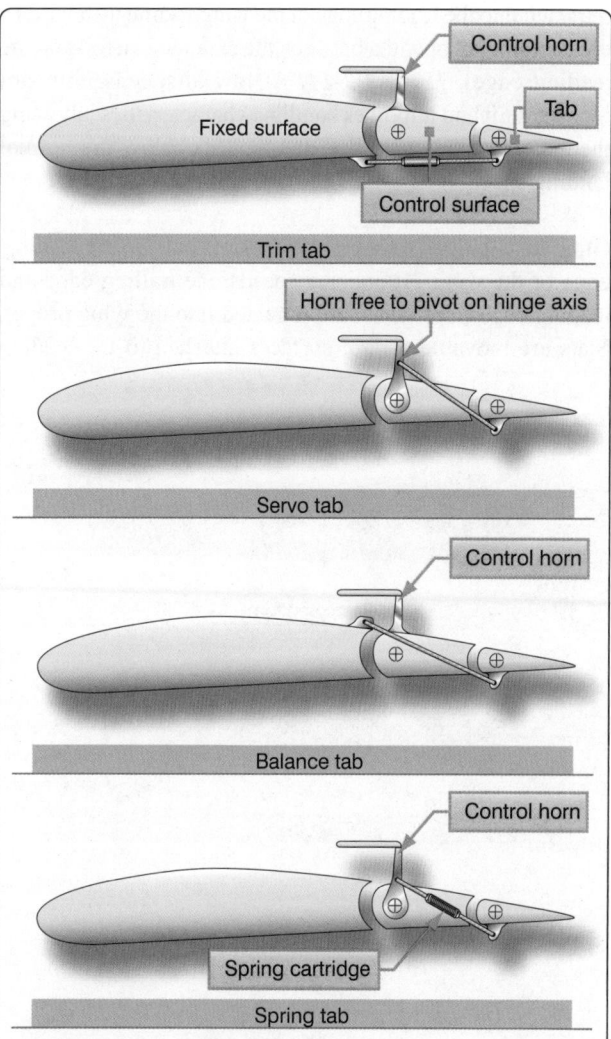

Figure 2-13. *Types of trim tabs.*

Auxiliary Lift Devices

Included in the auxiliary lift devices group of flight control surfaces are the wing flaps, spoilers, speed brakes, slats, leading edge flaps, and slots.

The auxiliary groups may be divided into two subgroups: those whose primary purpose is lift augmenting and those whose primary purpose is lift decreasing. In the first group are the flaps, both trailing edge and leading edge (slats), and slots. The lift decreasing devices are speed brakes and spoilers.

Lift Augmenting

Flaps are located on the trailing edge of the wing and are moveable to increase the wing area, thereby increasing lift on takeoff, and decreasing the speed during landing. These airfoils are retractable and fair into the wing contour. Others are simply a portion of the lower skin which extends into the airstream, thereby slowing the aircraft. Leading edge flaps, also referred to as slats, are airfoils extended from and

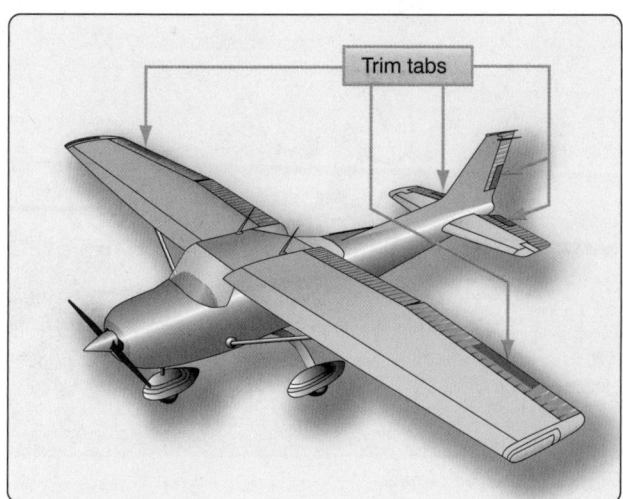

Figure 2-12. *Trim tabs.*

retracted into the leading edge of the wing. Some installations create a slot (an opening between the extended airfoil and the leading edge). *[Figure 2-14]* At low airspeeds, this slot increases lift and improves handling characteristics, allowing the aircraft to be controlled at airspeeds below the normal landing speed.

Other installations have permanent slots built in the leading edge of the wing. At cruising speeds, the trailing edge and leading edge flaps (slats) are retracted into the wing proper. Slats are movable control surfaces attached to the leading edges of the wings. When the slat is closed, it forms the leading edge of the wing. When in the open position (extended forward), a slot is created between the slat and the wing leading edge. At low airspeeds, this increases lift and improves handling characteristics, allowing the aircraft to be controlled at airspeeds below the normal landing speed. *[Figure 2-15]*

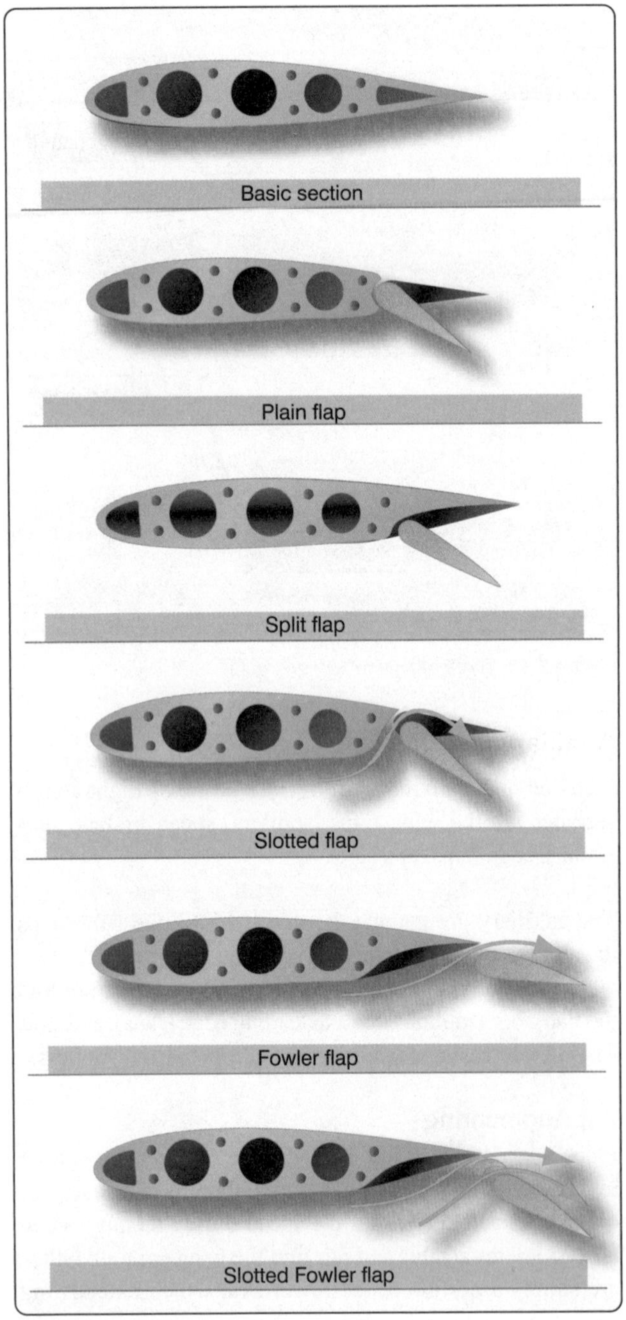

Lift Decreasing

Lift decreasing devices are the speed brakes (spoilers). In some installations, there are two types of spoilers. The ground spoiler is extended only after the aircraft is on the ground, thereby assisting in the braking action. The flight spoiler assists in lateral control by being extended whenever the aileron on that wing is rotated up. When actuated as speed brakes, the spoiler panels on both wings raise up. In-flight spoilers may also be located along the sides, underneath the fuselage, or back at the tail. *[Figure 2-16]* In some aircraft designs, the wing panel on the up aileron side rises more than the wing panel on the down aileron side. This provides speed brake operation and lateral control simultaneously.

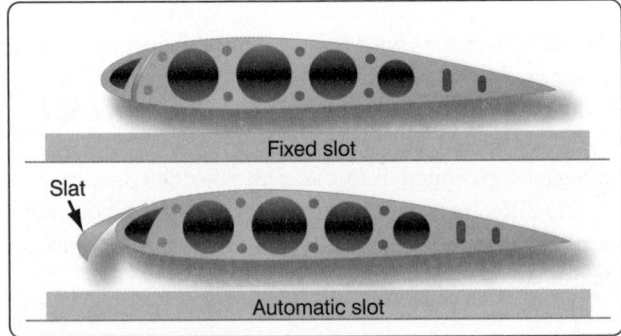

Figure 2-15. *Wing slots.*

Figure 2-16. *Speed brake.*

Figure 2-14. *Types of wing flaps.*

Basic section

Plain flap

Split flap

Slotted flap

Fowler flap

Slotted Fowler flap

Winglets

Winglets are the near-vertical extension of the wingtip that reduces the aerodynamic drag associated with vortices that develop at the wingtips as the airplane moves through the air. By reducing the induced drag at the tips of the wings, fuel consumption goes down and range is extended. *Figure 2-17* shows an example of a Learjet 60 with winglets.

Canard Wings

A canard wing aircraft is an airframe configuration of a fixed-wing aircraft in which a small wing or horizontal airfoil is ahead of the main lifting surfaces, rather than behind them as in a conventional aircraft. The canard may be fixed, movable, or designed with elevators. Good examples of aircraft with canard wings are the Rutan VariEze and Beechcraft 2000 Starship. *[Figures 2-18* and *2-19]*

Wing Fences

Wing fences are flat metal vertical plates fixed to the upper surface of the wing. They obstruct spanwise airflow along the wing, and prevent the entire wing from stalling at once. They are often attached on swept-wing aircraft to prevent the spanwise movement of air at high AOA. Their purpose is to provide better slow speed handling and stall characteristics. *[Figure 2-20]*

Control Systems for Large Aircraft

Mechanical Control

This is the basic type of system that was used to control early aircraft and is currently used in smaller aircraft where aerodynamic forces are not excessive. The controls are mechanical and manually operated.

The mechanical system of controlling an aircraft can include cables, push-pull tubes, and torque tubes. The cable system is the most widely used because deflections of the structure to which it is attached do not affect its operation. Some aircraft incorporate control systems that are a combination of all three. These systems incorporate cable assemblies, cable guides, linkage, adjustable stops, and control surface snubber or mechanical locking devices. These surface locking devices, usually referred to as a gust lock, limits the external wind forces from damaging the aircraft while it is parked or tied down.

Hydromechanical Control

As the size, complexity, and speed of aircraft increased, actuation of controls in flight became more difficult. It soon became apparent that the pilot needed assistance to overcome

Figure 2-18. *Canard wings on a Rutan VariEze.*

Figure 2-19. *The Beechcraft 2000 Starship has canard wings.*

Figure 2-20. *Aircraft stall fence.*

Figure 2-17. *Winglets on a Bombardier Learjet 60.*

the aerodynamic forces to control aircraft movement. Spring tabs, which were operated by the conventional control system, were moved so that the airflow over them actually moved the primary control surface. This was sufficient for the aircraft operating in the lowest of the high speed ranges (250–300 mph). For higher speeds, a power-assisted (hydraulic) control system was designed.

Conventional cable or push-pull tube systems link the flight deck controls with the hydraulic system. With the system activated, the pilot's movement of a control causes the mechanical link to open servo valves, thereby directing hydraulic fluid to actuators, which convert hydraulic pressure into control surface movements.

Because of the efficiency of the hydromechanical flight control system, the aerodynamic forces on the control surfaces cannot be felt by the pilot, and there is a risk of overstressing the structure of the aircraft. To overcome this problem, aircraft designers incorporated artificial feel systems into the design that provided increased resistance to the controls at higher speeds. Additionally, some aircraft with hydraulically powered control systems are fitted with a device called a stick shaker, which provides an artificial stall warning to the pilot.

Fly-By-Wire Control

The fly-by-wire (FBW) control system employs electrical signals that transmit the pilot's actions from the flight deck through a computer to the various flight control actuators. The FBW system evolved as a way to reduce the system weight of the hydromechanical system, reduce maintenance costs, and improve reliability. Electronic FBW control systems can respond to changing aerodynamic conditions by adjusting flight control movements so that the aircraft response is consistent for all flight conditions. Additionally, the computers can be programmed to prevent undesirable and dangerous characteristics, such as stalling and spinning.

Many of the new military high-performance aircraft are not aerodynamically stable. This characteristic is designed into the aircraft for increased maneuverability and responsive performance. Without the computers reacting to the instability, the pilot would lose control of the aircraft.

The Airbus A-320 was the first commercial airliner to use FBW controls. Boeing used them in their 777 and newer design commercial aircraft. The Dassault Falcon 7X was the first business jet to use a FBW control system.

High-Speed Aerodynamics

High-speed aerodynamics, often called compressible aerodynamics, is a special branch of study of aeronautics. It is utilized by aircraft designers when designing aircraft capable of speeds approaching Mach 1 and above. Because it is beyond the scope and intent of this handbook, only a brief overview of the subject is provided.

In the study of high-speed aeronautics, the compressibility effects on air must be addressed. This flight regime is characterized by the Mach number, a special parameter named in honor of Ernst Mach, the late 19th century physicist who studied gas dynamics. Mach number is the ratio of the speed of the aircraft to the local speed of sound and determines the magnitude of many of the compressibility effects.

As an aircraft moves through the air, the air molecules near the aircraft are disturbed and move around the aircraft. The air molecules are pushed aside much like a boat creates a bow wave as it moves through the water. If the aircraft passes at a low speed, typically less than 250 mph, the density of the air remains constant. But at higher speeds, some of the energy of the aircraft goes into compressing the air and locally changing the density of the air. The bigger and heavier the aircraft, the more air it displaces and the greater effect compression has on the aircraft.

This effect becomes more important as speed increases. Near and beyond the speed of sound, about 760 mph (at sea level), sharp disturbances generate a shockwave that affects both the lift and drag of an aircraft and flow conditions downstream of the shockwave. The shockwave forms a cone of pressurized air molecules which move outward and rearward in all directions and extend to the ground. The sharp release of the pressure, after the buildup by the shockwave, is heard as the sonic boom. *[Figure 2-21]*

Listed below are a range of conditions that are encountered by aircraft as their designed speed increases.

- Subsonic conditions occur for Mach numbers less than one (100–350 mph). For the lowest subsonic conditions, compressibility can be ignored.

Figure 2-21. *Breaking the sound barrier.*

- As the speed of the object approaches the speed of sound, the flight Mach number is nearly equal to one, M = 1 (350–760 mph), and the flow is said to be transonic. At some locations on the object, the local speed of air exceeds the speed of sound. Compressibility effects are most important in transonic flows and lead to the early belief in a sound barrier. Flight faster than sound was thought to be impossible. In fact, the sound barrier was only an increase in the drag near sonic conditions because of compressibility effects. Because of the high drag associated with compressibility effects, aircraft are not operated in cruise conditions near Mach 1.

- Supersonic conditions occur for numbers greater than Mach 1, but less then Mach 3 (760–2,280 mph). Compressibility effects of gas are important in the design of supersonic aircraft because of the shockwaves that are generated by the surface of the object. For high supersonic speeds, between Mach 3 and Mach 5 (2,280–3,600 mph), aerodynamic heating becomes a very important factor in aircraft design.

- For speeds greater than Mach 5, the flow is said to be hypersonic. At these speeds, some of the energy of the object now goes into exciting the chemical bonds which hold together the nitrogen and oxygen molecules of the air. At hypersonic speeds, the chemistry of the air must be considered when determining forces on the object. When the space shuttle re-enters the atmosphere at high hypersonic speeds, close to Mach 25, the heated air becomes an ionized plasma of gas, and the spacecraft must be insulated from the extremely high temperatures.

Additional technical information pertaining to high-speed aerodynamics can be found at bookstores, libraries, and numerous sources on the Internet. As the design of aircraft evolves and the speeds of aircraft continue to increase into the hypersonic range, new materials and propulsion systems will need to be developed. This is the challenge for engineers, physicists, and designers of aircraft in the future.

Rotary-Wing Aircraft Assembly & Rigging

The flight control units located in the flight deck of all helicopters are very nearly the same. All helicopters have either one or two of each of the following: collective pitch control, throttle grip, cyclic pitch control, and directional control pedals. *[Figure 2-22]* Basically, these units do the same things, regardless of the type of helicopter on which they are installed; however, the operation of the control system varies greatly by helicopter model.

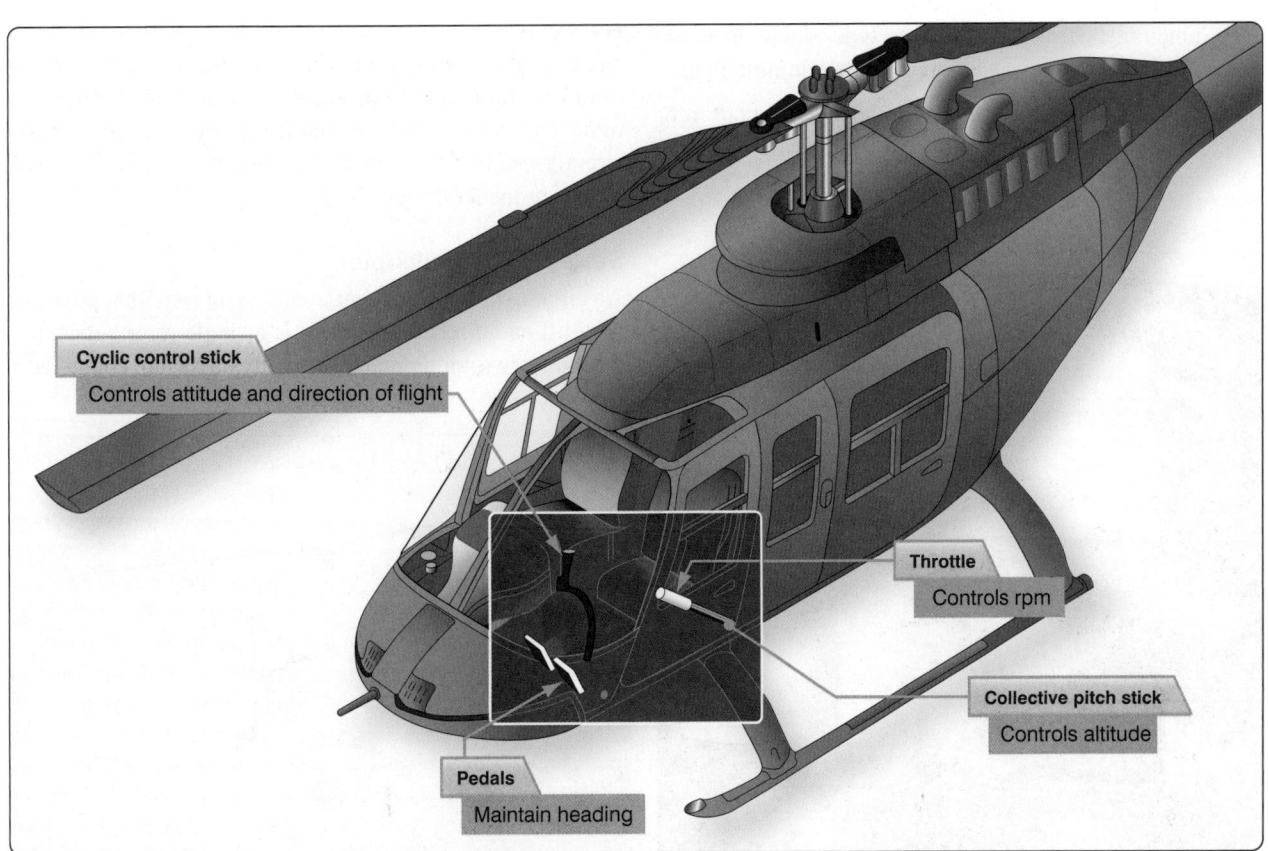

Figure 2-22. *Controls of a helicopter and the principal function of each.*

Rigging the helicopter coordinates the movements of the flight controls and establishes the relationship between the main rotor and its controls, and between the tail rotor and its controls. Rigging is not a difficult job, but it requires great precision and attention to detail. Strict adherence to rigging procedures described in the manufacturer's maintenance manuals and service instructions is a must. Adjustments, clearances, and tolerances must be exact.

Rigging of the various flight control systems can be broken down into the following three major steps:

1. Placing the control system in a specific position—holding it in position with pins, clamps, or jigs, then adjusting the various linkages to fit the immobilized control component.

2. Placing the control surfaces in a specific reference position—using a rigging jig, a precision bubble protractor, or a spirit level to check the angular difference between the control surface and some fixed surface on the aircraft. *[Figure 2-23]*

3. Setting the maximum range of travel of the various components—this adjustment limits the physical movement of the control system.

After completion of the static rigging, a functional check of the flight control system must be accomplished. The nature of the functional check varies with the type of helicopter and system concerned, but usually includes determining that:

1. The direction of movement of the main and tail rotor blades is correct in relation to movement of the pilot's controls.

2. The operation of interconnected control systems (engine throttle and collective pitch) is properly coordinated.

3. The range of movement and neutral position of the pilot's controls are correct.

4. The maximum and minimum pitch angles of the main rotor blades are within specified limits. This includes checking the fore-and-aft and lateral cyclic pitch and collective pitch blade angles.

5. The tracking of the main rotor blades is correct.

6. In the case of multirotor aircraft, the rigging and movement of the rotor blades are synchronized.

7. When tabs are provided on main rotor blades, they are correctly set.

8. The neutral, maximum, and minimum pitch angles and coning angles of the tail rotor blades are correct.

9. When dual controls are provided, they function correctly and in synchronization.

Upon completion of rigging, a thorough check should be made of all attaching, securing, and pivot points. All bolts, nuts, and rod ends should be properly secured and safetied as specified in the manufacturers' maintenance and service instructions.

Configurations of Rotary-Wing Aircraft

Autogyro

An autogyro is an aircraft with a free-spinning horizontal rotor that turns due to passage of air upward through the rotor. This air motion is created from forward motion of the aircraft resulting from either a tractor or pusher configured engine/propeller design. *[Figure 2-24]*

Single Rotor Helicopter

An aircraft with a single horizontal main rotor that provides both lift and direction of travel is a single rotor helicopter. A secondary rotor mounted vertically on the tail counteracts

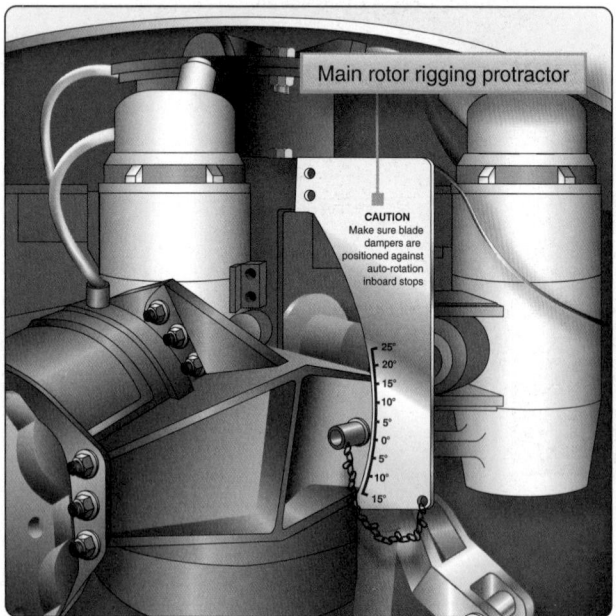

Figure 2-23. *A typical rigging protractor.*

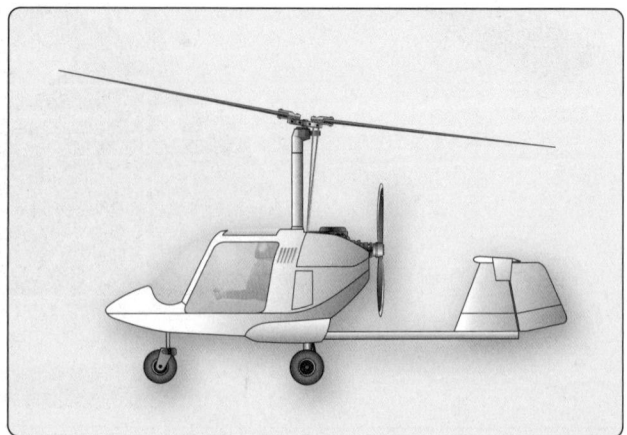

Figure 2-24. *An autogyro.*

the rotational force (torque) of the main rotor to correct yaw of the fuselage. *[Figure 2-25]*

Dual Rotor Helicopter

An aircraft with two horizontal rotors that provide both the lift and directional control is a dual rotor helicopter. The rotors are counterrotating to balance the aerodynamic torque and eliminate the need for a separate antitorque system. *[Figure 2-26]*

Types of Rotor Systems

Fully Articulated Rotor

A fully articulated rotor is found on aircraft with more than two blades and allows movement of each individual blade in three directions. In this design, each blade can rotate about the pitch axis to change lift; each blade can move back and forth in plane, lead and lag; and flap up and down through a hinge independent of the other blades. *[Figure 2-27]*

Semirigid Rotor

The semirigid rotor design is found on aircraft with two rotor blades. The blades are connected in a manner such that as one blade flaps up, the opposite blade flaps down.

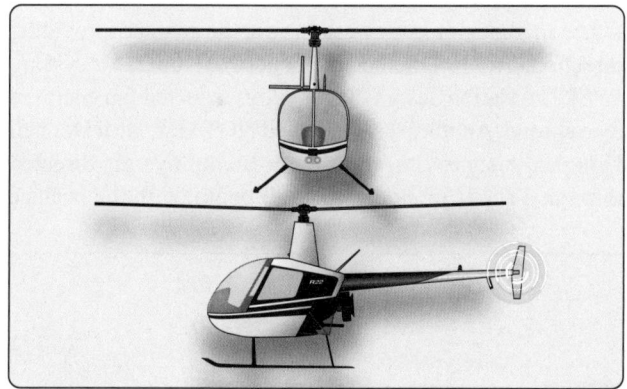

Figure 2-25. *Single rotor helicopter.*

Figure 2-26. *Dual rotor helicopter.*

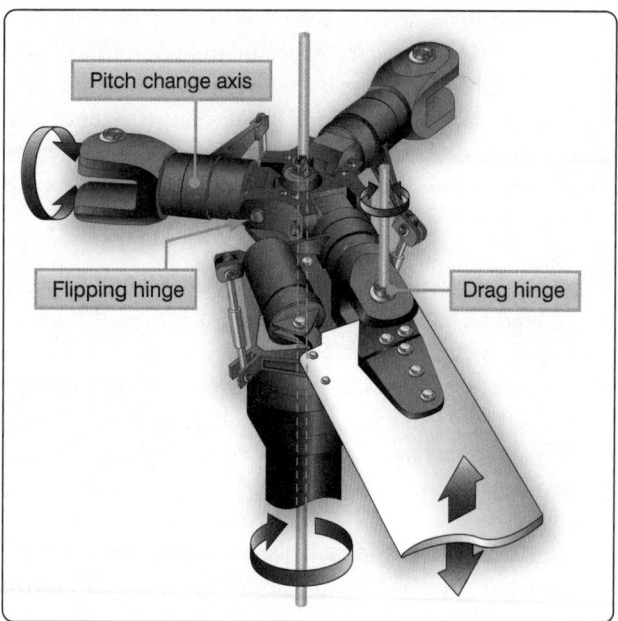

Figure 2-27. *Articulated rotor head.*

Rigid Rotor

The rigid rotor system is a rare design but potentially offers the best properties of both the fully articulated and semirigid rotors. In this design, the blade roots are rigidly attached to the rotor hub. The blades do not have hinges to allow lead-lag or flapping. Instead, the blades accommodate these motions by using elastomeric bearings. Elastomeric bearings are molded, rubber-like materials that are bonded to the appropriate parts. Instead of rotating like conventional bearings, they twist and flex to allow proper movement of the blades.

Forces Acting on the Helicopter

One of the differences between a helicopter and a fixed-wing aircraft is the main source of lift. The fixed-wing aircraft derives its lift from a fixed airfoil surface while the helicopter derives lift from a rotating airfoil called the rotor.

During hovering flight in a no-wind condition, the tip-path plane is horizontal, that is, parallel to the ground. Lift and thrust act straight up; weight and drag act straight down. The sum of the lift and thrust forces must equal the sum of the weight and drag forces in order for the helicopter to hover.

During vertical flight in a no-wind condition, the lift and thrust forces both act vertically upward. Weight and drag both act vertically downward. When lift and thrust equal weight and drag, the helicopter hovers; if lift and thrust are less than weight and drag, the helicopter descends vertically; if lift and thrust are greater than weight and drag, the helicopter rises vertically.

For forward flight, the tip-path plane is tilted forward, thus tilting the total lift-thrust force forward from the vertical. This resultant lift-thrust force can be resolved into two components: lift acting vertically upward and thrust acting horizontally in the direction of flight. In addition to lift and thrust, there is weight, the downward acting force, and drag, the rearward acting or retarding force of inertia and wind resistance.

In straight-and-level, unaccelerated forward flight, lift equals weight and thrust equals drag. (Straight-and-level flight is flight with a constant heading and at a constant altitude.) If lift exceeds weight, the helicopter climbs; if lift is less than weight, the helicopter descends. If thrust exceeds drag, the helicopter increases speed; if thrust is less than drag, it decreases speed.

In sideward flight, the tip-path plane is tilted sideward in the direction that flight is desired, thus tilting the total lift-thrust vector sideward. In this case, the vertical or lift component is still straight up, weight straight down, but the horizontal or thrust component now acts sideward with drag acting to the opposite side.

For rearward flight, the tip-path plane is tilted rearward and tilts the lift-thrust vector rearward. The thrust is then rearward and the drag component is forward, opposite that for forward flight. The lift component in rearward flight is straight up; weight, straight down.

Torque Compensation

Newton's third law of motion states "To every action there is an equal and opposite reaction." As the main rotor of a helicopter turns in one direction, the fuselage tends to rotate in the opposite direction. This tendency for the fuselage to rotate is called torque. Since torque effect on the fuselage is a direct result of engine power supplied to the main rotor, any change in engine power brings about a corresponding change in torque effect. The greater the engine power, the greater the torque effect. Since there is no engine power being supplied to the main rotor during autorotation, there is no torque reaction during autorotation.

The force that compensates for torque and provides for directional control can be produced by various means. The defining factor is dictated by the design of the helicopter, some of which do not have a torque issue. Single main rotor designs typically have an auxiliary rotor located on the end of the tail boom. This auxiliary rotor, generally referred to as a tail rotor, produces thrust in the direction opposite the torque reaction developed by the main rotor. *[Figure 2-25]* Foot pedals in the flight deck permit the pilot to increase or decrease tail rotor thrust, as needed, to neutralize torque effect.

Other methods of compensating for torque and providing directional control include the Fenestron® tail rotor system, an SUD Aviation design that employs a ducted fan enclosed by a shroud. Another design, called NOTAR®, a McDonnell Douglas design with no tail rotor, employs air directed through a series of slots in the tail boom, with the balance

Figure 2-28. *Aerospatiale Fenestron tail rotor system (left) and the McDonnell Douglas NOTAR® System (right).*

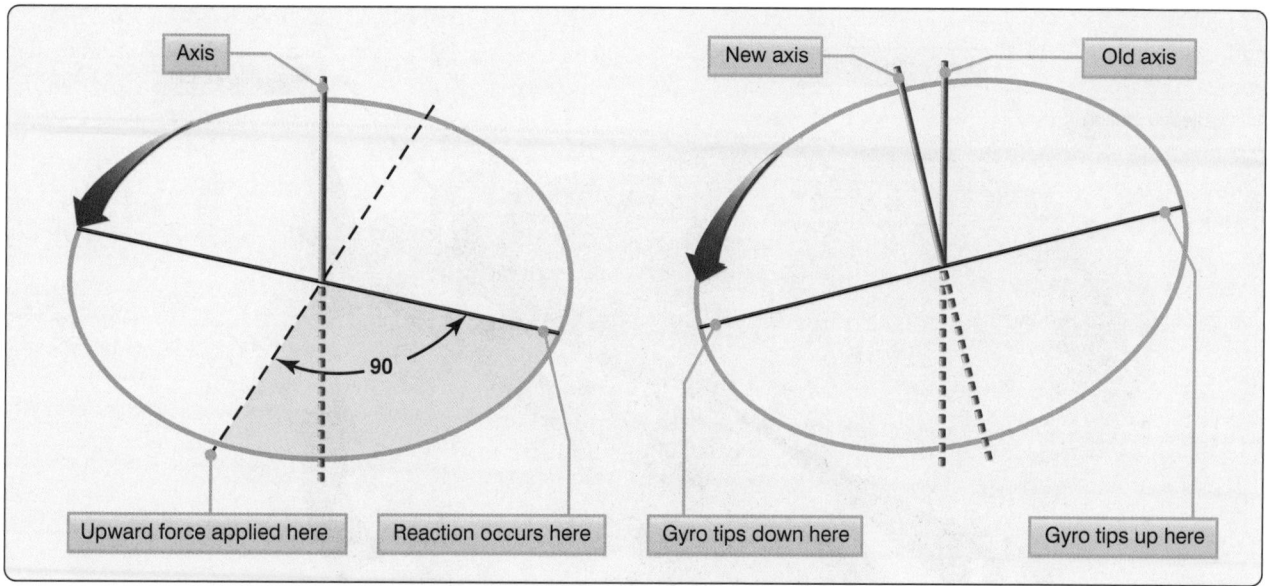

Figure 2-29. *Gyroscopic precession principle.*

exiting through a 90° duct located at the rear of the tail boom. *[Figure 2-28]*

Gyroscopic Forces

The spinning main rotor of a helicopter acts like a gyroscope. As such, it has the properties of gyroscopic action, one of which is precession. Gyroscopic precession is the resultant action or deflection of a spinning object when a force is applied to this object. This action occurs approximately 90° in the direction of rotation from the point where the force is applied. *[Figure 2-29]* Through the use of this principle, the tip-path plane of the main rotor may be tilted from the horizontal.

Examine a two-bladed rotor system to see how gyroscopic precession affects the movement of the tip-path plane. Moving the cyclic pitch control increases the AOA of one rotor blade with the result that a greater lifting force is applied at that point in the plane of rotation. This same control movement simultaneously decreases the AOA of the other blade the same amount, thus decreasing the lifting force applied at that point in the plane of rotation. The blade with the increased AOA tends to flap up; the blade with the decreased AOA tends to flap down. Because the rotor disc acts like a gyro, the blades reach maximum deflection at a point approximately 90° later in the plane of rotation. As shown in *Figure 2-30*, the retreating blade AOA is increased and the advancing blade AOA is decreased resulting in a tipping forward of the tip-path plane, since maximum deflection takes place 90° later when the blades are at the rear and front, respectively. In a rotor system using three or more blades, the movement of the cyclic pitch control changes the AOA of each blade an appropriate amount so that the end result is the same.

The movement of the cyclic pitch control in a two-bladed rotor system increases the AOA of one rotor blade with the result that a greater lifting force is applied at this point in the plane of rotation. This same control movement simultaneously decreases the AOA of the other blade a like amount, thus decreasing the lifting force applied at this point in the plane of rotation. The blade with the increased AOA tends to rise; the blade with the decreased AOA tends to lower. However, gyroscopic precession prevents the blades from rising or lowering to maximum deflection until a point approximately 90° later in the plane of rotation.

In a three-bladed rotor, the movement of the cyclic pitch control changes the AOA of each blade an appropriate amount so that the end result is the same, a tipping forward of the tip-path plane when the maximum change in AOA is made as each blade passes the same points at which the maximum increase and decrease are made for the two-bladed rotor as shown in *Figure 2-30*. As each blade passes the 90° position on the left, the maximum increase in AOA occurs. As each blade passes the 90° position to the right, the maximum decrease in AOA occurs. Maximum deflection takes place 90° later, maximum upward deflection at the rear and maximum downward deflection at the front; the tip-path plane tips forward.

Helicopter Flight Conditions

Hovering Flight

During hovering flight, a helicopter maintains a constant position over a selected point, usually a few feet above the ground. For a helicopter to hover, the lift and thrust produced by the rotor system act straight up and must equal the weight

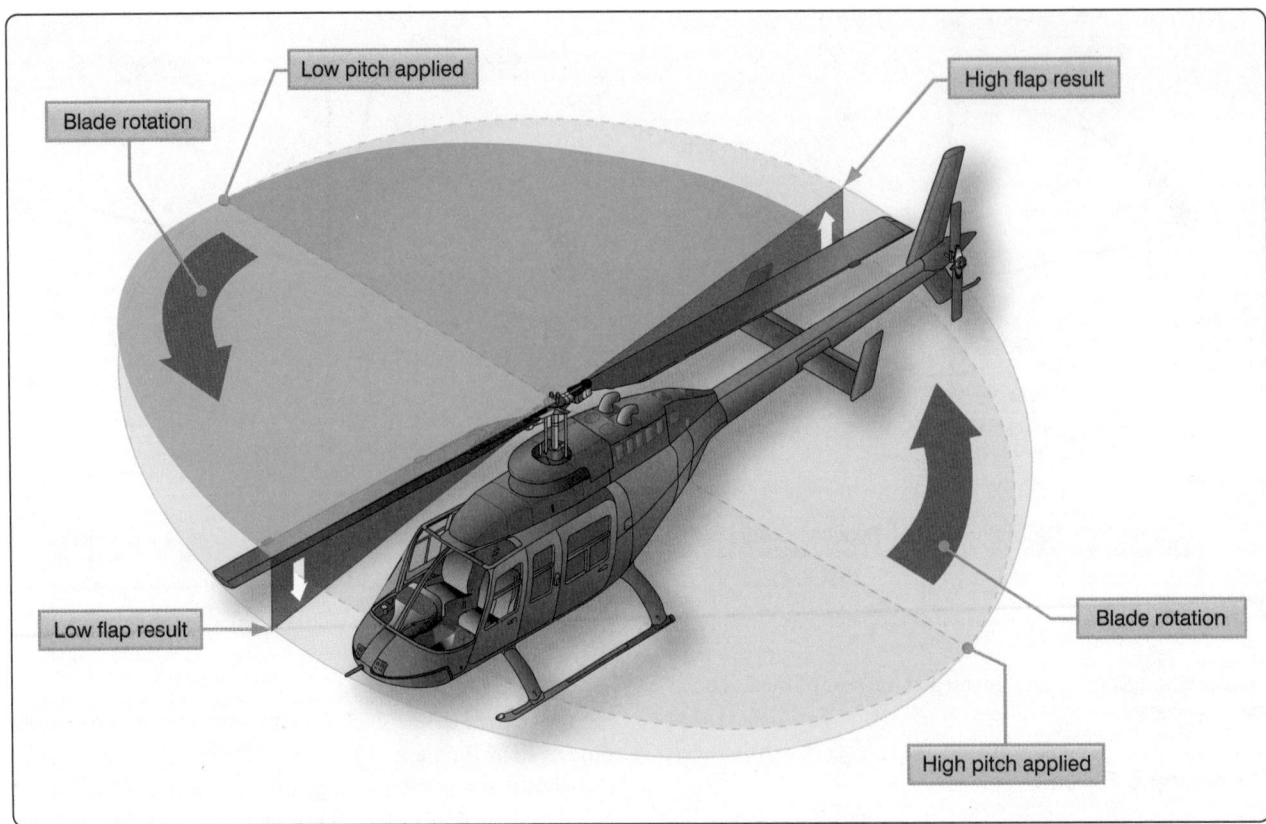

Figure 2-30. *Gyroscopic precession.*

and drag, which act straight down. *[Figure 2-31]* While hovering, the amount of main rotor thrust can be changed to maintain the desired hovering altitude. This is done by changing the angle of incidence (by moving the collective) of the rotor blades and hence the AOA of the main rotor blades. Changing the AOA changes the drag on the rotor blades, and the power delivered by the engine must change as well to keep the rotor speed constant.

The weight that must be supported is the total weight of the helicopter and its occupants. If the amount of lift is greater than the actual weight, the helicopter accelerates upwards until the lift force equals the weight gain altitude; if thrust is less than weight, the helicopter accelerates downward. When operating near the ground, the effect of the closeness to the ground changes this response.

The drag of a hovering helicopter is mainly induced drag incurred while the blades are producing lift. There is, however, some profile drag on the blades as they rotate through the air. Throughout the rest of this discussion, the term drag includes both induced and profile drag.

An important consequence of producing thrust is torque. As discussed earlier, Newton's third law states that for every action there is an equal and opposite reaction. Therefore, as the engine turns the main rotor system in a counterclockwise direction, the helicopter fuselage tends to turn clockwise. The amount of torque is directly related to the amount of engine power being used to turn the main rotor system. Remember, as power changes, torque changes.

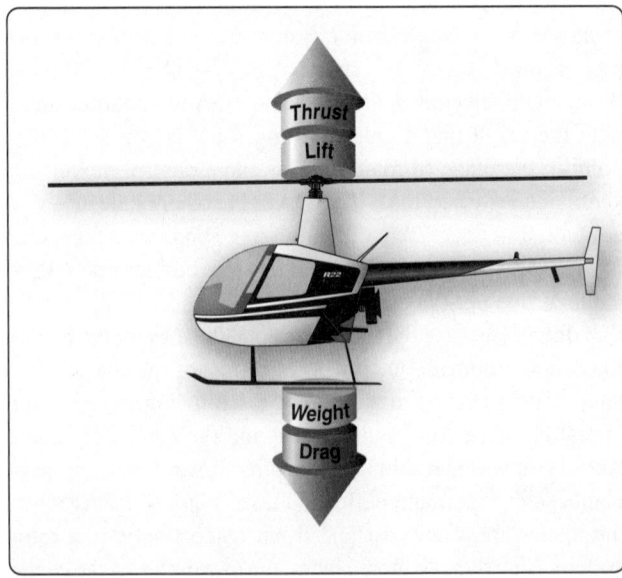

Figure 2-31. *To maintain a hover at a constant altitude, enough lift and thrust must be generated to equal the weight of the helicopter and the drag produced by the rotor blades.*

To counteract this torque-induced turning tendency, an antitorque rotor or tail rotor is incorporated into most helicopter designs. A pilot can vary the amount of thrust produced by the tail rotor in relation to the amount of torque produced by the engine. As the engine supplies more power to the main rotor, the tail rotor must produce more thrust to overcome the increased torque effect. This is done through the use of antitorque pedals.

Translating Tendency or Drift

During hovering flight, a single main rotor helicopter tends to drift or move in the direction of tail rotor thrust. This drifting tendency is called translating tendency. *[Figure 2-32]*

To counteract this drift, one or more of the following features may be used. All examples are for a counterclockwise rotating main rotor system.

- The main transmission is mounted at a slight angle to the left (when viewed from behind) so that the rotor mast has a built-in tilt to oppose the tail rotor thrust.

- Flight controls can be rigged so that the rotor disc is tilted to the right slightly when the cyclic is centered. Whichever method is used, the tip-path plane is tilted slightly to the left in the hover.

- If the transmission is mounted so that the rotor shaft is vertical with respect to the fuselage, the helicopter "hangs" left skid low in the hover. The opposite is true for rotor systems turning clockwise when viewed from above.

- In forward flight, the tail rotor continues to push to the right, and the helicopter makes a small angle with the wind when the rotors are level and the slip ball is in the middle. This is called inherent sideslip.

Ground Effect

When hovering near the ground, a phenomenon known as ground effect takes place. This effect usually occurs at heights between the surface and approximately one rotor diameter above the surface. The friction of the ground causes the downwash from the rotor to move outwards from the helicopter. This changes the relative direction of the downwash from a purely vertical motion to a combination of vertical and horizontal motion. As the induced airflow through the rotor disc is reduced by the surface friction, the lift vector increases. This allows a lower rotor blade angle for the same amount of lift, which reduces induced drag. Ground effect also restricts the generation of blade tip vortices due to the downward and outward airflow making a larger portion of the blade produce lift. When the helicopter gains altitude vertically, with no forward airspeed, induced airflow is no longer restricted, and the blade tip vortices increase with the decrease in outward airflow. As a result, drag increases

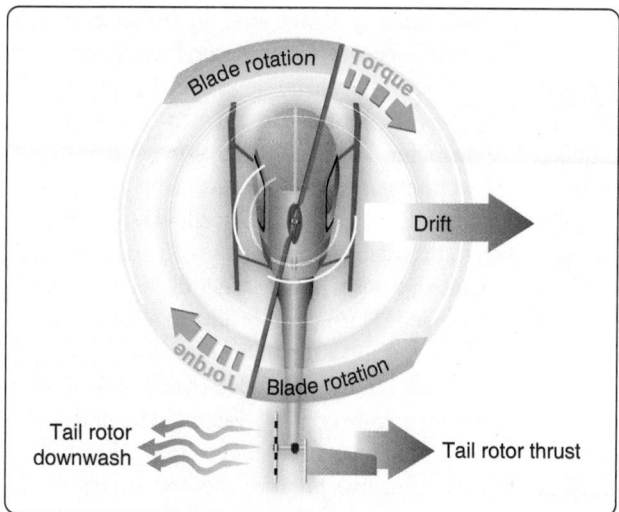

Figure 2-32. *A tail rotor is designed to produce thrust in a direction opposite torque. The thrust produced by the tail rotor is sufficient to move the helicopter laterally.*

which means a higher pitch angle, and more power is needed to move the air down through the rotor.

Ground effect is at its maximum in a no-wind condition over a firm, smooth surface. Tall grass, rough terrain, and water surfaces alter the airflow pattern, causing an increase in rotor tip vortices. *[Figure 2-33]*

Coriolis Effect (Law of Conservation of Angular Momentum)

The Coriolis effect is also referred to as the law of conservation of angular momentum. It states that the value of angular momentum of a rotating body does not change unless an external force is applied. In other words, a rotating body continues to rotate with the same rotational velocity until some external force is applied to change the speed of rotation. Angular momentum is moment of inertia (mass times distance from the center of rotation squared) multiplied by speed of rotation. Changes in angular velocity, known as angular acceleration and deceleration, take place as the mass of a rotating body is moved closer to or further away from the axis of rotation. The speed of the rotating mass increases or decreases in proportion to the square of the radius. An excellent example of this principle is a spinning ice skater. The skater begins rotation on one foot, with the other leg and both arms extended. The rotation of the skater's body is relatively slow. When a skater draws both arms and one leg inward, the moment of inertia (mass times radius squared) becomes much smaller and the body is rotating almost faster than the eye can follow. Because the angular momentum must remain constant (no external force applied), the angular velocity must increase. The rotor blade rotating about the rotor hub possesses angular momentum. As the rotor begins

2-21

to cone due to G-loading maneuvers, the diameter or the disc shrinks. Due to conservation of angular momentum, the blades continue to travel the same speed even though the blade tips have a shorter distance to travel due to reduced disc diameter. The action results in an increase in rotor rpm. Most pilots arrest this increase with an increase in collective pitch. Conversely, as G-loading subsides and the rotor disc flattens out from the loss of G-load induced coning, the blade tips now have a longer distance to travel at the same tip speed. This action results in a reduction of rotor rpm. However, if this drop in the rotor rpm continues to the point at which it attempts to decrease below normal operating rpm, the engine control system adds more fuel/power to maintain the specified engine rpm. If the pilot does not reduce collective pitch as the disc unloads, the combination of engine compensation for the rpm slow down and the additional pitch as G-loading increases may result in exceeding the torque limitations or power the engines can produce.

Vertical Flight

Hovering is actually an element of vertical flight. Increasing the AOA of the rotor blades (pitch) while keeping their rotation speed constant generates additional lift and the helicopter ascends. Decreasing the pitch causes the helicopter to descend. In a no wind condition, when lift and thrust are less than weight and drag, the helicopter descends vertically. If lift and thrust are greater than weight and drag, the helicopter ascends vertically. *[Figure 2-34]*

Forward Flight

In steady forward flight with no change in airspeed or vertical speed, the four forces of lift, thrust, drag, and weight must be in balance. Once the tip-path plane is tilted forward, the total lift-thrust force is also tilted forward. This resultant

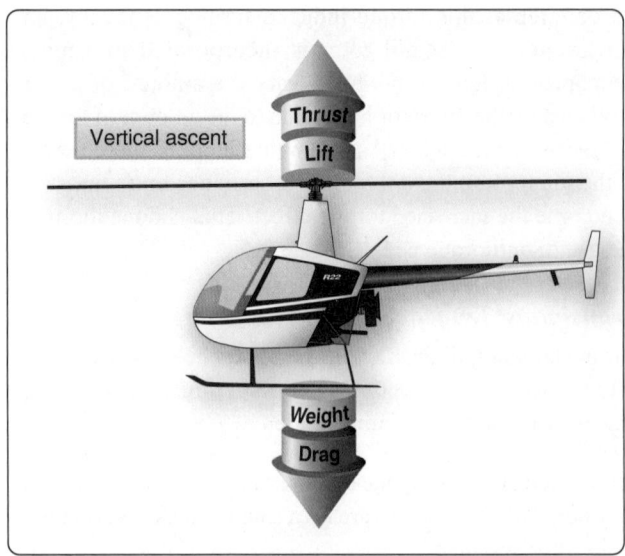

Figure 2-34. *To ascend vertically, more lift and thrust must be generated to overcome the forces of weight and drag.*

lift-thrust force can be resolved into two components—lift acting vertically upward and thrust acting horizontally in the direction of flight. In addition to lift and thrust, there is weight (the downward acting force) and drag (the force opposing the motion of an airfoil through the air). *[Figure 2-35]*

In straight-and-level (constant heading and at a constant altitude), unaccelerated forward flight, lift equals weight and thrust equals drag. If lift exceeds weight, the helicopter accelerates vertically until the forces are in balance; if thrust is less than drag, the helicopter slows until the forces are in balance. As the helicopter moves forward, it begins to lose altitude because lift is lost as thrust is diverted forward. However, as the helicopter begins to accelerate, the rotor system becomes more efficient due to the increased airflow. The result is excess power over that which is required to

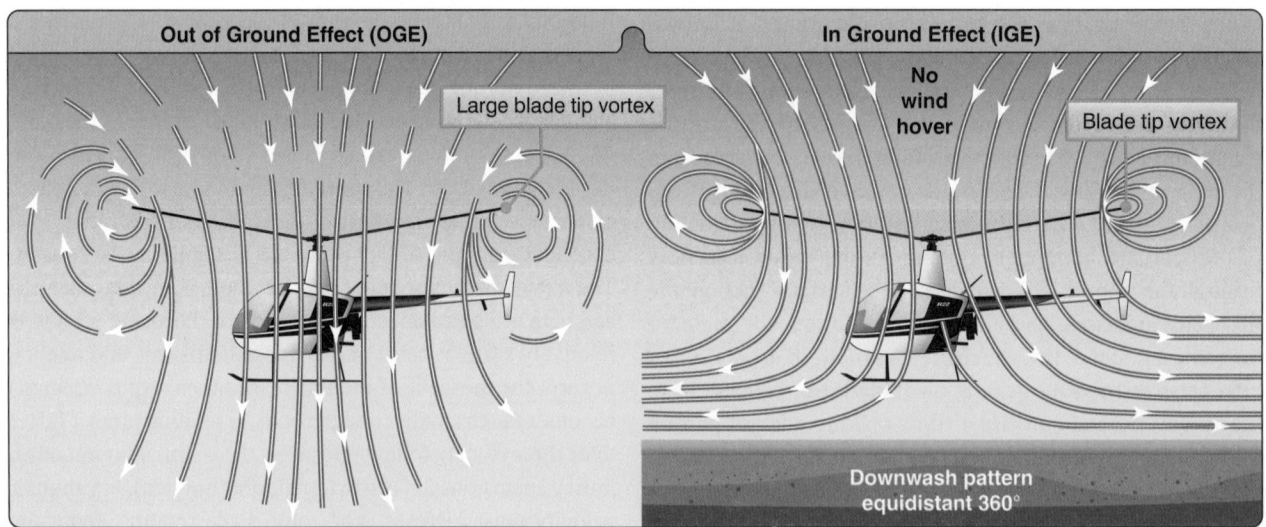

Figure 2-33. *Air circulation patterns change when hovering out of ground effect (OGE) and when hovering in ground effect (IGE).*

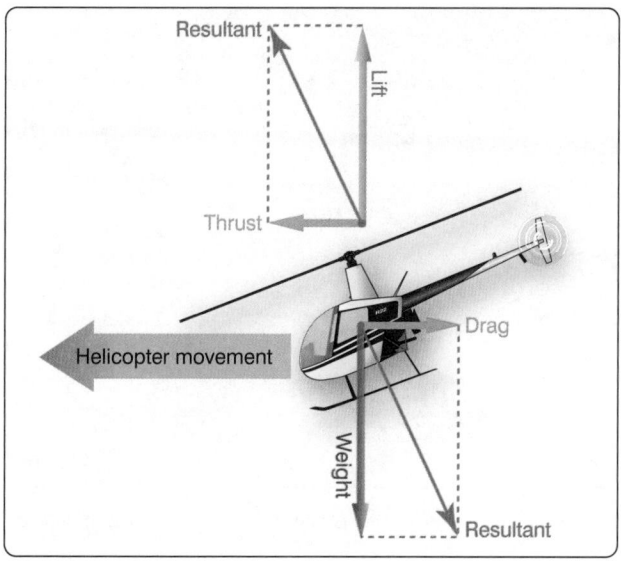

Figure 2-35. *The power required to maintain a straight-and-level flight and a stabilized airspeed.*

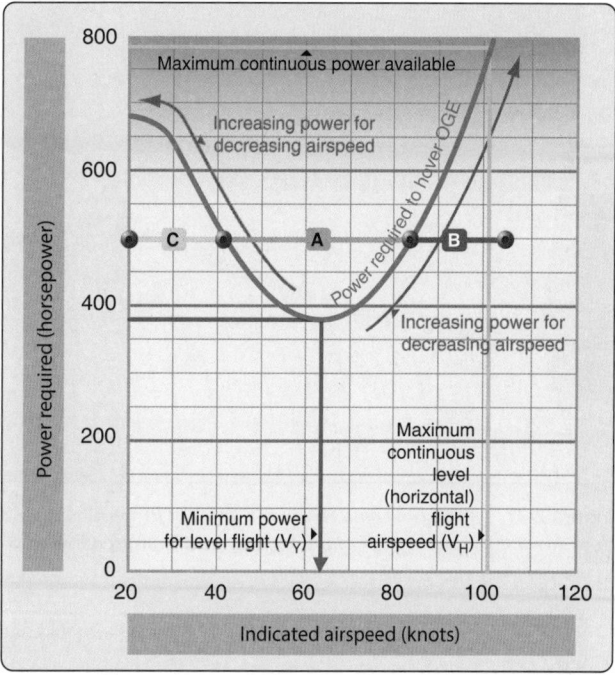

Figure 2-36. *Changing force vectors results in aircraft movement.*

hover. Continued acceleration causes an even larger increase in airflow through the rotor disc and more excess power. In order to maintain unaccelerated flight, the pilot must not make any changes in power or in cyclic movement. Any such changes would cause the helicopter to climb or descend. Once straight-and-level flight is obtained, the pilot should make note of the power (torque setting) required and not make major adjustments to the flight controls. *[Figure 2-36]*

Translational Lift

Improved rotor efficiency resulting from directional flight is called translational lift. The efficiency of the hovering rotor system is greatly improved with each knot of incoming wind gained by horizontal movement of the aircraft or surface wind. As incoming wind produced by aircraft movement or surface wind enters the rotor system, turbulence and vortices are left behind and the flow of air becomes more horizontal. In addition, the tail rotor becomes more aerodynamically efficient during the transition from hover to forward flight. Translational thrust occurs when the tail rotor becomes more aerodynamically efficient during the transition from hover to forward flight. As the tail rotor works in progressively less turbulent air, this improved efficiency produces more antitorque thrust, causing the nose of the aircraft to yaw left (with a main rotor turning counterclockwise) and forces the pilot to apply right pedal (decreasing the AOA in the tail rotor blades) in response. In addition, during this period, the airflow affects the horizontal components of the stabilizer found on most helicopters which tends to bring the nose of the helicopter to a more level attitude. *Figure 2-37* and *Figure 2-38* show airflow patterns at different speeds and how airflow affects the efficiency of the tail rotor.

Effective Translational Lift (ETL)

While transitioning to forward flight at about 16–24 knots, the helicopter experiences effective translational lift (ETL). As mentioned earlier in the discussion on translational lift, the rotor blades become more efficient as forward airspeed increases. Between 16–24 knots, the rotor system completely outruns the recirculation of old vortices and begins to work in relatively undisturbed air. The flow of air through the rotor system is more horizontal, therefore induced flow and induced drag are reduced. The AOA is subsequently increased, which makes the rotor system operate more efficiently. This increased efficiency continues with increased airspeed until the best climb airspeed is reached, and total drag is at its lowest point.

As speed increases, translational lift becomes more effective, the nose rises or pitches up, and the aircraft rolls to the right. The combined effects of dissymmetry of lift, gyroscopic precession, and transverse flow effect cause this tendency. It is important to understand these effects and anticipate correcting for them. Once the helicopter is transitioning through ETL, the pilot needs to apply forward and left lateral cyclic input to maintain a constant rotor-disc attitude. *[Figure 2-39]*

Dissymmetry of Lift

Dissymmetry of lift is the differential (unequal) lift between advancing and retreating halves of the rotor disc caused by the different wind flow velocity across each half. This difference in lift would cause the helicopter to be uncontrollable in any situation other than hovering in a calm wind. There must

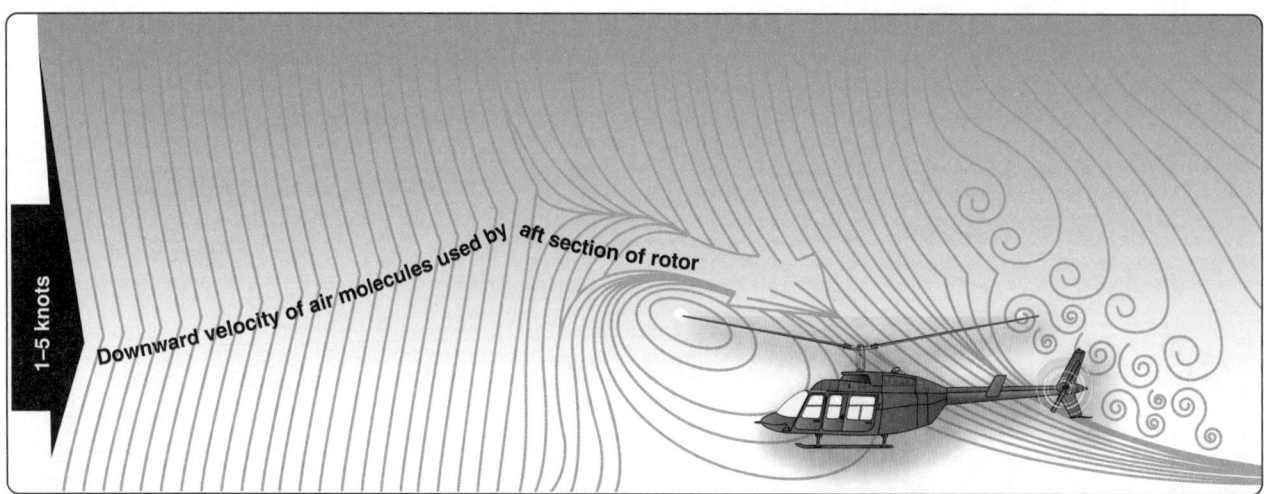

Figure 2-37. *The airflow pattern for 1–5 knots of forward airspeed. Note how the downwind vortex is beginning to dissipate and induced flow down through the rear of the rotor system is more horizontal.*

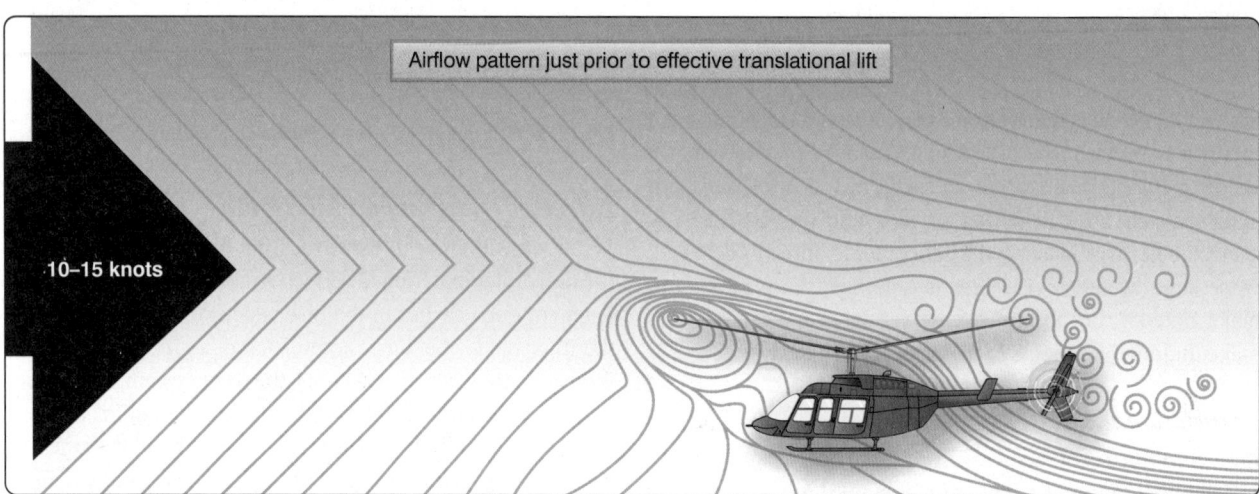

Figure 2-38. *An airflow pattern at a speed of 10–15 knots. At this increased airspeed, the airflow continues to become more horizontal. The leading edge of the downwash pattern is being overrun and is well back under the nose of the helicopter.*

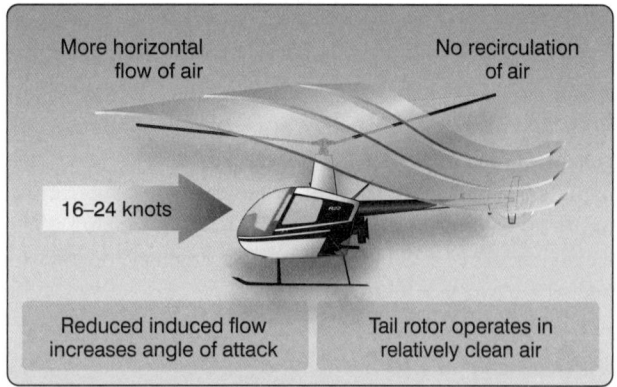

Figure 2-39. *Effective translational lift is easily recognized in actual flight by a transient induced aerodynamic vibration and increased performance of the helicopter.*

be a means of compensating, correcting, or eliminating this unequal lift to attain symmetry of lift.

When the helicopter moves through the air, the relative airflow through the main rotor disc is different on the advancing side than on the retreating side. The relative wind encountered by the advancing blade is increased by the forward speed of the helicopter; while the relative windspeed acting on the retreating blade is reduced by the helicopter's forward airspeed. Therefore, as a result of the relative windspeed, the advancing blade side of the rotor disc produces more lift than the retreating blade side. *[Figure 2-40]*

If this condition was allowed to exist, a helicopter with a counterclockwise main rotor blade rotation would roll to the left because of the difference in lift. In reality, the main rotor blades flap and feather automatically to equalize lift across the rotor disc. Articulated rotor systems, usually with three or

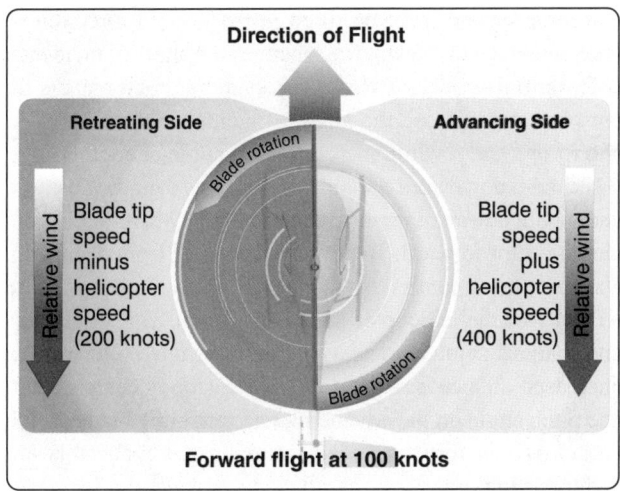

Figure 2-40. *The blade tip speed of this helicopter is approximately 300 knots. If the helicopter is moving forward at 100 knots, the relative windspeed on the advancing side is 400 knots. On the retreating side, it is only 200 knots. This difference in speed causes a dissymmetry of lift.*

more blades, incorporate a horizontal hinge (flapping hinge) to allow the individual rotor blades to move, or flap up and down as they rotate. A semirigid rotor system (two blades) utilizes a teetering hinge, which allows the blades to flap as a unit. When one blade flaps up, the other blade flaps down.

As the rotor blade reaches the advancing side of the rotor disc, it reaches its maximum upward flapping velocity. *[Figure 2-41A]* When the blade flaps upward, the angle between the chord line and the resultant relative wind decreases. This decreases the AOA, which reduces the amount of lift produced by the blade. At position C, the rotor blade is at its maximum downward flapping velocity. Due to downward flapping, the angle between the chord line and the resultant relative wind increases. This increases the AOA and thus the amount of lift produced by the blade.

The combination of blade flapping and slow relative wind acting on the retreating blade normally limits the maximum forward speed of a helicopter. At a high forward speed, the retreating blade stalls due to high AOA and slow relative wind speed. This situation is called "retreating blade stall" and is evidenced by a nose-up pitch, vibration, and a rolling tendency—usually to the left in helicopters with counterclockwise blade rotation. Pilots can avoid retreating blade stall by not exceeding the never-exceed speed. This speed is designated V_{NE} and is indicated on a placard and marked on the airspeed indicator by a red line.

During aerodynamic flapping of the rotor blades as they

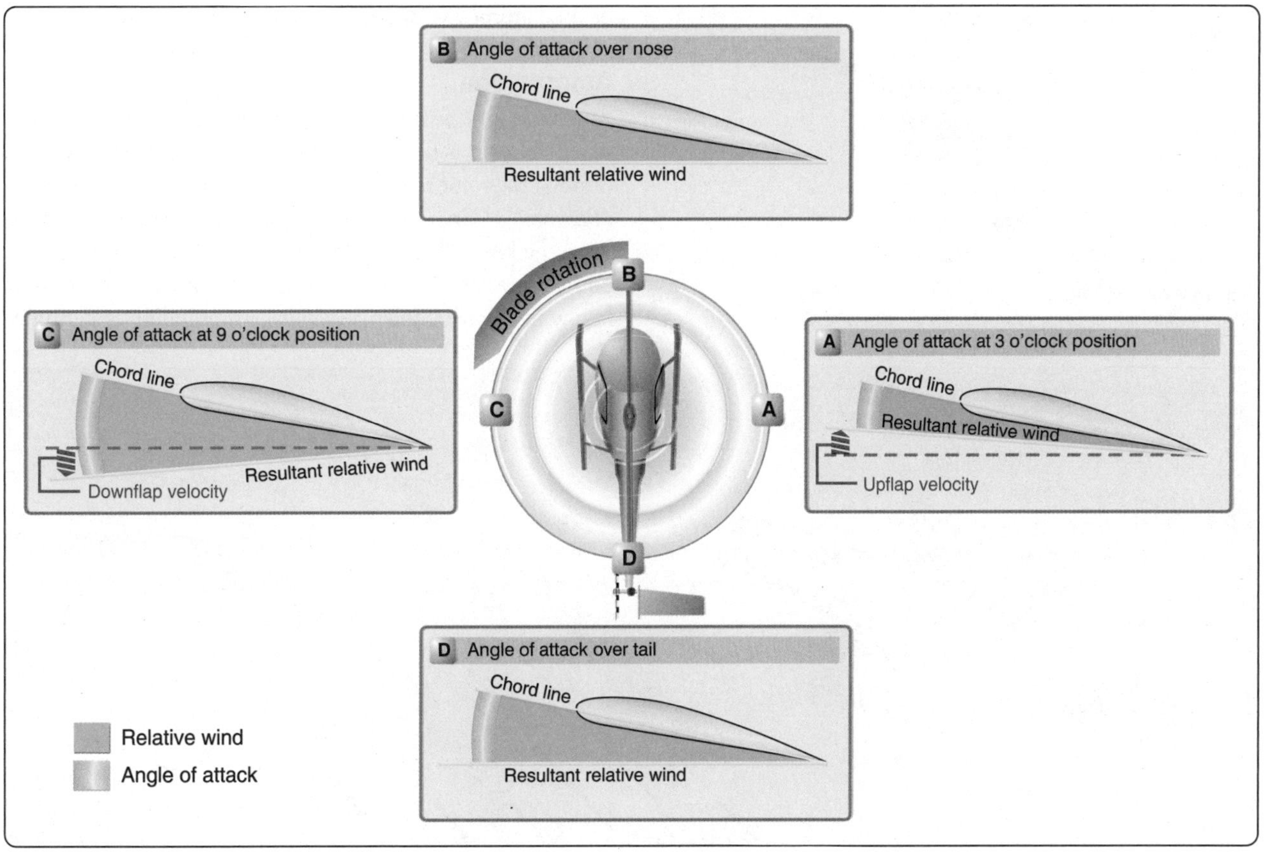

Figure 2-41. *The combined upward flapping (reduced lift) of the advancing blade and downward flapping (increased lift) of the retreating blade equalizes lift across the main rotor disc counteracting dissymmetry of lift.*

compensate for dissymmetry of lift, the advancing blade achieves maximum upward flapping displacement over the nose and maximum downward flapping displacement over the tail. This causes the tip-path plane to tilt to the rear and is referred to as blowback. *Figure 2-42* shows how the rotor disc is originally oriented with the front down following the initial cyclic input. As airspeed is gained and flapping eliminates dissymmetry of lift, the front of the disc comes up, and the back of the disc goes down. This reorientation of the rotor disc changes the direction in which total rotor thrust acts; the helicopter's forward speed slows, but can be corrected with cyclic input. The pilot uses cyclic feathering to compensate for dissymmetry of lift allowing them to control the attitude of the rotor disc.

Cyclic feathering compensates for dissymmetry of lift (changes the AOA) in the following way. At a hover, equal lift is produced around the rotor system with equal pitch and AOA on all the blades and at all points in the rotor system (disregarding compensation for translating tendency). The rotor disc is parallel to the horizon. To develop a thrust force, the rotor system must be tilted in the desired direction of movement. Cyclic feathering changes the angle of incidence differentially around the rotor system. Forward cyclic movements decrease the angle of incidence at one part on the rotor system while increasing the angle at another part. Maximum downward flapping of the blade over the nose and maximum upward flapping over the tail tilt both rotor disc and thrust vector forward. To prevent blowback from occurring, the pilot must continually move the cyclic forward as the velocity of the helicopter increases. *Figure 2-42* illustrates the changes in pitch angle as the cyclic is moved forward at increased airspeeds. At a hover, the cyclic is centered and the pitch angle on the advancing and retreating blades is the same. At low forward speeds, moving the cyclic forward reduces pitch angle on the advancing blade and increases pitch angle on the retreating blade. This causes a slight rotor tilt. At higher forward speeds, the pilot must continue to move the cyclic forward. This further reduces pitch angle on the advancing blade and further increases pitch angle on the retreating blade. As a result, there is even more tilt to the rotor than at lower speeds.

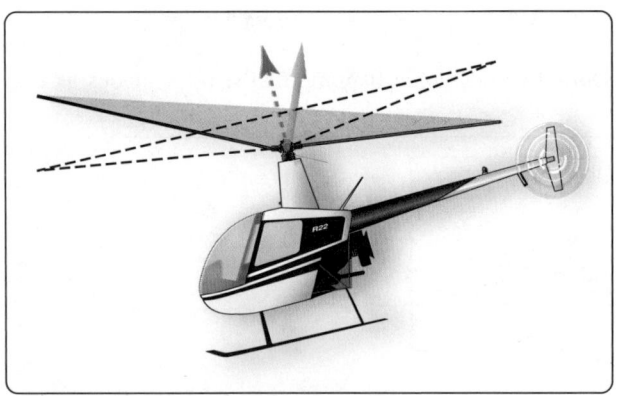

Figure 2-42. *To compensate for blowback, move the cyclic forward. Blowback is more pronounced with higher airspeeds.*

This horizontal lift component (thrust) generates higher helicopter airspeed. The higher airspeed induces blade flapping to maintain symmetry of lift. The combination of flapping and cyclic feathering maintains symmetry of lift and desired attitude on the rotor system and helicopter.

Autorotation

Autorotation is the state of flight in which the main rotor system of a helicopter is being turned by the action of air moving up through the rotor rather than engine power driving the rotor. *[Figure 2-43]* In normal, powered flight, air is drawn into the main rotor system from above and exhausted downward, but during autorotation, air moves up into the rotor system from below as the helicopter descends. Autorotation is permitted

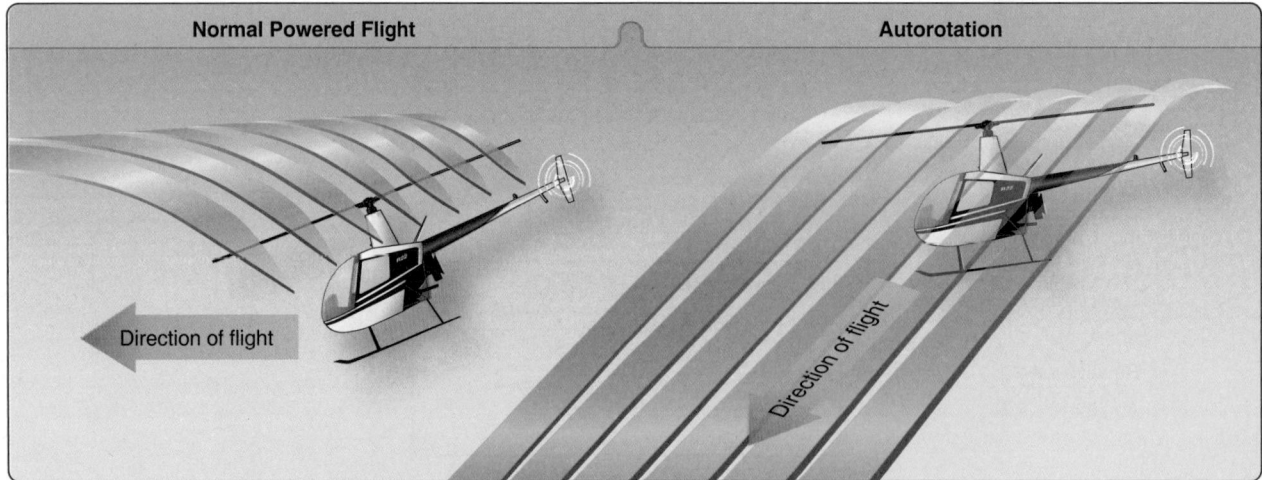

Figure 2-43. *During an autorotation, the upward flow of relative wind permits the main rotor blades to rotate at their normal speed. In effect, the blades are "gliding" in their rotational plane.*

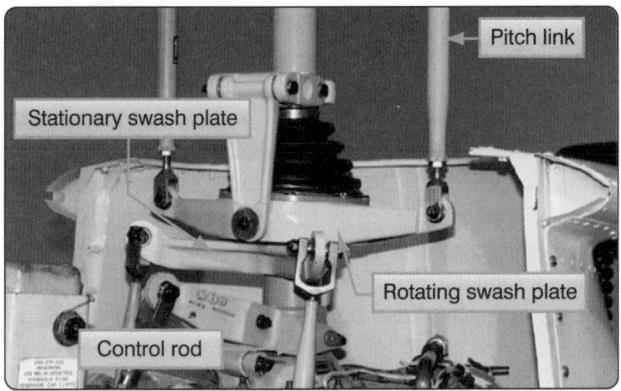

Figure 2-44. *Stationary and rotating swash plate.*

mechanically by a freewheeling unit, which is a special clutch mechanism that allows the main rotor to continue turning even if the engine is not running. If the engine fails, the freewheeling unit automatically disengages the engine from the main rotor allowing the main rotor to rotate freely. It is the means by which a helicopter can be landed safely in the event of an engine failure; consequently, all helicopters must demonstrate this capability in order to be certificated.

Rotorcraft Controls

Swash Plate Assembly

The purpose of the swash plate is to transmit control inputs from the collective and cyclic controls to the main rotor blades. It consists of two main parts: the stationary swash plate and the rotating swash plate. *[Figure 2-44]*

The stationary swash plate is mounted around the main rotor mast and connected to the cyclic and collective controls by a series of pushrods. It is restrained from rotating by an antidrive link but is able to tilt in all directions and move vertically. The rotating swash plate is mounted to the stationary swash plate by a uniball sleeve. It is connected to the mast by drive links and is allowed to rotate with the main rotor mast. Both swash plates tilt and slide up and down as one unit. The rotating swash plate is connected to the pitch horns by the pitch links.

There are three major controls in a helicopter that the pilot must use during flight. They are the collective pitch control, cyclic pitch control, and antitorque pedals or tail rotor control. In addition to these major controls, the pilot must also use the throttle control, which is mounted directly to the collective pitch control in order to fly the helicopter.

Collective Pitch Control

The collective pitch control is located on the left side of the pilot's seat and is operated with the left hand. The collective is used to make changes to the pitch angle of all the main rotor blades simultaneously, or collectively, as the name implies. As the collective pitch control is raised, there is a simultaneous and equal increase in pitch angle of all main rotor blades; as it is lowered, there is a simultaneous and equal decrease in pitch angle. *[Figure 2-45]* This is done through a series of mechanical linkages, and the amount

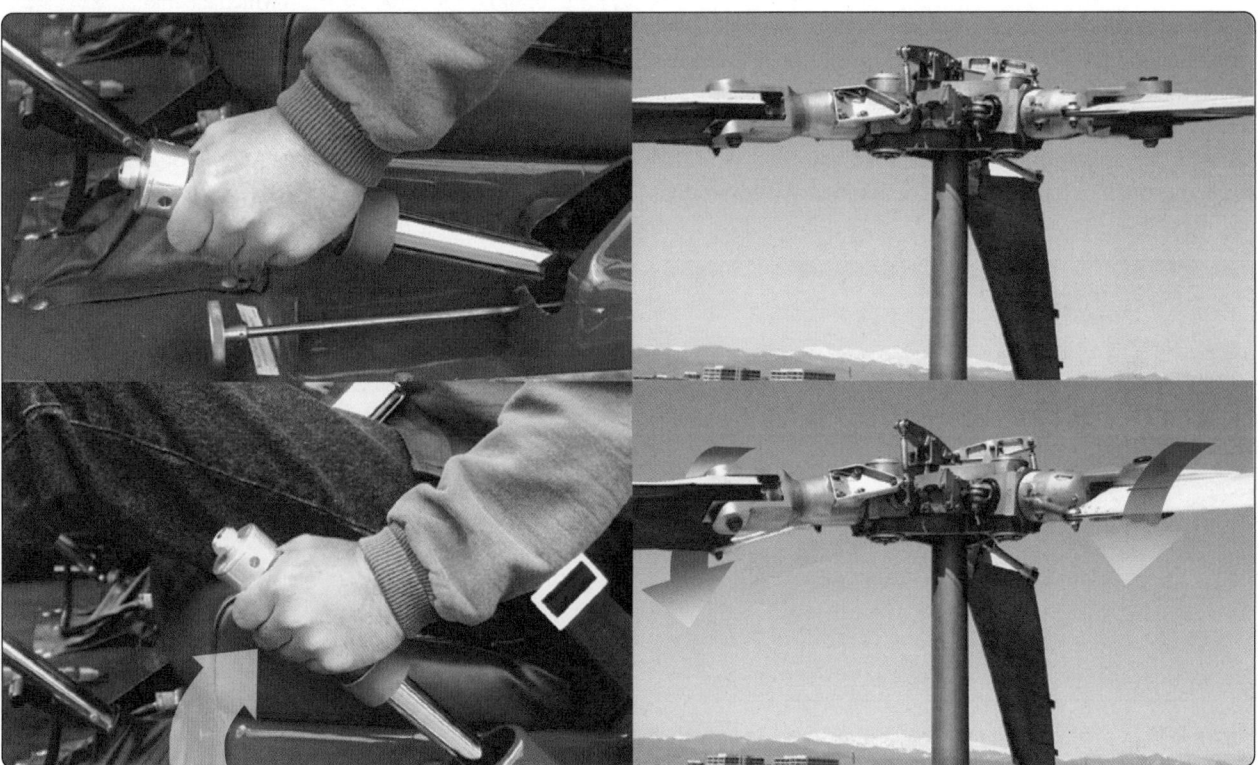

Figure 2-45. *Raising the collective pitch control increases the pitch angle by the same amount on all blades.*

of movement in the collective lever determines the amount of blade pitch change. An adjustable friction control helps prevent inadvertent collective pitch movement.

Throttle Control

The function of the throttle is to regulate engine rpm. If the correlator or governor system does not maintain the desired rpm when the collective is raised or lowered, or if those systems are not installed, the throttle must be moved manually with the twist grip to maintain rpm. The throttle control is much like a motorcycle throttle, and works almost the same way; twisting the throttle to the left increases rpm, twisting the throttle to the right decreases rpm. *[Figure 2-46]*

Governor/Correlator

A governor is a sensing device that senses rotor and engine rpm and makes the necessary adjustments in order to keep rotor rpm constant. Once the rotor rpm is set in normal operations, the governor keeps the rpm constant, and there is no need to make any throttle adjustments. Governors are common on all turbine helicopters (as it is a function of the fuel control system of the turbine engine), and used on some piston-powered helicopters.

A correlator is a mechanical connection between the collective lever and the engine throttle. When the collective lever is raised, power is automatically increased and when lowered, power is decreased. This system maintains rpm close to the desired value, but still requires adjustment of the throttle for fine tuning.

Some helicopters do not have correlators or governors and require coordination of all collective and throttle movements. When the collective is raised, the throttle must be increased; when the collective is lowered, the throttle must be decreased. As with any aircraft control, large adjustments of either collective pitch or throttle should be avoided. All corrections should be made with smooth pressure.

In piston helicopters, the collective pitch is the primary control for manifold pressure, and the throttle is the primary control for rpm. However, the collective pitch control also influences rpm, and the throttle also influences manifold pressure; therefore, each is considered to be a secondary control of the other's function. Both the tachometer (rpm indicator) and the manifold pressure gauge must be analyzed to determine which control to use. *Figure 2-47* illustrates this relationship.

Cyclic Pitch Control

The cyclic pitch control is mounted vertically from the flight deck floor, between the pilot's legs or, in some models, between the two pilot seats. *[Figure 2-48]* This primary flight control allows the pilot to fly the helicopter in any horizontal direction; fore, aft, and sideways. The total lift force is always perpendicular to the tip-path place of the main rotor. The purpose of the cyclic pitch control is to tilt the tip-path plane in the direction of the desired horizontal direction. The cyclic control changes the direction of this force and controls the attitude and airspeed of the helicopter.

The rotor disc tilts in the same direction the cyclic pitch control is moved. If the cyclic is moved forward, the rotor disc tilts forward; if the cyclic is moved aft, the disc tilts aft, and so on. Because the rotor disc acts like a gyro, the mechanical linkages for the cyclic control rods are rigged in such a way that they decrease the pitch angle of the rotor blade approximately 90° before it reaches the direction of cyclic displacement, and increase the pitch angle of the rotor blade approximately 90° after it passes the direction of displacement. An increase in pitch angle increases AOA; a decrease in pitch angle decreases AOA. For example, if the cyclic is moved forward, the AOA decreases as the rotor blade passes the right side of the helicopter and increases on the left side. This results in maximum downward deflection of the

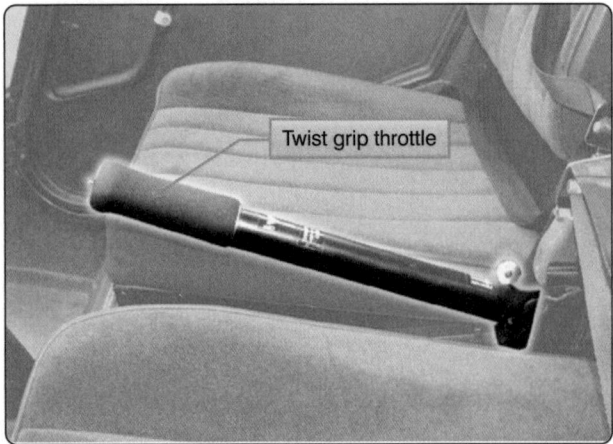

Figure 2-46. *A twist grip throttle is usually mounted on the end of the collective lever. The throttles on some turbine helicopters are mounted on the overhead panel or on the floor in the flight deck.*

If manifold pressure is	and rpm is	Solution
LOW	LOW	Increasing the throttle increases manifold pressure and rpm
HIGH	LOW	Lowering the collective pitch decreases manifold pressure and increases rpm
LOW	HIGH	Raising the collective pitch increases manifold pressure and decreases rpm
HIGH	HIGH	Reducing the throttle decreases manifold pressure and rpm

Figure 2-47. *Relationship between manifold pressure, rpm, collective, and throttle.*

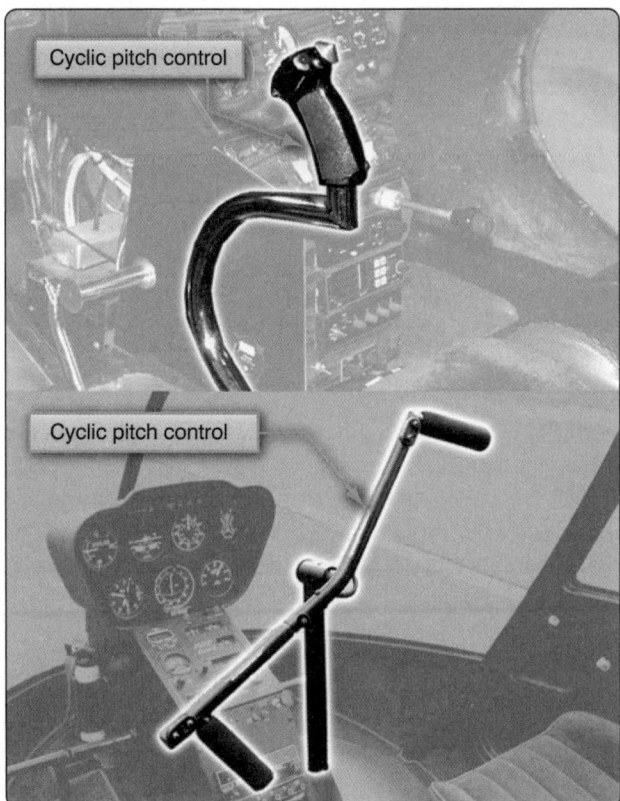

Figure 2-48. *The cyclic pitch control may be mounted vertically between the pilot's knees or on a teetering bar from a single cyclic located in the center of the helicopter. The cyclic can pivot in all directions.*

rotor blade in front of the helicopter and maximum upward deflection behind it, causing the rotor disc to tilt forward.

Antitorque Pedals

The antitorque pedals are located on the cabin floor by the pilot's feet. They control the pitch and, therefore, the thrust of the tail rotor blades. *[Figure 2-49]* Newton's third law applies to the helicopter fuselage and how it rotates in the opposite direction of the main rotor blades unless counteracted and controlled. To make flight possible and to compensate for this torque, most helicopter designs incorporate an antitorque rotor or tail rotor. The antitorque pedals allow the pilot to control the pitch angle of the tail rotor blades which in forward flight puts the helicopter in longitudinal trim and while at a hover, enables the pilot to turn the helicopter 360°. The antitorque pedals are connected to the pitch change mechanism on the tail rotor gearbox and allow the pitch angle on the tail rotor blades to be increased or decreased.

Helicopters that are designed with tandem rotors do not have an antitorque rotor. These helicopters are designed with both rotor systems rotating in opposite directions to counteract the torque, rather than using a tail rotor. Directional antitorque pedals are used for directional control of the aircraft while in

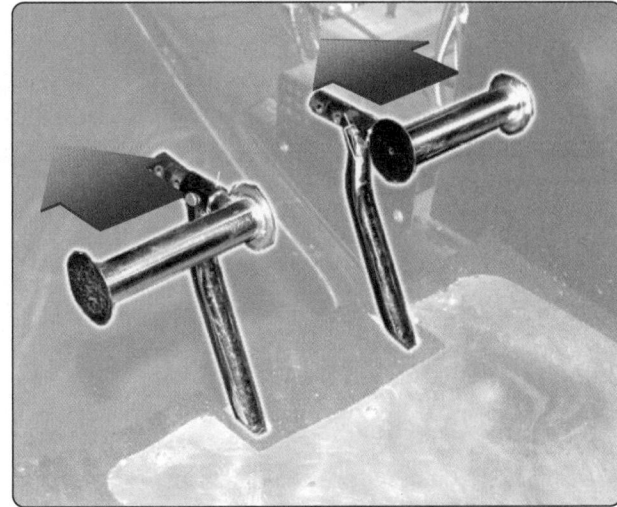

Figure 2-49. *Antitorque pedals compensate for changes in torque and control heading in a hover.*

flight, as well as while taxiing with the forward gear off the ground. With the right pedal displaced forward, the forward rotor disc tilts to the right, while the aft rotor disc tilts to the left. The opposite occurs when the left pedal is pushed forward; the forward rotor disc inclines to the left, and the aft rotor disc tilts to the right. Differing combinations of pedal and cyclic application can allow the tandem rotor helicopter to pivot about the aft or forward vertical axis, as well as pivoting about the center of mass.

Stabilizer Systems

Bell Stabilizer Bar System

Arthur M. Young discovered that stability could be increased significantly with the addition of a stabilizer bar perpendicular to the two blades. The stabilizer bar has weighted ends, which cause it to stay relatively stable in the plane of rotation. The stabilizer bar is linked with the swash plate in a manner that reduces the pitch rate. The two blades can flap as a unit and, therefore, do not require lag-lead hinges (the whole rotor slows down and accelerates per turn). Two-bladed systems require a single teetering hinge and two coning hinges to permit modest coning of the rotor disc as thrust is increased. The configuration is known under multiple names, including Hiller panels, Hiller system, Bell-Hiller system, and flybar system.

Offset Flapping Hinge

The offset flapping hinge is offset from the center of the rotor hub and can produce powerful moments useful for controlling the helicopter. The distance of the hinge from the hub (the offset) multiplied by the force produced at the hinge produces a moment at the hub. Obviously, the larger the offset, the greater the moment for the same force produced by the blade.

The flapping motion is the result of the constantly changing balance between lift, centrifugal, and inertial forces. This rising and falling of the blades is characteristic of most helicopters and has often been compared to the beating of a bird's wing. The flapping hinge, together with the natural flexibility found in most blades, permits the blade to droop considerably when the helicopter is at rest and the rotor is not turning over. During flight, the necessary rigidity is provided by the powerful centrifugal force that results from the rotation of the blades. This force pulls outward from the tip, stiffening the blade, and is the only factor that keeps it from folding up.

Stability Augmentation Systems (SAS)

Some helicopters incorporate stability augmentation systems (SAS) to help stabilize the helicopter in flight and in a hover. The simplest of these systems is a force trim system, which uses a magnetic clutch and springs to hold the cyclic control in the position at which it was released. More advanced systems use electric actuators that make inputs to the hydraulic servos. These servos receive control commands from a computer that senses helicopter attitude. Other inputs, such as heading, speed, altitude, and navigation information may be supplied to the computer to form a complete autopilot system. The SAS may be overridden or disconnected by the pilot at any time. SAS reduces pilot workload by improving basic aircraft control harmony and decreasing disturbances. These systems are very useful when the pilot is required to perform other duties, such as sling loading and search and rescue operations.

Helicopter Vibration

The following paragraphs describe the various types of vibrations. *Figure 2-50* shows the general levels into which frequencies are divided.

Extreme Low Frequency Vibration

Extreme low frequency vibration is pretty well limited to pylon rock. Pylon rocking (two to three cycles per second) is inherent with the rotor, mast, and transmission system. To keep the vibration from reaching noticeable levels, transmission mount dampening is incorporated to absorb the rocking.

Low Frequency Vibration

Low frequency vibrations (1/rev and 2/rev) are caused by the rotor itself. 1/rev vibrations are of two basic types: vertical

Helicopter Vibration Types	
Frequency Level	Vibration
Extreme low frequency	Less than 1/rev PYLON ROCK
Low frequency	1/rev or 2/rev type vibration
Medium frequency	Generally 4, 5, or 6/rev
High frequency	Tail rotor speed or faster

Figure 2-50. *Various helicopter vibration types.*

or lateral. A 1/rev is caused simply by one blade developing more lift at a given point than the other blade develops at the same point.

Medium Frequency Vibration

Medium frequency vibration (4/rev and 6/rev) is another vibration inherent in most rotors. An increase in the level of these vibrations is caused by a change in the capability of the fuselage to absorb vibration, or a loose airframe component, such as the skids, vibrating at that frequency.

High Frequency Vibration

High frequency vibrations can be caused by anything in the helicopter that rotates or vibrates at extremely high speeds. A high frequency vibration typically occurs when the tail rotor gears, tail drive shaft or the tail rotor engine, fan or shaft assembly vibrates or rotates at an equal or greater speed than the tail rotor.

Rotor Blade Tracking

Blade tracking is the process of determining the positions of the tips of the rotor blade relative to each other while the rotor head is turning, and of determining the corrections necessary to hold these positions within certain tolerances. The blades should all track one another as closely as possible. The purpose of blade tracking is to bring the tips of all blades into the same tip path throughout their entire cycle of rotation. Various methods of blade tracking are explained below.

Flag & Pole

The flag and pole method, as shown in *Figure 2-51,* shows the relative positions of the rotor blades. The blade tips are marked with chalk or a grease pencil. Each blade tip should be marked with a different color so that it is easy to determine the relationship of the other tips of the rotor blades to each other. This method can be used on all types of helicopters that do not have jet propulsion at the blade tips. Refer to the applicable maintenance manual for specific procedures.

Electronic Blade Tracker

The most common electronic blade tracker consists of a Balancer/Phazor, Strobex tracker, and Vibrex tester. *[Figures 2-52 through 2-54]* The Strobex blade tracker permits blade tracking from inside or outside the helicopter while on the ground or inside the helicopter in flight. The system uses a highly concentrated light beam flashing in sequence with the rotation of the main rotor blades so that a fixed target at the blade tips appears to be stopped. Each blade is identified by an elongated retroreflective number taped or attached to the underside of the blade in a uniform location. When viewed at an angle from inside the helicopter, the taped numbers will appear normal. Tracking can be accomplished with tracking tip cap reflectors and a strobe light. The tip

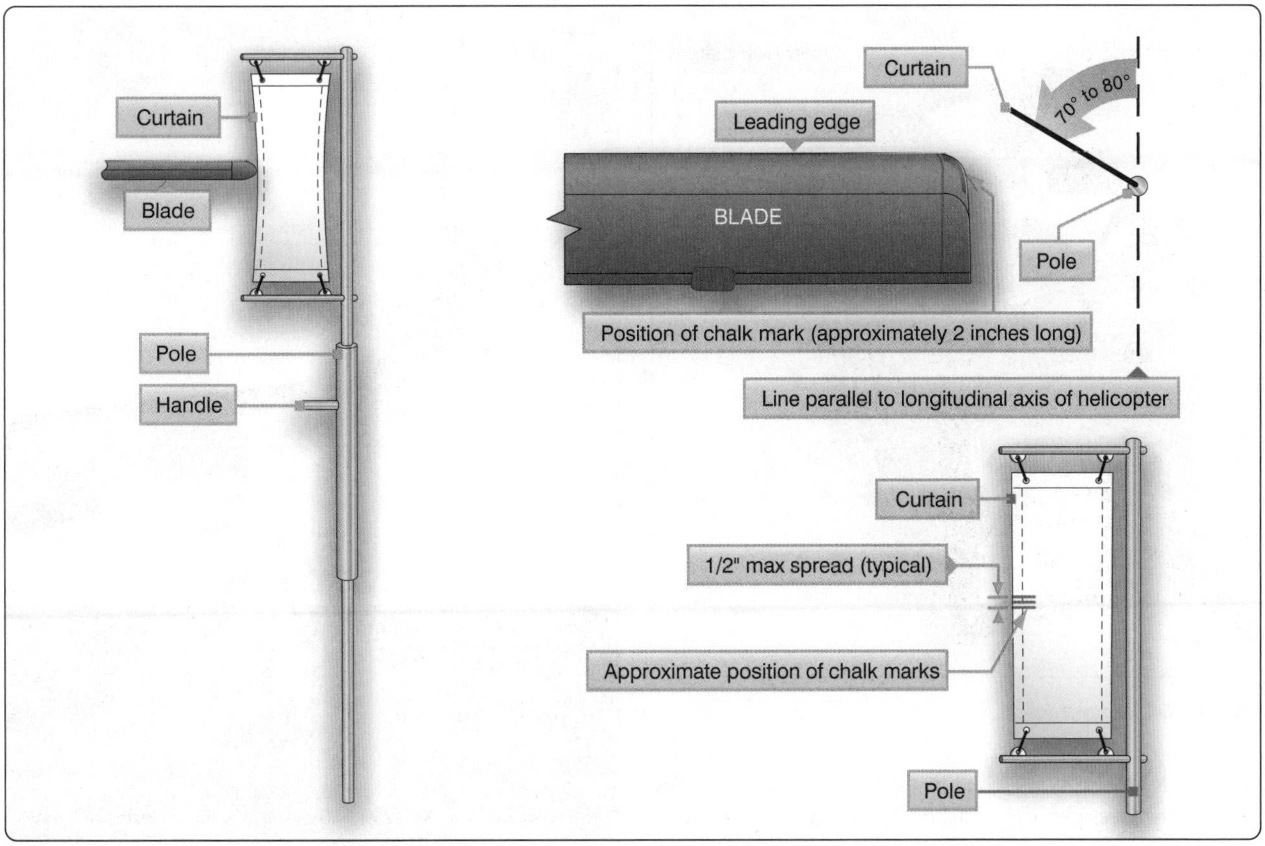

Curtain

Blade

Pole

Handle

Curtain

Leading edge

70° to 80°

BLADE

Pole

Position of chalk mark (approximately 2 inches long)

Line parallel to longitudinal axis of helicopter

Curtain

1/2" max spread (typical)

Approximate position of chalk marks

Pole

Figure 2-51. *Flag and pole blade tracking.*

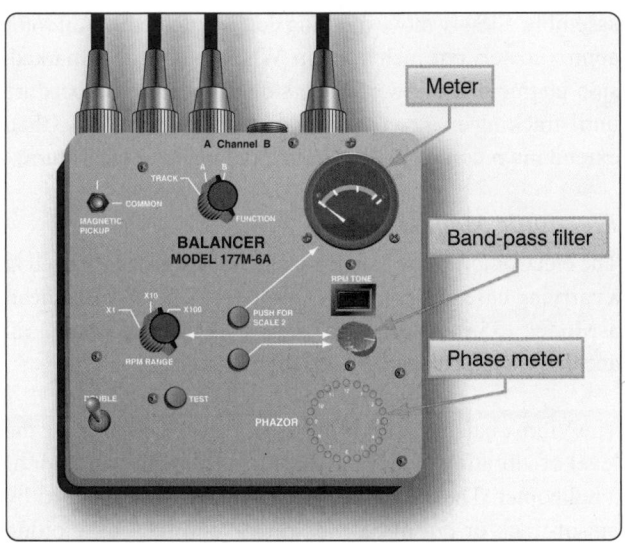

Meter

A Channel B

TRACK

COMMON

MAGNETIC
PICKUP

FUNCTION

BALANCER
MODEL 177M-6A

Band-pass filter

RPM TONE

X10

X1   X100

PUSH FOR
SCALE 2

Phase meter

RPM RANGE

TROUBLE   TEST

PHAZOR

Figure 2-52. *Balancer/Phazor.*

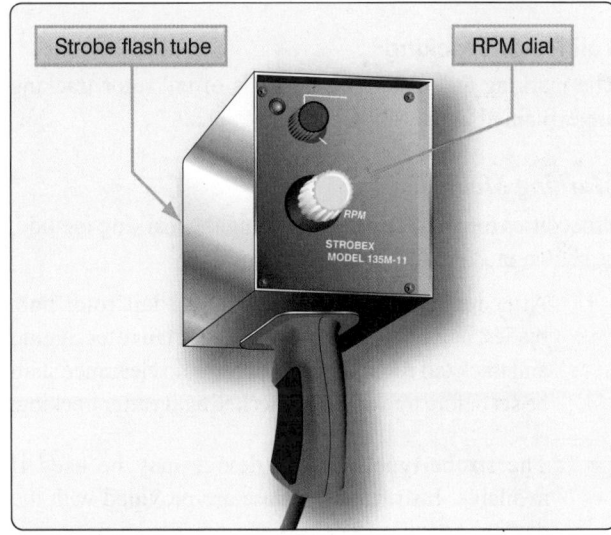

Strobe flash tube

RPM dial

RPM

STROBEX
MODEL 135M-11

Figure 2-53. *Strobex tracker.*

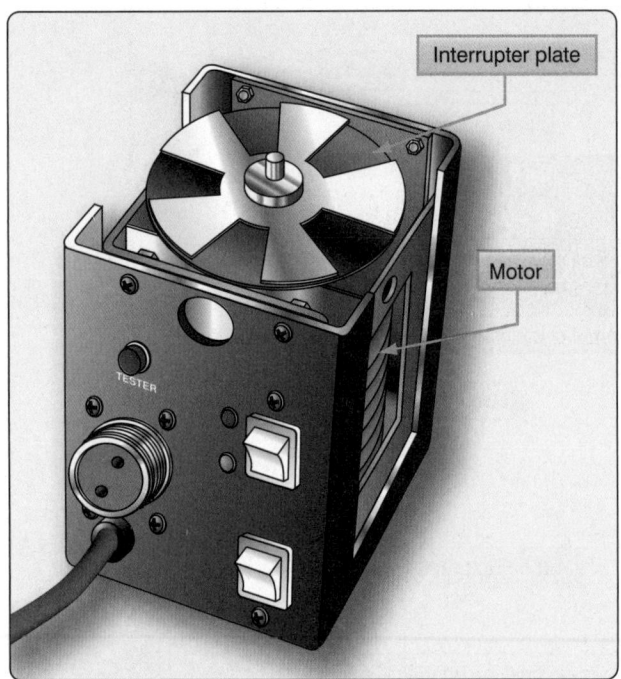

Figure 2-54. *Vibrex tracker.*

caps are temporarily attached to the tip of each blade. The high-intensity strobe light flashes in time with the rotating blades. The strobe light operates from the aircraft electrical power supply. By observing the reflected tip cap image, it is possible to view the track of the rotating blades. Tracking is accomplished in a sequence of four separate steps: ground tracking, hover verification, forward flight tracking, and autorotation rpm adjustment.

Tail Rotor Tracking

The marking and electronic methods of tail rotor tracking are explained in the following paragraphs.

Marking Method

Procedures for tail rotor tracking using the marking method, as shown in *Figure 2-55*, are as follows:

- After replacement or installation of tail rotor hub, blades, or pitch change system, check tail rotor rigging and track tail rotor blades. Tail rotor tip clearance shall be set before tracking and checked again after tracking.

- The strobe-type tracking device may be used if available. Instructions for use are provided with the device. Attach a piece of soft rubber hose six inches long on the end of a ½ × ½ inch pine stick or other flexible device. Cover the rubber hose with Prussian blue or similar type of coloring thinned with oil.

Note: Ground run-up shall be performed by authorized personnel only. Start engine in accordance with applicable

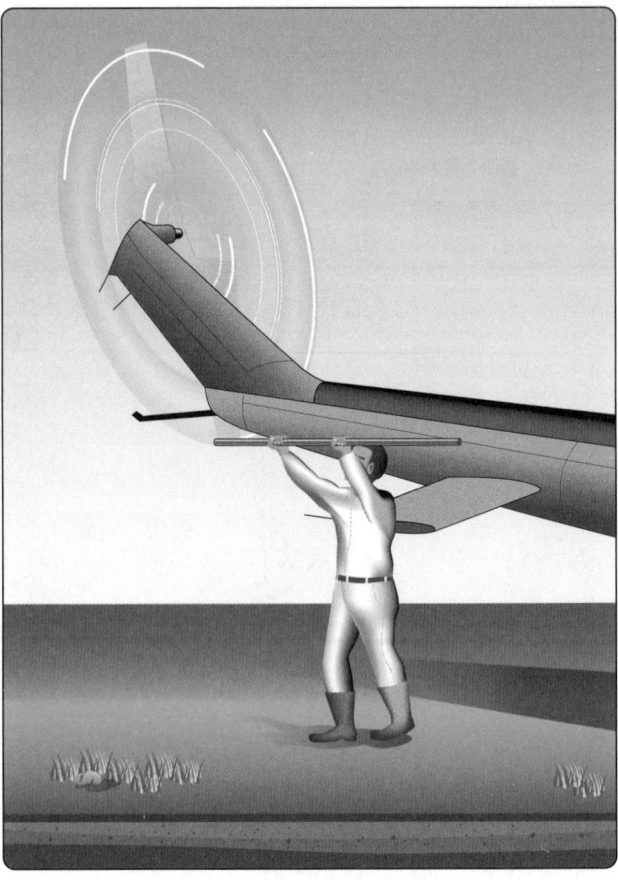

Figure 2-55. *Tail rotor tracking.*

maintenance manual. Run engine with pedals in neutral position. Reset marking device on underside of tail boom assembly. Slowly move marking device into disc of tail rotor approximately one inch from tip. When near blade is marked, stop engine and allow rotor to stop. Repeat this procedure until tracking mark crosses over to the other blade, then extend pitch control link of unmarked blade one half turn.

Electronic Method

The electronic Vibrex balancing and tracking kit is housed in a carrying case and consists of a Model 177M-6A Balancer, a Model 135M-11 Strobex, track and balance charts, an accelerometer, cables, and attaching brackets.

The Vibrex balancing kit is used to measure and indicate the level of vibration induced by the main rotor and tail rotor of a helicopter. The Vibrex analyzes the vibration induced by out-of-track or out-of-balance rotors, and then by plotting vibration amplitude and clock angle on a chart the amount and location of rotor track or weight change is determined. In addition, the Vibrex is used in troubleshooting by measuring the vibration levels and frequencies or rpm of unknown disturbances.

Rotor Blade Preservation & Storage

Accomplish the following requirements for rotor blade preservation and storage:

- Condemn, demilitarize, and dispose of locally any blade which has incurred nonrepairable damage.

- Tape all holes in the blade, such as tree damage, or foreign object damage (FOD) to protect the interior of the blade from moisture and corrosion.

- Thoroughly remove foreign matter from the entire exterior surface of blade with mild soap and water.

- Protect blade outboard eroded surfaces with a light coating of corrosion preventive or primer coating.

- Protect blade main bolt hole bushing, drag brace retention bolt hole bushing, and any exposed bare metal (i.e., grip and drag pads) with a light coating of corrosion preventive.

- Secure blade to shock-mounted support and secure container lid.

- Place copy of manufacturer's blade records, containing information required by Title 14 of the Code of Federal Regulations (14 CFR) section 91.417(a)(2)(ii), and any other blade records in a waterproof bag and insert into container record tube.

- Obliterate old markings from the container that pertained to the original shipment or to the original item it contained. Annotate the blade model, part number (P/N) and serial number, as applicable, on the outside of the container.

Helicopter Power Systems

Powerplant

The two most common types of engines used in helicopters are the reciprocating engine and the turbine engine. Reciprocating engines, also called piston engines, are generally used in smaller helicopters. Most training helicopters use reciprocating engines because they are relatively simple and inexpensive to operate. Turbine engines are more powerful and are used in a wide variety of helicopters. They produce a tremendous amount of power for their size but are generally more expensive to operate.

Reciprocating Engine

The reciprocating engine consists of a series of pistons connected to a rotating crankshaft. As the pistons move up and down, the crankshaft rotates. The reciprocating engine gets its name from the back-and-forth movement of its internal parts. The four-stroke engine is the most common type, and refers to the four different cycles the engine undergoes to produce power. *[Figure 2-56]*

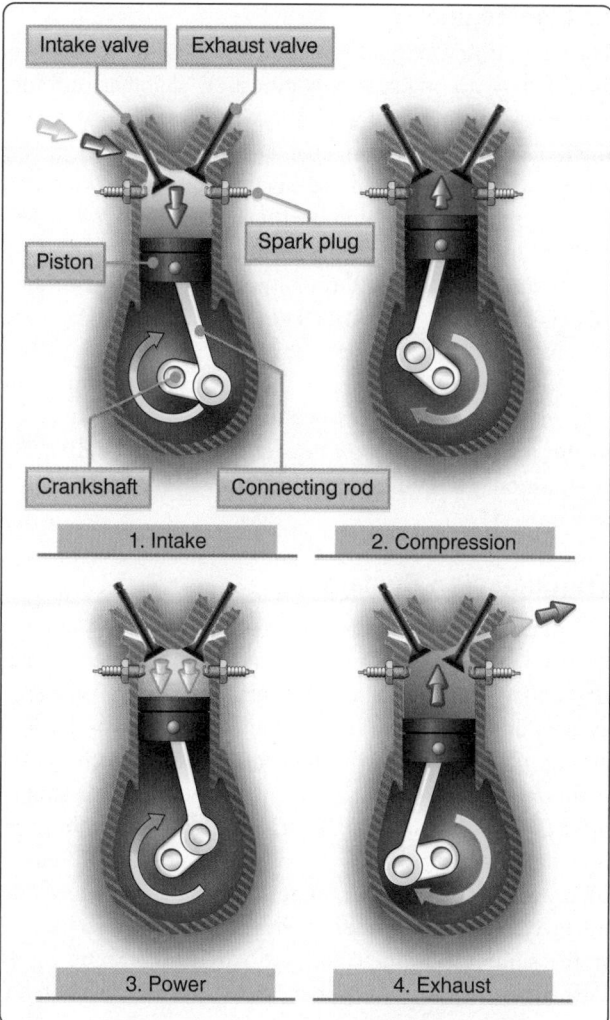

Figure 2-56. *The arrows indicate the direction of motion of the crankshaft and piston during the four-stroke cycle.*

When the piston moves away from the cylinder head on the intake stroke, the intake valve opens and a mixture of fuel and air is drawn into the combustion chamber. As the cylinder moves back toward the cylinder head, the intake valve closes, and the air-fuel mixture is compressed. When compression is nearly complete, the spark plugs fire and the compressed mixture is ignited to begin the power stroke. The rapidly expanding gases from the controlled burning of the air-fuel mixture drive the piston away from the cylinder head, thus providing power to rotate the crankshaft. The piston then moves back toward the cylinder head on the exhaust stroke where the burned gases are expelled through the opened exhaust valve. Even when the engine is operated at a fairly low speed, the four-stroke cycle takes place several hundred times each minute. In a four-cylinder engine, each cylinder operates on a different stroke. Continuous rotation of a crankshaft is maintained by the precise timing of the power strokes in each cylinder.

Turbine Engine

The gas turbine engine mounted on most helicopters is made up of a compressor, combustion chamber, turbine, and accessory gearbox assembly. The compressor draws filtered air into the plenum chamber and compresses it. The compressed air is directed to the combustion section through discharge tubes where atomized fuel is injected into it. The air-fuel mixture is ignited and allowed to expand. This combustion gas is then forced through a series of turbine wheels causing them to turn. These turbine wheels provide power to both the engine compressor and the accessory gearbox. Power is provided to the main rotor and tail rotor systems through the freewheeling unit which is attached to the accessory gearbox power output gear shaft. The combustion gas is finally expelled through an exhaust outlet. *[Figure 2-57]*

Transmission System

The transmission system transfers power from the engine to the main rotor, tail rotor, and other accessories during normal flight conditions. The main components of the transmission system are the main rotor transmission, tail rotor drive system, clutch, and freewheeling unit. The freewheeling unit, or autorotative clutch, allows the main rotor transmission to drive the tail rotor drive shaft during autorotation. Helicopter transmissions are normally lubricated and cooled with their own oil supply. A sight gauge is provided to check the oil level. Some transmissions have chip detectors located in the sump. These detectors are wired to warning lights located on the pilot's instrument panel that illuminate in the event of an internal problem. The chip detectors on modern

helicopters have a "burn off" capability and attempt to correct the situation without pilot action. If the problem cannot be corrected on its own, the pilot must refer to the emergency procedures for that particular helicopter.

Main Rotor Transmission

The primary purpose of the main rotor transmission is to reduce engine output rpm to optimum rotor rpm. This reduction is different for the various helicopters. As an example, suppose the engine rpm of a specific helicopter is 2,700. A rotor speed of 450 rpm would require a 6:1 reduction. A 9:1 reduction would mean the rotor would turn at 300 rpm. Most helicopters use a dual-needle tachometer or a vertical scale instrument to show both engine and rotor rpm or a percentage of engine and rotor rpm. The rotor rpm indicator normally is used only during clutch engagement to monitor rotor acceleration, and in autorotation to maintain rpm within prescribed limits. *[Figure 2-58]*

In helicopters with horizontally mounted engines, another purpose of the main rotor transmission is to change the axis of rotation from the horizontal axis of the engine to the vertical axis of the rotor shaft. *[Figure 2-59]*

Clutch

In a conventional airplane, the engine and propeller are directly connected. However, in a helicopter there is a different relationship between the engine and the rotor. Because of the greater weight of a rotor in relation to the power of the engine, as compared to the weight of a propeller and the power in an airplane, the rotor must be disconnected from the engine when the starter is engaged. A clutch allows

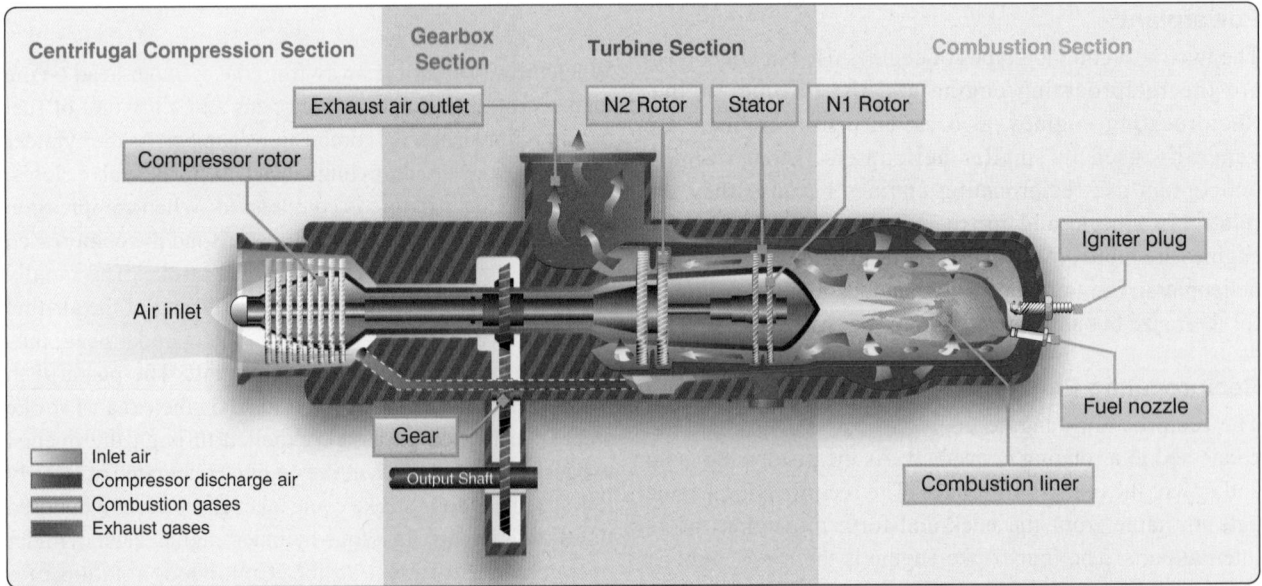

Figure 2-57. *Many helicopters use a turboshaft engine as shown above to drive the main transmission and rotor systems. The main difference between a turboshaft and a turbojet engine is that most of the energy produced by the expanding gases is used to drive a turbine rather than producing thrust through the expulsion of exhaust gases.*

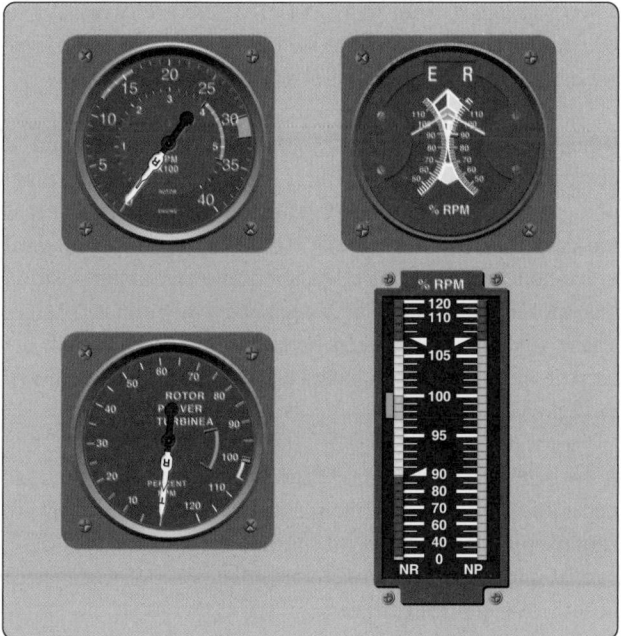

Figure 2-58. *There are various types of dual-needle tachometers; however, when the needles are superimposed, or married, the ratio of the engine rpm is the same as the gear reduction ratio.*

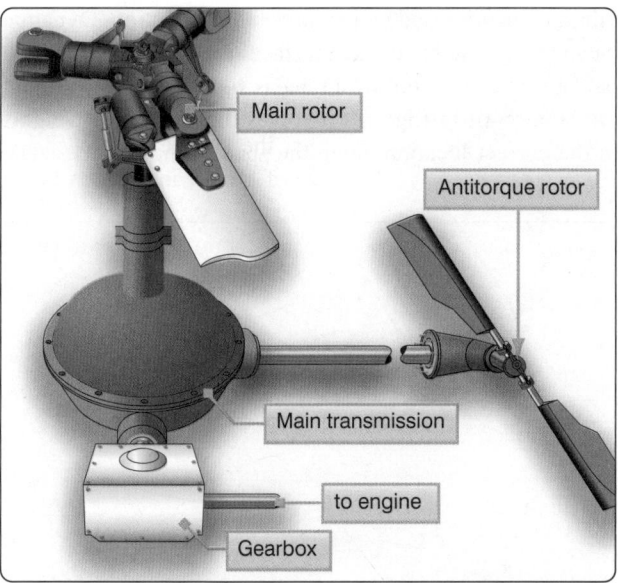

Figure 2-59. *The main rotor transmission and gearbox reduce engine output rpm to optimum rotor rpm and change the axis of rotation of the engine output shaft to the vertical axis for the rotor shaft.*

the engine to be started and then gradually pick up the load of the rotor.

On free turbine engines, no clutch is required, as the gas producer turbine is essentially disconnected from the power turbine. When the engine is started, there is little resistance from the power turbine. This enables the gas producer turbine to accelerate to normal idle speed without the load of the

transmission and rotor system dragging it down. As the gas pressure increases through the power turbine, the rotor blades begin to turn, slowly at first and then gradually accelerate to normal operating rpm.

On reciprocating helicopters, the two main types of clutches are the centrifugal clutch and the belt drive clutch.

Centrifugal Clutch

The centrifugal clutch is made up of an inner assembly and an outer drum. The inner assembly, which is connected to the engine driveshaft, consists of shoes lined with material similar to automotive brake linings. At low engine speeds, springs hold the shoes in, so there is no contact with the outer drum, which is attached to the transmission input shaft. As engine speed increases, centrifugal force causes the clutch shoes to move outward and begin sliding against the outer drum. The transmission input shaft begins to rotate, causing the rotor to turn, slowly at first, but increasing as the friction increases between the clutch shoes and transmission drum. As rotor speed increases, the rotor tachometer needle shows an increase by moving toward the engine tachometer needle. When the two needles are superimposed, the engine and the rotor are synchronized, indicating the clutch is fully engaged and there is no further slippage of the clutch shoes.

Belt Drive Clutch

Some helicopters utilize a belt drive to transmit power from the engine to the transmission. A belt drive consists of a lower pulley attached to the engine, an upper pulley attached to the transmission input shaft, a belt or a series of V-belts, and some means of applying tension to the belts. The belts fit loosely over the upper and lower pulley when there is no tension on the belts. This allows the engine to be started without any load from the transmission. Once the engine is running, tension on the belts is gradually increased. When the rotor and engine tachometer needles are superimposed, the rotor and the engine are synchronized, and the clutch is then fully engaged. Advantages of this system include vibration isolation, simple maintenance, and the ability to start and warm up the engine without engaging the rotor.

Freewheeling Unit

Since lift in a helicopter is provided by rotating airfoils, these airfoils must be free to rotate if the engine fails. The freewheeling unit automatically disengages the engine from the main rotor when engine rpm is less than main rotor rpm. This allows the main rotor and tail rotor to continue turning at normal in-flight speeds. The most common freewheeling unit assembly consists of a one-way sprag clutch located between the engine and main rotor transmission. This is usually in the upper pulley in a piston helicopter or mounted on the accessory gearbox in a turbine helicopter. When the engine is driving the rotor, inclined surfaces in the sprag clutch force rollers against an outer drum. This prevents the engine from exceeding transmission rpm. If the engine fails, the rollers move inward, allowing the outer drum to exceed the speed of the inner portion. The transmission can then exceed the speed of the engine. In this condition, engine speed is less than that of the drive system, and the helicopter is in an autorotative state.

Airplane Assembly & Rigging

The primary assembly of a type certificated aircraft is normally performed by the manufacturer at the factory. The assembly includes putting together the major components, such as the fuselage, empennage, wing sections, nacelles, landing gear, and installing the powerplant. Attached to the wing and empennage are primary flight control surfaces including ailerons, elevators, and rudder. Additionally, installation of auxiliary flight control surfaces may include wing flaps, spoilers, speed brakes, slats, and leading edge flaps.

The assembly of other aircraft outside of a manufacturer's facility is usually limited to smaller size and experimental amateur-built aircraft. Typically, after a major overhaul, repair, or alteration, the reassembly of an aircraft may include reattaching wings to the fuselage, balancing of and installation of flight control surfaces, installation of the landing gear, and installation of the powerplant(s).

Rebalancing of Control Surfaces

This section is presented for familiarization purposes only. Explicit instructions for the balancing of control surfaces are given in the manufacturer's service and overhaul manuals for the specific aircraft and must be followed closely.

Any time repairs on a control surface add weight fore or aft of the hinge center line, the control surface must be rebalanced. When an aircraft is repainted, the balance of the control surfaces must be checked. Any control surface that is out of balance is unstable and does not remain in a streamlined position during normal flight. For example, an aileron that is trailing edge heavy moves down when the wing deflects upward, and up when the wing deflects downward. Such a condition can cause unexpected and violent maneuvers of the aircraft. In extreme cases, fluttering and buffeting may develop to a degree that could cause the complete loss of the aircraft.

Rebalancing a control surface concerns both static and dynamic balance. A control surface that is statically balanced is also dynamically balanced.

Static Balance

Static balance is the tendency of an object to remain stationary

when supported from its own CG. There are two ways in which a control surface may be out of static balance. They are called underbalance and overbalance.

When a control surface is mounted on a balance stand, a downward travel of the trailing edge below the horizontal position indicates underbalance. Some manufacturers indicate this condition with a plus (+) sign. An upward movement of the trailing edge, above the horizontal position indicates overbalance. This is designated by a minus (–) sign. These signs show the need for more or less weight in the correct area to achieve a balanced control surface, as shown in *Figure 2-60*.

A tail-heavy condition (static underbalance) causes undesirable flight performance and is not usually allowed. Better flight operations are gained by nose-heavy static overbalance. Most manufacturers advocate the existence of nose-heavy control surfaces.

Dynamic Balance

Dynamic balance is that condition in a rotating body wherein all rotating forces are balanced within themselves so that no vibration is produced while the body is in motion. Dynamic balance as related to control surfaces is an effort to maintain balance when the control surface is submitted to movement on the aircraft in flight. It involves the placing of weights in the correct location along the span of the surfaces. The

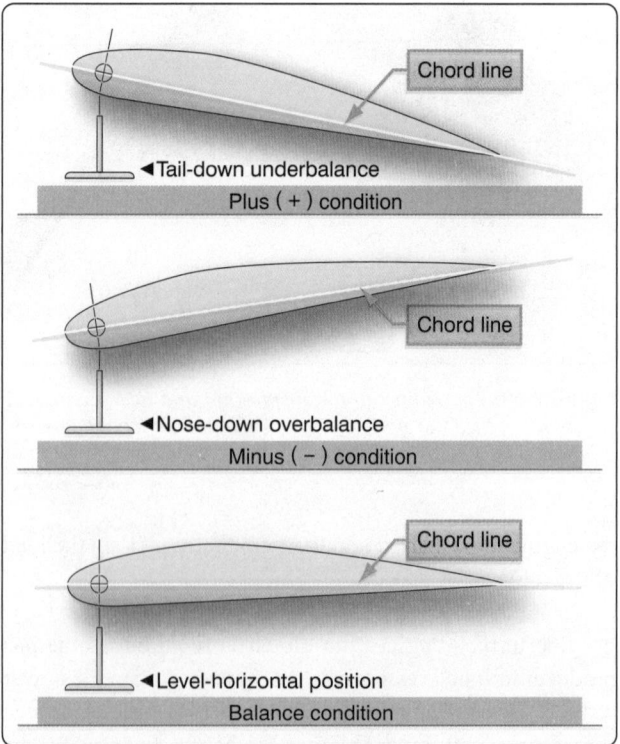

Figure 2-60. *Control surface static balance.*

location of the weights are, in most cases, forward of the hinge center line.

Rebalancing Procedures

Repairs to a control surface or its tabs generally increase the weight aft of the hinge center line, requiring static rebalancing of the control surface system, as well as the tabs. Control surfaces to be rebalanced should be removed from the aircraft and supported, from their own points, on a suitable stand, jig, or fixture. *[Figure 2-61]*

Trim tabs on the surface should be secured in the neutral position when the control surface is mounted on the stand. The stand must be level and be located in an area free of air currents. The control surface must be permitted to rotate freely about the hinge points without binding. Balance condition is determined by the behavior of the trailing edge when the surface is suspended from its hinge points. Any excessive friction would result in a false reaction as to the overbalance or underbalance of the surface.

When installing the control surface in the stand or jig, a neutral position should be established with the chord line of the surface in a horizontal position. Use a bubble protractor to determine the neutral position before continuing balancing procedures. *[Figure 2-62]*

Sometimes a visual check is all that is needed to determine whether the surface is balanced or unbalanced. Any trim tabs or other assemblies that are to remain on the surface during balancing procedures should be in place. If any assemblies or parts must be removed before balancing, they should be removed.

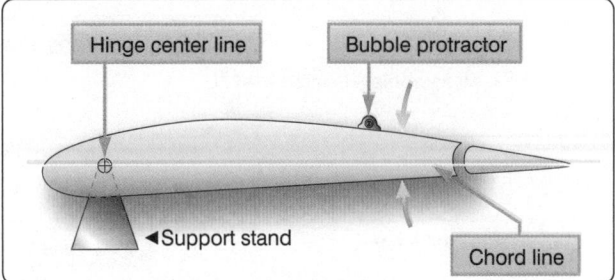

Figure 2-62. *Establishing a neutral position of the control surface.*

Rebalancing Methods

Several methods of balancing (rebalancing) control surfaces are in use by the various manufacturers of aircraft. The most common are the calculation method, scale method, and the balance beam method.

The calculation method of balancing a control surface has one advantage over the other methods in that it can be performed without removing the surface from the aircraft. In using the calculation method, the weight of the material from the repair area and the weight of the materials used to accomplish the repair must be known. Subtract the weight removed from the weight added to get the resulting net gain in the amount added to the surface. The distance from the hinge center line to the center of the repair area is then measured in inches. This distance must be determined to the nearest one-hundredth of an inch. *[Figure 2-63]*

The next step is to multiply the distance times the net weight of the repair. This results in an inch-pounds (in-lb) answer. If the in-lb result of the calculations is within specified tolerances, the control surface is considered balanced. If

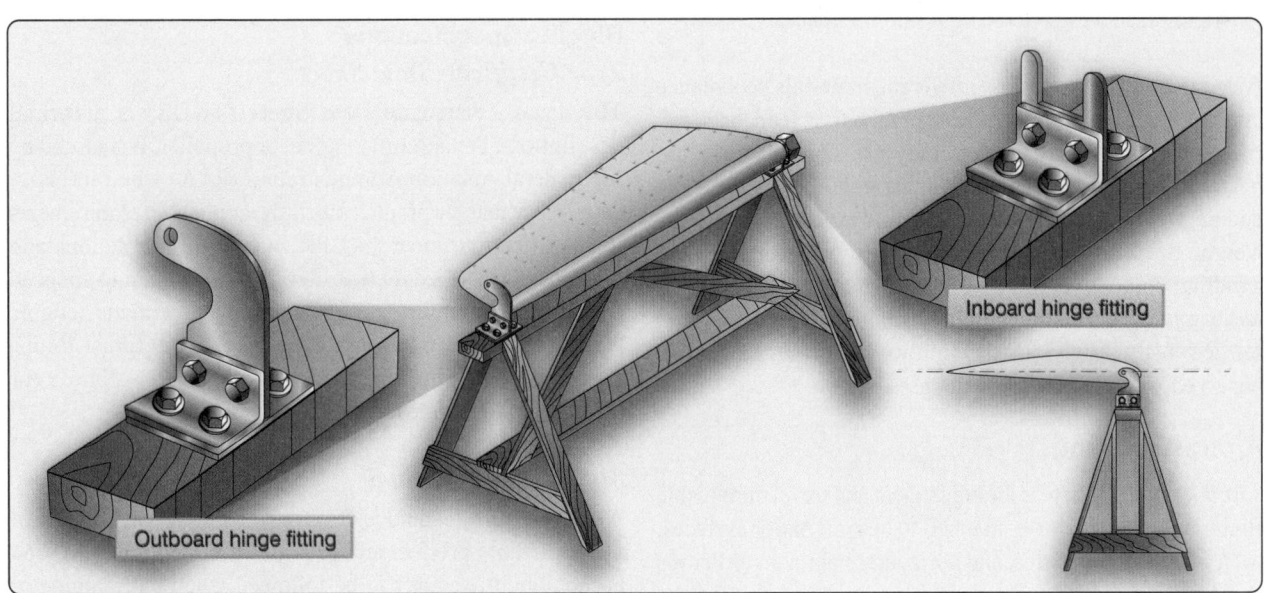

Figure 2-61. *Locally fabricated balancing fixture.*

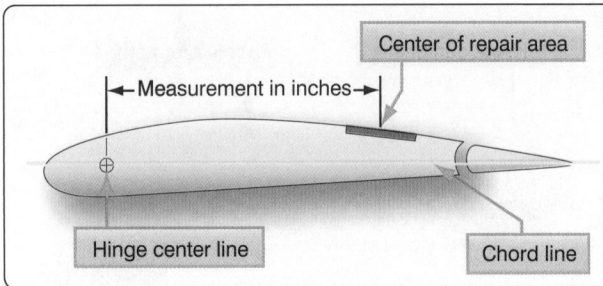

Figure 2-63. *Calculation method measurement.*

it is not within specified limits, consult the manufacturer's service manuals for the needed weights, material to use for weights, design for manufacture, and installation locations for addition of the weights.

The scale method of balancing a control surface requires the use of a scale that is graduated in hundredths of a pound. A support stand and balancing jigs for the surface are also required. *Figure 2-64* illustrates a control surface mounted for rebalancing purposes. Use of the scale method requires the removal of the control surface from the aircraft.

The balance beam method is used by the Cessna and Piper Aircraft companies. This method requires that a specialized tool be locally fabricated. The manufacturer's maintenance manual provides specific instructions and dimensions to fabricate the tool.

Once the control surface is placed on level supports, the weight required to balance the surface is established by moving the sliding weight on the beam. The maintenance manual indicates where the balance point should be. If the surface is found to be out of tolerance, the manual explains where to place weight to bring it into tolerance.

Aircraft manufacturers use different materials to balance control surfaces, the most common being lead or steel. Larger aircraft manufacturers may use depleted uranium because it has a heavier mass than lead. This allows the counterweights to be made smaller and still retain the same weight. Specific safety precautions must be observed when handling counterweights of depleted uranium because it is radioactive. The manufacturer's maintenance manual and service instructions must be followed and all precautions observed when handling the weights.

Aircraft Rigging

Aircraft rigging involves the adjustment and travel of movable flight controls which are attached to aircraft major surfaces, such as wings and vertical and horizontal stabilizers. Ailerons are attached to the wings, elevators are attached to the horizontal stabilizer, and the rudder is attached to the vertical

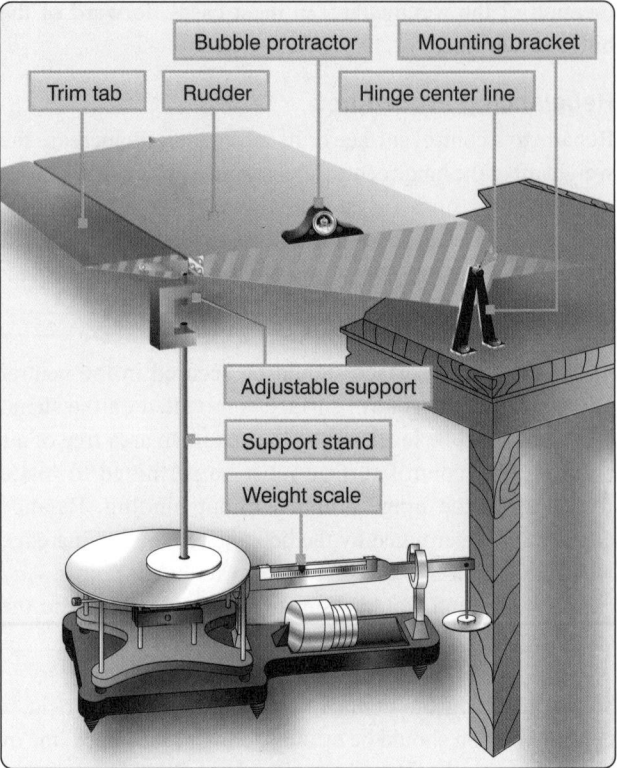

Figure 2-64. *Balancing setup.*

stabilizer. Rigging involves setting cable tension, adjusting travel limits of flight controls, and setting travel stops.

In addition to the flight controls, rigging is also performed on various components to include engine controls, flight deck controls, and retractable landing gear component parts. Rigging also includes the safetying of the attaching hardware using various types of cotter pins, locknuts, or safety wire.

Rigging Specifications
Type Certificate Data Sheet

The Type Certificate Data Sheet (TCDS) is a formal description of an aircraft, engine, or propeller. It is issued by the Federal Aviation Administration (FAA) when the FAA determines that the product meets the applicable requirements for certification under 14 CFR. It lists the limitations and information required for type certification, including airspeed limits, weight limits, control surface movements, engine make and model, minimum crew, fuel type, thrust limits, rpm limits, etc., and the various components eligible for installation on the product.

Maintenance Manual

A maintenance manual is developed by the manufacturer of the applicable product and provides the recommended and acceptable procedures to be followed when maintaining or repairing that product. Maintenance personnel are required by regulation to follow the applicable instructions set forth by the manufacturer. The Limitations section of the manual lists "life limits" of the product or its components that must be complied with during inspections and maintenance.

Structural Repair Manual (SRM)

The structural repair manual is developed by the manufacturer's engineering department to be used as a guideline to assist in the repair of common damage to a specific aircraft structure. It provides information for acceptable repairs of specific sections of the aircraft.

Manufacturer's Service Information

Information from the manufacturer may be in the form of information bulletins, service instructions, service bulletins, service letters, etc., that the manufacturer publishes to provide instructions for product improvement. Service instructions may include a recommended modification or repair that precedes the issuance of an Airworthiness Directive (AD). Service letters may provide more descriptive procedures or revise sections of the maintenance manuals. They may also include instructions for the installation and repair of optional equipment, not listed in the Type Certificate Data Sheet (TCDS).

Airplane Assembly

Aileron Installation

The manufacturer's maintenance and illustrated parts book must be followed to ensure the correct procedures and hardware are being used for installation of the control surfaces. All of the control surfaces require specific hardware, spacers, and bearings be installed to ensure the surface does not jam or become damaged during movement. After the aileron is connected to the flight deck controls, the control system must be inspected to ensure the cables/push-pull rods are routed properly. When a balance cable is installed, check for correct attachment and operation to determine the ailerons are moving in the proper direction and opposite each other.

Flap Installation

The design, installation, and systems that operate flaps are as varied as the models of airplanes on which they are installed. As with any system on a specific aircraft, the manufacturer's maintenance manual and the illustrated parts book must be followed to ensure the correct procedures and parts are used. Simple flap systems are usually operated manually by cables and/or torque tubes. Typically, many of the smaller manufactured airplane designs have flaps that are actuated by torque tubes and chains through a gear box driven by an electric motor.

Empennage Installation

The empennage, consisting of the horizontal and vertical stabilizer, is not normally removed and installed, unless the aircraft was damaged. Elevators, rudders, and stabilators are rigged the same as any other control surface, using the instructions provided in the manufacturer's maintenance manuals.

Control Operating Systems

Cable Systems

There are various types of cable:

*   Material—aircraft control cables are fabricated from carbon steel or stainless (corrosion resistant) steel. Additionally, some manufacturers use a nylon coated cable that is produced by extruding a flexible nylon coating over corrosion-resistant steel (CRES) cable. By adding the nylon coating to the corrosion resistant steel cable, it increases the service life by protecting the cable strands from friction wear, keeping dirt and grit out, and dampening vibration which can work-harden the wires in long runs of cable.

*   Cable construction—the basic component of a cable is a wire. The diameter of the wire determines the total diameter of the cable. A number of wires are preformed into a helical or spiral shape and then formed into a strand. These preformed strands are laid around a straight center strand to form a cable.

*   Cable designations—based on the number of strands and wires in each strand. The 7 × 19 cable is made up of seven strands of 19 wires each. Six of these strands are laid around the center strand. This cable is very flexible and is used in primary control systems and in other locations where operation over pulleys is frequent. The 7 × 7 cable consists of seven strands of seven wires each. Six of these strands are laid around the center strand. This cable is of medium flexibility and is used for trim tab controls, engine controls, and indicator controls. *[Figure 2-65]*

Types of control cable termination include:

*   Woven splice—a hand-woven 5-tuck splice used on aircraft cable. The process is very time consuming and produces only about 75 percent of the original cable

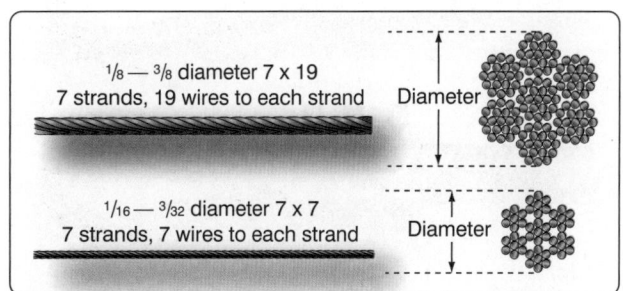

Figure 2-65. *Cable construction and cross-section.*

strength. The splice is rarely used except on some antique aircraft where the effort is made to keep all parts in their original configuration.

• Nicopress® process—a patented process using copper sleeves and may be used up to the full rated strength of the cable when the cable is looped around a thimble. *[Figure 2-66]* This process may also be used in place of the 5-tuck splice on cables up to and including $\frac{3}{8}$-inch diameter. Whenever this process is used for cable splicing, it is imperative that the tools, instructions, and data supplied by Nicopress® be followed exactly to ensure the desired cable function and strength is attained. The use of sleeves that are fabricated of material other than copper requires engineering approval for the specific application by the FAA.

• Swage-type terminals—manufactured in accordance with Army-Navy (AN) and Military Standards (MS), are suitable for use in civil aircraft up to, and including, maximum cable loads. *[Figure 2-67]*

When swaging tools are used, it is imperative that all the manufacturer's instructions, including 'go' and 'no-go' dimensions, be followed exactly to avoid defective and inferior swaging. Compliance with all of the instructions should result in the terminal developing the full-rated strength of the cable. The following basic procedures are used when swaging terminals onto cable ends:

• Cut the cable to length, allowing for growth during swaging. Apply a preservative compound to the cable end before insertion into the terminal barrel. Measure the internal length of the terminal end/barrel of the fitting to determine the proper length of the cable to be inserted. Transfer that measurement to the end of the cable and mark it with a piece of masking tape wrapped around the cable. This provides a positive mark to ensure the cable did not slip during the swaging process.

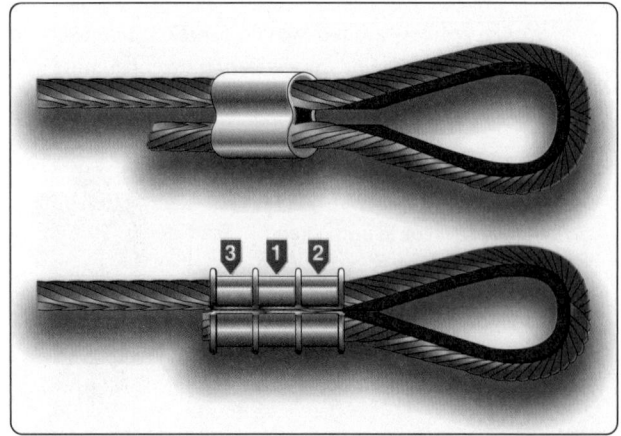

Figure 2-66. *Typical Nicopress® thimble-eye splice.*

2-40

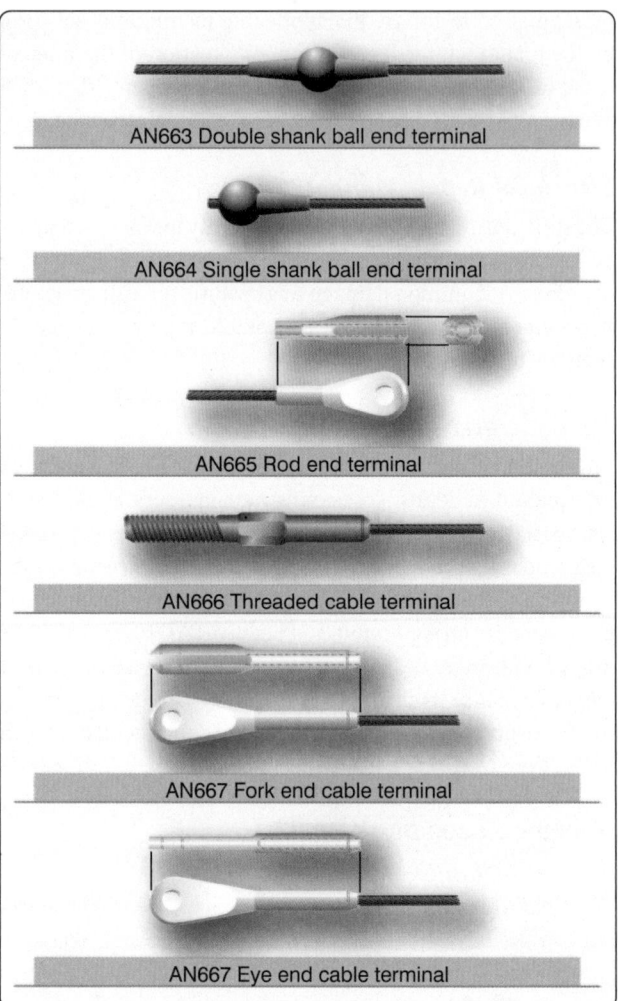

AN663 Double shank ball end terminal

AN664 Single shank ball end terminal

AN665 Rod end terminal

AN666 Threaded cable terminal

AN667 Fork end cable terminal

AN667 Eye end cable terminal

Figure 2-67. *Swage-type terminal fittings.*

Note: Never solder the cable ends to prevent fraying since the solder greatly increases the tendency of the cable to pull out of the terminal.

• Insert the cable into the terminal approximately one inch and bend it toward the terminal. Then, push the cable end all the way into the terminal. The bending action puts a slight kink in the cable end and provides enough friction to hold the terminal in place until the swaging operation is performed. *[Figure 2-68]*

• Accomplish the swaging operation in accordance with the instructions furnished by the manufacturer of the swaging equipment.

• Inspect the terminal after swaging to determine that it is free of die marks and splits and is not out of round. Check the cable for slippage at the masking tape and for cut and broken wire strands.

• Using a go/no-go gauge supplied by the swaging tool manufacturer or a micrometer and swaging chart, check the terminal shank diameter for proper dimension. *[Figures 2-69 and 2-70]*

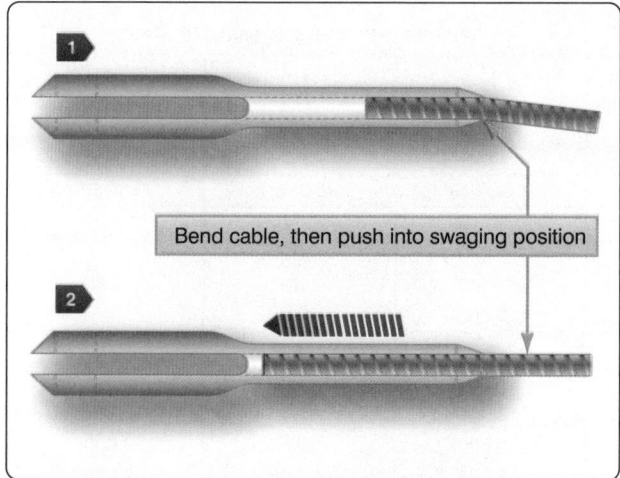

Figure 2-68. *Insertion of cable into terminal.*

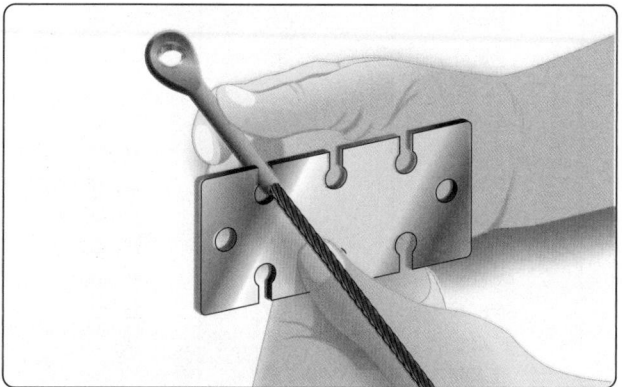

Figure 2-69. *Gauging terminal shank dimension after swaging.*

- Test the cable by proof-loading locally fabricated splices and newly installed swage terminal cable fittings for proper strength before installation. This is conducted by slowly applying a test load equal to 60 percent of the rated breaking strength of the cable listed in *Figure 2-71*.

This load should be held for at least 3 minutes. Any testing of this type can be dangerous. Suitable guards should be placed over the cable during the test to prevent injury to personnel in the event of cable failure. If a proper test fixture is not available, the load test should be contracted out and performed by a properly equipped facility.

Cable Inspection

Aircraft cable systems are subject to a variety of environmental conditions and deterioration. Wire or strand breakage is easy to recognize visually. Other kinds of deterioration, such as wear, corrosion, and distortion, are not easily seen. Special attention should be given to areas where cables pass through battery compartments, lavatories, and wheel wells. These are prime areas for corrosion. Special attention should be given to critical fatigue areas. Those areas are defined as anywhere the cable runs over, under, or around a pulley, sleeve, or through a fairlead; or any section where the cable is flexed, rubbed, or within 1 foot of a swaged-on fitting. Close inspection in these critical fatigue areas can be performed by rubbing a rag along the cable. If there are any broken strands, the rag snags on the cable. A more detailed inspection can be performed in areas that may be corroded or indicate a fatigue failure by loosing or removing the cable and bending it. This technique reveals internal broken strands not readily apparent from the outside. *[Figure 2-72]*

Cable System Installation

Cable Guides

Pulleys are used to guide cables and also to change the direction of cable movement. Pulley bearings are sealed and need no lubrication other than the lubrication done at the factory. Brackets fastened to the structure of the aircraft support the pulleys. Cables passing over pulleys are kept in place by guards. The guards are close fitting to prevent jamming or

| Cable size (inches) | Wire strands | Before Swaging | | | | After Swaging | |
		Outside diameter	Bore diameter	Bore length	Swaging length	Minimum breaking strength (pounds)	Shank diameter *
1/16	7 x 7	0.160	0.078	1.042	0.969	480	0.138
3/32	7 x 7	0.218	0.109	1.261	1.188	920	0.190
1/8	7 x 19	0.250	0.141	1.511	1.438	2,000	0.219
5/32	7 x 19	0.297	0.172	1.761	1.688	2,800	0.250
3/16	7 x 19	0.359	0.203	2.011	1.938	4,200	0.313
7/32	7 x 19	0.427	0.234	2.261	2.188	5,600	0.375
1/4	7 x 19	0.494	0.265	2.511	2.438	7,000	0.438
9/32	7 x 19	0.563	0.297	2.761	2.688	8,000	0.500
5/16	7 x 19	0.635	0.328	3.011	2.938	9,800	0.563
3/8	7 x 19	0.703	0.390	3.510	3.438	14,400	0.625
*Use gauges in kit for checking diameters.							

Figure 2-70. *Straight shank terminal dimensions.*

Nominal diameter of wire rope cable	Construction	Tolerance on diameter (plus only)	Allowable increase of diameter at cut end	Minimum Breaking Strength (Pounds)		
				MIL-W-83420 COMP A	MIL-W-83420 COMP B (CRES)	MIL-C-18375 (CRES)
INCHES		INCHES	INCHES	POUNDS	POUNDS	POUNDS
1/32	3 x 7	0.006	0.006	110	110	
3/64	7 x 7	0.008	0.008	270	270	
1/16	7 x 7	0.010	0.009	480	480	360
1/16	7 x 19	0.010	0.009	480	480	
3/32	7 x 7	0.012	0.010	920	920	700
3/32	7 x 19	0.012	0.010	1,000	920	
1/8	7 x 19	0.014	0.011	2,000	1,760	1,300
5/32	7 x 19	0.016	0.017	2,800	2,400	2,000
3/16	7 x 19	0.018	0.019	4,200	3,700	2,900
7/32	7 x 19	0.018	0.020	5,000	5,000	3,800
1/4	7 x 19	0.018	0.021	6,400	6,400	4,900
9/32	7 x 19	0.020	0.023	7,800	7,800	6,100
5/16	7 x 19	0.022	0.024	9,800	9,000	7,600
11/32	7 x 19	0.024	0.025	12,500		
3/8	7 x 19	0.026	0.027	14,400	12,000	11,000
7/16	6 x 19 IWRC	0.030	0.030	17,600	16,300	14,900
1/2	6 x 19 IWRC	0.033	0.033	22,800	22,800	19,300
9/16	6 x 19 IWRC	0.036	0.036	28,500	28,500	24,300
5/8	6 x 19 IWRC	0.039	0.039	35,000	35,000	30,100
3/4	6 x 19 IWRC	0.045	0.045	49,600	49,600	42,900
7/8	6 x 19 IWRC	0.048	0.048	66,500	66,500	58,000
1	6 x 19 IWRC	0.050	0.050	85,400	85,400	75,200
1 - 1/8	6 x 19 IWRC	0.054	0.054	106,400	106,400	
1 - 1/4	6 x 19 IWRC	0.057	0.057	129,400	129,400	
1 - 3/8	6 x 19 IWRC	0.060	0.060	153,600	153,600	
1 - 1/2	6 x 19 IWRC	0.062	0.062	180,500	180,500	

Figure 2-71. *Flexible cable construction.*

to prevent the cables from slipping off when they slacken due to temperature variations. Pulleys should be examined to ensure proper lubrication; smooth rotation and freedom from

Figure 2-72. *Cable inspection technique.*

abnormal cable wear patterns which can provide an indication of other problems in the cable system. *[Figure 2-73]*

Fairleads may be made from a nonmetallic material, such as phenolic, or a metallic material, such as soft aluminum. The fairlead completely encircles the cable where it passes through holes in bulkheads or other metal parts. Fairleads are used to guide cables in a straight line through or between structural members of the aircraft. Fairleads should never deflect the alignment of a cable more than 3° from a straight line.

Pressure seals are installed where cables (or rods) move through pressure bulkheads. The seal grips tightly enough to prevent excess air pressure loss but not enough to hinder movement of the cable. Pressure seals should be inspected at regular intervals to determine that the retaining rings are in place. If a retaining ring comes off, it may slide along the cable and cause jamming of a pulley. *[Figure 2-74]*

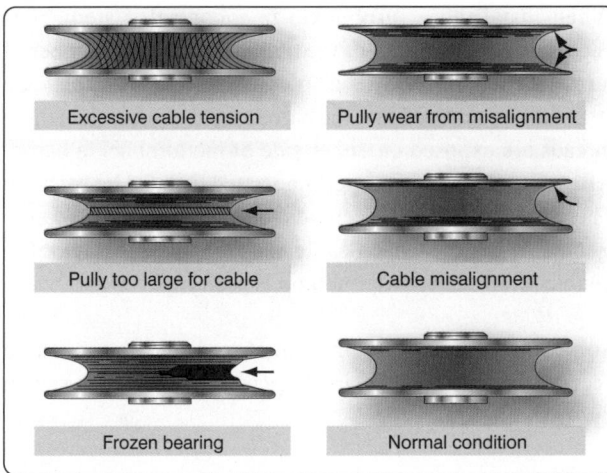

Figure 2-73. *Pulley wear patterns.*

Travel Adjustment

Control surfaces should move a certain distance in either direction from the neutral position. These movements must be synchronized with the movement of the flight deck controls. The flight control system must be adjusted (rigged) to obtain these requirements. The tools for measuring surface travel primarily include protractors, rigging fixtures, contour templates, and rulers. These tools are used when rigging flight control systems to assure that the desired travel has been obtained. Generally speaking, the rigging consists of

the following:

1. Positioning the flight control system in neutral and temporarily locking it there with rig pins or blocks;

2. Adjusting system cable tension and maintaining rudder, elevator, and ailerons in the neutral position; and

3. Adjusting the control stops to the aircraft manufacturer's specifications.

Cable Tension

For the aircraft to operate as it was designed, the cable tension for the flight controls must be correct. To determine the amount of tension on a cable, a tensiometer is used. When properly maintained, a tensiometer is 98 percent accurate. Cable tension is determined by measuring the amount of force needed to make an offset in the cable between two hardened steel blocks called anvils. A riser or plunger is pressed against the cable to form the offset. Several manufacturers make a variety of tensiometers, each type designed for different kinds of cable, cable sizes, and cable tensions. *[Figure 2-75]*

Rigging Fixtures

Rigging fixtures and templates are special tools (gauges) designed by the manufacturer to measure control surface travel. Markings on the fixture or template indicate desired control surface travel.

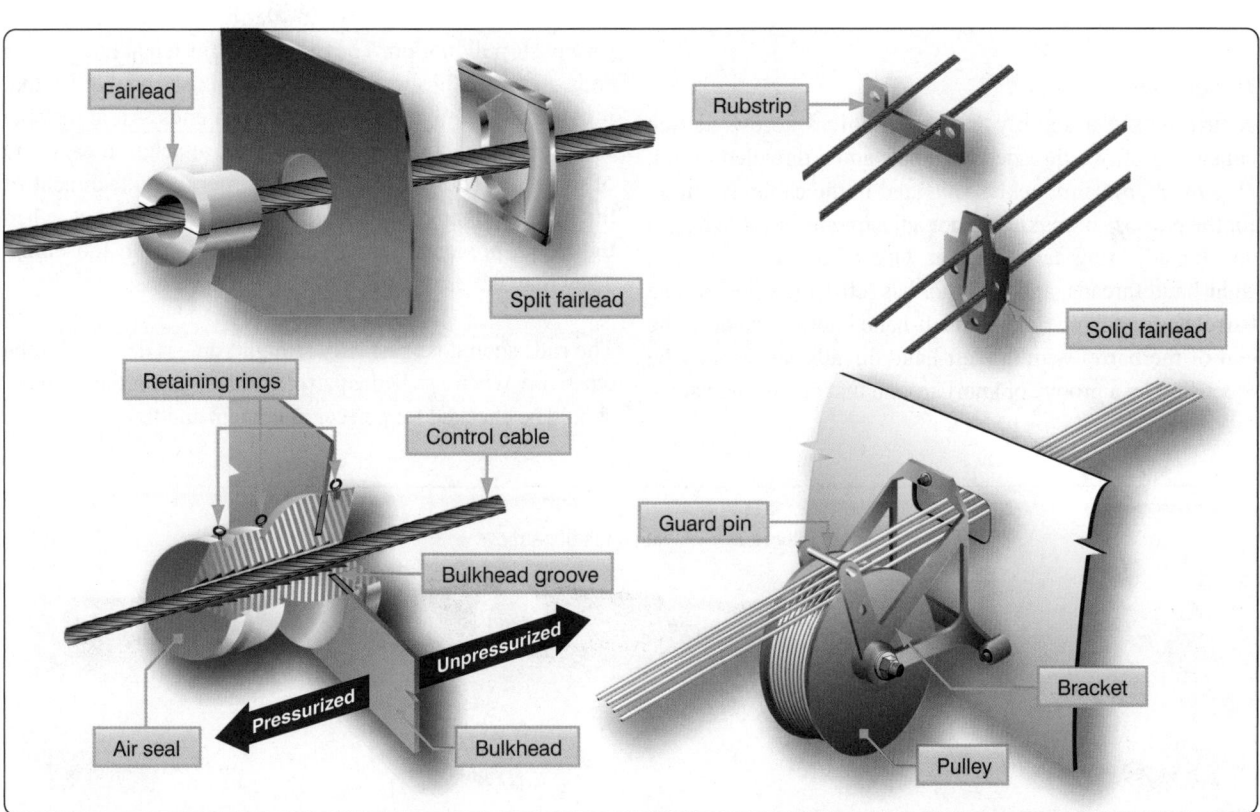

Figure 2-74. *Cable guides.*

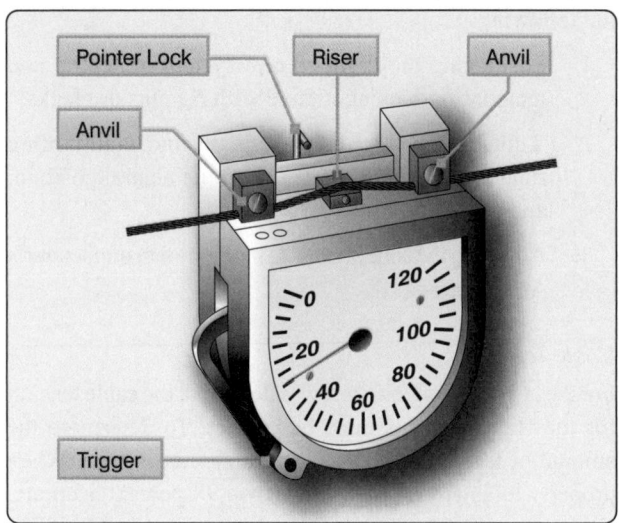

Figure 2-75. *Tensiometer.*

Tension Regulators

Cable tension regulators are used in some flight control systems because there is considerable difference in temperature expansion of the aluminum aircraft structure and the steel control cables. Some large aircraft incorporate tension regulators in the control cable systems to maintain a given cable tension automatically. The unit consists of a compression spring and a locking mechanism that allows the spring to make correction in the system only when the cable system is in neutral.

Turnbuckles

A turnbuckle assembly is a mechanical screw device consisting of two threaded terminals and a threaded barrel. *[Figure 2-76]* Turnbuckles are fitted in the cable assembly for the purpose of making minor adjustments in cable length and for adjusting cable tension. One of the terminals has right-hand threads, and the other has left-hand threads. The barrel has matching right- and left-hand internal threads. The end of the barrel with the left-hand threads can usually be identified by a groove or knurl around that end of the barrel.

When installing a turnbuckle in a control system, it is necessary to screw both of the terminals an equal number of turns into the barrel. It is also essential that all turnbuckle terminals be screwed into the barrel until not more than three threads are exposed on either side of the turnbuckle barrel. After a turnbuckle is properly adjusted, it must be safetied. There are a number of methods to safety a turnbuckle and/ or other types of swaged cable ends that are satisfactory. A double-wrap safety wire method is preferred.

Some turnbuckles are manufactured and designed to accommodate special locking devices. A typical unit is shown in *Figure 2-77*.

Cable Connectors

In addition to turnbuckles, cable connectors are used in some systems. These connectors enable a cable length to be quickly connected or disconnected from a system. *Figure 2-78* illustrates one type of cable connector in use.

Spring-Back

With a control cable properly rigged, the flight control should hit its stops at both extremes prior to the flight deck control. The spring-back is the small extra push that is needed for the flight deck control to hit its mechanical stop.

Push Rods (Control Rods)

Push rods are used as links in the flight control system to give push-pull motion. They may be adjusted at one or both ends. *Figure 2-79* shows the parts of a push rod. Notice that it consists of a tube with threaded rod ends. An adjustable antifriction rod end, or rod end clevis, attaches at each end of the tube. The rod end, or clevis, permits attachment of the tube to flight control system parts. The checknut, when tightened, prevents the rod end or clevis from loosening. They may have adjustments at one or both ends.

The rods should be perfectly straight, unless designed to be otherwise. When installed as part of a control system, the assembly should be checked for correct alignment and free movement.

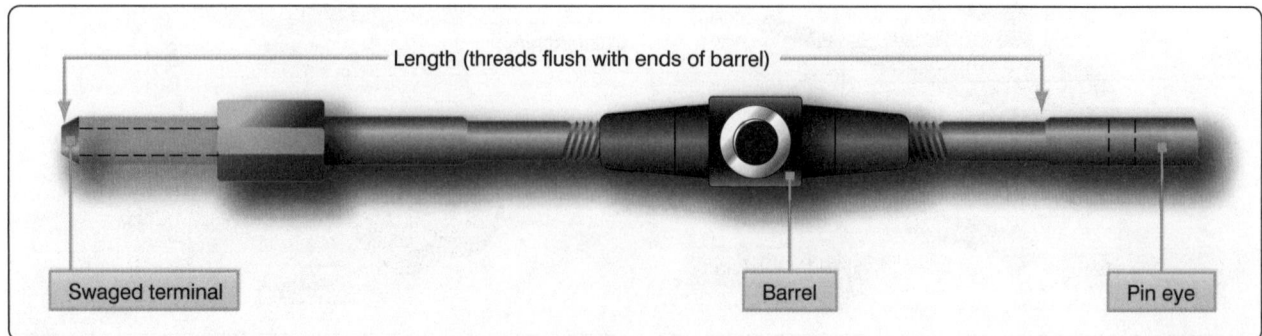

Figure 2-76. *Typical turnbuckle assembly.*

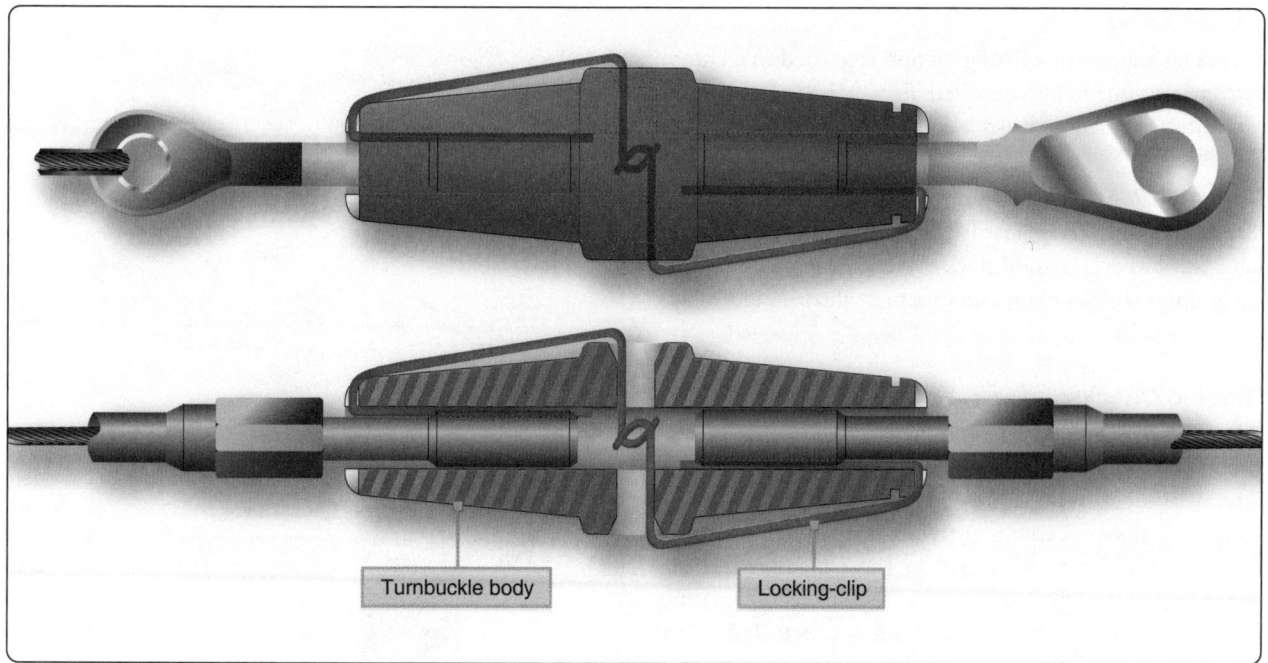

Figure 2-77. *Clip-type locking device and assembling in turnbuckle.*

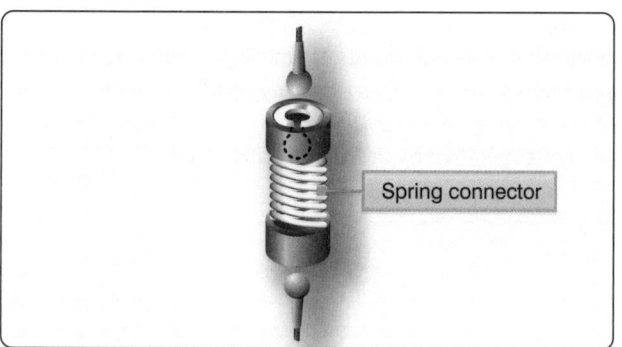

Figure 2-78. *Spring-type connector.*

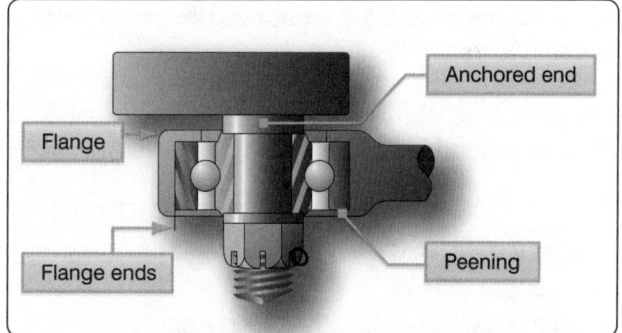

Figure 2-80. *Attached rod end.*

It is possible for control rods fitted with bearings to become disconnected because of failure of the peening that retains the ball races in the rod end. This can be avoided by installing the control rods so that the flange of the rod end is interposed between the ball race and the anchored end of the attaching pin or bolt as shown in *Figure 2-80*.

Another alternative is to place a washer, having a larger diameter than the hole in the flange, under the retaining nut on the end of the attaching pin or bolt. This retains the rod on the bolt in the event of a bearing failure.

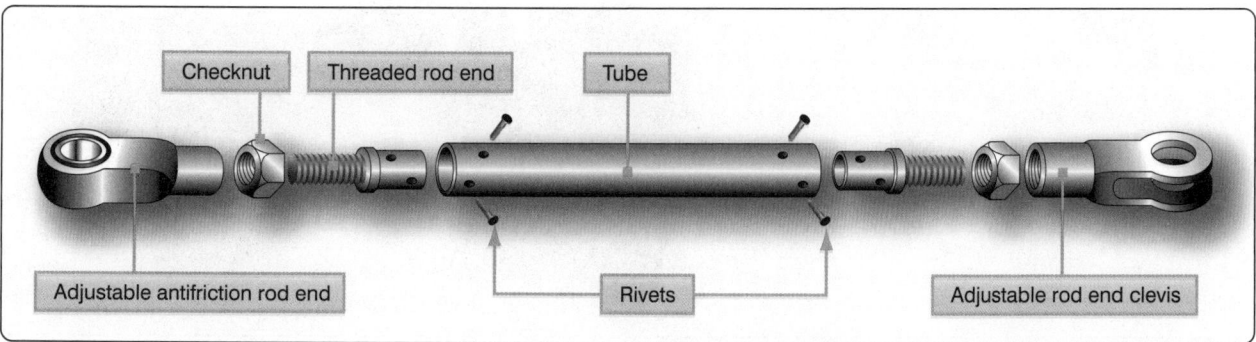

Figure 2-79. *Push rod.*

Torque Tubes

Where an angular or twisting motion is needed in a control system, a torque tube is installed. *Figure 2-81* shows how a torque tube is used to transmit motion in opposite directions.

Cable Drums

Cable drums are used primarily in trim tab systems. As the trim tab control wheel is moved clockwise or counterclockwise, the cable drum winds or unwinds to actuate the trim tab cables. *[Figure 2-82]*

Rigging Checks

All aircraft assembly and rigging must be performed in accordance with the requirements prescribed by the specific aircraft and/or aircraft component manufacturer. Correctly following the procedures provides for proper operation of the components in regard to their mechanical and aerodynamic function and ensures the structural integrity of the aircraft. Rigging procedures are detailed in the applicable manufacturer's maintenance or service manuals and applicable structural repair manuals. Additionally, aircraft specification or TCDS also provide information regarding control surface movement and weight and balance limits.

The purpose of this section is to explain the methods of checking the relative alignment and adjustment of an aircraft's

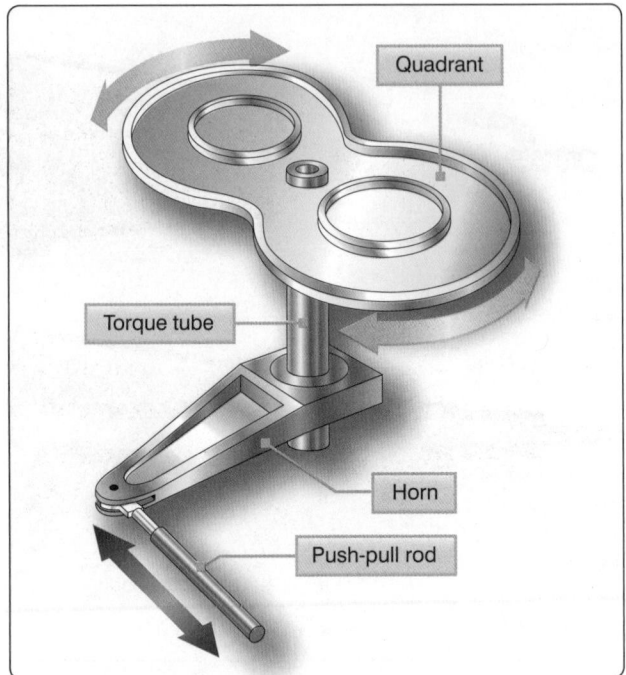

Figure 2-81. *Torque tube.*

main structural components. It is not intended to imply that the procedures are exactly as they may be in a particular aircraft. When rigging an aircraft, always follow the procedures and methods specified by the aircraft manufacturer.

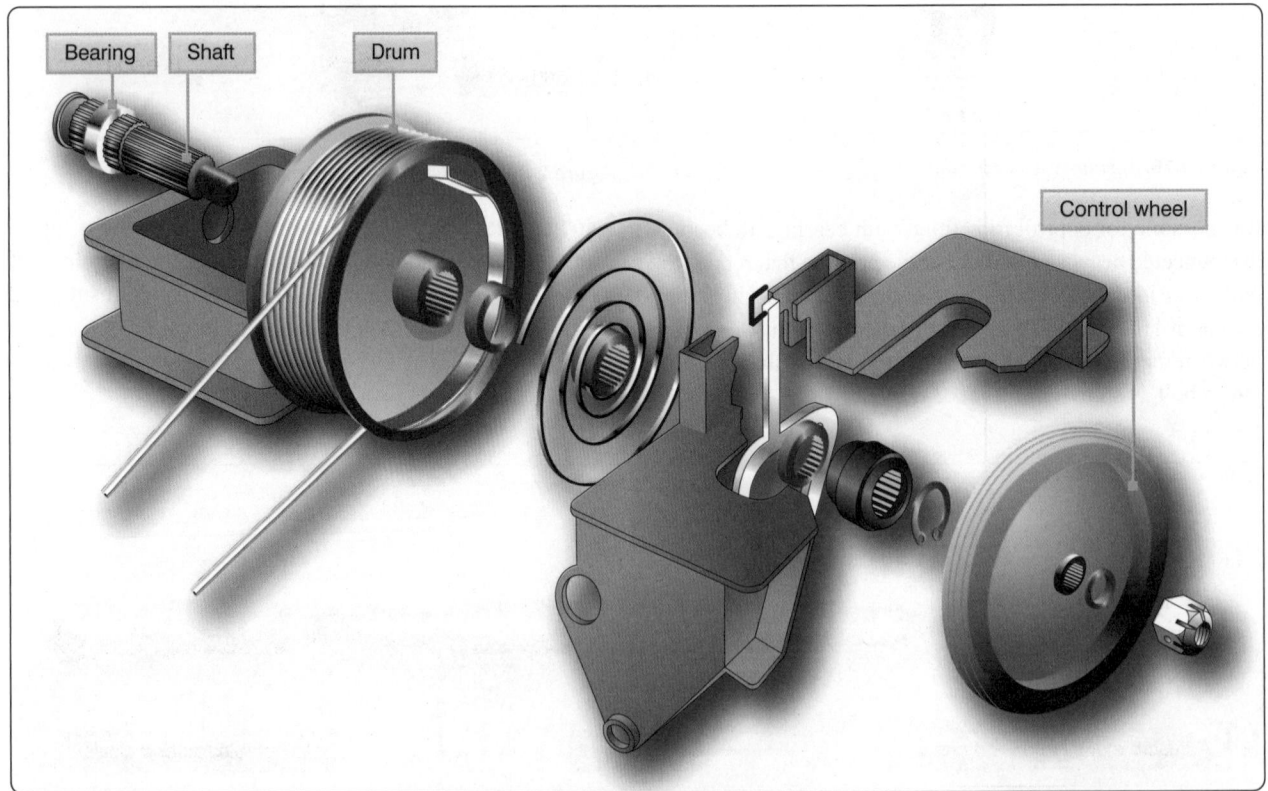

Figure 2-82. *Trim tab cable drum.*

Structural Alignment

The position or angle of the main structural components is related to a longitudinal datum line parallel to the aircraft center line and a lateral datum line parallel to a line joining the wing tips. Before checking the position or angle of the main components, the aircraft must be jacked and leveled.

Small aircraft usually have fixed pegs or blocks attached to the fuselage parallel to or coincident with the datum lines. A spirit level and a straight edge are rested across the pegs or blocks to check the level of the aircraft. This method of checking aircraft level also applies to many of the larger types of aircraft. However, the grid method is sometimes used on large aircraft. The grid plate is a permanent fixture installed on the aircraft floor or supporting structure. *[Figure 2-83]*

When the aircraft is to be leveled, a plumb bob is suspended from a predetermined position in the ceiling of the aircraft over the grid plate. The adjustments to the jacks necessary to level the aircraft are indicated on the grid scale. The aircraft is level when the plumb bob is suspended over the center point of the grid.

Certain precautions must be observed in all instances when jacking an aircraft. Normally, rigging and alignment checks should be performed in an enclosed hangar. If this cannot be accomplished, the aircraft should be positioned with the nose into the wind.

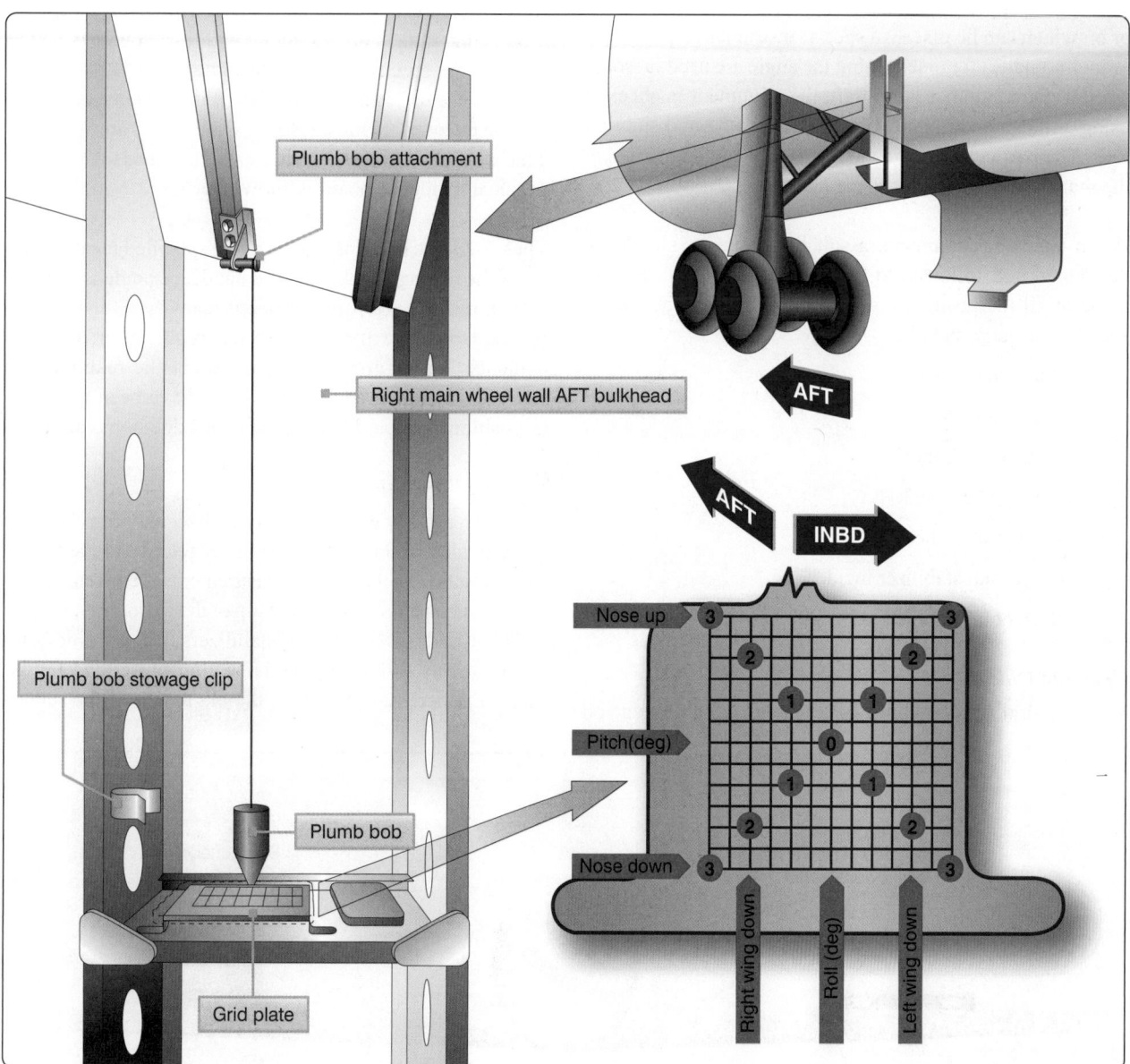

Figure 2-83. *Grid plate installed.*

The weight and loading of the aircraft should be exactly as described in the manufacturer's manual. In all cases, the aircraft should not be jacked until it is determined that the maximum jacking weight (if applicable) specified by the manufacturer is not exceeded.

With a few exceptions, the dihedral and incidence angles of conventional modern aircraft cannot be adjusted. Some manufacturers permit adjusting the wing angle of incidence to correct for a wing-heavy condition. The dihedral and incidence angles should be checked after hard landings or after experiencing abnormal flight loads to ensure that the components are not distorted and that the angles are within the specified limits.

There are several methods for checking structural alignment and rigging angles. Special rigging boards that incorporate, or on which can be placed, a special instrument (spirit level or inclinometer) for determining the angle are used on some aircraft. On a number of aircraft, the alignment is checked using a transit and plumb bobs or a theodolite and sighting rods. The particular equipment to use is usually specified in the manufacturer's maintenance manual.

When checking alignment, a suitable sequence should be developed and followed to be certain that the checks are made at all the positions specified. The alignment checks specified usually include:

- Wing dihedral angle
- Wing incidence angle
- Verticality of the fin
- Engine alignment
- A symmetry check
- Horizontal stabilizer incidence
- Horizontal stabilizer dihedral

Checking Dihedral

The dihedral angle should be checked in the specified

positions using the special boards provided by the aircraft manufacturer. If no such boards are available, a straight edge and a inclinometer can be used. Dihedral is normally checked using the front spar. The methods for checking dihedral are shown in *Figure 2-84*.

It is important that the dihedral be checked at the positions specified by the manufacturer. Certain portions of the wings or horizontal stabilizer may sometimes be horizontal or, on rare occasions, anhedral angles may be present.

Checking Incidence

Incidence is usually checked in at least two specified positions on the surface of the wing to ensure that the wing is free from twist. A variety of incidence boards are used to check the incidence angle. Some have stops at the forward edge, which must be placed in contact with the leading edge of the wing. Others are equipped with location pegs which fit into some specified part of the structure. The purpose in either case is to ensure that the board is fitted in exactly the position intended. In most instances, the boards are kept clear of the wing contour by short extensions attached to the board. A typical incidence board is shown in *Figure 2-85*.

When used, the board is placed at the specified locations on the surface being checked. If the incidence angle is correct, a inclinometer on top of the board reads zero, or within a specified tolerance of zero. Modifications to the areas where incidence boards are located can affect the reading. For example, if leading edge deicer boots have been installed, the position of a board having a leading edge stop is affected.

Checking Fin Verticality

After the rigging of the horizontal stabilizer has been checked, the verticality of the vertical stabilizer relative to the lateral datum can be checked. The measurements are taken from a given point on either side of the top of the fin to a given point on the left and right horizontal stabilizers. *[Figure 2-86]* The measurements should be similar within prescribed limits. When it is necessary to check the alignment of the rudder

Figure 2-84. *Checking dihedral.*

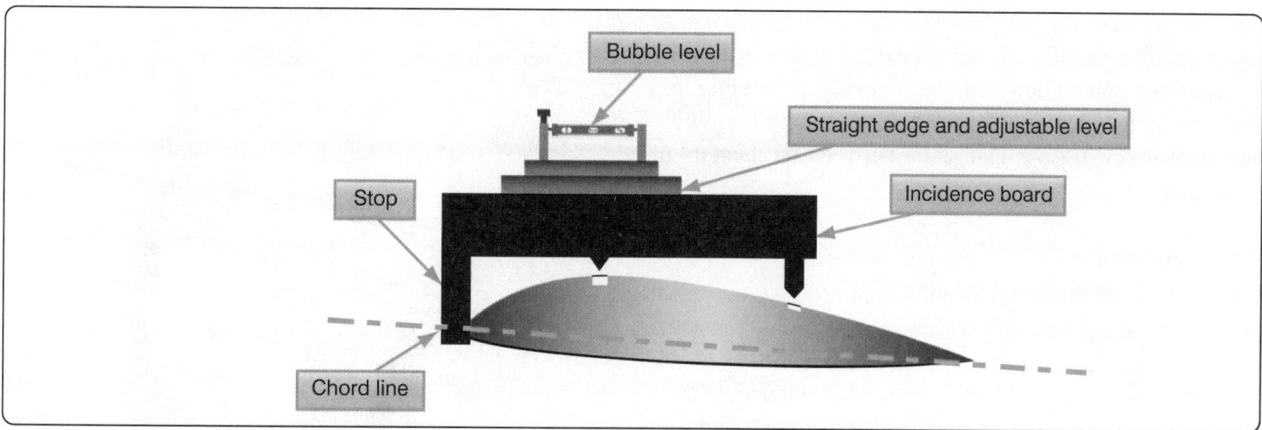

Figure 2-85. *A typical incidence board.*

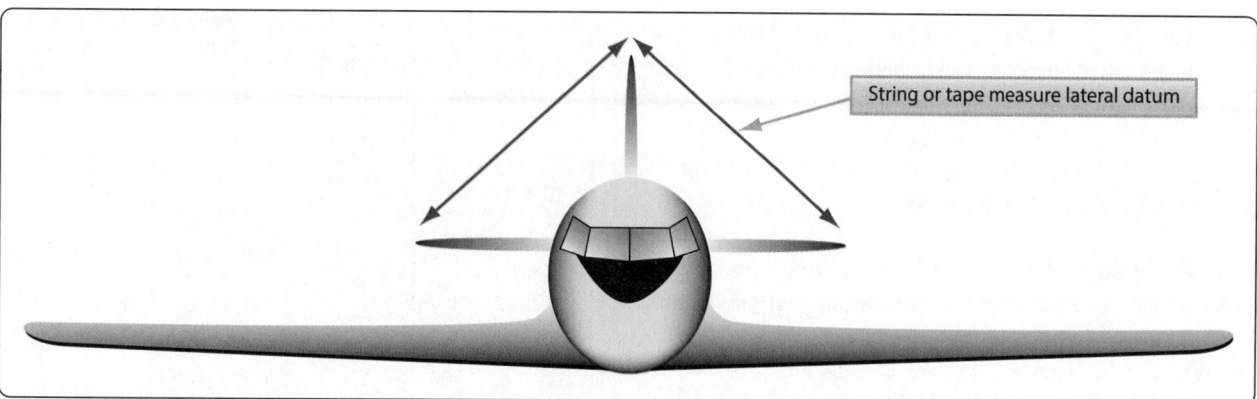

Figure 2-86. *Checking fin verticality.*

hinges, remove the rudder and pass a plumb bob line through the rudder hinge attachment holes. The line should pass centrally through all the holes. It should be noted that some aircraft have the leading edge of the vertical fin offset to the longitudinal center line to counteract engine torque.

Checking Engine Alignment

Engines are usually mounted with the thrust line parallel to the horizontal longitudinal plane of symmetry. However, this is not always true when the engines are mounted on the wings. Checking to ensure that the position of the engines, including any degree of offset is correct, depends largely on the type of mounting. Generally, the check entails a measurement from the center line of the mounting to the longitudinal center line of the fuselage at the point specified in the applicable manual. *[Figure 2-87]*

Symmetry Check

The principle of a typical symmetry check is illustrated in *Figure 2-87*. The precise figures, tolerances, and checkpoints for a particular aircraft are found in the applicable service or maintenance manual.

On small aircraft, the measurements between points are usually

taken using a steel tape. When measuring long distances, it is suggested that a spring scale be used with the tape to obtain equal tension. A five-pound pull is usually sufficient.

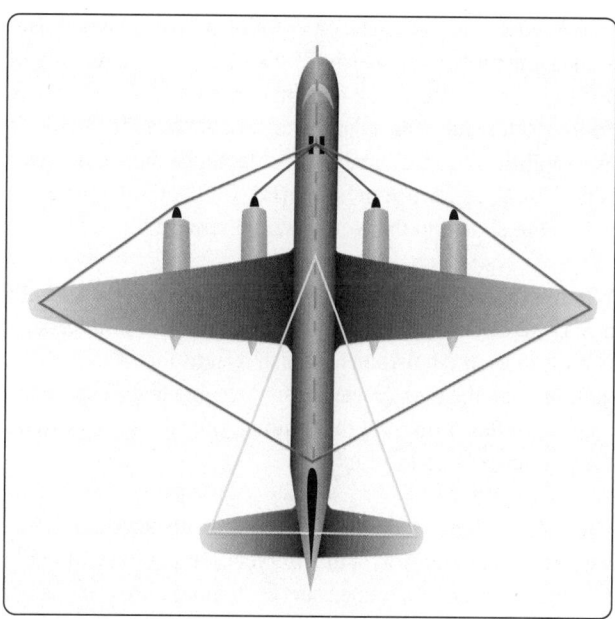

Figure 2-87. *Typical measurements used to check aircraft symmetry.*

On large aircraft, the positions at which the dimensions are to be taken are usually chalked on the floor. This is done by suspending a plumb bob from the checkpoints and marking the floor immediately under the point of each plumb bob. The measurements are then taken between the centers of each marking.

Cable Tension

When it has been determined that the aircraft is symmetrical and structural alignment is within specifications, the cable tension and control surface travel can be checked. To determine the amount of tension on a cable, a tensiometer is used. When properly maintained, a tensiometer is 98 percent accurate. Tensiometers are calibrated to maintain accuracy. Cable tension is determined by measuring the amount of force needed to make an offset in the cable between two hardened steel blocks called anvils. A riser or plunger is pressed against the cable to form the offset. Several manufacturers make a variety of tensiometers, each type designed for different kinds of cable, cable sizes, and cable tensions. One type of tensiometer is illustrated in *Figure 2-88*.

Following the manufacturer's instructions, lower the trigger. Then, place the cable to be tested under the two anvils and close the trigger (move it up). Movement of the trigger pushes up the riser, which pushes the cable at right angles to the two clamping points under the anvils. The force that is required to do this is indicated by the dial pointer. As the sample chart beneath the illustration shows, different numbered risers are used with different size cables. Each riser has an identifying number and is easily inserted into the tensiometer.

Included with each tensiometer is a conversion chart, which is used to convert the dial reading to pounds. The dial reading is converted to pounds of tension as follows. Using a No. 2 riser to measure the tension of a 5/32" diameter cable, a reading of 30 is obtained. The actual tension (see chart) of the cable is 70 lbs. Referring to the chart, also notice that a No. 1 riser is used with 1/16", 3/32", and 1/8" cable. Since the tensiometer is not designed for use in measuring 7/32" or 1/4" cable, no values are shown in the No. 3 riser column of the chart.

When actually taking a reading of cable tension in an aircraft, it may be difficult to see the dial. Therefore, a pointer lock is built in on the tensiometer. Push it in to lock the pointer, then remove the tensiometer from the cable and observe the reading. After observing the reading, pull the lock out and the pointer returns to zero.

Another variable that must be taken into account when adjusting cable tension is the ambient temperature of cable and the aircraft. To compensate for temperature variations, cable rigging charts are used when establishing cable tensions

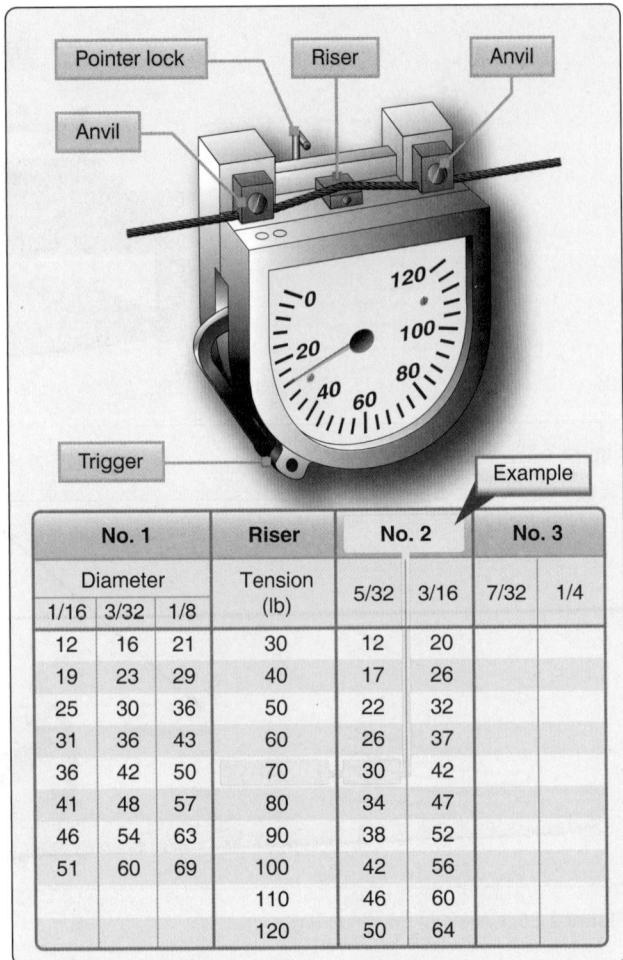

No. 1			Riser	No. 2		No. 3	
Diameter			Tension (lb)	5/32	3/16	7/32	1/4
1/16	3/32	1/8					
12	16	21	30	12	20		
19	23	29	40	17	26		
25	30	36	50	22	32		
31	36	43	60	26	37		
36	42	50	70	30	42		
41	48	57	80	34	47		
46	54	63	90	38	52		
51	60	69	100	42	56		
			110	46	60		
			120	50	64		

Figure 2-88. *Cable tensiometer and sample conversion chart.*

in flight control, landing gear, and other cable-operated systems. *[Figure 2-89]*

To use the chart, determine the size of the cable that is to be adjusted and the ambient air temperature. For example, assume that the cable size is 1/8" diameter, which is a 7-19 cable and the ambient air temperature is 85 °F. Follow the 85 °F line upward to where it intersects the curve for 1/8" cable. Extend a horizontal line from the point of intersection to the right edge of the chart. The value at this point indicates the tension (rigging load in pounds) to establish on the cable. The tension for this example is 70 pounds.

Control Surface Travel

In order for a control system to function properly, it must be correctly adjusted. Correctly rigged control surfaces move through a prescribed arc (surface-throw) and are synchronized with the movement of the flight deck controls. Rigging any control system requires that the aircraft manufacturer's instructions be followed as outlined in their maintenance manual.

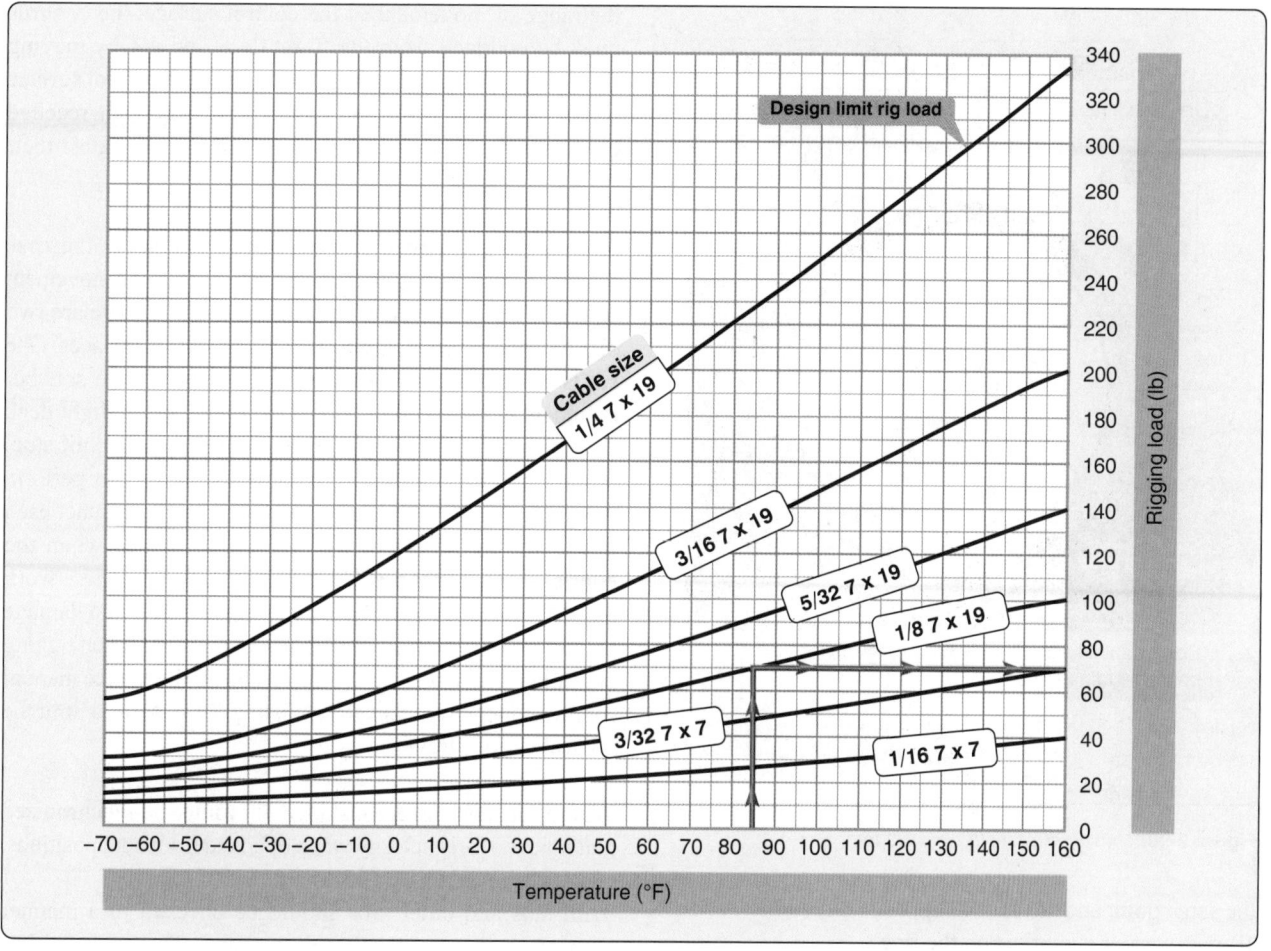

Figure 2-89. *Typical cable rigging chart.*

Therefore, the explanations in this chapter are limited to the three general steps listed below:

1. Lock the flight deck control, bellcranks, and the control surfaces in the neutral position.

2. Adjust the cable tension, maintaining the rudder, elevators, or ailerons in the neutral position.

3. Adjust the control stops to limit the control surface travel to the dimensions given for the aircraft being rigged.

The range of movement of the controls and control surfaces should be checked in both directions from neutral. There are various tools used for measuring surface travel, including protractors, rigging fixtures, contour templates, and rulers. These tools are used when rigging flight control systems to ensure that the aircraft is properly rigged and the manufacturer's specifications have been complied with.

Rigging fixtures and contour templates are special tools (gauges) designed by the manufacturer to measure control

surface travel. Markings on the fixture or template indicate desired control surface travel. In many instances, the aircraft manufacturer gives the travel of a particular control surface in degrees and inches. If the travel in inches is provided, a ruler can be used to measure surface travel in inches.

Protractors are tools for measuring angles in degrees. Various types of protractors are used to determine the travel of flight control surfaces. One protractor that can be used to measure aileron, elevator, or wing flap travel is the universal propeller protractor shown in *Figure 2-90*.

This protractor is made up of a frame, disc, ring, and two spirit levels. The disc and ring turn independently of each other and of the frame. (The center spirit level is used to position the frame vertically when measuring propeller blade angle.) The center spirit level is used to position the disc when measuring control surface travel. A disc-to-ring lock is provided to secure the disc and ring together when the zero on the ring vernier scale and the zero on the disc degree scale align. The ring-to-frame lock prevents the ring from moving when the disc is moved. Note that they start at

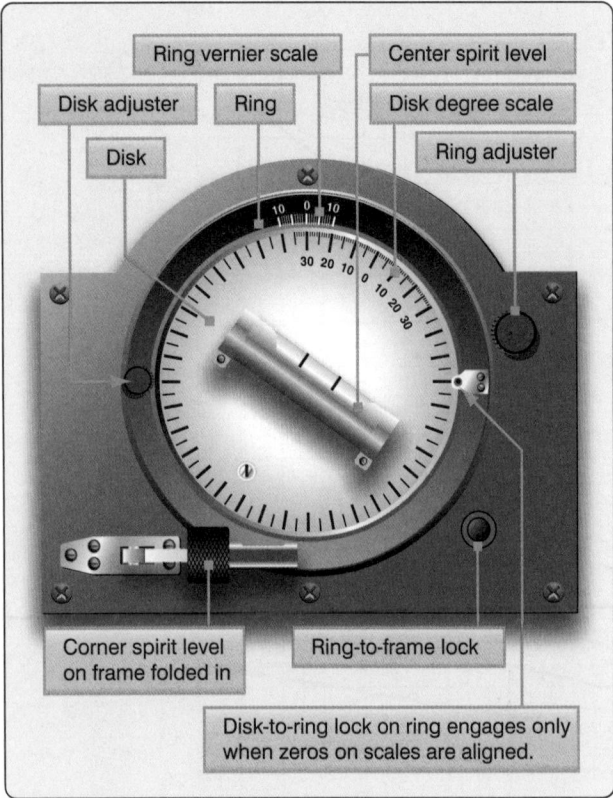

Ring vernier scale

Center spirit level

Disk adjuster   Ring

Disk degree scale

Disk

Ring adjuster

Corner spirit level
on frame folded in

Ring-to-frame lock

Disk-to-ring lock on ring engages only
when zeros on scales are aligned.

Figure 2-90. *Universal propeller protractor.*

the same point and advance in opposite directions. A double 10-part vernier is marked on the ring.

The rigging of the trim tab systems is performed in a similar manner. The trim tab control is set to the neutral (no trim) position, and the surface tab is usually adjusted to streamline with the control surface. However, on some aircraft, the specifications may require that the trim tabs be offset a degree or two from streamline when in the neutral position. After the tab and tab control are in the neutral position, adjust the control cable tension.

Pins, usually called rig pins, are sometimes used to simplify the setting of pulleys, levers, bellcranks, etc., in their neutral positions. A rig pin is a small metallic pin or clip. When rig pins are not provided, the neutral positions can be established by means of alignment marks, by special templates, or by taking linear measurements.

If the final alignment and adjustment of a system are correct, it should be possible to withdraw the rigging pins easily. Any undue tightness of the pins in the rigging holes indicates incorrect tensioning or misalignment of the system.

After a system has been adjusted, the full and synchronized movement of the controls should be checked. When checking

the range of movement of the control surface, the controls must be operated from the flight deck and not by moving the control surfaces. During the checking of control surface travel, ensure that chains, cables, etc., have not reached the limit of their travel when the controls are against their respective stops.

Adjustable and nonadjustable stops (whichever the case requires) are used to limit the throw-range or travel movement of the ailerons, elevator, and rudder. Usually there are two sets of stops for each of the three main control surfaces. One set is located at the control surface, either in the snubber cylinders or as structural stops; the other, at the flight deck control. Either of these may serve as the actual limit stop. However, those situated at the control surface usually perform this function. The other stops do not normally contact each other, but are adjusted to a definite clearance when the control surface is at the full extent of its travel. These work as override stops to prevent stretching of cables and damage to the control system during violent maneuvers. When rigging control systems, refer to the applicable maintenance manual for the sequence of steps for adjusting these stops to limit the control surface travel.

Where dual controls are installed, they must be synchronized and function satisfactorily when operated from both positions.

Trim tabs and other tabs should be checked in a manner similar to the main control surfaces. The tab position indicator must be checked to see that it functions correctly. If jackscrews are used to actuate the trim tab, check to see that they are not extended beyond the specified limits when the tab is in its extreme positions.

After determining that the control system functions properly and is correctly rigged, it should be thoroughly inspected to determine that the system is correctly assembled and operates freely over the specified range of movement.

Checking & Safetying the System
Whenever rigging is performed on any aircraft, it is good practice to have a second set of eyes inspect the control system to make certain that all turnbuckles, rod ends, and attaching nuts and bolts are correctly safetied.

As a general rule, all fasteners on an aircraft are safetied in some manner. Safetying is defined as securing by various means any nut, bolt, turnbuckle, etc., on the aircraft so that vibration does not cause it to loosen during operation.

Most aircraft manufacturers have a Standard Practices section in their maintenance manuals. These are the methods that should be used when working on a particular system of a

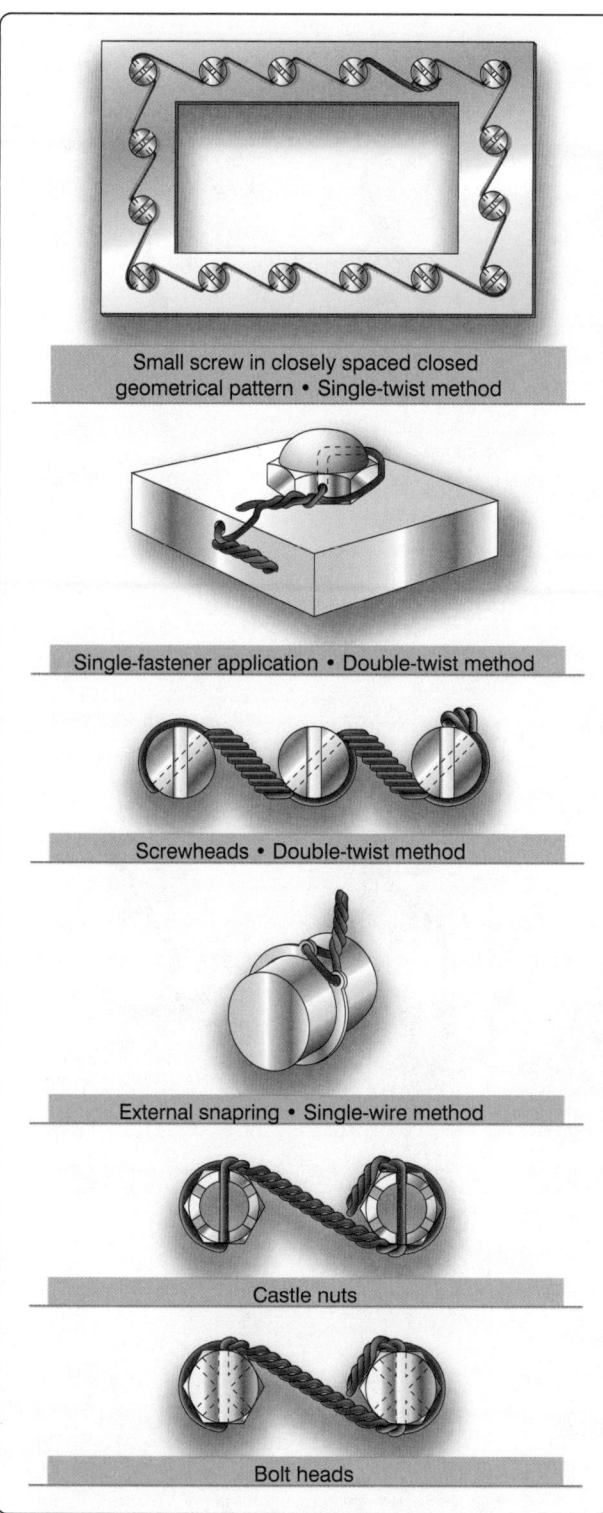

Small screw in closely spaced closed geometrical pattern • Single-twist method

Single-fastener application • Double-twist method

Screwheads • Double-twist method

External snapring • Single-wire method

Castle nuts

Bolt heads

Figure 2-91. *Double-wrap and single safety wire methods for nuts, bolts, and snap rings.*

specific aircraft. However, most standard aircraft hardware has a standard method of being safetied. The following information provides some of the most common methods used in aircraft safetying.

The most commonly used safety wire method is the double-twist, utilizing stainless steel or Monel wire in the .032 to .040-inch diameter range. This method is used on studs, cable turnbuckles, flight controls, and engine accessory attaching bolts. A single-wire method is used on smaller screws, bolts, and/or nuts when they are located in a closely spaced or closed geometrical pattern. The single-wire method is also used on electrical components and in places that are difficult to reach. *[Figure 2-91]*

Safety-of-flight emergency equipment, such as portable fire extinguishers, oxygen regulators, emergency valves, firewall shut-offs, and seals on first-aid kits, are safetied using a single copper wire (.020-inch diameter) or aluminum wire (.031-inch diameter). The wire on this emergency equipment is installed only to indicate the component is sealed or has not been actuated. It must be possible to break the wire seal by hand, without the use of any tools.

The use of safety wire pliers, or wire twisters, makes the job of safetying much easier on the mechanic's hands and produces a better finished product. *[Figure 2-92]*

The wire should have six to eight twists per inch of wire and be pulled taut while being installed. Where practicable, install the safety wire around the head of the fastener and twist it in such a manner that the loop of the wire is pulled close to the contour of the unit being safety wired, and in the direction that would have the tendency to tighten the fastener. *[Figure 2-93]*

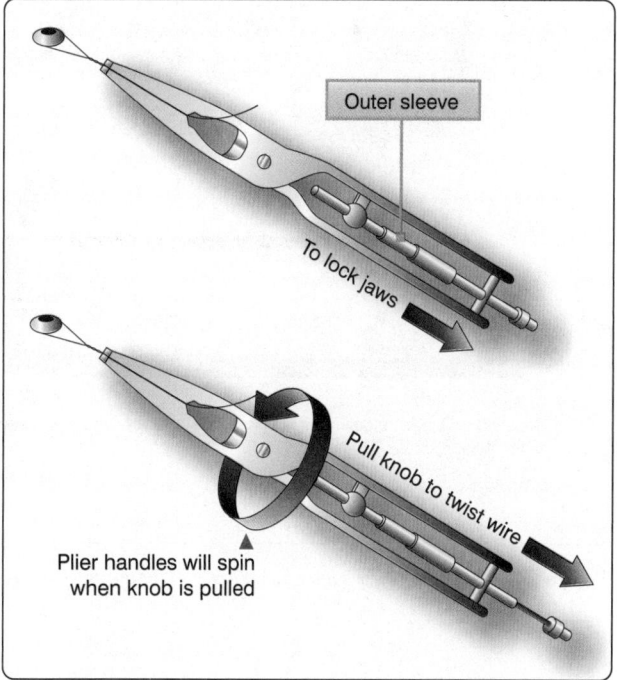

Figure 2-92. *Use of safety-wire pliers or wire twisters.*

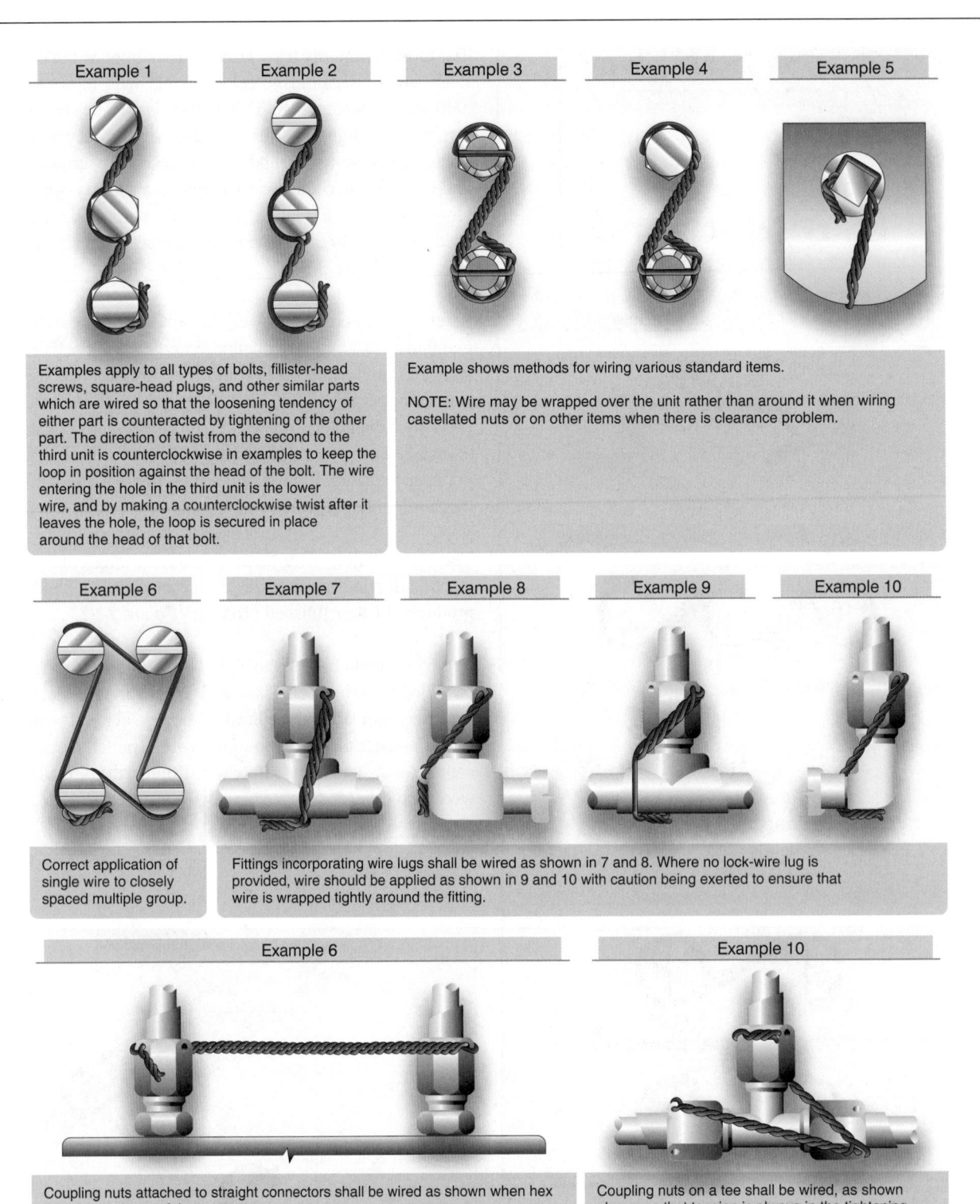

Example 1

Example 2

Example 3

Example 4

Example 5

Examples apply to all types of bolts, fillister-head screws, square-head plugs, and other similar parts which are wired so that the loosening tendency of either part is counteracted by tightening of the other part. The direction of twist from the second to the third unit is counterclockwise in examples to keep the loop in position against the head of the bolt. The wire entering the hole in the third unit is the lower wire, and by making a counterclockwise twist after it leaves the hole, the loop is secured in place around the head of that bolt.

Example shows methods for wiring various standard items.

NOTE: Wire may be wrapped over the unit rather than around it when wiring castellated nuts or on other items when there is clearance problem.

Example 6

Example 7

Example 8

Example 9

Example 10

Correct application of single wire to closely spaced multiple group.

Fittings incorporating wire lugs shall be wired as shown in 7 and 8. Where no lock-wire lug is provided, wire should be applied as shown in 9 and 10 with caution being exerted to ensure that wire is wrapped tightly around the fitting.

Example 6

Example 10

Coupling nuts attached to straight connectors shall be wired as shown when hex is an integral part of the connector.

Coupling nuts on a tee shall be wired, as shown above, so that tension is always in the tightening direction.

Figure 2-93. *Examples of various fasteners and methods of safetying.*

Cotter pins are used to secure such items as bolts, screws, pins, and shafts. They are used at any location where a turning or actuating movement takes place. The diameter of the cotter pin selected for any application should be the largest size that will fit consistent with the diameter of the cotter pin hole and/or the slots in the castellated nut. Cotter pins, like safety wire, should never be re-used on aircraft. [Figure 2-94]

Self-locking nuts are used in applications where they are not removed often. There are two types of self-locking nuts currently in use. One is all metal and the other has an insert, usually of fiber or nylon.

It is extremely important that the manufacturer's Illustrated Parts Book (IPB) be consulted for the correct type and grade of lock nut for various locations on the aircraft. The finish or plating color of the nut identifies the type of application and environment in which it can be used. For example, a cadmium-plated nut is gold in color and provides exceptionally good protection against corrosion, but should not be used in applications where the temperature may exceed 450 °F.

Repeated removal and installation causes the self-locking nut to lose its locking feature. They should be replaced when they are no longer capable of maintaining the minimum prevailing torque. [Figure 2-95]

Lock washers may be used with bolts and machine screws whenever a self-locking nut or castellated nut is not applicable. They may be of the split washer spring type, or a multi-serrated internal or external star washer.

Pal nuts may be a second nut tightened against the first and used to force the primary nut thread against the bolt or screw thread. They may also be of the type that are made of stamped spring steel and are to be used only once and replaced with new ones when removed.

Biplane Assembly & Rigging

Biplanes were some of the very first aircraft designs. The first powered heavier-than-air aircraft, the Wright Brothers' Wright Flyer, successfully flown on December 17, 1903, was a biplane.

The first biplanes were designed with very thin wing sections and, consequently, the wing structure needed to be strengthened by external bracing wires. The biplane configuration allowed the two wings to be braced against one another, increasing the structural strength. When the assembly and rigging of a biplane is accomplished in accordance with the approved instructions, a stable airworthy aircraft is the

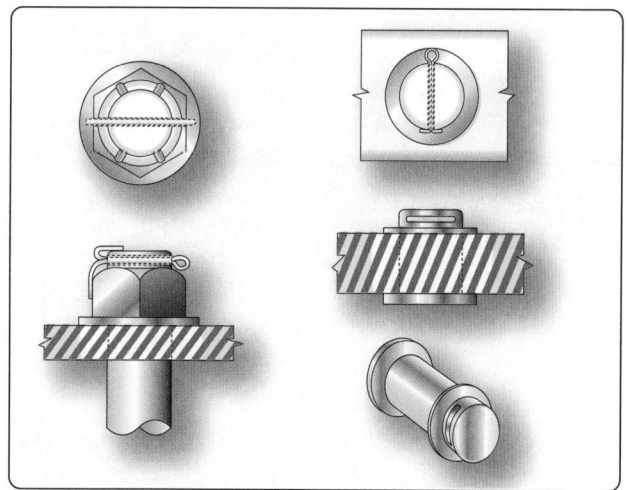

Figure 2-94. *Securing hardware with cotter pins.*

Fine Thread Series	
Thread Size	Minimum Prevailing Torque
7/16 - 20	8 inch-pounds
1/2 - 20	10 inch-pounds
9/16 - 18	13 inch-pounds
5/8 - 18	18 inch-pounds
3/4 - 16	27 inch-pounds
7/8 - 14	40 inch-pounds
1 - 14	55 inch-pounds
1-1/8 - 12	73 inch-pounds
1-1/4 - 12	94 inch-pounds

Coarse Thread Series	
Thread Size	Minimum Prevailing Torque
7/16 - 14	8 inch-pounds
1/2 - 13	10 inch-pounds
9/16 - 12	14 inch-pounds
5/8 - 11	20 inch-pounds
3/4 - 10	27 inch-pounds
7/8 - 9	40 inch-pounds
1 - 8	51 inch-pounds
1-1/8 - 8	68 inch-pounds
1-1/4 - 8	68 inch-pounds

Figure 2-95. *Minimum prevailing torque values for reused self-locking nuts.*

result.

Whether assembling an early model vintage aircraft that may have been disassembled for repair and restoration, or constructing and assembling a new aircraft, the following are some basic alignment procedures to follow.

To start, the fuselage must be level, fore and aft and laterally. The aircraft usually has specific leveling points designated

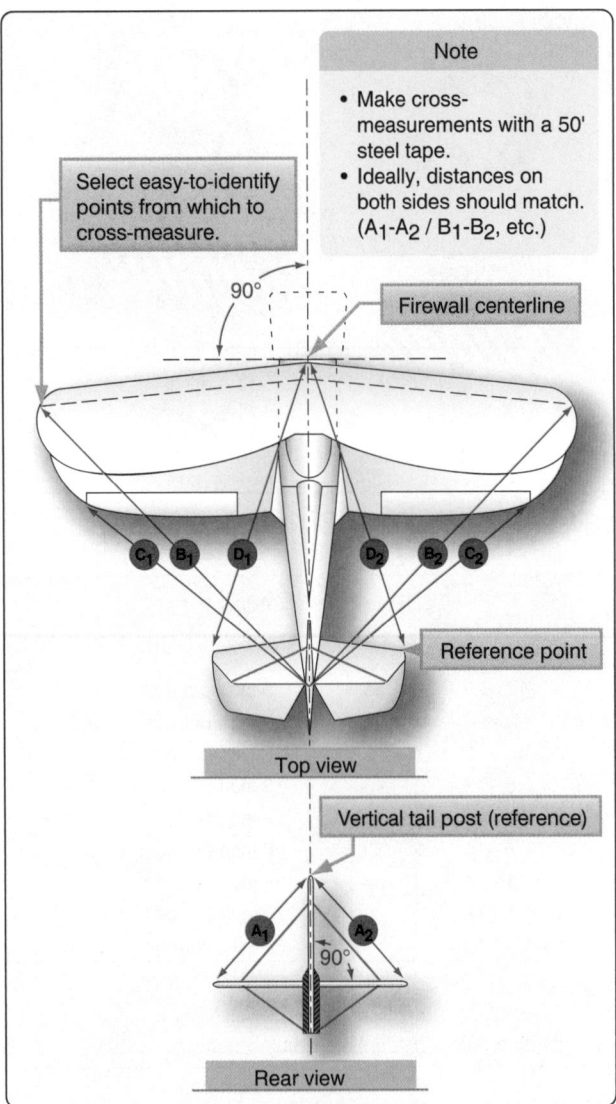

Figure 2-96. *Checking aircraft symmetry.*

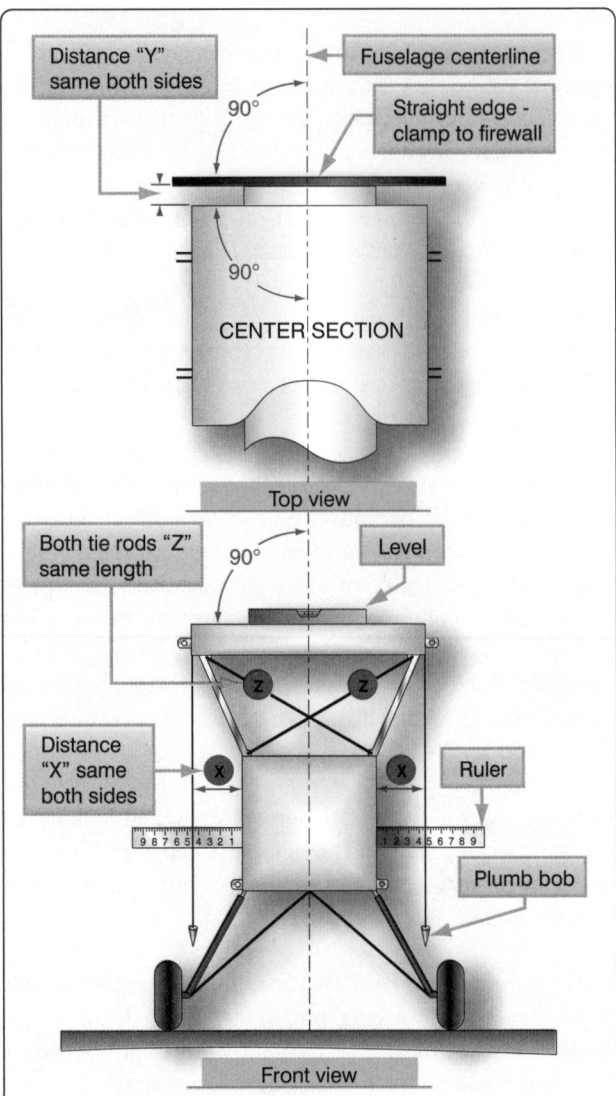

Figure 2-97. *Center section alignment.*

by the manufacturer or indicated on the plans. The fuselage should be blocked up off the landing gear so it is stable. A center line should be drawn on the floor the length of the fuselage and another line perpendicular to it at the firewall, for use as an additional alignment reference.

With the horizontal and vertical tail surfaces installed, the incident angle for the horizontal stabilizer should be set. The tail brace wires should be connected and tightened until the slack is removed. Alignment measurements should be checked as shown in *Figure 2-96.*

Install the elevator and rudder and clamp them in a neutral position. Verify the neutral position of the control stick and rudder pedals in the flight deck and secure them in order to simplify the connecting and final tensioning of the control cables.

If the biplane has a center section for the upper wing, it must be aligned as accurately as possible, because even the smallest error is compounded at the wing tip. Applicable cables and turnbuckles should be connected and the tension set as specified. *[Figure 2-97]* The stagger measurement can be checked as shown in *Figure 2-98.*

The lower wing sections should be individually attached to the fuselage and blocked up for support while the landing wires are connected and adjusted to obtain the dihedral called for in the specifications or plans. *[Figure 2-99]*

Next, connect the outer "N" struts to the left and right sections of the lower wing. Now, the upper wing can be attached and the flying wires installed. The slave struts can be installed and the ailerons connected using the same alignment and adjustment procedures used for the elevator and rudder. The incidence angle can be checked, as shown in *Figure 2-100.*

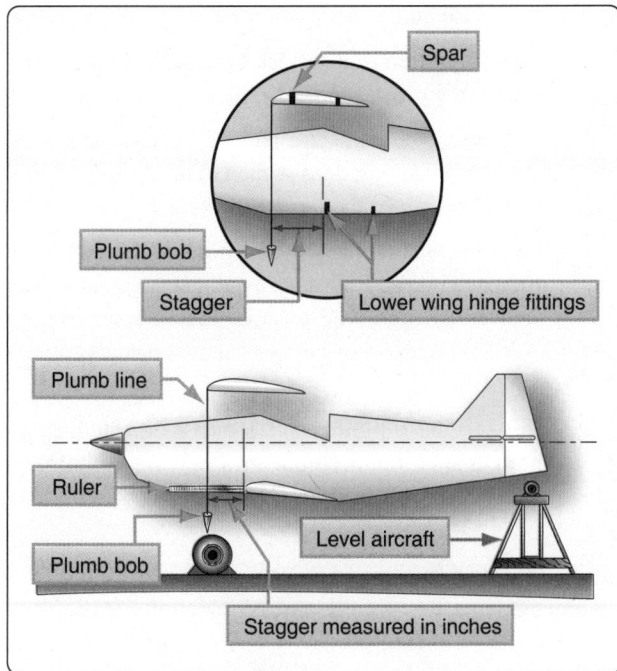

Figure 2-98. *Measuring stagger.*

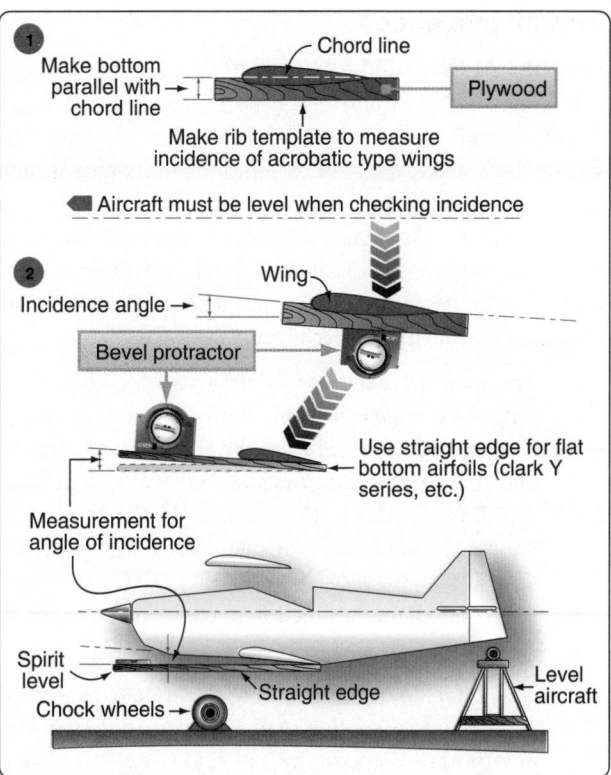

Figure 2-100. *Checking incidence.*

Once this point is reached, it is a matter of measuring, checking angles, and adjusting the various components to obtain the overall aircraft symmetry and desired alignment, as shown in *Figure 2-96.*

Also, remember that care should be used when tightening the wing wires because extra stress can be inadvertently induced into the wings. Always loosen one wire before tightening the opposite wire. Flying and landing wires are typically set at about 600 pounds and tail brace wires at about 300 pounds of tension.

When convinced the aircraft is properly rigged, move away from it and take a good look at the finished product. Are the wings symmetrical? Does the dihedral look even? Is the tail section square with the fuselage? Are the wing attaching hardware, flying wires, and control cables safetied? And the final task, before the first flight, is to complete the maintenance record entries.

As with any aircraft maintenance or repair, the instructions and specifications from the manufacturer, or the procedures and recommendations found in the construction plans, should be the primary method to perform the assembly and rigging of the aircraft.

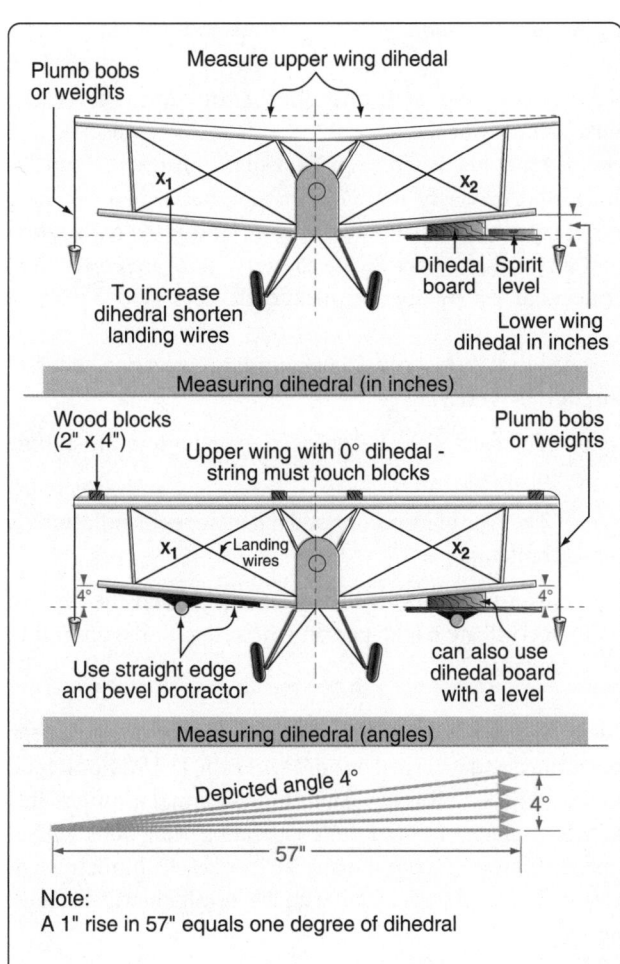

Figure 2-99. *Measuring dihedral.*

Aircraft Inspection

Purpose of Inspection Programs

The purpose of an aircraft inspection program is to ensure that the aircraft is airworthy. Per 14 CFR part 3, section 3.5, "airworthy" means the aircraft conforms to its type design and is in a condition for safe operation. By this definition and according to subsequent case law relating to the term and regulations for the issuance of a standard airworthiness certificate, there are two conditions that must be met for the aircraft to be considered airworthy:

1. *The aircraft must conform to its type design or properly altered condition.* Conformity to type design is considered attained when the aircraft configuration and the components installed are consistent with the drawings, specifications, and other data that are part of the type certificate (TC), which includes any supplemental type certificate (STC) and field approved alterations incorporated into the aircraft.

2. *The aircraft must be in a condition for safe operation.* This refers to the condition of the aircraft relative to wear and deterioration (e.g., skin corrosion, window delamination/crazing, fluid leaks, and tire wear beyond specified limits).

When flight hours and calendar time are accumulated into the life of an aircraft, some components wear out and others deteriorate. Inspections are developed to find these items, and repair or replace them before they affect the airworthiness of the aircraft.

Performing an Airframe Conformity & Airworthiness Inspection

To establish conformity of an aircraft product, start with a TCDS. This document is a formal description of the aircraft, the engine, or the propeller. It is issued by the Federal Aviation Administration (FAA) when they find that the product meets the applicable requirements for certification under 14 CFR.

The TCDS lists the limitations and information required for type certification of aircraft. It includes the certification basis and eligible serial numbers for the product. It lists airspeed limits, weight limits, control surface movements, engine make and models, minimum crew, fuel type, etc.; the horsepower and rpm limits, thrust limitations, size and weight for engines; and blade diameter, pitch, etc., for propellers. Additionally, it provides all the various components by make and model, eligible for installation on the applicable product.

A manufacturer's maintenance information may be in the form of service instructions, service bulletins, or service letters that the manufacturer publishes to provide instructions for product improvement or to revise and update maintenance

manuals. Service bulletins are not regulatory unless:

1. All or a portion of a service bulletin is incorporated as part of an airworthiness directive.

2. The service bulletins are part of the FAA-approved airworthiness limitations section of the manufacturer's manual or part of the type certificate.

3. The service bulletins are incorporated directly or by reference into an FAA-approved inspection program, such as an approved aircraft inspection program (AAIP) or continuous aircraft maintenance program (CAMP).

4. The service bulletins are listed as an additional maintenance requirement in a certificate holder's operations specifications (OpSpecs).

Airworthiness directives (ADs) are published by the FAA as amendments to 14 CFR part 39, section 39.13. They apply to the following products: aircraft, aircraft engines, propellers, and appliances. The FAA issues airworthiness directives when an unsafe condition exists in a product, and the condition is likely to exist or develop in other products of the same type design.

To perform the airframe conformity and verify the airworthiness of the aircraft, records must be checked and the aircraft inspected. The data plate on the airframe is inspected to verify its make, model, serial number, type certificate, or production certificate. Check the registration and airworthiness certificate to verify they are correct and reflect the "N" number on the aircraft.

Inspect aircraft records. Check current inspection status of aircraft, by verifying:

- The date of the last inspection and aircraft total time in service.

- The type of inspection and if it includes manufacturer's bulletins.

- The signature, certificate number, and the type of certificate of the person who returned the aircraft to service.

Identify if any major alterations or major repairs have been performed and recorded on an FAA Form 337, Major Repair and Alteration. Review any flight manual supplements (FMS) included in the Pilot's Operating Handbook (POH) and determine if there are any airworthiness limitations or required placards associated with the installation(s) that must be inspected.

Check for a current weight and balance report, and the current equipment list, current status of airworthiness directives for airframe, engine, propeller, and appliances. Also, check the limitations section of the manufacturer's manual to verify the status of any life-limited components.

Obtain the latest revision of the airframe TCDS and use it as a verification document to inspect and ensure the correct engines, propellers, and components are installed on the airframe.

Required Inspections

Preflight

Preflight for the aircraft is described in the POH for that specific aircraft and should be followed with the same attention given to the checklists for takeoff, inflight, and landing checklists.

Periodic Maintenance Inspections

Annual Inspection

With few exceptions, no person may operate an aircraft unless, within the preceding 12 calendar months, it has had an annual inspection in accordance with 14 CFR part 43 and was approved for return to service by a person authorized under section 43.7. (A certificated mechanic with an Airframe and Powerplant (A&P) rating must hold an inspection authorization (IA) to perform an annual inspection.) A checklist must be used and include as a minimum, the scope and detail of items (as applicable to the particular aircraft) in 14 CFR part 43, Appendix D.

100-hour Inspection

This inspection is required when an aircraft is operated under 14 CFR part 91 and used for hire, such as flight training. It is required to be performed every 100 hours of service in addition to the annual inspection. (The inspection may be performed by a certificated mechanic with an A&P rating.) A checklist must be used and as a minimum, the inspection must include the scope and detail of items (as applicable to the particular aircraft) in 14 CFR part 43, Appendix D.

Progressive Inspection

This inspection program can be performed under 14 CFR part 91, section 91.409(d), as an alternative to an annual inspection. However, the program requires that a written request be submitted by the registered owner or operator of an aircraft desiring to use a progressive inspection to the local FAA Flight Standards District Office (FSDO). It shall provide:

1. The name of a certificated mechanic holding an inspection authorization, a certificated airframe repair station, or the manufacturer of the aircraft to supervise or conduct the inspection.

2. A current inspection procedures manual available and readily understandable to the pilot and maintenance personnel containing in detail:

    • An explanation of the progressive inspection, including the continuity of inspection responsibility, the making of reports, and the keeping of records and technical reference material.

    • An inspection schedule, specifying the intervals in hours or days when routine and detailed inspections will be performed, and including instructions for exceeding an inspection interval by not more than 10 hours while en route, and for changing an inspection interval because of service experience.

    • Sample routine and detailed inspection forms and instructions for their use.

    • Sample reports and records and instructions for their use.

3. Enough housing and equipment for necessary disassembly and proper inspection of the aircraft.

4. Appropriate current technical information for the aircraft.

The frequency and detail of the progressive inspection program shall provide for the complete inspection of the aircraft within each 12 calendar months and be consistent with the manufacturer's recommendations and kind of operation in which the aircraft is engaged. The progressive inspection schedule must ensure that the aircraft will be airworthy at all times. A certificated A&P mechanic may perform a progressive inspection, as long as they are being supervised by a mechanic holding an Inspection Authorization.

If the progressive inspection is discontinued, the owner or operator must immediately notify the local FAA FSDO in writing. After discontinuance, the first annual inspection will be due within 12 calendar months of the last complete inspection of the aircraft under the progressive inspection.

Large Airplanes (over 12,500 lb)

Inspection requirements of 14 CFR part 91, section 91.409, to include paragraphs (e) and (f).

Paragraph (e) applies to large airplanes (to which 14 CFR part 125 is not applicable), turbojet multiengine airplanes, turbo propeller powered multiengine airplanes, and turbine-powered rotorcraft. Paragraph (f) lists the inspection programs that can be selected under paragraph (e).

The additional inspection requirements for these aircraft are placed on the operator because the larger aircraft typically are more complex and require a more detailed inspection program than is provided for in 14 CFR part 43, Appendix D.

An inspection program must be selected from one of the following four options by the owner or operator of the aircraft:

1. A continuous airworthiness inspection program that is part of a continuous airworthiness maintenance program currently in use by a person holding an air carrier operating certificate or an operating certificate issued under 14 CFR part 121 or 135.

2. An approved aircraft inspection program approved under 14 CFR part 135, section 135.419, and currently in use by a person holding an operating certificate issued under 14 CFR part 135.

3. A current inspection program recommended by the manufacturer.

4. Any other inspection program established by the registered owner or operator of the airplane or turbine-powered rotorcraft and approved by the FAA. This program must be submitted to the local FAA FSDO having jurisdiction of the area in which the aircraft is based. The program must be in writing and include at least the following information:

    (a) Instructions and procedures for the conduct of inspections for the particular make and model airplane or turbine-powered rotorcraft, including the necessary tests and checks. The instructions and procedures must set forth in detail the parts and areas of the airframe, engines, propellers, rotors, and appliances, including survival and emergency equipment, required to be inspected.

    (b) A schedule for performing the inspections that must be performed under the program expressed in terms of the time in service, calendar time, number of system operations (cycles), or any combination of these.

    This FAA approved owner/operator program can be revised at a future date by the FAA, if they find that revisions are necessary for the continued adequacy of the program. The owner/operator can petition the FAA within 30 days of notification to reconsider the notice to make changes.

Manufacturer's Inspection Program

This is a program developed by the manufacturer for their product. It is contained in the "Instructions for Continued Airworthiness" required under 14 CFR part 23, section 23.1529 and part 25, section 25.1529. It is in the form of a manual, or manuals as appropriate, for the quantity of data to be provided

and including, but not limited to, the following content:

- A description of the airplane and its systems and installations, including its engines, propellers, and appliances.

- Basic information describing how the airplane components and systems are controlled and operated, including any special procedures and limitations that apply.

- Servicing information that covers servicing points, capacities of tanks, reservoirs, types of fluids to be used, pressures applicable to the various systems, lubrication points, lubricants to be used, equipment required for servicing, tow instructions, mooring, jacking, and leveling information.

- Maintenance instructions with scheduling information for the airplane and each component that provides the recommended periods at which they should be cleaned, inspected, adjusted, tested, and lubricated, and the degree of inspection and work recommended at these periods.

- The recommended overhaul periods and necessary cross references to the airworthiness limitations section of the manual.

- The inspection program that details the frequency and extent of the inspections necessary to provide for the continued airworthiness of the airplane.

- Diagrams of structural access plates and information needed to gain access for inspections when access plates are not provided.

- Details for the application of special inspection techniques, including radiographic and ultrasonic testing where such processes are specified.

- A list of special tools needed.

- An Airworthiness Limitations section that is segregated and clearly distinguishable from the rest of the document. This section must set forth:

    1. Each mandatory replacement time, structural inspection interval, and related structural inspection procedures required for type certification or approved under 14 CFR part 23 or part 25.

    2. Each mandatory replacement time, inspection interval, related inspection procedure, and all critical design configuration control limitations approved under 14 CFR part 23 or part 25, for the fuel tank system.

The Airworthiness Limitations section must contain a legible statement in a prominent location that reads: "The Airworthiness Limitations section is FAA-approved and specifies maintenance required under 14 CFR part 43, sections 43.16 and part 91, section 91.403, unless an alternative program has been FAA-approved."

Any operator who wishes to adopt a manufacturers' inspection program should first contact their local FAA Flight Standards District Office, for further guidance.

Altimeter & Static System Inspections in Accordance with 14 CFR Part 91, Section 91.411

Any person operating an airplane or helicopter in controlled airspace under instrument flight rules (IFR) must have had, within the preceding 24 calendar months, each static pressure system, each altimeter instrument, and each automatic pressure altitude reporting system tested and inspected and found to comply with 14 CFR part 43, Appendix E. Those tests and inspections must be conducted by appropriately rated persons under 14 CFR.

Air Traffic Control (ATC) Transponder Inspections

Any person using an air traffic control (ATC) transponder must have had, within the preceding 24 calendar months, that transponder tested and inspected and found to comply with 14 CFR part 43, Appendix F, and part 91, section 91.411. Additionally, following any installation or maintenance on an ATC transponder where data correspondence error could be introduced, the integrated system must be tested and inspected and found to comply with 14 CFR part 43, Appendix E, and part 91, section 91.411 by an appropriately rated person under 14 CFR.

Emergency Locator Transmitter (ELT) Operational & Maintenance Practices in Accordance with Advisory Circular (AC) 91-44

This AC combined and updated several ACs on the subject of ELTs and receivers for airborne service.

Under the operating rules of 14 CFR part 91, most small U.S. registered civil airplanes equipped to carry more than one person must have an ELT attached to the airplane. 14 CFR part 91, section 91.207 defines the requirements of what type aircraft and when the ELT must be installed. It also states that an ELT that meets the requirements of Technical Standard Order (TSO)-C91 may not be used for new installations.

The pilot-in-command of an aircraft equipped with an ELT is responsible for its operation and, prior to engine shutdown at the end of each flight, should tune the VHF receiver to 121.5 MHz and listen for ELT activations. Maintenance personnel are responsible for accidental activation during the actual period of their work.

Maintenance of ELTs is subject to 14 CFR part 43 and part 91, section 91.413 and should be included in the required inspections. It is essential that the impact switch operation and the transmitter output be checked using the manufacturer's instructions. Testing of an ELT prior to installation or for maintenance reasons, should be conducted in a metal enclosure in order to avoid outside radiation by the transmitter. If this is not possible, the test should be conducted only within the first 5 minutes after any hour.

Manufacturers of ELTs are required to mark the expiration date of the battery, based on 50 percent of the useful life, on the outside of the transmitter. The batteries are required to be replaced on that date or when the transmitter has been in use for more than 1 cumulative hour. Water activated batteries, have virtually unlimited shelf life. They are not usually marked with an expiration date. They must be replaced after activation regardless of how long they were in service.

The battery replacement can be accomplished by a pilot on a portable type ELT that is readily accessible and can be removed and reinstalled in the aircraft by a simple operation. That would be considered preventive maintenance under 14 CFR part 43, section 43.3(g). Replacement batteries should be approved for the specific model of ELT and the installation performed in accordance with section 43.13.

AC 91-44 also contains additional information on:

- Airborne homing and alerting equipment for use with ELTs.

- Search and rescue responsibility.

- Alert and search procedures including various flight procedures for locating an ELT.

- The FAA Frequency Management Offices, for contacting by manufacturers when they are demonstrating and testing ELTs.

Although there is no regulatory requirement to install a 406 ELT, the benefits are numerous, regardless of regulatory minimums. All new installations must be a 406 MHz digital ELT. It must meet the standards of TSO C126. When installed, the new 406 MHz ELT should be registered so that if the aircraft were to go down, search and rescue could take full advantage of the benefits the system offers. The digital circuitry of the 406 MHz ELT can be coded with information about the aircraft type, base location, ownership, etc. This coding allows the search and rescue (SAR) coordinating centers to contact the registered owner or operator if a signal is detected to determine if the aircraft is flying or parked. This type of identification permits a rapid SAR response in the event of an accident, and will save valuable resources from a false alarm search.

Annual & 100-Hour Inspections

Preparation

An owner/operator bringing an aircraft into a maintenance facility for an annual or 100-hour inspection may not know what is involved in the process. This is the point at which the person who performs the inspection sits down with the customer to review the records and discuss any maintenance issues, repairs needed, or additional work the customer may want done. Moreover, the time spent on these items before starting the inspection usually saves time and money before the work is completed.

The work order describes the work that will be performed and the fee that the owner pays for the service. It is a contract that includes the parts, materials, and labor to complete the inspection. It may also include additional maintenance and repairs requested by the owner or found during the inspection.

Additional materials such as ADs, manufacturer's service bulletins and letters, and vendor service information must be researched to include the avionics and emergency equipment on the aircraft. The TCDS provides all the components eligible for installation on the aircraft.

The review of the aircraft records is one of the most important parts of any inspection. Those records provide the history of the aircraft. The records to be kept and how they are to be maintained are listed in 14 CFR part 91, section 91.417. Among those records that must be tracked are records of maintenance, preventive maintenance, and alteration, records of the last 100-hour, annual, or other required or approved inspections for the airframe, engine propeller, rotor, and appliances of an aircraft. The records must include:

- A description (or reference to data acceptable to the FAA) of the work performed.

- The date of completion of the work performed and the signature and certificate number of the person approving the aircraft for return to service.

- The total time in service and the current status of life-limited parts of the airframe, each engine, each propeller, and each rotor.

- The time since the last overhaul of all items installed on the aircraft which are required to be overhauled on a specified time basis.

- The current inspection status of the aircraft, including the time since the last inspection required by the program under which the aircraft and its appliances are maintained.

- The current status of applicable ADs including for each, the method of compliance, the AD number, and revision date. If the AD involves recurring action, the time and date when the next action is required.

- Copies of the forms prescribed by 14 CFR part 43, section 43.9, for each major alteration to the airframe and currently installed components.

The owner/operator is required to retain the records of inspection until the work is repeated, or for 1 year after the work is performed. Most of the other records that include total times and current status of life-limited parts, overhaul times, and AD status must be retained and transferred with the aircraft when it is sold.

14 CFR part 43, section 43.15 requires that each person performing a 100-hour or annual inspection shall use a checklist while performing the inspection. The checklist may be one developed by the person, one provided by the manufacturer of the equipment being inspected, or one obtained from another source. The checklist must include the scope and detail of the items contained in part 43, Appendix D.

The inspection checklist provided by the manufacturer is the preferred one to use. The manufacturer separates the areas to inspect such as engine, cabin, wing, empennage and landing gear. They typically list Service Bulletins and Service Letters for specific areas of the aircraft and the appliances that are installed.

Initial run-up provides an assessment to the condition of the engine prior to performing the inspection. The run-up should include full power and idle rpm, magneto operation, including positive switch grounding, fuel mixture check, oil and fuel pressure, and cylinder head and oil temperatures. After the engine run, check it for fuel, oil, and hydraulic leaks.

Following the checklist, the entire aircraft shall be opened by removing all necessary inspection plates, access doors, fairings, and cowling. The entire aircraft must then be cleaned to uncover hidden cracks or defects that may have been missed because of the dirt.

Following in order and using the checklist, visually inspect each item, or perform the checks or tests necessary to verify the condition of the component or system. Record discrepancies when they are found. The entire aircraft should be inspected and a list of discrepancies be presented to the owner.

A typical inspection following a checklist, on a small single-engine airplane may include in part, as applicable:

- The fuselage for damage, corrosion, and attachment of fittings, antennas, and lights; for "smoking rivets" especially in the landing gear area indicating the possibility of structural movement or hidden failure.

- The flight deck and cabin area for loose equipment that could foul the controls; seats and seat belts for defects and TSO tags; windows and windshields for deterioration; instruments for condition, markings, and operation; flight and engine controls for proper operation.

- The engine and attached components for visual evidence of leaks; studs and nuts for improper torque and obvious defects; engine mount and vibration dampeners for cracks, deterioration, and looseness; engine controls for defects, operation, and safetying; the internal engine for cylinder compression; spark plugs for operation; oil screens and filters for metal particles or foreign matter; exhaust stacks and mufflers for leaks, cracks, and missing hardware; cooling baffles for deterioration, damage, and missing seals; and engine cowling for cracks and defects.

- The landing gear group for condition and attachment; shock absorbing devices for leaks and fluid levels; retracting and locking mechanism for defects, damage, and operation; hydraulic lines for leakage; electrical system for chafing and switches for operation; wheels and bearings for condition; tires for wear and cuts; and brakes for condition and adjustment.

- The wing and center section assembly for condition, skin deterioration, distortion, structural failure, and attachment.

- The empennage assembly for condition, distortion, skin deterioration, evidence of failure (smoking rivets), secure attachment, and component operation and installation.

- The propeller group and system components for torque and proper safetying; the propeller for nicks, cracks, and oil leaks; the anti-icing devices for defects and operation; and the control mechanism for operation, mounting, and restricted movement.

- The radios and electronic equipment for improper installation and mounting; wiring and conduits for improper routing, insecure mounting, and obvious defects; bonding and shielding for installation and condition; and all antennas for condition, mounting, and operation. Additionally, if not already inspected and serviced, the main battery inspected for condition, mounting, corrosion, and electrical charge.

- Any and all installed miscellaneous items and components that are not otherwise covered by this listing for condition and operation.

With the aircraft inspection checklist completed, the list of discrepancies should be transferred to the work order. As part of the annual and 100-hour inspections, the engine oil is drained and replaced because new filters and/or clean screens have been installed in the engine. The repairs are then completed and all fluid systems serviced.

Before approving the aircraft for return to service after the annual or 100-hour inspection, 14 CFR states that the engine must be run to determine satisfactory performance in accordance with the manufacturers recommendations. The run must include:

- Power output (static and idle rpm)

- Magnetos (for drop and switch ground)

- Fuel and oil pressure

- Cylinder and oil temperature

After the run, the engine is inspected for fluid leaks and the oil level is checked a final time before close up of the cowling.

With the aircraft inspection completed, all inspections plates, access doors, fairing and cowling that were removed, must be reinstalled. It is a good practice to visually check inside the inspection areas for tools, shop rags, etc., prior to close up. Using the checklist and discrepancy list to review areas that were repaired will help ensure the aircraft is properly returned to service.

Upon completion of the inspection, the records for each airframe, engine, propeller, and appliance must be signed off. The record entry in accordance with 14 CFR part 43, section 43.11, must include the following information:

- The type inspection and a brief description of the extent of the inspection.

- The date of the inspection and aircraft total time in service.

- The signature, the certificate number, and kind of certificate held by the person approving or disapproving for return to service the aircraft, airframe, aircraft engine, propeller, appliance, component part, or portions thereof.

- For the annual and 100-hour inspection, if the aircraft is found to be airworthy and approved for return to service, enter the following statement: "I certify that this aircraft has been inspected in accordance with a (insert type) inspection and was determined to be in airworthy condition."

- If the aircraft is not approved for return to service because of necessary maintenance, noncompliance with applicable specifications, airworthiness directives, or other approved data, enter the following statement: "I certify that this aircraft has been inspected in accordance with a (insert type) inspection and a list of discrepancies and unairworthy items has been provided to the aircraft owner or operator."

If the owner or operator did not want the discrepancies and/ or unairworthy items repaired at the location where the inspection was accomplished, they may have the option of flying the aircraft to another location with a Special Flight Permit (Ferry Permit). An application for a Special Flight Permit can be made at the local FAA FSDO.

Other Aircraft Inspection & Maintenance Programs

Aircraft operating under 14 CFR part 135, Commuter and On Demand, have additional rules for maintenance that must be followed beyond those in 14 CFR parts 43 and 91.

14 CFR part 135, section 135.411 describes the applicable sections for maintaining aircraft that are type certificated for a passenger seating configuration, excluding any pilot seat, of nine seats or less, and which sections are applicable to maintaining aircraft with 10 or more passenger seats. The following sections apply to aircraft with nine seats or less:

- Section 135.415—requires each certificate holder to submit a Service Difficulty Report, whenever they have an occurrence, failure, malfunction, or defect in an aircraft concerning the list detailed in this section of the regulation.

- Section 135.417—requires each certificate holder to mail or deliver a Mechanical Interruption Report, for occurrences in multi-engine aircraft, concerning unscheduled flight interruptions, and the number of propeller featherings in flight, as detailed in this section of the regulation.

- Section 135.421—requires each certificate holder to comply with the manufacturer's recommended maintenance programs, or a program approved by the FAA for each aircraft, engine, propeller, rotor, and each item of emergency required by 14 CFR part 135. This section also details requirements for single-engine IFR passenger-carrying operations.

- Section 135.422—this section applies to multi-engine airplanes and details requirements for Aging Airplane Inspections and Records review. It excludes airplanes in schedule operations between any point within the State of Alaska.

Any certified operator using aircraft with ten or more passenger seats must have the required organization and maintenance programs, along with competent and knowledgeable people to ensure a safe operation. Title 14 of the CFR, sections 135.423 through 135.443 are numerous and complex, and compliance is required; however, they are not summarized in this handbook. It is the responsibility of the certificated operator to know and comply with these and all other applicable equirements of 14 CFR, and they should contact their local FAA FSDO for further guidance.

The approved aircraft inspection program (AAIP) is an FAA-approved inspection program for aircraft of nine or less passenger seats operated under 14 CFR part 135. The AAIP is an operator developed program tailored to their particular needs to satisfy aircraft inspection requirements. This program allows operators to develop procedures and time intervals for the accomplishment of inspection tasks in accordance with the needs of the aircraft, rather than repeat all the tasks at each 100-hour interval.

The operator is responsible for the AAIP. The program must encompass the total aircraft; including all avionics equipment, emergency equipment, cargo provisions, etc. FAA Advisory Circular 135-10 (as revised) provides detailed guidance to develop an approved aircraft inspection program. The following is a summary, in part, of elements that the program should include:

- A schedule of individual tasks (inspections) or groups of tasks, as well as the frequency for performing those tasks.

- Work forms designating those tasks with a signoff provision for each. The forms may be developed by the operator or obtained from another source.

- Instructions for accomplishing each task. These tasks must satisfy 14 CFR part 43, section 43.13(a), regarding methods, techniques, practices, tools, and equipment. The instructions should include adequate information in a form suitable for use by the person performing the work.

- Provisions for operator-developed revisions to referenced instructions should be incorporated in the operator's manual.

- A system for recording discrepancies and their correction.

- A means for accounting for work forms upon completion of the inspection. These forms are used to satisfy the requirements of 14 CFR part 91, section 91.417, so they must be complete, legible, and identifiable as to the aircraft and specific inspection to which they relate.

- Accommodation for variations in equipment and configurations between aircraft in the fleet.

- Provisions for transferring an aircraft from another program to the AAIP.

The development of the AAIP may come from one of the

following sources:

- An adoption of an aircraft manufacturer's inspection in its entirety. However, many aircraft manufacturers' programs do not encompass avionics, emergency equipment, appliances, and related installations that must be incorporated into the AAIP. The inspection of these items and systems will require additions to the program to ensure they comply with the air carrier's operation specifications and as applicable to 14 CFR.

- A modified manufacturer's program. The operator may modify a manufacturer's inspection program to suit its needs. Modifications should be clearly identified and provide an equivalent level of safety to those in the manufacturer's approved program.

- An operator-developed program. This type of program is developed in its entirety by the operator. It should include methods, techniques, practices, and standards necessary for proper accomplishment of the program.

- An existing progressive inspection program (14 CFR part 91.409(d)) may be used as a basis for the development of an AAIP.

As part of this inspection program, the FAA strongly recommends that a Corrosion Protection Control Program and a supplemental structural inspection type program be included.

A program revision procedure should be included so that an evaluation of any revision can be made by the operator prior to submitting them to the FAA for approval.

Procedures for administering the program should be established. These should include: defining the duties and responsibilities for all personnel involved in the program, scheduling inspections, recording their accomplishment, and maintaining a file of completed work forms.

The operator's manual should include a section that clearly describes the complete program, including procedures for program scheduling, recording, and accountability for continuing accomplishment of the program. This section serves to facilitate administration of the program by the certificate holder and to direct its accomplishment by mechanics or repair stations. The operator's manual should include instructions to accomplish the maintenance/inspections tasks. It should also contain a list of the necessary tools and equipment needed to perform the maintenance and inspections.

The FAA FSDO will provide each operator with computer-generated Operations Specifications when they approve the program.

Continuous Airworthiness Maintenance Program (CAMP)

The definition of maintenance in 14 CFR part 1 includes inspection. The inspection program required for 14 CFR part 121 and part 135 air carriers is part of the Continuous Airworthiness Maintenance Program (CAMP). CAMP is not required of every part 135 carrier; it depends on aircraft being operated. It is a complex program that requires an organization of experienced and knowledgeable aviation personnel to implement it.

The FAA has developed an Advisory Circular, AC 120-16 (as revised) Air Carrier Maintenance Programs, which explains the background as well as the FAA regulatory requirements for these programs. The AC applies to air carriers subject to 14 CFR parts 119, 121, and 135. For part 135, it applies only to aircraft type certificated with ten or more passenger seats.

Any person wanting to place their aircraft on this type of program should contact their local FAA FSDO for guidance.

Title 14 CFR part 125, section 125.247, Inspection Programs & Maintenance

This regulation applies to airplanes having a seating capacity of 20 or more passengers or a maximum payload capacity of 6,000 pounds or more when the aircraft is not required to be operated under 14 CFR parts 121, 129, 135, and 137. Inspection programs which may be approved for use under this 14 CFR part include, but are not limited to:

1. A continuous inspection program which is part of a current continuous airworthiness program approved for use by a certificate holder under 14 CFR part 121 or part 135;

2. Inspection programs currently recommended by the manufacturer of the airplane, airplane engines, propellers, appliances, or survival and emergency equipment; or

3. An inspection program developed by a certificate holder under 14 CFR part 125.

    The airplane subject to this part may not be operated unless:

    - The replacement times for life-limited parts specified in the aircraft type certificate data sheets, or other documents approved by the FAA are complied with;

    - Defects disclosed between inspections, or as a result of inspection, have been corrected in accordance with 14 CFR part 43; and

    - The airplane, including airframe, aircraft

engines, propellers, appliances, and survival and emergency equipment, and their component parts, is inspected in accordance with an inspection program approved by the FAA. These inspections must include at least the following:

- o Instructions, procedures and standards for the particular make and model of airplane, including tests and checks. The instructions and procedures must set forth in detail the parts and areas of the airframe, aircraft engines, propellers, appliances, and survival and emergency equipment required to be inspected.

- o A schedule for the performance of the inspections that must be performed under the program, expressed in terms of the time in service, calendar time, number of system operations, or any combination of these.

- o The person used to perform the inspections required by 14 CFR part 125, must be authorized to perform maintenance under 14 CFR part 43. The airplane subject to part 125 may not be operated unless the installed engines have been maintained in accordance with the overhaul periods recommended by the manufacturer or a program approved by the FAA; the engine overhaul periods are specified in the inspection programs required by 14 CFR part 125, section 125.247.

Piston-Engine & Turbine-Powered Helicopter Inspections

A piston-engine helicopter must be inspected in accordance with the scope and detail of 14 CFR part 43, Appendix D for an Annual Inspection. However, there are additional performance rules for inspections under 14 CFR part 43, section 43.15, requiring that each person performing an inspection under 14 CFR part 91 on a rotorcraft shall inspect these additional components in accordance with the maintenance manual or Instructions for Continued Airworthiness of the manufacturer concerned:

1. The drive shaft or similar systems.

2. The main rotor transmission gear box for obvious defects.

3. The main rotor and center section (or the equivalent area).

4. The auxiliary rotor.

The operator of a turbine-powered helicopter can elect to have

it inspected under 14 CFR part 91, section 91.409:

1. Annual inspection.

2. 100-hour inspection, when being used for compensation or hire.

3. A progressive inspection, when authorized by the FAA.

4. An inspection program listed under 14 CFR part 91, section 91.409 (f), when selected by the owner/operator and the selection is recorded in the aircraft maintenance records (14 CFR part 91, section 91.409(e)).

When performing any of the above inspections, the additional performance rules under 14 CFR part 43, section 43.15, for rotorcraft must be complied with.

Light Sport Aircraft & Aircraft Certificated as Experimental

Light sport aircraft and aircraft that are certificated in the experimental category are issued a Special Airworthiness Certificate by the FAA. Operating limitations are issued to these aircraft as a part of the Special Airworthiness Certificate that specify the required inspections and inspection intervals for the aircraft.

Typically, the operating limitations issued to these aircraft require that a condition inspection be performed once every 12 months. If the aircraft is used for compensation or hire (e.g., towing a glider, flight training), then it must also be inspected each 100 hours. A condition inspection is equivalent to the scope and detail of an annual inspection, the requirements of which are outlined in 14 CFR part 43, Appendix D.

An A&P or an appropriately rated repair station can perform the condition inspection on any of these aircraft. The FAA issues repairman certificates to individuals who are the builder of an amateur-built aircraft, which authorizes performance of the condition inspection. Additionally, repairman certificates can be issued to individuals for conducting inspections on light sport aircraft. There are two ratings available for light sport repairman certificate, each with different privileges as described in 14 CFR part 65, section 65.107, but both ratings authorize the repairman to conduct the annual condition inspection.

The operating limitations issued to the aircraft also require that the condition inspection be recorded in the aircraft maintenance records. The following or similarly worded statement is used:

"I certify that this aircraft has been inspected on [insert date] per the [insert either: scope and detail of 14 CFR part 43, Appendix D; or manufacturer's inspection procedures] and was found to be in a condition for safe operation." The entry will include the aircraft's total time-in-service (cycles if appropriate), and the name, signature, certificate number, and type of certificate held by the person performing the inspection.

Chapter 3
Aircraft Fabric Covering

General History

Fabric-covered aircraft play an important role in the history of aviation. The famous Wright Flyer utilized a fabric-covered wood frame in its design, and fabric covering continued to be used by many aircraft designers and builders during the early decades of production aircraft. The use of fabric covering on an aircraft offers one primary advantage: light weight. In contrast, fabric coverings have two disadvantages: flammability and lack of durability.

Finely woven organic fabrics, such as Irish linen and cotton, were the original fabrics used for covering airframes, but their tendency to sag left the aircraft structure exposed to the elements. To counter this problem, builders began coating the fabrics with oils and varnishes. In 1916, a mixture of cellulose dissolved in nitric acid, called nitrate dope, came into use as an aircraft fabric coating. Nitrate dope protected the fabric, adhered to it well, and tautened it over the airframe. It also gave the fabric a smooth, durable finish when dried. The major drawback to nitrate dope was its extreme flammability.

To address the flammability issue, aircraft designers tried a preparation of cellulose dissolved in butyric acid called butyrate dope. This mixture protected the fabric from dirt and moisture, but it did not adhere as well to the fabric as nitrate dope. Eventually, a system combining the two dope coatings was developed. First, the fabric was coated with nitrate dope for its adhesion and protective qualities. Then, subsequent coats of butyrate dope were added. Since the butyrate dope coatings reduced the overall flammability of the fabric covering, this system became the standard fabric treatment system.

The second problem, lack of durability, stems from the eventual deterioration of fabric from exposure to the elements that results in a limited service life. Although the mixture of nitrate dope and butyrate dope kept out dirt and water, solving some of the degradation issue, it did not address deterioration caused by ultraviolet (UV) radiation from the sun. Ultraviolet radiation passed through the dope and degraded not only the fabric, but also the aircraft structure underneath. Attempts to paint the coated fabric proved unsuccessful, because paint does not adhere well to nitrate dope. Eventually, aluminum solids were added to the butyrate coatings. This mixture reflected the sun's rays, prevented harmful UV rays from penetrating the dope, and protected the fabric, as well as the aircraft structure.

Regardless of treatments, organic fabrics have a limited lifespan; cotton or linen covering on an actively flown aircraft lasts only about 5–10 years. Furthermore, aircraft cotton has not been available for over 25 years. As the aviation industry developed more powerful engines and more aerodynamic aircraft structures, aluminum became the material of choice. Its use in engines, aircraft frames, and coverings revolutionized aviation. As a covering, aluminum protected the aircraft structure from the elements, was durable, and was not flammable.

Although aluminum and composite aircraft dominate modern aviation, advances in fabric coverings continue to be made because gliders, home-built, and light sport aircraft, as well as some standard and utility certificated aircraft, are still produced with fabric coverings. *[Figure 3-1]* The nitrate/butyrate dope process works well, but does not mitigate the short lifespan of organic fabrics. It was not until the introduction of polyester fabric as an aircraft covering in the 1950s that the problem of the limited lifespan of fabric covering was solved. The transition to polyester fabric had some problems because the nitrate and butyrate dope coating

Figure 3-1. *Examples of aircraft produced using fabric skin.*

process is not as suitable for polyester as it is for organic fabrics. Upon initial application of the dopes to polyester, good adhesion and protection occurred; as the dopes dried, they would eventually separate from the fabric. In other words, the fabric outlasted the coating.

Eventually, dope additives were developed that minimized the separation problem. For example, plasticizers keep the dried dope flexible and nontautening dope formulas eliminate separation of the coatings from the fabric. Properly protected and coated, polyester lasts indefinitely and is stronger than cotton or linen. Today, polyester fabric coverings are the standard and use of cotton and linen on United States certificated aircraft has ceased. In fact, the long staple cotton from which grade-A cotton aircraft fabric is made is no longer produced in this country.

Re-covering existing fabric aircraft is an accepted maintenance procedure. Not all aircraft covering systems include the use of dope coating processes. Modern aircraft covering systems that include the use of nondope fabric treatments show no signs of deterioration even after decades of service. In this chapter, various fabrics and treatment systems are discussed, as well as basic covering techniques.

Fabric Terms

To facilitate the discussion of fabric coverings for aircraft, the following definitions are presented. *Figure 3-2* illustrates some of these items.

- Warp—the direction along the length of fabric.
- Fill or weave—the direction across the width of the fabric.
- Count—the number of threads per inch in warp or filling.
- Ply—the number of yarns making up a thread.
- Bias—a cut, fold, or seam made diagonally to the warp or fill threads.
- Pinked edge—an edge which has been cut by machine or special pinking shears in a continuous series of Vs to prevent raveling.
- Selvage edge—the edge of cloth, tape, or webbing woven to prevent raveling.
- Greige—condition of polyester fabric upon completion of the production process before being heat shrunk.
- Cross-coat—brushing or spraying where the second coat is applied 90° to the direction the first coat was applied. The two coats together make a single cross coat. *[Figure 3-3]*

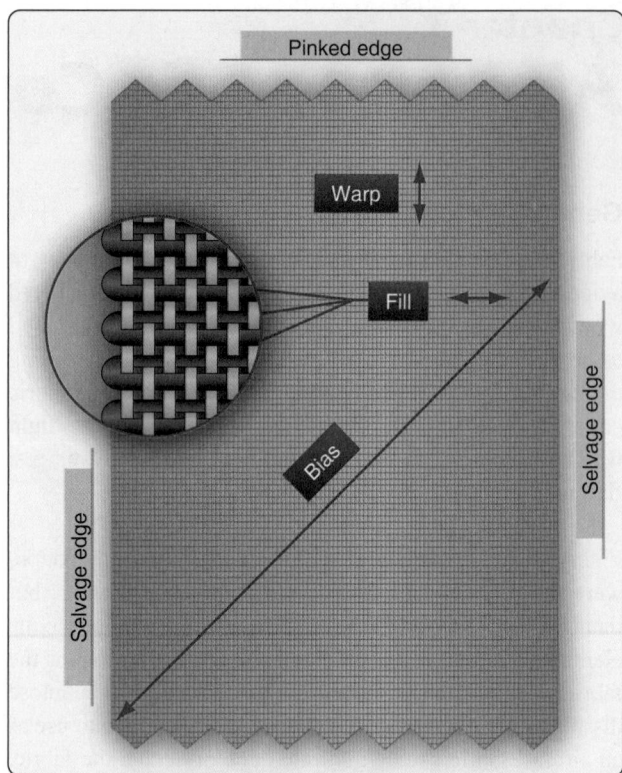

Figure 3-2. *Aircraft fabric nomenclature.*

Legal Aspects of Fabric Covering

When a fabric-covered aircraft is certificated, the aircraft manufacturer uses materials and techniques to cover the aircraft that are approved under the type certificate issued for that aircraft. The same materials and techniques must be used by maintenance personnel when replacing the aircraft fabric. Descriptions of these materials and techniques are in the manufacturer's service manual. For example, aircraft originally manufactured with cotton fabric can only be re-covered with cotton fabric unless the Federal Aviation Administration (FAA) approves an exception. Approved exceptions for alternate fabric-covering materials and procedures are common. Since polyester fabric coverings deliver performance advantages, such as lighter weight, longer life, additional strength, and lower cost, many older aircraft originally manufactured with cotton fabric have received approved alteration authority and have been re-covered with polyester fabric.

There are three ways to gain FAA approval to re-cover an aircraft with materials and processes other than those with which it was originally certificated. One is to do the work in accordance with an approved supplemental type certificate (STC). The STC must specify that it is for the particular aircraft model in question. It states in detail exactly what alternate materials must be used and what procedure(s) must be followed. Deviation from the STC data in any way renders

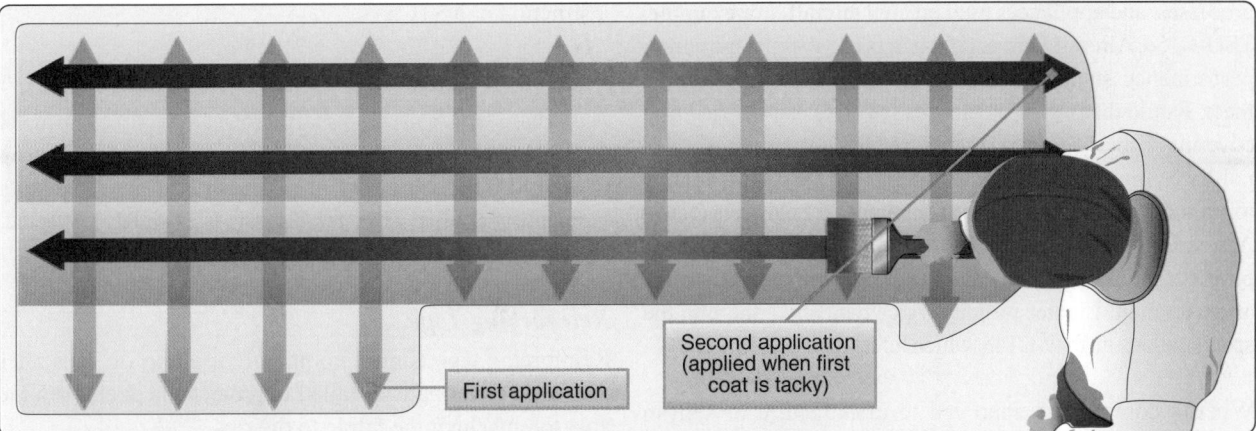

Figure 3-3. *A single cross coat is made up of two coats of paint applied 90° to each other.*

the aircraft unairworthy. The holder of the STC typically sells the materials and the use of the STC to the person wishing to re-cover the aircraft.

The second way to gain approval to re-cover an aircraft with different materials and processes is with a field approval. A field approval is a one-time approval issued by the FAA Flight Standards District Office (FSDO) permitting the materials and procedures requested to replace those of the original manufacturer. A field approval request is made on FAA Form 337. A thorough description of the materials and processes must be submitted with proof that, when the alteration is completed, the aircraft meets or exceeds the performance parameters set forth by the original type certificate.

The third way is for a manufacturer to secure approval through the Type Certificate Data Sheet (TCDS) for a new process. For example, Piper Aircraft Co. originally covered their PA-18s in cotton. Later, they secured approval to recover their aircraft with Dacron fabric. Recovering an older PA-18 with Dacron in accordance with the TCDS would be a major repair, but not an alteration as the TCDS holder has current approval for the fabric.

Advisory Circular (AC) 43.13-1, Acceptable Methods, Techniques, and Practices—Aircraft Inspection and Repair, contains acceptable practices for covering aircraft with fabric. It is a valuable source of general and specific information on fabric and fabric repair that can be used on Form 337 to justify procedures requested for a field approval. Submitting an FAA Form 337 does not guarantee a requested field approval. The FSDO inspector considers all aspects of the procedures and their effect(s) on the aircraft for which the request is being filed. Additional data may be required for approval.

Title 14 of the Code of Federal Regulations (14 CFR) part 43, Appendix A, states which maintenance actions are considered major repairs and which actions are considered major alterations. Fabric re-covering is considered a major repair and FAA Form 337 is executed whenever an aircraft is re-covered with fabric. Appendix A also states that changing parts of an aircraft wing, tail surface, or fuselage when not listed in the aircraft specifications issued by the FAA is a major alteration. This means that replacing cotton fabric with polyester fabric is a major alteration. A properly executed FAA Form 337 also needs to be approved in order for this alteration to be legal.

FAA Form 337, which satisfies the documentation requirements for major fabric repairs and alterations, requires participation of an FAA-certificated Airframe and Powerplant (A&P) mechanic with an Inspection Authorization (IA) in the re-covering process. Often the work involved in re-covering a fabric aircraft is performed by someone else, but under the supervision of the IA (IA certification requires A&P certification). This typically means the IA inspects the aircraft structure and the re-cover job at various stages to be sure STC or field approval specifications are being followed. The signatures of the IA and the FSDO inspector are required on the approved FAA Form 337. The aircraft logbook also must be signed by the FAA-certificated A&P mechanic. It is important to contact the local FSDO before making any major repair or alteration.

Approved Materials

There are a variety of approved materials used in aircraft fabric covering and repair processes. In order for the items to legally be used, the FAA must approve the fabric, tapes, threads, cords, glues, dopes, sealants, coatings, thinners, additives, fungicides, rejuvenators, and paints for the manufacturer, the holder of an STC, or a field approval.

Fabric

A Technical Standard Order (TSO) is a minimum performance standard issued by the FAA for specified materials, parts, processes, and appliances used on civil aircraft. For example, TSO-C15d, Aircraft Fabric, Grade A, prescribes the minimum performance standards that approved aircraft fabric must meet. Fabric that meets or exceeds the TSO can be used as a covering. Fabric approved to replace Grade-A cotton, such as polyester, must meet the same criteria. TSO-C15d also refers to another document, Society of Automotive Engineers (SAE) Aerospace Material Specification (AMS) 3806D, which details properties a fabric must contain to be an approved fabric for airplane cloth. Lighter weight fabrics typically adhere to the specifications in TSO-C14b, which refers to SAE AMS 3804C.

When a company is approved to manufacture or sell an approved aviation fabric, it applies for and receives a Parts Manufacturer Approval (PMA). Currently, only a few approved fabrics are used for aircraft coverings, such as the polyester fabrics Ceconite™, Stits/Polyfiber™, and Superflite™. These fabrics and some of their characteristics are shown in *Figure 3-4*. The holders of the PMA for these fabrics have also developed and gained approval for the various tapes, chords, threads, and liquids that are used in the covering process. These approved materials, along with the procedures for using them, constitute the STCs for each particular fabric covering process. Only the approved materials can be used. Substitution of other materials is forbidden and results in the aircraft being unairworthy.

Other Fabric Covering Materials

The following is an introduction to the supplemental materials used to complete a fabric covering job per manufacturer's instruction or a STC.

Anti-Chafe Tape

Anti-chafe tape is used on sharp protrusions, rib caps, metal seams, and other areas to provide a smoother surface to keep the fabric from being torn. It is usually self-adhesive cloth tape and is applied after the aircraft is cleaned, inspected, and primed, but before the fabric is installed.

Reinforcing Tape

Reinforcing tape is most commonly used on rib caps after the fabric covering is installed to protect and strengthen the area for attaching the fabric to the ribs.

Rib Bracing

Rib bracing tape is used on wing ribs before the fabric is installed. It is applied spanwise and alternately wrapped around a top rib cap and then a bottom rib cap progressing from rib to rib until all are braced. *[Figure 3-5]* Lacing the ribs in this manner holds them in the proper place and alignment during the covering process.

Surface Tape

Surface tape, made of polyester material and often pre-shrunk, is obtained from the STC holder. This tape, also known as finishing tape, is applied after the fabric is installed. It is used over seams, ribs, patches, and edges. Surface tape can have straight or pinked edges and comes in various widths. For curved surfaces, bias cut tape is available, which allows the tape to be shaped around a radius.

Approved Aircraft Fabrics					
Fabric Name or Type	Weight (oz/sq yd)	Count (warp x fill)	New Breaking Strength (lb) (warp, fill)	Minimum Deteriorated Breaking Strength	TSO
Ceconite™ 101	3.5	69 × 63	125,116	70% of original specified fabric	C-15d
Ceconite™ 102	3.16	60 × 60	106,113	70% of original specified fabric	C-15d
Polyfiber™ Heavy Duty-3	3.5	69 × 63	125,116	70% of original specified fabric	C-15d
Polyfiber™ Medium-3	3.16	60 × 60	106,113	70% of original specified fabric	C-15d
Polyfiber™ Uncertified Light	1.87	90 × 76	66,72	uncertified	
Superflight™ SF 101	3.7	70 × 51	80,130	70% of original specified fabric	C-15d
Superflight™ SF 102	2.7	72 × 64	90,90	70% of original specified fabric	C-15d
Superflight™ SF 104	1.8	94 × 91	75,55	uncertified	
Grade A Cotton	4.5	80 × 84	80,80	56 lb/in (70% of New)	C-15d

Figure 3-4. *Approved fabrics for covering aircraft.*

Figure 3-5. *Inter-rib bracing holds the ribs in place during the covering process.*

Rib Lacing Cord

Rib lacing cord is used to lace the fabric to the wing ribs. It must be strong and applied as directed to safely transfer in-flight loads from the fabric to the ribs. Rib lacing cord is available in a round or flat cross-section. The round cord is easier to use than the flat lacing, but if installed properly, the flat lacing results in a smoother finish over the ribs.

Sewing Thread

Sewing of polyester fabric is rare and mostly limited to the creation of prefitted envelopes used in the envelope method covering process. When a fabric seam must be made with no structure underneath it, a sewn seam could be used. Polyester threads of various specifications are used on polyester fabric. Different thread is specified for hand sewing versus machine sewing. For hand sewing, the thread is typically a three-ply, uncoated polyester thread with a 15-pound tensile strength. Machine thread is typically four-ply polyester with a 10-pound tensile strength.

Special Fabric Fasteners

Each fabric covering job involves a method of attaching the fabric to wing and empennage ribs. The original manufacturer's method of fastening should be used. In addition to lacing the fabric to the ribs with approved rib lacing cord, special clips, screws, and rivets are employed on some aircraft. *[Figure 3-6]* The first step in using any of these fasteners is to inspect the holes into which they fit. Worn holes may have to be enlarged or re-drilled according to the manufacturer's instructions. Use of approved fasteners is mandatory. Use of unapproved fasteners can render the covering job unairworthy if substituted. Screws and rivets often incorporate the use of a plastic or aluminum washer. All fasteners and rib lacing are covered with finishing tape once installed to provide a smooth finish and airflow.

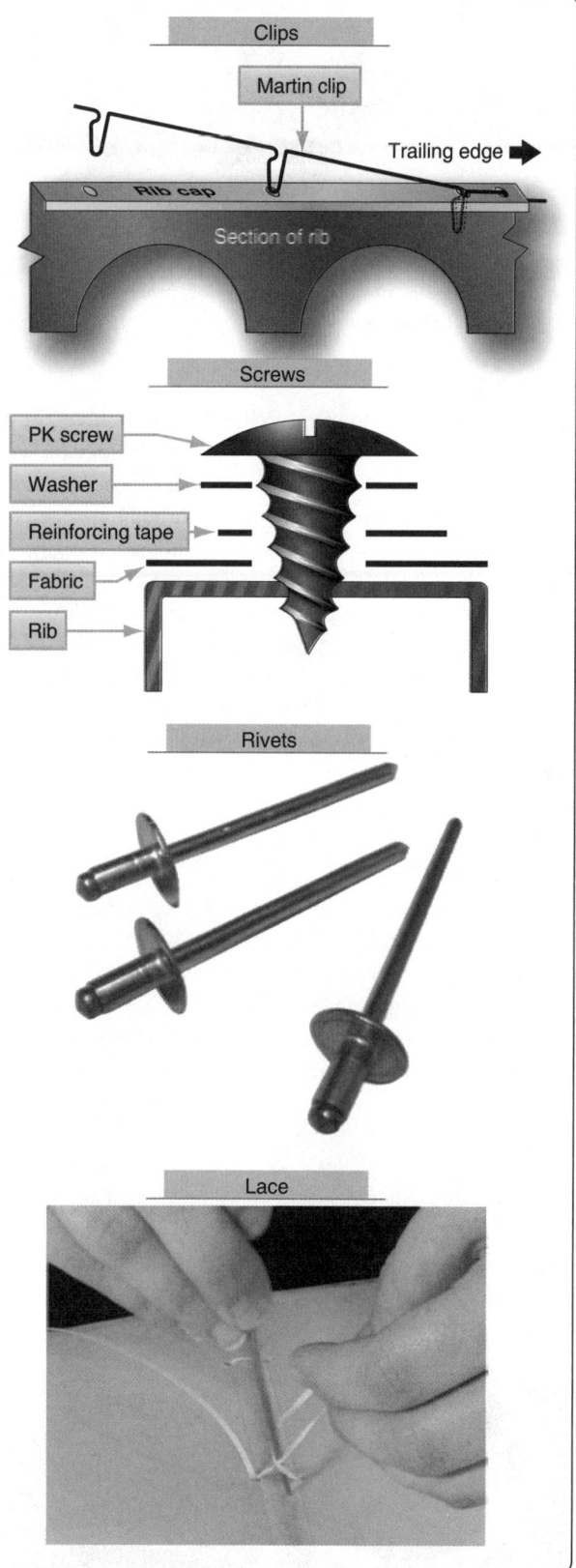

Figure 3-6. *Clips, screws, rivets, or lace are used to attach the fabric to wing and empennage ribs.*

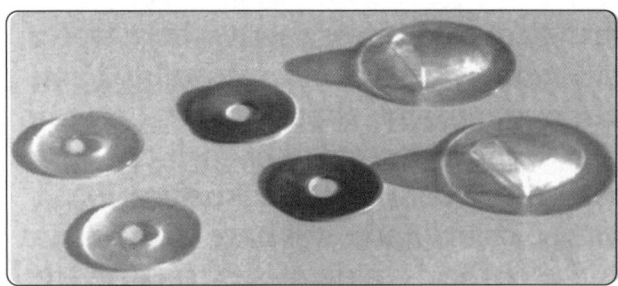

Figure 3-7. *Plastic, aluminum, and seaplane grommets are used to reinforce drain holes in the fabric covering.*

Grommets

Grommets are used to create reinforced drain holes in the aircraft fabric. Usually made of aluminum or plastic, they are glued or doped into place on the fabric surface. Once secured, a hole is created in the fabric through the center of the grommet. Often, this is done with a hot soldering pencil that also heat seals the fabric edge to prevent raveling. Seaplane grommets have a shield over the drain hole to prevent splashed water from entering the interior of the covered structure and to assist in siphoning out any water from within. *[Figure 3-7]* Drain holes using these grommets must be made before the grommets are put in place. Note that some drain holes do not require grommets if they are made through two layers of fabric.

Inspection Rings

The structure underneath an aircraft covering must be inspected periodically. To facilitate this in fabric-covered aircraft, inspection rings are glued or doped to the fabric. They provide a stable rim around an area of fabric that can be cut to allow viewing of the structure underneath. The fabric remains uncut until an inspection is desired. The rings are typically plastic or aluminum with an approximately three-inch inside diameter. Spring clip metal panel covers can be fitted to close the area once the fabric inside the inspection ring has been cut for access. *[Figure 3-8]* The location of the inspection rings are specified by the manufacturer. Additional rings are sometimes added to permit access to important areas that may not have been fitted originally with inspection access.

Figure 3-8. *Inspection rings and an inspection cover.*

Primer

The airframe structure of a fabric covered aircraft must be cleaned, inspected, and prepared before the fabric covering process begins. The final preparation procedure involves priming the structure with a treatment that works with the adhesive and first coats of fabric sealant that are to be utilized. Each STC specifies which primers, or if a wood structure, which varnishes are suitable. Most often, two-part epoxy primers are used on metal structure and two-part epoxy varnishes are used on wood structure. Utilize the primer specified by the manufacturer's or STC's instructions.

Fabric Cement

Modern fabric covering systems utilize special fabric cement to attach the fabric to the airframe. There are various types of cement. *[Figure 3-9]* In addition to good adhesion qualities, flexibility, and long life, fabric cements must be compatible with the primer and the fabric sealer that are applied before and after the cement.

Fabric Sealer

Fabric sealer surrounds the fibers in the fabric with a protective coating to provide adhesion and keep out dirt and moisture. The sealer is the first coat applied to the polyester fabric after it is attached to the airframe and heat shrunk to fit snugly. Dope-based fabric coating systems utilize non-tautening nitrate dope as the primary fabric sealant. The application of tautening dope may cause the fabric to become too taut resulting in excess stress on the airframe that could damage it. Nondope coating systems use proprietary sealers that are also nontautening. *[Figure 3-9]*

Fillers

After the fabric sealer is applied, a filler is used. It is sprayed on in a number of cross coats as required by the manufacturer or the fabric covering process STC. The filler contains solids or chemicals that are included to block UV light from reaching the fabric. Proper fill coating is critical because UV light is the single most destructive element that causes polyester fabric to deteriorate. Dope-based processes use butyrate dope fillers while other processes have their own proprietary formulas. When fillers and sealers are combined, they are known as fabric primers. Aluminum pastes and powders, formerly added to butyrate dope to provide the UV protection, have been replaced by premixed formulas.

Topcoats

Once the aircraft fabric has been installed, sealed, and fill-coat protected, finishing or topcoats are applied to give the aircraft its final appearance. Colored butyrate dope is common in dope-based processes, but various polyurethane topcoats are also available. It is important to use the topcoat

Aircraft Covering Systems							
APPROVED PROPRIETARY PRODUCT NAME							
Covering System	STC #	Allowable Fabrics	Base	Cement	Filler	UV Block	Topcoats
Air-Tech	SA7965SW	Ceconite™ Poly-Fiber™ Superflite™	Urethane	UA-55	PFU 1020 PFU 1030	PFU 1020 PFU 1030	CHSM Color Coat
			Water		PFUW 1050	PFUW 1050	
Ceconite™/ Randolph System	SA4503NM	Ceconite™	Dope	New Super Seam	Nitrate Dope	Rand-O-Fill	Colored Butyrate Dope Ranthane Polyurethane
Stits/Poly-Fiber™	SA1008WE	Poly-Fiber™	Vinyl	Poly-tak	Poly-brush	Poly-spray	Vinyl Poly-tone, Aero-Thane, or Ranthane Polyurethane
Stewart System	SA01734SE	Ceconite™ Poly-Fiber™	Water-borne	EkoBond	EkoFill	EkoFIll	EkoPoly
Superflite™ • System1 • System VI	SA00478CH and others	Superflite™ 101,102	Dope	U-500	Dacproofer	SrayFil	Tinted Butyrate Dope
		Superflite™ 101,102	Urethane	U-500	SF6500	SF6500	Superflite™ CAB

Figure 3-9. *Examples of FAA-approved fabric covering processes.*

products and procedures specified in the applicable STC to complete an airworthy fabric re-covering job.

The use of various additives is common at different stages when utilizing the above products. The following is a short list of additional products that facilitate the proper application of the fabric coatings. Note again that only products approved under a particular STC can be used. Substitution of similar products, even though they perform the same basic function, is not allowed.

- A catalyst accelerates a chemical reaction. Catalysts are specifically designed for each product with which they are mixed. They are commonly used with epoxies and polyurethanes.

- A thinner is a solvent or mixture of solvents added to a product to give it the proper consistency for application, such as when spraying or brushing.

- A retarder is added to a product to slow drying time. Used mostly in dope processes and topcoats, a retarder allow more time for a sprayed coating to flow and level, resulting in a deeper, glossier finish. It is used when the working temperature is elevated slightly above the ideal temperature for a product. It also can be used to prevent blushing of a dope finish when high humidity conditions exist.

- An accelerators contains solvents that speed up the drying time of the product with which it is mixed. It is typically used when the application working temperature is below that of the ideal working temperature. It can also be used for faster drying when airborne contaminants threaten a coating finish.

- Rejuvenator, used on dope finishes only, contains solvents that soften coatings and allow them to flow slightly. Rejuvenator also contains fresh plasticizers that mix into the original coatings. This increases the overall flexibility and life of the coatings.

- Fungicide and mildewicide additives are important for organic fabric covered aircraft because fabrics, such as cotton and linen, are hosts for fungus and mildew. Since fungus and mildew are not concerns when using polyester fabric, these additives are not required. Modern coating formulas contain premixed anti-fungal agents, providing sufficient insurance against the problem of fungus or mildew.

Available Covering Processes

The covering processes that utilize polyester fabric are the primary focus of this chapter. Examples of FAA-approved aircraft covering processes are listed in *Figure 3-9*. The processes can be distinguished by the chemical nature of the glue and coatings that are used. A dope-based covering process has been refined out of the cotton fabric era, with excellent results on polyester fabric. In particular, plasticizers added to the nitrate and butyrate dopes minimize the shrinking and tautening effects of the dope, establish flexibility, and allow esthetically pleasing tinted butyrate dope finishes that last indefinitely. Durable polyurethane-based processes integrate well with durable polyurethane topcoat finishes. Vinyl is the key ingredient in the popular Poly-Fiber covering system. Air Tech uses an acetone thinned polyurethane compatible system.

The most recent entry into the covering systems market is the Stewart Finishing System that uses waterborne technology to apply polyurethane coatings to the fabric. The glue used in the system is water-based and nonvolatile. The Stewart Finishing System is Environmental Protection Agency (EPA) compliant and STC approved. Both the Stewart and Air Tech systems operate with any of the approved polyester fabrics as stated in their covering system STCs.

All the modern fabric covering systems listed in *Figure 3-9* result in a polyester fabric covered aircraft with an indefinite service life. Individual preferences exist for working with the different approved processes. A description of basic covering procedures and techniques common to most of these systems follows later in this chapter.

Ceconite™, Polyfiber™, and Superflight™ are STC-approved fabrics with processes used to install polyester fabric coverings. Two companies that do not manufacturer their own fabric have gained STC approval for covering accessories and procedures to be used with these approved fabrics. The STCs specify the fabrics and the proprietary materials that are required to legally complete the re-covering job.

The aircraft fabric covering process is a three-step process. First, select an approved fabric. Second, follow the applicable STC steps to attach the fabric to the airframe and to protect it from the elements. Third, apply the approved topcoat to give the aircraft its color scheme and final appearance.

Although Grade-A cotton can be used on all aircraft originally certificated to be covered with this material, approved aircraft cotton fabric is no longer available. Additionally, due to the shortcomings of cotton fabric coverings, most of these aircraft have been re-covered with polyester fabric. In the rare instance the technician encounters a cotton fabric covered aircraft that is still airworthy, inspection and repair procedures specified in AC 43.13-1, Chapter 2, Fabric Covering, should be followed.

Determining Fabric Condition—Repair or Recover?

Re-covering an aircraft with fabric is a major repair and should only be undertaken when necessary. Often a repair to the present fabric is sufficient to keep the aircraft airworthy. The original manufacturer's recommendations or the covering process STC should be consulted for the type of repair required for the damage incurred by the fabric covering. AC 43.13-1 also gives guidelines and acceptable practices for repairing cotton fabric, specifically when stitching is concerned.

Often a large area that needs repair is judged in reference to the overall remaining lifespan of the fabric on the aircraft. For example, if the fabric has reached the limit of its durability, it is better to re-cover the entire aircraft than to replace a large damaged area when the remainder of the aircraft would soon need to be re-covered.

On aircraft with dope-based covering systems, continued shrinkage of the dope can cause the fabric to become too tight. Overly tight fabric may require the aircraft to be re-covered rather than repaired because excess tension on fabric can cause airframe structural damage. Loose fabric flaps in the wind during flight, affecting weight distribution and unduly stressing the airframe. It may also need to be replaced because of damage to the airframe.

Another reason to re-cover rather than repair occurs when dope coatings on fabric develop cracks. These cracks could expose the fabric beneath to the elements that can weaken it. Close observation and field testing must be used to determine if the fabrics are airworthy. If not, the aircraft must be re-covered. If the fabric is airworthy and no other problems exist, a rejuvenator can be used per manufacturer's instructions. This product is usually sprayed on and softens the coatings with very powerful solvents. Plasticizers in the rejuvenator become part of the film that fills in the cracks. After the rejuvenator dries, additional coats of aluminum-pigmented dope must be added and then final topcoats applied to finish the job. While laborious, rejuvenating a dope finish over strong fabric can save a great deal of time and money. Polyurethane-based finishes cannot be rejuvenated.

Fabric Strength

Deterioration of the strength of the present fabric covering is the most common reason to re-cover an aircraft. The strength of fabric coverings must be determined at every 100-hour and annual inspection. Minimum fabric breaking strength is used to determine if an aircraft requires re-covering.

Fabric strength is a major factor in the airworthiness of an aircraft. Fabric is considered to be airworthy until it deteriorates to a breaking strength less than 70 percent of the strength of the new fabric required for the aircraft. For example, if an aircraft was certificated with Grade-A cotton fabric that has a new breaking strength of 80 pounds, it becomes unairworthy when the fabric strength falls to 56 pounds, which is 70 percent of 80 pounds. If polyester fabric, which has a higher new breaking strength, is used to re-cover this same aircraft, it would also need to exceed 56 pounds breaking strength to remain airworthy.

In general, an aircraft is certified with a certain fabric based on its wing loading and its never exceed speed (V_{NE}). The higher the wing loading and V_{NE}, the stronger the fabric must be. On aircraft with wing loading of 9 pounds per square foot and over, or a V_{NE} of 160 miles per hour (mph) or higher, fabric equaling or exceeding the strength of Grade A cotton is required. This means the new fabric breaking strength must be at least 80 pounds and the minimum fabric breaking strength at which the aircraft becomes unairworthy is 56 pounds.

On aircraft with wing loading of 9 pounds per square foot or less, or a V_{NE} of 160 mph or less, fabric equaling or exceeding the strength of intermediate grade cotton is required. This means the new fabric breaking strength must be at least 65 pounds and the minimum fabric breaking strength at which the aircraft becomes unairworthy is 46 pounds.

Lighter weight fabric may be found to have been certified on gliders or sailplanes and may be used on many uncertificated aircraft or aircraft in the Light Sport Aircraft (LSA) category. For aircraft with wing loading less than 8 pounds per square foot or less, or V_{NE} of 135 mph or less, the fabric is considered unairworthy when the breaking strength has deteriorated to below 35 pounds (new minimum strength of 50 pounds). *Figure 3-10* summarizes these parameters.

How Fabric Breaking Strength is Determined

Manufacturer's instructions should always be consulted first for fabric strength inspection methodology. These instructions are approved data and may not require removal of a test strip to determine airworthiness of the fabric. In some cases, the manufacturer's information does not include any fabric inspection methods. It may refer the IA to AC 43.13-1, Chapter 2, Fabric Covering, which contains the approved FAA test strip method for breaking strength.

The test strip method for the breaking strength of aircraft covering fabrics uses standards published by the American Society for Testing and Materials (ASTM) for the testing of various materials. Breaking strength is determined by cutting a 1¼ inch by 4–6 inch strip of fabric from the aircraft covering. This sample should be taken from an area that is exposed to the elements—usually an upper surface. It is also wise to take the sample from an area that has a dark colored finish since this has absorbed more of the sun's UV rays and degraded faster. All coatings are then removed and the edges raveled to leave a 1-inch width. One end of the strip is clamped into a secured clamp and the other end is clamped such that a suitable container may be suspended from it. Weight is added to the container until the fabric breaks. The breaking strength of the fabric is equal to the weight of the lower clamp, the container, and the weight added to it. If the breaking strength is still in question, a sample should be sent to a qualified testing laboratory and breaking strength tests made in accordance with ASTM publication D5035.

Note that the fabric test strip must have all coatings removed from it for the test. Soaking and cleaning the test strip in methyl ethyl ketone (MEK) usually removes all the coatings.

Properly installed and maintained polyester fabric should give years of service before appreciable fabric strength degradation occurs. Aircraft owners often prefer not to have test strips cut out of the fabric, especially when the aircraft or the fabric covering is relatively new, because removal of a test strip damages the integrity of an airworthy

Fabric Performance Criteria				
IF YOUR PERFORMANCE IS...		FABRIC STRENGTH MUST BE...		
Loading	V_{NE} Speed	Type	New Breaking Strength	Minimum Breaking Strength
> 9 lb/sq ft	> 160 mph	≥ Grade A	> 80 lb	> 56
< 9 lb/sq ft	< 160 mph	≥ Intermediate	> 65 lb	> 46
< 8 lb/sq ft	< 135 mph	≥ Lightweight	> 50 lb	> 35

Figure 3-10. *Aircraft performance affects fabric selection.*

component if the fabric passes. The test strip area then must be repaired, costing time and money. To avoid cutting a strip out of airworthy fabric, the IA makes a decision based on knowledge, experience, and available nondestructive techniques as to whether removal of a test strip is warranted to ensure that the aircraft can be returned to service.

An aircraft made airworthy under an STC is subject to the instructions for continued airworthiness in that STC. Most STCs refer to AC 43.13-1 for inspection methodology. Poly-Fiber™ and Ceconite™ re-covering process STCs contain their own instructions and techniques for determining fabric strength and airworthiness. Therefore, an aircraft covered under those STCs may be inspected in accordance with this information. In most cases, the aircraft can be approved for return to service without cutting a strip from the fabric covering.

The procedures in the Poly-Fiber™ and Ceconite™ STCs outlined in the following paragraphs are useful when inspecting any fabric covered aircraft as they add to the information gathered by the IA to determine the condition of the fabric. However, following these procedures alone on aircraft not re-covered under these STCs does not make the aircraft airworthy. The IA must add their own knowledge, experience, and judgment to make a final determination of the strength of the fabric and whether it is airworthy.

Exposure to UV radiation appreciably reduces the strength of polyester fabric and forms the basis of the Poly-Fiber™ and Ceconite™ fabric evaluation process. All approved covering systems utilize fill coats applied to the fabric to protect it from UV. If installed according to the STC, these coatings should be sufficient to protect the fabric from the sun and should last indefinitely. Therefore, most of the evaluation of the strength of the fabric is actually an evaluation of the condition of its protective coating(s).

Upon a close visual inspection, the fabric coating(s) should be consistent, contain no cracks, and be flexible, not brittle. Pushing hard against the fabric with a knuckle should not damage the coating(s). It is recommended the inspector check in several areas, especially those most exposed to the sun. Coatings that pass this test can move to a simple test that determines whether or not UV light is passing through the coatings.

This test is based on the assumption that if visible light passes through the fabric coatings, then UV light can also. To verify whether or not visible light passes through the fabric coating, remove an inspection panel from the wing, fuselage, or empennage. Have someone hold an illuminated 60-watt lamp one foot away from the exterior of the fabric. No light should be visible through the fabric. If no light is visible, the fabric has not been weakened by UV rays and can be assumed to be airworthy. There is no need to perform the fabric strip strength test. If light is visible through the coatings, further investigation is required.

Fabric Testing Devices

Mechanical devices used to test fabric by pressing against or piercing the finished fabric are not FAA approved and are used at the discretion of the FAA-certificated mechanic to form an opinion on the general fabric condition. Punch test accuracy depends on the individual device calibration, total coating thickness, brittleness, and types of coatings and fabric. If the fabric tests in the lower breaking strength range with the mechanical punch tester or if the overall fabric cover conditions are poor, then more accurate field tests may be made.

The test should be performed on exposed fabric where there is a crack or chip in the coatings. If there is no crack or chip, coatings should be removed to expose the fabric wherever the test is to be done.

The Maule punch tester, a spring-loaded device with its scale calibrated in breaking strength, tests fabric strength by pressing against it while the fabric is still on the aircraft. It roughly equates strength in pounds per square inch (psi) of resistance to breaking strength. The tester is pushed squarely against the fabric until the scale reads the amount of maximum allowable degradation. If the tester does not puncture the fabric, it may be considered airworthy. Punctures near the breaking strength should be followed with further testing, specifically the strip breaking strength test described above. Usually, a puncture indicates the fabric is in need of replacement.

A second type of punch tester, the Seyboth, is not as popular as the Maule because it punctures a small hole in the fabric when the mechanic pushes the shoulder of the testing unit against the fabric. A pin with a color-coded calibrated scale protrudes from the top of the tester and the mechanic reads this scale to determine fabric strength. Since this device requires a repair regardless of the strength of the fabric indicated, it is not widely used.

Seyboth and Maule fabric strength testers designed for cotton- and linen-covered aircraft, not to be used on modern Dacron fabrics. Mechanical devices, combined with other information and experience, help the FAA-certificated mechanic judge the strength of the fabric. *[Figure 3-11]*

General Fabric Covering Process

It is required to have an IA involved in the process of re-covering a fabric aircraft because re-covering is a major repair

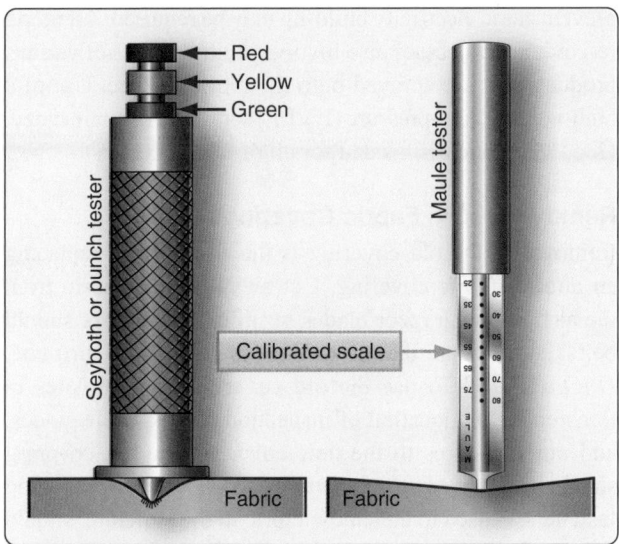

Figure 3-11. *Seyboth and Maule fabric strength testers.*

Figure 3-12. *Laying out fabric during a blanket method re-covering job.*

or major alteration. Signatures are required on FAA Form 337 and in the aircraft logbook. To ensure work progresses as required, the IA should be involved from the beginning, as well as at various stages throughout the process.

This section describes steps common to various STC and manufacturer covering processes, as well as the differences of some processes. To aid in proper performance of fabric covering and repair procedures, STC holders produce illustrated, step-by-step instructional manuals and videos that demonstrate the correct covering procedures. These training aids are invaluable to the inexperienced technician.

Since modern fabric coverings last indefinitely, a rare opportunity to inspect the aircraft exists during the re-covering process. Inspectors and owner-operators should use this opportunity to perform a thorough inspection of the aircraft before new fabric is installed.

The method of fabric attachment should be identical, as far as strength and reliability are concerned, to the method used by the manufacturer of the aircraft being recovered or repaired. Carefully remove the old fabric from the airframe, noting the location of inspection covers, drain grommets, and method of attachment. Either the envelope method or blanket method of fabric covering is acceptable, but a choice must be made prior to beginning the re-covering process.

Blanket Method vs. Envelope Method

In the blanket method of re-covering, multiple flat sections of fabric are trimmed and attached to the airframe. Certified greige polyester fabric for covering an aircraft can be up to 70 inches in width and used as it comes off the bolt. Each aircraft must be considered individually to determine the

size and layout of blankets needed to cover it. A single blanket cut for each small surface (i.e., stabilizers and control surfaces) is common. Wings may require two blankets that overlap. Fuselages are covered with multiple blankets that span between major structural members, often with a single blanket for the bottom. Very large wings may require more than two blankets of fabric to cover the entire top and bottom surfaces. In all cases, the fabric is adhered to the airframe using the approved adhesives, following specific rules for the covering process being employed. *[Figure 3-12]*

An alternative method of re-covering, the envelope method, saves time by using precut and pre-sewn envelopes of fabric to cover the aircraft. The envelopes must be sewn with approved machine sewing thread, edge distance, fabric fold, etc., such as those specified in AC 43.13-1 or an STC. Patterns are made and fabric is cut and stitched so that each major surface, including the fuselage and wings, can be covered with a single, close-fitting envelope. Since envelopes are cut to fit, they are slid into position, oriented with the seams in the proper place, and attached with adhesive to the airframe. Envelope seams are usually located over airframe structure in inconspicuous places, such as the trailing edge structures and the very top and bottom of the fuselage, depending on airframe construction. Follow the manufacturer's or STC's instructions for proper location of the sewn seams of the envelope when using this method. *[Figure 3-13]*

Preparation for Fabric Covering Work

Proper preparation for re-covering a fabric aircraft is essential. First, assemble the materials and tools required to complete the job. The holder of the STC usually supplies a materials and tools list either separately or in the STC manual. Control of temperature, humidity, and ventilation is needed in the work environment. If ideal environmental conditions cannot be met, additives are available that compensate for this for most re-covering products.

Rotating work stands for the fuselage and wings provide

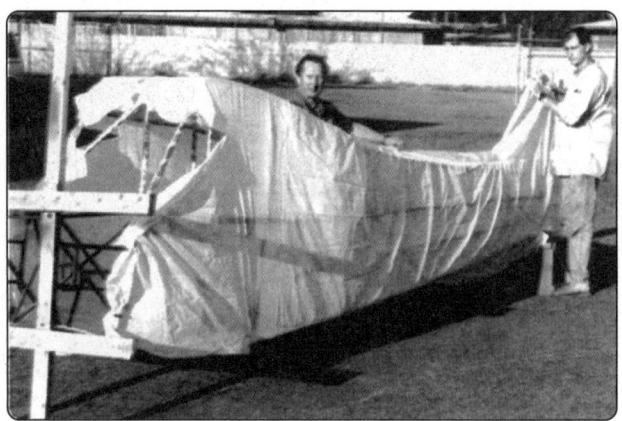

Figure 3-13. *A custom-fit presewn fabric envelope is slid into position over a fuselage for the envelope method of fabric covering. Other than fitting, most steps in the covering process are the same as with the blanket covering method.*

easy, alternating access to the upper and lower surfaces while the job is in progress. *[Figure 3-14]* They can be used with sawhorses or sawhorses can be used alone to support the aircraft structure while working. A workbench or table, as well as a rolling cart and storage cabinet, are also recommended. *Figure 3-15* shows a well conceived fabric covering workshop. A paint spray booth for sprayed-on coatings and space to store components awaiting work is also recommended.

Many of the substances used in most re-covering processes are highly toxic. Proper protection must be used to avoid serious short- and long-term adverse health effects. Eye protection, a proper respirator, and skin protection are vital. As mentioned in the beginning of this chapter, nitrate dope is very flammable. Proper ventilation and a rated fire extinguisher should be on hand when working with this and other covering process materials. Grounding of work to

prevent static electricity build-up may be required. All fabric re-covering processes also involve multiple coats of various products that are sprayed onto the fabric surface. Use of a high-volume, low-pressure (HVLP) sprayer is recommended. Good ventilation is needed for all of the processes.

Removal of Old Fabric Coverings

Removal of the old covering is the first step in replacing an aircraft fabric covering. Cut away the old fabric from the airframe with razor blades or utility knife. Care should be taken to ensure that no damage is done to the airframe. *[Figure 3-16]* To use the old covering for templates in transferring the location of inspection panels, cable guides, and other features to the new covering, the old covering should be removed in large sections. NOTE: any rib stitching fasteners, if used to attach the fabric to the structure, should be removed before the fabric is pulled free of the airframe. If fasteners are left in place, damage to the structure may occur during fabric removal.

Preparation of the Airframe Before Covering

Once the old fabric has been removed, the exposed airframe structure must be thoroughly cleaned and inspected. The IA collaborating on the job should be involved in this step of the process. Details of the inspection should follow the manufacturer's guidelines, the STC, or AC 43.13-1. All of the old adhesive must be completely removed from the airframe with solvent, such as MEK. A thorough inspection must be done and various components may be selected to be removed for cleaning, inspection, and testing. Any repairs that are required, including the removal and treatment of all corrosion, must be done at this time. If the airframe is steel tubing, many technicians take the opportunity to grit blast the entire airframe at this stage.

The leading edge of a wing is a critical area where airflow

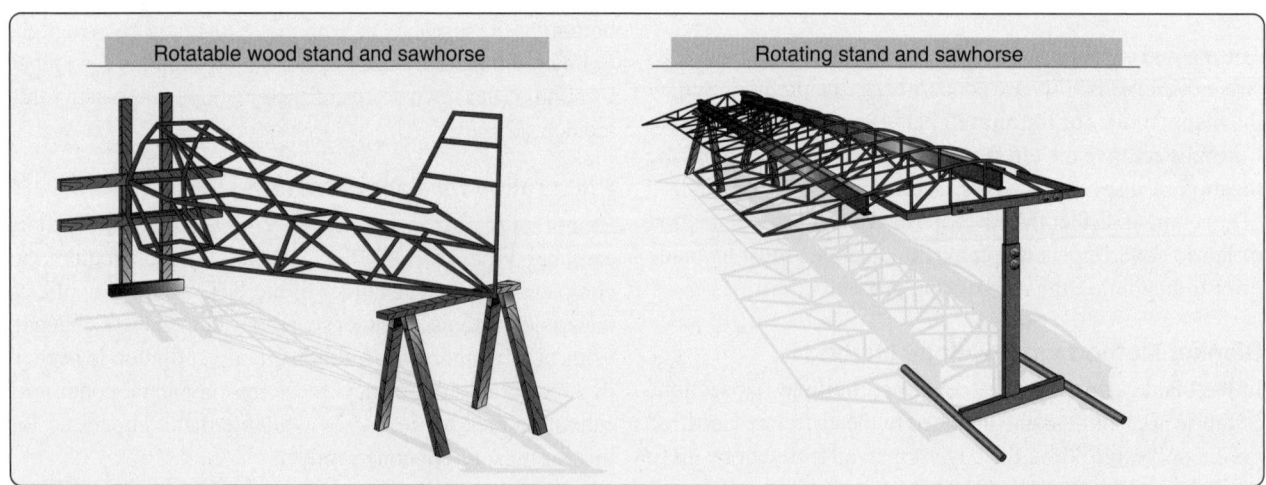

Rotatable wood stand and sawhorse

Rotating stand and sawhorse

Figure 3-14. *Rotating stands and sawhorses facilitate easy access to top and bottom surfaces during the fabric covering process.*

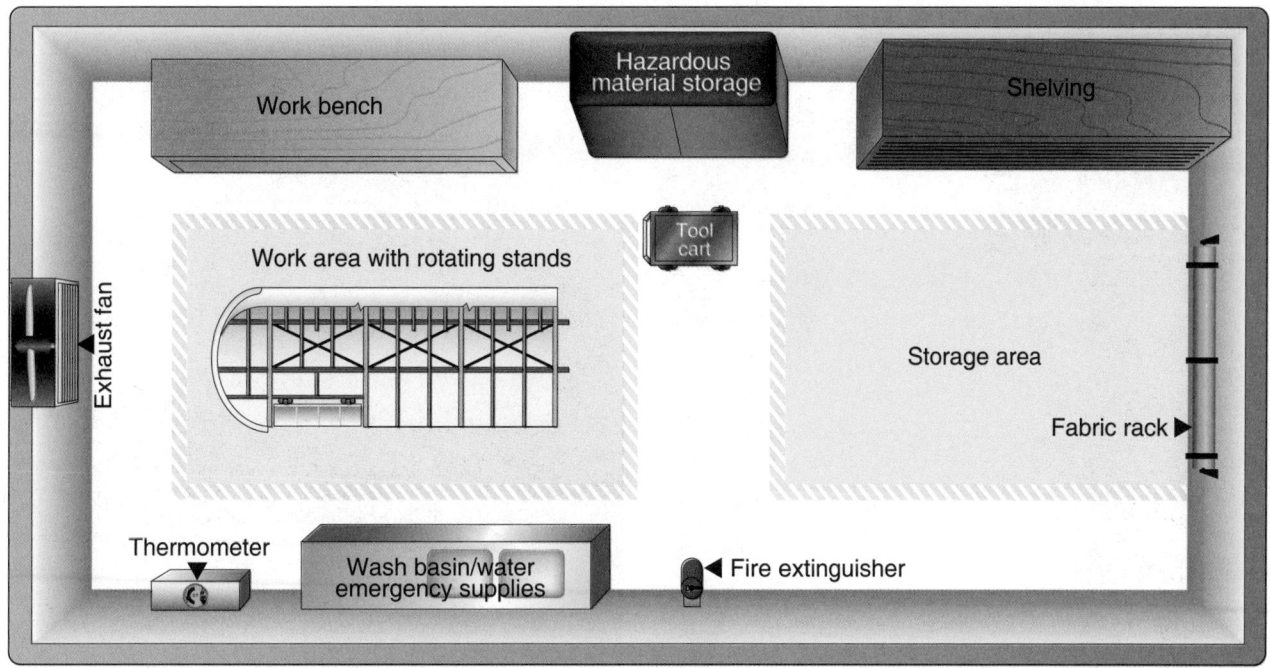

Figure 3-15. *Some components of a work area for covering an aircraft with fabric.*

diverges and begins its laminar flow over the wing's surfaces, which results in the generation of lift. It is beneficial to have a smooth, regular surface in this area. Plywood leading edges must be sanded until smooth, bare wood is exposed. If oil or grease spots exist, they must be cleaned with naphtha or other specified cleaners. If there are any chips, indentations, or irregularities, approved filler may be spread into these areas and sanded smooth. The entire leading edge should be cleaned before beginning the fabric covering process.

To obtain a smooth finish on fabric-covered leading edges of aluminum wings, a sheet of felt or polyester padding may be applied before the fabric is installed. This should only be done with the material specified in the STC under which

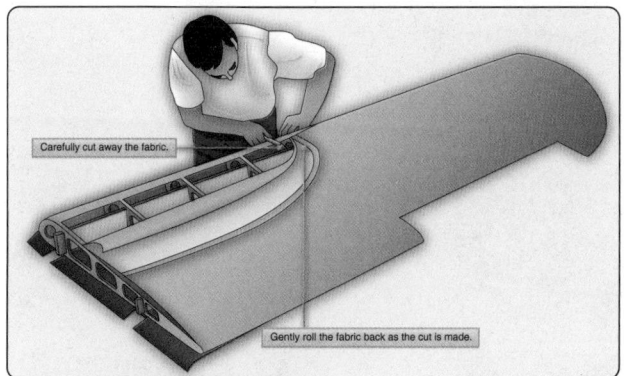

Figure 3-16. *Old fabric coverings are cut off in large pieces to preserve them as templates for locating various airframe features. Sharp blades and care must be used to avoid damaging the structure.*

the technician is working. The approved padding ensures compatibility with the adhesives and first coatings of the covering process. When a leading edge pad is used, check the STC process instructions for permission to make a cemented fabric seam over the padding. *[Figure 3-17]*

When completely cleaned, inspected, and repaired, an approved primer, or varnish if it is a wood structure, should be applied to the airframe. This step is sometimes referred to as dope proofing. Exposed aluminum must first be acid etched. Use the product(s) specified by the manufacturer or in the STC to prepare the metal before priming. Two part epoxy primers and varnishes, which are not affected by the fabric adhesive and subsequent coatings, are usually specified. One part primers, such as zinc chromate and spar varnish, are typically not acceptable. The chemicals in the adhesives dissolve the primers, and adhesion of the fabric to the airframe is lost.

Sharp edges, metal seams, the heads of rivets, and any other feature on the aircraft structure that might cut or wear through the fabric should be covered with anti-chafe tape. As described above, this cloth sticky-back tape is approved and should not be substituted with masking or any other kind of tape. Sometimes, rib cap strips need to have anti-chafe tape applied when the edges are not rounded over. *[Figure 3-18]*

Inter-rib bracing must also be accomplished before the fabric is installed. It normally does not have an adhesive attached to it and is wrapped only once around each rib. The single wrap

Figure 3-17. *The use of specified felt or padding over the wing leading edges before the fabric is installed results in a smooth regular surface.*

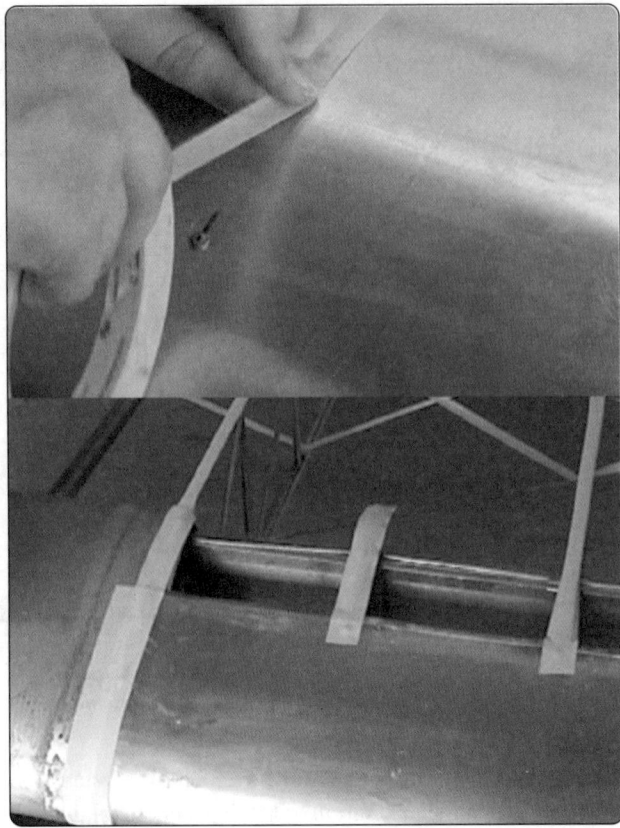

Figure 3-18. *Anti-chafe tape is applied to all features that might cut or wear through the fabric.*

around each rib is enough to hold the ribs in place during the covering process but allows small movements during the fabric shrinking process. *[Figure 3-19]*

Attaching Polyester Fabric to the Airframe

Inexperienced technicians are encouraged to construct a test panel upon which they can practice with the fabric and various substances and techniques to be used on the aircraft. It is often suggested to cover smaller surfaces first, such as the empennage and control surfaces. Mistakes on these can be corrected and are less costly if they occur. The techniques employed for all surfaces, including the wings and fuselage, are basically the same. Once dexterity has been established, the order in which one proceeds is often a personal choice.

When the airframe is primed and ready for fabric installation, it must receive a final inspection by an A&P with IA. When approved, attachment of the fabric may begin. The manufacturer's or STC's instructions must be followed without deviation for the job to be airworthy. The following are the general steps taken. Each approved process has its own nuances.

Seams

During installation, the fabric is overlapped and seamed

Figure 3-19. *Inter-rib bracing holds the ribs in place during the re-covering process.*

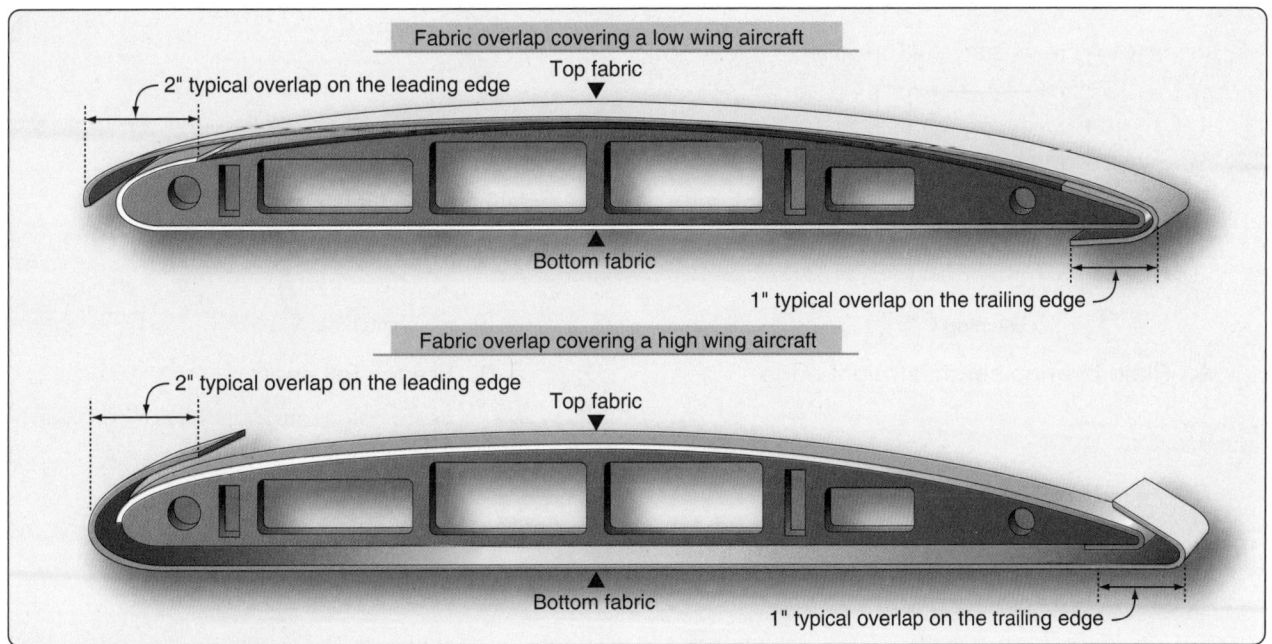

Figure 3-20. *For appearance, fabric can be overlapped differently on high wing and low wing aircraft.*

together. Primary concerns for fabric seams are strength, elasticity, durability, and good appearance. Whether using the blanket method or envelope method, position all fabric seams over airframe structure to which the fabric is to be adhered during the covering process, whenever possible. Unlike the blanket method, fabric seam overlap is predetermined in the envelope method. Seams sewn to the specifications in AC 43.13-1, the STC under which the work is being performed, or the manufacturer's instructions should perform adequately.

Most covering procedures for polyester fabric rely on doped or glued seams as opposed to sewn seams. They are simple and easy to make and provide excellent strength, elasticity, durability, and appearance. When using the blanket method, seam overlap is specified in the covering instructions and the FAA-certificated A&P mechanic must adhere to these specifications. Typically, a minimum of two to four inches of fabric overlap seam is required where ends of fabric are joined in areas of critical airflow, such as the leading edge of a wing. One to two inches of overlap is often the minimum in other areas.

When using the blanket method, options exist for deciding where to overlap the fabric for coverage. Function and the final appearance of the covering job should be considered. For example, fabric seams made on the wing's top surface of a high wing aircraft are not visible when approaching the aircraft. Seams on low wing aircraft and many horizontal stabilizers are usually made on the bottom of the wing for the same reason. *[Figure 3-20]*

Fabric Seams

Seams parallel to the line of flight are preferable; however, spanwise seams are acceptable.

Sewn Seams

Machine-sewn seams should be double stitched using any of the styles illustrated in *Figure 3-21* A, B, C, or D. A machine-sewn seam used to close an envelope at a wingtip, wing trailing edge, empennage and control surface trailing edge, and a fuselage longeron may be made with a single stitch when the seam will be positioned over a structure. *[Figure 3-21E]* The envelope size should accommodate fittings or other small protrusions with minimum excess for installation. Thick or protruding leading edge sewn seams should be avoided on thin airfoils with a sharp leading edge radius because they may act as a stall strip.

Hand sew, with plain overthrow or baseball stitches at a minimum of four stitches per inch, or permanent tacking, to the point where uncut fabric or a machine-sewn seam is reached. Lock hand sewing at a maximum of 10 stitch intervals with a double half hitch, and tie off the end stitch with a double half hitch. At the point where the hand-sewing or permanent tacking is necessary, cut the fabric so that it can be doubled under a minimum of 3/8 inch before sewing or permanent tacking is performed.

After hand sewing is complete, any temporary tacks used to secure the fabric over wood structures may be removed.

Cover a sewn spanwise seam on a wing's leading edge with

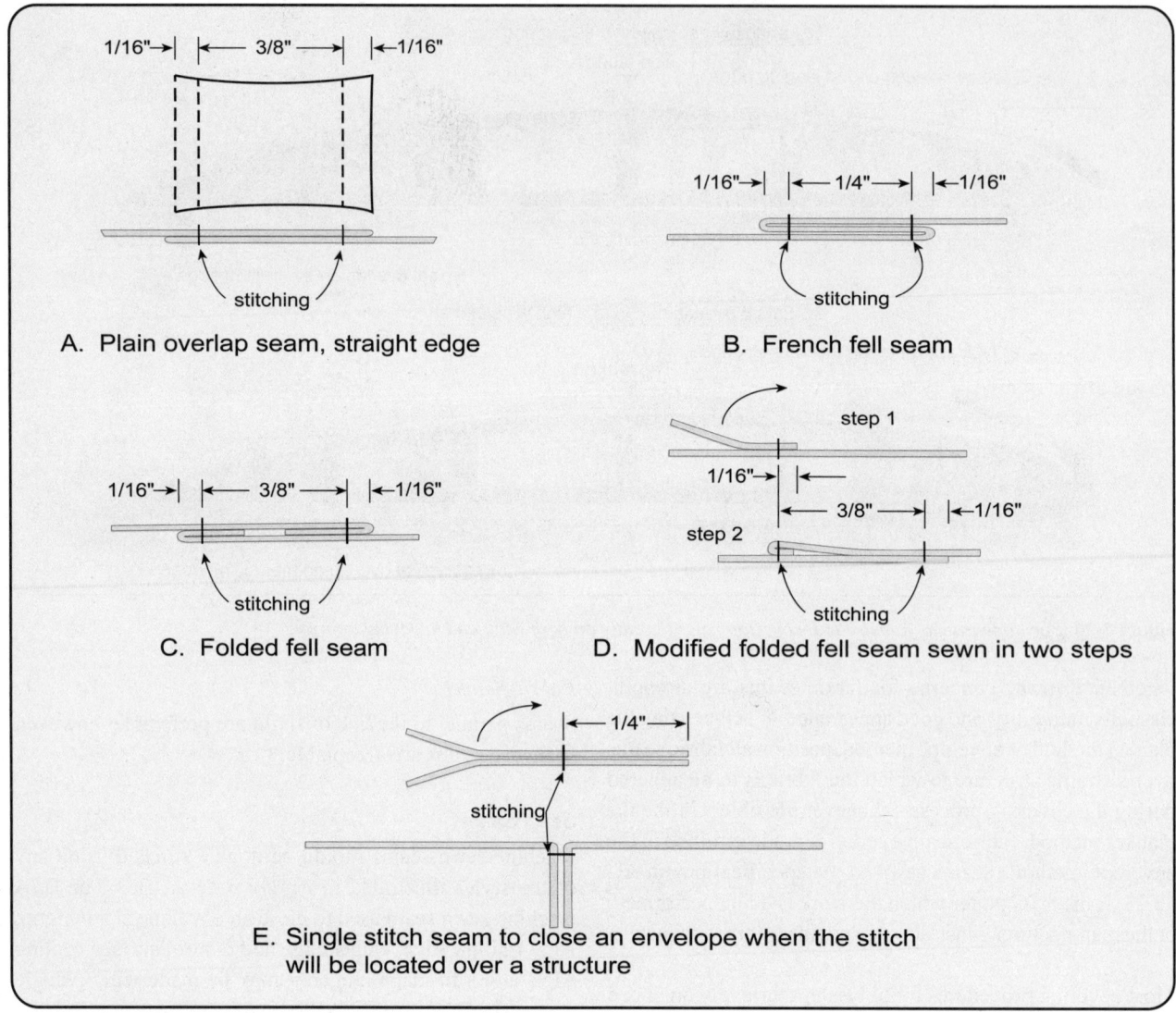

Figure 3-21. *Fabric Seams.*

a minimum 4-inch wide pinked-edged surface tape with the tape centered on the seam.

Cover a spanwise-sewn seam at the wing trailing edge with pinked-edge surface tape that is at least 3 inches wide. For aircraft with never-exceed speeds in excess of 200 mph, cut V notches at least 1 inch in depth and 1/4 inch in width in both edges of the surface tape when used to cover spanwise seams on trailing edges of control surfaces. Space notches at intervals not exceeding 6 inches. On tape less than 3 inches wide, the notches should be 1/3 the tape width. In the event the surface tape begins to separate because of poor adhesion or other causes, the tape will tear at a notched section, thus preventing progressive loosening of the entire length of the tape which could seriously affect the controllability of the aircraft. A loose tape acts as a trim tab only on a movable surface. It becomes a spoiler on a fixed surface and has no effect at the trailing edge other than drag.

Make spanwise-sewn seams on the wing's upper or lower surfaces in a manner that will minimize any protrusions. Cover the seams with finishing tape at least 3 inches wide, centering the tape on the seam.

Sewn seams parallel to the line of flight (chordwise) may be located over ribs. However, careful attention must be given to avoid damage to the seam threads by rib lace needles, screws, rivets, or wire clips that are used to attach the fabric to the rib. Cover chordwise seams with a finishing tape at least 3 inches wide with the tape centered on the seam.

Doped Seams

For an overlapped and doped span-wise seam on a wing's leading edge, overlap the fabric at least 4 inches and cover with finishing tape at least 4 inches wide, with the tape centered at the outside edge of the overlap seam.

For an overlapped and doped span-wise seam at the trailing edge, lap the fabric at least 3 inches and cover with pinked-edge surface tape at least 4 inches wide, with the tape centered on the outside edge of the overlap seam.

For an overlapped and doped seam on wingtips, wing butts, perimeters of wing control surfaces, perimeters of empennage surfaces, and all fuselage areas, overlap the fabric 2 inches and cover with a finishing tape that is at least 3 inches wide, centered on the outside edge of the overlap seam.

For an overlapped and doped seam on a wing's leading edge, on aircraft with a velocity never exceed speed (V_{NE}) up to and including 150 mph, overlap the fabric 2 inches and cover with a finishing tape that is at least 3 inches wide, with the tape centered on the outside edge of the overlap seam.

For an overlapped and doped seam on the perimeter of a wing (except a leading edge), perimeters of wing control surfaces, perimeters of empennage surfaces, and all areas of a fuselage, on aircraft with a V_{NE} speed up to and including 150 mph, overlap the fabric 1 inch and cover with a finishing tape that is at least 3 inches wide, centered on the outside edge of the overlap seam.

Fabric Cement

A polyester fabric covering is cemented or glued to the airframe structure at all points where it makes contact. Special formula adhesives have replaced nitrate dope for adhesion in most covering processes. The adhesive (as well as all subsequent coating materials) should be mixed for optimum characteristics at the temperature at which the work is being performed. Follow the manufacturer's or STC's guidance when mixing.

To attach the fabric to the airframe, first pre-apply two coats of adhesive to the structure at all points that will come in contact with the fabric. (It is important to follow the manufacturer's or STC's guidance as all systems are different.) Allow these to dry. The fabric is then spread over the surface and clamped into position. It should not be pulled tighter than the relaxed but not wrinkled condition it assumes when lying on the structure. Clamps or clothespins are used to attach the fabric completely around the perimeter. The Stewart System STC does not need clamps because the glue assumes a tacky condition when precoated and dried. There is sufficient adhesion in the precoat to position the fabric.

The fabric should be positioned in all areas before undertaking final adhesion. Final adhesion often involves lifting the fabric, applying a wet bed of cement, and pressing the fabric into the bed. An additional coat of cement over the top of the fabric is common. Depending on the process, wrinkles

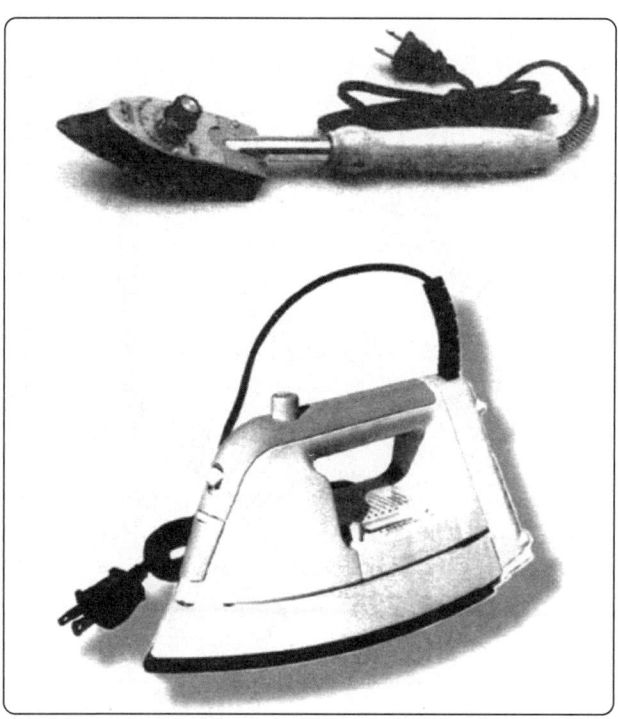

Figure 3-22. *Irons used during the fabric covering process.*

and excess cement are smoothed out with a squeegee or are ironed out. The Stewart System calls for heat activation of the cement precoats through the fabric with an iron while the fabric is in place. Follow the approved instructions for the covering method being used.

Fabric Heat Shrinking

Once the fabric has been glued to the structure, it can be made taut by heat shrinking. This process is done with an ordinary household iron that the technician calibrates before use. A smaller iron is also used to iron in small or tight places. *[Figure 3-22]* The iron is run over the entire surface of the fabric. Follow the instructions for the work being performed. Some processes avoid ironing seams while other processes begin ironing over structure and move to spanned fabric or vice-versa. It is important to shrink the fabric evenly. Starting on one end of a structure and progressing sequentially to the other end is not recommended. Skipping from one end to the other, and then to the middle, is more likely to evenly draw the fabric tight. *[Figure 3-23]*

The amount polyester fabric shrinks is directly related to the temperature applied. Polyester fabric can shrink nearly 5 percent at 250 °F and 10 percent at 350 °F. It is customary to shrink the fabric in stages, using a lower temperature first, before finishing with the final temperature setting. The first shrinking is used to remove wrinkles and excess fabric. The final shrinking gives the finished tautness desired. Each process has its own temperature regime for the stages of

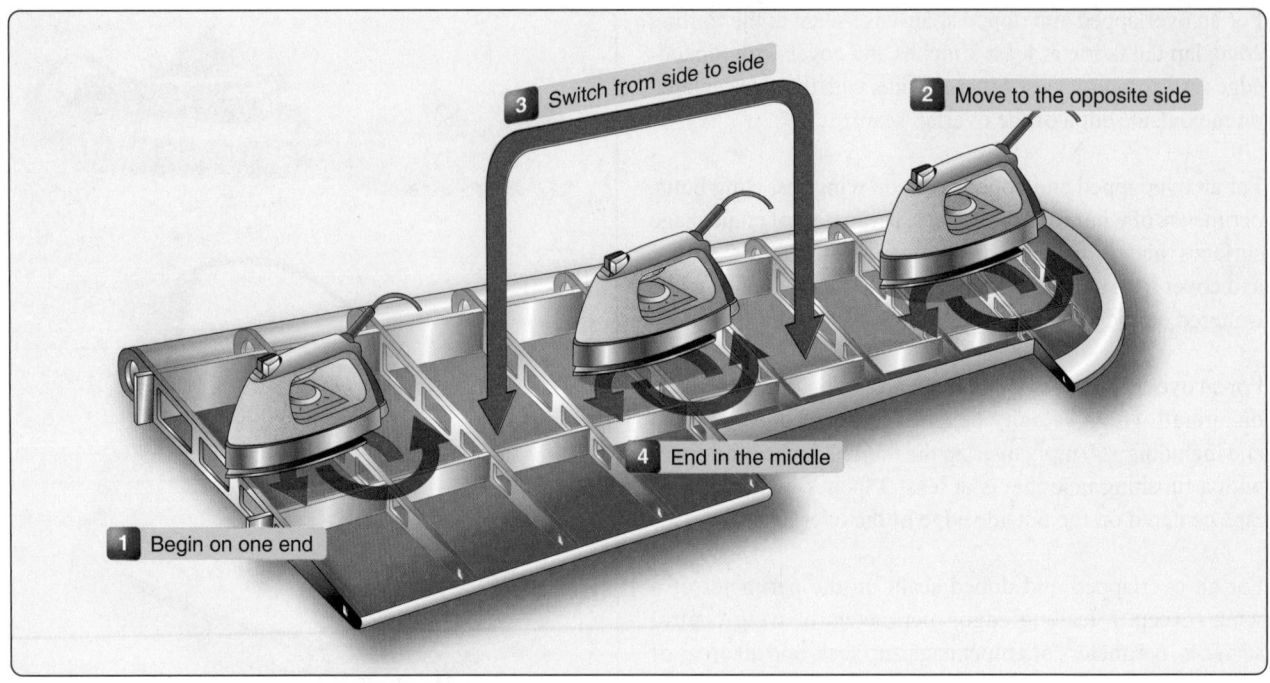

Figure 3-23. *An example of a wing fabric ironing procedure designed to evenly tauten the fabric.*

tautening. Typically ranging from 225 °F to 350 °F, it is imperative to follow the process instructions. Not all fabric covering processes use the same temperature range and maximum temperature. Ensure irons are calibrated to prevent damage at high temperature settings.

Attaching Fabric to the Wing Ribs

Once the fabric has been tautened, covering processes vary. Some require a sealing coat be applied to the fabric at this point. It is usually put on by brush to ensure the fibers are saturated. Other processes seal the fabric later. Whatever the process, the fabric on wings must be secured to the wing ribs with more than just cement. The forces caused by the airflow over the wings are too great for cement alone to hold the fabric in place. As described in the materials section, screws, rivets, clips and lacing hold the fabric in place on manufactured aircraft. Use the same attach method as used by the original aircraft manufacturer. Deviation requires a field approval. Note that fuselage and empennage attachments may be used on some aircraft. Follow the methodology for wing rib lacing described below and the manufacturer's instructions for attach point locations and any possible variations to what is presented here.

Care must always be taken to identify and eliminate any sharp edges that might wear through the fabric. Reinforcing tape of the exact same width as the rib cap is installed before any of the fasteners. This approved sticky-back tape helps prevent the fabric from tearing. *[Figure 3-24]* Then, screws, rivets, and clips simply attach into the predrilled holes in the rib caps

to hold the fabric to the caps. Rib lacing is a more involved process whereby the fabric is attached to the ribs with cord.

Rib Lacing

There are two kinds of rib lacing cord. One has a round cross-section and the other flat. Which to use is a matter of preference based on ease of use and final appearance. Only approved rib lacing cord can be used. Unless a rib is unusually deep from top to bottom, rib lacing uses a single length of cord that passes completely through the wing from the upper surface to the lower surface thereby attaching the top and bottom skin to the rib simultaneously.

Holes are laid out and pre-punched through the skin as close to the rib caps as possible to accept the lacing cord. *[Figure 3-25]* This minimizes leverage the fabric could develop while trying to pull away from the structure and prevents tearing. The location of the holes is not arbitrary. The spacing between lacing holes and knots must adhere to manufacturer's instructions, if available. STC lacing guidance refers to manufacturer's instructions or to that shown on the chart in *Figure 3-26* which is taken from AC 43.13-1. Notice that because of greater turbulence in the area of the propeller wash, closer spacing between the lacing is required there. This slipstream is considered to be the width of the propeller plus one additional rib. Ribs are normally laced from the leading edge to the trailing edge of the wing.

Rib lacing is done with a long curved needle to guide the cord in and out of holes and through the depth of the rib.

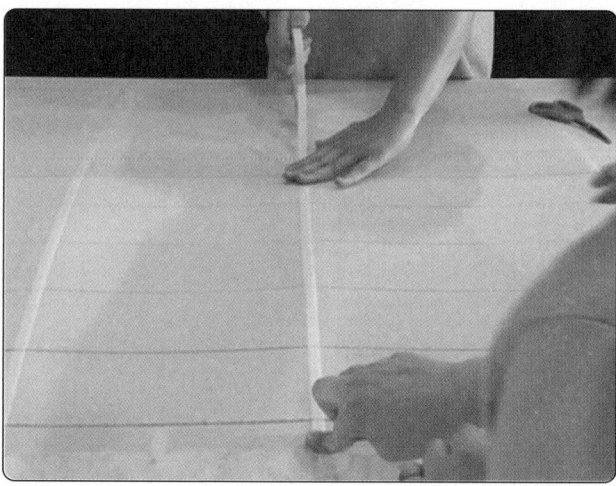

Figure 3-24. *Reinforcing tape the same width as the wing ribs is applied over all wing ribs.*

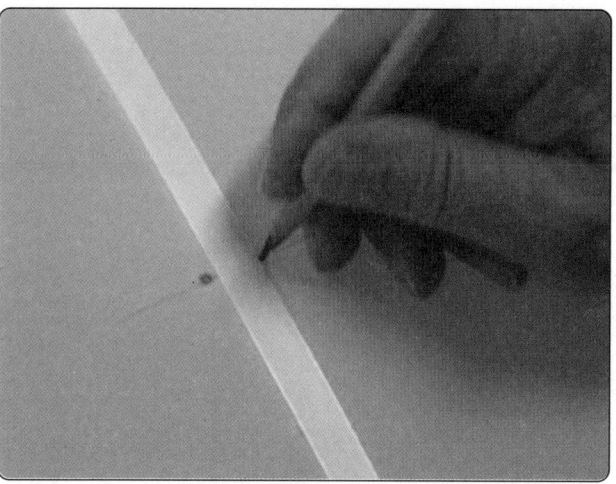

Figure 3-25. *A premarked location for a lacing hole, which is punched through the fabric with a pencil.*

The knots are designed not to slip under the forces applied and can be made in a series out of a single strand of lacing. Stitching can begin at the leading edge or trailing edge. A square knot with a half hitch on each side is typically used for the first knot when lacing a rib. *[Figure 3-27]* This is followed by a series of modified seine knots until the final knot is made and secured with a half hitch. *[Figure 3-28]* Hidden modified seine knots are also used. These knots are placed below the fabric surface so only a single strand of lacing is visible across the rib cap. *[Figure 3-29]*

Structure and accessories within the wing may prevent a continuous lacing. Ending the lacing and beginning again can avoid these obstacles. Lacing that is not long enough to complete the rib may be ended and a new starting knot can be initiated at the next set of holes. The lacing can also be extended by joining it with another piece of lacing using the splice knot shown in *Figure 3-30.*

Occasionally, lacing to just the rib cap is employed without lacing entirely through the wing and incorporating the cap on the opposite side. This is done where ribs are exceptionally deep or where through lacing is not possible, such as in an area where a fuel tank is installed. Changing to a needle with a tighter radius facilitates threading the lacing cord in these areas. Knotting procedures remain unchanged.

Technicians inexperienced at rib lacing should seek assistance to ensure the correct knots are being tied. STC holder videos are invaluable in this area. They present repeated close-up visual instruction and guidance to ensure airworthy lacing. AC 43.13-1, Chapter 2, Fabric Covering, also has in-depth instructions and diagrams as do some manufacturer's manuals and STC's instructions.

Rings, Grommets, & Gussets

When the ribs are laced and the fabric covering completely attached, the various inspection rings, drain grommets, reinforcing patches, and finishing tapes are applied. Inspection rings aid access to critical areas of the structure (pulleys, bell cranks, drag/anti-drag wires, etc.) once the fabric skin is in place. They are plastic or aluminum and normally cemented to the fabric using the approved cement and procedures. The area inside the ring is left intact. It is removed only when inspection or maintenance requires access through that ring. Once removed, preformed inspection panels are used to close the opening. The rings should be positioned as specified by the manufacturer. Lacking that information, they should be positioned as they were on the previous covering fabric. Additional rings should be installed by the technician if it is determined a certain area would benefit from access in the future. *[Figure 3-31]*

Water from rain and condensation can collect under the fabric covering and needs a way to escape. Drain grommets serve this purpose. There are a few different types as described in the materials section above. All are cemented into position in accordance with the approved process under which the work is being performed. Locations for the drain grommets should be ascertained from manufacturer's data. If not specified, AC 43.13-1 has acceptable location information. Each fabric covering STC may also give recommendations. Typically, drain grommets are located at the lowest part of each area of the structure (e.g., bottom of the fuselage, wings, empennage). *[Figure 3-32]* Each rib bay of the wings is usually drained with one or two grommets on the bottom of the trailing edge. Note that drain holes without grommets are sometimes approved in reinforced fabric.

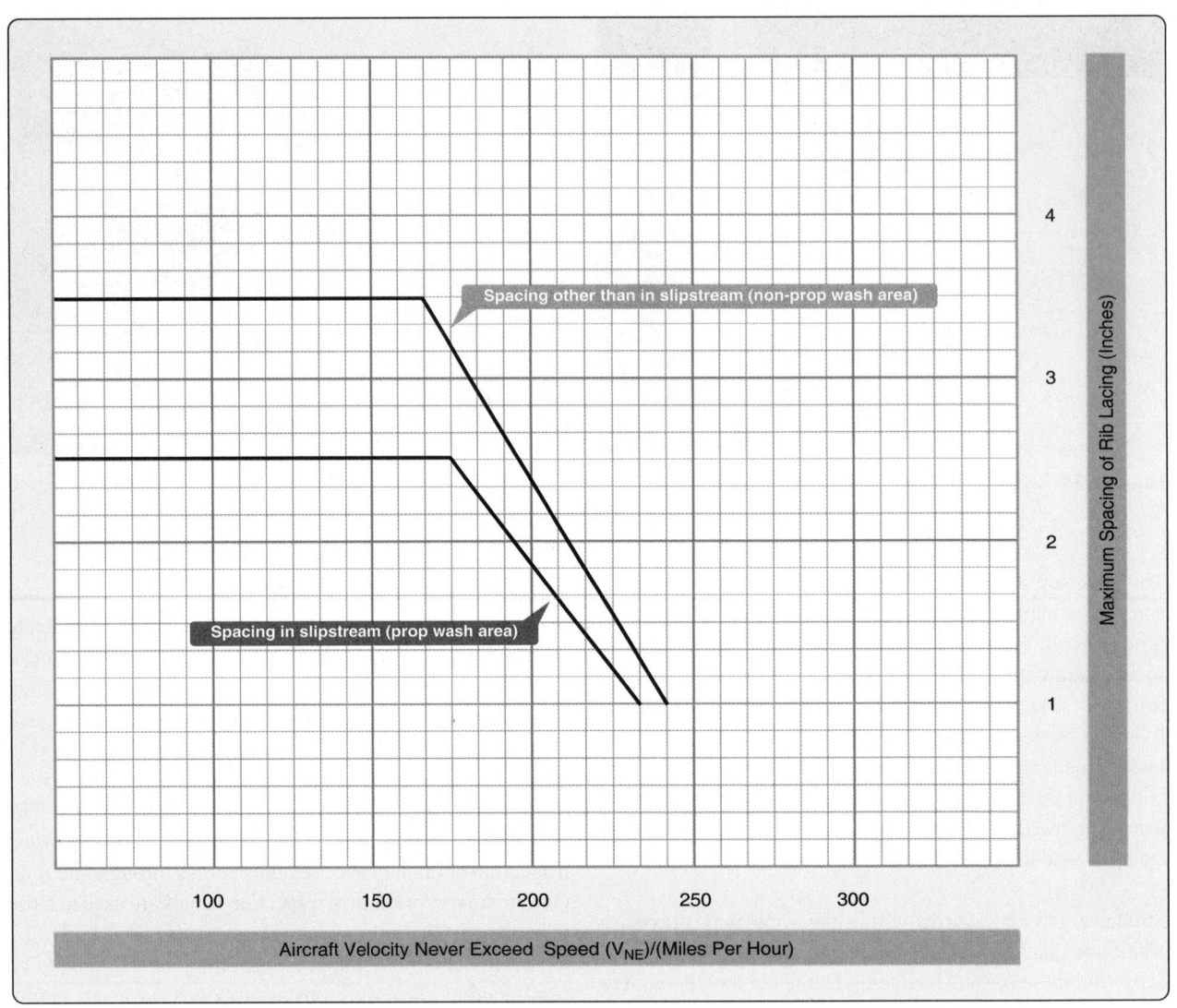

Figure 3-26. *A rib lacing spacing chart. Unless manufacturer data specifies otherwise, use the spacing indicated.*

It is possible that additional inspection rings and drain grommets have been specified after the manufacture of the aircraft. Check the Airworthiness Directives (ADs) and Service Bulletins for the aircraft being re-covered to ensure required rings and grommets have been installed.

Cable guide openings, strut-attach fitting areas, and similar features, as well as any protrusions in the fabric covering, are reinforced with fabric gussets. These are installed as patches in the desired location. They should be cut to fit exactly around the feature they reinforce to support the original opening made in the covering fabric. *[Figure 3-33]* Gussets made to keep protrusions from coming through the fabric should overlap the area they protect. Most processes call for the gusset material to be preshrunk and cemented into place using the approved covering process cementing procedures.

Finishing Tapes

Finishing tapes are applied to all seams, edges, and over the

ribs once all of the procedures above have been completed. They are used to protect these areas by providing smooth aerodynamic resistance to abrasion. The tapes are made from the same polyester material as the covering fabric. Use of lighter weight tapes is approved in some STCs. Preshrunk tapes are preferred because they react to exposure to the environment in the same way the as the fabric covering. This minimizes stress on the adhesive joint between the two. Straight edged and pinked tapes are available. The pinking provides greater surface area for adhesion of the edges and a smoother transition into the fabric covering. Only tapes approved in the STC under which work is being accomplished may be used to be considered airworthy.

Finishing tapes from one to six inches in width are used. Typically, two inch tapes cover the rib lacing and fuselage seams. Wing leading edges usually receive the widest tape with four inches being common. *[Figure 3-34]* Bias cut tapes are often used to wrap around the curved surfaces of the

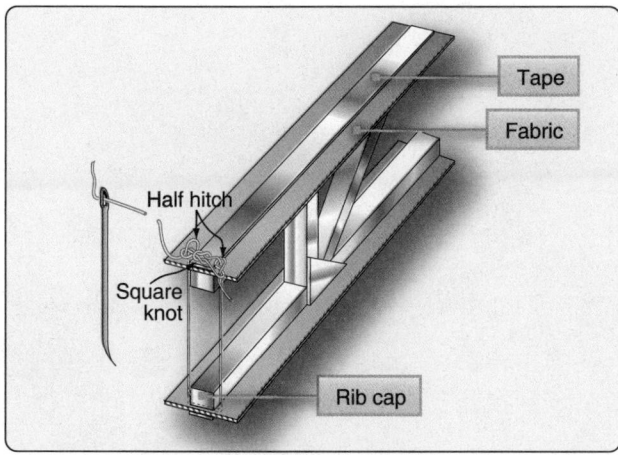

Figure 3-27. *A starter knot for rib lacing can be a square knot with a half hitch on each side.*

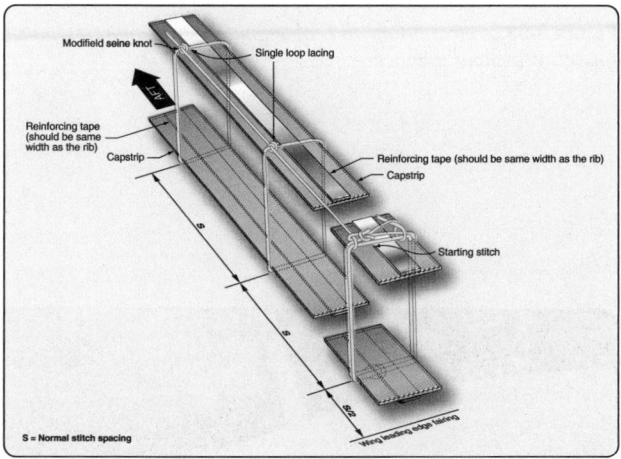

Figure 3-28. *In this example of rib lacing, modified seine knots are used and shown above the fabric surface. Hidden modified seine knots are common. They are made so that the knots are pushed or pulled below the fabric surface.*

airframe, such as the wing tips and empennage surface edges. They lay flat around the curves and do not require notching. Finishing tapes are attached with the process adhesive or the nitrate dope sealer when using a dope-based process. Generally, all chordwise tapes are applied first followed by the span-wise tapes at the leading and trailing edges. Follow the manufacturer's STC or AC 43.13-1 instructions.

Coating the Fabric

The sealer coat in most fabric covering processes is applied after all finishing tapes have been installed unless it was applied prior to rib lacing as in a dope-based finishing process. This coat saturates and completely surrounds the fibers in the polyester fabric, forming a barrier that keeps water and contaminants from reaching the fabric during its life. It is also used to provide adhesion of subsequent

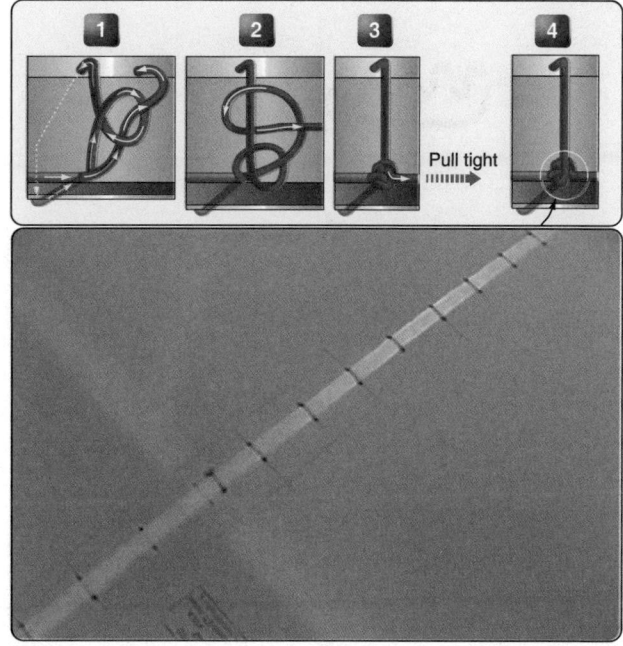

Figure 3-29. *Hiding rib lacing knots below the fabric surface results in a smooth surface.*

coatings. Usually brushed on in a cross coat application for thorough penetration, two coats of sealer are commonly used but processes vary on how many coats and whether spray coating is permitted.

With the sealer coats installed and dried, the next step provides protection from UV light, the only significant cause of deterioration of polyester fabric. Designed to prevent UV light from reaching the fabric and extend the life of the fabric indefinitely, these coating products, or fill coats, contain aluminum solids premixed into them that block the UV rays. They are sprayed on in the number of cross coats as specified in the manufacturer's STC or AC 43.13-1 instructions under which work is done. Two to four cross coats is common. Note that some processes may require coats of clear butyrate before the blocking formula is applied.

Fabric primer is a coating used in some approved covering processes that combines the sealer and fill coatings into one. Applied to fabric after the finishing tapes are installed, these fabric primers surround and seal the fabric fibers, provide good adhesion for all of the following coatings, and contain UV blocking agents. One modern primer contains carbon solids and others use chemicals that work similarly to sun block for human skin. Typically, two to four coats of fabric primer are sufficient before the top coatings of the final finish are applied. *[Figure 3-35]*

The FAA-certificated mechanic must strictly adhere to all instructions for thinning, drying times, sanding, and cleaning.

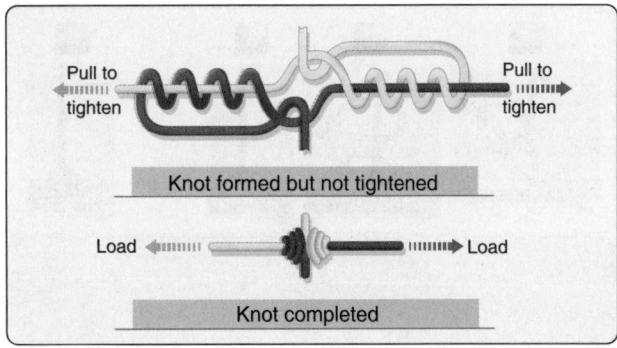

Figure 3-30. *The splice knot can be used to join two pieces of rib lacing cord.*

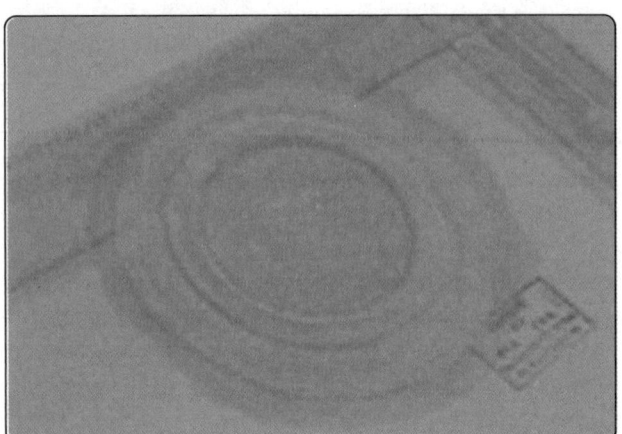

Figure 3-31. *This inspection ring was cemented into place on the fabric covering. The approved technique specifies the use of a fabric overlay that is cemented over the ring and to the fabric.*

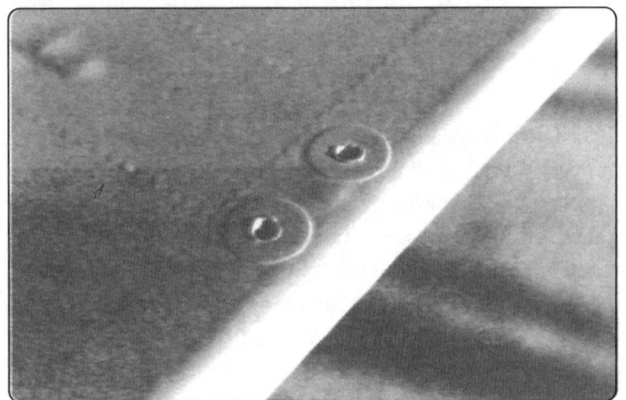

Figure 3-32. *Drain grommets cemented into place on the bottom side of a control surface.*

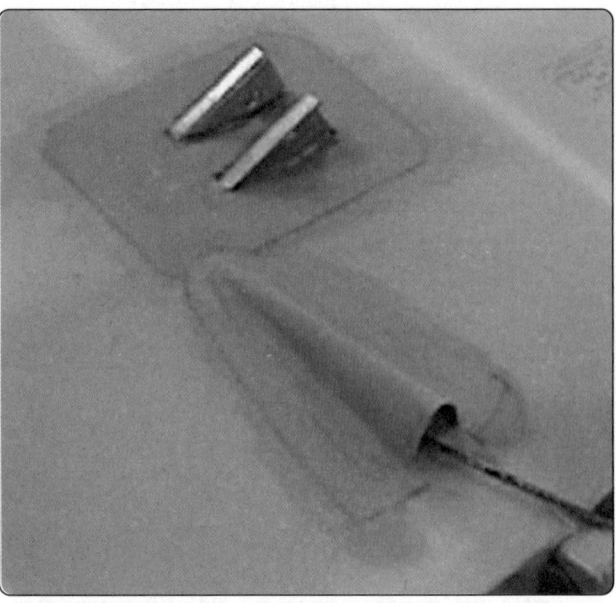

Figure 3-33. *A strut fitting and cable guide with reinforcing fabric gussets cemented in place.*

Figure 3-34. *Cement is brushed through a four-inch tape during installation over the fabric seam on a wing leading edge. Two-inch tapes cover the wing ribs and rib lacing.*

Figure 3-35. *Applying a primer with UV blocking by spraying cross coats.*

Small differences in the various processes exist and what works in one process may not be acceptable and could ruin the finish of another process. STCs are issued on the basis of the holder having successfully proven the effectiveness of both the materials and the techniques involved.

When the fill coats have been applied, the final appearance of the fabric covering job is crafted with the application of various topcoats. Due to the chemical nature of the fill coating upon which topcoats are sprayed, only specified materials can be used for top coating to ensure compatibility. Colored butyrate dope and polyurethane paint finishes are most common. They are sprayed on according to instructions.

Once the topcoats are dry, the trim (N numbers, stripes, etc.) can be added. Strict observation of drying times and instructions for buffing and waxing are critical to the quality of the final finish. Also, note that STC instructions may include insight on finishing the nonfabric portions of the airframe to best match the fabric covering finish.

Polyester Fabric Repairs

Applicable Instructions

Repairs to aircraft fabric coverings are inevitable. Always inspect a damaged area to ensure the damage is confined to the fabric and does not involve the structure below. A technician who needs to make a fabric repair must first identify which approved data was used to install the covering that needs to be repaired. Consult the logbook where an entry and reference to manufacturer data, an STC, or a field approval possibly utilizing practices from AC 43.13-1 should be recorded. The source of approved data for the covering job is the same source of approved data used for a repair.

This section discusses general information concerning repairs to polyester fabric. Thorough instructions for repairs made to cotton covered aircraft can be found in AC 43.13-1. It is the responsibility of the holder of an STC to provide maintenance instructions for the STC alteration in addition to materials specifications required to do the job.

Repair Considerations

The type of repair performed depends on the extent of the damage and the process under which the fabric was installed. The size of the damaged area is often a reference for whether a patch is sufficient to do the repair or whether a new panel should be installed. Repair size may also dictate the amount of fabric-to-fabric overlap required when patching and whether finishing tapes are required over the patch. Many STC repair procedures do not require finishing tapes. Some repairs in AC 43.13-1 require the use of tape up to six inches wide.

While many cotton fabric repairs involve sewing, nearly all repairs of polyester fabric are made without sewing. It is possible to apply the sewing repair techniques outlined in AC 43.13-1 to polyester fabric, but they were developed primarily for cotton and linen fabrics. STC instructions for repairs to polyester fabric are for cemented repairs which most technicians prefer as they are generally considered easier than sewn repairs. There is no compromise to the strength of the fabric with either method.

Patching or replacing a section of the covering requires prepping the fabric area around the damage where new fabric is to be attached. Procedures vary widely. Dope-based covering systems tend toward stripping off all coatings to cement raw fabric to raw fabric when patching or seaming in a new panel. From this point, the coatings are reapplied and finished as in the original covering process. Some polyurethane-based coating processes require only a scuffing of the topcoat with sandpaper before adhering small patches that are then refinished. *[Figure 3-36]* Still, other processes may remove the topcoats and cement a patch into the sealer or UV blocking coating. In some repair processes, preshrunk fabric is used and in others, the fabric is shrunk after it is in place. Varying techniques and temperatures for shrinking and gluing the fabric into a repair also exist.

These deviations in procedures underscore the critical nature of identifying and strictly adhering to the correct instructions from the approved data for the fabric covering in need of repair. A patch or panel replacement technique for one covering system could easily create an unairworthy repair if used on fabric installed with a different covering process.

Large section panel repairs use the same proprietary adhesives

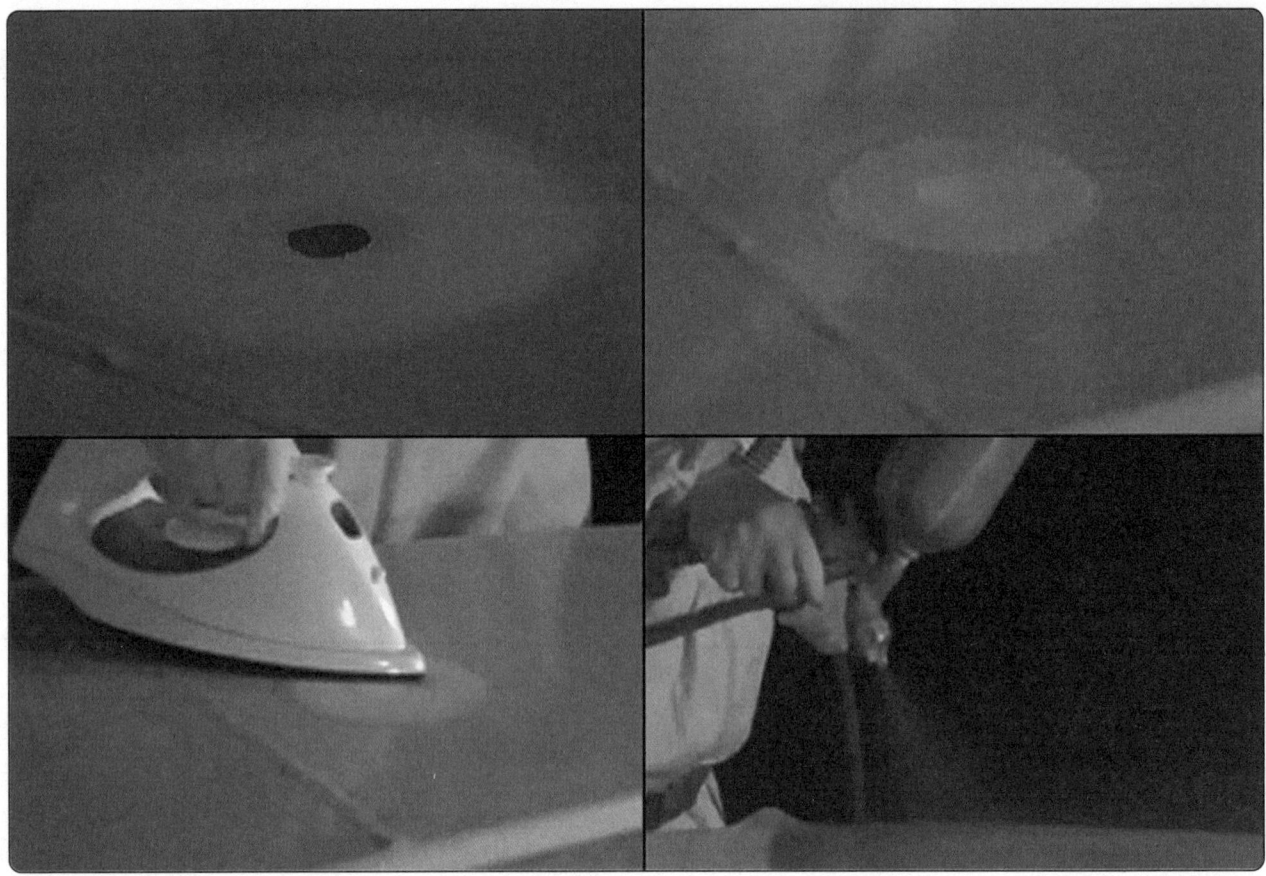

Figure 3-36. *A patch over this small hole on a polyurethane top coat is repaired in accordance with the repair instructions in the STC under which the aircraft was re-covered. It requires only a two-inch fabric overlap and scuffing into the top coat before cementing and refinishing. Other STC repair instructions may not allow this repair.*

and techniques and are only found in the instructions for the process used to install the fabric covering. A common technique for replacing any large damaged area is to replace all of the fabric between two adjacent structural members (e.g., two ribs, two longerons, between the forward and rear spars). Note that this is a major repair and carries with it the requirement to file an FAA Form 337.

Cotton-Covered Aircraft

You may encounter a cotton fabric-covered aircraft. In addition to other airworthiness criterion, the condition of the fabric under the finished surface is paramount as the cotton can deteriorate even while the aircraft is stored in a hangar. Inspection, in accordance with the manufacturer maintenance manual or AC 43.13-1, should be diligent. If the cotton covering is found to be airworthy, repairs to the fabric can be made under those specifications. This includes sewn-in and doped-in patches, as well as sewn-in and doped-in panel repairs. Due to the very limited number of airworthy aircraft that may still be covered with cotton, this handbook does not cover specific information on re-covering with cotton or cotton fabric maintenance and repair procedures. Refer to AC 43.13-1, Chapter 2, Fabric Covering, which thoroughly

addresses these issues.

Fiberglass Coverings

References to fiberglass surfaces in aircraft covering STCs, AC 43.13-1, and other maintenance literature address techniques for finishing and maintaining this kind of surface. However, this is typically limited to fiberglass radomes and fiberglass reinforced plywood surfaces and parts that are still in service. Use of dope-based processes on fiberglass is well established. Repair and apply coatings and finishes on fiberglass in accordance with manufacturer data, STC instructions, or AC 43.13-1 acceptable practices. Mildew, moisture, chemicals, or acids have no effect on glass fabric when used as a structure material. For more information on glass fabric, refer to AC 43.13-1(as revised).

Chapter 4
Aircraft Metal Structural Repair

Aircraft Metal Structural Repair

The satisfactory performance of an aircraft requires continuous maintenance of aircraft structural integrity. It is important that metal structural repairs be made according to the best available techniques because improper repair techniques can pose an immediate or potential danger. The reliability of an aircraft depends on the quality of the design, as well as the workmanship used in making the repairs. The design of an aircraft metal structural repair is complicated by the requirement that an aircraft be as light as possible. If weight were not a critical factor, repairs could be made with a large margin of safety. In actual practice, repairs must be strong enough to carry all of the loads with the required factor of safety, but they must not have too much extra strength. For example, a joint that is too weak cannot be tolerated, but a joint that is too strong can create stress risers that may cause cracks in other locations.

As discussed in Chapter 3, Aircraft Fabric Covering, sheet metal aircraft construction dominates modern aviation. Generally, sheet metal made of aluminum alloys is used in airframe sections that serve as both the structure and outer aircraft covering, with the metal parts joined with rivets or other types of fasteners. Sheet metal is used extensively in many types of aircraft from airliners to single engine airplanes, but it may also appear as part of a composite airplane, such as in an instrument panel. Sheet metal is obtained by rolling metal into flat sheets of various thicknesses ranging from thin (leaf) to plate (pieces thicker than 6 mm or 0.25 inch). The thickness of sheet metal, called gauge, ranges from 8 to 30 with the higher gauge denoting thinner metal. Sheet metal can be cut and bent into a variety of shapes.

Damage to metal aircraft structures is often caused by corrosion, erosion, normal stress, and accidents and mishaps. Sometimes aircraft structure modifications require extensive structural rework. For example, the installation of winglets on aircraft not only replaces a wing tip with a winglet, but also requires extensive reinforcing of the wing structure to carry additional stresses.

Numerous and varied methods of repairing metal structural portions of an aircraft exist, but no set of specific repair patterns applies in all cases. The problem of repairing a damaged section is usually solved by duplicating the original part in strength, kind of material, and dimensions. To make a

structural repair, the aircraft technician needs a good working knowledge of sheet metal forming methods and techniques. In general, forming means changing the shape by bending and forming solid metal. In the case of aluminum, this is usually done at room temperature. All repair parts are shaped to fit in place before they are attached to the aircraft or component.

Forming may be a very simple operation, such as making a single bend or a single curve, or it may be a complex operation, requiring a compound curvature. Before forming a part, the aircraft technician must give some thought to the complexity of the bends, the material type, the material thickness, the material temper, and the size of the part being fabricated. In most cases, these factors determine which forming method to use. Types of forming discussed in this chapter include bending, brake forming, stretch forming, roll forming, and spinning. The aircraft technician also needs a working knowledge of the proper use of the tools and equipment used in forming metal.

In addition to forming techniques, this chapter introduces the airframe technician to the tools used in sheet metal construction and repair, structural fasteners and their installation, how to inspect, classify, and assess metal structural damage, common repair practices, and types of repairs.

The repairs discussed in this chapter are typical of those used in aircraft maintenance and are included to introduce some of the operations involved. For exact information about specific repairs, consult the manufacturer's maintenance or structural repair manuals (SRM). General repair instructions are also discussed in Advisory Circular (AC) 43.13.1, Acceptable Methods, Techniques, and Practices—Aircraft Inspection and Repair.

Stresses in Structural Members

An aircraft structure must be designed so that it accepts all of the stresses imposed upon it by the flight and ground loads without any permanent deformation. Any repair made must accept the stresses, carry them across the repair, and then transfer them back into the original structure. These stresses are considered as flowing through the structure, so there must be a continuous path for them, with no abrupt changes in cross-sectional areas along the way. Abrupt changes in cross-sectional areas of aircraft structure that are subject to cycle loading or stresses result in a stress concentration that may induce fatigue cracking and eventual failure. A scratch or gouge in the surface of a highly stressed piece of metal causes a stress concentration at the point of damage and could lead to failure of the part. Forces acting on an aircraft, whether it is on the ground or in flight, introduce pulling, pushing, or twisting forces within the various members of the aircraft structure. While the aircraft is on the ground, the weight of the wings, fuselage, engines, and empennage causes forces to act downward on the wing and stabilizer tips, along the spars and stringers, and on the bulkheads and formers. These forces are passed from member to member causing bending, twisting, pulling, compression, and shearing forces.

As the aircraft takes off, most of the forces in the fuselage continue to act in the same direction; because of the motion of the aircraft, they increase in intensity. The forces on the wingtips and the wing surfaces, however, reverse direction; instead of being downward forces of weight, they become upward forces of lift. The forces of lift are exerted first against the skin and stringers, then are passed on to the ribs, and finally are transmitted through the spars to be distributed through the fuselage. The wings bend upward at their ends and may flutter slightly during flight. This wing bending cannot be ignored by the manufacturer in the original design and construction and cannot be ignored during maintenance. It is surprising how an aircraft structure composed of structural members and skin rigidly riveted or bolted together, such as a wing, can bend or act so much like a leaf spring.

The six types of stress in an aircraft are described as tension, compression, shear, bearing, bending, and torsion (or twisting). The first four are commonly called basic stresses; the last two, combination stresses. Stresses usually act in combinations rather than singly. [Figure 4-1]

Tension

Tension is the stress that resists a force that tends to pull apart. The engine pulls the aircraft forward, but air resistance tries to hold it back. The result is tension, which tends to stretch the aircraft. The tensile strength of a material is measured in pounds per square inch (psi) and is calculated by dividing the load (in pounds) required to pull the material apart by its cross-sectional area (in square inches).

The strength of a member in tension is determined on the basis of its gross area (or total area), but calculations involving tension must take into consideration the net area of the member. Net area is defined as the gross area minus that removed by drilling holes or by making other changes in the section. Placing rivets or bolts in holes makes no appreciable difference in added strength, as the rivets or bolts will not transfer tensional loads across holes in which they are inserted.

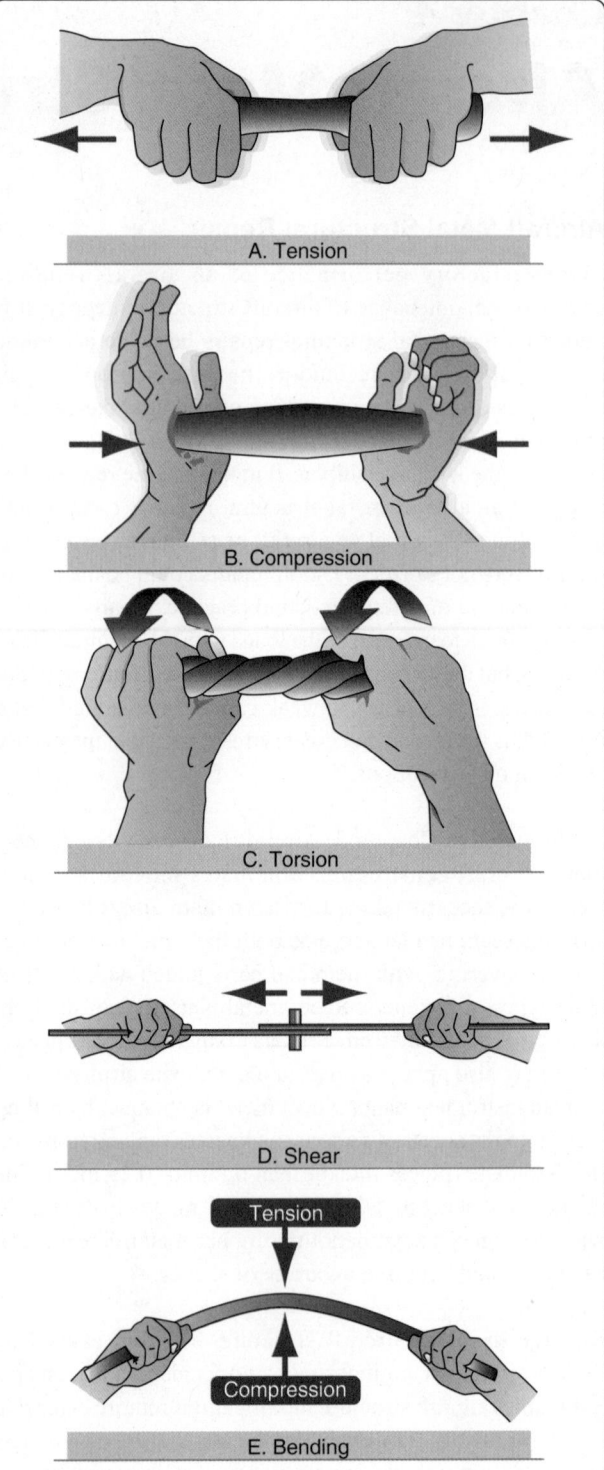

Figure 4-1. *Stresses in aircraft structures.*

Compression

Compression, the stress that resists a crushing force, tends to shorten or squeeze aircraft parts. The compressive strength of a material is also measured in psi. Under a compressive load, an undrilled member is stronger than an identical member with holes drilled through it. However, if a plug of equivalent or stronger material is fitted tightly in a drilled member, it transfers compressive loads across the hole, and the member carries approximately as large a load as if the hole were not there. Thus, for compressive loads, the gross or total area may be used in determining the stress in a member if all holes are tightly plugged with equivalent or stronger material.

Shear

Shear is the stress that resists the force tending to cause one layer of a material to slide over an adjacent layer. Two riveted plates in tension subject the rivets to a shearing force. Usually, the shear strength of a material is either equal to or less than its tensile or compressive strength. Shear stress concerns the aviation technician chiefly from the standpoint of the rivet and bolt applications, particularly when attaching sheet metal, because if a rivet used in a shear application gives way, the riveted or bolted parts are pushed sideways.

Bearing

Bearing stress resists the force that the rivet or bolt places on the hole. As a rule, the strength of the fastener should be such that its total shear strength is approximately equal to the total bearing strength of the sheet material. *[Figure 4-2]*

Torsion

Torsion is the stress that produces twisting. While moving the aircraft forward, the engine also tends to twist it to one side, but other aircraft components hold it on course. Thus, torsion is created. The torsional strength of a material is its resistance to twisting or torque (twisting stress). The stresses arising from this action are shear stresses caused by the rotation of adjacent planes past each other around a common reference axis at right angles to these planes. This action may be illustrated by a rod fixed solidly at one end and twisted by a weight placed on a lever arm at the other, producing the equivalent of two equal and opposite forces acting on the rod at some distance from each other. A shearing action is set up

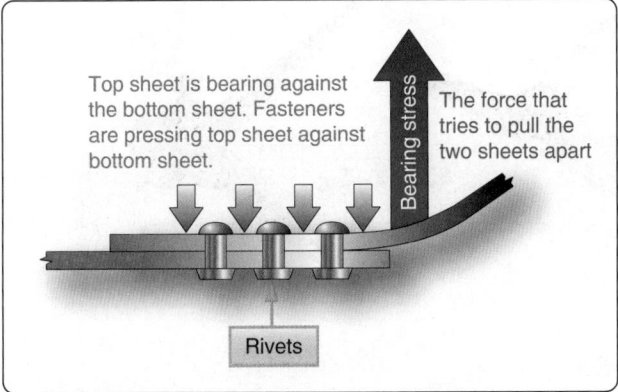

Figure 4-2. *Bearing stress.*

all along the rod, with the center line of the rod representing the neutral axis.

Bending

Bending (or beam stress) is a combination of compression and tension. The rod in *Figure 4-1E* has been shortened (compressed) on the inside of the bend and stretched on the outside of the bend. Note that the bending stress causes a tensile stress to act on the upper half of the beam and a compressive stress on the lower half. These stresses act in opposition on the two sides of the center line of the member, which is called the neutral axis. Since these forces acting in opposite directions are next to each other at the neutral axis, the greatest shear stress occurs along this line, and none exists at the extreme upper or lower surfaces of the beam.

Tools for Sheet Metal Construction & Repair

Without modern metalworking tools and machines, the job of the airframe technician would be more difficult and tiresome, and the time required to finish a task would be much greater. These specialized tools and machines help the airframe technician construct or repair sheet metal in a faster, simpler, and better manner than possible in the past. Powered by human muscle, electricity, or compressed air, these tools are used to lay out, mark, cut, sand, or drill sheet metal.

Layout Tools

Before fitting repair parts into an aircraft structure, the new sections must be measured and marked, or laid out to the dimensions needed to make the repair part. Tools utilized for this process are discussed in this section.

Scales

Scales are available in various lengths, with the 6-inch and 12-inch scales being the most common and affordable. A scale with fractions on one side and decimals on the other side is very useful. To obtain an accurate measurement, measure with the scale held on edge from the 1-inch mark instead of the end. Use the graduation marks on the side to set a divider or compass. *[Figure 4-3]*

Combination Square

A combination square consists of a steel scale with three heads that can be moved to any position on the scale and locked in place. The three heads are a stock head that measures 90° and 45° angles, a protractor head that can measure any angle between the head and the blade, and a center head that uses one side of the blade as the bisector of a 90° angle. The center of a shaft can be found by using the center head. Place the end of the shaft in the V of the head and scribe a line along the edge of the scale. Rotate the head about 90° and scribe another line along the edge of the scale. The two lines will cross at the center of the shaft. *[Figure 4-4]*

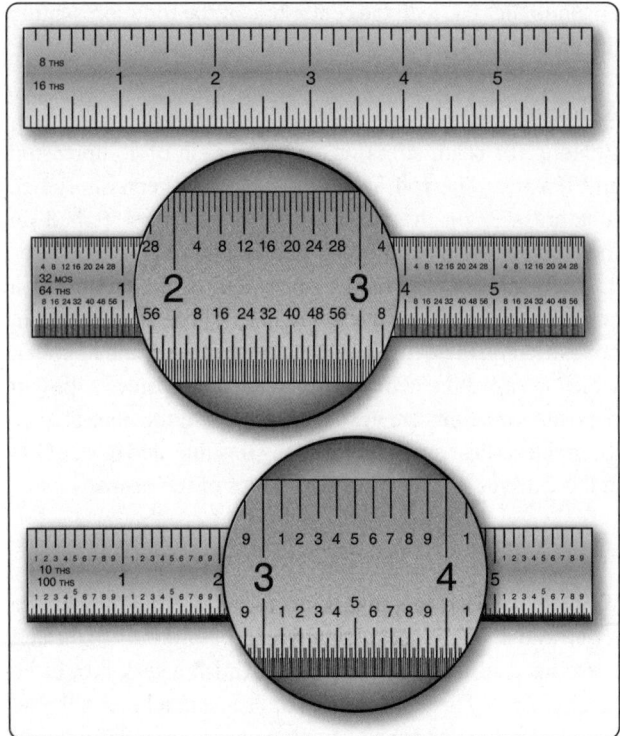

Figure 4-3. *Scales.*

Dividers

Dividers are used to transfer a measurement from a device to a scale to determine its value. Place the sharp points at the locations from which the measurement is to be taken. Then, place the points on a steel machinist's scale, but put one of the points on the 1-inch mark and measure from there. *[Figure 4-5]*

Rivet Spacers

A rivet spacer is used to make a quick and accurate rivet pattern layout on a sheet. On the rivet spacer, there are alignment marks for ½-inch, ¾-inch, 1-inch and 2-inch rivet spacing. *[Figure 4-6]*

Marking Tools

Pens

Fiber-tipped pens are the preferred method of marking lines and hole locations directly on aluminum, because the graphite in a No. 2 pencil can cause corrosion when used on aluminum. Make the layout on the protective membrane if it is still on the material, or mark directly on the material with a fiber-tipped pen, such as a fine-point Sharpie®, or cover the material with masking tape and then mark on the tape.

Scribes

A scribe is a pointed instrument used to mark or score metal to show where it is to be cut. A scribe should only be used when marks will be removed by drilling or cutting because it makes scratches that weaken the material and could cause corrosion. *[Figure 4-7]*

Punches

Punches are usually made of carbon steel that has been hardened and tempered. Generally classified as solid or hollow, punches are designed according to their intended use. A solid punch is a steel rod with various shapes at the end for different uses. For example, it is used to drive bolts out of holes, loosen frozen or tight pins and keys, knock out rivets, pierce holes in a material, etc. The hollow punch is sharp edged and used most often for cutting out blanks. Solid

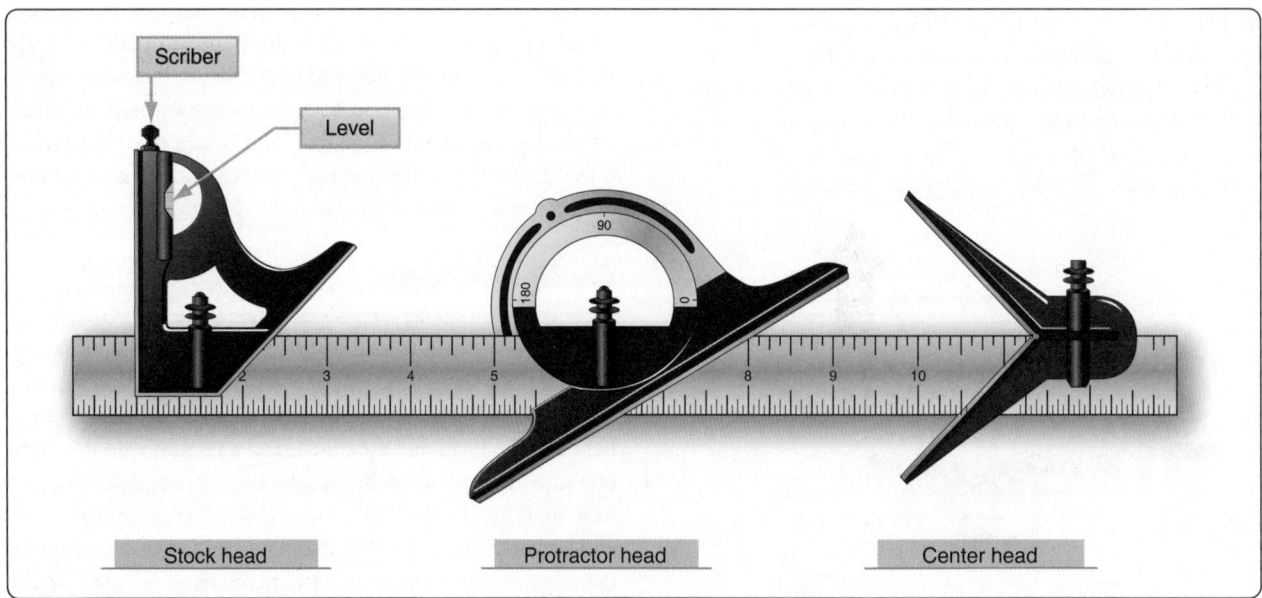

Figure 4-4. *Combination square.*

Figure 4-5. *Divider.*

Figure 4-6. *Rivet spacer.*

Figure 4-7. *Scribe.*

punches vary in both size and point design, while hollow punches vary in size.

Prick Punch

A prick punch is primarily used during layout to place reference marks on metal because it produces a small indentation. *[Figure 4-8]* After layout is finished, the indentation is enlarged with a center punch to allow for drilling. The prick punch can also be used to transfer dimensions from a paper pattern directly onto the metal. Take the following precautions when using a prick punch:

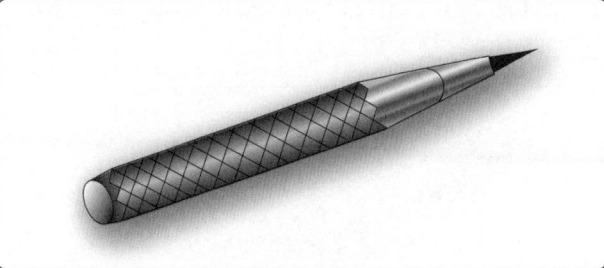

Figure 4-8. *Prick punch.*

- Never strike a prick punch a heavy blow with a hammer because it could bend the punch or cause excessive damage to the item being worked.

- Do not use a prick punch to remove objects from holes because the point of the punch spreads the object and causes it to bind even more.

Center Punch

A center punch is used to make indentations in metal as an aid in drilling. *[Figure 4-9]* These indentations help the drill, which has a tendency to wander on a flat surface, stay on the mark as it goes through the metal. The traditional center punch is used with a hammer, has a heavier body than the prick punch, and has a point ground to an angle of about 60°. Take the following precautions when using a center punch:

- Never strike the center punch with enough force to dimple the item around the indentation or cause the metal to protrude through the other side of the sheet.

- Do not use a center punch to remove objects from holes because the point of the punch spreads the object and causes it to bind even more.

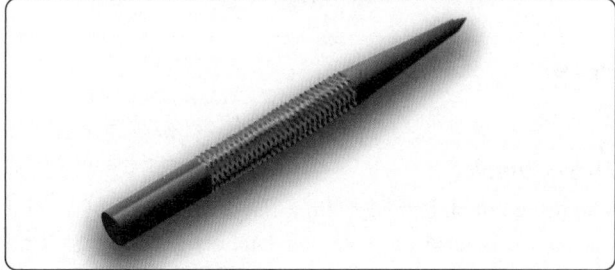

Figure 4-9. *Center punch.*

Automatic Center Punch

The automatic center punch performs the same function as an ordinary center punch, but uses a spring tension mechanism to create a force hard enough to make an indentation without the need for a hammer. The mechanism automatically strikes a blow of the required force when placed where needed and pressed. This punch has an adjustable cap for regulating the stroke; the point can be removed for replacement or sharpening. Never strike an automatic center punch with a hammer. *[Figure 4-10]*

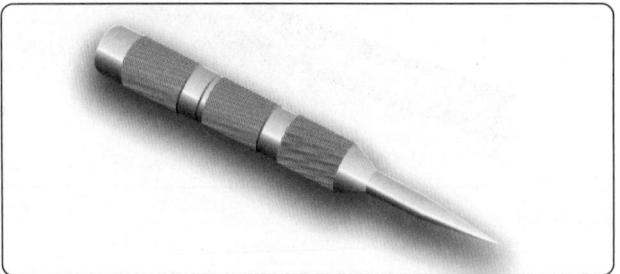

Figure 4-10. *Automatic center punch.*

Transfer Punch

A transfer punch uses a template or existing holes in the structure to mark the locations of new holes. The punch is centered in the old hole over the new sheet and lightly tapped with a mallet. The result should be a mark that serves to locate the hole in the new sheet. *[Figure 4-11]*

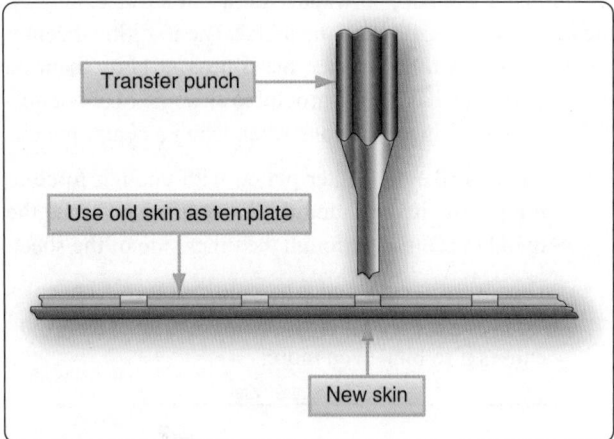

Transfer punch

Use old skin as template

New skin

Figure 4-11. *Transfer punch.*

Drive Punch

The drive punch is made with a flat face instead of a point because it is used to drive out damaged rivets, pins, and bolts that sometimes bind in holes. The size of the punch is determined by the width of the face, usually ⅛-inch to ¼-inch. *[Figure 4-12]*

Pin Punch

The pin punch typically has a straight shank characterized by a hexagonal body. Pin punch points are sized in ¹⁄₃₂-inch increments of an inch and range from ¹⁄₁₆-inch to ⅜-inch in diameter. The usual method for driving out a pin or bolt is to start working it out with a drive punch until the shank of the punch is touching the sides of the hole. Then use a pin

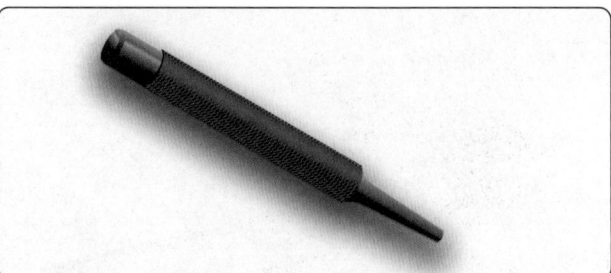

Figure 4-12. *Drive punch.*

punch to drive the pin or bolt the rest of the way out of the hole. *[Figure 4-13]*

Chassis Punch

A chassis punch is used to make holes in sheet metal parts for the installation of instruments and other avionics appliance, as well as lightening holes in ribs and spars. Sized in ¹⁄₁₆ of an inch, they are available in sizes from ½ inch to 3 inches. *[Figure 4-14]*

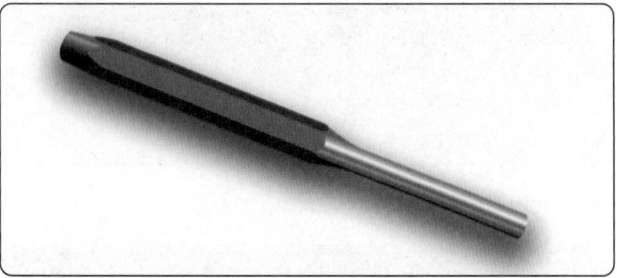

Figure 4-13. *Pin punch.*

Figure 4-14. *Chassis punch.*

Awl

A pointed tool for marking surfaces or for punching small holes, an awl is used in aircraft maintenance to place scribe marks on metal and plastic surfaces and to align holes, such as in the installation of a deicer boot. *[Figure 4-15]*

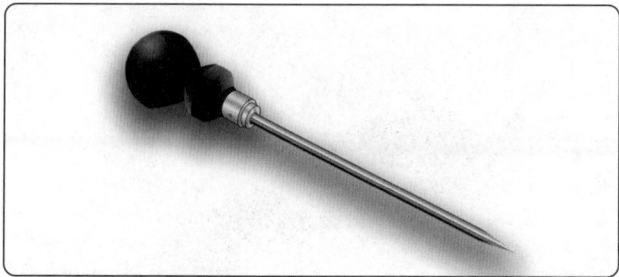

Figure 4-15. *Awl.*

Procedures for one use of an awl:

1. Place the metal to be scribed on a flat surface. Place a ruler or straightedge on the guide marks already measured and placed on the metal.

2. Remove the protective cover from the awl.

3. Hold the straightedge firmly. Hold the awl, as shown in *Figure 4-16*, and scribe a line along the straightedge.

4. Replace the protective cover on the awl.

Hole Duplicator

Available in a variety of sizes and styles, hole duplicators, or hole finders, utilize the old covering as a template to locate and match existing holes in the structure. Holes in a replacement sheet or in a patch must be drilled to match existing holes in the structure and the hole duplicator simplifies this process. *Figure 4-17* illustrates one type of hole duplicator. The peg on the bottom leg of the duplicator fits into the existing rivet hole. To make the hole in the replacement sheet or patch, drill through the bushing on the top leg. If the duplicator is properly made, holes drilled in this manner are in perfect alignment. A separate duplicator must be used for each diameter of rivet.

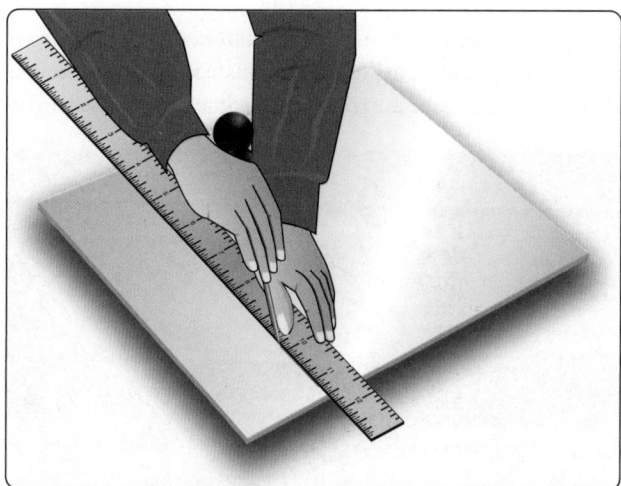

Figure 4-16. *Awl usage.*

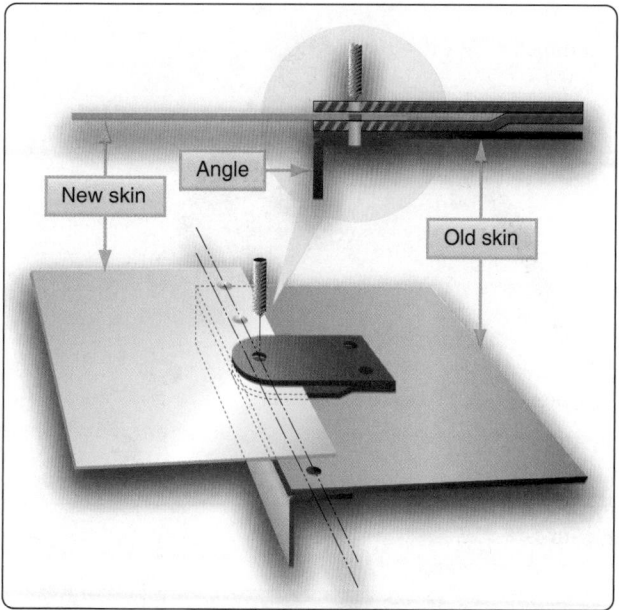

Figure 4-17. *Hole duplicator.*

Cutting Tools

Powered and nonpowered metal cutting tools available to the aviation technician include various types of saws, nibblers, shears, sanders, notchers, and grinders.

Circular-Cutting Saws

The circular-cutting saw cuts with a toothed, steel disc that rotates at high speed. Handheld or table mounted and powered by compressed air, this power saw cuts metal or wood. To prevent the saw from grabbing the metal, keep a firm grip on the saw handle at all times. Check the blade carefully for cracks prior to installation because a cracked blade can fly apart during use, possibly causing serious injury.

Kett Saw

The Kett saw is an electrically operated, portable circular cutting saw that uses blades of various diameters. *[Figure 4-18]* Since the head of this saw can be turned to any desired angle, it is useful for removing damaged sections on a stringer. The advantages of a Kett saw include:

1. Can cut metal up to $\frac{3}{16}$-inch in thickness.

2. No starting hole is required.

3. A cut can be started anywhere on a sheet of metal.

4. Can cut an inside or outside radius.

Pneumatic Circular-Cutting Saw

The pneumatic circular-cutting saw, useful for cutting out damage, is similar to the Kett saw. *[Figure 4-19]*

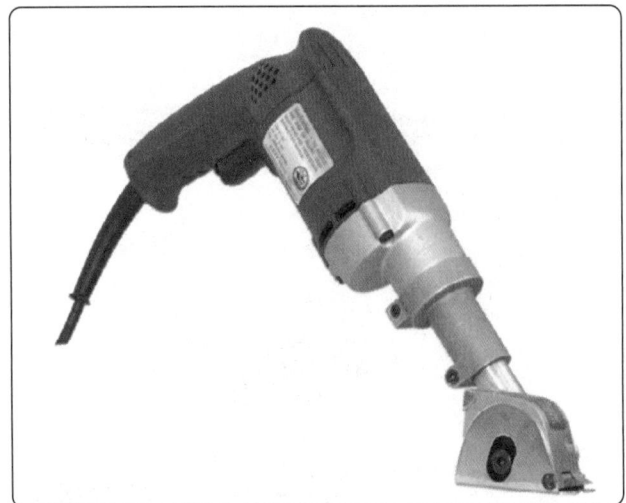

Figure 4-18. *Kett saw.*

Figure 4-19. *Pneumatic circular saw.*

Reciprocating Saw

The versatile reciprocating saw achieves cutting action through a push and pull (reciprocating) motion of the blade. This saw can be used right sideup or upside down, a feature that makes it handier than the circular saw for working in tight or awkward spots. A variety of blade types are available for reciprocating saws; blades with finer teeth are used for cutting through metal. The portable, air-powered reciprocating saw uses a standard hacksaw blade and can cut a 360° circle or a square or rectangular hole. Unsuited for fine precision work, this saw is more difficult to control than the pneumatic circular-cutting saw. A reciprocating saw should be used in such a way that at least two teeth of the saw blade are cutting at all times. Avoid applying too much downward pressure on the saw handle because the blade may break. *[Figure 4-20]*

Cut-off Wheel

A cut-off wheel is a thin abrasive disc driven by a high-speed pneumatic die-grinder and used to cut out damage on aircraft skin and stringers. The wheels come in different thicknesses and sizes. *[Figure 4-21]*

Figure 4-20. *Reciprocating saw.*

Figure 4-21. *Die grinder and cut-off wheel.*

Nibblers

Usually powered by compressed air, the nibbler is another tool for cutting sheet metal. Portable nibblers utilize a high speed blanking action (the lower die moves up and down and meets the upper stationary die) to cut the metal. *[Figure 4-22]* The shape of the lower die cuts out small pieces of metal approximately $\frac{1}{16}$ inch wide.

The cutting speed of the nibbler is controlled by the thickness of the metal being cut. Nibblers satisfactorily cut through sheets of metal with a maximum thickness of $\frac{1}{16}$ inch. Too

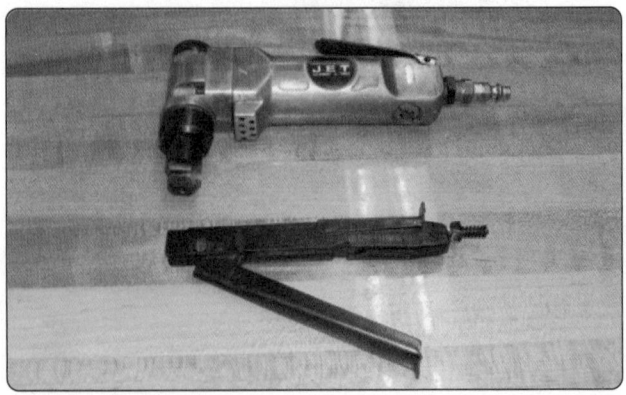

Figure 4-22. *Nibbler.*

much force applied to the metal during the cutting operation clogs the dies (shaped metal), causing them to fail or the motor to overheat. Both electric and hand nibblers are available.

Shop Tools

Due to size, weight, and/or power source, shop tools are usually in a fixed location, and the airframe part to be constructed or repaired is brought to the tool.

Squaring Shear

The squaring shear provides the airframe technician with a convenient means of cutting and squaring sheet metal. Available as a manual, hydraulic, or pneumatic model, this shear consists of a stationary lower blade attached to a bed and a movable upper blade attached to a crosshead. *[Figure 4-23]*

Two squaring fences, consisting of thick strips of metal used for squaring metal sheets, are placed on the bed. One squaring fence is placed on the right side and one on the left to form a 90° angle with the blades. A scale graduated in fractions of an inch is scribed on the bed for ease in placement.

To make a cut with a foot shear, move the upper blade down by placing the foot on the treadle and pushing downward. Once the metal is cut and foot pressure removed, a spring raises the blade and treadle. Hydraulic or pneumatic models utilize remote foot pedals to ensure operator safety.

The squaring shear performs three distinctly different operations:

1. Cutting to a line.
2. Squaring.
3. Multiple cutting to a specific size.

Figure 4-23. *Power squaring shear.*

When cutting to a line, place the sheet on the bed of the shears in front of the cutting blade with the cutting line even with the cutting edge of the bed. To cut the sheet with a foot shear, step on the treadle while holding the sheet securely in place.

Squaring requires several steps. First, one end of the sheet is squared with an edge (the squaring fence is usually used on the edge). Then, the remaining edges are squared by holding one squared end of the sheet against the squaring fence and making the cut, one edge at a time, until all edges have been squared.

When several pieces must be cut to the same dimensions, use the backstop, located on the back of the cutting edge on most squaring shears. The supporting rods are graduated in fractions of an inch and the gauge bar may be set at any point on the rods. Set the gauge bar the desired distance from the cutting blade of the shears and push each piece to be cut against the gauge bar. All the pieces can then be cut to the same dimensions without measuring and marking each one separately.

Foot-operated shears have a maximum metal cutting capacity of 0.063 inch of aluminum alloy. Use powered squaring shears for cutting thicker metals. *[Figure 4-24]*

Figure 4-24. *Foot-operated squaring shear.*

Throatless Shear

Airframe technicians use the throatless shear to cut aluminum sheets up to 0.063 inches. This shear takes its name from the fact that metal can be freely moved around the cutting blade during cutting because the shear lacks a "throat" down which metal must be fed. *[Figure 4-25]* This feature allows great flexibility in what shapes can be cut because the metal can be turned to any angle for straight, curved, and irregular cuts. Also, a sheet of any length can be cut.

A hand lever operates the cutting blade which is the top blade.

Figure 4-25. *Throatless shears.*

Throatless shears made by the Beverly Shear Manufacturing Corporation, called Beverly™ shears, are often used.

Scroll Shears

Scroll shears are used for cutting irregular lines on the inside of a sheet without cutting through to the edge. *[Figure 4-26]* The upper cutting blade is stationary while the lower blade is movable. A handle connected to the lower blade operates the machine.

Rotary Punch Press

Used in the airframe repair shop to punch holes in metal parts, the rotary punch can cut radii in corners, make washers, and perform many other jobs where holes are required. *[Figure 4-27]* The machine is composed of two cylindrical turrets, one mounted over the other and supported by the

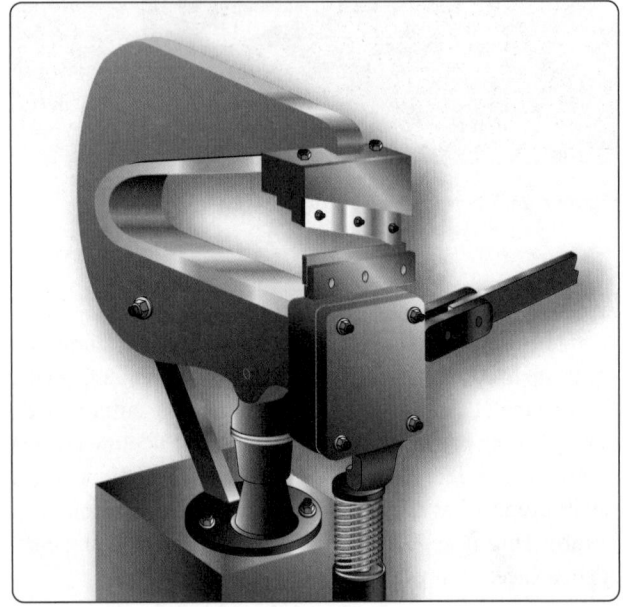

Figure 4-26. *Scroll shears.*

Figure 4-27. *Rotary punch press.*

frame, with both turrets synchronized to rotate together. Index pins, which ensure correct alignment at all times, may be released from their locking position by rotating a lever on the right side of the machine. This action withdraws the index pins from the tapered holes and allows an operator to turn the turrets to any size punch desired.

When rotating the turret to change punches, release the index lever when the desired die is within 1 inch of the ram, and continue to rotate the turret slowly until the top of the punch holder slides into the grooved end of the ram. The tapered index locking pins will then seat themselves in the holes provided and, at the same time, release the mechanical locking device, which prevents punching until the turrets are aligned. To operate the machine, place the metal to be worked between the die and punch. Pull the lever on the top of the machine toward the operator, actuating the pinion shaft, gear segment, toggle link, and the ram, forcing the punch through the metal. When the lever is returned to its original position, the metal is removed from the punch.

The diameter of the punch is stamped on the front of each die holder. Each punch has a point in its center that is placed in the center punch mark to punch the hole in the correct location.

Band Saw

A band saw consists of a toothed metal band coupled to, and continuously driven around, the circumferences of two wheels. It is used to cut aluminum, steel, and composite parts. *[Figure 4-28]* The speed of the band saw and the type and style of the blade depends on the material to be cut. Band saws are often designated to cut one type of material, and if a different material is to be cut, the blade is changed. The speed is controllable and the cutting platform can be tilted to cut angled pieces.

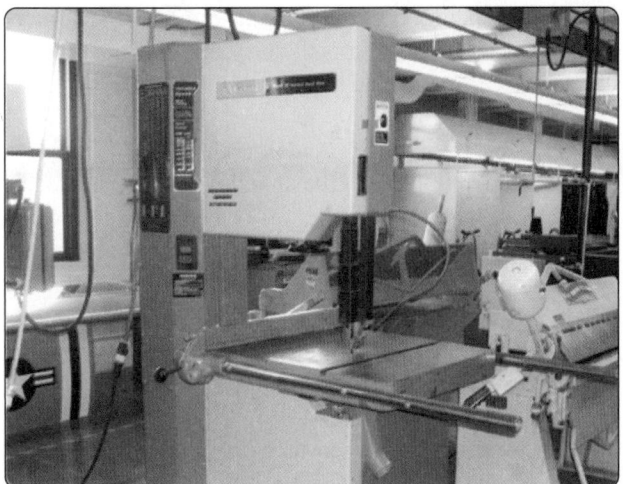

Figure 4-28. *Band saw.*

Disc Sander

Disc sanders have a powered abrasive-covered disc or belt and are used for smoothing or polishing surfaces. The sander unit uses abrasive paper of different grits to trim metal parts. It is much quicker to use a disc sander than to file a part to the correct dimension. The combination disc and belt sander has a vertical belt sander coupled with a disc sander and is often used in a metal shop. *[Figure 4-29]*

Belt Sander

The belt sander uses an endless abrasive belt driven by an electric motor to sand down metal parts much like the disc sander unit. The abrasive paper used on the belt comes in different degrees of grit or coarseness. The belt sander is available as a vertical or horizontal unit. The tension and tracking of the abrasive belt can be adjusted so the belt runs in the middle. *[Figure 4-30]*

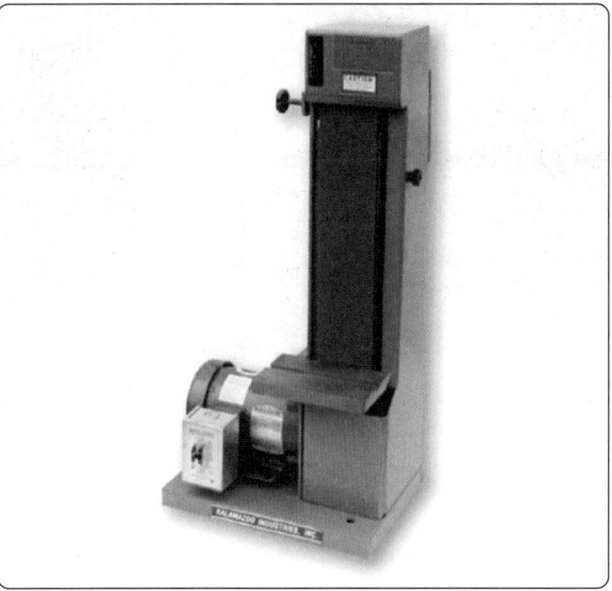

Figure 4-30. *Belt sander.*

Notcher

The notcher is used to cut out metal parts, with some machines capable of shearing, squaring, and trimming metal. *[Figure 4-31]* The notcher consists of a top and bottom die and most often cuts at a 90° angle, although some machines can cut metal into angles up to 180°. Notchers are available in manual and pneumatic models able to cut various thicknesses of mild steel and aluminum. This is an excellent tool for quickly removing corners from sheet metal parts. *[Figure 4-32]*

Wet or Dry Grinder

Grinding machines come in a variety of types and sizes, depending upon the class of work for which they are to be used. Dry and/or wet grinders are found in airframe repair shops. Grinders can be bench or pedestal mounted. A dry grinder usually has a grinding wheel on each end of a shaft

Figure 4-29. *Combination disc and belt sander.*

Figure 4-31. *Notcher.*

Figure 4-32. *Power notcher.*

that runs through an electric motor or a pulley operated by a belt. The wet grinder has a pump to supply a flow of water on a single grinding wheel. The water acts as a lubricant for faster grinding while it continuously cools the edge of the metal, reducing the heat produced by material being ground against the wheel. It also washes away any bits of metal or abrasive removed during the grinding operation. The water returns to a tank and can be re-used.

Grinders are used to sharpen knives, tools, and blades as well as grinding steel, metal objects, drill bits, and tools. *Figure 4-33* illustrates a common type bench grinder found in most airframe repair shops. It can be used to dress mushroomed heads on chisels and points on chisels, screwdrivers, and drills, as well as for removing excess metal from work and smoothing metal surfaces.

The bench grinder is generally equipped with one medium-grit and one fine-grit abrasive wheel. The medium-grit wheel is usually used for rough grinding where a considerable quantity of material is to be removed or where a smooth finish is unimportant. The fine-grit wheel is used for sharpening

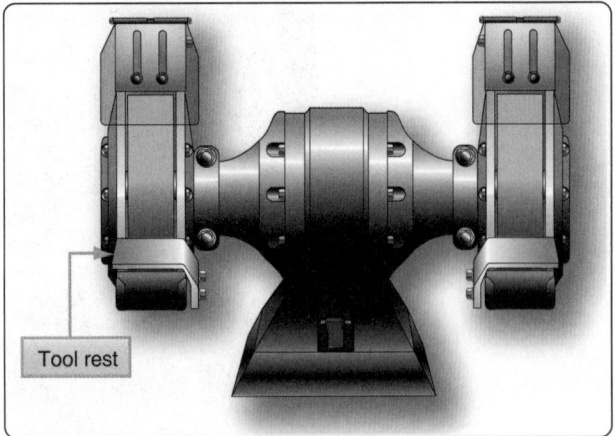

Tool rest

Figure 4-33. *Grinder.*

tools and grinding to close limits. It removes metal more slowly, gives the work a smooth finish, and does not generate enough heat to anneal the edges of cutting tools.

Before using any type of grinder, ensure that the abrasive wheels are firmly held on the spindles by the flange nuts. An abrasive wheel that comes off or becomes loose could seriously injure the operator in addition to ruining the grinder. A loose tool rest could cause the tool or piece of work to be "grabbed" by the abrasive wheel and cause the operator's hand to come in contact with the wheel, possibly resulting in severe wounds.

Always wear goggles when using a grinder, even if eye shields are attached to the grinder. Goggles should fit firmly against the face and nose. This is the only way to protect the eyes from the fine pieces of steel. Goggles that do not fit properly should be exchanged for ones that do fit. Be sure to check the abrasive wheel for cracks before using the grinder. A cracked abrasive wheel is likely to fly apart when turning at high speeds. Never use a grinder unless it is equipped with wheel guards that are firmly in place.

Grinding Wheels

A grinding wheel is made of a bonded abrasive and provides an efficient way to cut, shape, and finish metals. Available in a wide variety of sizes and numerous shapes, grinding wheels are also used to sharpen knives, drill bits, and many other tools, or to clean and prepare surfaces for painting or plating.

Grinding wheels are removable and a polishing or buffing wheel can be substituted for the abrasive wheel. Silicon carbide and aluminum oxide are the kinds of abrasives used in most grinding wheels. Silicon carbide is the cutting agent for grinding hard, brittle material, such as cast iron. It is also used in grinding aluminum, brass, bronze, and copper. Aluminum oxide is the cutting agent for grinding steel and other metals of high tensile strength.

Hand Cutting Tools

Many types of hand cutting tools are available to cut light gauge sheet metal. Four cutting tools commonly found in the air frame repair shop are straight hand snips, aviation snips, files, and burring tools.

Straight Snips

Straight snips, or sheet metal shears, have straight blades with cutting edges sharpened to an 85° angle. *[Figure 4-34]* Available in sizes ranging from 6 to 14 inches, they cut aluminum up to $\frac{1}{16}$ of an inch. Straight snips can be used for straight cutting and large curves, but aviation snips are better for cutting circles or arcs.

Figure 4-34. *Straight snips.*

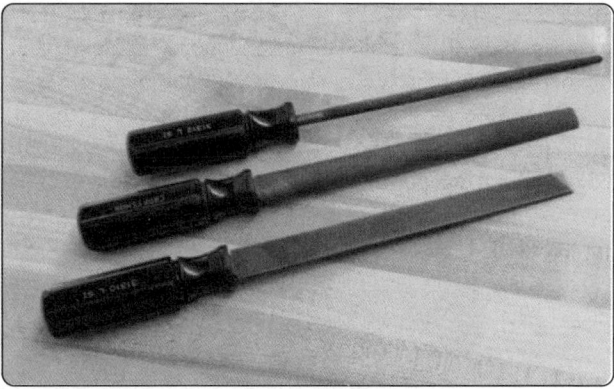

Figure 4-36. *Files.*

Aviation Snips

Aviation snips are used to cut holes, curved parts, round patches, and doublers (a piece of metal placed under a part to make it stiffer) in sheet metal. Aviation snips have colored handles to identify the direction of the cuts: yellow aviation snips cut straight, green aviation snips curve right, and red aviation snips curve left. *[Figure 4-35]*

Files

The file is an important but often overlooked tool used to shape metal by cutting and abrasion. Files have five distinct properties: length, contour, the form in cross section, the kind of teeth, and the fineness of the teeth. Many different types of files are available and the sizes range from 3 to 18 inches. *[Figure 4-36]*

The portion of the file on which the teeth are cut is called the face. The tapered end that fits into the handle is called the tang. The part of the file where the tang begins is the heel. The length of a file is the distance from the point or tip to the heel and does not include the tang. The teeth of the file do the cutting. These teeth are set at an angle across the face of the file. A file with a single row of parallel teeth is called a single-cut file. The teeth are cut at an angle of 65°–85° to the

centerline, depending on the intended use of the file. Files that have one row of teeth crossing another row in a crisscross pattern are called double-cut files. The angle of the first set usually is 40°–50° and that of the crossing teeth 70°–80°. Crisscrossing produces a surface that has a very large number of little teeth that slant toward the tip of the file. Each little tooth looks like an end of a diamond point cold chisel.

Files are graded according to the tooth spacing; a coarse file has a small number of large teeth, and a smooth file has a large number of fine teeth. The coarser the teeth, the more metal is removed on each stroke of the file. The terms used to indicate the coarseness or fineness of a file are rough, coarse, bastard, second cut, smooth, and dead smooth, and the file may be either single cut or double cut. Files are further classified according to their shape. Some of the more common types are: flat, triangle, square, half round, and round.

There are several filing techniques. The most common is to remove rough edges and slivers from the finished part before it is installed. Crossfiling is a method used for filing the edges of metal parts that must fit tightly together. Crossfiling involves clamping the metal between two strips of wood and filing the edge of the metal down to a preset line. Draw filing is used when larger surfaces need to be smoothed and squared. It is done by drawing the file over the entire surface of the work.

To protect the teeth of a file, files should be stored separately in a plastic wrap or hung by their handles. Files kept in a toolbox should be wrapped in waxed paper to prevent rust from forming on the teeth. File teeth can be cleaned with a file card.

Die Grinder

A die grinder is a handheld tool that turns a mounted cutoff wheel, rotary file, or sanding disc at high speed. *[Figure 4-37]* Usually powered by compressed air, electric die grinders are also used. Pneumatic die grinders run at

Figure 4-35. *Aviation snips.*

12,000 to 20,000 revolutions per minute (rpm) with the rotational speed controlled by the operator who uses a hand- or foot-operated throttle to vary the volume of compressed air. Available in straight, 45°, and 90° models, the die grinder is excellent for weld breaking, smoothing sharp edges, deburring, porting, and general high-speed polishing, grinding, and cutting.

Figure 4-37. *Die grinder.*

Burring Tool

This type of tool is used to remove a burr from an edge of a sheet or to deburr a hole. *[Figure 4-38]*

Hole Drilling

Drilling holes is a common operation in the airframe repair shop. Once the fundamentals of drills and their uses are learned, drilling holes for rivets and bolts on light metal is not difficult. While a small portable power drill is usually the most practical tool for this common operation in airframe metalwork, sometimes a drill press may prove to be the better piece of equipment for the job.

Portable Power Drills

Portable power drills operate by electricity or compressed air. Pneumatic drill motors are recommended for use on repairs around flammable materials where potential sparks from an

electric drill motor might become a fire hazard.

When using the portable power drill, hold it firmly with both hands. Before drilling, be sure to place a backup block of wood under the hole to be drilled to add support to the metal structure. The drill bit should be inserted in the chuck and tested for trueness or vibration. This may be visibly checked by running the motor freely. A drill bit that wobbles or is slightly bent should not be used since such a condition causes enlarged holes. The drill should always be held at right angles to the work regardless of the position or curvatures. Tilting the drill at any time when drilling into or withdrawing from the material may cause elongation (egg shape) of the hole. When drilling through sheet metal, small burrs are formed around the edge of the hole. Burrs must be removed to allow rivets or bolts to fit snugly and to prevent scratching. Burrs may be removed with a bearing scraper, a countersink, or a drill bit larger than the hole. If a drill bit or countersink is used, it should be rotated by hand. Always wear safety goggles while drilling.

Pneumatic Drill Motors

Pneumatic drill motors are the most common type of drill motor for aircraft repair work. *[Figure 4-39]* They are lightweight and have sufficient power and good speed control. Drill motors are available in many different sizes and models. Most drill motors used for aircraft sheet metal work are rated at 3,000 rpm, but if drilling deep holes or drilling in hard materials, such as corrosion resistant steel or titanium, a drill motor with more torque and lower rpm should be selected to prevent damage to tools and materials.

Right Angle & 45° Drill Motors

Right angle and 45° drill motors are used for positions that are not accessible with a pistol grip drill motor. Most right angle drill motors use threaded drill bits that are available in several lengths. Heavy-duty right angle drills are equipped with a chuck similar to the pistol grip drill motor. *[Figure 4-40]*

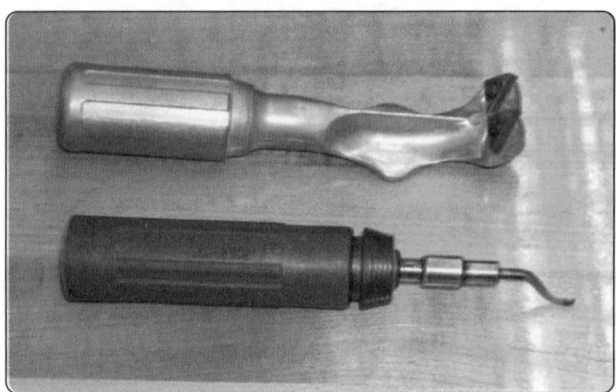

Figure 4-38. *Burring tools.*

Figure 4-39. *Drill motors.*

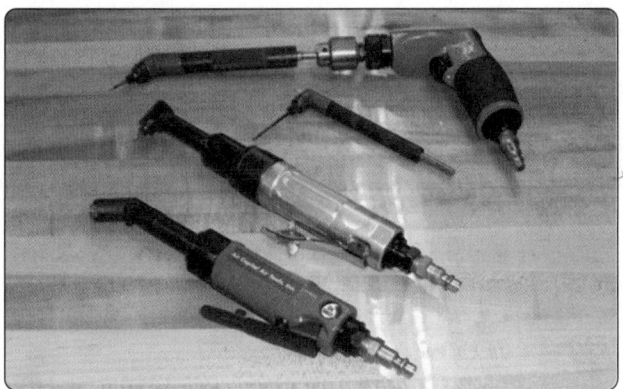

Figure 4-40. *Angle drill motors.*

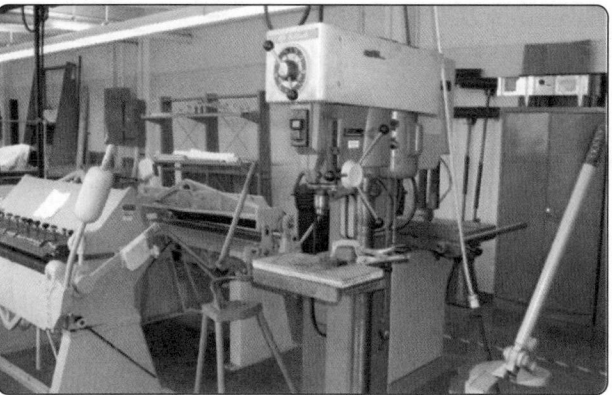

Figure 4-42. *Drill press.*

Two Hole

Special drill motors that drill two holes at the same time are used for the installation of nutplates. By drilling two holes at the same time, the distance between the holes is fixed and the holes line up perfectly with the holes in the nutplate. *[Figure 4-41]*

Drill Press

The drill press is a precision machine used for drilling holes that require a high degree of accuracy. It serves as an accurate means of locating and maintaining the direction of a hole that is to be drilled and provides the operator with a feed lever that makes the task of feeding the drill into the work easier. The upright drill press is the most common of the variety of drill presses available. *[Figure 4-42]*

When using a drill press, the height of the drill press table is adjusted to accommodate the height of the part to be drilled. When the height of the part is greater than the distance between the drill and the table, the table is lowered. When the height of the part is less than the distance between the drill and the table, the table is raised.

After the table is properly adjusted, the part is placed on the table and the drill is brought down to aid in positioning the metal so that the hole to be drilled is directly beneath the point of the drill. The part is then clamped to the drill press table to prevent it from slipping during the drilling operation. Parts not properly clamped may bind on the drill and start spinning, causing serious cuts on the operator's arms or body, or loss of fingers or hands. Always make sure the part to be drilled is properly clamped to the drill press table before starting the drilling operation.

The degree of accuracy that it is possible to attain when using the drill press depends to a certain extent on the condition of the spindle hole, sleeves, and drill shank. Therefore, special care must be exercised to keep these parts clean and free from nicks, dents, and warpage. Always be sure that the sleeve is securely pressed into the spindle hole. Never insert a broken drill in a sleeve or spindle hole. Be careful never to use the sleeve-clamping vise to remove a drill since this may cause the sleeve to warp.

The drill speed on a drill press is adjustable. Always select the optimum drill speed for the material to be drilled. Technically, the speed of a drill bit means its speed at the circumference, in surface feet per minute (sfm). The recommended speed for drilling aluminum alloy is from 200 to 300 sfm, and for mild steel is 30 to 50 sfm. In practice, this must be converted into rpm for each size drill. Machinist and mechanic handbooks include drill rpm charts or drill rpm may be computed by use of the formula:

$$\frac{CS \times 4}{D} = rpm$$

CS = The recommended cutting speed in sfm

D = The diameter of the drill bit in inches

Drill Extensions & Adapters

When access to a place where drilling is difficult or impossible with a straight drill motor, various types of drill extensions and adapters are used.

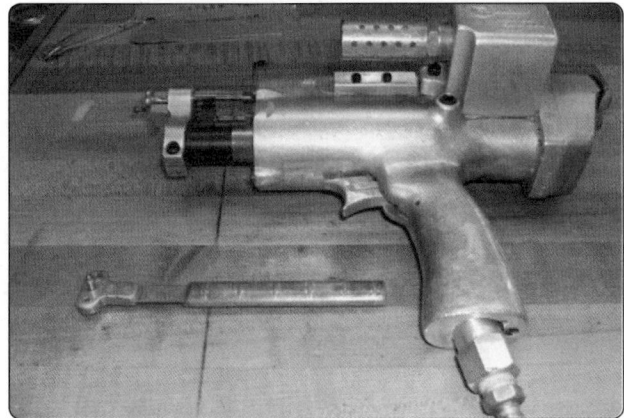

Figure 4-41. *Nutplate drill.*

Extension Drill Bits

Extension drill bits are widely used for drilling holes in locations that require reaching through small openings or past projections. These drill bits, which come in 6- to 12-inch lengths, are high speed with spring-tempered shanks. Extension drill bits are ground to a special notched point, which reduces end thrust to a minimum. When using extension drill bits always:

1. Select the shortest drill bit that will do the job. It is easier to control.

2. Check the drill bit for straightness. A bent drill bit makes an oversized hole and may whip, making it difficult to control.

3. Keep the drill bit under control. Extension drills smaller than ¼-inch must be supported by a drill guard made from a piece of tubing or spring to prevent whipping.

Straight Extension

A straight extension for a drill can be made from an ordinary piece of drill rod. The drill bit is attached to the drill rod by shrink fitting, brazing, or silver soldering.

Angle Adapters

Angle adapters can be attached to an electric or pneumatic drill when the location of the hole is inaccessible to a straight drill. Angle adapters have an extended shank fastened to the chuck of the drill. The drill is held in one hand and the adapter in the other to prevent the adapter from spinning around the drill chuck.

Snake Attachment

The snake attachment is a flexible extension used for drilling in places inaccessible to ordinary drills. Available for electric and pneumatic drill motors, its flexibility permits drilling around obstructions with minimum effort. *[Figure 4-43]*

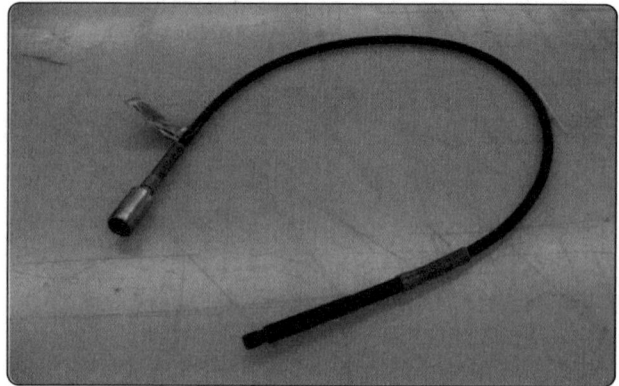

Figure 4-43. *Snake attachment.*

Types of Drill Bits

A wide variety of drill bits including specialty bits for specific jobs are available. *Figure 4-44* illustrates the parts of the drill bit and *Figure 4-45* shows some commonly used drill bits. High speed steel (HSS) drill bits come in short shank or standard length, sometimes called jobbers length. HSS drill bits can withstand temperatures nearing the critical range of 1,400 °F (dark cherry red) without losing their hardness. The industry standard for drilling metal (aluminum, steel, etc.), these drill bits stay sharper longer.

Step Drill Bits

Typically, the procedure for drilling holes larger than ³⁄₁₆ inch in sheet metal is to drill a pilot hole with a No. 40 or No. 30 drill bit and then to oversize with a larger drill bit to the correct size. The step drill combines these two functions into one step. The step drill bit consists of a smaller pilot drill point that drills the initial small hole. When the drill bit is advanced further into the material, the second step of the drill bit enlarges the hole to the desired size.

Step drill bits are designed to drill round holes in most metals,

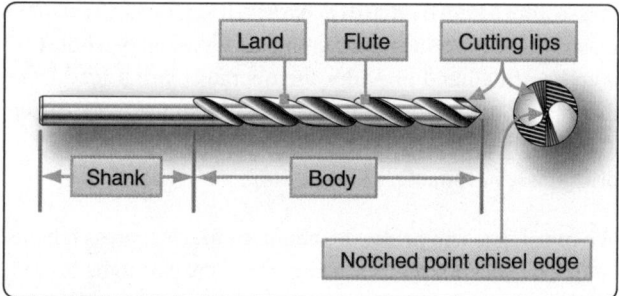

Figure 4-44. *Parts of a drill.*

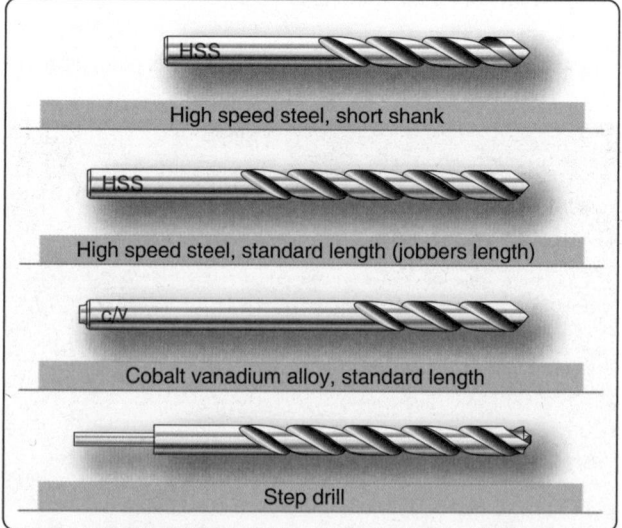

Figure 4-45. *Types of drill bits.*

plastic, and wood. Commonly used in general construction and plumbing, they work best on softer materials, such as plywood, but can be used on very thin sheet metal. Step drill bits can also be used to deburr holes left by other bits.

Cobalt Alloy Drill Bits

Cobalt alloy drill bits are designed for hard, tough metals like corrosion-resistant steel and titanium. It is important for the aircraft technician to note the difference between HSS and cobalt, because HSS drill bits wear out quickly when drilling titanium or stainless. Cobalt drill bits are excellent for drilling titanium or stainless steel, but do not produce a quality hole in aluminum alloys. Cobalt drill bits can be recognized by thicker webs and a taper at the end of the drill shank.

Twist Drill Bits

Easily the most popular drill bit type, the twist drill bit has spiral grooves or flutes running along its working length. *[Figure 4-46]* This drill bit comes in a single-fluted, two-fluted, three-fluted, and four-fluted styles. Single-fluted and two-fluted drill bits (most commonly available) are used for originating holes. Three-fluted and four-fluted drill bits are used interchangeably to enlarge existing holes. Twist drill bits are available in a wide choice of tooling materials and lengths with the variations targeting specific projects.

The standard twist drill bits used for drilling aluminum are made from HSS and have a 135° split point. Drill bits for titanium are made from cobalt vanadium for increased wear resistance.

Drill Bit Sizes

Drill diameters are grouped by three size standards: number, letter, and fractional. The decimal equivalents of standard drill are shown in *Figure 4-47*.

Drill Lubrication

Normal drilling of sheet material does not require lubrication, but lubrication should be provided for all deeper drilling

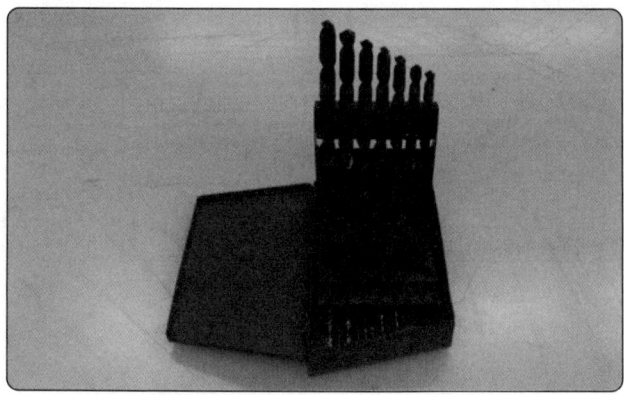

Figure 4-46. *Twist drill bits.*

Lubricants serve to assist in chip removal, which prolongs drill life and ensures a good finish and dimensional accuracy of the hole. It does not prevent overheating. The use of a lubricant is always a good practice when drilling castings, forgings, or heavy gauge stock. A good lubricant should be thin enough to help in chip removal but thick enough to stick to the drill. For aluminum, titanium, and corrosion-resistant steel, a cetyl alcohol based lubricant is the most satisfactory. Cetyl alcohol is a nontoxic fatty alcohol chemical produced in liquid, paste, and solid forms. The solid stick and block forms quickly liquefy at drilling temperatures. For steel, sulfurized mineral cutting oil is superior. Sulfur has an affinity for steel, which aids in holding the cutting oil in place. In the case of deep drilling, the drill should be withdrawn at intervals to relieve chip packing and to ensure the lubricant reaches the point. As a general rule, if the drill is large or the material hard, use a lubricant.

Reamers

Reamers, used for enlarging holes and finishing them smooth to a required size, are made in many styles. They can be straight or tapered, solid or expansive, and come with straight or helical flutes. *Figure 4-48* illustrates three types of reamers:

1. Three or four fluted production bullet reamers are customarily used where a finer finish and/or size is needed than can be achieved with a standard drill bit.

2. Standard or straight reamer.

3. Piloted reamer, with the end reduced to provide accurate alignment.

The cylindrical parts of most straight reamers are not cutting edges, but merely grooves cut for the full length of the reamer body. These grooves provide a way for chips to escape and a channel for lubricant to reach the cutting edge. Actual cutting is done on the end of the reamer. The cutting edges are normally ground to a bevel of 45° ± 5°.

Reamer flutes are not designed to remove chips like a drill. Do not attempt to withdraw a reamer by turning it in the reverse direction because chips can be forced into the surface, scarring the hole.

Drill Stops

A spring drill stop is a wise investment. *[Figure 4-49]* Properly adjusted, it can prevent excessive drill penetration that might damage underlying structure or injure personnel and prevent the drill chuck from marring the surface. Drill stops can be made from tubing, fiber rod, or hard rubber.

Drill Bushings & Guides

There are several types of tools available that aid in holding the drill perpendicular to the part. They consist of a hardened bushing anchored in a holder. *[Figure 4-50]*

Drill Size	Decimal (Inches)	Drill Size	Decimal (Inches)	Drill Size	Decimal (Inches)	Drill Size	Decimal (Inches)	Drill Size	Decimal (Inches)
80	.0135	50	.0700	22	.1570	G	.2610	31/64	.4844
79	.0145	49	.0730	21	.1590	17/64	.2656	1/2	.5000
1/54	.0156	48	.0760	20	.1610	H	.2660	33/64	.5156
78	.0160	5/64	.0781	19	.1660	I	.2720	17/32	.5312
77	.0180	47	.0785	18	.1695	J	.2770	35/64	.5469
76	.0200	46	.0810	11/64	.1718	K	.2810	9/16	.5625
75	.0210	45	.0820	17	.1730	9/32	.2812	37/64	.5781
74	.0225	44	.0860	16	.1770	L	.2900	19/32	.5937
73	.0240	43	.0890	15	.1800	M	.2950	39/84	.6094
72	.0250	42	.0935	14	.1820	19/64	.2968	5/8	.6250
71	.0260	3/32	.0937	13	.1850	N	.3020	41/64	.6406
70	.0280	41	.0960	3/16	.1875	5/16	.3125	21/32	.6562
69	.0293	40	.0980	12	.1890	O	.3160	43/64	.6719
68	.0310	39	.0995	11	.1910	P	.3230	11/16	.6875
1/32	.0312	38	.1015	10	.1935	21/64	.3281	45/64	.7031
67	.0320	37	.1040	9	.1960	Q	.3320	23/32	.7187
66	.0330	36	.1065	8	.1990	R	.3390	47/64	.7344
65	.0350	7/64	.1093	7	.2010	11/32	.3437	3/4	.7500
64	.0360	35	.1100	13/64	.2031	S	.3480	49/64	.7656
63	.0370	34	.1110	6	.2040	T	.3580	25/32	.7812
62	.0380	33	.1130	5	.2055	23/64	.3593	51/64	.7969
61	.0390	32	.1160	4	.2090	U	.3680	13/16	.8125
60	.0400	31	.1200	3	.2130	3/8	.3750	53/64	.8281
59	.0410	1/8	.1250	7/32	.2187	V	.3770	27/32	.8437
58	.0420	30	.1285	2	.2210	W	.3860	55/64	.8594
57	.0430	29	.1360	1	.2280	25/64	.3906	7/8	.8750
56	.0465	28	.1405	A	.2340	X	.3970	57/64	.8906
3/64	.0468	9/64	.1406	15/64	.2343	Y	.4040	29/32	.9062
55	.0520	27	.1440	B	.2380	13/32	.4062	59/64	.9219
54	.0550	26	.1470	C	.2420	Z	.4130	15/16	.9375
53	.0595	25	.1495	D	.2460	27/64	.4219	61/64	.9531
1/16	.0625	24	.1520	1/4	.2500	7/16	.4375	31/32	.9687
52	.0635	23	.1540	E	.2500	29/64	.4531	63/64	.9844
51	.0670	5/32	.1562	F	.2570	15/32	.4687	1	1.0000

Figure 4-47. *Drill sizes and decimal equivalents.*

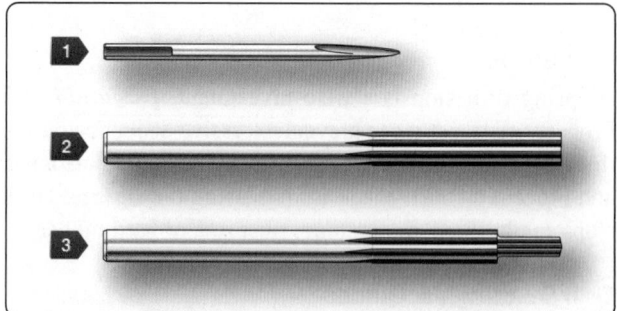

Figure 4-48. *Reamers.*

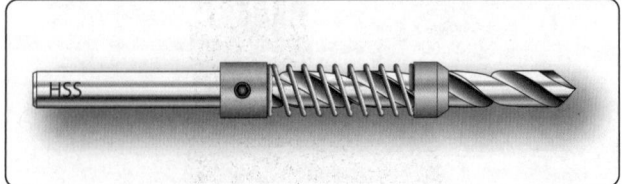

Figure 4-49. *Drill stop.*

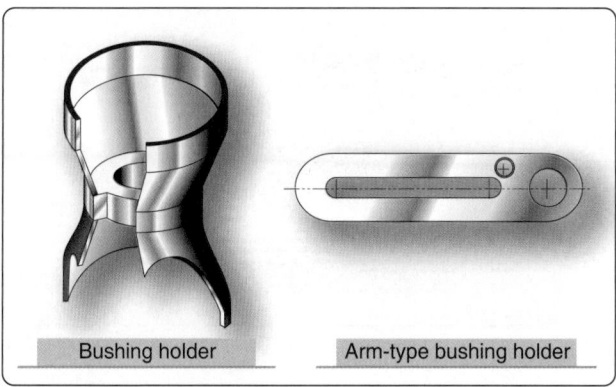

Figure 4-50. *Drill bushings.*

Drill bushing types:

1. Tube—hand-held in an existing hole

2. Commercial—twist lock

3. Commercial—threaded

Drill Bushing Holder Types

There are four types of drill bushing holder:

1. Standard—fine for drilling flat stock or tubing/rod; uses insert-type bushings.

2. Egg cup—improvement on standard tripod base; allows drilling on both flat and curved material; interchangeable bushings allows flexibility. *[Figure 4-51]*

3. Plate—used primarily for interchangeable production components; uses commercial bushings and self-feeding drills.

4. Arm—used when drilling critical structure; can be locked in position; uses interchangeable commercial bushings.

Hole Drilling Techniques

Precise location of drilled holes is sometimes required. When locating holes to close tolerances, accurately located punch marks need to be made. If a punch mark is too small, the chisel edge of the drill bit may bridge it and "walk off" the exact location before starting. If the punch mark is too heavy, it may deform the metal and/or result in a local strain hardening where the drill bit is to start cutting. The best size for a punch mark is about the width of the chisel edge of the drill bit to be used. This holds the drill point in place while starting. The procedure that ensures accurate holes follows: *[Figure 4-52]*

1. Measure and lay out the drill locations carefully and mark with crossed lines.

    Note: The chisel edge is the least efficient operating

Figure 4-51. *Bushing holder.*

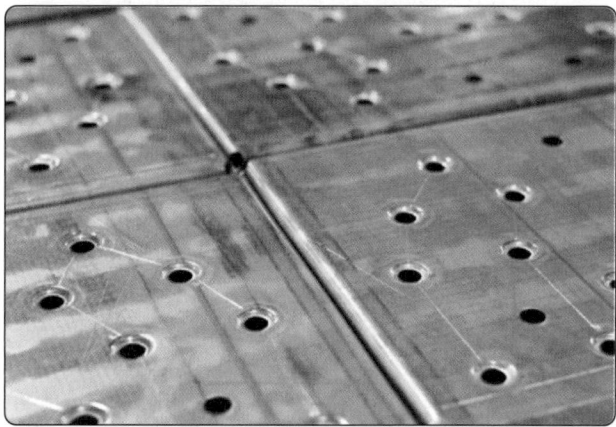

Figure 4-52. *Drilled sheet metal.*

surface element of the twist drill bit because it does not cut, but actually squeezes or extrudes the work material.

2. Use a sharp prick punch or spring-loaded center punch and magnifying glass to further mark the holes.

3. Seat a properly ground center punch (120°–135°) in the prick punch mark and, holding the center punch perpendicular to the surface, strike a firm square blow with a hammer.

4. Mark each hole with a small drill bit ($\frac{1}{16}$-inch recommended) to check and adjust the location prior to pilot drilling.

5. For holes $\frac{3}{16}$-inch and larger, pilot drilling is recommended. Select a drill bit equal to the width of the chisel edge of the final drill bit size. Avoid using a pilot drill bit that is too large because it would cause the corners and cutting lips of the final drill bit to be dulled, burned, or chipped. It also contributes to chattering and drill motor stalling. Pilot drill at each mark.

6. Place the drill point at the center of the crossed lines,

perpendicular to the surface, and, with light pressure, start drilling slowly. Stop drilling after a few turns and check to see if the drill bit is starting on the mark. It should be; if not, it is necessary to walk the hole a little by pointing the drill in the direction it should go, and rotating it carefully and intermittently until properly lined up.

7. Enlarge each pilot drilled hole to final size.

Drilling Large Holes

The following technique can be used to drill larger holes. Special tooling has been developed to drill large holes to precise tolerances. *[Figure 4-53]*

1. Pilot drill using a drill bushing. Bushings are sized for ⅛, ³⁄₁₆, or ¼ drill bits.

2. Step drill bits are used to step the hole to approximately ¹⁄₆₄-inch smaller than the final hole size. The aligning step diameter matches the pilot drill bit size.

3. Finish ream to size using a step reamer. The aligning step diameter matches the core drill bit size. Reamers should be available for both clearance and interference fit hole sizes.

    Note: Holes can also be enlarged by using a series of step reamers.

Chip Chasers

The chip chaser is designed to remove chips and burrs lodged between sheets of metal after drilling holes for riveting. *[Figure 4-54]* Chip chasers have a plastic molded handle and a flexible steel blade with a hook in the end.

Forming Tools

Sheet metal forming dates back to the days of the blacksmith who used a hammer and hot oven to mold metal into the desired form. Today's aircraft technician relies on a wide variety of powered and hand-operated tools to precisely bend and fold sheet metal to achieve the perfect shape. Forming tools include straight line machines, such as the bar folder and press brake, as well as rotary machines, such as the slip roll former. Forming sheet metal requires a variety of tools and equipment (both powered and manual), such as the piccolo former, shrinking and stretching tools, form blocks, and specialized hammers and mallets. *[Figure 4-55]*

Tempered sheet stock is used in forming operations whenever possible in typical repairs. Forming that is performed in the tempered condition, usually at room temperature, is known as cold-forming. Cold forming eliminates heat treatment and the straightening and checking operations required to remove the warp and twist caused by the heat treating process. Cold-

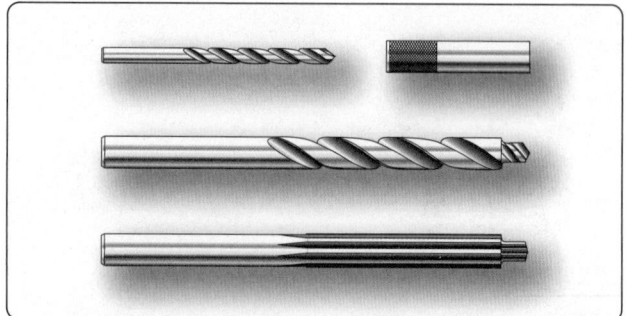

Figure 4-53. *Drilling large holes.*

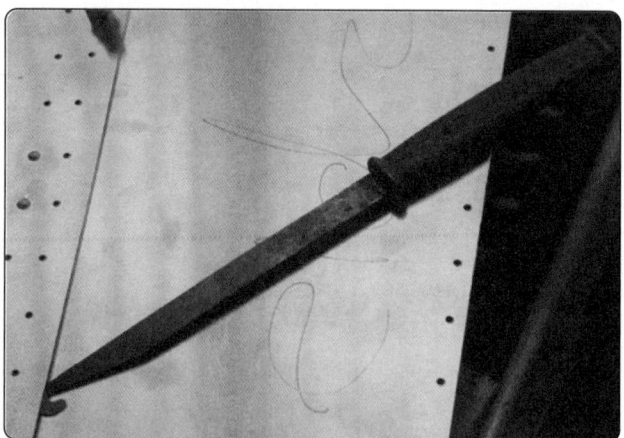

Figure 4-54. *Chip chaser.*

Figure 4-55. *Hammer and mallet forming.*

formed sheet metal experiences a phenomenon known as spring-back, which causes the worked piece to spring back slightly when the deforming force is removed. If the material shows signs of cracking during cold forming over small radii, the material should be formed in the annealed condition.

Annealing, the process of toughening steel by gradually heating and cooling it, removes the temper from metal, making it softer and easier to form. Parts containing small radii or compound curvatures must be formed in the annealed condition. After forming, the part is heat treated to a tempered condition before use on the aircraft.

Construction of interchangeable structural and nonstructural parts is achieved by forming flat sheet stock to make channel, angle, zee, and hat section members. Before a sheet metal part is formed, a flat pattern is made to show how much material is required in the bend areas, at what point the sheet must be inserted into the forming tool, or where bend lines are located. Determination of bend lines and bend allowances is discussed in greater detail in the section on layout and forming.

Bar Folding Machine

The bar folder is designed for use in making bends or folds along edges of sheets. *[Figure 4-56]* This machine is best suited for folding small hems, flanges, seams, and edges to be wired. Most bar folders have a capacity for metal up to 22 gauge in thickness and 42 inches in length. Before using the bar folder, several adjustments must be made for thickness of material, width of fold, sharpness of fold, and angle of fold. The adjustment for thickness of material is made by adjusting the screws at each end of the folder. As this adjustment is made, place a piece of metal of the desired thickness in the folder and raise the operating handle until the small roller rests on the cam. Hold the folding blade in this position and adjust the setscrews until the metal is clamped securely and evenly the full length of the folding blade. After the folder has been adjusted, test each end of the machine separately with a small piece of metal by actually folding it.

There are two positive stops on the folder, one for 45° folds or bends and the other for 90° folds or bends. A collar is provided that can be adjusted to any degree of bend within the capacity of the machine.

For forming angles of 45° or 90°, the appropriate stop is moved into place. This allows the handle to be moved forward to the correct angle. For forming other angles, the adjustable collar is used. This is accomplished by loosening the setscrew and setting the stop at the desired angle. After setting the stop, tighten the setscrew and complete the bend. To make the fold, adjust the machine correctly and then insert the metal. The metal goes between the folding blade and the jaw. Hold the metal firmly against the gauge and pull the operating handle toward the body. As the handle is brought forward, the jaw automatically raises and holds the metal until the desired fold is made. When the handle is returned to its original position, the jaw and blade return to their original positions and release the metal.

Cornice Brake

A brake is similar to a bar folder because it is also used for turning or bending the edges of sheet metal. The cornice brake is more useful than the bar folder because its design allows the sheet metal to be folded or formed to pass through the jaws from front to rear without obstruction. *[Figure 4-57]* In contrast, the bar folder can form a bend or edge only as wide as the depth of its jaws. Thus, any bend formed on a bar folder can also be made on the cornice brake.

In making ordinary bends with the cornice brake, the sheet is placed on the bed with the sight line (mark indicating line of bend) directly under the edge of the clamping bar. The clamping bar is then brought down to hold the sheet firmly in place. The stop at the right side of the brake is set for the proper angle or amount of bend and the bending leaf is raised until it strikes the stop. If other bends are to be made, the clamping bar is lifted and the sheet is moved to the correct position for bending.

The bending capacity of a cornice brake is determined by the manufacturer. Standard capacities of this machine are from 12- to 22-gauge sheet metal, and bending lengths are from 3 to 12 feet. The bending capacity of the brake is determined by the bending edge thickness of the various bending leaf bars.

Most metals have a tendency to return to their normal

Figure 4-56. *Bar folder.*

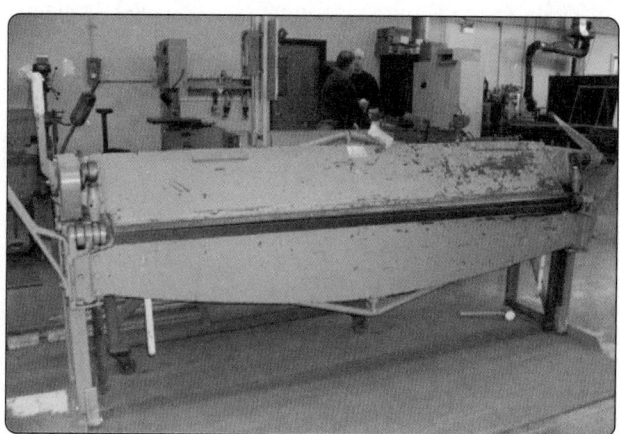

Figure 4-57. *Cornice brake.*

shape—a characteristic known as spring-back. If the cornice brake is set for a 90° bend, the metal bent probably forms an angle of about 87° to 88°. Therefore, if a bend of 90° is desired, set the cornice brake to bend an angle of about 93° to allow for spring-back.

Box & Pan Brake (Finger Brake)

The box and pan brake, often called the finger brake because it is equipped with a series of steel fingers of varying widths, lacks the solid upper jaw of the cornice brake. *[Figure 4-58]* The box and pan brake can be used to do everything that the cornice brake can do, as well as several things the cornice brake cannot do.

The box and pan brake is used to form boxes, pans, and other similar shaped objects. If these shapes were formed on a cornice brake, part of the bend on one side of the box would have to be straightened in order to make the last bend. With a finger brake, simply remove the fingers that are in the way and use only the fingers required to make the bend. The fingers are secured to the upper leaf by thumbscrews. All the fingers not removed for an operation must be securely seated and firmly tightened before the brake is used. The radius of the nose on the clamping fingers is usually rather small and frequently requires nose radius shims to be custom made for the total length of the bend.

Press Brake

Since most cornice brakes and box and pan brakes are limited to a maximum forming capacity of approximately 0.090-inch annealed aluminum, 0.063-inch 7075T6, or 0.063-inch stainless steel, operations that require the forming of thicker and more complex parts use a press brake. *[Figure 4-59]* The press brake is the most common machine tool used to bend sheet metal and applies force via mechanical and/or hydraulic components to shape the sheet metal between the punch and die. Narrow U-channels (especially with long legs) and hat channel stringers can be formed on the press brake by using special gooseneck or offset dies. Special urethane lower dies are useful for forming channels and stringers.

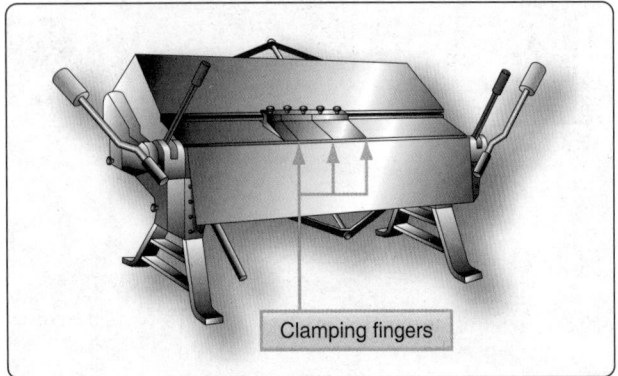

Figure 4-58. *Box and pan brake.*

Figure 4-59. *Press brake.*

Power press brakes can be set up with back stops (some are computer controlled) for high volume production. Press brake operations are usually done manually and require skill and knowledge of safe use.

Slip Roll Former

With the exception of the brake, the slip roll is probably used more than any other machine in the shop. *[Figure 4-60]* This machine is used to form sheets into cylinders or other straight curved surfaces. It consists of right and left end frames with three solid rolls mounted in between. Gears, which are operated by either a hand crank or a power drive, connect the two gripping rolls. These rolls can be adjusted to the thickness of the metal by using the two adjusting screws located on the bottom of each frame. The two most common of these forming machines are the slip roll former and the rotary former. Available in various sizes and capabilities, these machines come in manual or powered versions.

The slip roll former in *Figure 4-60* is manually operated and consists of three rolls, two housings, a base, and a handle. The handle turns the two front rolls through a system of gears enclosed in the housing. The front rolls serve as feeding, or gripping, rolls. The rear roll gives the proper curvature to the work. When the metal is started into the machine, the rolls grip the metal and carry it to the rear roll, which curves it. The desired radius of a bend is obtained by the rear roll. The bend radius of the part can be checked as the forming operation progresses by using a circle board or radius gauge. The gauges can be made by cutting a piece of material to the required finished radius and comparing it to the radius being formed by the rolling operation. On some material, the forming operation must be performed by passing the material through the rolls several times with progressive settings on the forming roll. On most machines, the top roll can be released on one end, permitting the formed sheet to be removed from the machine without distortion.

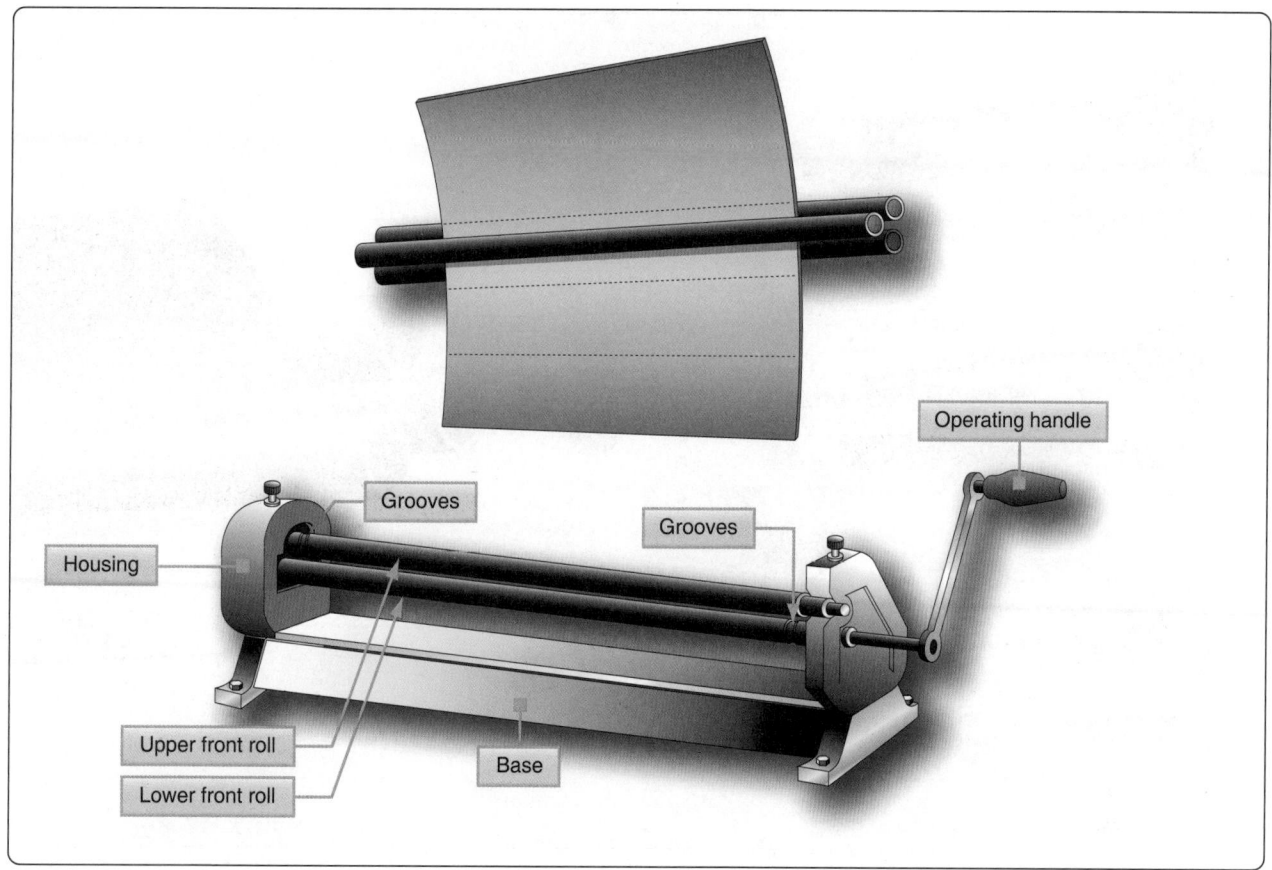

Figure 4-60. *Slip roll former.*

The front and rear rolls are grooved to permit forming of objects that have wired edges. The upper roll is equipped with a release that permits easy removal of the metal after it has been formed. When using the slip roll former, the lower front roll must be raised or lowered before inserting the sheet of metal. If the object has a folded edge, there must be enough clearance between the rolls to prevent flattening the fold. If a metal requiring special care (such as aluminum) is being formed, the rolls must be clean and free of imperfections.

The rear roll must be adjusted to give the proper curvature to the part being formed. There are no gauges that indicate settings for a specific diameter; therefore, trial and error settings must be used to obtain the desired curvature. The metal should be inserted between the rolls from the front of the machine. Start the metal between the rolls by rotating the operating handle in a clockwise direction. A starting edge is formed by holding the operating handle firmly with the right hand and raising the metal with the left hand. The bend of the starting edge is determined by the diameter of the part being formed. If the edge of the part is to be flat or nearly flat, a starting edge should not be formed.

Ensure that fingers and loose clothing are clear of the rolls before the actual forming operation is started. Rotate the operating handle until the metal is partially through the rolls and change the left hand from the front edge of the sheet to the upper edge of the sheet. Then, roll the remainder of the sheet through the machine. If the desired curvature is not obtained, return the metal to its starting position by rotating the handle counterclockwise. Raise or lower the rear roll and roll the metal through the rolls again. Repeat this procedure until the desired curvature is obtained, then release the upper roll and remove the metal. If the part to be formed has a tapered shape, the rear roll should be set so that the rolls are closer together on one end than on the opposite end. The amount of adjustment must be determined by experimentation. If the job being formed has a wired edge, the distance between the upper and lower rolls and the distance between the lower front roll and the rear roll should be slightly greater at the wired end than at the opposite end. *[Figure 4-61]*

Rotary Machine

The rotary machine is used on cylindrical and flat sheet metal to shape the edge or to form a bead along the edge. *[Figure 4-62]* Various shaped rolls can be installed on the rotary machine to perform these operations. The rotary machine works best with thinner annealed materials.

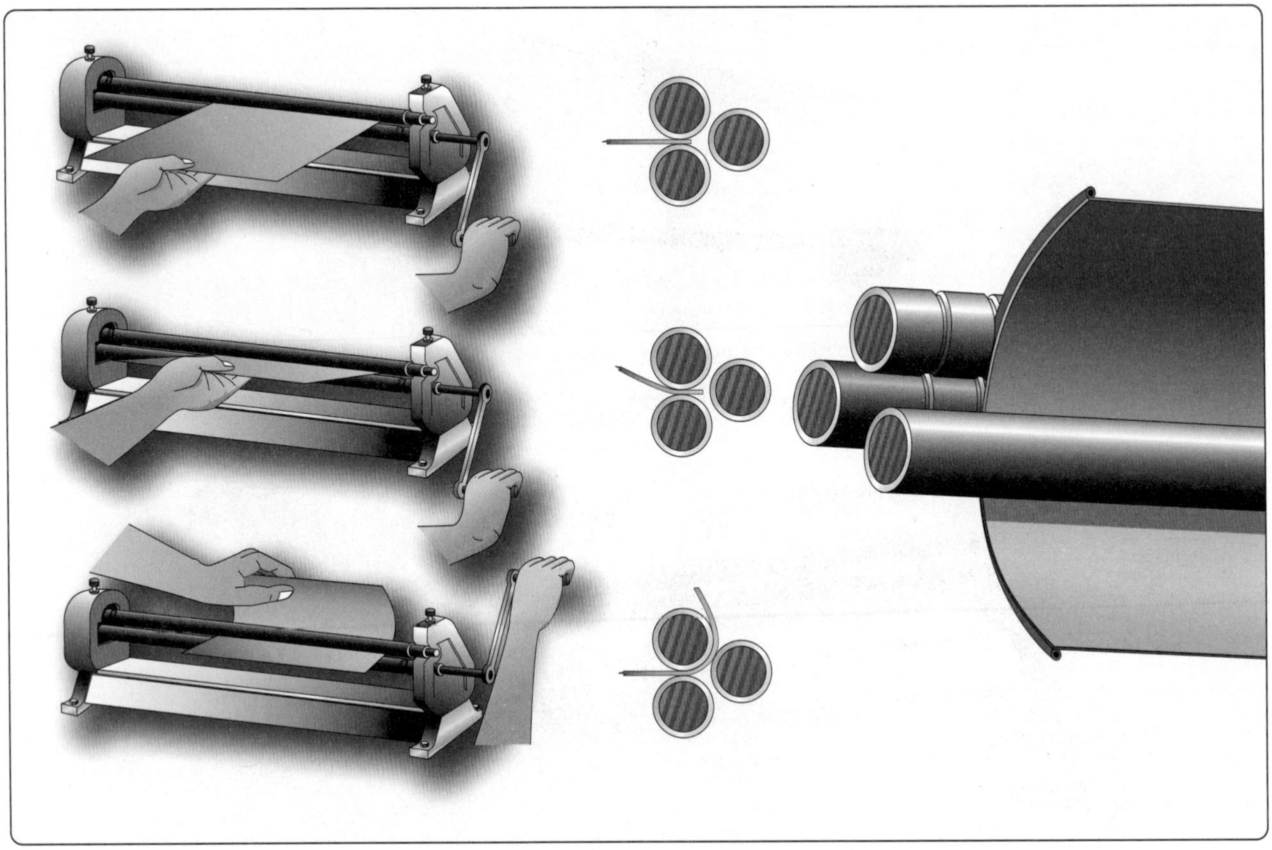

Figure 4-61. *Slip roll operation.*

Figure 4-62. *Rotary machine.*

Stretch Forming

In the process of stretch forming, a sheet of metal is shaped by stretching it over a formed block to just beyond the elastic limit where permanent set takes place with a minimum amount of spring-back. To stretch the metal, the sheet is rigidly clamped at two opposite edges in fixed vises. Then, the metal is stretched by moving a ram that carries the form block against the sheet with the pressure from the ram causing the material to stretch and wrap to the contour of the form block.

Stretch forming is normally restricted to relatively large parts with large radii of curvature and shallow depth, such as contoured skin. Uniform contoured parts produced at a faster speed give stretch forming an advantage over hand formed parts. Also, the condition of the material is more uniform than that obtained by hand forming.

Drop Hammer

The drop hammer forming process produces shapes by the progressive deformation of sheet metal in matched dies under the repetitive blows of a gravity-drop hammer or a power-drop hammer. The configurations most commonly formed by the process include shallow, smoothly contoured double-curvature parts, shallow-beaded parts, and parts with irregular and comparatively deep recesses. Small quantities of cup-shaped and box-shaped parts, curved sections, and contoured flanged parts are also formed. Drop hammer forming is not a precision forming method and cannot provide tolerances as close as 0.03-inch to 0.06-inch. Nevertheless, the process is often used for sheet metal parts, such as aircraft components, that undergo frequent design changes, or for which there is a short run expectancy.

Hydropress Forming

The rubber pad hydropress can be utilized to form many varieties of parts from aluminum and its alloys with relative ease. Phenolic, masonite, kirksite, and some types of hard setting moulding plastic have been used successfully as form blocks to press sheet metal parts, such as ribs, spars, fans, etc. To perform a press forming operation:

1. Cut a sheet metal blank to size and deburr edges.

2. Set the form block (normally male) on the lower press platen.

3. Place the prepared sheet metal blank (with locating pins to prevent shifting of the blank when the pressure is applied).

4. Lower or close the rubber pad-filled press head over the form block and the rubber envelope.

5. The form block forces the blank to conform to its contour.

Hydropress forming is usually limited to relatively flat parts with flanges, beads, and lightening holes. However, some types of large radii contoured parts can be formed by a combination of hand forming and pressing operations.

Spin Forming

In spin forming, a flat circle of metal is rotated at a very high speed to shape a seamless, hollow part using the combined forces of rotation and pressure. For example, a flat circular blank such as an aluminum disc, is mounted in a lathe in conjunction with a form block (usually made of hardwood). As the aircraft technician revolves the disc and form block together at high speeds, the disc is molded to the form block by applying pressure with a spinning stick or tool. It provides an economical alternative to stamping, casting, and many other metal forming processes. Propeller spinners are sometimes fabricated with this technique.

Aluminum soap, tallow, or ordinary soap can be used as a lubricant. The best adapted materials for spinning are the softer aluminum alloys, but other alloys can be used if the shape to be spun is not excessively deep or if the spinning is done in stages utilizing intermediate annealing to remove the effect of strain hardening that results from the spinning operation. Hot forming is used in some instances when spinning thicker and harder alloys. [Figure 4-63]

Figure 4-63. *Spin forming.*

Forming with an English Wheel

The English wheel, a popular type of metal forming tool used to create double curves in metal, has two steel wheels between which metal is formed. [Figure 4-64] Keep in mind that the English wheel is primarily a stretching machine, so it stretches and thins the metal before forming it into the desired shape. Thus, the operator must be careful not to over-stretch the metal.

To use the English wheel, place a piece of sheet metal between the wheels (one above and one below the metal). Then, roll the wheels against one another under a pre-adjusted

Figure 4-64. *English wheel.*

pressure setting. Steel or aluminum can be shaped by pushing the metal back and forth between the wheels. Very little pressure is needed to shape the panel, which is stretched or raised to the desired shape. It is important to work slowly and gradually curve the metal into the desired shape. Monitor the curvature with frequent references to the template.

The English wheel is used for shaping low crowns on large panels and polishing or planishing (to smooth the surface of a metal by rolling or hammering it) parts that have been formed with power hammers or hammer and shot bag.

Piccolo Former

The Piccolo former is used for cold forming and rolling sheet metal and other profile sections (extrusions). *[Figure 4-65]* The position of the ram is adjustable in height by means of either a handwheel or a foot pedal that permits control of the working pressure. Be sure to utilize the adjusting ring situated in the machine head to control the maximum working pressure. The forming tools are located in the moving ram and the lower tool holder. Depending on the variety of forming tools included, the operator can perform such procedures as forming edges, bending profiles, removing wrinkles, spot shrinking to remove buckles and dents, or expanding dome sheet metal. Available in either fiberglass (to prevent marring the surface) or steel (for working harder materials) faces, the tools are the quick-change type.

Shrinking & Stretching Tools

Shrinking Tools

Shrinking dies repeatedly clamp down on the metal, then shift inward. *[Figure 4-66]* This compresses the material between the dies, which actually slightly increases the thickness of the metal. Strain hardening takes place during this process, so it is best to set the working pressure high enough to complete the shape rather quickly (eight passes could be considered excessive).

Caution: Avoid striking a die on the radius itself when forming a curved flange. This damages the metal in the radius

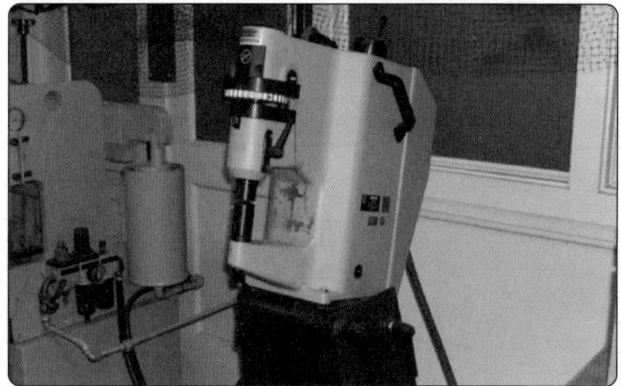

Figure 4-65. *Piccolo former.*

Figure 4-66. *Shrinking and stretching tools.*

and decreases the angle of bend.

Stretching Tools

Stretching dies repeatedly clamp down on the surface and then shift outward. This stretches the metal between the dies, which decreases the thickness in the stretched area. Striking the same point too many times weakens and eventually cracks the part. It is advantageous to deburr or even polish the edges of a flange that must undergo even moderate stretching to avoid crack formation. Forming flanges with existing holes causes the holes to distort and possibly crack or substantially weaken the flange.

Manual Foot-Operated Sheet Metal Shrinker

The manual foot-operated sheet metal shrinker operates very similarly to the Piccolo former though it only has two primary functions: shrinking and stretching. The only dies available are steel faced and therefore tend to mar the surface of the metal. When used on aluminum, it is necessary to gently blend out the surface irregularities (primarily in the cladding), then treat and paint the part.

Since this is a manual machine, it relies on leg power, as the operator repeatedly steps on the foot pedal. The more force is applied, the more stresses are concentrated at that single point. It yields a better part with a series of smaller stretches (or shrinks) than with a few intense ones. Squeezing the dies over the radius damages the metal and flattens out some of the bend. It may be useful to tape a thick piece of plastic or micarta to the opposite leg to shim the radius of the angle away from the clamping area of the dies.

Note: Watch the part change shape while slowly applying pressure. A number of small stretches works more effectively than one large one. If applying too much pressure, the metal has the tendency to buckle.

Hand-Operated Shrinker & Stretcher

The hand-operated shrinker and stretcher is similar to the manual foot-operated unit, except a handle is used to apply force to shrinking and stretching blocks. The dies are all metal and leave marks on aluminum that need to be blended out after the shrinking or stretching operation. *[Figure 4-67]*

Dollies & Stakes

Sheet metal is often formed or finished (planished) over anvils, available in a variety of shapes and sizes, called dollies and stakes. These are used for forming small, odd-shaped parts, or for putting on finishing touches for which a large machine may not be suited. Dollies are meant to be held in the hand, whereas stakes are designed to be supported by a flat cast iron bench plate fastened to the workbench. *[Figure 4-68]*

Most stakes have machined, polished surfaces that have been hardened. Use of stakes to back up material when chiseling, or when using any similar cutting tool, defaces the surface of the stake and makes it useless for finish work.

Hardwood Form Blocks

Hardwood form blocks can be constructed to duplicate practically any aircraft structural or nonstructural part. The wooden block or form is shaped to the exact dimensions and contour of the part to be formed.

V-Blocks

V-blocks made of hardwood are widely used in airframe metalwork for shrinking and stretching metal, particularly angles and flanges. The size of the block depends on the work being done and on personal preference. Although any type of hardwood is suitable, maple and ash are recommended for best results when working with aluminum alloys.

Figure 4-68. *Dollies and stakes.*

Shrinking Blocks

A shrinking block consists of two metal blocks and some device for clamping them together. One block forms the base and the other is cut away to provide space where the crimped material can be hammered. The legs of the upper jaw clamp the material to the base block on each side of the crimp to prevent the material from creeping away, but remains stationary while the crimp is hammered flat (being shrunk). This type of crimping block is designed to be held in a bench vise.

Shrinking blocks can be made to fit any specific need. The basic form and principle remain the same, even though the blocks may vary considerably in size and shape.

Sandbags

A sandbag is generally used as a support during the bumping process. A serviceable bag can be made by sewing heavy canvas or soft leather to form a bag of the desired size, and filling it with sand which has been sifted through a fine mesh screen.

Before filling canvas bags with sand, use a brush to coat the inside of the bag with softened paraffin or beeswax, which forms a sealing layer and prevents the sand from working

Figure 4-67. *Hand-operated shrinker and stretcher unit.*

through the pores of the canvas. Bags can also be filled with shot as an alternative to sand.

Sheet Metal Hammers & Mallets

The sheet metal hammer and the mallet are metal fabrication hand tools used for bending and forming sheet metal without marring or indenting the metal. The hammer head is usually made of high carbon, heat-treated steel, while the head of the mallet, which is usually larger than that of the hammer, is made of rubber, plastic, wood, or leather. In combination with a sandbag, V-blocks, and dies, sheet metal body hammers and mallets are used to form annealed metal. *[Figure 4-69]*

Sheet Metal Holding Devices

In order to work with sheet metal during the fabrication process, the aviation technician uses a variety of holding devices, such as clamps, vises, and fasteners to hold the work together. The type of operation being performed and the type of metal being used determine what type of the holding device is needed.

Clamps & Vises

Clamps and vises hold materials in place when it is not possible to handle a tool and the workpiece at the same time. A clamp is a fastening device with movable jaws that has opposing, often adjustable, sides or parts. An essential fastening device, it holds objects tightly together to prevent movement or separation. Clamps can be either temporary or permanent. Temporary clamps, such as the carriage clamp (commonly called the C-clamp), are used to position components while fixing them together.

C-Clamps

The C-clamp is shaped like a large C and has three main parts: threaded screw, jaw, and swivel head. *[Figure 4-70]* The swivel plate or flat end of the screw prevents the end from turning directly against the material being clamped. C-clamp

Figure 4-69. *Sheet metal mallet and hammers.*

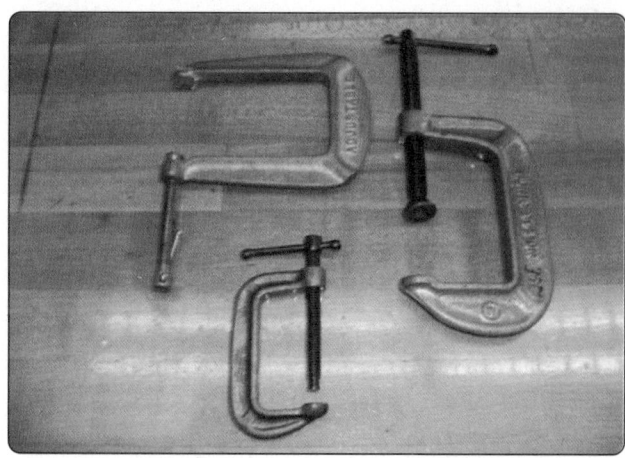

Figure 4-70. *C-clamps.*

size is measured by the dimension of the largest object the frame can accommodate with the screw fully extended. The distance from the center line of the screw to the inside edge of the frame or the depth of throat is also an important consideration when using this clamp. C-clamps vary in size from two inches upward. Since C-clamps can leave marks on aluminum, protect the aircraft covering with masking tape at the places where the C-clamp is used.

Vises

Vises are another clamping device that hold the workpiece in place and allow work to be done on it with tools such as saws and drills. The vise consists of two fixed or adjustable jaws that are opened or closed by a screw or a lever. The size of a vise is measured by both the jaw width and the capacity of the vise when the jaws are fully open. Vises also depend on a screw to apply pressure, but their textured jaws enhance gripping ability beyond that of a clamp.

Two of the most commonly used vises are the machinist's vise and the utility vise. *[Figure 4-71]* The machinist's vise has flat jaws and usually a swivel base, whereas the utility bench vise has scored, removable jaws and an anvil-faced back jaw. This vise holds heavier material than the machinist's vise and also grips pipe or rod firmly. The back jaw can be used as an anvil if the work being done is light. To avoid marring metal in the vise jaws, add some type of padding, such as a ready-made rubber jaw pad.

Reusable Sheet Metal Fasteners

Reusable sheet metal fasteners temporarily hold drilled sheet metal parts accurately in position for riveting or drilling. If sheet metal parts are not held tightly together, they separate while being riveted or drilled. The Cleco (also spelled Cleko) fastener is the most commonly used sheet metal holder. *[Figure 4-72]*

Figure 4-71. *A utility vise with swivel base and anvil.*

Figure 4-72. *Cleco fastener.*

Cleco Fasteners

The Cleco fastener consists of a steel cylinder body with a plunger on the top, a spring, a pair of step-cut locks, and a spreader bar. These fasteners come in six different sizes: $\frac{3}{32}$, $\frac{1}{8}$, $\frac{5}{32}$, $\frac{3}{16}$, $\frac{1}{4}$, and $\frac{3}{8}$-inch in diameter with the size stamped on the fastener. Color coding allows for easy size recognition. A special type of plier fits the six different sizes. When installed correctly, the reusable Cleco fastener keeps the holes in the separate sheets aligned.

Hex Nut & Wing Nut Temporary Sheet Fasteners

Hex nut and wing nut fasteners are used to temporarily fasten sheets of metal when higher clamp up pressure is required. *[Figure 4-73]* Hex nut fasteners provide up to 300 pounds of clamping force with the advantage of quick installation and removal with a hex nut runner. Wing nut sheet metal fasteners, characterized by wing shaped protrusions, not only provide

Figure 4-73. *Hex nut fastener.*

a consistent clamping force from 0 to 300 pounds, but the aircraft technician can turn and tighten these fasteners by hand. Cleco hex nut fasteners are identical to Cleco wing nut fasteners, but the Cleco hex nut can be used with pneumatic Cleco installers.

Aluminum Alloys

Aluminum alloys are the most frequently encountered type of sheet metal in aircraft repair. AC 43.13-1 Chapter 4, Metal Structure, Welding, and Brazing; Section 1, Identification of Metals (as revised) provides an in-depth discussion of all metal types. This section describes the aluminum alloys used in the forming processes discussed in the remainder of the chapter.

In its pure state, aluminum is lightweight, lustrous, and corrosion resistant. The thermal conductivity of aluminum is very high. It is ductile, malleable, and nonmagnetic. When combined with various percentages of other metals (generally copper, manganese, and magnesium), aluminum alloys that are used in aircraft construction are formed. Aluminum alloys are lightweight and strong. They do not possess the corrosion resistance of pure aluminum and are usually treated to prevent deterioration. Alclad™ aluminum is an aluminum alloy with a protective cladding of aluminum to improve its corrosion resistance.

To provide a visual means for identifying the various grades of aluminum and aluminum alloys, aluminum stock is usually marked with symbols such as a Government Specification Number, the temper or condition furnished, or the commercial code marking. Plate and sheet are usually marked with specification numbers or code markings in rows approximately five inches apart. Tubes, bars, rods, and extruded shapes are marked with specification numbers or code markings at intervals of three to five feet along the length of each piece.

The commercial code marking consists of a number that identifies the particular composition of the alloy. Additionally, letter suffixes designate the basic temper designations and subdivisions of aluminum alloys.

The aluminum and various aluminum alloys used in aircraft repair and construction are as follows:

- Aluminum designated by the symbol 1100 is used where strength is not an important factor, but where weight economy and corrosion resistance are desired. This aluminum is used for fuel tanks, cowlings, and oil tanks. It is also used for repairing wingtips and tanks. This material is weldable.

- Alloy 3003 is similar to 1100 and is generally used for the same purposes. It contains a small percentage of magnesium and is stronger and harder than 1100 aluminum.

- Alloy 2014 is used for heavy-duty forgings, plates, extrusions for aircraft fittings, wheels, and major structural components. This alloy is often used for applications requiring high strength and hardness, as well as for service at elevated temperatures.

- Alloy 2017 is used for rivets. This material is now in limited use.

- Alloy 2024, with or without Alclad™ coating, is used for aircraft structures, rivets, hardware, machine screw products, and other miscellaneous structural applications. In addition, this alloy is commonly used for heat-treated parts, airfoil and fuselage skins, extrusions, and fittings.

- Alloy 2025 is used extensively for propeller blades.

- Alloy 2219 is used for fuel tanks, aircraft skin, and structural components. This material has high fracture toughness and is readily weldable. Alloy 2219 is also highly resistant to stress corrosion cracking.

- Alloy 5052 is used where good workability, very good corrosion resistance, high fatigue strength, weldability, and moderate static strength are desired. This alloy is used for fuel, hydraulic, and oil lines.

- Alloy 5056 is used for making rivets and cable sheeting and in applications where aluminum comes into contact with magnesium alloys. Alloy 5056 is generally resistant to the most common forms of corrosion.

- Cast aluminum alloys are used for cylinder heads, crankcases, fuel injectors, carburetors, and landing wheels.

- Various alloys, including 3003, 5052, and 1100 aluminum, are hardened by cold working rather than by heat treatment. Other alloys, including 2017 and 2024, are hardened by heat treatment, cold working, or a combination of the two. Various casting alloys are hardened by heat treatment.

- Alloy 6061 is generally weldable by all commercial procedures and methods. It also maintains acceptable toughness in many cryogenic applications. Alloy 6061 is easily extruded and is commonly used for hydraulic and pneumatic tubing.

- Although higher in strength than 2024, alloy 7075 has a lower fracture toughness and is generally used in tension applications where fatigue is not critical. The T6 temper of 7075 should be avoided in corrosive environments. However, the T7351 temper of 7075 has excellent stress corrosion resistance and better fracture toughness than the T6 temper. The T76 temper is often used to improve the resistance of 7075 to exfoliate corrosion.

Structural Fasteners

Structural fasteners, used to join sheet metal structures securely, come in thousands of shapes and sizes with many of them specialized and specific to certain aircraft. Since some structural fasteners are common to all aircraft, this section focuses on the more frequently used fasteners. For the purposes of this discussion, fasteners are divided into two main groups: solid shank rivets and special purpose fasteners that include blind rivets.

Solid Shank Rivet

The solid shank rivet is the most common type of rivet used in aircraft construction. Used to join aircraft structures, solid shank rivets are one of the oldest and most reliable types of fastener. Widely used in the aircraft manufacturing industry, solid shank rivets are relatively low-cost, permanently installed fasteners. They are faster to install than bolts and nuts since they adapt well to automatic, high-speed installation tools. Rivets should not be used in thick materials or in tensile applications, as their tensile strengths are quite low relative to their shear strength. The longer the total grip length (the total thickness of sheets being joined), the more difficult it becomes to lock the rivet.

Riveted joints are neither airtight nor watertight unless special seals or coatings are used. Since rivets are permanently installed, they must be removed by drilling them out, a laborious task.

Description

Before installation, the rivet consists of a smooth cylindrical shaft with a factory head on one end. The opposite end is called the bucktail. To secure two or more pieces of sheet metal together, the rivet is placed into a hole cut just a bit larger in diameter than the rivet itself. Once placed in this predrilled hole, the bucktail is upset or deformed by any of several methods from hand-held hammers to pneumatically driven squeezing tools. This action causes the rivet to expand about 1½ times the original shaft diameter, forming a second head that firmly holds the material in place.

Rivet Head Shape

Solid rivets are available in several head shapes, but the universal (also known as protruding head) and the 100° countersunk head are the most commonly used in aircraft structures. Universal head rivets were developed specifically for the aircraft industry and designed as a replacement for both the round and brazier head rivets. These rivets replaced all protruding head rivets and are used primarily where the protruding head has no aerodynamic significance. They have a flat area on the head, a head diameter twice the shank diameter, and a head height approximately 42.5 percent of the shank diameter. *[Figure 4-74]*

The countersunk head angle can vary from 60° to 120°, but the 100° has been adopted as standard because this head style provides the best possible compromise between tension/shear strength and flushness requirements. This rivet is used where flushness is required because the rivet is flat-topped and undercut to allow the head to fit into a countersunk or dimpled hole. The countersunk rivet is primarily intended for use when aerodynamics smoothness is critical, such as on the external surface of a high-speed aircraft.

Typically, rivets are fabricated from aluminum alloys, such as 2017-T4, 2024-T4, 2117-T4, 7050, and 5056. Titanium, nickel-based alloys, such as Monel® (corrosion-resistant steel), mild steel or iron, and copper rivets are also used for rivets in certain cases.

Rivets are available in a wide variety of alloys, head shapes, and sizes and have a wide variety of uses in aircraft structure. Rivets that are satisfactory for one part of the aircraft are often unsatisfactory for another part. Therefore, it is important that an aircraft technician know the strength and driving properties of the various types of rivets and how to identify them, as well as how to drive or install them.

Solid rivets are classified by their head shape, by the material

from which they are manufactured, and by their size. Identification codes used are derived from a combination of the Military Standard (MS) and National Aerospace Standard (NAS) systems, as well as an older classification system known as AN for Army/Navy. For example, the prefix MS identifies hardware that conforms to written military standards. A letter or letters following the head-shaped code identify the material or alloy from which the rivet was made. The alloy code is followed by two numbers separated by a dash. The first number is the numerator of a fraction, which specifies the shank diameter in thirty-seconds of an inch. The second number is the numerator of a fraction in sixteenths of an inch and identifies the length of the rivet. Rivet head shapes and their identifying code numbers are shown in *Figure 4-75*.

The most frequently used repair rivet is the AD rivet because it can be installed in the received condition. Some rivet alloys, such as DD rivets (alloy 2024-T4), are too hard to drive in the received condition and must be annealed before they can be installed. Typically, these rivets are annealed and stored in a freezer to retard hardening, which has led to the nickname "ice box rivets." They are removed from the freezer just prior to use. Most DD rivets have been replaced by E-type rivets which can be installed in the received condition.

The head type, size, and strength required in a rivet are governed by such factors as the kind of forces present at the point riveted, the kind and thickness of the material to be riveted, and the location of the part on the aircraft. The type of head needed for a particular job is determined by where it is to be installed. Countersunk head rivets should be used where a smooth aerodynamic surface is required. Universal head rivets may be used in most other areas.

The size (or diameter) of the selected rivet shank should correspond in general to the thickness of the material being riveted. If an excessively large rivet is used in a thin material, the force necessary to drive the rivet properly causes an undesirable bulging around the rivet head. On the other hand,

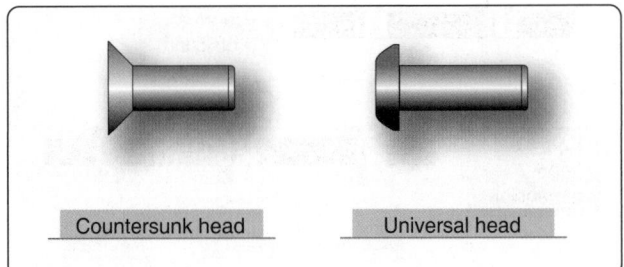

Figure 4-74. *Solid shank rivet styles.*

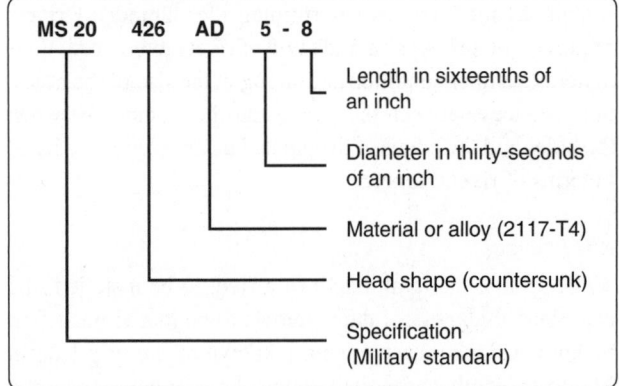

Figure 4-75. *Rivet head shapes and their identifying code numbers.*

if an excessively small rivet diameter is selected for thick material, the shear strength of the rivet is not great enough to carry the load of the joint. As a general rule, the rivet diameter should be at least two and a half to three times the thickness of the thicker sheet. Rivets most commonly chosen in the assembly and repair of aircraft range from ³⁄₃₂-inch to ³⁄₈-inch in diameter. Ordinarily, rivets smaller than ³⁄₃₂-inch in diameter are never used on any structural parts that carry stresses.

The proper sized rivets to use for any repair can also be determined by referring to the rivets (used by the manufacturer) in the next parallel row inboard on the wing or forward on the fuselage. Another method of determining the size of rivets to be used is to multiply the skin's thickness by 3 and use the next larger size rivet corresponding to that figure. For example, if the skin is 0.040 inch thick, multiply 0.040 inch by 3 to get 0.120 inch and use the next larger size of rivet, ⅛-inch (0.125 inch).

When rivets are to pass completely through tubular members, select a rivet diameter equivalent to at least ⅛ the outside diameter of the tube. If one tube sleeves or fits over another, take the outside diameter of the outside tube and use one-eighth of that distance as the minimum rivet diameter. A good practice is to calculate the minimum rivet diameter and then use the next larger size rivet.

Whenever possible, select rivets of the same alloy number as the material being riveted. For example, use 1100 and 3003 rivets on parts fabricated from 1100 and 3003 alloys, and 2117-1 and 2017-T rivets on parts fabricated from 2017 and 2024 alloys.

The size of the formed head is the visual standard of a proper rivet installation. The minimum and maximum sizes, as well as the ideal size, are shown in *Figure 4-76*.

Installation of Rivets

Repair Layout
Repair layout involves determining the number of rivets required, the proper size and style of rivets to be used, their material, temper condition and strength, the size of the holes, the distances between the holes, and the distance between the holes and the edges of the patch. Distances are measured in terms of rivet diameter.

Rivet Length
To determine the total length of a rivet to be installed, the combined thickness of the materials to be joined must first be known. This measurement is known as the grip length. The total length of the rivet equals the grip length plus the amount of rivet shank needed to form a proper shop head.

The properly formed shop head equals one and a half times the diameter of the rivet shank. Where A is total rivet length, B is grip length, and C is the length of the material needed to form a shop head, this formula can be represented as A = B + C. *[Figure 4-76]*

Rivet Strength
For structural applications, the strength of the replacement rivets is of primary importance. *[Figure 4-77]* Replace rivets with those of the same size and strength whenever possible. If the rivet hole becomes enlarged, deformed, or otherwise damaged; drill or ream the hole for the next larger size rivet. However, make sure that the edge distance and spacing is not less than minimums listed in the next paragraph. Rivets may not be replaced by a type having lower strength properties, unless the lower strength is adequately compensated by an increase in size or a greater number of rivets. For example, it is acceptable to replace 2017 rivets of 3/16 inch diameter or less, and 2024 rivets of 5/32 inch diameter or less with 2117 rivets for general repairs, provided the replacement rivets are 1/32 inch greater in diameter than the rivets they replace.

The 2117-T rivet is used for general repair work, since it requires no heat treatment, is fairly soft and strong, and is highly corrosion resistant when used with most types of alloys. Always consult the maintenance manual for correct rivet type and material. The type of rivet head to select for a particular repair job can be determined by referring to the

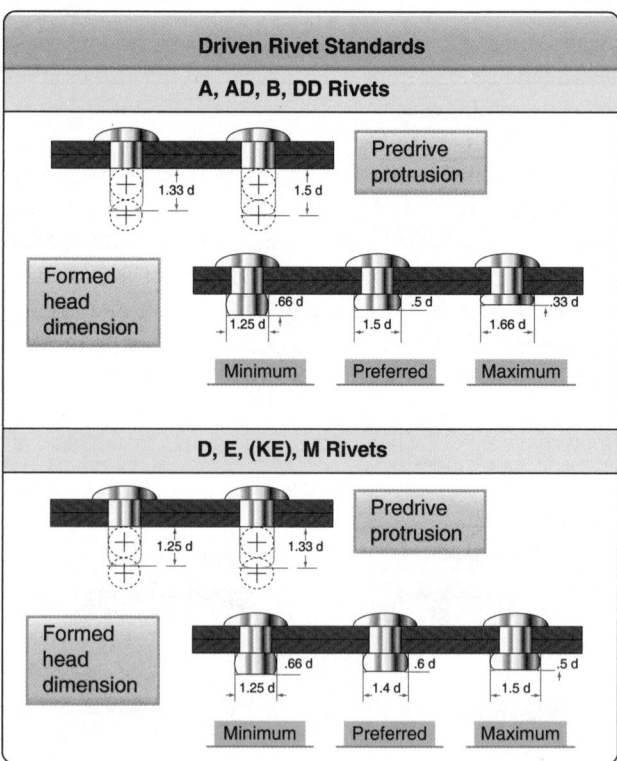

Figure 4-76. *Rivet formed head dimensions.*

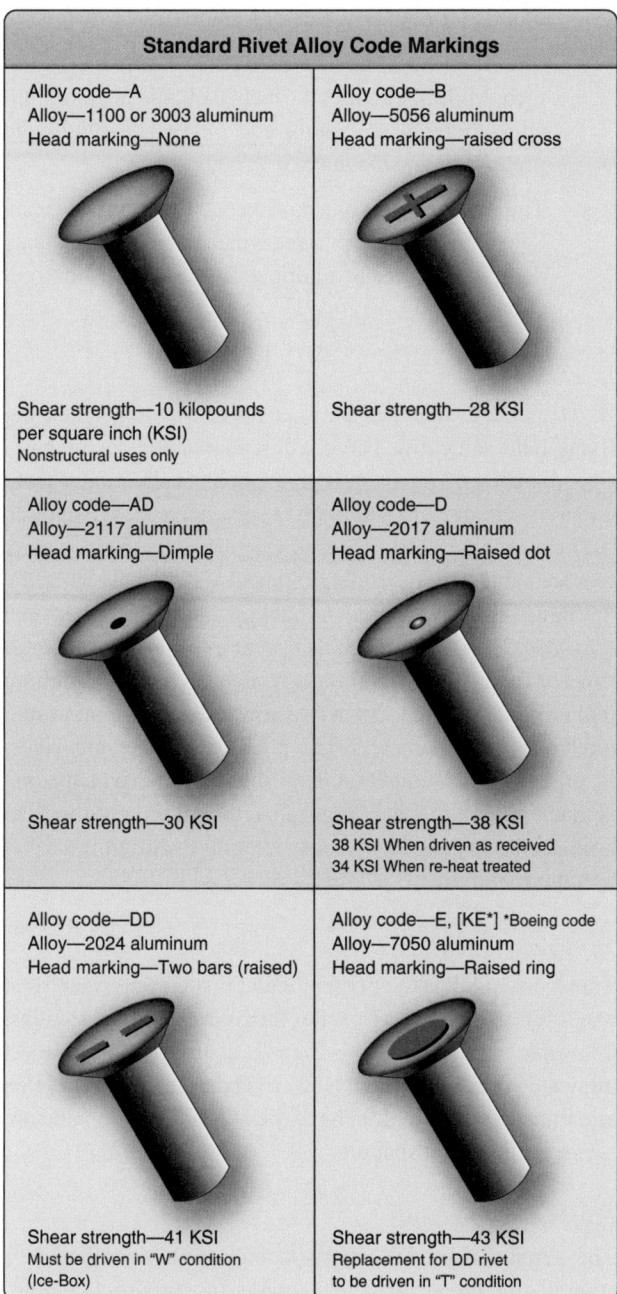

Standard Rivet Alloy Code Markings

Alloy code—A Alloy—1100 or 3003 aluminum Head marking—None	Alloy code—B Alloy—5056 aluminum Head marking—raised cross
Shear strength—10 kilopounds per square inch (KSI) Nonstructural uses only	Shear strength—28 KSI
Alloy code—AD Alloy—2117 aluminum Head marking—Dimple	Alloy code—D Alloy—2017 aluminum Head marking—Raised dot
Shear strength—30 KSI	Shear strength—38 KSI 38 KSI When driven as received 34 KSI When re-heat treated
Alloy code—DD Alloy—2024 aluminum Head marking—Two bars (raised)	Alloy code—E, [KE*] *Boeing code Alloy—7050 aluminum Head marking—Raised ring
Shear strength—41 KSI Must be driven in "W" condition (Ice-Box)	Shear strength—43 KSI Replacement for DD rivet to be driven in "T" condition

Figure 4-77. *Rivet alloy strength.*

type used within the surrounding area by the manufacturer. A general rule to follow on a flush-riveted aircraft is to apply flush rivets on the upper surface of the wing and stabilizers, on the lower leading edge back to the spar, and on the fuselage back to the high point of the wing. Use universal head rivets in all other surface areas. Whenever possible, select rivets of the same alloy number as the material being riveted.

Stresses Applied to Rivets

Shear is one of the two stresses applied to rivets. The shear strength is the amount of force required to cut a rivet that holds two or more sheets of material together. If the rivet holds two parts, it is under single shear; if it holds three sheets or parts, it is under double shear. To determine the shear strength, the diameter of the rivet to be used must be found by multiplying the thickness of the skin material by 3. For example, a material thickness of 0.040 inch multiplied by 3 equals 0.120 inch. In this case, the rivet diameter selected would be ⅛ (0.125) inch.

Tension is the other stress applied to rivets. The resistance to tension is called bearing strength and is the amount of tension required to pull a rivet through the edge of two sheets riveted together or to elongate the hole.

Rivet Spacing

Rivet spacing is measured between the centerlines of rivets in the same row. The minimum spacing between protruding head rivets shall not be less than 3½ times the rivet diameter. The minimum spacing between flush head rivets shall not be less than 4 times the diameter of the rivet. These dimensions may be used as the minimum spacing except when specified differently in a specific repair procedure or when replacing existing rivets.

On most repairs, the general practice is to use the same rivet spacing and edge distance (distance from the center of the hole to the edge of the material) that the manufacturer used in the area surrounding the damage. The SRM for the particular aircraft may also be consulted. Aside from this fundamental rule, there is no specific set of rules that governs spacing of rivets in all cases. However, there are certain minimum requirements that must be observed.

- When possible, rivet edge distance, rivet spacing, and distance between rows should be the same as that of the original installation.

- When new sections are to be added, the edge distance measured from the center of the rivet should never be less than 2 times the diameter of the shank; the distance between rivets or pitch should be at least 3 times the diameter; and the distance between rivet rows should never be less than 2½ times the diameter.

Figure 4-78 illustrates acceptable ways of laying out a rivet pattern for a repair.

Edge Distance

Edge distance, also called edge margin by some manufacturers, is the distance from the center of the first rivet to the edge of the sheet. It should not be less than 2 or more than 4 rivet diameters and the recommended edge distance is about 2½ rivet diameters. The minimum edge distance for universal rivets is 2 times the diameter of the rivet; the minimum edge distance for countersunk rivets is 2½ times the diameter of the

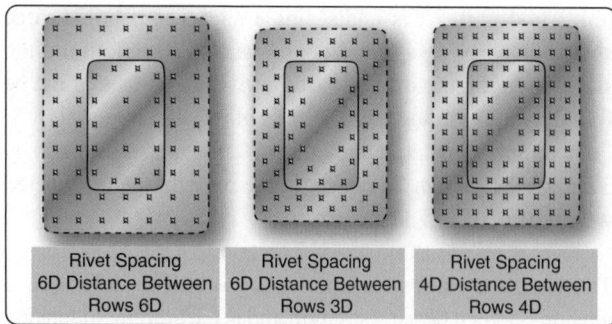

Figure 4-78. *Acceptable rivet patterns.*

rivet. If rivets are placed too close to the edge of the sheet, the sheet may crack or pull away from the rivets. If they are spaced too far from the edge, the sheet is likely to turn up at the edges. *[Figure 4-79]*

It is good practice to lay out the rivets a little further from the edge so that the rivet holes can be oversized without violating the edge distance minimums. Add $\frac{1}{16}$-inch to the minimum edge distance or determine the edge distance using the next size of rivet diameter.

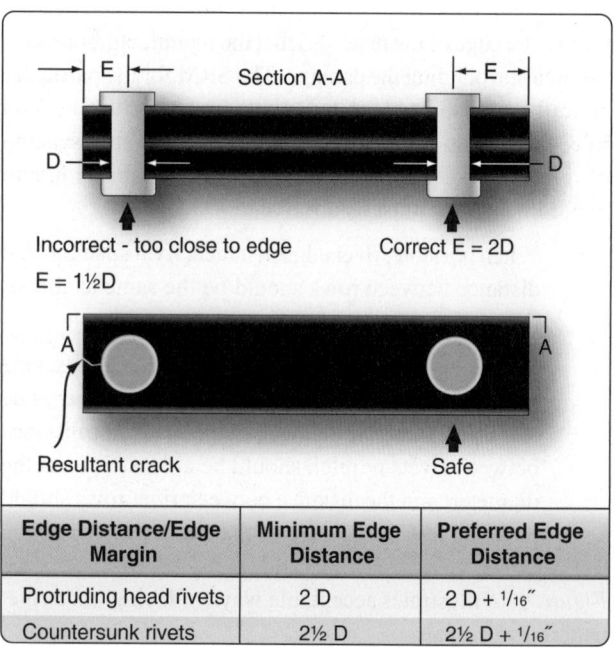

Figure 4-79. *Minimum edge distance.*

Edge Distance/Edge Margin	Minimum Edge Distance	Preferred Edge Distance
Protruding head rivets	2 D	2 D + $\frac{1}{16}$″
Countersunk rivets	2½ D	2½ D + $\frac{1}{16}$″

Two methods for obtaining edge distance:

- The rivet diameter of a protruding head rivet is $\frac{3}{32}$-inch. Multiply 2 times $\frac{3}{32}$-inch to obtain the minimum edge distance, $\frac{3}{16}$-inch, add $\frac{1}{16}$-inch to yield the preferred edge distance of $\frac{1}{4}$-inch.

- The rivet diameter of a protruding head rivet is $\frac{3}{32}$-inch. Select the next size of rivet, which is $\frac{1}{8}$-inch. Calculate the edge distance by multiplying 2 times $\frac{1}{8}$-inch to get $\frac{1}{4}$-inch.

Rivet Pitch

Rivet pitch is the distance between the centers of neighboring rivets in the same row. The smallest allowable rivet pitch is 3 rivet diameters. The average rivet pitch usually ranges from 4 to 6 rivet diameters, although in some instances rivet pitch could be as large as 10 rivet diameters. Rivet spacing on parts that are subjected to bending moments is often closer to the minimum spacing to prevent buckling of the skin between the rivets. The minimum pitch also depends on the number of rows of rivets. One-and three-row layouts have a minimum pitch of 3 rivet diameters, a two-row layout has a minimum pitch of 4 rivet diameters. The pitch for countersunk rivets is larger than for universal head rivets. If the rivet spacing is made at least $\frac{1}{16}$-inch larger than the minimum, the rivet hole can be oversized without violating the minimum rivet spacing requirement. *[Figure 4-80]*

Transverse Pitch

Transverse pitch is the perpendicular distance between rivet rows. It is usually 75 percent of the rivet pitch. The smallest allowable transverse pitch is 2½ rivet diameters. The smallest allowable transverse pitch is 2½ rivet diameters. Rivet pitch and transverse pitch often have the same dimension and are simply called rivet spacing.

Rivet Layout Example

The general rules for rivet spacing, as it is applied to a straight-row layout, are quite simple. In a one-row layout, find the edge distance at each end of the row and then lay off the rivet pitch (distance between rivets), as shown in *Figure 4-81*. In a two-row layout, lay off the first row, place the second row a distance equal to the transverse pitch from the first row, and then lay off rivet spots in the second row so that they fall midway between those in the first row. In the

Rivet Spacing	Minimum Spacing	Preferred Spacing
1 and 3 rows protruding head rivet layout	3D	3D + 1/16″
2 row protruding head rivet layout	4D	4D + 1/16″
1 and 3 rows countersunk head rivet layout	3/1/2D	3/1/2D + 1/16″
2 row countersunk head rivet layout	4/1/2D	4/1/2D + 1/16″

Figure 4-80. *Rivet spacing.*

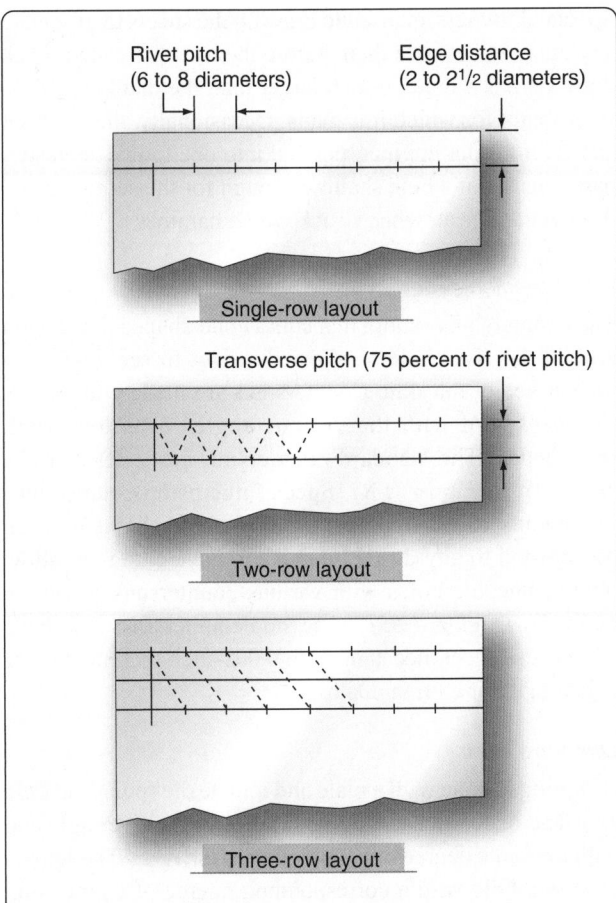

Rivet pitch
(6 to 8 diameters)

Edge distance
(2 to 2½ diameters)

Single-row layout

Transverse pitch (75 percent of rivet pitch)

Two-row layout

Three-row layout

Figure 4-81. *Rivet layout.*

three-row layout, first lay off the first and third rows, then use a straightedge to determine the second row rivet spots.

When splicing a damaged tube, and the rivets pass completely through the tube, space the rivets four to seven rivet diameters apart if adjacent rivets are at right angles to each other, and space them five to seven rivet diameters apart if the rivets are parallel to each other. The first rivet on each side of the joint should be no less than 2½ rivet diameters from the end of the sleeve.

Rivet Installation Tools

The various tools needed in the normal course of driving and upsetting rivets include drills, reamers, rivet cutters or nippers, bucking bars, riveting hammers, draw sets, dimpling dies or other types of countersinking equipment, rivet guns, and squeeze riveters. C-clamps, vises, and other fasteners used to hold sheets together when riveting were discussed earlier in the chapter. Other tools and equipment needed in the installation of rivets are discussed in the following paragraphs.

Hand Tools

A variety of hand tools are used in the normal course of

driving and upsetting rivets. They include rivet cutters, bucking bars, hand riveters, countersinks, and dimpling tools.

Rivet Cutter

The rivet cutter is used to trim rivets when rivets of the required length are unavailable. *[Figure 4-82]* To use the rotary rivet cutter, insert the rivet in the correct hole, place the required number of shims under the rivet head, and squeeze the cutter as if it were a pair of pliers. Rotation of the discs cuts the rivet to give the right length, which is determined by the number of shims inserted under the head. When using a large rivet cutter, place it in a vise, insert the rivet in the proper hole, and cut by pulling the handle, which shears off the rivet. If regular rivet cutters are not available, diagonal cutting pliers can be used as a substitute cutter.

Bucking Bar

The bucking bar, sometimes called a dolly, bucking iron, or bucking block, is a heavy chunk of steel whose countervibration during installation contributes to proper rivet installation. They come in a variety of shapes and sizes, and their weights ranges from a few ounces to 8 or 10 pounds, depending upon the nature of the work. Bucking bars are most often made from low-carbon steel that has been case hardened or alloy bar stock. Those made of better grades of steel last longer and require less reconditioning.

Bucking faces must be hard enough to resist indentation and remain smooth, but not hard enough to shatter. Sometimes, the more complicated bars must be forged or built up by welding. The bar usually has a concave face to conform to the shape of the shop head to be made. When selecting a bucking bar, the first consideration is shape. *[Figure 4-83]* If the bar does not have the correct shape, it deforms the rivet head; if the bar is too light, it does not give the necessary bucking weight, and the material may become bulged toward the shop head. If the bar is too heavy, its weight and the bucking force may cause the material to bulge away from the shop head.

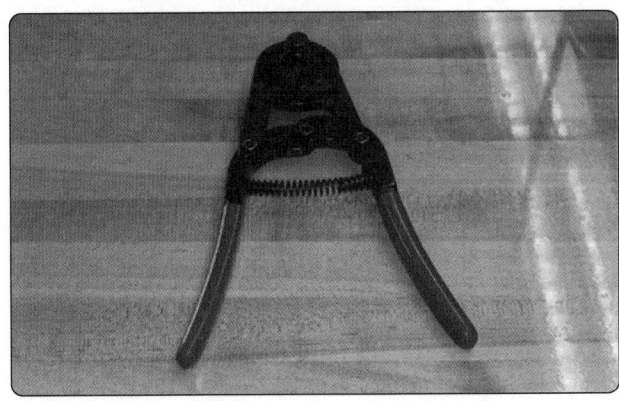

Figure 4-82. *Rivet cutters.*

Figure 4-83. *Bucking bars.*

This tool is used by holding it against the shank end of a rivet while the shop head is being formed. Always hold the face of the bucking bar at right angles to the rivet shank. Failure to do so causes the rivet shank to bend with the first blows of the rivet gun and causes the material to become marred with the final blows. The bucker must hold the bucking bar in place until the rivet is completely driven. If the bucking bar is removed while the gun is in operation, the rivet set may be driven through the material. Allow the weight of the bucking bar to do most of the work and do not bear down too heavily on the shank of the rivet. The operator's hands merely guide the bar and supply the necessary tension and rebound action. Coordinated bucking allows the bucking bar to vibrate in unison with the gun set. With experience, a high degree of skill can be developed.

Defective rivet heads can be caused by lack of proper vibrating action, the use of a bucking bar that is too light or too heavy, and failure to hold the bucking bar at right angles to the rivet. The bars must be kept clean, smooth, and well polished. Their edges should be slightly rounded to prevent marring the material surrounding the riveting operation.

Hand Rivet Set
A hand rivet set is a tool equipped with a die for driving a particular type rivet. Rivet sets are available to fit every size and shape of rivet head. The ordinary set is made of ½-inch carbon tool steel about 6 inches in length and is knurled to prevent slipping in the hand. Only the face of the set is hardened and polished.

Sets for universal rivets are recessed (or cupped) to fit the rivet head. In selecting the correct set, be sure it provides the proper clearance between the set and the sides of the rivet head and between the surfaces of the metal and the set. Flush or flat sets are used for countersunk and flathead rivets. To seat flush rivets properly, be sure that the flush sets are at least 1 inch in diameter.

Special draw sets are used to draw up the sheets to eliminate any opening between them before the rivet is bucked. Each draw set has a hole ¹⁄₃₂-inch larger than the diameter of the rivet shank for which it is made. Occasionally, the draw set and rivet header are incorporated into one tool. The header part consists of a hole shallow enough for the set to expand the rivet and head when struck with a hammer.

Countersinking Tool
The countersink is a tool that cuts a cone-shaped depression around the rivet hole to allow the rivet to set flush with the surface of the skin. Countersinks are made with angles to correspond with the various angles of countersunk rivet heads. The standard countersink has a 100° angle, as shown in *Figure 4-84*. Special microstop countersinks (commonly called stop countersinks) are available that can be adjusted to any desired depth and have cutters to allow interchangeable holes with various countersunk angles to be made. *[Figure 4-85]* Some stop countersinks also have a micrometer set mechanism, in 0.001-inch increments, for adjusting their cutting depths.

Dimpling Dies
Dimpling is done with a male and female die (punch and die set). The male die has a guide the size of the rivet hole and with the same degree of countersink as the rivet. The female die has a hole with a corresponding degree of countersink into which the male guide fits.

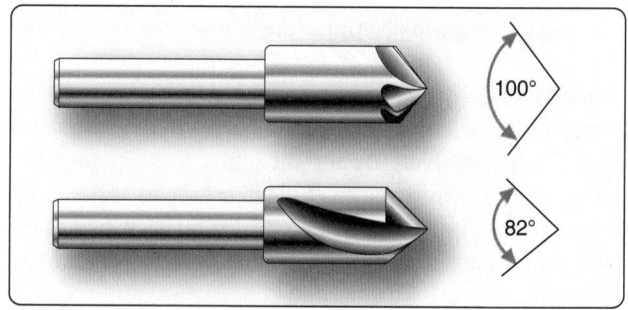

Figure 4-84. *Countersinks.*

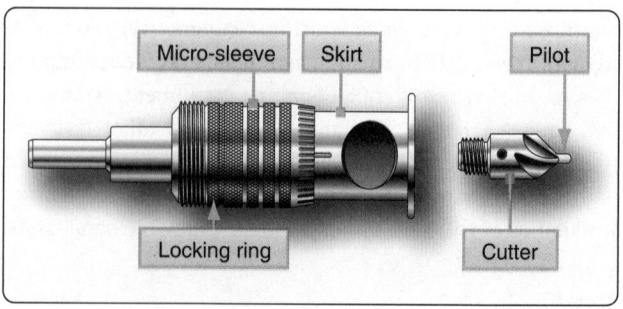

Figure 4-85. *Microstop countersink.*

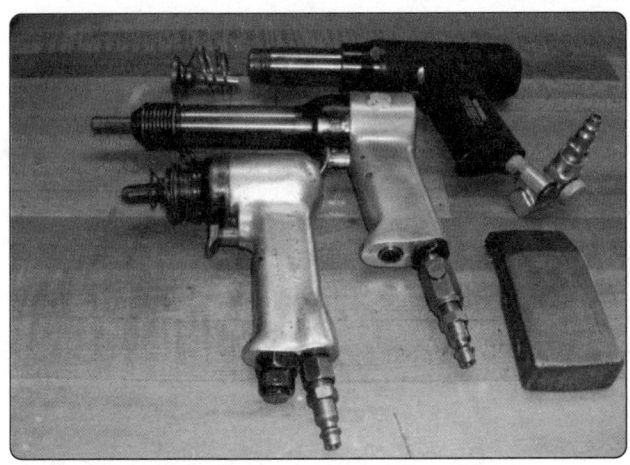

Figure 4-86. *Rivet guns.*

Power Tools

The most common power tools used in riveting are the pneumatic rivet gun, rivet squeezers, and the microshaver.

Pneumatic Rivet Gun

The pneumatic rivet gun is the most common rivet upsetting tool used in airframe repair work. It is available in many sizes and types. *[Figure 4-86]* The manufacturer's recommended capacity for each gun is usually stamped on the barrel. Pneumatic guns operate on air pressure of 90 to 100 pounds per square inch and are used in conjunction with interchangeable rivet sets. Each set is designed to fit the specific type of rivet and the location of the work. The shank of the set is designed to fit into the rivet gun. An air-driven hammer inside the barrel of the gun supplies force to buck the rivet.

Slow hitting rivet guns that strike from 900 to 2,500 blows per minute are the most common type. *[Figure 4-87]* These blows are slow enough to be easily controlled and heavy enough to do the job. These guns are sized by the largest rivet size continuously driven with size often based on the Chicago Pneumatic Company's old "X" series. A 4X gun (dash 8 or ¼ rivet) is used for normal work. The less powerful 3X gun is used for smaller rivets in thinner structure. 7X guns are used for large rivets in thicker structures. A rivet gun should upset a rivet in 1 to 3 seconds. With practice, an aircraft technician learns the length of time needed to hold down the trigger.

A rivet gun with the correct header (rivet set) must be held snugly against the rivet head and perpendicular to the surface

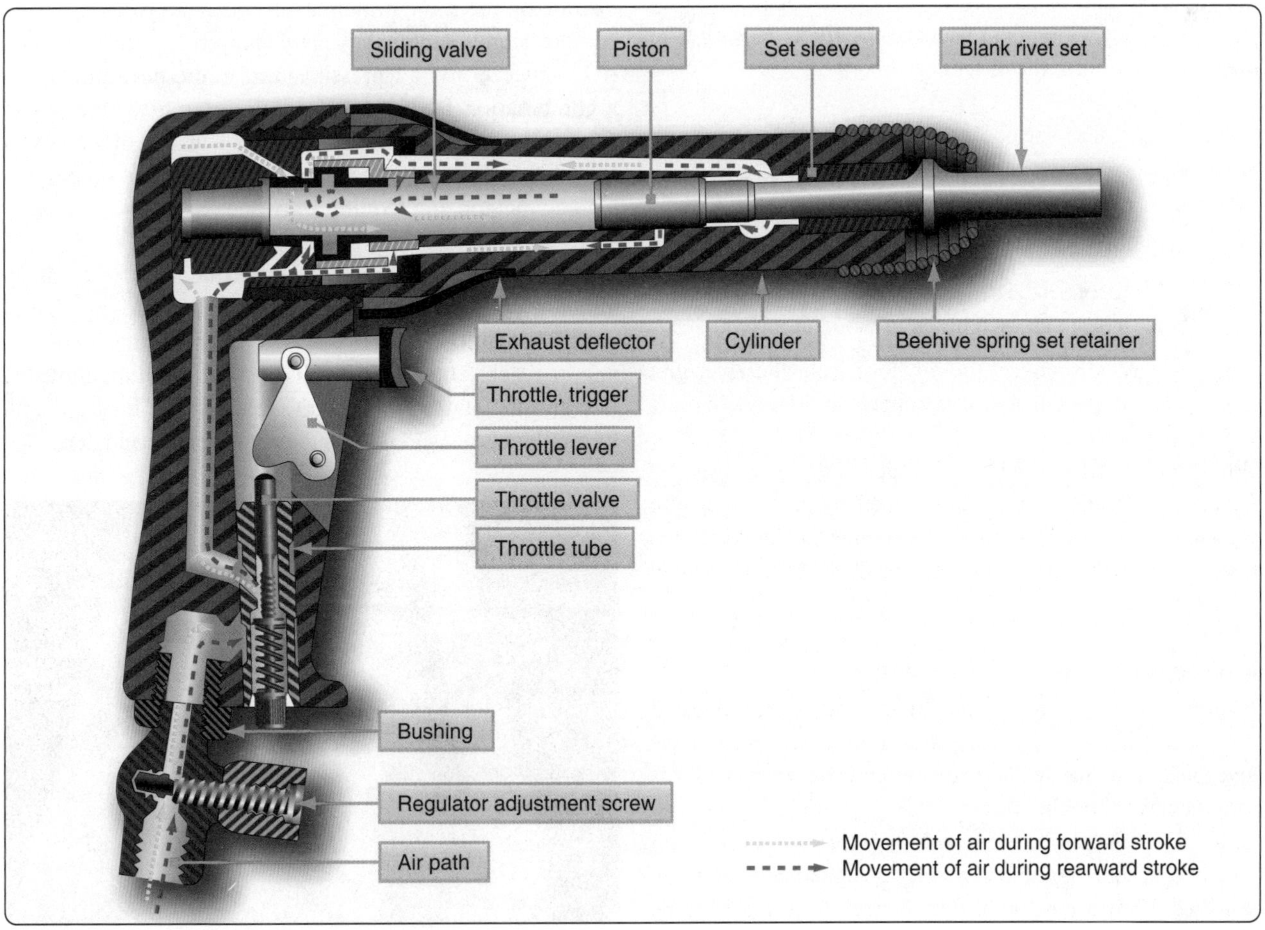

Sliding valve	Piston	Set sleeve	Blank rivet set

Exhaust deflector

Throttle, trigger

Throttle lever

Throttle valve

Throttle tube

Cylinder

Beehive spring set retainer

Bushing

Regulator adjustment screw

Air path

···········▷ Movement of air during forward stroke
- - - - ▶ Movement of air during rearward stroke

Figure 4-87. *Components of a rivet gun.*

while a bucking bar of the proper weight is held against the opposite end. The force of the gun must be absorbed by the bucking bar and not the structure being riveted. When the gun is triggered, the rivet is driven.

Always make sure the correct rivet header and the retaining spring are installed. Test the rivet gun on a piece of wood and adjust the air valve to a setting that is comfortable for the operator. The driving force of the rivet gun is adjusted by a needle valve on the handle. Adjustments should never be tested against anything harder than a wooden block to avoid header damage. If the adjustment fails to provide the best driving force, a different sized gun is needed. A gun that is too powerful is hard to control and may damage the work. On the other hand, if the gun is too light, it may work to harden the rivet before the head can be fully formed.

The riveting action should start slowly and be one continued burst. If the riveting starts too fast, the rivet header might slip off the rivet and damage the rivet (smiley) or damage the skin (eyebrow). Try to drive the rivets within 3 seconds, because the rivet will work harden if the driving process takes too long. The dynamic of the driving process has the gun hitting, or vibrating, the rivet and material, which causes the bar to bounce, or countervibrate. These opposing blows (low frequency vibrations) squeeze the rivet, causing it to swell and then form the upset head.

Some precautions to be observed when using a rivet gun are:

1. Never point a rivet gun at anyone at any time. A rivet gun should be used for one purpose only: to drive or install rivets.

2. Never depress the trigger mechanism unless the set is held tightly against a block of wood or a rivet.

3. Always disconnect the air hose from the rivet gun when it is not in use for any appreciable length of time.

While traditional tooling has changed little in the past 60 years, significant changes have been made in rivet gun ergonomics. Reduced vibration rivet guns and bucking bars have been developed to reduce the incidence of carpal tunnel syndrome and enhance operator comfort.

Rivet Sets/Headers
Pneumatic guns are used in conjunction with interchangeable rivet sets or headers. Each is designed to fit the type of rivet and location of the work. The shank of the rivet header is designed to fit into the rivet gun. An appropriate header must be a correct match for the rivet being driven. The working face of a header should be properly designed and smoothly polished. They are made of forged steel, heat treated to be tough but not too brittle. Flush headers come in various sizes.

Smaller ones concentrate the driving force in a small area for maximum efficiency. Larger ones spread the driving force over a larger area and are used for the riveting of thin skins.

Nonflush headers should fit to contact about the center two-thirds of the rivet head. They must be shallow enough to allow slight upsetting of the head in driving and some misalignment without eyebrowing the riveted surface. Care must be taken to match the size of the rivet. A header that is too small marks the rivet; while one too large marks the material.

Rivet headers are made in a variety of styles. *[Figure 4-88]* The short, straight header is best when the gun can be brought close to the work. Offset headers may be used to reach rivets in obstructed places. Long headers are sometimes necessary when the gun cannot be brought close to the work due to structural interference. Rivet headers should be kept clean.

Compression Riveting
Compression riveting (squeezing) is of limited value because this method of riveting can be used only over the edges of sheets or assemblies where conditions permit, and where the reach of the rivet squeezer is deep enough. The three types of rivet squeezers—hand, pneumatic, and pneudraulic—operate on the same principles. In the hand rivet squeezer, compression is supplied by hand pressure; in the pneumatic rivet squeezer, by air pressure; and in the pneudraulic, by a combination of air and hydraulic pressure. One jaw is stationary and serves as a bucking bar, the other jaw is movable and does the upsetting. Riveting with a squeezer is a quick method and requires only one operator.

These riveters are equipped with either a C-yoke or an alligator yoke in various sizes to accommodate any size of rivet. The working capacity of a yoke is measured by its gap and its reach. The gap is the distance between the movable jaw and the stationary jaw; the reach is the inside length of the throat measured from the center of the end sets. End

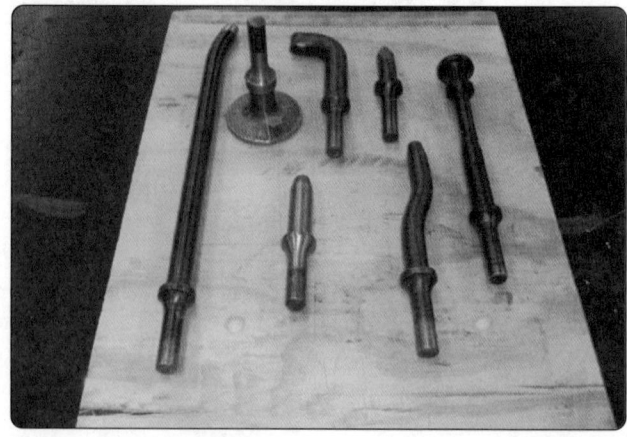

Figure 4-88. *Rivet headers.*

sets for rivet squeezers serve the same purpose as rivet sets for pneumatic rivet guns and are available with the same type heads, which are interchangeable to suit any type of rivet head. One part of each set is inserted in the stationary jaw, while the other part is placed in the movable jaws. The manufactured head end set is placed on the stationary jaw whenever possible. During some operations, it may be necessary to reverse the end sets, placing the manufactured head end set on the movable jaw.

Microshavers

A microshaver is used if the smoothness of the material (such as skin) requires that all countersunk rivets be driven within a specific tolerance. *[Figure 4-89]* This tool has a cutter, a stop, and two legs or stabilizers. The cutting portion of the microshaver is inside the stop. The depth of the cut can be adjusted by pulling outward on the stop and turning it in either direction (clockwise for deeper cuts). The marks on the stop permit adjustments of 0.001 inch. If the microshaver is adjusted and held correctly, it can cut the head of a countersunk rivet to within 0.002 inch without damaging the surrounding material.

Adjustments should always be made first on scrap material. When correctly adjusted, the microshaver leaves a small round dot about the size of a pinhead on the microshaved rivet. It may occasionally be necessary to shave rivets, normally restricted to MS20426 head rivets, after driving to obtain the required flushness. Shear head rivets should never be shaved.

Riveting Procedure

The riveting procedure consists of transferring and preparing the hole, drilling, and driving the rivets.

Hole Transfer

Accomplish transfer of holes from a drilled part to another part by placing the second part over first and using established holes as a guide. Using an alternate method, scribe hole location through from drilled part onto part to be drilled, spot with a center punch, and drill.

Hole Preparation

It is very important that the rivet hole be of the correct size and shape and free from burrs. If the hole is too small, the protective coating is scratched from the rivet when the rivet is driven through the hole. If the hole is too large, the rivet does not fill the hole completely. When it is bucked, the joint does not develop its full strength, and structural failure may occur at that spot.

If countersinking is required, consider the thickness of the metal and adopt the countersinking method recommended for that thickness. If dimpling is required, keep hammer blows or dimpling pressures to a minimum so that no undue work hardening occurs in the surrounding area.

Drilling

Rivet holes in repair may be drilled with either a light power drill or a hand drill. The standard shank twist drill is most commonly used. Drill bit sizes for rivet holes should be the smallest size that permits easy insertion of the rivet, approximately 0.003-inch greater than the largest tolerance of the shank diameter. The recommended clearance drill bits for the common rivet diameters are shown in *Figure 4-90*. Hole sizes for other fasteners are normally found on work documents, prints, or in manuals.

Before drilling, center punch all rivet locations. The center punch mark should be large enough to prevent the drill from slipping out of position, yet it must not dent the surface surrounding the center punch mark. Place a bucking bar behind the metal during punching to help prevent denting. To make a rivet hole the correct size, first drill a slightly undersized hole (pilot hole). Ream the pilot hole with a twist drill of the appropriate size to obtain the required dimension.

To drill, proceed as follows:

1. Ensure the drill bit is the correct size and shape.

Figure 4-89. *Microshaver.*

Rivet Diameter (in)	Drill Size	
	Pilot	Final
3/32	3/32 (0.0937)	#40 (0.098)
1/8	1/8 (0.125)	#30 (0.1285)
5/32	5/32 (0.1562)	#21 (0.159)
3/16	3/16 (0.1875)	#11 (0.191)
1/4	1/4 (0.250)	F (0.257)

Figure 4-90. *Drill sizes for standard rivets.*

2. Place the drill in the center-punched mark. When using a power drill, rotate the bit a few turns before starting the motor.

3. While drilling, always hold the drill at a 90° angle to the work or the curvature of the material.

4. Avoid excessive pressure, let the drill bit do the cutting, and never push the drill bit through stock.

5. Remove all burrs with a metal countersink or a file.

6. Clean away all drill chips.

When holes are drilled through sheet metal, small burrs are formed around the edge of the hole. This is especially true when using a hand drill because the drill speed is slow and there is a tendency to apply more pressure per drill revolution. Remove all burrs with a burr remover or larger size drill bit before riveting.

Driving the Rivet

Although riveting equipment can be either stationary or portable, portable riveting equipment is the most common type of riveting equipment used to drive solid shank rivets in airframe repair work.

Before driving any rivets into the sheet metal parts, be sure all holes line up perfectly, all shavings and burrs have been removed, and the parts to be riveted are securely fastened with temporary fasteners. Depending on the job, the riveting process may require one or two people. In solo riveting, the riveter holds a bucking bar with one hand and operates a riveting gun with the other.

If the job requires two aircraft technicians, a shooter, or gunner, and a bucker work together as a team to install rivets. An important component of team riveting is an efficient signaling system that communicates the status of the riveting process. This signaling system usually consists of tapping the bucking bar against the work and is often called the tap code. One tap may mean not fully seated, hit it again, while two taps may mean good rivet, and three taps may mean bad rivet, remove and drive another. Radio sets are also available for communication between the technicians.

Once the rivet is installed, there should be no evidence of rotation of rivets or looseness of riveted parts. After the trimming operation, examine for tightness. Apply a force of 10 pounds to the trimmed stem. A tight stem is one indication of an acceptable rivet installation. Any degree of looseness indicates an oversize hole and requires replacement of the rivet with an oversize shank diameter rivet. A rivet installation is assumed satisfactory when the rivet head is seated snugly against the item to be retained (0.005-inch feeler gauge should not go under rivet head for more than one-half the circumference) and the stem is proved tight.

Countersunk Rivets

An improperly made countersink reduces the strength of a flush-riveted joint and may even cause failure of the sheet or the rivet head. The two methods of countersinking commonly used for flush riveting in aircraft construction and repair are:

- Machine or drill countersinking.
- Dimpling or press countersinking.

The proper method for any particular application depends on the thickness of the parts to be riveted, the height and angle of the countersunk head, the tools available, and accessibility.

Countersinking

When using countersunk rivets, it is necessary to make a conical recess in the skin for the head. The type of countersink required depends upon the relation of the thickness of the sheets to the depth of the rivet head. Use the proper degree and diameter countersink and cut only deep enough for the rivet head and metal to form a flush surface.

Countersinking is an important factor in the design of fastener patterns, as the removal of material in the countersinking process necessitates an increase in the number of fasteners to assure the required load-transfer strength. If countersinking is done on metal below a certain thickness, a knife edge with less than the minimum bearing surface or actual enlarging of the hole may result. The edge distance required when using countersunk fasteners is greater than when universal head fasteners are used.

The general rule for countersinking and flush fastener installation procedures has been reevaluated in recent years because countersunk holes have been responsible for fatigue cracks in aircraft pressurized skin. In the past, the general rule for countersinking held that the fastener head must be contained within the outer sheet. A combination of countersinks too deep (creating a knife edge), number of pressurization cycles, fatigue, deterioration of bonding materials, and working fasteners caused a high stress concentration that resulted in skin cracks and fastener failures. In primary structure and pressurized skin repairs, some manufacturers are currently recommending the countersink depth be no more than ⅔ the outer sheet thickness or down to 0.020-inch minimum fastener shank depth, whichever is greater. Dimple the skin if it is too thin for machine countersinking. *[Figure 4-91]*

Keep the rivet high before driving to ensure the force of riveting is applied to the rivet and not to the skin. If the rivet is driven while it is flush or too deep, the surrounding skin is work hardened.

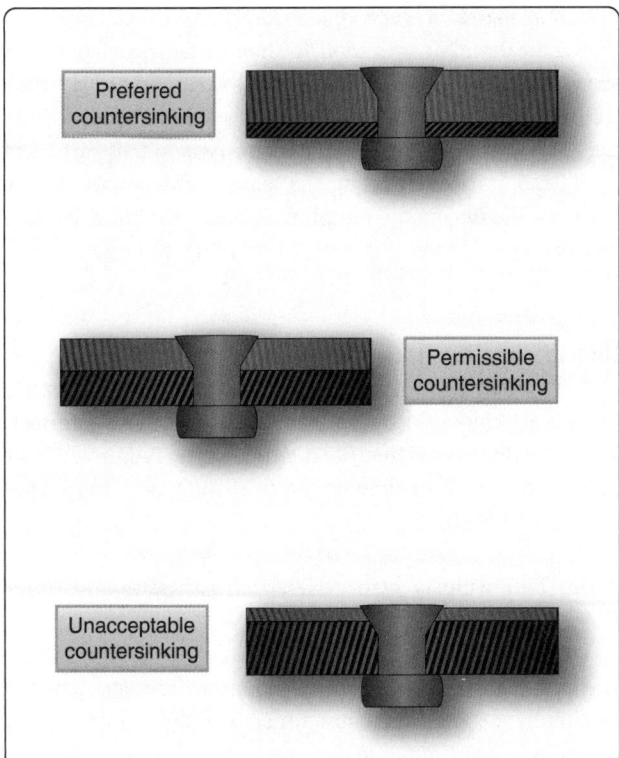

Figure 4-91. *Countersinking dimensions.*

Countersinking Tools

While there are many types of countersink tools, the most commonly used has an included angle of 100°. Sometimes types of 82° or 120° are used to form countersunk wells. *[Figure 4-84]* A six-fluted countersink works best in aluminum. There are also four- and three-fluted countersinks, but those are harder to control from a chatter standpoint. A single-flute type, such as those manufactured by the Weldon Tool Company®, works best for corrosion-resistant steel. *[Figure 4-92]*

The microstop countersink is the preferred countersinking tool. *[Figure 4-85]* It has an adjustable-sleeve cage that functions as a limit stop and holds the revolving countersink in a vertical position. Its threaded and replaceable cutters may have either a removable or an integral pilot that keeps the

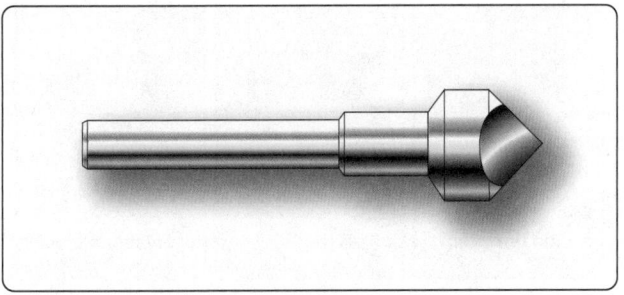

Figure 4-92. *Single-flute countersink.*

cutter centered in the hole. The pilot should be approximately 0.002-inch smaller than the hole size. It is recommended to test adjustments on a piece of scrap material before countersinking repair or replacement parts.

Freehand countersinking is needed where a microstop countersink cannot fit. This method should be practiced on scrap material to develop the required skill. Holding the drill motor steady and perpendicular is as critical during this operation as when drilling.

Chattering is the most common problem encountered when countersinking. Some precautions that may eliminate or minimize chatter include:

- Use sharp tooling.
- Use a slow speed and steady firm pressure.
- Use a piloted countersink with a pilot approximately 0.002-inch smaller than the hole.
- Use back-up material to hold the pilot steady when countersinking thin sheet material.
- Use a cutter with a different number of flutes.
- Pilot drill an undersized hole, countersink, and then enlarge the hole to final size.

Dimpling

Dimpling is the process of making an indentation or a dimple around a rivet hole to make the top of the head of a countersunk rivet flush with the surface of the metal. Dimpling is done with a male and female die, or forms, often called punch and die set. The male die has a guide the size of the rivet hole and is beveled to correspond to the degree of countersink of the rivet head. The female die has a hole into which the male guide fits and is beveled to a corresponding degree of countersink.

When dimpling, rest the female die on a solid surface. Then, place the material to be dimpled on the female die. Insert the male die in the hole to be dimpled and, with a hammer, strike the male die until the dimple is formed. Two or three solid hammer blows should be sufficient. A separate set of dies is necessary for each size of rivet and shape of rivet head. An alternate method is to use a countersunk head rivet instead of the regular male punch die, and a draw set instead of the female die, and hammer the rivet until the dimple is formed.

Dimpling dies for light work can be used in portable pneumatic or hand squeezers. *[Figure 4-93]* If the dies are used with a squeezer, they must be adjusted accurately to the thickness of the sheet being dimpled. A table riveter is also used for dimpling thin skin material and installing rivets. *[Figure 4-94]*

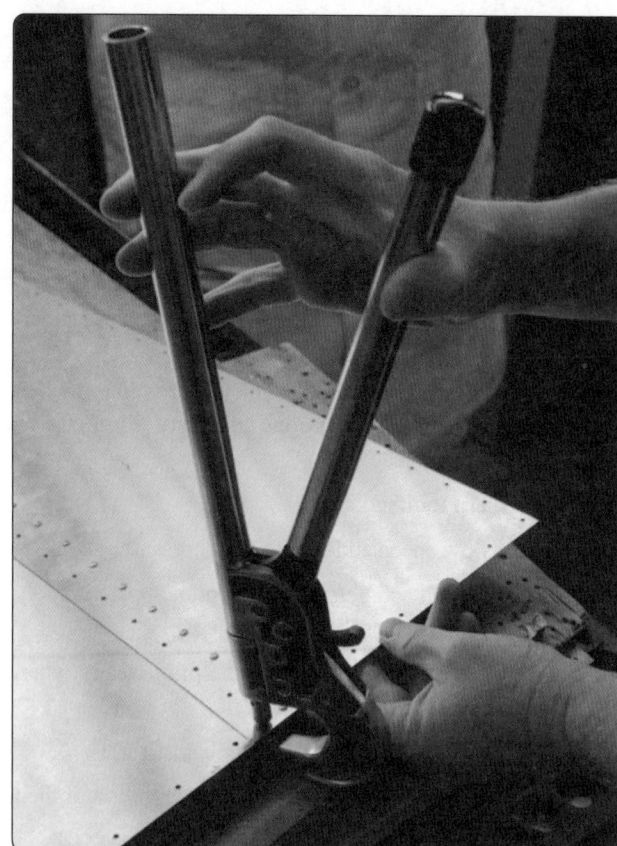

Figure 4-93. *Hand squeezers.*

Figure 4-94. *Table riveter.*

Coin Dimpling

The coin dimpling, or coin pressing, method uses a countersink rivet as the male dimpling die. Place the female die in the usual position and back it with a bucking bar. Place the rivet of the required type into the hole and strike the rivet with a pneumatic riveting hammer. Coin dimpling should be used only when the regular male die is broken or not available. Coin pressing has the distinct disadvantage of the rivet hole needing to be drilled to correct rivet size before the

dimpling operation is accomplished. Since the metal stretches during the dimpling operation, the hole becomes enlarged and the rivet must be swelled slightly before driving to produce a close fit. Because the rivet head causes slight distortions in the recess, and these are characteristic only to that particular rivet head, it is wise to drive the same rivet that was used as the male die during the dimpling process. Do not substitute another rivet, either of the same size or a size larger.

Radius Dimpling

Radius dimpling uses special die sets that have a radius and are often used with stationary or portable squeezers. Dimpling removes no metal and, due to the nestling effect, gives a stronger joint than the non-flush type. A dimpled joint reduces the shear loading on the rivet and places more load on the riveted sheets.

Note: Dimpling is also done for flush bolts and other flush fasteners.

Dimpling is required for sheets that are thinner than the minimum specified thickness for countersinking. However, dimpling is not limited to thin materials. Heavier parts may be dimpled without cracking by specialized hot dimpling equipment. The temper of the material, rivet size, and available equipment are all factors to be considered in dimpling. *[Figure 4-95]*

Hot Dimpling

Hot dimpling is the process that uses heated dimpling dies to

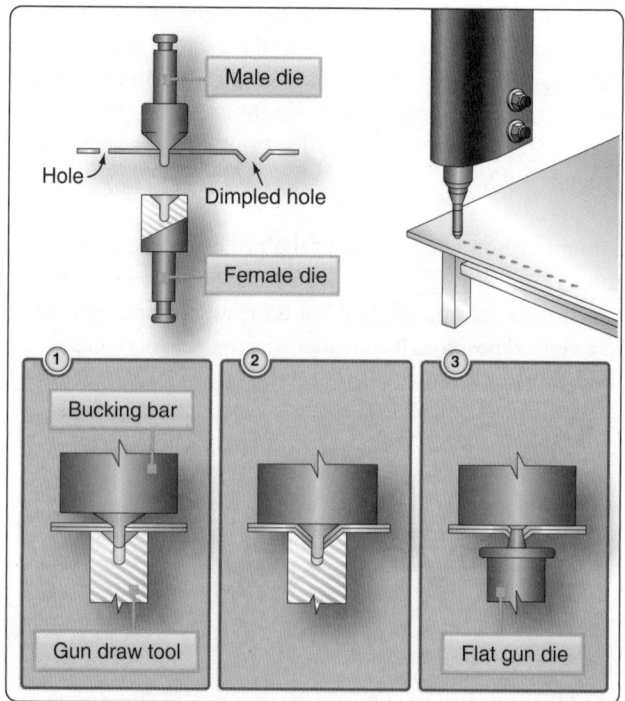

Figure 4-95. *Dimpling techniques.*

ensure the metal flows better during the dimpling process. Hot dimpling is often performed with large stationary equipment available in a sheet metal shop. The metal being used is an important factor because each metal presents different dimpling problems. For example, 2024-T3 aluminum alloy can be satisfactorily dimpled either hot or cold, but may crack in the vicinity of the dimple after cold dimpling because of hard spots in the metal. Hot dimpling prevents such cracking.

7075-T6 aluminum alloys are always hot dimpled. Magnesium alloys also must be hot dimpled because, like 7075-T6, they have low formability qualities. Titanium is another metal that must be hot dimpled because it is tough and resists forming. The same temperature and dwell time used to hot dimple 7075-T6 is used for titanium.

100° Combination Predimple & Countersink Method

Metals of different thicknesses are sometimes joined by a combination of dimpling and countersinking. *[Figure 4-96]* A countersink well made to receive a dimple is called a subcountersink. These are most often seen where a thin web is attached to heavy structure. It is also used on thin gap seals, wear strips, and repairs for worn countersinks.

Dimpling Inspection

To determine the quality of a dimple, it is necessary to make a close visual inspection. Several features must be checked. The rivet head should fit flush and there should be a sharp break from the surface into the dimple. The sharpness of the break is affected by dimpling pressure and metal thickness. Selected dimples should be checked by inserting a fastener to make sure that the flushness requirements are met. Cracked dimples are caused by poor dies, rough holes, or improper heating. Two types of cracks may form during dimpling:

- Radial cracks—start at the edge and spread outward as the metal within the dimple stretches. They are most common in 2024-T3. A rough hole or a dimple that is too deep causes such cracks. A small tolerance is usually allowed for radial cracks.

- Circumferential cracks—downward bending into the

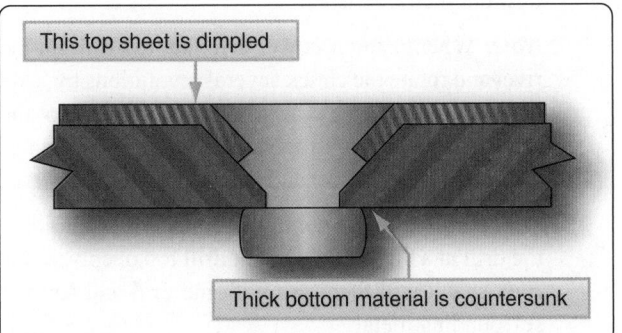

Figure 4-96. *Predimple and countersink method.*

This top sheet is dimpled

Thick bottom material is countersunk

draw die causes tension stresses in the upper portion of the metal. Under some conditions, a crack may be created that runs around the edge of the dimple. Such cracks do not always show since they may be underneath the cladding. When found, they are cause for rejection. These cracks are most common in hot-dimpled 7075 T6 aluminum alloy material. The usual cause is insufficient dimpling heat.

Evaluating the Rivet

To obtain high structural efficiency in the manufacture and repair of aircraft, an inspection must be made of all rivets before the part is put in service. This inspection consists of examining both the shop and manufactured heads and the surrounding skin and structural parts for deformities. A scale or rivet gauge can be used to check the condition of the upset rivet head to see that it conforms to the proper requirements. Deformities in the manufactured head can be detected by the trained eye alone. *[Figure 4-97]*

Some common causes of unsatisfactory riveting are improper bucking, rivet set slipping off or being held at the wrong angle, and rivet holes or rivets of the wrong size. Additional causes for unsatisfactory riveting are countersunk rivets not flush with the well, work not properly fastened together during riveting, the presence of burrs, rivets too hard, too much or too little driving, and rivets out of line.

Occasionally, during an aircraft structural repair, it is wise to examine adjacent parts to determine the true condition of neighboring rivets. In doing so, it may be necessary to remove the paint. The presence of chipped or cracked paint around the heads may indicate shifted or loose rivets. Look for tipped or loose rivet heads. If the heads are tipped or if rivets are loose, they show up in groups of several consecutive rivets and probably tipped in the same direction. If heads that appear to be tipped are not in groups and are not tipped in the same direction, tipping may have occurred during some previous installation.

Inspect rivets known to have been critically loaded, but that show no visible distortion, by drilling off the head and carefully punching out the shank. If, upon examination, the shank appears joggled and the holes in the sheet misaligned, the rivet has failed in shear. In that case, try to determine what is causing the shearing stress and take the necessary corrective action. Flush rivets that show head slippage within the countersink or dimple, indicating either sheet bearing failure or rivet shear failure, must be removed for inspection and replacement.

Joggles in removed rivet shanks indicate partial shear failure. Replace these rivets with the next larger size. Also, if the rivet

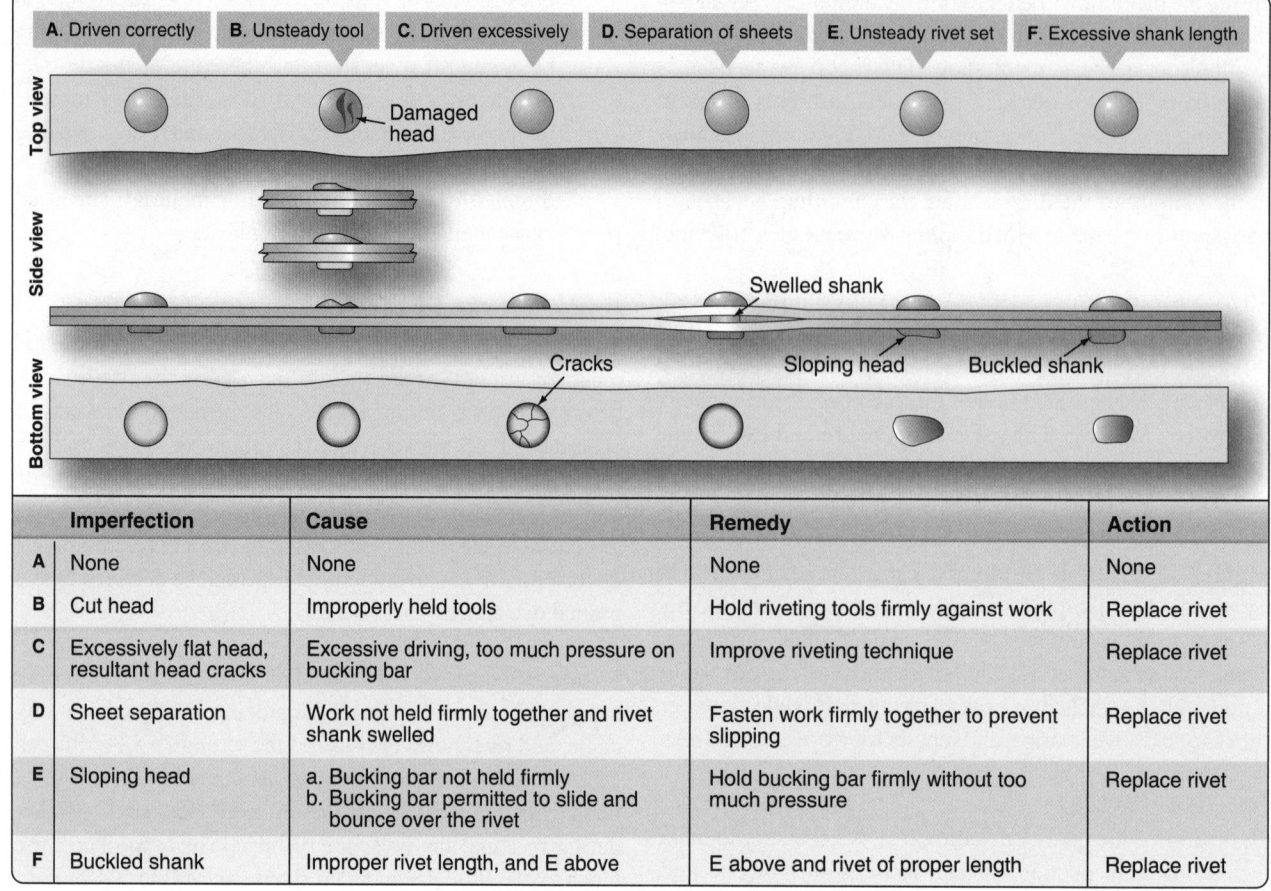

Figure 4-97. *Rivet defects.*

	Imperfection	Cause	Remedy	Action
A	None	None	None	None
B	Cut head	Improperly held tools	Hold riveting tools firmly against work	Replace rivet
C	Excessively flat head, resultant head cracks	Excessive driving, too much pressure on bucking bar	Improve riveting technique	Replace rivet
D	Sheet separation	Work not held firmly together and rivet shank swelled	Fasten work firmly together to prevent slipping	Replace rivet
E	Sloping head	a. Bucking bar not held firmly b. Bucking bar permitted to slide and bounce over the rivet	Hold bucking bar firmly without too much pressure	Replace rivet
F	Buckled shank	Improper rivet length, and E above	E above and rivet of proper length	Replace rivet

holes show elongation, replace the rivets with the next larger size. Sheet failures such as tear-outs, cracks between rivets, and the like usually indicate damaged rivets. The complete repair of the joint may require replacement of the rivets with the next larger size.

The general practice of replacing a rivet with the next larger size (¹⁄₃₂-inch greater diameter) is necessary to obtain the proper joint strength of rivet and sheet when the original rivet hole is enlarged. If the rivet in an elongated hole is replaced by a rivet of the same size, its ability to carry its share of the shear load is impaired and joint weakness results.

Removal of Rivets

When a rivet has to be replaced, remove it carefully to retain the rivet hole's original size and shape. If removed correctly, the rivet does not need to be replaced with one of the next larger size. Also, if the rivet is not removed properly, the strength of the joint may be weakened and the replacement of rivets made more difficult.

When removing a rivet, work on the manufactured head. It is more symmetrical about the shank than the shop head, and there is less chance of damaging the rivet hole or the material

around it. To remove rivets, use hand tools, a power drill, or a combination of both.

The procedure for universal or protruding head rivet removal is as follows:

1. File a flat area on the head of the rivet and center punch the flat surface for drilling.

   Note: On thin metal, back up the rivet on the upset head when center punching to avoid depressing the metal.

2. Use a drill bit one size smaller than the rivet shank to drill out the rivet head.

   Note: When using a power drill, set the drill on the rivet and rotate the chuck several revolutions by hand before turning on the power. This procedure helps the drill cut a good starting spot and eliminates the chance of the drill slipping off and tracking across the metal.

3. Drill the rivet to the depth of its head, while holding the drill at a 90° angle. Do not drill too deeply, as the rivet shank will then turn with the drill and tear the surrounding metal.

   Note: The rivet head often breaks away and climbs

the drill, which is a signal to withdraw the drill.

4. If the rivet head does not come loose of its own accord, insert a drift punch into the hole and twist slightly to either side until the head comes off.

5. Drive the remaining rivet shank out with a drift punch slightly smaller than the shank diameter.

On thin metal or unsupported structures, support the sheet with a bucking bar while driving out the shank. If the shank is unusually tight after the rivet head is removed, drill the rivet about two-thirds through the thickness of the material and then drive the rest of it out with a drift punch. *Figure 4-98* shows the preferred procedure for removing universal rivets.

The procedure for the removal of countersunk rivets is the same as described above except no filing is necessary. Be careful to avoid elongation of the dimpled or the countersunk holes. The rivet head should be drilled to approximately one-half the thickness of the top sheet. The dimple in 2117–T rivets usually eliminates the necessity of filing and center punching the rivet head.

To remove a countersunk or flush head rivet, you must:

1. Select a drill about 0.003-inch smaller than the rivet shank diameter.

2. Drill into the exact center of the rivet head to the approximate depth of the head.

3. Remove the head by breaking it off. Use a punch as a lever.

4. Punch out the shank. Use a suitable backup, preferably wood (or equivalent), or a dedicated backup block. If the shank does not come out easily, use a small drill and drill through the shank. Be careful not to elongate the hole.

Replacing Rivets

Replace rivets with those of the same size and strength whenever possible. If the rivet hole becomes enlarged, deformed, or otherwise damaged, drill or ream the hole for the next larger size rivet. Do not replace a rivet with a type having lower strength properties, unless the lower strength is adequately compensated by an increase in size or a greater number of rivets. It is acceptable to replace 2017 rivets of $\frac{3}{16}$-inch diameter or less, and 2024 rivets of $\frac{5}{32}$-inch diameter or less with 2117 rivets for general repairs, provided the replacement rivets are $\frac{1}{32}$-inch greater in diameter than the rivets they replace.

National Advisory Committee for Aeronautics (NACA) Method of Double Flush Riveting

A rivet installation technique known as the National Advisory Committee for Aeronautics (NACA) method has primary applications in fuel tank areas. *[Figure 4-99]* To make a NACA rivet installation, the shank is upset into a 82° countersink. In driving, the gun may be used on either the head or shank side. The upsetting is started with light blows, then the force increased and the gun or bar moved on the shank end so as to form a head inside the countersink well. If desired, the upset head may be shaved flush after driving. If utilizing this method, it is important to reference the manufacturer's instructions for repair or replacement.

Special Purpose Fasteners

Special purpose fasteners are designed for applications in which fastener strength, ease of installation, or temperature properties of the fastener require consideration. Solid shank rivets have been the preferred construction method for metal aircraft for many years because they fill up the hole, which results in good load transfer, but they are not always ideal. For example, the attachment of many nonstructural parts (aircraft interior furnishings, flooring, deicing boots, etc.) do not need the full strength of solid shank rivets.

To install solid shank rivets, the aircraft technician must have access to both sides of a riveted structure or structural part. There are many places on an aircraft where this access is impossible or where limited space does not permit the use of a bucking bar. In these instances, it is not possible to use solid shank rivets, and special fasteners have been designed that can be bucked from the front. *[Figure 4-100]* There are also areas of high loads, high fatigue, and bending on aircraft. Although the shear loads of riveted joints are very good, the tension, or clamp-up, loads are less than ideal.

Special purpose fasteners are sometimes lighter than solid shank rivets, yet strong enough for their intended use. These fasteners are manufactured by several corporations and have unique characteristics that require special installation tools, special installation procedures, and special removal procedures. Because these fasteners are often inserted in locations where one head, usually the shop head, cannot be seen, they are called blind rivets or blind fasteners.

Typically, the locking characteristics of a blind rivet are not as good as a driven rivet. Therefore, blind rivets are usually not used when driven rivets can be installed. Blind rivets shall not be used:

1. In fluid-tight areas.

2. On aircraft in air intake areas where rivet parts may be ingested by the engine.

3. On aircraft control surfaces, hinges, hinge brackets, flight control actuating systems, wing attachment fittings, landing gear fittings, on floats or amphibian

Rivet Removal

Remove rivets by drilling off the head and punching out the shank as illustrated.
1. File a flat area on the manufactured head of non-flush rivets.
2. Place a block of wood or a bucking bar under both flush and nonflush rivets when center punching the manufactured head.
3. Use a drill that is $^1/_{32}$ (0.0312) inch smaller than the rivet shank to drill through the head of the rivet. Ensure the drilling operation does not damage the skin or cut the sides of the rivet hole.
4. Insert a drift punch into the hole drilled in the rivet and tilt the punch to break off the rivet head.
5. Using a drift punch and hammer, drive out the rivet shank. Support the opposite side of the structure to prevent structural damage.

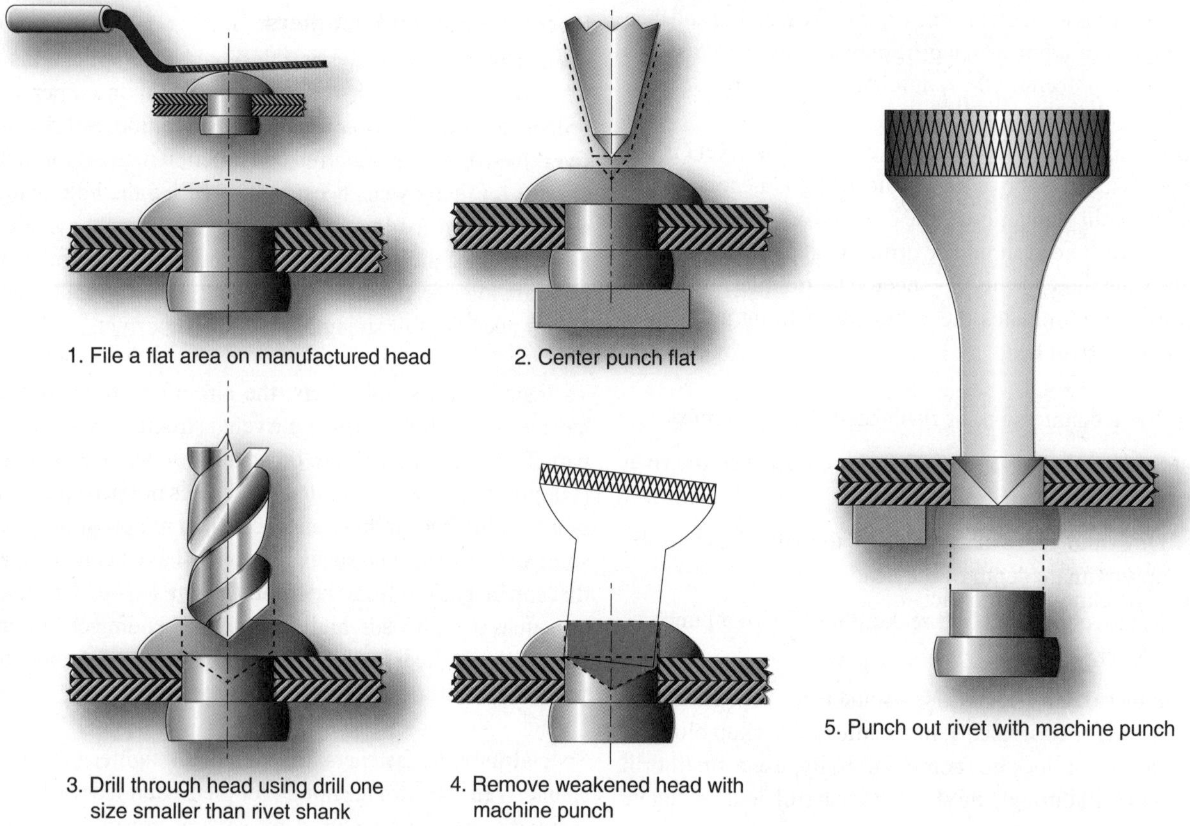

1. File a flat area on manufactured head

2. Center punch flat

3. Drill through head using drill one size smaller than rivet shank

4. Remove weakened head with machine punch

5. Punch out rivet with machine punch

Figure 4-98. *Rivet removal.*

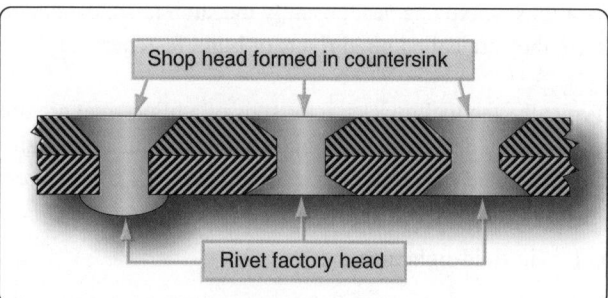

Shop head formed in countersink

Rivet factory head

Figure 4-99. *NACA riveting method.*

Figure 4-100. *Assorted fasteners.*

hulls below the water level, or other heavily stressed locations on the aircraft.

Note: For metal repairs to the airframe, the use of blind rivets must be specifically authorized by the airframe manufacturer or approved by a representative of the Federal Aviation Administration (FAA).

Blind Rivets

The first blind fasteners were introduced in 1940 by the Cherry Rivet Company (now Cherry® Aerospace), and the aviation industry quickly adopted them. The past decades have seen a proliferation of blind fastening systems based on the original concept, which consists of a tubular rivet with a fixed head and a hollow sleeve. Inserted within the rivet's core is a stem that is enlarged or serrated on its exposed end when activated by a pulling-type rivet gun. The lower end of the stem extends beyond the inner sheet of metal. This portion contains a tapered joining portion and a blind head that has a larger diameter than the stem or the sleeve of the tubular rivet.

When the pulling force of the rivet gun forces the blind head upward into the sleeve, its stem upsets or expands the lower end of the sleeve into a tail. This presses the inner sheet upward and closes any space that might have existed between it and the outer sheet. Since the exposed head of the rivet is held tightly against the outer sheet by the rivet gun, the sheets of metal are clamped, or clinched, together.

Note: Fastener manufacturers use different terminology to describe the parts of the blind rivet. The terms "mandrel," "spindle," and "stem" are often used interchangeably. For clarity, the word "stem" is used in this handbook and refers to the piece that is inserted into the hollow sleeve.

Friction-Locked Blind Rivets

Standard self-plugging blind rivets consist of a hollow sleeve and a stem with increased diameter in the plug section. The blind head is formed as the stem is pulled into the sleeve. Friction-locked blind rivets have a multiple-piece construction and rely on friction to lock the stem to the sleeve. As the stem is drawn up into the rivet shank, the stem portion upsets the shank on the blind side, forming a plug in the hollow center of the rivet. The excess portion of the stem breaks off at a groove due to the continued pulling action of the rivet gun. Metals used for these rivets are 2117-T4 and 5056-F aluminum alloy. Monel® is used for special applications.

Many friction-locked blind rivet center stems fall out due to vibration, which greatly reduces its shear strength. To combat that problem, most friction-lock blind rivets are replaced by

the mechanical-lock, or stem-lock, type of blind fasteners. However, some types, such as the Cherry SPR® ³⁄₃₂-inch Self-Plugging Rivet, are ideal for securing nutplates located in inaccessible and hard-to-reach areas where bucking or squeezing of solid rivets is unacceptable. *[Figure 4-101]*

Friction-lock blind rivets are less expensive than mechanical-lock blind rivets and are sometimes used for nonstructural applications. Inspection of friction-lock blind rivets is visual. A more detailed discussion on how to inspect riveted joints can be found later in this chapter. Removal of friction-lock blind rivets consists of punching out the friction-lock stem and then treating it like any other rivet.

Mechanical-Lock Blind Rivets

The self-plugging, mechanical-lock blind rivet was developed to prevent the problem of losing the center stem due to vibration. This rivet has a device on the puller or rivet head that locks the center stem into place when installed. Bulbed, self-plugging, mechanically-locked blind rivets form a large, blind head that provides higher strength in thin sheets when installed. They may be used in applications where the blind head is formed against a dimpled sheet.

Manufacturers such as Cherry® Aerospace (CherryMAX®, CherryLOCK®, Cherry SST®) and Alcoa Fastening Systems (Huck-Clinch®, HuckMax®, Unimatic®) make many variations of this of blind rivet. While similar in design, the tooling for these rivets is often not interchangeable.

The CherryMAX® Bulbed blind rivet is one of the earlier types of mechanical-lock blind rivets developed. Their main

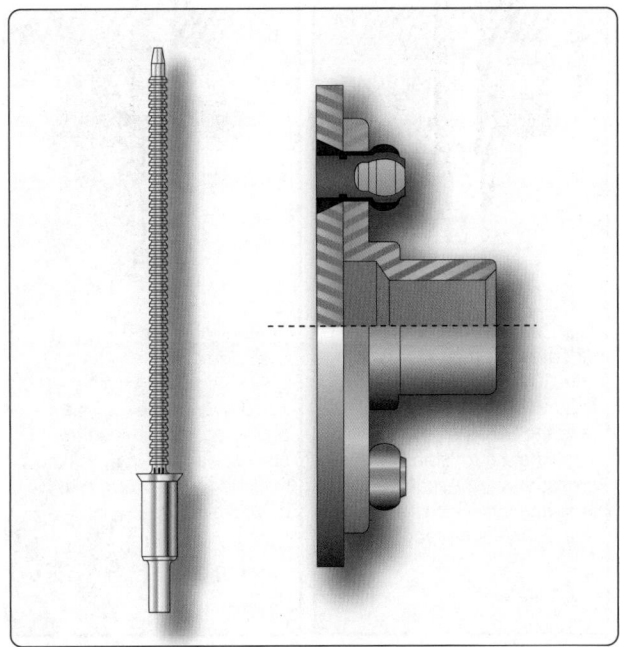

Figure 4-101. *Friction-lock blind rivet.*

advantage is the ability to replace a solid shank rivet size for size. The CherryMAX® Bulbed blind rivet consists of four parts:

1. A fully serrated stem with break notch, shear ring, and integral grip adjustment cone.

2. A driving anvil to ensure a visible mechanical lock with each fastener installation.

3. A separate, visible, and inspectable locking collar that mechanically locks the stem to the rivet sleeve.

4. A rivet sleeve with recess in the head to receive the locking collar.

It is called a bulbed fastener due to its large blind side bearing surface, developed during the installation process. These rivets are used in thin sheet applications and for use in materials that may be damaged by other types of blind rivets. This rivet features a safe-lock locking collar for more reliable joint integrity. The rough end of the retained stem in the center on the manufactured head must never be filed smooth because it weakens the strength of the lockring, and the center stem could fall out.

CherryMAX® bulbed rivets are available in three head styles: universal, 100° countersunk, and 100° reduced shear head styles. Their lengths are measured in increments of ¹⁄₁₆ inch. It is important to select a rivet with a length related to the grip length of the metal being joined. This blind rivet can be installed using either the Cherry® G750A or the newly released Cherry® G800 hand riveters, or either the pneumatic-hydraulic G704B or G747 CherryMAX® power tools. For installation, please refer to *Figure 4-102*.

The CherryMAX® mechanical-lock blind rivet is popular with general aviation repair shops because it features the one tool concept to install three standard rivet diameters and their oversize counterparts. *[Figure 4-103]* CherryMAX® rivets are available in four nominal diameters: ¹⁄₈, ⁵⁄₃₂, ³⁄₁₆, and ¹⁄₄-inch and three oversized diameters and four head styles: universal, 100° flush head, 120° flush head, and NAS1097 flush head. This rivet consists of a blind header, hollow rivet shell, locking (foil) collar, driving anvil, and pulling stem complete with wrapped locking collar. The rivet sleeve and the driving washer blind bulbed header takes up the extended shank and forms the bucktail.

The stem and rivet sleeve work as an assembly to provide radial expansion and a large bearing footprint on the blind side of the fastened surface. The lock collar ensures that the stem and sleeve remain assembled during joint loading and unloading. Rivet sleeves are made from 5056 aluminum, Monel® and INCO 600. The stems are made from alloy steel, CRES, and INCO® X-750. CherryMAX® rivets have an ultimate shear strength ranging from 50 KSI to 75 KSI.

Removal of Mechanically-Locked Blind Rivets
Mechanically-locked blind rivets are a challenge to remove because they are made from strong, hard metals. Lack of

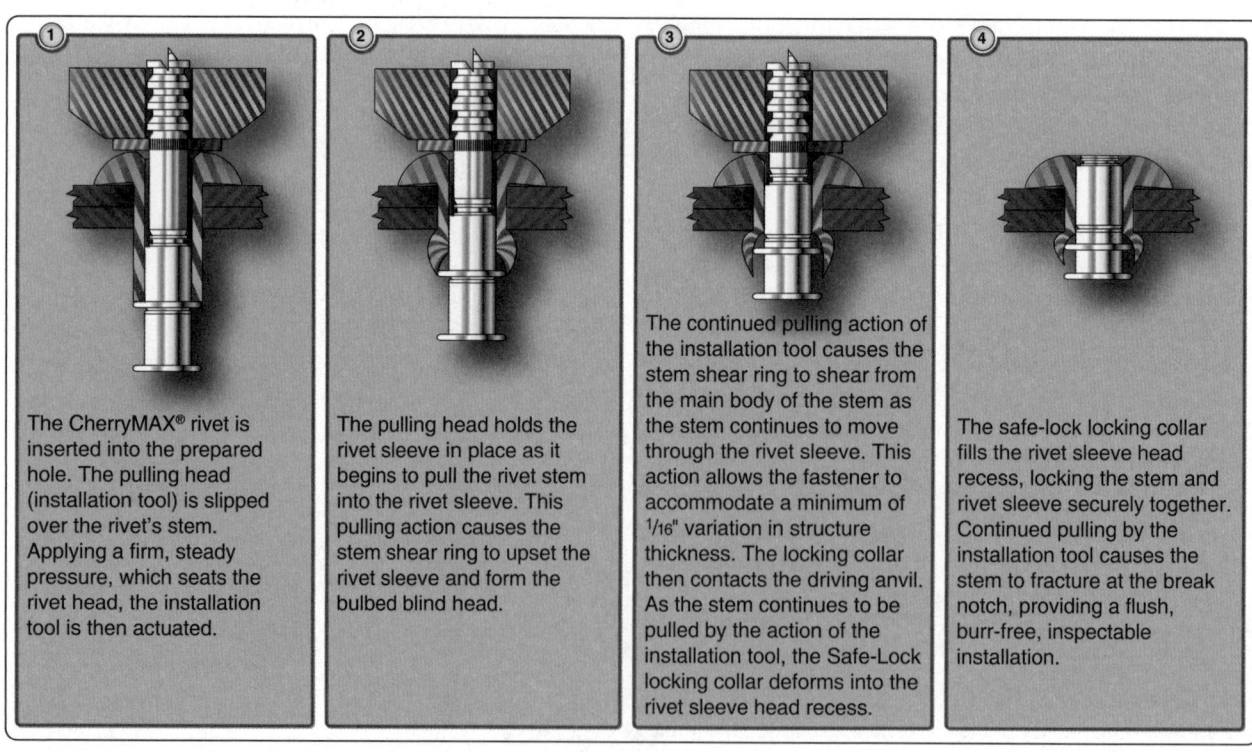

The CherryMAX® rivet is inserted into the prepared hole. The pulling head (installation tool) is slipped over the rivet's stem. Applying a firm, steady pressure, which seats the rivet head, the installation tool is then actuated.

The pulling head holds the rivet sleeve in place as it begins to pull the rivet stem into the rivet sleeve. This pulling action causes the stem shear ring to upset the rivet sleeve and form the bulbed blind head.

The continued pulling action of the installation tool causes the stem shear ring to shear from the main body of the stem as the stem continues to move through the rivet sleeve. This action allows the fastener to accommodate a minimum of ¹⁄₁₆" variation in structure thickness. The locking collar then contacts the driving anvil. As the stem continues to be pulled by the action of the installation tool, the Safe-Lock locking collar deforms into the rivet sleeve head recess.

The safe-lock locking collar fills the rivet sleeve head recess, locking the stem and rivet sleeve securely together. Continued pulling by the installation tool causes the stem to fracture at the break notch, providing a flush, burr-free, inspectable installation.

Figure 4-102. *CherryMAX® installation procedure.*

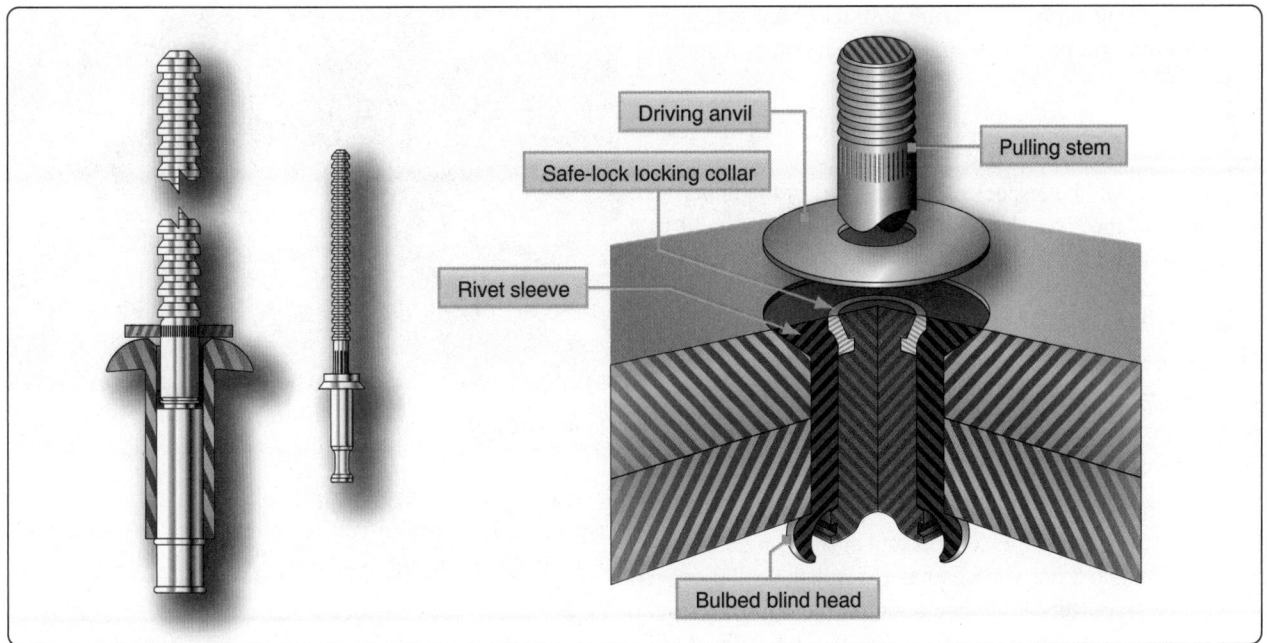

Figure 4-103. *CherryMAX® rivet.*

access poses yet another problem for the aviation technician. Designed for and used in difficult to reach locations means there is often no access to the blind side of the rivet or any way to provide support for the sheet metal surrounding the rivet's location when the aviation technician attempts removal.

The stem is mechanically locked by a small lock ring that needs to be removed first. Use a small center drill to provide a guide for a larger drill on top of the rivet stem and drill away the upper portion of the stem to destroy the lock. Try to remove the lock ring or use a prick punch or center punch to drive the stem down a little and remove the lock ring. After the lock ring is removed, the stem can be driven out with a drive punch. After the stem is removed, the rivet can be drilled out in the same way as a solid rivet. If possible, support the back side of the rivet with a backup block to prevent damage to the aircraft skin.

Pin Fastening Systems (High-Shear Fasteners)

A pin fastening system, or high-shear pin rivet, is a two-piece fastener that consists of a threaded pin and a collar. The metal collar is swaged onto the grooved end, effecting a firm tight fit. They are essentially threadless bolts.

High-shear rivets are installed with standard bucking bars and pneumatic riveting hammers. They require the use of a special gun set that incorporates collar swaging and trimming and a discharge port through which excess collar material is discharged. A separate size set is required for each shank diameter.

Installation of High-Shear Fasteners

Prepare holes for pin rivets with the same care as for other close tolerance rivets or bolts. At times, it may be necessary to spot-face the area under the head of the pin to ensure the head of the rivet fits tightly against the material. The spot-faced area should be ¹⁄₁₆-inch larger in diameter than the head diameter. Pin rivets may be driven from either end. Procedures for driving a pin rivet from the collar end are:

1. Insert the rivet in the hole.

2. Place a bucking bar against the rivet head.

3. Slip the collar over the protruding rivet end.

4. Place previously selected rivet set and gun over the collar. Align the gun until it is perpendicular to the material.

5. Depress the trigger on the gun, applying pressure to the rivet collar. This action causes the rivet collar to swage into the groove on the rivet end.

6. Continue the driving action until the collar is properly formed and excess collar material is trimmed off.

Procedures for driving a pin rivet from the head end are:

1. Insert the rivet in the hole.

2. Slip the collar over the protruding end of rivet.

3. Insert the correct size gun rivet set in a bucking bar and place the set against the collar of the rivet.

4. Apply pressure against the rivet head with a flush rivet set and pneumatic riveting hammer.

5. Continue applying pressure until the collar is formed in the groove and excess collar material is trimmed off.

Inspection

Pin rivets should be inspected on both sides of the material. The head of the rivet should not be marred and should fit tightly against the material.

Removal of Pin Rivets

The conventional method of removing rivets by drilling off the head may be utilized on either end of the pin rivet. Center punching is recommended prior to applying drilling pressure. In some cases, alternate methods may be needed:

- Grind a chisel edge on a small pin punch to a blade width of ⅛-inch. Place this tool at right angles to the collar and drive with a hammer to split the collar down one side. Repeat the operation on the opposite side. Then, with the chisel blade, pry the collar from the rivet. Tap the rivet out of the hole.

- Use a special hollow punch having one or more blades placed to split the collar. Pry the collar from the groove and tap out the rivet.

- Sharpen the cutting blades of a pair of nippers. Cut the collar in two pieces or use nippers at right angles to the rivet and cut through the small neck.

- A hollow-mill collar cutter can be used in a power hand drill to cut away enough collar material to permit the rivet to be tapped out of the work.

The high-shear pin rivet family includes fasteners, such as the Hi-Lok®, Hi-Tigue®, and Hi-Lite® made by Hi-Shear Corporation and the CherryBUCK® 95 KSI One-Piece Shear Pin and Cherry E-Z Buck® Shear Pin made by Cherry® Aerospace.

Hi-Lok® Fastening System

The threaded end of the Hi-Lok® two-piece fastener contains a hexagonal shaped recess. *[Figure 4-104]* The hex tip of an Allen wrench engages the recess to prevent rotation of the pin while the collar is being installed. The pin is designed in two basic head styles. For shear applications, the pin is made in countersunk style and in a compact protruding head style. For tension applications, the MS24694 countersunk and regular protruding head styles are available.

The self-locking, threaded Hi-Lok® collar has an internal counterbore at the base to accommodate variations in material thickness. At the opposite end of the collar is a wrenching device that is torqued by the driving tool until it shears off during installation, leaving the lower portion of the collar

Figure 4-104. *Hi-Lok®.*

seated with the proper torque without additional torque inspection. This shear-off point occurs when a predetermined preload or clamp-up is attained in the fastener during installation.

The advantages of Hi-Lok® two-piece fastener include its lightweight, high fatigue resistance, high strength, and its inability to be overtorqued. The pins, made from alloy steel, corrosion-resistant steel, or titanium alloy, come in many standard and oversized shank diameters. The collars are made of aluminum alloy, corrosion-resistant steel, or alloy steel. The collars have wrenching flats, fracture point, threads, and a recess. The wrenching flats are used to install the collar. The fracture point has been designed to allow the wrenching flats to shear when the proper torque has been reached. The threads match the threads of the pins and have been formed into an ellipse that is distorted to provide the locking action. The recess serves as a built-in washer. This area contains a portion of the shank and the transition area of the fastener.

The hole shall be prepared so that the maximum interference fit does not exceed 0.002-inch. This avoids build up of excessive internal stresses in the work adjacent to the hole. The Hi-Lok® pin has a slight radius under its head to increase fatigue life. After drilling, deburr the edge of the hole to allow the head to seat fully in the hole. The Hi-Lok® is installed in interference fit holes for aluminum structure and a clearance fit for steel, titanium, and composite materials.

Hi-Tigue® Fastening System

The Hi-Tigue® fastener offers all of the benefits of the Hi-Lok® fastening system along with a unique bead design that enhances the fatigue performance of the structure making it ideal for situations that require a controlled interference fit. The Hi-Tigue® fastener assembly consists of a pin and collar. These pin rivets have a radius at the transition area. During installation in an interference fit hole, the radius area will "cold work" the hole. These fastening systems can be easily confused, and visual reference should not be used for identification. Use part numbers to identify these fasteners.

Hi-Lite® Fastening System

The Hi-Lite® fastener is similar in design and principle to the Hi-Lok® fastener, but the Hi-Lite® fastener has a shorter transition area between the shank and the first load-bearing thread. Hi-Lite® has approximately one less thread. All Hi-Lite® fasteners are made of titanium.

These differences reduce the weight of the Hi-Lite® fastener without lessening the shear strength, but the Hi-Lite® clamping forces are less than that of a Hi-Lok® fastener. The Hi-Lite® collars are also different and thus are not interchangeable with Hi-Lok® collars. Hi-Lite® fasteners can be replaced with Hi-Lok® fasteners for most applications, but Hi-Loks® cannot be replaced with Hi-Lites®.

CherryBUCK® 95 KSI One-Piece Shear Pin

The CherryBUCK® is a bimetallic, one-piece fastener that combines a 95 KSI shear strength shank with a ductile, titanium-columbium tail. Theses fasteners are functionally interchangeable with comparable 6AI-4V titanium alloy two-piece shear fasteners, but with a number of advantages. Their one piece design means no foreign object damage (FOD), it has a 600 °F allowable temperature, and a very low backside profile.

Lockbolt Fastening Systems

Also pioneered in the 1940s, the lockbolt is a two-piece fastener that combines the features of a high-strength bolt and a rivet with advantages over each. *[Figure 4-105]* In general, a lockbolt is a nonexpanding fastener that has either a collar swaged into annular locking grooves on the pin shank or a type of threaded collar to lock it in place. Available with either countersunk or protruding heads, lockbolts are permanent type fasteners assemblies and consist of a pin and a collar.

A lockbolt is similar to an ordinary rivet in that the locking collar, or nut, is weak in tension and it is difficult to remove once installed. Some of the lockbolts are similar to blind

rivets and can be completely installed from one side. Others are fed into the workpiece with the manufactured head on the far side. The installation is completed on the near side with a gun similar to blind rivet gun. The lockbolt is easier and more quickly installed than the conventional rivet or bolt and eliminates the use of lockwashers, cotter pins, and special nuts. The lockbolt is generally used in wing splice fittings, landing gear fittings, fuel cell fittings, longerons, beams, skin splice plates, and other major structural attachment.

Often called huckbolts, lockbolts are manufactured by companies such as Cherry® Aerospace (Cherry® Lockbolt), Alcoa Fastening Systems (Hucktite® Lockbolt System), and SPS Technologies. Used primarily for heavily stressed structures that require higher shear and clamp-up values than can be obtained with rivets, the lockbolt and Hi-lok® are often used for similar applications. Lockbolts are made in various head styles, alloys, and finishes.

The lockbolt requires a pneumatic hammer or pull gun for installation. Lockbolts have their own grip gauge and an installation tool is required for their installation. *[Figure 4-106]* When installed, the lockbolt is rigidly and permanently locked in place. Three types of lockbolts are commonly used: pull-type, stump-type, and blind-type.

The pull-type lockbolt is mainly used in aircraft and primary and secondary structure. It is installed very rapidly and has approximately one-half the weight of equivalent AN steel bolts and nuts. A special pneumatic pull gun is required for installation of this type lockbolt, which can be performed by one operator since buckling is not required.

The stump-type lockbolt, although not having the extended stem with pull grooves, is a companion fastener to the pull-type lockbolt. It is used primarily where clearance does not permit effective installation of the pull-type lockbolt. It is driven with a standard pneumatic riveting hammer, with a hammer set attached for swaging the collar into the pin locking grooves, and a bucking bar.

The blind-type lockbolt comes as a complete unit or assembly and has exceptional strength and sheet pull-together characteristics. Blind-type lockbolts are used where only one side of the work is accessible and generally where it is difficult to drive a conventional rivet. This type lockbolt is installed in a manner similar to the pull-type lockbolt.

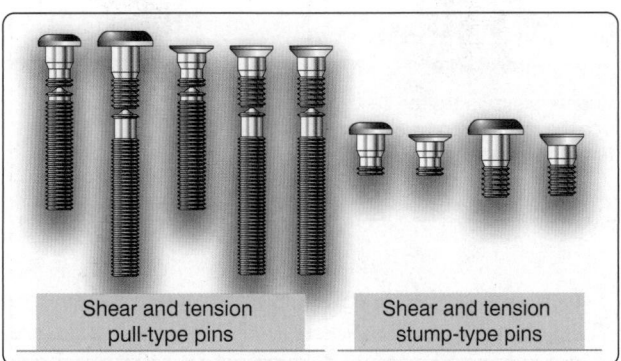

Figure 4-105. *Lockbolts.*

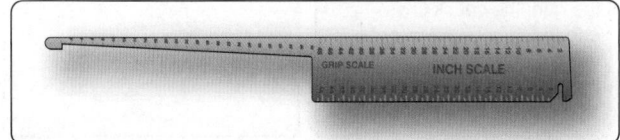

Figure 4-106. *Lockbolt grip gauge.*

The pins of pull- and stump-type lockbolts are made of heat-treated alloy steel or high-strength aluminum alloy. Companion collars are made of aluminum alloy or mild steel. The blind-type lockbolt consists of a heat-treated alloy steel pin, blind sleeve, filler sleeve, mild steel collar, and carbon steel washer.

These fasteners are used in shear and tension applications. The pull-type is more common and can be installed by one person. The stump type requires a two-person installation. An assembly tool is used to swage the collar onto the serrated grooves in the pin and break the stem flush to the top of the collar.

The easiest way to differentiate between tension and shear pins is the number of locking grooves. Tension pins normally have four locking grooves and shear pins have two locking grooves. The installation tooling preloads the pin while swaging the collar. The surplus end of the pin, called the pintail, is then fractured.

Installation Procedure

Installation of lockbolts involves proper drilling. The hole preparation for a lockbolt is similar to hole preparation for a Hi-Lok®. An interference fit is typically used for aluminum and a clearance fit is used for steel, titanium, and composite materials. [Figure 4-107]

Lockbolt Inspection

After installation, a lockbolt needs to be inspected to determine if installation is satisfactory. [Figure 4-108]

Inspect the lockbolt as follows:

1. The head must be firmly seated.

2. The collar must be tight against the material and have the proper shape and size.

3. Pin protrusion must be within limits.

Lockbolt Removal

The best way to remove a lockbolt is to remove the collar and

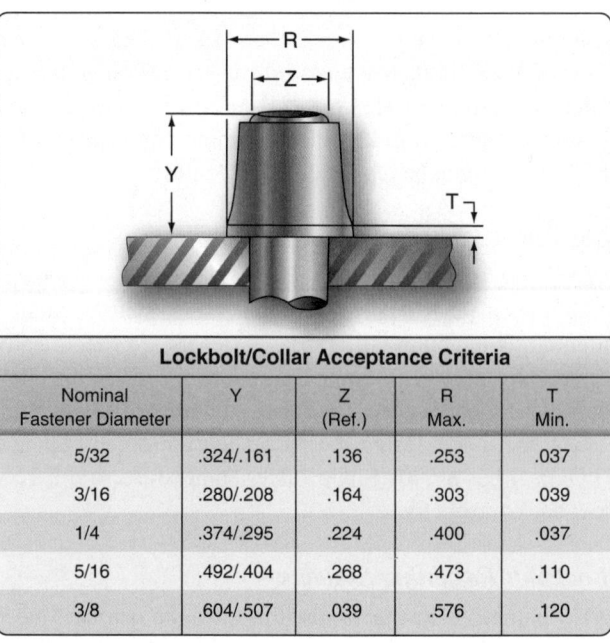

Lockbolt/Collar Acceptance Criteria				
Nominal Fastener Diameter	Y	Z (Ref.)	R Max.	T Min.
5/32	.324/.161	.136	.253	.037
3/16	.280/.208	.164	.303	.039
1/4	.374/.295	.224	.400	.037
5/16	.492/.404	.268	.473	.110
3/8	.604/.507	.039	.576	.120

Figure 4-108. *Lockbolt inspection.*

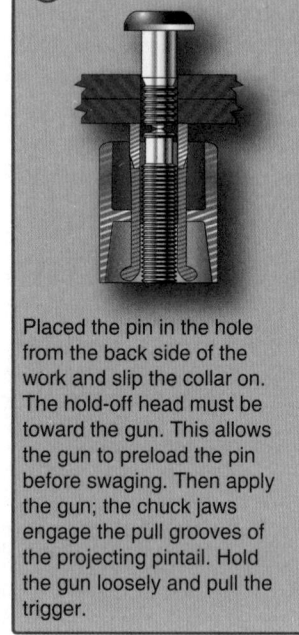

①	Placed the pin in the hole from the back side of the work and slip the collar on. The hold-off head must be toward the gun. This allows the gun to preload the pin before swaging. Then apply the gun; the chuck jaws engage the pull grooves of the projecting pintail. Hold the gun loosely and pull the trigger.

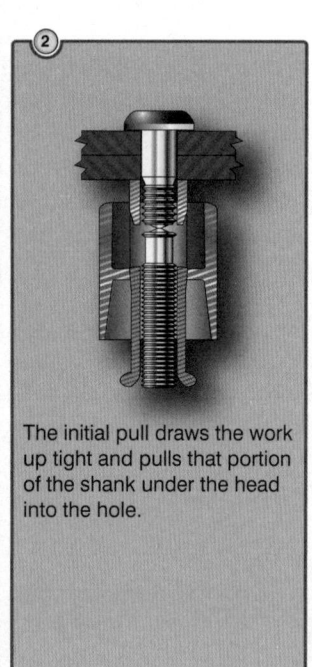

② The initial pull draws the work up tight and pulls that portion of the shank under the head into the hole.

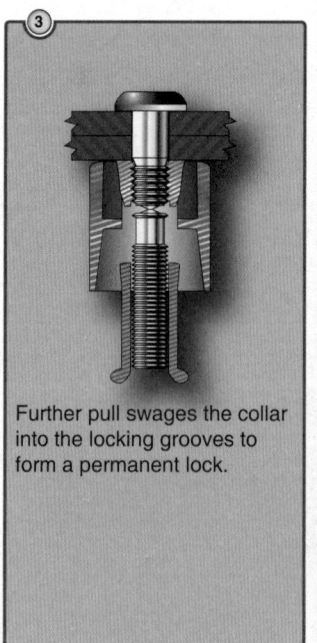

③ Further pull swages the collar into the locking grooves to form a permanent lock.

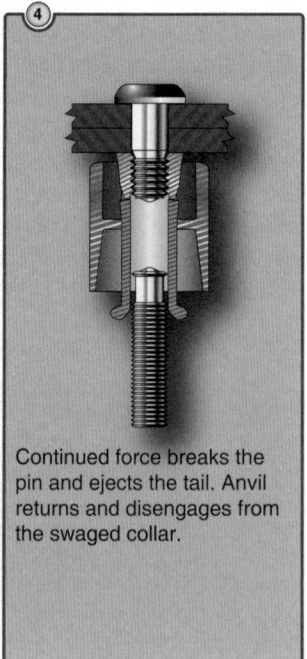

④ Continued force breaks the pin and ejects the tail. Anvil returns and disengages from the swaged collar.

Figure 4-107. *Lockbolt installation procedure.*

drive out the pin. The collar can be removed with a special collar cutter attached to a drill motor that mills off the collar without damaging the skin. If this is not possible, a collar splitter or small chisel can be used. Use a backup block on the opposite side to prevent elongation of the hole.

The Eddie-Bolt® 2 Pin Fastening System

The Eddie-Bolt® 2 looks similar to the Hi-Lok®, but has five flutes, equally spaced along a portion of the pin thread area. A companion threaded collar deforms into the flutes at a predetermined torque and locks the collar in place. The collar can be unscrewed using special tooling. This fastening system can be used in either clearance or interference-fit holes.

Blind Bolts

Bolts are threaded fasteners that support loads through pre-drilled holes. Hex, close-tolerance, and internal wrenching bolts are used in aircraft structural applications. Blind bolts have a higher strength than blind rivets and are used for joints that require high strength. Sometimes, these bolts can be direct replacements for the Hi-Lok® and lockbolt. Many of the new generation blind bolts are made from titanium and rated at 90 KSI shear strength, which is twice as much as most blind rivets.

Determining the correct length of the fastener is critical to correct installation. The grip length of a bolt is the distance from the underhead bearing surface to the first thread. The grip is the total thickness of material joined by the bolt. Ideally, the grip length should be a few thousands of an inch less than the actual grip to avoid bottoming the nut. Special grip gauges are inserted in the hole to determine the length of the blind bolt to be used. Every blind bolt system has its own grip gauge and is not interchangeable with other blind bolt or rivet systems.

Blind bolts are difficult to remove due to the hardness of the core bolt. A special removal kit is available from the manufacturer for removing each type of blind bolt. These kits make it easier to remove the blind bolt without damaging the hole and parent structure. Blind bolts are available in a pull-type and a drive-type.

Pull-Type Blind Bolt

Several companies manufacture the pull-type of blind bolt fastening systems. They may differ in some design aspects, but in general they have a similar function. The pull-type uses the drive nut concept and is composed of a nut, sleeve, and a draw bolt. Frequently used blind bolt systems include but are not limited to the Cherry Maxibolt® Blind Bolt system and the HuckBolt® fasteners which includes the Ti-Matic® Blind Bolt and the Unimatic® Advanced Bolt (UAB) blind bolt systems.

Cherry Maxibolt® Blind Bolt System

The Cherry Maxibolt® blind bolt, available in alloy steel and A-286 CRES materials, comes in four different nominal and oversized head styles. [Figure 4-109] One tool and pulling head installs all three diameters. The blind bolts create a larger blind side footprint and they provide excellent performance in thin sheet and nonmetallic applications. The flush breaking stem eliminates shaving while the extended grip range accommodates different application thicknesses. Cherry Maxibolts® are primarily used in structures where higher loads are required. The steel version is 112 KSI shear. The A286 version is 95 KSI shear. The Cherry® G83, G84, or G704 installation tools are required for installation.

Huck Blind Bolt System

The Huck Blind Bolt is a high strength vibration-resistant fastener. [Figure 4-110] These bolts have been used successfully in many critical areas, such as engine inlets and leading edge applications. All fasteners are installed with a combination of available hand, pneumatic, pneudraulic, or hydraulic pull-type tools (no threads) for ease of installation.

Huck Blind Bolts can be installed on blind side angle

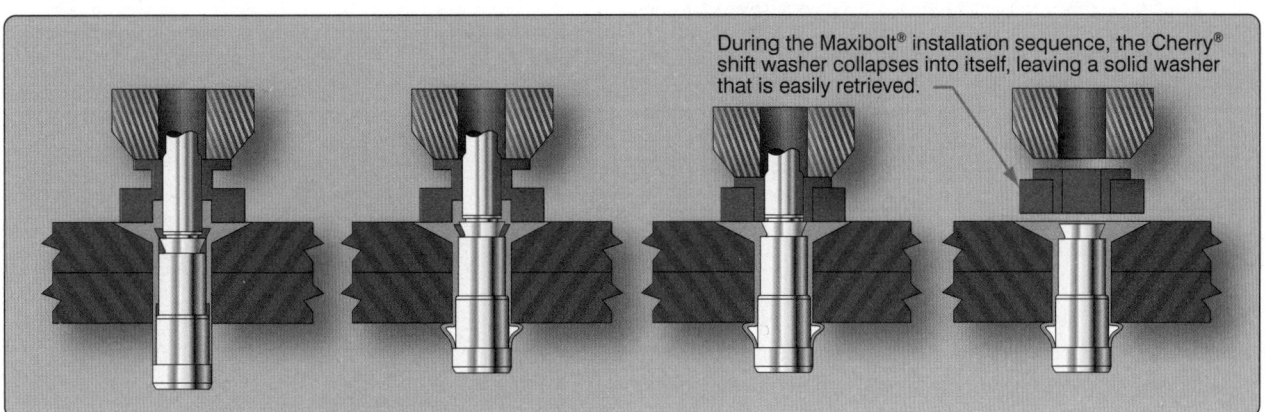

During the Maxibolt® installation sequence, the Cherry® shift washer collapses into itself, leaving a solid washer that is easily retrieved.

Figure 4-109. *Maxibolt® Blind Bolt System installation.*

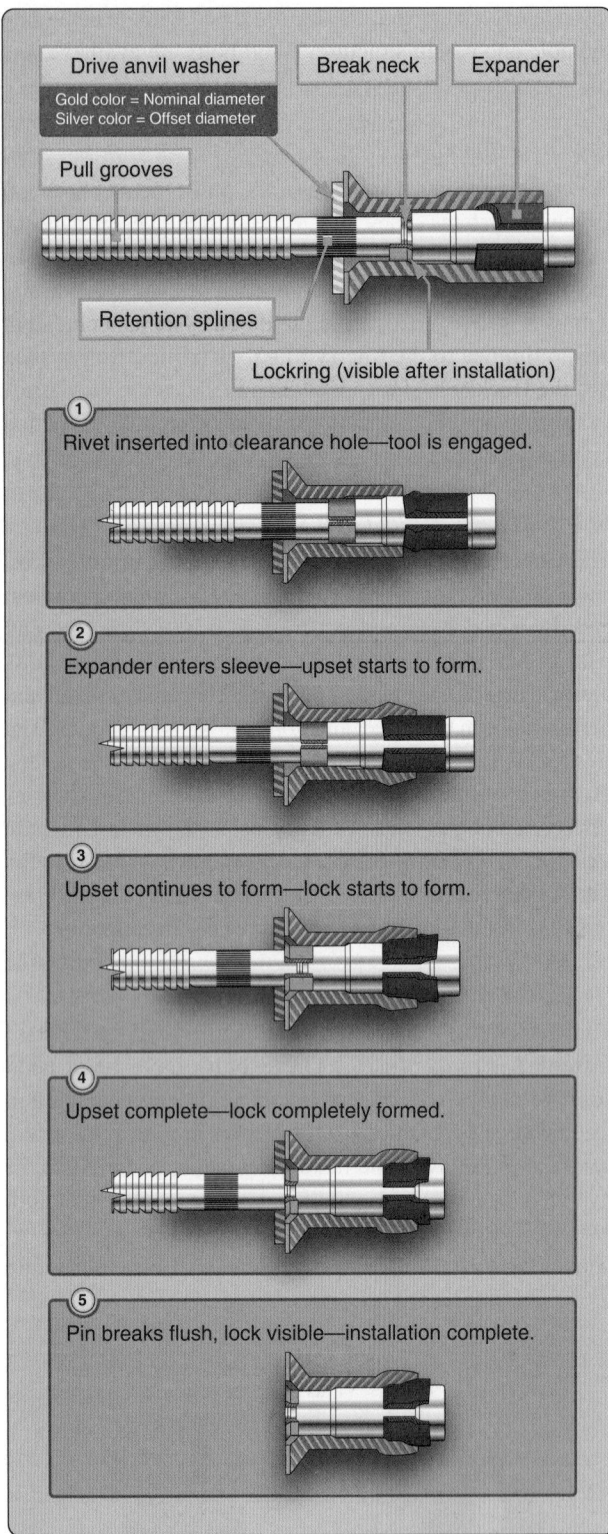

Figure 4-110. *Huck Blind Bolt system.*

surfaces up to 5° without loss of performance. The stem is mechanically locked to provide vibration-resistant FOD-free installations. The locking collar is forced into a conical pocket between stem and sleeve, creating high tensile capability. The lock collar fills the sleeve lock pocket to prevent leakage or corrosion pockets (crevice corrosion).

Flush head blind bolts are designed to install with a flush stem break that often requires no trimming for aerodynamic surfaces. The Huck Blind Bolt is available in high-strength A286 CRES at 95KSI shear strength in 5/32-inch through 3/8-inch diameters in 100° flush tension and protruding head. Also available are shear flush heads in 3/16-inch diameter. A286 CRES Huck Blind Bolts are also available in 1/64-inch oversize diameters for repair applications.

Drive Nut-Type of Blind Bolt

Jo-bolts, Visu-lok®, Composi-Lok®, OSI Bolt®, and Radial-Lok® fasteners use the drive nut concept and are composed of a nut, sleeve, and a draw bolt. *[Figure 4-111]* These types of blind bolts are used for high strength applications in metals and composites when there is no access to the blind side. Available in steel and titanium alloys, they are installed with special tooling. Both powered and hand tooling are available. During installation, the nut is held stationary while the core bolt is rotated by the installation tooling. The rotation of the core bolt draws the sleeve into the installed position and continues to retain the sleeve for the life of the fastener. The bolt has left hand threads and driving flats on the threaded end. A break-off relief allows the driving portion of the bolt to break off when the sleeve is properly seated. These types of bolts are available in many different head styles, including protruding head, 100° flush head, 130° flush head, and hex head.

Use the grip gauge available for the type of fastener and select the bolt grip after careful determination of the material thickness. The grip of the bolt is critical for correct installation. *[Figure 4-112]*

Installation procedure:

1. Install the fastener into the hole, and place the installation tooling over the screw (stem) and nut.

Figure 4-111. *Drive nut blind bolt.*

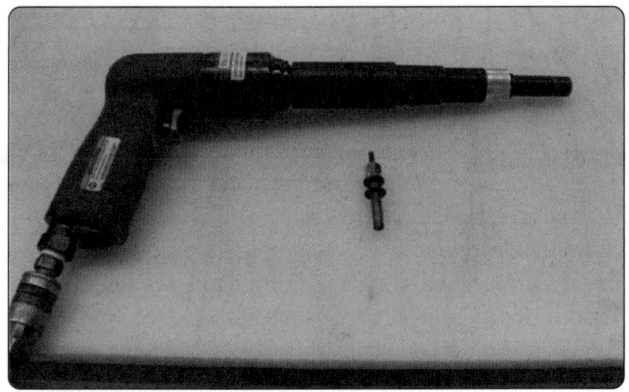

Figure 4-112. *Drive nut blind bolt installation tool.*

2. Apply torque to the screw with the installation tool while keeping the drive nut stationary. The screw continues to advance through the nut body causing the sleeve to be drawn up over the tapered nose of the nut. When the sleeve forms tightly against the blind side of the structure, the screw fractures in the break groove. The stem of Jo-bolts, Visu-lok®, and Composi-Lok® II fasteners does not break off flush with the head. A screw break-off shaver tool must be used if a flush installation is required. The stem of the newer Composi-Lok3® and OSI Bolt® break off flush.

Tapered Shank Bolt

Tapered shank bolts, such as the Taper-Lok®, are lightweight, high strength shear or tension bolts. This bolt has a tapered shank designed to provide an interference fit upon installation. Tapered shank bolts can be identified by a round head (rather than a screwdriver slot or wrench flats) and a threaded shank. The Taper-Lok® is comprised of a tapered, conical-shank fastener, installed into a precision tapered hole. The use of tapered shank bolts is limited to special applications such as high stress areas of fuel tanks. It is important that a tapered bolt not be substituted for any other type of fastener in repairs. It is equally as important not to substitute any other type of fastener for a tapered bolt.

Tapered shank bolts look similar to Hi-Lok® bolts after installation, but the tapered shank bolts do not have the hex recess at the threaded end of the bolt. Tapered shank bolts are installed in precision-reamed holes, with a controlled interference fit. The interference fit compresses the material around the hole that results in excellent load transfer, fatigue resistance, and sealing. The collar used with the tapered shank bolts has a captive washer, and no extra washers are required. New tapered shank bolt installation or rework of tapered shank bolt holes needs to be accomplished by trained personnel. Properly installed, these bolts become tightly wedged and do not turn while torque is applied to the nut.

Sleeve Bolts

Sleeve bolts are used for similar purposes as tapered shank bolts, but are easier to install. Sleeve bolts, such as the two piece SLEEVbolt®, consist of a tapered shank bolt in an expandable sleeve. The sleeve is internally tapered and externally straight. The sleeve bolt is installed in a standard tolerance straight hole. During installation, the bolt is forced into the sleeve. This action expands the sleeve which fills the hole. It is easier to drill a straight tolerance hole than it is to drill the tapered hole required for a tapered shank bolt.

Rivet Nut

The rivet nut is a blind installed, internally-threaded rivet invented in 1936 by the Goodrich Rubber Company for the purpose of attaching a rubber aircraft wing deicer extrusion to the leading edge of the wing. The original rivet nut is the Rivnut® currently manufactured by Bollhoff Rivnut Inc. The Rivnut® became widely used in the military and aerospace markets because of its many design and assembly advantages.

Rivet nuts are used for the installation of fairings, trim, and lightly loaded fittings that must be installed after an assembly is completed. *[Figure 4-113]* Often used for parts that are removed frequently, the rivet nut is available in two types: countersunk or flat head. Installed by crimping from one side, the rivet nut provides a threaded hole into which machine screws can be installed. Where a flush fit is required, the countersink style can be used. Rivet nuts made of alloy steel are used when increased tensile and shear strength is required.

Hole Preparation

Flat head rivet nuts require only the proper size of hole while flush installation can be made into either countersunk or dimpled skin. Metal thinner than the rivet nut head requires a dimple. The rivet nut size is selected according to the thickness of the parent material and the size of screw to be used. The part number identifies the type of rivet nut and the maximum grip length. Recommended hole sizes are shown in *Figure 4-114*.

Figure 4-113. *Rivet nut installation.*

Rivnut® Size	Drill Size	Hole Tolerance
No. 4	5/32	.155–.157
No. 6	#12	.189–.193
No. 8	#2	.221–.226

Figure 4-114. *Recommended hole sizes for rivet nut.*

Correct installation requires good hole preparation, removal of burrs, and holding the sheets in contact while heading. Like any sheet metal fastener, a rivet nut should fit snugly into its hole.

Blind Fasteners (Nonstructural)

Pop Rivets

Common pull-type pop rivets, produced for non-aircraft-related applications, are not approved for use on certificated aircraft structures or components. However, some homebuilt noncertificated aircraft use pull-type rivets for their structure. These types of rivets are typically made of aluminum and can be installed with hand tools.

Pull-Through Nutplate Blind Rivet

Nutplate blind rivets are used where the high shear strength of solid rivets is not required or if there is no access to install a solid rivet. The ³⁄₃₂-inch diameter blind rivet is most often used. The nut plate blind rivet is available with the pull-through and self-plugging locked spindle. *[Figure 4-115]*

The new Cherry® Rivetless Nut Plate, which replaces standard riveted nutplates, features a retainer that does not require flaring. This proprietary design eliminates the need for two additional rivet holes, as well as reaming, counterboring, and countersinking steps.

Forming Process

Before a part is attached to the aircraft during either manufacture or repair, it has to be shaped to fit into place.

This shaping process is called forming and may be a simple process, such as making one or two holes for attaching; it may be a complex process, such as making shapes with complex curvatures. Forming, which tends to change the shape or contour of a flat sheet or extruded shape, is accomplished by either stretching or shrinking the material in a certain area to produce curves, flanges, and various irregular shapes. Since the operation involves altering the shape of the stock material, the amount of shrinking and stretching almost entirely depends on the type of material used. Fully annealed (heated and cooled) material can withstand considerably more stretching and shrinking and can be formed at a much smaller bend radius than when it is in any of the tempered conditions.

When aircraft parts are formed at the factory, they are made on large presses or by drop hammers equipped with dies of the correct shape. Factory engineers, who designate specifications for the materials to be used to ensure the finished part has the correct temper when it leaves the machines, plan every part. Factory draftsmen prepare a layout for each part. *[Figure 4-116]*

Forming processes used on the flight line and those practiced in the maintenance or repair shop cannot duplicate a manufacturer's resources, but similar techniques of factory metal working can be applied in the handcrafting of repair parts.

Forming usually involves the use of extremely light-gauge alloys of a delicate nature that can be readily made useless by coarse and careless workmanship. A formed part may seem outwardly perfect, yet a wrong step in the forming procedure may leave the part in a strained condition. Such a defect may hasten fatigue or may cause sudden structural failure.

Of all the aircraft metals, pure aluminum is the most easily formed. In aluminum alloys, ease of forming varies with

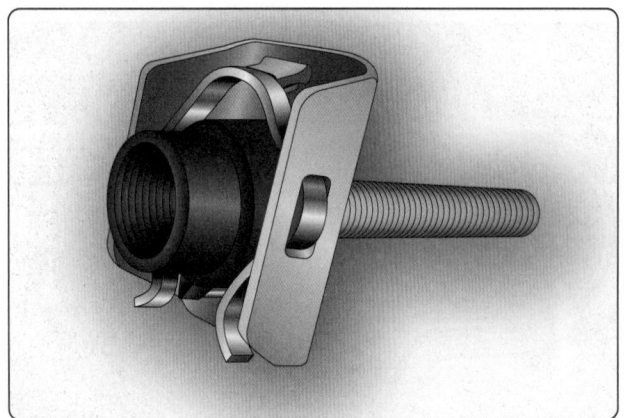

Figure 4-115. *Rivetless pull-through nutplate.*

Figure 4-116. *Aircraft formed at a factory.*

the temper condition. Since modern aircraft are constructed chiefly of aluminum and aluminum alloys, this section deals with the procedures for forming aluminum or aluminum alloy parts with a brief discussion of working with stainless steel, magnesium, and titanium.

Most parts can be formed without annealing the metal, but if extensive forming operations, such as deep draws (large folds) or complex curves, are planned, the metal should be in the dead soft or annealed condition. During the forming of some complex parts, operations may need to be stopped and the metal annealed before the process can be continued or completed. For example, alloy 2024 in the "0" condition can be formed into almost any shape by the common forming operations, but it must be heat treated afterward.

Forming Operations & Terms

Forming requires either stretching or shrinking the metal, or sometimes doing both. Other processes used to form metal include bumping, crimping, and folding.

Stretching

Stretching metal is achieved by hammering or rolling metal under pressure. For example, hammering a flat piece of metal causes the material in the hammered area to become thinner in that area. Since the amount of metal has not been decreased, the metal has been stretched. The stretching process thins, elongates, and curves sheet metal. It is critical to ensure the metal is not stretched too much, making it too thin, because sheet metal does not rebound easily. *[Figure 4-117]*

Stretching one portion of a piece of metal affects the surrounding material, especially in the case of formed and extruded angles. For example, hammering the metal in the horizontal flange of the angle strip over a metal block causes its length to increase (stretched), making that section longer than the section near the bend. To allow for this difference in length, the vertical flange, which tends to keep the material near the bend from stretching, would be forced to curve away from the greater length.

Shrinking

Shrinking metal is much more difficult than stretching it. During the shrinking process, metal is forced or compressed into a smaller area. This process is used when the length of a piece of metal, especially on the inside of a bend, is to be reduced. Sheet metal can be shrunk in by hammering on a V-block or by crimping and then using a shrinking block.

To curve the formed angle by the V-block method, place the angle on the V-block and gently hammer downward against the upper edge directly over the "V." While hammering, move the angle back and forth across the V-block to compress the material along the upper edge. Compression of the material along the upper edge of the vertical flange will cause the formed angle to take on a curved shape. The material in the horizontal flange will merely bend down at the center, and the length of that flange will remain the same. *[Figure 4-118]*

To make a sharp curve or a sharply bent flanged angle, crimping and a shrinking block can be used. In this process, crimps are placed in the one flange, and then by hammering the metal on a shrinking block, the crimps are driven, or shrunk, one at a time.

Cold shrinking requires the combination of a hard surface, such as wood or steel, and a soft mallet or hammer because a steel hammer over a hard surface stretches the metal, as opposed to shrinking it. The larger the mallet face is, the better.

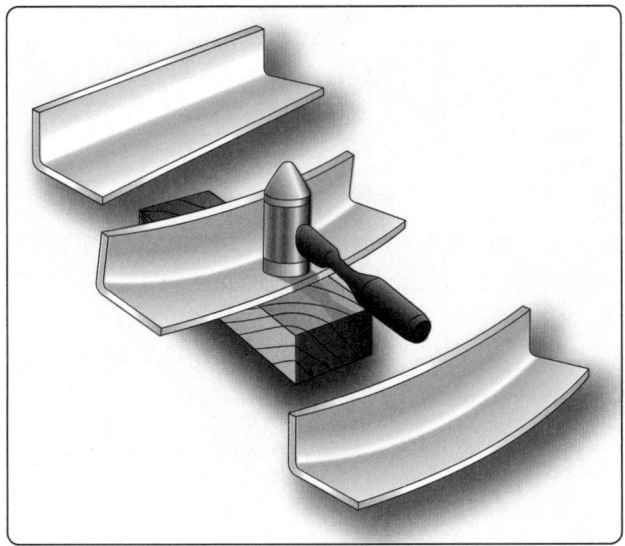

Figure 4-117. *Stretch forming metal.*

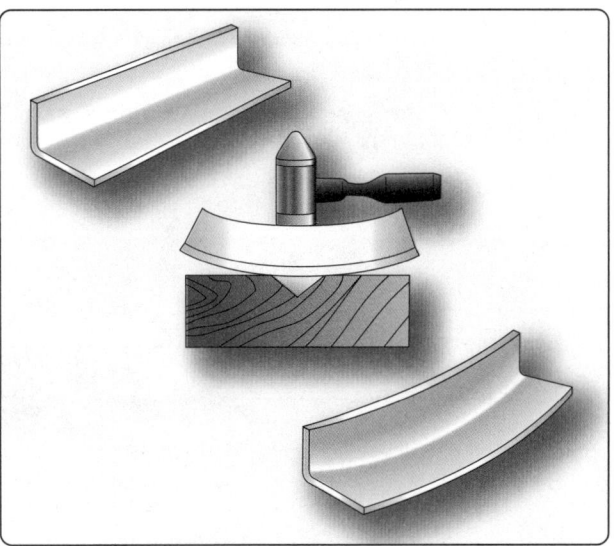

Figure 4-118. *Shrink forming metal.*

Bumping

Bumping involves shaping or forming malleable metal by hammering or tapping—usually with a rubber, plastic, or rawhide mallet. During this process, the metal is supported by a dolly, a sandbag, or a die. Each contains a depression into which hammered portions of the metal can sink. Bumping can be done by hand or by machine.

Crimping

Crimping is folding, pleating, or corrugating a piece of sheet metal in a way that shortens it or turning down a flange on a seam. It is often used to make one end of a piece of stove pipe slightly smaller so that one section may be slipped into another. Crimping one side of a straight piece of angle iron with crimping pliers causes it to curve. *[Figure 4-119]*

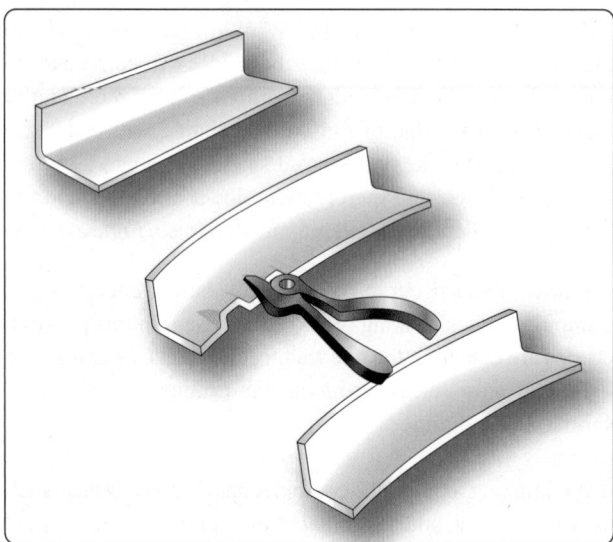

Figure 4-119. *Crimping metal.*

Folding Sheet Metal

Folding sheet metal is to make a bend or crease in sheets, plates, or leaves. Folds are usually thought of as sharp, angular bends and are generally made on folding machines such as the box and pan brake discussed earlier in this chapter.

Layout & Forming
Terminology

The following terms are commonly used in sheet metal forming and flat pattern layout. Familiarity with these terms aids in understanding how bend calculations are used in a bending operation. *Figure 4-120* illustrates most of these terms.

Base measurement—the outside dimensions of a formed part. Base measurement is given on the drawing or blueprint or may be obtained from the original part.

Leg—the longer part of a formed angle.

Flange—the shorter part of a formed angle—the opposite of leg. If each side of the angle is the same length, then each is known as a leg.

Grain of the metal—natural grain of the material is formed as the sheet is rolled from molten ingot. Bend lines should be made to lie at a 90° angle to the grain of the metal if possible.

Bend allowance (BA)—refers to the curved section of metal within the bend (the portion of metal that is curved in bending). The bend allowance may be considered as being the length of the curved portion of the neutral line.

Bend radius—the arc is formed when sheet metal is bent. This arc is called the bend radius. The bend radius is measured

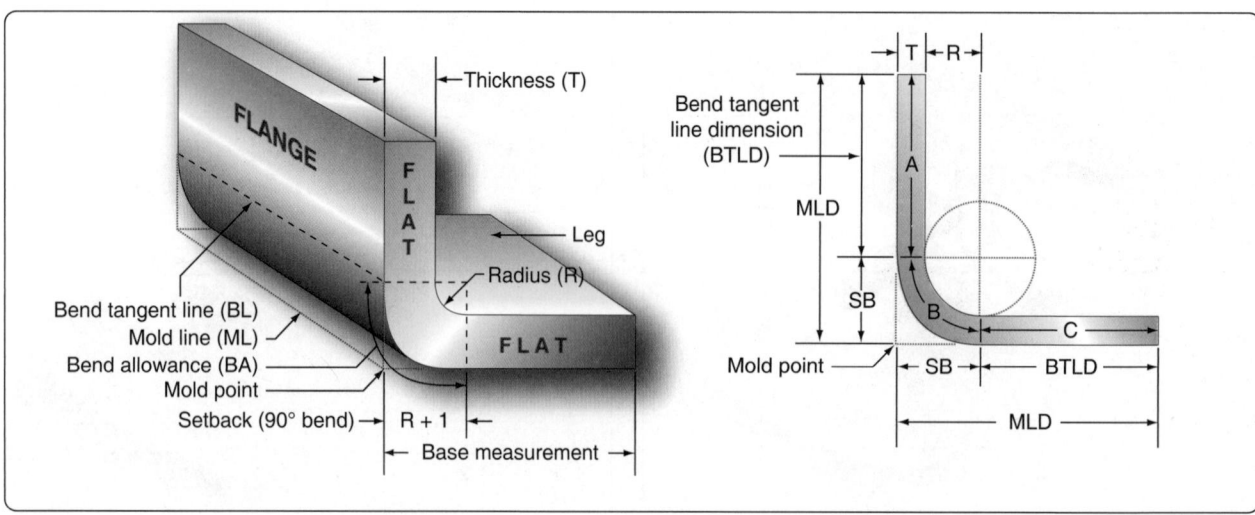

Figure 4-120. *Bend allowance terminology.*

from a radius center to the inside surface of the metal. The minimum bend radius depends on the temper, thickness, and type of material. Always use a Minimum Bend Radius Table to determine the minimum bend radius for the alloy that is going to be used. Minimum bend radius charts can be found in manufacturer's maintenance manuals.

Bend tangent line (BL)—the location at which the metal starts to bend and the line at which the metal stops curving. All the space between the bend tangent lines is the bend allowance.

Neutral axis—an imaginary line that has the same length after bending as it had before bending. *[Figure 4-121]* After bending, the bend area is 10 to 15 percent thinner than before bending. This thinning of the bend area moves the neutral line of the metal in towards the radius center. For calculation purposes, it is often assumed that the neutral axis is located at the center of the material, although the neutral axis is not exactly in the center of the material. However, the amount of error incurred is so slight that, for most work, assuming it is at the center is satisfactory.

Mold line (ML)—an extension of the flat side of a part beyond the radius.

Mold line dimension (MLD)—the dimension of a part made by the intersection of mold lines. It is the dimension the part would have if its corners had no radius.

Mold point—the point of intersection of the mold lines. The mold point would be the outside corner of the part if there were no radius.

K-Factor—the percentage of the material thickness where there is no stretching or compressing of the material, such as the neutral axis. This percentage has been calculated and is one of 179 numbers on the K chart corresponding to one of the angles between 0° and 180° to which metal can be bent. *[Figure 4-122]* Whenever metal is to be bent to any angle other than 90° (K-factor of 90° equal to 1), the corresponding K-factor number is selected from the chart and is multiplied by the sum of the radius (R) and the thickness (T) of the metal. The product is the amount of setback (see next paragraph) for the bend. If no K chart is available, the K-factor can be calculated with a calculator by using the following formula: the K value is the tangent of one-half the bend angle.

Setback (SB)—the distance the jaws of a brake must be setback from the mold line to form a bend. In a 90° bend, SB = R + T (radius of the bend plus thickness of the metal). The setback dimension must be determined prior to making the bend because setback is used in determining the location of the beginning bend tangent line. When a part has more than one bend, setback must be subtracted for each bend. The majority of bends in sheet metal are 90° bends. The K-factor must be used for all bends that are smaller or larger than 90°.

$$SB = K(R+T)$$

Sight line—also called the bend or brake line, it is the layout line on the metal being formed that is set even with the nose of the brake and serves as a guide in bending the work.

Flat—that portion of a part that is not included in the bend. It is equal to the base measurement (MLD) minus the setback. Flat = MLD – SB

Closed angle—an angle that is less than 90° when measured between legs, or more than 90° when the amount of bend is measured.

Open angle—an angle that is more than 90° when measured between legs, or less than 90° when the amount of bend is measured.

Total developed width (TDW)—the width of material measured around the bends from edge to edge. Finding the TDW is necessary to determine the size of material to be cut. The TDW is less than the sum of mold line dimensions since the metal is bent on a radius and not to a square corner as mold line dimensions indicate.

Layout or Flat Pattern Development

To prevent any waste of material and to get a greater degree of accuracy in the finished part, it is wise to make a layout or flat pattern of a part before forming it. Construction of interchangeable structural and nonstructural parts is achieved by forming flat sheet stock to make channel, angle, zee, or hat section members. Before a sheet metal part is formed, make a flat pattern to show how much material is required in the bend areas, at what point the sheet must be inserted into the forming tool, or where bend lines are located. Bend lines must be determined to develop a flat pattern for sheet metal forming.

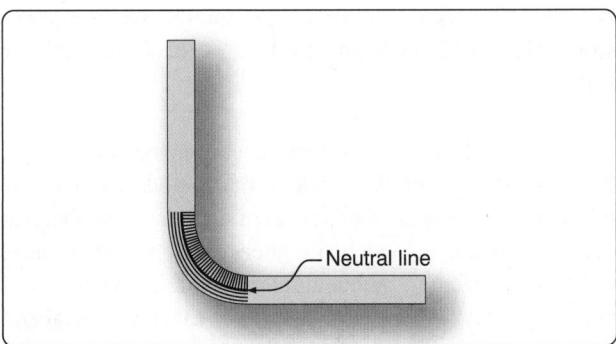

Figure 4-121. *Neutral line.*

Degree	K	Degree	K	Degree	K	Degree	K	Degree	K
1	0.0087	37	0.3346	73	0.7399	109	1.401	145	3.171
2	0.0174	38	0.3443	74	0.7535	110	1.428	146	3.270
3	0.0261	39	0.3541	75	0.7673	111	1.455	147	3.375
4	0.0349	40	0.3639	76	0.7812	112	1.482	148	3.487
5	0.0436	41	0.3738	77	0.7954	113	1.510	149	3.605
6	0.0524	42	0.3838	78	0.8097	114	1.539	150	3.732
7	0.0611	43	0.3939	79	0.8243	115	1.569	151	3.866
8	0.0699	44	0.4040	80	0.8391	116	1.600	152	4.010
9	0.0787	45	0.4142	81	0.8540	117	1.631	153	4.165
10	0.0874	46	0.4244	82	0.8692	118	1.664	154	4.331
11	0.0963	47	0.4348	83	0.8847	119	1.697	155	4.510
12	0.1051	48	0.4452	84	0.9004	120	1.732	156	4.704
13	0.1139	49	0.4557	85	0.9163	121	1.767	157	4.915
14	0.1228	50	0.4663	86	0.9324	122	1.804	158	5.144
15	0.1316	51	0.4769	87	0.9489	123	1.841	159	5.399
16	0.1405	52	0.4877	88	0.9656	124	1.880	160	5.671
17	0.1494	53	0.4985	89	0.9827	125	1.921	161	5.975
18	0.1583	54	0.5095	90	1.000	126	1.962	162	6.313
19	0.1673	55	0.5205	91	1.017	127	2.005	163	6.691
20	0.1763	56	0.5317	92	1.035	128	2.050	164	7.115
21	0.1853	57	0.5429	93	1.053	129	2.096	165	7.595
22	0.1943	58	0.5543	94	1.072	130	2.144	166	8.144
23	0.2034	59	0.5657	95	1.091	131	2.194	167	8.776
24	0.2125	60	0.5773	96	1.110	132	2.246	168	9.514
25	0.2216	61	0.5890	97	1.130	133	2.299	169	10.38
26	0.2308	62	0.6008	98	1.150	134	2.355	170	11.43
27	0.2400	63	0.6128	99	1.170	135	2.414	171	12.70
28	0.2493	64	0.6248	100	1.191	136	2.475	172	14.30
29	0.2586	65	0.6370	101	1.213	137	2.538	173	16.35
30	0.2679	66	0.6494	102	1.234	138	2.605	174	19.08
31	0.2773	67	0.6618	103	1.257	139	2.674	175	22.90
32	0.2867	68	0.6745	104	1.279	140	2.747	176	26.63
33	0.2962	69	0.6872	105	1.303	141	2.823	177	38.18
34	0.3057	70	0.7002	106	1.327	142	2.904	178	57.29
35	0.3153	71	0.7132	107	1.351	143	2.988	179	114.59
36	0.3249	72	0.7265	108	1.376	144	3.077	180	Inf.

Figure 4-122. *K-factor.*

When forming straight angle bends, correct allowances must be made for setback and bend allowance. If shrinking or stretching processes are to be used, allowances must be made so that the part can be turned out with a minimum amount of forming.

Making Straight Line Bends

When forming straight bends, the thickness of the material, its alloy composition, and its temper condition must be considered. Generally speaking, the thinner the material is, the more sharply it can be bent (the smaller the radius of bend), and the softer the material is, the sharper the bend

is. Other factors that must be considered when making straight line bends are bend allowance, setback, and brake or sight line.

The radius of bend of a sheet of material is the radius of the bend as measured on the inside of the curved material. The minimum radius of bend of a sheet of material is the sharpest curve, or bend, to which the sheet can be bent without critically weakening the metal at the bend. If the radius of bend is too small, stresses and strains weaken the metal and may result in cracking.

A minimum radius of bend is specified for each type of aircraft sheet metal. The minimum bend radius is affected by the kind of material, thickness of the material, and temper condition of the material. Annealed sheet can be bent to a radius approximately equal to its thickness. Stainless steel and 2024-T3 aluminum alloy require a fairly large bend radius.

Bending a U-Channel

To understand the process of making a sheet metal layout, the steps for determining the layout of a sample U-channel

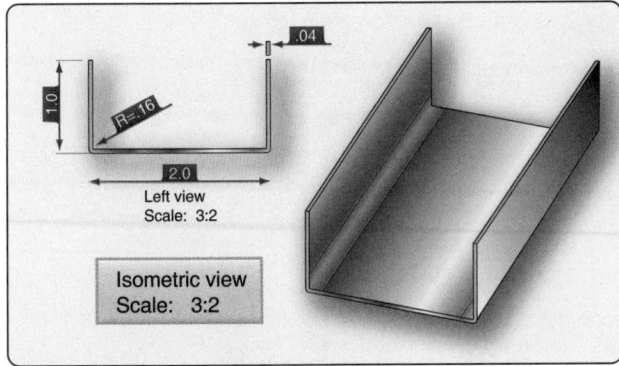

Figure 4-123. *U-channel example.*

will be discussed. *[Figure 4-123]* When using bend allowance calculations, the following steps for finding the total developed length can be computed with formulas, charts, or computer-aided design (CAD) and computer-aided manufacturing (CAM) software packages. This channel is made of 0.040-inch 2024-T3 aluminum alloy.

Step 1: Determine the Correct Bend Radius

Minimum bend radius charts are found in manufacturers' maintenance manuals. A radius that is too sharp cracks the material during the bending process. Typically, the drawing indicates the radius to use, but it is a good practice to double check. For this layout example, use the minimum radius chart in *Figure 4-124* to choose the correct bend radius for the alloy, temper, and the metal thickness. For 0.040, 2024-T3 the minimum allowable radius is 0.16-inch or ⁵⁄₃₂-inch.

Step 2: Find the Setback

The setback can be calculated with a formula or can be found in a setback chart available in aircraft maintenance manuals or Source, Maintenance, and Recoverability books (SMRs). *[Figure 4-125]*

CHART 204 MINIMUM BEND RADIUS FOR ALUMINUM ALLOYS						
Thickness	**5052-0 6061-0 5052-H32**	**7178-0 2024-0 5052-H34 6061-T4 7075-0**	**6061-T6**	**7075-T6**	**2024-T3 2024-T4**	**2024-T6**
.012	.03	.03	.03	.03	.06	.06
.016	.03	.03	.03	.03	.09	.09
.020	.03	.03	.03	.12	.09	.09
.025	.03	.03	.06	.16	.12	.09
.032	.03	.03	.06	.19	.12	.12
.040	.06	.06	.09	.22	.16	.16
.050	.06	.06	.12	.25	.19	.19
.063	.06	.09	.16	.31	.22	.25
.071	.09	.12	.16	.38	.25	.31
.080	.09	.16	.19	.44	.31	.38
.090	.09	.19	.22	.50	.38	.44
.100	.12	.22	.25	.62	.44	.50
.125	.12	.25	.31	.88	.50	.62
.160	.16	.31	.44	1.25	.75	.75
.190	.19	.38	.56	1.38	1.00	1.00
.250	.31	.62	.75	2.00	1.25	1.25
.312	.44	1.25	1.38	2.50	1.50	1.50
.375	.44	1.38	1.50	2.50	1.88	1.88
Bend radius is designated to the inside of the bend. All dimensions are in inches.						

Figure 4-124. *Minimum bend radius (from the Raytheon Aircraft Structural Inspection and Repair Manual).*

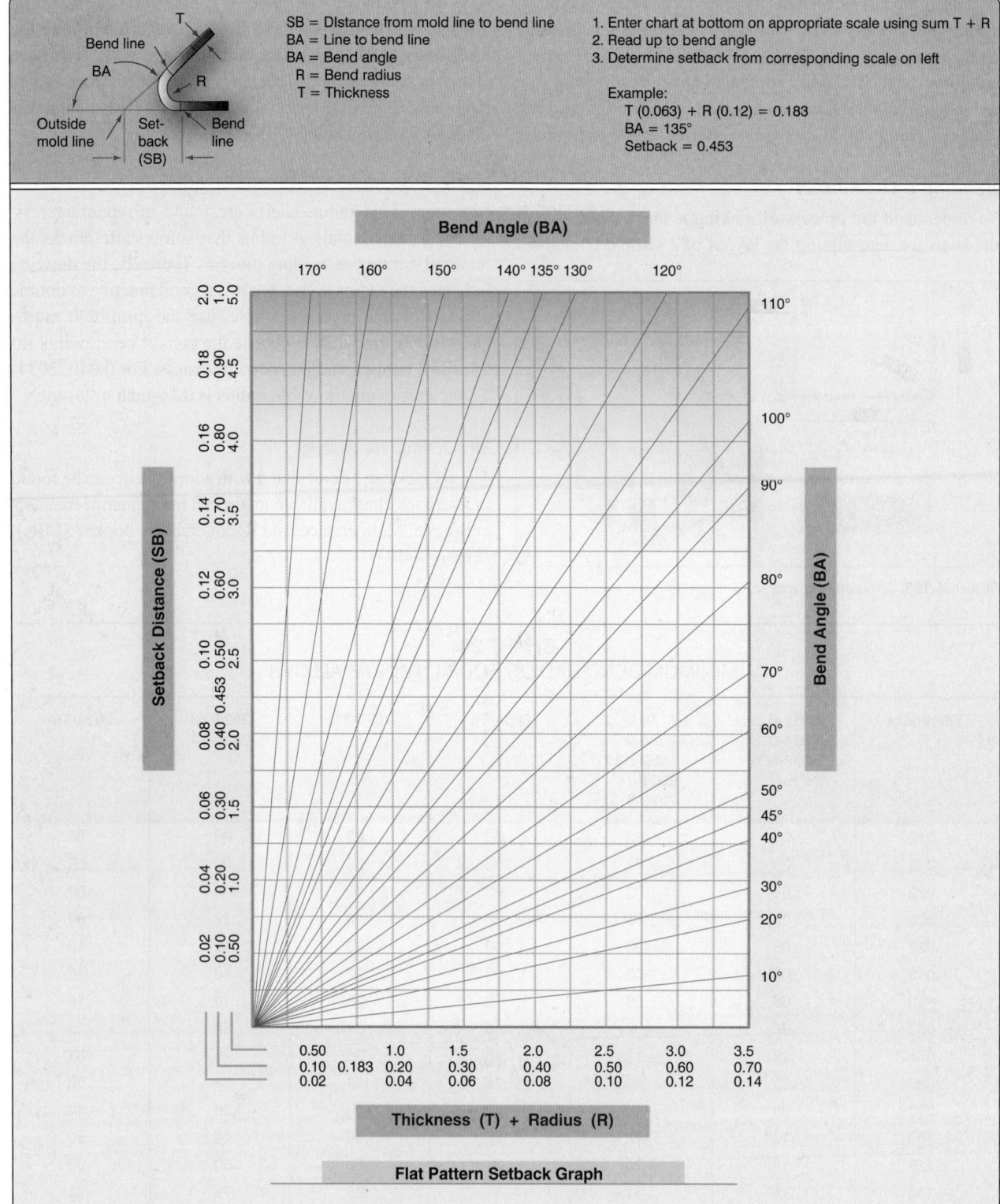

The chart contains the following labels and values:

SB = DIstance from mold line to bend line
BA = Line to bend line
BA = Bend angle
R = Bend radius
T = Thickness

1. Enter chart at bottom on appropriate scale using sum T + R
2. Read up to bend angle
3. Determine setback from corresponding scale on left

Example:
T (0.063) + R (0.12) = 0.183
BA = 135°
Setback = 0.453

Bend line
BA
Outside mold line
Set-back (SB)
Bend line
T
R

Bend Angle (BA)

170° 160° 150° 140° 135° 130° 120°

110°
100°
90°
80°
70°
60°
50°
45°
40°
30°
20°
10°

Setback Distance (SB)

2.0 1.0 5.0
0.18 0.90 4.5
0.16 0.80 4.0
0.14 0.70 3.5
0.12 0.60 3.0
0.10 0.50 2.5
0.08 0.453 2.0 0.40
0.06 0.30 1.5
0.04 0.20 1.0
0.02 0.10 0.50

Bend Angle (BA)

0.50		1.0	1.5	2.0	2.5	3.0	3.5
0.10	0.183	0.20	0.30	0.40	0.50	0.60	0.70
0.02		0.04	0.06	0.08	0.10	0.12	0.14

Thickness (T) + Radius (R)

Flat Pattern Setback Graph

Figure 4-125. *Setback chart.*

Using a Formula to Calculate the Setback

SB = setback

K = K-factor (K is 1 for 90° bends)

R = inside radius of the bend

T = material thickness

Since all of the angles in this example are 90° angles, the setback is calculated as follows:

SB = K(R+T) = 0.2 inches

Note: K = 1 for a 90° bend. For other than a 90° bend, use a K-factor chart.

Using a Setback Chart to Find the Setback

The setback chart is a quick way to find the setback and is useful for open and closed bends, because there is no need to calculate or find the K-factor. Several software packages and online calculators are available to calculate the setback. These programs are often used with CAD/CAM programs. *[Figure 4-125]*

- Enter chart at the bottom on the appropriate scale with the sum of the radius and material thickness.

- Read up to the bend angle.

- Find the setback from corresponding scale on the left.

Example:

- Material thickness is 0.063-inch.

- Bend angle is 135°.

- R + T = 0.183-inch.

Find 0.183 at the bottom of the graph. It is found in the middle scale.

- Read up to a bend angle of 135°.

- Locate the setback at the left hand side of the graph in the middle scale (0.435-inch). *[Figure 4-125]*

Step 3: Find the Length of the Flat Line Dimension

The flat line dimension can be found using the formula:

Flat = MLD – SB

MLD = mold line dimension

SB = setback

The flats, or flat portions of the U-channel, are equal to the mold line dimension minus the setback for each of the sides, and the mold line length minus two setbacks for the center flat. Two setbacks need to be subtracted from the center flat because this flat has a bend on either side.

The flat dimension for the sample U-channel is calculated in the following manner:

Flat dimension = MLD – SB

Flat 1 = 1.00-inch – 0.2-inch = 0.8-inch

Flat 2 = 2.00-inch – (2 × 0.2-inch) = 1.6-inch

Flat 3 = 1.00-inch – 0.2-inch = 0.8-inch

Step 4: Find the Bend Allowance

When making a bend or fold in a piece of metal, the bend allowance or length of material required for the bend must be calculated. Bend allowance depends on four factors: degree of bend, radius of the bend, thickness of the metal, and type of metal used.

The radius of the bend is generally proportional to the thickness of the material. Furthermore, the sharper the radius of bend, the less the material that is needed for the bend. The type of material is also important. If the material is soft, it can be bent very sharply; but if it is hard, the radius of bend is greater, and the bend allowance is greater. The degree of bend affects the overall length of the metal, whereas the thickness influences the radius of bend.

Bending a piece of metal compresses the material on the inside of the curve and stretches the material on the outside of the curve. However, at some distance between these two extremes lies a space which is not affected by either force. This is known as the neutral line or neutral axis and occurs at a distance approximately 0.445 times the metal thickness ($0.445 \times T$) from the inside of the radius of the bend. *[Figure 4-126]*

The length of this neutral axis must be determined so that sufficient material can be provided for the bend. This is called the bend allowance. This amount must be added to the overall length of the layout pattern to ensure adequate material for the bend. To save time in calculation of the bend allowance, formulas and charts for various angles, radii of bends, material thicknesses, and other factors have been developed.

Formula 1: Bend Allowance for a 90° Bend

To the radius of bend (R) add ½ the thickness of the metal (½T). This gives R + ½T, or the radius of the circle of the neutral axis. *[Figure 4-127]* Compute the circumference of this circle by multiplying the radius of the neutral line (R + ½T) by 2π (*Note:* π = 3.1416): 2π (R + ½T). Since a 90° bend is a quarter of the circle, divide the circumference by 4. This gives:

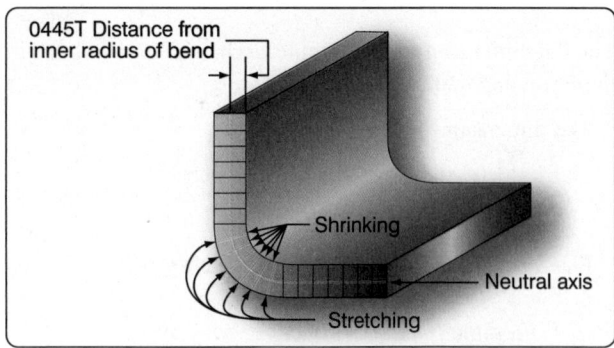

Figure 4-126. *Neutral axis and stresses resulting from bending.*

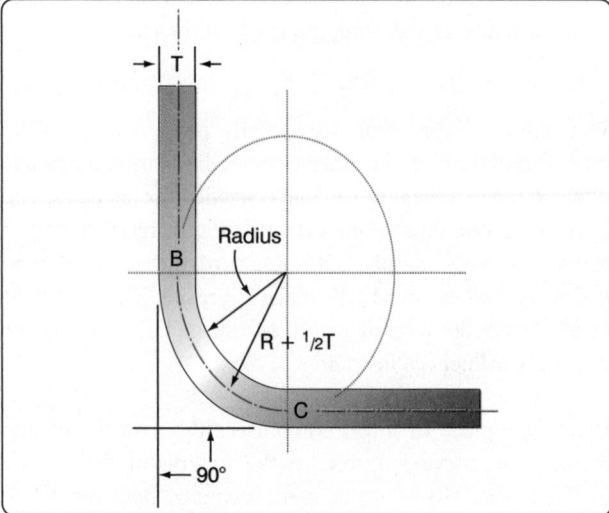

Figure 4-127. *Bend allowance for a 90° bend.*

$$\frac{2\pi\,(R + \tfrac{1}{2}T)}{4}$$

This is the bend allowance for a 90° bend. To use the formula for a 90° bend having a radius of ¼ inch for material 0.051-inch thick, substitute in the formula as follows.

$$\text{Bend allowance} = \frac{(2 \times 3.1416)(0.250 + \tfrac{1}{2}(0.051))}{4}$$

$$= \frac{6.2832(0.250 + 0.0255)}{4}$$

$$= \frac{6.2832(0.2755)}{4}$$

$$= 0.4327$$

The bend allowance, or the length of material required for the bend, is 0.4327 or ⁷⁄₁₆-inch.

Formula 2: Bend Allowance for a 90° Bend

This formula uses two constant values that have evolved over a period of years as being the relationship of the degrees in the bend to the thickness of the metal when determining the bend allowance for a particular application. By experimentation with actual bends in metals, aircraft engineers have found that accurate bending results could be obtained by using the following formula for any degree of bend from 1° to 180°.

Bend allowance = (0.01743R + 0.0078T)N where:

R = the desired bend radius

T = the thickness of the metal

N = number of degrees of bend

To use this formula for a 90° bend having a radius of .16-inch for material 0.040-inch thick, substitute in the formula as follows:

Bend allowance =
(0.01743 × 0.16) + (0.0078 × 0.040) × 90 = 0.27 inches

Use of Bend Allowance Chart for a 90° Bend

In *Figure 4-128,* the radius of bend is shown on the top line, and the metal thickness is shown on the left hand column. The upper number in each cell is the bend allowance for a 90° bend. The lower number in the cell is the bend allowance per 1° of bend. To determine the bend allowance for a 90° bend, simply use the top number in the chart.

Example: The material thickness of the U-channel is 0.040-inch and the bend radius is 0.16-inch.

Reading across the top of the bend allowance chart, find the column for a radius of bend of .156-inch. Now, find the block in this column that is opposite the material thickness (gauge) of 0.040 in the column at the left. The upper number in the cell is (0.273), the correct bend allowance in inches for a 90° bends.

Several bend allowance calculation programs are available online. Just enter the material thickness, radius, and degree of bend and the computer program calculates the bend allowance.

Use of Chart for Other Than a 90° Bend

If the bend is to be other than 90°, use the lower number in the block (the bend allowance for 1°) and compute the bend allowance.

Example:
The L-bracket shown in *Figure 4-129* is made from 2024-T3 aluminum alloy and the bend is 60° from flat. Note that the bend angle in the figure indicates 120°, but that is the number of degrees between the two flanges and not the bend angle

Metal Thickness	RADIUS OF BEND, IN INCHES													
	1/32 .031	1/16 .063	3/32 .094	1/8 .125	5/32 .156	3/16 .188	7/32 .219	1/4 .250	9/32 .281	5/16 .313	11/32 .344	3/8 .375	7/16 .438	1/2 .500
.020	.062 .000693	.113 .001251	.161 .001792	.210 .002333	.259 .002874	.309 .003433	.358 .003974	.406 .004515	.455 .005056	.505 .005614	.554 .006155	.603 .006695	.702 .007795	.799 .008877
.025	.066 .000736	.116 .001294	.165 .001835	.214 .002376	.263 .002917	.313 .003476	.362 .004017	.410 .004558	.459 .005098	.509 .005657	.558 .006198	.607 .006739	.705 .007838	.803 .008920
.028	.068 .000759	.119 .001318	.167 .001859	.216 .002400	.265 .002941	.315 .003499	.364 .004040	.412 .004581	.461 .005122	.511 .005680	.560 .006221	.609 .006762	.708 .007862	.805 .007862
.032	.071 .000787	.121 .001345	.170 .001886	.218 .002427	.267 .002968	.317 .003526	.366 .004067	.415 .004608	.463 .005149	.514 .005708	.562 .006249	.611 .006789	.710 .007889	.807 .008971
.038	.075 .00837	.126 .001396	.174 .001937	.223 .002478	.272 .003019	.322 .003577	.371 .004118	.419 .004659	.468 .005200	.518 .005758	.567 .006299	.616 .006840	.715 .007940	.812 .009021
.040	.077 .000853	.127 .001411	.176 .001952	.224 .002493	.273 .003034	.323 .003593	.372 .004134	.421 .004675	.469 .005215	.520 .005774	.568 .006315	.617 .006856	.716 .007955	.813 .009037
.051		.134 .001413	.183 .002034	.232 .002575	.280 .003116	.331 .003675	.379 .004215	.428 .004756	.477 .005297	.527 .005855	.576 .006397	.624 .006934	.723 .008037	.821 .009119
.064		.144 .001595	.192 .002136	.241 .002676	.290 .003218	.340 .003776	.389 .004317	.437 .004858	.486 .005399	.536 .005957	.585 .006498	.634 .007039	.732 .008138	.830 .009220
.072			.198 .002202	.247 .002743	.296 .003284	.436 .003842	.394 .004283	.443 .004924	.492 .005465	.542 .006023	.591 .006564	.639 .007105	.738 .008205	836 .009287
.078			.202 .002249	.251 .002790	.300 .003331	.350 .003889	.399 .004430	.447 .004963	.496 .005512	.546 .006070	.595 .006611	.644 .007152	.745 .008252	.840 .009333
.081			.204 .002272	.253 .002813	.302 .003354	.352 .003912	.401 .004453	.449 .004969	.498 .005535	.548 .006094	.598 .006635	.646 .007176	.745 .008275	.842 .009357
.091			.212 .002350	.260 .002891	.309 .003432	.359 .003990	.408 .004531	.456 .005072	.505 .005613	.555 .006172	.604 .006713	.653 .007254	.752 .008353	.849 .009435
.094			.214 .002374	.262 .002914	.311 .003455	.361 .004014	.410 .004555	.459 .005096	.507 .005637	.558 .006195	.606 .006736	.655 .007277	.754 .008376	.851 .009458
.102				.268 .002977	.317 .003518	.367 .004076	.416 .004617	.464 .005158	.513 .005699	.563 .006257	.612 .006798	.661 .007339	.760 .008439	.857 .009521
.109				.273 .003031	.321 .003572	.372 .004131	.420 .004672	.469 .005213	.518 .005754	.568 .006312	.617 .006853	.665 .008394	.764 .008493	.862 .009575
.125				.284 .003156	.333 .003697	.383 .004256	.432 .004797	.480 .005338	.529 .005678	.579 .006437	.628 .006978	.677 .007519	.776 .008618	.873 .009700
.156					.355 .003939	.405 .004497	.453 .005038	.502 .005579	.551 .006120	.601 .006679	.650 .007220	.698 .007761	.797 .008860	.895 .009942
.188						.417 .004747	.476 .005288	.525 .005829	.573 .006370	.624 .006928	.672 .007469	.721 .008010	.820 .009109	.917 .010191
.250								.568 .006313	.617 .006853	.667 .007412	.716 .007953	.764 .008494	.863 .009593	.961 .010675

Figure 4-128. *Bend allowance.*

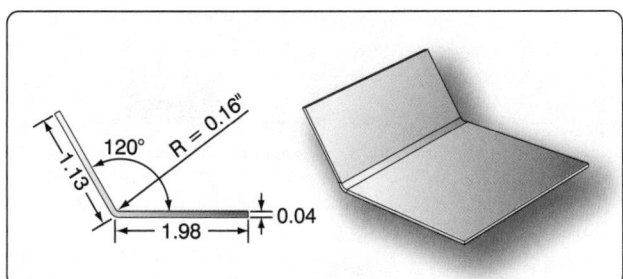

Figure 4-129. *Bend allowance for bends less than 90°.*

from flat. To find the correct bend angle, use the following formula:

Bend Angle = 180° − Angle between flanges

The actual bend is 60°. To find the correct bend radius for a 60° bend of material 0.040-inches thick, use the following procedure.

1. Go to the left side of the table and find 0.040-inch.

2. Go to the right and locate the bend radius of 0.16-inch (0.156-inch).

3. Note the bottom number in the block (0.003034).

4. Multiply this number by the bend angle:
   0.003034 × 60 = 0.18204

Step 5: Find the Total Developed Width of the Material

The total developed width (TDW) can be calculated when the dimensions of the flats and the bend allowance are found. The following formula is used to calculate TDW:

TDW = Flats + (bend allowance × number of bends)

For the U-channel example, this gives:

TDW = Flat 1 + Flat 2 + Flat 3 + (2 × BA)

TDW = 0.8 + 1.6 + 0.8 + (2 × 0.27)

TDW = 3.74-inches

Note that the amount of metal needed to make the channel is less than the dimensions of the outside of the channel (total of mold line dimensions is 4 inches). This is because the metal follows the radius of the bend rather than going from mold line to mold line. It is good practice to check that the calculated TDW is smaller than the total mold line dimensions. If the calculated TDW is larger than the mold line dimensions, the math was incorrect.

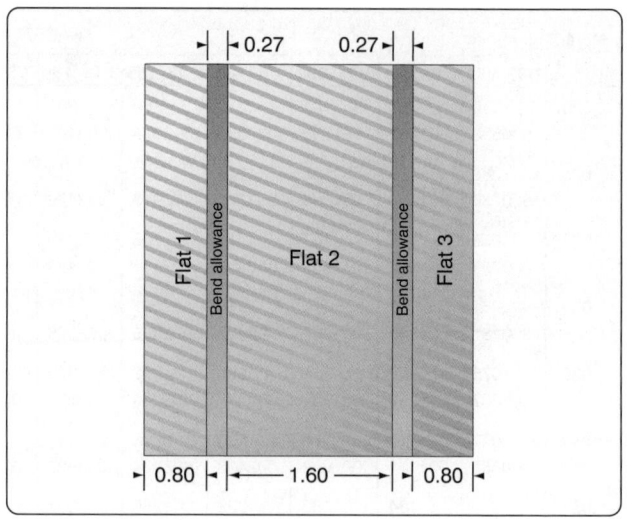

Figure 4-130. *Flat pattern layout.*

Step 6: Flat Pattern Lay Out

After a flat pattern layout of all relevant information is made, the material can be cut to the correct size, and the bend tangent lines can be drawn on the material. *[Figure 4-130]*

Step 7: Draw the Sight Lines on the Flat Pattern

The pattern laid out in *Figure 4-130* is complete, except for a sight line that needs to be drawn to help position the bend tangent line directly at the point where the bend should start. Draw a line inside the bend allowance area that is one bend radius away from the bend tangent line that is placed under the brake nose bar. Put the metal in the brake under the clamp and adjust the position of the metal until the sight line is directly below the edge of the radius bar. *[Figure 4-131]* Now, clamp the brake on the metal and raise the leaf to make the bend. The bend begins exactly on the bend tangent line.

Note: A common mistake is to draw the sight line in the middle of the bend allowance area, instead of one radius away from the bend tangent line that is placed under the brake nose bar.

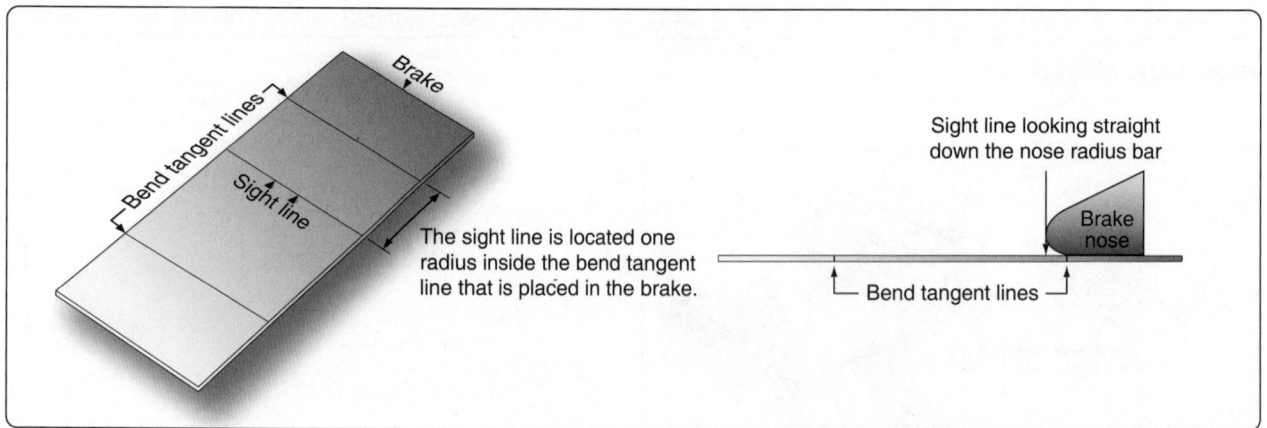

Figure 4-131. *Sight line.*

Using a J-Chart to Calculate Total Developed Width

The J-chart, often found in the SRM, can be used to determine bend deduction or setback and the TDW of a flat pattern layout when the inside bend radius, bend angle, and material thickness are known. *[Figure 4-132]* While not as accurate as the traditional layout method, the J-chart provides sufficient information for most applications. The J-chart does not require difficult calculations or memorized formulas because the required information can be found in the repair drawing or can be measured with simple measuring tools.

When using the J-chart, it is helpful to know whether the angle is open (greater than 90°) or closed (less than 90°) because the lower half of the J-chart is for open angles and the upper half is for closed angles.

To find the total developed width using a J-chart:

- Place a straightedge across the chart and connect the bend radius on the top scale with the material thickness on the bottom scale. *[Figure 4-132]*

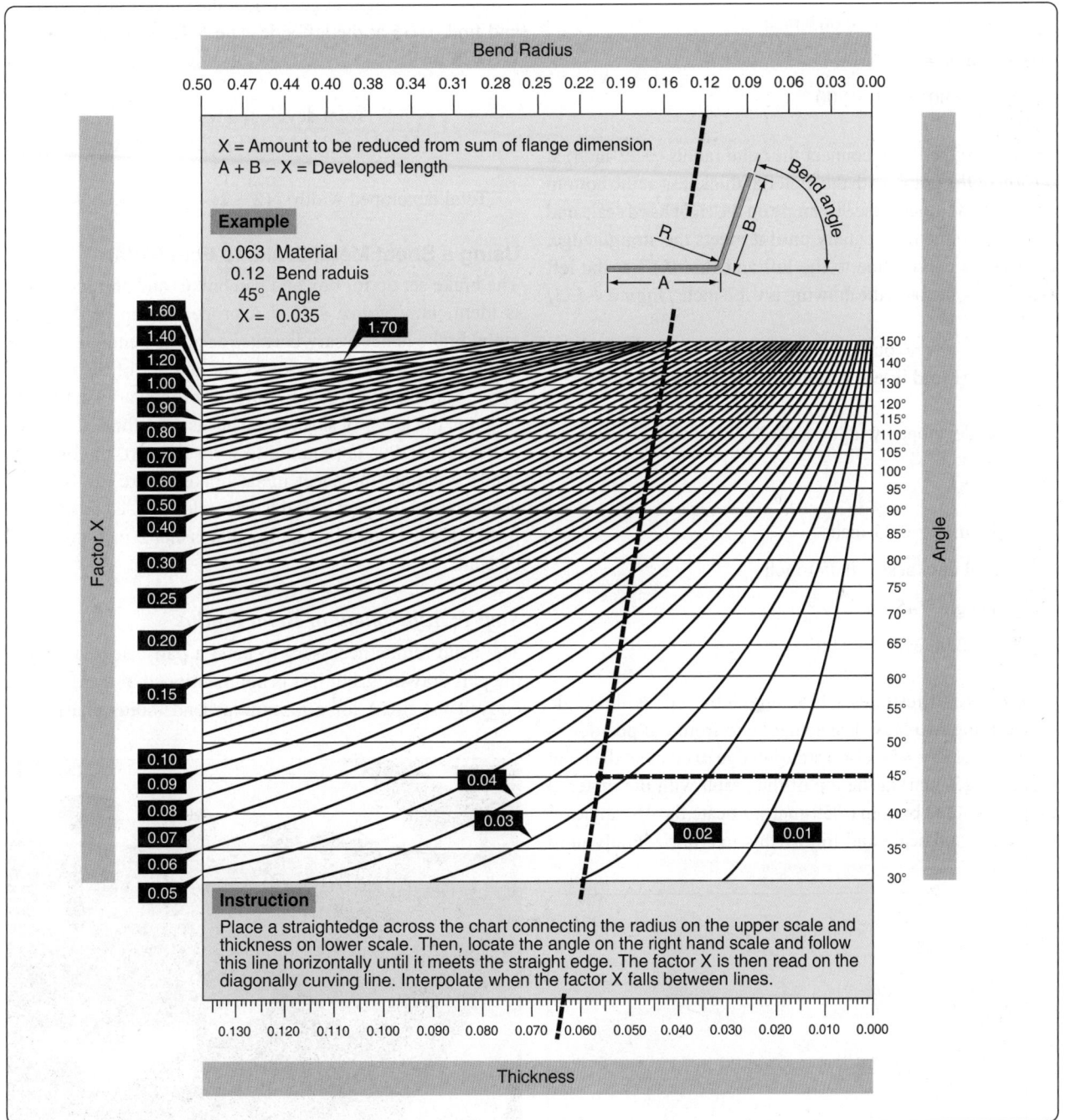

Figure 4-132. *J chart.*

- Locate the angle on the right hand scale and follow this line horizontally until it meets the straight edge.

- The factor X (bend deduction) is then read on the diagonally curving line.

- Interpolate when the X factor falls between lines.

- Add up the mold line dimensions and subtract the X factor to find the TDW.

Example 1

Bend radius = 0.22-inch

Material thickness = 0.063-inch

Bend angle = 90°

ML 1 = 2.00/ML 2 = 2.00

Use a straightedge to connect the bend radius (0.22-inch) at the top of the graph with the material thickness at the bottom (0.063-inch). Locate the 90° angle on the right hand scale and follow this line horizontally until it meets the straightedge. Follow the curved line to the left and find 0.17 at the left side. The X factor in the drawing is 0.17-inch. *[Figure 4-133]*

Total developed width =
(Mold line 1 + Mold line 2) – X factor

Total developed width = (2 + 2) – .17 = 3.83-inches

Example 2

Bend radius = 0.25-inch

Material thickness = 0.050-inch

Bend angle = 45°

ML 1 = 2.00/ML 2 = 2.00

Figure 4-134 illustrates a 135° angle, but this is the angle between the two legs. The actual bend from flat position is 45° (180 – 135 = 45). Use a straightedge to connect the bend radius (0.25-inch) at the top of the graph with the material thickness at the bottom (.050-inch). Locate the 45° angle on the right hand scale and follow this line horizontally until

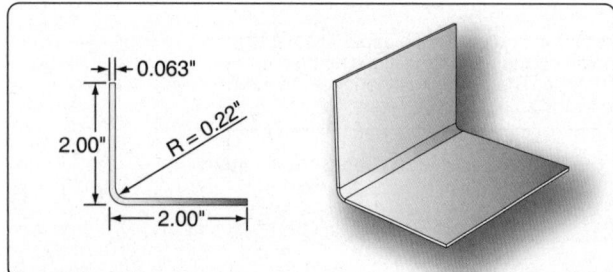

Figure 4-133. *Example 1 of J chart.*

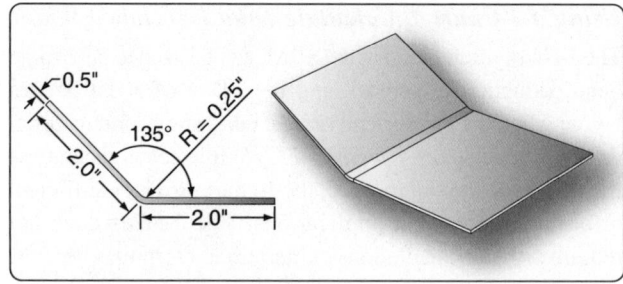

Figure 4-134. *Example 2 of J chart.*

it meets the straight edge. Follow the curved line to the left and find 0.035 at the left side. The X factor in the drawing is 0.035 inch.

Total developed width =
(Mold line 1 + Mold line 2) – X factor

Total developed width = (2 + 2) – .035 = 3.965-inch

Using a Sheet Metal Brake to Fold Metal

The brake set up for box and pan brakes and cornice brakes is identical. *[Figure 4-135]* A proper set up of the sheet metal brake is necessary because accurate bending of sheet metal depends on the thickness and temper of the material to be formed and the required radius of the part. Any time a different thickness of sheet metal needs to be formed or when a different radius is required to form the part, the operator needs to adjust the sheet metal brake before the brake is used to form the part. For this example, an L-channel made from 2024 –T3 aluminum alloy that is 0.032-inch thick will be bent.

Step 1: Adjustment of Bend Radius

The bend radius necessary to bend a part can be found in the part drawings, but if it is not mentioned in the drawing, consult the SRM for a minimum bend radius chart. This

Figure 4-135. *Brake radius nosepiece adjustment.*

chart lists the smallest radius allowable for each thickness and temper of metal that is normally used. To bend tighter than this radius would jeopardize the integrity of the part. Stresses left in the area of the bend may cause it to fail while in service, even if it does not crack while bending it.

The brake radius bars of a sheet metal brake can be replaced with another brake radius bar with a different diameter. *[Figure 4-136]* For example, a 0.032-inch 2024-T3 L channel needs to be bent with a radius of ⅛-inch and a radius bar with a ⅛-inch radius must be installed. If different brake radius bars are not available, and the installed brake radius bar is smaller than required for the part, it is necessary to bend some nose radius shims. *[Figure 4-137]*

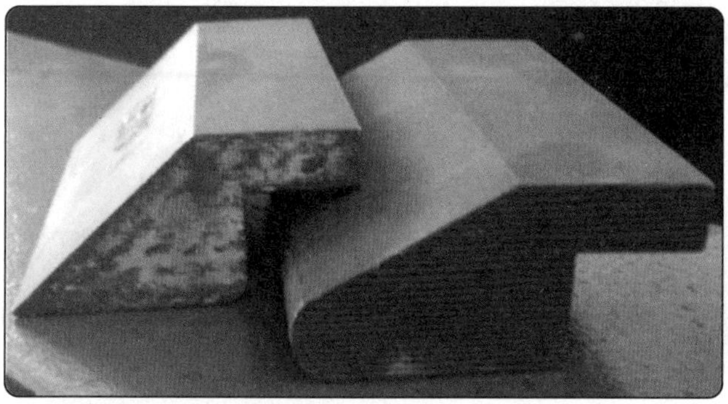

Figure 4-136. *Interchangeable brake radius bars.*

If the radius is so small that it tends to crack annealed aluminum, mild steel is a good choice of material. Experimentation with a small piece of scrap material is necessary to manufacture a thickness that increases the radius to precisely 1⁄16-inch or ⅛-inch. Use radius and fillet gauges to check this dimension. From this point on, each additional shim is added to the radius before it. *[Figure 4-138]*

Example: If the original nose was 1⁄16-inch and a piece of .063-inch material (1⁄16-inch) was bent around it, the new outside radius is ⅛-inch. If another .063-inch layer (1⁄16-inch) is added, it is now a 3⁄16-inch radius. If a piece of .032-inch (1⁄32-inch) instead of .063-inch material (1⁄16-inch) is bent around the ⅛-inch radius, a 5⁄32-inch radius results.

Step 2: Adjusting Clamping Pressure

The next step is setting clamping pressure. Slide a piece of the material with the same thickness as the part to be bent under the brake radius piece. Pull the clamping lever toward the operator to test the pressure. This is an over center type clamp and, when properly set, will not feel springy or spongy when pulled to its fully clamped position. The operator must be able to pull this lever over center with a firm pull and have it bump its limiting stops. On some brakes, this adjustment has to be made on both sides of the brake.

Place test strips on the table 3 inches from each end and one in the center between the bed and the clamp, adjust clamp

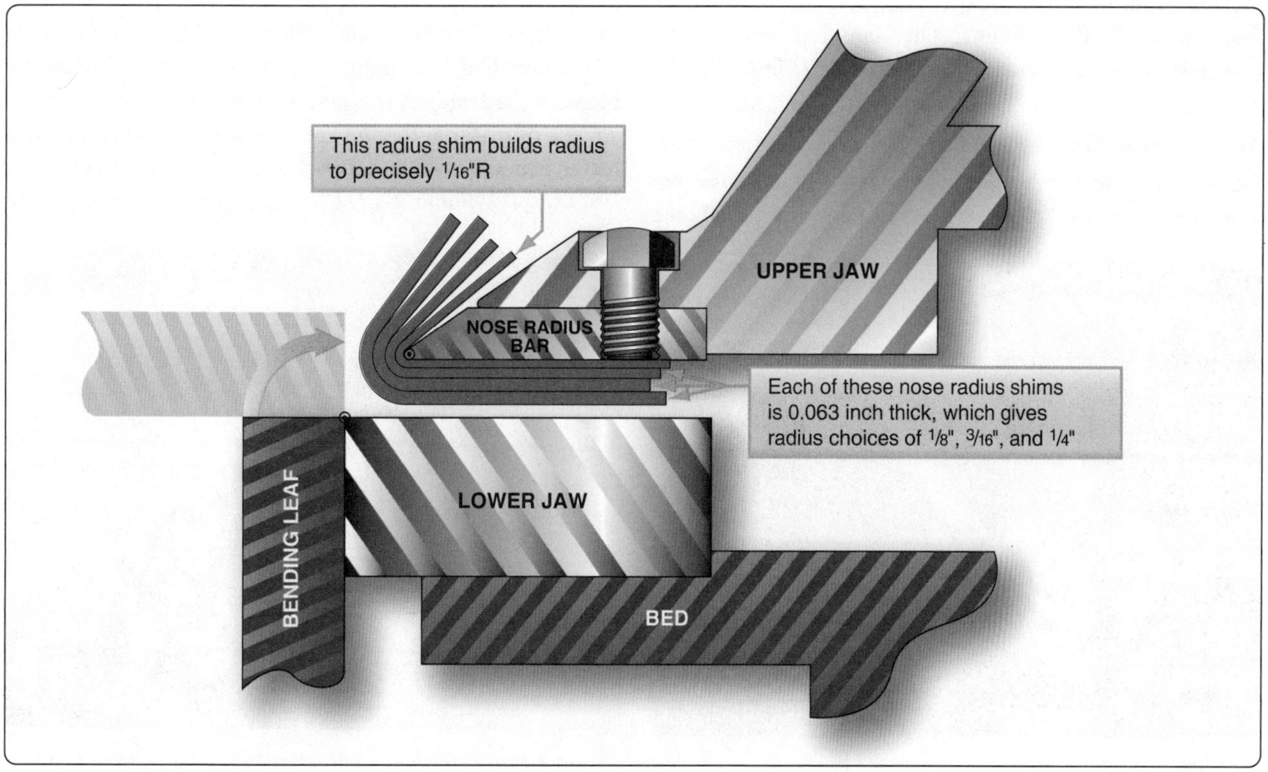

Figure 4-137. *Nose radius shims may be used when the brake radius bar is smaller than required.*

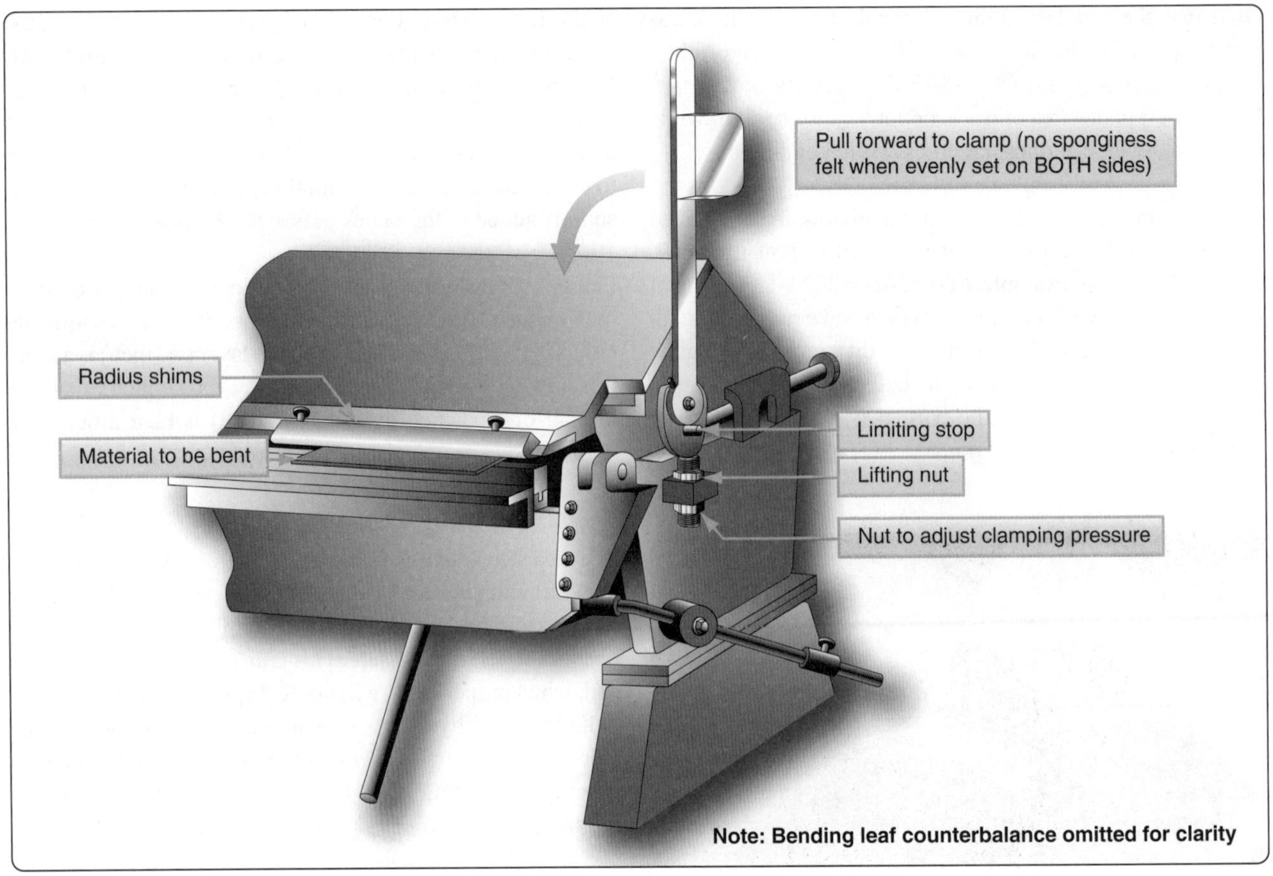

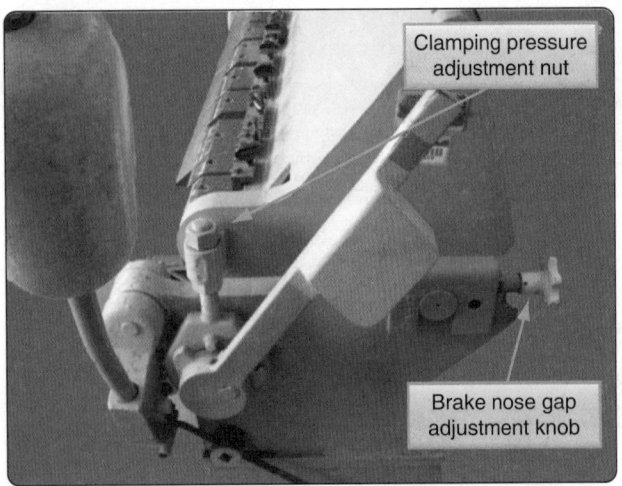

Figure 4-138. *General brake overview including radius shims.*

pressure until it is tight enough to prevent the work pieces from slipping while bending. The clamping pressure can be adjusted with the clamping pressure nut. *[Figure 4-139]*

Step 3: Adjusting the Nose Gap

Adjust the nose gap by turning the large brake nose gap adjustment knobs at the rear of the upper jaw to achieve its proper alignment. *[Figure 4-140]* The perfect setting is obtained when the bending leaf is held up to the angle of the finished bend and there is one material thickness between the bending leaf and the nose radius piece. Using a piece of material the thickness of the part to be bent as a feeler gauge can help achieve a high degree of accuracy. *[Figures 4-140* and *4-141]* It is essential this nose gap be

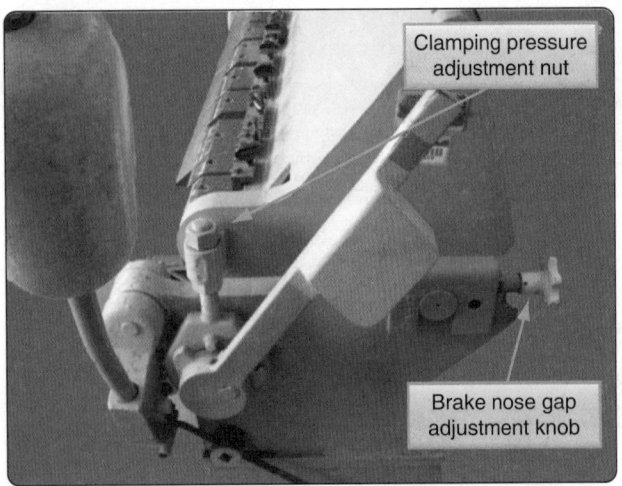

Figure 4-139. *Adjust clamping pressure with the clamping pressure nut.*

Figure 4-140. *Brake nose gap adjustment with piece of material same thickness as part to be formed.*

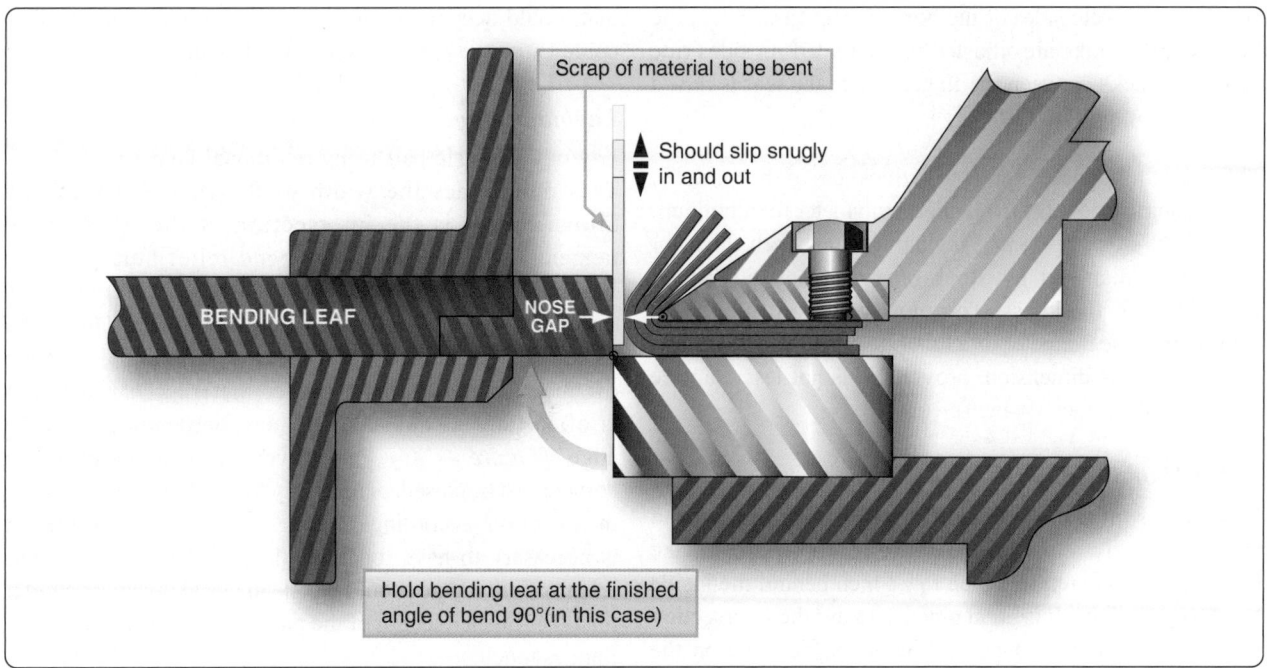

Figure 4-141. *Profile illustration of brake nose gap adjustment.*

perfect, even across the length of the part to be bent. Check by clamping two test strips between the bed and the clamp 3 inches from each end of the brake. *[Figure 4-142]* Bend 90° *[Figure 4-143]*, remove test strips, and place one on top of the other; they should match. *[Figure 4-144]* If they do not match, adjust the end with the sharper bend back slightly.

Folding a Box

A box can be formed the same way as the U-channel described on in the previous paragraphs, but when a sheet metal part has intersecting bend radii, it is necessary to remove material to make room for the material contained in the flanges. This is done by drilling or punching holes at the intersection of the inside bend tangent lines. These holes, called relief holes and whose diameter is approximately twice the bend radius, relieve stresses in the metal as it is bent and prevent the metal from tearing. Relief holes also provide a neatly trimmed corner from which excess material may be trimmed.

The larger and smoother the relief hole is, the less likely it will be that a crack will form in the corner. Generally, the radius of the relief hole is specified on the drawing. A box and pan brake, also called a finger brake, is used to bend the

Figure 4-143. *Brake alignment with two test strips bent at 90°.*

Figure 4-142. *Brake alignment with two test strips 3 inches from each end.*

Figure 4-144. *Brake alignment by comparing test strips.*

box. Two opposite sides of the box are bent first. Then, the fingers of the brake are adjusted so the folded-up sides ride up in the cracks between the fingers when the leaf is raised to bend the other two sides.

The size of relief holes varies with thickness of the material. They should be no less than ⅛-inch in diameter for aluminum alloy sheet stock up to and including 0.064-inch thick, or ³⁄₁₆₀-inch in diameter for stock ranging from 0.072-inch to 0.128-inch thickness. The most common method of determining the diameter of a relief hole is to use the radius of bend for this dimension, provided it is not less than the minimum allowance (⅛-inch).

Relief Hole Location

Relief holes must touch the intersection of the inside bend tangent lines. To allow for possible error in bending, make the relief holes extend ¹⁄₃₂-inch to ¹⁄₁₆-inch behind the inside bend tangent lines. It is good practice to use the intersection of these lines as the center for the holes. The line on the inside of the curve is cut at an angle toward the relief holes to allow for the stretching of the inside flange.

The positioning of the relief hole is important. *[Figure 4-145]* It should be located so its outer perimeter touches the intersection of the inside bend tangent lines. This keeps any material from interfering with the bend allowance area of the other bend. If these bend allowance areas intersected with each other, there would be substantial compressive stresses

that would accumulate in that corner while bending. This could cause the part to crack while bending.

Layout Method

Lay out the basic part using traditional layout procedures. This determines the width of the flats and the bend allowance. It is the intersection of the inside bend tangent lines that index the bend relief hole position. Bisect these intersected lines and move outward the distance of the radius of the hole on this line. This is the center of the hole. Drill at this point and finish by trimming off the remainder of the corner material. The trim out is often tangent to the radius and perpendicular to the edge. *[Figure 4-146]* This leaves an open corner. If the corner must be closed, or a slightly longer flange is necessary, then trim out accordingly. If the corner is to be welded, it is necessary to have touching flanges at the corners. The length of the flange should be one material thickness shorter than the finished length of the part so only the insides of the flanges touch.

Open & Closed Bends

Open and closed bends present unique problems that require more calculations than 90° bends. In the following 45° and a 135° bend examples, the material is 0.050-inch thick and the bend radius is ³⁄₁₆-inch.

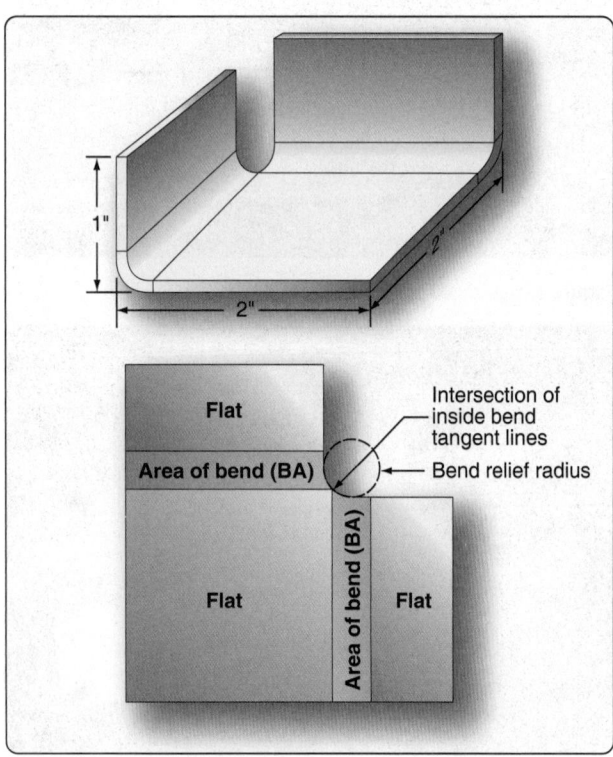

Figure 4-145. *Relief hole location.*

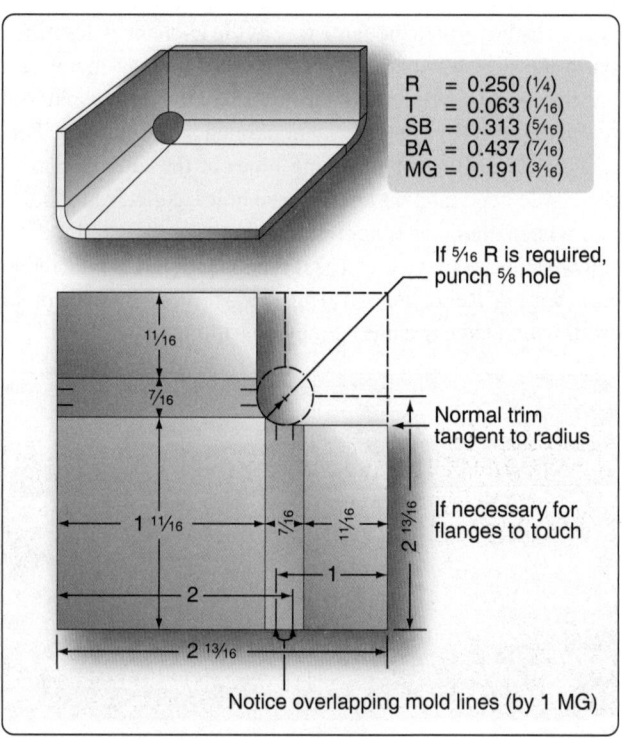

Figure 4-146. *Relief hole layout.*

Open End Bend (Less Than 90°)

Figure 4-147 shows an example for a 45° bend.

1. Look up K-factor in K chart. K-factor for 45° is 0.41421-inch.

2. Calculate setback.

   SB = K(R + T)

   SB = 0.41421-inch(0.1875-inch + 0.050-inch) = 0.098-inch

3. Calculate bend allowance for 45°. Look up bend allowance for 1° of bend in the bend allowance chart and multiply this by 45.

   0.003675-inch × 45 = 0.165-inch

4. Calculate flats.

   Flat = Mold line dimension – SB

   Flat 1 = .77-inch – 0.098-inch = 0.672-inch

   Flat 2 = 1.52-inch – 0.098-inch = 1.422-inch

5. Calculate TDW

   TDW = Flats + Bend allowance

   TDW = 0.672-inch + 1.422-inch + 0.165-inch = 2.259-inch.

Observe that the brake reference line is still located one radius from the bend tangent line.

Closed End Bend (More Than 90°)

Figure 4-148 shows an example of a 135° bend.

1. Look up K-factor in K chart. K-factor for 135° is 2.4142-inch.

2. Calculate SB.

   SB = K(R + T)

   SB = 2.4142-inch(0.1875-inch + 0.050-inch) = 0.57-inch

3. Calculate bend allowance for 135°. Look up bend allowance for 1° of bend in the bend allowance chart and multiply this by 135.

   0.003675-inch × 135 = 0.496-inch

4. Calculate flats.

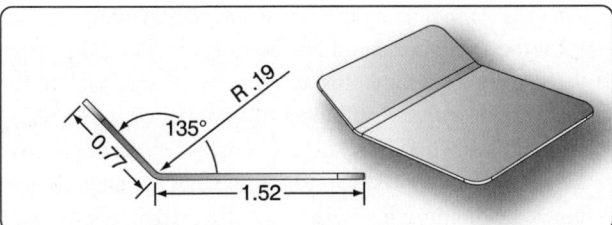

Figure 4-147. *Open bend.*

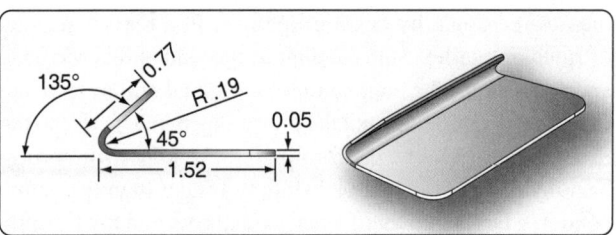

Figure 4-148. *Closed bend.*

Flat = Mold line dimension – SB

Flat 1 = 0.77-inch – 0.57-inch = 0.20-inch

Flat 2 = 1.52-inch – 0.57-inch = 0.95-inch

5. Calculate TDW.

   TDW = Flats + Bend allowance

   TDW = 0.20-inch + 0.95-inch + 0.496-inch = 1.65-inch

It is obvious from both examples that a closed bend has a smaller TDW than an open-end bend and the material length needs to be adjusted accordingly.

Hand Forming

All hand forming revolves around the processes of stretching and shrinking metal. As discussed earlier, stretching means to lengthen or increase a particular area of metal while shrinking means to reduce an area. Several methods of stretching and shrinking may be used, depending on the size, shape, and contour of the part being formed.

For example, if a formed or extruded angle is to be curved, either stretch one leg or shrink the other, whichever makes the part fit. In bumping, the material is stretched in the bulge to make it balloon, and in joggling, the material is stretched between the joggles. Material in the edge of lightening holes is often stretched to form a beveled reinforcing ridge around them. The following paragraphs discuss some of these techniques.

Straight Line Bends

The cornice brake and bar folder are ordinarily used to make straight bends. Whenever such machines are not available, comparatively short sections can be bent by hand with the aid of wooden or metal bending blocks.

After a blank has been laid out and cut to size, clamp it along the bend line between two wooden forming blocks held in a vise. The wooden forming blocks should have one edge rounded as needed for the desired radius of bend. It should also be curved slightly beyond 90° to allow for spring-back.

Bend the metal that protrudes beyond the bending block to

the desired angle by tapping lightly with a rubber, plastic, or rawhide mallet. Start tapping at one end and work back and forth along the edge to make a gradual and even bend. Continue this process until the protruding metal is bent to the desired angle against the forming block. Allow for spring-back by driving the material slightly farther than the actual bend. If a large amount of metal extends beyond the forming blocks, maintain hand pressure against the protruding sheet to prevent it from bouncing. Remove any irregularities by holding a straight block of hardwood edgewise against the bend and striking it with heavy blows of a mallet or hammer. If the amount of metal protruding beyond the bending blocks is small, make the entire bend by using the hardwood block and hammer.

Formed or Extruded Angles

Both formed and extruded types of angles can be curved (not bent sharply) by stretching or shrinking either of the flanges. Curving by stretching one flange is usually preferred since the process requires only a V-block and a mallet and is easily accomplished.

Stretching with V-Block Method

In the stretching method, place the flange to be stretched in the groove of the V-block. *[Figure 4-149]* (If the flange is to be shrunk, place the flange across the V-block.) Using a round, soft-faced mallet, strike the flange directly over the V portion with light, even blows while gradually forcing it downward into the V.

Begin at one end of the flange and form the curve gradually and evenly by moving the strip slowly back and forth,

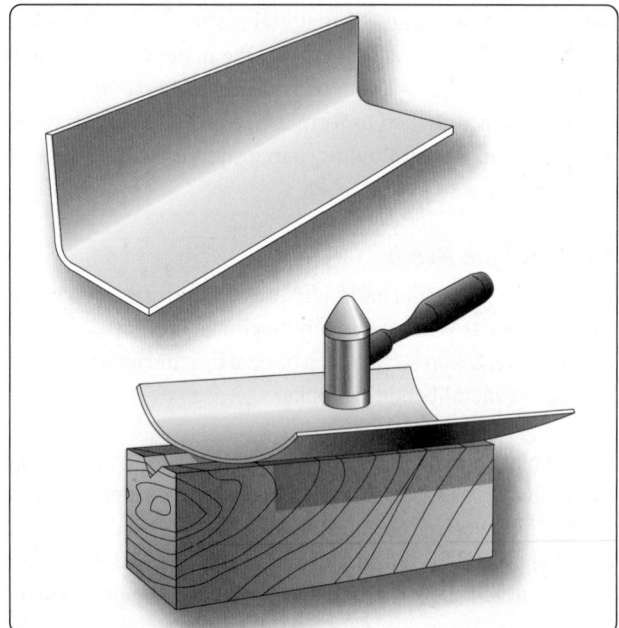

Figure 4-149. *V-block forming.*

distributing the hammer blows at equal spaces on the flange. Hold the strip firmly to keep it from bouncing when hammered. An overly heavy blow buckles the metal, so keep moving the flange across the V-block, but always lightly strike the spot directly above the V.

Lay out a full-sized, accurate pattern on a sheet of paper or plywood and periodically check the accuracy of the curve. Comparing the angle with the pattern determines exactly how the curve is progressing and just where it needs to be increased or decreased. It is better to get the curve to conform roughly to the desired shape before attempting to finish any one portion, because the finishing or smoothing of the angle may cause some other portion of the angle to change shape. If any part of the angle strip is curved too much, reduce the curve by reversing the angle strip on the V-block, placing the bottom flange up, and striking it with light blows of the mallet.

Try to form the curve with a minimum amount of hammering, for excessive hammering work hardens the metal. Work-hardening can be recognized by a lack of bending response or by springiness in the metal. It can be recognized very readily by an experienced worker. In some cases, the part may have to be annealed during the curving operation. If so, be sure to heat treat the part again before installing it on the aircraft.

Shrinking With V-Block & Shrinking Block Methods

Curving an extruded or formed angle strip by shrinking may be accomplished by either the previously discussed V-block method or the shrinking block method. While the V-block is more satisfactory because it is faster, easier, and affects the metal less, good results can be obtained by the shrinking block method.

In the V-block method, place one flange of the angle strip flat on the V-block with the other flange extending upward. Using the process outlined in the stretching paragraphs, begin at one end of the angle strip and work back and forth making light blows. Strike the edge of the flange at a slight angle to keep the vertical flange from bending outward.

Occasionally, check the curve for accuracy with the pattern. If a sharp curve is made, the angle (cross-section of the formed angle) closes slightly. To avoid such closing of the angle, clamp the angle strip to a hardwood board with the hammered flange facing upward using small C-clamps. The jaws of the C-clamps should be covered with masking tape. If the angle has already closed, bring the flange back to the correct angle with a few blows of a mallet or with the aid of a small hardwood block. If any portion of the angle strip is curved too much, reduce it by reversing the angle on the V-block and hammering with a suitable mallet, as explained in the previous paragraph on stretching. After obtaining the proper curve, smooth the entire angle by planishing with a soft-faced mallet.

If the curve in a formed angle is to be quite sharp or if the flanges of the angle are rather broad, the shrinking block method is generally used. In this process, crimp the flange that is to form the inside of the curve.

When making a crimp, hold the crimping pliers so that the jaws are about ⅛-inch apart. By rotating the wrist back and forth, bring the upper jaw of the pliers into contact with the flange, first on one side and then on the other side of the lower jaw. Complete the crimp by working a raised portion into the flange, gradually increasing the twisting motion of the pliers. Do not make the crimp too large because it will be difficult to work out. The size of the crimp depends upon the thickness and softness of the material, but usually about ¼-inch is sufficient. Place several crimps spaced evenly along the desired curve with enough space left between each crimp so that jaws of the shrinking block can easily be attached.

After completing the crimping, place the crimped flange in the shrinking block so that one crimp at a time is located between the jaws. *[Figure 4-150]* Flatten each crimp with light blows of a soft-faced mallet, starting at the apex (the closed end) of the crimp and gradually working toward the edge of the flange. Check the curve of the angle with the pattern periodically during the forming process and again after all the crimps have been worked out. If it is necessary to increase the curve, add more crimps and repeat the process. Space the additional crimps between the original ones so that the metal does not become unduly work hardened at any one point. If the curve needs to be increased or decreased slightly at any point, use the V-block.

After obtaining the desired curve, planish the angle strip over a stake or a wooden form.

Flanged Angles

The forming process for the following two flanged angles is slightly more complicated than the previously discussed angles because the bend is shorter (not gradually curved) and necessitates shrinking or stretching in a small or concentrated area. If the flange is to point toward the inside of the bend, the material must be shrunk. If it is to point toward the outside, it must be stretched.

Shrinking

In forming a flanged angle by shrinking, use wooden forming blocks similar to those shown in *Figure 4-151* and proceed as follows:

1. Cut the metal to size, allowing for trimming after forming. Determine the bend allowance for a 90° bend and round the edge of the forming block accordingly.

2. Clamp the material in the form blocks as shown in *Figure 4-151,* and bend the exposed flange against the block. After bending, tap the blocks slightly. This induces a setting process in the bend.

3. Using a soft-faced shrinking mallet, start hammering near the center and work the flange down gradually toward both ends. The flange tends to buckle at the bend because the material is made to occupy less space. Work the material into several small buckles instead of one large one and work each buckle

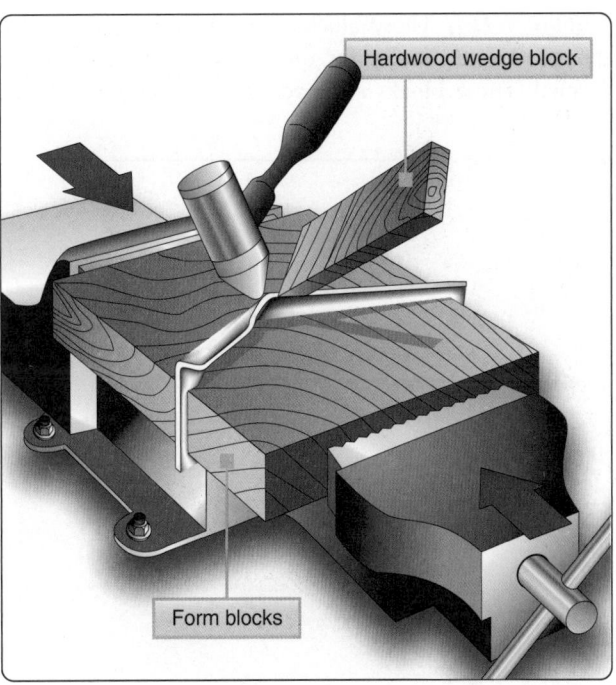

Figure 4-151. *Forming a flanged angle using forming blocks.*

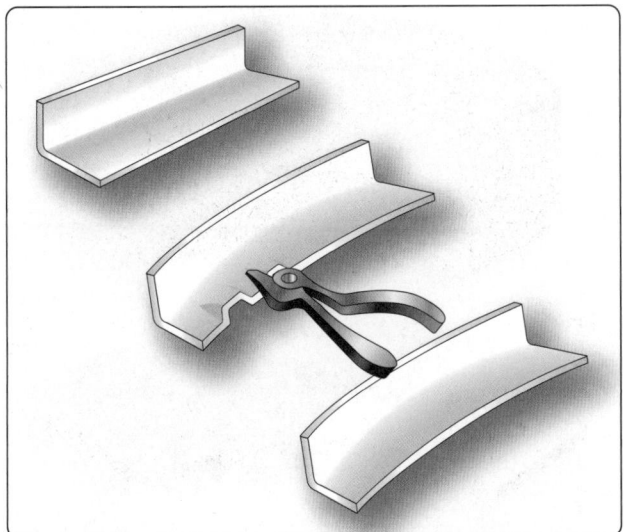

Figure 4-150. *Crimping a metal flange in order to form a curve.*

out gradually by hammering lightly and gradually compressing the material in each buckle. The use of a small hardwood wedge block aids in working out the buckles. *[Figure 4-152]*

4. Planish the flange after it is flattened against the form block and remove small irregularities. If the form blocks are made of hardwood, use a metal planishing hammer. If the forms are made of metal, use a soft-faced mallet. Trim the excess material away and file and polish.

Stretching

To form a flanged angle by stretching, use the same forming blocks, wooden wedge block, and mallet as used in the shrinking process and proceed as follows:

1. Cut the material to size (allowing for trim), determine bend allowance for a 90° bend, and round off the edge of the block to conform to the desired radius of bend.

2. Clamp the material in the form blocks. *[Figure 4-153]*

3. Using a soft-faced stretching mallet, start hammering near the ends and work the flange down smoothly and gradually to prevent cracking and splitting. Planish the flange and angle as described in the previous procedure, and trim and smooth the edges, if necessary.

Curved Flanged Parts

Curved flanged parts are usually hand formed with a concave flange, the inside edge, and a convex flange, the outside edge. The concave flange is formed by stretching, while the convex flange is formed by shrinking. Such parts are shaped with the aid of hardwood or metal forming blocks. *[Figure 4-154]* These blocks are made in pairs and are designed specifically for the shape of the area being formed. These blocks are made in pairs similar to those

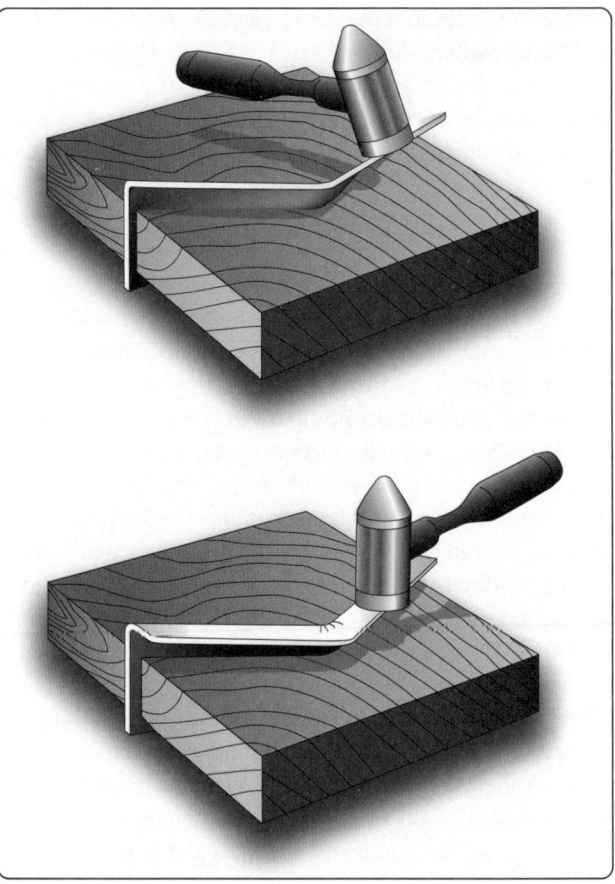

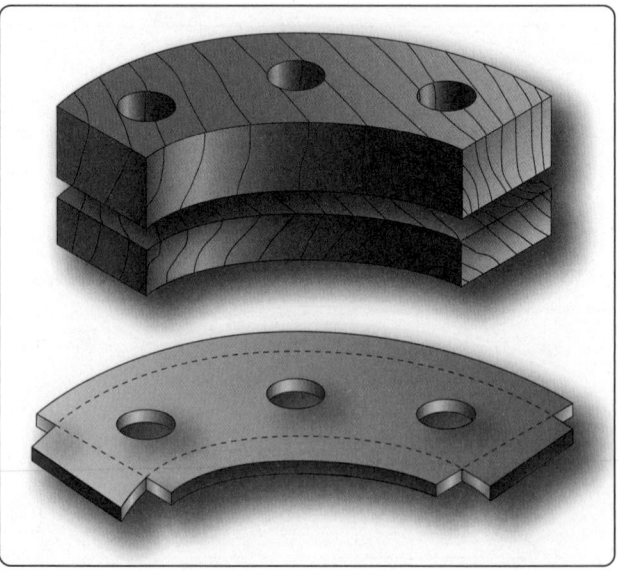

Figure 4-153. *Stretching a flanged angle.*

used for straight angle bends and are identified in the same manner. They differ in that they are made specifically for the particular part to be formed, they fit each other exactly, and they conform to the actual dimensions and contour of the finished article.

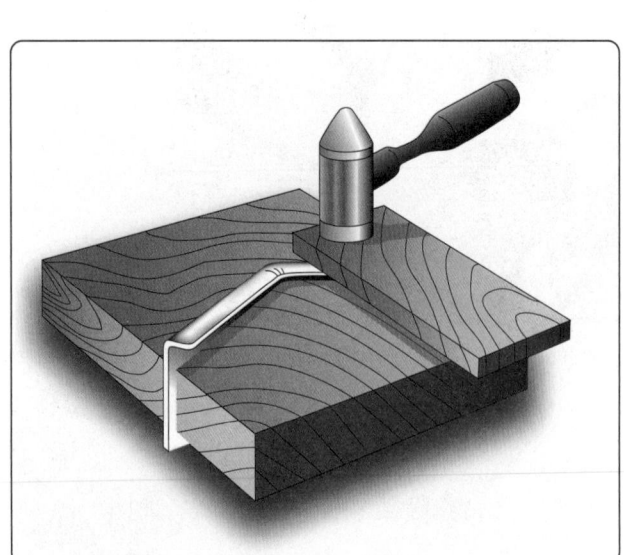

Figure 4-152. *Shrinking.*

Figure 4-154. *Forming blocks.*

The forming blocks may be equipped with small aligning pins to help line up the blocks and to hold the metal in place or they may be held together by C-clamps or a vise. They also may be held together with bolts by drilling through form blocks and the metal, provided the holes do not affect the strength of the finished part. The edges of the forming block are rounded to give the correct radius of bend to the part, and are undercut approximately 5° to allow for spring-back of the metal. This undercut is especially important if the material is hard or if the bend must be accurate.

The nose rib offers a good example of forming a curved flange because it incorporates both stretching and shrinking (by crimping). They usually have a concave flange, the inside edge, and a convex flange, the outside edge. Note the various types of forming represented in the following figures. In the plain nose rib, only one large convex flange is used. [Figure 4-155] Because of the great distance around the part and the likelihood of buckles in forming, it is rather difficult to form. The flange and the beaded (raised ridge on sheet metal used to stiffen the piece) portion of this rib provide sufficient strength to make this a good type to use. In Figure 4-156, the concave flange is difficult to form, but the outside flange is broken up into smaller sections by relief holes. In Figure 4-157, note that crimps are placed at equally spaced intervals to absorb material and cause curving, while also giving strength to the part.

In Figure 4-158, the nose rib is formed by crimping, beading, putting in relief holes, and using a formed angle riveted on each end. The beads and the formed angles supply strength to the part. The basic steps in forming a curved flange follow: [Figures 4-159 and 160]

1. Cut the material to size, allowing about ¼-inch excess material for trim and drill holes for alignment pins.

2. Remove all burrs (jagged edges). This reduces the possibility of the material cracking at the edges during the forming process.

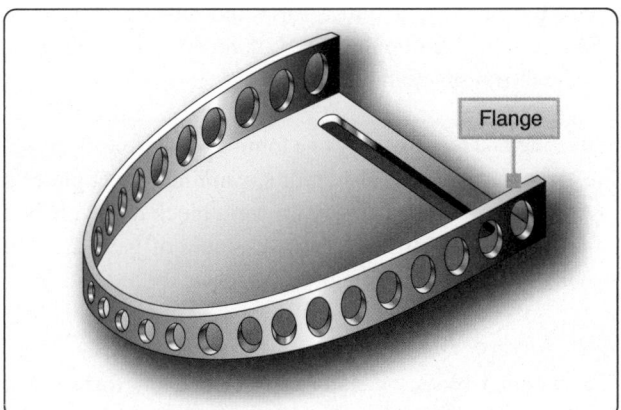

Figure 4-155. *Plain nose rib.*

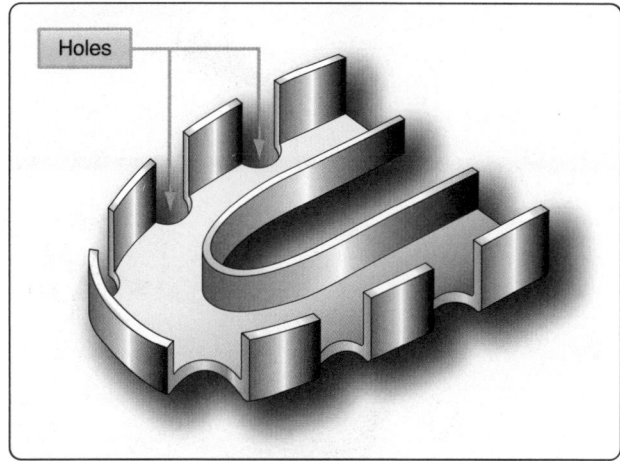

Figure 4-156. *Nose rib with relief holes.*

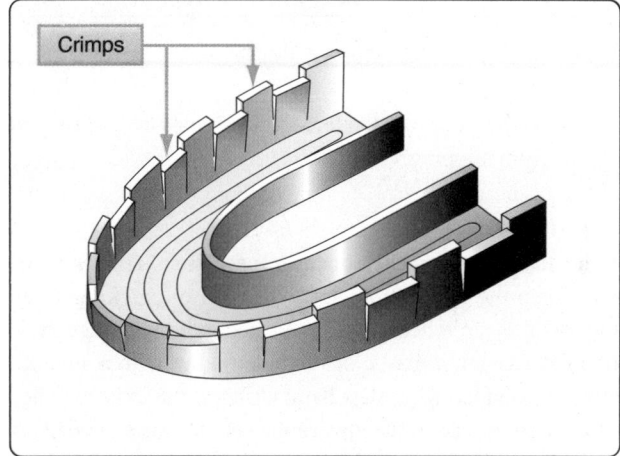

Figure 4-157. *Nose rib with crimps.*

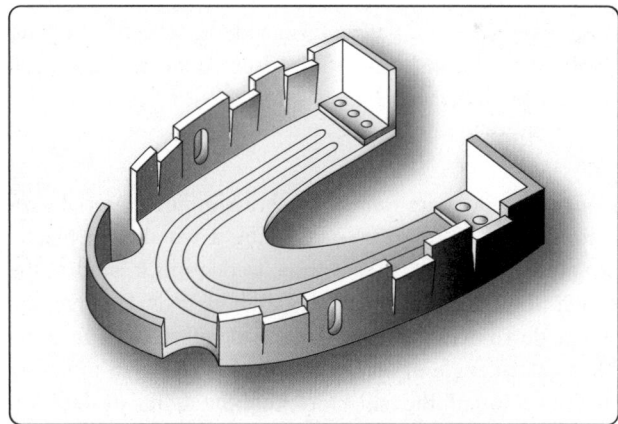

Figure 4-158. *Nose rib using a combination of forms.*

3. Locate and drill holes for alignment pins.

4. Place the material between the form blocks and clamp blocks tightly in a vise to prevent the material from moving or shifting. Clamp the work as closely as possible to the particular area being hammered to

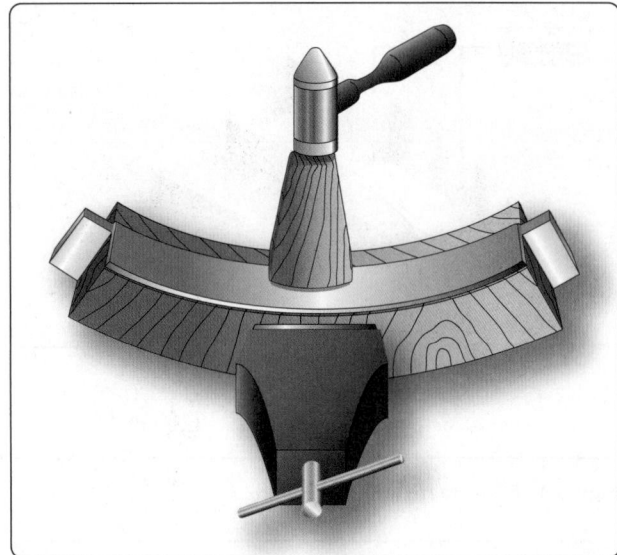

Figure 4-159. *Forming a concave flange.*

prevent strain on the form blocks and to keep the metal from slipping.

Concave Surfaces

Bend the flange on the concave curve first. This practice may keep the flange from splitting open or cracking when the metal is stretched. Should this occur, a new piece must be made. Using a plastic or rawhide mallet with a smooth, slightly rounded face, start hammering at the extreme ends of the part and continue toward the center of the bend. This procedure permits some of the metal at the ends of the part to be worked into the center of the curve where it is needed. Continue hammering until the metal is gradually worked down over the entire flange, flush with the form block. After the flange is formed, trim off the excess material and check the part for accuracy. *[Figure 4-159]*

Convex Surfaces

Convex surfaces are formed by shrinking the material over a form block. *[Figure 4-160]* Using a wooden or plastic shrinking mallet and a backup or wedge block, start at the center of the curve and work toward both ends. Hammer the flange down over the form, striking the metal with glancing blows at an angle of approximately 45° and with a motion that tends to pull the part away from the radius of the form block. Stretch the metal around the radius bend and remove the buckles gradually by hammering on a wedge block. Use the backup block to keep the edge of the flange as nearly perpendicular to the form block as possible. The backup block also lessens the possibility of buckles, splits, or cracks. Finally, trim the flanges of excess metal, planish, remove burrs, round the corners (if any), and check the part for accuracy.

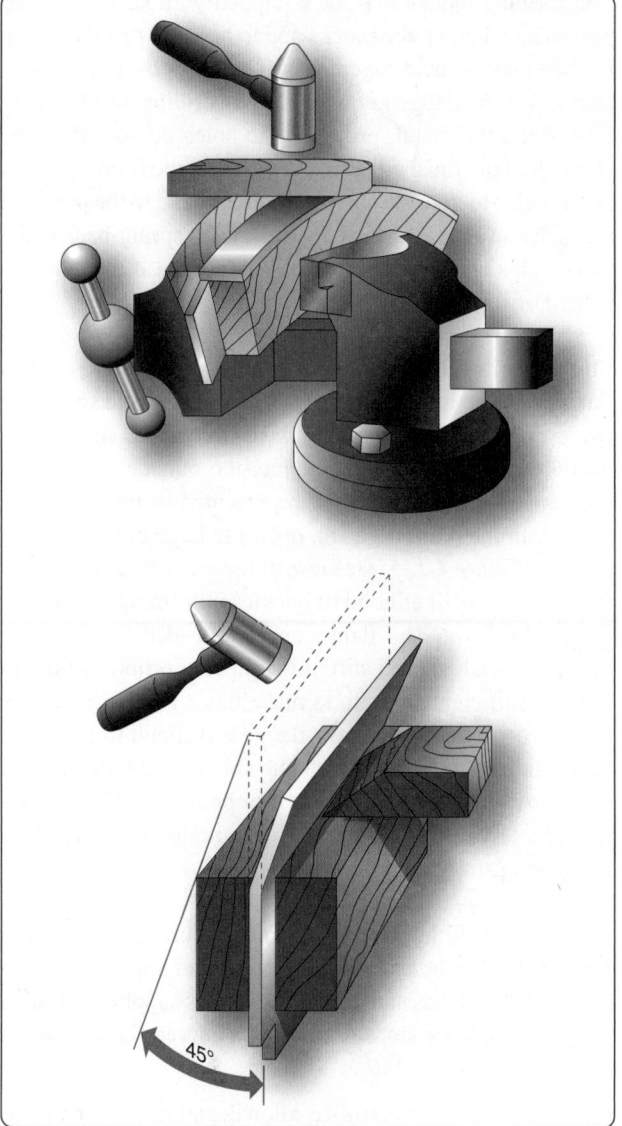

Figure 4-160. *Forming a convex flange.*

Forming by Bumping

As discussed earlier, bumping involves stretching the sheet metal by bumping it into a form and making it balloon. *[Figure 4-161]* Bumping can be done on a form block or female die, or on a sandbag.

Either method requires only one form: a wooden block, a lead die, or a sandbag. The blister, or streamlined cover plate, is an example of a part made by the form block or die method of bumping. Wing fillets are an example of parts that are usually formed by bumping on a sandbag.

Form Block or Die

The wooden block or lead die designed for form block bumping must have the same dimensions and contour as the outside of the blister. To provide enough bucking weight

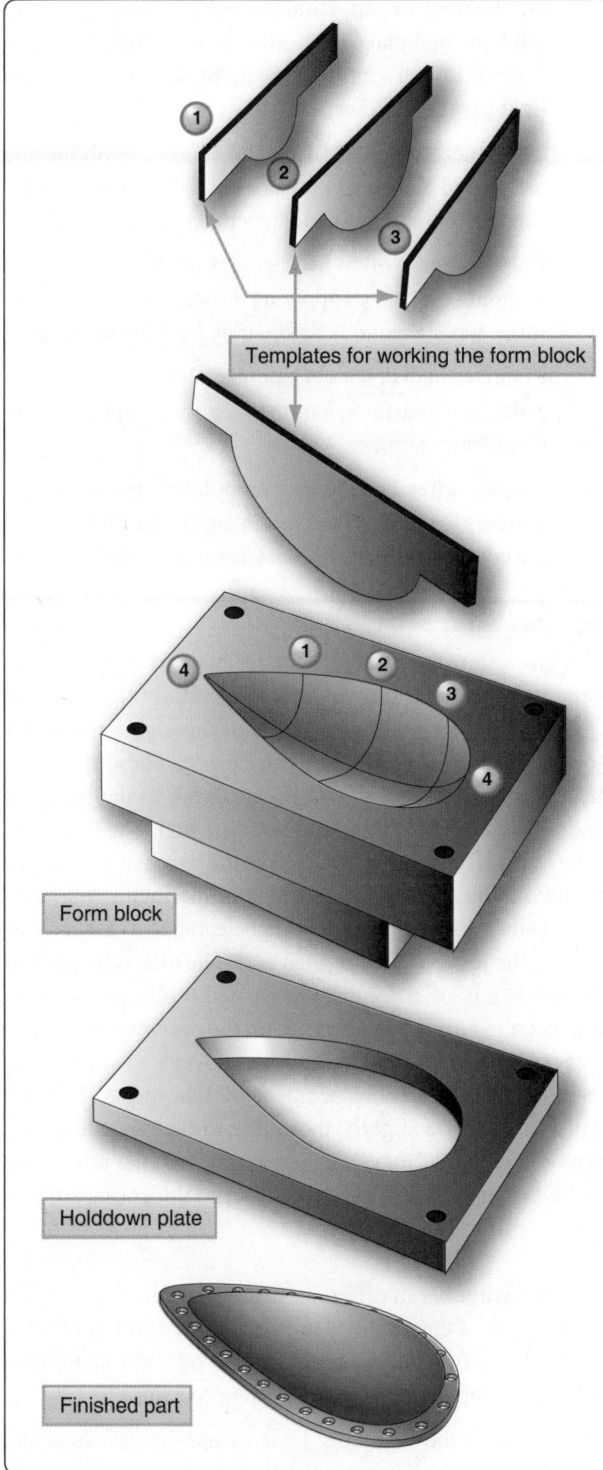

Templates for working the form block

Form block

Holddown plate

Finished part

Figure 4-161. *Form block bumping.*

and bearing surface for fastening the metal, the block or die should be at least one inch larger in all dimensions than the form requires.

Follow these procedures to create a form block:

1. Hollow the block out with tools, such as saws, chisels, gouges, files, and rasps.

2. Smooth and finish the block with sandpaper. The inside of the form must be as smooth as possible, because the slightest irregularity shows up on the finished part.

3. Prepare several templates (patterns of the cross-section), as shown in *Figure 4-161* so that the form can be checked for accuracy.

4. Shape the contour of the form at points 1, 2, and 3.

5. Shape the areas between the template checkpoints to conform the remaining contour to template 4. Shaping of the form block requires particular care because the more nearly accurate it is, the less time it takes to produce a smooth, finished part.

After the form is prepared and checked, perform the bumping as follows:

1. Cut a metal blank to size allowing an extra ½ to 1-inch to permit drawing.

2. Apply a thin coat of light oil to the block and the aluminum to prevent galling (scraping on rough spots).

3. Clamp the material between the block and steel plate. Ensure it is firmly supported yet it can slip a little toward the inside of the form.

4. Clamp the bumping block in a bench vise. Use a soft-faced rubber mallet, or a hardwood drive block with a suitable mallet, to start the bumping near the edges of the form.

5. Work the material down gradually from the edges with light blows of the mallet. Remember, the purpose of bumping is to work the material into shape by stretching rather than forcing it into the form with heavy blows. Always start bumping near the edge of the form. Never start near the center of the blister.

6. Before removing the work from the form, smooth it as much as possible by rubbing it with the rounded end of either a maple block or a stretching mallet.

7. Remove the blister from the bumping block and trim to size.

Sandbag Bumping

Sandbag bumping is one of the most difficult methods of hand forming sheet metal because there is no exact forming block to guide the operation. *[Figure 4-162]* In this method, a depression is made into the sandbag to take the shape of the hammered portion of the metal. The depression or pit has a tendency to shift from the hammering, which necessitates periodic readjustment during the bumping process. The degree of shifting depends largely on the contour or shape of the piece being formed, and whether glancing blows must be struck to stretch, draw, or shrink the metal. When forming

Figure 4-162. *Sandbag bumping.*

by this method, prepare a contour template or some sort of a pattern to serve as a working guide and to ensure accuracy of the finished part. Make the pattern from ordinary kraft or similar paper, folding it over the part to be duplicated. Cut the paper cover at the points where it would have to be stretched to fit, and attach additional pieces of paper with masking tape to cover the exposed portions. After completely covering the part, trim the pattern to exact size.

Open the pattern and spread it out on the metal from which the part is to be formed. Although the pattern does not lie flat, it gives a fairly accurate idea of the approximate shape of the metal to be cut, and the pieced-in sections indicate where the metal is to be stretched. When the pattern has been placed on the material, outline the part and the portions to be stretched using a felt-tipped pen. Add at least one inch of excess metal when cutting the material to size. Trim off the excess metal after bumping the part into shape.

If the part to be formed is radially symmetrical, it is fairly easy to shape since a simple contour template can be used as a working guide. The procedure for bumping sheet metal parts on a sandbag follows certain basic steps that can be applied to any part, regardless of its contour or shape.

1. Lay out and cut the contour template to serve as a working guide and to ensure accuracy of the finished part. (This can be made of sheet metal, medium to heavy cardboard, kraft paper, or thin plywood.)

2. Determine the amount of metal needed, lay it out, and cut it to size, allowing at least ½-inch in excess.

3. Place a sandbag on a solid foundation capable of supporting heavy blows and make a pit in the bag with a smooth-faced mallet. Analyze the part to determine the correct radius the pit should have for the forming operation. The pit changes shape with the hammering it receives and must be readjusted accordingly.

4. Select a soft round-faced or bell-shaped mallet with a contour slightly smaller than the contour desired on

the sheet metal part. Hold one edge of the metal in the left hand and place the portion to be bumped near the edge of the pit on the sandbag. Strike the metal with light glancing blows.

5. Continue bumping toward the center, revolving the metal, and working gradually inward until the desired shape is obtained. Shape the entire part as a unit.

6. Check the part often for accuracy of shape during the bumping process by applying the template. If wrinkles form, work them out before they become too large.

7. Remove small dents and hammer marks with a suitable stake and planishing hammer or with a hand dolly and planishing hammer.

8. Finally, after bumping is completed, use a pair of dividers to mark around the outside of the object. Trim the edge and file it smooth. Clean and polish the part.

Joggling

A joggle, often found at the intersection of stringers and formers, is the offset formed on a part to allow clearance for a sheet or another mating part. Use of the joggle maintains the smooth surface of a joint or splice. The amount of offset is usually small; therefore, the depth of the joggle is generally specified in thousandths of an inch. The thickness of the material to be cleared governs the depth of the joggle. In determining the necessary length of the joggle, allow an extra ¹⁄₁₆-inch to give enough added clearance to assure a fit between the joggled, overlapped part. The distance between the two bends of a joggle is called the allowance. This dimension is normally called out on the drawing. However, a general rule of thumb for figuring allowance is four times the thickness of the displacement of flat sheets. For 90° angles, it must be slightly more due to the stress built up at the radius while joggling. For extrusions, the allowance can be as much as 12 times the material thickness, so, it is important to follow the drawing.

There are a number of different methods of forming joggles. For example, if the joggle is to be made on a straight flange or flat piece of metal, it can be formed on a cornice brake. To form the joggle, use the following procedure:

1. Lay out the boundary lines of the joggle where the bends are to occur on the sheet.

2. Insert the sheet in the brake and bend the metal up approximately 20° to 30°.

3. Release the brake and remove the part.

4. Turn the part over and clamp it in the brake at the second bend line.

5. Bend the part up until the correct height of the joggle is attained.

6. Remove the part from the brake and check the joggle for correct dimensions and clearance.

When a joggle is necessary on a curved part or a curved flange, forming blocks or dies made of hardwood, steel, or aluminum alloy may be used. The forming procedure consists of placing the part to be joggled between the two joggle blocks and squeezing them in a vice or some other suitable clamping device. After the joggle is formed, the joggle blocks are turned over in the vice and the bulge on the opposite flange is flattened with a wooden or rawhide mallet. *[Figure 4-163]*

Since hardwood is easily worked, dies made of hardwood are satisfactory when the die is to be used only a few times. If a number of similar joggles are to be produced, use steel or aluminum alloy dies. Dies of aluminum alloy are preferred since they are easier to fabricate than those of steel and wear about as long. These dies are sufficiently soft and resilient to permit forming aluminum alloy parts on them without marring, and nicks and scratches are easily removed from their surfaces.

When using joggling dies for the first time, test them for accuracy on a piece of waste stock to avoid the possibility of ruining already fabricated parts. *[Figure 4-164]* Always keep the surfaces of the blocks free from dirt, filings, and the like, so that the work is not marred.

Lightening Holes

Lightening holes are cut in rib sections, fuselage frames, and other structural parts to decrease weight. To avoid weakening the member by removal of the material, flanges are often pressed around the holes to strengthen the area from which the material was removed.

Lightening holes should never be cut in any structural part unless authorized. The size of the lightening hole and the width of the flange formed around the hole are determined by design specifications. Margins of safety are considered in the specifications so that the weight of the part can be decreased and still retain the necessary strength. Lightening holes may be cut with a hole saw, a punch, or a fly cutter. The edges are filed smooth to prevent them from cracking or tearing.

Flanging Lightening Holes

Form the flange by using a flanging die, or hardwood or metal form blocks. Flanging dies consist of two matching parts: a female and a male die. For flanging soft metal, dies can be of hardwood, such as maple. For hard metal or for more permanent use, they should be made of steel. The pilot guide should be the same size as the hole to be flanged, and the shoulder should be the same width and angle as the desired flange.

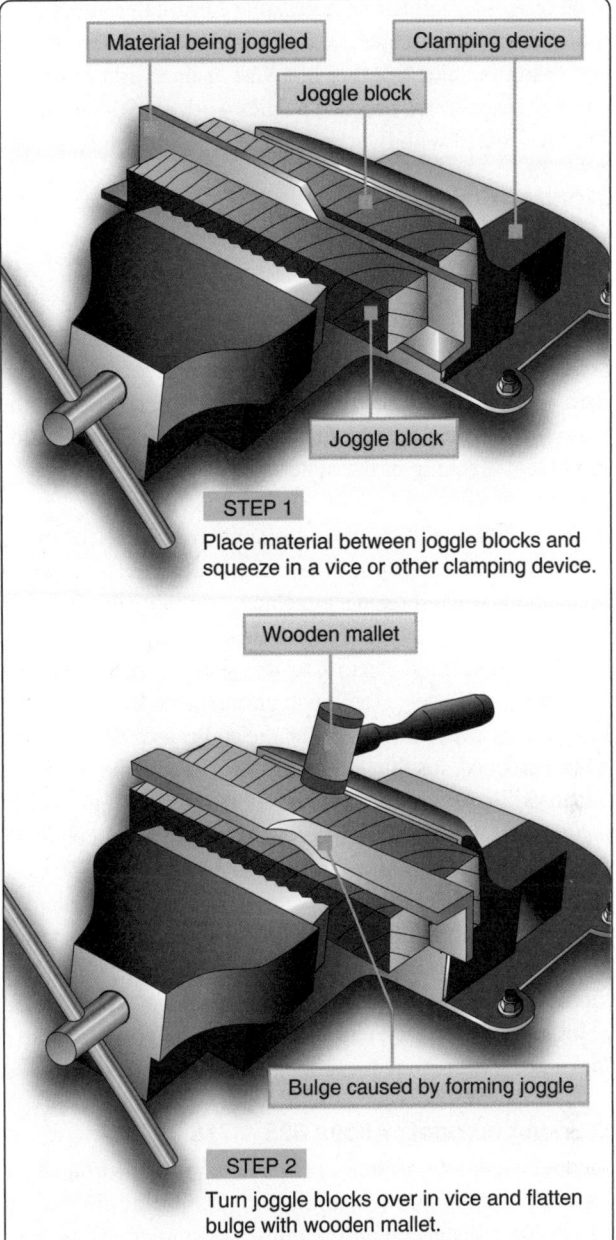

STEP 1

Place material between joggle blocks and squeeze in a vice or other clamping device.

STEP 2

Turn joggle blocks over in vice and flatten bulge with wooden mallet.

Figure 4-163. *Forming joggle using joggle blocks.*

Figure 4-164. *Samples of joggled metal.*

When flanging lightening holes, place the material between the mating parts of the die and form it by hammering or squeezing the dies together in a vise or in an arbor press (a small hand operated press). The dies work more smoothly if they are coated with light machine oil. *[Figure 4-165]*

Working Stainless Steel

Corrosion-resistant-steel (CRES) sheet is used on some parts of the aircraft when high strength is required. CRES causes magnesium, aluminum, or cadmium to corrode when it touches these metals. To isolate CRES from magnesium and aluminum, apply a finish that gives protection between their mating surfaces. It is important to use a bend radius that is larger than the recommended minimum bend radius to prevent cracking of the material in the bend area.

When working with stainless steel, make sure that the metal does not become unduly scratched or marred. Also, take special precautions when shearing, punching, or drilling this metal. It takes about twice as much pressure to shear or punch stainless steel as it does mild steel. Keep the shear or punch and die adjusted very closely. Too much clearance permits the metal to be drawn over the edge of the die and causes it to become work hardened, resulting in excessive strain on the machine. When drilling stainless steel, use an HSS drill bit ground to an included angle of 135°. Keep the drill speed about one-half that required for drilling mild steel, but never exceed 750 rpm. Keep a uniform pressure on the drill so the feed is constant at all times. Drill the material on a backing plate, such as cast iron, which is hard enough to permit the drill bit to cut completely through the stock without pushing the metal away from the drill point. Spot the drill bit before turning on the power and also make sure that pressure is exerted when the power is turned on.

Working Inconel® Alloys 625 & 718

Inconel® refers to a family of nickel-chromium-iron super alloys typically used in high-temperature applications. Corrosion resistance and the ability to stay strong in high temperatures led to the frequent use of these Inconel® alloys in aircraft powerplant structures. Inconel® alloys 625 and 718 can be cold formed by standard procedures used for steel and stainless steel.

Normal drilling into Inconel® alloys can break drill bits sooner and cause damage to the edge of the hole when the drill bit goes through the metal. If a hand drill is used to drill Inconel® alloys 625 and 718, select a 135° cobalt drill bit. When hand drilling, push hard on the drill, but stay at a constant chip rate. For example, with a No. 30 hole, push the drill with approximately 50 pounds of force. Use the maximum drill rpm as illustrated in *Figure 4-166*. A cutting fluid is not necessary when hand drilling.

The following drilling procedures are recommended:

- Drill pilot holes in loose repair parts with power feed equipment before preassembling them.

- Preassemble the repair parts and drill the pilot holes in the mating structure.

- Enlarge the pilot holes to their completed hole dimension.

When drilling Inconel®, autofeed-type drilling equipment is preferred.

Working Magnesium

Warning: Keep magnesium particles away from sources of ignition. Small particles of magnesium burn very easily. In sufficient concentration, these small particles can cause an explosion. If water touches molten magnesium, a steam explosion could occur. Extinguish magnesium fires with dry talc, calcium carbonate, sand, or graphite. Apply the powder on the burning metal to a depth of ½-inch or more. Do not use foam, water, carbon tetrachloride, or carbon dioxide. Magnesium alloys must not touch methyl alcohol.

Magnesium is the world's lightest structural metal. Like many other metals, this silvery-white element is not used in its pure state for stressed application. Instead, magnesium is alloyed with certain other metals (aluminum, zinc, zirconium, manganese, thorium, and rare earth metals) to obtain the strong, lightweight alloys needed for structural uses. When alloyed with these other metals, magnesium, yields alloys with excellent properties and high strength-

Figure 4-165. *Lightening hole die set.*

Drill Size	Maximum RPM
80–30	500
29–U	300
3/8	150

Figure 4-166. *Drill size and speed for drilling Inconel®.*

to-weight ratios. Proper combination of these alloying constituents provide alloys suitable for sand, permanent mold and die castings, forging, extrusions, rolled sheet, and plate with good properties at room temperature, as well as at elevated temperatures.

Lightweight is the best known characteristic of magnesium, an important factor in aircraft design. In comparison, aluminum weighs one and one half times more, iron and steel weigh four times more, and copper and nickel alloys weigh five times more. Magnesium alloys can be cut, drilled, and reamed with the same tools that are used on steel or brass, but the cutting edges of the tool must be sharp. Type B rivets (5056-F aluminum alloy) are used when riveting magnesium alloy parts. Magnesium parts are often repaired with clad 2024-T3 aluminum alloy.

While magnesium alloys can usually be fabricated by methods similar to those used on other metals, remember that many of the details of shop practice cannot be applied. Magnesium alloys are difficult to fabricate at room temperature; therefore, most operations must be performed at high temperatures. This requires preheating of the metal or dies, or both. Magnesium alloy sheets may be cut by blade shears, blanking dies, routers, or saws. Hand or circular saws are usually used for cutting extrusions to length. Conventional shears and nibblers should never be used for cutting magnesium alloy sheet because they produce a rough, cracked edge.

Shearing and blanking of magnesium alloys require close tool tolerances. A maximum clearance of 3 to 5 percent of the sheet thickness is recommended. The top blade of the shears should be ground with an included angle of 45° to 60°. The shear angle on a punch should be from 2° to 3°, with a 1° clearance angle on the die. For blanking, the shear angle on the die should be from 2° to 3° with a 1° clearance angle on the punch. Hold-down pressures should be used when possible. Cold shearing should not be accomplished on a hard-rolled sheet thicker than 0.064-inch or annealed sheet thicker than ⅛-inch. Shaving is used to smooth the rough, flaky edges of a magnesium sheet that has been sheared. This operation consists of removing approximately ¹⁄₃₂-inch by a second shearing.

Hot shearing is sometimes used to obtain an improved sheared edge. This is necessary for heavy sheet and plate stock. Annealed sheet may be heated to 600 °F, but hard-rolled sheet must be held under 400 °F, depending on the alloy used. Thermal expansion makes it necessary to allow for shrinkage after cooling, which entails adding a small amount of material to the cold metal dimensions before fabrication.

Sawing is the only method used in cutting plate stock more than ½-inch thick. Bandsaw raker-set blades of 4- to 6-tooth pitch are recommended for cutting plate stock or heavy extrusions. Small and medium extrusions are more easily cut on a circular cutoff saw having six teeth per inch. Sheet stock can be cut on handsaws having raker-set or straight-set teeth with an 8-tooth pitch. Bandsaws should be equipped with nonsparking blade guides to eliminate the danger of sparks igniting the magnesium alloy filings.

Cold working most magnesium alloys at room temperature is very limited, because they work harden rapidly and do not lend themselves to any severe cold forming. Some simple bending operations may be performed on sheet material, but the radius of bend must be at least 7 times the thickness of the sheet for soft material and 12 times the thickness of the sheet for hard material. A radius of 2 or 3 times the thickness of the sheet can be used if the material is heated for the forming operation.

Since wrought magnesium alloys tend to crack after they are cold-worked, the best results are obtained if the metal is heated to 450 °F before any forming operations are attempted. Parts formed at the lower temperature range are stronger because the higher temperature range has an annealing effect on the metal.

The disadvantages of hot working magnesium are:

1. Heating the dies and the material is expensive and troublesome.

2. There are problems in lubricating and handling materials at these temperatures.

The advantages to hot working magnesium are:

1. It is more easily formed when hot than are other metals.

2. Spring-back is reduced, resulting in greater dimensional accuracy.

When heating magnesium and its alloys, watch the temperature carefully as the metal is easily burned. Overheating also causes small molten pools to form within the metal. In either case, the metal is ruined. To prevent burning, magnesium must be protected with a sulfur dioxide atmosphere while being heated.

Proper bending around a short radius requires the removal of sharp corners and burrs near the bend line. Layouts should be made with a carpenter's soft pencil because any marring of the surface may result in fatigue cracks.

Press brakes can be used for making bends with short radii. Die and rubber methods should be used where bends are to be made at right angles, which complicate the use of a brake. Roll forming may be accomplished cold on equipment designed for forming aluminum. The most common method of forming and shallow drawing of magnesium is to use a rubber pad as the female die. This rubber pad is held in an inverted steel pan that is lowered by a hydraulic press ram. The press exerts pressure on the metal and bends it to the shape of the male die.

The machining characteristics of magnesium alloys are excellent, making possible the use of maximum speeds of the machine tools with heavy cuts and high feed rates. Power requirements for machining magnesium alloys are about one-sixth of those for mild steel.

Filings, shavings, and chips from machining operations should be kept in covered metal containers because of the danger of combustion. Do not use magnesium alloys in liquid deicing and water injection systems or in the integral fuel tank areas.

Working Titanium

Keep titanium particles away from sources of ignition. Small particles of titanium burn very easily. In sufficient concentration, these small particles can cause an explosion. If water touches molten titanium, a steam explosion could occur. Extinguish titanium fires with dry talc, calcium carbonate, sand, or graphite. Apply the powder on the burning metal to a depth of ½-inch or more. Do not use foam, water, carbon tetrachloride, or carbon dioxide.

Description of Titanium

Titanium in its mineral state, is the fourth most abundant structural metal in the earth's crust. It is lightweight, nonmagnetic, strong, corrosion resistant, and ductile. Titanium lies between the aluminum alloys and stainless steel in modulus, density, and strength at intermediate temperatures. Titanium is 30 percent stronger than steel, but is nearly 50 percent lighter. It is 60 percent heavier than aluminum, but twice as strong.

Titanium and its alloys are used chiefly for parts that require good corrosion resistance, moderate strength up to 600 °F (315 °C), and lightweight. Commercially pure titanium sheet may be formed by hydropress, stretch press, brake roll forming, drop hammer, or other similar operations. It is more difficult to form than annealed stainless steel. Titanium can also be worked by grinding, drilling, sawing, and the types of working used on other metals. Titanium must be isolated from magnesium, aluminum, or alloy steel because galvanic corrosion or oxidation of the other metals occurs upon contact.

Monel® rivets or standard close-tolerance steel fasteners should be used when installing titanium parts. The alloy sheet can be formed, to a limited extent, at room temperature. The forming of titanium alloys is divided into three classes:

- Cold forming with no stress relief
- Cold forming with stress relief
- Elevated temperature forming (built-in stress relief)

Over 5 percent of all titanium in the United States is produced in the form of the alloy Ti 6Al-4V, which is known as the workhorse of the titanium industry. Used in aircraft turbine engine components and aircraft structural components, Ti 6Al-4V is approximately 3 times stronger than pure titanium. The most widely used titanium alloy, it is hard to form.

The following are procedures for cold forming titanium 6Al-4V annealed with stress relief (room temperature forming):

1. It is important to use a minimum radius chart when forming titanium because an excessively small radius introduces excess stress to the bend area.

2. Stress relieves the part as follows: heat the part to a temperature above 1,250 °F (677 °C), but below 1,450 °F (788 °C). Keep the part at this temperature for more than 30 minutes but less than 10 hours.

3. A powerful press brake is required to form titanium parts. Regular hand-operated box and pan brakes cannot form titanium sheet material.

4. A power slip roller is often used if the repair patch needs to be curved to fit the contour of the aircraft.

Titanium can be difficult to drill, but standard high-speed drill bits may be used if the bits are sharp, if sufficient force is applied, and if a low-speed drill motor is used. If the drill bit is dull, or if it is allowed to ride in a partially drilled hole, an overheated condition is created, making further drilling extremely difficult. Therefore, keep holes as shallow as possible; use short, sharp drill bits of approved design; and flood the area with large amounts of cutting fluid to facilitate drilling or reaming.

When working titanium, it is recommended that you use carbide or 8 percent cobalt drill bits, reamers, and countersinks. Ensure the drill or reamer is rotating to prevent scoring the side of the hole when removing either of them from a hole. Use a hand drill only when positive-power-feed drills are not available.

The following guidelines are used for drilling titanium:

- The largest diameter hole that can be drilled in a single step is 0.1563-inch because a large force is required.

Larger diameter drill bits do not cut satisfactorily when much force is used. Drill bits that do not cut satisfactorily cause damage to the hole.

- Holes with a diameter of 0.1875-inch and larger can be hand drilled if the operator:
  - Starts with a hole with a diameter of 0.1563-inch.
  - Increases the diameter of the hole in 0.0313-inch or 0.0625-inch increments.

- Cobalt vanadium drill bits last much longer than HSS bits.

- The recommended drill motor rpm settings for hand drilling titanium are listed in *Figure 4-167*.

- The life of a drill bit is shorter when drilling titanium than when drilling steel. Do not use a blunt drill bit or let a drill bit rub the surface of the metal and not cut it. If one of these conditions occurs, the titanium surface becomes work hardened, and it is very difficult to start the drill again.

- When hand drilling two or more titanium parts at the same time, clamp them together tightly. To clamp them together, use temporary bolts, Cleco clamps, or tooling clamps. Put the clamps around the area to drill and as near the area as possible.

- When hand drilling thin or flexible parts, put a support (such as a block of wood) behind the part.

- Titanium has a low thermal conductivity. When it becomes hot, other metals become easily attached to it. Particles of titanium often become welded to the sharp edges of the drill bit if the drill speed is too high. When drilling large plates or extrusions, use a water soluble coolant or sulphurized oil.

Note: The intimate metal-to-metal contact in the metal working process creates heat and friction that must be reduced or the tools and the sheet metal used in the process are quickly damaged and/or destroyed. Coolants, also called cutting fluids, are used to reduce the friction at the interface of the tool and sheet metal by transferring heat away from the tool and sheet metal. Thus, the use of cutting fluids increases productivity, extends tool life, and results in a higher quality of workmanship.

Hole Size (inches)	Drill Speed (rpm)
0.0625	920 to 1830 rpm
0.125	460 to 920 rpm
0.1875	230 to 460 rpm

Figure 4-167. *Hole size and drill speed for drilling titanium.*

Basic Principles of Sheet Metal Repair

Aircraft structural members are designed to perform a specific function or to serve a definite purpose. The primary objective of aircraft repair is to restore damaged parts to their original condition. Very often, replacement is the only way this can be done effectively. When repair of a damaged part is possible, first study the part carefully to fully understand its purpose or function.

Strength may be the principal requirement in the repair of certain structures, while others may need entirely different qualities. For example, fuel tanks and floats must be protected against leakage; cowlings, fairings, and similar parts must have such properties as neat appearance, streamlined shape, and accessibility. The function of any damaged part must be carefully determined to ensure the repair meets the requirements.

An inspection of the damage and accurate estimate of the type of repair required are the most important steps in repairing structural damage. The inspection includes an estimate of the best type and shape of repair patch to use; the type, size, and number of rivets needed; and the strength, thickness, and kind of material required to make the repaired member no heavier (or only slightly heavier) and just as strong as the original.

When investigating damage to an aircraft, it is necessary to make an extensive inspection of the structure. When any component or group of components has been damaged, it is essential that both the damaged members and the attaching structure be investigated, since the damaging force may have been transmitted over a large area, sometimes quite remote from the point of original damage. Wrinkled skin, elongated or damaged bolt or rivet holes, or distortion of members usually appears in the immediate area of such damage, and any one of these conditions calls for a close inspection of the adjacent area. Check all skin, dents, and wrinkles for any cracks or abrasions.

Nondestructive inspection methods (NDI) are used as required when inspecting damage. NDI methods serve as tools of prevention that allow defects to be detected before they develop into serious or hazardous failures. A trained and experienced technician can detect flaws or defects with a high degree of accuracy and reliability. Some of the defects found by NDI include corrosion, pitting, heat/stress cracks, and discontinuity of metals.

When investigating damage, proceed as follows:

- Remove all dirt, grease, and paint from the damaged and surrounding areas to determine the exact condition of each rivet, bolt, and weld.

- Inspect skin for wrinkles throughout a large area.

- Check the operation of all movable parts in the area.

- Determine if repair would be the best procedure.

In any aircraft sheet metal repair, it is critical to:
- Maintain original strength,

- Maintain original contour, and

- Minimize weight.

Maintaining Original Strength

Certain fundamental rules must be observed if the original strength of the structure is to be maintained.

Ensure that the cross-sectional area of a splice or patch is at least equal to or greater than that of the damaged part. Avoid abrupt changes in cross-sectional area. Eliminate dangerous stress concentration by tapering splices. To reduce the possibility of cracks starting from the corners of cutouts, try to make cutouts either circular or oval in shape. Where it is necessary to use a rectangular cutout, make the radius of curvature at each corner no smaller than ½-inch. If the member is subjected to compression or bending loads, the patch should be placed on the outside of the member to obtain a higher resistance to such loads. If the patch cannot be placed there, material one gauge thicker than the original shall be used for the repair.

Replace buckled or bent members or reinforce them by attaching a splice over the affected area. A buckled part of the structure shall not be depended upon to carry its load again, no matter how well the part may be strengthened.

The material used in all replacements or reinforcements must be similar to that used in the original structure. If an alloy weaker than the original must be substituted for it, a heavier thickness must be used to give equivalent cross-sectional strength. A material that is stronger, but thinner, cannot be substituted for the original because one material can have greater tensile strength but less compressive strength than another, or vice versa. Also, the buckling and torsional strength of many sheet metal and tubular parts depends primarily on the thickness of the material rather than its allowable compressive and shear strengths. The manufacturer's SRM often indicates what material can be used as a substitution and how much thicker the material needs to be. *Figure 4-168* is an example of a substitution table found in an SRM.

Care must be taken when forming. Heat-treated and cold-worked aluminum alloys stand very little bending without cracking. On the other hand, soft alloys are easily formed, but they are not strong enough for primary structure. Strong alloys can be formed in their annealed (heated and allowed to cool slowly) condition, and heat treated before assembling to develop their strength.

The size of rivets for any repair can be determined by referring to the rivets used by the manufacturer in the next parallel rivet row inboard on the wing or forward on the fuselage. Another method of determining the size of rivets to be used is to multiply the thickness of the skin by three and use the next larger size rivet corresponding to that figure. For example, if the skin thickness is 0.040-inch, multiply 0.040-inch by 3, which equals 0.120-inch; use the next larger size rivet, ⅛-inch (0.125-inch). The number of rivets to be used for a repair can be found in tables in manufacturer's SRMs or in Advisory Circular (AC) 43.13-1 (as revised), Acceptable Methods, Techniques, and Practices—Aircraft Inspection and Repair. *Figure 4-169* is a table from AC 43.13-1 that is used to calculate the number of rivets required for a repair.

Extensive repairs that are made too strong can be as undesirable as repairs weaker than the original structure. All aircraft structure must flex slightly to withstand the forces imposed during takeoff, flight, and landing. If a repaired area is too strong, excessive flexing occurs at the edge of the completed repair, causing acceleration of metal fatigue.

Shear Strength & Bearing Strength

Aircraft structural joint design involves an attempt to find the optimum strength relationship between being critical in shear and critical in bearing. These are determined by the failure mode affecting the joint. The joint is critical in shear if less than the optimum number of fasteners of a given size are installed. This means that the rivets will fail, and not the sheet, if the joint fails. The joint is critical in bearing if more than the optimum number of fasteners of a given size are installed; the material may crack and tear between holes, or fastener holes may distort and stretch while the fasteners remain intact.

Maintaining Original Contour

Form all repairs in such a manner to fit the original contour perfectly. A smooth contour is especially desirable when making patches on the smooth external skin of high-speed aircraft.

Keeping Weight to a Minimum

Keep the weight of all repairs to a minimum. Make the size of the patches as small as practicable and use no more rivets than are necessary. In many cases, repairs disturb the original balance of the structure. The addition of excessive weight in each repair may unbalance the aircraft, requiring adjustment of the trim-and-balance tabs. In areas such as the spinner on the propeller, a repair requires application of balancing patches in order to maintain a perfect balance of the propeller.

Shape	Initial Material	Replacement Material
Sheet 0.016 to 0.125	Clad 2024–T42 (F)	Clad 2024–T3 2024–T3 Clad 7075–T6 (A) 7075–T6 (A)
	Clad 2024–T3	2024–T3 Clad 7075–T6 (A) 7075–T6 (A)
	Clad 7075–T6	7075–T6
Formed or Extruded Section	2024–T42 (F)	7075–T6 (A) (B)

Sheet Material to be Replaced	Material Replacement Factor									
	7075–T6	Clad 7075–T6	2024–T3		Clad 2024–T3		(F) 2024–T4 2024–T42		(F) Clad 2024–T4 Clad 2024–T42	
	(C)	(C) (H)	(D)	(E)	(D)	(E)	(D)	(E)	(D)	(E)
7075–T6	1.00	1.10	1.20	1.78	1.30	1.83	1.20	1.78	1.24	1.84
Clad 7075–T6	1.00	1.00	1.13	1.70	1.22	1.76	1.13	1.71	1.16	1.76
2024–T3	1.00 (A)	1.00 (A)	1.00	1.00	1.09	1.10	1.00	1.10	1.03	1.14
Clad 2024–T3	1.00 (A)	1.00 (A)	1.00	1.00	1.00	1.00	1.00	1.00	1.03	1.00
2024–T42	1.00 (A)	1.00 (A)	1.00	1.00	1.00	1.00	1.00	1.00	1.00	1.14
Clad 2024–T42	1.00 (A)	1.00 (A)	1.00	1.00	1.00	1.00	1.00	1.00	1.00	1.00
7178–T6	1.28	1.28	1.50	1.90	1.63	2.00	1.86	1.90	1.96	1.98
Clad 7178–T6	1.08	1.18	1.41	1.75	1.52	1.83	1.75	1.75	1.81	1.81
5052–H34 (G) (H)	1.00 (A)	1.00 (A)	1.00	1.00	1.00	1.00	1.00	1.00	1.00	1.00

Notes

- All dimensions are in inches, unless given differently.

- It is possible that more protection from corrosion will be necessary when bare mineral is used to replace Clad material. Refer to 51-10-2.

- It is possible for the material replacement factor to be a lower value for a specific location on the airplane. To get that value, contact Boeing for a case by case analysis.

- Refer to Figure 3 for minimun bend radii.

- Example:
  To refer 0.040 thick 7075–T6 with Clad 7075–T6, multiply the gage by the material replacement factor to get the replacement gage 0. 040 x 1.10 = 0.045.

(A) These materials cannot be used as replacements for the initial material in areas that are pressured.

(B) They also cannot be used in the wing interspar structure at the wing center section structure.

(C) Use the next thicker standard gage when you use a formed section as a replacement for an extrusion.

(D) For all gages of flat sheet and formed sections.

(E) For flat sheet less than 0.071 thick.

(F) For flat sheet 0.071 thick and thicker, and for formed sections.

(G) 2024–T4 and 2024–T42 are equivalent.

(H) A compound to give protection from corrosion must be applied to bare material that is used to replace 5052–H34.

Figure 4-168. *Material substitution.*

When flight controls are repaired and weight is added, it is very important to perform a balancing check to determine if the flight control is still within its balance limitations. Failure to do so could result in flight control flutter.

Flutter & Vibration Precautions

To prevent severe vibration or flutter of flight control surfaces during flight, precautions must be taken to stay within the design balance limitations when performing maintenance or repair. The importance of retaining the proper balance and rigidity of aircraft control surfaces cannot be overemphasized.

| Thickness "t" in inches | No. of 2117–T4 (AD) protruding head rivets required per inch of width "W" | | | | | No. of Bolts |
| | Rivet size | | | | | |
	3/32	1/8	5/32	3/16	1/4	AN–3
.016	6.5	4.9	- -	- -	- -	- -
.020	6.5	4.9	3.9	- -	- -	- -
.025	6.9	4.9	3.9	- -	- -	- -
.032	8.9	4.9	3.9	3.3	- -	- -
.036	10.0	5.6	3.9	3.3	2.4	- -
.040	11.1	6.2	4.0	3.3	2.4	- -
.051	- -	7.9	5.1	3.6	2.4	3.3
.064	- -	9.9	6.5	4.5	2.5	3.3
.081	- -	12.5	8.1	5.7	3.1	3.3
.091	- -	- -	9.1	6.3	3.5	3.3
.102	- -	- -	10.3	7.1	3.9	3.3
.128	- -	- -	12.9	8.9	4.9	3.3

Notes
   a. For stringer in the upper surface of a wing, or in a fuselage, 80 percent of the number of rivets shown in the table may be used.
   b. For intermediate frames, 60 percent of the number shown may be used.
   c. For single lap sheet joints, 75 percent of the number shown may be used.

Engineering Notes
   a. The load per inch of width of material was calculated by assuming a strip 1 inch wide in tension.
   b. Number of rivets required was calculated for 2117–T4 (AD) rivets, based on a rivet allowable shear stress equal to percent of the sheet allowable tensile stress, and a sheet allowable bearing stress equal to 160 percent of the sheet allowable tensile stress, using nominal hole diameters for rivets.
   c. Combinations of shoot thickness and rivet size above the underlined numbers are critical in (i.e., will fail by) bearing on the sheet; those below are critical in shearing of the rivets.
   d. The number of AN–3 bolts required below the underlined number was calculated based on a sheet allowable tensile stress of 55.000 psi and a bolt allowable single shear load of 2.126 pounds.

Figure 4-169. *Rivet calculation table.*

The effect of repair or weight change on the balance and CG is proportionately greater on lighter surfaces than on the older heavier designs. As a general rule, repair the control surface in such a manner that the weight distribution is not affected in any way, in order to preclude the occurrence of flutter of the control surface in flight. Under certain conditions, counterbalance weight is added forward of the hinge line to maintain balance. Add or remove balance weights only when necessary in accordance with the manufacturer's instructions. Flight testing must be accomplished to ensure flutter is not a problem. Failure to check and retain control surface balance within the original or maximum allowable value could result in a serious flight hazard.

Aircraft manufacturers use different repair techniques and repairs designed and approved for one type of aircraft are not automatically approved for other types of aircraft. When repairing a damaged component or part, consult the applicable section of the manufacturer's SRM for the aircraft. Usually the SRM contains an illustration for a similar repair along with a list of the types of material, rivets and rivet spacing, and the methods and procedures to be used. Any additional knowledge needed to make a repair is also detailed. If the necessary information is not found in the SRM, attempt to find a similar repair or assembly installed by the manufacturer of the aircraft.

Inspection of Damage

When visually inspecting damage, remember that there may be other kinds of damage than that caused by impact from foreign objects or collision. A rough landing may overload one of the landing gear, causing it to become sprung; this would be classified as load damage. During inspection and sizing up of the repair job, consider how far the damage caused by the sprung shock strut extends to supporting structural members.

A shock occurring at one end of a member is transmitted throughout its length; therefore, closely inspect all rivets, bolts, and attaching structures along the complete member for any evidence of damage. Make a close examination for rivets that have partially failed and for holes that have been elongated.

Whether specific damage is suspected or not, an aircraft structure must occasionally be inspected for structural integrity. The following paragraphs provide general guidelines for this inspection.

When inspecting the structure of an aircraft, it is very important to watch for evidence of corrosion on the inside. This is most likely to occur in pockets and corners where moisture and salt spray may accumulate; therefore, drain holes must always be kept clean.

While an injury to the skin covering caused by impact with an object is plainly evident, a defect, such as distortion or failure of the substructure, may not be apparent until some evidence develops on the surface, such as canted, buckled or wrinkled covering, and loose rivets or working rivets. A working rivet is one that has movement under structural stress, but has not loosened to the extent that movement can be observed. This situation can sometimes be noted by a dark, greasy residue or deterioration of paint and primers around rivet heads. External indications of internal injury must be watched for and correctly interpreted. When found, an investigation of the substructure in the vicinity should be made and corrective action taken.

Warped wings are usually indicated by the presence of parallel skin wrinkles running diagonally across the wings and extending over a major area. This condition may develop from unusually violent maneuvers, extremely rough air, or extra hard landings. While there may be no actual rupture of any part of the structure, it may be distorted and weakened. Similar failures may also occur in fuselages. Small cracks in the skin covering may be caused by vibration and they are frequently found leading away from rivets.

Aluminum alloy surfaces having chipped protective coating, scratches, or worn spots that expose the surface of the metal should be recoated at once, as corrosion may develop rapidly. The same principle is applied to aluminum clad (Alclad™) surfaces. Scratches, which penetrate the pure aluminum surface layer, permit corrosion to take place in the alloy beneath.

A simple visual inspection cannot accurately determine if suspected cracks in major structural members actually exist or the full extent of the visible cracks. Eddy current and ultrasonic inspection techniques are used to find hidden damage.

Types of Damage & Defects

Types of damage and defects that may be observed on aircraft parts are defined as follows:

- Brinelling—occurrence of shallow, spherical depressions in a surface, usually produced by a part having a small radius in contact with the surface under high load.

- Burnishing—polishing of one surface by sliding contact with a smooth, harder surface. Usually there is no displacement or removal of metal.

- Burr—a small, thin section of metal extending beyond a regular surface, usually located at a corner or on the edge of a hole.

- Corrosion—loss of metal from the surface by chemical or electrochemical action. The corrosion products generally are easily removed by mechanical means. Iron rust is an example of corrosion.

- Crack—a physical separation of two adjacent portions of metal, evidenced by a fine or thin line across the surface caused by excessive stress at that point. It may extend inward from the surface from a few thousandths of an inch to completely through the section thickness.

- Cut—loss of metal, usually to an appreciable depth over a relatively long and narrow area, by mechanical means, as would occur with the use of a saw blade, chisel, or sharp-edged stone striking a glancing blow.

- Dent—indentation in a metal surface produced by an object striking with force. The surface surrounding the indentation is usually slightly upset.

- Erosion—loss of metal from the surface by mechanical action of foreign objects, such as grit or fine sand. The eroded area is rough and may be lined in the direction in which the foreign material moved relative to the surface.

- Chattering—breakdown or deterioration of metal surface by vibratory or chattering action. Although chattering may give the general appearance of metal loss or surface cracking, usually, neither has occurred.

- Galling—breakdown (or build-up) of metal surfaces due to excessive friction between two parts having relative motion. Particles of the softer metal are torn loose and welded to the harder metal.

- Gouge—groove in, or breakdown of, a metal surface from contact with foreign material under heavy pressure. Usually it indicates metal loss but may be largely the displacement of material.

- Inclusion—presence of foreign or extraneous material wholly within a portion of metal. Such material is introduced during the manufacture of rod, bar or tubing by rolling or forging.

- Nick—local break or notch on an edge. Usually it involves the displacement of metal rather than loss.

- Pitting—sharp, localized breakdown (small, deep cavity) of metal surface, usually with defined edges.

- Scratch—slight tear or break in metal surface from light, momentary contact by foreign material.

- Score—deeper (than scratch) tear or break in metal surface from contact under pressure. May show discoloration from temperature produced by friction.

- Stain—a change in color, locally causing a noticeably different appearance from the surrounding area.

- Upsetting—a displacement of material beyond the normal contour or surface (a local bulge or bump). Usually it indicates no metal loss.

Classification of Damage

Damages may be grouped into four general classes. In many cases, the availabilities of repair materials and time are the most important factors in determining if a part should be repaired or replaced.

Negligible Damage

Negligible damage consists of visually apparent, surface damage that do not affect the structural integrity of the component involved. Negligible damage may be left as is or may be corrected by a simple procedure without restricting flight. In both cases, some corrective action must be taken to keep the damage from spreading. Negligible or minor damage areas must be inspected frequently to ensure the damage does not spread. Permissible limits for negligible damage vary for different components of different aircraft and should be carefully researched on an individual basis. Failure to ensure that damages within the specified limit of negligible damage may result in insufficient structural strength of the affected support member for critical flight conditions.

Small dents, scratches, cracks, and holes that can be repaired by smoothing, sanding, stop drilling, or hammering out, or otherwise repaired without the use of additional materials, fall in this classification. *[Figure 4-170]*

Damage Repairable by Patching

Damage repairable by patching is any damage exceeding negligible damage limits that can be repaired by installing splice members to bridge the damaged portion of a structural part. The splice members are designed to span the damaged areas and to overlap the existing undamaged surrounding structure. The splice or patch material used in internal riveted and bolted repairs is normally the same type of material as

the damaged part, but one gauge heavier. In a patch repair, filler plates of the same gauge and type of material as that in the damaged component may be used for bearing purposes or to return the damaged part to its original contour. Structural fasteners are applied to members and the surrounding structure to restore the original load-carrying characteristics of the damaged area. The use of patching depends on the extent of the damage and the accessibility of the component to be repaired.

Damage Repairable by Insertion

Damage must be repaired by insertion when the area is too large to be patched or the structure is arranged such that repair members would interfere with structural alignment (e.g., in a hinge or bulkhead). In this type of repair, the damaged portion is removed from the structure and replaced by a member identical in material and shape. Splice connections at each end of the insertion member provide for load transfer to the original structure.

Damage Necessitating Replacement of Parts

Components must be replaced when their location or extent of damage makes repair impractical, when replacement is more economical than repair, or when the damaged part is relatively easy to replace. For example, replacing damaged castings, forgings, hinges, and small structural members, when available, is more practical than repairing them. Some highly stressed members must be replaced because repair would not restore an adequate margin of safety.

Repairability of Sheet Metal Structure

The following criteria can be used to help an aircraft technician decide upon the repairability of a sheet metal structure:

- Type of damage.

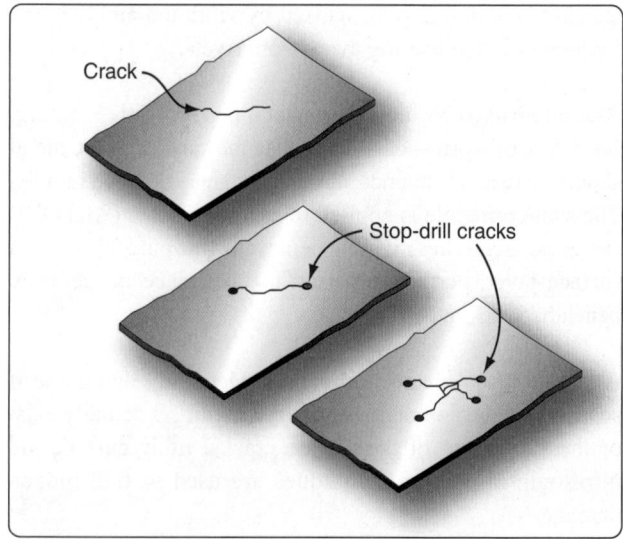

Figure 4-170. *Repair of cracks by stop-drilling.*

- Type of original material.

- Location of the damage.

- Type of repair required.

- Tools and equipment available to make the repair.

The following methods, procedures, and materials are only typical and should not be used as the authority for a repair.

Structural Support During Repair

During repair, the aircraft must be adequately supported to prevent further distortion or damage. It is also important that the structure adjacent to the repair is supported when it is subject to static loads. The aircraft structure can be supported adequately by the landing gear or by jacks where the work involves a repair, such as removing the control surfaces, wing panels, or stabilizers. Cradles must be prepared to hold these components while they are removed from the aircraft.

When the work involves extensive repair of the fuselage, landing gear, or wing center section, a jig (a device for holding parts in position to maintain their shape) may be constructed to distribute the loads while repairs are being accomplished. *Figure 4-171* shows a typical aircraft jig. Always check the applicable aircraft maintenance manual for specific support requirements.

Assessment of Damage

Before starting any repair, the extent of damage must be fully evaluated to determine if repair is authorized or even practical. This evaluation should identify the original material used and the type of repair required. The assessment of the damage begins with an inspection of riveted joints and an inspection for corrosion.

Inspection of Riveted Joints

Inspection consists of examining both the shop and manufactured heads and the surrounding skin and structural parts for deformities.

During the repair of an aircraft structural part, examine adjacent parts to determine the condition of neighboring rivets. The presence of chipped or cracked paint around the heads may indicate shifted or loose rivets. If the heads are tipped or if rivets are loose, they show up in groups of several consecutive rivets and are probably tipped in the same direction. If heads that appear to be tipped are not in groups and are not tipped in the same direction, tipping may have occurred during some previous installation.

Inspect rivets that are known to have been critically loaded, but that show no visible distortion, by drilling off the head and carefully punching out the shank. If upon examination, the

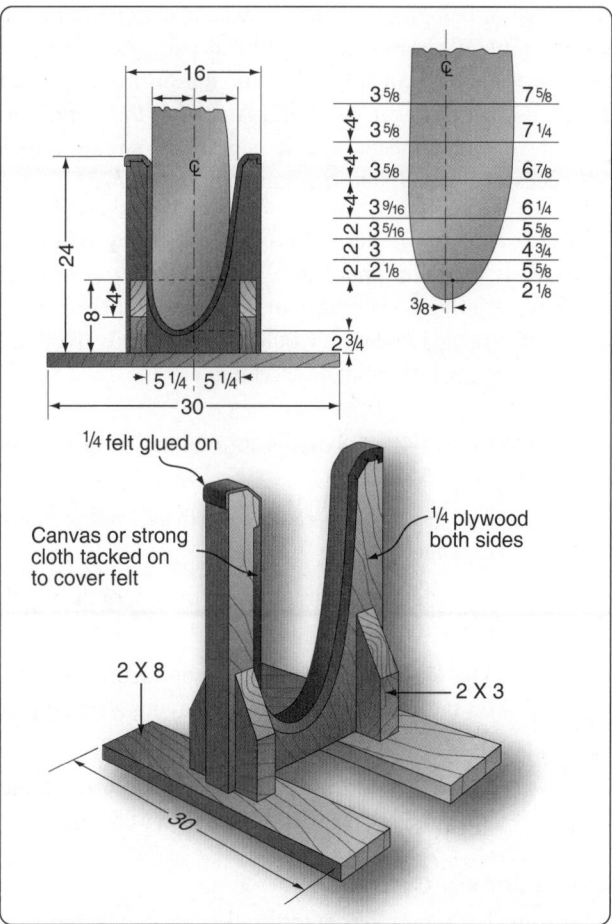

Figure 4-171. *Aircraft jig used to hold components during repairs.*

shank appears joggled and the holes in the sheet misaligned, the rivet has failed in shear. In that case, determine what is causing the stress and take necessary corrective action. Countersunk rivets that show head slippage within the countersink or dimple, indicating either sheet bearing failure or rivet shear failure, must be replaced.

Joggles in removed rivet shanks indicate partial shear failure. Replace these rivets with the next larger size. Also, if the rivet holes show elongation, replace the rivets with the next larger size. Sheet failures, such as tearouts, cracks between rivets, and the like, usually indicate damaged rivets, and the complete repair of the joint may require replacement of the rivets with the next larger size.

The presence of a black residue around the rivets is not an indication of looseness, but it is an indication of movement (fretting). The residue, which is aluminum oxide, is formed by a small amount of relative motion between the rivet and the adjacent surface. This is called fretting corrosion, or smoking, because the aluminum dust quickly forms a dark, dirty looking trail, like a smoke trail. Sometimes, the thinning of the moving pieces can propagate a crack. If a rivet is suspected of being defective, this residue may be removed with a general purpose abrasive hand pad, such as those manufactured by Scotch Brite™, and the surface inspected for signs of pitting or cracking. Although the condition indicates the component is under significant stress, it does not necessarily precipitate cracking. *[Figure 4-172]*

Airframe cracking is not necessarily caused by defective rivets. It is common practice in the industry to size rivet patterns assuming one or more of the rivets is not effective. This means that a loose rivet would not necessarily overload adjacent rivets to the point of cracking.

Rivet head cracking is acceptable under the following conditions:

- The depth of the crack is less than ⅛ of the shank diameter.
- The width of the crack is less than 1/16 of the shank diameter.
- The length of the crack is confined to an area on the head within a circle having a maximum diameter of 1¼ times the shank diameter.
- Cracks should not intersect, which creates the potential for the loss of a portion of a head.

Inspection for Corrosion

Corrosion is the gradual deterioration of metal due to a chemical or electrochemical reaction with its environment. The reaction can be triggered by the atmosphere, moisture, or other agents. When inspecting the structure of an aircraft, it is important to watch for evidence of corrosion on both the outside and inside. Corrosion on the inside is most likely to occur in pockets and corners where moisture and salt spray may accumulate; therefore, drain holes must always be kept clean. Also inspect the surrounding members for evidence of corrosion.

Damage Removal

To prepare a damaged area for repair:

1. Remove all distorted skin and structure in damaged area.

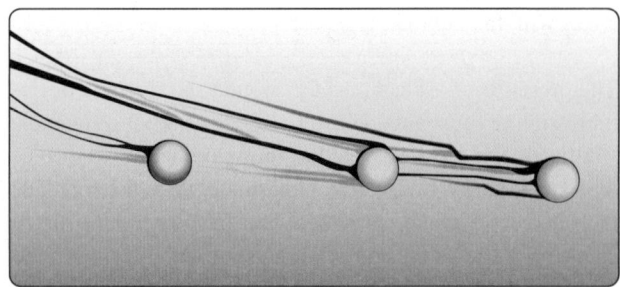

Figure 4-172. *Smoking rivet.*

2. Remove damaged material so that the edges of the completed repair match existing structure and aircraft lines.

3. Round all square corners.

4. Smooth out any abrasions and/or dents.

5. Remove and incorporate into the new repair any previous repairs joining the area of the new repair.

Repair Material Selection

The repair material must duplicate the strength of the original structure. If an alloy weaker than the original material has to be used, a heavier gauge must be used to give equivalent cross-sectional strength. A lighter gauge material should not be used even when using a stronger alloy.

Repair Parts Layout

All new sections fabricated for repairing or replacing damaged parts in a given aircraft should be carefully laid out to the dimensions listed in the applicable aircraft manual before fitting the parts into the structure.

Rivet Selection

Normally, the rivet size and material should be the same as the original rivets in the part being repaired. If a rivet hole has been enlarged or deformed, the next larger size rivet must be used after reworking the hole. When this is done, the proper edge distance for the larger rivet must be maintained. Where access to the inside of the structure is impossible and blind rivets must be used in making the repair, always consult the applicable aircraft maintenance manual for the recommended type, size, spacing, and number of rivets needed to replace either the original installed rivets or those that are required for the type of repair being performed.

Rivet Spacing & Edge Distance

The rivet pattern for a repair must conform to instructions in the applicable aircraft manual. The existing rivet pattern is used whenever possible.

Corrosion Treatment

Prior to assembly of repair or replacement parts, make certain that all existing corrosion has been removed in the area and that the parts are properly insulated one from the other.

Approval of Repair

Once the need for an aircraft repair has been established, Title 14 of the Code of Federal Regulations (14 CFR) defines the approval process. 14 CFR part 43, section 43.13(a) states that each person performing maintenance, alteration, or preventive maintenance on an aircraft, engine, propeller, or appliance shall use the methods, techniques, and practices prescribed in the current manufacturer's maintenance manual or instructions for continued airworthiness prepared by its manufacturer, or other methods, techniques, or practices acceptable to the Administrator. AC 43.13-1 contains methods, techniques, and practices acceptable to the Administrator for the inspection and repair of nonpressurized areas of civil aircraft, only when there are no manufacturer repair or maintenance instructions. This data generally pertains to minor repairs. The repairs identified in this AC may only be used as a basis for FAA approval for major repairs. The repair data may also be used as approved data, and the AC chapter, page, and paragraph listed in block 8 of FAA Form 337 when:

a. The user has determined that it is appropriate to the product being repaired;

b. It is directly applicable to the repair being made; and

c. It is not contrary to manufacturer's data.

Engineering support from the aircraft manufacturer is required for repair techniques and methods that are not described in the aircraft maintenance manual or SRM.

FAA Form 337, Major Repair and Alteration, must be completed for repairs to the following parts of an airframe and repairs of the following types involving the strengthening, reinforcing, splicing, and manufacturing of primary structural members or their replacement, when replacement is by fabrication, such as riveting or welding. *[Figure 4-173]*

- Box beams
- Monocoque or semimonocoque wings or control surfaces
- Wing stringers or chord members
- Spars
- Spar flanges
- Members of truss-type beams
- Thin sheet webs of beams
- Keel and chine members of boat hulls or floats
- Corrugated sheet compression members that act as flange material of wings or tail surfaces
- Wing main ribs and compression members
- Wing or tail surface brace struts, fuselage longerons
- Members of the side truss, horizontal truss or bulkheads
- Main seat support braces and brackets
- Landing gear brace struts
- Repairs involving the substitution of material
- Repair of damaged areas in metal or plywood stressed

covering exceeding six inches in any direction

- Repair of portions of skin sheets by making additional seams
- Splicing of thin sheets
- Repair of three or more adjacent wing or control surface ribs or the leading edge of wings and control surfaces between such adjacent ribs

For major repairs made in accordance with a manual or specifications acceptable to the Administrator, a certificated repair station may use the customer's work order upon which the repair is recorded in place of the FAA Form 337.

Repair of Stressed Skin Structure

In aircraft construction, stressed skin is a form of construction in which the external covering (skin) of an aircraft carries part or all of the main loads. Stressed skin is made from high strength rolled aluminum sheets. Stressed skin carries a large portion of the load imposed upon an aircraft structure. Various specific skin areas are classified as highly critical, semicritical, or noncritical. To determine specific repair requirements for these areas, refer to the applicable aircraft maintenance manual.

Minor damage to the outside skin of the aircraft can be repaired by applying a patch to the inside of the damaged sheet. A filler plug must be installed in the hole made by the removal of the damaged skin area. It plugs the hole and forms a smooth outside surface necessary for aerodynamic smoothness of the aircraft. The size and shape of the patch is determined in general by the number of rivets required in the repair. If not otherwise specified, calculate the required number of rivets by using the rivet formula. Make the patch plate of the same material as the original skin and of the same thickness or of the next greater thickness.

Patches

Skin patches may be classified as two types:

- Lap or scab patch
- Flush patch

Lap or Scab Patch

The lap or scab type of patch is an external patch where the edges of the patch and the skin overlap each other. The overlapping portion of the patch is riveted to the skin. Lap patches may be used in most areas where aerodynamic smoothness is not important. *Figure 4-174* shows a typical patch for a crack and or for a hole.

When repairing cracks or small holes with a lap or scab patch, the damage must be cleaned and smoothed. In repairing cracks, a small hole must be drilled in each end and sharp

MAJOR REPAIR AND ALTERATION
(Airframe, Powerplant, Propeller, or Appliance)

US Department
of Transportation
Federal Aviation
Administration

OMB No. 2120-0020 Exp: 01/31/2023	Electronic Tracking Number
	For FAA Use Only

INSTRUCTIONS: Print or type all entries. See Title 14 CFR §43.9, Part 43 Appendix B, and AC 43.9-1 (or subsequent revision thereof) for instructions and disposition of this form. This report is required by law (49 U.S.C. §44701). Failure to report can result in a civil penalty for each such violation. (49 U.S.C. §46301(a))

1. Aircraft

Nationality and Registration Mark	Serial No.	
Make	Model	Series

2. Owner

Name *(As shown on registration certificate)*	Address *(As shown on registration certificate)*
	Address
	City State
	Zip Country

3. For FAA Use Only

4. Type		5. Unit Identification			
Repair	Alteration	Unit	Make	Model	Serial No.
☐	☐	AIRFRAME	——————	*(As described in Item 1 above)*	——————
☐	☐	POWERPLANT			
☐	☐	PROPELLER			
☐	☐	APPLIANCE	Type		
			Manufacturer		

6. Conformity Statement

A. Agency's Name and Address	B. Kind of Agency	
Name	U. S. Certificated Mechanic	Manufacturer
Address	Foreign Certificated Mechanic	C. Certificate No.
City State	Certificated Repair Station	
Zip Country	Certificated Maintenance Organization	

D. I certify that the repair and/or alteration made to the unit(s) identified in item 5 above and described on the reverse or attachments hereto have been made in accordance with the requirements of Part 43 of the U.S. Federal Aviation Regulations and that the information furnished herein is true and correct to the best of my knowledge.

Extended range fuel per 14 CFR Part 43 App. B ☐	Signature/Date of Authorized Individual

7. Approval for Return to Service

Pursuant to the authority given persons specified below, the unit identified in item 5 was inspected in the manner prescribed by the Administrator of the Federal Aviation Administration and is ☐ Approved ☐ Rejected

BY	FAA Flt. Standards Inspector	Manufacturer	Maintenance Organization	Persons Approved by Canadian Department of Transport
	FAA Designee	Repair Station	Inspection Authorization	Other *(Specify)*

Certificate or Designation No.	Signature/Date of Authorized Individual

Figure 4-173. *FAA Form 337, Major Repair and Alteration (Airframe, Powerplant, Propeller, or Appliance).*

NOTICE

Weight and balance or operating limitation changes shall be entered in the appropriate aircraft record. An alteration must be compatible with all previous alterations to assure continued conformity with the applicable airworthiness requirements.

8. Description of Work Accomplished
(If more space is required, attach additional sheets. Identify with aircraft nationality and registration mark and date work completed.)

_____   _____

_____   _____
Nationality and Registration Mark   Date

Additional Sheets Are Attached

Figure 4-173. *FAA Form 337, Major Repair and Alteration (Airframe, Powerplant, Propeller, or Appliance) continued.*

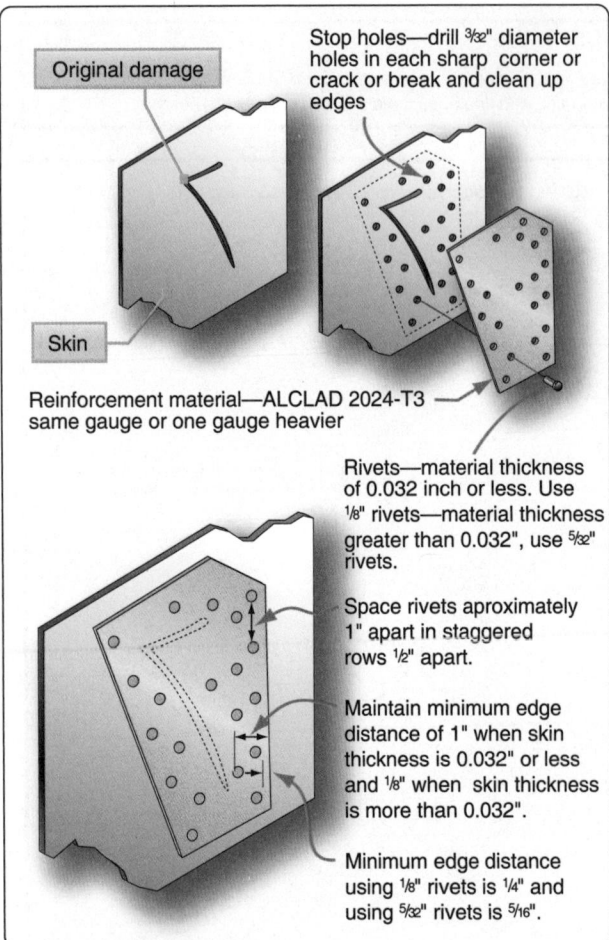

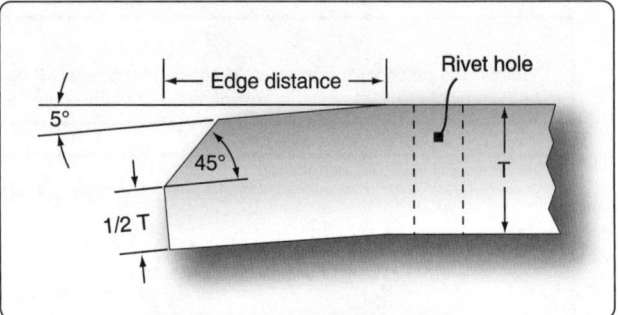

Figure 4-175. *Lap patch edge preparation.*

Figure 4-174. *Lap or scab patch (crack).*

bend of the crack before applying the patch. These holes relieve the stress at these points and prevent the crack from spreading. The patch must be large enough to install the required number of rivets. It may be cut circular, square, or rectangular. If it is cut square or rectangular, the corners are rounded to a radius no smaller than ¼-inch. The edges must be chamfered to an angle of 45° for ½ the thickness of the material, and bent down 5° over the edge distance to seal the edges. This reduces the chance that the repair is affected by the airflow over it. These dimensions are shown in *Figure 4-175*.

Flush Patch

A flush patch is a filler patch that is flush to the skin when applied it is supported by and riveted to a reinforcement plate which is, in turn, riveted to the inside of the skin. *Figure 4-176* shows a typical flush patch repair. The doubler is inserted through the opening and rotated until it slides in place under the skin. The filler must be of the same gauge and material as the original skin. The doubler should be of material one gauge heavier than the skin.

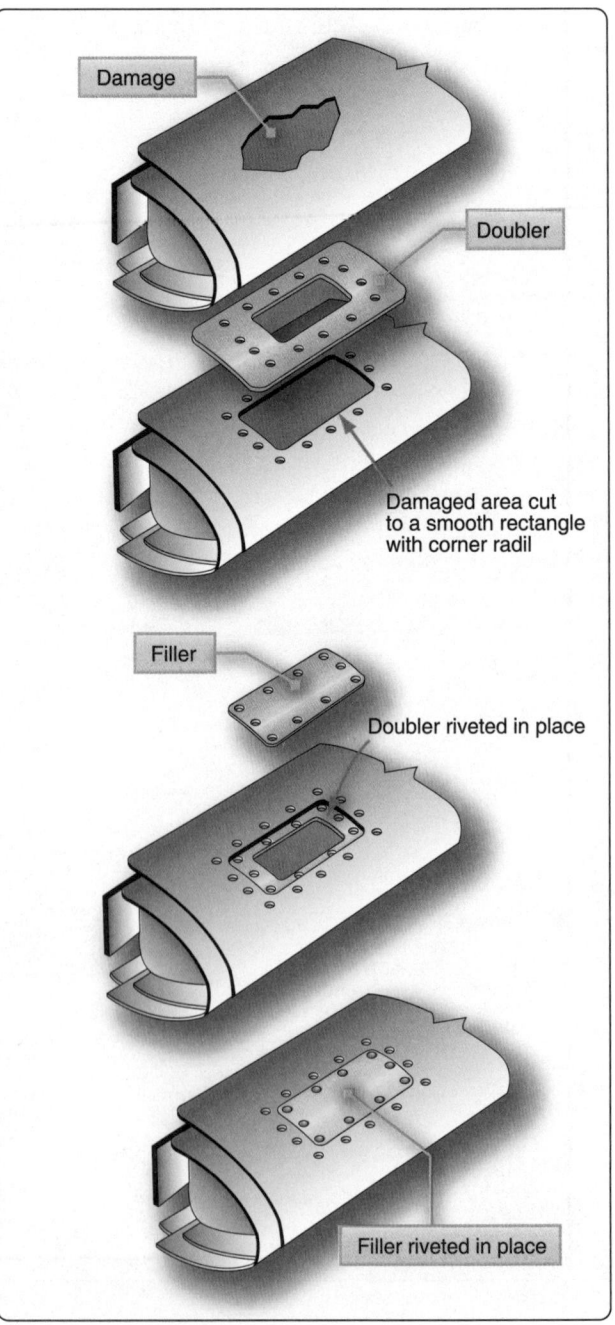

Figure 4-176. *Typical flush patch repair.*

Open & Closed Skin Area Repair

The factors that determine the methods to be used in skin repair are accessibility to the damaged area and the instructions found in the aircraft maintenance manual. The skin on most areas of an aircraft is inaccessible for making the repair from the inside and is known as closed skin. Skin that is accessible from both sides is called open skin.

Usually, repairs to open skin can be made in the conventional manner using standard rivets, but in repairing closed skin, some type of special fastener must be used. The exact type to be used depends on the type of repair being made and the recommendations of the aircraft manufacturer.

Design of a Patch for a Non-pressurized Area

Damage to the aircraft skin in a non-pressurized area can be repaired by a flush patch if a smooth skin surface is required or by an external patch in noncritical areas. *[Figure 4-177]* The first step is to remove the damage. Cut the damage to a round, oval, or rectangular shape. Round all corners of a rectangular patch to a minimum radius of 0.5-inch. The minimum edge distance used is 2 times the diameter and the rivet spacing is typically between 4-6 times the diameter. The size of the doubler depends on the edge distance and rivet spacing. The doubler material is of the same material as the damaged skin, but of one thickness greater than the damaged skin. The size of the doubler depends on the edge distance and rivet spacing. The insert is made of the same material and thickness as the damaged skin. The size and type of rivets should be the same as rivets used for similar joints on the aircraft. The SRM indicates what size and type of rivets to use.

Typical Repairs for Aircraft Structures

This section describes typical repairs of the major structural parts of an airplane. When repairing a damaged component or part, consult the applicable section of the manufacturer's SRM for the aircraft. Normally, a similar repair is illustrated, and the types of material, rivets, and rivet spacing and the methods and procedures to be used are listed. Any additional knowledge needed to make a repair is also detailed. If the necessary information is not found in the SRM, attempt to find a similar repair or assembly installed by the manufacturer of the aircraft.

Floats

To maintain the float in an airworthy condition, periodic and frequent inspections should be made because of the rapidity of corrosion on metal parts, particularly when the aircraft is operated in salt water. Inspection of floats and hulls involves examination for damage due to corrosion, collision with other objects, hard landings, and other conditions that may lead to failure.

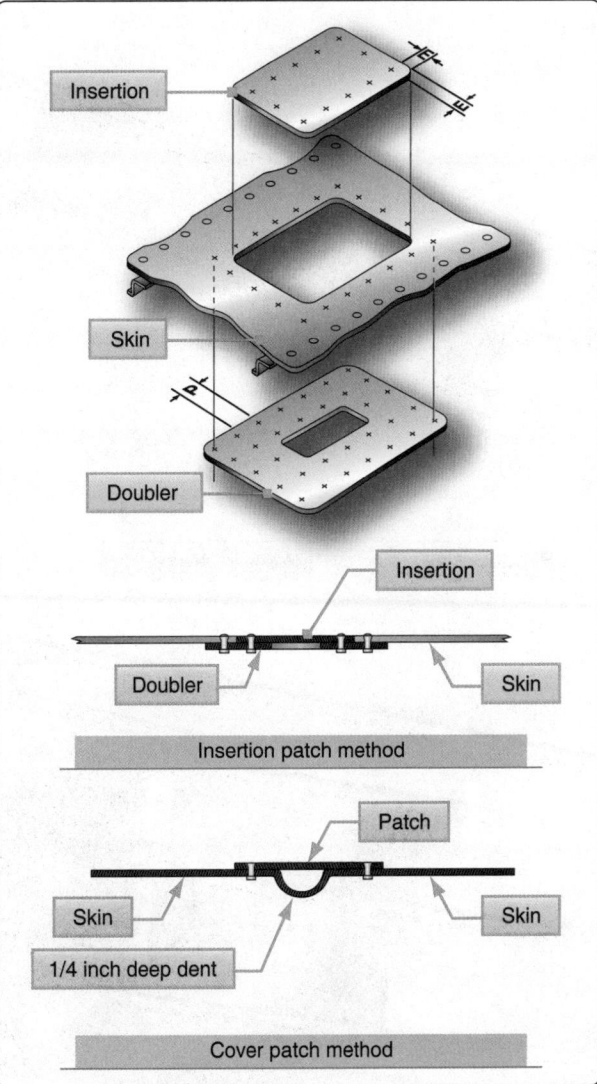

Figure 4-177. *Repair patch for a non-pressurized area.*

Note: Blind rivets should not be used on floats or amphibian hulls below the water line.

Sheet-metal floats should be repaired using approved practices; however, the seams between sections of sheet metal should be waterproofed with suitable fabric and sealing compound. A float that has undergone hull repairs should be tested by filling it with water and allowing it to stand for at least 24 hours to see if any leaks develop. *[Figure 4-178]*

Corrugated Skin Repair

Some of the flight controls of smaller general aviation aircraft have beads in their skin panels. The beads give some stiffness to the thin skin panels. The beads for the repair patch can be formed with a rotary former or press brake. *[Figure 4-179]*

Replacement of a Panel

Damage to metal aircraft skin that exceeds repairable limits

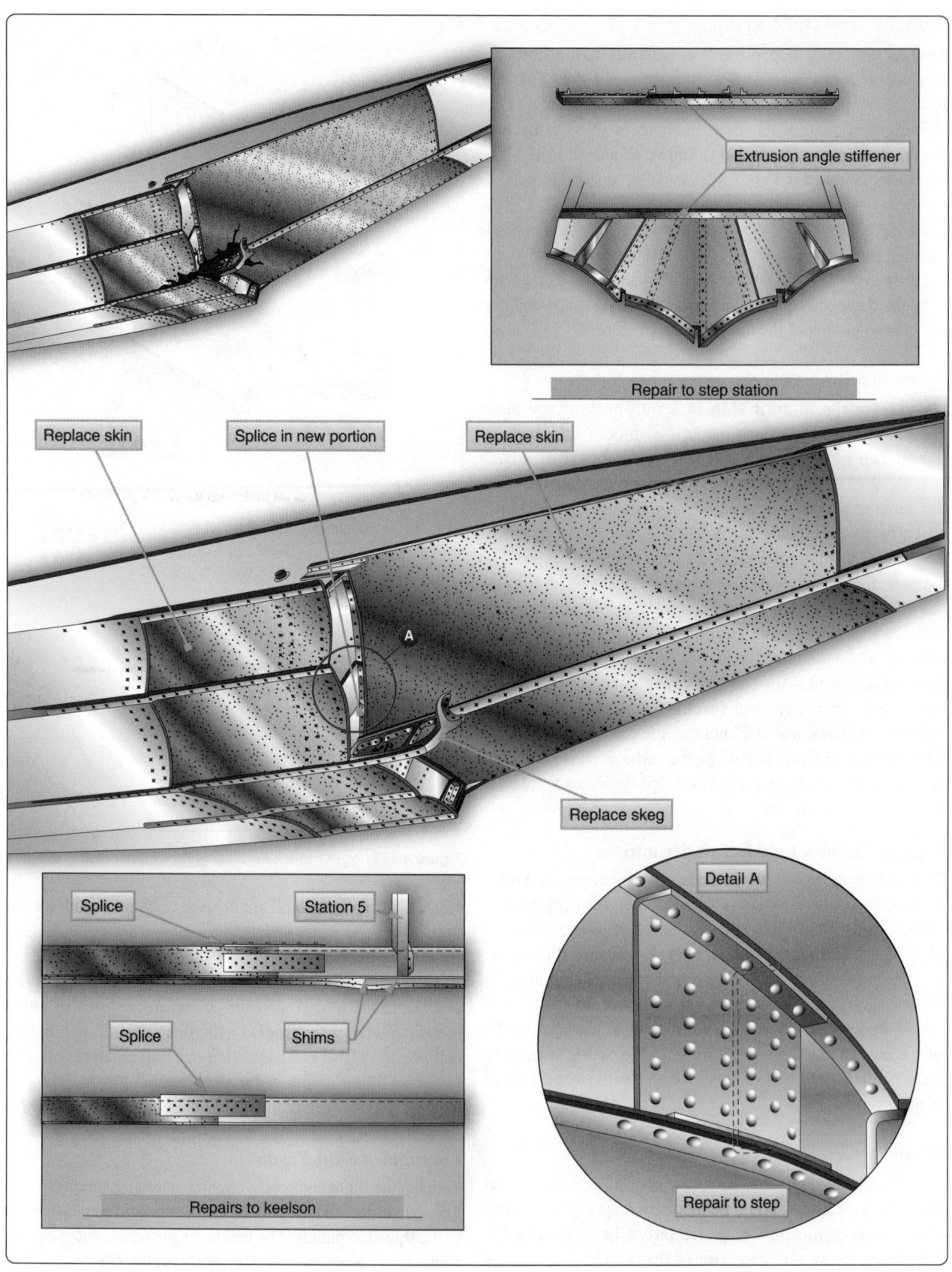

Figure 4-178. *Float repair.*

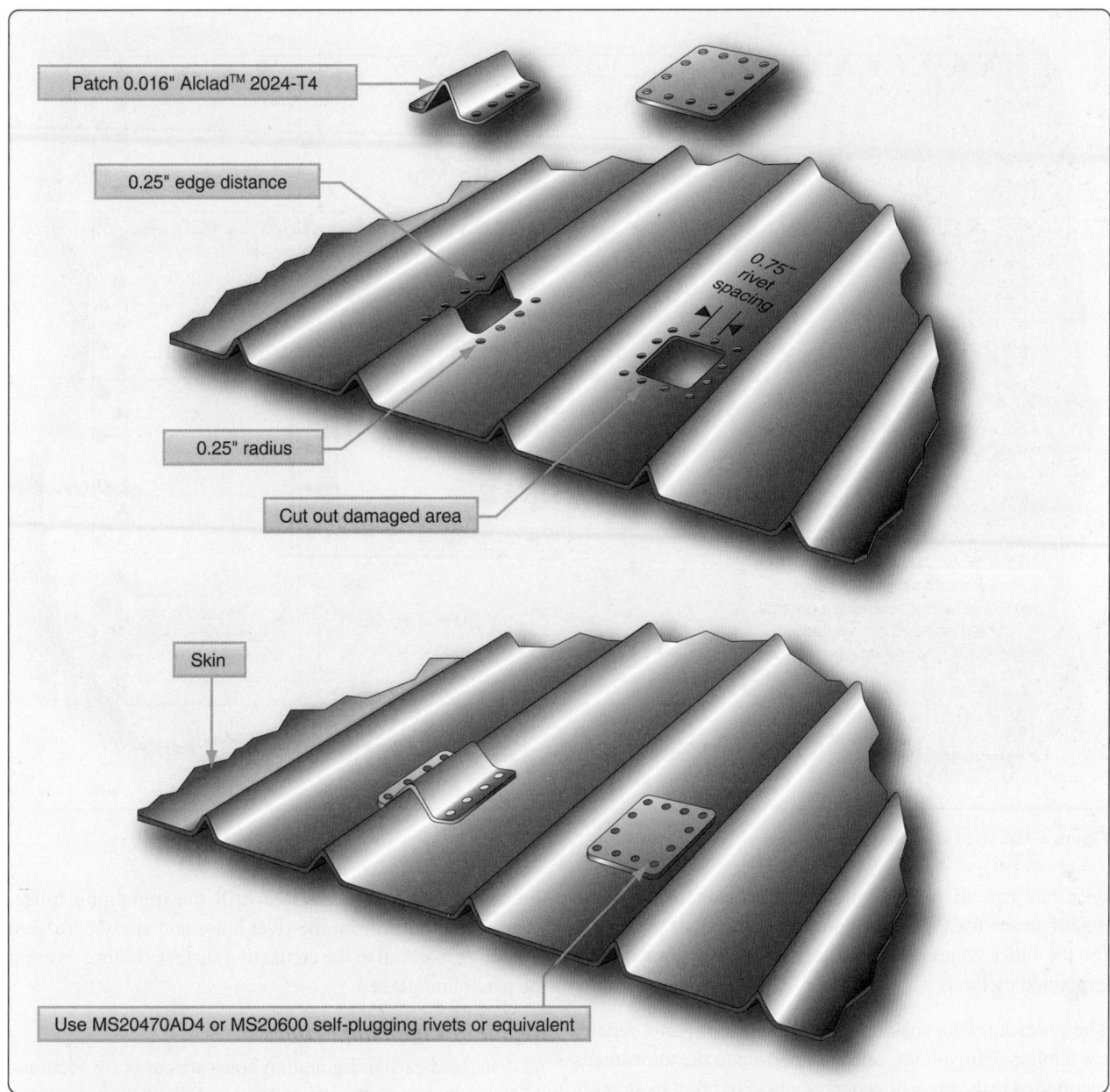

Patch 0.016" Alclad™ 2024-T4

0.25" edge distance

0.75" rivet spacing

0.25" radius

Cut out damaged area

Skin

Use MS20470AD4 or MS20600 self-plugging rivets or equivalent

Figure 4-179. *Beaded skin repair on corrugated surfaces.*

requires replacement of the entire panel. *[Figure 4-180]* A panel must also be replaced when there are too many previous repairs in a given section or area.

In aircraft construction, a panel is any single sheet of metal covering. A panel section is the part of a panel between adjacent stringers and bulk heads. Where a section of skin is damaged to such an extent that it is impossible to install a standard skin repair, a special type of repair is necessary. The particular type of repair required depends on whether the damage is repairable outside the member, inside the member, or to the edges of the panel.

Outside the Member

For damage that, after being trimmed, has 8½ rivet diameters or more of material, extend the patch to include the manufacturer's row of rivets and add an extra row inside the members.

Inside the Member

For damage that, after being trimmed, has less than 8½ manufacturer's rivet diameters of material inside the members, use a patch that extends over the members and an extra row of rivets along the outside of the members.

Edges of the Panel

For damage that extends to the edge of a panel, use

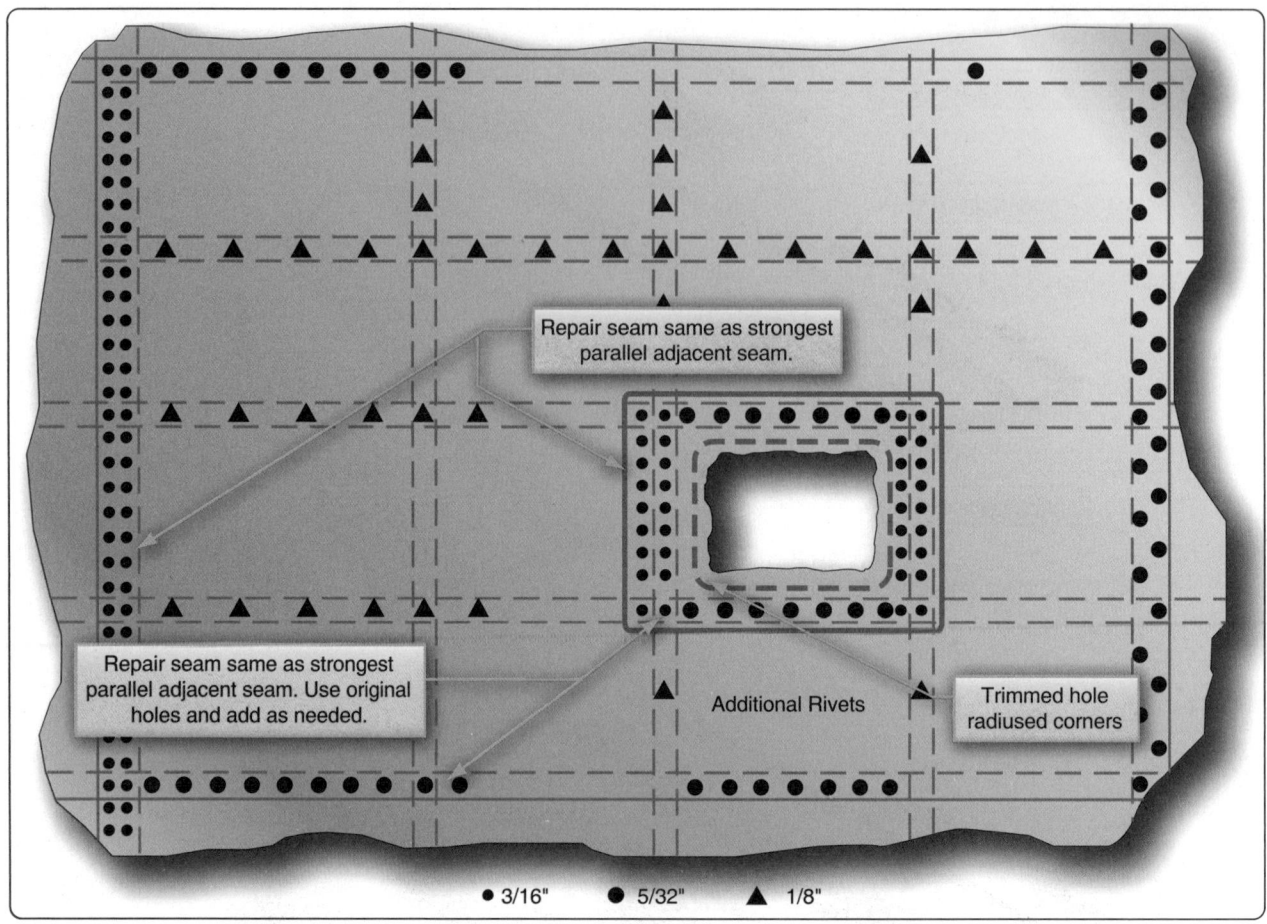

Repair seam same as strongest parallel adjacent seam.

Repair seam same as strongest parallel adjacent seam. Use original holes and add as needed.

Additional Rivets

Trimmed hole radiused corners

● 3/16"   ● 5/32"   ▲ 1/8"

Figure 4-180. *Replacement of an entire panel.*

only one row of rivets along the panel edge, unless the manufacturer used more than one row. The repair procedure for the other edges of the damage follows the previously explained methods.

The procedures for making all three types of panel repairs are similar. Trim out the damaged portion to the allowances mentioned in the preceding paragraphs. For relief of stresses at the corners of the trim-out, round them to a minimum radius of ½-inch. Lay out the new rivet row with a transverse pitch of approximately five rivet diameters and stagger the rivets with those put in by the manufacturer. Cut the patch plate from material of the same thickness as the original or the next greater thickness, allowing an edge distance of 2½ rivet diameters. At the corners, strike arcs having the radius equal to the edge distance.

Chamfer the edges of the patch plate for a 45° angle and form the plate to fit the contour of the original structure. Turn the edges downward slightly so that the edges fit closely. Place the patch plate in its correct position, drill one rivet hole, and temporarily fasten the plate in place with a fastener. Using a hole finder, locate the position of a second hole, drill it, and insert a second fastener. Then, from the back side and through

the original holes, locate and drill the remaining holes. Remove the burrs from the rivet holes and apply corrosion protective material to the contacting surfaces before riveting the patch into place.

Repair of Lightening Holes

As discussed earlier, lightening holes are cut in rib sections, fuselage frames, and other structural parts to reduce the weight of the part. The holes are flanged to make the web stiffer. Cracks can develop around flanged lightening holes, and these cracks need to be repaired with a repair plate. The damaged area (crack) needs to be stop drilled or the damage must be removed. The repair plate is made of the same material and thickness as the damaged part. Rivets are the same as in surrounding structure and the minimum edge distance is 2 times the diameter and spacing is between four to six times the diameter. *Figure 4-181* illustrates a typical lightening hole repair.

Repairs to a Pressurized Area

The skin of aircraft that are pressurized during flight is highly stressed. The pressurization cycles apply loads to the skin, and the repairs to this type of structure requires more rivets

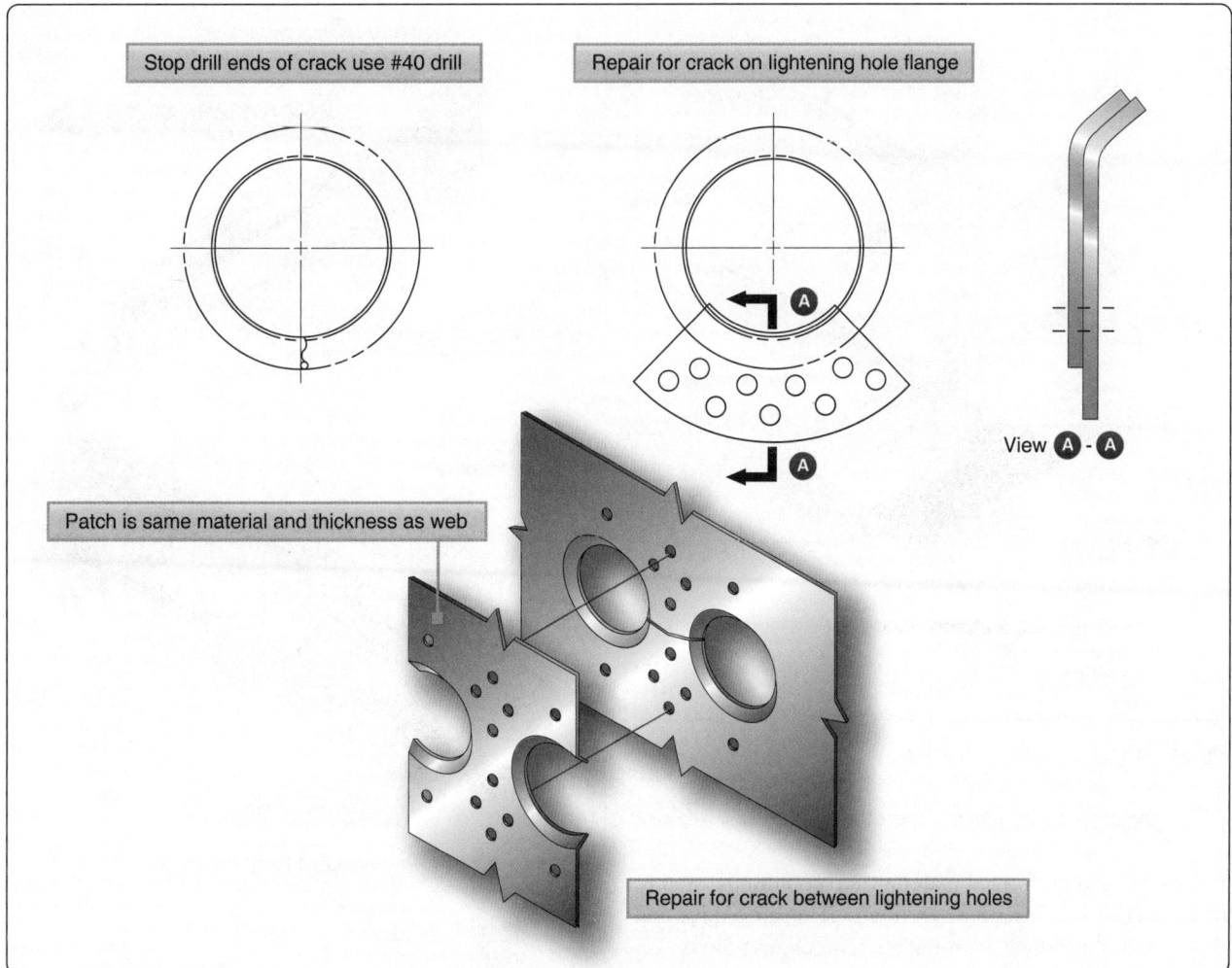

Figure 4-181. *Repair of lightening holes.*

than a repair to a nonpressurized skin. *[Figure 4-182]*

1. Remove the damaged skin section.

2. Radius all corners to 0.5-inch.

3. Fabricate a doubler of the same type of material as, but of one size greater thickness than, the skin. The size of the doubler depends on the number of rows, edge distance, and rivets spacing.

4. Fabricate an insert of the same material and same thickness as the damaged skin. The skin to insert clearance is typically 0.015-inch to 0.035-inch.

5. Drill the holes through the doubler, insertion, and original skin.

6. Spread a thin layer of sealant on the doubler and secure the doubler to the skin with Clecos.

7. Use the same type of fastener as in the surrounding area, and install the doubler to the skin and the insertion to the doubler. Dip all fasteners in the sealant before installation.

Stringer Repair

The fuselage stringers extend from the nose of the aircraft to the tail, and the wing stringers extend from the fuselage to the wing tip. Surface control stringers usually extend the length of the control surface. The skin of the fuselage, wing, or control surface is riveted to stringers.

Stringers may be damaged by vibration, corrosion, or collision. Because stringers are made in many different shapes, repair procedures differ. The repair may require the use of preformed or extruded repair material, or it may require material formed by the airframe technician. Some repairs may need both kinds of repair material. When repairing a stringer, first determine the extent of the damage and remove the rivets from the surrounding area. *[Figure 4-183]* Then, remove the damaged area by using a hacksaw, keyhole saw, drill, or file. In most cases, a stringer repair requires the use of insert and splice angle. When locating the splice angle on the stringer during repair, be sure to consult the applicable structural repair manual for the repair piece's position. Some stringers are repaired by placing the splice angle on the inside, whereas

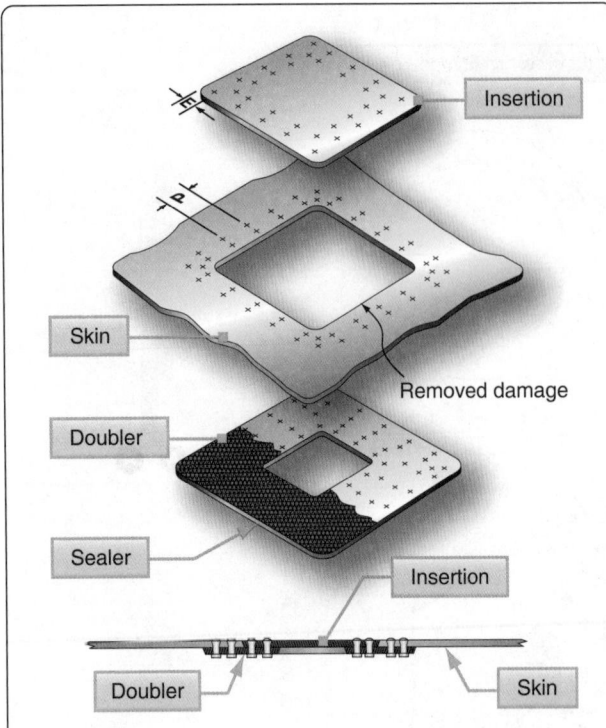

Figure 4-182. *Pressurized skin repair.*

If damage has been cut away from center section of stringer length, both ends of new portion must be attached as shown below.

Use AN470 or AN456 AD3 rivets

0.064" 245-T4 Alclad™ strip

0.58" 0.10" rad.

0.58"

3.35" 0.40" 0.90"

3.35" 0.90"

0.90" 0.90"

0.90" 0.20"

0.04" 245-T4 Alclad™

Stringer CS-14 and CS-15

0.50"

0.35"

0.35" A - A

Original structure

Repair parts

Repair parts in cross section

Figure 4-183. *Stringer repair.*

others are repaired by placing it on the outside.

Extrusions and preformed materials are commonly used to repair angles and insertions or fillers. If repair angles and fillers must be formed from flat sheet stock, use the brake. It may be necessary to use bend allowance and sight lines when making the layout and bends for these formed parts. For repairs to curved stringers, make the repair parts so that they fit the original contour.

Figure 4-184 shows a stringer repair by patching. This repair is permissible when the damage does not exceed two-thirds of the width of one leg and is not more than 12 inches long. Damage exceeding these limits can be repaired by one of the following methods.

Figure 4-185 illustrates repair by insertion where damage exceeds two-thirds of the width of one leg and after a portion of the stringer is removed. *Figure 4-186* shows repair by insertion when the damage affects only one stringer and exceeds 12 inches in length. *Figure 4-187* illustrates repair by an insertion when damage affects more than one stringer.

Former or Bulkhead Repair

Bulkheads are the oval-shaped members of the fuselage that give form to and maintain the shape of the structure. Bulkheads or formers are often called forming rings, body frames, circumferential rings, belt frames, and other similar names. They are designed to carry concentrated stressed loads.

There are various types of bulkheads. The most common type is a curved channel formed from sheet stock with stiffeners added. Others have a web made from sheet stock with extruded angles riveted in place as stiffeners and flanges. Most of these members are made from aluminum alloy. Corrosion-resistant steel formers are used in areas that are exposed to high temperatures.

Bulkhead damages are classified in the same manner as other damages. Specifications for each type of damage are established by the manufacturer and specific information is given in the maintenance manual or SRM for the aircraft. Bulkheads are identified with station numbers that are very

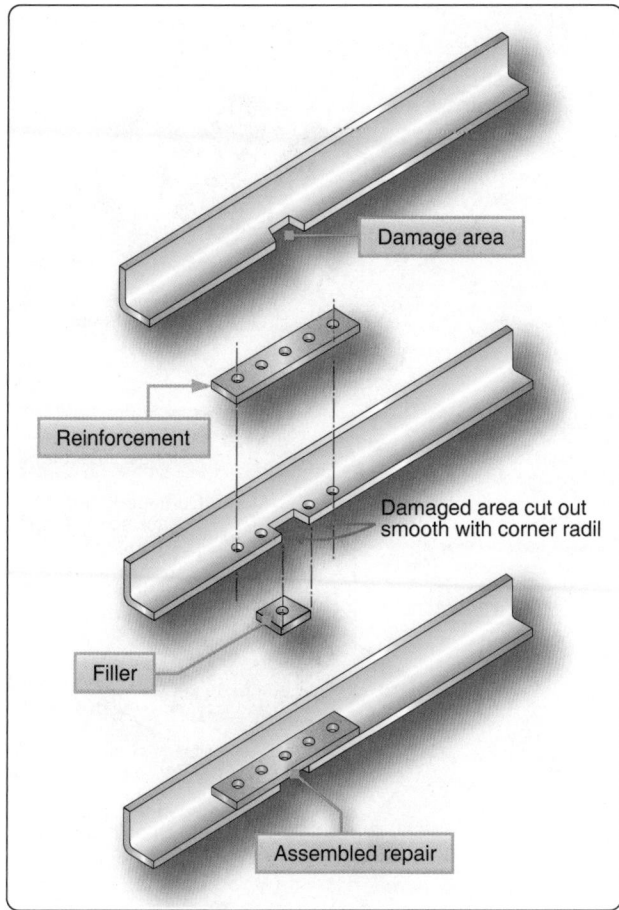

Figure 4-184. *Stringer repair by patching.*

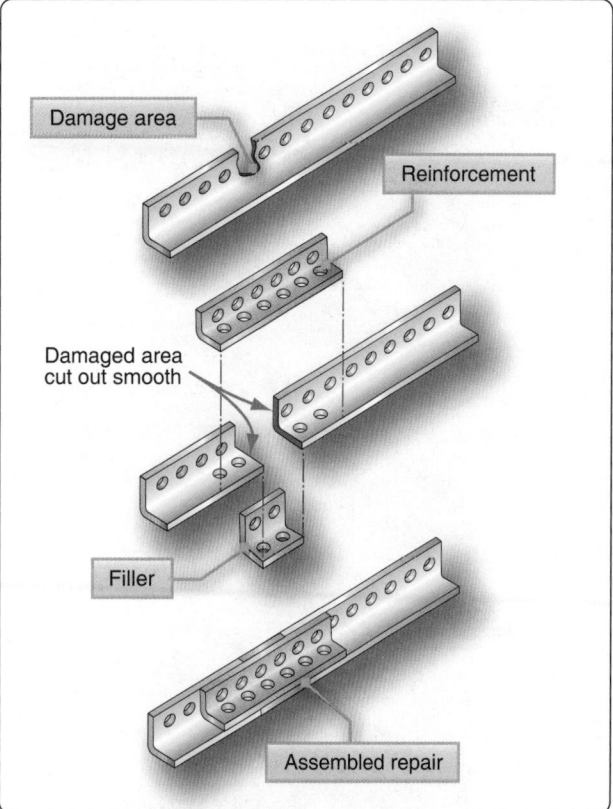

Figure 4-185. *Stringer repair by insertion when damage exceeds two-thirds of one leg in width.*

helpful in locating repair information. *Figure 4-188* is an example of a typical repair for a former, frame section, or bulkhead repair.

1. Stop drill the crack ends with a No. 40 size drill.

2. Fabricate a doubler of the same material but one size thicker than the part being repaired. The doubler should be of a size large enough to accommodate ⅛-inch rivet holes spaced one inch apart, with a minimum edge distance of 0.30-inch and 0.50-inch spacing between staggered rows. *[Figure 4-189]*

3. Attach the doubler to the part with clamps and drill holes.

4. Install rivets.

Most repairs to bulkheads are made from flat sheet stock if spare parts are not available. When fabricating the repair from flat sheet, remember the substitute material must provide cross-sectional tensile, compressive, shear, and bearing strength equal to the original material. Never substitute material that is thinner or has a cross-sectional area less than the original material. Curved repair parts made from flat sheet

must be in the "0" condition before forming, and then must be heat treated before installation.

Longeron Repair

Generally, longerons are comparatively heavy members that serve approximately the same function as stringers. Consequently, longeron repair is similar to stringer repair. Because the longeron is a heavy member and more strength is needed than with a stringer, heavy rivets are used in the repair. Sometimes bolts are used to install a longeron repair, due to the need for greater accuracy, they are not as suitable as rivets. Also, bolts require more time for installation.

If the longeron consists of a formed section and an extruded angle section, consider each section separately. A longeron repair is similar to a stringer repair, but keep the rivet pitch between 4 and 6 rivet diameters. If bolts are used, drill the bolt holes for a light drive fit.

Spar Repair

The spar is the main supporting member of the wing. Other components may also have supporting members called spars that serve the same function as the spar does in the wing. Think of spars as the hub, or base, of the section in which they are located, even though they are not in the center. The spar

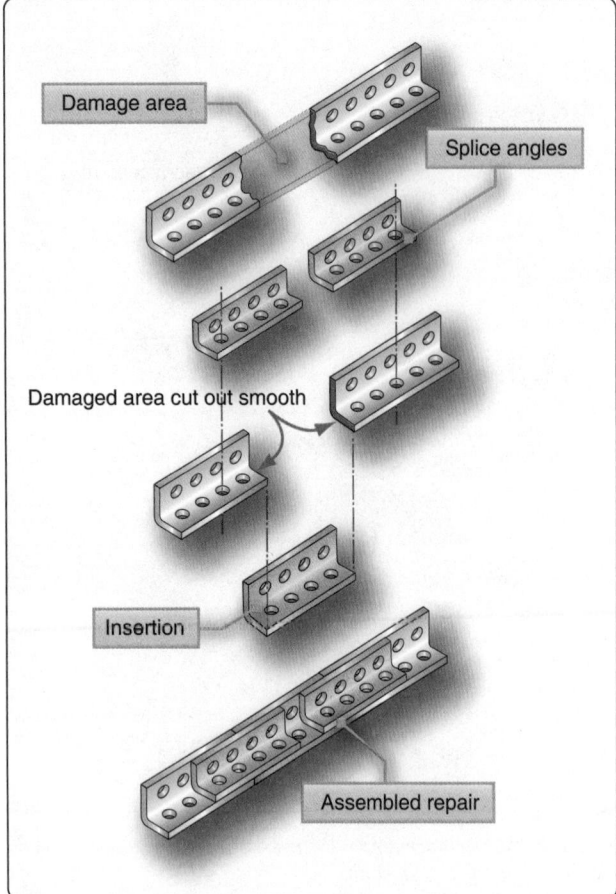

Figure 4-186. *Stringer repair by insertion when damage affects only one stringer.*

is usually the first member located during the construction of the section, and the other components are fastened directly or indirectly to it. Because of the load the spar carries, it is very important that particular care be taken when repairing this member to ensure the original strength of the structure is not impaired. The spar is constructed so that two general classes of repairs, web repairs and cap strip repairs, are usually necessary.

Figures 4-189 and *4-190* are examples of typical spar repairs. The damage to the spar web can be repaired with a round or rectangular doubler. Damage smaller than 1-inch is typically repaired with a round doubler and larger damage is repaired with a rectangular doubler.

1. Remove the damage and radius all corners to 0.5-inch.

2. Fabricate doubler; use same material and thickness. The doubler size depends on edge distance (minimum of 2D) and rivet spacing (4-6D).

3. Drill through the doubler and the original skin and secure doubler with Clecos.

4. Install rivets.

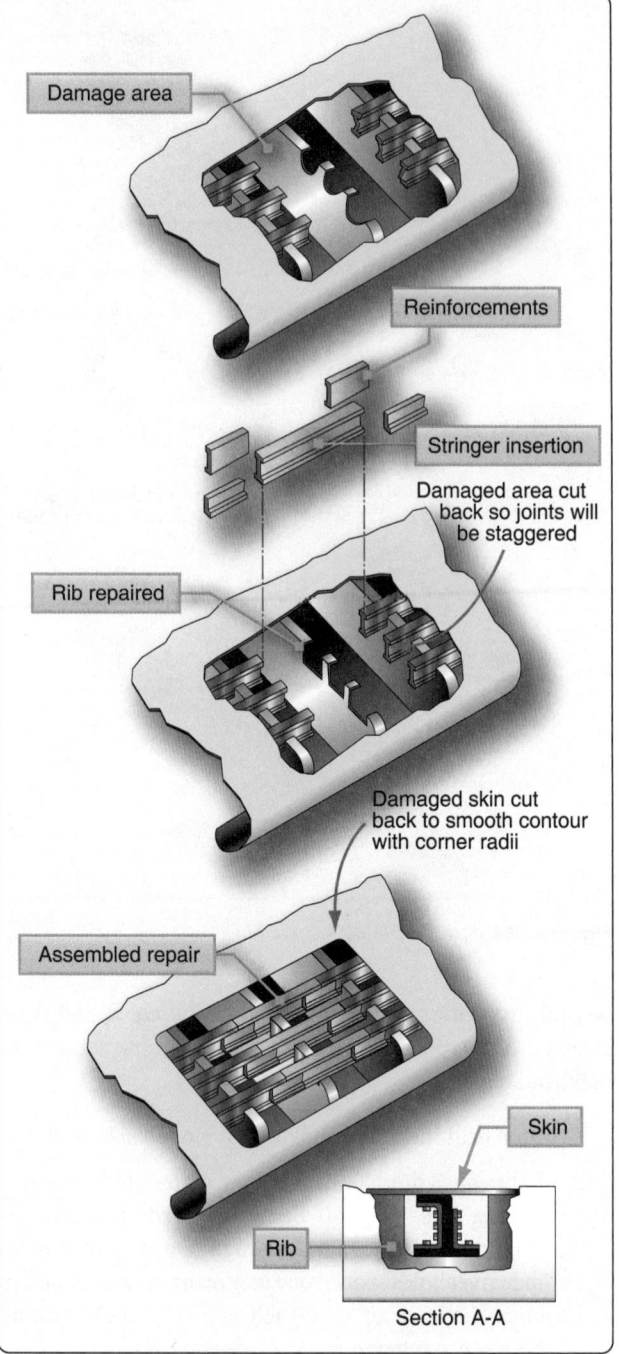

Figure 4-187. *Stringer repair by insertion when damage affects more than one stringer.*

Rib & Web Repair

Web repairs can be classified into two types:

1. Those made to web sections considered critical, such as those in the wing ribs.

2. Those considered less critical, such as those in elevators, rudders, flaps, and the like.

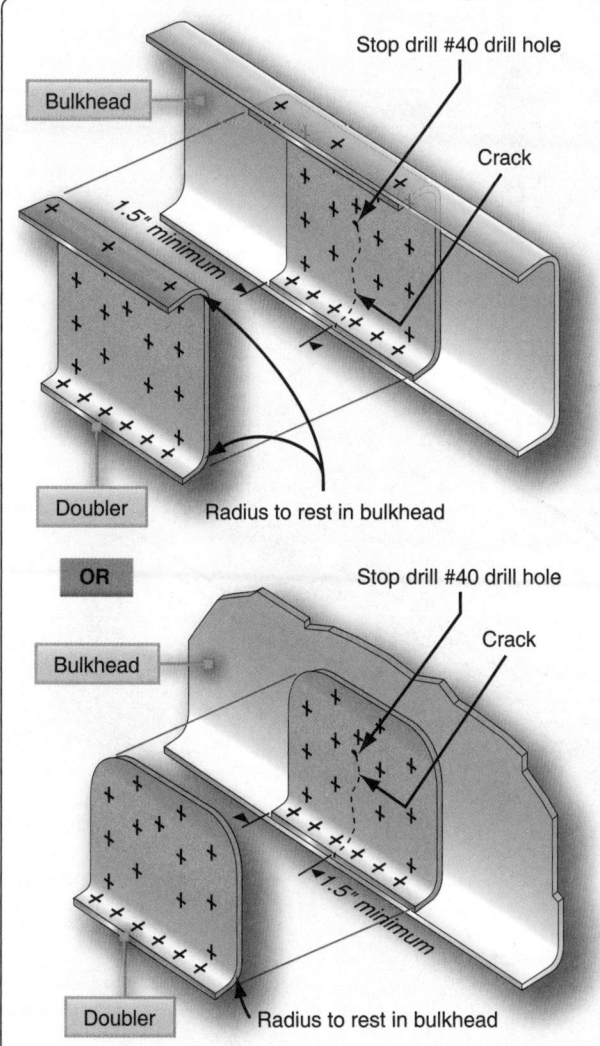

Figure 4-188. *Bulkhead repair.*

Web sections must be repaired in such a way that the original strength of the member is restored. In the construction of a member using a web, the web member is usually a light gauge aluminum alloy sheet forming the principal depth of the member. The web is bounded by heavy aluminum alloy extrusions known as cap strips. These extrusions carry the loads caused by bending and also provide a foundation for attaching the skin. The web may be stiffened by stamped beads, formed angles, or extruded sections riveted at regular intervals along the web.

The stamped beads are a part of the web itself and are stamped in when the web is made. Stiffeners help to withstand the compressive loads exerted upon the critically stressed web members. Often, ribs are formed by stamping the entire piece from sheet stock. That is, the rib lacks a cap strip, but does have a flange around the entire piece, plus lightening holes in the web of the rib. Ribs may be formed with stamped beads for stiffeners, or they may have extruded angles riveted on the web for stiffeners.

Most damages involve two or more members, but only one member may be damaged and need repairing. Generally, if the web is damaged, cleaning out the damaged area and installing a patch plate are all that is required.

The patch plate should be of sufficient size to ensure room for at least two rows of rivets around the perimeter of the damage that includes proper edge distance, pitch, and transverse pitch for the rivets. The patch plate should be of material having the same thickness and composition as the original member. If any forming is necessary when making the patch plate, such as fitting the contour of a lightening hole, use material in the "0" condition and then heat treat it after forming.

Damage to ribs and webs that requires a repair larger than a simple plate probably needs a patch plate, splice plates, or angles and an insertion. *[Figure 4-191]*

Leading Edge Repair

The leading edge is the front section of a wing, stabilizer, or other airfoil. The purpose of the leading edge is to streamline the forward section of the wings or control surfaces to ensure effective airflow. The space within the leading edge is sometimes used to store fuel. This space may also house extra equipment, such as landing lights, plumbing lines, or thermal anti-icing systems.

The construction of the leading edge section varies with the type of aircraft. Generally, it consists of cap strips, nose ribs, stringers, and skin. The cap strips are the main lengthwise extrusions, and they stiffen the leading edges and furnish a base for the nose ribs and skin. They also fasten the leading edge to the front spar.

The nose ribs are stamped from aluminum alloy sheet or machined parts. These ribs are U-shaped and may have their web sections stiffened. Regardless of their design, their purpose is to give contour to the leading edge. Stiffeners are used to stiffen the leading edge and supply a base for fastening the nose skin. When fastening the nose skin, use only flush rivets.

Leading edges constructed with thermal anti-icing systems consist of two layers of skin separated by a thin air space. The inner skin, sometimes corrugated for strength, is perforated to conduct the hot air to the nose skin for anti-icing purposes.

Damage can be caused by contact with other objects, namely, pebbles, birds, and hail. However, the major cause of damage is carelessness while the aircraft is on the ground.

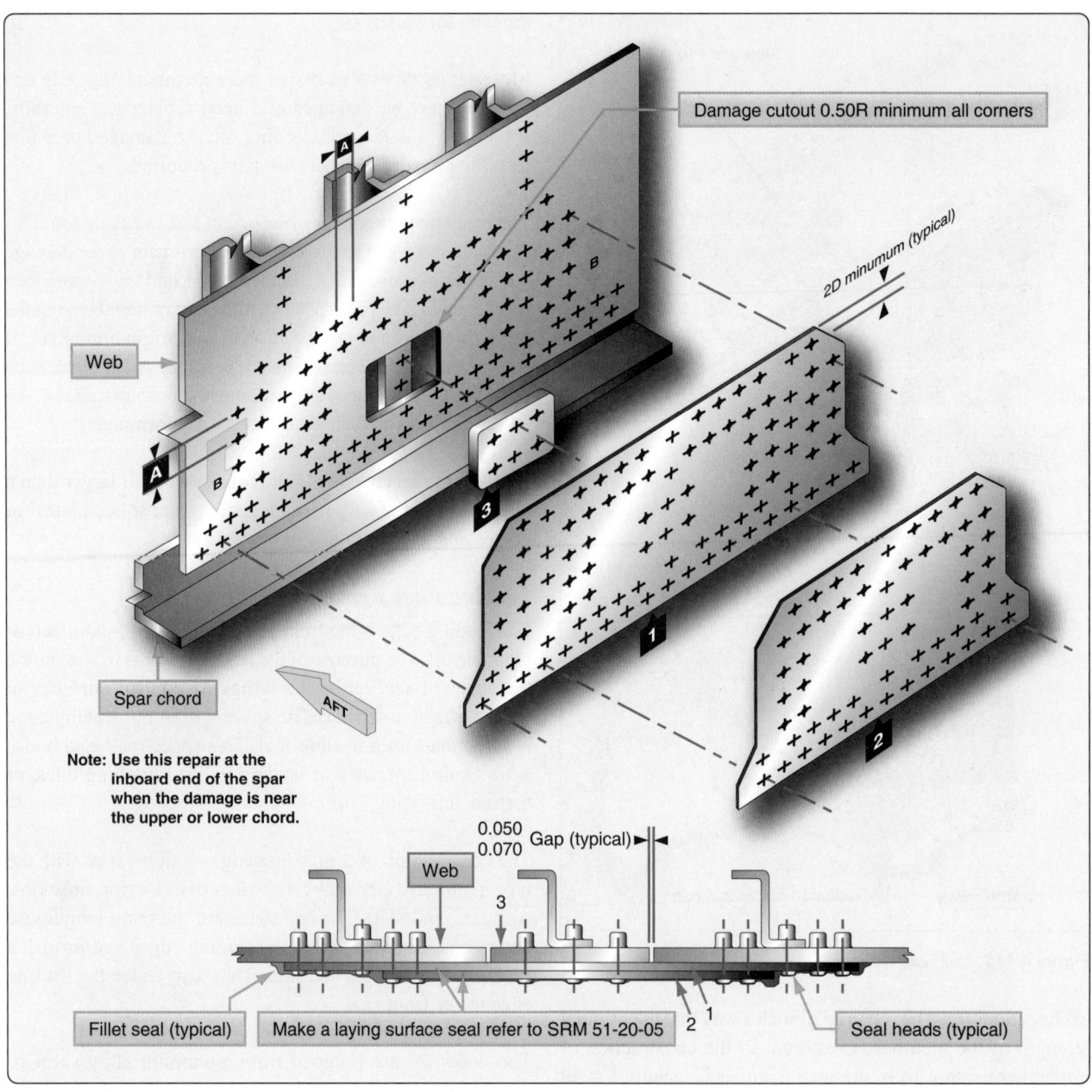

Figure 4-189. *Wing spar repair.*

A damaged leading edge usually involves several structural parts. FOD probably involves the nose skin, nose ribs, stringers, and possibly the cap strip. Damage involving all of these members necessitates installing an access door to make the repair possible. First, the damaged area has to be removed and repair procedures established. The repair needs insertions and splice pieces. If the damage is serious enough, it may require repair of the cap strip and stringer, a new nose rib, and a skin panel. When repairing a leading edge, follow the procedures prescribed in the appropriate repair manual for this type of repair. *[Figure 4-192]* Repairs to leading edges are more difficult to accomplish than repairs to flat and straight structures because the repair parts need to be

formed to fit the existing structure.

Trailing Edge Repair

A trailing edge is the rear-most part of an airfoil found on the wings, ailerons, rudders, elevators, and stabilizers. It is usually a metal strip that forms the shape of the edge by tying the ends of a rib section together and joining the upper and lower skins. Trailing edges are not structural members, but they are considered to be highly stressed in all cases.

Damage to a trailing edge may be limited to one point or extended over the entire length between two or more rib sections. Besides damage resulting from collision and

Upper flange
Spar web
Damage
Lower flange
Case A
Case B
Patch
Patch
Same material and thickness

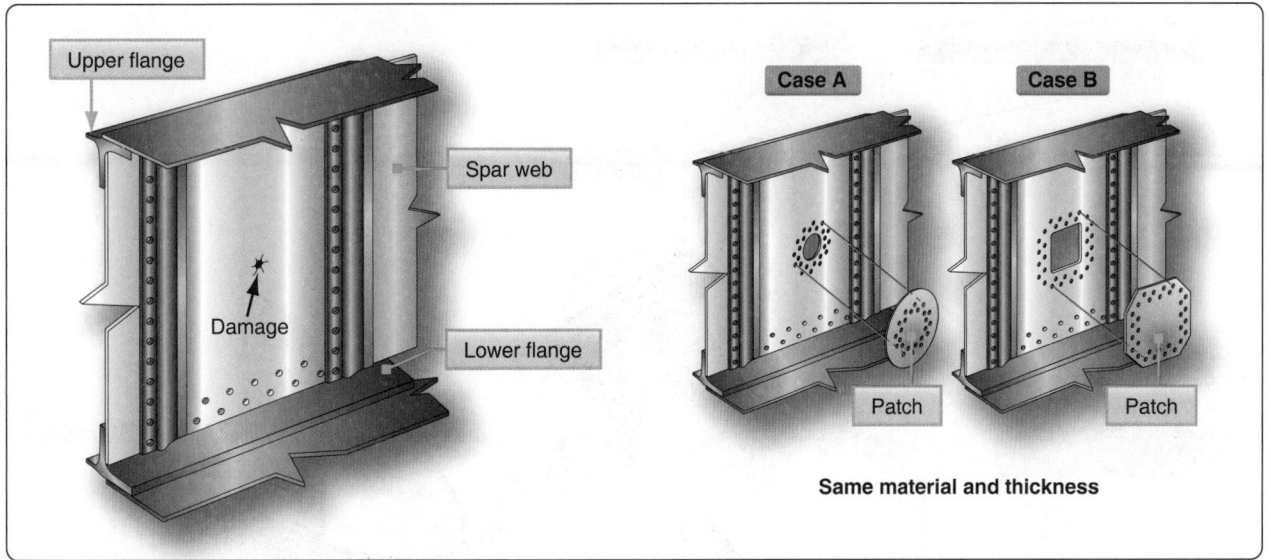

Figure 4-190. *Wing spar repair.*

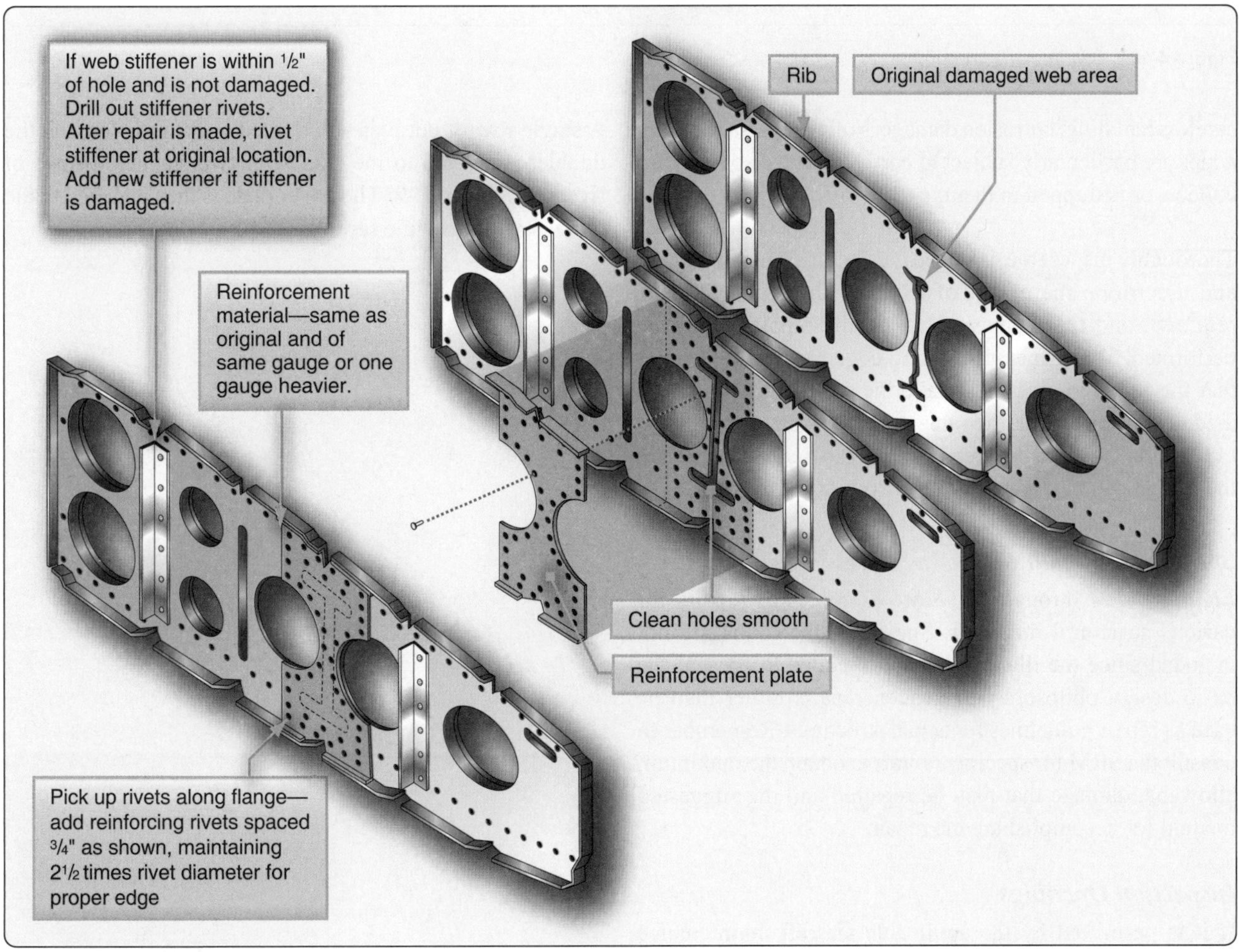

If web stiffener is within 1/2" of hole and is not damaged. Drill out stiffener rivets. After repair is made, rivet stiffener at original location. Add new stiffener if stiffener is damaged.
Reinforcement material—same as original and of same gauge or one gauge heavier.
Rib
Original damaged web area
Clean holes smooth
Reinforcement plate
Pick up rivets along flange—add reinforcing rivets spaced 3/4" as shown, maintaining 2 1/2 times rivet diameter for proper edge

Figure 4-191. *Wing rib repair.*

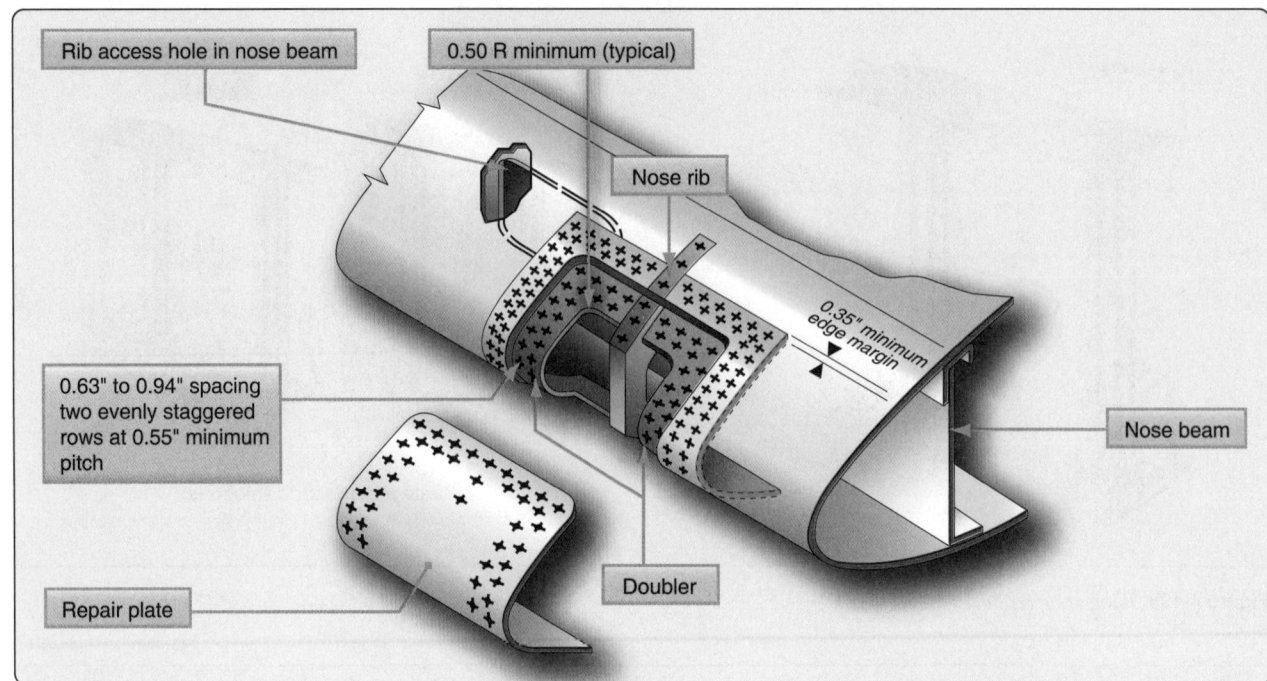

Figure 4-192. *Leading edge repair.*

careless handling, corrosion damage is often present. Trailing edges are particularly subject to corrosion because moisture collects or is trapped in them.

Thoroughly inspect the damaged area before starting repairs, and determine the extent of damage, the type of repair required, and the manner in which the repair should be performed. When making trailing edge repairs, remember that the repaired area must have the same contour and be made of material with the same composition and temper as the original section. The repair must also be made to retain the design characteristics of the airfoil. *[Figure 4-193]*

Specialized Repairs

Figures 4-194 through *4-198* are examples of repairs for various structural members. Specific dimensions are not included since the illustrations are intended to present the basic design philosophy of general repairs rather than be used as repair guidelines for actual structures. Remember to consult the SRM for specific aircraft to obtain the maximum allowable damage that may be repaired and the suggested method for accomplishing the repair.

Inspection Openings

If it is permitted by the applicable aircraft maintenance manual, installation of a flush access door for inspection purposes sometimes makes it easier to repair the internal structure as well as damage to the skin in certain areas. This installation consists of a doubler and a stressed cover plate.

A single row of nut plates is riveted to the doubler, and the doubler is riveted to the skin with two staggered rows of rivets. *[Figure 4-199]* The cover plate is then attached to the doubler with machine screws.

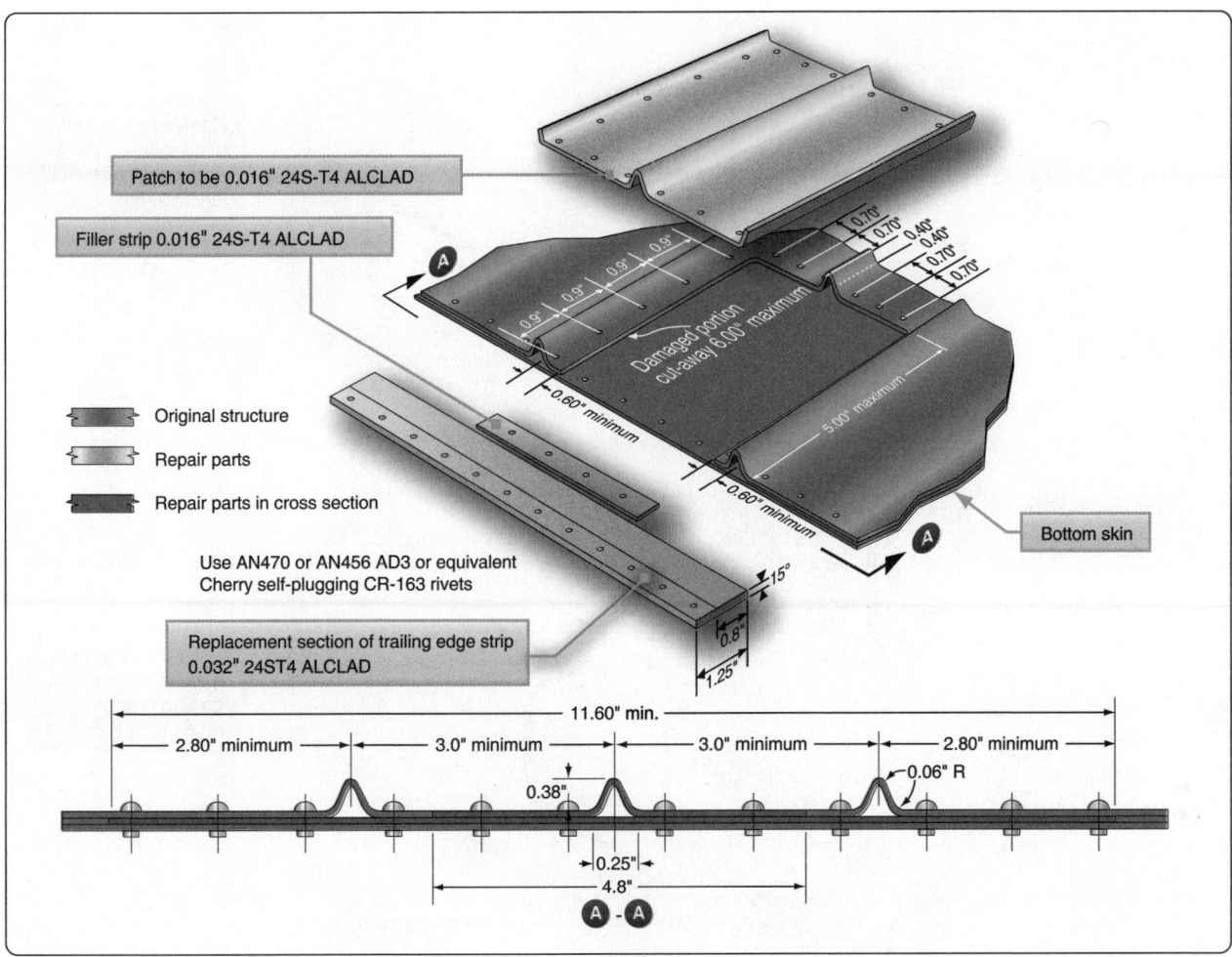

Patch to be 0.016" 24S-T4 ALCLAD

Filler strip 0.016" 24S-T4 ALCLAD

Original structure

Repair parts

Repair parts in cross section

Use AN470 or AN456 AD3 or equivalent
Cherry self-plugging CR-163 rivets

Replacement section of trailing edge strip
0.032" 24ST4 ALCLAD

0.9"
0.9"
0.9"
0.9"

Damaged portion
cut-away 6.00" maximum

5.00" maximum

0.70"
0.70"
0.40"
0.40"
0.70"
0.70"

0.60" minimum

0.60" minimum

Bottom skin

A

A

15°

0.8"

1.25"

11.60" min.

2.80" minimum

3.0" minimum

3.0" minimum

2.80" minimum

0.38"

0.06" R

0.25"

4.8"

A - A

Figure 4-193. *Trailing edge repair.*

4-109

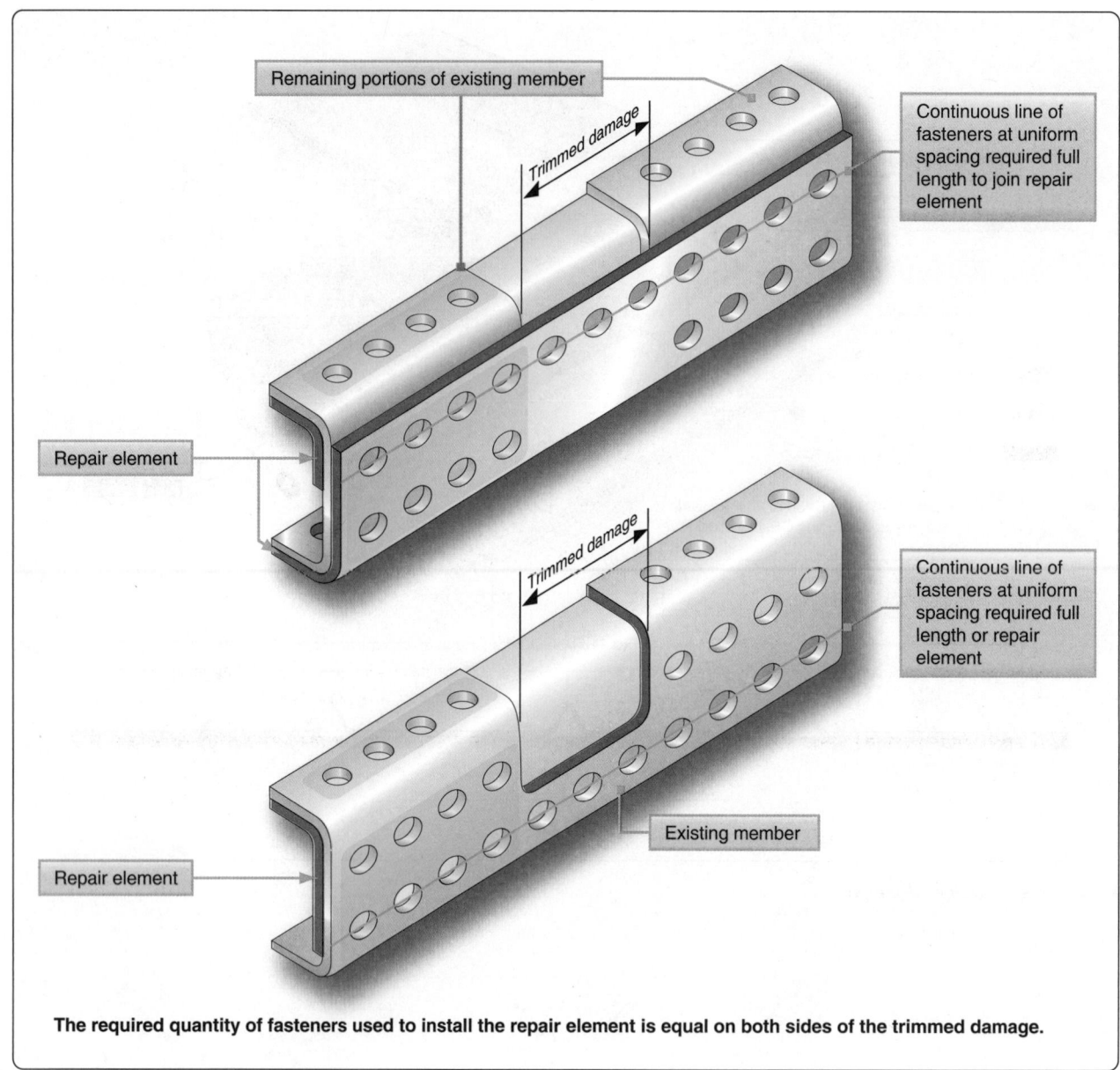

Remaining portions of existing member

Trimmed damage

Continuous line of fasteners at uniform spacing required full length to join repair element

Repair element

Trimmed damage

Continuous line of fasteners at uniform spacing required full length or repair element

Existing member

Repair element

The required quantity of fasteners used to install the repair element is equal on both sides of the trimmed damage.

Figure 4-194. *C-channel repair.*

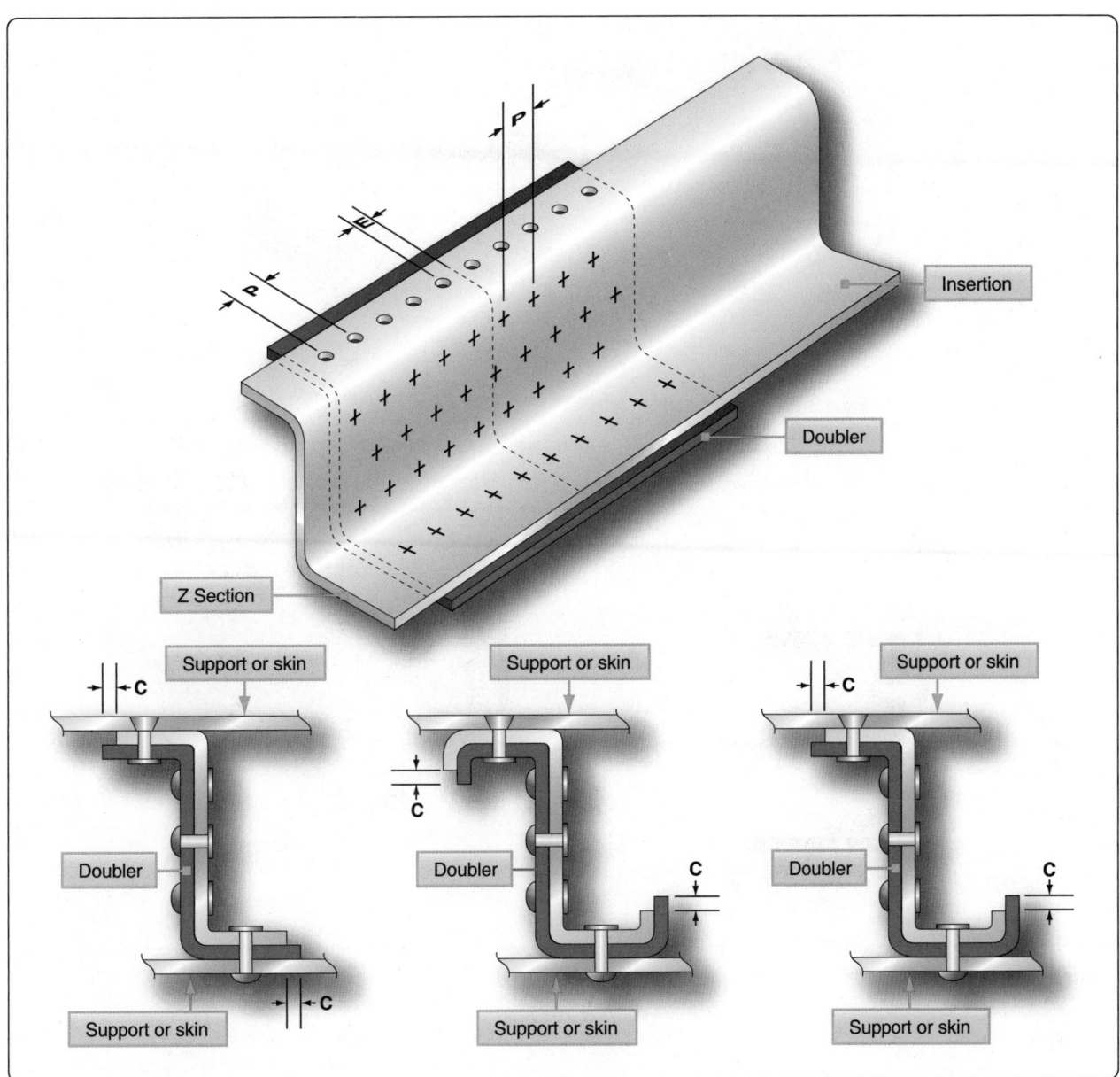

Figure 4-195. *Primary Z-section repair.*

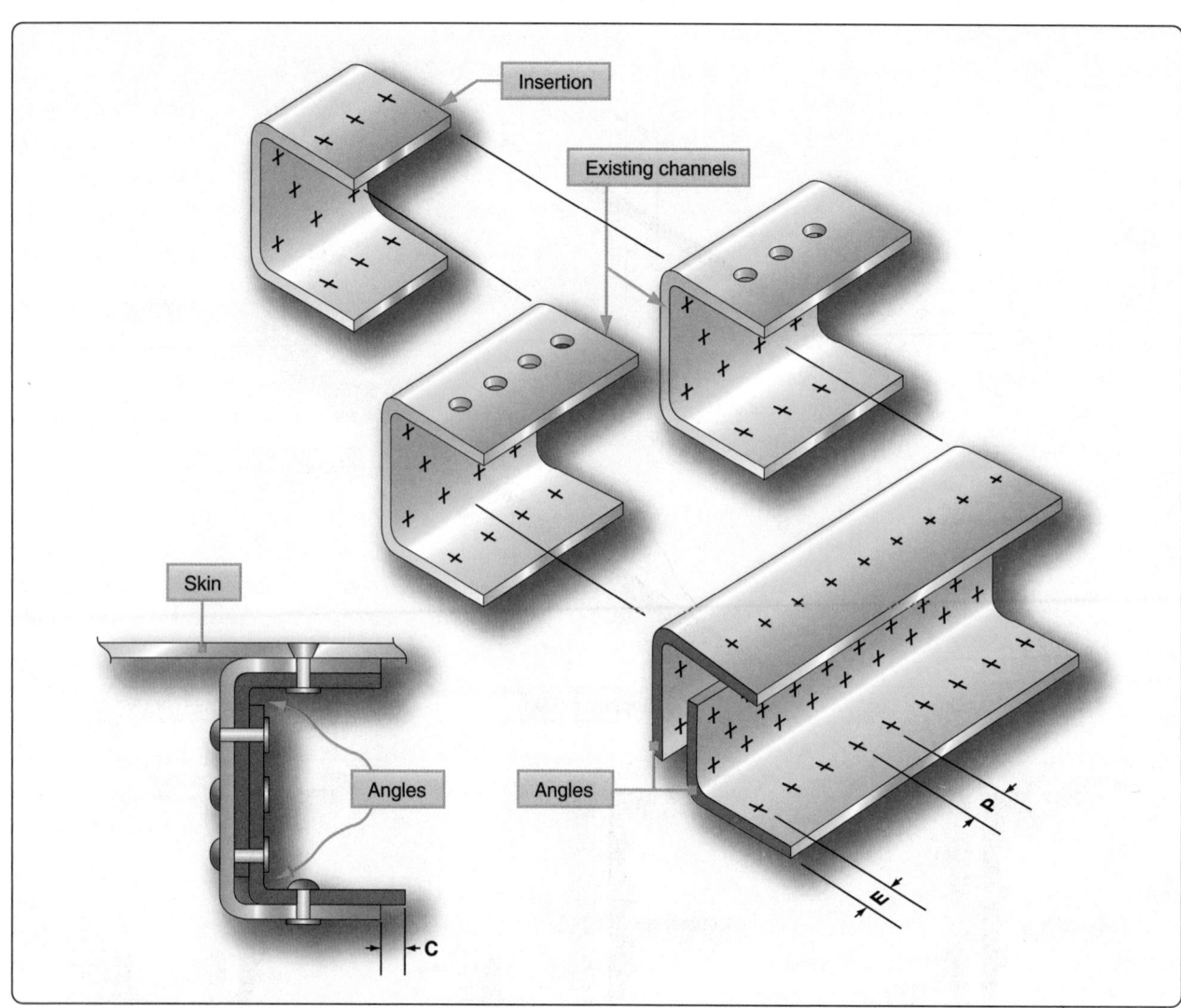

Figure 4-196. *U-channel repair.*

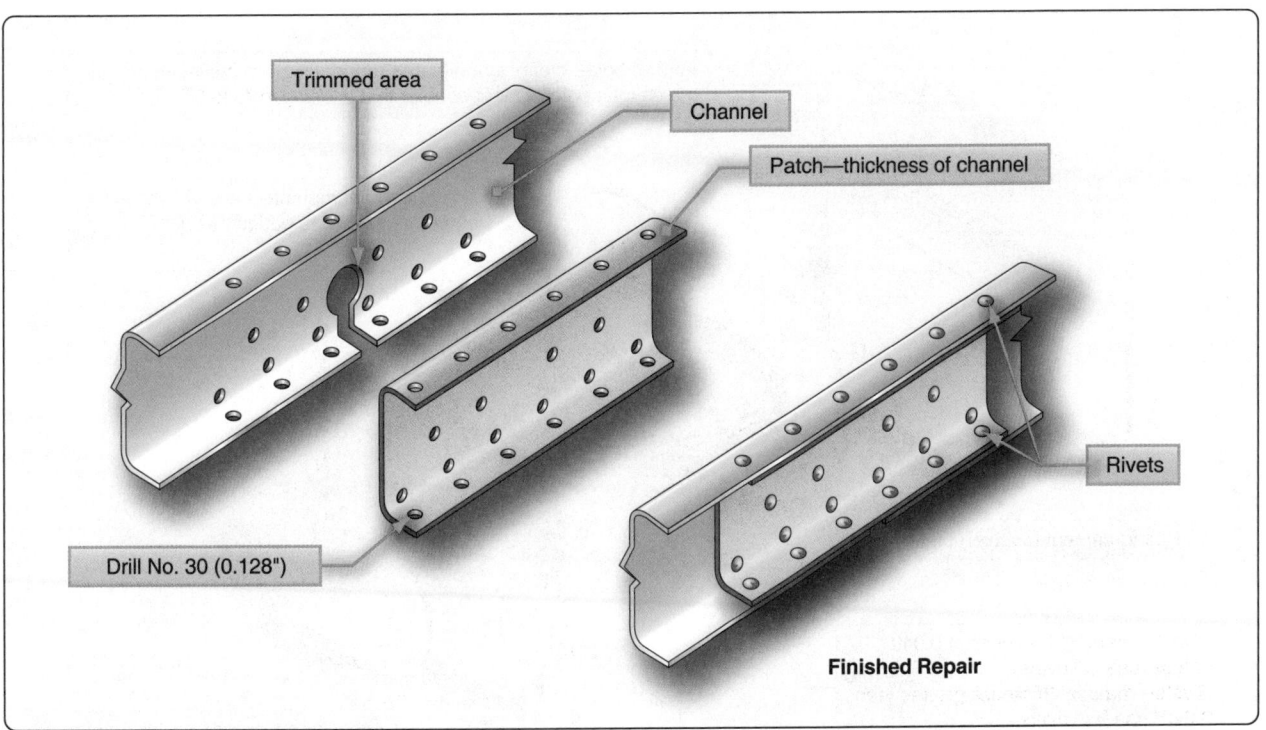

Figure 4-197. *Channel repair by patching.*

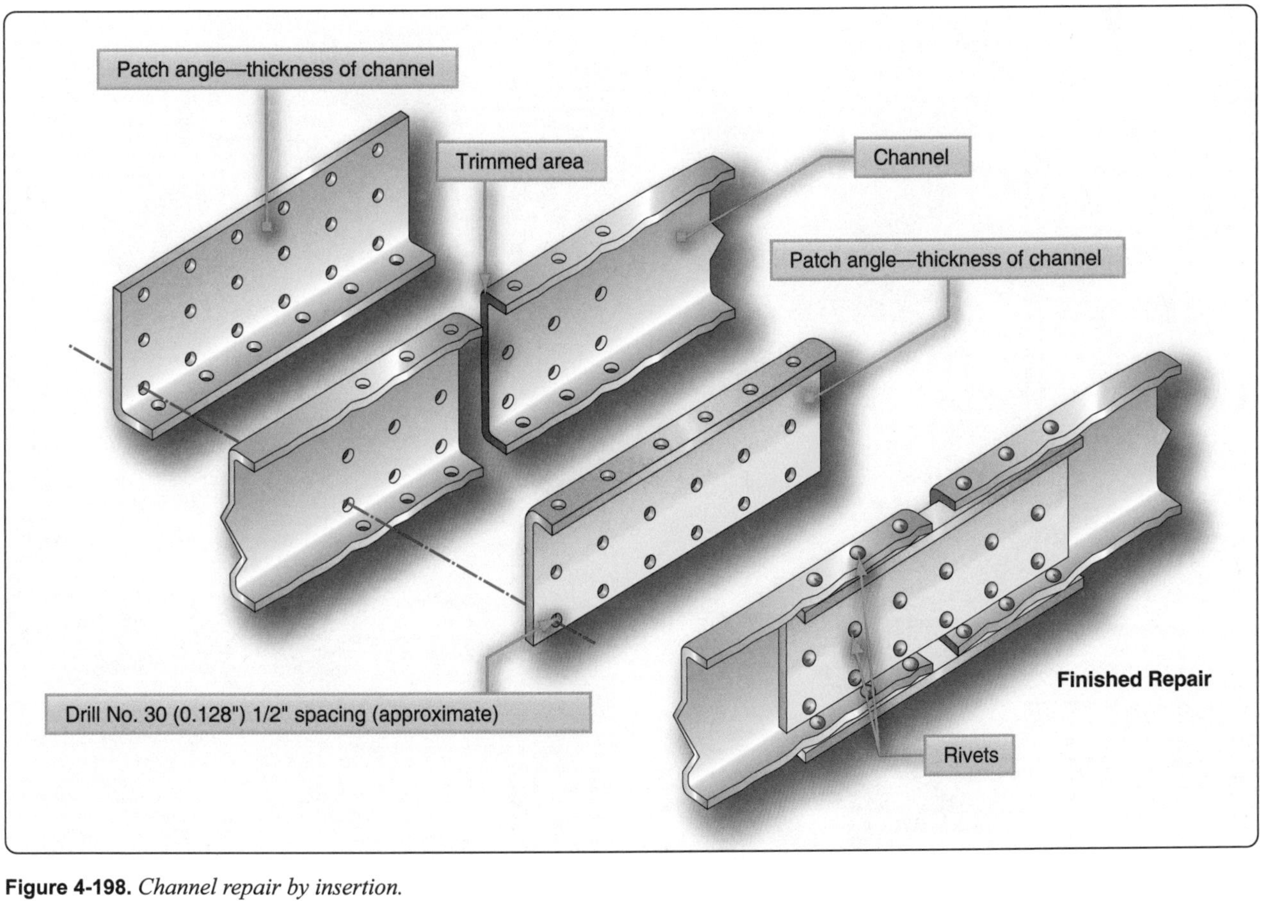

Figure 4-198. *Channel repair by insertion.*

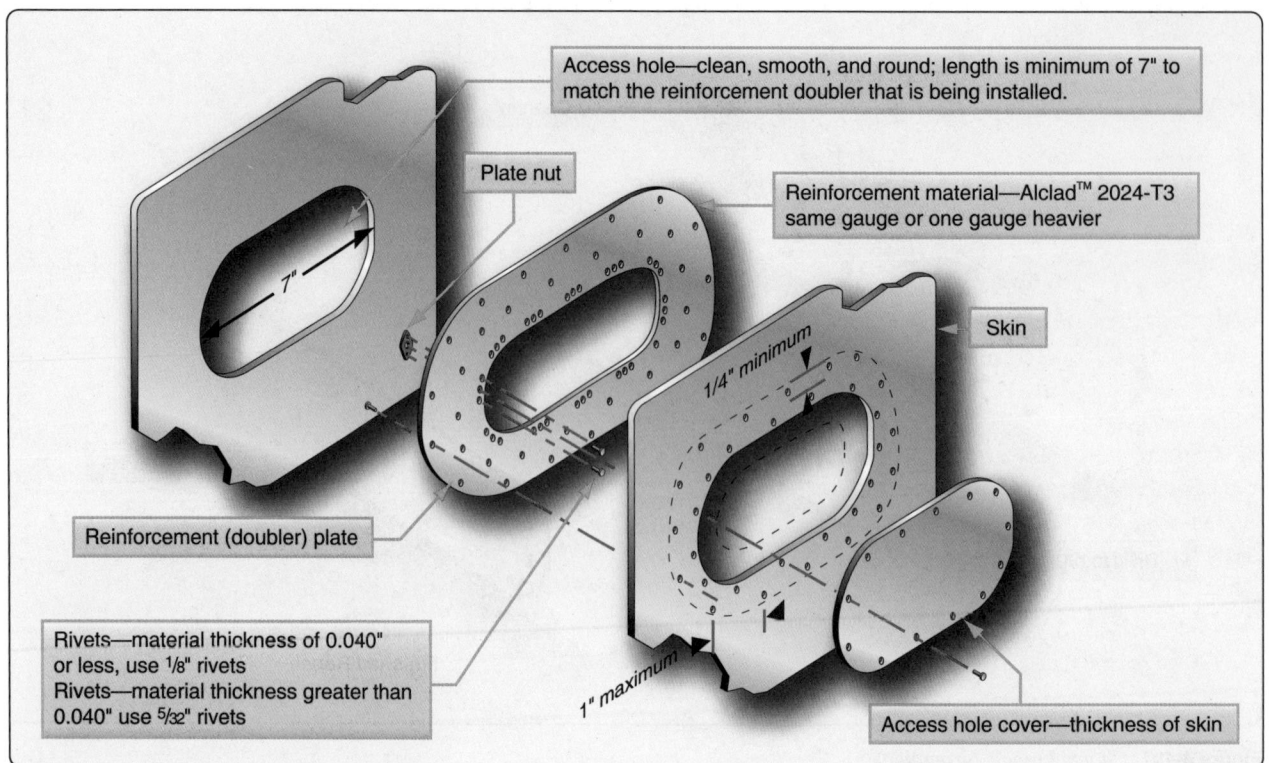

Access hole—clean, smooth, and round; length is minimum of 7" to match the reinforcement doubler that is being installed.

Plate nut

Reinforcement material—Alclad™ 2024-T3 same gauge or one gauge heavier

Skin

7"

1/4" minimum

Reinforcement (doubler) plate

Rivets—material thickness of 0.040" or less, use ⅛" rivets
Rivets—material thickness greater than 0.040" use ⁵⁄₃₂" rivets

1" maximum

Access hole cover—thickness of skin

Figure 4-199. *Inspection hole.*

Chapter 5
Aircraft Welding

Introduction

Welding can be traced back to the Bronze Age, but it was not until the 19th century that welding as we know it today was invented. Some of the first successful commercially manufactured aircraft were constructed from welded steel tube frames.

As the technology and manufacturing processes evolved in the aircraft and aerospace industry, lighter metals, such as aluminum, magnesium, and titanium, were used in their construction. New processes and methods of welding these metals were developed. This chapter provides some of the basic information needed to understand and initiate the various welding methods and processes.

Traditionally, welding is defined as a process that joins metal by melting or hammering the work pieces until they are united together. With the right equipment and instruction, almost anyone with some basic mechanical skill, dexterity, and practice can learn to weld.

There are three general types of welding: gas, electric arc, and electric resistance. Each type of welding has several variations, some of which are used in the construction of aircraft. Additionally, there are some new welding processes that have been developed in recent years that are highlighted for the purpose of information.

This chapter addresses the welding equipment, methods, and various techniques used during the repair of aircraft and fabrication of component parts, including the processes of brazing and soldering of various metals.

Types of Welding

Gas Welding

Gas welding is accomplished by heating the ends or edges of metal parts to a molten state with a high temperature flame. The oxy-acetylene flame, with a temperature of approximately 6,300 °Fahrenheit (F), is produced with a torch burning acetylene and mixing it with pure oxygen. Hydrogen may be used in place of acetylene for aluminum welding, but the heat output is reduced to about 4,800 °F. Gas welding was the method most commonly used in production on aircraft materials under $\frac{3}{16}$-inch in thickness until the mid 1950s, when it was replaced by electric welding for economic (not engineering) reasons. Gas welding continues to be a very popular and proven method for repair operations.

Nearly all gas welding in aircraft fabrication is performed with oxy-acetylene welding equipment consisting of:

- Two cylinders, acetylene and oxygen.

- Acetylene and oxygen pressure regulators and cylinder pressure gauges.

- Two lengths of colored hose (red for acetylene and green for oxygen) with adapter connections for the regulators and torch.

- A welding torch with an internal mixing head, various size tips, and hose connections.

- Welding goggles fitted with appropriate colored lenses.

- A flint or spark lighter.

- Special wrench for acetylene tank valve if needed.

- An appropriately-rated fire extinguisher.

The equipment may be permanently installed in a shop, but most welding outfits are of the portable type. *[Figure 5-1]*

Electric Arc Welding

Electric arc welding is used extensively by the aircraft industry in both the manufacture and repair of aircraft. It can be used

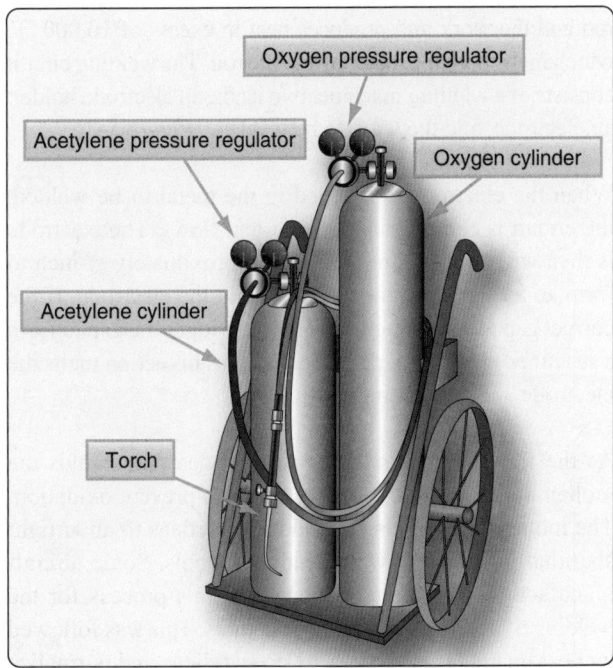

Figure 5-1. *Portable oxy-acetylene welding outfit.*

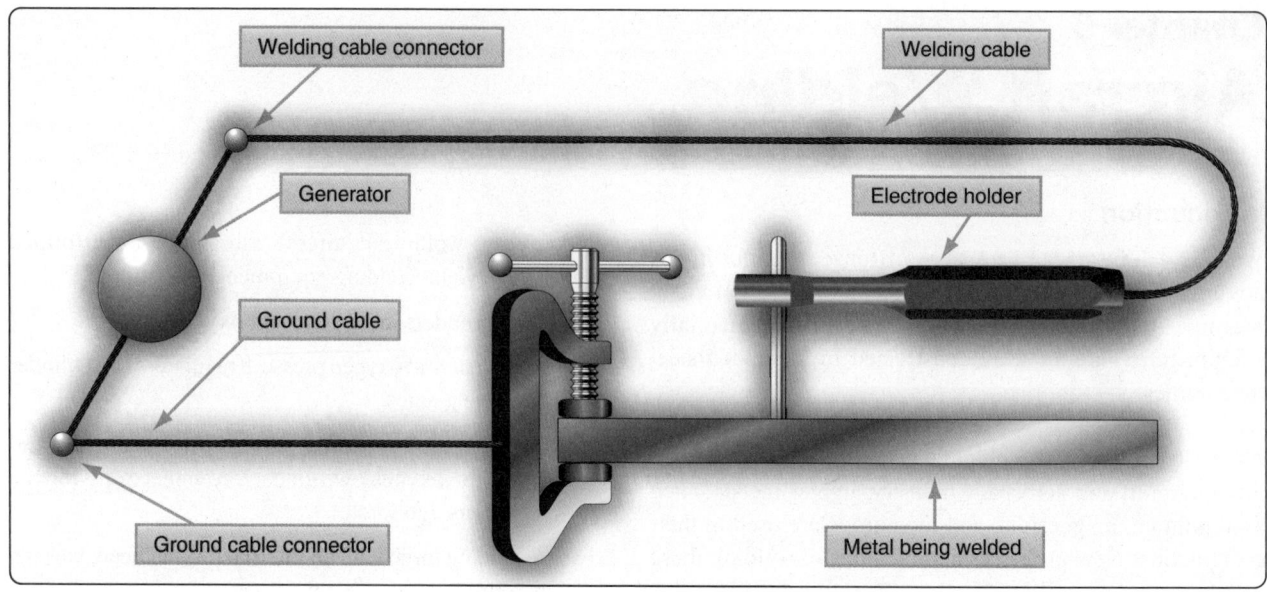

Figure 5-2. *Typical arc welding circuit.*

satisfactorily to join all weldable metals, provided that the proper processes and materials are used. The four types of electric arc welding are addressed in the following paragraphs.

Shielded Metal Arc Welding (SMAW)

Shielded metal arc welding (SMAW) is the most common type of welding and is usually referred to as "stick" welding. The equipment consists of a metal wire rod coated with a welding flux that is clamped in an electrode holder that is connected by a heavy electrical cable to a low voltage and high current in either alternating current (AC) or direct current (DC), depending on the type of welding being done. An arc is struck between the rod and the work and produces heat in excess of 10,000 °F, which melts both the material and the rod. The welding circuit consists of a welding machine, two leads, an electrode holder, an electrode, and the work to be welded. *[Figure 5-2]*

When the electrode is touched to the metal to be welded, the circuit is complete and the current flows. The electrode is then withdrawn from the metal approximately ¼-inch to form an air gap between the metal and the electrode. If the correct gap is maintained, the current bridges the gap to form a sustained electric spark called the arc. This action melts the electrode and the coating of flux.

As the flux melts, it releases an inert gas that shields the molten puddle from oxygen in the air to prevent oxidation. The molten flux covers the weld and hardens to an airtight slag that protects the weld bead as it cools. Some aircraft manufacturers, such as Stinson, used this process for the welding of 4130 steel fuselage structures. This was followed by heat treatment in an oven to stress relieve and normalize the structure. Shown in *Figure 5-3* is a typical arc welding

machine with cables, ground clamp, and electrode holder.

Gas Metal Arc Welding (GMAW)

Gas metal arc welding (GMAW) was formerly called metal inert gas (MIG) welding. It is an improvement over stick welding because an uncoated wire electrode is fed into and through the torch and an inert gas, such as argon, helium, or carbon dioxide, flows out around the wire to protect the puddle from oxygen. The power supply is connected to the torch and the work, and the arc produces the intense heat needed to melt the work and the electrode. *[Figure 5-4]*

Low-voltage, high-current DC is typically used with GMAW welding. *Figure 5-5* shows the equipment required for a typical MIG welding setup.

This method of welding can be used for large volume

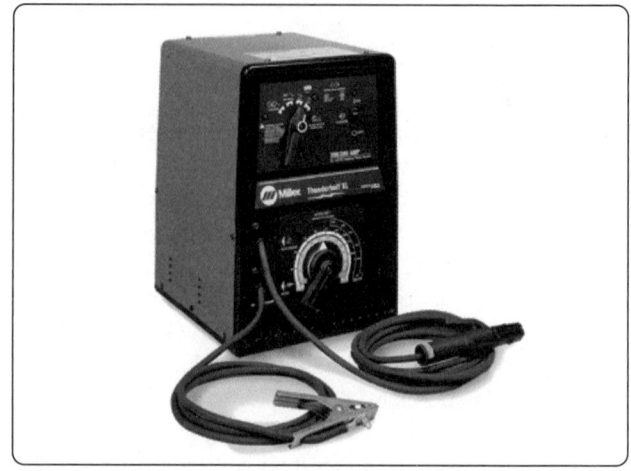

Figure 5-3. *Stick welder–Shielded Metal Arc Welder (SMAW).*

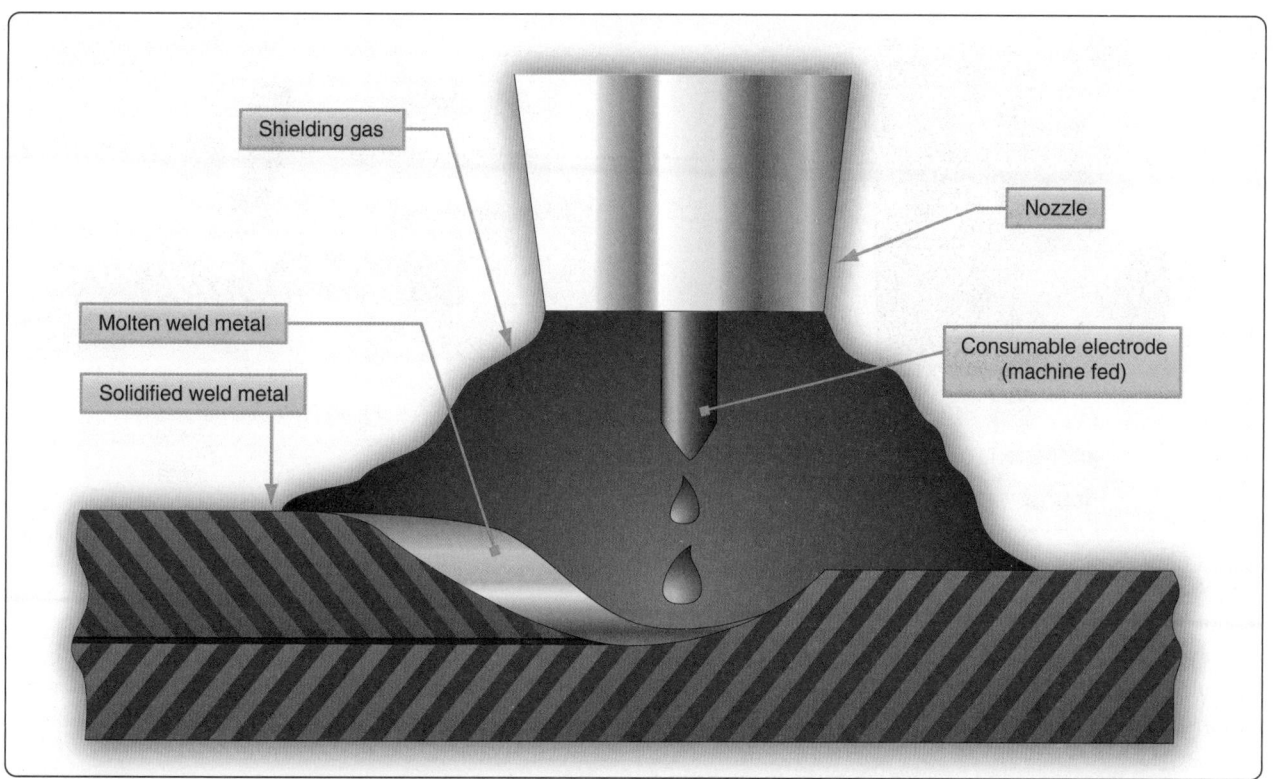

Figure 5-4. *Metal inert gas (MIG) welding process.*

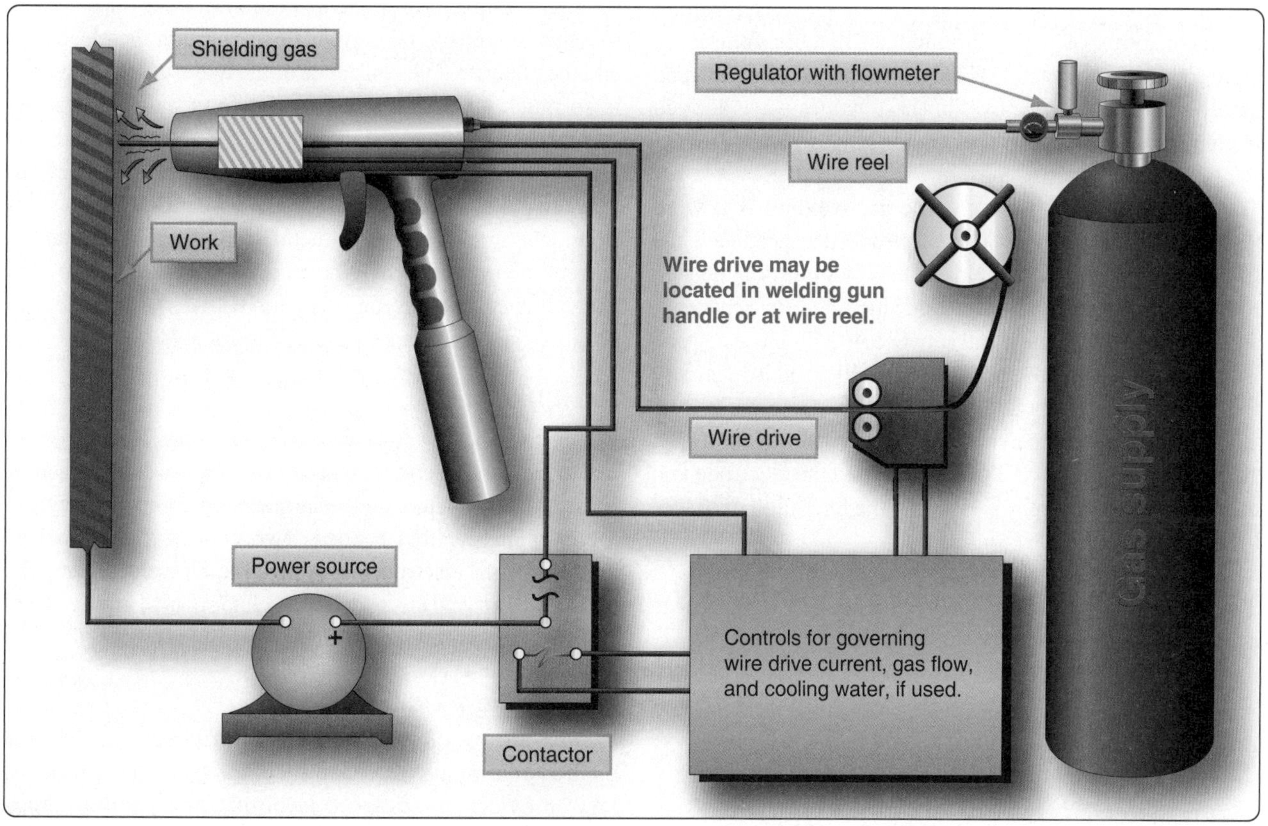

Figure 5-5. *MIG welding equipment.*

Figure 5-6. *MIG welder–gas metal arc welder (GMAW).*

manufacturing and production work; it is not well suited to repair work because weld quality cannot be easily determined without destructive testing. *Figure 5-6* depicts a typical power source used for MIG welding.

Gas Tungsten Arc Welding (GTAW)

Gas tungsten arc welding (GTAW) is a method of electric arc welding that fills most of the needs in aircraft maintenance and repair when proper procedures and materials are used. It is the preferred method to use on stainless steel, magnesium, and most forms of thick aluminum. It is more commonly known as Tungsten Inert Gas (TIG) welding and by the trade names of Heliarc or Heliweld. These names were derived from the inert helium gas that was originally used.

The first two methods of electric arc welding that were addressed used a consumable electrode that produced the filler for the weld. In TIG welding, the electrode is a tungsten rod that forms the path for the high amperage arc between it and the work to melt the metal at over 5,400 °F. The electrode is not consumed and used as filler so a filler rod is manually fed into the molten puddle in almost the same manner as when using an oxy-acetylene torch. A stream of inert gas, such as argon or helium, flows out around the electrode and envelopes the arc thereby preventing the formation of oxides in the molten puddle. *[Figure 5-7]*

The versatility of a TIG welder is increased by the choice of the power supply being used. DC of either polarity or AC may be used. *[Figure 5-8]*

- Either select the welder setting to DC straight polarity (the work being the positive and the torch being negative) when welding mild steel, stainless steel, and titanium; or
- Select AC for welding aluminum and magnesium.

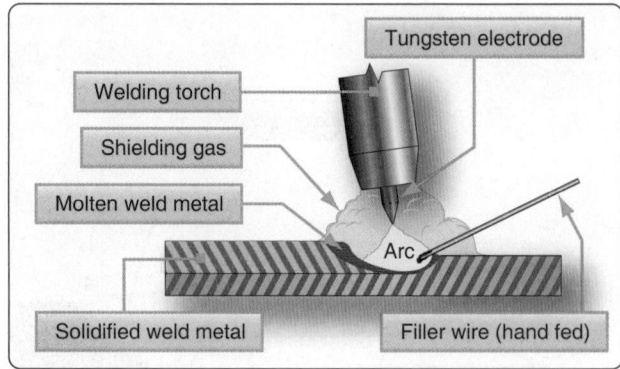

Figure 5-7. *Tungsten inert gas (TIG) welding process.*

Figure 5-9 is a typical power source for TIG welding along with a torch, foot operated current control, regulator for inert gas, and assorted power cables.

Electric Resistance Welding

Electric resistance welding, either spot welding or seam welding, is typically used to join thin sheet metal components during the manufacturing process.

Spot Welding

Two copper electrodes are held in the jaws of the spot welding machine, and the material to be welded is clamped between them. Pressure is applied to hold the electrodes tightly together and electrical current flows through the electrodes and the material. The resistance of the material being welded is so much higher than that of the copper electrodes that enough heat is generated to melt the metal. The pressure on the electrodes forces the molten spots in the two pieces of metal to unite, and this pressure is held after the current stops flowing long enough for the metal to solidify. The amount of current, pressure, and dwell time are all carefully controlled and matched to the type of material and the thickness to produce the correct spot welds. *[Figure 5-10]*

Seam Welding

Rather than having to release the electrodes and move the material to form a series of spot welds, a seam-welding machine is used to manufacture fuel tanks and other components where a continuous weld is needed. Two copper wheels replace the bar-shaped electrodes. The metal to be welded is moved between them, and electric pulses create spots of molten metal that overlap to form the continuous seam.

Plasma Arc Welding (PAW)

Plasma arc welding (PAW) was developed in 1964 as a method of bringing better control to the arc welding process. PAW provides an advanced level of control and accuracy using automated equipment to produce high quality welds

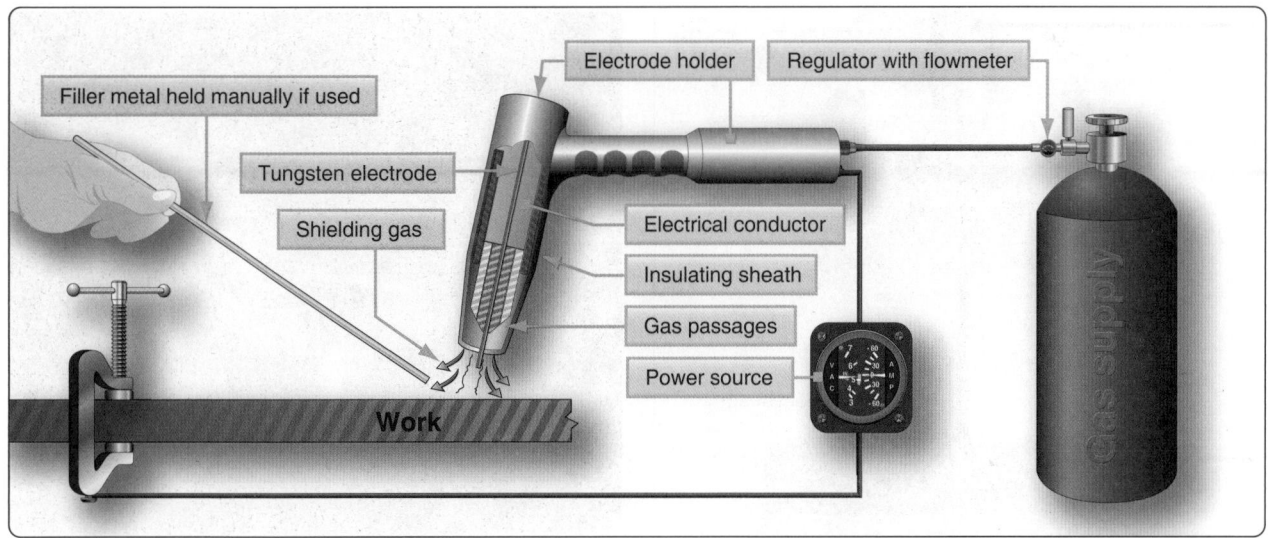

Figure 5-8. *Typical setup for TIG welding.*

Figure 5-9. *TIG welder–gas tungsten arc welder (GTAW).*

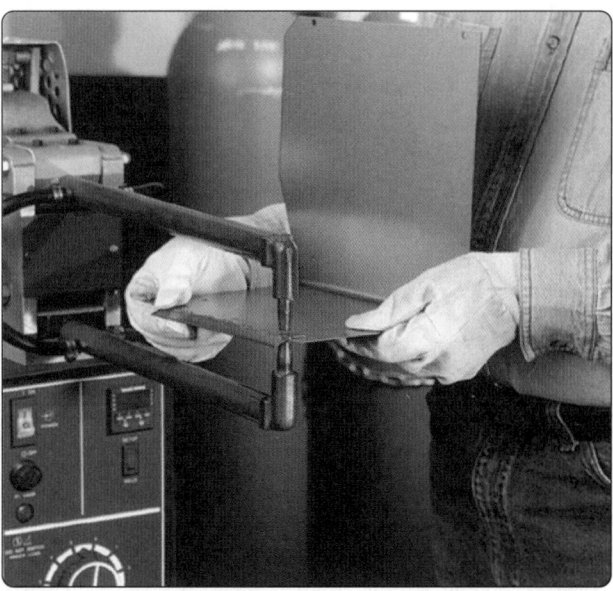

Figure 5-10. *Spot welding thin sheet metal.*

in miniature and precision applications. Furthermore, PAW is equally suited to manual operation and can be performed by a person using skills similar to those for GTAW.

In the plasma welding torch, a nonconsumable tungsten electrode is located within a fine-bore copper nozzle. A pilot arc is initiated between the torch electrode and nozzle tip. This arc is then transferred to the metal being welded. *[Figure 5-11]*

By forcing the plasma gas and arc through a constricted orifice, the torch delivers a high concentration of heat to a small area. The plasma process produces exceptionally high quality welds. *[Figure 5-12]*

Plasma gas is normally argon. The torch also uses a secondary gas, such as argon/helium or argon/nitrogen, that assists in shielding the molten weld puddle and minimizing oxidation of the weld.

Like GTAW, the PAW process can be used to weld most commercial metals, and it can be used for a wide variety of metal thicknesses. On thin material, from foil to ⅛-inch, the process is desirable because of the low heat input. The process provides relatively constant heat input because arc length variations are not very critical. On material thicknesses greater than ⅛-inch and using automated equipment, a keyhole technique is often used to produce full penetration single-path welds. In the keyhole technique, the plasma

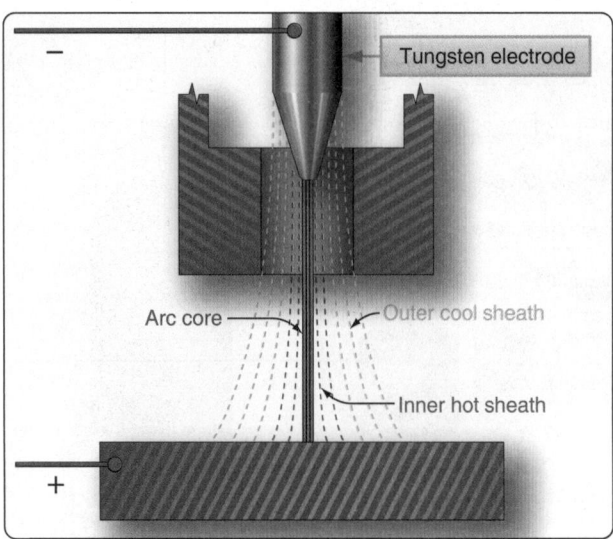

Figure 5-11. *The plasma welding process.*

Figure 5-12. *Plasma arc.*

completely penetrates the work piece. The molten weld metal flows to the rear of the keyhole and solidifies as the torch moves on. The high quality welds produced are characterized by deep, narrow penetration and a small weld face.

When PAW is performed manually, the process requires a high degree of welding skills similar to that required for GTAW. However, the equipment is more complex and requires a high degree of knowledge to set up and use. The equipment required for PAW includes a welding machine, a special plasma arc control system, the plasma welding torch (water-cooled), the source of plasma and shielding gas, and filler material, when required. Because of the cost associated with this equipment, this process is very limited outside of manufacturing facilities.

Plasma Arc Cutting

When a plasma cutting torch is used, the gas is usually compressed air. The plasma cutting machine works by constricting an electrical arc in a nozzle and forcing the ionized gas through it. This heats the gas that melts the metal which is blown away by the air pressure. By increasing air pressure and intensifying the arc with higher voltages, the cutter is capable of blasting through thicker metals and blowing away the dross with minimal cleanup.

Plasma arc systems can cut all electrically conductive metals, including aluminum and stainless steel. These two metals cannot be cut by oxy-fuel cutting systems because they have an oxide layer that prevents oxidation from occurring. Plasma cutting works well on thin metals and can successfully cut brass and copper in excess of two inches thick.

Plasma cutting machines can rapidly and precisely cut

through, gouge, or pierce any electrically conductive metal without preheating. The plasma cutter produces a precise kerf (cut) width and a small heat-affected zone (HAZ) that prevents warping and damage.

Gas Welding & Cutting Equipment

Welding Gases

Acetylene

This is the primary fuel for oxy-fuel welding and cutting. It is chemically very unstable, and is stored in special cylinders designed to keep the gas dissolved. The cylinders are packed with a porous material and then saturated with acetone. When the acetylene is added to the cylinder, it dissolves; in this solution, it becomes stable. Pure acetylene stored in a free state explodes from a slight shock at 29.4 pounds per square inch (psi). The acetylene pressure gauge should never be set higher than 15 psi for welding or cutting.

Argon

Argon is a colorless, odorless, tasteless, and non-toxic inert gas. Inert gas cannot combine with other elements. It has a very low chemical reactivity and low thermal conductivity. It is used as a gas shield for the electrode in MIG, TIG, and plasma welding equipment.

Helium

Helium is a colorless, odorless, tasteless, and non-toxic inert gas. Its boiling and melting points are the lowest of the elements and it normally exists only in gas form. It is used as a protective gas shield for many industrial uses including electric arc welding.

Figure 5-13. *Single-stage acetylene regulator. Note the maximum 15-psi working pressure. The notched groove cylinder connection nut indicates a left-hand thread.*

Figure 5-14. *Two-stage oxygen regulator. No groove on the cylinder connection nut indicates a right-hand thread.*

Hydrogen

Hydrogen is a colorless, odorless, tasteless, and highly flammable gas. It can be used at a higher pressure than acetylene and is used for underwater welding and cutting. It also can be used for aluminum welding using the oxy-hydrogen process.

Oxygen

Oxygen is a colorless, odorless, and nonflammable gas. It is used in the welding process to increase the combustion rate which increases the flame temperature of flammable gas.

Pressure Regulators

A pressure regulator is attached to a gas cylinder and is used to lower the cylinder pressure to the desired working pressure. Regulators have two gauges, one indicating the pressure in the cylinder and the second showing the working pressure. By turning the adjustment knob in or out, a spring operating a flexible diaphragm opens or closes a valve in the regulator. Turning the knob in causes the flow and pressure to increase; backing it out decreases the flow and pressure.

There are two types of regulators: single stage and two stage. They perform the same function but the two-stage regulator maintains a more constant outlet pressure and flow as the cylinder volume and pressure drops. Two-stage regulators can be identified by a larger, second pressure chamber under the regulator knob. *[Figures 5-13 and 5-14]*

Welding Hose

A welding hose connects the regulators to the torch. It is typically a double hose joined together during manufacture. The acetylene hose is red and has left hand threads indicated by a groove cut into the connection nut. The oxygen hose is green and has right hand threads indicated by the absence of a groove on the connection nut.

Welding hoses are produced in different sizes from ¼-inch to ½-inch inside diameter (ID). The hose should be marked for light, standard, and heavy duty service plus a grade indicating whether it has an oil- and/or flame-resistant cover. The hose should have the date of manufacture, maximum working pressure of 200 psi, and indicate that it meets specification IP-90 of the Rubber Manufacturers Association and the Compressed Gas Association for rubber welding hoses. Grade-R hose should only be used with acetylene gas. A T-grade hose must be used with propane, MAPP®, and all other fuel gases.

Check Valves & Flashback Arrestors

The check valve stops the reverse flow of the gas and can be installed either between the regulator and the hose or the hose and the torch. *[Figure 5-15]* Excessive overheating of cutting, welding, and heating tips can cause flashback conditions. A flashback can be caused when a tip is overheated and the gas ignites before passing out of the tip. The flame is then burning internally rather than on the outside of the tip and is usually identified by a shrill hissing or squealing noise.

A flashback arrestor installed on each hose prevents a high pressure flame or oxygen-fuel mixture from being pushed back into either cylinder causing an explosion. The flashback arrestors incorporate a check valve that stops the reverse flow of gas and the advancement of a flashback fire. *[Figure 5-16]*

Figure 5-15. *Check valves.*

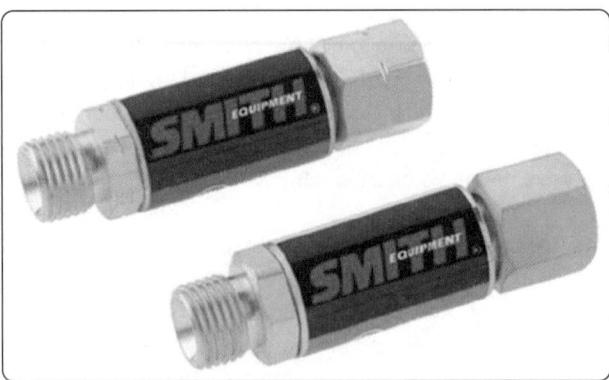

Figure 5-16. *Flashback arrestors.*

Torches

Equal Pressure Torch

The equal pressure torch is the most commonly used torch for oxy-acetylene welding. It has a mixing chamber and uses acetylene fuel at 1–15 psi. The flame is easy to adjust and there is less chance of flashback with this torch. There are several small lightweight torches of this type that are ideal for aviation welding projects. The Smith Airline™ and the Meco Midget™ torches are small enough to be used in close confined areas, lightweight enough to reduce fatigue during long welding sessions yet, with the appropriate tips, are capable of welding 0.250-inch steel.

Injector Torch

The injector torch uses fuel gas at pressures between just above 0 and 2 psi. This torch is typically used with propane and propylene gas. High-pressure oxygen comes through a small nozzle inside the torch head and pulls the fuel gas along with it via a venturi effect. The low-pressure injector torch is more prone to flashback.

Cutting Torch

The cutting torch is an attachment added to the torch handle that allows the cutting of metal. The cutting process is fundamentally the rapid burning or oxidizing of the metal in a localized area. The metal is heated to a bright red color (1,400 °F to 1,600 °F), which is the kindling temperature, using only the preheat jets. Then, a jet of high-pressure oxygen released by the lever on the cutting attachment is directed against the heated metal. This oxygen blast combines with the hot metal and forms an intensely hot oxide. The molten oxide is blown down the sides of the cut, heating the metal in its path to the kindling temperature as the torch is moved along the line of the desired cut. The heated metal also burns to an oxide that is blown away on the underside of the piece. *[Figure 5-17]*

Torch Tips

The torch tip delivers and controls the final flow of gases. It is important that you use the correct tip with the proper gas pressures for the work to be welded satisfactorily. The size of the tip opening—not the temperature—determines the amount of heat applied to the work. If an excessively small tip is used, the heat provided is insufficient to produce penetration to the proper depth. If the tip is too large, the heat is too great, and holes are burned in the metal.

Torch tip sizes are designated by numbers. The manufacturer can provide a chart with recommended sizes for welding specific thicknesses of metal. With use, a torch tip becomes clogged with carbon deposits. If it is allowed to contact the molten pool, particles of slag may clog the tip. This may cause a backfire, which is a momentary backward flow of the gases at the torch tip. A backfire is rarely dangerous, but molten metal may be splattered when the flame pops. Tips should be cleaned with the proper size tip cleaner to avoid enlarging the tip opening.

Welding Eyewear

Protective eyewear for use with oxy-fuel welding outfits is available in several styles and must be worn to protect the welder's eyes from the bright flame and flying sparks. This eyewear is not for use with arc welding equipment.

Some of the styles available have individual lenses and include goggles that employ a head piece and/or an elastic head strap to keep them snug around the eyes for protection from the occasional showering spark. *[Figure 5-18]* Another popular style is the rectangular eye shield that takes a standard 2-inch by 4.25-inch lens. This style is available with an elastic strap but is far more comfortable and better fitting when attached to a proper fitting adjustable headgear. It can be worn over prescription glasses, provides protection from flying sparks, and accepts a variety of standard shade and

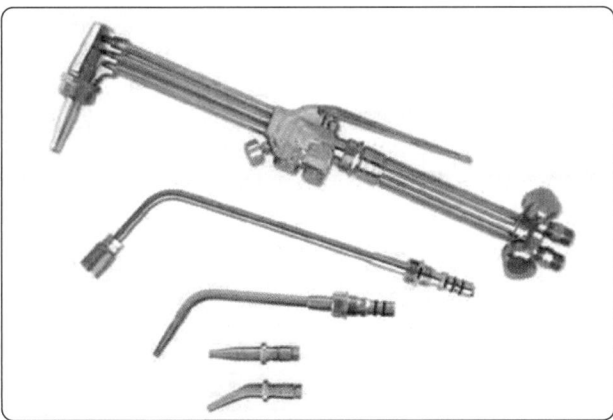

Figure 5-17. *Torch handle with cutting, heating, and welding tips.*

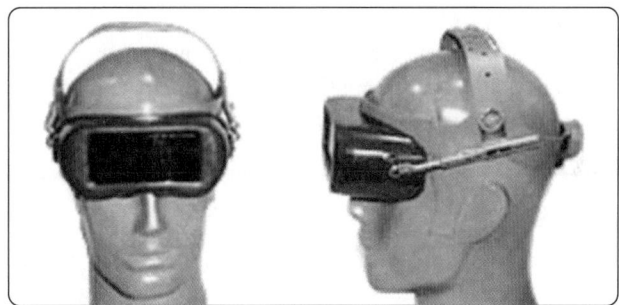

Figure 5-19. *Gas welding eye shield attached to adjustable headgear.*

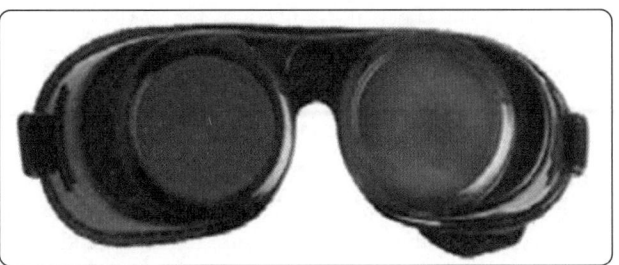

Figure 5-18. *Welding goggles.*

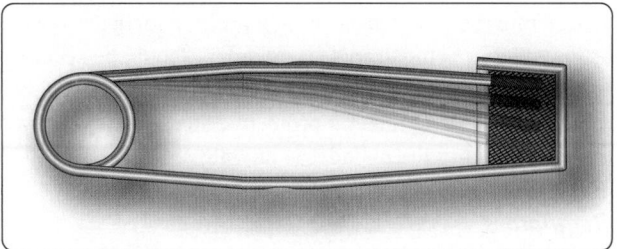

Figure 5-20. *Torch lighter.*

color lenses. A clear safety glass lens is added in front of the shaded lens to protect it from damage. *[Figure 5-19]*

It was standard practice in the past to select a lens shade for gas welding based on the brightness of flame emitting from the torch. The darkest shade of lens showing a clear definition of the work was normally the most desirable. However, when flux was used for brazing and welding, the torch heat caused the sodium in the flux to give off a brilliant yellow-orange flare, hiding a clear view of the weld area and causing many eye problems.

Various types of lens and colors were tried for periods of time without much success. It was not until the late 1980s that TM Technologies developed and patented a new green glass designed especially for aluminum oxy-fuel welding. It not only eliminated the sodium orange flare completely, but also provided the necessary protection from ultraviolet, infrared, and blue light, and impact to meet the requirements of the American National Standards Institute (ANSI) Z87-1989 Safety Standards for a special purpose lens. This lens can be used for welding and brazing all metals using an oxy-fuel torch.

Torch Lighters

Torch lighters are called friction lighters or flint strikers. The lighter consists of a file-shaped piece of steel, usually recessed in a cuplike device, and a replaceable flint, which when drawn across the steel produces a shower of sparks to light the fuel

gas. An open flame or match should never be used to light a torch, because accumulated gas may envelop the hand and when ignited cause a severe burn. *[Figure 5-20]*

Filler Rod

The use of the proper type of filler rod is very important for oxy-acetylene welding. This material adds not only reinforcement to the weld area, but also desired properties to the finished weld. By selecting the proper rod, tensile strength or ductility can be secured in a weld. Similarly, the proper rod can help retain the desired amount of corrosion resistance. In some cases, a suitable rod with a lower melting point helps to avoid cracks caused by expansion and contraction.

Welding rods may be classified as ferrous or nonferrous. Ferrous rods include carbon and alloy steel rods, as well as cast-iron rods. Nonferrous rods include brass, aluminum, magnesium, copper, silver, and their various alloys.

Welding rods are manufactured in standard 36-inch lengths and in diameters from $\frac{1}{16}$-inch to $\frac{3}{8}$-inch. The diameter of the rod to be used is governed by the thickness of the metals to be joined. If the rod is too small, it cannot conduct heat away from the puddle rapidly enough, and a burned hole results. A rod too large in diameter draws heat away and chills the puddle, resulting in poor penetration of the joined metal. All filler rods should be cleaned prior to use.

Equipment Setup

Setting up acetylene welding equipment in preparation for welding should be accomplished in a systematic and definite order to avoid costly damage to equipment and compromising the safety of personnel.

Gas Cylinders

All cylinders should be stored and transported in the upright position, especially acetylene cylinders, because they contain an absorbent material saturated with liquid acetone. If the cylinder were laid on its side, allowing the acetone to enter and contaminate the regulator, hose, and torch, fuel starvation and a resultant flashback in the system could result. If an acetylene cylinder must be placed on its side for a period of time, it must be stored in the upright position for at least twice as long before being used. Gas cylinders should be secured, usually with a chain, in a permanent location or in a suitable mobile cart. The cylinder's protective steel cap should not be removed until the cylinder is put into service.

Regulators

Prior to installing the regulator on a gas cylinder, open the cylinder shutoff valve for an instant to blow out any foreign material that may be lodged in the outlet. Close the valve and wipe off the connection with a clean oil-free cloth. Connect the acetylene pressure regulator to the acetylene cylinder and tighten the left-hand nut. Connect the oxygen pressure regulator to the oxygen cylinder and tighten the right-hand nut. The connection fittings are brass and do not require a lot of torque to prevent them from leaking. At this time, check to ensure the adjusting screw on each pressure regulator is backed out by turning counterclockwise until it turns freely.

Hoses

Connect the red hose with the left-hand threads to the acetylene pressure regulator and the green hose with the right-hand threads to the oxygen pressure regulator. This is the location, between the regulator and hose, in which flashback arrestors should be installed. Again, because the fittings are brass and easily damaged, tighten only enough to prevent leakage.

Stand off to the side away from the face of the gauges. Now, very slowly open the oxygen cylinder valve and read the cylinder gauge to check the contents in the tank. The oxygen cylinder shutoff valve has a double seat valve and should be opened fully against its stop to seat the valve and prevent a leak. The acetylene cylinder shutoff valve should be slowly opened just enough to get the cylinder pressure reading on the regulator and then one half of a turn more. This allows a quick shutoff, if needed.

Note: As a recommended safety practice, the cylinders should not be depleted in content below 20 psi. This prevents the possible reverse flow of gas from the opposite tank.

Both hoses should be blown out before attaching to the torch. This is accomplished for each cylinder by turning the pressure adjusting screw in (clockwise) until the gas escapes, and then quickly backing the screw out (counterclockwise) to shut off the flow. This should be done in a well ventilated open space, free from sparks, flames, or other sources of ignition.

Connecting Torch

Connect the red hose with the left-hand thread connector nut to the left-hand thread fitting on the torch. Connect the green hose with the right-hand thread connector nut to the right-hand thread fitting on the torch. Close the valves on the torch handle and check all connections for leaks as follows:

- Turn in the adjusting screw on the oxygen pressure regulator until the working pressure indicates 10 psi. Turn in the adjusting screw on the acetylene pressure regulator until the working pressure indicates 5 psi.

- Back out both adjusting screws on the regulators and verify that the working pressure remains steady. If it drops and pressure is lost, a leak is indicated between the regulator and the torch.

- A general tightening of all connections should fix the leak. Repeat a check of the system.

- If a leak is still indicated by a loss in working pressure, a mixture of soapy water on all the connections reveals the source of the leak. Never check for a leak with a flame because a serious explosion could occur.

Select the Tip Size

Welding and cutting tips are available in a variety of sizes for almost any job, and are identified by number. The higher the number, the bigger the hole in the tip, allowing more heat to be directed onto the metal and allowing thicker metal to be welded or cut.

Welding tips have one hole and cutting tips have a number of holes. The cutting tip has one large hole in the center for the cutting oxygen and a number of smaller holes around it that supply fuel, gas, and oxygen for the preheating flame. The selection of the tip size is very important, not only for the quality of the weld and/or the efficiency of the cutting process, but for the overall operation of the welding equipment and safety of the personnel using it.

Starvation occurs if torch tips are operated at less than the required volume of gas, leading to tip overheating and possible flashbacks. Incorrect tip size and obstructed tip orifices can also cause overheating and/or flashback conditions.

Welding Tip Size Conversion Chart									
Wire Drill	Decimal Inch	Metric Equiv. (mm)	Smiths™ AW1A	Henrob/ Dillion	Harris 15	Victor J Series	Meco N Midget™	Aluminum Thickness (in)	Steel Thickness (in)
97	0.0059	0.150						Foil	Foil
85	0.0110	0.279							
80	0.0135	0.343		#00			#00		
76	0.0200	0.508	AW200				#0		
75	0.0220	0.559		#0	#0	#000			.015
74	0.0225	0.572	AW20					.025	
73	0.0240	0.610					0.5		
72	0.0250	0.635		0.5					
71	0.0260	0.660	AW201		1				
70	0.0280	0.711				#00	1		.032
69	0.0292	0.742	AW202						
67	0.0320	0.813	AW203				1.5	.040	
66	0.0340	0.864		1					
65	0.0350	0.889			2	#0	2	.050	.046
63	0.0370	0.940	AW204				2.5		
60	0.0400	1.016				1			
59	0.0410	1.041		1.5					
58	0.0420	1.067			3		3		.062
57	0.0430	1.092	AW205						
56	0.0465	1.181	AW206			2	4	.063	
55	0.0520	1.321		2	4				.093
54	0.0550	1.397	AW207				4.5		
53	0.0595	1.511			5	3			.125
52	0.0635	1.613	AW208				5	.100	
51	0.0670	1.702			6				.187
49	0.0730	1.854	AW209	2.5		4	5.5		
48	0.0760	1.930			7			.188	.250
47	0.0780	1.981					6		
45	0.0820	2.083			8				.312
44	0.0860	2.184	AW210				6.5	.25	
43	0.0890	2.261			9	5	7		.375
42	0.0930	2.362		3					
40	0.0980	2.489			10				
36	0.1060	2.692				6			
35	0.1100	2.794			13				

Figure 5-21. *Chart of recommended tip sizes for welding various thicknesses of metal.*

All fuel cylinders have a limited capacity to deliver gas to the tip. That capacity is further limited by the gas contents remaining in the cylinder and the temperature of the cylinder.

The following provides some recommended procedures to guard against overheating and flashbacks:

- Refer to the manufacturer's recommendations for tip size based on the metal's thickness.

- Use the recommended gas pressure settings for the tip size being used.

- Provide the correct volume of gas as recommended for each tip size.

- Do not use an excessively long hose, one with multiple splices, or one that may be too small in diameter and restrict the flow of gas.

Note: Acetylene is limited to a maximum continuous withdrawal rate of one-seventh of the cylinder's rated capacity when full. For example, an acetylene cylinder that has a capacity of 330 cubic feet has a maximum withdrawal of 47 cubic feet per hour. This is determined by dividing 330 (cylinder capacity) by 7 (one-seventh of the cylinder capacity). As a safety precaution, it is recommended that flashback arrestors be installed between the regulators and the gas supply hoses of all welding outfits. *Figure 5-21* shows recommended tip sizes of different manufacturers, for welding various thickness of metals.

Adjusting the Regulator Working Pressure

The working pressure should be set according to the manufacturer's recommendation for the tip size that is being used to weld or cut. This is a recommended method that works for most welding and cutting operations.

In a well ventilated area, open the acetylene valve on the torch and turn the adjusting screw on the acetylene pressure regulator clockwise until the desired pressure is set. Close the acetylene valve on the torch. Then, set the oxygen pressure in the same manner by opening the oxygen valve on the torch and turning the adjusting screw clockwise on the oxygen regulator until desired pressure is set. Then, close the oxygen valve on the torch handle. With the working pressures set, the welding or cutting operation can be initiated.

Lighting & Adjusting the Torch

With the proper working pressures set for the acetylene and oxygen, open the torch acetylene valve a quarter to a half turn. Direct the torch away from the body and ignite the acetylene gas with the flint striker. Open the acetylene valve until the black sooty smoke disappears from the flame. The pure acetylene flame is long, bushy, and has a yellowish color. Open the torch oxygen valve slowly and the flame shortens and turns to a bluish-white color that forms a bright inner luminous cone surrounded by an outer flame envelope. This is a neutral flame that should be set before either a carburizing or oxidizing flame mixture is set.

Different Flames

The three types of flame commonly used for welding are neutral, carburizing, and oxidizing. Each serves a specific purpose. *[Figure 5-22]*

Neutral Flame

The neutral flame burns at approximately 5,850 °F at the tip of the inner luminous cone and is produced by a balanced mixture of acetylene and oxygen supplied by the torch. The neutral flame is used for most welding because it does not alter the composition of the base metal. When using this flame on steel, the molten metal puddle is quiet and clear, and the metal flows to give a thoroughly fused weld without burning or sparking.

Carburizing Flame

The carburizing flame burns at approximately 5,700 °F at

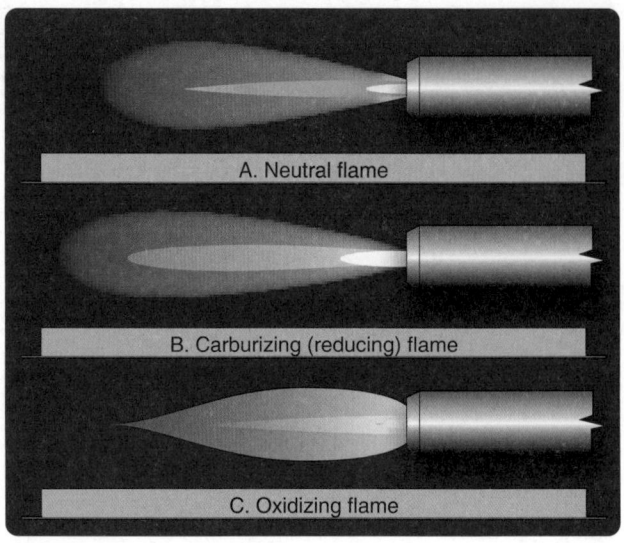

Figure 5-22. *Oxy-acetylene flames.*

the tip of the inner core. It is also referred to as a reducing flame because it tends to reduce the amount of oxygen in the iron oxides. The flame burns with a coarse rushing sound, and has a bluish-white inner cone, a white center cone, and a light blue outer cone.

The flame is produced by burning more acetylene than oxygen, and can be recognized by the greenish feathery tip at the end of the cone. The longer the feather, the more acetylene is in the mix. For most welding operations, the length of the feather should be about twice the length of the inner cone.

The carburizing flame is best used for welding high-carbon steels, for hard facing, and for welding such nonferrous alloys as aluminum, nickel, and Monel.

Oxidizing Flame

The oxidizing flame burns at approximately 6,300 °F and is produced by burning an excess of oxygen. It takes about two parts of oxygen to one part acetylene to produce this flame. It can be identified by the shorter outer flame and the small, white, inner cone. To obtain this flame, start with a neutral flame and then open the oxygen valve until the inner cone is about one-tenth of its original length. The oxidizing flame makes a hissing sound, and the inner cone is somewhat pointed and purplish in color at the tip.

The oxidizing flame does have some specific uses. A slightly oxidizing flame is used for bronze welding (brazing) of steel and cast iron. A stronger oxidizing flame is used for fusion welding of brass and bronze. If an oxidizing flame is used on steel, it causes the molten metal to foam, give off sparks, and burn.

Soft or Harsh Flames

With each size of tip, a neutral, carburizing, or oxidizing flame can be obtained. It is also possible to obtain a soft or harsh flame by decreasing or increasing the working pressure of both gases (observing the maximum working pressure of 15 psi for acetylene gas).

For some work, it may be desirable to have a soft or low velocity flame without a reduction of thermal output. This can be achieved by reducing the working pressure using a larger tip and closing the torch valves until the neutral flame is quiet and steady. It is especially desirable to use a soft flame when welding aluminum to avoid blowing holes in the metal when the puddle is formed.

Handling of the Torch

It should be cautioned that improper adjustment or handling of the torch may cause the flame to backfire or, in rare cases, to flashback. A backfire is a momentary backward flow of gases at the torch tip that causes the flame to go out. A backfire may be caused by touching the tip against the work, overheating the tip, by operating the torch at other than recommended pressures, by a loose tip or head, or by dirt or slag in the end of the tip, and may cause molten metal to be splattered when the flame pops.

A flashback is dangerous because it is the burning of gases within the torch. It is usually caused by loose connections, improper pressures, or overheating of the torch. A shrill hissing or squealing noise accompanies a flashback, and unless the gases are turned off immediately, the flame may burn back through the hose and regulators causing great damage and personal injury. The cause of the flashback should always be determined and the problem corrected before relighting the torch. All gas welding outfits should have a flashback arrestor.

Oxy-acetylene Cutting

Cutting ferrous metals by the oxy-acetylene process is primarily the rapid burning or oxidizing of the metal in a localized area. This is a quick and inexpensive way to cut iron and steel where a finished edge is not required.

Figure 5-23 shows an example of a cutting torch. It has the conventional oxygen and acetylene valves in the torch handle that control the flow of the two gases to the cutting head. It also has an oxygen valve below the oxygen lever on the cutting head so that a finer adjustment of the flame can be obtained.

The size of the cutting tip is determined by the thickness of the metal to be cut. Set the regulators to the recommended working pressures for the cutting torch based on the tip size

Figure 5-23. *Cutting torch with additional tools.*

selected. Before beginning any cutting operation, the area should be clear of all combustible material and the proper protective equipment should be worn by personnel engaged in the cutting operation.

The flame for the torch in *Figure 5-23* is set by first closing the oxygen valve below the cutting lever and fully opening the oxygen valve on the handle. (This supplies the high-pressure oxygen blast when the cutting lever is actuated.) The acetylene valve on the handle is then opened and the torch is lit with a striker. The acetylene flame is increased until the black soot is gone. Then, open the oxygen valve below the cutting lever and adjust the flame to neutral. If more heat is needed, open the valves to add more acetylene and oxygen. Actuate the cutting lever and readjust the preheat flame to neutral if necessary. The metal is heated to a bright red color (1,400 °F–1,600 °F, which is the kindling or ignition temperature) by the preheat orifices in the tip of the cutting torch. Then, a jet of high-pressure oxygen is directed against it by pressing the oxygen lever on the torch. This oxygen blast combines with the red-hot metal and forms an intensely hot molten oxide that is blown down the sides of the cut. As the torch is moved along the intended cut line, this action continues heating the metal in its path to the kindling temperature. The metal, thus heated, also burns to an oxide that is blown away to the underside of the piece.

Proper instruction and practice provides the knowledge and skill to become proficient in the technique needed to cut with a torch. Hold the torch in either hand, whichever is most comfortable. Use the thumb of that hand to operate the oxygen cutting lever. Use the other hand to rest the torch on and steady it along the cut line.

Begin at the edge of the metal and hold the tip perpendicular to the surface, preheating until the spot turns bright red. Lightly depress the cutting lever to allow a shower of sparks and molten metal to blow through the cut. Fully depress the cutting lever and move the torch slowly in the direction of the intended cut.

Practice and experience allow the technician to learn how to judge the speed at which to move the torch. It should be just fast enough to allow the cut to penetrate completely without excessive melting around the cut. If the torch is moved too fast, the metal will not be preheated enough, and the cutting action stops. If this happens, release the cutting lever, preheat the cut to bright red, depress the lever, and continue with the cut.

Shutting Down the Gas Welding Equipment

Shutting down the welding equipment is fairly simple when some basic steps are followed:

- Turn off the flame by closing the acetylene valve on the torch first. This shuts the flame off quickly. Then, close the oxygen valve on the torch handle. Also, close oxygen valve on cutting torch, if applicable.

- If the equipment is not used in the immediate future (approximately the next 30 minutes), the valves on the acetylene and oxygen cylinders should be closed and pressure relieved from the hoses.

- In a well-ventilated area, open the acetylene valve on the torch and allow the gas to escape to the outside atmosphere, and then close the valve.

- Open the oxygen valve on the torch, allow the gas to escape, and then close the valve.

- Close both the acetylene and oxygen regulators by backing out the adjusting screw counterclockwise until loose.

- Carefully coil the hose to prevent kinking and store it to prevent damage to the torch and tip.

Gas Welding Procedures & Techniques

The material to be welded, the thickness of the metal, the type of joint, and the position of the weld dictates the procedure and technique to be used.

When light-gauge metal is welded, the torch is usually held with the hose draped over the wrist. *[Figure 5-24]* To weld heavy materials, the more common grip may provide better control of the torch. *[Figure 5-25]*

The torch should be held in the most comfortable position that allows the tip to be in line with the joint to be welded, and inclined between 30° and 60° from the perpendicular. This position preheats the edges just ahead of the molten puddle. The best angle depends on the type of weld, the amount of preheating required, and the thickness and type of metal. The thicker the metal, the more vertical the torch must be for proper heat penetration. The white cone of the flame should be held about ⅛-inch from the surface of the metal.

Welding can be performed by pointing the torch flame in the direction that the weld is progressing. This is referred to as

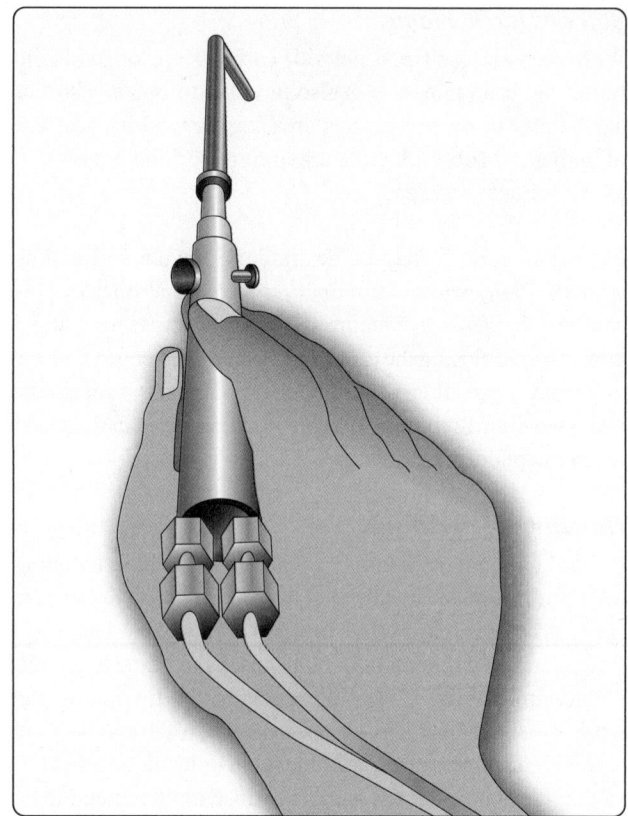

Figure 5-24. *Hand position for light-gauge materials.*

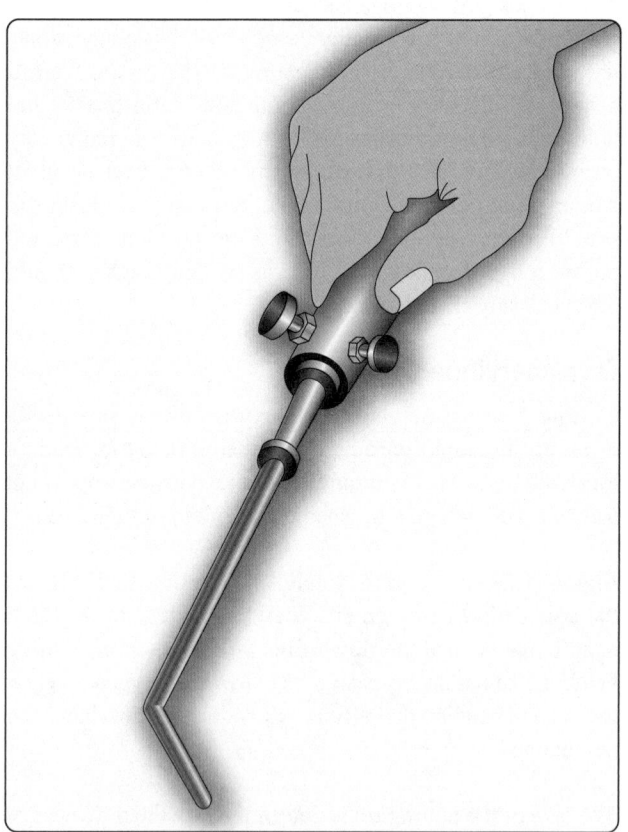

Figure 5-25. *Hand position for heavy-gauge materials.*

forehand welding, and is the most commonly used method for lighter tubing and sheet metal. The filler rod is kept ahead of the tip in the direction the weld is going and is added to the puddle.

For welding thick metals or heavy plate, a technique called backhand welding can be used. In this method, the torch flame is pointed back toward the finished weld and the filler rod is added between the flame and the weld. This method provides a greater concentration of heat for welding thicker metals and would rarely be used in aircraft maintenance.

Puddle

If the torch is held in the correct position, a small puddle of molten metal forms. The puddle should be centered in the joint and composed of equal parts of those pieces being welded. After the puddle appears, the tip should be moved in a semicircular arc or circular motion equally between the pieces to ensure an even distribution of heat.

Adding Filler Rod to the Puddle

As the metal melts and the puddle forms, filler rod is needed to replace the metal that flows out from around the joint. The rod is added to the puddle in the amount that provides for the completed fillet to be built up about one-fourth the thickness of the base metal. The filler rod selected should be compatible with the base metal being welded.

Correct Forming of a Weld

The form of the weld metal has considerable bearing upon the strength and fatigue resistance of a joint. The strength of an improperly made weld is usually less than the strength for which the joint was designed. Low-strength welds are generally the result of insufficient penetration; undercutting of the base metal at the toe of the weld; poor fusion of the weld metal with the base metal; trapped oxides, slag, or gas pockets in the weld; overlap of the weld metal on the base metal; too much or too little reinforcement; or overheating of the weld.

Characteristics of a Good Weld

A completed weld should have the following characteristics:

1. The seam should be smooth, the bead ripples evenly spaced, and of a uniform thickness.

2. The weld should be built up, slightly convex, thus providing extra thickness at the joint.

3. The weld should taper off smoothly into the base metal.

4. No oxide should be formed on the base metal close to the weld.

5. The weld should show no signs of blowholes, porosity, or projecting globules.

6. The base metal should show no signs of burns, pits, cracks, or distortion.

Although a clean, smooth weld is desirable, this characteristic does not necessarily mean that the weld is a good one; it may be dangerously weak inside. However, when a weld is rough, uneven, and pitted, it is almost always unsatisfactory inside. Welds should never be filed to give them a better appearance, since filing deprives the weld of part of its strength. Welds should never be filled with solder, brazing material, or filler of any sort.

When it is necessary to reweld a joint, all old weld material must be removed before the operation is begun. It must be remembered that reheating the area may cause the base metal to lose some of its strength and become brittle. This should not be confused with a post-weld heat treatment that does not raise the metal to a high enough temperature to cause harm to the base material.

Oxy-Acetylene Welding of Ferrous Metals

Steel (Including SAE 4130)

Low-carbon steel, low-alloy steel (e.g., 4130), cast steel, and wrought iron are easily welded with the oxy-acetylene flame. Low-carbon and low-alloy steels are the ferrous materials that are gas welded most frequently. As the carbon content of steel increases, it may be repaired by welding using specific procedures for various alloy types. Factors involved are the carbon content and hardenability. For corrosion-resistant and heat-resistant nickel chromium steels, the allowed weldability depends upon their stability, carbon content, and reheat treatment.

The Society of Automotive Engineers (SAE) and the American Iron and Steel Institute (AISI) provide a designation system that is an accepted standard for the industry. SAE 4130 is an alloy steel that is an ideal material for constructing fuselages and framework on small aircraft; it is also used for motorcycle and high-end bicycle frames and race car frames and roll cages. The tubing has high tensile strength, malleability, and is easy to weld.

The number '4130' is also an AISI 4-digit code that defines the approximate chemical composition of the steel. The '41' indicates a low-alloy steel containing chromium and molybdenum (chromoly) and the '30' designates a carbon content of 0.3 percent. 4130 steel also contains small amounts of manganese, phosphorus, sulfur, and silicon, but like all steels, it contains mostly iron.

In order to make a good weld, the carbon content of the

steel must not be altered to any appreciable degree, nor can other atmospheric chemical constituents be added to or subtracted from the base metal without seriously altering the properties of the metal. However, many welding filler wires do contain constituents different from the base material for specific reasons, which is perfectly normal and acceptable if approved materials are used. Molten steel has a great affinity for carbon, oxygen, and nitrogen combining with the molten puddle to form oxides and nitrates, both of which lower the strength of steel. When welding with an oxy-acetylene flame, the inclusion of impurities can be minimized by observing the following precautions:

- Maintain an exact neutral flame for most steels and a slight excess of acetylene when welding alloys with a high nickel or chromium content, such as stainless steel.

- Maintain a soft flame and control the puddle.

- Maintain a flame sufficient to penetrate the metal and manipulate it so that the molten metal is protected from the air by the outer envelope of flame.

- Keep the hot end of the welding rod in the weld pool or within the flame envelope.

- When the weld is complete and still in the red heat, circle the outer envelope of the torch around the entire weldment to bring it evenly to a dull red. Slowly back the torch away from the weldment to ensure a slow cooling rate.

Chrome Molybdenum

The welding technique for chrome molybdenum (chrome-moly) is practically the same as that for carbon steels, except for sections over ¾6-inch thick. The surrounding area must be preheated to a temperature between 300 °F and 400 °F before beginning to weld. If this is not done, the sudden quenching of the weld area after the weld is complete may cause a brittle grain structure of untempered martensite that must be eliminated with post-weld heat treatments. Untempered martensite is a glass-like structure that takes the place of the normally ductile steel structure and makes the steel prone to cracking, usually near the edge of the weld. This preheating also helps to alleviate some of the distortion caused by welding along with using proper practices found in other sections of this chapter.

A soft neutral flame should be used for welding and must be maintained during the process. If the flame is not kept neutral, an oxidizing flame may cause oxide inclusions and fissures. A carburizing flame makes the metal more hardenable by raising the carbon content. The volume of the flame must be sufficient to melt the base metal, but not hot enough to overheat the base metal and cause oxide inclusions or a loss

of metal thickness. The filler rod should be compatible with the base metal. If the weld requires high strength, special low-alloy filler is used, and the piece is heat treated after welding.

It may be advantageous to TIG weld 4130 chrome-moly sections over 0.093-inch thickness followed by a proper post-weld heat treatment as this can result in less overall distortion. However, do not eliminate the post-weld heat treatment as doing so could severely limit the fatigue life of the weldment due to the formed martensitic grain structure.

Stainless Steel

The procedure for welding stainless steel is basically the same as that for carbon steels. There are, however, some special precautions you must take to obtain the best results.

Only stainless steel used for nonstructural members of aircraft can be welded satisfactorily. The stainless steel used for structural components is cold worked or cold rolled and, if heated, loses some of its strength. Nonstructural stainless steel is obtained in sheet and tubing form and is often used for exhaust collectors, stacks, or manifolds. Oxygen combines very readily with this metal in the molten state, and you must take extreme care to prevent this from occurring.

A slightly carburizing flame is recommended for welding stainless steel. The flame should be adjusted so that a feather of excess acetylene, about ¼6-inch long, forms around the inner cone. Too much acetylene, however, adds carbon to the metal and causes it to lose its resistance to corrosion. The torch tip size should be one or two sizes smaller than that prescribed for a similar gauge of low-carbon steel. The smaller tip lessens the chances of overheating and subsequent loss of the corrosion-resistant qualities of the metal.

To prevent the formation of chromium oxide, a specially compounded flux for stainless steel, should be used. The flux, when mixed with water, can be spread on the underside of the joint and on the filler rod. Since oxidation must be avoided as much as possible, use sufficient flux. The filler rod used should be of the same composition as the base metal.

When welding, hold the filler rod within the envelope of the torch flame so that the rod is melted in place or melted at the same time as the base metal. Add the filler rod by allowing it to flow into the molten pool. Do not stir the weld pool, because air enters the weld and increases oxidation. Avoid rewelding any portion or welding on the reverse side of the weld, which results in warping and overheating of the metal.

Another method used to keep oxygen from reaching the metal is to surround the weld with a blanket of inert gas. This is done by using a TIG welder to perform welding of stainless steel. It is a recommended method for excellent weld results and does not require the application of flux and its subsequent cleanup.

Oxy-Acetylene Welding of Nonferrous Metals

Nonferrous metals are those that contain no iron. Examples of nonferrous metals are lead, copper, silver, magnesium, and the most important in aircraft construction, aluminum. Some of these metals are lighter than the ferrous metals, but in most cases, they are not as strong. Aluminum manufacturers have compensated for the lack of strength of pure aluminum by alloying it with other metals or by cold working it. For still greater strength, some aluminum alloys are also heat treated.

Aluminum Welding

Gas welding of certain aluminum alloys can be accomplished successfully, but it requires some practice and the appropriate equipment to produce a successful weld. Before attempting to weld aluminum for the first time, become familiar with how the metal reacts under the welding flame.

A good example for practice and to see how aluminum reacts to a welding flame, heat a piece of aluminum sheet on a welding bench. Hold a torch with a neutral flame perpendicular to the sheet and bring the tip of the inner cone almost in contact with the metal. Observe that the metal suddenly melts away, almost without any indication, and leaves a hole in the metal. Now repeat the operation, only this time hold the torch at an angle of about 30° to the surface. This allows for better control of the heat and allows the surface metal to melt without forming a hole. Practice by slowly moving the flame along the surface until the puddle can be controlled without melting holes. Once that is mastered, practice on flanged joints by tacking and welding without filler rod. Then, try welding a butt joint using flux and filler rod. Practice and experience provides the visual indication of the melting aluminum so that a satisfactory weld can be performed.

Aluminum gas welding is usually confined to material between 0.031-inch and 0.125-inch in thickness. The weldable aluminum alloys used in aircraft construction are 1100, 3003, 4043, and 5052. Alloy numbers 6053, 6061, and 6151 can also be welded, but since these alloys are in the heat-treated condition, welding should not be done unless the parts can be reheat treated.

Proper preparation prior to welding any metal is essential to produce a satisfactory weld. This preparation is especially critical during oxy-acetylene welding of aluminum. Select the proper torch tip for the thickness of metal being welded. Tip selection for aluminum is always one size larger than one would normally choose for the same thickness in a steel sheet. A rule of thumb: ¾ metal thickness = tip orifice.

Set the proper regulator pressure using the following method for oxy-acetylene welding of aluminum. This method has been used by all aircraft factories since World War II. Start by slowly opening the valve on the oxygen cylinder all the way until it stops to seat the upper packing. Now, barely crack open the acetylene cylinder valve until the needle on the gauge jumps up, then open one-quarter turn more. Check the regulators to ensure the adjusting screws are turned counterclockwise all the way out and loose. Now, open both torch valves wide open, about two full turns (varies with the torch model). Turn the acetylene regulator by adjusting the screw until the torch blows a light puff at a two-inch distance. Now, hold the torch away from the body and light it with the striker, adjusting the flame to a bright yellow bushy flame with the regulator screw. Add oxygen by slowly turning in the oxygen regulator screw to get a loud blue flame with a bright inner cone, perhaps a bit of the "fuel-rich" feather or carburizing secondary cone. By alternately turning in each of the torch valves a little bit, the flame setting can be lowered to what is needed to either tack or weld.

Special safety eyewear must also be used to protect the welder and provide a clear view through the yellow-orange flare given off by the incandescing flux. Special purpose green-glass lens have been designed and patented especially for aluminum oxy-fuel welding by TM Technologies. These lenses cut the sodium orange flare completely and provide the necessary protection from ultraviolet, infrared, blue light, and impact. They meet safety standard ANSI Z87-1989 for a special-purpose lens.

Apply flux either to the material, the filler, or both if needed. The aluminum welding flux is a white powder mixed one part powder to two parts clean spring or mineral water. (Do not use distilled water.) Mix a paste that can be brushed on the metal. Heating the filler or the part with the torch before applying the flux helps the flux dry quickly and not pop off when the torch heat approaches. Proper safety precautions, such as eye protection, adequate ventilation, and avoiding the fumes are recommended.

The material to be welded must be free of oil or grease. It should be cleaned with a solvent; the best being denatured isopropyl (rubbing) alcohol. A stainless toothbrush should be used to scrub off the invisible aluminum oxide film just prior to welding but after cleaning with alcohol. Always clean the filler rod or filler wire prior to use with alcohol and a clean cloth.

Make the best possible fit-up for joints to avoid large gaps and select the appropriate filler metal that is compatible with the base metal. The filler should not be a larger diameter than the pieces to be welded. *[Figure 5-26]*

Begin by tacking the pieces. The tacks should be applied 1–1½-inches apart. Tacks are done hot and fast by melting the edges of the metal together, if they are touching, or by adding filler to the melting edges when there is a gap. Tacking requires a hotter flame than welding. So, if the thickness of the metal being welded is known, set the length of the inner cone of the flame roughly three to four metal thicknesses in length for tacking. (Example: .063 aluminum sheet = ³⁄₁₆–¼ inch inner cone.)

Once the edges are tacked, begin welding by either starting at the second tack and continuing on, or starting the weld one inch in from the end and then welding back to the edge of the sheet. Allow this initial skip-weld to chill and solidify. Then, begin to weld from the previous starting point and continue all the way to the end. Decrease the heat at the end of the seam to allow the accumulated heat to dissipate. The last inch or so is tricky and must be dabbed to prevent blow-through. (Dabbing is the adding of filler metal in the molten pool while controlling the heat on the metal by raising and lowering the torch.)

Weld bead appearance, or making ringlets, is caused by the movement of the torch and dabbing the filler metal. If the torch and add filler metal is moved at the same time, the ringlet is more pronounced. A good weld has a bead that is not too proud and has penetration that is complete.

Immediately after welding, the flux must be cleaned by using hot (180 °F) water and the stainless steel brush, followed by liberal rinsing with fresh water. If only the filler was fluxed, the amount of cleanup is minimal. All flux residues must be removed from voids and pinholes. If any particular area is suspect to hidden flux, pass a neutral flame over it and a yellow-orange incandescence will betray hiding residues.

Proper scrubbing with an etching solution and waiting no longer than 20 minutes to prime and seal avoids the lifting, peeling, or blistering of the finished topcoat.

Magnesium Welding

Gas welding of magnesium is very similar to welding aluminum using the same equipment. Joint design also follows similar practice to aluminum welding. Care must be taken to avoid designs that may trap flux after the welding is completed, with butt and edge welds being preferred. Of special interest is the high expansion rate of magnesium-based alloys, and the special attention that must be given to avoid stresses being set up in the parts. Rigid fixtures should

Filler Metal Selection Chart					
Base Metals	1100 3003	5005	5052	5086 DO NOT GAS WELD	6061
6061	4043 (a) 4047	4043 (a) 5183 5356 5556 5554 (d) 5654 (c)	5356 5183 5554 5556 5654 (d) 4043 (a)	5356 5183 5356 5556 5654 (c)	4043 (a) 4047 5556 5183 5554 (d)
5086 DO NOT GAS WELD	5356 4043 (a)	5356 5183 5556	5356 5183 5556	5356 5183 5556	
5052	5183 5356 5556 4043 (a,b)	4043 (a) 5183 5356 5556 4047	5654 (c) 5183 5356 5556 5554 (d) 4043 (a)		
5005	5183 5356 5556 4043 (a,b)	5183 5356 4043 (a,b)			
1100 3003	1110 4043 (a)				
For explanation of (a. b. c. d) see below					

Copyright © 1997 TM Technologies

(a) 4043, because of its silicon (Si) content, is less susceptible to hot cracking but has less ductility and may crack when planished.
(b) For applications at sustained temperatures above ISOF because of intergranular corrosion.
(c) Low temperature service at ISOF and below.
(d) 5554 is suitable for elevated temperatures.

Note: *When choosing between 5356, 5183, 5556, be aware that 5356 is the weakest and 5556 is the strongest, with 5183 in between. Also, 4047 has more Si than 4043, therefore less sensitivity to hot cracking, slightly higher weld shear strength, and less ductility.*

Figure 5-26. *Filler metal selection chart.*

be avoided; use careful planning to eliminate distortion.

In most cases, filler material should match the base material in alloy. When welding two different magnesium alloys together, the material manufacturer should be consulted for recommendations. Aluminum should never be welded to magnesium. As in aluminum welding, a flux is required to break down the surface oxides and ensure a sound weld. Fluxes sold specifically for the purpose of fusion welding magnesium are available in powder form and are mixed with water in the same manner as for aluminum welding. Use the minimum amount of flux necessary to reduce the corrosive effects and cleaning time required after the weld is finished. The sodium-flare reducing eye protection used for aluminum welding is of the same benefit on magnesium welding.

Welding is done with a neutral flame setting using the same tip size for aluminum welding. The welding technique follows the same pattern as aluminum with the welding being completed in a single pass on sheet gauge material. Generally, the TIG process has replaced gas welding of magnesium due to the elimination of the corrosive flux and its inherent limitations on joint design.

Brazing & Soldering

Torch Brazing of Steel

The definition of joining two pieces of metal by brazing typically meant using brass or bronze as the filler metal. However, that definition has been expanded to include any metal joining process in which the bonding material is a nonferrous metal or alloy with a melting point higher than 800 °F, but lower than that of the metals being joined.

Brazing requires less heat than welding and can be used to join metals that may be damaged by high heat. However, because the strength of a brazed joint is not as great as that of a welded joint, brazing is not used for critical structural repairs on aircraft. Also, any metal part that is subjected to a sustained high temperature should not be brazed.

Brazing is applicable for joining a variety of metals, including brass, copper, bronze and nickel alloys, cast iron, malleable iron, wrought iron, galvanized iron and steel, carbon steel, and alloy steels. Brazing can also be used to join dissimilar metals, such as copper to steel or steel to cast iron.

When metals are joined by brazing, the base metal parts are not melted. The brazing metal adheres to the base metal by molecular attraction and intergranular penetration; it does not fuse and amalgamate with them.

In brazing, the edges of the pieces to be joined are usually beveled as in welding steel. The surrounding surfaces must be cleaned of dirt and rust. Parts to be brazed must be securely fastened together to prevent any relative movement. The strongest brazed joint is one in which the molten filler metal is drawn in by capillary action, requiring a close fit.

A brazing flux is necessary to obtain a good union between the base metal and the filler metal. It destroys the oxides and floats them to the surface, leaving a clean metal surface free from oxidation. A brazing rod can be purchased with a flux coating already applied, or any one of the numerous fluxes available on the market for specific application may be used. Most fluxes contain a mixture of borax and boric acid.

The base metal should be preheated slowly with a neutral soft flame until it reaches the flowing temperature of the filler metal. If a filler rod that is not precoated with flux is used,

heat about 2 inches of the rod end with the torch to a dark purple color and dip it into the flux. Enough flux adheres to the rod that it is unnecessary to spread it over the surface of the metal. Apply the flux-coated rod to the red-hot metal with a brushing motion, using the side of the rod; the brass flows freely into the steel. Keep the torch heat on the base metal to melt the filler rod. Do not melt the rod with the torch. Continue to add the rod as the brazing progresses, with a rhythmic dipping action so that the bead is built to a uniform width and height. The job should be completed rapidly and with the fewest possible passes of the rod and torch.

Notice that some metals are good conductors of heat and dissipate the heat more rapidly away from the joint. Other metals are poor conductors that tend to retain the heat and overheat readily. Controlling the temperature of the base metal is extremely important. The base metal must be hot enough for the brazing filler to flow, but never overheated to the filler boiling point. This causes the joint to be porous and brittle.

The key to even heating of the joint area is to watch the appearance of the flux. The flux should change appearance uniformly when even heat is being applied. This is especially important when joining two metals of different mass or conductivity.

The brazing rod melts when applied to the red-hot base metal and runs into the joint by capillary attraction. (Note that molten brazing filler metal tends to flow toward the area of higher temperature.) In a torch heated assembly, the outer metal surfaces are slightly hotter than the interior joint surfaces. The filler metal should be deposited directly adjacent to the joint. Where possible, the heat should be applied to the assembly on the side opposite to where the filler is applied because the filler metal tends to flow toward the source of greater heat.

After the brazing is complete, the assembly or component must be cleaned. Since most brazing fluxes are water soluble, a hot water rinse (120 °F or hotter) and a wire brush remove the flux. If the flux was overheated during the brazing process, it usually turns green or black. In this case, the flux needs to be removed with a mild acid solution recommended by the manufacturer of the flux in use.

Torch Brazing of Aluminum

Torch brazing of aluminum is done using similar methods as brazing of other materials. The brazing material itself is an aluminum/silicon alloy having a slightly lower melting temperature than the base material. Aluminum brazing occurs at temperatures over 875 °F, but below the melting point of the parent metal. This is performed with a specific aluminum brazing flux. Brazing is best suited to joint configurations that have large surface areas in contact, such as the lap, or for fitting fuel tank bungs and fittings. Either acetylene or hydrogen may be used as fuel gas, both being used for production work for many years. Using eye protection that reduces the sodium flare, such as the TM2000 lens, is recommended.

When using acetylene, the tip size is usually the same, or one size smaller than that used for welding of aluminum. A 1–2X reducing flame is used to form a slightly cooler flame, and the torch is held back at a greater distance using the outer envelope as the heat source rather than the inner cone. Prepare the flux and apply in the same manner as the aluminum welding flux, fluxing both the base metal and filler material. Heat the parts with the outer envelope of the flame, watching for the flux to begin to liquefy; the filler may be applied at that point. The filler should flow easily. If the part gets overheated, the flux turns brown or grey. If this happens, reclean and re-flux the part before continuing. Brazing is more easily accomplished on 1100, 3003, and 6061 aluminum alloys. 5052 alloy is more difficult; proper cleaning and practice are vital. There are brazing products sold that have the flux contained in hollow spaces in the filler metal itself, which typically work only on 1100, 3003, and 6061 alloys as the flux is not strong enough for use on 5052. Cleaning after brazing is accomplished the same as with oxy-fuel welding of aluminum, using hot water and a clean stainless brush. The flux is corrosive, so every effort should be made to remove it thoroughly and quickly after the brazing is completed.

Soldering

Soldering is a method of thermally joining metal parts with a molten nonferrous alloy that melts at a temperature below 800 °F. The molten alloy is pulled up between close-fitting parts by capillary action. When the alloy cools and hardens, it forms a strong, leak-proof connection.

Soft solder is chiefly used to join copper and brass where a leak proof joint is desired, and sometimes for fitting joints to promote rigidity and prevent corrosion. Soft soldering is generally performed only in minor repair jobs. Soft solder is also used to join electrical connections. It forms a strong union with low electrical resistance.

Soft soldering does not require the heat of an oxy-fuel gas torch and can be performed using a small propane or MAPP® torch, an electrical soldering iron, or in some cases, a soldering copper, that is heated by an outside source, such as an oven or torch. The soft solders are chiefly alloys of tin and lead. The percentages of tin and lead vary considerably in the various solders with a corresponding change in their melting points ranging from 293 °F to 592 °F. Half-and-half (50/50) is the most common general-purpose solder. It contains equal portions of tin and lead and melts at approximately 360 °F.

To get the best results for heat transfer when using an electrical soldering iron or a soldering copper, the tip must be clean and have a layer of solder on it. This is usually referred to as being tinned. The hot iron or copper should be fluxed and the solder wiped across the tip to form a bright, thin layer of solder.

Flux is used with soft solder for the same reasons as with brazing. It cleans the surface area to be joined and promotes the flow by capillary action into the joint. Most fluxes should be cleaned away after the job is completed because they cause corrosion. Electrical connections should be soldered only with soft solder containing rosin. Rosin does not corrode the electrical connection.

Aluminum Soldering

The soldering of aluminum is much like the soldering of other metals. The use of special aluminum solders is required, along with the necessary flux. Aluminum soldering occurs at temperatures below 875 °F. Soldering can be accomplished using the oxy-acetylene, oxy-hydrogen, or even an air-propane torch setup. A neutral flame is used in the case of either oxy-acetylene or oxy-hydrogen. Depending on the solder and flux type, most common aluminum alloys can be soldered. Being of lower melting temperature, a tip one or two sizes smaller than required for welding is used, along with a soft flame setting.

Joint configurations for aluminum soldering follow the same guidelines as any other base material. Lap joints are preferred to tee or butt joints due to the larger surface contact area. However parts, such as heat exchanger tubes, are a common exception to this.

Normally, the parts are cleaned as for welding or brazing, and the flux is applied according to manufacturer's instructions. The parts are evenly heated with the outer envelope of the flame to avoid overheating the flux, and the solder is applied in a fashion similar to that for other base metals. Cleaning after soldering may not be required to prevent oxidation because some fluxes are not corrosive. However, it is always advisable to remove all flux residues after soldering.

Aluminum soldering is commonly used in such applications as the repair of heat exchanger or radiator cores originally using a soldered joint. It is not, however, to be used as a direct replacement repair for brazing or welding.

Silver Soldering

The principle use of silver solder in aircraft work is in the fabrication of high-pressure oxygen lines and other parts that must withstand vibration and high temperatures.

Silver solder is used extensively to join copper and its alloys, nickel and silver, as well as various combinations of these metals and thin steel parts. Silver soldering produces joints of higher strength than those produced by other brazing processes.

Flux must be used in all silver soldering operations to ensure the base metal is chemically clean. The flux removes the film of oxide from the base metal and allows the silver solder to adhere to it.

All silver solder joints must be physically, as well as chemically, clean. The joint must be free of dirt, grease, oil, and/or paint. After removing the dirt, grease, etc., any oxide (rust and/or corrosion) should be removed by grinding or filing the piece until bright metal can be seen. During the soldering operation, the flux continues to keep the oxide away from the metal and aid in the flow of the solder.

The three recommended types of joint for silver soldering are lap, flanged, and edge. With these, the metal is formed to furnish a seam wider than the base metal thickness and provide the type of joint that holds up under all types of loads. *[Figure 5-27]*

The oxy-acetylene flame for silver soldering should be a soft neutral or slightly reducing flame. That is, a flame with a slight excess of acetylene. During both preheating and application of the solder, the tip of the inner cone of the flame should be held about ½-inch from the work. The flame should be kept moving so that the metal does not overheat.

When both parts of the base metal are at the correct temperature, the flux flows and solder can be applied directly adjacent to the edge of the seam. It is necessary to simultaneously direct the flame over the seam and keep it moving so that the base metal remains at an even temperature.

Gas Tungsten Arc Welding (TIG Welding)

The TIG process as it is known today is a combination of the work done by General Electric in the 1920s to develop the basic process, the work done by Northrop in the 1940s to develop the torch itself, and the use of helium-shielding gas and a tungsten electrode. The process was developed for welding magnesium in the Northrop XP-56 flying wing to eliminate the corrosion and porosity issues with the atomic hydrogen process they had been using with a boron flux. It was not readily used on other materials until the late 1950s when it found merit in welding space-age super alloys. It was also later used on other metals, such as aluminum and steel, to a much greater degree.

Modern TIG welding machines are offered in DC, AC, or with AC/DC configurations, and use either transformer or inverter-based technology. Typically, a machine capable of AC output is required for aluminum. The TIG torch itself has changed little since the first Northrop patent. TIG welding is similar to oxy-fuel welding in that the heat source (torch) is manipulated with one hand, and the filler, if used, is manipulated with the other. A distinct difference is to control the heat input to the metal. The heat control may be preset and fixed by a machine setting or variable by use of a foot pedal or torch-mounted control.

Several types of tungsten electrode are used with the TIG welder. Thoriated and zirconiated electrodes have better electron emission characteristics than pure tungsten, making them more suitable for DC operations on transformer-based machines, or either AC or DC with the newer inverter-based machines. Pure tungsten provides a better current balance with AC welding with a transformer based machine, which is advantageous when welding aluminum and magnesium. The equipment manufacturers' suggestions for tungsten type and form should be followed as this is an ever changing part of the TIG technology.

The shape of the electrode used in the TIG welding torch is an important factor in the quality and penetration of the weld. The tip of the electrode should be shaped on a dedicated grinding stone or a special-purpose tungsten grinder to avoid contaminating the electrode. The grinding should be done longitudinally, not radially, with the direction of stone travel away from the tip. *Figure 5-28* shows the effects of a sharp versus blunt electrode with transformer-based machines.

When in doubt, consult the machine manufacturer for the

Sharper Electrode	Blunter Electrode
Easy arc starting	Usually harder to start the arc
Handles less amperage	Handles more amperage
Wider arc shape	Narrower arc shape
Good arc stability	Potential for arc wander
Less weld penetration	Better weld penetration
Shorter electrode life	Longer electrode life

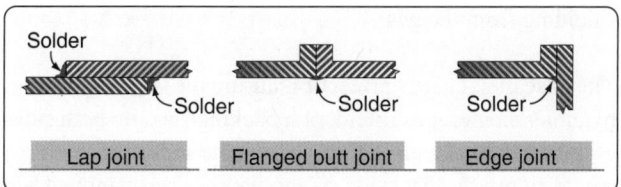

Figure 5-27. *Silver solder joints.*

Figure 5-28. *Effects of sharp and blunt electrodes.*

latest up-to-date suggestions on tungsten preparation or if problems arise.

The general guidelines for weld quality, joint fit prior to welding, jigging, and controlling warp all apply to this process in the same regard as any other welding method. Of particular note are the additional process steps that sometimes must be taken to perform a quality weld; these are dealt within their appropriate sections.

TIG Welding 4130 Steel Tubing

Welding 4130 with TIG is not much different than welding other steels as far as technique is concerned. The following information generally addresses material under 0.120-inch thick.

Clean the steel of any oil or grease and use a stainless steel wire brush to clean the work piece prior to welding. This is to prevent porosity and hydrogen embrittlement during the welding process. The TIG process is highly susceptible to these problems, much more so than oxy-acetylene welding, so care must be taken to ensure all oils and paint are removed from all surfaces of the parts to be welded.

Use a TIG welder with high-frequency starting to eliminate arc strikes. Do not weld where there is any breeze or draft; the welds should be allowed to cool slowly. Preheating is not necessary for tubing of less than 0.120-inch wall thickness; however, post-weld tempering (stress relieving) is still recommended to prevent the possible brittleness of the area surrounding the weld due to the untempered martensite formations caused by the rapid cooling of the weld inherent to the TIG process.

If you use 4130 filler rod, preheat the work before welding and heat treat afterward to avoid cracking. In a critical situation such as this, engineering should be done to determine preheat and post-weld heat treatment needed for the particular application.

Weld at a slower speed, make sufficiently large fillets, and make them flat or slightly convex, not concave. After the welding is complete, allow the weldment to cool to room temperature. Using an oxy-acetylene torch set to a neutral flame, heat the entire weldment evenly to 1,100 °F–1,200 °F; hold this temperature for about 45 minutes per inch of metal thickness. The temperature is generally accepted to be a dull red in ambient lighting. Note that for most tubing sections, the temperature needs to be held for only a minute or two. This process is found in most materials engineering handbooks written by the Materials Information Society (ASM) and other engineering sources. When working on a critical component, seek engineering help if there is any doubt.

TIG Welding Stainless Steel

Stainless steels, or more precisely, corrosion-resisting steels, are a family of iron-based metals that contain chromium in amounts ranging from 10 percent to about 30 percent. Nickel is added to some of the stainless steels, which reduces the thermal conductivity and decreases the electrical conductivity. The chromium-nickel steels belong in the AISI 300 series of stainless steels. They are nonmagnetic and have austenitic microstructure. These steels are used extensively in aircraft in which strength or resistance to corrosion at high temperature is required.

All of the austenitic stainless steels are weldable with most welding processes, with the exception of AISI 303, which contains high sulfur, and AISI 303Se, which contains selenium to improve its machinability.

The austenitic stainless steels are slightly more difficult to weld than mild-carbon steel. They have lower melting temperatures, and a lower coefficient of thermal conductivity, so welding current can be lower. This helps on thinner materials because these stainless steels have a higher coefficient of thermal expansion, requiring special precautions and procedures to be used to reduce warping and distortion. Any of the distortion-reducing techniques, such as skip welding or back-step welding, should be used. Fixtures and/or jigs should be used where possible. Tack welds should be applied twice as often as normal.

The selection of the filler metal alloy for welding the stainless steel is based on the composition of the base metal. Filler metal alloys for welding austenitic type stainless include AISI No. 309, 310, 316, 317, and 347. It is possible to weld several different stainless base metals with the same filler metal alloy. Follow the manufacturer's recommendations.

Clean the base metal just prior to welding to prevent the formation of oxides. Clean the surface and joint edges with a nonchlorinated solvent, and brush with a stainless steel wire brush to remove the oxides. Clean the filler material in the same manner.

To form a weld bead, move the torch along the joint at a steady speed using the forehand method. Dip the filler metal into the center of the weld puddle to ensure adequate shielding from the gas.

The base metal needs protection during the welding process by either an inert gas shield, or a backing flux, on both sides of the weld. Back purging uses a separate supply of shielding gas to purge the backside of the weld of any ambient air. Normally, this requires sealing off the tubular structures or using other various forms of shields and tapes to contain the shielding gas. A special flux may also be used on the inside of tubular structures in place of a back purge. This is especially advantageous with exhaust system repairs in which sealing off the entire system is time consuming. The flux is the same as is used for the oxy-acetylene welding process on stainless materials.

TIG Welding Aluminum

TIG welding of aluminum uses similar techniques and filler materials as oxy-fuel welding. Consult with the particular welding machine manufacturer for recommendations on tungsten type and size, as well as basic machine settings for a particular weldment because this varies with specific machine types. Typically, the machine is set to an AC output waveform because it causes a cleaning action that breaks up surface oxides. Argon or helium shielding gas may be used, but argon is preferred because it uses less by volume than helium. Argon is a heavier gas than helium, providing better cover, and it provides a better cleaning action when welding aluminum.

Filler metal selection is the same as used with the oxy-fuel process; however, the use of a flux is not needed as the shielding gas prevents the formation of aluminum oxide on the surface of the weld pool, and the AC waveform breaks up any oxides already on the material. Cleaning of the base metal and filler follows the same guidelines as for oxy-fuel welding. When welding tanks of any kind, it is a good practice to back-purge the inside of the tank with a shielding gas. This promotes a sound weld with a smooth inner bead profile that can help lessen pinhole leaks and future fatigue failures.

Welding is done with similar torch and filler metal angles as in oxy-fuel welding. The tip on the tungsten is held a short distance ($1/16$–$1/8$-inch) from the surface of the material, taking care not to ever let the molten pool contact the tungsten and contaminate it. Contamination of the tungsten must be dealt with by removal of the aluminum from the tungsten and re-grinding the tip to the factory recommended profile.

TIG Welding Magnesium

Magnesium alloys can be welded successfully using the same type joints and preparation that are used for steel or aluminum. However, because of its high thermal conductivity and coefficient of thermal expansion, which combine to cause severe stresses, distortion, and cracking, extra precautions must be taken. Parts must be clamped in a fixture or jig. Smaller welding beads, faster welding speed, and the use of a lower melting point and lower shrinkage filler rods are recommended.

DC, both straight or reverse polarity, and AC, with superimposed high frequency for arc stabilization, are commonly used for welding magnesium. DC reverse polarity provides better cleaning action of the metal and is preferred for manual welding operations.

AC power sources should be equipped with a primary contactor operated by a control switch on the torch or a foot control for starting or stopping the arc. Otherwise, the arcing that occurs while the electrode approaches or draws away from the work piece may result in burned spots on the work.

Argon is the most common used shielding gas for manual welding operations. Helium is the preferred gas for automated welding because it produces a more stable arc than argon and permits the use of slightly longer arc lengths. Zirconiated, thoriated, and pure tungsten electrodes are used for TIG welding magnesium alloys.

The welding technique for magnesium is similar to that used for other non-ferrous metals. The arc should be maintained at about $5/16$-inch. Tack welds should be used to maintain fit and prevent distortion. To prevent weld cracking, weld from the middle of a joint towards the end, and use starting and run off plates to start and end the weld. Minimize the number of stops during welding. After a stop, the weld should be restarted about $1/2$-inch from the end of the previous weld. When possible, make the weld in one uninterrupted pass.

TIG Welding Titanium

The techniques for welding titanium are similar to those required for nickel-based alloys and stainless steels. To produce a satisfactory weld, emphasis is placed on the surface cleanliness and the use of inert gas to shield the weld area. A clean environment is one of the requirements to weld titanium.

TIG welding of titanium is performed using DC straight polarity. A water-cooled torch, equipped with a $3/4$-inch ceramic cup and a gas lens, is recommended. The gas lens provides a uniform, nonturbulent inert gas flow. Thoriated tungsten electrodes are recommended for TIG welding of titanium. The smallest diameter electrode that can carry the required current should be used. A remote contactor controlled by the operator should be employed to allow the arc to be broken without removing the torch from the cooling weld metal, allowing the shielding gas to cover the weld until the temperature drops.

Most titanium welding is performed in an open fabrication shop. Chamber welding is still in use on a limited basis, but field welding is common. A separate area should be set aside and isolated from any dirt producing operations, such as grinding or painting. Additionally, the welding area should be free of air drafts and the humidity should be controlled.

Molten titanium weld metal must be totally shielded from contamination by air. Molten titanium reacts readily with oxygen, nitrogen, and hydrogen; exposure to these elements in air or in surface contaminants during welding can adversely affect titanium weld properties and cause weld embrittlement. Argon is preferred for manual welding because of better arc stability characteristics. Helium is used in automated welding and when heavier base metals or deeper penetration is required.

Care must be taken to ensure that the heat affected zones and the root side of the titanium welds are shielded until the weld metal temperature drops below 800 °F. This can be accomplished using shielding gas in three separate gas streams during welding.

1. The first shielding of the molten puddle and adjacent surfaces is provided by the flow of gas through the torch. Manufacturer recommendations should be followed for electrodes, tip grinding, cup size, and gas flow rates.

2. The secondary, or trailing, shield of gas protects the solidified weld metal and the heat affected zone until the temperature drops. Trailing shields are custom-made to fit a specific torch and a particular welding operation.

3. The third, or backup, flow is provided by a shielding device that can take many forms. On straight seam welds, it may be a grooved copper backing bar clamped behind the seam allowing the gas flow in the groove and serving as a heat sink. Irregular areas may be enclosed with aluminum tents taped to the backside of welds and purged with the inert gas.

Titanium weld joints are similar to those employed with other metals. Before welding, the weld joint surfaces must be cleaned and remain free of any contamination during the welding operation. Detergent cleaners and nonchlorinated cleaners, such as denatured isopropyl alcohol, may be used. The same requirements apply to the filler rod, it too must be cleaned and free of all contaminates. Welding gloves, especially the one holding the filler, must be contaminate free.

A good indication and measure of weld quality for titanium is the weld color. A bright silver weld indicates that the shielding is satisfactory and the heat affected zone and backup was properly purged until weld temperatures dropped. Straw-colored films indicate slight contamination, unlikely to affect mechanical properties; dark blue films or white powdery oxide on the weld would indicate a seriously deficient purge. A weld in that condition must be completely removed and rewelded.

Arc Welding Procedures, Techniques, & Welding Safety Equipment

Arc welding, also referred to as stick welding, has been performed successfully on almost all types of metals. This section addresses the procedures as they may apply to fusion welding of steel plate and provides the basic steps and procedures required to produce an acceptable arc weld. Additional instruction and information pertaining to arc welding of other metals can be obtained from training institutions and the various manufacturers of the welding equipment.

The first step in preparing to arc weld is to make certain that the necessary equipment is available and that the welding machine is properly connected and in good working order. Particular attention should be given to the ground connection, since a poor connection results in a fluctuating arc, that is difficult to control.

When using a shielded electrode, the bare end of the electrode should be clamped in its holder at a 90° angle to the jaws. (Some holders allow the electrode to be inserted at a 45° angle when needed for various welding positions.)

Before starting to weld, the following typical list of items should be checked:

•   Is the proper personal safety equipment being used, including a welding helmet, welding gloves, protective clothing, and footwear; if not, in an adequately ventilated area, appropriate breathing equipment?

•   Has the ground connection been properly made to the work piece and is it making a good connection?

•   Has the proper type and size electrode been selected for the job?

•   Is the electrode properly secured in the holder?

•   Does the polarity of the machine coincide with that of the electrode?

•   Is the machine in good working order and is it adjusted to provide the necessary current for the job?

The welding arc is established by touching the base metal plate with the electrode and immediately withdrawing it a short distance. At the instant the electrode touches the plate, a rush of current flows through the point of contact. As the electrode is withdrawn, an electric arc is formed, melting a spot on the plate and at the end of the electrode.

Correctly striking an arc takes practice. The main difficulty in confronting a beginner in striking the arc is sticking the electrode to the work. If the electrode is not withdrawn promptly upon contact with the metal, the high amperage

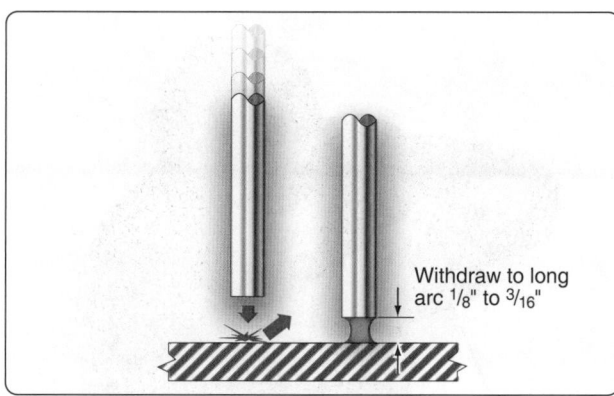

Figure 5-29. *Touch method of starting an arc.*

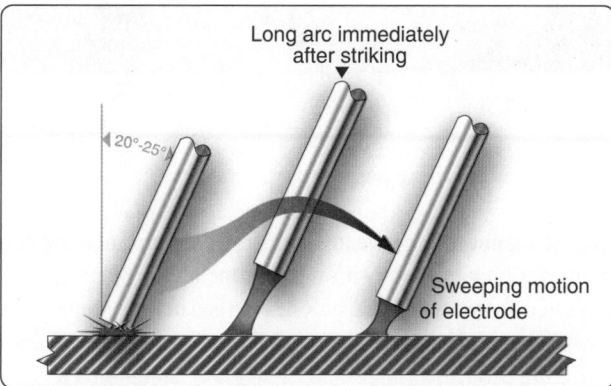

Figure 5-30. *Scratch/sweeping method of starting the arc.*

flows through the electrode causing it to stick or freeze to the plate and practically short circuits the welding machine. A quick roll of the wrist, either right or left, usually breaks the electrode loose from the work piece. If that does not work, quickly unclamp the holder from the electrode, and turn off the machine. A small chisel and hammer frees the electrode from the metal so it can be regripped in the holder. The welding machine can then be turned back on.

There are two essentially similar methods of striking the arc. One is the touch or tapping method. When using this method, the electrode should be held in a vertical position and lowered until it is an inch or so above the point where the arc is to be struck. Then, the electrode is lightly tapped on the work piece and immediately lifted to form an arc approximately ¼-inch in length. *[Figure 5-29]*

The second (and usually easier to master) is a scratch or sweeping method. To strike the arc by the scratch method, the electrode is held just above the plate at an angle of 20°–25°. The arc should be struck by sweeping the electrode with a wrist motion and lightly scratching the plate. The electrode is then lifted immediately to form an arc. *[Figure 5-30]*

Either method takes some practice, but with time and

experience, it becomes easy. The key is to raise the electrode quickly, but only about ¼-inch from the base or the arc is lost. If it is raised too slowly, the electrode sticks to the plate.

To form a uniform bead, the electrode must be moved along the plate at a constant speed in addition to the downward feed of the electrode. If the rate of advance is too slow, a wide overlapping bead forms with no fusion at the edges. If the rate is too fast, the bead is too narrow and has little or no fusion at the plate.

The proper length of the arc cannot be judged by looking at it. Instead, depend on the sound that the short arc makes. This is a sharp cracking sound, and it should be heard during the time the arc is being moved down to and along the surface of the plate.

A good weld bead on a flat plate should have the following characteristics:

*   Little or no splatter on the surface of the plate.

*   An arc crater in the bead of approximately ¹⁄₁₆-inch when the arc has been broken.

*   The bead should be built up slightly, without metal overlap at the top surface.

*   The bead should have a good penetration of approximately ¹⁄₁₆-inch into the base metal.

Figure 5-31 provides examples of operator's technique and welding machine settings.

When advancing the electrode, it should be held at an angle of about 20° to 25° in the direction of travel moving away from the finished bead. *[Figure 5-32]*

If the arc is broken during the welding of a bead and the electrode is removed quickly, a crater is formed at the point where the arc ends. This shows the depth of penetration or fusion that the weld is getting. The crater is formed by the pressure of the gases from the electrode tip forcing the weld metal toward the edges of the crater. If the electrode is removed slowly, the crater is filled.

If you need to restart an arc of an interrupted bead, start just ahead of the crater of the previous weld bead, as shown in position 1, *Figure 5-33*. Then, the electrode should be returned to the back edge of the crater (step 2). From this point, the weld may be continued by welding right through the crater and down the line of weld as originally planned (step 3).

Once a bead has been formed, every particle of slag must be removed from the area of the crater before restarting the arc. This is accomplished with a pick hammer and wire brush and prevents the slag from becoming trapped in the weld.

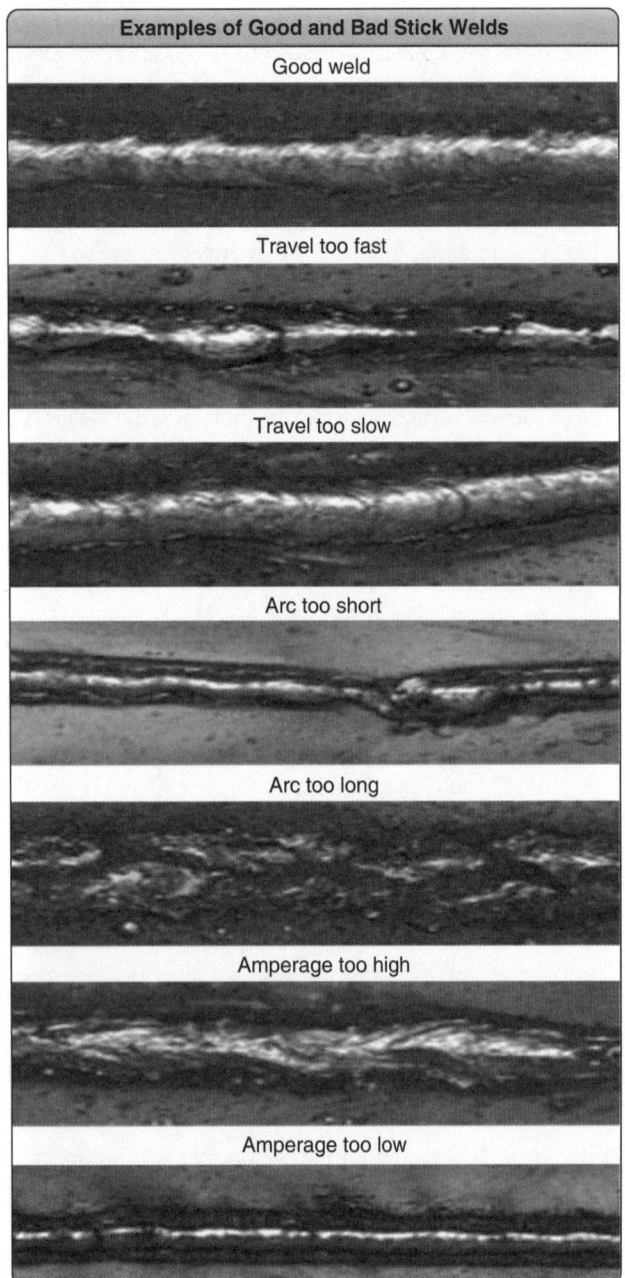

Examples of Good and Bad Stick Welds

Good weld

Travel too fast

Travel too slow

Arc too short

Arc too long

Amperage too high

Amperage too low

Figure 5-31. *Examples of good and bad stick welds.*

Multiple Pass Welding

Groove and fillet welds in heavy metals often require the deposit of a number of beads to complete a weld. It is important that the beads be deposited in a predetermined sequence to produce the soundest welds with the best proportions. The number of beads is determined by the thickness of the metal being welded.

Plates from ⅛-inch to ¼-inch can be welded in one pass, but they should be tacked at intervals to keep them aligned. Any weld on a plate thicker than ¼-inch should have the edges

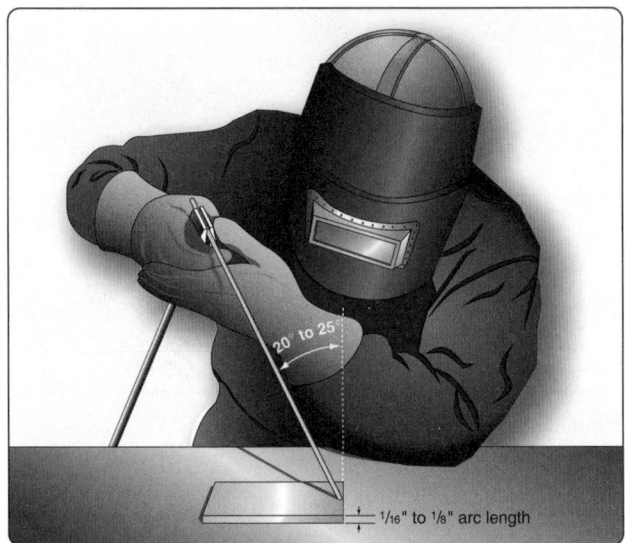

Figure 5-32. *Angle of electrode.*

beveled and multiple passes.

The sequence of the bead deposits is determined by the kind of joint and the position of the metal. All slag must be removed from each bead before another bead is deposited. Typical multiple-pass groove welding of butt joints is shown in *Figure 5-34.*

Techniques of Position Welding

Each time the position of a welded joint or the type of joint is changed, it may be necessary to change any one or a combination of the following:

- Current value
- Electrode
- Polarity
- Arc length
- Welding technique

Current values are determined by the electrode size, as well as the welding position. Electrode size is governed by the thickness of the metal and the joint preparation. The electrode type is determined by the welding position. Manufacturers

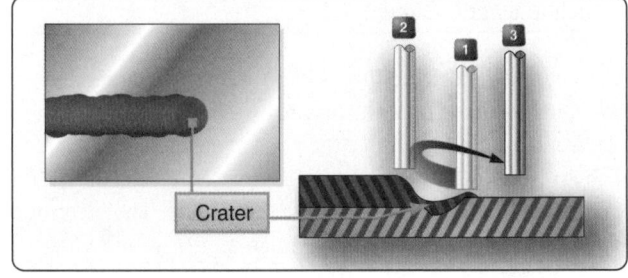

Figure 5-33. *Restarting the arc.*

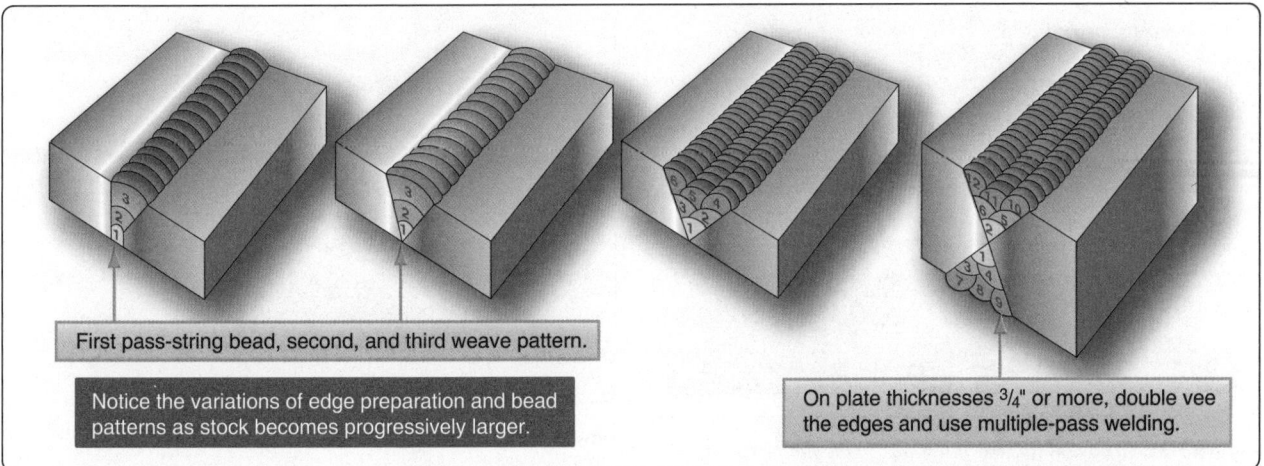

Figure 5-34. *Multiple-pass groove welding of butt joints.*

specify the polarity to be used with each electrode. Arc length is controlled by a combination of the electrode size, welding position, and welding current.

Since it is impractical to cite every possible variation occasioned by different welding conditions, only the information necessary for the commonly used positions and welds is discussed here.

Flat Position Welding

There are four types of welds commonly used in flat position welding: bead, groove, fillet, and lap joint. Each type is discussed separately in the following paragraphs.

Bead Weld

The bead weld utilizes the same technique that is used when depositing a bead on a flat metal surface. *[Figure 5-35]* The only difference is that the deposited bead is at the butt joint of two steel plates, fusing them together. Square butt joints may be welded in one or multiple passes. If the thickness of the metal is such that complete fusion cannot be obtained by welding from one side, the joint must be welded from both sides. Most joints should first be tack-welded to ensure alignment and reduce warping.

Groove Weld

Groove welding may be performed on a butt joint or an outside corner joint. Groove welds are made on butt joints where the metal to be welded is ¼-inch or more in thickness. The butt joint can be prepared using either a single or double groove depending on the thickness of the plate. The number of passes required to complete a weld is determined by the thickness of the metal being welded and the size of the electrode being used.

Any groove weld made in more than one pass must have

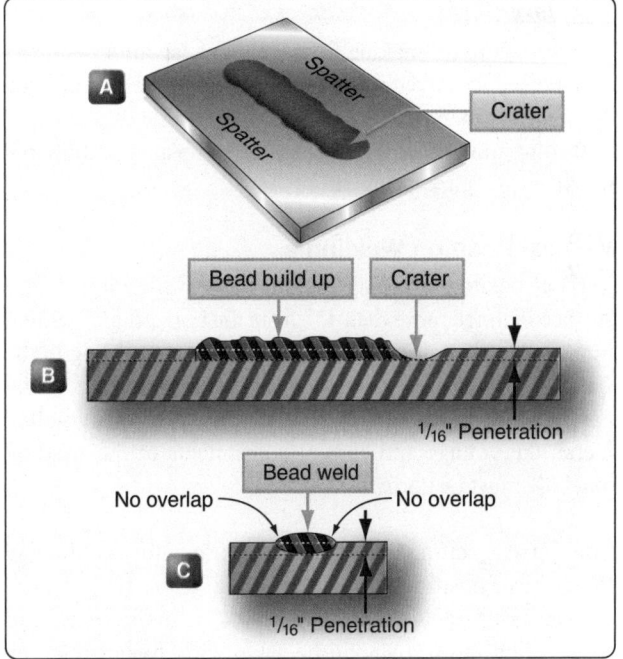

Figure 5-35. *Proper bead weld.*

the slag, spatter, and oxide carefully removed from all previous weld deposits before welding over them. Some of the common types of groove welds performed on butt joints in the flat position are shown in *Figure 5-36*.

Fillet Weld

Fillet welds are used to make tee and lap joints. The electrode should be held at an angle of 45° to the plate surface. The electrode should be tilted at an angle of about 15° in the direction of welding. Thin plates should be welded with little or no weaving motion of the electrode and the weld is made in one pass. Fillet welding of thicker plates may require two or more passes using a semicircular weaving motion of the electrode. *[Figure 5-37]*

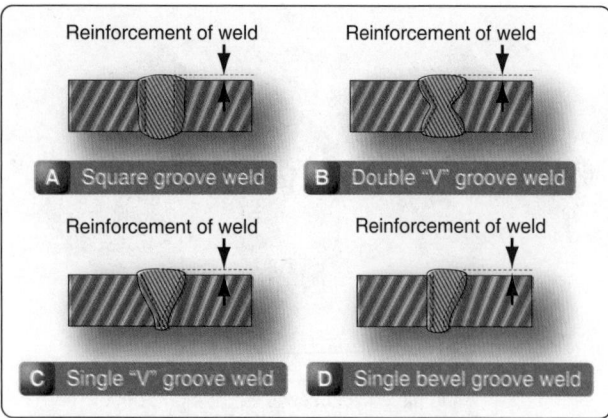

Figure 5-36. *Groove welds on butt joints in the flat position.*

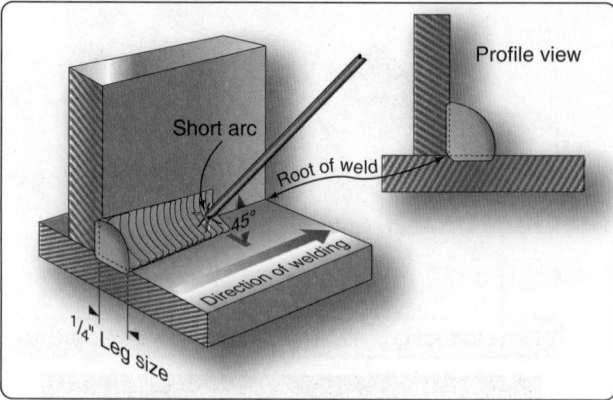

Figure 5-37. *Tee joint fillet weld.*

Lap Joint Weld

The procedure for making fillet weld in a lap joint is similar to that used in the tee joint. The electrode is held at about a 30° angle to the vertical and tilted to an angle of about 15° in the direction of welding when joining plates of the same thickness. *[Figure 5-38]*

Vertical Position Welding

Vertical positing welding includes any weld applied to a surface inclined more than 45° from the horizontal. Welding in the vertical position is more difficult than welding in the flat position because of the force of gravity. The molten metal has the tendency to run down. To control the flow of molten metal, the voltage and current adjustments of the welding machine must be correct.

The current setting, or amperage, is less for welding in the vertical position than for welding in the flat position for similar size electrodes. Additionally, the current used for welding upward should be set slightly higher than the current used for welding downward on the same work piece. When welding up, hold the electrode 90° to the vertical, and weld moving the bead upward. Focus on welding the sides of the joint and the middle takes care of itself. In welding downward, with the hand below the arc and the electrode tilted about 15° upward, the weld should move downward.

Overhead Position Welding

Overhead position welding is one of the most difficult in welding since a very short arc must be constantly maintained to control the molten metal. The force of gravity tends to cause the molten metal to drop down or sag from the plate, so it is important that protective clothing and head gear be worn at all times when performing overhead welding.

For bead welds in an overhead position, the electrode should

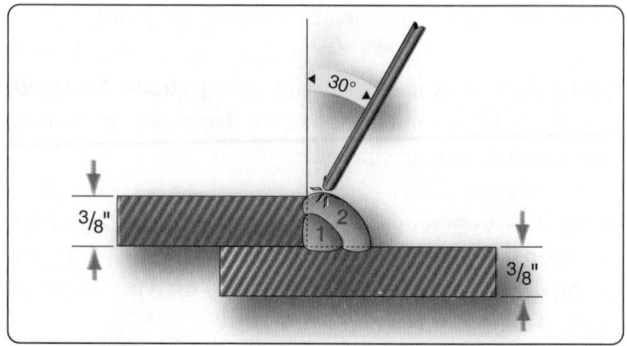

Figure 5-38. *Typical lap joint fillet weld.*

be held at an angle of 90° to the base metal. In some cases where it is desirable to observe the arc and the crater of the weld, the electrode may be held at an angle of 15° in the direction of welding.

When making fillet welds on overhead tee or lap joints, a short arc should be held, and there should be no weaving of the electrode. The arc motion should be controlled to secure good penetration to the root of the weld and good fusion to the plates. If the molten metal becomes too fluid and tends to sag, the electrode should be whipped away quickly from the center ahead of the weld to lengthen the arc and allow the metal to solidify. The electrode should then be returned immediately to the crater of the weld and the welding continued.

Anyone learning or engaged in arc welding should always have a good view of the weld puddle. Otherwise there is no way to ensure that the welding is in the joint and keeping the arc on the leading edge of the puddle. For the best view, the welder should keep their head off to the side and out of the fumes so they can see the puddle.

Expansion & Contraction of Metals

The expansion and contraction of metal is a factor taken into consideration during the design and manufacturing of all aircraft. It is equally important to recognize and allow for the dimensional changes and metal stress that may occur during any welding process.

Heat causes metals to expand; cooling causes them to contract. Therefore, uneven heating causes uneven expansion, and uneven cooling causes uneven contraction. Under such conditions, stresses are set up within the metal. These forces must be relieved, and unless precautions are taken, warping or buckling of the metal takes place. Likewise, on cooling, if nothing is done to take up the stress set up by the contraction forces, further warping may result; or if the metal is too heavy to permit this change in shape, the stresses remain within the metal itself.

The coefficient of linear expansion of a metal is the amount in inches that a one inch piece of metal expands when its temperature is raised 1 °F. The amount that a piece of metal expands when heat is applied is found by multiplying the coefficient of linear expansion by the temperature rise and multiplying that product by the length of the metal in inches.

Expansion and contraction have a tendency to buckle and warp thin sheet metal ⅛-inch or thinner. This is the result of having a large surface area that spreads heat rapidly and dissipates it soon after the source of heat is removed. The most effective method of alleviating this situation is to remove the heat from the metal near the weld, preventing it from spreading across the whole surface area. This can be done by placing heavy pieces of metal, known as chill bars, on either side of the weld; to absorb the heat and prevent it from spreading. Copper is most often used for chill bars because of its ability to absorb heat readily. Welding fixtures sometimes use this same principle to remove heat from the base metal. Expansion can also be controlled by tack welding at intervals along the joint.

The effect of welding a seam longer than 10 or 12 inches is to draw the seam together as the weld progresses. If the edges of the seam are placed in contact with each other throughout their length before welding starts, the far ends of the seam actually overlap before the weld is completed. This tendency can be overcome by setting the pieces to be welded with the seam spaced correctly at one end and increasing the space at the opposite end. *[Figure 5-39]*

The amount of space allowed depends on the type of material, the thickness of the material, the welding process being used, and the shape and size of the pieces to be welded. Instruction and/or welding experience dictates the space needed to produce a stress-free joint.

The weld is started at the correctly spaced end and proceeds toward the end that has the increased gap. As the seam is

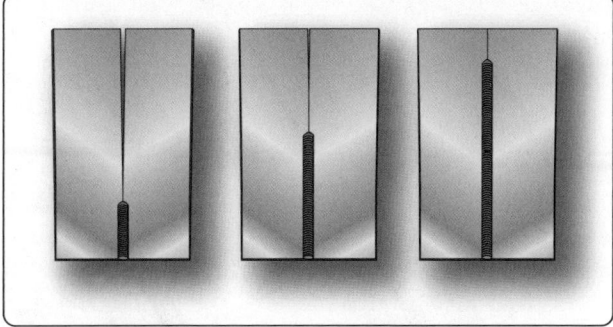

Figure 5-39. *Allowance for a straight butt weld when joining steel sheets.*

welded, the space closes and should provide the correct gap at the point of welding. Sheet metal under ¹⁄₁₆-inch can be handled by flanging the edges, tack welding at intervals, and then by welding between the tacks.

There are fewer tendencies for plate stock over ⅛-inch to warp and buckle when welded because the greater thickness limits the heat to a narrow area and dissipates it before it travels far on the plate.

Preheating the metal before welding is another method of controlling expansion and contraction. Preheating is especially important when welding tubular structures and castings. Great stress can be set up in tubular welds by contraction. When welding two members of a tee joint, one tube tends to draw up because of the uneven contraction. If the metal is preheated before the welding operation begins, contraction still takes place in the weld, but the accompanying contraction in the rest of the structure is at almost the same rate, and internal stress is reduced.

Welded Joints Using Oxy-Acetylene Torch

Figure 5-40 shows various types of basic joints.

Butt Joints

A butt joint is made by placing two pieces of material edge to edge, without overlap, and then welding. A plain butt joint is used for metals from ¹⁄₁₆-inch to ⅛-inch in thickness. A filler rod is used when making this joint to obtain a strong weld.

The flanged butt joint can be used in welding thin sheets, ¹⁄₁₆-inch or less. The edges are prepared for welding by turning up a flange equal to the thickness of the metal. This type of joint is usually made without the use of a filler rod.

If the metal is thicker than ⅛-inch, it may be necessary to bevel the edges so that the heat from the torch can completely penetrate the metal. These bevels may be either single or double-bevel type or single or double-V type. A filler rod is used to add strength and reinforcement to the weld.

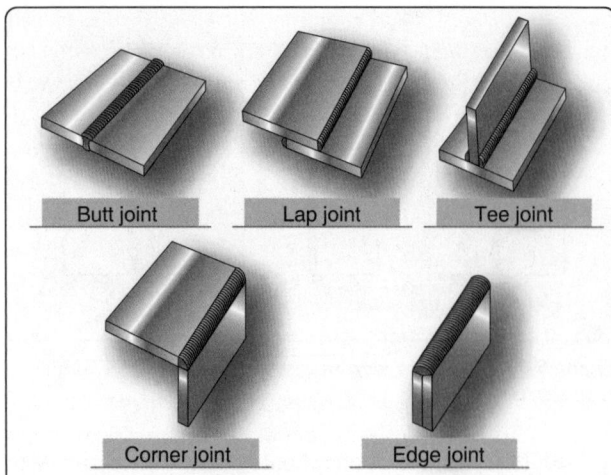

Figure 5-40. *Basic joints.*

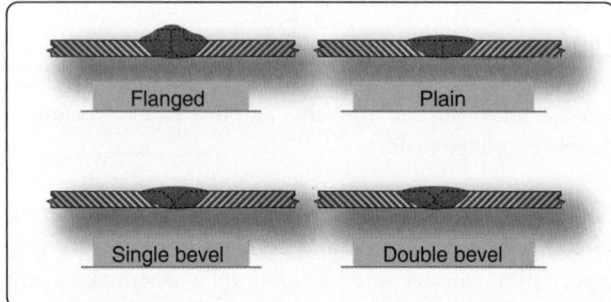

Figure 5-41. *Types of butt joints.*

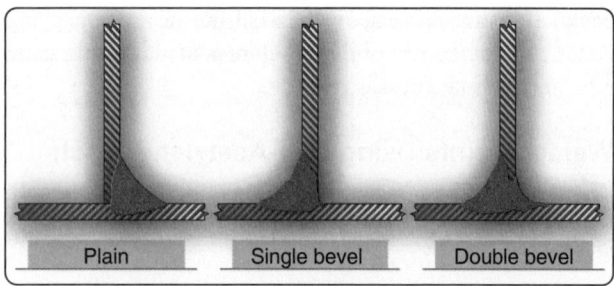

Figure 5-42. *Types of tee joints showing filler penetration.*

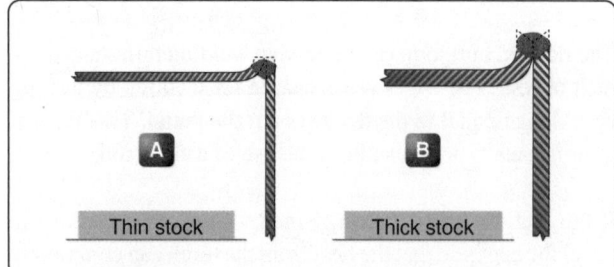

Figure 5-43. *Edge joints.*

[Figure 5-41]

Repair of cracks by welding may be considered just another type of butt joint. The crack should be stop drilled at either end and then welded like a plain butt joint using filler rod. In most cases, the welding of the crack does not constitute a complete repair and some form of reinforcement is still required, as described in following sections.

Tee Joints

A tee joint is formed when the edge or end of one piece is welded to the surface of another. *[Figure 5-42]* These joints are quite common in aircraft construction, particularly in tubular structures. The plain tee joint is suitable for most thicknesses of metal used in aircraft, but heavier thicknesses require the vertical member to be either single or double-beveled to permit the heat to penetrate deeply enough. The dark areas in *Figure 5-42* show the depth of heat penetration and fusion required. It is a good practice to leave a gap between the parts, about equal to the metal thickness to aid full penetration of the weld. This is common when welding from only one side with tubing clusters. Tight fitment of the parts prior to welding does not provide for a proper weldment unless full penetration is secured, and this is much more difficult with a gapless fitment.

Edge Joints

An edge joint is used when two pieces of sheet metal must be fastened together and load stresses are not important. Edge joints are usually made by bending the edges of one or both parts upward, placing the two ends parallel to each other, and welding along the outside of the seam formed by the two joined edges. The joint shown in *Figure 5-43A* requires no filler rod since the edges can be melted down to fill the seam. The joint shown in *Figure 5-43B,* being thicker material, must be beveled for heat penetration; filler rod is added for reinforcement.

Corner Joints

A corner joint is made when two pieces of metal are brought together so that their edges form a corner of a box or enclosure. *[Figure 5-44]* The corner joint shown in *Figure 5-44A* requires no filler rod, since the edges fuse to make the weld. It is used where the load stress is not important. The type shown in *Figure 5-44B* is used on heavier metals, and filler rod is added for roundness and strength. If a higher stress is to be placed on the corner, the inside is reinforced with another weld bead. *[Figure 5-44C]*

Lap Joints

The lap joint is seldom used in aircraft structures when welding with oxy-acetylene, but is commonly used and joined by spot welding. The single lap joint has very little resistance

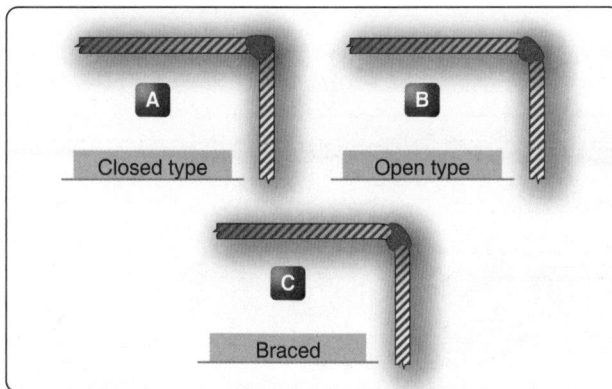

Figure 5-44. *Corner joints.*

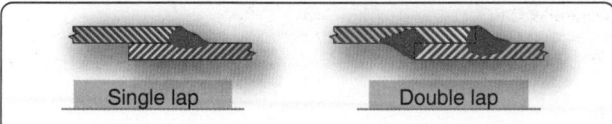

Figure 5-45. *Single and double lap joints.*

to bending, and cannot withstand the shearing stress to which the weld may be subjected under tension or compression loads. The double lap joint offers more strength, but requires twice the amount of welding required on the simpler, more efficient butt weld. *[Figure 5-45]*

Repair of Steel Tubing Aircraft Structure by Welding

Dents at a Cluster Weld

Dents at a cluster weld can be repaired by welding a formed steel patch plate over the dented area and surrounding tubes. Remove any existing finish on the damaged area and thoroughly clean prior to welding.

To prepare the patch plate, cut a section from a steel sheet of the same material and thickness as the heaviest tube damaged. Fashion the reinforcement plate so that the fingers extend over the tubes a minimum of 1½ times the respective tube diameter. The plate may be cut and formed prior to welding or cut and tack welded to the cluster, then heated and formed around the joint to produce a snug smooth contour. Apply sufficient heat to the plate while forming so there is a gap of no more than 1⁄16-inch from the contour of the joint to the plate.

In this operation, avoid unnecessary heating and exercise care to prevent damage at the point of the angle formed by any two adjacent fingers of the plate. After the plate is formed and tack welded to the joint, weld all the plate edges to the cluster joint. *[Figure 5-46]*

Dents Between Clusters

A damaged tubular section can be repaired using welded split sleeve reinforcement. The damaged member should be carefully straightened and should be stop drilled at the ends of any cracks with a No. 40 drill bit.

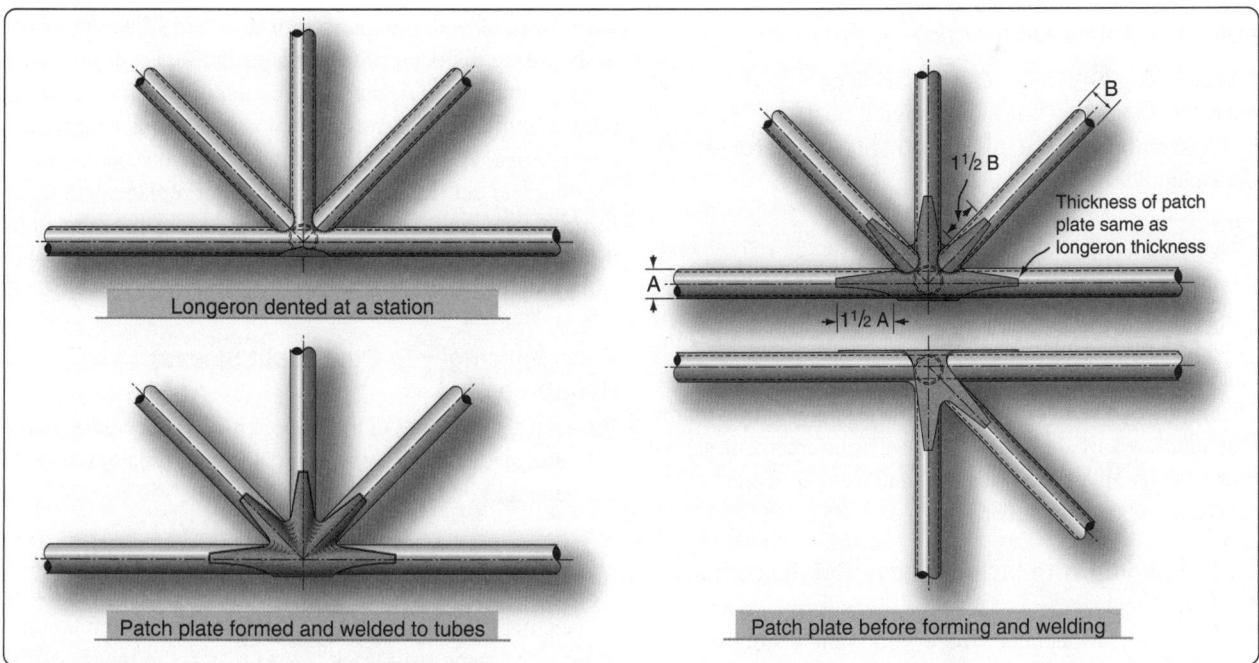

Figure 5-46. *Repair of tubing dented at a cluster.*

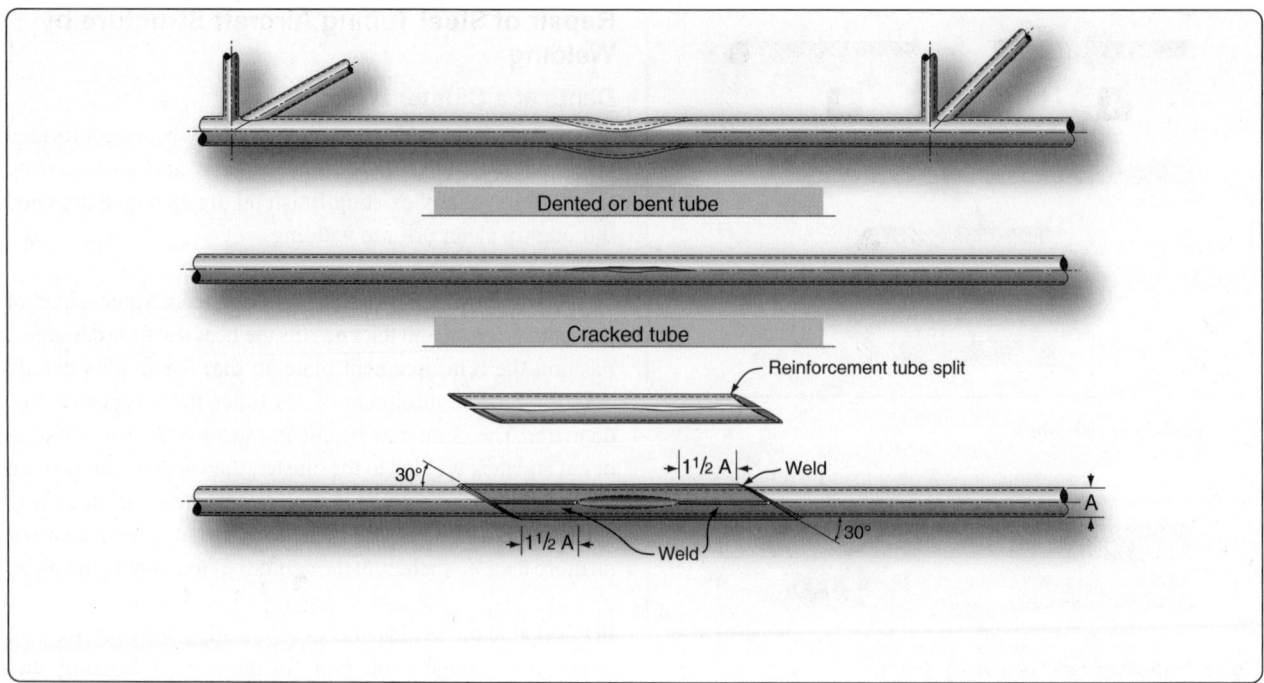

Figure 5-47. *Repair using welded sleeve.*

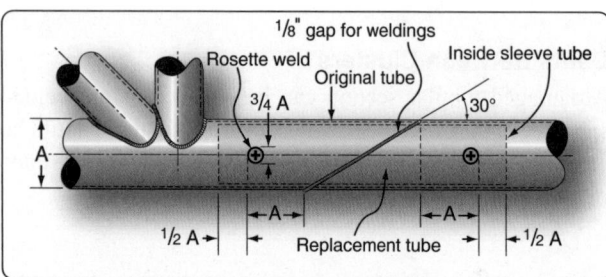

Figure 5-48. *Splicing with inner sleeve method.*

Select a length of steel tube of the same material and at least the same wall thickness having an inside diameter approximately equal to the outside diameter of the damaged tube.

Diagonally cut the selected piece at a 30° angle on both ends so the minimum distance of the sleeve from the edge of the crack or dent is not less than 1½ times the diameter of the damaged tube. Then, cut through the entire length of the sleeve and separate the half sections as shown in *Figure 5-47*. Clamp the two sleeve sections in the proper position on the damaged area of the tube. Weld the reinforcement sleeve along the length of the two sides, and weld both ends of the sleeve to the damaged tube.

Tube Splicing with Inside Sleeve Reinforcement

If a partial replacement of the tube is necessary, do an inner sleeve splice, especially where you want a smooth tube surface.

Make a diagonal cut to remove the damaged section of the tube, and remove the burrs from the inner and outer cut edges with a file or similar means. Diagonally cut a replacement steel tube of the same material, diameter, and wall thickness to match the length of the removed section of the damaged tube. The replacement tube should allow a ⅛-inch gap for welding at each end to the stubs of the original tube.

Select a length of steel tubing of the same material and at least the same wall thickness with an outside diameter equal to the inside diameter of the damaged tube. From this inner sleeve tube material, cut two sections of tubing, each of such a length that the ends of the inner sleeve is a minimum distance of 1½ times the tube diameter from the nearest end of the diagonal cut. Tack the outer and inner replacement tubes using rosette welds. Weld the inner sleeve to the tube stubs through the ⅛-inch gap forming a weld bead over the gap and joining with the new replacement section. *[Figure 5-48]*

Tube Splicing with Outer Split Sleeve Reinforcement

If partial replacement of a damaged tube is necessary, make the outer sleeve splice using a replacement tube of the same diameter and material. *[Figures 5-49 and 5-50]*

To perform the outer sleeve repair, remove the damaged section of the tube, utilizing a 90° cut at either end. Cut a replacement steel tube of the same material, diameter, and at least the same wall thickness to match the length of the removed portion of the damaged tube. The replacement

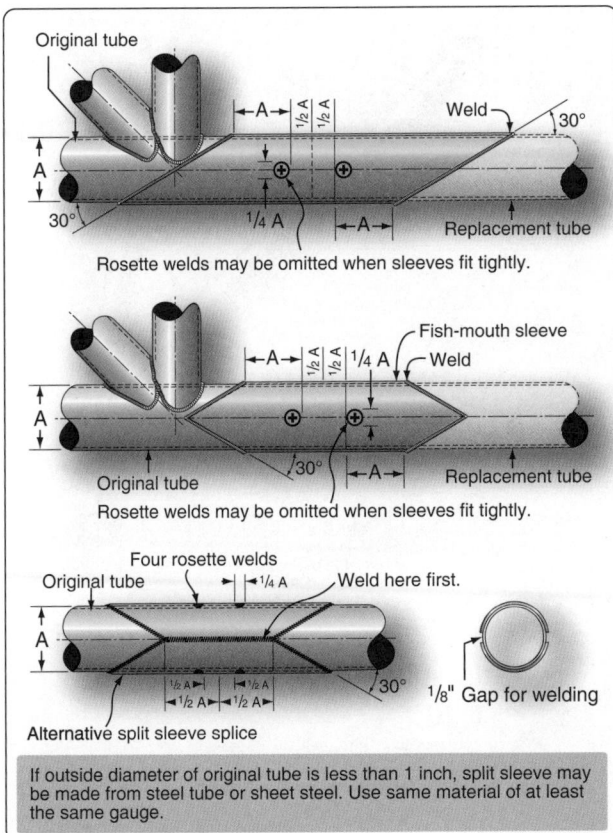

Original tube

Weld

30°

A

30°

←A→|½ A|½ A

¼ A ←A→

Replacement tube

Rosette welds may be omitted when sleeves fit tightly.

Fish-mouth sleeve

Weld

←A→|½ A|½ A| ¼ A

A

Original tube

30° ←A→

Replacement tube

Rosette welds may be omitted when sleeves fit tightly.

Four rosette welds

Original tube   ►|◄¼ A   Weld here first.

A

½ A►|◄ |►½ A◄|
½ A►|◄   ◄½ A►

30°   ⅛" Gap for welding

Alternative split sleeve splice

If outside diameter of original tube is less than 1 inch, split sleeve may be made from steel tube or sheet steel. Use same material of at least the same gauge.

Figure 5-49. *Splicing by the outer sleeve method.*

tube must bear against the stubs of the original tube with a tolerance of ±¹⁄₆₄-inch. The material selected for the outer sleeve must be of the same material and at least the same wall thickness as the original tube. The clearance between the inside diameter of the sleeve and the outside diameter of the original tube may not exceed ¹⁄₁₆-inch. From this outer sleeve tube material, either cut diagonally or fishmouth two sections

of tubing, each of such a length that the nearest end of the outer sleeve is a minimum distance of 1½ tube diameters from the end of the cut on the original tube. Use the fish mouth sleeve wherever possible. Remove all burrs from the edges of the replacement tube, sleeves, and the original tube stubs.

Slip the two sleeves over the replacement tube, align the replacement tube with the original tube stubs, and slip the sleeves over the center of each joint. Adjust the sleeves to the area to provide maximum reinforcement.

Tack weld the two sleeves to the replacement tube in two places before welding ends. Apply a uniform weld around both ends of one of the reinforcement sleeves and allow the weld to cool. Then, weld around both ends of the remaining reinforcement tube. Allow one sleeve weld to cool before welding the remaining tube to prevent undue warping.

Landing Gear Repairs

Some components of a landing gear may be repaired by welding while others, when damaged, may require replacement. Representative types of repairable and nonrepairable landing gear assemblies are shown in *Figure 5-51*.

The landing gear types shown in A, B, and C of this figure are repairable axle assemblies. They are formed from steel tubing and may be repaired by any of the methods described in this chapter or in FAA Advisory Circular (AC) 43.13-1, Acceptable Methods, Techniques, and Practices—Aircraft Inspection and Repair. However, it must be determined if the assemblies were heat treated. Assemblies originally heat treated must be reheat treated after a welding repair.

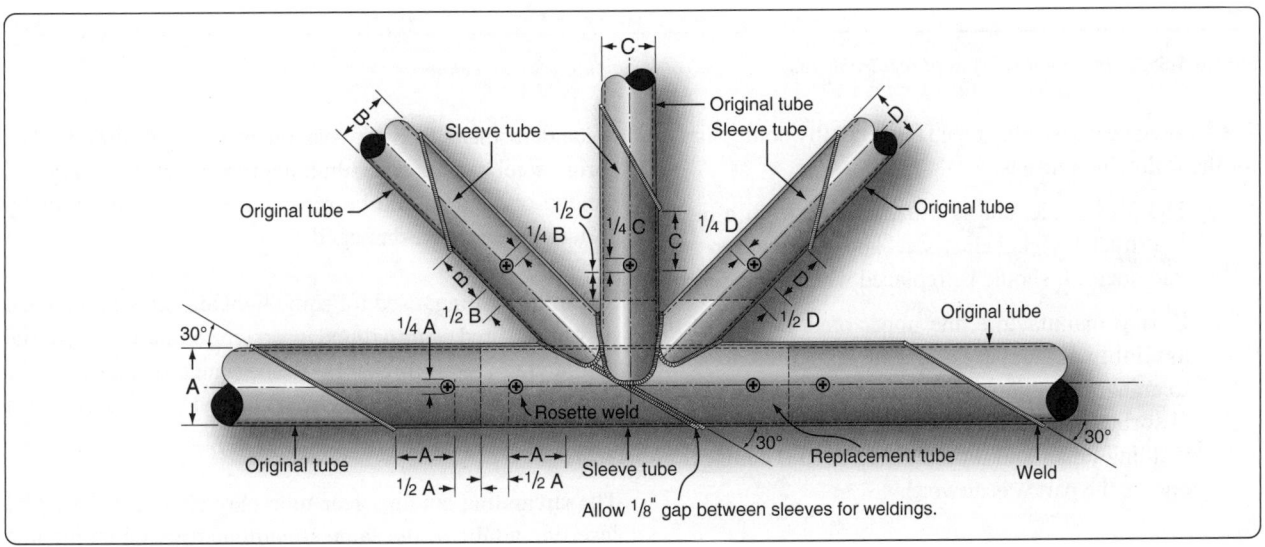

Figure 5-50. *Tube replacement at a cluster by outer sleeve method.*

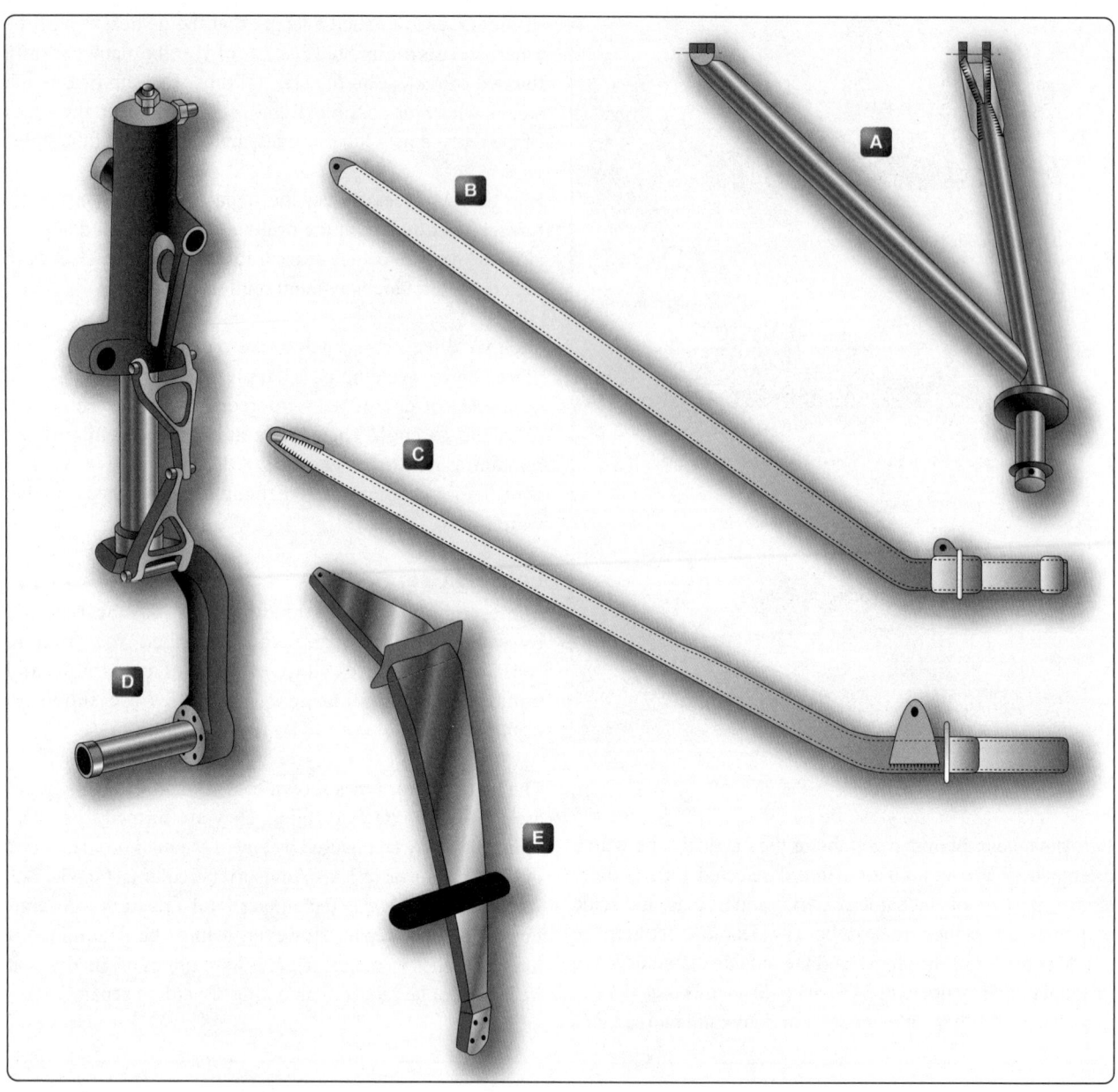

Figure 5-51. *Representative types of repairable and nonrepairable landing gear assemblies.*

The landing gear assembly type D is generally nonrepairable for the following reasons:

1. The lower axle stub is usually made from a highly heat-treated nickel alloy steel and machined to close tolerances. It should be replaced when damaged.

2. During manufacture, the upper oleo section of the assembly is heat treated and machined to close tolerances to assure proper functioning of the shock absorber. These parts would be distorted by any welding repair and should be replaced if damaged to ensure the part was airworthy.

The spring-steel leaf, shown as type E, is a component of a standard main landing gear on many light aircraft. The spring-steel part is, in general, nonrepairable, should not be welded on, and should be replaced when it is excessively sprung or otherwise damaged.

Streamline tubing, used for some light aircraft landing gear, may be repaired using a round insert tube of the same material and having a wall thickness of one gauge thicker than the original streamline tube and inserting and welding as shown in *Figure 5-52.*

The streamline landing gear tube may also be repaired by inserting a tube of the same streamline original tubing and welding. This can be accomplished by cutting off the trailing

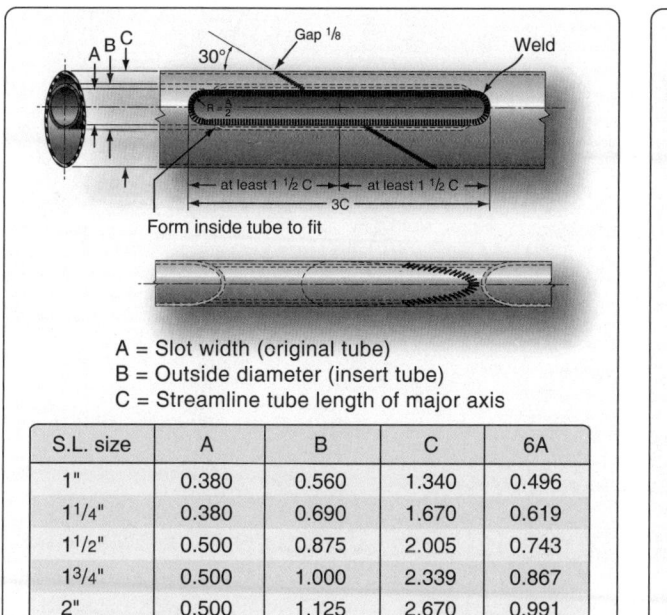

A = Slot width (original tube)
B = Outside diameter (insert tube)
C = Streamline tube length of major axis

S.L. size	A	B	C	6A
1"	0.380	0.560	1.340	0.496
1¹/₄"	0.380	0.690	1.670	0.619
1¹/₂"	0.500	0.875	2.005	0.743
1³/₄"	0.500	1.000	2.339	0.867
2"	0.500	1.125	2.670	0.991
2¹/₄"	0.500	1.250	3.008	1.115
2¹/₂"	0.500	1.380	3.342	1.239

Round insert tube (B) should be of same material and one gauge thicker than original streamline tube (C).

Figure 5-52. *Streamline landing gear repair using round tube.*

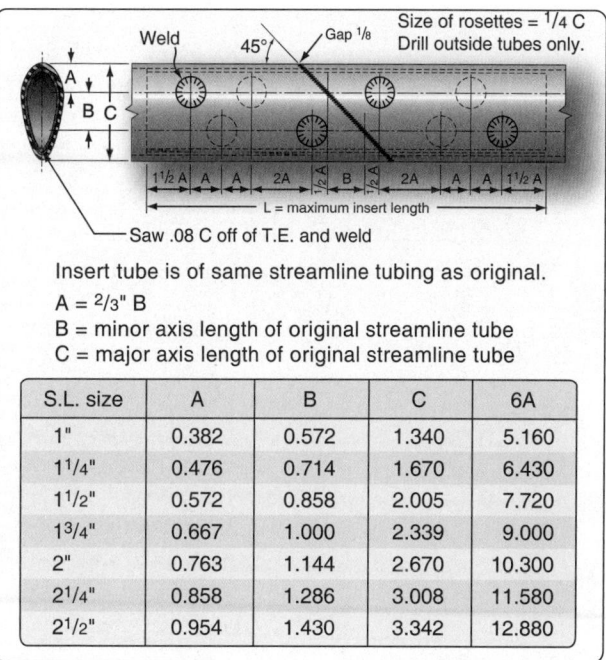

Insert tube is of same streamline tubing as original.
A = ²/₃" B
B = minor axis length of original streamline tube
C = major axis length of original streamline tube

S.L. size	A	B	C	6A
1"	0.382	0.572	1.340	5.160
1¹/₄"	0.476	0.714	1.670	6.430
1¹/₂"	0.572	0.858	2.005	7.720
1³/₄"	0.667	1.000	2.339	9.000
2"	0.763	1.144	2.670	10.300
2¹/₄"	0.858	1.286	3.008	11.580
2¹/₂"	0.954	1.430	3.342	12.880

Figure 5-53. *Streamline tube splice using split insert.*

edge of the insert and fitting it into the original tube. Once fitted, remove the insert, weld the trailing edge back together, and reinsert into the original tube. Use the figures and weld as indicated in *Figure 5-53*.

Engine Mount Repairs

All welding on an engine mount should be performed by an experienced welder and be of the highest quality, since vibration tends to accentuate any minor defect.

The preferred method to repair an engine mount member is by using a larger diameter replacement tube telescoped over the stub of the original member using fish-mouth and rosette welds. 30° scarf welds are also acceptable in place of the fish-mouth welds.

One of the most important aspects to keep in mind when repairing an engine mount is that the alignment of the structure must be maintained. This can be accomplished by attaching to a fixture designed for that purpose, or bolting the mount to an engine and/or airframe before welding.

All cracked welds should be ground out and only high-grade filler rod of the appropriate material should be used.

If all members of the mount are out of alignment, the mount should be replaced with one supplied by the manufacturer or with one built to conform to the manufacturer's drawings

and specifications.

Minor damage, such as a crack adjacent to an engine attachment lug, can be repaired by rewelding the ring and extending a gusset or a mounting lug past the damaged area. Engine mount rings that are extensively damaged must not be repaired unless the method of repair is specifically approved by FAA Engineering, a Designated Engineering Representative (DAR), or the repair is accomplished in accordance with FAA-approved instructions.

If the manufacturer stress relieved the engine mount after welding, the engine mount should again be stress relieved after weld repairs are made.

Rosette Welding

Rosette welds are used on many of the type repairs that were previously discussed. They are holes, typically one-fourth the diameter of the original tube, drilled in the outer splice and welded around the circumference for attachment to the inner replacement tube or original tube structure.

Chapter 6
Aircraft Wood & Structural Repair

Introduction

Wood was among the first materials used to construct aircraft. Most of the airplanes built during World War I (WWI) were constructed of wood frames with fabric coverings. Wood was the material of choice for aircraft construction into the 1930s. Part of the reason was the slow development of strong, lightweight metal aircraft structures and the lack of suitable corrosion-resistant materials for all-metal aircraft.

In the late 1930s, the British airplane company DeHavilland designed and developed a bomber named the Mosquito. Well into the late 1940s, DeHavilland produced more than 7,700 airplanes made of spruce, birch plywood, and balsa wood. *[Figure 6-1]*

During the early part of WWII, the U.S. government put out a contract to build three flying boats. Hughes Aircraft ultimately won the contract with the mandate to use only materials not critical to the war, such as aluminum and steel. Hughes designed the aircraft to be constructed out of wood.

After many delays and loss of government funding, Howard Hughes continued construction, using his own money and completing one aircraft. On November 2, 1947, during taxi tests in the harbor at Long Beach, California, Hughes piloted the Spruce Goose for over a mile at an altitude of 70 feet, proving it could fly.

This was the largest seaplane and the largest wooden aircraft ever constructed. Its empty weight was 300,000 pounds with a maximum takeoff weight of 400,000 pounds. The entire airframe, surface structures, and flaps were composed of laminated wood with fabric covered primary control surfaces. It was powered by eight Pratt & Whitney R-4360 radial engines, each producing 3,000 horsepower. *[Figure 6-2]*

As aircraft design and manufacturing evolved, the development of lightweight metals and the demand for increased production moved the industry away from aircraft constructed entirely of wood. Some general aviation aircraft were produced with wood spars and wings, but today only a limited number of wood aircraft are produced. Most of those are built by their owners for education or recreation and not for production.

Quite a number of airplanes in which wood was used as

the primary structural material still exist and are operating, including certificated aircraft that were constructed during the 1930s and later. With the proper maintenance and repair procedures, these older aircraft can be maintained in an airworthy condition and kept operational for many years.

Wood Aircraft Construction & Repairs

The information presented in this chapter is general in nature and should not be regarded as a substitute for specific instructions contained in the aircraft manufacturer's maintenance and repair manuals. Methods of construction vary greatly with different types of aircraft, as do the various repair and maintenance procedures required to keep them airworthy.

When specific manufacturer's manuals and instructions are

Figure 6-1. *British DeHavilland Mosquito bomber.*

Figure 6-2. *Hughes Flying Boat, H-4 Hercules named the Spruce Goose.*

not available, the Federal Aviation Administration (FAA) Advisory Circular (AC) 43.13-1, Acceptable Methods, Techniques, and Practices—Aircraft Inspection and Repair, can be used as reference for inspections and repairs. The AC details in the first paragraph, Purpose, the criteria necessary for its use. In part, it stipulates that the use of the AC is acceptable to the FAA for the inspection and minor repair of nonpressurized areas of civil aircraft.

It also specifies that the repairs identified in the AC may also be used as a basis for FAA approval of major repairs when listed in block 8 of FAA Form 337, Major Repair and Alteration, when:

1. The user has determined that it is appropriate to the product being repaired;

2. It is directly applicable to the repair being made; and

3. It is not contrary to manufacturer's data.

Certificated mechanics that have the experience of working on wooden aircraft are becoming rare. Title 14 of the Code of Federal Regulations (14 CFR) part 65 states in part that a certificated mechanic may not perform any work for which they are rated unless they have performed the work concerned at an earlier date. This means that if an individual does not have the previous aviation woodworking experience performing the repair on an aircraft, regulation requires a certificated and appropriately rated mechanic or repairman who has had previous experience in the operation concerned to supervise that person.

The ability to inspect wood structures and recognize defects (dry rot, compression failures, etc.) can be learned through experience and instruction from knowledgeable certificated mechanics and appropriately qualified technical instructors.

Inspection of Wood Structures

To properly inspect an aircraft constructed or comprised of wood components, the aircraft must be dry. It should be placed in a dry, well-ventilated hangar with all inspection covers, access panels, and removable fairings opened and removed. This allows interior sections and compartments to thoroughly dry. Wet, or even damp, wood causes swelling and makes it difficult to make a proper determination of the condition of the glue joints.

If there is any doubt that the wood is dry, a moisture meter should be utilized to verify the percentage of moisture in the structure. Nondestructive meters are available that check moisture without making holes in the surface. The ideal range is 8–12 percent, with any reading over 20 percent providing an environment for the growth of fungus in the wood.

External & Internal Inspection

The inspection should begin with an examination of the external surface of the aircraft. This provides a general assessment of the overall condition of the wood and structure. The wings, fuselage, and empennage should be inspected for undulation, warping, or any other disparity from the original shape. Where the wings, fuselage, or empennage structure and skins form stressed structures, no departure from the original contour or shape is permissible. *[Figure 6-3]*

Where light structures using single plywood covering are concerned, some slight sectional undulation or bulging between panels may be permissible if the wood and glue are sound. However, where such conditions exist, a careful check must be made of the attachment of the plywood to its supporting structure. A typical example of a distorted single plywood structure is illustrated in *Figure 6-4*.

The contours and alignment of leading and trailing edges are of particular importance. A careful check should be made for any deviation from the original shape. Any distortion of these light plywood and spruce structures is indicative of deterioration, and a detailed internal inspection has to be made for security of these parts to the main wing structure. If deterioration is found in these components, the main wing structure may also be affected.

Splits in the fabric covering on plywood surfaces must be investigated to ascertain whether the plywood skin beneath

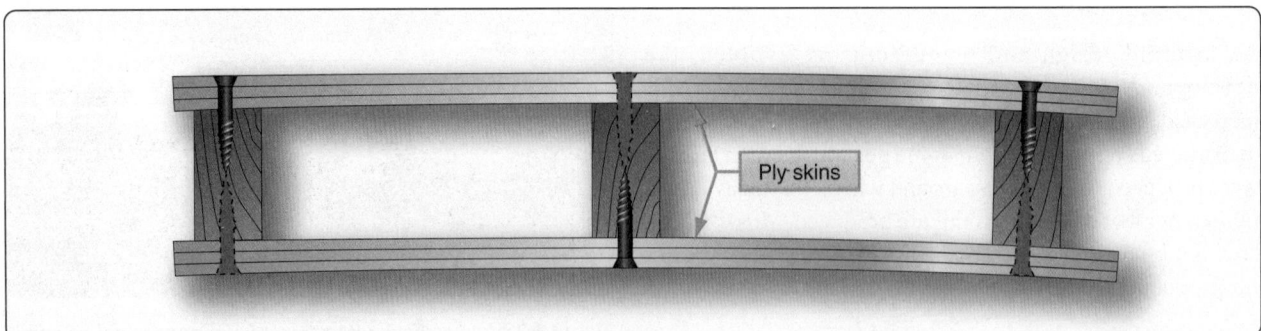

Figure 6-3. *Cross sectional view of a stressed skin structure.*

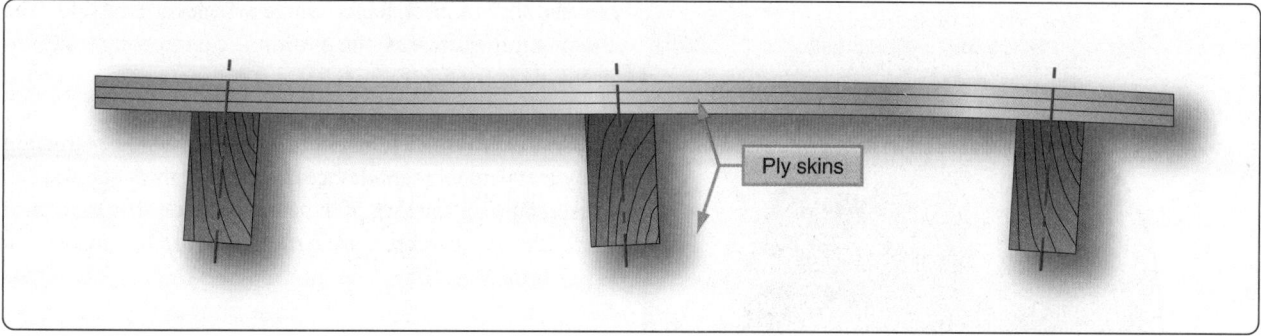

Figure 6-4. *A distorted single plywood structure.*

is serviceable. In all cases, remove the fabric and inspect the plywood, since it is common for a split in the plywood skin to initiate a similar defect in the protective fabric covering.

Although a preliminary inspection of the external structure can be useful in assessing the general condition of the aircraft, note that wood and glue deterioration can often take place inside a structure without any external indications. Where moisture can enter a structure, it seeks the lowest point, where it stagnates and promotes rapid deterioration. A musty or moldy odor apparent as you remove the access panels during the initial inspection is a good indication of moisture, fungal growth, and possible decay.

Glue failure and wood deterioration are often closely related, and the inspection of glued joints must include an examination of the adjacent wood structure. *Note:* Water need not be present for glue deterioration to take place.

The inspection of a complete aircraft for glue or wood deterioration requires scrutiny of parts of the structure that may be known, or suspected, trouble spots. In many instances, these areas are boxed in or otherwise inaccessible. Considerable dismantling may be required. It may be necessary to cut access holes in some of the structures to facilitate the inspection. Do such work only in accordance with approved drawings or instructions in the maintenance manual for the aircraft concerned. If drawings and manuals are not available, engineering review may be required before cutting access holes.

Glued Joint Inspection

The inspection of glued joints in wooden aircraft structures presents considerable difficulties. Even where access to the joint exists, it is still difficult to positively assess the integrity of the joint. Keep this in mind when inspecting any glue joint.

Some common factors in premature glue deterioration include:

- Chemical reactions of the glue caused by aging or moisture, extreme temperatures, or a combination of these factors.

- Mechanical forces caused mainly by wood shrinkage.

- Development of fungal growths.

An aircraft painted in darker colors experiences higher skin temperatures and heat buildup within its structure. Perform a more detailed inspection on a wooden aircraft structure immediately beneath the upper surfaces for signs of deteriorating adhesives.

Aircraft that are exposed to large cyclic changes of temperature and humidity are especially prone to wood shrinkage that may lead to glue joint deterioration. The amount of movement of a wooden member due to these changes varies with the size of each member, the rate of growth of the tree from which it was cut, and the way the wood was converted in relation to the grain.

This means that two major structural members joined to each other by glue are not likely to have identical characteristics. Over a period of time, differential loads are transmitted across the glue joint because the two members do not react identically. This imposes stresses in the glue joint that can normally be accommodated when the aircraft is new and for some years afterwards. However, glue tends to deteriorate with age, and stresses at the glued joints may cause failure of the joints. This is a fact even when the aircraft is maintained under ideal conditions.

The various cuts of lumber from a tree have tendency to shrink and warp in the direction(s) indicated in the yellow area around each cut in *Figure 6-5*.

When checking a glue line (the edge of the glued joint) for condition, all protective coatings of paint should be removed by careful scraping. It is important to ensure that the wood is not damaged during the scraping operation. Scraping should

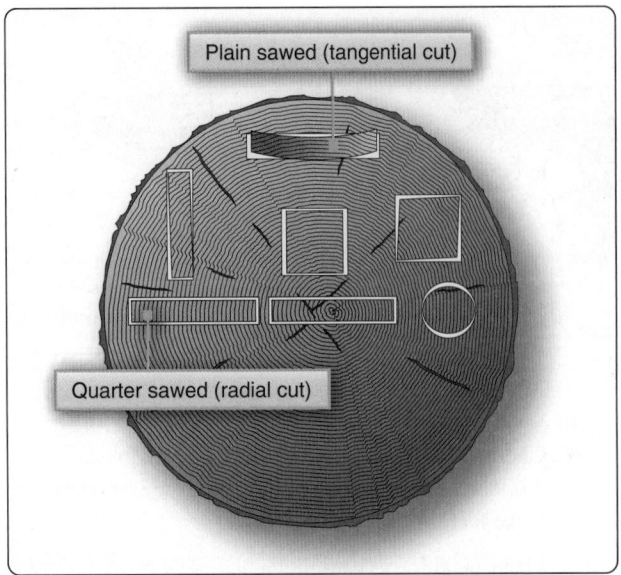

Figure 6-5. *Effects of shrinkage on the various shapes during drying from the green condition.*

cease immediately when the wood is revealed in its natural state and the glue line is clearly discernible. At this point in the inspection, it is important that the surrounding wood is dry; otherwise, you will get a false indication of the integrity of the glue line due to swelling of the wood and subsequent closing of the joint.

Inspect the glue line using a magnifying glass. Where the glue line tends to part, or where the presence of glue cannot be detected or is suspect, probe the glue line with a thin feeler gauge. If any penetration is observed, the joint is defective. The structure usually dictates the feeler gauge thickness,

but use the thinnest feeler gauge whenever possible. The illustration indicates the points a feeler gauge should probe. *[Figure 6-6]*

Pressure exerted on a joint either by the surrounding structure or by metal attachment devices, such as bolts or screws, can cause a false appearance of the glue condition. The joint must be relieved of this pressure before the glue line inspection is performed.

A glued joint may fail in service as a result of an accident or because of excessive mechanical loads having been imposed upon it. Glued joints are generally designed to take shear loads. If a joint is expected to take tension loads, it is secured by a number of bolts or screws in the area of tension loading. In all cases of glued joint failure, whatever the direction of loading, there should be a fine layer of wood fibers adhering to the glue. The presence of fibers usually indicates that the joint itself is not at fault.

Examination of the glue under magnification that does not reveal any wood fibers, but shows an imprint of the wood grain, indicates that the cause of the failure was the predrying of the glue before applying pressure during the manufacture of the joint. If the glue exhibits an irregular appearance with star-shaped patterns, this is an indication that precuring of the glue occurred before pressure was applied, or that pressure had been incorrectly applied or maintained on the joint. If there is no evidence of wood fiber adhesion, there may also be glue deterioration.

Wood Condition

Wood decay and dry rot are usually easy to detect. Decay

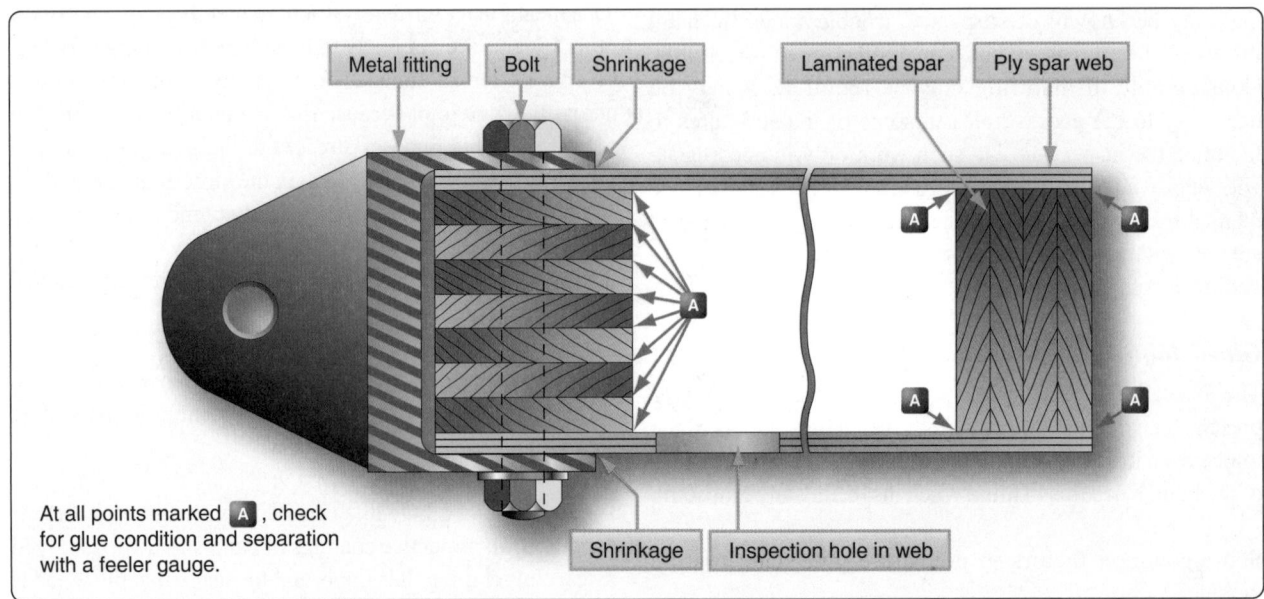

Figure 6-6. *Inspection points for laminated glue joints.*

may be evident as either a discoloration or a softening of the wood. Dry rot is a term loosely applied to many types of decay, but especially to a condition that, in an advanced stage, permits the wood to be crushed to a dry powder. The term is actually a misnomer for any decay, since all fungi require considerable moisture for growth.

Dark discolorations of the wood or gray stains running along the grain are indicative of water penetration. If such discoloration cannot be removed by light scraping, replace the part. Disregard local staining of the wood by dye from a synthetic adhesive hardener.

In some instances where water penetration is suspected, a few screws removed from the area in question reveal, by their degree of corrosion, the condition of the surrounding joint. *[Figure 6-7]*

Another method of detecting water penetration is to remove the bolts holding the fittings at spar root-end joints, aileron hinge brackets, etc. Corrosion on the surface of such bolts and wood discoloration provide a useful indication of water penetration.

Plain brass screws are normally used for reinforcing glued wooden members. For hardwoods, such as mahogany or ash, steel screws may be used. Unless specified by the aircraft manufacturer, replace removed screws with new screws of identical length, but one gauge larger in diameter.

Inspection experience with a particular type of aircraft provides insight to the specific areas most prone to water penetration and moisture entrapment. Wooden aircraft are more prone to the damaging effects of water, especially without the protection of covered storage. Control system openings, fastener holes, cracks or breaks in the finish, and the interfaces of metal fittings and the wood structure are points that require additional attention during an inspection. Additionally, windshield and window frames, the area under the bottom of entrance and cargo doors, and the lower sections of the wing and fuselage are locations that require detailed inspections for water damage and corrosion on all aircraft.

The condition of the fabric covering on plywood surfaces provides an indication of the condition of the wood underneath. If there is any evidence of poor adhesion, cracks in the fabric, or swelling of the wood, remove the fabric to allow further inspection. The exposed surface shows water penetration by the existence of dark gray streaks along the grain and dark discoloration at ply joints or screw holes.

Cracks in wood spars are often hidden under metal fittings or metal rib flanges and leading edge skins. Any time a

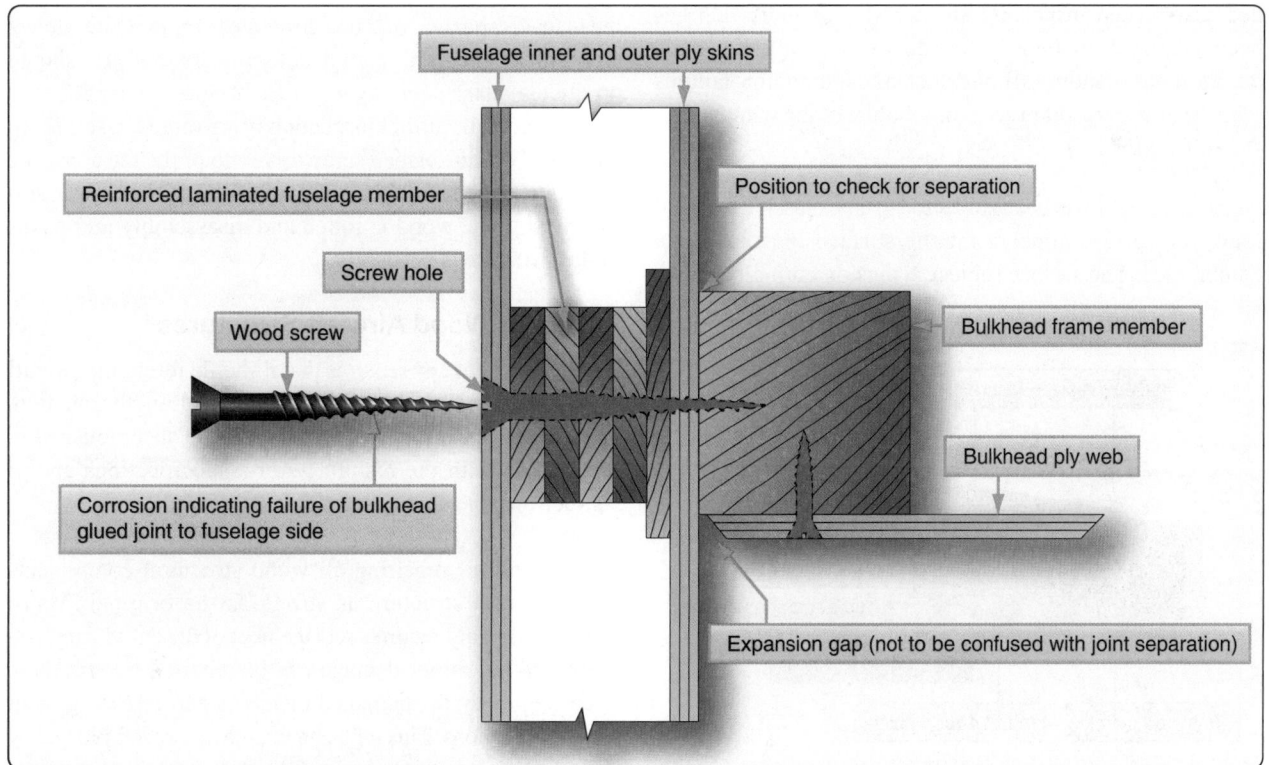

Figure 6-7. *Checking a glued joint for water penetration.*

reinforcement plate exists that is not feathered out on its ends, a stress riser exists at the ends of the plate. A failure of the primary structure can be expected at this point. *[Figure 6-8]*

As part of the inspection, examine the structure for other defects of a mechanical nature, including any location where bolts secure fittings that take load-carrying members, or where the bolts are subject to landing or shear loads. Remove the bolts and examine the holes for elongation or surface crushing of the wood fibers. It is important to ensure the bolts are a good fit in the holes. Check for evidence of bruises or crushing of the structural member, which can be caused by overtorquing of the bolts.

Check all metal fittings that are attached to a wood structure for looseness, corrosion, cracks, or bending. Areas of particular concern are strut attach fittings, spar butt fittings, aileron and flap hinges, jury strut fittings, compression struts, control cable pulley brackets, and landing gear fittings. All exposed end grain wood, particularly the spar butts, should be inspected for cracking or checking.

Inspect structural members for compression failures, which is indicated by rupture across the wood fibers. This is a serious defect that can be difficult to detect. If a compression failure is suspected, a flashlight beam shown along the member and running parallel to the grain, will assist in revealing it. The surface will appear to have minute ridges or lines running across the grain. Particular attention is necessary when inspecting any wooden member that has been subjected to abnormal bending or compression loads during a hard landing. If undetected, compression failures of the spar may result in structural failure of the wing during flight. *[Figure 6-9]*

When a member has been subjected to an excessive bending load, the failure appears on the surface that has been compressed. The surface subject to tension normally shows

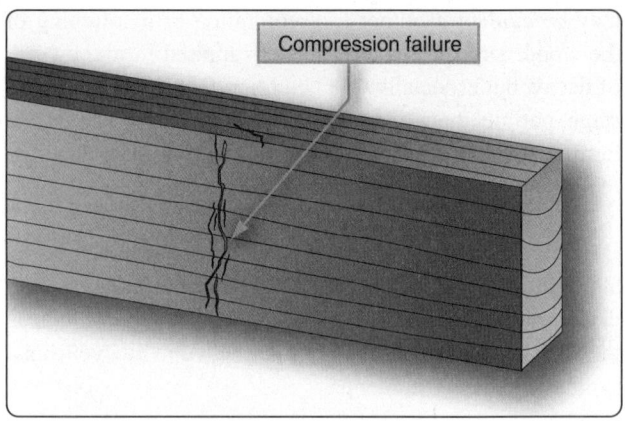

Figure 6-9. *Pronounced compression failure in wood beam.*

no defects. In the case of a member taking an excessive direct compression load, the failure is apparent on all surfaces.

The front and rear spars should be checked for longitudinal cracks at the ends of the plywood reinforcement plates where the lift struts attach. *[Figure 6-8]* Check the ribs on either side of the strut attach points for cracks where the cap strips pass over and under the spars, and for missing or loose rib-to-spar attach nails. All spars, those in the wing(s) and empennage, should be inspected on the face and top surface for compression cracks. A borescope can be utilized by accessing existing inspection holes.

Various mechanical methods can be employed to enhance the visual inspection of wood structures. Tapping the subject area with a light plastic hammer or screwdriver handle should produce a sharp solid sound. If the suspected area sounds hollow and dull, further inspection is warranted. Use a sharp metal awl or thin-bladed screwdriver to probe the area. The wood structure should be solid and firm. If the area is soft and mushy, the wood is rotted and disassembly and repair of the structure is necessary.

Repair of Wood Aircraft Structures

The standard for any repair is that it should return the aircraft or component to its original condition in strength, function, and aerodynamic shape. It should also be accomplished in accordance with the manufacturer's specifications and/or instructions, or other approved data.

The purpose of repairing all wood structural components is to obtain a structure as strong as the original. Major damage probably requires replacement of the entire damaged assembly, but minor damage can be repaired by removing or cutting away the damaged members and replacing them with new sections. This replacement may be accomplished by gluing, glue and nails, or glue and screw-reinforced splicing.

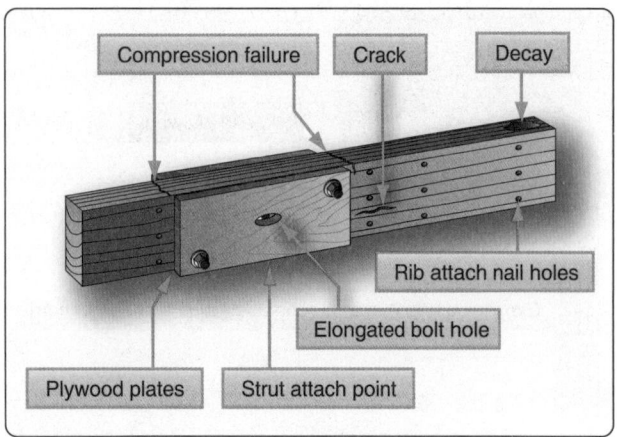

Figure 6-8. *Areas likely to incur structural damage.*

Materials

Several forms of wood are commonly used in aircraft.

- Solid wood or the adjective "solid" used with such nouns as "beam" or "spar" refers to a member consisting of one piece of wood.

- Laminated wood is an assembly of two or more layers of wood that have been glued together with the grain of all layers or laminations approximately parallel.

- Plywood is an assembled product of wood and glue that is usually made of an odd number of thin plies, or veneers, with the grain of each layer placed 90° with the adjacent ply or plies.

- High-density material includes compreg, impreg, or similar commercially made products, heat-stabilized wood, or any of the hardwood plywoods commonly used as bearing or reinforcement plates.

Suitable Wood

The various species of wood listed in *Figure 6-10* are acceptable for structural purposes when used for the repair of aircraft. Spruce is the preferred choice and the standard by which the other wood is measured. *Figure 6-10* provides a comparison of other wood that may be suitable for aircraft repair. It lists the strength and characteristics of the wood in comparison to spruce. The one item common to all the species is that the slope of the grain cannot be steeper than 1:15.

All solid wood and plywood used for the construction and repair of aircraft should be of the highest quality and grade. For certificated aircraft, the wood should have traceability to a source that can provide certification to a military specification (MIL-SPEC). The term "aircraft quality" or "aircraft grade" is referred to and specified in some repair documents, but that grade wood cannot be purchased from a local lumber company. To purchase the material, contact one of the specialty aircraft supply companies and request a certification document with the order. The MIL-SPEC for solid spruce is MIL-S-6073 and for plywood it is MIL-P-6070B.

When possible, fabricated wood components should be purchased from the aircraft manufacturer, or someone who may have a Parts Manufacturer Approval (PMA) to produce replacement parts for the aircraft. With either of these sources supplying the wood components, the mechanic can be assured of installing approved material. At the completion of the repair, as always, it is the responsibility of the person returning the aircraft to service to determine the quality of the replacement wood and the airworthiness of the subsequent repair.

To help determine the suitability of the wood, inspect it for defects that would make it unsuitable material to repair or construct an aircraft. The type, location, and amount or size of the defects grade the wood for possible use. All woods used for structural repair of aircraft are classified as softwood. Softwood is typically used for construction and is graded based on strength, load carrying ability, and safety. Hardwoods, on the other hand, are typically appearance woods and are graded based on the number and size of clear cuttings from the tree.

Defects Permitted

The following defects are permitted in the wood species used for aircraft repair that are identified in *Figure 6-10*:

1. Cross grain—Spiral grain, diagonal grain, or a combination of the two is acceptable if the grain does not diverge from the longitudinal axis of the material more than specified in *Figure 6-10* column 3. A check of all four faces of the board is necessary to determine the amount of divergence. The direction of free-flowing ink frequently assists in determining grain direction.

2. Wavy, curly, and interlocked grain—Acceptable, if local irregularities do not exceed limitations specified for spiral and diagonal grain.

3. Hard knots—Sound, hard knots up to ⅜-inch in diameter are acceptable if: (1) they are not projecting portions of I-beams, along the edges of rectangular or beveled unrouted beams, or along the edges of flanges of box beams (except in portions of low stress); (2) they do not cause grain divergence at the edges of the board or in the flanges of a beam more than specified in *Figure 6-10* column 3; and (3) they are in the center third of the beam and not closer than 20-inches to another knot or other defect (pertains to ⅜-inch knots; smaller knots may be proportionately closer). Knots greater than ¼-inch must be used with caution.

4. Pin knot clusters—Small clusters are acceptable if they produce only a small effect on grain direction.

5. Pitch pockets—Acceptable in center portion of a beam if they are at least 14-inches apart when they lie in the same growth ring and do not exceed 1½-inches in length by ⅛-inch width by ⅛-inch depth, and if they are not along the projecting portions of I-beams, along the edges of rectangular or beveled unrouted beams, or along the edges of the flanges of box beams.

6. Mineral streaks—Acceptable if careful inspection fails to reveal any decay.

Defects Not Permitted

The following defects are not permitted in wood used for aircraft repair. If a defect is listed as unacceptable, please refer to the previous section, Defects Permitted, for acceptable conditions.

Species of Wood	Strength Properties (as compared to spruce)	Maximum Permissible Grain Deviation (slope of grain)	Remarks
1	2	3	4
Spruce (Picea) Sitka (P. sitchensis) Red (P. rubra) White (P. glauca)	100%	1.15	Excellent for all uses. Considered standard for this table.
Douglas fir (Pseudotsuga taxifolia)	Exceeds spruce	1.15	May be used as substitute for spruce in same sizes or in slightly reduced sizes if reductions are substantiated. Difficult to work with hand tools. Some tendency to split and splinter during fabrication and much greater care in manufacture is necessary. Large solid pieces should be avoided due to inspection difficulties. Satisfactory for gluing .
Noble fir (Abies procera, also known as Abies nobilis)	Slightly exceeds spruce except 8% deficient in shear	1.15	Satisfactory characteristics of workability, warping, and splitting. May be used as direct substitute for spruce in same sizes if shear does not become critical. Hardness somewhat less than spruce. Satisfactory for gluing.
Western hemlock (Tsuga heterophylla)	Slightly exceeds spruce	1.15	Less uniform in texture than spruce. May be used as direct substitute for spruce. Upland growth superior to lowland growth. Satisfactory for gluing.
Northern white pine, also known as Eastern white pine (Pinus strobus)	Properties between 85% and 96% those of spruce	1.15	Excellent working qualities and uniform in properties, but somewhat low in hardness and shock-resistance. Cannot be used as substitute for spruce without increase in sizes to compensate for lesser strength. Satisfactory for gluing.
Port Orford white cedar (Chamaecyparis lawsoniana)	Exceeds spruce	1.15	May be used as substitute for spruce in same sizes or in slightly reduced sizes if reductions are substantiated. Easy to work with hand tools. Gluing is difficult, but satisfactory joints can be obtained if suitable precautions are taken.
Yellow poplar (Liriodendron tulipifera)	Slightly less than spruce except in compression (crushing) and shear	1.15	Excellent working qualities. Should not be used as a direct substitute for spruce without carefully accounting for slightly reduced strength properties. Somewhat low in shock-resistance. Satisfactory for gluing.

Figure 6-10. *Selection and properties of wood for aircraft repairs.*

1. Cross grain—unacceptable.

2. Wavy, curly, and interlocked grain – unacceptable.

3. Hard knots—unacceptable.

4. Pin knot clusters—unacceptable, if they produce large effect on grain direction.

5. Spike knots—knots running completely through the depth of a beam perpendicular to the annual rings and appear most frequently in quarter-sawed lumber. Reject wood containing this defect.

6. Pitch pockets—unacceptable.

7. Mineral streaks—unacceptable, if accompanied by decay.

8. Checks, shakes, and splits—checks are longitudinal cracks extending, in general, across the annual rings. Shakes are longitudinal cracks usually between two annual rings. Splits are longitudinal cracks caused by artificially induced stress. Reject wood containing these defects.

9. Compression—very detrimental to strength and is difficult to recognize readily, compression wood is characterized by high specific gravity, has the appearance of an excessive growth of summer wood, and in most species shows little contrast in color between spring wood and summer wood. If in doubt, reject the material or subject samples to toughness machine test to establish the quality of the wood.

Reject all material containing compression wood.

10. Compression failures—caused from overstress in compression due to natural forces during the growth of the tree, felling trees on rough or irregular ground, or rough handling of logs or lumber. Compression failures are characterized by a buckling of the fibers that appears as streaks substantially at right angles to the grain on the surface of the piece, and vary from pronounced failures to very fine hairlines that require close inspection to detect. Reject wood containing obvious failures. If in doubt, reject the wood or make a further inspection in the form of microscopic examination or toughness test, the latter being more reliable.

11. Tension—forming on the upper side of branches and leaning trunks of softwood trees, tension wood is caused by the natural overstressing of trying to pull the branches and leaning trunk upright. It is typically harder, denser, and may be darker in color than normal wood, and is a serious defect, having higher than usual longitudinal shrinkage that may break down due to uneven shrinkage. When in doubt, reject the wood.

12. Decay—rot, dote, red heart, purple heart, etc., must not appear on any piece. Examine all stains and discoloration carefully to determine whether or not they are harmless or in a stage of preliminary or advanced decay.

Glues (Adhesives)

Because adhesives play a critical role in the bonding of aircraft structure, the mechanic must employ only those types of adhesives that meet all of the performance requirements necessary for use in certified aircraft. The product must be used strictly in accordance with the aircraft and adhesive manufacturer's instructions. All instructions must be followed exactly, including the mixing ratios, the ambient and surface temperatures, the open and closed assembly times, the gap-filling ability, or glue line thickness, the spread of the adhesive, whether one or two surfaces, and the amount of clamping pressure and time required for full cure of the adhesive.

AC 43.13-1 provides information on the criteria for identifying adhesives that are acceptable to the FAA. It stipulates the following:

1. Refer to the aircraft maintenance or repair manual for specific instructions on acceptable adhesive selection for use on that type aircraft.

2. Adhesives meeting the requirements of a MIL-SPEC, Aerospace Material Specification (AMS), or Technical Standard Order (TSO) for wooden aircraft

structures are satisfactory, provided they are found to be compatible with existing structural materials in the aircraft and fabrication methods to be used in the repair.

New adhesives have been developed in recent years, and some of the older ones are still in use. Some of the more common adhesives that have been used in aircraft construction and repair include casein glue, plastic resin glue, resorcinol glue, and epoxy adhesives.

Casein glue should be considered obsolete for all aircraft repairs. The adhesive deteriorates when exposed to moisture and temperature variations that are part of the normal operating environment of any aircraft.

Note: Some modern adhesives are incompatible with casein adhesive. If a joint that has previously been bonded with casein is to be reglued using another type adhesive, all traces of the casein must be scraped off before a new adhesive is applied. If any casein adhesive is left, residual alkalinity may cause the new adhesive to fail to cure properly.

Plastic resin glue, also known as a urea-formaldehyde adhesive, came on the market in the middle to late 1930s. Tests and practical applications have shown that exposure to moist conditions, and particularly to a warm humid environment, under swell-shrink stress, leads to deterioration and eventual failure of the bond. For these reasons, plastic resin glue should be considered obsolete for all aircraft repairs. Discuss any proposed use of this type adhesive on aircraft with FAA engineering prior to use.

Resorcinol glue, or resorcinol-formaldehyde glue, is a two-component synthetic adhesive consisting of resin and a catalyst. It was first introduced in 1943 and almost immediately found wide application in the wood boat-building and wood aircraft industry in which the combination of high durability and moderate-temperature curing was extremely important. It has better wet-weather and ultraviolet (UV) resistance than other adhesives. This glue meets all strength and durability requirements if the fit of the joint and proper clamping pressure results in a very thin and uniform bond line.

The manufacturer's product data sheets must be followed regarding mixing, usable temperature range, and the open and close assembly times. It is very important that this type of glue is used at the recommended temperatures because the full strength of the joint cannot be relied on if assembly and curing temperatures are below 70 °F. With that in mind, higher temperatures shorten the working life because of a faster cure rate, and open and closed assembly times must be shortened.

Epoxy adhesive is a two-part synthetic resin product that depends less on joint quality and clamping pressure. However, many epoxies have not exhibited joint durability in the presence of moisture and elevated temperatures and are not recommended for structural aircraft bonding unless they meet the acceptable standards set forth by the FAA in AC 43.13-1, as referenced earlier in this chapter.

Definition of Terms Used in the Glue Process

- Close contact adhesive—a non-gap-filling adhesive (e.g., resorcinol-formaldehyde glue) suitable for use only in those joints where the surfaces to be joined can be brought into close contact by means of adequate pressure, to allow a glue line of no more than 0.005-inch gap.

- Gap-filling adhesive—an adhesive suitable for use in those joints in which the surfaces to be joined may not be close or in continuous contact (e.g., cpoxy adhesives) due either to the impracticability of applying adequate pressure or to the slight inaccuracies of fabricating the joint.

- Glue line—resultant layer of adhesive joining any two adjacent wood layers in the assembly.

- Single spread—spread of adhesive to one surface only.

- Double spread—spread of adhesive to both surfaces and equally divided between the two surfaces to be joined.

- Open assembly time—period of time between the application of the adhesive and the assembly of the joint components.

- Closed assembly time—time elapsing between the assembly of the joints and the application of pressure.

- Pressing or clamping time—time during which the components are pressed tightly together under recommended pressure until the adhesive cures (may vary from 10 to 150 pounds per square inch (psi) for softwoods, depending on the viscosity of the glue).

- Caul—a clamping device, usually two rigid wooden bars, to keep an assembly of flat panel boards aligned during glue-up. It is assembled with long bolts and placed on either side of the boards, one on top and another below, and parallel with the pipe/bar clamps. A caul is usually finished and waxed before each use to keep glue from adhering to it.

- Adhesive pot life—time elapsed from the mixing of the adhesive components until the mixture must be discarded, because it no longer performs to its specifications. The manufacturer's product data sheet may define this as working time or useful life; once

expired, the adhesive must not be used. It lists the specific temperature and quantity at which the sample amount can be worked. Pot life is a product of time and temperature. The cooler the mix is kept, within the recommended temperature range, the longer it is usable.

Preparation of Wood for Gluing

Satisfactory glue joints in aircraft should develop the full strength of the wood under all conditions of stress. To produce this result, the conditions involved in the gluing operation must be carefully controlled to obtain a continuous, thin, uniform film of solid glue in the joint with adequate adhesion to both surfaces of the wood. The following conditions are required:

1. Proper and equal moisture content of wood to be joined (8 to 12 percent).

2. Properly prepared wood surfaces that are machined or planed, and not sanded or sawed.

3. Selection of the proper adhesive for the intended task, which is properly prepared and of good quality.

4. The application of good gluing techniques, including fitment, recommended assembly times, and adequate equal pressure applied to the joint.

5. Performing the gluing operation under the recommended temperature conditions.

The surfaces to be joined must be clean, dry, and free from grease, oil, wax, paint, etc. Keep large prepared surfaces covered with a plastic sheet or masking paper prior to the bonding operation. It is advisable to clean all surfaces with a vacuum cleaner just prior to adhesive application.

Smooth even surfaces produced on planers and joiners with sharp knives and correct feed adjustments are the best surfaces for gluing solid wood. The use of sawn surfaces for gluing has been discouraged for aircraft component assembly because of the difficulty in producing a surface free of crushed fibers. Glue joints made on surfaces that are covered with crushed fibers do not develop the normal full strength of the wood.

Some of the surface changes in plywood, such as glazing and bleed-through, that occur in manufacture and may interfere with the adhesion of glue in secondary gluing are easily recognized. A light sanding of the surface with 220-grit sandpaper in the direction of the grain restores the surface fibers to their original condition, removes the gloss, and improves the adhesion of the glue. In contrast to these recognized surface conditions, wax deposits from cauls used during hot pressing produce unfavorable gluing surfaces that are not easily detected.

Wetting tests are a useful means of detecting the presence of wax. A finely sprayed mist or drops of water on the surface of wax-coated plywood bead and do not wet the wood. This test may also give an indication of the presence of other materials or conditions that would degrade a glue joint. Only a proper evaluation of the adhesion properties, using gluing tests, determines the gluing characteristics of the plywood surfaces.

Preparing Glues for Use

The manufacturer's directions should be followed for the preparation of any glue or adhesive. Unless otherwise specified by the glue manufacturer, clear, cool water should be used with glues that require mixing with water. The recommended proportions of glue, catalyst, and water or other solvent should be determined by the weight of each component. Mixing can be either by hand or machine. Whatever method is used, the glue should be thoroughly mixed and free of air bubbles, foam, and lumps of insoluble material.

Applying the Glue/Adhesive

To make a satisfactorily bonded joint, it is generally desirable to apply adhesive to both surfaces and join in a thin even layer. The adhesive can be applied with a brush, glue spreader, or a grooved rubber roller. Follow the adhesive manufacturer's application instructions for satisfactory results.

Be careful to ensure the surfaces make good contact and the joint is positioned correctly before applying the adhesive. Keep the open assembly time as short as possible and do not exceed the recommended times indicated in the product data sheet.

Pressure on the Joint

To ensure the maximum strength of the bonded surfaces, apply even force to the joint. Non-uniform gluing pressure commonly results in weak areas and strong areas in the same joint. The results of applied pressure are illustrated in *Figure 6-11.*

Use pressure to squeeze the glue out into a thin continuous film between the wood layers, to force air from the joint, to bring the wood surfaces into intimate contact with the glue, and to hold them in this position during the setting of the glue. Pressure may be applied by means of clamps, elastic straps, weight, vacuum bags, or other mechanical devices. Other methods used to apply pressure to joints in aircraft gluing operations range from the use of brads, nails, and screws to the use of electric and hydraulic power presses.

The amount of pressure required to produce strong joints in aircraft assembly operations may vary from 10 to 150 psi for softwoods and as high as 200 psi for hardwoods. Insufficient

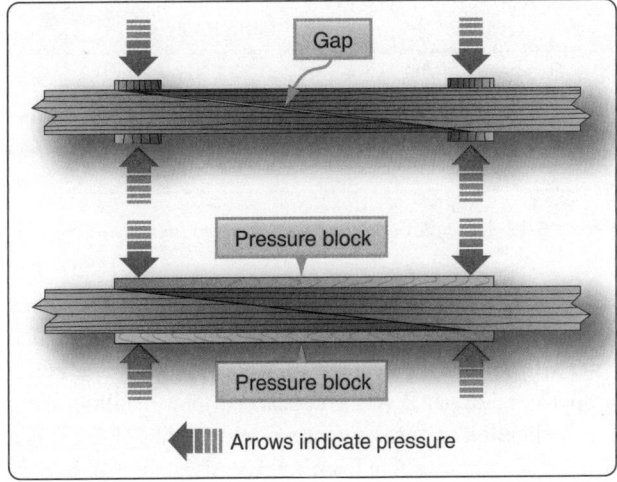

Figure 6-11. *Even distribution of gluing pressure creates a strong, gap-free joint.*

pressure to poorly machined or fitted wood joints usually results in a thick glue line, indicating a weak joint, and should be carefully avoided.

High clamping pressure is neither essential nor desirable, provided good contact between the surfaces being joined is obtained. When pressure is applied, a small quantity of glue should be squeezed from the joint. This excess should be removed before it sets. It is important that full pressure be maintained on the joint for the entire cure time of the adhesive because the adhesive does not chemically relink and bond if it is disturbed before it is fully cured.

The full curing time of the adhesive is dependent on the ambient temperature; therefore, it is very important to follow the manufacturer's product data sheets for all phases of the gluing operation from the shelf life to the moisture content of the wood to the proper mixing of the adhesive to the application, and especially to the temperature. The successful assembly and fabrication depends on the workmanship and quality of the joints and following the glue manufacturer's instructions.

All gluing operations should be performed above 70 °F for proper performance of the adhesive. Higher temperatures shorten the assembly times, as does coating the pieces of wood with glue and exposing openly to the air. This open assembly promotes a more rapid thickening of the glue than pieces being mated together as soon as the spreading of the glue is completed.

Figure 6-12 provides an example of resorcinol resin glue and the allowable assembly times and gluing pressure when in the open and closed assembly condition. All examples are for an ambient temperature of 75 °F.

Glue	Gluing Pressure	Type of Assembly	Maximum Assembly Time
Resorcinol resins	100–250 psi	Closed	Up to 50 minutes
	100–250 psi	Open	Up to 12 minutes
	Less than 100 psi	Closed	Up to 40 minutes
	Less than 100 psi	Open	Up to 10 minutes

Figure 6-12. *Examples of differences for open and closed assembly times.*

Figure 6-13 provides examples of strong and weak glue joints resulting from different gluing conditions. A is a well-glued joint with a high percentage of wood failure made under proper conditions; B is a glue-starved joint resulting from the application of excessive pressure with thin glues; C is a dried glue joint resulting from an excessively long assembly time and/or insufficient pressure.

Testing Glued Joints

Satisfactory glue joints in aircraft should develop the full strength of the wood under all conditions of stress. Tests should be made by the mechanic prior to gluing a joint of a major repair, such as a wing spar. Whenever possible, perform tests using pieces cut from the actual wood used for the repair under the same mechanical and environmental conditions that the repair will undergo.

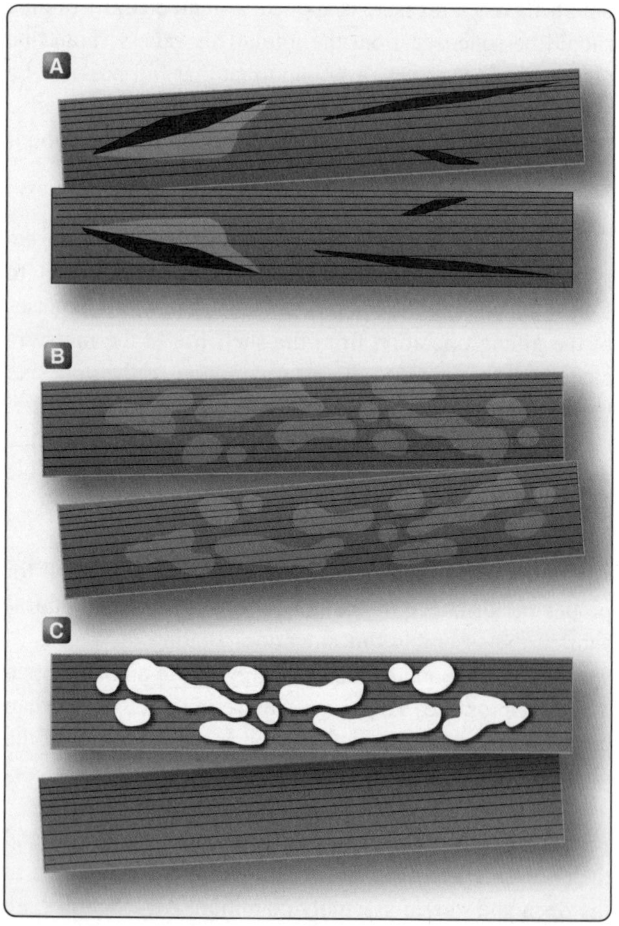

Perform a sample test using two pieces of scrap wood from the intended repair, each cut approximately 1" × 2" × 4". The pieces should be joined by overlapping each approximately 2 inches. The type of glue, pressure, and curing time should be the same as used for the actual repair. After full cure, place the test sample in a bench vise and break the joint by exerting pressure on the overlapping member. The fractured glue faces should show a high percentage of at least 75 percent of the wood fibers evenly distributed over the fractured glue surface. *[Figure 6-14]*

Repair of Wood Aircraft Components
Wing Rib Repairs

Ribs that have sustained damage may be repaired or replaced, depending upon the type of damage and location in the aircraft. If new parts are available from the aircraft manufacturer or the holder of a PMA for the part, it is advisable to replace the part rather than to repair it.

If you make a repair to a rib, do the work in such a manner and using materials of such quality that the completed repair is at least equal to the original part in aerodynamic function, structural strength, deterioration, and other qualities affecting airworthiness, such as fit and finish. When manufacturer's repair manuals or instructions are not available, acceptable methods of repairing damaged ribs are described in AC 43.13-1 under Wood Structure Repairs.

When necessary, a rib can be fabricated and installed using the same materials and dimensions from a manufacturer-approved drawing or by reference to an original rib. However, if you fabricated it from an existing rib, you must provide evidence to verify that the dimensions are accurate and the materials are correct for the replacement part.

You can repair a cap strip of a wood rib using a scarf splice. The repair is reinforced on the side opposite the wing covering by a spruce block that extends beyond the scarf joint not less than three times the thickness of the strips being repaired. Reinforce the entire splice, including the spruce reinforcing block, on each side with a plywood side plate.

Figure 6-13. *Strong and weak glue joints.*

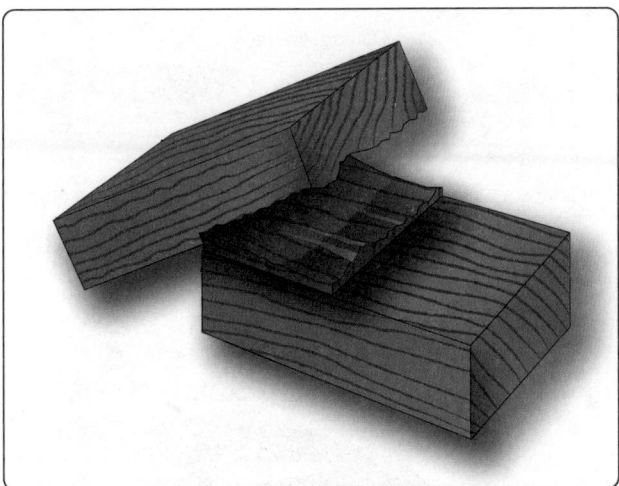

Figure 6-14. *An example of good glue joint.*

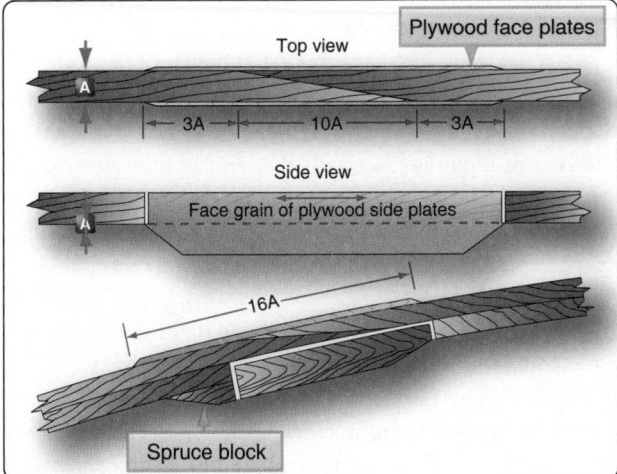

Figure 6-15. *A rib cap strip repair.*

The scarf length bevel is 10 times dimension A (thickness of the rib cap strip) with the spruce reinforcement block being 16 times dimension A (the scarf length plus extension on either end of the scarf). The plywood splice plates should be of the same material and thickness as the original plates used to fabricate the rib. The spruce block should have a 5:1 bevel on each end. *[Figure 6-15]*

These specific rib repairs describing the use of one scarf splice implies that either the entire forward or aft portion of the cap strip beyond the damage can be replaced to complete the repair and replace the damaged section. Otherwise, replacement of the damaged section may require a splice repair at both ends of the replaced section of the cap strip using the indicated dimensions for cutting and reinforcing of each splice.

When a cap strip is to be repaired at a point where there is a joint between it and cross members of the rib, make the

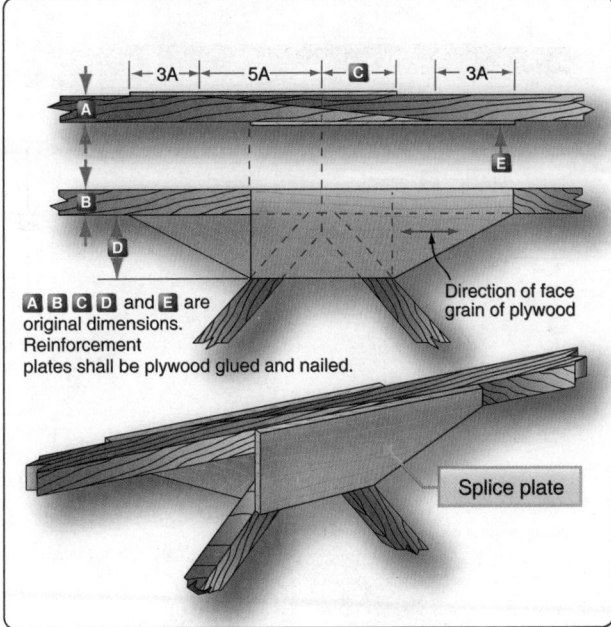

Figure 6-16. *Cap strip repair at cross member.*

repair by reinforcing the scarf joint with plywood gussets, as shown in *Figure 6-16*.

If a cap strip must be repaired where it crosses a spar, reinforce the joint with a continuous gusset extending over the spar, as shown in *Figure 6-17*.

The scarf joints referred to in the rib repairs are the most satisfactory method of fabricating an end joint between two solid wood members. When the scarf splice is used to repair a solid wood component, the mechanic must be aware of the direction and slope of the grain. To ensure the full strength of the joint, the scarf cut is made in the general direction of the grain on both connecting ends of the wood and then correctly oriented to each other when glued. *[Figure 6-18]*

The trailing edge of a rib can be replaced and repaired by removing the damaged portion of the cap strip and inserting a softwood block of white pine, spruce, or basswood. The entire repair is then reinforced with plywood gussets and nailed and glued, as shown in *Figure 6-19*.

Compression ribs are of many different designs, and the proper method of repairing any part of this type of rib is specified by the manufacturer. All repairs should be performed using recommended or approved practices, materials, and adhesives.

Figure 6-20A illustrates the repair of a compression rib of the I section type (i.e., wide, shallow cap strips, and a center plywood web with a rectangular compression member on each side of the web). The rib damage suggests that the upper

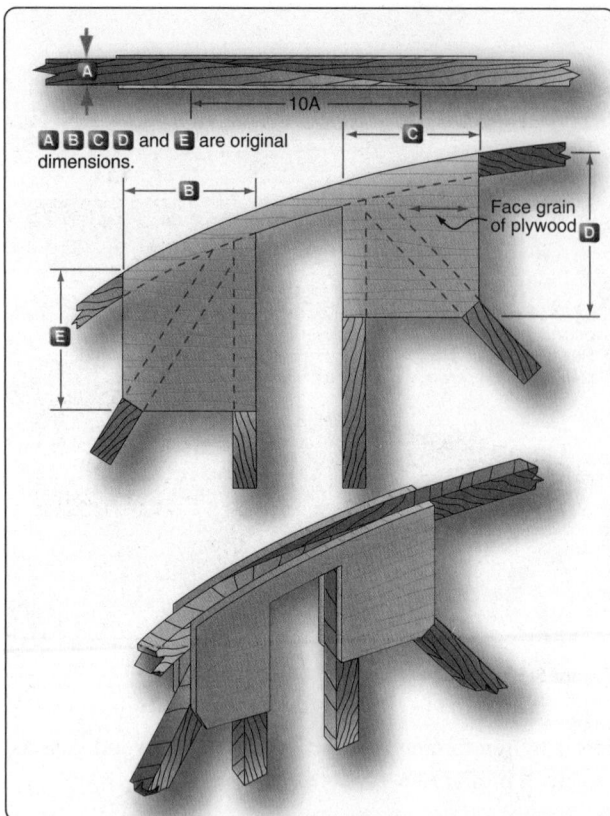

Figure 6-17. *Cap strip repair at a spar.*

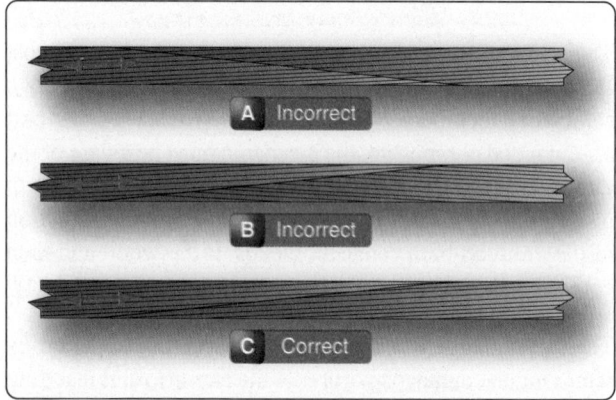

Figure 6-18. *Relationship of scarf slope to grain slope.*

and lower cap strips, the web member, and the compression members are cracked completely through. To facilitate this repair, cut the compression members as shown in *Figure 6-20D* and repair as recommended using replacement sections to the rear spar. Cut the damaged cap strips and repair as shown in *Figure 6-20*, replacing the aft section of the cap strips. Plywood side plates are then bonded on each side diagonally to reinforce the damaged web as shown in *Figure 6-20, A-A.*

Figure 6-20B illustrates a compression rib of the type that is

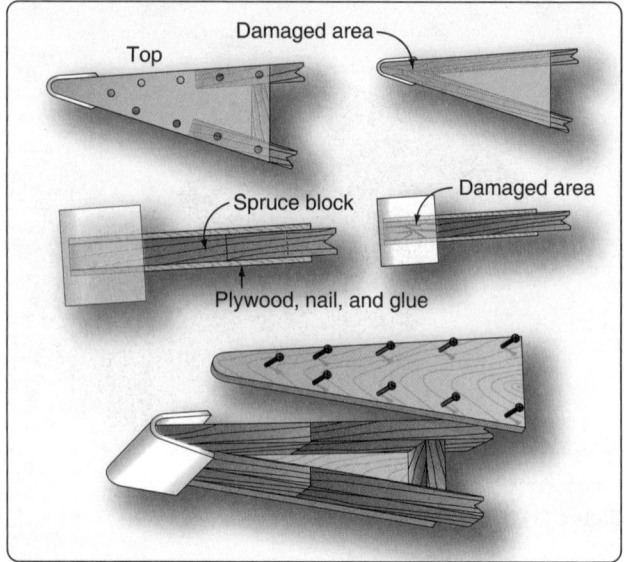

Figure 6-19. *Rib trailing edge repair.*

a standard rib with rectangle compression members added to one side and a plywood web to the other side. The method used in this repair is essentially the same as in *Figure 6-20A*, except that the plywood reinforcement plate, shown in *Figure 6-20B-B*, is continued the full distance between the spars.

Figure 6-20C illustrates a compression rib of the I type with a rectangular vertical member on each side of the web. The method of repair is essentially the same as in *Figure 6-20A*, except the plywood reinforcement plates on each side, shown in *Figure 6-20C-C*, are continued the full distance between the spars.

Wing Spar Repairs

Wood wing spars are fabricated in various designs using solid wood, plywood, or a combination of the two. *[Figure 6-21]*

When a spar is damaged, the method of repair must conform to the manufacturer's instructions and recommendations. In the absence of manufacturer's instructions, contact the FAA for advice and approval before making repairs to the spar and following recommendations in AC 43.13-1. If instructions are not available for a specific type of repair, it is highly recommended that you request appropriate engineering assistance to evaluate and provide guidance for the intended repair.

Shown in *Figure 6-22* is a recommended method to repair either a solid or laminated rectangle spar. The slope of the scarf in any stressed part, such as a spar, should not be steeper than 15 to 1.

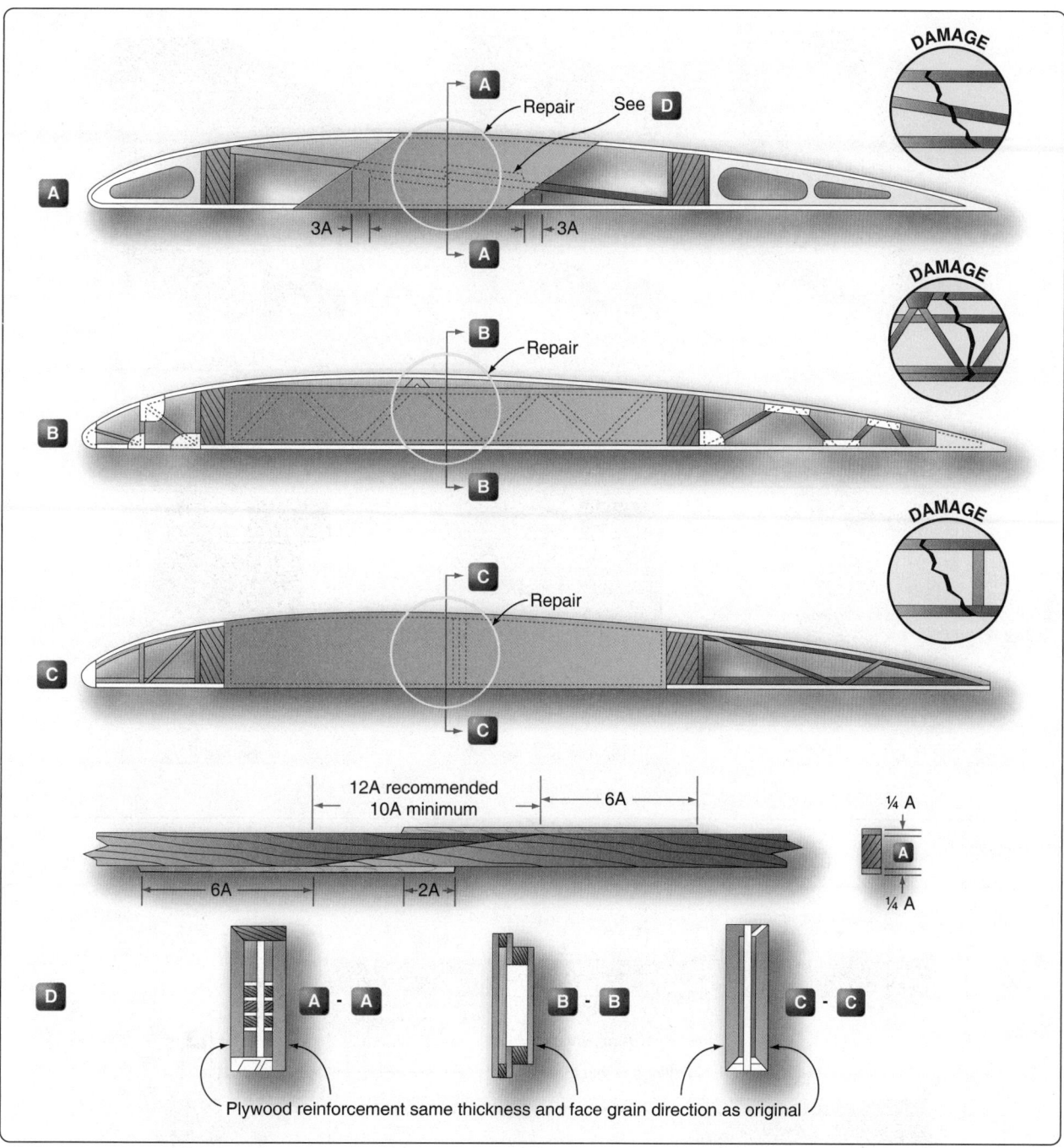

Figure 6-20. *Typical compression rib repair.*

Unless otherwise specified by the aircraft manufacturer, a damaged spar may be spliced at almost any point except at wing attachment fittings, landing gear fittings, engine mount fittings, or lift-and-interplane strut fittings. These fittings may not overlap any part of the splice. The reinforcement plates of the splice should not interfere with the proper attachment or alignment of the fittings. Taper reinforcement plates on the ends at a 5:1 slope *[Figure 6-23]*.

The use of a scarf joint to repair a spar or any other component

of an aircraft is dependent on the accessibility to the damaged section. It may not be possible to utilize a scarf repair where recommended, so the component may have to be replaced. A scarf must be precisely cut on both adjoining pieces to ensure an even thin glue line; otherwise, the joint may not achieve full strength. The primary difficulty encountered in making this type of joint is obtaining the same bevel on each piece. *[Figure 6-24]*

The mating surfaces of the scarf must be smooth. You can

Box
I
Double I
C
Plain rectangular
Routed

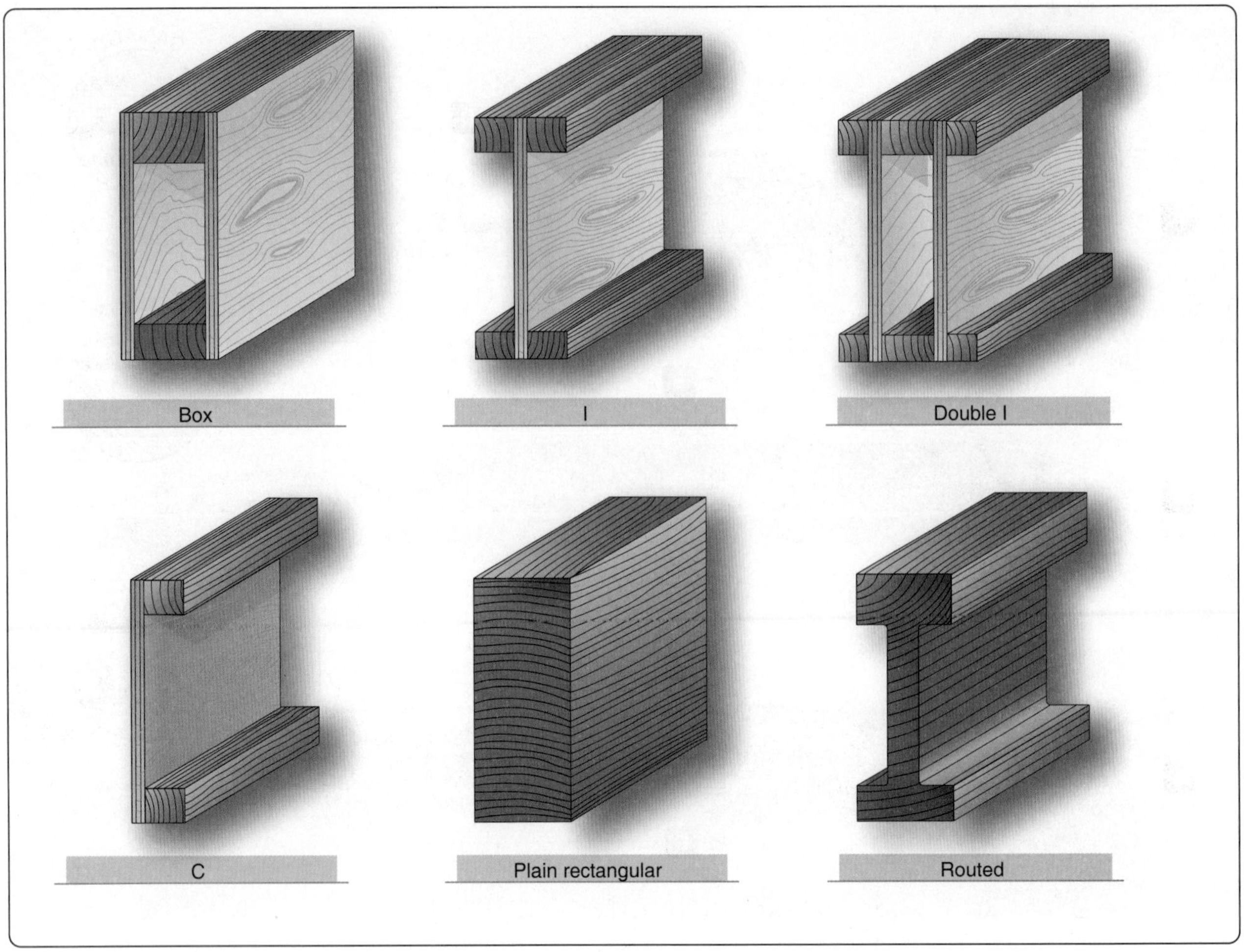

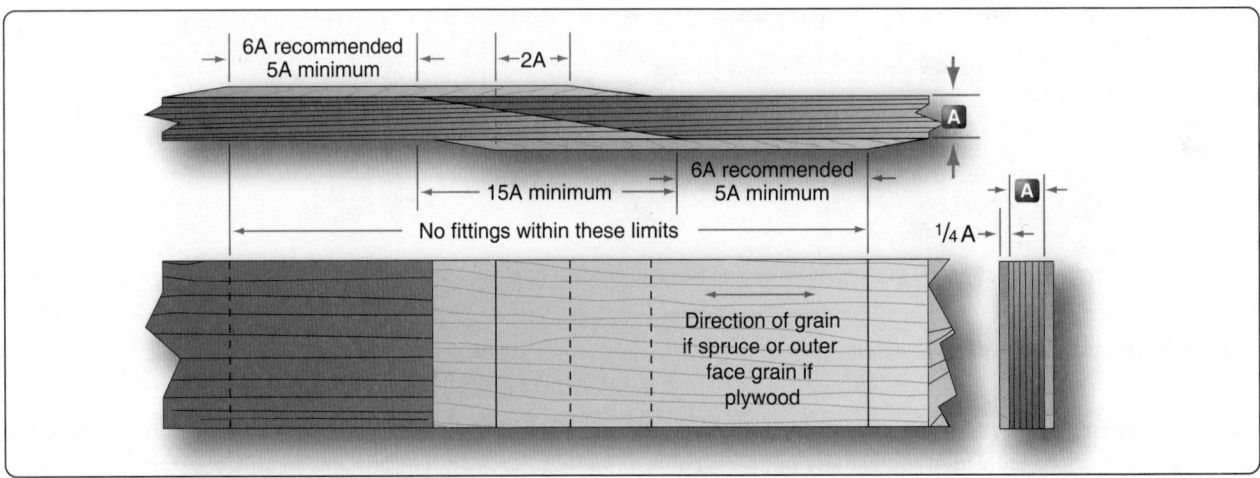

Figure 6-21. *Typical splice repair of solid rectangular spar.*

6A recommended
5A minimum
2A
A
15A minimum
6A recommended
5A minimum
A
No fittings within these limits
1/4 A
Direction of grain
if spruce or outer
face grain if
plywood

Figure 6-22. *Typical splice repair of solid rectangular spar.*

machine smooth a saw cut using any of a variety of tools, such as a plane, a joiner, or a router. For most joints, you need a beveled fixture set at the correct slope to complete the cut. *Figure 6-25* illustrates one method of producing an accurate scarf joint.

Once the two bevels are cut for the intended splice, clamp the pieces to a flat guide board of similar material. Then, work a sharp, fine-tooth saw all the way through the joint. Remove the saw, decrease pressure, and tap one of the pieces on the end to close the gap. Work the saw again through the joint. Continue this procedure until the joint is perfectly parallel

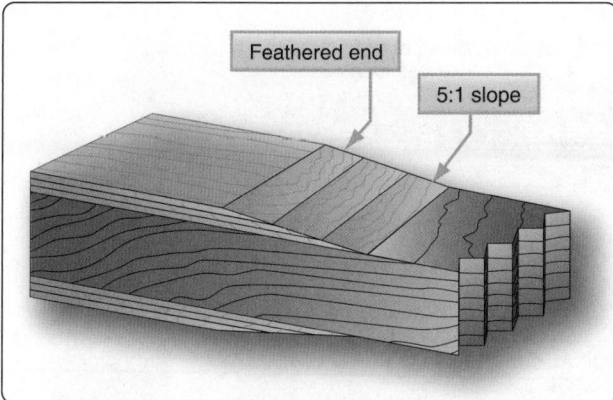

Figure 6-23. *Tapered faceplate.*

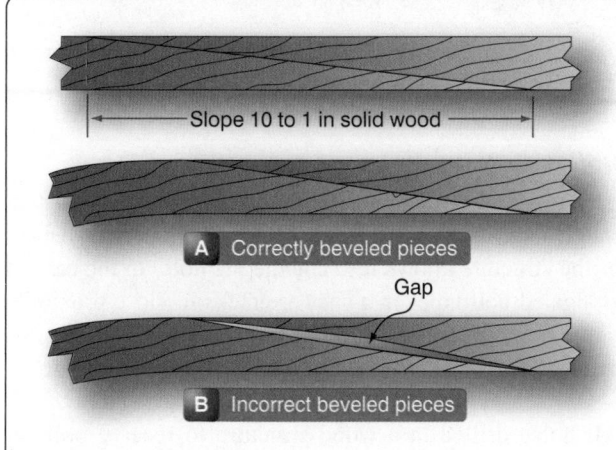

Figure 6-24. *Beveled scarf joint.*

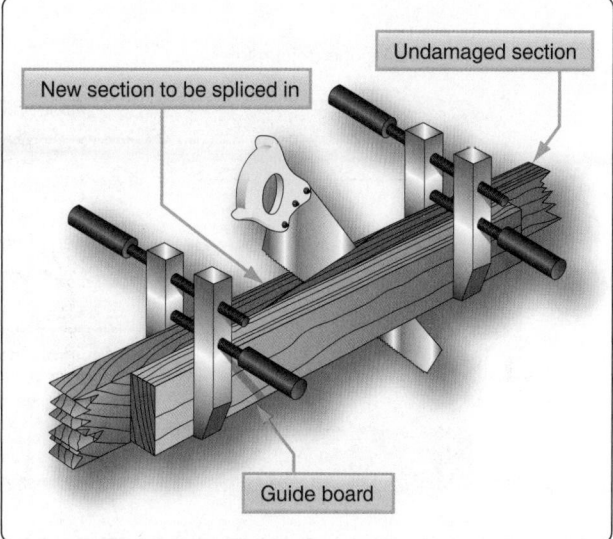

Figure 6-25. *Making a scarf joint.*

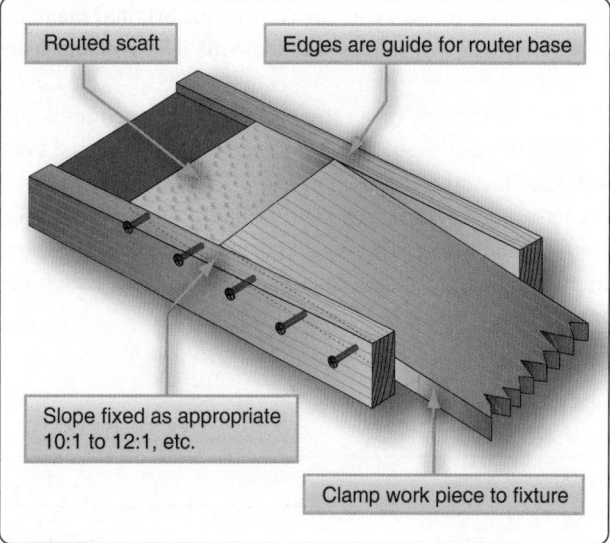

Figure 6-26. *Scarf cutting fixture.*

with matching surfaces. Then, make a light cut with the grain, using a sharp plane, to smooth both mating surfaces.

Another method of cutting a scarf uses a simple scarf-cutting fixture that you can also fabricate for use with a router. Extend the work piece beyond the edge so the finished cut results in a feathered edge across the end of the scarf. *[Figure 6-26]* There are numerous tools made by individuals, and there are commercial plans for sale with instructions for building scarf-cutting tools. Most of them work, but some are better than others. The most important requirement for the tool is that it produces a smooth, repeatable cut at the appropriate angle.

Local damage to the top or bottom edge of a solid spar may be repaired by removing the damaged portion and fabricating a replacement filler block of the same material as the spar. Full width doublers are fabricated as shown and then all three pieces are glued and clamped to the spar. Nails or screws should not be used in spar repairs. A longitudinal crack in a solid spar may be repaired using doublers made from the proper thickness plywood. Care must be taken to ensure the doublers extend the minimum distance beyond the crack.

[Figure 6-27]

A typical repair to a built-up I spar is illustrated using plywood reinforcement plates with solid wood filler blocks. As with all repairs, the reinforcement plate ends should be feathered out to a 5:1 slope. *[Figure 6-28]*

Repair methods for the other types of spar illustrated at the start of this section all follow the basic steps of repair. The wood used should be of the same type and size as the original spar. Always splice and reinforce plywood webs with the same type of plywood as the original. Do not use solid wood to replace plywood webs because plywood is stronger in shear than solid wood of the same thickness. The splices and scarf cuts must be of the correct slope for the repair with the face

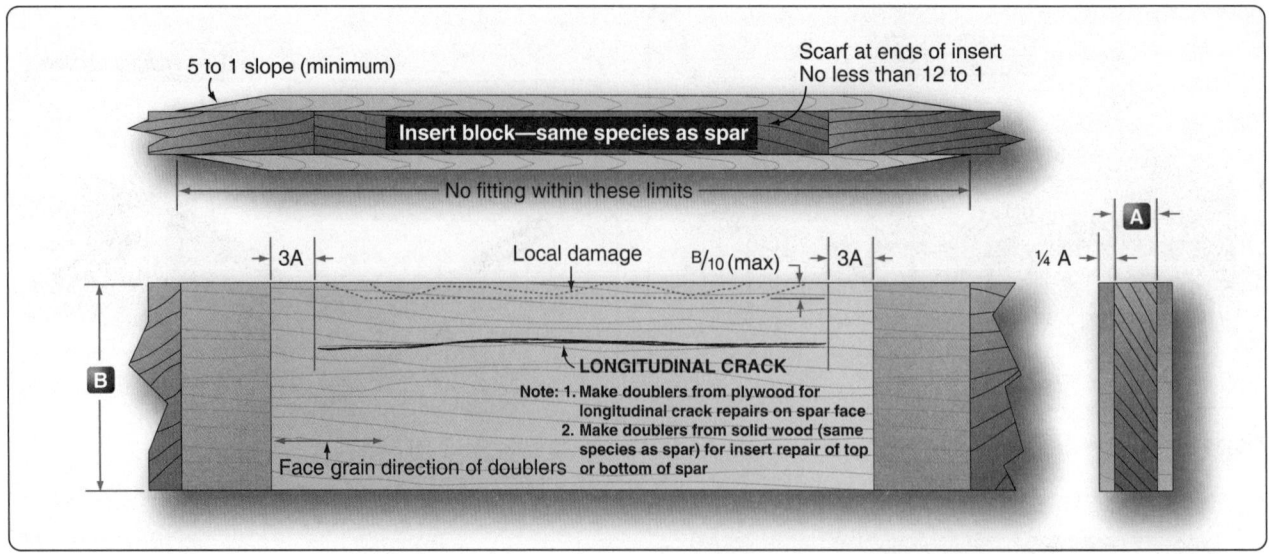

Figure 6-27. *A method to repair damage to solid spar.*

grain running in the same direction as the original member. Not more than two splices should be made in any one spar.

When a satisfactory repair to a spar cannot be accomplished, the spar should be replaced. New spars may be obtained from the manufacturer or the holder of a PMA for that part. An owner-produced spar may be installed provided it is made from a manufacturer-approved drawing. Care should be taken to ensure that any replacement spars accurately match the manufacturer's original design.

Bolt & Bushing Holes

All bolts and bushings used in aircraft structures must fit snugly into the holes. If the bolt or bushing is loose, movement of the structure allows it to enlarge the hole. In the case of elongated bolt holes in a spar or cracks in close proximity to the bolt holes, the repair may require a new section to be spliced in the spar, or replacement of the entire spar.

All holes drilled in a wood structure to receive bolts or bushings should be of such size that inserting the bolt or bushing requires a light tapping with a wood or rawhide mallet. If the hole is so tight that heavy blows are necessary,

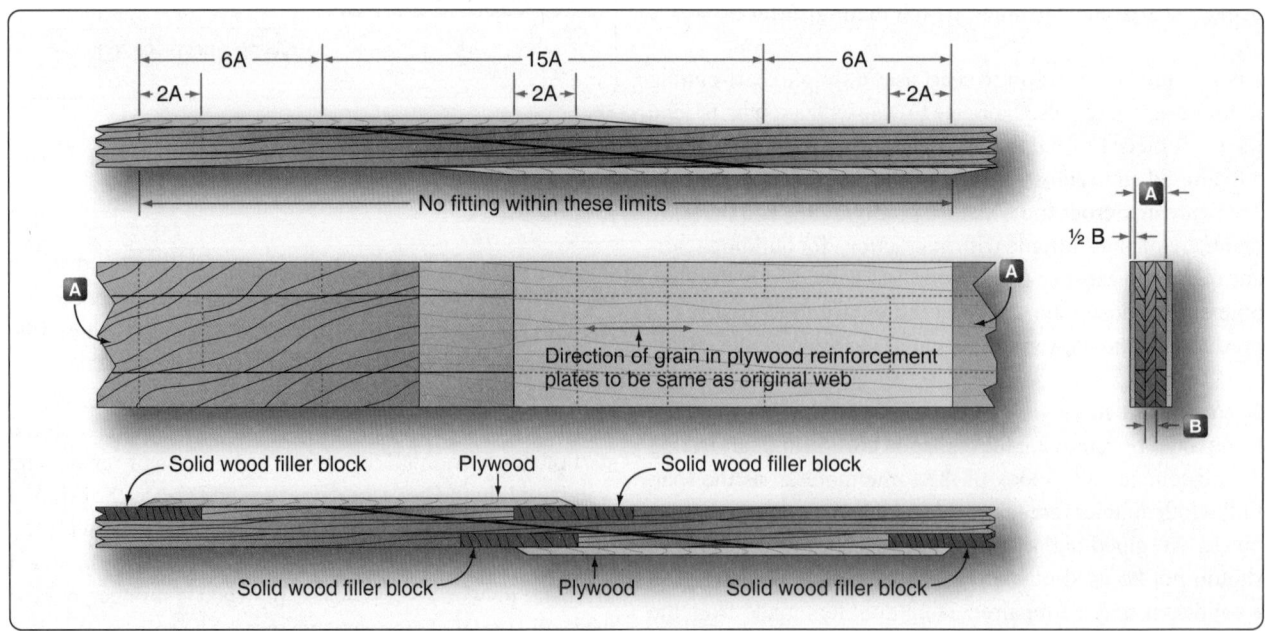

Figure 6-28. *Repairs to a built-up I spar.*

deformation of the wood may cause splitting or unequal load distribution.

For boring accurate smooth holes, it is recommended that a drill press be utilized where possible. Holes should be drilled with sharp bits using slow steady pressure. Standard twist drills can be used in wood when sharpened to a 60° angle. However, a better designed drill was developed for wood boring called a lip and spur or brad point. The center of the drill has a spur with a sharp point and four sharp corners to center and cut rather than walk as a conventional drill sometimes does. It has the outside corner of the cutting edges leading, so that it cuts the periphery of the hole first and maximizes the chance that the wood fibers cut cleanly, leaving a smooth bore.

Forstner bits bore precise, flat bottomed holes in wood, in any orientation with respect to the wood grain. They must be used in a drill press because more force is needed for their cutting action. Also, they are not designed to clear chips from the hole and must be pulled out periodically to do this. A straight, accurate bore-through hole can be completed by drilling through the work piece and into a piece of wood backing the work piece.

All holes bored for bolts that are to hold fittings in place should match the hole diameter in the fitting. Bushings made of steel, aluminum, or plastic are sometimes used to prevent crushing the wood when bolts are tightened. Holes drilled in the wood structure should be sealed after being drilled. This can be accomplished by application of varnish or other acceptable sealer into the open hole. The sealer must be allowed to dry or cure thoroughly prior to the bolts or bushings being installed.

Plywood Skin Repairs

Plywood skin can be repaired using a number of different methods depending on the size of the hole and its location on the aircraft. Manufacturer's instructions, when available, should be the first source of a repair scheme. AC 43.13-1 provides other acceptable methods of repair. Some of those are featured in the following section.

Fabric Patch

A fabric patch is the simplest method to repair a small hole in plywood. This repair is used on holes not exceeding 1-inch in diameter after being trimmed to a smooth outline. The edges of the trimmed hole should first be sealed, preferably with a two-part epoxy varnish. This varnish requires a long cure time, but it provides the best seal on bare wood.

The fabric used for the patch should be of an approved material using the cement recommended by the manufacturer of the fabric system. The fabric patch should be cut with pinking shears and overlap the plywood skin by at least 1-inch. A fabric patch should not be used to repair holes in the leading edge of a wing, in the frontal area of the fuselage, or nearer than 1-inch to any frame member.

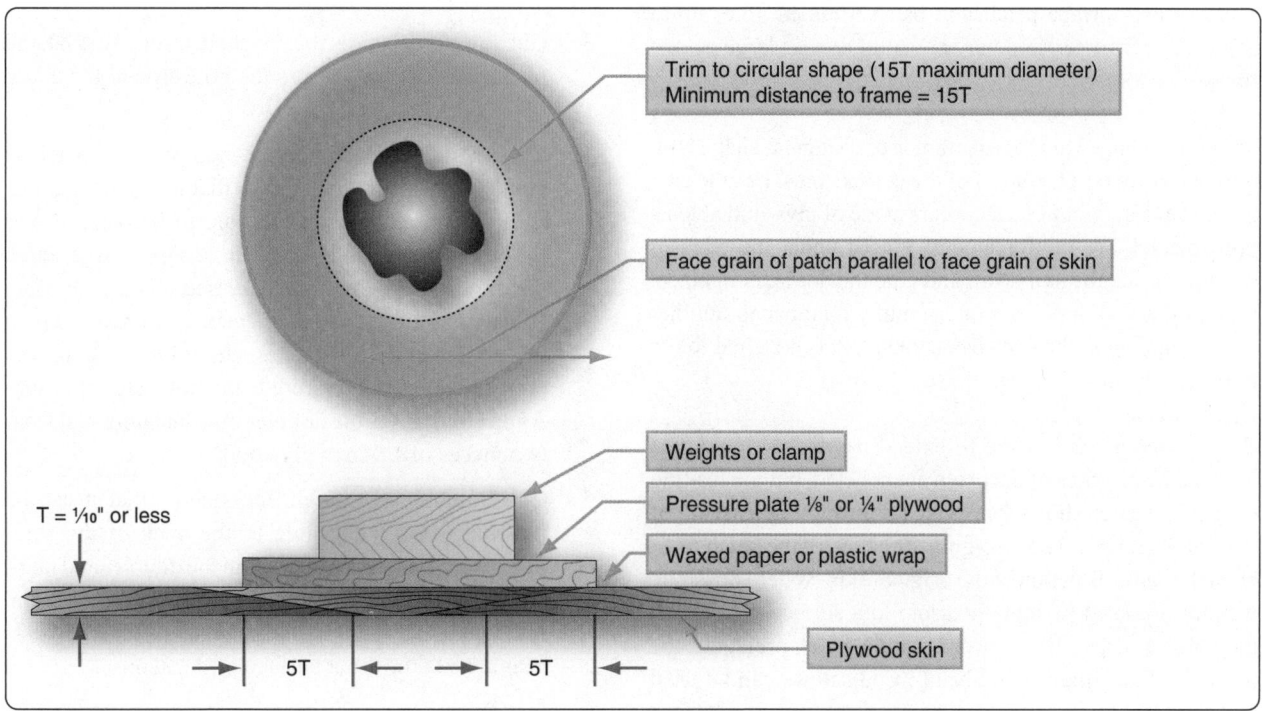

Figure 6-29. *Splayed patch.*

Splayed Patch

A splayed patch is a flush patch. The term splayed denotes that the edges of the patch are tapered, with the slope cut at a 5:1 ratio to the thickness of the skin. This may be used for small holes where the largest dimension of the hole to be repaired is not more than 15 times the skin thickness and the skin is not more than ⅒-inch thick. This calculates to nothing larger than a 1½-inch trimmed hole in very thin plywood.

Using the sample ⅒-inch thick plywood and a maximum trimmed hole size of 1½-inches, and cutting a 5:1 scarf, results in a 2½-inches round section to be patched. The patch should be fabricated with a 5:1 scarf, from the same type and thickness plywood as the surface being repaired.

Glue is applied to the beveled edges and the patch is set with the grain parallel to the surface being repaired. A pressure plate of thicker plywood cut to the exact size of the patch is centered over the patch covered with waxed paper. A suitable weight is used for pressure until the glue has set. The repair is then sanded and finished to match the original surface. *[Figure 6-29]*

Surface Patch

Plywood skins not over ⅛-inch thick that are damaged between or along framing members may be repaired with a surface or overlay patch. Surface patches located aft of the 10 percent chord line, or which wrap around the leading edge and terminate aft of the 10 percent chord line, are permissible. You can use surface patches to patch trimmed holes up to a 50-inch perimeter, and may cover an area as large as one frame or rib space.

Trim the damaged area to a rectangle or triangular shape with rounded corners. The radius of the corners must be at least 5 times the skin thickness. Doublers made of plywood at least ¼-inch thick are reinforcements placed under the edge of the hole inside the skin. Nail and glue the doublers in place. Extend the doublers from one framing member to another and strengthen at the ends by saddle gussets attached to the framing members. *[Figure 6-30]*

The surface patch is sized to extend beyond the cutout as indicated. All edges of the patch are beveled, but the leading edge of the patch should be beveled at an angle at least 4:1 of the skin thickness. The face-grain direction of the patch must be in the same direction of the original skin. Where possible, weights are used to apply pressure to a surface patch until the glue has dried. If the location of the patch precludes the use of weight, small round head wood screws can be used to apply glue pressure to secure the patch. After a surface patch has dried, the screws can be removed and the holes

filled. The patch should be covered with fabric that overlaps the original surface by at least 2-inches. The fabric should be from one of the approved fabric covering systems using the procedures recommended by the manufacturer to cement and finish the fabric.

Plug Patch

Two types of plug patch, oval and round, may be used on plywood skins. Because the plug patch is only a skin repair, use it only for damage that does not involve the supporting structure under the skin.

Cut the edges of a plug patch at right angles to the surface of the skin. Cut the skin also to a clean round or oval hole with edges at right angles to the surface. Cut the patch to the exact size of the hole; when installed, the edge of the patch forms a butt joint with the edge of the hole.

You can use a round plug patch where the cutout repair is no larger than 6-inches in diameter. Sample dimensions for holes of 4-inches and 6-inches in diameter appear in *Figure 6-31.*

The following steps provide a method for making a round plug patch:

1. Cut a round patch large enough to cover the intended repair. If applicable for size, use the sample dimensions in *Figure 6-31.* The patch must be of the same material and thickness as the original skin.

2. Place the patch over the damaged spot and mark a circle of the same size as the patch.

3. Cut the skin inside the marked circle so that the plug patch fits snugly into the hole around the entire perimeter.

4. Cut a doubler of soft quarter-inch plywood, such as poplar. A small patch is cut so that its outside radius is 5⁄8-inch greater than the hole to be patched and the inside radius is 5⁄8-inch less. For a large patch the dimensions would be increased to 7⁄8-inch each. If the curvature of the skin surface is greater than a rise of 1⁄8-inch in 6-inches, the doubler should be preformed to the curvature using hot water or steam. As an alternative, the doubler may be laminated from two pieces of 1⁄8-inch plywood.

5. Cut the doubler through one side so that it can be inserted through the hole to the back of the skin. Place the patch plug centered on the doubler and mark around its perimeter. Apply a coat of glue outside the line to the outer half of the doubler surface that will bear against the inner surface of the skin.

6. Install the doubler by slipping it through the cutout

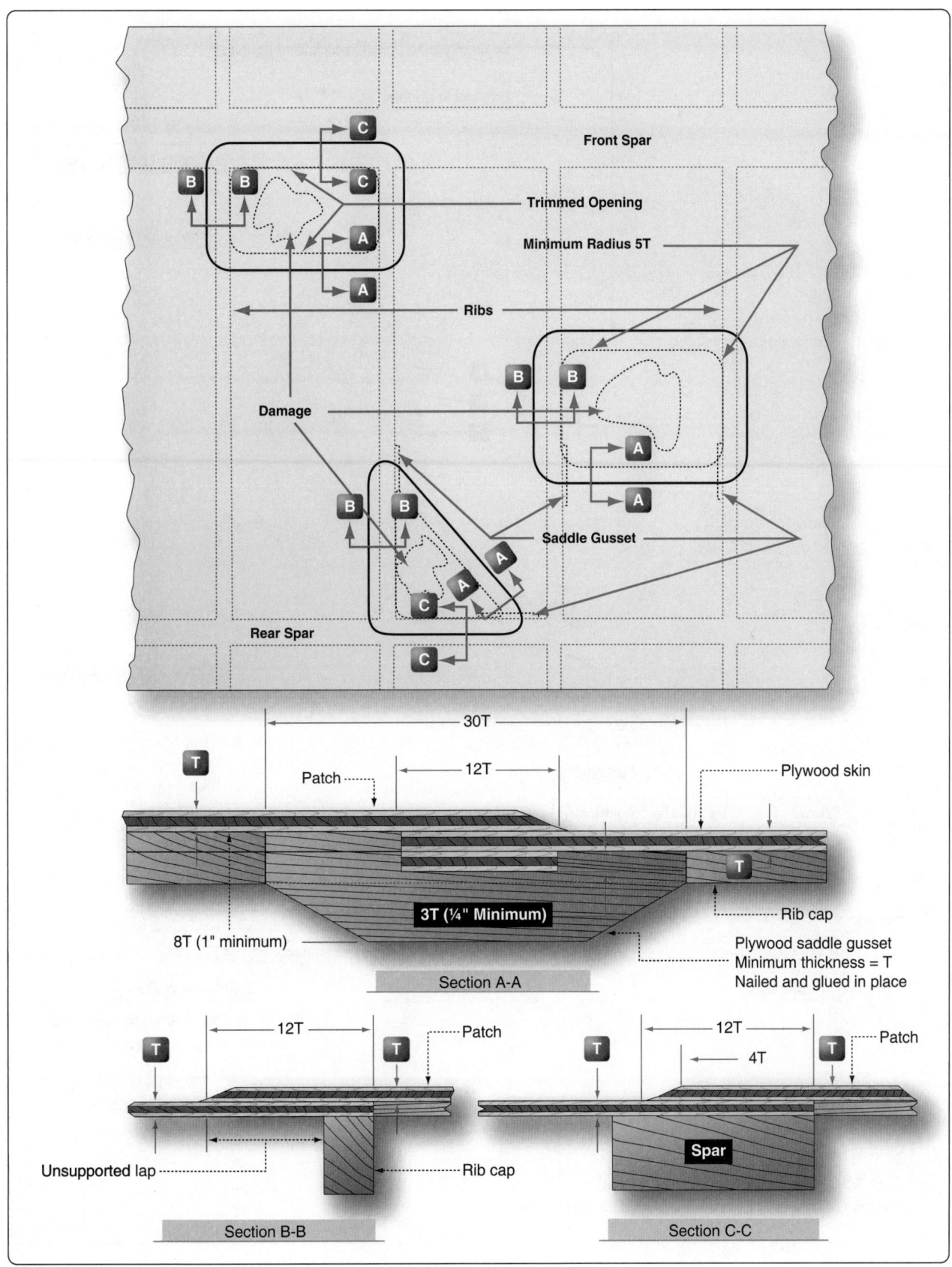

Figure 6-30. *Surfaces patches.*

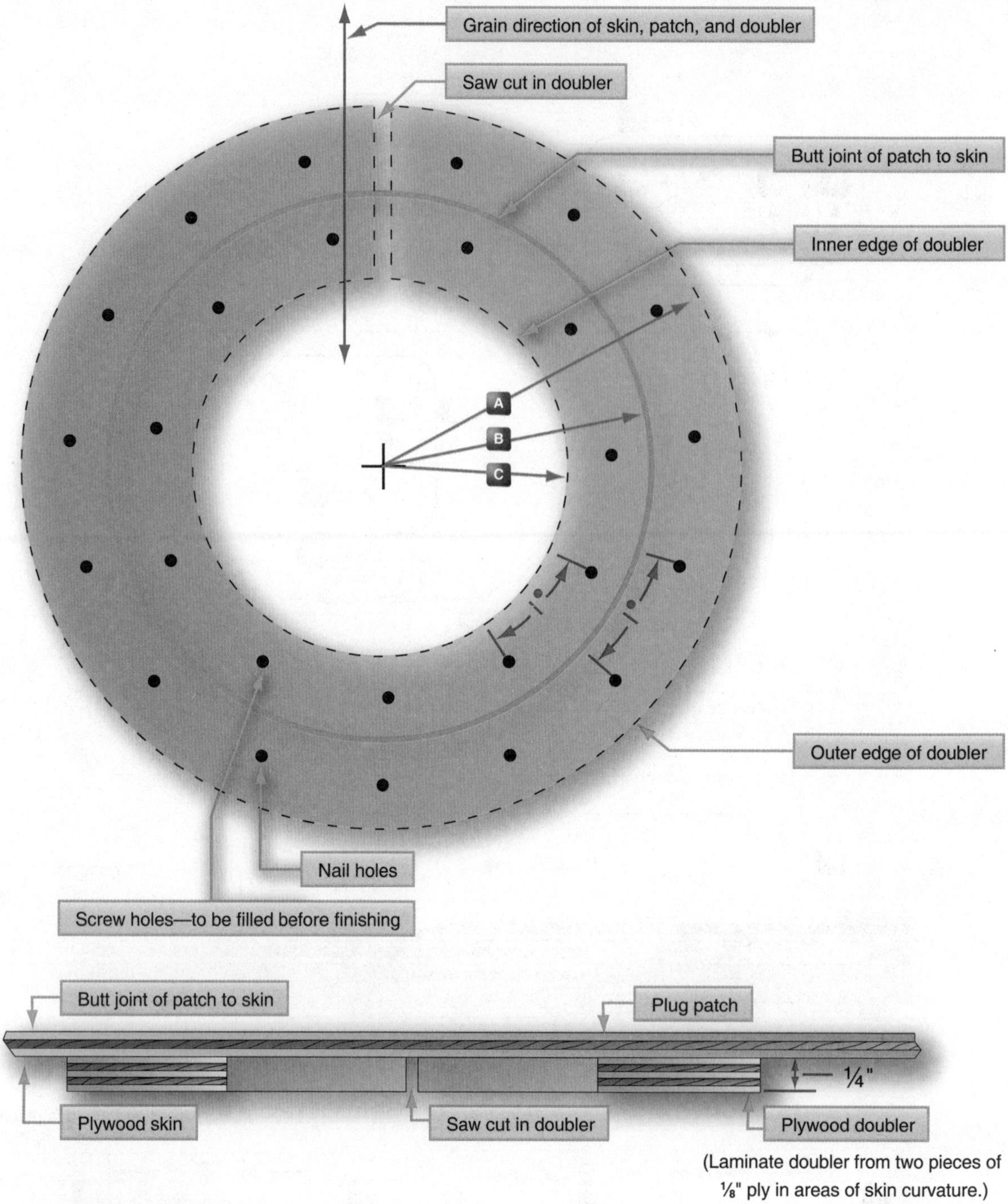
Grain direction of skin, patch, and doubler
Saw cut in doubler
Butt joint of patch to skin
Inner edge of doubler
A
B
C
Outer edge of doubler
Nail holes
Screw holes—to be filled before finishing
Butt joint of patch to skin
Plug patch
1/4"
Plywood skin
Saw cut in doubler
Plywood doubler
(Laminate doubler from two pieces of 1/8" ply in areas of skin curvature.)

DIMENSIONS

	A	B	C
Small circular plug patch	2 5/8"	2"	1 3/8"
Large circular plug patch	3 7/8"	3"	2 1/8"

(Two rows of screws and nails are required for a large patch.)

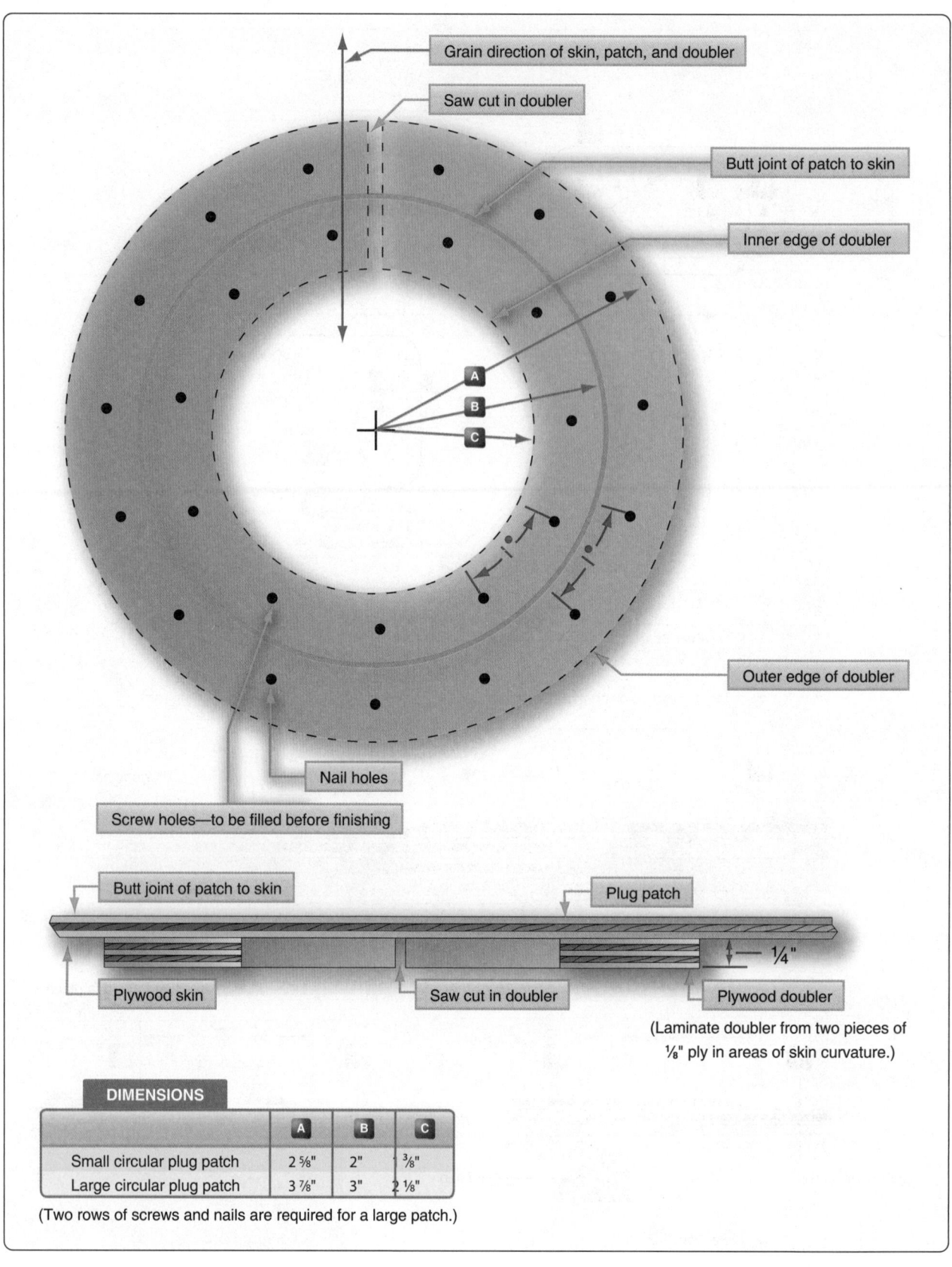

Figure 6-31. *Round plug patch assembly.*

hole and place it so that the mark is concentric with the hole. Nail it in place with nailing strips, while holding a bucking bar or similar object under the doubler for backup. Place waxed paper between the nailing strips and the skin. Cloth webbing under the nailing strips facilitates removal of the strips and nails after the glue dries.

7. After the glue has set for the installed doubler, and you have removed the nail strips, apply glue to the inner half of the doubler and to the patch plug. Drill holes around the plug's circumference to accept No. 4 round head wood screws. Insert the plug with the grain aligned to the surface wood.

8. Apply the pressure to the patch by means of the wood screws. No other pressure is necessary.

9. After the glue has set, remove the screws and fill the nail and screw holes. Sand and finish to match the original surface.

The steps for making an oval plug patch are identical to those for making the round patch. The maximum dimensions for large oval patches are 7-inches long and 5-inches wide. Oval patches must be cut, so when installed, the face grain matches the direction of the original surface. *[Figure 6-32]*

Scarf Patch

A properly prepared and installed scarf patch is the best repair for damaged plywood and is preferred for most skin repairs. The scarf patch has edges beveled at a 12:1 slope; the splayed patch is beveled at a 5:1 slope. The scarf patch also uses reinforcements under the patch at the glue joints.

Much of the outside surface of a plywood aircraft is curved. If the damaged plywood skin has a radius of curvature not greater than 100 times the skin thickness, you can install a scarf patch. However, it may be necessary to soak or steam the patch, to preform it prior to gluing it in place. Shape backing blocks or other reinforcements to fit the skin curvature.

You can make scarf cuts in plywood with various tools, such as a hand plane, spoke shave, a sharp scraper, or sanding block. Sawn or roughly filed surfaces are not recommended because they are normally inaccurate and do not form the best glue joint.

The Back of the Skin is Accessible for Repair

When the back of a damaged plywood skin is accessible, such as a fuselage skin, repair it with scarf patches cut and

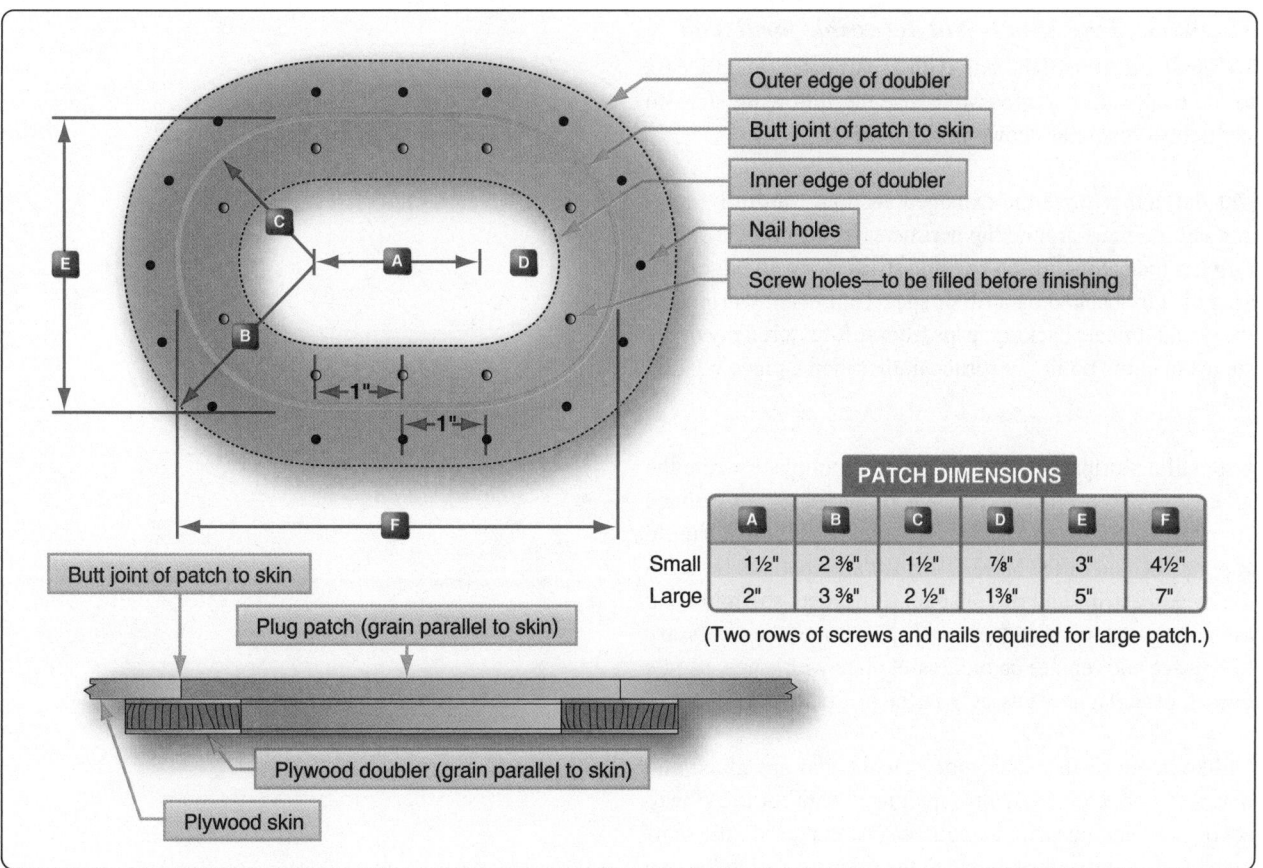

PATCH DIMENSIONS						
	A	B	C	D	E	F
Small	1½"	2⅜"	1½"	⅞"	3"	4½"
Large	2"	3⅜"	2½"	1⅜"	5"	7"

(Two rows of screws and nails required for large patch.)

Figure 6-32. *An oval plug patch.*

installed with the grain parallel to the surface skin. Details for this type of repair are shown in *Figure 6-33*.

Figure 6-33, Section A-A, shows methods of support for a scarf between frame members using permanent backing and gussets. When the damage follows or extends to a framing member, support the scarf as shown in section B-B. When the scarf does not quite extend to a frame member, support the patch as shown in section C-C.

Damage that does not exceed 25 times the skin thickness (3⅛-inches for ⅛-inch thick skin) after being trimmed to a circular shape can be repaired as shown in section D-D, provided the trimmed opening is not nearer than 15 times the skin thickness to a frame member (1⅞-inches for ⅛-inch thick skin).

A temporary backing block is carefully shaped from solid wood and fitted to the inside surface of the skin. A piece of waxed paper or plastic wrap is placed between the block and the underside of the skin. The scarf patch is installed and temporarily attached to the backing block, being held together in place with nailing strips. When the glue sets, remove the nails and block, leaving a flush surface on both sides of the repaired skin.

The Back of the Skin Is Not Accessible for Repair

To repair a section of the skin with a scarf patch when access to the back side is not possible, use the following steps to facilitate a repair, as shown in *Figure 6-34*.

Cut out and remove the damaged section. Carefully mark and cut the scarf around the perimeter of the hole. Working through the cutout, install backing strips along all edges that are not fully backed by a rib or spar. To prevent warping of the skin, fabricate backing strips from soft-textured plywood, such as yellow poplar or spruce, rather than a piece of solid wood.

Use nailing strips to hold backing strips in place while the glue sets. Use a bucking bar, where necessary, to provide support for nailing. A saddle gusset of plywood should support the end of the backing strip at all junctions between the backing strips and ribs or spars. If needed, nail and bond the new gusset plate to the rib or spar. It may be necessary to remove and replace an old gusset plate with a new saddle gusset, or nail a new gusset over the original.

Unlike some of the other type patches that are glued and installed as one process, this repair must wait for the glue to set on the backing strips and gussets. At that point, the scarf patch can be cut and fit to match the grain, and glued, using weight for pressure on the patch as appropriate. When dry, fill and finish the repair to match the original surface.

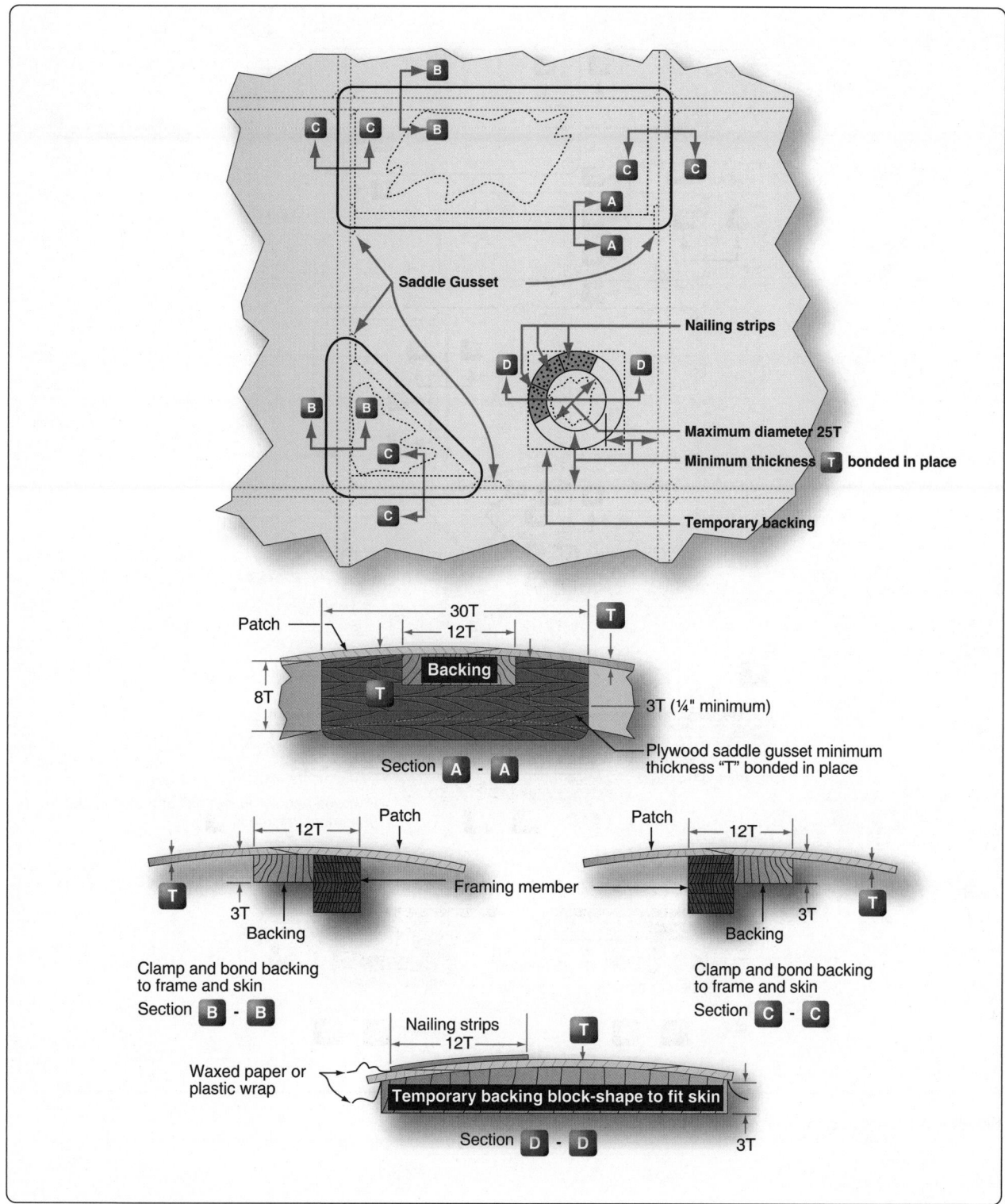

Figure 6-33. *Scarf patches, back of skin accessible.*

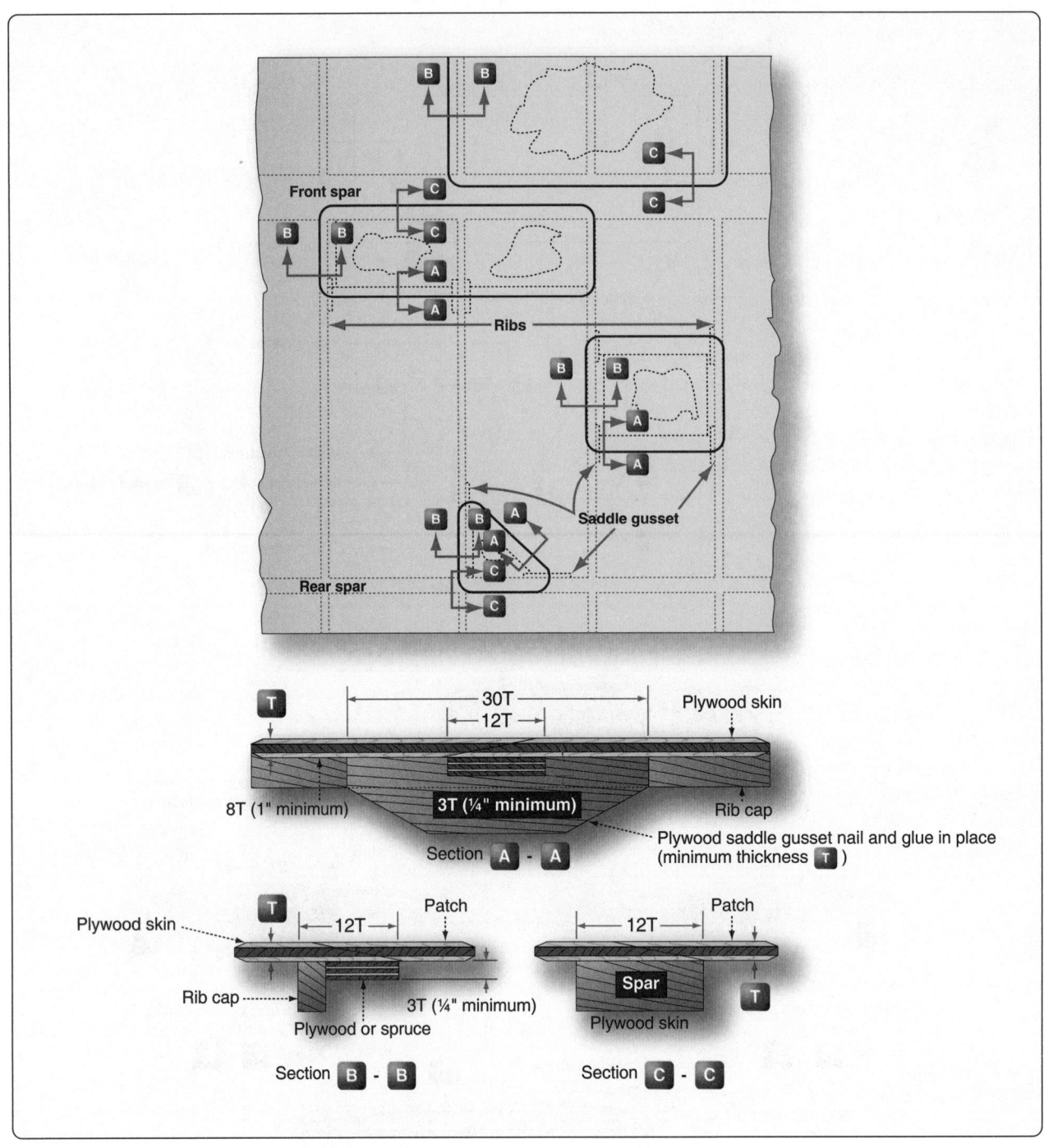

Figure 6-34. *Scarf patches, back of skin not accessible.*

Chapter 7
Advanced Composite Materials

Description of Composite Structures

Introduction

Composite materials are becoming more important in the construction of aerospace structures. Aircraft parts made from composite materials, such as fairings, spoilers, and flight controls, were developed during the 1960s for their weight savings over aluminum parts. New generation large aircraft are designed with all composite fuselage and wing structures, and the repair of these advanced composite materials requires an in-depth knowledge of composite structures, materials, and tooling. The primary advantages of composite materials are their high strength, relatively low weight, and corrosion resistance.

Laminated Structures

Composite materials consist of a combination of materials that are mixed together to achieve specific structural properties. The individual materials do not dissolve or merge completely in the composite, but they act together as one. Normally, the components can be physically identified as they interface with one another. The properties of the composite material are superior to the properties of the individual materials from which it is constructed.

An advanced composite material is made of a fibrous material embedded in a resin matrix, generally laminated with fibers oriented in alternating directions to give the material strength and stiffness. Fibrous materials are not new; wood is the most common fibrous structural material known to man.

Applications of composites on aircraft include:

- Fairings
- Flight control surfaces
- Landing gear doors
- Leading and trailing edge panels on the wing and stabilizer
- Interior components
- Floor beams and floor boards
- Vertical and horizontal stabilizer primary structure on large aircraft
- Primary wing and fuselage structure on new generation large aircraft

- Turbine engine fan blades
- Propellers

Major Components of a Laminate

An isotropic material has uniform properties in all directions. The measured properties of an isotropic material are independent of the axis of testing. Metals such as aluminum and titanium are examples of isotropic materials.

A fiber is the primary load carrying element of the composite material. The composite material is only strong and stiff in the direction of the fibers. Unidirectional composites have predominant mechanical properties in one direction and are said to be anisotropic, having mechanical and/or physical properties that vary with direction relative to natural reference axes inherent in the material. Components made from fiber-reinforced composites can be designed so that the fiber orientation produces optimum mechanical properties, but they can only approach the true isotropic nature of metals, such as aluminum and titanium.

A matrix supports the fibers and bonds them together in the composite material. The matrix transfers any applied loads to the fibers, keeps the fibers in their position and chosen orientation, gives the composite environmental resistance, and determines the maximum service temperature of a composite.

Strength Characteristics

Structural properties, such as stiffness, dimensional stability, and strength of a composite laminate, depend on the stacking sequence of the plies. The stacking sequence describes the distribution of ply orientations through the laminate thickness. As the number of plies with chosen orientations increases, more stacking sequences are possible. For example, a symmetric eight-ply laminate with four different ply orientations has 24 different stacking sequences.

Fiber Orientation

The strength and stiffness of a composite buildup depends on the orientation sequence of the plies. The practical range of strength and stiffness of carbon fiber extends from values as low as those provided by fiberglass to as high as those provided by titanium. This range of values is determined by the orientation of the plies to the applied load. Proper selection of ply orientation in advanced composite materials is necessary to provide a structurally efficient design. The part might require 0° plies to react to axial loads, ±45° plies to react to shear loads, and 90° plies to react to side loads. Because the strength design requirements are a function of the applied load direction, ply orientation and ply sequence have to be correct. It is critical during a repair to replace each damaged ply with a ply of the same material and ply orientation.

The fibers in a unidirectional material run in one direction and the strength and stiffness is only in the direction of the fiber. Pre-impregnated (prepreg) tape is an example of a unidirectional ply orientation.

The fibers in a bidirectional material run in two directions, typically 90° apart. A plain weave fabric is an example of a bidirectional ply orientation. These ply orientations have strength in both directions but not necessarily the same strength. *[Figure 7-1]*

The plies of a quasi-isotropic layup are stacked in a 0°, –45°, 45°, and 90° sequence or in a 0°, –60°, and 60° sequence. *[Figure 7-2]* These types of ply orientation simulate the properties of an isotropic material. Many aerospace composite structures are made of quasi-isotropic materials.

Warp Clock

Warp indicates the longitudinal fibers of a fabric. The warp is the high strength direction due to the straightness of the fibers. A warp clock is used to describe direction of fibers on a diagram, spec sheet, or manufacturer's sheets. If the warp clock is not available on the fabric, the orientation is defaulted to zero as the fabric comes off the roll. Therefore, 90° to zero is the width of the fabric across. *[Figure 7-3]*

Fiber Forms

All product forms generally begin with spooled unidirectional raw fibers packaged as continuous strands. An individual fiber is called a filament. The word strand is also used to identify an individual glass fiber. Bundles of filaments are identified as tows, yarns, or rovings. Fiberglass yarns are twisted, while Kevlar® yarns are not. Tows and rovings do not have any twist. Most fibers are available as dry fiber that needs to be impregnated (impreg) with a resin before use or prepreg materials where the resin is already applied to the fiber.

Roving

A roving is a single grouping of filament or fiber ends, such

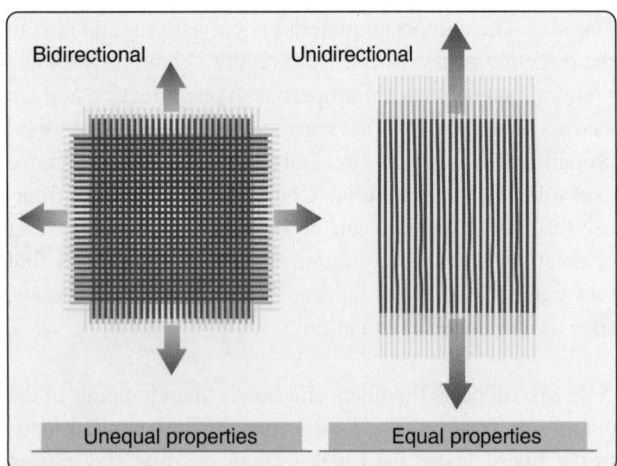

Figure 7-1. *Bidirectional and unidirectional material properties.*

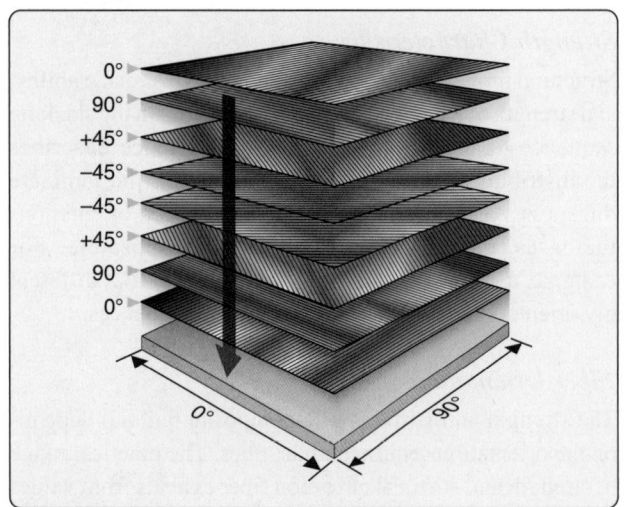

Figure 7-2. *Quasi-isotropic material layup.*

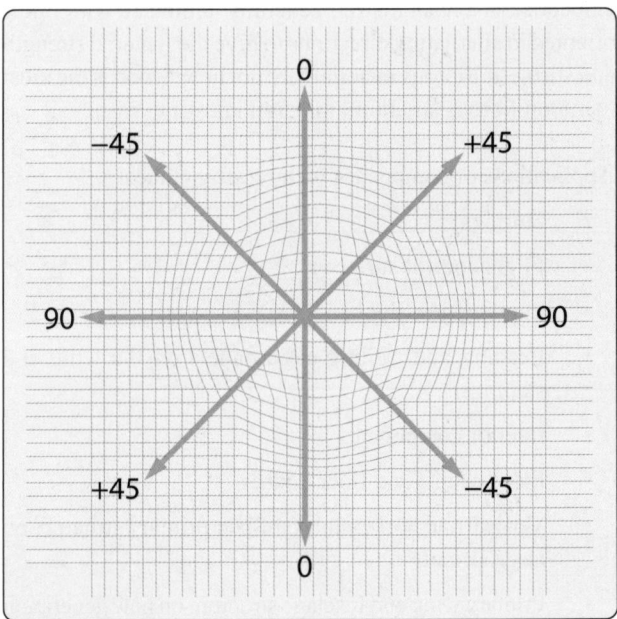

Figure 7-3. *A warp clock.*

as 20-end or 60-end glass rovings. All filaments are in the same direction and they are not twisted. Carbon rovings are usually identified as 3K, 6K, or 12K rovings, K meaning 1,000 filaments. Most applications for roving products utilize mandrels for filament winding and then resin cure to final configuration.

Unidirectional (Tape)

Unidirectional prepreg tapes have been the standard within the aerospace industry for many years, and the fiber is typically impregnated with thermosetting resins. The most common method of manufacture is to draw collimated raw (dry) strands into the impregnation machine where hot melted resins are combined with the strands using heat and pressure. Tape products have high strength in the fiber direction and virtually no strength across the fibers. The fibers are held in place by the resin. Tapes have a higher strength than woven fabrics. *[Figure 7-4]*

Bidirectional (Fabric)

Most fabric constructions offer more flexibility for layup of complex shapes than straight unidirectional tapes offer. Fabrics offer the option for resin impregnation either by solution or the hot melt process. Generally, fabrics used for structural applications use like fibers or strands of the same weight or yield in both the warp (longitudinal) and fill (transverse) directions. For aerospace structures, tightly woven fabrics are usually the choice to save weight, minimizing resin void size, and maintaining fiber orientation during the fabrication process.

Woven structural fabrics are usually constructed with reinforcement tows, strands, or yarns interlocking upon themselves with over/under placement during the weaving process. The more common fabric styles are plain or satin weaves. The plain weave construction results from each fiber alternating over and then under each intersecting strand (tow, bundle, or yarn). With the common satin weaves, such as 5 harness or 8 harness, the fiber bundles traverse both in warp and fill directions changing over/under position less frequently.

These satin weaves have less crimp and are easier to distort than a plain weave. With plain weave fabrics and most 5 or 8 harness woven fabrics, the fiber strand count is equal in both warp and fill directions. For example, 3K plain weave often has an additional designation, such as 12 x 12, meaning there are twelve tows per inch in each direction. This count designation can be varied to increase or decrease fabric weight or to accommodate different fibers of varying weight. *[Figure 7-5]*

Nonwoven (Knitted or Stitched)

Knitted or stitched fabrics can offer many of the mechanical advantages of unidirectional tapes. Fiber placement can be straight or unidirectional without the over/under turns of woven fabrics. The fibers are held in place by stitching with fine yarns or threads after preselected orientations of one or more layers of dry plies. These types of fabrics offer a wide range of multi-ply orientations. Although there may be some added weight penalties or loss of some ultimate reinforcement fiber properties, some gain of interlaminar shear and toughness properties may be realized. Some common stitching yarns are polyester, aramid, or thermoplastics. *[Figure 7-6]*

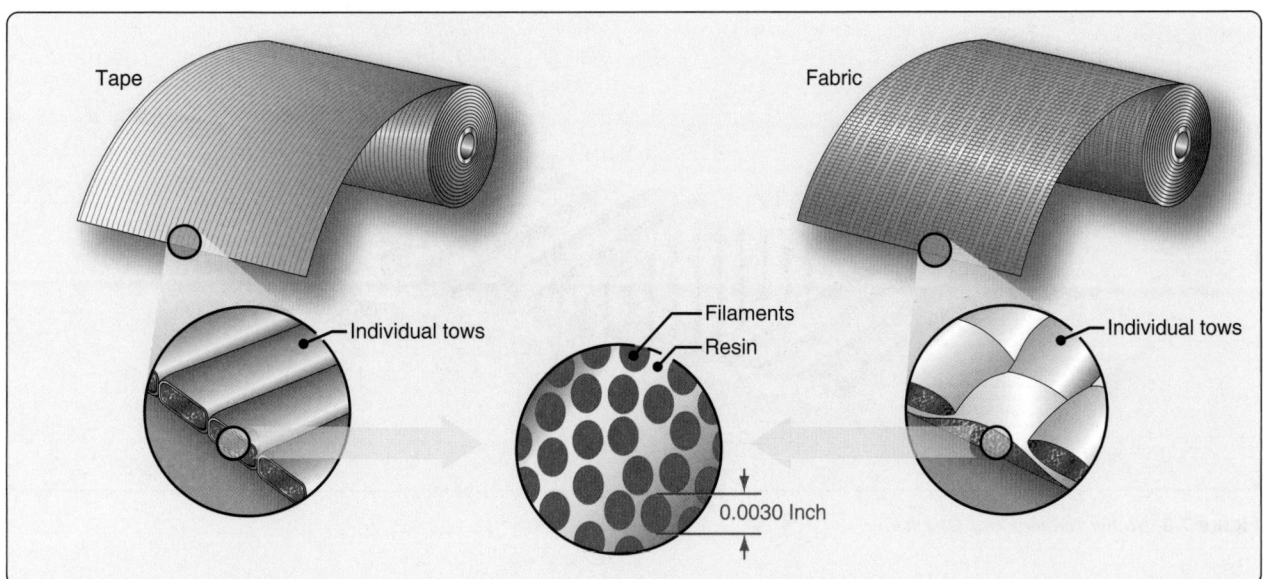

Figure 7-4. *Tape and fabric products.*

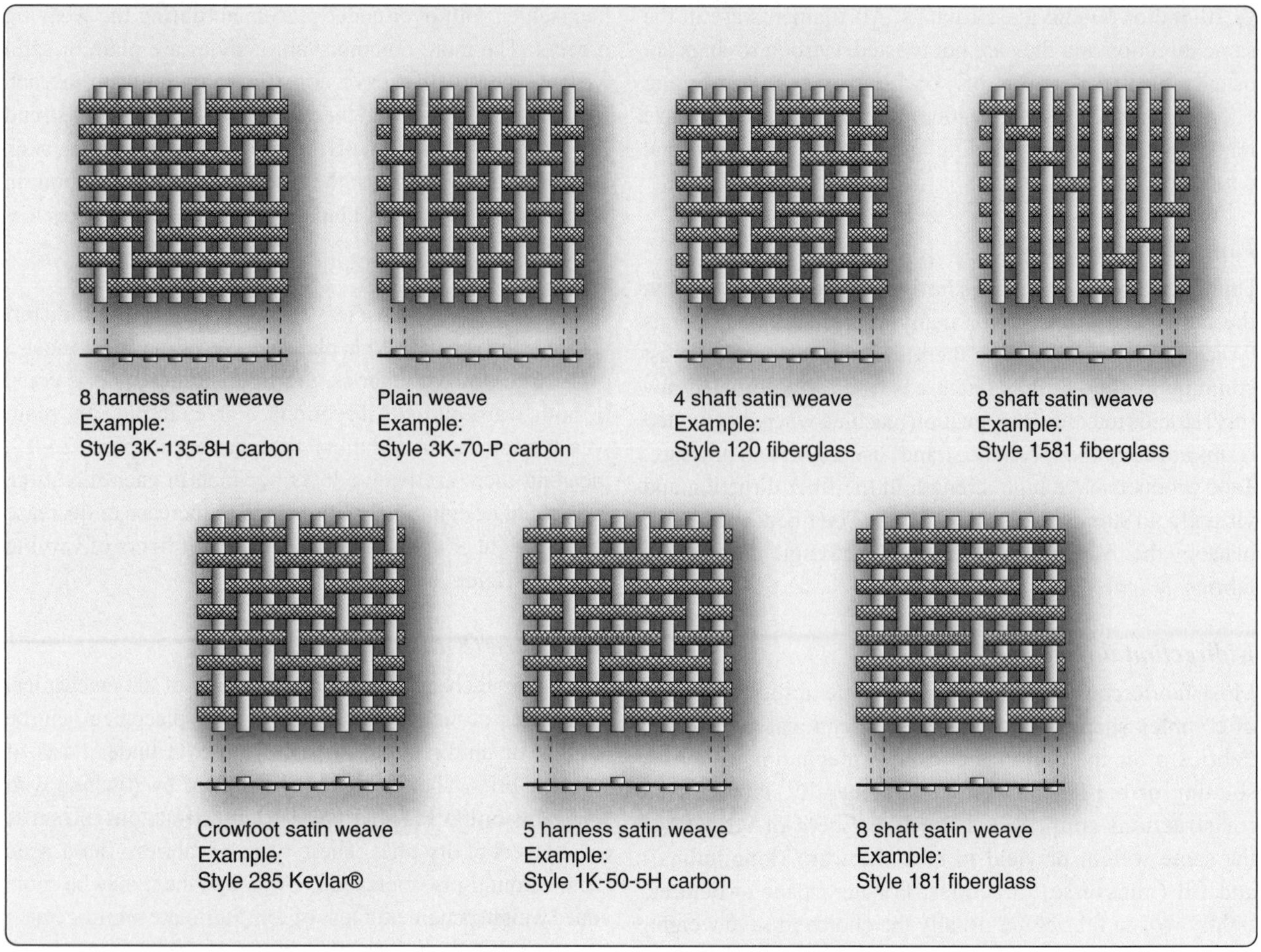

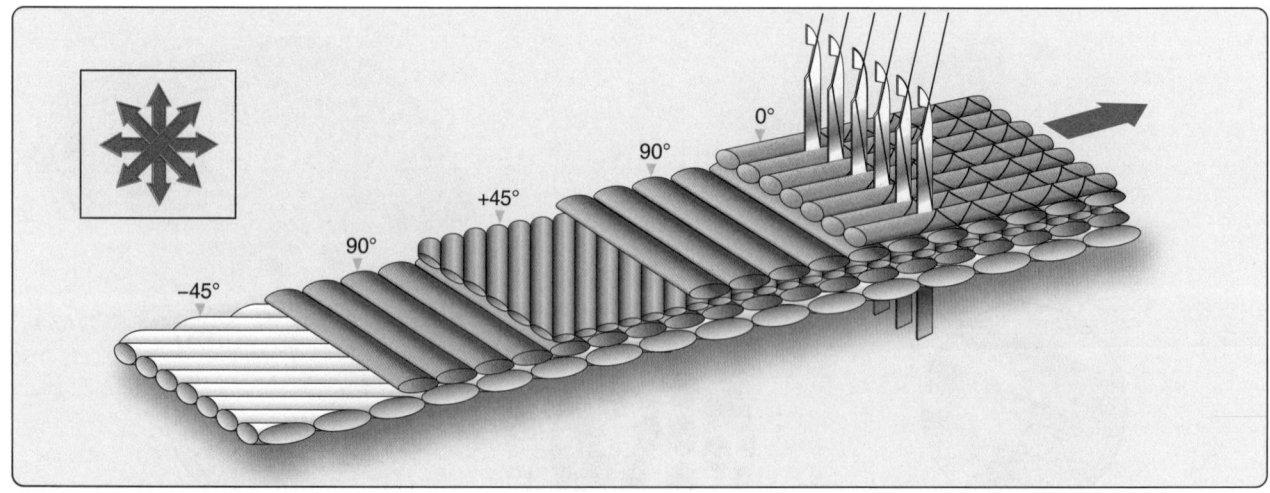

Figure 7-5. *Typical fabric weave styles.*

8 harness satin weave
Example:
Style 3K-135-8H carbon

Plain weave
Example:
Style 3K-70-P carbon

4 shaft satin weave
Example:
Style 120 fiberglass

8 shaft satin weave
Example:
Style 1581 fiberglass

Crowfoot satin weave
Example:
Style 285 Kevlar®

5 harness satin weave
Example:
Style 1K-50-5H carbon

8 shaft satin weave
Example:
Style 181 fiberglass

Figure 7-6. *Nonwoven material (stitched).*

−45°   90°   +45°   90°   0°

Types of Fiber

Fiberglass

Fiberglass is often used for secondary structure on aircraft, such as fairings, radomes, and wing tips. Fiberglass is also used for helicopter rotor blades. There are several types of fiberglass used in the aviation industry. Electrical glass, or E-glass, is identified as such for electrical applications. It has high resistance to current flow. E-glass is made from borosilicate glass. S-glass and S2-glass identify structural fiberglass that have a higher strength than E-glass. S-glass is produced from magnesia-alumina-silicate. Advantages of fiberglass are lower cost than other composite materials, chemical or galvanic corrosion resistance, and electrical properties (fiberglass does not conduct electricity). Fiberglass has a white color and is available as a dry fiber fabric or prepreg material.

Kevlar

Kevlar® is DuPont's name for aramid fibers. Aramid fibers are light weight, strong, and tough. Two types of aramid fiber are used in the aviation industry. Kevlar® 49 has a high stiffness and Kevlar® 29 has a low stiffness. An advantage of aramid fibers is their high resistance to impact damage, so they are often used in areas prone to impact damage. The main disadvantage of aramid fibers is their general weakness in compression and hygroscopy. Service reports have indicated that some parts made from Kevlar® absorb up to 8 percent of their weight in water. Therefore, parts made from aramid fibers need to be protected from the environment. Another disadvantage is that Kevlar® is difficult to drill and cut. The fibers fuzz easily and special scissors are needed to cut the material. Kevlar® is often used for military ballistic and body armor applications. It has a natural yellow color and is available as dry fabric and prepreg material. Bundles of aramid fibers are not sized by the number of fibers like carbon or fiberglass but by the weight.

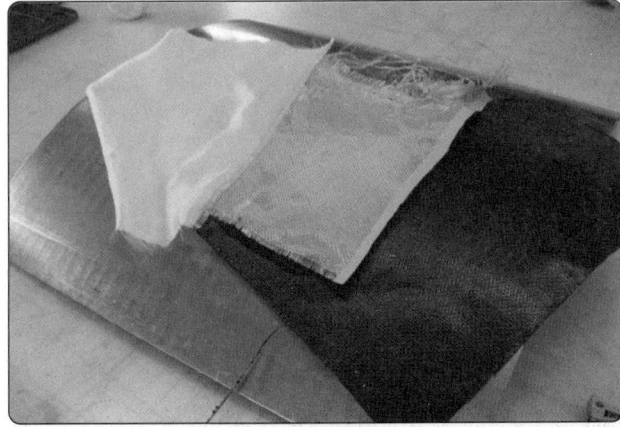

Figure 7-7. *Fiberglass (left), Kevlar® (middle), and carbon fiber material (right).*

Carbon/Graphite

One of the first distinctions to be made among fibers is the difference between carbon and graphite fibers, although the terms are frequently used interchangeably. Carbon and graphite fibers are based on graphene (hexagonal) layer networks present in carbon. If the graphene layers, or planes, are stacked with three dimensional order, the material is defined as graphite. Usually extended time and temperature processing is required to form this order, making graphite fibers more expensive. Bonding between planes is weak. Disorder frequently occurs such that only two-dimensional ordering within the layers is present. This material is defined as carbon.

Carbon fibers are very stiff and strong, 3 to 10 times stiffer than glass fibers. Carbon fiber is used for structural aircraft applications, such as floor beams, stabilizers, flight controls, and primary fuselage and wing structure. Advantages include its high strength and corrosion resistance. Disadvantages include lower conductivity than aluminum; therefore, a lightning protection mesh or coating is necessary for aircraft parts that are prone to lightning strikes. Another disadvantage of carbon fiber is its high cost. Carbon fiber is gray or black in color and is available as dry fabric and prepreg material. Carbon fibers have a high potential for causing galvanic corrosion when used with metallic fasteners and structures. *[Figure 7-7]*

Boron

Boron fibers are very stiff and have a high tensile and compressive strength. The fibers have a relatively large diameter and do not flex well; therefore, they are available only as a prepreg tape product. An epoxy matrix is often used with the boron fiber. Boron fibers are used to repair cracked aluminum aircraft skins, because the thermal expansion of boron is close to aluminum and there is no galvanic corrosion potential. The boron fiber is difficult to use if the parent material surface has a contoured shape. The boron fibers are very expensive and can be hazardous for personnel. Boron fibers are used primarily in military aviation applications.

Ceramic Fibers

Ceramic fibers are used for high-temperature applications, such as turbine blades in a gas turbine engine. The ceramic fibers can be used to temperatures up to 2,200 °F.

Lightning Protection Fibers

An aluminum airplane is quite conductive and is able to dissipate the high currents resulting from a lightning strike. Carbon fibers are 1,000 times more resistive than aluminum to current flow, and epoxy resin is 1,000,000 times more resistive (i.e., perpendicular to the skin). The surface of an external composite component often consists of a ply or layer of conductive material for lightning strike protection because composite materials are less conductive than aluminum. Many different types of conductive materials are used ranging from nickel-coated graphite cloth to metal meshes to aluminized fiberglass to conductive paints. The materials are available for wet layup and as prepreg.

In addition to a normal structural repair, the technician must also recreate the electrical conductivity designed into the part. These types of repair generally require a conductivity test to be performed with an ohmmeter to verify minimum electrical resistance across the structure. When repairing these types of structures, it is extremely important to use only the approved materials from authorized vendors, including such items as potting compounds, sealants, adhesives, and so forth. *[Figures 7-8 and 7-9]*

Matrix Materials
Thermosetting Resins
Resin is a generic term used to designate the polymer. The resin, its chemical composition, and physical properties fundamentally affect the processing, fabrication, and ultimate properties of a composite material. Thermosetting resins are the most diverse and widely used of all man-made materials. They are easily poured or formed into any shape, are compatible with most other materials, and cure readily (by heat or catalyst) into an insoluble solid. Thermosetting resins are also excellent adhesives and bonding agents.

Polyester Resins
Polyester resins are relatively inexpensive, fast processing resins used generally for low cost applications. Low smoke producing polyester resins are used for interior parts of the aircraft. Fiber-reinforced polyesters can be processed by many methods. Common processing methods include matched metal molding, wet layup, press (vacuum bag) molding, injection molding, filament winding, pultrusion, and autoclaving.

Vinyl Ester Resin
The appearance, handling properties, and curing characteristics of vinyl ester resins are the same as those of conventional polyester resins. However, the corrosion resistance and mechanical properties of vinyl ester composites are much improved over standard polyester resin composites.

Phenolic Resin
Phenol-formaldehyde resins were first produced commercially in the early 1900s for use in the commercial market. Urea-formaldehyde and melamine-formaldehyde appeared in the 1920–1930s as a less expensive alternative for lower temperature use. Phenolic resins are used for interior components because of their low smoke and flammability characteristics.

Epoxy
Epoxies are polymerizable thermosetting resins and are available in a variety of viscosities from liquid to solid. There are many different types of epoxy, and the technician should use the maintenance manual to select the correct type for a specific repair. Epoxies are used widely in resins for prepreg materials and structural adhesives. The advantages of epoxies are high strength and modulus, low levels of volatiles, excellent adhesion, low shrinkage, good chemical resistance, and ease of processing. Their major disadvantages are brittleness and the reduction of properties in the presence of moisture. The processing or curing of epoxies is slower than polyester resins. Processing techniques include autoclave molding, filament winding, press molding, vacuum bag

Figure 7-8. *Copper mesh lightning protection material.*

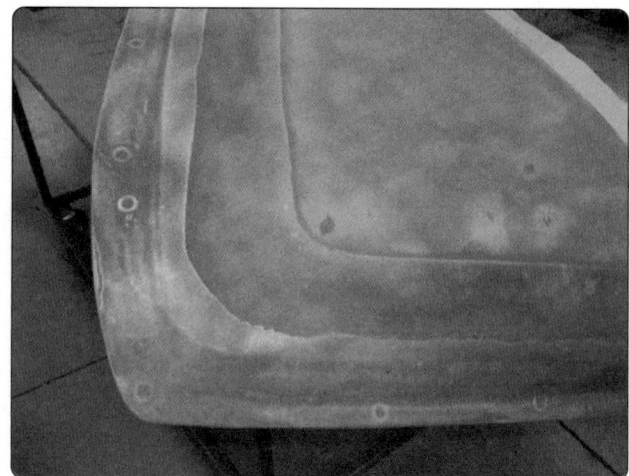

Figure 7-9. *Aluminum mesh lightning protection material.*

Figure 7-10. *Two-part wet layup epoxy resin system with pump dispenser.*

molding, resin transfer molding, and pultrusion. Curing temperatures vary from room temperature to approximately 350 °F (180 °C). The most common cure temperatures range between 250° and 350 °F (120–180 °C). *[Figure 7-10]*

Polyimides

Polyimide resins excel in high-temperature environments where their thermal resistance, oxidative stability, low coefficient of thermal expansion, and solvent resistance benefit the design. Their primary uses are circuit boards and hot engine and airframe structures. A polyimide may be either a thermoset resin or a thermoplastic. Polyimides require high cure temperatures, usually in excess of 550 °F (290 °C). Consequently, normal epoxy composite bagging materials are not usable, and steel tooling becomes a necessity. Polyimide bagging and release films, such as Kapton® are used. It is extremely important that Upilex® replace the lower cost nylon bagging and polytetrafluoroethylene (PTFE) release films common to epoxy composite processing. Fiberglass fabrics must be used for bleeder and breather materials instead of polyester mat materials due to the low melting point of polyester.

Polybenzimidazoles (PBI)

Polybenzimidazole resin is extremely high temperature resistant and is used for high-temperature materials. These resins are available as adhesive and fiber.

Bismaleimides (BMI)

Bismaleimide resins have a higher temperature capability and higher toughness than epoxy resins, and they provide excellent performance at ambient and elevated temperatures. The processing of bismaleimide resins is similar to that for epoxy resins. BMIs are used for aero engines and high temperature components. BMIs are suitable for standard autoclave processing, injection molding, resin transfer molding, and sheet molded compound (SMC) among others.

Thermoplastic Resins

Thermoplastic materials can be softened repeatedly by an increase of temperature and hardened by a decrease in temperature. Processing speed is the primary advantage of thermoplastic materials. Chemical curing of the material does not take place during processing, and the material can be shaped by molding or extrusion when it is soft.

Semicrystalline Thermoplastics

Semicrystalline thermoplastics possess properties of inherent flame resistance, superior toughness, good mechanical properties at elevated temperatures and after impact, and low moisture absorption. They are used in secondary and primary aircraft structures. Combined with reinforcing fibers, they are available in injection molding compounds, compression-moldable random sheets, unidirectional tapes, prepregs fabricated from tow (towpreg), and woven prepregs. Fibers impregnated in semicrystalline thermoplastics include carbon, nickel-coated carbon, aramid, glass, quartz, and others.

Amorphous Thermoplastics

Amorphous thermoplastics are available in several physical forms, including films, filaments, and powders. Combined with reinforcing fibers, they are also available in injection molding compounds, compressive moldable random sheets, unidirectional tapes, woven prepregs, etc. The fibers used are primarily carbon, aramid, and glass. The specific advantages of amorphous thermoplastics depend upon the polymer. Typically, the resins are noted for their processing ease and speed, high temperature capability, good mechanical properties, excellent toughness and impact strength, and chemical stability. The stability results in unlimited shelf life, eliminating the cold storage requirements of thermoset prepregs.

Polyether Ether Ketone (PEEK)

Polyether ether ketone, better known as PEEK, is a high-temperature thermoplastic. This aromatic ketone material offers outstanding thermal and combustion characteristics and resistance to a wide range of solvents and proprietary fluids. PEEK can also be reinforced with glass and carbon.

Curing Stages of Resins

Thermosetting resins use a chemical reaction to cure. There are three curing stages, which are called A, B, and C.

- A stage: The components of the resin (base material and hardener) have been mixed but the chemical reaction has not started. The resin is in the A stage during a wet layup procedure.

- B stage: The components of the resin have been mixed

and the chemical reaction has started. The material has thickened and is tacky. The resins of prepreg materials are in the B stage. To prevent further curing the resin is placed in a freezer at 0 °F. In the frozen state, the resin of the prepreg material stays in the B stage. The curing starts when the material is removed from the freezer and warmed again.

- C stage: The resin is fully cured. Some resins cure at room temperature and others need an elevated temperature cure cycle to fully cure.

Pre-Impregnated Products (Prepregs)

Prepreg material consists of a combination of a matrix and fiber reinforcement. It is available in unidirectional form (one direction of reinforcement) and fabric form (several directions of reinforcement). All five of the major families of matrix resins can be used to impregnate various fiber forms. The resin is then no longer in a low-viscosity stage, but has been advanced to a B stage level of cure for better handling characteristics. The following products are available in prepreg form: unidirectional tapes, woven fabrics, continuous strand rovings, and chopped mat. Prepreg materials must be stored in a freezer at a temperature below 0 °F to retard the curing process. Prepreg materials are cured with an elevated temperature. Many prepreg materials used in aerospace are impregnated with an epoxy resin and they are cured at either 250 °F or 350 °F. Prepreg materials are cured with an autoclave, oven, or heat blanket. They are typically purchased and stored on a roll in a sealed plastic bag to avoid moisture contamination. *[Figure 7-11]*

Dry Fiber Material

Dry fiber materials, such as carbon, glass, and Kevlar® are used for many aircraft repair procedures. The dry fabric is impregnated with a resin just before the repair work starts. This process is often called wet layup. The main advantage of using the wet layup process is that the fiber and resin can

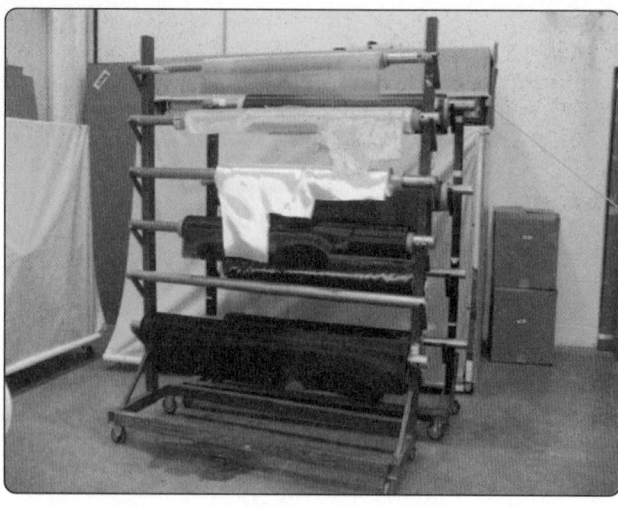

Figure 7-12. *Dry fabric materials (top to bottom: aluminum lightning protection mess, Kevlar®, fiberglass, and carbon fiber).*

be stored for a long time at room temperature. The composite can be cured at room temperature or an elevated temperature cure can be used to speed up the curing process and increase the strength. The disadvantage is that the process is messy and reinforcement properties are less than prepreg material properties. *[Figure 7-12]*

Thixotropic Agents

Thixotropic agents are gel-like at rest but become fluid when agitated. These materials have high static shear strength and low dynamic shear strength at the same time to lose viscosity under stress.

Adhesives
Film Adhesives

Structural adhesives for aerospace applications are generally supplied as thin films supported on a release paper and stored under refrigerated conditions (–18 °C, or 0 °F). Film adhesives are available using high-temperature aromatic

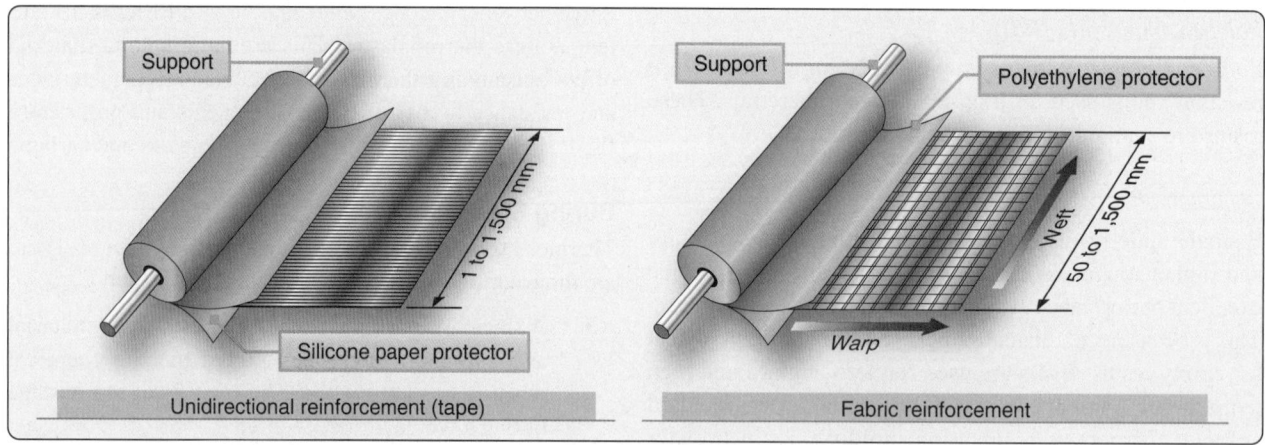

Figure 7-11. *Tape and fabric prepreg materials.*

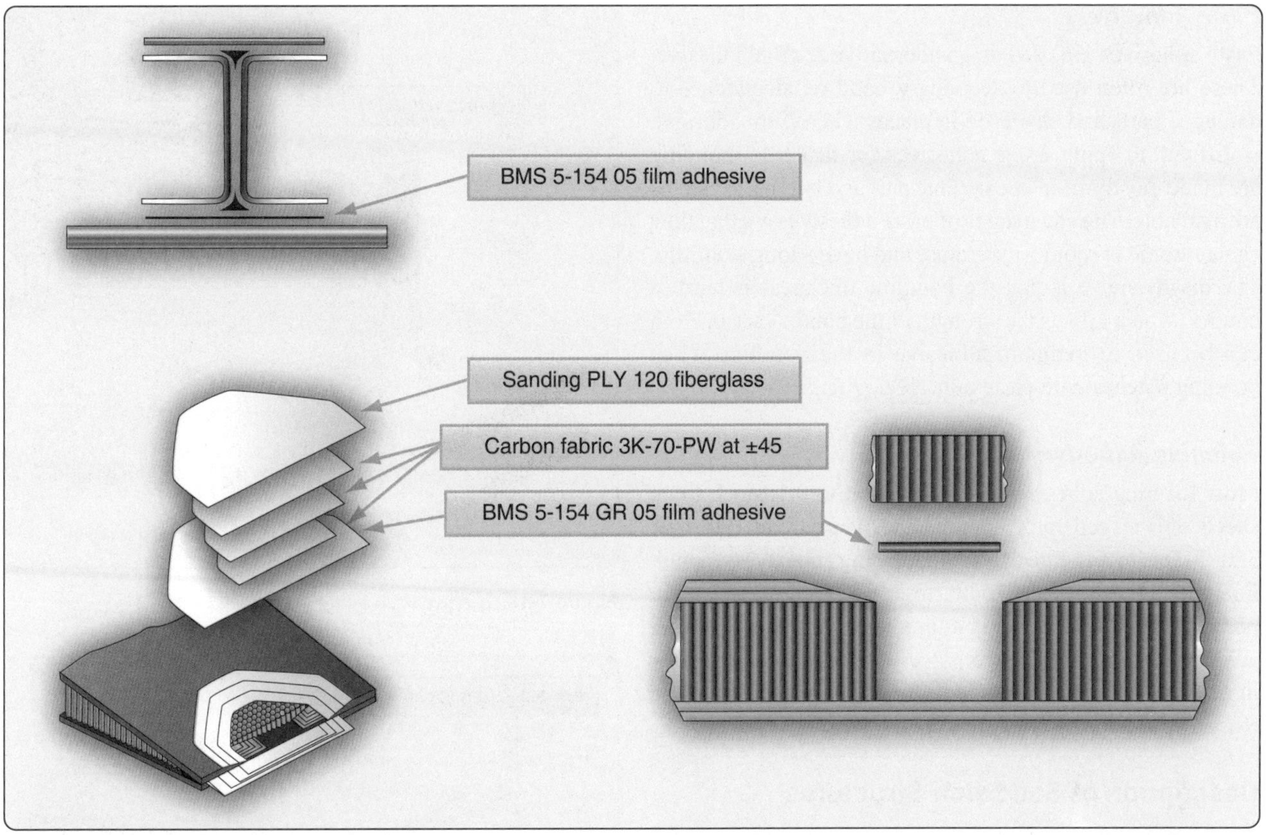

Figure 7-13. *The use of film adhesive mess, Kevlar®, fiberglass, and carbon fiber.*

Labels in figure:
- BMS 5-154 05 film adhesive
- Sanding PLY 120 fiberglass
- Carbon fabric 3K-70-PW at ±45
- BMS 5-154 GR 05 film adhesive

Figure 7-14. *A roll of film adhesive.*

Figure 7-15. *Two-part paste adhesive.*

amine or catalytic curing agents with a wide range of flexibilizing and toughening agents. Rubber-toughened epoxy film adhesives are widely used in aircraft industry. The upper temperature limit of 121–177 °C (250–350 °F) is usually dictated by the degree of toughening required and by the overall choice of resins and curing agents. In general, toughening of a resin results in a lower usable service temperature. Film materials are frequently supported by fibers that serve to improve handling of the films prior to cure, control adhesive flow during bonding, and assist in bond line thickness control. Fibers can be incorporated as short-fiber mats with random orientation or as woven cloth. Commonly encountered fibers are polyesters, polyamides (nylon), and glass. Adhesives containing woven cloth may have slightly degraded environmental properties because of wicking of water by the fiber. Random mat scrim cloth is not as efficient for controlling film thickness as woven cloth because the unrestricted fibers move during bonding. Spun-bonded nonwoven scrims do not move and are, therefore, widely used. *[Figures 7-13 and 7-14]*

Paste Adhesives

Paste adhesives are used as an alternative to film adhesive. These are often used to secondary bond repair patches to damaged parts and also used in places where film adhesive is difficult to apply. Paste adhesives for structural bonding are made mostly from epoxy. One part and two part systems are available. The advantages of paste adhesives are that they can be stored at room temperature and have a long shelf life. The disadvantage is that the bondline thickness is hard to control, which affects the strength of the bond. A scrim cloth can be used to maintain adhesive in the bondline when bonding patches with paste adhesive. *[Figure 7-15]*

Foaming Adhesives

Most foaming adhesives are 0.025-inch to 0.10-inch thick sheets of B staged epoxy. Foam adhesives cure at 250 °F or 350 °F. During the cure cycle, the foaming adhesives expand. Foaming adhesives need to be stored in the freezer just like prepregs, and they have only a limited storage life. Foaming adhesives are used to splice pieces of honeycomb together in a sandwich construction and to bond repair plugs to the existing core during a prepreg repair. *[Figure 7-16]*

Description of Sandwich Structures

Theory A sandwich construction is a structural panel concept that consists in its simplest form of two relatively thin, parallel face sheets bonded to and separated by a relatively thick, lightweight core. The core supports the face sheets against buckling and resists out-of-plane shear loads. The core must have high shear strength and compression stiffness. Composite sandwich construction is most often fabricated using autoclave cure, press cure, or vacuum bag cure. Skin laminates may be precured and subsequently bonded to core, co-cured to core in one operation, or a combination of the two methods. Examples of honeycomb structure are: wing spoilers, fairings, ailerons, flaps, nacelles, floor boards, and rudders. *[Figure 7-17]*

Properties

Sandwich construction has high bending stiffness at minimal weight in comparison to aluminum and composite laminate construction. Most honeycombs are anisotropic; that is, properties are directional. *Figure 7-18* illustrates the advantages of using a honeycomb construction. Increasing the core thickness greatly increases the stiffness of the honeycomb construction, while the weight increase is minimal. Due to the high stiffness of a honeycomb construction, it is not necessary to use external stiffeners, such as stringers and frames. *[Figure 7-18]*

Facing Materials

Most honeycomb structures used in aircraft construction have aluminum, fiberglass, Kevlar®, or carbon fiber face sheets.

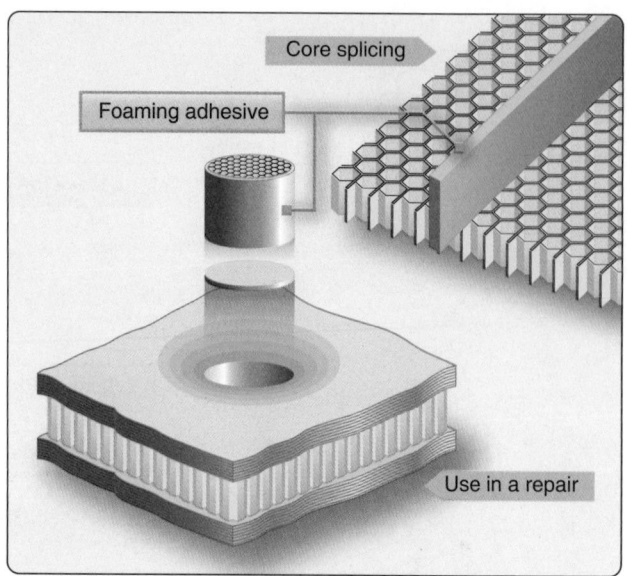

Figure 7-16. *The use of foaming adhesives.*

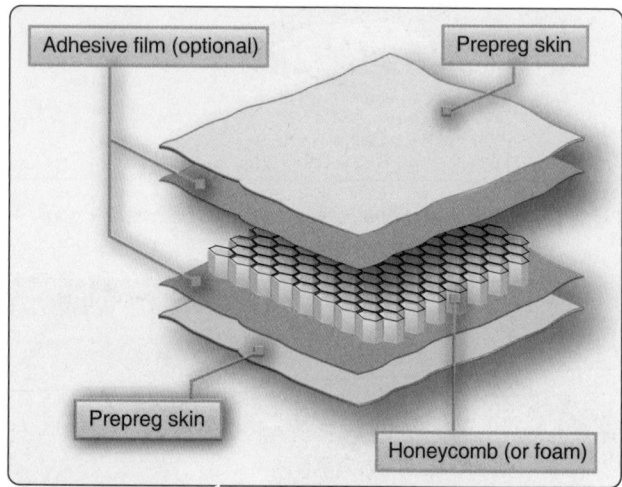

Figure 7-17. *Honeycomb sandwich construction.*

	Solid Material	Core Thickness t	Core Thickness 3t
Thickness	1.0	7.0	37.0
Flexural Strength	1.0	3.5	9.2
Weight	1.0	1.03	1.06

Figure 7-18. *Strength and stiffness of honeycomb sandwich material compared to a solid laminate.*

Carbon fiber face sheets cannot be used with aluminum honeycomb core material, because it causes the aluminum to corrode. Titanium and steel are used for specialty applications in high-temperature constructions. The face sheets of many components, such as spoilers and flight controls, are very thin—sometimes only 3 or 4 plies. Field reports have indicated that these face sheets do not have a good impact resistance.

Core Materials

Honeycomb

Each honeycomb material provides certain properties and has specific benefits. [Figure 7-19] The most common core material used for aircraft honeycomb structures is aramid paper (Nomex® or Korex®). Fiberglass is used for higher strength applications.

- Kraft paper—relatively low strength, good insulating properties, is available in large quantities, and has a low cost.

- Thermoplastics—good insulating properties, good energy absorption and/or redirection, smooth cell walls, moisture and chemical resistance, are environmentally compatible, aesthetically pleasing, and have a relatively low cost.

- Aluminum—best strength-to-weight ratio and energy absorption, has good heat transfer properties, electromagnetic shielding properties, has smooth, thin cell walls, is machinable, and has a relatively low cost.

- Steel—good heat transfer properties, electromagnetic shielding properties, and heat resistant.

- Specialty metals (titanium)—relatively high strength-to-weight ratio, good heat transfer properties, chemical resistance, and heat resistant to very high temperatures.

- Aramid paper—flame resistant, fire retardant, good insulating properties, low dielectric properties, and good formability.

- Fiberglass—tailorable shear properties by layup, low dielectric properties, good insulating properties, and good formability.

- Carbon—good dimensional stability and retention, high-temperature property retention, high stiffness, very low coefficient of thermal expansion, tailorable thermal conductivity, relatively high shear modulus, and very expensive.

- Ceramics—heat resistant to very high temperatures, good insulating properties, is available in very small cell sizes, and very expensive. [Figure 7-19]

Honeycomb core cells for aerospace applications are usually hexagonal. The cells are made by bonding stacked sheets

Figure 7-19. *Honeycomb core materials.*

at special locations. The stacked sheets are expanded to form hexagons. The direction parallel to the sheets is called ribbon direction.

Bisected hexagonal core has another sheet of material cutting across each hexagon. Bisected hexagonal honeycomb is stiffer and stronger than hexagonal core. Overexpanded core is made by expanding the sheets more than is needed to make hexagons. The cells of overexpanded core are rectangular. Overexpanded core is flexible perpendicular to the ribbon direction and is used in panels with simple curves. Bell-shaped core, or flexicore, has curved cell walls, that make it flexible in all directions. Bell-shaped core is used in panels with complex curves.

Honeycomb core is available with different cell sizes. Small sizes provide better support for sandwich face sheets. Honeycomb is also available in different densities. Higher density core is stronger and stiffer than lower density core. [Figure 7-20]

Foam

Foam cores are used on homebuilts and lighter aircraft to give strength and shape to wing tips, flight controls, fuselage sections, wings, and wing ribs. Foam cores are not commonly used on commercial type aircraft. Foams are typically heavier than honeycomb and not as strong. A variety of foams can be used as core material including:

- Polystyrene (better known as styrofoam)—aircraft grade styrofoam with a tightly closed cell structure and no voids between cells; high compressive strength and good resistance to water penetration; can be cut with a hot wire to make airfoil shapes.

- Phenolic—very good fire-resistant properties and can have very low density, but relatively low mechanical

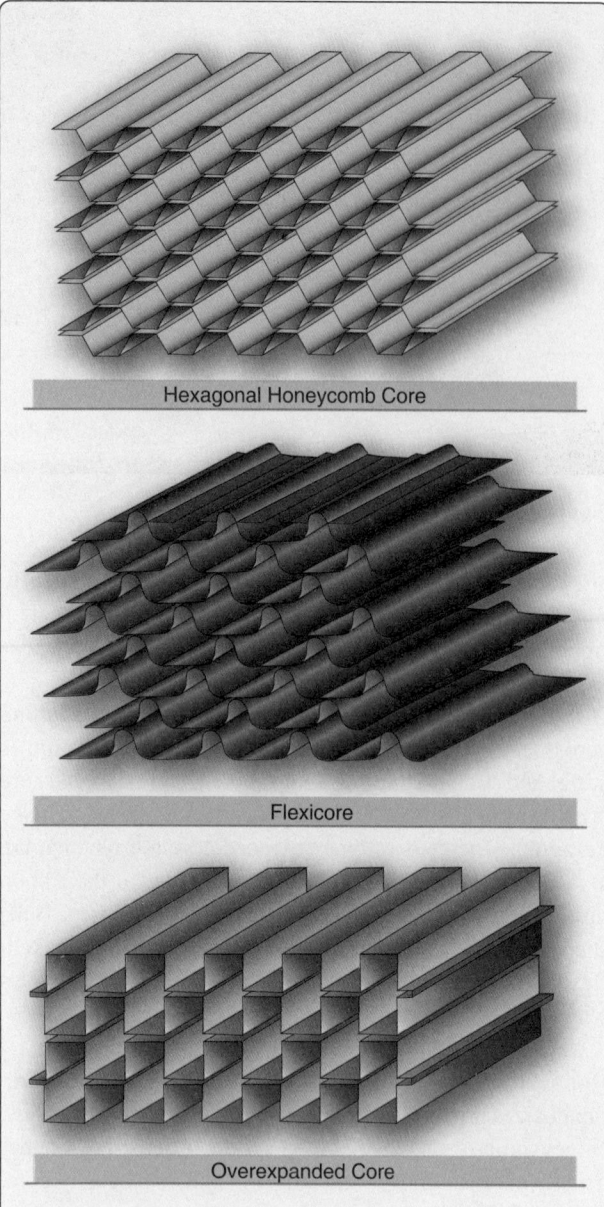

Figure 7-20. *Honeycomb density.*

properties.

- Polyurethane—used for producing the fuselage, wing tips, and other curved parts of small aircraft; relatively inexpensive, fuel resistant, and compatible with most adhesives; do not use a hot wire to cut polyurethane foam; easily contoured with a large knife and sanding equipment.

- Polypropylene—used to make airfoil shapes; can be cut with a hot wire; compatible with most adhesives and epoxy resins; not for use with polyester resins, dissolves in fuels and solvents.

- Polyvinyl chloride (PVC) (Divinycell, Klegecell, and Airex)—a closed cell medium- to high-density foam with high compression strength, durability, and excellent fire resistance; can be vacuum formed to compound shapes and be bent using heat; compatible with polyester, vinyl ester, and epoxy resins.

- Polymethacrylimide (Rohacell)—a closed-cell foam used for lightweight sandwich construction; excellent mechanical properties, high-dimensional stability under heat, good solvent resistance, and outstanding creep compression resistance; more expensive than the other types of foams, but has greater mechanical properties.

Balsa Wood

Balsa is a natural wood product with elongated closed cells; it is available in a variety of grades that correlate to the structural, cosmetic, and physical characteristics. The density of balsa is less than one-half of the density of conventional wood products. However, balsa has a considerably higher density than the other types of structural cores.

Manufacturing & In-Service Damage

Manufacturing Defects

Manufacturing defects include:

- Delamination

- Resin starved areas

- Resin rich areas

- Blisters, air bubbles

- Wrinkles

- Voids

- Thermal decomposition

Manufacturing damage includes anomalies, such as porosity, microcracking, and delaminations resulting from processing discrepancies. It also includes such items as inadvertent edge cuts, surface gouges and scratches, damaged fastener holes, and impact damage. Examples of flaws occurring in manufacturing include a contaminated bondline surface or inclusions, such as prepreg backing paper or separation film, that is inadvertently left between plies during layup. Inadvertent (nonprocess) damage can occur in detail parts or components during assembly or transport or during operation.

A part is resin rich if too much resin is used, for nonstructural applications this is not necessarily bad, but it adds weight. A part is called resin starved if too much resin is bled off during the curing process or if not enough resin is applied during the wet layup process. Resin-starved areas are indicated by fibers that show to the surface. The ratio of 60:40 fiber to resin ratio is considered optimum. Sources of manufacturing defects include:

- Improper cure or processing
- Improper machining
- Mishandling
- Improper drilling
- Tool drops
- Contamination
- Improper sanding
- Substandard material
- Inadequate tooling
- Mislocation of holes or details

Damage can occur at several scales within the composite material and structural configuration. This ranges from damage in the matrix and fiber to broken elements and failure of bonded or bolted attachments. The extent of damage controls repeated load life and residual strength and is critical to damage tolerance.

Fiber Breakage

Fiber breakage can be critical because structures are typically designed to be fiber dominant (i.e., fibers carry most of the loads). Fortunately, fiber failure is typically limited to a zone near the point of impact and is constrained by the impact object size and energy. Only a few of the service-related events listed in the previous section could lead to large areas of fiber damage.

Matrix Imperfections

Matrix imperfections usually occur on the matrix-fiber interface or in the matrix parallel to the fibers. These imperfections can slightly reduce some of the material properties but are seldom critical to the structure, unless the matrix degradation is widespread. Accumulation of matrix cracks can cause the degradation of matrix-dominated properties. For laminates designed to transmit loads with their fibers (fiber dominant), only a slight reduction of properties is observed when the matrix is severely damaged. Matrix cracks, or microcracks, can significantly reduce properties dependent on the resin or the fiber-resin interface, such as interlaminar shear and compression strength. Micro-cracking can have a very negative effect on properties of high-temperature resins. Matrix imperfections may develop into delaminations, which are a more critical type of damage.

Delamination & Debonds

Delaminations form on the interface between the layers in the laminate. Delaminations may form from matrix cracks that grow into the interlaminar layer or from low-energy impact. Debonds can also form from production nonadhesion along the bondline between two elements and initiate delamination in adjacent laminate layers. Under certain conditions, delaminations or debonds can grow when subjected to repeated loading and can cause catastrophic failure when the laminate is loaded in compression. The criticality of delaminations or debonds depend on:

- Dimensions.
- Number of delaminations at a given location.
- Location—in the thickness of laminate, in the structure, proximity to free edges, stress concentration region, geometrical discontinuities, etc.
- Loads—behavior of delaminations and debonds depend on loading type. They have little effect on the response of laminates loaded in tension. Under compression or shear loading, however, the sublaminates adjacent to the delaminations or debonded elements may buckle and cause a load redistribution mechanism that leads to structural failure.

Combinations of Damages

In general, impact events cause combinations of damages. High-energy impacts by large objects (e.g., turbine blades) may lead to broken elements and failed attachments. The resulting damage may include significant fiber failure, matrix cracking, delamination, broken fasteners, and debonded elements. Damage caused by low-energy impact is more contained, but may also include a combination of broken fibers, matrix cracks, and multiple delaminations.

Flawed Fastener Holes

Improper hole drilling, poor fastener installation, and missing fasteners may occur in manufacturing. Hole elongation can occur due to repeated load cycling in service.

In-Service Defects

In-service defects include:

- Environmental degradation
- Impact damage
- Fatigue
- Cracks from local overload
- Debonding
- Delamination
- Fiber fracturing
- Erosion

Many honeycomb structures, such as wing spoilers, fairings, flight controls, and landing gear doors, have thin face

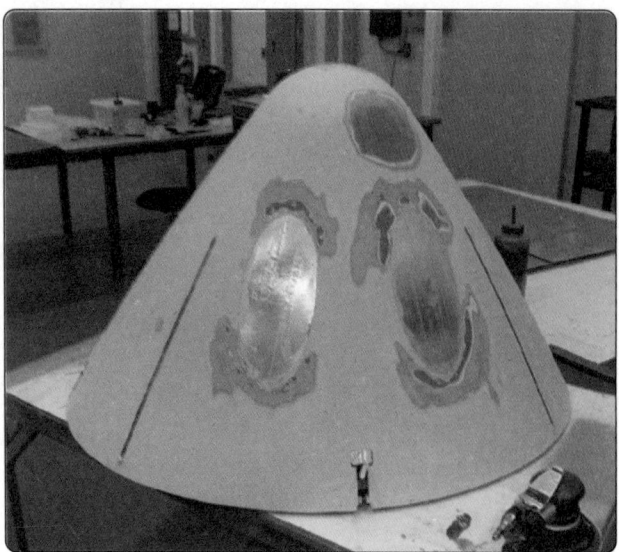

Figure 7-21. *Damage to radome honeycomb sandwich structure.*

The repair of parts due to liquid ingression can vary depending on the liquid, most commonly water or Skydrol (hydraulic fluid). Water tends to create additional damage in repaired parts when cured unless all moisture is removed from the part. Most repair material systems cure at temperatures above the boiling point of water, which can cause a disbond at the skin-to-core interface wherever trapped water resides. For this reason, core drying cycles are typically included prior to performing any repair. Some operators take the extra step of placing a damaged but unrepaired part in the autoclave to dry to preclude any additional damage from occurring during the cure of the repair. Skydrol presents a different problem. Once the core of a sandwich part is saturated, complete removal of Skydrol is almost impossible. The part continues to weep the liquid even in cure until bondlines can become contaminated and full bonding does not occur. Removal of contaminated core and adhesive as part of the repair is highly recommended. *[Figure 7-21]*

sheets which have experienced durability problems that could be grouped into three categories: low resistance to impact, liquid ingression, and erosion. These structures have adequate stiffness and strength but low resistance to a service environment in which parts are crawled over, tools dropped, and service personnel are often unaware of the fragility of thin-skinned sandwich parts. Damages to these components, such as core crush, impact damages, and disbonds, are quite often easy to detect with a visual inspection due to their thin face sheets. However, they are sometimes overlooked or damaged by service personnel who do not want to delay aircraft departure or bring attention to their accidents, which might reflect poorly on their performance record. Therefore, damages are sometimes allowed to go unchecked, often resulting in growth of the damage due to liquid ingression into the core. Nondurable design details (e.g., improper core edge close-outs) also lead to liquid ingression.

Erosion capabilities of composite materials have been known to be less than that of aluminum and, as a result, their application in leading-edge surfaces has been generally avoided. However, composites have been used in areas of highly complex geometry, but generally with an erosion coating. The durability and maintainability of some erosion coatings are less than ideal. Another problem, not as obvious as the first, is that edges of doors or panels can erode if they are exposed to the air stream. This erosion can be attributed to improper design or installation/fit-up. On the other hand, metal structures in contact or in the vicinity of these composite parts may show corrosion damage due to inappropriate choice of aluminum alloy, damaged corrosion sealant of metal parts during assembly or at splices, or insufficient sealant and/or lack of glass fabric isolation plies at the interfaces of spars, ribs, and fittings. *[Figure 7-22]*

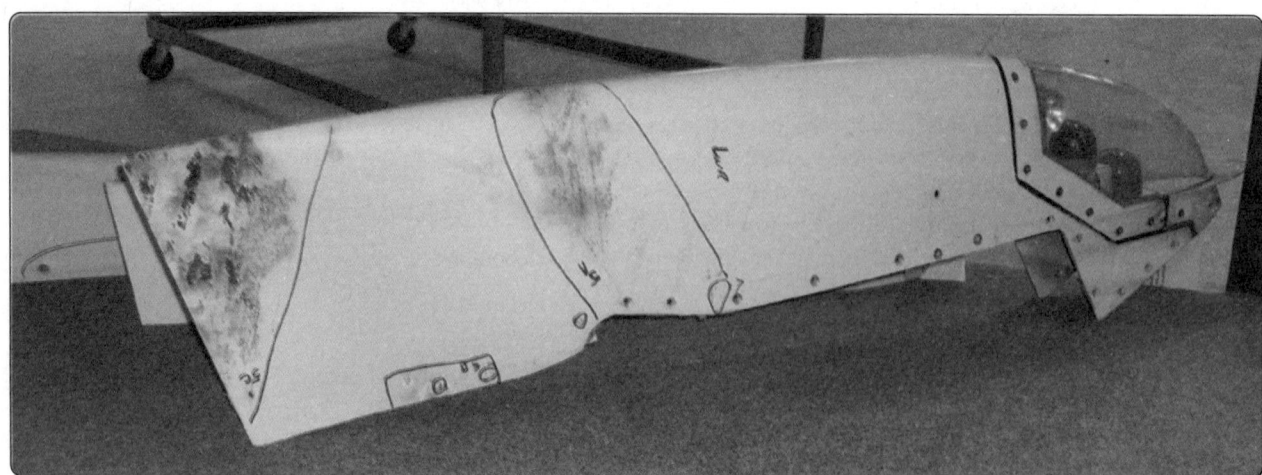

Figure 7-22. *Erosion damage to wingtip.*

Corrosion

Many fiberglass and Kevlar® parts have a fine aluminum mesh for lightning protection. This aluminum mesh often corrodes around the bolt or screw holes. The corrosion affects the electrical bonding of the panel, and the aluminum mesh needs to be removed and new mesh installed to restore the electrical bonding of the panel. *[Figure 7-23]*

Ultraviolet (UV) light affects the strength of composite materials. Composite structures need to be protected by a top coating to prevent the effects of UV light. Special UV primers and paints have been developed to protect composite materials.

Nondestructive Inspection (NDI) of Composites

Visual Inspection

A visual inspection is the primary inspection method for in-service inspections. Visible damage may include scorches, stains, dents, penetration, abrasions, or chips in the composite surface. Once damage is detected, the affected area needs to be inspected closer using flashlights, magnifying glasses, mirrors, and borescopes. These tools are used to magnify defects that otherwise might not be seen easily and to allow visual inspection of areas that are not readily accessible. Resin starvation, resin richness, wrinkles, ply bridging, discoloration (due to overheating, lightning strike, etc.), impact damage by any cause, foreign matter, blisters, and disbonding are some of the discrepancies that can be detected with a visual inspection. Visual inspection cannot find internal flaws in the composite, such as delaminations, disbonds, and matrix crazing. More sophisticated NDI techniques are needed to detect these types of defects.

Audible Sonic Testing (Coin Tapping)

Sometimes referred to as audio, sonic, or coin tap, this technique makes use of frequencies in the audible range (10 Hz to 20 Hz). A surprisingly accurate method in the hands of experienced personnel, tap testing is perhaps the most common technique used for the detection of delamination and/or disbond. The method is accomplished by tapping the inspection area with a solid round disc or lightweight hammer-like device and listening to the response of the structure to the hammer. *[Figure 7-24]* A clear, sharp, ringing sound is indicative of a well-bonded solid structure, while a dull or thud-like sound indicates a discrepant area.

The tapping rate needs to be rapid enough to produce enough sound for any difference in sound tone to be discernable to the ear. Tap testing is effective on thin skin to stiffener bondlines, honeycomb sandwich with thin face sheets, or even near the surface of thick laminates, such as rotorcraft blade supports. Again, inherent in the method is the possibility that changes within the internal elements of the structure might produce pitch changes that are interpreted as defects, when in fact they are present by design. This inspection should be accomplished in as quiet an area as possible and by experienced personnel familiar with the part's internal configuration. This method is not reliable for structures with more than four plies. It is often used to map out the damage on thin honeycomb facesheets. *[Figure 7-24]*

Automated Tap Test

This test is very similar to the manual tap test except that a solenoid is used instead of a hammer. The solenoid produces multiple impacts in a single area. The tip of the impactor has a transducer that records the force versus time signal of the impactor. The magnitude of the force depends on the impactor, the impact energy, and the mechanical properties of the structure. The impact duration (period) is not sensitive

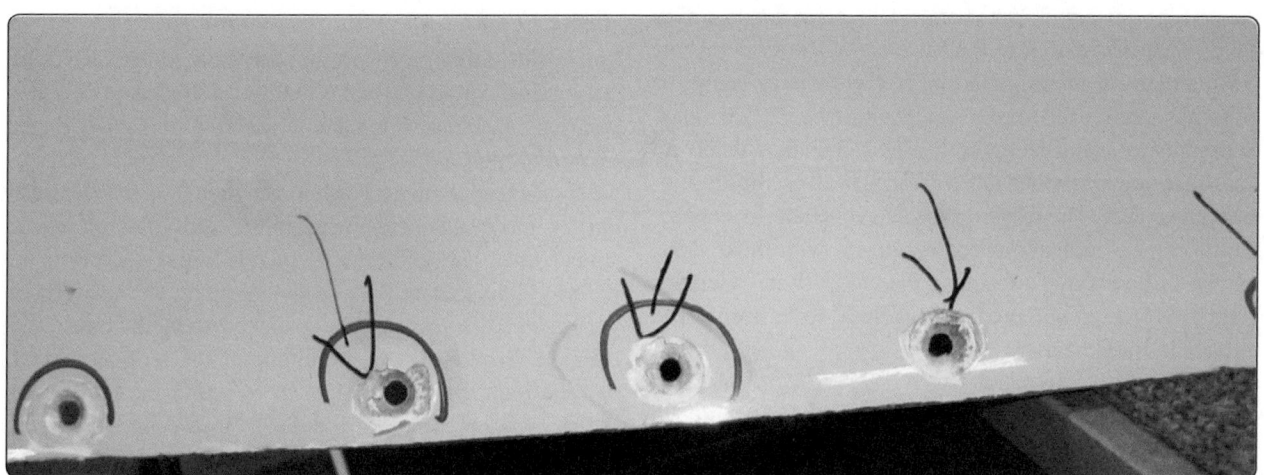

Figure 7-23. *Corrosion of aluminum lightning protection mesh.*

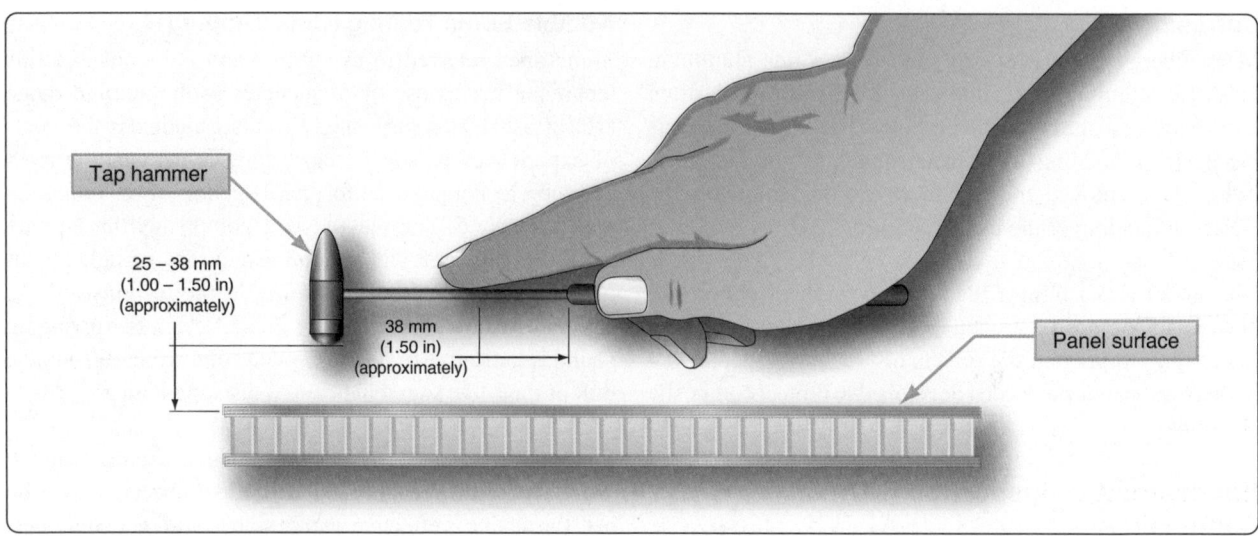

Figure 7-24. *Tap test with tap hammer.*

to the magnitude of the impact force; however, this duration changes as the stiffness of the structure is altered. Therefore, the signal from an unflawed region is used for calibration, and any deviation from this unflawed signal indicates the existence of damage.

Ultrasonic Inspection

Ultrasonic inspection has proven to be a very useful tool for the detection of internal delaminations, voids, or inconsistencies in composite components not otherwise discernable using visual or tap methodology. There are many ultrasonic techniques; however, each technique uses sound wave energy with a frequency above the audible range. *[Figure 7-25]* A high-frequency (usually several MHz) sound wave is introduced into the part and may be directed to travel normal to the part surface, or along the surface of the part, or at some predefined angle to the part surface. You may need to try different directions to locate the flow. The introduced sound is then monitored as it travels its assigned route through the part for any significant change. Ultrasonic sound waves have properties similar to light waves. When an ultrasonic wave strikes an interrupting object, the wave or energy is either absorbed or reflected back to the surface. The disrupted or diminished sonic energy is then picked up by a receiving transducer and converted into a display on an oscilloscope or a chart recorder. The display allows the operator to evaluate the discrepant indications comparatively with those areas known to be good. To facilitate the comparison, reference standards are established and utilized to calibrate the ultrasonic equipment.

The repair technician must realize that the concepts outlined here work fine in the repetitive manufacturing environment, but are likely to be more difficult to implement in a repair environment given the vast number of different composite components installed on the aircraft and the relative complexity of their construction. The reference standards would also have to take into account the transmutations that take place when a composite component is exposed to an in-service environment over a prolonged period or has been the subject of repair activity or similar restorative action. The four most common ultrasonic techniques are discussed next.

Through Transmission Ultrasonic Inspection

Through transmission ultrasonic inspection uses two transducers, one on each side of the area to be inspected. The ultrasonic signal is transmitted from one transducer to the other transducer. The loss of signal strength is then measured by the instrument. The instrument shows the loss as a percent of the original signal strength or the loss in decibels. The signal loss is compared to a reference standard. Areas with a greater loss than the reference standard indicate a defective area.

Pulse Echo Ultrasonic Inspection

Single-side ultrasonic inspection may be accomplished using pulse echo techniques. In this method, a single search unit is working as a transmitting and a receiving transducer that is excited by high voltage pulses. Each electrical pulse activates the transducer element. This element converts the electrical energy into mechanical energy in the form of an ultrasonic sound wave. The sonic energy travels through a Teflon® or methacrylate contact tip into the test part. A waveform is generated in the test part and is picked up by the transducer element. Any change in amplitude of the received signal, or time required for the echo to return to the transducer, indicates the presence of a defect. Pulse echo inspections are used to find delaminations, cracks, porosity, water, and disbonds of bonded components. Pulse echo does not find

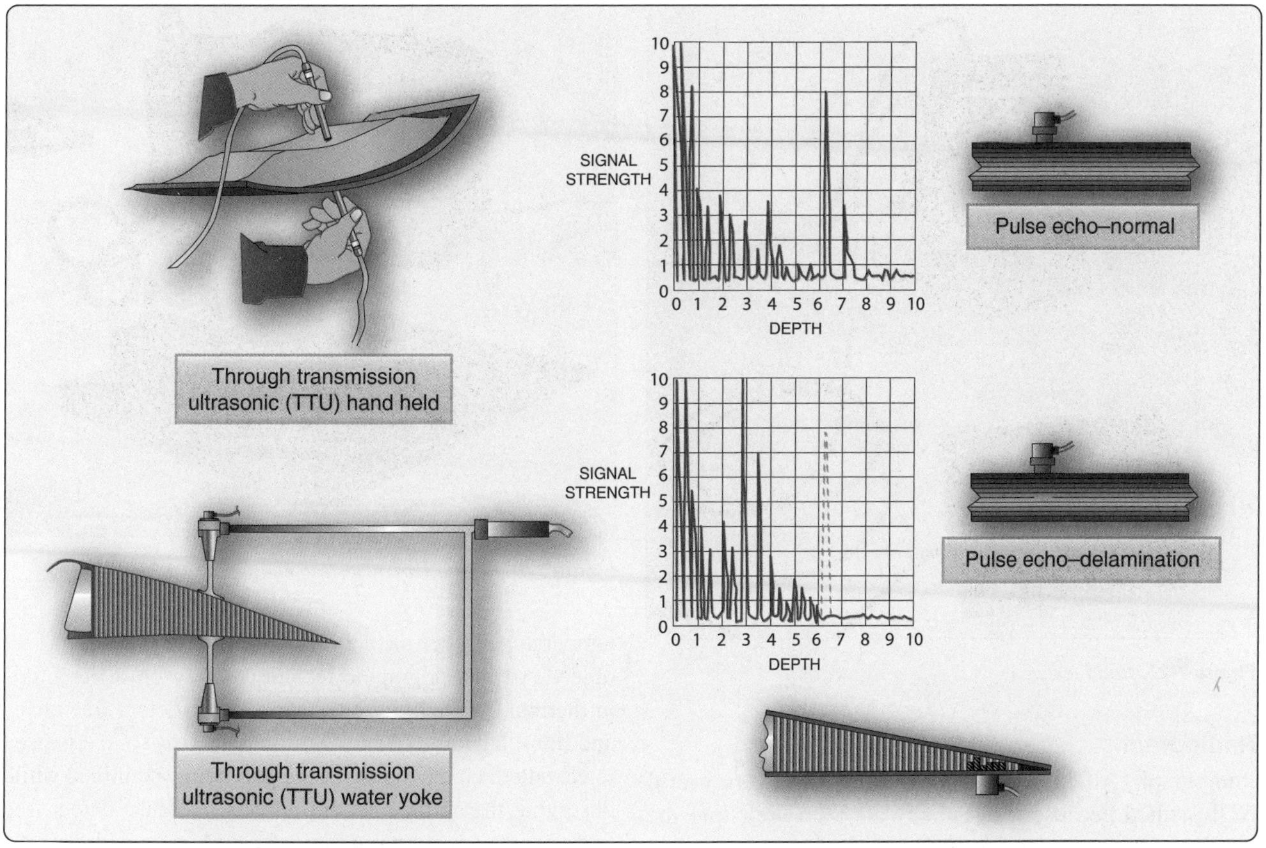

Figure 7-25. *Ultrasonic testing methods.*

Through transmission
ultrasonic (TTU) hand held

SIGNAL
STRENGTH

DEPTH

Pulse echo–normal

SIGNAL
STRENGTH

DEPTH

Pulse echo–delamination

Through transmission
ultrasonic (TTU) water yoke

Figure 7-26. *Pulse echo test equipment.*

disbonds or defects between laminated skins and honeycomb core. *[Figure 7-26]*

Ultrasonic Bond Tester Inspection

Low-frequency and high-frequency bond testers are used for ultrasonic inspections of composite structures. These bond testers use an inspection probe that has one or two transducers. The high-frequency bond tester is used to detect delaminations and voids. It cannot detect a skin-to-honeycomb core disbond or porosity. It can detect defects as small as 0.5-inch in diameter. The low-frequency bond tester uses two transducers and is used to detect delamination, voids, and skin to honeycomb core disbands. This inspection method does not detect which side of the part is damaged, and cannot detect defects smaller than 1.0-inch. *[Figure 7-27]*

Phased Array Inspection

Phased array inspection is one of the latest ultrasonic instruments to detect flaws in composite structures. It operates under the same principle of operation as pulse echo, but it uses 64 sensors at the same time, which speeds up the process. *[Figure 7-28]*

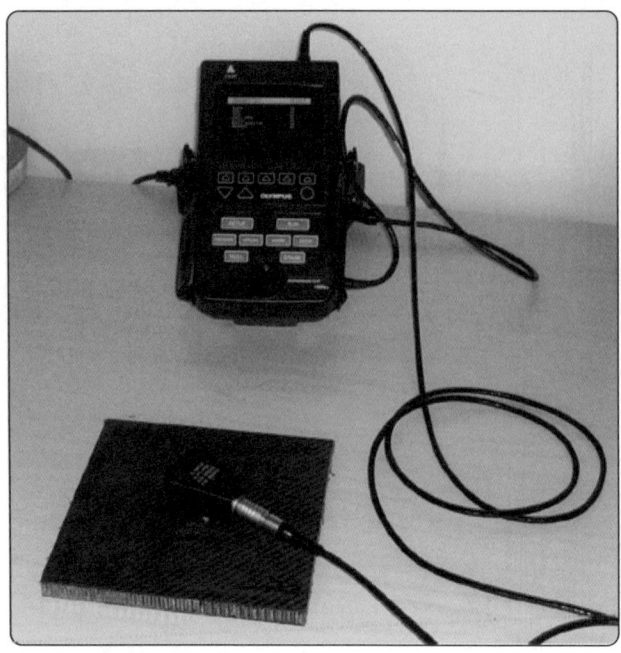

Figure 7-27. *Bond tester.*

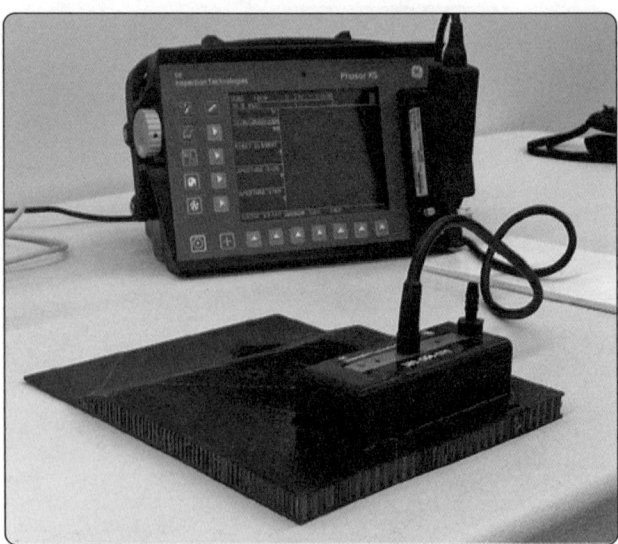

Figure 7-28. *Phased array testing equipment.*

Radiography

Radiography, often referred to as X-ray, is a very useful NDI method because it essentially allows a view into the interior of the part. This inspection method is accomplished by passing X-rays through the part or assembly being tested while recording the absorption of the rays onto a film sensitive to X-rays. The exposed film, when developed, allows the inspector to analyze variations in the opacity of the exposure recorded onto the film, in effect creating a visualization of the relationship of the component's internal details. Since the method records changes in total density through its thickness, it is not a preferred method for detecting defects such as delaminations that are in a plane that is normal to the ray direction. It is a most effective method, however, for detecting flaws parallel to the X-ray beam's centerline. Internal anomalies, such as delaminations in the corners, crushed core, blown core, water in core cells, voids in foam adhesive joints, and relative position of internal details, can readily be seen via radiography. Most composites are nearly transparent to X-rays, so low energy rays must be used. Because of safety concerns, it is impractical to use around aircraft. Operators should always be protected by sufficient lead shields, as the possibility of exposure exists either from the X-ray tube or from scattered radiation. Maintaining a minimum safe distance from the X-ray source is always essential.

Thermography

Thermal inspection comprises all methods in which heat-sensing devices are used to measure temperature variations for parts under inspection. The basic principle of thermal inspection consists of measuring or mapping of surface temperatures when heat flows from, to, or through a test object. All thermographic techniques rely on differentials in thermal conductivity between normal, defect free areas, and those having a defect. Normally, a heat source is used to elevate the temperature of the part being examined while observing the surface heating effects. Because defect free areas conduct heat more efficiently than areas with defects, the amount of heat that is either absorbed or reflected indicates the quality of the bond. The type of defects that affect the thermal properties include debonds, cracks, impact damage, panel thinning, and water ingress into composite materials and honeycomb core. Thermal methods are most effective for thin laminates or for defects near the surface.

Neutron Radiography

Neutron radiography is a nondestructive imaging technique that is capable of visualizing the internal characteristics of a sample. The transmission of neutrons through a medium is dependent upon the neutron cross sections for the nuclei in the medium. Differential attenuation of neutrons through a medium may be measured, mapped, and then visualized. The resulting image may then be utilized to analyze the internal characteristics of the sample. Neutron radiography is a complementary technique to X-ray radiography. Both techniques visualize the attenuation through a medium. The major advantage of neutron radiography is its ability to reveal light elements such as hydrogen found in corrosion products and water.

Moisture Detector

A moisture meter can be used to detect water in sandwich honeycomb structures. A moisture meter measures the radio frequency (RF) power loss caused by the presence of water. The moisture meter is often used to detect moisture in nose

Figure 7-29. *Moisture tester equipment.*

radomes. *[Figure 7-29] Figure 7-30* provides a comparison of NDI testing equipment.

Composite Repairs

Layup Materials

Hand Tools

Prepreg and dry fabrics can be cut with hand tools, such as scissors, pizza cutters, and knives. Materials made from Kevlar® are more difficult to cut than fiberglass or carbon and tools wear quicker. A squeegee and a brush are used to impregnate dry fibers with resin for wet layup. Markers, rulers, and circle templates are used to make a repair layout. *[Figure 7-31]*

Air Tools

Air-driven power tools, such as drill motors, routers, and grinders, are used for composite materials. Electric motors are not recommended, because carbon is a conductive material that can cause an electrical short circuit. If electric tools are used, they need to be of the totally enclosed type. *[Figure 7-32]*

Caul Plate

A caul plate made from aluminum is often used to support the part during the cure cycle. A mold release agent, or parting film, is applied to the caul plate so that the part does not attach to the caul plate. A thin caul plate is also used on top of the repair when a heat bonder is used. The caul plate provides a more uniform heated area and it leaves a smoother finish of the composite laminate.

Support Tooling & Molds

Certain repairs require tools to support the part and/or maintain surface contour during cure. A variety of materials can be used to manufacture these tools. The type of material depends on the type of repair, cure temperature, and whether it is a temporary or permanent tool. Support tooling is necessary for oven and autoclave cure due to the high cure temperature. The parts deform if support tooling is not used. There are many types of tooling material available. Some are molded to a specific part contour and others are used as rigid supports to maintain the contour during cure. Plaster is an inexpensive and easy material for contour tooling. It can be filled with fiberglass, hemp, or other material. Plaster is not very durable, but can be used for temporary tools. Often, a layer of fiberglass-reinforced epoxy is placed on the tool side surface to improve the finish quality. Tooling resins are used to impregnate fiberglass, carbon fiber, or other reinforcements to make permanent tools. Complex parts are made from metal or high-temperature tooling boards that are machined with 5-axis CNC equipment to make master tools that can be used to fabricate aircraft parts. *[Figures 7-33 and 7-34]*

Vacuum Bag Materials

Repairs of composite aircraft components are often performed with a technique known as vacuum bagging. A plastic bag is sealed around the repair area. Air is then removed from the bag, which allows repair plies to be drawn together with no air trapped in between. Atmospheric pressure bears on the repair and a strong, secure bond is created.

Several processing materials are used for vacuum bagging a part. These materials do not become part of the repair and are discarded after the repair process.

Release Agents

Release agents, also called mold release agents, are used so that the part comes off the tool or caul plate easily after curing.

Bleeder Ply

The bleeder ply creates a path for the air and volatiles to escape from the repair. Excess resin is collected in the bleeder. Bleeder material could be made of a layer of fiberglass, nonwoven polyester, or it could be a perforated Teflon® coated material. The structural repair manual (SRM) indicates what type and how many plies of bleeder are required. As a general rule, the thicker the laminate, the more bleeder plies are required.

Method of Inspection	Type of Defect							
	Disbond	Delamination	Dent	Crack	Hole	Water Ingestion	Overheat and Burns	Lightning Strike
Visual	X (1)	X (1)	X	X	X		X	X
X-Ray	X (1)	X (1)		X (1)		X		
Ultrasonic TTU	X	X						
Ultrasonic pulse echo		X				X		
Ultrasonic bondtester	X	X						
Tap test	X (2)	X (2)						
Infrared thermography	X (3)	X (3)				X		
Dye penetrant				X (4)				
Eddy current				X (4)				
Shearography	X (3)	X (3)						

Notes: (1) For defects that open to the surface
(2) For thin structure (3 plies or less)
(3) The procedures for this type of inspection are being developed
(4) This procedure is not recommended

Figure 7-30. *Comparison of NDI testing equipment.*

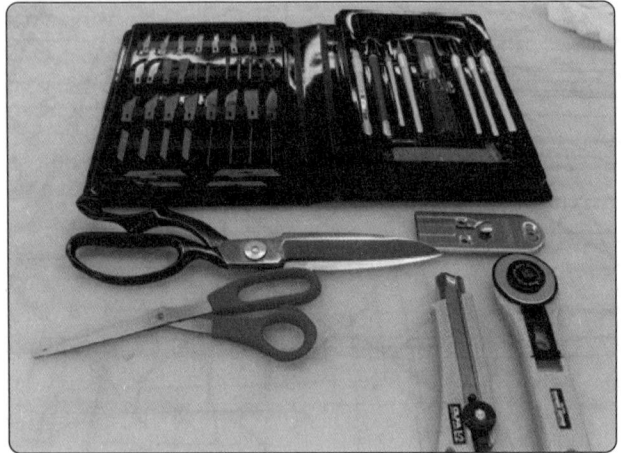

Figure 7-31. *Hand tools for layup.*

Figure 7-32. *Air tools used for composite repair.*

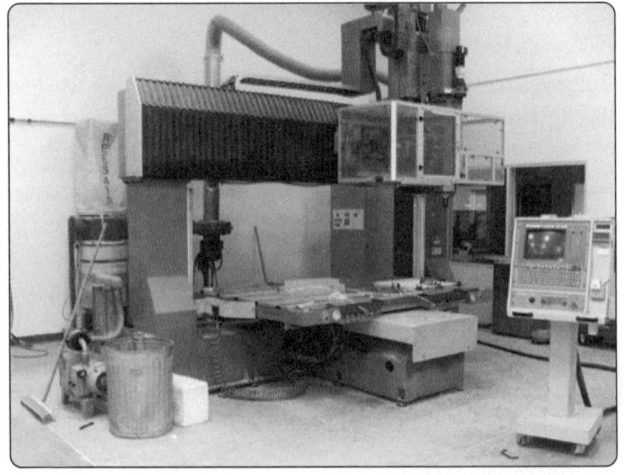

Figure 7-33. *Five-axis CNC equipment for tool and mold making.*

Figure 7-34. *A mold of an inlet duct.*

Peel Ply

Peel plies are often used to create a clean surface for bonding purposes. A thin layer of fiberglass is cured with the repair part. Just before the part is bonded to another structure, the peel ply is removed. The peel ply is easy to remove and leaves a clean surface for bonding. Peel plies are manufactured from polyester, nylon, flouronated ethylene propylene (FEP), or coated fiberglass. They can be difficult to remove if overheated. Some coated peel plies can leave an undesirable contamination on the surface. The preferred peel ply material is polyester that has been heat-set to eliminate shrinkage.

Layup Tapes

Vacuum bag sealing tape, also called sticky tape, is used to seal the vacuum bag to the part or tool. Always check the temperature rating of the tape before use to ensure that you use appropriately rated tape.

Perforated Release Film

Perforated parting film is used to allow air and volatiles out of the repair, and it prevents the bleeder ply from sticking to the part or repair. It is available with different size holes and hole spacing depending on the amount of bleeding required.

Solid Release Film

Solid release films are used so that the prepreg or wet layup plies do not stick to the working surface or caul plate. Solid release film is also used to prevent the resins from bleeding through and damaging the heat blanket or caul plate if they are used.

Breather Material

The breather material is used to provide a path for air to get out of the vacuum bag. The breather must contact the bleeder. Typically, polyester is used in either 4-ounce or 10-ounce weights. Four ounces is used for applications below 50 pounds per square inch (psi) and 10 ounces is used for 50–100 psi.

Vacuum Bag

The vacuum bag material provides a tough layer between the repair and the atmosphere. The vacuum bag material is available in different temperature ratings, so make sure that the material used for the repair can handle the cure temperature. Most vacuum bag materials are one time use, but material made from flexible silicon rubber is reusable. Two small cuts are made in the bagging material so that the vacuum probe valve can be installed. The vacuum bag is not very flexible and plies need to be made in the bag if complex

Figure 7-36. *Bagging of complex part.*

Figure 7-35. *Bagging materials.*

Figure 7-37. *Self-sealing vacuum bag with heater element.*

shapes are to be bagged. Sometimes, an envelope type bag is used, but the disadvantage of this method is that the vacuum pressure might crush the part. Reusable bags made from silicon rubber are available that are more flexible. Some have a built-in heater blanket that simplifies the bagging task. *[Figures 7-35, 7-36, and 7-37]*

Vacuum Equipment

A vacuum pump is used to evacuate air and volatiles from the vacuum bag so that atmospheric pressure consolidates the plies. A dedicated vacuum pump is used in a repair shop. For repairs on the aircraft, a mobile vacuum pump could be used. Most heat bonders have a built-in vacuum pump. Special air hoses are used as vacuum lines, because regular air hoses might collapse when a vacuum is applied. The vacuum lines that are used in the oven or autoclave need to be able to withstand the high temperatures in the heating device. A vacuum pressure regulator is sometimes used to lower the vacuum pressure during the bagging process.

Vacuum Compaction Table

A vacuum compaction table is a convenient tool for debulking composite layups with multiple plies. Essentially a reusable vacuum bag, a compaction table consists of a metal table surface with a hinged cover. The cover includes a solid frame, a flexible membrane, and a vacuum seal. Repair plies are laid up on the table surface and sealed beneath the cover with vacuum to remove entrapped air. Some compaction tables are heated but most are not.

Heat Sources
Oven

Composite materials can be cured in ovens using various pressure application methods. *[Figure 7-38]* Typically, vacuum bagging is used to remove volatiles and trapped air and utilizes atmospheric pressure for consolidation. Another method of pressure application for oven cures is the use of shrink wrapping or shrink tape. The oven uses heated air circulated at high speed to cure the material system. Typical oven cure temperatures are 250 °F and 350 °F. Ovens have a temperature sensor to feed temperature data back to the oven controller. The oven temperature can differ from the actual part temperature depending upon the location of the oven sensor and the location of the part in the oven. The thermal mass of the part in the oven is generally greater than the surrounding oven and during rise to temperature, the part temperature can lag the oven temperature by a considerable amount. To deal with these differences, at least two thermocouples must be placed on the part and connected to a temperature-sensing device (separate chart recorder, hot bonder, etc.) located outside the oven. Some oven controllers can be controlled by thermocouples placed on the repair part.

Figure 7-38. *Walk-in curing oven.*

Autoclave

An autoclave system allows a complex chemical reaction to occur inside a pressure vessel according to a specified time, temperature, and pressure profile in order to process a variety of materials. *[Figure 7-39]* The evolution of materials and processes has taken autoclave operating conditions from 120 °C (250 °F) and 275 kPa (40 psi) to well over 760 °C (1,400 °F) and 69,000 kPa (10,000 psi). Autoclaves that are operated at lower temperatures and pressures can be pressurized by air, but if higher temperatures and pressures are required for the cure cycle, a 50/50 mixture of air and nitrogen or 100 percent nitrogen should be used to reduce the change of an autoclave fire.

The major elements of an autoclave system are a vessel to contain pressure, sources to heat the gas stream and circulate it uniformly within the vessel, a subsystem to apply vacuum to parts covered by a vacuum bag, a subsystem to control operating parameters, and a subsystem to load the molds into the autoclave. Modern autoclaves are computer controlled and the operator can write and monitor all types of cure cycle programs. The most accurate way to control the cure cycle is to control the autoclave controller with thermocouples that are placed on the actual part.

Most parts processed in autoclaves are covered with a vacuum bag that is used primarily for compaction of laminates and to provide a path for removal of volatiles. The bag allows the part to be subjected to differential pressure in the autoclave without being directly exposed to the autoclave atmosphere.

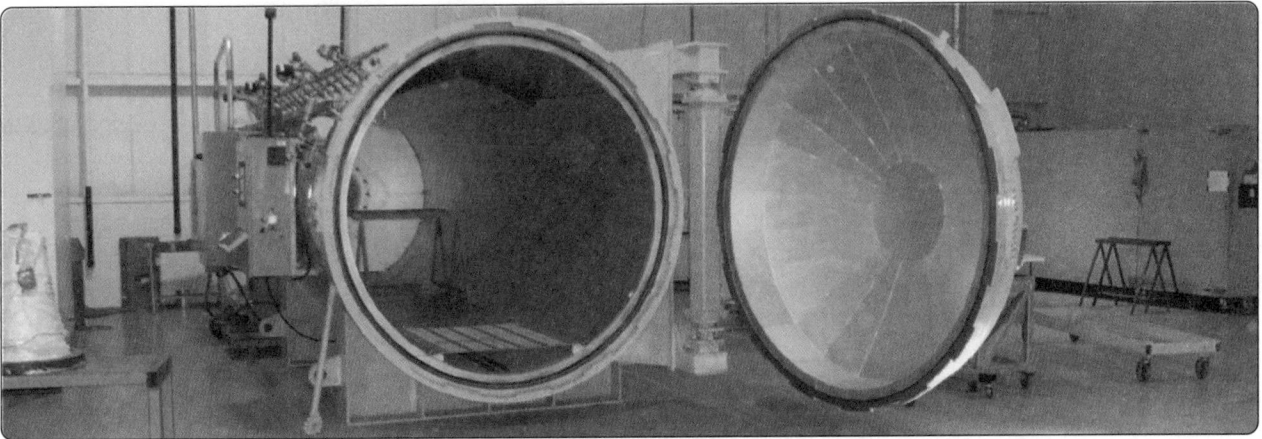

Figure 7-39. *Autoclave.*

The vacuum bag is also used to apply varying levels of vacuum to the part.

Heat Bonder & Heat Lamps

Typical on-aircraft heating methods include electrical resistance heat blankets, infrared heat lamps, and hot air devices. All heating devices must be controlled by some means so that the correct amount of heat can be applied. This is particularly important for repairs using prepreg material and adhesives, because controlled heating and cooling rates are usually prescribed.

Heat Bonder

A heat bonder is a portable device that automatically controls heating based on temperature feedback from the repair area. Heat bonders also have a vacuum pump that supplies and monitors the vacuum in the vacuum bag. The heat bonder controls the cure cycle with thermocouples that are placed near the repair. Some repairs require up to 10 thermocouples.

Modern heat bonders can run many different types of cure programs and cure cycle data can be printed out or uploaded to a computer. *[Figure 7-40]*

Heat Blanket

A heat blanket is a flexible heater. It is made of two layers of silicon rubber with a metal resistance heater between the two layers of silicon. Heat blankets are a common method of applying heat for repairs on the aircraft. Heat blankets may be controlled manually; however, they are usually used in conjunction with a heat bonder. Heat is transferred from the blanket via conduction. Consequently, the heat blanket must conform to and be in 100 percent contact with the part, which is usually accomplished using vacuum bag pressure. *[Figure 7-41]*

Heat Lamp

Infrared heat lamps can also be used for elevated temperature curing of composites if a vacuum bag is not utilized. However, they are generally not effective for producing curing temperatures above 150 °F, or for areas larger than two square feet. It is also difficult to control the heat applied with a lamp, and lamps tend to generate high-surface temperatures

Figure 7-40. *Heat bonder equipment.*

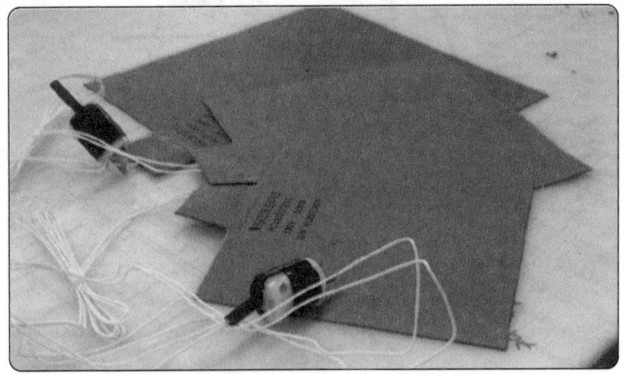

Figure 7-41. *Heat blankets.*

quickly. If controlled by thermostats, heat lamps can be useful in applying curing heat to large or irregular surfaces. Heat bonders can be used to control heat lamps.

Hot Air System

Hot air systems can be used to cure composite repairs, and are mainly restricted to small repairs and for drying the repair area. A heat generator supplies hot air that is directed into an insulated enclosure set up around the repair area after vacuum bagging has been deployed. The hot air surrounds the repair for even temperature rise.

Heat Press Forming

During the press forming process, flat stacked thermoplastic prepreg is heated to above melt temperature (340–430 °C, or 645–805 °F) in an oven, rapidly (1–10 seconds) shuttled to a forming die, pressed to shape, and consolidated and cooled under pressure (700–7,000 kPa, or 100–1,000 psi). *[Figure 7-42]* In production, press forming dies usually are matched male-female sets constructed of steel or aluminum. However, rubber, wood, phenolics, and so on can be used during prototyping. The die set can be maintained at room temperature throughout the forming-consolidation cycle. But, the use of a hot die (120–200 °C, or 250–390 °F) allows control of the cooling-down rate (avoiding part warpage and controlling morphology in semicrystalline thermoplastic prepreg, such as PEEK and polyphenylene sulfide) and extends the forming window promoting better ply slip.

The main disadvantage with this method is that the press only applies pressure in one direction, and hence, it is difficult to make complex-shaped (e.g., beads, closed corners) parts or parts with legs that approach vertical. Since the temperature of the die set need not be cycled with each part, rapid forming times of between 10 minutes and 2 hours are achievable with

Figure 7-42. *Heat press.*

press forming.

Thermocouples

A thermocouple (TC) is a thermoelectric device used to accurately measure temperatures. It may be connected to a simple temperature reading device, or connected to a hot bonder, oven, or other type of controller that regulates the amount of heat. TCs consist of a wire with two leads of dissimilar metals that are joined at one end. Heating the joint produces an electric current, which is converted to a temperature reading with a TC monitor. Select the type of wire (J or K) and the type of connector that are compatible with the local temperature monitoring equipment (hot bonder, oven, autoclave, etc.). TC wire is available with different types of insulation; check the manufacturer's product data sheets to ensure the insulation withstands the highest cure temperature. Teflon-insulated wire is generally good for 390 °F and lower cures; Kapton-insulated wire should be used for higher temperatures.

Thermocouple Placement

Thermocouple placement is the key in obtaining proper cure temperatures throughout the repair. In general, the thermocouples used for temperature control should be placed as close as possible to the repair material without causing it to become embedded in the repair or producing indentations in the repair. They should also be placed in strategic hot or cold locations to ensure the materials are adequately cured but not exposed to excessively high temperatures that could degrade the material structural properties. The thermocouples should be placed as close as practical to the area that needs to be monitored. The following steps should be taken when using thermocouples:

- Never use fewer than three thermocouples to monitor a heating cycle.

- If bonding a precured patch, place the thermocouple near the center of the patch.

- A control thermocouple may be centered over a low-temperature (200 °F or lower) co-cured patch as long as it is placed on top of a thin metallic sheet to prevent a thermocouple indentation onto the patch. This may allow for a more accurate control of the patch temperature.

- The thermocouples installed around the perimeter of the repair patch should be placed approximately 0.5-inch away from the edge of the adhesive line.

- Place flash tape below and above the thermocouple tips to protect them from resin flash and to protect the control unit from electrical shorts.

- Do not place the thermocouple under the vacuum port as the pressure may damage the lead and cause erroneous readings to occur.

- Do not place thermocouple wires adjacent to or crossing the heat blanket power cord to prevent erroneous temperature readings caused by magnetic flux lines.

- Do not place any control thermocouple beyond the heat blanket's two-inch overlap of the repair to prevent the controller from trying to compensate for the lower temperature.

- Always leave slack in the thermocouple wire under the vacuum bag to prevent the thermocouple from being pulled away from the area to be monitored as vacuum is applied.

Thermal Survey of Repair Area

In order to achieve maximum structural bonded composite repair, it is essential to cure these materials within the recommended temperature range. Failure to cure at the correct temperatures can produce weak patches and/or bonding surfaces and can result in a repair failure during service. A thermal survey should be performed prior to installing the repair to ensure proper and uniform temperatures can be achieved. The thermal survey determines the heating and insulation requirements, as well as TC locations for the repair area. The thermal survey is especially useful for determining the methods of heating (hot air modules, heat lamps, heat blanket method and monitoring requirements in cases where heat sinks (substructure for instance) exist in the repair area). It should be performed for all types of heating methods to preclude insufficient, excessive, or uneven heating of the repair area.

Temperature Variations in Repair Zone

Thermal variations in the repair area occur for many reasons. Primary among these are material type, material thickness, and underlying structure in the repair zone. For these reasons, it is important to know the structural composition of the area to be repaired. Substructure existing in the repair zone conducts heat away from the repair area, resulting in a cold spot directly above the structure. Thin skins heat quickly and can easily be overheated. Thick skin sections absorb heat slowly and take longer to reach soak temperature. The thermal survey identifies these problem areas and allows the technician to develop the heat and insulation setup required for even heating of the repair area.

Thermal Survey

During the thermal survey process, try to determine possible hot and cold areas in the repair zone. Temporarily attach a patch of the same material and thickness, several thermal couples, heating blanket, and a vacuum bag to the repair area. Heat the area and, after the temperature is stabilized,

record the thermocouple temperatures. Add insulation if the temperature of the thermocouple varies more than 10 degrees from average. The areas with a stringer and rib indicate a lower temperature than the middle of the patch because they act as a heat sink. Add insulation to these areas to increase the temperature. *[Figure 7-43]*

Solutions to Heat Sink Problems

Additional insulation can be placed over the repair area. This insulation can also be extended beyond the repair area to minimize heat being conducted away. Breather materials and fiberglass cloths work well, either on top of the vacuum bag or within the vacuum bag or on the accessible backside of the structure. Place more insulation over cool spots and less insulation over hot spots. If access is available to the backside of the repair area, additional heat blankets could be placed there to heat the repair area more evenly.

Types of Layups
Wet Layups

During the wet layup process, a dry fabric is impregnated with a resin. Mix the resin system just before making the repair. Lay out the repair plies on a piece of fabric and impregnate the fabric with the resin. After the fabric is impregnated, cut the repair plies, stack in the correct ply orientation, and vacuum bag. Wet layup repairs are often used with fiberglass for nonstructural applications. Carbon and Kevlar® dry fabric could also be used with a wet layup resin system. Many resin systems used with wet layup cure at room temperature, are easy to accomplish, and the materials can be stored at room temperature for long period of times. The disadvantage of room temperature wet layup is that it does not restore the strength and durability of the original structure and parts that were cured at 250 °F or 350 °F during manufacturing. Some wet layup resins use an elevated temperature cure and have improved properties. In general, wet layup properties are less than properties of prepreg material.

Epoxy resins may require refrigeration until they are used. This prevents the aging of the epoxy. The label on the container states the correct storage temperature for each component. The typical storage temperature is between 40 °F and 80 °F for most epoxy resins. Some resin systems require storage below 40 °F.

Prepreg

Prepreg is a fabric or tape that is impregnated with a resin during the manufacturing process. The resin system is already mixed and is in the B stage cure. Store the prepreg material in a freezer below 0 °F to prevent further curing of the resin. The material is typically placed on a roll and a backing material is placed on one side of the material so that the prepreg does not stick together. The prepreg material is

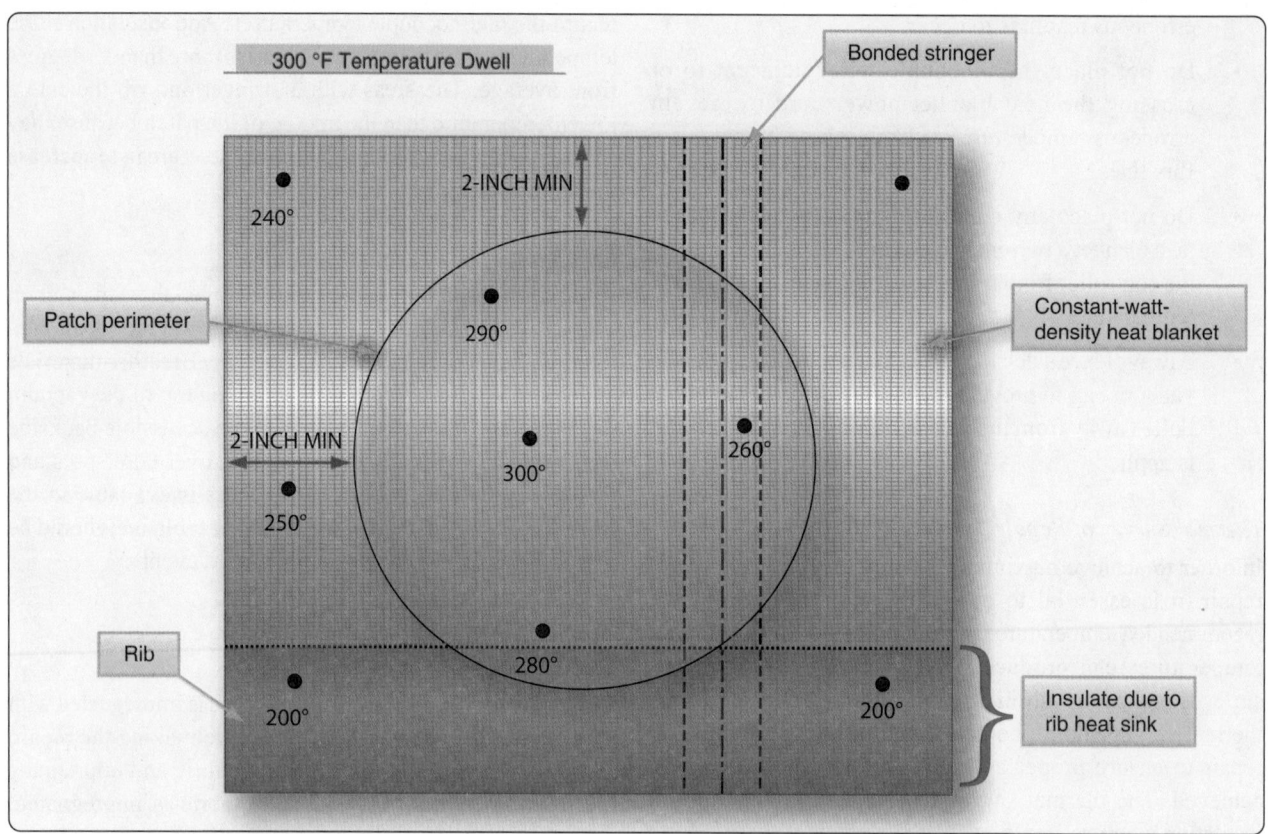

Figure 7-43. *Thermal survey example.*

sticky and adheres to other plies easily during the stack-up process. You must remove the prepreg from the freezer and let the material thaw, which might take 8 hours for a full roll. Store the prepreg materials in a sealed, moisture proof bag. Do not open these bags until the material is completely thawed, to prevent contamination of the material by moisture.

After the material is thawed and removed from the backing material, cut it in repair plies, stack in the correct ply orientation, and vacuum bag. Do not forget to remove the backing material when stacking the plies. Cure prepregs at an elevated cure cycle; the most common temperatures used are 250 °F and 350 °F. Autoclaves, curing ovens, and heat bonders can be used to cure the prepreg material.

Consolidation is necessary if parts are made from several layers of prepreg, because large quantities of air can be trapped between each prepreg layer. Remove this trapped air by covering the prepreg with a perforated release film and a breather ply, and apply a vacuum bag. Apply the vacuum for 10 to 15 minutes at room temperature. Typically, attach the first consolidated ply to the tool face and repeat this process after every 3 or 5 layers depending on the prepreg thickness and component shape.

Store prepreg, film adhesive, and foaming adhesives in

a freezer at a temperature below 0 °F. If these types of materials need to be shipped, place them in special containers filled with dry ice. The freezer must not be of the automatic defrost type; the auto-defrost cycle periodically warms the inside of the freezer, which can reduce the shelf life and consume the allowable out-time of the composite material. Freezers must be capable of maintaining 0 °F or below; most household freezers meet this level. Walk-in freezers can be used for large volume cold storage. If usage is small, a chest-type freezer may suffice. Refrigerators are used to store laminating and paste adhesives and should be kept near 40 °F. *[Figure 7-44]*

Uncured prepreg materials have time limits for storage and use. *[Figure 7-45]* The maximum time allowed for storing of a prepreg at low temperature is called the storage life, which is typically 6 months to a year. The material can be tested, and the storage life could be extended by the material manufacturer. The maximum time allowed for material at room temperature before the material cures is called the mechanical life. The recommended time at room temperature to complete layup and compaction is called the handling life. The handling life is shorter than the mechanical life. The mechanical life is measured from the time the material is removed from the freezer until the time the material is returned to the freezer. The operator must keep records of

Figure 7-44. *Walk-in freezer for storing prepreg materials.*

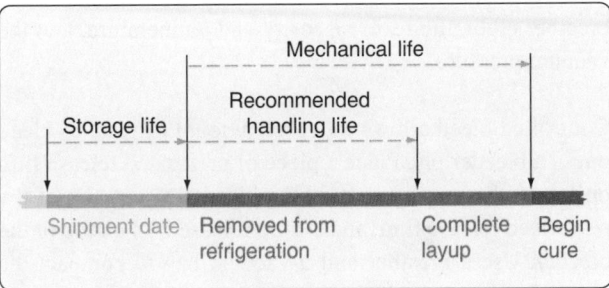

Figure 7-45. *Storage life for prepreg materials.*

the time in and out of the freezer. Material that exceeds the mechanical life needs to be discarded.

Many repair facilities cut the material in smaller kits and store them in moisture-proof bags that thaw quicker when removed from the freezer. This also limits the time out of the freezer for a big roll.

All frozen prepreg materials need to be stored in moisture-proof bag to avoid moisture contamination. All prepreg material should be protected from dust, oil, vapors, smoke, and other contaminants. A clean room for repair layup would be best, but if a clean room is not available, the prepreg should be protected by storing them in bags or keeping them covered with plastic. Before starting the layup, cover the unprotected sides of the prepreg with parting film, and clean the area being repaired immediately before laying up the repair plies.

Prepreg material is temperature sensitive. Excessively high temperatures cause the material to begin curing, and excessively low temperatures make the material difficult to handle. For repairs on aircraft in very cold or very hot climates, the area should be protected by a tent around the repair area. Prepare the prepreg repair plies in a controlled-

temperature environment and bring them to the repair area immediately before using them.

Co-curing

Co-curing is a process wherein two parts are simultaneously cured. The interface between the two parts may or may not have an adhesive layer. Co-curing often results in poor panel surface quality, which is prevented by using a secondary surfacing material co-cured in the standard cure cycle or a subsequent fill-and-fair operation. Co-cured skins may also have poorer mechanical properties, requiring the use of reduced design values.

A typical co-cure application is the simultaneous cure of a stiffener and a skin. Adhesive film is frequently placed into the interface between the stiffener and the skin to increase fatigue and peel resistance. Principal advantages derived from the co-cure process are excellent fit between bonded components and guaranteed surface cleanliness.

Secondary Bonding

Secondary bonding utilizes precured composite detail parts, and uses a layer of adhesive to bond two precured composite parts. Honeycomb sandwich assemblies commonly use a secondary bonding process to ensure optimal structural performance. Laminates co-cured over honeycomb core may have distorted plies that have dipped into the core cells. As a result, compressive stiffness and strength can be reduced as much as 10 and 20 percent, respectively.

Precured laminates undergoing secondary bonding usually have a thin nylon or fiberglass peel ply cured onto the bonding surfaces. While the peel ply sometimes hampers nondestructive inspection of the precured laminate, it has been found to be the most effective means of ensuring surface cleanliness prior to bonding. When the peel ply is stripped away, a pristine surface becomes available. Light scuff sanding removes high resin peak impressions produced by the peel ply weave which, if they fracture, create cracks in the bondline.

Composite materials can be used to structurally repair, restore, or enhance aluminum, steel, and titanium components. Bonded composite doublers have the ability to slow or stop fatigue crack growth, replace lost structural area due to corrosion grind-outs, and structurally enhance areas with small and negative margins. This technology has often been referred to as a combination of metal bonding and conventional on-aircraft composite bonded repair. Boron prepreg tape with an epoxy resin is most often used for this application.

Co-bonding

In the co-bonding process, one of the detail parts is precured with the mating part being cured simultaneously with the adhesive. Film adhesive is often used to improve peel strength.

Layup Process (Typical Laminated Wet Layup)
Layup Techniques

Read the SRM and determine the correct repair material, number of plies required for the repair, and the ply orientation. Dry the part, remove the damage, and taper sand the edges of damaged area. Use a piece of thin plastic, and trace the size of each repair ply from the damaged area. Indicate the ply orientation of each ply on the trace sheet. Copy the repair ply information to a piece of repair material that is large enough to cut all plies. Impregnate the repair material with resin, place a piece of transparent release film over the fabric, cut out the plies, and lay up the plies in the damaged area. The plies are usually placed using the smallest ply first taper layup sequence, but an alternative method is to use the largest ply first layup sequence. In this sequence, the first layer of reinforcing fabric completely covers the work area, followed by successively smaller layers, and then is finished with an extra outer layer or two extending over the patch and onto the sound laminate for some distance. Both methods are illustrated in *Figures 7-46* and *7-47*.

Bleedout Technique

The traditional bleedout using a vacuum bag technique places a perforated release film and a breather/bleeder ply on top of the repair. The holes in the release film allow air to breath and resin to bleed off over the entire repair area. The amount of resin bled off depends on the size and number of holes

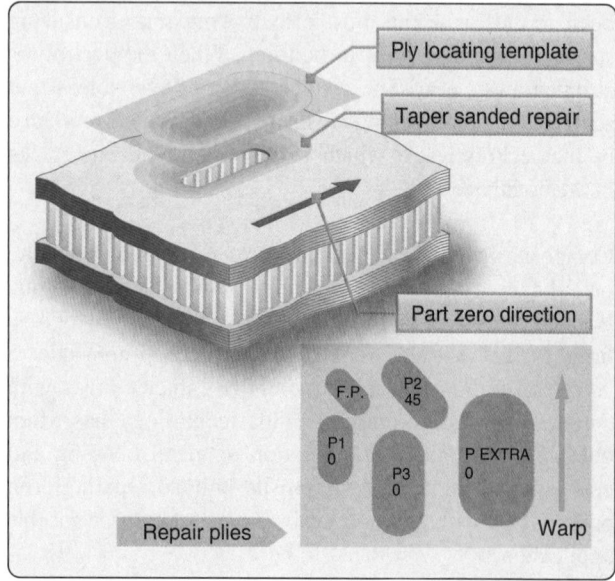

Figure 7-46. *Repair layup process.*

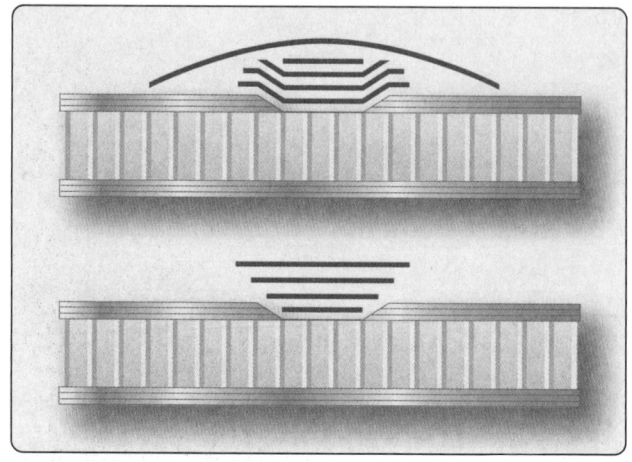

Figure 7-47. *Different layup techniques.*

in the perforated release film, the thickness of the bleeder/breather cloth, the resin viscosity and temperature, and the vacuum pressure.

Controlled bleed allows a limited amount of resin to bleed out in a bleeder ply. Place a piece of perforated release film on top of the prepreg material, a bleeder ply on top of the perforated release film, and a solid release film on top of the bleeder. Use a breather and a vacuum bag to compact the repair. The breather allows the air to escape. The bleeder can only absorb a limited amount of resin, and the amount of resin that is bled can be controlled by using multiple bleeder plies. Too many bleeder plies can result in a resin-starved repair. Always consult the maintenance manual or manufacturer tech sheets for correct bagging and bleeding techniques.

No Bleedout

Prepreg systems with 32 to 35 percent resin content are typically no-bleed systems. These prepregs contain exactly the amount of resin needed in the cured laminate; therefore, resin bleedoff is not desired. Bleedout of these prepregs results in a resin-starved repair or part. Many high-strength prepregs in use today are no-bleed systems. No bleeder is used, and the resin is trapped/sealed so that none bleeds away. Consult the maintenance manual to determine if bleeder plies are required for the repair. A sheet of solid release film (no holes) is placed on top of the prepreg and taped off at the edges with flash tape. Small openings are created at the edges of the tape so that air can escape. A breather and vacuum bag are installed to compact the prepreg plies. The air can escape on the edge of the repair but no resin can bleed out. *[Figure 7-48]*

Horizontal (or edge) bleedout is used for small room temperature wet layup repairs. A 2-inch strip of breather cloth is placed around the repair or part (edge breather). There is no need for a release film because there is no bleeder/breather

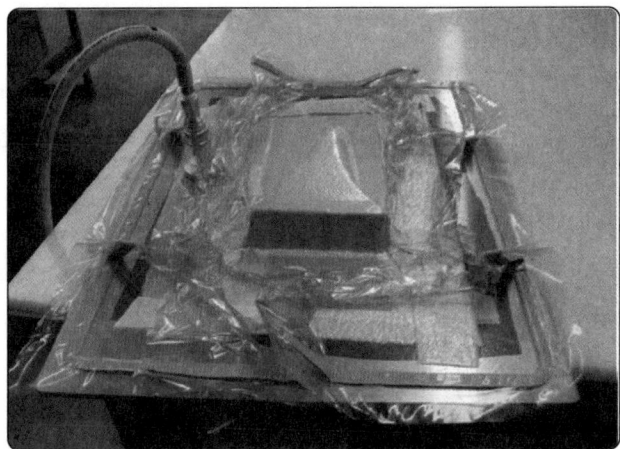

Figure 7-48. *Vacuum bagging of contoured part.*

Type	Example	Comments
Symmetrical, balanced	(+45, −45, 0, 0, −45, +45)	Flat, constant midplane stress
Nonsymmetrical, balanced	(90, +45, 0, 90, −45, 0)	Induces curvature
Symmetrical, nonbalanced	(−45, 0, 0, −45)	Induces twist
Nonsymmetrical, nonbalanced	(90, −45, 0, 90, −45, 0)	Induces twist and curvature

Figure 7-50. *Examples of the effects caused by nonsymmetrical laminates.*

cloth on top of the repair. The part is impregnated with resin, and the vacuum bag is placed over the repair. A vacuum is applied and a squeegee is used to remove air and excess resin to the edge breather.

Ply Orientation Warp Clock

In order to minimize any residual thermal stresses caused during cure of the resin, it is always good practice to design a symmetrical, or balanced, laminate. Examples of balance laminates are presented in *Figure 7-49*. The first example uses unidirectional tape, and examples 2 and 3 are typical quasi-isotropic laminates fabricated from woven cloth.

Figure 7-50 presents examples of the effects caused by nonsymmetrical laminates. These effects are most pronounced in laminates that are cured at high temperature in an autoclave or oven due to the thermal stresses developed in the laminate as the laminate cools down from the cure temperature to room temperature. Laminates cured at room temperature using typical wet layup do not exhibit the same degree of distortion due to the much smaller thermal stresses. The strength and stiffness of a composite buildup depends on the ply orientation. The practical range of strength and stiffness of carbon epoxy extends from values as low as those provided by fiberglass to as high as those provided by titanium. This range of values is determined by the orientation

Example	Lamina	Written as
1	±45°, −45°, 0°, 0°, −45°, +45°	(+45, −45, 0) S
2	±45°, 0°/90°, ±45°, 0°/90°, 0°/90°, ±45°, 0°/90°, ±45°	(±45, 0/90)2S
3	±45°, ±45°, 0°/90°, 0°/90°, ±45°, ±45°	([±45] 2, 0/90) S

Figure 7-49. *Examples of balance laminates.*

of the plies to the applied load. Because the strength design requirement is a function of the applied load direction, ply orientation and ply sequence must be correct. It is critical during a repair operation to replace each damaged ply with a ply of the same material and orientation or an approved substitute.

Warp is the longitudinal fibers of a fabric. The warp is the high-strength direction due to the straightness of the fibers. A warp clock is used to describe direction of fibers on a diagram, spec sheet, or manufacturer's sheets. If the warp clock is not available on the fabric, the orientation is defaulted to zero as the fabric comes off the roll. Therefore, 90° to zero is across the width of the fabric. 90° to zero is also called the fill direction.

Mixing Resins

Epoxy resins, like all multipart materials, must be thoroughly mixed. Some resin systems have a dye added to aid in seeing how well the material is mixed. Since many resin systems do not have a dye, the resin must be mixed slowly and fully for three minutes. Air enters into the mixture if the resin is mixed too fast. If the resin system is not fully mixed, the resin may not cure properly. Make sure to scrape the edges and bottom of the mixing cup to ensure that all resin is mixed correctly.

Do not mix large quantities of quick curing resin. These types of resins produce heat after they are mixed. Smoke can burn or poison you when the resin overheats. Mix only the amount of material that is required. Mix more than one batch if more material is needed than the maximum batch size.

Saturation Techniques

For wet layup repair, impregnate the fabric with resin. It is important to put the right amount of resin on the fabric. Too much or too little resin affects the strength of the repair. Air that is put into the resin or not removed from the fabric also reduces the repair strength.

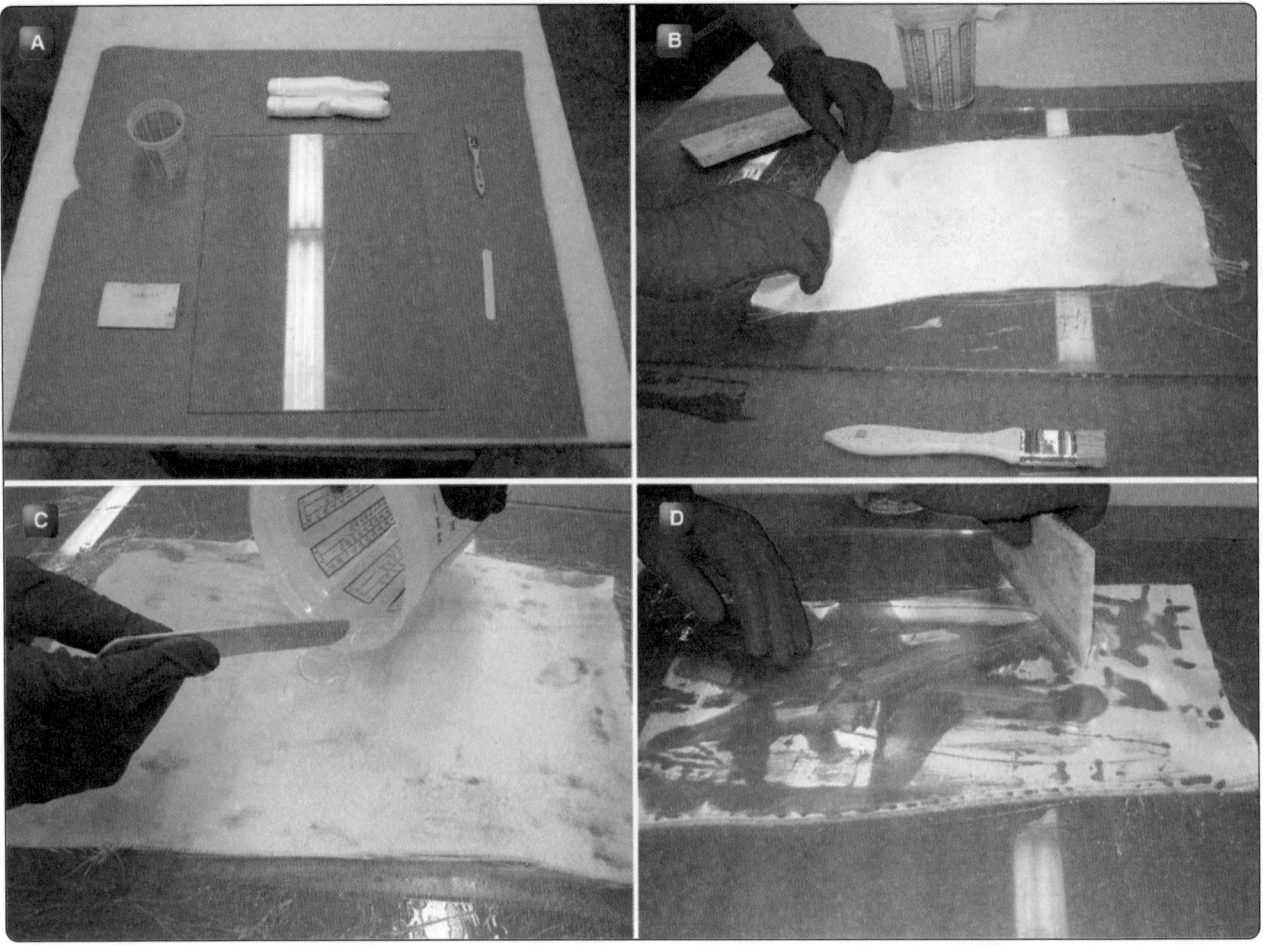

Figure 7-51. *Fabric impregnation with a brush or squeegee: A) wet layup materials; B) fabric placement; C) fabric impregnation; D) squeegee used to thoroughly wet the fabric.*

Fabric Impregnation With a Brush or Squeegee

The traditional way of impregnating the fabric is by using a brush or squeegee. The technician puts a mold release compound or a release film on a caul plate so that the plies will not adhere to the caul plate. Place a sheet of fabric on the caul plate and apply resin in the middle of the sheet. Use a brush or squeegee to thoroughly wet the fabric. More plies of fabric and resin are added and the process is repeated until all plies are impregnated. A vacuum bag will be used to consolidate the plies and to bleed off excess resin and volatiles. Most wet layup processes have a room temperature cure but extra heat, up to 150 °F, are used to speed up the curing process. *[Figure 7-51]*

Fabric Impregnation Using a Vacuum Bag

The vacuum-assisted impregnation method is used to impregnate repair fabric with a two-part resin while enclosed inside a vacuum bag. This method is preferred for tight-knit weaves and when near optimum resin-to-fiber ratio is required. Compared to squeegee impregnation, this process reduces the level of entrapped air within the fabric and offers a more controlled and contained configuration for completing the impregnation process.

Vacuum-assisted impregnation consists of the following steps:

1. Place vacuum bag sealing tape on the table surface around the area that is used to impregnate the material. The area should be at least 4 inches larger than the material to be impregnated.

2. Place an edge breather cloth next to the vacuum bag sealing tape. The edge breather should be 1–2 inches wide.

3. Place a piece of solid parting film on the table. The sheet should be 2-inches larger than the material to be impregnated.

4. Weigh the fabric to find the amount of resin mix that is necessary to impregnate the material.

5. Lay the fabric on the parting film.

6. Put a piece of breather material between the fabric and the edge breather to provide an air path.

7. Pour the resin onto the fabric. The resin should be a continuous pool in the center area of the fabric.

8. Put vacuum probes on the edge breather.

9. Place a second piece of solid parting film over the fabric. This film should be the same size or larger than the first piece.

10. Place and seal the vacuum bag, and apply vacuum to the bag.

11. Allow 2 minutes for the air to be removed from the fabric.

12. Sweep the resin into the fabric with a squeegee. Slowly sweep the resin from the center to the edge of the fabric. The resin should be uniformly distributed over all of the fabric.

13. Remove the fabric and cut the repair plies.

Vacuum Bagging Techniques

Vacuum bag molding is a process in which the layup is cured under pressure generated by drawing a vacuum in the space between the layup and a flexible sheet placed over it and sealed at the edges. In the vacuum bag molding process, the plies are generally placed in the mold by hand layup using prepreg or wet layup. High-flow resins are preferred for vacuum bag molding.

Single Side Vacuum Bagging

This is the preferred method if the repair part is large enough for a vacuum bag on one side of the repair. The vacuum bag is taped in place with tacky tape and a vacuum port is placed through the bag to create the vacuum.

Envelope Bagging

Envelope bagging is a process in which the part to be repaired is completely enclosed in a vacuum bag or the bag is wrapped around the end of the component to obtain an adequate seal. It is frequently used for removable aircraft parts, such as flight controls, access panels, etc., and when a part's geometry and/or the repair location makes it very difficult to properly vacuum bag and seal the area in a vacuum. In some cases, a part may be too small to allow installation of a single-side bag vacuum. Other times, the repair is located on the end of a large component that must have a vacuum bag wrapped around the ends and sealed all the way around. [Figure 7-52]

Alternate Pressure Application
Shrink Tape

Another method of pressure application for oven cures is the use of shrink wrapping or shrink tape. This method is

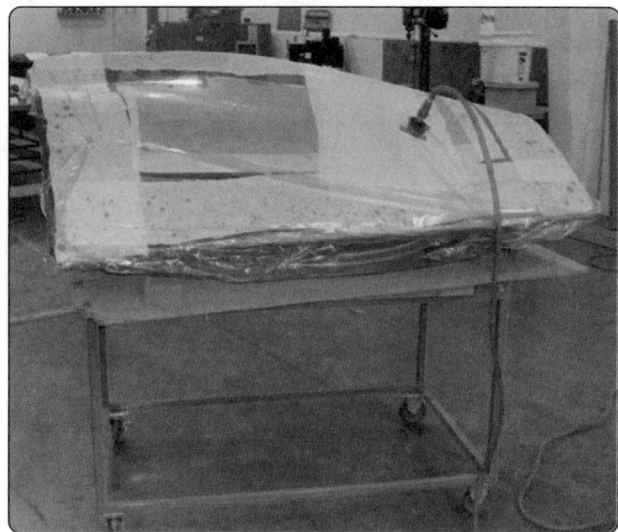

Figure 7-52. *Envelope bagging of repair.*

commonly used with parts that have been filament wound, because some of the same rules for application apply. The tape is wrapped around the completed layup, usually with only a layer of release material between the tape and the layup. Heat is applied to the tape, usually using a heat gun to make the tape shrink, a process that can apply a tremendous amount of pressure to the layup. After shrinking, the part is placed in the oven for cure. High quality parts can be made inexpensively using shrink tape.

C-Clamps

Parts can also be pressed together with clamps. This technique is used for solid laminate edges of honeycomb panels. Clamps (e.g., C-clamps and spring clamps) are used for pressing together the edges of components and/or repair details. Always use clamps with pressure distribution pads because damage to the part may occur if the clamping force is too high. Spring clamps can be used in applications where resin squeeze-out during cure would require C-clamps to be retightened periodically.

Shotbags & Weights

Shotbags and weights can be used also to provide pressure, but their use is limited due to the low level of pressure imposed.

Curing of Composite Materials

A cure cycle is the time/temperature/pressure cycle used to cure a thermosetting resin system or prepreg. The curing of a repair is as important as the curing of the original part material. Unlike metal repairs in which the materials are premanufactured, composite repairs require the technician to manufacture the material. This includes all storage, processing, and quality control functions. An aircraft repair's cure cycle starts with material storage. Materials that are stored incorrectly can begin to cure before they are used for a repair. All time and temperature requirements must be met and documented. Consult the aircraft structural repair manual to determine the correct cure cycle for the part that needs to be repaired.

Room Temperature Curing

Room temperature curing is the most advantageous in terms of energy savings and portability. Room temperature cure wet layup repairs do not restore either the strength or the durability of the original 250 °F or 350 °F cure components and are often used for wet layup fiberglass repairs for noncritical components. Room temperature cure repairs can be accelerated by the application of heat. Maximum properties are achieved at 150 °F. A vacuum bag can be used to consolidate the plies and to provide a path for air and volatiles to escape.

Elevated Temperature Curing

All prepreg materials are cured with an elevated temperature cure cycle. Some wet layup repairs use an elevated cure cycle as well to increase repair strength and to speed up the curing process. The curing oven and heat bonder uses a vacuum bag to consolidate the plies and to provide a path for air and volatiles to escape. The autoclave uses vacuum and positive pressure to consolidate the plies and to provide a path for air and volatiles to escape. Most heating devices use a programmable computer control to run the cure cycles. The operator can select from a menu of available cure cycles or write their own program. Thermocouples are placed near the repair, and they provide temperature feedback for the

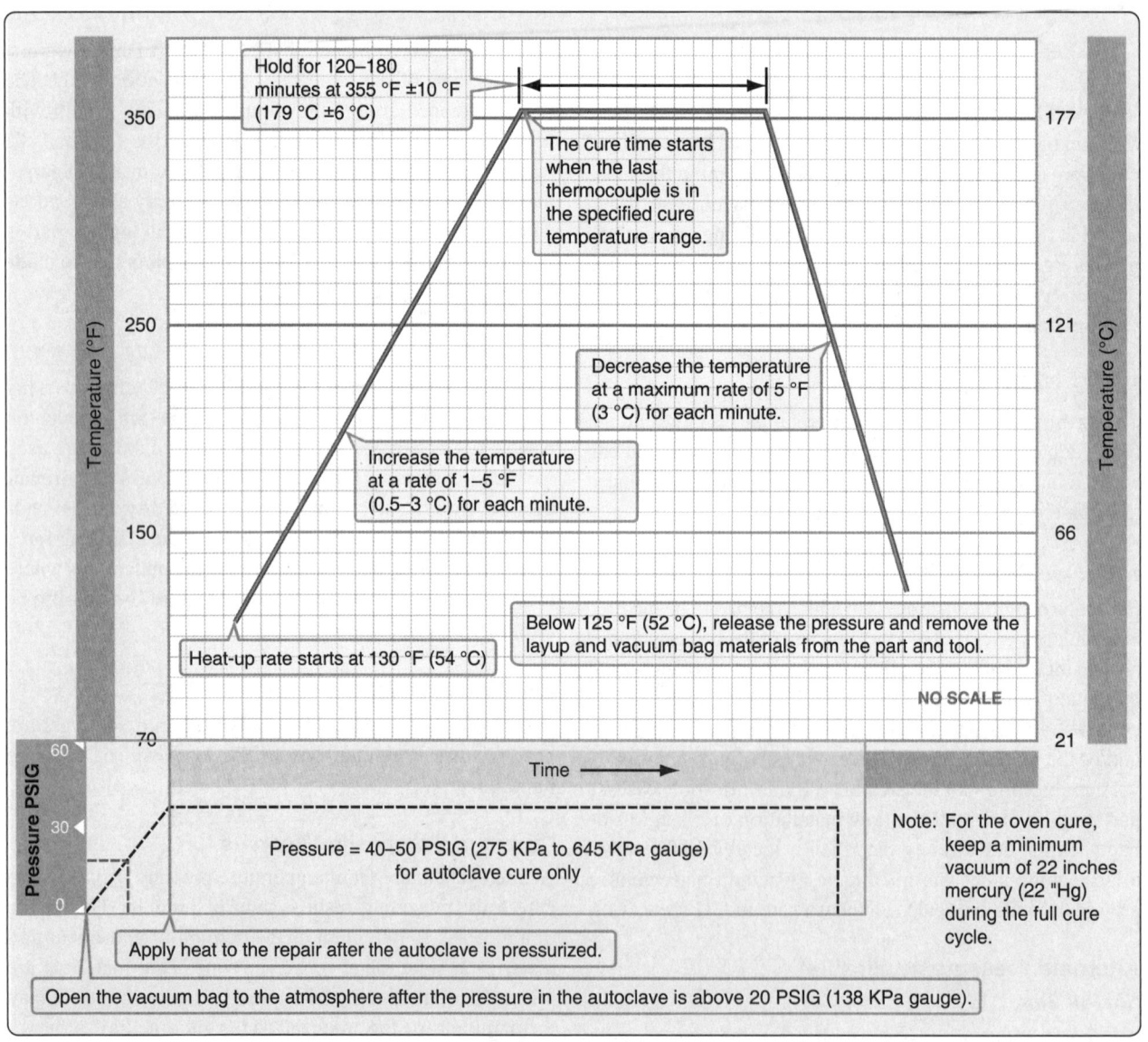

Figure 7-53. *Autoclave cure.*

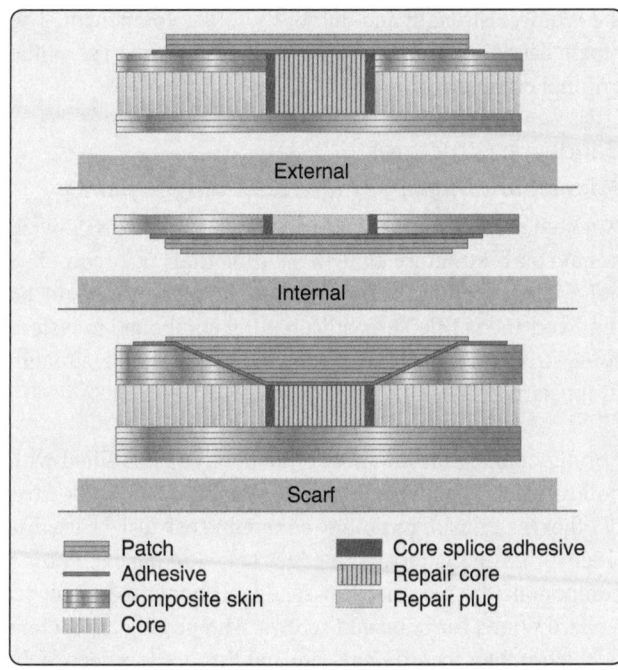

Figure 7-54. *Typical repairs for honeycomb sandwich structure.*

Patch	Core splice adhesive
Adhesive	Repair core
Composite skin	Repair plug
Core	

External

Internal

Scarf

• Hold or soak: The heating device maintains the temperature for a predetermined period.

• Cool down: The heating device cools down at a set temperature. Cool down temperatures are typically below 5 °F per minute. When the heating device is below 125 °F, the part can be removed. When an autoclave is used for curing parts, make sure that the pressure in the autoclave is relieved before the door is opened. *[Figure 7-53]*

The curing process is accomplished by the application of heat and pressure to the laminate. The resin begins to soften and flow as the temperature is increased. At lower temperatures, very little reaction occurs. Any volatile contaminants, such as air and/or water, are drawn out of the laminate with vacuum during this time. The laminate is compacted by applying pressure, usually vacuum (atmospheric pressure); autoclaves apply additional pressure, typically 50–100 psi. As the temperature approaches the final cure temperature, the rate of reaction greatly increases, and the resin begins to gel and harden. The hold at the final cure lets the resin finish curing and attain the desired structural properties.

Composite Honeycomb Sandwich Repairs

A large proportion of current aerospace composite components are light sandwich structures that are susceptible to damage and are easily damaged. Because sandwich structure is a bonded construction and the face sheets are thin, damage to sandwich structure is usually repaired by bonding. Repairs to sandwich honeycomb structure use similar techniques for the most common types of face sheet materials, such as fiberglass, carbon, and Kevlar®. Kevlar® is often repaired with fiberglass. *[Figure 7-54]*

Damage Classification

A temporary repair meets the strength requirements, but is limited by time or flight cycles. At the end of the repair's life, the repair must be removed and replaced. An interim repair

heating device. Typical curing temperature for composite materials is 250 °F or 350 °F. The temperature of large parts that are cured in an oven or autoclave might be different from that of an oven or autoclave during the cure cycle, because they act like a heat sink. The part temperature is most important for a correct cure, so thermocouples are placed on the part to monitor and control part temperature. The oven or autoclave air temperature probe that measures oven or autoclave temperature is not always a reliable device to determine part curing temperature. The oven temperature and the part temperature can be substantially different if the part or tool acts as a heat sink.

The elevated cure cycle consists of at least three segments:

• Ramp up: The heating device ramps up at a set temperature typically between 3 °F to 5 °F per minute.

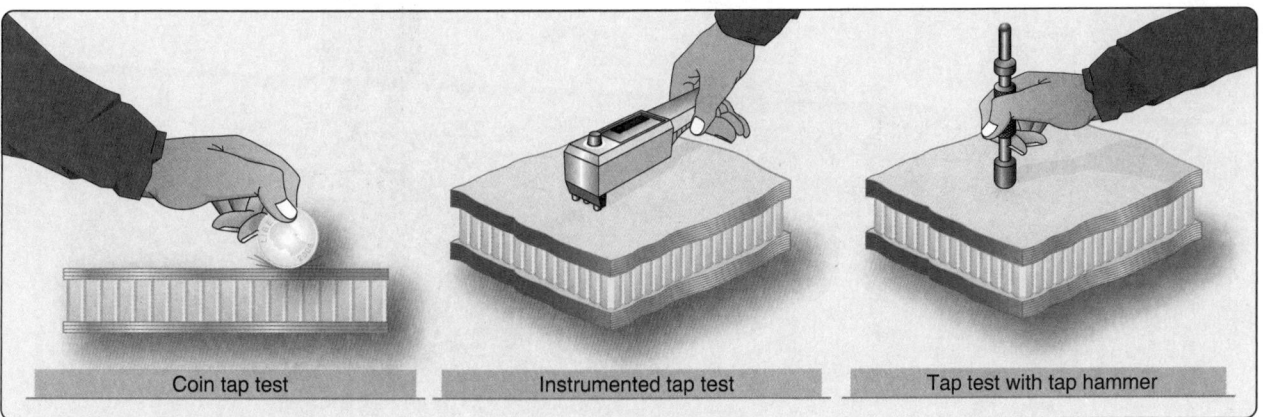

| Coin tap test | Instrumented tap test | Tap test with tap hammer |

Figure 7-55. *Tap testing techniques.*

Breather cloth
Heat blanket
Breather cloth
Thermocouple
Repair area

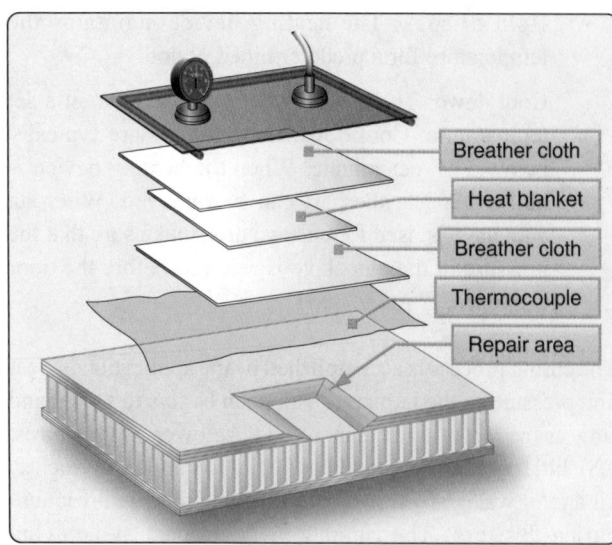

Figure 7-56. *Vacuum bag method for drying parts.*

restores the required strength to the component. However, this repair does not restore the required durability to the component. Therefore, it has a different inspection interval and/or method. A permanent repair is a repair that restores the required strength and durability to the component. The repair has the same inspection method and interval as the original component.

Sandwich Structures

Minor Core Damage (Filler & Potting Repairs)

A potted repair can be used to repair damage to a sandwich honeycomb structure that is smaller than 0.5 inch. The honeycomb material could be left in place or could be removed and is filled up with a potting compound to restore some strength. Potted repairs do not restore the full strength of the part.

Potting compounds are most often epoxy resins filled with hollow glass, phenolic or plastic microballoons, cotton, flox, or other materials. The potting compound can also be used as filler for cosmetic repairs to edges and skin panels. Potting compounds are also used in sandwich honeycomb panels as hard points for bolts and screws. The potting compound is heavier than the original core and this could affect flight control balance. The weight of the repair must be calculated and compared with flight control weight and balance limits set out in the SRM.

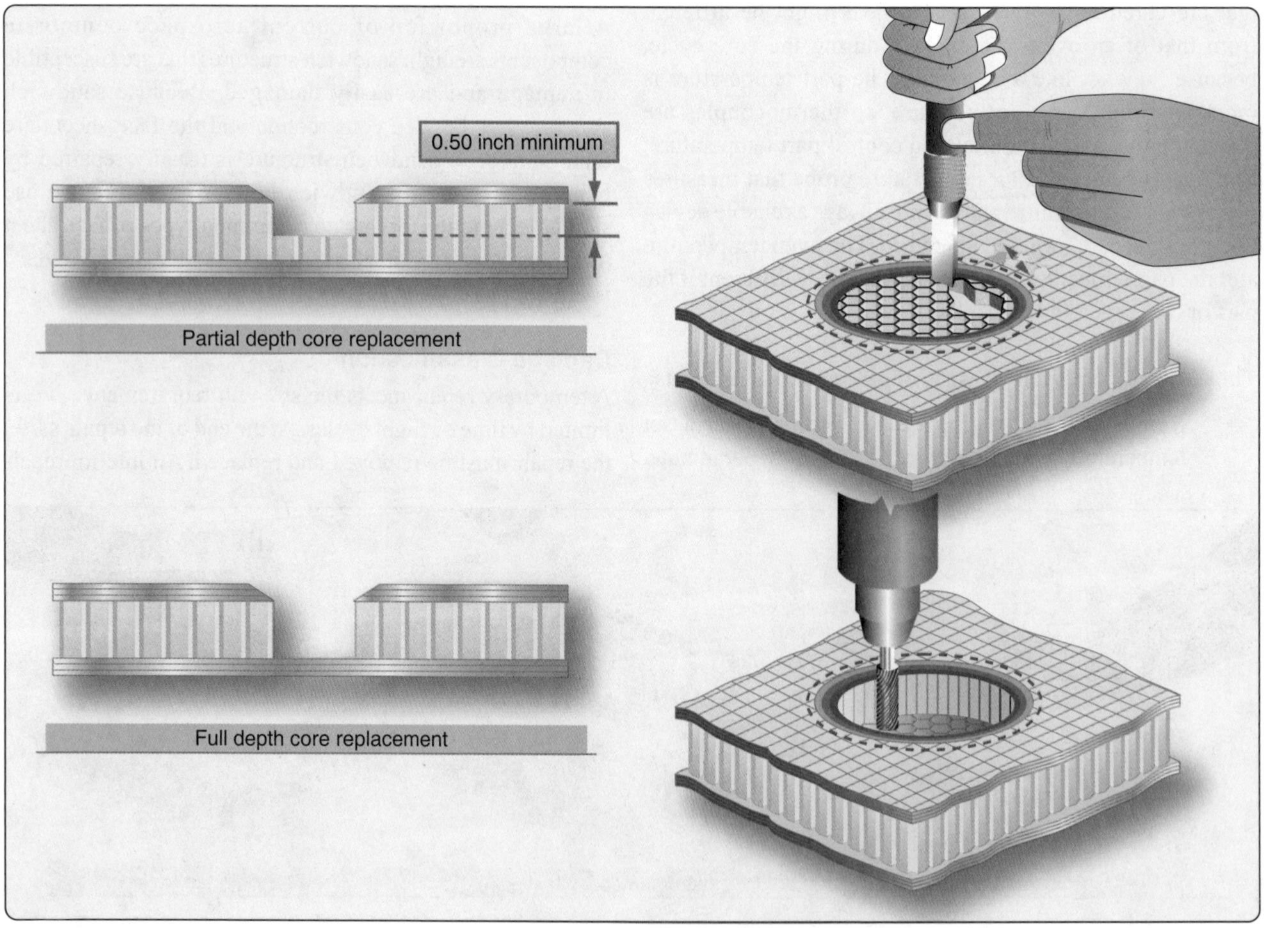

0.50 inch minimum
Partial depth core replacement
Full depth core replacement

Figure 7-57. *Core damage removal.*

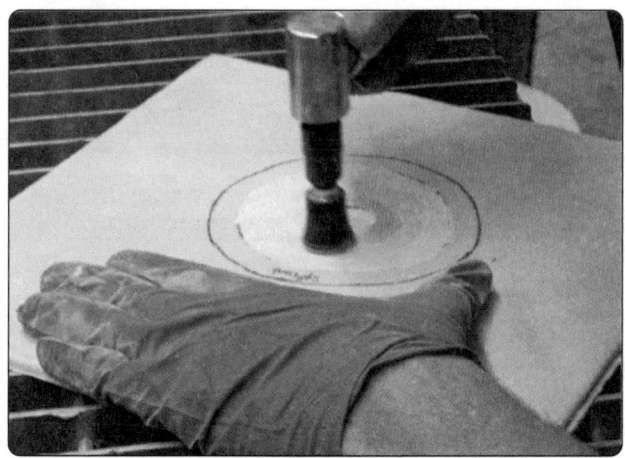

Figure 7-58. *Taper sanding of repair area.*

Damage Requiring Core Replacement & Repair to One or Both Faceplates

Note: the following steps are not a substitution for the aircraft specific Structural Repair Manual (SRM). Do not assume that the repair methods used by one manufacturer are applicable to another manufacturer.

Step 1: Inspect the Damage

Thin laminates can be visually inspected and tap tested to map out the damage. *[Figure 7-55]* Thicker laminates need more in-depth NDI methods, such as ultrasonic inspection. Check in the vicinity of the damage for entry of water, oil, fuel, dirt, or other foreign matter. Water can be detected with X-ray, back light, or a moisture detector.

Step 2: Remove Water From Damaged Area

Water needs to be removed from the core before the part is repaired. *[Figure 7-56]* If the water is not removed, it boils during the elevated temperature cure cycle and the face sheets blow off the core, resulting in more damage. Water in the honeycomb core could also freeze at the low temperatures that exist at high altitudes, which could result in disbonding of the face sheets.

Step 3: Remove the Damage

Trim out the damage to the face sheet to a smooth shape with rounded corners, or a circular or oval shape. Do not damage the undamaged plies, core, or surrounding material. If the core is damaged as well, remove the core by trimming to the same outline as the skin. *[Figure 7-57]*

Step 4: Prepare the Damaged Area

Use a flexible disc sander or a rotating pad sander to taper sand a uniform taper around the cleaned up damage. Some manufacturers give a taper ratio, such as 1:40, and others prescribe a taper distance like a 1-inch overlap for each existing ply of the face sheet. Remove the exterior finish, including conductive coating for an area that is at least 1 inch larger than the border of the taper. Remove all sanding dust with dry compressed air and a vacuum cleaner. Use a clean cloth moistened with approved solvent to clean the damaged area. *[Figure 7-58]*

Step 5: Installation of Honeycomb Core (Wet Layup)

Use a knife to cut the replacement core. The core plug must be of the same type, class, and grade of the original core. The direction of the core cells should line up with the honey comb of the surrounding material. The plug must be trimmed to the right length and be solvent washed with an approved cleaner.

For a wet layup repair, cut two plies of woven fabric that fit on the inside surface of the undamaged skin. Impregnate the fabric plies with a resin and place in the hole. Use potting compound around the core and place it in the hole. For a prepreg repair, cut a piece of film adhesive that fits the hole and use a foaming adhesive around the plug. The plug should touch the sides of the hole. Line up the cells of the plug with

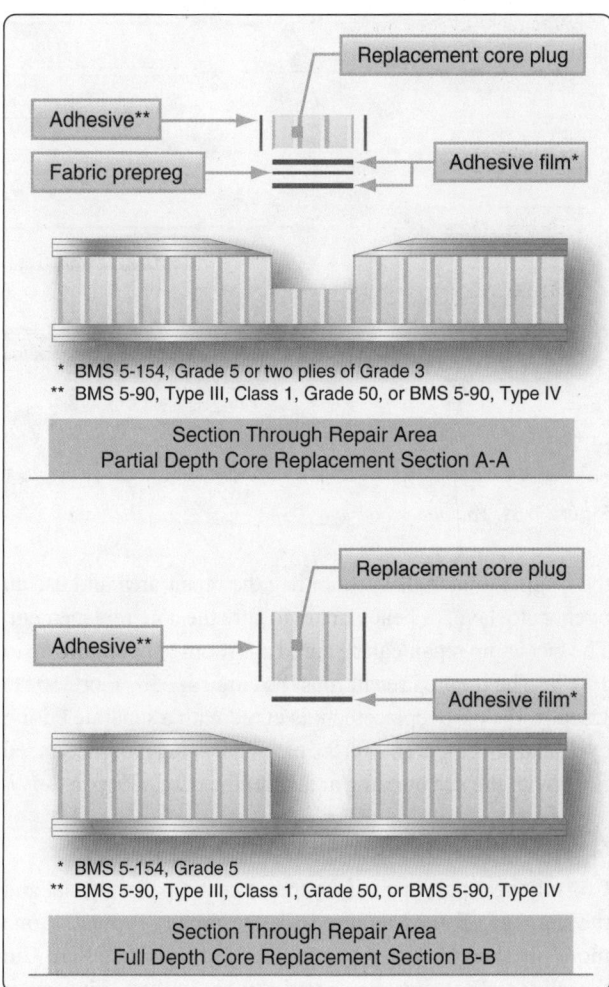

Figure 7-59. *Core replacement.*

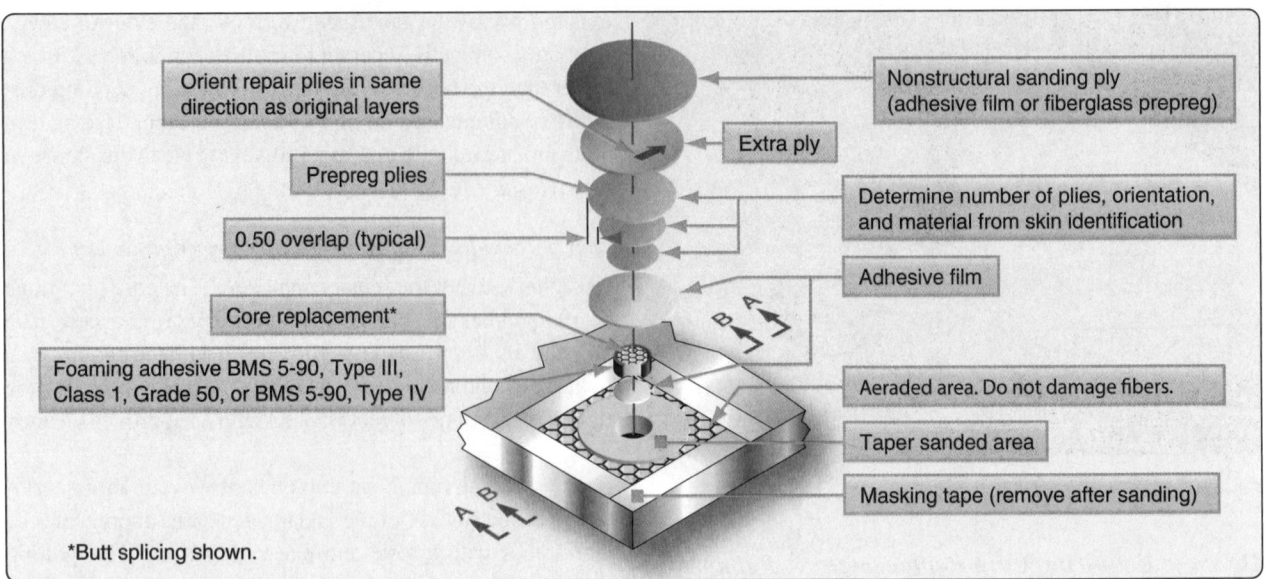

Figure 7-60. *Repair ply installation.*

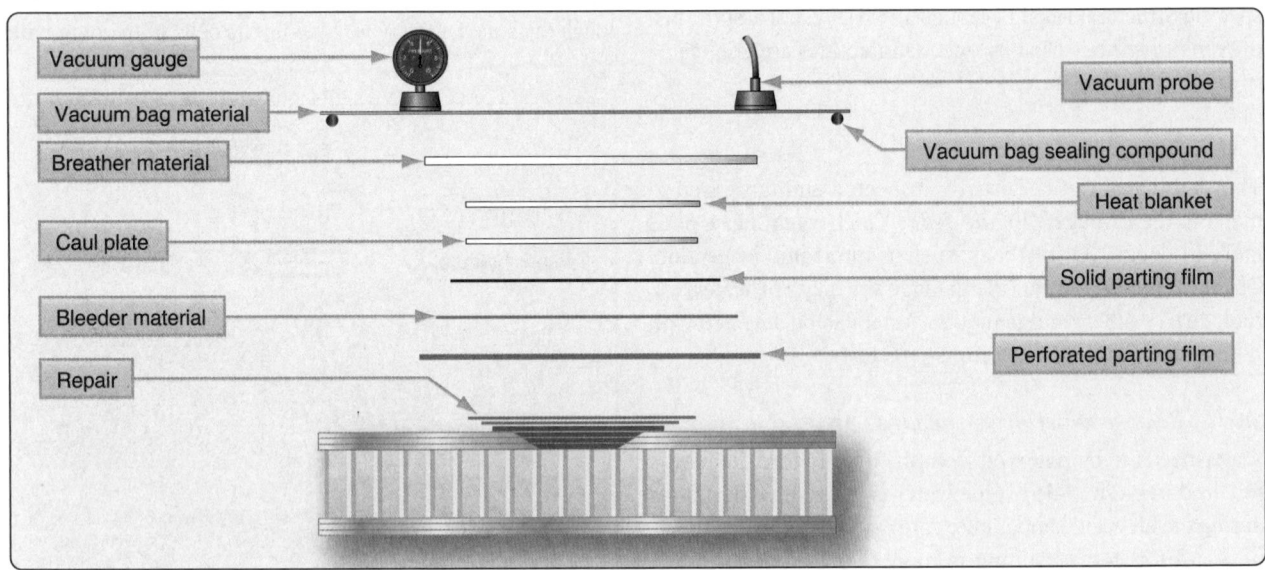

Figure 7-61. *Vacuum processing.*

the original material. Vacuum bag the repair area and use an oven, autoclave, or heat blanket to cure the core replacement. The wet layup repair can be cured at a room temperature up to 150 °F. The prepreg repair must be cured at 250 °F or 350 °F. Usually, the core replacement is cured with a separate curing cycle and not co-cured with the patch. The plug must be sanded flush with the surrounding area after the cure. *[Figure 7-59]*

Step 6: Prepare & Install the Repair Plies

Consult the repair manual for the correct repair material and the number of plies required for the repair. Typically, one more ply than the original number of plies is installed. Cut the plies to the correct size and ply orientation. The repair plies must be installed with the same orientation as that of the original plies being repaired. Impregnate the plies with resin for the wet layup repair, or remove the backing material from the prepreg material. The plies are usually placed using the smallest ply first taper layup sequence. *[Figure 7-60]*

Step 7: Vacuum Bag the Repair

Once the ply materials are in place, vacuum bagging is used to remove air and to pressurize the repair for curing. Refer to *Figure 7-61* for bagging instructions.

Step 8: Curing the Repair

The repair is cured at the required cure cycle. Wet layup repairs can be cured at room temperature. An elevated temperature up to 150 °F can be used to speed up the cure.

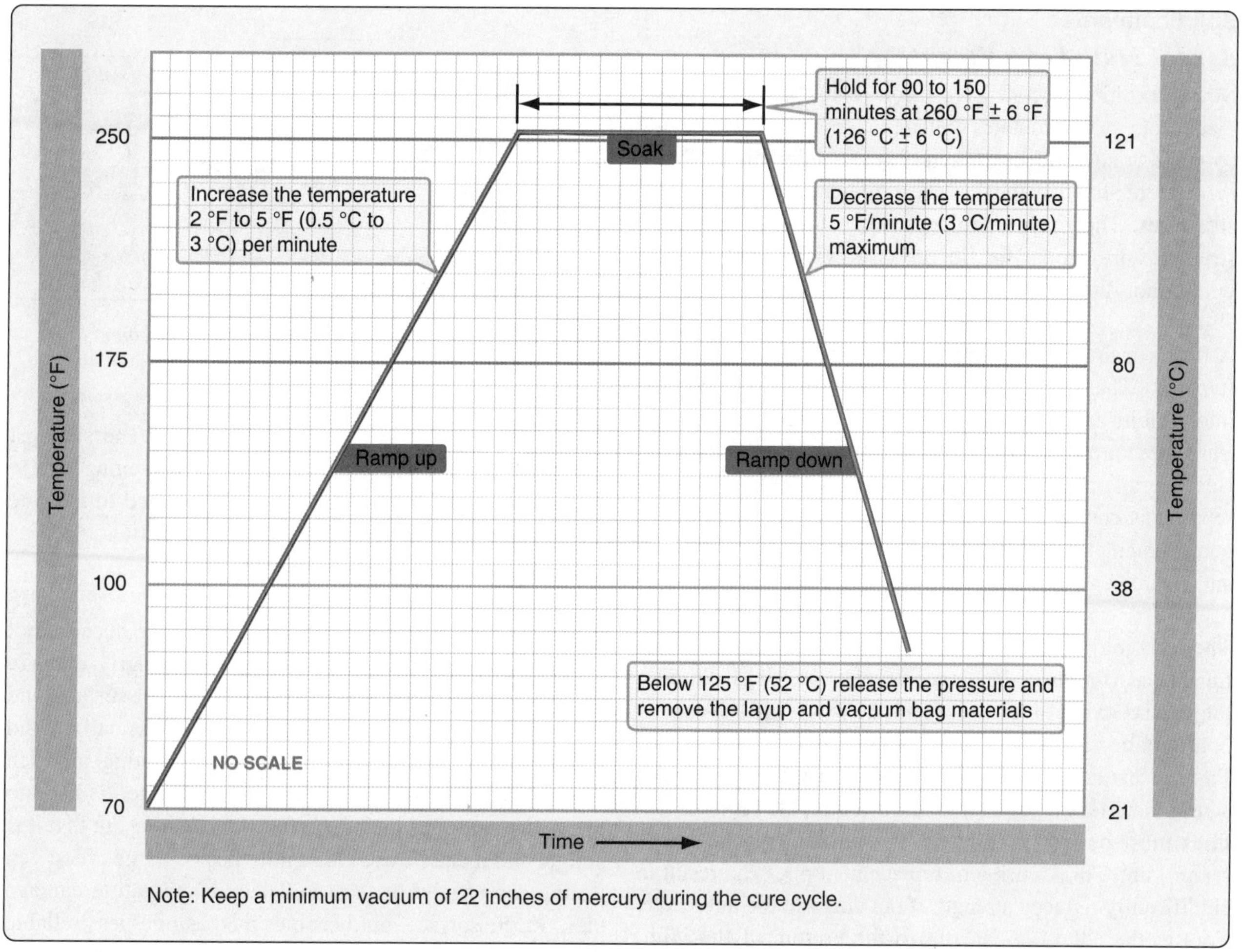

Figure 7-62. *Curing the repair.*

The graph (Figure 7-62) shows temperature in °F (left axis: 70, 100, 175, 250) and °C (right axis: 21, 38, 80, 121) versus Time.

Hold for 90 to 150 minutes at 260 °F ± 6 °F (126 °C ± 6 °C)

Increase the temperature 2 °F to 5 °F (0.5 °C to 3 °C) per minute

Decrease the temperature 5 °F/minute (3 °C/minute) maximum

Soak

Ramp up

Ramp down

Below 125 °F (52 °C) release the pressure and remove the layup and vacuum bag materials

NO SCALE

Time

Note: Keep a minimum vacuum of 22 inches of mercury during the cure cycle.

The prepreg repair needs to be cured at an elevated cure cycle. *[Figure 7-62]* Parts that can be removed from the aircraft could be cured in a hot room, oven, or autoclave. A heating blanket is used for on-aircraft repairs.

Remove the bagging materials after curing and inspect the repair. The repair should be free from pits, blisters, resin-rich and resin-starved areas. Lightly sand the repair patch to produce a smooth finish without damaging the fibers. Apply top finish and conductive coating (lightning protection).

Step 9: Post Repair Inspection
Use visual, tap, and/or ultrasonic inspection to inspect the repair. Remove the repair patch if defects are found. *[Figure 7-63]*

Perform a balance check if a repair to a flight control surface was made, and ensure that the repaired flight control is within limits of the SRM. Failure to do so could result in flight control flutter, and safety of flight could be affected.

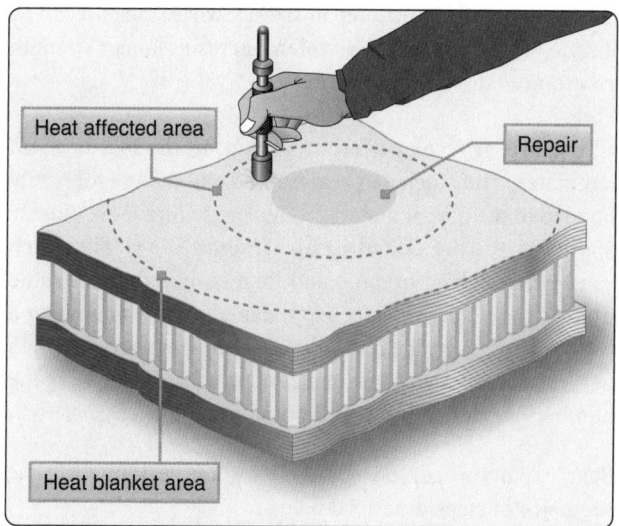

Heat affected area

Repair

Heat blanket area

Figure 7-63. *Post-repair inspection.*

Solid Laminates
Bonded Flush Patch Repairs

New generation aircraft have fuselage and wing structures made from solid laminates that are externally stiffened with co-cured or co-bonded stringers. These solid laminates have many more plies than the face sheets of honeycomb sandwich structures. The flush repair techniques for solid laminate structures are similar for fiberglass, Kevlar®, and graphite with minor differences.

A flush repair can be stepped or, more commonly, scarved (tapered). The scarf angles are usually small to ease the load into the joint and to prevent the adhesive from escaping. This translates into thickness-to-length ratios of 1:10 to 1:70. Because inspection of bonded repairs is difficult, bonded repairs, as contrasted with bolted repairs, require a higher commitment to quality control, better trained personnel, and cleanliness.

The scarf joint is more efficient from the viewpoint of load transfer as it reduces load eccentricity by closely aligning the neutral axis of the parent and the patch. However, this configuration has many drawbacks in making the repair. First, to maintain a small taper angle, a large quantity of sound material must be removed. Second, the replacement plies must be very accurately laid up and placed in the repair joint. Third, curing of replacement plies can result in significantly reduced strength if not cured in the autoclave. Fourth, the adhesive can run to the bottom of the joint, creating a nonuniform bond line. This can be alleviated by approximating the scarf with a series of small steps. For these reasons, unless the part is lightly loaded, this type of repair is usually performed at a repair facility where the part can be inserted into the autoclave, which can result in part strength as strong as the original part.

There are several different repair methods for solid laminates. The patch can be precured and then secondarily bonded to the parent material. This procedure most closely approximates the bolted repair. *[Figure 7-64]* The patch can be made from prepreg and then co-cured at the same time as the adhesive. The patch can also be made using a wet layup repair. The curing cycle can also vary in length of time, cure temperature, and cure pressure, increasing the number of possible repair combinations.

Scarf repairs of composite laminates are performed in the sequence of steps described below.

Step 1: Inspection & Mapping of Damage

The size and depth of damage to be repaired must be accurately surveyed using appropriate nondestructive evaluation (NDE) techniques. A variety of NDE techniques can be used to

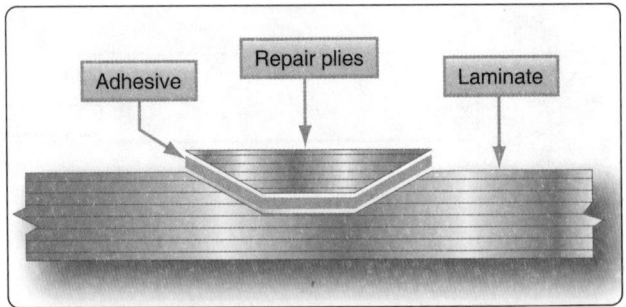

Figure 7-64. *A precured patch can be secondarily bound to the parent material.*

inspect for damage in composite structures. The simplest technique is visual inspection, in which whitening due to delamination and/or resin cracking can be used to indicate the damage area in semitransparent composites, such as glass-polyester and glass-vinyl ester laminates.

Visual inspection is not an accurate technique because not all damage is detectable to the eye, particularly damage hidden by paint, damage located deep below the surface, and damage in nontransparent composites, such as carbon and aramid laminates. A popular technique is tap testing, in which a lightweight object, such as a coin or hammer, is used to locate damage. The main benefits of tap testing are that it is simple and it can be used to rapidly inspect large areas. Tap testing can usually be used to detect delamination damage close to the surface, but becomes increasingly less reliable the deeper the delamination is located below the surface. Tap testing is not useful for detecting other types of damage, such as resin cracks and broken fibers.

More advanced NDE techniques for inspecting composites are impedance testing, x-ray radiography, thermography, and ultrasonics. Of these techniques, ultrasonics is arguably the most accurate and practical and is often used for surveying damage. Ultrasonics can be used to detect small delaminations located deep below the surface, unlike visual inspection and tap testing.

Step 2: Removal of Damaged Material

Once the scope of the damaged area to be repaired has been determined, the damaged laminate must be removed. The edges of the sound laminate are then tapered back to a shallow angle. The taper slope ratio, also known as the scarf angle, should be less than 12 to 1 (< 5°) to minimize the shear strains along the bond line after the repair patch is applied. The shallow angle also compensates for some errors in workmanship and other shop variables that might diminish patch adhesion. *[Figure 7-65]*

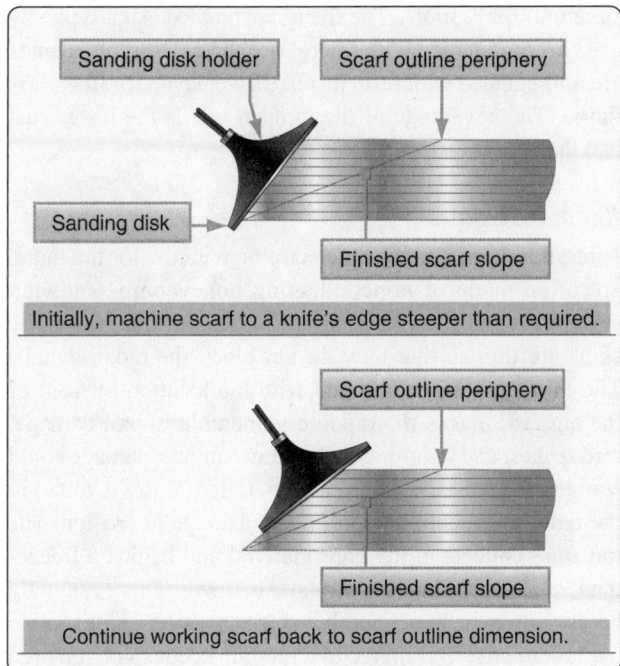

Figure 7-65. *Scarf patch of solid laminate.*

In figure 7-65:
- Sanding disk holder
- Scarf outline periphery
- Sanding disk
- Finished scarf slope
- Initially, machine scarf to a knife's edge steeper than required.
- Scarf outline periphery
- Finished scarf slope
- Continue working scarf back to scarf outline dimension.

Step 3: Surface Preparation

The laminate close to the scarf zone should be lightly abraded with sandpaper, followed by the removal of dust and contaminates. It is recommended that, if the scarf zone has been exposed to the environment for any considerable period of time, it should be cleaned with a solvent to remove contamination.

Step 4: Molding

A rigid backing plate having the original profile of the composite structure is needed to ensure the repair has the same geometry as the surrounding structure.

Step 5: Laminating

Laminated repairs are usually done using the smallest ply-first taper sequence. While this repair is acceptable, it produces relatively weak, resin-rich areas at each ply edge at the repair interface. The largest ply first laminate sequence, where the first layer of reinforcing fabric completely covers the work area, produces a stronger interface joint. Follow the manufacturer's SRM instructions.

Selection of the reinforcing material is critical to ensuring the repair has acceptable mechanical performance. The reinforcing fabric or tape should be identical to the reinforcement material used in the original composite. Also, the fiber orientation of the reinforcing layers within the repair laminate should match those of the original part laminate, so that the mechanical properties of the repair are as close to original as possible.

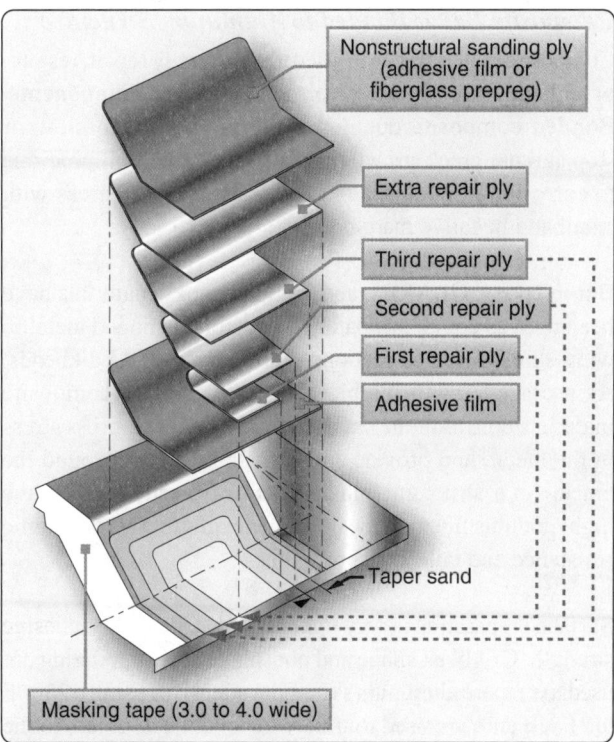

Figure 7-66. *Trailing edge repair.*

In figure 7-66:
- Nonstructural sanding ply (adhesive film or fiberglass prepreg)
- Extra repair ply
- Third repair ply
- Second repair ply
- First repair ply
- Adhesive film
- Taper sand
- Masking tape (3.0 to 4.0 wide)

Step 6: Finishing

After the patch has cured, a conducting mesh and finish coat should be applied if needed.

Trailing Edge & Transition Area Patch Repairs

Trailing edges of control panels are highly vulnerable to damage. The aft 4 inches are especially subject to ground collision and handling, as well as to lightning strike. Repairs in this region can be difficult because both the skins and the trailing edge reinforcement may be involved. The repairs to a honeycomb core on a damaged edge or panel are similar to the repair of a sandwich honeycomb structure discussed in the Damage Requiring Core Replacement and Repair to One or Both Faceplate Repair sections. Investigate the damage, remove damaged plies and core, dry the part, install new core, layup the repair plies, curing and post inspection. A typical trail edge repair is shown in *Figure 7-66*.

Resin Injection Repairs

Resin injection repairs are used on lightly loaded structures for small damages to a solid laminate due to delamination. Two holes are drilled on the outside of the delamination area and a low-viscosity resin is injected in one hole until it flows out the other hole. Resin injection repairs are sometimes used on sandwich honeycomb structure to repair a facesheet disbond. Disadvantages of the resin injection method are that the fibers are cut as a result of drilling holes, it is difficult to remove moisture from the damaged area, and it is difficult to achieve complete infusion of resin. *[Figure 7-67]*

Composite Patch Bonded to Aluminum Structure

Composite materials can be used to structurally repair, restore, or enhance aluminum, steel, and titanium components. Bonded composite doublers have the ability to slow or stop fatigue crack growth, replace lost structural area due to corrosion grindouts, and structurally enhance areas with small and negative margins.

Boron epoxy, GLARE®, and graphite epoxy materials have been used as composite patches to restore damaged metallic wing skins, fuselage sections, floor beams, and bulkheads. As a crack growth inhibitor, the stiff bonded composite materials constrain the cracked area, reduce the gross stress in the metal, and provide an alternate load path around the crack. As a structural enhancement or blendout filler, the high modulus fiber composites offer negligible aerodynamic resistance and tailorable properties.

Surface preparation is very important to achieve the adhesive strength. Grit blast silane and phosphoric acid anodizing are used to prepare aluminum skin. Film adhesives using a 250 °F (121 °C) cure are used routinely to bond the doublers to the metallic structure. Critical areas of the installation process include a good thermal cure control, having and maintaining water-free bond surfaces, and chemically and physically prepared bond surfaces.

Secondarily bonded precured doublers and in-situ cured doublers have been used on a variety of structural geometries ranging from fuselage frames to door cutouts to blade stiffeners. Vacuum bags are used to apply the bonding and curing pressure between the doubler and metallic surface.

Fiberglass Molded Mat Repairs

Fiberglass molded mats consists of short fibers, and the strength is much less than other composite products that use continuous fibers. Fiberglass molded mats are not used for structural repair applications, but could be used for non-structural applications. The fiberglass molded mat is typically used in combination with fiberglass fabric. The molded mats are impregnated with resin just like a wet layup for fiberglass fabric. The advantage of the molded mat is the lower cost and the ease of use.

Radome Repairs

Aircraft radomes, being an electronic window for the radar, are often made of nonconducting honeycomb sandwich structure with only three or four plies of fiberglass. The skins are thin so that they do not block the radar signals. The thin structure, combined with the location in front of the aircraft, makes the radome vulnerable to hail damage, bird strikes, and lightning strikes. Low-impact damage could lead to disbonds and delamination. Often, water is found in the radome structure due to impact damage or erosion. The moisture collects in the core material and begins a freeze-thaw cycle each time the airplane is flown. This eventually breaks down the honeycomb material causing a soft spot on the radome itself. Damage to a radome needs to be repaired quickly to avoid further damage and radar signal obstructions. Trapped water or moisture can produce a shadow on the radar image and severely degrade the performance of the radar. To detect water ingression in radomes, the available NDE techniques include x-ray radiography, infrared thermography, and a radome moisture meter that measures the RF power loss caused by the presence of water. The repairs to radomes are similar to repairs to other honeycomb structures, but the technician needs to realize that repairs could affect the radar performance. A special tool is necessary to repair severely damaged radomes. *[Figure 7-68]*

Transmissivity testing after radome repair ensures that the radar signal is transmitted properly through the radome. Radomes have lightning protections strips bonded to the outside of the radome to dissipate the energy of a lightning

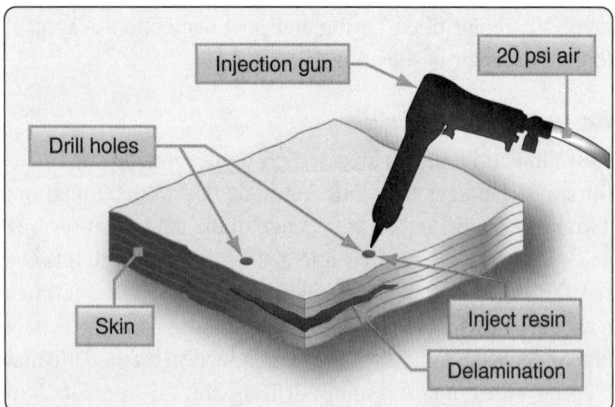

Figure 7-67. *Resin injection repair.*

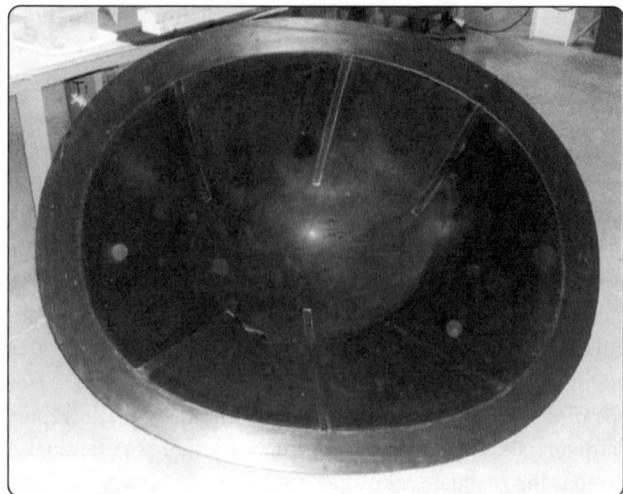

Figure 7-68. *Radome repair tool.*

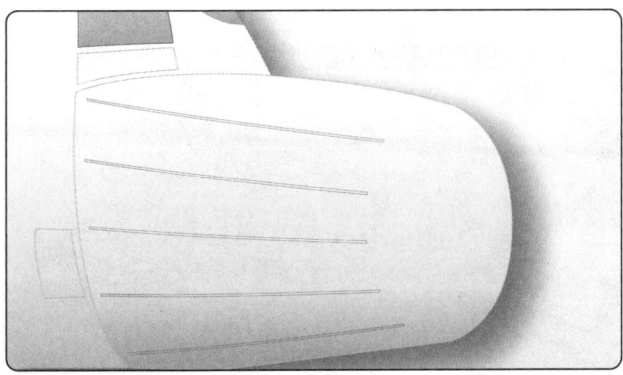

Figure 7-69. *Lightning protection strips on a radome.*

strike. It is important that these lightning protection strips are in good condition to avoid damage to the radome structure. Typical failures of lightning protection strips that are found during inspection are high resistance caused by shorts in the strips or attaching hardware and disbonding of the strips from the radome surface. *[Figures 7-69]*

External Bonded Patch Repairs

Repairs to damaged composite structures can be made with an external patch. The external patch repair could be made with prepreg, a wet layup, or a precured patch. External patches are usually stepped to reduce the stress concentration at the edge of the patch. The disadvantages of the external patch are the eccentricity of the loading that causes peel stresses and the protrusion of the patch in the air stream. The advantage of the external patch is that it is easier to accomplish than a flush scarf-type repair.

External Bonded Repair With Prepreg Plies

The repair methods for carbon, fiberglass, and Kevlar® are similar. Fiberglass is sometimes used to repair Kevlar® material. The main steps in repairing damage with an external patch are investigating and mapping the damage, removal of the damage, layup of the repair plies, vacuum bagging, curing, and finish coating.

Step 1: Investigating & Mapping the Damage
Use the tap test or ultrasonic test to map out the damage.

Step 2: Damage Removal
Trim out the damage to a smooth round or oval shape. Use scotch or sand paper to rough up the parent surface at least 1 inch larger than the patch size. Clean the surface with an approved solvent and cheese cloth.

Step 3: Layup of the Repair Plies
Use the SRM to determine the number, size, and orientation of the repair plies. The repair ply material and orientation must be the same as the orientation of the parent structure.

The repair can be stepped to reduce peel stresses at the edges.

Step 4: Vacuum Bagging
A film adhesive is placed over the damaged area and the repair layup is placed on top of the repair. The vacuum bagging materials are placed on top of the repair (see Prepreg Layup and Controlled Bleed Out) and a vacuum is applied.

Step 5: Curing the Repair
The prepreg patch can be cured with a heater blanket that is placed inside the vacuum bag, oven, or autoclave when the part can be removed from the aircraft. Most prepregs and film adhesives cure at either 250 °F or 350 °F. Consult the SRM for the correct cure cycle.

Step 6: Applying Top Coat
Remove the vacuum bag from the repair after the cure and inspect the repair, remove the patch if the repair is not satisfactory. Lightly sand the repair and apply a protective topcoating.

External Repair Using Wet Layup & Double Vacuum Debulk Method (DVD)

Generally, the properties of a wet layup repair are not as good as a repair with prepreg material; but by using a DVD method, the properties of the wet layup process can be improved. The DVD process is a technique to remove entrapped air that causes porosity in wet layup laminates. The DVD process is often used to make patches for solid laminate structures for complex contoured surfaces. The wet layup patch is prepared in a DVD tool and then secondary bonded to the aircraft structure. *[Figure 7-70]* The laminating process is similar to a standard wet layup process. The difference is how the patch is cured.

Double Vacuum Debulk Principle

The double vacuum bag process is used to fabricate wet layup or prepreg repair laminates. Place the impregnated fabric within the debulking assembly, shown in *Figure 7-70*. To begin the debulking process, evacuate the air within the inner flexible vacuum bag. Then, seal the rigid outer box onto the inner vacuum bag, and evacuate the volume of air between the rigid outer box and inner vacuum bag. Since the outer box is rigid, the second evacuation prevents atmospheric pressure from pressing down on the inner vacuum bag over the patch. This subsequently prevents air bubbles from being pinched off within the laminate and facilitates air removal by the inner vacuum. Next, heat the laminate to a predetermined debulking temperature in order to reduce the resin viscosity and further improve the removal of air and volatiles from the laminate. Apply the heat through a heat blanket that is controlled with thermocouples placed directly on the heat

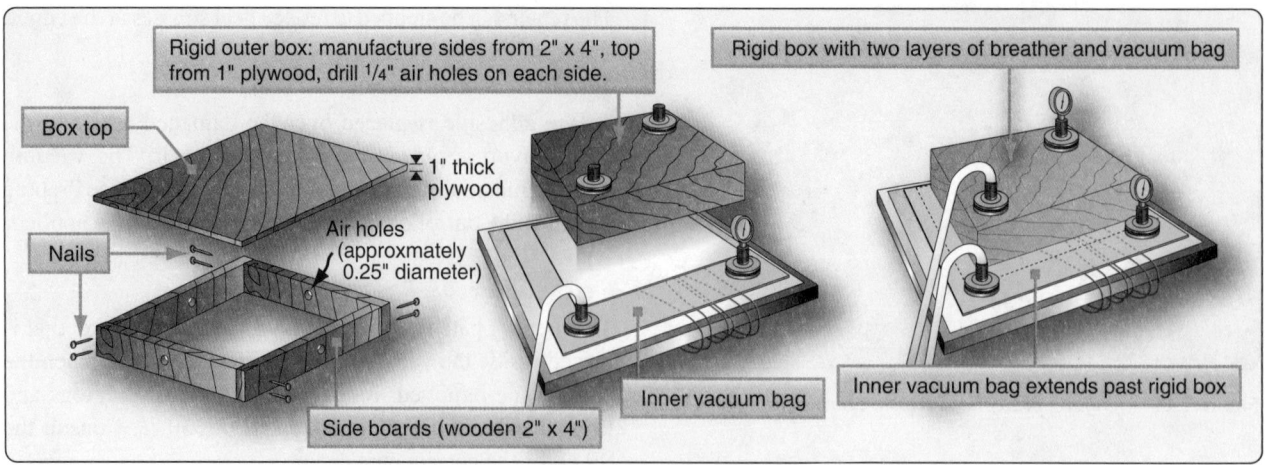

Figure 7-70. *DVD tool made from wood two by fours and plywood.*

blanket. Once the debulking cycle is complete, compact the laminate to consolidate the plies by venting the vacuum source attached to the outer rigid box, allowing atmospheric pressure to reenter the box and provide positive pressure against the inner vacuum bag. Upon completion of the compaction cycle, remove the laminate from the assembly and prepare for cure.

DVD tools can be purchased commercially but can also be fabricated locally from wood two-by-fours and sheets of plywood, as illustrated in *Figure 7-70.*

Patch Installation on the Aircraft

After the patch comes out of the DVD tool, it is still possible to form it to the contour of the aircraft, but the time is typically limited to 10 minutes. Place a film adhesive, or paste adhesive, on the aircraft skin and place the patch on the aircraft. Use a vacuum bag and heater blanket to cure the adhesive. *[Figures 7-71 and 7-72]*

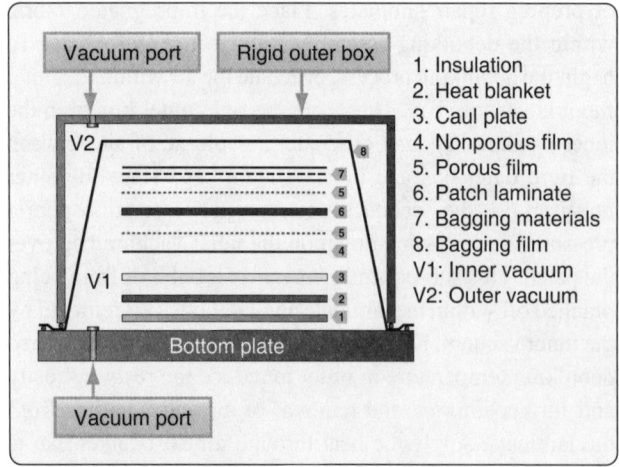

Figure 7-71. *Double vacuum debulk schematic.*

External Repair Using Precured Laminate Patches

Precured patches are not very flexible and cannot be used on highly curved or compound curved surfaces. The repair steps are similar as in External Bonded Repair With Prepreg Plies, except step 3 and 4 that follow.

Step 3: A Precured Patch

Consult the SRM for correct size, ply thickness, and orientation. You can laminate and cure the precured patch in the repair shop and secondary bond to the parent structure, or obtain standard precured patches. *[Figure 7-73]*

Step 4: For a Precured Patch

Apply film adhesive or paste adhesive to the damaged area and place the precured patch on top. Vacuum bag the repair and cure at the correct temperature for the film adhesive or paste adhesive. Most film adhesives cure at either 250 °F or 350 °F. Some paste adhesives cure at room temperature although an elevated temperature could be used to speed the curing process.

Bonded versus Bolted Repairs

Bonded repair concepts have found applicability in both types of manufacturing assembly methods. They have the advantage of not introducing stress concentrations by drilling fastener holes for patch installation and can be stronger than original part material. The disadvantage of bonded repairs is that most repair materials require special storage, handling, and curing procedures.

Bolted repairs are quicker and easier to fabricate than bonded repairs. They are normally used on composite skins thicker than 0.125-inch to ensure sufficient fastener bearing area is available for load transfer. They are prohibited in honeycomb sandwich assemblies due to the potential for moisture intrusion from the fastener holes and the resulting

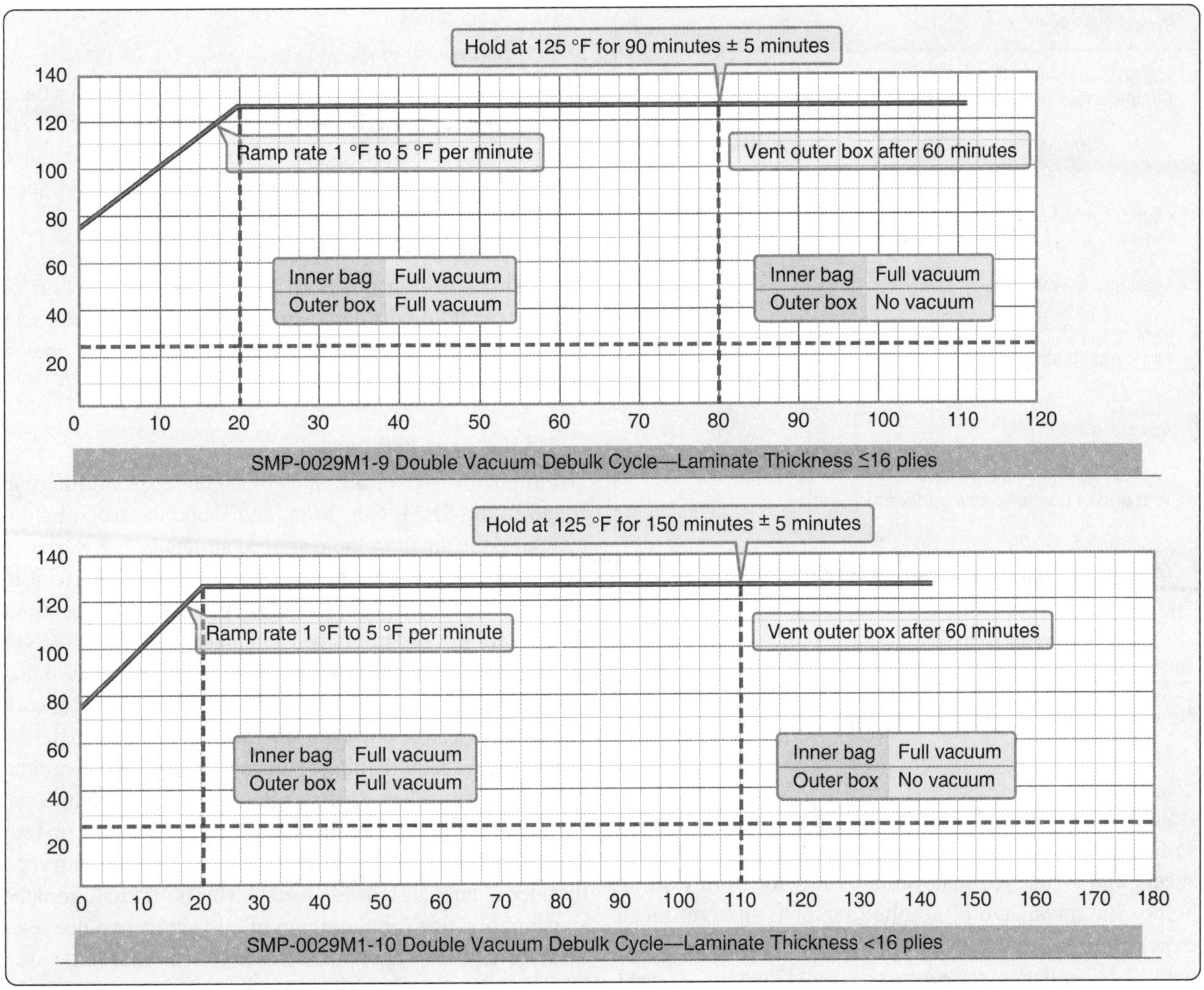
Hold at 125 °F for 90 minutes ± 5 minutes
Ramp rate 1 °F to 5 °F per minute
Vent outer box after 60 minutes
Inner bag Full vacuum
Outer box Full vacuum
Inner bag Full vacuum
Outer box No vacuum
140
120
100
80
60
40
20
0
10
20
30
40
50
60
70
80
90
100
110
120
SMP-0029M1-9 Double Vacuum Debulk Cycle—Laminate Thickness ≤16 plies
Hold at 125 °F for 150 minutes ± 5 minutes
Ramp rate 1 °F to 5 °F per minute
Vent outer box after 60 minutes
Inner bag Full vacuum
Outer box Full vacuum
Inner bag Full vacuum
Outer box No vacuum
140
120
100
80
60
40
20
0
10
20
30
40
50
60
70
80
90
100
110
120
130
140
150
160
170
180
SMP-0029M1-10 Double Vacuum Debulk Cycle—Laminate Thickness <16 plies

Figure 7-72. *DVD cure cycle.*

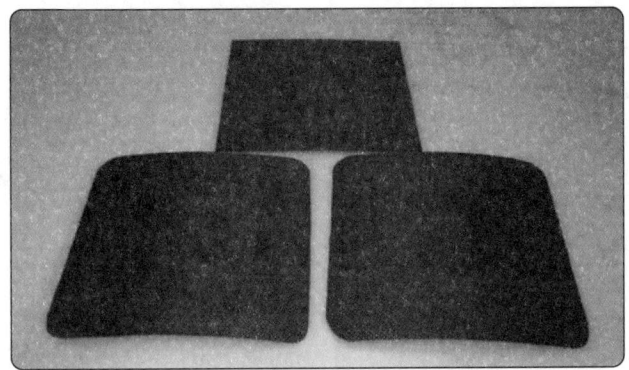

Figure 7-73. *Precured patches.*

core degradation. Bolted repairs are heavier than comparable bonded repairs, limiting their use on weight-sensitive flight control surfaces.

Honeycomb sandwich parts often have thin face sheets and are most effectively repaired by using a bonded scarf type repair. A bonded external step patch can be used as an alternative. Bolted repairs are not effective for thin laminates because of the low bearing stress of the composite laminate. Thicker solid laminates used on larger aircraft can be up to an inch thick in highly loaded areas and these types of laminates cannot be effectively repaired using a bonded scarf type repair. *[Figure 7-74]*

Bolted Repairs

Aircraft designed in the 1970s used composite sandwich honeycomb structure for lightly loaded secondary structure, but new large aircraft use thick solid laminates for primary structure instead of sandwich honeycomb. These thick solid laminate structures are quite different from the traditional sandwich honeycomb structures used for flight controls, landing gear doors, flaps, and spoilers of today's aircraft. They present a challenge to repair and are difficult to repair with a bonded repair method. Bolted repair methods have been developed to repair thicker solid laminates.

Bonded versus bolted repair	Bolted	Bonded
Lightly loaded structures – laminate thickness less than 0.1"		X
Highly loaded structures – laminate thickness between 0.125" – 0.5"	X	X
Highly loaded structures – laminate thickness larger than 0.5"	X	
High peeling stresses	X	
Honeycomb structure		X
Dry surfaces	X	X
Wet and/or contaminated surfaces	X	
Disassembly required	X	
Restore unnotched strength		X

Figure 7-74. *Bolted versus bonded repair.*

Bolted repairs are not desirable for honeycomb sandwich structure due to the limited bearing strength of the thin face sheets and weakened honeycomb structure from drilling holes. The advantage of a bolted repair is that you need to select only patch material and fasteners, and the repair method is similar to a sheet metal repair. There is no need for curing the repair and storing the prepreg repair material and film adhesives in a freezer. Patches may be made from aluminum, titanium, steel, or precured composite material. Composite patches are often made from carbon fiber with an epoxy resin or fiberglass with an epoxy resin.

You can repair a carbon fiber structure with an aluminum patch, but you must place a layer of fiberglass cloth between the carbon part and the aluminum patch to prevent galvanic corrosion. Titanium and precured composite patches are preferred for repair of highly loaded components. Precured carbon/epoxy patches have the same strength and stiffness as the parent material as they are usually cured similarly.

Titanium or stainless steel fasteners are used for bolted repairs of a carbon fiber structure. Aluminum fasteners corrode if used with carbon fiber. Rivets cannot be used because the installation of rivets using a rivet gun introduce damage to the hole and surrounding structure and rivets expand during installation, which is undesirable for composite structures because it could cause delimination of the composite material.

Repair Procedures

Step 1: Inspection of the Damage

The tap test is not effective to detect delamination in thick laminates unless the damage is close to the surface. An ultrasonic inspection is necessary to determine the damage area. Consult the SRM to find an applicable NDI procedure.

Step 2: Removal of the Damage

The damaged area needs to be trimmed to a round or rectangular hole with large smooth radii to prevent stress concentrations. Remove the damage with a sander, router, or similar tool.

Step 3: Patch Preparation

Determine the size of the patch based on repair information found in the SRM. Cut, form, and shape the patch before attaching the patch to the damaged structure. It is easier to make the patch a little bigger than calculated and trim to size after drilling all fastener holes. In some cases, the repair patches are stocked preshaped and predrilled. If cutting is to be performed, standard shop procedures should be used that are suitable for the patch material. Titanium is hard to work and requires a large powerful slip roller to curve the material. Metal patches require filing to prevent crack initiation around the cut edges. When drilling pilot holes in the composite, the holes for repair fasteners must be a minimum of four diameters from existing fasteners and have a minimum edge distance of three fastener diameters. This is different from the standard practice for aluminum of allowing a two diameter distance. Specific pilot hole sizes and drill types to be used should follow specific SRM instructions. *[Figure 7-75]*

Step 4: Hole Pattern Lay Out

To locate the patch on the damaged area, draw two perpendicular centerlines on the parent structure and on the patch material that define the principal load or geometric directions. Then, lay out hole pattern on the patch and drill pilot holes in the patch material. Align the two perpendicular centerlines of the patch with the lines on the parent structure and transfer the pilot holes to the parent material. Use clecos to keep the patch in place. Mark the edges of the patch so that it can be returned to the same location easily.

Step 5: Drilling & Reaming Holes in Patch & Parent Structure

Composite skins should be backed up to prevent splitting. Enlarge the pilot holes in the patch and parent materials with a drill $\frac{1}{64}$ undersize and then ream all holes to the correct size. A tolerance of +0.0025/–0.000-inch is usually recommended for aircraft parts. For composites, this means interference fasteners are not used.

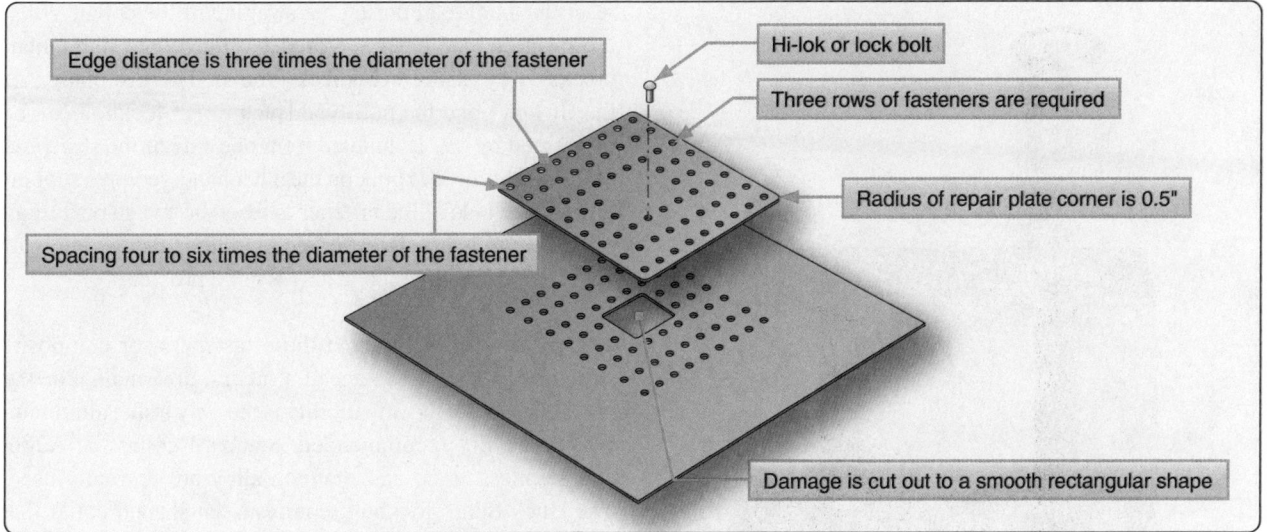

Figure 7-75. *Repair layout for bolted repair of composite structure.*

Step 6: Fastener Installation

Once fastener holes are drilled full size and reamed, permanent fasteners are installed. Before installation, measure the fastener grip length for each fastener using a grip length gauge. As different fasteners are required for different repairs, consult the SRM for permissible fastener type and installation procedure. However, install all fasteners wet with sealant and with proper torque for screws and bolts.

Step 7: Sealing of Fasteners & Patch

Sealants are applied to bolted repairs for prevention of water/moisture intrusion, chemical damage, galvanic corrosion, and fuel leaks. They also provide contour smoothness. The sealant must be applied to a clean surface. Masking tape is usually placed around the periphery of the patch, parallel with the patch edges and leaving a small gap between the edge of the patch and the masking tape. Sealing compound is applied into this gap.

Step 8: Application of Finish Coat & Lightning Protection Mesh

The repair needs to be sanded, primed, and painted with an approved paint system. A lightning protection mesh needs to be applied if composite patches are used in an area that is prone to lightning strikes.

Fasteners Used with Composite Laminates

Many companies make specialty fasteners for composite structures and several types of fasteners are commonly used: threaded fasteners, lock bolts, blind bolts, blind rivets, and specialty fasteners for soft structures, such as honeycomb panels. The main differences between fasteners for metal and composite structures are the materials and the footprint diameter of nuts and collars.

Corrosion Precautions

Neither fiberglass nor Kevlar® fiber-reinforced composites cause corrosion problems when used with most fastener materials. Composites reinforced with carbon fibers, however, are quite cathodic when used with materials, such as aluminum or cadmium, the latter of which is a common plating used on fasteners for corrosion protection.

Fastener Materials

Titanium alloy Ti-6Al-4V is the most common alloy for fasteners used with carbon fiber reinforced composite structures. Austenitic stainless steels, superalloys (e.g., A286), multiphase alloys (e.g., MP35N or MP159), and nickel alloys (e.g., alloy 718) also appear to be very compatible with carbon fiber composites.

Fastener System for Sandwich Honeycomb Structures (SPS Technologies Comp Tite)

The adjustable sustain preload (ASP) fastening system provides a simplified method of fastening composite, soft core, metallic or other materials, which are sensitive to fastener clamp-up or installation force conditions. Clamping force can be infinitely adjustable within maximum recommended torque limits and no further load is applied during installation of the lock collar. The fastener is available in two types. The ASP® has full shank and the 2ASP® has a pilot type shank. *[Figures 7-76 and 7-77]*

Hi-Lok® & Huck-Spin® Lockbolt Fasteners

Most composite primary structures for the aircraft industry are fastened with Hi-Loks® (Hi-Shear Corp.) or Huck-Spin® lockbolts for permanent installations. The Hi-Lok® is a threaded fastener that incorporates a hex key in the

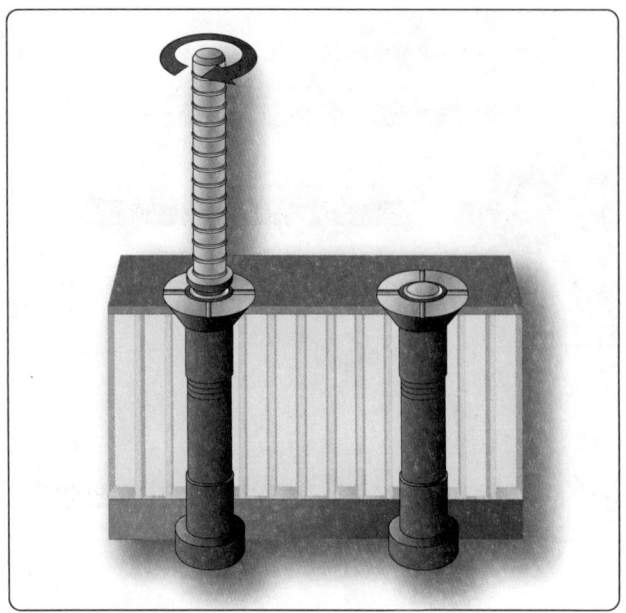

Figure 7-76. *ASP fastener system.*

react the axial load during the swaging of the collar. When the swaging load reaches a predetermined limit, the pintail breaks away at the breakneck groove. The installation of the Hi-Lok® and the pull-type Huck-Spin® lockbolt can be performed by one technician from one side of the structure. The stump-type lockbolt, on the other hand, requires support on the head side of the fastener to react the swage operation. This method is usually reserved for automated assembly of detail structure in which access is not a problem.

The specific differences in these fasteners for composite structure in contrast to metal structure are small. For the Hi-Lok®, material compatibility is the only issue; aluminum collars are not recommended. Standard collars of A286, 303 stainless steel, and titanium alloy are normally used. The Huck-Spin® lockbolt requires a hat-shaped collar that incorporates a flange to spread the high bearing loads during installation. The lockbolt pin designed for use in composite structure has six annular grooves as opposed to five for metal structure. *[Figures 7-79 and 7-80]*

Eddie-Bolt® Fasteners

Eddie-Bolt® fasteners (Alcoa) are similar in design to Hi-Loks® and are a natural choice for carbon fiber composite structures. The Eddie-Bolt® pin is designed with flutes in the threaded portion, which allow a positive lock to be made during installation using a specially designed mating nut or

threaded end to react to the torque applied to the collar during installation. The collar includes a frangible portion that separates at a predetermined torque value. *[Figure 7-78]*

The lockbolt incorporates a collar that is swaged into annular grooves. It comes in two types: pull and stump. The pull-type is the most common, where a frangible pintail is used to

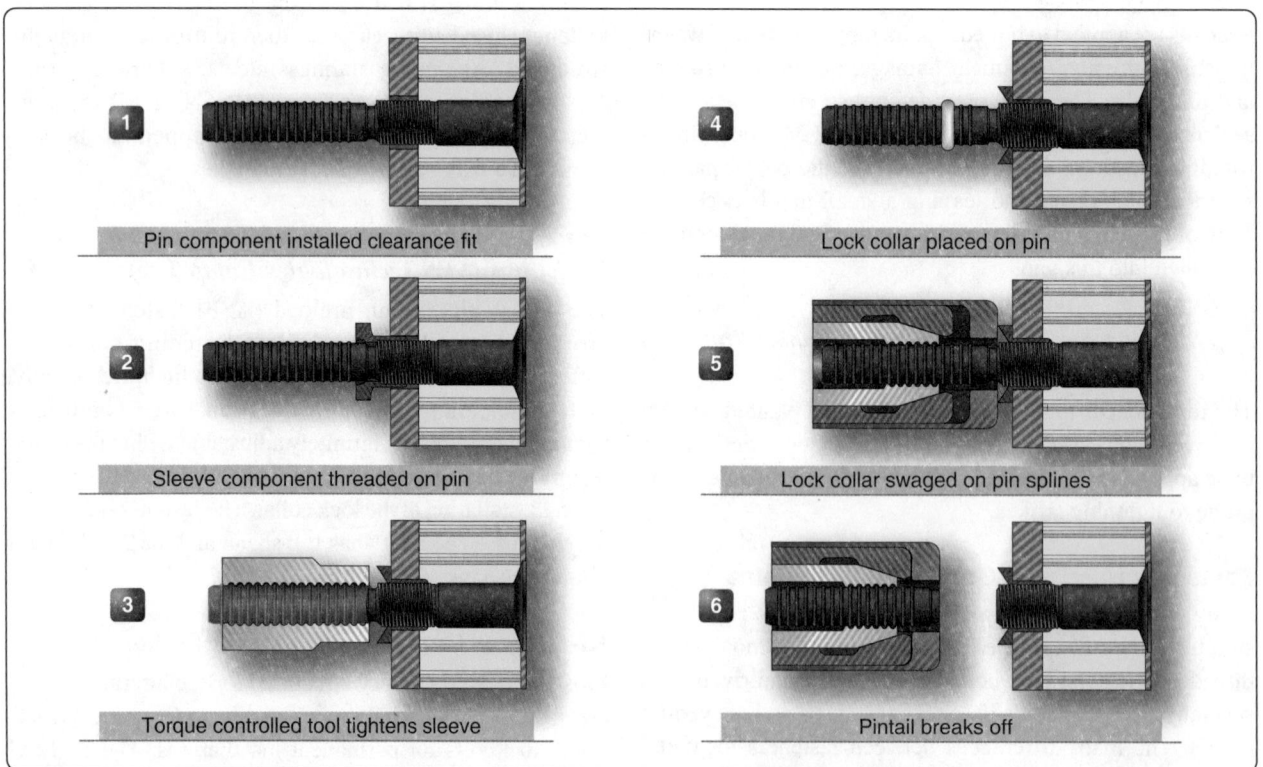

Figure 7-77. *ASP fastener system installation sequence.*

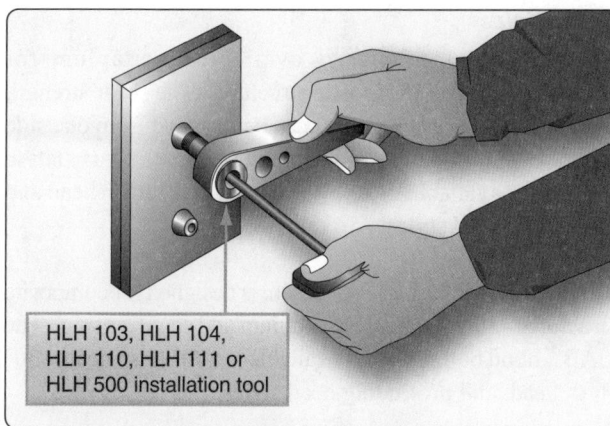

Figure 7-78. *Hi-Lok® installation.*

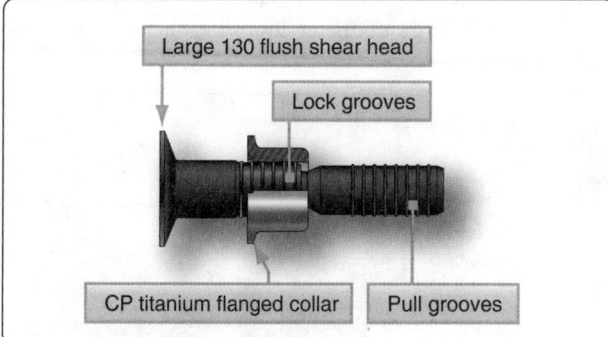

Figure 7-79. *Huck-Spin® lockbolt.*

collar. The mating nut has three lobes that serve as driving ribs. During installation, at a predetermined preload, the lobes compress the nut material into the flutes of the pin and form the locking feature. The advantage for composite structure is that titanium alloy nuts can be used for compatibility and weight saving without the fear of galling. The nuts spin on freely, and the locking feature is established at the end of the installation cycle. *[Figure 7-81]*

Cherry's E-Z Buck® (CSR90433) Hollow Rivet

The Cherry Hollow End E-Z Buck® rivet is made from titanium/columbium alloy and has a shear strength of 40 KSI. The E-Z Buck® rivet is designed to be used in a double flush application for fuel tanks. The main advantage of this type of rivet is that it takes less than half the force of a solid rivet of the same material. The rivets are installed with automated riveting equipment or a rivet squeezer. Special optional dies ensure that the squeezer is always centered during installation, avoiding damage to the structure. *[Figure 7-82]*

Blind Fasteners

Composite structures do not require as many fasteners as metal aircraft because stiffeners and doublers are co-cured with the skins, eliminating many fasteners. The size of panels on aircraft has increased in composite structures, which causes backside inaccessibility. Therefore, blind fasteners or screws and nutplates must be used in these areas. Many manufacturers make blind fasteners for composite structures; a few are discussed below.

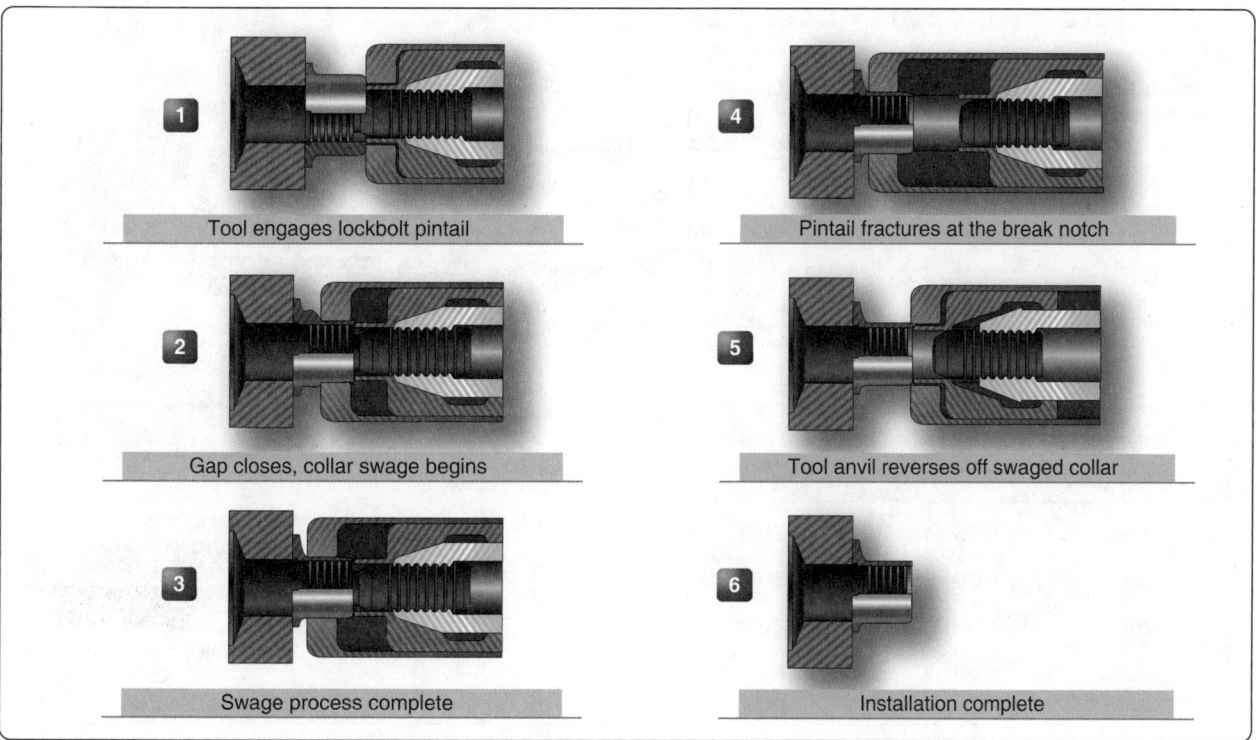

Figure 7-80. *Huck-Spin® installation sequence.*

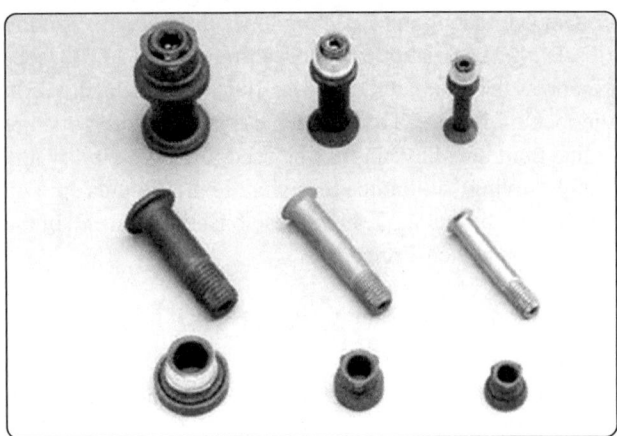

Figure 7-81. *Eddie-Bolts®.*

Blind Bolts

The Cherry Maxibolt® is available in titanium for compatibility with composite structures. The shear strength of the Maxibolt® is 95 KSI. It can be installed from one side with a G-83 or equivalent pneumatic-hydraulic installation tool, and is available in 100° flush head, 130° flush head and protruding head styles. *[Figure 7-83]*

The Alcoa UAB™ blind bolt system is designed for composite structures and is available in titanium and stainless steel. The UAB™ blind bolt system is available in 100° flush head, 130° flush head, and protruding head styles.

The Accu-Lok™ Blind Fastening System is designed specifically for use in composite structures in which access is limited to one side of the structure. It combines high joint preload with a large diameter footprint on the blind side.

Rivet Diameter	A REF	CSR 90433			
		B REF	C1 DIA	C2 DIA	D DIA
1/8 (–4)	0.028	0.028	0.195 0.189	0.195 0.189	0.132 0.129
5/32 (–5)	0.037	0.037	0.247 0.242	0.247 0.242	0.162 0.159
3/16 (–6)	0.046	0.046	0.302 0.297	0.302 0.297	0.195 0.191
7/32 (–7)	0.046	0.046	0.328 0.323	0.328 0.323	0.227 0.224

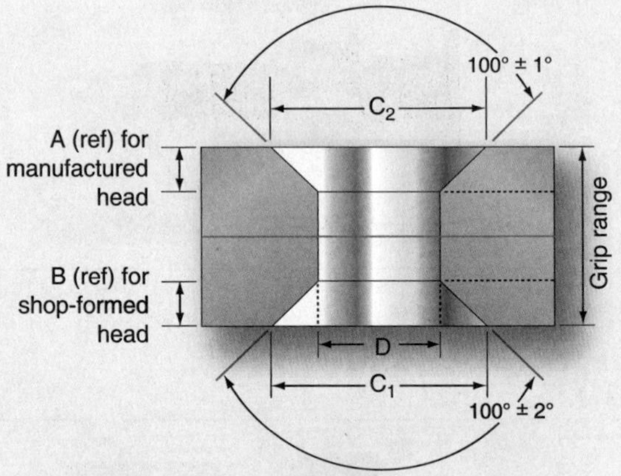

Hollow End E-Z Buck® Nominal Diameter	Upset Load (Lb) + 200 Lb
1/8" (–4)	2,500
5/32" (–5)	2,700
3/16" (–6)	3,000
7/32" (–7)	3,750

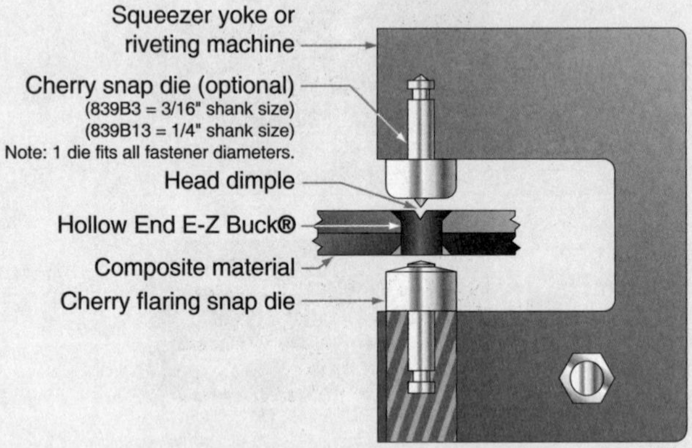

Cherry Flaring Snap Die Part Numbers

Rivet Diameter	3/16" Diameter Mount	1/4" Diameter Mount
1/8"	839B1-4	839B10-4
5/32"	839B1-5	839B10-5
3/16"	839B1-6	839B10-6
7/32"	839B1-7	839B10-7

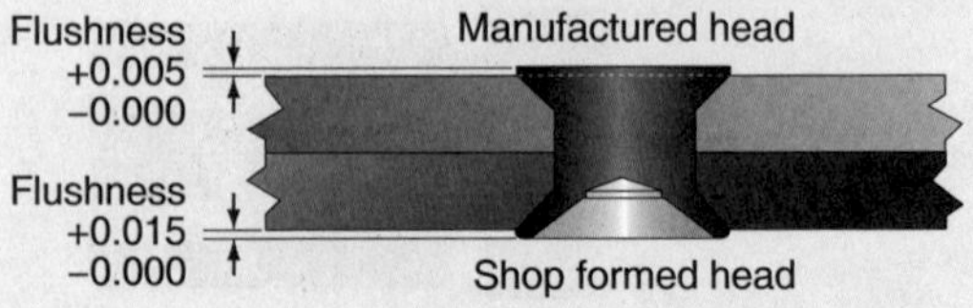

$100° \pm 1°$

C_2

A (ref) for manufactured head

B (ref) for shop-formed head

D

C_1

$100° \pm 2°$

Grip range

Squeezer yoke or riveting machine

Cherry snap die (optional)
(839B3 = 3/16" shank size)
(839B13 = 1/4" shank size)
Note: 1 die fits all fastener diameters.

Head dimple

Hollow End E-Z Buck®

Composite material

Cherry flaring snap die

Flushness +0.005 –0.000

Manufactured head

Flushness +0.015 –0.000

Shop formed head

Figure 7-82. *Cherry's E-Z Buck hollow rivet.*

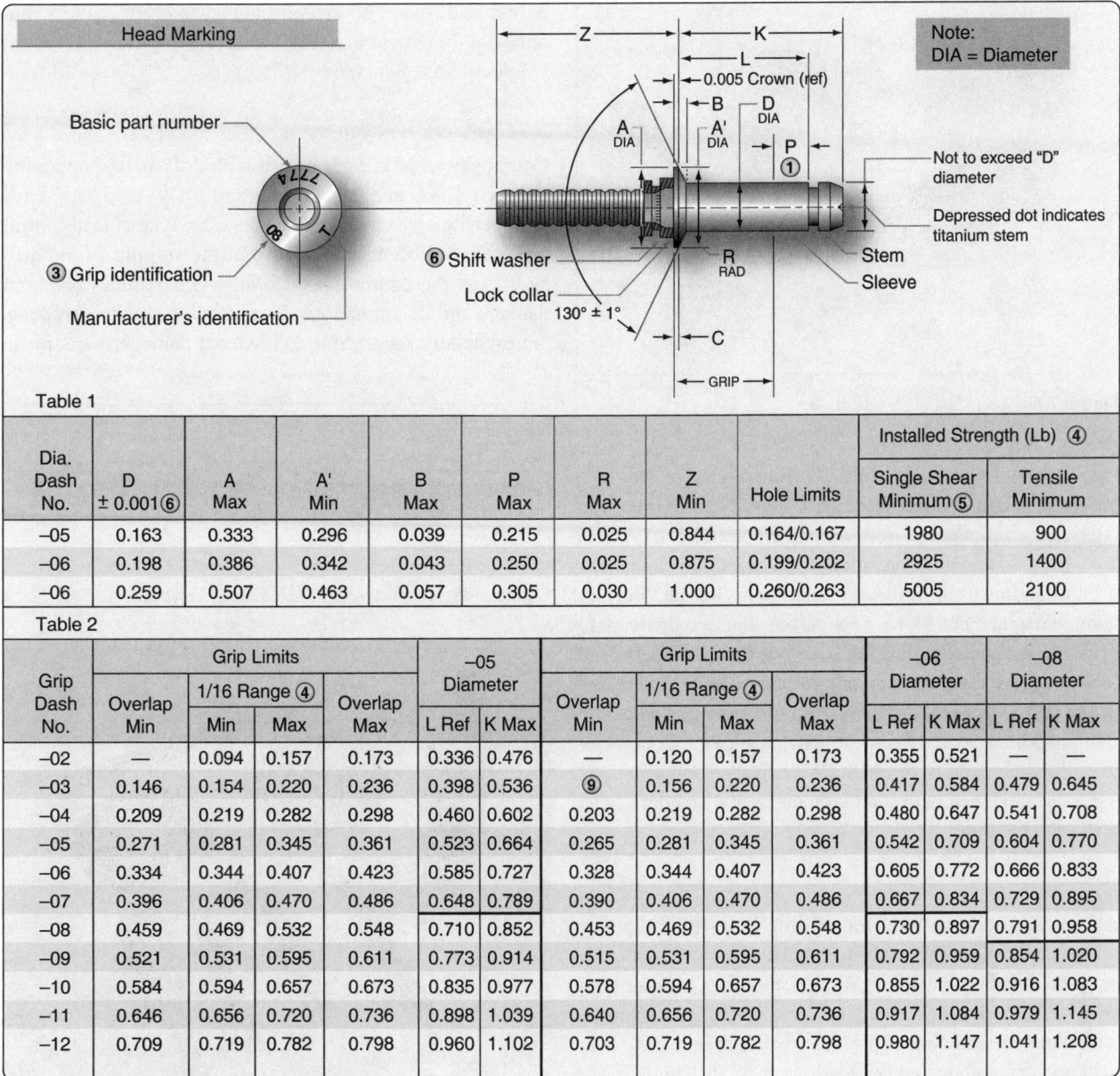

Table 1

Dia. Dash No.	D ± 0.001(6)	A Max	A' Min	B Max	P Max	R Max	Z Min	Hole Limits	Installed Strength (Lb) (4) Single Shear Minimum (5)	Tensile Minimum
–05	0.163	0.333	0.296	0.039	0.215	0.025	0.844	0.164/0.167	1980	900
–06	0.198	0.386	0.342	0.043	0.250	0.025	0.875	0.199/0.202	2925	1400
–08	0.259	0.507	0.463	0.057	0.305	0.030	1.000	0.260/0.263	5005	2100

Table 2

Grip Dash No.	Grip Limits Overlap Min	1/16 Range (4) Min	1/16 Range (4) Max	Overlap Max	–05 Diameter L Ref	–05 Diameter K Max	Grip Limits Overlap Min	1/16 Range (4) Min	1/16 Range (4) Max	Overlap Max	–06 Diameter L Ref	–06 Diameter K Max	–08 Diameter L Ref	–08 Diameter K Max
–02	—	0.094	0.157	0.173	0.336	0.476	—	0.120	0.157	0.173	0.355	0.521	—	—
–03	0.146	0.154	0.220	0.236	0.398	0.536	(9)	0.156	0.220	0.236	0.417	0.584	0.479	0.645
–04	0.209	0.219	0.282	0.298	0.460	0.602	0.203	0.219	0.282	0.298	0.480	0.647	0.541	0.708
–05	0.271	0.281	0.345	0.361	0.523	0.664	0.265	0.281	0.345	0.361	0.542	0.709	0.604	0.770
–06	0.334	0.344	0.407	0.423	0.585	0.727	0.328	0.344	0.407	0.423	0.605	0.772	0.666	0.833
–07	0.396	0.406	0.470	0.486	0.648	0.789	0.390	0.406	0.470	0.486	0.667	0.834	0.729	0.895
–08	0.459	0.469	0.532	0.548	0.710	0.852	0.453	0.469	0.532	0.548	0.730	0.897	0.791	0.958
–09	0.521	0.531	0.595	0.611	0.773	0.914	0.515	0.531	0.595	0.611	0.792	0.959	0.854	1.020
–10	0.584	0.594	0.657	0.673	0.835	0.977	0.578	0.594	0.657	0.673	0.855	1.022	0.916	1.083
–11	0.646	0.656	0.720	0.736	0.898	1.039	0.640	0.656	0.720	0.736	0.917	1.084	0.979	1.145
–12	0.709	0.719	0.782	0.798	0.960	1.102	0.703	0.719	0.782	0.798	0.980	1.147	1.041	1.208

Figure 7-83. *Cherry's titanium Maxibolt.*

The large footprint enables distribution of the joint preload over a larger area, virtually eliminating the possibility of delaminating the composite structure. The shear strength of the Accu-Lok™ is 95 KSI, and it is available in 100° flush head, 130° flush head, and protruding head styles. A similar fastener designed by Monogram is called the Radial-Lok®. *[Figure 7-84]*

Fiberlite

The fiberlite fastening system uses composite materials for a wide range of aerospace hardware. The strength of fiberlite fasteners is equivalent to aluminum at two-thirds the weight. The composite fastener provides good material compatibility with carbon fiber and fiberglass.

Screws & Nutplates in Composite Structures

The use of screws and nutplates in place of Hi-Loks® or blind fasteners is recommended if a panel must be removed periodically for maintenance. Nutplates used in composite structures usually require three holes: two for attachment of the nutplate and one for the removable screw, although rivetless nut plates and adhesive bonded nutplates are available that do not require drilling and countersinking two extra holes.

Machining Processes & Equipment
Drilling

Hole drilling in composite materials is different from drilling

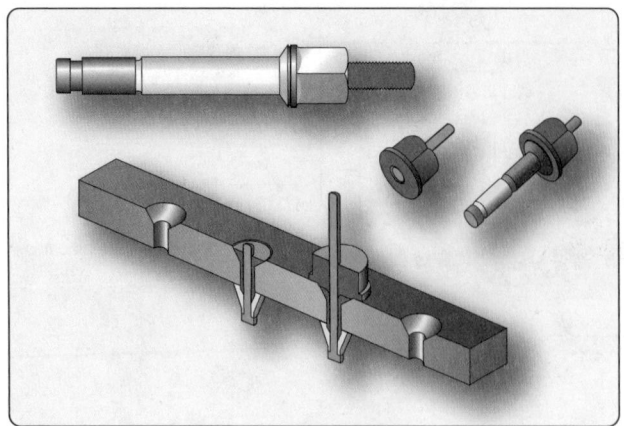

Figure 7-84. *Accu-Lok ™ installation.*

holes in metal aircraft structures. Different types of drill bits, higher speeds, and lower feeds are required to drill precision holes. Structures made from carbon fiber and epoxy resin are very hard and abrasive, requiring special flat flute drills or similar four-flute drills. Aramid fiber (Kevlar®)/epoxy composites are not as hard as carbon but are difficult to drill unless special cutters are used because the fibers tend to fray or shred unless they are cut clean while embedded in the epoxy. Special drill bits with clothes pin points and fish-tail points have been developed that slice the fibers prior to pulling them out of the drilled hole. If the Kevlar®/epoxy part is sandwiched between two metal parts, standard twist drills can be used.

Equipment

Air-driven tools are used for drilling holes in composite materials. Drill motors with free speed of up to 20,000 rpm are used. A general rule for drilling composites is to use high speed and a low feed rate (pressure). Drilling equipment with a power feed control produces better hole quality than drill motors without power feed control. Drill guides are recommended, especially for thicker laminates.

Do not use standard twist drill bits for drilling composite structures. Standard high-speed steel is unacceptable, because it dulls immediately, generates excessive heat, and causes ply delamination, fiber tear-out, and unacceptable hole quality.

Drill bits used for carbon fiber and fiberglass are made from diamond-coated material or solid carbide because the fibers are so hard that standard high-speed steel (HSS) drill bits do not last long. Typically, twist drills are used, but brad point drills are also available. The Kevlar® fibers are not as hard as carbon, and standard HSS drill bits can be used. The hole quality can be poor if standard drill bits are used and the preferred drill style is the sickle-shaped Klenk drill. This drill first pulls on the fibers and then shears them, which results in a better quality hole. Larger holes can

be cut with diamond-coated hole saws or fly cutters, but only use fly cutters in a drill press, and not in a drill motor. *[Figures 7-85, 7-86, and 7-87]*

Processes & Precautions

Composite materials are drilled with drill motors operating between 2,000 and 20,000 rpm and a low feed rate. Drill motors with a hydraulic dash pod or other type of feed control are preferred because they restrict the surging of the drill as it exits the composite materials. This reduces breakout damage and delaminations. Parts made from tape products are especially susceptible to breakout damage; parts made

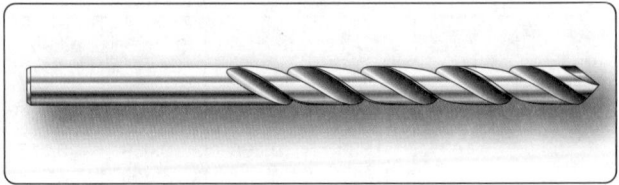

Figure 7-85. *Klenk-type drill for drilling Kevlar®.*

Figure 7-86. *Drilling and cutting tools for composite materials.*

Figure 7-87. *Autofeed drill.*

from fabric material have experienced less damage. The composite structure needs to be backed with a metal plate or sheet to avoid breakout. Holes in composite structures are often predrilled with a small pilot hole, enlarged with a diamond-coated or carbide drill bit and reamed with a carbide reamer to final hole size.

Back counterboring is a condition that can occur when carbon/epoxy parts mate metal substructure parts. The back edge of the hole in the carbon/epoxy part can be eroded or radiused by metal chips being pulled through the composite. The condition is more prevalent when there are gaps between the parts or when the metal debris is stringy rather than small chips. Back counterboring can be minimized or eliminated by changing feeds and speeds, cutter geometry, better part clamp-up adding a final ream pass, using a peck drill, or combination of these.

When drilling combinations of composite parts with metal parts, the metal parts may govern the drilling speed. For example, even though titanium is compatible with carbon/epoxy material from a corrosion perspective, lower drilling speeds are required in order to ensure no metallurgical damage occurs to the titanium. Titanium is drilled with low speed and high feed. Drill bits suitable for titanium might not be suitable for carbon or fiberglass. Drill bits that are used for drilling titanium are often made from cobalt-vanadium; drill bits used for carbon fiber are made from carbide or are diamond coated to increase drill life and to produce an accurate hole. Small-diameter high-speed steel drill bits, such as No. 40 drill, which are used to manually drill pilot holes, are typically used because carbide drills are relatively brittle and are easily broken. The relatively low cost of these small HSS drill bits offsets the limited life expectancy. High-speed steel drill bits may last for only one hole.

The most common problem with carbide cutters used in hand-drill operations is handling damage (chipped edges) to the cutters. A sharp drill with a slow constant feed can produce a 0.1 mm (0.004-inch) tolerance hole through carbon/epoxy plus thin aluminum, especially if a drill guide is used. With hard tooling, tighter tolerances can be maintained. When the structure under the carbon/epoxy is titanium, drills can pull titanium chips through the carbon/epoxy and enlarge the hole. In this case, a final ream operation may be required to hold tight hole tolerances. Carbide reamers are needed for holes through carbon/epoxy composite structure. In addition, the exit end of the hole needs good support to prevent splintering and delaminations when the reamer removes more than about 0.13 mm (0.005-inch) on the diameter. The support can be the substructure or a board held firmly against the back surface. Typical reaming speeds are about one-half of the drilling speed.

Cutting fluids are not normally used or recommended for drilling thin (less than 6.3 mm, or 0.25-inch thick) carbon/epoxy structures. It is good practice to use a vacuum while drilling into composite materials to prevent carbon dust from freely floating around the work area.

Countersinking

Countersinking a composite structure is required when flush head fasteners are to be installed in the assembly. For metallic structures, a 100° included angle shear or tension head fastener has been the typical approach. In composite structures, two types of fasteners are commonly used: a 100° included angle tension head fastener or a 130° included angle head fastener. The advantage of the 130° head is that the fastener head can have about the same diameter as a tension head 100° fastener with the head depth of a shear-type head 100° fastener. For seating flush fasteners in composite parts, it is recommended that the countersink cutters be designed to produce a controlled radius between the hole and the countersink to accommodate the head-to-shank fillet radius on the fasteners. In addition, a chamfer operation or a washer may be required to provide proper clearance for protruding head fastener head-to-shank radii. Whichever head style is used, a matching countersink/chamfer must be prepared in the composite structure.

Carbide cutters are used for producing a countersink in carbon/epoxy structure. These countersink cutters usually have straight flutes similar to those used on metals. For Kevlar® fiber/epoxy composites, S-shaped positive rake cutting flutes are used. If straight-fluted countersink cutters are used, a special thick tape can be applied to the surface to allow for a clean cutting of the Kevlar® fibers, but this is not as effective as the S-shaped fluted cutters. Use of a piloted countersink cutter is recommended because it ensures better concentricity between the hole and the countersink and decreases the possibility of gaps under the fasteners due to misalignment or delaminations of the part.

Use a microstop countersink gauge to produce consistent countersink wells. Do not countersink through more than 70 percent of the skin depth because a deeper countersink well reduces material strength. When a piloted countersink cutter is used, the pilot must be periodically checked for wear, as wear can cause reduction of concentricity between the hole and countersink. This is especially true for countersink cutters with only one cutting edge. For piloted countersink cutters, position the pilot in the hole and bring the cutter to full rpm before beginning to feed the cutter into the hole and preparing the countersink. If the cutter is in contact with the composite before triggering the drill motor, you may get splintering.

Cutting Processes & Precautions

Cutters that work well for metals would either have a short life or produce a poorly cut edge if used for composite materials. The cutters that are used for composites vary with the composite material that is being cut. The general rule for cutting composites is high speed and slow feed.

- Carbon fiber reinforced plastics: Carbon fiber is very hard and quickly wears out high speed steel cutters. For most trimming and cutting tasks, diamond grit cutters are best. Aluminum-oxide or silicon-carbide sandpaper or cloth is used for sanding. Silicon-carbide lasts longer then aluminum-oxide. Router bits can also be made from solid carbide or diamond coated.

- Glass fiber reinforced plastics: Glass fibers, like carbon, are very hard and quickly wear out high-speed steel cutters. Fiberglass is drilled with the same type and material drill bits as carbon fiber.

- Aramid (Kevlar®) fiber-reinforced plastics: Aramid fiber is not as hard as carbon and glass fiber, and cutters made from high-speed steel can be used. To prevent loose fibers at the edge of aramid composites, hold the part and then cut with a shearing action. Aramid composites need to be supported with a plastic backup plate. The aramid and backup plate are cut through at the same time. Aramid fibers are best cut by being held in tension and then sheared. There are specially shaped cutters that pull on the fibers and then shear them. When using scissors to cut aramid fabric or prepreg, they must have a shearing edge on one blade and a serrated or grooved surface on the other. These serrations hold the material from slipping. Sharp blades should always be used as they minimize fiber damage. Always clean the scissor serrations immediately after use so the uncured resin does not ruin the scissors.

Always use safety glasses and other protective equipment when using tools and equipment.

Cutting Equipment

The bandsaw is the equipment that is most often used in a repair shop for cutting composite materials. A toothless carbide or diamond-coated saw blade is recommended. A typical saw blade with teeth does not last long if carbon fiber or fiberglass is cut. *[Figure 7-88]* Air-driven hand tools, such as routers, saber saws, die grinders, and cut-off wheels can be used to trim composite parts. Carbide or diamond-coated cutting tools produce a better finish and they last much longer. Specialized shops have ultrasonic, waterjet, and laser cutters. These types of equipment are numerical controlled (NC) and produce superior edge and hole quality. A waterjet cutter cannot be used for honeycomb structure because it introduces water in the part. Do not cut anything else on equipment that is used for composites because other materials can contaminate the composite material.

Prepreg materials can be cut with a CNC Gerber table. The use of this equipment speeds up the cutting process and optimizes the use of the material. Design software is available that calculates how to cut plies for complex shapes. *[Figures 7-89]*

Repair Safety

Advanced composite materials including prepreg, resin systems, cleaning solvents, and adhesives could be

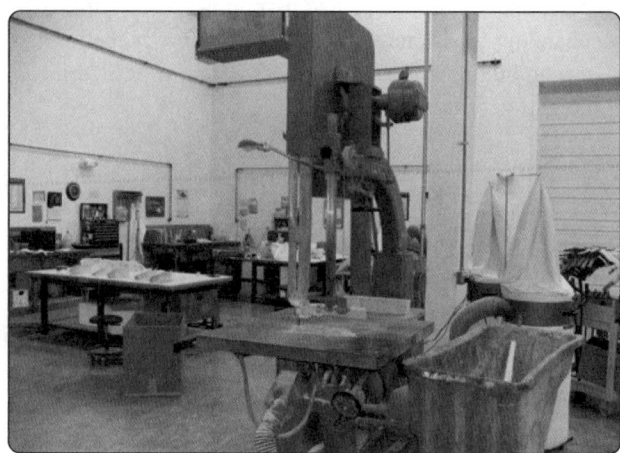

Figure 7-88. *Bandsaw.*

hazardous, and it is important that you use personal protection equipment. It is important to read and understand the Safety Data Sheets (SDS) and handle all chemicals, resins, and fibers correctly. The SDS lists the hazardous chemicals in the material system, and it outlines the hazards. The material could be a respiratory irritant or carcinogenic, or another kind of dangerous substance.

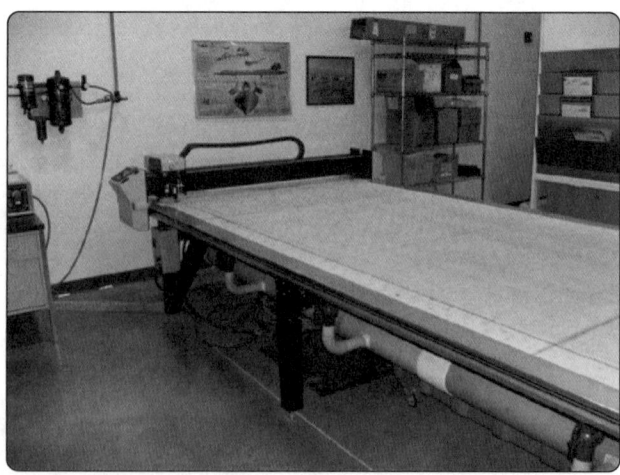

Figure 7-89. *Gerber cutting table.*

Eye Protection

Always protect eyes from chemicals and flying objects. Wear safety glasses at all times and, when mixing or pouring acids, wear a face shield. Never wear corrective contact lenses in a shop, even with safety glasses. Some of the chemical solvents can melt the lenses and damage eyes. Dust can also get under the lenses, causing damage.

Respiratory Protection

Do not breathe carbon fiber dust and always ensure that there is a good flow of air where the work is performed. Always use equipment to assist in breathing when working in a confined space. Use a vacuum near the source of the dust to remove the dust from the air. When sanding or applying paint, you need a dust mask or a respirator. A properly fitted dust mask provides the protection needed. For application of paints, a sealed respirator with the correct filters or a fresh air supply respirator is required.

Downdraft Tables

A downdraft table is an efficient and economical device for protecting workers from harmful dust caused by sanding and grinding operations. The tables are also useful housekeeping tools because the majority of particulate material generated by machining operations is immediately collected for disposal. Downdraft tables should be sized and maintained to have an average face velocity between 100 and 150 cubic feet per minute. The downdraft table draws contaminants like dust and fibers away from the operator's material. Downdraft tables should be monitored and filters changed on a regular basis to provide maximum protection and particulate collection.

Skin Protection

During composite repair work, protect your skin from hazardous materials. Chemicals could remain on hands that burn sensitive skin. Always wear gloves and clothing that offer protection against toxic materials. Use only approved gloves that protect skin and do not contaminate the composite material. Always wash hands prior to using the toilet or eating. Damaged composite components should be handled with care. Single fibers can easily penetrate the skin, splinter off, and become lodged in the skin.

Fire Protection

Most solvents are flammable. Close all solvent containers and store in a fireproof cabinet when not in use. Make sure that solvents are kept away from areas where static electricity can occur. Static electricity can occur during sanding operations or when bagging material is unrolled. It is preferable to use air-driven tools. If electric tools are used, ensure that they are the enclosed type. Do not mix too much resin. The resin could overheat and start smoking caused by the exothermic process. Ensure that a fire extinguisher is always nearby.

Transparent Plastics

Plastics cover a broad field of organic synthetic resin and may be divided into two main classifications: thermoplastics and thermosetting plastics.

a. Thermoplastics—may be softened by heat and can be dissolved in various organic solvents. Acrylic plastic is commonly used as a transparent thermoplastic material for windows, canopies, etc. Acrylic plastics are known by the trade names of Lucite® or Plexiglas® and by the British as Perspex®, and meet the military specifications of MIL-P-5425 for regular acrylic and MIL-P-8184 for craze-resistant acrylic.

b. Thermosetting plastics—do not soften appreciably under heat but may char and blister at temperatures of 240–260 °C (400–500 °F). Most of the molded products of synthetic resin composition, such as phenolic, urea-formaldehyde, and melamine formaldehyde resins, belong to the thermosetting group. Once the plastic becomes hard, additional heat does not change it back into a liquid as it would with a thermoplastic.

Optical Considerations

Scratches and other types of damage that obstruct the vision of the pilots are not acceptable. Some types of damage might be acceptable at the edges of the windshield.

Identification

Storage & Handling

Because transparent thermoplastic sheets soften and deform when they are heated, they must be stored where the temperature never becomes excessive. Store them in a cool, dry location away from heating coils, radiators, or steam pipes, and away from such fumes as are found in paint spray booths or paint storage areas.

Keep paper-masked transparent sheets out of the direct rays of the sun, because sunlight accelerates deterioration of the adhesive, causing it to bond to the plastic, and making it difficult to remove.

Store plastic sheets with the masking paper in place, in bins that are tilted at a 10° angle from the vertical to prevent buckling. If the sheets are stored horizontally, take care to avoid getting dirt and chips between them. Stacks of sheets must never be over 18 inches high, with the smallest sheets stacked on top of the larger ones so there is no unsupported overhang. Leave the masking paper on the sheets as long as possible, and take care not to scratch or gouge the sheets by sliding them against each other or across rough or dirty tables.

Store formed sections with ample support so they do not lose their shape. Vertical nesting should be avoided. Protect formed parts from temperatures higher than 120 °F (49 °C), and leave their protective coating in place until they are installed on the aircraft.

Forming Procedures & Techniques

Transparent acrylic plastics get soft and pliable when they are heated to their forming temperatures and can be formed to almost any shape. When they cool, they retain the shape to which they were formed. Acrylic plastic may be cold-bent into a single curvature if the material is thin and the bending radius is at least 180 times the thickness of the sheet. Cold bending beyond these limits impose so much stress on the surface of the plastic that tiny fissures or cracks, called crazing, form.

Heating

Wear cotton gloves when handling the plastic to eliminate finger marks on the soft surface. Before heating any transparent plastic material, remove all of the masking paper and adhesive from the sheet. If the sheet is dusty or dirty, wash it with clean water and rinse it well. Dry the sheet thoroughly by blotting it with soft absorbent paper towels.

For the best results when hot forming acrylics, adhere to the temperatures recommended by the manufacturer. Use a forced-air oven that can operate over a temperature range of 120–374 °F (49–190 °C). If the part gets too hot during the forming process, bubbles may form on the surface and impair the optical qualities of the sheet.

For uniform heating, it is best to hang the sheets vertically by grasping them by their edges with spring clips and suspending the clips in a rack. *[Figure 7-90]* If the piece is too small to hold with clips, or if there is not enough trim area, lay the sheets on shelves or racks covered with soft felt or flannel. Be sure there is enough open space to allow the air to circulate around the sheet and heat it evenly.

Small forming jobs, such as landing light covers, may be heated in a kitchen baking oven. Infrared heat lamps may be used if they are arranged on 7- to 8-inch centers and enough of them are used in a bank to heat the sheet evenly. Place the lamps about 18-inches from the material.

Never use hot water or steam directly on the plastic to heat it because this likely causes the acrylic to become milky or cloudy.

Forms

Heated acrylic plastic molds with almost no pressure, so the forms used can be of very simple construction. Forms made of pressed wood, plywood, or plaster are adequate to form simple curves, but reinforced plastic or plaster may be needed to shape complex or compound curves. Since hot plastic conforms to any waviness or unevenness, the form used must be completely smooth. To ensure this, sand the form and cover it with soft cloth, such as outing flannel or billiard felt. The mold should be large enough to extend beyond the trim line of the part, and provisions should be made for holding the hot plastic snug against the mold as it cools.

A mold can be made for a complex part by using the damaged part itself. If the part is broken, tape the pieces together, wax or grease the inside so the plaster does not stick to it, and support the entire part in sand. Fill the part with plaster and allow it to harden, and then remove it from the mold. Smooth out any roughness and cover it with soft cloth. It is now ready to use to form the new part.

Forming Methods

Simple Curve Forming

Heat the plastic material to the recommended temperature, remove it from the heat source, and carefully drape it over the prepared form. Carefully press the hot plastic to the form and either hold or clamp the sheet in place until it cools. This process may take from 10–30 minutes. Do not force cool it.

Compound Curve Forming

Compound curve forming is normally used for canopies or complex wingtip light covers, and it requires a great deal of specialized equipment. There are four commonly used methods, each having its advantages and disadvantages.

Stretch Forming

Preheated acrylic sheets are stretched mechanically over the form in much the same way as is done with the simple curved

Figure 7-90. *Hanging an acrylic sheet.*

piece. Take special care to preserve uniform thickness of the material, since some parts must stretch more than others.

Male & Female Die Forming

Male and female die forming requires expensive matching male and female dies. The heated plastic sheet is placed between the dies that are then mated. When the plastic cools, the dies are opened.

Vacuum Forming Without Forms

Many aircraft canopies are formed by this method. In this process, a panel, which has cut into it the outline of the desired shape, is attached to the top of a vacuum box. The heated and softened sheet of plastic is then clamped on top of the panel. When the air in the box is evacuated, the outside air pressure forces the hot plastic through the opening and forms the concave canopy. It is the surface tension of the plastic that shapes the canopy.

Vacuum Forming With a Female Form

If the shape needed is other than that which would be formed by surface tension, a female mold, or form must be used. It is placed below the plastic sheet and the vacuum pump is connected. When air from the form is evacuated, the outside air pressure forces the hot plastic sheet into the mold and fills it.

Sawing & Drilling
Sawing

Several types of saws can be used with transparent plastics; however, circular saws are the best for straight cuts. The blades should be hollow ground or have some set to prevent binding. After the teeth are set, they should be side dressed to produce a smooth edge on the cut. Band saws are recommended for cutting flat acrylic sheets when the cuts must be curved or where the sheet is cut to a rough dimension to be trimmed later. Close control of size and shape may be obtained by band sawing a piece to within 1⁄16-inch of the desired size, as marked by a scribed line on the plastic, and then sanding it to the correct size with a drum or belt sander.

Drilling

Unlike soft metal, acrylic plastic is a very poor conductor of heat. Make provisions for removing the heat when drilling. Deep holes need cooling, and water-soluble cutting oil is a satisfactory coolant since it has no tendency to attack the plastic.

The drill used on acrylics must be carefully ground and free from nicks and burrs that would affect the surface finish. Grind the drill with a greater included angle than would be used for soft metal. The rake angle should be zero in order to scrape, and not cut. The length of the cutting edge (and

hence the width of the lip) can be reduced by increasing the included angle of the drill. *[Figure 7-91]* Whenever holes are drilled completely through acrylic, the standard twist drills should be modified to a 60° tip angle, the cutting edge to a zero rake angle, and the back lip clearance angle increased to 12-15°. Drills specially modified for drilling acrylic are available from authorized distributors and dealers.

The patented Unibit® is good for drilling small holes in aircraft windshields and windows. *[Figure 7-92]* It can cut holes from 1⁄8 to 1⁄2-inch in 1⁄32-inch increments and produces good smooth holes with no stress cracks around their edges.

Cementing

Polymerizable cements are those in which a catalyst is added to an already thick monomer-polymer syrup to promote rapid hardening. Cement PS-30® and Weld-On 40® are polymerizable cements of this type. They are suitable for cementing all types of plexiglas acrylic cast sheet and parts molded from plexiglas molding pellets. At room temperature, the cements harden (polymerize) in the container in about 45 minutes after mixing the components. They harden more rapidly at higher temperatures. The cement joints are usually hard enough for handling within 4 hours after assembly. The joints may be machined within 4 hours after assembly, but it is better to wait 24 hours.

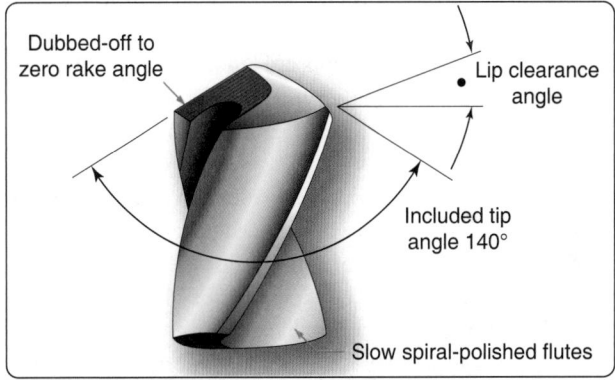

Figure 7-91. *Drill with an included angle of 140° is used to drill acrylic plastics.*

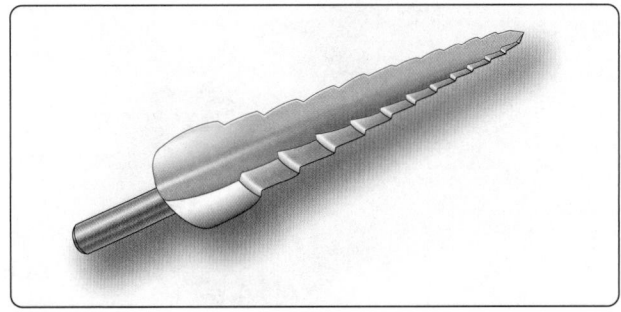

Figure 7-92. *Unibit® drill for drilling acrylic plastics.*

Application of Cement

PS-30® and Weld-On 40® joints retain excellent appearance and color stability after outdoor exposure. These cements produce clear, transparent joints and should be used when the color and appearance of the joints are important.

PS-30® and Weld-On 40® should be used at temperatures no lower than 65 °F. If cementing is done in a room cooler than 65 °F, it requires a longer time to harden and the joint strength is reduced.

The cement should be prepared with the correct proportions of components as given in the manufacturer's instructions and thoroughly mixed, making sure neither the mixing container nor mixing paddle adds color or effects the hardening of the cement. Clean glass or polyethylene mixing containers are preferred. Because of their short pot life (approximately 45 minutes), Cement PS-30® and Weld-On 40® must be used quickly once the components are mixed. Time consumed in preparation shortens the effective working time, making it necessary to have everything ready to be cemented before the cements are mixed. For better handling, pour cement within 20 minutes of mixing. For maximum joint strength, the final cement joint should be free of bubbles. It is usually sufficient to allow the mixed cement to stand for 10 minutes before cementing to allow bubbles to rise to the surface. The gap joint technique can only be used with colorless plexiglas acrylic or in cases where joints are hidden. If inconspicuous joints in colored plexiglas acrylic are needed, the parts must be fitted closely, using closed V-groove, butt, or arc joints.

Cement forms, or dams, may be made with masking tape as long as the adhesive surface does not contact the cement. This is easily done with a strip of cellophane tape placed over the masking tape adhesive. The tape must be chosen carefully. The adhesive on ordinary cellophane tape prevents the cure of PS-30® and Weld-On 40®. Before actual fabrication of parts,

sample joints should be tried to ensure that the tape system used does not harm the cement. Since it is important for all of the cement to remain in the gap, only contact pressure should be used.

Bubbles tend to float to the top of the cement bead in a gap joint after the cement is poured. These cause no problem if the bead is machined off. A small wire (not copper) or similar object may be used to lift some bubbles out of the joint; however, the cement joint should be disturbed as little as possible.

Polymerizable cements shrink as the cement hardens. Therefore, the freshly poured cement bead should be left above the surfaces being cemented to compensate for the shrinkage. If it is necessary for appearances, the bead may be machined off after the cement has set.

Repairs

Whenever possible, replace, rather than repair, extensively damaged transparent plastic. A carefully patched part is not the equal of a new section, either optically or structurally. At the first sign of crack development, drill a small hole with a #30 or a ⅛-inch drill at the extreme ends of the cracks. *[Figure 7-93]* This serves to localize the cracks and to prevent further splitting by distributing the strain over a large area. If the cracks are small, stopping them with drilled holes usually suffices until replacement or more permanent repairs can be made.

Cleaning

Plastics have many advantages over glass for aircraft use, but they lack the surface hardness of glass and care must be exercised while servicing the aircraft to avoid scratching or otherwise damaging the surface. Clean the plastic by washing it with plenty of water and mild soap, using a clean, soft, grit-free cloth, sponge, or bare hands. Do not use gasoline, alcohol, benzene, acetone, carbon tetrachloride, fire extinguisher or deicing fluids, lacquer thinners, or window

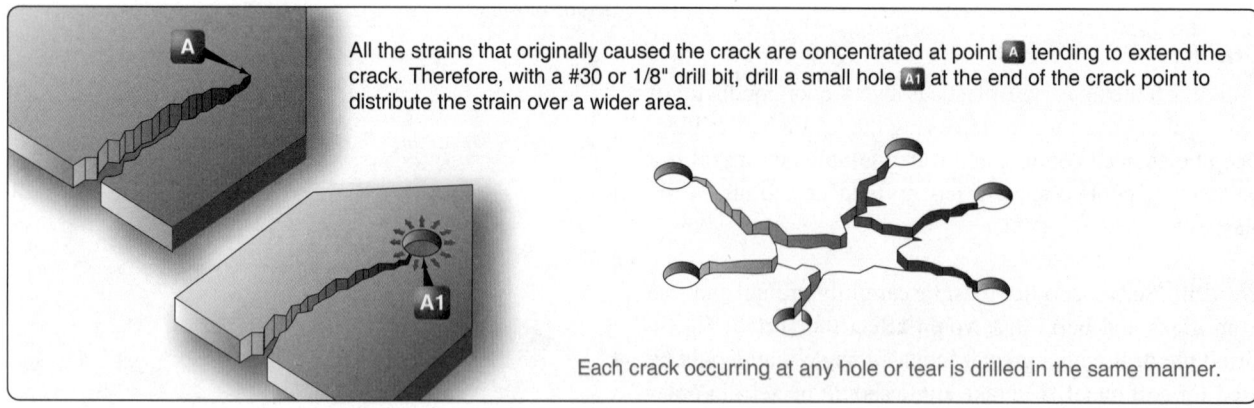

All the strains that originally caused the crack are concentrated at point **A** tending to extend the crack. Therefore, with a #30 or 1/8" drill bit, drill a small hole **A1** at the end of the crack point to distribute the strain over a wider area.

Each crack occurring at any hole or tear is drilled in the same manner.

Figure 7-93. *Stop drilling of cracks.*

cleaning sprays. These soften the plastic and cause crazing.

Plastics should not be rubbed with a dry cloth since it is likely to cause scratches and to build up an electrostatic charge that attracts dust particles to the surface. If, after removing dirt and grease, no great amount of scratching is visible, finish the plastic with a good grade of commercial wax. Apply the wax in a thin even coat and bring to a high polish by rubbing lightly with a soft cloth.

Polishing

Do not attempt hand polishing or buffing until the surface is clean. A soft, open-type cotton or flannel buffing wheel is suggested. Minor scratches may be removed by vigorously rubbing the affected area by hand, using a soft clean cloth dampened with a mixture of turpentine and chalk, or by applying automobile cleanser with a damp cloth. Remove the cleaner and polish with a soft, dry cloth. Acrylic and cellulose acetate plastics are thermoplastic. Friction created by buffing or polishing too long in one spot can generate sufficient heat to soften the surface. This condition produces visual distortion and should be avoided.

Windshield Installation

Use material equivalent to that originally used by the manufacturer of the aircraft for replacement panels. There are many types of transparent plastics on the market. Their properties vary greatly, particularly expansion characteristics, brittleness under low temperatures, resistance to discoloration when exposed to sunlight, surface checking, etc. Information on these properties is in MIL-HDBK-17, Plastics for Flight Vehicles, Part II Transparent Glazing Materials. These properties are considered by aircraft manufacturers in selecting materials to be used in their designs and the use of substitutes having different characteristics may result in subsequent difficulties.

Installation Procedures

When installing a replacement panel, use the same mounting method employed by the manufacturer of the aircraft. While the actual installation varies from one type of aircraft to another, consider the following major principles when installing any replacement panel.

1. Never force a plastic panel out of shape to make it fit a frame. If a replacement panel does not fit easily into the mounting, obtain a new replacement or heat the whole panel and re-form. When possible, cut and fit a new panel at ordinary room temperature.

2. In clamping or bolting plastic panels into their mountings, do not place the plastic under excessive compressive stress. It is easy to develop more than 1,000 psi on the plastic by overtorquing a nut and bolt. Tighten each nut to a firm fit, and then back the nut off one full turn (until they are snug and can still be rotated with the fingers).

3. In bolted installations, use spacers, collars, shoulders, or stop-nuts to prevent tightening the bolt excessively. Whenever such devices are used by the aircraft manufacturer, retain them in the replacement installation. It is important that the original number of bolts, complete with washers, spacers, etc., be used. When rivets are used, provide adequate spacers or other satisfactory means to prevent excessive tightening of the frame to the plastic.

4. Mount plastic panels between rubber, cork, or other gasket material to make the installation waterproof, to reduce vibration, and to help to distribute compressive stresses on the plastic.

5. Plastics expand and contract considerably more than the metal channels in which they are mounted. Mount windshield panels to a sufficient depth in the channel to prevent it from falling out when the panel contracts at low temperatures or deforms under load. When the manufacturer's original design permits, mount panels to a minimum depth of 1⅛-inches, and with a clearance of ⅛-inch between the plastic and bottom of the channel.

6. In installations involving bolts or rivets, make the holes through the plastic oversize by ⅛-inch and center so that the plastic does not bind or crack at the edge of the holes. The use of slotted holes is also recommended.

Chapter 8
Aircraft Painting & Finishing

Introduction

Paint, or more specifically its overall color and application, is usually the first impression that is transmitted to someone when they look at an aircraft for the first time. Paint makes a statement about the aircraft and the person who owns or operates it. The paint scheme may reflect the owner's ideas and color preferences for an amateur-built aircraft project, or it may be colors and identification for the recognition of a corporate or air carrier aircraft.

Paint is more than aesthetics; it affects the weight of the aircraft and protects the integrity of the airframe. The topcoat finish is applied to protect the exposed surfaces from corrosion and deterioration. Also, a properly painted aircraft is easier to clean and maintain because the exposed surfaces are more resistant to corrosion and dirt, and oil does not adhere as readily to the surface.

A wide variety of materials and finishes are used to protect and provide the desired appearance of the aircraft. The term "paint" is used in a general sense and includes primers, enamels, lacquers, and the various multipart finishing formulas. Paint has three components: resin as coating material, pigment for color, and solvents to reduce the mix to a workable viscosity.

Internal structure and unexposed components are finished to protect them from corrosion and deterioration. All exposed surfaces and components are finished to provide protection and to present a pleasing appearance. Decorative finishing includes trim striping, the addition of company logos and emblems, and the application of decals, identification numbers, and letters.

Finishing Materials

A wide variety of materials are used in aircraft finishing. Some of the more common materials and their uses are described in the following paragraphs.

Acetone

Acetone is a fast-evaporating colorless solvent. It is used as an ingredient in paint, nail polish, and varnish removers. It is a strong solvent for most plastics and is ideal for thinning fiberglass resin, polyester resins, vinyl, and adhesives. It is also used as a superglue remover. Acetone is a heavy-duty degreaser suitable for metal preparation and removing grease from fabric covering prior to doping. It should not be used

as a thinner in dope because of its rapid evaporation, which causes the doped area to cool and collect moisture. This absorbed moisture prevents uniform drying and results in blushing of the dope and a flat no-gloss finish.

Alcohol

Butanol, or butyl alcohol, is a slow-drying solvent that can be mixed with aircraft dope to retard drying of the dope film on humid days, thus preventing blushing. A mixture of dope solvent containing 5 to 10 percent of butyl alcohol is usually sufficient for this purpose. Butanol and ethanol alcohol are mixed together in ratios ranging from 1:1 to 1:3 to use to dilute wash coat primer for spray applications because the butyl alcohol retards the evaporation rate.

Ethanol or denatured alcohol is used to thin shellac for spraying and as a constituent of paint and varnish remover. It can also be used as a cleaner and degreaser prior to painting.

Isopropyl, or rubbing alcohol, can be used as a disinfectant. It is used in the formulation of oxygen system cleaning solutions. It can be used to remove grease pencil and permanent marker from smooth surfaces, or to wipe hand or fingerprint oil from a surface before painting.

Benzene

Benzene is a highly flammable, colorless liquid with a sweet odor. It is a product used in some paint and varnish removers. It is an industrial solvent that is regulated by the Environmental Protection Agency (EPA) because it is an extremely toxic chemical compound when inhaled or absorbed through the skin. It has been identified as a Class A carcinogen known to cause various forms of cancer. It should be avoided for use as a common cleaning solvent for paint equipment and spray guns.

Methyl Ethyl Ketone (MEK)

Methyl ethyl ketone (MEK), also referred to as 2-Butanone, is a highly flammable, liquid solvent used in paint and varnish removers, paint and primer thinners, in surface coatings, adhesives, printing inks, as a catalyst for polyester resin hardening, and as an extraction medium for fats, oils, waxes, and resins. Because of its effectiveness as a quickly evaporating solvent, MEK is used in formulating high solids coatings that help to reduce emissions from coating operations. Persons using MEK should use protective gloves and have adequate ventilation to avoid the possible irritation effects of skin contact and breathing of the vapors.

Methylene Chloride

Methylene chloride is a colorless, volatile liquid completely miscible with a variety of other solvents. It is widely used in paint strippers and as a cleaning agent/degreaser for metal parts. It has no flash point under normal use conditions and can be used to reduce the flammability of other substances.

Toluene

Referred to as toluol or methylbenzene, toluene is a clear, water-insoluble liquid with a distinct odor similar to that of benzene. It is a common solvent used in paints, paint thinners, lacquers, and adhesives. It has been used as a paint remover in softening fluorescent-finish, clear-topcoat sealing materials. It is also an acceptable thinner for zinc chromate primer. It has been used as an antiknocking additive in gasoline. Prolonged exposure to toluene vapors should be avoided because it may be linked to brain damage.

Turpentine

Turpentine is obtained by distillation of wood from certain pine trees. It is a flammable, water-insoluble liquid solvent used as a thinner and quick-drier for varnishes, enamels, and other oil-based paints. Turpentine can be used to clean paint equipment and paint brushes used with oil-based paints.

Mineral Spirits

Sometimes referred to as white spirit, Stoddard solvent, or petroleum spirits, mineral spirits is a petroleum distillate used as a paint thinner and mild solvent. The reference to the name Stoddard came from a dry cleaner who helped to develop it in the 1920s as a less volatile dry cleaning solvent and as an alternative to the more volatile petroleum solvents that were being used for cleaning clothes. It is the most widely used solvent in the paint industry, used in aerosols, paints, wood preservatives, lacquers, and varnishes. It is also commonly used to clean paint brushes and paint equipment. Mineral spirits are used in industry for cleaning and degreasing machine tools and parts because it is very effective in removing oils and greases from metal. It has low odor, is less flammable, and less toxic than turpentine.

Naphtha

Naphtha is one of a wide variety of volatile hydrocarbon mixtures that is sometimes processed from coal tar but more often derived from petroleum. Naphtha is used as a solvent for various organic substances, such as fats and rubber, and in the making of varnish. It is used as a cleaning fluid and is incorporated into some laundry soaps. Naphtha has a low flashpoint and is used as a fuel in portable stoves and lanterns. It is sold under different names around the world and is known as white gas, or Coleman fuel, in North America.

Linseed Oil

Linseed oil is the most commonly used carrier in oil paint. It makes the paint more fluid, transparent, and glossy. It is used to reduce semipaste oil colors, such as dull black stenciling paint and insignia colors, to a brushing consistency. Linseed oil is also used as a protective coating on the interior of metal tubing. Linseed oil is derived from pressing the dried ripe flax seeds of the flax plant to obtain the oil and then using a process called solvent extraction. Oil obtained without the solvent extraction process is marketed as flaxseed oil. The term "boiled linseed oil" indicates that it was processed with additives to shorten its drying time.

A note of caution is usually added to packaging of linseed oil with the statement, "Risk of Fire from Spontaneous Combustion Exists with this Product." Linseed oil generates heat as it dries. Oily materials and rags must be properly disposed after use to eliminate the possible cause of spontaneous ignition and fire.

Thinners

Thinners include a plethora of solvents used to reduce the viscosity of any one of the numerous types of primers, subcoats, and topcoats. The types of thinner used with the various coatings is addressed in other sections of this chapter.

Varnish

Varnish is a transparent protective finish primarily used for finishing wood. It is available in interior and exterior grades. The exterior grade does not dry as hard as the interior grade, allowing it to expand and contract with the temperature changes of the material being finished. Varnish is traditionally a combination of a drying oil, a resin, and a thinner or solvent. It has little or no color, is transparent, and has no added pigment. Varnish dries slower than most other finishes. Resin varnishes dry and harden when the solvents in them evaporate. Polyurethane and epoxy varnishes remain liquid after the evaporation of the solvent but quickly begin to cure through chemical reactions of the varnish components.

Primers

The importance of primers in finishing and protection is generally misunderstood and underestimated because it is invisible after the topcoat finish is applied. A primer is the foundation of the finish. Its role is to bond to the surface, inhibit corrosion of metal, and provide an anchor point for the finish coats. It is important that the primer pigments be either anodic to the metal surface or passivate the surface should moisture be present. The binder must be compatible with the finish coats. Primers on nonmetallic surfaces do not require sacrificial or passivating pigments. Some of the various primer types are discussed below.

Wash Primers

Wash primers are water-thin coatings of phosphoric acid in solutions of vinyl butyral resin, alcohol, and other ingredients. They are very low in solids with almost no filling qualities. Their functions are to passivate the surface, temporarily provide corrosion resistance, and provide an adhesive base for the next coating, such as a urethane or epoxy primer. Wash primers do not require sanding and have high-corrosion protection qualities. Some have a very small recoat time frame that must be considered when painting larger aircraft. The manufacturers' instructions must be followed for satisfactory results.

Red Iron Oxide

Red oxide primer is an alkyd resin-based coating that was developed for use over iron and steel located in mild environmental conditions. It can be applied over rust that is free of loose particles, oil, and grease. It has limited use in the aviation industry.

Gray Enamel Undercoat

This is a single component, nonsanding primer compatible with a wide variety of topcoats. It fills minor imperfections, dries fast without shrinkage, and has high-corrosion resistance. It is a good primer for composite substrates.

Urethane

This is a term that is misused or interchanged by painters and manufacturers alike. It is typically a two-part product that uses a chemical activator to cure by linking molecules together to form a whole new compound. Polyurethane is commonly used when referring to urethane, but not when the product being referred to is acrylic urethane.

Urethane primer, like the urethane paint, is also a two-part product that uses a chemical activator to cure. It is easy to sand and fills well. The proper film thickness must be observed, because it can shrink when applied too heavily. It is typically applied over a wash primer for best results. Special precautions must be taken by persons spraying because the activators contain isocyanates (discussed further in the Protective Equipment section at the end of this chapter).

Epoxy

Epoxy is a synthetic, thermosetting resin that produces tough, hard, chemical-resistant coatings and adhesives. It uses a catalyst to chemically activate the product, but it is not classified as hazardous because it contains no isocyanates. Epoxy can be used as a nonsanding primer/sealer over bare metal and it is softer than urethane, so it has good chip resistance. It is recommended for use on steel tube frame aircraft prior to installing fabric covering.

Zinc Chromate

Zinc chromate is a corrosion-resistant pigment that can be added to primers made of different resin types, such as epoxy, polyurethane, and alkyd. Older type zinc chromate is distinguishable by its bright yellow color when compared to the light green color of some of the current brand primers. Moisture in the air causes the zinc chromate to react with the metal surface, and it forms a passive layer that prevents corrosion. Zinc chromate primer was, at one time, the standard primer for aircraft painting. Environmental concerns and new formula primers have all but replaced it.

Identification of Paints

Dope

When fabric-covered aircraft ruled the sky, dope was the standard finish used to protect and color the fabric. The dope imparted additional qualities of increased tensile strength, airtightness, weather-proofing, ultraviolet (UV) protection, and tautness to the fabric cover. Aircraft dope is essentially a colloidal solution of cellulose acetate or nitrate combined with plasticizers to produce a smooth, flexible, homogeneous film.

Dope is still used on fabric covered aircraft as part of a covering process. However, the type of fabric being used to cover the aircraft has changed. Grade A cotton or linen was the standard covering used for years, and it still may be used if it meets the requirements of the Federal Aviation Administration (FAA), Technical Standard Order (TSO) C-15d/AMS 3806c.

Polyester fabric coverings now dominate in the aviation industry. These new fabrics have been specifically developed for aircraft and are far superior to cotton and linen. The protective coating and topcoat finishes used with the Ceconite® polyester fabric covering materials are part of a Supplemental Type Certificate (STC) and must be used as specified when covering any aircraft with a Standard Airworthiness Certificate. The Ceconite® covering procedures use specific brand name, nontautening nitrate and butyrate dope as part of the STC.

The Poly-Fiber® system also uses a special polyester fabric covering as part of its STC, but it does not use dope. All the liquid products in the Poly-Fiber® system are made from vinyl, not from cellulose dope. The vinyl coatings have several real advantages over dope: they remain flexible, they do not shrink, they do not support combustion, and they are easily removed from the fabric with MEK, which simplifies most repairs.

Synthetic Enamel

Synthetic enamel is an oil-based, single-stage paint (no clear coat) that provides durability and protection. It can be mixed with a hardener to increase the durability and shine while decreasing the drying time. It is one of the more economical types of finish.

Lacquers

The origin of lacquer dates back thousands of years to a resin obtained from trees indigenous to China. In the early 1920s, nitrocellulose lacquer was developed from a process using cotton and wood pulp.

Nitrocellulose lacquers produce a hard, semiflexible finish that can be polished to a high sheen. The clear variety yellows as it ages, and it can shrink over time to a point that the surface crazes. It is easy to spot repair because each new coat of lacquer softens and blends into the previous coat. This was one of the first coatings used by the automotive industry in mass production, because it reduced finishing times from almost two weeks to two days.

Acrylic lacquers were developed to eliminate the yellowing problems and crazing of the nitrocellulose lacquers. General Motors started using acrylic lacquer in the mid-1950s, and they used it into the 1960s on some of their premium model cars. Acrylics have the same working properties but dry to a less brittle and more flexible film than nitrocellulose lacquer.

Lacquer is one of the easiest paints to spray, because it dries quickly and can be applied in thin coats. However, lacquer is not very durable; bird droppings, acid rain, and gasoline spills actually eat down into the paint. It still has limited use on collector and show automobiles because they are usually kept in a garage, protected from the environment.

The current use of lacquer for an exterior coating on an aircraft is almost nonexistent because of durability and environmental concerns. Upwards of 85 percent of the volatile organic compounds (VOCs) in the spray gun ends up in the atmosphere, and some states have banned its use.

There are some newly developed lacquers that use a catalyst, but they are used mostly in the woodworking and furniture industry. They have the ease of application of nitrocellulose lacquer with much better water, chemical, and abrasion resistance. Additionally, catalyzed lacquers cure chemically, not solely through the evaporation of solvents, so there is a reduction of VOCs released into the atmosphere. It is activated when the catalyst is added to the base mixture.

Polyurethane

Polyurethane is at the top of the list when compared to other coatings for abrasion-, stain-, and chemical-resistant properties. Polyurethane was the coating that introduced the wet look. It has a high degree of natural resistance to the damaging effects of UV rays from the sun. Polyurethane is usually the first choice for coating and finishing the corporate and commercial aircraft in today's aviation environment.

Urethane Coating

The term urethane applies to certain types of binders used for paints and clear coatings. (A binder is the component that holds the pigment together in a tough, continuous film and provides film integrity and adhesion.) Typically, urethane is a two-part coating that consists of a base and catalyst that, when mixed, produces a durable, high-gloss finish that is abrasion- and chemical-resistant.

Acrylic Urethanes

Acrylic simply means plastic. It dries to a harder surface but is not as resistant to harsh chemicals as polyurethane. Most acrylic urethanes need additional UV inhibitors added when subject to the UV rays of the sun.

Methods of Applying Finish

There are several methods of applying aircraft finish. Among the most common are dipping, brushing, and spraying.

Dipping

The application of finishes by dipping is generally confined to factories or large repair stations. The process consists of dipping the part to be finished in a tank filled with the finishing material. Primer coats are frequently applied in this manner.

Brushing

Brushing has long been a satisfactory method of applying finishes to all types of surfaces. Brushing is generally used for small repair work and on surfaces where it is not practicable to spray paint.

The material to be applied should be thinned to the proper consistency for brushing. A material that is too thick has a tendency to pull or rope under the brush. If the materials are too thin, they are likely to run or not cover the surface adequately. Proper thinning and substrate temperature allows the finish to flow-out and eliminates the brush marks.

Spraying

Spraying is the preferred method for a quality finish. Spraying is used to cover large surfaces with a uniform layer of material, which results in the most cost effective method of application. All spray systems have several basic similarities. There must be an adequate source of compressed air, a reservoir or feed tank to hold a supply of the finishing material, and a device for controlling the combination of the air and finishing material ejected in an atomized cloud or spray against the surface to be coated.

A self-contained, pressurized spray can of paint meets the above requirements and satisfactory results can be obtained painting components and small areas of touchup. However, the aviation coating materials available in cans is limited, and this chapter addresses the application of mixed components through a spray gun.

There are two main types of spray equipment. A spray gun with an integral paint container is adequate for use when painting small areas. When large areas are painted, pressure-feed equipment is more desirable since a large supply of finishing material can be applied without the interruption of having to stop and refill a paint container. An added bonus is the lighter overall weight of the spray gun and the flexibility of spraying in any direction with a constant pressure to the gun.

The air supply to the spray gun must be entirely free of water or oil in order to produce the optimum results in the finished product. Water traps, as well as suitable filters to remove any trace of oil, must be incorporated in the air pressure supply line. These filters and traps must be serviced on a regular basis.

Finishing Equipment

Paint Booth
A paint booth may be a small room in which components of an aircraft are painted, or it can be an aircraft hangar big enough to house the largest aircraft. Whichever it is, the location must be able to protect the components or aircraft from the elements. Ideally, it would have temperature and humidity controls; but, in all cases, the booth or hangar must have good lighting, proper ventilation, and be dust free.

A simple paint booth can be constructed for a small aircraft by making a frame out of wood or polyvinyl chloride (PVC) pipe. It needs to be large enough to allow room to walk around and maneuver the spray gun. The top and sides can be covered with plastic sheeting stapled or taped to the frame. An exhaust fan can be added to one end with a large air-conditioning filter placed on the opposite end to filter incoming air. Lights should be large enough to be set up outside of the spray booth and shine through the sheeting or plastic windows. The ideal amount of light would be enough to produce a glare off of all the surfaces to be sprayed. This type of temporary booth can be set up in a hangar, a garage, or outside on a ramp, if the weather and temperature are favorable.

Normally, Environmental Protection Agency (EPA) regulations do not apply to a person painting one airplane. However, anyone planning to paint an aircraft should be aware that local clean air regulations may be applicable to an airplane painting project. When planning to paint an aircraft at an airport, it would be a good idea to check with the local airport authority before starting.

Air Supply
The air supply for paint spraying using a conventional siphon feed spray gun should come from an air compressor with a storage tank big enough to provide an uninterrupted supply of air with at least 90 pounds per square inch (psi) providing 10 cubic feet per minute (CFM) of air to the spray gun.

The compressor needs to be equipped with a regulator, water trap, air hose, and an adequate filter system to ensure that clean, dry, oil-free air is delivered to the spray gun.

If using one of the newer high-volume low-pressure (HVLP) spray guns and using a conventional compressor, it is better to use a two-stage compressor of at least a 5 horsepower (hp) that operates at 90 psi and provides 20 CFM to the gun. The key to the operation of the newer HVLP spray guns is the air volume, not the pressure.

If purchasing a new complete HVLP system, the air supply is from a turbine compressor. An HVLP turbine has a series of fans, or stages, that move a lot of air at low pressure. More stages provide greater air output (rated in CFM), which means better atomization of the coating being sprayed. The intake air is also the cooling air for the motor. This air is filtered from dirt and dust particles prior to entering the turbine. Some turbines also have a second filter for the air supply to the spray gun. The turbine does not produce oil or water to contaminate the air supply, but the air supply from the turbine heats up, causing the paint to dry faster, so you may need an additional length of hose to reduce the air temperature at the spray gun.

Spray Equipment
Air Compressors
Piston–type compressors are available with one-stage and multiple-stage compressors, various size motors, and various size supply tanks. The main requirement for painting is to ensure the spray gun has a continuous supplied volume of air. Piston-type compressors compress air and deliver it to a storage tank. Most compressors provide over 100 psi, but only the larger ones provide the volume of air needed for an uninterrupted supply to the gun. The multistage compressor is a good choice for a shop when a large volume of air is needed for pneumatic tools. When in doubt about the size of the compressor, compare the manufacturer's specifications and get the largest one possible. *[Figure 8-1]*

Large Coating Containers

For large painting projects, such as spraying an entire aircraft, the quantity of mixed paint in a pressure tank provides many advantages. The setup allows a greater area to be covered without having to stop and fill the cup on a spray gun. The painter is able to keep a wet paint line, and more material is applied to the surface with less overspray. It provides the flexibility of maneuvering the spray gun in any position without the restriction and weight of an attached paint cup. Remote pressure tanks are available in sizes from 2 quarts to over 60 gallons. *[Figure 8-2]*

System Air Filters

The use of a piston-type air compressor for painting requires that the air supply lines include filters to remove water and oil. A typical filter assembly is shown in *Figure 8-3*.

Miscellaneous Painting Tools & Equipment

Some tools that are available to the painter include:

- Masking paper/tape dispenser that accommodates various widths of masking paper. It includes a masking tape dispenser that applies the tape to one edge of the paper as it is rolled off to facilitate one person applying the paper and tape in a single step.

- Electronic and magnetic paint thickness gauges to measure dry paint thickness.

- Wet film gauges to measure freshly applied wet paint.

- Infrared thermometers to measure coating and substrate surfaces to verify that they fall in the recommended temperature range prior to spraying.

Figure 8-2. *Pressure paint tank.*

Spray Guns

A top quality spray gun is a key component in producing a quality finish in any coating process. It is especially important when painting an aircraft because of the large area and varied surfaces that must be sprayed.

When spray painting, it is of utmost importance to follow the manufacturer's recommendations for correct sizing of the air cap, fluid tip, and needle combinations. The right combination provides the best coverage and the highest quality finish in the shortest amount of time.

All of the following examples of the various spray guns

Figure 8-1. *Standard air compressor.*

Figure 8-3. *Air line filter assembly.*

(except the airless) are of the air atomizing type. They are the most capable of providing the highest quality finish.

Siphon-Feed Gun

The siphon-feed gun is a conventional spray gun familiar to most people, with a one quart paint cup located below the gun. Regulated air passes through the gun and draws (siphons) the paint from the supply cup. This is an external mix gun, which means the air and fluid mix outside the air cap. This gun applies virtually any type coating and provides a high quality finish. *[Figure 8-4]*

Gravity-Feed Gun

A gravity-feed gun provides the same high-quality finish as a siphon-feed gun, but the paint supply is located in a cup on top of the gun and supplied by gravity. The operator can make fine adjustments between the atomizing pressure and fluid flow and utilize all material in the cup. This also is an external mix gun. *[Figure 8-5]*

The HVLP production spray gun is an internal mix gun. The air and fluid is mixed inside the air cap. Because of the low pressure used in the paint application, it transfers at least 65 percent and upwards of 80 percent of the finish material to the surface. HVLP spray guns are available with a standard cup located underneath or in a gravity-feed model with the cup on top. The sample shown can be connected with hoses to a remote paint material container holding from 2 quarts to 60 gallons. *[Figure 8-6]*

Because of more restrictive EPA regulations, and the fact that more paint is being transferred to the surface with less waste from overspray, a large segment of the paint and coating

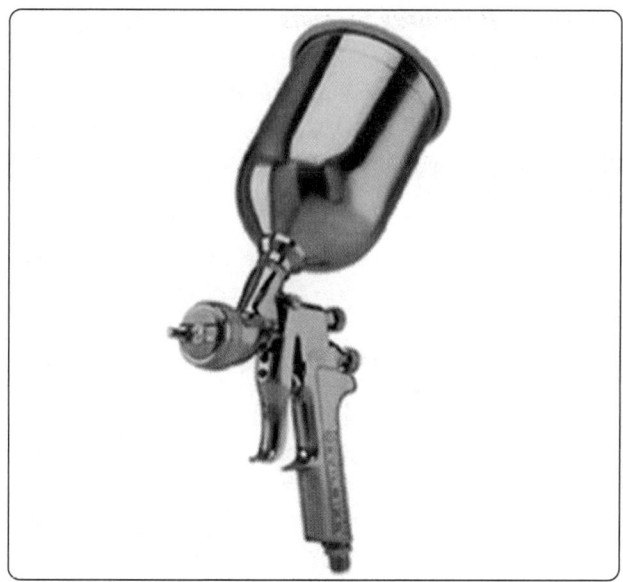

Figure 8-5. *Gravity-feed spray gun.*

Figure 8-6. *A high-volume, low-pressure (HVLP) spray gun.*

industry is switching to HVLP spray equipment.

Airless spraying does not directly use compressed air to atomize the coating material. A pump delivers paint to the spray gun under high hydraulic pressure (500 to 4,500 psi) to atomize the fluid. The fluid is then released through an orifice in the spray nozzle. This system increases transfer efficiency and production speed with less overspray than conventional air atomized spray systems. It is used for production work but does not provide the fine finish of air atomized systems. *[Figure 8-7]*

Figure 8-4. *Siphon-feed spray gun.*

Fresh Air Breathing Systems

Fresh air breathing systems should be used whenever coatings are being sprayed that contain isocyanides. This includes all polyurethane coatings. The system incorporates a high-capacity electric air turbine that provides a constant source of fresh air to the mask. The use of fresh air breathing systems is also highly recommended when spraying chromate primers and chemical stripping aircraft. The system provides cool filtered breathing air with up to 200 feet of hose, which allows the air pump intake to be placed in an area of fresh air, well outside of the spraying area. *[Figure 8-8]*

A charcoal-filtered respirator should be used for all other spraying and sanding operations to protect the lungs and respiratory tract. The respirator should be a double-cartridge, organic vapor type that provides a tight seal around the nose and mouth. The cartridges can be changed separately, and should be changed when detecting odor or experiencing nose or throat irritation. The outer prefilters should be changed if experiencing increased resistance to breathing. *[Figure 8-9]*

Viscosity Measuring Cup

This is a small cup with a long handle and a calibrated orifice in the bottom that allows the liquid in the cup to drain out at a specific timed rate. Coating manufacturers recommend spraying their product at a specific pressure and viscosity. That viscosity is determined by measuring the efflux (drain) time of the liquid coating through the cup orifice. The time (in seconds) is listed on most paint manufacturers' product/technical data pages. The measurement determines if the mixed coating meets the recommended viscosity for spraying.

There are different manufacturers of the viscosity measuring

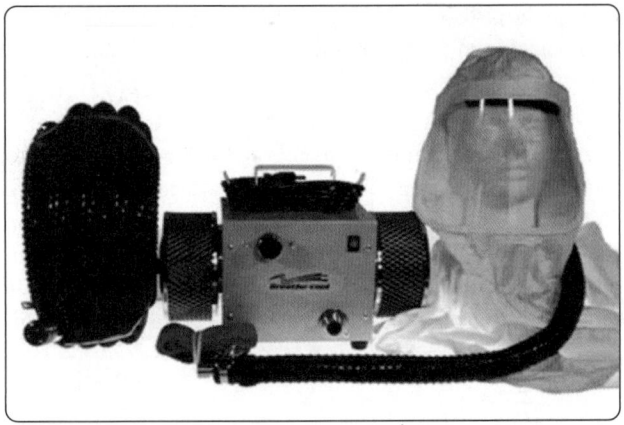

Figure 8-8. *Breathe-Cool II® supplied air respirator system with Tyvek® hood.*

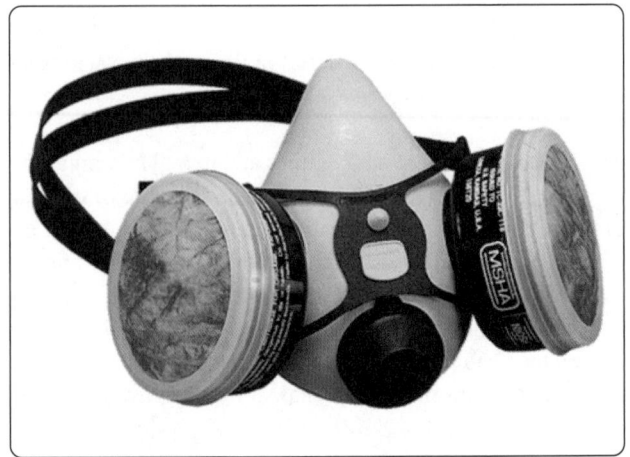

Figure 8-9. *Charcoal-filtered respirator.*

devices, but the most common one listed and used for spray painting is known as a Zahn cup. The orifice number must correspond to the one listed on the product/technical data sheet. For most primers and topcoats, the #2 or #3 Zahn cup is the one recommended. *[Figure 8-10]*

To perform an accurate viscosity measurement, it is very important that the temperature of the sample material be within the recommended range of 73.5 °F ± 3.5 °F (23 °C ± 2 °C), and then proceed as follows:

Figure 8-7. *Airless spray gun.*

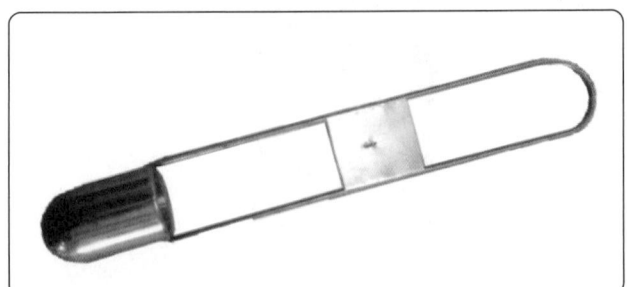

Figure 8-10. *A Zahn cup viscosity measuring cup.*

1. Thoroughly mix the sample with minimum bubbles.

2. Dip the Zahn cup vertically into the sample being tested, totally immersing the cup below the surface.

3. With a stopwatch in one hand, briskly lift the cup out of the sample. As the top edge of the cup breaks the surface, start the stopwatch.

4. Stop the stopwatch when the first break in the flow of the liquid is observed at the orifice exit. The number in seconds is referred to as the efflux time.

5. Record the time on the stopwatch and compare it to the coating manufacturer's recommendation. Adjust the viscosity, if necessary, but be aware not to thin the coating below recommendations that could result in the release of VOCs into the atmosphere above the regulated limitations.

Mixing Equipment

Use a paint shaker for all coatings within 5 days of application to ensure the material is thoroughly mixed. Use a mechanical paint stirrer to mix larger quantities of material. If a mechanical stirrer is driven by a drill, the drill should be pneumatic, instead of electric. The sparks from an electric drill can cause an explosion from the paint vapors.

Preparation

Surfaces

The most important part of any painting project is the preparation of the substrate surface. It takes the most work and time, but with the surface properly prepared, the results are a long-lasting, corrosion-free finish. Repainting an older aircraft requires more preparation time than a new paint job because of the additional steps required to strip the old paint, and then clean the surface and crevices of paint remover. Paint stripping is discussed in another section of this chapter.

It is recommended that all the following procedures be performed using protective clothing, rubber gloves, and goggles, in a well-ventilated area, at temperatures between 68 °F and 100 °F.

Aluminum surfaces are the most common on a typical aircraft. The surface should be scrubbed with Scotch-Brite® pads using an alkaline aviation cleaner. The work area should be kept wet and rinsed with clean water until the surface is water break free. This means that there are no beads or breaks in the water surface as it flows over the aluminum surface.

The next step is to apply an acid etch solution to the surface. Following manufacturers' suggestions, this is applied like a wash using a new sponge and covering a small area while keeping it wet and allowing it to contact the surface for between 1 and 2 minutes. It is then rinsed with clean water without allowing the solution to dry on the surface. Continue this process until all the aluminum surfaces are washed and rinsed. Extra care must be taken to thoroughly rinse this solution from all the hidden areas that it may penetrate. It provides a source for corrosion to form if not completely removed.

When the surfaces are completely dry from the previous process, the next step is to apply Alodine® or another type of an aluminum conversion coating. This coating is also applied like a wash, allowing the coating to contact the surface and keeping it wet for 2 to 5 minutes without letting it dry. It then must be thoroughly rinsed with clean water to remove all chemical salts from the surface. Depending on the brand, the conversion coating may color the aluminum a light gold or green, but some brands are colorless. When the surface is thoroughly dry, the primer should be applied as soon as possible as recommended by the manufacturer.

The primer should be one that is compatible with the topcoat finish. Two-part epoxy primers provide excellent corrosion resistance and adhesion for most epoxy and urethane surfaces and polyurethane topcoats. Zinc chromate should not be used under polyurethane paints.

Composite surfaces that need to be primed may include the entire aircraft if it is constructed from those materials, or they may only be components of the aircraft, such as fairings, radomes, antennas, and the tips of the control surfaces.

Epoxy sanding primers have been developed that provide an excellent base over composites and can be finish sanded with 320 grit using a dual action orbital sander. They are compatible with two-part epoxy primers and polyurethane topcoats.

Topcoats must be applied over primers within the recommended time window, or the primer may have to be scuff sanded before the finish coat is applied. Always follow the recommendations of the coating manufacturer.

Primer & Paint

Purchase aircraft paint for the aviation painting project. Paint manufacturers use different formulas for aircraft and automobiles because of the environments they operate in. The aviation coatings are formulated to have more flexibility and chemical resistance than the automotive paint.

It is also highly recommended that compatible paints of the same brand are used for the entire project. The complete system (of a particular brand) from etching to primers and reducers to the finish topcoat are formulated to work together. Mixing brands is a risk that may ruin the entire project.

When purchasing the coatings for a project, always request a manufacturer's technical or material data and safety data sheets, for each component used. Before starting to spray, read the sheets. If the manufacturer's recommendations are not followed, a less than satisfactory finish or a hazard to personal safety or the environment may result. It cannot be emphasized enough to follow the manufacturer's recommendations. The finished result is well worth the effort.

Before primer or paint is used for any type application, it must be thoroughly mixed. This is done so that any pigment that may have settled to the bottom of the container is brought into suspension and distributed evenly throughout the paint. Coatings now have shelf lives listed in their specification sheets. If a previously opened container is found to have a skin or film formed over the primer or paint, the film must be completely removed before mixing. The material should not be used if it has exceeded its shelf life and/or has become thick or jelled.

Mechanical shaking is recommended for all coatings within 5 days of use. After opening, a test with a hand stirrer should be made to ensure that all the pigment has been brought into suspension. Mechanical stirring is recommended for all two-part coatings. When mixing any two-part paint, the catalyst/activator should always be added to the base or pigmented component. The technical or material data sheet of the coating manufacturer should be followed for recommended times of induction (the time necessary for the catalyst to react with the base prior to application). Some coatings do not require any induction time after mixing, and others need 30 minutes of reaction time before being applied.

Thinning of the coating material should follow the recommendations of the manufacturer. The degree of thinning depends on the method of application. For spray application, the type of equipment, air pressure, and atmospheric conditions guide the selection and mixing ratios for the thinners. Because of the importance of accurate thinning to the finished product, use a viscosity measuring (flow) cup. Material thinned using this method is the correct viscosity for the best application results.

Thin all coating materials and mix in containers separate from the paint cup or pot. Then, filter the material through a paint strainer recommended for the type coating you are spraying as you pour it into the cup or supply pot.

Spray Gun Operation

Adjusting the Spray Pattern

To obtain the correct spray pattern, set the recommended air pressure on the gun, usually 40 to 50 psi for a conventional gun. Test the pattern of the gun by spraying a piece of masking paper taped to the wall. Hold the gun square to the wall approximately 8 to 10 inches from the surface. (With hand spread, it is the distance from the tip of the thumb to the tip of the little finger.)

All spray guns (regardless of brand name) have the same type of adjustments. The upper control knob proportions the air flow, adjusting the spray pattern of the gun. *[Figure 8-11]* The lower knob adjusts the fluid passing the needle, which in turn controls the amount or volume of paint being delivered through the gun.

Pull the trigger lever fully back. Move the gun across the paper, and alternately adjust between the two knobs to obtain a spray fan of paint that is wet from top to bottom (somewhat like the pattern at dial 10.) Turning in (to the right) on the lower, or fluid knob, reduces the amount of paint going through the gun. Turning out increases the volume of paint. Turning out (to the left) on the upper, or pattern control knob, widens the spray pattern. Turning in reduces it to a cone shape (as shown with dial set at 0).

Once the pattern is set on the gun, the next step is to follow the correct spraying technique for applying the coating to the surface.

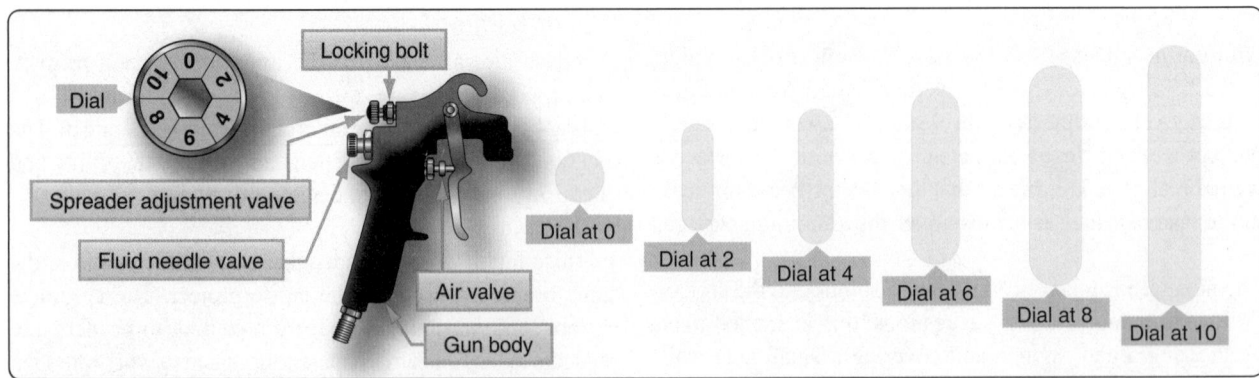

Figure 8-11. *Adjustable spray pattern.*

Applying the Finish

If the painter has never used a spray gun to apply a finish coat of paint, and the aircraft has been completely prepared, cleaned, primed, and ready for the topcoat, they may need to pause for some practice. Reading a book or an instruction manual is a good start as it provides the basic knowledge about the movement of the spray gun across the surface. Also, if available, the opportunity to observe an aircraft being painted is well worth the time.

At this point in the project, the aircraft has already received its primer coats. The difference between the primer and the finish topcoat is that the primer is flat (no gloss) and the finish coat has a glossy surface (some more than others, depending on the paint). The flat finish of the primer is obtained by paying attention to the basics of trigger control distance from the surface and consistent speed of movement of the spray gun across the surface.

Primer is typically applied using a crosscoat spray pattern. A crosscoat is one pass of the gun from left to right, followed by another pass moving up and down. The starting direction does not matter as long as the spraying is accomplished in two perpendicular passes. The primer should be applied in light coats as cross-coating is the application of two coats of primer.

Primer does not tend to run because it is applied in light coats. The gloss finish requires a little more experience with the gun. A wetter application produces the gloss, but

the movement of the gun, overlap of the spray pattern, and the distance from the surface all affect the final product. It is very easy to vary one or another, yielding runs or dry spots and a less than desirable finish. Practice not only provides some experience, but also provides the confidence needed to produce the desired finish.

Start the practice by spraying the finish coat material on a flat, horizontal panel. The spray pattern has been already adjusted by testing it on the masking paper taped to the wall. Hold the gun 8–10 inches away from and perpendicular to the surface. Pull the trigger enough for air to pass through the cap and start a pass with the gun moving across the panel. As it reaches the point to start painting, squeeze the trigger fully back and continue moving the gun about one foot per second across the panel until the end is reached. Then, release the trigger enough to stop the paint flow but not the air flow. *[Figure 8-12]*

The constant air flow through the gun maintains a constant pressure, rather than a buildup of pressure each time that the trigger is released. This would cause a buildup of paint at the end of each pass, causing runs and sags in the finish. Repeat the sequence of the application, moving back in the opposite direction and overlapping the first pass by 50 percent. This is accomplished by aiming the center of the spray pattern at the outer edge of the first pass and continuing the overlap with each successive pass of the gun.

Once the painter has mastered spraying a flat horizontal panel, practice next on a panel that is positioned vertically against a wall. This is the panel that shows the value of applying a light

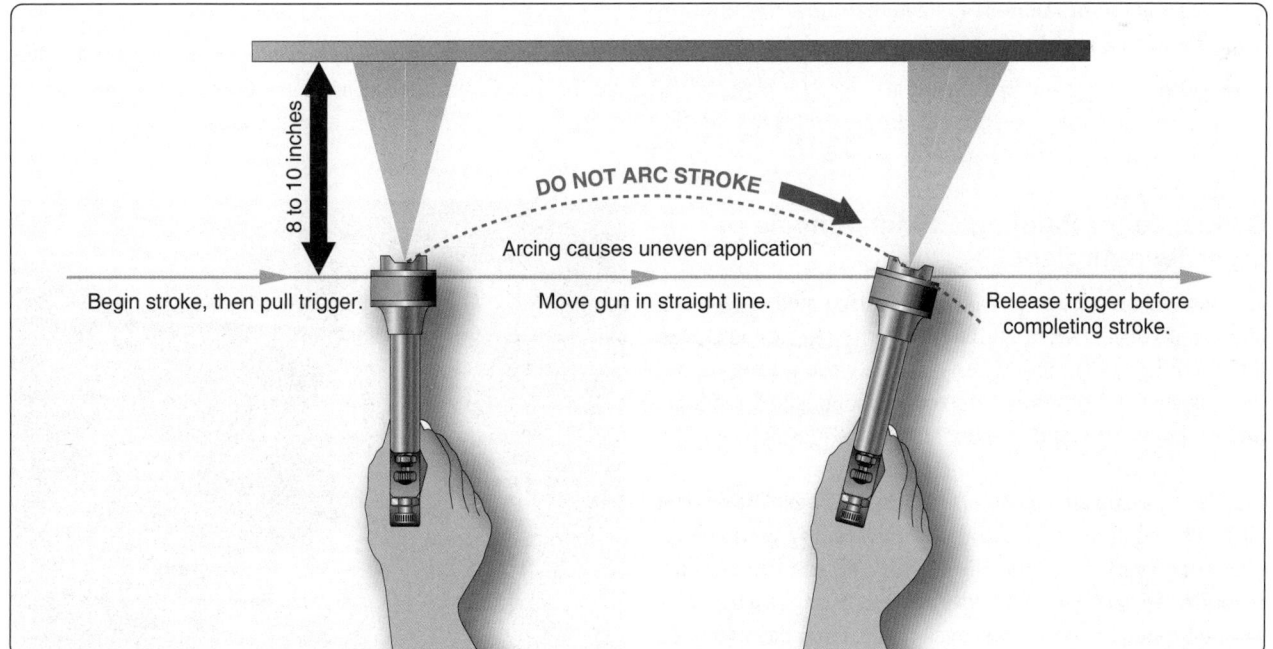

Figure 8-12. *Proper spray application.*

tack coat before spraying on the second coat. The tack coat holds the second coat from sagging and runs. Practice spraying this test panel both horizontally with overlapping passes and then rotate the air cap 90° on the gun and practice spraying vertically with the same 50 percent overlapping passes.

Practice cross-coating the paint for an even application. Apply two light spray passes horizontally, overlapping each by 50 percent, and allowing it to tack. Then, spray vertically with overlapping passes, covering the horizontal sprayed area. When practice results in a smooth, glossy, no-run application on the vertical test panel, you are ready to try your skill on the actual project.

Common Spray Gun Problems

A quick check of the spray pattern can be verified before using the gun by spraying some thinner or reducer, compatible with the finish used, through the gun. It is not of the same viscosity as the coating, but it indicates if the gun is working properly before the project is started.

If the gun is not working properly, use the following information to troubleshoot the problem:

- A pulsating, or spitting, fan pattern may be caused by a loose nozzle, clogged vent hole on the supply cup, or the packing may be leaking around the needle.

- If the spray pattern is offset to one side or the other, the air ports in the air cap or the ports in the horns may be plugged.

- If the spray pattern is heavy on the top or the bottom, rotate the air cap 180°. If the pattern reverses, the air cap is the problem. If it stays the same, the fluid tip or needle may be damaged.

- Other spray pattern problems may be a result of improper air pressure, improper reducing of the material, or wrong size spray nozzle.

Sequence for Painting a Single-Engine or Light Twin Airplane

As a general practice on any surface being painted, spray each application of coating in a different direction to facilitate even and complete coverage. After you apply the primer, apply the tack coat and subsequent top coats in opposite directions, one coat vertically and the next horizontally, as appropriate.

Start by spraying all the corners and gaps between the control surfaces and fixed surfaces. Paint the leading and trailing edges of all surfaces. Spray the landing gear and wheel wells, if applicable, and paint the bottom of the fuselage up the sides to a horizontal break, such as a seam line. Paint the underside of the horizontal stabilizer. Paint the vertical stabilizer and the rudder, and then move to the top of the horizontal stabilizer. Spray the top and sides of the fuselage down to the point of the break from spraying the underside of the fuselage. Then, spray the underside of the wings. Complete the job by spraying the top of the wings.

The biggest challenge is to control the overspray and keep the paint line wet. The ideal scenario would be to have another experienced painter with a second spray gun help with the painting. It is much easier to keep the paint wet and the job is completed in half the time.

Common Paint Troubles

Common problems that may occur during the painting of almost any project but are particularly noticeable and troublesome on the surfaces of an aircraft include poor adhesion, blushing, pinholes, sags and/or runs, "orange peel," fisheyes, sanding scratches, wrinkling, and spray dust.

Poor Adhesion

- Improper cleaning and preparation of the surface to be finished.

- Application of the wrong primer.

- Incompatibility of the topcoat with the primer. [Figure 8-13]

- Improper thinning of the coating material or selection of the wrong grade reducer.

- Improper mixing of materials.

- Contamination of the spray equipment and/or air supply.

Correction for poor adhesion requires a complete removal of the finish, a determination and correction of the cause, and a complete refinishing of the affected area.

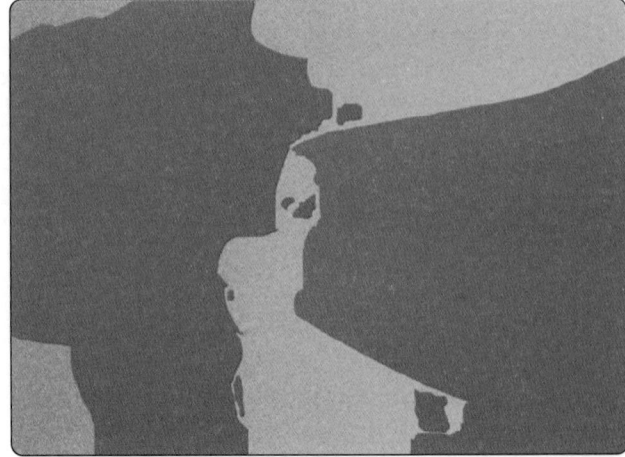

Figure 8-13. *Example of poor adhesion.*

Blushing

Blushing is the dull milky haze that appears in a paint finish. *[Figure 8-14]* It occurs when moisture is trapped in the paint. Blushing forms when the solvents quickly evaporate from the sprayed coating, causing a drop in temperature that is enough to condense the water in the air. It usually forms when the humidity is above 80 percent. Other causes include:

- Incorrect temperature (below 60 °F or above 95 °F).
- Incorrect reducer (fast drying) being used.
- Excessively high air pressure at the spray gun.

If blushing is noticed during painting, a slow-drying reducer can sometimes be added to the paint mixture, and then the area resprayed. If blushing is found after the finish has dried, the area must be sanded down and repainted.

Pinholes

Pinholes are tiny holes, or groups of holes, that appear in the surface of the finish as a result of trapped solvents, air, or moisture. *[Figure 8-15]* Examples include:

- Contaminants in the paint or air lines.
- Poor spraying techniques that allow excessively heavy or wet paint coats, which tend to trap moisture or solvent under the finish.
- Use of the wrong thinner or reducer, either too fast by quick drying the surface and trapping solvents or too slow and trapping solvents by subsequent topcoats.

If pinholes occur during painting, the equipment and painting technique must be evaluated before continuing. When dry, sand the surface smooth and then repaint.

Figure 8-15. *Example of pinholes.*

Sags & Runs

Sags and runs are usually caused by applying too much paint to an area, by holding the spray gun too close to the surface, or moving the gun too slowly across the surface. *[Figure 8-16]* Other causes include:

- Too much reducer in the paint (too thin).
- Incorrect spray gun setting of air-paint mixture.

Sags and runs can be avoided by following the recommended thinning instructions for the coatings being applied and taking care to use the proper spray gun techniques, especially on vertical surfaces and projected edges. Dried sags and runs must be sanded out and the surface repainted.

Orange Peel

"Orange peel" refers to the appearance of a bumpy surface, much like the skin of an orange. *[Figure 8-17]* It can be the result of a number of factors with the first being the improper adjustment of the spray gun. Other causes include:

Figure 8-14. *Example of blushing.*

Figure 8-16. *Example of sags and runs.*

Figure 8-17. *Example of orange peel.*

- Not enough reducer (too thick) or the wrong type reducer for the ambient temperature.

- Material not uniformly mixed.

- Forced drying method, either with fans or heat, is too quick.

- Too little flash time between coats.

- Spray painting when the ambient or substrate temperature is either too hot or too cold.

Light orange peel can be wet sanded or buffed out with polishing compound. In extreme cases, it has to be sanded smooth and resprayed.

Fisheyes

Fisheyes appear as small holes in the coating as it is being applied, which allows the underlying surface to be seen. *[Figure 8-18]* Usually, it is due to the surface not being cleaned of all traces of silicone wax. If numerous fisheyes appear when spraying a surface, stop spraying and clean off all the wet paint. Then, thoroughly clean the surface to remove all traces of silicone with a silicone wax remover.

The most effective way to eliminate fisheyes is to ensure that the surface about to be painted is clean and free from any type of contamination. A simple and effective way to check this is referred to as a water break test. Using clean water, spray, pour, or gently hose down the surface to be painted. If the water beads up anywhere on the surface, it is not clean. The water should flatten out and cover the area with an unbroken film.

If the occasional fisheye appears when spraying, wait until the first coat sets up and then add a recommended amount of fisheye eliminator to the subsequent finish coats. Fisheyes may appear during touchup of a repair. A coat of sealer may help, but a complete removal of the finish may be the only solution.

One last check before spraying is to ensure that the air compressor has been drained of water, the regulator cleaned, and the system filters are clean or have been replaced so that this source of contamination is eliminated.

Sanding Scratches

Sanding scratches appear in the finish paint when the surface has not been properly sanded and/or sealed prior to spraying the finish coats. *[Figure 8-19]* This usually shows up in nonmetal surfaces. Composite cowling, wood surfaces, and plastic fairings must be properly sanded and sealed before painting. The scratches may also appear if an overly rapid quick-drying thinner is used.

The only fix after the finish coat has set up is to sand down the affected areas using a finer grade of sandpaper, follow with a recommended sealer, and then repaint.

Wrinkling

Wrinkling is usually caused by trapped solvents and unequal drying of the paint finish due to excessively thick or solvent-

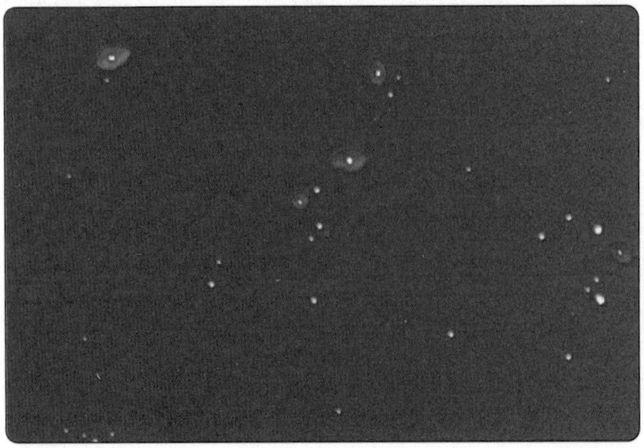

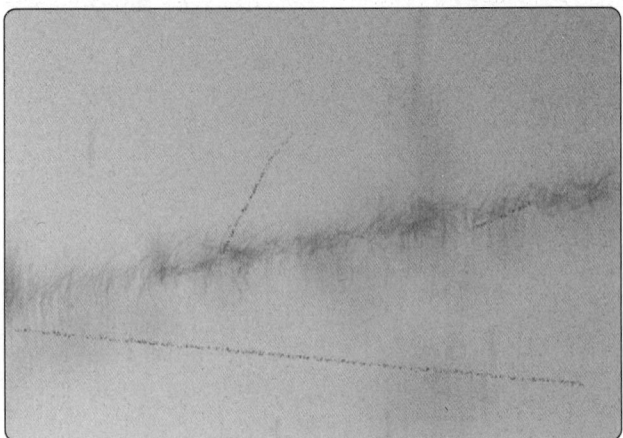

Figure 8-18. *Example of fisheyes.*

Figure 8-19. *Example of sanding scratches.*

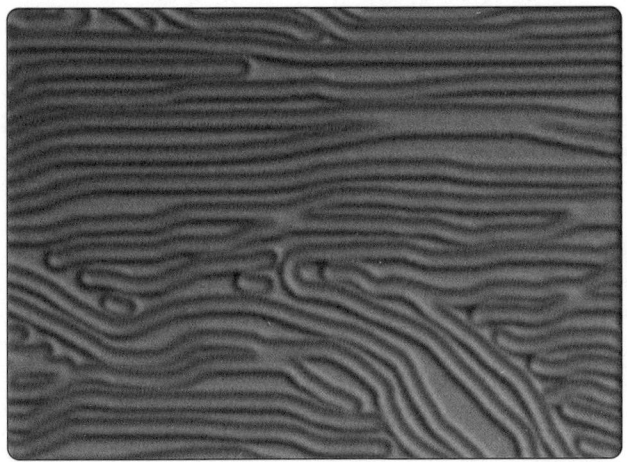

Figure 8-20. *Example of wrinkling.*

heavy paint coats. *[Figure 8-20]* Fast reducers can also contribute to wrinkling if the sprayed coat is not allowed to dry thoroughly. Thick coatings and quick-drying reducers allow the top surface of the coating to dry, trapping the solvents underneath. If another heavy coat is applied before the first one dries, wrinkles may result. It may also have the effect of lifting the coating underneath, almost with the same result as a paint stripper.

Rapid changes in ambient temperatures while spraying may cause an uneven release of the solvents, causing the surface to dry, shrink, and wrinkle. Making the mistake of using an incompatible thinner, or reducer, when mixing the coating materials may cause not only wrinkles but other problems as well. Wrinkled paint must be completely removed and the surface refinished.

Spray Dust

Spray dust is caused by the atomized spray particles from the gun becoming dry before reaching the surface being painted, thus failing to flow into a continuous film. *[Figure 8-21]* This may be caused by:

- Incorrect spray gun setting of air pressure, paint flow, or spray pattern.
- Spray gun being held too far from the surface.
- Material being improperly thinned or the wrong reducers being used with the finish coats.

The affected area needs to be sanded and recoated.

Painting Trim & Identification Marks

Masking & Applying the Trim

At this point in the project, the entire aircraft has been painted with the base color and all the masking paper and tape carefully removed. Refer again to the coating manufacturer's technical data sheet for "dry and recoat" times for the appropriate temperatures and "dry to tape" time that must elapse before safe application and removal of tape on new paint without it lifting.

Masking Materials

When masking for the trim lines, use 3M® Fine Line tape. It is solvent proof, available in widths of ⅛–1 inch and, when applied properly, produces a sharp edge paint line. A good quality masking tape should be used with masking paper to cover all areas not being trimmed to ensure the paper does not lift and allow overspray on the basecoat. Do not use newspaper to mask the work as paint penetrates newspaper. Using actual masking paper is more efficient, especially if used with a masking paper/tape dispenser as part of the finishing equipment.

Masking for the Trim

After the base color has dried and cured for the recommended time shown in the manufacturer's technical data sheet, the next step is to mask for the trim. The trim design can be simple, with one or two color stripes running along the fuselage, or it can be an elaborate scheme covering the entire aircraft. Whichever is chosen, the basic masking steps are the same.

If unsure of a design, there are numerous websites that provide the information and software to do a professional job. If electing to design a personalized paint scheme, the proposed design should be portrayed on a silhouette drawing of the aircraft as close to scale as possible. It is much easier to change a drawing than to remask the aircraft.

Start by identifying a point on the aircraft from which to

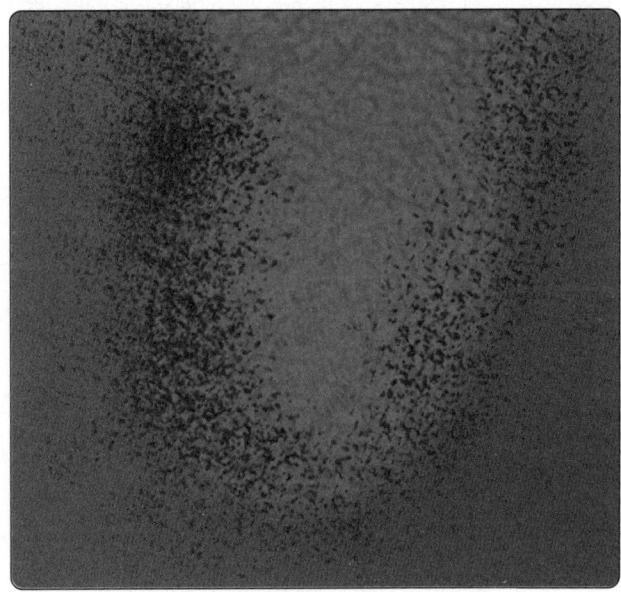

Figure 8-21. *Example of spray dust.*

initiate the trim lines using the Fine Line tape. If the lines are straight and/or have large radius curves, use ¾-inch or one-inch tape and keep it pulled tight. The wider tape is much easier to control when masking a straight line. Smaller radius curves may require ½-inch or even ¼-inch tape. Try and use the widest tape that lays flat and allows for a smooth curve. Use a small roller (like those used for wallpaper seams) to go back over and roll the tape edges firmly onto the surface to ensure they are flat.

Finish masking the trim lines on one side of the aircraft, to include the fuselage, vertical fin and rudder, the engine nacelles and wing(s). Once complete, examine the lines. If adjustments are needed to the placement or design, now is the time to correct it. With one side of the aircraft complete, the entire design and placement can be transferred to the opposite side.

Different methods can be employed to transfer the placement of the trim lines from one side of the aircraft to the other. One method is to trace the design on paper and then apply it to the other side, starting at the same point opposite the first starting point. Another method is to use the initial starting point and apply the trim tape using sheet metal or rivet lines as reference, along with measurements, to position the tape in the correct location.

When both sides are completed, a picture can be taken of each side and a comparison made to verify the tape lines on each side of the aircraft are identical.

With the Fine Line taping complete, some painters apply a sealing strip of ¾-inch or 1-inch masking tape covering half and extending over the outside edge of the Fine Line tape. This provides a wider area to apply the masking paper and adds an additional seal to the Fine Line tape. Now, apply the masking paper using 1-inch tape, placing half the width of the tape on the paper and half on the masked trim tape.

Use only masking paper made for painting and a comparable quality masking tape. With all the trim masking complete, cover the rest of the exposed areas of the aircraft to prevent overspray from landing on the base color. Tape the edges of the covering material to ensure the spray does not drift under it.

Now, scuff-sand all the area of trim to be painted to remove the gloss of the base paint. The use of 320-grit for the main area and a fine mesh Scotch-Brite pad next to the tape line should be sufficient. Then, blow all the dust and grit off the aircraft, and wipe down the newly sanded trim area with a degreaser and a tack cloth. Press or roll down the trim tape edges one more time before painting.

There are some various methods used by painters to ensure that a sharp defined tape line is attained upon removal of the tape. The basic step is to first use the 3M® Fine Line tape to mask the trim line. Some painters then spray a light coat of the base color or clear coat just prior to spraying the trim color. This will seal the tape edge line and ensure a clean sharp line when the tape is removed.

If multiple colors are used for the trim, cover the trim areas not to be sprayed with masking paper. When the first color is sprayed and dried, remove the masking paper from the next trim area to spray and cover the trim area that was first sprayed, taking care not to press the masking paper or tape into the freshly dried paint.

With all the trim completed, the masking paper should be removed as soon as the last trimmed area is dry to the touch. Carefully remove the Fine Line trim edge tape by slowly pulling it back onto itself at a sharp angle. Remove all trim and masking tape from the base coat as soon as possible to preclude damage to the paint.

As referenced previously, use compatible paint components from the same manufacturer when painting trim over the base color. This reduces the possibility of an adverse reaction between the base coat and the trim colors.

Display of Nationality & Registration Marks

The complete regulatory requirement for identification and marking of a U.S.-registered aircraft can be found in Title 14 of the Code of Federal Regulations (14 CFR), Part 45, Identification and Registration Marking.

In summary, the regulation states that the marks must:

- Be painted on the aircraft or affixed by other means to insure a similar degree of permanence;

- Have no ornamentation;

- Contrast in color with the background; and

- Be legible.

The letters and numbers may be taped off and applied at the same time and using the same methods as when the trim is applied, or they may be applied later as decals of the proper size and color.

Display of Marks

Each operator of an aircraft shall display on the aircraft marks consisting of the Roman capital letter "N" (denoting United States registration) followed by the registration number of the aircraft. Each suffix letter must also be a

Roman capital letter.

Location & Placement of Marks

On fixed-wing aircraft, marks must be displayed on either the vertical tail surfaces or the sides of the fuselage. If displayed on the vertical tail surfaces, they shall be horizontal on both surfaces of a single vertical tail or on the outer surfaces of a multivertical tail. If displayed on the fuselage surfaces, then horizontally on both sides of the fuselage between the trailing edge of the wing and the leading edge of the horizontal stabilizer. Exceptions to the location and size requirement for certain aircraft can be found in 14 CFR part 45.

On rotorcraft, marks must be displayed horizontally on both surfaces of the cabin, fuselage, boom, or tail. On airships, balloons, powered parachutes, and weight-shift control aircraft, display marks as required by 14 CFR part 45.

Size Requirements for Different Aircraft

Almost universally for U.S.-registered, standard certificated, fixed-wing aircraft, the marks must be at least 12 inches high. A glider may display marks at least 3 inches high.

In all cases, the marks must be of equal height, two-thirds as wide as they are high, and the characters must be formed by solid lines one-sixth as wide as they are high. The letters "M" and "W" may be as wide as they are high.

The spacing between each character may not be less than one-fourth of the character width. The marks required by 14 CFR part 45 for fixed-wing aircraft must have the same height, width, thickness, and spacing on both sides of the aircraft.

The marks must be painted or, if decalcomanias (decals), be affixed in a permanent manner. Other exceptions to the size and location of the marks are applicable to aircraft with Special Airworthiness certificates and those penetrating ADIZ and DEWIZ airspace. The current 14 CFR part 45 should be consulted for a complete copy of the rules.

Decals

Markings are placed on aircraft surfaces to provide servicing instructions, fuel and oil specifications, tank capacities, and to identify lifting and leveling points, walkways, battery locations, or any areas that should be identified. These markings can be applied by stenciling or by using decals.

Decals are used instead of painted instructions because they are usually less expensive and easier to apply. Decals used on aircraft are usually of three types: paper, metal, or vinyl film. These decals are suitable for exterior and interior surface application.

To assure proper adhesion of decals, clean all surfaces thoroughly with aliphatic naphtha to remove grease, oil, wax, or foreign matter. Porous surfaces should be sealed and rough surfaces sanded, followed by cleaning to remove any residue.

The instructions to be followed for applying decals are usually printed on the reverse side of each decal. A general application procedure for each type of decal is presented in the following paragraphs to provide familiarization with the techniques involved.

Paper Decals

Immerse paper decals in clean water for 1 to 3 minutes. Allowing decals to soak longer than 3 minutes causes the backing to separate from the decal while immersed. If decals are allowed to soak less than 1 minute, the backing does not separate from the decal.

Place one edge of the decal on the prepared receiving surface and press lightly, then slide the paper backing from beneath the decal. Perform any minor alignment with the fingers. Remove water by gently blotting the decal and adjacent area with a soft, absorbent cloth. Remove air or water bubbles trapped under the decal by wiping carefully toward the nearest edge of the decal with a cloth. Allow the decal to dry.

Metal Decals with Cellophane Backing

Apply metal decals with cellophane backing adhesive as follows:

1. Immerse the decal in clean, warm water for 1 to 3 minutes.

2. Remove it from the water and dry carefully with a clean cloth.

3. Remove the cellophane backing, but do not touch adhesive.

4. Position one edge of the decal on the prepared receiving surface. On large foil decals, place the center on the receiving surface and work outward from the center to the edges.

5. Remove all air pockets by rolling firmly with a rubber roller, and press all edges tightly against the receiving surface to ensure good adhesion.

Metal Decals With Paper Backing

Metal decals with a paper backing are applied similarly to those having a cellophane backing. However, it is not necessary to immerse the decal in water to remove the backing. It may be peeled from the decal without moistening. Follow the manufacturer's recommendation for activation of the adhesive, if necessary, before application. The decal should be positioned and smoothed out following the procedures given for cellophane-backed decals.

Metal Decals with No Adhesive

Apply decals with no adhesive in the following manner:

1. Apply one coat of cement, Military Specification MIL-A-5092, to the decal and prepared receiving surface.

2. Allow cement to dry until both surfaces are tacky.

3. Apply the decal and smooth it down to remove air pockets.

4. Remove excess adhesive with a cloth dampened with aliphatic naphtha.

Vinyl Film Decals

To apply vinyl film decals, separate the paper backing from the plastic film. Remove any paper backing adhering to the adhesive by rubbing the area gently with a clean cloth saturated with water. Remove small pieces of remaining paper with masking tape.

1. Place vinyl film, adhesive side up, on a clean porous surface, such as wood or blotter paper.

2. Apply recommended activator to the adhesive in firm, even strokes to the adhesive side of decal.

3. Position the decal in the proper location, while adhesive is still tacky, with only one edge contacting the prepared surface.

4. Work a roller across the decal with overlapping strokes until all air bubbles are removed.

Removal of Decals

Paper decals can be removed by rubbing the decal with a cloth dampened with lacquer thinner. If the decals are applied over painted or doped surfaces, use lacquer thinner sparingly to prevent removing the paint or dope.

Remove metal decals by moistening the edge of the foil with aliphatic naphtha and peeling the decal from the adhering surface. Work in a well-ventilated area.

Vinyl film decals are removed by placing a cloth saturated with MEK on the decal and scraping with a plastic scraper. Remove the remaining adhesive by wiping with a cloth dampened with a dry-cleaning solvent.

Paint System Compatibility

The use of several different types of paint, coupled with several proprietary coatings, makes repair of damaged and deteriorated areas particularly difficult. Paint finishes are not necessarily compatible with each other. The following general rules for coating compatibility are included for information

and are not necessarily listed in order of importance:

1. Old type zinc chromate primer may be used directly for touchup of bare metal surfaces and for use on interior finishes. It may be overcoated with wash primers if it is in good condition. Acrylic lacquer finishes do not adhere to this material.

2. Modified zinc chromate primer does not adhere satisfactorily to bare metal. It must never be used over a dried film of acrylic nitrocellulose lacquer.

3. Nitrocellulose coatings adhere to acrylic finishes, but the reverse is not true. Acrylic nitrocellulose lacquers may not be used over old nitrocellulose finishes.

4. Acrylic nitrocellulose lacquers adhere poorly to bare metal and both nitrocellulose and epoxy finishes. For best results, the lacquers must be applied over fresh, successive coatings of wash primer and modified zinc chromate. They also adhere to freshly applied epoxy coatings (dried less than 6 hours).

5. Epoxy topcoats adhere to any paint system that is in good condition, and may be used for general touchup, including touchup of defects in baked enamel coatings.

6. Old wash primer coats may be overcoated directly with epoxy finishes. A new second coat of wash primer must be applied if an acrylic finish is to be applied.

7. Old acrylic finishes may be refinished with new acrylic if the old coating is softened using acrylic nitrocellulose thinner before touchup.

8. Damage to epoxy finishes can best be repaired by using more epoxy, since neither of the lacquer finishes stick to the epoxy surface. In some instances, air-drying enamels may be used for touchup of epoxy coatings if edges of damaged areas are abraded with fine sandpaper.

Paint Touchup

Paint touchup may be required on an aircraft following repair to the surface substrate. Touchup may also be used to cover minor topcoat damage, such as scratches, abrasions, permanent stains, and fading of the trim colors. One of the first steps is to identify the paint that needs to be touched up.

Identification of Paint Finishes

Existing finishes on current aircraft may be any one of several types, a combination of two or more types, or combinations of general finishes with special proprietary coatings.

Any of the finishes may be present at any given time, and repairs may have been made using material from several different type coatings. Some detailed information for the identification of each finish is necessary to ensure the topcoat application does not react adversely with the undercoat. A simple test can be used to confirm the nature of the coatings present.

The following procedure aids in identification of the paint finish. Apply a coating of engine oil (MIL SPEC, MIL-PRF-7808, turbine oil, or equivalent) to a small area of the surface to be checked. Old nitrocellulose finishes soften within a period of a few minutes. Acrylic and epoxy finishes show no effects.

If still not identified, wipe a small area of the surface in question with a rag wet with MEK. The MEK picks up the pigment from an acrylic finish, but has no effect on an epoxy coating. Just wipe the surface, and do not rub. Heavy rubbing picks up even epoxy pigment from coatings that are not thoroughly cured. Do not use MEK on nitrocellulose finishes. *Figure 8-22* provides a solvent test to identify the coating on an aircraft.

Surface Preparation for Touchup

In the case of a repair and touchup, once the aircraft paint coating has been identified, the surface preparation follows some basic rules.

The first rule, as with the start of any paint project, is to wash and wipe down the area with a degreaser and silicone wax remover before starting to sand or abrade the area.

If a whole panel or section within a seam line can be refinished during a touchup, it eliminates having to match and blend the topcoat to an existing finish. The area of repair should be stripped to a seam line and the finish completely redone from wash primer to the topcoat, as applicable. The paint along the edge of the stripped area should be hand-sanded wet and feathered with a 320-grade paper.

For a spot repair that requires blending of the coating, an area about three times the area of the actual repair will need to be prepared for blending of the paint. If the damaged area is through the primer to the substrate, the repair area should be abraded with 320 aluminum oxide paper on a double-action (D/A) air sander. Then, the repair and the surrounding area should be wet sanded using the air sander fitted with 1500 wet paper. The area should then be wiped with a tack cloth prior to spraying.

Apply a crosscoat of epoxy primer to the bare metal area, following the material data sheet for drying and recoat times. Abrade the primer area lightly with 1500 wet or dry, and then abrade the unsanded area around the repair with cutting compound. Clean and wipe the area with a degreasing solvent, such as isopropyl alcohol, and then a tack cloth.

Mix the selected topcoat paint that is compatible for the repair. Apply two light coats over the sanded repair area, slightly extending the second coat beyond the first. Allow time for the first coat to flash before applying the second coat. Then, thin the topcoat by one-third to one-half with a compatible reducer and apply one more coat, extending beyond the first two coats. Allow to dry according to the material data sheet before buffing and polishing the blended area.

If the damage did not penetrate the primer, and only the topcoat is needed for the finish, complete the same steps that would follow a primer coat.

Paint touchup procedures generally are the same for almost

3–5 Minute Contact With Cotton Wad Saturated With Test Solvent									
Hitrate	Nitrate dope	Butyrate dope	Nitro-cellulose lacquer	Poly-tone Poly-brush Poly-spray	Synthetic enamel	Acrylic lacquer	Acrylic enamel	Urethane enamel	Epoxy paint
Methanol	S	IS	IS	IS	PS	IS	PS	IS	IS
Toluol (Toluene)	IS	IS	IS	S	IS	S	ISW	IS	IS
MEK (Methyl ethyl ketone)	S	S	S	S	ISW	S	ISW	IS	IS
Isopropanol	IS	IS	IS	IS	IS	S	IS	IS	IS
Methylene chloride	SS	VS	S	VS	ISW	S	ISW	ISW	ISW

IS – Insoluble   S – Soluble
ISW – Insoluble, film wrinkles   SS – Slightly Soluble
PS – Penetrate film, slight softening without wrinkling   VS – Very Soluble

Figure 8-22. *Chart for solvent testing of coating.*

any repair. The end result, however, is affected by numerous variables, which include the preparation, compatibility of the finishing materials, color match, selection of reducers and/or retarders based on temperature, and experience and expertise of the painter.

Stripping the Finish

The most experienced painter, the best finishing equipment, and newest coatings, do not produce the desired finish on an aircraft if the surface was not properly prepared prior to refinishing. Surface preparation for painting of an entire aircraft typically starts with the removal of the paint. This is done not only for the weight reduction that is gained by stripping the many gallons of topcoats and primers, but for the opportunity to inspect and repair corrosion or other defects uncovered by the removal of the paint.

Before any chemical stripping can be performed, all areas of the aircraft not being stripped must be protected. The stripper manufacturer can recommend protective material for this purpose. This normally includes all window material, vents and static ports, rubber seals and tires, and composite components that may be affected by the chemicals.

The removal of paint from an aircraft, even a small single-engine model, involves not only the labor but a concern for the environment. You should recognize the impact and regulatory requirements that are necessary to dispose of the water and coating materials removed from the aircraft.

Chemical Stripping

At one time, most chemical strippers contained methylene chloride, considered an environmentally acceptable chemical until 1990. It was very effective in removing multiple layers of paint. However, in 1990, it was listed as a toxic air contaminant that caused cancer and other medical problems and was declared a Hazardous Air Pollutant (HAP) by the EPA in the Clean Air Act Amendments of 1990.

Since then, other substitute chemical strippers were tested, from formic acid to benzyl alcohol. None of them were found to be particularly effective in removing multiple layers of paint. Most of them were not friendly to the environment.

One of the more recent entries into the chemical stripping business is an environmentally friendly product known as EFS-2500, which works by breaking the bond between the substrate and primer. This leads to a secondary action that causes the paint to lift both primer and top coat off the surface as a single film. Once the coating is lifted, it is easily removed with a squeegee or high-pressure water.

This product differs from conventional chemical strippers by not melting the coatings. Cleanup is easier, and the product complies with EPA rules on emissions. Additionally, it passed Boeing testing specifications related to sandwich corrosion, immersion corrosion, and hydrogen embrittlement. EFS-2500 has no chlorinated components, is non-acidic, nonflammable, nonhazardous, biodegradable, and has minimal to no air pollution potential.

The stripper can be applied using existing common methods, such as airless spraying, brushing, rolling, or immersion in a tank. It works on all metals, including aluminum, magnesium, cadmium plate, titanium, wood, fiberglass, ceramic, concrete, plaster, and stone.

Plastic Media Blasting (PMB)

Plastic media blasting (PMB) is one of the stripping methods that reduces and may eliminate a majority of environmental pollution problems that can be associated with the earlier formulations of some chemical stripping. PMB is a dry abrasive blasting process designed to replace chemical paint stripping operations. PMB is similar to conventional sand blasting except that soft, angular plastic particles are used as the blasting medium. The process has minimum effect on the surface under the paint because of the plastic medium and relatively low air pressure used in the process. The media, when processed through a reclamation system, can be reused up to 10 times before it becomes too small to effectively remove the paint.

PMB is most effective on metal surfaces, but it has been used successfully on composite surfaces after it was found to produce less visual damage than removing the paint by sanding.

New Stripping Methods

Various methods and materials for stripping paint and other coatings are under development and include:

- A laser stripping process used to remove coatings from composites.
- Carbon dioxide pellets (dry ice) used in conjunction with a pulsed flashlamp that rapidly heats a thin layer of paint, which is then blasted away by the ice pellets.

Safety in the Paint Shop

All paint booths and shops must have adequate ventilation systems installed that not only remove the toxic air but, when properly operating, reduce and/or eliminate overspray and dust from collecting on the finish. All electric motors used in the fans and exhaust system should be grounded and enclosed to eliminate sparks. The lighting systems and all bulbs should be covered and protected against breakage.

Proper respirators and fresh air breathing systems must be available to all personnel involved in the stripping and painting process. When mixing any paint or two-part coatings, eye protection and respirators should be worn.

An appropriate number and size of the proper class fire extinguishers should be available in the shop or hangar during all spraying operations. They should be weighed and certified, as required, to ensure they work in the event they are needed. Fireproof containers should be available for the disposal of all paint and solvent soaked rags.

Storage of Finishing Materials

All chemical components that are used to paint an aircraft burn in their liquid state. They should be stored away from all sources of heat or flames. The ideal place would be in fireproof metal cabinets located in a well-ventilated area.

Some of the finishing components have a shelf life listed in the material or technical data sheet supplied by the coating manufacturer. Those materials should be marked on the container, with a date of purchase, in the event that they are not used immediately.

Protective Equipment for Personnel

The process of painting, stripping, or refinishing an aircraft requires the use of various coatings, chemicals, and procedures that may be hazardous if proper precautions are not utilized to protect personnel involved in their use.

The most significant hazards are airborne chemicals inhaled either from the vapors of opened paint containers or atomized mist resulting from spraying applications. There are two types of devices available to protect against airborne hazards: respirators and forced-air breathing systems.

A respirator is a device worn over the nose and mouth to filter particles and organic vapors from the air being inhaled. The most common type incorporate double charcoal-filtered cartridges with replaceable dust filters that fits to the face over the nose and mouth with a tight seal. When properly used, this type of respirator provides protection against the inhalation of organic vapors, dust, mists of paints, lacquers, and enamels. A respirator does not provide protection against paints and coatings containing isocyanates (polyurethane paint).

A respirator must be used in an area of adequate ventilation. If breathing becomes difficult, there is a smell or taste of the contaminant(s), or an individual becomes dizzy or feels nauseous, they should leave the area and seek fresh air and assistance as necessary. Carefully read the warnings furnished with each respirator describing the limits and materials for which they provide protection.

A forced-air breathing system must be used when spraying any type of polyurethane or any coating that contains isocyanates. It is also recommended for all spraying and stripping of any type, whether chemical or media blasting. The system provides a constant source of fresh air for breathing, which is pumped into the mask through a hose from an electric turbine pump.

Protective clothing, such as Tyvek® coveralls, should be worn that not only protects personnel from the paint but also help keep dust off the painted surfaces. Rubber gloves must be worn when any stripper, etching solution, conversion coatings, and solvent is used.

When solvents are used for cleaning paint equipment and spray guns, the area must be free of any open flame or other heat source. Solvent should not be randomly sprayed into the atmosphere when cleaning the guns. Solvents should not be used to wash or clean paint and other coatings from bare hands and arms. Use protective gloves and clothing during all spraying operations.

In most states, there are Occupational Safety Hazard Administration (OSHA) regulations in effect that may require personnel to be protected from vapors and other hazards while on the job. In any hangar or shop, personnel must be vigilant and provide and use protection for safety.

Chapter 9
Aircraft Electrical System

Introduction

The satisfactory performance of any modern aircraft depends to a very great degree on the continuing reliability of electrical systems and subsystems. Improperly or carelessly installed or maintained wiring can be a source of both immediate and potential danger. The continued proper performance of electrical systems depends on the knowledge and technique of the mechanic who installs, inspects, and maintains the electrical system wires and cables.

Ohm's Law

Ohm's Law describes the basic mathematical relationships of electricity. The law was named after German Physicist George Simon Ohm (1789–1854). Basically, Ohm's Law states that the current (electron flow) through a conductor is directly proportional to the voltage (electrical pressure) applied to that conductor and inversely proportional to the resistance of the conductor. The unit used to measure resistance is called the ohm. The symbol for the ohm is the Greek letter omega (Ω). In mathematical formulas, the capital letter R refers to resistance. The resistance of a conductor and the voltage applied to it determine the number of amperes of current flowing through the conductor. Thus, 1 ohm of resistance limits the current flow to 1 ampere in a conductor to which a voltage of 1 volt is applied. The primary formula derived from Ohm's Law is: $E = I \times R$ (E = electromotive force measured in volts, I = current flow measured in amps, and R = resistance measured in ohms). This formula can also be written to solve for current or resistance:

$$I = \frac{E}{R}$$

$$R = \frac{E}{I}$$

Ohm's Law provides a foundation of mathematical formulas that predict how electricity responds to certain conditions. *[Figure 9-1]* For example, Ohm's Law can be used to calculate that a lamp of 12 Ohms (Ω) passes a current of 2 amps when connected to a 24-volt direct current (DC) power source.

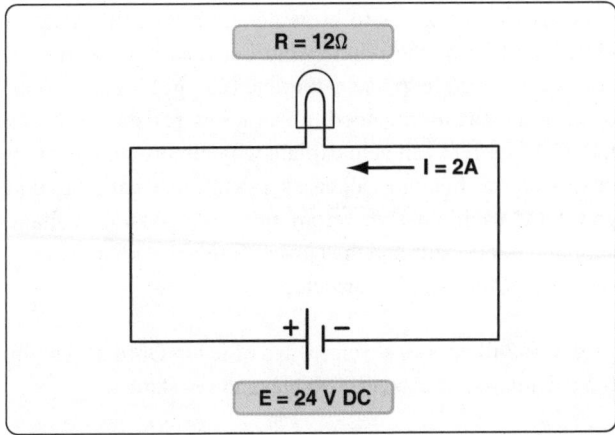

Figure 9-1. *Ohm's Law used to calculate how much current a lamp will pass when connected to a 24-volt DC power source.*

Example 1
A 28-volt landing light circuit has a lamp with 4 ohms of resistance. Calculate the total current of the circuit.

$$I = \frac{E}{R}$$

$$I = \frac{28 \text{ volts}}{4\Omega}$$

$$I = 7 \text{ amps}$$

Example 2
A 28-volt deice boot circuit has a current of 6.5 amps. Calculate the resistance of the deice boot.

$$R = \frac{E}{I}$$

$$R = \frac{28 \text{ volts}}{6.5 \text{ amps}}$$

$$R = 4.31\Omega$$

Example 3

A taxi light has a resistance of 4.9Ω and a total current of 2.85 amps. Calculate the system voltage.

$$E = I \times R$$

$$E = 2.85 \times 4.9Ω$$

$$E = 14 \text{ volts}$$

Whenever troubleshooting aircraft electrical circuits, it is always valuable to consider Ohm's Law. A good understanding of the relationship between resistance and current flow can help one determine if a circuit contains an open or a short. Remembering that a low resistance means increased current can help explain why circuit breakers pop or fuses blow. In almost all cases, aircraft loads are wired in parallel to each other; therefore, there is a constant voltage supplied to all loads and the current flow through a load is a function of that load's resistance.

Figure 9-2 illustrates several ways of using Ohm's Law for the calculation of current, voltage, and resistance.

Current

Electrical current is the movement of electrons. This electron movement is referred to as current, flow, or current flow. In practical terms, this movement of electrons must take place within a conductor (wire). Current is typically measured in amps. The symbol for current is I and the symbol for amps is A.

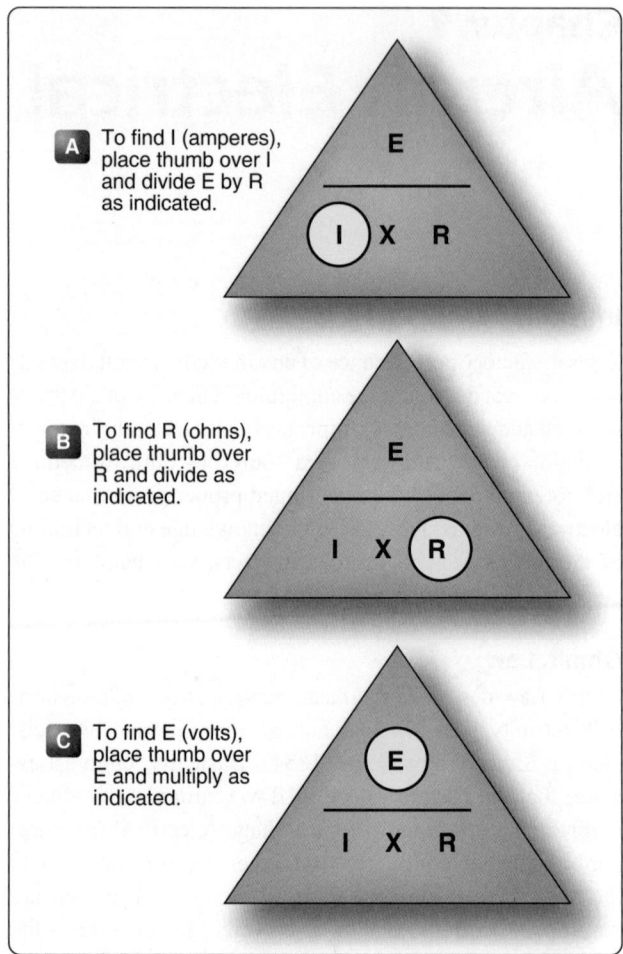

Figure 9-2. *Ohm's Law chart.*

The current flow is actually the movement of the free electrons found within conductors. Common conductors include copper, silver, aluminum, and gold. The term "free electron" describes a condition in some atoms where the outer electrons are loosely bound to their parent atom. These loosely bound electrons are easily motivated to move in a given direction when an external source, such as a battery, is applied to the circuit. These electrons are attracted to the positive terminal of the battery, while the negative terminal is the source of the electrons. So, the measure of current is actually the number of electrons moving through a conductor in a given amount of time.

The internationally accepted unit for current is the ampere (A). One ampere (A) of current is equivalent to 1 coulomb (C) of charge passing through a conductor in 1 second. One coulomb of charge equals 6.28×10^{18} electrons. Obviously, the unit of amperes is a much more convenient term to use than coulombs. The unit of coulombs is simply too small to be practical.

When current flow is in one direction, it is called direct current (DC). Later in the text, the form of current that periodically oscillates back and forth within the circuit is discussed. The present discussion is concerned only with the use of DC. It should be noted that as with the movement of any mass, electron movement (current flow) only occurs when there is a force present to push the electrons. This force is commonly called voltage (described in more detail in the next section). When a voltage is applied across the conductor, an electromotive force creates an electric field within the conductor, and a current is established. The electrons do not move in a straight direction, but undergo repeated collisions with other nearby atoms within a conductor. These collisions usually knock other free electrons from their atoms, and these electrons move on toward the positive end of the conductor with an average velocity called the drift velocity, which is relatively low speed. To understand the nearly instantaneous speed of the effect of the current, it is helpful to visualize a long tube filled with steel balls. *[Figure 9-3]*

It can be seen that a ball introduced in one end of the tube, which represents the conductor, immediately causes a ball to be emitted at the opposite end of the tube. Thus, electric

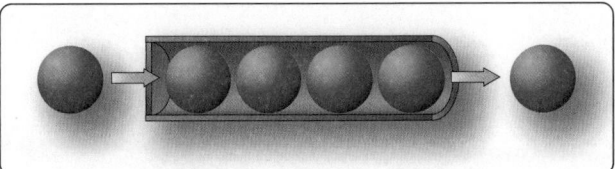

Figure 9-3. *Electron flow.*

current can be viewed as instantaneous, even though it is the result of a relatively slow drift of electrons.

Conventional Current Theory & Electron Theory

There are two competing schools of thought regarding the flow of electricity. The two explanations are the conventional current theory and the electron theory. Both theories describe the movement of electrons through a conductor. They simply explain the direction current moves. Typically during troubleshooting or the connection of electrical circuits, the use of either theory can be applied as long as it is used consistently. The Federal Aviation Administration (FAA) officially defines current flow using electron theory (negative to positive).

The conventional current theory was initially advanced by Benjamin Franklin, who reasoned that current flowed out of a positive source into a negative source or an area that lacked an abundance of charge. The notation assigned to the electric charges was positive (+) for the abundance of charge and negative (−) for a lack of charge. It then seemed natural to visualize the flow of current as being from the positive (+) to the negative (−). Later discoveries were made that proved that just the opposite is true. Electron theory describes what actually happens in the case of an abundance of electrons flowing out of the negative (−) source to an area that lacks electrons or the positive (+) source. Both conventional flow and electron flow are used in industry.

Electromotive Force (Voltage)

Voltage is most easily described as electrical pressure force. It is the electromotive force (EMF), or the push or pressure from one end of the conductor to the other, that ultimately moves the electrons. The symbol for EMF is the capital letter E. EMF is always measured between two points and voltage is considered a value between two points. For example, across the terminals of the typical aircraft battery, voltage can be measured as the potential difference of 12 volts or 24 volts. That is to say that between the two terminal posts of the battery, there is a voltage available to push current through a circuit. Free electrons in the negative terminal of the battery move toward the excessive number of positive charges in the positive terminal. The net result is a flow or current through a conductor. There cannot be a flow in a conductor unless there is an applied voltage from a battery,

generator, or ground power unit. The potential difference, or the voltage across any two points in an electrical system, can be determined by:

$$V_1 - V_2 = V_{Drop}$$

Example

The voltage at one point is 14 volts. The voltage at a second point in the circuit is 12.1 volts. To calculate the voltage drop, use the formula above to get a total voltage drop of 1.9 volts.

Figure 9-4 illustrates the flow of electrons of electric current. Two interconnected water tanks demonstrate that when a difference of pressure exists between the two tanks, water flows until the two tanks are equalized. *Figure 9-4* shows the level of water in tank A to be at a higher level, reading 10 pounds per square inch (psi) (higher potential energy), than the water level in tank B, reading 2 psi (lower potential energy). Between the two tanks, there is 8 psi potential difference. If the valve in the interconnecting line between the tanks is opened, water flows from tank A into tank B until the level of water (potential energy) of both tanks is equalized. It is important to note that it was not the pressure in tank A that caused the water to flow; rather, it was the difference in pressure between tank A and tank B that caused the flow.

This comparison illustrates the principle that electrons move, when a path is available, from a point of excess electrons (higher potential energy) to a point deficient in electrons (lower potential energy). The force that causes this movement is the potential difference in electrical energy between the two points. This force is called the electrical pressure (voltage), the potential difference, or the electromotive force (electron moving force).

Resistance

The two fundamental properties of current and voltage are related by a third property known as resistance. In any electrical circuit, when voltage is applied to it, a current results. The resistance of the conductor determines the amount of current that flows under the given voltage. In general, the greater the circuit resistance, the less the current.

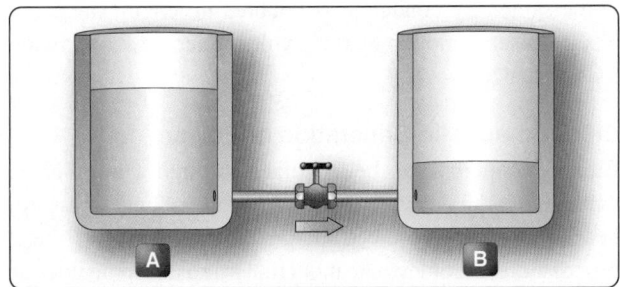

Figure 9-4. *Difference of pressure.*

If the resistance is reduced, then the current will increase. This relation is linear in nature and is known as Ohm's Law. An example would be if the resistance of a circuit is doubled, and the voltage is held constant, then the current through the resistor is cut in half.

There is no distinct dividing line between conductors and insulators; under the proper conditions, all types of material conduct some current. Materials offering a resistance to current flow midway between the best conductors and the poorest conductors (insulators) are sometimes referred to as semiconductors and find their greatest application in the field of transistors.

The best conductors are materials, chiefly metals, that possess a large number of free electrons. Conversely, insulators are materials having few free electrons. The best conductors are silver, copper, gold, and aluminum, but some nonmetals, such as carbon and water, can be used as conductors. Materials such as rubber, glass, ceramics, and plastics are such poor conductors that they are usually used as insulators. The current flow in some of these materials is so low that it is usually considered zero.

Factors Affecting Resistance

The resistance of a metallic conductor is dependent on the type of conductor material. It has been pointed out that certain metals are commonly used as conductors because of the large number of free electrons in their outer orbits. Copper is usually considered the best available conductor material, since a copper wire of a particular diameter offers a lower resistance to current flow than an aluminum wire of the same diameter. However, aluminum is much lighter than copper, and for this reason, as well as cost considerations, aluminum is often used when the weight factor is important.

The resistance of a metallic conductor is directly proportional to its length. The longer the length of a given size of wire, the greater the resistance. *Figure 9-5* shows two wire conductors of different lengths. If 1 volt of electrical pressure is applied across the two ends of the conductor that is 1 foot in length and the resistance to the movement of free electrons is assumed to be 1 ohm, the current flow is limited to 1 ampere. If the same size conductor is doubled in length, the same electrons set in motion by the 1 volt applied now find twice the resistance.

Electromagnetic Generation of Power

Electrical energy can be produced through a number of methods. Common methods include the use of light, pressure, heat, chemical, and electromagnetic induction. Of these processes, electromagnetic induction is most responsible for the generation of the majority of the electrical power used

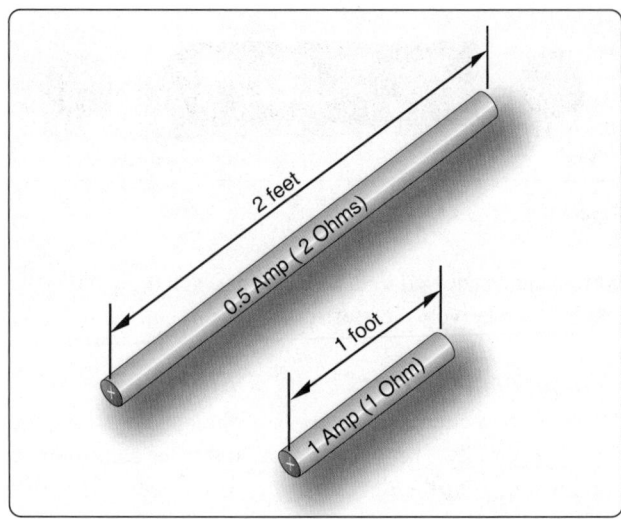

Figure 9-5. *Resistance varies with length of conductor.*

by humans. Virtually all mechanical devices (generators and alternators) that produce electrical power employ the process of electromagnetic induction. The use of light, pressure, heat, and chemical sources for electrical power is found on aircraft but produce a minimal amount of all the electrical power consumed during a typical flight.

In brief, light can produce electricity using a solar cell (photovoltaic cell). These cells contain a certain chemical that converts light energy into voltage/current.

Using pressure to generate electrical power is commonly known as the piezoelectric effect. The piezoelectric effect (piezo or piez taken from Greek: to press; pressure; to squeeze) is a result of the application of mechanical pressure on a dielectric or nonconducting crystal.

Chemical energy can be converted into electricity, most commonly in the form of a battery. A primary battery produces electricity using two different metals in a chemical solution like alkaline electrolyte. A chemical reaction exists between the metals which frees more electrons in one metal than in the other.

Heat used to produce electricity creates the thermoelectric effect. When a device called a thermocouple is subjected to heat, a voltage is produced. A thermocouple is a junction between two different metals that produces a voltage related to a temperature difference. If the thermocouple is connected to a complete circuit, a current also flows. Thermocouples are often found on aircraft as part of a temperature monitoring system, such as a cylinder head temperature gauge.

Electromagnetic induction is the process of producing a voltage (EMF) by moving a magnetic field in relationship to a

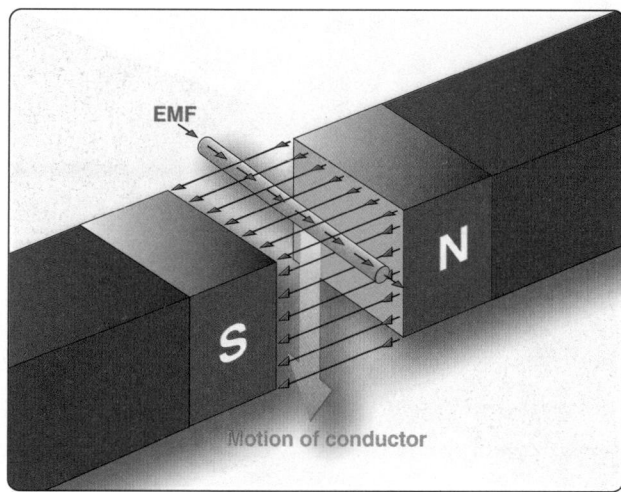

Figure 9-6. *Inducing an EMF in a conductor.*

conductor. As shown in *Figure 9-6,* when a conductor (wire) is moved through a magnetic field, an EMF is produced in the conductor. If a complete circuit is connected to the conductor, the voltage also produces a current flow.

One single conductor does not produce significant voltage/current via electromagnetic induction. *[Figure 9-6]* In practice, instead of a single wire, a coil of wire is moved through the magnetic field of a strong magnet. This produces a greater electrical output. In many cases, the magnetic field is created by using a powerful electromagnet. This allows for the production of a greater voltage/current due to the stronger magnetic field produced by the electromagnet when compared to an ordinary magnet.

Please note that this text often refers to voltage/current in regards to electrical power. Remember voltage (electrical pressure) must be present to produce a current (electron flow). Hence, the output energy generated through the process of electromagnetic induction always consists of voltage. Current also results when a complete circuit is connected to that voltage. Electrical power is produced when there is both electrical pressure E (EMF) and current (I). Power = Current × Voltage (P = I × E)

It is the relative motion between a conductor and a magnetic field that causes current to flow in the conductor. Either the conductor or magnet can be moving or stationary. When a magnet and its field are moved through a coiled conductor, as shown in *Figure 9-7,* a DC voltage with a specific polarity is produced. The polarity of this voltage depends on the direction in which the magnet is moved and the position of the north and south poles of the magnetic field. The generator left-hand rule can be used to determine the direction of current flow within the conductor. *[Figure 9-8]* Of course, the direction of current flow is a function of the polarity of

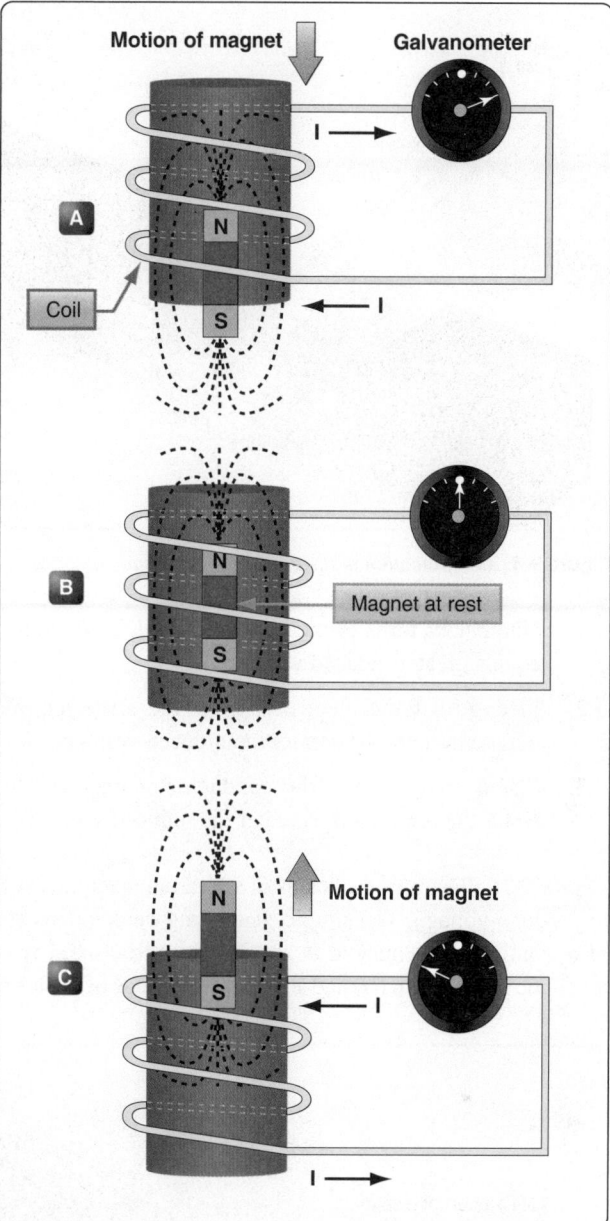

Figure 9-7. *Inducing a current flow.*

the voltage induced in to the conductor.

In practice, producing voltage/current using the process of electromagnetic induction requires a rotating machine. Generally speaking, on all aircraft, a generator or alternator employs the principles of electromagnetic induction to create electrical power for the aircraft. Either the magnetic field can rotate or the conductor can rotate. *[Figure 9-9]* The rotating component is driven by a mechanical device, such as an aircraft engine.

During the process of electromagnetic induction, the value of the induced voltage/current depends on three basic factors:

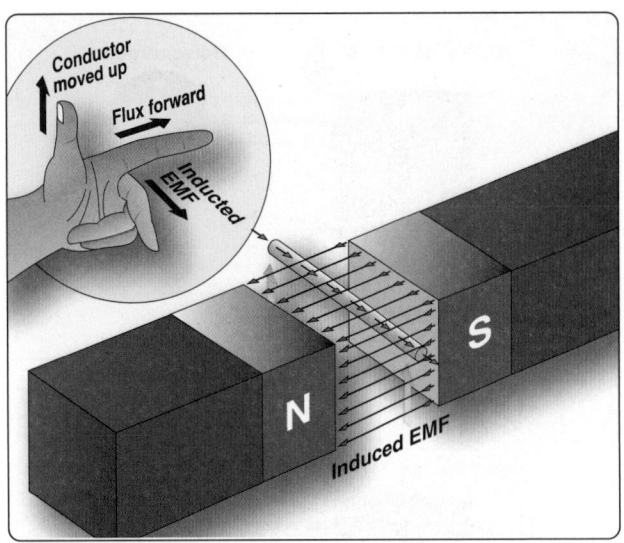

Figure 9-8. *An application of the generator left-hand rule.*

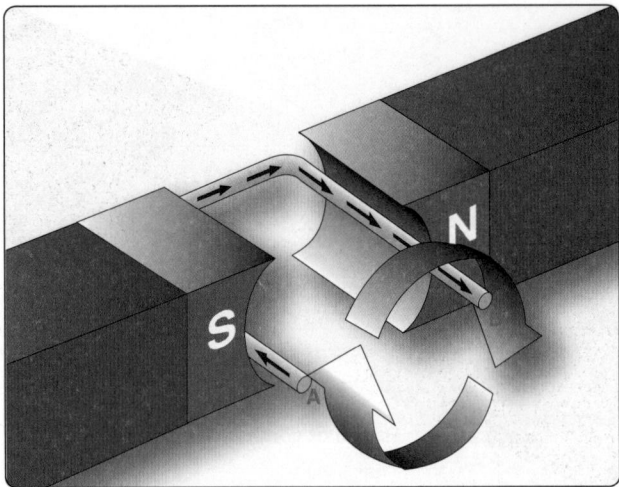

Figure 9-9. *Voltage induced in a loop.*

1. Number of turns in the conductor coil (more loops equals greater induced voltage).

2. Strength of the electromagnet (the stronger the magnetic field, the greater the induced voltage).

3. Speed of rotation of the conductor or magnet (the faster the rotation, the greater the induced voltage).

Figure 9-10 illustrates the basics of a rotating machine used to produce voltage. The simple generating device consists of a rotating loop, marked A and B, placed between two magnetic poles, north (N) and south (S). The ends of the loop

are connected to two metal slip rings (collector rings), C1 and C2. Current is taken from the collector rings by brushes. If the loop is considered as separate wires, A and B, and the left-hand rule for generators is applied, then it can be observed that as wire B moves up across the field, a voltage is induced that causes the current to flow toward the reader. As wire A moves down across the field, a voltage is induced that causes the current to flow away from the reader. When the wires are formed into a loop, the voltages induced in the two sides of the loop are combined. Therefore, for explanatory purposes, the action of either conductor, A or B, while rotating in the magnetic field is similar to the action of the loop.

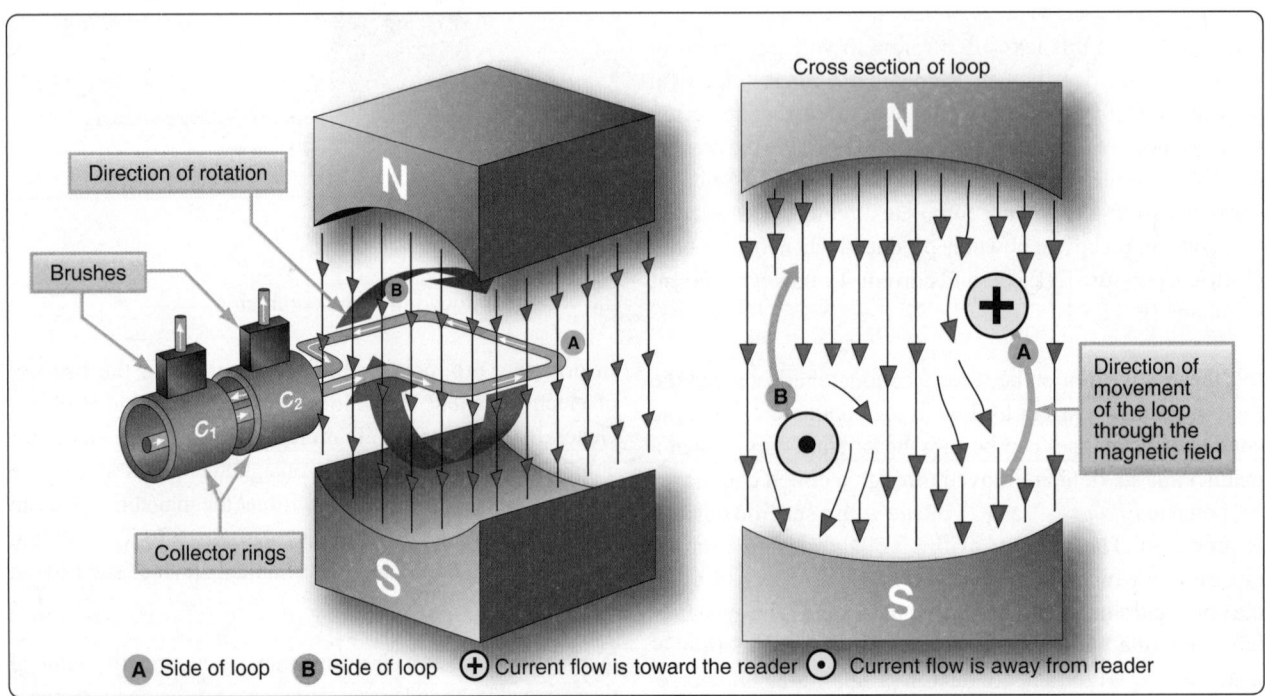

Figure 9-10. *Simple generator.*

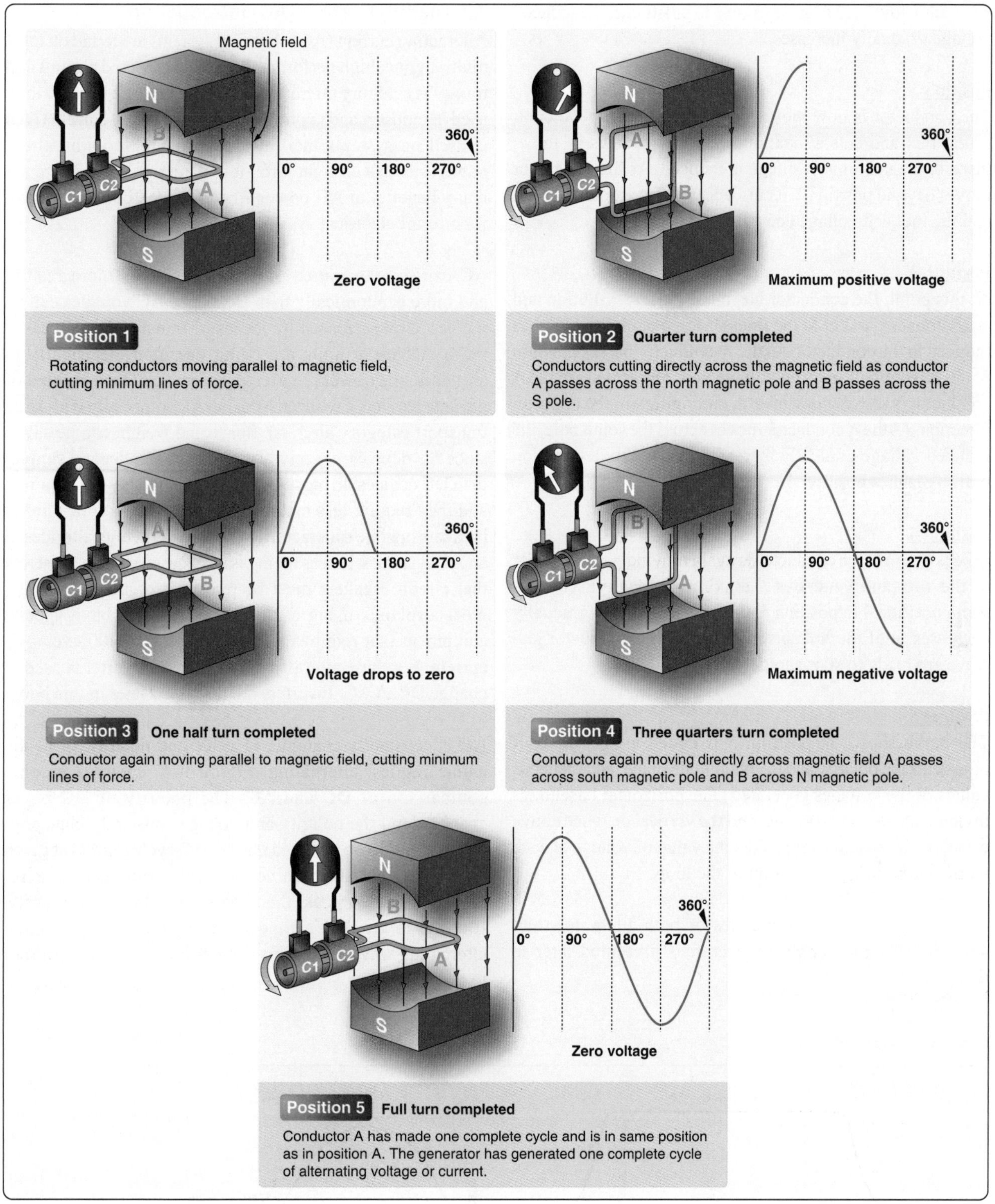
Magnetic field
N
S
A
B
C1
C2
0°
90°
180°
270°
360°
Zero voltage
Position 1
Rotating conductors moving parallel to magnetic field, cutting minimum lines of force.
Maximum positive voltage
Position 2
Quarter turn completed
Conductors cutting directly across the magnetic field as conductor A passes across the north magnetic pole and B passes across the S pole.
Voltage drops to zero
Position 3
One half turn completed
Conductor again moving parallel to magnetic field, cutting minimum lines of force.
Maximum negative voltage
Position 4
Three quarters turn completed
Conductors again moving directly across magnetic field A passes across south magnetic pole and B across N magnetic pole.
Zero voltage
Position 5
Full turn completed
Conductor A has made one complete cycle and is in same position as in position A. The generator has generated one complete cycle of alternating voltage or current.

Figure 9-11. *Generation of a sine wave.*

Figure 9-11 illustrates the generation of alternating current (AC) with a simple loop conductor rotating in a magnetic field. As it is rotated in a counterclockwise direction, varying voltages are induced in the conductive loop.

Position 1

The conductor A moves parallel to the lines of force. Since it cuts no lines of force, the induced voltage is zero. As the conductor advances from position 1 to position 2, the induced voltage gradually increases.

Position 2
The conductor is now moving in a direction perpendicular to the flux and cuts a maximum number of lines of force; therefore, a maximum voltage is induced. As the conductor moves beyond position 2, it cuts a decreasing amount of flux, and the induced voltage decreases.

Position 3
At this point, the conductor has made half a revolution and again moves parallel to the lines of force, and no voltage is induced in the conductor. As the A conductor passes position 3, the direction of induced voltage now reverses since the A conductor is moving downward, cutting flux in the opposite direction. As the A conductor moves across the south pole, the induced voltage gradually increases in a negative direction until it reaches position 4.

Position 4
Like position 2, the conductor is again moving perpendicular to the flux and generates a maximum negative voltage. From position 4 to position 5, the induced voltage gradually decreases until the voltage is zero, and the conductor and wave are ready to start another cycle.

Position 5
The curve shown at position 5 is called a sine wave. It represents the polarity and the magnitude of the instantaneous values of the voltages generated. The horizontal baseline is divided into degrees, or time, and the vertical distance above or below the baseline represents the value of voltage at each particular point in the rotation of the loop.

The specific operating principles of both alternators and generators as they apply to aircraft is presented later in this text.

Alternating Current (AC) Introduction
Alternating current (AC) electrical systems are found on most multi-engine, high performance turbine powered aircraft and transport category aircraft. AC is the same type of electricity used in industry and to power our homes. Direct current (DC) is used on systems that must be compatible with battery power, such as on light aircraft and automobiles. There are many benefits of AC power when selected over DC power for aircraft electrical systems.

AC can be transmitted over long distances more readily and more economically than DC, since AC voltages can be increased or decreased by means of transformers. Because more and more units are being operated electrically in airplanes, the power requirements are such that a number of advantages can be realized by using AC (especially with large transport category aircraft). Space and weight can be saved since AC devices, especially motors, are smaller and simpler than DC devices. In most AC motors, no brushes are required, and they require less maintenance than DC motors. Circuit breakers operate satisfactorily under loads at high altitudes in an AC system, whereas arcing is so excessive on DC systems that circuit breakers must be replaced frequently. Finally, most airplanes using a 24-volt DC system have special equipment that requires a certain amount of 400 cycle AC current. For these aircraft, a unit called an inverter is used to change DC to AC. Inverters are discussed later in this book.

AC is constantly changing in value and polarity, or as the name implies, alternating. *Figure 9-12* shows a graphic comparison of DC and AC. The polarity of DC never changes, and the polarity and voltage constantly change in AC. It should also be noted that the AC cycle repeats at given intervals. With AC, both voltage and current start at zero, increase, reach a peak, then decrease and reverse polarity. If one is to graph this concept, it becomes easy to see the alternating wave form. This wave form is typically referred to as a sine wave.

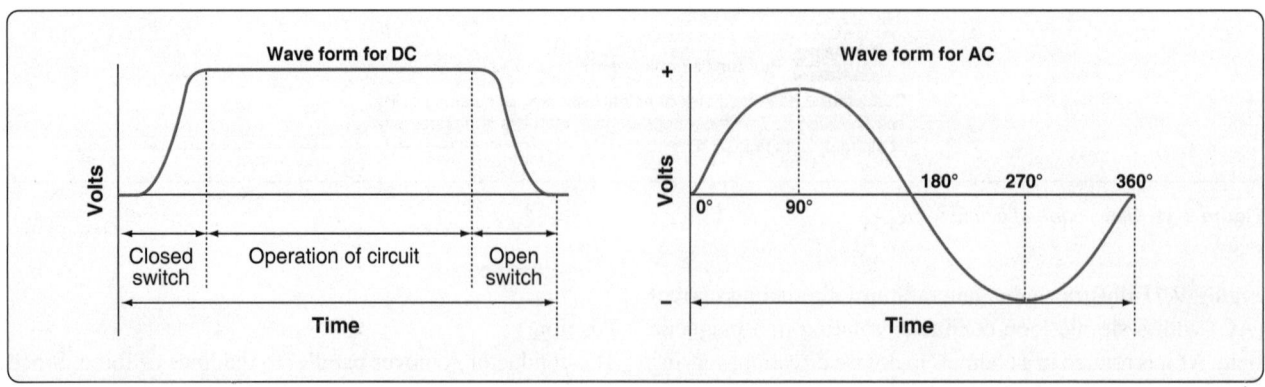

Figure 9-12. *DC and AC voltage curves.*

Definitions

Values of AC

There are three values of AC that apply to both voltage and current. These values help to define the sine wave and are called instantaneous, peak, and effective. It should be noted that during the discussion of these terms, the text refers to voltage. But remember, the values apply to voltage and current in all AC circuits.

Instantaneous

An instantaneous voltage is the value at any instant in time along the AC wave. The sine wave represents a series of these values. The instantaneous value of the voltage varies from zero at 0° to maximum at 90°, back to zero at 180°, to maximum in the opposite direction at 270°, and to zero again at 360°. Any point on the sine wave is considered the instantaneous value of voltage.

Peak

The peak value is the largest instantaneous value, often referred to as the maximum value. The largest single positive value occurs after a certain period of time when the sine wave reaches 90°, and the largest single negative value occurs when the wave reaches 270°. Although important in the understanding of the AC sine wave, peak values are seldom used by aircraft technicians.

Effective

The effective values for voltage are always less than the peak (maximum) values of the sine wave and approximate DC voltage of the same value. For example, an AC circuit of 24 volts and 2 amps should produce the same heat through a resistor as a DC circuit of 24 volts and 2 amps. The effective value is also known as the root mean square, or RMS value, which refers to the mathematical process by which the value is derived.

Most AC meters display the effective value of the AC. In almost all cases, the voltage and current ratings of a system or component are given in effective values. In other words, the industry ratings are based on effective values. Peak and instantaneous values, used only in very limited situations, would be stated as such. In the study of AC, any values given for current or voltage are assumed to be effective values unless otherwise specified. In practice, only the effective values of voltage and current are used.

The effective value is equal to .707 times the peak (maximum) value. Conversely, the peak value is 1.41 times the effective value. Thus, the 110 volt value given for AC is only 0.707 of the peak voltage of this supply. The maximum voltage is approximately 155 volts (110 × 1.41 = 155 volts maximum).

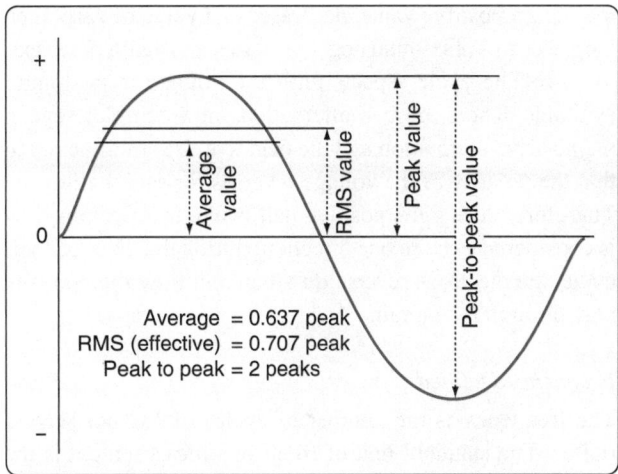

Figure 9-13. *Values of AC.*

How often the AC waveform repeats is known as the AC frequency. The frequency is typically measured in cycles per second (CPS) or hertz (Hz). One Hz equals one CPS. The time it takes for the sine wave to complete one cycle is known as period (P). Period is a value or time period and typically measured in seconds, milliseconds, or microseconds. It should be noted that the time period of a cycle can change from one system to another; it is always said that the cycle completes in 360° (related to the 360° of rotation of an AC alternator). *[Figure 9-13]*

Cycle Defined

A cycle is a completion of a pattern. Whenever a voltage or current passes through a series of changes, returns to the starting point, and then repeats the same series of changes, the series is called a cycle. When the voltage values are graphed, as in *Figure 9-14*, the complete AC cycle is displayed. One complete cycle is often referred to as the sine wave and said to be 360°. It is typical to start the sine wave where the voltage is zero. The voltage then increases to a

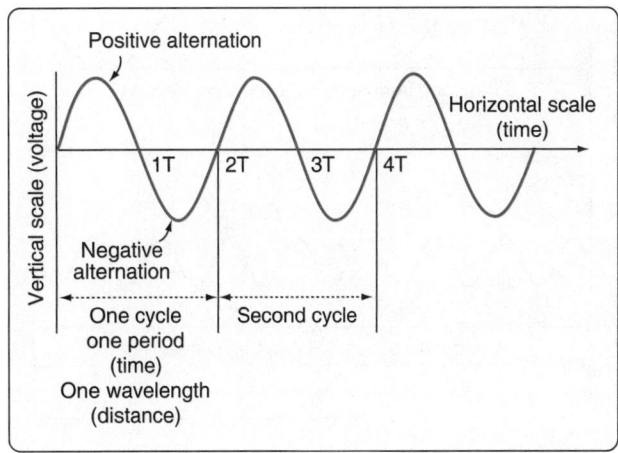

Figure 9-14. *Cycle of voltage.*

maximum positive value, decreases to a value of zero, then increases to a maximum negative value, and again decreases to zero. The cycle repeats until the voltage is no longer available. There are two alternations in a complete cycle: the positive alternation and the negative. It should be noted that the polarity of the voltage reverses for each half cycle. Therefore, during the positive half cycle, the electron flow is considered to be in one direction; during the negative half cycle, the electrons reverse direction and flow the opposite way through the circuit.

Frequency Defined

The frequency is the number of cycles of AC per second (CPS). The standard unit of frequency measurement is the Hz. *[Figure 9-15]* In a generator, the voltage and current pass through a complete cycle of values each time a coil or conductor passes under a north and south pole of the magnet. The number of cycles for each revolution of the coil or conductor is equal to the number of pairs of poles. The frequency, then, is equal to the number of cycles in one revolution multiplied by the number of revolutions per second.

Period Defined

The time required for a sine wave to complete one full cycle is called a period (P). A period is typically measured in seconds, milliseconds, or microseconds. *[Figure 9-14]* The period of a sine wave is inversely proportional to the frequency. That is to say that the higher the frequency, the shorter the period. The mathematical relationship between frequency and period is given as:

Period

$$P = \frac{1}{f}$$

Frequency

$$F = \frac{1}{P}$$

Wavelength Defined

The distance that a waveform travels during a period is commonly referred to as a wavelength and is indicated by the Greek letter lambda (λ). Wavelength is related to frequency by the formula:

$$\frac{wave\ speed}{frequency} = wavelength$$

The higher the frequency is, the shorter the wavelength is. The measurement of wavelength is taken from one point on the waveform to a corresponding point on the next waveform. *[Figure 9-14]* Since wavelength is a distance, common units of measure include meters, centimeters, millimeters, or nanometers. For example, a sound wave of frequency 20 Hz would have wavelength of 17 meters and a visible red light wave of $4.3 \times 10\ -12$ Hz would have a wavelength of roughly 700 nanometers. Keep in mind that the actual wavelength depends on the media through which the waveform must travel.

Phase Relationships

Phase is the relationship between two sine waves, typically measured in angular degrees. For example, if there are two different alternators producing power, it would be easy to compare their individual sine waves and determine their phase relationship. In *Figure 9-16B*, there is a 90° phase difference between the two voltage waveforms. A phase relationship can be between any two sine waves. The phase relationship can be measured between two voltages of different alternators or the current and voltage produced by the same alternator.

Figure 9-16A shows a voltage signal and a current signal superimposed on the same time axis. Notice that when the voltage increases in the positive alternation that the current also increases. When the voltage reaches its peak value, so does the current. Both waveforms then reverse and decrease back to a zero magnitude, then proceed in the same manner in the negative direction as they did in the positive direction. When two waves are exactly in step with each other, they are said to be in phase. To be in phase, the two waveforms must go through their maximum and minimum points at the same time and in the same direction.

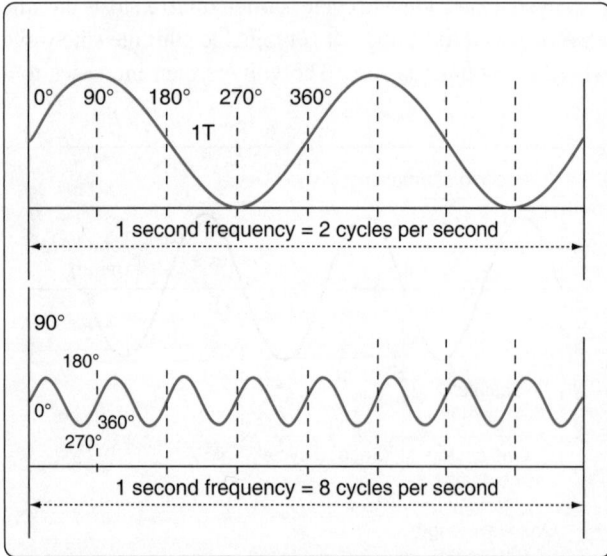

Figure 9-15. *Frequency in cycles per second.*

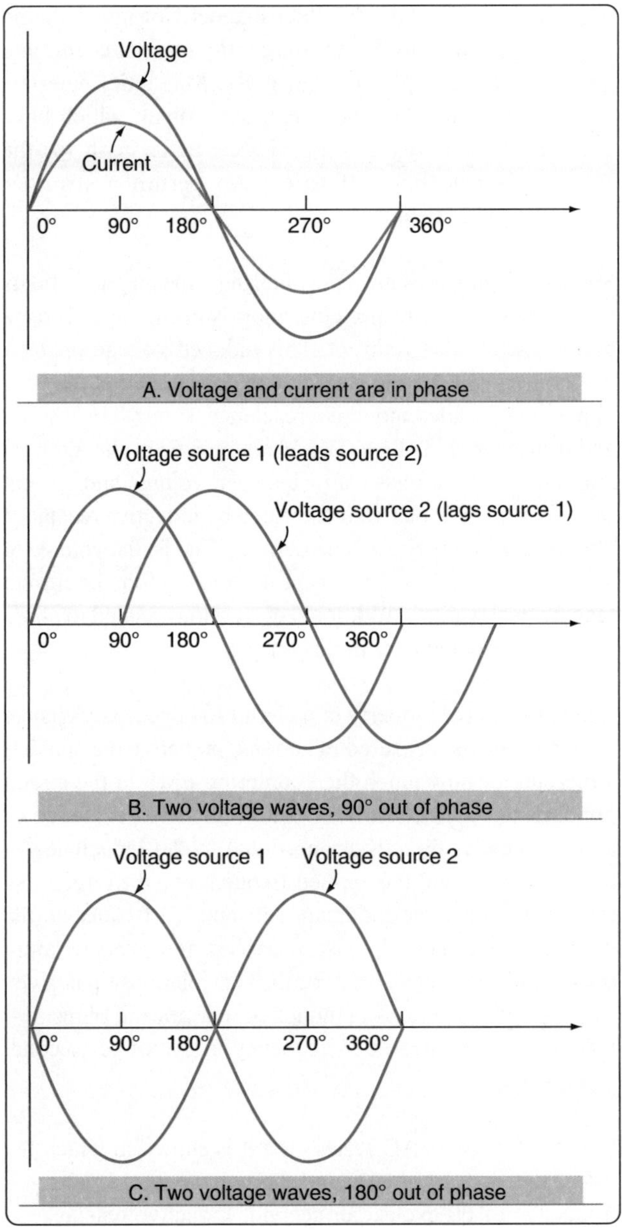
Voltage
Current
0°
90°
180°
270°
360°
A. Voltage and current are in phase
Voltage source 1 (leads source 2)
Voltage source 2 (lags source 1)
0°
90°
180°
270°
360°
B. Two voltage waves, 90° out of phase
Voltage source 1
Voltage source 2
0°
90°
180°
270°
360°
C. Two voltage waves, 180° out of phase

Figure 9-16. *In-phase and out-of-phase conditions.*

When two waveforms go through their maximum and minimum points at different times, a phase difference exists between the two. In this case, the two waveforms are said to be out of phase with each other. The terms lead and lag are often used to describe the phase difference between waveforms. The waveform that reaches its maximum or minimum value first is said to lead the other waveform. *Figure 9-16B* shows this relationship. On the other hand, the second waveform is said to be lagging the first source. When a waveform is said to be leading or lagging, the difference in degrees is usually stated. If the two waveforms differ by 360°, they are said to be in phase with each other. If there is a 180° difference between the two signals, then they are still out of phase even though they are both reaching their minimum and maximum values at the same time. *[Figure 9-16C]*

Opposition to Current Flow of AC

There are three factors that can create an opposition to the flow of electrons (current) in an AC circuit. Resistance, similar to resistance of DC circuits, is measured in ohms and has a direct influence on AC regardless of frequency. Inductive reactance and capacitive reactance, on the other hand, oppose current flow only in AC circuits, not in DC circuits. Since AC constantly changes direction and intensity, inductors and capacitors may also create an opposition to current flow in AC circuits. It should also be noted that inductive reactance and capacitive reactance may create a phase shift between the voltage and current in an AC circuit. Whenever analyzing an AC circuit, it is very important to consider the resistance, inductive reactance, and the capacitive reactance. All three have an effect on the current of that circuit.

Resistance

As mentioned, resistance creates an opposition to current in an AC circuit similar to the resistance of a DC circuit. The current through a resistive portion of an AC circuit is inversely proportional to the resistance and directly proportional to the voltage applied to that circuit or portion of the circuit. The equations I = E / R & E = I × R show how current is related to both voltage and resistance. It should be noted that resistance in an AC circuit does not create a phase shift between voltage and current.

Figure 9-17 shows how a circuit of 10 ohms allows 11.5 amps of current flow through an AC resistive circuit of 115 volts.

$$I = \frac{E}{R}$$

$$I = \frac{115V}{10\Omega}$$

$$I = 11.5 \text{ amps}$$

Inductive Reactance

When moving a magnet through a coil of wire, a voltage is induced across the coil. If a complete circuit is provided, then a current will also be induced. The amount of induced voltage is directly proportional to the rate of change of the magnetic field with respect to the coil. Conversely, current flowing through a coil of wire produces a magnetic field. When this wire is formed into a coil, it then becomes a basic inductor.

The primary effect of a coil is its property to oppose any change in current through it. This property is called inductance. When current flows through any conductor, a magnetic field starts to expand from the center of the wire. As the lines of magnetic

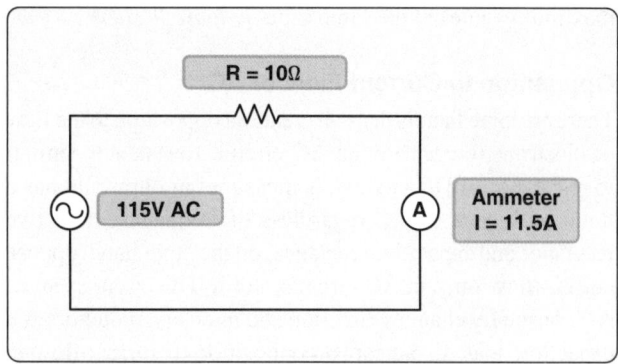

Figure 9-17. *Resistance.*

force grow outward through the conductor, they induce an EMF in the conductor itself. The induced voltage is always in the direction opposite to the direction of the applied current flow. The effects of this countering EMF are to oppose the applied current. This effect is only a temporary condition. Once the current reaches a steady value in the conductor, the lines of magnetic force are no longer expanding and the countering EMF is no longer present. Since AC is constantly changing in value, the inductance repeats in a cycle always opposite the applied voltage. It should be noted that the unit of measure for inductance is the henry (H).

The physical factors that affect inductance are:

1. Number of turns—doubling the number of turns in a coil produces a field twice as strong if the same current is used. As a general rule, the inductance varies with the square of the number of turns.

2. Cross-sectional area of the coil—the inductance of a coil increases directly as the cross-sectional area of the core increases. Doubling the radius of a coil increases the inductance by a factor of four.

3. Length of a coil—doubling the length of a coil, while keeping the same number of turns, reduces inductance by one-half.

4. Core material around which the coil is formed—

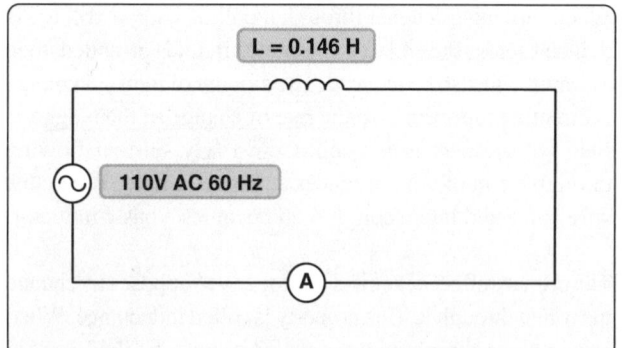

Figure 9-18. *AC circuit containing inductance.*

coils are wound on either magnetic or nonmagnetic materials. Some nonmagnetic materials include air, copper, plastic, and glass. Magnetic materials include nickel, iron, steel, and cobalt, which have a permeability that provides a better path for the magnetic lines of force and permit a stronger magnetic field.

Since AC is in a constant state of change, the magnetic fields within an inductor are also continuously changing and create an inducted voltage/current. This induced voltage opposes the applied voltage and is known as the counter EMF. This opposition is called inductive reactance, symbolized by X_L, and is measured in ohms. This characteristic of the inductor may also create a phase shift between voltage and current of the circuit. The phase shift created by inductive reactance always causes voltage to lead current. That is, the voltage of an inductive circuit reaches its peak values before the current reaches peak values. Additional discussions related to phase shift are presented later in this chapter.

Inductance is the property of a circuit to oppose any change in current and is measured in henries. Inductive reactance is a measure of how much the countering EMF in the circuit opposes the applied current. The inductive reactance of a component is directly proportional to the inductance of the component and the applied frequency to the circuit. By increasing either the inductance or applied frequency, the inductive reactance likewise increases and presents more opposition to current in the circuit. This relationship is given as $X_L = 2\pi fL$ Where X_L = inductive reactance in ohms, L = inductance in henries, f = frequency in cycles per second, and $\pi = 3.1416$.

In *Figure 9-18*, an AC series circuit is shown in which the inductance is 0.146 henry and the voltage is 110 volts at a frequency of 60 cycles per second. Inductive reactance is determined by the following method.

$$X_L = 2\pi \times f \times L$$

$$X_L = 6.28 \times 60 \times 0.146$$

$$X_L = 55\Omega$$

In AC series circuits, inductive reactance is added like resistances in series in a DC circuit. *[Figure 9-19]* The total reactance in the illustrated circuit equals the sum of the individual reactances.

$$X_L = X_{L1} + X_{L2}$$

$$X_L = 10\Omega + 15\Omega$$

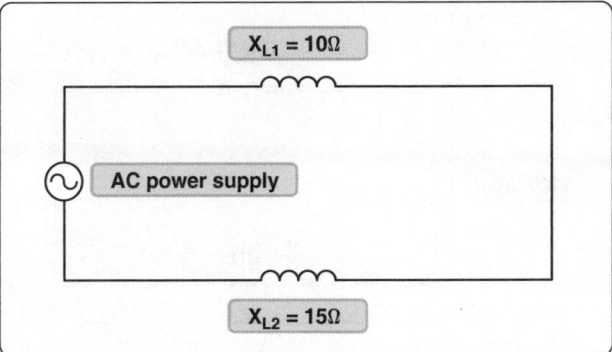

Figure 9-19. *Inductances in series.*

$X_{LT} = 25\Omega$

The total reactance of inductors connected in parallel is found the same way as the total resistance in a parallel circuit. *[Figure 9-20]* Thus, the total reactance of inductances connected in parallel, as shown, is expressed as:

$$X_{LT} = \cfrac{1}{\cfrac{1}{X_{L1}} + \cfrac{1}{X_{L2}} + \cfrac{1}{X_{L3}}}$$

$$X_{LT} = \cfrac{1}{\cfrac{1}{15} + \cfrac{1}{15} + \cfrac{1}{15}}$$

$X_{LT} = 5\Omega$

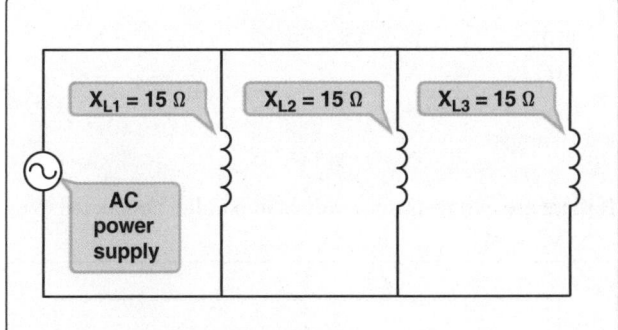

Figure 9-20. *Inductances in parallel.*

Capacitive Reactance

Capacitance is the ability of a body to hold an electric charge. In general, a capacitor is constructed of two parallel plates separated by an insulator. The insulator is commonly called the dielectric. The capacitor's plates have the ability to store

electrons when charged by a voltage source. The capacitor discharges when the applied voltage is no longer present and the capacitor is connected to a current path. In an electrical circuit, a capacitor serves as a reservoir or storehouse for electricity.

The basic unit of capacitance is the farad and is given by the letter F. By definition, one farad is one coulomb of charge stored with one volt across the plates of the capacitor. In practical terms, one farad is a large amount of capacitance. Typically, in electronics, much smaller units are used. The two more common smaller units are the microfarad (μF), which is 10^{-6} farad and the picofarad (pF), which is 10^{-12} farad.

Capacitance is a function of the physical properties of the capacitor:

1. The capacitance of parallel plates is directly proportional to their area. A larger plate area produces a larger capacitance, and a smaller area produces less capacitance. If we double the area of the plates, there is room for twice as much charge.

2. The capacitance of parallel plates is inversely proportional to the distance between the plates.

3. The dielectric material effects the capacitance of parallel plates. The dielectric constant of a vacuum is defined as 1, and that of air is very close to 1. These values are used as a reference, and all other materials have values relative to that of air (vacuum).

When an AC is applied in the circuit, the charge on the plates constantly changes. *[Figure 9-21]* This means that electricity must flow first from Y clockwise around to X, then from X counterclockwise around to Y, then from Y clockwise around to X, and so on. Although no current flows through the insulator between the plates of the capacitor, it constantly flows in the remainder of the circuit between X and Y. As this current alternates to and from the capacitor, a certain time lag is created. When a capacitor charges or discharges through a resistance, a certain amount of time is required for a full

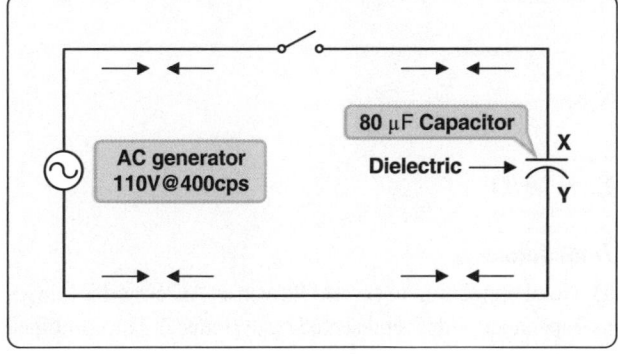

Figure 9-21. *Capacitor in an AC circuit.*

charge or discharge. The voltage across the capacitor does not change instantaneously. The rate of charging or discharging is determined by the time constant of the circuit. This rate of charge and discharge creates an opposition to current flow in AC circuits known as capacitive reactance. Capacitive reactance is symbolized by X_C and is measured in ohms. This characteristic of a capacitor may also create a phase shift between voltage and current of the circuit. The phase shift created by capacitive reactance always causes current to lead voltage. That is, the current of a capacitive circuit reaches its peak values before the voltage reaches peak values.

Capacitive reactance is a measure of how much the capacitive circuit opposes the applied current flow. Capacitive reactance is measured in ohms. The capacitive reactance of a circuit is indirectly proportional to the capacitance of the circuit and the applied frequency to the circuit. By increasing either the capacitance or applied frequency, the capacitive reactance decreases, and vice versa. This relationship is given as:

$$X_C = \frac{1}{2\pi fC}$$

Where: X_C = capacitive reactance in ohms, C = capacitance in farads, f = frequency in cycles per second, and π = 3.1416.

In *Figure 9-21,* a series circuit is shown in which the applied voltage is 110 volts at 400 cps, and the capacitance of a condenser is 80 mf. Find the capacitive reactance and the current flow.

To find the capacitive reactance, the following equation:

$$X_C = \frac{1}{2\pi fC}$$

First, the capacitance, 80 μf, is changed to farads by dividing 80 by 1,000,000, since 1 million microfarads is equal to 1 farad. This quotient equals 0.000080 farad. This is substituted in the equation:

$$X_C = \frac{1}{2\pi fC}$$

$$X_C = \frac{1}{2\pi(400)(0.000080)}$$

$$X_C = 4.97\Omega$$

Impedance

The total opposition to current flow in an AC circuit is known as impedance and is represented by the letter Z. The combined effects of resistance, inductive reactance, and capacitive

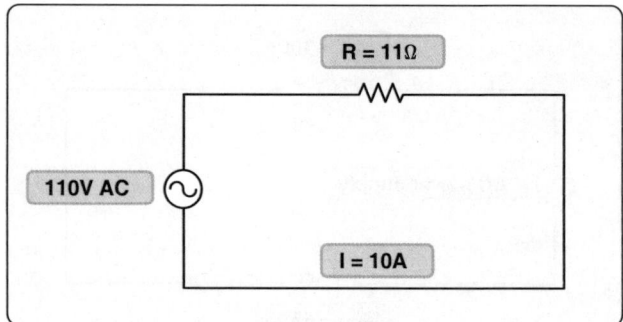

Figure 9-22. *Ohm's Law applies to AC circuit only when circuit consists of resistance only. Impedance (Z) = Resistance (R).*

reactance make up impedance (the total opposition to current flow in an AC circuit). In order to accurately calculate voltage and current in AC circuits, the effect of inductance and capacitance along with resistance must be considered. Impedance is measured in ohms.

The rules and equations for DC circuits apply to AC circuits only when that circuit contains resistance alone and no inductance or capacitance. In both series and parallel circuits, if an AC circuit consists of resistance only, the value of the impedance is the same as the resistance, and Ohm's Law for an AC circuit, I = E/Z, is exactly the same as for a DC circuit. *Figure 9-22* illustrates a series circuit containing a heater element with 11 ohms resistance connected across a 110-volt source. To find how much current flows if 110 volts AC is applied, the following example is solved:

$$I = \frac{E}{Z}$$

$$I = \frac{110V}{11\Omega}$$

I = 10 amps

If there are two resistance values in parallel connected to an

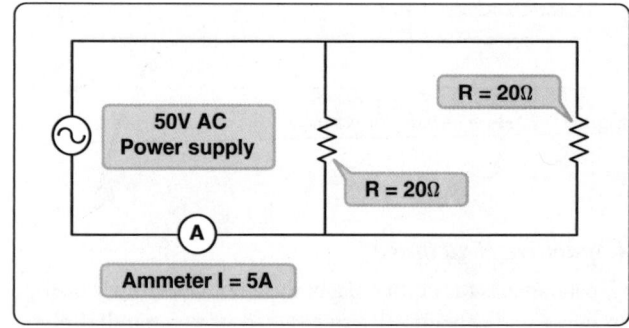

Figure 9-23. *Two resistance values in parallel connected to an AC voltage. Impedance is equal to the total resistance of the circuit.*

AC voltage, as seen in *Figure 9-23,* impedance is equal to the total resistance of the circuit. Once again, the calculations would be handled the same as if it were a DC circuit and the following would apply:

$$R_T = \cfrac{1}{\cfrac{1}{R_1} + \cfrac{1}{R_2}}$$

$$R_T = \cfrac{1}{\cfrac{1}{20} + \cfrac{1}{20}}$$

$$R_T = 10\Omega$$

Since this is a pure resistive circuit $R_T = Z$ (Resistance = Impedance)

$$Z_T = R_T$$

$$Z_T = 10\Omega$$

To determine the current flow in the circuit use the equation:

$$I = \frac{E}{Z}$$

$$I = \frac{50V}{10\Omega}$$

$$I = 5 \text{ amps}$$

Impedance is the total opposition to current flow in an AC circuit. If a circuit has inductance or capacitance, one must take into consideration resistance (R), inductive reactance (X_L), and/or capacitive reactance (X_C) to determine impedance (Z). In this case, Z does not equal R_T. Resistance and reactance (inductive or capacitive) cannot be added directly, but they can be considered as two forces acting at right angles to each other. Thus, the relation between resistance, reactance, and impedance may be illustrated by a right triangle. *[Figure 9-24]* Since these quantities may be related to the sides of a right triangle, the formula for finding the impedance can be found using the Pythagorean Theorem. It states that the square of the hypotenuse is equal to the sum of the squares of the other two sides. Thus, the value of any side of a right triangle can be found if the other two sides are known.

In practical terms, if a series AC circuit contains resistance and inductance, as shown in *Figure 9-25,* the relation between the sides can be stated as:

$$Z^2 = R^2 + (X_L - X_C)^2$$

The square root of both sides of the equation gives:

$$Z = \sqrt{R^2 + (X_L - X_C)^2}$$

This formula can be used to determine the impedance when the values of inductive reactance and resistance are known. It can be modified to solve for impedance in circuits containing capacitive reactance and resistance by substituting X_C in the formula in place of X_L. In circuits containing resistance with both inductive and capacitive reactance, the reactances can be combined; but because their effects in the circuit are exactly opposite, they are combined by subtraction (the smaller number is always subtracted from the larger):

$$Z = X_L - X_C$$

or

$$X = X_C - X_L$$

Figure 9-25 shows example 1. Here, a series circuit containing a resistor and an inductor are connected to a source of 110 volts at 60 cycles per second. The resistive element is a simple measuring 6 ohms, and the inductive element is a coil with an inductance of 0.021 henry. What is the value of the impedance and the current through the circuit?

Solution:
First, the inductive reactance of the coil is computed:

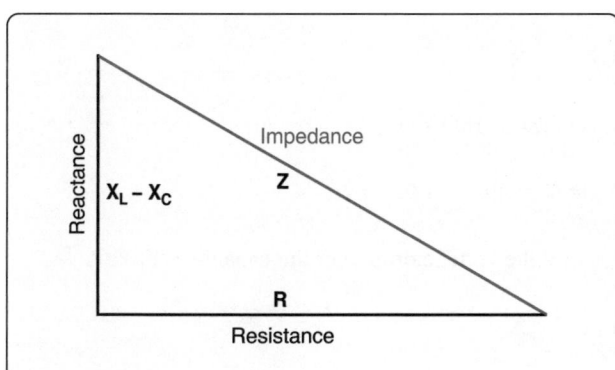

Figure 9-24. *Impedance triangle.*

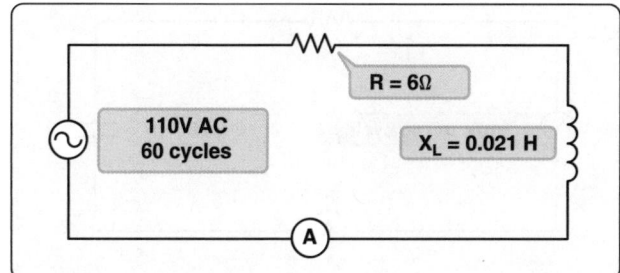

Figure 9-25. *A circuit containing resistance and inductance.*

$X_L = 2\pi \times f \times L$

$X_L = 6.28 \times 60 \times 0.021$

$X_L = 8$ ohms inductive reactance

Next, the total impedance is computed:

$Z = \sqrt{R^2 + X_L^2}$

$Z = \sqrt{6^2 + 8^2}$

$Z = \sqrt{36 + 64}$

$Z = \sqrt{100}$

$Z = 10\Omega$

Remember when making calculations for Z always use inductive reactance not inductance, and use capacitive reactance, not capacitance.

Once impedance is found, the total current can be calculated.

$I = \dfrac{E}{Z}$

$I = \dfrac{110V}{10\Omega}$

$I = 11$ amps

Since this circuit is resistive and inductive, there is a phase shift where voltage leads current.

Example 2 is a series circuit illustrated in which a capacitor of 200 μf is connected in series with a 10 ohm resistor. [Figure 9-26] What is the value of the impedance, the current flow, and the voltage drop across the resistor?

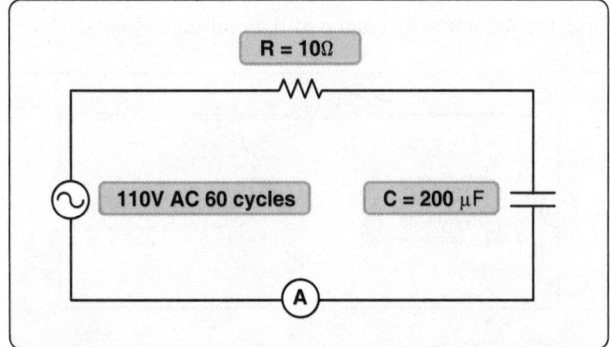

Figure 9-26. *A circuit containing resistance and capacitance.*

Solution:
First, the capacitance is changed from microfarads to farads. Since 1 million microfarads equal 1 farad, then 200 μf = 0.000200 farads.

Next solve for capacitive reactance:

$X_C = \dfrac{1}{2\pi fC}$

$X_C = \dfrac{1}{2\pi(60)(.00020)}$

$X_C = \dfrac{1}{0.07536}$

$X_C = 13\Omega$

To find the impedance,

$Z = \sqrt{R^2 + X_C^2}$

$Z = \sqrt{10^2 + 13^2}$

$Z = 16.4\Omega$

Since this circuit is resistive and capacitive, there is a phase shift where current leads voltage:

To find the current:

$I_T = \dfrac{E}{Z}$

$I_T = \dfrac{110V}{6.4\Omega}$

$I_T = 6.7$ amps

To find the voltage drop across the resistor (E_R):

$E_R = I \times R$

$E_R = 6.7A \times 10\Omega$

$E_R = 67$ volts

To find the voltage drop over the capacitor (E_C):

$E_C = I \times X_C$

$E_C = 6.7A \times 13\Omega$

$E_C = 86.1$ volts

The sum of these two voltages does not equal the applied voltage, since the current leads the voltage. Use the following formula to find the applied voltage:

$$E = \sqrt{(E_R)^2 + (E_C)2}$$

$$E = \sqrt{67^2 + 86.1^2}$$

$$E = \sqrt{4,489 + 7,413}$$

$$E = \sqrt{11,902}$$

$E = 110$ volts

When the circuit contains resistance, inductance, and capacitance, the following equation is used to find the impedance.

$$Z = \sqrt{R^2 + (X_L - X_C)^2}$$

Example 3: What is the impedance of a series circuit consisting of a capacitor with a capacitive reactance of 7 ohms, an inductor with an inductive reactance of 10 ohms, and a resistor with a resistance of 4 ohms? *[Figure 9-27]*

Solution:

$$Z = \sqrt{R^2 + (X_L - X_C)^2}$$

$$Z = \sqrt{4^2 + (10 - 7)^2}$$

$$Z = \sqrt{25}$$

$$Z = 5\Omega$$

To find total current:

$$I_T = \frac{E_T}{Z}$$

$$I_T = \frac{110V}{5\Omega}$$

$I_T = 22$ amps

Remember that inductive and capacitive reactances can cause a phase shift between voltage and current. In this example, inductive reactance is larger than capacitive reactance, so the voltage leads current.

It should be noted that since inductive reactance, capacitive reactance, and resistance affect each other at right angles, the voltage drops of any series AC circuit should be added using vector addition. *Figure 9-28* shows the voltage drops over the series AC circuit described in example 3 above.

To calculate the individual voltage drops, simply use the equations:

$$E_R = I \times R$$

$$E_{X_L} = I \times X_L$$

$$E_{X_C} = I \times X_C$$

To determine the total applied voltage for the circuit, each individual voltage drop must be added using vector addition.

$$E_T = \sqrt{E_R^2 + (E_L - E_C)^2}$$

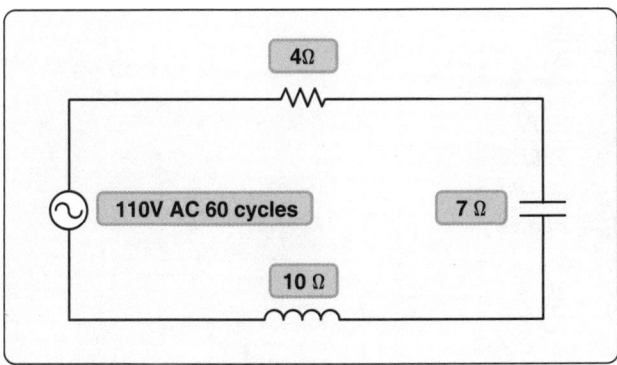

Figure 9-27. *A circuit containing resistance, inductance, and capacitance.*

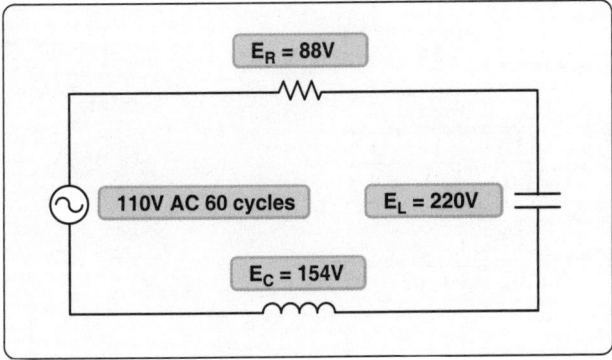

Figure 9-28. *Voltage drops.*

$$E_T = \sqrt{88^2 + (220 - 154)^2}$$

$$E_T = \sqrt{88^2 + 66^2}$$

$$E_T = \sqrt{12,100}$$

$$E_T = 110 \text{ volts}$$

Parallel AC Circuits

When solving parallel AC circuits, one must also use a derivative of the Pythagorean Theorem. The equation for finding impedance in an AC circuit is as follows:

$$Z = \sqrt{\left(\frac{1}{R}\right)^2 + \left(\frac{1}{X_L} - \frac{1}{X_C}\right)^2}$$

To determine the total impedance of the parallel circuit shown in *Figure 9-29*, one would first determine the capacitive and inductive reactances. (Remember to convert microfarads to farads.)

$$X_L = 2\pi FL$$

$$X_L = 2\pi(400)(0.02)$$

$$X_L = 50\Omega$$

$$X_C = \frac{1}{2\pi FC}$$

$$100\mu f = 0.0001F$$

$$X_C = \frac{1}{2\pi(400)(0.0001)}$$

$$X_C = 4\Omega$$

Next, the impedance can be found:

$$Z = \frac{1}{\sqrt{\left(\frac{1}{R}\right)^2 + \left(\frac{1}{X_L} - \frac{1}{X_C}\right)^2}}$$

$$Z = \frac{1}{\sqrt{\left(\frac{1}{50}\right)^2 + \left(\frac{1}{50} - \frac{1}{4}\right)^2}}$$

$$Z = \frac{1}{\sqrt{(.02)^2 + (.02 - .25)^2}}$$

$$Z = \frac{1}{\sqrt{.0004 + .0529}}$$

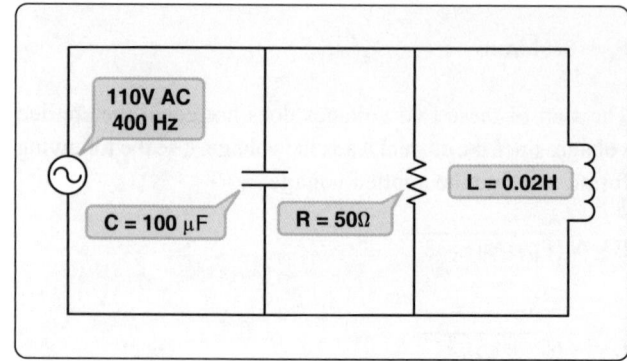

Figure 9-29. *Total impedance of parallel circuit.*

$$Z = \frac{1}{.23}$$

$$Z = 4.33\Omega$$

To determine the current flow in the circuit:

$$I_T = \frac{E_T}{Z}$$

$$I_T = \frac{100V}{4.33\Omega}$$

$$I_T = 23.09 \text{ amps}$$

To determine the current flow through each parallel path of the circuit, calculate I_R, I_L, and I_C.

$$I_R = \frac{E}{R}$$

$$I_R = \frac{100V}{50\Omega}$$

$$I_R = 2 \text{ amps}$$

$$I_L = \frac{E}{X_L}$$

$$I_L = \frac{100V}{50\Omega}$$

$$I_L = 2 \text{ amps}$$

$$I_C = \frac{E}{X_C}$$

$$I_C = \frac{100V}{4\Omega}$$

$I_C = 25$ amps

It should be noted that the total current flow of parallel circuits is found by using vector addition of the individual current flows as follows:

$$I_T = \sqrt{I_R^2 + (I_L - I_C)^2}$$

$$I_T = \sqrt{2^2 + (2 - 25)^2}$$

$$I_T = \sqrt{2^2 + 23^2}$$

$$I_T = \sqrt{4 + 529}$$

$$I_T = \sqrt{533}$$

$$I_T = 23 \text{ amps}$$

Power in AC Circuits

Since voltage and current determine power, there are similarities in the power consumed by both AC and DC circuits. In AC however, current is a function of both the resistance and the reactance of the circuit. The power consumed by any AC circuit is a function of the applied voltage and both circuit's resistance and reactance. AC circuits have two distinct types of power, one created by the resistance of the circuit and one created by the reactance of the circuit.

True Power

True power of any AC circuit is commonly referred to as the working power of the circuit. True power is the power consumed by the resistance portion of the circuit and is measured in watts (W). True power is symbolized by the letter P and is indicated by any wattmeter in the circuit. True power is calculated by the formula:

$$P = I^2 \times Z$$

Apparent Power

Apparent power in an AC circuit is sometimes referred to as the reactive power of a circuit. Apparent power is the power consumed by the entire circuit, including both the resistance and the reactance. Apparent power is symbolized by the letter S and is measured in volt-amps (VA). Apparent power is a product of the effective voltage multiplied by the effective current. Apparent power is calculated by the formula:

$$S = I^2 \times Z$$

Power Factor

As seen in *Figure 9-30*, the resistive power and the reactive power effect the circuit at right angles to each other. The power factor in an AC circuit is created by this right angle effect.

Power factor can be defined as the mathematical difference between true power and apparent power. Power factor (PF) is a ratio and always a measurement between 0 and 100. The power factor is directly related to the phase shift of a circuit. The greater the phase shift of a circuit the lower the power factor. For example, an AC circuit that is purely inductive (contains reactance only and no resistance) has a phase shift of 90° and a power factor of 0.0. An AC circuit that is purely resistive (has no reactance) has a phase shift of 0 and a power factor of 100. Power factor is calculated by using the following formula:

$$PF = \frac{\text{True Power (Watts)}}{\text{Apparent Power (VA)}} \times 100$$

Example of calculating PF: *Figure 9-31* shows an AC load connected to a 50 volt power supply. The current draw of the circuit is 5 amps and the total resistance of the circuit is 8 ohms. Determine the true power, the apparent power, and the power factor for this circuit.

Solution:

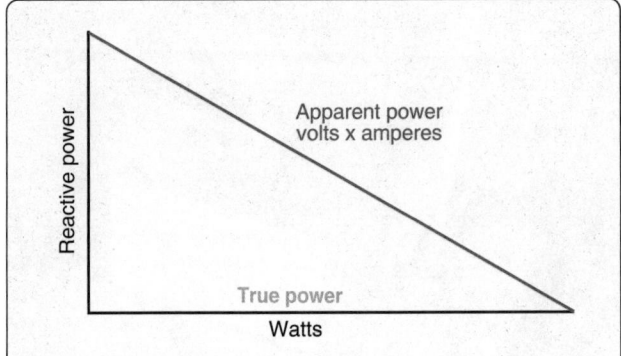

Figure 9-30. *Power relations in AC circuit.*

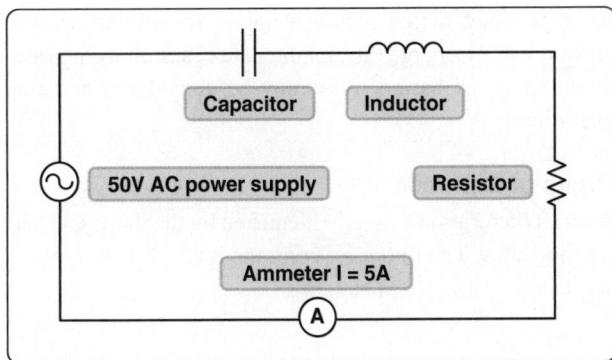

Figure 9-31. *AC load connected to a 50-volt power supply.*

$$P = I^2 \times R$$

$$P = 5^2 \times 8$$

$$P = 200 \text{ Watts}$$

$$S = E \times I$$

$$S = 50 \times 5$$

$$S = 250VA$$

$$PF = \frac{TP}{S} \times 100$$

$$PF = \frac{200}{250} \times 100$$

$$PF = 80$$

Power factor can also be represented as a percentage. Using a percentage to show power factor, the circuit in the previous example would have a power factor of 80 percent.

It should be noted that a low power factor is undesirable. Circuits with a lower power factor create excess load on the power supply and produce inefficiency in the system. Aircraft AC alternators must typically operate with a power factor between 90 percent and 100 percent. It is therefore very important to carefully consider power factor when designing the aircraft electrical system.

Aircraft Batteries

Aircraft batteries are used for many functions (e.g., ground power, emergency power, improving DC bus stability, and fault clearing). Most small private aircraft use lead-acid batteries. Most commercial and corporate aircraft use nickel-cadmium (NiCd) batteries. However, other lead acid types of batteries are becoming available, such as the valve-regulated lead-acid (VRLA) batteries. The battery best suited for a particular application depends on the relative importance of several characteristics, such as weight, cost, volume, service or shelf life, discharge rate, maintenance, and charging rate. Any change of battery type may be considered a major alteration.

Types of Batteries

Aircraft batteries are usually identified by the material used for the plates. The two most common types of battery used are lead-acid and NiCd batteries.

Lead-Acid Batteries

Dry Charged Cell Lead-Acid Batteries

Dry charged cell lead-acid batteries, also known as flooded or wet batteries, are assembled with electrodes (plates) that have been fully charged and dried. The electrolyte is added to the battery when it is placed in service, and battery life begins when the electrolyte is added. An aircraft storage battery consists of 6 or 12 lead-acid cells connected in series. The open circuit voltage of the 6 cell battery is approximately 12 volts, and the open circuit voltage of the 12-cell battery is approximately 24 volts. Open circuit voltage is the voltage of the battery when it is not connected to a load. When flooded (vented) batteries are on charge, the oxygen generated at the positive plates escapes from the cell. Concurrently, at the negative plates, hydrogen is generated from water and escapes from the cell. The overall result is the gassing of the cells and water loss. Therefore, flooded cells require periodic water replenishment. *[Figure 9-32]*

Valve-Regulated Lead-Acid (VRLA) Batteries

VRLA batteries contain all electrolyte absorbed in glass-mat

Figure 9-32. *Lead-acid battery installation.*

Figure 9-33. *Valve-regulated lead-acid battery (sealed battery).*

separators with no free electrolyte and are sometimes referred to as sealed batteries. *[Figure 9-33]* The electrochemical reactions for VRLA batteries are the same as flooded batteries, except for the gas recombination mechanism that is predominant in VRLA batteries. These types of battery are used in general aviation and turbine powered aircraft and are sometimes authorized replacements for NiCd batteries.

When VRLA batteries are on charge, oxygen combines chemically with the lead at the negative plates in the presence of H_2SO_4 to form lead sulfate and water. This oxygen recombination suppresses the generation of hydrogen at the negative plates. Overall, there is no water loss during charging. A very small quantity of water may be lost as a result of self-discharge reactions; however, such loss is so small that no provisions are made for water replenishment. The battery cells have a pressure relief safety valve that may vent if the battery is overcharged.

NiCd Batteries

A NiCd battery consists of a metallic box, usually stainless steel, plastic-coated steel, painted steel, or titanium containing a number of individual cells. *[Figure 9-34]* These cells are connected in series to obtain 12 volts or 24 volts. The cells are connected by highly conductive nickel copper links. Inside the battery box, the cells are held in place by partitions, liners, spacers, and a cover assembly. The battery has a ventilation system to allow the escape of the gases produced during an overcharge condition and provide cooling during normal operation.

NiCd cells installed in an aircraft battery are typical of the vented cell type. The vented cells have a vent or low pressure release valve that releases any generated oxygen and hydrogen gases when overcharged or discharged rapidly. This also means the battery is not normally damaged by

excessive rates of overcharge, discharge, or even negative charge. The cells are rechargeable and deliver a voltage of 1.2 volts during discharge.

Aircraft that are outfitted with NiCd batteries typically have a fault protection system that monitors the condition of the battery. The battery charger is the unit that monitors the condition of the battery and the following conditions are monitored:

1. Overheat condition,

2. Low temperature condition (below –40 °F),

3. Cell imbalance,

4. Open circuit, and

5. Shorted circuit.

If the battery charger finds a fault, it turns off and sends a fault signal to the Electrical Load Management System (ELMS).

NiCd batteries are capable of performing to their rated capacity when the ambient temperature of the battery is in the range of approximately 60–90 °F. An increase or decrease in temperature from this range results in reduced capacity. NiCd batteries have a ventilation system to control the temperature of the battery. A combination of high battery temperature (in excess of 160 °F) and overcharging can lead to a condition called thermal runaway. *[Figure 9-35]* The temperature of the battery has to be constantly monitored to ensure safe operation. Thermal runaway can result in a NiCd chemical fire and/or explosion of the NiCd battery under recharge by a constant-voltage source and is due to cyclical, ever-increasing temperature and charging current. One or more shorted cells or an existing high temperature and low charge can produce the following cyclical sequence of events:

1. Excessive current,

2. Increased temperature,

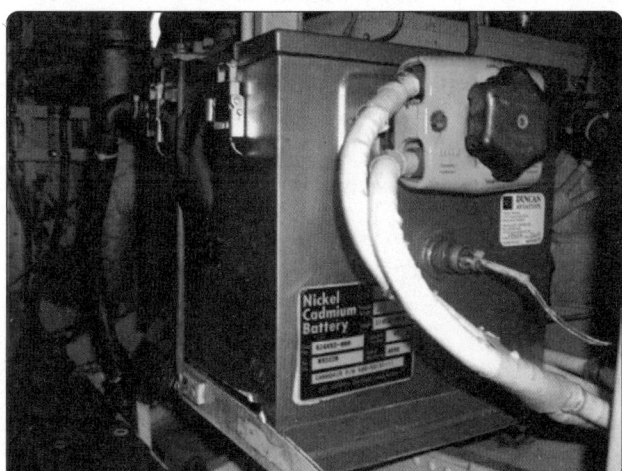

Figure 9-34. *NiCd battery installation.*

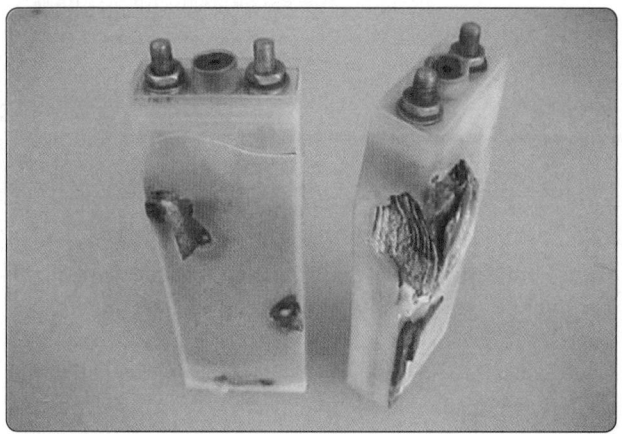

Figure 9-35. *Thermal runaway damage.*

3. Decreased cell(s) resistance,

4. Further increased current, and

5. Further increased temperature.

This does not become a self-sustaining thermal-chemical action if the constant-voltage charging source is removed before the battery temperature is in excess of 160 °F.

Capacity

Capacity is measured quantitatively in ampere-hours delivered at a specified discharge rate to a specified cut-off voltage at room temperature. The cut-off voltage is 1.0 volt per cell. Battery available capacity depends upon several factors including such items as:

1. Cell design (cell geometry, plate thickness, hardware, and terminal design govern performance under specific usage conditions of temperature, discharge rate, etc.).

2. Discharge rate (high current rates yield less capacity than low rates).

3. Temperature (capacity and voltage levels decrease as battery temperature moves away from the 60 °F (16 °C) to 90 °F (32 °C) range toward the high and low extremes).

4. Charge rate (higher charge rates generally yield greater capacity).

Aircraft Battery Ratings by Specification

The one-hour rate is the rate of discharge a battery can endure for 1 hour with the battery voltage at or above 1.67 volts per cell, or 20 volts for a 24-volt lead-acid battery, or 10 volts for a 12-volt lead-acid battery. The one-hour capacity, measured in ampere hours (Ah), is the product of the discharge rate and time (in hours) to the specified end voltage.

The emergency rate is the total essential load, measured in amperes, required to support the essential bus for 30 minutes. This is the rate of discharge a battery can endure for 30 minutes with the battery voltage at or above 1.67 volts per cell, or 20 volts for a 24 volt lead-acid battery, or 10 volts for a 12 volt lead-acid battery.

Storing & Servicing Facilities

Separate facilities for storing and/or servicing flooded electrolyte lead-acid and NiCd batteries must be maintained. Introduction of acid electrolyte into alkaline electrolyte causes permanent damage to vented (flooded electrolyte) NiCd batteries and vice versa. However, batteries that are sealed can be charged and capacity checked in the same area. Because the electrolyte in a valve-regulated lead-acid battery is absorbed in the separators and porous plates, it cannot contaminate a NiCd battery even when they are serviced in the same area.

Warning: It is extremely dangerous to store or service lead-acid and NiCd batteries in the same area. Introduction of acid electrolytes into alkaline electrolyte destroys the NiCd, and vice versa.

Battery Freezing

Discharged lead-acid batteries exposed to cold temperatures are subject to plate damage due to freezing of the electrolyte. To prevent freezing damage, maintain each cell's specific gravity at 1.275 or, for sealed lead-acid batteries, check open circuit voltage. *[Figure 9-36]* NiCd battery electrolyte is not as susceptible to freezing because no appreciable chemical change takes place between the charged and discharged states. However, the electrolyte freezes at approximately –75 °F.

Note: Only a load check determines overall battery condition.

Temperature Correction

U.S.-manufactured lead-acid batteries are considered fully charged when the specific gravity reading is between 1.275 and 1.300. A ⅓ discharged battery reads about 1.240 and a ⅔ discharged battery shows a specific gravity reading of about 1.200 when tested by a hydrometer at an electrolyte temperature of 80 °F. However, to determine precise specific gravity readings, a temperature correction should be applied to the hydrometer indication. *[Figure 9-37]* As an example, for a hydrometer reading of 1.260 and electrolyte temperature of 40 °F, the corrected specific gravity reading of the electrolyte is 1.244.

Battery Charging

Operation of aircraft batteries beyond their ambient temperature or charging voltage limits can result in excessive cell temperatures leading to electrolyte boiling, rapid

Specific Gravity	Freezing Point		State of Charge (SOC) for Sealed Lead-Acid Batteries at 70°		
	°C	°F	SOC	12 volt	24 volt
1.300	–70	–95	100%	12.9	25.8
1.275	–62	–80	75%	12.7	25.4
1.250	–52	–62	50%	12.4	24.8
1.225	–37	–35	25%	12.0	24.0
1.200	–26	–16			
1.175	–20	–04			
1.150	–15	+05			
1.125	–10	+13			
1.100	–08	+19			

Figure 9-36. *Lead-acid battery electrolyte freezing points.*

Electrolyte Temperature		Points to Subtract From or Add to Specific Gravity Readings
°C	°F	12 volt
+60	+140	+0.024
+55	+130	+0.020
+49	+120	+0.016
+43	+110	+0.012
+38	+100	+0.008
+33	+90	+0.004
+27	+80	0
+23	+70	−0.004
+15	+60	−0.008
+10	+50	−0.012
+05	+40	−0.016
−02	+30	−0.020
−07	+20	−0.024
−13	+10	−0.028
−18	0	−0.032
−23	−10	−0.036
−28	−20	−0.040
−35	−30	−0.044

Figure 9-37. *Sulfuric acid temperature correction.*

deterioration of the cells, and battery failure. The relationship between maximum charging voltage and the number of cells in the battery is also significant. This determines (for a given ambient temperature and state of charge) the rate at which energy is absorbed as heat within the battery. For lead-acid batteries, the voltage per cell must not exceed 2.35 volts. In the case of NiCd batteries, the charging voltage limit varies with design and construction. Values of 1.4 and 1.5 volts per cell are generally used. In all cases, follow the recommendations of the battery manufacturer.

Constant Voltage (CV) Charging
The battery charging system in an airplane is of the constant voltage type. An engine-driven generator, capable of supplying the required voltage, is connected through the aircraft electrical system directly to the battery. A battery switch is incorporated in the system so that the battery may be disconnected when the airplane is not in operation.

The voltage of the generator is accurately controlled by means of a voltage regulator connected in the field circuit of the generator. For a 12-volt system, the voltage of the generator is adjusted to approximately 14.25. On 24-volt systems, the adjustment should be between 28 and 28.5 volts. When these conditions exist, the initial charging current through the battery is high. As the state of charge increases, the battery voltage also increases, causing the current to taper down. When the battery is fully charged, its voltage is almost equal to the generator voltage, and very little current flows into the battery. When the charging current is low, the battery may remain connected to the generator without damage.

When using a constant-voltage system in a battery shop, a voltage regulator that automatically maintains a constant voltage is incorporated in the system. A higher capacity battery (e.g., 42 Ah) has a lower resistance than a lower capacity battery (e.g., 33 Ah). Hence, a high-capacity battery draws a higher charging current than a low-capacity battery when both are in the same state of charge and when the charging voltages are equal. The constant voltage method is the preferred charging method for lead-acid batteries.

Constant Current (CC) Charging
Constant current charging is the most convenient for charging batteries outside the airplane because several batteries of varying voltages may be charged at once on the same system. A constant current charging system usually consists of a rectifier to change the normal AC supply to DC. A transformer is used to reduce the available 110-volt or 220-volt AC supply to the desired level before it is passed through the rectifier. If a constant current charging system is used, multiple batteries may be connected in series, provided that the charging current is kept at such a level that the battery does not overheat or gas excessively.

The constant current charging method is the preferred method for charging NiCd batteries. Typically, a NiCd battery is constant current charged at a rate of 1CA until all the cells have reached at least 1.55V. Another charge cycle follows at 0.1CA, again until all cells have reached 1.55V. The charge is finished with an overcharge or top-up charge, typically for not less than 4 hours at a rate of 0.1CA. The purpose of the overcharge is to expel as much, if not all the gases collected on the electrodes, hydrogen on the anode, and oxygen on the cathode; some of these gases recombine to form water that, in turn, raises the electrolyte level to its highest level after which it is safe to adjust the electrolyte levels. During the overcharge or top-up charge, the cell voltages go beyond 1.6V and then slowly start to drop. No cell should rise above 1.71V (dry cell) or drop below 1.55V (gas barrier broken).

Charging is done with vent caps loosened or open. A stuck vent might increase the pressure in the cell. It also allows for refilling of water to correct levels before the end of the top-up charge while the charge current is still on. However, cells should be closed again as soon as the vents have been cleaned and checked since carbon dioxide dissolved from outside air carbonates the cells and ages the battery.

Battery Maintenance
Battery inspection and maintenance procedures vary with the type of chemical technology and the type of physical construction. Always follow the battery manufacturer's

approved procedures. Battery performance at any time in a given application depends upon the battery's age, state of health, state of charge, and mechanical integrity, which you can determine according to the following:

- To determine the life and age of the battery, record the install date of the battery on the battery. During normal battery maintenance, battery age must be documented either in the aircraft maintenance log or in the shop maintenance log.

- Lead-acid battery state of health may be determined by duration of service interval (in the case of vented batteries), by environmental factors (such as excessive heat or cold), and by observed electrolyte leakage (as evidenced by corrosion of wiring and connectors or accumulation of powdered salts). If the battery needs to be refilled often, with no evidence of external leakage, this may indicate a poor state of the battery, the battery charging system, or an overcharge condition.

- Use a hydrometer to determine the specific gravity of the lead-acid battery electrolyte, which is the weight of the electrolyte compared to the weight of pure water. Take care to ensure the electrolyte is returned to the cell from which it was extracted. When a specific gravity difference of 0.050 or more exists between cells of a battery, the battery is approaching the end of its useful life and replacement should be considered. Electrolyte level may be adjusted by the addition of distilled water. Do not add electrolyte.

- Battery state of charge is determined by the cumulative effect of charging and discharging the battery. In a normal electrical charging system, the aircraft generator or alternator restores a battery to full charge during a flight of 1 hour to 90 minutes.

- Proper mechanical integrity involves the absence of any physical damage, as well as assurance that hardware is correctly installed and the battery is properly connected. Battery and battery compartment venting system tubes, nipples, and attachments, when required, provide a means of avoiding the potential buildup of explosive gases, and should be checked periodically to ensure that they are securely connected and oriented in accordance with the maintenance manual's installation procedures. Always follow procedures approved for the specific aircraft and battery system to ensure that the battery system is capable of delivering specified performance.

Battery & Charger Characteristics

The following information is provided to acquaint the user with characteristics of the more common aircraft battery and battery charger types. *[Figure 9-38]* Products may vary

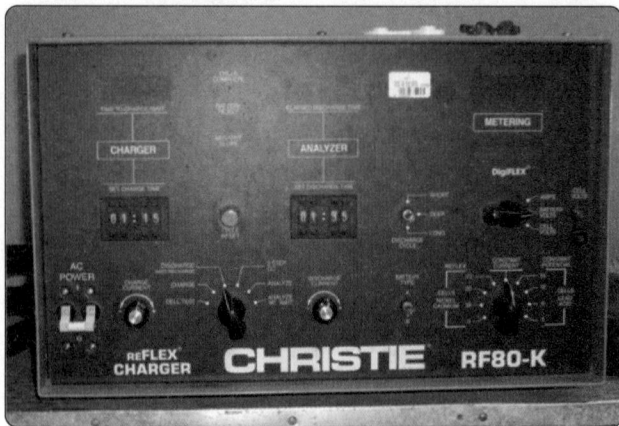

Figure 9-38. *Battery charger.*

from these descriptions due to different applications of available technology. Consult the manufacturer for specific performance data.

Note: Never connect a lead-acid battery to a charger, unless properly serviced.

Lead-Acid Batteries

Lead-acid vented batteries have a two volt nominal cell voltage. Batteries are constructed so that individual cells cannot be removed. Occasional addition of water is required to replace water loss due to overcharging in normal service. Batteries that become fully discharged may not accept recharge. Lead-acid sealed batteries are similar in most respects to lead-acid vented batteries, but do not require the addition of water.

The lead-acid battery is economical and has extensive application but is heavier than an equivalent performance battery of another type. The battery is capable of a high rate of discharge and low-temperature performance. However, maintaining a high rate of discharge for a period of time usually warps the cell plates, shorting out the battery. Its electrolyte has a moderate specific gravity, and state of charge can be checked with a hydrometer.

Lead-acid batteries are usually charged by regulated DC voltage sources. This allows maximum accumulation of charge in the early part of recharging.

NiCd Batteries

NiCd vented batteries have a 1.2-volt nominal cell voltage. Occasional addition of distilled water is required to replace water loss due to overcharging in normal service. Cause of failure is usually shorting or weakening of a cell. After replacing the bad cell with a good cell, the battery's life can be extended for 5 or more years. Full discharge is not harmful to this type of battery.

NiCd sealed batteries are similar in most respects to NiCd vented batteries, but do not normally require the addition of water. Fully discharging the battery (to zero volts) may cause irreversible damage to one or more cells, leading to eventual battery failure due to low capacity.

The state of charge of a NiCd battery cannot be determined by measuring the specific gravity of the potassium hydroxide electrolyte. The electrolyte specific gravity does not change with the state of charge. The only accurate way to determine the state of charge of a NiCd battery is by a measured discharge with a NiCd battery charger and following the manufacturer's instructions. After the battery has been fully charged and allowed to stand for at least 2 hours, the fluid level may be adjusted, if necessary, using distilled or demineralized water. Because the fluid level varies with the state of charge, water should never be added while the battery is installed in the aircraft. Overfilling the battery results in electrolyte spewage during charging. This causes corrosive effects on the cell links, self-discharge of the battery, dilution of the electrolyte density, possible blockage of the cell vents, and eventual cell rupture.

Constant current battery chargers are usually provided for NiCd batteries because the NiCd cell voltage has a negative temperature coefficient. With a constant voltage charging source, a NiCd battery having a shorted cell might overheat due to excessive overcharge and undergo a thermal runaway, destroying the battery and creating a possible safety hazard to the aircraft. Pulsed-current battery chargers are sometimes provided for NiCd batteries.

Caution: It is important to use the proper charging procedures for batteries under test and maintenance. These charging regimes for reconditioning and charging cycles are defined by the aircraft manufacturer and should be closely followed.

Aircraft Battery Inspection

Aircraft battery inspection consists of the following items:

1. Inspect battery sump jar and lines for condition and security.

2. Inspect battery terminals and quickly disconnect plugs and pins for evidence of corrosion, pitting, arcing, and burns. Clean as required.

3. Inspect battery drain and vent lines for restriction, deterioration, and security.

4. Routine preflight and postflight inspection procedures should include observation for evidence of physical damage, loose connections, and electrolyte loss.

Ventilation Systems

Modern airplanes are equipped with battery ventilating systems. The ventilating system removes gasses and acid fumes from the battery in order to reduce fire hazards and to eliminate damage to airframe parts. Air is carried from a scoop outside the airplane through a vent tube to the interior of the battery case. After passing over the top of the battery, air, battery gasses, and acid fumes are carried through another tube to the battery sump. This sump is a glass or plastic jar of at least one pint capacity. In the jar is a felt pad about 1 inch thick saturated with a 5-percent solution of bicarbonate of soda and water. The tube carrying fumes to the sump extends into the jar to within about ¼ inch of the felt pad. An overboard discharge tube leads from the top of the sump jar to a point outside the airplane. The outlet for this tube is designed so there is negative pressure on the tube whenever the airplane is in flight. This helps to ensure a continuous flow of air across the top of the battery through the sump and outside the airplane. The acid fumes going into the sump are neutralized by the action of the soda solution, thus preventing corrosion of the aircraft's metal skin or damage to a fabric surface.

Installation Practices

- External surface—Clean the external surface of the battery prior to installation in the aircraft.

- Replacing lead-acid batteries—When replacing lead-acid batteries with NiCd batteries, a battery temperature or current monitoring system must be installed. Neutralize the battery box or compartment and thoroughly flush with water and dry. A flight manual supplement must also be provided for the NiCd battery installation. Acid residue can be detrimental to the proper functioning of a NiCd battery, as alkaline is to a lead-acid battery.

- Battery venting—Battery fumes and gases may cause an explosive mixture or contaminated compartments and should be dispersed by adequate ventilation. Venting systems often use ram pressure to flush fresh air through the battery case or enclosure to a safe overboard discharge point. The venting system pressure differential should always be positive and remain between recommended minimum and maximum values. Line runs should not permit battery overflow fluids or condensation to be trapped and prevent free airflow.

- Battery sump jars—A battery sump jar installation may be incorporated in the venting system to dispose of battery electrolyte overflow. The sump jar should be of adequate design and the proper neutralizing agent used. The sump jar must be located only on the

Trouble	Probable Cause	Corrective Action
Apparent loss of capacity	Very common when recharging on a constant potential bus, as in aircraft	Reconditioning will alleviate this condition.
	Usually indicates imbalance between cells because of difference in temperature, charge efficiency, self-discharge rate, etc., in the cells	
	Electrolyte level too low Battery not fully charged	Charge. Adjust electrolyte level. Check aircraft voltage regulator. If OK, reduce maintenance interval.
Complete failure to operate	Defective connection in equipment circuitry in which battery is installed, such as broken lead, inoperative relay, or improper receptacle installation	Check and correct external circuitry.
	End terminal connector loose or disengaged Poor intercell connections	Clean and retighten hardware using proper torque values.
	Open circuit or dry cell	Replace defective cell.
Excessive spewage of electrolyte	High charge voltage High temperature during charge Electrolyte level too high	Clean battery, charge, and adjust electrolyte level.
	Loose or damaged vent cap	Clean battery, tighten or replace cap, charge, and adjust electrolyte level.
	Damaged cell and seal	Short out all cells to 0 volts, clean battery, replace defective cell, charge, and adjust electrolyte level.
Failure of one or more cells to rise to the required 1.55 volts at the end of charge	Negative electrode not fully charged Cellophane separator damage	Discharge battery and recharge. If the cell still fails to rise to 1.55 volts or if the cell's voltage rises to 1.55 volts or above and then drops, remove cell and replace.
Distortion of cell case to cover	Overcharged, overdischarged, or overheated cell with internal short	Discharge battery and disassemble. Replace defective cell. Recondition battery.
	Plugged vent cap	Replace vent cap.
	Overheated battery	Check voltage regulator: treat battery as above, replacing battery case and cover and all other defective parts.
Foreign material within the cell case	Introduced into cell through addition of impure water or water contaminated with acid	Discharge battery and disassemble, remove cell and replace, recondition battery.
Frequent addition of water	Cell out of balance	Recondition battery.
	Damaged "O" ring, vent cap	Replace damaged parts.
	Leaking cell	Discharge battery and disassemble. Replace defective cell, recondition battery.
	Charge voltage too high	Adjust voltage regulator.
Corrosion of top hardware	Acid flumes or spray or other corrosive atmosphere	Replace parts. Battery should be kept clean and kept away from such environments.
Discolored or burned end connectors or intercell connectors.	Dirty connections. Loose connection. Improper mating of parts.	Clean parts: replace if necessary. Retighten hardware using proper torque values. Check to see that parts are properly mated.

Distortion of battery case and/or cover.	Explosion caused by:
	   Dry cells
	   Charger failure
	   High charge voltage
	   Plugged vent caps
	   Loose intercell connectors
		Discharge battery and disassemble. Replace damaged parts and recondition.

Figure 9-39. *Battery troubleshooting guide.*

discharge side of the battery venting system.

- Installing batteries—When installing batteries in an aircraft, exercise care to prevent inadvertent shorting of the battery terminals. Serious damage to the aircraft structure (frame, skin and other subsystems, avionics, wire, fuel, etc.) can be sustained by the resultant high discharge of electrical energy. This condition may normally be avoided by insulating the terminal posts during the installation process. Remove the grounding lead first for battery removal, then the positive lead. Connect the grounding lead of the battery last to minimize the risk of shorting the hot terminal of the battery during installation.

- Battery hold down devices—Ensure that the battery hold down devices are secure, but not so tight as to exert excessive pressure that may cause the battery to buckle causing internal shorting of the battery.

- Quick-disconnect type battery—If a quick-disconnect type of battery connector that prohibits crossing the battery lead is not employed, ensure that the aircraft wiring is connected to the proper battery terminal. Reverse polarity in an electrical system can seriously damage a battery and other electrical components. Ensure that the battery cable connections are tight to prevent arcing or a high resistance connection.

Troubleshooting

See *Figure 9-39* for a troubleshooting chart.

DC Generators & Controls

DC generators transform mechanical energy into electrical energy. As the name implies, DC generators produce direct current and are typically found on light aircraft. In many cases, DC generators have been replaced with DC alternators. Both devices produce electrical energy to power the aircraft's electrical loads and charge the aircraft's battery. Even though they share the same purpose, the DC alternator and DC generator are very different. DC generators require a control circuit in order to ensure the generator maintains the correct voltage and current for the current electrical conditions of the aircraft. Typically, aircraft generators maintain a nominal output voltage of approximately 14 volts or 28 volts.

Generators

The principles of electromagnetic induction were discussed earlier in this chapter. These principles show that voltage is induced in the armature of a generator throughout the entire 360° rotation of the conductor. The armature is the rotating portion of a DC generator. As shown, the voltage being induced is AC. *[Figure 9-40]*

Since the conductor loop is constantly rotating, some means

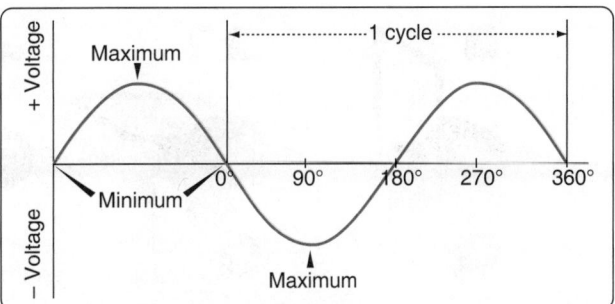

Figure 9-40. *Output of an elementary generator.*

Figure 9-41. *Generator slip rings and loop rotate; brushes are stationary.*

must be provided to connect this loop of wire to the electrical loads. As shown in *Figure 9-41,* slip rings and brushes can be used to transfer the electrical energy from the rotating loop to the stationary aircraft loads. The slip rings are connected to the loop and rotate; the brushes are stationary and allow a current path to the electrical loads. The slip rings are typically a copper material and the brushes are a soft carbon substance.

It is important to remember that the voltage being produced by this basic generator is AC, and AC voltage is supplied to the slip rings. Since the goal is to supply DC loads, some means must be provided to change the AC voltage to a DC voltage. Generators use a modified slip ring arrangement, known as a commutator, to change the AC produced in the generator loop into a DC voltage. The action of the commutator allows the generator to produce a DC output.

By replacing the slip rings of the basic AC generator with two half cylinders (the commutator), a basic DC generator is obtained. In *Figure 9-42*, the red side of the coil is connected to the red segment and the amber side of the coil to the amber segment. The segments are insulated from each other. The two stationary brushes are placed on opposite sides of the

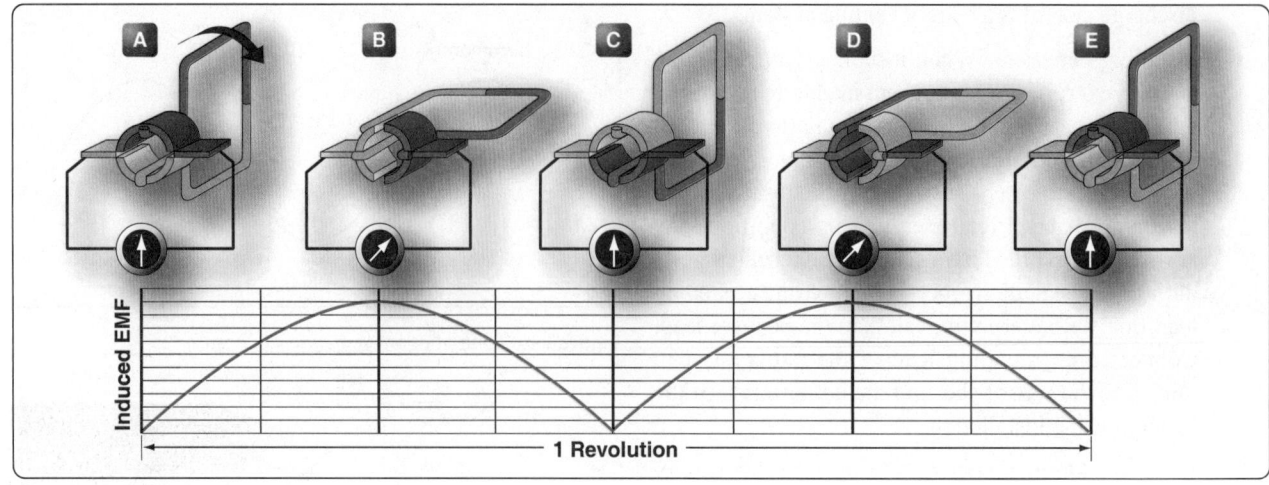

Figure 9-42. *A two-piece slip ring, or commutator, allows brushes to transfer current that flows in a single direction (DC).*

commutator and are so mounted that each brush contacts each segment of the commutator as the commutator revolves simultaneously with the loop. The rotating parts of a DC generator (coil and commutator) are called an armature.

As seen in the very simple generator of *Figure 9-42*, as the loop rotates the brushes make contact with different segments of the commutator. In positions A, C, and E, the brushes touch the insulation between the brushes; when the loop is in these positions, no voltage is being produced. In position B, the positive brush touches the red side of the conductor loop. In position D, the positive brush touches the amber side of the armature conductor. This type of connection reversal changes the AC produced in the conductor coil into DC to power the aircraft. An actual DC generator is more complex, having several loops of wire and commutator segments.

Because of this switching of commutator elements, the red brush is always in contact with the coil side moving downward, and the amber brush is always in contact with the coil side moving upward. Though the current actually reverses its direction in the loop in exactly the same way as in the AC generator, commutator action causes the current to flow always in the same direction through the external circuit or meter.

The voltage generated by the basic DC generator in *Figure 9-42* varies from zero to its maximum value twice for each revolution of the loop. This variation of DC voltage is called ripple and may be reduced by using more loops, or coils, as shown in *Figure 9-43*.

As the number of loops is increased, the variation between maximum and minimum values of voltage is reduced *[Figure 9-43]*, and the output voltage of the generator approaches a steady DC value. For each additional loop in

the rotor, another two commutator segments is required. A photo of a typical DC generator commutator is shown in *Figure 9-44*.

Construction Features of DC Generators

The major parts, or assemblies, of a DC generator are a field frame, a rotating armature, and a brush assembly. The parts of a typical aircraft generator are shown in *Figure 9-45*.

Field Frame

The frame has two functions: to hold the windings needed to produce a magnetic field, and to act as a mechanical support for the other parts of the generator. The actual electromagnet

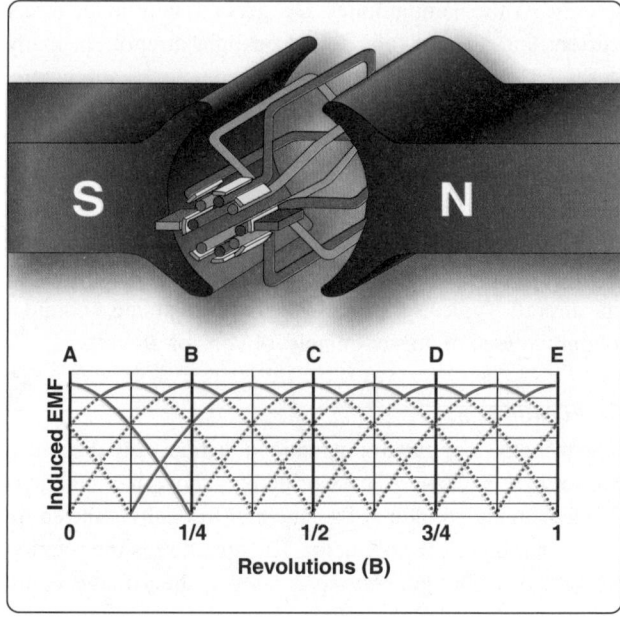

Figure 9-43. *Increasing the number of coils reduces the ripple in the voltage.*

Figure 9-44. *Typical DC generator commutator.*

conductor is wrapped around pieces of laminated metal called field poles. The poles are typically bolted to the inside of the frame and laminated to reduce eddy current losses and serve the same purpose as the iron core of an electromagnet; they concentrate the lines of force produced by the field coils. The field coils are made up of many turns of insulated wire and are usually wound on a form that fits over the iron core of the pole to which it is securely fastened. *[Figure 9-46]*

A DC current is fed to the field coils to produce an electromagnetic field. This current is typically obtained from an external source that provides voltage and current

regulation for the generator system. Generator control systems are discussed later in this chapter.

Armature

The armature assembly of a generator consists of two primary elements: the wire coils (called windings) wound around an iron core and the commutator assembly. The armature windings are evenly spaced around the armature and mounted on a steel shaft. The armature rotates inside the magnetic field produced by the field coils. The core of the armature acts as an iron conductor in the magnetic field and, for this reason, is laminated to prevent the circulation of eddy currents. A typical armature assembly is shown in *Figure 9-47*.

Commutators

Figure 9-48 shows a cross-sectional view of a typical commutator. The commutator is located at the end of an armature and consists of copper segments divided by a thin insulator. The insulator is often made from the mineral mica. The brushes ride on the surface of the commutator forming the electrical contact between the armature coils and the external circuit. A flexible, braided copper conductor, commonly called a pigtail, connects each brush to the external circuit. The brushes are free to slide up and down in their holders in order to follow any irregularities in the surface of the commutator. The constant making and breaking of electrical connections between the brushes and the commutator segments, along with

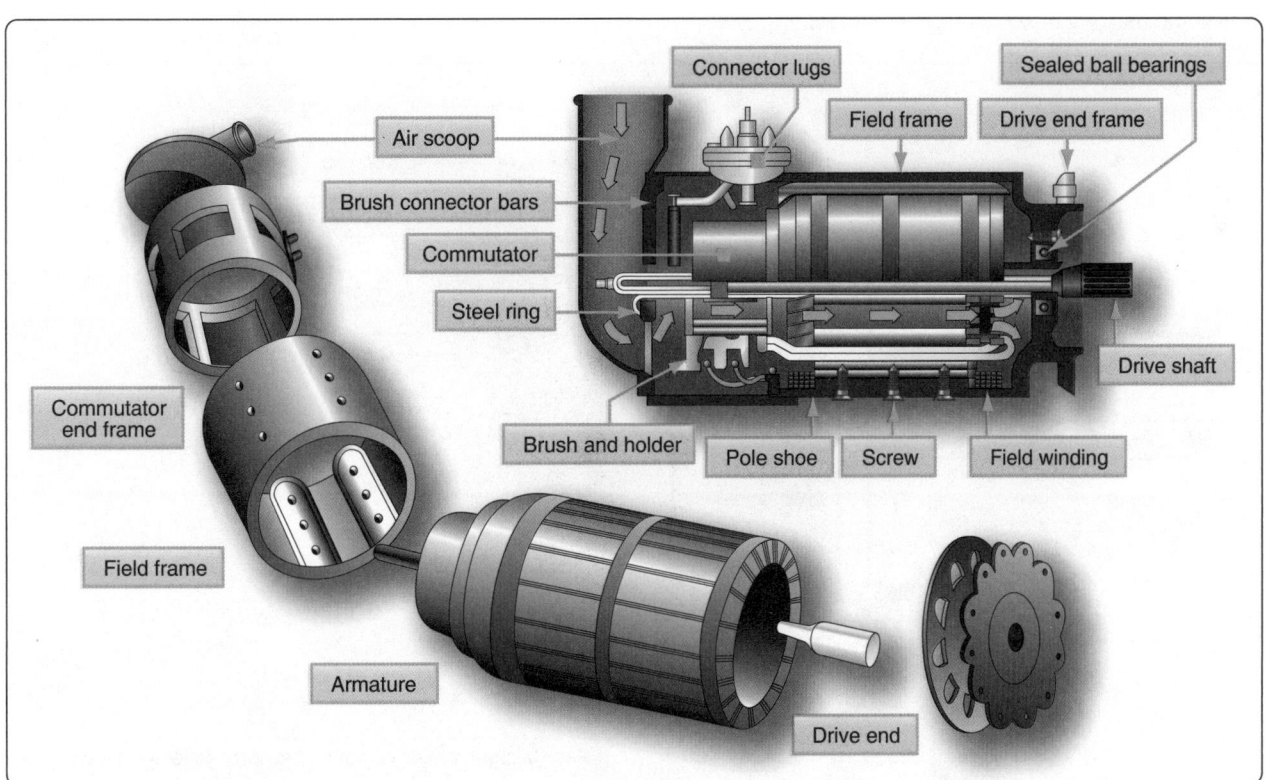

Figure 9-45. *Typical 24-volt aircraft generator.*

Figure 9-46. *Generator field frame.*

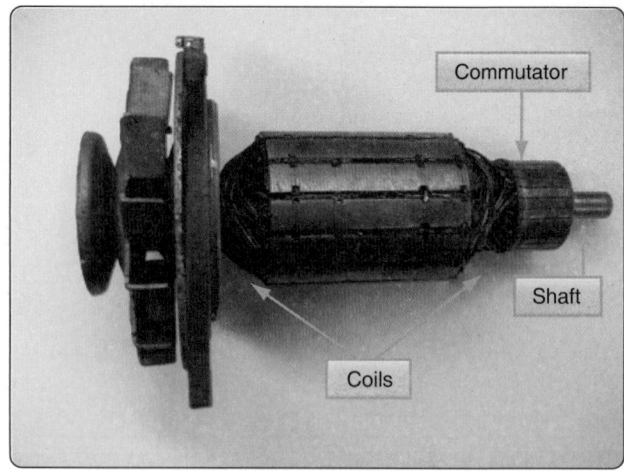

Figure 9-47. *A drum-type armature.*

the friction between the commutator and the brush, causes brushes to wear out and need regular attention or replacement. For these reasons, the material commonly used for brushes is high-grade carbon. The carbon must be soft enough to prevent undue wear of the commutator and yet hard enough to provide reasonable brush life. Since the contact resistance of carbon is fairly high, the brush must be quite large to provide a current path for the armature windings.

The commutator surface is highly polished to reduce friction as much as possible. Oil or grease must never be used on a commutator, and extreme care must be used when cleaning it to avoid marring or scratching the surface.

Types of DC Generators

There are three types of DC generators: series wound, parallel (shunt) wound, and series-parallel (or compound wound). The appropriate generator is determined by the connections to the armature and field circuits with respect to the external circuit. The external circuit is the electrical load powered by the generator. In general, the external circuit is used for charging the aircraft battery and supplying power to all electrical equipment being used by the aircraft. As their names imply, windings in series have characteristics different from windings in parallel.

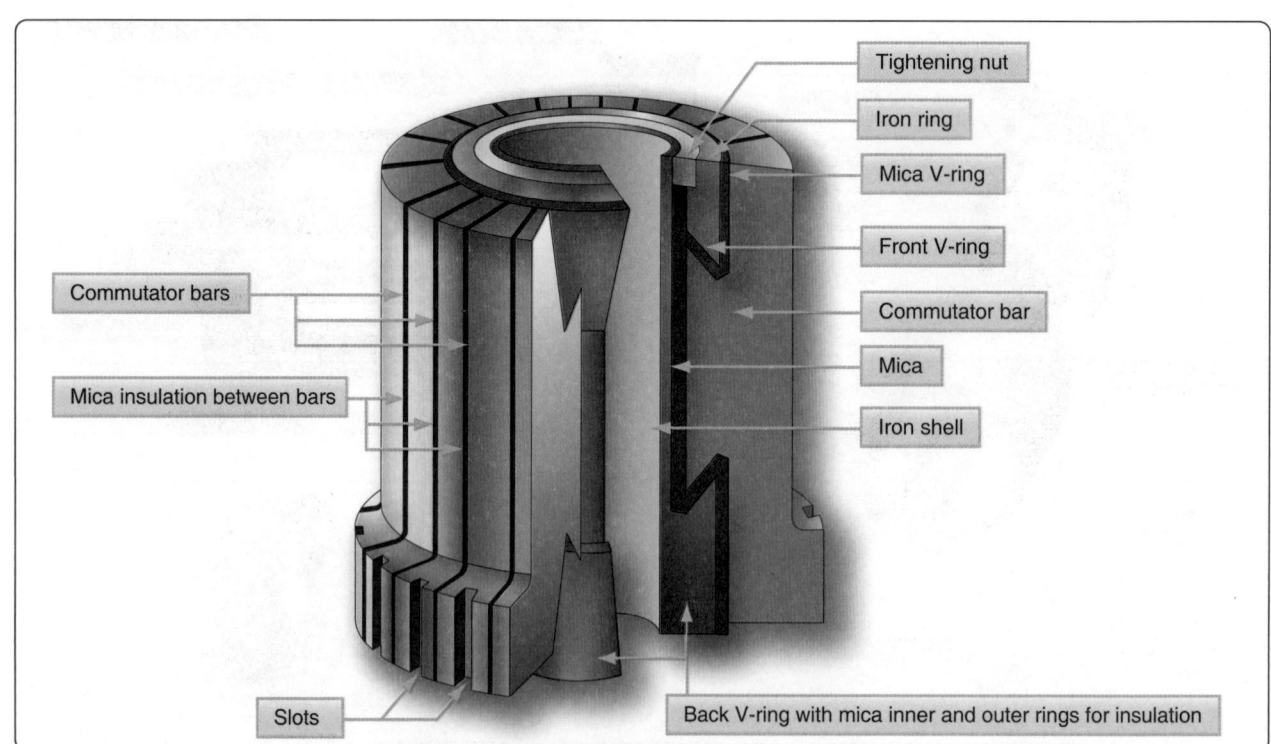

Figure 9-48. *Commutator with portion removed to show construction.*

Series Wound DC Generators

The series generator contains a field winding connected in series with the external circuit. *[Figure 9-49]* Series generators have very poor voltage regulation under changing load, since the greater the current is through the field coils to the external circuit, the greater the induced EMFs and the greater the output voltage is. When the aircraft electrical load is increased, the voltage increases; when the load is decreased, the voltage decreases.

Since the series wound generator has such poor voltage and current regulation, it is never employed as an airplane generator. Generators in airplanes have field windings, that are connected either in shunt or in compound formats.

Parallel (Shunt) Wound DC Generators

A generator having a field winding connected in parallel with the external circuit is called a shunt generator. *[Figure 9-50]* It should be noted that, in electrical terms, shunt means parallel. Therefore, this type of generator could be called either a shunt generator or a parallel generator.

In a shunt generator, any increase in load causes a decrease in the output voltage, and any decrease in load causes an increase output voltage. This occurs since the field winding is connected in parallel to the load and armature, and all the current flowing in the external circuit passes only through the armature winding (not the field).

As shown in *Figure 9-50A*, the output voltage of a shunt generator can be controlled by means of a rheostat inserted in series with the field windings. As the resistance of the field circuit is increased, the field current is reduced; consequently, the generated voltage is also reduced. As the field resistance

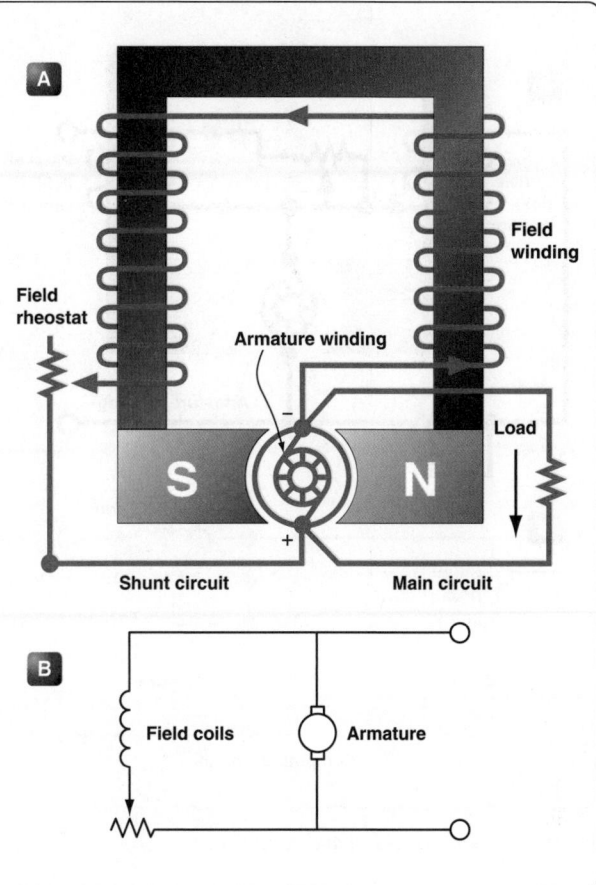

Figure 9-50. *Shunt wound generator.*

is decreased, the field current increases and the generator output increases. In the actual aircraft, the field rheostat would be replaced with an automatic control device, such as a voltage regulator.

Compound Wound DC Generators

A compound wound generator employs two field windings one in series and another in parallel with the load. *[Figure 9-51]* This arrangement takes advantage of both the series and parallel characteristics described earlier. The output of a compound wound generator is relatively constant, even with changes in the load.

Generator Ratings

A DC generator is typically rated for its voltage and power output. Each generator is designed to operate at a specified voltage, approximately 14 or 28 volts. It should be noted that aircraft electrical systems are designed to operate at one of these two voltage values. The aircraft's voltage depends on which battery is selected for that aircraft. Batteries are either 12 or 24 volts when fully charged. The generator selected must have a voltage output slightly higher than the battery voltage. Hence, the 14- or 28-volt rating is required for aircraft DC generators.

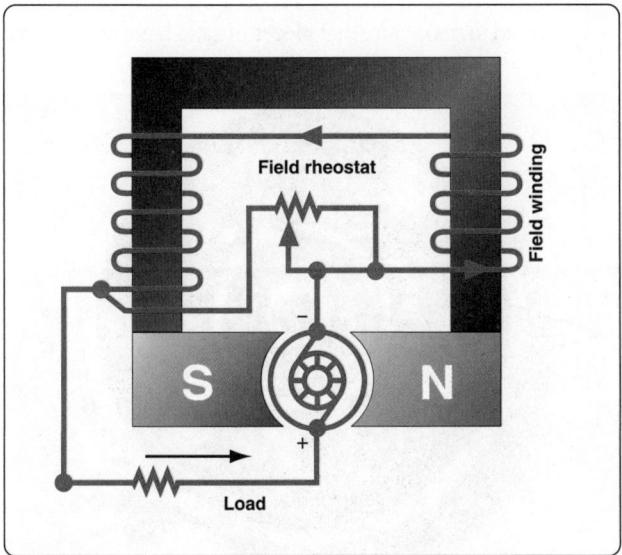

Figure 9-49. *Diagram of a series wound generator.*

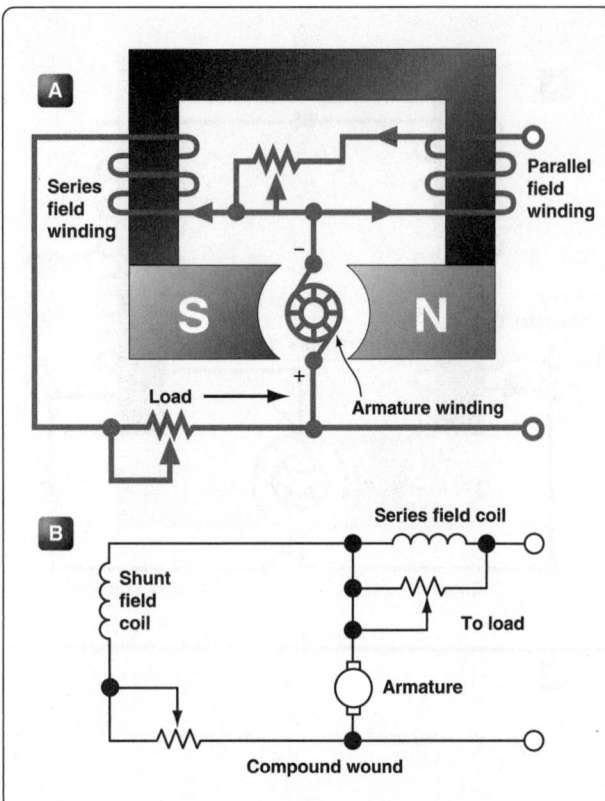

Figure 9-51. *Compound wound generator.*

The power output of any generator is given as the maximum number of amperes the generator can safely supply. Generator rating and performance data are stamped on the nameplate attached to the generator. When replacing a generator, it is important to choose one of the proper ratings.

The rotation of generators is termed either clockwise or counterclockwise, as viewed from the driven end. The direction of rotation may also be stamped on the data plate. It is important that a generator with the correct rotation be used; otherwise, the polarity of the output voltage is reversed. The speed of an aircraft engine varies from idle rpm to takeoff rpm; however, during the major portion of a flight, it is at a constant cruising speed. The generator drive is usually geared to turn the generator between 1⅛ and 1½ times the engine crankshaft speed. Most aircraft generators have a speed at which they begin to produce their normal voltage. Called the "coming in" speed, it is usually about 1,500 rpm.

DC Generator Maintenance

The following information about the inspection and maintenance of DC generator systems is general in nature because of the large number of differing aircraft generator systems. These procedures are for familiarization only. Always follow the applicable manufacturer's instructions for a given generator system. In general, the inspection of

the generator installed in the aircraft should include the following items:

1. Security of generator mounting.
2. Condition of electrical connections.
3. Dirt and oil in the generator. If oil is present, check engine oil seals. Blow out any dirt with compressed air.
4. Condition of generator brushes.
5. Generator operation.
6. Voltage regulator operation.

Sparking of brushes quickly reduces the effective brush area in contact with the commutator bars. The degree of such sparking should be determined. Excessive wear warrants a detailed inspection and possible replacement of various components. *[Figure 9-52]*

Manufacturers usually recommend the following procedures to seat brushes that do not make good contact with slip rings or commutators. Lift the brush sufficiently to permit the insertion of a strip of extra-fine 000 (triple aught) grit, or finer, sandpaper under the brush, rough side towards the carbon brush. *[Figure 9-53]*

Pull the sandpaper in the direction of armature rotation, being careful to keep the ends of the sandpaper as close to the slip ring or commutator surface as possible in order to avoid rounding the edges of the brush. When pulling the sandpaper back to the starting point, raise the brush so it does not ride on the sandpaper. Sand the brush only in the direction of rotation. Carbon dust resulting from brush sanding should be thoroughly cleaned from all parts of the generators after a sanding operation.

After the generator has run for a short period, brushes should be inspected to make sure that pieces of sand have not become

Figure 9-52. *Wear areas of commutator and brushes.*

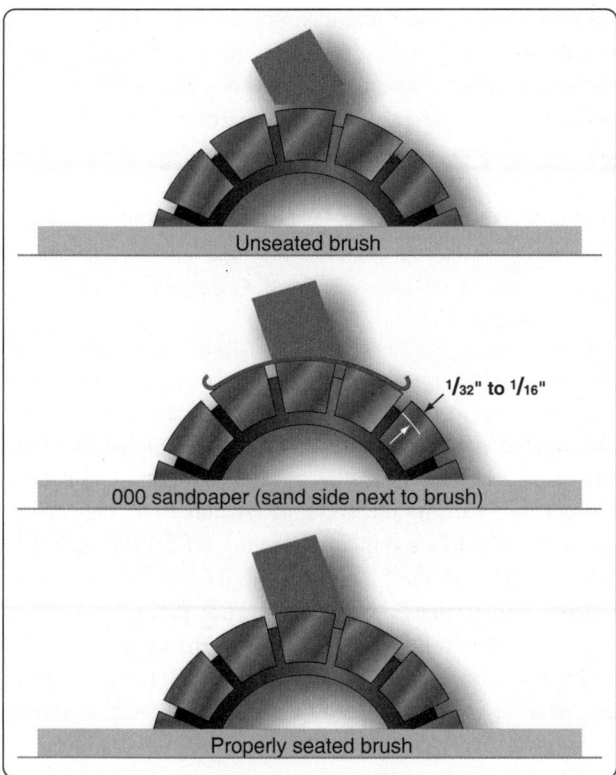

Figure 9-53. *Seating brushes with sandpaper.*

embedded in the brush. Under no circumstances should emery cloth or similar abrasives be used for seating brushes (or smoothing commutators), since they contain conductive materials that cause arcing between brushes and commutator bars. It is important that the brush spring pressure be correct. Excessive pressure causes rapid wear of brushes. Too little pressure, however, allows bouncing of the brushes, resulting in burned and pitted surfaces. The pressure recommended by the manufacturer should be checked by the use of a spring scale graduated in ounces. Brush spring tension on some generators can be adjusted. A spring scale is used to measure the pressure that a brush exerts on the commutator.

Flexible low-resistance pigtails are provided on most heavy current carrying brushes, and their connections should be securely made and checked at frequent intervals. The pigtails should never be permitted to alter or restrict the free motion of the brush. The purpose of the pigtail is to conduct the current from the armature, through the brushes, to the external circuit of the generator.

Generator Controls
Theory of Generator Control
All aircraft are designed to operate within a specific voltage range (for example 13.5–14.5 volts). And since aircraft operate at a variety of engine speeds (remember, the engine drives the generator) and with a variety of electrical demands,

all generators must be regulated by some control system. The generator control system is designed to keep the generator output within limits for all flight variables. Generator control systems are often referred to as voltage regulators or generator control units (GCU).

Aircraft generator output can easily be adjusted through control of the generator's magnetic field strength. Remember, the strength of the magnetic field has a direct effect on generator output. More field current means more generator output and vice versa. *Figure 9-54* shows a simple generator control used to adjust field current. When field current is controlled, generator output is controlled. Keep in mind, this system is manually adjusted and would not be suitable for aircraft. Aircraft systems must be automatic and are therefore a bit more complex.

There are two basic types of generator controls: electromechanical and solid-state (transistorized). The electromechanical type controls are found on older aircraft and tend to require regular inspection and maintenance. Solid-state systems are more modern and typically considered to have better reliability and more accurate generator output control.

Functions of Generator Control Systems
Most generator control systems perform a number of functions related to the regulation, sensing, and protection of the DC generation system. Light aircraft typically require a less complex generator control system than larger multiengine aircraft. Some of the functions listed below are not found on light aircraft.

Voltage Regulation
The most basic of the GCU functions is that of voltage regulation. Regulation of any kind requires the regulation unit to take a sample of a generator output and compare that sample to a known reference. If the generator's output

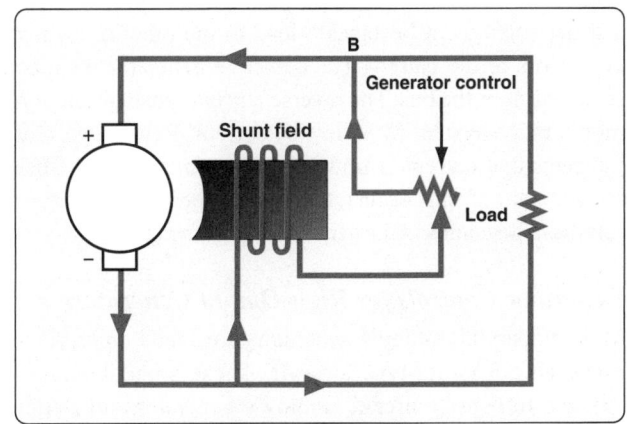

Figure 9-54. *Regulation of generator voltage by field rheostat.*

voltage falls outside of the set limits, then the regulation unit must provide an adjustment to the generator field current. Adjusting field current controls generator output.

Overvoltage Protection

The overvoltage protection system compares the sampled voltage to a reference voltage. The overvoltage protection circuit is used to open the relay that controls the field excitation current. It is typically found on more complex generator control systems.

Parallel Generator Operations

On multiengine aircraft, a paralleling feature must be employed to ensure all generators operate within limits. In general, paralleling systems compare the voltages between two or more generators and adjust the voltage regulation circuit accordingly.

Overexcitation Protection

When one generator in a paralleled system fails, one of the generators can become overexcited and tends to carry more than its share of the load, if not all of the loads. Basically, this condition causes the generator to produce too much current. If this condition is sensed, the overexcited generator must be brought back within limits, or damage occurs. The overexcitation circuit often works in conjunction with the overvoltage circuit to control the generator.

Differential Voltage

This function of a control system is designed to ensure all generator voltage values are within a close tolerance before being connected to the load bus. If the output is not within the specified tolerance, then the generator contactor is not allowed to connect the generator to the load bus.

Reverse Current Sensing

If the generator cannot maintain the required voltage level, it eventually begins to draw current instead of providing it. This situation occurs, for example, if a generator fails. When a generator fails, it becomes a load to the other operating generators or the battery. The defective generator must be removed from the bus. The reverse current sensing function monitors the system for a reverse current. Reverse current indicates that current is flowing to the generator not from the generator. If this occurs, the system opens the generator relay and disconnects the generator from the bus.

Generator Controls for High-Output Generators

Most modern high-output generators are found on turbine-powered corporate-type aircraft. These small business jets and turboprop aircraft employ a generator and starter combined into one unit. This unit is referred to as a starter-

generator. A starter-generator has the advantage of combining two units into one housing, saving space and weight. Since the starter-generator performs two tasks, engine starting and generation of electrical power, the control system for this unit is relatively complex.

A simple explanation of a starter-generator shows that the unit contains two sets of field windings. One field is used to start the engine and one used for the generation of electrical power. *[Figure 9-55]*

During the start function, the GCU must energize the series field and the armature causes the unit to act like a motor. During the generating mode, the GCU must disconnect the series field, energize the parallel field, and control the current produced by the armature. At this time, the starter-generator acts like a typical generator. Of course, the GCU must perform all the functions described earlier to control voltage and protect the system. These functions include voltage regulation, reverse current sensing, differential voltage, overexcitation protection, overvoltage protection, and parallel generator operations. A typical GCU is shown in *Figure 9-56*.

In general, modern GCUs for high-output generators employ solid-state electronic circuits to sense the operations of the generator or starter-generator. The circuitry then controls a series of relays and/or solenoids to connect and disconnect the unit to various distribution buses. One unit found in almost all voltage regulation circuitry is the zener diode. The zener diode is a voltage sensitive device that is used to monitor system voltage. The zener diode, connected in conjunction to the GCU circuitry, then controls the field current, which in turn controls the generator output.

Generator Controls for Low-Output Generators

A typical generator control circuit for low-output generators modifies current flow to the generator field to control

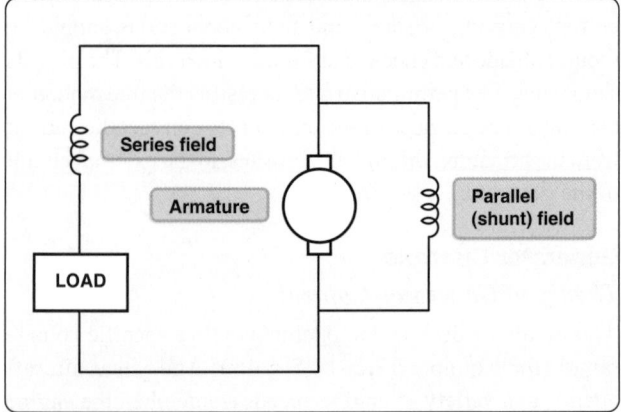

Figure 9-55. *Starter-generator.*

Figure 9-56. *Generator control unit (GCU).*

generator output power. As flight variables and electrical loads change, the GCU must monitor the electrical system and make the appropriate adjustments to ensure proper system voltage and current. The typical generator control is referred to as a voltage regulator or a GCU.

Since most low-output generators are found on older aircraft, the control systems for these systems are electromechanical devices. (Solid-state units are found on more modern aircraft that employ DC alternators and not DC generators.) The two most common types of voltage regulator are the carbon pile regulator and the three-unit regulator. Each of these units controls field current using a type of variable resistor. Controlling field current then controls generator output. A simplified generator control circuit is shown in *Figure 9-57.*

Carbon Pile Regulators

The carbon pile regulator controls DC generator output by sending the field current through a stack of carbon discs (the carbon pile). The carbon discs are in series with the generator field. If the resistance of the discs increases, the field current decreases and the generator output goes down. If the resistance of the discs decreases, the field current increases and generator output goes up. As seen in *Figure 9-58*, a voltage coil is installed in parallel with the generator output

leads. The voltage coil acts like an electromagnet that increases or decrease strength as generator output voltage changes. The magnetism of the voltage coil controls the pressure on the carbon stack. The pressure on the carbon stack controls the resistance of the carbon; the resistance of the carbon controls field current and the field current controls generator output.

Carbon pile regulators require regular maintenance to ensure accurate voltage regulation; therefore, most have been replaced on aircraft with more modern systems.

Three-Unit Regulators

The three-unit regulator used with DC generator systems is made of three distinct units. Each of these units performs a specific function vital to correct electrical system operation. A typical three-unit regulator consists of three relays mounted in a single housing. Each of the three relays monitors generator outputs and opens or closes the relay contact points according to system needs. A typical three-unit regulator is shown in *Figure 9-59*.

Voltage Regulator

The voltage regulator section of the three-unit regulator is used to control generator output voltage. The voltage

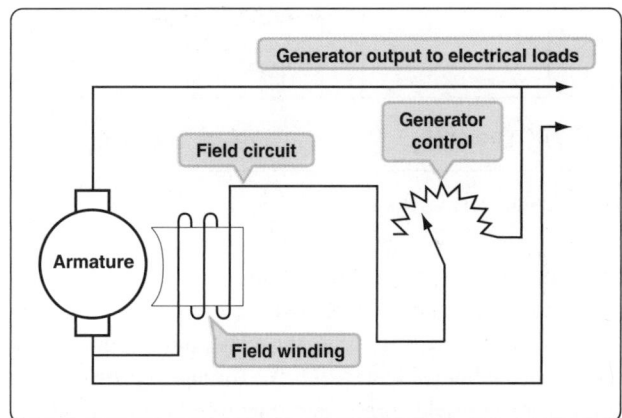

Figure 9-57. *Voltage regulator for low-output generator.*

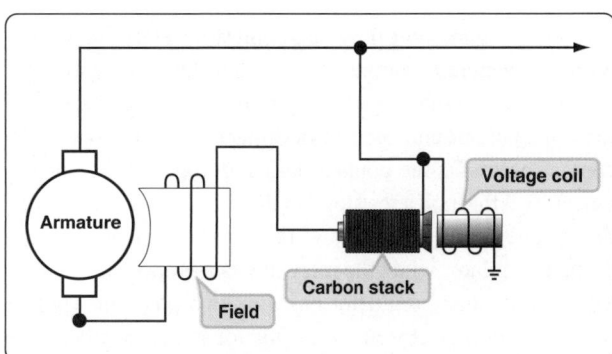

Figure 9-58. *Carbon pile regulator.*

Figure 9-59. *The three relays found on this regulator are used to regulate voltage, limit current, and prevent reverse current flow.*

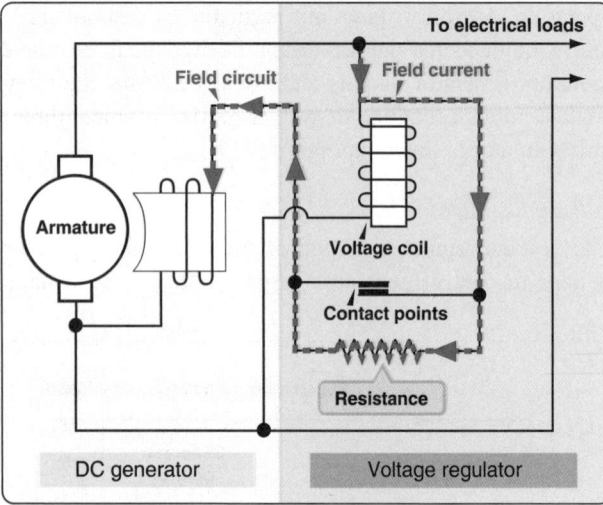

Figure 9-60. *Voltage regulator.*

regulator monitors generator output and controls the generator field current as needed. If the regulator senses that system voltage is too high, the relay points open and the current in the field circuit must travel through a resistor. This resistor lowers field current and therefore lowers generator output. Remember, generator output goes down whenever generator field current goes down.

As seen in *Figure 9-60*, the voltage coil is connected in parallel with the generator output, and it therefore measures the voltage of the system. If voltage gets beyond a predetermined limit, the voltage coil becomes a strong magnet and opens the contact points. If the contact points are open, field current must travel through a resistor and therefore field current goes down. The dotted arrow shows the current flow through the voltage regulator when the relay points are open.

Since this voltage regulator has only two positions (points open and points closed), the unit must constantly be in adjustment to maintain accurate voltage control. During

normal system operation, the points are opening and closing at regular intervals. The points are in effect vibrating. This type of regulator is sometimes referred to as a vibrating-type regulator. As the points vibrate, the field current raises and lowers and the field magnetism averages to a level that maintains the correct generator output voltage. If the system requires more generator output, the points remain closed longer and vice versa.

Current Limiter

The current limiter section of the three-unit regulator is designed to limit generator output current. This unit contains a relay with a coil wired in series with respect to the generator output. As seen in *Figure 9-61*, all the generator output current must travel through the current coil of the relay. This creates a relay that is sensitive to the current output of the generator. That is, if generator output current increases, the relay points open and vice versa. The dotted line shows the current flow to the generator field when the current limiter points are open. It should be noted that, unlike the voltage regulator relay, the current limiter is typically closed during normal flight. Only during extreme current loads must the current limiter points open; at that time, field current is lowered and generator output is kept within limits.

Reverse-Current Relay

The third unit of a three-unit regulator is used to prevent current from leaving the battery and feeding the generator. This type of current flow would discharge the battery and is opposite of normal operation. It can be thought of as a reverse current situation and is known as reverse-current relay. The simple reverse-current relay shown in *Figure 9-62* contains both a voltage coil and a current coil.

The voltage coil is wired in parallel to the generator output and is energized any time the generator output reaches its

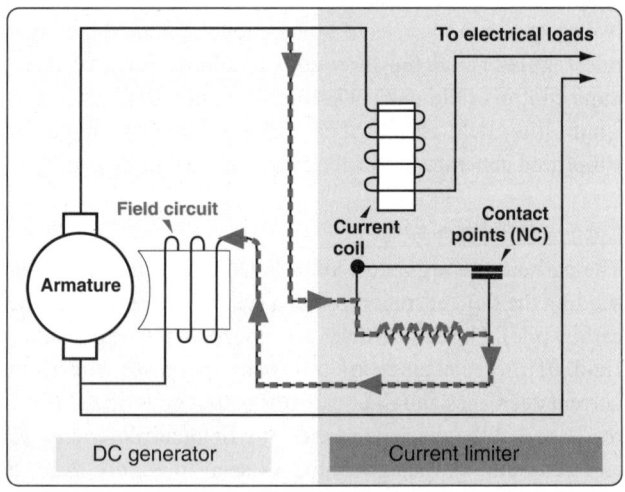

Figure 9-61. *Current limiter.*

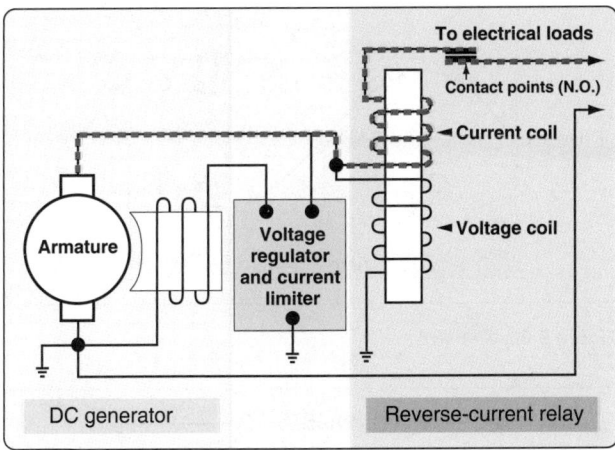

Figure 9-62. *Reverse-current relay.*

operational voltage. As the voltage coil is energized, the contact points close and the current is then allowed to flow to the aircraft electrical loads, as shown by the dotted lines. The diagram shows the reverse current relay in its normal operating position; the points are closed and current is flowing from the generator to the aircraft electrical loads. As current flows to the loads, the current coil is energized and the points remain closed. If there is no generator output due to a system failure, the contact points open because magnetism in the relay is lost. With the contact points open, the generator is automatically disconnected from the aircraft electrical system, which prevents reverse flow from the load bus to the generator. A typical three-unit regulator for aircraft generators is shown in *Figure 9-63*.

As seen in *Figure 9-63*, all three units of the regulator work together to control generator output. The regulator monitors generator output and controls power to the aircraft loads as needed for flight variables. Note that the vibrating regulator just described was simplified for explanation purposes. A typical vibrating regulator found on an aircraft would probably be more complex.

DC Alternators & Controls

DC alternators (like generators) change mechanical energy into electrical energy by the process of electromagnetic induction. In general, DC alternators are lighter and more efficient than DC generators. DC alternators and their related controls are found on modern, light, piston-engine aircraft. The alternator is mounted in the engine compartment driven by a v-belt, or drive gear mechanism, which receives power from the aircraft engine. *[Figure 9-64]* The control system of a DC alternator is used to automatically regulate alternator output power and ensure the correct system voltage for various flight parameters.

DC Alternators

DC alternators contain two major components: the armature winding and the field winding. The field winding (which produces a magnetic field) rotates inside the armature and, using the process of electromagnetic induction, the armature produces a voltage. This voltage produced by the armature is fed to the aircraft electrical bus and produces a current to power the electrical loads. *Figure 9-65* shows a basic diagram of a typical alternator.

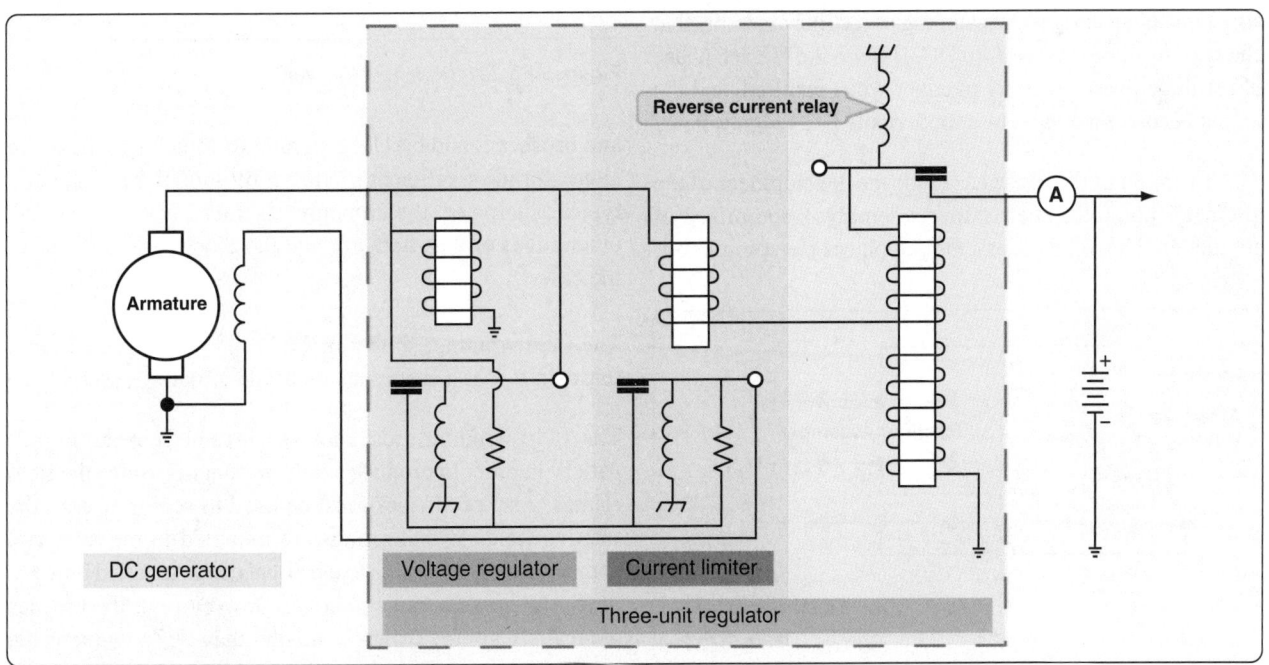

Figure 9-63. *Three-unit regulator for variable speed generators.*

Figure 9-64. *DC alternator installation.*

The armature used in DC alternators actually contains three coils of wire. Each coil receives current as the magnetic field rotates inside the armature. The resulting output voltage consists of three distinct AC sine waves, as shown in *Figure 9-66*. The armature winding is known as a three- phase armature, named after the three different voltage waveforms produced.

Figure 9-67 shows the two common methods used to connect the three phase armature windings: the delta winding and the Y winding. For all practical purposes, the two windings produce the same results in aircraft DC alternators.

Since the three-phase voltage produced by the alternators armature is AC, it is not compatible with typical DC electrical loads and must be rectified (changed to DC). Therefore, the armature output current is sent through a rectifier assembly that changes the three-phase AC to DC. *[Figure 9-67]* Each phase of the three-phase armature overlaps when rectified, and the output becomes a relatively smooth ripple DC. *[Figure 9-68]*

The invention of the diode has made the development of the alternator possible. The rectifier assembly is comprised of six diodes. This rectifier assembly replaces the commutator

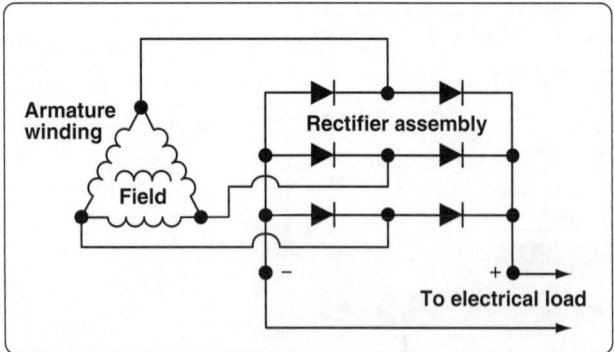

Figure 9-65. *Diagram of a typical alternator.*

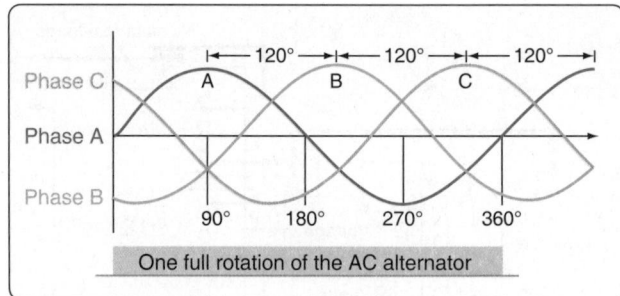

Figure 9-66. *Sine waves.*

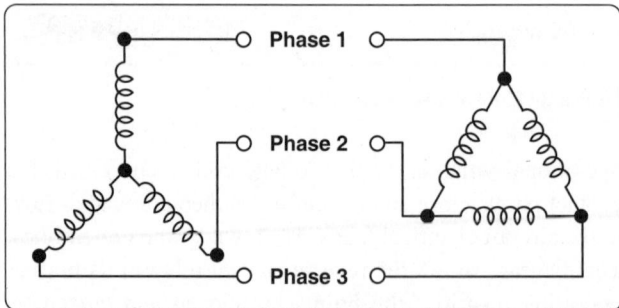

Figure 9-67. *Three-phase armature windings: Y on the left and delta winding on the right.*

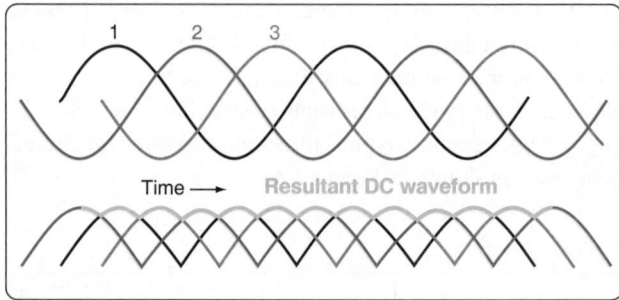

Figure 9-68. *Relatively smooth ripple DC.*

and brushes found on DC generators and helps to make the alternator more efficient. *Figure 9-69* shows the inside of a typical alternator; the armature assembly is located on the outer edges of the alternator and the diodes are mounted to the case.

The field winding, shown in *Figure 9-70*, is mounted to a rotor shaft so it can spin inside of the armature assembly.

The field winding must receive current from an aircraft battery in order to produce an electromagnet. Since the field rotates, a set of brushes must be used to send power to the rotating field. Two slip rings are mounted to the rotor and connect the field winding to electrical contacts called brushes. Since the brushes carry relatively low current, the brushes of an alternator are typically smaller than those found inside a DC generator. *[Figure 9-71]* DC alternator brushes last

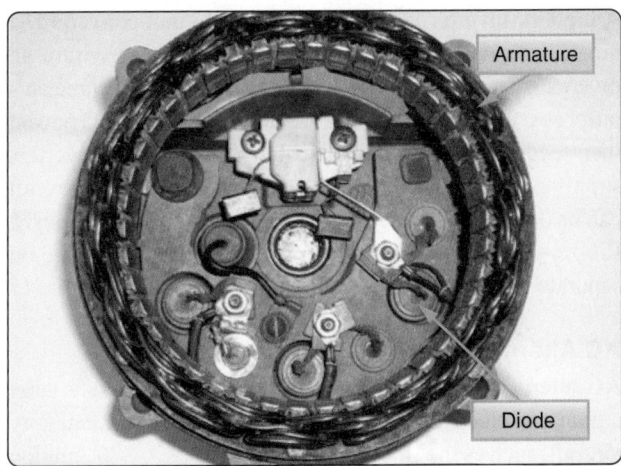

Figure 9-69. *Diode assembly.*

Figure 9-70. *Alternator field winding.*

longer and require less maintenance than those found in a DC generator.

The alternator case holds the alternator components inside a compact housing that mounts to the engine. Aircraft alternators either produce a nominal 14-volt output or a 26-volt output. The physical size of the alternator is typically a function of the alternator's amperage output. Common alternators for light aircraft range in output form 60–120 amps.

Alternator Voltage Regulators

Voltage regulators for DC alternators are similar to those found on DC generators. The general concepts are the same in that adjusting alternator field current controls alternator output. Regulators for most DC alternators are either the vibrating-relay type or solid-state regulators, which are found on most modern aircraft. Vibrating-relay regulators are similar to those discussed in the section on generator regulators. As the points of the relay open, the field current is lowered and alternator output is lowered and vice versa.

Figure 9-71. *Alternator brushes.*

Solid-State Regulators

Solid-state regulators for modern light aircraft are often referred to as alternator control units (ACUs). These units contain no moving parts and are generally considered to be more reliable and provide better system regulation than vibrating-type regulators. Solid-state regulators rely on transistor circuitry to control alternator field current and alternator output. The regulator monitors alternator output voltage/current and controls alternator field current accordingly. Solid-state regulators typically provide additional protection circuitry not found in vibrating-type regulators. Protection may include over- or under-voltage protection, overcurrent protection, as well as monitoring the alternator for internal defects, such as a defective diode. In many cases, the ACU also provides a warning indication to the pilot if a system malfunction occurs.

A key component of any solid-state voltage regulator is known as the zener diode. *Figure 9-72* shows the schematic diagram symbol of a zener diode, as well as one installed in an ACU.

The operation of a zener diode is similar to a common diode in that the zener only permits current flow in one direction. This is true until the voltage applied to the zener reaches a certain level. At that predetermined voltage level, the zener then permits current flow with either polarity. This is known as the breakdown or zener voltage.

As an ACU monitors alternator output, the zener diode is connected to system voltage. When the alternator output reaches the specific zener voltage, the diode controls a transistor in the circuit, which in turn controls the alternator field current. This is a simplified explanation of the complete circuitry of an ACU. *[Figure 9-73]* However, it is easy to see how the zener diode and transistor circuit are used in place of an electromechanical relay in a vibrating-type regulator. The use of solid-state components creates a more accurate regulator that requires very little maintenance. The solid-state ACU is, therefore, the control unit of choice for modern aircraft with DC alternators.

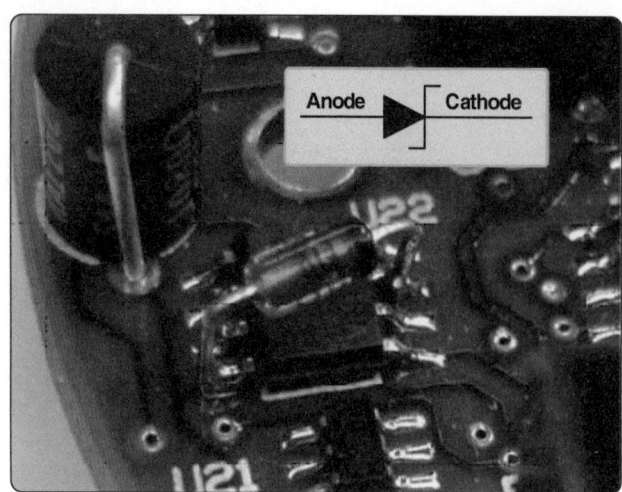

Figure 9-72. *Zener diode.*

Power Systems

Since certain electrical systems operate only on AC, many aircraft employ a completely AC electrical system, as well as a DC system. The typical AC system would include an AC alternator (generator), a regulating system for that alternator, AC power distribution buses, and related fuses and wiring. Note that when referring to AC systems, the terms "alternator" and "generator" are often used interchangeably. This chapter uses the term "AC alternator."

AC power systems are becoming more popular on modern aircraft. Light aircraft tend to operate most electrical systems using DC, therefore the DC battery can easily act as a backup power source. Some modern light aircraft also employ a small AC system. In this case, the light aircraft probably uses an AC inverter to produce the AC needed for this system.

Inverters are commonly used when only a small amount of AC is required for certain systems. Inverters may also be used as a backup AC power source on aircraft that employ an AC alternator. *Figure 9-74* shows a typical inverter that might be found on modern aircraft.

A modern inverter is a solid-state device that converts DC power into AC power. The electronic circuitry within an inverter is quite complex; however, for an aircraft technician's purposes, the inverter is simply a device that uses DC power, then feeds power to an AC distribution bus. Many inverters supply both 26-volt AC, as well as 115-volt AC. The aircraft can be designed to use either voltage or both simultaneously. If both voltages are used, the power must be distributed on separate 26-and 115-volt AC buses.

AC Alternators

AC alternators are found only on aircraft that use a large amount of electrical power. Virtually all transport category aircraft, such as the Boeing 757 or the Airbus A-380, employ one AC alternator driven by each engine. These aircraft also have an auxiliary AC alternator driven by the auxiliary power unit. In most cases, transport category aircraft also have at least one more AC backup power source, such as an AC inverter or a small AC alternator driven by a ram-air turbine (RAT).

AC alternators produce a three-phase AC output. For each revolution of the alternator, the unit produces three separate voltages. The sine waves for these voltages are separated by 120°. *[Figure 9-75]* This wave pattern is similar to those produced internally by a DC alternator; however, in this case, the AC alternator does not rectify the voltage and the output of the unit is AC.

The modern AC alternator does not utilize brushes or slip rings and is often referred to as a brushless AC alternator. This brushless design is extremely reliable and requires very little maintenance. In a brushless alternator, energy to or from the alternator's rotor is transferred using magnetic energy. In other words, energy from the stator to the rotor

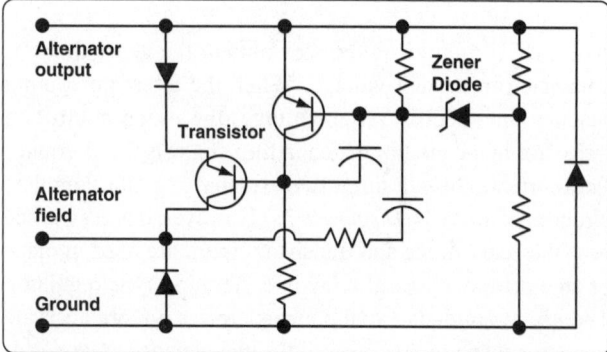

Figure 9-73. *ACU circuitry.*

Figure 9-74. *Inverter.*

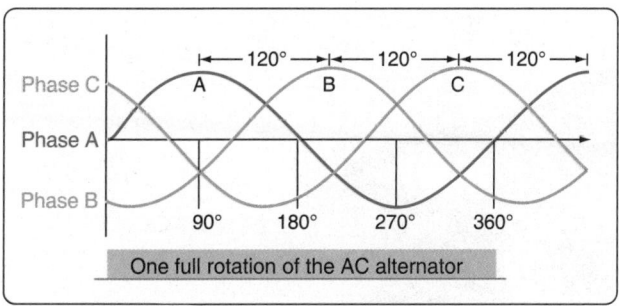

Figure 9-75. *AC alternator sine waves.*

Figure 9-76. *Large aircraft AC alternator.*

is transferred using magnetic flux energy and the process of electromagnetic induction. A typical large aircraft AC alternator is shown in *Figure 9-76*.

As seen in *Figure 9-77*, the brushless alternator actually contains three generators: the exciter generator (armature and permanent magnet field), the pilot exciter generator (armature and fields windings), and the main AC alternator (armature winding and field windings). The need for brushes is eliminated by using a combination of these three distinct generators.

The exciter is a small AC generator with a stationary field made of a permanent magnet and two electromagnets. The exciter armature is three phase and mounted on the rotor shaft. The exciter armature output is rectified and sent to the pilot exciter field and the main generator field.

The pilot exciter field is mounted on the rotor shaft and is connected in series with the main generator field. The pilot exciter armature is mounted on the stationary part of the assembly. The AC output of the pilot exciter armature is supplied to the generator control circuitry where it is rectified, regulated, and then sent to the exciter field windings. The current sent to the exciter field provides the voltage regulation for the main AC alternator. If greater AC alternator output is needed, there is more current sent to the exciter field and vice versa.

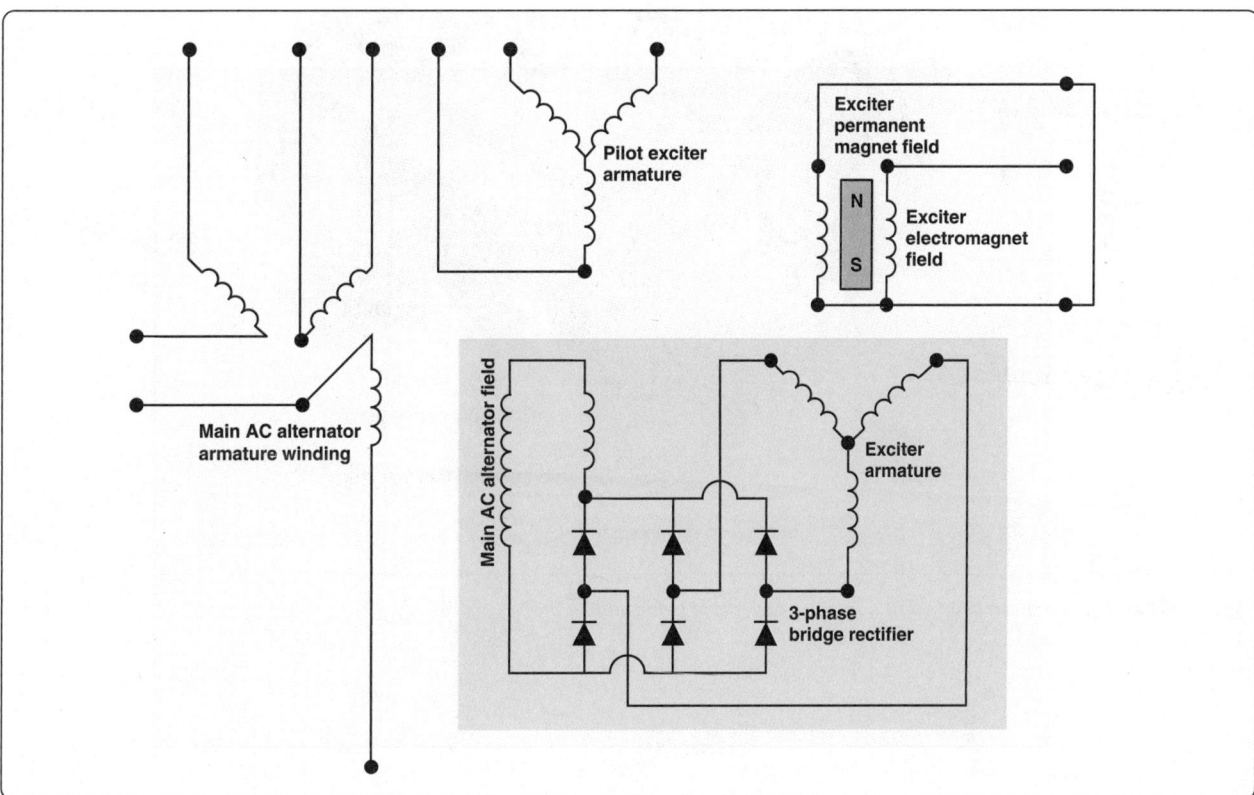

Figure 9-77. *Schematic of an AC alternator.*

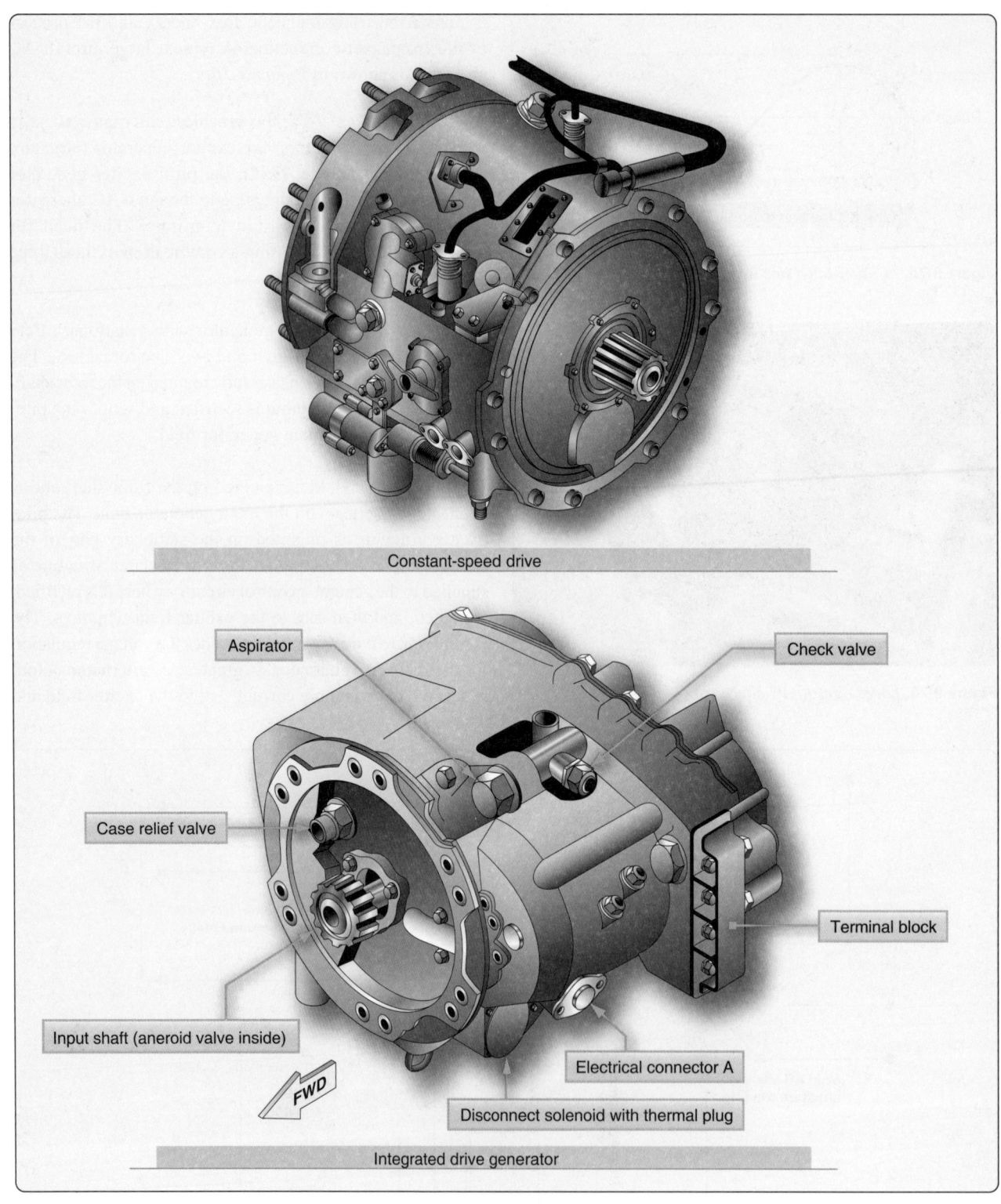

Constant-speed drive

Aspirator

Check valve

Case relief valve

Terminal block

Input shaft (aneroid valve inside)

FWD

Electrical connector A

Disconnect solenoid with thermal plug

Integrated drive generator

Figure 9-78. *Constant-speed drive (top) and integrated drive generator (bottom).*

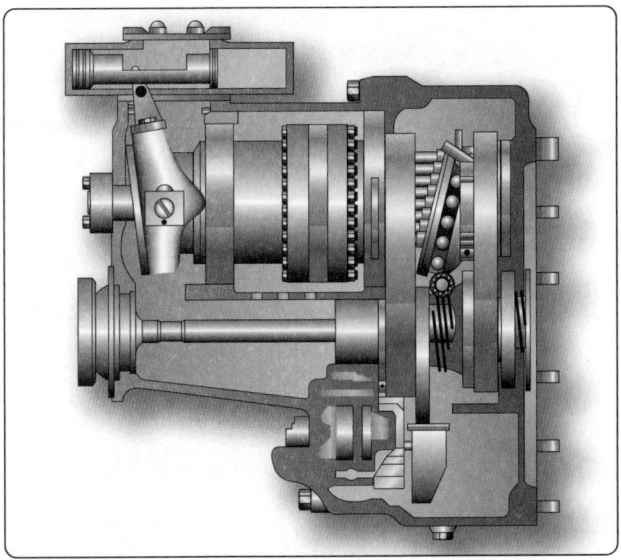

Figure 9-79. *A hydraulic constant speed drive for an AC alternator.*

In short, the exciter permanent magnet and armature starts the generation process, and the output of the exciter armature is rectified and sent to the pilot exciter field. The pilot exciter field creates a magnetic field and induces power in the pilot exciter armature through electromagnetic induction. The output of the pilot exciter armature is sent to the main alternator control unit and then sent back to the exciter field. As the rotor continues to turn, the main AC alternator field generates power into the main AC alternator armature, also using electromagnetic induction. The output of the main AC armature is three-phase AC and used to power the various electrical loads.

Some alternators are cooled by circulating oil through the internal components of the alternator. The oil used for cooling is supplied from the constant speed drive assembly and often cooled by an external oil cooler assembly. Located in the flange connecting the generator and drive assemblies, ports make oil flow between the constant speed drive and the generator possible. This oil level is critical and typically checked on a routine basis.

Alternator Drive

The unit shown in *Figure 9-78* contains an alternator assembly combined with an automatic drive mechanism. The automatic drive controls the alternator's rotational speed which allows the alternator to maintain a constant 400-Hz AC output.

All AC alternators must rotate at a specific rpm to keep the frequency of the AC voltage within limits. Aircraft AC alternators should produce a frequency of approximately 400 Hz. If the frequency strays more than 10 percent from this value, the electrical systems do not operate correctly. A

unit called a constant-speed drive (CSD) is used to ensure the alternator rotates at the correct speed to ensure a 400-Hz frequency. The CSD can be an independent unit or mounted within the alternator housing. When the CSD and the alternator are contained within one unit, the assembly is known as an integrated drive generator (IDG).

The CSD is a hydraulic unit similar to an automatic transmission found in a modern automobile. The engine of the automobile can change rpm while the speed of the car remains constant. This is the same process that occurs for an aircraft AC alternator. If the aircraft engine changes speed, the alternator speed remains constant. A typical hydraulic-type drive is shown in *Figure 9-79*. This unit can be controlled either electrically or mechanically. Modern aircraft employ an electronic system. The constant-speed drive enables the alternator to produce the same frequency at slightly above engine idle rpm as it does at maximum engine rpm.

The hydraulic transmission is mounted between the AC alternator and the aircraft engine. Hydraulic oil or engine oil is used to operate the hydraulic transmission, which creates a constant output speed to drive the alternator. In some cases, this same oil is used to cool the alternator as shown in the CSD cutaway view of *Figure 9-79*. The input drive shaft is powered by the aircraft engine gear case. The output drive shaft, on the opposite end of the transmission, engages the drive shaft of the alternator. The CSD employs a hydraulic pump assembly, a mechanical speed control, and a hydraulic drive. Engine rpm drives the hydraulic pump, the hydraulic drive turns the alternator. The speed control unit is made up of a wobble plate that adjusts hydraulic pressure to control output speed.

Figure 9-80 shows a typical electrical circuit used to control alternator speed. The circuit controls the hydraulic assembly found in a typical CSD. As shown, the alternator input speed is monitored by a tachometer (tach) generator. The tach generator signal is rectified and sent to the valve assembly. The valve assembly contains three electromagnetic coils that operate the valve. The AC alternator output is sent through a control circuit that also feeds the hydraulic valve assembly. By balancing the force created by the three electromagnets, the valve assembly controls the flow of fluid through the automatic transmission and controls the speed of the AC alternator.

It should be noted that an AC alternator also produces a constant 400 Hz if that alternator is driven directly by an engine that rotates at a constant speed. On many aircraft, the auxiliary power unit operates at a constant rpm. AC alternators driven by these APUs are typically driven directly by the engine, and there is no CSD required. For these units, the APU engine controls monitor the alternator output

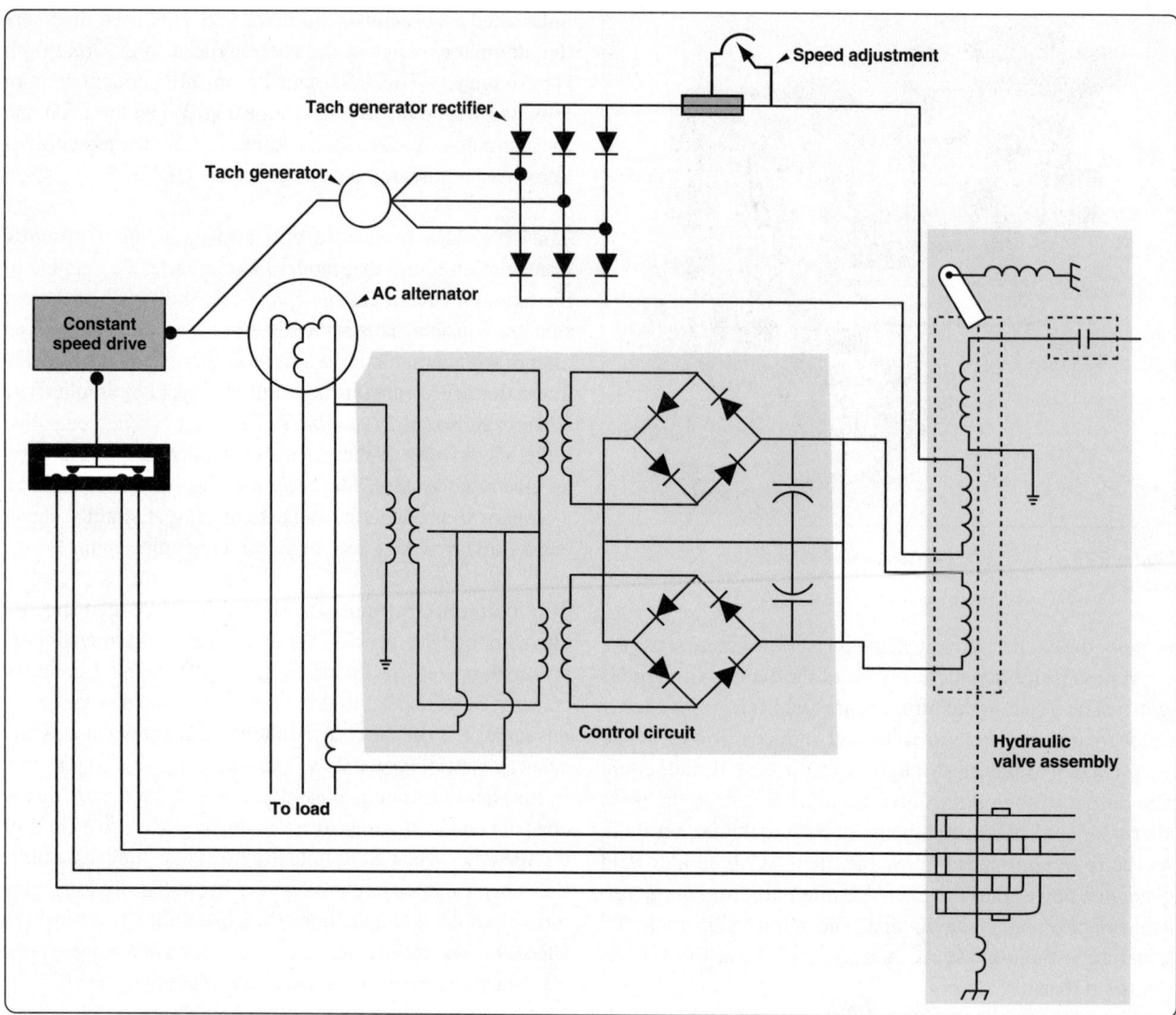

Figure 9-80. *Speed control circuit.*

frequency. If the alternator output frequency varies from 400 Hz, the APU speed control adjusts the engine rpm accordingly to keep the alternator output within limits.

AC Alternators Control Systems

Modern aircraft that employ AC alternators use several computerized control units, typically located in the aircraft's equipment bay for the regulation of AC power throughout the aircraft. *Figure 9-81* shows a photo of a typical equipment bay and computerized control units.

Since AC alternators are found on large transport category aircraft designed to carry hundreds of passengers, their control systems always have redundant computers that provide safety in the event of a system failure. Unlike DC systems, AC systems must ensure that the output frequency of the alternator stays within limits. If the frequency of an alternator varies from 400 Hz, or if two or more alternators

connected to the same bus are out of phase, damage occurs to the system. All AC alternator control units contain circuitry that regulates both voltage and frequency. These control units also monitor a variety of factors to detect any system failures and take protective measures to ensure the integrity of the electrical system. The two most common units used to control AC alternators are the bus power control unit (BPCU) and the generator control unit (GCU). In this case, the term "generator" is used, and not alternator, although the meaning is the same.

The GCU is the main computer that controls alternator functions. The BPCU is the computer that controls the distribution of AC power to the power distribution buses located throughout the aircraft. There is typically one GCU used to monitor and control each AC alternator, and there can be one or more BPCUs on the aircraft. BPCUs are described later in this chapter; however, please note that the

Figure 9-81. *Line replaceable units in an equipment rack.*

BPCU works in conjunction with the GCUs to control AC on modern aircraft.

A typical GCU ensures the AC alternator maintains a constant voltage, typically between 115 to 120 volts. The GCU ensures the maximum power output of the alternator is never exceeded. The GCU provides fault detection and circuit protection in the event of an alternator failure. The GCU monitors AC frequency and ensures the output if the alternator remains 400 Hz. The basic method of voltage regulation is similar to that found in all alternator systems; the output of the alternator is controlled by changing the strength of a magnetic field. As shown in *Figure 9-82*, the GCU controls the exciter field magnetism within the brushless alternator to control alternator output voltage. The frequency is controlled by the CDS hydraulic unit in conjunction with signals monitored by the GCU.

The GCU is also used to turn the AC alternator on or off. When the pilot selects the operation of an AC alternator, the GCU monitors the alternator's output to ensure voltage and frequency are within limits. If the GCU is satisfied with the alternator's output, the GCU sends a signal to an electrical contactor that connects the alternator to the appropriate AC distribution bus. The contactor, often call the generator breaker, is basically an electromagnetic solenoid that controls a set of large contact points. The large contact points are necessary in order to handle the large amounts of current produced by most AC alternators. This same contactor is activated in the event the GCU detects a fault in the alternator

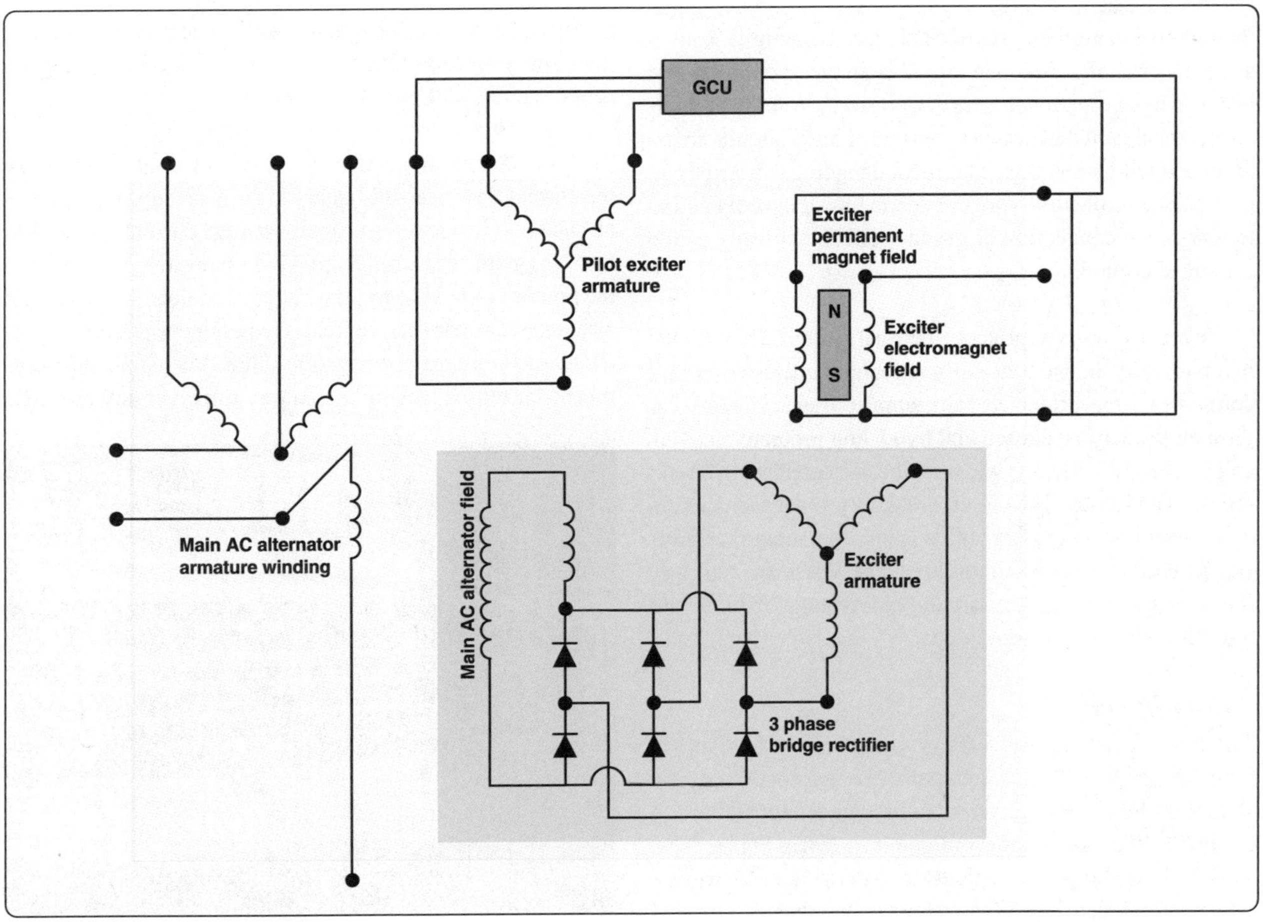

Figure 9-82. *Schematic GCU control of the exciter field magnetism.*

output; however, in this case the contactor would disconnect the alternator from the bus.

Aircraft Electrical Systems

Virtually all aircraft contain some form of an electrical system. The most basic aircraft must produce electricity for operation of the engine's ignition system. Modern aircraft have complex electrical systems that control almost every aspect of flight. In general, electrical systems can be divided into different categories according to the function of the system. Common systems include lighting, engine starting, and power generation.

Small Single-Engine Aircraft

Light aircraft typically have a relatively simple electrical system because simple aircraft generally require less redundancy and less complexity than larger transport category aircraft. On most light aircraft, there is only one electrical system powered by the engine-driven alternator or generator. The aircraft battery is used for emergency power and engine starting. Electrical power is typically distributed through one or more common points known as an electrical bus (or bus bar).

Almost all electrical circuits must be protected from faults that can occur in the system. Faults are commonly known as opens or shorts. An open circuit is an electrical fault that occurs when a circuit becomes disconnected. A short circuit is an electrical fault that occurs when one or more circuits create an unwanted connection. The most dangerous short circuit occurs when a positive wire creates an unwanted connection to a negative connection or ground. This is typically called a short to ground.

There are two ways to protect electrical systems from faults: mechanically and electrically. Mechanically, wires and components are protected from abrasion and excess wear through proper installation and by adding protective covers and shields. Electrically, wires can be protected using circuit breakers and fuses. The circuit breakers protect each system in the event of a short circuit. It should be noted that fuses can be used instead of circuit breakers. Fuses are typically found on older aircraft. A circuit breaker panel from a light aircraft is shown in *Figure 9-83*.

Battery Circuit

The aircraft battery and battery circuit is used to supply power for engine starting and to provide a secondary power supply in the event of an alternator (or generator) failure. A schematic of a typical battery circuit is shown in *Figure 9-84*. This diagram shows the relationship of the starter and external power circuits that are discussed later in this chapter. The bold lines found on the diagram represent large wire (see the wire

leaving the battery positive connection), which is used in the battery circuit due to the heavy current provided through these wires. Because batteries can supply large current flows, a battery is typically connected to the system through an electrical solenoid. At the start/end of each flight, the battery is connected/disconnected from the electrical distribution bus through the solenoid contacts. A battery master switch on the flight deck is used to control the solenoid.

Although they are very similar, there is often confusion between the terms "solenoid" and "relay." A solenoid is typically used for switching high current circuits and relays are used to control lower current circuits. To help illuminate the confusion, the term "contactor" is often used when describing a magnetically operated switch. For general purposes, an aircraft technician may consider the terms relay, solenoid, and contactor synonymous. Each of these three terms may be used on diagrams and schematics to describe electrical switches controlled by an electromagnet.

Here it can be seen that the battery positive wire is connected to the electrical bus when the battery master switch is active. A battery solenoid is shown in *Figure 9-85*. The battery switch is often referred to as the master switch since it turns off or on virtually all electrical power by controlling the battery connection. Note how the electrical connections of the battery solenoid are protected from electrical shorts by rubber covers at the end of each wire.

The ammeter shown in the battery circuit is used to monitor the current flow from the battery to the distribution bus. When all systems are operating properly, battery current should flow from the main bus to the battery giving a positive indication on the ammeter. In this case, the battery is being charged. If the aircraft alternator (or generator) experiences a malfunction, the ammeter indicates a negative value. A negative indication means current is leaving the battery to power any electrical

Figure 9-83. *Light aircraft circuit breaker panel.*

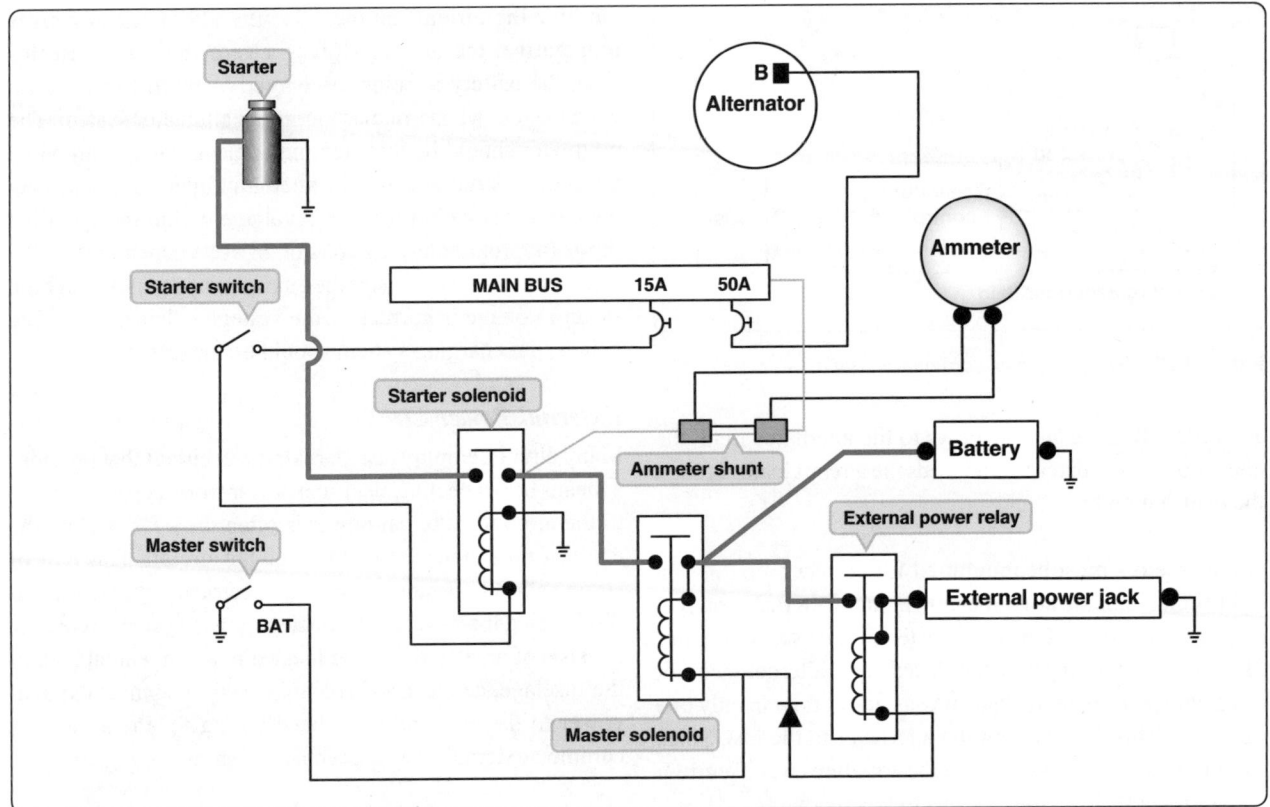

Figure 9-84. *Schematic of typical battery circuit.*

load connected to the bus. The battery is being discharged and the aircraft is in danger of losing all electrical power.

Generator Circuit

Generator circuits are used to control electrical power between the aircraft generator and the distribution bus. Typically, these circuits are found on older aircraft that have not upgraded to an alternator. Generator circuits control power to the field winding and electrical power from the generator to the electrical bus. A generator master switch is used to turn on the generator typically by controlling field current. If the generator is spinning and current is sent to

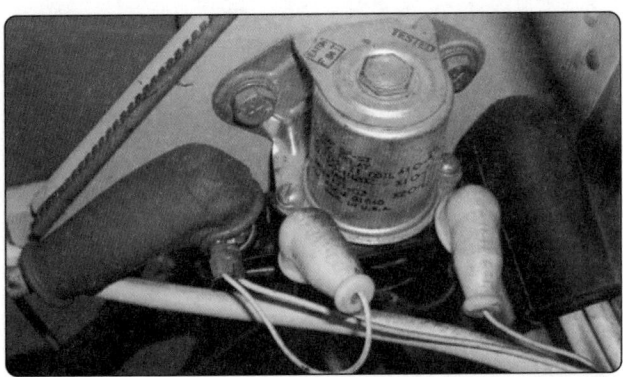

Figure 9-85. *Battery solenoid.*

the field circuit, the generator produces electrical power. The power output of the generator is controlled through the generator control unit (or voltage regulator). A simplified generator control circuit is shown in *Figure 9-86*.

As can be seen in *Figure 9-86,* the generator switch controls the power to the generator field (F terminal). The generator output current is supplied to the aircraft bus through the armature circuit (A terminal) of the generator.

Alternator Circuit

Alternator circuits, like generator circuits, must control power both to and from the alternator. The alternator is controlled by the pilot through the alternator master switch. The alternator master switch in turn operates a circuit within the alternator control unit (or voltage regulator) and sends current to the alternator field. If the alternator is powered by the aircraft engine, the alternator produces electrical power for the aircraft electrical loads. The alternator control circuit contains the three major components of the alternator circuit: alternator, voltage regulator, and alternator master switch. *[Figure 9-87]*

The voltage regulator controls the generator field current according to aircraft electrical load (alternator). If the aircraft engine is running and the alternator master switch is on, the voltage regulator adjusts current to the alternator field

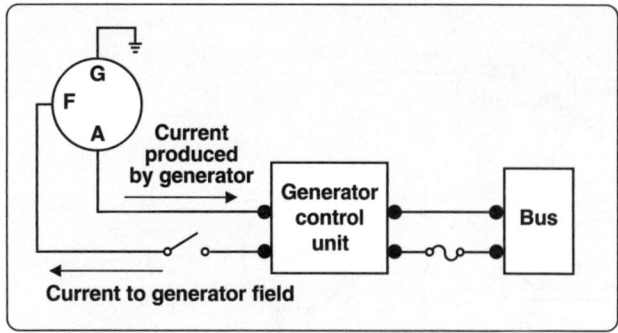

Figure 9-86. *Simplified generator control circuit.*

as needed. If more current flows to the alternator field, the alternator output increases and feeds the aircraft loads through the distribution bus.

All alternators must be monitored for correct output. Most light aircraft employ an ammeter to monitor alternator output. *Figure 9-88* shows a typical ammeter circuit used to monitor alternator output. An ammeter placed in the alternator circuit is a single polarity meter that shows current flow in only one direction. This flow is from the alternator to the bus. Since the alternator contains diodes in the armature circuit, current cannot reverse flow from the bus to the alternator.

When troubleshooting an alternator system, be sure to monitor the aircraft ammeter. If the alternator system is inoperative, the ammeter gives a zero indication. In this case, the battery is being discharged. A voltmeter is also a valuable tool when troubleshooting an alternator system. The voltmeter should be installed in the electrical system while the engine is running and the alternator operating. A system operating normally produces a voltage within the specified limits (approximately 14 volts or 28 volts depending on the electrical system). Consult the aircraft manual and verify the system voltage is correct. If the voltage is below specified values, the charging system should be inspected.

External Power Circuit

Many aircraft employ an external power circuit that provides a means of connecting electrical power from a ground source to the aircraft. External power is often used for starting the engine or maintenance activities on the aircraft. This type of system allows operation of various electrical systems without discharging the battery. The external power systems typically consists of an electrical plug located in a convenient area of the fuselage, an electrical solenoid used to connect external power to the bus, and the related wiring for the system. A common external power receptacle is shown in *Figure 9-89*.

Figure 9-90 shows how the external power receptacle connects to the external power solenoid through a reverse polarity diode. This diode is used to prevent any accidental

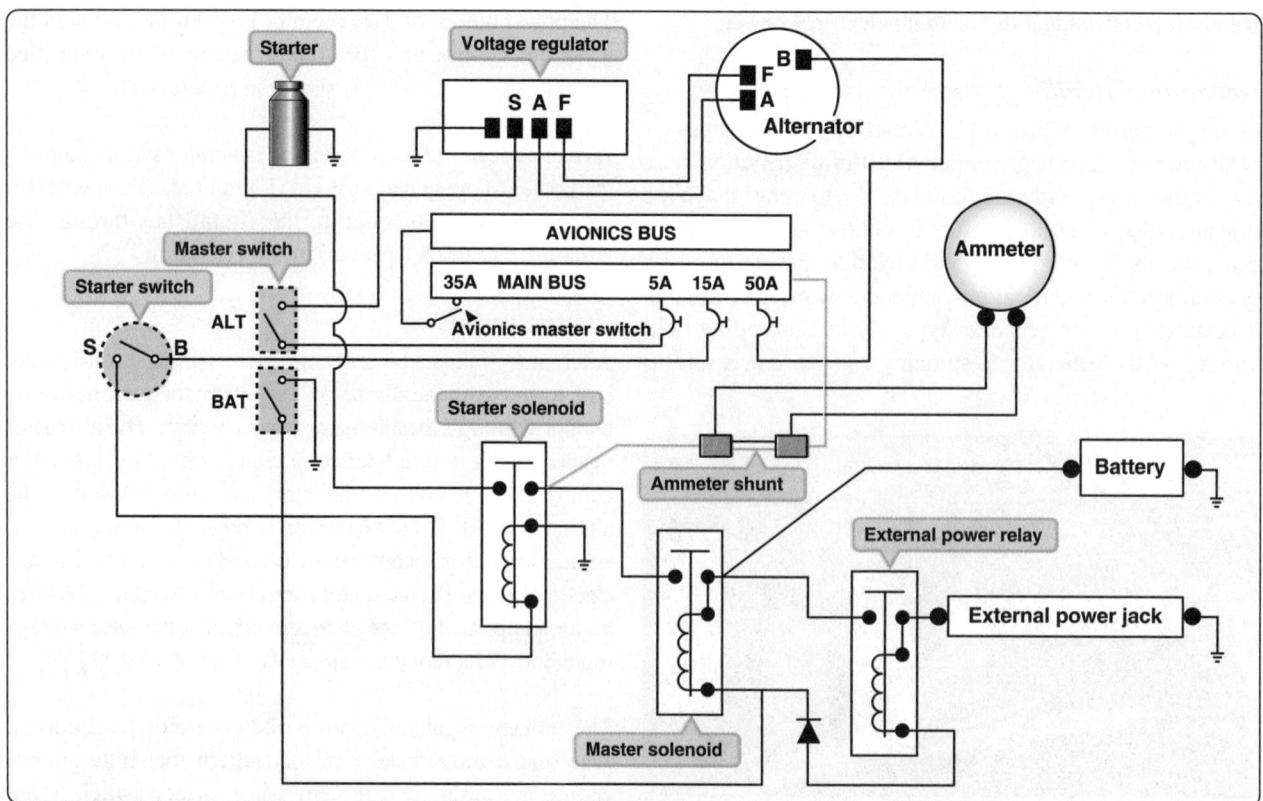

Figure 9-87. *Alternator control circuit.*

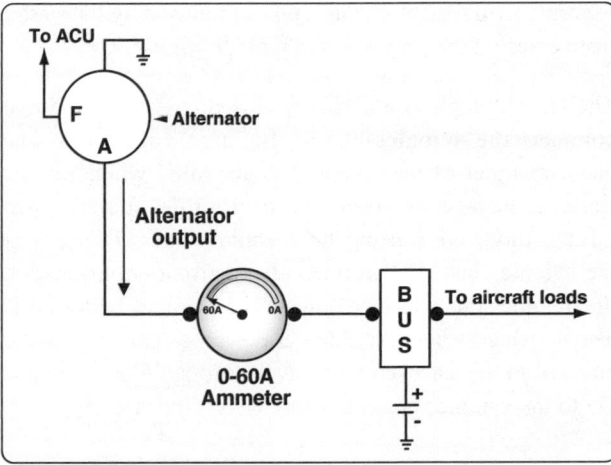

Figure 9-88. *Typical ammeter circuit used to monitor alternator output.*

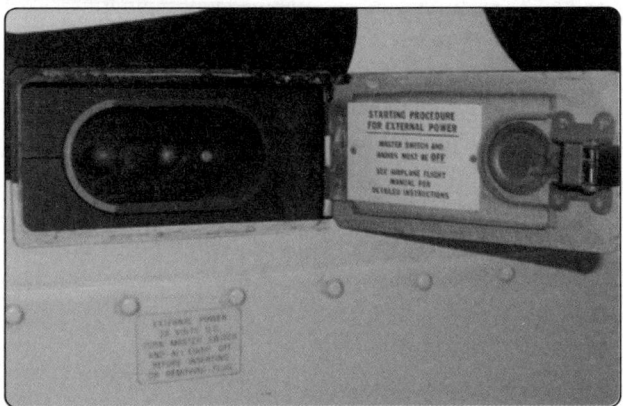

Figure 9-89. *External power receptacle.*

connection in the event the external power supply has the incorrect polarity (i.e., a reverse of the positive and negative electrical connections). A reverse polarity connection could be catastrophic to the aircraft's electrical system. If a ground power source with a reverse polarity is connected, the diode

blocks current and the external power solenoid does not close. This diagram also shows that external power can be used to charge the aircraft battery or power the aircraft electrical loads. For external power to start the aircraft engine or power electrical loads, the battery master switch must be closed.

Starter Circuit

Virtually all modern aircraft employ an electric motor to start the aircraft engine. Since starting the engine requires several horsepower, the starter motor can often draw 100 or more amperes. For this reason, all starter motors are controlled through a solenoid. *[Figure 9-91]*

The starter circuit must be connected as close as practical to the battery since large wire is needed to power the starter motor and weight savings can be achieved when the battery and the starter are installed close to each other in the aircraft. As shown in the starter circuit diagram, the start switch can be part of a multifunction switch that is also used to control the engine magnetos. *[Figure 9-92]*

The starter can be powered by either the aircraft battery or the external power supply. Often when the aircraft battery is weak or in need of charging, the external power circuit is used to power the starter. During most typical operations, the starter is powered by the aircraft battery. The battery master must be on and the master solenoid closed in order to start the engine with the battery.

Avionics Power Circuit

Many aircraft contain a separate power distribution bus specifically for electronics equipment. This bus is often referred to as an avionics bus. Since modern avionics equipment employs sensitive electronic circuits, it is often advantageous to disconnect all avionics from electrical power to protect their circuits. For example, the avionics bus is often depowered when the starter motor is activated. This helps to

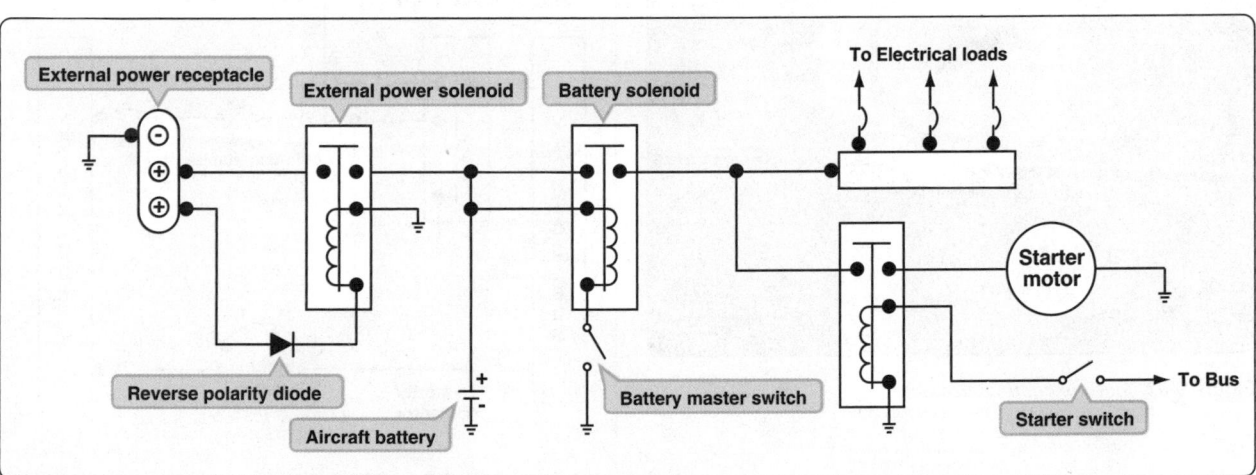

Figure 9-90. *A simple external power circuit diagram.*

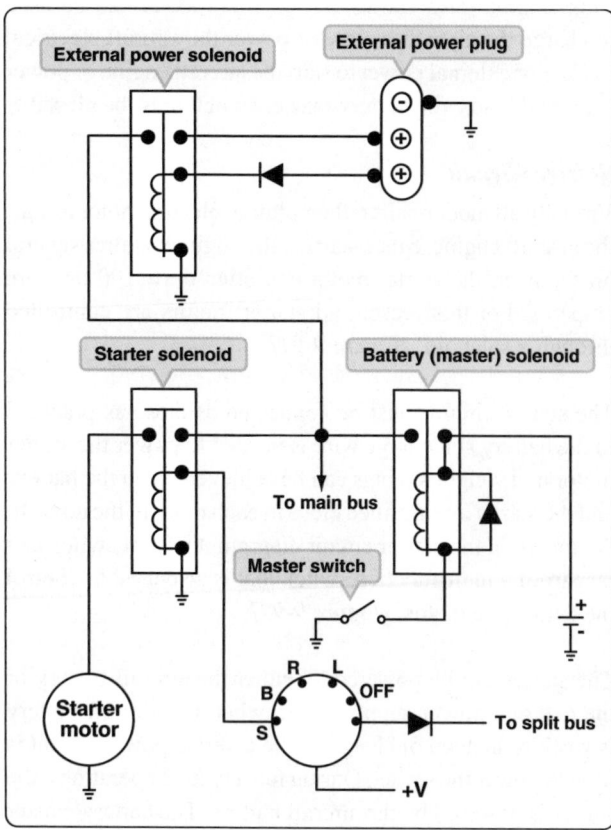

Figure 9-91. *Starter circuit.*

Figure 9-92. *Multifunction starter switch.*

prevent any transient voltage spikes produced by the starter from entering the sensitive avionics. *[Figure 9-93]*

The circuit employs a normally closed (NC) solenoid that connects the avionics bus to the main power bus. The electromagnet of the solenoid is activated whenever the starter is engaged. Current is sent from the starter switch through diode D1, causing the solenoid to open and depower the avionics bus. At that time, all electronics connected to the avionics bus will lose power. The avionics contactor is also activated whenever external power is connected to the aircraft. In this case, current travels through diodes D2 and D3 to the avionics bus contactor.

A separate avionics power switch may also be used to disconnect the entire avionics bus. A typical avionics power switch is shown wired in series with the avionics power bus. In some cases, this switch is combined with a circuit breaker and performs two functions (called a circuit breaker switch). It should also be noted that the avionics contactor is often referred to as a split bus relay, since the contactor separates (splits) the avionics bus from the main bus.

Landing Gear Circuit

Another common circuit found on light aircraft operates the retractable landing gear systems on high-performance

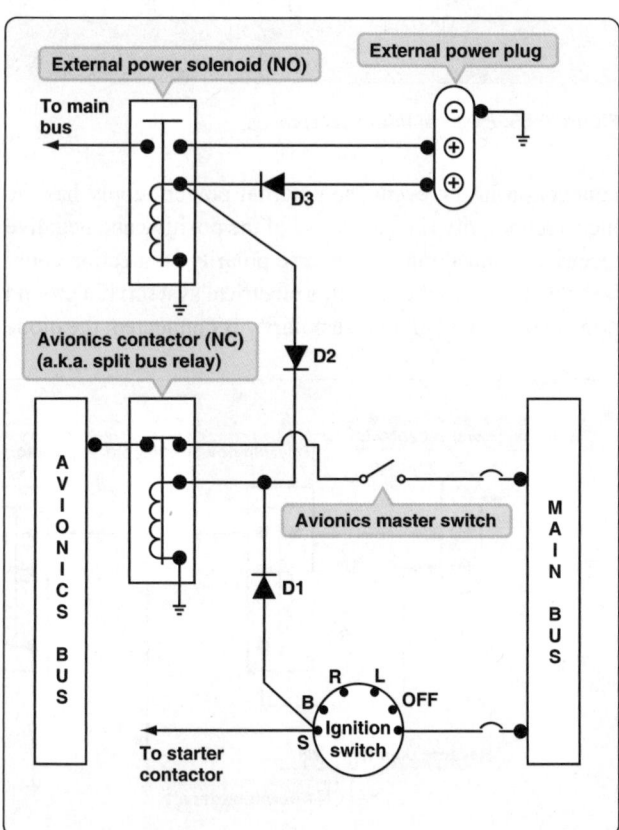

Figure 9-93. *Avionics power circuit.*

light aircraft. These airplanes typically employ a hydraulic system to move the gear. After takeoff, the pilot moves the gear position switch to the retract position, starting an electric motor. The motor operates a hydraulic pump, and the hydraulic system moves the landing gear. To ensure correct operation of the system, the landing gear electrical system is relatively complex. The electrical system must detect the position of each gear (right, left, nose) and determine when each reaches full up or down; the motor is then controlled accordingly. There are safety systems to help prevent accidental actuation of the gear.

A series of limit switches are needed to monitor the position of each gear during the operation of the system. (A limit switch is simply a spring-loaded, momentary contact switch that is activated when a gear reaches its limit of travel.) Typically, there are six limit switches located in the landing gear wheel wells. The three up-limit switches are used to detect when the gear reaches the full retract (UP) position. Three down-limit switches are used to detect when the gear reaches the full extended (DOWN) position. Each of these switches is mechanically activated by a component of the landing gear assembly when the appropriate gear reaches a given limit.

The landing gear system must also provide an indication to the pilot that the gear is in a safe position for landing. Many aircraft employ a series of three green lights when all three gears are down and locked in the landing position. These three lights are activated by the up- and down-limit switches found in the gear wheel well. A typical instrument panel showing the landing gear position switch and the three gears down indicators is shown in *Figure 9-94*.

The hydraulic motor/pump assembly located in the upper left corner of *Figure 9-95* is powered through either the UP or DOWN solenoids (top left). The solenoids are controlled by the gear selector switch (bottom left) and the six landing gear limit switches (located in the center of *Figure 9-95*). The three gear DOWN indicators are individual green lights (center of *Figure 9-95*) controlled by the three gear DOWN switches. As each gear reaches its DOWN position, the limit switch moves to the DOWN position, and the light is illuminated.

Figure 9-95 shows the landing gear in the full DOWN position. It is always important to know gear position when reading landing gear electrical diagrams. Knowing gear position helps the technician to analyze the diagram and understand correct operation of the circuits. Another important concept is that more than one circuit is used to operate the landing gear. On this system, there is a low current control circuit fused at 5 amps (CB2, top right of

Figure 9-94. *Instrument panel showing the landing gear position switch and the three gear down indicators.*

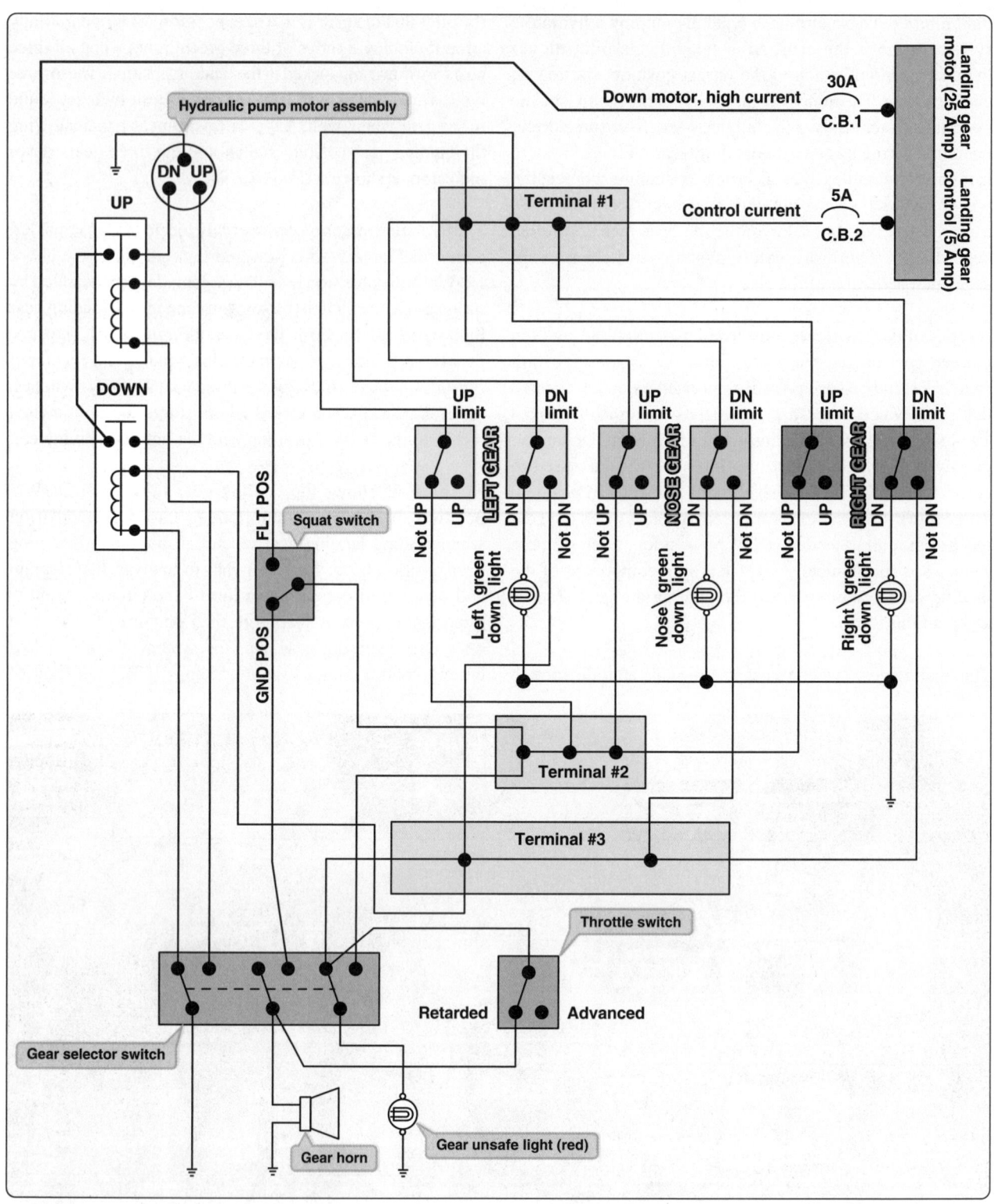

Figure 9-95. *Aircraft landing gear schematic while gear is in the DOWN and locked position.*

Figure 9-95). This circuit is used for indicator lights and the control of the gear motor contactors. There is a separate circuit to power the gear motor fused at 30 amps (CB3, top right of *Figure 9-95*). Since this circuit carries a large current flow, the wires would be as short as practical and carefully protected with rubber boots or nylon insulators.

The following paragraphs describe current flow through the landing gear circuit as the system moves the gear up and down. Be sure to refer to *Figure 9-96* often during the following discussions. *Figure 9-96* shows current flow when the gear is traveling to the extend (DOWN) position. Current flow is highlighted in red for each description.

To run the gear DOWN motor, current must flow in the control circuit leaving CB2 through terminal 1 to the NOT DOWN contacts of the DOWN limit switches, through terminal 3, to the DOWN solenoid positive terminal (upper left). The negative side of the DOWN solenoid coil is connected to ground through the gear selector switch. Remember, the gear DOWN switches are wired in parallel and activated when the gear reach the full-DOWN position. All three gears must reach full-DOWN to shut off the gear DOWN motor. Also note that the gear selector switch controls the negative side of the gear solenoids. The selector switch has independent control of the gear UP and DOWN motors through control of the ground circuit to both the UP and DOWN solenoids.

When the landing gear control circuit is sending a positive voltage to the DOWN solenoid, and the gear selector switch is sending negative voltage, the solenoid magnet is energized. When the gear-DOWN solenoid is energized, the high-current gear motor circuit sends current from CB1 through the down solenoid contact points to the gear DOWN motor. When the motor runs, the hydraulic pump produces pressure and the gear begins to move. When all three gears reach the DOWN position, the gear-DOWN switches move to the DOWN position, the three green lights illuminate, and the gear motor turns off completing the gear-DOWN cycle.

Figure 9-97 shows the landing gear electrical diagram with the current flow path shown in red as the gear moves to the retract (UP) position. Starting in the top right corner of the diagram, current must flow through CB2 in the control circuit through terminal 1 to each of the three gear-UP switches. With the gear-UP switches in the not UP position, current flows to terminal 2 and eventually through the squat switch to the UP solenoid electromagnet coil. The UP solenoid coil receives negative voltage through the gear selector switch. With the UP solenoid coil activated, the UP solenoid closes and power travels through the motor circuit. To power the motor, current leaves the bus through CB1 to the terminal at the DOWN solenoid onward through the UP solenoid to the UP motor. (Remember, current cannot travel through the DOWN solenoid at this time since the DOWN solenoid is not activated.) As the UP motor runs, each gear travels to the retract position. As this occurs, the gear UP switches move from the NOT UP position to the UP position. When the last gear reaches up, the current no longer travels to terminal 2 and the gear motor turns off. It should be noted that similar to DOWN, the gear switches are wired in parallel, which means the gear motor continues to run until all three gear reach the required position.

During both the DOWN and UP cycles of the landing gear operation, current travels from the limit switches to terminal 2. From terminal 2, there is a current path through the gear selector switch to the gear unsafe light. If the gear selector disagrees with the current gear position (e.g., gear is DOWN and pilot has selected UP), the unsafe light is illuminated. The gear unsafe light is shown at the bottom of *Figure 9-96*.

The squat switch (shown mid left of *Figure 9-96*) is used to determine if the aircraft is on the GROUND or in FLIGHT. This switch is located on a landing gear strut. When the weight of the aircraft compresses the strut, the switch is activated and moved to the GROUND position. When the switch is in the GROUND position, the gear cannot be retracted and a warning horn sounds if the pilot selects gear UP. The squat switch is sometimes referred to as the weight-on-wheels switch.

A throttle switch is also used in conjunction with landing gear circuits on most aircraft. If the throttle is retarded (closed) beyond a certain point, the aircraft descends and eventually lands. Therefore, many manufacturers activate a throttle switch whenever engine power is reduced. If engine power is reduced too low, a warning horn sounds telling the pilot to lower the landing gear. Of course, this horn need not sound if the gear is already DOWN or the pilot has selected the DOWN position on the gear switch. This same horn also sounds if the aircraft is on the ground, and the gear handle is moved to the UP position. *Figure 9-96* shows the gear warning horn in the bottom left corner.

AC Supply

Many modern light aircraft employ a low-power AC electrical system. Commonly, the AC system is used to power certain instruments and some lighting that operate only using AC. The electroluminescent panel has become a popular lighting system for aircraft instrument panels and requires AC. Electroluminescent lighting is very efficient and lightweight; therefore, excellent for aircraft installations. The electroluminescent material is a paste-like substance that glows when supplied with a voltage. This material is typically molded into a plastic panel and used for lighting.

A device called an inverter is used to supply AC when needed for light aircraft. Simply put, the inverter changes DC into AC. Two types of inverters may be found on aircraft: rotary inverters and static inverters. Rotary inverters are found only on older aircraft due to its poor reliability, excess weight, and inefficiency. The rotary inverters employee a DC motor that spins an AC generator. The unit is typically one unit and contains a voltage regulator circuit to ensure voltage stability. Most aircraft have a modern static inverter instead of a rotary inverter. Static inverters, as the name implies, contain no moving parts and use electronic circuitry to convert DC to

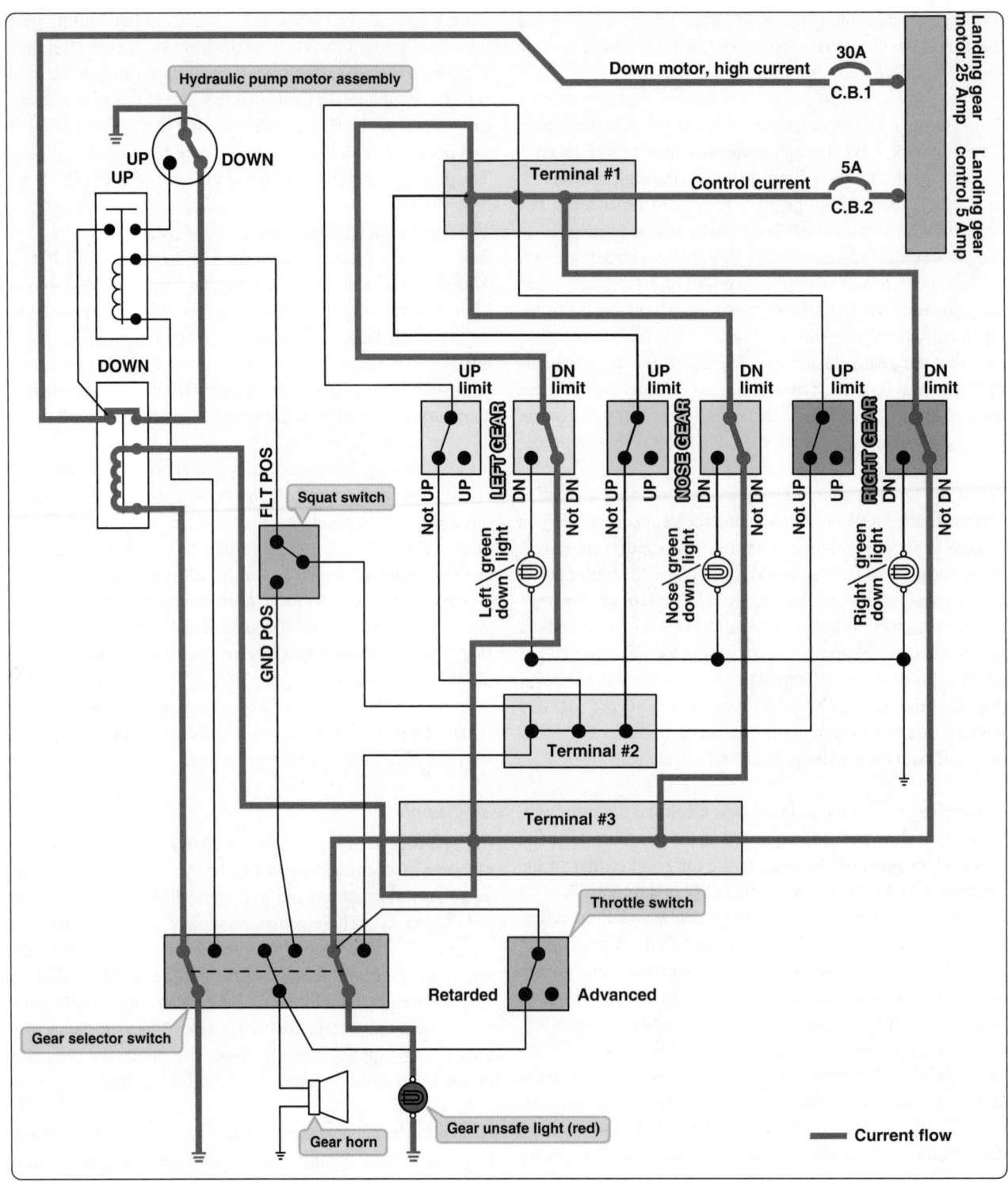

Figure 9-96. *Landing gear moving down diagram.*

AC. *Figure 9-98* shows a static inverter. Whenever AC is used on light aircraft, a distribution circuit separated from the DC system must be employed. *[Figure 9-99]*

Some aircraft use an inverter power switch to control AC power. Many aircraft simply power the inverter whenever the DC bus is powered and no inverter power switch is needed. On complex aircraft, more than one inverter may be used to provide a backup AC power source. Many inverters also offer more than one voltage output. Two common voltages found on aircraft inverters are 26VAC and 115VAC.

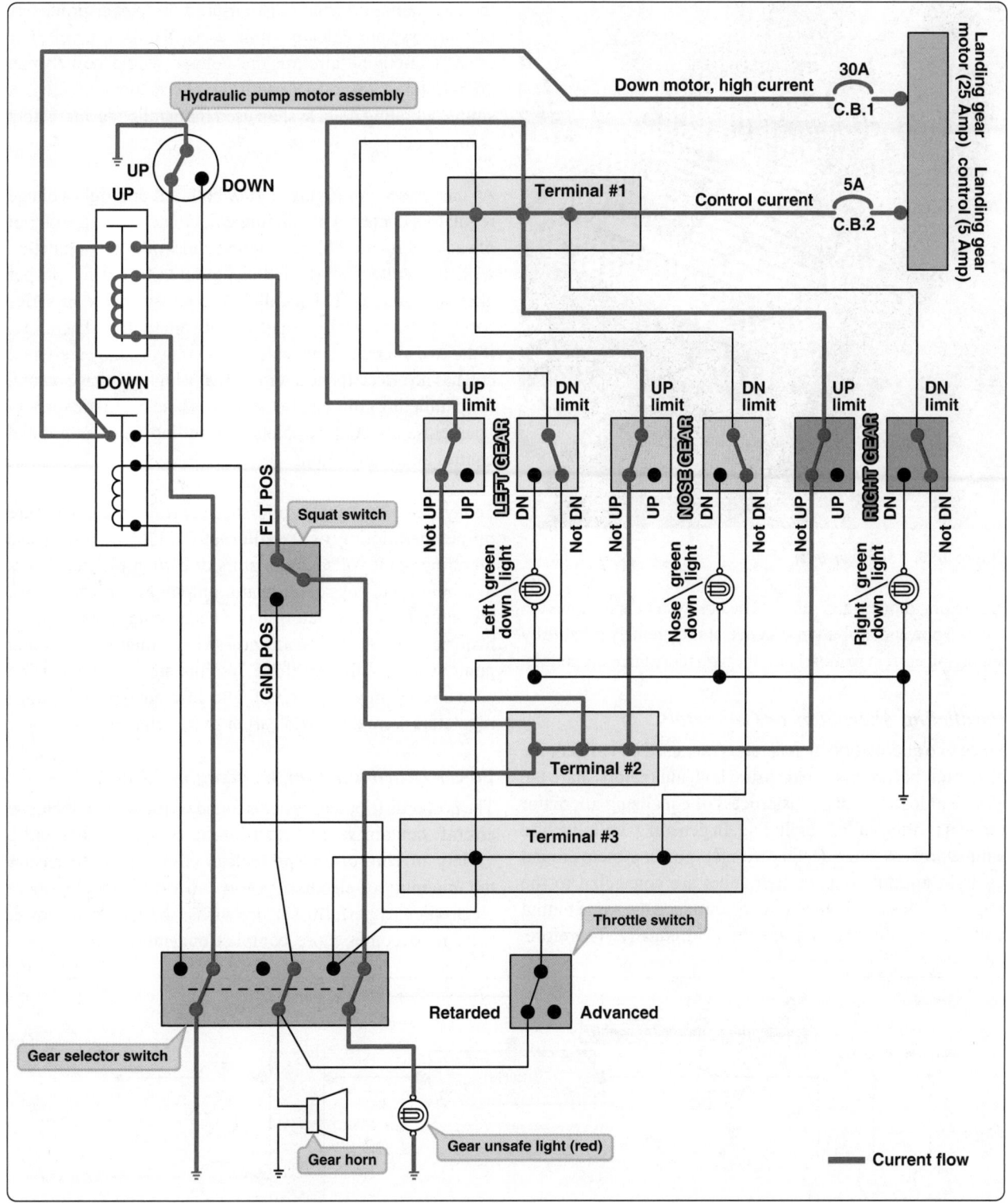

Figure 9-97. *Aircraft landing gear schematic while gear is moving to the UP position.*

Light Multiengine Aircraft

Multiengine aircraft typically fly faster, higher, and farther than single engine aircraft. Multiengine aircraft are designed for added safety and redundancy and, therefore, often contain a more complex power distribution system when compared to light single-engine aircraft. With two engines, these aircraft can drive two alternators (or generators) that supply current to the various loads of the aircraft. The electrical distribution bus system is also divided into two or more systems. These bus systems are typically connected through a series of circuit

Figure 9-98. *A static inverter.*

protectors, diodes, and relays. The bus system is designed to create a power distribution system that is extremely reliable by supplying current to most loads through more than one source.

Paralleling Alternators or Generators

Since two alternators (or generators) are used on twin engine aircraft, it becomes vital to ensure both alternators share the electrical load equally. This process of equalizing alternator outputs is often called paralleling. In general, paralleling is a simple process when dealing with DC power systems found on light aircraft. If both alternators are connected to the same load bus and both alternators produce the same output voltage, the alternators share the load equally. Therefore,

the paralleling systems must ensure both power producers maintain system voltage within a few tenths of a volt. For most twin-engine aircraft, the voltage would be between 26.5-volt and 28-volt DC with the alternators operating. A simple vibrating point system used for paralleling alternators is found in *Figure 9-100*.

As can be seen in *Figure 9-100*, both left and right voltage regulators contain a paralleling coil connected to the output of each alternator. This paralleling coil works in conjunction with the voltage coil of the regulator to ensure proper alternator output. The paralleling coils are wired in series between the output terminals of both alternators. Therefore, if the two alternators provide equal voltages, the paralleling coil has no effect. If one alternator has a higher voltage output, the paralleling coils create the appropriate magnetic force to open/close the contact points, controlling field current and control alternator output.

Today's aircraft employ solid-state control circuits to ensure proper paralleling of the alternators. Older aircraft use vibrating point voltage regulators or carbon-pile regulators to monitor and control alternator output. For the most part, all carbon-pile regulators have been replaced except on historic aircraft. Many aircraft still maintain a vibrating point system, although these systems are no longer being used on contemporary aircraft. The different types of voltage regulators were described earlier in this chapter.

Power Distribution on Multiengine Aircraft

The power distribution systems found on modern multiengine aircraft contain several distribution points (buses) and a variety of control and protection components to ensure the reliability of electrical power. As aircraft employ more electronics to perform various tasks, the electrical power systems becomes more complex and more reliable. One

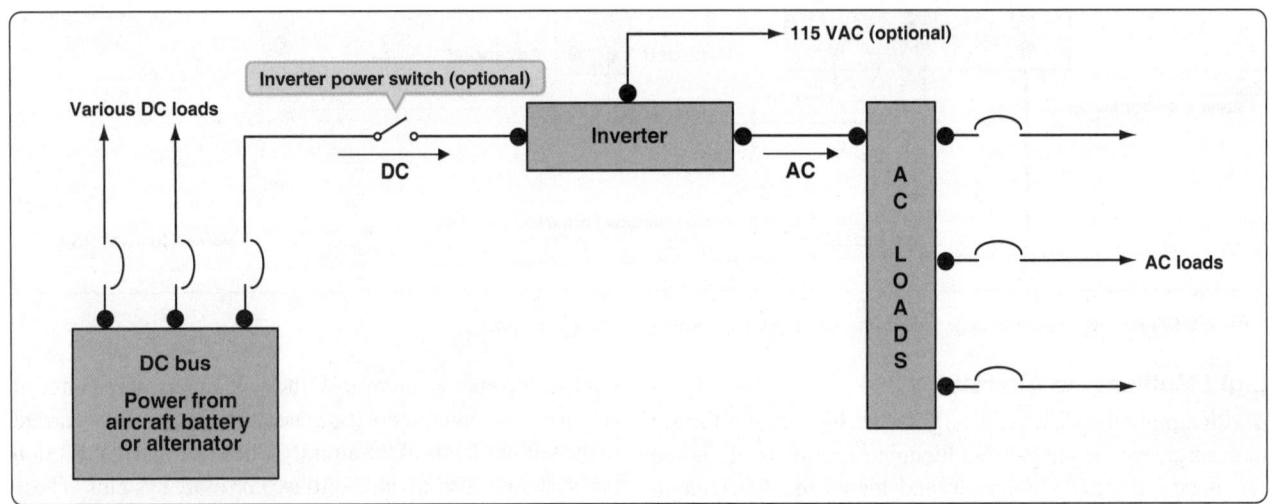

Figure 9-99. *Distribution circuit.*

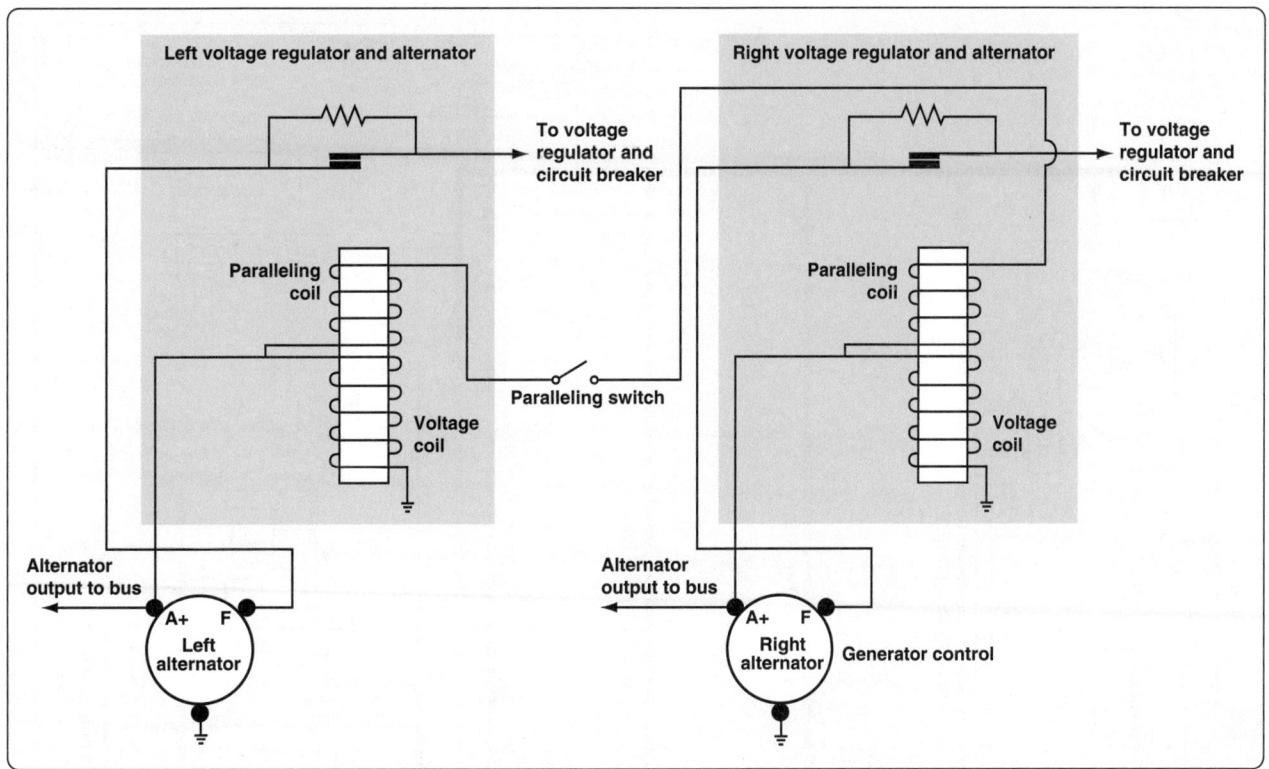

Figure 9-100. *Vibrating point system used for paralleling alternators.*

means to increase reliability is to ensure more than one power source can be used to power any given load. Another important design concept is to supply critical electrical loads from more than one bus. Twin-engine aircraft, such as a typical corporate jet or commuter aircraft, have two DC generators; they also have multiple distribution buses fed from each generator. *Figure 9-101* shows a simplified diagram of the power distribution system for a twin-engine turboprop aircraft.

This aircraft contains two starter generator units used to start the engines and generate DC electrical power. The system is typically defined as a split-bus power distribution system since there is a left and right generator bus that splits (shares) the electrical loads by connecting to each sub-bus through a diode and current limiter. The generators are operated in parallel and equally carry the loads.

The primary power supplied for this aircraft is DC, although small amounts of AC are supplied by two inverters. The aircraft diagram shows the AC power distribution at the top and mid left side of the diagram. One inverter is used for main AC power and the second is operated in standby and ready as a backup. Both inverters produce 26-volt AC and 115-volt AC. There is an inverter select relay operated by a pilot controlled switch used to choose which inverter is active.

The hot battery bus (right side of *Figure 9-101*) shows a direct connection to the aircraft battery. This bus is always hot if there is a charged battery in the aircraft. Items powered by this bus may include some basics like the entry door lighting and the aircraft clock, which should always have power available. Other items on this bus would be critical to flight safety, such as fire extinguishers, fuel shutoffs, and fuel pumps. During a massive system failure, the hot battery bus is the last bus on the aircraft that should fail.

If the battery switch is closed and the battery relay activated, battery power is connected to the main battery bus and the isolation bus. The main battery bus carries current for engine starts and external power. So the main battery bus must be large enough to carry the heaviest current loads of the aircraft. It is logical to place this bus as close as practical to the battery and starters and to ensure the bus is well protected from shorts to ground.

The isolation bus connects to the left and right buses and receives power whenever the main battery bus is energized. The isolation bus connects output of the left and right generators in parallel. The output of the two generators is then sent to the loads through additional buses. The generator buses are connected to the isolation bus through a fuse known as a current limiter. Current limiters are high amperage fuses that isolate buses if a short circuit occurs. There are several current limiters used in this system for protection

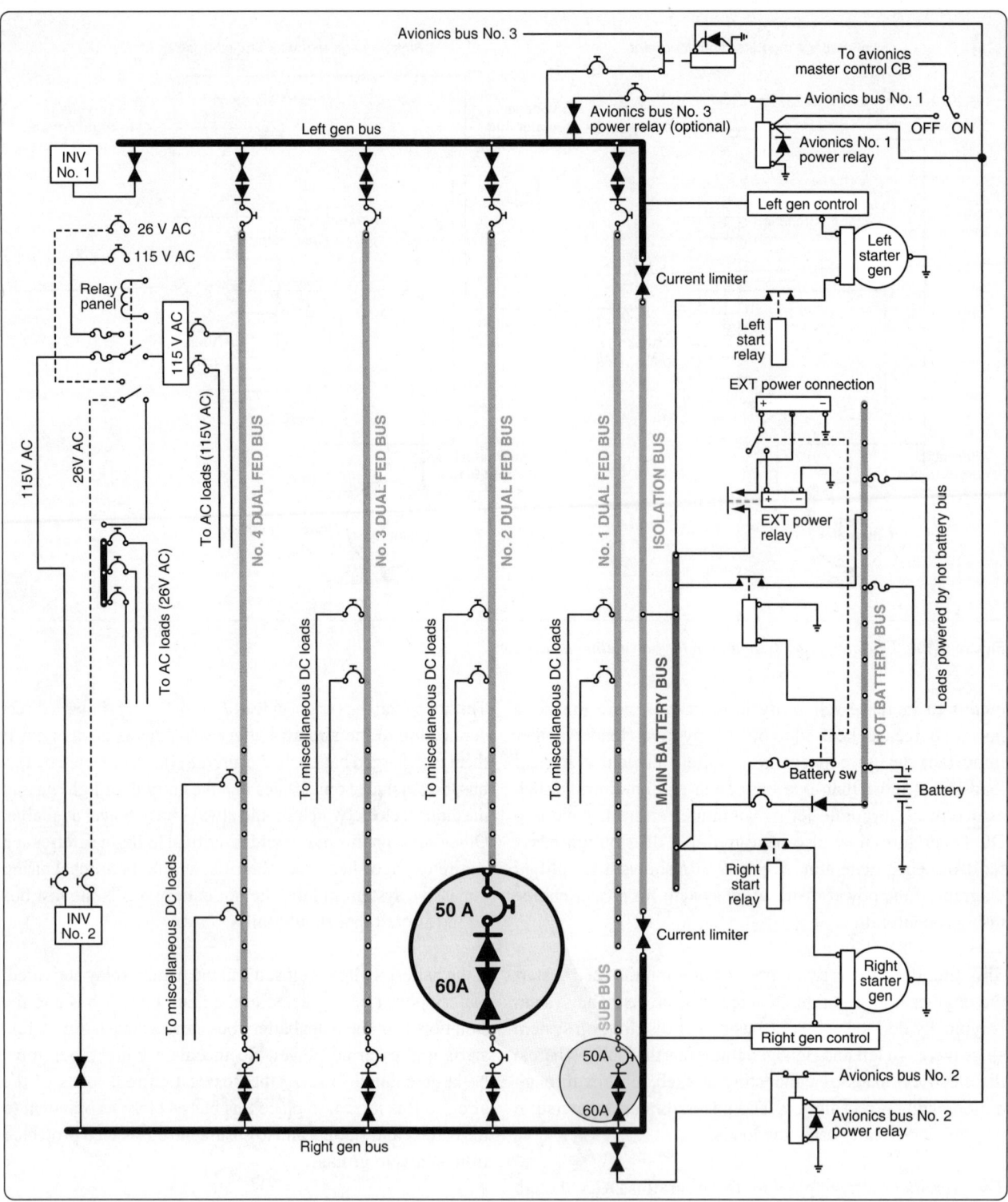

Figure 9-101. *Diagram of the power distribution system for a twin-engine turboprop aircraft.*

between buses. As can be seen in *Figure 9-101*, a current limiter symbol looks like two triangles pointed toward each other. The current limiter between the isolation bus and the main generator buses are rated at 325 amps and can only be replaced on the ground. Most current limiters are designed for ground replacement only and only after the malfunction that caused the excess current draw is repaired.

The left and right DC generators are connected to their respective main generator buses. Each generator feeds its respective bus, and since the buses are connected under normal circumstances, the generators operate in parallel.

Both generators feed all loads together. If one generator fails or a current limiter opens, the generators can operate independently. This design allows for redundancy in the event of failure and provides battery backup in the event of a dual generator failure.

In the center of *Figure 9-101* are four dual-feed electrical buses. These buses are considered dual-feed since they receive power from both the left and right generator buses. If a fault occurs, either generator bus can power any or all loads on a dual-feed bus. During the design phase of the aircraft, the electrical loads must be evenly distributed between each of the dual-feed buses. It is also important to power redundant systems from different buses. For example, the pilot's windshield heat would be powered by a different bus from the one that powers the copilot's windshield heat. If one bus fails, at least one windshield heat continues to work properly, and the aircraft can be landed safely in icing conditions.

Notice that the dual-feed buses are connected to the main generator buses through both a current limiter and a diode. Remember, a diode allows current flow in only one direction. *[Figure 9-102]*

The current can flow from the generator bus to the dual-feed bus, but the current cannot flow from the dual fed bus to the main generator bus. The diode is placed in the circuit so the main bus must be more positive than the sub bus for current flow. This circuit also contains a current limiter and a circuit breaker. The circuit breaker is located on the flight deck and can be reset by the pilot. The current limiter can only be replaced on the ground by a technician. The circuit breaker is rated at a slightly lower current value than the current limiter; therefore, the circuit breaker should open if a current overload exists. If the circuit breaker fails to open, the current limiter provides backup protection and disconnects the circuit.

Large Multiengine Aircraft

Transport category aircraft typically carry hundreds of passengers and fly thousands of miles each trip. Therefore, large aircraft require extremely reliable power distribution systems that are computer controlled. These aircraft have multiple power sources (AC generators) and a variety of distribution buses. A typical airliner contains two or more main AC generators driven by the aircraft turbine engines, as well as more than one backup AC generator. DC systems are also employed on large aircraft and the ship's battery is used to supply emergency power in case of a multiple failures.

The AC generator (sometimes called an alternator) produces three-phase 115-volt AC at 400 Hz. AC generators were discussed previously in this chapter. Since most modern transport category aircraft are designed with two engines, there are two main AC generators. The APU also drives an AC generator. This unit is available during flight if one of the main generators fails. The main and auxiliary generators are typically similar in output capacity and supply a maximum of 110 kilovolt amps (KVA). A fourth generator, driven by an emergency ram air turbine, is also available in the event the two main generators and one auxiliary generator fail. The emergency generator is typically smaller and produces less power. With four AC generators available on modern aircraft, it is highly unlikely that a complete power failure occurs. However, if all AC generators are lost, the aircraft battery will continue to supply DC electrical power to operate vital systems.

AC Power Systems

Transport category aircraft use large amounts of electrical power for a variety of systems. Passenger comfort requires power for lighting, audio visual systems, and galley power for food warmers and beverage coolers. A variety of electrical systems are required to fly the aircraft, such as flight control systems, electronic engine controls, communication, and navigation systems. The output capacity of one engine-driven AC generator can typically power all necessary electrical systems. A second engine-driven generator is operated during flight to share the electrical loads and provide redundancy.

The complexity of multiple generators and a variety of distribution buses requires several control units to maintain a constant supply of safe electrical power. The AC electrical system must maintain a constant output of 115 to 120 volts at

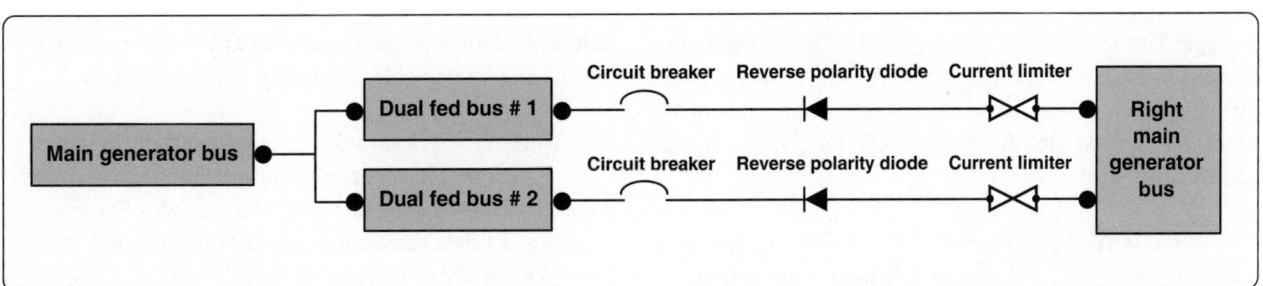

Figure 9-102. *Dual-feed bus system.*

a frequency of 400 Hz (±10 percent). The system must ensure power limits are not exceeded. AC generators are connected to the appropriate distribution buses at the appropriate time, and generators are in phase when needed. There is also the need to monitor and control any external power supplied to the aircraft, as well as control of all DC electrical power.

Two electronic line replaceable units are used to control the electrical power on a typical large aircraft. The generator control unit (GCU) is used for control of AC generator functions, such as voltage regulation and frequency control. The bus power control unit (BPCU) is used to control the distribution of electrical power between the various distribution buses on the aircraft. The GCU and BPCU work together to control electrical power, detect faults, take corrective actions when needed, and report any defect to the pilots and the aircraft's central maintenance system. There is typically one GCU for each AC generator and at least one BPCU to control bus connections. These LRUs are located in the aircraft's electronics equipment bay and are designed for easy replacement.

When the pilot calls for generator power by activating the generator control switch on the flight deck, the GCU monitors the system to ensure correct operation. If all systems are operating within limits, the GCU energizes the appropriate generator circuits and provides voltage regulation for the system. The GCU also monitors AC output to ensure a constant 400-Hz frequency. If the generator output is within limits, the GCU then connects the electrical power to the main generator bus through an electrical contactor (solenoid). These contactors are often called generator breakers (GB) since they break (open) or make (close) the main generator circuit.

After generator power is available, the BPCU activates various contactors to distribute the electrical power. The BPCU monitors the complete electrical system and communicates with the GCU to ensure proper operation. The BPCU employs remote current sensors known as a current transformers (CT) to monitor the system. *[Figure 9-103]*

A CT is an inductive unit that surrounds the main power cables of the electrical distribution system. As AC power flows through the main cables, the CT receives an induced voltage. The amount of CT voltage is directly related to the current flowing through the cable. The CT connects to the BPCU, which allows accurate current monitoring of the system. A typical aircraft employs several CTs throughout the electrical system.

The BPCU is a dedicated computer that controls the electrical connections between the various distribution buses found on the aircraft. The BPCU uses contactors (solenoids) called bus

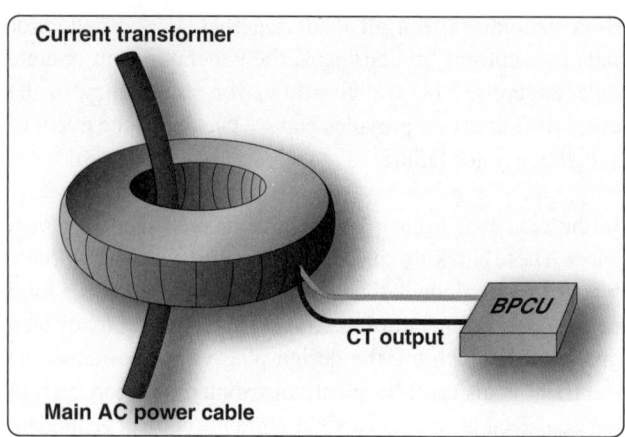

Figure 9-103. *Current transformer.*

tie breakers (BTB) for connection of various circuits. These BTBs open/close the connections between the buses as needed for system operation as called for by the pilots and the BPCU. This sounds like a simple task, yet to ensure proper operation under a variety of conditions, the bus system becomes very complex. There are three common types of distribution bus systems found on transport category aircraft: split bus, parallel bus, and split parallel.

Split-Bus Power Distribution Systems
Modern twin-engine aircraft, such as the Boeing 737, 757, 777, Airbus A-300, A-320, and A-310, employ a split-bus power distribution system. During normal conditions, each engine-driven AC generator powers only one main AC bus. The buses are kept split from each other, and two generators can never power the same bus simultaneously. This is very important since the generator output current is not phase regulated. (If two out-of-phase generators were connected to the same bus, damage to the system would occur.) The split-bus system does allow both engine-driven generators to power any given bus, but not at the same time. Generators must remain isolated from each other to avoid damage. The GCUs and BPCU ensures proper generator operation and power distribution.

On all modern split bus systems, the APU can be started and operated during flight. This allows the APU generator to provide back-up power in the event of a main generator failure. A fourth emergency generator powered by the ram air turbine is also available if the other generators fail.

The four AC generators are shown at the bottom of *Figure 9-104*. These generators are connected to their respective buses through the generator breakers. For example, generator 1 sends current through GB1 to AC bus 1. AC bus 1 feeds a variety of primary electrical loads, and also feeds sub-buses that in turn power additional loads.

With both generators operating and all systems normal, AC bus 1 and AC bus 2 are kept isolated. Typically during flight, the APB (bottom center of *Figure 9-104*) would be open and the APU generator off; the emergency generator (bottom right) would also be off and disconnected. If generator one should fail, the following happens:

1. The GB 1 is opened by the GCU to disconnect the failed generator.

2. The BPCU closes BTB 1 and BTB 2. This supplies AC power to AC bus 1 from generator 2.

3. The pilots start the APU and connect the APU generator. At that time, the BPCU and GCUs move the appropriate BTBs to correctly configure the system so the APU powers bus 1 and generator 2 powers bus 2. Once again, two AC generators operate independently to power AC bus 1 and 2.

If all generators fail, AC is also available through the static inverter (center of *Figure 9-104*). The inverter is powered from the hot battery bus and used for essential AC loads if all AC generators fail. Of course, the GCUs and BPCU take the appropriate actions to disconnect defective units and continue to feed essential AC loads using inverter power.

To produce DC power, AC bus 1 sends current to its transformer rectifier (TR), TR 1 (center left of *Figure 9-104*). The TR unit is used to change AC to DC. The TR contains a transformer to step down the voltage from 115-volt AC to 26-volt AC and a rectifier to change the 26-volt AC to 26-volt DC. The output of the TR is therefore compatible with the aircraft battery at 26-volt DC. Since DC power is not phase sensitive, the DC buses are connected during normal operation. In the event of a bus problem, the BPCU may isolate one or more DC buses to ensure correct distribution of DC power. This aircraft contains two batteries that are used to supply emergency DC power.

Parallel Systems

Multiengine aircraft, such as the Boeing 727, MD-11, and the early Boeing 747, employ a parallel power distribution system. During normal flight conditions, all engine-driven generators connect together and power the AC loads. In this configuration, the generators are operated in parallel; hence the name parallel power distribution system. In a parallel system, all generator output current must be phase regulated.

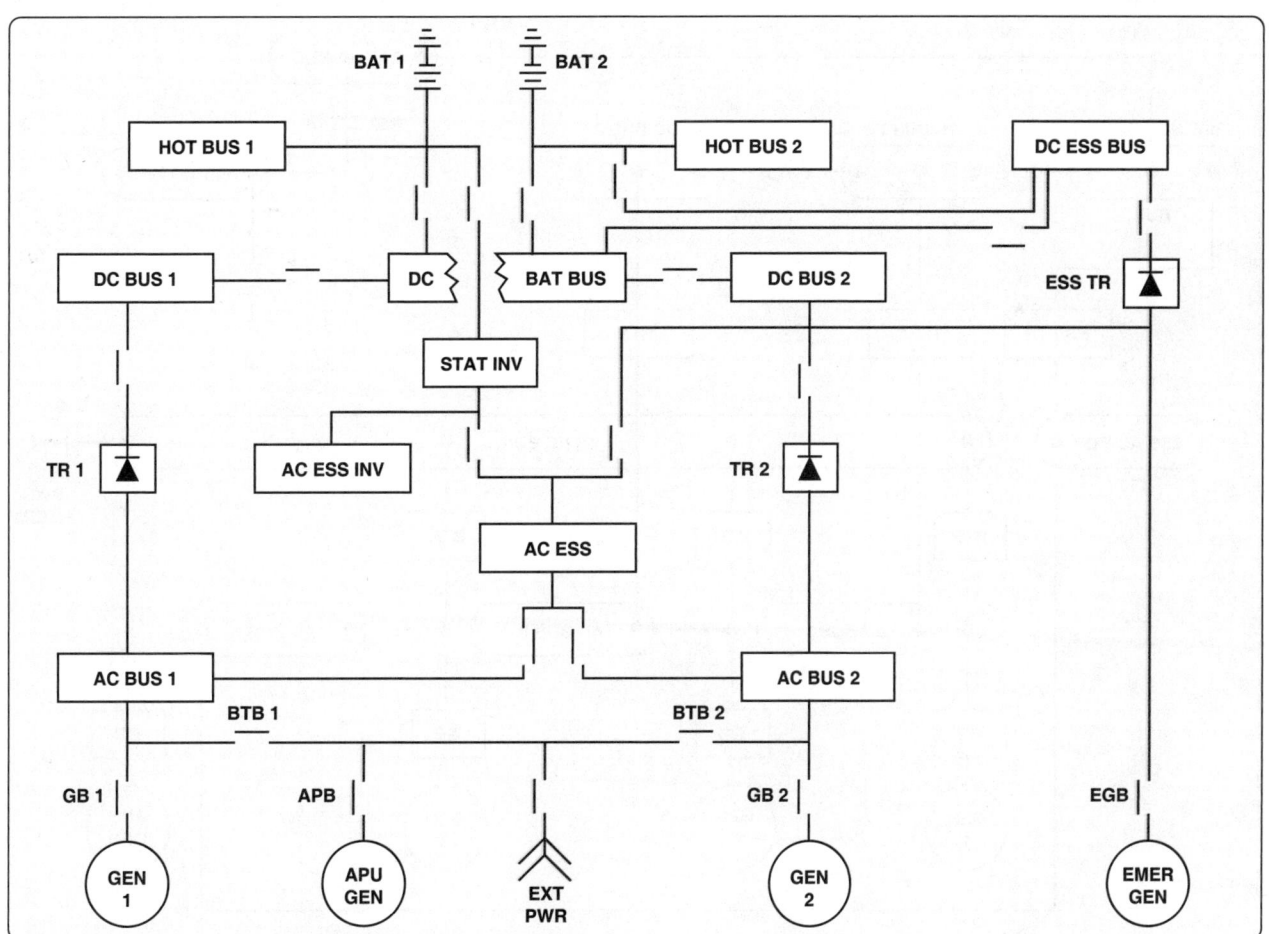

Figure 9-104. *Schematic of split-bus power distribution system.*

Before generators are connected to the same bus, their output frequency must be adjusted to ensure the AC output reaches the positive and negative peaks simultaneously. During the flight, generators must maintain this in-phase condition for proper operation.

One advantage of parallel systems is that in the event of a generator failure, the buses are already connected and the defective generator need only be isolated from the system. A paralleling bus, or synchronizing bus, is used to connect the generators during flight. The synchronizing bus is often referred to as the sync bus. Most of these systems are less automated and require that flight crew monitor systems and manually control bus contactors. BTBs are operated by the flight crew through the electrical control panel and used to connect all necessary buses. GBs are used to connect and disconnect the generators.

Figure 9-105 shows a simplified parallel power distribution system. This aircraft employs three main-engine driven generators and one APU generator. The APU (bottom right)

is not operational in flight and cannot provide backup power. The APU generator is for ground operations only. The three main generators (bottom of *Figure 9-105*) are connected to their respective AC bus through GBs one, two, and three. The AC buses are connected to the sync bus through three BTBs. In this manner, all three generators share the entire AC electrical loads. Keep in mind, all generators connected to the sync bus must be in phase. If a generator fails, the flight crew would simply isolate the defective generator and the flight would continue without interruption.

The number one and two DC buses (*Figure 9-105* top left) are used to feed the DC electrical loads of the aircraft. DC bus 1 receives power form AC bus 1 though TR1. DC bus 2 is fed in a similar manner from AC bus 2. The DC buses also connect to the battery bus and eventually to the battery. The essential DC bus (top left) can be fed from DC bus 1 or the essential TR. A diode prevents the essential DC bus from powcring DC bus 1. The essential DC bus receives power from the essential TR, which receives power from the essential AC bus. This provides an extra layer of redundancy

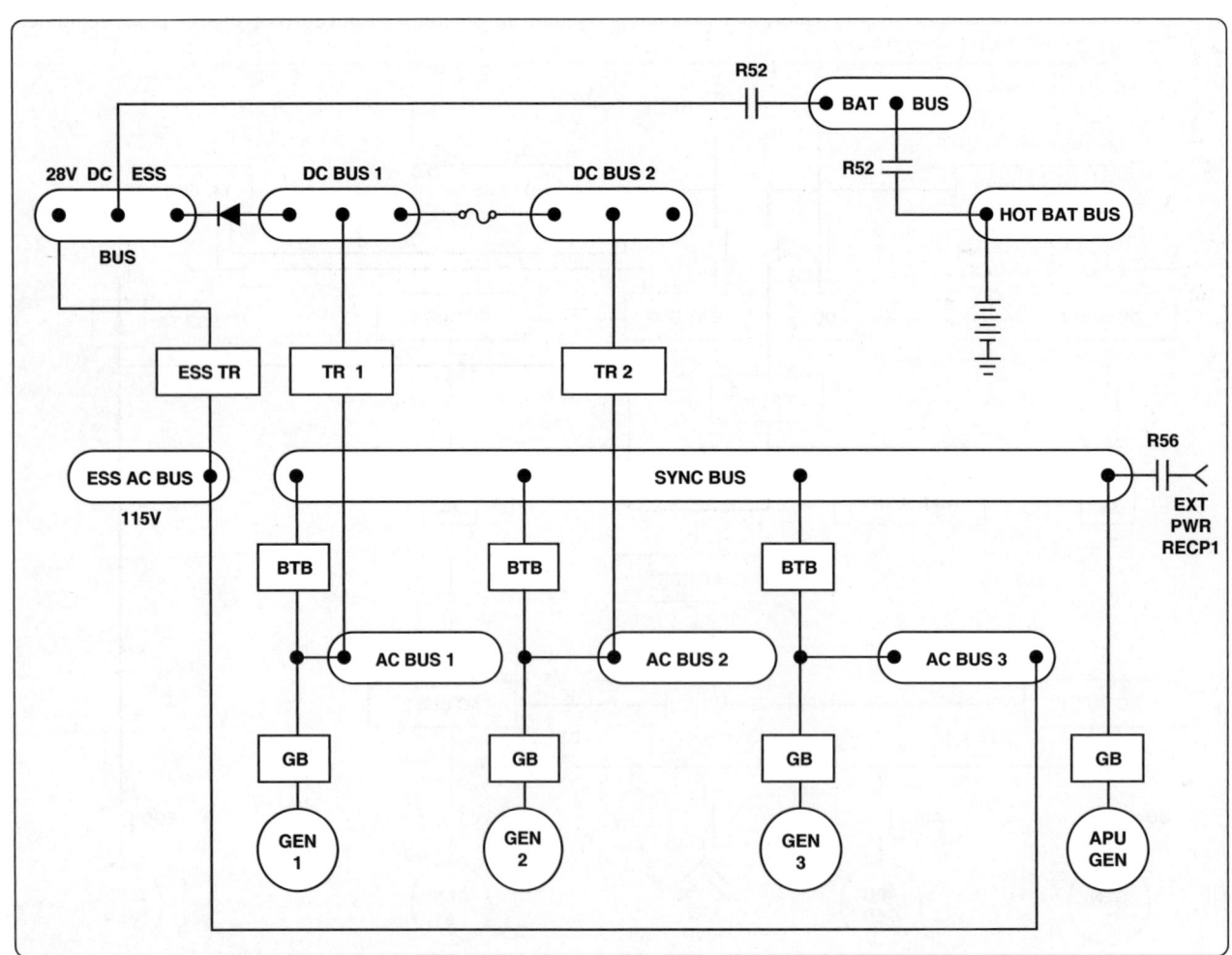

Figure 9-105. *Parallel power distribution system.*

since the essential AC bus can be isolated and fed from any main generator. *Figure 9-105* shows generator 3 powering the essential AC bus.

Split-Parallel Systems

A split-parallel bus basically employs the best of both split-bus and the parallel-bus systems. The split-parallel system is found on the Boeing 747-400 and contains four generators driven by the main engines and two APU-driven generators. The system can operate with all generators in parallel, or the generators can be operated independently as in a split-bus system. During a normal flight, all four engine-driven generators are operated in parallel. The system is operated in split-bus mode only under certain failure conditions or when using external power. The Boeing 747-400 split-parallel system is computer controlled using four GCUs and two BPCUs. There is one GCU controlling each generator; BPCU 1 controls the left side bus power distribution, and BPCU 2 controls the right side bus power. The GCUs and BPCUs operate similarly to those previously discussed under the split-bus system.

Figure 9-106 shows a simplified split-parallel power distribution system. The main generators (top of *Figure 9-106*) are driven by the main turbine engines. Each generator is connected to its load bus through a generator control breaker (GCB). The generator control unit closes the GCB when the pilot calls for generator power and all systems are operating normally. Each load bus is connected to various electrical

systems and additional sub-buses. The BTBs are controlled by the BPCU and connect each load bus to the left and right sync bus. A split systems breaker (SSB) is used to connect the left and right sync buses and is closed during a normal flight. With the SSB, GCBs, and BTBs, in the closed position the generators operate in parallel. When operating in parallel, all generators must be in phase.

If the aircraft electrical system experiences a malfunction, the control units make the appropriate adjustments to ensure all necessary loads receive electrical power. For example, if generator 1 fails, GCU 1 detects the fault and commands GCB 1 to open. With GCB 1 open, load bus 1 now feeds from the sync bus and the three operating generators. In another example, if load bus 4 should short to ground, BPCU 4 opens the GCB 4 and BTB 4. This isolates the shorted bus (load bus 4). All loads on the shorted bus are no longer powered, and generator 4 is no longer available. However, with three remaining generators operational, the flight continues safely.

As do all large aircraft, the Boeing 747-400 contains a DC power distribution system. The DC system is used for battery and emergency operations. The DC system is similar to those previously discussed, powered by TR units. The TRs are connected to the AC buses and convert AC into 26-volt DC. The DC power systems are the final backups in the event of a catastrophic electrical failure. The systems most critical to fly the aircraft can typically receive power from the battery. This aircraft also contains two static inverters to provide emergency AC power when needed.

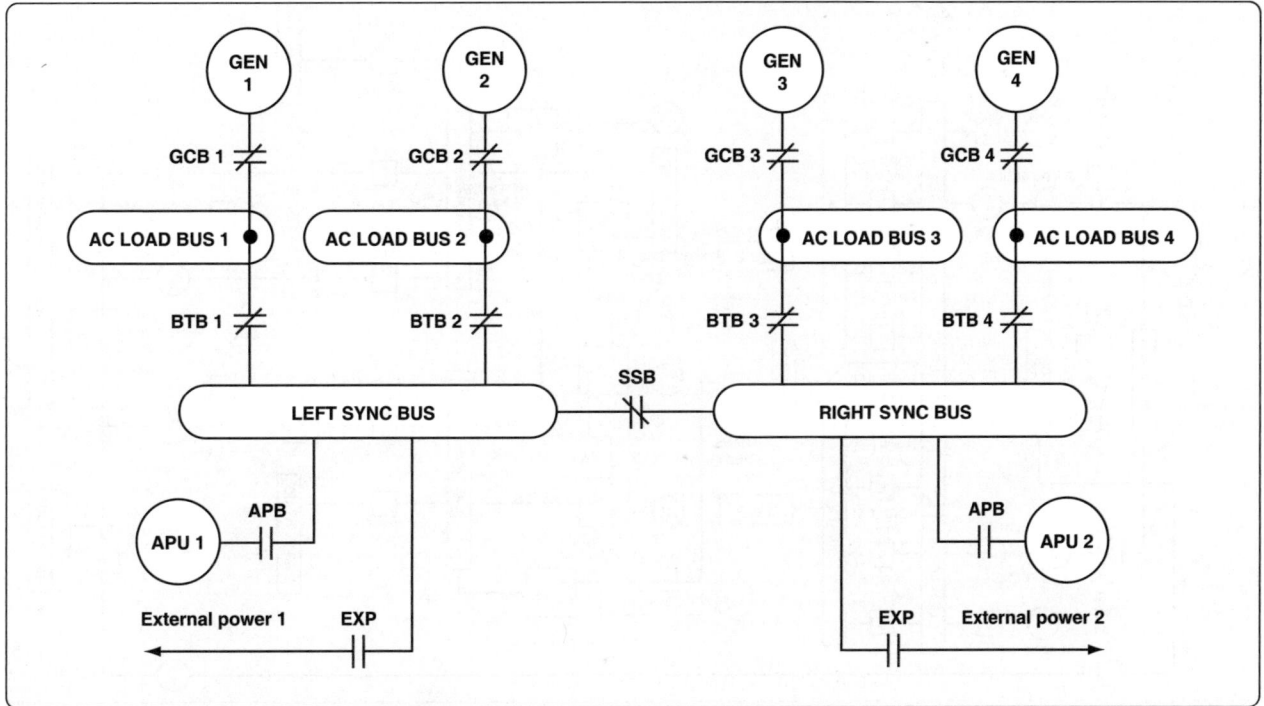

Figure 9-106. *Split-parallel distribution system.*

Wiring Installation

Wiring Diagrams

Electrical wiring diagrams are included in most aircraft service manuals and specify information, such as the size of the wire and type of terminals to be used for a particular application. Furthermore, wiring diagrams typically identify each component within a system by its part number and its serial number, including any changes that were made during the production run of an aircraft. Wiring diagrams are often used for troubleshooting electrical malfunctions.

Block Diagrams

A block diagram is used as an aid for troubleshooting complex electrical and electronic systems. A block diagram consists of individual blocks that represent several components, such as a printed circuit board or some other type of replaceable module. *Figure 9-107* is a block diagram of an aircraft electrical system.

Pictorial Diagrams

In a pictorial diagram, pictures of components are used instead of the conventional electrical symbols found in schematic diagrams. A pictorial diagram helps the maintenance technician visualize the operation of a system. *[Figure 9-108]*

Schematic Diagrams

A schematic diagram is used to illustrate a principle of operation, and therefore does not show parts as they actually appear or function. *[Figure 9-109]* However, schematic diagrams do indicate the location of components with respect to each other. Schematic diagrams are best utilized for troubleshooting.

Wire Types

The satisfactory performance of any modern aircraft depends to a very great degree on the continuing reliability of electrical systems and subsystems. Improperly or carelessly maintained wiring can be a source of both immediate and potential danger. The continued proper performance of electrical systems depends on the knowledge and techniques of the technician who installs, inspects, and maintains the electrical system wires and cables.

Procedures and practices outlined in this section are general recommendations and are not intended to replace the manufacturer's instructions and approved practices.

A wire is described as a single, solid conductor, or as a stranded conductor covered with an insulating material. *Figure 9-110* illustrates these two definitions of a wire. Because of in-flight vibration and flexing, conductor round wire should be stranded to minimize fatigue breakage.

The term "cable," as used in aircraft electrical installations, includes:

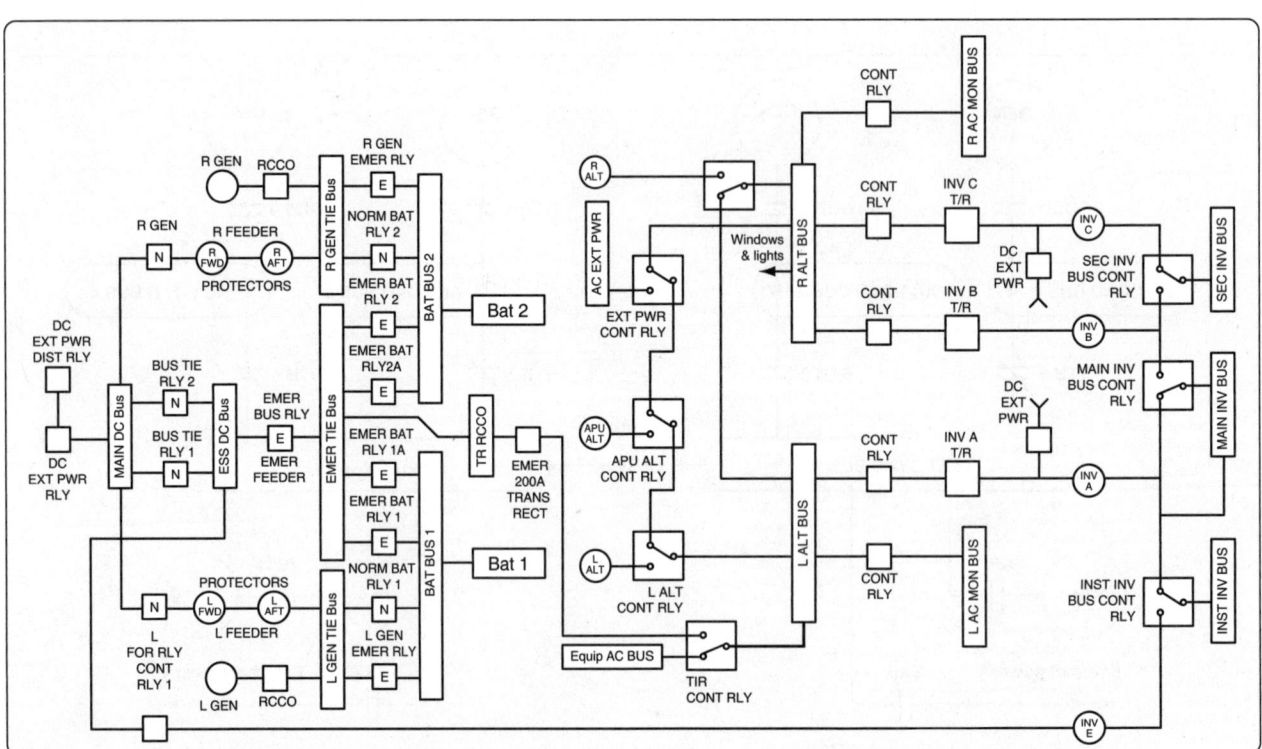

Figure 9-107. *Block diagram of an aircraft electrical system.*

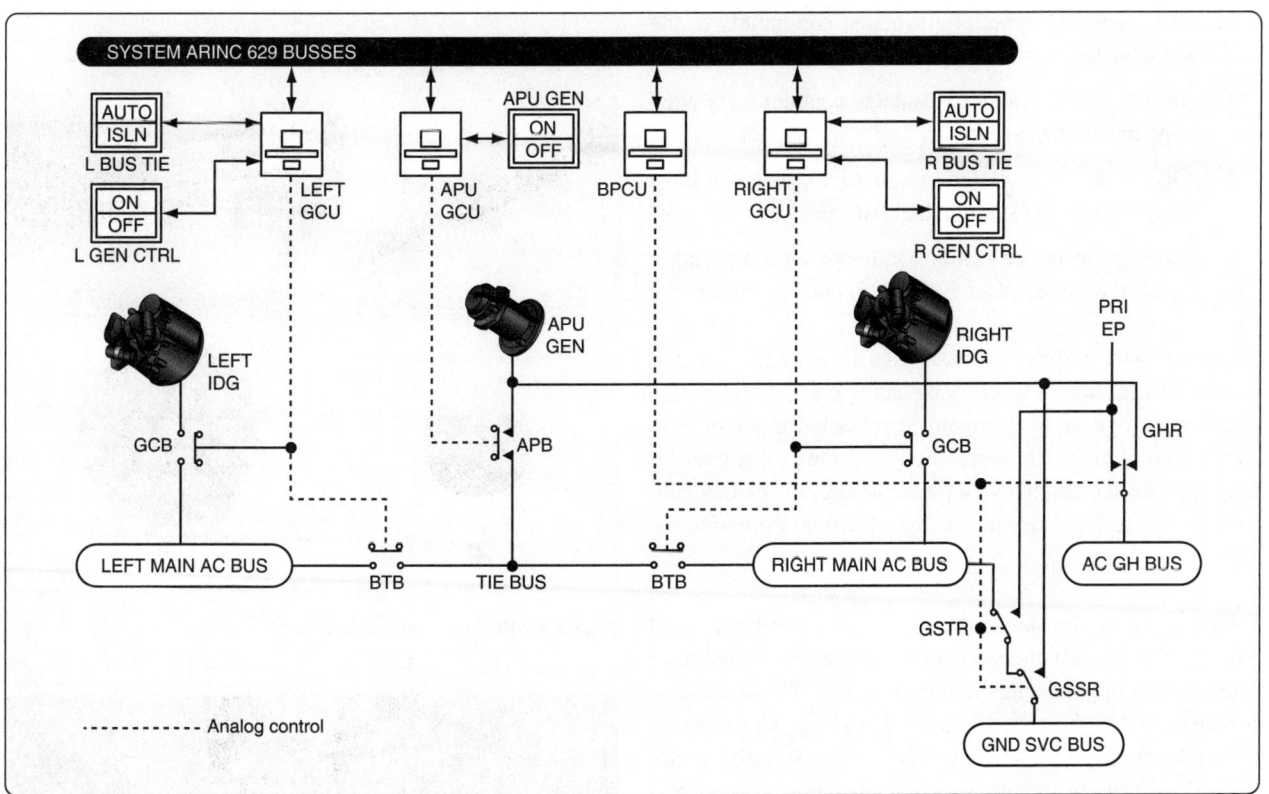

Figure 9-108. *Pictorial diagram of an aircraft electrical system.*

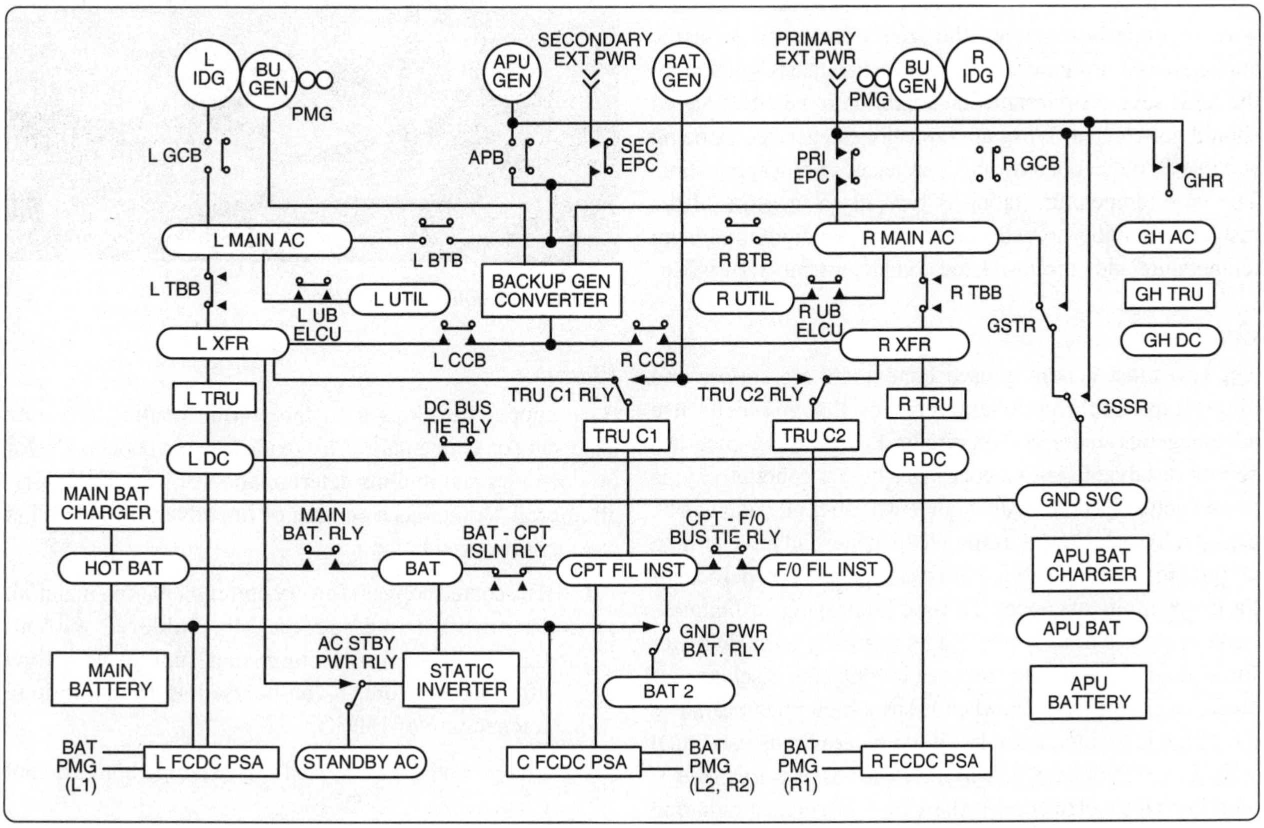

Figure 9-109. *Schematic diagram.*

1. Two or more separately insulated conductors in the same jacket.

2. Two or more separately insulated conductors twisted together (twisted pair).

3. One or more insulated conductors covered with a metallic braided shield (shielded cable).

4. A single insulated center conductor with a metallic braided outer conductor (radio frequency cable).

The term "wire harness" is used when an array of insulated conductors are bound together by lacing cord, metal bands, or other binding in an arrangement suitable for use only in specific equipment for which the harness was designed; it may include terminations. Wire harnesses are extensively used in aircraft to connect all the electrical components. *[Figure 9-111]*

For many years, the standard wire in light aircraft has been MIL-W-5086A, which uses a tin-coated copper conductor rated at 600 volts and temperatures of 105 °C. This basic wire is then coated with various insulating coatings. Commercial and military aircraft use wire that is manufactured under MIL-W-22759 specification, which complies with current military and FAA requirements.

The most important consideration in the selection of aircraft wire is properly matching the wire's construction to the application environment. Wire construction that is suitable for the most severe environmental condition to be encountered should be selected. Wires are typically categorized as being suitable for either open wiring or protected wiring application. The wire temperature rating is typically a measure of the insulation's ability to withstand the combination of ambient temperature and current-related conductor temperature rise.

Conductor

The two most generally used conductors are copper and aluminum. Each has characteristics that make its use advantageous under certain circumstances. Also, each has certain disadvantages. Copper has a higher conductivity; is more ductile; has relatively high tensile strength; and can be easily soldered. Copper is more expensive and heavier than aluminum. Although aluminum has only about 60 percent of the conductivity of copper, it is used extensively. Its lightness makes possible long spans, and its relatively large diameter for a given conductivity reduces corona (the discharge of electricity from the wire when it has a high potential). The discharge is greater when small diameter wire is used than when large diameter wire is used. Some bus bars are made of aluminum instead of copper where there is a greater radiating surface for the same conductance. The characteristics of copper and aluminum are compared in *Figure 9-112.*

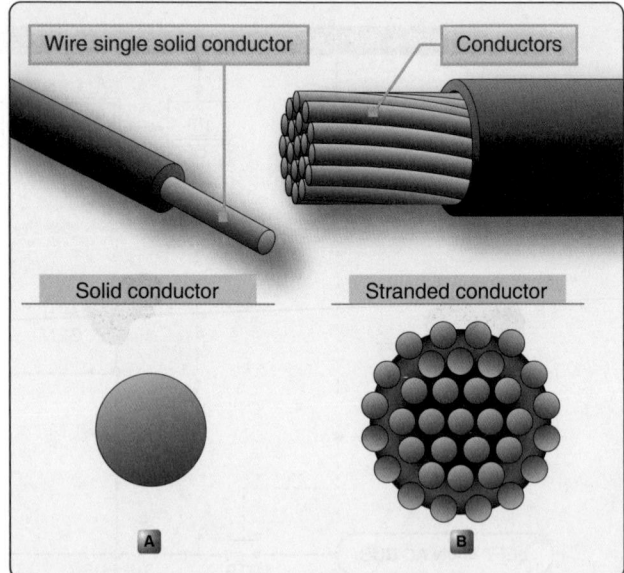

Figure 9-110. *Aircraft electrical cable.*

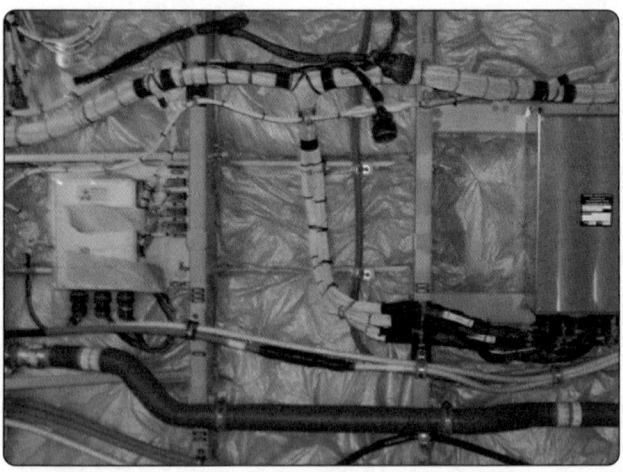

Figure 9-111. *Shielded wire harness.*

Plating

Bare copper develops a surface oxide coating at a rate dependent on temperature. This oxide film is a poor conductor of electricity and inhibits determination of wire. Therefore, all aircraft wiring has a coating of tin, silver, or nickel that has far slower oxidation rates.

1. Tin-coated copper is a very common plating material. Its ability to be successfully soldered without highly active fluxes diminishes rapidly with time after manufacture. It can be used up to the limiting temperature of 150 °C.

2. Silver-coated wire is used where temperatures do not exceed 200 °C (392 °F).

3. Nickel-coated wire retains its properties beyond 260 °C, but most aircraft wire using such coated

Characteristic	Copper	Aluminum
Tensile strength (lb-in)	55,000	25,000
Tensile strength for same conductivity (lb)	55,000	40,000
Weight for same conductivity (lb)	100	48
Cross section for same conductivity (CM)	100	160
Specific resistance (ohm/mil ft)	10.6	17

Figure 9-112. *Aircraft electrical cable.*

strands has insulation systems that cannot exceed that temperature on long-term exposure. Soldered terminations of nickel-plated conductor require the use of different solder sleeves or flux than those used with tin- or silver-plated conductor.

Insulation

Two fundamental properties of insulation materials are insulation resistance and dielectric strength. These are entirely different and distinct properties.

Insulation resistance is the resistance to current leakage through and over the surface of insulation materials. Insulation resistance can be measured with a megohmmeter/ insulation tester without damaging the insulation, and data so obtained serves as a useful guide in determining the general condition of the insulation. However, the data obtained in this manner may not give a true picture of the condition of the insulation. Clean, dry insulation having cracks or other faults might show a high value of insulation resistance but would not be suitable for use.

Dielectric strength is the ability of the insulator to withstand potential difference and is usually expressed in terms of the voltage at which the insulation fails because of the electrostatic stress. Maximum dielectric strength values can be measured by raising the voltage of a test sample until the insulation breaks down.

The type of conductor insulation material varies with the type of installation. Characteristics should be chosen based on environment, such as abrasion resistance, arc resistance, corrosion resistance, cut-through strength, dielectric strength, flame resistant, mechanical strength, smoke emission, fluid resistance, and heat distortion. Such types of insulation materials (e.g., PVC/nylon, Kapton®, and Teflon®) are no longer used for new aircraft designs, but might still be installed on older aircraft. Insulation materials for new aircraft designs are made of Tefzel®, Teflon®/Kapton®/Teflon® and PTFE/Polyimide/PTFE. The development of better and safer insulation materials is ongoing.

Since electrical wire may be installed in areas where

inspection is infrequent over extended periods of time, it is necessary to give special consideration to heat-aging characteristics in the selection of wire. Resistance to heat is of primary importance in the selection of wire for aircraft use, as it is the basic factor in wire rating. Where wire may be required to operate at higher temperatures due either to high ambient temperatures, high current loading, or a combination of the two, selection should be made on the basis of satisfactory performance under the most severe operating conditions.

Wire Shielding

With the increase in number of highly sensitive electronic devices found on modern aircraft, it has become very important to ensure proper shielding for many electric circuits. Shielding is the process of applying a metallic covering to wiring and equipment to eliminate electromagnetic interference (EMI). EMI is caused when electromagnetic fields (radio waves) induce high frequency (HF) voltages in a wire or component. The induced voltage can cause system inaccuracies or even failure.

Use of shielding with 85 percent coverage or greater is recommended. Coaxial, triaxial, twinaxial, or quadraxial cables should be used, wherever appropriate, with their shields connected to ground at a single point or multiple points, depending upon the purpose of the shielding. *[Figure 9-113]* The airframe grounded structure may also be used as an EMI shield.

Wire Substitutions

When a replacement wire is required in the repair and modification of existing aircraft, the maintenance manual for that aircraft must first be reviewed to determine if the original aircraft manufacturer (OAM) has approved any substitution. If not, then the manufacturer must be contacted for an acceptable replacement.

Figure 9-113. *Shielded wire harness for flight control.*

Areas Designated as Severe Wind & Moisture Problem (SWAMP)

SWAMP areas differ from aircraft to aircraft but are usually wheel wells, near wing flaps, wing folds, pylons, and other exterior areas that may have a harsh environment. Wires in these areas have often an exterior jacket to protect them from the environment. Wires for these applications often have design features incorporated into their construction that may make the wire unique; therefore, an acceptable substitution may be difficult, if not impossible, to find. It is very important to use the wire type recommended in the aircraft manufacturer's maintenance handbook. Insulation or jacketing varies according to the environment. *[Figure 9-114]*

Wire Size Selection

Wire is manufactured in sizes according to a standard known as the American wire gauge (AWG). As shown in *Figure 9-115*, the wire diameters become smaller as the gauge numbers become larger. Typical wire sizes range from a number 40 to number 0000.

Gauge numbers are useful in comparing the diameter of wires, but not all types of wire or cable can be measured accurately with a gauge. A wire gauge tool may be used to determine the size of an unmarked wire. Larger wires are usually stranded to increase their flexibility. In such cases, the total area can be determined by multiplying the area of one strand (usually computed in circular mils when diameter or gauge number is known) by the number of strands in the wire or cable.

Several factors must be considered in selecting the size of wire for transmitting and distributing electric power.

1. Wires must have sufficient mechanical strength to allow for service conditions.

2. Allowable power loss (I2 R loss) in the line represents electrical energy converted into heat. The use of large conductors reduces the resistance and therefore the I2 R loss. However, large conductors are more expensive,

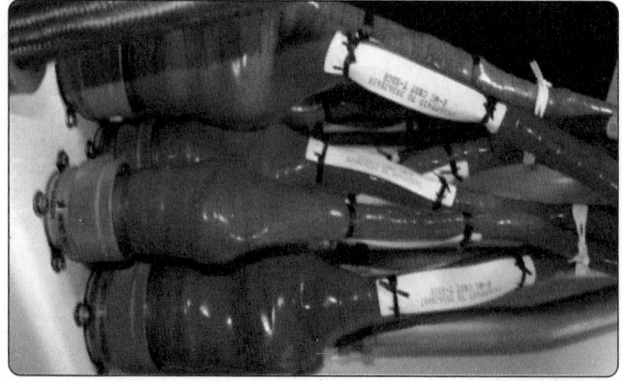

Figure 9-114. *Wire harness with protective jacket.*

heavier, and need more substantial support.

3. If the source maintains a constant voltage at the input to the lines, any variation in the load on the line causes a variation in line current and a consequent variation in the IR drop in the line. A wide variation in the IR drop in the line causes poor voltage regulation at the load. The obvious remedy is to reduce either current or resistance. A reduction in load current lowers the amount of power being transmitted, whereas a reduction in line resistance increases the size and weight of conductors required. A compromise is generally reached whereby the voltage variation at the load is within tolerable limits and the weight of line conductors is not excessive.

4. When current is drawn through the conductor, heat is generated. The temperature of the wire rises until the heat radiated, or otherwise dissipated, is equal to the heat generated by the passage of current through the line. If the conductor is insulated, the heat generated in the conductor is not so readily removed as it would be if the conductor were not insulated. Thus, to protect the insulation from too much heat, the current through the conductor must be maintained below a certain value. When electrical conductors are installed in locations where the ambient temperature is relatively high, the heat generated by external sources constitutes an appreciable part of the total conductor heating. Allowance must be made for the influence of external heating on the allowable conductor current, and each case has its own specific limitations. The maximum allowable operating temperature of insulated conductors varies with the type of conductor insulation being used.

If it is desirable to use wire sizes smaller than #20, particular attention should be given to the mechanical strength and installation handling of these wires (e.g., vibration, flexing, and termination). Wires containing less than 19 strands must not be used. Consideration should be given to the use of high-strength alloy conductors in small-gauge wires to increase mechanical strength. As a general practice, wires smaller than size #20 should be provided with additional clamps and be grouped with at least three other wires. They should also have additional support at terminations, such as connector grommets, strain relief clamps, shrinkable sleeving, or telescoping bushings. They should not be used in applications where they are subjected to excessive vibration, repeated bending, or frequent disconnection from screw termination. *[Figure 9-116]*

Current Carrying Capacity

In some instances, the wire may be capable of carrying more

Cross Section			Ohms per 1,000 ft		
Gauge Number	Diameter (mils)	Circular (mils)	Square inches	25 °C (77 °F)	65 °C (149 °F)
0000	460.0	212,000.0	0.166	0.0500	0.0577
000	410.0	168,000.0	0.132	0.0630	0.0727
00	365.0	133,000.0	0.105	0.0795	0.0917
0	325.0	106,000.0	0.0829	0.100	0.166
1	289.0	83,700.0	0.0657	0.126	0.146
2	258.0	66,400.0	0.0521	0.159	0.184
3	229.0	52,600.0	0.0413	0.201	0.232
4	204.0	41,700.0	0.0328	0.253	0.292
5	182.0	33,100.0	0.0260	0.319	0.369
6	162.0	26,300.0	0.0206	0.403	0.465
7	144.0	20,800.0	0.0164	0.508	0.586
8	128.0	16,500.0	0.0130	0.641	0.739
9	114.0	13,100.0	0.0103	0.808	0.932
10	102.0	10,400.0	0.00815	1.02	1.18
11	91.0	8,230.0	0.00647	1.28	1.48
12	81.0	6,530.0	0.00513	1.62	1.87
13	72.0	5,180.0	0.00407	2.04	2.36
14	64.0	4,110.0	0.00323	2.58	2.97
15	57.0	3,260.0	0.00256	3.25	3.75
16	51.0	2,580.0	0.00203	4.09	4.73
17	45.0	2,050.0	0.00161	5.16	5.96
18	40.0	1,620.0	0.00128	6.51	7.51
19	36.0	1,290.0	0.00101	8.21	9.48
20	32.0	1,020.0	0.000802	10.40	11.90
21	28.5	810.0	0.000636	13.10	15.10
22	25.3	642.0	0.000505	16.50	19.00
23	22.6	509.0	0.000400	20.80	24.00
24	20.1	404.0	0.000317	26.20	30.20
25	17.9	320.0	0.000252	33.00	38.10
26	15.9	254.0	0.000200	41.60	48.00
27	14.2	202.0	0.000158	52.50	60.60
28	12.6	160.0	0.000126	66.20	76.40
29	11.3	127.0	0.0000995	83.40	96.30
30	10.0	101.0	0.0000789	105.00	121.00
31	8.9	79.7	0.0000626	133.00	153.00
32	8.0	63.2	0.0000496	167.00	193.00
33	7.1	50.1	0.0000394	211.00	243.00
34	6.3	39.8	0.0000312	266.00	307.00
35	5.6	31.5	0.0000248	335.00	387.00
36	5.0	25.0	0.0000196	423.00	488.00
37	4.5	19.8	0.0000156	533.00	616.00
38	4.0	15.7	0.0000123	673.00	776.00
39	3.5	12.5	0.0000098	848.00	979.00
40	3.1	9.9	0.0000078	1,070.00	1,230.00

Figure 9-115. *American wire gauge for standard annealed solid copper wire.*

current than is recommended for the contacts of the related connector. In this instance, it is the contact rating that dictates the maximum current to be carried by a wire. Wires of larger gauge may need to be used to fit within the crimp range of connector contacts that are adequately rated for the current being carried. *Figure 9-117* gives a family of curves whereby the bundle derating factor may be obtained.

Maximum Operating Temperature

The current that causes a temperature steady state condition equal to the rated temperature of the wire should not be exceeded. Rated temperature of the wire may be based upon the ability of either the conductor or the insulation to withstand continuous operation without degradation.

Single Wire in Free Air

Determining a wiring system's current-carrying capacity begins with determining the maximum current that a given-sized wire can carry without exceeding the allowable temperature difference (wire rating minus ambient °C). The curves are based upon a single copper wire in free air. [Figure 9-117]

Wires in a Harness

When wires are bundled into harnesses, the current derived for a single wire must be reduced, as shown in *Figure 9-118*. The amount of current derating is a function of the number of wires in the bundle and the percentage of the total wire bundle capacity that is being used.

Harness at Altitude

Since heat loss from the bundle is reduced with increased altitude, the amount of current should be derated. *Figure 9-119* gives a curve whereby the altitude-derating factor may be obtained.

Aluminum Conductor Wire

When aluminum conductor wire is used, sizes should be selected on the basis of current ratings shown in *Figure 9-120*. The use of sizes smaller than #8 is discouraged. Aluminum wire should not be attached to engine mounted accessories or used in areas having corrosive fumes, severe vibration, mechanical stresses, or where there is a need for frequent disconnection. Use of aluminum wire is also discouraged for runs of less than 3 feet. Termination hardware should be of the type specifically designed for use with aluminum conductor wiring.

Computing Current Carrying Capacity

The following section presents some examples on how to calculate the load carrying capacity of aircraft electrical wire. The calculation is a step by step approach and several graphs are used to obtain information to compute the current carrying capacity of a particular wire.

Example 1

Assume a harness (open or braided) consisting of 10 wires, size 20, 200 °C rated copper, and 25 wires size 22, 200 °C rated copper, is installed in an area where the ambient temperature is 60 °C and the aircraft is capable of operating at a 35,000 foot altitude. Circuit analysis reveals that 7 of the 35 wires in the bundle ($^7/_{35}$ = 20 percent) are carrying power currents near or up to capacity.

Step 1—Refer to the single wire in free air graph in *Figure 9-117*. Determine the change of temperature of the wire to determine free air ratings. Since the wire is in an ambient temperature of 60 °C and rated at 200 °C, the change of the temperature is 200 °C – 60 °C = 140 °C. Follow the 140 °C temperature difference horizontally until it intersects with wire size line on *Figure 9-117*. The free air rating for size 20 is 21.5 amps, and the free air rating for size 22 is 16.2 amps.

Step 2—Refer to the bundle derating curves in *Figure 9-118*. The 20 percent curve is selected since circuit analysis indicate that 20 percent or less of the wire in the harness would be carrying power currents and less than 20 percent of the bundle capacity would be used. Find 35 (on the horizontal axis), since there are 35 wires in the bundle, and determine a derating factor of 0.52 (on the vertical axis) from the 20 percent curve.

Step 3—Derate the size 22 free air rating by multiplying 16.2 by 0.52 to get 8.4 amps in harness rating. Derate the size 20 free air rating by multiplying 21.5 by 0.52 to get 11.2 amps in-harness rating.

Step 4—Refer to the altitude derating curve in *Figure 9-119*. Look for 35,000 feet (on the horizontal axis) since that is the altitude at which the aircraft is operating. Note that the wire must be derated by a factor of 0.86 (found on the vertical axis). Derate the size 22 harness rating by multiplying 8.4 amps by 0.86 to get 7.2 amps. Derate the size 20 harness rating by multiplying 11.2 amps by 0.86 to get 9.6 amps.

Step 5—To find the total harness capacity, multiply the total number of size 22 wires by the derated capacity (25 × 7.2 = 180.0 amps) and add to that the number of size 20 wires multiplied by the derated capacity (10 × 9.6 = 96.8 amps) and multiply the sum by the 20 percent harness capacity factor. Thus, the total harness capacity is (180.0 + 96.0) × 0.20 = 55.2 amps. It has been determined that the total harness current should not exceed 55.2 A, size 22 wire should not carry more than 7.2 amps and size 20 wire should not carry more than 9.6 amps.

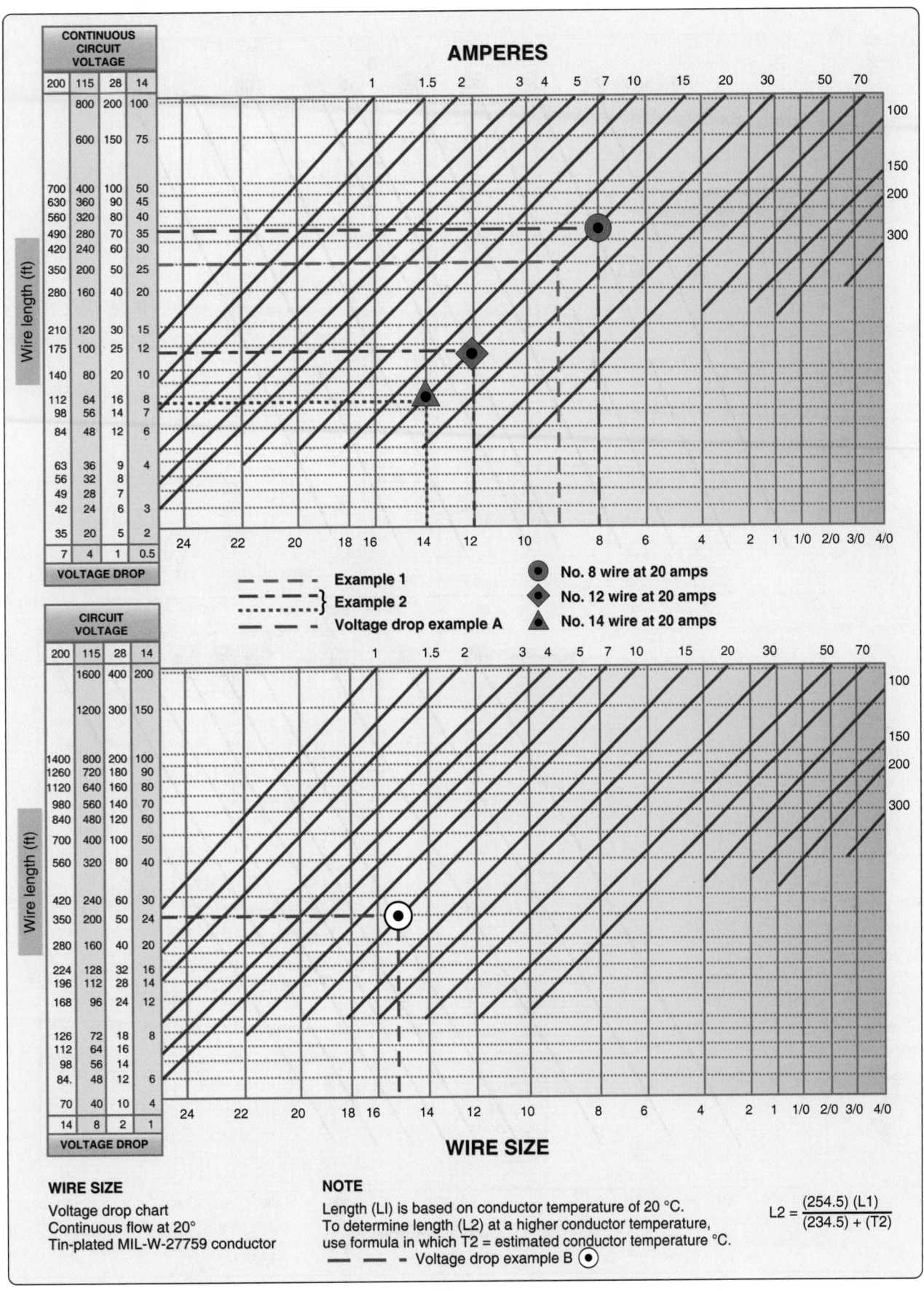

Figure 9-116. *Conductor chart, continuous (top) and intermittent flow (bottom).*

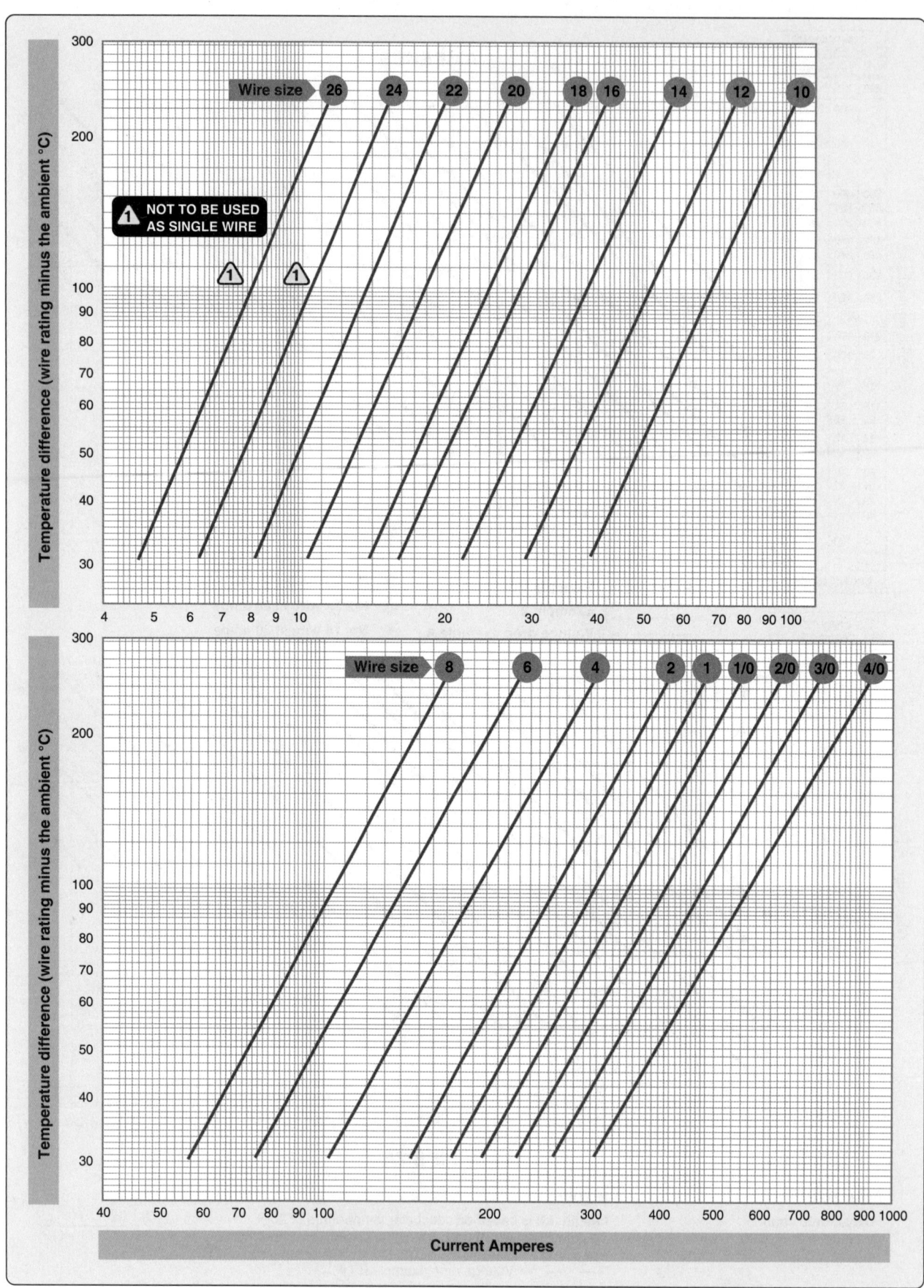

Figure 9-117. *Single copper wire in free air.*

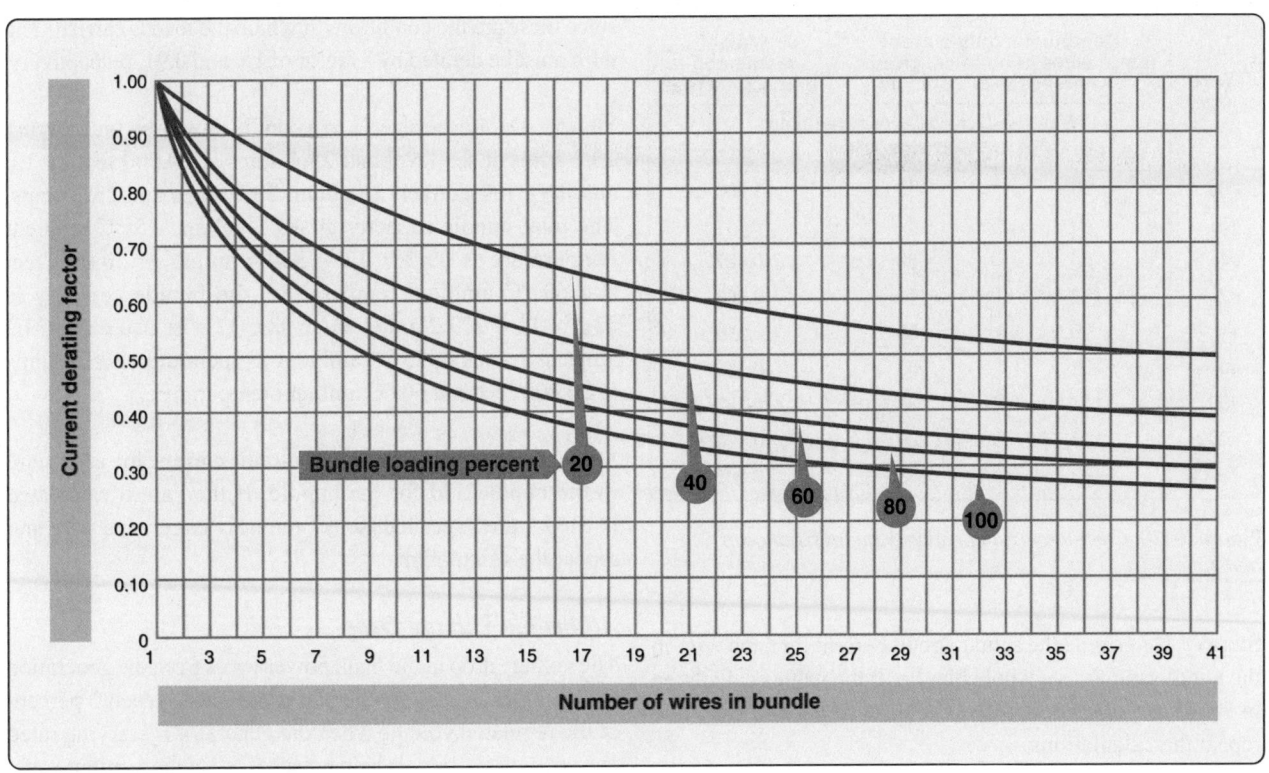

Figure 9-118. *Bundle derating curve.*

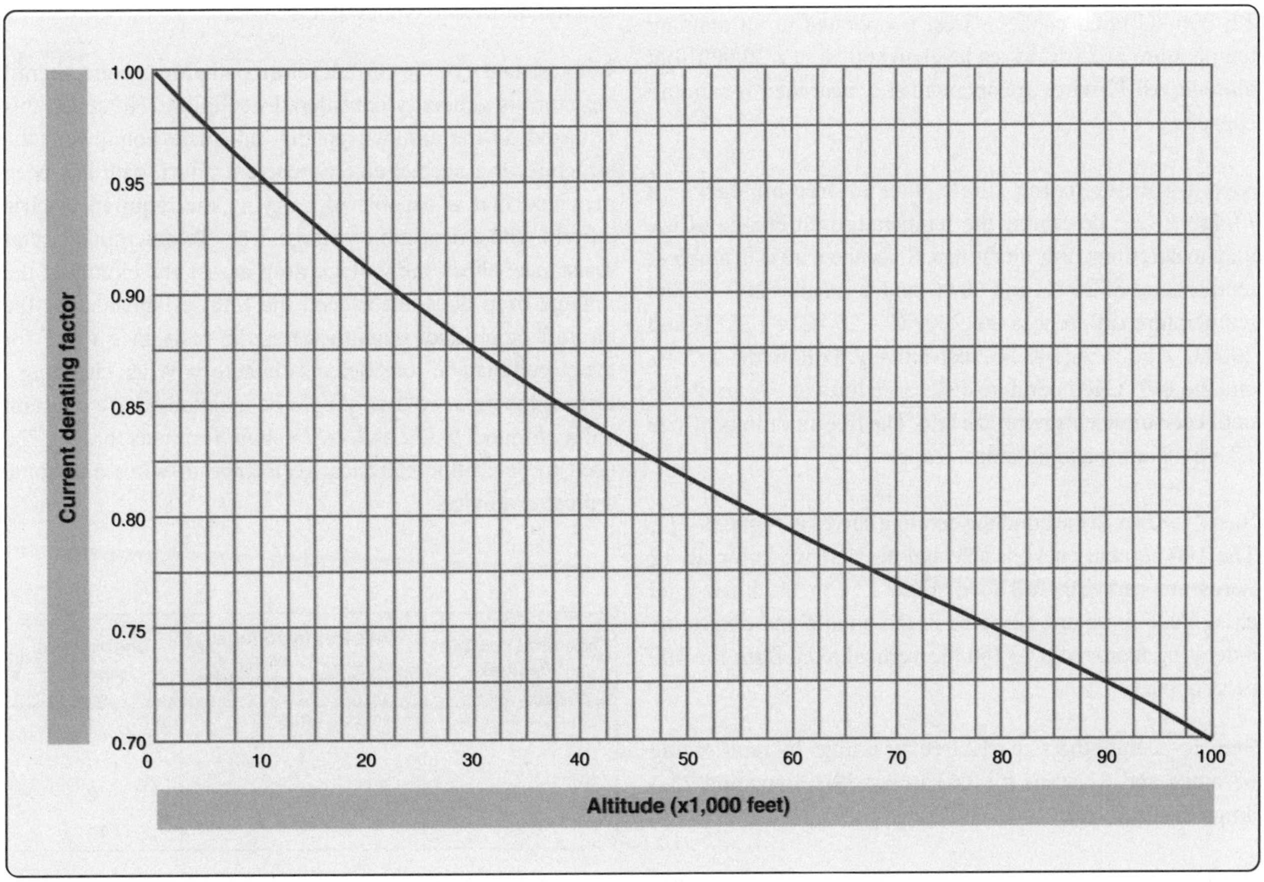

Figure 9-119. *Altitude derating curve.*

Wire size	Continuous duty current (amp) wires in bundles, groups, or harnesses or conduits		Max. resistance ohms/1000 feet
	Wire conductor temperature rating		
	@ 105 °C	@ 150 °C	@ 20 °C
#8	30	45	1.093
#6	40	61	0.641
#4	54	82	0.427
#2	76	113	0.268
#1	90	133	0.214
#0	102	153	0.169
#00	117	178	0.133
#000	138	209	0.109
#0000	163	248	0.085

Figure 9-120. *Current-carrying capacity and resistance of aluminum wire.*

Step 6—Determine the actual circuit current for each wire in the bundle and for the whole bundle. If the values calculated in step 5 are exceeded, select the next larger size wire and repeat the calculations.

Example 2

Assume a harness (open or braided), consisting of 12 size 12, 200 °C rated copper wires, is operated in an ambient temperature of 25 °C at sea level and 60 °C at a 20,000-foot altitude. All 12 wires are operated at or near their maximum capacity.

Step 1—Refer to the single wire in free air curve in *Figure 9-117*, determine the temperature difference of the wire to determine free air ratings. Since the wire is in ambient temperature of 25 °C and 60 °C and is rated at 200 °C, the temperature differences are 200 °C – 25 °C = 175 °C and 200 °C – 60 °C = 140 °C, respectively. Follow the 175 °C and the 140 °C temperature difference lines on *Figure 9-116* until each intersects wire size line. The free air ratings of size 12 are 68 amps and 59 amps, respectively.

Step 2—Refer to the bundling derating curves in *Figure 9-118*. The 100 percent curve is selected because we know all 12 wires are carrying full load. Find 12 (on the horizontal axis) since there are 12 wires in the bundle and determine a derating factor of 0.43 (on the vertical axis) from the 100 percent curve.

Step 3—Derate the size #12 free air ratings by multiplying 68 amps and 61 amps by 0.43 to get 29.2 amps and 25.4 amps, respectively.

Step 4—Refer to the altitude derating curve of *Figure 9-119*, look for sea level and 20,000 feet (on the horizontal axis)

since these are the conditions at which the load is carried. The wire must be derated by a factor of 1.0 and 0.91, respectively.

Step 5—Derate the size 12 in a bundle ratings by multiplying 29.2 amps at sea level and 25.4 amps at 20,000 feet by 1.0 and 0.91, respectively to obtain 29.2 amps and 23.1 amps. The total bundle capacity at sea level and 25 °C ambient temperature is 29.2 × 12 = 350.4 amps. At 20,000 feet and 60 °C ambient temperature, the bundle capacity is 23.1 × 12 = 277.2 amps. Each size 12 wire can carry 29.2 amps at sea level, 25 °C ambient temperature or 23.1 amps at 20,000 feet and 60 °C ambient temperature.

Step 6—Determine the actual circuit current for each wire in the bundle and for the bundle. If the values calculated in Step 5 are exceeded, select the next larger size wire and repeat the calculations.

Allowable Voltage Drop

The voltage drop in the main power wires from the generation source or the battery to the bus should not exceed 2 percent of the regulated voltage when the generator is carrying rated current or the battery is being discharged at the 5-minute rate. The tabulation shown in *Figure 9-121* defines the maximum acceptable voltage drop in the load circuits between the bus and the utilization equipment ground.

The resistance of the current return path through the aircraft structure is generally considered negligible. However, this is based on the assumption that adequate bonding to the structure or a special electric current return path has been provided that is capable of carrying the required electric current with a negligible voltage drop. To determine circuit resistance, check the voltage drop across the circuit. If the voltage drop does not exceed the limit established by the aircraft or product manufacturer, the resistance value for the circuit may be considered satisfactory. When checking a circuit, the input voltage should be maintained at a constant value. *Figures 9-122* and *9-123* show formulas that may be used to determine electrical resistance in wires and some typical examples.

Nominal system voltage	Allowable voltage drop during continuous operation	Intermittent operation
14	0.5	1
28	1	2
115	4	8
200	7	14

Figure 9-121. *Tabulation chart (allowable voltage drop between bus and utilization equipment ground).*

The following formula can be used to check the voltage drop. The resistance/ft can be found in *Figures 9-122* and *9-123* for the wire size.

Calculated voltage drop (VD) = resistance/ft × length × current

Electric Wire Chart Instructions

To select the correct size of electrical wire, two major requirements must be met:

1. The wire size should be sufficient to prevent an excessive voltage drop while carrying the required current over the required distance. *[Figure 9-121]*

2. The size should be sufficient to prevent overheating of the wire carrying the required current. (See Maximum Operating Temperature earlier in this chapter for computing current carrying capacity methods.)

To meet the two requirements for selecting the correct wire

Voltage drop	Run lengths (feet)	Circuit current (amps)	Wire size from chart	Check calculated voltage drop (VD) = (resistance/feet) (length) (current)
1	107	20	No. 6	VD = (0.00044 ohms/feet) (107 x 20) = 0.942
0.5	90	20	No. 4	VD = (0.00028 ohms/feet) (90 x 20) = 0.504
4	88	20	No. 12	VD = (0.00202 ohms/feet) (88 x 20) = 3.60
7	100	20	No. 14	VD = (0.00306 ohms/feet) (100 x 20) = 6.12

Figure 9-122. *Determining required tin-plated copper wire size and checking voltage drop.*

Maximum Voltage drop	Wire size	Circuit current (amps)	Maximum wire run length (feet)	Check calculated voltage drop (VD) = (resistance/feet) (length) (current)
1	No. 10	20	39	VD = (0.00126 ohms/feet) (39 x 20) = 0.98
0.5	---		19.5	VD = (0.00126 ohms/feet) (19.5 x 20) = 0.366
4	---		156	VD = (0.00126 ohms/feet) (156 x 20) = 3.93
7	---		273	VD = (0.00126 ohms/feet) (273 x 20) = 6.88

Figure 9-123. *Determining maximum tin-plated copper wire length and checking voltage drop.*

size using *Figure 9-116*, the following must be known:

1. The wire length in feet.

2. The number of amperes of current to be carried.

3. The allowable voltage drop permitted.

4. The required continuous or intermittent current.

5. The estimated or measured conductor temperature.

6. Is the wire to be installed in conduit and/or bundle?

7. Is the wire to be installed as a single wire in free air?

Example A

Find the wire size in *Figure 9-116* using the following known information:

1. The wire run is 50 feet long, including the ground wire.

2. Current load is 20 amps.

3. The voltage source is 28 volts from bus to equipment.

4. The circuit has continuous operation.

5. Estimated conductor temperature is 20 °C or less. The scale on the left of the chart represents maximum wire length in feet to prevent an excessive voltage drop for a specified voltage source system (e.g., 14V, 28V, 115V, 200V). This voltage is identified at the top of scale and the corresponding voltage drop limit for continuous operation at the bottom. The scale (slant lines) on top of the chart represents amperes. The scale at the bottom of the chart represents wire gauge.

Step 1—From the left scale, find the wire length 50 feet under the 28V source column.

Step 2—Follow the corresponding horizontal line to the right until it intersects the slanted line for the 20-amp load.

Step 3—At this point, drop vertically to the bottom of the chart. The value falls between No. 8 and No. 10. Select the next larger size wire to the right, in this case No. 8. This is the smallest size wire that can be used without exceeding the voltage drop limit expressed at the bottom of the left scale. This example is plotted on the wire chart in *Figure 9-116*. Use *Figure 9-116 (top)* for continuous flow and *Figure 9-116 (bottom)* for intermittent flow.

Example B

Find the wire size in *Figure 9-116* using the following known information:

1. The wire run is 200 feet long, including the ground wire.

2. Current load is 10 amps.

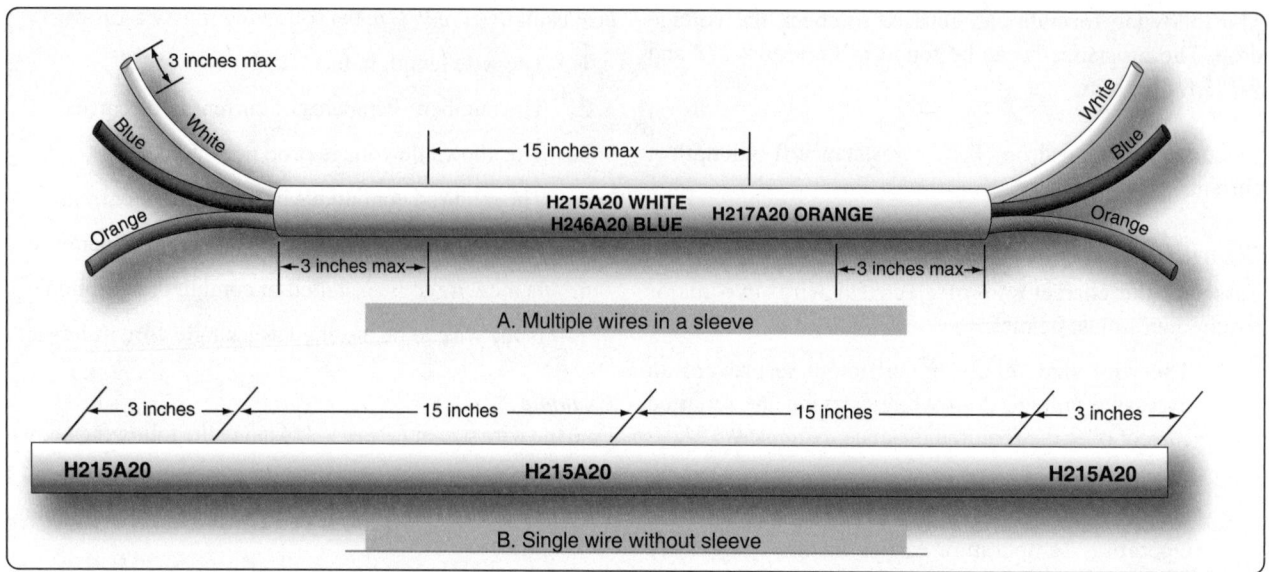

Figure 9-124. *Wire markings for single wire without sleeve.*

3. The voltage source is 115 volts from bus to equipment.

4. The circuit has intermittent operation.

Step 1—From the left scale, find the wire length of 200 feet under the 115V source column.

Step 2—Follow the corresponding horizontal line to the right until it intersects the slanted line for the 10-amp load.

Step 3—At this point, drop vertically to the bottom of the chart. The value falls between No. 16 and No. 14. Select the next larger size wire to the right—in this case, No. 14. This is the smallest size wire that can be used without exceeding the voltage drop limit expressed at the bottom of the left scale.

Wire Identification

The proper identification of electrical wires and cables with their circuits and voltages is necessary to provide safety of operation, safety to maintenance personnel, and ease of maintenance. All wire used on aircraft must have its type identification imprinted along its length. It is common practice to follow this part number with the five digit/letter Commercial and Government Entity (CAGE) code identifying the wire manufacturer. You can identify the performance capabilities of existing installed wire you need to replace, and avoid the inadvertent use of a lower performance and unsuitable replacement wire.

Placement of Identification Markings

Identification markings should be placed at each end of the wire and at 15-inch maximum intervals along the length of the wire. Wires less than 3 inches in length need not be identified. Wires 3 to 7 inches in length should be identified approximately at the center. Added identification marker sleeves should be located so that ties, clamps, or supporting devices need not be removed to read the identification. The wire identification code must be printed to read horizontally (from left to right) or vertically (from top to bottom). The two methods of marking wire or cable are as follows:

1. Direct marking is accomplished by printing the cable's outer covering. *[Figure 9-124B]*

2. Indirect marking is accomplished by printing a heat-shrinkable sleeve and installing the printed sleeve on the wire or cables outer covering. Indirectly-marked wire or cable should be identified with printed sleeves at each end and at intervals not longer than 6 feet. *[Figure 9-125]* The individual wires inside a cable should be identified within 3 inches of their termination. *[Figure 9-124A]*

Types of Wire Markings

The preferred method is to mark directly on the wire without causing insulation degradation. Teflon-coated wires, shielded wiring, multiconductor cable, and thermocouple wires usually require special sleeves to carry identification marks. There are some special wire marking machines available that can be used to stamp directly on the type wires mentioned

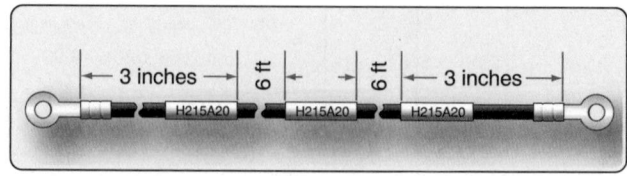

Figure 9-125. *Spacing of printed identification marks (indirect marking).*

above. Whatever method of marking is used, the marking should be legible and the color should contrast with the wire insulation or sleeve.

Several different methods can be used to mark directly on the wire: hot stamp marking, ink jet printers, and laser jet printers. *[Figure 9-126]* The hot stamp method can damage the insulation of a newer type of wire that utilizes thin insulators. Fracture of the insulation wall and penetration to the conductor of these materials by the stamping dies have occurred. Later in service, when these openings have been wetted by various fluids or moisture, serious arcing and surface tracking have damaged wire bundles.

Identification sleeves can be used if the direct marking on the wire is not possible. *[Figure 9-127]*

Flexible sleeving, either clear or opaque, is satisfactory for general use. When color-coded or striped component wire is used as part of a cable, the identification sleeve should specify which color is associated with each wire identification code. Identification sleeves are normally used for identifying the following types of wire or cable: unjacketed shielded wire, thermocouple wire, coaxial cable, multiconductor cable, and high temperature wire. In most cases, identification tape can be used in place of sleeving. For sleeving exposed to high temperatures (over 400 °F), materials, such as silicone fiberglass, should be used. Polyolefin sleeving should be used in areas where resistance to solvent and synthetic hydraulic fluids is necessary. Sleeves may be secured in place with cable ties or by heat shrinking. The identification sleeving for various sizes of wire is shown in *Figure 9-128*.

Wire Installation & Routing
Open Wiring
Interconnecting wire is used in point-to-point open harnesses, normally in the interior or pressurized fuselage, with each

Figure 9-126. *Laser wire printer.*

Figure 9-127. *Alternate method of identifying wire bundles.*

wire providing enough insulation to resist damage from handling and service exposure. Electrical wiring is often installed in aircraft without special enclosing means. This practice is known as open wiring and offers the advantages of ease of maintenance and reduced weight.

Wire Groups & Bundles & Routing
Wires are often installed in bundles to create a more organized installation. These wire bundles are often called wire harnesses. Wire harnesses are often made in the factory or electrical shop on a jig board so that the wire bundles could be

Wire size		Sleeving size	
AN #	AL #	No.	Nominal ID (inch)
24		12	0.085
22		11	0.095
20		10	0.106
18		9	0.118
16		8	0.113
14		7	0.148
12		6	0.166
10		4	0.208
8	8	2	0.263
6	6	0	0.330
4	4	3/8 inch	0.375
2	2	1/2 inch	0.500
1	1	1/2 inch	0.500
0	0	5/8 inch	0.625
00	00	5/8 inch	0.625
000	000	3/4 inch	0.750
0000	0000	3/4 inch	0.750

Figure 9-128. *Recommended size of identification sleeving.*

Figure 9-129. *Cable harness jig board.*

preformed to fit into the aircraft. *[Figure 9-129]* As a result, each harness for a particular aircraft installation is identical in shape and length. The wiring harness could be covered by a shielding (metal braid) to avoid EMI. Grouping or bundling certain wires, such as electrically unprotected power wiring and wiring going to duplicate vital equipment, should be avoided. Wire bundles should generally be less than 75 wires, or 1½ to 2 inches in diameter where practicable. When several wires are grouped at junction boxes, terminal blocks, panels, etc., identity of the groups within a bundle can be retained.

Slack in Wire Bundles

Wiring should be installed with sufficient slack so that bundles and individual wires are not under tension. Wires connected to movable or shock-mounted equipment should have sufficient length to allow full travel without tension on the bundle. Wiring at terminal lugs or connectors should have sufficient slack to allow two reterminations without replacement of wires. This slack should be in addition to the drip loop and the allowance for movable equipment. Normally, wire groups or bundles should not exceed ½ inch deflection between support points. *[Figure 9-130]* This measurement may be exceeded if there is no possibility of the wire group or bundle touching a surface that may cause abrasion. Sufficient slack should be provided at each end to permit replacement of terminals and ease of maintenance; prevent mechanical strain on the wires, cables, junctions, and supports; permit free movement of shock- and vibration-mounted equipment; and allow shifting of equipment, as necessary, to perform alignment, servicing, tuning, removal of dust covers, and changing of internal components while installed in aircraft.

Twisting Wires

When specified on the engineering drawing, or when accomplished as a local practice, parallel wires must sometimes be twisted. The following are the most common examples:

1. Wiring in the vicinity of magnetic compass or flux valve

2. Three-phase distribution wiring

3. Certain other wires (usually radio wiring) as specified on engineering drawings

Twist the wires so they lie snugly against each other, making approximately the number of twists per foot as shown in *Figure 9-131.* Always check wire insulation for damage after twisting. If the insulation is torn or frayed, replace the wire.

Spliced Connections in Wire Bundles

Splicing is permitted on wiring as long as it does not affect the reliability and the electromechanical characteristics of the wiring. Splicing of power wires, coaxial cables, multiplex bus, and large-gauge wire must have approved data. Splicing of electrical wire should be kept to a minimum and avoided

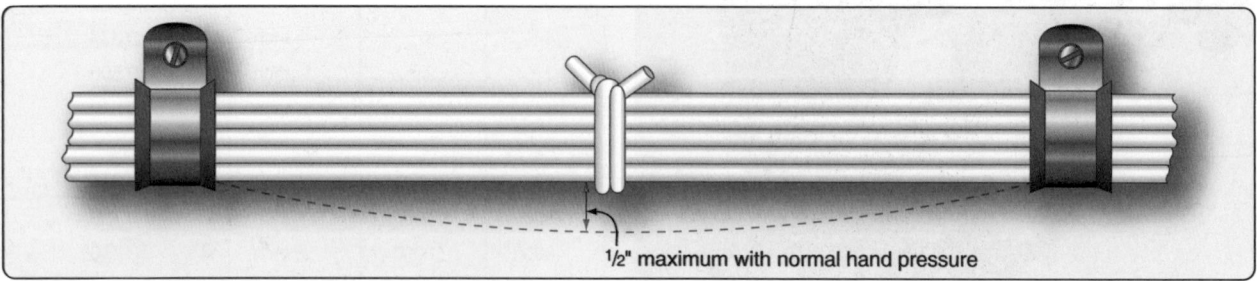

½" maximum with normal hand pressure

Figure 9-130. *Slack between supports of a cable harness.*

Gauge #	22	20	18	16	14	12	10	8	6	4
2 Wires	10	10	9	8	7 1/2	7	6 1/2	6	5	4
3 Wires	10	10	8 1/2	7	6 1/2	6	5 1/2	5	4	3

Figure 9-131. *Recommended number of wire twists per foot.*

entirely in locations subject to extreme vibrations. Splicing of individual wires in a group or bundle should have engineering approval, and the splice(s) should be located to allow periodic inspection.

Many types of aircraft splice connector are available for use when splicing individual wires. Use of a self-insulated splice connector is preferred; however, a non-insulated splice connector may be used provided the splice is covered with plastic sleeving that is secured at both ends. Environmentally sealed splices that conform to MIL-T-7928 provide a reliable means of splicing in SWAMP areas. However, a non-insulated splice connector may be used, provided the splice is covered with dual-wall shrink sleeving of a suitable material.

There should be no more than one splice in any one wire segment between any two connectors or other disconnect points. Exceptions include when attaching to the spare pigtail lead of a potted connector, when splicing multiple wires to a single wire, when adjusting wire size to fit connector contact crimp barrel size, and when required to make an approved repair.

Splices in bundles must be staggered to minimize any increase in the size of the bundle, preventing the bundle from fitting into its designated space or causing congestion that adversely affects maintenance. *[Figure 9-132]*

Splices should not be used within 12 inches of a termination device, except when attaching to the pigtail spare lead of a potted termination device, to splice multiple wires to a single wire, or to adjust the wire sizes so that they are compatible with the contact crimp barrel sizes.

Bend Radii

The minimum radius of bends in wire groups or bundles must not be less than 10 times the outside diameter of the largest wire or cable, except that at the terminal strips where wires break out at terminations or reverse direction in a bundle. Where the wire is suitably supported, the radius may be three times the diameter of the wire or cable. Where it is not practical to install wiring or cables within the radius requirements, the bend should be enclosed in insulating tubing. The radius for thermocouple wire should be done in accordance with the manufacturer's recommendation and shall be sufficient to avoid excess losses or damage to the cable. Ensure that RF cables (e.g., coaxial and triaxial) are bent at a radius of no less than six times the outside diameter of the cable.

Protection Against Chafing

Wires and wire groups should be protected against chafing or abrasion in those locations where contact with sharp surfaces or other wires would damage the insulation, or chafing could occur against the airframe or other components. Damage to the insulation can cause short circuits, malfunction, or inadvertent operation of equipment.

Protection Against High Temperature

Wiring must be routed away from high-temperature equipment and lines to prevent deterioration of insulation. Wires must be rated so the conductor temperature remains within the wire specification maximum when the ambient temperature and heat rise related to current-carrying capacity are taken into account. The residual heating effects caused by exposure to sunlight when aircraft are parked for extended periods should also be taken into account. Wires, such as those used in fire detection, fire extinguishing, fuel shutoff, and fly-by-wire flight control systems that must operate during and after a fire, must be selected from types that are qualified to provide circuit integrity after exposure to fire for a specified period. Wire insulation deteriorates rapidly when subjected to high temperatures.

Separate wires from high-temperature equipment, such as resistors, exhaust stacks, heating ducts, to prevent insulation

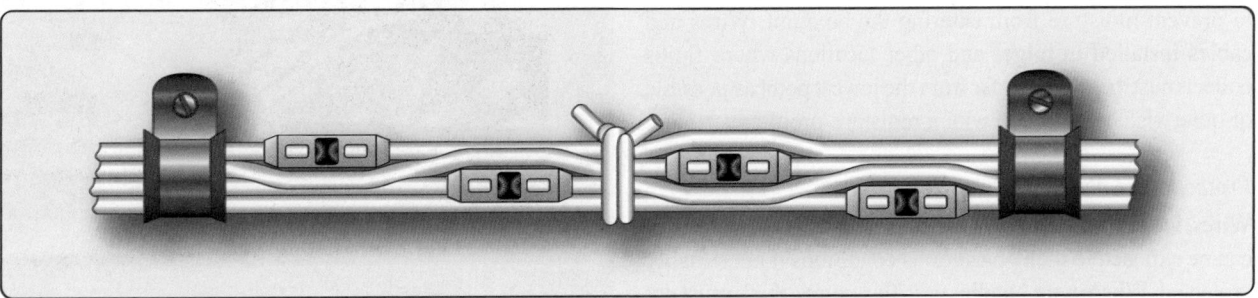

Figure 9-132. *Staggered splices in wire bundle.*

breakdown. Insulate wires that must run through hot areas with a high-temperature insulation material, such as fiberglass or PTFE. Avoid high-temperature areas when using cables with soft plastic insulation, such as polyethylene, because these materials are subject to deterioration and deformation at elevated temperatures. Many coaxial cables have this type of insulation.

Protection Against Solvents & Fluids

An arcing fault between an electrical wire and a metallic flammable fluid line may puncture the line and result in a fire. Every effort must be made to avoid this hazard by physical separation of the wire from lines and equipment containing oxygen, oil, fuel, hydraulic fluid, or alcohol. Wiring must be routed above these lines and equipment with a minimum separation of 6 inches or more whenever possible. When such an arrangement is not practicable, wiring must be routed so that it does not run parallel to the fluid lines. A minimum of 2 inches must be maintained between wiring and such lines and equipment, except when the wiring is positively clamped to maintain at least ½-inch separation, or when it must be connected directly to the fluid-carrying equipment. Install clamps as shown in *Figure 9-133*. These clamps should not be used as a means of supporting the wire bundle. Additional clamps should be installed to support the wire bundle and the clamps fastened to the same structure used to support the fluid line(s) to prevent relative motion.

Wires, or groups of wires, should enter a junction box, or terminate at a piece of equipment in an upward direction where practicable. Ensure that a trap, or drip loop, is provided to prevent fluids or condensation from running into wire or cable ends that slope downward toward a connector, terminal block, panel, or junction block. A drip loop is an area where the wire(s) are made to travel downward and then up to the connector. *[Figure 9-134]* Fluids and moisture will flow along the wires to the bottom of the loop and be trapped there to drip or evaporate without affecting electrical conductivity in the wire, junction, or connected device.

Where wires must be routed downwards to a junction box or electrical unit and a drip loop is not possible, the entrance should be sealed according to manufacturer's specifications to prevent moisture from entering the box/unit. Wires and cables installed in bilges and other locations where fluids collect must be routed as far from the lowest point as possible or otherwise be provided with a moisture-proof covering.

Protection of Wires in Wheel Well Areas

Wires located on landing gear and in the wheel well area can be exposed to many hazardous conditions if not suitably protected. Where wire bundles pass flex points, there must not be any strain on attachments or excessive slack when parts are fully extended or retracted. The wiring and protective tubing must be inspected frequently and replaced at the first sign of wear.

Wires should be routed so that fluids drain away from the connectors. When this is not practicable, connectors must be potted. Wiring which must be routed in wheel wells or other external areas must be given extra protection in the form of harness jacketing and connector strain relief. Conduits or flexible sleeving used to protect wiring must be equipped with drain holes to prevent entrapment of moisture.

The technician should check during inspections that wires and cables are adequately protected in wheel wells and other areas where they may be exposed to damage from impact of

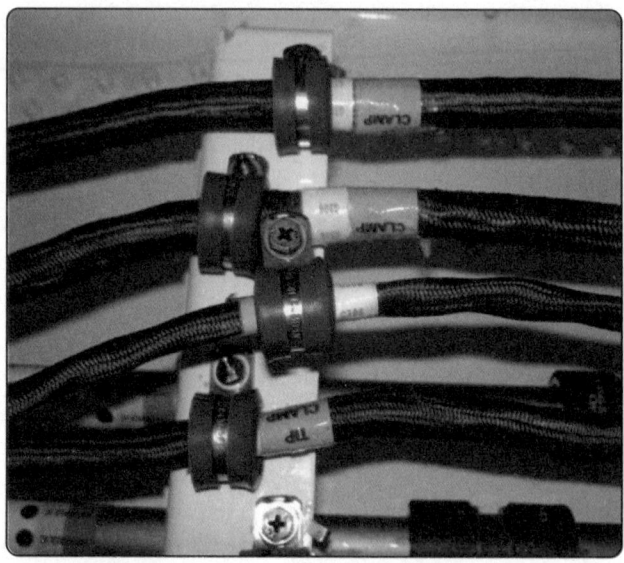

Figure 9-133. *Positive separation of wire and fluid lines and wire clamps.*

Figure 9-134. *Drip loop.*

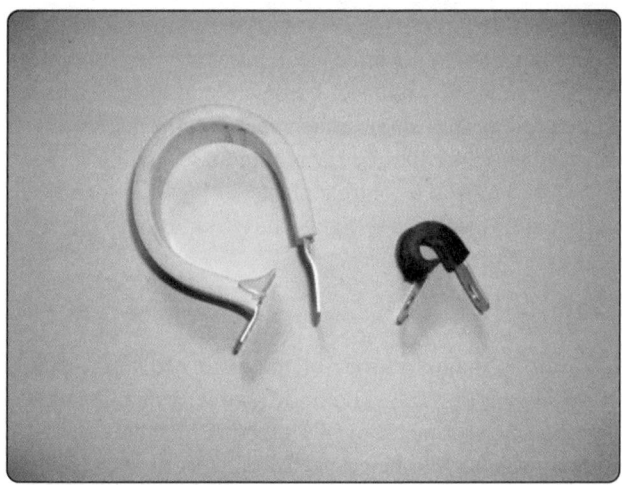

Figure 9-135. *Wire clamps.*

rocks, ice, mud, etc. (If rerouting of wires or cables is not practical, protective jacketing may be installed). This type of installation must be held to a minimum.

Clamp Installation

Wires and wire bundles must be supported by clamps or plastic cable straps. *[Figure 9-135]* Clamps and other primary support devices must be constructed of materials that are compatible with their installation and environment, in terms of temperature, fluid resistance, exposure to ultraviolet (UV) light, and wire bundle mechanical loads. They should be spaced at intervals not exceeding 24 inches. Clamps on wire bundles should be selected so that they have a snug fit without pinching wires. *[Figures 9-136 through 9-138]*

Caution: The use of metal clamps on coaxial RF cables may cause problems, if clamp fit is such that RF cable's original cross section is distorted.

Clamps on wire bundles should not allow the bundle to move through the clamp when a slight axial pull is applied. Clamps on RF cables must fit without crushing and must

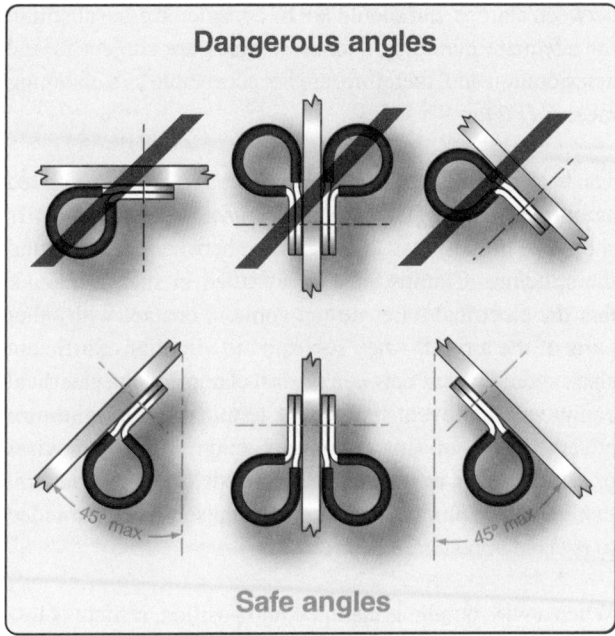

Figure 9-136. *Safe angle for cable clamps.*

be snug enough to prevent the cable from moving freely through the clamp, but may allow the cable to slide through the clamp when a light axial pull is applied. The cable or wire bundle may be wrapped with one or more turns of electrical tape when required to achieve this fit. Plastic clamps or cable ties must not be used where their failure could result in interference with movable controls, wire bundle contact with movable equipment, or chafing damage to essential or unprotected wiring. They must not be used on vertical runs where inadvertent slack migration could result in chafing or other damage. Clamps must be installed with their attachment hardware positioned above them, wherever practicable, so that they are unlikely to rotate as the result of wire bundle weight or wire bundle chafing. *[Figure 9-136]*

Clamps lined with nonmetallic material should be used to support the wire bundle along the run. Tying may be used

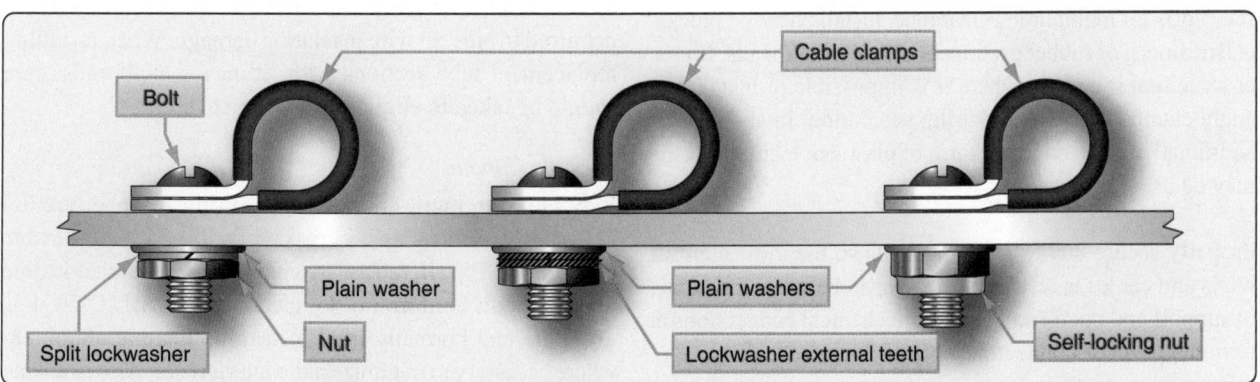

Figure 9-137. *Typical mounting hardware for MS-21919 cable clamps.*

between clamps, but should not be considered as a substitute for adequate clamping. Adhesive tapes are subject to age deterioration and, therefore, are not acceptable as a clamping means. *[Figure 9-137]*

The back of the clamp, whenever practical, should be rested against a structural member. *[Figure 9-138]* Stand-offs should be used to maintain clearance between the wires and the structure. Clamps must be installed in such a manner that the electrical wires do not come in contact with other parts of the aircraft when subjected to vibration. Sufficient slack should be left between the last clamp and the electrical equipment to prevent strain at the terminal and to minimize adverse effects on shock-mounted equipment. Where wires or wire bundles pass through bulkheads or other structural members, a grommet or suitable clamp should be provided to prevent abrasion.

When a wire bundle is clamped into position, if there is less than ⅜-inch of clearance between the bulkhead cutout and the wire bundle, a suitable grommet should be installed as indicated in *Figure 9-139*. The grommet may be cut at a 45° angle to facilitate installation, provided it is cemented in place and the slot is located at the top of the cutout.

Wire & Cable Clamp Inspection

Inspect wire and cable clamps for proper tightness. Where cables pass through structure or bulkheads, inspect for proper clamping and grommets. Inspect for sufficient slack between the last clamp and the electronic equipment to prevent strain at the cable terminals and to minimize adverse effects on shock-mounted equipment. Wires and cables are supported by suitable clamps, grommets, or other devices at intervals of not more than 24 inches, except when contained in troughs, ducts, or conduits. The supporting devices should be of a suitable size and type, with the wires and cables held securely in place without damage to the insulation.

Use metal stand-offs to maintain clearance between wires and structure. Tape or tubing is not acceptable as an alternative to stand-offs for maintaining clearance. Install phenolic blocks, plastic liners, or rubber grommets in holes, bulkheads, floors, or structural members where it is impossible to install off-angle clamps to maintain wiring separation. In such cases, additional protection in the form of plastic or insulating tape may be used.

Properly secure clamp retaining bolts so the movement of wires and cables is restricted to the span between the points of support and not on soldered or mechanical connections at terminal posts or connectors.

Movable Controls Wiring Precautions

Clamping of wires routed near movable flight controls must be attached with steel hardware and must be spaced so that failure of a single attachment point cannot result in interference with controls. The minimum separation between wiring and movable controls must be at least ½ inch when the bundle is displaced by light hand pressure in the direction of the controls.

Conduit

Conduit is manufactured in metallic and nonmetallic materials and in both rigid and flexible forms. Primarily, its purpose is for mechanical protection of cables or wires. Conduit size should be selected for a specific wire bundle application to allow for ease in maintenance, and possible future circuit expansion, by specifying the conduit inner diameter (ID) about 25 percent larger than the maximum diameter of the wire bundle. *[Figure 9-140]*

Conduit problems can be avoided by following these guidelines:

- Do not locate conduit where passengers or maintenance personnel might use it as a handhold or footstep.

- Provide drain holes at the lowest point in a conduit run. Drilling burrs should be carefully removed.

- Support conduit to prevent chafing against structure and to avoid stressing its end fittings.

Rigid Conduit

Damaged conduit sections should be repaired to preclude injury to the wires or wire bundle that may consume as much as 80 percent of the tube area. Minimum acceptable tube bend radii for rigid conduit are shown in *Figure 9-141*. Kinked or wrinkled bends in rigid conduits are not recommended and should be replaced. Tubing bends that have been flattened into an ellipse and have a minor diameter of less than 75 percent of the nominal tubing diameter should be replaced, because the tube area has been reduced by at least 10 percent. Tubing that has been formed and cut to final length should be deburred to prevent wire insulation damage. When installing replacement tube sections with fittings at both ends, care should be taken to eliminate mechanical strain.

Flexible Conduit

Flexible aluminum conduit conforming to specification MIL-C-6136 is available in two types: Type I, bare flexible conduit, and Type II, rubber-covered flexible conduit. Flexible brass conduit conforming to specification MIL-C-7931 is available and normally used instead of flexible aluminum where necessary to minimize radio interference. Also available is a plastic flexible tubing. (Reference MIL-T-8191A.) Flexible conduit may be used where it is impractical to use rigid conduit,

Cable clamps
Incorrect
Angle bracket
Correct
Wire is pinched in clamp
"Angle" member
"Z" member

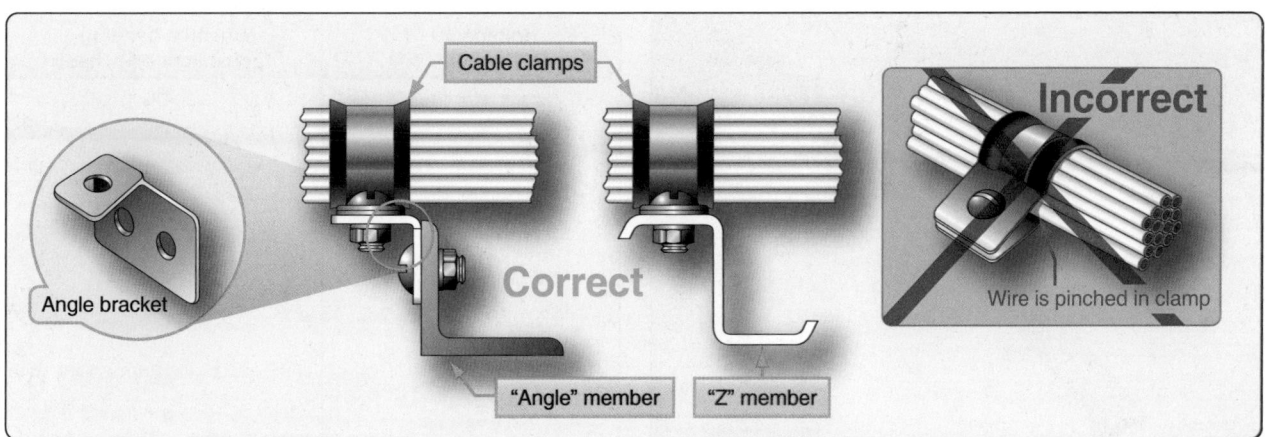

Figure 9-138. *Installing cable clamp to structure.*

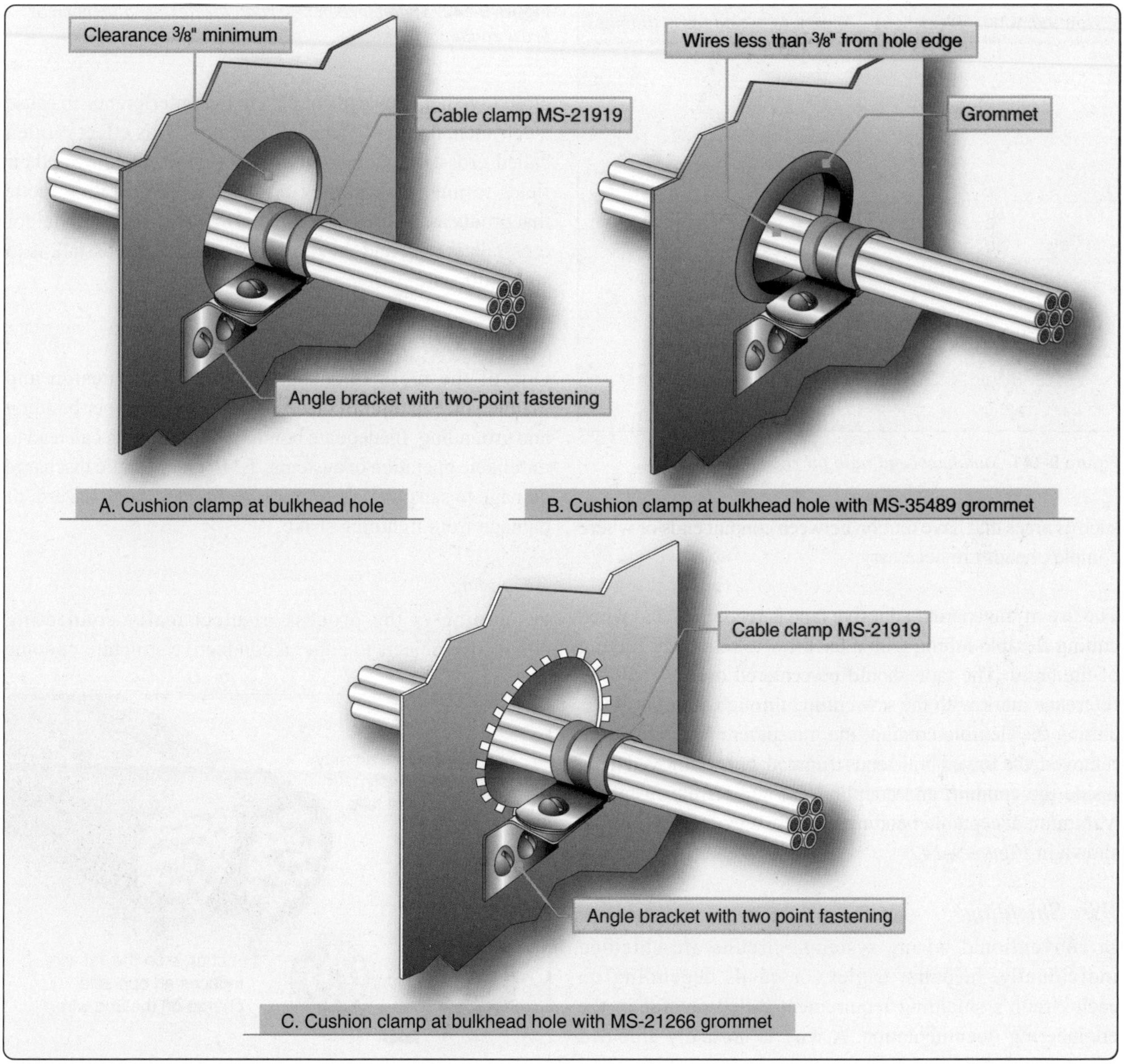

Clearance 3/8" minimum
Cable clamp MS-21919
Angle bracket with two-point fastening
A. Cushion clamp at bulkhead hole
Wires less than 3/8" from hole edge
Grommet
B. Cushion clamp at bulkhead hole with MS-35489 grommet
Cable clamp MS-21919
Angle bracket with two point fastening
C. Cushion clamp at bulkhead hole with MS-21266 grommet

Figure 9-139. *Clamping at a bulkhead hole.*

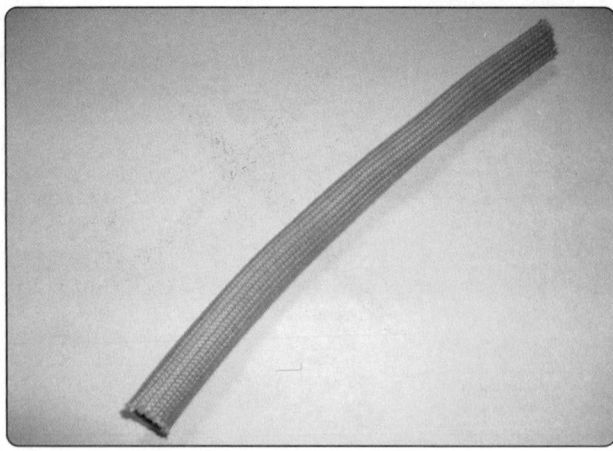

Figure 9-140. *Flexible conduit.*

Nominal tube OD (inches)	Minimum bend radius (inches)
$1/8$	$3/8$
$3/16$	$7/16$
$1/4$	$9/16$
$3/8$	$15/16$
$1/2$	$1 1/4$
$5/8$	$1 1/2$
$3/4$	$1 3/4$
1	3
$1 1/4$	$3 3/4$
$1 1/2$	5
$1 3/4$	7
2	8

Figure 9-141. *Minimum bend radii for rigid conduit.*

such as areas that have motion between conduit ends or where complex bends are necessary.

The use of transparent adhesive tape is recommended when cutting flexible tubing with a hacksaw to minimize fraying of the braid. The tape should be centered over the cutting reference mark with the saw cutting through the tape. After cutting the flexible conduit, the transparent tape should be removed, the frayed braid ends trimmed, burrs removed from inside the conduit, and coupling nut and ferrule installed. Minimum acceptable bending radii for flexible conduit are shown in *Figure 9-142.*

Wire Shielding

In conventional wiring systems, circuits are shielded individually, in pairs, triples, or quads depending on each circuit's shielding requirement called out for in the engineering documentation. A wire is normally shielded when it is anticipated that the circuit can be affected by another circuit in the wire harness. When the wires come

Nominal ID of conduit (inches)	Minimum bending radius inside (inches)
$3/16$	$2 1/4$
$1/4$	$2 3/4$
$3/8$	$3 3/4$
$1/2$	$3 3/4$
$5/8$	$3 3/4$
$3/4$	$4 1/4$
1	$5 3/4$
$1 1/4$	8
$1 1/2$	$8 1/4$
$1 3/4$	9
2	$9 3/4$
$2 1/2$	10

Figure 9-142. *Minimum bending radii for flexible aluminum or brass conduit.*

close together, they can couple enough interference to cause a detrimental upset to attached circuitry. This effect is often called crosstalk. Wires must come close enough for their fields to interact, and they must be in an operating mode that produces the crosstalk effect. However, the potential for crosstalk is real, and the only way to prevent crosstalk is to shield the wire. *[Figure 9-143]*

Bonding & Grounding

One of the more important factors in the design and maintenance of aircraft electrical systems is proper bonding and grounding. Inadequate bonding or grounding can lead to unreliable operation of systems, EMI, electrostatic discharge damage to sensitive electronics, personnel shock hazard, or damage from lightning strike.

Grounding

Grounding is the process of electrically connecting conductive objects to either a conductive structure or some

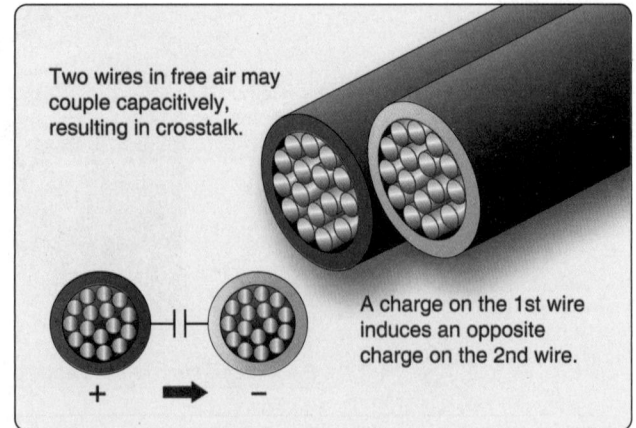

Two wires in free air may couple capacitively, resulting in crosstalk.

A charge on the 1st wire induces an opposite charge on the 2nd wire.

Figure 9-143. *Crosstalk.*

other conductive return path for the purpose of safely completing either a normal or fault circuit. *[Figure 9-144]*

If wires carrying return currents from different types of sources, such as signals of DC and AC generators, are connected to the same ground point or have a common connection in the return paths, an interaction of the currents occurs. Mixing return currents from various sources should be avoided because noise is coupled from one source to another and can be a major problem for digital systems. To minimize the interaction between various return currents, different types of ground should be identified and used. As a minimum, the design should use three ground types: (1) AC returns, (2) DC returns, and (3) all others.

For distributed power systems, the power return point for an alternative power source would be separated. For example, in a two-AC generator (one on the right side and the other on the left side) system, if the right AC generator was supplying backup power to equipment located on the left side, (left equipment rack) the backup AC ground return should be labeled "AC Right." The return currents for the left generator should be connected to a ground point labeled "AC Left."

The design of the ground return circuit should be given as much attention as the other leads of a circuit. A requirement for proper ground connections is that they maintain an impedance that is essentially constant. Ground return circuits should have a current rating and voltage drop adequate for satisfactory operation of the connected electrical and electronic equipment. EMI problems that can be caused by a system's power wire can be reduced substantially by locating the associated ground return near the origin of the power wiring (e.g., circuit breaker panel) and routing the power wire and its ground return in a twisted pair. Special care should be exercised to ensure replacement on ground return leads. The use of numbered insulated wire leads instead of

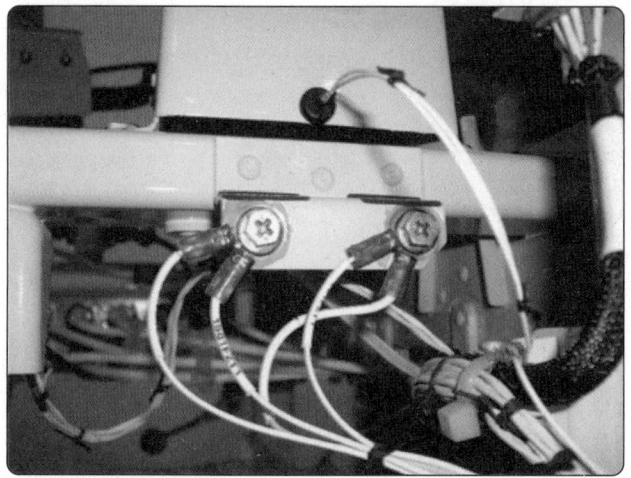

Figure 9-144. *Ground wires.*

bare grounding jumpers may aid in this respect. In general, equipment items should have an external ground connection, even when internally grounded. Direct connections to a magnesium structure must not be used for ground return because they may create a fire hazard.

Power ground connections for generators, transformer rectifiers, batteries, external power receptacles, and other heavy-current loads must be attached to individual grounding brackets that are attached to aircraft structure with a proper metal-to-metal bonding attachment. This attachment and the surrounding structure must provide adequate conductivity to accommodate normal and fault currents of the system without creating excessive voltage drop or damage to the structure. At least three fasteners, located in a triangular or rectangular pattern, must be used to secure such brackets in order to minimize susceptibility to loosening under vibration. If the structure is fabricated of a material, such as carbon fiber composite (CFC), that has a higher resistivity than aluminum or copper, it is necessary to provide an alternative ground path(s) for power return current. Special attention should be considered for composite aircraft.

Power return or fault current ground connections within flammable vapor areas must be avoided. If they must be made, make sure these connections do not arc, spark, or overheat under all possible current flow or mechanical failure conditions, including induced lightning currents. Criteria for inspection and maintenance to ensure continued airworthiness throughout the expected life of the aircraft should be established. Power return fault currents are normally the highest currents flowing in a structure. These can be the full generator current capacity. If full generator fault current flows through a localized region of the carbon fiber structure, major heating and failure can occur. CFC and other similar low-resistive materials must not be used in power return paths. Additional voltage drops in the return path can cause voltage regulation problems. Likewise, repeated localized material heating by current surges can cause material degradation. Both problems may occur without warning and cause no repeatable failures or anomalies.

The use of common ground connections for more than one circuit or function should be avoided except where it can be shown that related malfunctions that could affect more than one circuit do not result in a hazardous condition. Even when the loss of multiple systems does not, in itself, create a hazard, the effect of such failure can be quite distracting to the crew.

Bonding

Bonding is the electrical connecting of two or more conducting objects not otherwise adequately connected.

The following bonding requirements must be considered:

- Equipment bonding—low-impedance paths to aircraft structure are normally required for electronic equipment to provide radio frequency return circuits and for most electrical equipment to facilitate reduction in EMI. The cases of components that produce electromagnetic energy should be grounded to structure. To ensure proper operation of electronic equipment, it is particularly important to conform the system's installation specification when interconnections, bonding, and grounding are being accomplished.

- Metallic surface bonding—all conducting objects on the exterior of the airframe must be electrically connected to the airframe through mechanical joints, conductive hinges, or bond straps capable of conducting static charges and lightning strikes. Exceptions may be necessary for some objects, such as antenna elements, whose function requires them to be electrically isolated from the airframe. Such items should be provided with an alternative means to conduct static charges and/or lightning currents, as appropriate.

- Static bonds—all isolated conducting parts inside and outside the aircraft, having an area greater than 3 square inches and a linear dimension over 3 inches, that are subjected to appreciable electrostatic charging due to precipitation, fluid, or air in motion, should have a mechanically secure electrical connection to the aircraft structure of sufficient conductivity to dissipate possible static charges. A resistance of less than 1 ohm when clean and dry generally ensures such dissipation on larger objects. Higher resistances are permissible in connecting smaller objects to airframe structure.

Testing of Bonds & Grounds

The resistance of all bond and ground connections should be tested after connections are made before re-finishing. The resistance of each connection should normally not exceed 0.003 ohm. A high quality test instrument, an AN/USM-21A or equivalent, is required to accurately measure the very low resistance values.

Bonding Jumper Installation

Bonding jumpers should be made as short as practicable, and installed in such a manner that the resistance of each connection does not exceed .003 ohm. The jumper should not interfere with the operation of movable aircraft elements, such as surface controls, nor should normal movement of these elements result in damage to the bonding jumper. *[Figure 9-145]*

- Bonding connections—to ensure a low-resistance connection, nonconducting finishes, such as paint and anodizing films, should be removed from the

attachment surface to be contacted by the bonding terminal. Electrical wiring should not be grounded directly to magnesium parts.

- Corrosion protection—one of the more frequent causes of failures in electrical system bonding and grounding is corrosion. The areas around completed connections should be post-finished quickly with a suitable finish coating.

- Corrosion prevention—electrolytic action may rapidly corrode a bonding connection if suitable precautions are not taken. Aluminum alloy jumpers are recommended for most cases; however, copper jumpers should be used to bond together parts made of stainless steel, cadmium-plated steel, copper, brass, or bronze. Where contact between dissimilar metals cannot be avoided, the choice of jumper and hardware should be such that corrosion is minimized; the part likely to corrode should be the jumper or associated hardware.

- Bonding jumper attachment—the use of solder to attach bonding jumpers should be avoided. Tubular members should be bonded by means of clamps to which the jumper is attached. Proper choice of clamp material should minimize the probability of corrosion.

- Ground return connection—when bonding jumpers carry substantial ground return current, the current rating of the jumper should be determined to be adequate, and a negligible voltage drop is produced. *[Figure 9-146]*

Lacing & Tying Wire Bundles

Ties, lacing, and straps are used to secure wire groups or bundles to provide ease of maintenance, inspection, and installation. Straps may not be used in areas of SWAMP, such as wheel wells, near wing flaps, or wing folds. They

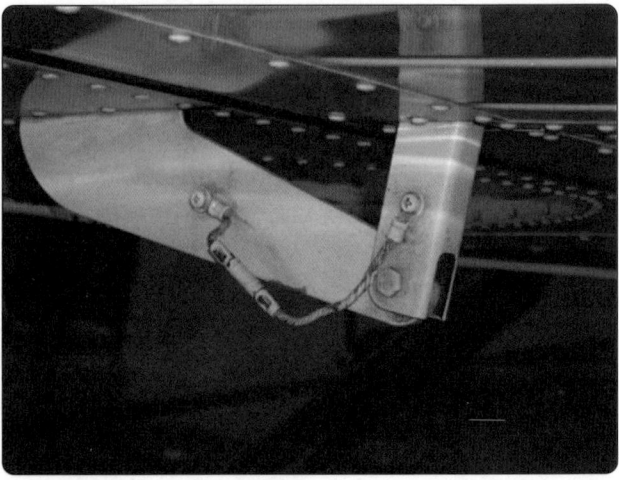

Figure 9-145. *Bonding jumpers.*

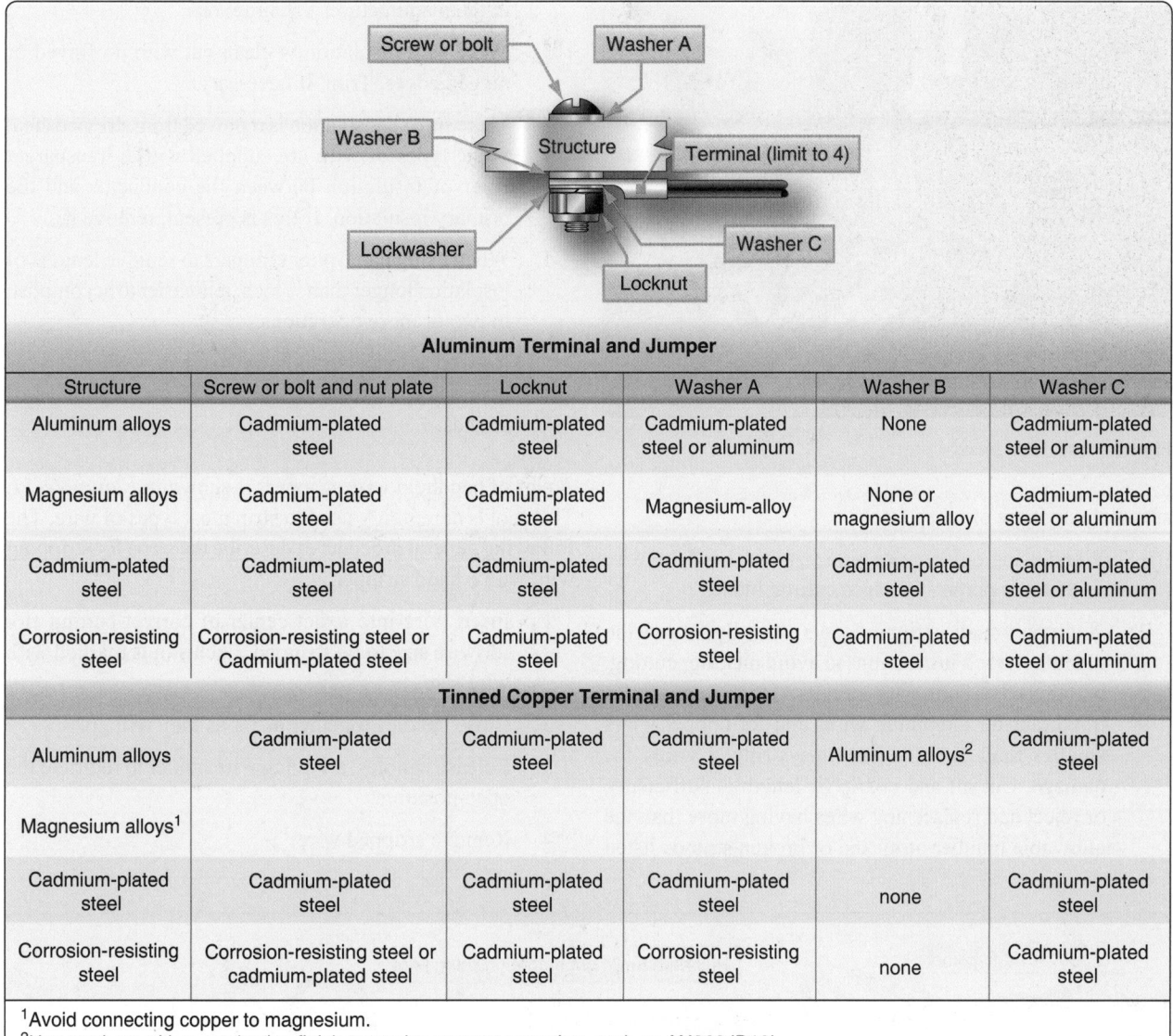

Structure	Screw or bolt and nut plate	Locknut	Washer A	Washer B	Washer C
Aluminum Terminal and Jumper					
Aluminum alloys	Cadmium-plated steel	Cadmium-plated steel	Cadmium-plated steel or aluminum	None	Cadmium-plated steel or aluminum
Magnesium alloys	Cadmium-plated steel	Cadmium-plated steel	Magnesium-alloy	None or magnesium alloy	Cadmium-plated steel or aluminum
Cadmium-plated steel	Cadmium-plated steel	Cadmium-plated steel	Cadmium-plated steel	Cadmium-plated steel	Cadmium-plated steel or aluminum
Corrosion-resisting steel	Corrosion-resisting steel or Cadmium-plated steel	Cadmium-plated steel	Corrosion-resisting steel	Cadmium-plated steel	Cadmium-plated steel or aluminum
Tinned Copper Terminal and Jumper					
Aluminum alloys	Cadmium-plated steel	Cadmium-plated steel	Cadmium-plated steel	Aluminum alloys[2]	Cadmium-plated steel
Magnesium alloys[1]					
Cadmium-plated steel	Cadmium-plated steel	Cadmium-plated steel	Cadmium-plated steel	none	Cadmium-plated steel
Corrosion-resisting steel	Corrosion-resisting steel or cadmium-plated steel	Cadmium-plated steel	Corrosion-resisting steel	none	Cadmium-plated steel

[1] Avoid connecting copper to magnesium.
[2] Use washers with a conductive finish treated to prevent corrosion, such as AN960JD10L.

Figure 9-146. *Bolt and nut bonding or grounding to flat surface.*

may not be used in high vibration areas where failure of the strap would permit wiring to move against parts that could damage the insulation and foul mechanical linkages or other moving mechanical parts. They also may not be used where they could be exposed to UV light, unless the straps are resistant to such exposure. *[Figure 9-147]*

The single cord-lacing method and tying tape may be used for wire groups of bundles 1 inch in diameter or less. The recommended knot for starting the single cord-lacing method is a clove hitch secured by a double-looped overhand knot. *[Figure 9-148, step A]* Use the double cord-lacing method on wire bundles 1 inch in diameter or larger. When using the double cord-lacing method, employ a bowline-on-a-bight as the starting knot. *[Figure 9-149, step A]*

Tying

Use wire group or bundle ties where the supports for the wire are more than 12 inches apart. A tie consists of a clove hitch around the wire group or bundle, secured by a square knot. *[Figure 9-150]*

Wire Termination
Stripping Wire

Before wire can be assembled to connectors, terminals, splices, etc., the insulation must be stripped from connecting ends to expose the bare conductor. Copper wire can be stripped in a number of ways depending on the size and insulation.

Aluminum wire must be stripped using extreme care, since individual strands break very easily after being nicked. The following general precautions are recommended when

Figure 9-147. *Wire lacing.*

stripping any type of wire:

1. When using any type of wire stripper, hold the wire so that it is perpendicular to cutting blades.

2. Adjust automatic stripping tools carefully; follow the manufacturer's instructions to avoid nicking, cutting, or otherwise damaging strands. This is especially important for aluminum wires and for copper wires smaller than No. 10. Examine stripped wires for damage. Cut off and restrip (if length is sufficient), or reject and replace any wires having more than the allowable number of nicked or broken strands listed

in the manufacturer's instructions.

3. Make sure insulation is clean-cut with no frayed or ragged edges. Trim, if necessary.

4. Make sure all insulation is removed from stripped area. Some types of wire are supplied with a transparent layer of insulation between the conductor and the primary insulation. If this is present, remove it.

5. When using hand-plier strippers to remove lengths of insulation longer than ¾ inch, it is easier to accomplish in two or more operations.

6. Retwist copper strands by hand or with pliers, if necessary, to restore natural lay and tightness of strands.

A pair of handheld wire strippers is shown in *Figure 9-151.* This tool is commonly used to strip most types of wire. The following general procedures describe the steps for stripping wire with a hand stripper.

1. Insert wire into exact center of correct cutting slot for wire size to be stripped. Each slot is marked with wire size.

2. Close handles together as far as they will go.

3. Release handles, allowing wire holder to return to the open position.

4. Remove stripped wire.

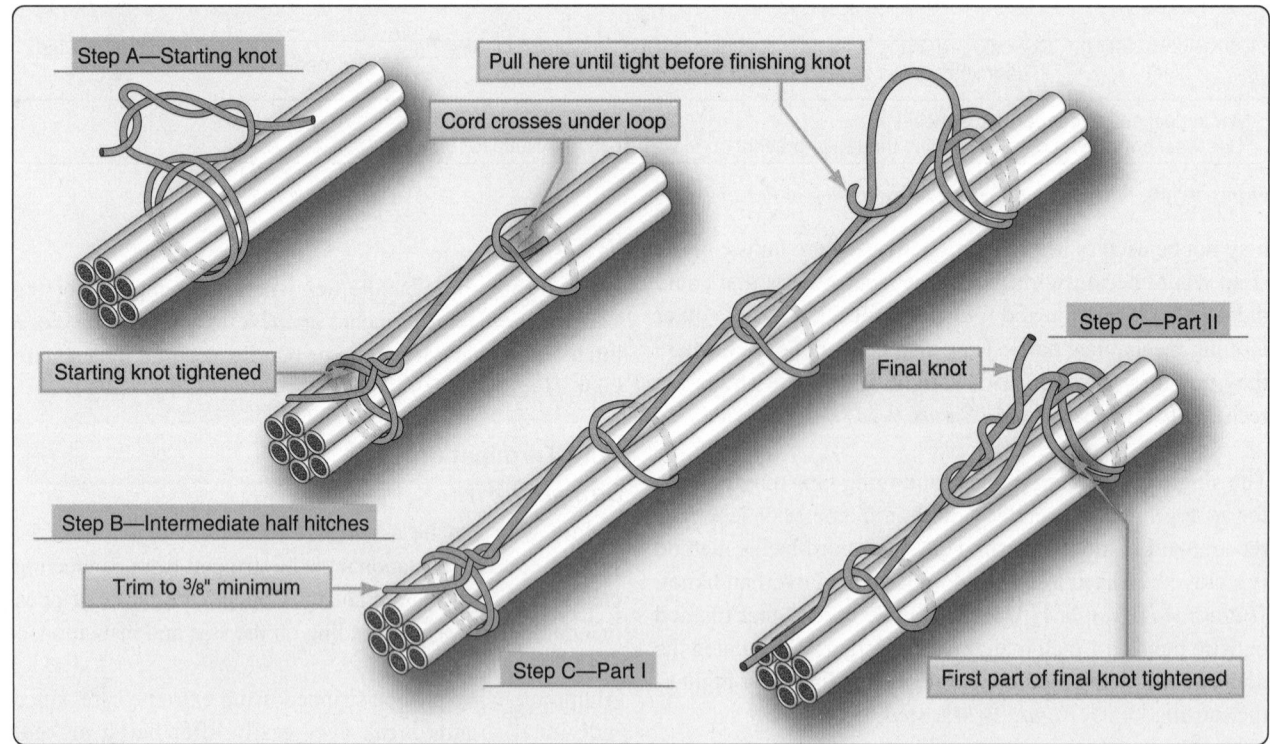

Figure 9-148. *Single cord-lacing method.*

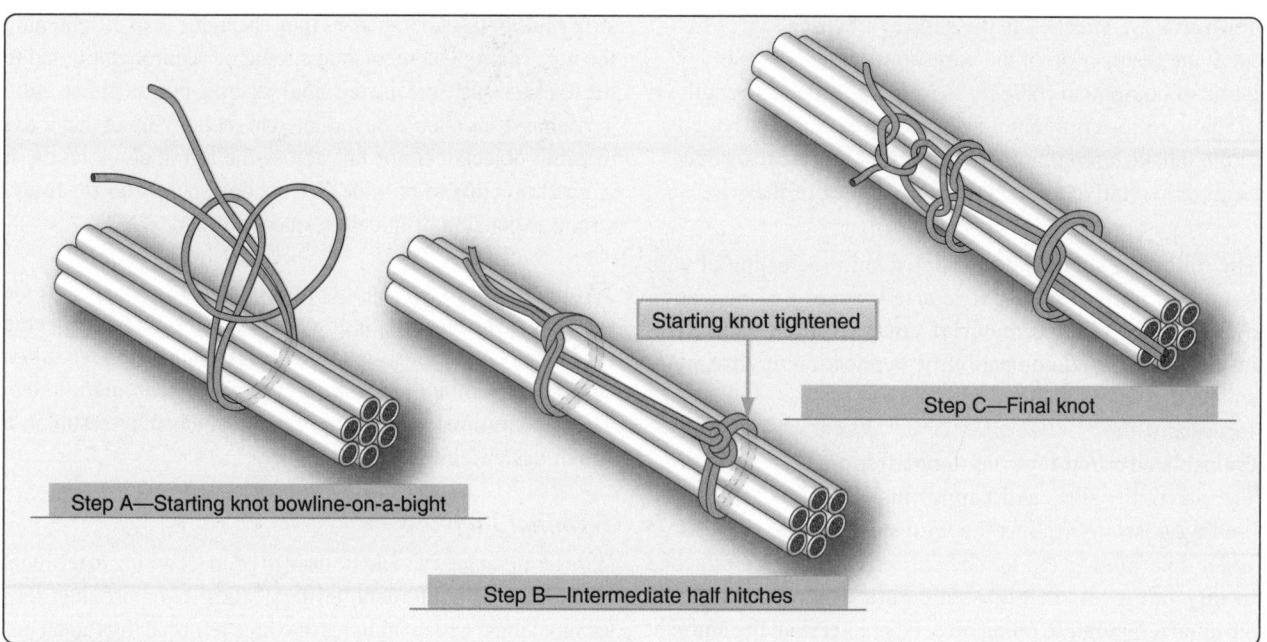

Step A—Starting knot bowline-on-a-bight

Step B—Intermediate half hitches

Starting knot tightened

Step C—Final knot

Figure 9-149. *Double cord-lacing.*

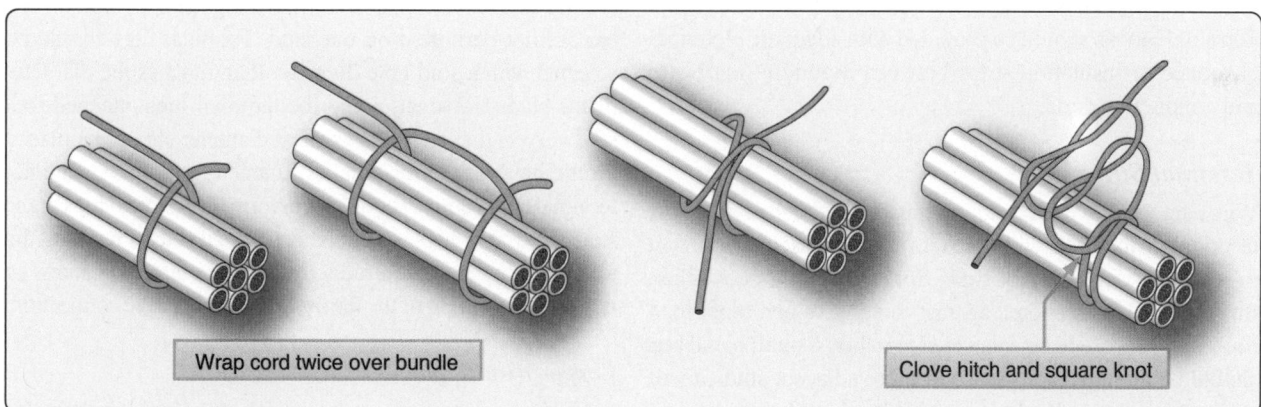

Wrap cord twice over bundle

Clove hitch and square knot

Figure 9-150. *Tying.*

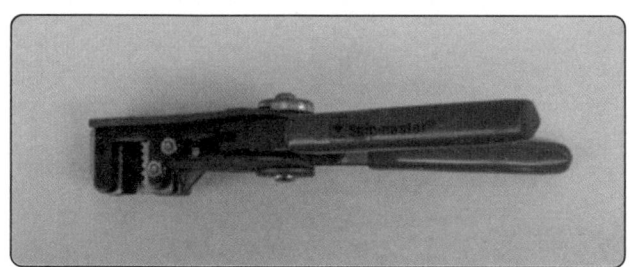

Figure 9-151. *Wire strippers.*

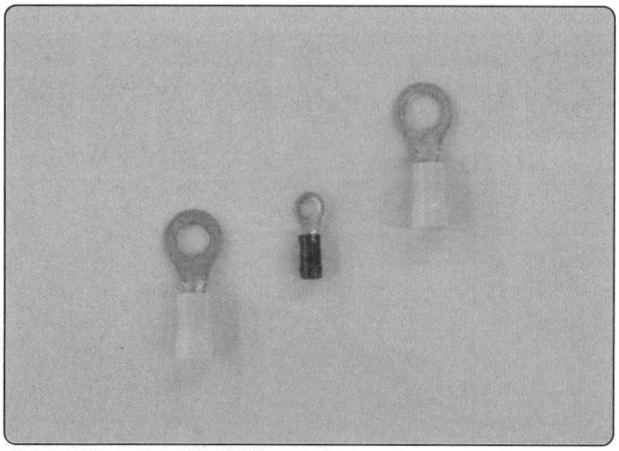

Figure 9-152. *Ring-tongue terminals.*

Terminals are attached to the ends of electrical wires to facilitate connection of the wires to terminal strips or items of equipment. *[Figure 9-152]* The tensile strength of the wire-to-terminal joint should be at least equivalent to the tensile strength of the wire itself, and its resistance negligible relative to the normal resistance of the wire.

The following should be considered in the selection of wire terminals: current rating, wire size (gauge) and insulation diameter, conductor material compatibility, stud size, insulation material compatibility, application environment, and solder versus solderless.

Preinsulated crimp-type ring-tongue terminals are preferred. The strength, size, and supporting means of studs and binding posts, as well as the wire size, may be considered when determining the number of terminals to be attached to any one post. In high-temperature applications, the terminal temperature rating must be greater than the ambient temperature plus current related temperature rise. Use of nickel-plated terminals and of uninsulated terminals with high-temperature insulating sleeves should be considered. Terminal blocks should be provided with adequate electrical clearance or insulation strips between mounting hardware and conductive parts.

Terminal Strips

Wires are usually joined at terminal strips. *[Figure 9-153]* A terminal strip fitted with barriers may be used to prevent the terminals on adjacent studs from contacting each other. Studs should be anchored against rotation. When more than four terminals are to be connected together, a small metal bus should be mounted across two or more adjacent studs. In all cases, the current should be carried by the terminal contact surfaces and not by the stud itself. Defective studs should be replaced with studs of the same size and material since terminal strip studs of the smaller sizes may shear due to overtightening the nut. The replacement stud should be securely mounted in the terminal strip and the terminal securing nut should be tight. Terminal strips should be mounted in such a manner that loose metallic objects cannot fall across the terminals or studs. It is good practice to provide at least one spare stud for future circuit expansion or in case a stud is broken.

Terminal strips that provide connection of radio and electronic systems to the aircraft electrical system should be inspected for loose connections, metallic objects that may have fallen across the terminal strip, dirt and grease accumulation, etc. These conditions can cause arcing, which may result in a fire or system failures.

Terminal Lugs

Wire terminal lugs should be used to connect wiring to terminal block studs or equipment terminal studs. No more than four terminal lugs, which includes the three terminal lugs and a bus bar, should be connected to any one stud. The total number of terminal lugs per stud always includes a common bus bar joining adjacent studs. Four terminal lugs plus a common bus bar are not permitted on one stud. Terminal lugs should be selected with a stud hole diameter that matches the diameter of the stud. However, when the terminal lugs attached to a stud vary in diameter, the greatest diameter should be placed on the bottom and the smallest diameter on top. Tightening terminal connections should not deform the terminal lugs or the studs. Terminal lugs should be positioned so that bending the terminal lug is not required to remove the fastening screw or nut, and movement of the terminal lugs tighten the connection.

Copper Wire Terminals

Solderless crimp-style, copper wire, terminal lugs may be used which conform to MIL-T-7928. Spacers or washers should not be used between the tongues of terminal lugs. *[Figure 9-154]*

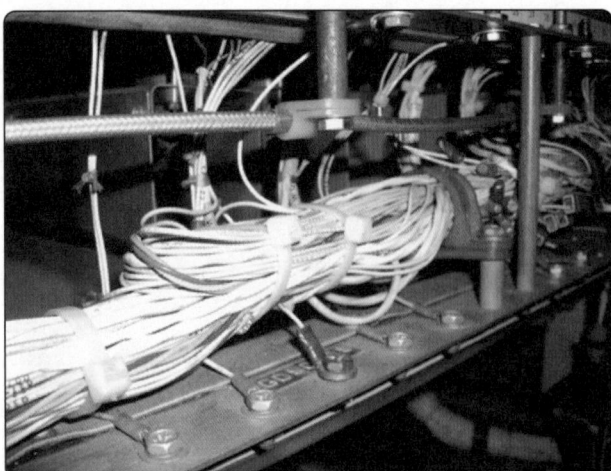

Figure 9-153. *Terminal strip.*

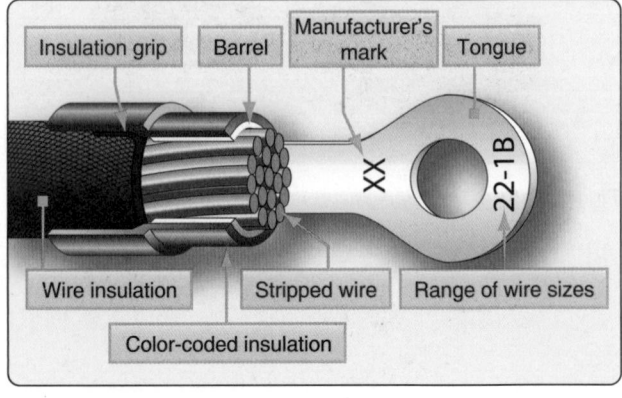

Figure 9-154. *Wire terminal.*

Aluminum Wire Terminals

The aluminum terminal lugs should be crimped to aluminum wire only. The tongue of the aluminum terminal lugs, or the total number of tongues of aluminum terminal lugs when stacked, should be sandwiched between two flat washers when terminated on terminal studs. Spacers or washers should not be used between the tongues of terminal lugs. Special attention should be given to aluminum wire and cable installations to guard against conditions that would result in excessive voltage drop and high resistance at junctions that may ultimately lead to failure of the junction. Examples of such conditions are improper installation of terminals and washers, improper torsion (torquing of nuts), and inadequate terminal contact areas.

Pre-Insulated Splices

Pre-insulated terminal lugs and splices must be installed using a high-quality crimping tool. Such tools are provided with positioners for the wire size and are adjusted for each wire size. It is essential that the crimp depth be appropriate for each wire size. If the crimp is too deep, it may break or cut individual strands. If the crimp is not deep enough, it may not be tight enough to retain the wire in the terminal or connector. Crimps that are not tight enough are also susceptible to high resistance due to corrosion buildup between the crimped terminal and the wire. *[Figure 9-155]*

Crimping Tools

Hand, portable, and stationary power tools are available for crimping terminal lugs. These tools crimp the barrel to the conductor, and simultaneously form the insulation support to the wire insulation. *[Figure 9-156]*

Emergency Splicing Repairs

Broken wires can be repaired by means of crimped splices, by using terminal lugs from which the tongue has been cut off, or by soldering together and potting broken strands. These repairs are applicable to copper wire. Damaged aluminum wire must not be temporarily spliced. These repairs are for temporary emergency use only and should

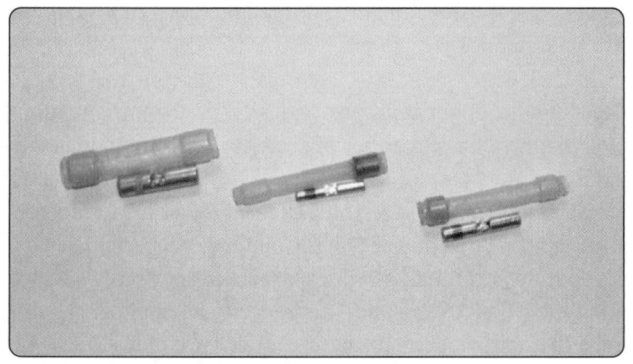

Figure 9-155. *Terminal splices.*

be replaced as soon as possible with permanent repairs. Since some manufacturers prohibit splicing, the applicable manufacturer's instructions should always be consulted.

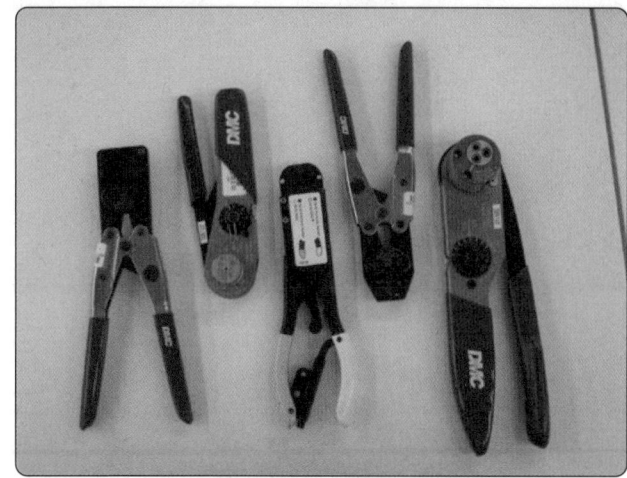

Figure 9-156. *Crimping pliers.*

Junction Boxes

Junction boxes are used for collecting, organizing, and distributing circuits to the appropriate harnesses that are attached to the equipment. *[Figure 9-157]* Junction boxes are also used to conveniently house miscellaneous components, such as relays and diodes. Junction boxes that are used in high-temperature areas should be made of stainless steel. Replacement junction boxes should be fabricated using the same material as the original or from a fire-resistant, nonabsorbent material, such as aluminum, or an acceptable plastic material. Where fireproofing is necessary, a stainless steel junction box is recommended. Rigid construction prevents oil-canning of the box sides that could result in internal short circuits. In all cases, drain holes should be provided in the lowest portion of the box. Cases of electrical power equipment must be insulated from metallic structure to avoid ground fault related fires.

The junction box arrangement should permit easy access to any installed items of equipment, terminals, and wires. Where marginal clearances are unavoidable, an insulating material should be inserted between current carrying parts and any grounded surface. It is not good practice to mount equipment on the covers or doors of junction boxes, since inspection for internal clearance is impossible when the door or cover is in the closed position.

Junction boxes should be securely mounted to the aircraft structure in such a manner that the contents are readily accessible for inspection. When possible, the open side should face downward or at an angle so that loose metallic objects,

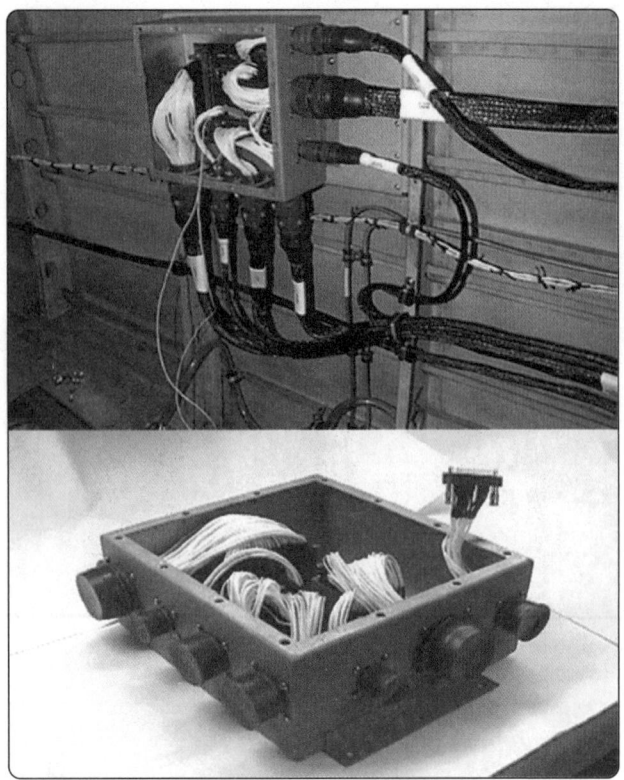

Figure 9-157. *Junction boxes.*

such as washers or nuts, tend to fall out of the junction box rather than wedge between terminals.

Junction box layouts should take into consideration the necessity for adequate wiring space and possible future additions. Electrical wire bundles should be laced or clamped inside the box so that cables do not touch other components, prevent ready access, or obscure markings or labels. Cables at entrance openings should be protected against chafing by using grommets or other suitable means.

AN/MS Connectors

Connectors (plugs and receptacles) facilitate maintenance when frequent disconnection is required. There is a multitude of types of connectors. The connector types that use crimped contacts are generally used on aircraft. Some of the more common types are the round cannon type, the rectangular, and the module blocks. Environmentally resistant connectors should be used in applications subject to fluids, vibration, heat, mechanical shock, and/or corrosive elements.

When HIRF/lightning protection is required, special attention should be given to the terminations of individual or overall shields. The number and complexity of wiring systems have resulted in an increased use of electrical connectors. *[Figure 9-158]* The proper choice and application of connectors is a significant part of the aircraft wiring

Figure 9-158. *Electrical connectors.*

system. Connectors must be kept to a minimum, selected, and installed to provide the maximum degree of safety and reliability to the aircraft. For the installation of any particular connector assembly, the specification of the manufacturer or the appropriate governing agency must be followed.

Types of Connector

Connectors must be identified by an original identification number derived from MIL Specification (MS) or OEM specification. *Figure 9-159* provides information about MS style connectors.

Environment-resistant connectors are used in applications where they are probably subjected to fluids, vibration, heat, mechanical shock, corrosive elements, etc. Firewall class connectors incorporating these same features should, in addition, be able to prevent the penetration of the fire through the aircraft firewall connector opening and continue to function without failure for a specified period of time when exposed to fire. Hermetic connectors provide a pressure seal for maintaining pressurized areas. When EMI/RFI protection is required, special attention should be given to the termination of individual and overall shields. Backshell adapters designed for shield termination, connectors with conductive finishes, and EMI grounding fingers are available for this purpose.

Rectangular connectors are typically used in applications where a very large number of circuits are accommodated in a single mated pair. *[Figure 9-160]* They are available with a great variety of contacts, which can include a mix of standard, coaxial, and large power types. Coupling is accomplished by various means. Smaller types are secured with screws which hold their flanges together. Larger ones have integral guide pins that ensure correct alignment, or jackscrews that both align and lock the connectors. Rack and panel connectors

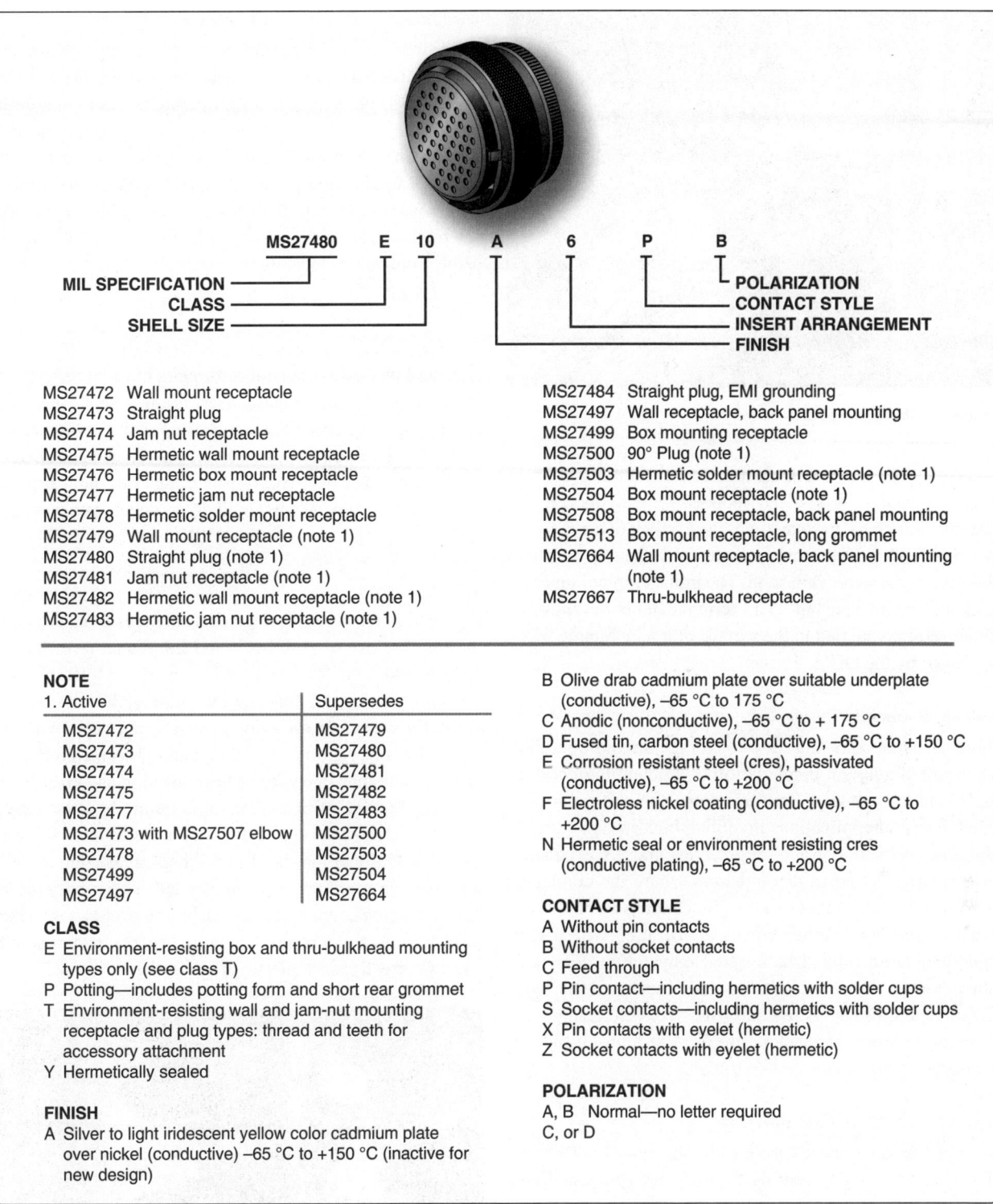

MS27480 E 10 A 6 P B

MIL SPECIFICATION
CLASS
SHELL SIZE
POLARIZATION
CONTACT STYLE
INSERT ARRANGEMENT
FINISH

MS27472 Wall mount receptacle
MS27473 Straight plug
MS27474 Jam nut receptacle
MS27475 Hermetic wall mount receptacle
MS27476 Hermetic box mount receptacle
MS27477 Hermetic jam nut receptacle
MS27478 Hermetic solder mount receptacle
MS27479 Wall mount receptacle (note 1)
MS27480 Straight plug (note 1)
MS27481 Jam nut receptacle (note 1)
MS27482 Hermetic wall mount receptacle (note 1)
MS27483 Hermetic jam nut receptacle (note 1)

MS27484 Straight plug, EMI grounding
MS27497 Wall receptacle, back panel mounting
MS27499 Box mounting receptacle
MS27500 90° Plug (note 1)
MS27503 Hermetic solder mount receptacle (note 1)
MS27504 Box mount receptacle (note 1)
MS27508 Box mount receptacle, back panel mounting
MS27513 Box mount receptacle, long grommet
MS27664 Wall mount receptacle, back panel mounting (note 1)
MS27667 Thru-bulkhead receptacle

NOTE

1. Active

Active	Supersedes
MS27472	MS27479
MS27473	MS27480
MS27474	MS27481
MS27475	MS27482
MS27477	MS27483
MS27473 with MS27507 elbow	MS27500
MS27478	MS27503
MS27499	MS27504
MS27497	MS27664

CLASS

E Environment-resisting box and thru-bulkhead mounting types only (see class T)
P Potting—includes potting form and short rear grommet
T Environment-resisting wall and jam-nut mounting receptacle and plug types: thread and teeth for accessory attachment
Y Hermetically sealed

FINISH

A Silver to light iridescent yellow color cadmium plate over nickel (conductive) –65 °C to +150 °C (inactive for new design)

B Olive drab cadmium plate over suitable underplate (conductive), –65 °C to 175 °C
C Anodic (nonconductive), –65 °C to + 175 °C
D Fused tin, carbon steel (conductive), –65 °C to +150 °C
E Corrosion resistant steel (cres), passivated (conductive), –65 °C to +200 °C
F Electroless nickel coating (conductive), –65 °C to +200 °C
N Hermetic seal or environment resisting cres (conductive plating), –65 °C to +200 °C

CONTACT STYLE

A Without pin contacts
B Without socket contacts
C Feed through
P Pin contact—including hermetics with solder cups
S Socket contacts—including hermetics with solder cups
X Pin contacts with eyelet (hermetic)
Z Socket contacts with eyelet (hermetic)

POLARIZATION

A, B Normal—no letter required
C, or D

Figure 9-159. *MS connector information sheet.*

use integral or rack-mounted pins for alignment and box mounting hardware for couplings.

Module blocks are types of junctions that accept crimped contacts similar to those on connectors. Some use internal busing to provide a variety of circuit arrangements. They are useful where a number of wires are connected for power or signal distribution. When used as grounding modules, they save and reduce hardware installation on the aircraft. Standardized modules are available with wire end grommet seals for environmental applications and are track mounted. Function module blocks are used to provide an easily

Figure 9-160. *Rectangular connectors.*

wired package for environment-resistant mounting of small resistors, diodes, filters, and suppression networks. Inline terminal junctions are sometimes used in lieu of a connector when only a few wires are terminated and when the ability to disconnect the wires is desired. The inline terminal junction is environment resistant. The terminal junction splice is small and may be tied to the surface of a wire bundle when approved by the OEM.

Voltage & Current Rating

Selected connectors must be rated for continuous operation under the maximum combination of ambient temperature and circuit current load. Hermetic connectors and connectors used in circuit applications involving high-inrush currents should be derated. It is good engineering practice to conduct preliminary testing in any situation where the connector is to operate with most or all of its contacts at maximum rated current load. When wiring is operating with a high conductor temperature near its rated temperature, connector contact sizes should be suitably rated for the circuit load. This may require an increase in wire size. Voltage derating is required when connectors are used at high altitude in non-pressurized areas.

Spare Contacts for Future Wiring

To accommodate future wiring additions, spare contacts are normally provided. Locating the unwired contacts along the outer part of the connector facilitates future access. A good practice is to provide two spares on connectors with 25 or fewer contacts; 4 spares on connectors with 26 to 100 contacts; and 6 spares on connectors with more than 100 contacts. Spare contacts are not normally provided on receptacles of components that are unlikely to have added wiring. Connectors must have all available contact cavities filled with wired or unwired contacts. Unwired contacts should be provided with a plastic grommet sealing plug.

Wire Installation into the Connector

Wires that perform the same function in redundant systems must be routed through separate connectors. On systems critical to flight safety, system operation wiring should be routed through separate connectors from the wiring used for system failure warning. It is also good practice to route a system's indication wiring in separate connectors from its failure warning circuits to the extent practicable. These steps can reduce an aircraft's susceptibility to incidents that might result from connector failures.

Adjacent Locations

Mating of adjacent connectors should not be possible. In order to ensure this, adjacent connector pairs must be different in shell size, coupling means, insert arrangement, or keying arrangement. When such means are impractical, wires should be routed and clamped so that incorrectly mated pairs cannot reach each other. Reliance on markings or color stripes is not recommended as they are likely to deteriorate with age. *[Figure 9-161]*

Sealing

Connectors must be of a type that excludes moisture entry through the use of peripheral and interfacial seal that are compressed when the connector is mated. Moisture entry through the rear of the connector must be avoided by correctly matching the wire's outside diameter with the connector's rear grommet sealing range. It is recommended that no more than one wire be terminated in any crimp style contact. The use of heat-shrinkable tubing to build up the wire diameter, or the application of potting to the wire entry area as additional means of providing a rear compatibility with the rear grommet is recommended. These extra means have inherent penalties and should be considered only where other means cannot be used. Unwired spare contacts should have a correctly sized plastic plug installed.

Figure 9-161. *Connector arrangement to avoid wrong connection.*

Drainage

Connectors must be installed in a manner that ensures moisture and fluids drain out of and not into the connector when unmated. Wiring must be routed so that moisture accumulated on the bundle drains away from connectors. When connectors must be mounted in a vertical position, as through a shelf or floor, the connectors must be potted or environmentally sealed. In this situation, it is better to have the receptacle faced downward so that it is less susceptible to collecting moisture when unmated.

Wire Support

A rear accessory back shell must be used on connectors that are not enclosed. Connectors with very small size wiring, or subject to frequent maintenance activity, or located in high-vibration areas must be provided with a strain-relief-type back shell. The wire bundle should be protected from mechanical damage with suitable cushion material where it is secured by the clamp. Connectors that are potted or have molded rear adapters do not normally use a separate strain relief accessory. Strain relief clamps should not impart tension on wires between the clamp and contact. *[Figure 9-162]*

Sufficient wire length must be provided at connectors to ensure a proper drip loop and that there is no strain on termination after a complete replacement of the connector and its contacts.

Coaxial Cable

All wiring needs to be protected from damage. However, coaxial and triaxial cables are particularly vulnerable to certain types of damage. Personnel should exercise care while handling or working around coaxial. *[Figure 9-163]* Coaxial damage can occur when clamped too tightly, or when they are bent sharply (normally at or near connectors). Damage can also be incurred during unrelated maintenance actions around the coaxial cable. Coaxial cable can be severely damaged on the inside without any evidence of damage on the outside. Coaxial cables with solid center conductors should not be used. Stranded center coaxial cables can be used as a direct replacement for solid center coaxial. *[Figure 9-164]* Coaxial cable precautions include:

- Never kink coaxial cable.
- Never drop anything on coaxial cable.
- Never step on coaxial cable.
- Never bend coaxial cable sharply.
- Never loop coaxial cable tighter than the allowable bend radius.
- Never pull on coaxial cable except in a straight line.
- Never use coaxial cable for a handle, lean on it, or hang things on it (or any other wire).

Figure 9-162. *Backshells with strain relief.*

Figure 9-163. *Coaxial cables.*

Wire Inspection

Aircraft service imposes severe environmental condition on electrical wire. To ensure satisfactory service, inspect wire annually for abrasions, defective insulation, condition of terminations, and potential corrosion. Grounding connections for power, distribution equipment, and electromagnetic shielding must be given particular attention to ensure that electrical bonding resistance has not been significantly increased by the loosening of connections or corrosion.

Electrical System Components

Switches

Switches are devices that open and close circuits. They consist of one or more pair of contacts. The current in the circuit flows when the contacts are closed. Switches with momentary contacts actuate the circuit temporarily, and they return to the normal position with an internal spring when the switch is released. Switches with continuous contacts remain in position when activated. Hazardous errors in switch operation can be avoided by logical and consistent installation. Two-position on/off switches should be mounted

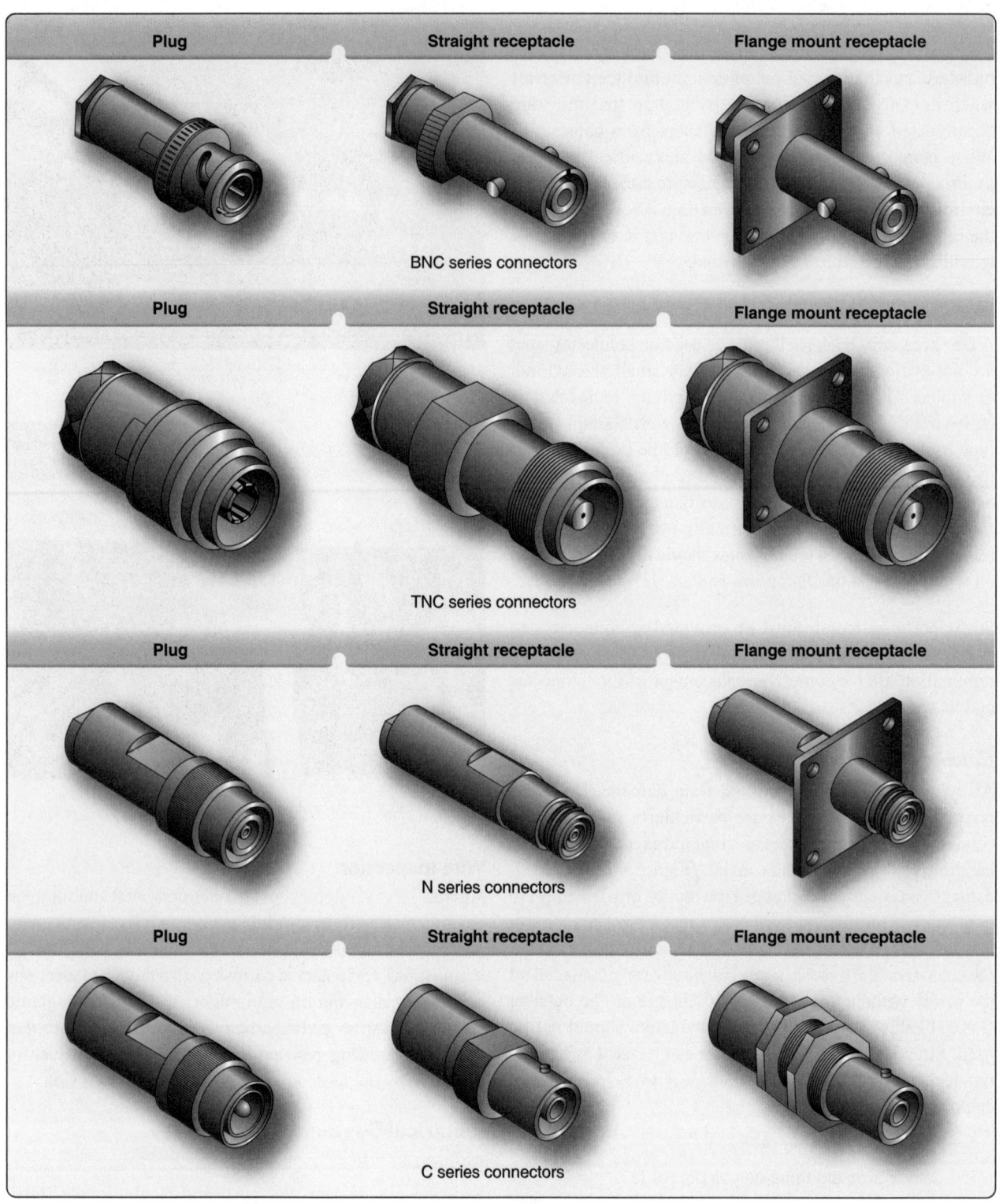

Plug	Straight receptacle	Flange mount receptacle

BNC series connectors

Plug	Straight receptacle	Flange mount receptacle

TNC series connectors

Plug	Straight receptacle	Flange mount receptacle

N series connectors

Plug	Straight receptacle	Flange mount receptacle

C series connectors

Figure 9-164. *Coaxial cable connectors.*

so that the on position is reached by an upward or forward movement of the toggle. When the switch controls movable aircraft elements, such as landing gear or flaps, the toggle should move in the same direction as the desired motion. Inadvertent operation of a switch can be prevented by mounting a suitable guard over the switch. *[Figure 9-165]*

A specifically designed switch should be used in all circuits where a switch malfunction would be hazardous. Such switches are of rugged construction and have sufficient contact capacity to break, make, and carry continuously the connected load current. Snap action design is generally preferred to obtain rapid opening and closing of contacts regardless of the speed of the operating toggle or plunger, thereby minimizing contact arcing. The nominal current rating of the conventional aircraft switch is usually stamped on the switch housing. This rating represents the continuous current rating with the contacts closed. Switches should be derated from their nominal current rating for the following types of circuits:

1. High rush-in circuits—contain incandescent lamps that can draw an initial current 15 times greater than the continuous current. Contact burning or welding may occur when the switch is closed.

2. Inductive circuits—magnetic energy stored in solenoid coils or relays is released and appears as an arc when the control switch is opened.

3. Motors—DC motors draw several times their rated current during starting, and magnetic energy stored in their armature and field coils is released when the control switch is opened.

Figure 9-166 is used for selecting the proper nominal switch rating when the continuous load current is known. This selection is essentially a derating to obtain reasonable switch efficiency and service life.

Type of Switches

Single-pole single-throw (SPST)—opens and closes a single circuit. Pole indicates the number of separate circuits that can be activated, and throw indicates the number of current paths.

Double-pole single-throw (DPST)—turn two circuits on and off with one lever.

Single-pole double-throw (SPDT)—route circuit current to either of two paths. The switch is ON in both positions. For example, switch turns on red lamp in one position and turns on green lamp in the other position.

Double-pole double-throw (DPDT)—activates two separate circuits at the same time.

Double-throw switches—have either two or three positions.

Two-position switch—pole always connected to one of the two throws. Three-position switches have a center OFF position that disconnects the pole from both throws.

Spring-loaded switches—available in two types: 1) normally open (NO) and 2) normally closed (NC). The contacts of the NO switch are disconnected in the normal position and become closed when the switch is activated. The switch returns to the normal position when the applied force to the switch is released. The contacts of the NC switch are connected in the normal position and become open when the switch is activated. The switch returns to the normal position when the applied force to the switch is released.

Toggle & Rocker Switches

Toggle and rocker switches control most of aircraft's electrical components. *[Figure 9-167]* Aircraft that are outfitted with a glass flight deck often use push buttons to control electrical components.

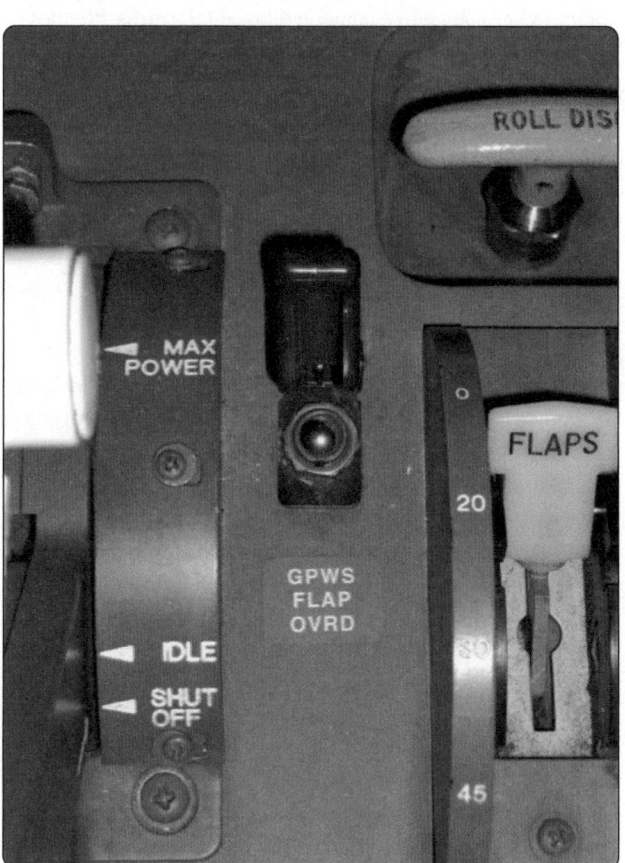

Figure 9-165. *Switch guard.*

Nominal system voltage (DC)	Type of load	Derating factor
28V	Lamp	8
28V	Inductive	4
28V	Resistive	2
28V	Motor	3
12V	Lamp	5
12V	Inductive	2
12V	Resistive	1
12V	Motor	2

Figure 9-166. *Derating table for switches.*

Figure 9-167. *Toggle and rocker switches.*

Rotary Switches

Rotary switches are activated by twisting a knob or shaft and are commonly found on radio control panels. Rotary switches are utilized for controlling more than two circuits.

Precision (Micro) Switches

Micro switches require very little pressure to activate. These types of switches are spring loaded, once the pressure is removed, the contacts return to the normal position. These types of switches are typically single-pole double-throw (SPDT) or double-pole double-throw (DPDT) and have three contacts: normally open, normally closed, and common. Micro switches are used to detect position or to limit travel of moving parts, such as landing gear, flaps, spoilers, etc. *[Figure 9-168]*

Relays & Solenoids (Electromagnetic Switches)

Relays are used to control the flow of large currents using a small current. A low-power DC circuit is used to activate the relay and control the flow of large AC currents. They are used to switch motors and other electrical equipment on and off and to protect them from overheating. A solenoid is a special

Figure 9-168. *A micro switch.*

type of relay that has a moving core. The electromagnet core in a relay is fixed. Solenoids are mostly used as mechanical actuators but can also be used for switching large currents. Relays are only used to switch currents.

Solenoids

Solenoids are used as switching devices where a weight reduction can be achieved or electrical controls can be simplified. The foregoing discussion of switch ratings is generally applicable to solenoid contact ratings. Solenoids have a movable core/armature that is usually made of steel or iron, and the coil is wrapped around the armature. The solenoid has an electromagnetic tube and the armature moves in and out of the tube. *[Figure 9-169]*

Relays

The two main types of relays are electromechanical and solid state. Electromechanical relays have a fixed core and a moving plate with contacts on it, while solid-state relays work similar to transistors and have no moving parts. Current flowing through the coil of an electromechanical relay creates a magnetic field that attracts a lever and changes the switch contacts. The coil current can be on or off so relays have two switch positions. These can be made as a single throw or double throw switch. Residual magnetism is a common problem and the contacts may stay closed or are opened by a slight amount of residual magnetism. A relay is an electrically operated switch and is therefore subject to dropout under low system voltage conditions. Relays allow one circuit to switch a second circuit that can be completely separate from the first. For example, a low voltage DC battery circuit can use a relay to switch a 110-volt three-phase AC circuit. There is no electrical connection inside the relay between the two circuits; the link is magnetic and mechanical. *[Figure 9-170]*

Current Limiting Devices

Conductors should be protected with circuit breakers or fuses located as close as possible to the electrical power source bus. Normally, the manufacturer of the electrical equipment specifies the fuse or circuit breaker to be used when installing

Figure 9-169. *Solenoid.*

Wire AN gauge copper	Circuit breaker amperage	Fuse amperage
22	5	5
20	7.5	5
18	10	10
16	15	10
14	20	15
12	30	20
10	40	30
8	50	50
6	80	70
4	100	70
2	125	100
1		150
0		150

Figure 9-171. *Wired and circuit protection chart.*

equipment. The circuit breaker or fuse should open the circuit before the conductor emits smoke. To accomplish this, the time current characteristic of the protection device must fall below that of the associated conductor. Circuit protector characteristics should be matched to obtain the maximum utilization of the connected equipment. *Figure 9-171* shows a chart used in selecting the circuit breaker and fuse protection for copper conductors. This limited chart is applicable to a specific set of ambient temperatures and wire bundle sizes and is presented as typical only. It is important to consult such guides before selecting a conductor for a specific purpose. For example, a wire run individually in the open air may be protected by the circuit breaker of the next higher rating to that shown on the chart.

Fuses

A fuse is placed in series with the voltage source and all current must flow through it. *[Figure 9-172]* The fuse consists of a strip of metal that is enclosed in a glass or plastic housing. The metal strip has a low melting point and is usually made of lead, tin, or copper. When the current exceeds the capacity of the fuse the metal strip heats up and breaks. As a result of this, the flow of current in the circuit stops.

There are two basic types of fuses: fast acting and slow blow. The fast-acting type opens very quickly when their particular current rating is exceeded. This is important for electric devices that can quickly be destroyed when too much current flows through them for even a very small amount of time. Slow blow fuses have a coiled construction inside. They are designed to open only on a continued overload, such as a short circuit.

Circuit Breakers

A circuit breaker is an automatically operated electrical switch designed to protect an electrical circuit from damage caused by an overload or short circuit. Its basic function is to detect a fault condition and immediately discontinue electrical flow. Unlike a fuse that operates once and then has to be replaced, a circuit breaker can be reset to resume normal operation. All resettable circuit breakers should open the circuit in which they are installed regardless of the position of the operating control when an overload or circuit fault exists. Such circuit breakers are referred to as trip-free. Automatic reset circuit breakers automatically reset themselves. They should not be used as circuit protection devices in aircraft. When a circuit breaker trips, the electrical circuit should be checked and the fault removed before the circuit breaker is reset. Sometimes circuit breakers trip for no apparent

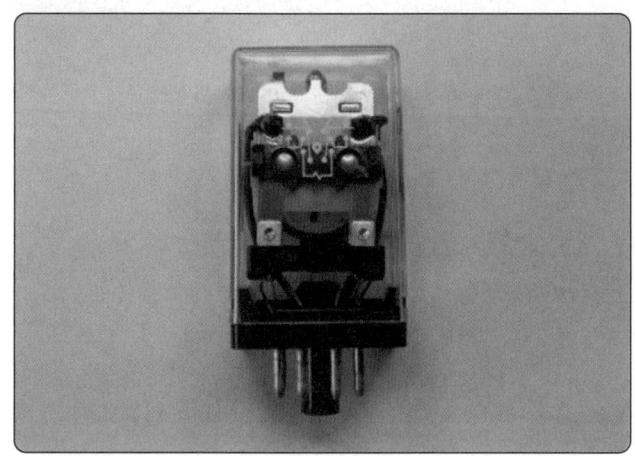

Figure 9-170. *Relay.*

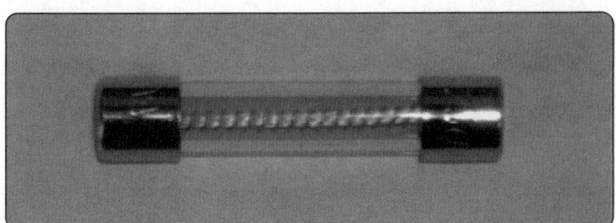

Figure 9-172. *A fuse.*

reason, and the circuit breaker can be reset one time. If the circuit breaker trips again, there exists a circuit fault and the technician must troubleshoot the circuit before resetting the circuit breaker. *[Figure 9-173]*

Some new aircraft designs use a digital circuit protection architecture. This system monitors the amperage through a particular circuit. When the maximum amperage for that circuit is reached, the power is rerouted away from the circuit. This system reduces the use of mechanical circuit breakers. The advantages are weight savings and the reduction of mechanical parts.

Aircraft Lighting Systems

Aircraft lighting systems provide illumination for both exterior and interior use. Lights on the exterior provide illumination for such operations as landing at night, inspection of icing conditions, and safety from midair collision. Interior lighting provides illumination for instruments, flight decks, cabins, and other sections occupied by crewmembers and passengers. Certain special lights, such as indicator and warning lights, indicate the operation status of equipment.

Exterior Lights

Position, anticollision, landing, and taxi lights are common examples of aircraft exterior lights. Some lights are required for night operations. Other types of exterior lights, such as wing inspection lights, are of great benefit for specialized flying operations.

Position Lights

Aircraft operating at night must be equipped with position lights that meet the minimum requirements specified by Title 14 of the Code of Federal Regulations. A set of position lights consist of one red, one green, and one white light. *[Figures 9-174 and 9-175]*

On some types of installations, a switch in the flight deck provides for steady or flashing operation of the position lights. On many aircraft, each light unit contains a single lamp mounted on the surface of the aircraft. Other types of position light units contain two lamps and are often streamlined into the surface of the aircraft structure. The green light unit is always mounted at the extreme tip of the right wing. The red unit is mounted in a similar position on the left wing. The white unit is usually located on the vertical stabilizer in a position where it is clearly visible through a wide angle from the rear of the aircraft. *Figure 9-176* illustrates a schematic diagram of a position light circuit. Position lights are also known as navigation lights.

There are, of course, many variations in the position light circuits used on different aircraft. All circuits are protected by fuses or circuit breakers, and many circuits include flashing and dimming equipment. Small aircraft are usually equipped

Figure 9-174. *A left wing tip position light (red) and a white strobe light.*

Figure 9-173. *Circuit breaker panel.*

Figure 9-175. *A right wing tip position light, also known as a navigation light.*

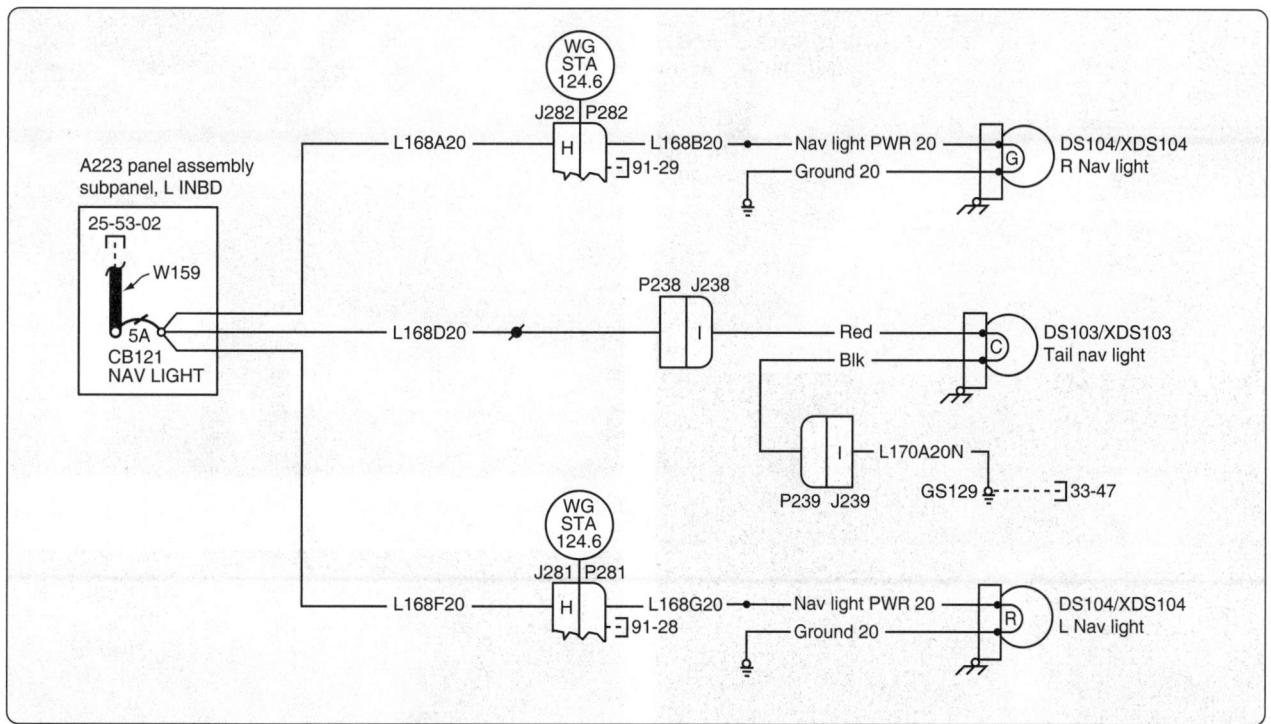

Figure 9-176. *Navigation light system schematic.*

with a simplified control switch and circuitry. In some cases, one control knob or switch is used to turn on several sets of lights; for example, one type utilizes a control knob, the first movement of which turns on the position lights and the instrument panel lights. Further rotation of the control knob increases the intensity of only the panel lights. A flasher unit is seldom included in the position light circuitry of very light aircraft but is used in small twin-engine aircraft. Traditional position lights use incandescent light bulbs. LED lights have been introduced on modern aircraft because of their good visibility, high reliability, and low power consumption.

Anticollision Lights

An anticollision light system may consist of one or more lights. They are rotating beam lights that are usually installed on top of the fuselage or tail in such a location that the light does not affect the vision of the crewmember or detract from the visibility of the position lights. Large transport type aircraft use an anticollision light on top and one on the bottom of the aircraft. *Figure 9-177* shows a typical anticollision light installation in a vertical stabilizer.

An anticollision light unit usually consists of one or two rotating lights operated by an electric motor. The light may be fixed but mounted under rotating mirrors inside a protruding red glass housing. The mirrors rotate in an arc, and the resulting flash rate is between 40 and 100 cycles per minute. Newer aircraft designs use a LED type of anticollision light. The anticollision light is a safety light to warn other aircraft,

especially in congested areas.

A white strobe light is a second type of anti-collision light that is also common. Usually mounted at the wing tips and, possibly, at empennage extremities, strobe lights produce an extremely bright intermittent flash of white light that is highly visible. The light is produced by a high voltage discharge of a capacitor. A dedicated power pack houses the capacitor and supplies voltage to a sealed xenon-filled tube. The xenon ionizes with a flash when the voltage is applied. A strobe light is shown in *Figure 9-174*.

Landing & Taxi Lights

Landing lights are installed in aircraft to illuminate runways during night landings. These lights are very powerful and are directed by a parabolic reflector at an angle providing a maximum range of illumination. Landing lights of smaller aircraft are usually located midway in the leading edge of each wing or streamlined into the aircraft surface. Landing lights for larger transport category aircraft are usually located in the leading edge of the wing close to the fuselage. Each light may be controlled by a relay, or it may be connected directly into the electric circuit. On some aircraft, the landing light is mounted in the same area with a taxi light. *[Figure 9-178]* A sealed beam, halogen, or high intensity xenon discharge lamp is used.

Taxi lights are designed to provide illumination on the ground while taxiing or towing the aircraft to or from a runway, taxi

Figure 9-177. *Anticollision lights.*

Figure 9-178. *Landing lights.*

strip, or in the hangar area. *[Figure 9-179]* Taxi lights are not designed to provide the degree of illumination necessary for landing lights. On aircraft with tricycle landing gear, either single or multiple taxi lights are often mounted on the non-steerable part of the nose landing gear. They are positioned at an oblique angle to the center line of the aircraft to provide illumination directly in front of the aircraft and also some illumination to the right and left of the aircraft's path. On some aircraft, the dual taxi lights are supplemented by wingtip clearance lights controlled by the same circuitry. Taxi lights are also mounted in the recessed areas of the wing leading edge, often in the same area with a fixed landing light.

Many small aircraft are not equipped with any type of taxi light, but rely on the intermittent use of a landing light to illuminate taxiing operations. Still other aircraft utilize a dimming resistor in the landing light circuit to provide reduced illumination for taxiing. A typical circuit for taxi lights is shown in *Figure 9-180*.

Some large aircraft are equipped with alternate taxi lights located on the lower surface of the aircraft, aft of the nose radome. These lights, operated by a separate switch from the main taxi lights, illuminate the area immediately in front of and below the aircraft nose.

Figure 9-179. *Taxi lights.*

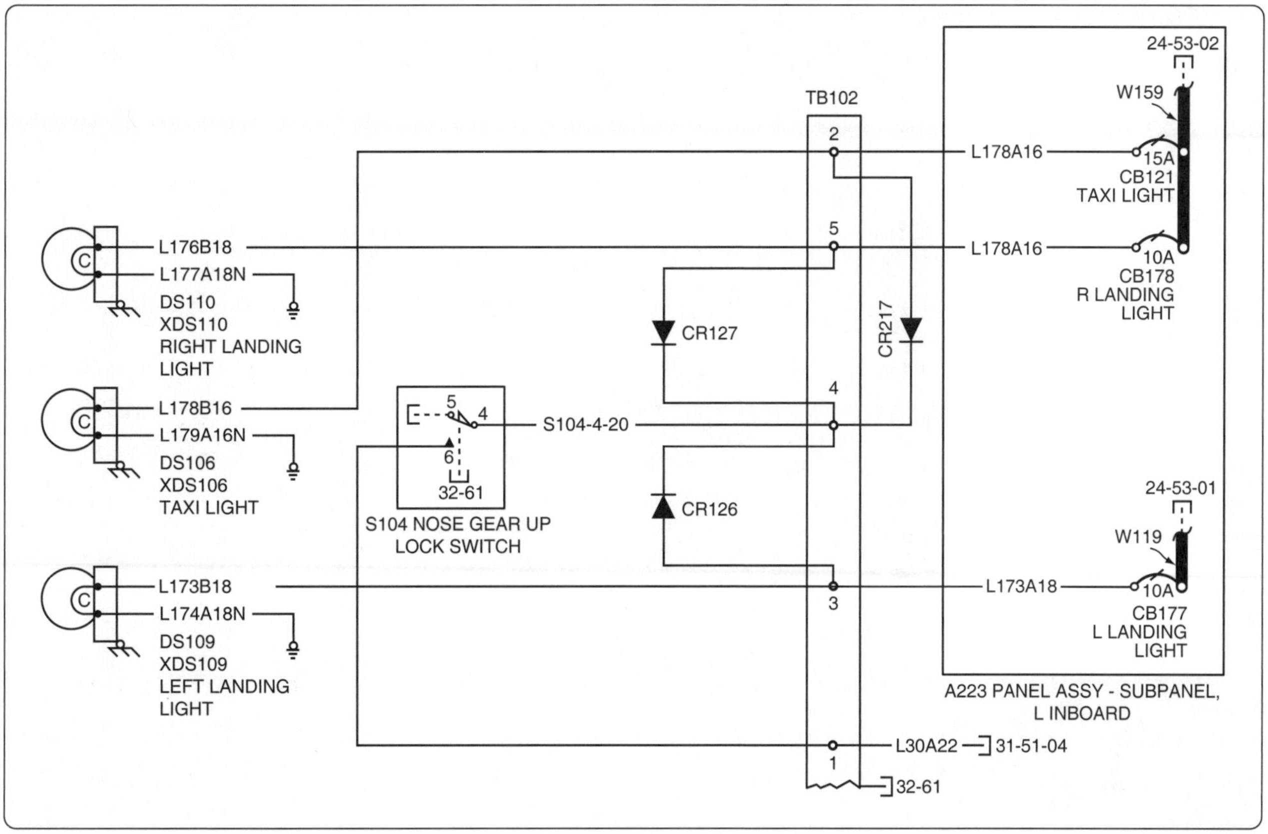

Figure 9-180. *Taxi light circuit.*

The figure shows a wiring diagram with the following labels:

24-53-02
W159
L178A16
15A
CB121
TAXI LIGHT

TB102
2

5
L178A16
10A
CB178
R LANDING
LIGHT

CR217

L176B18
L177A18N
DS110
XDS110
RIGHT LANDING
LIGHT

CR127

L178B16
L179A16N
DS106
XDS106
TAXI LIGHT

5 4
6
32-61
S104-4-20
S104 NOSE GEAR UP
LOCK SWITCH

4

CR126

24-53-01
W119
L173A18
10A
CB177
L LANDING
LIGHT

L173B18
L174A18N
DS109
XDS109
LEFT LANDING
LIGHT

3

A223 PANEL ASSY - SUBPANEL,
L INBOARD

L30A22 — 31-51-04
1
32-61

Wing Inspection Lights

Some aircraft are equipped with wing inspection lights to illuminate the leading edge of the wings to permit observation of icing and general condition of the   se areas in flight. These lights permit visual detection of ice formation on wing leading edges while flying at night. They are usually controlled through a relay by an on/off toggle switch in the flight deck. Some wing inspection light systems may include or be supplemented by additional lights, sometimes called nacelle lights, that illuminate adjacent areas, such a cowl flaps or the landing gear. These are normally the same type of lights and can be controlled by the same circuits.

Interior Lights

Aircraft are equipped with interior lights to illuminate the cabin. *[Figure 9-181]* Often white and red light settings are provided. Commercial aircraft have a lighting system that illuminates the main cabin, an independent lighting system so that passengers can read when the cabin lights are off, and an emergency lighting system on the floor of the aircraft to aid passengers during an emergency.

Maintenance & Inspection of Lighting Systems

Inspection of an aircraft's lighting system normally includes checking the condition and security of all visible wiring, connections, terminals, fuses, and switches. A continuity light or meter can be used in making these checks, since the cause of many troubles can often be located by systematically testing each circuit for continuity.

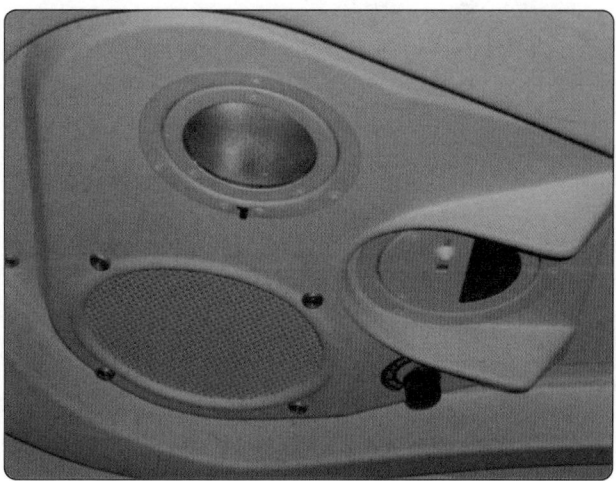

Figure 9-181. *Interior flight deck and cabin light system.*

Chapter 10
Aircraft Systems

Introduction

Since the beginning of manned flight, it has been recognized that supplying the pilot with information about the aircraft and its operation could be useful and lead to safer flight. The Wright Brothers had very few instruments on their Wright Flyer, but they did have an engine tachometer, an anemometer (wind meter), and a stop watch. They were obviously concerned about the aircraft's engine and the progress of their flight. From that simple beginning, a wide variety of instruments have been developed to inform flight crews of different parameters. Instrument systems now exist to provide information on the condition of the aircraft, engine, components, the aircraft's attitude in the sky, weather, cabin environment, navigation, and communication. *Figure 10-1* shows various instrument panels from the Wright Flyer to a modern jet airliner.

The ability to capture and convey all of the information a pilot may want, in an accurate, easily understood manner, has been a challenge throughout the history of aviation. As the range of desired information has grown, so too have the size and complexity of modern aircraft, thus expanding even further the need to inform the flight crew without sensory overload or overcluttering the flight deck. As a result, the old flat panel in the front of the flight deck with various individual instruments attached to it has evolved into a sophisticated computer-controlled digital interface with flat-panel display screens and prioritized messaging. A visual comparison between a conventional flight deck and a glass flight deck is shown in *Figure 10-2*.

There are usually two parts to any instrument or instrument system. One part senses the situation and the other part displays it. In analog instruments, both of these functions often take place in a single unit or instrument (case). These are called direct-sensing instruments. Remote-sensing requires the information to be sensed, or captured, and then sent to a separate display unit in the flight deck. Both analog and digital instruments make use of this method. *[Figure 10-3]*

The relaying of important bits of information can be done in various ways. Electricity is often used by way of wires that carry sensor information into the flight deck. Sometimes pneumatic lines are used. In complex, modern aircraft, this can lead to an enormous amount of tubing and wiring terminating behind the instrument display panel. More efficient information transfer has been accomplished via the

Figure 10-1. *From top to bottom: instruments of the Wright Flyer, instruments on a World War I era aircraft, a late 1950s/early 1960s Boeing 707 airliner flight deck, and an Airbus A380 glass flight deck.*

Figure 10-2. *A conventional instrument panel of the C-5A Galaxy (top) and the glass flight deck of the C-5B Galaxy (bottom).*

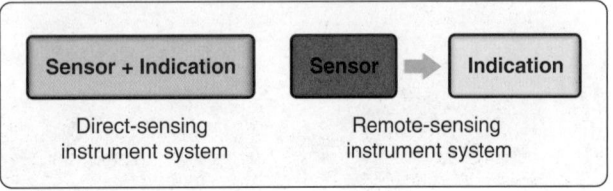

Sensor + Indication	Sensor ➡ Indication
Direct-sensing instrument system	Remote-sensing instrument system

Figure 10-3. *There are two parts to any instrument system—the sensing mechanism and the display mechanism.*

use of digital data buses. Essentially, these are wires that share message carrying for many instruments by digitally encoding the signal for each. This reduces the number of wires and weight required to transfer remotely sensed information for the pilot's use. Flat-panel computer display screens that can be controlled to show only the information desired are also lighter in weight than the numerous individual gauges it would take to display the same information simultaneously. An added bonus is the increased reliability inherent in these solid-state systems.

It is the job of the aircraft technician to understand and maintain all aircraft, including these various instrument systems. Accordingly, in this chapter, discussions begin

with analog instruments and refer to modern digital instrumentation when appropriate.

Classifying Instruments

There are three basic kinds of instruments classified by the job they perform: flight instruments, engine instruments, and navigation instruments. There are also miscellaneous gauges and indicators that provide information that do not fall into these classifications, especially on large complex aircraft. Flight control position, cabin environmental systems, electrical power, and auxiliary power units (APUs), for example, are all monitored and controlled from the flight deck via the use of instrument systems.

Flight Instruments

The instruments used in controlling the aircraft's flight attitude, altitude, speed, and direction are known as the flight instruments. There are basic flight instruments, such as the altimeter that displays aircraft altitude; the airspeed indicator; and the magnetic direction indicator, a form of compass. Additionally, an artificial horizon, turn coordinator, and vertical speed indicator are flight instruments present in most aircraft. Much variation exists for these instruments, which is explained throughout this chapter. Over the years, flight instruments have come to be situated similarly on the instrument panels in most aircraft. This basic T arrangement for flight instruments is shown in *Figure 10-4*. The top center position directly in front of the pilot and copilot is the basic display position for the artificial horizon even in modern glass flight decks (those with solid-state, flat-panel screen indicating systems).

Original analog flight instruments are operated by air pressure and the use of gyroscopes. This avoids the use of

Figure 10-4. *The basic T arrangement of analog flight instruments. At the bottom of the T is a heading indicator that functions as a compass but is driven by a gyroscope and not subject to the oscillations common to magnetic direction indicators.*

electricity, which could put the pilot in a dangerous situation if the aircraft lost electrical power. Development of sensing and display techniques, combined with advanced aircraft electrical systems, has made it possible for reliable primary and secondary instrument systems that are electrically operated. Nonetheless, often a pneumatic altimeter, a gyro artificial horizon, and a magnetic direction indicator are retained somewhere in the instrument panel for redundancy. *[Figure 10-5]*

Figure 10-5. *This electrically operated flat screen display instrument panel, or glass flight deck, retains an analog airspeed indicator, a gyroscope-driven artificial horizon, and an analog altimeter as a backup should electric power be lost, or a display unit fails.*

Engine Instruments

Engine instruments are those designed to measure operating parameters of the aircraft's engine(s). These are usually quantity, pressure, speed, and temperature indications. The most common engine instruments are the fuel and oil quantity and pressure gauges, tachometers, and temperature gauges. *Figure 10-6* contains various engine instruments found on reciprocating and turbine-powered aircraft.

Engine instrumentation is often displayed in the center of the flight deck where it is easily visible to the pilot and copilot. *[Figure 10-7]* On light aircraft requiring only one flight crewmember, this may not be the case. Multiengine aircraft often use a single gauge for a particular engine parameter, but it displays information for all engines through the use of multiple pointers on the same dial face.

Navigation Instruments

Navigation instruments are those that contribute information used by the pilot to guide the aircraft along a definite course. This group includes compasses of various kinds, some of which incorporate the use of radio signals to define a specific course while flying the aircraft en route from one airport to another. Other navigational instruments are designed specifically to direct the pilot's approach to landing at an airport. Traditional navigation instruments include a clock and a magnetic compass. Along with the airspeed indicator and wind information, these can be used to calculate navigational progress. Radios and instruments sending locating information via radio waves have replaced these manual efforts in modern aircraft. Global position systems (GPS) use satellites to pinpoint the location of the aircraft via geometric triangulation. This technology is built into some aircraft instrument packages for navigational purposes. Many of these aircraft navigational systems are discussed in chapter 11 of this handbook. *[Figure 10-8]*

Instruments can also be classified according to the principle upon which they operate. Some use mechanical methods to measure pressure and temperature. Some utilize magnetism and electricity to sense and display a parameter. Others depend on the use of gyroscopes in their primary workings. Still others utilize solid state sensors and computers to process and display important information. In the following sections, the different operating principles for sensing parameters are explained. Then, an overview of many of the engine, flight, and navigation instruments is given.

Pressure Measuring Instruments

A number of instruments inform the pilot of the aircraft's condition and flight situations through the measurement of

Reciprocating engines	Turbine engines
Oil pressure	Oil pressure
Oil temperature	Exhaust gas temperature (EGT)
Cylinder head temperature (CHT)	Turbine inlet temperature (TIT) or turbine gas temperature (TGT)
Manifold pressure	Engine pressure ratio (EPR)
Fuel quantity	Fuel quantity
Fuel pressure	Fuel pressure
	Fuel flow
Tachometer	Tachometer (percent calibrated)
	N_1 and N_2 compressor speeds
Carburetor temperature	Torquemeter (on turboprop and turboshaft engines)

Figure 10-6. *Common engine instruments. Note: For example purposes only. Some aircraft may not have these instruments or may be equipped with others.*

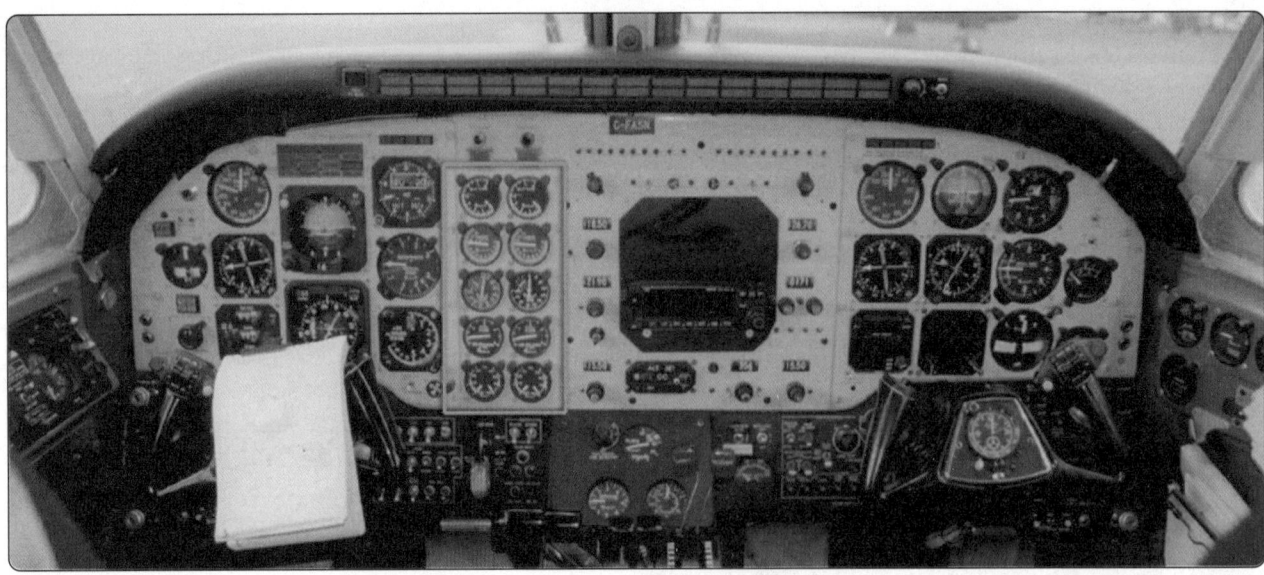

Figure 10-7. *An engine instrumentation located in the middle of the instrument panel is shared by the pilot and co-pilot.*

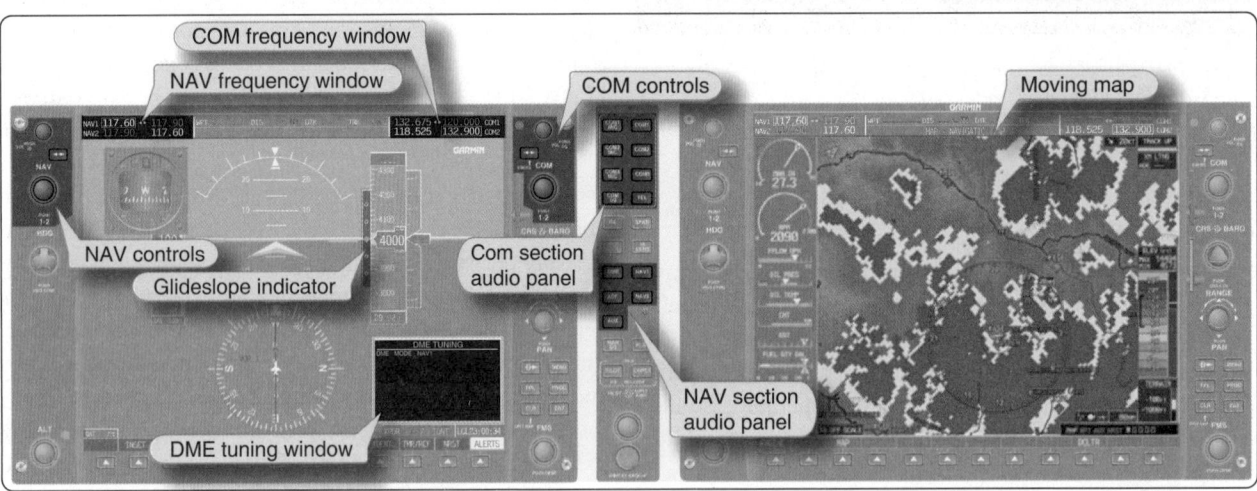

Figure 10-8. *Navigation instruments.*

pressure. Pressure-sensing instruments can be found in the flight group and the engine group. They can be either direct reading or remote sensing. These are some of the most critical instruments on the aircraft and must accurately inform the pilot to maintain safe operations. Pressure measurement involves some sort of mechanism that can sense changes in pressure. A technique for calibration and displaying the information is then added to inform the pilot. The type of pressure needed to be measured often makes one sensing mechanism more suited for use in a particular instance. The three fundamental pressure-sensing mechanisms used in aircraft instrument systems are the Bourdon tube, the diaphragm or bellows, and the solid-state sensing device.

A Bourdon tube is used to measure relatively high pressures and is illustrated in *Figure 10-9*. The open end of this coiled tube is fixed in place and the other end is sealed and free to move. When pressure is directed into the open end of the

tube, the unfixed portion of the coiled tube tends to straighten out. The higher the pressure, the more the tube straightens. When the pressure is reduced, the tube recoils. A pointer is attached to this moving end of the tube, usually through a linkage of small shafts and gears. By calibrating this motion of the straightening tube, a face or dial of the instrument can be created. Thus, by observing the pointer movement along the scale of the instrument face positioned behind it, pressure increases and decreases are communicated to the pilot.

The Bourdon tube is the internal mechanism for many pressure gauges used on aircraft. Most Bourdon tubes are made from brass, bronze, or copper. Alloys of these metals can be made to coil and uncoil the tube consistently numerous times.

Bourdon tube gauges are simple and reliable. Some of the instruments that use a Bourdon tube mechanism include the

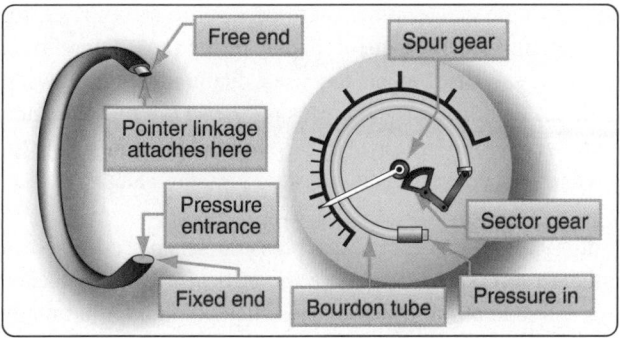

Figure 10-9. *The Bourdon tube is one of the basic mechanisms for sensing pressure.*

engine oil pressure gauge, hydraulic pressure gauge, oxygen tank pressure gauge, and deice boot pressure gauge. Since the pressure of a heated liquid or gas increases as temperature increases, Bourdon tube mechanisms can also be used to measure temperature. This is done by calibrating the pointer connecting linkage and relabeling the face of the gauge with a temperature scale. Oil temperature gauges often employ Bourdon tube mechanisms. *[Figure 10-10]*

Since the sensing and display of pressure or temperature information using a Bourdon tube mechanism usually occurs in a single instrument housing, they are most often direct reading gauges. But the Bourdon tube sensing device can also be used remotely. Regardless, it is necessary to direct the fluid to be measured into the Bourdon tube. For example, a common direct-reading gauge measuring engine oil pressure and indicating it to the pilot in the flight deck is mounted in the instrument panel. A length of small tube connects a pressurized oil port on the engine, runs though the firewall, and into the back of the gauge. This setup is especially functional on light, single-engine aircraft in which the engine is mounted just forward of the instrument panel in the forward end of the fuselage. However, a remote sensing unit can be more practical on twin-engine aircraft where the engines are a long distance from the flight deck pressure display. Here, the Bourdon tube's motion is converted to an electrical signal and carried to the flight deck display via a wire. This is lighter and more efficient, eliminating the possibility of leaking fluids into the passenger compartment of the aircraft.

The diaphragm and bellows are two other basic sensing mechanisms employed in aircraft instruments for pressure measurement. They are most often used to measure relatively low pressures. The diaphragm is a hollow, thin-walled metal disc, usually corrugated. When pressure is introduced through an opening on one side of the disc, the entire disc expands. By placing linkage in contact against the other side of the disc, the movement of the pressurized diaphragm can be transferred to a pointer that registers the movement against the scale on the instrument face. *[Figure 10-11]*

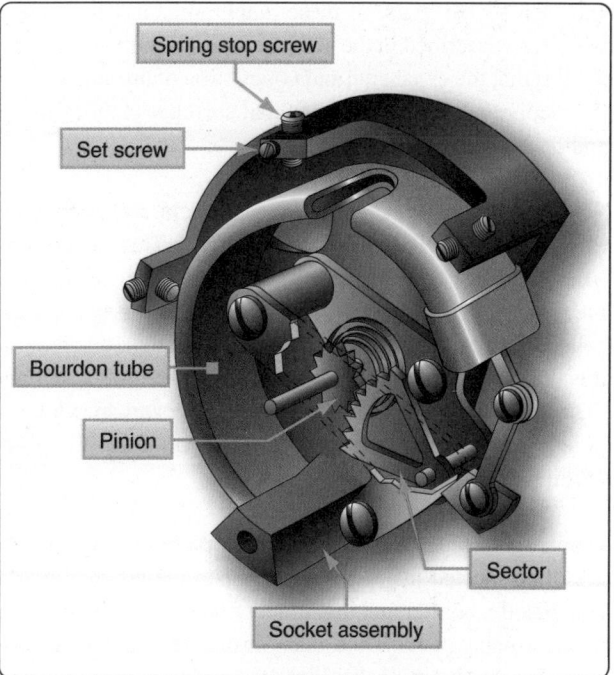

Figure 10-10. *The Bourdon tube mechanism can be used to measure pressure or temperature by recalibrating the pointer's connecting linkage and scaling instrument face to read in degrees Celsius or Fahrenheit.*

Diaphragms can also be sealed. The diaphragm can be evacuated before sealing, retaining absolutely nothing inside. When this is done, the diaphragm is called an aneroid. Aneroids are used in many flight instruments. A diaphragm can also be filled with a gas to standard atmospheric pressure

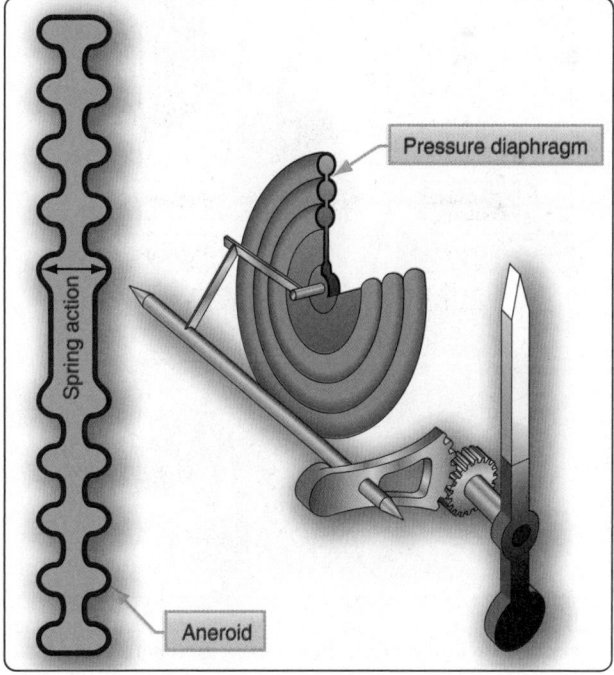

Figure 10-11. *A diaphragm used for measuring pressure. An evacuated sealed diaphragm is called an aneroid.*

and then sealed. Each of these diaphragms has their uses, which are described in the next section. The common factor in all is that the expansion and contraction of the side wall of the diaphragm is the movement that correlates to increasing and decreasing pressure.

When a number of diaphragm chambers are connected together, the device is called a bellows. This accordion-like assembly of diaphragms can be very useful when measuring the difference in pressure between two gases, called differential pressure. Just as with a single diaphragm, it is the movement of the side walls of the bellows assembly that correlates with changes in pressure and to which a pointer linkage and gearing is attached to inform the pilot. [Figure 10-12]

Diaphragms, aneroids, and bellows pressure sensing devices are often located inside the single instrument housing that contains the pointer and instrument dial read by the pilot on the instrument panel. Thus, many instruments that make use of these sensitive and reliable mechanisms are direct reading gauges. But, many remote sensing instrument systems also make use of the diaphragm and bellows. In this case, the sensing device containing the pressure sensitive diaphragm or bellows is located remotely on the engine or airframe. It is part of a transducer that converts the pressure into an electrical signal. The transducer, or transmitter, sends the

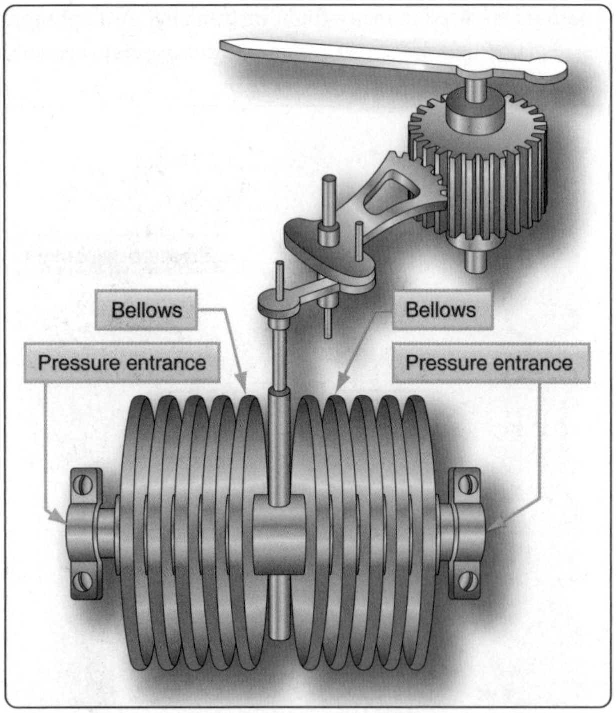

Figure 10-12. *A bellows unit in a differential pressure gauge compares two different pressure values. End movement of the bellows away from the side with the highest pressure input occurs when the pressures in the bellows are not equal. The indicator linkage is calibrated to display the difference.*

signal to the gauge in the flight deck, or to a computer, for processing and subsequent display of the sensed condition. Examples of instruments that use a diaphragm or bellows in a direct reading or remote sensing gauge are the altimeter, vertical speed indicator, cabin differential pressure gauge (in pressurized aircraft), and manifold pressure gauge.

Solid-state microtechnology pressure sensors are used in modern aircraft to determine the critical pressures needed for safe operation. Many of these have digital output ready for processing by electronic flight instrument computers and other onboard computers. Some sensors produce small electric signals that are converted to digital format for use by computers. As with the analog sensors described above, the key to the function of solid-state sensors is their consistent property changes as pressure changes.

The solid-state sensors used in most aviation applications exhibit varying electrical output or resistance changes when pressure changes occur. Crystalline piezoelectric, piezoresistor, and semiconductor chip sensors are most common. In the typical sensor, tiny wires are embedded in the crystal or pressure-sensitive semiconductor chip. When pressure deflects the crystal(s), a small amount of electricity is created or, in the case of a semiconductor chip and some crystals, the resistance changes. Since the current and resistance changes vary directly with the amount of deflection, outputs can be calibrated and used to display pressure values.

Nearly all of the pressure information needed for engine, airframe, and flight instruments can be captured and/or calculated through the use of solid-state pressure sensors in combination with temperature sensors. But continued use of aneroid devices for comparisons involving absolute pressure is notable. Solid-state pressure-sensing systems are remote sensing systems. The sensors are mounted on the aircraft at convenient and effective locations.

Types of Pressure

Pressure is a comparison between two forces. Absolute pressure exists when a force is compared to a total vacuum, or absolutely no pressure. It is necessary to define absolute pressure, because the air in the atmosphere is always exerting pressure on everything. Even when it seems there is no pressure being applied, like when a balloon is deflated, there is still atmospheric pressure inside and outside of the balloon. To measure that atmospheric pressure, it is necessary to compare it to a total absence of pressure, such as in a vacuum. Many aircraft instruments make use of absolute pressure values, such as the altimeter, the rate-of-climb indicator, and the manifold pressure gauge. As stated, this is usually done with an aneroid.

The most common type of pressure measurement is gauge pressure. This is the difference between the pressure to be measured and the atmospheric pressure. The gauge pressure inside the deflated balloon is therefore 0 pounds per square inch (psi) because the pressure inside the balloon is equal to the pressure outside the balloon. Gauge pressure is easily measured and is obtained by ignoring the fact that the atmosphere is always exerting its pressure on everything. For example, a tire is filled with air to 32 psi at a sea level location and checked with a gauge to read 32 psi, which is the gauge pressure. The approximately 14.7 psi of air pressing on the outside of the tire is ignored. The absolute pressure in the tire is 32 psi plus the 14.7 psi that is needed to balance the 14.7 psi on the outside of the tire. So, the tire's absolute pressure is approximately 46.7 psi. If the same tire is inflated to 32 psi at a location 10,000 feet above sea level, the air pressure on the outside of the tire would only be approximately 10 psi, due to the thinner atmosphere. The pressure inside the tire required to balance this would be 32 psi plus 10 psi, making the absolute pressure of the tire 42 psi. So, the same tire with the same amount of inflation and performance characteristics has different absolute pressure values. Gauge pressure, however, remains the same, indicating the tires are inflated identically. It this case, gauge pressure is more useful in informing us of the condition of the tire.

Gauge pressure measurements are simple and widely useful. They eliminate the need to measure varying atmospheric pressure to indicate or monitor a particular pressure situation. Gauge pressure should be assumed, unless otherwise indicated, or unless the pressure measurement is of a type known to require absolute pressure.

In many instances in aviation, it is desirable to compare the pressures of two different elements to arrive at useful information for operating the aircraft. When two pressures are compared in a gauge, the measurement is known as differential pressure and the gauge is a differential pressure gauge. An aircraft's airspeed indicator is a differential pressure gauge. It compares ambient air pressure with ram air pressure to determine how fast the aircraft is moving through the air. A turbine's engine pressure ratio (EPR) gauge is also a differential pressure gauge. It compares the pressure at the inlet of the engine with that at the outlet to indicate the thrust developed by the engine. Both of these differential pressure gauges and others are discussed further in this chapter and throughout this handbook.

In aviation, there is also a commonly used pressure known as standard pressure. Standard pressure refers to an established or standard value that has been created for atmospheric pressure. This standard pressure value is 29.92 inches of mercury ("Hg), 1,013.2 hectopascal (hPa), or 14.7 psi. It is

part of a standard day that has been established that includes a standard temperature of 15 °C at sea level. Specific standard day values have also been established for air density, volume, and viscosity. All of these values are developed averages since the atmosphere is continuously fluctuating. They are used by engineers when designing instrument systems and are sometimes used by technicians and pilots. Often, using a standard value for atmospheric pressure is more desirable than using the actual value. For example, at 18,000 feet and above, all aircraft use 29.92 "Hg as a reference pressure for their instruments to indicate altitude. This results in altitude indications in all flight decks being identical. Therefore, an accurate means is established for maintaining vertical separation of aircraft flying at these high altitudes.

Pressure Instruments
Engine Oil Pressure

The most important instrument used by the pilot to perceive the health of an engine is the engine oil pressure gauge. *[Figure 10-13]* Oil pressure is usually indicated in pounds per square inch (psi). The normal operating range is typically represented by a green arc on the circular gauge. For exact acceptable operating range, consult the manufacturer's operating and maintenance data.

In reciprocating and turbine engines, oil is used to lubricate and cool bearing surfaces where parts are rotating or sliding past each other at high speeds. A loss of pressurized oil to these areas would rapidly cause excessive friction and over-temperature conditions, leading to catastrophic engine

Figure 10-13. *An analog oil pressure gauge is driven by a Bourdon tube. Oil pressure is vital to engine health and must be monitored by the pilot.*

failure. As mentioned, aircraft using analog instruments often use direct reading Bourdon tube oil pressure gauges. *Figure 10-13* shows the instrument face of a typical oil pressure gauge of this type. Digital instrument systems use an analog or digital remote oil pressure sensing unit that sends output to the computer, driving the display of oil pressure value(s) on the aircraft's flight deck display screens. Oil pressure may be displayed in a circular or linear gauge fashion and may even include a numerical value on screen. Often, oil pressure is grouped with other engine parameter displays on the same page or portion of a page on the display. *Figure 10-14* shows this grouping on a Garmin G1000 digital instrument display system for general aviation aircraft.

Manifold Pressure

In reciprocating engine aircraft, the manifold pressure gauge indicates the pressure of the air in the engine's induction manifold. This is an indication of power being developed by the engine. The higher the pressure of the fuel air mixture going into the engine, the more power it can produce. For normally aspirated engines, maximum manifold pressure would be slightly less than the ambient atmospheric pressure.

Turbocharged or supercharged engines pressurize the air being mixed with the fuel, so full power indications are above the ambient atmospheric pressure.

Most manifold pressure gauges are calibrated in inches of mercury, although digital displays may have the option to display in a different scale. A typical analog gauge makes use of an aneroid described above. When atmospheric pressure acts on the aneroid inside the gauge, the connected pointer indicates the current air pressure. A line running from the intake manifold into the gauge presents intake manifold air pressure to the aneroid, so the gauge indicates the absolute pressure in the intake manifold. An analog manifold pressure gauge, along with its internal workings, is shown in *Figure 10-15*. The digital presentation of manifold pressure is at the top of the engine instruments displayed on the Garmin G1000 multifunctional display in *Figure 10-14*. The aircraft's operating manual contains data on managing manifold pressure in relation to fuel flow and propeller pitch and for achieving various performance profiles during different phases of run-up and flight.

Figure 10-14. *Oil pressure indication with other engine-related parameters shown in a column on the left side of this digital flight deck display panel.*

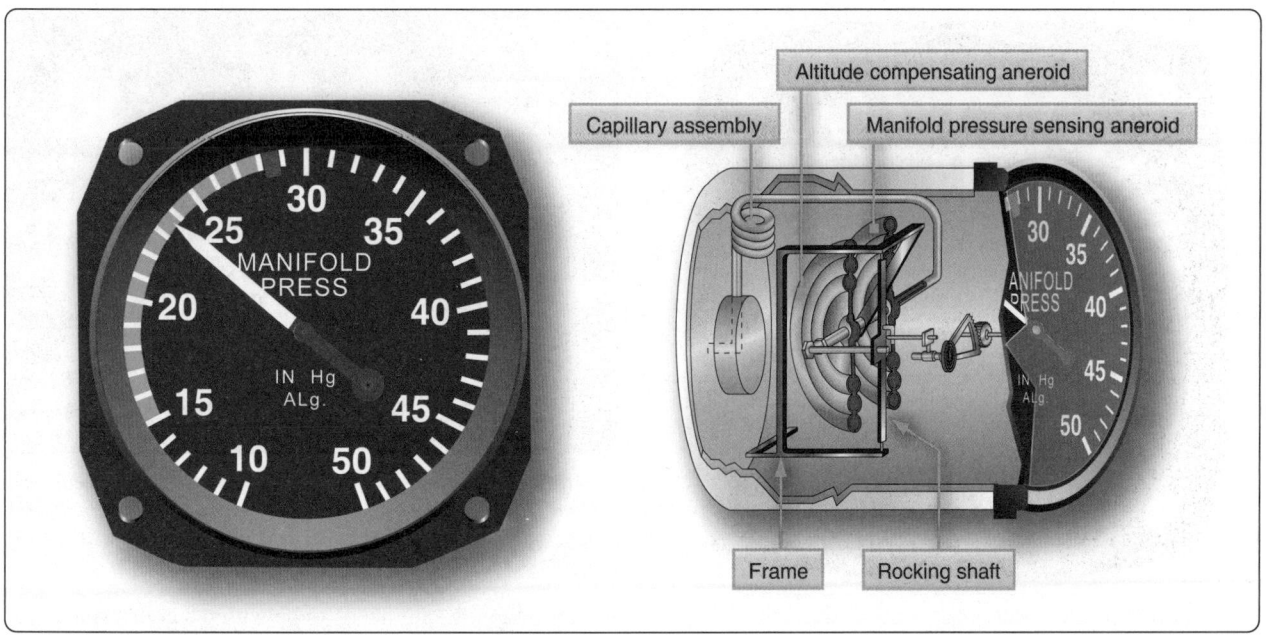

Figure 10-15. *An analog manifold pressure indicator instrument dial calibrated in inches of mercury (left). The internal workings of an analog manifold pressure gauge are shown on the right. Air from the intake manifold surrounds the aneroid causing it to deflect and indicate pressure on the dial through the use of linkage to the pointer (right).*

Engine Pressure Ratio (EPR)

Turbine engines have their own pressure indication that relates the power being developed by the engine. It is called the engine pressure ratio (EPR) indicator (EPR gauge). This gauge compares the total exhaust pressure to the pressure of the ram air at the inlet of the engine. With adjustments for temperature, altitude, and other factors, the EPR gauge presents an indication of the thrust being developed by the engine. Since the EPR gauge compares two pressures, it is a differential pressure gauge. It is a remote-sensing instrument that receives its input from an engine pressure ratio transmitter or, in digital instrument systems displays, from a computer. The pressure ratio transmitter contains the bellows arrangement that compares the two pressures and converts the ratio into an electric signal used by the gauge for indication. *[Figure 10-16]*

Fuel Pressure

Fuel pressure gauges also provide critical information to the pilot. *[Figure 10-17]* Typically, fuel is pumped out of various fuel tanks on the aircraft for use by the engines. A malfunctioning fuel pump, or a tank that has been emptied beyond the point at which there is sufficient fuel entering the pump to maintain desired output pressure, is a condition that requires the pilot's immediate attention. While direct-sensing fuel pressure gauges using Bourdon tubes, diaphragms,

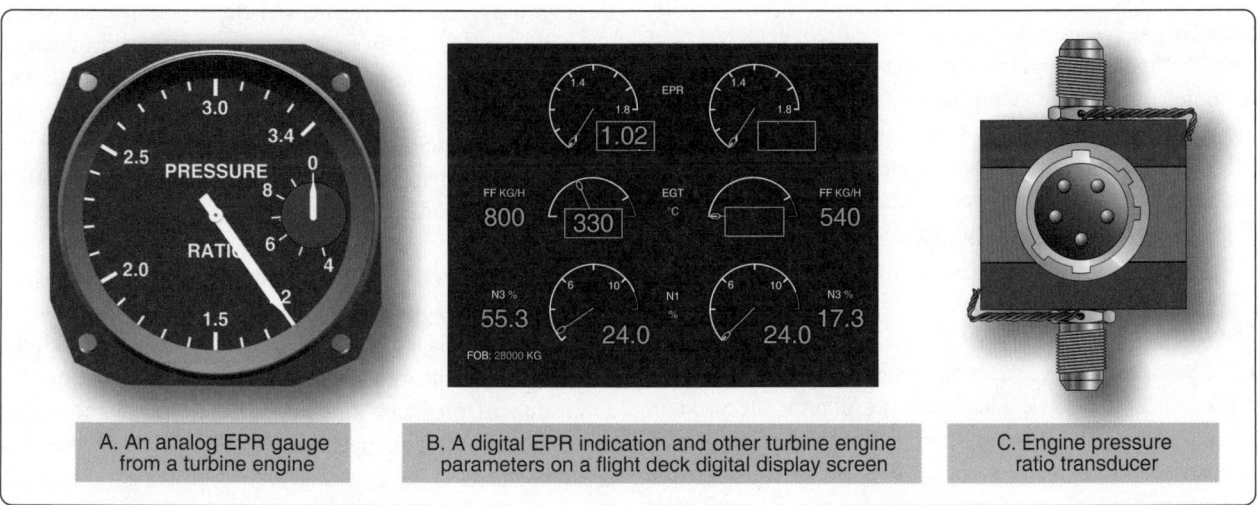

A. An analog EPR gauge from a turbine engine

B. A digital EPR indication and other turbine engine parameters on a flight deck digital display screen

C. Engine pressure ratio transducer

Figure 10-16. *Engine pressure ratio gauges.*

Figure 10-17. *A typical analog fuel pressure gauge.*

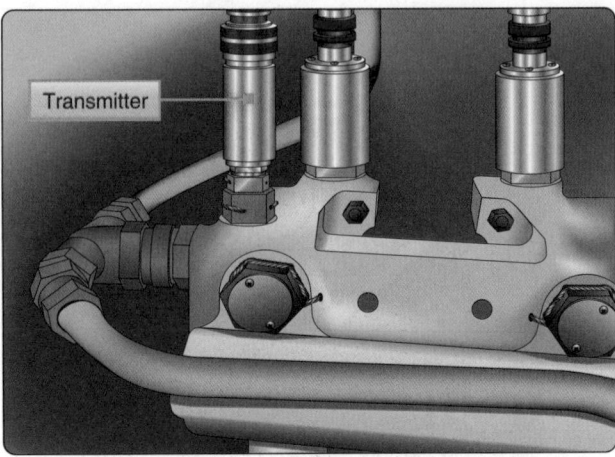

Figure 10-18. *A hydraulic pressure transmitter senses and converts pressure into an electrical output for indication by the flight deck gauge or for use by a computer that analyzes and displays the pressure in the flight deck when requested or required.*

and bellows sensing arrangements exist, it is particularly undesirable to run a fuel line into the flight deck, due to the potential for fire should a leak develop. Therefore, the preferred arrangement is to have whichever sensing mechanism that is used be part of a transmitter device that uses electricity to send a signal to the indicator in the flight deck. Sometimes, indications monitoring the fuel flow rate are used instead of fuel pressure gauges. Fuel flow indications are discussed in the fuel system chapter of this handbook.

Hydraulic Pressure

Numerous other pressure monitoring gauges are used on complex aircraft to indicate the condition of various support systems not found on simple light aircraft. Hydraulic systems are commonly used to raise and lower landing gear, operate flight controls, apply brakes, and more. Sufficient pressure in the hydraulic system developed by the hydraulic pump(s) is required for normal operation of hydraulic devices. Hydraulic pressure gauges are often located in the flight deck and at or near the hydraulic system servicing point on the airframe. Remotely located indicators used by maintenance personnel are almost always direct reading Bourdon tube type gauges. Flight deck gauges usually have system pressure transmitted from sensors or computers electrically for indication. *Figure 10-18* shows a hydraulic pressure transmitter in place in a high-pressure aircraft hydraulic system.

Vacuum Pressure

Gyro pressure gauge, vacuum gauge, or suction gauge are all terms for the same gauge used to monitor the vacuum developed in the system that actuates the air driven gyroscopic flight instruments. Air is pulled through the instruments, causing the gyroscopes to spin. The speed at

which the gyros spin needs to be within a certain range for correct operation. This speed is directly related to the suction pressure that is developed in the system. The suction gauge is extremely important in aircraft relying solely on vacuum-operated gyroscopic flight instruments.

Vacuum is a differential pressure indication, meaning the pressure to be measured is compared to atmospheric pressure through the use of a sealed diaphragm or capsule. The gauge is calibrated in inches of mercury. It shows how much less pressure exists in the system than in the atmosphere. *Figure 10-19* shows a suction gauge calibrated in inches of mercury.

Pressure Switches

In aviation, it is often sufficient to simply monitor whether the pressure developed by a certain operating system is too high or too low, so that an action can take place should one of these conditions occur. This is often accomplished through the use of a pressure switch. A pressure switch is a simple device usually made to open or close an electric circuit when a certain pressure is reached in a system. It can be manufactured so that the electric circuit is normally open and can then close when a certain pressure is sensed, or the circuit can be closed and then opened when the activation pressure is reached. [*Figure 10-20*]

Pressure switches contain a diaphragm to which the pressure being sensed is applied on one side. The opposite side of the diaphragm is connected to a mechanical switching mechanism for an electric circuit. Small fluctuations or a buildup of pressure against the diaphragm move the diaphragm, but not enough to throw the switch. Only when pressure meets or exceeds a preset level designed into the structure of the switch does the diaphragm move far enough

Figure 10-19. *Vacuum suction gauge.*

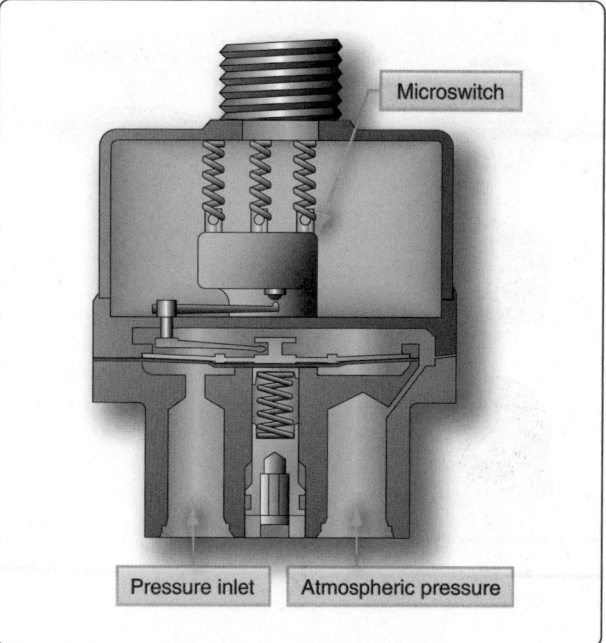

Figure 10-21. *A normally open pressure switch positioned in an electrical circuit causes the circuit to be open as well. The switch closes, allowing electricity to flow when pressure is applied beyond the switch's preset activation point. Normally, closed pressure switches allow electricity to flow through the switch in a circuit but open when pressure reaches a preset activation point, thus opening the electrical circuit.*

given in the flight deck. Should a loss of oil pressure occur, the pressure against the diaphragm becomes insufficient to hold the switched contacts open. When the contacts close, they close the circuit to the low oil pressure indicator, usually a light, to warn the pilot of the situation.

Pressure gauges for various components or systems work similarly to those mentioned above. Some sort of sensing device, appropriate for the pressure being measured or monitored, is matched with an indicating display system. If appropriate, a properly rated pressure switch is installed in the system and wired into an indicating circuit. Further discussion of specific instruments occurs throughout this handbook as the operation of various systems and components are discussed.

Pitot-Static Systems

Some of the most important flight instruments derive their indications from measuring air pressure. Gathering and distributing various air pressures for flight instrumentation is the function of the pitot-static system.

Pitot Tubes & Static Vents

On simple aircraft, this may consist of a pitot-static system head or pitot tube with impact and static air pressure ports and leak-free tubing connecting these air pressure pick-up points to the instruments that require the air for their

Figure 10-20. *A pressure switch can be used in addition to, or instead of, a pressure gauge.*

for the mechanical device on the opposite side to close the switch contacts and complete the circuit. *[Figure 10-21]* Each switch is rated to close (or open) at a certain pressure and must only be installed in the proper location.

A low oil pressure indication switch is a common example of how pressure switches are employed. It is installed in an engine so pressurized oil can be applied to the switch's diaphragm. Upon starting the engine, oil pressure increases and the pressure against the diaphragm is sufficient to hold the contacts in the switch open. As such, current does not flow through the circuit and no indication of low oil pressure is

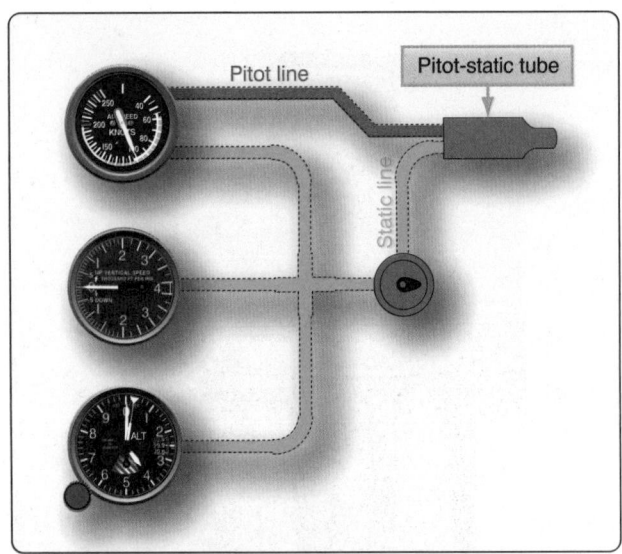

Figure 10-22. *A simple pitot-static system is connected to the primary flight instruments.*

pressure system. The altimeter is connected to the static pressure system.

A pitot tube is open and faces into the airstream to receive the full force of the impact air pressure as the aircraft moves forward. This air passes through a baffled plate designed to protect the system from moisture and dirt entering the tube. Below the baffle, a drain hole is provided, allowing moisture to escape. The ram air is directed aft to a chamber in the shark fin of the assembly. An upright tube, or riser, leads this pressurized air out of the pitot assemble to the airspeed indicator.

The aft section of the pitot tube is equipped with small holes on the top and bottom surfaces that are designed to collect air pressure that is at atmospheric pressure in a static, or still, condition. The static section also contains a riser tube and the air is run out the pitot assembly through tubes and is connected to the altimeter, the airspeed indicator, and the vertical speed indicator. *[Figure 10-23]*

indications. The altimeter, airspeed indicator, and vertical speed indicator are the three most common pitot-static instruments. *Figure 10-22* illustrates a simple pitot-static system connected to these three instruments. All three instruments are connected to the static pressure system. The airspeed indicator is additionally connected to the pitot

Many pitot-static tube heads contain heating elements to prevent icing during flight. The pilot can send electric current to the element with a switch in the flight deck when ice-forming conditions exist. The pitot tube heat switch may be wired so that when the ignition switch is turned off when

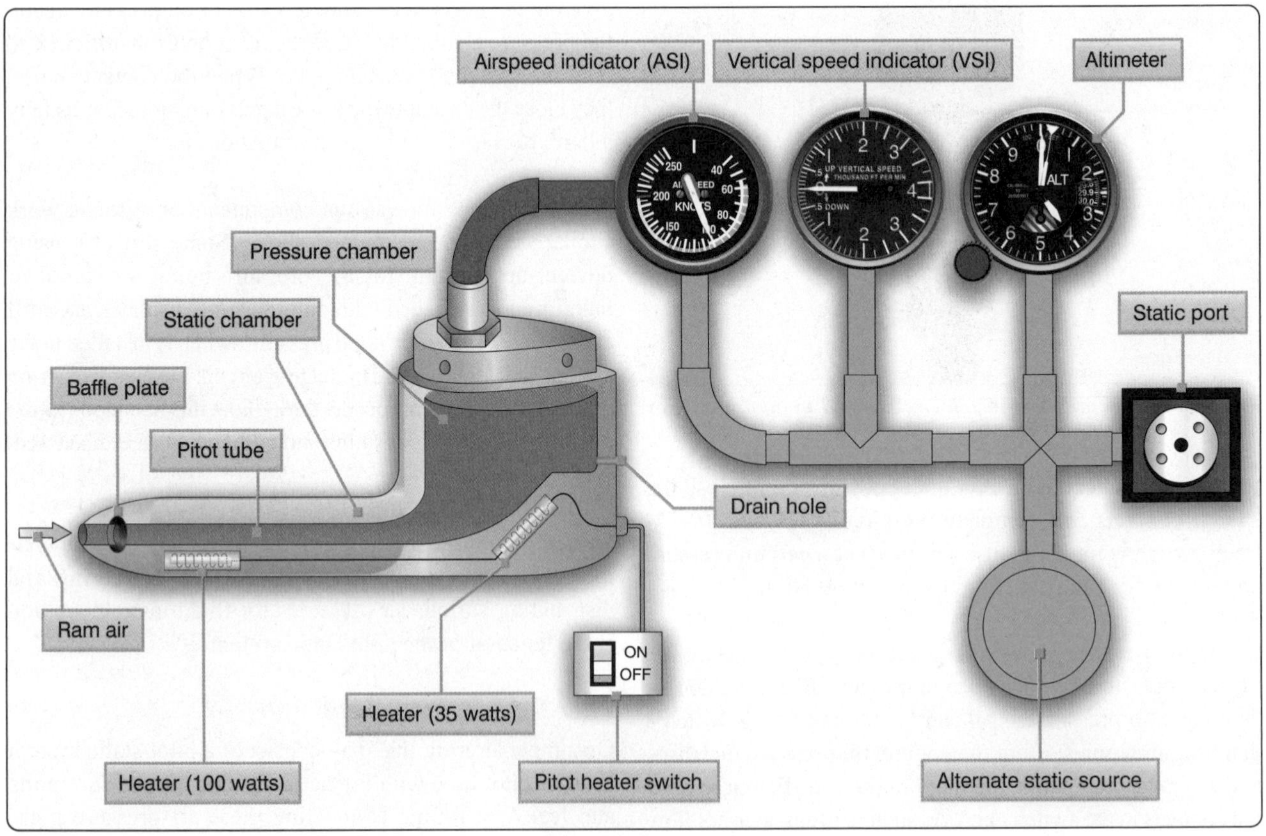

Figure 10-23. *A typical pitot-static system head, or pitot tube, collects ram air and static pressure for use by the flight instruments.*

the aircraft is shut down, a pitot tube heater inadvertently left on does not continue to draw current and drain the battery. Caution should be exercised when near the pitot tube, as these heating elements make the tube too hot to be touched without receiving a burn.

The pitot-static tube is mounted on the outside of the aircraft at a point where the air is least likely to be turbulent. It is pointed in a forward direction parallel to the aircraft's line of flight. The location may vary. Some are on the nose of the fuselage and others may be located on a wing. A few may even be found on the empennage. Various designs exist but the function remains the same, to capture impact air pressure and static air pressure and direct them to the proper instruments. *[Figure 10-24]*

Most aircraft equipped with a pitot-static tube have an alternate source of static air pressure provided for emergency use. The pilot may select the alternate with a switch in the flight deck should it appear the flight instruments are not providing accurate indications. On low-flying unpressurized aircraft, the alternate static source may simply be air from the cabin. *[Figure 10-25]* On pressurized aircraft, cabin air pressure may be significantly different than the outside ambient air pressure. If used as an alternate source for static air, instrument indications would be grossly inaccurate. In this case, multiple static vent pickup points are employed. All are located on the outside of the aircraft and plumbed so the pilot can select which source directs air into the instruments.

Figure 10-25. *On unpressurized aircraft, an alternate source of static air is cabin air.*

On electronic flight displays, the choice is made for which source is used by the computer or by the flight crew.

Another type of pitot-static system provides for the location of the pitot and static sources at separate positions on the aircraft. The pitot tube in this arrangement is used only to gather ram air pressure. Separate static vents are used to collect static air pressure information. Usually, these are located flush on the side of the fuselage. *[Figure 10-26]* There may be two or more vents. A primary and alternate source vent is typical, as well as separate dedicated vents for the pilot and first officer's instruments. Also, two primary vents may be located on opposite sides of the fuselage and connected with Y tubing for input to the instruments. This is done to compensate for any variations in static air pressure on the vents due to the aircraft's attitude. Regardless of the number and location of separate static vents, they may be heated as well as the separate ram air pitot tube to prevent icing.

Figure 10-24. *Pitot-static system heads, or pitot tubes, can be of various designs and locations on airframes.*

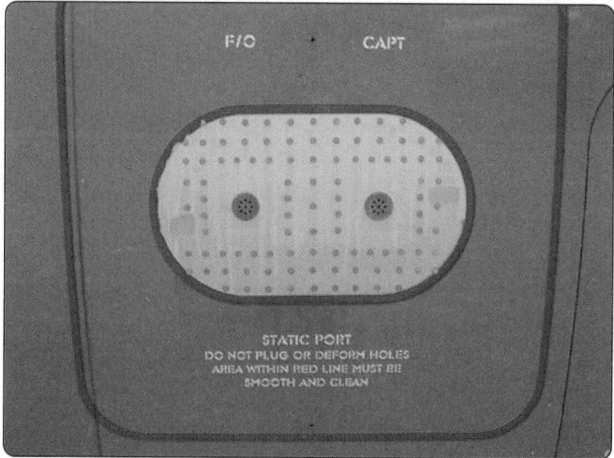

Figure 10-26. *Heated primary and alternate static vents located on the sides of the fuselage.*

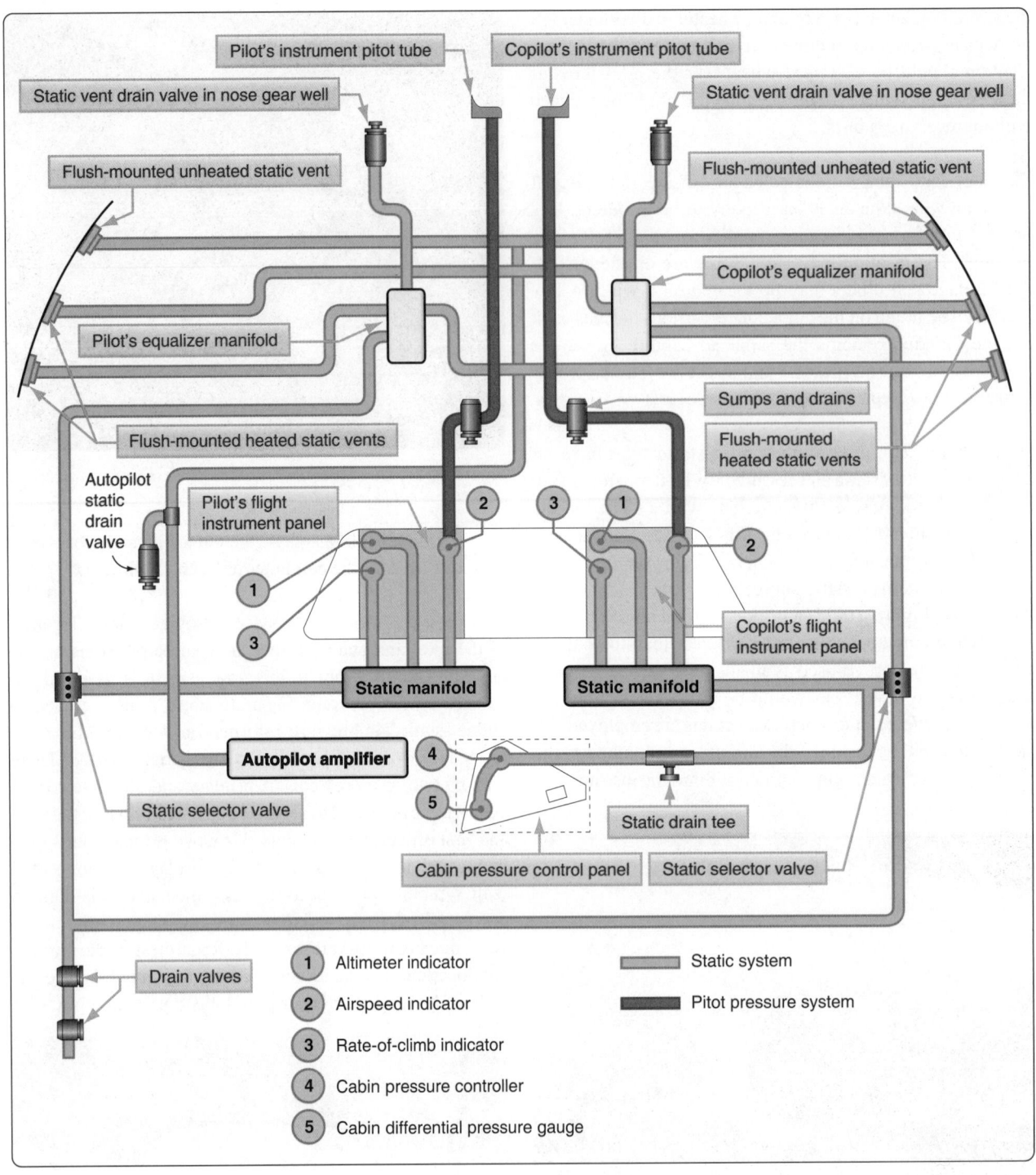

Figure 10-27. *Schematic of a typical pitot-static system on a pressurized multiengine aircraft.*

The pitot-static systems of complex, multiengine, and pressurized aircraft can be elaborate. Additional instruments, gauges, the autopilot system, and computers may need pitot and static air information. *Figure 10-27* shows a pitot-static system for a pressurized multiengine aircraft with dual analog instrument panels in the flight deck. The additional set of flight instruments for the copilot alters and complicates the pitot-static system plumbing. Additionally, the autopilot system requires static pressure information, as does the cabin pressurization unit. Separate heated sources for static air pressure are taken from both sides of the airframe to feed independent static air pressure manifolds; one each for the pilot's flight instruments and the copilot's flight instruments. This is designed to ensure that there is always one set of flight instruments operable in case of a malfunction.

Air Data Computers (ADC) & Digital Air Data Computers (DADC)

High performance and jet transport category aircraft pitot-static systems may be more complicated. These aircraft frequently operate at high altitude where the ambient temperature can exceed 50 °F below zero. The compressibility of air is also altered at high speeds and at high altitudes. Airflow around the fuselage changes, making it difficult to pick up consistent static pressure inputs. The pilot must compensate for all factors of air temperature and density to obtain accurate indications from instruments. While many analog instruments have compensating devices built into them, the use of an air data computer (ADC) is common for these purposes on high-performance aircraft. Moreover, modern aircraft utilize digital air data computers (DADC). The conversion of sensed air pressures into digital values makes them more easily manipulated by the computer to output accurate information that has compensated for the many variables encountered. *[Figure 10-28]*

Essentially, all pressures and temperatures captured by sensors are fed into the ADC. Analog units utilize transducers to convert these to electrical values and manipulate them in various modules containing circuits designed to make the proper compensations for use by different instruments and systems. A DADC usually receives its data in digital format. Systems that do not have digital sensor outputs will first convert inputs into digital signals via an analog-to-digital converter. Conversion can take place inside the computer or in a separate unit designed for this function. Then, all calculation and compensations are performed digitally by the computer. Outputs from the ADC are electric to drive servo motors or for use as inputs in pressurization systems, flight control units, and other systems. DADC outputs are distributed to these same systems and the flight deck display using a digital data bus.

There are numerous benefits of using ADCs. Simplification of pitot-static plumbing lines creates a lighter, simpler, system with fewer connections, so it is less prone to leaks and easier to maintain. One-time compensation calculations can be done inside the computer, eliminating the need to build compensating devices into numerous individual instruments or units of the systems using the air data. DADCs can run a number of checks to verify the plausibility of data received from any source on the aircraft. Thus, the crew can be alerted automatically of a parameter that is out of the ordinary. Change to an alternate data source can also be automatic so accurate flight deck and systems operations are continuously maintained. In general, solid-state technology is more reliable and modern units are small and lightweight. *Figure 10-29* shows a schematic of how a DADC is connected into the aircraft's pitot-static and other systems.

Pitot-Static Pressure-Sensing Flight Instruments

The basic flight instruments are directly connected to the pitot-static system on many aircraft. Analog flight instruments primarily use mechanical means to measure and indicate various flight parameters. Digital flight instrument systems use electricity and electronics to do the same. Discussion of the basic pitot-static flight instruments begins with analog instruments to which further information about modern digital instrumentation is added.

Altimeters & Altitude

An altimeter is an instrument that is used to indicate the height of the aircraft above a predetermined level, such as sea level or in the case of a radio/radar altimeter, the height of terrain beneath the aircraft. The most common way to measure this distance is rooted in discoveries made by scientists centuries ago. Seventeenth century work proving that the air in the atmosphere exerted pressure on the things around us led Evangelista Torricelli to the invention of the barometer. Also in that century, using the concept of this first atmospheric air pressure measuring instrument, Blaise Pascal was able to show that a relationship exists between altitude and air pressure. As altitude increases, air pressure decreases. The amount that it decreases is measurable and consistent for any given altitude change. Therefore, by measuring air pressure, altitude can be determined. *[Figure 10-30]*

Altimeters that measure the aircraft's altitude by measuring the pressure of the atmospheric air are known as pressure altimeters. A pressure altimeter is made to measure the ambient air pressure at any given location and altitude. In aircraft, it is connected to the static vent(s) via tubing in the pitot-static system. The relationship between the measured

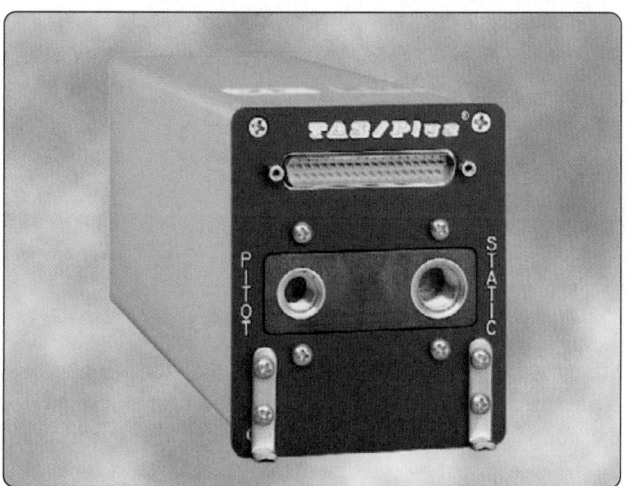

Figure 10-28. *Teledyne's 90004 TAS/Plus air data computer (ADC) computes air data information from the pitot-static pneumatic system, aircraft temperature probe, and barometric correction device to help create a clear indication of flight conditions.*

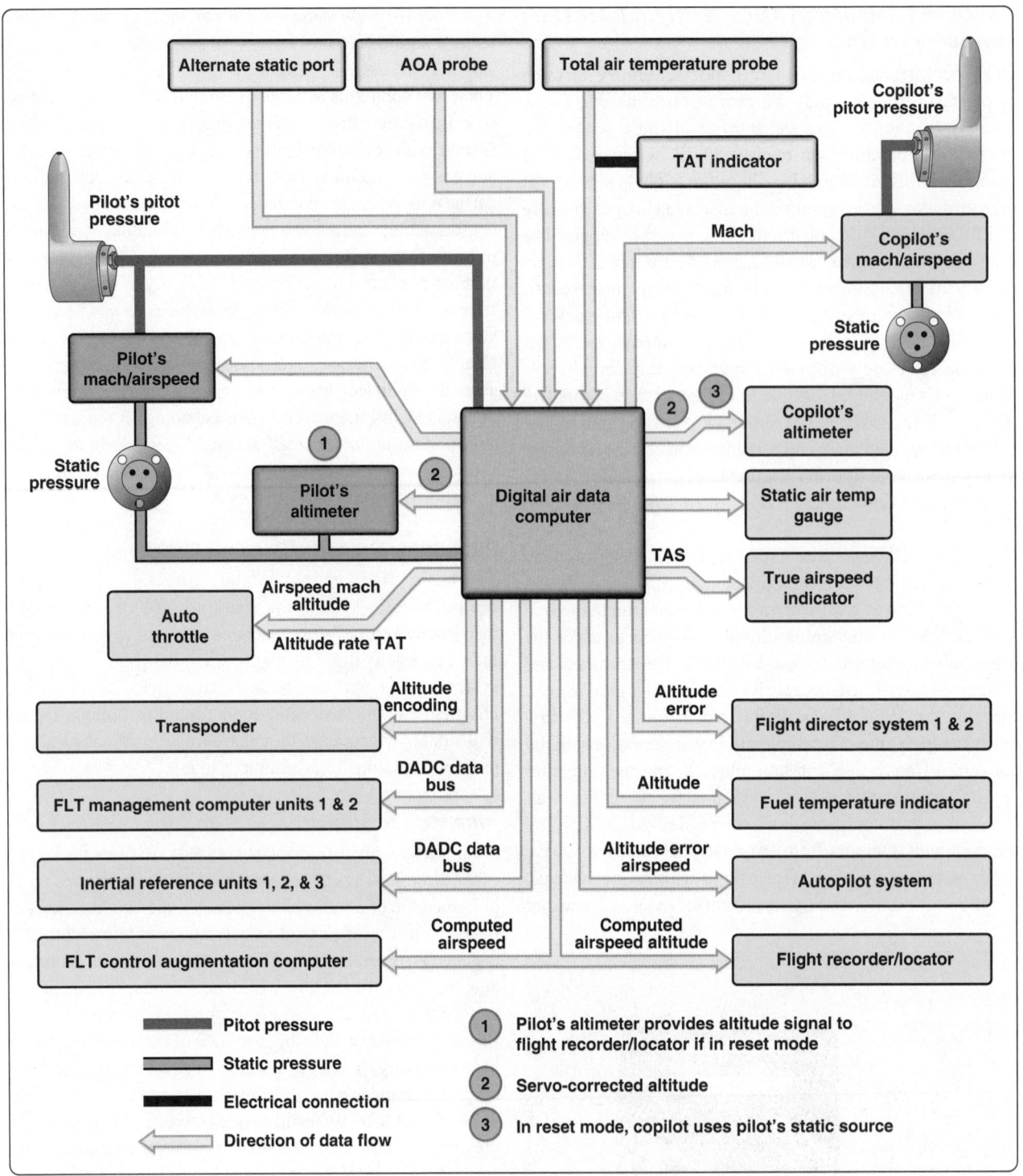

Figure 10-29. *ADCs receive input from the pitot-static sensing devices and process them for use by numerous aircraft systems.*

pressure and the altitude is indicated on the instrument face, which is calibrated in feet. These devices are direct-reading instruments that measure absolute pressure. An aneroid or aneroid bellows is at the core of the pressure altimeter's inner workings. Attached to this sealed diaphragm are the linkages and gears that connect it to the indicating pointer. Static air pressure enters the airtight instrument case and surrounds the aneroid. At sea level, the altimeter indicates zero when

this pressure is exerted by the ambient air on the aneroid. As air pressure is reduced by moving the altimeter higher in the atmosphere, the aneroid expands and displays altitude on the instrument by rotating the pointer. As the altimeter is lowered in the atmosphere, the air pressure around the aneroid increases and the pointer moves in the opposite direction. *[Figure 10-31]*

Atmosphere pressure	
Altitude (ft)	Pressure (psi)
Sea level	14.69
2,000	13.66
4,000	12.69
6,000	11.77
8,000	10.91
10,000	10.10
12,000	9.34
14,000	8.63
16,000	7.96
18,000	7.34
20,000	6.75
22,000	6.20
24,000	5.69
26,000	5.22
28,000	4.77
30,000	4.36
32,000	3.98
34,000	3.62
36,000	3.29
38,000	2.99
40,000	2.72
42,000	2.47
44,000	2.24
46,000	2.04
48,000	1.85
50,000	1.68

Figure 10-30. *Air pressure is inversely related to altitude. This consistent relationship is used to calibrate the pressure altimeter.*

The face, or dial, of an analog altimeter is read similarly to a clock. As the longest pointer moves around the dial, it is registering the altitude in hundreds of feet. One complete revolution of this pointer indicates 1,000 feet of altitude.

The second-longest point moves more slowly. Each time it reaches a numeral, it indicates 1,000 feet of altitude. Once around the dial for this pointer is equal to 10,000 feet. When the longest pointer travels completely around the dial one time, the second-longest point moves only the distance between two numerals—indicating 1,000 feet of altitude has been attained. If so equipped, a third, shortest or thinnest pointer registers altitude in 10,000 foot increments. When this pointer reaches a numeral, 10,000 feet of altitude has been attained. Sometimes a black-and-white or red-and-white cross-hatched area is shown on the face on the instrument until the 10,000 foot level has been reached. *[Figure 10-32]*

Many altimeters also contain linkages that rotate a numerical counter in addition to moving pointers around the dial. This quick reference window allows the pilot to simply read the numerical altitude in feet. The motion of the rotating digits or drum-type counter during rapid climb or descent makes it difficult or impossible to read the numbers. Reference can then be directed to the classic clock-style indication. *Figure 10-33* illustrates the inner workings behind this type of mechanical digital display of pressure altitude.

True digital instrument displays can show altitude in numerous ways. Use of a numerical display rather than a reproduction of the clock-type dial is most common. Often a

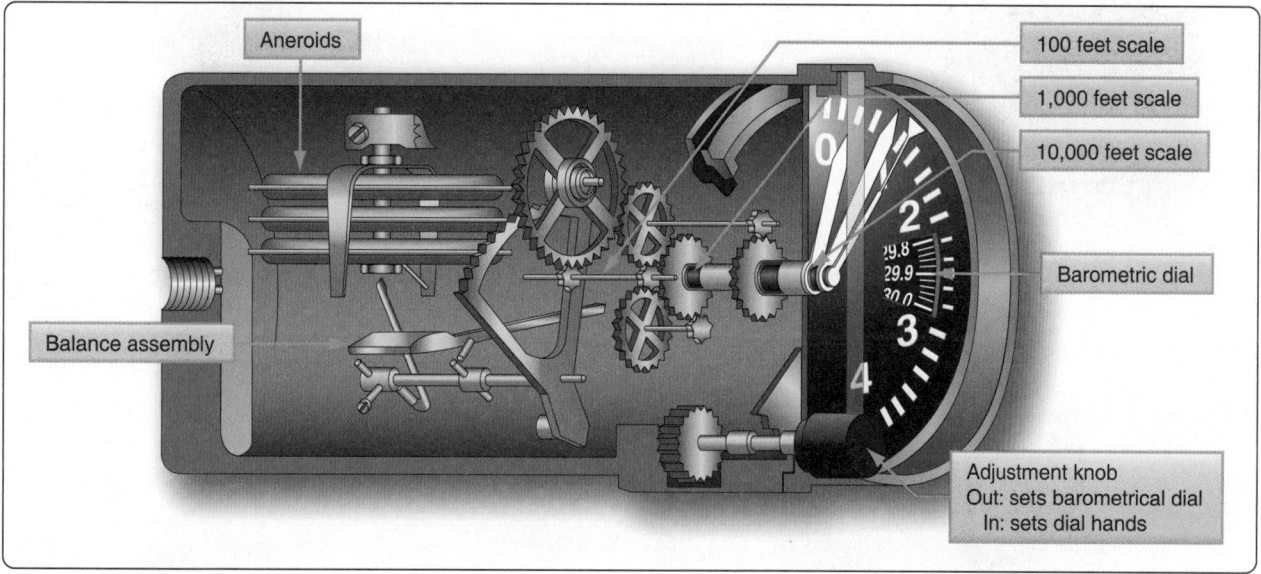

Figure 10-31. *The internal arrangement of a sealed diaphragm pressure altimeter. At sea level and standard atmospheric conditions, the linkage attached to the expandable diaphragm produces an indication of zero. When altitude increases, static pressure on the outside of the diaphragm decreases and the aneroid expands, producing a positive indication of altitude. When altitude decreases, atmospheric pressure increases. The static air pressure on the outside of the diaphragm increases and the pointer moves in the opposite direction, indicating a decrease in altitude.*

Figure 10-32. *A sensitive altimeter with three pointers and a cross-hatched area displayed during operation below 10,000 feet.*

digital numeric display of altitude is given on the electronic primary flight display near the artificial horizon depiction. A linear vertical scale may also be presented to put this hard numerical value in perspective. An example of this type of display of altitude information is shown in *Figure 10-34.*

Accurate measurement of altitude is important for numerous reasons. The importance is magnified in instrument flight rules (IFR) conditions. For example, avoidance of tall obstacles and rising terrain relies on precise altitude indication, as does flying at a prescribed altitude assigned by air traffic control (ATC) to avoid colliding with other aircraft. Measuring altitude with a pressure measuring device is fraught with complications. Steps are taken to refine pressure altitude indication to compensate for factors that may cause an inaccurate display.

A major factor that affects pressure altitude measurements is the naturally occurring pressure variations throughout the atmosphere due to weather conditions. Different air masses develop and move over the earth's surface, each with inherent pressure characteristics. These air masses cause the weather we experience, especially at the boundary areas between air masses known as fronts. Accordingly, at sea level, even if the temperature remains constant, air pressure rises and falls as weather system air masses come and go. The values in *Figure 10-30,* therefore, are averages for theoretical purposes.

To maintain altimeter accuracy despite varying atmospheric pressure, a means for setting the altimeter was devised. An adjustable pressure scale visible on the face of an analog altimeter known as a barometric or Kollsman window is set to read the existing atmospheric pressure that has been

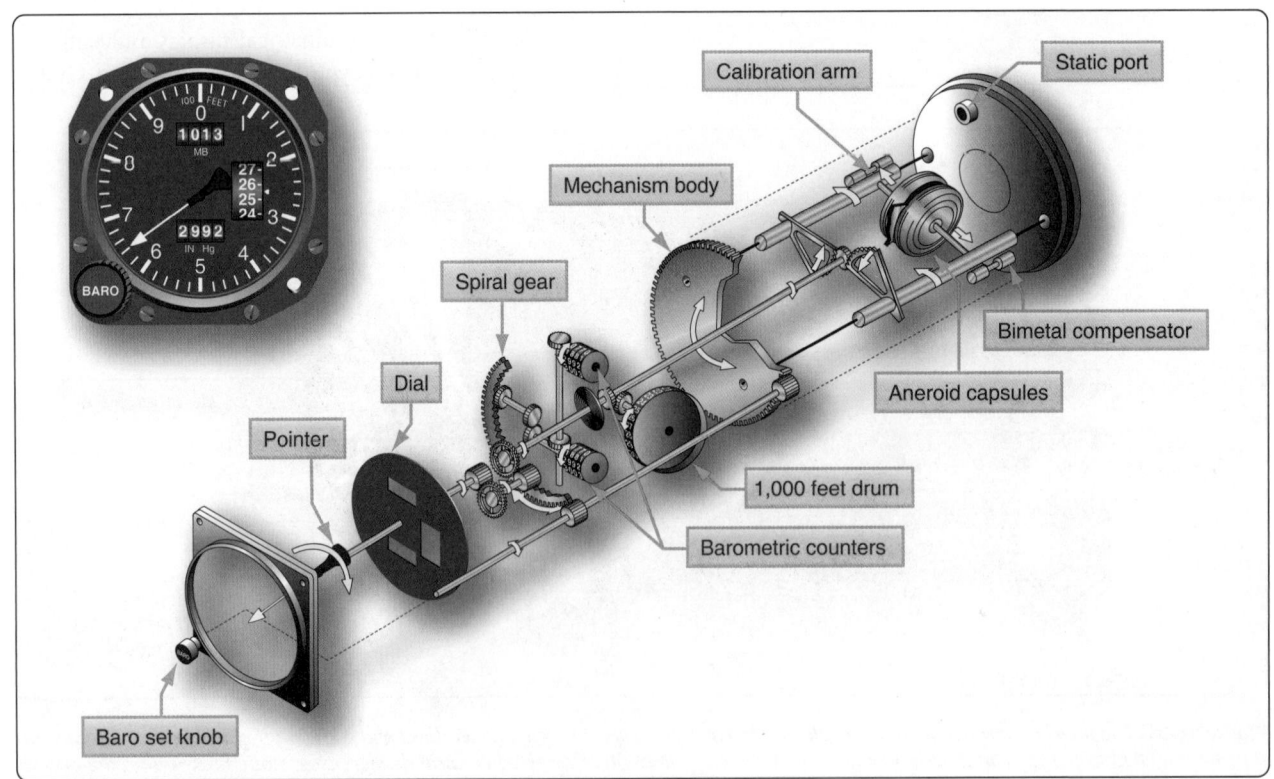

Figure 10-33. *A drum-type counter can be driven by the altimeter's aneroid for numerical display of altitude. Drums can also be used for the altimeter's setting indications.*

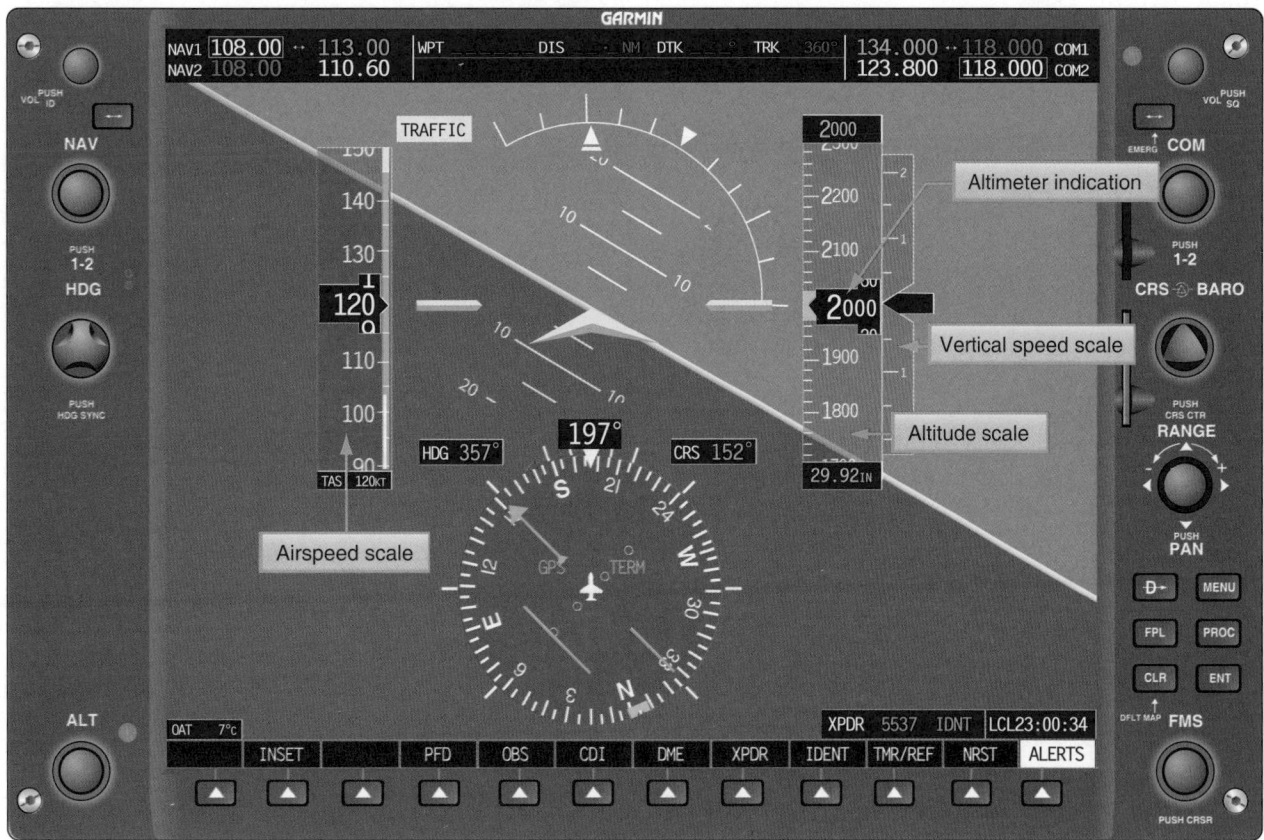

Figure 10-34. *This primary flight display unit of a Garmin 1000 series glass flight deck instrumentation package for light aircraft indicates altitude using a vertical linear scale and a numerical counter. As the aircraft climbs or descends, the scale behind the black numerical altitude readout changes.*

corrected to mean sea level (MSL). This tells the altimeter what barometric pressure is considered zero feet in altitude. The instrument will then indicate the altitude of the aircraft above mean sea level. This altitude, adjusted for atmospheric pressure changes due to weather and air mass pressure inconsistency, is known as the indicated altitude.

It must be noted that in flight below 18,000 feet, the altimeter setting is changed to match that of the closest available weather reporting station or airport. This keeps the altimeter accurate as the flight progresses.

While there was little need for exact altitude measurement in early fixed wing aviation, knowing one's altitude provided the pilot with useful references while navigating in the three dimensions of the atmosphere. As air traffic grew and the desire to fly in any weather conditions increased, exact altitude measurement became more important and the altimeter was refined. In 1928, Paul Kollsman invented the means for adjusting an altimeter to reflect variations in air pressure from standard atmospheric pressure. The very next year, Jimmy Doolittle made his successful flight demonstrating the feasibility of instrument flight with no visual references outside of the flight deck using a Kollsman sensitive altimeter.

The term "pressure altitude" is used to describe the indication an altimeter gives when 29.92 is set in the Kollsman window. When flying in U.S. airspace above 18,000 feet mean sea level (MSL), pilots are required to set their altimeters to 29.92. With all aircraft referencing this standard pressure level, vertical separation between aircraft assigned to different altitudes by ATC should be assured. This is the case if all altimeters are functioning properly and pilots hold their assigned altitudes. The actual, or true altitude, is less important than keeping aircraft from colliding, which is accomplished by all aircraft above 18,000 feet referencing the same barometric pressure (29.92 "Hg) on their altimeters. *[Figure 10-35]*

Temperature also affects the accuracy of an altimeter. The aneroid diaphragms used in altimeters are usually made of metal. Their elasticity changes as their temperature changes. This can lead to a false indication, especially at high altitudes when the ambient air is very cold. A bimetallic compensating device is built into many sensitive altimeters to correct for varying temperature. *Figure 10-33* shows one such device on a drum-type altimeter.

Temperature also affects air density, which has great impact on the performance of an aircraft. Although this does not

Figure 10-35. *Above 18,000 feet MSL, all aircraft are required to set 29.92 as the reference pressure in the Kollsman window. The altimeter then reads pressure altitude. Depending on the atmospheric pressure that day, the true or actual altitude of the aircraft may be above or below what is indicated (pressure altitude).*

cause the altimeter to produce an errant reading, flight crews must be aware that performance changes with temperature variations in the atmosphere. The term density altitude describes altitude corrected for nonstandard temperature. That is, the density altitude is the standard day altitude (pressure altitude) at which an aircraft would experience similar performance as it would on the non-standard day currently being experienced. For example, on a very cold day, the air is denser than on a standard day, so an aircraft performs as though it is at a lower altitude. The density altitude is lower that day. On a very hot day, the reverse is true, and an aircraft performs as though it were at a higher elevation where the air is less dense. The density altitude is higher that day.

Conversion factors and charts have been produced so pilots can calculate the density altitude on any particular day. Inclusion of nonstandard air pressure due to weather systems and humidity can also be factored. So, while the effects of temperature on aircraft performance do not cause an altimeter to indicate falsely, an altimeter indication can be misleading in terms of aircraft performance if these effects are not considered. *[Figure 10-36]*

Other factors can cause an inaccurate altimeter indication. Scale error is a mechanical error whereby the scale of the instrument is not aligned so the altimeter pointers indicate correctly. Periodic testing and adjustment by trained technicians using calibrated equipment ensures scale error is kept to a minimum.

The pressure altimeter is connected to the pitot-static system and must receive an accurate sample of ambient air pressure to indicate the correct altitude. Position error, or installation error, is that inaccuracy caused by the location of the static

vent that supplies the altimeter. While every effort is made to place static ports in undisturbed air, airflow over the airframe changes with the speed and attitude of the aircraft. The amount of this air pressure collection error is measured in test flights, and a correction table showing the variances can be included with the altimeter for the pilot's use. Normally, location of the static vents is adjusted during these test flights so that the position error is minimal. *[Figure 10-37]* Position error can be removed by the ADC in modern aircraft, so the pilot need not be concerned about this inaccuracy.

Static system leaks can affect the static air input to the altimeter or ADC resulting in inaccurate altimeter indications. It is for this reason that static system maintenance includes leak checks every 24 months, regardless of whether any discrepancy has been noticed. See the instrument maintenance section toward the end of this chapter for further information on this mandatory check. It should also be understood that analog mechanical altimeters are mechanical devices that often reside in a hostile environment. The significant vibration and temperature range swings encountered by the instruments and the pitot static system (i.e., the tubing connections and fittings) can sometime create damage or a leak, leading to instrument malfunction. Proper care upon installation is the best preventive action. Periodic inspection and testing can also assure integrity.

The mechanical nature of the analog altimeter's diaphragm pressure measuring apparatus has limitations. The diaphragm itself is only so elastic when responding to static air pressure changes. Hysteresis is the term for when the material from which the diaphragm is made takes a set during long periods of level flight. If followed by an abrupt altitude change, the indication lags or responds slowly while expanding or contracting during a rapid altitude change. While temporary,

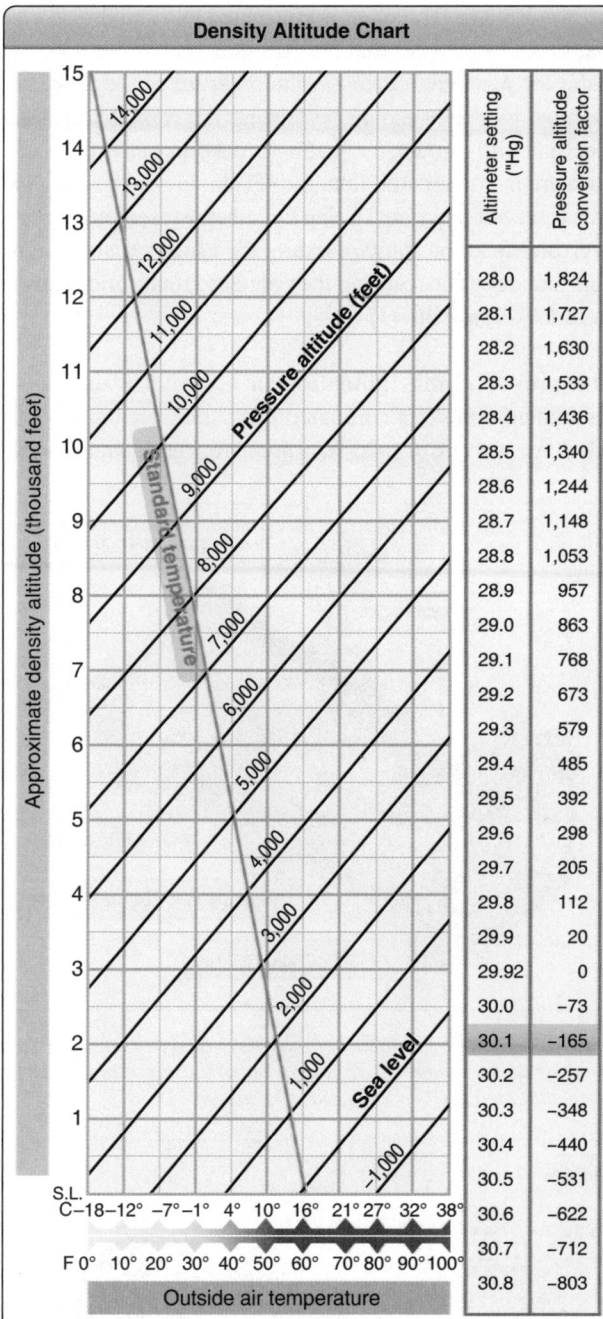

Density Altitude Chart

Altimeter setting ("Hg)	Pressure altitude conversion factor
28.0	1,824
28.1	1,727
28.2	1,630
28.3	1,533
28.4	1,436
28.5	1,340
28.6	1,244
28.7	1,148
28.8	1,053
28.9	957
29.0	863
29.1	768
29.2	673
29.3	579
29.4	485
29.5	392
29.6	298
29.7	205
29.8	112
29.9	20
29.92	0
30.0	-73
30.1	-165
30.2	-257
30.3	-348
30.4	-440
30.5	-531
30.6	-622
30.7	-712
30.8	-803

Figure 10-36. *The effect of air temperature on aircraft performance is expressed as density altitude.*

this limitation does cause an inaccurate altitude indication.

It should be noted that many modern altimeters are constructed to integrate into flight control systems, autopilots, and altitude monitoring systems, such as those used by ATC. The basic pressure-sensing operation of these altimeters is the same, but a means for transmitting the information is added.

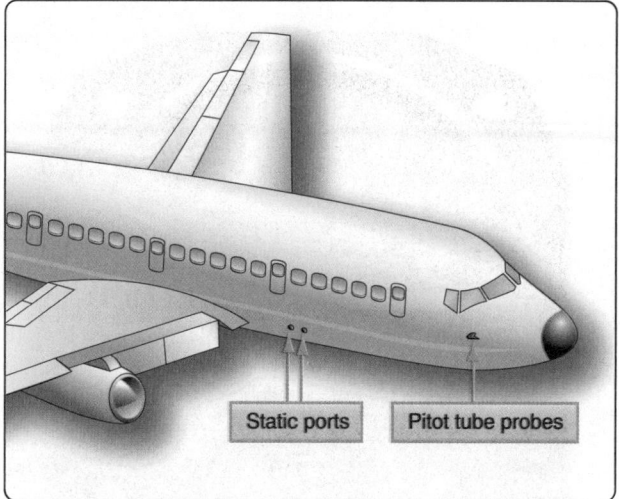

Figure 10-37. *The location of the static ports is selected to keep altimeter position error to a minimum.*

Vertical Speed Indicator

An analog vertical speed indicator (VSI) may also be referred to as a vertical velocity indicator (VVI), or rate-of-climb indicator. It is a direct reading, differential pressure gauge that compares static pressure from the aircraft's static system directed into a diaphragm with static pressure surrounding the diaphragm in the instrument case. Air is free to flow unrestricted in and out of the diaphragm but is made to flow in and out of the case through a calibrated orifice. A pointer attached to the diaphragm indicates zero vertical speed when the pressure inside and outside the diaphragm are the same. The dial is usually graduated in 100s of feet per minute. A zeroing adjustment screw, or knob, on the face of the instrument is used to center the pointer exactly on zero while the aircraft is on the ground. *[Figure 10-38]*

As the aircraft climbs, the unrestricted air pressure in the diaphragm lowers as the air becomes less dense. The case air pressure surrounding the diaphragm lowers more slowly, having to pass through the restriction created by the orifice. This causes unequal pressure inside and outside the diaphragm, which in turn causes the diaphragm to contract a bit and the pointer indicates a climb. The process works in reverse for an aircraft in a descent. If a steady climb or descent is maintained, a steady pressure differential is established between the diaphragm and case pressure surrounding it, resulting in an accurate indication of the rate of climb via graduations on the instrument face. *[Figure 10-39]*

A shortcoming of the rate-of-climb mechanism as described is that there is a lag of six to nine seconds before a stable differential pressure can be established that indicates the actual climb or descent rate of the aircraft. An instantaneous vertical speed indicator (IVSI) has a built-in mechanism to reduce this lag. A small, lightly sprung dashpot, or piston,

Figure 10-38. *A typical vertical speed indicator.*

reacts to the direction change of an abrupt climb or descent. As this small accelerometer does so, it pumps air into or out of the diaphragm, hastening the establishment of the pressure differential that causes the appropriate indication. *[Figure 10-40]*

Gliders and lighter-than-air aircraft often make use of a variometer. This is a differential VSI that compares static pressure with a known pressure. It is very sensitive and gives an instantaneous indication. It uses a rotating vane with a pointer attached to it. The vane separates two chambers. One is connected to the aircraft's static vent or is open to the atmosphere. The other is connected to a small reservoir inside the instrument that is filled to a known pressure. As static air pressure increases, the pressure in the static air chamber increases and pushes against the vane. This rotates the vane and pointer, indicating a descent since the static pressure is now greater than the set amount in the chamber with reservoir pressure. During a climb, the reservoir pressure is greater than the static pressure; the vane is pushed in the opposite direction, causing the pointer to rotate and indicate a climb. *[Figure 10-41]*

The rate-of-climb indication in a digitally displayed instrument system is computed from static air input to the ADC. An aneroid, or solid-state pressure sensor, continuously

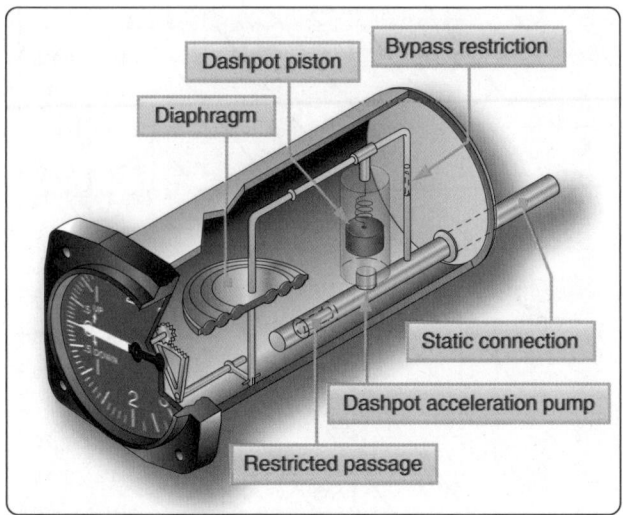

Figure 10-40. *The small dashpot in this IVSI reacts abruptly to a climb or descent pumping air into or out of the diaphragm causing an instantaneously vertical speed indication.*

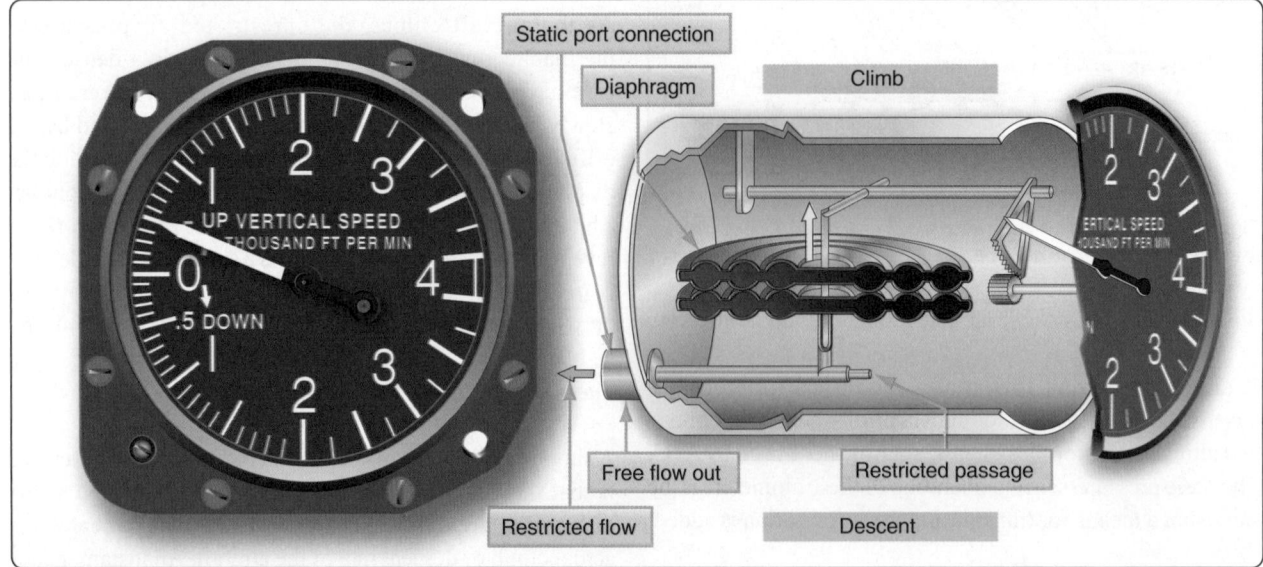

Figure 10-39. *The VSI is a differential pressure gauge that compares free-flowing static air pressure in the diaphragm with restricted static air pressure around the diaphragm in the instrument case.*

reacts to changes in static pressure. The digital clock within the computer replaces the calibrated orifice found on an analog instrument. As the static pressure changes, the computer's clock can be used to develop a rate for the change. Using the known lapse rate conversion for air pressure as altitude increases or decreases, a figure for climb or descent in fpm can be calculated and sent to the flight deck. The vertical speed is often displayed near the altimeter information on the primary flight display. *[Figure 10-34]*

Airspeed Indicators

The airspeed indicator is another primary flight instrument that is also a differential pressure gauge. Ram air pressure from the aircraft's pitot tube is directed into a diaphragm in an analog airspeed instrument case. Static air pressure from the aircraft static vent(s) is directed into the case surrounding the diaphragm. As the speed of the aircraft varies, the ram air pressure varies, expanding or contracting the diaphragm. Linkage attached to the diaphragm causes a pointer to move over the instrument face, which is calibrated in knots or miles per hour (mph). *[Figure 10-42]*

The relationship between the ram air pressure and static air pressure produces the indication known as indicated airspeed (IAS). As with the altimeter, there are other factors that must be considered in measuring airspeed throughout all phases of flight. These can cause inaccurate readings or indications that are not useful to the pilot in a particular situation. In analog airspeed indicators, the factors are often compensated for with ingenious mechanisms inside the case and on the instrument dial face. Digital flight instruments can have calculations performed in the ADC so the desired accurate indication is displayed.

While the relationship between ram air pressure and static air

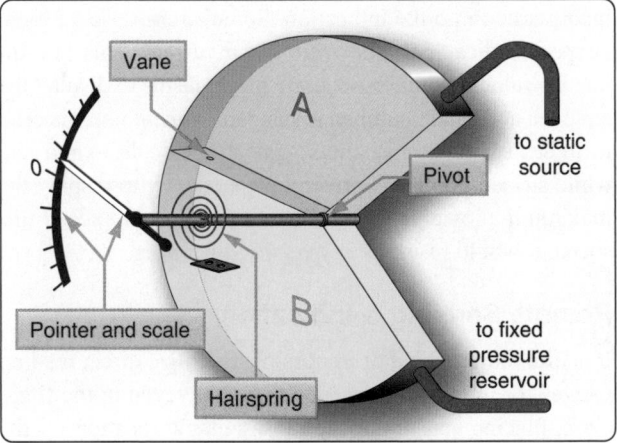

Figure 10-41. *A variometer uses differential pressure to indicate vertical speed. A rotating vane separating two chambers (one with static pressure, the other with a fixed pressure reservoir), moves the pointer as static pressure changes.*

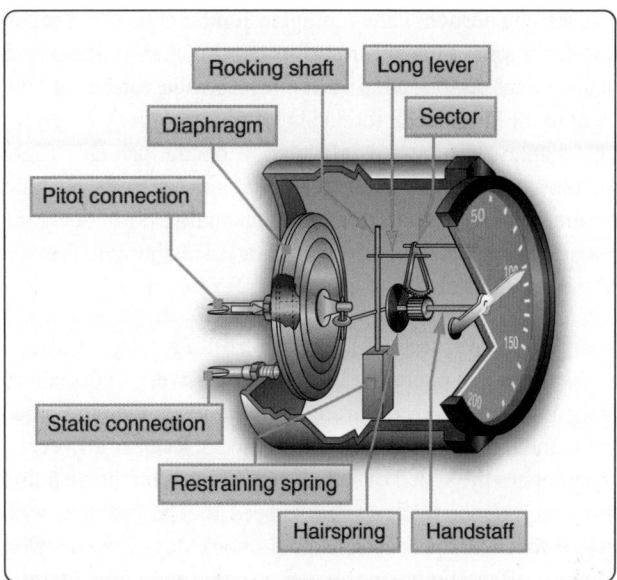

Figure 10-42. *An airspeed indicator is a differential pressure gauge that compares ram air pressure with static pressure.*

pressure is the basis for most airspeed indications, it can be more accurate. Calibrated airspeed (CAS) takes into account errors due to position error of the pitot static pickups. It also corrects for the nonlinear nature of the pitot static pressure differential when it is displayed on a linear scale. Analog airspeed indicators come with a correction chart that allows cross-referencing of indicated airspeed to calibrated airspeed for various flight conditions. These differences are typically very small and often are ignored. Digital instruments have these corrections performed in the ADC.

More importantly, indicated airspeed does not take into account temperature and air pressure differences needed to indicate true airspeed (TAS). These factors greatly affect airspeed indication. True airspeed, therefore, is the same as indicated airspeed when standard day conditions exist. But when atmospheric temperature or pressure varies, the relationship between the ram air pressure and static pressure alters. Analog airspeed instruments often include bimetallic temperature compensating devices that can alter the linkage movement between the diaphragm and the pointer movement. There can also be an aneroid inside the airspeed indicator case that can compensate for non-standard pressures. Alternatively, true airspeed indicators exist that allow the pilot to set temperature and pressure variables manually with external knobs on the instrument dial. The knobs rotate the dial face and internal linkages to present an indication that compensates for non-standard temperature and pressure, resulting in a true airspeed indication. *[Figure 10-43]*

Digital flight instrument systems perform all of the calculations for true airspeed in the ADC. Ram air from the pitot tube and static air from the static vent(s) are run into the sensing portion of the computer. Temperature information is also input. This information can be manipulated and calculations performed so a true airspeed value can be digitally sent to the flight deck for display. Refer to *Figure 10-34* for the display of airspeed information on the primary flight display on a light aircraft. Note that similar to its position in the standard T configuration of an analog flight deck, the airspeed indication is just left of the artificial horizon display. Refer to *Figure 10-4* showing T configuration.

Complications continue when considering airspeed indications and operating limitations. It is very important to keep high-speed aircraft from traveling faster than the speed of sound if they are not designed to do so. Even as an aircraft approaches the speed of sound, certain parts on the airframe may experience airflows that exceed it. The problem with this is that near the speed of sound, shock waves can develop that can affect flight controls and, in some cases, can literally tear the aircraft apart if not designed for supersonic airflow. A further complication is that the speed of sound changes with altitude and temperature. So, a safe true airspeed at sea level could put the aircraft in danger at altitude due to the lower speed of sound. *[Figure 10-44]*

In order to safeguard against these dangers, pilots monitor airspeed closely. A maximum allowable speed is established for the aircraft during certification flight testing. This speed is known the critical Mach number or Mcrit. Mach is a term for the speed of sound. The critical Mach number is expressed as a decimal of Mach such as 0.8 Mach. This means 8/10 of the speed of sound, regardless of what the actual speed of sound is at any particular altitude.

Standard Altitude, Temperature, and the Speed of Sound		
Altitude (feet)	Temperature (°F)	Speed of sound (knots)
Sea level	59	661
2,000	52	657
4,000	48	652
6,000	38	648
8,000	30	643
10,000	23	638
12,000	16	633
14,000	9	629
16,000	2	624
18,000	−5	619
20,000	−12	614
22,000	−19	609
24,000	−27	604
26,000	−34	599
28,000	−41	594
30,000	−48	589
32,000	−55	584
34,000	−62	579
36,000	−69	574
38,000	−70	574
40,000	−70	574
42,000	−70	574
44,000	−70	574
46,000	−70	574
48,000	−70	574
50,000	−70	574

Figure 10-44. *As temperatures fall at higher altitudes, the speed of sound is reduced.*

Many high performance aircraft are equipped with a Machmeter for monitoring Mcrit. The Machmeter is essentially an airspeed instrument that is calibrated in relation to Mach on the dial. Various scales exist for subsonic and supersonic aircraft. *[Figure 10-45]* In addition to the ram air/static air diaphragm arrangement, Machmeters also contain an altitude sensing diaphragm. It adjusts the input to the pointer so changes in the speed of sound due to altitude are incorporated into the indication. Some aircraft use a Mach/airspeed indicator as shown in *Figure 10-46*. This two-in-one instrument contains separate mechanisms to display the airspeed and Mach number. A standard white pointer is used to indicate airspeed in knots against one scale. A red and white striped pointer is driven independently to display the maximum allowable speed. Should the aircraft exceed this speed, it would result in an overspeed warning.

Remote Sensing & Indication

It is often impractical or impossible to utilize direct reading gauges for information needed to be conveyed in the flight deck. Placing sensors at the most suitable location on the airframe or engine and transmitting the collected data electrically through wires to the displays in the flight deck is a widely used method of remote-sensing and indicating on aircraft. Many remote sensing instrument systems

Figure 10-43. *An analog true airspeed indicator. The pilot manually aligns the outside air temperature with the pressure altitude scale, resulting in an indication of true airspeed.*

Figure 10-45. *A Machmeter indicates aircraft speed relative to the speed of sound.*

Figure 10-46. *A combination Mach/airspeed indicator shows airspeed with a white pointer and Mach number with a red and white striped pointer. Each pointer is driven by separate internal mechanisms.*

consist simply of the sensing device, transmitter unit, and the flight deck indicator unit connected to each other by wires. For pressure flight instruments, the ADC and pickup devices (pitot tubes, static vents, etc.) comprise the sensing and transmitter unit. Many aircraft collect sensed data in dedicated engine and airframe computers. There, the information can be processed. An output section of the computer then transmits it electrically or digitally to the flight deck for display. Remote-sensing instrument systems operate with high reliability and accuracy. They are powered by the aircraft's electrical system.

Small electric motors inside the instrument housings are used to position the pointers, instead of direct-operating mechanical linkages. They receive electric current from the output section of the ADC or other computers. They also receive input from sensing transmitters or transducers that are remotely located on the aircraft. By varying the electric signal, the motors are turned to the precise location needed to reflect the correct indication. Direct electric transmission of information from different types of sensors is accomplished with a few reliable and relatively simple techniques. Note that digital flight deck displays receive all of their input from a digital air data computer (DADC) and other computers, via a digital data bus and do not use electric motors. The data packages transmitted via the bus contain the instructions on how to illuminate the display screen.

Synchro-Type Remote-Indicating Instruments

A synchro system is an electric system used for transmitting information from one point to another. The word "synchro" is a shortened form of the word "synchronous," and refers to any one of a number of similarly operating two-unit electrical systems capable of measuring, transmitting, and indicating a certain parameter on the aircraft. Most position-indicating instruments are designed around a synchro system, such as the flap position indicator. Fluid pressure indicators also commonly use synchro systems. Synchro systems are used as remote position indicators for landing gear, autopilot systems, radar, and many other remote-indicating applications. The most common types of synchro system are the autosyn, selsyn, and magnesyn synchro systems.

These systems are similar in construction, and all operate by exploiting the consistent relationship between electricity and magnetism. The fact that electricity can be used to create magnetic fields that have definite direction, and that magnetic fields can interact with magnets and other electromagnetic fields, is the basis of their operation.

DC Selsyn Systems

On aircraft with direct current (DC) electrical systems, the DC selsyn system is widely used. As mentioned, the selsyn system consists of a transmitter, an indicator, and connecting wires. The transmitter consists of a circular resistance winding and a rotatable contact arm. The rotatable contact arm turns on a shaft in the center of the resistance winding. The two ends of the arm are brushes and always touch the winding on opposite sides. *[Figure 10-47]* On position indicating systems, the shaft to which the contact arm is fastened protrudes through the end of transmitter housing and is attached to the unit whose position is to be transmitted (e.g., flaps, landing gear). The transmitter is often connected to the moving unit through a mechanical linkage. As the unit moves, it causes the transmitter shaft to turn. The arm is turned so that voltage is applied through the brushes to any two points around the circumference of the resistance winding. The rotor shaft of DC selsyn systems, measuring other kinds of data, operates the same way, but may not protrude outside of the housing. The sensing device, which imparts rotary motion to the shaft, could be located inside the transmitter housing.

Referring to *Figure 10-47*, note that the resistance winding of the transmitter is tapped off in three fixed places, usually 120° apart. These taps distribute current through the toroidal windings of the indicator motor. When current flows through these windings, a magnetic field is created. Like all magnetic fields, a definite north and south direction to the field exists.

As the transmitter rotor shaft is turned, the voltage-supplying contact arm moves. Because it contacts the transmitter resistance winding in different positions, the resistance between the supply arm and the various tapoffs changes. This causes the voltage flowing through the tapoffs to change as the resistance of sections of the winding become longer or shorter. The result is that varied current is sent via the tapoffs to the three windings in the indicator motor.

The resultant magnetic field created by current flowing through the indicator coils changes as each receives varied current from the tapoffs. The direction of the magnetic field also changes. Thus, the direction of the magnetic field across the indicating element corresponds in position to the moving arm in the transmitter. A permanent magnet is attached to the centered rotor shaft in the indicator, as is the indicator pointer. The magnet aligns itself with the direction of the magnetic field and the pointer does as well. Whenever the magnetic field changes direction, the permanent magnet and pointer realign with the new position of the field. Thus, the position of the aircraft device is indicated.

Landing gear contain mechanical devices that lock the gear up, called an up-lock, or down, called a down-lock. When the DC selsyn system is used to indicate the position of the landing gear, the indicator can also show that the up-lock or down-lock is engaged. This is done by again varying the current flowing through the indicator's coils. Switches located on the actual locking devices close when the locks engage. Current from the selsyn system described above flows through the switch and a small additional circuit. The circuit adds an additional resistor to one of the transmitter winding sections created by the rotor arm and a tapoff. This changes the total resistance of that section. The result is a change in the current flowing through one of the indicator's motor coils. This, in turn, changes the magnetic field around that coil. Therefore, the combined magnetic field created by all three motor coils is also affected, causing a shift in the direction of the indicator's magnetic field. The permanent magnet and pointer align with the new direction and shift to the locked position on the indicator dial. *Figure 10-48* shows a simplified diagram of a lock switch in a three-wire selsyn system and an indicator dial.

AC Synchro Systems

Aircraft with alternating current (AC) electrical power systems make use of autosyn or magnasysn synchro remote indicating systems. Both operate in a similar way to the DC selsyn system, except that AC power is used. Thus, they make use of electric induction, rather than resistance current flows defined by the rotor brushes. Magnasyn systems use permanent magnet rotors such as those found in the DC selsyn system. Usually, the transmitter magnet is larger than the indicator magnet, but the electromagnetic response of the indicator rotor magnet and pointer remains the same. It aligns with the magnetic field set up by the coils, adopting the same angle of deflection as the transmitter rotor. *[Figure 10-49]*

Autosyn systems are further distinguished by the fact that the transmitter and indicator rotors used are electro-magnets rather than permanent magnets. Nonetheless, like a permanent magnet, an electro-magnet aligns with the direction of the magnetic field created by current flowing through the stator coils in the indicator. Thus, the indicator pointer position mirrors the transmitter rotor position. *[Figure 10-50]*

AC synchro systems are wired differently than DC systems. The varying current flows through the transmitter and indicator stator coils are induced as the AC cycles through zero and the rotor magnetic field flux is allowed to flow. The important characteristic of all synchro systems is maintained by both the autosyn and magnasyn systems. That is, the position of the transmitter rotor is mirrored by the rotor in the indicator. These systems are used in many of the same applications as the DC systems and more. Since they are usually part of instrumentation for high performance aircraft, adaptations of autosyn and magnasyn synchro systems are frequently used in directional indicators and in autopilot systems.

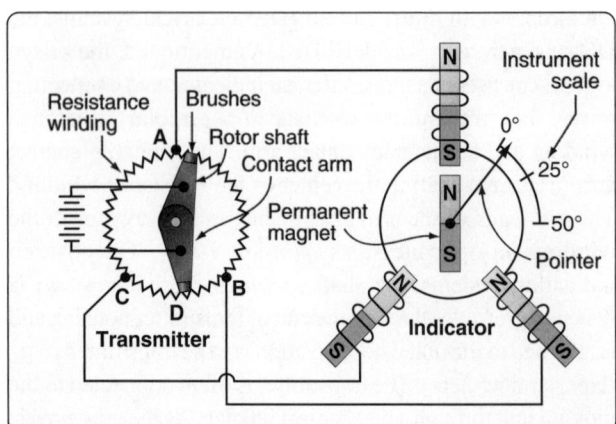

Figure 10-47. *A schematic of a DC selsyn synchro remote indicating system.*

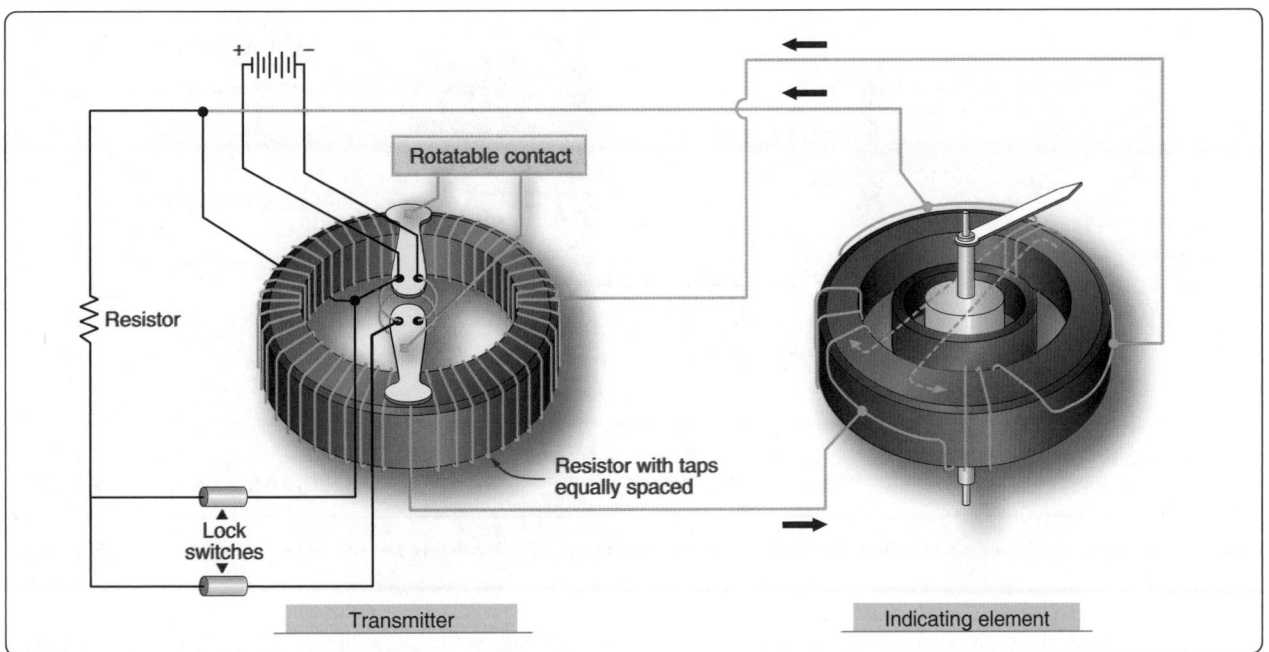

Figure 10-48. *A lock switch circuit can be added to the basic DC selsyn synchro system when used to indicate landing gear position and up- and down-locked conditions on the same indicator.*

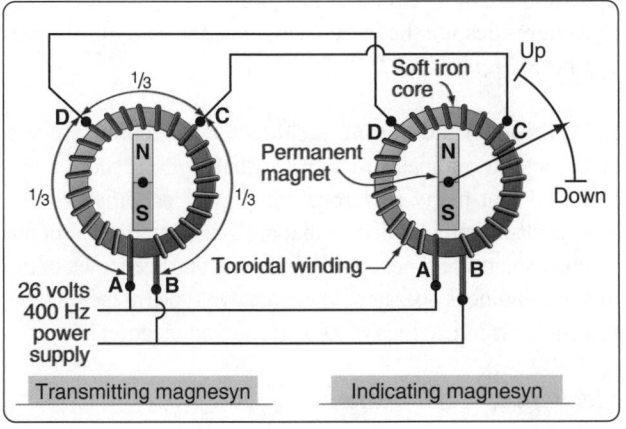

Figure 10-49. *A magnasysn synchro remote-indicating system uses AC. It has permanent magnet rotors in the transmitter and indictor.*

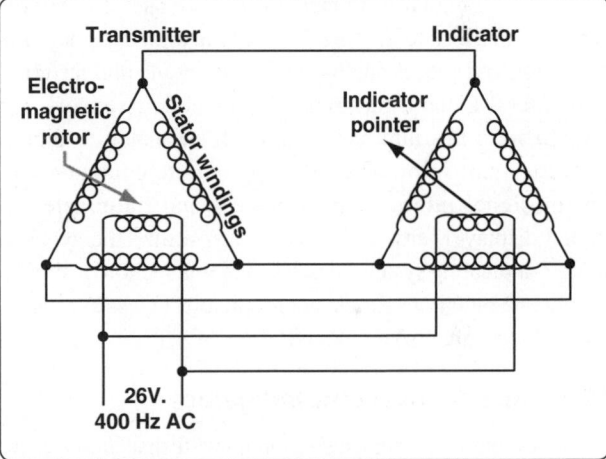

Figure 10-50. *An autosyn remote-indicating system utilizes the interaction between magnetic fields set up by electric current flow to position the indicator pointer.*

Remote Indicating Fuel & Oil Pressure Gauges

Fuel and oil pressure indications can be conveniently obtained through the use of synchro systems. As stated previously, running fuel and oil lines into the cabin to direct reading gauges is not desirable due to the possibility of leaking fluid. Additionally, there is an increased risk of fire in the cabin.

By locating the transmitter of a synchro system remotely, fluid pressure can be directed into it without a long tubing run. Inside the transmitter, the motion of a pressure bellows can be geared to the transmitter rotor in such a way as to make the rotor turn. *[Figure 10-51]* As in all synchros, the transmitter rotor turns proportional to the pressure sensed, which varies the voltages set up in the resistor windings of the synchro

stator. These voltages are transmitted to the indicator coils that develop the magnetic field that positions the pointer.

Often on twin-engine aircraft, synchro mechanisms for each engine can be used to drive separate pointers on the same indicator. By placing the coils one behind the other, the pointer shaft from the rear indicator motor can be sent through the hollow shaft of the forward indicator motor. Thus, each pointer responds with the magnet's alignment in its own motor's magnetic field while sharing the same gauge housing. Labeling the pointer's engine 1 or 2 removes any doubt about which indicator pointer is being observed. A similar principle is employed in an indicator that has side-by-side indications

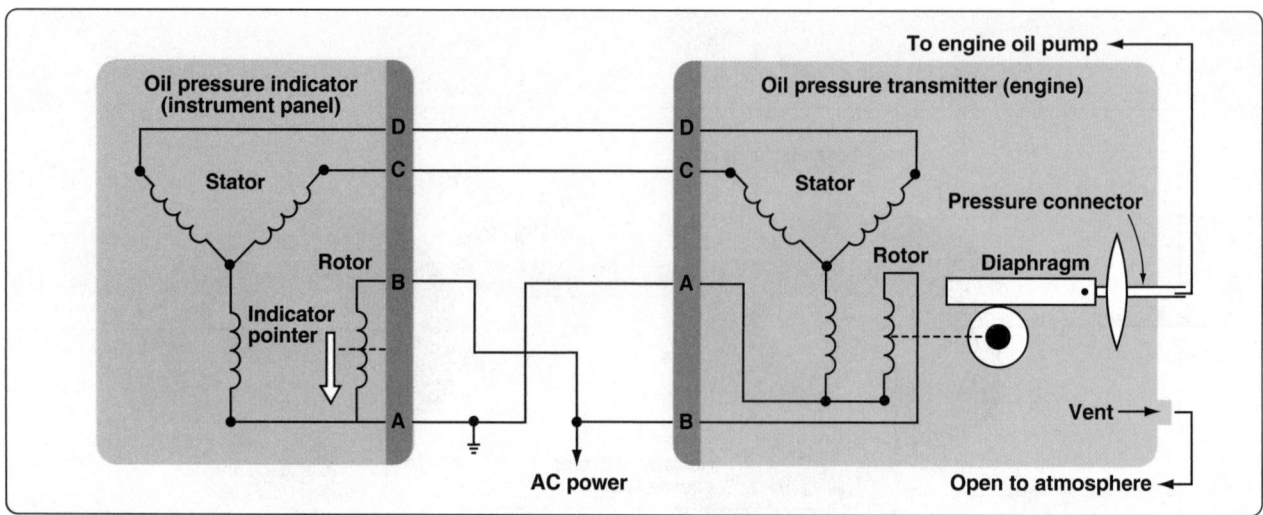

Figure 10-51. *Remote pressure sensing indicators change linear motion to rotary motion in the sensing mechanism part of the synchro transmitter.*

for different parameters, such as oil pressure and fuel pressure in the same indicator housing. Each parameter has its own synchro motor for positioning its pointer.

Aircraft with digital instrumentation make use of pressure-sensitive solid-state sensors that output digital signals for collection and processing by dedicated engine and airframe computers. Others may retain their analog sensors but may forward this information through an analog to digital converter unit from which the appropriate computer can obtain digital information to process and illuminate the digital display. Many more instruments utilize the synchro remote-indicating systems described in this section or similar synchros. Sometimes simple, more suitable, or less expensive technologies are also employed.

Mechanical Movement Indicators

There are many instruments on an aircraft that indicate the mechanical motion of a component, or even the aircraft itself. Some utilize the synchro remote-sensing and indicating systems described above. Other means for capturing and displaying mechanical movement information are also used. This section discusses some unique mechanical motion indicators and groups instruments by function. All give valuable feedback to the pilot on the condition of the aircraft in flight.

Tachometers

The tachometer, or tach, is an instrument that indicates the speed of the crankshaft of a reciprocating engine. It can be a direct- or remote-indicating instrument, the dial of which is calibrated to indicate revolutions per minutes (rpm). On reciprocating engines, the tach is used to monitor engine power and to ensure the engine is operated within certified limits.

Gas turbine engines also have tachometers. They are used to monitor the speed(s) of the compressor section(s) of the engine. Turbine engine tachometers are calibrated in percentage of rpm with 100 percent corresponding to optimum turbine speed. This allows similar operating procedures despite the varied actual engine rpm of different engines. *[Figure 10-52]*

In addition to the engine tachometer, helicopters use a tachometer to indicator main rotor shaft rpm. It should also be noted that many reciprocating-engine tachometers also have built-in numeric drums that are geared to the rotational mechanism inside. These are hour meters that keep track of the time the engine is operated. There are two types of tachometer system in wide use today: mechanical and electrical.

Mechanical Tachometers

Mechanical tachometer indicating systems are found on small, single-engine light aircraft in which a short distance exists between the engine and the instrument panel. They consist of an indicator connected to the engine by a flexible drive shaft. The drive shaft is geared into the engine so that when the engine turns, so does the shaft. The indicator contains a flyweight assembly coupled to a gear mechanism that drives a pointer. As the drive shaft rotates, centrifugal force acts on the flyweights and moves them to an angular position. This angular position varies with the rpm of the engine. The amount of movement of the flyweights is transmitted through the gear mechanism to the pointer. The pointer rotates to indicate this movement on the tachometer indicator, which is directly related to the rpm of the engine. *[Figure 10-53]*

A more common variation of this type of mechanical tachometer uses a magnetic drag cup to move the pointer in

Figure 10-52. *A tachometer for a reciprocating engine is calibrated in rpm. A tachometer for a turbine engine is calculated in percent of rpm.*

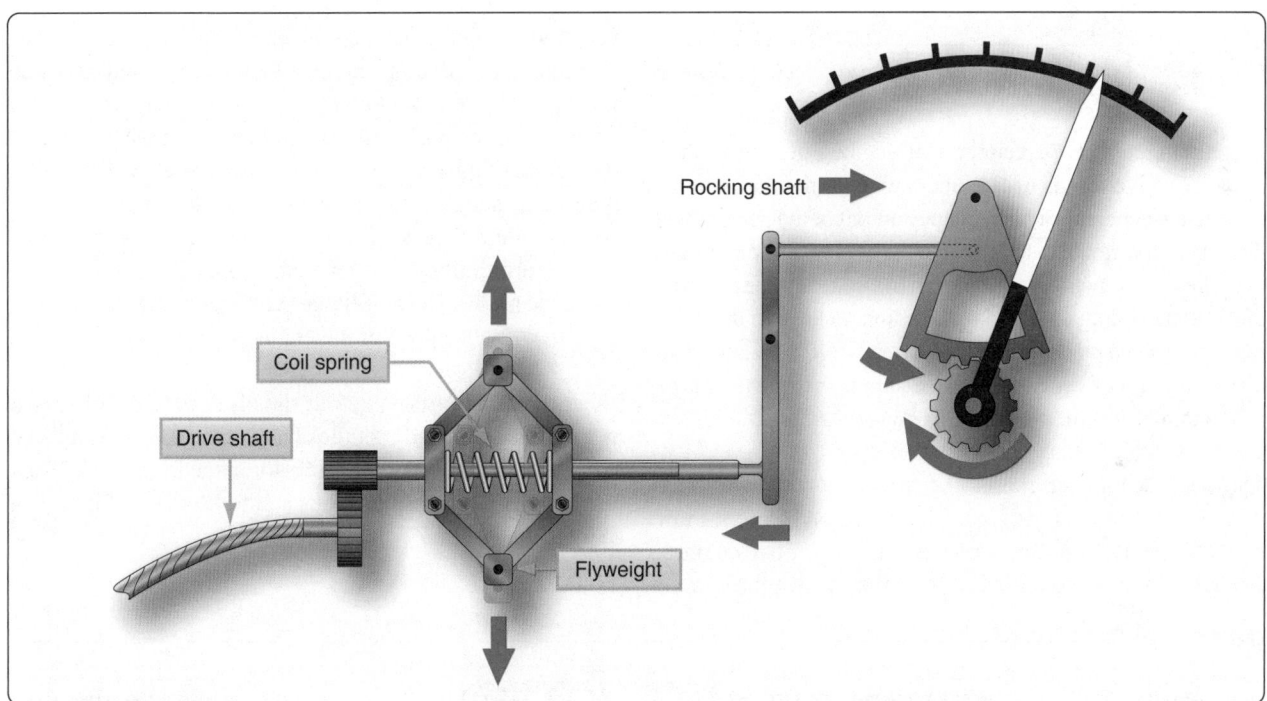

Figure 10-53. *The simplified mechanism of a flyweight type mechanical tachometer.*

the indicator. As the drive shaft turns, it rotates a permanent magnet in a close-tolerance aluminum cup. A shaft attached to the indicating point is attached to the exterior center of the cup. As the magnet is rotated by the engine flex drive cable, its magnetic field cuts through the conductor surrounding it, creating eddy currents in the aluminum cup. This current flow creates its own magnetic field, which interacts with the rotating magnet's flux field. The result is that the cup tends to rotate, and with it, the indicating pointer. A calibrated restraining spring limits the cup's rotation to the arc of motion of the pointer across the scale on the instrument face. *[Figure 10-54]*

Electric Tachometers

It is not practical to use a mechanical linkage between the engine and the rpm indicator on aircraft with engines not mounted in the fuselage just forward of the instrument panel. Greater accuracy with lower maintenance is achieved through the use of electric tachometers. A wide variety of electric tachometer systems can be employed, so manufacturer's instructions should be consulted for details of each specific tachometer system.

A popular electric tachometer system makes use of a small AC

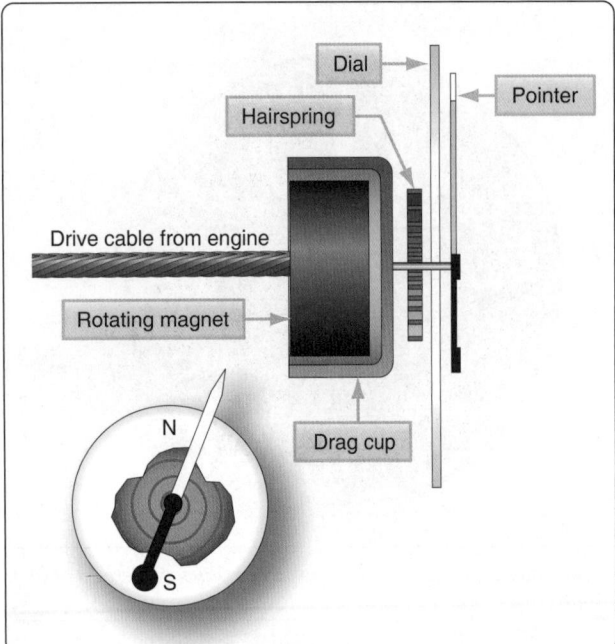

Figure 10-54. *A simplified magnetic drag cup tachometer indicating device.*

generator mounted to a reciprocating engine's gear case or the accessory drive section of a turbine engine. As the engine turns, so does the generator. The frequency output of the generator is directly proportional to the speed of the engine. It is connected via wires to a synchronous motor in the indicator that mirrors this output. A drag cup, or drag disc link, is used to drive the indicator as in a mechanical tachometer. *[Figure 10-55]* Two different types of generator units, distinguished by their type of mounting system, are shown in *Figure 10-56.*

The dual tachometer consists of two tachometer indicator units housed in a single case. The indicator pointers show simultaneously, on one or two scales, the rpm of two engines. A dual tachometer on a helicopter often shows the rpm of

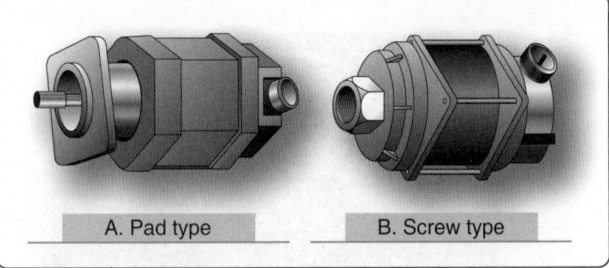

Figure 10-56. *Different types of tach generators.*

the engine and the rpm of the main rotor. A comparison of the voltages produced by the two tach generators of this type of helicopter indicator gives information concerning clutch slippage. A third indication showing this slippage is sometimes included in the helicopter tachometer. *[Figure 10-57]*

Some turbine engines use tachometer probes for rpm indication, rather than a tach generator system. They provide a great advantage in that there are no moving parts. The probes are sealed units that are mounted on a flange and protrude into the compressor section of the engine. A magnetic field is set up inside the probe that extends through pole pieces and out the end of the probe. A rotating gear wheel, which moves at the same speed as the engine compressor shaft, alters the magnetic field flux density as it moves past the pole pieces at close proximity. This generates voltage signals in coils inside the probe. The amplitude (voltage) of the signals vary directly with the speed of the engine.

The tachometer probe's output signals need to be processed in a remotely located module. They must also be amplified to drive a servo motor type indicator in the flight deck. They may also be used as input for an autothrottle or flight data acquisition system. *[Figure 10-58]*

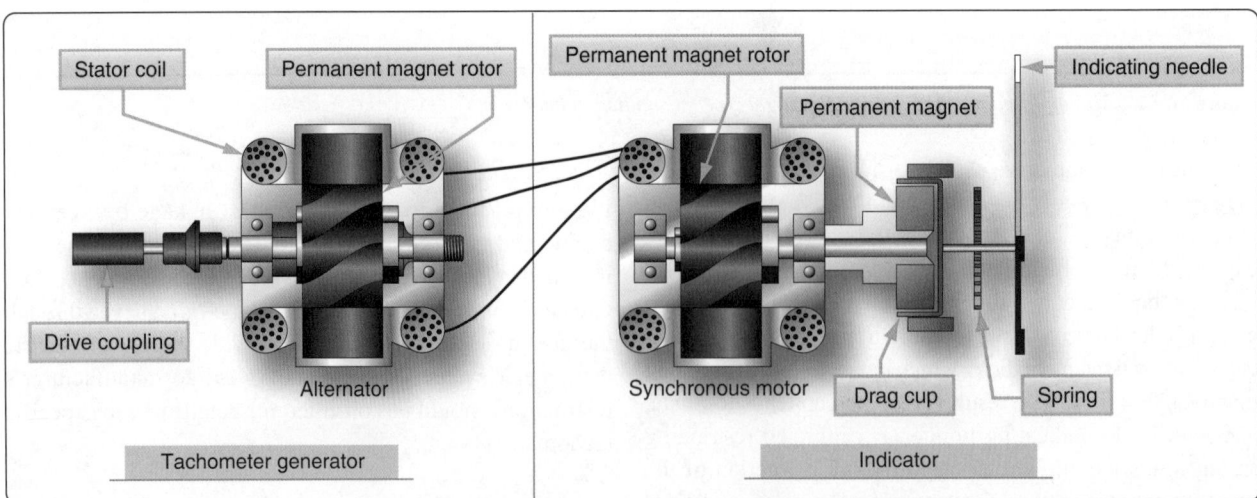

Figure 10-55. *An electric tachometer system with synchronous motors and a drag cup indicator.*

Figure 10-57. *A helicopter tachometer with engine rpm, rotor rpm, and slippage indications.*

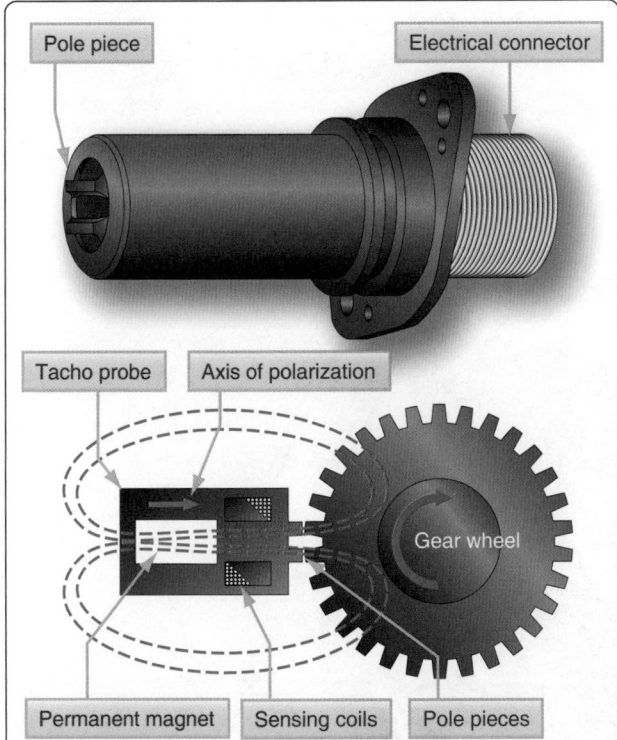

Figure 10-58. *A tacho probe has no moving parts. The rate of magnetic flux field density change is directly related to engine speed.*

Synchroscope

The synchroscope is an instrument that indicates whether two or more rotating devices, such as engines, are synchronized. Since synchroscopes compare rpm, they utilize the output from tachometer generators. The instrument consists of a small electric motor that receives electrical current from the generators of both engines. Current from the faster running engine controls the direction in which the synchroscope motor rotates.

If both engines are operating at exactly the same speed, the synchroscope motor does not operate. If one engine operates faster than the other, its tach generator signal causes the synchroscope motor to turn in a given direction. Should the speed of the other engine then become greater than that of the first engine, the signal from its tach generator causes the synchroscope motor to reverse itself and turn in the opposite direction. The pilot makes adjustments to steady the pointer so it does not move.

One use of synchroscope involve designating one of the engines as a master engine. The rpm of the other engine(s) is always compared to the rpm of this master engine. The dial face of the synchroscope indicator looks like *Figure 10-59.* "Slow" and "fast" represent the other engine's rpm relative to the master engine, and the pilot makes adjustments accordingly.

Accelerometers

An accelerometer is an instrument that measures acceleration. It is used to monitor the forces acting upon an airframe. Accelerometers are also used in inertial reference navigation systems. The installation of accelerometers is usually limited to high-performance and aerobatic aircraft.

Simple accelerometers are mechanical, direct-reading instruments calibrated to indicate force in Gs. One G is equal to one times the force of gravity. The dial face of an accelerometer is scaled to show positive and negative forces. When an aircraft initiates a rapid climb, positive G force tends to push one back into one's seat. Initiating a rapid decent causes a force in the opposite direction, resulting in a negative G force.

Most accelerometers have three pointers. One is continuously

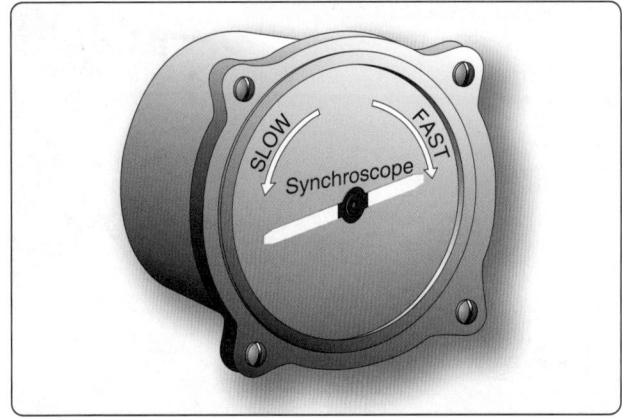

Figure 10-59. *This synchroscope indicates the relative speed of the slave engine to the master.*

indicating the acceleration force experienced. The other two contain ratcheting devices. The positive G pointer follows the continuous pointer and stay at the location on the dial where the maximum positive force is indicated. The negative G pointer does the same for negative forces experienced. Both max force pointers can be reset with a knob on the instrument face.

The accelerometer operates on the principle of inertia. A mass, or weight, inside is free to slide along a shaft in response to the slightest acceleration force. When a maneuver creates an accelerating force, the aircraft and instrument move, but inertia causes the weight to stay at rest in space. As the shaft slides through the weight, the relative position of the weight on the shaft changes. This position corresponds to the force experienced. Through a series of pulleys, springs, and shafts, the pointers are moved on the dial to indicate the relative strength of the acceleration force. [*Figure 10-60*] Forces can act upon an airframe along the three axes of flight. Single and multi-axis accelerometers are available, although most flight deck gauges are of the single-axis type. Inertial reference navigation systems make use of multi-axis accelerometers to continuously, mathematically calculate the location of the aircraft in a three dimensional plane.

Electric and digital accelerometers also exist. Solid-state sensors are employed, such as piezoelectric crystalline devices. In these instruments, when an accelerating force is applied, the amount of resistance, current flow, or capacitance changes in direct relationship to the size of the force. Micro-electric signals integrate well with digital computers designed to process and display information in the flight deck.

Stall Warning & Angle of Attack (AOA) Indicators

An aircraft's angle of attack (AOA) is the angle formed between the wing cord centerline and the relative wind. At a certain angle, airflow over the wing surfaces is insufficient to create enough lift to keep the aircraft flying, and a stall occurs. An instrument that monitors the AOA allows the pilot to avoid such a condition.

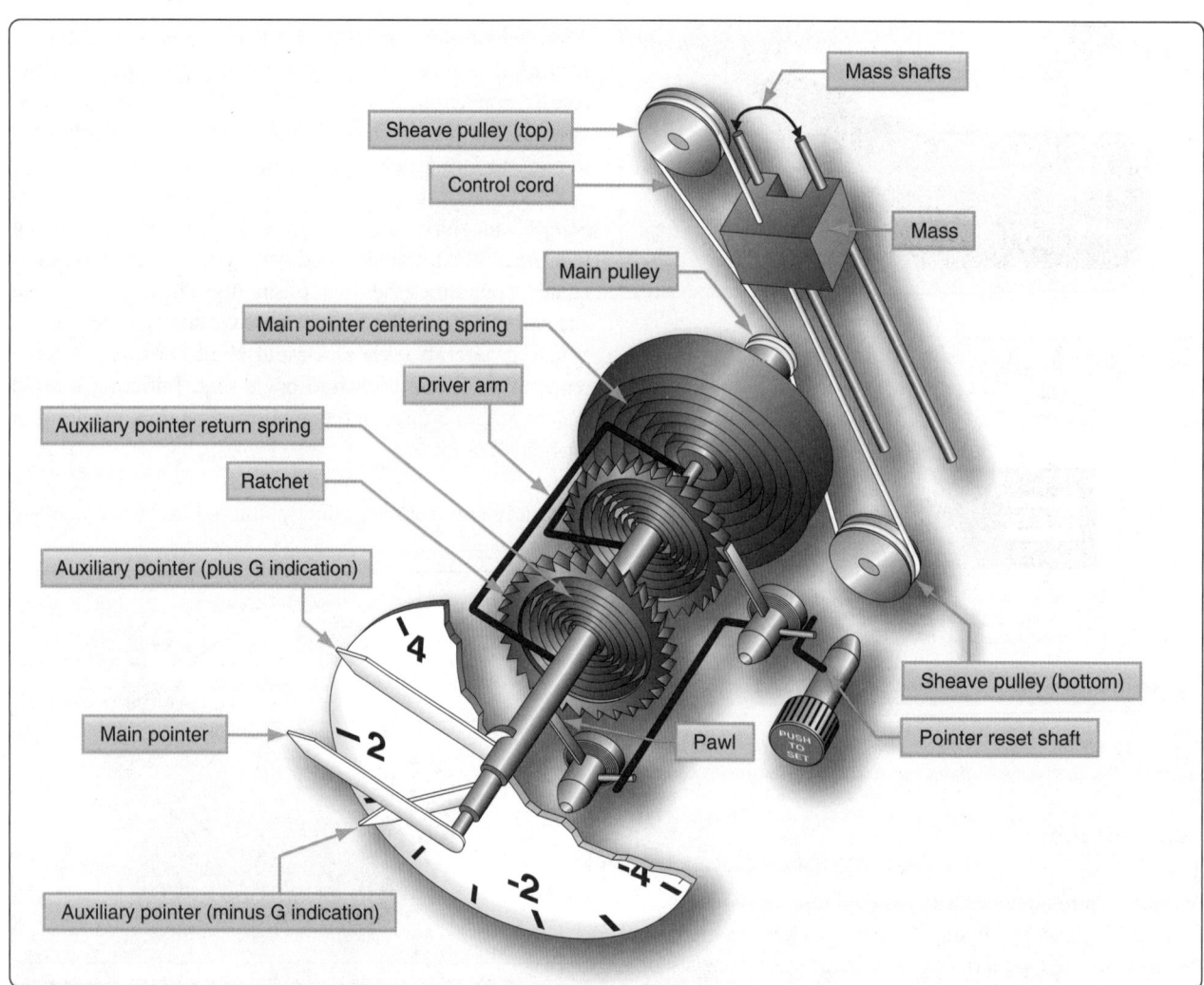

Figure 10-60. *The inner workings of a mass-type accelerometer.*

The simplest form of AOA indicator is a stall warning device that does not have a gauge located in the flight deck. It uses an aural tone to warn of an impending stall due to an increase in AOA. This is done by placing a reed in a cavity just aft of the leading edge of the wing. The cavity has an open passage to a precise point on the leading edge.

In flight, air flows over and under a wing. The point on the wing leading edge where the oncoming air diverges is known as the point of stagnation. As the AOA of the wing increases, the point of stagnation moves down below the open passage that leads inside the wing to the reed. Air flowing over the curved leading edge speeds up and causes a low pressure. This causes air to be sucked out of the inside of the wing through the passage. The reed vibrates as the air rushes by making a sound audible in the flight deck. *[Figure 10-61]*

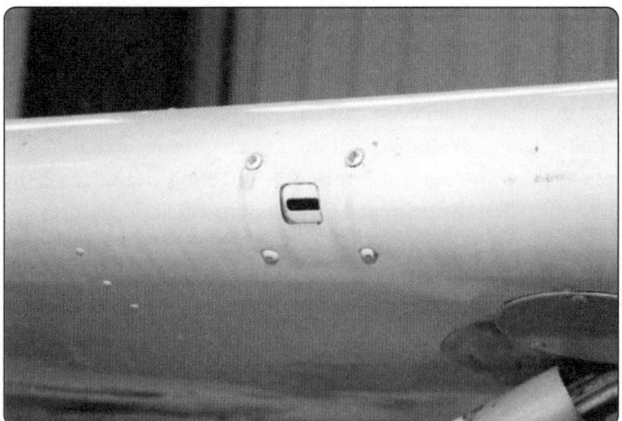

Figure 10-61. *A reed-type stall warning device is located behind this opening in the leading edge of the wing. When the angle of attack increases to near the point of a stall, low-pressure air flowing over the opening causes a suction, which audibly vibrates the reed.*

Another common device makes use of an audible tone as the AOA increases to near the point where the aircraft will stall. This stall warning device includes an electric switch that opens and closes a circuit to a warning horn audible in the flight deck. It may also be wired into a warning light circuit. The switch is located near the point of stagnation on the wing leading edge. A small lightly sprung tab activates the switch. At normal AOA, the tab is held down by air that diverges at the point of stagnation and flows under the wing. This holds the switch open so the horn does not sound nor the warning light illuminate. As the AOA increases, the point of stagnation moves down. The divergent air that flows up and over the wing now pushes the tab upward to close the switch and complete the circuit to the horn or light. *[Figure 10-62]*

A true AOA indicating system detects the local AOA of the aircraft and displays the information on a flight deck indicator. It also may be designed to furnish reference information to other systems on high-performance aircraft. The sensing mechanism and transmitter are usually located on the forward side of the fuselage. It typically contains a heating element to ensure ice-free operation. Signals are sent from the sensor to the flight deck or computer(s) as required. An AOA indicator may be calibrated in actual angle degrees, arbitrary units, percentage of lift used, symbols, or even fast/slow. *[Figure 10-63]*

There are two main types of AOA sensors in common use. Both detect the angular difference between the relative wind and the fuselage, which is used as a reference plane. One uses a vane, known as an alpha vane, externally mounted to the outside of the fuselage. It is free to rotate in the wind. As the AOA changes, air flowing over the vane changes its

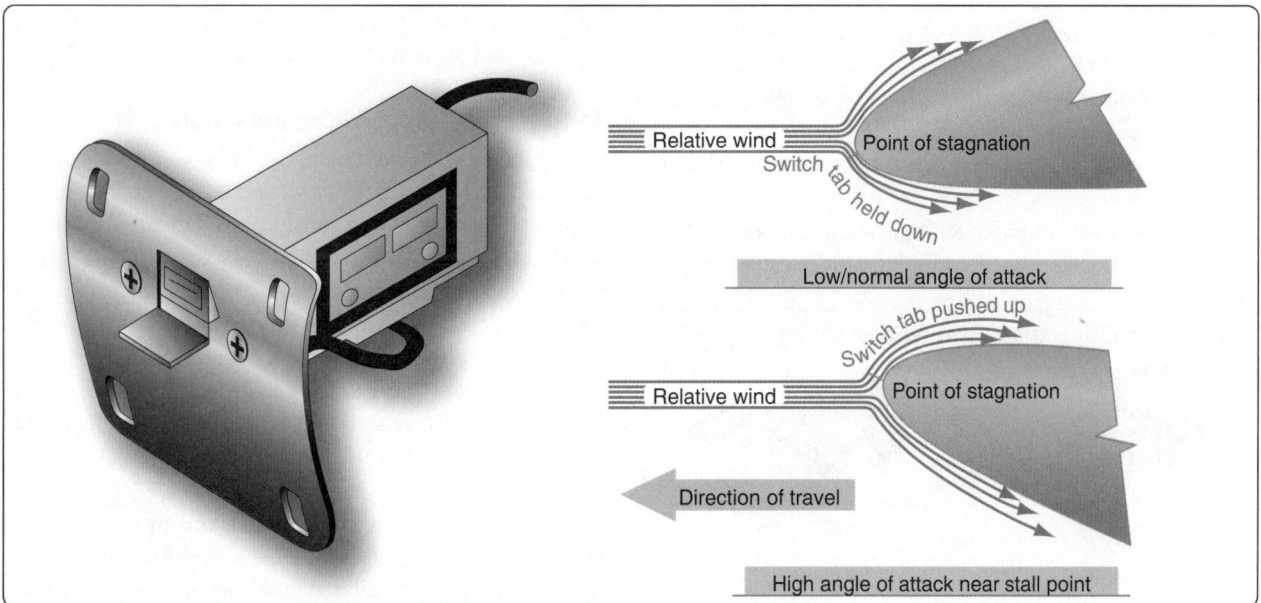

Figure 10-62. *A popular stall warning switch located in the wing leading edge.*

Figure 10-63. *Angle of attack indicator.*

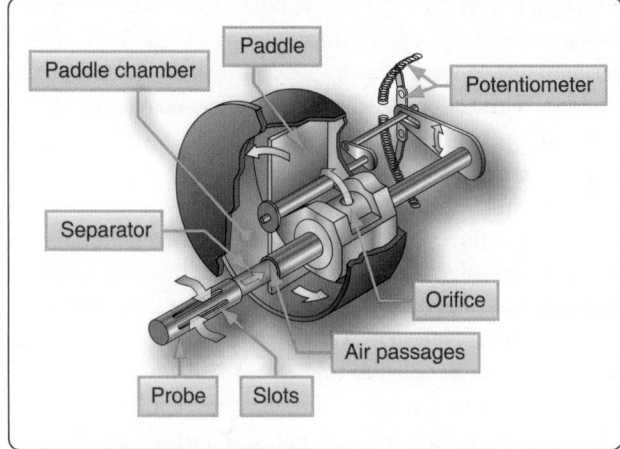

Figure 10-65. *The internal structure of a slotted probe airstream direction detector.*

angle. The other uses two slots in a probe that extends out of the side of the fuselage into the airflow. The slots lead to different sides of movable paddles in a chamber of the unit just inside the fuselage skin. As the AOA varies, the air pressure ported by each of the slots changes and the paddles rotate to neutralize the pressures. The shaft upon which the paddles rotate connects to a potentiometer wiper contact that is part of the unit. The same is true of the shaft of the alpha vane. The changing resistance of the potentiometer is used in a balanced bridge circuit to signal a motor in the indicator to move the pointer proportional to the AOA. *[Figures 10-64 and 10-65]*

Modern aircraft AOA sensor units send output signals to the ADC. There, the AOA data is used to create an AOA indication, usually on the primary flight display. AOA information can also be integrated with flap and slat position information to better determine the point of stall. Additionally, AOA sensors of the type described are subject to position error since airflow around the alpha vane and slotted probe changes somewhat with airspeed and aircraft attitude. The errors are small but can be corrected in the ADC.

To incorporate a warning of an impending stall, many AOA systems signal a stick shaker motor that literally shakes the

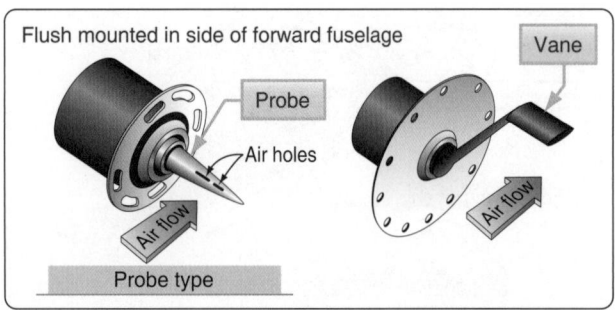

Figure 10-64. *A slotted AOA probe and an alpha vane.*

control column to warn the pilot as the aircraft approaches a stall condition. Electrical switches are actuated in the AOA indicator at various preset AOA to activate the motor that drives an unbalanced weighted ring, causing the column to shake. Some systems include a stick pusher actuator that pushes the control yoke forward, lowering the nose of the aircraft when the critical AOA is approached. Regardless of the many existing variations for warning of an impending stall, the AOA system triggers all stall warnings in high performance aircraft.

Temperature Measuring Instruments

The temperature of numerous items must be known for an aircraft to be operated properly. Engine oil, carburetor mixture, inlet air, free air, engine cylinder heads, heater ducts, and exhaust gas temperature of turbine engines are all items requiring temperature monitoring. Many other temperatures must also be known. Different types of thermometers are used to collect and present temperature information.

Non-Electric Temperature Indicators

The physical characteristics of most materials change when exposed to changes in temperature. The changes are consistent, such as the expansion or contraction of solids, liquids, and gases. The coefficient of expansion of different materials varies and it is unique to each material. Most everyone is familiar with the liquid mercury thermometer. As the temperature of the mercury increases, it expands up a narrow passage that has a graduated scale upon it to read the temperature associated with that expansion. The mercury thermometer has no application in aviation.

A bimetallic thermometer is very useful in aviation. The temperature sensing element of a bimetallic thermometer is made of two dissimilar metals strips bonded together. Each metal expands and contracts at a different rate when temperature changes. One end of the bimetallic strip is fixed, the other end is coiled. A pointer is attached to the coiled end which is set in the instrument housing. When the bimetallic strip is heated, the two metals expand. Since their expansion rates differ and they are attached to each other, the effect is that the coiled end tries to uncoil as the one metal expands faster than the other. This moves the pointer across the dial face of the instrument. When the temperature drops, the metals contract at different rates, which tends to tighten the coil and move the pointer in the opposite direction.

Direct reading bimetallic temperature gauges are often used in light aircraft to measure free air temperature or outside air temperature (OAT). In this application, a collecting probe protrudes through the windshield of the aircraft to be exposed to the atmospheric air. The coiled end of the bimetallic strip in the instrument head is just inside the windshield where it can be read by the pilot. *[Figures 10-66 and 10-67]*

A bourdon tube is also used as a direct reading non-electric temperature gauge in simple, light aircraft. By calibrating the dial face of a bourdon tube gauge with a temperature scale, it can indicate temperature. The basis for operation is the consistent expansion of the vapor produced by a volatile liquid in an enclosed area. This vapor pressure changes directly with temperature. By filling a sensing bulb with such a volatile liquid and connecting it to a bourdon tube, the tube causes an indication of the rising and falling vapor pressure due to temperature change. Calibration of the dial face in degrees Fahrenheit or Celsius, rather than psi, provides a temperature reading. In this type of gauge, the sensing bulb is placed in the area needing to have temperature measured. A long capillary tube connects the bulb to the bourdon tube in the instrument housing. The narrow diameter of the capillary tube ensures that the volatile liquid is lightweight and stays primarily in the sensor bulb. Oil temperature is sometimes measured this way.

Electrical Temperature Measuring Indication

The use of electricity in measuring temperature is very common in aviation. The following measuring and indication

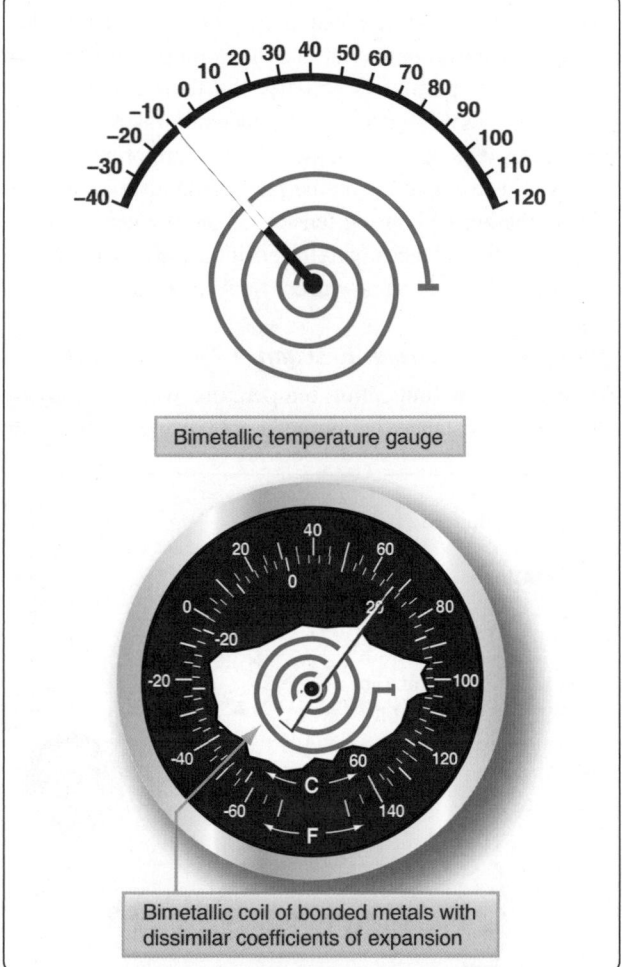

Figure 10-66. *A bimetallic temperature gauge works because of the dissimilar coefficients of expansion of two metals bonded together. When bent into a coil, cooling or heating causes the dissimilar metal coil to tighten, or unwind, moving the pointer across the temperature scale on the instrument dial face.*

Figure 10-67. *A bimetallic outside air temperature gauge and its installation on a light aircraft.*

10-35

systems can be found on many types of aircraft. Certain temperature ranges are more suitably measured by one or another type of system.

Electrical Resistance Thermometer

The principle parts of the electrical resistance thermometer are the indicating instrument, the temperature-sensitive element (or bulb), and the connecting wires and plug connectors. Electrical resistance thermometers are used widely in many types of aircraft to measure carburetor air, oil, free air temperatures, and more. They are used to measure low and medium temperatures in the –70 °C to 150 °C range.

For most metals, electrical resistance changes as the temperature of the metal changes. This is the principle upon which a resistance thermometer operates. Typically, the electrical resistance of a metal increases as the temperature rises. Various alloys have a high temperature-resistance coefficient, meaning their resistance varies significantly with temperature. This can make them suitable for use in temperature sensing devices. The metal resistor is subjected to the fluid or area in which temperature needs to be measured. It is connected by wires to a resistance measuring device inside the flight deck indicator. The instrument dial is calibrated in degrees Fahrenheit or Celsius as desired rather than in ohms. As the temperature to be measured changes, the resistance of the metal changes and the resistance measuring indicator shows to what extent.

A typical electrical resistance thermometer looks like any other temperature gauge. Indicators are available in dual form for use in multiengine aircraft. Most indicators are self-compensating for changes in flight deck temperature. The heat-sensitive resistor is manufactured so that it has a definite resistance for each temperature value within its working range. The temperature-sensitive resistor element is a length or winding made of a nickel/manganese wire or other suitable alloy in an insulating material. The resistor is protected by a closed-end metal tube attached to a threaded plug with a hexagonal head. [Figure 10-68] The two ends of the winding are brazed, or welded, to an electrical receptacle

Figure 10-68. *An electric resistance thermometer sensing bulb.*

designed to receive the prongs of the connector plug.

The indicator contains a resistance-measuring instrument. Sometimes it uses a modified form of the Wheatstone-bridge circuit. The Wheatstone-bridge meter operates on the principle of balancing one unknown resistor against other known resistances. A simplified form of a Wheatstone-bridge circuit is shown in *Figure 10-69*. Three equal values of resistance *[Figure 10-69A, B, and C]* are connected into a diamond shaped bridge circuit. A resistor with an unknown value *[Figure 10-69D]* is also part of the circuit. The unknown resistance represents the resistance of the temperature bulb of the electrical resistance thermometer system. A galvanometer is attached across the circuit at points X and Y.

When the temperature causes the resistance of the bulb to equal that of the other resistances, no potential difference exists between points X and Y in the circuit. Therefore, no current flows in the galvanometer leg of the circuit. If the temperature of the bulb changes, its resistance also changes, and the bridge becomes unbalanced, causing current to flow through the galvanometer in one direction or the other. The galvanometer pointer is actually the temperature gauge pointer. As it moves against the dial face calibrated in degrees, it indicates temperature. Many indicators are provided with a zero adjustment screw on the face of the instrument. This adjusts the zeroing spring tension of the pointer when the bridge is at the balance point (the position at which the bridge circuit is balanced and no current flows through the meter).

Ratiometer Electrical Resistance Thermometers

Another way of indicating temperature when employing an electric resistance thermometer is by using a ratiometer.

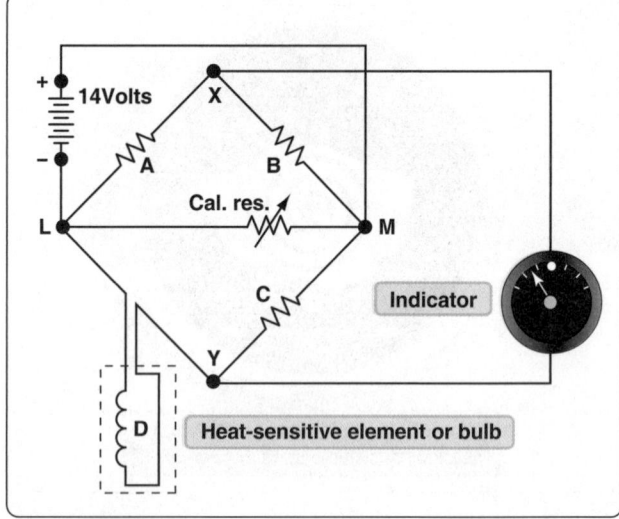

Figure 10-69. *The internal structure of an electric resistance thermometer indicator features a bridge circuit, galvanometer, and variable resistor, which is outside the indicator in the form of the temperature sensor.*

The Wheatstone-bridge indicator is subject to errors from line voltage fluctuation. The ratiometer is more stable and can deliver higher accuracy. As its name suggests, the ratiometer electrical resistance thermometer measures a ratio of current flows.

The resistance bulb sensing portion of the ratiometer electric resistance thermometer is essentially the same as described above. The circuit contains a variable resistance and a fixed resistance to provide the indication. It contains two branches for current flow. Each has a coil mounted on either side of the pointer assembly that is mounted within the magnetic field of a large permanent magnet. Varying current flow through the coils causes different magnetic fields to form, which react with the larger magnetic field of the permanent magnet. This interaction rotates the pointer against the dial face that is calibrated in degrees Fahrenheit or Celsius, giving a temperature indication. *[Figure 10-70]*

The magnetic pole ends of the permanent magnet are closer at the top than they are at the bottom. This causes the magnetic field lines of flux between the poles to be more concentrated at the top. As the two coils produce their magnetic fields, the stronger field interacts and pivots downward into the weaker, less concentrated part of the permanent magnet field, while the weaker coil magnetic field shifts upward toward the more concentrated flux field of the large magnet. This provides a balancing effect that changes but stays in balance as the coil field strengths vary with temperature and the resultant current flowing through the coils.

For example, if the resistance of the temperature bulb is equal to the value of the fixed resistance (R), equal values

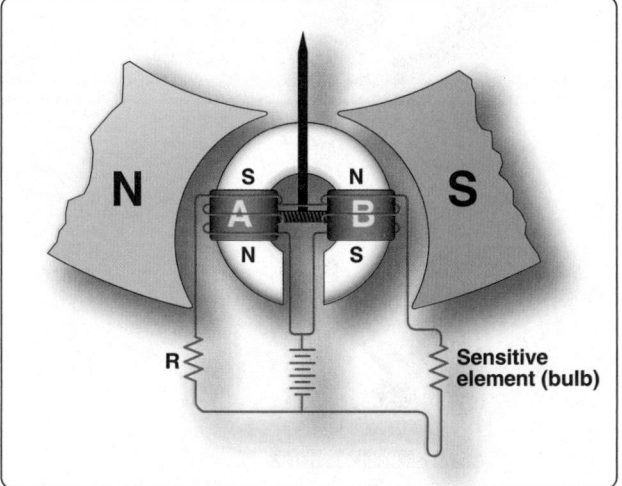

Figure 10-70. *A ratiometer temperature measuring indicator has two coils. As the sensor bulb resistance varies with temperature, different amounts of current flow through the coils. This produces varying magnetic fields. These fields interact with the magnetic field of a large permanent magnet, resulting in an indication of temperature.*

of current flow through the coils. The torques, caused by the magnetic field each coil creates, are the same and cancel any movement in the larger magnetic field. The indicator pointer will be in the vertical position. If the bulb temperature increases, its resistance also increases. This causes the current flow through coil A circuit branch to increase. This creates a stronger magnetic field at coil A than at coil B. Consequently, the torque on coil A increases, and it is pulled downward into the weaker part of the large magnetic field. At the same time, less current flows through the sensor bulb resistor and coil B, causing coil B to form a weaker magnetic field that is pulled upward into the stronger flux area of the permanent magnet's magnetic field. The pointer stops rotating when the fields reach a new balance point that is directly related to the resistance in the sensing bulb. The opposite of this action would take place if the temperature of the heat-sensitive bulb should decrease.

Ratiometer temperature measuring systems are used to measure engine oil, outside air, carburetor air, and other temperatures in many types of aircraft. They are especially in demand to measure temperature conditions where accuracy is important, or large variations of supply voltages are encountered.

Thermocouple Temperature Indicators

A thermocouple is a circuit or connection of two unlike metals. The metals are touching at two separate junctions. If one of the junctions is heated to a higher temperature than the other, an electromotive force is produced in the circuit. This voltage is directly proportional to the temperature. So, by measuring the amount of electromotive force, temperature can be determined. A voltmeter is placed across the colder of the two junctions of the thermocouple. It is calibrated in degrees Fahrenheit or Celsius, as needed. The hotter the high-temperature junction (hot junction) becomes, the greater the electromotive force produced, and the higher the temperature indication on the meter. *[Figure 10-71]*

Thermocouples are used to measure high temperatures. Two common applications are the measurement of cylinder head temperature (CHT) in reciprocating engines and exhaust gas temperature (EGT) in turbine engines. Thermocouple leads are made from a variety of metals, depending on the maximum temperature to which they are exposed. Iron and constantan, or copper and constantan, are common for CHT measurement. Chromel and alumel are used for turbine EGT thermocouples.

The amount of voltage produced by the dissimilar metals when heated is measured in millivolts. Therefore, thermocouple leads are designed to provide a specific amount of resistance in the thermocouple circuit (usually very little). Their material, length, or cross-sectional size cannot be altered

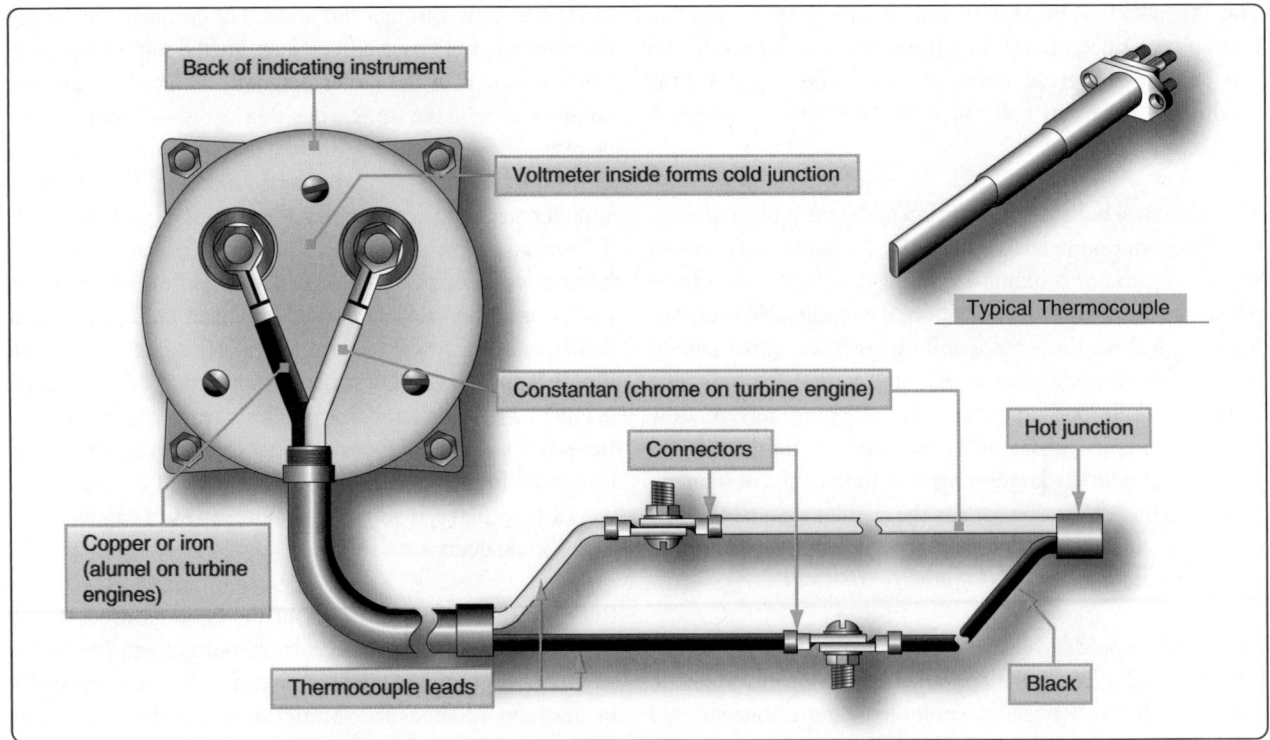

Figure 10-71. *Thermocouples combine two unlike metals that cause current flow when heated.*

without compensation for the change in total resistance that would result. Each lead that makes a connection back to the voltmeter must be made of the same metal as the part of the thermocouple to which it is connected. For example, a copper wire is connected to the copper portion of the hot junction and a constantan wire is connected to the constantan part.

The hot junction of a thermocouple varies in shape depending on its application. Two common types are the gasket and the bayonet. In the gasket type, two rings of the dissimilar metals are pressed together to form a gasket that can be installed under a spark plug or cylinder hold down nut. In the bayonet type, the metals come together inside a perforated protective sheath. Bayonet thermocouples fit into a hole or well in a cylinder head. On turbine engines, they are found mounted on the turbine inlet or outlet case and extend through the case into the gas stream. Note that for CHT indication, the cylinder chosen for the thermocouple installation is the one that runs the hottest under most operating conditions. The location of this cylinder varies with different engines. *[Figure 10-72]*

The cold junction of the thermocouple circuit is inside the instrument case. Since the electromotive force set up in the circuit varies with the difference in temperature between the hot and cold junctions, it is necessary to compensate the indicator mechanism for changes in flight deck temperature which affect the cold junction. This is accomplished by using a bimetallic spring connected to the indicator mechanism. This actually works the same as the bimetallic thermometer described previously. When the leads are disconnected

from the indicator, the temperature of the flight deck area around the instrument panel can be read on the indicator dial. *[Figure 10-73]* Numeric LED indictors for CHT are also common in modern aircraft.

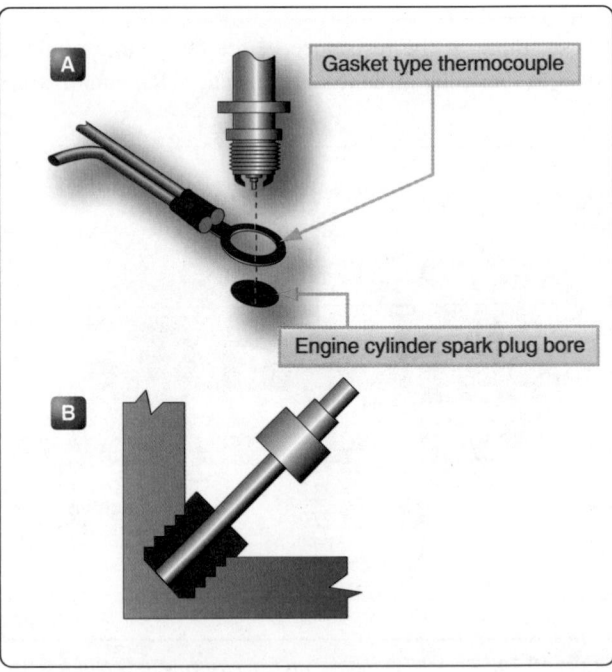

Figure 10-72. *A cylinder head temperature thermocouple with a gasket type hot junction is made to be installed under the spark plug or a cylinder hold down nut of the hottest cylinder (A). A bayonet type thermocouple is installed in a bore in the cylinder wall (B).*

Figure 10-73. *Typical thermocouple temperature indicators.*

Turbine Gas Temperature Indicating Systems

EGT is a critical variable of turbine engine operation. The EGT indicating system provides a visual temperature indication in the flight deck of the turbine exhaust gases as they leave the turbine unit. In certain turbine engines,

the temperature of the exhaust gases is measured at the entrance to the turbine unit. This is referred to as a turbine inlet temperature (TIT) indicating system.

Several thermocouples are used to measure EGT or TIT. They are spaced at intervals around the perimeter of the engine turbine casing or exhaust duct. The tiny thermocouple voltages are typically amplified and used to energize a servomotor that drives the indicator pointer. Gearing a digital drum indication off of the pointer motion is common. *[Figure 10-74]* The EGT indicator shown is a hermetically sealed unit. The instrument's scale ranges from 0 °C to 1,200 °C, with a vernier dial in the upper right-hand corner and a power off warning flag located in the lower portion of the dial.

A TIT indicating system provides a visual indication at the instrument panel of the temperature of gases entering the turbine. Numerous thermocouples can be used with the average voltage representing the TIT. Dual thermocouples exist containing two electrically independent junctions within a single probe. One set of these thermocouples is paralleled to transmit signals to the flight deck indicator. The other set of parallel thermocouples provides temperature signals to engine monitoring and control systems. Each circuit is electrically independent, providing dual system reliability.

A schematic for the turbine inlet temperature system for one engine of a four-engine turbine aircraft is shown in *Figure 10-75.* Circuits for the other three engines are identical to this system. The indicator contains a bridge circuit, a chopper circuit, a two-phase motor to drive the pointer, and a feedback potentiometer. Also included are a voltage reference circuit, an amplifier, a power-off flag, a power supply, and an over-temperature warning light. Output of the amplifier energizes the variable field of the two-phase motor that positions the indicator main pointer and a digital indicator. The motor also drives the feedback potentiometer to provide a humming signal to stop the drive motor when

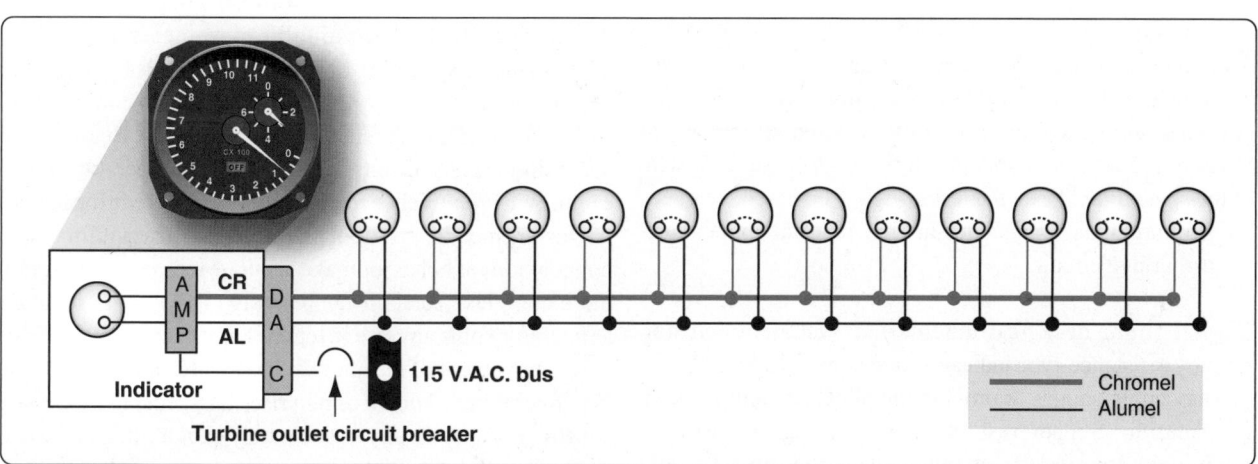

Figure 10-74. *A typical exhaust gas temperature thermocouple system.*

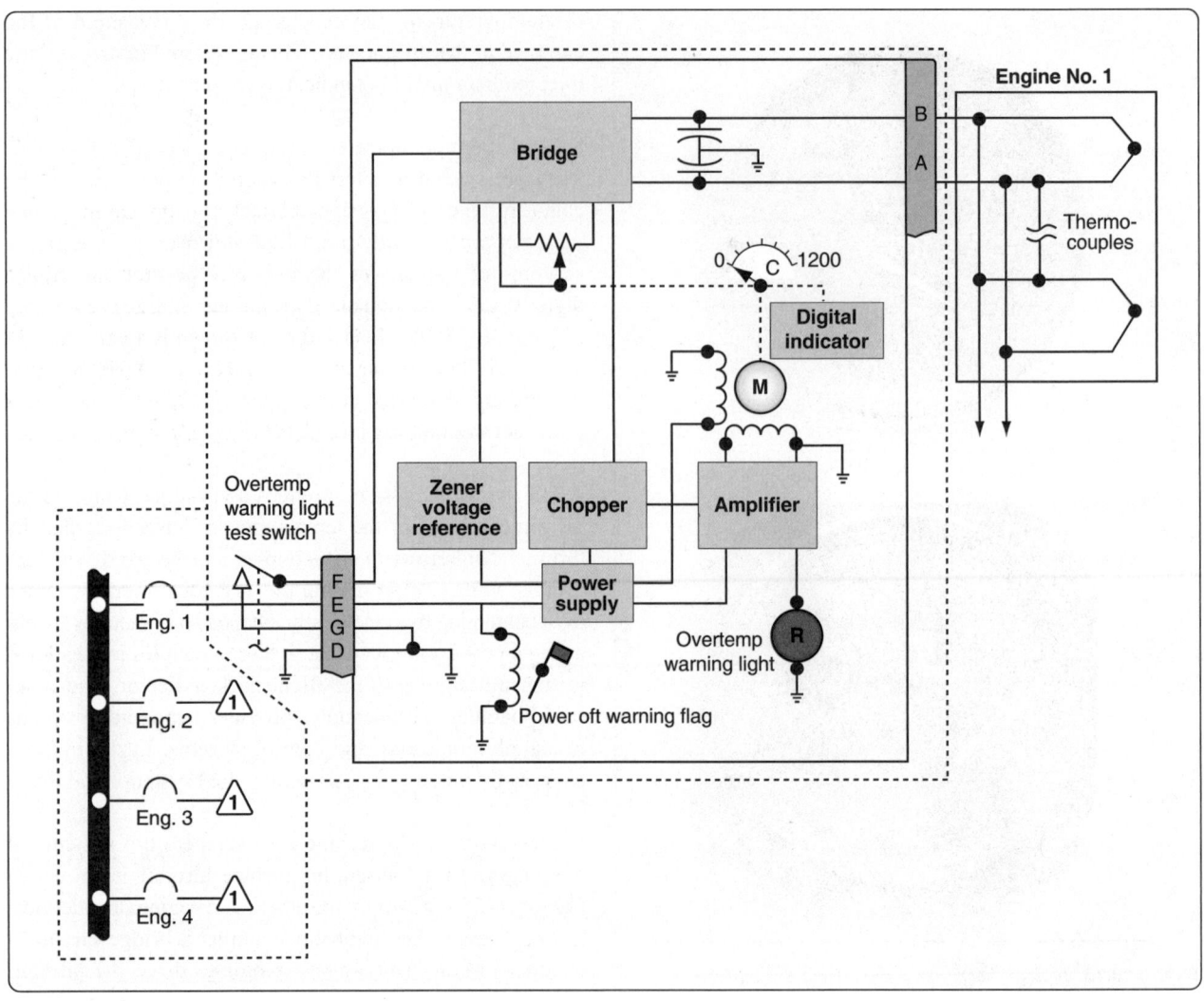

Figure 10-75. *A typical analog turbine inlet temperature indicating system.*

the correct pointer position, relative to the temperature signal, has been reached. The voltage reference circuit provides a closely regulated reference voltage in the bridge circuit to preclude error from input voltage variation to the indicator power supply.

The over-temperature warning light in the indicator illuminates when the TIT reaches a predetermined limit. An external test switch is usually installed so that over-temperature warning lights for all the engines can be tested at the same time. When the test switch is operated, an over-temperature signal is simulated in each indicator temperature control bridge circuit.

Digital flight deck instrumentation systems need not employ resistance-type indicators and adjusted servo-driven thermocouple gauges to provide the pilot with temperature information. Sensor resistance and voltage values are input to the appropriate computer, where they are adjusted, processed, monitored, and output for display on flight deck

display panels. They are also sent for use by other computers requiring temperature information for the control and monitoring of various integrated systems.

Total Air Temperature Measurement

Air temperature is a valuable parameter that many performance monitoring and control variables depend on. During flight, static air temperature changes continuously and accurate measurement presents challenges. Below Mach 0.2, a simple resistance-type or bimetallic temperature gauge can provide relatively accurate air temperature information. At faster speeds, friction, the air's compressibility, and boundary layer behavior make accurate temperature capture more complex. Total air temperature (TAT) is the static air temperature plus any rise in temperature caused by the high-speed movement of the aircraft through the air. The increase in temperature is known as ram rise. TAT-sensing probes are constructed specifically to accurately capture this value and transmit signals for flight deck indication, as well as for use in various engine and aircraft systems.

Simple TAT systems include a sensor and an indicator with a built-in resistance balance circuit. Air flow through the sensor is designed so that air with the precise temperature impacts a platinum alloy resistance element. The sensor is engineered to capture temperature variations in terms of varying the resistance of the element. When placed in the bridge circuit, the indicator pointer moves in response to the imbalance caused by the variable resistor.

More complex systems use signal correction technology and amplified signals sent to a servo motor to adjust the indicator in the flight deck. These systems include closely regulated power supply and failure monitoring. They often use numeric drum type readouts but can also be sent to an LCD driver to illuminate LCD displays. Many LCD displays are multifunctional, capable of displaying static air temperature and true airspeed. In fully digital systems, the correction signals are input into the ADC. There, they can be manipulated appropriately for flight deck display or for whichever system requires temperature information. *[Figure 10-76]*

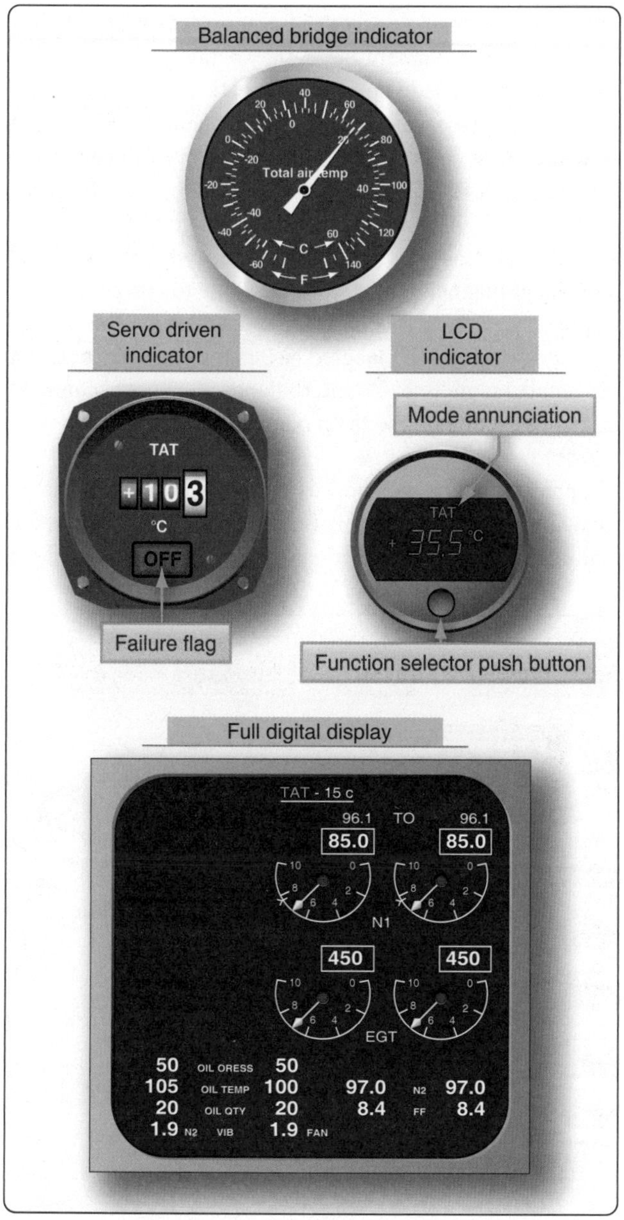

TAT sensor/probe design is complicated by the potential of ice forming during icing conditions. Left unheated, a probe may cease to function properly. The inclusion of a heating element threatens accurate data collection. Heating the probe must not affect the resistance of the sensor element. *[Figure 10-77]*

Close attention is paid to airflow and materials conductivity during the design phase. Some TAT sensors channel bleed air through the units to affect the flow of outside air, so that it flows directly onto the platinum sensor without gaining added energy from the probe heater.

Direction Indicating Instruments

A myriad of techniques and instruments exist to aid the pilot in navigation of the aircraft. An indication of direction is part of this navigation. While the next chapter deals with communication and navigation, this section discusses some of the magnetic direction indicating instruments. Additionally, a common, reliable gyroscopic direction indicator is discussed in the gyroscopic instrument section of this chapter.

Magnetic Compass

Having an instrument on board an aircraft that indicates direction can be invaluable to the pilot. 14 CFR part 91, section 91.205 requires that aircraft with standard category airworthiness certificates have a magnetic direction indicator for VFR flight during the day. The magnetic compass is a direction finding instrument that has been used for navigation for hundreds of years. It is a simple instrument that takes advantage of the earth's magnetic field.

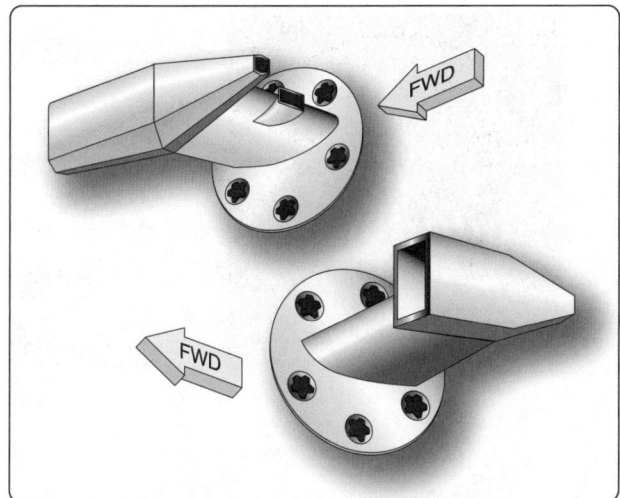

Figure 10-76. *Different flight deck TAT displays.*

Figure 10-77. *Total air temperature (TAT) probes.*

Figure 10-78 shows the earth and the magnetic field that surrounds it. The magnetic north pole is very close to the geographic North Pole of the globe, but they are not the same. An ordinary permanent magnet that is free to do so, aligns itself with the direction of the earth's magnetic field. Upon this principle, an instrument is constructed that the pilot can reference for directional orientation. Permanent magnets are attached under a float that is mounted on a pivot so it is free to rotate in the horizontal plane. As such, the magnets align with the earth's magnetic field. A numerical compass card, usually graduated in 5° increments, is constructed around the perimeter of the float. It serves as the instrument dial. The entire assembly is enclosed in a sealed case that is filled with a liquid similar to kerosene. This dampens vibration and oscillation of the moving float assembly and decreases friction.

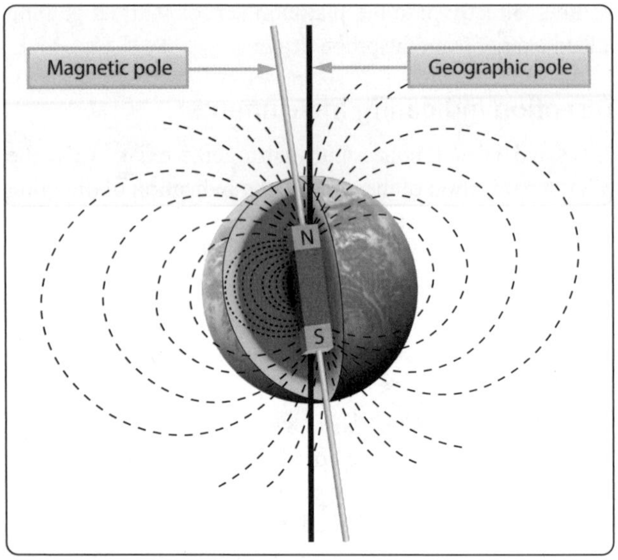

On the front of the case, a glass face allows the numerical compass card to be referenced against a vertical lubber line. The magnetic heading of the aircraft is read by noting the graduation on which the lubber line falls. Thus, direction in any of 360° can be read off the dial as the magnetic float compass card assembly holds its alignment with magnetic north, while the aircraft changes direction.

The liquid that fills the compass case expands and contracts as altitude changes and temperature fluctuates. A bellows diaphragm expands and contracts to adjust the volume of the space inside the case so it remains full. *[Figure 10-79]*

There are accuracy issues associated with using a magnetic compass. The main magnets of a compass align not only with the earth's magnetic field, they actually align with the composite field made up of all magnetic influences around them, meaning local electromagnetic influence from metallic structures near the compass and operation of the aircraft's electrical system. This is called magnetic deviation. It causes a magnet's alignment with the earth's magnetic field to be altered. Compensating screws are turned, which move small permanent magnets in the compass case to correct for this magnetic deviation. The two set-screws are on the face of the instrument and are labeled N-S and E-W. They position the small magnets to counterbalance the local magnetic influences acting on the main compass magnets.

The process for knowing how to adjust for deviation is known as swinging the compass. It is described in the instrument maintenance pages near the end of this chapter. Magnetic

Figure 10-78. *The earth and its magnetic field.*

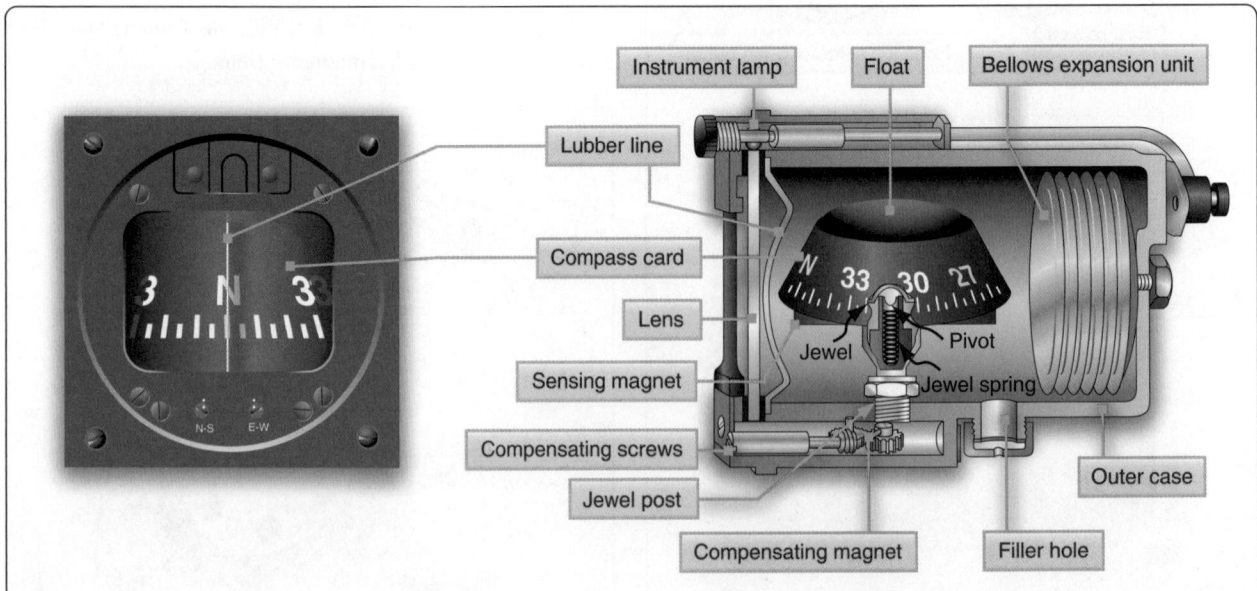

Figure 10-79. *The parts of a typical magnetic compass.*

deviation cannot be overlooked. It should never be more than 10°. Using nonferrous mounting screws and shielding or twisting the wire running to the compass illuminating lamp are additional steps taken to keep deviation to a minimum.

Another compass error is called magnetic variation. It is caused by the difference in location between the earth's magnetic poles and the geographic poles. There are only a few places on the planet where a compass pointing to magnetic north is also pointing to geographic North. A line drawn through these locations is called the Agonic line. At all other points, there is some variation between that which a magnetic compass indicates is north and geographic (true) North. Isogonic lines drawn on aeronautical charts indicate points of equal variation. Depending on the location of the aircraft, airmen must add or subtract degrees from the magnetic indication to obtain true geographic location information. *[Figure 10-80]*

The earth's magnetic field exits the poles vertically and arches around to extend past the equator horizontally or parallel to the earth's surface. *[Figure 10-78]* Operating an aircraft near the magnetic poles causes what is known as dip error. The compass magnets pull downward toward the pole, rather than horizontally, as is the case near the equator. This downward motion causes inaccuracy in the indication. Although the compass float mechanism is weighted to compensate, the closer the aircraft is to the north or south magnetic poles, the more pronounced the errors.

Dip errors manifest themselves in two ways. The first is called acceleration error. If an aircraft is flying on an east-west path and simply accelerates, the inertia of the float mechanism causes the compass to swing to the north. Rapid deceleration causes it to swing southward. Second, if flying toward the North Pole and a banked turn is made, the downward pull

of the magnetic field initially pulls the card away from the direction of the turn. The opposite is true if flying south from the North Pole and a banked turn is initiated. In this case, there is initially a pull of the compass indicator toward the direction of the turn. These kinds of movements are called turning errors.

Another peculiarity exists with the magnetic compass that is not dip error. Look again at the magnetic compass in *Figure 10-79*. If flying north or toward any indicated heading, turning the aircraft to the left causes a steady decrease in the heading numbers. But, before the turn is made, the numbers to the left on the compass card are actually increasing. The numbers to the right of the lubber line rotate behind it on a left turn. So, the compass card rotates opposite to the direction of the intended turn. This is because, from the pilot's seat, you are actually looking at the back of the compass card. While not a major problem, it is more intuitive to see the 360° of direction oriented as they are on an aeronautical chart or a hand-held compass.

Vertical Magnetic Compass

Solutions to the shortcomings of the simple magnetic compass described above have been engineered. The vertical magnetic compass is a variation of the magnetic compass that eliminates the reverse rotation of the compass card just described. By mounting the main indicating magnets of the compass on a shaft rather than a float, through a series of gears, a compass card can be made to turn about a horizontal axis. This allows the numbers for a heading, towards which the pilot wants to turn, to be oriented correctly on the indicating card. In other words, when turning right, increasing numbers are to the right; when turning left, decreasing numbers rotate in from the left. *[Figure 10-81]*

Many vertical magnetic compasses have also replaced the liquid-filled instrument housing with a dampening cup that uses eddy currents to dampen oscillations. Note that a vertical magnetic compass and a directional gyro look very similar and are often in the lower center position of the instrument panel basic T. Both use the nose of an aircraft as the lubber line against which a rotating compass card is read. Vertical magnetic compasses are characterized by the absence of the hand adjustment knob found on DGs, which is used to align the gyro with a magnetic indication.

Remote Indicating Compass

Magnetic deviation is compensated for by swinging the compass and adjusting compensating magnets in the instrument housing. A better solution to deviation is to remotely locate the magnetic compass in a wing tip or vertical stabilizer where there is very little interference with the earth's magnetic field. By using a synchro remote indicating system, the magnetic compass float assembly can act as the rotor of the synchro system. As the float mechanism rotates to

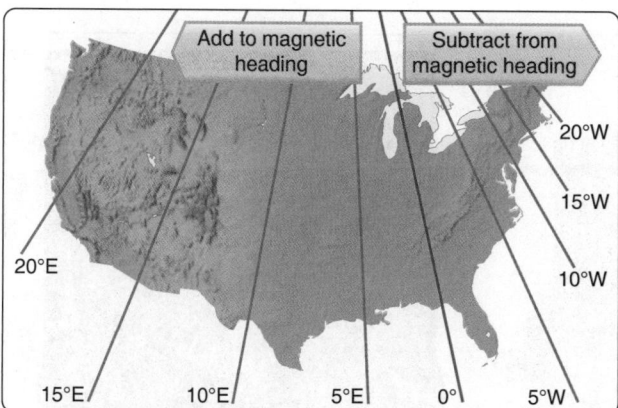

Figure 10-80. *Aircraft located along the agonic line have 0° of variation between magnetic north and true north. Locations on and between the isogonic lines require addition or subtraction, as shown, to magnetic indications to arrive at a true geographic direction.*

Figure 10-81. *A vertical magnetic direction indicator provides a realistic reference of headings.*

section of this chapter assists in understanding this device.

A gyroscopic direction indicator is augmented by magnetic direction information from a remotely located compass. The type of compass used is called a flux valve or flux gate compass. It consists of a very magnetically permeable circular segmented core frame called a spider. The earth's magnetic field flows through this iron core and varies its distribution through segments of the core as the flux valve is rotated via the movement of the aircraft. Pickup coil windings are located on each of the core's spider legs that are positioned 120° apart. *[Figure 10-83]*

The distribution of earth's magnetic field flowing through the legs is unique for every directional orientation of the aircraft. A coil is placed in the center of the core and is energized by AC current. As the AC flow passes through zero while changing direction, the earth's magnetic field is allowed to flow through the core. Then, it is blocked or gated as the magnetic field of the core current flow builds to its peak again. The cycle is repeated at the frequency of the AC supplied to the excitation coil. The result is repeated flow and nonflow of the earth's flux across the pickup coils. During each cycle, a unique voltage is induced in each of the pickup coils reflecting the orientation of the aircraft in the earth's magnetic field.

align with magnetic north in the remotely located compass, a varied electric current can be produced in the transmitter. This alters the magnetic field produced by the coils of the indicator in the flight deck, and a magnetic indication relatively free from deviation is displayed. Many of these systems are of the magnesyn type.

Remote Indicating Slaved Gyro Compass (Flux Gate Compass)

An elaborate and very accurate method of direction indication has been developed that combines the use of a gyro, a magnetic compass, and a remote indicating system. *[Figure 10-82]* It is called the slaved gyro compass or flux gate compass system. A study of the gyroscopic instruments

The electricity that flows from each of the pickup coils is transmitted out of the flux valve via wires into a second unit. It contains an autosyn transmitter, directional gyro, an amplifier, and a triple wound stator that is similar to that found in the indicator of a synchro system. Unique voltage is induced in the center rotor of this stator which reflects the voltage received from the flux valve pickup coils sent through the stator coils. It is amplified and used to augment the position of the DG. The gyro is wired to be the rotor of an autosyn synchro system, which transmits the position

| Flux valve or flux gate | DG/Amplifier or slaved gyro | Direction indicator |

Figure 10-82. *Components used to provide direction indication.*

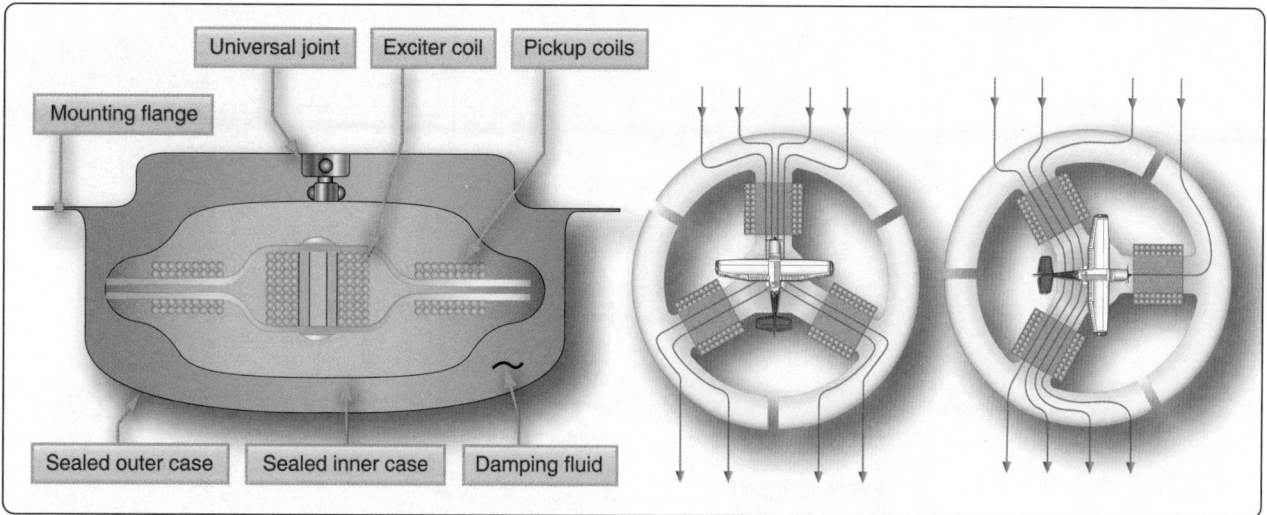

Figure 10-83. *As the aircraft turns in the earth's magnetic field, the lines of flux flow lines vary through the permeable core of flux gate, creating variable voltages at the three pickoffs.*

of the gyro into an indicator unit located in the flight deck. *[Figure 10-84]* In the indicator, a vertical compass card is rotated against a small airplane type lubber line like that in a vertical magnetic compass.

Further enhancements to direction finding systems of this type involving the integration of radio navigation aids are common. The radio magnetic indicator (RMI) is one such variation. *[Figure 10-85]* In addition to the rotating direction indicator of the slaved gyro compass, it contains two pointers. Each pointer can show the bearing to either a very high frequency (VHF) omnidirectional range (VOR) station or to a non-directional automatic direction finder (ADF) beacon. These and other radio navigation aids are discussed further in the communications and navigation chapter of this handbook. It should also be noted that integration of slaved gyro direction indicating system information into auto-pilot systems is also possible.

Solid State Magnetometers

Solid state magnetometers are used on many modern aircraft. They have no moving parts and are extremely accurate. Tiny layered structures react to magnetism on a molecular level resulting in variations in electron activity. These low power consuming devices can sense not only the direction to the earth's magnetic poles, but also the angle of the flux field. They are free from oscillation that plagues a standard magnetic compass. They feature integrated processing algorithms and easy integration with digital systems. *[Figure 10-86]*

Sources of Power for Gyroscopic Instruments

Gyroscopic instruments are essential instruments used on all aircraft. They provide the pilot with critical attitude and directional information and are particularly important while flying under IFR. The sources of power for these instruments can vary. The main requirement is to spin the gyroscopes at a high rate of speed. Originally, gyroscopic instruments were strictly vacuum driven. A vacuum source pulled air across the gyro inside the instruments to make the gyros spin. Later, electricity was added as a source of power. The turning armature of an electric motor doubles as the gyro rotor. In some aircraft, pressure, rather than vacuum, is used to induce the gyro to spin. Various systems and powering configurations have been developed to provide reliable operation of the gyroscopic instruments.

Vacuum Systems

Vacuum systems are very common for driving gyro instruments. In a vacuum system, a stream of air directed against the rotor vanes turns the rotor at high speed. The action is similar to a water wheel. Air at atmospheric pressure is first drawn through a filter(s). It is then routed into the instrument and directed at vanes on the gyro rotor. A suction line leads from the instrument case to the vacuum source. From there, the air is vented overboard. Either a venturi or a vacuum pump can be used to provide the vacuum required to spin the rotors of the gyro instruments.

The amount of vacuum required for instrument operation is usually between 3½ inches to 4½ inches of mercury. It is usually adjusted by a vacuum relief valve located in the supply line. Some turn-and-bank indicators require a lower vacuum setting. This can be obtained through the use of an additional regulating valve in the turn and bank vacuum supply line.

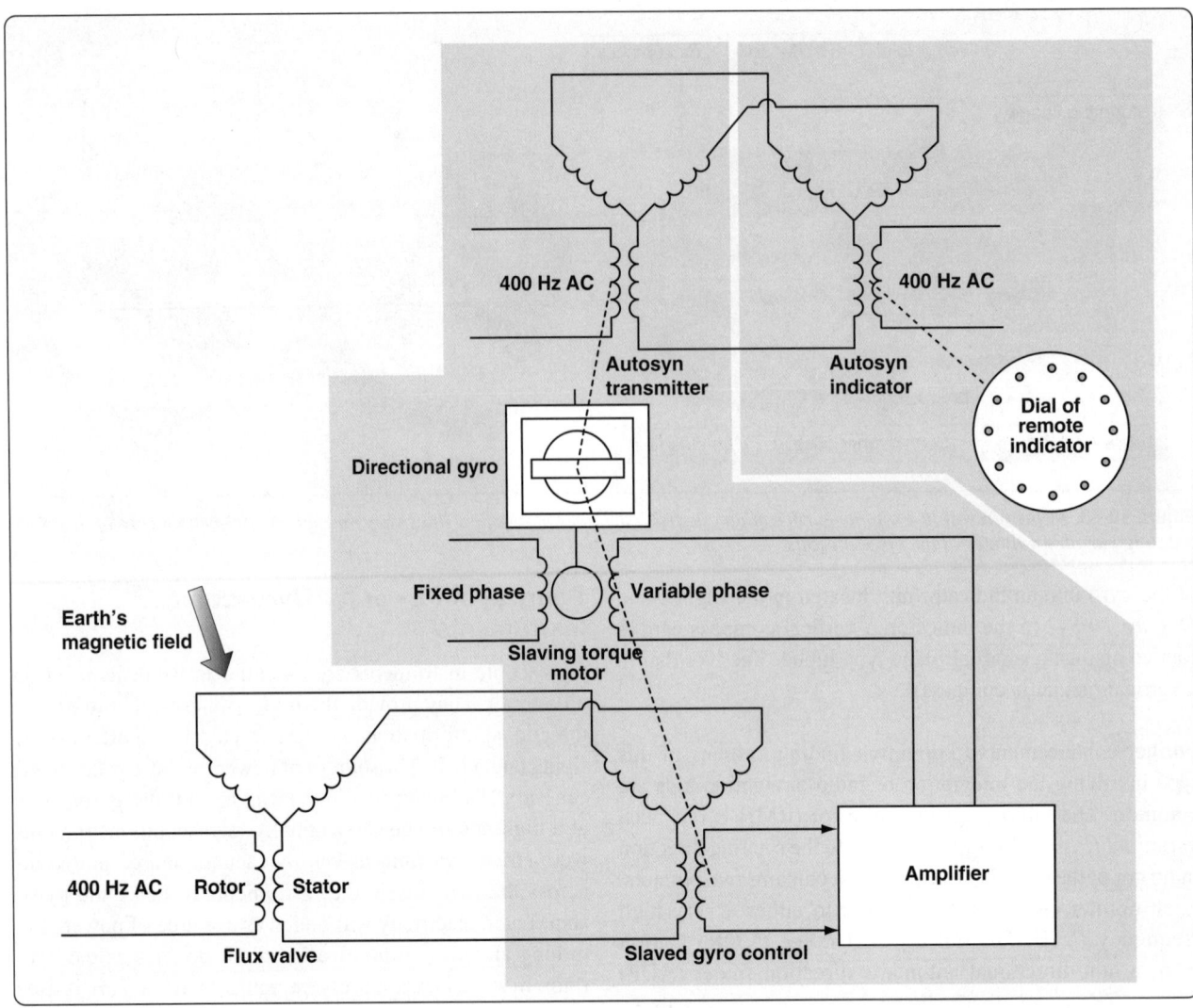

Figure 10-84. *A simplified schematic of a flux gate, or slaved gyro, compass system.*

Venturi Tube Systems

The velocity of the air rushing through a venturi can create sufficient suction to spin instrument gyros. A line is run from the gyro instruments to the throat of the venturi mounted on the outside of the airframe. The low pressure in the venturi tube pulls air through the instruments, spins the gyros, and expels the air overboard through the venturi. This source of gyro power is used on many simple, early aircraft.

A light, single-engine aircraft can be equipped with a 2-inch venturi (2 inches of mercury vacuum capacity) to operate the turn and bank indicator. It can also have a larger 8-inch venturi to power the attitude and heading indicators. Simplified illustrations of these venturi vacuum systems are shown in *Figure 10-87*. Normally, air going into the instruments is filtered.

The advantages of a venturi as a suction source are its relatively low cost and its simplicity of installation and operation. It also requires no electric power. But there are serious limitations. A venturi is designed to produce the desired vacuum at approximately 100 mph at standard sea level conditions. Wide variations in airspeed or air density cause the suction developed to fluctuate. Airflow can also be hampered by ice that can form on the venturi tube. Additionally, since the rotor does not reach normal operating speed until after takeoff, preflight operational checks of venturi powered gyro instruments cannot be made. For these reasons, alternate sources of vacuum power were developed.

Engine-Driven Vacuum Pump

The vane-type engine-driven pump is the most common source of vacuum for gyros installed in general aviation, light aircraft. One type of engine-driven pump is geared to the engine and is connected to the lubricating system to seal, cool, and lubricate the pump. Another commonly used pump is a dry vacuum pump. It operates without external lubrication and installation requires no connection to the

Figure 10-85. *A radio magnetic indicator (RMI) combines a slaved gyro heading indication (red triangle at top of gauge) with magnetic bearing information to a VOR station (solid pointer) and an ADF station (hollow pointer).*

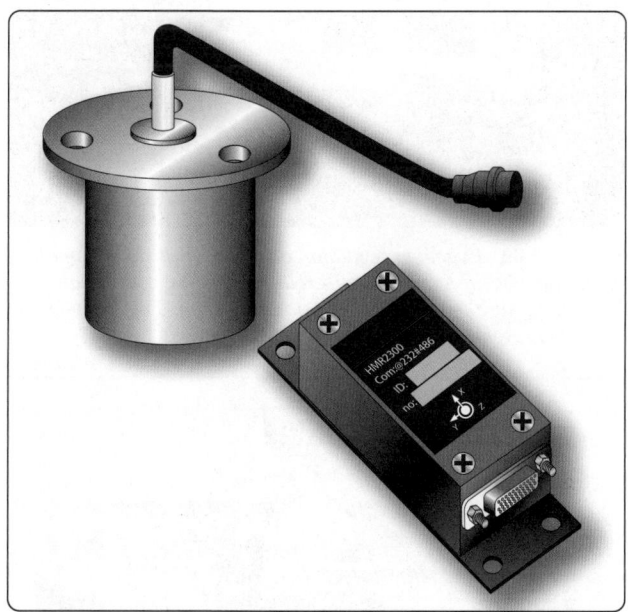

Figure 10-86. *Solid state magnetometer units.*

engine oil supply. It also does not need the air oil separator or gate check valve found in wet pump systems. In many other respects, the dry pump system and oil lubricated system are the same. *[Figure 10-88]*

When a vacuum pump develops a vacuum (negative pressure), it also creates a positive pressure at the outlet of the pump. This pressure is compressed air. Sometimes,

it is utilized to operate pressure gyro instruments. The components for pressure systems are much the same as those for a vacuum system as listed below. Other times, the pressure developed by the vacuum pump is used to inflate de-ice boots or inflatable seals or it is vented overboard.

An advantage of engine-driven pumps is their consistent performance on the ground and in flight. Even at low engine rpm, they can produce more than enough vacuum so that a regulator in the system is needed to continuously provide the correct suction to the vacuum instruments. As long as the engine operates, the relatively simple vacuum system adequately spins the instrument gyros for accurate indications. However, engine failure, especially on single-engine aircraft, could leave the pilot without attitude and directional information at a critical time. To thwart this shortcoming, often the turn and bank indicator operates with an electrically driven gyro that can be driven by the battery for a short time. Thus, when combined with the aircraft's magnetic compass, sufficient attitude and directional information is still available.

Multiengine aircraft typically contain independent vacuum systems for the pilot and copilot instruments driven by separate vacuum pumps on each of the engines. Should an engine fail, the vacuum system driven by the still operating engine supplies a full complement of gyro instruments. An interconnect valve may also be installed to connect the failed instruments to the still operational pump.

Typical Pump-Driven System

The following components are found in a typical vacuum system for gyroscopic power supply. A brief description is given of each. Refer to the figures for detailed illustrations.

Air-oil separator—oil and air in the vacuum pump are exhausted through the separator, which separates the oil from the air; the air is vented overboard, and the oil is returned to the engine sump. This component is not present when a dry-type vacuum pump is used. The self-lubricating nature of the pump vanes requires no oil.

Vacuum regulator or suction relief valve—since the system capacity is more than is needed for operation of the instruments, the adjustable vacuum regulator is set for the vacuum desired for the instruments. Excess suction in the instrument lines is reduced when the spring-loaded valve opens to atmospheric pressure. *[Figure 10-89]*

Gate check valve—prevents possible damage to the instruments by engine backfire that would reverse the flow of air and oil from the pump. *[Figure 10-90]*

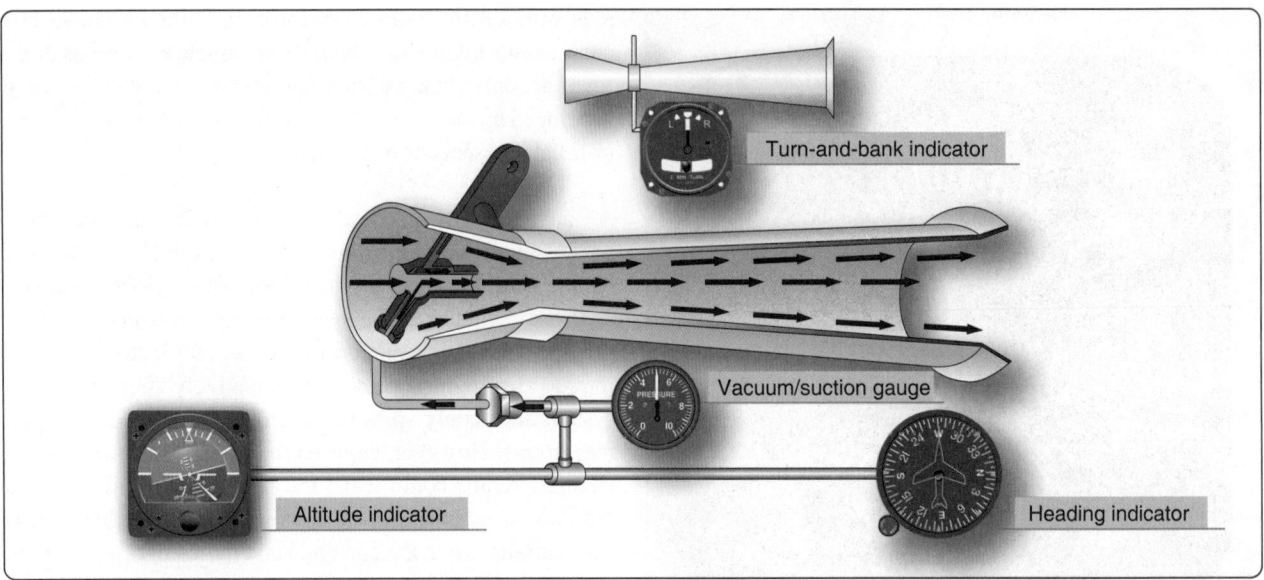

Figure 10-87. *Simple venturi tube systems for powering gyroscopic instruments.*

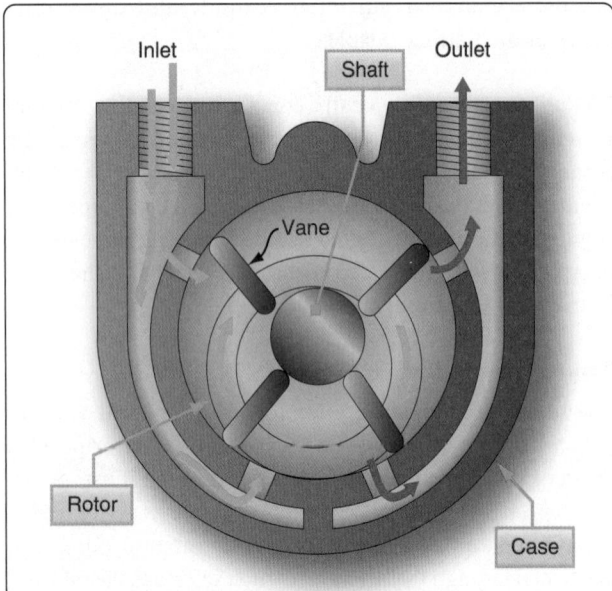

Figure 10-88. *Cutaway view of a vane-type engine-driven vacuum pump used to power gyroscopic instruments.*

Figure 10-89. *A vacuum regulator, also known as a suction relief valve, includes a foam filter. To relieve vacuum, outside air of a higher pressure must be drawn into the system. This air must be clean to prevent damage to the pump.*

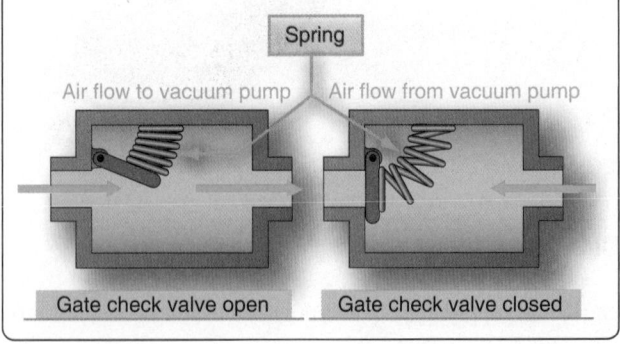

Figure 10-90. *Gate check valve used to prevent vacuum system damage from engine backfire.*

Pressure relief valve—since a reverse flow of air from the pump would close both the gate check valve and the suction relief valve, the resulting pressure could rupture the lines. The pressure relief valve vents positive pressure into the atmosphere.

Selector valve—In twin-engine aircraft having vacuum pumps driven by both engines, the alternate pump can be selected to provide vacuum in the event of either engine or pump failure, with a check valve incorporated to seal off the failed pump.

Restrictor valve—Since the turn needle of the turn and bank indicator operates on less vacuum than that required by the other instruments, the vacuum in the main line must be reduced for use by this instrument. An inline restrictor valve performs this function. This valve is either a needle valve or a spring-loaded regulating valve that maintains a constant, reduced vacuum for the turn-and-bank indicator.

Air filter—A master air filter screens foreign matter from the air flowing through all the gyro instruments. It is an extremely important filter requiring regular maintenance. Clogging of the master filter reduces airflow and causes a lower reading on the suction gauge. Each instrument is also provided with individual filters. In systems with no master filter that rely only upon individual filters, clogging of a filter does not necessarily show on the suction gauge.

Suction gauge—a pressure gauge which indicates the difference between the pressure inside the system and atmospheric or flight deck pressure. It is usually calibrated in inches of mercury. The desired vacuum and the minimum and maximum limits vary with gyro system design. If the desired vacuum for the attitude and heading indicators is 5 inches and the minimum is 4.6 inches, a reading below the latter value indicates that the airflow is not spinning the gyros fast enough for reliable operation. In many aircraft, the system provides a suction gauge selector valve permitting the pilot to check the vacuum at several points in the system.

Suction/vacuum pressures discussed in conjunction with the operation of vacuum systems are actually negative pressures, indicated as inches of mercury below that of atmospheric pressure. The minus sign is usually not presented, as the importance is placed on the magnitude of the vacuum developed. In relation to an absolute vacuum (0 psi or 0 "Hg), instrument vacuum systems have positive pressure.

Figure 10-91 shows a typical engine-driven pump vacuum system containing the above components. A pump capacity of approximately 10 "Hg at engine speeds above 1,000 rpm is normal. Pump capacity and pump size vary in different aircraft, depending on the number of gyros to be operated.

Twin-Engine Aircraft Vacuum System Operation

Twin-engine aircraft vacuum systems are more complicated. They contain an engine-driven vacuum pump on each engine. The associated lines and components for each pump are isolated from each other and act as two independent vacuum systems. The vacuum lines are routed from each vacuum pump through a vacuum relief valve and through a check valve to the vacuum four-way selector valve. The four-way valve permits either pump to supply a vacuum manifold. From the manifold, flexible hoses connect the vacuum-operated instruments into the system. To reduce the vacuum for the turn and bank indicators, needle valves are included in both lines to these units. Lines to the artificial horizons and the directional gyro receive full vacuum. From the instruments, lines are routed to the vacuum gauge through a turn and bank selector valve. This valve has three positions: main, left turn and bank (T&B), and right T&B. In the main position, the vacuum gauge indicates the vacuum in the lines

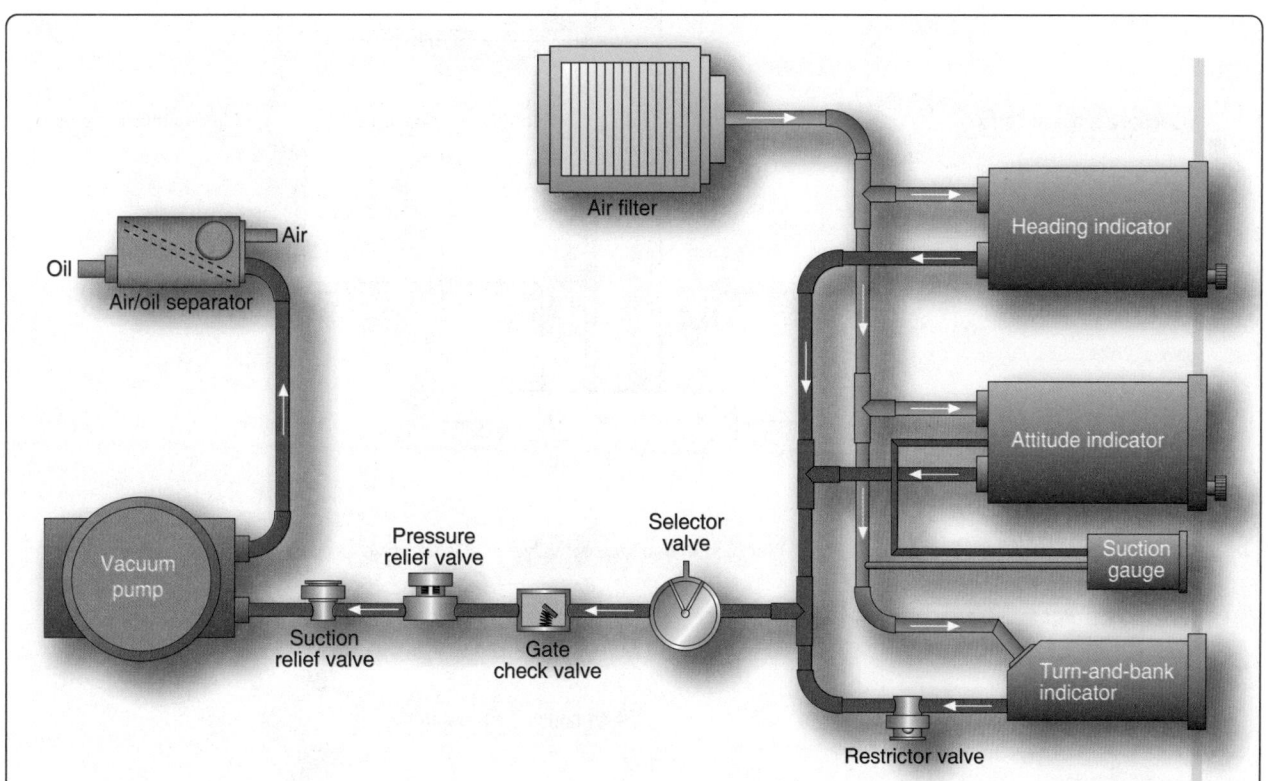

Figure 10-91. *A typical pump-driven vacuum system for powering gyroscopic instruments.*

of the artificial horizons and directional gyro. In the other positions, the lower value of vacuum for the turn and bank indicators can be read.

A schematic of this twin-engine aircraft vacuum system is shown in *Figure 10-92*. Note the following components: two engine-driven pumps, two vacuum relief valves, two flapper type check valves, a vacuum manifold, a vacuum restrictor for each turn and bank indicator, an engine four-way selector valve, one vacuum gauge, and a turn-and-bank selector valve. Not shown are system and individual instrument filters. A drain line may also be installed at the low point in the system.

Pressure-Driven Gyroscopic Instrument Systems

Gyroscopic instruments are finely balanced devices with jeweled bearings that must be kept clean to perform properly. When early vacuum systems were developed, only oil-lubricated pumps were available. Even with the use of air-oil separators, the pressure outputs of these pumps contain traces of oil and dirt. As a result, it was preferred to draw clean air through the gyro instruments with a vacuum system, rather than using pump output pressure that presented the risk of contamination. The development of self-lubricated dry pumps greatly reduced pressure output contaminates. This made pressure gyro systems possible.

At high altitudes, the use of pressure-driven gyros is more efficient. Pressure systems are similar to vacuum systems and make use of the same components, but they are designed for pressure instead of vacuum. Thus, a pressure regulator is used instead of a suction relief valve. Filters are still extremely important to prevent damage to the gyros. Normally, air is filtered at the inlet and outlet of the pump in a pressure gyro system.

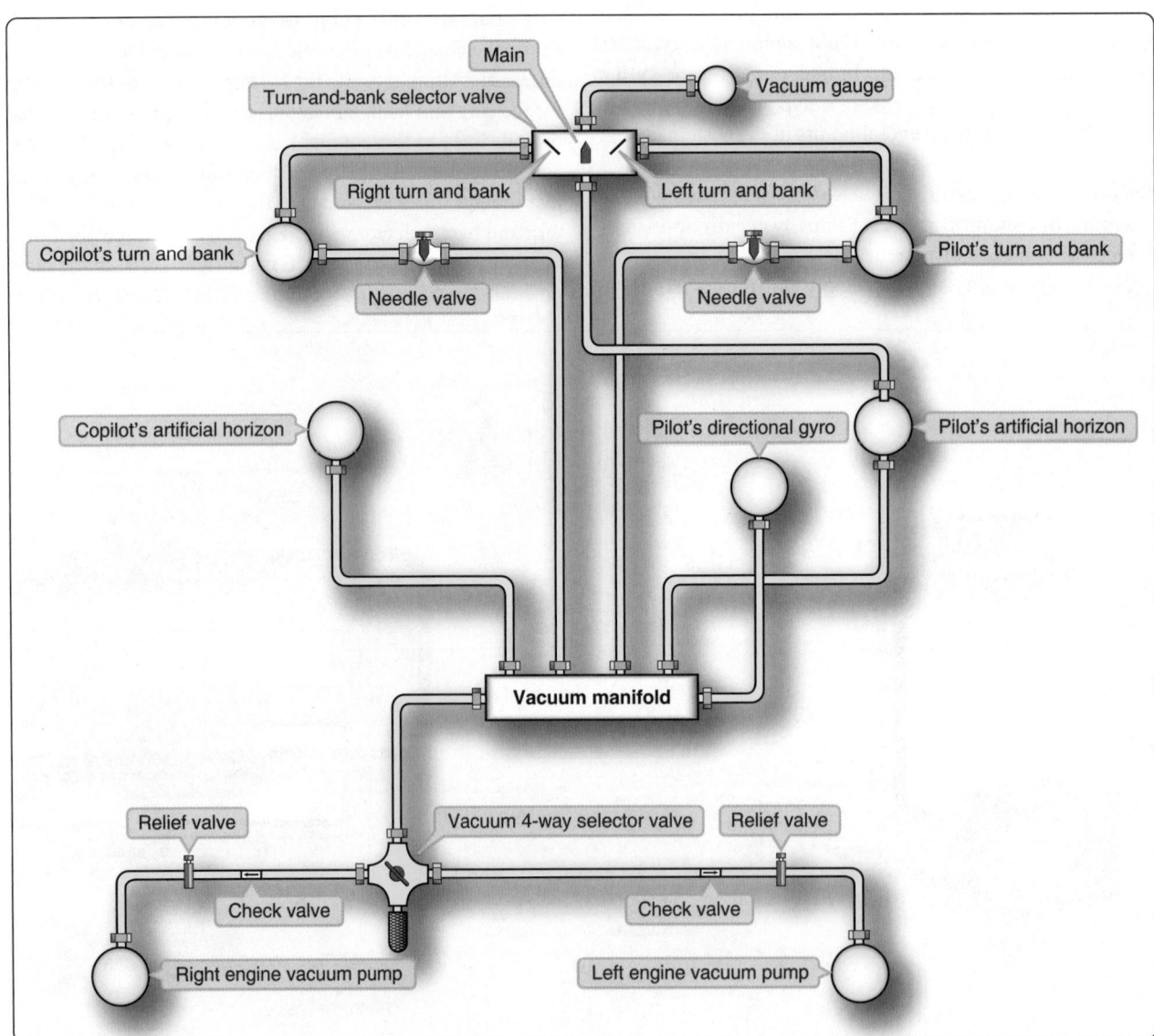

Figure 10-92. *An example of a twin-engine instrument vacuum system.*

Electrically-Driven Gyroscopic Instrument Systems

A spinning motor armature can act as a gyroscope. This is the basis for electrically driven gyroscopic instruments in which the gyro rotor spin is powered by an electric motor.

Electric gyros have the advantage of being powered by battery for a limited time if a generator fails or an engine is lost. Since air is not sent through the gyro to spin the rotor, contamination worries are also reduced. Also, elimination of vacuum pumps, plumbing, and vacuum system components saves weight.

On many small, single-engine aircraft, electric turn-and-bank or turn coordinators are combined with vacuum-powered attitude and directional gyro instruments as a means for redundancy. The reverse is also possible. By combining both types of instruments in the instrument panel, the pilot has more options. On more complex multiengine aircraft, reliable, redundant electrical systems make use of all electric-powered gyro instruments possible.

It should be noted that electric gyro instruments have a failure flag. The flag will come in if the gyro is not powered or is spinning at too slow of a speed. Usually, this is in the form of a red flag with the word "off" or "gyro" written on it.

Principles of Gyroscopic Instruments

Mechanical Gyros

Three of the most common flight instruments, the attitude indicator, heading indicator, and turn needle of the turn-and-bank indicator, are controlled by gyroscopes. To understand how these instruments operate, knowledge of gyroscopic principles and instrument power systems is required.

A mechanical gyroscope, or gyro, is comprised of a wheel or rotor with its mass concentrated around its perimeter. The rotor has bearings to enable it to spin at high speeds. [Figure 10-93A]

Different mounting configurations are available for the rotor and axle, which allow the rotor assembly to rotate about one or two axes perpendicular to its axis of spin. To suspend the rotor for rotation, the axle is first mounted in a supporting ring called a gimbal. [Figure 10-93B] If brackets are attached 90° around the supporting ring from where the spin axle attached, the supporting ring and rotor can both move freely 360°. When in this configuration, the gyro is said to be a captive gyro. It can rotate about only one axis that is perpendicular to the axis of spin. [Figure 10-93C]

The supporting ring can also be mounted inside an outer ring. The bearing points are the same as the bracket just described, 90° around the supporting ring from where the spin axle attached. Attachment of a bracket to this outer ring allows the rotor to rotate in two planes while spinning. Both of these are perpendicular to the spin axis of the rotor. The plane that the rotor spins in due to its rotation about its axle is not counted as a plane of rotation.

A gyroscope with this configuration, two rings plus the mounting bracket, is said to be a free gyro because it is free to rotate about two axes that are both perpendicular to the rotor's spin axis. [Figure 10-93D] As a result, the supporting ring with spinning gyro mounted inside is free to turn 360° inside the outer ring.

Unless the rotor of a gyro is spinning, it has no unusual properties; it is simply a wheel universally mounted. When the rotor is rotated at a high speed, the gyro exhibits a couple of unique characteristics. The first is called gyroscopic rigidity, or rigidity in space. This means that the rotor of a free gyro always points in the same direction no matter which way the base of the gyro is positioned. [Figure 10-94]

Gyroscopic rigidity depends upon several design factors:

1. Weight—for a given size, a heavy mass is more

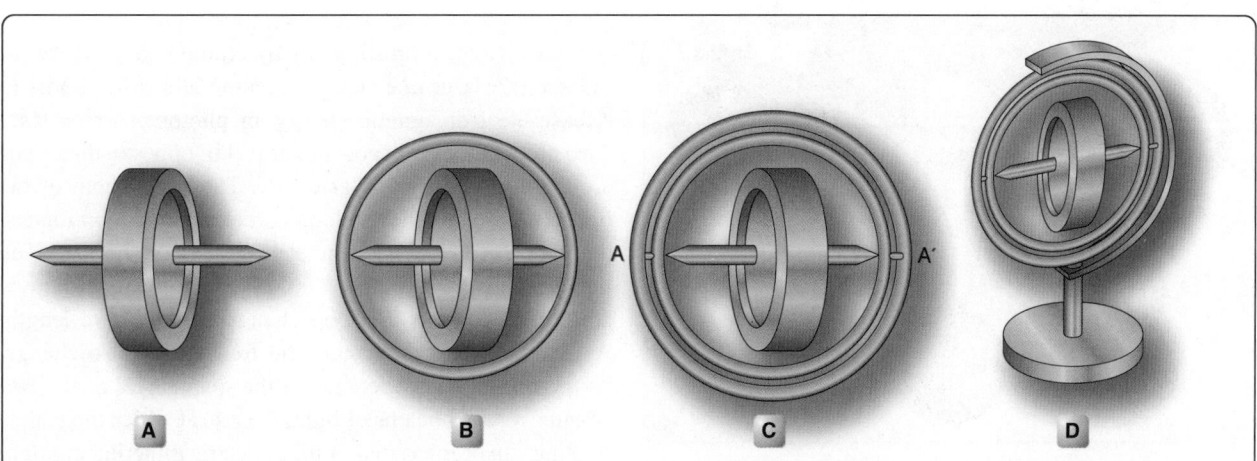

Figure 10-93. *Gyroscopes.*

resistant to disturbing forces than a light mass.

2. Angular velocity—the higher the rotational speed, the greater the rigidity or resistance is to deflection.

3. Radius at which the weight is concentrated—maximum effect is obtained from a mass when its principal weight is concentrated near the rim, rotating at high speed.

4. Bearing friction—any friction applies a deflecting force to a gyro. Minimum bearing friction keeps deflecting forces at a minimum.

This characteristic of gyros to remain rigid in space is exploited in the attitude-indicating instruments and the directional indicators that use gyros. The spinning gyro remains fixed in space and the airplane moves around it. It acts as a reference to allow measurement of changes in attitude or direction.

Precession is a second important characteristic of gyroscopes. By applying a force to the horizontal axis of the gyro, a unique phenomenon occurs. The applied force is resisted. Instead of responding to the force by moving about the horizontal axis, the gyro moves in response about its vertical axis. Stated another way, an applied force to the axis of the spinning gyro does not cause the axis to tilt. Rather, the gyro responds as though the force was applied 90° around in the direction of rotation of the gyro rotor. The gyro rotates rather than tilts. *[Figure 10-95]* This predictable controlled precession of a gyroscope is utilized in a turn and bank instrument.

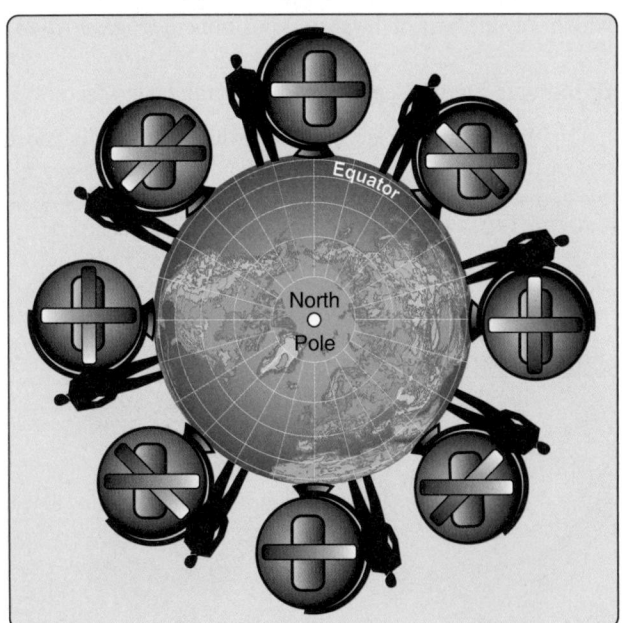

Figure 10-94. *Once spinning, a free gyro rotor stays oriented in the same position in space despite the position or location of its base.*

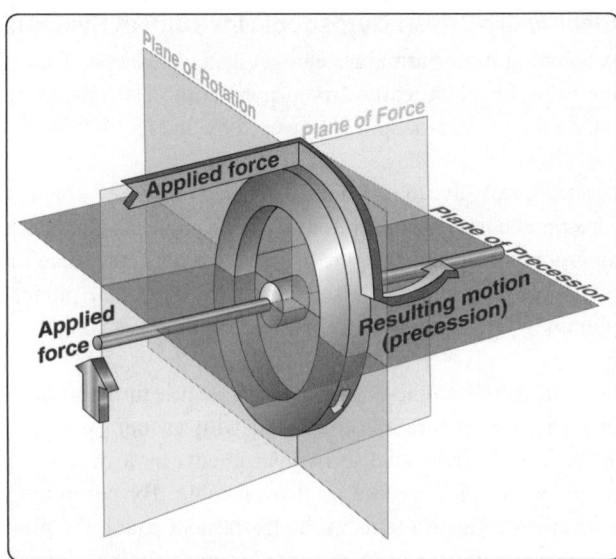

Figure 10-95. *When a force is applied to a spinning gyroscope, it reacts as though the force came from 90° further around the rotor in the direction it is spinning. The plane of the applied force, the plane of the rotation, and the plane in which the gyro responds (known as the plane of precession), are all perpendicular to each other.*

Solid State Gyros & Related Systems

Improved attitude and direction information is always a goal in aviation. Modern aircraft make use of highly accurate solid-state attitude and directional devices with no moving parts. This results in very high reliability and low maintenance.

Ring Laser Gyros (RLG)

The ring laser gyro (RLG) is widely used in commercial aviation. The basis for RLG operation is that it takes time for light to travel around a stationary, nonrotating circular path. Light takes longer to complete the journey if the path is rotating in the same direction as the light is traveling. And, it takes less time for the light to complete the loop if the path is rotating in the direction opposite to that of the light. Essentially, the path is made longer or shorter by the rotation of the path. *[Figure 10-96]* This is known as the Sagnac effect.

A laser is light amplification by stimulated emission of radiation. A laser operates by exciting atoms in plasma to release electromagnetic energy, or photons. A ring laser gyro produces laser beams that travel in opposite directions around a closed triangular cavity. The wavelength of the light traveling around the loop is fixed. As the loop rotates, the path the lasers must travel lengthens or shortens. The light wavelengths compress or expand to complete travel around the loop as the loop changes its effective length. As the wavelengths change, the frequencies also change. By examining the difference in the frequencies of the two counter-rotating beams of light, the rate at which the path is rotating can be measured. A piezoelectric dithering motor in the center of the unit vibrates to prevent lock-in of the output

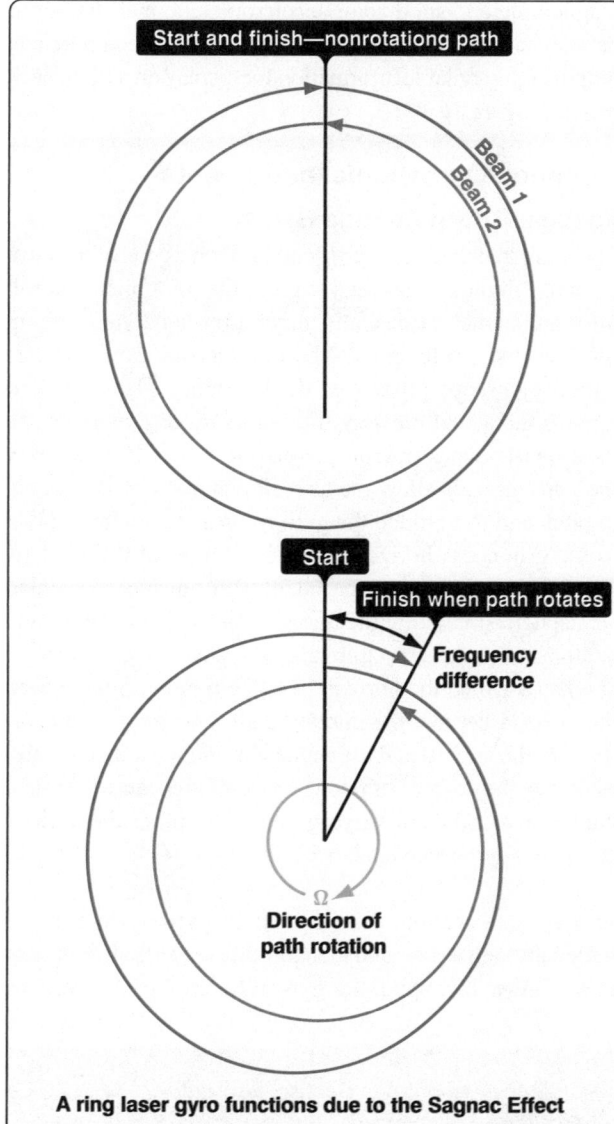

Figure 10-96. *Light traveling in opposite directions around a non-rotating path arrives at the end of the loop at the same time (top). When the path rotates, light traveling with the rotation must travel farther to complete one loop. Light traveling against the rotation completes the loop sooner (bottom).*

signal at low rotational speeds. It causes units installed on aircraft to hum when operating. *[Figure 10-97]*

An RLG is remotely mounted so the cavity path rotates around one of the axes of flight. The rate of frequency phase shift detected between the counter-rotating lasers is proportional to the rate that the aircraft is moving about that axis. On aircraft, an RLG is installed for each axis of flight. Output can be used in analog instrumentation and autopilot systems. It is also easily made compatible for use by digital display computers and for digital autopilot computers.

RLGs are very rugged and have a long service life with

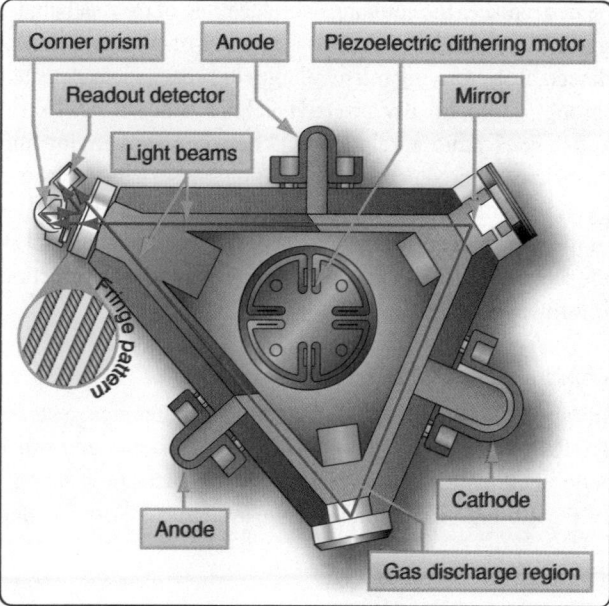

Figure 10-97. *The ring laser gyro is rugged, accurate, and free of friction.*

virtually no maintenance due to their lack of moving parts. They measure movement about an axis extremely quickly and provide continuous output. They are extremely accurate and generally are considered superior to mechanical gyroscopes.

Microelectromechanical-Based Attitude & Directional Systems

On aircraft, microelectromechanical systems (MEMS) devices save space and weight. Through the use of solid-state MEMS devices, reliability is increased primarily due to the lack of moving parts. The development of MEMS technology for use in aviation instrumentation integrates with the use of ADCs. This newest improvement in technology is low cost and promises to proliferate through all forms of aviation.

MEMS for gyroscopic applications are used in small, general aviation aircraft, as well as larger commercial aircraft. Tiny vibration-based units with resistance and capacitance measuring pick-offs are accurate and reliable and only a few millimeters in length and width. They are normally integrated into a complete micro-electronic solid-state chip designed to yield an output after various conditioning processes are performed. The chips, which are analogous to tiny circuit boards, can be packaged for installation inside a dedicated computer or module that is installed on the aircraft.

While a large mechanical gyroscope spins in a plane, its rigidity in space is used to observe and measure the movement of the aircraft. The basis of operation of many MEMS gyroscopes is the same despite their tiny size. The difference is that a vibrating or oscillating piezoelectric device replaces the spinning, weighted ring of the mechanical gyro. Still, once set in motion, any out-of-plane motion is detectable by varying microvoltages or capacitances detected through geometrically arranged pickups. Since piezoelectric substances have a relationship between movement and electricity, microelectrical stimulation can set a piezoelectric gyro in motion and the tiny voltages produced via the movement in the piezo can be extracted. They can be input as the required variables needed to compute attitude or direction information. *[Figure 10-98]*

Other Attitude & Directional Systems

In modern aircraft, attitude heading and reference systems (AHRS) have taken the place of the gyroscope and other individual instruments. While MEMS devices provide part of the attitude information for the system, GPS, solid state

magnetometers, solid state accelerometers, and digital air data signals are all combined in an AHRS to compute and output highly reliable information for display on a flight deck panel. *[Figure 10-99]*

Common Gyroscopic Instruments

Vacuum-Driven Attitude Gyros

The attitude indicator, or artificial horizon, is one of the most essential flight instruments. It gives the pilot pitch and roll information that is especially important when flying without outside visual references. The attitude indicator operates with a gyroscope rotating in the horizontal plane. Thus, it mimics the actual horizon through its rigidity in space. As the aircraft pitches and rolls in relation to the actual horizon, the gyro gimbals allow the aircraft and instrument housing to pitch and roll around the gyro rotor that remains parallel to the ground. A horizontal representation of the airplane in miniature is fixed to the instrument housing. A painted semisphere simulating the horizon, the sky, and the ground is attached to the gyro gimbals. The sky and ground meet at what is called the horizon bar. The relationship between the horizon bar and the miniature airplane are the same as those of the aircraft and the actual horizon. Graduated scales reference the degrees of pitch and roll. Often, an adjustment knob allows pilots of varying heights to place the horizon bar at an appropriate level. *[Figure 10-100]*

In a typical vacuum-driven attitude gyro system, air is pulled through a filter and then through the attitude indicator in a manner that spins the gyro rotor inside. An erecting

Figure 10-98. *The relative scale size of a MEMS gyro.*

Figure 10-99. *Instrumentation displayed within a glass flight deck using an attitude heading and reference system (AHRS) computer.*

Figure 10-100. *A typical vacuum-driven attitude indicator shown with the aircraft in level flight (left) and in a climbing right turn (right).*

mechanism is built into the instrument to assist in keeping the gyro rotor rotating in the intended plane. Precession caused by bearing friction makes this necessary. After air engages the scalloped drive on the rotor, it flows from the instrument to the vacuum pump through four ports. These ports all exhaust the same amount of air when the gyro is rotating in plane. When the gyro rotates out of plane, air tends to port out of one side more than another. Vanes close to prevent this, causing more air to flow out of the opposite side. The force from this unequal venting of the air re-erects the gyro rotor. *[Figure 10-101]*

Early vacuum-driven attitude indicators were limited in how far the aircraft could pitch or roll before the gyro gimbals contacted stops, causing abrupt precession and tumbling of the gyro. Many of these gyros include a caging device. It is used to erect the rotor to its normal operating position prior to flight or after tumbling. A flag indicates that the gyro must be uncaged before use. More modern gyroscopic instruments are built so they do not tumble, regardless of the angular movement of the aircraft about its axes.

In addition to the contamination potential introduced by the air-drive system, other shortcomings exist in the performance of vacuum-driven attitude indicators. Some are induced by the erection mechanism. The pendulous vanes that move to direct airflow out of the gyro respond not only to forces caused by a deviation from the intended plane of rotation, but centrifugal force experienced during turns also causes the vanes to allow asymmetric porting of the gyro vacuum air. The result is inaccurate display of the aircraft's attitude, especially

in skids and steep banked turns. Also, abrupt acceleration and deceleration imposes forces on the gyro rotor. Suspended in its gimbals, it acts similar to an accelerometer, resulting in a false nose-up or nose-down indication. Pilots must learn to recognize these errors and adjust accordingly.

Electric Attitude Indicators

Electric attitude indicators are very similar to vacuum-driven gyro indicators. The main difference is in the drive mechanism. Inside the gimbals of an electric gyro, a small squirrel cage electric motor is the rotor. It is typically driven by 115-volt, 400-cycle AC. It turns at approximately 21,000 rpm.

Other characteristics of the vacuum-driven gyro are shared by the electric gyro. The rotor is still oriented in the horizontal plane. The free gyro gimbals allow the aircraft and instrument case to rotate around the gyro rotor that remains rigid in space. A miniature airplane fixed to the instrument case indicates the aircraft's attitude against the moving horizon bar behind it.

Electric attitude indicators address some of the shortcomings of vacuum-driven attitude indicators. Since there is no air flowing through an electric attitude indicator, air filters, regulators, plumbing lines and vacuum pump(s) are not needed. Contamination from dirt in the air is not an issue, resulting in the potential for longer bearing life and less precession. Erection mechanism ports are not employed, so pendulous vanes responsive to centrifugal forces are eliminated.

It is still possible that the gyro may experience precession and need to be erected. This is done with magnets rather than

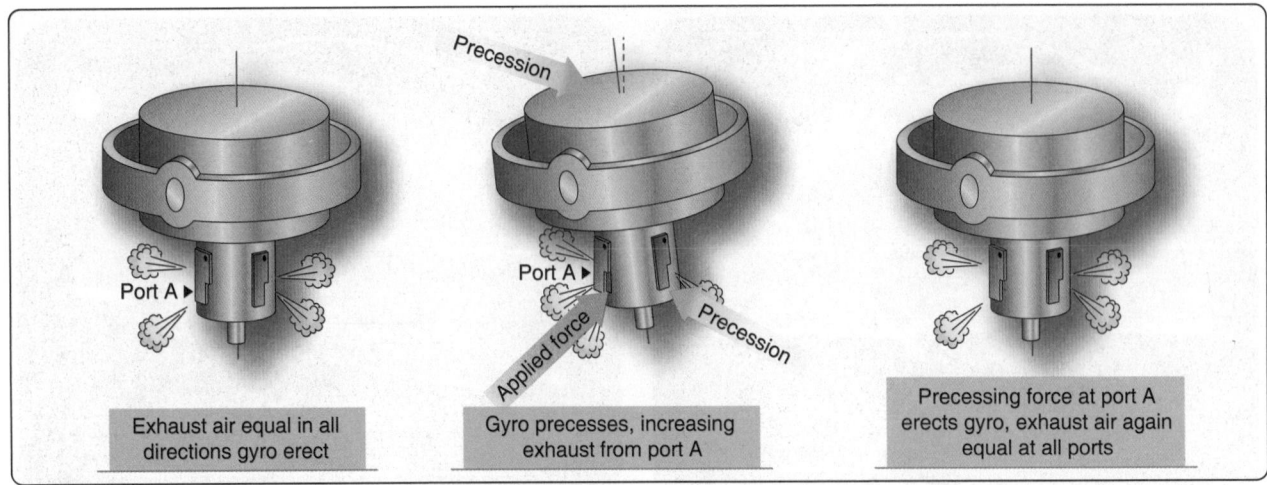

Figure 10-101. *The erecting mechanism of a vacuum-driven attitude indicator.*

vent ports. A magnet attached to the top of the gyro shaft spins at approximately 21,000 rpm. Around this magnet, but not attached to it, is a sleeve that is rotated by magnetic attraction at approximately 44 to 48 rpm. Steel balls are free to move around the sleeve. If the pull of gravity is not aligned with the axis of the gyro, the balls fall to the low side. The resulting precession re-aligns the axis of rotation vertically.

Typically, electric attitude indicator gyros can be caged manually by a lever and cam mechanism to provide rapid erection. When the instrument is not getting sufficient power for normal operation, an off flag appears in the upper right hand face of the instrument. *[Figure 10-102]*

Gyroscopic Direction Indicator or Directional Gyro (DG)

The gyroscopic direction indicator or directional gyro (DG) is often the primary instrument for direction. Because a magnetic compass fluctuates so much, a gyro aligned with the magnetic compass gives a much more stable heading indication. Gyroscopic direction indicators are located at the center base of the instrument panel basic T.

A vacuum-powered DG is common on many light aircraft. Its basis for operation is the gyro's rigidity in space. The gyro rotor spins in the vertical plane and stays aligned with the direction to which it is set. The aircraft and instrument case moves around the rigid gyro. This causes a vertical compass card that is geared to the rotor gimbal to move. It is calibrated in degrees, usually with every 30 degrees labeled. The nose of a small, fixed airplane on the instrument glass indicates the aircraft's heading. *[Figure 10-103]*

Vacuum-driven direction indicators have many of the same basic gyroscopic instrument issues as attitude indicators. Built-in compensation for precession varies and a caging device is usually found. Periodic manual realignment with the magnetic compass by the pilot is required during flight.

Turn Coordinators

Many aircraft make use of a turn coordinator. The rotor of the gyro in a turn coordinator is canted upwards 30°. As

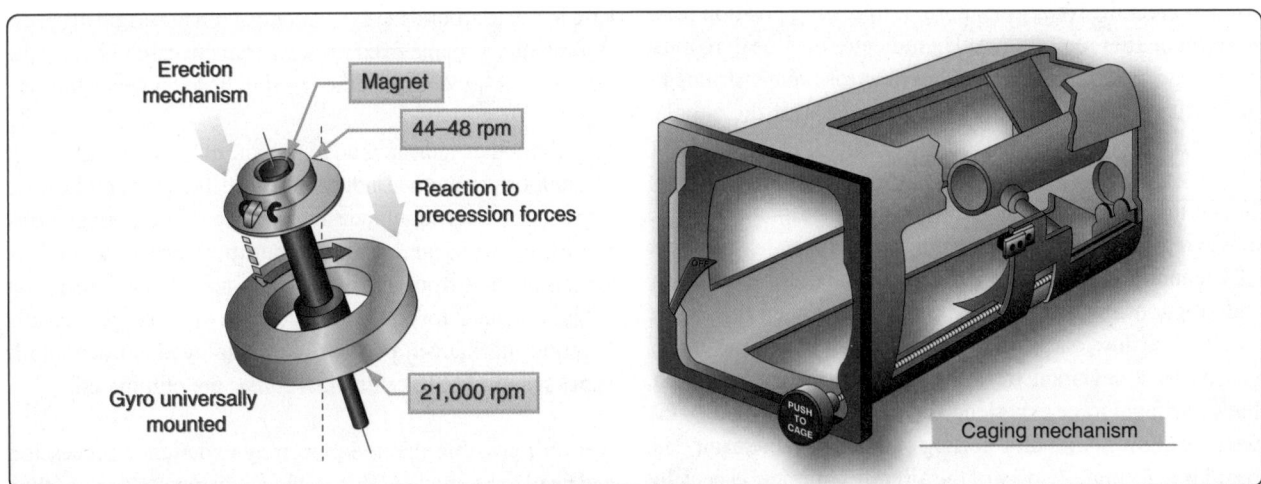

Figure 10-102. *Erecting and caging mechanisms of an electric attitude indicator.*

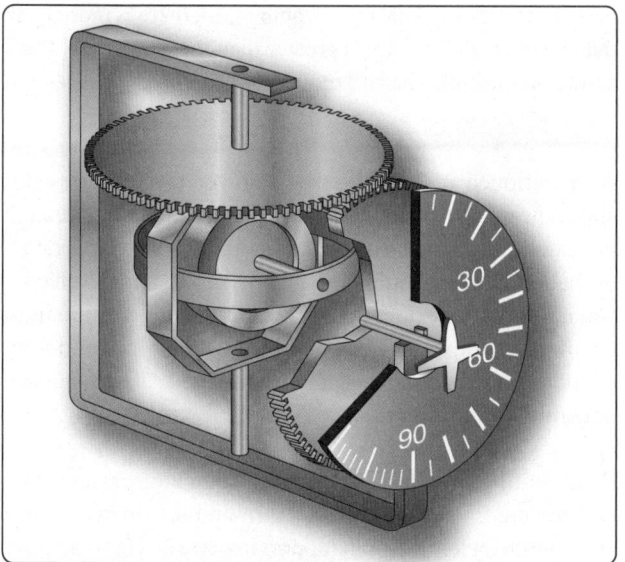

Figure 10-103. *A typical vacuum-powered gyroscopic direction indicator, also known as a directional gyro.*

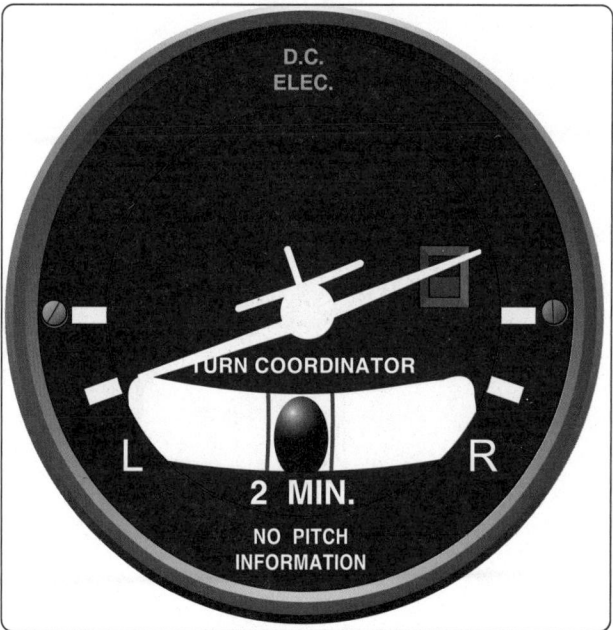

Figure 10-104. *A turn coordinator senses and indicates the rate of both roll and yaw.*

such, it responds not only to movement about the vertical axis, but also to roll movements about the longitudinal axis. This is useful because it is necessary to roll an aircraft to turn it about the vertical axis. Instrument indication of roll, therefore, is the earliest possible warning of a departure from straight-and-level flight.

Typically, the face of the turn coordinator has a small airplane symbol. The wing tips of the airplane provide the indication of level flight and the rate at which the aircraft is turning. *[Figure 10-104]*

Turn-and-Slip Indicator

The turn-and-slip indicator may also be referred to as the turn-and-bank indicator, or needle-and-ball indicator. Regardless, it shows the correct execution of a turn while banking the aircraft and indicates movement about the vertical axis of the aircraft (yaw). Most turn-and-slip indicators are located below the airspeed indicator of the instrument panel basic T, just to the left of the direction indicator.

The turn-and-slip indicator is actually two separate devices built into the same instrument housing: a turn indicator pointer and slip indicator ball. The turn pointer is operated by a gyro that can be driven by a vacuum, air pressure, or by electricity. The ball is a completely independent device. It is a round agate, or steel ball, in a glass tube filled with dampening fluid. It moves in response to gravity and centrifugal force experienced in a turn. The gyro turn-and-slip indicator may be driven electrically or pneumatically (by vacuum or air pressure).

Turn indicators vary. They all indicate the rate at which the

aircraft is turning. Three degrees of turn per second cause an aircraft to turn 360° in 2 minutes. This is considered a standard turn. This rate can be indicated with marks right and left of the pointer, which normally rests in the vertical position. Sometimes, no marks are present, and the width of the pointer is used as the calibration device. In this case, one pointer width deflection from vertical is equal to the 3° per second standard 2-minute turn rate. Faster aircraft tend to turn more slowly and have graduations or labels that indicate 4-minute turns. In other words, a pointer's width or alignment with a graduation mark on this instrument indicates that the aircraft is turning a 1½° per second and completes a 360° turn in 4 minutes. It is customary to placard the instrument face with words indicating whether it is a 2-or 4-minute turn indicator. *[Figure 10-105]*

The turn pointer indicates the rate at which an aircraft is turning about its vertical axis. It does so by using the precession of a gyro to tilt a pointer. The gyro spins in a vertical plane aligned with the longitudinal axis of the aircraft. When the aircraft rotates about its vertical axis during a turn, the force experienced by the spinning gyro is exerted about the vertical axis. Due to precession, the reaction of the gyro rotor is 90° further around the gyro in the direction of spin. This means the reaction to the force around the vertical axis is movement around the longitudinal axis of the aircraft. This causes the top of the rotor to tilt to the left or right. The pointer is attached with linkage that makes the pointer deflect in the opposite direction, which matches the direction of turn. So, the aircraft's turn around the vertical axis is indicated around the longitudinal axis on the gauge. This is intuitive

Figure 10-105. *Turn-and-slip indicator.*

outside of the turn. During a slipping turn, there is more bank than needed, and gravity is greater than the centrifugal force acting on the ball. The ball moves in the curved glass toward the inside of the turn.

As mentioned previously, often power for the turn-and-slip indicator gyro is electrical if the attitude and direction indicators are vacuum powered. This allows limited operation off battery power should the vacuum system and the electric generator fail. The directional and attitude information from the turn-and-slip indicator, combined with information from the pitot static instruments, allow continued safe emergency operation of the aircraft.

Electrically powered turn-and-slip indicators are usually DC powered. Vacuum-powered turn-and-slip indicators are usually run on less vacuum (approximately 2 "Hg) than fully gimbaled attitude and direction indicators. Regardless, proper vacuum must be maintained for accurate turn rate information to be displayed.

Autopilot Systems

An aircraft automatic pilot system controls the aircraft without the pilot directly maneuvering the controls. The autopilot maintains the aircraft's attitude and/or direction and returns the aircraft to that condition when it is displaced from it. Automatic pilot systems are capable of keeping aircraft stabilized laterally, vertically, and longitudinally.

The primary purpose of an autopilot system is to reduce the work strain and fatigue of controlling the aircraft during long flights. Most autopilots have both manual and automatic modes of operation. In the manual mode, the pilot selects

to the pilot when regarding the instrument, since the pointer indicates in the same direction as the turn. *[Figure 10-106]*

The slip indicator (ball) part of the instrument is an inclinometer. The ball responds only to gravity during coordinated straight-and-level flight. Thus, it rests in the lowest part of the curved glass between the reference wires. When a turn is initiated, and the aircraft is banked, both gravity and the centrifugal force of the turn act upon the ball. If the turn is coordinated, the ball remains in place. Should a skidding turn exist, the centrifugal force exceeds the force of gravity on the ball and it moves in the direction of the

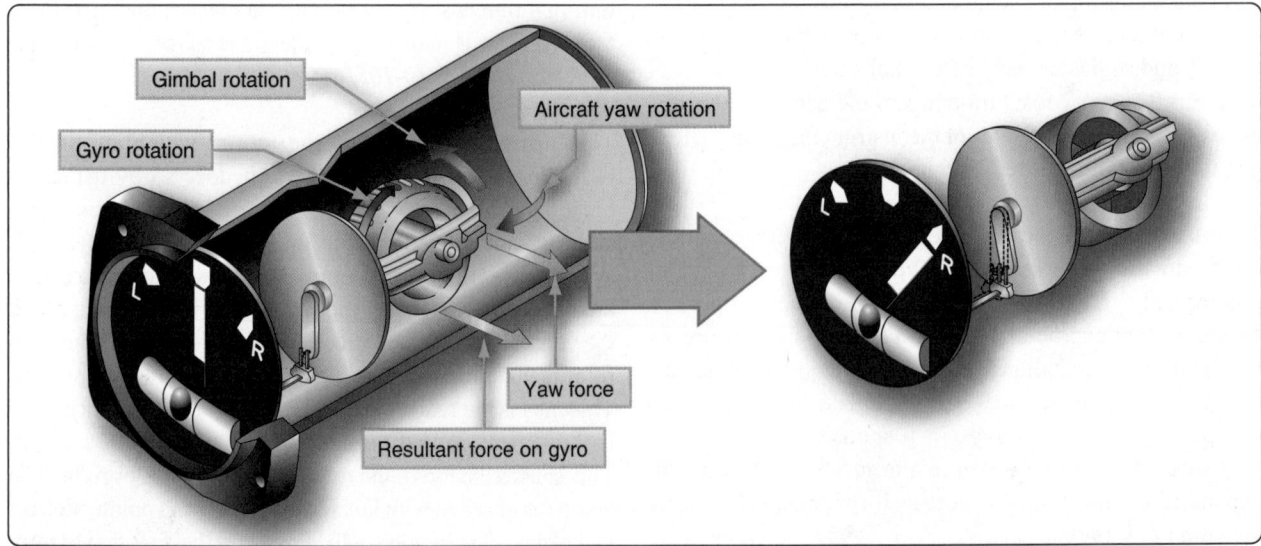

Figure 10-106. *The turn-and-slip indicator's gyro reaction to the turning force in a right hand turn. The yaw force results in a force on the gyro 90° around the rotor in the direction it is turning due to precession. This causes the top of the rotor to tilt to the left. Through connecting linkage, the pointer tilts to the right.*

each maneuver and makes small inputs into an autopilot controller. The autopilot system moves the control surfaces of the aircraft to perform the maneuver. In automatic mode, the pilot selects the attitude and direction desired for a flight segment. The autopilot then moves the control surfaces to attain and maintain these parameters.

Autopilot systems provide for one-, two-, or three-axis control of an aircraft. Those that manage the aircraft around only one axis control the ailerons. They are single-axis autopilots, known as wing leveler systems, usually found on light aircraft. [Figure 10-107] Other autopilots are two-axis systems that control the ailerons and elevators. Three-axis autopilots control the ailerons, elevators, and the rudder. Two-and three-axis autopilot systems can be found on aircraft of all sizes.

There are many autopilot systems available. They feature a wide range of capabilities and complexity. Light aircraft typically have autopilots with fewer capabilities than high-performance and transport category aircraft. Integration of navigation functions is common, even on light aircraft autopilots. As autopilots increase in complexity, they not only manipulate the flight control surfaces, but other flight parameters as well.

Some modern small aircraft, high-performance, and transport category aircraft have very elaborate autopilot systems known as automatic flight control systems (AFCS). These three-axis systems go far beyond steering the airplane. They control the aircraft during climbs, descents, cruise, and approach to landing. Some even integrate an auto-throttle function that automatically controls engine thrust that makes auto-landings possible.

For further automatic control, flight management systems (FMS) have been developed. Through the use of computers, an entire flight profile can be programmed ahead of time allowing the pilot to supervise its execution. An FMS computer coordinates nearly every aspect of a flight, including the autopilot and auto throttle systems, navigation route selection, fuel management schemes, and more.

Basis for Autopilot Operation

The basis for autopilot system operation is error correction. When an aircraft fails to meet the conditions selected, an error is said to have occurred. The autopilot system automatically corrects that error and restores the aircraft to the flight attitude desired by the pilot. There are two basic ways modern autopilot systems do this. One is position based and the other is rate based. A position based autopilot manipulates the aircraft's controls so that any deviation from the desired attitude of the aircraft is corrected. This is done by memorizing the desired aircraft attitude and moving the control surfaces so that the aircraft returns to that attitude. Rate based autopilots use information about the rate of movement of the aircraft and move control surfaces to counter the rate of change that

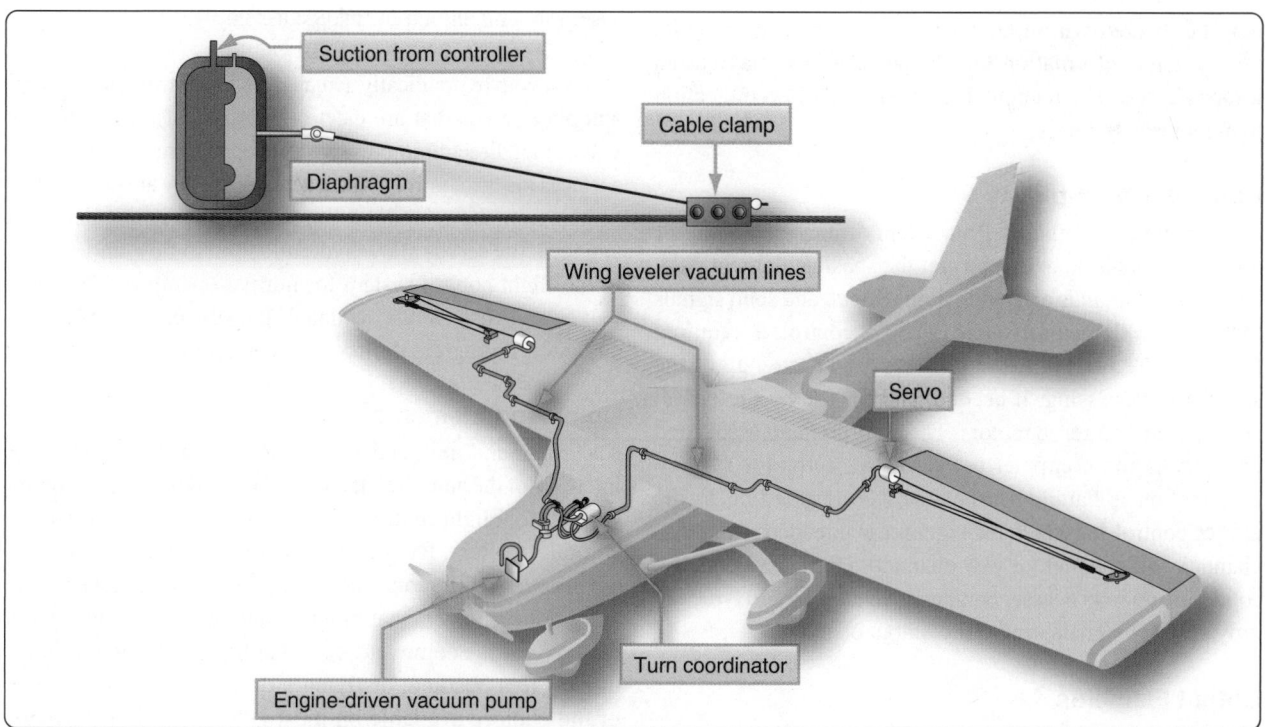

Figure 10-107. *The wing leveler system on a small aircraft is a vacuum-operated single-axis autopilot. Only the ailerons are controlled. The aircraft's turn coordinator is the sensing element. Vacuum from the instrument vacuum system is metered to the diaphragm cable servo to move the ailerons when the turn coordinator senses roll.*

causes the error. Most large aircraft use rate-based autopilot systems. Small aircraft may use either.

Autopilot Components

Most autopilot systems consist of four basic elements, plus various switches and auxiliary units. The four basic elements are: sensing elements, computing element, output elements, and command elements. Many advanced autopilot systems contain a fifth element: feedback or follow-up. This refers to signals sent as corrections are being made by the output elements to advise the autopilot of the progress being made. [Figure 10-108]

Sensing Elements

The attitude (or vertical) and directional gyros, the turn coordinator, and an altitude control are the autopilot sensing elements. These units sense the movements of the aircraft. They generate electric signals that are used by the autopilot to automatically take the required corrective action needed to keep the aircraft flying as intended. The sensing gyros can be located in the flight deck mounted instruments. They can also be remotely mounted. Remote gyro sensors drive the servo displays in the flight deck panel, as well as provide the input signals to the autopilot computer.

Modern digital autopilots may use a variety of different sensors. MEMS gyros may be used or accompanied by the use solid state accelerometers and magnetometers. Rate based systems may not use gyros at all. Various input sensors may be located within the same unit or in separate units that transfer information via digital data bus. Navigation information is also integrated via digital data bus connection to avionics computers.

Computer & Amplifier

The computing element of an autopilot may be analog or digital. Its function is to interpret the sensing element data, integrate commands and navigational input, and send signals to the output elements to move the flight controls as required to control the aircraft. An amplifier is used to strengthen the signal for processing, if needed, and for use by the output devices, such as servo motors. The amplifier and associated circuitry is the computer of an analog autopilot system. Information is handled in channels corresponding to the axis of control for which the signals are intended (i.e., pitch channel, roll channel, or yaw channel). Digital systems use solid state microprocessor computer technology and typically only amplify signals sent to the output elements.

Output Elements

The output elements (components) of an autopilot system are the servos that cause actuation of the flight control surfaces. They are independent devices for each of the control channels that integrate into the regular flight control system. Autopilot servo designs vary widely depending on the method of actuation of the flight controls. Cable-actuated systems typically utilize electric servo motors or electro-pneumatic servos. Hydraulic actuated flight control systems use electro-hydraulic autopilot servos. Digital fly-by-wire aircraft utilize the same actuators for carrying out manual and autopilot maneuvers. When the autopilot is engaged, the actuators respond to commands from the autopilot rather than exclusively from the pilot. Regardless, autopilot servos must allow unimpeded control surface movement when the autopilot is not operating.

Aircraft with cable actuated control surfaces use two basic types of electric motor-operated servos. In one, a motor is connected to the servo output shaft through reduction gears. The motor starts, stops, and reverses direction in response to the commands of autopilot computer. The other type of electric servo uses a constantly running motor geared to the output shaft through two magnetic clutches. The clutches are arranged so that energizing one clutch transmits motor torque to turn the output shaft in one direction; energizing the other clutch turns the shaft in the opposite direction. [Figure 10-109] Electro-pneumatic servos can also be used to drive cable flight controls in some autopilot systems. They are controlled by electrical signals from the autopilot amplifier and actuated by an appropriate air pressure source. The source may be a vacuum system pump or turbine engine bleed air. Each servo consists of an electromagnetic valve assembly and an output linkage assembly.

Aircraft with hydraulically actuated flight control systems have autopilot servos that are electro-hydraulic. They are control valves that direct fluid pressure as needed to move the control surfaces via the control surface actuators. They are powered by signals from the autopilot computer. When the autopilot is not engaged, the servos allow hydraulic fluid to flow unrestricted in the flight control system for normal operation. The servo valves can incorporate feedback transducers to update the autopilot of progress during error correction.

Command Elements

The command unit, called a flight controller, is the human interface of the autopilot. It allows the pilot to tell the autopilot what to do. Flight controllers vary with the complexity of the autopilot system. By pressing the desired function buttons, the pilot causes the controller to send instruction signals to the autopilot computer, enabling it to activate the proper servos to carry out the command(s). Level flight, climbs, descents, turning to a heading, or flying a desired heading are some of the choices available on most autopilots. Many aircraft make use of a multitude of radio navigational aids. These can be selected to issue commands directly to the autopilot computer. [Figure 10-110]

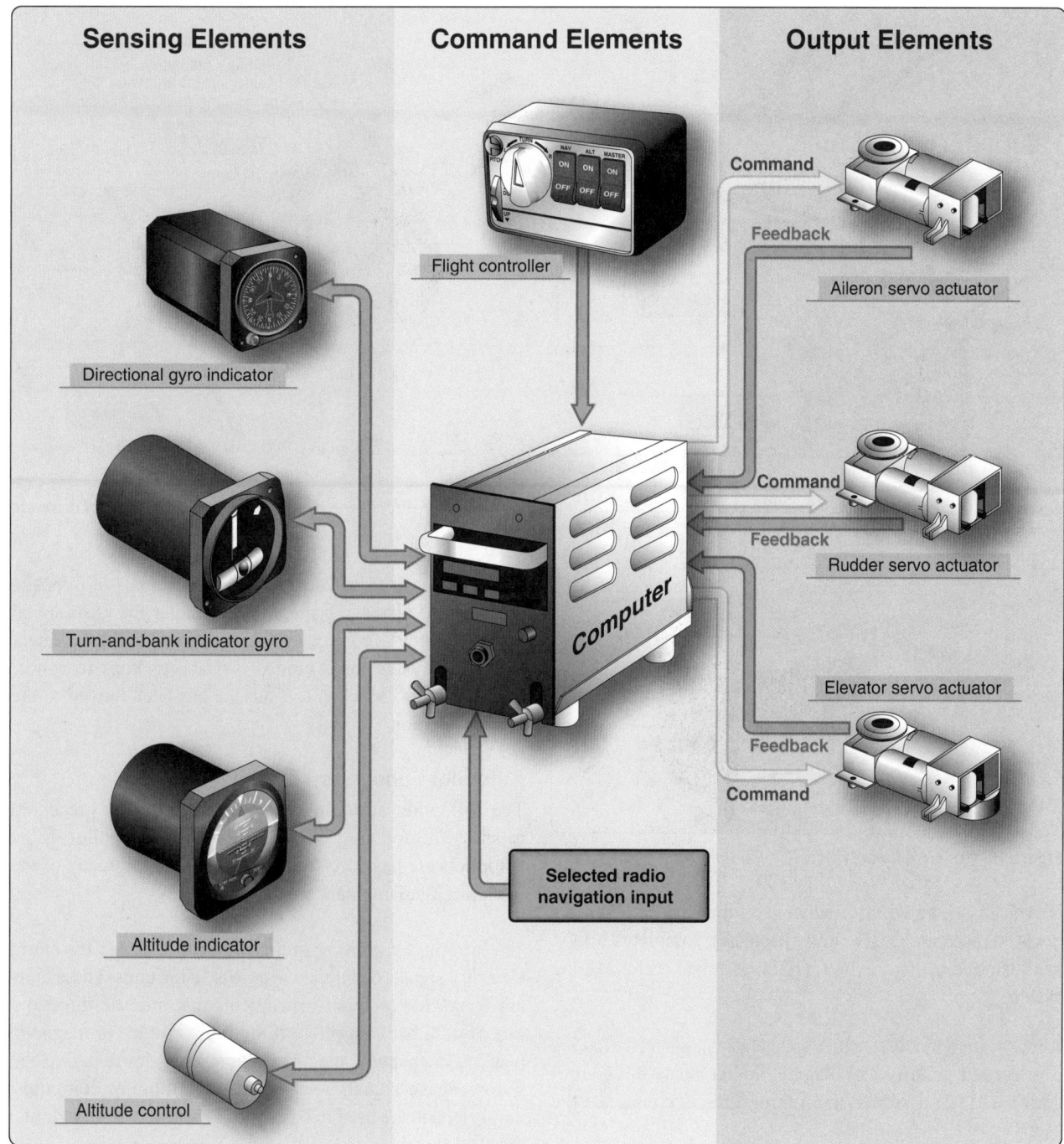

Figure 10-108. *Typical analog autopilot system components.*

In addition to an on/off switch on the autopilot controller, most autopilots have a disconnect switch located on the control wheel(s). This switch, operated by thumb pressure, can be used to disengage the autopilot system should a malfunction occur in the system or any time the pilot wishes to take manual control of the aircraft.

Feedback or Follow-up Element

As an autopilot maneuvers the flight controls to attain a desired flight attitude, it must reduce control surface correction as the desired attitude is nearly attained so the controls and aircraft come to rest on course. Without doing so, the system would continuously overcorrect. Surface deflection would occur until the desired attitude is attained. But movement would still occur as the surface(s) returned to pre-error position. The attitude sensor would once again detect an error and begin the correction process all over again. Various electric feedback, or follow-up signals, are generated to progressively reduce the error message in the autopilot so that continuous over correction does not take place. This

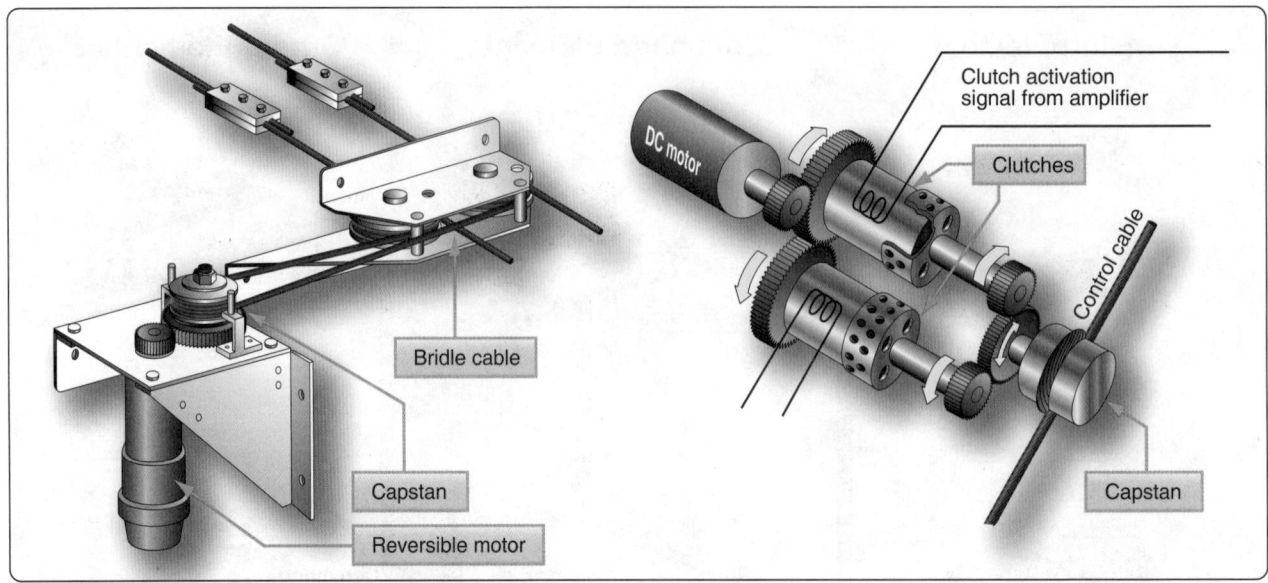

Figure 10-109. *A reversible motor with capstan and bridle cable (left), and a single-direction constant motor with clutches that drive the output shafts and control cable in opposite directions (right).*

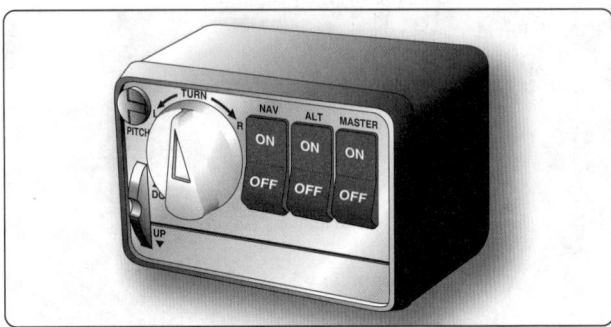

Figure 10-110. *An autopilot controller of a simple autopilot system.*

is typically done with transducers on the surface actuators or in the autopilot servo units. Feedback completes a loop as illustrated in *Figure 10-111*. This is called a closed-loop system.

A rate system receives error signals from a rate gyro that are of a certain polarity and magnitude that cause the control surfaces to be moved. As the control surfaces counteract the

error and move to correct it, follow-up signals of opposite polarity and increasing magnitude counter the error signal until the aircraft's correct attitude is restored. A displacement follow-up system uses control surface pickups to cancel the error message when the surface has been moved to the correct position.

Autopilot Functions

The following autopilot system description is presented to show the function of a simple analog autopilot. Most autopilots are far more sophisticated; however, many of the operating fundamentals are similar.

The automatic pilot system flies the aircraft by using electrical signals developed in gyro-sensing units. These units are connected to flight instruments that indicate direction, rate of turn, bank, or pitch. If the flight attitude or magnetic heading is changed, electrical signals are developed in the gyros. These signals are sent to the autopilot computer/amplifier and are used to control the operation of servo units.

A servo for each of the three control channels converts electrical signals into mechanical force, which moves the control surface in response to corrective signals or pilot commands. The rudder channel receives two signals that determine when and how much the rudder moves. The first signal is a course signal derived from a compass system. As long as the aircraft remains on the magnetic heading it was on when the autopilot was engaged, no signal develops. But, any deviation causes the compass system to send a signal to the rudder channel that is proportional to the angular displacement of the aircraft from the preset heading.

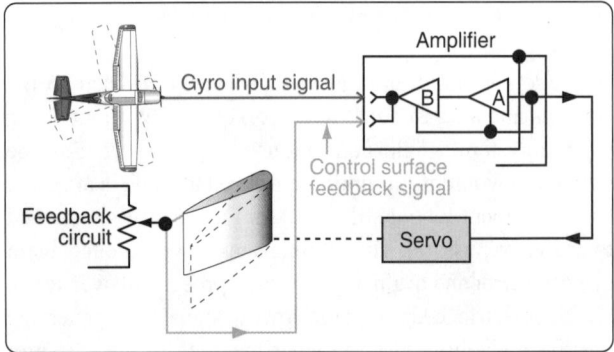

Figure 10-111. *Basic function of an analog autopilot system including follow-up or feedback signal.*

The second signal received by the rudder channel is the rate signal that provides information anytime the aircraft is turning about the vertical axis. This information is provided by the turn-and-bank indicator gyro. When the aircraft attempts to turn off course, the rate gyro develops a signal proportional to the rate of turn, and the course gyro develops a signal proportional to the amount of displacement. The two signals are sent to the rudder channel of the amplifier, where they are combined, and their strength is increased. The amplified signal is then sent to the rudder servo. The servo turns the rudder in the proper direction to return the aircraft to the selected magnetic heading.

As the rudder surface moves, a follow-up signal is developed that opposes the input signal. When the two signals are equal in magnitude, the servo stops moving. As the aircraft arrives on course, the course signal reaches a zero value, and the rudder is returned to the streamline position by the follow-up signal.

The aileron channel receives its input signal from a transmitter located in the gyro horizon indicator. Any movement of the aircraft about its longitudinal axis causes the gyro-sensing unit to develop a signal to correct for the movement. This signal is amplified, phase detected, and sent to the aileron servo, which moves the aileron control surfaces to correct for the error. As the aileron surfaces move, a follow-up signal builds up in opposition to the input signal. When the two signals are equal in magnitude, the servo stops moving. Since the ailerons are displaced from the streamline, the aircraft now starts moving back toward level flight with the input signal becoming smaller and the follow-up signal driving the control surfaces back toward the streamline position. When the aircraft has returned to level flight roll attitude, the input signal is again zero. At the same time, the control surfaces are streamlined, and the follow-up signal is zero.

The elevator channel circuits are similar to those of the aileron channel, with the exception that the elevator channel detects and corrects changes in pitch attitude of the aircraft. For altitude control, a remotely mounted unit containing an altitude pressure diaphragm is used. Similar to the attitude and directional gyros, the altitude unit generates error signals when the aircraft has moved from a preselected altitude. This is known as an altitude hold function. The signals control the pitch servos, which move to correct the error. An altitude select function causes the signals to continuously be sent to the pitch servos until a preselected altitude has been reached. The aircraft then maintains the preselected altitude using altitude hold signals.

Yaw Dampening

Many aircraft have a tendency to oscillate around their vertical axis while flying a fixed heading. Near continuous rudder input is needed to counteract this effect. A yaw damper is used to correct this motion. It can be part of an autopilot system or a completely independent unit. A yaw damper receives error signals from the turn coordinator rate gyro. Oscillating yaw motion is counteracted by rudder movement, which is made automatically by the rudder servo(s) in response to the polarity and magnitude of the error signal.

Automatic Flight Control System (AFCS)

An aircraft autopilot with many features and various autopilot related systems integrated into a single system is called an automatic flight control system (AFCS). These were formerly found only on high-performance aircraft. Currently, due to advances in digital technology for aircraft, modern aircraft of any size may have AFCS.

AFCS capabilities vary from system to system. Some of the advances beyond ordinary autopilot systems are the extent of programmability, the level of integration of navigational aids, the integration of flight director and autothrottle systems, and combining of the command elements of these various systems into a single integrated flight control human interface. *[Figure 10-112]*

It is at the AFCS level of integration that an autothrottle system is integrated into the flight director and autopilot systems with glide scope modes so that auto landings are

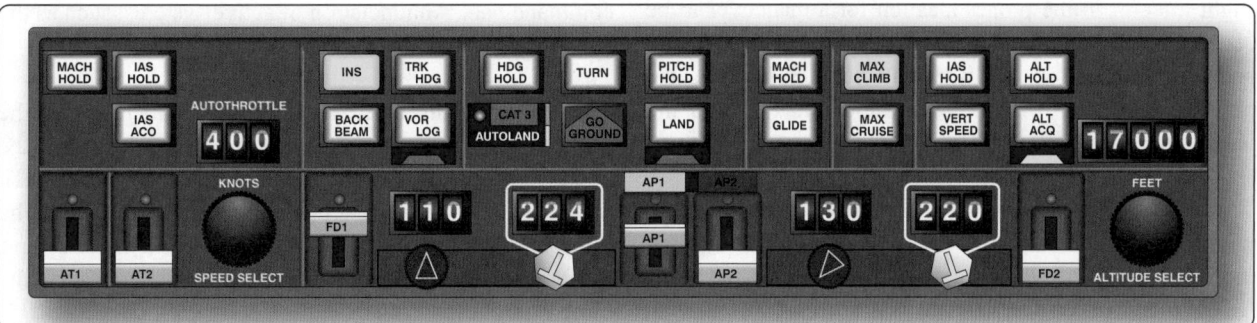

Figure 10-112. *The AFCS control panel commands several integrated systems from a single panel including: flight directors, autopilots, autothrottles, autoland, and navigational aids. Mode selections for many features are made from this single interface.*

possible. Small general aviation aircraft being produced with AFCS may lack the throttle-dependent features.

Modern general aviation AFCS are fully integrated with digital attitude heading and reference systems (AHRS) and navigational aids including glideslope. They also contain modern computer architecture for the autopilot (and flight director systems) that is slightly different than described above for analog autopilot systems. Functionality is distributed across a number of interrelated computers and includes the use of intelligent servos that handle some of the error correction calculations. The servos communicate with dedicated avionics computers and display unit computers through a control panel, while no central autopilot computer exists. *[Figure 10-113]*

Flight Director Systems

A flight director system is an instrument system consisting of electronic components that compute and indicate the aircraft attitude required to attain and maintain a preselected flight condition. A command bar on the aircraft's attitude indicator shows the pilot how much and in what direction the attitude of the aircraft must be changed to achieve the desired result. The computed command indications relieve the pilot of many of the mental calculations required for instrument flights, such as interception angles, wind drift correction, and rates of climb and descent.

Essentially, a flight director system is an autopilot system without the servos. All of the same sensing and computations are made, but the pilot controls the airplane and makes maneuvers by following the commands displayed on the instrument panel. Flight director systems can be part of an autopilot system or exist on aircraft that do not possess full autopilot systems. Many autopilot systems allow for the option of engaging or disengaging a flight director display.

Flight director information is displayed on the instrument that displays the aircraft's attitude. The process is accomplished with a visual reference technique. A symbol representing the aircraft is fit into a command bar positioned by the flight director in the proper location for a maneuver to be accomplished. The symbols used to represent the aircraft and the command bar vary by manufacturer. Regardless, the object is always to fly the aircraft symbol into the command bar symbol. *[Figure 10-114]*

The instrument that displays the flight director commands is known as a flight director indicator (FDI), attitude director indicator (ADI), or electronic attitude director indicator (EADI). It may even be referred to as an artificial horizon with flight director. This display element combines with the other primary components of the flight director system. Like an autopilot, these consist of the sensing elements, a computer, and an interface panel.

Integration of navigation features into the attitude indicator is highly useful. The flight director contributes to this usefulness by indicating to the pilot how to maneuver the airplane to navigate a desired course. Selection of the VOR function on the flight director control panel links the computer to the omnirange receiver. The pilot selects a desired course and the flight director displays the bank attitude necessary to intercept and maintain this course. Allocations for wind drift and calculation of the intercept angle is performed automatically.

Flight director systems vary in complexity and features. Many have altitude hold, altitude select, pitch hold, and other features. But flight director systems are designed to offer the greatest assistance during the instrument approach phase of flight. ILS localizer and glideslope signals are transmitted through the receivers to the computer and are presented as command indications. This allows the pilot to fly the airplane down the optimum approach path to the runway using the flight director system.

With the altitude hold function engaged, level flight can be maintained during the maneuvering and procedure turn phase of an approach. Altitude hold automatically disengages when the glideslope is intercepted. Once inbound on the localizer, the command signals of the flight director are maintained in a centered or zero condition. Interception of the glideslope causes a downward indication of the command pitch indicator. Any deviation from the proper glideslope path causes a fly-up or fly-down command indication. The pilot needs only to keep the airplane symbol fit into the command bar.

Electronic Instruments

Electronic Attitude Director Indicator (EADI)

The EADI is an advanced version of attitude and electric attitude indicators previously discussed. In addition to displaying the aircraft's attitude, numerous other situational flight parameters are displayed. Most notable are those that relate to instrument approaches and the flight director command bars. Annunciation of active systems, such as the AFCS and navigation systems, is typical.

The concept behind an EADI is to put all data related to the flight situation in close proximity for easy observation by the pilot. *[Figure 10-115]* Most EADIs can be switched between different display screens depending on the preference of the pilot and the phase of flight. EADIs vary from manufacturer to manufacturer and aircraft to aircraft. However, most of the same information is displayed.

EADIs can be housed in a single instrument housing or

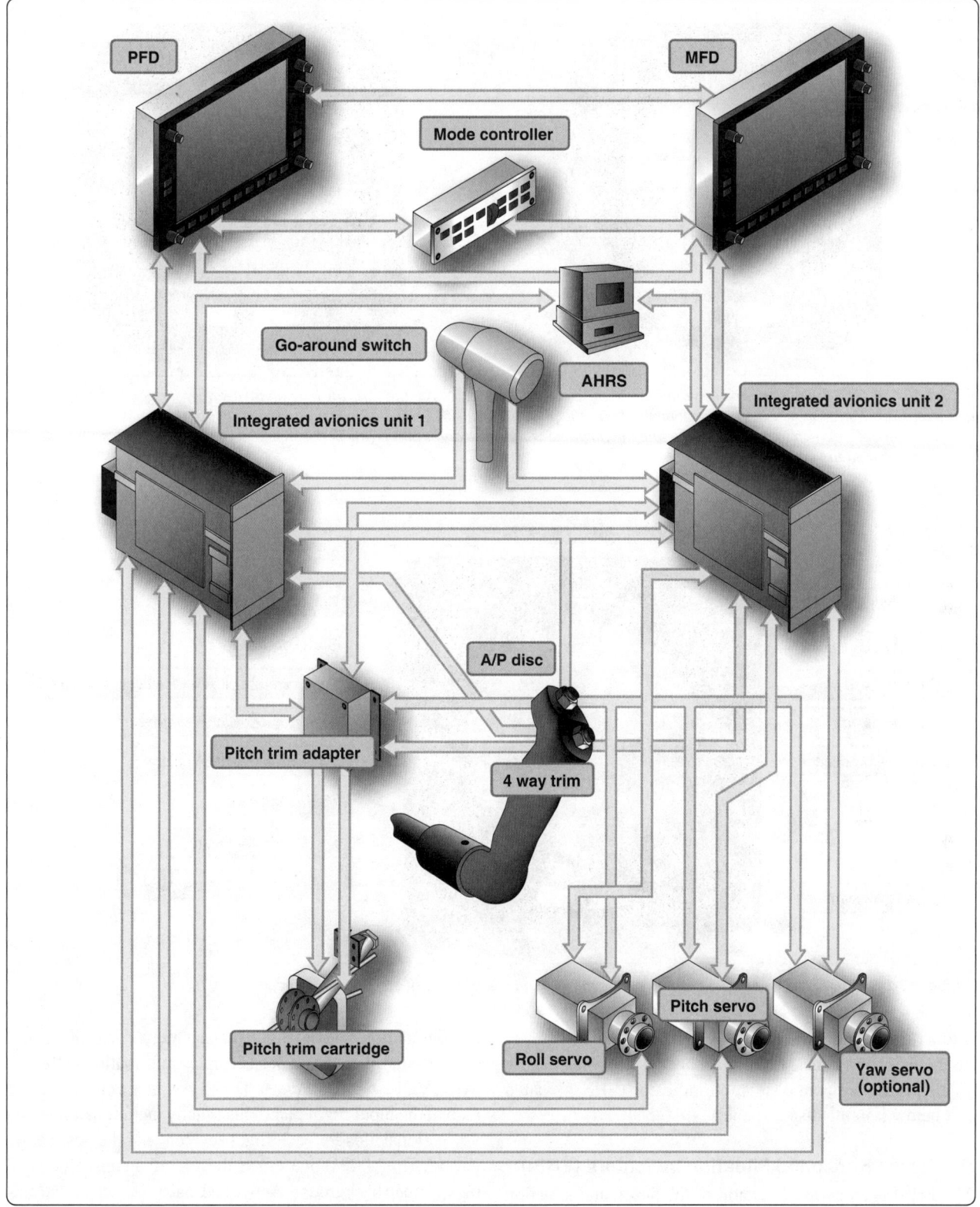

Figure 10-113. *Automatic flight control system (AFCS) of a Garmin G1000 glass flight deck instrument system for a general aviation aircraft.*

can be part of an electronic instrument display system. One such system, the electronic flight instrument system (EFIS), uses a cathode ray tube EADI display driven by a signal generator. Large-screen glass flight deck displays use LCD technology to display EADI information as part of an entire situational display directly in front of the pilot in the middle of the instrument panel. Regardless, the EADI is the primary flight instrument used for aircraft attitude

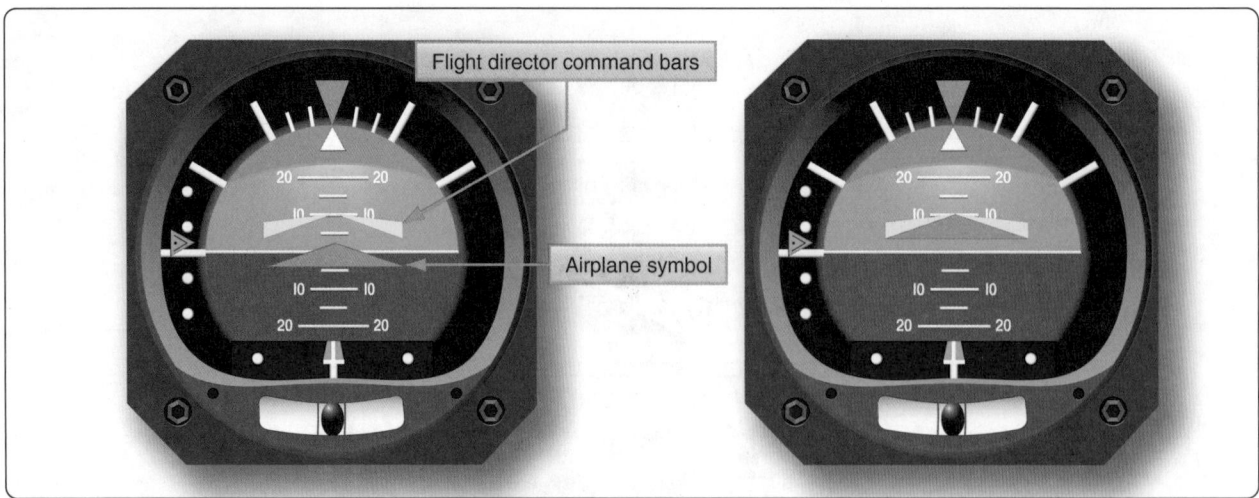

Figure 10-114. *The flight director command bar signals the pilot how to steer the aircraft for a maneuver. By flying the aircraft so the triangular airplane symbol fits into the command bar, the pilot performs the maneuver calculated by the flight director. The instrument shown on the left is commanding a climb while the airplane is flying straight and level. The instrument on the right shows that the pilot has accomplished the maneuver.*

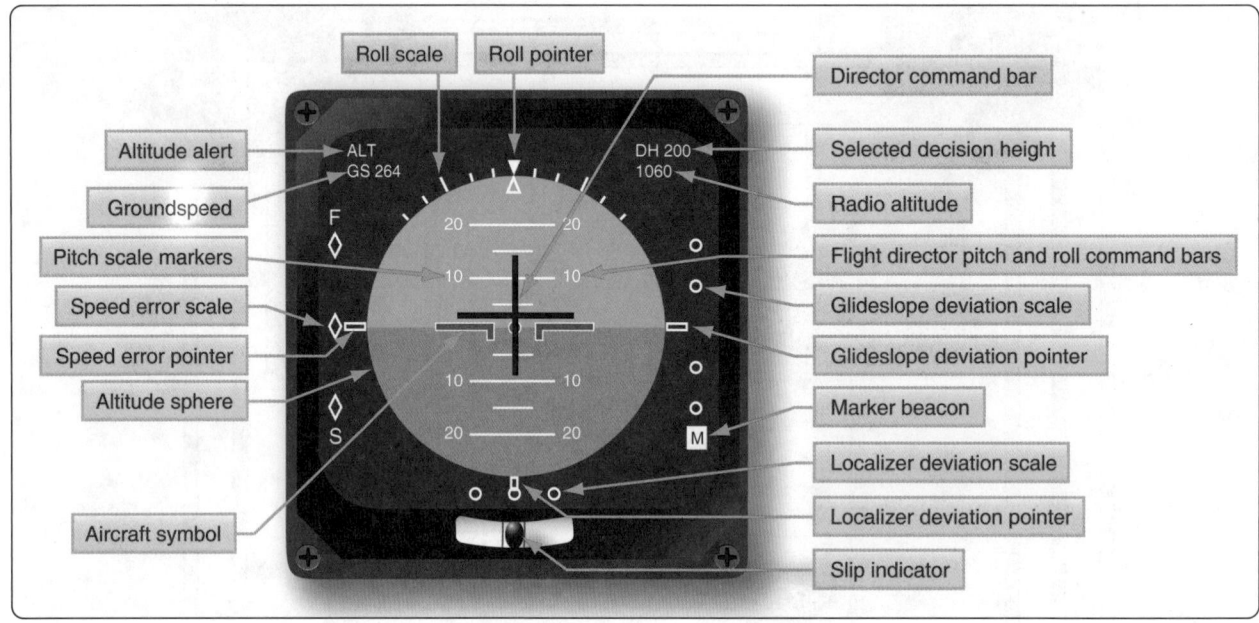

Figure 10-115. *Some of the many parameters and features of an electronic attitude director indicator (EADI).*

information during instrument flying and especially during instrument approaches. It is almost always accompanied by an electronic horizontal situation indicator (EHSI) located just below it in the display panel.

Electronic Horizontal Situation Indicators (EHSI)

The EHSI is an evolved version of the horizontal situation indicator (HSI), which was born from the gyroscopic direction indicator or directional gyro. The HSI incorporates directional information to two different navigational aids, as well as the heading of the aircraft. The EHSI does this and more. Its primary purpose is to display as much useful navigational information as possible.

In conjunction with a flight management computer and a display controller, an EHSI can display information in PLAN, MAP, VOR, and ILS modes. The PLAN mode shows a fixed map of the input flight plan. This usually includes all selected navigational aids for each flight segment and the destination airport. The MAP mode shows the aircraft against a detailed moving map background. Active and inactive navigational aids are shown, as well as other airports and waypoints. Weather radar information may be selected to be shown in scale as a background. Some HSIs can depict other air traffic when integrated with the TCAS system. Unlike a standard HSI, an EHSI may show only the pertinent portion of the compass rose. Annunciation of active mode and selected features appear with other pertinent information, such as distance and arrival time to the next waypoint, airport designators, wind direction and speed, and more. *[Figure 10-116]* There are many different displays that vary by manufacturer.

The VOR view of an EHSI presents a more traditional focus on a selected VOR, or other navigational station being used, during a particular flight segment. The entire compass rose, the traditional lateral deviation pointer, to/from information, heading, and distance information are standard. Other information may also be displayed. *[Figure 10-117]* The ILS mode of an EHSI shows the aircraft in relation to the ILS approach aids and selected runway with varying degrees of details. With this information displayed, the pilot need not consult printed airport approach information, allowing full attention to flying the aircraft.

Electronic Flight Information Systems

In an effort to increase the safety of operating complicated aircraft, computers and computer systems have been

incorporated. Flight instrumentation and engine and airframe monitoring are areas particularly well suited to gain advantages from the use of computers. They contribute by helping to reduce instrument panel clutter and focusing the pilot's attention only on matters of imminent importance.

"Glass flight deck" is a term that refers to the use of flat-panel display screens in flight deck instrumentation. In reality, it also refers to the use of computer-produced images that have replaced individual mechanical gauges. Moreover, computers and computer systems monitor the processes and components of an operating aircraft beyond human ability while relieving the pilot of the stress from having to do so.

Computerized electronic flight instrument systems have additional benefits. The solid-state nature of the components increases reliability. Also, microprocessors, data buses, and LCDs all save space and weight. The following systems have been developed and utilized on aircraft for a number of years. New systems and computer architecture are sure to come in the future.

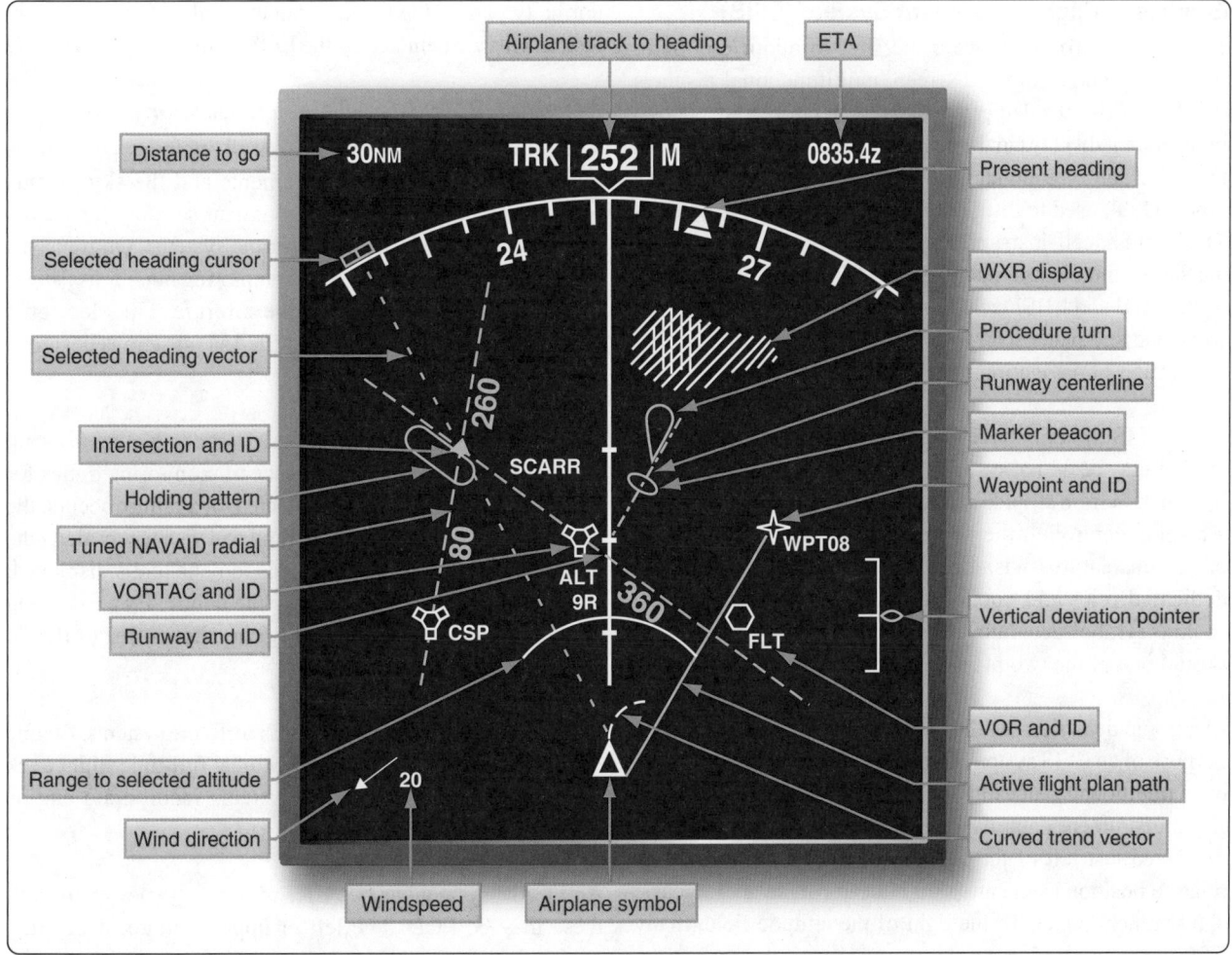

Figure 10-116. *An EHSI presents navigational information for the entire flight. The pilot selects the mode most useful for a particular phase of flight, ranging from navigational planning to instrument approach to landing. The MAP mode is used during most of the flight.*

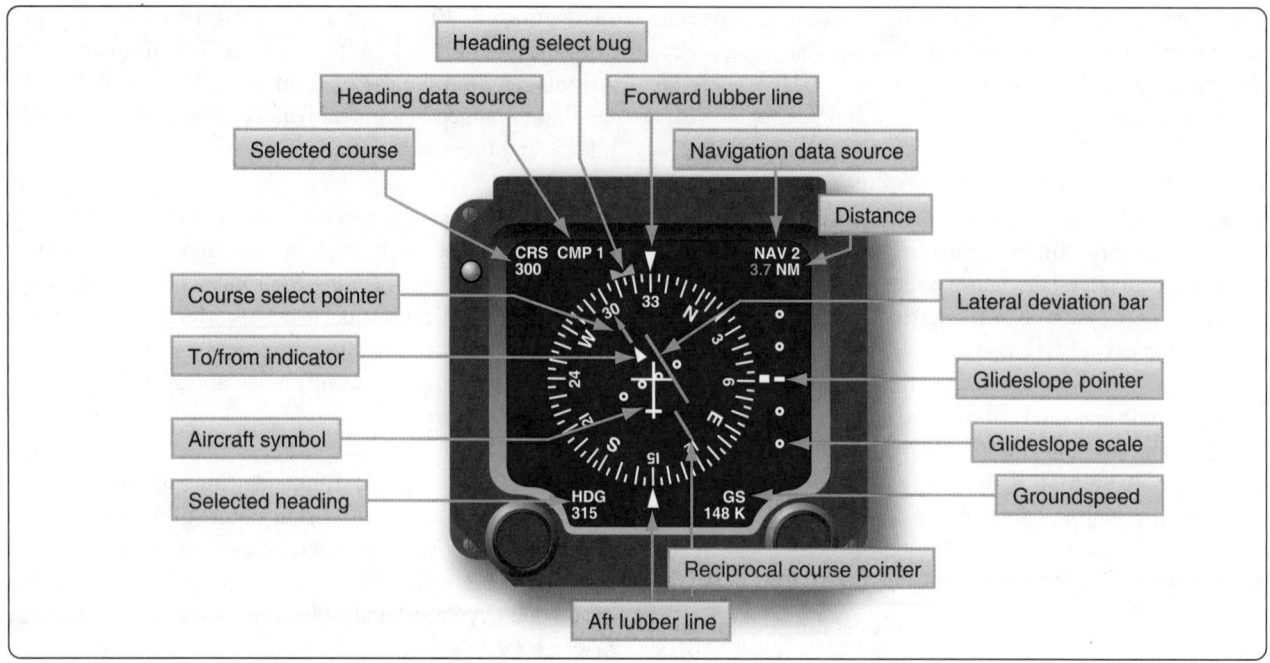

Figure 10-117. *Approach and VOR mode presentation of an electronic horizontal situation indicator.*

Electronic Flight Instrument System (EFIS)

The flight instruments were the first to adopt computer technology and utilize flat screen, multifunctional displays (MFD). EFIS uses dedicated signal generators to drive two independent displays in the center of the basic T. The attitude indicator and directional gyro are replaced by cathode ray tubes (CRT) used to display EADI and EHSI presentations. These enhanced instruments operate alongside ordinary mechanic and electric instruments with limited integration. Still, EADI and EHSI technology is very desirable, reducing workload and panel scan with the added safety provided by integration of navigation information as described.

Early EFIS systems have analog technology, while newer models may be digital systems. The signal generators receive information from attitude and navigation equipment. Through a display controller, the pilot can select the various mode or screen features wishing to be displayed. Independent dedicated pilot and copilot systems are normal. A third, backup symbol generator is available to assume operation should one of the two primary units fail. *[Figure 10-118]*

Electronic depiction of ADI and HSI information is the core purpose of an EFIS system. Its expanded size and capabilities over traditional gauges allow for integration of even more flight instrument data. A vertical airspeed scale is typically displayed just left of the attitude field. This is in the same relative position as the airspeed indicator in an analog basic T instrument panel. To the right of the attitude field, many EFIS systems display an altitude and vertical speed scale. Since most EFIS EADI depictions include the inclinometer,

normally part of the turn coordinator, all of the basic flight instruments are depicted by the EFIS display. *[Figure 10-119]*

Electronic Centralized Aircraft Monitor (ECAM)

The pilot's workload on all aircraft includes continuous monitoring of the flight instruments and the sky outside of the aircraft. It also includes vigilant scrutiny for proper operation of the engine and airframe systems. On transport category aircraft, this can mean monitoring numerous gauges in addition to maneuvering the aircraft. The electronic centralized aircraft monitoring (ECAM) system is designed to assist with this duty.

The basic concept behind ECAM (and other monitoring systems) is automatic performance of monitoring duties for the pilot. When a problem is detected, or a failure occurs, the primary display, along with an aural and visual cue, alerts the pilot. Corrective action that needs to be taken is displayed, as well as suggested action due to the failure. By performing system monitoring automatically, the pilot is free to fly the aircraft until a problem occurs.

Early ECAM systems only monitor airframe systems. Engine parameters are displayed on traditional full-time flight deck gauges. Later model ECAM systems incorporate engine displays, as well as airframe.

An ECAM system has two CRT monitors. In newer aircraft, these may be LCD. The left or upper monitor, depending on the aircraft panel layout, displays information on system status and any warnings associated corrective actions. This

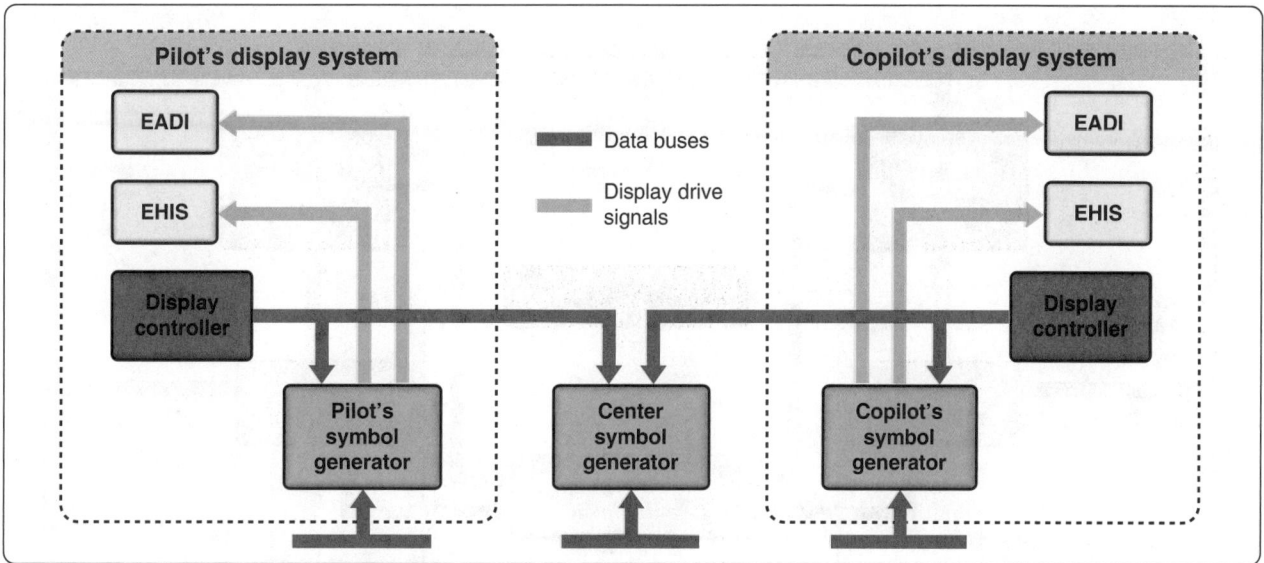

Figure 10-118. *A simplified diagram of an EFIS system. The EADI and EHSI displays are CRT units in earlier systems. Modern systems use digital displays, sometimes with only one multifunctional display unit replacing the two shown. Independent digital processors can also be located in a single unit to replace the three separate symbol generators.*

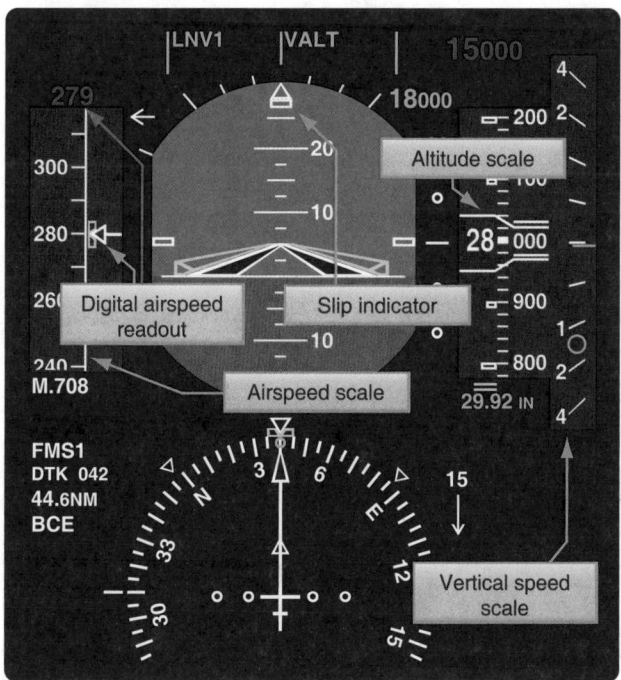

Figure 10-119. *An EFIS EADI displays an airspeed scale to the left of the horizon sphere and an altimeter and vertical speed scale to the right. The slip indicator is the small rectangle under the direction triangles at the top. This EFIS display presents all of the flight information in the conventional flight deck basic T.*

is done in a checklist format. The right or lower monitor displays accompanying system information in a pictorial form, such as a diagram of the system being referred to on the primary monitor.

The ECAM monitors are typically powered by separate signal generators. Aircraft data inputs are fed into two flight warning computers. Analog inputs are first fed through a system data analog converter and then into the warning computers. The warning computers process the information and forward information to the signal generators to illuminate the monitors. *[Figure 10-120]*

There are four basic modes to the ECAM system: flight phase, advisory, failure related, and manual. The flight phase mode is normally used. The phases are: preflight, takeoff, climb, cruise, descent, approach, and post landing. Advisory and failure–related modes will appear automatically as the situation requires. When an advisory is shown on the primary monitor, the secondary monitor will automatically display the system schematic with numerical values. The same is true for the failure-related mode, which takes precedent over all other modes regardless of which mode is selected at the time of the failure. Color coding is used on the displays to draw attention to matters in order of importance. Display modes are selected via a separate ECAM control panel shown in *Figure 10-121.*

The manual mode of an ECAM is set by pressing one of the synoptic display buttons on the control panel. This allows the display of system diagrams. A failure warning or advisory event will cancel this view. *[Figure 10-122]*

ECAM flight warning computers self-test upon startup. The signal generators are also tested. A maintenance panel allows for testing annunciation and further testing upon demand. BITE stands for built-in test equipment. It is standard for monitoring systems to monitor themselves as well as the aircraft systems. All of the system inputs to the flight warning computers can also be tested for continuity from this panel, as

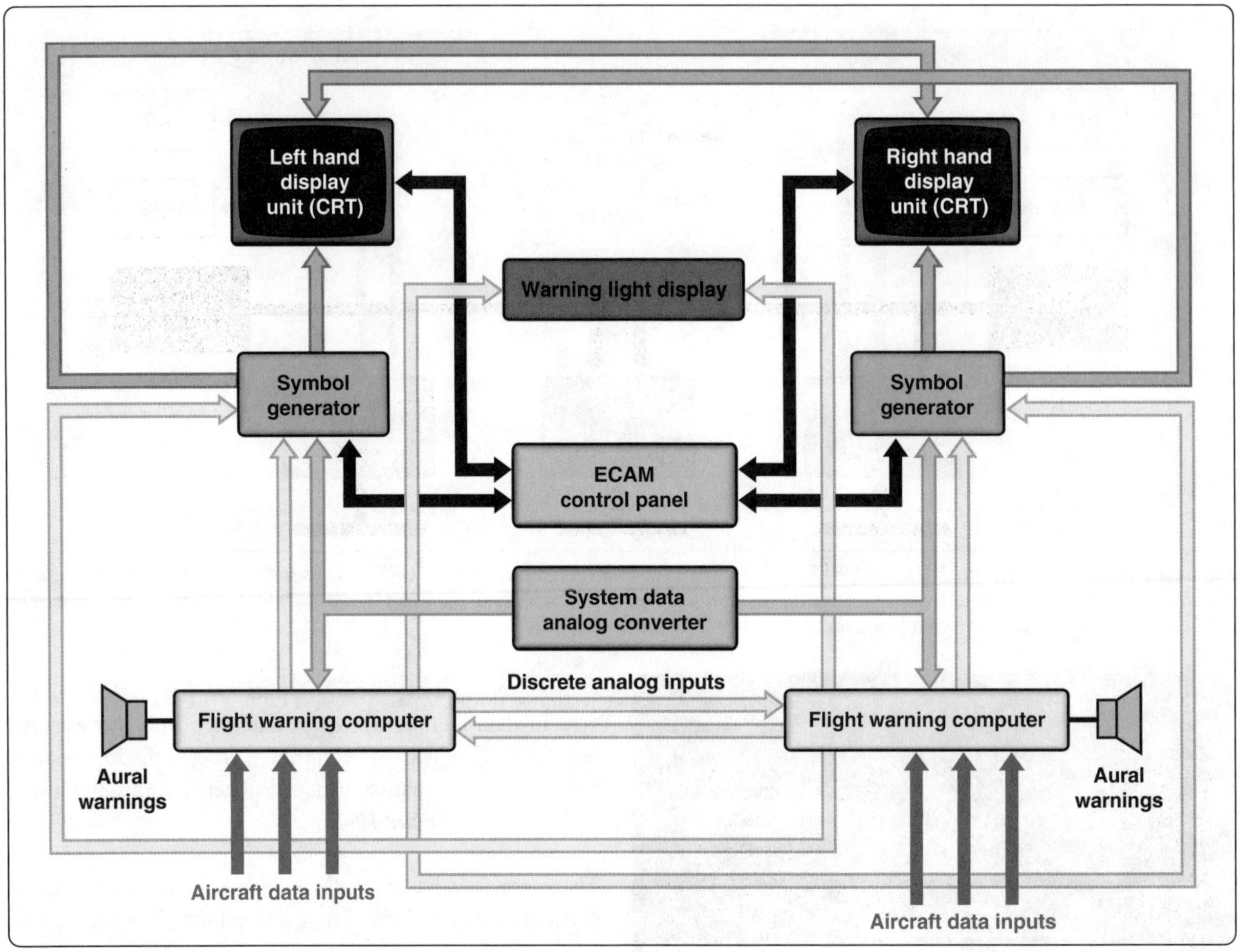

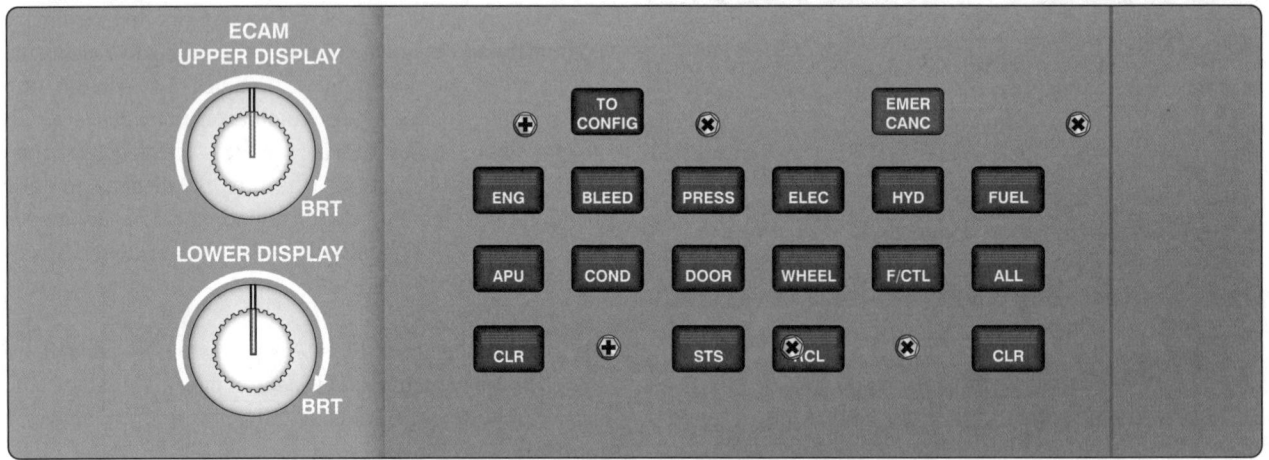

Figure 10-120. *An electronic centralized aircraft monitor (ECAM) system displays aircraft system status, checklists, advisories, and warnings on a pair of controllable monitors.*

Figure 10-121. *An ECAM display control panel.*

well as inputs and outputs of the system data analog converter. Any individual system faults will be listed on the primary display as normal. Faults in the flight warning computers and signal generators will annunciate on the maintenance panel. *[Figure 10-123]* Follow the manufacturer's guidelines when testing ECAM and related systems.

Engine Indicating & Crew Alerting System (EICAS)

An engine indicating and crew alerting system (EICAS) performs many of the same functions as an ECAM system. The objective is still to monitor the aircraft systems for the pilot. All EICASs display engine, as well as airframe, parameters.

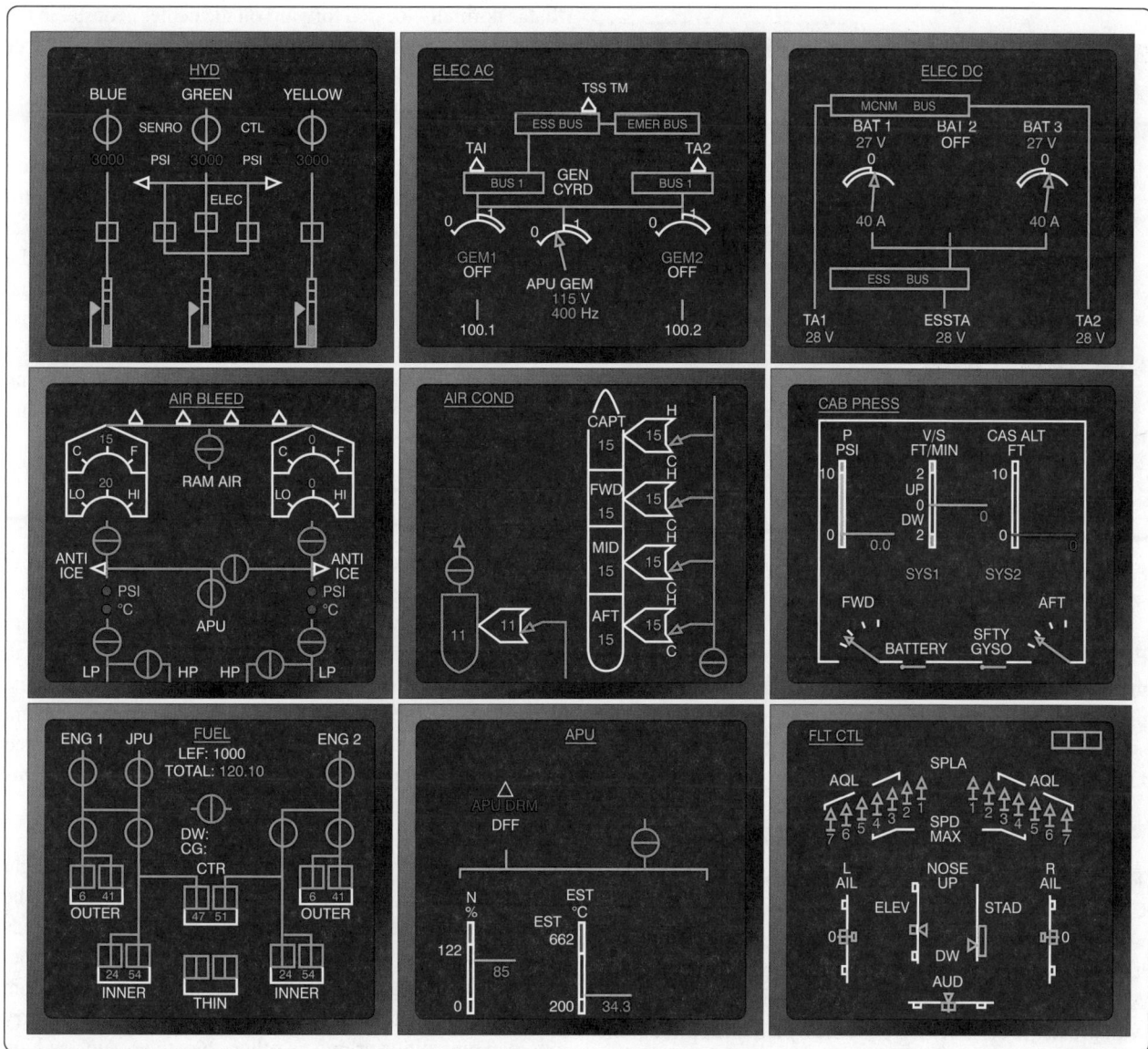

Figure 10-122. *Nine of the 12 available system diagrams from the ECAM manual mode.*

Traditional gauges are not utilized, other than as a standby combination engine gauge in case of total system failure.

EICAS is also a two-monitor, two-computer system with a display select panel. Both monitors receive information from the same computer. The second computer serves as a standby. Digital and analog inputs from the engine and airframe systems are continuously monitored. Caution and warning lights, as well as aural tones, are incorporated. *[Figure 10-124]*

EICAS provides full time primary engine parameters (EPR, N_1, EGT) on the top, primary monitor. Advisories and warning are also shown there. Secondary engine parameters and nonengine system status are displayed on the bottom screen. The lower screen is also used for maintenance

diagnosis when the aircraft is on the ground. Color coding is used, as well as message prioritizing.

The display select panel allows the pilot to choose which computer is actively supplying information. It also controls the display of secondary engine information and system status displays on the lower monitor. EICAS has a unique feature that automatically records the parameters of a failure event to be reviewed afterwards by maintenance personnel. Pilots that suspect a problem may be occurring during flight can press the event record button on the display select panel. This also records the parameters for that flight period to be studied later by maintenance. Hydraulic, electrical, environmental, performance, and APU data are examples of what may be recorded.

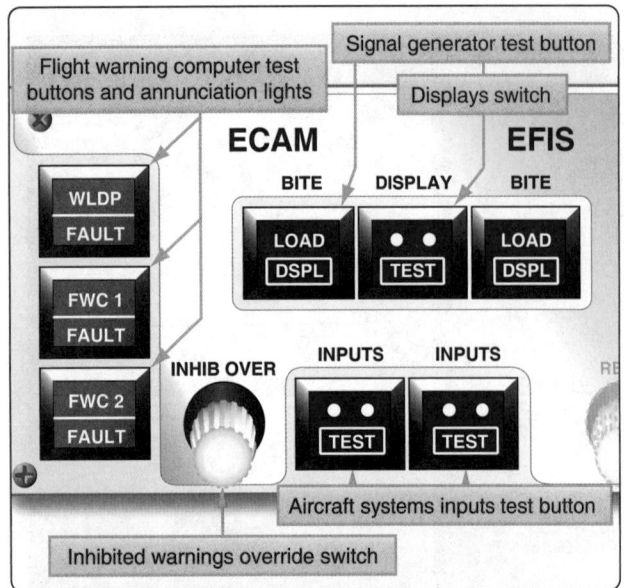

Figure 10-123. *An ECAM maintenance panel used for testing and annunciating faults in the ECAM system.*

EICAS uses BITE for systems and components. A maintenance panel is included for technicians. From this panel, when the aircraft is on the ground, push-button switches display information pertinent to various systems for analysis. *[Figure 10-125]*

Flight Management System (FMS)

The highest level of automated flight system is the flight FMS. Companies flying aircraft for hire have special results they wish to achieve. On-time performance, fuel conservation, and long engine and component life all contribute to profitability. An FMS helps achieve these results by operating the aircraft with greater precision than possible by a human pilot alone.

A FMS can be thought of as a master computer system that has control over all other systems, computerized and otherwise. As such, it coordinates the adjustment of flight, engine, and airframe parameters either automatically or by instructing the pilot how to do so. Literally, all aspects of the flight are considered, from preflight planning to pulling up to the jet-way upon landing, including in-flight amendments to planned courses of action.

The main component of an FMS is the flight management computer (FMC). It communicates with the EICAS or ECAM, the ADC, the thrust management computer that controls the autothrottle functions, the EIFIS symbol generators, the automatic flight control system, the inertial reference system, collision avoidance systems, and all of the radio navigational aids via data busses. *[Figure 10-126]*

The interface to the system is a control display unit (CDU)

that is normally located forward on the center pedestal in the flight deck. It contains a full alphanumeric keypad, a CRT or LCD display/work screen, status and condition annunciators, and specialized function keys. *[Figure 10-127]*

The typical FMS uses two FMS FMCs that operate independently as the pilot's unit and the copilot's unit. However, they do crosstalk through the data busses. In normal operation, the pilot and copilot divide the workload, with the pilot's CDU set to supervise and interface with operational parameters and the copilot's CDU handling navigational chores. This is optional at the flightcrew's discretion. If a main component fails (e.g., an FMC or a CDU), the remaining operational units continue to operate with full control and without system compromise.

Each flight of an aircraft has vertical, horizontal, and navigational components, which are maintained by manipulating the engine and airframe controls. While doing so, numerous options are available to the pilot. Rate of climb, thrust settings, EPR levels, airspeed, descent rates, and other terms can be varied. Commercial air carriers use the FMC to establish guidelines by which flights can be flown. Usually, these promote the company's goals for fuel and equipment conservation. The pilot need only enter variables as requested and respond to suggested alternatives as the FMC presents them.

The FMC has stored in its database literally hundreds of flight plans with predetermined operational parameters that can be selected and implemented. Integration with NAV-COM aids allows the FMS to change radio frequencies as the flight plan is enacted. Internal computations using direct input from fuel flow and fuel quantity systems allow the FMC to carry out lean operations or pursue other objectives, such as high performance operations if making up time is paramount on a particular flight. Weather and traffic considerations are also integrated. The FMS can handle all variables automatically but communicates via the CDU screen to present its planned action, gain consensus, or ask for an input or decision.

As with the monitoring systems, FMS includes BITE. The FMC continuously monitors its entire systems and inputs for faults during operation. Maintenance personal can retrieve system generated and pilot recorded fault messages. They may also access maintenance pages that call out line replaceable units (LRUs) to which faults have been traced by the BITE system. Follow manufacturers' procedures for interfacing with maintenance data information.

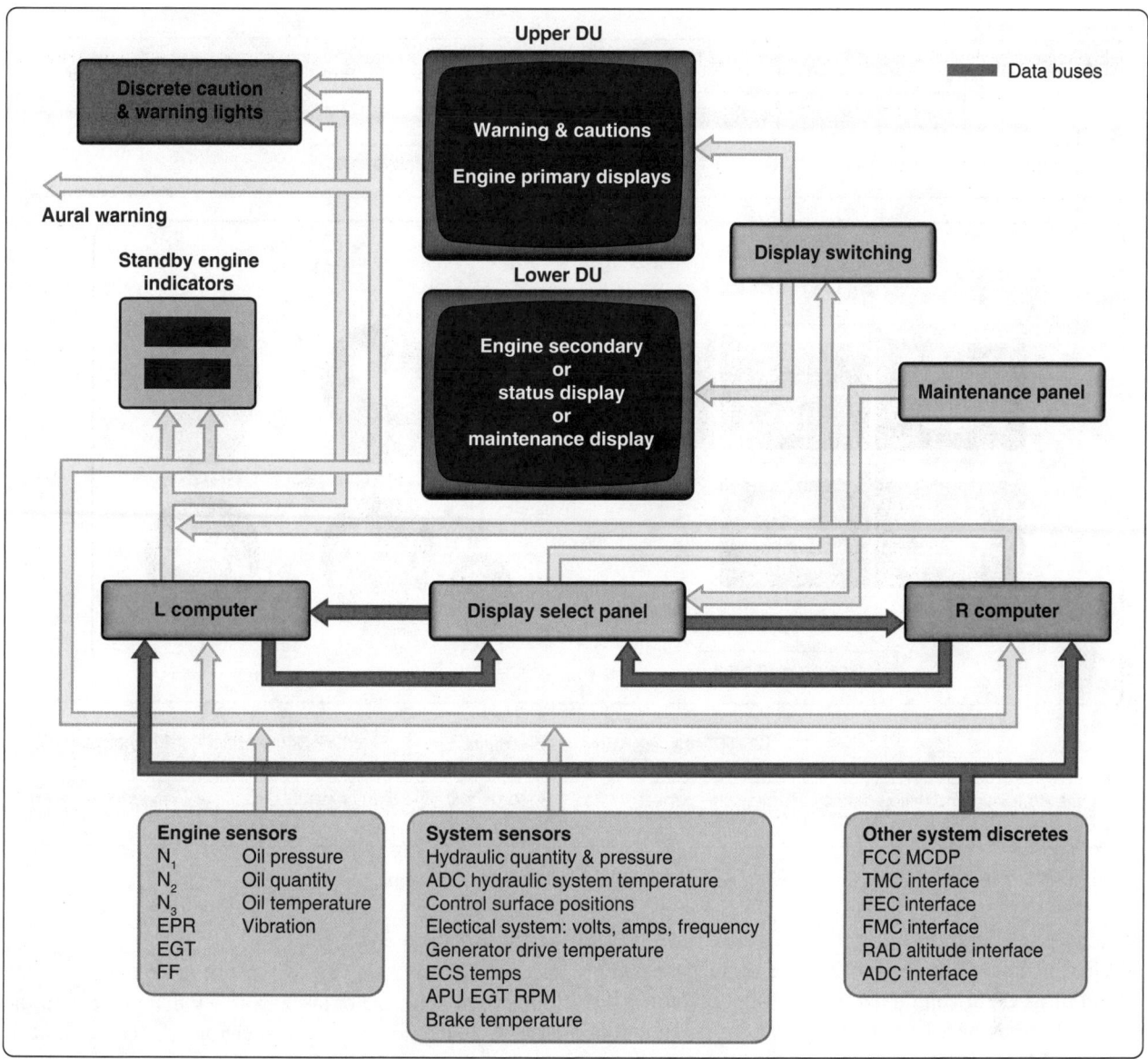

Figure 10-124. *Schematic of an engine indicating and crew alerting system (EICAS).*

Warnings & Cautions

Annunciator Panels

An annunciator panel is a panel of warning lights in plain sight of the pilot. These lights are identified by the name of the system they represent and are usually covered with colored lenses to show the meaning of the condition they announce. Annunciator panels are installed for two purposes: to display current conditions and to notify of unsatisfactory conditions. Standardized colors are used to differentiate between visual messages. For example, the color green indicates a satisfactory condition. Yellow is used to caution of a serious condition that requires further monitoring. Red is the color for an unsatisfactory condition which needs immediate attention or action. Whether part of the instrument face or of a visual warning system, these colors give quick-reference information to the pilot.

Most aircraft include annunciator lights that illuminate when an event demanding attention occurs. These use the aforementioned colors in a variety of presentations. Individual lights near the associated flight deck instrument or a collective display of lights for various systems in a central location are common. Words label each light or are part of the light itself to identify any problem quickly and plainly.

On complex aircraft, the status of numerous systems and components must be known and maintained. Centralized warning systems have been developed to annunciate critical messages concerning a multitude of systems and components in a simplified, organized manner. Often, this will be done by locating a single annunciator panel somewhere on the instrument panel. These analog aircraft warning systems may look different in various aircraft and depend on manufacturer

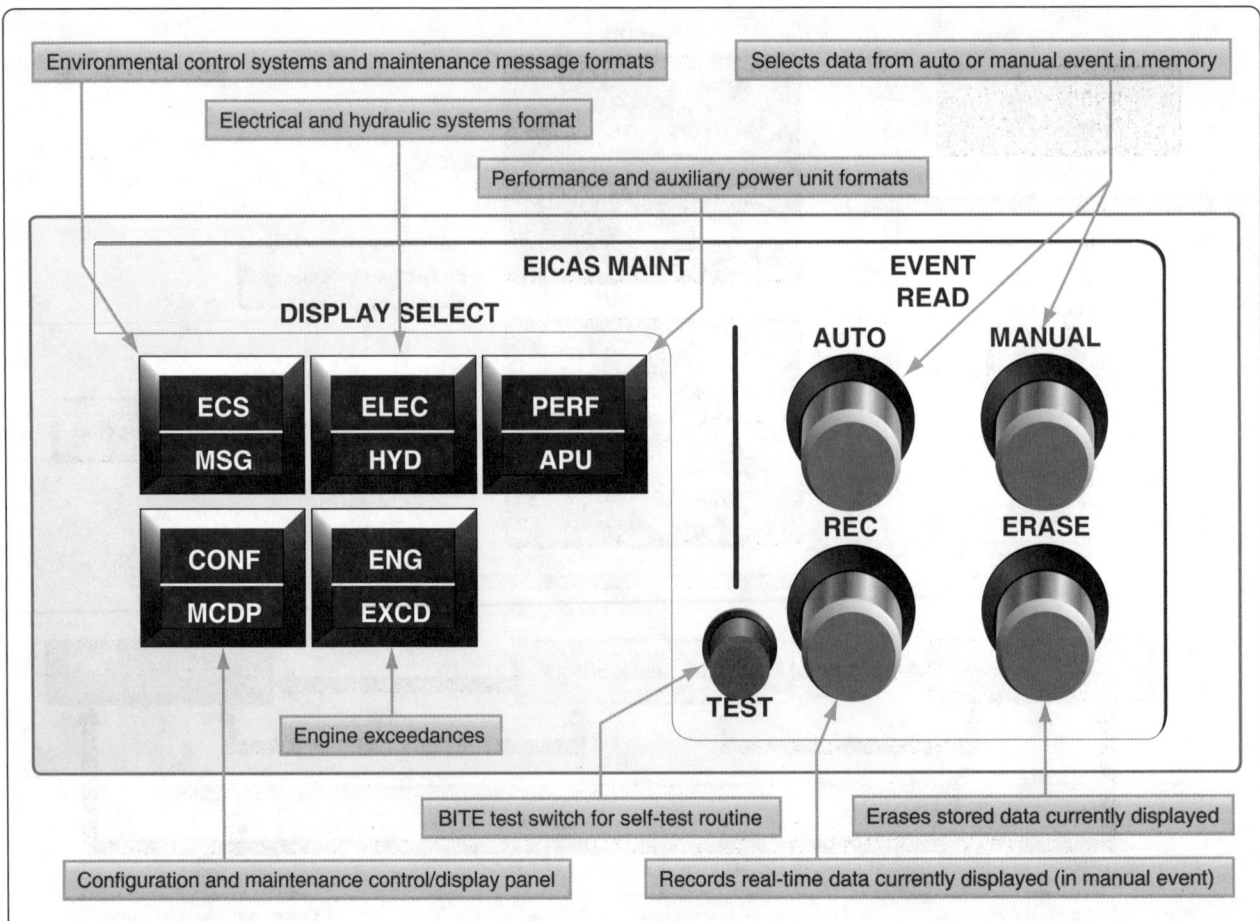

Environmental control systems and maintenance message formats

Electrical and hydraulic systems format

Performance and auxiliary power unit formats

Selects data from auto or manual event in memory

EICAS MAINT

EVENT READ

DISPLAY SELECT

ECS	ELEC	PERF
MSG	HYD	APU

CONF	ENG
MCDP	EXCD

AUTO

MANUAL

REC

ERASE

TEST

Engine exceedances

BITE test switch for self-test routine

Erases stored data currently displayed

Configuration and maintenance control/display panel

Records real-time data currently displayed (in manual event)

Figure 10-125. *The EICAS maintenance control panel is for the exclusive use of technicians.*

preference and the systems installed. *[Figure 10-128]* EFIS provide for annunciation of advisory and warning messages as part of its flight control and monitoring capabilities, as previously described. Usually, the primary display unit is designated as the location to display annunciations.

Master caution lights are used to draw the attention of the crew to a critical situation in addition to an annunciator that describes the problem. These master caution lights are centrally wired and illuminate whenever any of the participating systems or components require attention. Once notified, the pilot may cancel the master caution, but a dedicated system or component annunciator light stays illuminated until the situation that caused the warning is rectified. Cancelling resets the master caution lights to warn of a subsequent fault event even before the initial fault is corrected. *[Figure 10-129]* Press to test is available for the entire annunciator system, which energizes all warning circuitry and lights to confirm readiness. Often, this test exposes the need to replace the tiny light bulbs that are used in the system.

Aural Warning Systems

Aircraft aural warning systems work in conjunction with illuminated annunciator systems. They audibly inform the pilot of a situation requiring attention. Various tones and phrases sound in the flight deck to alert the crew when certain conditions exist. For example, an aircraft with retractable landing gear uses an aural warning system to alert the crew to an unsafe condition. A bell sounds if the throttle is retarded and the landing gear is not in a down and locked condition.

A typical transport category aircraft has an aural warning system that alerts the pilot with audio signals for the following: abnormal takeoff, landing, pressurization, mach airspeed conditions, an engine or wheel well fire, calls from the crew call system, collision avoidance recommendations, and more. *Figure 10-130* shows some of the problems that trigger aural warnings and the action to be taken to correct the situation.

Clocks

Whether called a clock or a chronometer, an FAA-approved time indicator is required in the flight deck of IFR-certified

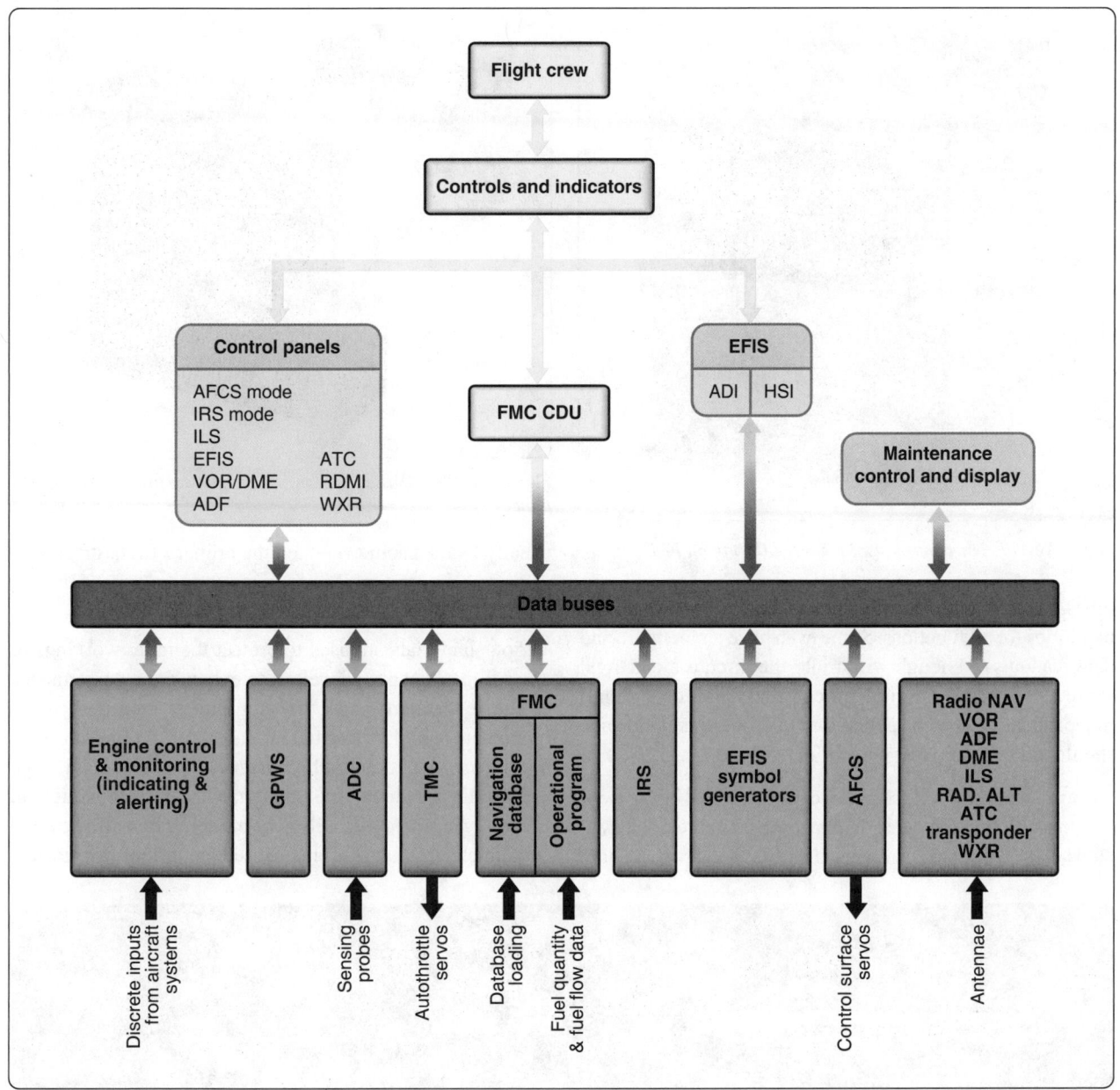

Figure 10-126. *A flight management system (FMS) integrates numerous engine, aircraft, and navigational systems to provide overall management of the flight.*

aircraft. Pilots use a clock during flight to time maneuvers and for navigational purposes. The clock is usually mounted near the flight instrument group, often near the turn coordinator. It indicates hours, minutes, and seconds.

For many years, the mechanical 8-day clock was the standard aircraft timekeeping device largely because it continues to run without electrical power as long as it has been hand wound. The mechanical 8-day clock is reliable and accurate enough for its intended use. Some mechanical aircraft clocks feature a push-button elapsed time feature. *[Figure 10-131]*

As electrical systems developed into the reliable, highly redundant systems that exist today, use of an electric clock to replace the mechanical clock began. An electric clock is an analog devise that may also have an elapsed time feature. It can be wired to the battery or battery bus. Thus, it continues to operate in the event of a power failure. Electric aircraft clocks are often used in multiengine aircraft where complete loss of electrical power is unlikely.

Many modern aircraft have a digital electronic clock with LED readout. This device comes with the advantages of low power consumption and high reliability due to the lack of moving parts. It is also very accurate. Solid-state electronics allow for expanded features, such as elapsed time, flight

Figure 10-127. *The control display unit (CDU) of an FMS.*

Figure 10-129. *A master caution switch removed from the instrument panel.*

time that starts automatically upon takeoff, a stop watch, and memories for all functions. Some even have temperature and date readouts. Although wired into the aircraft's electrical system, electronic digital clocks may include a small independent battery inside the unit that operates the device should aircraft electrical power fail. *[Figure 10-132]*

On aircraft with fully digital computerized instrument systems utilizing flat panel displays, the computer's internal clock, or a GPS clock, can be used with a digital time readout

usually located somewhere on the primary flight display.

Instrument Housings & Handling

Various materials are used to protect the inner workings of aircraft instruments, as well as to enhance the performance of the instrument and other equipment mounted in the immediate vicinity. Instrument cases can be one piece or multipiece. Aluminum alloy, magnesium alloy, steel, iron, and plastic are all common materials for case construction. Electric instruments usually have a steel or iron alloy case to contain electromagnetic flux caused by current flow inside.

Figure 10-128. *The centralized analog annunciator panel has indicator lights from systems and components throughout the aircraft. It is supported by the master caution system.*

Examples of Aircraft Aural Warnings				
Stage of Operation	Warning System	Warning Signal	Cause of Warning Signal Activation	Corrective action
Takeoff	Flight control	Intermittent horn	Throttles are advanced and any of the following conditions exist: 1. Speed brakes are not down 2. Flaps are not in takeoff range 3. Auxiliary power exhaust door is open 4. Stabilizer is not in the takeoff setting	Correct the aircraft to proper takeoff conditions
In flight	Mach warning	Clacker	Equivalent airspeed or mach number exceeds limits	Decrease aircraft speed
In flight	Pressurization	Intermittent horn	If cabin pressure becomes equal to atmospheric pressure at the specific altitude (altitude at time of occurrence)	Correct the condition
Landing	Landing gear	Continuous horn	Landing gear is not down and locked when flaps are less than full up and throttle is retarded to idle	Raise flaps; advance throttle
Any stage	Fire warning	Continuous bell	Any overheat condition or fire in any engine or nacelle, or main wheel or nose wheel well, APU engine, or any compartment having fire warning system installed Whenever the fire warning system is tested	1. Lower the heat in the the area where in the F/W was activated 2. Signal may be silenced pushing the F/W bell cutout switch or the APU cutout switch
Any stage	Communications	High chime	Any time captain's call button is pressed at external power panel forward or rearward cabin attendant's panel	Release button; if button remains locked in, pull button out

Figure 10-130. *Aircraft aural warnings.*

Figure 10-131. *A typical mechanical 8-day aircraft clock.*

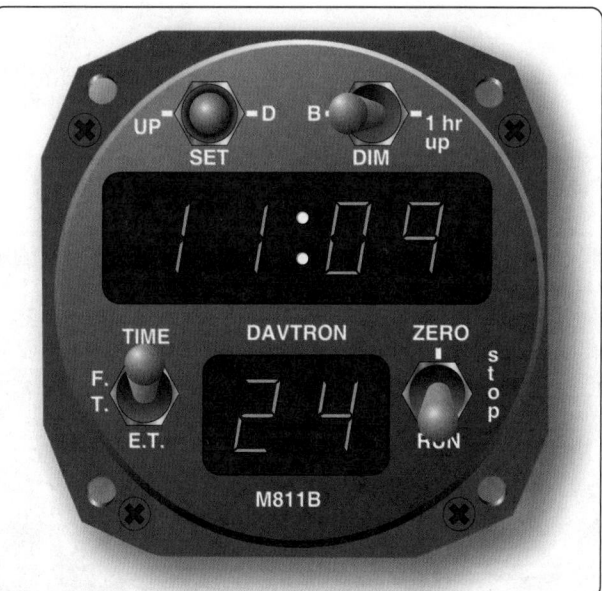

Figure 10-132. *A typical aircraft electronic clock.*

Despite their rugged outward appearance, all instruments, especially analog mechanical instruments, should be handled with special care and should never be dropped. A crack in an airtight instrument case renders it unairworthy. Ports should never be blown into and should be plugged until the instrument is installed. Cage all gyro instruments until mounted in the instrument panel. Observe all cautions written on the instrument housing and follow the manufacturer's instruction for proper handling and shipping, as well as installation.

Instrument Installations & Markings

Instrument Panels

Instrument panels are usually made from sheet aluminum alloy and are painted a dark, nonglare color. They sometimes contain subpanels for easier access to the backs of instruments during maintenance. Instrument panels are usually shock-mounted to absorb low-frequency, high-amplitude shocks. The mounts absorb most of the vertical and horizontal vibration but permit the instruments to operate under conditions of minor vibration. Bonding straps are used to ensure electrical continuity from the panel to the airframe. *[Figure 10-133]*

The type and number of shock mounts to be used for instrument panels are determined by the weight of the unit. Shock-mounted instrument panels should be free to move in all directions and have sufficient clearance to avoid striking the supporting structure. When a panel does not have adequate clearance, inspect the shock mounts for looseness, cracks, or deterioration.

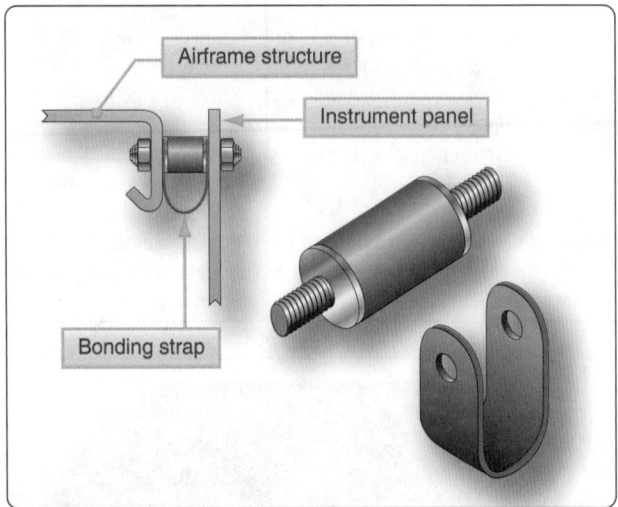

Figure 10-133. *Instrument panel shock mounts.*

Instrument panel layout is seemingly random on older aircraft. The advent of instrument flight made the flight instruments of critical importance when flying without outside reference to the horizon or ground. As a result, the basic T arrangement for flight instruments was adopted, as mentioned in the beginning of this chapter. *[Figure 10-4]* Electronic flight instrument systems and digital flight deck displays have kept the same basic T arrangement for flight instrument and data presentations. The flight instruments and basic T are located directly in front of the pilot and copilot's seats. Some light aircraft have only one full set of flight instruments that are located in front of the left seat.

The location of engine instruments and navigation instruments varies. Ideally, they should be accessible to both the pilot and copilot. Numerous variations exist to utilize the limited space in the center of the instrument panel and still provide accessibility by the flight crew to all pertinent instruments. On large aircraft, a center pedestal and overhead panels help create more space. On small aircraft, the engine instruments are often moved to allow navigation instruments and radios to occupy the center of the instrument panel. *[Figure 10-134]* On modern aircraft, EFIS and digital flight information systems reduce panel clutter and allow easier access to all instruments by both crewmembers. Controllable display panels provide the ability to select from pages of information that, when not displayed, are completely gone from view and use no instrument panel space.

Figure 10-134. *Flight instruments directly in front of the pilot, engine instruments to the left and right, and navigation instruments and radios primarily to the right, which is the center of the instrument panel. This arrangement is commonly on light aircraft to be flown by a single pilot.*

Instrument Mounting

The method of mounting instruments in their respective panels depends on the design of the instrument case. In one design, the bezel is flanged in such a manner that the instrument can be flush mounted in its cutout from the rear of the panel. Integral, self-locking nuts are provided at the rear faces of the flange corners to receive mounting screws from the front of the panel. The flanged-type instrument can also be mounted to the front of the panel. In this case, nut-plates are usually installed in the panel itself. Nonferrous screws are usually used to mount the instruments.

There are also instrument mounting systems where the instruments are flangeless. A special clamp, shaped and dimensioned to fit the instrument case, is permanently secured to the rear face of the panel. The instrument is slid into the panel from the front and into the clamp. The clamp's tightening screw is accessible from the front side of the panel. *[Figure 10-135]* Regardless of how an instrument is mounted, it should not be touching or be so close as to touch another instrument during the shock of landing.

Instrument Power Requirements

Many aircraft instruments require electric power for operation. Even nonelectric instruments may include electric lighting. Only a limited amount of electricity is produced by the aircraft's electric generator(s). It is imperative that the electric load of the instruments, radios, and other equipment on board the aircraft does not exceed this amount.

Electric devices, including instruments, have power ratings. These show what voltage is required to correctly operate the unit and the amount of amperage it draws when operating to capacity. The rating must be checked before installing any component. Replacement of a component with one that has the same power rating is recommended to ensure the potential electric load of the installed equipment remains within the limits the aircraft manufacturer intended. Adding a component with a different rating or installing a completely new component may require a load check be performed. This is essentially an on the ground operational check to ensure the electrical system can supply all of the electricity consuming devices installed on the aircraft. Follow the manufacturer's instructions on how to perform this check.

Instrument Range Markings

Many instruments contain colored markings on the dial face to indicate, at a glance, whether a particular system or component is within a range of operation that is safe and desirable or if an undesirable condition exists. These markings are put on the instrument by the original equipment manufacturer in accordance with the Aircraft Specifications in the Type Certificate Data Sheet. Data describing these

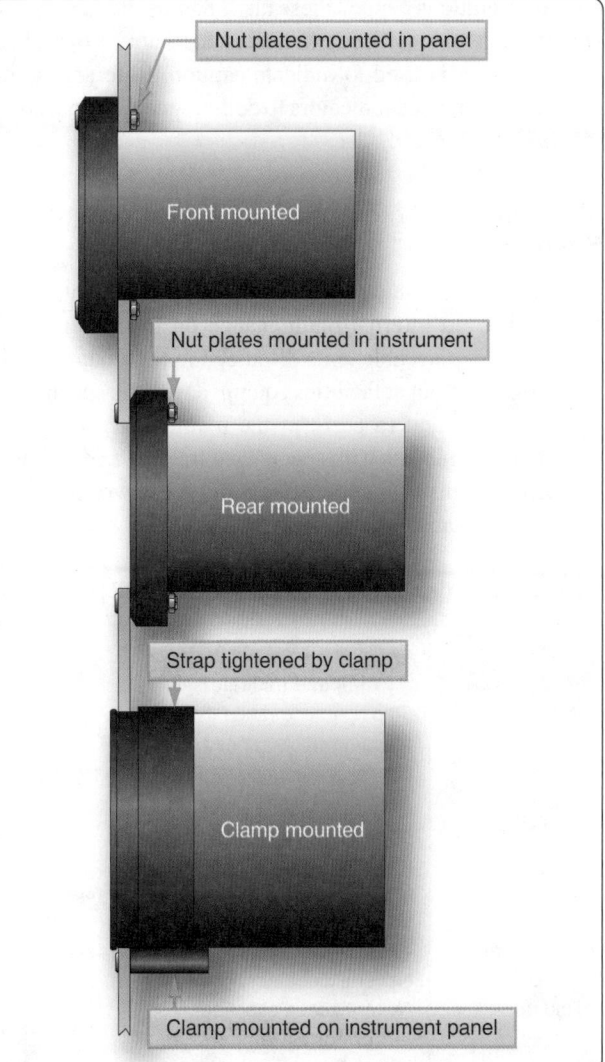

Figure 10-135. *Instrument mounts— flanged (top and middle) and flangeless (bottom).*

limitations can also sometimes be found in the aircraft manufacturer's operating and maintenance manuals and in the flight manual.

Occasionally, the aircraft technician may find it necessary to apply these marking to an approved replacement instrument on which they do not appear. It is crucial that the instrument be marked correctly and only in accordance with approved data. The marking may be placed on the cover glass of the instrument with paint or decals. A white slippage mark is made to extend from the glass to the instrument case. Should the glass rotate in the bezel, the marking will no longer be aligned properly with the calibrated instrument dial. The broken slippage mark indicates this to the pilot or technician.

The colors used as range markings are red, yellow, green, blue, or white. The markings can be in the form of an arc or a radial line. Red is used to indicate maximum and minimum ranges; operations beyond these markings are dangerous and should be avoided. Green indicates the normal operating range. Yellow is used to indicate caution. Blue and white are used on airspeed indicators to define specific conditions. [Figures 10-136 and 10-137]

Maintenance of Instruments & Instrument Systems

An FAA airframe and powerplant (A&P) technician is not qualified to do internal maintenance on instruments and related line replaceable units discussed in this chapter. This must be carried out at facilities equipped with the specialized equipment needed to perform the maintenance properly. Qualified technicians with specialized training and intimate knowledge of instruments perform this type of work, usually under repair station certification.

However, certified airframe technicians and A&P technicians are charged with a wide variety of maintenance functions related to instruments and instrument systems. Installation, removal, inspection, troubleshooting, and functional checks are all performed in the field by certified personnel. It is also a responsibility of the certified technician holding an airframe rating to know what maintenance is required and to access the approved procedures for meeting those requirements.

In the following paragraphs, various maintenance and servicing procedures and suggestions are given. The discussion follows the order in which the various instruments and instrument systems were presented throughout this chapter. This is not meant to represent all of the maintenance required by any of the instruments or instruments systems. The aircraft manufacturer's and instrument manufacturer's approved maintenance documents should always be consulted for required maintenance and servicing instructions. FAA regulations must also be observed.

Instrument	Range marking	Instrument	Range marking
Airspeed indicator		**Oil temperature gauge**	
White arc	Flap operating range	Green arc	Normal operating range
bottom	Flaps-down stall speed	Yellow arc	Precautionary range
top	Maximum airspeed for flaps-down flight	Red radial line	Maximum and/or minimum permissible oil temperature
Green arc	Normal operating range	**Tachometer (reciprocating engine)**	
bottom	Flaps-up stall speed	Green arc	Normal operating range
top	Maximum airspeed for rough air	Yellow arc	Precautionary range
Blue radial line	Best single-engine rate-of-climb airspeed	Red arc	Restricted operating range
Yellow arc	Structural warning area	Red radial line	Maximum permissible rotational speed
bottom	Maximum airspeed for rough air	**Tachometer (turbine engine)**	
top	Never-exceed airspeed	Green arc	Normal operating range
Red radial line	Never-exceed airspeed	Yellow arc	Precautionary range
Carburetor air temperature		Red radial line	Maximum permissible rotational speed
Green arc	Normal operating range	**Tachometer (helicopter)**	
Yellow arc	Range in which carburetor ice is most likely to form	Engine tachometer	
Red radial line	Maximum allowable inlet air temperature	Green arc	Normal operating range
Cylinder head temperature		Yellow arc	Precautionary range
Green arc	Normal operating range	Red radial line	Maximum permissible rotational speed
Yellow arc	Operation approved for limited time	Rotor tachometer	
Red radial line	Never-exceed temperature	Green arc	Normal operating range
Manifold pressure gauge		Red radial line	Maximum and minimum rotor speed for power-off operational conditions
Green arc	Normal operating range	**Torque indicator**	
Yellow arc	Precautionary range	Green arc	Normal operating range
Red radial line	Maximum permissible manifold absolute pressure	Yellow arc	Precautionary range
Fuel pressure gauge		Red radial line	Maximum permissible torque pressure
Green arc	Normal operating range	**Exhaust gas temperature indicator (turbine engine)**	
Yellow arc	Precautionary range	Green arc	Normal operating range
Red radial line	Maximum and/or minimum permissible fuel pressure	Yellow arc	Precautionary range
Oil pressure gauge		Red radial line	Maximum permissible gas temperature
Green arc	Normal operating range	**Gas producer N1 tachometer (turboshaft helicopter)**	
Yellow arc	Precautionary range	Green arc	Normal operating range
Red radial line	Maximum and/or minimum permissible oil pressure	Yellow arc	Precautionary range
		Red radial line	Maximum permissible rotational speed

Figure 10-136. *Instrument range markings.*

Figure 10-137. *An airspeed indicator makes extensive use of range markings.*

Altimeter Tests

When an aircraft is to be operated under IFR, an altimeter test must have been performed within the previous 24 months. Title 14 of the Code of Federal Regulations (14 CFR) part 91, section 91.411, requires this test, as well as tests on the pitot-static system and on the automatic pressure altitude reporting system. The certified airframe or A&P mechanic is not qualified to perform the altimeter inspections. They must be conducted by either the manufacturer or a certified repair station. 14 CFR part 43, Appendix E details the requirements for these tests.

Pitot-Static System Maintenance & Tests

Water trapped in a pitot static system may cause inaccurate or intermittent indications on the pitot-static flight instruments. This is especially a problem if the water freezes in flight. Many systems are fitted with drains at the low points in the system to remove any moisture during maintenance. Lacking this, dry compressed air or nitrogen may be blown through the lines of the system. Always disconnect all pitot-static instruments before doing so and always blow from the instrument end of the system towards the pitot and static ports. This procedure must be followed by a leak check described below. Systems with drains can be drained without requiring a leak check. Upon completion, the technician must ensure that the drains are closed and made secure in accordance with approved maintenance procedures.

Aircraft pitot-static systems must be tested for leaks after the installation of any component parts or when system malfunction is suspected. It must also be tested every 24 months if on an IFR certified aircraft intended to be flown as such as called out in 14 CFR part 91, section 91.411. Certified airframe and A&P technicians may perform this test.

The method of leak testing depends on the type of aircraft, its pitot-static system, and the testing equipment available. *[Figure 10-138]* Essentially, a testing device is connected into the static system at the static vent end, and pressure is reduced in the system by the amount required to indicate 1,000 feet on the altimeter. Then, the system is sealed and observed for 1 minute. A loss of altitude of more than 100 feet is not permissible. If a leak exists, a systematic check of portions of the system is conducted until the leak is isolated. Most leaks occur at fittings. The pitot portion of the pitot-static system is checked in a similar fashion. Follow the manufacturer's instructions when performing all pitot-static system checks. In all cases, pressure and suction must be applied and released slowly to avoid damage to the aircraft instruments. Pitot-static system leak check units usually have their own built-in altimeters. This allows a functional cross-check of the aircraft's altimeter with the calibrated test unit's altimeter while performing the static system check. However, this does not meet the requirements of 14 CFR part 91, section 91.411 for altimeter tests.

Upon completion of the leak test, be sure that the system is returned to the normal flight configuration. If it is necessary to block off various portions of a system, check to be sure that all blanking plugs, adaptors, or pieces of adhesive tape have been removed.

Tachometer Maintenance

Tachometer indicators should be checked for loose glass, chipped scale markings, or loose pointers. The difference in indications between readings taken before and after lightly tapping the instrument should not exceed approximately 15 rpm. This value may vary, depending on the tolerance established by the indicator manufacturer. Both tachometer generator and indicator should be inspected for tightness of mechanical and electrical connections, security of mounting, and general condition. For detailed maintenance procedures, the manufacturer's instructions should always be consulted.

When an engine equipped with an electrical tachometer is running at idle rpm, the tachometer indicator pointers may fluctuate and read low. This is an indication that the synchronous motor is not synchronized with the generator output. As the engine speed is increased, the motor should synchronize and register the rpm correctly. The rpm at which synchronization occurs varies with the design of the tachometer system. If the instrument pointer(s) oscillate(s) at speeds above the synchronizing value, determine that the total oscillation does not exceed the allowable tolerance.

Pointer oscillation can also occur with a mechanical indication system if the flexible drive is permitted to whip. The drive shaft should be secured at frequent intervals to prevent it from whipping. When installing mechanical type indicators, be sure that the flexible drive has adequate clearance behind the panel. Any bends necessary to route the drive should not cause strain on the instrument when it is secured to the panel. Avoid sharp bends in the drive. An improperly installed drive can cause the indicator to fail to read or to read incorrectly.

Magnetic Compass Maintenance & Compensation

The magnetic compass is a simple instrument that does not require setting or a source of power. A minimum of maintenance is necessary, but the instrument is delicate and should be handled carefully during inspection. The following items are usually included in an inspection:

1. The compass indicator should be checked for correct readings on various cardinal headings and re-compensated if necessary.

2. Moving parts of the compass should work easily.

3. The compass bowl should be correctly suspended on an antivibration device and should not touch any part of the metal container.

4. The compass bowl should be filled with liquid. The liquid should not contain any bubbles or have any discoloration. Airframe mechanics cannot refill the fluid of a whiskey compass.

5. The scale should be readable and be well lit.

Compass magnetic deviation is caused by electromagnetic interference from ferrous materials and operating electrical components in the flight deck. Deviation can be reduced by swinging the compass and adjusting its compensating magnets. An example of how to perform this calibration process is given below. The results are recorded on a compass correction card which is placed near the compass in the flight deck. *[Figure 10-139]*

There are various ways to swing a compass. The following is meant as a representative method. Follow the aircraft manufacturer's instructions for method and frequency of swinging the magnetic compass. This is usually accomplished at flight hour or calendar intervals. Compass calibration is also performed when a new electric component is added to the flight deck, such as a new radio. A complete list of conditions requiring a compass swing and procedure can be found in FAA Advisory Circular (AC) 43.13-1 (as revised), Chapter 12, Section 3, paragraph 12-37.

To swing a compass, a compass rose is required. Most airports have one painted on the tarmac in a low-traffic area where maintenance personnel can work. One can also be made with chalk and a good compass. The area where the compass rose is laid out should be far from any possible electromagnetic disturbances, including those underground, and should remain clear of any ferrous vehicles or large equipment while the procedure takes place. *[Figure 10-140]*

The aircraft should be in level flight attitude for the compass swing procedure. Tail draggers need to have the aft end of the fuselage propped up, preferably with wood, aluminum, or some other nonferrous material. The aircraft interior and baggage compartments should be free from miscellaneous items that might interfere with the compass. All normal equipment should be on board and turned on to simulate a flight condition. The engine(s) should be running.

The basic idea when swinging a compass is to note the deviation along the north-south radial and the east-west radial. Then, adjust the compensating magnets of the compass to eliminate as much deviation as possible. Begin

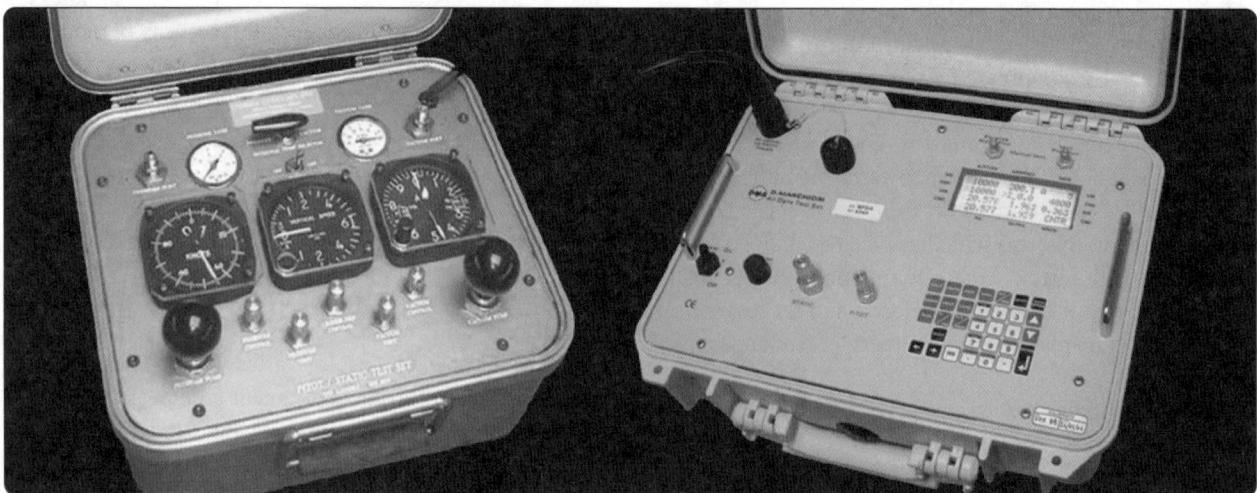

Figure 10-138. *An analog pitot-static system test unit (left) and a digital pitot static test unit (right).*

by centering or zeroing the compass' compensating magnets with a non-ferrous screw driver. Align the longitudinal axis of the aircraft on the N-S radial facing north. Adjust the N-S compensating screw so the indication is 0°. Next, align the longitudinal axis of the aircraft on the E-W radial facing east. Adjust the E-W compensating screw so that the compass indicates 90°. Now, move the aircraft to be aligned with the N-S radial facing south. If the compass indicates 180°, there is no deviation while the aircraft is heading due north or due south. However, this is unlikely. Whatever the south-facing indication is, adjust the N-S compensating screw to eliminate half of the deviation from 180°. Continue around to face the aircraft west on the E-W radial and use the E-W compensating screw to eliminate half of the west-facing deviation from 270°.

Once this is done, return the aircraft to alignment with the N-S radial facing north and record the indication. Up to 10° deviation is allowed. If the deviation cannot be corrected to within 10°, the compass should be replaced if further inspection reveals that there is no other reasons for this deviation. Align the aircraft with the radials every 30° around the compass rose and record each indication on the compass compensation card. Date and sign the card and place it in full view of the pilot near the compass in the flight deck.

Vacuum System Maintenance

Errors in the indication presented on a vacuum gyroscopic instrument could be the result of any factor that prevents the vacuum system from operating within the design suction limits. Errors can also be caused by problems within the instrument, such as friction, worn parts, or broken parts. Any source that disturbs the free rotation of the gyro at design speed is undesirable resulting in excessive precession and failure of the instruments to maintain accurate indication. The aircraft technician is responsible for the prevention or correction of vacuum system malfunctions. Usually this consists of cleaning or replacing filters, checking and correcting insufficient vacuum, or removing and replacing the vacuum pump or instruments. A list of the most common malfunctions, together with their correction, is included in *Figure 10-141.*

Autopilot System Maintenance

The information in this section does not apply to any particular autopilot system but gives general information that relates to all autopilot systems. Maintenance of an autopilot system consists of visual inspection, replacement of components, cleaning, lubrication, and an operational checkout of the system. Consult the manufacturer's maintenance manual for all of these procedures.

With the autopilot disengaged, the flight controls should function smoothly. The resistance offered by the autopilot servos should not affect the control of the aircraft. The interconnecting mechanisms between the autopilot system and the flight control system should be correctly aligned and smooth in operation. When applicable, the operating cables should be checked for tension.

An operational check is important to assure that every circuit is functioning properly. An autopilot operational check should be performed on new installations, after replacement of an autopilot component, or whenever a malfunction in the autopilot is suspected.

After the aircraft's main power switch has been turned on, allow the gyros to come up to speed and the amplifier to warm up before engaging the autopilot. Some systems are designed with safeguards that prevent premature autopilot

Figure 10-139. *A magnetic compass with a deviation correction card attached, on which the results of swinging the compass should be recorded.*

Figure 10-140. *The compass rose on this airport ramp can be used to swing an aircraft magnetic compass.*

engagement. While holding the control column in the normal flight position, engage the autopilot system using the switch on the autopilot controller.

After the system is engaged, perform the operational checks specified for the particular aircraft. In general, the checks are as follows:

1. Rotate the turn knob to the left; the left rudder pedal should move forward, and the control column wheel should move to the left and slightly aft.

2. Rotate the turn knob to the right; the right rudder pedal should move forward, and the control column wheel should move to the right and slightly aft. Return the turn knob to the center position; the flight controls should return to the level-flight position.

3. Rotate the pitch-trim knob forward; the control column should move forward.

4. Rotate the pitch-trim knob aft; the control column should move aft.

If the aircraft has a pitch-trim system installed, it should function to add down-trim as the control column moves forward and add up-trim as the column moves aft. Many pitch-trim systems have an automatic and a manual mode of operation. The above action occurs only in the automatic mode.

Check to see if it is possible to manually override or overpower the autopilot system in all control positions. Center all the controls when the operational checks have been completed. Disengage the autopilot system and check for freedom of the control surfaces by moving the control columns and rudder pedals. Then, reengage the system and check the emergency disconnect release circuit. The autopilot should disengage each time the release button on the control yoke is actuated.

When performing maintenance and operational checks on a specific autopilot system, always follow the procedure recommended by the aircraft or equipment manufacturer.

LCD Display Screens

Electronic and digital instrument systems utilizing LCD technology may have special considerations for the care of the display screens. Antireflective coatings are sometimes used to reduce glare and make the displays more visible. These treatments can be degraded by human skin oils and certain cleaning agents, such as those containing ammonia. It is very important to clean the display lens using a clean, lint-free cloth and a cleaner that is specified as safe for antireflective coatings, preferable one recommended by the aircraft manufacturer.

Problem and Potential Causes	Isolation Procedure	Correction
1. No vacuum pressure or insufficient pressure		
Defective vacuum gauge	Check opposite engine system on the gauge	Replace faulty vacuum gauge
Vacuum relief valve incorrectly adjusted	Change valve adjustment	Make final adjustment to correct setting
Vacuum relief valve installed backward	Visually inspect	Install lines properly
Broken lines	Visually inspect	Replace line
Lines crossed	Visually inspect	Install lines properly
Obstruction in vacuum line	Check for collapsed line	Clean & test line; replace defective part(s)
Vacuum pump failure	Remove and inspect	Replace faulty pump
Vacuum regulator valve incorrectly adjusted	Make valve adjustment and note pressure	Adjust to proper pressure
Vacuum relief valve dirty	Clean and adjust relief valve	Replace valve if adjustment fails
2. Excessive vacuum		
Relief valve improperly adjusted		Adjust relief valve to proper setting
Inaccurate vacuum gauge	Check calibration of gauge	Replace faulty gauge
3. Gyro horizon bar fails to respond		
Instrument caged	Visually inspect	Uncage instrument
Instrument filter dirty	Check filter	Replace or clean as necessary
Insufficient vacuum	Check vacuum setting	Adjust relief valve to proper setting
Instrument assembly worn or dirty		Replace instrument
4. Turn-and-bank indicator fails to respond		
No vacuum supplied to instrument	Check lines and vacuum system	Clean and replace lines and components
Instrument filter clogged	Visually inspect	Replace filter
Defective instrument	Test with properly functioning instrument	Replace faulty instrument
5. Turn-and-bank pointer vibrates		
Defective instrument	Test with properly functioning instrument	Replace defective instrument

Figure 10-141. *Vacuum system troubleshooting guide.*

Chapter 11
Communication & Navigation

Introduction

With the mechanics of flight secured, early aviators began the tasks of improving operational safety and functionality of flight. These were developed in large part through the use of reliable communication and navigation systems. Today, with thousands of aircraft aloft at any one time, communication and navigation systems are essential to safe, successful flight. Continuing development is occurring. Smaller, lighter, and more powerful communication and navigation devices increase situational awareness on the flight deck. Coupled with improved displays and management control systems, the advancement of aviation electronics is relied upon to increase aviation safety.

Clear radio voice communication was one of the first developments in the use of electronics in aviation. Navigational radios soon followed. Today, numerous electronic navigation and landing aids exist. Electronic devices also exist to assist with weather, collision avoidance, automatic flight control, flight recording, flight management, public address, and entertainment systems.

Avionics in Aviation Maintenance

Avionics is a conjunction of the words aviation and electronics. It is used to describe the electronic equipment found in modern aircraft. The term "avionics" was not used until the 1970s. For many years, aircraft had electrical devices, but true solid-state electronic devices were only introduced in large numbers in the 1960s.

Airframe and engine maintenance is required on all aircraft and is not likely to ever go away. Aircraft instrument maintenance and repair also has an inevitable part in aviation maintenance. The increased use of avionics in aircraft over the past 50 years has increased the role of avionics maintenance in aviation. However, modern, solid-state, digital avionics are highly reliable. Mean times between failures are high, and maintenance rates of avionics systems compared to mechanical systems are likely to be lower.

The first decade of avionics proliferation saw a greater increase in the percent of cost of avionics compared to the overall cost of an aircraft. In some military aircraft with highly refined navigation, weapons targeting, and monitoring

systems, it hit a high estimate of 80 percent of the total cost of the aircraft. Currently, the ratio of the cost of avionics to the cost of the total aircraft is beginning to decline. This is due to advances in digital electronics and numerous manufacturers offering highly refined instrumentation, communication, and navigation systems that can be fitted to nearly any aircraft. New aircraft of all sizes are manufactured with digital glass flight decks, and many owners of older aircraft are retrofitting digital avionics to replace analog instrumentation and radio navigation equipment.

The airframe and powerplant (A&P) maintenance technician needs to be familiar with the general workings of various avionics. Maintenance of the actual avionics devices is often reserved for the avionics manufacturers or certified repair stations. However, the installation and proper operation of these devices and systems remains the responsibility of the field technician. This chapter discusses some internal components used in avionics devices. It also discusses a wide range of common communication and navigational aids found on aircraft. The breadth of avionics is so wide that discussion of all avionics devices is not possible.

History of Avionics

The history of avionics is the history of the use of electronics in aviation. Both military and civil aviation requirements contributed to the development. The First World War brought about an urgent need for communications. Voice communications from ground-to-air and from aircraft-to-aircraft were established. *[Figure 11-1]* The development of aircraft reliability and use for civilian purposes in the 1920s led to increased instrumentation and set in motion the need to conquer blind flight—flight without the ground being visible. Radio beacon direction finding was developed for en route navigation. Toward the end of the decade, instrument navigation combined with rudimentary radio use to produce the first safe blind landing of an aircraft.

In the 1930s, the first all radio-controlled blind-landing was accomplished. At the same time, radio navigation using ground-based beacons expanded. Instrument navigation certification for airline pilots began. Low and medium frequency radio waves were found to be problematic at

Figure 11-1. *Early voice communication radio tests in 1917. Courtesy of AT&T Archives and History Center.*

night and in weather. By the end of the decade, use of high-frequency radio waves was explored and included the advent of high-frequency radar.

In the 1940s, after two decades of development driven by mail carrier and passenger airline requirements, World War II injected urgency into the development of aircraft radio communication and navigation. Communication radios, despite their size, were essential on board aircraft. *[Figure 11-2]* Very high frequencies were developed for communication and navigational purposes. Installation of the first instrument landing systems for blind landings began mid-decade and, by the end of the decade, the very high-frequency omni-directional range (VOR) navigational network was instituted. It was also in the 1940s that the first transistor was developed, paving the way for modern, solid-state electronics.

Civilian air transportation increased over the ensuing decades. Communication and navigation equipment was refined. Solid-state radio development, especially in the 1960s, produced a wide range of small, rugged radio and navigational equipment for aircraft. The space program began and added a higher level of communication and navigational necessity. Communication satellites were also launched. The Cold War military build-up caused developments in guidance and navigation and gave birth to the concept of using satellites for positioning.

In the 1970s, concept-validation of satellite navigation was introduced for the military and Block I global positioning system (GPS) satellites were launched well into the 1980s. Back on earth, the long-range navigation system (LORAN) was constructed. Block II GPS satellites were commissioned in the mid-80s and GPS became operational in 1990 with the full 24-satellite system operational in 1994.

In the new millennium, the Federal Aviation Administration (FAA) assessed the national airspace system (NAS) and traffic projections for the future. Gridlock is predicted by 2022. Therefore, a complete overhaul of the NAS, including communication and navigational systems, has been developed and undertaken. The program is called NextGen. It uses the latest technologies to provide a more efficient and effective system of air traffic management. Heavily reliant on global satellite positioning of aircraft in flight and on the ground, NextGen combines GPS technology with automatic dependent surveillance broadcast technology (ADS-B) for traffic separation. A large increase in air system capacity is the planned result. Overhauled ground facilities accompany the technology upgrades mandated for aircraft. NextGen implementation has started and is currently scheduled through the year 2025.

For the past few decades, avionics development has increased at a faster pace than that of airframe and powerplant development. This is likely to continue in the near future. Improvements to solid-state electronics in the form of micro-

Figure 11-2. *Bomber onboard radio station.*

and nano-technologies continue to this day. Trends are toward lighter, smaller devices with remarkable capability and reliability. Integration of the wide range of communication and navigational aids is a focus.

Fundamentals of Electronics

Analog Versus Digital Electronics

Electronic devices represent and manipulate real-world phenomenon through the use of electrical signals. Electronic circuits are designed to perform a wide array of manipulations. Analog representations are continuous. Some aspect of an electric signal is modified proportionally to the real-world item that is being represented. For example, a microphone has electricity flowing through it that is altered when sound is applied. The type and strength of the modification to the electric signal is characteristic of the sound that is made into the microphone. The result is that sound, a real-world phenomenon, is represented electronically. It can then be moved, amplified, and reconverted from an electrical signal back into sound and broadcast from a speaker across the room or across the globe.

Since the flow of electricity through the microphone is continuous, the sound continuously modifies the electric signal. On an oscilloscope, an analog signal is a continuous curve. *[Figure 11-3]* An analog electric signal can be modified by changing the signal's amplitude, frequency, or phase.

A digital electronic representation of a real-world event is discontinuous. The essential characteristics of the continuous event are captured as a series of discrete incremental values. Electronically, these representative samplings are successive chains of voltage and non-voltage signals. They can be transported and manipulated in electronic circuits. When the samples are sufficiently small and occur with high frequency, real-world phenomenon can be represented to appear continuous.

Noise

A significant advantage of digital electronics over analog electronics is the control of noise. Noise is any alteration of the represented real-world phenomenon that is not intended or desired. Consider the operation of a microphone when understanding noise. A continuous analog voltage is modified by a voice signal that results in the continuous voltage varying in proportion to the volume and tone of the input sound. However, the voltage responds and modifies to any input. Thus, background sounds also modify the continuous voltage as will electrostatic activity and circuitry imperfections. This alteration by phenomenon that are not the intended modifier is noise.

During the processing of digitized data, there is little or no signal degradation. The real-world phenomenon is represented in a string of binary code. A series of ones and zeros are electronically created as a sequence of voltage or no voltage and carried through processing stages. It is relatively immune to outside alteration once established. If a signal is close to the set value of the voltage, it is considered to be that voltage. If the signal is close to zero, it is considered to be no voltage. Small variations or modifications from undesired phenomenon are ignored. *Figure 11-4* illustrates an analog sine wave and a digital sine wave. Any unwanted voltage will modify the analog curve. The digital steps are not modified by small foreign inputs. There is either voltage or no voltage.

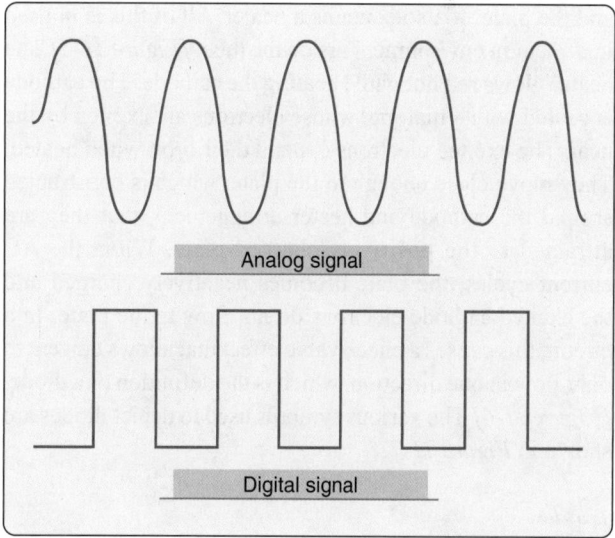

Figure 11-4. *Analog signals are continuous voltage modified by all external events including those that are not desired called noise. Digital signals are a series of voltage or no voltage that represent a desired event.*

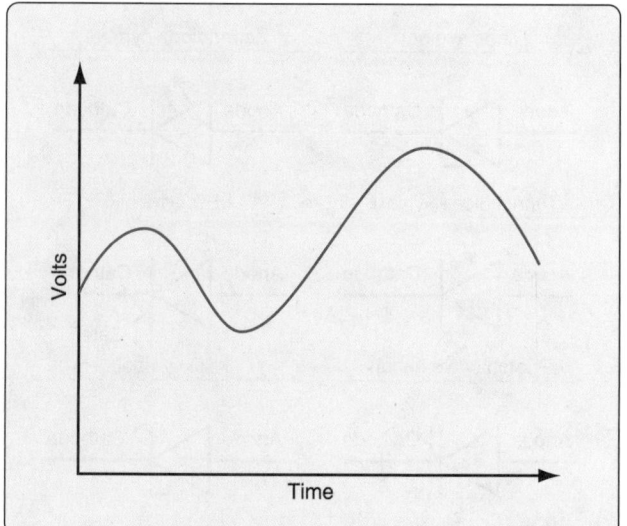

Figure 11-3. *An analog signal displayed on an oscilloscope is a continuous curve.*

Analog Electronics

Early aircraft were equipped with radio communication and navigational devices that were constructed with analog electronic circuits. They used vacuum tubes that functioned as electron control valves. These were later replaced by solid-state devices. Today, digital electronic circuits dominate modern avionics. A brief look at various electron control valves used on aircraft follows.

Electron Control Valves

Electron control valves are an essential part of an electronic circuit. Control of electron flow enables the circuit to produce the desired outcome. Early aircraft made use of vacuum tubes to control electron flow. Later, transistors replaced vacuum tubes. Semiconductors used in transistors and integrated circuits have enabled the solid-state digital electronics found in aircraft today.

Vacuum Tubes

Electron control valves found in the analog circuits of early aircraft electronics are constructed of vacuum tubes. Only antique aircraft retain radios with these devices due to their size and inability to withstand the harsh vibration and shock of the aircraft operating environment. However, they do function, and a description is included here as a foundation for the study of more modern electronic circuits and components.

Diodes

A diode acts as a check valve in an alternating current (AC) circuit. It allows current to flow during half of the AC cycle but not the other half. In this manner, it creates a pulsating direct current (DC) with current that drops to zero in between pulses. A diode tube has two active electrodes: the cathode and the plate. It also contains a heater. All of this is housed in a vacuum environment inside the tube. *[Figure 11-5]* The heater glows red hot while heating the cathode. The cathode is coated with a material whose electrons are excited by the heat. The excited electrons expand their orbit when heated. They move close enough to the plate, which is constructed around the cathode and heater arrangement, that they are attracted to the positively-charged plate. When the AC current cycles, the plate becomes negatively charged and the excited cathode electrons do not flow to the plate. In a circuit, this causes a check valve effect that allows current to only flow in one direction, which is the definition of a diode. *[Figure 11-6]* The various symbols used to depict diodes are shown in *Figure 11-7*.

Triodes

A triode is an electron control valve containing three elements. It is often used to control a large amount of current with a smaller current flow. In addition to the cathode, plate, and

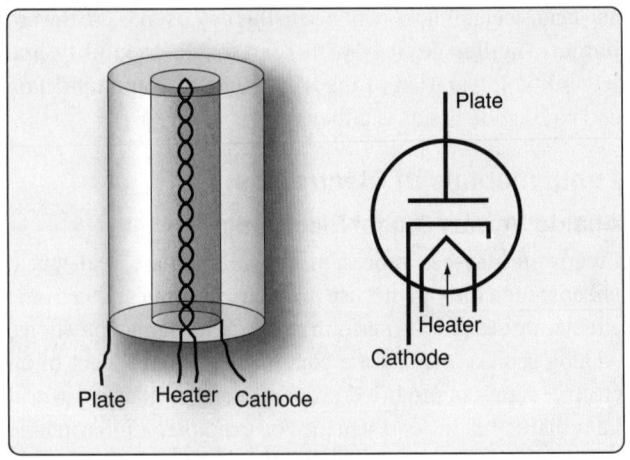

Figure 11-5. *A vacuum tube diode contains a cathode, heater, and plate. Note that the arrow formed in the symbol for the heater points to the direction of electron flow.*

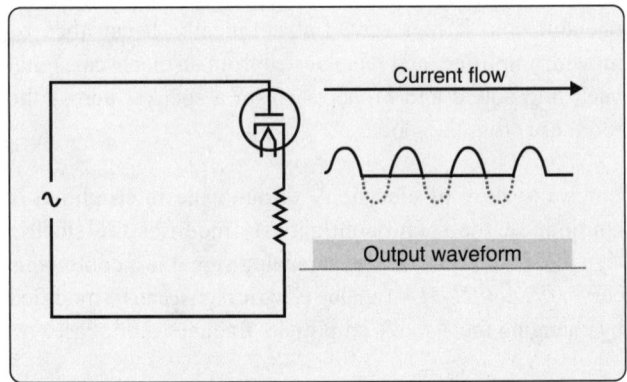

Figure 11-6. *A vacuum tube diode in a circuit allows current to flow in one direction only. The output waveform illustrates the lack of current flow as the AC cycles.*

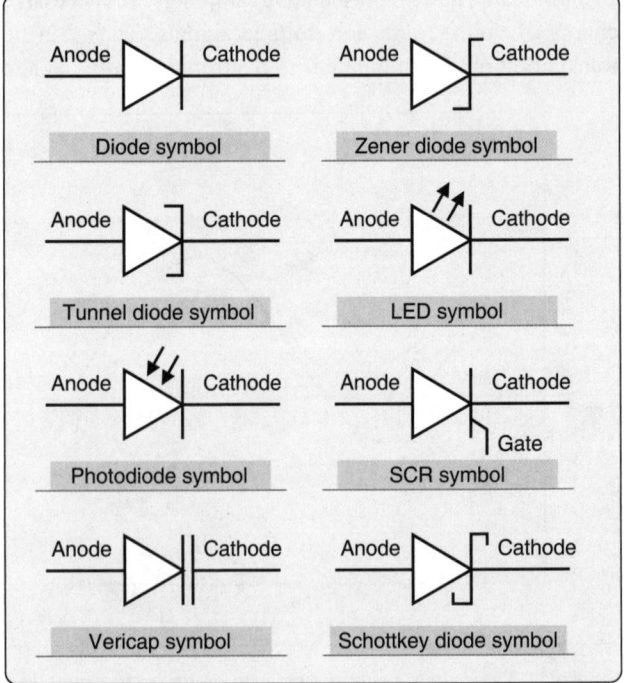

Figure 11-7. *Diode symbols.*

heater present in the diode, a triode also contains a grid. The grid is composed of fine wire spiraled between the cathode and the plate but closer to the cathode. Applying voltage to the grid can influence the cathode's electrons, which normally flow to the plate when the cathode is heated. Changes in the relatively small amount of current that flows through the grid can greatly impact the flow of electrons from the cathode to the plate. *[Figure 11-8]*

Figure 11-9 illustrates a triode in a simple circuit. AC voltage input is applied to the grid. A high-resistance resistor is used so that only minimum voltage passes through to the grid. As this small AC input voltage varies, the amount of DC output in the cathode-plate circuit also varies. When the input signal is positive, the grid is positive. This aids in drawing electrons from the cathode to the plate. However, when the AC input signal cycles to negative, the grid becomes negatively charged and flow from the cathode to the plate is cut off with the help of the negatively charged grid that repels the electrons on the cathode.

Tetrodes

A tetrode vacuum tube electron control valve has four elements. In addition to the cathode, plate, and grid found in a triode, a tetrode also contains a screen grid. The cathode and plate of a vacuum tube electron control valve can act as a capacitor. At high frequencies, the capacitance is so low that feedback occurs. The output in the plate circuit feeds back into the control grid circuit. This causes an oscillation generating AC voltage that is unwanted. By placing a screen grid between the anode and the control grid windings, this feedback and the inter-electrode capacitive effect of the anode and cathode are neutralized. *[Figure 11-10]*

Figure 11-11 illustrates a tetrode in a circuit. The screen grid is powered by positive DC voltage. The inter-electrode capacitance is now between the screen grid and the plate. A capacitor is located between the screen grid and ground. AC feedback generated in the screen grid goes to ground and

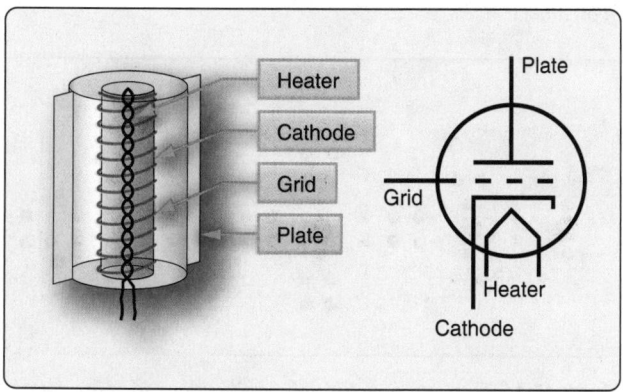

Figure 11-8. *A triode has three elements: the cathode, plate, and a grid.*

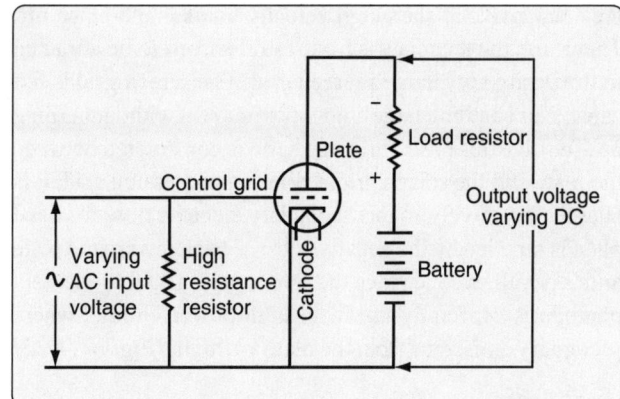

Figure 11-9. *Varying AC input voltage to the grid circuit in a triode produces a varying DC output.*

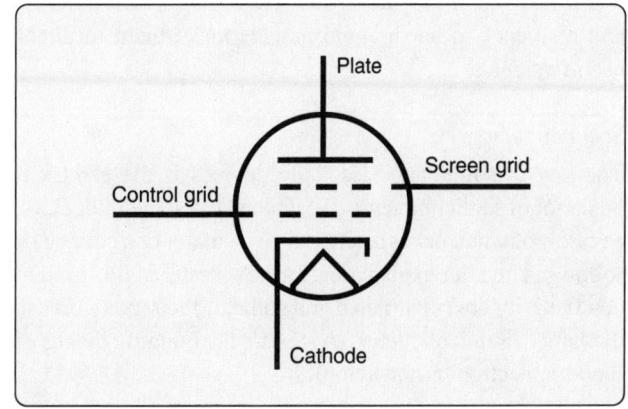

Figure 11-10. *A tetrode is a four element electron control valve vacuum tube including a cathode, a plate, a control grid, and a screen grid.*

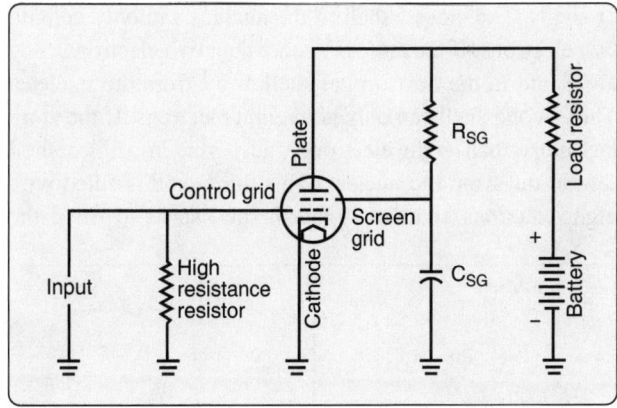

Figure 11-11. *To enable a triode to be used at high frequencies, a screen grid is constructed between the plate and the control grid.*

does not oscillate. This allows use of the tetrode at higher frequencies than a triode.

Pentodes

The plate in a vacuum tube can have a secondary emission that must be controlled. When electrons flow from the cathode through the control grid and screen grid to the plate, they can arrive at such high velocity that some bounce off. Therefore, the tendency is for those electrons to be attracted to the positively charged screen grid. The screen grid is not capable of handling large amounts of current without burning up. To solve this problem, a third grid is constructed between the plate and the screen grid. Called a suppression grid, it is charged negatively so that secondary electron flow from the plate is repelled by the negative charge back toward the plate and is not allowed to reach the screen grid. The five-element pentode is especially useful in high-power circuits where secondary emissions from the plate are high. *[Figure 11-12]*

Solid-State Devices

Solid-state devices began replacing vacuum tube electron control valves in the late 1950s. Their long life, reliability, and resilience in harsh environments make them ideal for use in avionics.

Semiconductors

The key to solid-state electronic devices is the electrical behavior of semiconductors. To understand semiconductors, a review of what makes a material an insulator or a conductor follows. Then, an explanation for how materials of limited conductivity are constructed and some of their many uses is explained. Semiconductor devices are the building blocks of modern electronics and avionics.

An atom of any material has a characteristic number of electrons orbiting the nucleus of the atom. The arrangement of the electrons occurs in somewhat orderly orbits called rings or shells. The closest shell to the nucleus can only contain two electrons. If the atom has more than two electrons, they are found in the next orbital shell away from the nucleus. This second shell can only hold eight electrons. If the atom has more than eight electrons, they orbit in a third shell farther out from the nucleus. This third shell is filled with eight electrons and then a fourth shell starts to fill if the

element still has more electrons. However, when the fourth shell contains eight electrons, the number of electrons in the third shell begins to increase again until a maximum of 18 is reached. *[Figure 11-13]*

The outer most orbital shell of any atom's electrons is called the valence shell. The number of electrons in the valence shell determines the chemical properties of the material. When the valence shell has the maximum number of electrons, it is complete, and the electrons tend to be bound strongly to the nucleus. Materials with this characteristic are chemically stable. It takes a large amount of force to move the electrons in this situation from one atom valence shell to that of another. Since the movement of electrons is called electric current, substances with complete valence shells are known as good insulators because they resist the flow of electrons (electricity). *[Figure 11-14]*

In atoms with an incomplete valence shell, that is, those without the maximum number of electrons in their valence shell, the electrons are bound less strongly to the nucleus. The material is chemically disposed to combine with other materials or other identical atoms to fill in the unstable valence configuration and bring the number of electrons in the valence shell to maximum. Two or more substances may share the electrons in their valence shells and form a covalent bond. A covalent bond is the method by which atoms complete their valence shells by sharing valence electrons with other atoms.

Electrons in incomplete valence shells may also move freely from valence shell to valence shell of different atoms or compounds. In this case, these are known as free electrons. As stated, the movement of electrons is known as electric

Shell or Orbit Number	1	2	3	4	5
Maximum number of electrons	2	8	18	32	50

Figure 11-13. *Maximum number of electrons in each orbital shell of an atom.*

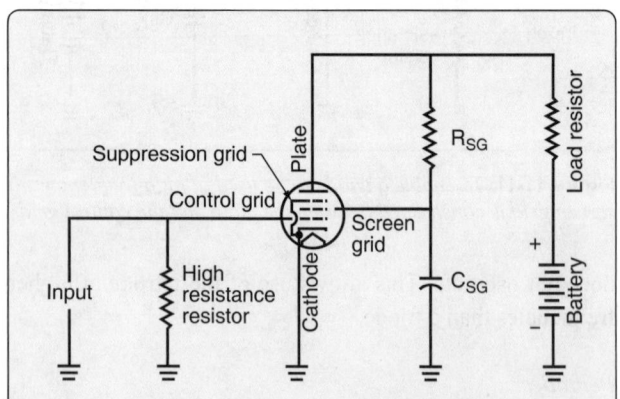

Figure 11-12. *A pentode contains a suppression grid that controls secondary electron emissions from the plate at high power. This keeps the current in the screen grid from becoming too high.*

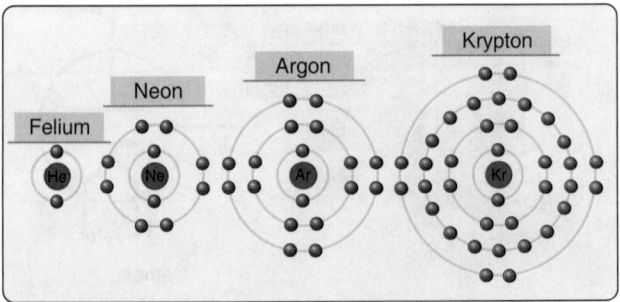

Figure 11-14. *Elements with full valence shells are good insulators. Most insulators used in aviation are compounds of two or more elements that share electrons to fill their valence shells.*

current or current flow. When electrons move freely from atom to atom or compound to compound, the substance is known as a conductor. *[Figure 11-15]*

Not all materials are pure elements, that is, substances made up of one kind of atom. Compounds occur when two or more different types of atoms combine. They create a new substance with different characteristics than any of the component elements. When compounds form, valence shells and their maximum number of electrons remain the rule of physics. The new compound molecule may either share electrons to fill the valence shell or free electrons may exist to make it a good conductor.

Silicon is an atomic element that contains four electrons in its valence shell. It tends to combine readily with itself and form a lattice of silicon atoms in which adjacent atoms share electrons to fill out the valance shell of each to the maximum of eight electrons. *[Figure 11-16]* This unique symmetric alignment of silicon atoms results in a crystalline structure.

Once bound together, the valence shells of each silicon atom are complete. In this state, movement of electrons does not occur easily. There are no free electrons to move to another atom and no space in the valence shells to accept a free electron. Therefore, silicon in this form is a good insulator.

Silicon is a primary material used in the manufacture of semiconductors. Germanium and a few other materials are also used.

Since silicon is an insulator, it must be modified to become a semiconductor. The process often used is called doping. Starting with ultra-pure silicon crystal, arsenic, phosphorus, or some other element with five valence electrons in each atom is mixed into the silicon. The result is a silicon lattice with flaws. *[Figure 11-17]* The elements bond, but numerous free electrons are present in the material from the 5th electron that is part of the valence shell of the doping element atoms. These free electrons can now flow under certain conditions. Thus, the silicon becomes semiconductive. The conditions required for electron flow in a semiconductor are discussed in the following paragraphs.

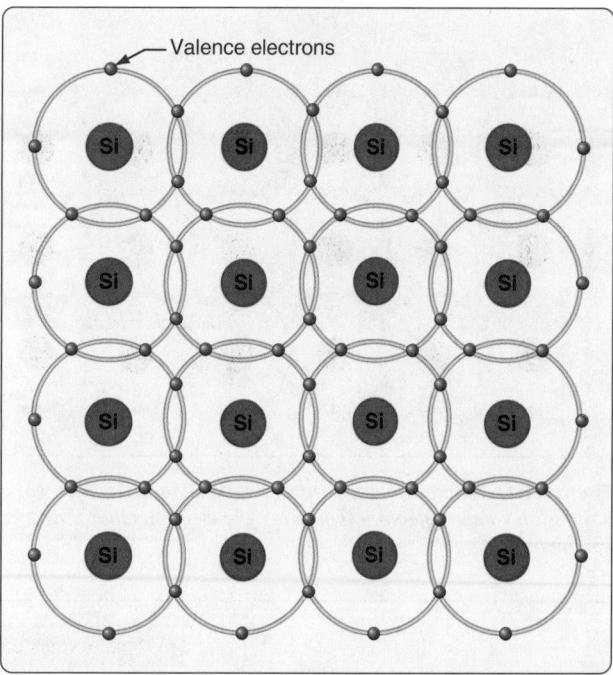

Figure 11-16. *The silicon atoms with just the valence shell electrons share these valence electrons with each other. By sharing with four other silicon atoms, the number of electrons in each silicon atom valence shell becomes eight, which is the maximum number. This makes the substance stable and it resists any flow of electrons.*

When silicon is doped with an element or compound containing five electrons in its valence shell, the result is a negatively charged material due to the excess free electrons, and the fact that electrons are negatively charged. This is known as an N-type semiconductor material. It is also known as a donor material because, when it is used in electronics, it donates the extra electrons to current flow.

Doping silicon can also be performed with an element that has only three valence electrons, such as boron, gallium, or indium. Valence electron sharing still occurs, and the silicon atoms with interspersed doping element atoms form a lattice molecular structure. However, in this case, there are many valence shells where there are only seven electrons and not eight. This greatly changes the properties of the material. The absence of the electrons, called holes, encourages electron flow due to the preference to have eight electrons in all valence shells. Therefore, this type of doped silicon is also semiconductive. It is known as P-type material or as an acceptor since it accepts electrons in the holes under certain conditions. *[Figure 11-18]*

Combining N- and P-type semiconductor material in certain ways can produce very useful results. A look at various semiconductor devices follows.

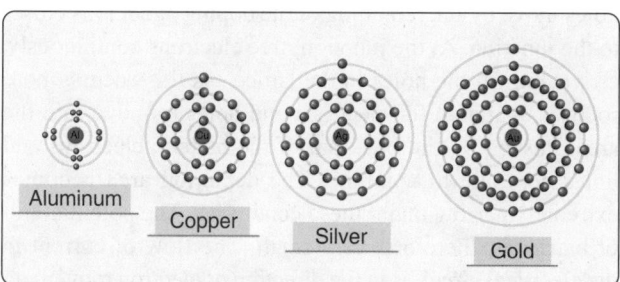

Figure 11-15. *The valence shells of elements that are common conductors have one (or three) electrons.*

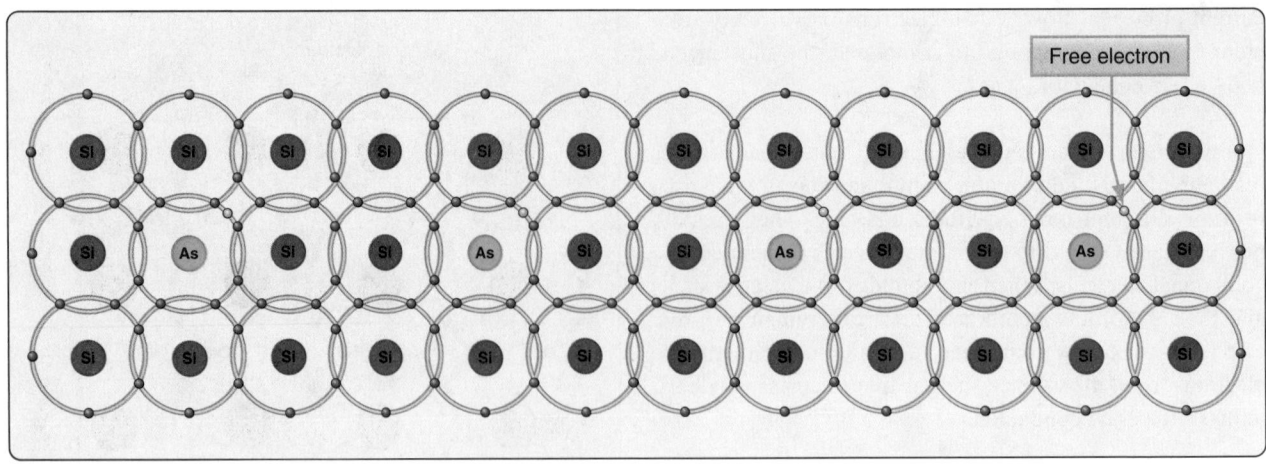

Figure 11-17. *Silicon atoms doped with arsenic form a lattice work of covalent bonds. Free electrons exist in the material from the arsenic atom's 5th valence electron. These are the electrons that flow when the semiconductor material, known as N-type or donor material, is conducting.*

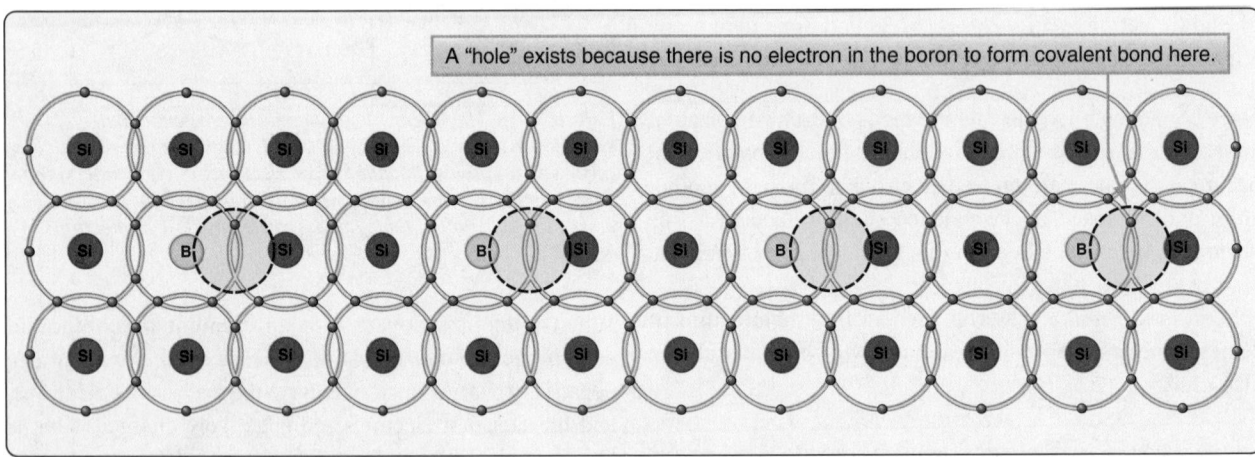

Figure 11-18. *The lattice of boron doped silicon contains holes where the three boron valence shell electrons fail to fill in the combined valence shells to the maximum of eight electrons. This is known as P-type semiconductor material or acceptor material.*

Semiconductor Diodes

A diode is an electrical device that allows current to flow in one direction through the device but not the other. A simple device that can be made from N- and P-type semiconductors is a semiconductor diode. When joined, the junction of these two materials exhibits unique properties. Since there are holes in the P-type material, free electrons from the N-type material are attracted to fill these holes. Once combined, the area at the junction of the two materials where this happens is said to be depleted. There are no longer free electrons or holes. However, having given up some electrons, the N-type material next to the junction becomes slightly positively charged, and having received electrons, the P-type material next to the junction becomes slightly negatively charged. The depletion area at the junction of the two semiconductor materials constitutes a barrier or potential hill. The intensity of the potential hill is proportional to the width of the depletion area (where the electrons from the N-type material have filled holes in the P-type material). *[Figure 11-19]*

The two semiconductors joined in this manner form a diode that can be used in an electrical circuit. A voltage source is attached to the diode. When the negative terminal of the battery is attached to the N-type semiconductor material and the positive terminal is attached to the P-type material, electricity can flow in the circuit. The negative potential of the battery forces free electrons in the N-type material toward the junction. The positive potential of the battery forces holes in the P-type material toward the other side of the junction. The holes move by the rebinding of the doping agent ions closer to the junction. At the junction, free electrons continuously arrive and fill the holes in the lattice. As this occurs, more room is available for electrons and holes to move into the area. Pushed by the potential of the battery, electrons and holes continue to combine. The depletion area becomes extremely narrow under these conditions. The potential hill or barrier is, therefore, very small. The flow of current in the electrical circuit is in the direction of electron movement shown in *Figure 11-20*.

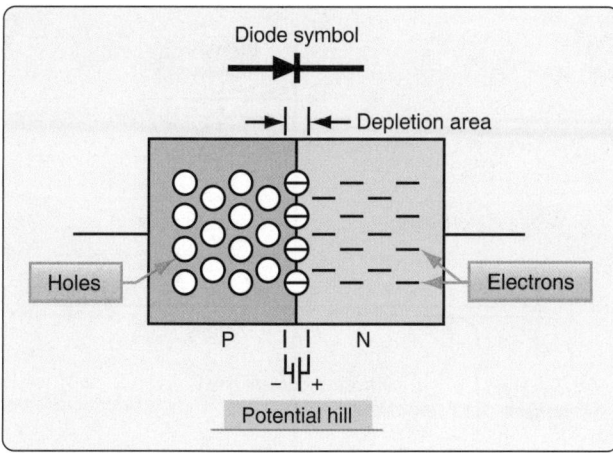

Figure 11-19. *A potential hill.*

In similar circuits where the negative battery terminal is attached to the N-type semiconductor material and the positive terminal is attached to the P-type material, current flows from N-type, or donor material, to P-type receptor material. This is known as a forward-biased semiconductor. A voltage of approximately 0.7 volts is needed to begin the current flow over the potential hill. Thereafter, current flow is linear with the voltage. However, temperature affects the ease at which electrons and holes combine given a specific voltage.

If the battery terminals are reversed, the semiconductor diode circuit is said to be reversed biased. *[Figure 11-21]* Attaching the negative terminal of the battery to the P-type material

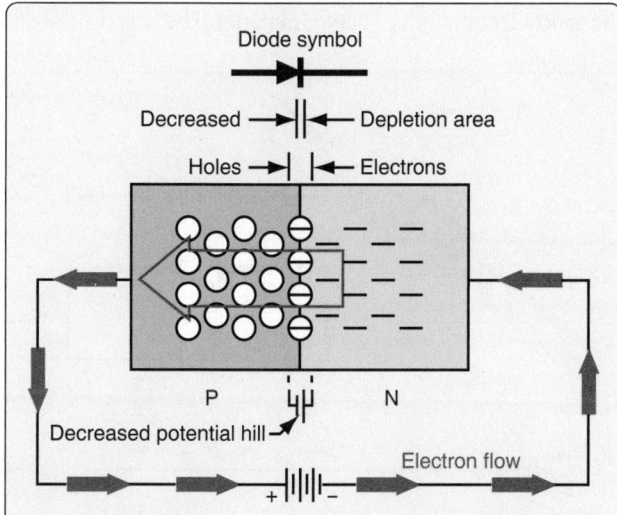

Figure 11-20. *The flow of current and the P-N junction of a semiconductor diode attached to a battery in a circuit.*

attracts the holes in the P-type material away from the junction in the diode. The positive battery terminal attached to the N-type material attracts the free electrons from the junction in the opposite direction. In this way, the width of the area of depletion at the junction of the two materials

increases. The potential hill is greater. Current cannot climb the hill; therefore, no current flows in the circuit. The semiconductors do not conduct.

Semiconductor diodes are used often in electronic circuits. When AC current is applied to a semiconductor diode, current flows during one cycle of the AC but not during the other cycle. The diode, therefore, becomes a rectifier. When it is forward biased, electrons flow; when the AC cycles, electrons do not flow. A simple AC rectifier circuit containing a semiconductor diode and a load resistor is illustrated in *Figure 11- 22*. Semiconductor diode symbols and examples of semiconductor diodes are shown in *Figure 11-23*.

Note: Electron flow is typically discussed in this text. The conventional current flow concept where electricity is thought to flow from the positive terminal of the battery through a circuit to the negative terminal is sometimes used in the field.

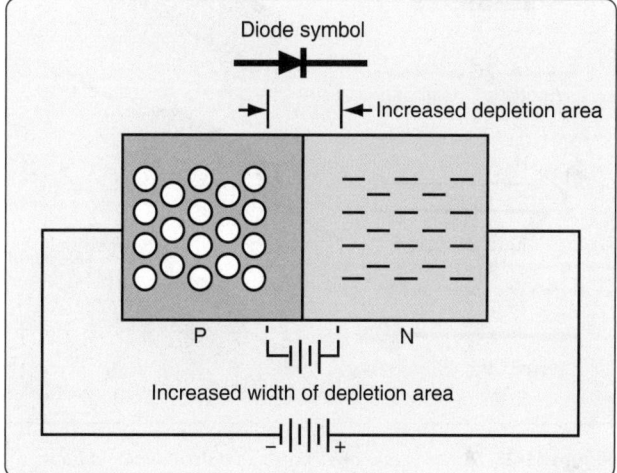

Figure 11-21. *A reversed biased condition.*

Semiconductor diodes have limitations. They are rated for a range of current flow. Above a certain level, the diode overheats and burns up. The amount of current that passes through the diode when forward biased is directly proportional to the amount of voltage applied. But, as mentioned, it is affected by temperature.

Figure 11-24 indicates the actual behavior of a semiconductor diode. In practice, a small amount of current does flow through a semiconductor diode when reversed biased. This is known as leakage current and it is in the micro amperage range. However, at a certain voltage, the blockage of current flow in a reversed biased diode breaks down completely. This voltage is known as the avalanche voltage because the diode can no longer hold back the current and the diode fails.

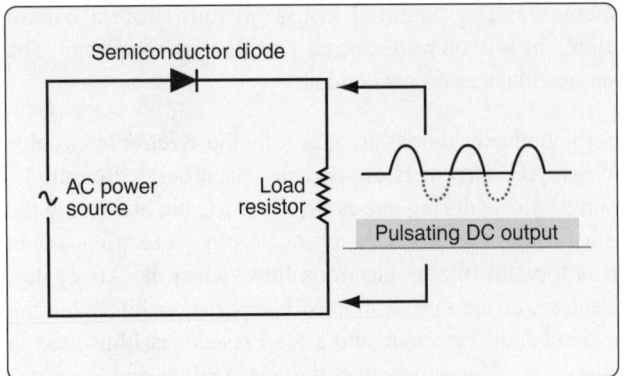

Figure 11-22. *A semiconductor diode acts as a check valve in an AC circuit resulting in a pulsating DC output.*

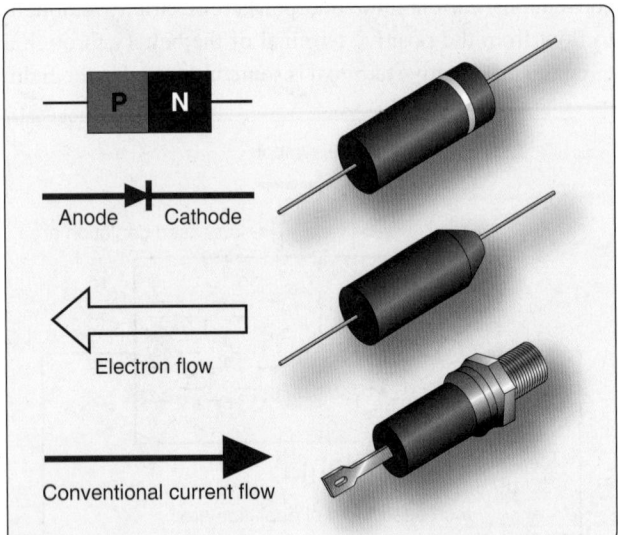

Figure 11-23. *Symbols and drawings of semiconductor diodes.*

Zener Diodes

Diodes can be designed with a zener voltage. This is similar to avalanche flow. When reversed biased, only leakage current flows through the diode. However, as the voltage is increased, the zener voltage is reached. The diode lets current flow freely through the diode in the direction in which it is normally blocked. The diode is constructed to be able to handle the zener voltage and the resulting current, whereas avalanche voltage burns out a diode. A zener diode can be used as means of dropping voltage or voltage regulation. It can be used to step down circuit voltage for a particular application but only when certain input conditions exist. Zener diodes are constructed to handle a wide range of voltages. *[Figure 11-25]*

Transistors

While diodes are very useful in electronic circuits, semiconductors can be used to construct true control valves known as transistors. A transistor is little more than a sandwich

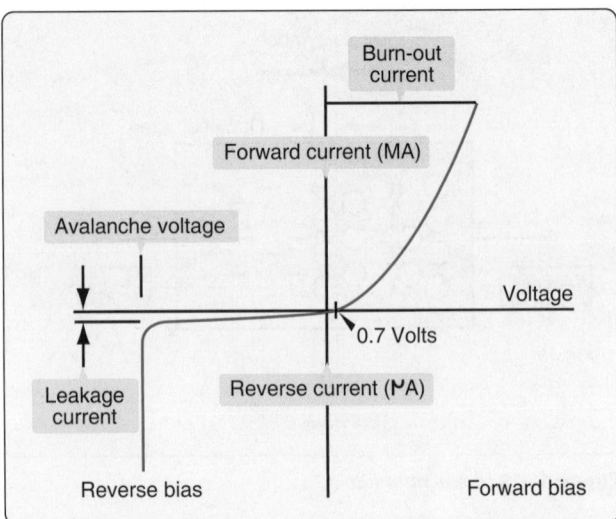

Figure 11-24. *A semiconductor diode.*

of N-type semiconductor material between two pieces of P-type semiconductor material or vice versa. However, a transistor exhibits some remarkable properties and is the building block of all things electronic. *[Figure 11-26]* As with any union of dissimilar types of semiconductor materials, the junctions of the P- and N- materials in a transistor have depletion areas that create potential hills for the flow of electrical charges.

Like a vacuum tube triode, the transistor has three electrodes or terminals, one each for the three layers of semiconductor material. The emitter and the collector are on the outside of the sandwiched semiconductor material. The center material

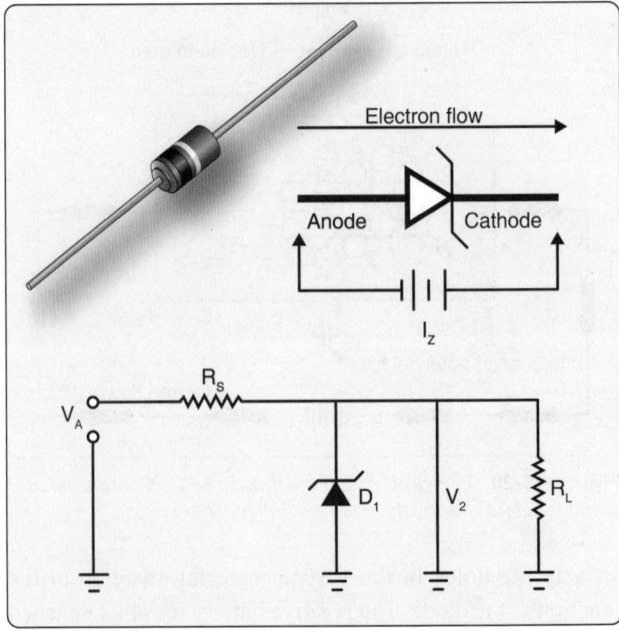

Figure 11-25. *A zener diode, when reversed biased, will break down and allow a prescribed voltage to flow in the direction normally blocked by the diode.*

is known as the base. A change in a relatively small amount of voltage applied to the base of the transistor allows a relatively large amount of current to flow from the collector to the emitter. In this way, the transistor acts as a switch with a small input voltage controlling a large amount of current.

If a transistor is put into a simple battery circuit, such as the one shown in *Figure 11-27*, voltage from the battery (EB) forces free electrons and holes toward the junction between the base and the emitter just as it does in the junction of a semiconductor diode. The emitter-base depletion area becomes narrow as free electrons combine with the holes at the junction. Current (IB) (solid arrows) flows through the junction in the emitter-base battery circuit. At the same time, an emitter-collector circuit is constructed with a battery (EC) of much higher voltage in its circuit. Because of the narrow depletion area at the emitter-base junction, current IC is able to cross the collector base junction, flow through emitter-base junction, and complete the collector-emitter battery circuit (hollow arrows).

To some extent, varying the voltage to the base material can increase or decrease the current flow through the transistor as the emitter-base depletion area changes width in response to the base voltage. If base voltage is removed, the emitter-base depletion area becomes too wide and all current flow through the transistor ceases.

Current in the transistor circuit illustrated has a relationship as follows: $I_E = I_B + I_C$. It should be remembered that it is the voltage applied to the base that turns the collector-emitter transistor current on or off.

Controlling a large amount of current flow with a small independent input voltage is very useful when building electronic circuits. Transistors are the building blocks from which all electronic devices are made, including Boolean gates that are used to create microprocessor chips. As production techniques have developed, the size of reliable transistors has shrunk. Now, hundreds of millions and even billions of transistors may be used to construct a single chip such as the one that powers your computer and various avionic devices.

Silicon Controlled Rectifiers

Combination of semiconductor materials is not limited to a two-type, three-layer sandwich transistor. By creating a four-layer sandwich of alternating types of semiconductor material (i.e., PNPN or NPNP), a slightly different semiconductor diode is created. As is the case in a two-layer diode, circuit current is either blocked or permitted to flow through the diode in a single direction.

Within a four-layer diode, sometimes known as a Shockley diode, there are three junctions. The behavior of the junctions and the entire four-layer diode can be understood by considering it to be two interconnected three-layer transistors. *[Figure 11-28]* Transistor behavior includes no current flow until the base material receives an applied voltage to narrow the depletion area at the base-emitter junction. The base materials in the four-layer diode transistor model receive charge from the other transistor's collector. With no other means of reducing any of the depletion areas at the junctions, it appears that current does not flow in either direction in this device. However, if a large voltage is applied to forward bias the anode or cathode, at some point the ability to block flow breaks down. Current flows through whichever transistor is charged. Collector current then charges the base of the other transistor and current flows through the entire device.

Some caveats are necessary with this explanation. The transistors that comprise this four-layer diode must be constructed of material similar to that described in a zener diode. That is, it must be able to endure the current flow without burning out. In this case, the voltage that causes the diode to conduct is known as breakover voltage rather than breakdown voltage. Additionally, this diode has the unique characteristic of allowing current flow to continue until the applied voltage is reduced significantly, in most cases, until it is reduced to zero. In AC circuits, this would occur when the AC cycles.

While the four-layer, Shockley diode is useful as a switching device, a slight modification to its design creates a silicon-controlled rectifier (SCR). To construct a SCR, an additional terminal known as a gate is added. It provides more control and utility. In the four-layer semiconductor construction, there are always two junctions forward biased and one junction reversed biased. The added terminal allows the momentary application of voltage to the reversed biased junction. All three junctions then become forward biased and current at the anode flows through the device. Once voltage is applied to the gate, the SCR become latched or locked on. Current continues to flow through it until the level drops off significantly, usually to zero. Then, another applied voltage through the gate is needed to reactivate the current flow. *[Figures 11-29 and 11-30]*

SCRs are often used in high voltage situations, such as power switching, phase controls, battery chargers, and inverter circuits. They can be used to produce variable DC voltages for motors and are found in welding power supplies. Often, lighting dimmer systems use SCRs to reduce the average voltage applied to the lights by only allowing current flow during part of the AC cycle. This is controlled by controlling the pulses to the SCR gate and eliminating the massive heat

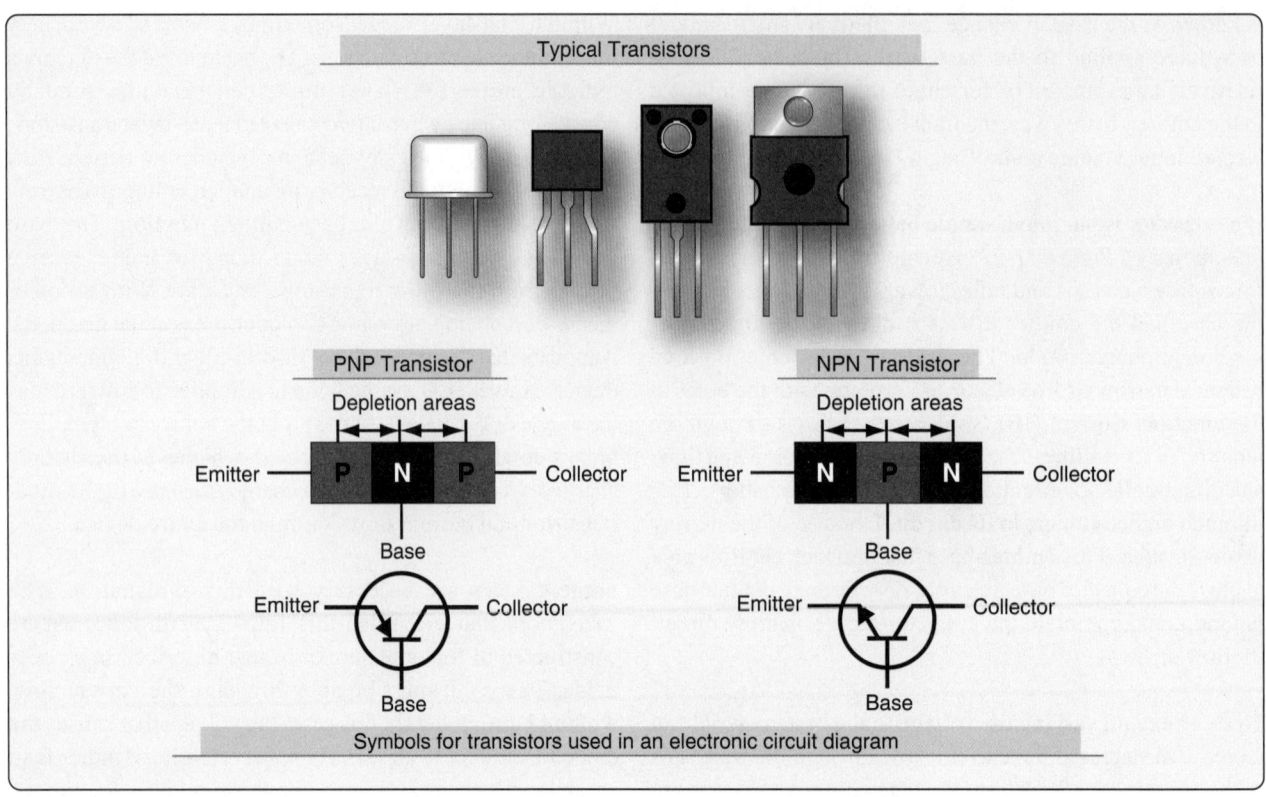

Figure 11-26. *Typical transistors, diagrams of a PNP and NPN transistor, and the symbol for those transistors when depicted in an electronic circuit diagram.*

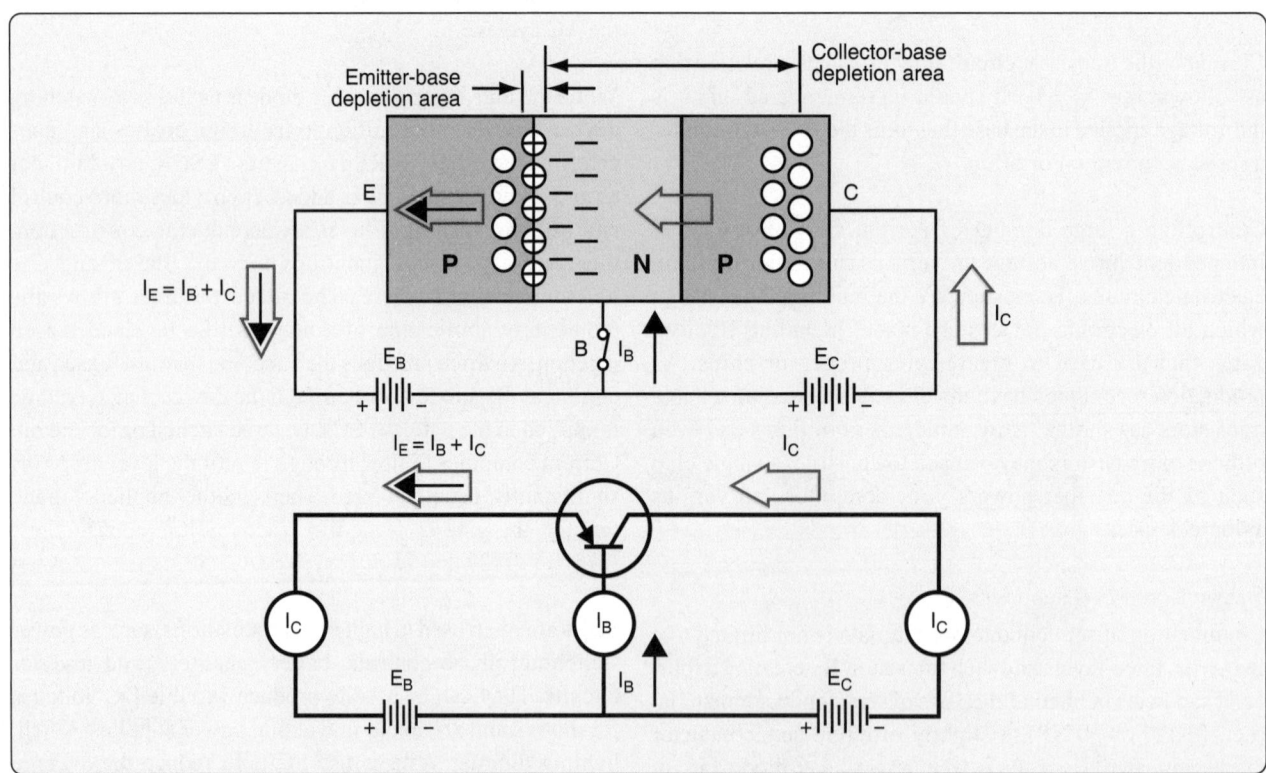

Figure 11-27. *The effect of applying a small voltage to bias the emitter-base junction of a transistor (top). A circuit diagram for this same transistor (bottom).*

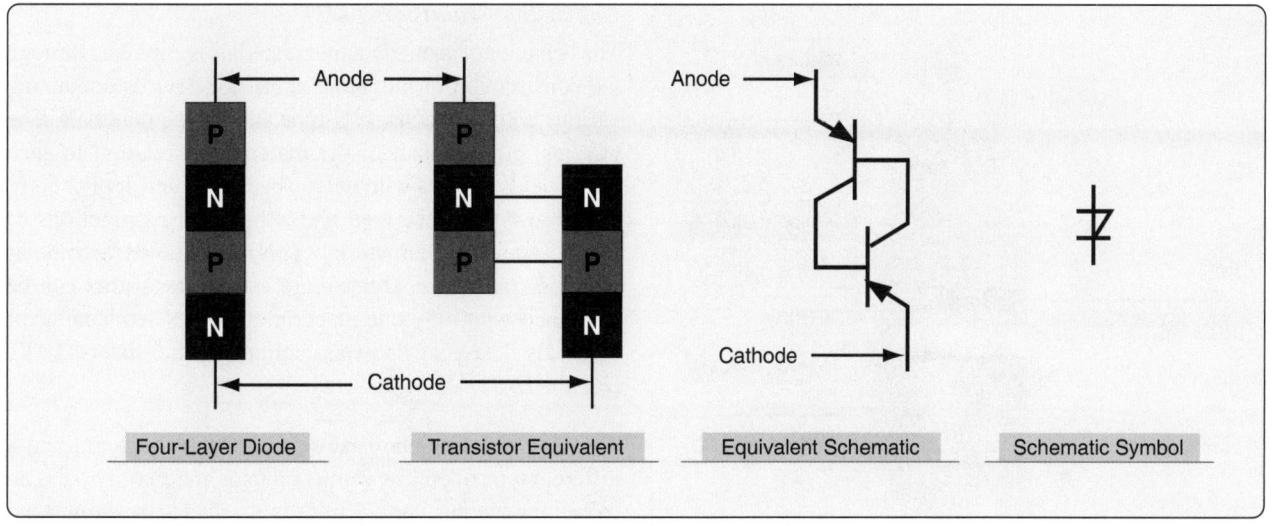

Figure 11-28. *A four-layer semiconductor diode behaves like two transistors. When breakover voltage is reached, the device conducts current until the voltage is removed.*

dissipation caused when using resistors to reduce voltage. *Figure 11-31* graphically depicts the timing of the gate pulse that limits full cycle voltage to the load. By controlling the phase during which, time the SCR is latched, a reduced average voltage is applied.

Triacs

SCRs are limited to allowing current flow in one direction only. In AC circuitry, this means only half of the voltage cycle can be used and controlled. To access the voltage in the reverse cycle from an AC power source, a triac can be used. A triac is also a four-layer semiconductor device. It differs from an SCR in that it allows current flow in both directions. A triac has a gate that works the same way as in a SCR; however, a positive or negative pulse to the gate triggers current flow in a triac. The pulse polarity determines the direction of the current flow through the device.

Figure 11-32 illustrates a triac and shows a triac in a simple circuit. It can be triggered with a pulse of either polarity and remains latched until the voltage declines, such as when the AC cycles. Then, it needs to be triggered again. In many ways, the triac acts as though it is two SCRs connected side by side only in opposite directions. Like an SCR, the timing of gate pulses determines the amount of the total voltage that is allowed to pass. The output waveform if triggered at 90° is shown in *Figure 11-32*. Because a triac allows current to flow in both directions, the reverse cycle of AC voltage can also be used and controlled.

When used in actual circuits, triacs do not always maintain the same phase firing point in reverse as they do when fired with a positive pulse. This problem can be regulated somewhat through the use of a capacitor and a diac in the gate circuit. However, as a result, where precise control is

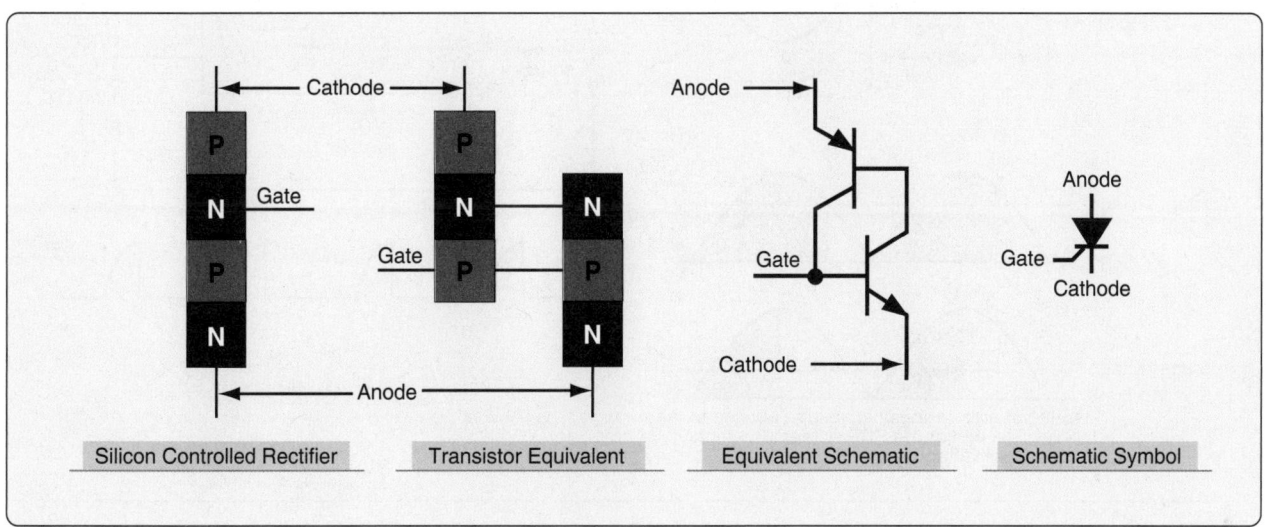

Figure 11-29. *A silicon controlled rectifier (SCR) allows current to pass in one direction when the gate receives a positive pulse to latch the device in the on position. Current ceases to flow when it drops below holding current, such as when AC current reverses cycle.*

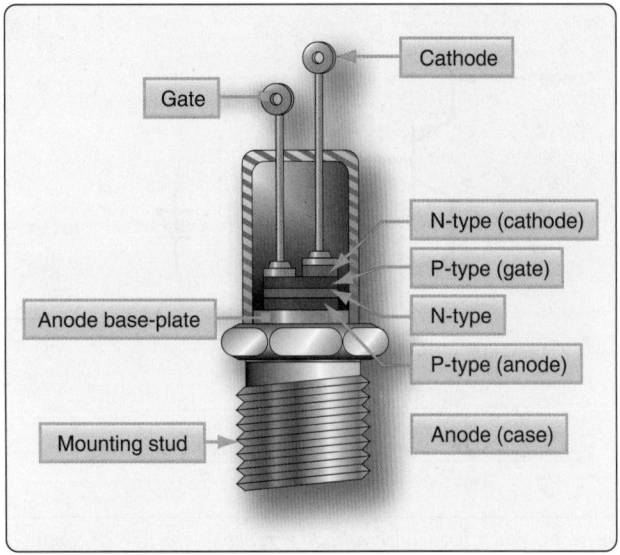

Figure 11-30. *Cross-section of a medium-power SCR.*

required, two SCRs in reverse of each other are often used instead of the triac. Triacs do perform well in lower voltage circuits. *Figure 11-33* illustrates the semiconductor layering in a triac.

Note: The four layers of N- and P-type materials are not uniform as they were in previously described semiconductor devices. None the less, gate pulses affect the depletion areas at the junctions of the materials in the same way allowing current to flow when the areas are narrowed.

Unijunction Transistors (UJT)

The behavior of semiconductor materials is exploited through the construction of numerous transistor devices containing various configurations of N-type and P-type materials. The physical arrangement of the materials in relation to each other yields devices with unique behaviors and applications. The transistors described above having two junctions of P-type and N-type materials (PN) are known as bipolar junction transistors. Other more simple transistors can be fashioned with only one junction of the PN semiconductor materials. These are known as unijunction transistors (UJT). *[Figure 11-34]*

The UJT contains one base semiconductor material and a different type of emitter semiconductor material. There is no collector material. One electrode is attached to the emitter and two electrodes are attached to the base material at opposite ends. These are known as base 1 (B1) and base 2 (B2). The electrode configuration makes the UJT appear physically the same as a bipolar junction transistor. However, there is only one PN junction in the UJT and it behaves differently.

The base material of a UJT behaves like a resistor between the electrodes. With B2 positive with respect to B1, voltage gradually drops as it flows through the base. *[Figure 11-35]* By placing the emitter at a precise location along the base material gradient, the amount of voltage needed to be applied to the emitter electrode to forward bias the UJT base-emitter junction is determined. When the applied emitter voltage

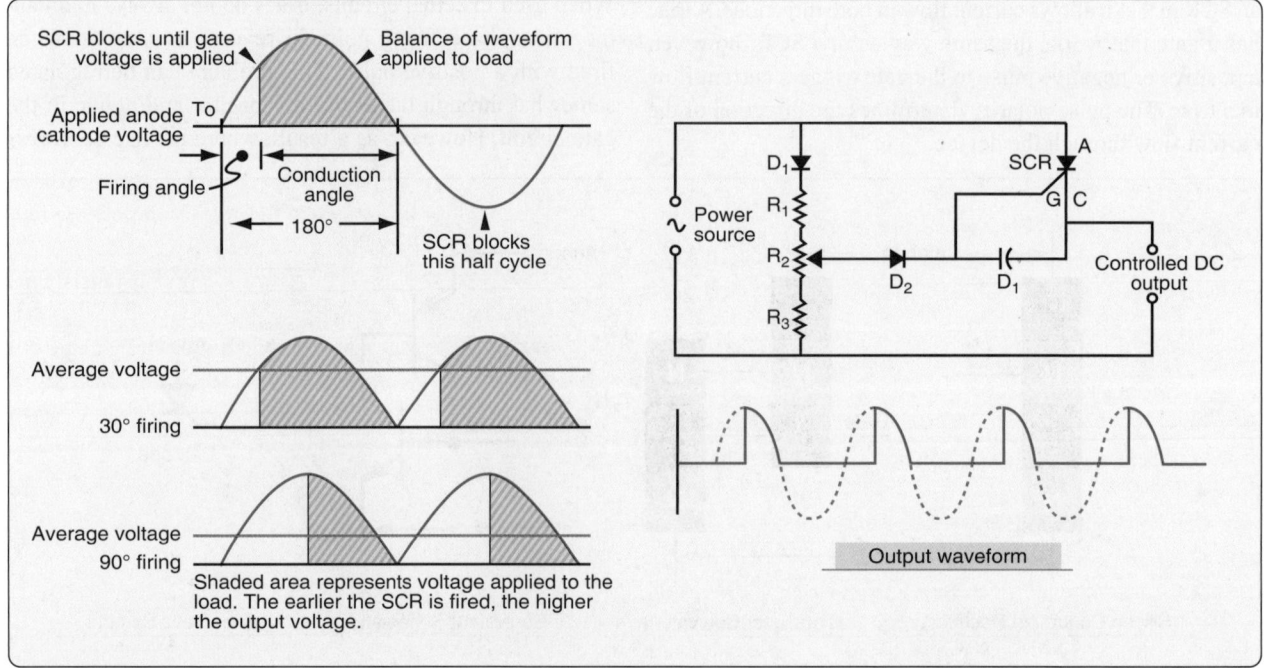

Figure 11-31. *Phase control is a key application for SCR. By limiting the percentage of a full cycle of AC voltage that is applied to a load, a reduced voltage results. The firing angle or timing of a positive voltage pulse through the SCR's gate latches the device open allowing current flow until it drops below the holding current, which is usually at or near zero voltage as the AC cycle reverses.*

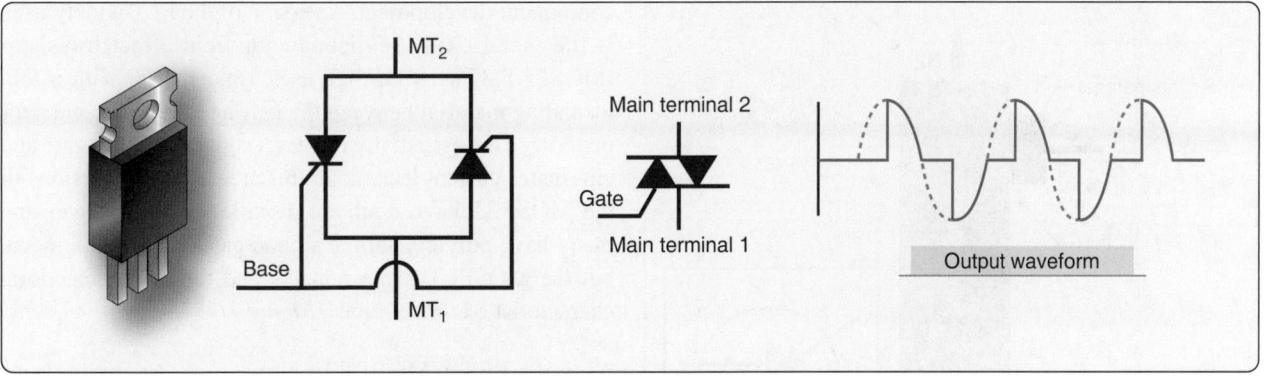

Figure 11-32. *A triac is a controlled semiconductor device that allows current flow in both directions.*

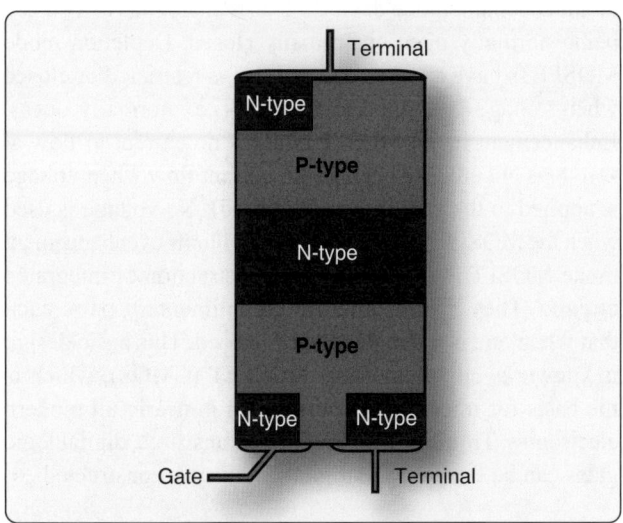

Figure 11-33. *The semiconductor layering in a triac. A positive or negative gate pulse with respect to the upper terminal allows current to flow through the devise in either direction.*

exceeds the voltage at the gradient point where the emitter is attached, the junction is forward biased and current flows freely from the B1 electrode to the E electrode. Otherwise, the junction is reversed biased and no significant current flows although there is some leakage. By selecting a UJT with the correct bias level for a particular circuit, the applied emitter voltage can control current flow through the device.

UJTs of a wide variety of designs and characteristics exist. A description of all of them is beyond the scope of this discussion. In general, UJTs have some advantages over bipolar transistors. They are stable in a wide range of temperatures. In some circuits, use of UJTs can reduce the overall number of components used, which saves money and potentially increases reliability. They can be found in switching circuits, oscillators, and wave shaping circuits. However, four-layered semiconductor thyristors that function the same as the UJT just described are less expensive and most often used.

Field Effect Transistors (FET)

As shown in the triac and the UJT, creative arrangement of semiconductor material types can yield devices with a variety of characteristics. The field effect transistor (FET) is another such device which is commonly used in electronic circuits. Its N- and P-type material configuration is shown in *Figure 11-36*. A FET contains only one junction of the two types of semiconductor material. It is located at the gate where it contacts the main current carrying portion of the device. Because of this, when an FET has a PN junction, it is known as a junction field effect transistor (JFET). All FETs operate by expanding and contracting the depletion area at the junction of the semiconductor materials.

One of the materials in a FET or JFET is called the channel. It is usually the substrate through which the current needing to be controlled flows from a source terminal to a drain terminal. The other type of material intrudes into the channel and acts as the gate. The polarity and amount of voltage applied to the gate can widen or narrow the channel due to expansion or shrinking of the depletion area at the junction of the semiconductors. This increases or decreases the amount of current that can flow through the channel. Enough reversed biased voltage can be applied to the gate to prevent the flow of current through the channel. This allows the FET to act as a switch. It can also be used as a voltage-controlled resistance.

FETs are easier to manufacture than bipolar transistors and have the advantage of staying on once current flow begins without continuous gate voltage applied. They have higher impedance than bipolar transistors and operate cooler. This makes their use ideal for integrated circuits where millions of FETs may be in use on the same chip. FETs come in N-channel and P-channel varieties.

Metal Oxide Semiconductor Field Effect Transistors (MOSFETs) & Complementary Metal Oxide Semiconductor (CMOS)

The basic FET has been modified in numerous ways and continues to be at the center of faster and smaller electronic

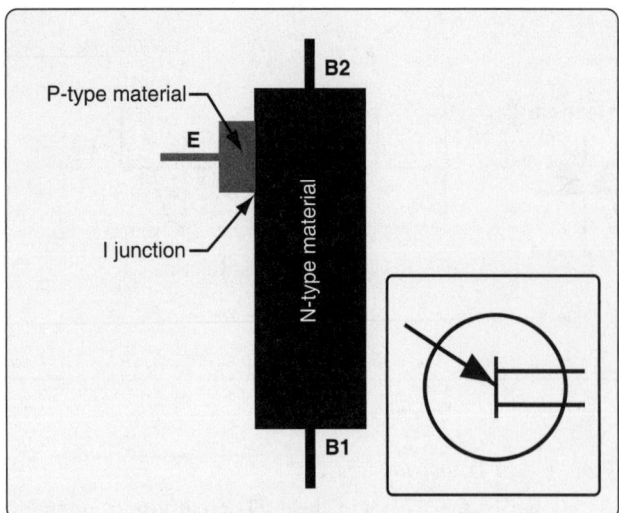

Figure 11-34. *A unijunction transistor (UJT).*

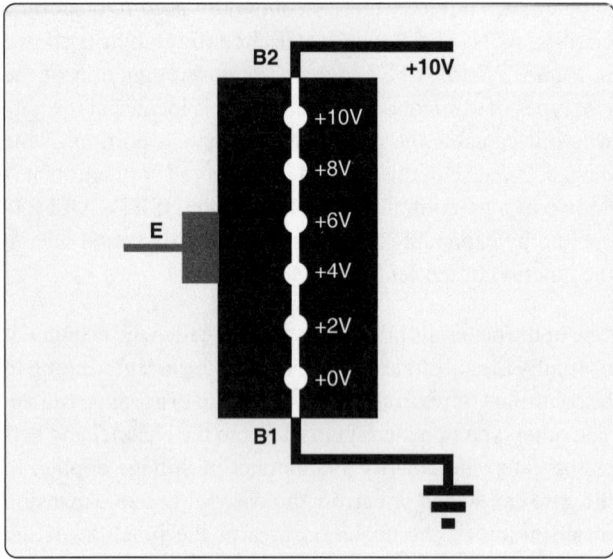

Figure 11-35. *The voltage gradient in a UJT.*

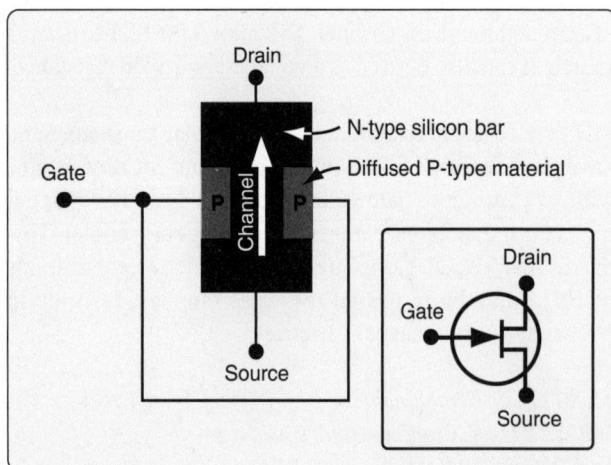

Figure 11-36. *The basic structure of a field effect transistor and its electronic symbol.*

component development. A version of the FET widely used is the metal oxide semiconductor field effect transistor (MOSFET). The MOSFET uses a metal gate with a thin insulating material between the gate and the semiconductor material. This essentially creates a capacitor at the gate and eliminates current leakage in this area. Modern versions of the MOSFET have a silicon dioxide insulating layer and many have poly-crystalline silicon gates rather than metal, but the MOSFET name remains and the basic behavioral characteristic are the same. *[Figure 11-37]*

As with FETs, MOSFETs come with N-channels or P-channels. They can also be constructed as depletion mode or enhancement mode devices. This is analogous to a switch being normally open or normally closed. Depletion mode MOSFETs have an open channel that is restricted or closed when voltage is applied to the gate (i.e., normally open). Enhancement mode MOSFETs allow no current to flow at zero bias but create a channel for current flow when voltage is applied to the gate (normally closed). No voltage is used when the MOSFETs are at zero bias. Millions of enhancement mode MOSFETs are used in the construction of integrated circuits. They are installed in complimentary pairs such that when one is open, the other is closed. This basic design is known as complementary MOSFET (CMOS), which is the basis for integrated circuit design in nearly all modern electronics. Through the use of these transistors, digital logic gates can be formed, and digital circuitry is constructed.

Other more specialized FETs exist. Some of their unique characteristics are owed to design alterations and others to material variations. The transistor devices discussed above use silicon-based semiconductors. But the use of other semiconductor materials can yield variations in performance. Metal semiconductor FETs (MESFETS) for example, are often used in microwave applications. They have a combined metal and semiconductor material at the gate and are typically made from gallium arsenide or indium phosphide. MESFETs are used for their quickness when starting and stopping current flows especially in opposite directions. High electron mobility transistors (HEMT) and pseudomorphic high electron mobility transistors (PHEMT) are also constructed from gallium arsenide semiconductor material and are used for high-frequency applications.

Photodiodes & Phototransistors

Light contains electromagnetic energy that is carried by photons. The amount of energy depends on the frequency of light of the photon. This energy can be very useful in the operation of electronic devices since all semiconductors are affected by light energy. When a photon strikes a semiconductor atom, it raises the energy level above what is needed to hold its electrons in orbit. The extra energy frees an electron enabling it to flow as current. The vacated position

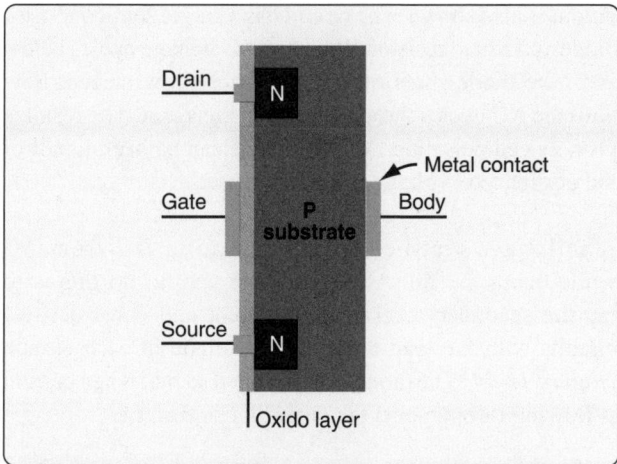

Figure 11-37. *A MOSFET has a metal gate and an oxide layer between it and the semiconductor material to prevent current leakage.*

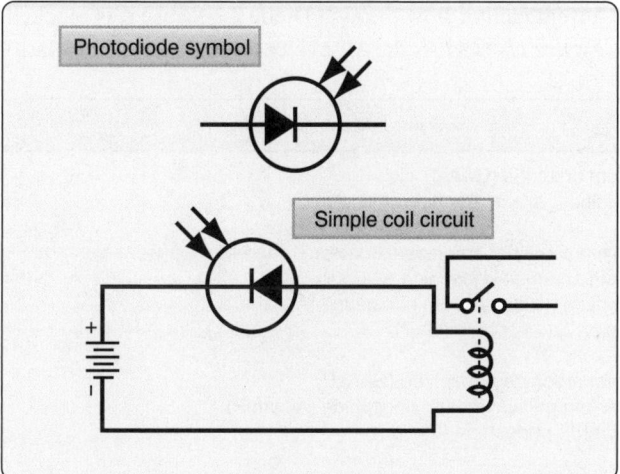

Figure 11-38. *The symbol for a photodiode and a photodiode in a simple coil circuit.*

of the electron becomes a hole. In photodiodes, this occurs in the depletion area of the reversed biased PN junction turning on the device and allowing current to flow.

Figure 11-38 illustrates a photodiode in a coil circuit. In this case, the light striking the photodiode causes current to flow in the circuit whereas the diode would have otherwise blocked it. The result is the coil energizes and closes another circuit enabling its operation.

A photon activated transistor could be used to carry even more current than a photodiode. In this case, the light energy is focused on a collector-base junction. This frees electrons in the depletion area and starts a flow of electrons from the base that turns on the transistor. Once on, heavier current flows from the emitter to the collector. *[Figure 11-39]* In practice, engineers have developed numerous ways to use the energy in light photons to trigger semiconductor devices in electronic circuits. *[Figure 11-40]*

Light Emitting Diodes

Light emitting diodes (LEDs) have become so commonly used in electronics that their importance may tend to be overlooked. Numerous avionics displays and indicators use LEDs for indicator lights, digital readouts, and backlighting of liquid crystal display (LCD) screens.

LEDs are simple and reliable. They are constructed of semiconductor material. When a free electron from a semiconductor drops into a semiconductor hole, energy is given off. This is true in all semiconductor materials. However, the energy released when this happens in certain materials is in the frequency range of visible light. *Figure 11-41* is a table that illustrates common LED colors and the semiconductor material that is used in the construction of the diode.

Note: When the diode is reversed biased, no light is given off. When the diode is forward biased, the energy given off is visible in the color characteristic for the material being used. *Figure 11-42* illustrates the anatomy of a single LED, the symbol of an LED, and a graphic depiction of the LED process.

Basic Analog Circuits

The solid-state semiconductor devices described in the previous section of this chapter can be found in both analog and digital electronic circuits. As digital electronics evolve, analog circuitry is being replaced. However, many aircraft still make use of analog electronics in radio and navigation equipment, as well as in other aircraft systems. A brief look at some of the basic analog circuits follows.

Rectifiers

Rectifier circuits change AC voltage into DC voltage and are one of the most commonly used type of circuits in aircraft electronics. *[Figure 11-43]* The resulting DC waveform

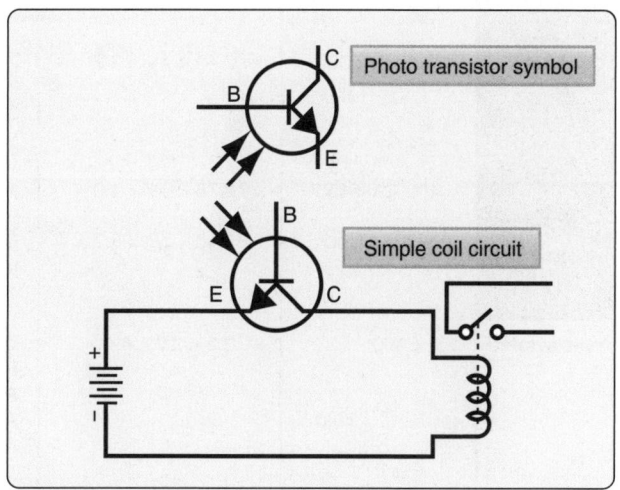

Figure 11-39. *A photo transistor in a simple coil circuit (bottom) and the symbol for a phototransistor (top).*

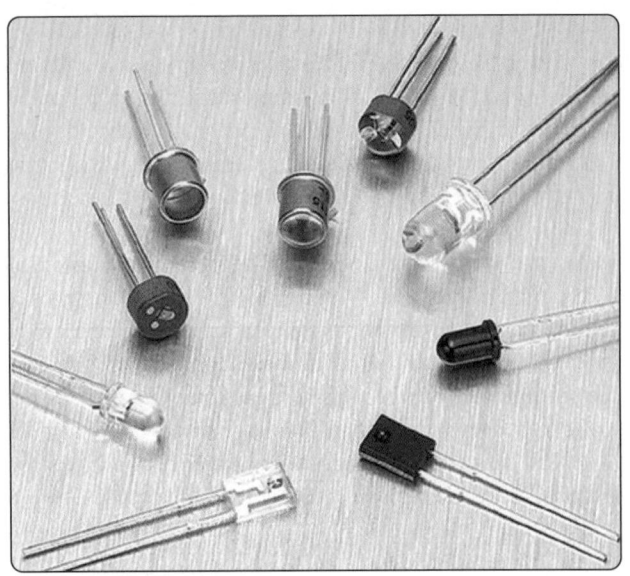

Figure 11-40. *Phototransistors.*

output is also shown. The circuit has a single semiconductor diode and a load resistor. When the AC voltage cycles below zero, the diode shuts off and does not allow current flow until the AC cycles through zero voltage again. The result is pronounced pulsating DC. While this can be useful, half of the original AC voltage is not being used.

A full wave rectifier creates pulsating DC from AC while using the full AC cycle. One way to do this is to tap the secondary coil at its midpoint and construct two circuits with the load resistor and a diode in each circuit. *[Figure 11-44]* The diodes are arranged so that when current is flowing through one, the other blocks current.

When the AC cycles so the top of the secondary coil of the transformer is positive, current flows from ground, through the load resistor (V_{RL}), Diode 1, and the upper half of the coil. Current cannot flow through Diode 2 because it is blocked. *[Figure 11-44A]* As the AC cycles through zero, the polarity

Color	Wavelength (nm)	Voltage (V)	Semiconductor Material
Infrared	$\lambda > 760$	$\Delta V < 1.9$	Gallium arsenide (GaAs) Aluminium gallium arsenide (AlGaAs)
Red	$610 < \lambda < 760$	$1.63 < \Delta V < 2.03$	Aluminium gallium arsenide (AlGaAs) Gallium arsenide phosphide (GaAsP) Aluminium gallium indium phosphide (AlGaInP) Gallium(III) phosphide (GaP)
Orange	$590 < \lambda < 610$	$2.03 < \Delta V < 2.10$	Gallium arsenide phosphide (GaAsP) Aluminium gallium indium phosphide (AlGaInP) Gallium(III) phosphide (GaP)
Yellow	$570 < \lambda < 590$	$2.10 < \Delta V < 2.18$	Gallium arsenide phosphide (GaAsP) Aluminium gallium indium phosphide (AlGaInP) Gallium(III) phosphide (GaP)
Green	$500 < \lambda < 570$	$1.9[32] < \Delta V < 4.0$	Indium gallium nitride (InGaN) / Gallium(III) nitride (GaN) Gallium(III) phosphide (GaP) Aluminium gallium indium phosphide (AlGaInP) Aluminium gallium phosphide (AlGaP)
Blue	$450 < \lambda < 500$	$2.48 < \Delta V < 3.7$	Zinc selenide (ZnSe) Indium gallium nitride (InGaN) Silicon carbide (SiC) as substrate Silicon (Si) as substrate — (under development)
Violet	$400 < \lambda < 450$	$2.76 < \Delta V < 4.0$	Indium gallium nitride (InGaN)
Purple	multiple types	$2.48 < \Delta V < 3.7$	Dual blue/red LEDs, blue with red phosphor, or white with purple plastic
Ultraviolet	$\lambda < 400$	$3.1 < \Delta V < 4.4$	Diamond (235 nm)[33] Boron nitride (215 nm)[34][35] Aluminium nitride (AlN) (210 nm)[36] Aluminium gallium nitride (AlGaN) Aluminium gallium indium nitride (AlGaInN) — (down to 210 nm)[37]
White	Broad spectrum	$\Delta V = 3.5$	Blue/UV diode with yellow phosphor

Figure 11-41. *LED colors and the materials used to construct them as well as their wavelength and voltages.*

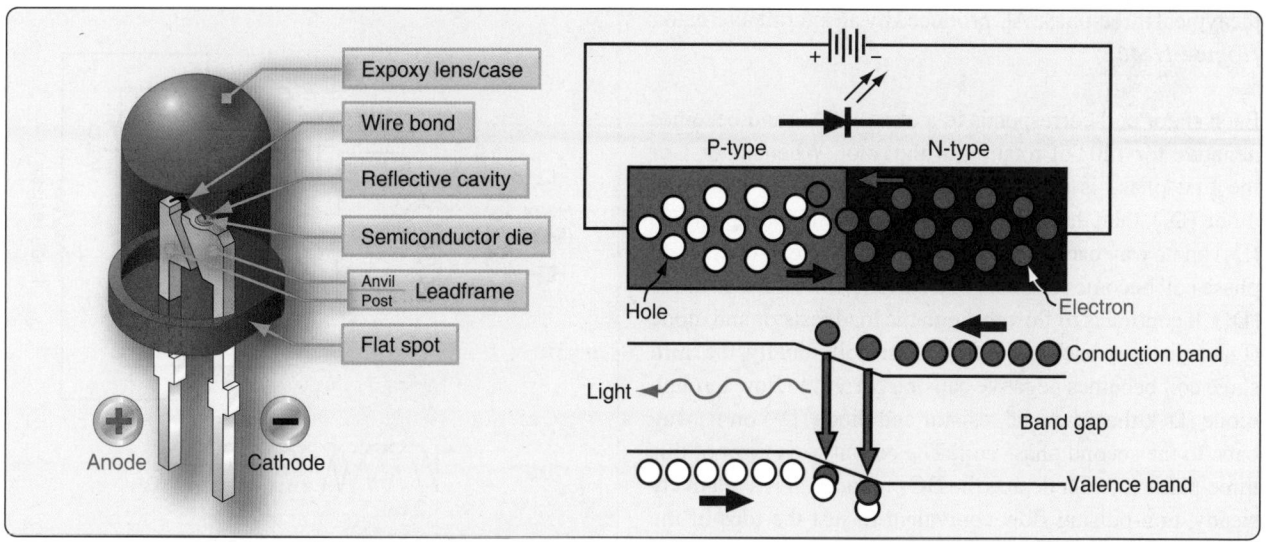

Figure 11-42. *A close up of a single LED (left) and the process of a semi-conductor producing light by electrons dropping into holes and giving off energy (right). The symbol for a light emitting diode is the diode symbol with two arrows pointing away from the junction.*

of the secondary coil changes. *[Figure 11-44B]* Current then flows from ground, through the load resistor, Diode 2, and the bottom half of the secondary coil. Current flow through Diode 1 is blocked. This arrangement yields positive DC from cycling AC with no wasted current.

Another way to construct a full wave rectifier uses four semiconductor diodes in a bridge circuit. Because the secondary coil of the transformer is not tapped at the center, the resultant DC voltage output is twice that of the two-diode full wave rectifier. *[Figure 11-45]* During the first half of the AC cycle, the bottom of the secondary coil is negative. Current flows from it through diode (D_1), then through the load resistor, and through diode (D_2) on its way back to the top of the secondary coil. When the AC reverses its cycle, the polarity of the secondary coil changes. Current flows

from the top of the coil through diode (D_3), then through the load resistor, and through diode (D_4) on its way back to the bottom of the secondary coil. The output waveform reflects the higher voltage achieved by rectifying the full AC cycle through the entire length of the secondary coil.

Use and rectification of three-phase AC is also possible on aircraft with a specific benefit. The output DC is very smooth and does not drop to zero. A six-diode circuit is built to rectify

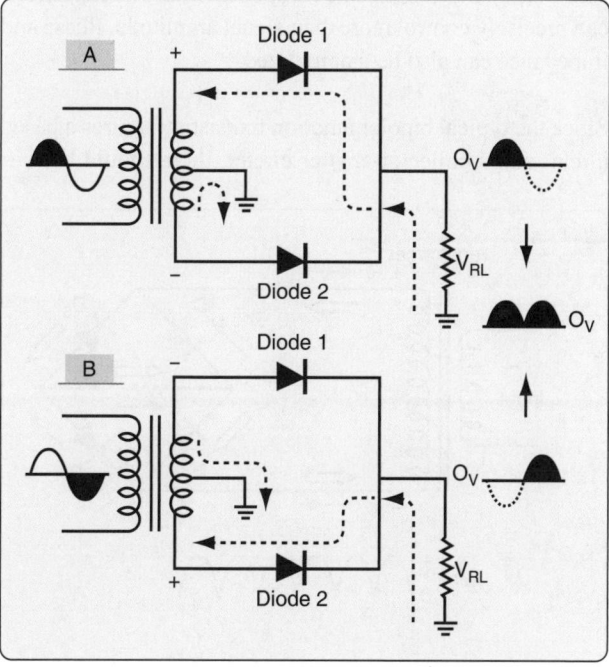

Figure 11-44. *A full wave rectifier can be built by center tapping the secondary coil of the transformer and using two diodes in separate circuits. This rectifies the entire AC input into a pulsating DC with twice the frequency of a half wave rectifier.*

Figure 11-43. *A half wave rectifier uses one diode to produce pulsating DC current from AC. Half of the AC cycle is wasted when the diode blocks the current flow as the AC cycles below zero.*

the typical three-phase AC produced by an aircraft alternator. [Figure 11-46]

Each stator coil corresponds to a phase of AC and becomes negative for 120° of rotation of the rotor. When stator 1 or the first phase is negative, current flows from it through diode (D_1), then through the load resistor and through diode (D_2) on its way back to the third phase coil. Next, the second phase coil becomes negative and current flows through diode (D_3). It continues to flow through the load resistor and diode (D_4) on its way back to the first phase coil. Finally, the third stage coil becomes negative causing current to flow through diode (D_5), then the load resistor and diode (D_6) on its way back to the second phase coil. The output waveform of this three-phase rectifier depicts the DC produced. It is a relatively steady, non-pulsing flow equivalent to just the tops of the individual curves. The phase overlap prevents voltage from falling to zero producing smooth DC from AC.

Amplifiers

An amplifier is a circuit that changes the amplitude of an electric signal. This is done through the use of transistors. As mentioned, a transistor that is forward biased at the base-emitter junction and reversed biased at the collector-base junction is turned on. It can conduct current from the collector to the emitter. Because a small signal at the base can cause a large current to flow from collector to emitter, a transistor in itself can be said to be an amplifier. However, a transistor properly wired into a circuit with resistors, power sources, and other electronic components, such as capacitors, can precisely control more than signal amplitude. Phase and impedance can also be manipulated.

Since the typical bipolar junction transistor requires a based circuit and a collector-emitter circuit, there should be four

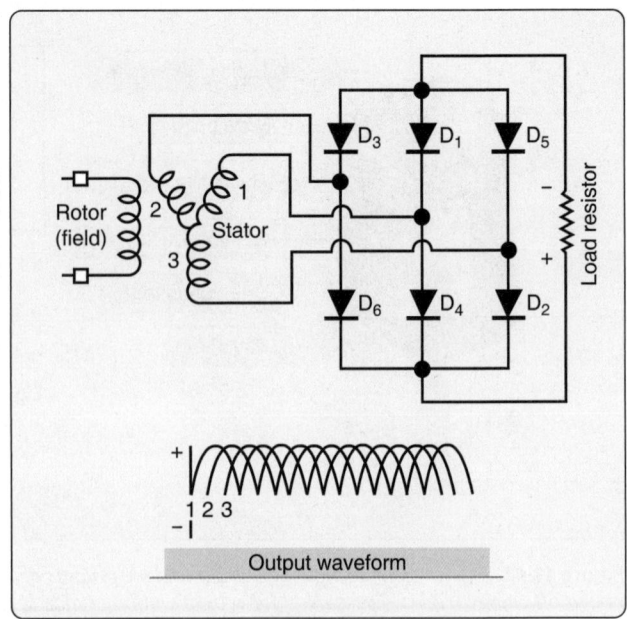

Figure 11-46. *A six-diode, three-phase AC rectifier.*

terminals, two for each circuit. However, the transistor only has three terminals (i.e., the base, the collector, and the emitter). Therefore, one of the terminals must be common to both transistor circuits. The selection of the common terminal affects the output of the amplifier.

The three basic amplifier types, named for which terminal of the transistor is the common terminal to both transistor circuits, include:

1. Common-emitter amplifier.
2. Common-collector amplifier.
3. Common-base amplifier.

Common-Emitter Amplifier

The common-emitter amplifier controls the amplitude of an electric signal and inverts the phase of the input signal. *Figure 11-47* illustrates a common-emitter amplifier for AC using a NPN transistor and its output signal graph. Common emitter circuits are characterized by high current gain and a 180° voltage phase shift from input to output. It is for the amplification of a microphone signal to drive a speaker. As always, adequate voltage of the correct polarity to the base puts the transistor in the active mode or turns it on. Then, as the base input current fluctuates, the current through the transistor fluctuates proportionally. However, AC cycles through positive and negative polarity. Every 180°, the transistor shuts off because the polarity to the base-emitter junction of the transistor is not correct to forward bias the junction. To keep the transistor on, a DC biasing voltage of the correct polarity (shown as a 2.3 volts (V) battery) is placed in series with the input signal in the base circuit to hold the transistor in the active mode as the AC polarity changes. This way the transistor stays

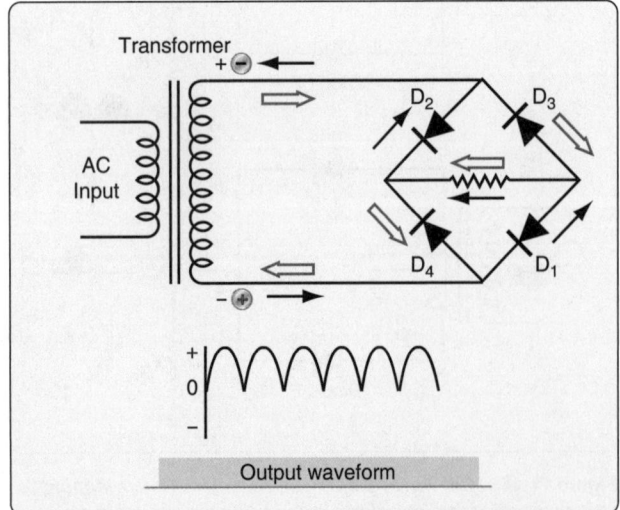

Figure 11-45. *The bridge-type four-diode full wave rectifier circuit is most commonly used to rectify single-phase AC into DC.*

in the active mode to amplify an entire AC signal.

Transistors are rated by ratio of the collector current to the base current, or Beta (β). This is established during the manufacture of the unit and cannot be changed. A 100 β transistor can handle 100 times more current through a collector-emitter circuit than the base input signal. This current in *Figure 11-47* is provided from the 15V battery, V_1. So, the amplitude of amplification is a factor of the beta of the transistor and any inline resistors used in the circuits. The fluctuations of the output signal, however, are entirely controlled by the fluctuations of current input to the transistor base.

If measurements of input and output voltages are made, it is shown that as the input voltage increases, the output voltage decreases. This accounts for the inverted phase produced by a common-emitter circuit. *[Figure 11-47]*

Common-Collector Amplifier

Another basic type of amplifier circuit is the common-collector amplifier. Common-collector circuits are characterized by high current gain, but virtually no voltage gain. The input circuit and the load circuit in this amplifier share the collector terminal of the transistor used. Because the load is in series with the emitter, both the input current and output current run through it. This causes a directly proportional relationship between the input and the output. The current gain in this circuit configuration is high. A small amount of input current can control a large amount of current to flow from the collector to the emitter. A common collector amplifier circuit

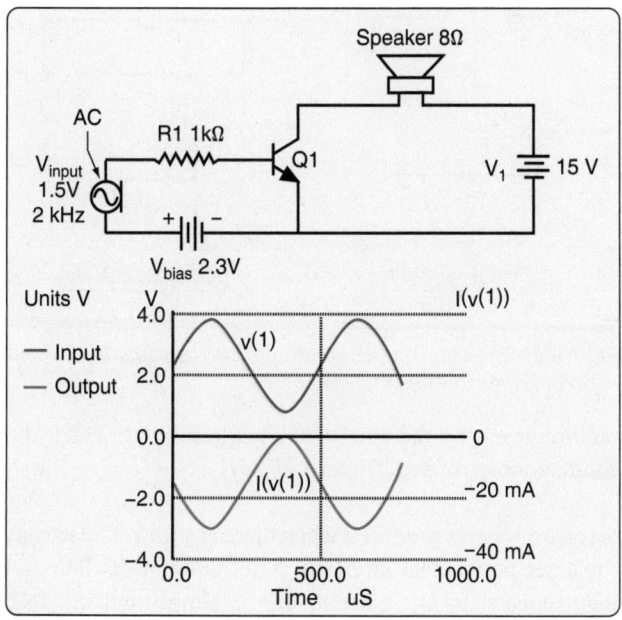

Figure 11-47. *A common-emitter amplifier circuit for amplifying an AC microphone signal to drive a speaker (top) and the graph of the output signal showing a 180 degree shift in phase (bottom).*

is illustrated in *Figure 11-48*. The base current needs to flow through the PN junction of the transistor, which has about a 0.7V threshold to be turned on. The output current of the amplifier is the beta value of the transistor plus 1.

During AC amplification, the common-collector amplifier has the same problem that exists in the common-emitter amplifier. The transistor must stay on or in the active mode regardless of input signal polarity. When the AC cycles through zero, the transistor turns off because the minimum amount of current to forward bias the transistor is not available. The addition of a DC biasing source (battery) in series with the AC signal in the input circuit keeps the transistor in the active mode throughout the full AC cycle. *[Figure 11-49]*

A common-collector amplifier can also be built with a PNP transistor. *[Figure 11-50]* It has the same characteristics as the NPN common-collector amplifier shown in *Figure 11-50*. When arranged with a high resistance in the input circuit and a small resistance in the load circuit, the common-collector amplifier can be used to step down the impedance of a signal. *[Figure 11-51]*

Common-Base Amplifier

A third type of amplifier circuit using a bipolar transistor is the common-base amplifier. In this circuit, the shared transistor terminal is the base terminal. *[Figure 11-52]* This causes a unique situation in which the base current is actually larger than the collector or emitter current. As such, the common-base amplifier does not boost current as the other amplifiers do. It attenuates current but causes a high gain in voltage. A very small fluctuation in base voltage in the input circuit causes a large variation in output voltage. The effect on the circuit output is direct, so the output voltage phase is the same as the input signal but much greater in amplitude.

As with the other amplifier circuits, when amplifying an AC signal with a common-base amplifier circuit, the input signal to the base must include a DC source to forward bias the transistor's base-emitter junction. This allows current to flow from the collector to the emitter during both cycles of the AC. A circuit for AC amplification is illustrated in *Figure 11-53* with a graph of the output voltage showing the large increase produced. The common-base amplifier is limited in its use since it does not increase current flow. This makes it the least used configuration. However, it is used in radio frequency amplification because of the low input Z. *Figure 11-54* summarizes the characteristics of the bipolar amplifier circuits discussed above.

Note: There are many variations in circuit design. JFETs and MOSFETS are also used in amplifier circuits, usually in small signal amplifiers due to their low noise outputs.

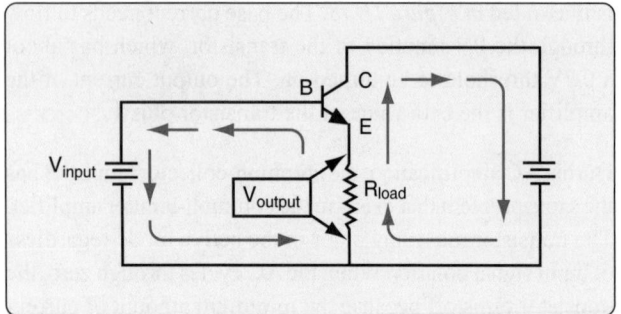

Figure 11-48. *A basic common-collector amplifier circuit. Both the input and output circuits share a path through the load and the emitter. This causes a direct relationship of the output current to the input current.*

Oscillator Circuits

Oscillators function to make AC from DC. They can produce various waveforms as required by electronic circuits. There are many different types of oscillators and oscillator circuits. Some of the most common types are discussed below.

A sine wave is produced by generators when a conductor is rotated in a uniform magnetic field. The typical AC sine wave is characterized by a gradual build-up and decline of voltage in one direction, followed by a similar smooth build-up to peak voltage and decline to zero again in the opposite direction. The value of the voltage at any given time in the cycle can be calculated by taking the peak voltage and

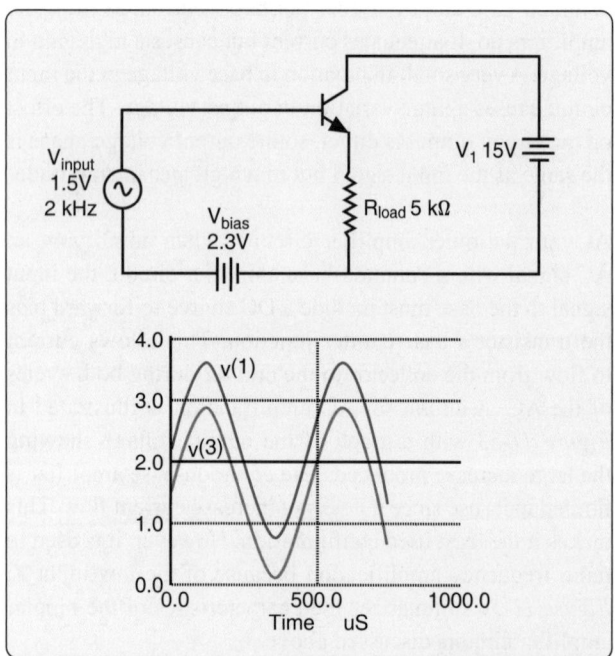

Figure 11-49. *A DC biasing current is used to keep the transistor of a common-collector amplifier in the active mode when amplifying AC (top). The output of this amplifier is in phase and directly proportional to the input (bottom). The difference in amplitude between the two is the 0.7V used to bias the PN junction of the transistor in the input circuit.*

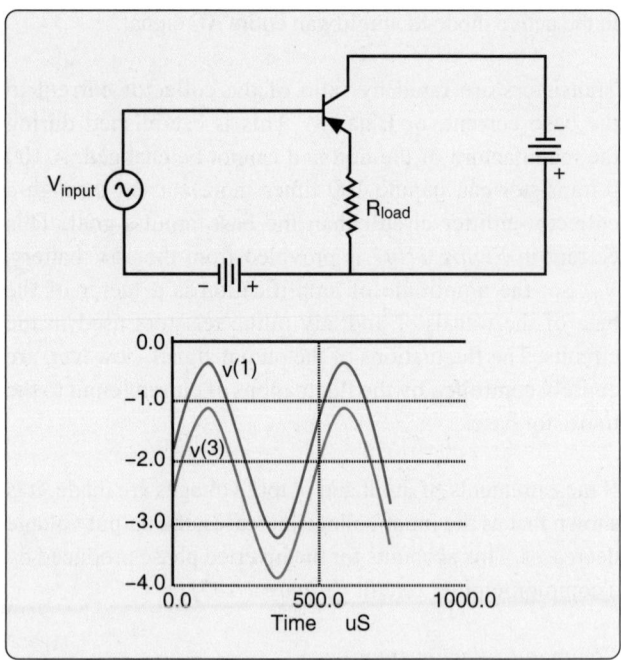

Figure 11-50. *A common-collector amplifier circuit with a PNP transistor has the same characteristics as that of a common-collector amplifier with a NPN transistor except for reversed voltage polarities and current direction.*

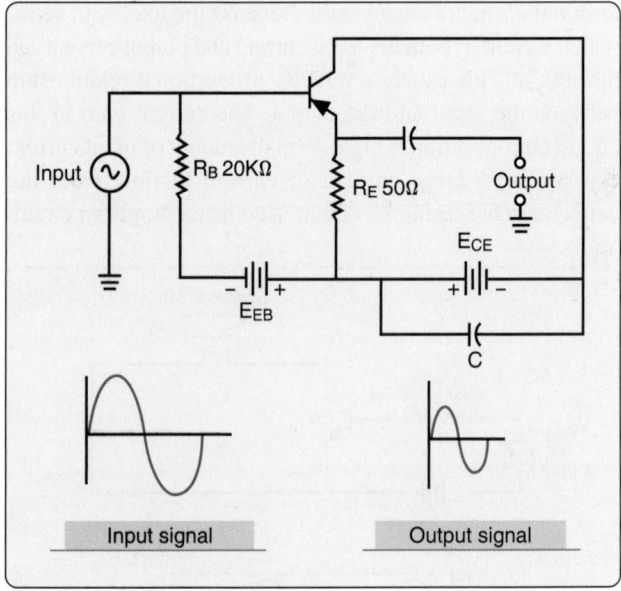

Figure 11-51. *This common-collector circuit has high input impedance and low output impedance.*

multiplying it by the sine of the angle through which the conductor has rotated. *[Figure 11-55]*

A square wave is produced when there is a flow of electrons for a set period that stops for a set amount of time and then repeats. In DC current, this is simply pulsing DC. *[Figure 11-56]* This same wave form can be of opposite polarities when passed through a transformer to produce AC. Certain oscillators produce square waves.

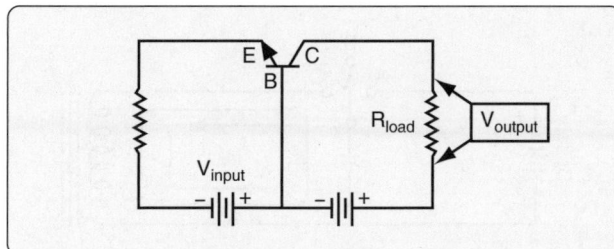

Figure 11-52. *A common-base amplifier circuit for DC current.*

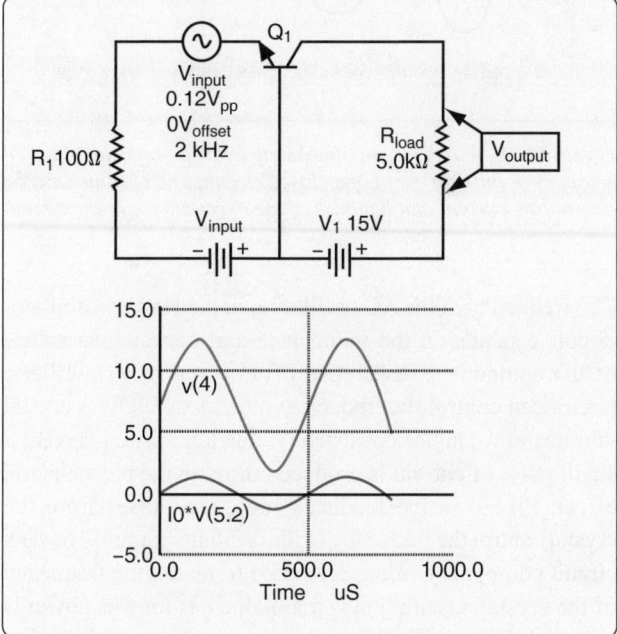

Figure 11-53. *In a common-base amplification circuit for AC (top), output voltage amplitude is greatly increased in phase with the input signal (bottom).*

An oscillator known as a relaxation oscillator produces another kind of wave form, a sawtooth wave. A slow rise from zero to peak voltage is followed by a rapid drop-off of voltage back to almost zero. Then it repeats. *[Figure 11-57]* In the circuit, a capacitor slowly charges through a resistor. A neon bulb is wired across the capacitor. When its ignition voltage is reached, the bulb conducts. This short-circuits the charged capacitor, which causes the voltage to drop to nearly zero and the bulb goes out. Then, the voltage rises again as the cycle repeats.

Electronic Oscillation

Oscillation in electronic circuits is accomplished by combining a transistor and a tank circuit. A tank circuit is comprised of a capacitor and coil parallel to each other. *[Figure 11-58]* When attached to a power source by closing switch A, the capacitor charges to a voltage equal to the battery voltage. It stays charged, even when the circuit to the battery is open (switch in position B). When the switch is put in position C, the capacitor and coil are in a closed circuit. The capacitor discharges through the coil. While receiving the energy from the capacitor, the coil stores it by building up an electromagnetic field. When the capacitor is fully discharged, the coil stops conducting. The magnetic field collapses, which induces current flow. The current charges the opposite plate of the capacitor. When completely charged, the capacitor discharges into the coil again. The magnetic field builds again and stops when the capacitor is fully discharged. The magnetic field collapses again, which induces current that charges the original plate of the capacitor and the cycle repeats.

This oscillation of charging and discharging the capacitor through the coil would continue indefinitely if a circuit could be built with no resistance. This is not possible. However, a circuit can be built using a transistor that restores losses due to resistance. There are various ways to accomplish this. The Hartley oscillator circuit in *Figure 11-59* is one. The circuit can oscillate indefinitely as long as it is connected to power.

When the switch is closed, current begins to flow in the oscillator circuit. The transistor base is supplied with biasing current through the voltage divider R_A and R_B. This allows current to flow through the transistor from the collector to the emitter, through R_E and through the lower portion of the center tapped coil that is labeled L_1. The current increasing through this coil builds a magnetic field that induces current in the upper half of the coil labeled L_2. The current from L_2 charges capacitor C_2, which increases the forward bias of the transistor. This allows an increasing flow of current through the transistor, R_E, and L_1 until the transistor is saturated and capacitor C_1 is fully charged. Without force to add electrons to capacitor C_1, it discharges and begins

Type of Amplifier	Impedance	Voltage Gain	Current Gain	Power Gain	Phase
Common-emitter	Input: fairly high Output: fairly high	Relatively large	Relatively large	Large	Inverts phase
Common-collector	Input: high Output: low	Always less than one	Relatively large	Relatively large	Output same as input
Common-base	Input: low Output: high	Large	Always less than one	Relatively large	Output same as input

Figure 11-54. *PN junction transistor amplifier characteristics.*

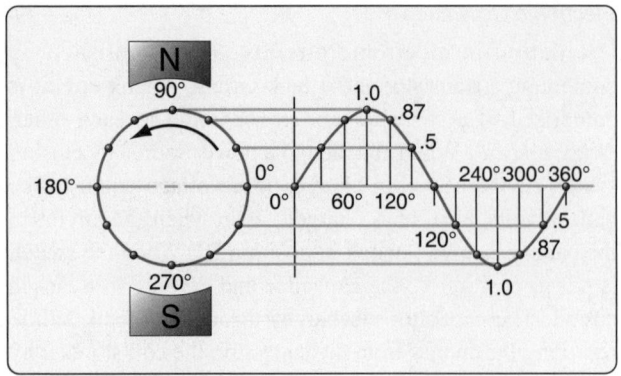

Figure 11-55. *Voltage over time of sine waveform electricity created when a conductor is rotated through a uniform magnetic field, such as in a generator.*

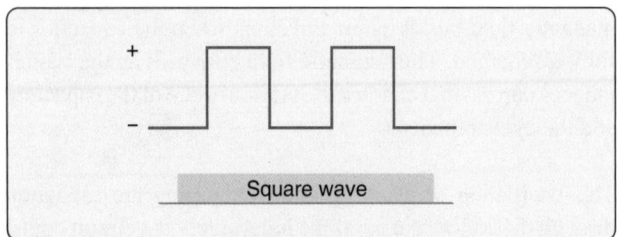

Figure 11-56. *The waveform of pulsing DC is a square wave.*

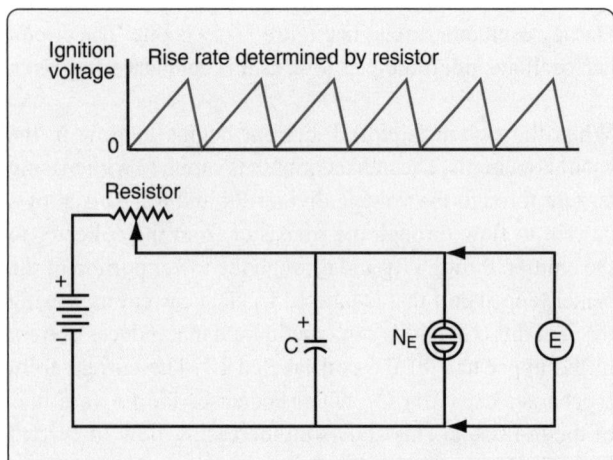

Figure 11-57. *A relaxation oscillator produces a sawtooth wave output.*

the oscillation in the tank circuit described in the previous section. As C_1 becomes fully charged, current to charge C_2 reduces and C_2 also discharges. This adds the energy needed to the tank circuit to compensate for resistance losses. As C_2 is discharging, it reduces forward biasing and eventually the transistor becomes reversed biased and cuts off. When the opposite plate of capacitor C_1 is fully charged, it discharges, and the oscillation is in progress. The transistor base becomes forward biased again, allowing for current flow through the resistor R_E, coil L_1, etc.

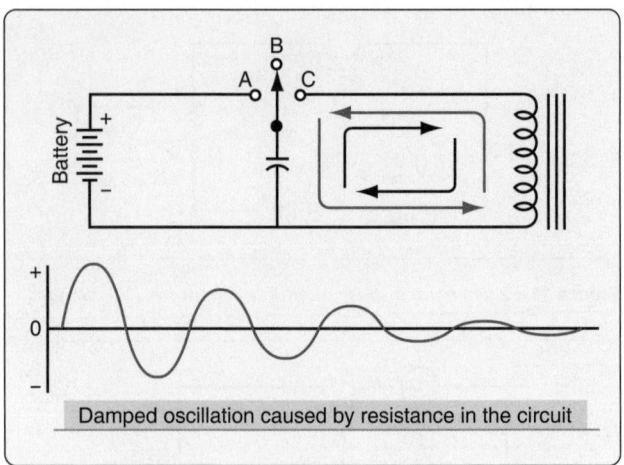

Damped oscillation caused by resistance in the circuit

Figure 11-58. *A tank circuit alternately charges opposite plates of a capacitor through a coil in a closed circuit. The oscillation is an alternating current that diminishes due to resistance in the circuit.*

The frequency of the AC oscillating in the Hartley oscillator circuit depends on the inductance and capacitance values of the components used. Use of a crystal in an oscillator circuit can control the frequency more accurately. A crystal vibrates at a single, consistent frequency. When flexed, a small pulse of current is produced through the piezoelectric effect. Placed in the feedback loop, the pulses from the crystal control the frequency of the oscillator circuit. The tank circuit component values are tuned to match the frequency of the crystal. Oscillation is maintained as long as power is supplied. *[Figure 11-60]*

Other types of oscillator circuits used in electronics and computers have two transistors that alternate being in the active mode. They are called multi-vibrators. The choice of oscillator in an electronic device depends on the exact type of manipulation of electricity required to permit the device to function as desired.

Digital Electronics

The above discussion of semiconductors, semiconductor devices, and circuitry is only an introduction to the electronics found in communications and navigation avionics. In-depth maintenance of the interior electronics on most avionics devices is performed only by certified repair stations and trained avionics technicians. The airframe technician is responsible for installation, maintenance, inspection, and proper performance of avionics in the aircraft.

Modern aircraft increasingly employs digital electronics in avionics rather than analog electronics. Transistors are used in digital electronics to construct circuits that act as digital logic gates. The purpose and task of a device is achieved by manipulating electric signals through the logic gates. Thousands, and even millions, of tiny transistors can be

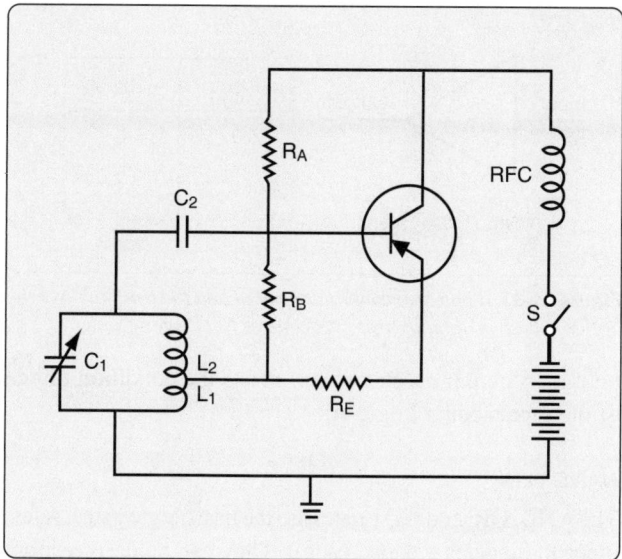

Figure 11-59. *A Hartley oscillator uses a tank circuit and a transistor to maintain oscillation whenever power is applied.*

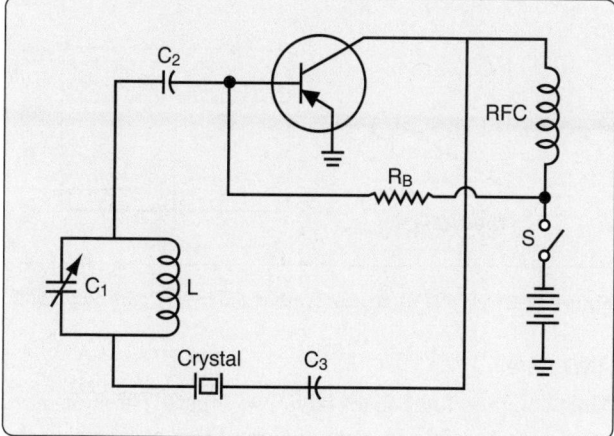

Figure 11-60. *A crystal in an electronic oscillator circuit is used to tune the frequency of oscillation.*

placed on a chip to create the digital logic landscape through which a component's signals are processed.

Digital Building Blocks

Digital logic is based on the binary number system. There are two conditions that may exist, 1 or 0. In a digital circuit, these are equivalent to voltage or no voltage. Within the binary system, these two conditions are called Logic 1 and Logic 0. Using just these two conditions, gates can be constructed to manipulate information. There are a handful of common logic gates that are used. By combining any number of these tiny solid-state gates, significant memorization, manipulation, and calculation of information can be performed.

The NOT Gate

The NOT gate is the simplest of all gates. If the input to the gate is Logic 1, then the output is NOT Logic 1. This means that it is Logic 0, since there are only two conditions in the binary world. In an electronic circuit, a NOT gate would invert the input signal. In other words, if there was voltage at the input to the gate, there would be no output voltage. The gate can be constructed with transistors and resistors to yield this electrical logic every time. (The gate or circuit would also have to invert an input of Logic 0 into an output of Logic 1.)

To understand logic gates, truth tables are often used. A truth table gives all of the possibilities in binary terms for each gate containing a characteristic logic function. For example, a truth table for a NOT gate is illustrated in *Figure 11-61*. Any input (A) is NOT present at the output (B). This is simple, but it defines this logic situation. A tiny NOT gate circuit can be built using transistors that produce these results. In other words, a circuit can be built such that if voltage arrives at

the gate, no voltage is output or vice-versa.

When using transistors to build logic gates, the primary concern is to operate them within the circuits so the transistors are either OFF (not conducting) or fully ON (saturated). In this manner, reliable logic functions can be performed. The variable voltage and current situations present during the active mode of the transistor are of less importance.

Figure 11-62 illustrates an electronic circuit diagram that performs the logic NOT gate function. Any input, either a no voltage or voltage condition, yields the opposite output. This gate is built with bipolar junction transistors, resistors, and a few diodes. Other designs exist that may have different components.

When examining and discussing digital electronic circuits, the electronic circuit design of a gate is usually not presented. The symbol for the logic gate is most often used. *[Figure 11-61]* The technician can then concentrate on the configuration of the logic gates in relation to each other. A brief discussion of the other logic gates, their symbols, and truth tables follow.

Buffer Gate

Another logic gate with only one input and one output is the buffer. It is a gate with the same output as the input. While this may seem redundant or useless, an amplifier may be considered a buffer in a digital circuit because if there is voltage present at the input, there is an output voltage. If there is no voltage at the input, there is no output voltage. When used as an amplifier, the buffer can change the values of a signal. This is often done to stabilize a weak or varying signal. All gates are amplifiers subject to output fluctuations. The buffer steadies the output of the upstream device while maintaining its basic characteristic. Another application of a buffer that is two NOT gates, is to use it to isolate a portion of a circuit. *[Figure 11-63]*

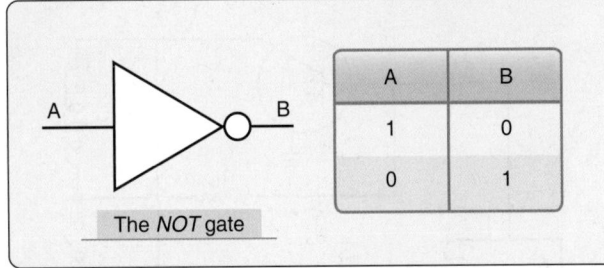

Figure 11-61. *A NOT logic gate symbol and a NOT gate truth table.*

A	B
1	0
0	1

The *NOT* gate

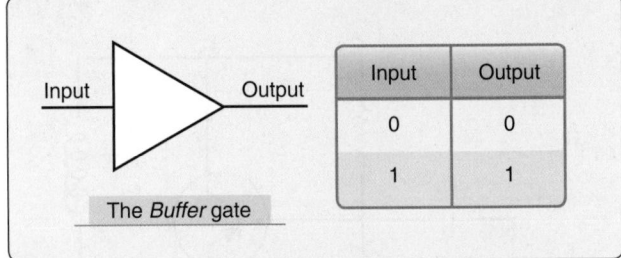

Figure 11-63. *A buffer or amplifier symbol and the truth table of the buffer, which is actually two consecutive NOT gates.*

Input	Output
0	0
1	1

The *Buffer* gate

AND Gate

Most common logic gates have two inputs. Three or more inputs are possible on some gates. When considering the characteristics of any logic gate, an output of Logic 1 is sought and a condition for the inputs is stated or examined. For example, *Figure 11-64* illustrates an AND gate. For an AND gate to have a Logic 1 output, both inputs have to be Logic 1. In an actual electronic circuit, this means that for a voltage to be present at the output, the AND gate circuit has to receive voltage at both of its inputs. As pointed out, there are different arrangements of electronic components that yield this result. Whichever is used is summarized and presented as the AND gate symbol. The truth table in *Figure 11-64* illustrates that there is only one way to have an output of Logic 1 or voltage when using an AND gate.

OR Gate

Another useful and common logic gate is the OR gate. In an OR gate, to have an output of Logic 1 (voltage present), one of the inputs must be Logic 1. As seen in *Figure 11-65,* only one of the inputs needs to be Logic 1 for there to be an output of Logic 1. When both inputs are Logic 1, the OR gate has

a Logic 1 output because it still meets the condition of one of the inputs being Logic 1.

NAND Gate

The AND, OR, and NOT gates are the basic logic gates. A few other logic gates are also useful. They can be derived from combining the AND, OR, and NOT gates. The NAND gate is a combination of an AND gate and a NOT gate. This means that AND gate conditions must be met and then inverted. So, the NAND gate is an AND gate followed by a NOT gate. The truth table for a NAND gate is shown in *Figure 11-66* along with its symbol. If a Logic 1 output is to exist from a NAND gate, inputs A and B must not both be Logic 1. Or, if a NAND gate has both inputs Logic 1, the output is Logic 0. Stated in electronic terms, if there is to be an output voltage, then the inputs cannot both have voltage or, if both inputs have voltage, there is no output voltage.

Note: The values in the output column of the NAND gate table are exactly the opposite of the output values in the AND gate truth table.

NOR Gate

A NOR gate is similarly arranged except that it is an inverted OR gate. If there is to be a Logic 1 output, or output voltage, then neither input can be Logic 1 or have input voltage. This is the same as satisfying the OR gate conditions and then putting output through a NOT gate. The NOR gate truth table in *Figure 11-67* shows that the NOR gate output values are exactly the opposite of the OR gate output values.

The NAND gate and the NOR gate have a unique distinction. Each one can be the only gate used in circuitry to produce the same output as any of the other logic gates. While it may be inefficient, it is testimonial to the flexibility that designers have when working with logic gates, the NAND and NOR gates in particular.

EXCLUSIVE OR Gate

Another common logic gate is the EXCLUSIVE OR gate. It is the same as an OR gate except for the condition where both inputs are Logic 1. In an OR gate, there would be Logic

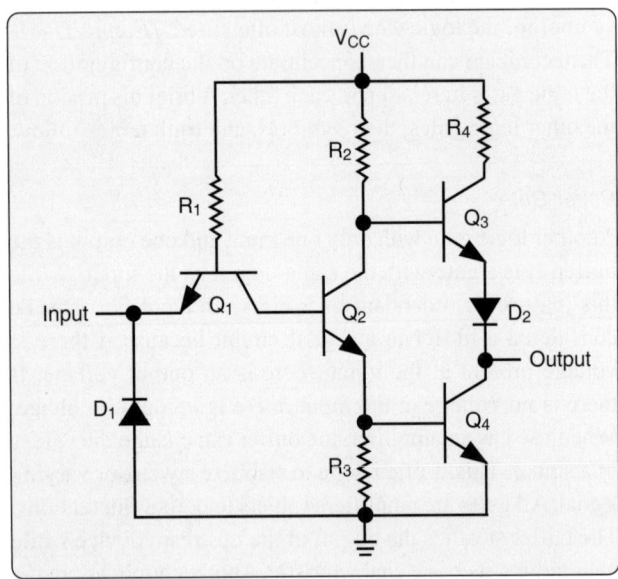

Figure 11-62. *An electronic circuit that reliably performs the NOT logic function.*

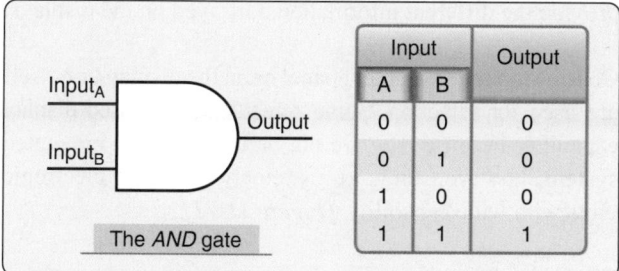

Figure 11-64. *An AND gate symbol and its truth table.*

Input		Output
A	B	
0	0	0
0	1	0
1	0	0
1	1	1

The AND gate

1 output when both inputs are Logic 1. This is not allowed in an EXCLUSIVE OR gate. When either of the inputs is Logic 1, the output is Logic 1. But, if both inputs are logic 1, the Logic 1 output is excluded or Logic 0. *[Figure 11-68]*

Negative Logic Gates

There are also negative logic gates. The negative OR and the negative AND gates are gates wherein the inputs are inverted rather than inverting the output. This creates a unique set of outputs as seen in the truth tables in *Figure 11-69*. The negative OR gate is not the same as the NOR gate as is sometimes misunderstood. Neither is the negative AND gate the same as the NAND gate. However, as the truth tables reveal, the output of a negative AND gate is the same as a NOR gate, and the output of a negative OR gate is the same as a NAND gate.

In summary, electronic circuits use transistors to construct logic gates that produce outputs related to the inputs shown in the truth tables for each kind of gate. The gates are then assembled with other components to manipulate data in digital circuits. The electronic digital signals used are voltage or no voltage representations of Logic 1 or Logic 0 conditions. By using a series of voltage output or no voltage output gates, manipulation, computation, and storage of data takes place.

Digital Aircraft Systems

Digital aircraft systems are the present and future of aviation. From communication and navigation to engine and flight controls, increased proliferation of digital technology increases reliability and performance. Processing, storing, and transferring vital information for the operation of an aircraft in digital form provides a usable common language

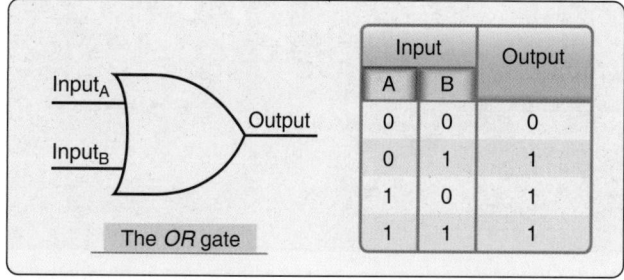

Input		Output
A	B	
0	0	0
0	1	1
1	0	1
1	1	1

The OR gate

Figure 11-65. *An OR gate symbol and its truth table.*

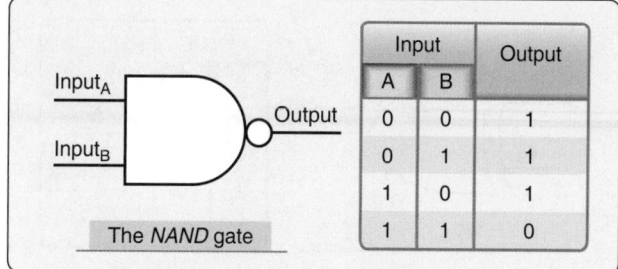

Figure 11-66. *A NAND gate symbol and its truth table illustrating that the NAND gate is an inverted AND gate.*

Input		Output
A	B	
0	0	1
0	1	1
1	0	1
1	1	0

The NAND gate

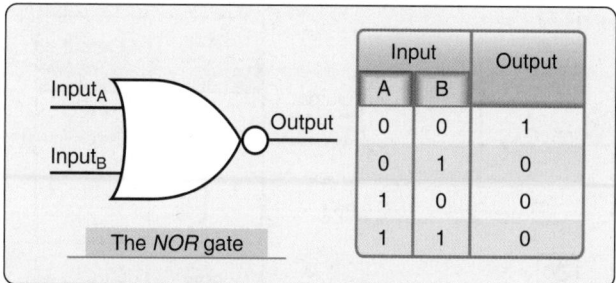

Figure 11-67. *A NOR gate symbol and its truth table illustrating that the NOR gate is an inverted OR gate.*

Input		Output
A	B	
0	0	1
0	1	0
1	0	0
1	1	0

The NOR gate

for monitoring, control, and safety. Integration of information from different systems is simplified. Self-monitoring, built-in test equipment (BITE) and air-to-ground data links increase maintenance efficiency. Digital buss networking allows aircraft system computers to interact for a coordinated comprehensive approach to flight operations.

Digital Data Displays

Modern digital data displays are the most visible features of digital aircraft systems. They extend the functional advantages of state of the art digital communication and navigation avionics and other digital aircraft systems via the use of an enhanced interface with the pilot. The result is an increase in situational awareness and overall safety of flight. Digital data displays are the glass of the glass flight deck. They expand the amount, clarity, and proximity of the information presented to the pilot. *[Figure 11-70]*

Many digital data displays are available from numerous manufacturers as original equipment in new aircraft, or as retrofit components or complete retrofit systems for older aircraft. Approval for retrofit displays is usually accomplished through supplementary type certificate (STC) awarded to the equipment manufacturer.

Early digital displays presented scale indication in digital or integer format readouts. Today's digital data displays are analogous to computer screen presentations. Numerous aircraft and flight instrument readouts and symbolic presentations are combined with communication and navigational information on multifunctional displays (MFD).

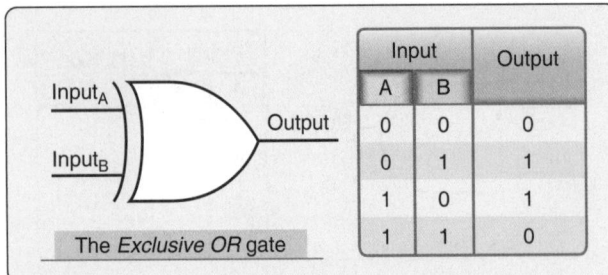

Figure 11-68. *An EXCLUSIVE OR gate symbol and its truth table, which is similar to an OR gate but excludes output when both inputs are the same.*

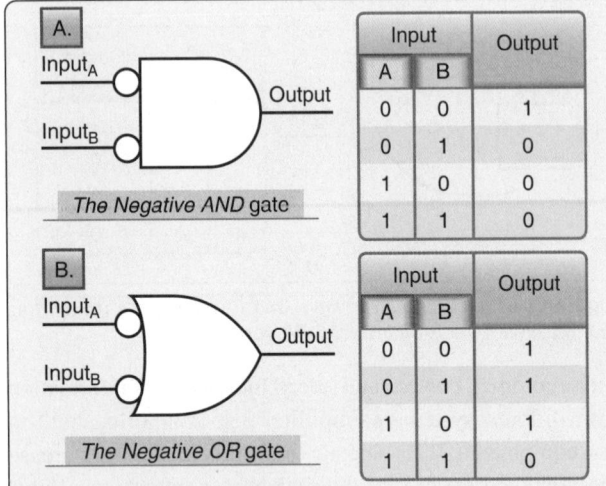

Figure 11-69. *The NEGATIVE AND gate symbol and its truth table (A) and the NEGATIVE OR gate symbol and truth table (B). The inputs are inverted in the NEGATIVE gates.*

Often a display has a main function with potential to back-up another display should it fail. Names, such as primary flight display (PFD), secondary flight display, navigational display (ND), etc., are often used to describe a display by its primary use. The hardware composition of the displays is essentially the same. Avionics components and computers combine to provide the different information portrayed on the displays.

Controls on the instrument panel or on the display unit itself are used for selection. Some screens have limited display capability because they are not part of a totally integrated system; however, they are extremely powerful electronic units with wide capability. *[Figure 11-71]*

The basis of the information displayed on what is known as a PFD, is usually an electronic flight instrument system (EFIS) like representation of the aircraft attitude indicator in the upper half of the display, and an electronic horizontal situation indicator display on the lower half. Numerous ancillary readouts are integrated or surround the electronic attitude indicator and the horizontal situation indicator (HSI). On full glass flight deck PFDs, all of the basic T instrument indications are presented and much more, such as communication and navigation information, weather data, terrain features, and approach information. Data displays for engine parameters, hydraulics, fuel, and other airframe systems are often displayed on the secondary flight display or on an independent display made for this purpose. *[Figure 11-72]*

As with other avionics components, repair and maintenance of the internal components of digital data displays is reserved for licensed repair stations only.

Digital Tuners & Audio Panels

Numerous communication and navigation devices are described in the following sections of this chapter. Many of these use radio waves and must be tuned to a desired frequency for operation. As a flight progresses, retuning and changing from one piece of equipment to another can occur frequently. An audio panel or digital tuner consolidates various communication and navigation radio selection controls into a single unit. The pilot can select and use, or

Figure 11-70. *A modern glass flight deck on a general aviation aircraft. Digital data displays replace many older instruments and indicators of the past.*

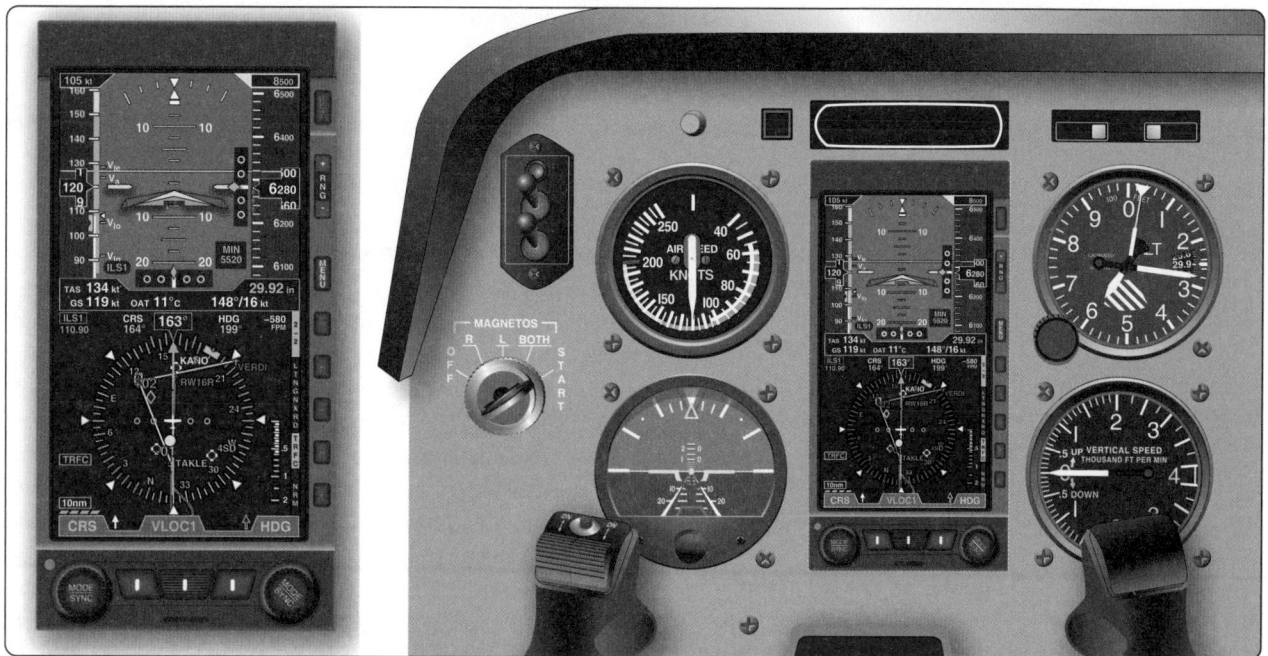

Figure 11-71. *A retrofit digital data display.*

select and tune, most of the aircraft's avionics from this one control interface. *[Figure 11-73]*

Radio Communication

Much of aviation communication and navigation is accomplished through the use of radio waves. Communication by radio was the first use of radio frequency transmissions in aviation.

Radio Waves

A radio wave is invisible to the human eye. It is electromagnetic in nature and part of the electronic spectrum of wave activity that includes gamma rays, x-rays, ultraviolet rays, infrared waves, and visible light rays, as well all radio waves. *[Figure 11-74]* The atmosphere is filled with these waves. Each wave occurs at a specific frequency and has a corresponding wavelength. The relationship between frequency and wavelength is inversely proportional. A high frequency wave has a short wavelength and a low frequency wave has a long wavelength.

In aviation, a variety of radio waves are used for communication. *Figure 11-75* illustrates the radio spectrum that includes the range of common aviation radio frequencies and their applications.

Note: A wide range of frequencies are used from low frequency (LF) at 100 kHz (100,000 cycles per second) to super high frequency (SHF) at nearly 10gHz (10,000,000,000 cycles per second). The Federal Communications Commission (FCC) controls the assignment of frequency usage.

AC power of a particular frequency has a characteristic length of conductor that is resonant at that frequency. This length is the wavelength of the frequency that can be seen on an oscilloscope. Fractions of the wavelength also resonate, especially half of a wavelength, which is the same as half of the AC sine wave or cycle.

The frequency of an AC signal is the number of times the AC cycles every second. AC applied to the center of a radio antenna, a conductor half the wavelength of the AC frequency, travels the length of the antenna, collapses, and travels the

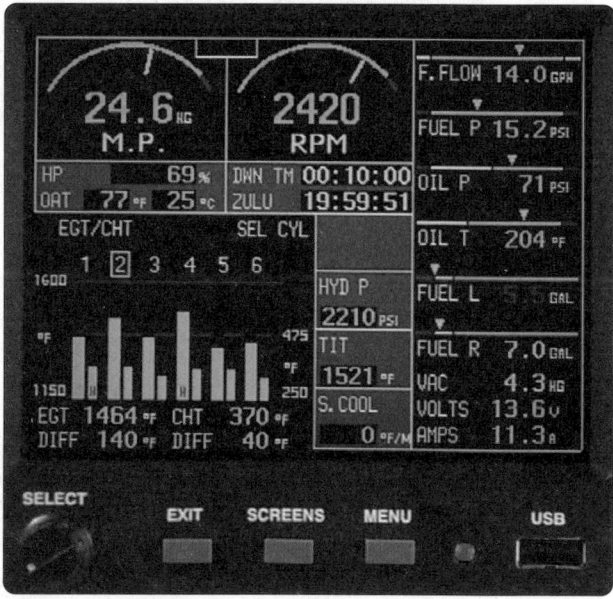

Figure 11-72. *A digital data display dedicated to the depiction of engine and airframe system parameter status.*

length of the antenna in the opposite direction. The number of times it does this every second is known as the radio wave signal frequency or radio frequency as shown in *Figure 11-75*. As the current flows through the antenna, corresponding electromagnetic and electric fields build, collapse, build in the opposite direction, and collapse again. *[Figure 11-76]*

To transmit radio waves, an AC generator is placed at the midpoint of an antenna. As AC current builds and collapses in the antenna, a magnetic field also builds and collapses around it. An electric field also builds and subsides as the voltage shifts from one end of the antenna to the other. Both fields, the magnetic and the electric, fluctuate around the antenna at the same time. The antenna is half the wavelength of the AC signal received from the generator. At any one point along the antenna, voltage and current vary inversely to each other.

Because of the speed of the AC, the electromagnetic fields and electric fields created around the antenna do not have time to completely collapse as the AC cycles. Each new current flow creates new fields around the antenna that force the not-totally-collapsed fields from the previous AC cycle out into space. These are the radio waves. The process is continuous as long as AC is applied to the antenna. Thus, steady radio waves of a frequency determined by the input AC frequency propagate out into space.

Radio waves are directional and propagate out into space at 186,000 miles per second. The distance they travel depends on the frequency and the amplification of the signal AC sent to the antenna. The electric field component and the electromagnetic field component are oriented at 90° to each other, and at 90° to the direction that the wave is traveling. *[Figure 11-77]*

Types of Radio Waves

Radio waves of different frequencies have unique characteristics as they propagate through the atmosphere. Very low frequency (VLF), LF, and medium frequency (MF) waves have relatively long wavelengths and utilize correspondingly long antennas. Radio waves produced at these frequencies ranging from 3 kHz to 3 mHz are known as ground waves or surface waves. This is because they follow the curvature of the earth as they travel from the broadcast antenna to the receiving antenna. Ground waves are particularly useful for long distance transmissions. Automatic direction finders (ADF) and LORAN navigational aids use these frequencies. *[Figure 11-78]*

High frequency (HF) radio waves travel in a straight line and do not curve to follow the earth's surface. This would limit transmissions from the broadcast antenna to receiving antennas only in the line-of-sight of the broadcast antenna except for a unique characteristic. HF radio waves bounce off of the ionosphere layer of the atmosphere. This refraction

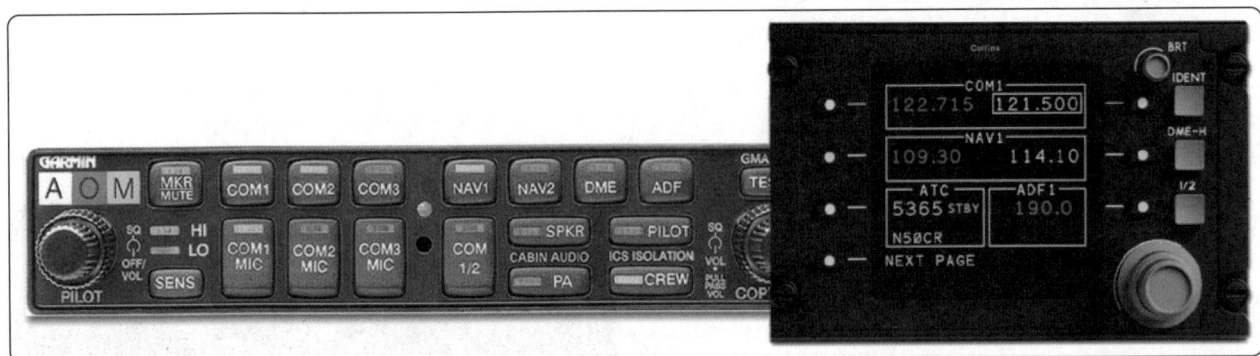

Figure 11-73. *An audio panel in a general aviation aircraft integrates the selection of several radio-based communication and navigational aids into a single control panel (left). A digital tuner (right) does the same on a business class aircraft and allows the frequency of each device to be tuned from the same panel as well.*

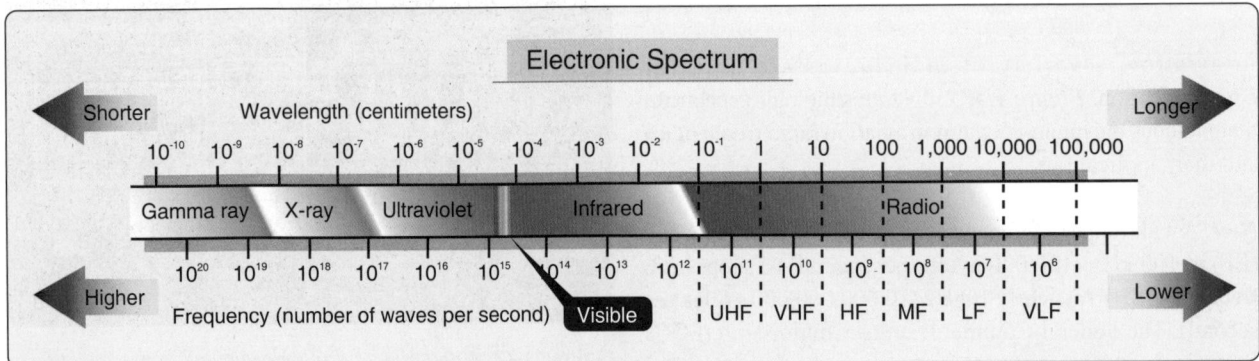

Figure 11-74. *Radio waves are just some of the electromagnetic waves found in space.*

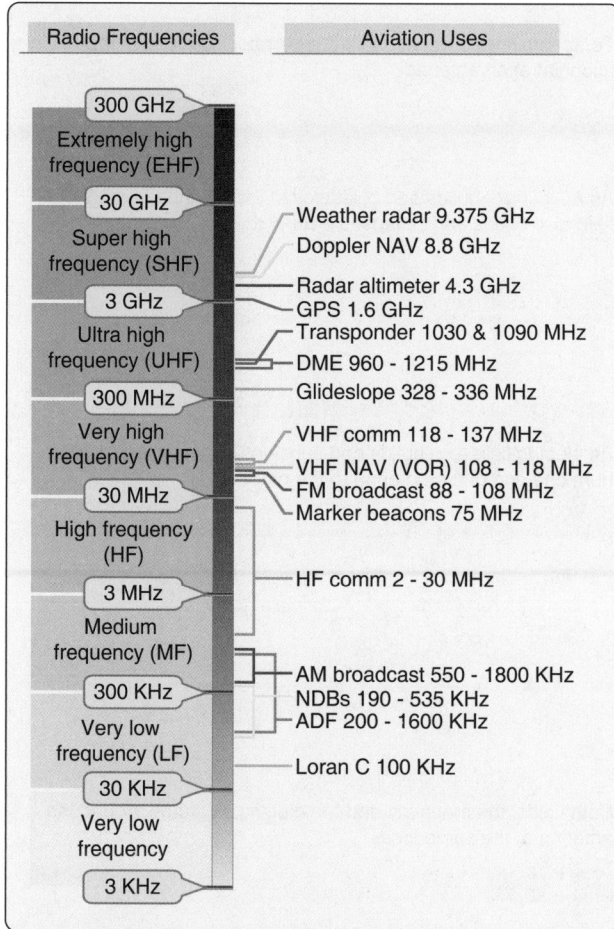

Radio Frequencies	Aviation Uses

300 GHz
Extremely high frequency (EHF)
30 GHz
Super high frequency (SHF) — Weather radar 9.375 GHz — Doppler NAV 8.8 GHz
3 GHz — Radar altimeter 4.3 GHz — GPS 1.6 GHz
Ultra high frequency (UHF) — Transponder 1030 & 1090 MHz — DME 960 - 1215 MHz
300 MHz — Glideslope 328 - 336 MHz
Very high frequency (VHF) — VHF comm 118 - 137 MHz — VHF NAV (VOR) 108 - 118 MHz — FM broadcast 88 - 108 MHz
30 MHz — Marker beacons 75 MHz
High frequency (HF)
3 MHz — HF comm 2 - 30 MHz
Medium frequency (MF)
300 KHz — AM broadcast 550 - 1800 KHz — NDBs 190 - 535 KHz — ADF 200 - 1600 KHz
Very low frequency (LF) — Loran C 100 KHz
30 KHz
Very low frequency
3 KHz

Figure 11-75. *There is a wide range of radio frequencies. Only the very low frequencies and the extremely high frequencies are not used in aviation.*

extends the range of HF signals beyond line-of-sight. As a result, transoceanic aircraft often use HF radios for voice communication. The frequency range is between 2 to 25 MHz. These kinds of radio waves are known as sky waves. *[Figure 11-78]*

Above HF transmissions, radio waves are known as space waves. They are only capable of line-of-sight transmission and do not refract off of the ionosphere. *[Figure 11-78]* Most aviation communication and navigational aids operate with space waves. This includes VHF (30 - 300 MHz), UHF (300 MHz - 3 GHz), and super high frequency (SHF) (3 - 30 Ghz) radio waves.

VHF communication radios are the primary communication radios used in aviation. They operate in the frequency range from 118.0 MHz to 136.975 MHz. Seven hundred and twenty separate and distinct channels have been designated in this range with 25 kilohertz spacing between each channel. Further division of the bandwidth is possible, such as in Europe where 8.33 kilohertz separate each

VHF communication channel. VHF radios are used for communications between aircraft and air traffic control (ATC), as well as air-to-air communication between aircraft. When using VHF, each party transmits and receives on the same channel. Only one party can transmit at any one time.

Loading Information onto a Radio Wave

The production and broadcast of radio waves does not convey any significant information. The basic radio wave discussed above is known as a carrier wave. To transmit and receive useful information, this wave is altered or modulated by an information signal. The information signal contains the unique voice or data information desired to be conveyed. The modulated carrier wave then carries the information from the transmitting radio to the receiving radio via their respective antennas. Two common methods of modulating carrier waves are amplitude modulation and frequency modulation.

Amplitude Modulation (AM)

A radio wave can be altered to carry useful information by modulating the amplitude of the wave. A DC signal, for example from a microphone, is amplified and then superimposed over the AC carrier wave signal. As the varying DC information signal is amplified, the amplifier output current varies proportionally. The oscillator that creates the carrier wave does so with this varying current. The oscillator frequency output is consistent because it is built into the oscillator circuit. But the amplitude of the oscillator output varies in relation to the fluctuating current input. *[Figure 11-79]*

When the modulated carrier wave strikes the receiving antenna, voltage is generated that is the same as that which was applied to the transmitter antenna. However, the signal is weaker. It is amplified so that it can be demodulated. Demodulation is the process of removing the original information signal from the carrier wave. Electronic circuits containing capacitors, inductors, diodes, filters, etc., remove all but the desired information signal identical to the original input signal. Then, the information signal is typically amplified again to drive speakers or other output devices. *[Figure 11-80]*

AM has limited fidelity. Atmospheric noises or static alter the amplitude of a carrier wave making it difficult to separate the intended amplitude modulation caused by the information signal and that which is caused by static. It is used in aircraft VHF communication radios.

Frequency Modulation (FM)

Frequency modulation (FM) is widely considered superior to AM for carrying and deciphering information on radio waves. A carrier wave modulated by FM retains its constant amplitude. However, the information signal alters the

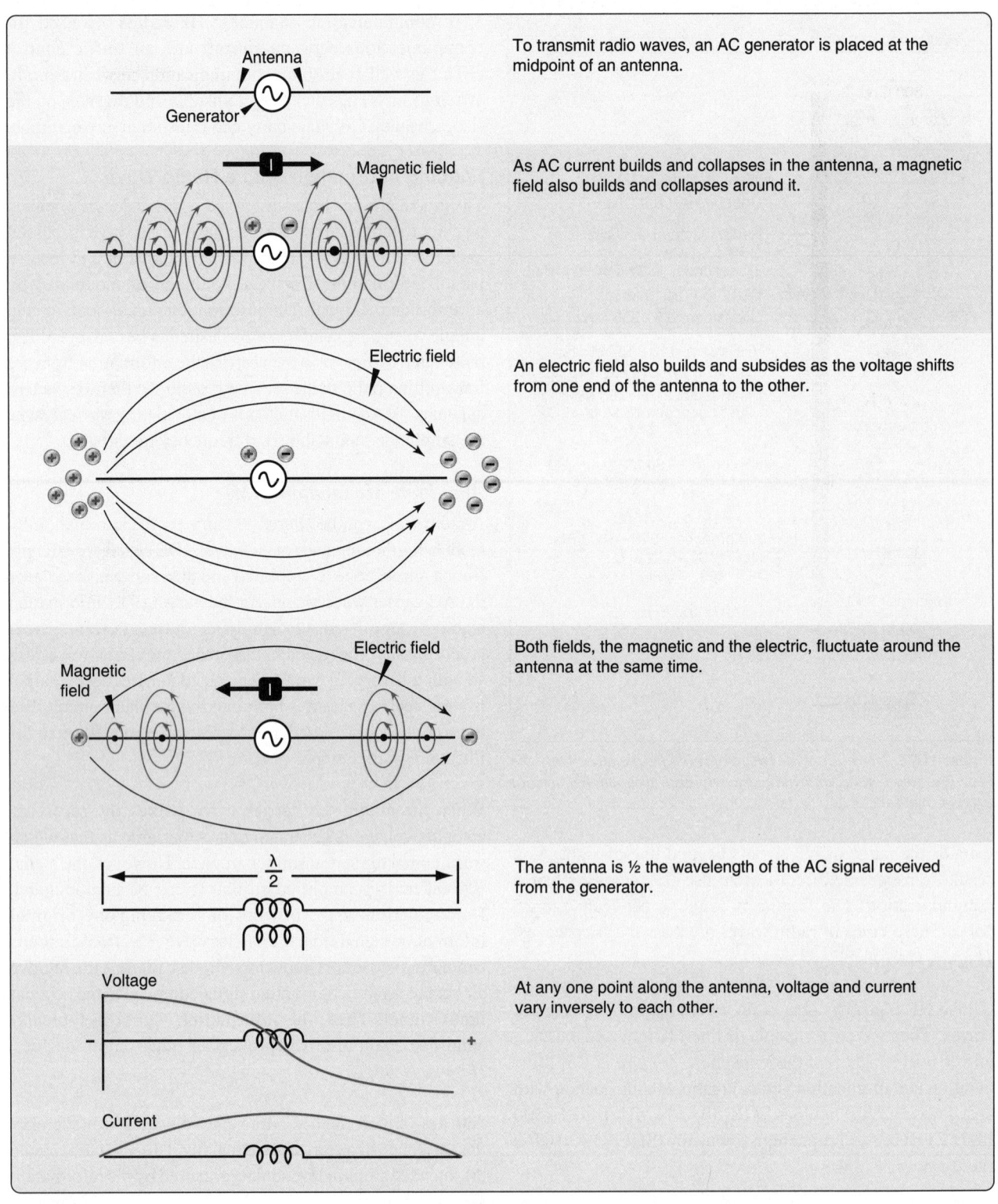

To transmit radio waves, an AC generator is placed at the midpoint of an antenna.

As AC current builds and collapses in the antenna, a magnetic field also builds and collapses around it.

An electric field also builds and subsides as the voltage shifts from one end of the antenna to the other.

Both fields, the magnetic and the electric, fluctuate around the antenna at the same time.

The antenna is ½ the wavelength of the AC signal received from the generator.

At any one point along the antenna, voltage and current vary inversely to each other.

Figure 11-76. *Radio waves are produced by applying an AC signal to an antenna. This creates a magnetic and electric field around the antenna. They build and collapse as the AC cycles. The speed at which the AC cycles does not allow the fields to completely collapse before the next fields build. The collapsing fields are then forced out into space as radio waves.*

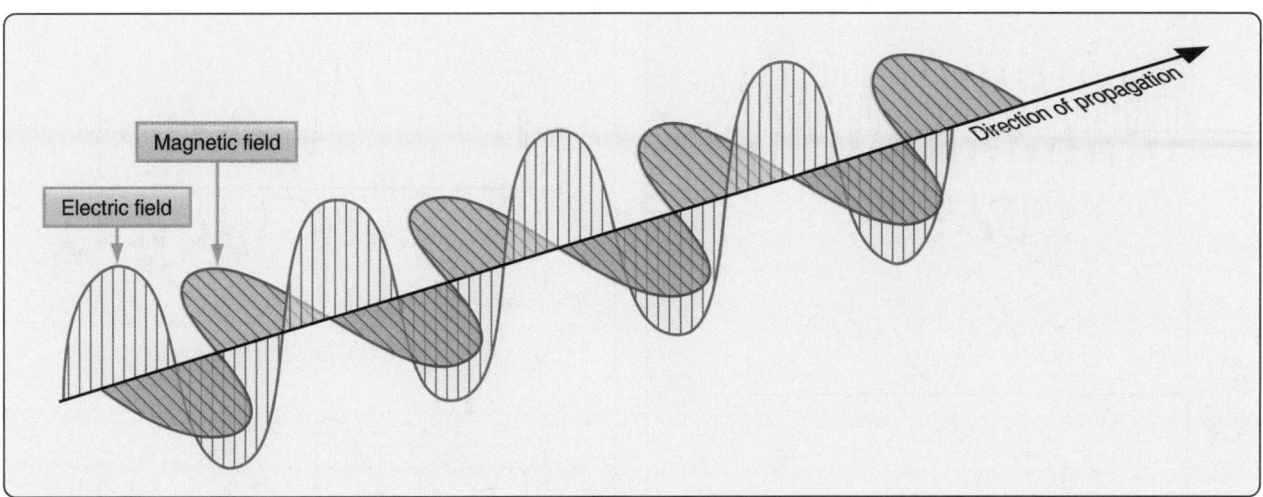

Figure 11-77. *The electric field and the magnetic field of a radio wave are perpendicular to each other and to the direction of propagation of the wave.*

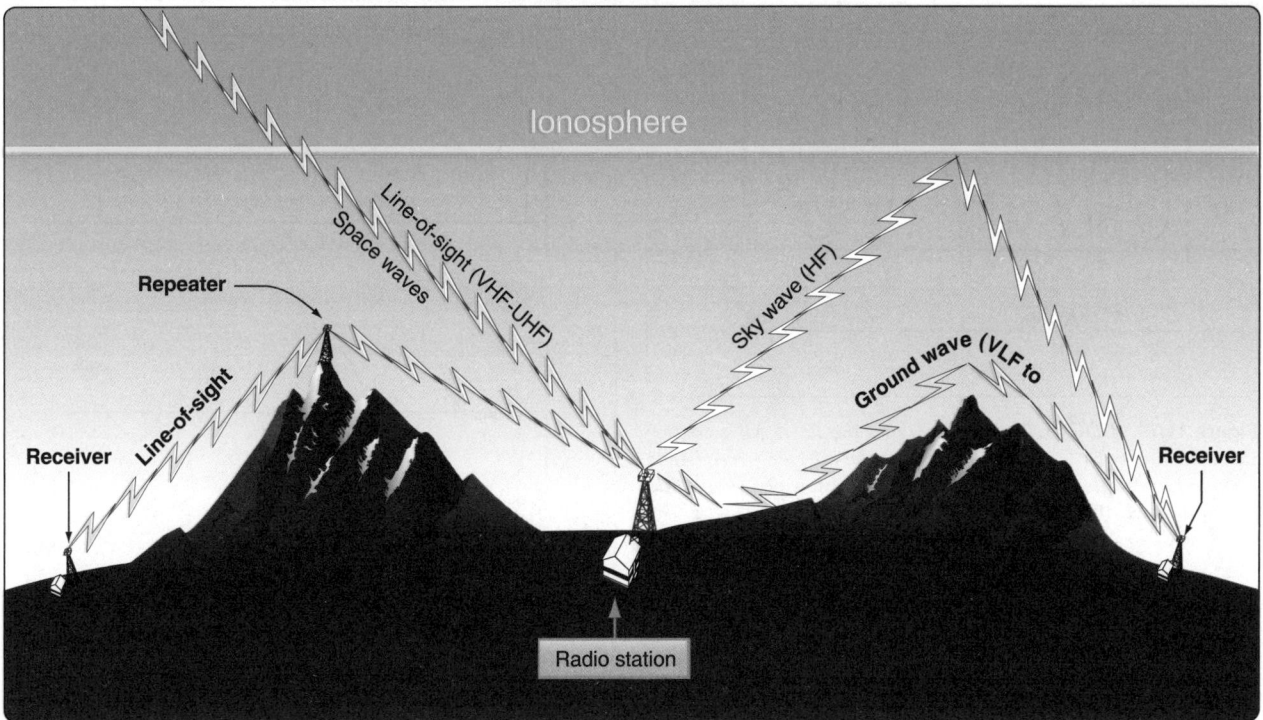

Figure 11-78. *Radio waves behave differently in the atmosphere depending in their frequency.*

frequency of the carrier wave in proportion to the strength of the signal. Thus, the signal is represented as slight variations to the normally consistent timing of the oscillations of the carrier wave. *[Figure 11-81]*

Since the transmitter oscillator output fluctuates during modulation to represent the information signal, FM bandwidth is greater than AM bandwidth. This is overshadowed by the ease with which noise and static can be removed from the FM signal. FM has a steady current flow and requires less power to produce since modulating an oscillator producing a carrier wave takes less power than modulating the amplitude of a signal using an amplifier.

Demodulation of an FM signal is similar to that of an AM receiver. The signal captured by the receiving antenna is usually amplified immediately since signal strength is lost as the wave travels through the atmosphere. Numerous circuits are used to isolate, stabilize, and remove the information from the carrier wave. The result is then amplified to drive the output device.

Single Sideband (SSB)

When two AC signals are mixed together, such as when a carrier wave is modulated by an information signal, three main frequencies result:

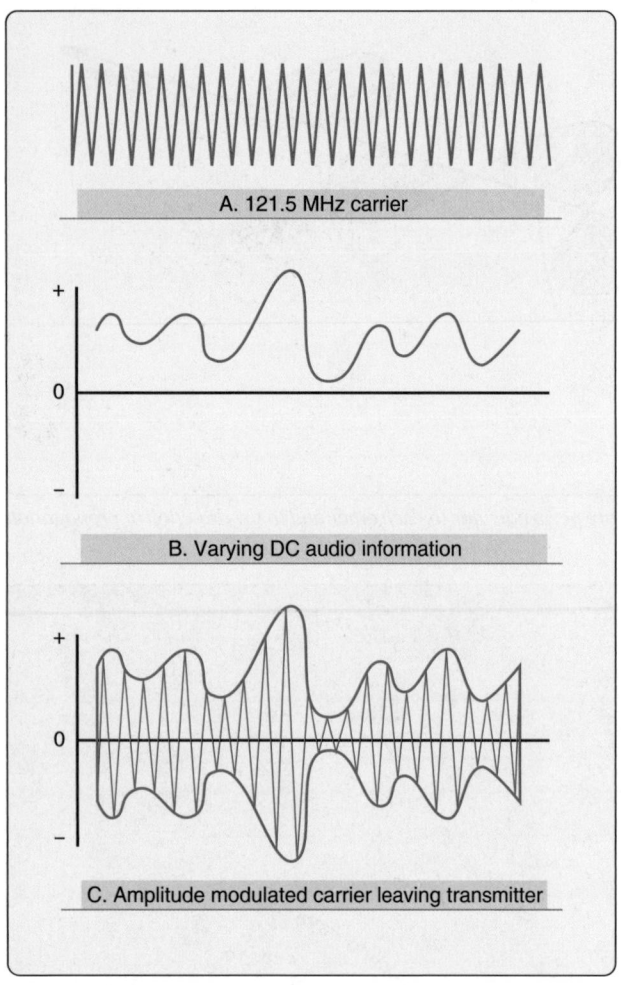

Figure 11-79. *A DC audio signal modifies the 121.5 MHz carrier wave as shown in C. The amplitude of the carrier wave (A) is changed in relation to modifier (B). This is known as amplitude modulation (AM).*

1. Original carrier wave frequency;
2. Carrier wave frequency plus the modulating frequency; and
3. Carrier wave frequency minus the modulating frequency.

Due to the fluctuating nature of the information signal, the modulating frequency varies from the carrier wave up or down to the maximum amplitude of the modulating frequency during AM. These additional frequencies on either side of the carrier wave frequency are known as sidebands. Each sideband contains the unique information signal desired to be conveyed. The entire range of the lower and upper sidebands including the center carrier wave frequency is known as bandwidth. *[Figure 11-82]*

There are a limited number of frequencies within the usable frequency ranges (i.e., LF, HF, and VHF). If different

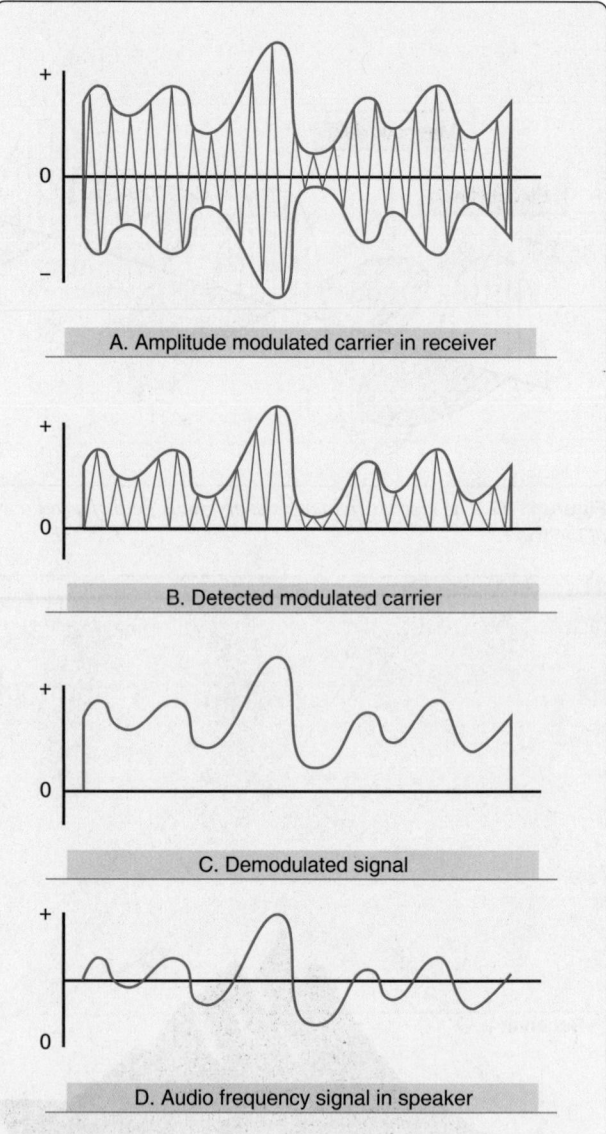

Figure 11-80. *Demodulation of a received radio signal involves separating the carrier wave from the information signal.*

broadcasts are made on frequencies that are too close together, some of the broadcast from one frequency interferes with the adjacent broadcast due to overlapping sidebands. The FCC divides the various frequency bands and issues rules for their use. Much of this allocation is to prevent interference. The spacing between broadcast frequencies is established so that a carrier wave can expand to include the upper and lower sidebands and still not interfere with a signal on an adjacent frequency.

As use of the radio frequencies increases, more efficient allocation of bandwidth is imperative. Sending information via radio waves using the narrowest bandwidth possible is the focus of engineering moving forward. At the same time, fully representing all of the desired information or increasing

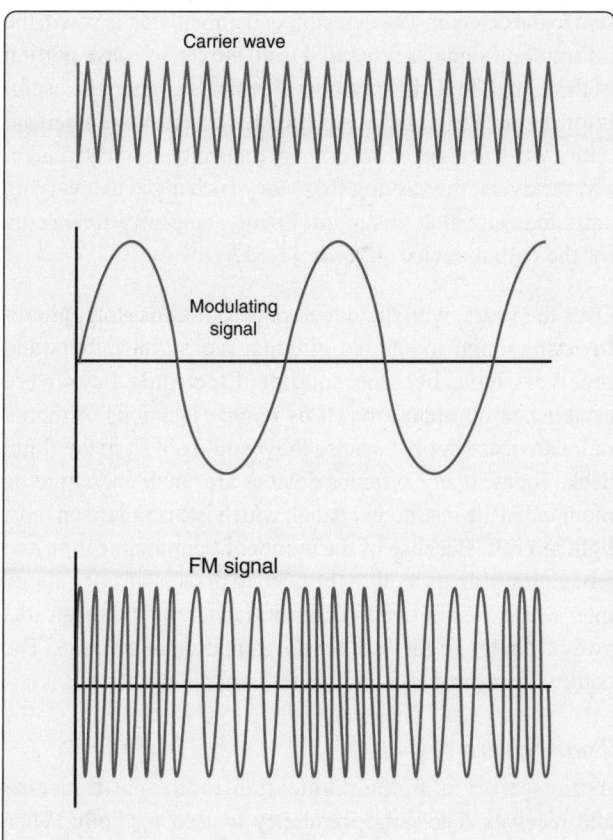

Figure 11-81. *A frequency modulated (FM) carrier wave retains the consistent amplitude of the AC sine wave. It encodes the unique information signal with slight variations to the frequency of the carrier wave. These variations are shown as space variations between the peaks and valleys of the wave on an oscilloscope.*

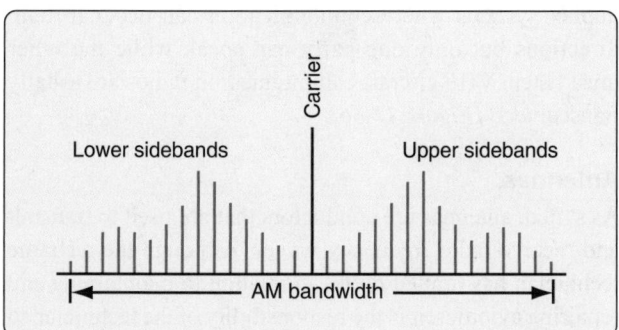

Figure 11-82. *The bandwidth of an AM signal contains the carrier wave, the carrier wave plus the information signal frequencies, and the carrier wave minus the information signal frequencies.*

the amount of information conveyed is also desired. Various methods are employed to keep bandwidth to a minimum, many of which restrict the quality or quantity of information able to be transmitted.

In lower frequency ranges, such as those used for ground wave and some sky wave broadcasts, SSB transmissions are a narrow bandwidth solution. Each sideband represents

the initial information signal in its entirety. Therefore, in an SSB broadcast, the carrier wave and either the upper or lower sidebands are filtered out. Only one sideband with its frequencies is broadcast since it contains all of the needed information. This cuts the bandwidth required in half and allows more efficient use of the radio spectrum. SSB transmissions also use less power to transmit the same amount of information over an equal distance. Many HF long-distance aviation communications are SSB. *[Figure 11-83]*

Radio Transmitters & Receivers

Radio transmitters and receivers are electronic devices that manipulate electricity resulting in the transmission of useful information through the atmosphere or space.

Transmitters

A transmitter consists of a precise oscillating circuit or oscillator that creates an AC carrier wave frequency. This is combined with amplification circuits or amplifiers. The distance a carrier wave travels is directly related to the amplification of the signal sent to the antenna.

Other circuits are used in a transmitter to accept the input information signal and process it for loading onto the carrier wave. Modulator circuits modify the carrier wave with the processed information signal. Essentially, this is all there is to a radio transmitter.

Note: Modern transmitters are highly refined devices with extremely precise frequency oscillation and modulation. The circuitry for controlling, filtering, amplifying, modulating, and oscillating electronic signals can be complex.

A transmitter prepares and sends signals to an antenna that, in the process described above, radiates the waves out into the atmosphere. A transmitter with multiple channel (frequency) capability contains tuning circuitry that enables the user to select the frequency upon which to broadcast. This adjusts

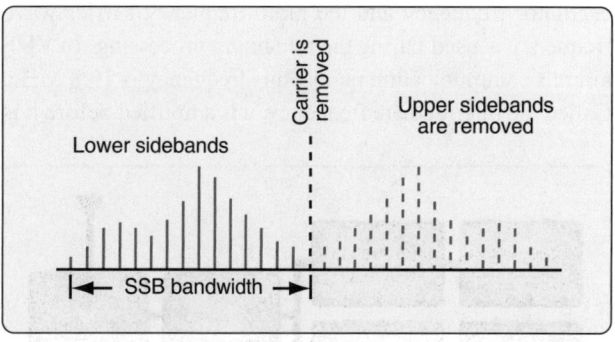

Figure 11-83. *The additional frequencies above and below the carrier wave produced during modulation with the information signal are known as sidebands. Each sideband contains the unique information of the information signal and can be transmitted independent of the carrier wave and the other sideband.*

the oscillator output to the precise frequency desired. It is the oscillator frequency that is being tuned. *[Figure 11-84]*

As shown in *Figure 11-84*, most radio transmitters generate a stable oscillating frequency and then use a frequency multiplier to raise the AC to the transmitting frequency. This allows oscillation to occur at frequencies that are controllable and within the physical working limits of the crystal in crystal-controlled oscillators.

Receivers

Antennas are simply conductors of lengths proportional to the wavelength of the oscillated frequency put out by the transmitter. An antenna captures the desired carrier wave as well as many other radio waves that are present in the atmosphere. A receiver is needed to isolate the desired carrier wave with its information. The receiver also has circuitry to separate the information signal from the carrier wave. It prepares it for output to a device, such as speakers or a display screen. The output is the information signal originally introduced into the transmitter.

A common receiver is the super heterodyne receiver. As with any receiver, it must amplify the desired radio frequency captured by the antenna since it is weak from traveling through the atmosphere. An oscillator in the receiver is used to compare and select the desired frequency out of all of the frequencies picked up by the antenna. The undesired frequencies are sent to ground.

A local oscillator in the receiver produces a frequency that is different than the radio frequency of the carrier wave. These two frequencies are mixed in the mixer. Four frequencies result from this mixing. They are the radio frequency, the local oscillator frequency, and the sum and difference of these two frequencies. The sum and difference frequencies contain the information signal.

The frequency that is the difference between the local oscillator frequency and the radio frequency carrier wave frequency is used during the remaining processing. In VHF aircraft communication radios, this frequency is 10.8 MHz. Called the intermediate frequency, it is amplified before it is sent to the detector. The detector, or demodulator, is where the information signal is separated from the carrier wave portion of the signal. In AM, since both sidebands contain the useful information, the signal is rectified leaving just one sideband with a weak version of the original transmitter input signal. In FM receivers, the varying frequency is changed to a varying amplitude signal at this point. Finally, amplification occurs for the output device. *[Figure 11-85]*

Over the years, with the development of transistors, micro-transistors, and integrated circuits, radio transmitters and receivers have become smaller. Electronic bays were established on older aircraft as remote locations to mount radio devices simply because they would not fit in the flight deck. Today, many avionics devices are small enough to be mounted in the instrument panel, which is customary on most light aircraft. Because of the number of communication and navigation aids, as well as the need to present an uncluttered interface to the pilot, most complicated aircraft retain an area away from the flight deck for the mounting of avionics. The control heads of these units remain on the flight deck.

Transceivers

A transceiver is a communication radio that transmits and receives. The same frequency is used for both. When transmitting, the receiver does not function. The push to talk (PTT) switch blocks the receiving circuitry and allows the transmitter circuitry to be active. In a transceiver, some of the circuitry is shared by the transmitting and receiving functions of the device. So is the antenna. This saves space and the number of components used. Transceivers are half duplex systems where communication can occur in both directions but only one party can speak while the other must listen. VHF aircraft communication radios are usually transceivers. *[Figure 11-86]*

Antennas

As stated, antennas are conductors that are used to transmit and receive radio frequency waves. Although the airframe technician has limited duties in relation to maintaining and repairing avionics, it is the responsibility of the technician to install, inspect, repair, and maintain aircraft radio antennas.

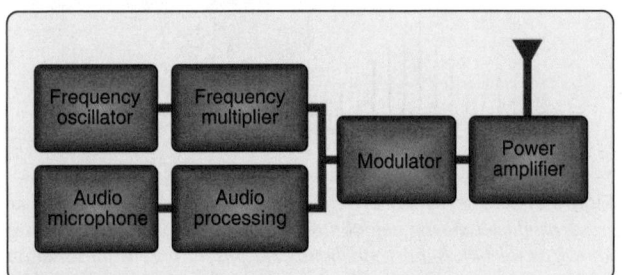

Figure 11-84. *Block diagram of a basic radio transmitter.*

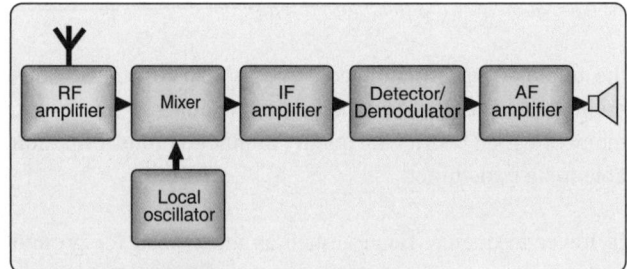

Figure 11-85. *The basic stages used in a receiver to produce an output from a radio wave.*

Three characteristics are of major concern when considering antennas:

1. Length.
2. Polarization.
3. Directivity.

The exact shape and material from which an antenna is made can alter its transmitting and receiving characteristics. Also note that some non-metallic aircraft have antennas imbedded into the composite material as it is built up.

Length

When an AC signal is applied to an antenna, it has a certain frequency. There is a corresponding wavelength for that frequency. An antenna that is half the length of this wavelength is resonant. During each phase of the applied AC, all voltage and current values experience the full range of their variability. As a result, an antenna that is half the wavelength of the corresponding AC frequency is able to allow full voltage and full current flow for the positive phase of the AC signal in one direction. The negative phase of the full AC sine wave is accommodated by the voltage and current simply changing direction in the conductor. Thus, the applied AC frequency flows through its entire wavelength, first in one direction and then in the other. This produces the strongest signal to be radiated by the transmitting antenna. It also facilitates capture of the wave and maximum induced voltage in the receiving antenna. *[Figure 11-87]*

Most radios, especially communication radios, use the same antenna for transmitting and receiving. Multichannel radios could use a different length antenna for each frequency, however, this is impractical. Acceptable performance can exist from a single antenna half the wavelength of a median frequency. This antenna can be made effectively shorter by placing a properly rated capacitor in series with the transmission line from the transmitter or receiver. This electrically shortens the resonant circuit of which the antenna is a part. An antenna may be electrically lengthened by adding an inductor in the circuit. Adjusting antenna length in this fashion allows the use of a single antenna for multiple frequencies in a narrow frequency range.

Many radios use a tuning circuit to adjust the effective length of the antenna to match the wavelength of the desired frequency. It contains a variable capacitor and an inductor connected in parallel in a circuit. Newer radios use a more efficient tuning circuit. It uses switches to combine frequencies from crystal controlled circuits to create a resonant frequency that matches the desired frequency. Either way, the physical antenna length is a compromise when using a multichannel communication or navigation device that must be electronically tuned for the best performance.

A formula can be used to find the ideal length of a half wavelength antenna required for a particular frequency as

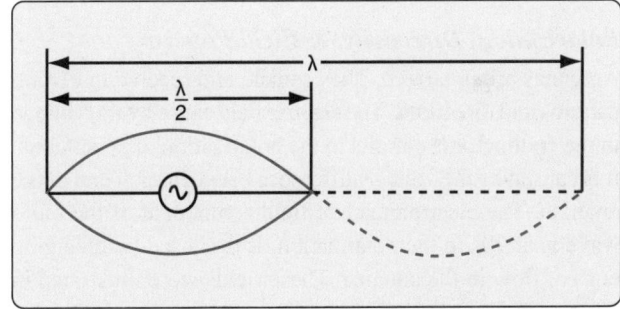

Figure 11-87. *An antenna equal to the full length of the applied AC frequency wavelength would have the negative cycle current flow along the antenna as shown by the dotted line. An antenna that is ½ wavelength allows current to reverse its direction in the antenna during the negative cycle. This results in low current at the ends of the ½ wavelength antenna and high current in the center. As energy radiates into space, the field is strongest 90° to the antenna where the current flow is strongest.*

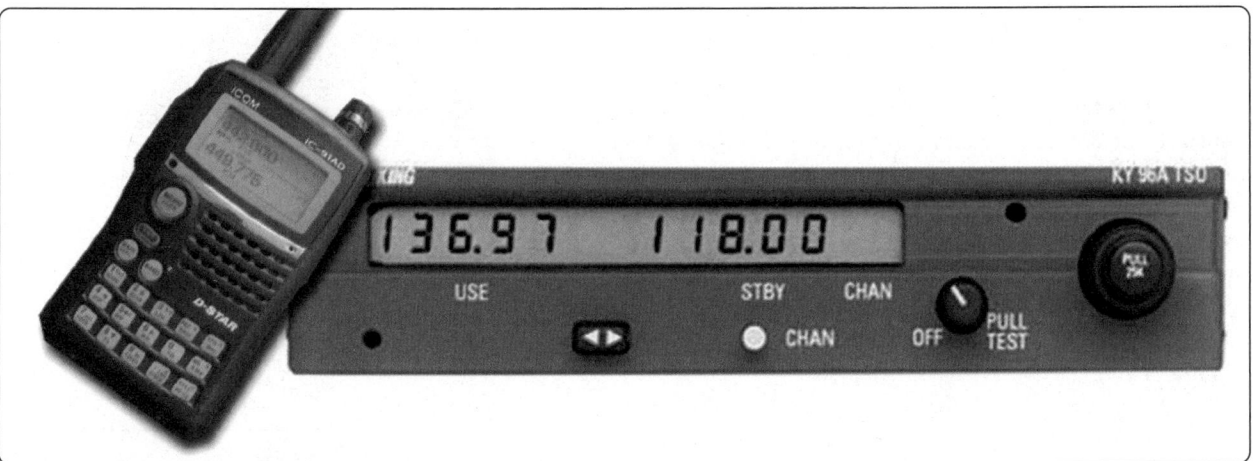

Figure 11-86. *VHF aircraft communication transceivers.*

follows:

$$\text{Antenna Length (feet)} = \frac{468}{\text{F MHz}}$$

The formula is derived from the speed of propagation of radio waves, which is approximately 300 million meters per second. It takes into account the dielectric effect of the air at the end of an antenna that effectively shortens the length of the conductor required.

VHF radio frequencies used by aircraft communication radios are 118–136.975 MHz. The corresponding half wavelengths of these frequencies are 3.96 – 3.44 feet (47.5–41.2 inches). Therefore, VHF antennas are relatively long. Antennas one-quarter of the wavelength of the transmitted frequency are often used. This is possible because when mounted on a metal fuselage, a ground plane is formed and the fuselage acts as the missing one-quarter length of the half wavelength antenna. This is further discussed in the following antenna types section.

Polarization, Directivity, & Field Pattern

Antennas are polarized. They radiate and receive in certain patterns and directions. The electric field cause by the voltage in the conductor is parallel to the polarization of an antenna. It is caused by the voltage difference between each end of the antenna. The electromagnetic field component of the radio wave is at 90° to the polarization. It is caused by changing current flow in the antenna. These fields were illustrated in *Figure 11-76* and *11-77*. As radio waves radiate out from the antenna they propagate in a specific direction and in a specific pattern. This is the antenna field. The orientation of the electric and electromagnetic fields remains at 90° to each other but radiate from antenna with varying strength in different directions. The strength of the radiated field varies depending on the type of antenna and the angular proximity to it. All antennas, even those that are omnidirectional, radiate a stronger signal in some direction compared to other directions. This is known as the antenna field directivity.

Receiving antennas with the same polarization as the transmitting antenna generate the strongest signal. A vertically polarized antenna is mounted up and down. It radiates waves out from it in all directions. To receive the strongest signal from these waves, the receiving antenna should also be positioned vertically so the electromagnetic component of the radio wave can cross it at as close to a 90° angle as possible for most of the possible proximities. *[Figure 11-88]*

Horizontally polarized antennas are mounted side to side (horizontally). They radiate in a donut-like field. The

strongest signals come from, or are received at, 90° to the length of the antenna. There is no field generated off of the end of the antenna. *Figure 11-89* illustrates the field produced by a horizontally polarized antenna.

Many vertical and horizontal antennas on aircraft are mounted at a slight angle off plane. This allows the antenna to receive a weak signal rather than no signal at all when the polarization of the receiving antenna is not identical to the transmitting antenna. *[Figure 11-90]*

Types

There are three basic types of antennas used in aviation:

1. Dipole antenna.

2. Marconi antenna.

3. Loop antenna.

Dipole Antenna

The dipole antenna is the type of antenna referred to in the discussion of how a radio wave is produced. It is a conductor, the length of which is approximately equal to half the wavelength of the transmission frequency. This sometimes is referred to as a Hertz antenna. The AC transmission current is fed to a dipole antenna in the center. As the current alternates, current flow is greatest in the middle of the antenna and gradually less as it approaches the ends. Then, it changes direction and flows the other way. The result is that the largest electromagnetic field is in the middle of the antenna and the strongest radio wave field is perpendicular to the length of the antenna. Most dipole antennas in aviation are horizontally polarized.

A common dipole antenna is the V-shaped VHF navigation

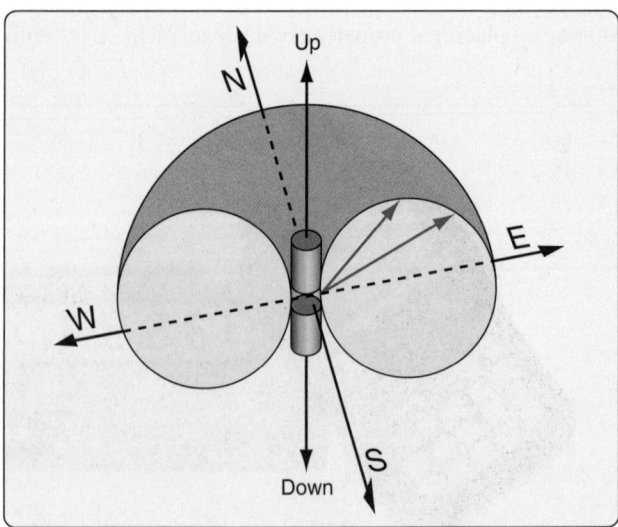

Figure 11-88. *A vertically polarized antenna radiates radio waves in a donut-like pattern in all directions.*

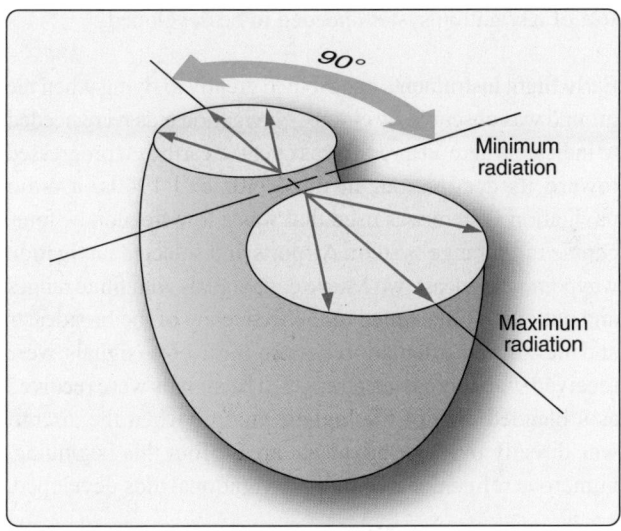

Figure 11-89. *A horizontally polarized antenna radiates in a donut-like pattern. The strongest signal is at 90° to the length of the conductor.*

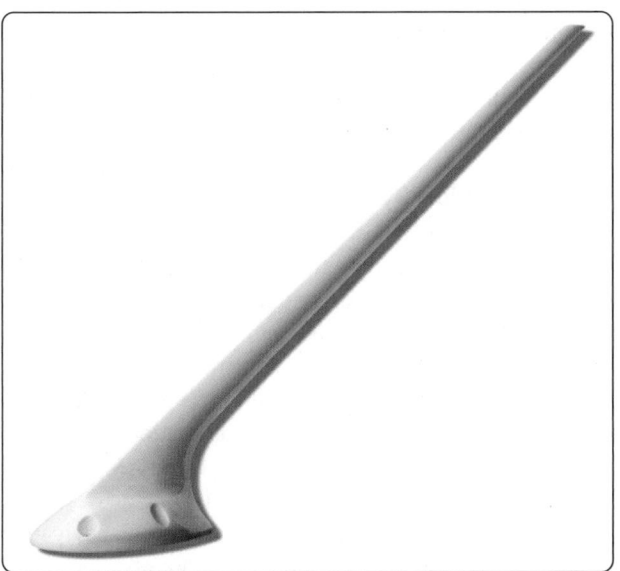

Figure 11-90. *Many antenna are canted for better reception.*

antenna, known as a VOR antenna, found on numerous aircraft. Each arm of the V is one-fourth wavelength creating a half wave antenna which is fed in the center. This antenna is horizontally polarized. For a dipole receiving antenna, this means it is most sensitive to signals approaching the antenna from the sides rather than head-on in the direction of flight. *[Figure 11-91]*

Marconi Antenna

A Marconi antenna is a one-fourth wave antenna. It achieves the efficiency of a half wave antenna by using the mounting surface of the conductive aircraft skin to create the second one-fourth wavelength. Most aircraft VHF communications antennas are Marconi antennas. They are vertically polarized

and create a field that is omnidirectional. On fabric skinned aircraft, the ground plane that makes up the second one-fourth wavelength of the antenna must be fashioned under the skin where the Marconi antenna is mounted. This can be done with thin aluminum or aluminum foil. Sometimes four or more wires are extended under the skin from the base of the vertical antenna that serve as the ground plane. This is enough to give the antenna the proper conductive length. The same practice is also utilized on ground-based antennas. *[Figure 11-92]*

Loop Antenna

The third type of antenna commonly found on aircraft is the loop antenna. When the length of an antenna conductor is fashioned into a loop, its field characteristics are altered significantly from that of a straight-half wavelength antenna. It also makes the antenna more compact and less prone to damage.

Used as a receiving antenna, the loop antenna's properties are highly direction-sensitive. A radio wave intercepting the loop directly broadside causes equal current flow in both sides of the loop. However, the polarity of the current flows is opposite each other. This causes them to cancel out and produce no signal. When a radio wave strikes the loop antenna in line with the plane of the loop, current is generated first in one side, and then in the other side. This causes the current flows to have different phases and the strongest signal can be generated from this angle. The phase difference (and strength) of the generated current varies proportionally to the angle at which the radio wave strikes the antenna loop. This is useful and is discussed further in the section on automatic direction finder (ADF) navigational aids. *[Figure 11-93]*

Transmission Lines

Transmitters and receivers must be connected to their antenna(s) via conductive wire. These transmission lines

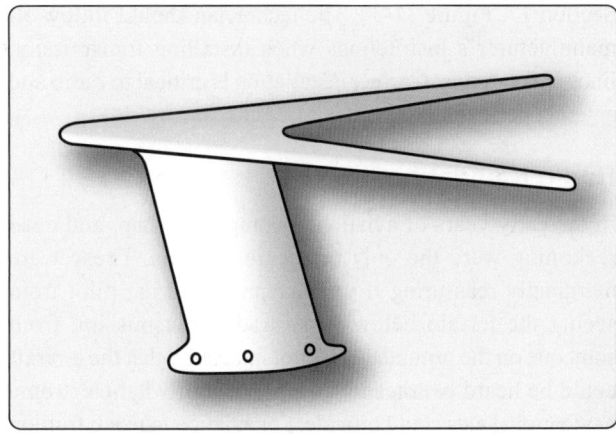

Figure 11-91. *The V-shaped VOR navigation antenna is a common dipole antenna.*

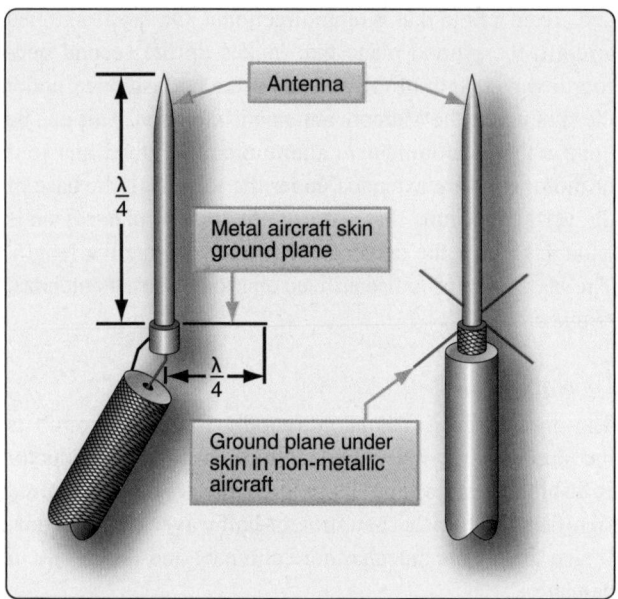

Figure 11-92. *On a metal-skinned aircraft, a ¼ wavelength Marconi antenna is used. The skin is the ground plane that creates the 2nd quarter of the antenna required for resonance (left). On a non-metallic-skinned aircraft, wires, conductive plates or strips equal in length to the antenna must be installed under the skin to create the ground plane (right).*

are coaxial cable, also known as coax. Coax consists of a center wire conductor surrounded by a semirigid insulator. Surrounding the wire and insulator material is a conductive, braided cover that runs the length of the cable. Finally, a waterproof covering is set around the braided shield to protect the entire assembly from the elements. The braided cover in the coax shields the inner conductor from any external fields. It also prevents the fields generated by the internal conductor from radiating. For optimum performance, the impedance of the transmission line should be equal to the impedance of the antenna. In aviation antenna applications, this is often approximately 50 ohms. *[Figure 11-94]* Special connectors are used for coaxial cable. A variety can be seen in Advisory Circular (AC) 43.13-1b, Chapter 11, Section 17, Figure 11-37. The technician should follow all manufacturer's instructions when installing transmission lines and antenna. Correct installation is critical to radio and antenna performance.

Radio Navigation

In the early years of aviation, a compass, a map, and dead reckoning were the only navigational tools. These were marginally reassuring if weather prevented the pilot from seeing the terrain below. Voice radio transmission from someone on the ground to the pilot indicating that the aircraft could be heard overhead was a preview of what electronic navigational aids could provide. For aviation to reach fruition as a safe, reliable, consistent means of transportation, some

sort of navigation system needed to be developed.

Early flight instruments contributed greatly to flying when the ground was obscured by clouds. Navigation aids were needed to indicate where an aircraft was over the earth as it progressed toward its destination. In the 1930s and 1940s, a radio navigation system was used that was a low frequency, four-course radio range system. Airports and selected navigation waypoints broadcast two Morse code signals with finite ranges and patterns. Pilots tuned to the frequency of the broadcasts and flew in an orientation pattern until both signals were received with increasing strength. The signals were received as a blended tone of the highest volume when the aircraft was directly over the broadcast area. From this beginning, numerous refinements to radio navigational aids developed.

Radio navigation aids supply the pilot with intelligence that maintains or enhances the safety of flight. As with communication radios, navigational aids are avionics devices, the repair of which must be carried out by trained technicians at certified repair stations. However, installation, maintenance and proper functioning of the electronic units, as well as their antennas, displays, and any other peripheral devices, are the responsibilities of the airframe technician.

VOR Navigation System

One of the oldest and most useful navigational aids is the very high frequency omni-directional range (VOR) navigation system. The four main components of a typical system are: a receiver, a visual indicator, a frequency selector (controller or control panel), and antennas. The system was constructed after WWII and is still in use today. It consists of thousands of land-based transmitter stations, or VORs, that communicate with radio receiving equipment on board aircraft. Many of the VORs are located along airways. The Victor airway system is built around the VOR navigation system. Ground VOR transmitter units are also located at airports where they are known as TVOR (terminal VOR). The U.S. Military has a navigational system known as TACAN that operates similarly to the VOR system. Sometimes VOR and TACAN transmitters share a location. These sites are known as VORTACs.

The position of all VORs, TVORs, and VORTACs are marked on aeronautical charts along with the name of the station, the frequency to which an airborne receiver must be tuned to use the station, and a Morse code designation for the station. Some VORs also broadcast a voice identifier on a separate frequency that is included on the chart. *[Figure 11-95]*

VOR uses VHF radio waves (108–117.95 MHz) with 50 kHz separation between each channel. This keeps atmospheric interference to a minimum but limits the VOR to line-of-

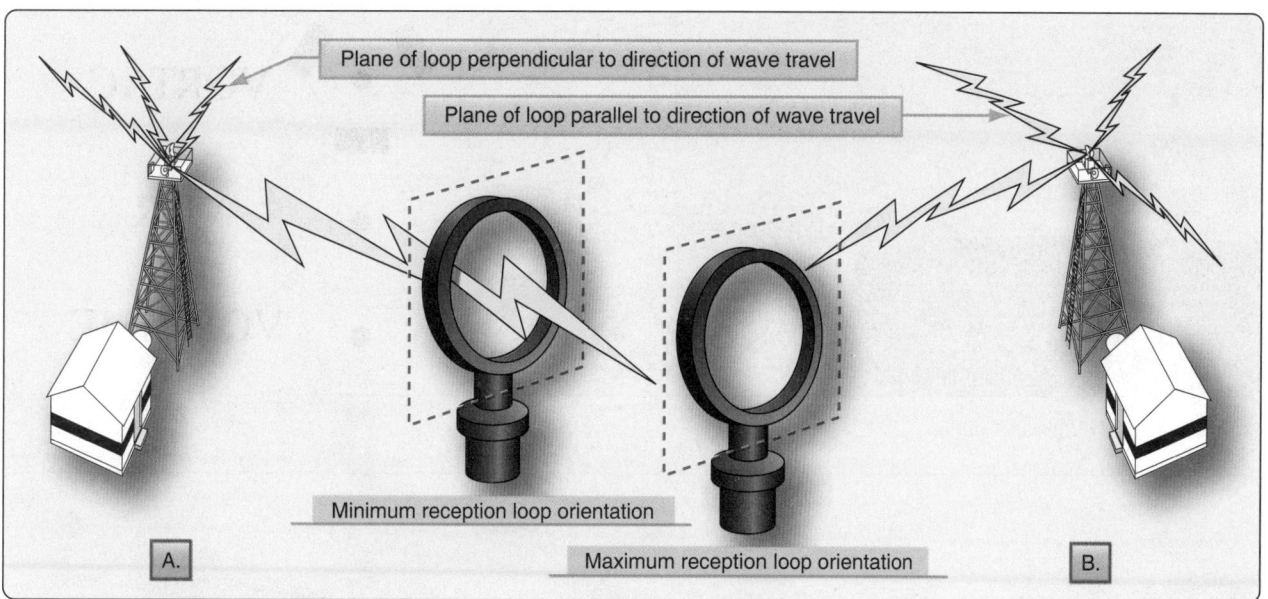

Figure 11-93. *A loop antenna is highly direction-sensitive. A signal origin perpendicular or broadside to the loop creates a weak signal (A). A signal origin parallel or in the plain of the loop creates a strong signal (B).*

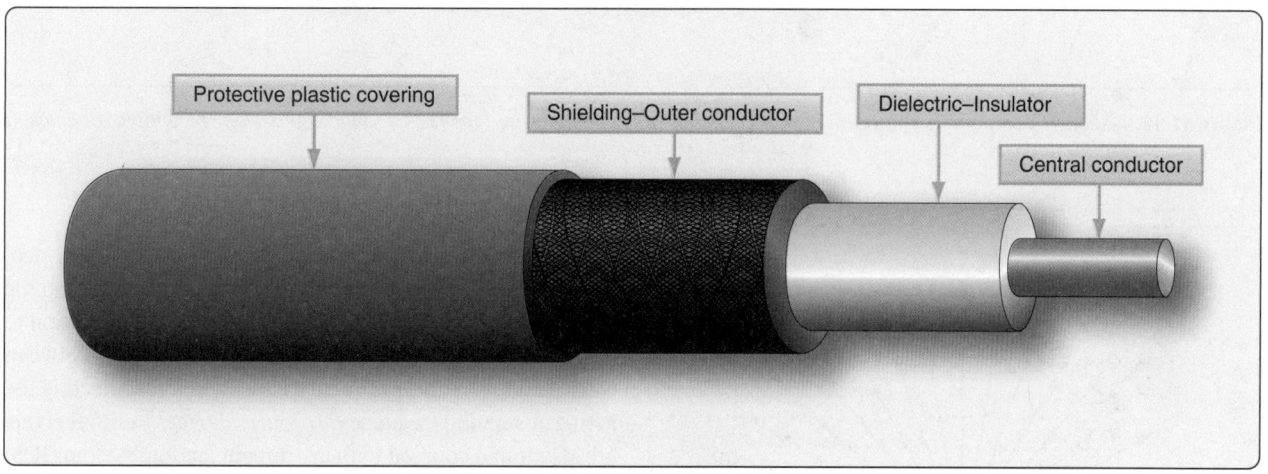

Figure 11-94. *Coaxial cable is used as the transmission line between an antenna and its transmitters and/or receiver.*

sight usage. To receive VOR VHF radio waves, generally a V-shaped, horizontally polarized, bi-pole antenna is used. Other type antennas are also certified. Follow the manufacturer's instructions for installation location. *[Figure 11-96]*

The signals produced by a VOR transmitter propagate 360° from the unit and are used by aircraft to navigate to and from the station with the help of an onboard VOR receiver and display instruments. A pilot is not required to fly a pattern to intersect the signal from a VOR station since it propagates out in every direction. The radio waves are received as long as the aircraft is in range of the ground unit and regardless of the aircraft's direction of travel. *[Figure 11-97]*

A VOR transmitter produces two signals that a receiver

on board an aircraft uses to locate itself in relation to the ground station. One signal is a reference signal. The second is produced by electronically rotating a variable signal. The variable signal is in phase with the reference signal when at magnetic north but becomes increasingly out of phase as it is rotated to 180°. As it continues to rotate to 360° (0°), the signals become increasingly in phase until they are in phase again at magnetic north. The receiver in the aircraft deciphers the phase difference and determines the aircraft's position in degrees from the VOR ground based unit. *[Figure 11-98]*

Most aircraft carry a dual VOR receiver. Sometimes, the VOR receivers are part of the same avionics unit as the VHF communication transceiver(s). These are known as NAV/COM radios. Internal components are shared since frequency bands for each are adjacent. *[Figure 11-99]*

11-41

Figure 11-95. *A VOR ground station.*

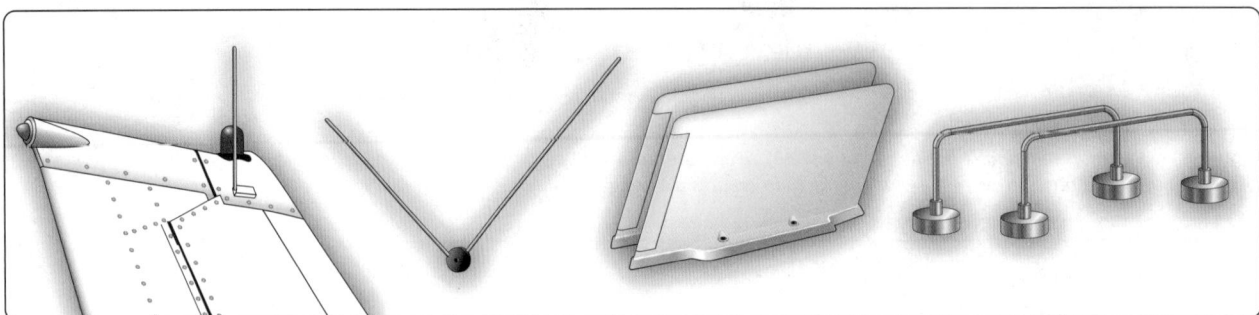

Figure 11-96. *V-shaped, horizontally polarized, bi-pole antennas are commonly used for VOR and VOR/glideslope reception. All antenna shown are VOR/glideslope antenna.*

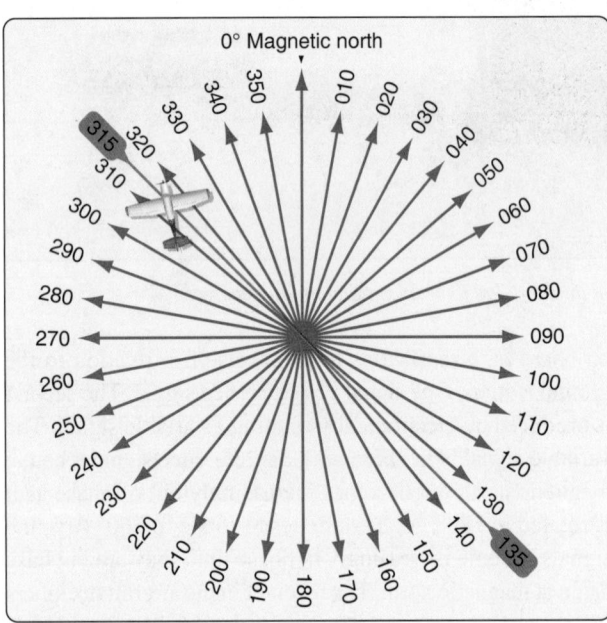

Figure 11-97. *A VOR transmitter produces signals for 360° radials that an airborne receiver uses to indicate the aircraft's location in relation to the VOR station regardless of the aircraft's direction of flight. The aircraft shown is on the 315° radial even though it does not have a heading of 315°.*

Large aircraft may have two dual receivers and even dual antennas. Normally, one receiver is selected for use and the second is tuned to the frequency of the next VOR station to be encountered en route. A means for switching between NAV 1 and NAV 2 is provided as is a switch for selecting the active or standby frequency. *[Figure 11-100]* VOR receivers are also found coupled with instrument landing system (ILS) receivers and glideslope receivers.

A VOR receiver interprets the bearing in degrees to (or from) the VOR station where the signals are generated. It also produces DC voltage to drive the display of the deviation from the desired course centerline to (or from) the selected station. Additionally, the receiver decides whether or not the aircraft is flying toward the VOR or away from it. These items can be displayed a number of different ways on various instruments. Older aircraft are often equipped with a VOR gauge dedicated to display only VOR information. This is also called an omni-bearing selector (OBS) or a course deviation indicator (CDI). *[Figure 11-101]*

The CDI linear indicator remains essentially vertical but moves left and right across the graduations on the instrument face to show deviation from being on course. Each graduation represents 2°. The OBS knob rotates the azimuth ring. When

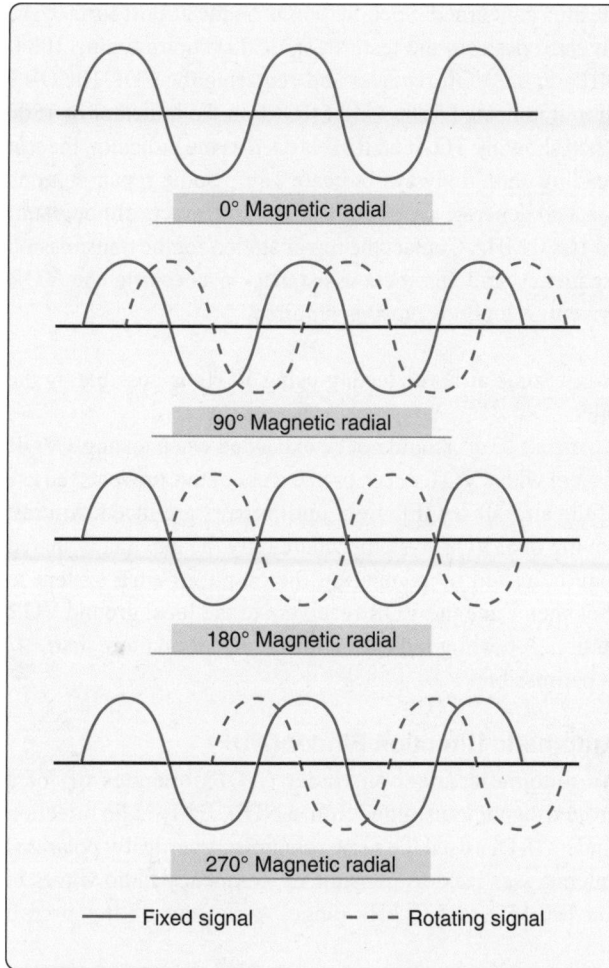

Figure 11-100. *An airliner VOR control head with two independent NAV receivers each with an active and standby tuning circuit controlled by a toggle switch.*

Figure 11-98. *The phase relationship of the two broadcast VOR signals.*

Figure 11-99. *A NAV/COM receiver typically found in light aircraft.*

in range of a VOR, the pilot rotates the OBS until the course deviation indicator centers. For each location of an aircraft, the OBS can be rotated to two positions where the CDI will center. One produces an arrow in the TO window of the gauge indicating that the aircraft is traveling toward the VOR station. The other selectable bearing is 180° from this. When chosen, the arrow is displayed in the FROM window indicating the aircraft is moving away from the VOR on the course selected. The pilot must steer the aircraft to the heading with the CDI centered to fly directly to or from the VOR. The displayed VOR information is derived from deciphering the phase relationship between the two simultaneously transmitted signals from the VOR ground station. When power is lost or the VOR signal is weak or interrupted, a NAV

warning flag comes into view. *[Figure 11-101]*

A separate gauge for the VOR information is not always used. As flight instruments and displays have evolved, VOR navigation information has been integrated into other instruments displays, such as the radio magnetic indicator (RMI), the HSI, an EFIS display or an electronic attitude director indicator (EADI). Flight management systems and automatic flight control systems are also made to integrate VOR information to automatically control the aircraft on its planned flight segments. Flat panel MFDs integrate VOR information into moving map presentations and other selected displays. The basic information of the radial bearing in degrees, course deviation indication, and to/from information remains unchanged however. *[Figure 11-102]*

At large airports, an instrument landing system (ILS) guides the aircraft to the runway while on an instrument landing approach. The aircraft's VOR receiver is used to interpret the radio signals. It produces a more sensitive course deviation indication on the same instrument display as the VOR CDI display. This part of the ILS is known as the localizer and is discussed below. While tuned to the ILS localizer frequency, the VOR circuitry of the VOR/ILS receiver is inactive.

It is common at VOR stations to combine the VOR transmitter with distance measuring equipment (DME) or a nondirectional beacon (NDB) such as an ADF transmitter and antenna. When used with a DME, pilots can gain an exact fix on their location using the VOR and DME together. Since the VOR indicates the aircraft's bearing to the VOR transmitter and a co-located DME indicates how far away the station is, this relieves the pilot from having to fly over the station to

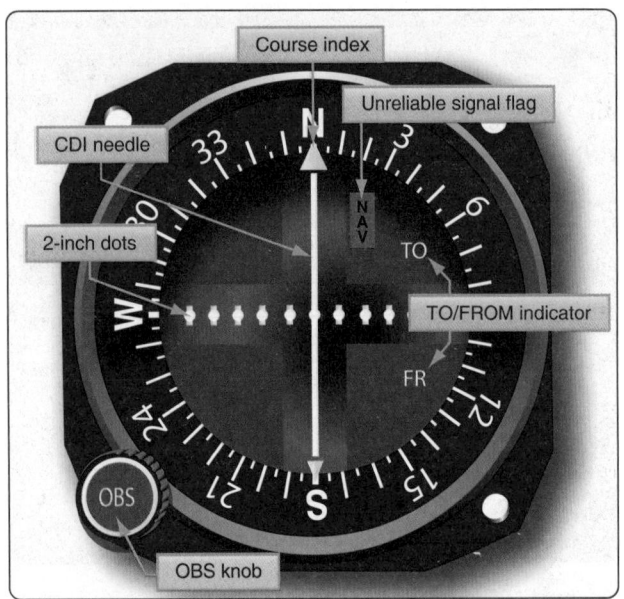

Figure 11-101. *A traditional VOR gauge, also known as a course deviation indicator (CDI) or an omni-bearing selector (OBS).*

the area concerned. Specific points on the airport surface are given to perform the test. Most VOTs require tuning 108.0 MHz on the VOR receiver and centering the CDI. The OBS should indicate 0° showing FROM on the indicator or 180° when showing TO. If an RMI is used as the indicator, the test heading should always indicate 180°. Some repair stations can also generate signals to test VOR receivers although not on 108.0 MHz. Contact the repair station for the transmission frequency and for their assistance in checking the VOR system. A logbook entry is required.

Note: Some airborne testing using VOTs is possible by the pilot.

An error of ±4° should not be exceeded when testing a VOR system with a VOT. An error in excess of this prevents the use of the aircraft for IFR fight until repairs are made. Aircraft having dual VOR systems where only the antenna is shared may be tested by comparing the output of each system to the other. Tune the VOR receivers to the local ground VOR station. A bearing indication difference of no more than ±4° is permissible.

Automatic Direction Finder (ADF)

An automatic direction finder (ADF) operates off of a ground signal transmitted from a NDB. Early radio direction finders (RDF) used the same principle. A vertically polarized antenna was used to transmit LF frequency radio waves in the 190 kHz to 535 kHz range. A receiver on the aircraft

know with certainty their location. These navigational aids are discussed separately in the following sections.

Functional accuracy of VOR equipment is critical to the safety of flight. VOR receivers are operationally tested using VOR test facilities (VOT). These are located at numerous airports that can be identified in the Airport Facilities Directory for

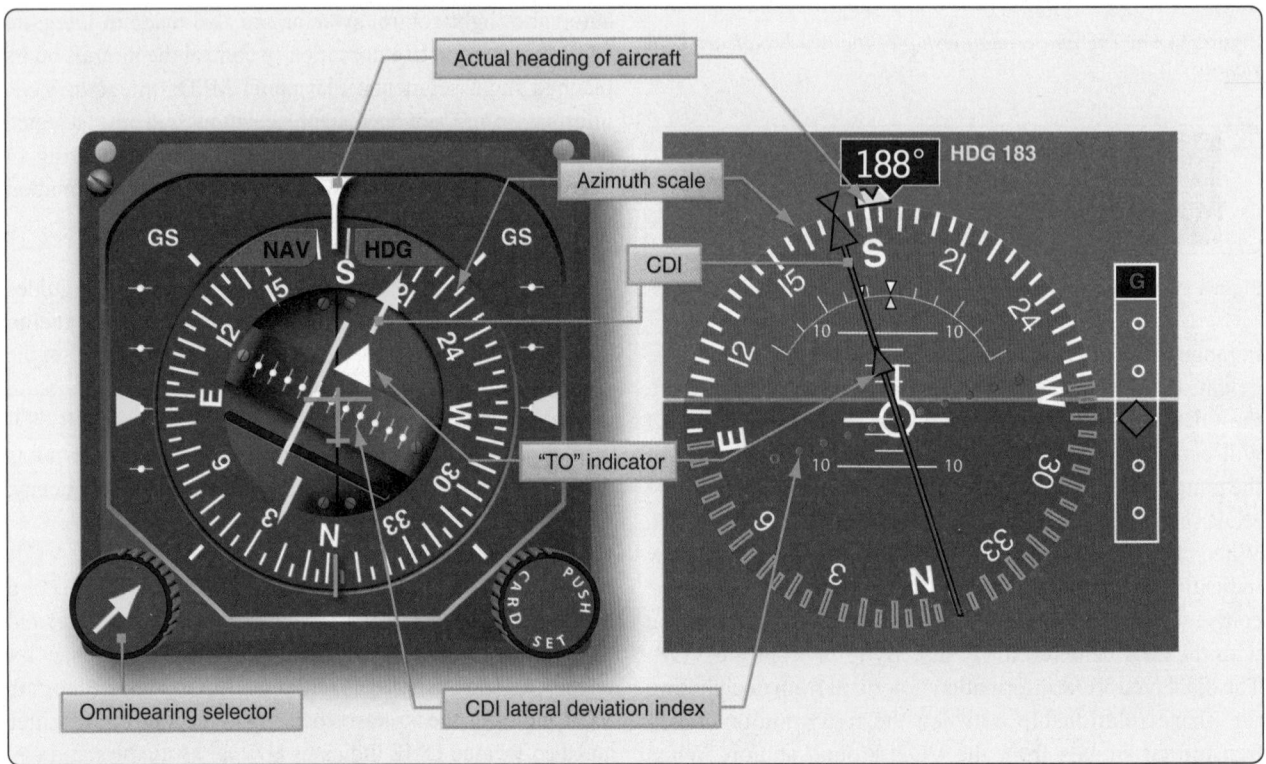

Figure 11-102. *A mechanical HSI (left) and an electronic HSI (right) both display VOR information.*

was tuned to the transmission frequency of the NDB. Using a loop antenna, the direction to (or from) the antenna could be determined by monitoring the strength of the signal received. This was possible because a radio wave striking a loop antenna broadside induces a null signal. When striking it in the plane of the loop, a much stronger signal is induced. The NDB signals were modulated with unique Morse code pulses that enabled the pilot to identify the beacon to which they were navigating.

With RDF systems, a large rigid loop antenna was installed inside the fuselage of the aircraft. The broadside of the antenna was perpendicular to the aircraft's longitudinal axis. The pilot listened for variations in signal strength of the LF broadcast and maneuvered the aircraft so a gradually increasing null signal was maintained. This took them to the transmitting antenna. When over flown, the null signal gradually faded as the aircraft became farther from the station. The increasing or decreasing strength of the null signal was the only way to determine if the aircraft was flying to or from the NDB. A deviation left or right from the course caused the signal strength to sharply increase due to the loop antenna's receiving properties.

The ADF improved on this concept. The broadcast frequency range was expanded to include MF up to about 1800 kHz. The heading of the aircraft no longer needed to be changed to locate the broadcast transmission antenna. In early model ADFs, a rotatable antenna was used instead. The antenna rotated to seek the position in which the signal was null. The direction to the broadcast antenna was shown on an azimuth scale of an ADF indicator in the flight deck. This type of instrument is still found in use today. It has a fixed card with 0° always at the top of a non-rotating dial. A pointer indicates the relative bearing to the station. When the indication is 0°, the aircraft is on course to (or from) the station. [Figure 11-103]

As ADF technology progressed, indicators with rotatable azimuth cards became the norm. When an ADF signal is received, the pilot rotates the card so that the present heading is at the top of the scale. This results in the pointer indicating the magnetic bearing to the ADF transmitter. This is more intuitive and consistent with other navigational practices. [Figure 11-104]

In modern ADF systems, an additional antenna is used to remove the ambiguity concerning whether the aircraft is heading to or from the transmitter. It is called a sense antenna. The reception field of the sense antenna is omnidirectional. When combined with the fields of the loop antenna, it forms a field with a single significant null reception area on one side. This is used for tuning and produces an indication in the

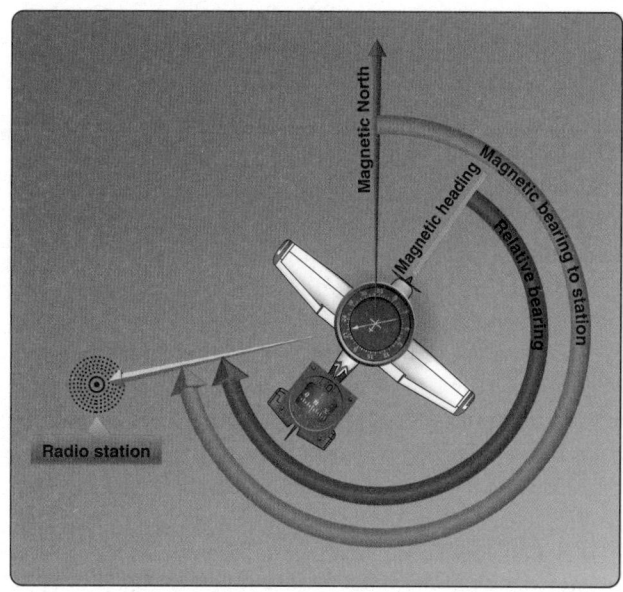

Figure 11-103. *Older ADF indicators have nonrotating azimuth cards. 0° is fixed at the top of the instrument and the pointer always indicates the relative bearing to the ADF transmission antenna. To fly to the station, the pilot turns the aircraft until the ADF pointer indicates 0°.*

direction toward the ADF station at all times. The onboard ADF receiver needs only to be tuned to the correct frequency of the broadcast transmitter for the system to work. The loop and sense antenna are normally housed in a single, low profile antenna housing. [Figure 11-105]

Any ground antenna transmitting LF or MF radio waves in

Figure 11-104. *A movable card ADF indicator can be rotated to put the aircraft's heading at the top of the scale. The pointer then points to the magnetic bearing the ADF broadcast antenna.*

range of the aircraft receiver's tuning capabilities can be used for ADF. This includes those from AM radio stations. Audible identifier tones are loaded on the NDB carrier waves. Typically, a two-character Morse code designator is used. With an AM radio station transmission, the AM broadcast is heard instead of a station identifier code. The frequency for an NDB transmitter is given on an aeronautical chart next to a symbol for the transmitter. The identifying designator is also given. *[Figure 11-106]*

ADF receivers can be mounted in the flight deck with the controls accessible to the user. This is found on many general aviation aircraft. Alternately, the ADF receiver is mounted in a remote avionics bay with only the control head in the flight deck. Dual ADF receivers are common. ADF information can be displayed on the ADF indicators mentioned or it can be digital. Modern, flat, multipurpose electronic displays usually display the ADF digitally. *[Figure 11-107]* When ANT is selected on an ADF receiver, the loop antenna is cut out and only the sense antenna is active. This provides better multi-directional reception of broadcasts in the ADF frequency range, such as weather or AWAS broadcasts.

When the best frequency oscillator (BFO) is selected on an

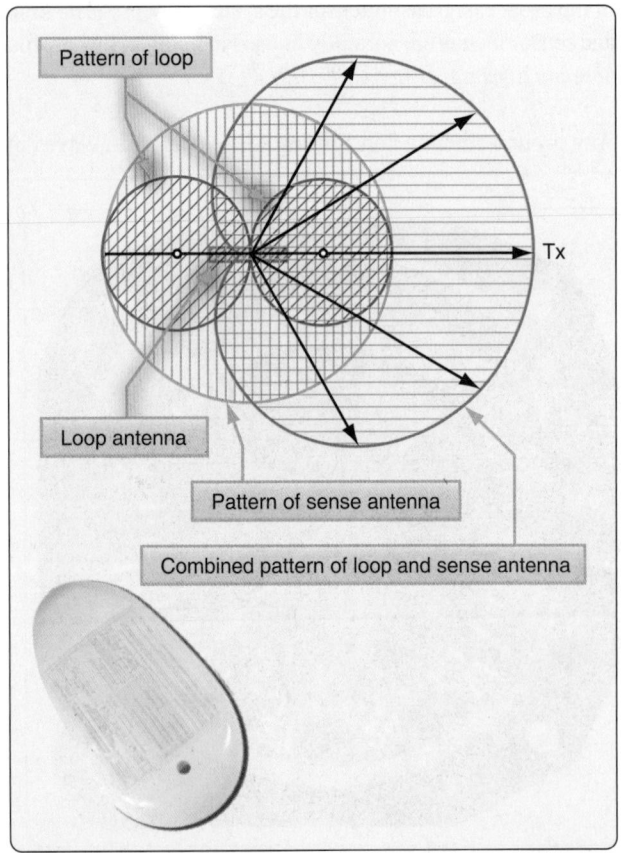

Figure 11-105. *The reception fields of a loop and sense antenna combine to create a field with a sharp null on just one side. This removes directional ambiguity when navigating to an ADF station.*

Figure 11-106. *Nondirectional broadcast antenna in the LF and medium frequency range are used for ADF navigation.*

ADF receiver/controller, an internal beat frequency oscillator is connected to the IF amplifier inside the ADF receiver. This is used when an NDB does not transmit a modulated signal.

Continued refinements to ADF technology has brought it to its current state. The rotating receiving antenna is replaced by a fixed loop with a ferrite core. This increases sensitivity and allows a smaller antenna to be used. The most modern ADF systems have two loop antennas mounted at 90° to each other. The received signal induces voltage that is sent to two stators in a resolver or goniometer. The goniometer stators induce voltage in a rotor that correlates to the signal of the fixed loops. The rotor is driven by a motor to seek the null. The same motor rotates the pointer in the flight deck indicator to show the relative or magnetic bearing to the station. *[Figure 11-108]*

Technicians should note that the installation of the ADF antenna is critical to a correct indication since it is a directional device. Calibration with the longitudinal axis of the fuselage or nose of the aircraft is important. A single null reception area must exist in the correct direction. The antenna

Figure 11-107. *A flight deck mountable ADF receiver used on general aviation aircraft.*

must be oriented so the ADF indicates station location when the aircraft is flying toward it rather than away. Follow all manufacturer's instructions.

Radio Magnetic Indicator (RMI)

To save space in the instrument panel and to consolidate related information into one easy to use location, the radio magnetic indicator (RMI) has been developed. It is widely used. The RMI combines indications from a

magnetic compass, VOR, and ADF into one instrument. [Figure 11-109]

The azimuth card of the RMI is rotated by a remotely located flux gate compass. Thus, the magnetic heading of the aircraft is always indicated. The lubber line is usually a marker or triangle at the top of the instrument dial. The VOR receiver drives the solid pointer to indicate the magnetic direction TO a tuned VOR station. When the ADF is tuned to an NDB, the double, or hollow pointer, indicates the magnetic bearing to the NDB.

Since the flux gate compass continuously adjusts the azimuth card so that the aircraft heading is at the top of the instrument, pilot workload is reduced. The pointers indicate where the VOR and ADF transmission stations are located in relationship to where the aircraft is currently positioned. Push buttons allow conversion of either pointer to either ADF or VOR for navigation involving two of one type of station and none of the other.

Instrument Landing Systems (ILS)

An ILS is used to land an aircraft when visibility is poor. This radio navigation system guides the aircraft down a slope to the touch down area on the runway. Multiple radio transmissions are used that enable an exact approach to landing with an ILS. A localizer is one of the radio transmissions. It is used to provide horizontal guidance to the center line of the runway. A separate glideslope broadcast provides vertical guidance of the aircraft down the proper slope to the touch down point. Compass locator transmissions for outer and middle approach marker beacons aid the pilot in intercepting the

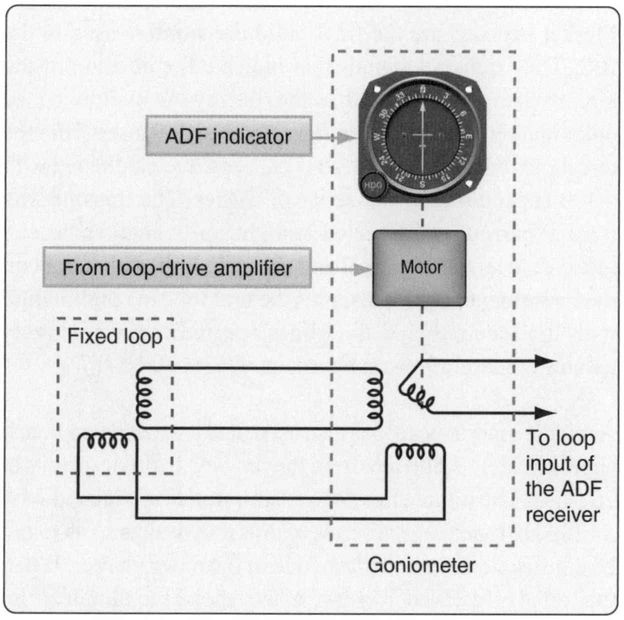

Figure 11-108. *In modern ADF, a rotor in a goniometer replaces the rotating loop antenna used in earlier models.*

Figure 11-109. *A radio magnetic indicator (RMI) combines a magnetic compass, VOR, and ADF indications.*

approach navigational aid system. Marker beacons provide distance-from-the-runway information. Together, all of these radio signals make an ILS a very accurate and reliable means for landing aircraft. *[Figure 11-110]*

Localizer

The localizer broadcast is a VHF broadcast in the lower range of the VOR frequencies (108 MHz–111.95 MHz) on odd frequencies only. Two modulated signals are produced from a horizontally polarized antenna complex beyond the far end of the approach runway. They create an expanding field that is 2½° wide (about 1,500 feet) 5 miles from the runway. The field tapers to runway width near the landing threshold. The left side of the approach area is filled with a VHF carrier wave modulated with a 90 Hz signal. The right side of the approach contains a 150 MHz modulated signal. The aircraft's VOR receiver is tuned to the localizer VHF frequency that can be found on published approach plates and aeronautical charts.

The circuitry specific to standard VOR reception is inactive while the receiver uses localizer circuitry and components common to both. The signals received are passed through filters and rectified into DC to drive the course deviation indicator. If the aircraft receives a 150 Hz signal, the CDI of the VOR/ILS display deflects to the left. This indicates that the runway is to the left. The pilot must correct course with a turn to the left. This centers the course deviation indicator on the display and centers the aircraft with the centerline of the runway. If the 90 Hz signal is received by the VOR receiver, the CDI deflects to the right. The pilot must turn toward the right to center the CDI and the aircraft with the runway center line. *[Figure 11-111]*

Glideslope

The vertical guidance required for an aircraft to descend for a landing is provided by the glideslope of the ILS. The glideslope provides vertical guidance for correct angle of descent. Radio signals funnel the aircraft down to the touchdown point on the runway at an angle of approximately 3°. The transmitting glideslope antenna is located off to the side of the approach runway approximately 1,000 feet from the threshold. It transmits in a wedge-like pattern with the field narrowing as it approaches the runway. *[Figure 11-112]*

The glideslope transmitter antenna is horizontally polarized. The transmitting frequency range is UHF between 329.3 MHz and 335.0 MHz. The frequency is paired to the localizer frequency of the ILS. When the VOR/ILS receiver is tuned for the approach, the glideslope receiver is automatically tuned. Like the localizer, the glideslope transmits two signals, one modulated at 90 Hz and the other modulated at 150 Hz. The aircraft's glideslope receiver deciphers the signals similar to

the method of the localizer receiver. It drives a vertical course deviation indicator known as the glideslope indicator. The glideslope indicator operates identically to the localizer CDI only 90° to it. The VOR/ILS localizer CDI and the glideslope are displayed together on whichever kind of instrumentation is in the aircraft. *[Figure 11-113]*

The UHF antenna for aircraft reception of the glideslope signals comes in many forms. A single dipole antenna mounted inside the nose of the aircraft is a common option. Antenna manufacturers have also incorporated glideslope reception into the same dipole antenna used for the VHS VOR/ILS localizer reception. Blade type antennas are also used. *[Figures 11-114]* Figure 11-115 shows a VOR and a glideslope receiver for a GA aircraft ILS.

Compass Locators

It is imperative that a pilot be able to intercept the ILS to enable its use. A compass locator is a transmitter designed for this purpose. There is typically one located at the outer marker beacon 4–7 miles from the runway threshold. Another may be located at the middle marker beacon about 3,500 feet from the threshold. The outer marker compass locator is a 25 watt NDB with a range of about 15 miles. It transmits omnidirectional LF radio waves (190 Hz to 535 Hz) keyed with the first two letters of the ILS identifier. The ADF receiver is used to intercept the locator so no additional equipment is required. If a middle marker compass locator is in place, it is similar but is identified with the last two letters of the ILS identifier. Once located, the pilot maneuvers the aircraft to fly down the glidepath to the runway.

Marker Beacons

Marker beacons are the final radio transmitters used in the ILS. They transmit signals that indicate the position of the aircraft along the glidepath to the runway. As mentioned, an outer marker beacon transmitter is located 4–7 miles from the threshold. It transmits a 75 MHz carrier wave modulated with a 400 Hz audio tone in a series of dashes. The transmission is very narrow and directed straight up. A marker beacon receiver receives the signal and uses it to light a blue light on the instrument panel. This, plus the oral tone in combination with the localizer and the glideslope indicator, positively locates the aircraft on an approach. *[Figure 11-115]*

A middle marker beacon is also used. It is located on approach approximately 3,500 feet from the runway. It also transmits at 75 MHz. The middle marker transmission is modulated with a 1300 Hz tone that is a series of dots and dashes so as to not be confused with the all dash tone of the outer marker. When the signal is received, it is used in the receiver to illuminate an amber-colored light on the instrument panel. *[Figure 11-116]*

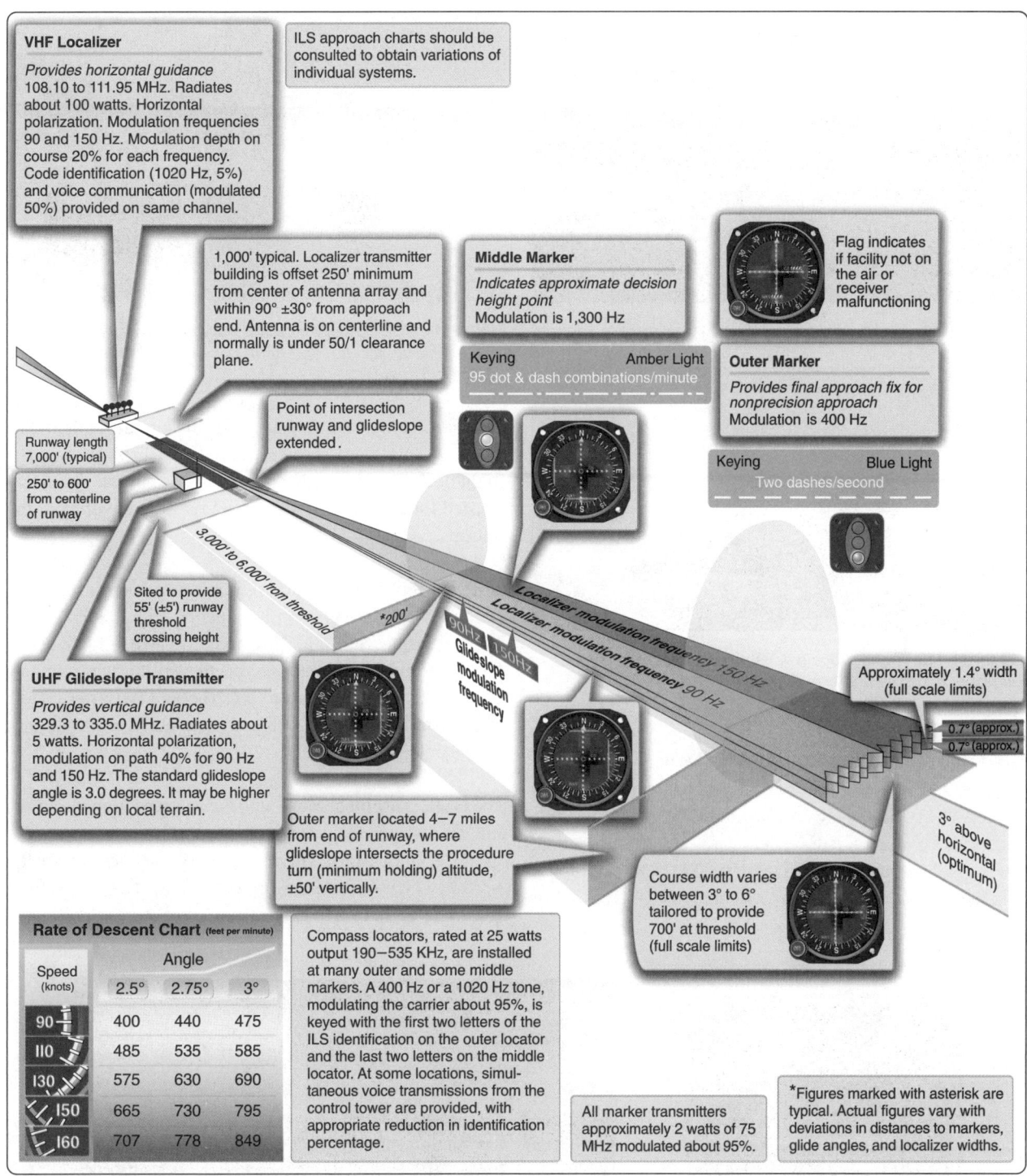

VHF Localizer

Provides horizontal guidance
108.10 to 111.95 MHz. Radiates about 100 watts. Horizontal polarization. Modulation frequencies 90 and 150 Hz. Modulation depth on course 20% for each frequency. Code identification (1020 Hz, 5%) and voice communication (modulated 50%) provided on same channel.

ILS approach charts should be consulted to obtain variations of individual systems.

1,000' typical. Localizer transmitter building is offset 250' minimum from center of antenna array and within 90° ±30° from approach end. Antenna is on centerline and normally is under 50/1 clearance plane.

Middle Marker

Indicates approximate decision height point
Modulation is 1,300 Hz

Flag indicates if facility not on the air or receiver malfunctioning

Keying   Amber Light
95 dot & dash combinations/minute

Outer Marker

Provides final approach fix for nonprecision approach
Modulation is 400 Hz

Keying   Blue Light
Two dashes/second

Point of intersection runway and glideslope extended.

Runway length 7,000' (typical)

250' to 600' from centerline of runway

3,000' to 6,000' from threshold

Sited to provide 55' (±5') runway threshold crossing height

*200'

Glideslope modulation frequency

90Hz 150Hz

Localizer modulation frequency 150 Hz
Localizer modulation frequency 90 Hz

Approximately 1.4° width (full scale limits)

0.7° (approx.)
0.7° (approx.)

UHF Glideslope Transmitter

Provides vertical guidance
329.3 to 335.0 MHz. Radiates about 5 watts. Horizontal polarization, modulation on path 40% for 90 Hz and 150 Hz. The standard glideslope angle is 3.0 degrees. It may be higher depending on local terrain.

Outer marker located 4–7 miles from end of runway, where glideslope intersects the procedure turn (minimum holding) altitude, ±50' vertically.

3° above horizontal (optimum)

Course width varies between 3° to 6° tailored to provide 700' at threshold (full scale limits)

Rate of Descent Chart (feet per minute)

Speed (knots)	Angle		
	2.5°	2.75°	3°
90	400	440	475
110	485	535	585
130	575	630	690
150	665	730	795
160	707	778	849

Compass locators, rated at 25 watts output 190–535 KHz, are installed at many outer and some middle markers. A 400 Hz or a 1020 Hz tone, modulating the carrier about 95%, is keyed with the first two letters of the ILS identification on the outer locator and the last two letters on the middle locator. At some locations, simultaneous voice transmissions from the control tower are provided, with appropriate reduction in identification percentage.

All marker transmitters approximately 2 watts of 75 MHz modulated about 95%.

*Figures marked with asterisk are typical. Actual figures vary with deviations in distances to markers, glide angles, and localizer widths.

Figure 11-110. *Components of an instrument landing system (ILS).*

Some ILS approaches have an inner marker beacon that transmits a signal modulated with 3000 Hz in a series of dots only. It is placed at the land-or-go-around decision point of the approach close to the runway threshold. If present, the signal when received is used to illuminate a white light on the instrument panel. The three marker beacon lights are usually incorporated into the audio panel of a general aviation aircraft or may exist independently on a larger aircraft. Electronic display aircraft usually incorporate marker lights or indicators close to the glideslope display near attitude director indicator. *[Figure 11-117]*

ILS radio components can be tested with an ILS test unit. Localizer, glideslope, and marker beacon signals are generated to ensure proper operation of receivers and correct display on flight deck instruments. *[Figure 11-118]*

Figure 11-111. *An ILS localizer antenna.*

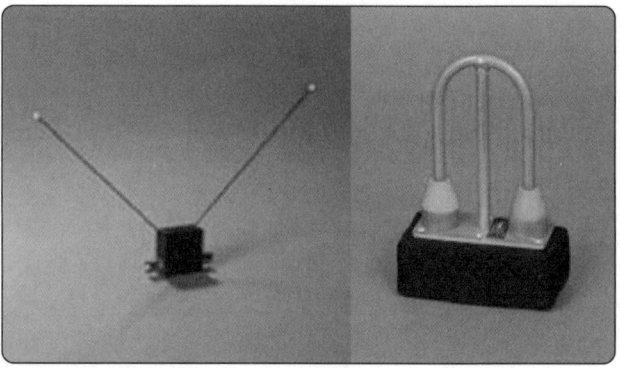

Figure 11-114. *Glideslope antennas—designed to be mounted inside a non-metallic aircraft nose (left), and mounted inside or outside the aircraft (right).*

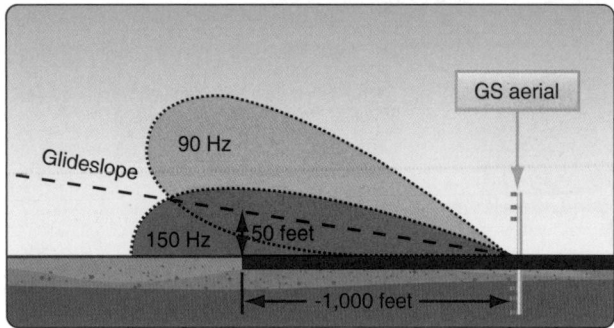

Figure 11-112. *A glideslope antenna broadcasts radio signals to guide an aircraft vertically to the runway.*

Distance Measuring Equipment (DME)

Many VOR stations are co-located with the military version of the VOR station, which is known as TACAN. When this occurs, the navigation station is known as a VORTAC station. Civilian aircraft make use of one of the TACAN features not originally installed at civilian VOR stations–DME. A DME

system calculates the distance from the aircraft to the DME unit at the VORTAC ground station and displays it on the flight deck. It can also display calculated aircraft speed and elapsed time for arrival when the aircraft is traveling to the station.

DME ground stations have subsequently been installed at civilian VORs, as well as in conjunction with ILS localizers. These are known as VOR/DME and ILS/DME or LOC/DME. The latter aid in approach to the runway during landings. The DME system consists of an airborne DME transceiver, display, and antenna, as well as the ground based DME unit and its antenna. *[Figure 11-119]*

The DME is useful because with the bearing (from the VOR) and the distance to a known point (the DME antenna at the VOR), a pilot can positively identify the location of the aircraft. DME operates in the UHF frequency range from 962 MHz to 1213 MHz. A carrier signal transmitted from

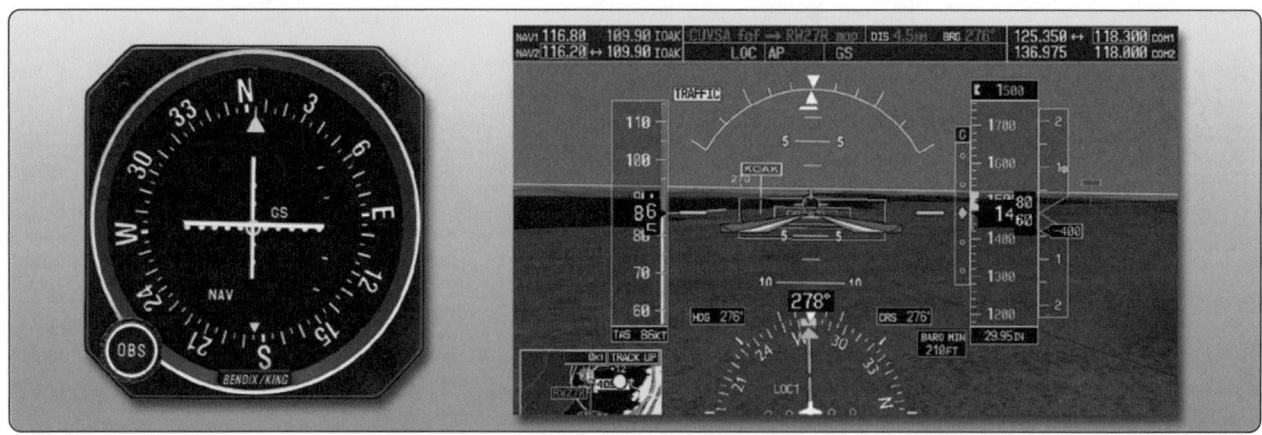

Figure 11-113. *A traditional course deviation indicator is shown on the left. The horizontal white line is the deviation indicator for the glideslope. The vertical line is for the localizer. On the right, a Garmin G-1000 PFD illustrates an aircraft during an ILS approach. The narrow vertical scale on the right of the attitude indicator with the "G" at the top is the deviation scale for the glideslope. The green diamond moves up and down to reflect the aircraft being above or below the glidepath. The diamond is shown centered indicating the aircraft is on course vertically. The localizer CDI can be seen at the bottom center of the display. It is the center section of the vertical green course indicator. LOC1 is displayed to the left of it.*

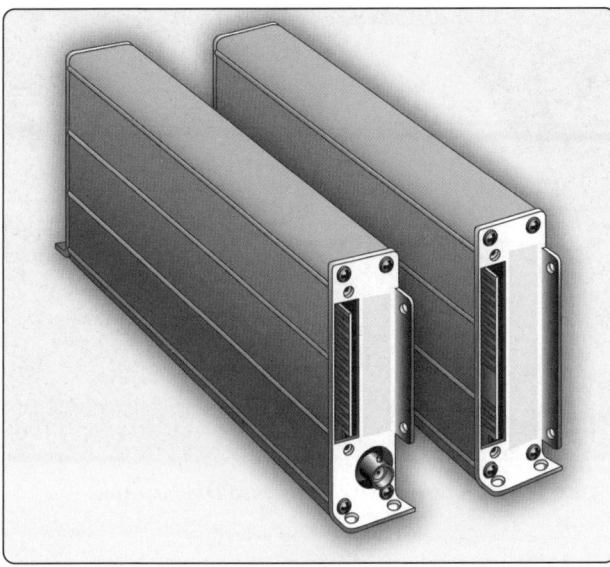

Figure 11-115. *A localizer and glideslope receiver for a general aviation aircraft ILS.*

Figure 11-116. *Various marker beacon instrument panel display lights.*

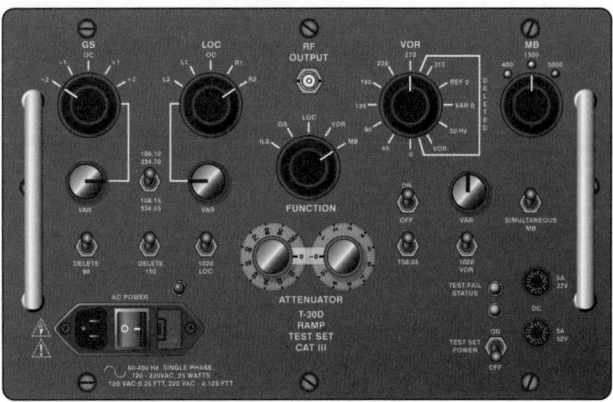

Figure 11-118. *An ILS test unit.*

the aircraft is modulated with a string of integration pulses. The ground unit receives the pulses and returns a signal to the aircraft. The time that transpires for the signal to be sent and returned is calculated and converted into nautical miles for display. Time to station and speed are also calculated and displayed. DME readout can be on a dedicated DME display or it can be part of an EHSI, EADI, EFIS, or on the primary flight display in a glass flight deck. *[Figure 11-120]*

The DME frequency is paired to the co-located VOR or VORTAC frequency. When the correct frequency is tuned for the VOR signal, the DME is tuned automatically. Tones are broadcast for the VOR station identification and then for the DME. The hold selector on a DME panel keeps the DME tuned in while the VOR selector is tuned to a different VOR. In most cases, the UHF of the DME is transmitted and received via a small blade-type antenna mounted to the underside of the fuselage centerline. *[Figure 11-121]*

A traditional DME displays the distance from the DME transmitter antenna to the aircraft. This is called the slant distance. It is very accurate. However, since the aircraft is at altitude, the distance to the DME ground antenna from a

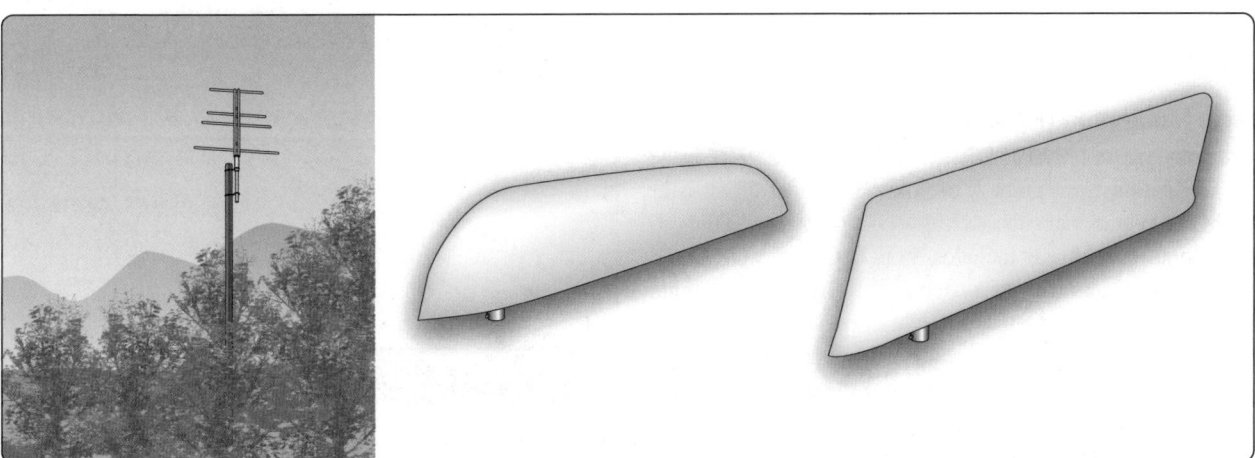

Figure 11-117. *An outer marker transmitter antenna 4–7 miles from the approach runway transmits a 75 MHz signal straight up (left). Aircraft mounted marker beacon receiver antennas are shown (center and right).*

Figure 11-119. *A VOR with DME ground station.*

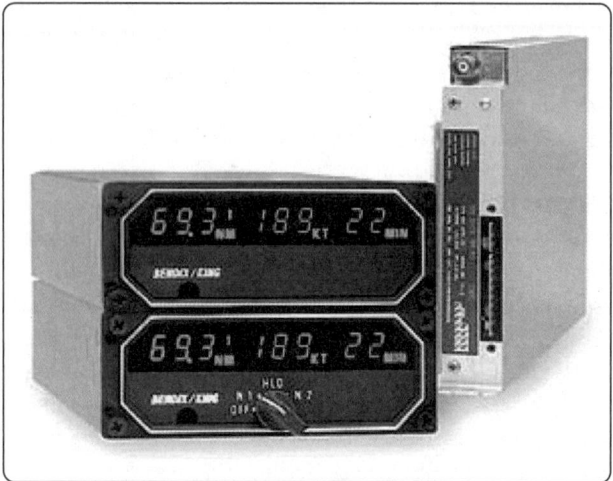

Figure 11-120. *Distance information from the DME can be displayed on a dedicated DME instrument or integrated into any of the electronic navigational displays found on modern aircraft. A dual display DME is shown with its remote mounted receiver.*

point directly beneath the aircraft is shorter. Some modern DMEs are equipped to calculate this ground distance and display it. *[Figure 11-122]*

Area Navigation (RNAV)

Area navigation (RNAV) is a general term used to describe the navigation from point A to point B without direct over flight of navigational aids, such as VOR stations or ADF non-directional beacons. It includes VORTAC and VOR/DME based systems, as well as systems of RNAV based around LORAN, GPS, INS, and the FMS of transport category aircraft. However, until recently, the term RNAV was most commonly used to describe the area navigation or the process of direct flight from point A to point B using VORTAC and VOR/DME based references which are discussed in this section.

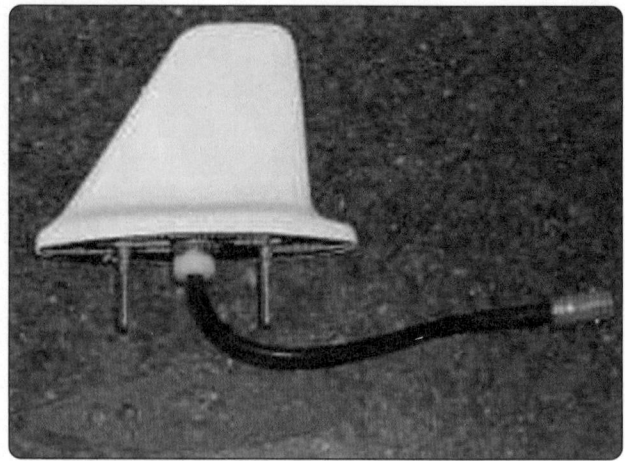

Figure 11-121. *A typical aircraft-mounted DME antenna.*

All RNAV systems make use of waypoints. A waypoint is a designated geographical location or point used for route definition or progress-reporting purposes. It can be defined or described by using latitude/longitude grid coordinates or, in the case of VOR based RNAV, described as a point on a VOR radial followed by that point's distance from the VOR station (i.e., 200/25 means a point 25 nautical miles from the VOR station on the 200° radial).

Figure 11-123 illustrates an RNAV route of flight from airport A to airport B. The VOR/DME and VORTAC stations shown are used to create phantom waypoints that are overflown rather than the actual stations. This allows a more direct route to be taken. The phantom waypoints are entered into the RNAV course-line computer (CLC) as a radial and distance number pair. The computer creates the waypoints and causes the aircraft's CDI to operate as though they are actual VOR stations. A mode switch allows the choice between standard VOR navigation and RNAV.

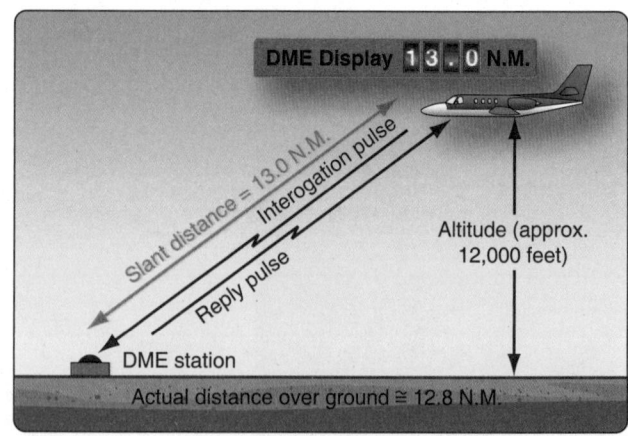

Figure 11-122. *Many DME's only display the slant distance, which is the actual distance from the aircraft to the DME station. This is different than the ground distance due to the aircraft being at altitude. Some DMEs compute the ground distance for display.*

VOR based RNAV uses the VOR receiver, antenna, and VOR display equipment, such as the CDI. The computer in the RNAV unit uses basic geometry and trigonometry calculations to produce heading, speed, and time readouts for each waypoint. VOR stations need to be within line-of-sight and operational range from the aircraft for RNAV use. *[Figure 11-124]*

RNAV has increased in flexibility with the development of GPS. Integration of GPS data into a planned VOR RNAV flight plan is possible as is GPS route planning without the use of any VOR stations.

Radar Beacon Transponder

A radar beacon transponder, or simply, a transponder, provides positive identification and location of an aircraft on the radar screens of ATC. For each aircraft equipped with an altitude encoder, the transponder also provides the pressure altitude of the aircraft to be displayed adjacent to the on-screen blip that represents the aircraft. *[Figure 11-125]*

Radar capabilities at airports vary. Generally, two types of radar are used by air traffic control (ATC). The primary radar transmits directional UHF or SHF radio waves sequentially in all directions. When the radio waves encounter an aircraft, part of those waves reflect back to a ground antenna. Calculations are made in a receiver to determine the direction

Figure 11-124. *RNAV unit from a general aviation aircraft.*

and distance of the aircraft from the transmitter. A blip or target representing the aircraft is displayed on a radar screen also known as a plan position indicator (PPI). The azimuth direction and scaled distance from the tower are presented giving controllers a two dimensional fix on the aircraft. *[Figure 11-126]*

A secondary surveillance radar (SSR) is used by ATC to verify the aircraft's position and to add the third dimension of altitude to its location. SSD radar transmits coded pulse trains that are received by the transponder on board the aircraft. Mode 3/A pulses, as they are known, aid in confirming the location of the aircraft. When verbal communication is established with ATC, a pilot is instructed to select one of 4,096 discrete codes on the transponder. These are digital octal codes. The ground station transmits a pulse of energy at 1030 MHz and the transponder transmits a reply with

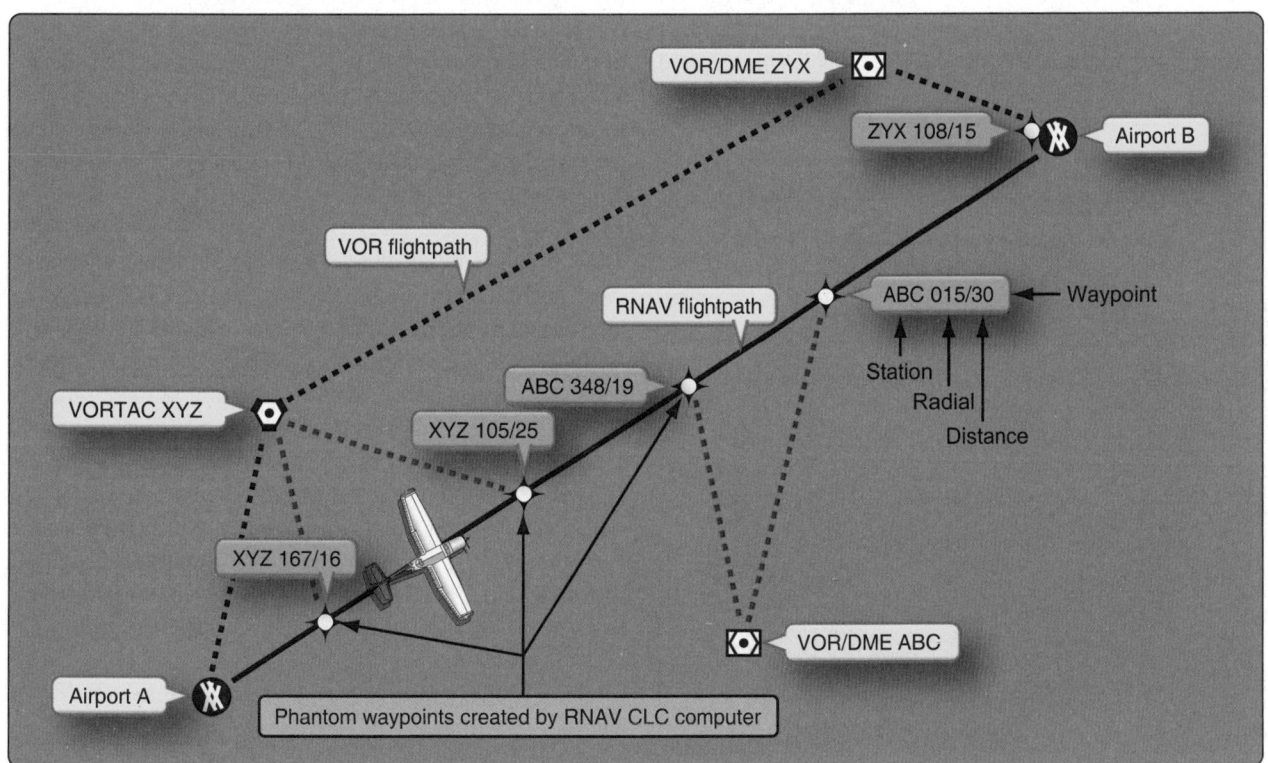

Figure 11-123. *The pilot uses the aircraft's course deviation indicator to fly to and from RNAV phantom waypoints created by computer. This allows direct routes to be created and flown rather than flying from VOR to VOR.*

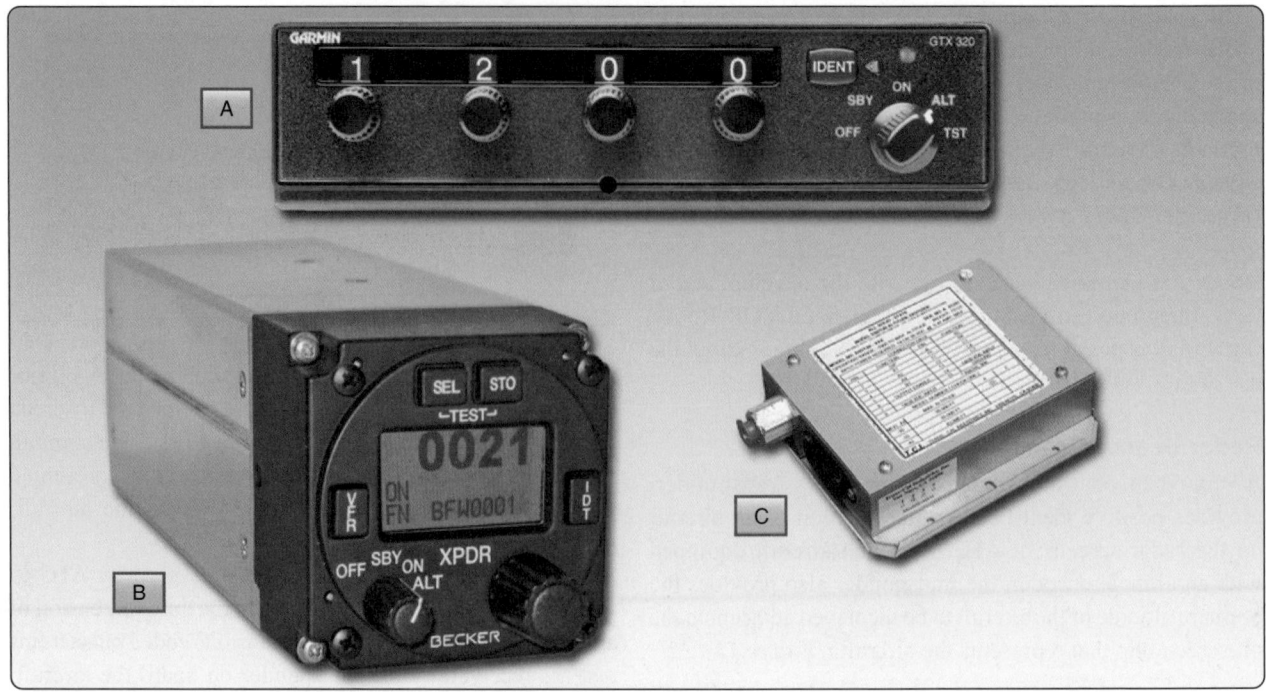

Figure 11-125. *A traditional transponder control head (A), a lightweight digital transponder (B), and a remote altitude encoder (C) that connects to a transponder to provide ATC with an aircraft's altitude displayed on a PPI radar screen next to the target that represents the aircraft.*

the assigned code attached at 1090 MHz. This confirms the aircraft's location typically by altering its target symbol on the radar screen. As the screen may be filled with many confirmed aircraft, ATC can also ask the pilot to ident. By pressing the IDENT button on the transponder, it transmits in such a way that the aircraft's target symbol is highlighted on the PPI to be distinguishable.

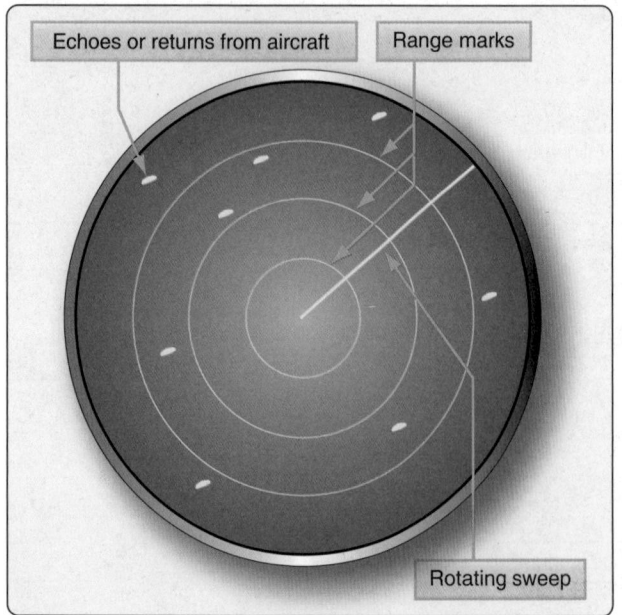

Figure 11-126. *A plan position indicator (PPI) for ATC primary radar locates target aircraft on a scaled field.*

To gain altitude clarification, the transponder control must be placed in the ALT or Mode C position. The signal transmitted back to ATC in response to pulse interrogation is then modified with a code that places the pressure altitude of the aircraft next to the target symbol on the radar screen. The transponder gets the pressure altitude of the aircraft from an altitude encoder that is electrically connected to the transponder. Typical aircraft transponder antennas are illustrated in *Figure 11-127.*

The ATC/aircraft transponder system described is known as Air Traffic Control Radar Beacon System (ATCRBS). To increase safety, Mode S altitude response has been developed. With Mode S, each aircraft is pre-assigned a unique identity code that displays along with its pressure altitude on ATC radar when the transponder responds to SSR interrogation. Since no other aircraft respond with this code, the chance of two pilots selecting the same response code on the transponder is eliminated. A modern flight data processor computer (FDP) assigns the beacon code and searches flight plan data for useful information to be displayed on screen next to the target in a data block for each aircraft. *[Figure 11-128]*

Mode S is sometimes referred to as mode select. It is a data packet protocol that is also used in onboard collision avoidance systems. When used by ATC, Mode S interrogates one aircraft at a time. Transponder workload is reduced by not having to respond to all interrogations in an airspace.

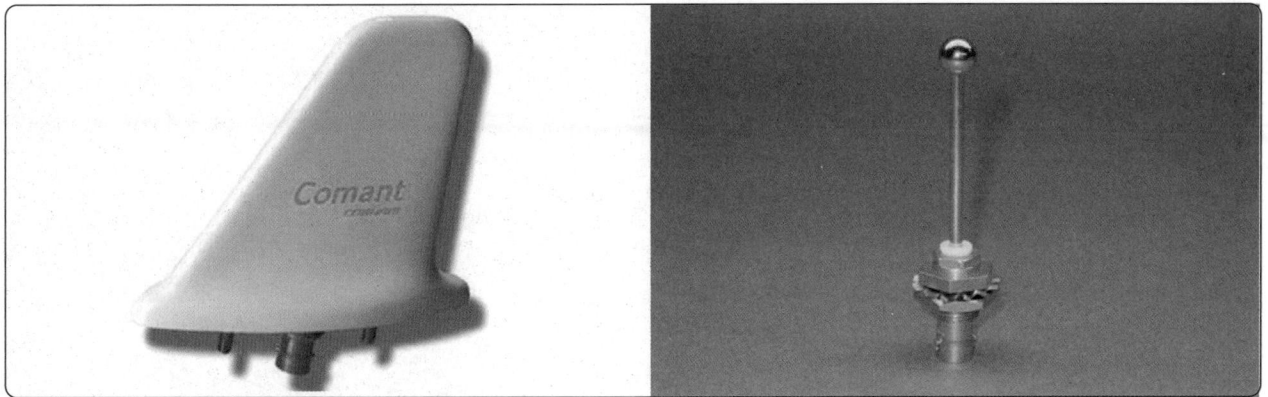

Figure 11-127. *Aircraft radar beacon transponder antennas transmit and receive UHF and SHF radio waves.*

Additionally, location information is more accurate with Mode S. A single reply in which the phase of the transponder reply is used to calculate position, called monopulse, is sufficient to locate the aircraft. Mode S also contains capacity for a wider variety of information exchange that is untapped potential for the future. At the same time, compatibility with older radar and transponder technology has been maintained.

Transponder Tests & Inspections

Title 14 of the Code of Federal Regulations (CFR) part 91, section 91.413 states that all transponders on aircraft flown into controlled airspace are required to be inspected and tested in accordance with 14 CFR part 43, Appendix F, every 24 calendar months. Installation or maintenance that may introduce a transponder error is also cause for inspection and test in accordance with Appendix F. Only an appropriately rated repair station, the aircraft manufacturer (if it installed transponder), and holders of a continuous airworthy program are approved to conduct the procedures. As with many radio-electronic devices, test equipment exists to test airworthy operation of a transponder. *[Figure 11-129]*

Operating a transponder in a hangar or on the ramp does not immunize it from interrogation and reply. Transmission of certain codes reserved for emergencies or military activity must be avoided. The procedure to select a code during ground operation is to do so with the transponder in the OFF or STANDBY mode to avoid inadvertent transmission. Code 0000 is reserved for military use and is a transmittable code. Code 7500 is used in a hijack situation and 7600 and 7700 are also reserved for emergency use. Even the inadvertent transmission of code 1200 reserved for VFR flight not under ATC direction could result in evasion action. All signals received from a radar beacon transponder are taken seriously by ATC.

Altitude Encoders

Altitude encoders convert the aircraft's pressure altitude into a code sent by the transponder to ATC. Increments of 100 feet are usually reported. Encoders have varied over the years. Some are built into the altimeter instrument used in the instrument panel and connected by wires to the transponder. Others are mounted out of sight on an avionics rack or similar out of the way place. These are known as blind encoders. On transport category aircraft, the altitude encoder may be a large black box with a static line connection to an internal aneroid. Modern general aviation encoders are smaller and more lightweight, but still often feature an internal aneroid and static line connection. Some encoders use microtransistors and are completely solid-state including the pressure sensing device from which the altitude is derived. No static port connection is required. Data exchange with GPS and other systems is becoming common. *[Figure 11-130]*

When a transponder selector is set on ALT, the digital pulse message sent in response to the secondary surveillance radar interrogation becomes the digital representation of the pressure altitude of the aircraft. There are 1280 altitude codes, one for each 100 feet of altitude between 1200 feet mean sea level (MSL) and 126,700 feet MSL. Each altitude increment is assigned a code. While these would be 1280 of the same codes used for location and IDENT, the Mode C (or S) interrogation deactivates the 4096 location codes and causes the encoder to become active. The correct altitude code is sent to the transponder that replies to the interrogation. The SSR receiver recognized this as a response to a Mode C (or S) interrogation and interprets the code as altitude code.

Collision Avoidance Systems

The ever increasing volume of air traffic has caused a corresponding increase in concern over collision avoidance. Ground-based radar, traffic control, and visual vigilance are no longer adequate in today's increasingly crowded skies. Onboard collision avoidance equipment, long a staple in larger aircraft, is now common in general aviation aircraft.

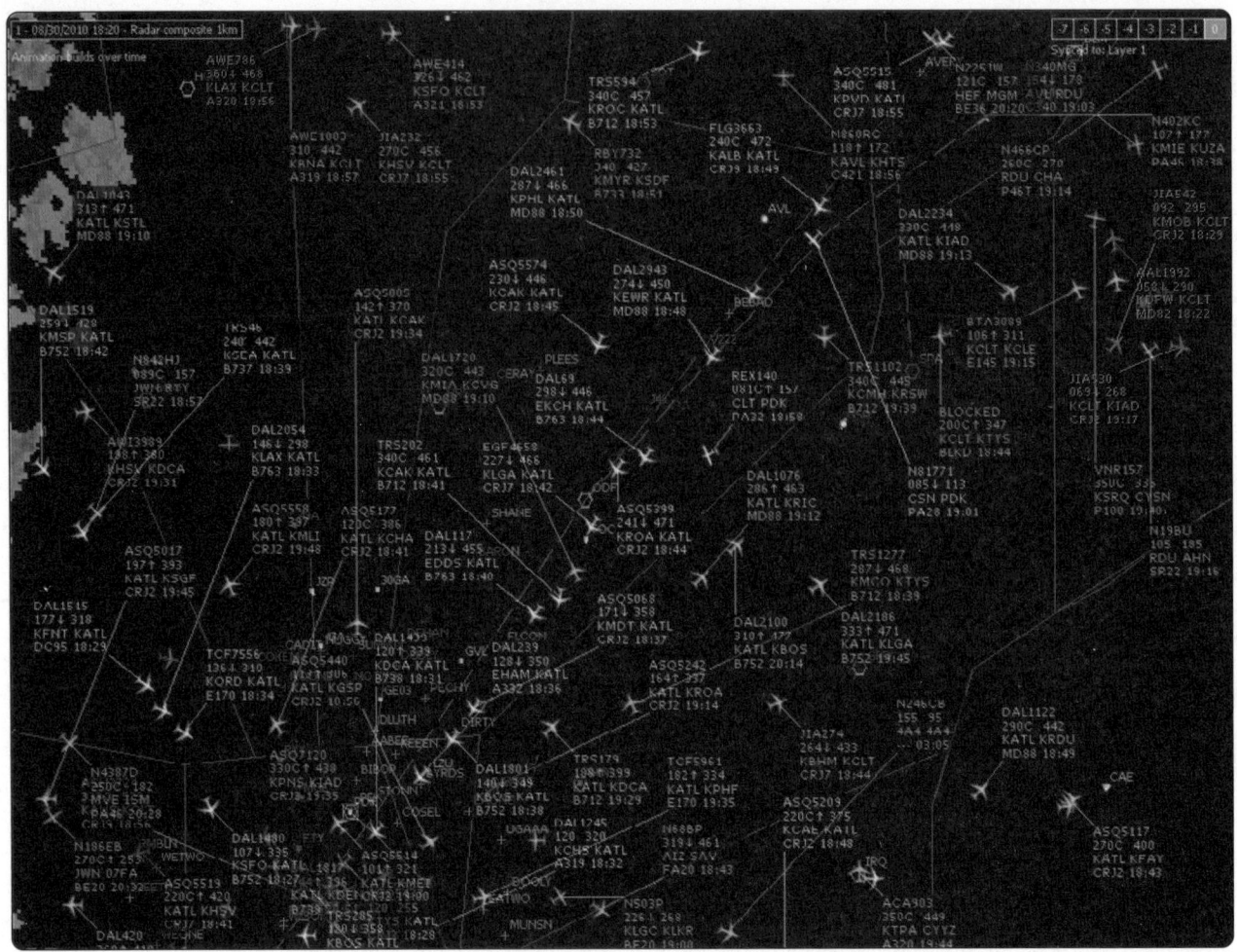

Figure 11-128. *Air traffic control radar technology and an onboard radar beacon transponder work together to convey and display air traffic information on a PPI radar screen. A modern approach ATC PPI is shown. Targets representing aircraft are shown as little aircraft on the screen. The nose of the aircraft indicates the direction of travel. Most targets shown above are airliners. The data block for each target includes the following information either transmitted by the transponder or matched and loaded from flight plans by a flight data processor computer: call sign, altitude/speed, origination/destination, and aircraft type/ETA (ZULU time). A "C" after the altitude indicates the information came from a Mode C equipped transponder. The absence of a C indicates Mode S is in use. An arrow up indicates the aircraft is climbing. An arrow down indicates a descent. White targets are arrivals, light blue targets are departures, all other colors are for arrivals and departures to different airports in the area.*

New applications of electronic technology combined with lower costs make this possible.

Traffic Collision Avoidance Systems (TCAS)

Traffic collision avoidance systems (TCAS) are transponder based air-to-air traffic monitoring and alerting systems. There are two classes of TCAS. TCAS I was developed to accommodate the general aviation community and regional airlines. This system identifies traffic in a 35–40 mile range of the aircraft and issues Traffic Advisories (TA) to assist pilots in visual acquisition of intruder aircraft. TCAS I is mandated on aircraft with 10 to 30 seats.

TCAS II is a more sophisticated system. It is required internationally in aircraft with more than 30 seats or weighing more than 15,000 kg. TCAS II provides the information of TCAS I, but also analyzes the projected flightpath of approaching aircraft. If a collision or near miss is imminent, the TCAS II computer issues a Resolution Advisory (RA). This is an aural command to the pilot to take a specific evasive action (i.e., DESCEND). The computer is programmed such that the pilot in the encroaching aircraft receives an RA for evasive action in the opposite direction (if it is TCAS II equipped). *[Figure 11-131]*

The transponder of an aircraft with TCAS is able to interrogate the transponders of other aircraft nearby using SSR technology (Mode C and Mode S). This is done with a 1030 MHz signal. Interrogated aircraft transponders reply with an encoded 1090 MHz signal that allows the TCAS computer to display the position and altitude of each aircraft. Should the aircraft come within the horizontal or vertical distances shown in *Figure 11-131*, an audible TA is

Figure 11-129. *A handheld transponder test unit.*

announced. The pilot must decide whether to take action and what action to take. TCAS II equipped aircraft use continuous reply information to analyze the speed and trajectory of target aircraft in close proximity. If a collision is calculated to be imminent, an RA is issued.

TCAS target aircraft are displayed on a screen on the flight deck. Different colors and shapes are used to depict approaching aircraft depending on the imminent threat level. Since RAs are currently limited to vertical evasive maneuvers, some stand-alone TCAS displays are electronic vertical speed indicators. Most aircraft use some version of an electronic HSI on a navigational screen or page to display TCAS information. *[Figure 11-132]* A multifunction display may depict TCAS and weather radar information on the same screen. *[Figure 11-133]* A TCAS control panel *[Figure 11-134]* and computer are required to work with a compatible transponder and its antenna(s). Interface with EFIS or other previously installed or selected display(s) is also required.

TCAS may be referred to as airborne collision avoidance system (ACAS), which is the international name for the same system. TCAS II with the latest revisions is known as Version 7. The accuracy and reliability of this TCAS information is such that pilots are required to follow a TCAS RA over an ATC command.

ADS-B

Collision avoidance is a significant part of the FAA's NextGen plan for transforming the National Airspace

Figure 11-130. *Modern altitude encoders for general aviation aircraft.*

System (NAS). Increasing the number of aircraft using the same quantity of airspace and ground facilities requires the implementation of new technologies to maintain a high level of performance and safety. The successful proliferation of global navigation satellite systems (GNSS), such as GPS, has led to the development of a collision avoidance system known as Automatic Dependent Surveillance-Broadcast (ADS-B). ADS-B is an integral part of NextGen program. The implementation of its ground and airborne infrastructure is currently underway. ADS-B is active in parts of the United States and around the world. *[Figure 11-135]*

ADS-B is considered in two segments: ADS-B OUT and ADS-B IN. ADS-B OUT combines the positioning information available from a GPS receiver with on-board flight status information, i.e., location including altitude, velocity, and time. It then broadcasts this information to other ADS-B equipped aircraft and ground stations. *[Figure 11-136]*

Two different frequencies are used to carry these broadcasts with data link capability. The first is an expanded use of the 1090 MHz Mode-S transponder protocol known as 1090 ES. The second, largely being introduced as a new broadband solution for general aviation implementation of ADS-B, is at 978 MHz. A 978 universal access transceiver (UAT) is used to accomplish this. An omni-directional antenna is required in addition to the GPS antenna and receiver. Airborne receivers of an ADS-B use the information to plot the location and movement of the transmitting aircraft on a flight deck display similar to TCAS. *[Figure 11-137]*

Inexpensive ground stations (compared to radar) are constructed in remote and obstructed areas to proliferate ADS-B. Ground stations share information from airborne ADS-Bs with other ground stations that are part of the air traffic management system (ATMS). Data is transferred with no need for human acknowledgement. Microwave and satellite transmissions are used to link the network.

For traffic separation and control, ADS-B has several advantages over conventional ground-based radar. The first is the entire airspace can be covered with a much lower expense. The aging ATC radar system that is in place is expensive to maintain and replace. Additionally, ADS-B provides more accurate information since the vector state is generated from the aircraft with the help of GPS satellites. Weather is a greatly reduced factor with ADS-B. Ultra-high frequency GPS transmissions are not affected. Increased positioning

accuracy allows for higher density traffic flow and landing approaches, an obvious requirement to operate more aircraft in and out of the same number of facilities. The higher degree of control available also enables routing for fewer weather delays and optimal fuel burn rates. Collision avoidance is expanded to include runway incursion from other aircraft and support vehicles on the surface of an airport.

ADS-B IN offers features not available in TCAS. Equipped aircraft are able to receive abundant data to enhance situational awareness. Traffic information services-broadcast (TIS-B) supply traffic information from non-ADS-B aircraft and ADS-B aircraft on a different frequency. Ground radar

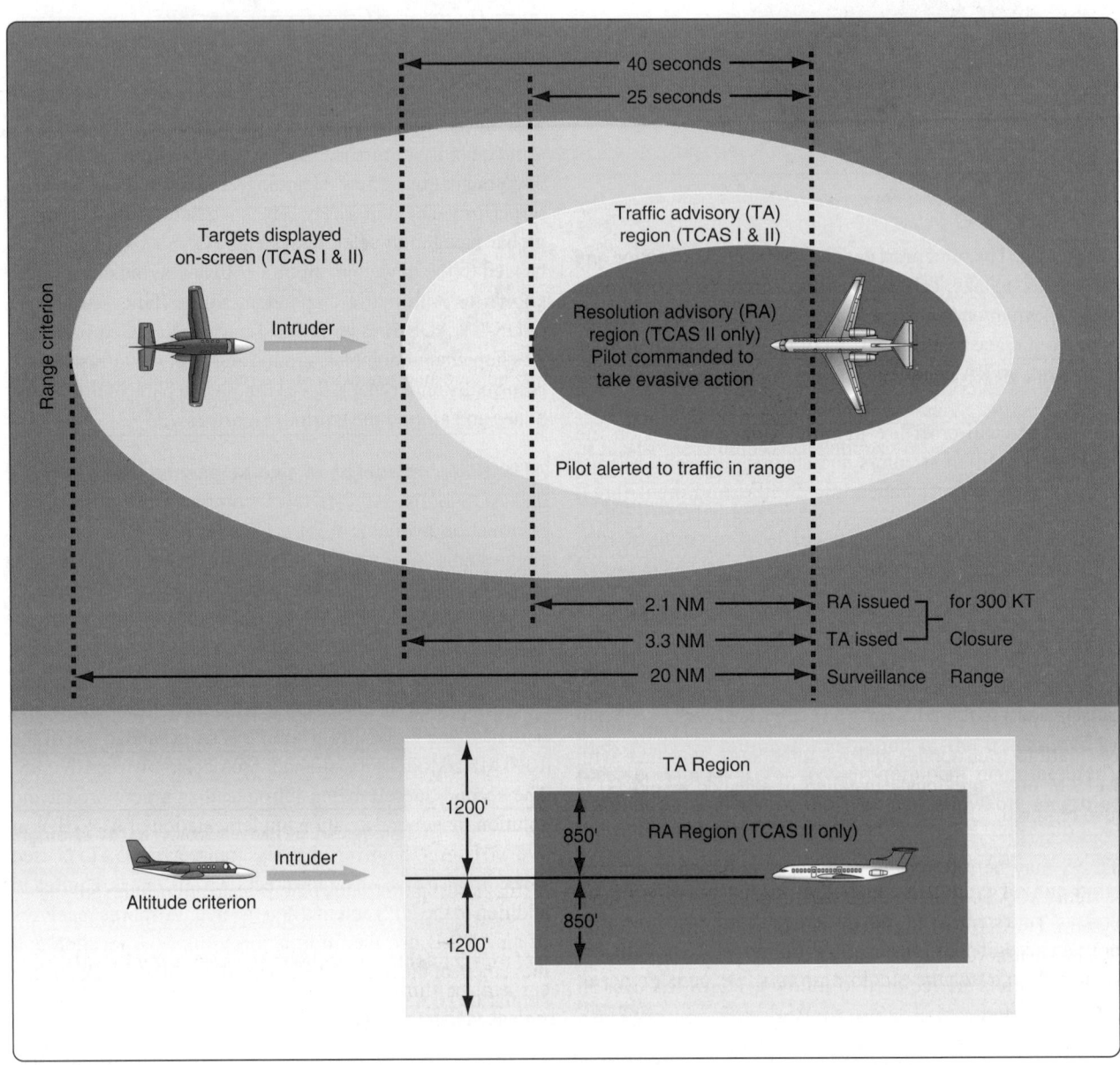

Figure 11-131. *Traffic collision and avoidance system (TCAS) uses an aircraft's transponder to interrogate and receive replies from other aircraft in close proximity. The TCAS computer alerts the pilot as to the presence of an intruder aircraft and displays the aircraft on a screen in the flight deck. Additionally, TCAS II equipped aircraft receive evasive maneuver commands from the computer that calculates trajectories of the aircraft to predict potential collisions or near misses before they become unavoidable.*

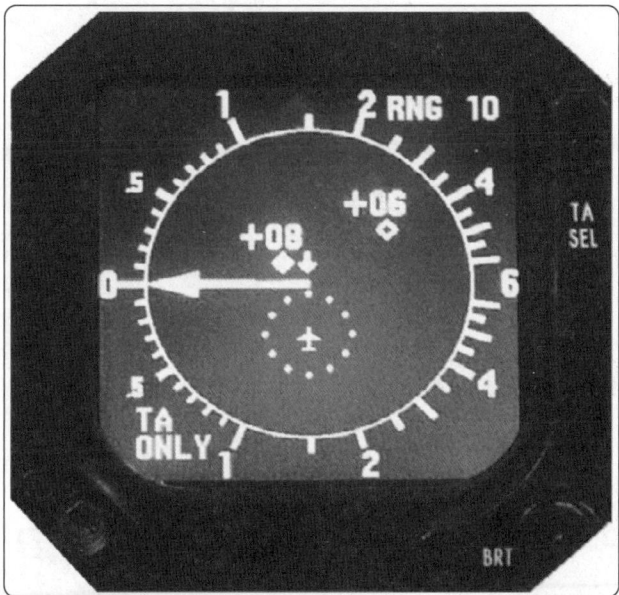

Figure 11-132. *TCAS information displayed on an electronic vertical speed indicator.*

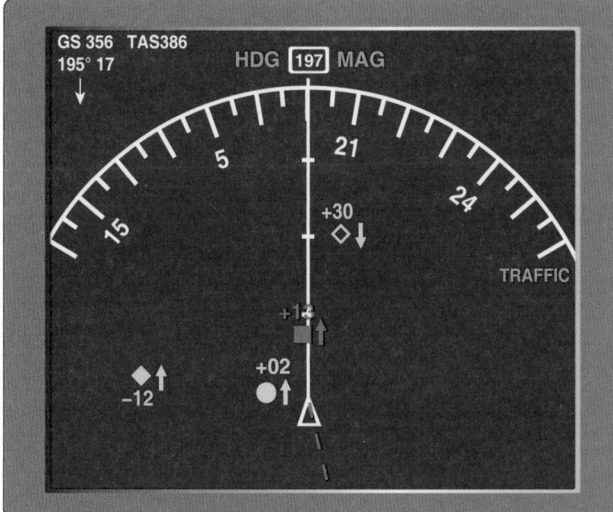

Figure 11-133. *TCAS information displayed on a multifunction display. An open diamond indicates a target; a solid diamond represents a target that is within 6 nautical miles of 1,200 feet vertically. A yellow circle represents a target that generates a TA (25-48 seconds before contact). A red square indicates a target that generates an RA in TCAS II (contact within 35 seconds). A (+) indicates the target aircraft is above and a (-) indicates it is below. The arrows show if the target is climbing or descending.*

monitoring of surface targets, and any traffic data in the linked network of ground stations is sent via ADS-B IN to the flight deck. This provides a more complete picture than air-to-air only collision avoidance. Flight information services-broadcast (FIS-B) is also received by ADS-B IN. Weather text and graphics, ATIS information, and NOTAMs are able to be received in aircraft that have 987 UAT capability. *[Figure 11-138]*

Figure 11-134. *This control panel from a Boeing 767 controls the transponder for ATC use and TCAS.*

ADS-B test units are available for trained maintenance personnel to verify proper operation of ADS-B equipment. This is critical since close tolerance of air traffic separation depends on accurate data from each aircraft and throughout all components of the ADS-B system. *[Figure 11-139]*

Radio Altimeter

A radio altimeter, or radar altimeter, is used to measure the distance from the aircraft to the terrain directly beneath it. It is used primarily during instrument approach and low level or night flight below 2,500 feet. The radio altimeter supplies the primary altitude information for landing decision height. It incorporates an adjustable altitude bug that creates a visual or aural warning to the pilot when the aircraft reaches that altitude. Typically, the pilot will abort a landing if the decision height is reached and the runway is not visible.

Using a transceiver and a directional antenna, a radio altimeter broadcasts a carrier wave at 4.3 GHz from the aircraft directly toward the ground. The wave is frequency

Figure 11-135. *Low power requirements allow remote ADS-B stations with only solar or propane support. This is not possible with ground radar due to high power demands which inhibit remote area radar coverage for air traffic purposes.*

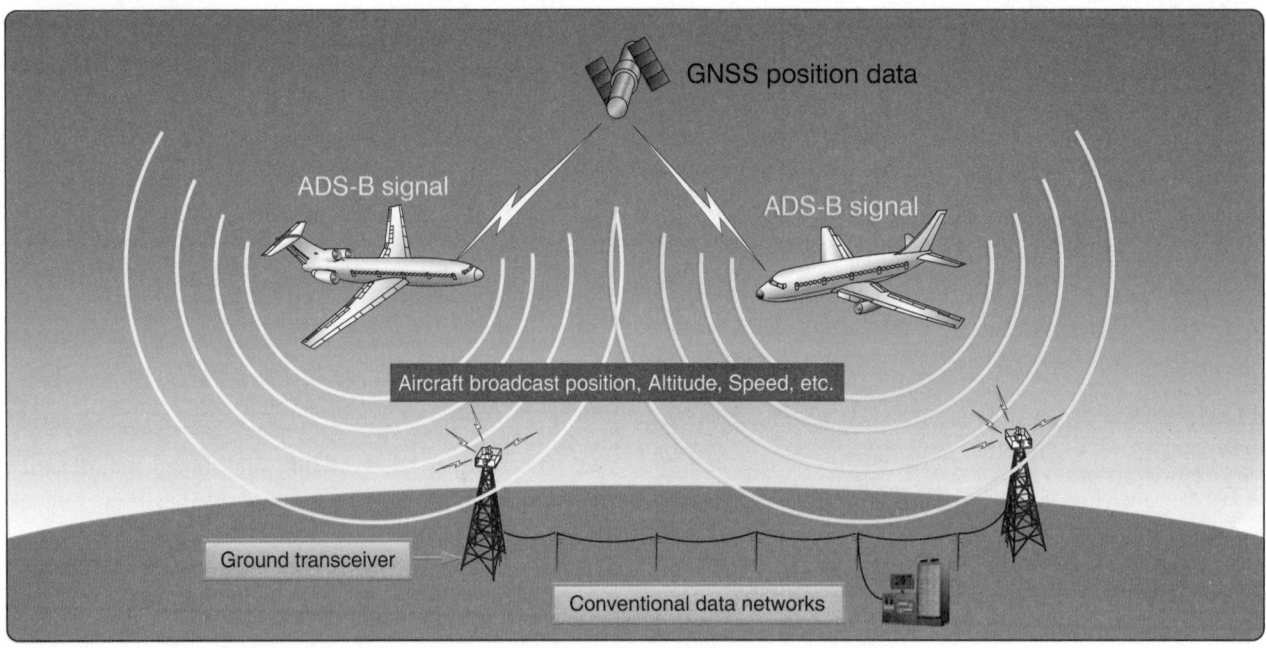

Figure 11-136. *ADS-B OUT uses satellites to identify the position aircraft. This position is then broadcast to other aircraft and to ground stations along with other flight status information.*

modulated at 50 MHz and travels at a known speed. It strikes surface features and bounces back toward the aircraft where a second antenna receives the return signal. The transceiver processes the signal by measuring the elapsed time the signal traveled and the frequency modulation that occurred. The display indicates height above the terrain also known as above ground level (AGL). *[Figure 11-140]*

A radar altimeter is more accurate and responsive than an air pressure altimeter for AGL information at low altitudes. The transceiver is usually located remotely from the indicator. Multifunctional and glass flight deck displays typically integrate decision height awareness from the radar altimeter as a digital number displayed on the screen with a bug, light, or color change used to indicate when that altitude is reached.

Large aircraft may incorporate radio altimeter information into a ground proximity warning system (GPWS) which aurally alerts the crew of potentially dangerous proximity to the terrain below the aircraft. A decision height (DH) window displays the radar altitude on the EADI in *Figure 11-141*.

Weather Radar

There are three common types of weather aids used in an aircraft flight deck that are often referred to as weather radar:

1. Actual on-board radar for detecting and displaying weather activity;

2. Lightning detectors; and

3. Satellite or other source weather radar information that is uploaded to the aircraft from an outside source.

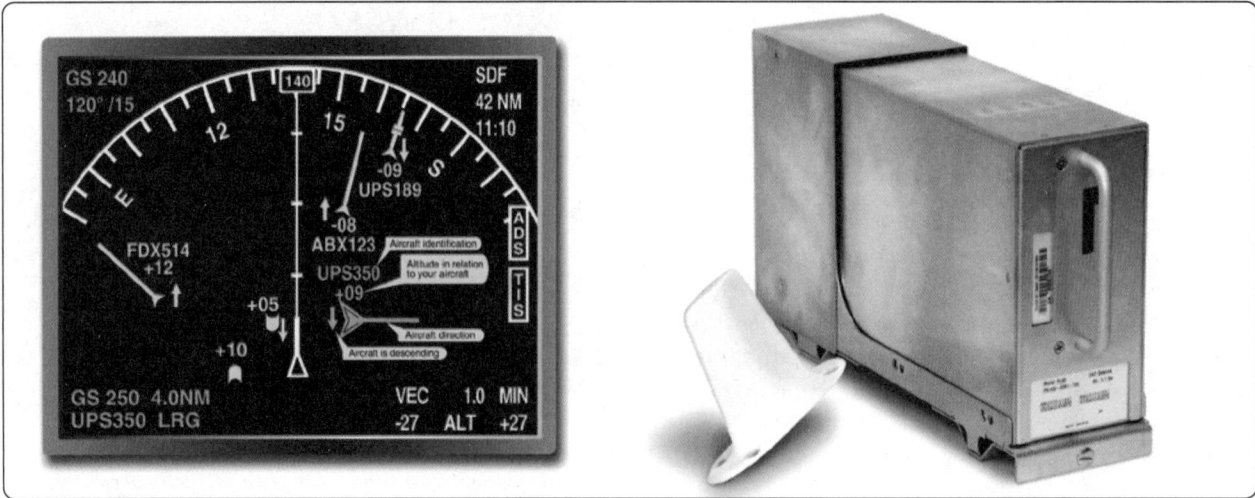

Figure 11-137. *A flight deck display of ADS-B generated targets (left) and an ADS-B airborne receiver with antenna (right).*

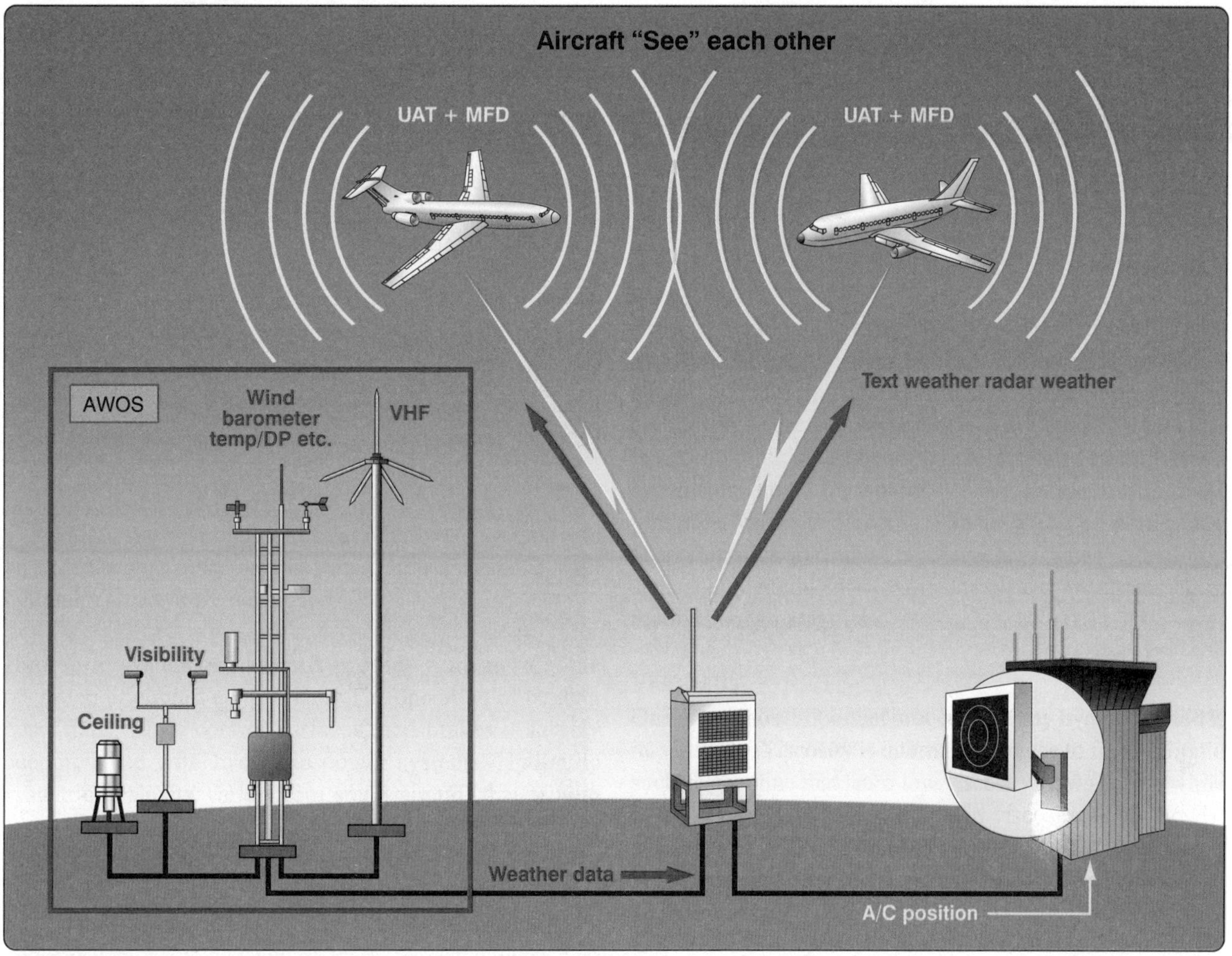

Aircraft "See" each other
UAT + MFD
UAT + MFD
Text weather radar weather
AWOS
Wind
barometer
temp/DP etc.
VHF
Visibility
Ceiling
Weather data
A/C position

Figure 11-138. *ADS-B IN enables weather and traffic information to be sent into the flight deck. In addition to AWOS weather, NWS can also be transmitted.*

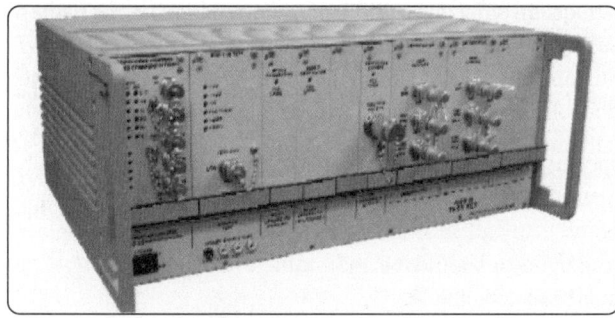

Figure 11-139. *An ADS-B test unit.*

On-board weather radar systems can be found in aircraft of all sizes. They function similar to ATC primary radar except the radio waves bounce off of precipitation instead of aircraft. Dense precipitation creates a stronger return than light precipitation. The on-board weather radar receiver is set up to depict heavy returns as red, medium returns as yellow and light returns as green on a display in the flight deck. Clouds do not create a return. Magenta is reserved to depict intense or extreme precipitation or turbulence. Some aircraft have a dedicated weather radar screen. Most modern

GEAR
DH
TRI
40
DH
SET
10
200
TEST
ALT AGL
TERRA
DO NOT PAINT
DO NOT PAINT

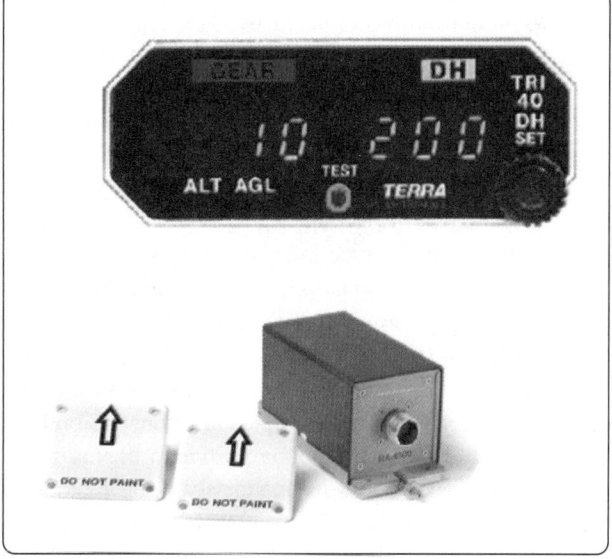

Figure 11-140. *A digital display radio altimeter (top), and the two antennas and transceiver for a radio/radar altimeter (bottom).*

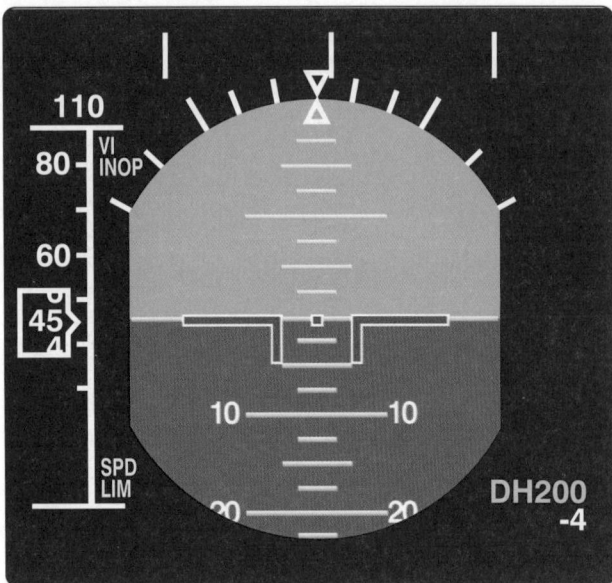

Figure 11-141. *The decision height, DH200, in the lower right corner of this EADI display uses the radar altimeter as the source of altitude information.*

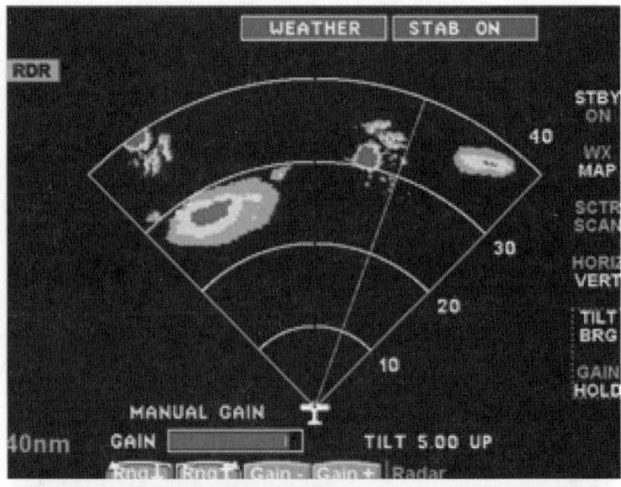

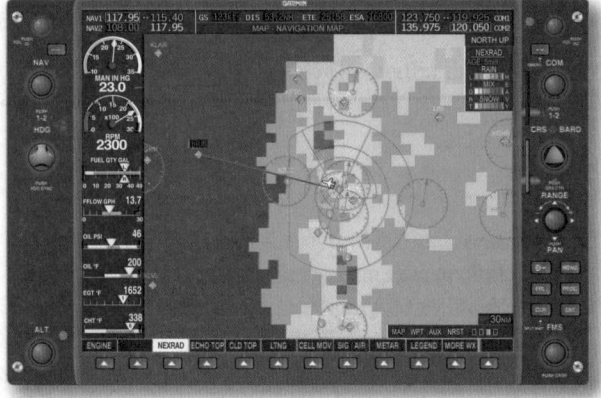

Figure 11-142. *A dedicated weather radar display (top) and a multifunctional navigation display with weather radar overlay (bottom).*

aircraft integrate weather radar display into the navigation display(s). *Figure 11-142* illustrates weather radar displays found on aircraft.

Radio waves used in weather radar systems are in the SHF range such as 5.44 GHz or 9.375 GHz. They are transmitted forward of the aircraft from a directional antenna usually located behind a non-metallic nose cone. Pulses of approximately 1 micro-second in length are transmitted. A duplexer in the radar transceiver switches the antenna to receive for about 2,500 micro seconds after a pulse is transmitted to receive and process any returns. This cycle repeats and the receiver circuitry builds a two dimensional image of precipitation for display. Gain adjustments control the range of the radar. A control panel facilitates this and other adjustments. *[Figure 11-143]*

Severe turbulence, windshear, and hail are of major concern to the pilot. While hail provides a return on weather radar, windshear and turbulence must be interpreted from the movement of any precipitation that is detected. An alert is annunciated if this condition occurs on a weather radar system so equipped. Dry air turbulence is not detectable. Ground clutter must also be attenuated when the radar sweep includes any terrain features. The control panel facilitates this.

Special precautions must be followed by the technician during maintenance and operation of weather radar systems. The radome covering the antenna must only be painted with approved paint to allow the radio signals to pass unobstructed. Many radomes also contain grounding strips to conduct lightning strikes and static away from the dome.

When operating the radar, it is important to follow all manufacturer instructions. Physical harm is possible from the high energy radiation emitted, especially to the eyes and testes. Do not look into the antenna of a transmitting radar. Operation of the radar should not occur in hangars unless special radio wave absorption material is used. Additionally, operation of radar should not take place while the radar is pointed toward a building or when refueling takes place. Radar units should be maintained and operated only by qualified personnel.

Lightning detection is a second reliable means for identifying potentially dangerous weather. Lightning gives off its own electromagnetic signal. The azimuth of a lightning strike can be calculated by a receiver using a loop type antenna such as that used in ADF. *[Figure 11-144]* Some lightning detectors make use of the ADF antenna. The range of the lightning strike is closely associated with its intensity. Intense strikes are plotted as being close to the aircraft.

Stormscope is a proprietary name often associated with lightning detectors. There are others that work in a similar

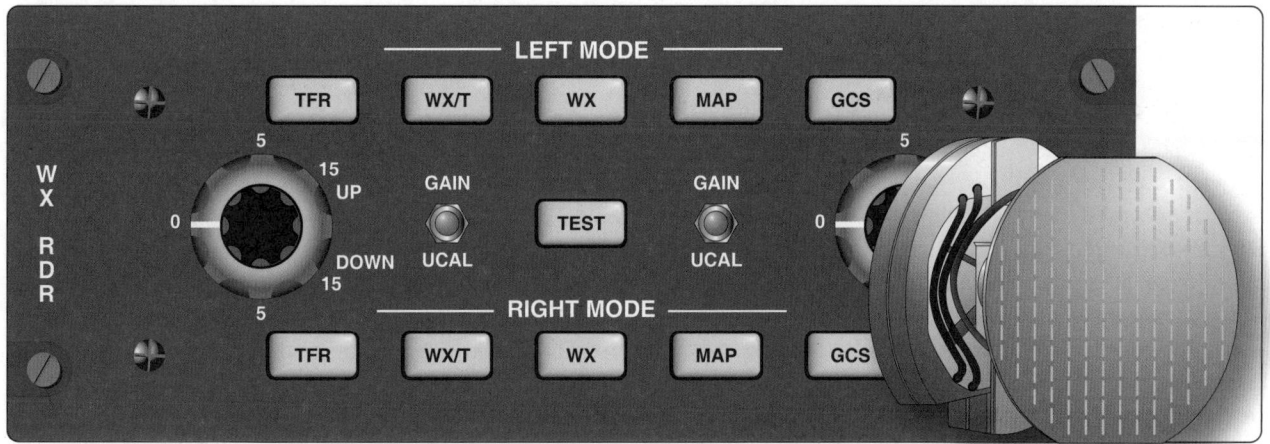

Figure 11-143. *A typical on-board weather radar system for a high performance aircraft uses a nose-mounted antenna that gimbals. It is usually controlled by the inertial reference system (IRS) to automatically adjust for attitude changes during maneuvers so that the radar remains aimed at the desired weather target. The pilot may also adjust the angle and sweep manually as well as the gain. A dual mode control panel allows separate control and display on the left or right HSI or navigational display.*

manner. A dedicated display plots the location of each strike within a 200 mile range with a small mark on the screen. As time progresses, the marks may change color to indicate their age. Nonetheless, a number of lightning strikes in a small area indicates a storm cell, and the pilot can navigate around it. Lightning strikes can also be plotted on a multifunctional navigation display. *[Figure 11-145]*

A third type of weather radar is becoming more common in all classes of aircraft. Through the use of orbiting satellite systems and/or ground up-links, such as described with ADS-B IN, weather information can be sent to an aircraft in flight virtually anywhere in the world. This includes text data as well as real-time radar information for overlay on an aircraft's navigational display(s). Weather radar data produced remotely and sent to the aircraft is refined through consolidation of various radar views from different angles and satellite imagery. This produces more accurate depictions of actual weather conditions. Terrain databases are integrated to eliminate ground clutter. Supplemental data includes the entire range of intelligence available from

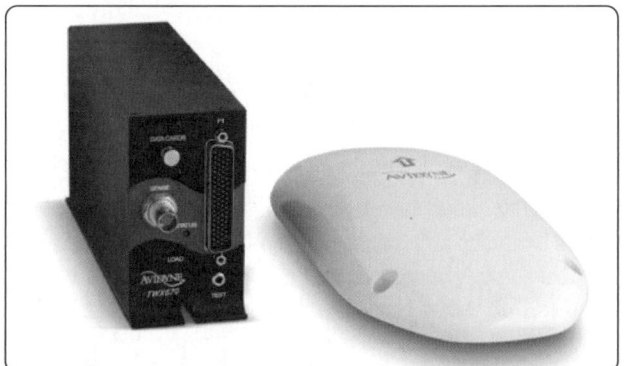

Figure 11-144. *A receiver and antenna from a lightning detector system.*

the National Weather Service (NWS) and the National Oceanographic and Atmospheric Administration (NOAA). *Figure 11-146* illustrates a plain language weather summary received in an aircraft along with a list of other weather information available through satellite or ground link weather information services.

As mentioned, to receive an ADS-B weather signal, a 1090 ES or 970 UAT transceiver with associated antenna needs to be installed on board the aircraft. Satellite weather services are received by an antenna matched to the frequency of the service. Receivers are typically located remotely and interfaced with existing navigational and multifunction displays. Handheld GPS units also may have satellite weather capability. *[Figure 11-147]*

Emergency Locator Transmitter (ELT)

An emergency locator transmitter (ELT) is an independent battery powered transmitter activated by the excessive G-forces experienced during a crash. It transmits a digital signal every 50 seconds on a frequency of 406.025 MHz at 5 watts for at least 24 hours. The signal is received anywhere in the world by satellites in the Cospas-Sarsat satellite system. Two types of satellites, low earth orbiting satellites (LEOSATs) and geostationary satellites (GEOSATs), are used with different, complimentary capability. The signal is partially processed and stored in the satellites and then relayed to ground stations known as local user terminals (LUTs). Further deciphering of a signal takes place at the LUTs, and appropriate search and rescue operations are notified through mission control centers (MCCs) set up for this purpose.

Note: Maritime vessel emergency locating beacons (EPIRBs) and personal locator beacons (PLBs) use the exact same system. The United States portion of the Cospas-Sarsat system is

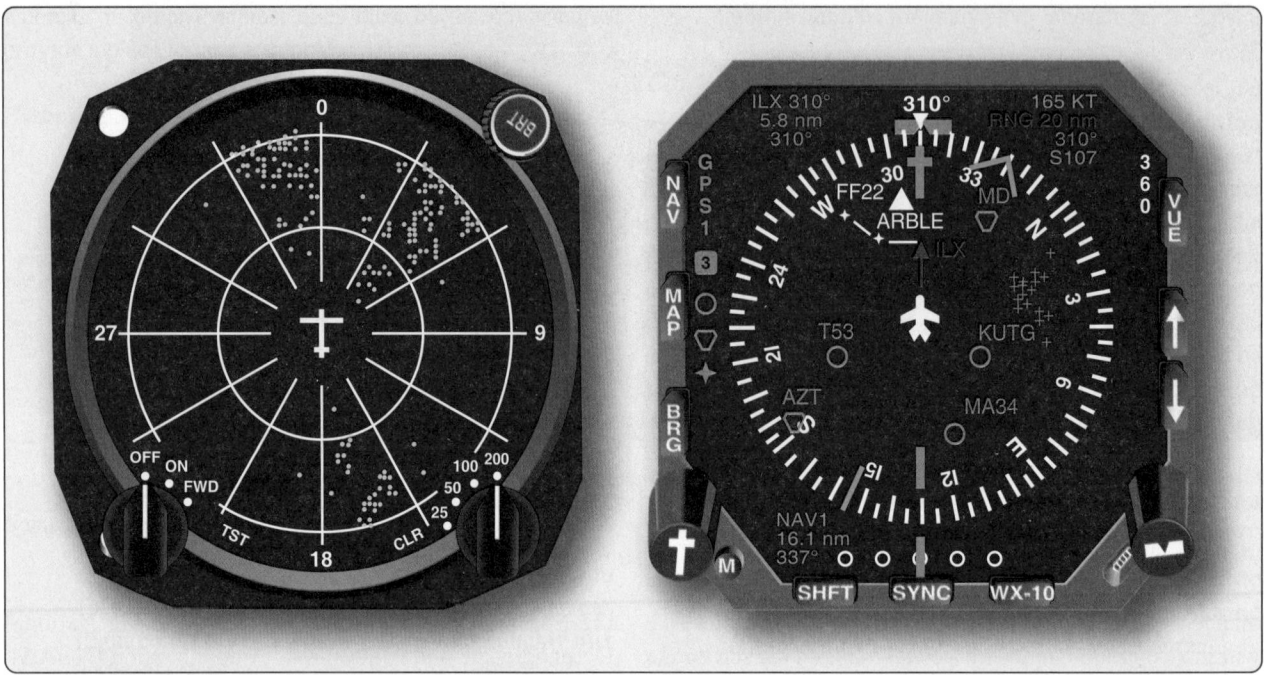

Figure 11-145. *A dedicated stormscope lightning detector display (left), and an electronic navigational display with lightning strikes overlaid in the form of green "plus" signs (right).*

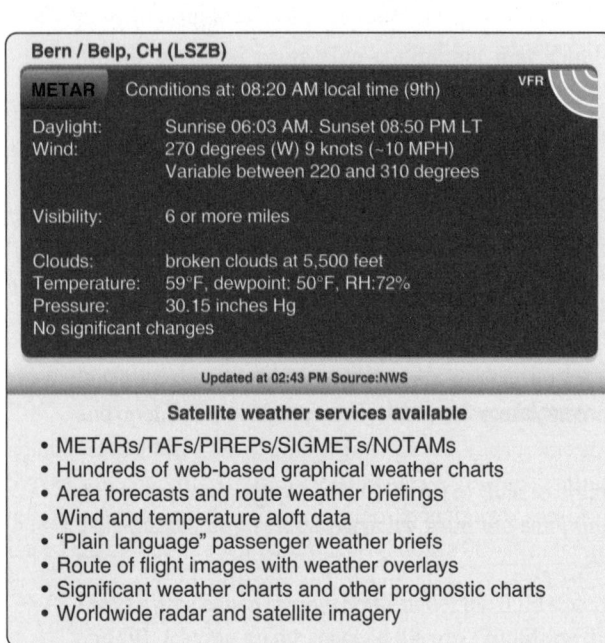

Figure 11-146. *A plain language METAR weather report received in the flight deck from a satellite weather service for aircraft followed by a list of various weather data that can be radioed to the flight deck from a satellite weather service.*

maintained and operated by NOAA. *Figure 11-148* illustrates the basic components in the Cospas-Sarsat system.

ELTs are required to be installed in aircraft according to FAR 91.207. This encompasses most general aviation aircraft not operating under parts 135 or 121. ELTs must be inspected within 12 months of previous inspection for proper

installation, battery corrosion, operation of the controls and crash sensor, and the presence of a sufficient signal at the antenna. Built-in test equipment facilitates testing without transmission of an emergency signal. The remainder of the inspection is visual. Technicians are cautioned to not activate the ELT and transmit an emergency distress signal. Inspection must be recorded in maintenance records including the new expiration date of the battery. This must also be recorded on the outside of the ELT.

ELTs are typically installed as far aft in the fuselage of an aircraft as is practicable just forward of the empennage. The built-in G-force sensor is aligned with the longitudinal axis of the aircraft. Helicopter ELTs may be located elsewhere on the airframe. They are equipped with multidirectional activation devices. Follow ELT and airframe manufacturer's instructions for proper installation, inspection, and maintenance of all ELTs. *Figure 11-149* illustrates ELTs mounted locations.

Use of Doppler technology enables the origin of the 406 MHz ELT signal to be calculated within 2 to 5 kilometers. Second generation 406 MHz ELT digital signals are loaded with GPS location coordinates from a receiver inside the ELT unit or integrated from an outside unit. This reduces the location accuracy of the crash site to within 100 meters. The digital signal is also loaded with unique registration information. It identifies the aircraft, the owner, and contact information, etc. When a signal is received, this is used to immediately research the validity of the alert to ensure it is a true emergency transmission so that rescue resources are not deployed needlessly.

Figure 11-147. *A satellite weather receiver and antenna enable display of real-time textual and graphic weather information beyond that of airborne weather radar. A handheld GPS can also be equipped with these capabilities. A built-in multifunctional display with satellite weather overlays and navigation information can be found on many aircraft.*

ELTs with automatic G-force activation mounted in aircraft are easily removable. They often contain a portable antenna so that crash victims may leave the site and carry the operating ELT with them. A flight deck mounted panel is required to alert the pilot if the ELT is activated. It also allows the ELT to be armed, tested, and manually activated if needed. *[Figure 11-150]*

Modern ELTs may also transmit a signal on 121.5 MHz. This is an analog transmission that can be used for homing. Prior to 2009, 121.5 MHz was a worldwide emergency frequency monitored by the Cospas-Sarsat satellites. However, it has been replaced by the 406 MHz standard. Transmissions on 121.5 MHz are no longer received and relayed via satellite.

The use of a 406 MHz ELT has not been mandated by the FAA. An older 121.5 MHz ELT satisfies the requirements of FAR Part 91.207 in all except new aircraft. Thousands of aircraft registered in the United States remain equipped with ELTs that transmit a .75 watt analog 121.5 MHz emergency signal when activated. The 121.5 MHz frequency is still an emergency frequency and is not monitored by over-flying aircraft and control towers.

Technicians are required to perform an inspection/test of 121.5 MHz ELTs within 12 months of the previous one and inspect for the same integrity as required for the 406 MHz ELTs mentioned above. However, older ELTs often lack the built-in test circuitry of modern ELTs certified to TSO C-126. Therefore, a true operational test may include activating the signal. This can be done by removing the antenna and installing a dummy load. Any activation of an ELT signal is required to only be done between the top of each hour and 5 minutes after the hour. The duration of activation must be no longer than three audible sweeps. Contact of the local control tower or flight service station before testing is recommended.

It must be noted that older 121.5 MHz analog signal ELTs often also transmit an emergency signal on a frequency of 243.0 MHz. This has long been the military emergency frequency. Its use is being phased out in favor of digital ELT signals and satellite monitoring. Testing the functionality of a 121.5 MHz transmitter should be accomplished per manufacturer's instructions. Improvements in coverage, location accuracy, identification of false alerts, and shortened response times are so significant with 406 MHz ELTs, they are currently the service standard worldwide.

Testing Considerations for 406 MHz ELTs

Care should be taken to prevent accidentally triggering a SAR (search and rescue) response. Accidental activation of an ELT will generate an emergency signal that cannot be distinguished from that of an actual emergency and could lead to expensive and frustrating searches. Moreover, the unwarranted ELT signal could tie up the emergency frequencies such that a genuine emergency signal would not be picked up. In addition, if an ELT signal is transmitted on or near an airport, it may render some radio communications channels unusable.

Regardless of where the ELT is, or the duration of activation, a 406 MHz beacon broadcast will be detected by at least one Geostationary Local User Terminal (GEOLUT) and possibly every Low Earth Orbit Local User Terminal (LEOLUT) in the Cospas-Sarsat System. Alert messages will be routed to every Mission Control Center (MCC) in the Cospas-Sarsat System for coordination around the world and a response will be made (unless prior coordination is made with Cospas-Sarsat and local authorities).

Direct connect testing is preferred to prevent inadvertent activation of the SAR response system. Over-air testing should always be avoided if possible. Use of an antenna boot

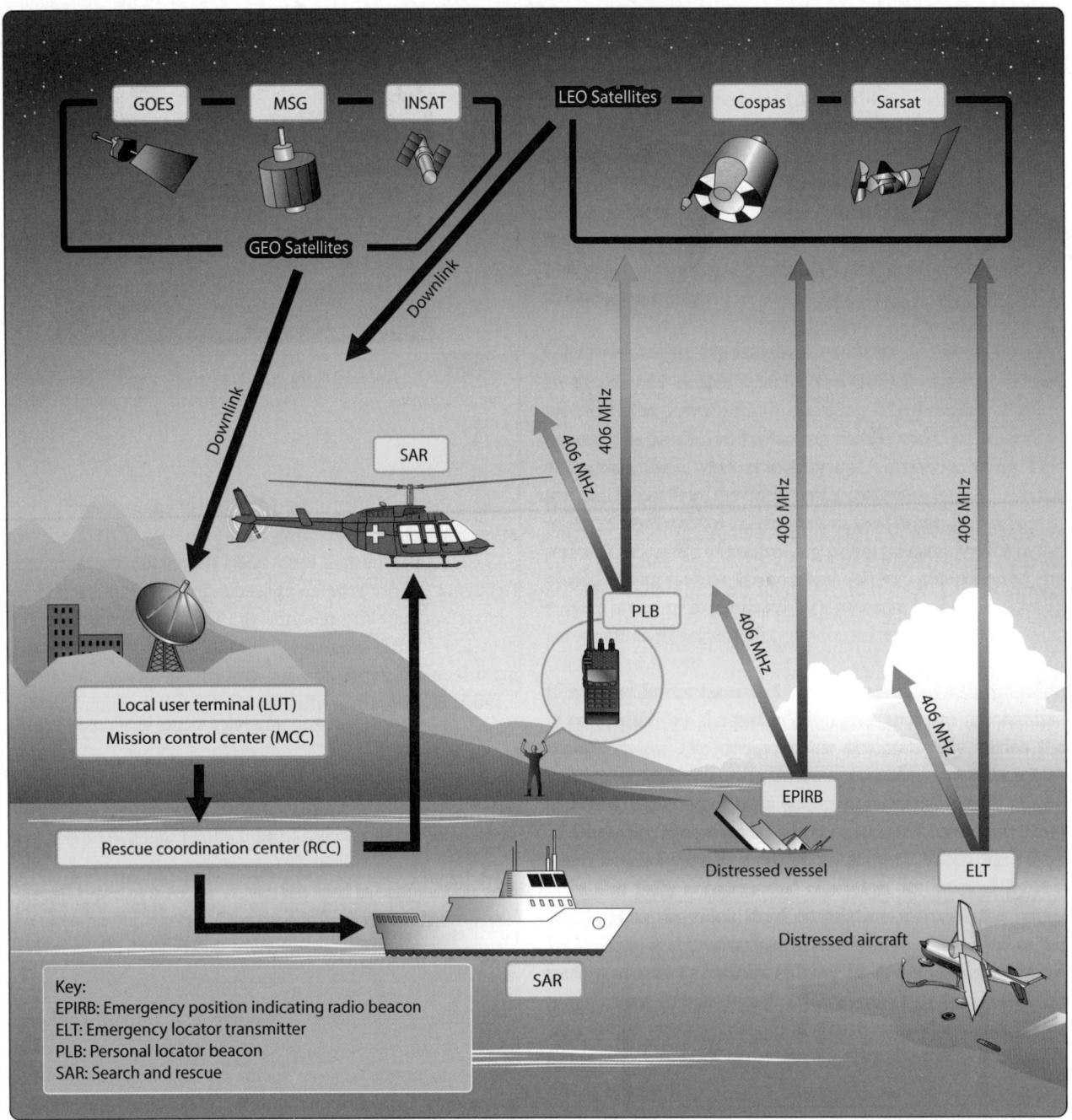

Figure 11-148. *The basic operating components of the satellite-based Cospas-Sarsat rescue system of which aircraft ELTs are a part.*

or a direct connection from test equipment to the antenna port is preferred.

Testing an ELT system in a metal hangar will not guarantee the radiated signal will not be detected by the Cospas-Sarsat System. Technicians testing ELT devices in a hangar should treat the test as if they were testing outside.

When testing an ELT, a 50-ohm dummy load or antenna boot should be used to prevent the signal from being radiated into space. The signal must be attenuated to less than -51 dBW (a power flux density of -37.4 dB (W/m^2) or a field intensity of -11.6 dB (V/m).

If over-air testing must be accomplished, technicians should carefully follow Cospas-Sarsat instructions and use the built-in test message on the ELT device. The ELT test message is different from messages transmitted during an emergency, but is still detectable by the Cospas-Sarsat System. Cospas-Sarsat should be contacted prior to performing over-air testing. Cospas-Sarsat can be contacted at www.cospas-sarsat.int/en/. If over-air testing must be accomplished, the local air traffic control (ATC) facility should be contacted in advance.

Follow test set instructions or place the test set a minimum

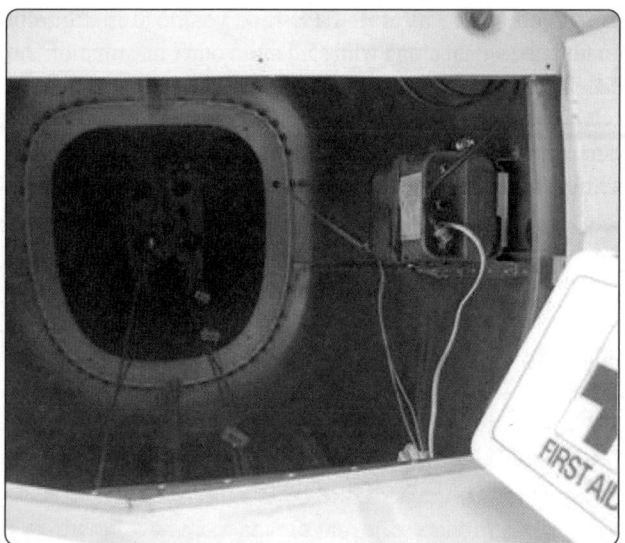

Figure 11-149. *An emergency locator transmitter (ELT) mounting location is generally far aft in a fixed-wing aircraft fuselage in line with the longitudinal axis. Helicopter mounting location and orientation varies.*

Figure 11-150. *An ELT and its components including a flight deck-mounted panel, the ELT, a permanent mount antenna, and a portable antenna.*

of 12 meters, (39.4 feet) from the ELT antenna. Test in each mode and frequency the ELT unit transmits.

406 MHz Testing

Verify the device is outputting a signal of not less than 17 dBm (50 mW) and not greater than 26 dBm (400 mW).

Verify the device is transmitting on the correct frequency. This can be done by running the ELT self-test and detecting the signal with an ELT test set. Receiving and decoding a

test message is an indication the unit is transmitting on a correct frequency.

Using appropriate test equipment and shielding, note the ELT code transmitted and verify that the ELT code is registered with Cospas-Sarsat. The testing technician or aircraft owner should verify the information on file with Cospas-Sarsat is accurate and up to date.

Determine that the ELT aural indicator can be heard in the flight deck with the aircraft engine(s) off, and that the visual indicator can be seen from the crew's normal sitting position. If possible, this should be performed in a way that will prevent a SAR response (e.g., with a dummy load installed).

Perform an operational check of the g switch. This should be performed in a way that will prevent a SAR response (e.g., with a dummy load installed). Replace if the g switch fails to activate.

Ensure all cables except the 406 MHz transmitter output are reconnected. Ensure the 406 MHz transmitter is connected to a test set if possible. Activate the ELT (use the remote switch if installed), and determine if the system is radiating a strong 406 MHz signal. Ensure the system is reset if necessary.

If equipped with a water-activated circuit, connect the ELT to a test set if possible. Activate the ELT by shorting the water-sensing leads and determine if the system is radiating a strong 406 MHz signal. Ensure the system is reset if necessary.

Long Range Aid to Navigation System (LORAN)

Long range aid to navigation system (LORAN) is a type of RNAV that is no longer available in the United States. It was developed during World War II, and the most recent edition, LORAN-C, has been very useful and accurate to aviators as well as maritime sailors. LORAN uses radio wave pulses from a series of towers and an on-board receiver/computer to positively locate an aircraft amid the tower network. There are twelve LORAN transmitter tower "chains" constructed across North America. Each chain has a master transmitter tower and a handful of secondary towers. All broadcasts from the transmitters are at the same frequency, 100 KHz. Therefore, a LORAN receiver does not need to be tuned. Being in the low frequency range, the LORAN transmissions travel long distances and provide good coverage from a small number of stations.

Precisely-timed, synchronized pulse signals are transmitted from the towers in a chain. The LORAN receiver measures the time to receive the pulses from the master tower and two other towers in the chain. It calculates the aircraft's position based on the intersection of parabolic curves representing elapsed signal times from each of these known points.

The accuracy and proliferation of GPS navigation has caused the U.S. Government to cease support for the LORAN navigation system citing redundancy and expense of operating the towers as reasons. Panel-mounted LORAN navigation units will likely be removed and replaced by GPS units in aircraft that have not already done so. *[Figure 11-151]*

Global Positioning System (GPS)

Global positioning system (GPS) navigation is the fastest growing type of navigation in aviation. It is accomplished through the use of NAVSTAR satellites set and maintained in orbit around the earth by the U.S. Government. Continuous coded transmissions from the satellites facilitate locating the position of an aircraft equipped with a GPS receiver with extreme accuracy. GPS can be utilized on its own for en route navigation, or it can be integrated into other navigation systems, such as VOR/RNAV, inertial reference, or flight management systems.

There are three segments of GPS: the space segment, the control segment, and the user segment. Aircraft technicians are only involved with user segment equipment such as GPS receivers, displays, and antennas.

Twenty-four satellites (21 active, 3 spares) in six separate planes of orbit 12,625 miles above the planet comprise what is known as the space segment of the GPS system. The satellites are positioned such that in any place on earth at any one time, at least four will be a minimum of 15° above the horizon. Typically, between five and eight satellites are in view. *[Figure 11-152]*

Two signals loaded with digitally coded information are transmitted from each satellite. The L1 channel transmission on a 1575.42 MHz carrier frequency is used in civilian aviation. Satellite identification, position, and time are conveyed to the aircraft GPS receiver on this digitally modulated signal along with status and other information. An L2 channel 1227.60 MHz transmission is used by the military.

The amount of time it takes for signals to reach the aircraft GPS receiver from transmitting satellites is combined with each satellite's exact location to calculate the position of an aircraft. The control segment of the GPS monitors each satellite to ensure its location and time are precise. This control is accomplished with five ground-based receiving stations, a master control station, and three transmitting antennas. The receiving stations forward status information received from the satellites to the master control station. Calculations are made, and corrective instructions are sent to the satellites via the transmitters.

The user segment of the GPS is comprised of the thousands of receivers installed in aircraft as well as every other receiver that uses the GPS transmissions. Specifically, for the aircraft technician, the user section consists of a control panel/display, the GPS receiver circuitry, and an antenna. The control, display and receiver are usually located in a single unit which also may include VOR/ILS circuitry and a VHF communications transceiver. GPS intelligence is integrated into the multifunctional displays of glass flight deck aircraft. *[Figure 11-153]*

The GPS receiver measures the time it takes for a signal to arrive from three transmitting satellites. Since radio waves travel at 186,000 miles per second, the distance to each satellite can be calculated. The intersection of these ranges provides a two dimensional position of the aircraft. It is expressed in latitude/longitude coordinates. By incorporating

Figure 11-151. *Panel-mounted LORAN units are now obsolete as LORAN signals are no longer generated from the tower network.*

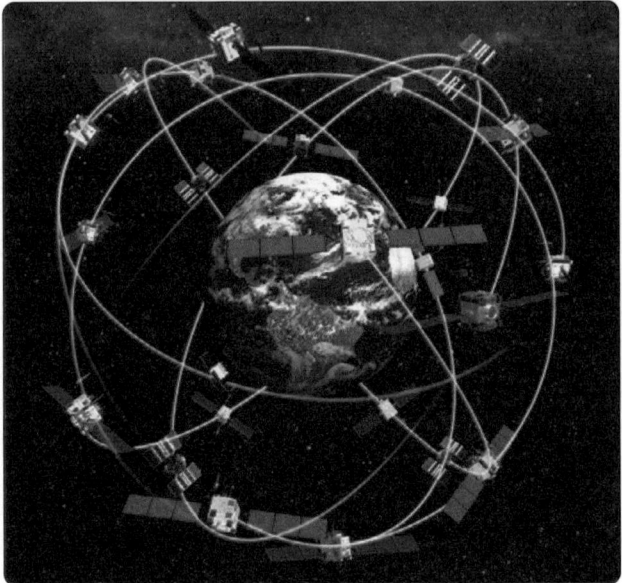

Figure 11-152. *The space segment of GPS consists of 24 NAVSTAR satellites in six different orbits around the earth.*

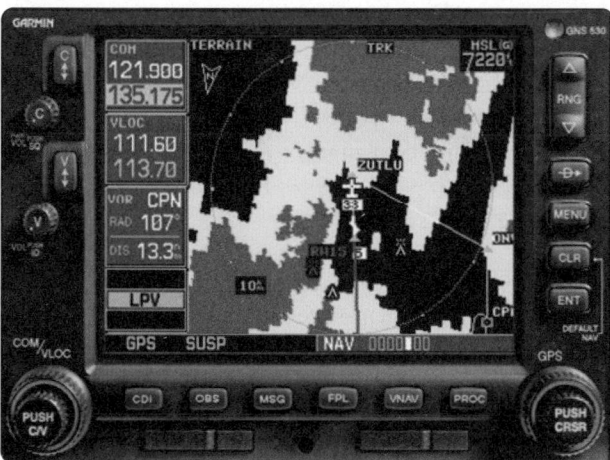

Figure 11-153. *A GPS unit integrated with NAV/COM circuitry.*

the distance to a fourth satellite, the altitude above the surface of the earth can be calculated as well. This results in a three dimensional fix. Additional satellite inputs refine the accuracy of the position.

Having deciphered the position of the aircraft, the GPS unit processes many useful navigational outputs such as speed, direction, bearing to a waypoint, distance traveled, time of arrival, and more. These can be selected to display for use. Waypoints can be entered and stored in the unit's memory. Terrain features, airport data, VOR/RNAV and approach information, communication frequencies, and more can also be loaded into a GPS unit. Most modern units come with moving map display capability.

A main benefit of GPS use is immunity from service disruption due to weather. Errors are introduced while the carrier waves travel through the ionosphere; however, these are corrected and kept to a minimum. GPS is also relatively inexpensive. GPS receivers for IFR navigation in aircraft must be built to TSO-129A. This raises the price above that of handheld units used for hiking or in an automobile. But the overall cost of GPS is low due to its small infrastructure. Most of the inherent accuracy is built into the space and control segments permitting reliable positioning with inexpensive user equipment.

The accuracy of current GPS is within 20 meters horizontally and a bit more vertically. This is sufficient for en route navigation with greater accuracy than required. However, departures and approaches require more stringent accuracy. Integration of the wide area augmentation system (WAAS) improves GPS accuracy to within 7.6 meters and is discussed below. The future of GPS calls for additional accuracy by adding two new transmissions from each satellite. An L2C channel is for general use in non-safety critical application. An aviation dedicated L5 channel provides the accuracy

required for category I, II, and III landings. It enables the NextGen NAS plan along with ADS-B.

Wide Area Augmentation System (WAAS)

To increase the accuracy of GPS for aircraft navigation, the wide area augmentation system (WAAS) was developed. It consists of approximately 25 precisely surveyed ground stations that receive GPS signals and ultimately transmit correction information to the aircraft. An overview of WAAS components and its operation is shown in *Figure 11-154*.

WAAS ground stations receive GPS signals and forward position errors to two master ground stations. Time and location information is analyzed, and correction instructions are sent to communication satellites in geostationary orbit over the NAS. The satellites broadcast GPS-like signals that WAAS enabled GPS receivers use to correct position information received from GPS satellites.

A WAAS-enabled GPS receiver is required to use the wide area augmentation system. If equipped, an aircraft qualifies to perform precision approaches into thousands of airports without any ground-based approach equipment. Separation minimums are also able to be reduced between aircraft that are WAAS equipped. The WAAS system is known to reduce position errors to 1–3 meters laterally and vertically.

Inertial Navigation System (INS)/Inertial Reference System (IRS)

An inertial navigation system (INS) is used on some large aircraft for long range navigation. This may also be

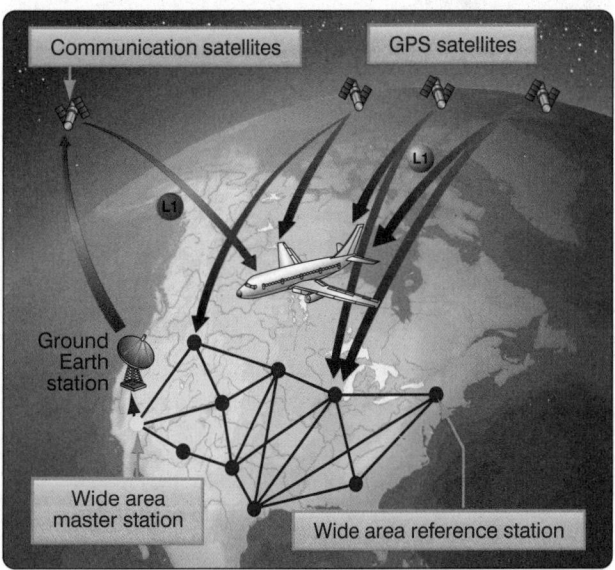

Figure 11-154. *The wide area augmentation system (WAAS) is used to refine GPS positions to a greater degree of accuracy. A WAAS enabled GPS receiver is required for its use as corrective information is sent from geostationary satellites directly to an aircraft's GPS receiver for use.*

identified as an inertial reference system (IRS), although the IRS designation is generally reserved for more modern systems. An INS/IRS is a self-contained system that does not require input radio signals from a ground navigation facility or transmitter. The system derives attitude, velocity, and direction information from measurement of the aircraft's accelerations given a known starting point. The location of the aircraft is continuously updated through calculations based on the forces experienced by INS accelerometers. A minimum of two accelerometers is used, one referenced to north, and the other referenced to east. In older units, they are mounted on a gyro-stabilized platform. This averts the introduction of errors that may result from acceleration due to gravity.

An INS uses complex calculation made by an INS computer to convert applied forces into location information. An interface control head is used to enter starting location position data while the aircraft is stationary on the ground. This is called initializing. *[Figure 11-155]* From then on, all motion of the aircraft is sensed by the built-in accelerometers and run through the computer. Feedback and correction loops are used to correct for accumulated error as flight time progresses. The amount an INS is off in one hour of flight time is a reference point for determining performance. Accumulated error of less than one mile after one hour of flight is possible. Continuous accurate adjustment to the gyro-stabilized platform to keep it parallel to the Earth's surface is a key requirement to reduce accumulated error. A latitude/longitude coordinate system is used when giving the location output.

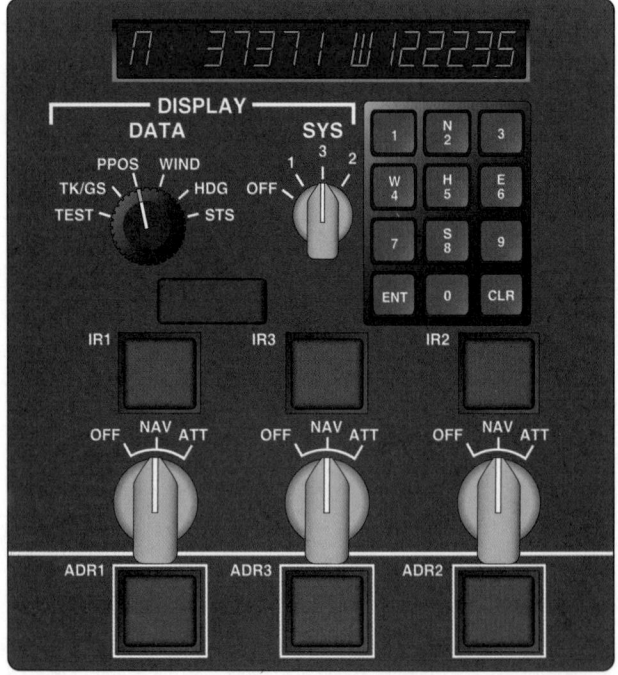

Figure 11-155. *An interface panel for three air data and inertial reference systems on an Airbus. The keyboard is used to initialize the system. Latitude and longitude position is displayed at the top.*

INS is integrated into an airliner's flight management system and automatic flight control system. Waypoints can be entered for a predetermined flightpath and the INS will guide the aircraft to each waypoint in succession. Integration with other NAV aids is also possible to ensure continuous correction and improved accuracy but is not required.

Modern INS systems are known as IRS. They are completely solid-state units with no moving parts. Three-ring, laser gyros replace the mechanical gyros in the older INS platform systems. This eliminates precession and other mechanical gyro shortcomings. The use of three solid-state accelerometers, one for each plane of movement, also increases accuracy. The accelerometer and gyro output are input to the computer for continuous calculation of the aircraft's position.

The most modern IRS integration is the satellite GPS. The GPS is extremely accurate in itself. When combined with IRS, it creates one of the most accurate navigation systems available. The GPS is used to initialize the IRS so the pilot no longer needs to do so. GPS also feeds data into the IRS computer to be used for error correction. Occasional service interruptions and altitude inaccuracies of the GPS system pose no problem for IRS/GPS. The IRS functions continuously and is completely self-contained within the IRS unit. Should the GPS falter, the IRS portion of the system continues without it. The latest electronic technology has reduced the size and weight of INS/IRS avionics units significantly. *Figure 11-156* shows a modern micro-IRS unit that measures approximately six inches on each side.

Aircraft Communication Addressing & Reporting System (ACARS)

ACARS is a two-way communication link between an airliner in flight and the airline's main ground facilities. Data is collected in the aircraft by digital sensors and is transmitted to the ground facilities. Replies from the ground may be printed out so the appropriate flight crewmember can have a hard copy of the response.

Installation of Communication & Navigation Equipment

Approval of New Avionics Equipment Installations

Most of the avionics equipment discussed in this chapter is only repairable by the manufacturer or FAA-certified repair stations that are licensed to perform specific work. The airframe technician; however, must competently remove, install, inspect, maintain, and troubleshoot these ever increasingly complicated electronic devices and systems. It is imperative to follow all equipment and airframe manufacturers' instruction

Figure 11-156. *A modern micro-IRS with built-in GPS.*

when dealing with an aircraft's avionics.

The revolution to GPS navigation and the pace of modern electronic development results in many aircraft owner operators upgrading flight decks with new avionics. The aircraft technician must only perform airworthy installations. The avionics equipment to be installed must be a TSO'd device that is approved for installation in the aircraft in question. The addition of a new piece of avionics equipment and/or its antenna is a minor alteration if previously approved by the airframe manufacturer. A certificated airframe technician is qualified to perform the installation and return the aircraft to service. The addition of new avionics not on the aircraft's approved equipment list is considered a major alteration and requires an FAA Form 337 to be completed. A technician with an inspection authorization is required to complete the approval for return to service following a major alteration and to sign the corresponding approval for return to service block on Form 337.

Most new avionics installations are approved and performed under an STC. The equipment manufacturer supplies a list of aircraft on which the equipment has been approved for installation. The STC includes thorough installation and maintenance instructions which the technician must follow. Regardless, if not on the aircraft's original equipment list, the STC installation is considered a major alteration and an FAA Form 337 must be filed. The STC is referenced as the required approved data.

Occasionally, an owner/operator or technician wishes to install an electronic device in an aircraft that has no STC for the model aircraft in question. A field approval and a Form 337 must be filed on which it must be shown that the installation

will be performed in accordance with approved data.

Considerations

There are many factors which the technician must consider prior to altering an aircraft by the addition of avionics equipment. These factors include the space available, the size and weight of the equipment, and previously accomplished alterations. The power consumption of the added equipment must be considered to calculate and determine the maximum continuous electrical load on the aircraft's electrical system. Each installation should also be planned to allow easy access for inspection, maintenance, and exchange of units.

The installation of avionics equipment is partially mechanical, involving sheet metal work to mount units, racks, antennas, and controls. Routing of the interconnecting wires, cables, antenna leads, etc. is also an important part of the installation process. When selecting a location for the equipment, use the area(s) designated by the airframe manufacturer or the STC. If such information is not available, select a location for installation that will carry the loads imposed by the weight of the equipment, and which is capable of withstanding the additional inertia forces.

If an avionics device is to be mounted in the instrument panel and no provisions have been made for such an installation, ensure that the panel is not a primary structure prior to making any cutouts. To minimize the load on a stationary instrument panel, a support bracket may be installed between the rear of the electronics case or rack and a nearby structural member of the aircraft. *[Figure 11-157]*

Avionics radio equipment must be securely mounted to the aircraft. All mounting bolts must be secured by locking devices to prevent loosening from vibration. Adequate clearance between all units and adjacent structure must be provided to prevent mechanical damage to electric wiring or to the avionic equipment from vibration, chafing, or landing shock.

Do not locate avionics equipment and wiring near units containing combustible fluids. When separation is impractical, install baffles or shrouds to prevent contact of the combustible fluids with any electronic equipment in the event of plumbing failure.

Cooling & Moisture

The performance and service life of most avionics equipment is seriously limited by excessive ambient temperatures. High performance aircraft with avionics equipment racks typically route air-conditioned air over the avionics to keep them cool. It is also common for non-air conditioned aircraft to use a blower or scooped ram air to cool avionics installations. When adding a unit to an aircraft, the installation should

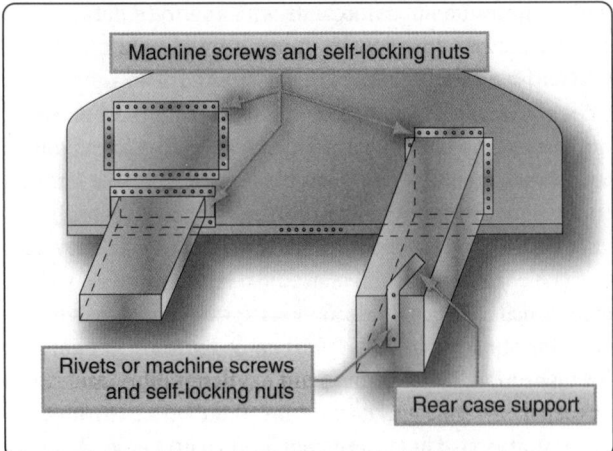

Figure 11-157. *An avionics installation in a stationary instrument panel may include a support for the avionics case.*

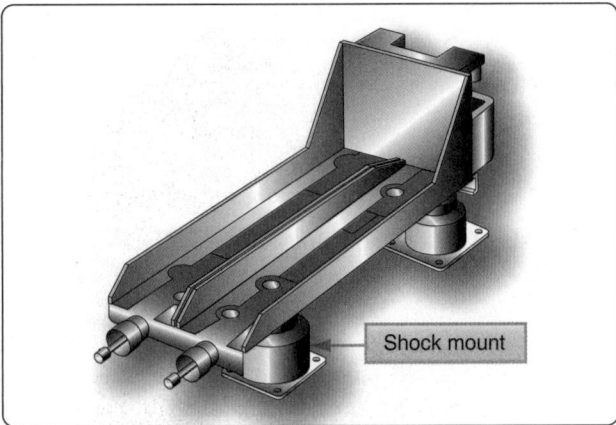

Figure 11-158. *A shock mounted equipment rack is often used to install avionics.*

be planned so that it can dissipate heat readily. In some installations, it may be necessary to produce airflow over the new equipment either with a blower or through the use of routed ram air. Be sure that proper baffling is used to prevent water from reaching any electronics when ducting outside air. The presence of water in avionics equipment areas promotes rapid deterioration of the exposed components and could lead to failure.

Vibration Isolation

Vibration is a continued motion by an oscillating force. The amplitude and frequency of vibration of the aircraft structure will vary considerably with the type of aircraft. Avionics equipment is sensitive to mechanical shock and vibration and is normally shock mounted to provide some protection against in-flight vibration and landing shock.

Special shock mounted racks are often used to isolate avionics equipment from vibrating structure. *[Figure 11-158]* Such mounts should provide adequate isolation over the entire range of expected vibration frequencies. When installing shock mounts, assure that the equipment weight does not exceed the weight-carrying capabilities of the mounts. Radio equipment installed on shock mounts must have sufficient clearance from surrounding equipment and structure to allow for normal swaying of the equipment.

Radios installed in instrument panels do not ordinarily require vibration protection since the panel itself is usually shock mounted. However, make certain that the added weight of any added equipment can be safely carried by the existing mounts. In some cases, it may be necessary to install larger capacity mounts or to increase the number of mounting points.

Periodic inspection of the shock mounts is required, and defective mounts should be replaced with the proper type. The following factors to observe during the inspection are:

1. Deterioration of the shock-absorbing material;

2. Stiffness and resiliency of the material; and

3. Overall rigidity of the mount.

If the mount is too stiff, it may not provide adequate protection against the shock of landing. If the shock mount is not stiff enough, it may allow prolonged vibration following an initial shock.

Shock-absorbing materials commonly used in shock mounts are usually electrical insulators. For this reason, each electronic unit mounted with shock mounts must be electrically bonded to a structural member of the aircraft to provide a current path to ground. This is accomplished by secure attachment of a tinned copper wire braid from the component, across the mount, to the aircraft structure as shown in *Figure 11-159*. Occasional bonding is accomplished with solid aluminum or copper material where a short flexible strap is not possible.

Reducing Radio Interference

Suppression of unwanted electromagnetic fields and electrostatic interference is essential on all aircraft. In communication radios, this is noticeable as audible noise. In other components, the effects may not be audible but pose a threat to proper operation. Large discharges of static electricity can permanently damage the sensitive solid-state microelectronics found in nearly all modern avionics.

Shielding

Many components of an aircraft are possible sources of electrical interference which can deteriorate the performance and reliability of avionics components. Rotating electrical devices, switching devices, ignition systems, propeller control systems, AC power lines, and voltage regulators all produce potential damaging fields. Shielding wires to

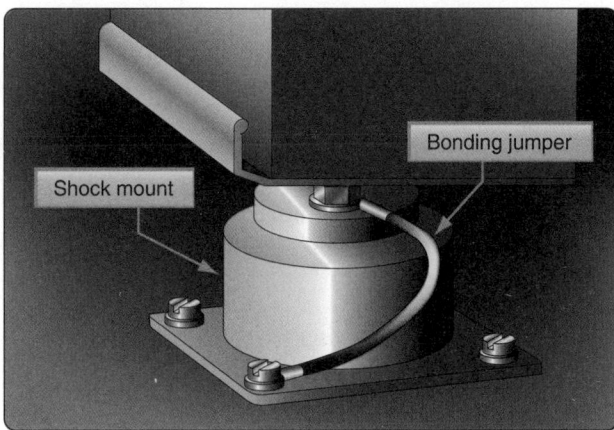

Figure 11-159. *A bonding jumper is used to ground an equipment rack and avionics chassis around the non-conductive shock mount material.*

electric components and ignition systems dissipates radio frequency noise energy. Instead of radiating into space, the braided conductive shielding guides unwanted current flows to ground. To prevent the build-up of electrical potential, all electrical components should also be bonded to the aircraft structure (ground).

Isolation

Isolation is another practical method of radio frequency suppression to prevent interference. This involves separating the source of the noise from the input circuits of the affected equipment. In some cases, noise in a receiver may be entirely eliminated simply by moving the antenna lead-in wire just a few inches away from a noise source. On other occasions, when shielding and isolation are not effective, a filter may need to be installed in the input circuit of an affected component.

Bonding

The aircraft surface can become highly charged with static electricity while in flight. Measures are required to eliminate the build-up and radiation of unwanted electrical charges. One of the most important measures taken to eliminate unwanted electrical charges which may damage or interfere with avionics equipment is bonding. Charges flowing in paths of variable resistance due to such causes as intermittent contact from vibration or the movement of a control surface produce electrical disturbances (noise) in avionics. Bonding provides the necessary electric connection between metallic parts of an aircraft to prevent variable resistance in the airframe. It provides a low-impedance ground return which minimizes interference from static electricity charges.

All metal parts of the aircraft should be bonded to prevent the development of electrical potential build-up. Bonding also provides the low resistance return path for single-wire electrical systems. Bonding jumpers and clamps are examples of bonding connectors. Jumpers should be as short

as possible. Be sure finishes are removed in the contact area of a bonding device so that metal-to-metal contact exists. Resistance should not exceed .003 ohm. When a jumper is used only to reduce radio frequency noise and is not for current carrying purposes, a resistance of 0.01 ohm is satisfactory.

Static Discharge Wicks

Static dischargers, or wicks, are installed on aircraft to reduce radio receiver interference. This interference is caused by corona discharge emitted from the aircraft as a result of precipitation static. Corona occurs in short pulses which produce noise at the radio frequency spectrum. Static dischargers are normally mounted on the trailing edges of the control surfaces, wing tips and the vertical stabilizer. They discharge precipitation static at points a critical distance away from avionics antennas where there is little or no coupling of the static to cause interference or noise.

Flexible and semi-flexible dischargers are attached to the aircraft structure by metal screws, rivets, or epoxy. The connections should be checked periodically for security. A resistance measurement from the mount to the airframe should not exceed 0.1 ohm. Inspect the condition of all static dischargers in accordance with manufacturer's instructions. *Figure 11-160* illustrates examples of static dischargers.

Installation of Aircraft Antenna Systems

Knowledge of antenna installation and maintenance is especially important as these tasks are performed by the aircraft technician. Antennas take many forms and sizes dependent upon the frequency of the transmitter and receiver to which they are connected. Airborne antennas must be mechanically secure. The air loads on an antenna are significant and must be considered. Antennas must be electrically matched to the receiver and transmitter that they serve. They must also be mounted in interference free locations and in areas where signals can be optimally transmitted and received. Antennas must also have the same polarization as the ground station.

The following procedures describe the installation of a typical rigid antenna. They are presented as an example only. Always follow the manufacturer's instructions when installing any antenna. An incorrect antenna installation could cause equipment failure.

1. Place a template similar to that shown in *Figure 11-161* on the fore-and-aft centerline at the desired location. Drill the mounting holes and correct diameter hole for the transmission line cable in the fuselage skin.

2. Install a reinforcing doubler of sufficient thickness to reinforce the aircraft skin. The length and width of

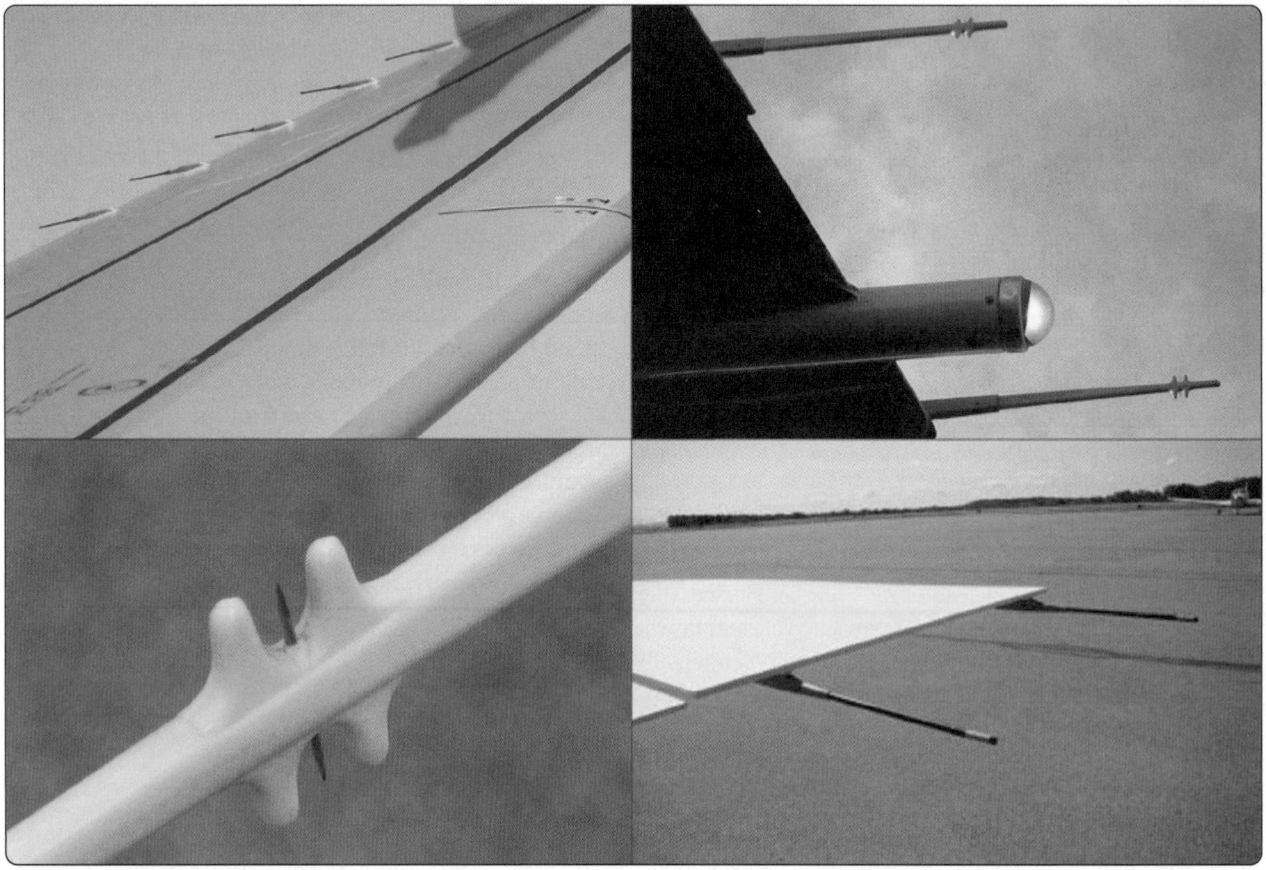

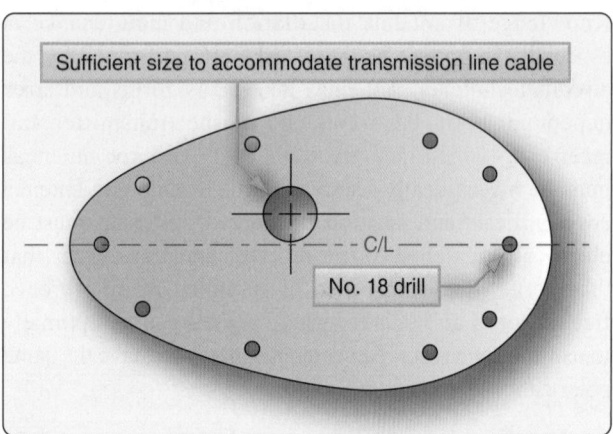

Figure 11-160. *Static dischargers or wicks dissipate built up static energy in flight at points a safe distance from avionics antennas to prevent radio frequency interference.*

Sufficient size to accommodate transmission line cable

C/L

No. 18 drill

Figure 11-161. *A typical antenna mounting template.*

the reinforcing plate should approximate the example shown in *Figure 11-162*.

3. Install the antenna on the fuselage, making sure that the mounting bolts are tightened firmly against the reinforcing doubler, and the mast is drawn tight against the gasket. If a gasket is not used, seal between the mast and the fuselage with a suitable sealer, such as zinc chromate paste or equivalent.

The mounting bases of antennas vary in shape and sizes; however, the aforementioned installation procedure is typical of mast-type antenna installations.

Transmission Lines

A transmitting or receiving antenna is connected directly to its associated transmitter or receiver by a transmission line. This is a shielded wire also known as coax. Transmission lines may vary from only a few feet to many feet in length. They must transfer energy with minimal loss. Transponders, DME and other pulse type transceivers require transmission lines that are precise in length. The critical length of transmission lines provides minimal attenuation of the transmitted or received signal. Refer to the equipment manufacturer's installation manual for the type and allowable length of transmission lines.

To provide the proper impedance matching for the most efficient power transfer, a balun may be used in some antenna installations. It is formed in the transmission line connection to the antenna. A balun in a dipole antenna installation is illustrated in *Figure 11-163*.

Coax connectors are usually used with coax cable to ensure a secure connection. Many transmission lines are part of the equipment installation kit with connectors previously installed. The aircraft technician is also able to install these

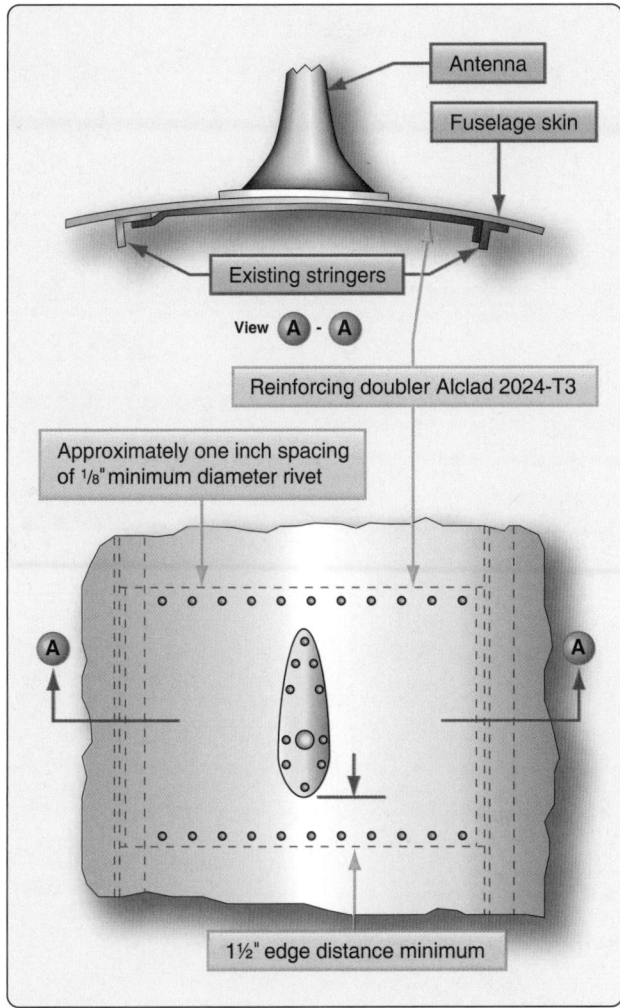

Figure 11-162. *A typical antenna installation on a skin panel including a doubler.*

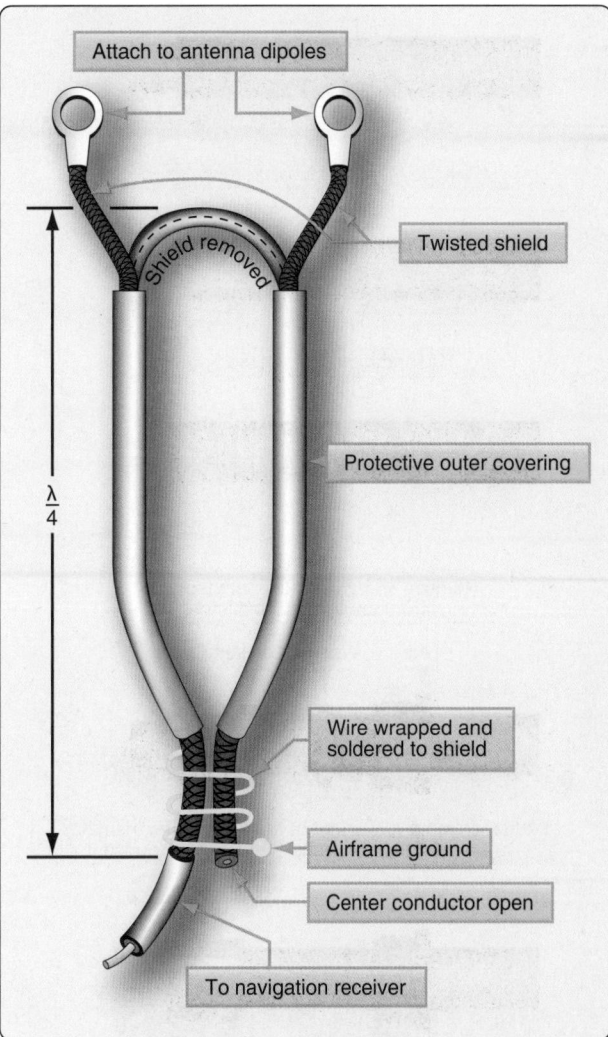

Figure 11-163. *A balun in a dipole antenna installation provides the proper impedance for efficient power transfer.*

connectors on coax. *Figure 11-164* illustrates the basic steps used when installing a coax cable connector.

When installing coaxial cable, secure the cables firmly along their entire length at intervals of approximately 2 feet. To assure optimum operation, coaxial cables should not be routed or tied to other wire bundles. When bending coaxial cable, be sure that the bend is at least 10 times the size of the cable diameter. In all cases, follow the equipment manufacturer's instructions.

Maintenance Procedures

Detailed instructions, procedures, and specifications for the servicing of avionics equipment are contained in the manufacturer's operating manuals. Additional instructions for removal and installation of the units are contained in the maintenance manual for the aircraft in which the equipment is installed. Although an installation may appear to be a simple procedure, many avionics troubles are attributed to

careless oversights during equipment replacement. Loose cable connections, switched cable terminations, improper bonding, worn shock mounts, improper safety wiring, and failure to perform an operational check after installation may result in poor performance or inoperative avionics.

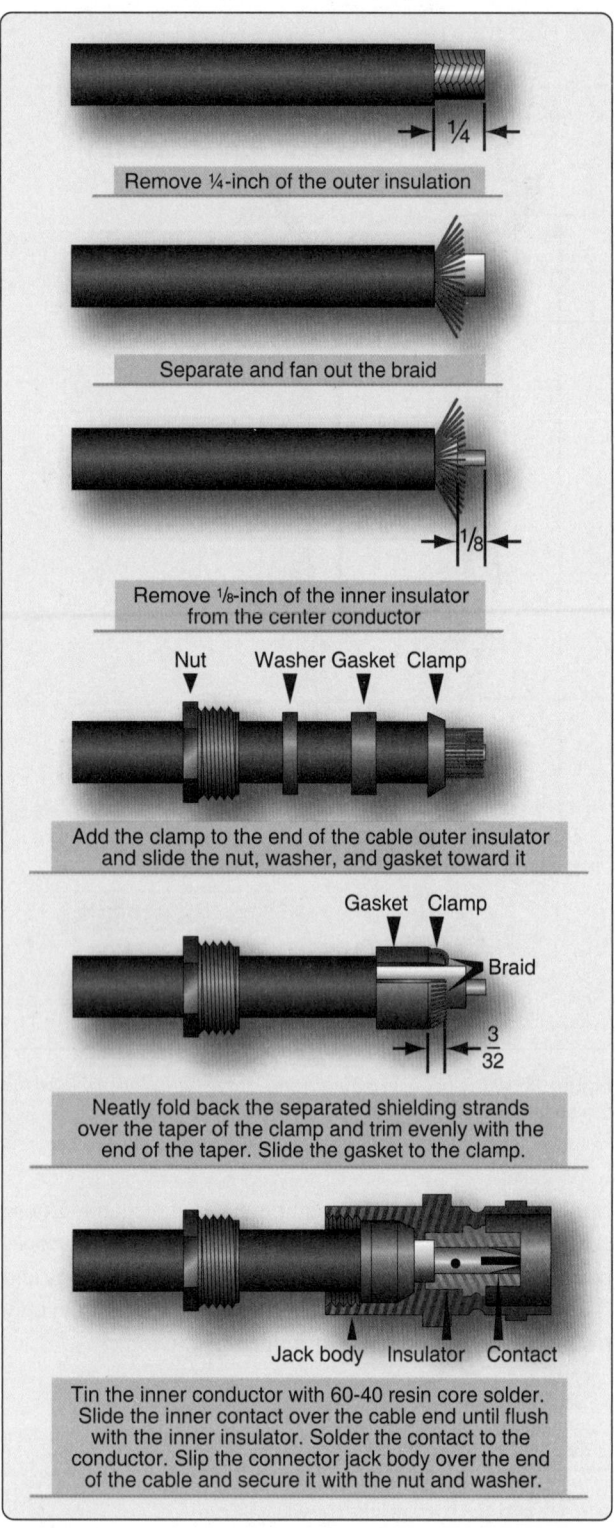

Remove ¼-inch of the outer insulation

Separate and fan out the braid

Remove ⅛-inch of the inner insulator
from the center conductor

Nut   Washer Gasket Clamp

Add the clamp to the end of the cable outer insulator
and slide the nut, washer, and gasket toward it

Gasket   Clamp

Braid

$\frac{3}{32}$

Neatly fold back the separated shielding strands
over the taper of the clamp and trim evenly with the
end of the taper. Slide the gasket to the clamp.

Jack body   Insulator   Contact

Tin the inner conductor with 60-40 resin core solder.
Slide the inner contact over the cable end until flush
with the inner insulator. Solder the contact to the
conductor. Slip the connector jack body over the end
of the cable and secure it with the nut and washer.

Figure 11-164. *Steps in attaching a connector to coax cable used as antenna transmission lines.*

Chapter 12
Hydraulic & Pneumatic Power Systems

Aircraft Hydraulic Systems

The word "hydraulics" is based on the Greek word for water and originally meant the study of the physical behavior of water at rest and in motion. Today, the meaning has been expanded to include the physical behavior of all liquids, including hydraulic fluid. Hydraulic systems are not new to aviation. Early aircraft had hydraulic brake systems. As aircraft became more sophisticated, newer systems with hydraulic power were developed.

Hydraulic systems in aircraft provide a means for the operation of aircraft components. The operation of landing gear, flaps, flight control surfaces, and brakes is largely accomplished with hydraulic power systems. Hydraulic system complexity varies from small aircraft that require fluid only for manual operation of the wheel brakes to large transport aircraft where the systems are large and complex. To achieve the necessary redundancy and reliability, the system may consist of several subsystems. Each subsystem has a power generating device (pump), reservoir, accumulator, heat exchanger, filtering system, etc. System operating pressure may vary from a couple hundred pounds per square inch (psi) in small aircraft and rotorcraft to 5,000 psi in large transports. Hydraulic systems have many advantages as power sources for operating various aircraft units; they combine the advantages of light weight, ease of installation, simplification of inspection, and minimum maintenance requirements. Hydraulic operations are also almost 100 percent efficient, with only negligible loss due to fluid friction.

Hydraulic Fluid

Hydraulic system liquids are used primarily to transmit and distribute forces to various units to be actuated. Liquids are able to do this because they are almost incompressible. Pascal's Law states that pressure applied to any part of a confined liquid is transmitted with undiminished intensity to every other part. Thus, if a number of passages exist in a system, pressure can be distributed through all of them by means of the liquid.

Manufacturers of hydraulic devices usually specify the type of liquid best suited for use with their equipment in view of the working conditions, the service required, temperatures expected inside and outside the systems, pressures the liquid must withstand, the possibilities of corrosion, and other conditions that must be considered. If incompressibility and fluidity were the only qualities required, any liquid that is not too thick could be used in a hydraulic system. But a satisfactory liquid for a particular installation must possess a number of other properties. Some of the properties and characteristics that must be considered when selecting a satisfactory liquid for a particular system are discussed in the following paragraphs.

Viscosity

One of the most important properties of any hydraulic fluid is its viscosity. Viscosity is internal resistance to flow. A liquid such as gasoline that has a low viscosity flows easily, while a liquid such as tar that has a high viscosity flows slowly. Viscosity increases as temperature decreases. A satisfactory liquid for a given hydraulic system must have enough body to give a good seal at pumps, valves, and pistons, but it must not be so thick that it offers resistance to flow, leading to power loss and higher operating temperatures. These factors add to the load and to excessive wear of parts. A fluid that is too thin also leads to rapid wear of moving parts or of parts that have heavy loads. The instruments used to measure the viscosity of a liquid are known as viscometers or viscosimeters. Several types of viscometers are in use today. The Saybolt viscometer measures the time required, in seconds, for 60 milliliters of the tested fluid at 100 °F to pass through a standard orifice. The time measured is used to express the fluid's viscosity, in Saybolt universal seconds or Saybolt FUROL seconds. *[Figure 12-1]*

Chemical Stability

Chemical stability is another property that is exceedingly important in selecting a hydraulic liquid. It is the liquid's ability to resist oxidation and deterioration for long periods. All liquids tend to undergo unfavorable chemical changes under severe operating conditions. This is the case, for example, when a system operates for a considerable period of time at high temperatures. Excessive temperatures have a great effect on the life of a liquid. It should be noted that the temperature of the liquid in the reservoir of an operating hydraulic system does not always represent a true state of operating conditions. Localized hot spots occur on bearings,

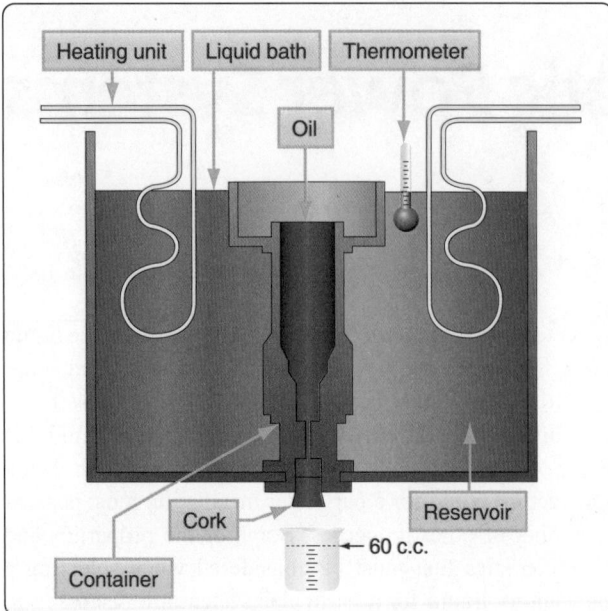

Figure 12-1. *Saybolt viscosimeter.*

gear teeth, or at the point where liquid under pressure is forced through a small orifice. Continuous passage of a liquid through these points may produce local temperatures high enough to carbonize or sludge the liquid, yet the liquid in the reservoir may not indicate an excessively high temperature.

Liquids with a high viscosity have a greater resistance to heat than light or low-viscosity liquids that have been derived from the same source. The average hydraulic liquid has a low viscosity. Fortunately, there is a wide choice of liquids available for use within the viscosity range required of hydraulic liquids.

Liquids may break down if exposed to air, water, salt, or other impurities, especially if they are in constant motion or subject to heat. Some metals, such as zinc, lead, brass, and copper, have an undesirable chemical reaction on certain liquids. These chemical processes result in the formation of sludge, gums, and carbon or other deposits that clog openings, cause valves and pistons to stick or leak, and give poor lubrication to moving parts. As soon as small amounts of sludge or other deposits are formed, the rate of formation generally increases more rapidly. As they are formed, certain changes in the physical and chemical properties of the liquid take place. The liquid usually becomes darker in color, higher in viscosity, and acids are formed.

Flash Point

Flash point is the temperature at which a liquid gives off vapor in sufficient quantity to ignite momentarily or flash when a flame is applied. A high flash point is desirable for hydraulic liquids because it indicates good resistance to combustion and

a low degree of evaporation at normal temperatures.

Fire Point

Fire point is the temperature at which a substance gives off vapor in sufficient quantity to ignite and continue to burn when exposed to a spark or flame. Like flash point, a high fire point is required of desirable hydraulic liquids.

Types of Hydraulic Fluids

To assure proper system operation and to avoid damage to nonmetallic components of the hydraulic system, the correct fluid must be used. When adding fluid to a system, use the type specified in the aircraft manufacturer's maintenance manual or on the instruction plate affixed to the reservoir or unit being serviced.

The three principal categories of hydraulic fluids are:

1. Minerals.
2. Polyalphaolefins.
3. Phosphate esters.

When servicing a hydraulic system, the technician must be certain to use the correct category of replacement fluid. Hydraulic fluids are not necessarily compatible. For example, contamination of the fire-resistant fluid MIL-H-83282 with MIL-H-5606 may render the MIL-H-83282 non-fire-resistant.

Mineral-Based Fluids

Mineral oil-based hydraulic fluid (MIL-H-5606) is the oldest, dating back to the 1940s. It is used in many systems, especially where the fire hazard is comparatively low. MIL-H-6083 is simply a rust-inhibited version of MIL-H-5606. They are completely interchangeable. Suppliers generally ship hydraulic components with MIL-H-6083. Mineral-based hydraulic fluid (MIL–H-5606) is processed from petroleum. It has an odor similar to penetrating oil and is dyed red. Some synthetic hydraulic fluids are dyed purple and even green, depending on the identity of the fluid. Synthetic rubber seals are used with petroleum-based fluids.

Polyalphaolefin-Based Fluids

MIL-H-83282 is a fire-resistant hydrogenated polyalphaolefin-based fluid developed in the 1960s to overcome the flammability characteristics of MIL-H-5606. MIL-H-83282 is significantly more flame resistant than MIL-H-5606, but a disadvantage is the high viscosity at low temperature. It is generally limited to –40 °F. However, it can be used in the same system and with the same seals, gaskets, and hoses as MIL-H-5606. MIL-H-46170 is the rust-inhibited version of MIL-H-83282. Small aircraft predominantly use MIL-H-5606, but some have switched to MIL-H-83282 if they can accommodate the high viscosity at low temperature.

Phosphate Ester-Based Fluid

These fluids are used in most commercial transport category aircraft and are extremely fire-resistant. However, they are not fireproof and under certain conditions, they burn. In addition, these fluids are very susceptible to contamination from water in the atmosphere. The earliest generation of these fluids was developed after World War II as a result of the growing number of aircraft hydraulic brake fires that drew the collective concern of the commercial aviation industry. Progressive development of these fluids occurred as a result of performance requirements of newer aircraft designs. The airframe manufacturers dubbed these new generations of hydraulic fluid, such as Skydrol® and Hyjet®, as types based on their performance.

Today, types IV and V fluids are used. Two distinct classes of type IV fluids exist based on their density: class I fluids are low density and class II fluids are standard density. The class I fluids provide weight savings advantages versus class II. In addition to the type IV fluids that are currently in use, type V fluids are being developed in response to industry demands for a more thermally stable fluid at higher operating temperatures. Type V fluids will be more resistant to hydrolytic and oxidative degradation at high temperature than the type IV fluids.

Intermixing of Fluids

Due to the difference in composition, petroleum-based and phosphate ester-based fluids will not mix; neither are the seals for any one fluid usable with or tolerant of any of the other fluids. Should an aircraft hydraulic system be serviced with the wrong type fluid, immediately drain and flush the system and maintain the seals according to the manufacturer's specifications.

Compatibility with Aircraft Materials

Aircraft hydraulic systems designed around phosphate ester-based fluids should be virtually trouble-free if properly serviced. Phosphate ester-based fluids do not appreciably affect common aircraft metals—aluminum, silver, zinc, magnesium, cadmium, iron, stainless steel, bronze, chromium, and others—as long as the fluids are kept free of contamination. Thermoplastic resins, including vinyl compositions, nitrocellulose lacquers, oil-based paints, linoleum, and asphalt may be softened chemically due to phosphate ester-based fluids. However, this chemical action usually requires longer than just momentary exposure, and spills that are wiped up with soap and water do not harm most of these materials. Paints that are resistant to phosphate ester-based fluids include epoxies and polyurethanes. Today, polyurethanes are the standard of the aircraft industry because of their ability to keep a bright, shiny finish for long periods of time and for the ease with which they can be removed.

Hydraulic systems require the use of special accessories that are compatible with the hydraulic fluid. Appropriate seals, gaskets, and hoses must be specifically designated for the type of fluid in use. Care must be taken to ensure that the components installed in the system are compatible with the fluid. When gaskets, seals, and hoses are replaced, positive identification should be made to ensure that they are made of the appropriate material. Phosphate ester-based type V fluid is compatible with natural fibers and with a number of synthetics, including nylon and polyester, which are used extensively in most aircraft. Petroleum oil hydraulic system seals of neoprene or Buna-N are not compatible with phosphate ester-based fluids and must be replaced with seals of butyl rubber or ethylene-propylene elastomers.

Hydraulic Fluid Contamination

Experience has shown that trouble in a hydraulic system is inevitable whenever the liquid is allowed to become contaminated. The nature of the trouble, whether a simple malfunction or the complete destruction of a component, depends to some extent on the type of contaminant. Two general contaminants are:

- Abrasives, including such particles as core sand, weld spatter, machining chips, and rust.

- Nonabrasives, including those resulting from oil oxidation and soft particles worn or shredded from seals and other organic components.

Contamination Check

Whenever it is suspected that a hydraulic system has become contaminated or the system has been operated at temperatures in excess of the specified maximum, a check of the system should be made. The filters in most hydraulic systems are designed to remove most foreign particles that are visible to the naked eye. Hydraulic liquid that appears clean to the naked eye may be contaminated to the point that it is unfit for use. Thus, visual inspection of the hydraulic liquid does not determine the total amount of contamination in the system. Large particles of impurities in the hydraulic system are indications that one or more components are being subjected to excessive wear. Isolating the defective component requires a systematic process of elimination. Fluid returned to the reservoir may contain impurities from any part of the system. To determine which component is defective, liquid samples should be taken from the reservoir and various other locations in the system. Samples should be taken in accordance with the applicable manufacturer's instructions for a particular hydraulic system. Some hydraulic systems are equipped with permanently installed bleed valves for taking liquid samples, whereas on other systems, lines must be disconnected to provide a place to take a sample.

Hydraulic Sampling Schedule

- Routine sampling—each system should be sampled at least once a year, or every 3,000 flight hours, or whenever the airframe manufacturer suggests.

- Unscheduled maintenance—when malfunctions may have a fluid related cause, samples should be taken.

- Suspicion of contamination—if contamination is suspected, fluids should be drained and replaced, with samples taken before and after the maintenance procedure.

Sampling Procedure

- Pressurize and operate hydraulic system for 10 to 15 minutes. During this period, operate various flight controls to activate valves and thoroughly mix hydraulic fluid.

- Shut down and depressurize the system.

- Before taking samples, always be sure to wear the proper personal protective equipment that should include, at the minimum, safety glasses and gloves.

- Wipe off sampling port or tube with a lint-free cloth. Do not use shop towels or paper products that could produce lint. Generally speaking, the human eye can see particles down to about 40 microns in size. Since we are concerned with particles down to 5 microns in size, it is easy to contaminate a sample without ever knowing it.

- Place a waste container under the reservoir drain valve and open valve so that a steady, but not forceful, stream is running.

- Allow approximately 1 pint (250 ml) of fluid to drain. This purges any settled particles from the sampling port.

- Insert a precleaned sample bottle under the fluid stream and fill, leaving an air space at the top. Withdraw the bottle and cap immediately.

- Close drain valve.

- Fill out sample identification label supplied in sample kit, making sure to include customer name, aircraft type, aircraft tail number, hydraulic system sampled, and date sampled. Indicate on the sample label under remarks if this is a routine sample or if it is being taken due to a suspected problem.

- Service system reservoirs to replace the fluid that was removed.

- Submit samples for analysis to laboratory.

Contamination Control

Filters provide adequate control of the contamination problem during all normal hydraulic system operations. Control of the size and amount of contamination entering the system from any other source is the responsibility of the people who service and maintain the equipment. Therefore, precautions should be taken to minimize contamination during maintenance, repair, and service operations. If the system becomes contaminated, the filter element should be removed and cleaned or replaced. As an aid in controlling contamination, the following maintenance and servicing procedures should be followed at all times:

- Maintain all tools and the work area (workbenches and test equipment) in a clean, dirt-free condition.

- A suitable container should always be provided to receive the hydraulic liquid that is spilled during component removal or disassembly procedures.

- Before disconnecting hydraulic lines or fittings, clean the affected area with dry-cleaning solvent.

- All hydraulic lines and fittings should be capped or plugged immediately after disconnecting.

- Before assembly of any hydraulic components, wash all parts in an approved dry-cleaning solvent, i.e., Stoddard solvent.

- After cleaning the parts in the dry-cleaning solution, dry the parts thoroughly and lubricate them with the recommended preservative or hydraulic liquid before assembly. Use only clean, lint-free cloths to wipe or dry the component parts.

- All seals and gaskets should be replaced during the reassembly procedure. Use only those seals and gaskets recommended by the manufacturer.

- All parts should be connected with care to avoid stripping metal slivers from threaded areas. All fittings and lines should be installed and torqued in accordance with applicable technical instructions.

- All hydraulic servicing equipment should be kept clean and in good operating condition.

Contamination, both particulate and chemical, is detrimental to the performance and life of components in the aircraft hydraulic system. Contamination enters the system through normal wear of components, by ingestion through external seals during servicing, or maintenance, when the system is opened to replace/repair components, etc. To control the particulate contamination in the system, filters are installed in the pressure line, in the return line, and in the pump case drain line of each system. The filter rating is given in microns as an indication of the smallest particle size that is filtered out. The replacement interval of these filters is established by the manufacturer and is included in the maintenance manual. In the absence of specific replacement instructions, a recommended service life of the filter elements is:

- Pressure filters—3,000 hours.

- Return Filters—1,500 hours.

- Case drain filters—600 hours.

Hydraulic System Flushing

When inspection of hydraulic filters or hydraulic fluid evaluation indicates that the fluid is contaminated, flushing the system may be necessary. This must be done according to the manufacturer's instructions; however, a typical procedure for flushing is as follows:

1. Connect a ground hydraulic test stand to the inlet and outlet test ports of the system. Verify that the ground unit fluid is clean and contains the same fluid as the aircraft.

2. Change the system filters.

3. Pump clean, filtered fluid through the system, and operate all subsystems until no obvious signs of contamination are found during inspection of the filters. Dispose of contaminated fluid and filter. Note: A visual inspection of hydraulic filters is not always effective.

4. Disconnect the test stand and cap the ports.

5. Ensure that the reservoir is filled to the full line or proper service level.

It is very important to check if the fluid in the hydraulic test stand, or mule, is clean before the flushing operation starts. A contaminated hydraulic test stand can quickly contaminate other aircraft if used for ground maintenance operations.

Health & Handling

Some phosphate ester-based fluids are blended with performance additives. Phosphate esters are good solvents and dissolve away some of the fatty materials of the skin. Repeated or prolonged exposure may cause drying of the skin, which if unattended, could result in complications, such as dermatitis or even secondary infection from bacteria. Phosphate ester-based fluids could cause itching of the skin but have not been known to cause allergic-type skin rashes. Always use the proper gloves and eye protection when handling any type of hydraulic fluid. When phosphate ester-based mist or vapor exposure is possible, a respirator capable of removing organic vapors and mists must be worn. Ingestion of any hydraulic fluid should be avoided. Although

small amounts do not appear to be highly hazardous, any significant amount should be tested in accordance with manufacturer's direction, followed with hospital supervised stomach treatment.

Basic Hydraulic Systems

Regardless of its function and design, every hydraulic system has a minimum number of basic components in addition to a means through which the fluid is transmitted. A basic system consists of a pump, reservoir, directional valve, check valve, pressure relieve valve, selector valve, actuator, and filter. [Figure 12-2]

Open Center Hydraulic Systems

An open center system is one having fluid flow, but no pressure in the system when the actuating mechanisms are idle. The pump circulates the fluid from the reservoir, through the selector valves, and back to the reservoir. [Figure 12-3] The open center system may employ any number of subsystems, with a selector valve for each subsystem. Unlike the closed center system, the selector valves of the open center system are always connected in series with each other. In this arrangement, the system pressure line goes through each selector valve. Fluid is always allowed free passage through each selector valve and back to the reservoir until one of the selector valves is positioned to operate a mechanism.

When one of the selector valves is positioned to operate an actuating device, fluid is directed from the pump through one of the working lines to the actuator. [Figure 12-3B] With the selector valve in this position, the flow of fluid through the valve to the reservoir is blocked. The pressure builds up in the system to overcome the resistance and moves the piston of the actuating cylinder; fluid from the opposite end of the

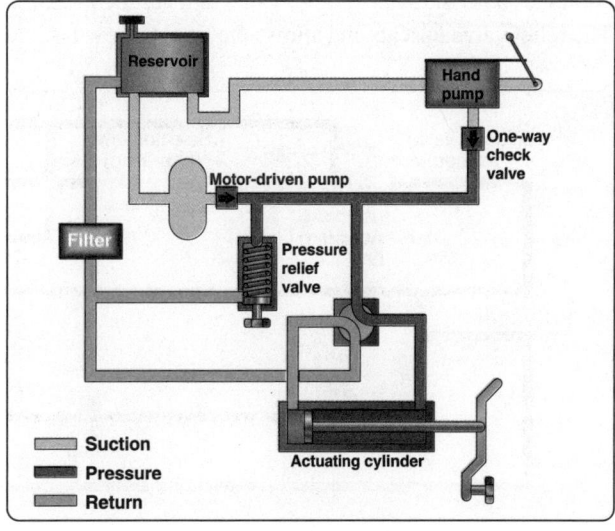

Figure 12-2. *Basic hydraulic system.*

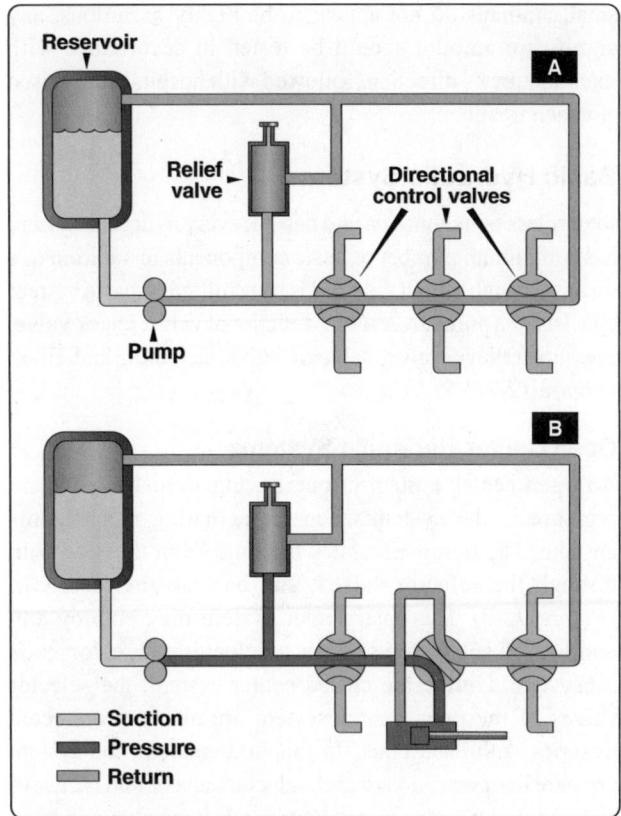

Figure 12-3. *Open center hydraulic system.*

the reservoir. The system pressure remains at the relief valve set pressure until the selector valve is manually returned to the neutral position. This action reopens the open center flow and allows the system pressure to drop to line resistance pressure.

The manually engaged and pressure disengaged type of selector valve is similar to the valve previously discussed. When the actuating mechanism reaches the end of its cycle, the pressure continues to rise to a predetermined pressure. The valve automatically returns to the neutral position and to open center flow.

Closed-Center Hydraulic Systems

In the closed-center system, the fluid is under pressure whenever the power pump is operating. The three actuators are arranged in parallel and actuating units B and C are operating at the same time, while actuating unit A is not operating. This system differs from the open-center system in that the selector or directional control valves are arranged in parallel and not in series. The means of controlling pump pressure varies in the closed-center system. If a constant delivery pump is used, the system pressure is regulated by a pressure regulator. A relief valve acts as a backup safety device in case the regulator fails.

If a variable displacement pump is used, system pressure is controlled by the pump's integral pressure mechanism compensator. The compensator automatically varies the volume output. When pressure approaches normal system pressure, the compensator begins to reduce the flow output of the pump. The pump is fully compensated (near zero flow) when normal system pressure is attained. When the pump is in this fully compensated condition, its internal bypass mechanism provides fluid circulation through the pump for cooling and lubrication. A relief valve is installed in the system as a safety backup. *[Figure 12-4]* An advantage of the open-center system over the closed-center system is that the continuous pressurization of the system is eliminated. Since the pressure is built up gradually after the selector valve is

actuator returns to the selector valve and flows back to the reservoir. Operation of the system following actuation of the component depends on the type of selector valve being used.

Several types of selector valves are used in conjunction with the open center system. One type is both manually engaged and manually disengaged. First, the valve is manually moved to an operating position. Then, the actuating mechanism reaches the end of its operating cycle, and the pump output continues until the system relief valve relieves the pressure. The relief valve unseats and allows the fluid to flow back to

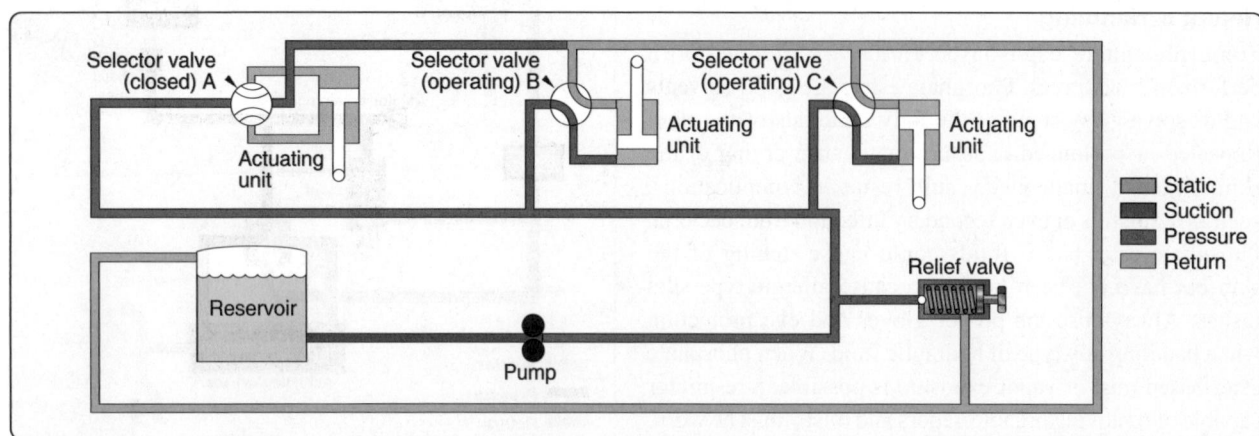

Figure 12-4. *A basic closed-center hydraulic system with a variable displacement pump.*

moved to an operating position, there is very little shock from pressure surges. This action provides a smoother operation of the actuating mechanisms. The operation is slower than the closed-center system, in which the pressure is available the moment the selector valve is positioned. Since most aircraft applications require instantaneous operation, closed-center systems are the most widely used.

Hydraulic Power Systems

Evolution of Hydraulic Systems

Smaller aircraft have relatively low flight control surface loads, and the pilot can operate the flight controls by hand.

Hydraulic systems were utilized for brake systems on early aircraft. When aircraft started to fly faster and got larger in size, the pilot was not able to move the control surfaces by hand anymore, and hydraulic power boost systems were introduced. Power boost systems assist the pilot in overcoming high control forces, but the pilot still actuates the flight controls by cable or push rod.

Many modern aircraft use a power supply system and fly-by-wire flight control. The pilot input is electronically sent to the flight control servos. Cables or push rods are not used. Small power packs are the latest evolution of the hydraulic system. They reduce weight by eliminating hydraulic lines and large quantities of hydraulic fluid. Some manufacturers are reducing hydraulic systems in their aircraft in favor of electrically controlled systems. The Boeing 787 is the first aircraft designed with more electrical systems than hydraulic systems.

Hydraulic Power Pack System

A hydraulic power pack is a small unit that consists of an electric pump, filters, reservoir, valves, and pressure relief valve. *[Figure 12-5]* The advantage of the power pack is that there is no need for a centralized hydraulic power supply system and long stretches of hydraulic lines, which reduces weight. Power packs could be driven by either an engine gearbox or electric motor. Integration of essential valves, filters, sensors, and transducers reduces system weight, virtually eliminates any opportunity for external leakage, and simplifies troubleshooting. Some power pack systems have an integrated actuator. These systems are used to control the stabilizer trim, landing gear, or flight control surfaces directly, thus eliminating the need for a centralized hydraulic system.

Hydraulic System Components

Figure 12-6 is a typical example of a hydraulic system in a large commercial aircraft. The following sections discuss the components of such system in more detail.

Figure 12-5. *Hydraulic power pack.*

Reservoirs

The reservoir is a tank in which an adequate supply of fluid for the system is stored. Fluid flows from the reservoir to the pump, where it is forced through the system and eventually returned to the reservoir. The reservoir not only supplies the operating needs of the system, but it also replenishes fluid lost through leakage. Furthermore, the reservoir serves as an overflow basin for excess fluid forced out of the system by thermal expansion (the increase of fluid volume caused by temperature changes), the accumulators, and by piston and rod displacement.

The reservoir also furnishes a place for the fluid to purge itself of air bubbles that may enter the system. Foreign matter picked up in the system may also be separated from the fluid in the reservoir or as it flows through line filters. Reservoirs are either pressurized or nonpressurized.

Baffles and/or fins are incorporated in most reservoirs to keep the fluid within the reservoir from having random movement, such as vortexing (swirling) and surging. These conditions can cause fluid to foam and air to enter the pump along with the fluid. Many reservoirs incorporate strainers in the filler neck to prevent the entry of foreign matter during servicing. These strainers are made of fine mesh screening and are usually referred to as finger strainers because of their shape. Finger strainers should never be removed or punctured as a means of speeding up the pouring of fluid into the reservoir. Reservoirs could have an internal trap to make sure fluid goes to the pumps during negative-G conditions.

Most aircraft have emergency hydraulic systems that take over if main systems fail. In many such systems, the pumps of both systems obtain fluid from a single reservoir. Under such circumstances, a supply of fluid for the emergency pump is ensured by drawing the hydraulic fluid from the bottom of the reservoir. The main system draws its fluid through a standpipe

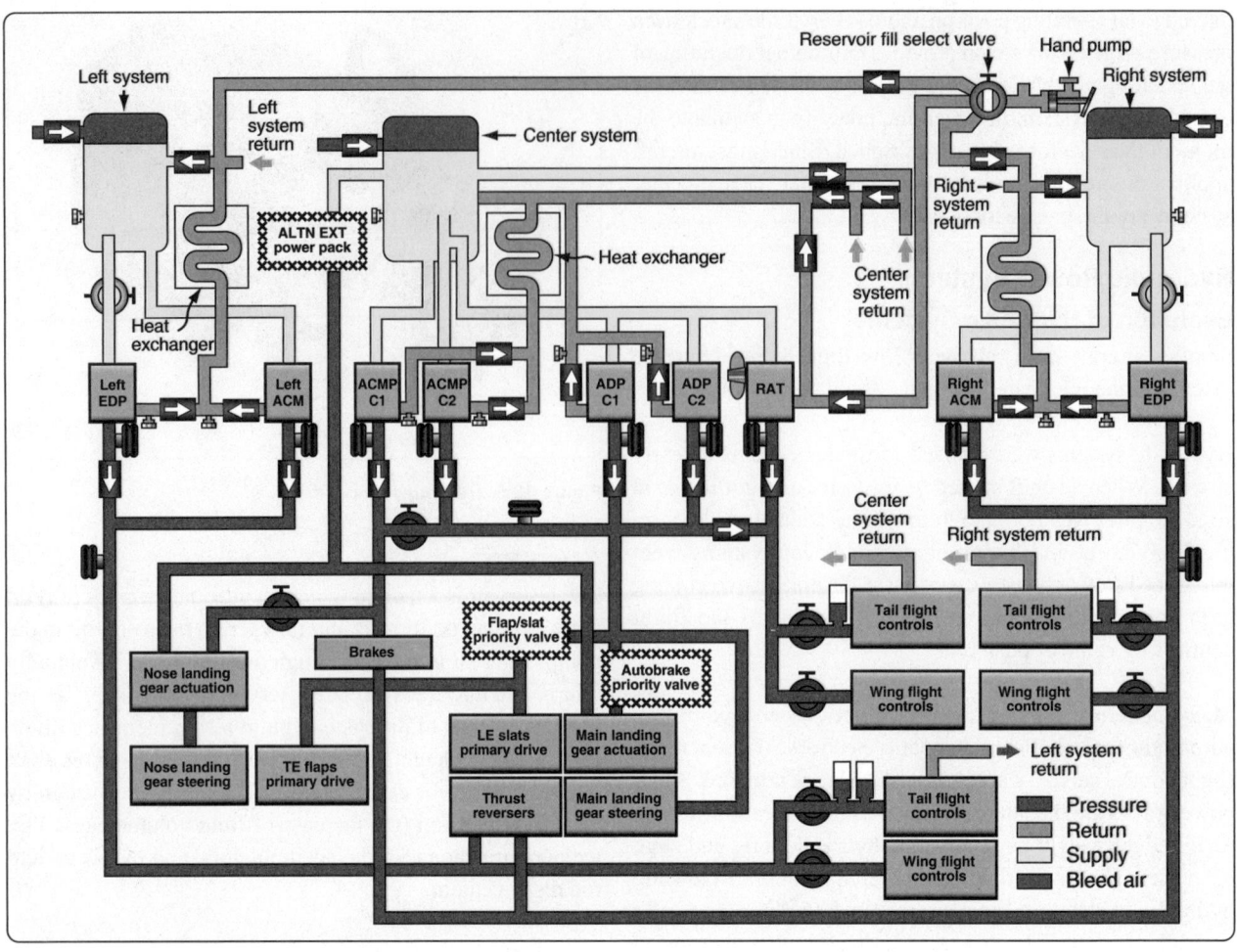

Figure 12-6. *Large commercial aircraft hydraulic system.*

located at a higher level. With this arrangement, should the main system's fluid supply become depleted, adequate fluid is left for operation of the emergency system. *Figure 12-7* illustrates that the engine-driven pump (EDP) is not able to draw fluid any more if the reservoir gets depleted below the standpipe. The alternating current motor-driven pump (ACMP) still has a supply of fluid for emergency operations.

Nonpressurized Reservoirs

Nonpressurized reservoirs are used in aircraft that are not designed for violent maneuvers, do not fly at high altitudes, or in which the reservoir is located in the pressurized area of the aircraft. High altitude in this situation means an altitude where atmospheric pressure is inadequate to maintain sufficient flow of fluid to the hydraulic pumps. Most nonpressurized reservoirs are constructed in a cylindrical shape. The outer housing is manufactured from a strong corrosion-resistant metal. Filter elements are normally installed within the reservoir to clean returning system hydraulic fluid.

In some of the older aircraft, a filter bypass valve is incorporated to allow fluid to bypass the filter in the event the filter becomes clogged. Reservoirs can be serviced by pouring fluid directly into the reservoir through a filler strainer (finger strainer) assembly incorporated within the filler well to strain out impurities as the fluid enters the reservoir. Generally, nonpressurized reservoirs use a visual gauge to indicate the fluid quantity. Gauges incorporated on or in the reservoir may be a direct reading glass tube-type or a float-type rod that is visible through a transparent dome. In some cases, the fluid quantity may also be read in the flight deck through the use of quantity transmitters. A typical nonpressurized reservoir is shown in *Figure 12-8.* This reservoir consists of a welded body and cover assembly clamped together. Gaskets are incorporated to seal against leakage between assemblies.

Nonpressurized reservoirs are slightly pressurized due to thermal expansion of fluid and the return of fluid to the reservoir from the main system. This pressure ensures that there is a positive flow of fluids to the inlet ports of the hydraulic pumps. Most reservoirs of this type are vented directly to the atmosphere or cabin with only a check valve and filter to control the outside air source. The reservoir system includes a pressure and vacuum relief valve. The purpose of

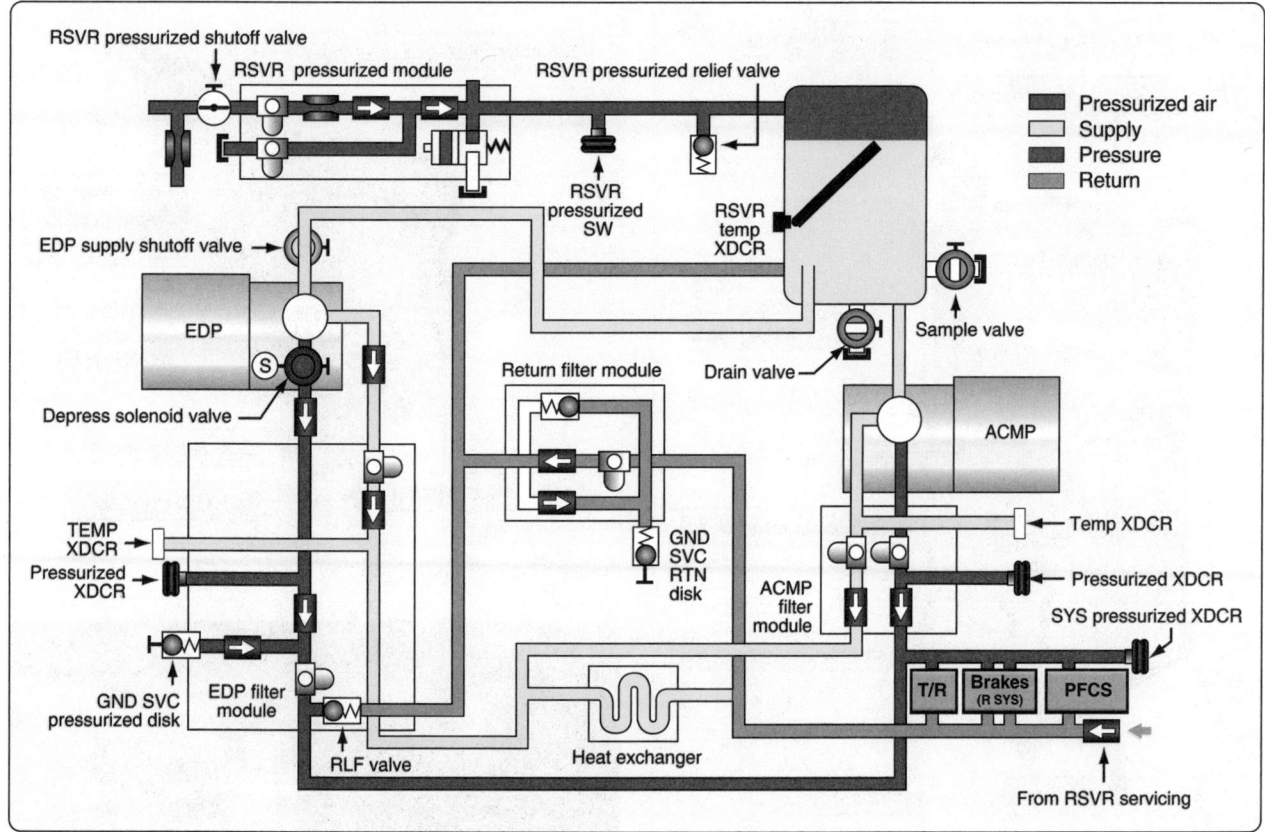

Figure 12-7. *Hydraulic reservoir standpipe for emergency operations.*

the valve is to maintain a differential pressure range between the reservoir and cabin. A manual air bleed valve is installed on top of the reservoir to vent the reservoir. The valve is connected to the reservoir vent line to allow depressurization of the reservoir. The valve is actuated prior to servicing the reservoir to prevent fluid from being blown out of the filler as the cap is being removed. The manual bleed valve also needs to be actuated if hydraulic components need to be replaced.

Pressurized Reservoirs

Reservoirs on aircraft designed for high-altitude flight are usually pressurized. Pressurizing assures a positive flow of fluid to the pump at high altitudes when low atmospheric pressures are encountered. On some aircraft, the reservoir is pressurized by bleed air taken from the compressor section of the engine. On others, the reservoir may be pressurized by hydraulic system pressure.

Air-Pressurized Reservoirs

Air-pressurized reservoirs are used in many commercial transport-type aircraft. *[Figures 12-9 and 12-10]* Pressurization of the reservoir is required because the reservoirs are often located in wheel wells or other non-pressurized areas of the aircraft and at high altitude there is not enough atmospheric pressure to move the fluid to

the pump inlet. Engine bleed air is used to pressurize the reservoir. The reservoirs are typically cylindrical in shape. The following components are installed on a typical reservoir:

- Reservoir pressure relief valve—prevents over pressurization of the reservoir. Valve opens at a preset value.

- Sight glasses (low and overfull)—provides visual indication for flight crews and maintenance personnel that the reservoir needs to be serviced.

- Reservoir sample valve—used to draw a sample of hydraulic fluid for testing.

- Reservoir drain valve—used to drain the fluids out of the reservoir for maintenance operation.

- Reservoir temperature transducer—provides hydraulic fluid temperature information for the flight deck. *[Figure 12-11]*

- Reservoir quantity transmitter—transmits fluid quantity to the flight deck so that the flight crew can monitor fluid quantity during flight. *[Figure 12-11]*

A reservoir pressurization module is installed close to the reservoir. *[Figure 12-12]* The reservoir pressurization module supplies airplane bleed air to the reservoirs. The module

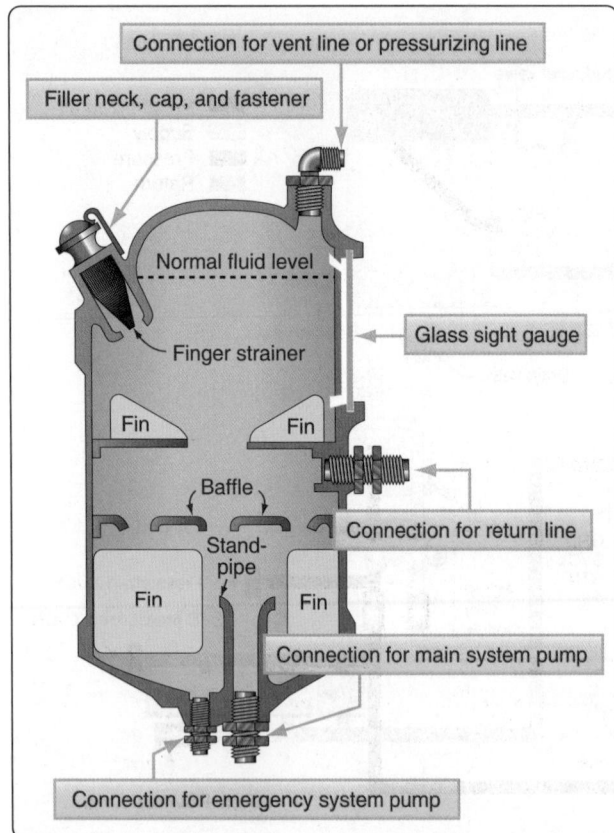

Figure 12-8. *Nonpressurized reservoir.*

Figure 12-9. *Air-pressurized reservoir.*

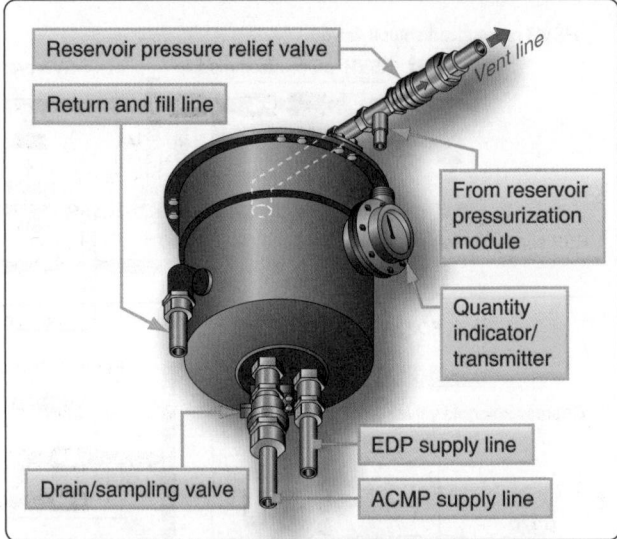

Figure 12-10. *Components of an air-pressurized reservoir.*

Figure 12-11. *Temperature and quantity sensors.*

consists of the following parts:

- Filters (2).

- Check valves (2).

- Test port.

- Manual bleed valve.

- Gauge port.

A manual bleeder valve is incorporated into the module. During hydraulic system maintenance, it is necessary to relieve reservoir air pressure to assist in the installation and removal of components, lines, etc. This type of valve is small in size and has a push button installed in the outer case. When the bleeder valve push button is pushed, pressurized air from the reservoir flows through the valve to an overboard vent until the air pressure is depleted or the button is released. When the button is released, the internal spring causes the

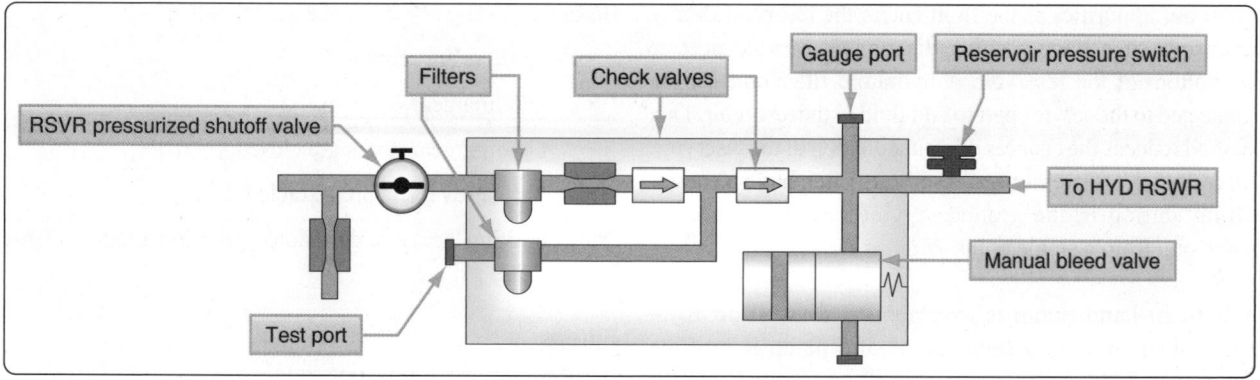

Figure 12-12. *Reservoir pressurization module.*

poppet to return to its seat. Some hydraulic fluid can escape from the manual bleed valve when the button is depressed.

Caution: Put a rag around the air bleed valve on the reservoir pressurization module to catch hydraulic fluid spray. Hydraulic fluid spray can cause injuries to persons.

Fluid-Pressurized Reservoirs

Some aircraft hydraulic system reservoirs are pressurized by hydraulic system pressure. Regulated hydraulic pump output pressure is applied to a movable piston inside the cylindrical reservoir. This small piston is attached to and moves a larger piston against the reservoir fluid. The reduced force of the small piston when applied by the larger piston is adequate to provide head pressure for high altitude operation. The small piston protrudes out of the body of the reservoir. The amount exposed is used as a reservoir fluid quantity indicator. *Figure 12-13* illustrates the concept behind the fluid-pressurized hydraulic reservoir.

The reservoir has five ports: pump suction, return, pressurizing, overboard drain, and bleed port. Fluid is supplied to the pump through the pump suction port. Fluid returns to the reservoir from the system through the return port. Pressure from the pump enters the pressurizing cylinder in the top of the reservoir through the pressurizing port. The overboard drain port drains the reservoir, when necessary, while performing maintenance. The bleed port is used as an aid in servicing the reservoir. When servicing a system equipped with this type of reservoir, place a container under the bleed drain port. The fluid should then be pumped into the reservoir until air-free fluid flows through the bleed drain port.

The reservoir fluid level is indicated by the markings on the part of the pressurizing cylinder that moves through the reservoir dust cover assembly. There are three fluid level markings indicated on the cover: full at zero system pressure (FULL ZERO PRESS), full when system is pressurized (FULL SYS PRESS), and REFILL. When the system is unpressurized and the pointer on the reservoir lies between the two full marks, a marginal reservoir fluid level is indicated. When the system is pressurized, and the pointer lies between REFILL and FULL SYS PRESS, a marginal reservoir fluid level is also indicated.

Reservoir Servicing

Nonpressurized reservoirs can be serviced by pouring fluid directly into the reservoir through a filler strainer (finger strainer) assembly incorporated within the filler well to

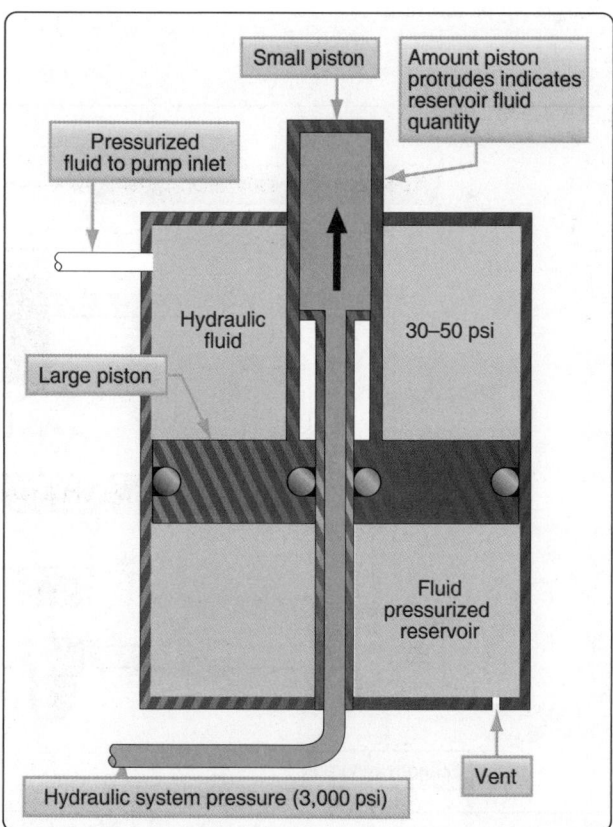

Figure 12-13. *Operating principle behind a fluid-pressurized hydraulic reservoir.*

strain out impurities as the fluid enters the reservoir. Many reservoirs also have a quick disconnect service port at the bottom of the reservoir. A hydraulic filler unit can be connected to the service port to add fluid to the reservoir. This method reduces the chances of contamination of the reservoir. Aircraft that use pressurized reservoirs often have a central filling station in the ground service bay to service all reservoirs from a single point. *[Figure 12-14]*

A built-in hand pump is available to draw fluid from a container through a suction line and pump it into the reservoirs. Additionally, a pressure fill port is available for attachment of a hydraulic mule or serving cart, which uses an external pump to push fluid into the aircraft hydraulic system. A check valve keeps the hand pump output from exiting the pressure fill port. A single filter is located downstream of both the pressure fill port and the hand pump to prevent the introduction of contaminants during fluid servicing.

It is very important to follow the maintenance instructions when servicing the reservoir. To get the correct results when the hydraulic fluid quantities are checked, or the reservoirs are to be filled, the airplane should be in the correct configuration. Failure to do so could result in overservicing of the reservoir. This configuration could be different for each aircraft. The following service instructions are an example of a large transport-type aircraft.

Before servicing always make sure that the:

- Spoilers are retracted,

- Landing gear is down,

- Landing gear doors are closed,

- Thrust reversers are retracted, and

- Parking brake accumulator pressure reads at least 2,500 psi.

Filters

A filter is a screening or straining device used to clean the hydraulic fluid, preventing foreign particles and contaminating substances from remaining in the system. *[Figure 12-15]* If such objectionable material were not removed, the entire hydraulic system of the aircraft could fail through the breakdown or malfunctioning of a single unit of the system.

The hydraulic fluid holds in suspension tiny particles of metal that are deposited during the normal wear of selector valves, pumps, and other system components. Such minute particles of metal may damage the units and parts through which they pass if they are not removed by a filter. Since tolerances within the hydraulic system components are quite small, it is apparent that the reliability and efficiency of the entire system depends upon adequate filtering.

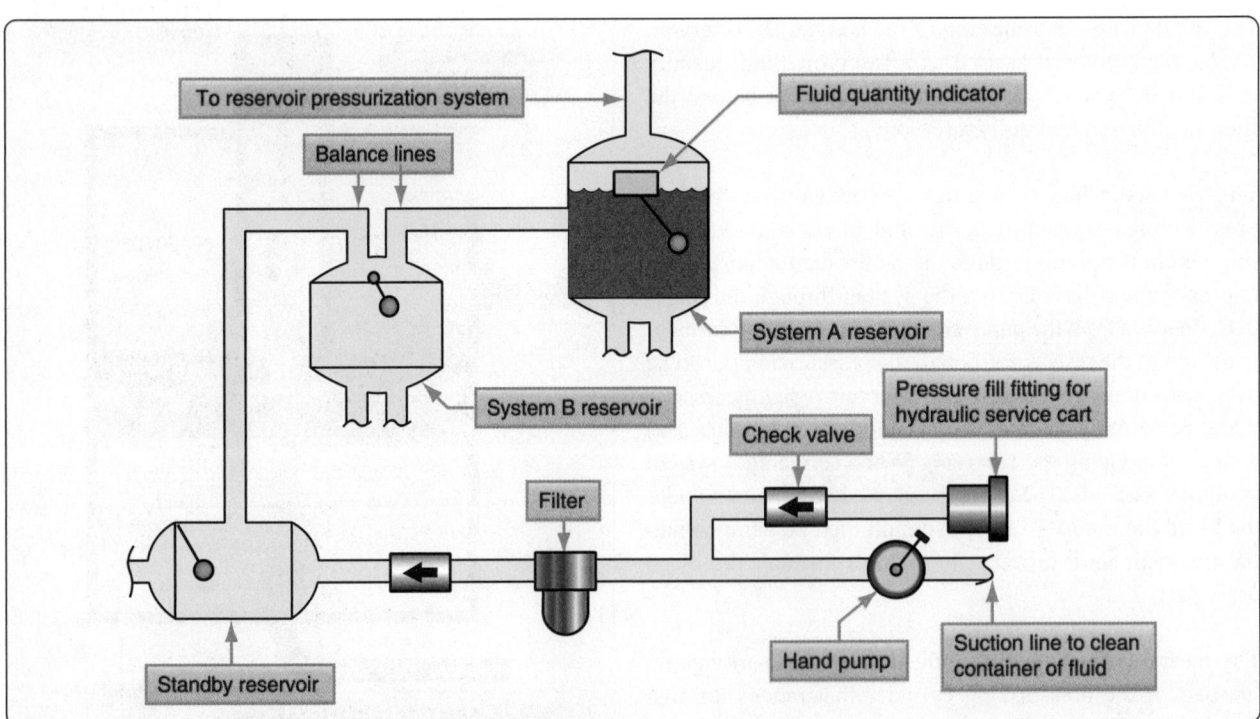

Figure 12-14. *The hydraulic ground serive station on a Boeing 737 provides for hydraulic fluid servicing with a hand pump or via an external pressure fluid source. All three reservoirs are serviced from the same location.*

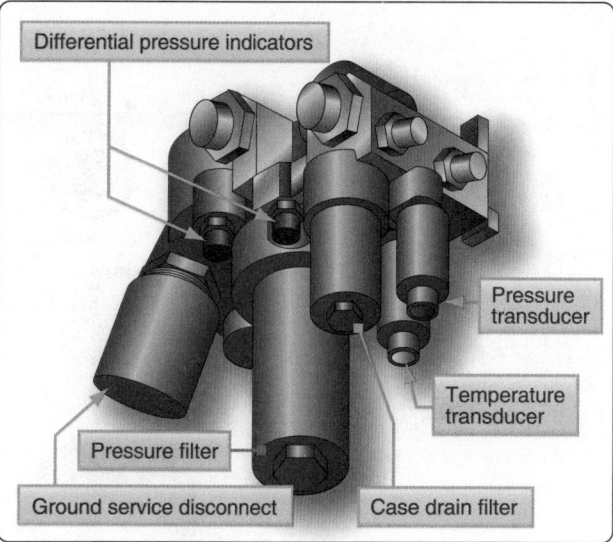

Figure 12-15. *Filter module components.*

Filters may be located within the reservoir, in the pressure line, in the return line, or in any other location the designer of the system decides that they are needed to safeguard the hydraulic system against impurities. Modern design often uses a filter module that contains several filters and other components. *[Figure 12-16]* There are many models and styles of filters. Their position in the aircraft and design requirements determine their shape and size. Most filters used in modern aircraft are of the inline type. The inline filter assembly is comprised of three basic units: head assembly, bowl, and element. The head assembly is secured to the aircraft structure and connecting lines. Within the head, there is a bypass valve that routes the hydraulic fluid directly from the inlet to the outlet port if the filter element becomes clogged with foreign matter. The bowl is the housing that holds the element to the filter head and is removed when element removal is required.

The element may be a micron, porous metal, or magnetic type. The micron element is made of a specially treated paper and is normally thrown away when removed. The porous metal and magnetic filter elements are designed to be cleaned by various methods and replaced in the system.

Micron-Type Filters

A typical micron-type filter assembly utilizes an element made of specially treated paper that is formed in vertical convolutions (wrinkles). An internal spring holds the elements in shape. The micron element is designed to prevent the passage of solids greater than 10 microns (0.000394 inch) in size. *[Figure 12-17]* In the event that the filter element becomes clogged, the spring-loaded relief valve in the filter head bypasses the fluid after a differential pressure of 50 psi has been built up. Hydraulic fluid enters the filter through the inlet port in the filter body and flows around the element inside the bowl. Filtering takes place as the fluid passes through the element into the hollow core, leaving the foreign material on the outside of the element.

Maintenance of Filters

Maintenance of filters is relatively easy. It mainly involves cleaning the filter and element or cleaning the filter and replacing the element. Filters using the micron-type element should have the element replaced periodically according to applicable instructions. Since reservoir filters are of the micron type, they must also be periodically changed or cleaned. For filters using other than the micron-type element, cleaning the filter and element is usually all that is necessary. However, the element should be inspected very closely to ensure that it is completely undamaged. The methods and materials used in cleaning all filters are too numerous to be included in this text. Consult the manufacturer's instructions for this information.

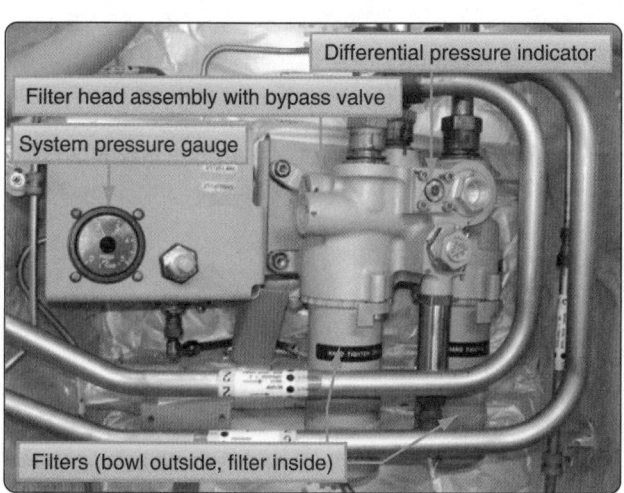

Figure 12-16. *A transport category filter module with two filters.*

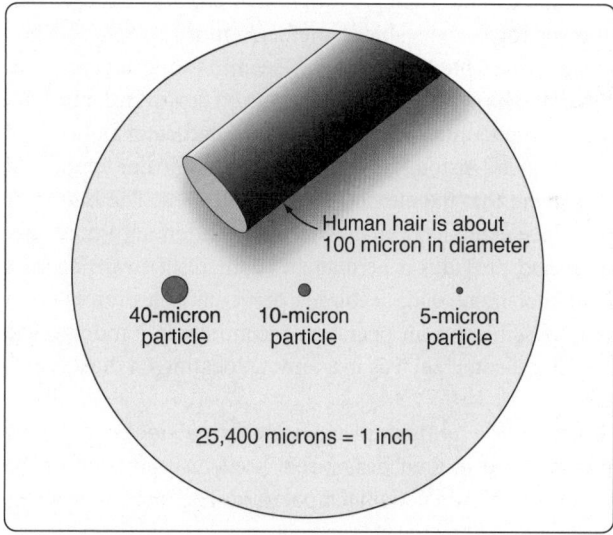

Figure 12-17. *Size comparison in microns.*

When replacing filter elements, be sure that there is no pressure on the filter bowl. Protective clothing and a face shield must be used to prevent fluid from contacting the eye. Replace the element with one that has the proper rating. After the filter element has been replaced, the system must be pressure tested to ensure that the sealing element in the filter assembly is intact.

In the event of a major component failure, such as a pump, consideration must be given to replacing the system filter elements, as well as the failed component.

Filter Bypass Valve

Filter modules are often equipped with a bypass relief valve. The bypass relief valve opens if the filter clogs, permitting continued hydraulic flow and operation of aircraft systems. Dirty oil is preferred over no flow at all. *Figure 12-18* shows the principle of operation of a filter bypass valve. The ball valve opens when the filter becomes clogged and the pressure over the filter increases.

Filter Differential Pressure Indicators

The extent to which a filter element is loaded can be determined by measuring the drop in hydraulic pressure across the element under rated flow conditions. This drop, or differential pressure, provides a convenient means of monitoring the condition of installed filter elements and is the operating principle used in the differential pressure or loaded-filter indicators found on many filter assemblies.

Differential pressure indicating devices have many configurations, including electrical switches, continuous-reading visual indicators (gauges), and visual indicators with memory. Visual indicators with memory usually take the form of magnetic or mechanically latched buttons or pins that extend when the differential pressure exceeds that allowed for a serviceable element. *[Figure 12-18, top]* When this increased pressure reaches a specific value, inlet pressure forces the spring-loaded magnetic piston downward, breaking the magnetic attachment between the indicator button and the magnetic piston. This allows the red indicator to pop out, signifying that the element must be cleaned. The button or pin, once extended, remains in that position until manually reset and provides a permanent (until reset) warning of a loaded element. This feature is particularly useful where it is impossible for an operator to continuously monitor the visual indicator, such as in a remote location on the aircraft.

Some button indicators have a thermal lockout device incorporated in their design that prevents operation of the indicator below a certain temperature. The lockout prevents the higher differential pressure generated at cold temperatures by high fluid viscosity from causing a false indication of a

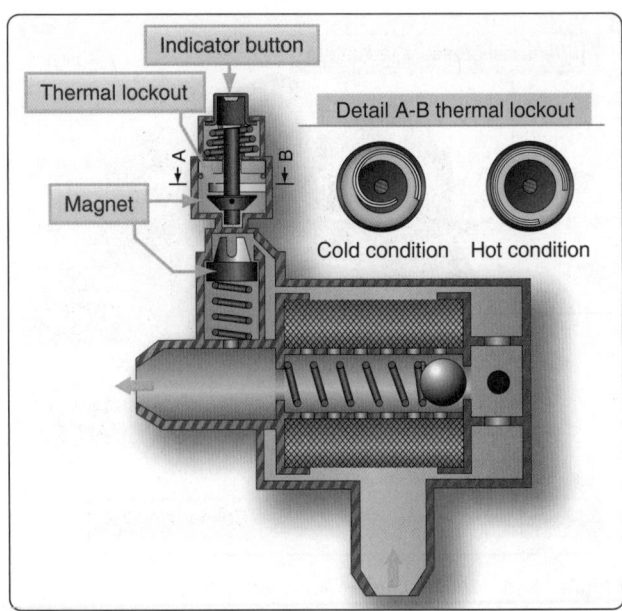

Figure 12-18. *Filter bypass valve.*

loaded filter element.

Differential pressure indicators are a component part of the filter assembly in which they are installed and are normally tested and overhauled as part of the complete assembly. With some model filter assemblies, however, it is possible to replace the indicator itself without removal of the filter assembly if it is suspected of being inoperative or out of calibration. It is important that the external surfaces of button-type indicators be kept free of dirt or paint to ensure free movement of the button. Indications of excessive differential pressure, regardless of the type of indicator employed, should never be disregarded. All such indications must be verified and action taken, as required, to replace the loaded filter element. Failure to replace a loaded element can result in system starvation, filter element collapse, or the loss of filtration where bypass assemblies are used. Verification of loaded filter indications is particularly important with button-type indicators as they may have been falsely triggered by mechanical shock, vibration, or cold start of the system. Verification is usually obtained by manually resetting the indicator and operating the system to create a maximum flow demand ensuring that the fluid is at near normal operating temperatures.

Pumps

All aircraft hydraulic systems have one or more power-driven pumps and may have a hand pump as an additional unit when the engine-driven pump is inoperative. Power-driven pumps are the primary source of energy and may be either engine driven, electric motor driven, or air driven. As a general rule, electrical motor pumps are installed for use in emergencies or during ground operations. Some aircraft can deploy a ram air turbine (RAT) to generate hydraulic power.

Hand Pumps

The hydraulic hand pump is used in some older aircraft for the operation of hydraulic subsystems and in a few newer aircraft systems as a backup unit. Hand pumps are generally installed for testing purposes, as well as for use in emergencies. Hand pumps are also installed to service the reservoirs from a single refilling station. The single refilling station reduces the chances for the introduction of fluid contamination.

Several types of hand pumps are used: single action, double action, and rotary. A single action hand pump draws fluid into the pump on one stroke and pumps that fluid out on the next stroke. It is rarely used in aircraft due to this inefficiency.

Double-action hand pumps produce fluid flow and pressure on each stroke of the handle. *[Figure 12-19]* The double-action hand pump consists essentially of a housing that has a cylinder bore and two ports, a piston, two spring-loaded check valves, and an operating handle. An O-ring on the piston seals against leakage between the two chambers of the piston cylinder bore. An O-ring in a groove in the end of the pump housing seals against leakage between the piston rod and housing.

When the piston is moved to the right, the pressure in the chamber left of the piston is lowered. The inlet port ball check valve opens, and hydraulic fluid is drawn into the chamber. At the same time, the rightward movement of the piston forces the piston ball check valve against its seat. Fluid in the

chamber to the right of the piston is forced out of the outlet port into the hydraulic system. When the piston is moved to the left, the inlet port ball check valve seats. Pressure in the chamber left of the piston rises, forcing the piston ball check valve off of its seat. Fluid flows from the left chamber through the piston to the right chamber. The volume in the chamber right of the piston is smaller than that of the left chamber due to the displacement created by the piston rod. As the fluid from the left chamber flows into the smaller right chamber, the excess volume of fluid is forced out of the outlet port to the hydraulic system.

A rotary hand pump may also be employed. It produces continuous output while the handle is in motion. *Figure 12-20* shows a rotary hand pump in a hydraulic system.

Power-Driven Pumps

Many of the power driven hydraulic pumps of current aircraft are of variable delivery, compensator-controlled type. Constant delivery pumps are also in use. Principles of operation are the same for both types of pumps. Modern aircraft use a combination of engine-driven power pumps, electrical-driven power pumps, air-driven power pumps, power transfer units (PTU), and pumps driven by a RAT. For example, large aircraft, such as the Airbus A380, have

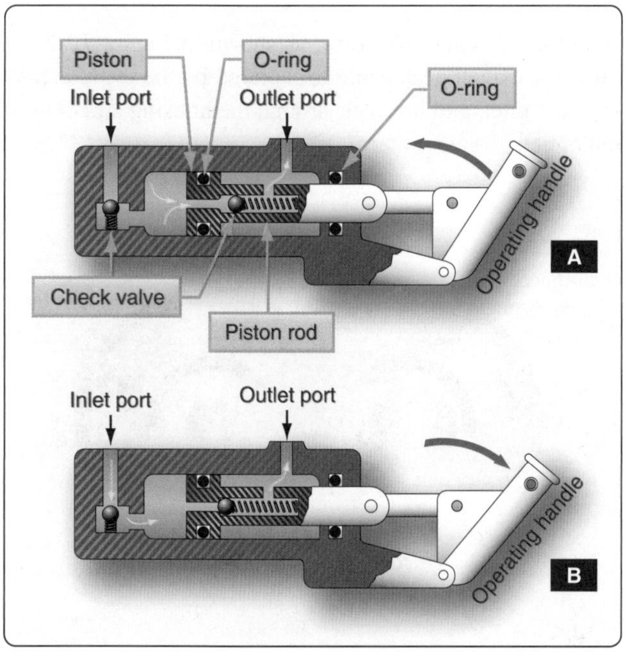

Figure 12-20. *Rotary hand pump.*

Figure 12-19. *Double action hand pump.*

two hydraulic systems, eight engine-driven pumps, and three electrical driven pumps. The Boeing 777 has three hydraulic systems with two engine driven pumps, four electrical driven pumps, two air driven pumps, and a hydraulic pump motor driven by the RAT. [Figure 12-21 and 12-22]

Figure 12-21. *Engine-driven pump.*

Figure 12-22. *Electrically-driven pump.*

Classification of Pumps

All pumps may be classified as either positive-displacement (constant displacement) or nonpositive-displacement (variable displacement). Most pumps used in hydraulic systems are positive displacement. Engine driven pumps may be either constant displacement or variable displacement. A nonpositive-displacement pump produces a continuous flow. However, because it does not provide a positive internal seal against slippage, its output varies considerably as pressure varies. Centrifugal and impeller pumps are examples of nonpositive-displacement pumps. If the output port of a nonpositive-displacement pump was blocked off, the pressure would rise, and output would decrease to zero.

Although the pumping element would continue moving, flow would stop because of slippage inside the pump. In a positive-displacement pump, slippage is negligible compared to the pump's volumetric output flow. If the output port were plugged, pressure would increase instantaneously to the point that the pump pressure relief valve opens.

Constant-Displacement Pumps

A constant-displacement pump, regardless of pump rotations per minute, forces a fixed or unvarying quantity of fluid through the outlet port during each revolution of the pump. Constant-displacement pumps are sometimes called constant-volume or constant-delivery pumps. They deliver a fixed quantity of fluid per revolution, regardless of the pressure demands. Since the constant-delivery pump provides a fixed quantity of fluid during each revolution of the pump, the quantity of fluid delivered per minute depends upon pump rotations per minute. When a constant-displacement pump is used in a hydraulic system in which the pressure must be kept at a constant value, a pressure regulator is required.

Gear-Type Power Pump

A gear-type power pump is a constant-displacement pump. It consists of two meshed gears that revolve in a housing. [Figure 12-23] The driving gear is driven by the aircraft engine or some other power unit. The driven gear meshes with, and is driven by, the driving gear. Clearance between the teeth as they mesh and between the teeth and the housing is very small. Excessive clearance will result in lower output pressure. The inlet port of the pump is connected to the reservoir, and the outlet port is connected to the pressure line.

When the driving gear turns, as shown in *Figure 12-23,* it turns the driven gear. Fluid is captured by the teeth as they pass the inlet, and it travels around the housing and exits at the outlet.

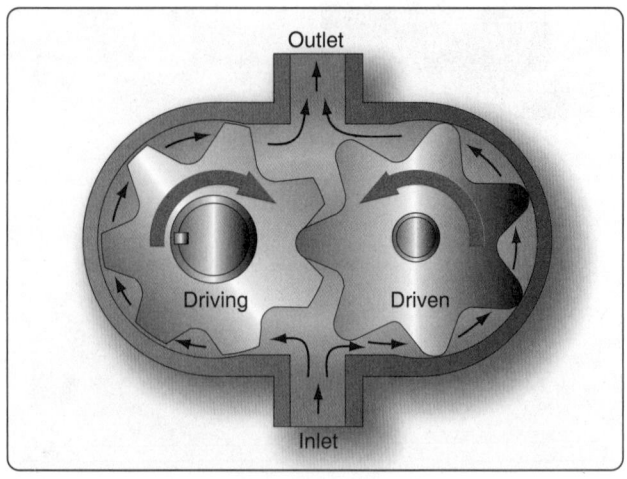

Figure 12-23. *Gear-type power pump.*

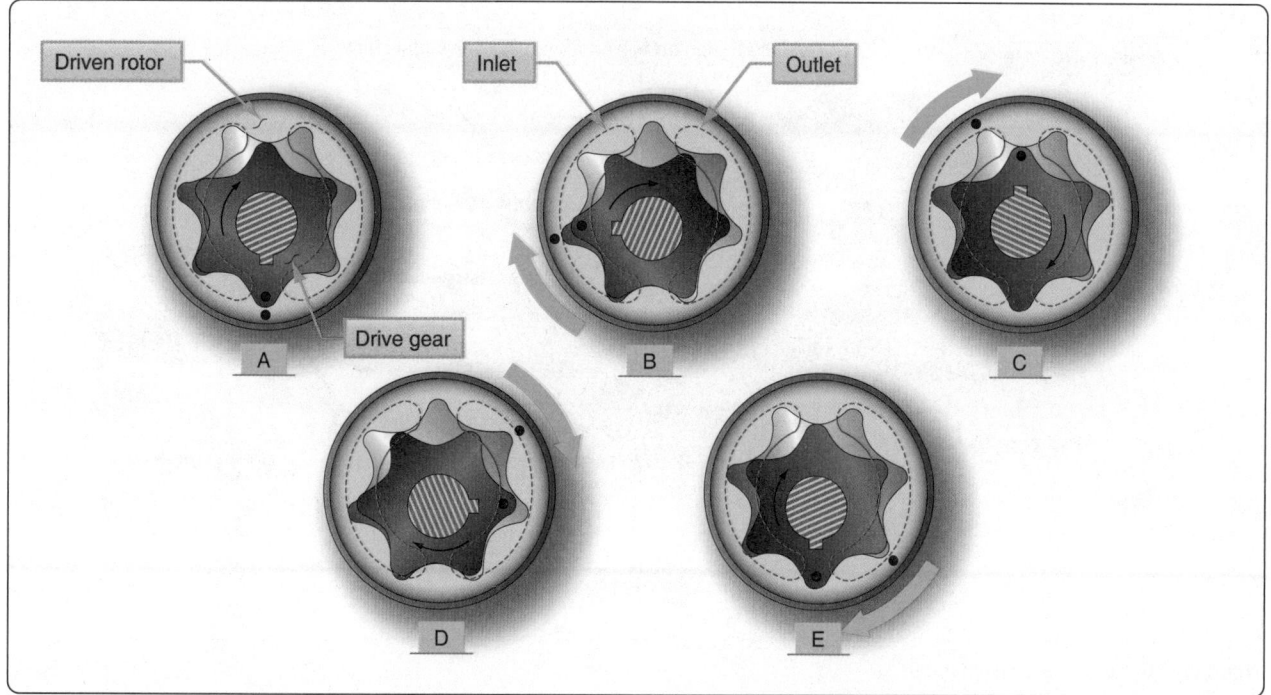

Figure 12-24. *Gerotor pump.*

Gerotor Pump

A gerotor-type power pump consists essentially of a housing containing an eccentric-shaped stationary liner, an internal gear rotor having seven wide teeth of short height, a spur driving gear having six narrow teeth, and a pump cover that contains two crescent-shaped openings. *[Figure 12-24]* One opening extends into an inlet port and the other extends into an outlet port. During the operation of the pump, the gears turn clockwise together. As the pockets between the gears on the left side of the pump move from a lowermost position toward a topmost position, the pockets increase in size, resulting in the production of a partial vacuum within these pockets. Since the pockets enlarge while over the inlet port crescent, fluid is drawn into them. As these same pockets (now full of fluid) rotate over to the right side of the pump, moving from the topmost position toward the lowermost position, they decrease in size. This results in the fluid being expelled from the pockets through the outlet port crescent.

Piston Pump

Piston pumps can be constant-displacement or variable-displacement pumps. The common features of design and operation that are applicable to all piston-type hydraulic pumps are described in the following paragraphs. Piston-type power-driven pumps have flanged mounting bases for the purpose of mounting the pumps on the accessory drive cases of aircraft engines. A pump drive shaft, which turns the mechanism, extends through the pump housing slightly beyond the mounting base. Torque from the driving unit is transmitted to the pump drive shaft by a drive coupling. The drive coupling is a short shaft with a set of male splines on both ends. The splines on one end engage with female splines in a driving gear; the splines on the other end engage with female splines in the pump drive shaft. Pump drive couplings are designed to serve as safety devices. The shear section of the drive coupling, located midway between the two sets of splines, is smaller in diameter than the splines. If the pump becomes unusually hard to turn or becomes jammed, this section shears, preventing damage to the pump or driving unit. *[Figure 12-25]* The basic pumping mechanism of piston-type pumps consists of a multiple-bore cylinder block, a piston for each bore, and a valve plate with inlet and outlet slots. The purpose of the valve plate slots is to let fluid into and out of the bores as the pump operates. The cylinder bores lie parallel to and symmetrically around the pump axis. All aircraft axial-piston pumps have an odd number of pistons. *[Figure 12-26]*

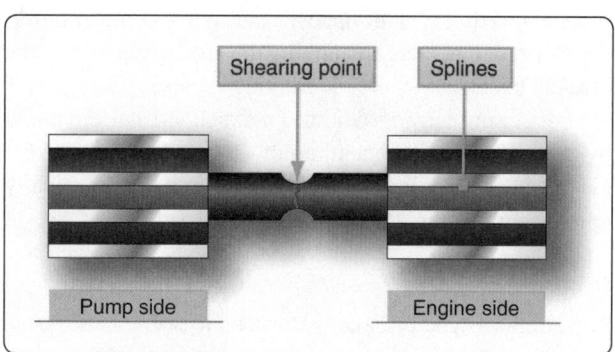

Figure 12-25. *Hydraulic pump shear shaft.*

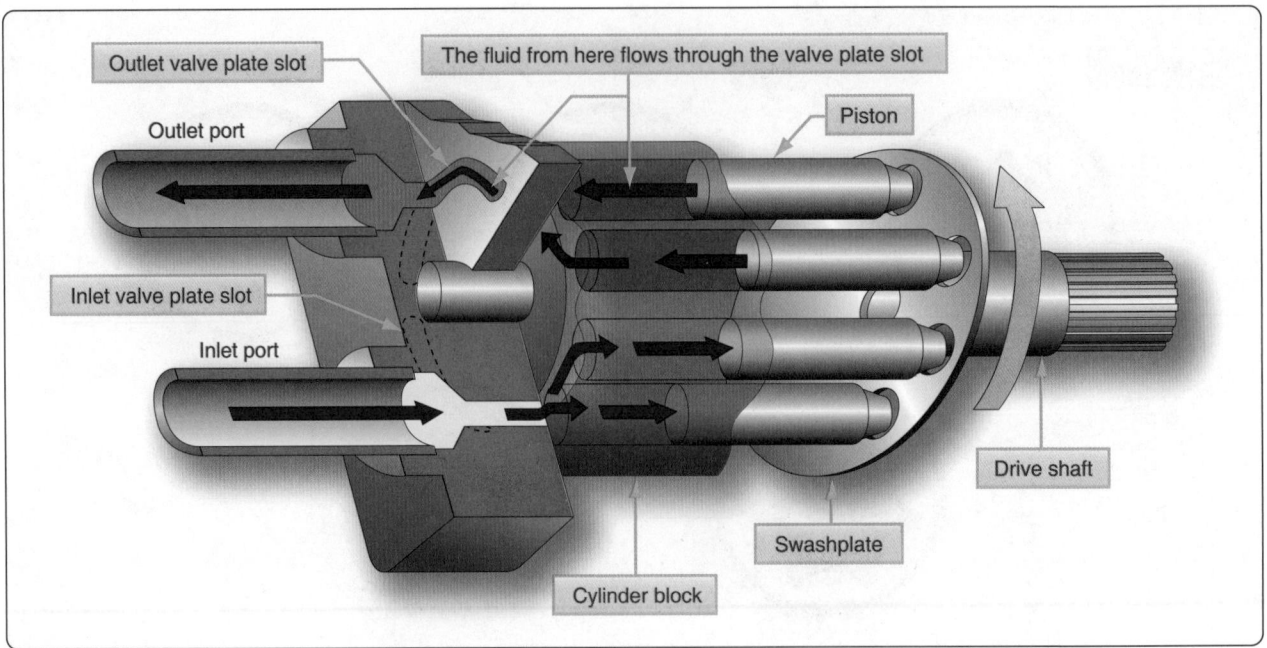

Figure 12-26. *Axial inline piston pump.*

Bent Axis Piston Pump

A typical constant-displacement axial-type pump is shown in *Figure 12-27.* The angular housing of the pump causes a corresponding angle to exist between the cylinder block and the drive shaft plate to which the pistons are attached. It is this angular configuration of the pump that causes the pistons to stroke as the pump shaft is turned. When the pump operates, all parts within the pump (except the outer races of the bearings that support the drive shaft, the cylinder bearing pin on which the cylinder block turns, and the oil seal) turn together as a rotating group. At one point of rotation of the rotating group, a minimum distance exists between the top of the cylinder block and the upper face of the drive shaft plate. Because of the angled housing at a point of rotation 180° away, the distance between the top of the cylinder block and the upper face of the drive shaft plate is at a maximum. At any given moment of operation, three of the pistons are moving away from the top face of the cylinder block, producing a partial vacuum in the bores in which these pistons operate. This occurs over the inlet port, so fluid is drawn into these bores at this time. On the opposite side of the cylinder block, three different pistons are moving toward the top face of the block. This occurs while the rotating group is passing over the outlet port causing fluid to be expelled from the pump by these pistons. The continuous and rapid action of the pistons is overlapping in nature and results in a practically nonpulsating pump output.

Inline Piston Pump

The simplest type of axial piston pump is the swash plate design in which a cylinder block is turned by the drive shaft.

Pistons fitted to bores in the cylinder block are connected through piston shoes and a retracting ring so that the shoes bear against an angled swash plate. As the block turns, the piston shoes follow the swash plate, causing the pistons to reciprocate. The ports are arranged in the valve plate so that the pistons pass the inlet as they are pulled out, and pass the outlet as they are forced back in. In these pumps, displacement is determined by the size and number of pistons, as well as their stroke length, which varies with the swash plate angle. This constant-displacement pump is illustrated in *Figure 12-26.*

Vane Pump

The vane-type power pump is also a constant-displacement pump. It consists of a housing containing four vanes (blades), a hollow steel rotor with slots for the vanes, and a coupling to turn the rotor. *[Figure 12-28]* The rotor is positioned off center within the sleeve. The vanes, which are mounted in the slots in the rotor, together with the rotor, divide the bore of the sleeve into four sections. As the rotor turns, each section passes one point where its volume is at a minimum and another point where its volume is at a maximum. The volume gradually increases from minimum to maximum during the first half of a revolution and gradually decreases from maximum to minimum during the second half of the revolution. As the volume of a given section increases, that section is connected to the pump inlet port through a slot in the sleeve. Since a partial vacuum is produced by the increase in volume of the section, fluid is drawn into the section through the pump inlet port and the slot in the sleeve. As the rotor turns through the second half of the revolution and the

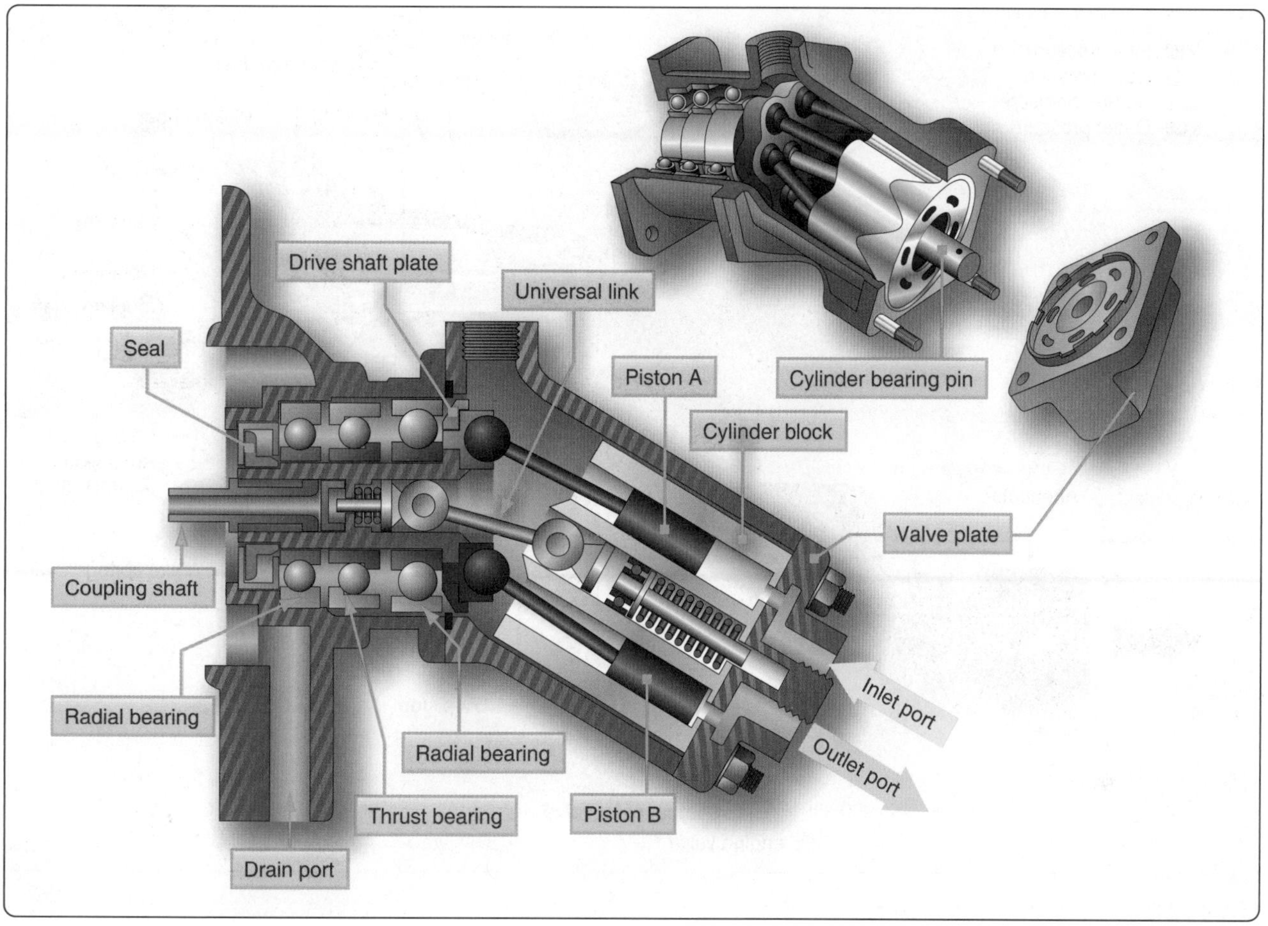

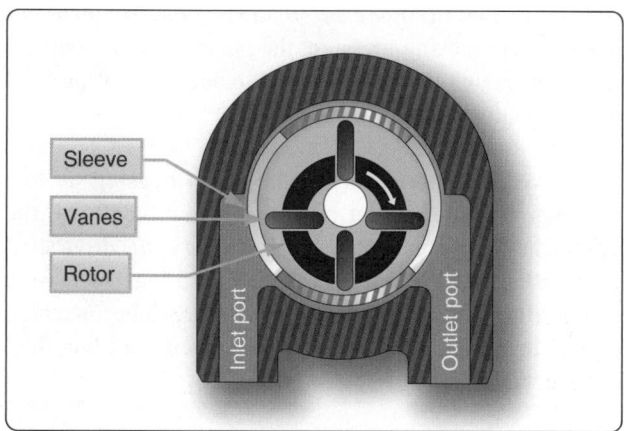

Figure 12-27. *Bent axis piston pump.*

Figure 12-28. *Vane-type power pump.*

volume of the given section is decreasing, fluid is displaced out of the section, through the slot in the sleeve aligned with the outlet port, and out of the pump.

Variable-Displacement Pump

A variable-displacement pump has a fluid output that is varied to meet the pressure demands of the system. The pump output is changed automatically by a pump compensator within the pump. The following paragraph discusses a two-stage Vickers variable displacement pump. The first stage of the pump consists of a centrifugal pump that boosts the pressure before the fluid enters the piston pump. *[Figure 12-29]*

Basic Pumping Operation

The aircraft's engine rotates the pump drive shaft, cylinder block, and pistons via a gearbox. Pumping action is generated by piston shoes that are restrained and slide on the shoe bearing plate in the yoke assembly. Because the yoke is at an angle to the drive shaft, the rotary motion of the shaft is converted to piston reciprocating motion.

As the piston begins to withdraw from the cylinder block, system inlet pressure forces fluid through a porting arrangement in the valve plate into the cylinder bore. The piston shoes are restrained in the yoke by a piston shoe retaining plate and a shoe plate during the intake stroke. As the drive shaft continues to turn the cylinder block, the piston shoe continues following the yoke bearing surface. This begins to return the piston into its bore (i.e., toward the valve block).

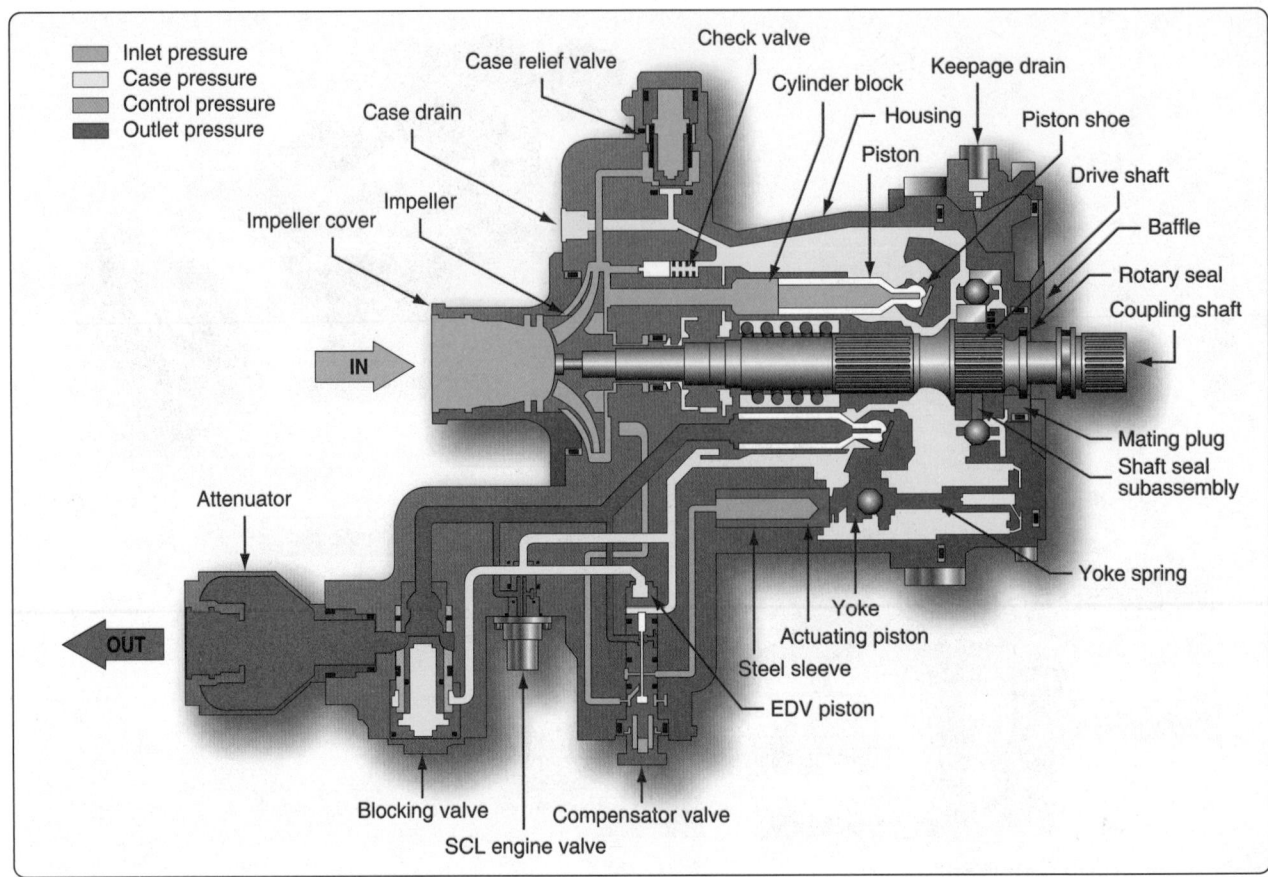

Figure 12-29. *Variable displacement pump.*

The fluid contained in the bore is precompressed, then expelled through the outlet port. Discharge pressure holds the piston shoe against the yoke bearing surface during the discharge stroke and also provides the shoe pressure balance and fluid film through an orifice in the piston and shoe subassembly.

With each revolution of the drive shaft and cylinder block, each piston goes through the pumping cycle described above, completing one intake and one discharge stroke. High-pressure fluid is ported out through the valve plate, past the blocking valve, to the pump outlet. The blocking valve is designed to remain open during normal pump operation. Internal leakage keeps the pump housing filled with fluid for lubrication of rotating parts and cooling. The leakage is returned to the system through a case drain port. The case valve relief valve protects the pump against excessive case pressure, relieving it to the pump inlet.

Normal Pumping Mode
The pressure compensator is a spool valve that is held in the closed position by an adjustable spring load. *[Figure 12-30]* When pump outlet pressure (system pressure) exceeds the pressure setting (2,850 psi for full flow), the spool

moves to admit fluid from the pump outlet against the yoke actuator piston. In *Figure 12-30,* the pressure compensator is shown at cracking pressure; the pump outlet pressure is just high enough to move the spool to begin admitting fluid to the actuator piston.

The yoke is supported inside the pump housing on two bearings. At pump outlet pressures below 2,850 psi, the yoke is held at its maximum angle relative to the drive shaft centerline by the force of the yoke return spring. Decreasing system flow demand causes outlet pressure to become high enough to crack the compensator valve open and admit fluid to the actuator piston.

This control pressure overcomes the yoke return spring force and strokes the pump yoke to a reduced angle. The reduced angle of the yoke results in a shorter stroke for the pistons and reduced displacement. *[Figure 12-31]*

The lower displacement results in a corresponding reduction in pump flow. The pump delivers only that flow required to maintain the desired pressure in the system. When there is no demand for flow from the system, the yoke angle decreases to nearly zero degrees stroke angle. In this mode, the unit

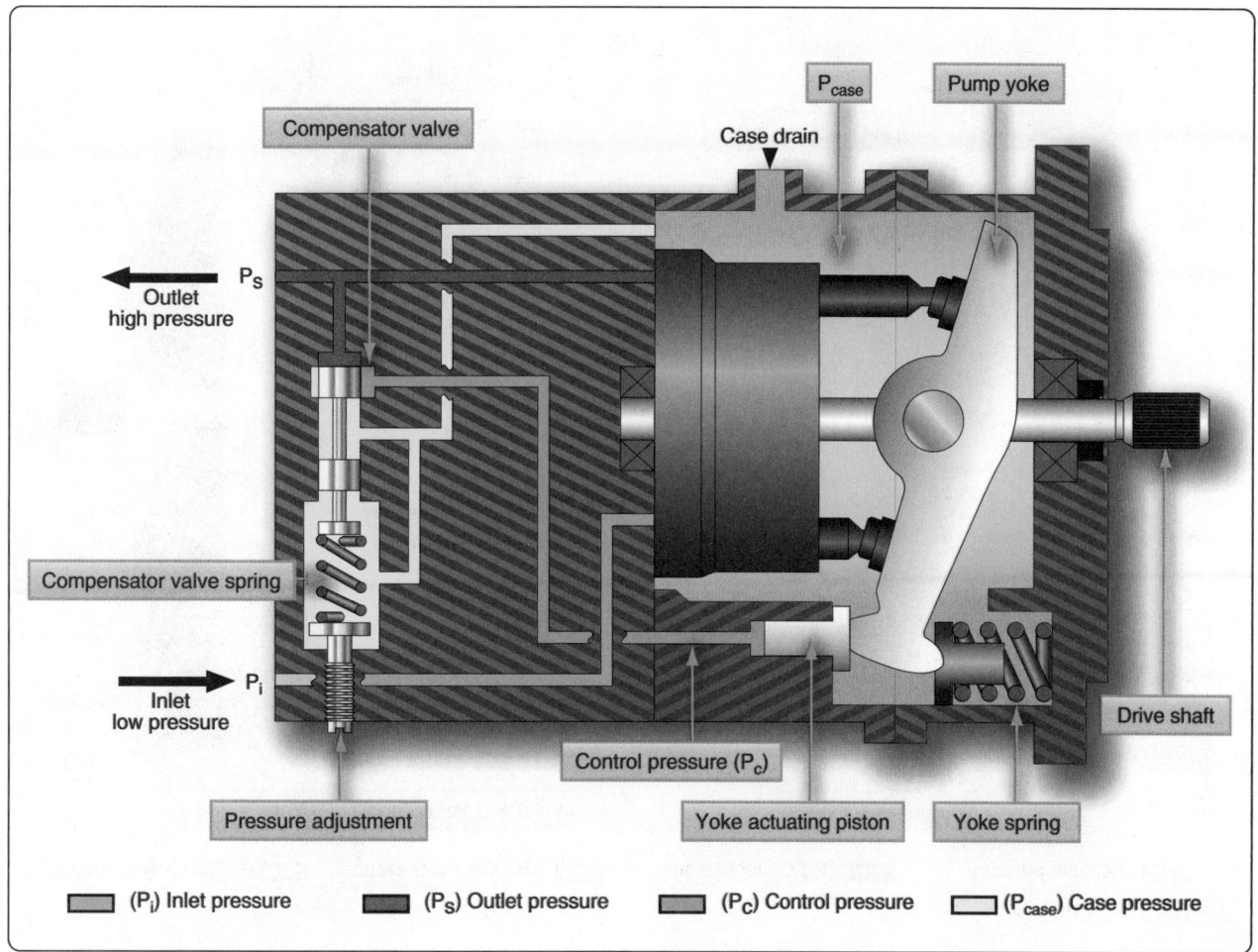

Figure 12-30. *Normal pumping mode.*

pumps only its own internal leakage. Thus, at pump outlet pressures above 2,850 psi, pump displacement decreases as outlet pressure rises. At system pressures below this level, no fluid is admitted through the pressure compensator valve to the actuator piston and the pump remains at full displacement, delivering full flow. Pressure is then

determined by the system demand. The unit maintains zero flow at system pressure of 3,025 psi.

Depressurized Mode

When the solenoid valve is energized, the electrical depressurization valve (EDV) solenoid valve moves up against the spring force and the outlet fluid is ported to the EDV control piston on the top of the compensator (depressurizing piston). *[Figure 12-32]* The high-pressure fluid pushes the compensator spool beyond its normal metering position. This removes the compensator valve from the circuit and connects the actuator piston directly to the pump outlet. Outlet fluid is also ported to the blocking valve spring chamber, which equalizes pressure on both sides of its plunger. The blocking valve closes due to the force of the blocking valve spring and isolates the pump from the external hydraulic system. The pump strokes itself to zero delivery at an outlet pressure that is equal to the pressure required on the actuator piston to reduce the yoke angle to nearly zero, approximately 1,100 psi. This depressurization and blocking feature can be used to reduce the load on the engine during startup and, in a multiple pump system, to isolate one pump at a time and check for proper system pressure output.

Figure 12-31. *Yoke angle.*

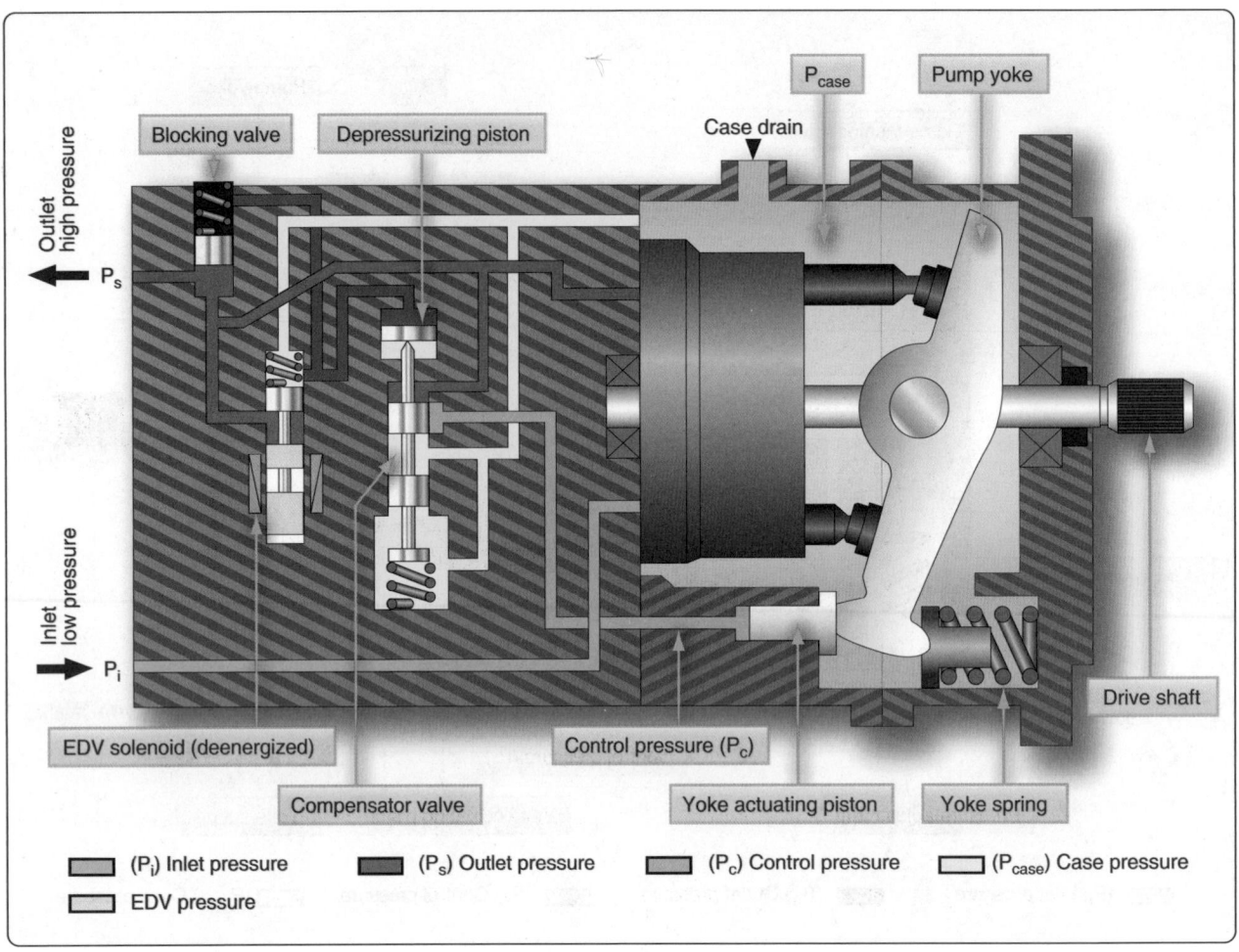

Figure 12-32. *Depressurized mode.*

Valves

Flow Control Valves

Flow control valves control the speed and/or direction of fluid flow in the hydraulic system. They provide for the operation of various components when desired and the speed at which the component operates. Examples of flow control valves include: selector valves, check valves, sequence valves, priority valves, shuttle valves, quick disconnect valves, and hydraulic fuses.

Selector Valves

A selector valve is used to control the direction of movement of a hydraulic actuating cylinder or similar device. It provides for the simultaneous flow of hydraulic fluid both into and out of the unit. Hydraulic system pressure can be routed with the selector valve to operate the unit in either direction and a corresponding return path for the fluid to the reservoir is provided. There are two main types of selector valves: open-center and closed-center. An open center valve allows a continuous flow of system hydraulic fluid through the valve even when the selector is not in a position to actuate a unit. A closed-center selector valve blocks the flow of fluid through the valve when it is in the NEUTRAL or OFF position. *[Figure 12-33A]*

Selector valves may be poppet-type, spool-type, piston-type, rotary-type, or plug-type. *[Figure 12-34]* Regardless, each selector valve has a unique number of ports. The number of ports is determined by the particular requirements of the system in which the valve is used. Closed-centered selector valves with four ports are most common in aircraft hydraulic systems. These are known as four-way valves. *Figure 12-33* illustrates how this valve connects to the pressure and return lines of the hydraulic system, as well as to the two ports on a common actuator. Most selector valves are mechanically controlled by a lever or electrically controlled by solenoid or servo. *[Figure 12-35]*

The four ports on a four-way selector valve always have the same function. One port receives pressurized fluid from the system hydraulic pump. A second port always returns fluid to the reservoir. The third and fourth ports are used to connect the selector valve to the actuating unit. There are two ports

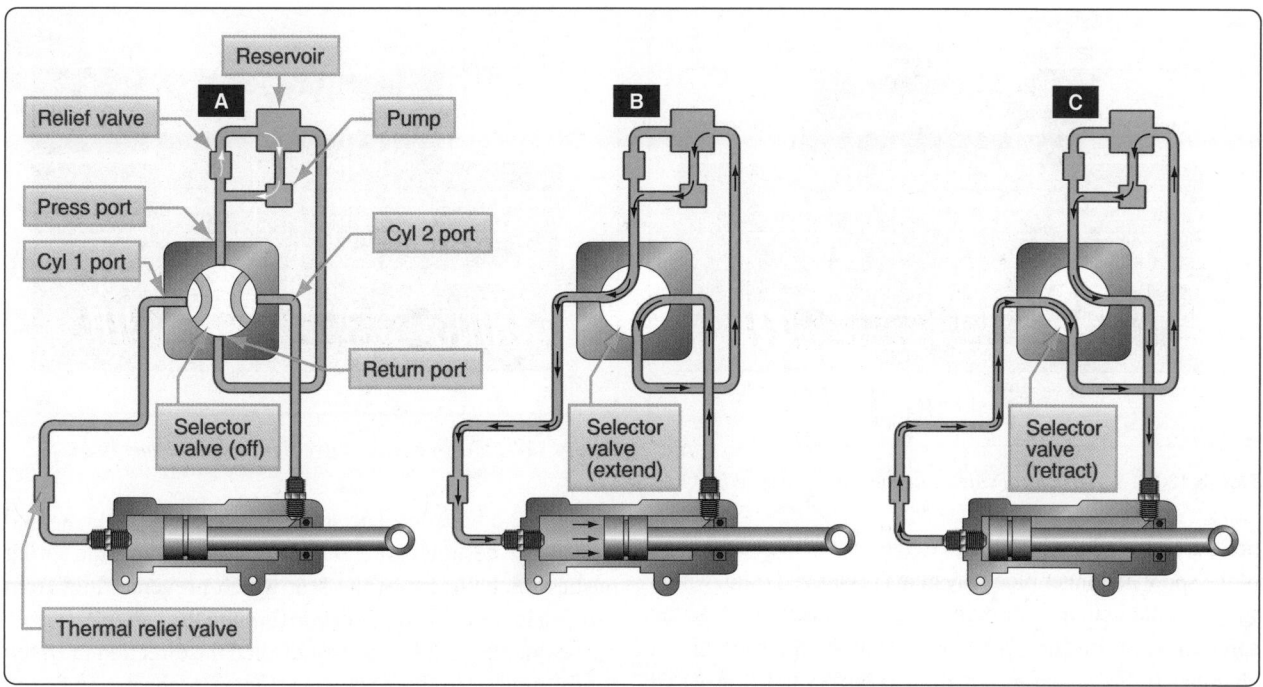

Figure 12-33. *Operation of a closed-center four-way selector valve, which controls an actuator.*

on the actuating unit. When the selector valve is positioned to connect pressure to one port on the actuator, the other actuator port is simultaneously connected to the reservoir return line through selector valve. *[Figure 12-33B]* Thus, the unit operates in a certain direction. When the selector valve is positioned to connect pressure to the other port on the actuating unit, the original port is simultaneously connected to the return line through the selector valve and the unit operates in the opposite direction. *[Figure 12-33C]*

Figure 12-36 illustrates the internal flow paths of a solenoid operated selector valve. The closed center valve is shown in the NEUTRAL or OFF position. Neither solenoid is energized. The pressure port routes fluid to the center lobe on the spool, which blocks the flow. Fluid pressure flows through the pilot valves and applies equal pressure on both ends of the spool. The actuator lines are connected around the spool to the return line.

When selected via a switch in the flight deck, the right solenoid is energized. The right pilot valve plug shifts left, which blocks pressurized fluid from reaching the right end of the main spool. The spool slides to the right due to greater

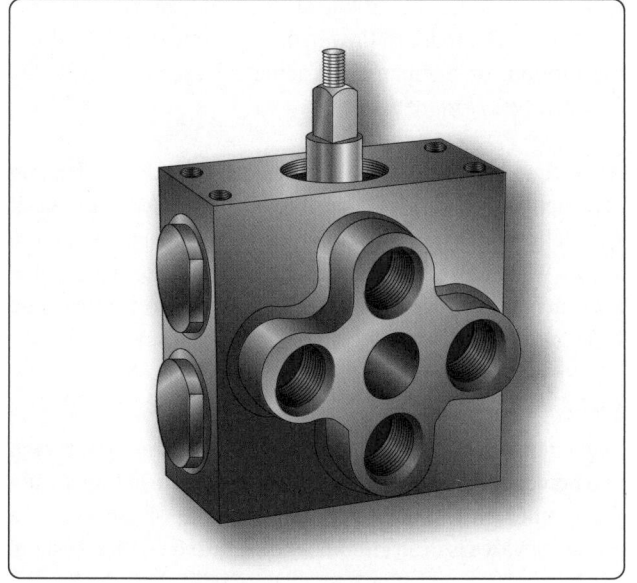

Figure 12-34. *A poppet-type four-way selector valve.*

Figure 12-35. *Four-way servo control valve.*

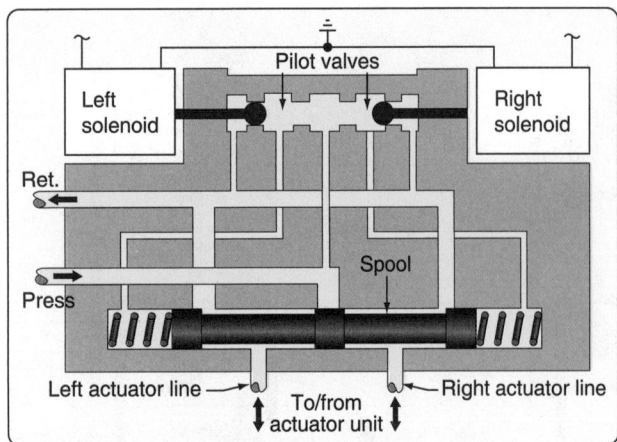

Figure 12-36. *Servo control valve solenoids not energized.*

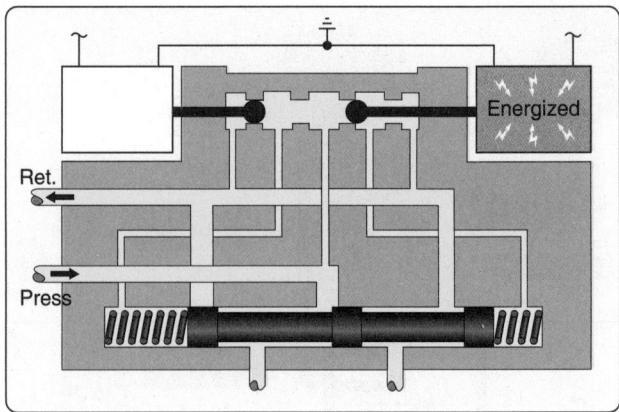

Figure 12-37. *Servo control valve right solenoid energized.*

pressure applied on the left end of the spool. The center lobe of the spool no longer blocks system pressurized fluid, which flows to the actuator through the left actuator line. At the same time, return flow is blocked from the main spool left chamber so the actuator (not shown) moves in the selected direction. Return fluid from the moving actuator flows through the right actuator line past the spool and into the return line. *[Figure 12-37]*

Typically, the actuator or moving device contacts a limit switch when the desired motion is complete. The switch causes the right solenoid to de-energize and the right pilot valve reopens. Pressurized fluid can once again flow through the pilot valve and into the main spool right end chamber. There, the spring and fluid pressure shift the spool back to the left into the NEUTRAL or OFF position shown in *Figure 12-36.*

To make the actuator move in the opposite direction, the flight deck switch is moved in the opposite direction. All motion inside the selector valve is the same as described above but in the opposite direction. The left solenoid is energized. Pressure is applied to the actuator through the right port and return fluid from the left actuator line is connected to the return port through the motion of the spool to the left.

Check Valve

Another common flow control valve in aircraft hydraulic systems is the check valve. A check valve allows fluid to flow unimpeded in one direction but prevents or restricts fluid flow in the opposite direction. A check valve may be an independent component situated inline somewhere in the hydraulic system or it may be built-in to a component. When part of a component, the check valve is said to be an integral check valve.

A typical check valve consists of a spring-loaded ball and

seat inside a housing. The spring compresses to allow fluid flow in the designed direction. When flow stops, the spring pushes the ball against the seat which prevents fluid from flowing in the opposite direction through the valve. An arrow on the outside of the housing indicated the direction in which fluid flow is permitted. *[Figure 12-38]* A check valve may also be constructed with spring loaded flapper or coned shape piston instead of a ball.

Orifice-Type Check Valve

Some check valves allow full fluid flow in one direction and restricted flow in the opposite direction. These are known as orifice-type check valves, or damping valves. The valve contains the same spring, ball, and seat combination as a normal check valve, but the seat area has a calibrated orifice machined into it. Thus, fluid flow is unrestricted in the designed direction while the ball is pushed off of its seat. The downstream actuator operates at full speed. When fluid back flows into the valve, the spring forces the ball against the seat which limits fluid flow to the amount that can pass through the orifice. The reduced flow in this opposite direction slows the motion, or dampens, the actuator associated with the check valve. *[Figure 12-38]*

An orifice check valve may be included in a hydraulic landing gear actuator system. When the gear is raised, the check valve allows full fluid flow to lift the heavy gear at maximum speed. When lowering the gear, the orifice in the check valve prevents the gear from violently dropping by restricting fluid flow out of the actuating cylinder.

Sequence Valves

Sequence valves control the sequence of operation between two branches in a circuit; they enable one unit to automatically set another unit into motion. An example of the use of a sequence valve is in an aircraft landing gear actuating system. In a landing gear actuating system, the landing gear doors must open before the landing gear starts to extend. Conversely,

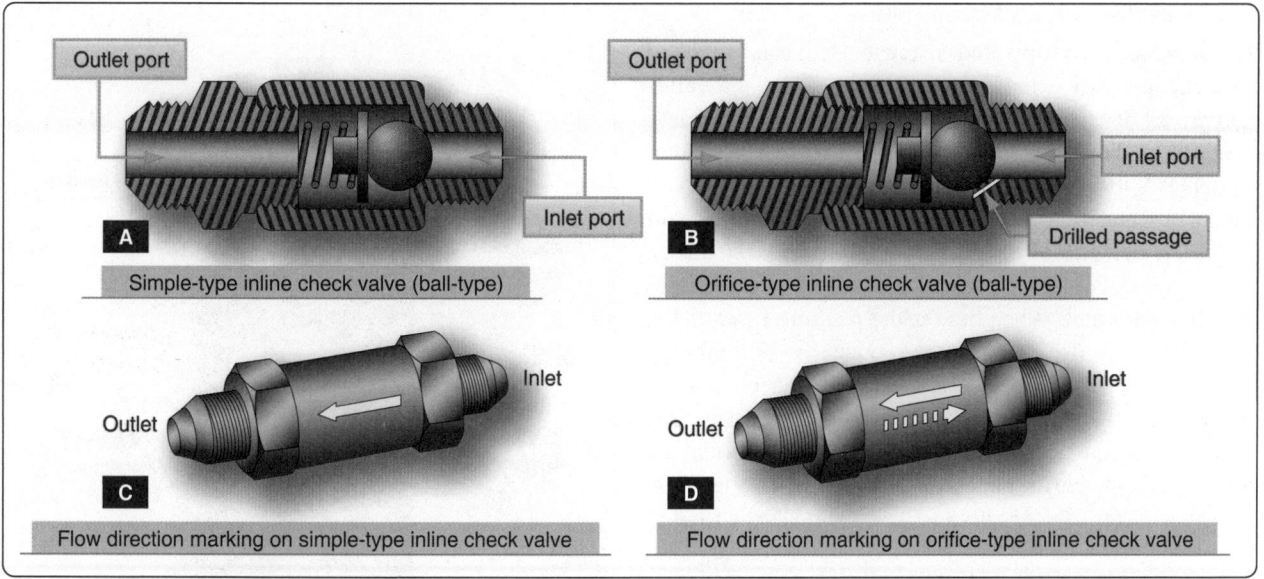

Figure 12-38. *An inline check valve and orifice type inline check valve.*

the landing gear must be completely retracted before the doors close. A sequence valve installed in each landing gear actuating line performs this function. A sequence valve is somewhat similar to a relief valve except that, after the set pressure has been reached, the sequence valve diverts the fluid to a second actuator or motor to do work in another part of the system. There are various types of sequence valves. Some are controlled by pressure, some are controlled mechanically, and some are controlled by electric switches.

Pressure-Controlled Sequence Valve

The operation of a typical pressure-controlled sequence valve is illustrated in *Figure 12-36.* The opening pressure is obtained by adjusting the tension of the spring that normally holds the piston in the closed position. (Note that the top part of the piston has a larger diameter than the lower part.) Fluid

enters the valve through the inlet port, flows around the lower part of the piston and exits the outlet port, where it flows to the primary (first) unit to be operated. *[Figure 12-39A]* This fluid pressure also acts against the lower surface of the piston.

When the primary actuating unit completes its operation, pressure in the line to the actuating unit increases sufficiently to overcome the force of the spring, and the piston rises. The valve is then in the open position. *[Figure 12-39B]* The fluid entering the valve takes the path of least resistance and flows to the secondary unit. A drain passage is provided to allow any fluid leaking past the piston to flow from the top of the valve. In hydraulic systems, this drain line is usually connected to the main return line.

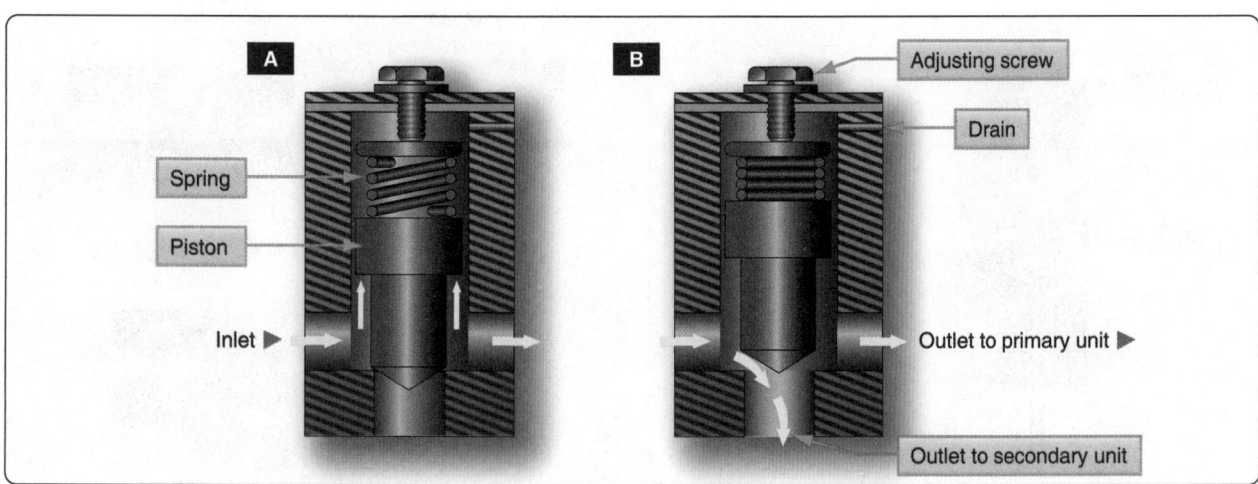

Figure 12-39. *A pressure-controlled sequence valve.*

Mechanically-Operated Sequence Valve

The mechanically-operated sequence valve is operated by a plunger that extends through the body of the valve. *[Figure 12-40]* The valve is mounted so that the plunger is operated by the primary unit. A check valve, either a ball or a poppet, is installed between the fluid ports in the body. It can be unseated by either the plunger or fluid pressure. Port A and the actuator of the primary unit are connected by a common line. Port B is connected by a line to the actuator of the secondary unit. When fluid under pressure flows to the primary unit, it also flows into the sequence valve through port A to the seated check valve in the sequence valve. In order to operate the secondary unit, the fluid must flow through the sequence valve. The valve is located so that the primary unit moves the plunger as it completes its operation. The plunger unseats the check valve and allows the fluid to flow through the valve, out port B, and to the secondary unit.

Priority Valves

A priority valve gives priority to the critical hydraulic subsystems over noncritical systems when system pressure is low. For instance, if the pressure of the priority valve is set for 2,200 psi, all systems receive pressure when the pressure is above 2,200 psi. If the pressure drops below 2,200 psi, the priority valve closes, and no fluid pressure flows to the noncritical systems. *[Figure 12-41]* Some hydraulic designs use pressure switches and electrical shutoff valves to assure that the critical systems have priority over noncritical systems when system pressure is low.

Quick Disconnect Valves

Quick disconnect valves are installed in hydraulic lines to prevent loss of fluid when units are removed. Such valves are installed in the pressure and suction lines of the system immediately upstream and downstream of the power pump. In addition to pump removal, a power pump can be disconnected

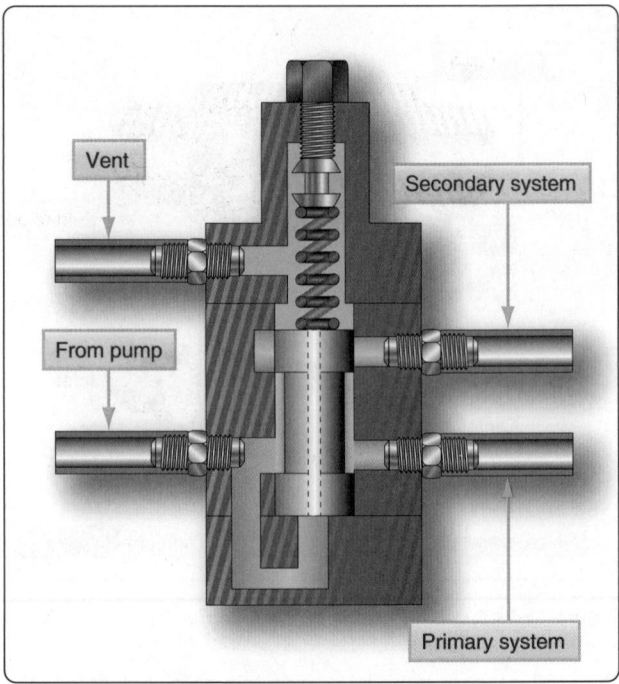

Figure 12-41. *Priority valve.*

from the system and a hydraulic test stand connected in its place. These valve units consist of two interconnecting sections coupled together by a nut when installed in the system. Each valve section has a piston and poppet assembly. These are spring loaded to the closed position when the unit is disconnected. *[Figure 12-42]*

Hydraulic Fuses

A hydraulic fuse is a safety device. Fuses may be installed at strategic locations throughout a hydraulic system. They detect a sudden increase in flow, such as a burst downstream,

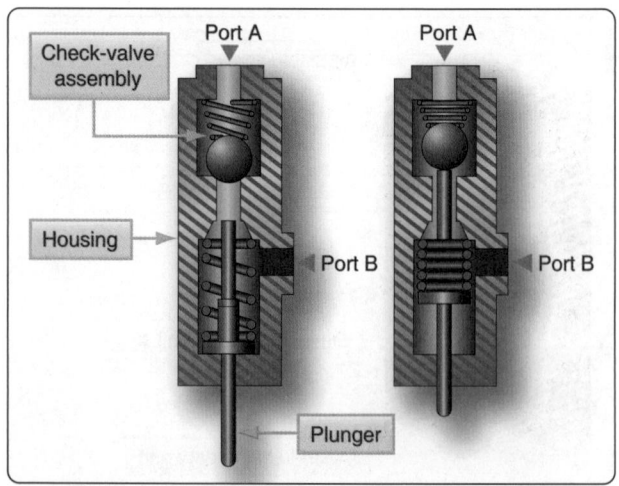

Figure 12-40. *Mechanically operated sequence valve.*

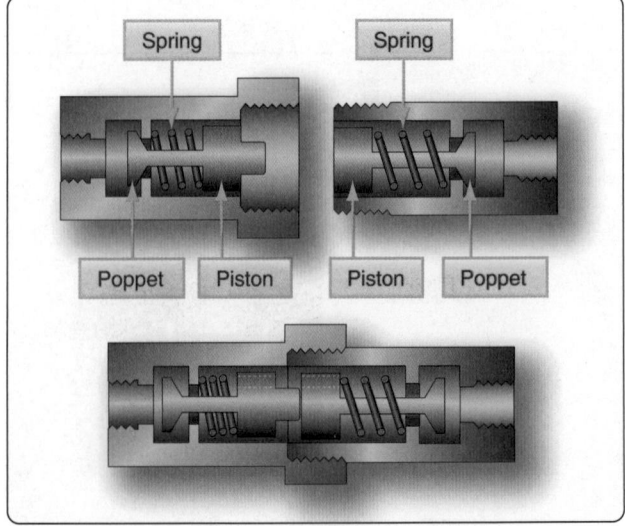

Figure 12-42. *A hydraulic quick-disconnect valve.*

and shut off the fluid flow. By closing, a fuse preserves hydraulic fluid for the rest of the system. Hydraulic fuses are fitted to the brake system, leading edge flap and slat extend and retract lines, nose landing gear up and down lines, and the thrust reverser pressure and return lines. One type of fuse, referred to as the automatic resetting type, is designed to allow a certain volume of fluid per minute to pass through it. If the volume passing through the fuse becomes excessive, the fuse closes and shuts off the flow. When the pressure is removed from the pressure supply side of the fuse, it automatically resets itself to the open position. Fuses are usually cylindrical in shape, with an inlet and outlet port at opposite ends. [Figure 12-43]

Pressure Control Valves

The safe and efficient operation of fluid power systems, system components, and related equipment requires a means of controlling pressure. There are many types of automatic pressure control valves. Some of them are an escape for pressure that exceeds a set pressure; some only reduce the pressure to a lower pressure system or subsystem; and some keep the pressure in a system within a required range.

Relief Valves

Hydraulic pressure must be regulated in order to use it to perform the desired tasks. A pressure relief valve is used to limit the amount of pressure being exerted on a confined liquid. This is necessary to prevent failure of components or rupture of hydraulic lines under excessive pressures. The pressure relief valve is, in effect, a system safety valve.

The design of pressure relief valves incorporates adjustable spring-loaded valves. They are installed in such a manner as to discharge fluid from the pressure line into a reservoir return line when the pressure exceeds the predetermined maximum for which the valve is adjusted. Various makes and designs of pressure relief valves are in use, but, in general, they all

employ a spring-loaded valving device operated by hydraulic pressure and spring tension. [Figure 12-44] Pressure relief valves are adjusted by increasing or decreasing the tension on the spring to determine the pressure required to open the valve. They may be classified by type of construction or uses in the system. The most common types of valve are:

1. Ball type—in pressure relief valves with a ball-type valving device, the ball rests on a contoured seat. Pressure acting on the bottom of the ball pushes it off its seat, allowing the fluid to bypass.

2. Sleeve type—in pressure relief valves with a sleeve-type valving device, the ball remains stationary and a sleeve-type seat is moved up by the fluid pressure. This allows the fluid to bypass between the ball and the sliding sleeve-type seat.

3. Poppet type—in pressure relief valves with a poppet-type valving device, a cone-shaped poppet may have any of several design configurations; however, it is basically a cone and seat machined at matched angles to prevent leakage. As the pressure rises to its predetermined setting, the poppet is lifted off its seat, as in the ball-type device. This allows the fluid to pass through the opening created and out the return port.

Pressure relief valves cannot be used as pressure regulators in large hydraulic systems that depend on engine-driven pumps for the primary source of pressure because the pump is constantly under load and the energy expended in holding the pressure relief valve off its seat is changed into heat. This

Figure 12-43. *Hydraulic fuse.*

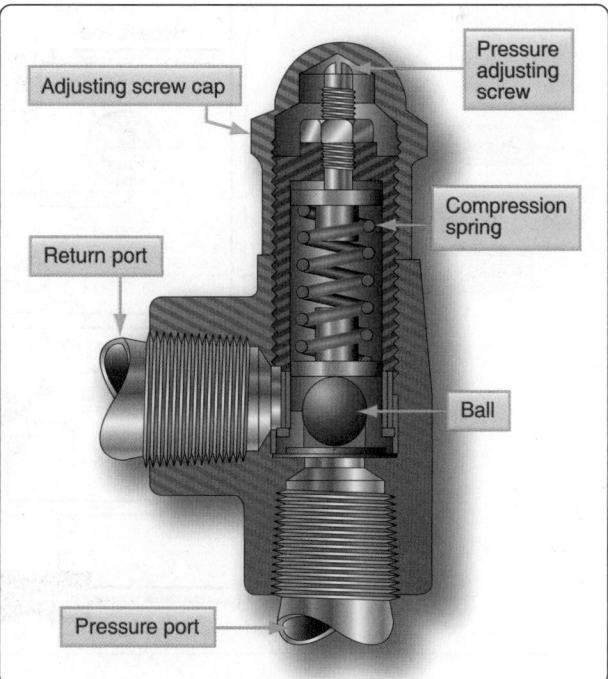

Figure 12-44. *Pressure relief valve.*

heat is transferred to the fluid and, in turn, to the packing rings, causing them to deteriorate rapidly. Pressure relief valves, however, may be used as pressure regulators in small, low-pressure systems or when the pump is electrically driven and is used intermittently.

Pressure relief valves may be used as:

1. System relief valve—the most common use of the pressure relief valve is as a safety device against the possible failure of a pump compensator or other pressure regulating device. All hydraulic systems that have hydraulic pumps incorporate pressure relief valves as safety devices.

2. Thermal relief valve—the pressure relief valve is used to relieve excessive pressures that may exist due to thermal expansion of the fluid. They are used where a check valve or selector valve prevents pressure from being relieved through the main system relief valve. Thermal relief valves are usually smaller than system relief valves. As pressurized fluid in the line in which

it is installed builds to an excessive amount, the valve poppet is forced off its seat. This allows excessive pressurized fluid to flow through the relief valve to the reservoir return line. When system pressure decreases to a predetermined pressure, spring tension overcomes system pressure and forces the valve poppet to the closed position.

Pressure Regulators

The term pressure regulator is applied to a device used in hydraulic systems that are pressurized by constant-delivery-type pumps. One purpose of the pressure regulator is to manage the output of the pump to maintain system operating pressure within a predetermined range. The other purpose is to permit the pump to turn without resistance (termed unloading the pump) at times when pressure in the system is within normal operating range. The pressure regulator is located in the system so that pump output can get into the system pressure circuit only by passing through the regulator. The combination of a constant-delivery-type pump and the

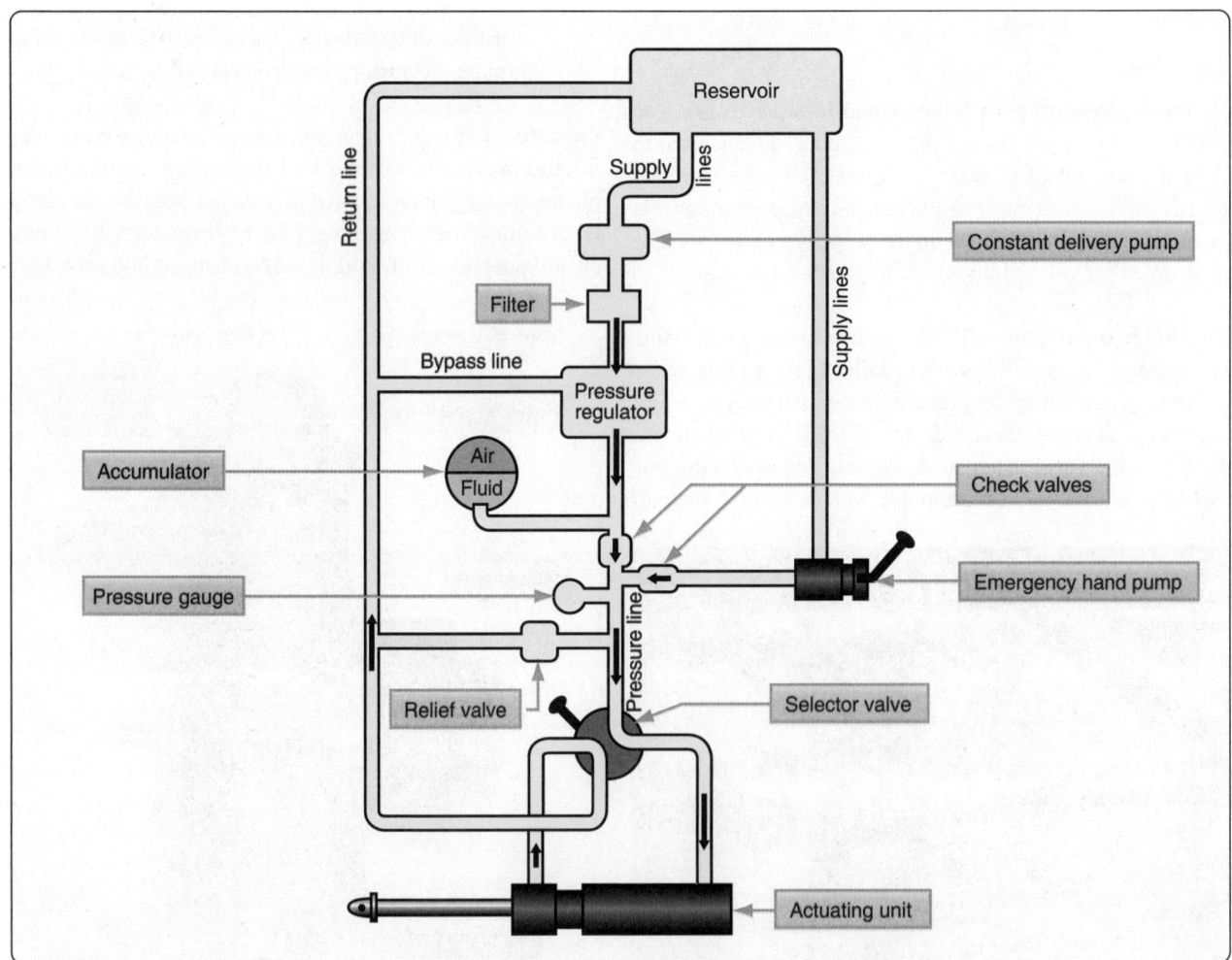

Figure 12-45. *The location of a pressure regulator in a basic hydraulic system. The regulator unloads the constant delivery pump by bypassing fluid to the return line when the predetermined system pressure is reached.*

pressure regulator is virtually the equivalent of a compensator-controlled, variable-delivery-type pump. *[Figure 12-45]*

Pressure Reducers

Pressure reducing valves are used in hydraulic systems where it is necessary to lower the normal system operating pressure by a specified amount. Pressure reducing valves provide a steady pressure into a system that operates at a lower pressure than the supply system. A reducing valve can normally be set for any desired downstream pressure within the design limits of the valve. Once the valve is set, the reduced pressure is maintained regardless of changes in supply pressure (as long as the supply pressure is at least as high as the reduced pressure desired) and regardless of the system load, if the load does not exceed the designed capacity of the reducer. *[Figure 12-46]*

Shuttle Valves

In certain fluid power systems, the supply of fluid to a subsystem must be from more than one source to meet system requirements. In some systems, an emergency system is provided as a source of pressure in the event of normal system failure. The emergency system usually actuates only essential components. The main purpose of the shuttle valve is to isolate the normal system from an alternate or emergency system. It is small and simple; yet, it is a very important component. *[Figure 12-47]* The housing contains three ports—normal system inlet, alternate or emergency system inlet, and outlet. A shuttle valve used to operate more than one actuating unit may contain additional unit outlet ports.

Enclosed in the housing is a sliding part called the shuttle. Its purpose is to seal off one of the inlet ports. There is a shuttle seat at each inlet port. When a shuttle valve is in the normal

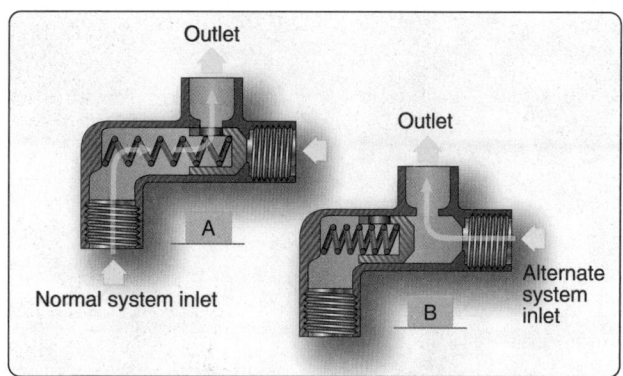

Figure 12-47. *A spring-loaded piston-type shuttle valve in normal configuration (A) and with alternate/emergency supply (B).*

operation position, fluid has a free flow from the normal system inlet port, through the valve, and out through the outlet port to the actuating unit. The shuttle is seated against the alternate system inlet port and held there by normal system pressure and by the shuttle valve spring. The shuttle remains in this position until the alternate system is activated. This action directs fluid under pressure from the alternate system to the shuttle valve and forces the shuttle from the alternate system inlet port to the normal system inlet port. Fluid from the alternate system then has a free flow to the outlet port but is prevented from entering the normal system by the shuttle, which seals off the normal system port.

The shuttle may be one of four types:

1. Sliding plunge.
2. Spring-loaded piston.
3. Spring-loaded ball.
4. Spring-loaded poppet.

In shuttle valves that are designed with a spring, the shuttle is normally held against the alternate system inlet port by the spring.

Shutoff Valves

Shutoff valves are used to shut off the flow of fluid to a particular system or component. In general, these types of valves are electrically powered. Shutoff valves are also used to create a priority in a hydraulic system and are controlled by pressure switches. *[Figure 12-48]*

Accumulators

The accumulator is a steel sphere divided into two chambers by a synthetic rubber diaphragm. The upper chamber contains fluid at system pressure, while the lower chamber is charged with nitrogen or air. Cylindrical types are also used in high-pressure hydraulic systems. Many aircraft have several accumulators in the hydraulic system. There may be a main system accumulator and an emergency system accumulator.

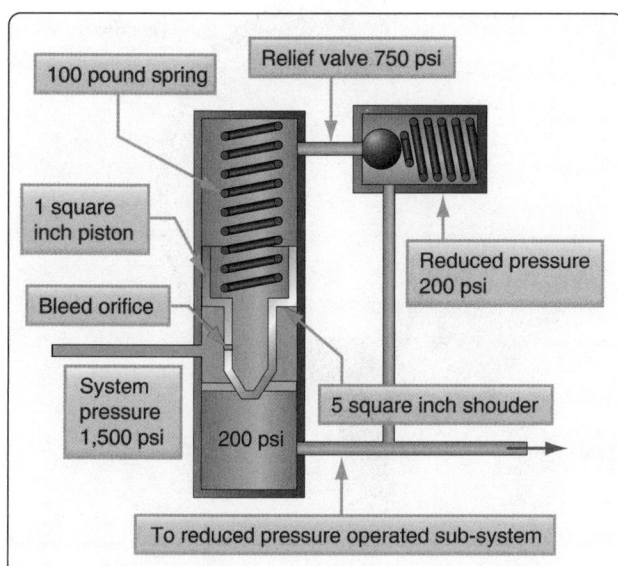

Figure 12-46. *Operating mechanism of a pressure reducing valve.*

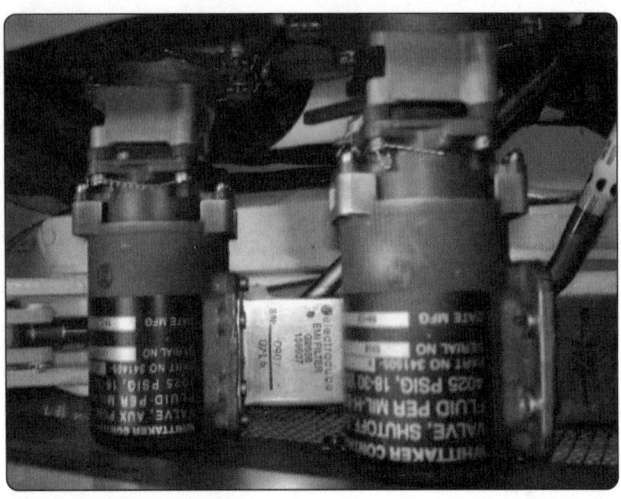

Figure 12-48. *Shutoff valves.*

There may also be auxiliary accumulators located in various sub-systems.

The function of an accumulator is to:

- Dampen pressure surges in the hydraulic system caused by actuation of a unit and the effort of the pump to maintain pressure at a preset level.

- Aid or supplement the power pump when several units are operating at once by supplying extra power from its accumulated, or stored, power.

- Store power for the limited operation of a hydraulic unit when the pump is not operating.

- Supply fluid under pressure to compensate for small internal or external (not desired) leaks that would cause the system to cycle continuously by action of the pressure switches continually kicking in.

Types of Accumulators

There are two general types of accumulators used in aircraft hydraulic systems: spherical and cylindrical.

Spherical

The spherical-type accumulator is constructed in two halves that are fastened and threaded, or welded, together. Two threaded openings exist. The top port accepts fittings to connect to the pressurized hydraulic system to the accumulator. The bottom port is fitted with a gas servicing valve, such as a Schrader valve. A synthetic rubber diaphragm, or bladder, is installed in the sphere to create two chambers. Pressurized hydraulic fluid occupies the upper chamber and nitrogen or air charges the lower chamber. A screen at the fluid pressure port keeps the diaphragm, or bladder, from extruding through the port when the lower chamber is charged, and hydraulic fluid pressure is zero. A rigid button or disc may also be attached to the diaphragm, or bladder, for this purpose. *[Figure 12-49]* The bladder is installed through a large opening in the bottom

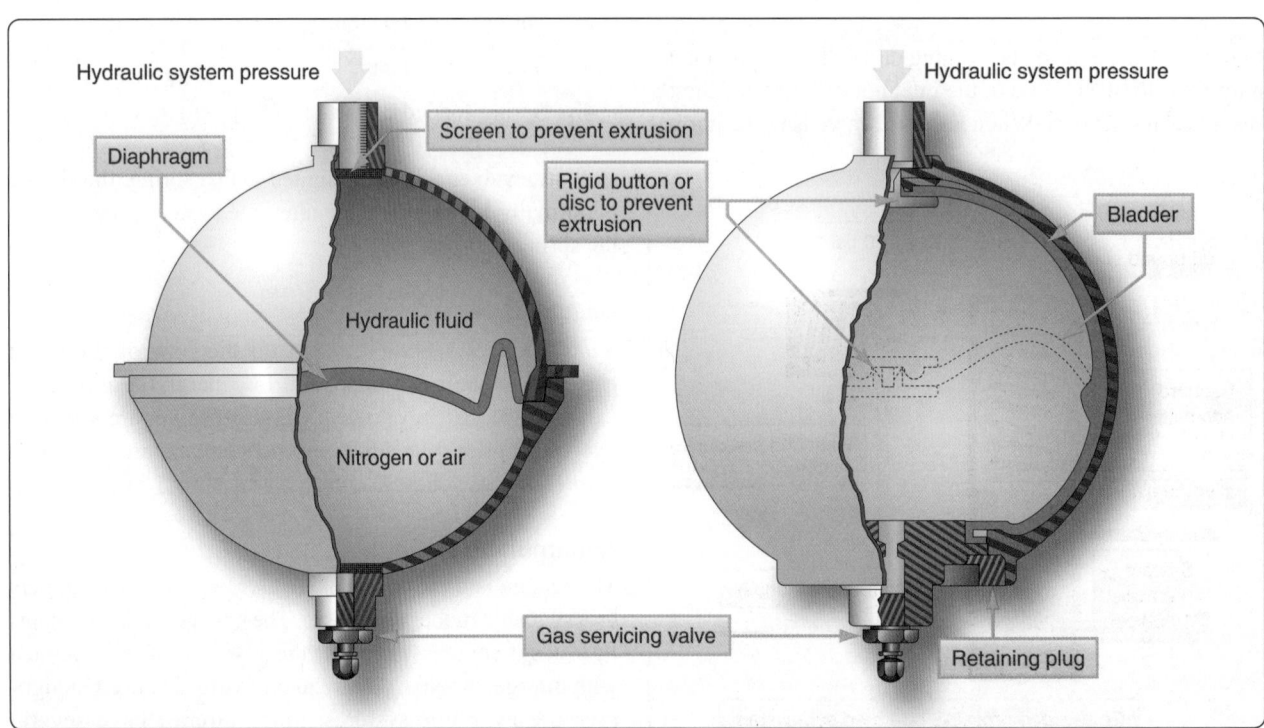

Figure 12-49. *A spherical accumulator with diaphragm (left) and bladder (right). The dotted lines in the right drawing depict the bladder when the accumulator is charged with both hydraulic system fluid and nitrogen preload.*

of the sphere and is secured with a threaded retainer plug. The gas servicing valve mounts into the retainer plug.

Cylindrical

Cylindrical accumulators consist of a cylinder and piston assembly. End caps are attached to both ends of the cylinder. The internal piston separates the fluid and air/nitrogen chambers. The end caps and piston are sealed with gaskets and packings to prevent external leakage around the end caps and internal leakage between the chambers. In one end cap, a hydraulic fitting is used to attach the fluid chamber to the hydraulic system. In the other end cap, a filler valve is installed to perform the same function as the filler valve installed in the spherical accumulator. *[Figure 12-50]*

Operation

In operation, the compressed-air chamber is charged to a predetermined pressure that is somewhat lower than the system operating pressure. This initial charge is referred to as the accumulator preload. As an example of accumulator operation, let us assume that the cylindrical accumulator is designed for a preload of 1,300 psi in a 3,000-psi system. When the initial charge of 1,300 psi is introduced into the unit, hydraulic system pressure is zero. As air pressure is applied through a gas servicing valve, it moves the piston toward the opposite end until it bottoms. If the air behind the piston has a pressure of 1,300 psi, the hydraulic system pump has to create a pressure within the system greater than 1,300 psi before the hydraulic fluid can actuate the piston. At 1,301 psi the piston starts to move within the cylinder, compressing the air as it moves. At 2,000 psi, it has backed up several inches. At 3,000 psi, the piston has backed up to its normal operating position, compressing the air until it occupies a space less than one-half the length of the cylinder. When actuation of hydraulic units lowers the system pressure, the compressed air expands against the piston, forcing fluid

from the accumulator. This supplies an instantaneous supply of fluid to the hydraulic system component. The charged accumulator may also supply fluid pressure to actuate a component(s) briefly in case of pump failure.

Maintenance of Accumulators

Maintenance consists of inspections, minor repairs, replacement of component parts, and testing. There is an element of danger in maintaining accumulators. Therefore, proper precautions must be strictly observed to prevent injury and damage.

Before disassembling any accumulator, ensure that all preload air (or nitrogen) pressure has been discharged. Failure to release the preload could result in serious injury to the technician. Before making this check, be certain you know the type of high-pressure air valve used. When you know that all air pressure has been removed, you can take the unit apart. Be sure to follow manufacturer's instructions for the specific unit you have.

Heat Exchangers

Transport-type aircraft use heat exchangers in their hydraulic power supply system to cool the hydraulic fluid from the hydraulic pumps. This extends the service life of the fluid and the hydraulic pumps. They are located in the fuel tanks of the aircraft. The heat exchangers use aluminum finned

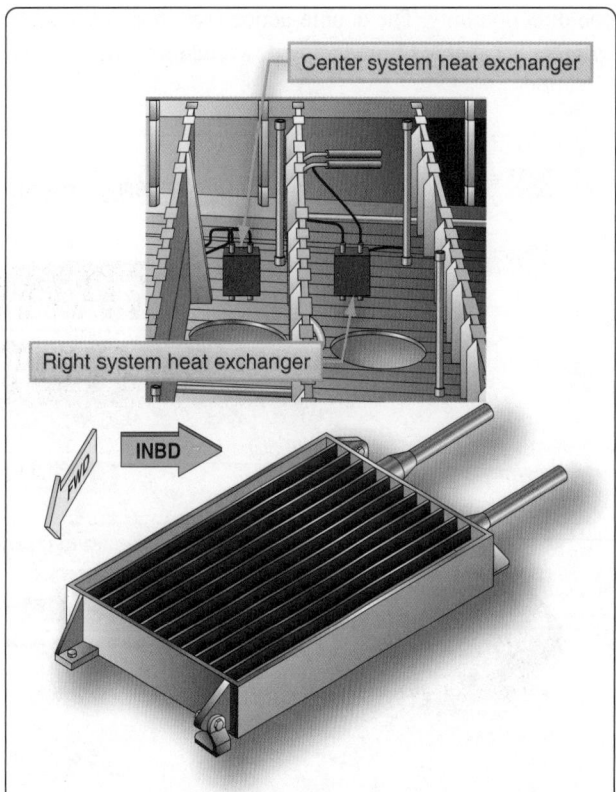

Figure 12-51. *Hydraulic heat exchanger.*

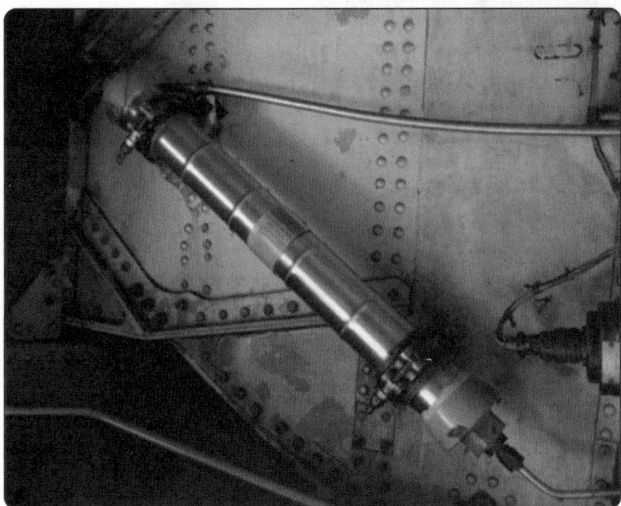

Figure 12-50. *Cylindrical accumulator.*

tubes to transfer heat from the fluid to the fuel. The fuel in the tanks that contain the heat exchangers must be maintained at a specific level to ensure adequate cooling of the fluid. *[Figure 12-51]*

Actuators

An actuating cylinder transforms energy in the form of fluid pressure into mechanical force, or action, to perform work. It is used to impart powered linear motion to some movable object or mechanism. A typical actuating cylinder consists of a cylinder housing, one or more pistons and piston rods, and some seals. The cylinder housing contains a polished bore in which the piston operates, and one or more ports through which fluid enters and leaves the bore. The piston and rod form an assembly. The piston moves forward and backward within the cylinder bore, and an attached piston rod moves into and out of the cylinder housing through an opening in one end of the cylinder housing.

Seals are used to prevent leakage between the piston and the cylinder bore and between the piston rod and the end of the cylinder. Both the cylinder housing and the piston rod have provisions for mounting and for attachment to an object or mechanism that is to be moved by the actuating cylinder.

Actuating cylinders are of two major types: single action and double action. The single-action (single port) actuating cylinder is capable of producing powered movement in one direction only. The double-action (two ports) actuating cylinder is capable of producing powered movement in two directions.

Linear Actuators

A single-action actuating cylinder is illustrated in *Figure 12-52A*. Fluid under pressure enters the port at the left and pushes against the face of the piston, forcing the piston to the right. As the piston moves, air is forced out of the spring chamber through the vent hole, compressing the spring. When pressure on the fluid is released to the point it exerts less force than is present in the compressed spring, the spring pushes the piston toward the left. As the piston moves to the left, fluid is forced out of the fluid port. At the same time, the moving piston pulls air into the spring chamber through the vent hole. A three-way control valve is normally used for controlling the operation of a single-action actuating cylinder.

A double-action (two ports) actuating cylinder is illustrated in *Figure 12-52B*. The operation of a double-action actuating cylinder is usually controlled by a four-way selector valve. *Figure 12-53* shows an actuating cylinder interconnected with a selector valve. Operation of the selector valve and actuating cylinder is discussed below.

When the selector valve is placed in the ON or EXTEND position, fluid is admitted under pressure to the left-hand chamber of the actuating cylinder. *[Figure 12-53]* This results in the piston being forced toward the right. As the piston moves toward the right, it pushes return fluid out of the right-hand chamber and through the selector valve to the reservoir. When the selector valve is placed in its RETRACT position, as illustrated in *Figure 12-50,* fluid pressure enters

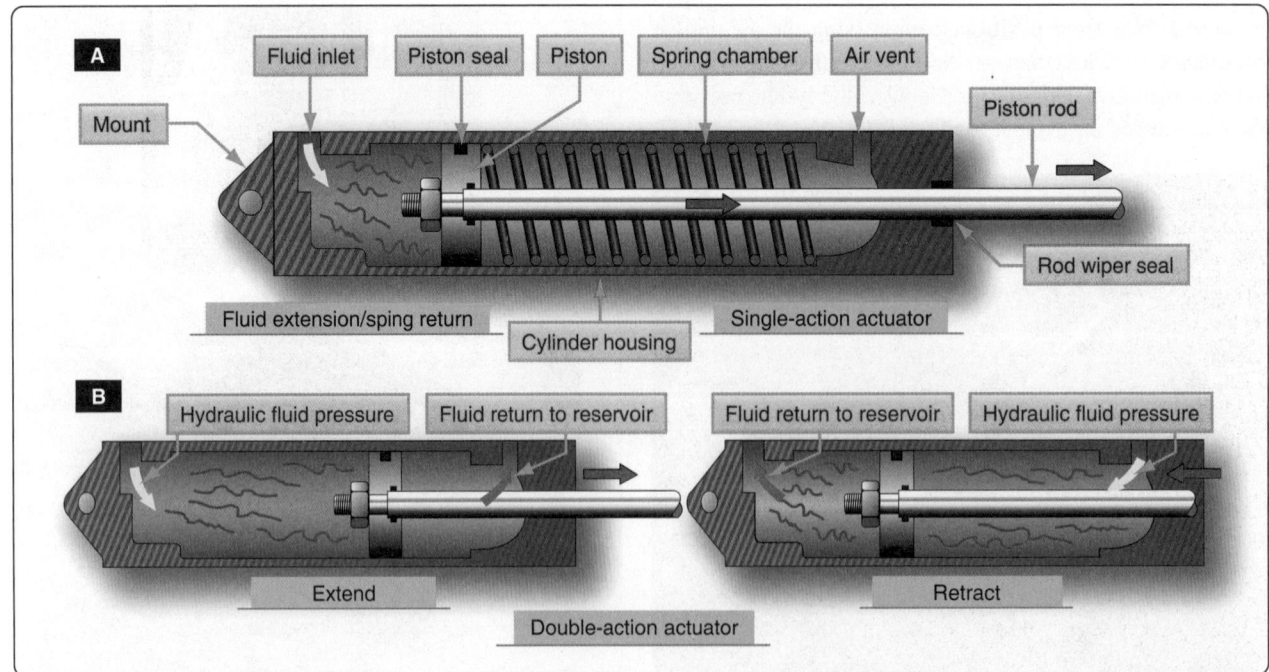

Figure 12-52. *Linear actuator.*

the right chamber, forcing the piston toward the left. As the piston moves toward the left, it pushes return fluid out of the left chamber and through the selector valve to the reservoir.

Besides having the ability to move a load into position, a double-acting cylinder also has the ability to hold a load in position. This capability exists because when the selector valve used to control operation of the actuating cylinder is placed in the off position, fluid is trapped in the chambers on both sides of the actuating cylinder piston. Internal locking actuators also are used in some applications.

Rotary Actuators

Rotary actuators can mount right at the part without taking up the long stroke lengths required of cylinders. Rotary actuators are not limited to the 90° pivot arc typical of cylinders; they can achieve arc lengths of 180°, 360°, or even 720° or more, depending on the configuration. An often-used type of rotary actuator is the rack and pinion actuator used for many nose wheel steering mechanisms. In a rack-and-pinion actuator, a long piston with one side machined into a rack engages a pinion to turn the output shaft. *[Figure 12-54]* One side of the piston receive fluid pressure while the other side is connected to the return. When the piston moves, it rotates the pinion.

Hydraulic Motor

Piston-type motors are the most commonly used in hydraulic systems. *[Figure 12-55]* They are basically the same as hydraulic pumps except they are used to convert hydraulic energy into mechanical (rotary) energy. Hydraulic motors

are either of the axial inline or bent-axis type. The most commonly used hydraulic motor is the fixed-displacement bent-axis type. These types of motors are used for the activation of trailing edge flaps, leading edge slats, and stabilizer trim. Some equipment uses a variable-displacement piston motor where very wide speed ranges are desired. Although some piston-type motors are controlled by directional control valves, they are often used in combination with variable-displacement pumps. This pump-motor combination is used to provide a transfer of power between a driving element and a driven element. Some applications for which hydraulic transmissions may be used are speed reducers, variable speed drives, constant speed or constant torque drives, and torque converters.

Some advantages of hydraulic transmission of power over mechanical transmission of power are as follows:

- Quick, easy speed adjustment over a wide range while the power source is operating at a constant (most efficient) speed.

- Rapid, smooth acceleration or deceleration.

- Control over maximum torque and power.

- Cushioning effect to reduce shock loads.

- Smoother reversal of motion.

Ram Air Turbine (RAT)

The RAT is installed in the aircraft to provide electrical and hydraulic power if the primary sources of aircraft power are

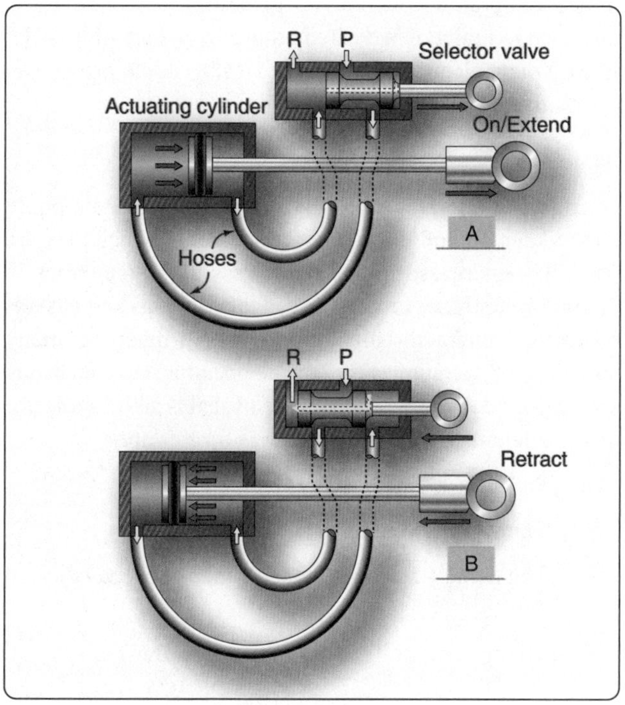

Figure 12-53. *Linear actuator operation.*

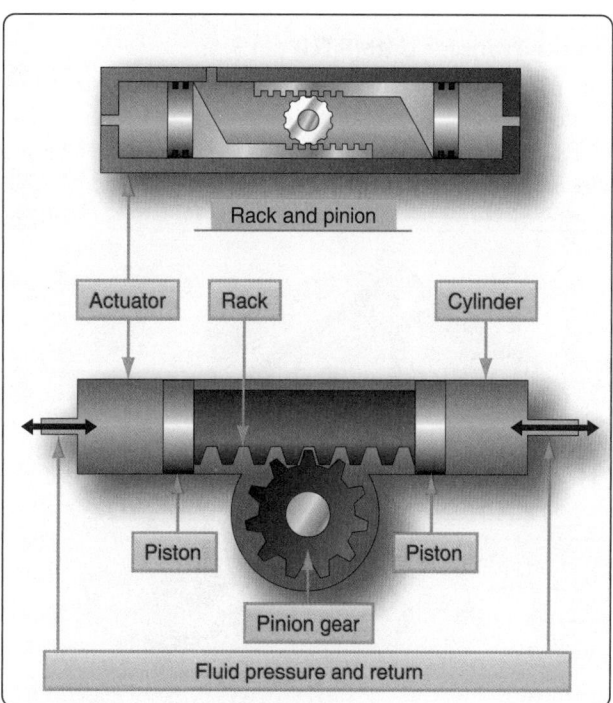

Figure 12-54. *Rack and pinion gear.*

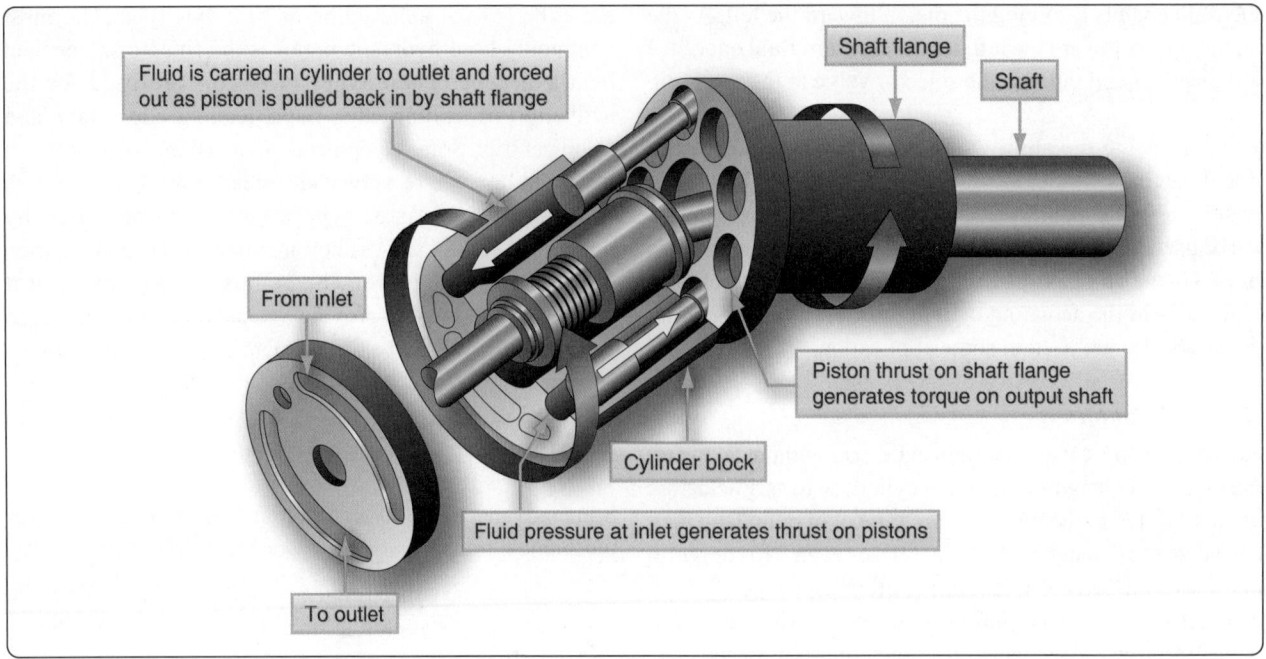

Figure 12-55. *Bent axis piston motor.*

lost. Ram air is used to turn the blades of a turbine that, in turn, operates a hydraulic pump and generator. The turbine and pump assembly is generally installed on the inner surface of a door installed in the fuselage. The door is hinged, allowing the assembly to be extended into the slipstream by pulling a manual release in the flight deck. In some aircraft, the RAT automatically deploys when the main hydraulic pressure system fails, and/or electrical system malfunction occurs. *[Figure 12-56]*

Power Transfer Unit (PTU)

The PTU is able to transfer power but not fluid. It transfers power between two hydraulic systems. Different types of PTUs are in use; some can only transfer power in one direction while others can transfer power both ways. Some

PTUs have a fixed displacement, while others use a variable displacement hydraulic pump. The two units, hydraulic pump and hydraulic motor, are connected via a single drive shaft so that power can be transferred between the two systems. Depending on the direction of power transfer, each unit in turn works either as a motor or a pump. *[Figure 12-57]*

Hydraulic Motor-Driven Generator (HMDG)

The HMDG is a servo-controlled variable displacement motor integrated with an AC generator. The HMDG is designed to maintain a desired output frequency of 400 Hz. In case of an electrical failure, the HMDG could provide an alternative source of electrical power.

Seals

Seals are used to prevent fluid from passing a certain point, and to keep air and dirt out of the system in which they are used. The increased use of hydraulics and pneumatics in aircraft systems has created a need for packings and gaskets of varying characteristics and design to meet the many variations of operating speeds and temperatures to which they are subjected. No one style or type of seal is satisfactory for all installations. Some of the reasons for this are:

- Pressure at which the system operates.

- The type fluid used in the system.

- The metal finish and the clearance between adjacent parts.

- The type motion (rotary or reciprocating), if any.

Seals are divided into three main classes: packings, gaskets,

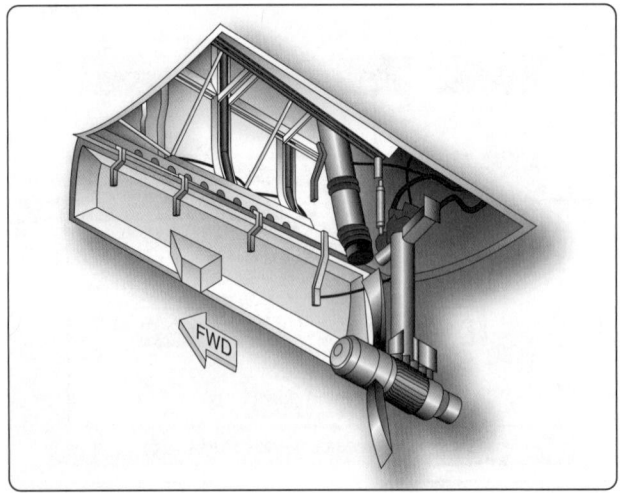

Figure 12-56. *Ram air turbine.*

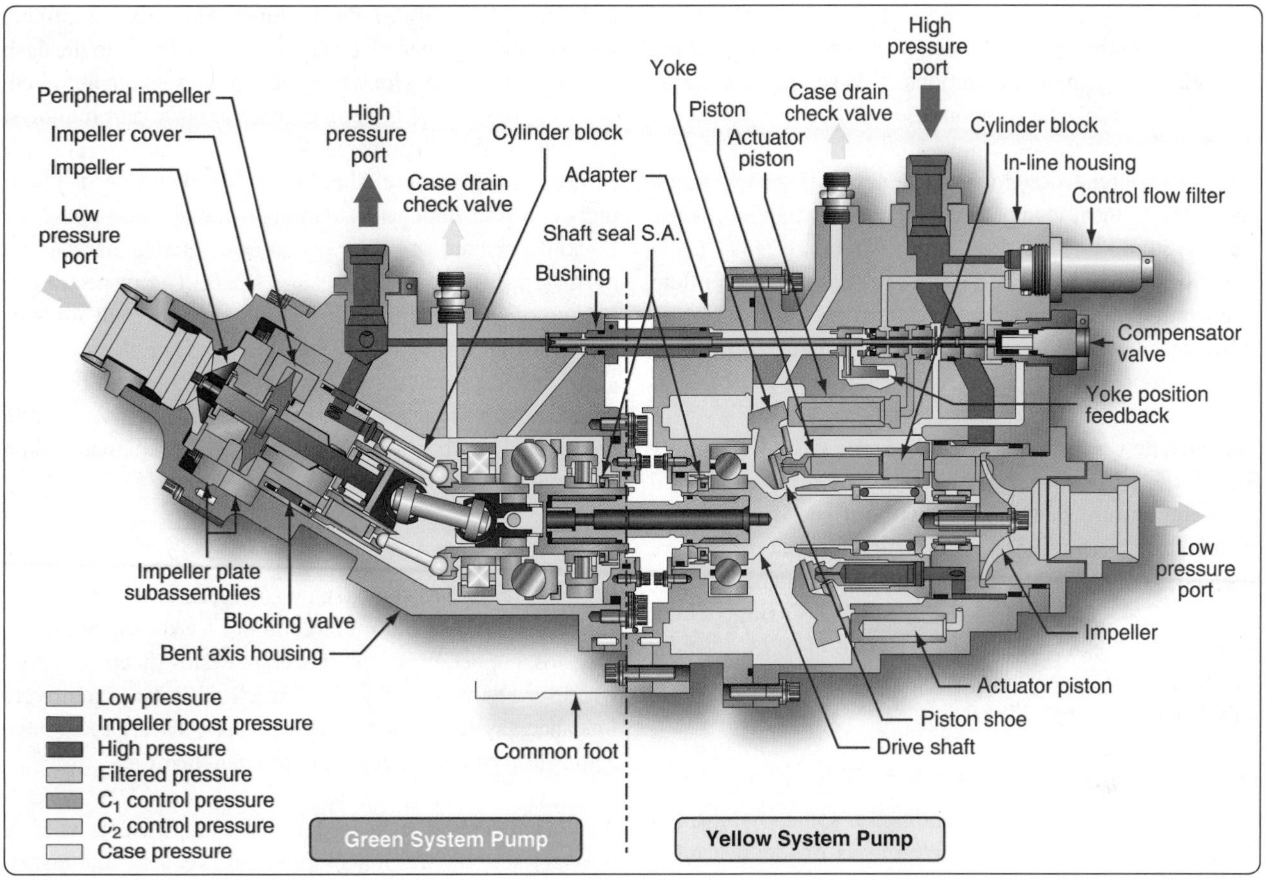

Figure 12-57. *Power transfer unit.*

and wipers. A seal may consist of more than one component, such as an O-ring and a backup ring, or possibly an O-ring and two backup rings. Hydraulic seals used internally on a sliding or moving assembly are normally called packings. *[Figure 12-58]* Hydraulic seals used between nonmoving fittings and bosses are normally called gaskets.

V-Ring Packings

V-ring packings (AN6225) are one-way seals and are always installed with the open end of the V facing the pressure. V-ring packings must have a male and female adapter to hold them in the proper position after installation. It is also necessary to torque the seal retainer to the value specified by the manufacturer of the component being serviced, or the seal may not give satisfactory service.

U-Ring

U-ring packings (AN6226) and U-cup packings are used in brake assemblies and brake master cylinders. The U-ring and U-cup seals pressure in only one direction; therefore, the lip of the packings must face toward the pressure. U-ring packings are primarily low-pressure packings to be used with pressures of less than 1,000 psi.

O-Rings

Most packings and gaskets used in aircraft are manufactured in the form of O-rings. An O-ring is circular in shape, and its cross-section is small in relation to its diameter.

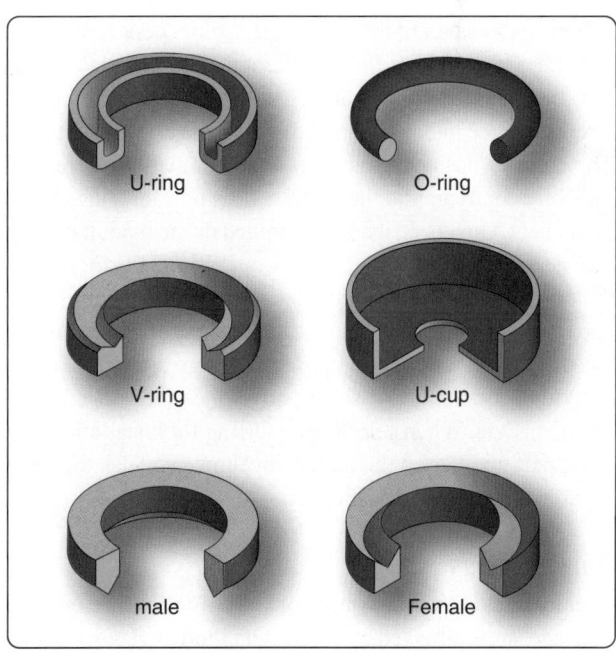

Figure 12-58. *Packings.*

The cross-section is truly round and has been molded and trimmed to extremely close tolerances. The O-ring packing seals effectively in both directions. This sealing is done by distortion of its elastic compound.

Advances in aircraft design have made new O-ring composition necessary to meet changing conditions. Hydraulic O-rings were originally established under Air Force-Navy (AN) specification numbers 6227, 6230, and 6290 for use in fluid at operating temperatures ranging from −65 °F to +160 °F. When new designs raised operating temperatures to a possible +275 °F, more compounds were developed and perfected.

Recently, newer compounds were developed under Military Standard (MS) specifications that offered improved low-temperature performance without sacrificing high-temperature performance. These superior materials were adopted in the MS28775 O-ring, which is replacing AN6227 and AN6230 O-rings, and the MS28778 O-ring, which is replacing the AN6290 O-ring. These O-rings are now standard for systems where the operating temperatures may vary from −65 °F to +275 °F.

O-Ring Color Coding

Manufacturers provide color coding on some O-rings, but this is not a reliable or complete means of identification. The color coding system does not identify sizes, but only system fluid or vapor compatibility and, in some cases, the manufacturer. Color codes on O-rings that are compatible with MIL-H-5606 fluid always contains blue but may also contain red or other colors. Packings and gaskets suitable for use with phosphate ester-based fluids are always coded with a green stripe, but may also have a blue, grey, red, green, or yellow dot as a part of the color code. Color codes on O-rings that are compatible with hydrocarbon fluid always contain red, but never contain blue. A colored stripe around the circumference indicates that the O-ring is a boss gasket seal. The color of the stripe indicates fluid compatibility: red for fuel, blue for hydraulic fluid. The coding on some rings is not permanent. On others, it may be omitted due to manufacturing difficulties or interference with operation. Furthermore, the color coding system provides no means to establish the age of the O-ring or its temperature limitations. Because of the difficulties with color coding, O-rings are available in individual hermetically sealed envelopes labeled with all pertinent data. When selecting an O-ring for installation, the basic part number on the sealed envelope provides the most reliable compound identification.

Backup Rings

Backup rings (MS28782) made of Teflon™ do not deteriorate with age, are unaffected by any system fluid or vapor, and can tolerate temperature extremes in excess of those encountered in high pressure hydraulic systems. Their dash numbers indicate not only their size but also relate directly to the dash number of the O-ring for which they are dimensionally suited. They are procurable under a number of basic part numbers, but they are interchangeable; any Teflon™ backup ring may be used to replace any other Teflon™ backup ring if it is of proper overall dimension to support the applicable O-ring. Backup rings are not color coded or otherwise marked and must be identified from package labels. The inspection of backup rings should include a check to ensure that surfaces are free from irregularities, that the edges are clean cut and sharp, and that scarf cuts are parallel. When checking Teflon™ spiral backup rings, make sure that the coils do not separate more than ¼ inch when unrestrained. Be certain that backup rings are installed downstream of the O-ring. *[Figure 12-59]*

Gaskets

Gaskets are used as static (stationary) seals between two flat surfaces. Some of the more common gasket materials are asbestos, copper, cork, and rubber. Asbestos sheeting is used wherever a heat resistant gasket is needed. It is used extensively for exhaust system gaskets. Most asbestos exhaust gaskets have a thin sheet of copper edging to prolong their life.

A solid copper washer is used for spark plug gaskets where it is essential to have a noncompressible, yet semisoft gasket. Cork gaskets can be used as an oil seal between the engine crankcase and accessories, and where a gasket is required that is capable of occupying an uneven or varying space caused by a rough surface or expansion and contraction. Rubber sheeting can be used where there is a need for a compressible gasket. It should not be used in any place where it may come in contact with gasoline or oil because the rubber deteriorates very rapidly when exposed to these substances. Gaskets are used in fluid systems around the end caps of actuating cylinders, valves, and other units. The gasket generally used for this purpose is in the shape of an O-ring, similar to O-ring packings.

Seal Materials

Most seals are made from synthetic materials that are compatible with the hydraulic fluid used. Seals used for MIL-H-5606 hydraulic fluid are not compatible with phosphate ester-based fluids and servicing the hydraulic system with the wrong fluid could result in leaks and system malfunctions. Seals for systems that use MIl-H-5606 are made of neoprene or Buna-N. Seals for phosphate ester-based fluids are made from butyl rubber or ethylene-propylene elastomers.

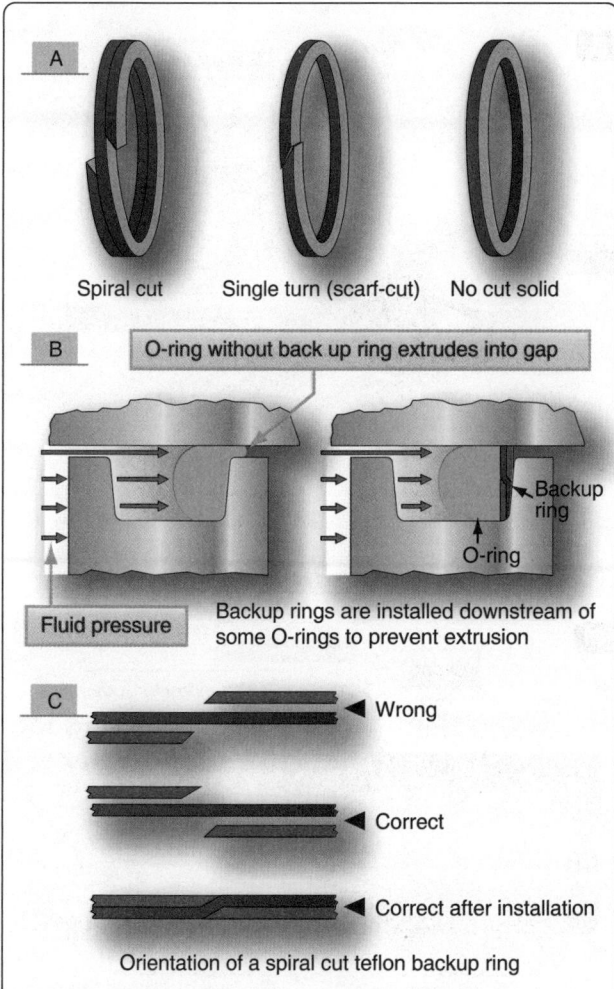

A
Spiral cut   Single turn (scarf-cut)   No cut solid

B
O-ring without back up ring extrudes into gap

Backup ring

O-ring

Fluid pressure

Backup rings are installed downstream of some O-rings to prevent extrusion

C
Wrong

Correct

Correct after installation

Orientation of a spiral cut teflon backup ring

Figure 12-59. *Backup O-rings installed downstream.*

O-Ring Installation

When removing or installing O-rings, avoid using pointed or sharp-edged tools that might cause scratching or marring of hydraulic component surfaces or cause damage to the O-rings. Special tooling for the installation of O-rings is available. While using the seal removal and the installation tools, contact with cylinder walls, piston heads, and related precision components is not desirable.

After the removal of all O-rings, the parts that receive new O-rings have to be cleaned and inspected to make sure that they are free from all contamination. Each replacement O-ring should be removed from its sealed package and inspected for defects, such as blemishes, abrasions, cuts, or punctures. Although an O-ring may appear perfect at first glance, slight surface flaws may exist. These flaws are often capable of preventing satisfactory O-ring performance under the variable operating pressures of aircraft systems; therefore, O-rings should be rejected for flaws that affect their performance. Such flaws are difficult to detect, and

one aircraft manufacturer recommends using a four-power magnifying glass with adequate lighting to inspect each ring before it is installed. By rolling the ring on an inspection cone or dowel, the inner diameter surface can also be checked for small cracks, particles of foreign material, or other irregularities that cause leakage or shorten the life of the O-ring. The slight stretching of the ring when it is rolled inside out helps to reveal some defects not otherwise visible.

After inspection and prior to installation, immerse the O-ring in clean hydraulic fluid. During the installation, avoid rolling and twisting the O-ring to maneuver it into place. If possible, keep the position of the O-ring's mold line constant. When the O-ring installation requires spanning or inserting through sharply threaded areas, ridges, slots, and edges, use protective measures, such as O-ring entering sleeves, as shown in *Figure 12-60A*. After the O-ring is placed in the cavity provided, gently roll the O-ring with the fingers to remove any twist that might have occurred during installation. *[Figure 12-61]*

Wipers

Wipers are used to clean and lubricate the exposed portions of piston shafts. They prevent dirt from entering the system and help protect the piston shaft against scoring. Wipers may be either metallic or felt. They are sometimes used together, a felt wiper installed behind a metallic wiper.

Large Aircraft Hydraulic Systems

Figure 12-62 provides an overview of hydraulic components in large aircraft.

Boeing 737 Next Generation Hydraulic System

The Boeing 737 Next Generation has three 3,000 psi hydraulic systems: system A, system B, and standby. The standby system is used if system A and/or B pressure is lost. The hydraulic systems power the following aircraft systems:

- Flight controls.
- Leading edge flaps and slats.
- Trailing edge flaps.
- Landing gear.
- Wheel brakes.
- Nose wheel steering.
- Thrust reversers.
- Autopilots.

Reservoirs

The system A, B, and standby reservoirs are located in the wheel well area. The reservoirs are pressurized by bleed air

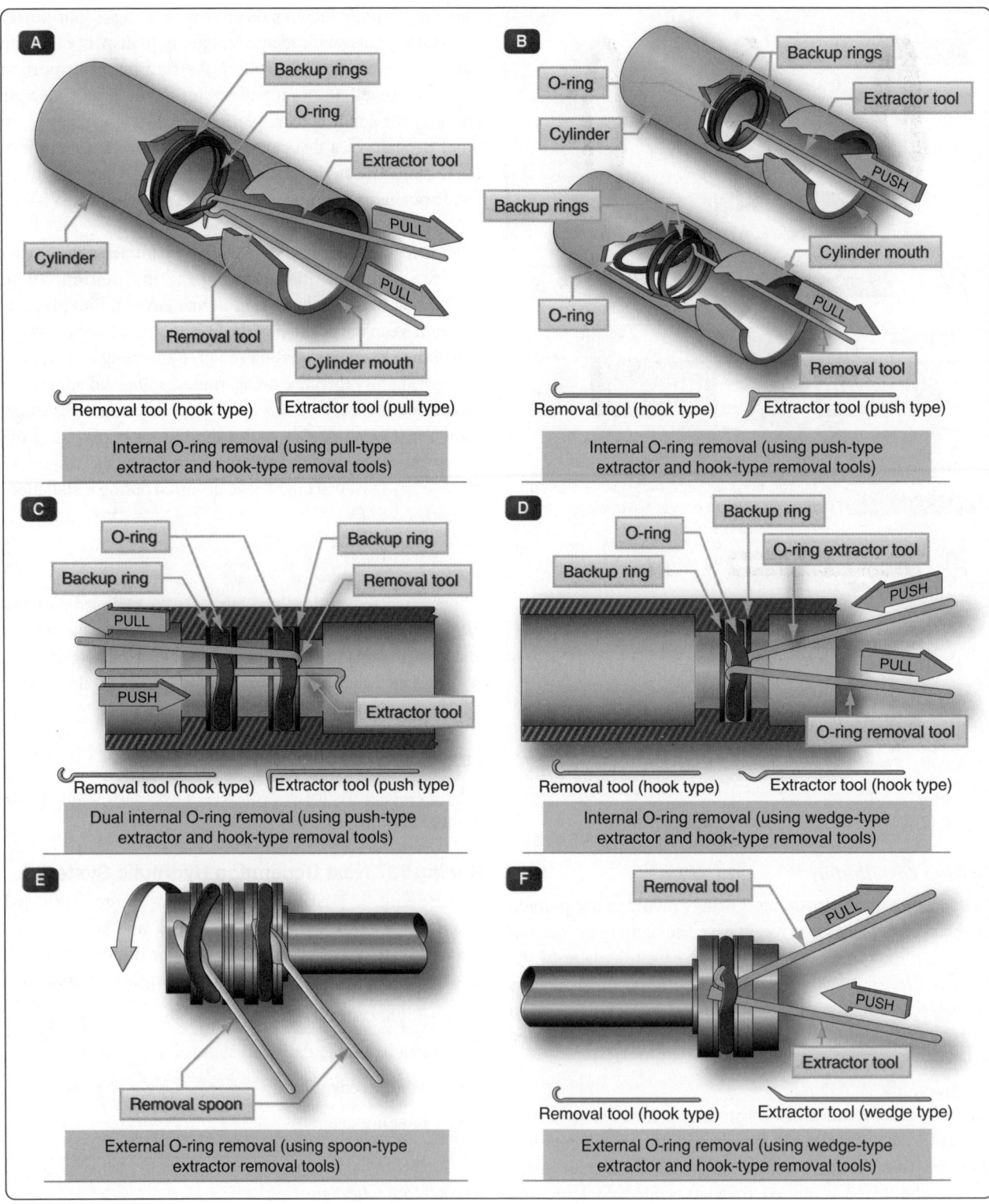

Figure 12-60. *O-ring installation techniques.*

through a pressurization module. The standby reservoir is connected to the system B reservoir for pressurization and servicing. The positive pressure in the reservoir ensures a positive flow of fluid to the pumps. The reservoirs have a standpipe that prevents the loss of all hydraulic fluid if a leak develops in the engine-driven pump or its related lines.

The engine-driven pump draws fluid through a standpipe in the reservoir and the AC motor pump draws fluid from the bottom of the reservoir. *[Figure 12-63]*

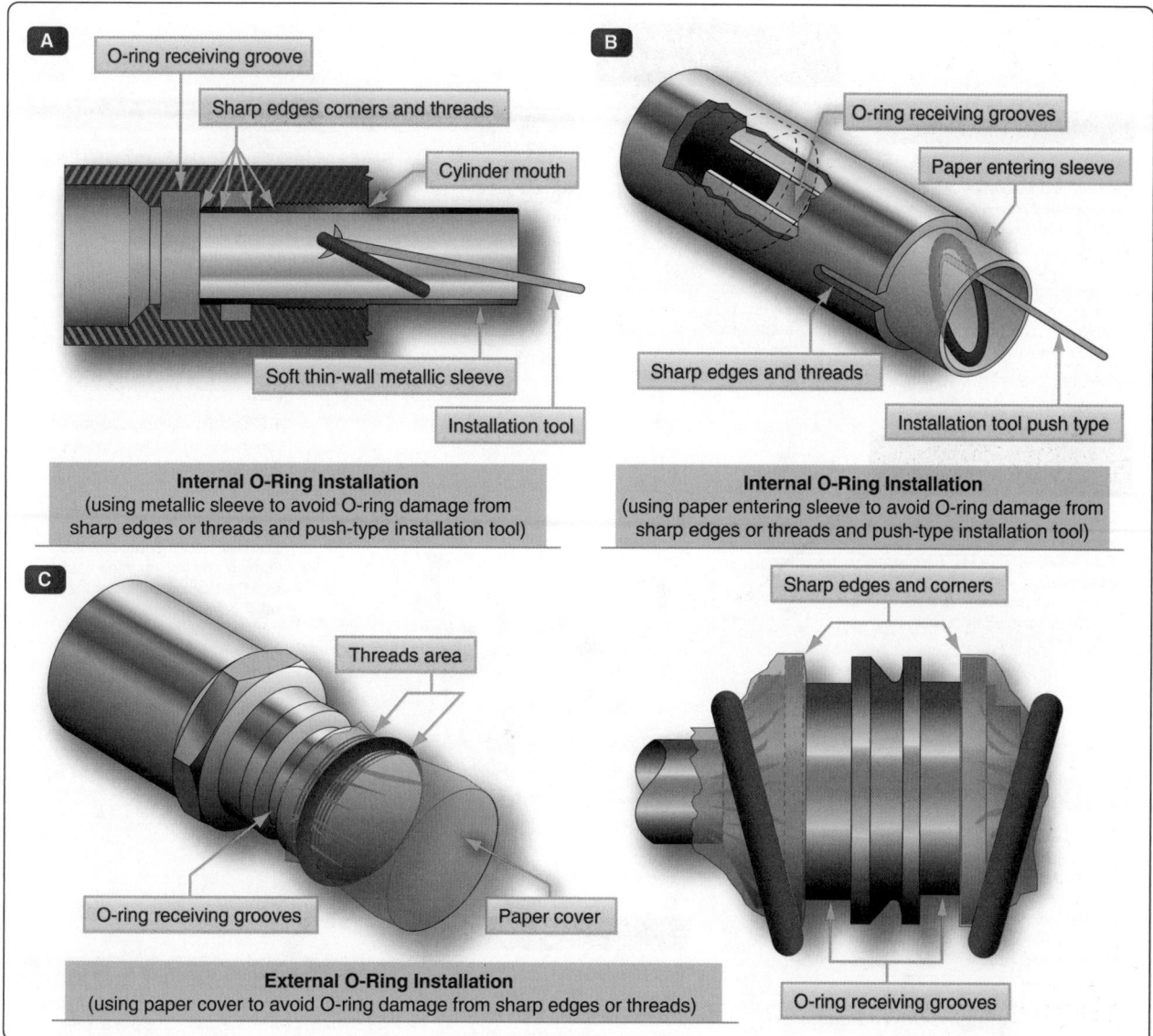

Figure 12-61. *More O-ring installation techniques.*

Pumps

Refer to *Figure 12-64* for the following description. Both A and B hydraulic systems have an engine-driven pump (EDP) and an ACMP. The system A engine-driven pump is installed on the number 1 engine and the system B engine-driven pump is installed on the number 2 engine. The AC pumps are controlled by a switch on the flight deck. The hydraulic case drain fluid that lubricates and cools the pumps return to the reservoir through a heat exchanger. *[Figure 12-65]* The heat exchanger for the A system is installed in the main fuel tank No. 1, and the heat exchanger for the B system is installed in the main fuel tank No. 2. Minimum fuel for ground operation of electric motor-driven pumps is 1,675 pounds in the related main tank. Pressure switches, located in the EDP and ACMP pump output lines, send signals to illuminate the related LOW PRESSURE light if pump output

pressure is low. The related system pressure transmitter sends the combined pressure of the EDP and ACMP to the related hydraulic system pressure indicator.

Filter Units

Filter modules are installed in the pressure, case drain, and return lines to clean the hydraulic fluid. Filters have a differential pressure indicator that pops out when the filter is dirty and needs to be replaced.

Power Transfer Unit (PTU)

The purpose of the PTU is to supply the additional volume of hydraulic fluid needed to operate the autoslats and leading-edge flaps and slats at the normal rate when system B EDP malfunctions. The PTU unit consists of a hydraulic motor and hydraulic pump that are connected through a shaft. The

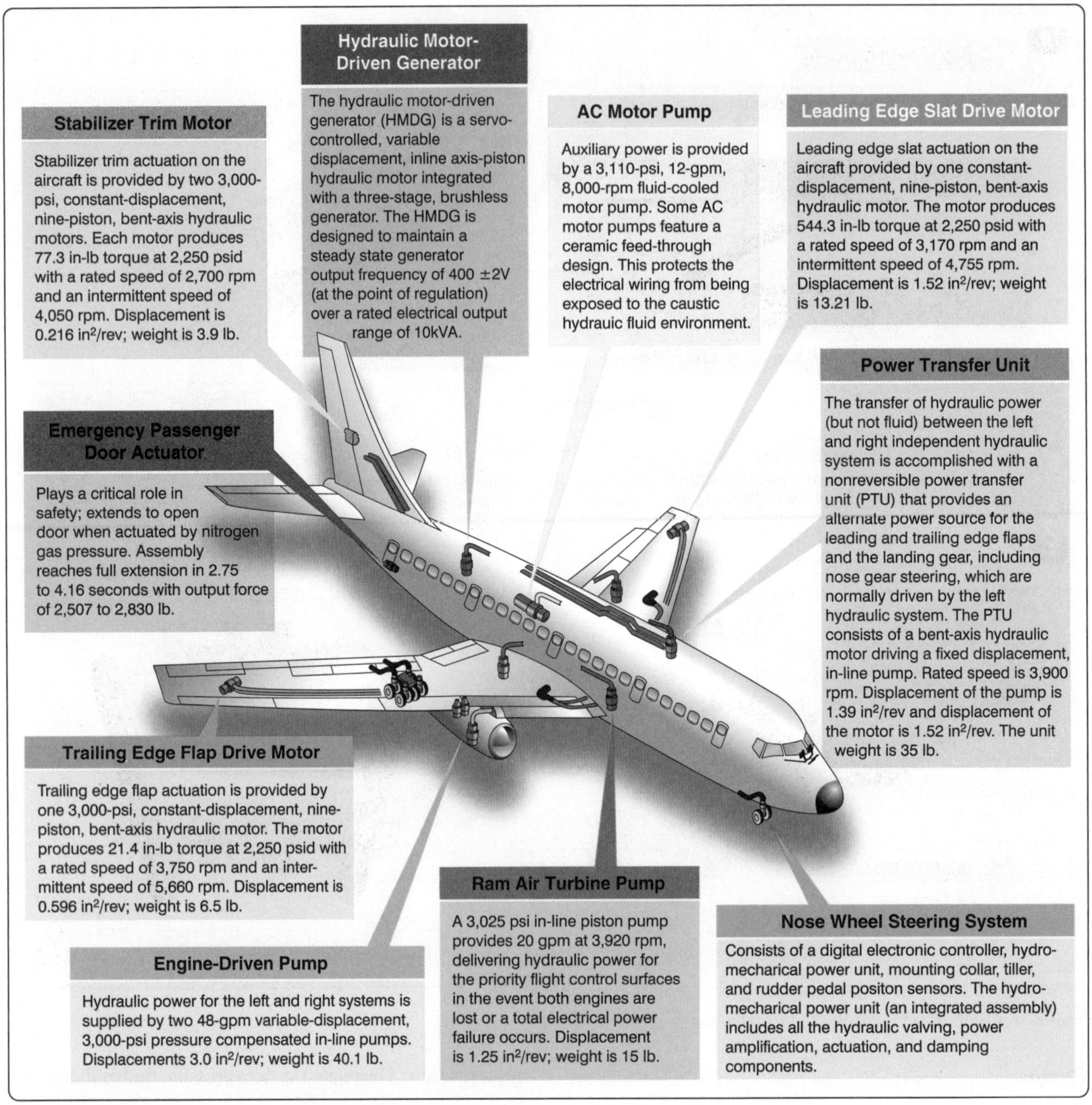

Stabilizer Trim Motor

Stabilizer trim actuation on the aircraft is provided by two 3,000-psi, constant-displacement, nine-piston, bent-axis hydraulic motors. Each motor produces 77.3 in-lb torque at 2,250 psid with a rated speed of 2,700 rpm and an intermittent speed of 4,050 rpm. Displacement is 0.216 in²/rev; weight is 3.9 lb.

Hydraulic Motor-Driven Generator

The hydraulic motor-driven generator (HMDG) is a servo-controlled, variable displacement, inline axis-piston hydraulic motor integrated with a three-stage, brushless generator. The HMDG is designed to maintain a steady state generator output frequency of 400 ±2V (at the point of regulation) over a rated electrical output range of 10kVA.

AC Motor Pump

Auxiliary power is provided by a 3,110-psi, 12-gpm, 8,000-rpm fluid-cooled motor pump. Some AC motor pumps feature a ceramic feed-through design. This protects the electrical wiring from being exposed to the caustic hydrauic fluid environment.

Leading Edge Slat Drive Motor

Leading edge slat actuation on the aircraft provided by one constant-displacement, nine-piston, bent-axis hydraulic motor. The motor produces 544.3 in-lb torque at 2,250 psid with a rated speed of 3,170 rpm and an intermittent speed of 4,755 rpm. Displacement is 1.52 in²/rev; weight is 13.21 lb.

Power Transfer Unit

The transfer of hydraulic power (but not fluid) between the left and right independent hydraulic system is accomplished with a nonreversible power transfer unit (PTU) that provides an alternate power source for the leading and trailing edge flaps and the landing gear, including nose gear steering, which are normally driven by the left hydraulic system. The PTU consists of a bent-axis hydraulic motor driving a fixed displacement, in-line pump. Rated speed is 3,900 rpm. Displacement of the pump is 1.39 in²/rev and displacement of the motor is 1.52 in²/rev. The unit weight is 35 lb.

Emergency Passenger Door Actuator

Plays a critical role in safety; extends to open door when actuated by nitrogen gas pressure. Assembly reaches full extension in 2.75 to 4.16 seconds with output force of 2,507 to 2,830 lb.

Trailing Edge Flap Drive Motor

Trailing edge flap actuation is provided by one 3,000-psi, constant-displacement, nine-piston, bent-axis hydraulic motor. The motor produces 21.4 in-lb torque at 2,250 psid with a rated speed of 3,750 rpm and an intermittent speed of 5,660 rpm. Displacement is 0.596 in²/rev; weight is 6.5 lb.

Ram Air Turbine Pump

A 3,025 psi in-line piston pump provides 20 gpm at 3,920 rpm, delivering hydraulic power for the priority flight control surfaces in the event both engines are lost or a total electrical power failure occurs. Displacement is 1.25 in²/rev; weight is 15 lb.

Nose Wheel Steering System

Consists of a digital electronic controller, hydro-mecharical power unit, mounting collar, tiller, and rudder pedal positon sensors. The hydro-mechanical power unit (an integrated assembly) includes all the hydraulic valving, power amplification, actuation, and damping components.

Engine-Driven Pump

Hydraulic power for the left and right systems is supplied by two 48-gpm variable-displacement, 3,000-psi pressure compensated in-line pumps. Displacements 3.0 in²/rev; weight is 40.1 lb.

Figure 12-62. *Large aircraft hydraulic systems.*

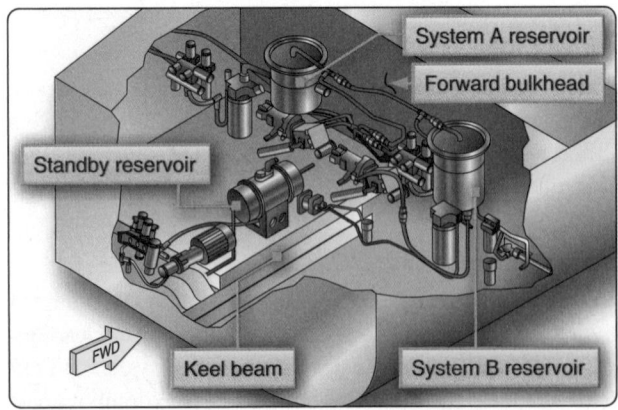

Figure 12-63. *Hydraulic reservoirs on a Boeing 737.*

PTU uses system A pressure to drive a hydraulic motor. The hydraulic motor of the PTU unit is connected through a shaft with a hydraulic pump that can draw fluid from the system B reservoir. The PTU can only transfer power and cannot transfer fluid. The PTU operates automatically when all of the following conditions are met:

- System B EDP pressure drops below limits.
- Aircraft airborne.
- Flaps are less than 15° but not up.

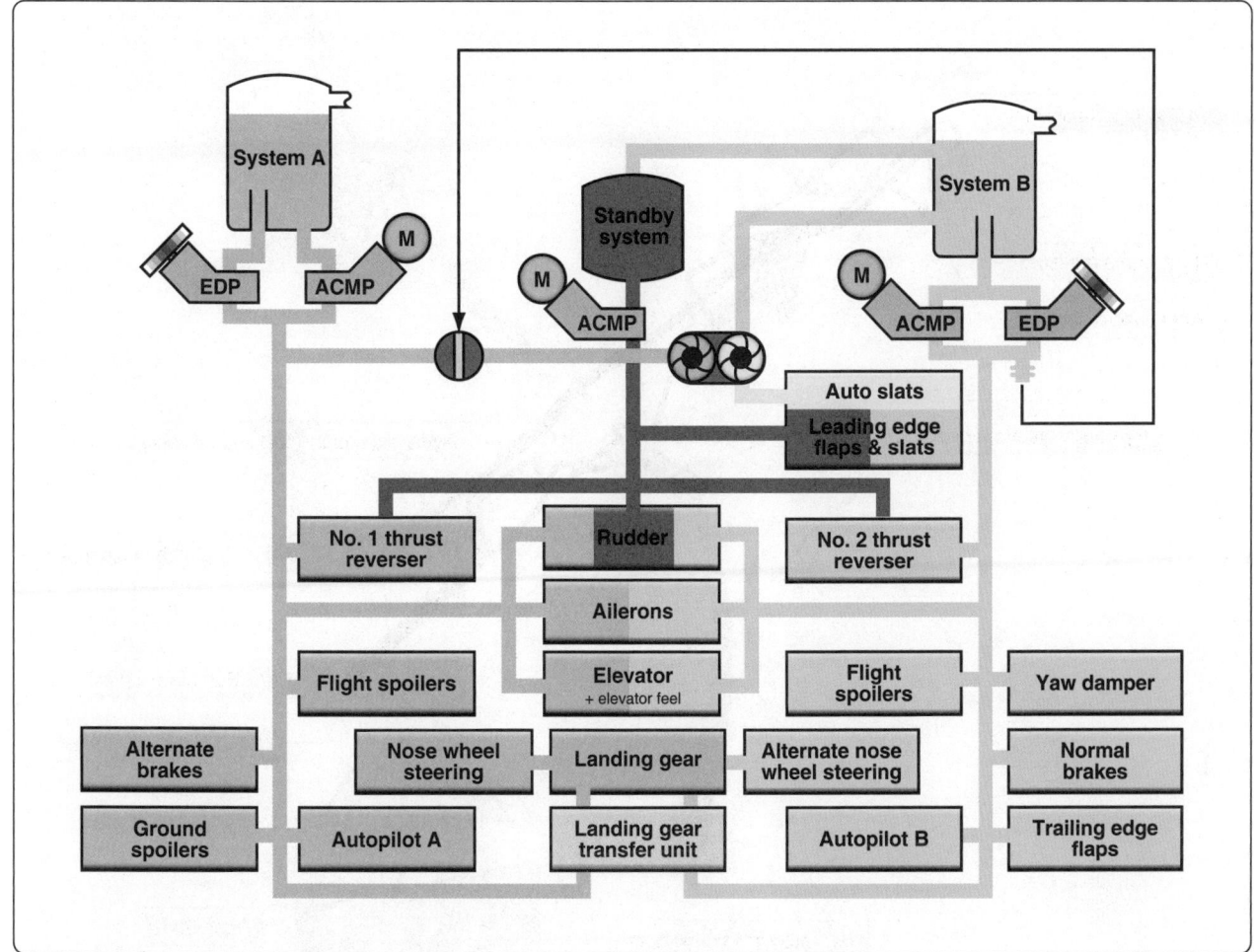

Figure 12-64. *Boeing 737 hydraulic system (simplified).*

Landing Gear Transfer Unit

The purpose of the landing gear transfer unit is to supply the volume of hydraulic fluid needed to raise the landing gear at the normal rate when system A EDP is lost. The system B EDP supplies the volume of hydraulic fluid needed to operate the landing gear transfer unit when all of the following conditions are met:

- Aircraft airborne.
- No. 1 engine rpm drops below a limit value.
- Landing gear lever is up.
- Either or both main landing gear not up and locked.

Standby Hydraulic System

The standby hydraulic system is provided as a backup if system A and/or B pressure is lost. The standby system can be activated manually or automatically and uses a single electric ACMP to power:

- Thrust reversers.
- Rudder.

- Leading edge flaps and slats (extend only).
- Standby yaw damper.

Indications

A master caution light illuminates if an overheat or low pressure is detected in the hydraulic system. An overheat light on the flight deck illuminates if an overheat is detected in either system A or B and a low-pressure light illuminates if a low pressure is detected in system A and B.

Boeing 777 Hydraulic System

The Boeing 777 is equipped with three hydraulic systems. The left, center, and right systems deliver hydraulic fluid at a rated pressure of 3,000 psi (207 bar) to operate flight controls, flap systems, actuators, landing gear, and brakes. Primary hydraulic power for the left and right systems is provided by two EDPs and supplemented by two on-demand ACMPs. Primary hydraulic power for the center system is provided by two electric motor pumps (ACMP) and supplemented by two on-demand air turbine-driven pumps. The center system provides hydraulic power for the engine thrust reversers,

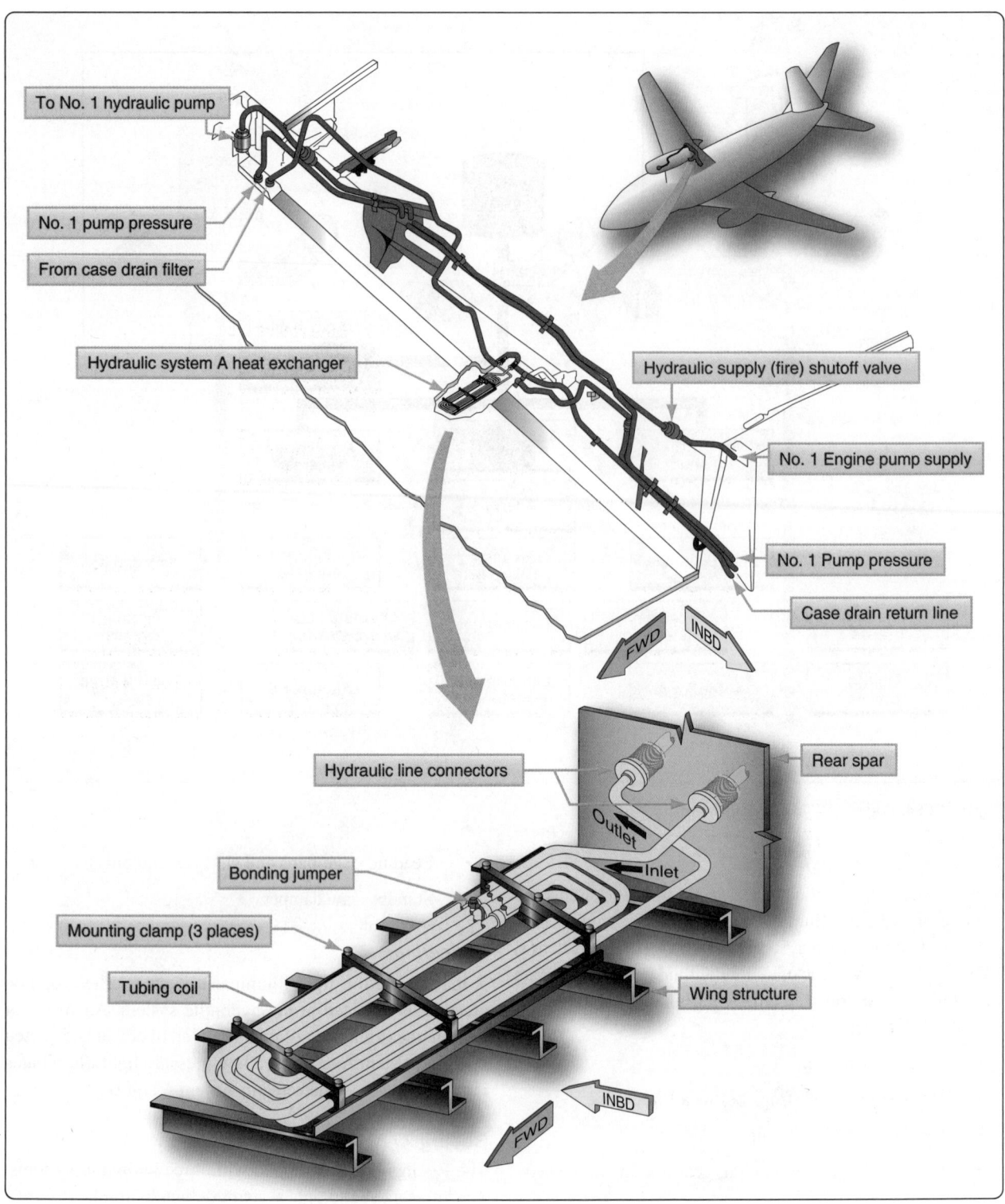

Figure 12-65. *Boeing 737 hydraulic case drain fluid heat exchanger installed in the fuel tank.*

primary flight controls, landing gear, and flaps/slats. Under emergency conditions, hydraulic power is generated by the ram air turbine (RAT), which is deployed automatically and drives a variable displacement inline pump. The RAT pump provides flow to the center system flight controls. *[Figure 12-66]*

Left & Right System Description

The left and right hydraulic systems are functionally the same. The left hydraulic system supplies pressurized hydraulic fluid to operate the left thrust reverser and the flight control systems. The right hydraulic system supplies pressurized

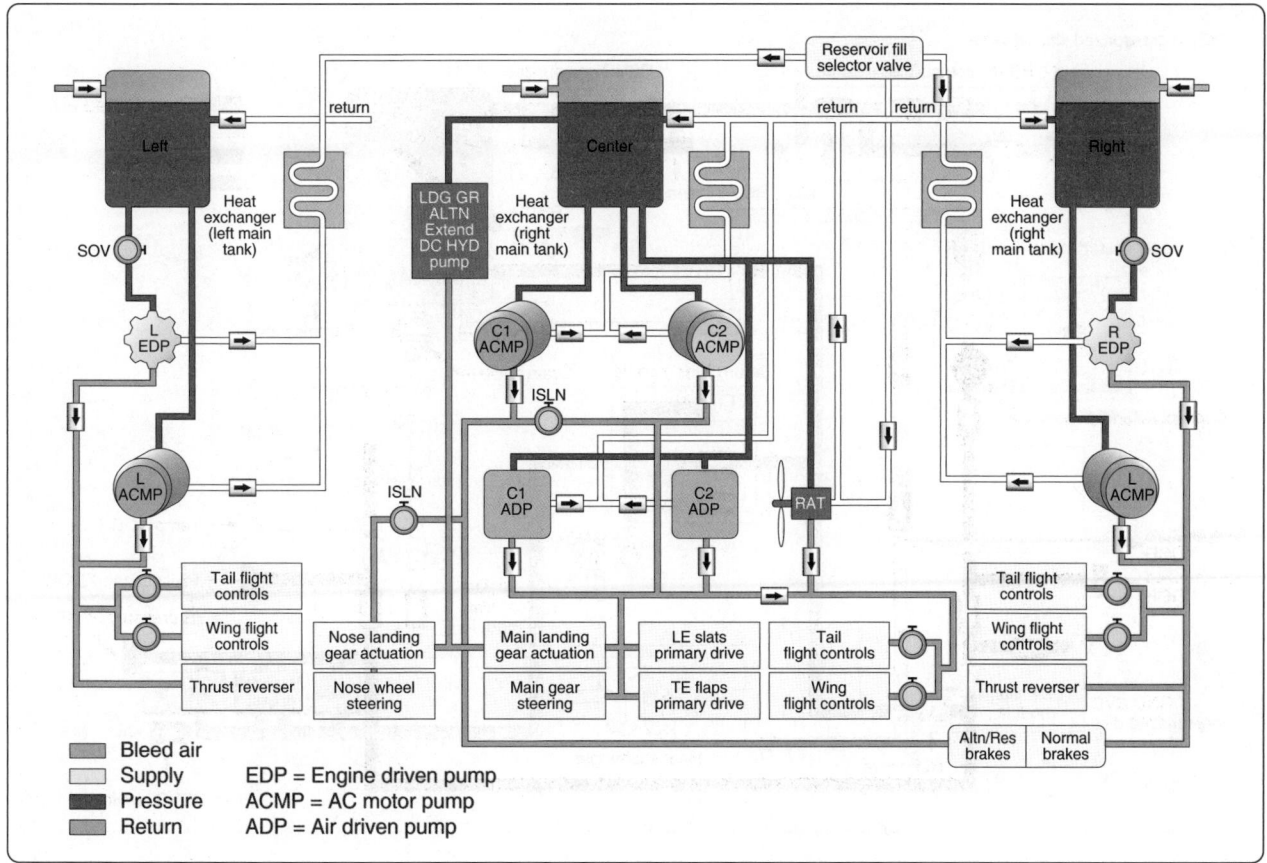

Figure 12-66. *A Boeing 777 hydraulic system.*

hydraulic fluid to operate the right thrust reverser, flight control systems, and the normal brake system. *[Figure 12-67]*

Reservoir

The hydraulic system reservoirs of the left and right system contain the hydraulic fluid supply for the hydraulic pumps. The reservoir is pressurized by bleed air through a reservoir pressurization module. The EDP draws fluid through a standpipe. The ACMP draws fluid from the bottom of the reservoir. If the fluid level in the reservoir gets below the standpipe, the EDP cannot draw any fluid any longer, and the ACMP is the only source of hydraulic power. The reservoir can be serviced through a center servicing point in the fuselage of the aircraft. The reservoir has a sample valve for contamination testing purposes, a temperature transmitter for temperature indication on the flight deck, a pressure transducer for reservoir pressure, and a drain valve for reservoir draining.

Pumps

The EDPs are the primary pumps for the left and right hydraulic systems. The EDPs get reservoir fluid through the EDP supply shutoff valves. The EDPs operates whenever the engines operate. A solenoid valve in each EDP controls the

pressurization and depressurization of the pump. The pumps are variable displacement inline piston pumps consisting of a first stage impeller pump and a second stage piston pump. The impeller pump delivers fluid under pressure to the piston pump. The ACMPs are the demand pumps for the left and right hydraulic systems. The ACMPs normally operate only when there is high hydraulic system demand.

Filter Module

Pressure and case drain filter modules clean the pressure flows and the case drain flows of the hydraulic pumps. A return filter module cleans the return flow of hydraulic fluid from the user systems. The module can be bypassed if the filter clogs, and a visible indicator pops to indicate a clogged filter. The heat exchanger, which is installed in the wing fuel tanks, cools the hydraulic fluid from ACMP and EDP case drain lines before the fluid goes back to the reservoir.

Indication

The hydraulic system sensors send pressure, temperature, and quantity signals to the flight deck. A reservoir quantity transmitter and temperature transducer are installed on each of the reservoirs, and a hydraulic reservoir pressure switch is located on the pneumatic line between the reservoir

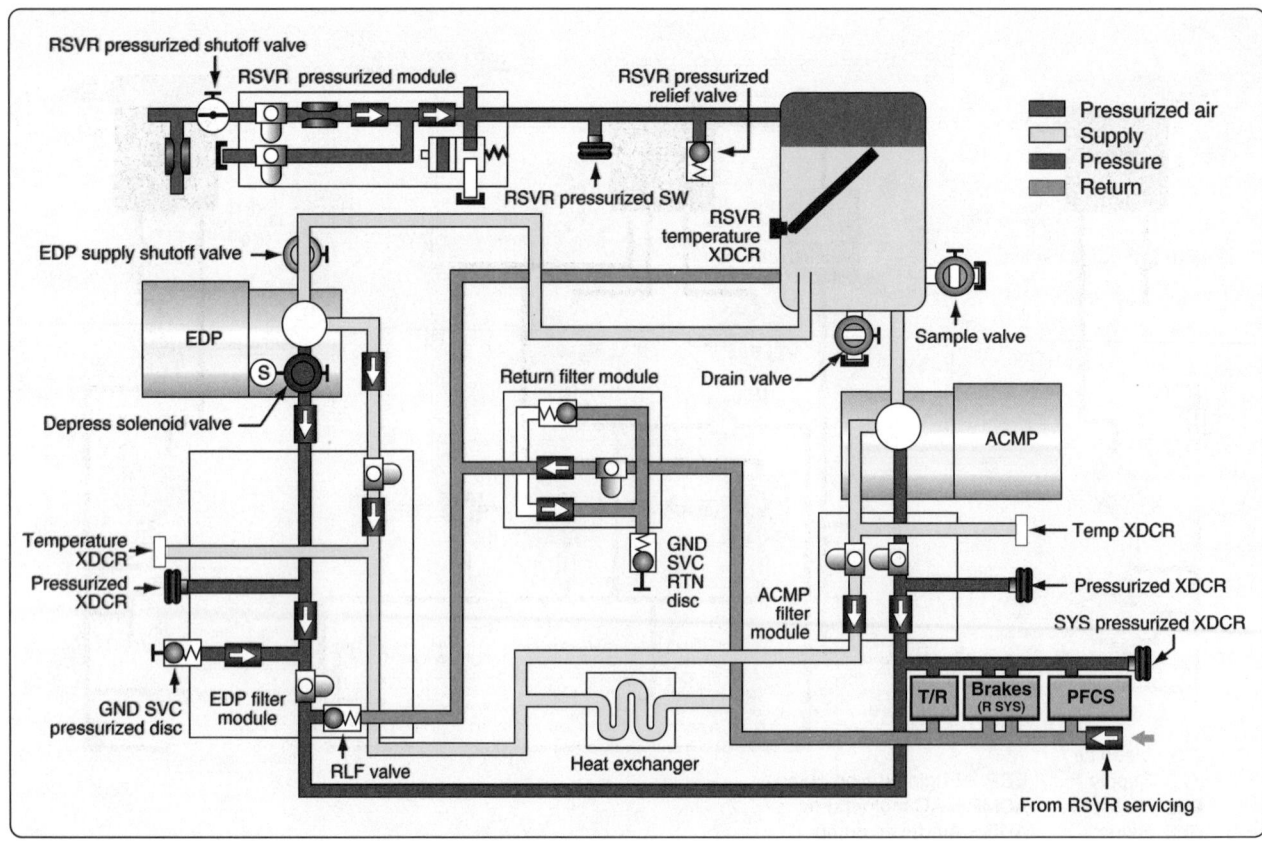

Figure 12-67. *Right hydraulic system of a Boeing 777. A left system is similar.*

pressurization module and the reservoir. The ACMP and EDP filter modules each have a pressure transducer to measure pump output pressure. A temperature transducer is installed in the case drain line of each filter module and measures pump case drain fluid temperature. A system pressure transducer measures hydraulic system pressure. A pressure relief valve on the EDP filter module protects the system against overpressurization. *[Figure 12-67]*

Center Hydraulic System

The center hydraulic system supplies pressurized hydraulic fluid to operate these systems. *[Figure 12-68]*

- Nose landing gear actuation.

- Nose landing gear steering.

- Alternate brakes.

- Main landing gear actuation.

- Main landing gear steering.

- Trailing edge flaps.

- Leading edge slat.

- Flight controls.

Reservoir

The hydraulic system reservoir of the center system contains the hydraulic fluid supply for the hydraulic pumps. The reservoir is pressurized by bleed air through a reservoir pressurization module. The reservoir supplies fluid to the ADPs, the RAT, and one of the ACMPs through a standpipe. The other ACMP gets fluid from the bottom of the reservoir. The reservoir also supplies hydraulic fluid to the landing gear alternate extension system.

The ACMPs are the primary pumps in the center hydraulic system and are normally turned on. The ADPs are the demand pumps in the center system. They normally operate only when the center system needs more hydraulic flow capacity. The RAT system supplies an emergency source of hydraulic power to the center hydraulic system flight controls. A reservoir quantity transmitter and temperature transducer are installed on the reservoir. A hydraulic reservoir pressure switch is installed on the pneumatic line between the reservoir and the reservoir pressurization module.

Filter

Filter modules clean the pressure and case drain output of the hydraulic pumps. A return filter module cleans the return flow of hydraulic fluid from the user systems. The module

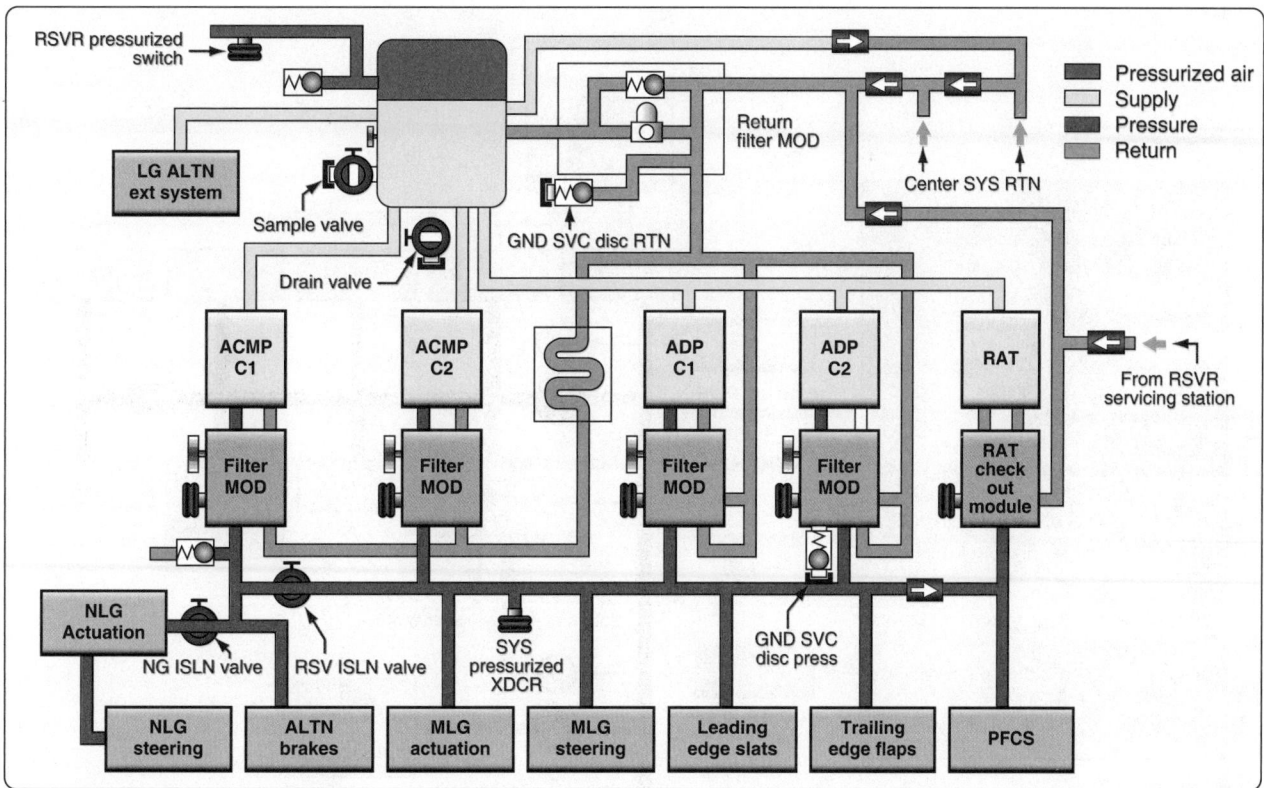

Figure 12-68. *Center hydraulic system.*

can be bypassed. The heat exchanger cools the hydraulic fluid from the ACMP case drains before the fluid goes back to the reservoir. ADP case drain fluid does not go through the heat exchangers.

The ACMP and ADP filter modules each have a pressure transducer to measure pump output pressure. A temperature transducer in each filter module measures the pump case drain temperature. A system pressure transducer measures hydraulic system pressure.

Pressure relief valves in each ADP filter module prevent system overpressurization. A pressure relief valve near ACMP C1 supplies overpressure protection for the center hydraulic isolation system (CHIS).

Center Hydraulic Isolation System (CHIS)

The CHIS supplies engine burst protection and a reserve brakes and steering function. CHIS operation is fully automatic. Relays control the electric motors in the reserve and nose gear isolation valves. When the CHIS system is operational, it prevents hydraulic operation of the leading-edge slats.

ACMP C1 gets hydraulic fluid from the bottom of the center system reservoir. All other hydraulic pumps in the center system get fluid through a standpipe in the reservoir. This gives ACMP C1 a 1.2-gallon (4.5 liter) reserve supply of hydraulic fluid.

The reserve and nose gear isolation valves are normally open. Both valves close if the quantity in the center system reservoir is low (less than 0.40) and the airspeed is more than 60 knots for more than one second. When CHIS is active, this divides the center hydraulic system into different parts. The NLG actuation and steering and the leading-edge slat hydraulic lines are isolated from center system pressure. The output of ACMP C1 goes only to the alternate brake system.

The output of the other center hydraulic system pumps goes to the trailing edge flaps, the MLG actuation and steering, and the flight controls. If there is a leak in the NLG actuation and steering or LE slat lines, there is no further loss of hydraulic fluid. The alternate brakes, the trailing edge flaps, the MLG actuation and steering, and the PFCS continue to operate normally.

If there is a leak in the trailing edge flaps, the MLG actuation and steering, or the flight control lines, the reservoir loses fluid down to the standpipe level (0.00 indication). This causes a loss of these systems, but the alternate brake system continues to get hydraulic power from ACMP C1. If there is a leak in the lines between ACMP C1 and the alternate brake system, all center hydraulic system fluid is lost.

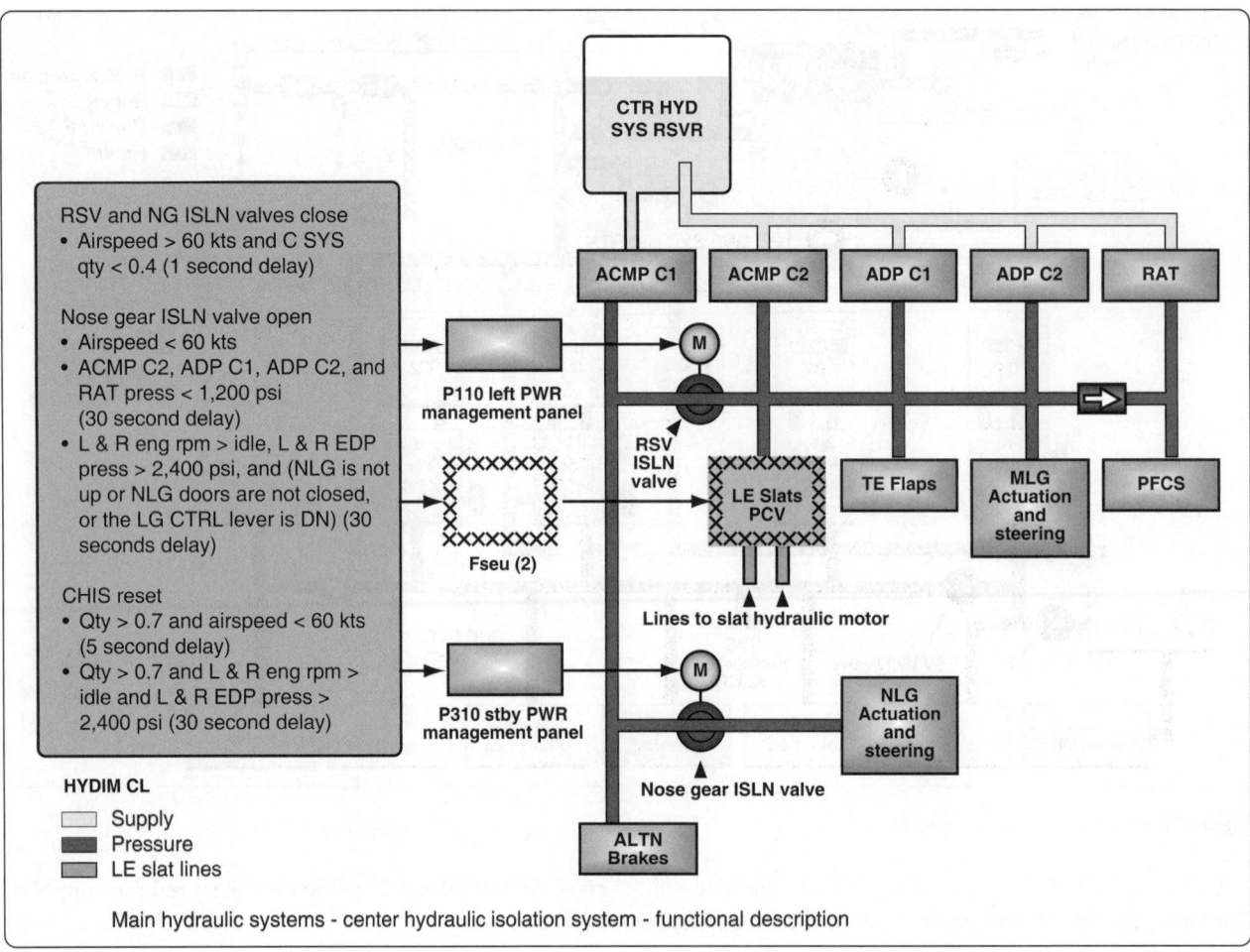

Figure 12-69. *Center hydraulic isolation system.*

Nose Gear Isolation Valve

The nose gear isolation valve opens for any of these conditions:

- Airspeed is less than 60 knots.

- Pump pressures for ACMP C2, ADP C1, ADP C2, and the RAT is less than 1,200 psi for 30 seconds.

- Left and right engine rpm is above idle, left and right EDP pressure is more than 2,400 psi, and the NLG is not up, the NLG doors are not closed, or the landing gear lever is not up for 30 seconds.

The first condition permits the flight crew to operate the NLG steering when airspeed is less than 60 knots (decreased rudder control authority during taxi). The second condition permits operation of the NLG actuation and steering if the hydraulic leak is in the part of the center hydraulic system isolated by the reserve isolation valve. The third condition permits operation of the NLG actuation and steering if there has not been an engine burst and the other hydraulic systems are pressurized. The nose gear isolation valve opens when pressure is necessary at the NLG. If the NLG is not fully retracted or the NLG doors are not closed, the nose gear isolation valve opens to let the NLG complete the retraction. When the landing gear lever is moved to the down position, the nose gear isolation valve opens to let the NLG extend with center system pressure.

Central Hydraulic System Reset

Both valves open again automatically when the center system quantity is more than 0.70 and airspeed is less than 60 knots for 5 seconds. Both valves also reset when the center system quantity is more than 0.70 and both engines and both engine-driven pumps operate normally for 30 seconds. [Figure 12-69]

Aircraft Pneumatic Systems

Some aircraft manufacturers have equipped their aircraft with a high pressure pneumatic system (3,000 psi) in the past. The last aircraft to utilize this type of system was the Fokker F27. Such systems operate a great deal like hydraulic systems, except they employ air instead of a liquid for transmitting power. Pneumatic systems are sometimes used for:

- Brakes.
- Opening and closing doors.
- Driving hydraulic pumps, alternators, starters, water injection pumps, etc.
- Operating emergency devices.

Both pneumatic and hydraulic systems are similar units and use confined fluids. The word confined means trapped or completely enclosed. The word fluid implies such liquids as water, oil, or anything that flows. Since both liquids and gases flow, they are considered as fluids; however, there is a great deal of difference in the characteristics of the two. Liquids are practically incompressible; a quart of water still occupies about a quart of space regardless of how hard it is compressed. But gases are highly compressible; a quart of air can be compressed into a thimbleful of space. In spite of this difference, gases and liquids are both fluids and can be confined and made to transmit power. The type of unit used to provide pressurized air for pneumatic systems is determined by the system's air pressure requirements.

High-Pressure Systems

For high-pressure systems, air is usually stored in metal bottles at pressures ranging from 1,000 to 3,000 psi,

depending on the particular system. *[Figure 12-70]* This type of air bottle has two valves, one of which is a charging valve. A ground-operated compressor can be connected to this valve to add air to the bottle. The other valve is a control valve. It acts as a shutoff valve, keeping air trapped inside the bottle until the system is operated. Although the high-pressure storage cylinder is light in weight, it has a definite disadvantage. Since the system cannot be recharged during flight, operation is limited by the small supply of bottled air. Such an arrangement cannot be used for the continuous operation of a system. Instead, the supply of bottled air is reserved for emergency operation of such systems as the landing gear or brakes. The usefulness of this type of system is increased, however, if other air-pressurizing units are added to the aircraft. *[Figure 12-71]*

Pneumatic System Components

Pneumatic systems are often compared to hydraulic systems, but such comparisons can only hold true in general terms. Pneumatic systems do not utilize reservoirs, hand pumps, accumulators, regulators, or engine-driven or electrically driven power pumps for building normal pressure. But similarities do exist in some components.

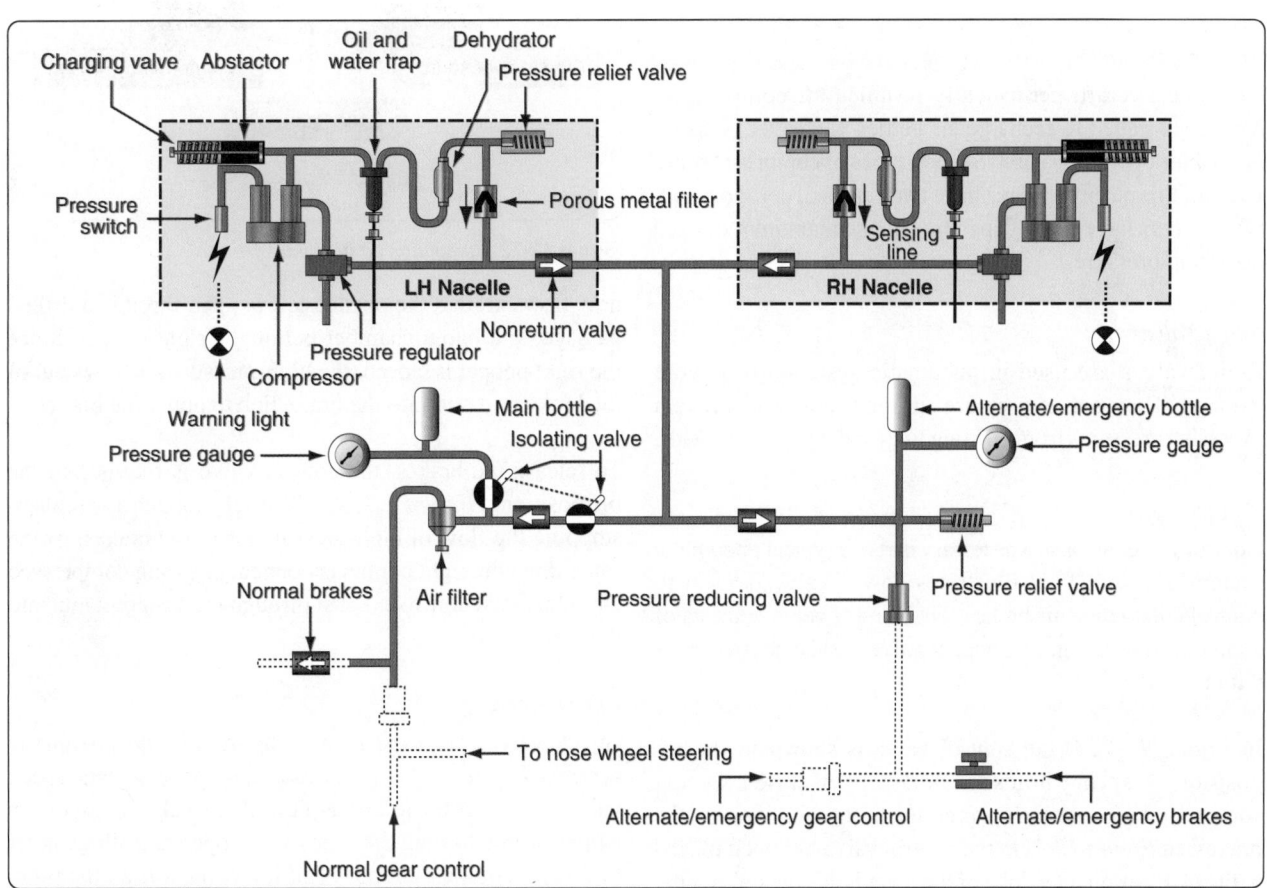

Figure 12-70. *High-pressure pneumatic system.*

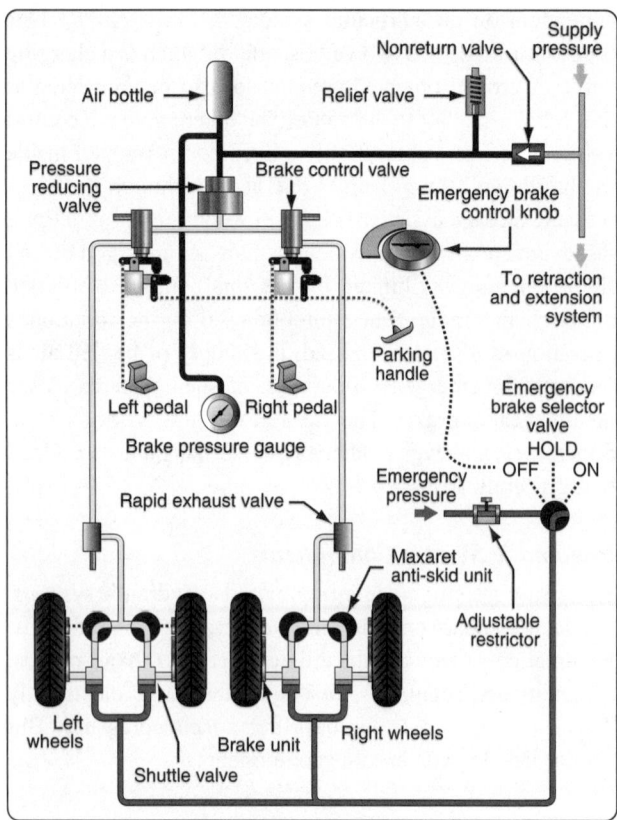

Figure 12-71. *Pneumatic brake system.*

Air Compressors

On some aircraft, permanently installed air compressors have been added to recharge air bottles whenever pressure is used for operating a unit. Several types of compressors are used for this purpose. Some have two stages of compression, while others have three, depending on the maximum desired operating pressure.

Relief Valves

Relief valves are used in pneumatic systems to prevent damage. They act as pressure limiting units and prevent excessive pressures from bursting lines and blowing out seals.

Control Valves

Control valves are also a necessary part of a typical pneumatic system. *Figure 12-72* illustrates how a valve is used to control emergency air brakes. The control valve consists of a three-port housing, two poppet valves, and a control lever with two lobes.

In *Figure 12-72A*, the control valve is shown in the off position. A spring holds the left poppet closed so that compressed air entering the pressure port cannot flow to the brakes. In *Figure 12-72B*, the control valve has been placed in the on position. One lobe of the lever holds the left poppet open, and a spring closes the right poppet. Compressed air

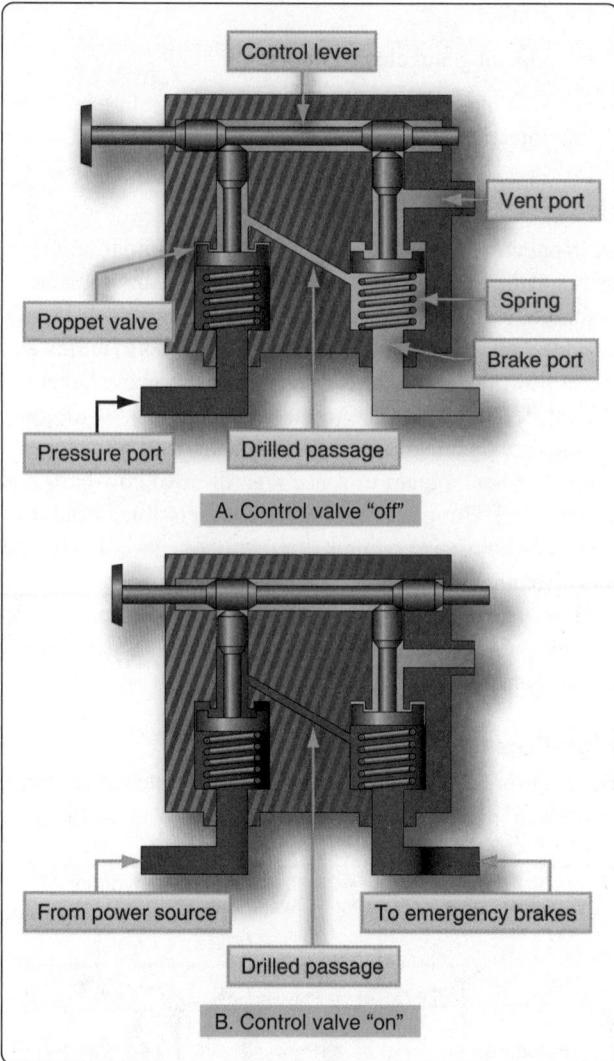

Figure 12-72. *Pneumatic control valve.*

now flows around the opened left poppet, through a drilled passage, and into a chamber below the right poppet. Since the right poppet is closed, the high-pressure air flows out of the brake port and into the brake line to apply the brakes.

To release the brakes, the control valve is returned to the off position. *[Figure 12-72A]* The left poppet now closes, stopping the flow of high-pressure air to the brakes. At the same time, the right poppet is opened, allowing compressed air in the brake line to exhaust through the vent port and into the atmosphere.

Check Valves

Check valves are used in both hydraulic and pneumatic systems. *Figure 12-73* illustrates a flap-type pneumatic check valve. Air enters the left port of the check valve, compresses a light spring, forcing the check valve open and allowing air to flow out of the right port. But if air enters from the right, air pressure closes the valve, preventing a flow of air out of

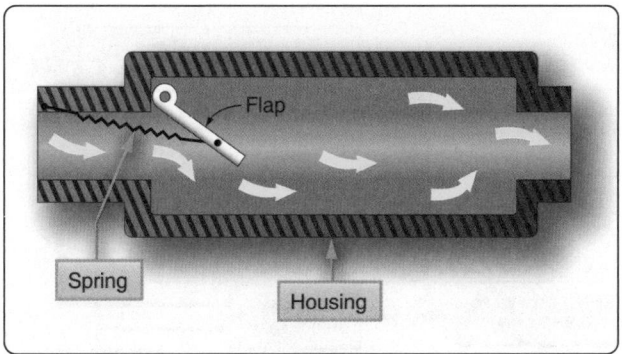

Figure 12-73. *Flap-type pneumatic check valve.*

Figure 12-74. *Pneumatic orifice valve.*

the left port. Thus, a pneumatic check valve is a one-direction flow control valve.

Restrictors

Restrictors are a type of control valve used in pneumatic systems. *Figure 12-74* illustrates an orifice-type restrictor with a large inlet port and a small outlet port. The small outlet port reduces the rate of airflow and the speed of operation of an actuating unit.

Variable Restrictor

Another type of speed-regulating unit is the variable restrictor. *[Figure 12-75]* It contains an adjustable needle valve, which has threads around the top and a point on the lower end. Depending on the direction turned, the needle valve moves the sharp point either into or out of a small opening to decrease or increase the size of the opening. Since air entering the inlet port must pass through this opening before reaching the outlet port, this adjustment also determines the rate of airflow through the restrictor.

Filters

Pneumatic systems are protected against dirt by means of various types of filters. A micronic filter consists of a housing

with two ports, a replaceable cartridge, and a relief valve. Normally, air enters the inlet, circulates around the cellulose cartridge, and flows to the center of the cartridge and out the outlet port. If the cartridge becomes clogged with dirt, pressure forces the relief valve open and allows unfiltered air to flow out the outlet port.

A screen-type filter is similar to the micron filter but contains a permanent wire screen instead of a replaceable cartridge. In the screen filter, a handle extends through the top of the housing and can be used to clean the screen by rotating it against metal scrapers.

Desiccant/Moisture Separator

The moisture separator in a pneumatic system is always located downstream of the compressor. Its purpose is to remove any moisture caused by the compressor. A complete moisture separator consists of a reservoir, a pressure switch, a dump valve, and a check valve. It may also include a regulator and a relief valve. The dump valve is energized and deenergized by the pressure switch. When deenergized, it completely purges the separator reservoir and lines up to the compressor. The check valve protects the system against pressure loss during the dumping cycle and prevents reverse flow through the separator.

Chemical Drier

Chemical driers are incorporated at various locations in a pneumatic system. Their purpose is to absorb any moisture that may collect in the lines and other parts of the system. Each drier contains a cartridge that should be blue in color. If otherwise noted, the cartridge is to be considered contaminated with moisture and should be replaced.

Emergency Backup Systems

Many aircraft use a high-pressure pneumatic back-up source of power to extend the landing gear or actuate the brakes,

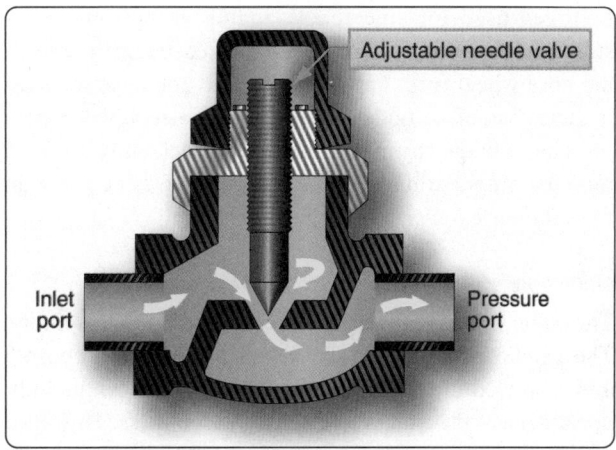

Figure 12-75. *Variable pneumatic restrictor.*

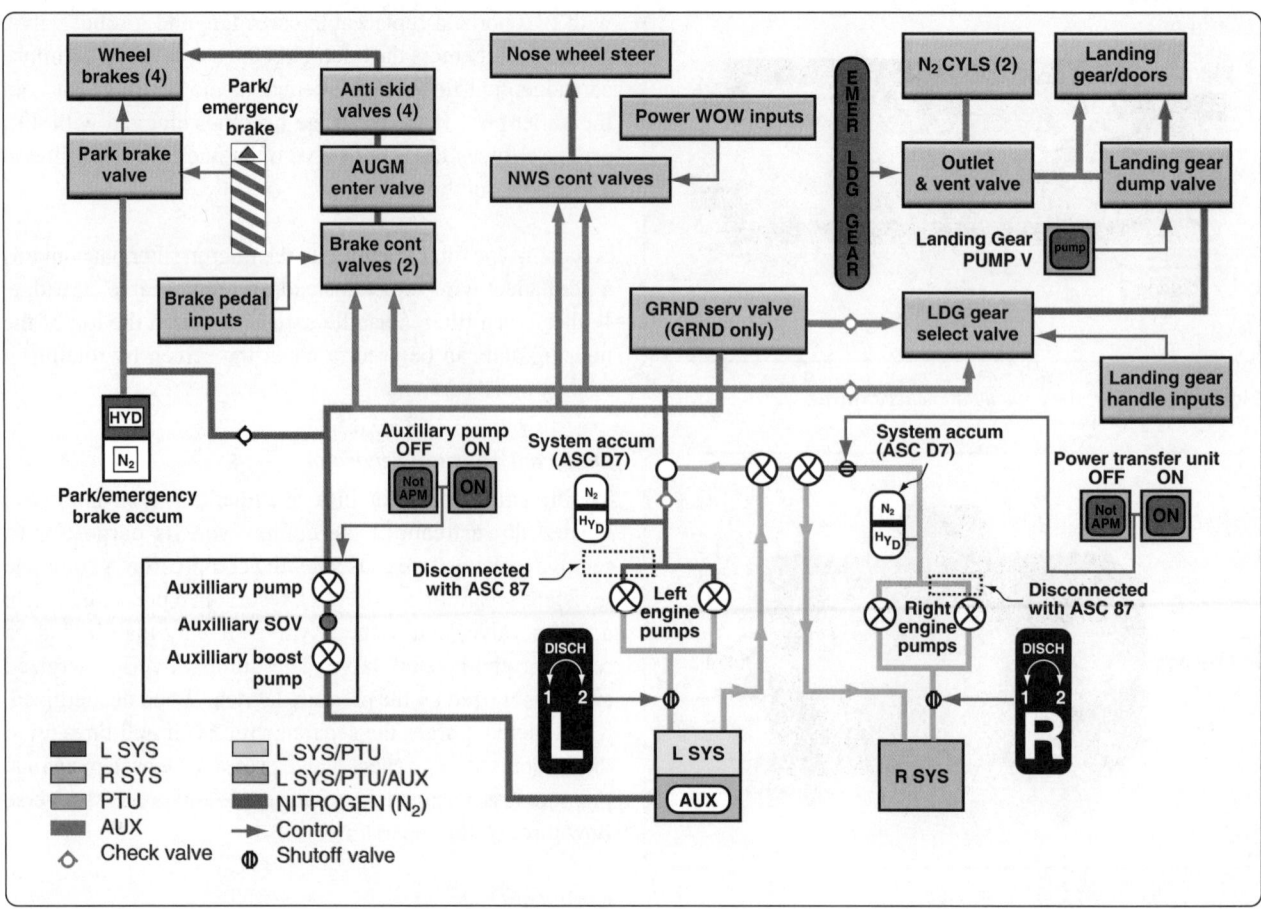

Figure 12-76. *Pneumatic emergency landing gear extension system.*

if the main hydraulic braking system fails. The nitrogen is not directly used to actuate the landing gear actuators or brake units but, instead, it applies the pressurized nitrogen to move hydraulic fluid to the actuator. This process is called pneudraulics. The following paragraph discusses the components and operation of an emergency pneumatic landing gear extension system used on a business jet. *[Figure 12-76]*

Nitrogen Bottles

Nitrogen used for emergency landing gear extension is stored in two bottles, one bottle located on each side of the nose wheel well. Nitrogen from the bottles is released by actuation of an outlet valve. Once depleted, the bottles must be recharged by maintenance personnel. Fully serviced pressure is approximately 3,100 psi at 70 °F/21 °C, enough for only one extension of the landing gear.

Gear Emergency Extension Cable & Handle

The outlet valve is connected to a cable and handle assembly. The handle is located on the side of the copilot's console and is labeled EMER LDG GEAR. Pulling the handle fully upward opens the outlet valve, releasing compressed nitrogen into the landing gear extension system. Pushing the handle

fully downward closes the outlet valve and allows any nitrogen present in the emergency landing gear extension system to be vented overboard. The venting process takes approximately 30 seconds.

Dump Valve

As compressed nitrogen is released to the landing gear selector/dump valve during emergency extension, the pneudraulic pressure actuates the dump valve portion of the landing gear selector/dump valve to isolate the landing gear system from the remainder of hydraulic system. When activated, a blue DUMP legend is illuminated on the LDG GR DUMP V switch, located on the flight deck overhead panel. A dump valve reset switch is used to reset the dump valve after the system has been used and serviced.

Emergency Extension Sequence:

1. Landing gear handle is placed in the DOWN position.

2. Red light in the landing gear control handle is illuminated.

3. EMER LDG GEAR handle is pulled fully outward.

4. Compressed nitrogen is released to the landing gear

selector/dump valve.

5. Pneudraulic pressure actuates the dump valve portion of the landing gear selector/dump valve.

6. Blue DUMP legend is illuminated on the LDG GR DUMP switch.

7. Landing gear system is isolated from the remainder of hydraulic system.

8. Pneudraulic pressure is routed to the OPEN side of the landing gear door actuators, the UNLOCK side of the landing gear uplock actuators, and the EXTEND side of the main landing gear sidebrace actuators and nose landing gear extend/retract actuator.

9. Landing gear doors open.

10. Uplock actuators unlock.

11. Landing gear extends down and locks.

12. Three green DOWN AND LOCKED lights on the landing gear control panel are illuminated.

13. Landing gear doors remain open.

Medium-Pressure Systems

A medium-pressure pneumatic system (50–150 psi) usually does not include an air bottle. Instead, it generally draws air from the compressor section of a turbine engine. This process is often called bleed air and is used to provide pneumatic power for engine starts, engine deicing, wing deicing, and in some cases, it provides hydraulic power to the aircraft systems (if the hydraulic system is equipped with an air-driven hydraulic pump). Engine bleed air is also used to pressurize the reservoirs of the hydraulic system. Bleed air systems are discussed in more detail in the powerplant handbook.

Low-Pressure Systems

Many aircraft equipped with reciprocating engines obtain a supply of low-pressure air from vane-type pumps. These pumps are driven by electric motors or by the aircraft engine. *Figure 12-77* shows a schematic view of one of these pumps, which consists of a housing with two ports, a drive shaft, and two vanes. The drive shaft and the vanes contain slots, so the vanes can slide back and forth through the drive shaft. The shaft is eccentrically mounted in the housing, causing the vanes to form four different sizes of chambers (A, B, C, and D). In the position shown, B is the largest chamber and is connected to the supply port. As depicted in *Figure 12-77,* outside air can enter chamber B of the pump. When the pump begins to operate, the drive shaft rotates and changes positions of the vanes and sizes of the chambers. Vane No. 1 then moves to the right, separating chamber B from the supply port. Chamber B now contains trapped air.

As the shaft continues to turn, chamber B moves downward

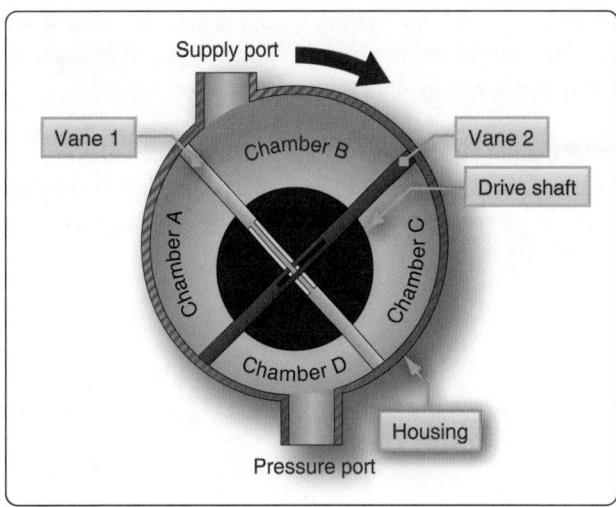

Figure 12-77. *Schematic of vane-type air pump.*

and becomes increasingly smaller, gradually compressing its air. Near the bottom of the pump, chamber B connects to the pressure port and sends compressed air into the pressure line. Then chamber B moves upward again becoming increasingly larger in area. At the supply port, it receives another supply of air. There are four such chambers in this pump and each goes through this same cycle of operation. Thus, the pump delivers to the pneumatic system a continuous supply of compressed air from 1 to 10 psi. Low-pressure systems are used for wing deicing boot systems.

Pneumatic Power System Maintenance

Maintenance of the pneumatic power system consists of servicing, troubleshooting, removal, and installation of components, and operational testing.

The air compressor's lubricating oil level should be checked daily in accordance with the applicable manufacturer's instructions. The oil level is indicated by means of a sight gauge or dipstick. When refilling the compressor oil tank, the oil (type specified in the applicable instructions manual) is added until the specified level. After the oil is added, ensure that the filler plug is torqued, and safety wire is properly installed.

The pneumatic system should be purged periodically to remove the contamination, moisture, or oil from the components and lines. Purging the system is accomplished by pressurizing it and removing the plumbing from various components throughout the system. Removal of the pressurized lines causes a high rate of airflow through the system, causing foreign matter to be exhausted from the system. If an excessive amount of foreign matter, particularly oil, is exhausted from any one system, the lines and components should be removed and cleaned or replaced.

Upon completion of pneumatic system purging and after

reconnecting all the system components, the system air bottles should be drained to exhaust any moisture or impurities that may have accumulated there.

After draining the air bottles, service the system with nitrogen or clean, dry compressed air. The system should then be given a thorough operational check and an inspection for leaks and security.

Chapter 13
Aircraft Landing Gear Systems

Landing Gear Types

Aircraft landing gear supports the entire weight of an aircraft during landing and ground operations. They are attached to primary structural members of the aircraft. The type of gear depends on the aircraft design and its intended use. Most landing gear have wheels to facilitate operation to and from hard surfaces, such as airport runways. Other gear feature skids for this purpose, such as those found on helicopters, balloon gondolas, and in the tail area of some tail dragger aircraft. Aircraft that operate to and from frozen lakes and snowy areas may be equipped with landing gear that have skis. Aircraft that operate to and from the surface of water have pontoon-type landing gear. Regardless of the type of landing gear utilized, shock absorbing equipment, brakes, retraction mechanisms, controls, warning devices, cowling, fairings, and structural members necessary to attach the gear to the aircraft are considered parts of the landing gear system. *[Figure 13-1]*

Numerous configurations of landing gear types can be found. Additionally, combinations of two types of gear are common. Amphibious aircraft are designed with gear that allow landings to be made on water or dry land. The gear features pontoons for water landing with extendable wheels for landings on hard surfaces. A similar system is used to allow the use of skis and wheels on aircraft that operate on both slippery, frozen surfaces and dry runways. Typically, the skis are retractable to allow use of the wheels when needed. *Figure 13-2* illustrates this type of landing gear.

Note: References to auxiliary landing gear refer to the nose gear, tail gear, or outrigger-type gear on any particular aircraft. Main landing gear are the two or more large gear located close to the aircraft's center of gravity.

Figure 13-1. *Basic landing gear types include those with wheels (a), skids (b), skis (c), and floats or pontoons (d).*

Figure 13-2. *An amphibious aircraft with retractable wheels (left) and an aircraft with retractable skis (right).*

Figure 13-3. *Tail wheel configuration landing gear on a DC-3 (left) and a STOL Maule MX-7-235 Super Rocket.*

Landing Gear Arrangement

Three basic arrangements of landing gear are used: tail wheel-type landing gear (also known as conventional gear), tandem landing gear, and tricycle-type landing gear.

Tail Wheel-Type Landing Gear

Tail wheel-type landing gear is also known as conventional gear because many early aircraft use this type of arrangement. The main gear are located forward of the center of gravity, causing the tail to require support from a third wheel assembly. A few early aircraft designs use a skid rather than a tail wheel. This helps slow the aircraft upon landing and provides directional stability. The resulting angle of the aircraft fuselage, when fitted with conventional gear, allows the use of a long propeller that compensates for older, underpowered engine design. The increased clearance of the forward fuselage offered by tail wheel-type landing gear is also advantageous when operating in and out of non-paved runways. Today, aircraft are manufactured with conventional gear for this reason and for the weight savings accompanying the relatively light tail wheel assembly. *[Figure 13-3]*

The proliferation of hard surface runways has rendered the tail skid obsolete in favor of the tail wheel. Directional control is maintained through differential braking until the speed of the aircraft enables control with the rudder. A steerable tail wheel, connected by cables to the rudder or rudder pedals, is also a common design. Springs are incorporated for dampening. *[Figure 13-4]*

Tandem Landing Gear

Few aircraft are designed with tandem landing gear. As the name implies, this type of landing gear has the main gear and tail gear aligned on the longitudinal axis of the aircraft. Sailplanes commonly use tandem gear, although many only have one actual gear forward on the fuselage with a skid under the tail. A few military bombers, such as the B-47 and the B-52, have tandem gear, as does the U2 spy plane. The VTOL Harrier has tandem gear but uses small outrigger gear under the wings for support. Generally, placing the gear only under the fuselage facilitates the use of very flexible wings. *[Figure 13-5]*

Figure 13-4. *The steerable tail wheel of a Pitts Special.*

Figure 13-5. *Tandem landing gear along the longitudinal axis of the aircraft permits the use of flexible wings on sailplanes (left) and select military aircraft like the B-52 (center). The VTOL Harrier (right) has tandem gear with outrigger-type gear.*

Figure 13-6. *Tricycle-type landing gear with dual main wheels on a Learjet (left) and a Cessna 172, also with tricycle gear (right).*

Tricycle-Type Landing Gear

The most commonly used landing gear arrangement is the tricycle-type landing gear. It is comprised of main gear and nose gear. *[Figure 13-6]*

Tricycle-type landing gear is used on large and small aircraft with the following benefits:

1. Allows more forceful application of the brakes without nosing over when braking, which enables higher landing speeds.

2. Provides better visibility from the flight deck, especially during landing and ground maneuvering.

3. Prevents ground-looping of the aircraft. Since the aircraft center of gravity is forward of the main gear, forces acting on the center of gravity tend to keep the aircraft moving forward rather than looping, such as with a tail wheel-type landing gear.

The nose gear of a few aircraft with tricycle-type landing gear is not controllable. It simply casters as steering is accomplished with differential braking during taxi. However, nearly all aircraft have steerable nose gear. On light aircraft,

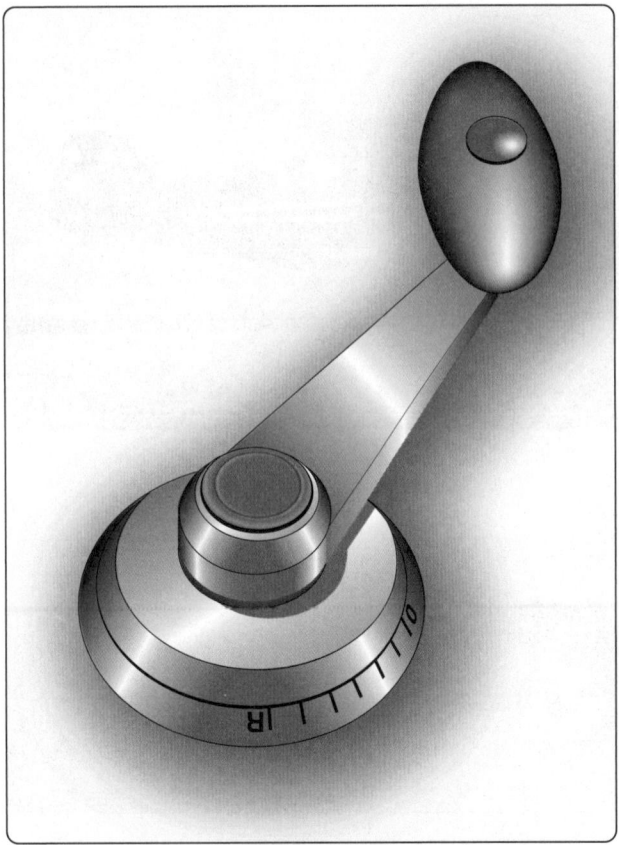

Figure 13-7. *A nose wheel steering tiller located on the flight deck.*

Figure 13-9. *Triple bogie main landing gear assembly on a Boeing 777.*

the nose gear is directed through mechanical linkage to the rudder pedals. Heavy aircraft typically utilize hydraulic power to steer the nose gear. Control is achieved through an independent tiller in the flight deck. *[Figure 13-7]*

The main gear on a tricycle-type landing gear arrangement is attached to reinforced wing structure or fuselage structure. The number and location of wheels on the main gear vary. Many main gear have two or more wheels. *[Figure 13-8]* Multiple wheels spread the weight of the aircraft over a larger area. They also provide a safety margin should one tire fail. Heavy aircraft may use four or more wheel assemblies on each main gear. When more than two wheels are attached to a landing gear strut, the attaching mechanism is known as a bogie. The number of wheels included in the bogie is a function of the gross design weight of the aircraft and the surface type on which the loaded aircraft is required to land. *Figure 13-9* illustrates the triple bogie main gear of a Boeing 777.

The tricycle-type landing gear arrangement consists of many parts and assemblies. These include air/oil shock struts, gear alignment units, support units, retraction and safety devices, steering systems, wheel and brake assemblies, etc. A main landing gear of a transport category aircraft is illustrated in *Figure 13-10* with many of the parts identified as an introduction to landing gear nomenclature.

Fixed & Retractable Landing Gear
Further classification of aircraft landing gear can be made into two categories: fixed and retractable. Many small, single-engine light aircraft have fixed landing gear, as do a few light twins. This means the gear is attached to the airframe and remains exposed to the slipstream as the aircraft is flown. As discussed in Chapter 2 of this handbook, as the speed of

Figure 13-8. *Dual main gear of a tricycle-type landing gear.*

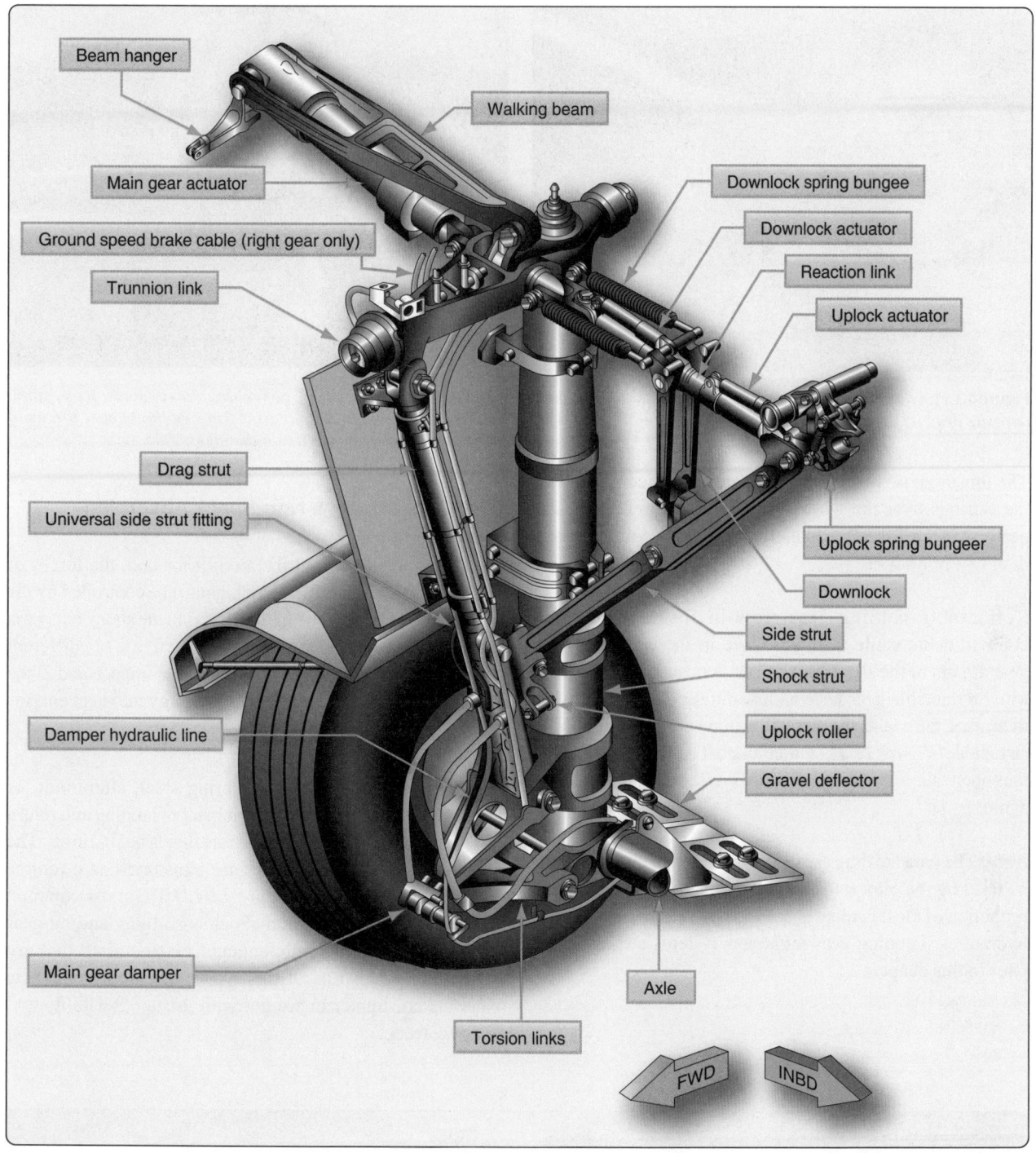

Figure 13-10. *Nomenclature of a main landing gear bogie truck.*

an aircraft increases, so does parasite drag. Mechanisms to retract and stow the landing gear to eliminate parasite drag add weight to the aircraft. On slow aircraft, the penalty of this added weight is not overcome by the reduction of drag, so fixed gear is used. As the speed of the aircraft increases, the drag caused by the landing gear becomes greater and a means to retract the gear to eliminate parasite drag is required, despite the weight of the mechanism.

A great deal of the parasite drag caused by light aircraft landing gear can be reduced by building gear as aerodynamically as possible and by adding fairings or wheel pants to streamline the airflow past the protruding assemblies. A small, smooth profile to the oncoming wind greatly reduces landing gear parasite drag. *Figure 13-11* illustrates a Cessna aircraft landing gear used on many of the manufacturer's light planes.

Figure 13-11. *Wheel fairings, or pants, and low profile struts reduce parasite drag on fixed gear aircraft.*

Figure 13-13. *Non-shock absorbing struts made from steel, aluminum, or composite material transfer the impact forces of landing to the airframe at a non-damaging rate.*

The thin cross section of the spring steel struts combine with the fairings over the wheel and brake assemblies to raise performance of the fixed landing gear by keeping parasite drag to a minimum.

Retractable landing gear stow in fuselage or wing compartments while in flight. Once in these wheel wells, gear are out of the slipstream and do not cause parasite drag. Most retractable gear have a close-fitting panel attached to them that fairs with the aircraft skin when the gear is fully retracted. *[Figure 13-12]* Other aircraft have separate doors that open, allowing the gear to enter or leave, and then close again.

Note: The parasite drag caused by extended landing gear can be used by the pilot to slow the aircraft. The extension and retraction of most landing gear is usually accomplished with hydraulics. Landing gear retraction systems are discussed later in this chapter.

Shock Absorbing & Non-Shock Absorbing Landing Gear

In addition to supporting the aircraft for taxi, the forces of impact on an aircraft during landing must be controlled by the landing gear. This is done in two ways: 1) the shock energy is altered and transferred throughout the airframe at a different rate and time than the single strong pulse of impact, and 2) the shock is absorbed by converting the energy into heat energy.

Leaf-Type Spring Gear

Many aircraft utilize flexible spring steel, aluminum, or composite struts that receive the impact of landing and return it to the airframe to dissipate at a rate that is not harmful. The gear flexes initially and forces are transferred as it returns to its original position. *[Figure 13-13]* The most common example of this type of non-shock absorbing landing gear are the thousands of single-engine Cessna aircraft that use it. Landing gear struts of this type made from composite materials are lighter in weight with greater flexibility and do not corrode.

Figure 13-12. *The retractable gear of a Boeing 737 fair into recesses in the fuselage. Panels attached to the landing gear provide smooth airflow over the struts. The wheel assemblies mate with seals to provide aerodynamic flow without doors.*

Figure 13-14. *Rigid steel landing gear is used on many early aircraft.*

Rigid

Before the development of curved spring steel landing struts, many early aircraft were designed with rigid, welded steel landing gear struts. Shock load transfer to the airframe is direct with this design. Use of pneumatic tires aids in softening the impact loads. *[Figure 13-14]* Modern aircraft that use skid-type landing gear make use of rigid landing gear with no significant ill effects. Rotorcraft, for example, typically experience low impact landings that are able to be directly absorbed by the airframe through the rigid gear (skids).

Bungee Cord

The use of bungee cords on non-shock absorbing landing gear is common. The geometry of the gear allows the strut assembly to flex upon landing impact. Bungee cords are positioned between the rigid airframe structure and the flexing gear assembly to take up the loads and return them to the airframe at a non-damaging rate. The bungees are made of many individual small strands of elastic rubber that must be inspected for condition. Solid, donut-type rubber cushions are also used on some aircraft landing gear. *[Figure 13-15]*

Shock Struts

True shock absorption occurs when the shock energy of landing impact is converted into heat energy, as in a shock strut landing gear. This is the most common method of landing shock dissipation in aviation. It is used on aircraft of all sizes. Shock struts are self-contained hydraulic units that support an aircraft while on the ground and protect the structure during landing. They must be inspected and serviced regularly to ensure proper operation.

There are many different designs of shock struts, but most operate in a similar manner. The following discussion is general in nature. For information on the construction, operation, and servicing of a specific aircraft shock, consult the manufacturer's maintenance instructions.

A typical pneumatic/hydraulic shock strut uses compressed air or nitrogen combined with hydraulic fluid to absorb and dissipate shock loads. It is sometimes referred to as an air/oil or oleo strut. A shock strut is constructed of two telescoping cylinders or tubes that are closed on the external ends. The upper cylinder is fixed to the aircraft and does not move. The lower cylinder is called the piston and is free to slide in and out of the upper cylinder. Two chambers are formed. The lower chamber is always filled with hydraulic fluid and the upper chamber is filled with compressed air or nitrogen. An orifice located between the two cylinders provides a passage for the fluid from the bottom chamber to enter the top cylinder chamber when the strut is compressed. *[Figure 13-16]*

Most shock struts employ a metering pin similar to that shown in *Figure 13-16* for controlling the rate of fluid flow

Figure 13-15. *Piper Cub bungee cord landing gear transfer landing loads to the airframe (left and center). Rubber, donut-type shock transfer is used on some Mooney aircraft (right).*

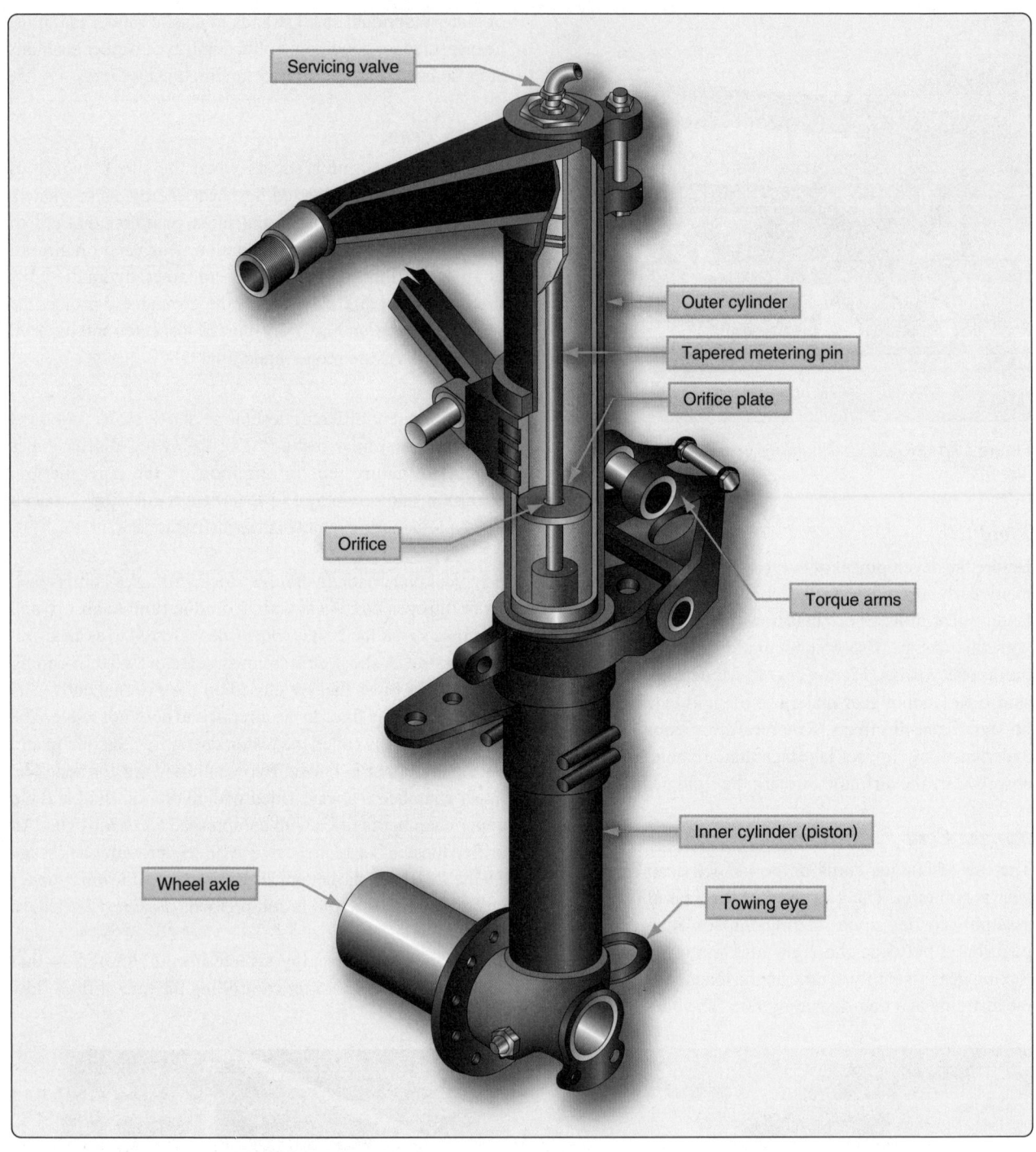

Figure 13-16. *A landing gear shock strut with a metering pin to control the flow of hydraulic fluid from the lower chamber to the upper chamber during compression.*

from the lower chamber into the upper chamber. During the compression stroke, the rate of fluid flow is not constant. It is automatically controlled by the taper of the metering pin in the orifice. When a narrow portion of the pin is in the orifice, more fluid can pass to the upper chamber. As the diameter of the portion of the metering pin in the orifice increases, less fluid passes. Pressure build-up caused by strut compression and the hydraulic fluid being forced through the metered

orifice causes heat. This heat is converted impact energy. It is dissipated through the structure of the strut.

On some types of shock struts, a metering tube is used. The operational concept is the same as that in shock struts with metering pins, except the holes in the metering tube control the flow of fluid from the bottom chamber to the top chamber during compression. *[Figure 13-17]*

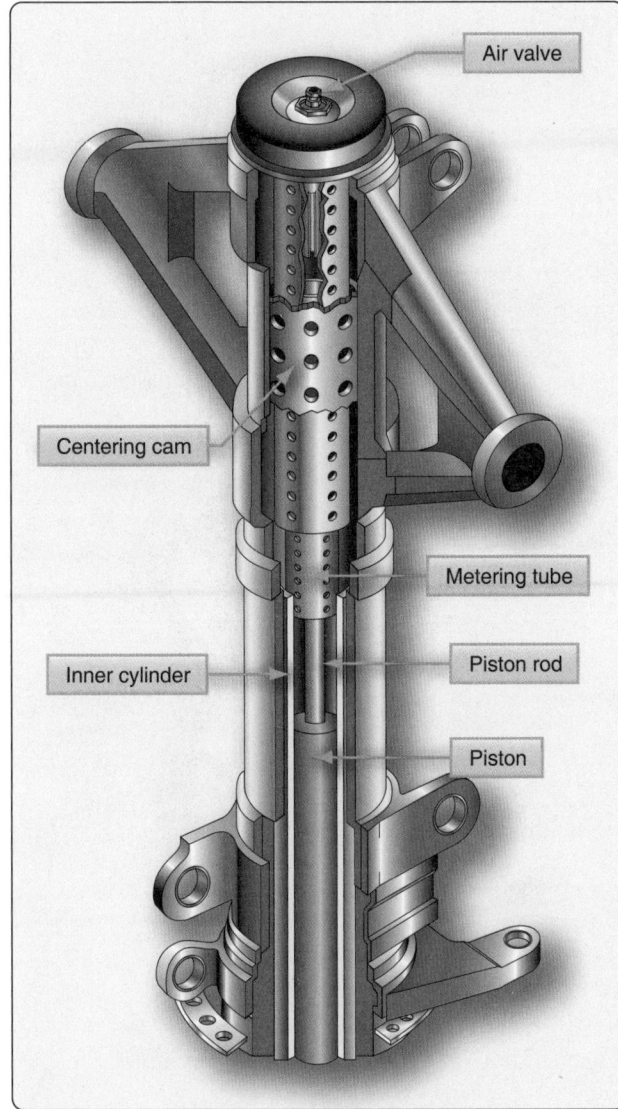

Figure 13-17. *Some landing gear shock struts use an internal metering tube rather than a metering pin to control the flow of fluid from the bottom cylinder to the top cylinder.*

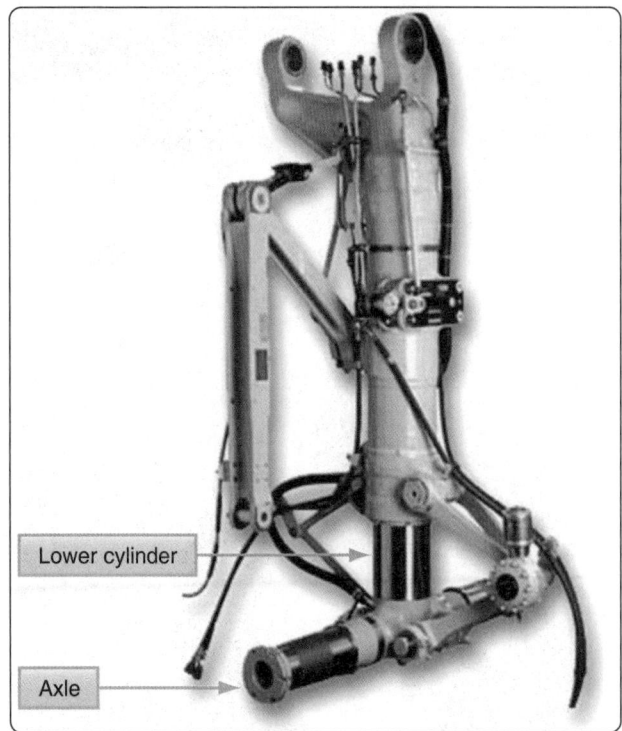

Figure 13-18. *Axles machined out of the same material as the landing gear lower cylinder.*

The upper cylinder of a shock strut typically contains a valve fitting assembly. It is located at or near the top of the cylinder. The valve provides a means of filling the strut with hydraulic fluid and inflating it with air or nitrogen as specified by the manufacturer. A packing gland is employed to seal the sliding joint between the upper and lower telescoping cylinders. It is installed in the open end of the outer cylinder. A packing gland wiper ring is also installed in a groove in the lower bearing or gland nut on most shock struts. It is designed to keep the sliding surface of the piston from carrying dirt, mud, ice, and snow into the packing gland and upper cylinder. Regular cleaning of the exposed portion of the strut piston helps the wiper do its job and decreases the possibility of damage to the packing gland, which could cause the strut to a leak.

To keep the piston and wheels aligned, most shock struts are equipped with torque links or torque arms. One end of the links is attached to the fixed upper cylinder. The other end is attached to the lower cylinder (piston), so it cannot rotate. This keeps the wheels aligned. The links also retain the piston in the end of the upper cylinder when the strut is extended, such as after takeoff. *[Figure 13-19]*

Nose gear shock struts are provided with a locating cam assembly to keep the gear aligned. A cam protrusion is attached to the lower cylinder, and a mating lower cam recess is attached to the upper cylinder. These cams line up

Upon lift off or rebound from compression, the shock strut tends to extend rapidly. This could result in a sharp impact at the end of the stroke and damage to the strut. It is typical for shock struts to be equipped with a damping or snubbing device to prevent this. A recoil valve on the piston or a recoil tube restricts the flow of fluid during the extension stroke, which slows the motion and prevents damaging impact forces.

Most shock struts are equipped with an axle as part of the lower cylinder to provide installation of the aircraft wheels. Shock struts without an integral axle have provisions on the end of the lower cylinder for installation of the axle assembly. Suitable connections are provided on all shock strut upper cylinders to attach the strut to the airframe. *[Figure 13-18]*

Figure 13-19. *Torque links align the landing gear and retain the piston in the upper cylinder when the strut is extended.*

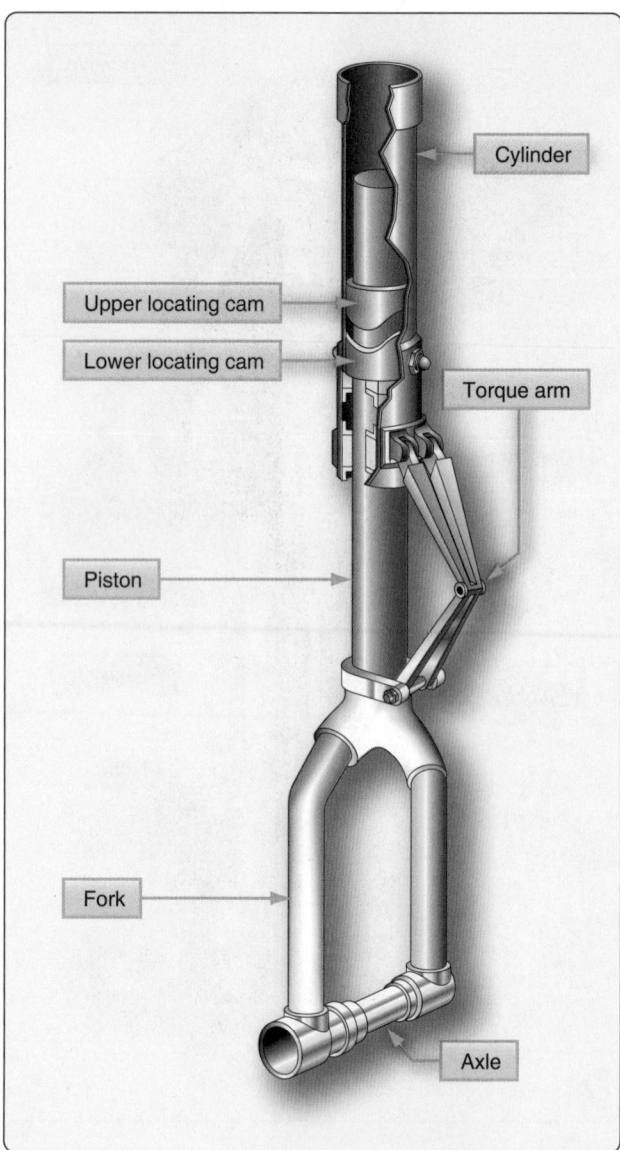

Figure 13-20. *An upper locating cam mates into a lower cam recess when the nose landing gear shock strut is extended before landing and before the gear is retracted into the wheel well.*

the wheel and axle assembly in the straight-ahead position when the shock strut is fully extended. This allows the nose wheel to enter the wheel well when the nose gear is retracted and prevents structural damage to the aircraft. It also aligns the wheels with the longitudinal axis of the aircraft prior to landing when the strut is fully extended. *[Figure 13-20]* Many nose gear shock struts also have attachments for the installation of an external shimmy damper. *[Figure 13-21]*

Nose gear struts are often equipped with a locking or disconnect pin to enable quick turning of the aircraft while towing or positioning the aircraft when on the ramp or in a hangar. Disengagement of this pin allows the wheel fork spindle on some aircraft to rotate 360°, thus enabling the aircraft to be turned in a tight radius. At no time should the nose wheel of any aircraft be rotated beyond limit lines marked on the airframe.

Nose and main gear shock struts on many aircraft are also equipped with jacking points and towing lugs. Jacks should always be placed under the prescribed points. When towing lugs are provided, the towing bar should be attached only to these lugs. *[Figure 13-22]*

Shock struts contain an instruction plate that gives directions for filling the strut with fluid and for inflating the strut. The instruction plate is usually attached near filler inlet and air valve assembly. It specifies the correct type of hydraulic fluid to use in the strut and the pressure to which the strut should be inflated. It is of utmost importance to become familiar with these instructions prior to filling a shock strut with hydraulic fluid or inflating it with air or nitrogen.

Shock Strut Operation

Figure 13-23 illustrates the inner construction of a shock strut. Arrows show the movement of the fluid during compression and extension of the strut. The compression stroke of the shock strut begins as the aircraft wheels touch the ground. As

Figure 13-21. *A shimmy damper helps control oscillations of the nose gear.*

Figure 13-22. *A towing lug on a landing gear is the designed means for attaching a tow bar.*

the center of mass of the aircraft moves downward, the strut compresses, and the lower cylinder or piston is forced upward into the upper cylinder. The metering pin is therefore moved up through the orifice. The taper of the pin controls the rate of fluid flow from the bottom cylinder to the top cylinder at all points during the compression stroke. In this manner, the greatest amount of heat is dissipated through the walls of the strut. At the end of the downward stroke, the compressed air in the upper cylinder is further compressed which limits the compression stroke of the strut with minimal impact. During taxi operations, the air in the tires and the strut combine to smooth out bumps.

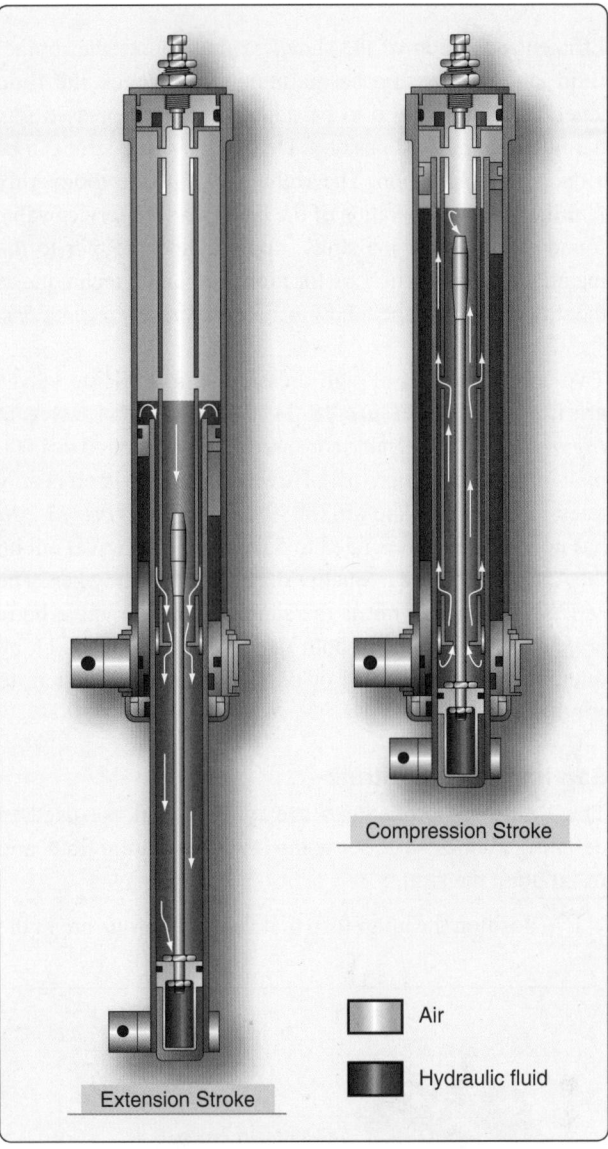

Compression Stroke

Air

Hydraulic fluid

Extension Stroke

Figure 13-23. *Fluid flow during shock strut operation is controlled by the taper of the metering pin in the shock strut orifice.*

Insufficient fluid, or air in the strut, cause the compression stroke to not be properly limited. The strut could bottom out, resulting in impact forces to be transferred directly to the airframe through the metallic structure of the strut. In a properly serviced strut, the extension stroke of the shock strut operation occurs at the end of the compression stroke. Energy stored in the compressed air in the upper cylinder causes the aircraft to start moving upward in relation to the ground and lower strut cylinder as the strut tries to rebound to its normal position. Fluid is forced back down into the lower cylinder through restrictions and snubbing orifices. The snubbing of fluid flow during the extension stroke dampens the strut rebound and reduces oscillation caused by the spring action of the compressed air. A sleeve, spacer, or bumper ring incorporated into the strut limits the extension stroke.

Efficient operation of the shock struts requires that proper fluid and air pressure be maintained. To check the fluid level, most struts need to be deflated and compressed into the fully compressed position. Deflating a shock strut can be a dangerous operation. The technician must be thoroughly familiar with the operation of the high-pressure service valve found at the top of the strut's upper cylinder. Refer to the manufacturer's instructions for proper deflating technique of the strut in question and follow all necessary safety precautions.

Two common types of high pressure strut servicing valves are illustrated in *Figure 13-24*. The AN6287-1 valve in *Figure 13-24A* has a valve core assembly and is rated to 3,000 pounds per square inch (psi). However, the core itself is only rated to 2,000 psi. The MS28889-1 valve in *Figure 13-24B* has no valve core. It is rated to 5,000 psi. The swivel nut on the AN6287-1 valve is smaller than the valve body hex. The MS28889-1 swivel nut is the same size as the valve body hex. The swivel nuts on both valves engage threads on an internal stem that loosens or draws tight the valve stem to a metal seat.

Servicing Shock Struts

The following procedures are typical of those used in deflating a shock strut, servicing it with hydraulic fluid, and re-inflating the strut.

1. Position the aircraft so that the shock struts are in the normal ground operating position. Make certain that personnel, work stands, and other obstacles are clear of the aircraft. If the maintenance procedures require, securely jack the aircraft.

2. Remove the cap from the air servicing valve. *[Figure 13-25A]*

3. Check the swivel nut for tightness.

4. If the servicing valve is equipped with a valve core, depress it to release any air pressure that may be trapped under the core in the valve body. *[Figure 13-25B]* Always be positioned to the side of the trajectory of any valve core in case it releases. Propelled by strut air pressure, serious injury could result.

5. Loosen the swivel nut. For a valve with a valve core (AN2687-1), rotate the swivel nut one turn (counter clockwise). Using a tool designed for the purpose, depress the valve core to release all of the air in the strut. For a valve without a valve core (MS28889), rotate the swivel nut sufficiently to allow the air to escape.

6. When all air has escaped from the strut, it should be compressed completely. Aircraft on jacks may need to have the lower strut jacked with an exerciser jack to achieve full compression of the strut. *[Figure 13-26]*

7. Remove the valve core of an AN6287 valve *[Figure 13-25D]* using a valve core removal tool.

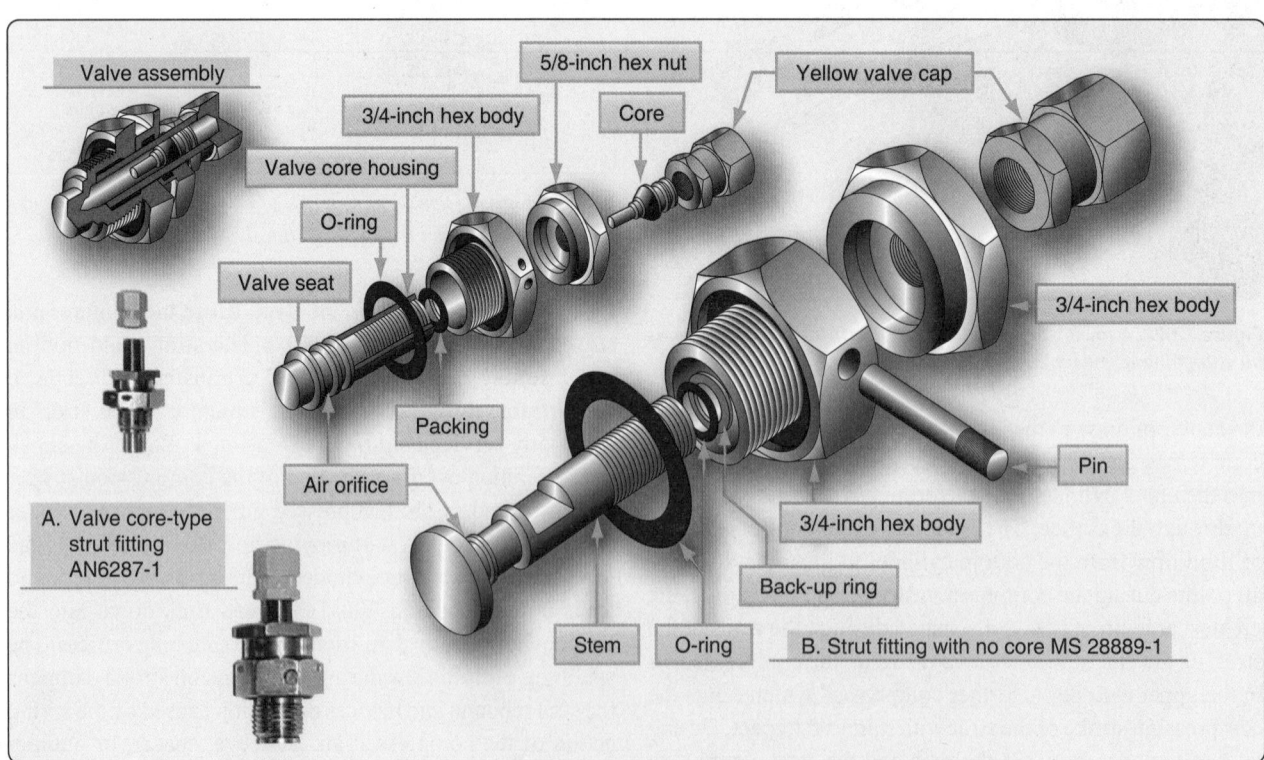

Figure 13-24. *Valve core-type (A) and core-free valve fittings (B) are used to service landing gear shock struts.*

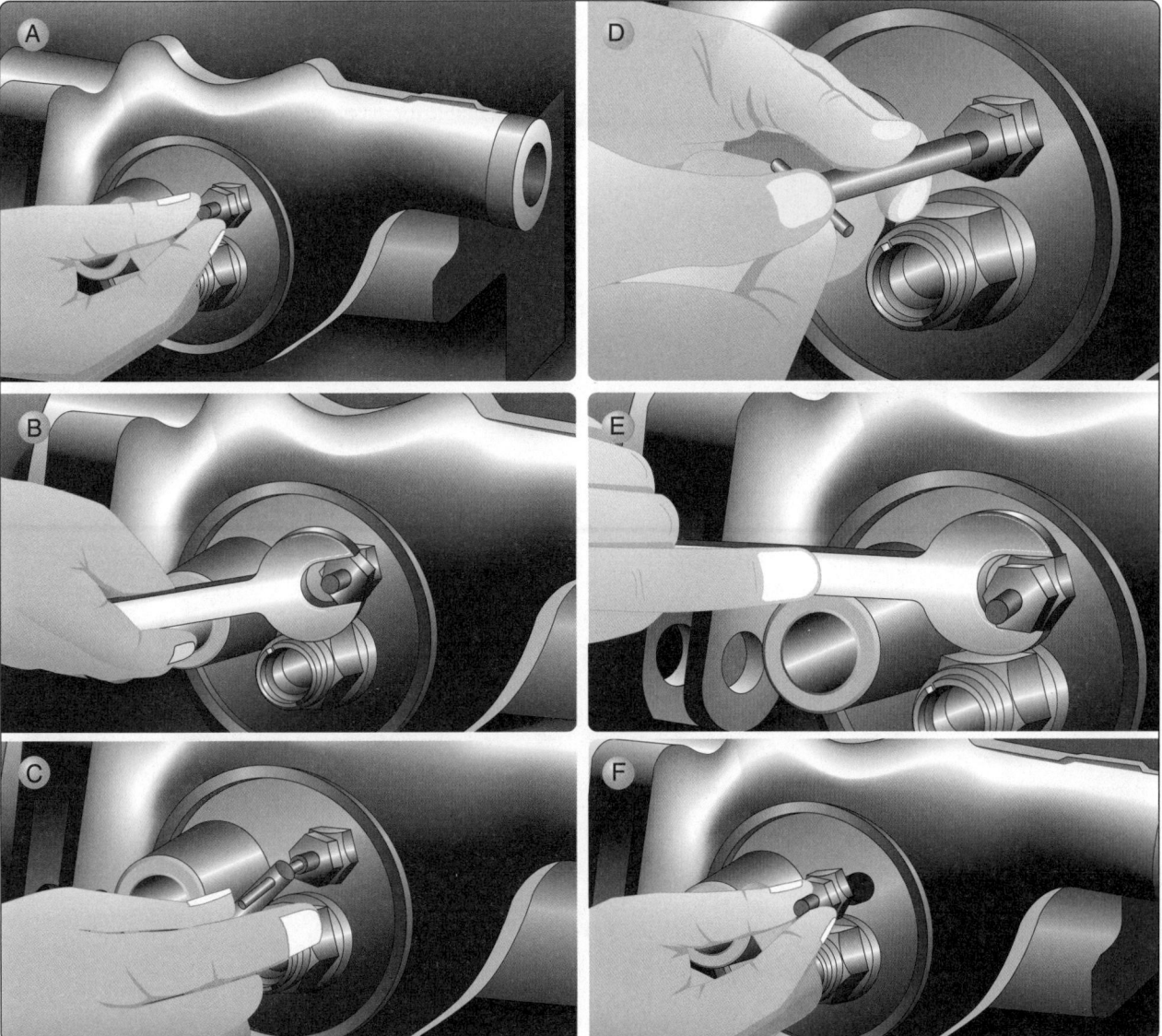

Figure 13-25. *Steps in servicing a landing gear shock strut include releasing the air from the strut and removing the service valve from the top of the strut to permit the introduction of hydraulic fluid. Note that the strut is illustrated horizontally. On an actual aircraft installation, the strut is serviced in the vertical position (landing gear down).*

[*Figure 13-27*] Then, remove the entire service valve by unscrewing the valve body from the strut. [*Figure 13-25E*]

8. Fill the strut with hydraulic fluid to the level of the service valve port with the approved hydraulic fluid.

9. Re-install the air service valve assembly using a new O-ring packing. Torque according to applicable manufacturer's specifications. If an AN2687-1 valve, install a new valve core.

10. Inflate the strut. A threaded fitting from a controlled source of high pressure air or nitrogen should be screwed onto the servicing valve. Control the flow with the service valve swivel nut. The correct amount of inflation is measured in psi on some struts. Other manufacturers specify struts to be inflated until extension of the lower strut is a certain measurement. Follow manufacturer's instructions. Shock struts should always be inflated slowly to avoid excess heating and over inflation.

11. Once inflated, tighten the swivel nut and torque as specified.

12. Remove the fill hose fitting and finger tighten the valve cap of the valve.

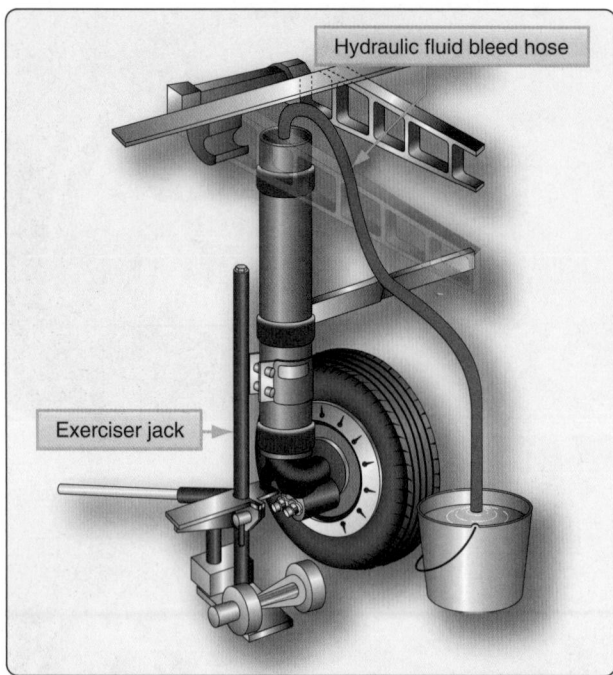

Figure 13-26. *Air trapped in shock strut hydraulic fluid is bled by exercising the strut through its full range of motion while the end of an air-tight bleed hose is submerged in a container of hydraulic fluid.*

Figure 13-27. *This valve tool features internal and external thread chasers, a notched valve core removal/installation tool, and a tapered end for depressing a valve core or clearing debris.*

Bleeding Shock Struts

It may be necessary to bleed a shock strut during the service operation or when air becomes trapped in the hydraulic fluid inside the strut. This can be caused by low hydraulic fluid quantity in the strut. Bleeding is normally done with the aircraft on jacks to facilitate repeated extension and

compression of the strut to expel the entrapped air. An example procedure for bleeding the shock strut follows.

1. Construct and attach a bleed hose containing a fitting suitable for making an airtight connection at the shock strut service valve port. Ensure a long enough hose to reach the ground while the aircraft is on jacks.

2. Jack the aircraft until the shock struts are fully extended.

3. Release any air pressure in the shock strut.

4. Remove the air service valve assembly.

5. Fill the strut to the level of the service port with approved hydraulic fluid.

6. Attach the bleed hose to the service port and insert the free end of the hose into a container of clean hydraulic fluid. The hose end must remain below the surface of the fluid.

7. Place an exerciser jack or other suitable jack under the shock strut jacking point. Compress and extend the strut fully by raising and lowering the jack. Continue this process until all air bubbles cease to form in the container of hydraulic fluid. Compress the strut slowly and allow it to extend by its own weight. Ensure that the hydraulic fluid is still at the appropriate level before moving on to the next step.

8. Remove the exerciser jack. Lower the aircraft and remove all other jacks.

9. Remove the bleed hose assembly and fitting from the service port of the strut.

10. Install the air service valve, torque, and inflate the shock strut to the manufacturer's specifications.

Landing Gear Alignment, Support, & Retraction

Retractable landing gear consist of several components that enable it to function. Typically, these are the torque links, trunnion and bracket arrangements, drag strut linkages, electrical and hydraulic gear retraction devices, as well as locking, sensing, and indicating components. Additionally, nose gear have steering mechanisms attached to the gear.

Alignment

As previously mentioned, a torque arm or torque links assembly keeps the lower strut cylinder from rotating out of alignment with the longitudinal axis of the aircraft. In some strut assemblies, it is the sole means of retaining the piston in the upper strut cylinder. The link ends are attached to the fixed upper cylinder and the moving lower cylinder with a hinge pin in the center to allow the strut to extend and compress.

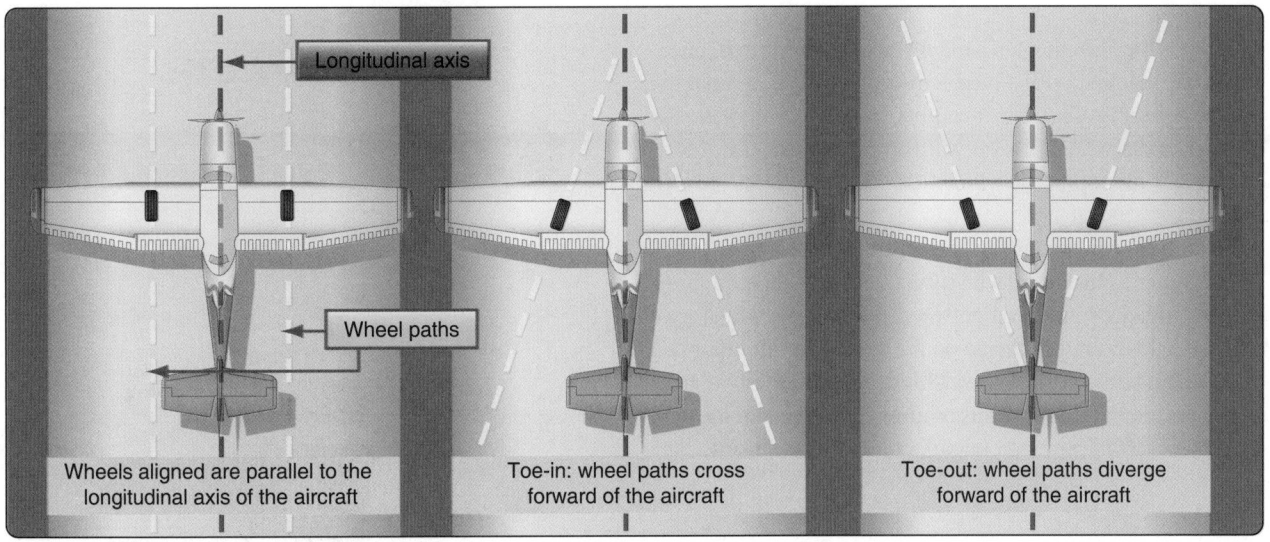

Figure 13-28. *Wheel alignment on an aircraft.*

Alignment of the wheels of an aircraft is also a consideration. Normally, this is set by the manufacturer and only requires occasional attention such as after a hard landing. The aircraft's main wheels must be inspected and adjusted, if necessary, to maintain the proper toe-in or toe-out and the correct camber. Toe-in and toe-out refer to the path a main wheel would take in relation to the airframe longitudinal axis or centerline if the wheel was free to roll forward. Three possibilities exist. The wheel would roll either: 1) parallel to the longitudinal axis (aligned); 2) converge on the longitudinal axis (toe-in); or 3) veer away from the longitudinal axis (toe-out). *[Figure 13-28]*

The manufacturer's maintenance instructions give the procedure for checking and adjusting toe-in or toe-out. A general procedure for checking alignment on a light aircraft follows. To ensure that the landing gear settle properly for a toe-in/toe-out test, especially on spring steel strut aircraft, two aluminum plates separated with grease are put under each wheel. Gently rock the aircraft on the plates to cause the gear to find the at rest position preferred for alignment checks.

A straight edge is held across the front of the main wheel tires just below axle height. A carpenter's square placed against the straight edge creates a perpendicular that is parallel to the longitudinal axis of the aircraft. Slide the square against the wheel assembly to see if the forward and aft sections of the tire touch the square. A gap in front indicates the wheel is toed-in. A gap in the rear indicates the wheel is toed-out. *[Figure 13-29]*

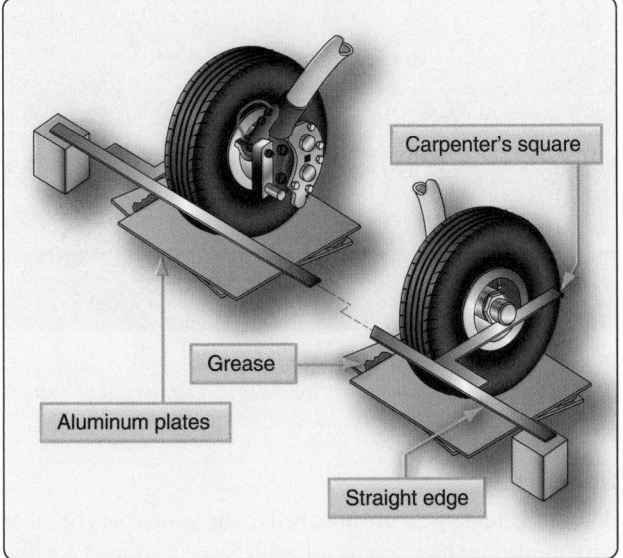

Figure 13-29. *Finding toe-in and toe-out on a light aircraft with spring steel struts.*

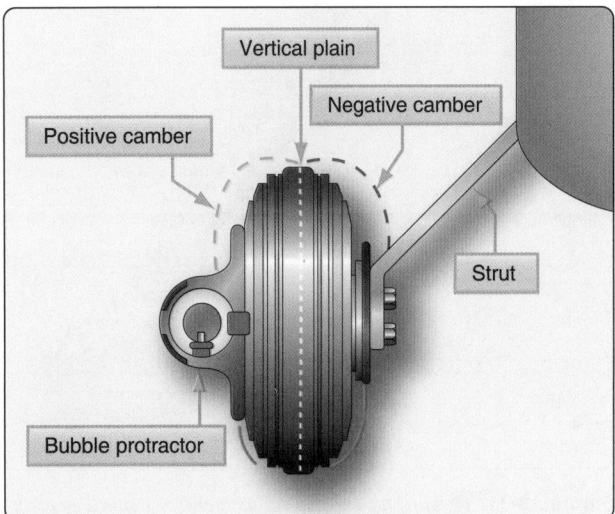

Figure 13-30. *Camber of a wheel is the amount the wheel is tilted out of the vertical plain. It can be measured with a bubble protractor.*

Camber is the alignment of a main wheel in the vertical plain. It can be checked with a bubble protractor held against the wheel assembly. The wheel camber is said to be positive if the top of the wheel tilts outward from vertical. Camber is negative if the top of the wheel tilts inward. *[Figure 13-30]*

Adjustments can be made to correct small amounts of wheel misalignment. On aircraft with spring steel gear, tapered shims can be added or removed between the bolt-on wheel axle and the axle mounting flange on the strut. Aircraft equipped with air/oil struts typically use shims between the two arms of the torque links as a means of aligning toe-in and toe-out. *[Figure 13-31]* Follow all manufacturer's instructions.

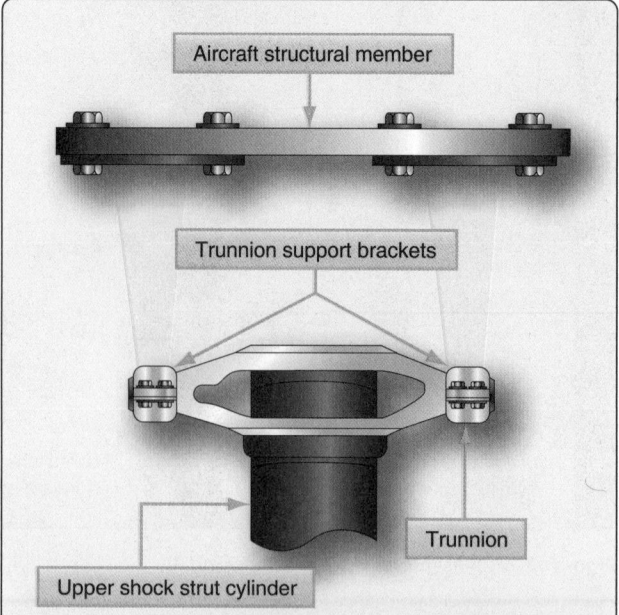

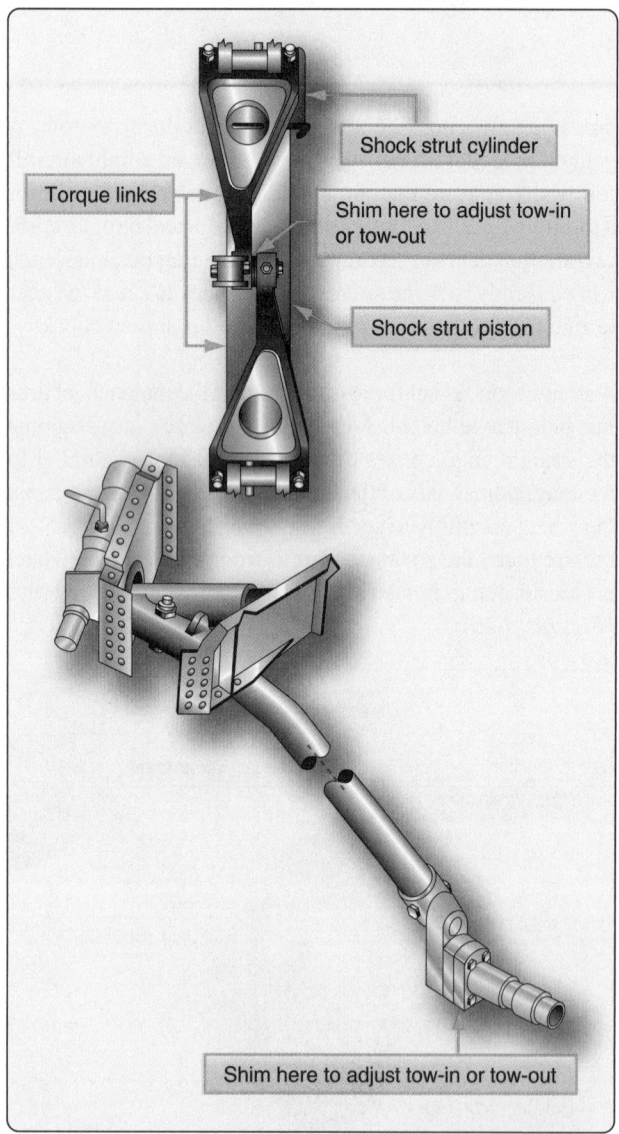

Figure 13-31. *Toe-in and toe-out adjustments on small aircraft with spring steel landing gear are made with shims behind the axle assembly. On shock strut aircraft, the shims are placed where the torque links couple.*

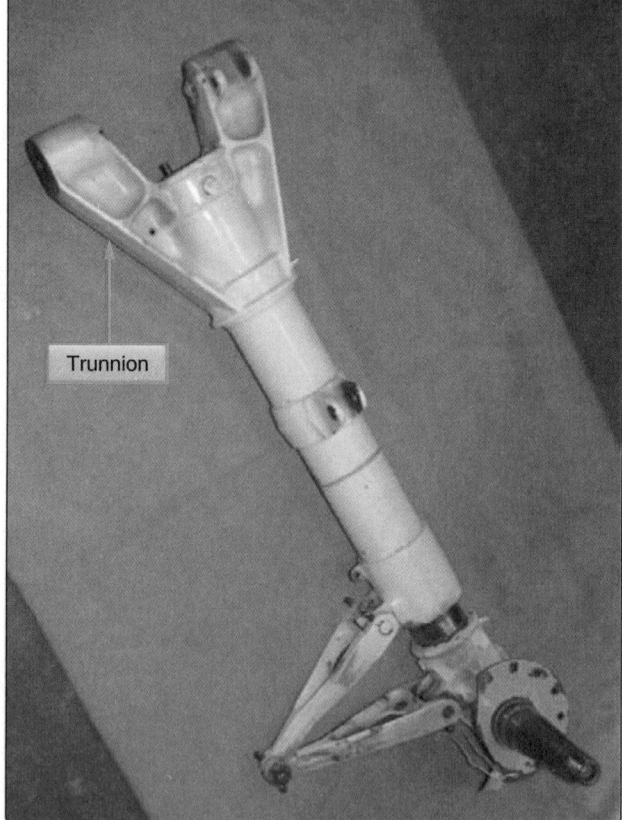

Figure 13-32. *The trunnion is a fixed structural support that is part of or attached to the upper strut cylinder of a landing gear strut. It contains bearing surfaces so the gear can retract.*

Support

Aircraft landing gear are attached to the wing spars or other structural members, many of which are designed for the specific purpose of supporting the landing gear. Retractable gear must be engineered in such a way as to provide strong

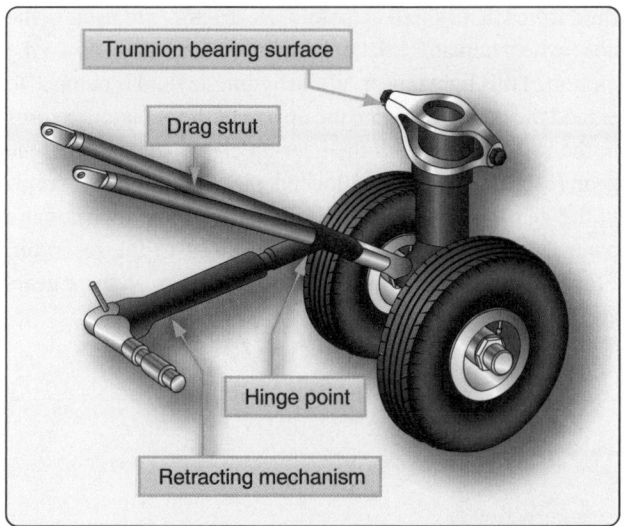

Figure 13-33. *A hinged drag strut holds the trunnion and gear firm for landing and ground operation. It folds at the hinge to allow the gear to retract.*

attachment to the aircraft and still be able to move into a recess or well when stowed. A trunnion arrangement is typical. The trunnion is a fixed structural extension of the upper strut cylinder with bearing surfaces that allow the entire gear assembly to move. It is attached to aircraft structure in such a way that the gear can pivot from the vertical position required for landing and taxi to the stowed position used during flight. *[Figure 13-32]*

While in the vertical gear down position, the trunnion is free to swing or pivot. Alone, it cannot support the aircraft without collapsing. A drag brace is used to restrain against the pivot action built into the trunnion attachment. The upper end of the two-piece drag brace is attached to the aircraft structure and the lower end to the strut. A hinge near the middle of the brace allows the brace to fold and permits the gear to retract. For ground operation, the drag brace is straightened over

center to a stop and locked into position so the gear remains rigid. *[Figure 13-33]* The function of a drag brace on some aircraft is performed by the hydraulic cylinder used to raise and lower the gear. Cylinder internal hydraulic locks replace the over-center action of the drag brace for support during ground maneuvers.

Small Aircraft Retraction Systems

As the speed of a light aircraft increases, there reaches a point where the parasite drag created by the landing gear in the wind is greater than the induced drag caused by the added weight of a retractable landing gear system. Thus, many light aircraft have retractable landing gear. There are many unique designs. The simplest contains a lever in the flight deck mechanically linked to the gear. Through mechanical advantage, the pilot extends and retracts the landing gear by operating the lever. Use of a roller chain, sprockets, and a hand crank to decrease the required force is common.

Electrically operated landing gear systems are also found on light aircraft. An all-electric system uses an electric motor and gear reduction to move the gear. The rotary motion of the motor is converted to linear motion to actuate the gear. This is possible only with the relatively lightweight gear found on smaller aircraft. An all-electric gear retraction system is illustrated in *Figure 13-34.*

A more common use of electricity in gear retraction systems is that of an electric/hydraulic system found in many Cessna and Piper aircraft. This is also known as a power pack system. A small lightweight hydraulic power pack contains several components required in a hydraulic system. These include the reservoir, a reversible electric motor-driven hydraulic pump, a filter, high-and-low pressure control valves, a thermal relief valve, and a shuttle valve. Some power packs incorporate an emergency hand pump. A hydraulic actuator for each gear is

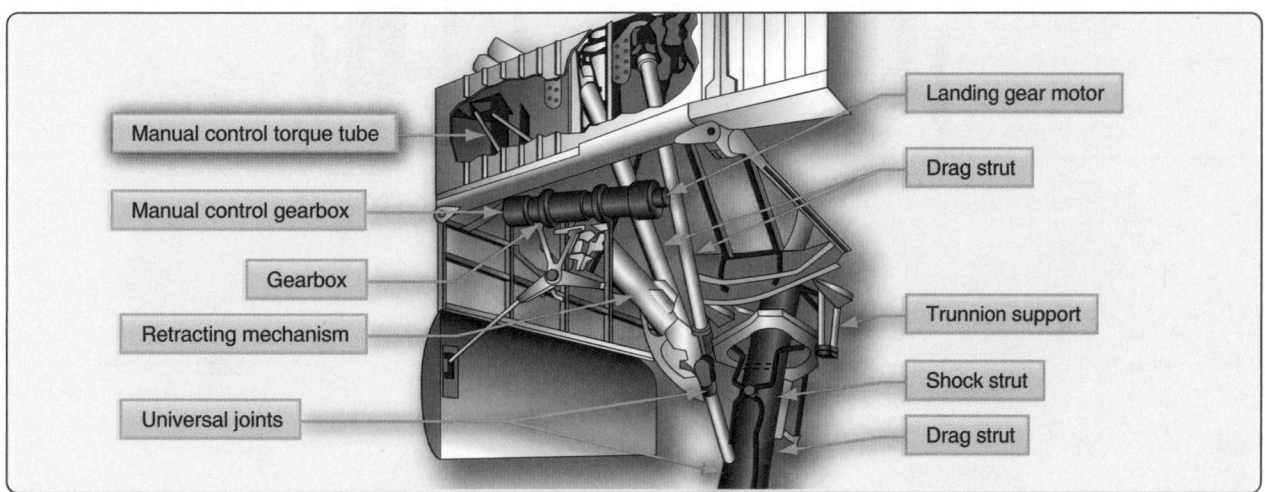

Figure 13-34. *A geared electric motor landing gear retraction system.*

driven to extend or retract the gear by fluid from the power pack. *Figure 13-35* illustrates a power pack system while gear is being lowered. *Figure 13-36* shows the same system while the gear is being raised.

When the flight deck gear selection handle is put in the gear-down position, a switch is made that turns on the electric motor in the power pack. The motor turns in the direction to rotate the hydraulic gear pump so that it pumps fluid to the gear-down side of the actuating cylinders. Pump pressure moves the spring-loaded shuttle valve to the left to allow

fluid to reach all three actuators. Restrictors are used in the nose wheel actuator inlet and outlet ports to slow down the motion of this lighter gear. While hydraulic fluid is pumped to extend the gear, fluid from the upside of the actuators returns to the reservoir through the gear-up check valve. When the gear reach the down and locked position, pressure builds in the gear-down line from the pump and the low-pressure control valve unseats to return the fluid to the reservoir. Electric limit switches turn off the pump when all three gears are down and locked.

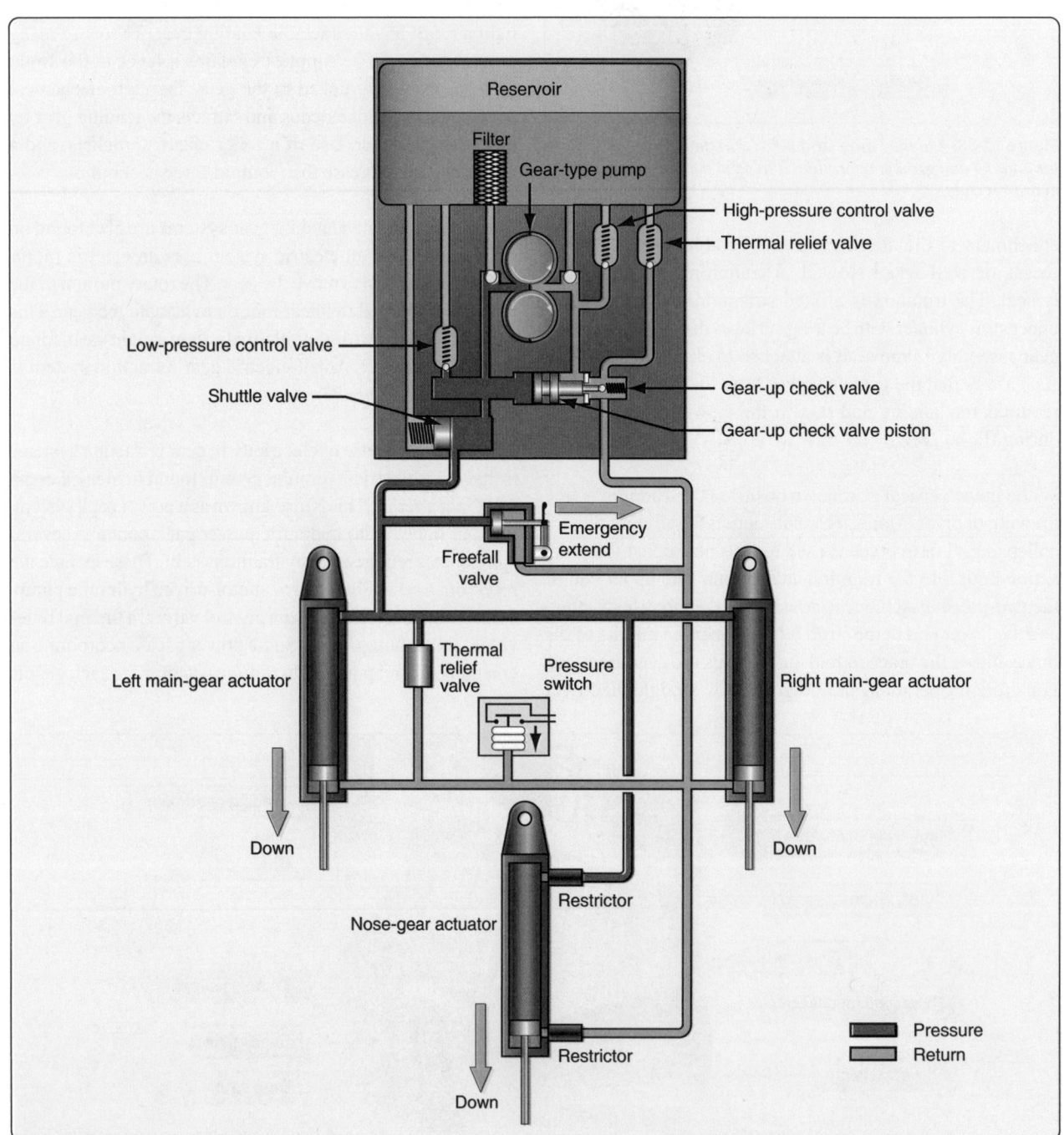

Figure 13-35. *A popular light aircraft gear retraction system that uses a hydraulic power pack in the gear down condition.*

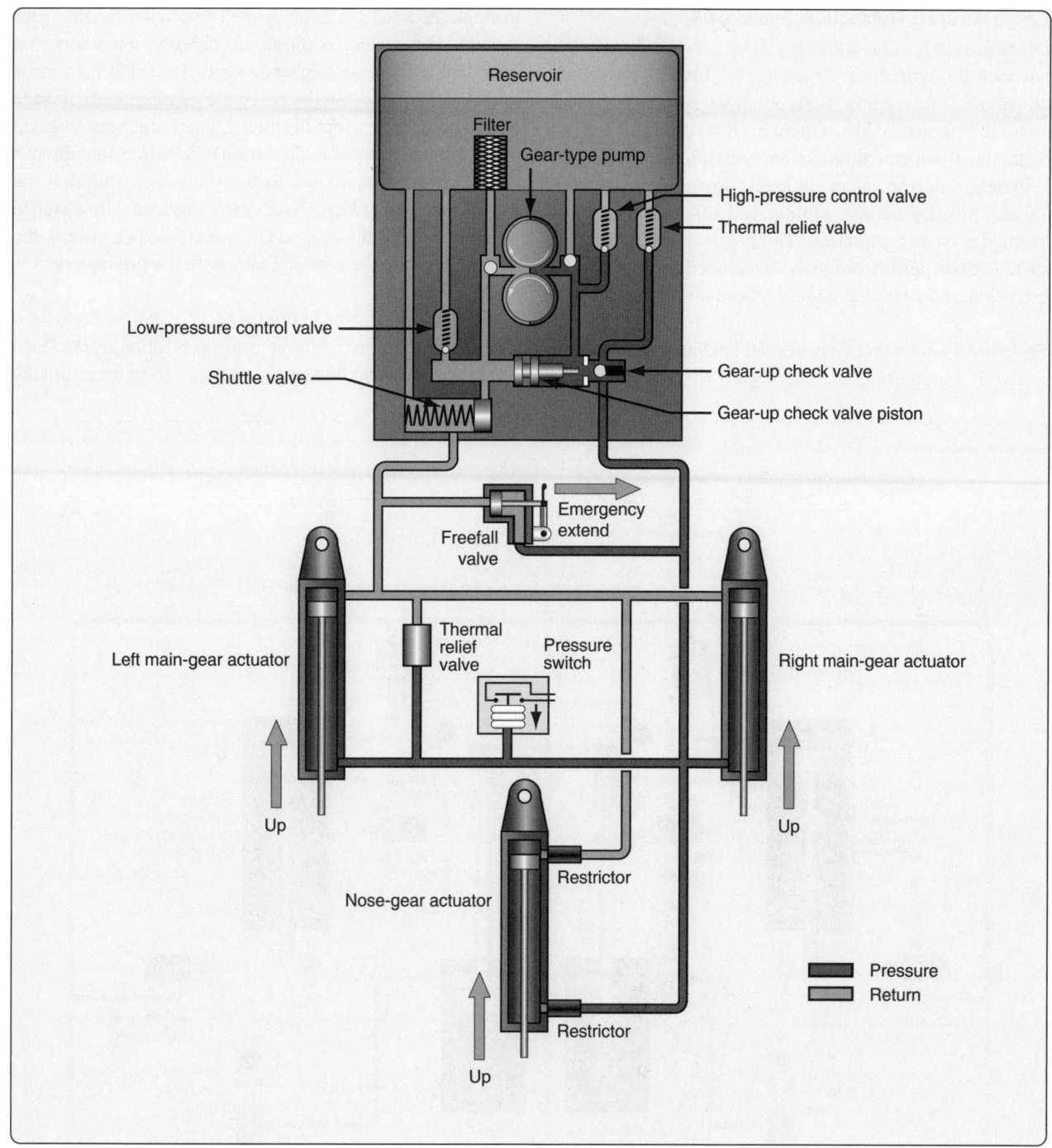

Figure 13-36. *A hydraulic power pack gear retraction system in the gear up condition.*

To raise the gear, the flight deck gear handle is moved to the gear-up position. This sends current to the electric motor, which drives the hydraulic gear pump in the opposite direction causing fluid to be pumped to the gear-up side of the actuators. In this direction, pump inlet fluid flows through the filter. Fluid from the pump flows thought the gear-up check valve to the gear-up sides of the actuating cylinders. As the cylinders begin to move, the pistons release the mechanical down locks that hold the gear rigid for ground operations. Fluid from the gear-down side of the actuators returns to the reservoir through the shuttle valve. When the three gears are fully retracted, pressure builds in the system, and a pressure switch is opened that cuts power to the electric pump motor. The gear are held in the retracted position with hydraulic pressure. If pressure declines, the pressure switch closes to run the pump and raise the pressure until the pressure switch opens again.

Large Aircraft Retraction Systems

Large aircraft retraction systems are nearly always powered by hydraulics. Typically, the hydraulic pump is driven off of the engine accessory drive. Auxiliary electric hydraulic pumps are also common. Other devices used in a hydraulically-operated retraction system include actuating cylinders, selector valves, uplocks, downlocks, sequence valves, priority valves, tubing, and other conventional hydraulic system components. These units are interconnected so that they permit properly sequenced retraction and extension of the landing gear and the landing gear doors.

The correct operation of any aircraft landing gear retraction system is extremely important. *Figure 13-37* illustrates an example of a simple large aircraft hydraulic landing gear system. The system is on an aircraft that has doors that open before the gear is extended and close after the gear is retracted. The nose gear doors operate via mechanical linkage and do not require hydraulic power. There are many gear and gear door arrangements on various aircraft. Some aircraft have gear doors that close to fair the wheel well after the gear is extended. Others have doors mechanically attached to the outside of the gear so that when it stows inward, the door stows with the gear and fairs with the fuselage skin.

In the system illustrated in *Figure 13-37*, when the flight deck gear selector is moved to the gear-up position, it positions a selector valve to allow pump pressure from the hydraulic

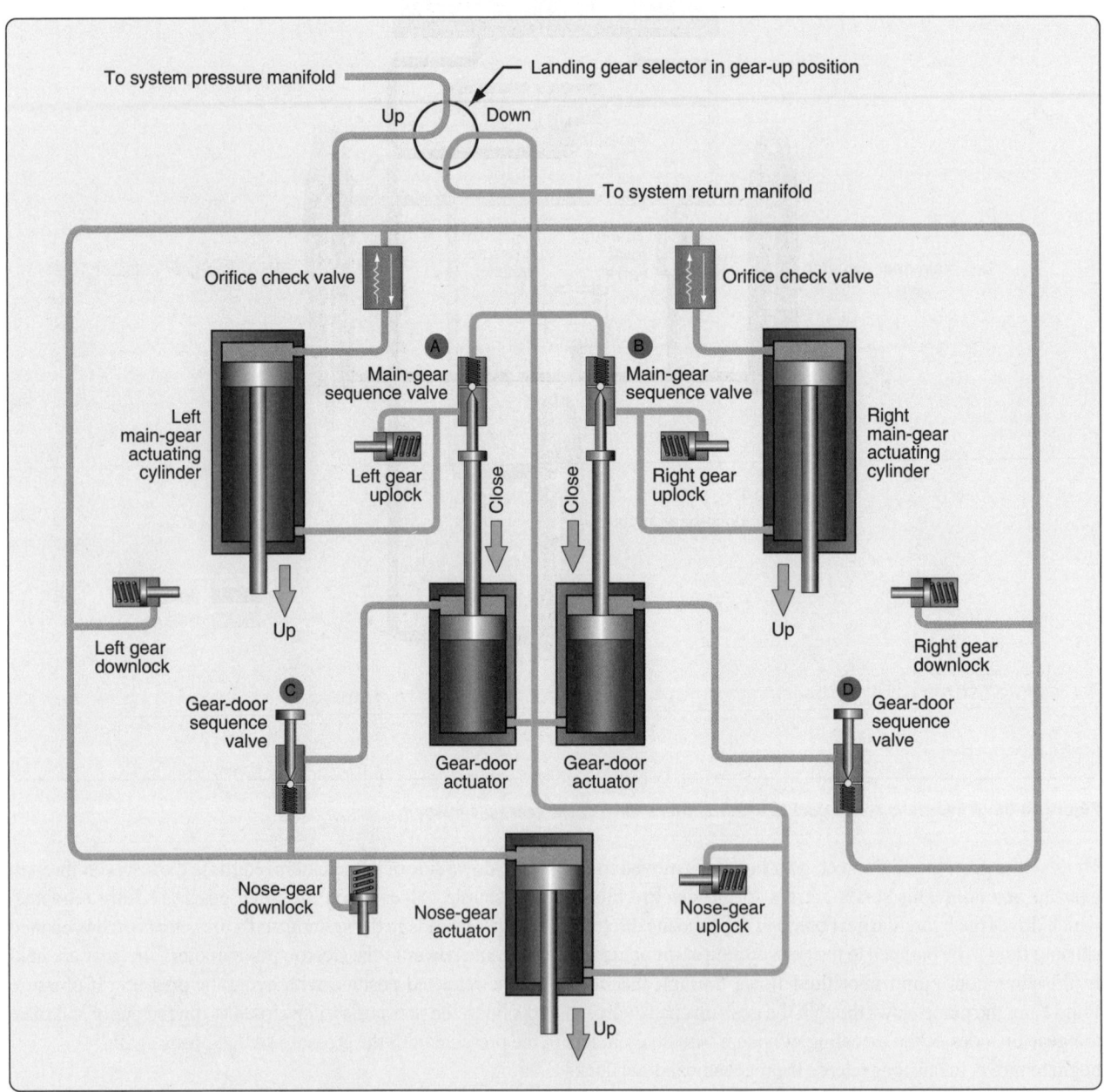

Figure 13-37. *A simple large aircraft hydraulic gear retraction system.*

system manifold to access eight different components. The three downlocks are pressurized and unlocked so the gear can be retracted. At the same time, the actuator cylinder on each gear also receives pressurized fluid to the gear-up side of the piston through an unrestricted orifice check valve. This drives the gear into the wheel well. Two sequence valves (C and D) also receive fluid pressure. Gear door operation must be controlled so that it occurs after the gear is stowed. The sequence valves are closed and delay flow to the door actuators. When the gear cylinders are fully retracted, they mechanically contact the sequence valve plungers that open the valves and allow fluid to flow into the close side of the door actuator cylinders. This closes the doors. Sequence valves A and B act as check valves during retraction. They allow fluid to flow one way from the gear-down side of the main gear cylinders back into the hydraulic system return manifold through the selector valve.

To lower the gear, the selector is put in the gear-down position. Pressurized hydraulic fluid flows from the hydraulic manifold to the nose gear uplock, which unlocks the nose gear. Fluid flows to the gear-down side of the nose gear actuator and extends it. Fluid also flows to the open side of the main gear door actuators. As the doors open, sequence valves A and B block fluid from unlocking the main gear uplocks and prevent fluid from reaching the down side of the main gear actuators. When the doors are fully open, the door actuator engages the plungers of both sequence valves to open the valves. The main gear uplocks, then receives fluid pressure and unlock. The main gear cylinder actuators receive fluid on the down side through the open sequence valves to extend the gear. Fluid from each main gear cylinder up-side flows to the hydraulic system return manifold through restrictors in the orifice check valves. The restrictors slow the extension of the gear to prevent impact damage.

There are numerous hydraulic landing gear retraction system designs. Priority valves are sometimes used instead of mechanically operated sequence valves. This controls some gear component activation timing via hydraulic pressure. Particulars of any gear system are found in the aircraft maintenance manual. The aircraft technician must be thoroughly familiar with the operation and maintenance requirements of this crucial system.

Emergency Extension Systems

The emergency extension system lowers the landing gear if the main power system fails. There are numerous ways in which this is done depending on the size and complexity of the aircraft. Some aircraft have an emergency release handle in the flight deck that is connected through a mechanical linkage to the gear uplocks. When the handle is operated, it releases the uplocks and allows the gear to free-fall to the

extended position under the force created by gravity acting upon the gear. Other aircraft use a non-mechanical back-up, such as pneumatic power, to unlatch the gear.

The popular small aircraft retraction system shown in *Figures 13-35* and *13-36* uses a free-fall valve for emergency gear extension. Activated from the flight deck, when the free-fall valve is opened, hydraulic fluid is allowed to flow from the gear-up side of the actuators to the gear-down side of the actuators, independent of the power pack. Pressure holding the gear up is relieved, and the gear extends due to its weight. Air moving past the gear aids in the extension and helps push the gear into the down-and-locked position.

Large and high-performance aircraft are equipped with redundant hydraulic systems. This makes emergency extension less common since a different source of hydraulic power can be selected if the gear does not function normally. If the gear still fails to extend, some sort of unlatching device is used to release the uplocks and allow the gear to free fall. *[Figure 13-38]*

In some small aircraft, the design configuration makes emergency extension of the gear by gravity and air loads alone impossible or impractical. Force of some kind must therefore be applied. Manual extension systems, wherein the pilot mechanically cranks the gear into position, are common. Consult the aircraft maintenance manual for all emergency landing gear extension system descriptions of operation, performance standards, and emergency extension tests as required.

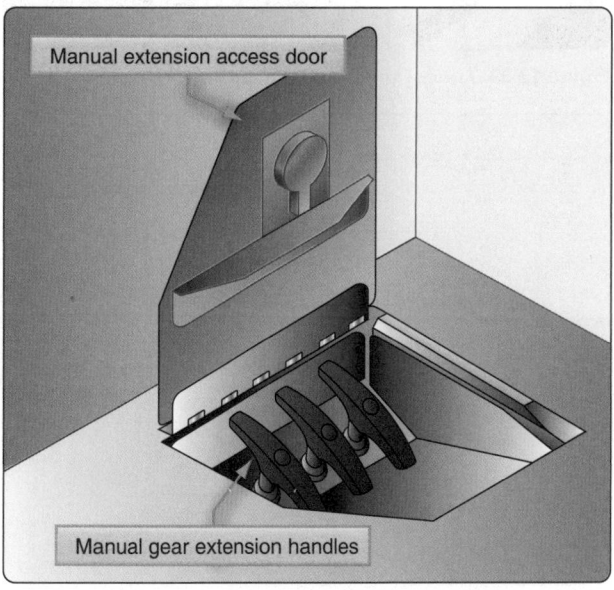

Figure 13-38. *These emergency gear extension handles in a Boeing 737 are located under a floor panel on the flight deck. Each handle releases the gear uplock via a cable system so the gear can freefall into the extended position.*

Landing Gear Safety Devices

There are numerous landing gear safety devices. The most common are those that prevent the gear from retracting or collapsing while on the ground. Gear indicators are another safety device. They are used to communicate to the pilot the position status of each individual landing gear at any time. A further safety device is the nose wheel centering device mentioned previously in this chapter.

Safety Switch

A landing gear squat switch, or safety switch, is found on most aircraft. This is a switch positioned to open and close depending on the extension or compression of the main landing gear strut. [Figure 13-39] The squat switch is wired into any number of system operating circuits. One circuit prevents the gear from being retracted while the aircraft is

Figure 13-39. *Typical landing gear squat switches.*

on the ground. There are different ways to achieve this lock-out. A solenoid that extends a shaft to physically disable the gear position selector is one such method found on many aircraft. When the landing gear is compressed, the squat safety switch is open, and the center shaft of the solenoid protrudes a hardened lock-pin through the landing gear control handle so that it cannot be moved to the up position. At takeoff, the landing gear strut extends. The safety switch closes and allows current to flow in the safety circuit. The solenoid energizes and retracts the lock-pin from the selector handle. This permits the gear to be raised. [Figure 13-40]

The use of proximity sensors for gear position safety switches is common in high-performance aircraft. An electromagnetic sensor returns a different voltage to a gear logic unit depending on the proximity of a conductive target to the switch. No physical contact is made. When the gear is in the designed position, the metallic target is close to the inductor in the sensor which reduces the return voltage. This type of sensing is especially useful in the landing gear environment where switches with moving parts can become contaminated with dirt and moisture from runways and taxi ways. The technician is required to ensure that sensor targets are installed the correct distance away from the sensor. Go–no go gauges are often used to set the distance. [Figure 13-41]

Ground Locks

Ground locks are commonly used on aircraft landing gear as extra insurance that the landing gear will remain down and locked while the aircraft is on the ground. They are external devices that are placed in the retraction mechanism to prevent its movement. A ground lock can be as simple as a pin placed into the pre-drilled holes of gear components that keep the gear from collapsing. Another commonly used ground lock clamps onto the exposed piston of the gear retraction cylinder

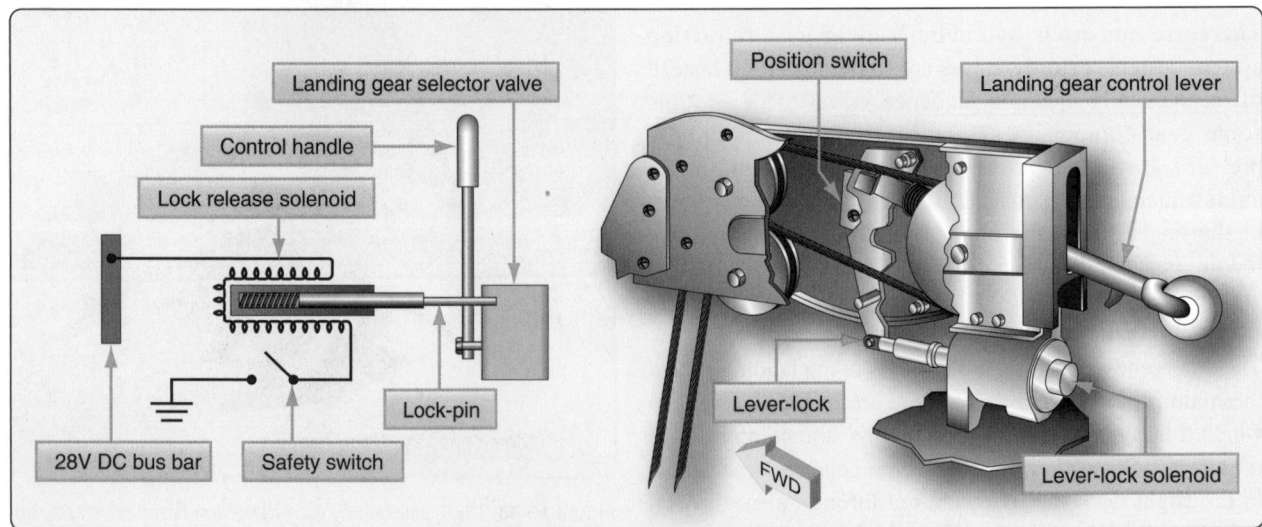

Figure 13-40. *A landing gear safety circuit with solenoid that locks the control handle and selector valve from being able to move into the gear up position when the aircraft is on the ground. The safety switch, or squat switch, is located on the aircraft landing gear.*

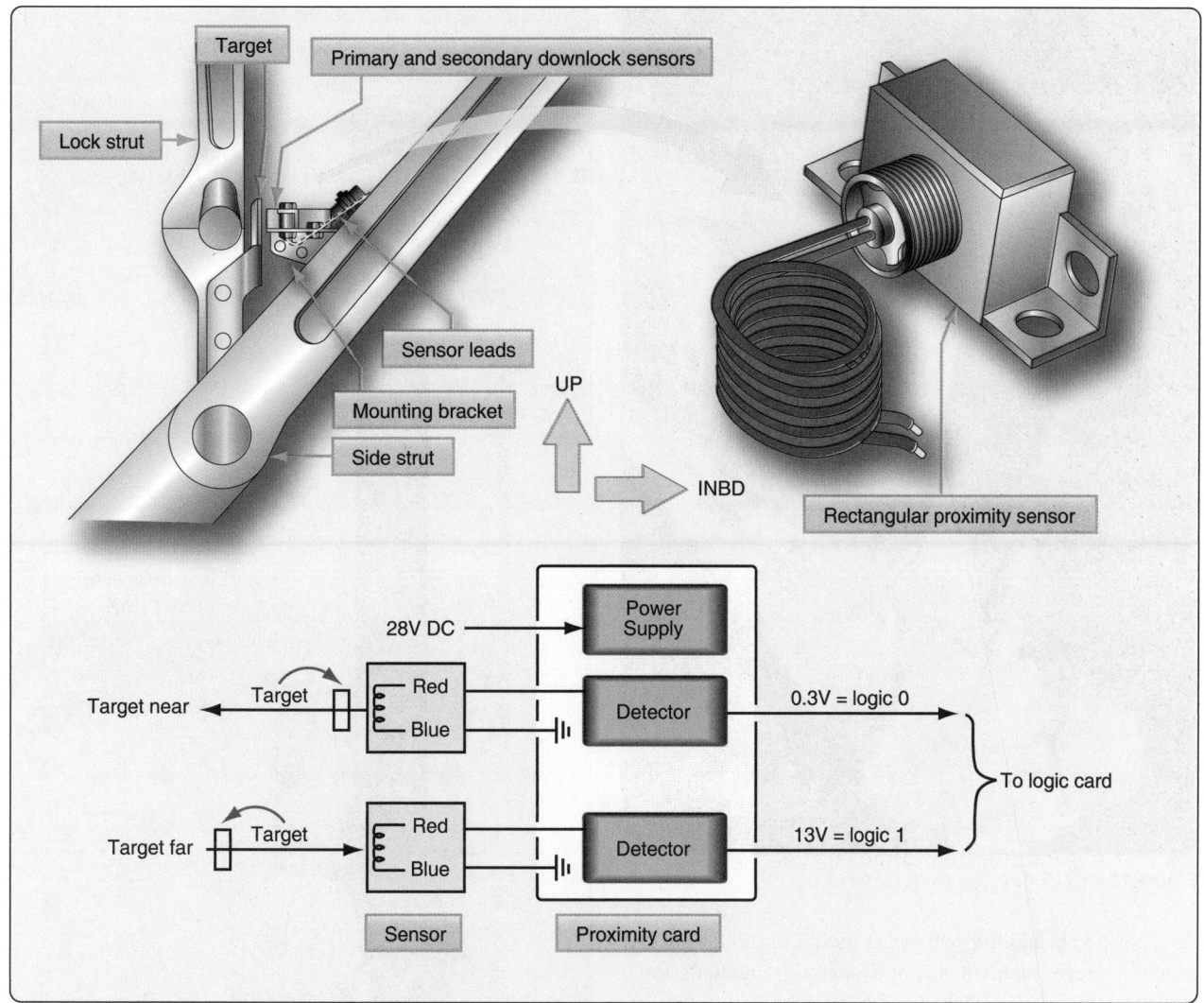

Figure 13-41. *Proximity sensors are used instead of contact switches on many landing gear.*

that prevents it from retracting. All ground locks should have red streamers attached to them, so they are visible and removed before flight. Ground locks are typically carried in the aircraft and put into place by the flight crew during the post landing walk-around. *[Figure 13-42]*

Landing Gear Position Indicators

To provide a visual indication of landing gear position, indicators are installed in the flight deck or flight compartment. Gear warning devices are incorporated on all retractable gear aircraft and usually consist of a horn or some other aural device and a red warning light. The horn blows and the light comes on when one or more throttles are retarded and the landing gear is in any position other than down and locked.

Position indicators are typically:

    Off = gear up and locked
    Red = unsafe

    Green = down and locked

There is usually an amber in transit light to indicate a gear that is in transition.

Landing gear position indicators are located on the instrument panel adjacent to the gear selector handle. They are used to inform the pilot of gear position status. There are many arrangements for gear indication. Usually, there is a dedicated light for each gear. The most common display for the landing gear being down and locked is an illuminated green light. Three green lights means it is safe to land. All lights out typically indicates that the gear is up and locked, or there may be gear up indicator lights. Gear in transit lights are used on some aircraft as are barber pole displays when a gear is not up or down and locked. Blinking indicator lights also indicate gear in transit. Some manufacturer's use a gear disagree annunciation when the landing gear is not in the same position as the selector. Many aircraft monitor gear

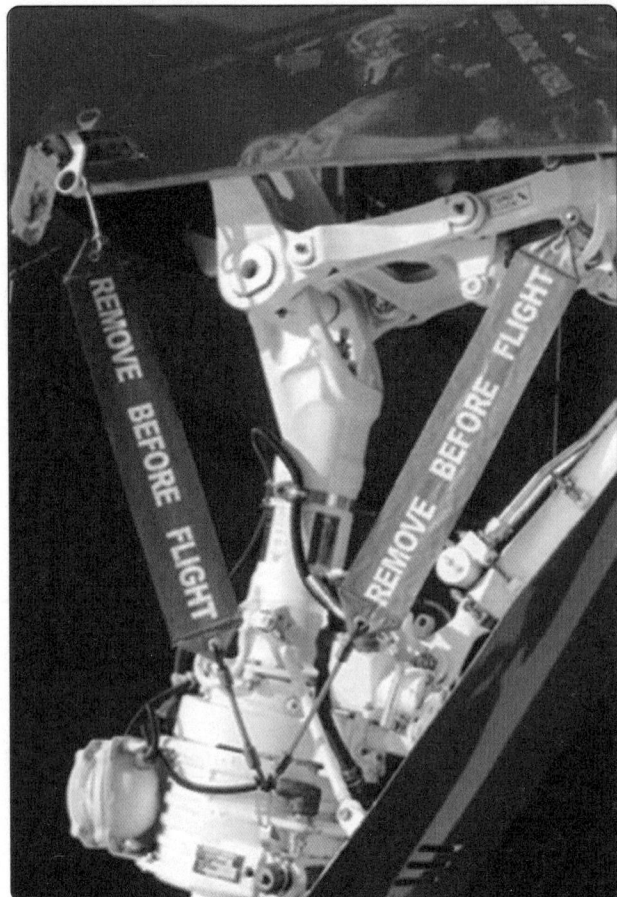

Figure 13-42. *Gear pin ground lock devices.*

door position in addition to the gear itself. Consult the aircraft manufacturer's maintenance and operating manuals for a complete description of the landing gear indication system. [*Figure 13-43 and Figure 13-44*]

Nose Wheel Centering

Since most aircraft have steerable nose wheel gear assemblies for taxiing, a means for aligning the nose gear before retraction is needed. Centering cams built into the shock strut structure accomplish this. An upper cam is free to mate into a lower cam recess when the gear is fully extended. This aligns the gear for retraction. When weight returns to the wheels after landing, the shock strut is compressed, and the centering cams separate allowing the lower shock strut (piston) to rotate in the upper strut cylinder. This rotation is controlled to steer the aircraft. [*Figure 13-45*] Small aircraft sometimes incorporate an external roller or guide pin on the strut. As the strut is folded into the wheel well during retraction, the roller or guide pin engages a ramp or track mounted to the wheel well structure. The ramp/track guides the roller or pin in such a manner that the nose wheel is straightened as it enters the wheel well.

Figure 13-43. *Landing gear selector panels with position indicator lights. The Boeing 737 panel illuminates red lights above the green lights when the gear is in transit.*

Landing Gear System Maintenance

The moving parts and dirty environment of the landing gear make this an area of regular maintenance. Because of the stresses and pressures acting on the landing gear, inspection, servicing, and other maintenance becomes a continuous process. The most important job in the maintenance of the aircraft landing gear system is thorough accurate inspections. To properly perform inspections, all surfaces should be cleaned to ensure that no trouble spots are undetected.

Periodically, it is necessary to inspect shock struts, trunnion and brace assemblies and bearings, shimmy dampers, wheels,

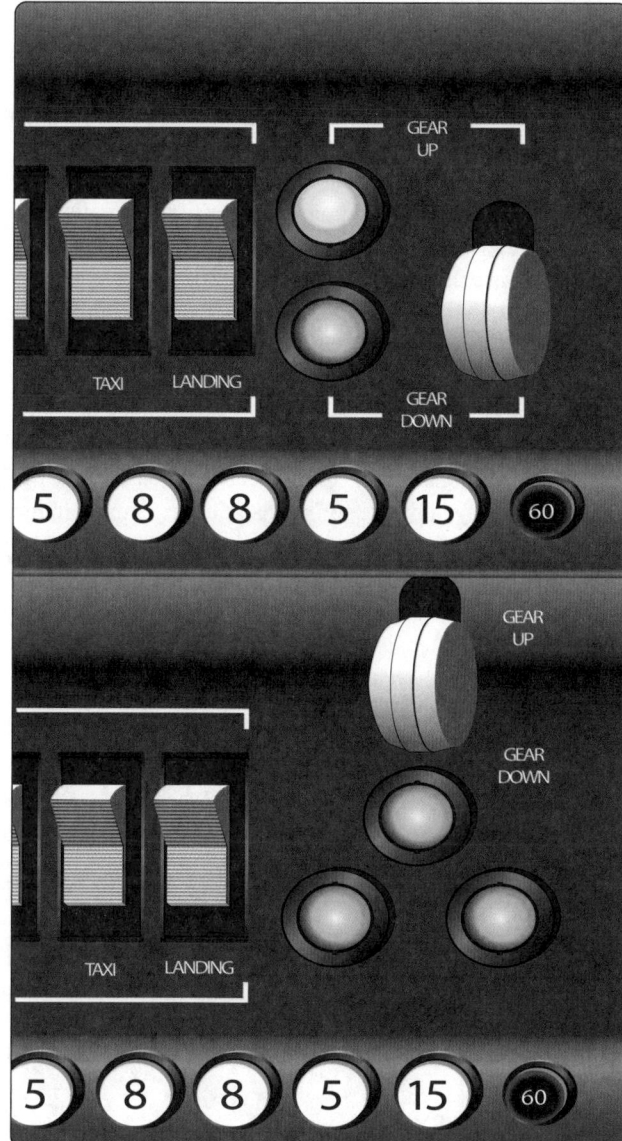

Figure 13-44. *Typical landing gear indicator lights and lever as found on small aircraft.*

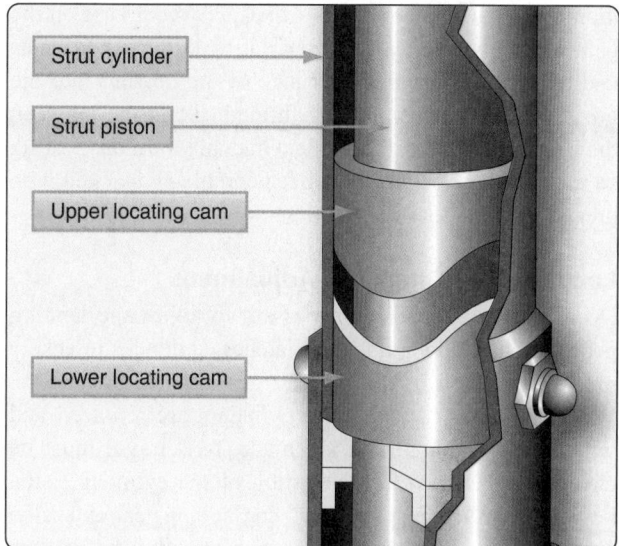

Figure 13-45. *A cutaway view of a nose gear internal centering cam.*

wheel bearings, tires, and brakes. Landing gear position indicators, lights, and warning horns must also be checked for proper operation. During all inspections and visits to the wheel wells, ensure all ground safety locks are installed.

Other landing gear inspection items include checking emergency control handles and systems for proper position and condition. Inspect landing gear wheels for cleanliness, corrosion, and cracks. Check wheel tie bolts for looseness. Examine anti-skid wiring for deterioration. Check tires for wear, cuts, deterioration, presence of grease or oil, alignment of slippage marks, and proper inflation. Inspect landing gear mechanism for condition, operation, and proper adjustment. Lubricate the landing gear, including the nose wheel steering. Check steering system cables for wear, broken strands,

alignment, and safetying. Inspect landing gear shock struts for such conditions as cracks, corrosion, breaks, and security. Where applicable, check brake clearances and wear.

Various types of lubricant are required to lubricate points of friction and wear on landing gear. Specific products to be used are given by the manufacturer in the maintenance manual. Lubrication may be accomplished by hand or with a grease gun. Follow manufacturer's instructions. Before applying grease to a pressure grease fitting, be sure the fitting is wiped clean of dirt and debris, as well as old hardened grease. Dust and sand mixed with grease produce a very destructive abrasive compound. Wipe off all excess grease while greasing the gear. The piston rods of all exposed strut cylinders and actuating cylinders should be clean at all times.

Periodically, wheel bearings must be removed, cleaned, inspected, and lubricated. When cleaning a wheel bearing, use the recommended cleaning solvent. Do not use gasoline or jet fuel. Dry the bearing by directing a blast of dry air between the rollers. Do not direct the air so that it spins the bearing as without lubrication, this could cause the bearing to fly apart resulting in injury. When inspecting the bearing, check for defects that would render it unserviceable, such as cracks, flaking, broken bearing surfaces, roughness due to impact pressure or surface wear, corrosion or pitting, discoloration from excessive heat, cracked or broken bearing cages, and scored or loose bearing cups or cones that would affect proper seating on the axle or wheel. If any discrepancies are found, replace the bearing with a serviceable unit. Bearings should be lubricated immediately after cleaning and inspection to prevent corrosion.

To lubricate a tapered roller bearing, use a bearing lubrication tool or place a small amount of the approved grease on the palm of the hand. Grasp the bearing with the other hands and press the larger diameter side of the bearing into the grease to force it completely through the space between the bearing rollers and the cone. Gradually turn the bearing so that all of the rollers have been completely packed with grease. *[Figure 13-46]*

Landing Gear Rigging & Adjustment

Occasionally, it becomes necessary to adjust the landing gear switches, doors, linkages, latches, and locks to ensure proper operation of the landing gear system and doors. When landing gear actuating cylinders are replaced and when length adjustments are made, over-travel must be checked. Over-travel is the action of the cylinder piston beyond the movement necessary for landing gear extension and retraction. The additional action operates the landing gear latch mechanisms.

A wide variety of aircraft types and landing gear system designs result in procedures for rigging and adjustment that vary from aircraft to aircraft. Uplock and downlock clearances, linkage adjustments, limit switch adjustments, and other adjustments must be confirmed by the technician in the manufacturer's maintenance data before taking action. The following examples of various adjustments are given to convey concepts, rather than actual procedures for any particular aircraft.

Adjusting Landing Gear Latches

The adjustment of various latches is a primary concern to the aircraft technician. Latches are generally used in landing gear systems to hold the gear up or down and/or to hold the gear doors open or closed. Despite numerous variations, all latches are designed to do the same thing. They must operate automatically at the proper time, and they must hold the unit in the desired position. A typical landing gear door latch is

examined below. Many gear up latches operate similarly. Clearances and dimensional measurements of rollers, shafts, bushings, pins, bolts, etc., are common.

On this particular aircraft, the landing gear door is held closed by two latches. To have the door locked securely, both latches must grip and hold the door tightly against the aircraft structure. The principle components of each latch mechanism are shown in *Figure 13-47*. They are a hydraulic latch cylinder, a latch hook, a spring-loaded crank-and-lever linkage with sector, and the latch roller.

When hydraulic pressure is applied, the cylinder operates the linkage to engage (or disengage) the hook with (or from) the roller on the gear door. In the gear-down sequence, the hook is disengaged by the spring load on the linkage. In the gear-up sequence, when the closing door is in contact with the latch hook, the cylinder operates the linkage to engage the latch hook with the door roller. Cables on the landing gear emergency extension system are connected to the sector to permit emergency release of the latch rollers. An uplock switch is installed on, and actuated by, each latch to provide a gear up indication in the flight deck.

With the gear up and the door latched, inspect the latch roller for proper clearance as shown in *Figure 13-48A*. On this installation, the required clearance is ⅛ ± 3/32-inch. If the roller is not within tolerance, it may be adjusted by loosening its mounting bolts and raising or lowering the latch roller support. This is accomplished via the elongated holes and serrated locking surfaces of the latch roller support and serrated plate. *[Figure 13-48B]*

Gear Door Clearances

Landing gear doors have specific allowable clearances between the doors and the aircraft structure that must be maintained. Adjustments are typically made at the hinge

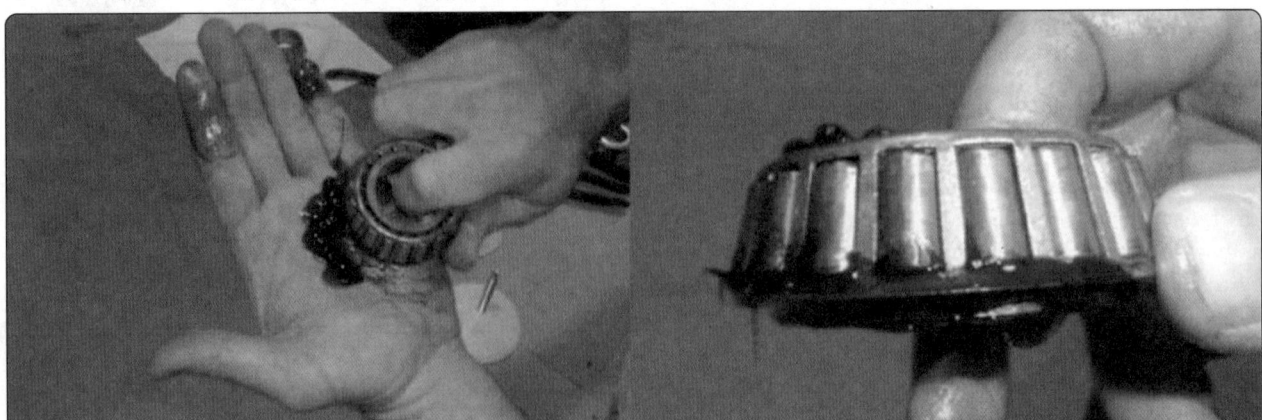

Figure 13-46. *Packing grease into a clean, dry bearing can be done by hand in the absence of a bearing grease tool. Press the bearing into the grease on the palm of the hand until it passes completely through the gap between the rollers and the inner race all the way around the bearing.*

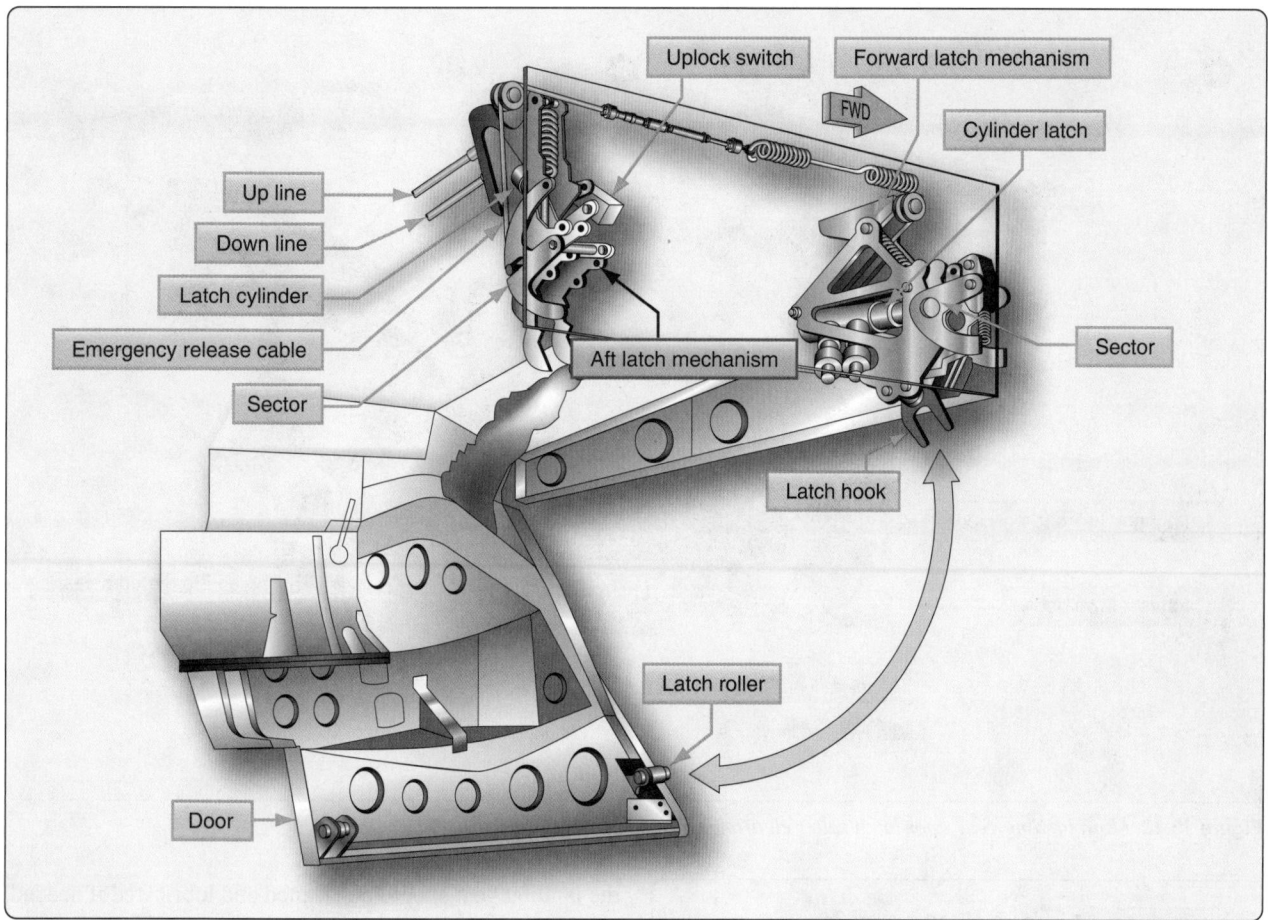

Figure 13-47. *An example of a main landing gear door latch mechanism.*

installations or to the connecting links that support and move the door. On some installations, door hinges are adjusted by placing a serrated hinge with an elongated mounting hole in the proper position in a hinge support fitting. Using serrated washers, the mounting bolt is torqued to hold the position. *Figure 13-49* illustrates this type of mounting, which allows linear adjustments via the elongated hole.

The distance landing gear doors open or close may depend upon the length of the door linkage. Rod end adjustments are common to fit the door. Adjustments to door stops are also a possibility. The manufacturer's maintenance manual specifies the length of the linkages and gives procedure for adjusting the stops. Follow all specified procedures that are accomplished with the aircraft on jacks and the gear retracted. Doors that are too tight can cause structural damage. Doors that are too loose catch wind in flight, which could cause wear and potential failure, as well as parasite drag.

Drag & Side Brace Adjustment

Each landing gear has specific adjustments and tolerances per the manufacturer that permit the gear to function as intended. A common geometry used to lock a landing gear in the down position involves a collapsible side brace that is extended and held in an over-center position through the use of a locking link. Springs and actuators may also contribute to the motion of the linkage. Adjustments and tests are needed to ensure proper operation.

Figure 13-50 illustrates a landing gear on a small aircraft with such a side brace. It consists of an upper and lower link hinged at the center that permits the brace to jackknife during retraction of the gear. The upper end pivots on a trunnion attached to structure in the wheel well overhead. The lower end is attached to the shock strut. A locking link is incorporated between the upper end of the shock strut and the lower drag link. It is adjustable to provide the correct amount of over-center travel of the side brace links. This locks the gear securely in the down position to prevent collapse of the gear.

To adjust the over-center position of the side brace locking link, the aircraft must be placed on jacks. With the landing gear in the down position, the lock link end fitting is adjusted so that the side brace links are held firmly over-center. When the gear is held inboard six inches from the down and locked position and then released, the gear must free fall into the locked down position.

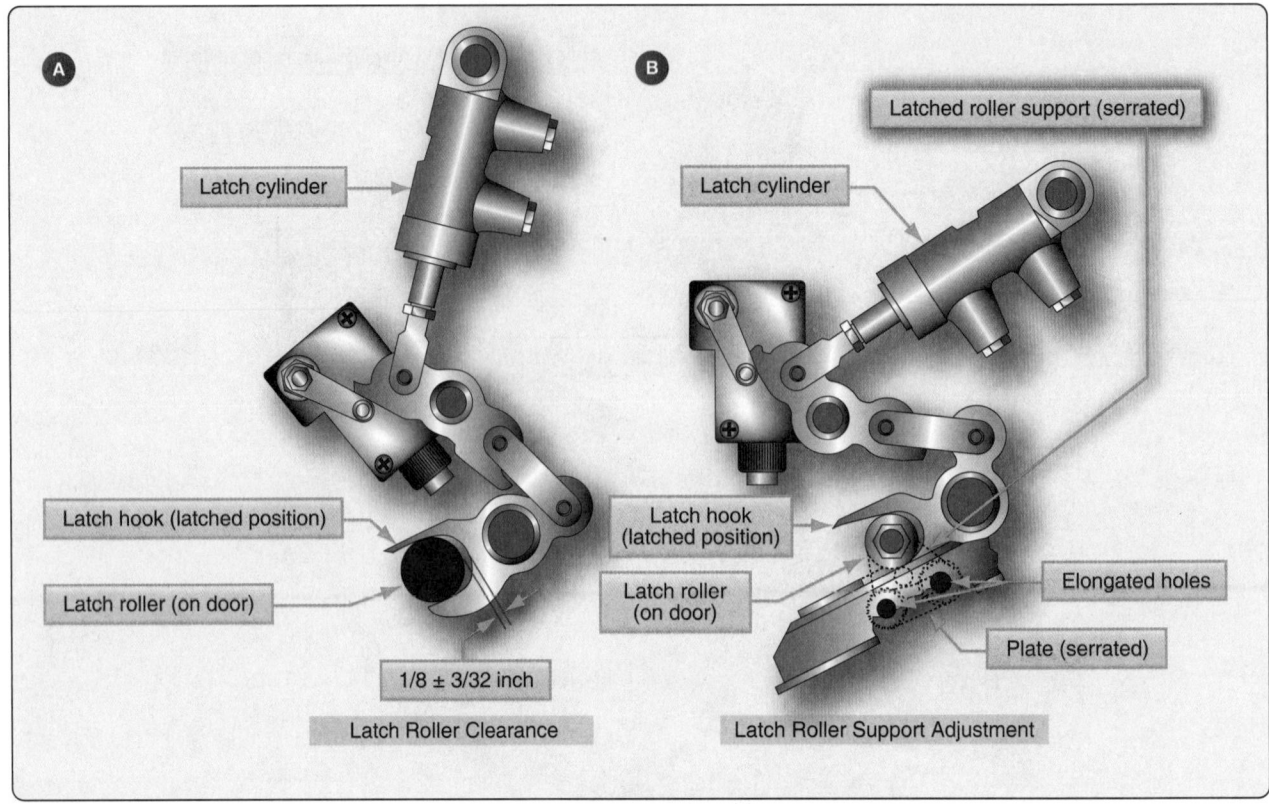

Figure 13-48. *Main landing gear door latch roller clearance measurement and adjustment.*

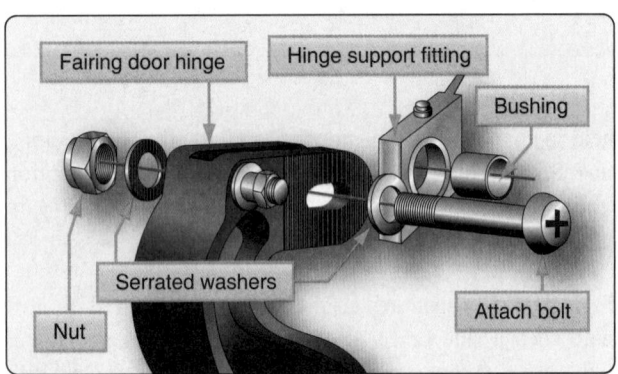

Figure 13-49. *An adjustable door hinge installation for setting door clearance.*

In addition to the amount the side brace links are adjusted to travel over center, down lock spring tension must also be checked. This is accomplished with a spring scale. The tension on this particular gear is between 40 and 60 pounds. Check the manufacturer's maintenance data for each aircraft to ensure correct tensions exist and proper adjustments are made.

Landing Gear Retraction Test

The proper functioning of a landing gear system and components can be checked by performing a landing gear retraction test. This is also known as swinging the gear. The aircraft is properly supported on jacks for this check, and the landing gear should be cleaned and lubricated if needed. The gear is then raised and lowered as though the aircraft were in flight while a close visual inspection is performed. All parts of the system should be observed for security and proper operation. The emergency back-up extension system should be checked whenever swinging the gear.

Retraction tests are performed at various times, such as during annual inspection. Any time a landing gear component is replaced that could affect the correct functioning of the landing gear system, a retraction test should follow when adjustments to landing gear linkages or components that affect gear system performance are made. It may be necessary to swing the gear after a hard or overweight landing. It is also common to swing the gear while attempting to locate a malfunction within the system. For all required retraction tests and the specific inspection points to check, consult the manufacturer's maintenance manual for the aircraft in question as each landing gear system is unique.

The following is a list of general inspection items to be performed while swinging the gear:

1. Check the landing gear for proper extension and retraction.

2. Check all switches, lights, and warning devices for proper operation.

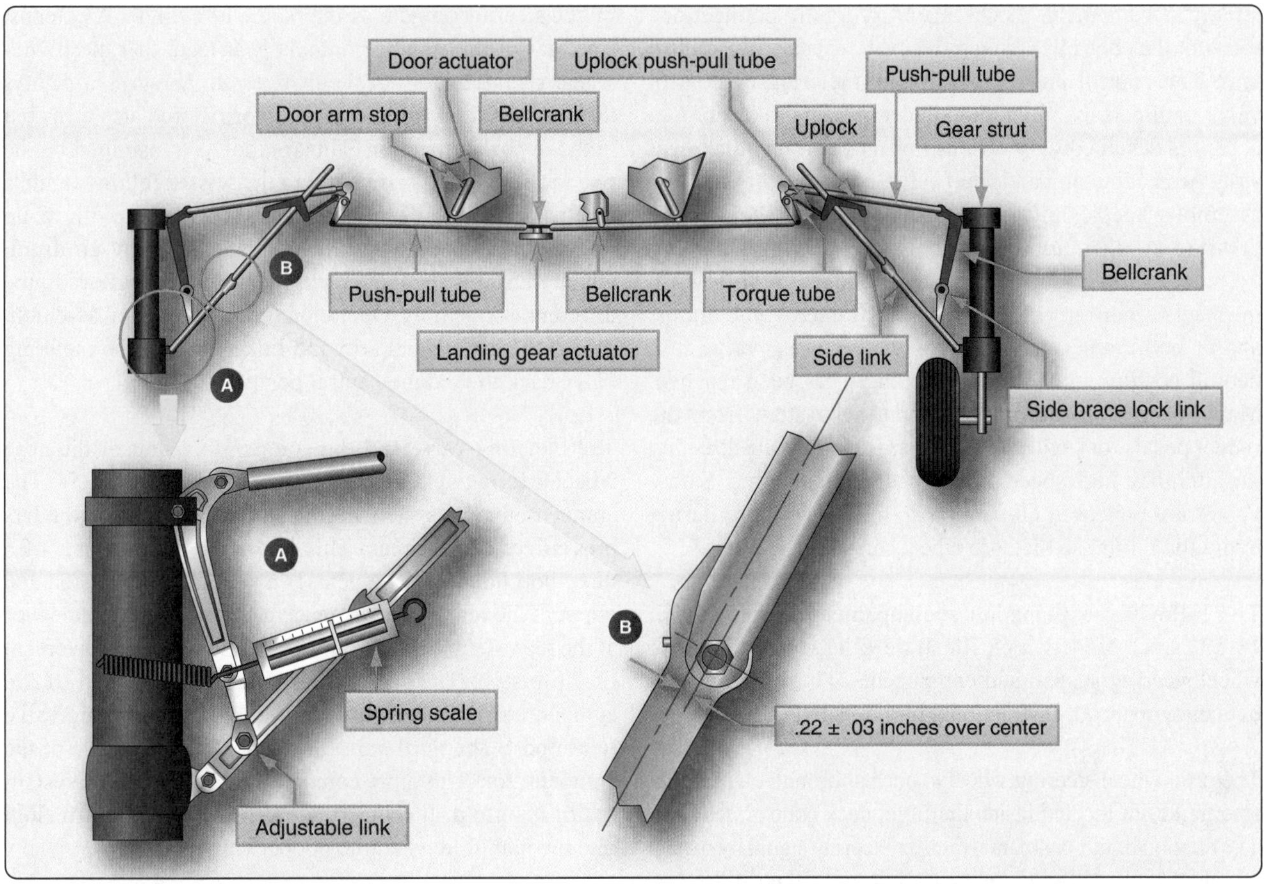

Figure 13-50. *Over-center adjustments on a small aircraft main gear.*

3. Check the landing gear doors for clearance and freedom from binding.

4. Check landing gear linkage for proper operation, adjustment, and general condition.

5. Check the alternate/emergency extension or retraction systems for proper operation.

6. Investigate any unusual sounds, such as those caused by rubbing, binding, chafing, or vibration.

Nose Wheel Steering Systems

The nose wheel on most aircraft is steerable from the flight deck via a nose wheel steering system. This allows the aircraft to be directed during ground operation. A few simple aircraft have nose wheel assemblies that caster. Such aircraft are steered during taxi by differential braking.

Small Aircraft

Most small aircraft have steering capabilities through the use of a simple system of mechanical linkages connected to the rudder pedals. Push-pull tubes are connected to pedal horns on the lower strut cylinder. As the pedals are depressed, the movement is transferred to the strut piston axle and wheel assembly which rotates to the left or right. *[Figure 13-51]*

Figure 13-51. *Nose wheel steering on a light aircraft often uses a push-pull rod system connected to the rudder pedals.*

Large Aircraft

Due to their mass and the need for positive control, large aircraft utilize a power source for nose wheel steering. Hydraulic power predominates. There are many different designs for large aircraft nose steering systems. Most share similar characteristics and components. Control of the steering is from the flight deck through the use of a small wheel, tiller, or joystick typically mounted on the left side wall. Switching the system on and off is possible on some aircraft. Mechanical, electrical, or hydraulic connections transmit the controller input movement to a steering control unit. The control unit is a hydraulic metering or control valve. It directs hydraulic fluid under pressure to one or two actuators designed with various linkages to rotate the lower strut. An accumulator and relief valve, or similar pressurizing assembly, keeps fluid in the actuators and system under pressure at all times. This permits the steering actuating cylinders to also act as shimmy dampers. A follow-up mechanism consists of various gears, cables, rods, drums, and/or bell-crank, etc. It returns the metering valve to a neutral position once the steering angle has been reached. Many systems incorporate an input subsystem from the rudder pedals for small degrees of turns made while directing the aircraft at high speed during takeoff and landing. Safety valves are typical in all systems to relieve pressure during hydraulic failure so the nose wheel can swivel.

The following explanation accompanies Figures 13-52, 13-53, and 13-54, which illustrate a large aircraft nose wheel steering system and components. These figures and explanation are for instructional purposes only.

The nose wheel steering wheel connects through a shaft to a steering drum located inside the flight deck control pedestal. The rotation of this drum transmits the steering signal by means of cables and pulleys to the control drum of the differential assembly. Movement of the differential assembly is transmitted by the differential link to the metering valve assembly where it moves the selector valve to the selected position. This provides the hydraulic power for turning the nose gear.

As shown in *Figure 13-53,* pressure from the aircraft hydraulic system is directed through the open safety shutoff valve into a line leading to the metering valve. The metering valve then routes the pressurized fluid out of port A, through the right turn alternating line, and into steering cylinder A. This is a one-port cylinder and pressure forces the piston to begin extension. Since the rod of this piston connects to the nose steering spindle on the nose gear shock strut which pivots at point X, the extension of the piston turns the steering spindle gradually toward the right. As the nose wheel turns, fluid is forced out of steering cylinder B through the left turn alternating line and into port B of the metering valve. The metering valve directs this return fluid into a compensator that routes the fluid into the aircraft hydraulic system return manifold.

As described, hydraulic pressure starts the nose gear turning. However, the gear should not be turned too far. The nose gear steering system contains devices to stop the gear at the selected angle of turn and hold it there. This is accomplished with follow-up linkage. As stated, the nose gear is turned

by the steering spindle as the piston of cylinder A extends. The rear of the spindle contains gear teeth that mesh with a gear on the bottom of the orifice rod. *[Figure 13-52]* As the nose gear and spindle turn, the orifice rod also turns but in the opposite direction. This rotation is transmitted by the two sections of the orifice rod to the scissor follow-up links located at the top of the nose gear strut. As the follow-up links return, they rotate the connected follow-up drum, which transmits the movement by cables and pulleys to the differential assembly. Operation of the differential assembly causes the differential arm and links to move the metering valve back toward the neutral position.

The metering valve and the compensator unit of the nose wheel steering system are illustrated in *Figure 13-54*. The compensator unit system keeps fluid in the steering cylinders pressurized at all times. This hydraulic unit consists of a three-port housing that encloses a spring-loaded piston and poppet. The left port is an air vent that prevents trapped air at the rear of the piston from interfering with the movement of the piston. The second port located at the top of the compensator connects through a line to the metering valve return port. The third port is located at the right side of the compensator. This port connects to the hydraulic system return manifold. It routes the steering system return fluid into the manifold when the poppet valve is open.

The compensator poppet opens when pressure acting on the piston becomes high enough to compress the spring. In this system, 100 psi is required. Therefore, fluid in the metering valve return line is contained under that pressure. The 100-psi pressure also exists throughout the metering valve and back through the cylinder return lines. This pressurizes the steering cylinders at all times and permits them to function as shimmy dampers.

Shimmy Dampers

Torque links attached from the stationary upper cylinder of a nose wheel strut to the bottom moveable cylinder or piston of the strut are not sufficient to prevent most nose gear from the tendency to oscillate rapidly, or shimmy, at certain speeds. This vibration must be controlled through the use of a shimmy damper. A shimmy damper controls nose wheel shimmy through hydraulic damping. The damper can be built integrally within the nose gear, but most often it is an external unit attached between the upper and lower shock struts. It is active during all phases of ground operation while permitting the nose gear steering system to function normally.

Steering Damper

As mentioned above, large aircraft with hydraulic steering hold pressure in the steering cylinders to provide the required damping. This is known as steering damping. Some older

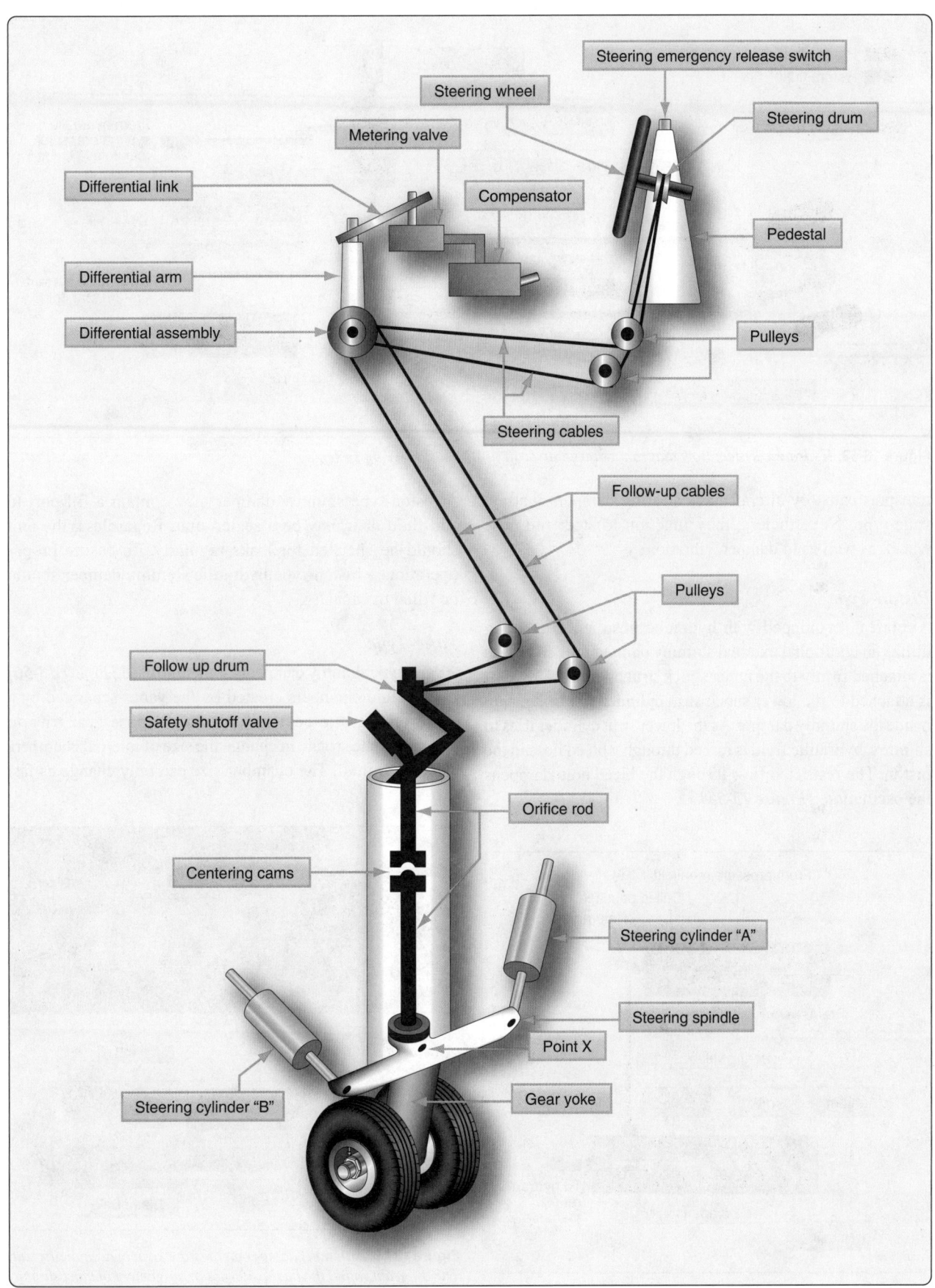

Figure 13-52. *Example of a large aircraft hydraulic nose wheel steering system with hydraulic and mechanical units.*

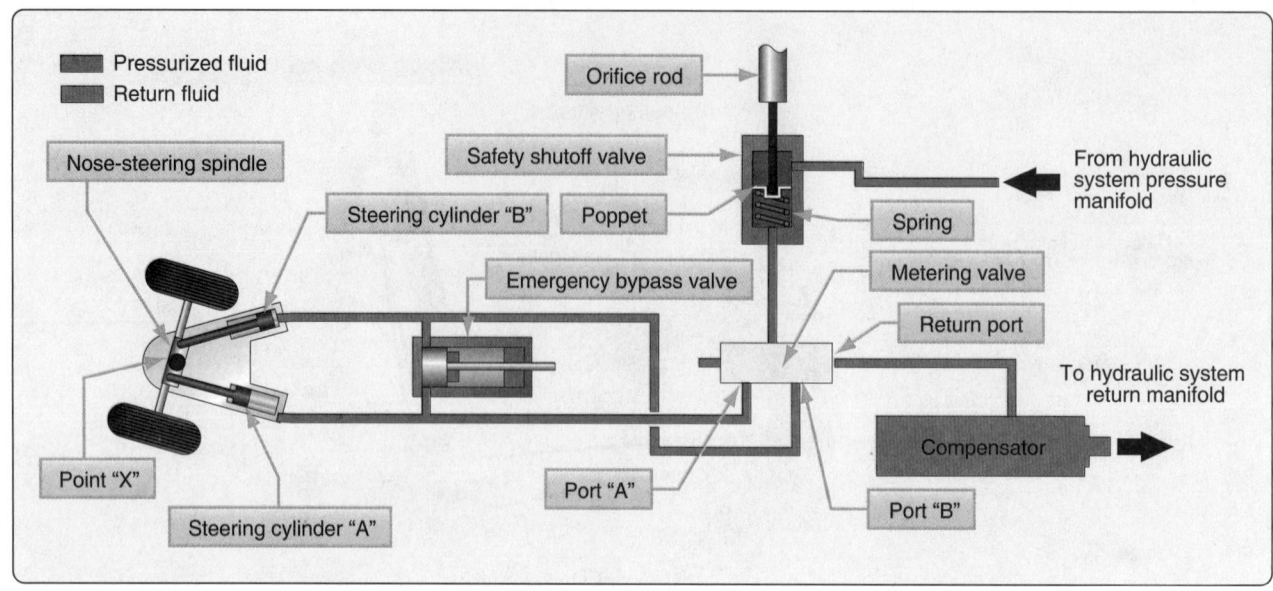

Figure 13-53. *Hydraulic system flow diagram of large aircraft nose wheel steering system.*

transport category aircraft have steering dampers that are vane-type. Nevertheless, they function to steer the nose wheel, as well as to dampen vibration.

Piston-Type

Aircraft not equipped with hydraulic nose wheel steering utilize an additional external shimmy damper unit. The case is attached firmly to the upper shock strut cylinder. The shaft is attached to the lower shock strut cylinder and to a piston inside the shimmy damper. As the lower strut cylinder tries to shimmy, hydraulic fluid is forced through a bleed hole in the piston. The restricted flow through the bleed hole dampens the oscillation. *[Figure 13-55]*

A piston-type shimmy damper may contain a fill port to add fluid or it may be a sealed unit. Regardless, the unit should be checked for leaks regularly. To ensure proper operation, a piston-type hydraulic shimmy damper should be filled to capacity.

Vane-Type

A vane-type shimmy damper is sometime used.*[Figure 13-56]* It uses fluid chambers created by the vanes separated by a valve orifice in a center shaft. As the nose gear tries to oscillate, vanes rotate to change the size of internal chambers filled with fluid. The chamber size can only change as fast

Figure 13-54. *Hydraulic system flow diagram of large aircraft nose wheel steering system.*

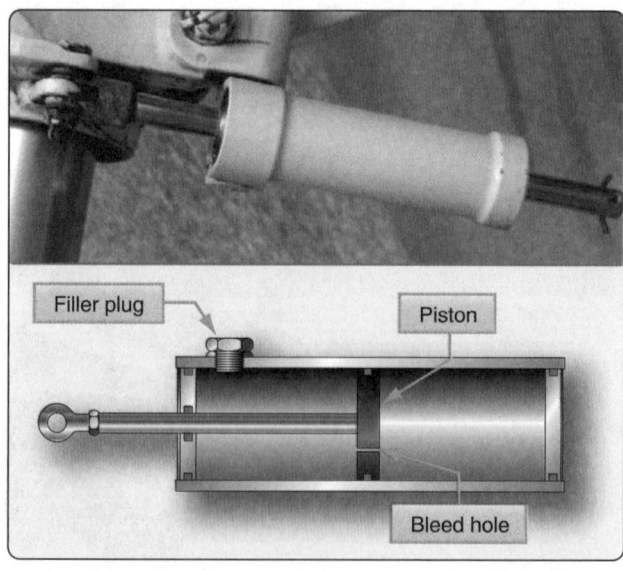

Figure 13-55. *A shimmy damper on the nose strut of a small aircraft. The diagram shows the basic internal arrangement of most shimmy dampers. The damper in the photo is essentially the same except the piston shaft extends through both ends of the damper cylinder body.*

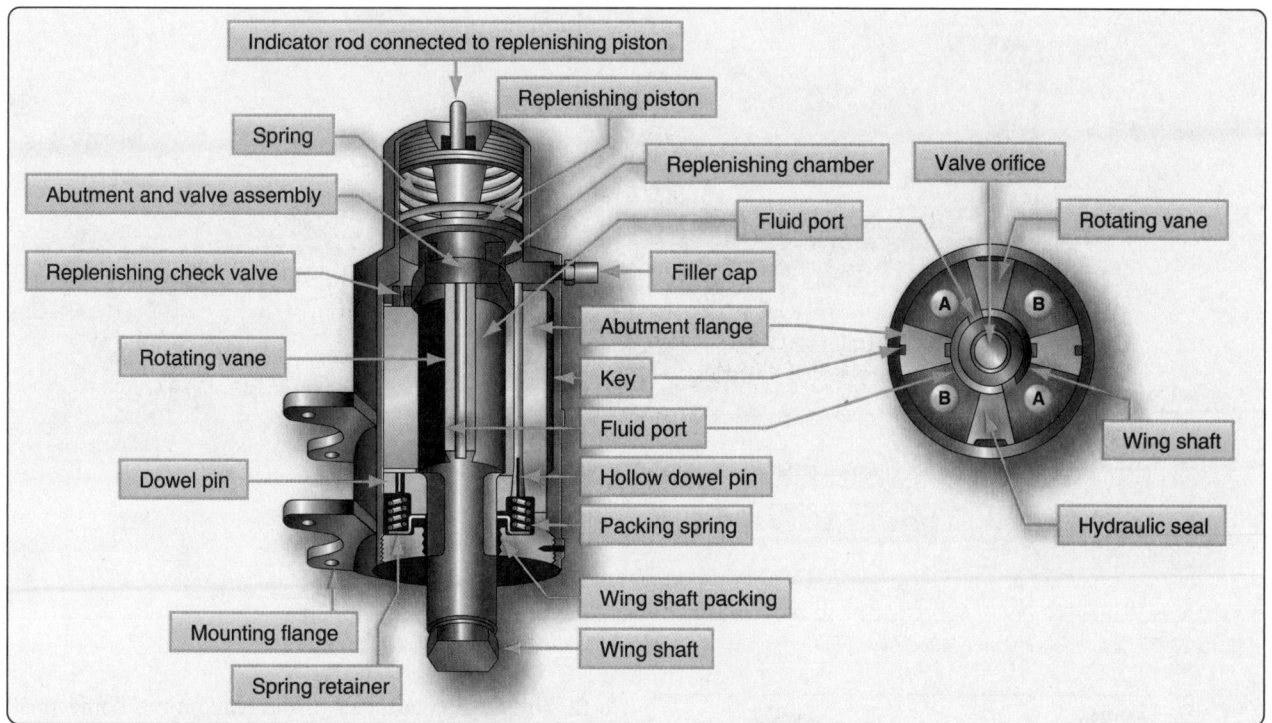

Figure 13-56. *A typical vane-type shimmy damper.*

as the fluid can be forced through the orifice. Thus, the gear oscillation is dissipated by the rate of fluid flow. An internal spring-loaded replenishing reservoir keeps pressurized fluid in the working chambers and thermal compensation of the orifice size is included. As with the piston type shimmy damper, the vane-type damper should be inspected for leaks and kept serviced. A fluid level indicator protrudes from the reservoir end of the unit.

Non-Hydraulic Shimmy Damper

Non-hydraulic shimmy dampers are currently certified for many aircraft. They look and fit similar to piston-type shimmy dampers but contain no fluid inside. In place of the metal piston, a rubber piston presses out against the inner diameter of the damper housing when the shimmy motion is received through the shaft. The rubber piston rides on a very thin film of grease and the rubbing action between the piston and the housing provides the damping.

This is known as surface-effect damping. The materials used to construct this type of shimmy damper provide a long service life without the need to ever add fluid to the unit. *[Figure 13-57]*

Aircraft Wheels

Aircraft wheels are an important component of a landing gear system. With tires mounted upon them, they support the entire weight of the aircraft during taxi, takeoff, and landing. The typical aircraft wheel is lightweight, strong, and made

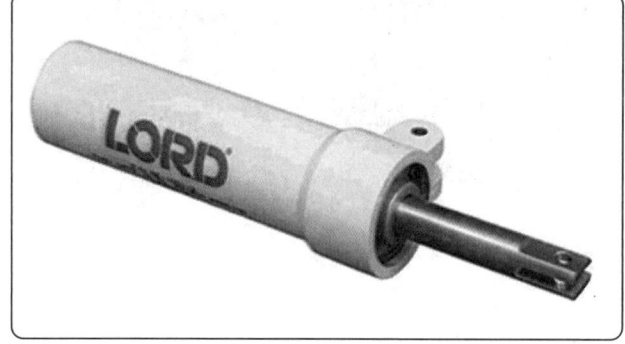

Figure 13-57. *A non-hydraulic shimmy damper uses a rubber piston with lubricant that dampens via motion against the inner diameter of the unit housing.*

from aluminum alloy. Some magnesium alloy wheels also exist. Early aircraft wheels were of single piece construction, much the same as the modern automobile wheel. As aircraft tires were improved for the purpose they serve, they were made stiffer to better absorb the forces of landing without blowing out or separating from the rim. Stretching such a tire over a single piece wheel rim was not possible. A two-piece wheel was developed. Early two-piece aircraft wheels were essentially one-piece wheels with a removable rim to allow mounting access for the tire. These are still found on older aircraft. *[Figure 13-58]* Later, wheels with two nearly symmetrical halves were developed. Nearly all modern aircraft wheels are of this two-piece construction. *[Figures 13-59* and *13-60]*

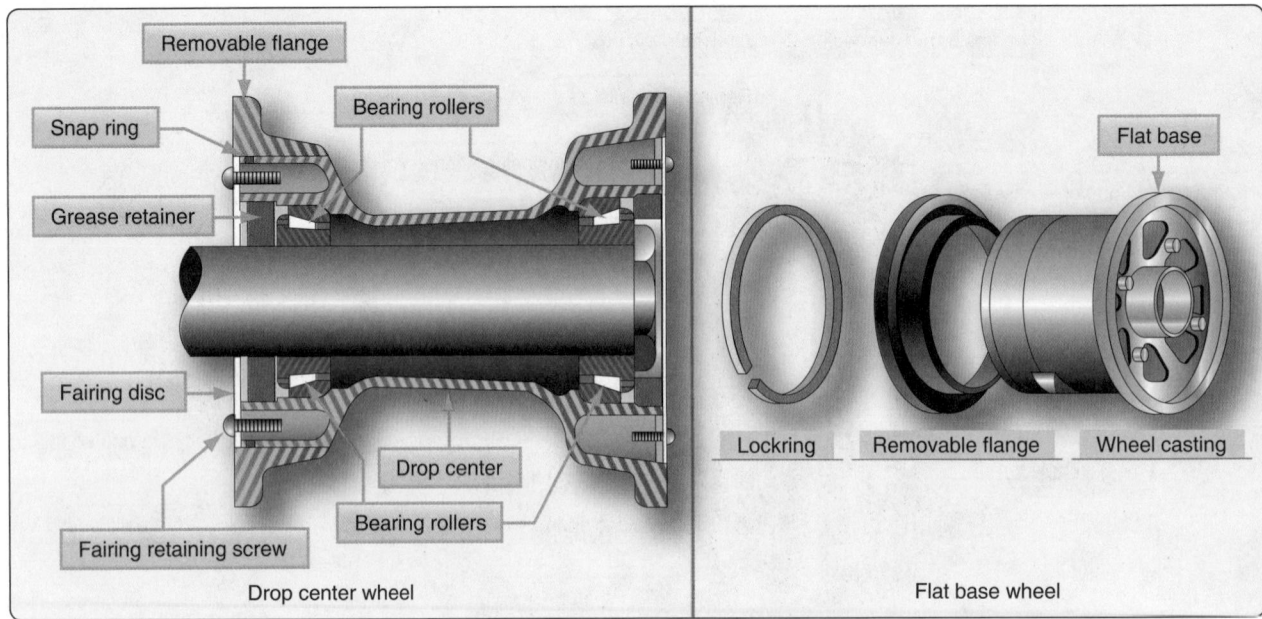

Figure 13-58. *Removable flange wheels found on older aircraft are either drop center or flat base types.*

Figure 13-59. *Two-piece split-wheel aircraft wheels found on modern light aircraft.*

Wheel Construction

The typical modern two-piece aircraft wheel is cast or forged from aluminum or magnesium alloy. The halves are bolted together and contain a groove at the mating surface for an o-ring, which seals the rim since most modern aircraft utilize tubeless tires. The bead seat area of a wheel is where the tire actually contacts the wheel. It is the critical area that accepts the significant tensile loads from the tire during landing. To strengthen this area during manufacturing, the bead seat area is typically rolled to prestress it with a compressive stress load.

Inboard Wheel Half

Wheel halves are not identical. The primary reason for this is that the inboard wheel half must have a means for accepting and driving the rotor(s) of the aircraft brakes that are mounted on both main wheels. Tangs on the rotor are fitted into steel reinforced keyways on many wheels. Other wheels have steel keys bolted to the inner wheel halves. These are made

to fit slots in the perimeter of the brake rotor. Some small aircraft wheels have provisions for bolting the brake rotor to the inner wheel half. Regardless, the inner wheel half is distinguishable from the outer wheel half by its brake mounting feature. *[Figure 13-61]*

Both wheel halves contain a bearing cavity formed into the center that accepts the polished steel bearing cup, tapered roller bearing, and grease retainer of a typical wheel bearing set-up. A groove may also be machined to accept a retaining clip to hold the bearing assembly in place when the wheel assembly is removed. The wheel bearings are a very important part of the wheel assembly and are discussed in a later section of this chapter.

The inner wheel half of a wheel used on a high-performance aircraft is likely to have one or more thermal plugs. *[Figure 13-62]* During heavy braking, temperatures can become so great that tire temperature and pressure rise to a level resulting in explosion of the wheel and tire assembly. The thermal plug core is filled with a low melting point alloy. Before tire and wheel temperatures reach the point of explosion, the core melts and deflates the tire. The tire must be removed from service, and the wheel must be inspected in accordance with the wheel manufacturer's instructions before return to service if a thermal plug melts. Adjacent wheel assemblies should also be inspected for signs of damage. A heat shield is commonly installed under the inserts designed to engage the brake rotor to assist in protecting the wheel and tire assembly from overheating.

An overinflation safety plug may also be installed in the inner wheel half. This is designed to rupture and release all

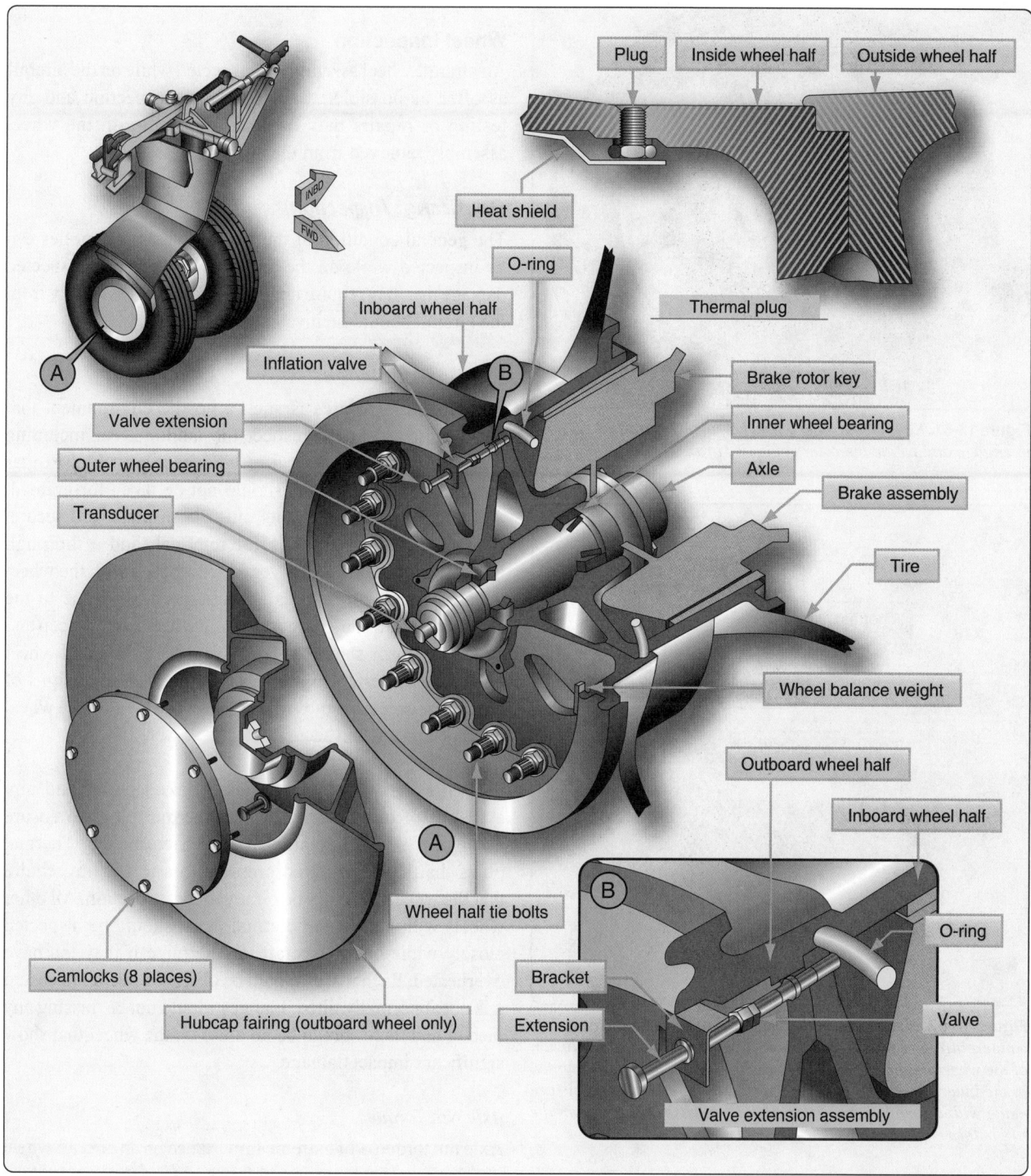

Figure 13-60. *Features of a two piece aircraft wheel found on a modern airliner.*

of the air in the tire should it be over inflated. The fill valve is also often installed in the inner wheel half with the stem extending through holes in the outer wheel half to permit access for inflation and deflation.

Outboard Wheel Half

The outboard wheel half bolts to the inboard wheel half to make up the wheel assembly upon which the tire is mounted.

The center boss is constructed to receive a bearing cup and bearing assembly as it does on the inboard wheel half. The outer bearing and end of the axle is capped to prevent contaminants from entering this area. Aircraft with anti-skid brake systems typically mount the wheel-spin transducer here. It is sealed and may also serve as a hub cap. The 737 outer wheel half illustrated in *Figure 13-60* also has a hub cap fairing over the entire wheel half. This is to fair it with

Figure 13-61. *Keys on the inner wheel half of an aircraft wheel used to engage and rotate the rotors of a disc brake.*

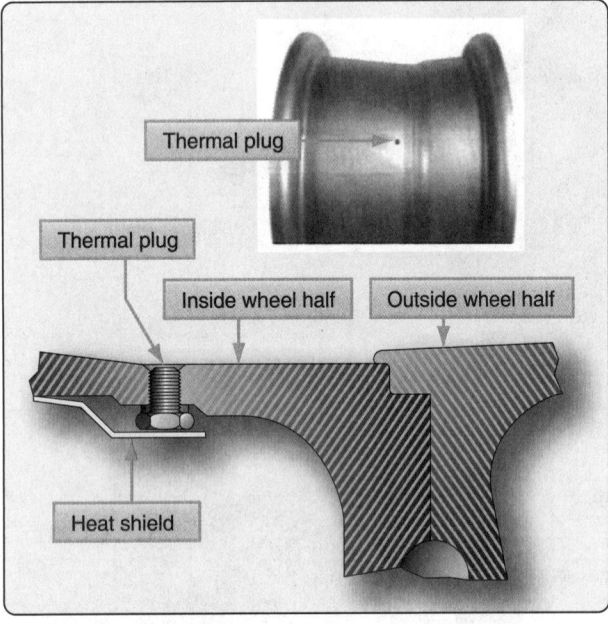

Figure 13-62. *Heavy use of the aircraft brakes can cause tire air temperature and pressure to rise to a level resulting in explosion of the wheel assembly. To alleviate this, thermal plug(s) mounted in the inner wheel half of a high performance aircraft wheels are made with a fusible core that melts and releases the air from the tire before explosion.*

the wind since the outer wheel half does not close behind a gear door on this aircraft. Hub caps may also be found on fixed gear aircraft.

The outboard wheel half provides a convenient location of the valve stem used to inflate and deflate tubeless tires. Alternately, it may contain a hole through which a valve stem extension may pass from the inner wheel half or the valve stem itself may fit through such a hole if a tube-type tire is used.

Wheel Inspection

An aircraft wheel assembly is inspected while on the aircraft as often as possible. A more detailed inspection and any testing or repairs may be accomplished with the wheel assembly removed from the aircraft.

On Aircraft Inspection

The general condition of the aircraft wheel assemblies can be inspected while on the aircraft. Any signs of suspected damage that may require removal of the wheel assembly from the aircraft should be investigated.

Proper Installation

The landing gear area is such a hostile environment that the technician should inspect the landing gear including the wheels, tires, and brakes whenever possible. Proper installation of the wheels should not be taken for granted. All wheel tie bolts and nuts must be in place and secure. A missing bolt is grounds for removal, and a thorough inspection of the wheel halves in accordance with the wheel manufacturer's procedures must be performed due to the stresses that may have occurred. The wheel hub dust cap and anti-skid sensor should also be secure. The inboard wheel half should interface with the brake rotor with no signs of chafing or excessive movement. All brake keys on the wheel must be present and secure.

Examine the wheels for cracks, flaked paint, and any evidence of overheating. Inspect thermal plugs to ensure no sign of the fusible alloy having been melted. Thermal plugs that have permitted pressure loss in the tire require that the wheel assembly be removed for inspection. All other wheels with brakes and thermal plugs should be inspected closely while on the aircraft to determine if they too have overheated. Each wheel should be observed overall to ensure it is not abnormally tilted. Flanges should not be missing any pieces, and there should be no areas on the wheel that show significant impact damage.

Axle Nut Torque

Axle nut torque is of extreme importance on an aircraft wheel installation. If the nut is too loose, the bearing and wheel assembly may have excessive movement. The bearing cup(s) could loosen and spin, which could damage the wheel. There could also be impact damage from the bearing rollers which leads to bearing failure. *[Figure 13-63]* An over-torqued axle nut prevents the bearing from properly accepting the weight load of the aircraft. The bearing spins without sufficient lubrication to absorb the heat caused by the higher friction level. This too leads to bearing failure. All aircraft axle nuts must be installed and torqued in accordance with the airframe manufacturer's maintenance procedures.

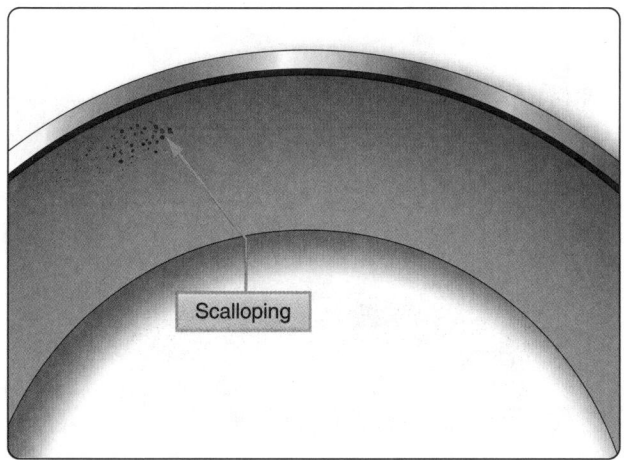

Figure 13-63. *Improper loose torque on the axle nut can cause excessive end play leading to bearing race damage known as scalloping. Eventually, this leads to bearing failure.*

Off Aircraft Wheel Inspection

Discrepancies found while inspecting a wheel mounted on the aircraft may require further inspection with the wheel removed from the aircraft. Other items such as bearing condition, can only be performed with the wheel assembly removed. A complete inspection of the wheel requires that the tire be removed from the wheel rim. Observe the following caution when removing a wheel assembly from an aircraft.

Caution: Deflate the tire before starting the procedure of removing the wheel assembly from the aircraft. Wheel assemblies have been known to explode while removing the axle nut, especially when dealing with high pressure, high performance tires. The torque of the nut can be the only force holding together a defective wheel or one with broken tie bolts. When loosened, the high internal pressure of the tire can create a catastrophic failure that could be lethal to the technician. It is also important to let aircraft tires cool before removal. Three hours or more is needed for cool down. Approach the wheel assembly from the front or rear, not broadside. Do not stand in the path of the released air and valve core trajectory when removing air from the tire as it could seriously injure the technician should it release from the valve stem.

Note: As a precautionary measure, remove only one tire and wheel assembly from a pair at a time. This leaves a tire and wheel assembly in place should the aircraft fall off its jack, resulting in less chance of damage to the aircraft and injury to personnel.

Loosening the Tire from the Wheel Rim

After inflation and usage, an aircraft tire has a tendency to adhere to the wheel, and the bead must be broken to remove the tire. There are mechanical and hydraulic presses designed for this purpose. In the absence of a device specifically made for the job, an arbor press can be used with patience working sequentially around the wheel as close as possible to the bead. *[Figure 13-64]* As stated above, there should be no air pressure in the tire while it is being pressed off of the wheel. Never pry a tire off of the rim with a screwdriver or other device. The wheels are relatively soft. Any nick or deformation causes a stress concentration that can easily lead to wheel failure.

Disassembly of the Wheel

Disassembly of the wheel should take place in a clean area on a flat surface, such as a table. Remove the wheel bearing first and set aside for cleaning and inspecting. The tie bolts can then be removed. Do not use an impact tool to disassemble the tie bolts. Aircraft wheels are made of relatively soft

Figure 13-64. *Tire beads must be broken from the wheel to remove the tire. A mechanical removal tool designed for breaking the bead is shown in (A); a hydraulic press designed with the capacity for large aircraft wheels is shown in (B); and an arbor press is shown in (C). All are tools available to the technician for this purpose.*

aluminum and magnesium alloys. They are not designed to receive the repeated hammering of an impact tool and will be damaged if used.

Cleaning the Wheel Assembly

Clean the wheel halves with the solvent recommended by the wheel manufacturer. Use of a soft brush helps this process. Avoid abrasive techniques, materials, and tools, such as scrapers, capable of removing the finish off of the wheel. Corrosion can quickly form and weaken the wheel if the finish is missing in an area. When the wheels are clean, they can be dried with compressed air.

Cleaning the Wheel Bearings

The bearings should be removed from the wheel to be cleaned with the recommended solvent, such as varsol, naptha, or Stoddard® solvent. Soaking the bearings in solvent is acceptable to loosen any dried-on grease. Bearings are brushed clean with a soft bristle brush and dried with compressed air. Never rotate the bearing while drying with compressed air. The high-speed metal to metal contact of the bearing rollers with the race causes heat that damages the metal surfaces. The bearing parts could also cause injury should the bearing come apart. Always avoid steam cleaning of bearings. The surface finish of the metals will be compromised leading to early failure.

Wheel Bearing Inspection

Once cleaned, the wheel bearing is inspected. There are many unacceptable conditions of the bearing and bearing cup, which are grounds for rejection. In fact, nearly any flaw detected in a bearing assembly is likely to be grounds for replacement.

Common conditions of a bearing that are cause for rejection are as follows:

Galling—caused by rubbing of mating surfaces. The metal gets so hot it welds, and the surface metal is destroyed as the motion continues and pulls the metal apart in the direction of motion. *[Figure 13-65]*

Spalling—a chipped away portion of the hardened surface of a bearing roller or race. *[Figure 13-66]*

Overheating—caused by lack of sufficient lubrication results in a bluish tint to the metal surface. The ends of the rollers shown were overheated causing the metal to flow and deform, as well as discolor. The bearing cup raceway is usually discolored as well. *[Figure 13-67]*

Brinelling—caused by excessive impact. It appears as indentations in the bearing cup raceways. Any static overload

Figure 13-65. *Galling is caused by rubbing of mating surfaces. The metal gets so hot it welds, and the surface metal is destroyed as the motion continues and pulls the metal apart in the direction of motion.*

Figure 13-66. *Spalling is a chipped away portion of the hardened surface of a bearing roller or race.*

Figure 13-67. *Overheating caused by lack of sufficient lubrication results in a bluish tint to the metal surface. The ends of the rollers shown were overheated causing the metal to flow and deform, as well as discolor. The bearing cup raceway is usually discolored as well.*

or severe impact can cause true brinelling that leads to vibration and premature bearing failure. *[Figure 13-68]*

False Brinelling—caused by vibration of the bearing while in a static state. Even with a static overload, lubricant can be forced from between the rollers and the raceway. Submicroscopic particles removed at the points of metal-to-metal contact oxidize. They work to remove more particles spreading the damage. This is also known as frictional corrosion. It can be identified by a rusty coloring of the lubricant. *[Figure 13-69]*

Staining and surface marks—located on the bearing cup as grayish black streaks with the same spacing as the rollers and caused by water that has gotten into the bearing. It is the first stage of deeper corrosion that follows. *[Figure 13-70]*

Etching and corrosion—caused when water and the damage caused by water penetrates the surface treatment of the bearing element. It appears as a reddish/brown discoloration. *[Figure 13-71]*

Bruising—caused by fine particle contamination possibly from a bad seal or improper maintenance of bearing cleanliness. It leaves a less than smooth surface on the bearing cup. *[Figure 13-72]*

The bearing cup does not require removal for inspection; however, it must be firmly seated in the wheel half boss. There should be no evidence that a cup is loose or able to spin. *[Figure 13-73]* The cup is usually removed by heating the wheel in a controlled oven and pressing it out or tapping it out with a non-metallic drift. The installation procedure is similar. The wheel is heated, and the cup is cooled with dry ice before it is tapped into place with a non-metallic hammer or drift. The outside of the race is often sprayed with

Figure 13-69. *False brinelling is caused by vibration of the bearing while in a static state. Even with a static overload, it can force the lubricant from between the rollers and the raceway. Submicroscopic particles removed at the points of metal-to-metal contact oxidize. They work to remove more particles spreading the damage. This is also known as frictional corrosion. It can be identified by a rusty coloring of the lubricant.*

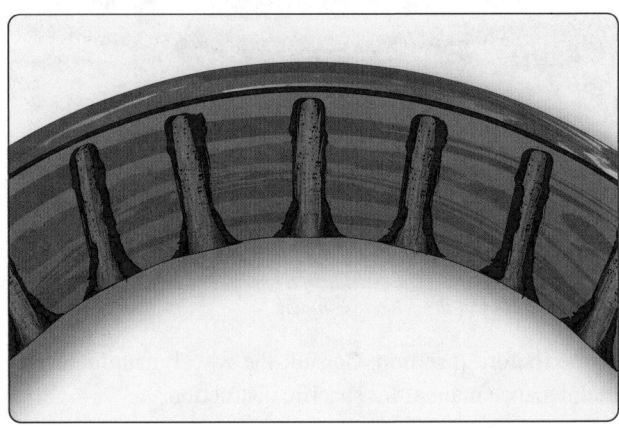

Figure 13-70. *Staining and surface marks on the bearing cup that are grayish black streaks with the same spacing as the rollers are caused by water that has gotten into the bearing. It is the first stage of deeper corrosion that will follow.*

Figure 13-68. *Brinelling is caused by excessive impact. It appears as indentations in the bearing cup raceways. Any static overload or severe impact can cause true brinelling, which leads to vibration and premature bearing failure.*

Figure 13-71. *Etching and corrosion is caused when water, and the damage caused by water, penetrates the surface treatment of the bearing element. It appears as a reddish/brown discoloration.*

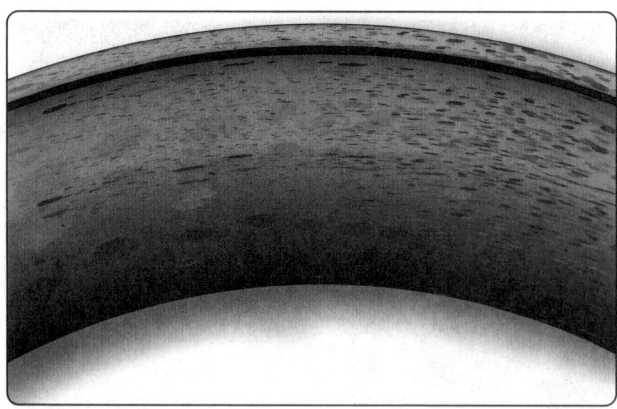

Figure 13-72. *Bruising is caused by fine particle contamination possibly from a bad seal or improper maintenance of bearing cleanliness. It leaves a less than smooth surface on the bearing cup.*

Figure 13-73. *Bearing cups should be tight in the wheel boss and should never rotate. The outside of a bearing cup that was spinning while installed in the wheel is shown.*

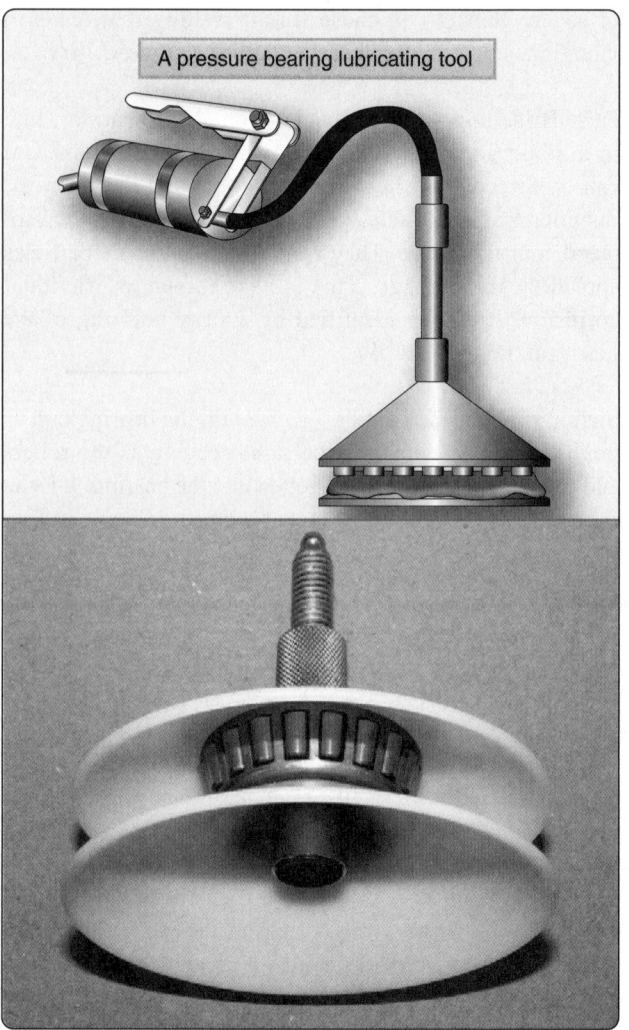

Figure 13-74. *A pressure bearing lubricating tool.*

primer before insertion. Consult the wheel manufacturer's maintenance manual for specific instructions.

Bearing Handling & Lubrication

Handling of bearings is of the utmost importance. Contamination, moisture, and vibration, even while the bearing is in a static state, can ruin a bearing. Avoid conditions where these may affect bearings and be sure to install and torque bearings into place according manufacturer's instructions.

Proper lubrication is a partial deterrent to negative environmental impacts on a bearing. Use the lubricant recommended by the manufacturer. Use of a pressure bearing packing tool or adapter is also recommended as the best method to remove any contaminants from inside the bearing that may have remained after cleaning. *[Figure 13-74]*

Inspection of the Wheel Halves

A thorough visual inspection of each wheel half should be conducted for discrepancies specified in the wheel manufacturer's maintenance data. Use of a magnifying glass is recommended. Corrosion is one of the most common problems encountered while inspecting wheels. Locations where moisture is trapped should be checked closely. It is possible to dress out some corrosion according to the manufacturer's instructions. An approved protective surface treatment and finish must be applied before returning the wheel to service. Corrosion beyond stated limits is cause for rejection of the wheel.

In addition to corrosion, cracks in certain areas of the wheel are particularly prevalent. One such area is the bead seat area. *[Figure 13-75]* The high stress of landing is transferred to the wheel by the tire in this contact area. Hard landings produce distortion or cracks that are very difficult to detect. This is a concern on all wheels and is most problematic in high-pressure, forged wheels. Dye penetrant inspection is generally ineffective when checking for cracks in the bead area. There is a tendency for cracks to close up tightly once the tire is dismounted, and the stress is removed from the metal. Eddy

Figure 13-75. *The bead seat areas of a light aircraft wheel set. Eddy current testing for cracks in the bead seat area is common.*

current inspection of the bead seat area is required. Follow the wheel manufacturer's instruction when performing the eddy current check.

The wheel brake disc drive key area is another area in which cracks are common. The forces experienced when the keys drive the disc against the stopping force of the brakes are high. Generally, a dye penetrant test is sufficient to reveal cracks in this area. All drive keys should be secure with no movement possible. No corrosion is permitted in this area. *[Figure 13-76]*

Wheel Tie Bolt Inspection

Wheel half tie bolts are under great stress while in service and require inspection. The tie bolts stretch and change dimension usually at the threads and under the bolt head. These are areas where cracks are most common. Magnetic particle inspection can reveal these cracks. Follow the maintenance manual procedures for inspecting tie bolts.

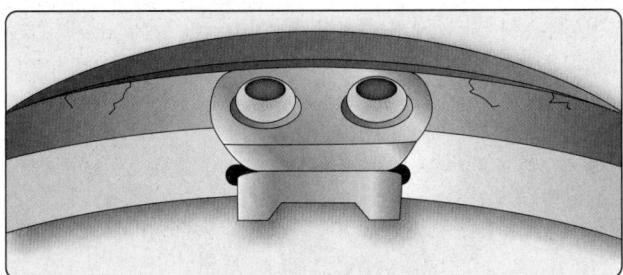

Figure 13-76. *Inspection for cracks in the wheel disc drive key area is performed with dye penetrant on many wheels.*

Key & Key Screw Inspection

On most aircraft inner wheel halves, keys are screwed or bolted to the wheel to drive the brake disc(s). The drive keys are subject to extreme forces when the brakes are applied. As mentioned, there should be no movement between the wheel and the keys. The bolts should be checked for security, and the area around the keys should be inspected for cracks. There is also a limitation on how worn the keys can be since too much wear allows excessive movement. The wheel manufacturer's maintenance instructions should be used to perform a complete inspection of this critical area.

Fusible Plug Inspection

Fusible plugs or thermal plugs must be inspected visually. These threaded plugs have a core that melts at a lower temperature than the outer part of the plug. This is to release air from the tire should the temperature rise to a dangerous level. A close inspection should reveal whether any core has experienced deformation that might be due to high temperature. If detected, all thermal plugs in the wheel should be replaced with new plugs. *[Figure 13-77]*

Balance Weights

The balance of an aircraft wheel assembly is important. When manufactured, each wheel set is statically balanced. Weights are added to accomplish this if needed. They are a permanent part of the wheel assembly and must be installed to use the wheel. The balance weights are bolted to the wheel halves and can be removed when cleaning and inspecting the wheel. They must be re-fastened in their original position. When a tire is mounted to a wheel, balancing of the wheel and tire assembly may require that additional weights be added. These are usually installed around the circumference of the outside of the wheel and should not be taken as substitutes for the factory wheel set balance weights. *[Figure 13-78]*

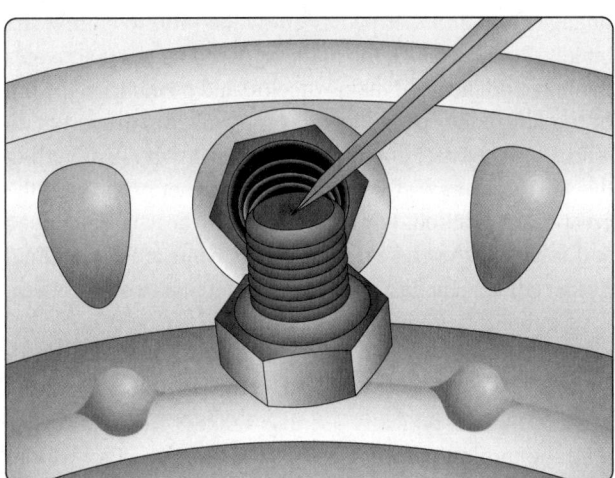

Figure 13-77. *Visually inspect the core of a thermal or fusible plug for deformation associated with heat exposure. Replace all of the plugs if any appear to have begun to deform.*

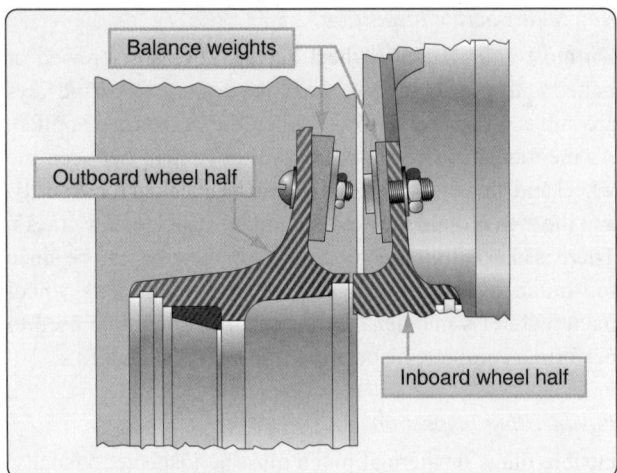

Figure 13-78. *Two piece aircraft wheels are statically balanced when manufactured and may include weights attached to each wheel half that must stay with the wheel during its entire serviceable life.*

Aircraft Brakes

Very early aircraft have no brake system to slow and stop the aircraft while it is on the ground. Instead, they rely on slow speeds, soft airfield surfaces, and the friction developed by the tail skid to reduce speed during ground operation. Brake systems designed for aircraft became common after World War I as the speed and complexity of aircraft increased, and the use of smooth, paved runway surfaces proliferated. All modern aircraft are equipped with brakes. Their proper functioning is relied upon for safe operation of the aircraft on the ground. The brakes slow the aircraft and stop it in a reasonable amount of time. They hold the aircraft stationary during engine run-up and, in many cases, steer the aircraft during taxi. On most aircraft, each of the main wheels is equipped with a brake unit. The nose wheel or tail wheel does not have a brake.

In the typical brake system, mechanical and/or hydraulic linkages to the rudder pedals allow the pilot to control the brakes. Pushing on the top of the right rudder pedal activates the brake on the right main wheel(s) and pushing on the top of the left rudder pedal operates the brake on the left main wheel(s). The basic operation of brakes involves converting the kinetic energy of motion into heat energy through the creation of friction. A great amount of heat is developed and forces on the brake system components are demanding. Proper adjustment, inspection, and maintenance of the brakes is essential for effective operation.

Types & Construction of Aircraft Brakes

Modern aircraft typically use disc brakes. The disc rotates with the turning wheel assembly while a stationary caliper resists the rotation by causing friction against the disc when the brakes are applied. The size, weight, and landing speed of the aircraft influence the design and complexity of the disc brake system. Single, dual, and multiple disc brakes are common types of brakes. Segmented rotor brakes are used on large aircraft. Expander tube brakes are found on older large aircraft. The use of carbon discs is increasing in the modern aviation fleet.

Single Disc Brakes

Small, light aircraft typically achieve effective braking using a single disc keyed or bolted to each wheel. As the wheel turns, so does the disc. Braking is accomplished by applying friction to both sides of the disc from a non-rotating caliper bolted to the landing gear axle flange. Pistons in the caliper housing under hydraulic pressure force wearable brake pads or linings against the disc when the brakes are applied. Hydraulic master cylinders connected to the rudder pedals supply the pressure when the upper halves of the rudder pedals are pressed.

Floating Disc Brakes

A floating disc brake is illustrated in *Figure 13-79*. A more detailed, exploded view of this type of brake is shown in *Figure 13-80*. The caliper straddles the disc. It has three cylinders bored through the housing, but on other brakes this number may vary. Each cylinder accepts an actuating piston assembly comprised mainly of a piston, a return spring, and an automatic adjusting pin. Each brake assembly has six brake linings or pucks. Three are located on the ends of the pistons, which are in the outboard side of the caliper. They are designed to move in and out with the pistons and apply pressure to the outboard side of the disc. Three more linings are located opposite of these pucks on the inboard side of the caliper. These linings are stationary.

The brake disc is keyed to the wheel. It is free to move laterally in the key slots. This is known as a floating disc. When the brakes are applied, the pistons move out from the outboard cylinders and their pucks contact the disc. The disc

Figure 13-79. *A single disc brake is a floating-disc, fixed caliper brake.*

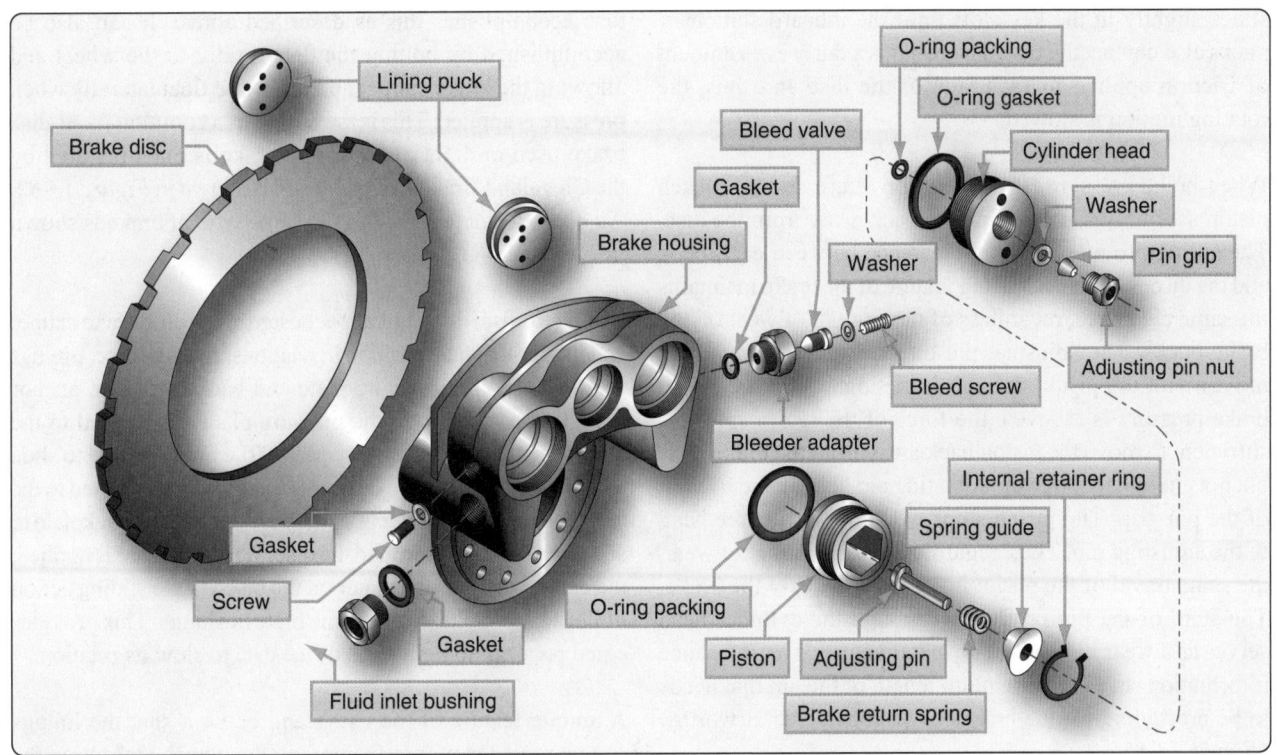

Figure 13-80. *An exploded view of a single-disc brake assembly found on a light aircraft.*

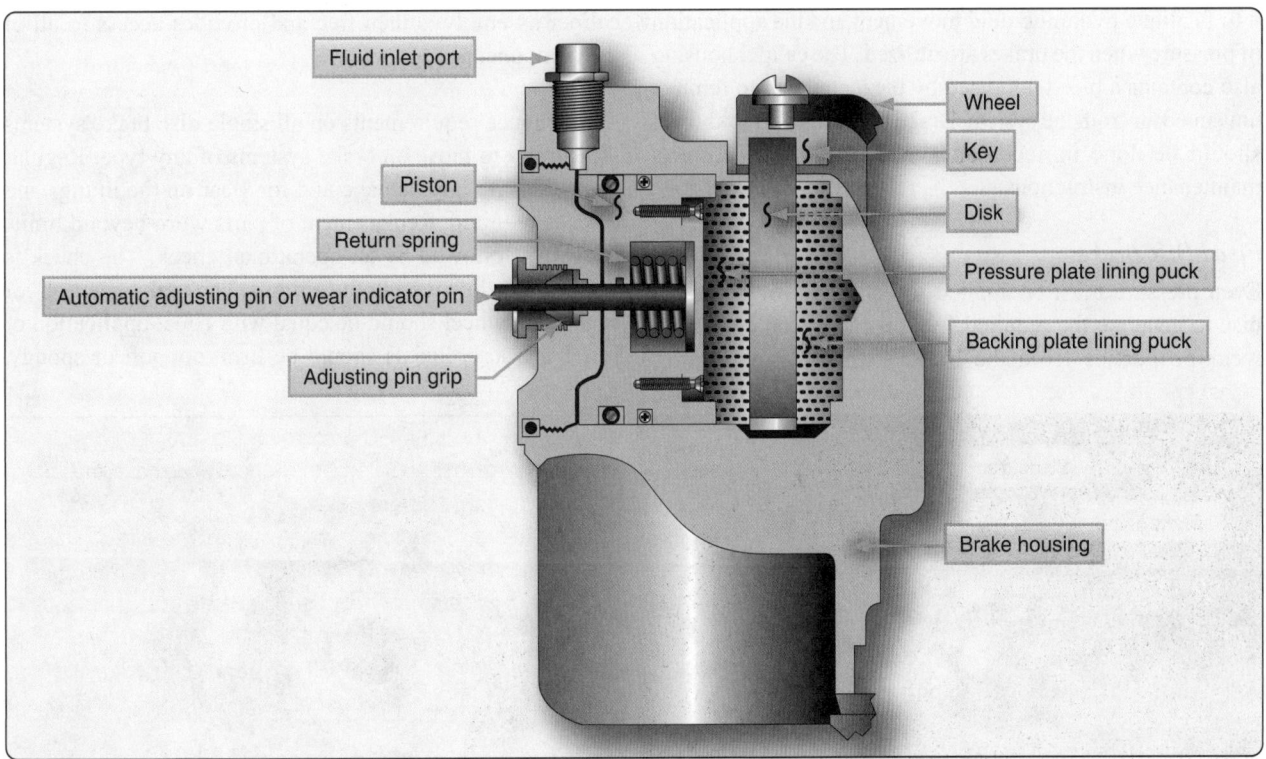

Figure 13-81. *A cross-sectional view of a Goodyear single-disc brake caliper illustrates the adjusting pin assembly that doubles as a wear indicator.*

slides slightly in the key slots until the inboard stationary pucks also contact the disc. The result is a fairly even amount of friction applied to each side of the disc and thus, the rotating motion is slowed.

When brake pressure is released, the return spring in each piston assembly forces the piston back away from the disc. The spring provides a preset clearance between each puck and the disc. The self-adjusting feature of the brake maintains the same clearance, regardless of the amount of wear on the brake pucks. The adjusting pin on the back of each piston moves with the piston through a frictional pin grip. When brake pressure is relieved, the force of the return spring is sufficient to move the piston back away from the brake disc, but not enough to move the adjusting pin held by the friction of the pin grip. The piston stops when it contacts the head of the adjusting pin. Thus, regardless of the amount of wear, the same travel of the piston is required to apply the brake. The stem of the pin protruding through the cylinder head serves as a wear indicator. The manufacturer's maintenance information states the minimum length of the pin that needs to be protruding for the brakes to be considered airworthy. [Figure 13-81]

The brake caliper has the necessary passages machined into it to facilitate hydraulic fluid movement and the application of pressure when the brakes are utilized. The caliper housing also contains a bleed port used by the technician to remove unwanted air from the system. Brake bleeding, as it is known, should be done in accordance with the manufacturer's maintenance instructions.

Fixed-Disc Brakes
Even pressure must be applied to both sides of the brake disc to generate the required friction and obtain consistent wear properties from the brake linings. The floating

disc accomplishes this as described above. It can also be accomplished by bolting the disc rigidly to the wheel and allowing the brake caliper and linings to float laterally when pressure is applied. This is the design of a common fixed-disc brake used on light aircraft. The brake is manufactured by the Cleveland Brake Company and is shown in *Figure 13-82*. An exploded detail view of the same type of brake is shown in *Figure 13-83*.

The fixed-disc, floating-caliper design allows the brake caliper and linings to adjust position in relationship to the disc. Linings are riveted to the pressure plate and backplate. Two anchor bolts that pass through the pressure plate are secured to the cylinder assembly. The other ends of the bolts are free to slide in and out of bushings in the torque plate, which is bolted to the axle flange. The cylinder assembly is bolted to the backplate to secure the assembly around the disc. When pressure is applied, the caliper and linings center on the disc via the sliding action of the anchor bolts in the torque plate bushings. This provides equal pressure to both sides of the disc to slow its rotation.

A unique feature of the Cleveland brake is that the linings can be replaced without removing the wheel. Unbolting the cylinder assembly from the backplate allows the anchor bolts to slide out of the torque plate bushings. The entire caliper assembly is then free and provides access to all of the components.

Maintenance requirements on all single disc brake systems are similar to those on brake systems of any type. Regular inspection for any damage and for wear on the linings and discs is required. Replacement of parts worn beyond limits is always followed by an operational check. The check is performed while taxiing the aircraft. The braking action for each main wheel should be equal with equal application of pedal pressure. Pedals should be firm, not soft or spongy,

Figure 13-82. *A Cleveland brake on a light aircraft is a fixed-disc brake. It allows the brake caliper to move laterally on anchor bolts to deliver even pressure to each side of the brake disc.*

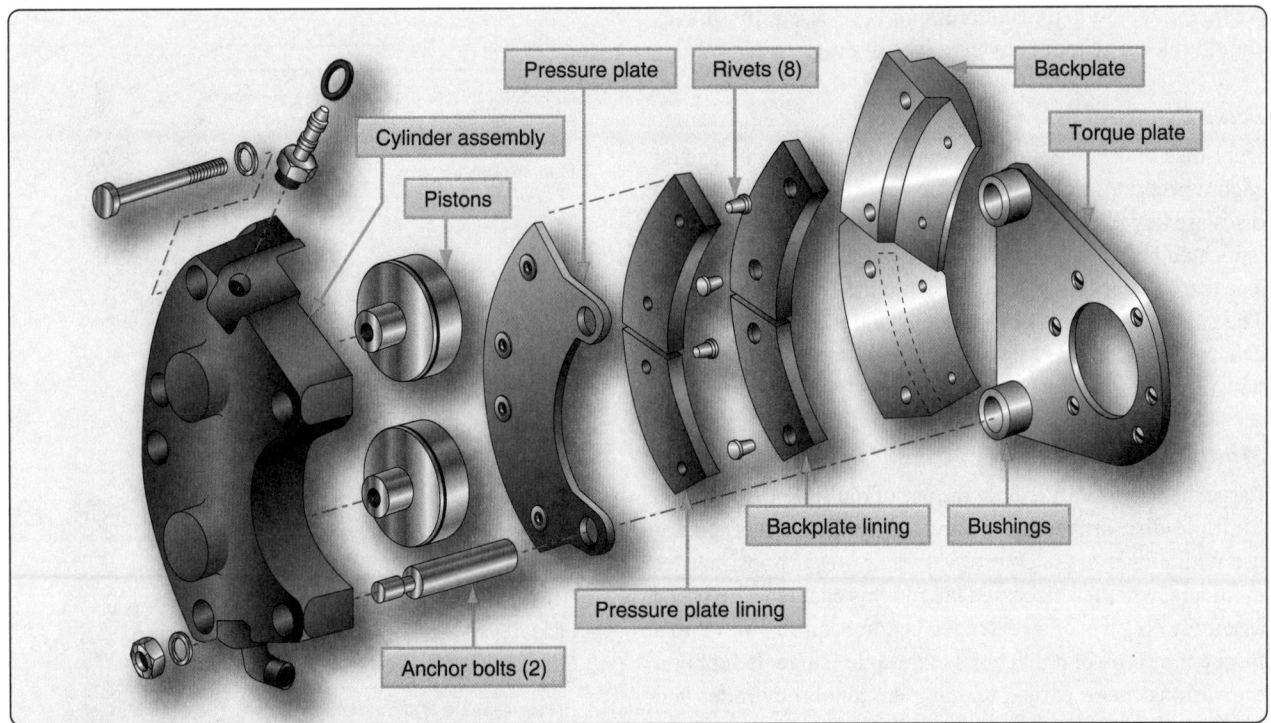

Figure 13-83. *An exploded view of a dual-piston Cleveland brake assembly.*

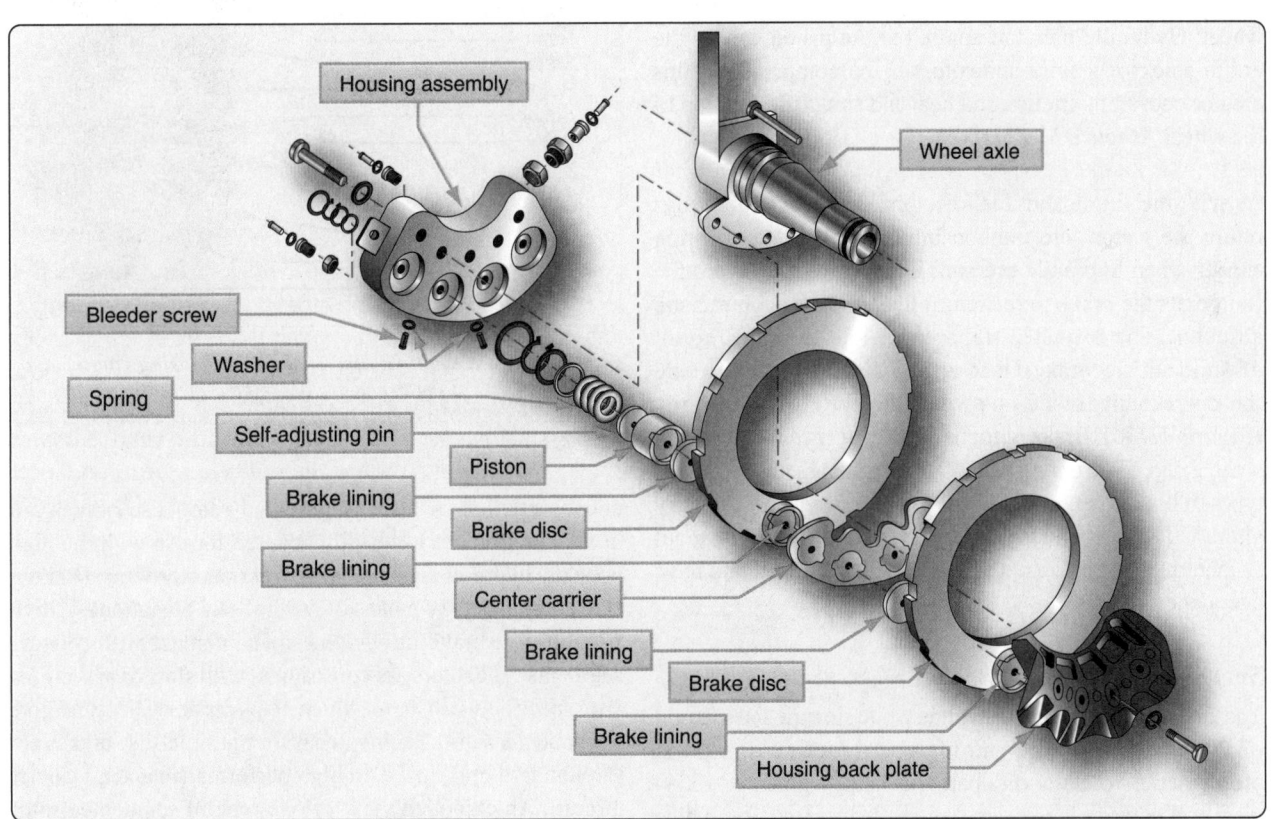

Figure 13-84. *A dual-disc brake is similar to a single-disc brake. It uses a center carrier to hold brake linings against each of the discs.*

when applied. When pedal pressure is released, the brakes should release without any evidence of drag.

Dual-Disc Brakes

Dual-disc brakes are used on aircraft where a single disc on each wheel does not supply sufficient braking friction. Two discs are keyed to the wheel instead of one. A center carrier is located between the two discs. It contains linings on each side that contact each of the discs when the brakes are applied. The caliper mounting bolts are long and mount through the center carrier, as well as the backplate which bolts to the housing assembly. [Figure 13-84]

Multiple-Disc Brakes

Large, heavy aircraft require the use of multiple-disc brakes. Multiple-disc brakes are heavy duty brakes designed for use with power brake control valves or power boost master cylinders, which is discussed later in this chapter. The brake assembly consists of an extended bearing carrier similar to a torque tube type unit that bolts to the axle flange. It supports the various brake parts, including an annular cylinder and piston, a series of steel discs alternating with copper or bronze-plated discs, a backplate, and a backplate retainer. The steel stators are keyed to the bearing carrier, and the copper or bronze plated rotors are keyed to the rotating wheel. Hydraulic pressure applied to the piston causes the entire stack of stators and rotors to be compressed. This creates enormous friction and heat and slows the rotation of the wheel. [Figure 13-85]

As with the single and dual-disc brakes, retracting springs return the piston into the housing chamber of the bearing carrier when hydraulic pressure is relieved. The hydraulic fluid exits the brake to the return line through an automatic adjuster. The adjuster traps a predetermined amount of fluid in the brakes that is just sufficient to provide the correct clearances between the rotors and stators. [Figure 13-86] Brake wear is typically measured with a wear gauge that is not part of the brake assembly. These types of brake are typically found on older transport category aircraft. The rotors and stators are relatively thin, only about ⅛-inch thick. They do not dissipate heat very well and have a tendency to warp.

Segmented Rotor-Disc Brakes

The large amount of heat generated while slowing the rotation of the wheels on large and high-performance aircraft is problematic. To better dissipate this heat, segmented rotor-disc brakes have been developed. Segmented rotor-disc brakes are multiple-disc brakes but of more modern design than the type discussed earlier. There are many variations. Most feature numerous elements that aid in the control

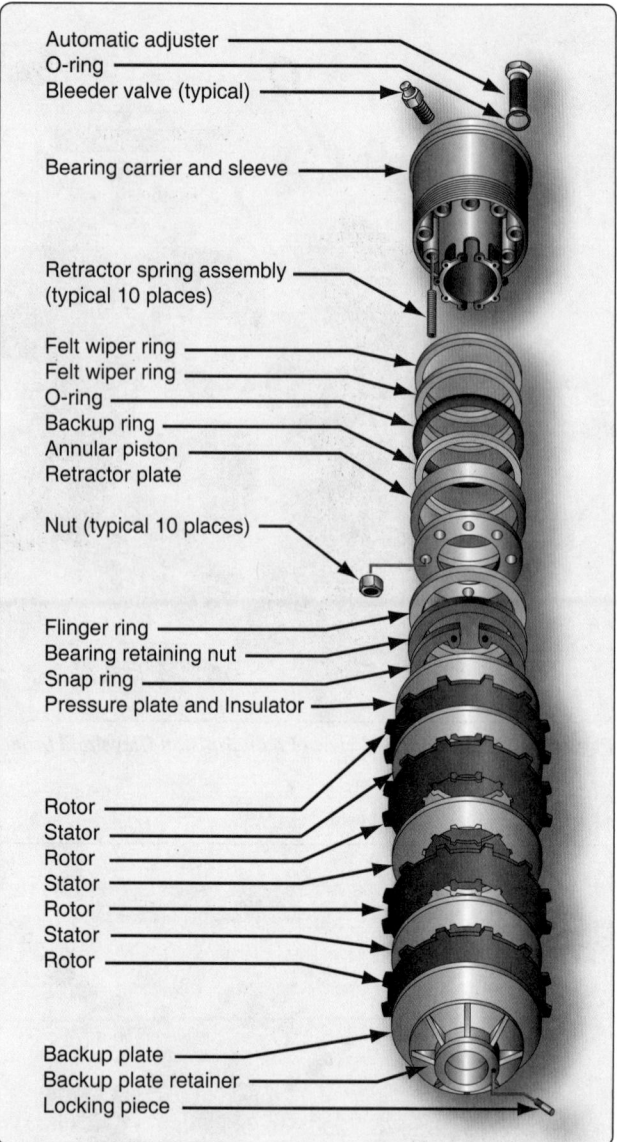

Figure 13-85. *A multiple disc brake with bearing carrier upon which the parts of the brake are assembled including an annular cylinder and piston assembly that apply pressure evenly to a stack of rotors and stators.*

and dissipation of heat. Segmented rotor-disc brakes are heavy-duty brakes especially adapted for use with the high pressure hydraulic systems of power brake systems. Braking is accomplished by means of several sets of stationary, high friction type brake linings that make contact with rotating segments. The rotors are constructed with slots or in sections with space between them, which helps dissipate heat and give the brake its name. Segmented rotor multiple-disc brakes are the standard brake used on high performance and air carrier aircraft. An exploded view of one type of segmented rotor brake assembly is shown in *Figure 13-87*.

The description of a segmented rotor brake is very similar to the multiple-disc type brake previously described. The brake

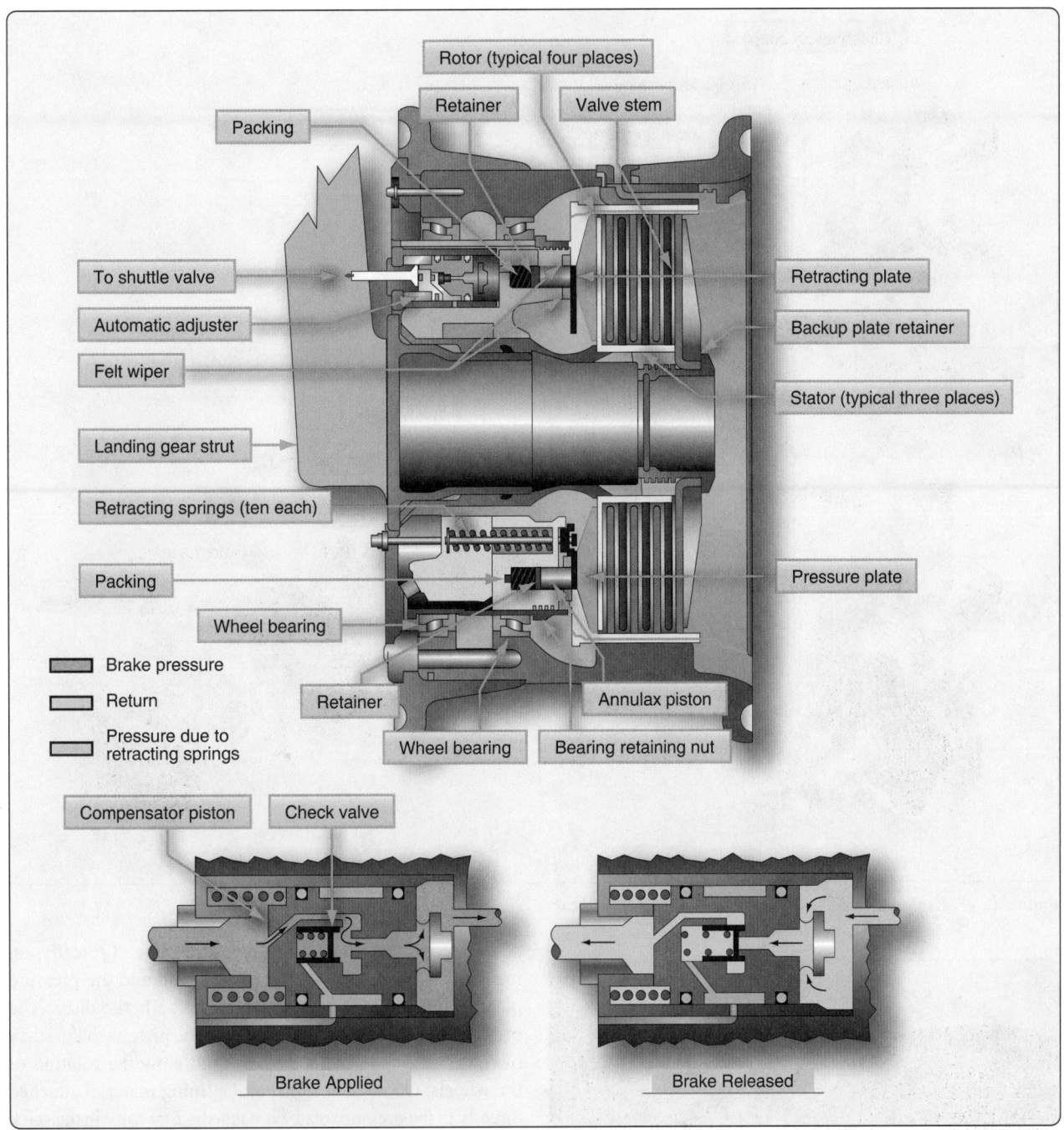

Labels in figure:
Rotor (typical four places)
Retainer
Valve stem
Packing
To shuttle valve
Automatic adjuster
Felt wiper
Landing gear strut
Retracting springs (ten each)
Packing
Wheel bearing
Brake pressure
Return
Pressure due to retracting springs
Retainer
Wheel bearing
Annulax piston
Bearing retaining nut
Retracting plate
Backup plate retainer
Stator (typical three places)
Pressure plate
Compensator piston
Check valve
Brake Applied
Brake Released

Figure 13-86. *A multiple-disc brake with details of the automatic adjuster.*

assembly consists of a carrier, a piston and piston cup seal, a pressure plate, an auxiliary stator plate, rotor segments, stator plates, automatic adjusters, and a backing plate.

The carrier assembly, or brake housing with torque tube, is the basic unit of the segmented rotor brake. It is the part that attaches to the landing gear shock strut flange upon which the other components of the brake are assembled. On some brakes, two grooves or cylinders are machined into the carrier to receive the piston cups and pistons. *[Figure 13-87]* Most segmented rotor-disc brakes have numerous individual

cylinders machined into the brake housing into which fit the same number of actuating pistons. Often, these cylinders are supplied by two different hydraulic sources, alternating every other cylinder from a single source. If one source fails, the brake still operates sufficiently on the other. *[Figure 13-88]* External fittings in the carrier or brake housing admit the hydraulic fluid. A bleed port can also be found.

A pressure plate is a flat, circular, high-strength steel, non-rotating plate notched on the inside circumference to fit over the stator drive sleeves or torque tube spines. The brake

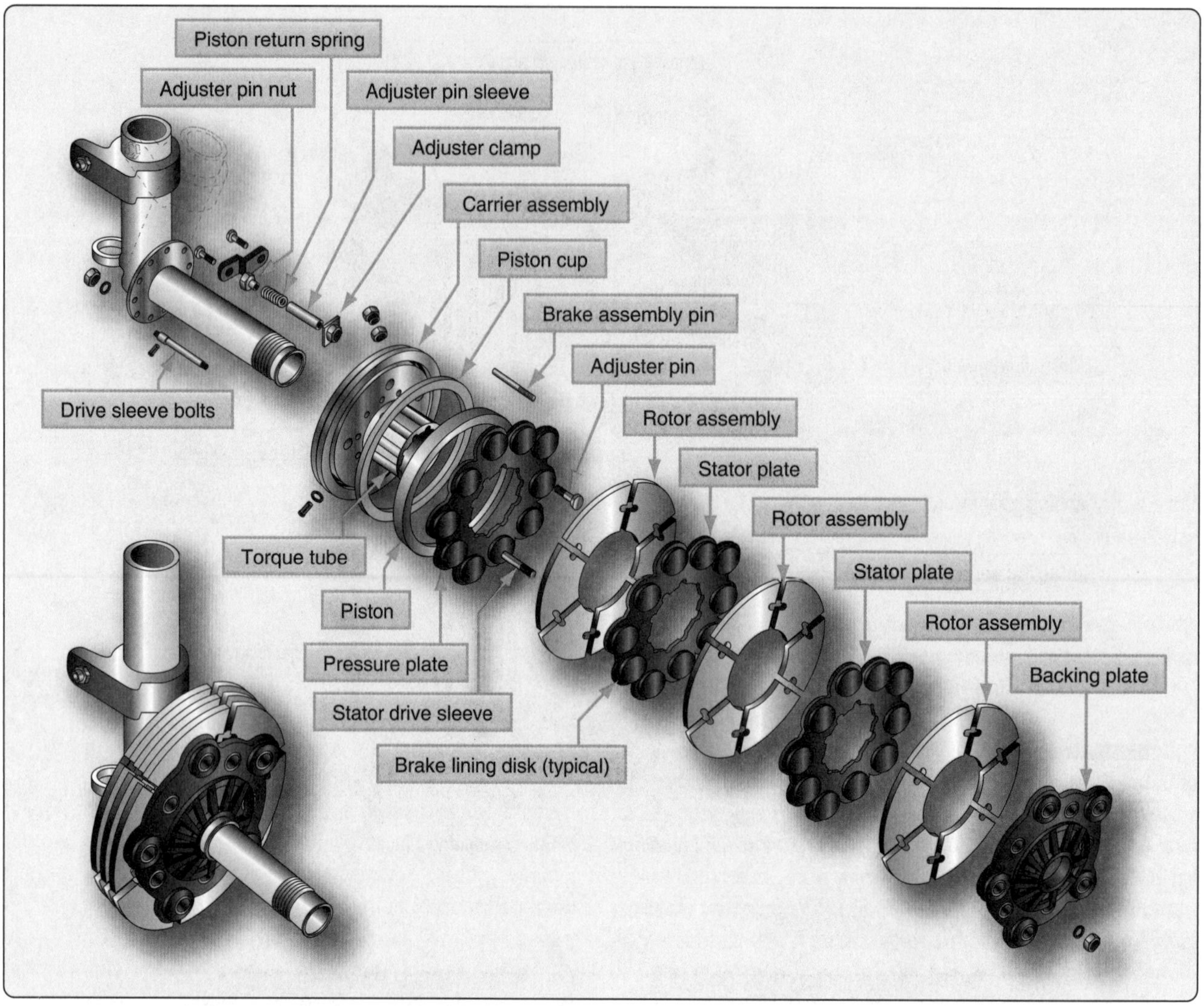

Figure 13-87. Exploded and detail views of segmented rotor brakes.

Figure 13-88. Many modern segmented rotor disc brakes use a housing machined to fit numerous individual actuating pistons.

actuating pistons contact the pressure plate. Typically, an insulator is used between the piston head and the pressure plate to impede heat conduction from the brake discs. The pressure plate transfers the motion of the pistons to the stack of rotors and stators that compress to slow the rotation of the wheels. On most designs, brake lining material attached directly to the pressure plate contacts the first rotor in the stack to transfer the motion of the piston(s). [Figure 13-87] An auxiliary stator plate with brake lining material on the side opposite the pressure plate can also be used.

Any number of alternating rotors and stators are sandwiched under hydraulic pressure against the backing plate of the brake assembly when the brakes are applied. The backing plate is a heavy steel plate bolted to the housing or torque tube at a fixed dimension from the carrier housing. In most cases, it has brake lining material attached to it and contacts the last rotor in the stack. [Figure 13-87]

Stators are flat plates notched on the internal circumference to be held stationary by the torque tube spines. They have wearable brake lining material riveted or adhered to each side to make contact with adjacent rotors. The liner is typically constructed of numerous isolated blocks. *[Figure 13-87]* The space between the liner blocks aids in the dissipation of heat. The composition of the lining materials vary. Steel is often used.

Rotors are slit or segmented discs that have notches or tangs in the external circumference that key to the rotating wheel. Slots or spaces between sections of the rotor create segments that allow heat to dissipate faster than it would if the rotor was solid. They also allow for expansion and prevent warping. *[Figure 13-87]* Rotors are usually steel to which a frictional surface is bonded to both sides. Typically, sintered metal is used in creating the rotor contact surface.

Segmented multiple-disc brakes use retraction spring assemblies with auto clearance adjusters to pull the backplate away from the rotor and stator stack when brake pressure is removed. This provides clearance so the wheel can turn unimpeded by contact friction between the brake parts but keeps the units in close proximity for rapid contact and braking when the brakes are applied. The number of retraction devices varies with brake design. *Figure 13-89* illustrates a brake assembly used on a Boeing 737 transport category aircraft. In the cutaway view, the number and locations of the auto adjustment retraction mechanisms can be seen. Details of the mechanisms are also shown.

Instead of using a pin grip assembly for auto adjustment, an adjuster pin, ball, and tube operate in the same manner. They move out when brake pressure is applied, but the ball in the tube limits the amount of the return to that equal to the brake lining wear. Two independent wear indicators are used on the brake illustrated. An indicator pin attached to the backplate protrudes through the carrier. The amount that it protrudes with the brakes applied is measured to ascertain if new linings are required.

Note: Other segmented multiple-disc brakes may use slightly different techniques for pressure plate retraction and wear indication. Consult the manufacturer's maintenance information to ensure wear indicators are read correctly.

Carbon Brakes

The segmented multiple-disc brake has given many years of reliable service to the aviation industry. It has evolved through time in an effort to make it lightweight and to dissipate the frictional heat of braking in a quick, safe manner. The latest iteration of the multiple-disc brake is the carbon-disc brake. It is currently found on high performance and air carrier aircraft. Carbon brakes are so named because carbon fiber materials

are used to construct the brake rotors. *[Figure 13-90]*

Carbon brakes are approximately forty percent lighter than conventional brakes. On a large transport category aircraft, this alone can save several hundred pounds in aircraft weight. The carbon fiber discs are noticeably thicker than sintered steel rotors but are extremely light. They are able to withstand temperatures fifty percent higher than steel component brakes. The maximum designed operating temperature is limited by the ability of adjacent components to withstand the high temperature. Carbon brakes have been shown to withstand two to three times the heat of a steel brake in non-aircraft applications. Carbon rotors also dissipate heat faster than steel rotors. A carbon rotor maintains its strength and dimensions at high temperatures. Moreover, carbon brakes last twenty to fifty percent longer than steel brakes, which results in reduced maintenance.

The only impediment to carbon brakes being used on all aircraft is the high cost of manufacturing. The price is expected to lower as technology improves and greater numbers of aircraft operators enter the market.

Expander Tube Brakes

An expander tube brake is a different approach to braking that is used on aircraft of all sizes produced in the 1930s–1950s. It is a lightweight, low pressure brake bolted to the axle flange that fits inside an iron brake drum. A flat, fabric-reinforced neoprene tube is fitted around the circumference of a wheel-like torque flange. The exposed flat surface of the expander tube is lined with brake blocks similar to brake lining material. Two flat frames bolt to the sides of the torque flange. Tabs on the frames contain the tube and allow evenly spaced torque bars to be bolted in place across the tube between each brake block. These prevent circumferential movement of the tube on the flange. *[Figure 13-91]*

The expander tube is fitted with a metal nozzle on the inner surface. Hydraulic fluid under pressure is directed through this fitting into the inside of the tube when the brakes are applied. The tube expands outward, and the brake blocks make contact with the wheel drum causing friction that slows the wheel. As hydraulic pressure is increased, greater friction develops. Semi-elliptical springs located under the torque bars return the expander tube to a flat position around the flange when hydraulic pressure is removed. The clearance between the expander tube and the brake drum is adjustable by rotating an adjuster on some expander tube brakes. Consult the manufacturer's maintenance manual for the correct clearance setting. *Figure 13-92* gives an exploded view of an expander tube brake, detailing its components.

Expander tube brakes work well but have some drawbacks.

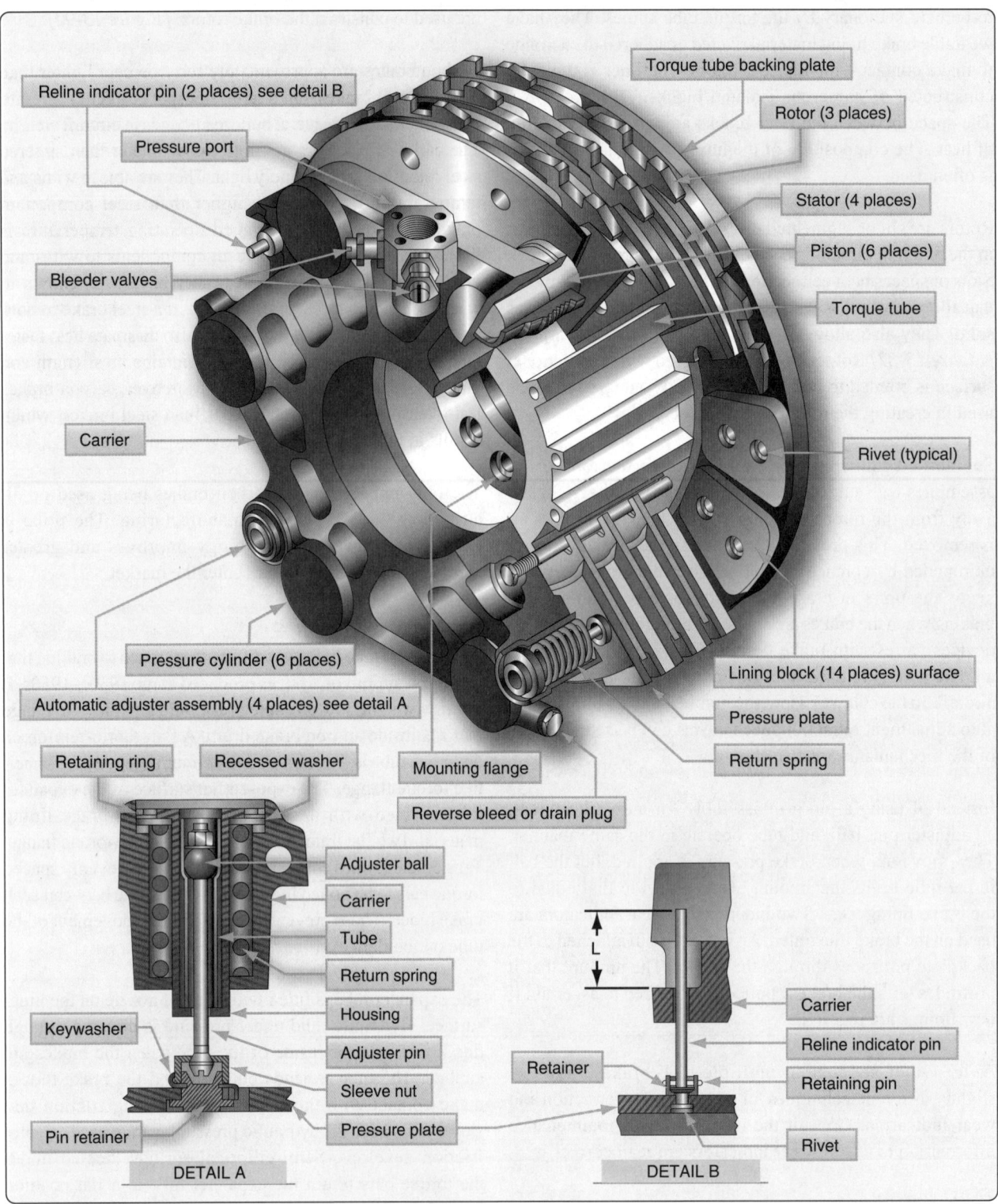

Figure 13-89. *The multiple-disc brake assembly and details from a Boeing 737.*

They tend to take a setback when cold. They also have a tendency to swell with temperature and leak. They may drag inside the drum if this occurs. Eventually, expander brakes were abandoned in favor of disc brake systems.

Brake Actuating Systems

The various brake assemblies, described in the previous section, all use hydraulic power to operate. Different means of delivering the required hydraulic fluid pressure to brake assemblies are discussed in this section. There are three basic actuating systems:

Figure 13-90. *A carbon brake for a Boeing 737.*

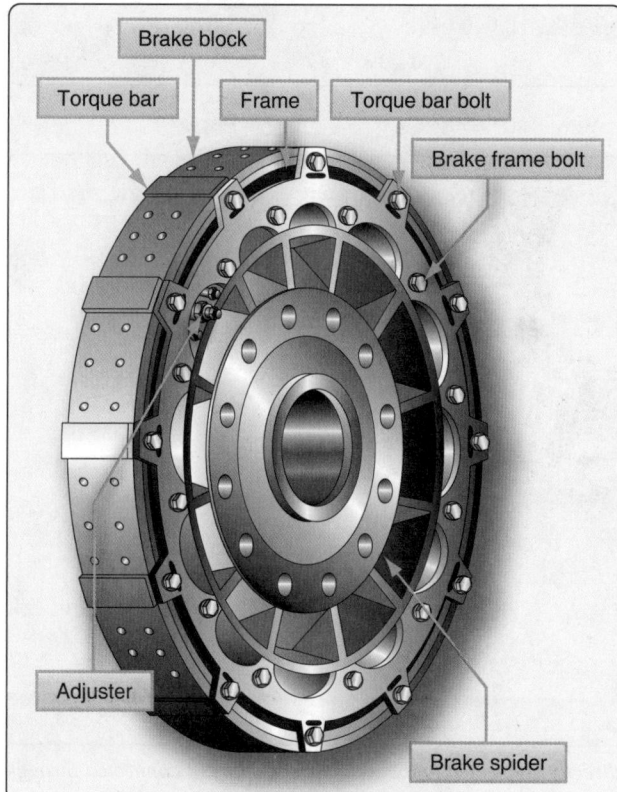

Brake block

Torque bar | Frame | Torque bar bolt

Brake frame bolt

Adjuster

Brake spider

Figure 13-91. *An expander tube brake assembly.*

1. An independent system not part of the aircraft main hydraulic system;

2. A booster system that uses the aircraft hydraulic system intermittently when needed; and

3. A power brake system that only uses the aircraft main. hydraulic system(s) as a source of pressure.

Systems on different aircraft vary, but the general operation is similar to those described.

Independent Master Cylinders

In general, small, light aircraft and aircraft without hydraulic systems use independent braking systems. An independent brake system is not connected in any way to the aircraft hydraulic system. Master cylinders are used to develop the necessary hydraulic pressure to operate the brakes. This is similar to the brake system of an automobile.

In most brake actuating systems, the pilot pushes on the tops of the rudder pedals to apply the brakes. A master cylinder for each brake is mechanically connected to the corresponding rudder pedal (i.e., right main brake to the right rudder pedal, left main brake to the left rudder pedal). *[Figure 13-93]* When the pedal is depressed, a piston inside a sealed fluid-filled chamber in the master cylinder forces hydraulic fluid through a line to the piston(s) in the brake assembly. The brake piston(s) push the brake linings against the brake rotor to create the friction that slows the wheel rotation. Pressure is increased throughout the entire brake systems and against the rotor as the pedal is pushed harder.

Many master cylinders have built-in reservoirs for the brake hydraulic fluid. Others have a single remote reservoir that services both of the aircraft's two master cylinders. *[Figure 13-94]* A few light aircraft with nose wheel steering have only one master cylinder that actuates both main wheel brakes. This is possible because steering the aircraft during taxi does not require differential braking. Regardless of the set-up, it is the master cylinder that builds up the pressure required for braking.

A master cylinder used with a remote reservoir is illustrated in *Figure 13-95*. This particular model is a Goodyear master cylinder. The cylinder is always filled with air-free, contaminant-free hydraulic fluid as is the reservoir and the line that connects the two together. When the top of the rudder pedal is depressed, the piston arm is mechanically moved forward into the master cylinder. It pushes the piston against the fluid, which is forced through the line to the brake. When pedal pressure is released, the return springs in the brake assembly retract the brake pistons back into the brake housing. The hydraulic fluid behind the pistons is displaced and must return to the master cylinder. As it does, a return spring in the master cylinder move the piston, piston rod and rudder pedal back to the original position (brake off, pedal not depressed). The fluid behind the master cylinder piston flows back into the reservoir. The brake is ready to be applied again.

Hydraulic fluid expands as temperature increases. Trapped

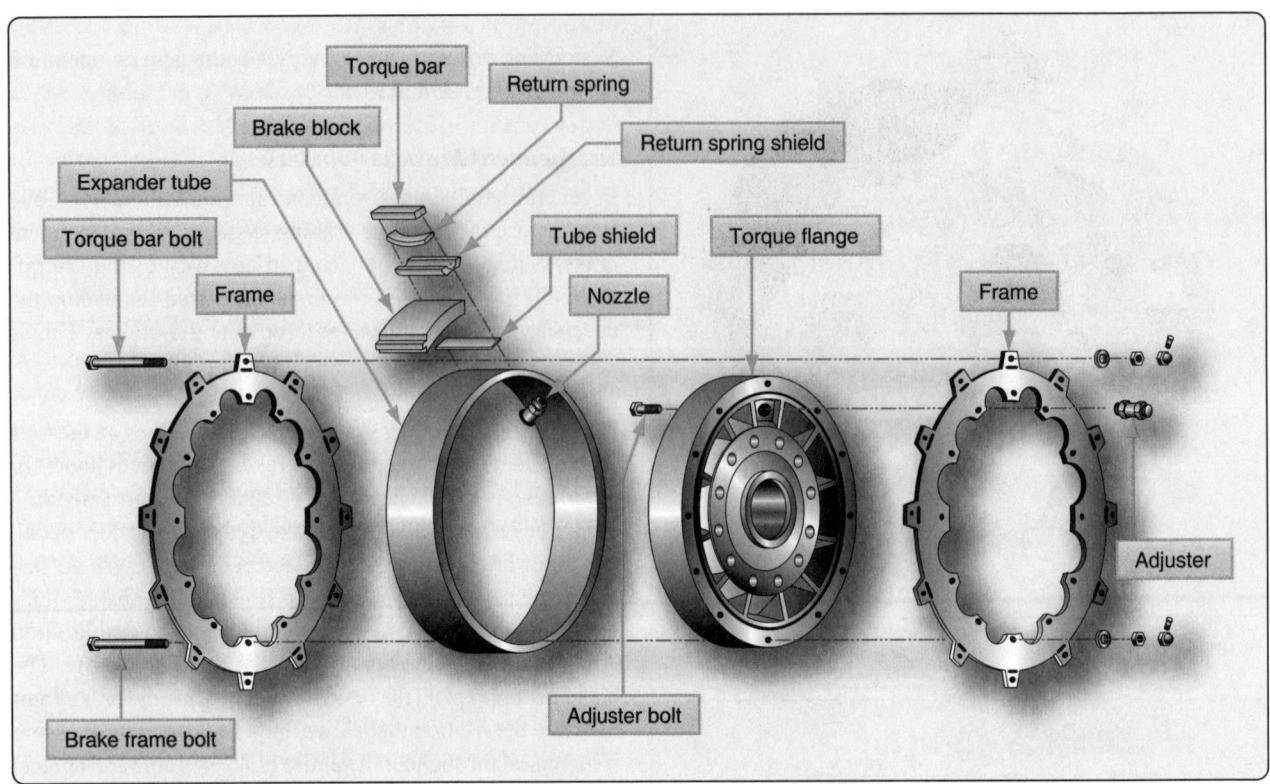

Figure 13-92. *An exploded view of an expander tube brake.*

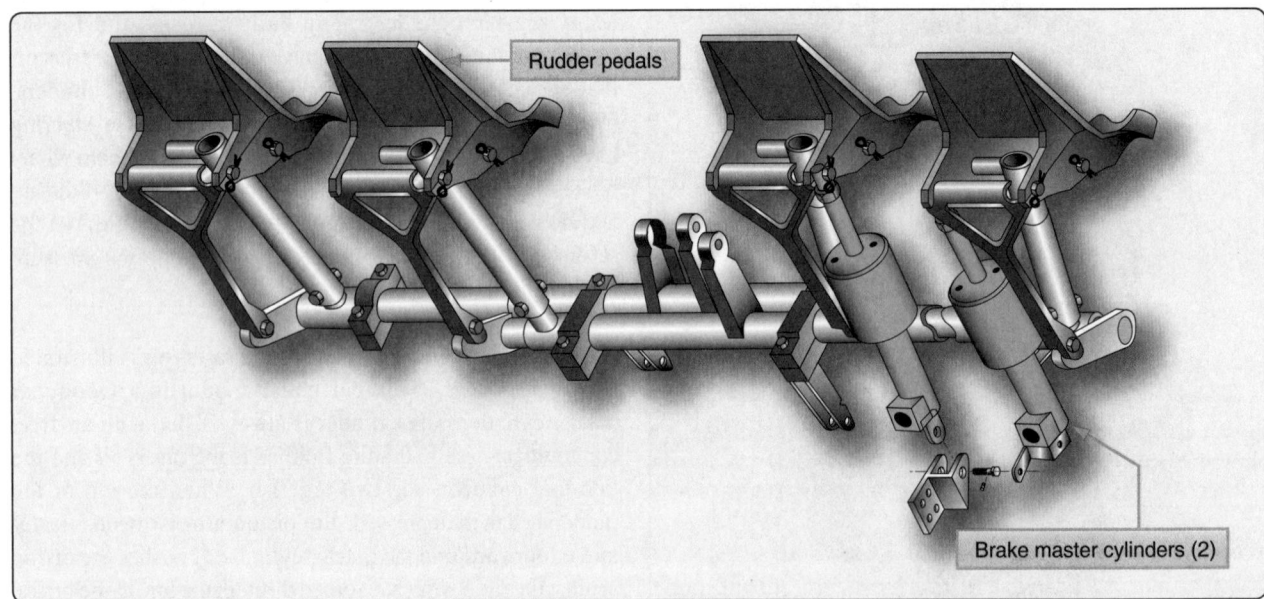

Figure 13-93. *Master cylinders on an independent brake system are directly connected to the rudder pedals or are connected through mechanical linkage.*

fluid can cause a brake to drag against the rotor(s). Leaks may also result. When the brakes are not applied, fluid must be allowed to expand safely without causing these issues. A compensating port is included in most master cylinders to facilitate this. In the master cylinder in *Figure 13-95*, this port is opened when the piston is fully retracted. Fluid in the brake system is allowed to expand into the reservoir, which

has the capacity to accept the extra fluid volume. The typical reservoir is also vented to the atmosphere to provide positive pressure on the fluid.

The forward side of the piston head contains a seal that closes off the compensating port when the brakes are applied so that pressure can build. The seal is only effective in the forward

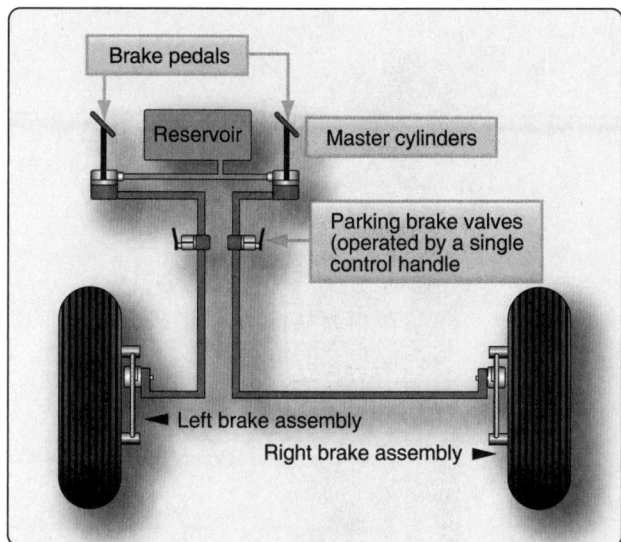

Figure 13-94. *A remote reservoir services both master cylinders on some independent braking systems.*

is relieved by a spring in the mechanical linkage.

A common requirement of all braking systems is for there to be no air mixed in with the hydraulic fluid. Since air is compressible and hydraulic fluid essentially is not, any air under pressure when the brakes are applied causes spongy brakes. The pedals do not feel firm when pushed down due to the air compressing. Brake systems must be bled to remove all air from the system. Instructions for bleeding the brakes are in the manufacturer's maintenance information. Brake systems equipped with Goodyear master cylinders must be bled from the top down to ensure any air trapped behind the master cylinder piston is removed.

An alternative common arrangement of independent braking systems incorporates two master cylinders, each with its own integral fluid reservoir. Except for the reservoir location, the brake system is basically the same as just described. The master cylinders are mechanically linked to the rudder pedals as before. Depressing the top of a pedal causes the piston rod to push the piston into the cylinder forcing the fluid out to the brake assembly. The piston rod rides in a compensator sleeve and contains an O-ring that seals the rod to the piston when the rod is moved forward. This blocks the compensating ports. When released, a spring returns the piston to its original position which refills the reservoir as it returns. The rod end seal retracts away from the piston head allowing a free flow of fluid from the cylinder through the compensating ports in the piston to the reservoir. *[Figure 13-96]*

The parking brake mechanism is a ratcheting type that operates as described. A servicing port is supplied at the top of the master cylinder reservoir. Typically, a vented plug is installed in the port to provide positive pressure on the fluid.

direction. When the piston is returning, or is fully retracted to the off position, fluid behind the piston is free to flow through piston head ports to replenish any fluid that may be lost downstream of the master cylinder. The aft end of the master cylinder contains a seal that prevents leakage at all times. A rubber boot fits over the piston rod and the aft end of the master cylinder to keep out dust.

A parking brake for this remote reservoir master cylinder brake system is a ratcheting mechanical device between the master cylinder and the rudder pedals. With the brakes applied, the ratchet is engaged by pulling the parking brake handle. To release the brakes, the rudder pedals are depressed further allowing the ratchet to disengage. With the parking brake set, any expansion of hydraulic fluid due to temperature

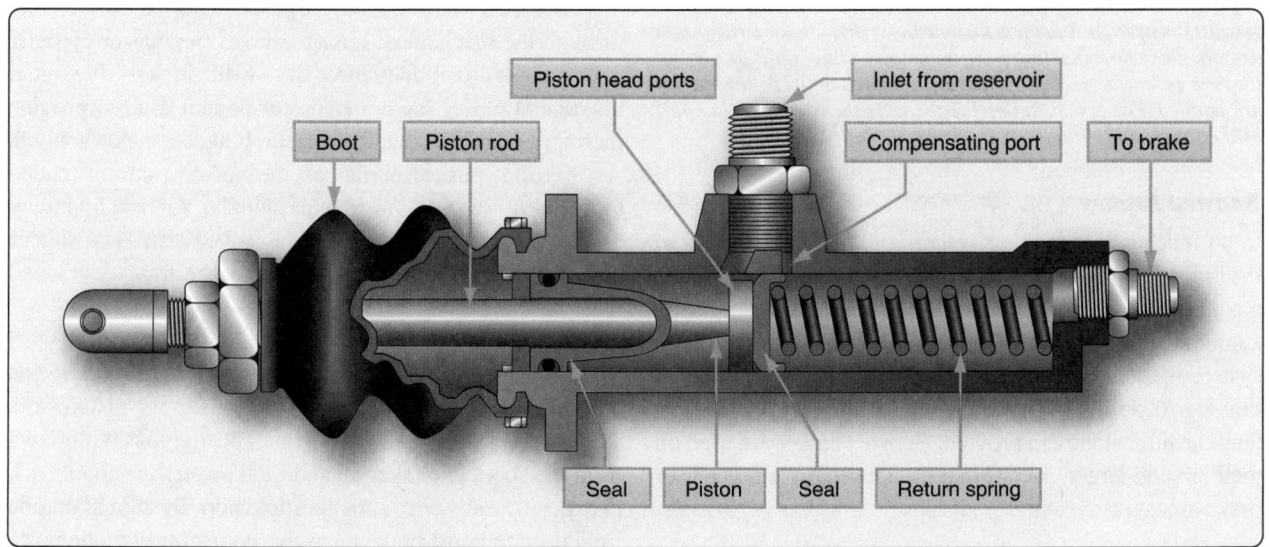

Figure 13-95. *A Goodyear brake master cylinder from an independent braking system with a remote reservoir.*

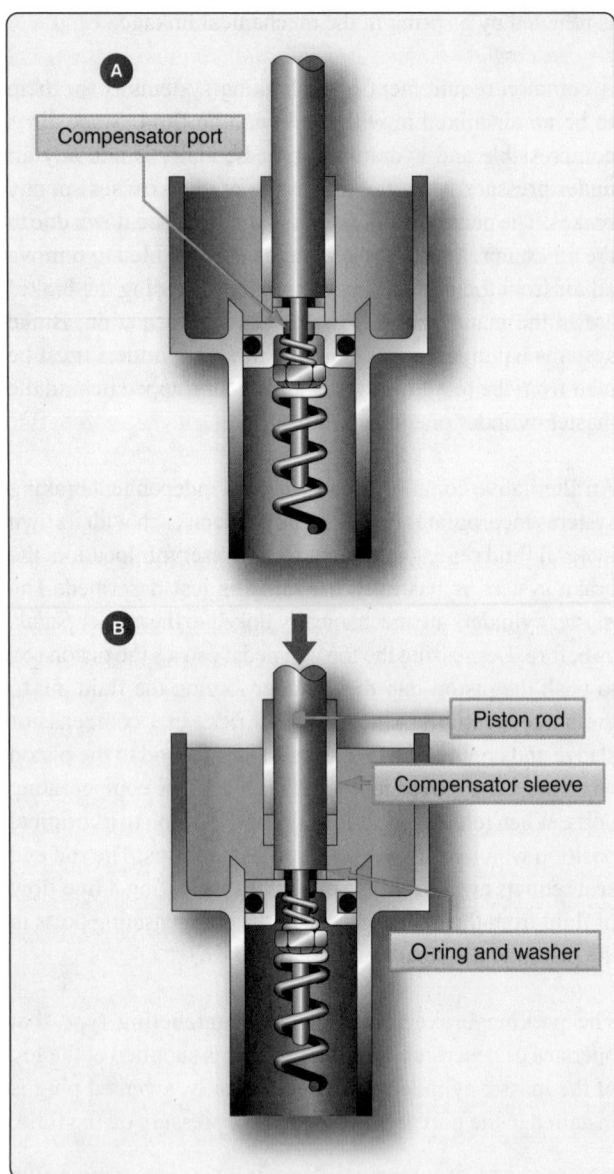

Figure 13-96. *A common master cylinder with built-in reservoir is shown. Illustration A depicts the master cylinder when the brakes are off. The compensating port is open to allow fluid to expand into the reservoir should temperature increase. In B, the brakes are applied. The piston rod-end seal covers the compensating port as it contacts the piston head.*

Boosted Brakes

In an independent braking system, the pressure applied to the brakes is only as great as the foot pressure applied to the top of the rudder pedal. Boosted brake actuating systems augment the force developed by the pilot with hydraulic system pressure when needed. The boost is only during heavy braking. It results in greater pressure applied to the brakes than the pilot alone can provide. Boosted brakes are used on medium and larger aircraft that do not require a full power brake actuating system.

A boosted brake master cylinder for each brake is mechanically

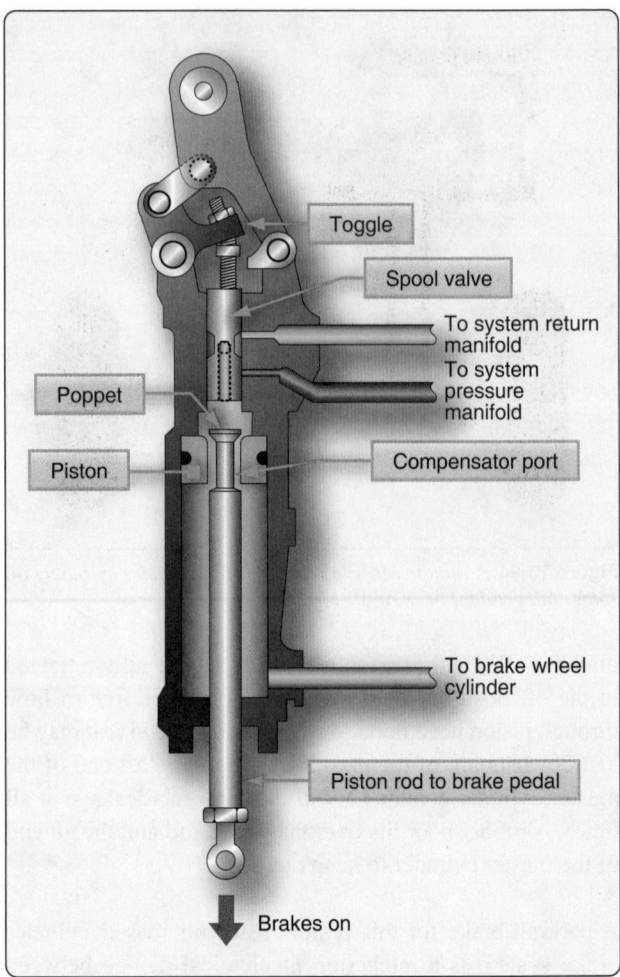

Figure 13-97. *A master cylinder for a boosted brake system augments foot pedal pressure with aircraft system hydraulic pressure during heavy braking.*

attached to the rudder pedals. However, the boosted brake master cylinder operates differently. *[Figure 13-97]*

When the brakes are applied, the pressure from the pilot's foot through the mechanical linkage moves the master cylinder piston in the direction to force fluid to the brakes. The initial movement closes the compensator poppet used to provide thermal expansion relief when the brakes are not applied. As the pilot pushes harder on the pedal, a spring-loaded toggle moves a spool valve in the cylinder. Aircraft hydraulic system pressure flows through the valve to the back side of the piston. Pressure is increased, as is the force developed to apply the brakes.

When the pedal is released, the piston rod travels in the opposite direction, and the piston returns to the piston stop. The compensating poppet reopens. The toggle is withdrawn from the spool via linkages, and fluid pushes the spool back to expose the system return manifold port. System hydraulic fluid used to boost brake pressure returns through the port.

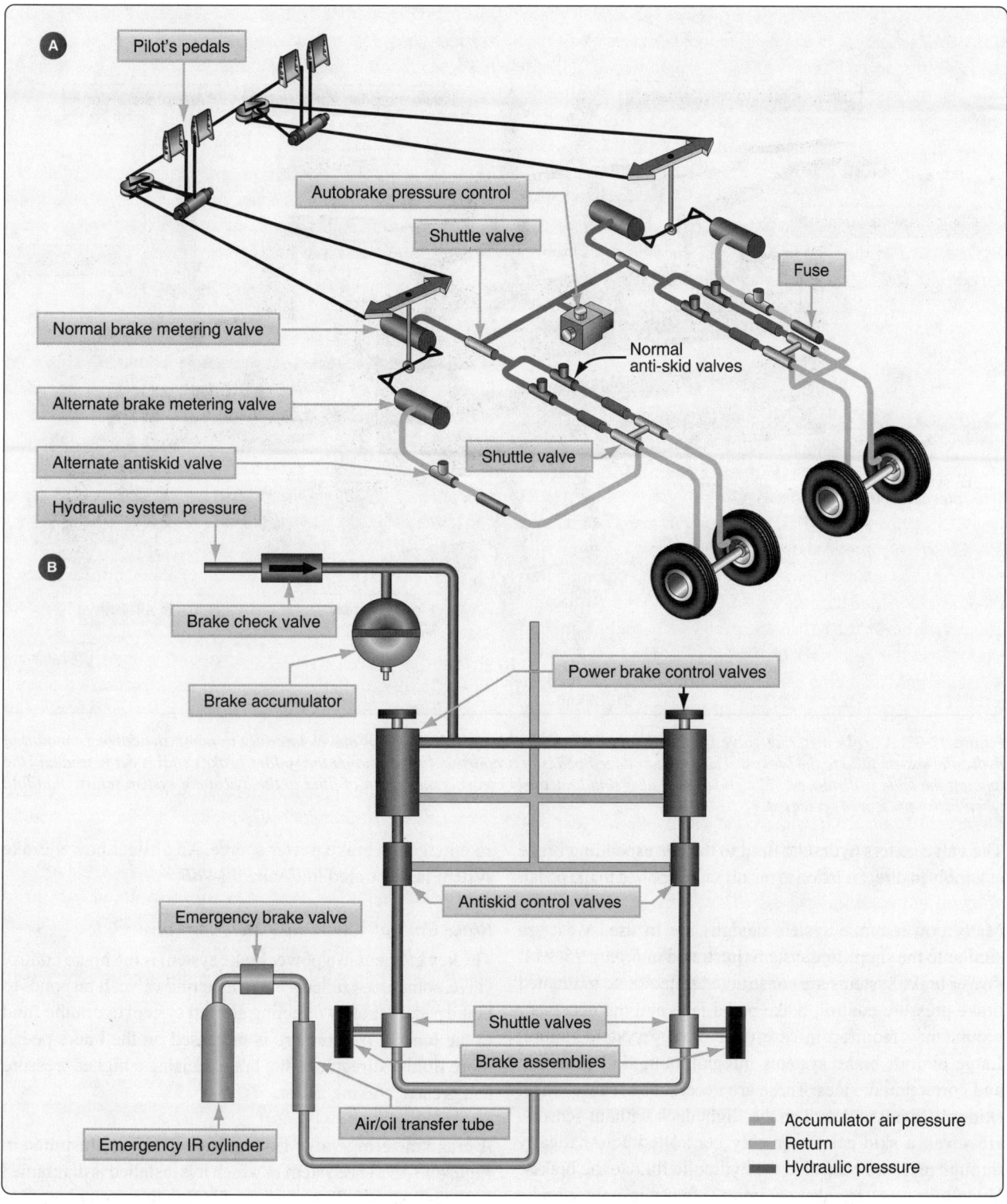

Figure 13-98. *The orientation of components in a basic power brake system is shown in A. The general layout of an airliner power brake system is shown in B.*

Power Brakes

Large and high-performance aircraft are equipped with power brakes to slow, stop, and hold the aircraft. Power brake actuating systems use the aircraft hydraulic system as the source of power to apply the brakes. The pilot presses on the top of the rudder pedal for braking as with the other actuating systems. The volume and pressure of hydraulic fluid required cannot be produced by a master cylinder. Instead, a power brake control valve or brake metering valve receives the brake pedal input either directly or through linkages.

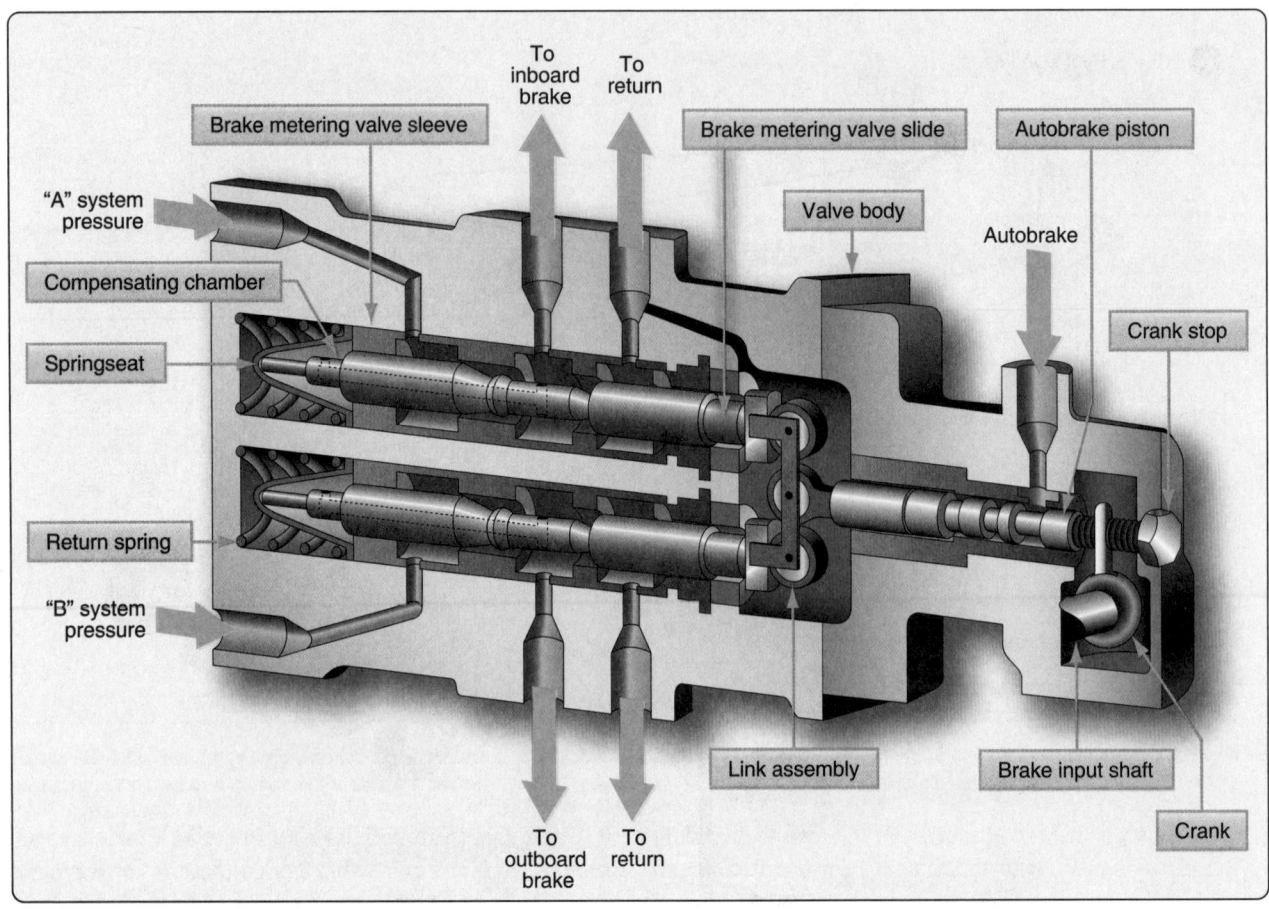

Figure 13-99. *A brake metering valve from a Boeing 737. A machined slide or spool moves laterally to admit the correct amount of hydraulic system fluid to the brakes. The pressure developed is in proportion to the amount the rudder/brake pedal is depressed and the amount the slide is displaced. The slide/spool also simultaneously controls the return of fluid to the hydraulic system return manifold when brake pressure is released.*

The valve meters hydraulic fluid to the corresponding brake assembly in direct relation to the pressure applied to the pedal.

Many power brake system designs are in use. Most are similar to the simplified system illustrated in *Figure 13-98A*. Power brake systems are constructed to facilitate graduated brake pressure control, brake pedal feel, and the necessary redundancy required in case of hydraulic system failure. Large aircraft brake systems integrate anti-skid detection and correction devices. These are necessary because wheel skid is difficult to detect on the flight deck without sensors. However, a skid can be quickly controlled automatically through pressure control of the hydraulic fluid to the brakes. Hydraulic fuses are also commonly found in power brake systems. The hostile environment around the landing gear increases the potential for a line to break or sever, a fitting to fail, or other hydraulic system malfunctions to occur where hydraulic fluid is lost en route to the brake assemblies. A fuse stops any excessive flow of fluid when detected by closing to retain the remaining fluid in the hydraulic system. Shuttle valves are used to direct flow from optional sources of fluid, such as in redundant systems or during the use of

an emergency brake power source. An airliner power brake system is illustrated in *Figure 13-98B*.

Brake Control Valve/Brake Metering Valve
The key element in a power brake system is the brake control valve, sometimes called a brake metering valve. It responds to brake pedal input by directing aircraft system hydraulic fluid to the brakes. As pressure is increased on the brake pedal, more fluid is directed to the brake causing a higher pressure and greater braking action.

A brake metering valve from a Boeing 737 is illustrated in *Figure 13-99*. The system in which it is installed is diagramed in *Figure 13-100*. Two sources of hydraulic pressure provide redundancy in this brake system. A brake input shaft, connected to the rudder/brake pedal through mechanical linkages, provides the position input to the metering valve. As in most brake control valves, the brake input shaft moves a tapered spool or slide in the valve so that it allows hydraulic system pressure to flow to the brakes. At the same time, the slide covers and uncovers access to the hydraulic system return port as required.

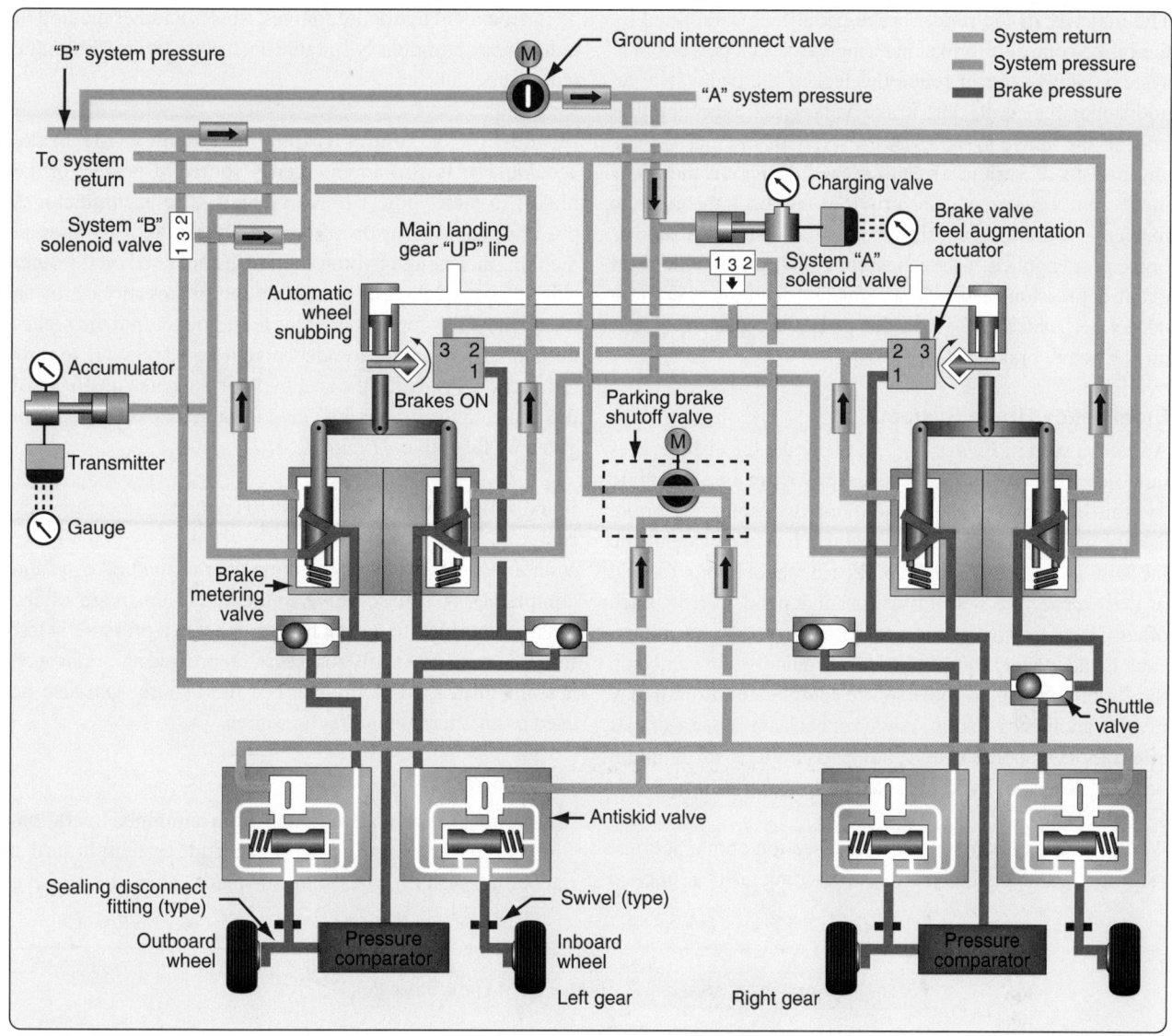

Figure 13-100. *The power brake system on a Boeing 737.*

When the rudder/brake pedal is depressed, the slide in the metering valve moves to the left. *[Figure 13-99]* It covers the return port so pressure can build in the brake system. The hydraulic supply pressure chamber is connected to the brake system pressure chamber by the movement of the slide, which due to its taper, unblocks the passage between these two. As the pedal is depressed further, the valve slide moves farther to the left. This enables more fluid to flow to the brakes due to the narrowing shape of the slide. Brake pressure increases with the additional fluid. A passage in the slide directs brake pressure fluid into a compensating chamber at the end of the slide. This acts on the end of the slide creating a return force that counters the initial slide movement and gives feel to the brake pedal. As a result, the pressure and return ports are closed and pressure proportional to the foot pressure on the pedal is held on the brakes. When the pedal is released, a return spring and compensating chamber pressure drive the slide to the right into its original position (return port

open, supply pressure chamber and brake pressure chambers blocked from each other).

The metering valve operates as described simultaneously for the inboard and the outboard brakes. *[Figure 13-99]* The design of the link assembly is such that a single side of the metering valve can operate even if the other fails. Most brake control valves and metering valves function in a similar manner, although many are single units that supply only one brake assembly.

The auto brake, referenced in the metering valve diagram, is connected into the landing gear retraction hydraulic line. Pressurized fluid enters this port and drives the slide slightly to the left to apply the brakes automatically after takeoff. This stops the wheels from rotating when retracted into the wheel wells. Auto brake pressure is withheld from this port when the landing gear is fully stowed since the retraction system is depressurized.

The majority of the rudder/brake pedal feel is supplied by the brake control or brake metering valve in a power brake system. Many aircraft refine the feel of the pedal with an additional feel unit. The brake valve feel augmentation unit, in the above system, uses a series of internal springs and pistons of various sizes to create a force on the brake input shaft movement. This provides feel back through the mechanical linkages consistent with the amount of rudder/brake pedal applied. The request for light braking with slight pedal depression results in a light feel to the pedal and a harder resistance feel when the pedals are pushed harder during heavy braking. [Figure 13-101]

Emergency Brake Systems

As can be seen in *Figure 13-100,* the brake metering valves not only receive hydraulic pressure from two separate hydraulic systems, they also feed two separate brake assemblies. Each main wheel assembly has two wheels. The inboard wheel brake and the outboard wheel brake, located in their respective wheel rims, are independent from each other. In case of hydraulic system failure or brake failure, each is independently supplied to adequately slow and stop the aircraft without the other. More complicated aircraft may involve another hydraulic system for back-up or use a similar alternation of sources and brake assemblies to maintain braking in case of hydraulic system or brake failure.

Note: In the segmented rotor brake section above, a brake assembly was described that had alternating pistons supplied by independent hydraulic sources. This is another method of redundancy particularly suitable on, but not limited to, single main wheel aircraft.

In addition to supply system redundancy, the brake accumulator is also an emergency source of power for the brakes in many power brake systems. The accumulator is pre-charged with air or nitrogen on one side of its internal diaphragm. Enough hydraulic fluid is contained on the other side of the diaphragm to operate the brakes in case of an emergency. It is forced out of the accumulator into the brakes through the system lines under enough stored pressure to slow the aircraft. Typically, the accumulator is located upstream of the brake control/metering valve to capitalize on the control given by the valve. [Figure 13-102]

Some simpler power brake systems may use an emergency source of brake power that is delivered directly to the brake assemblies and bypasses the remainder of the brake system completely. A shuttle valve immediately upstream of the brake units shifts to accept this source when pressure is lost from the primary supply sources. Compressed air or nitrogen is sometimes used. A pre-charged fluid source can also be used as an alternate hydraulic source.

Parking Brake

The parking brake system function is a combined operation. The brakes are applied with the rudder pedals and a ratcheting system holds them in place when the parking

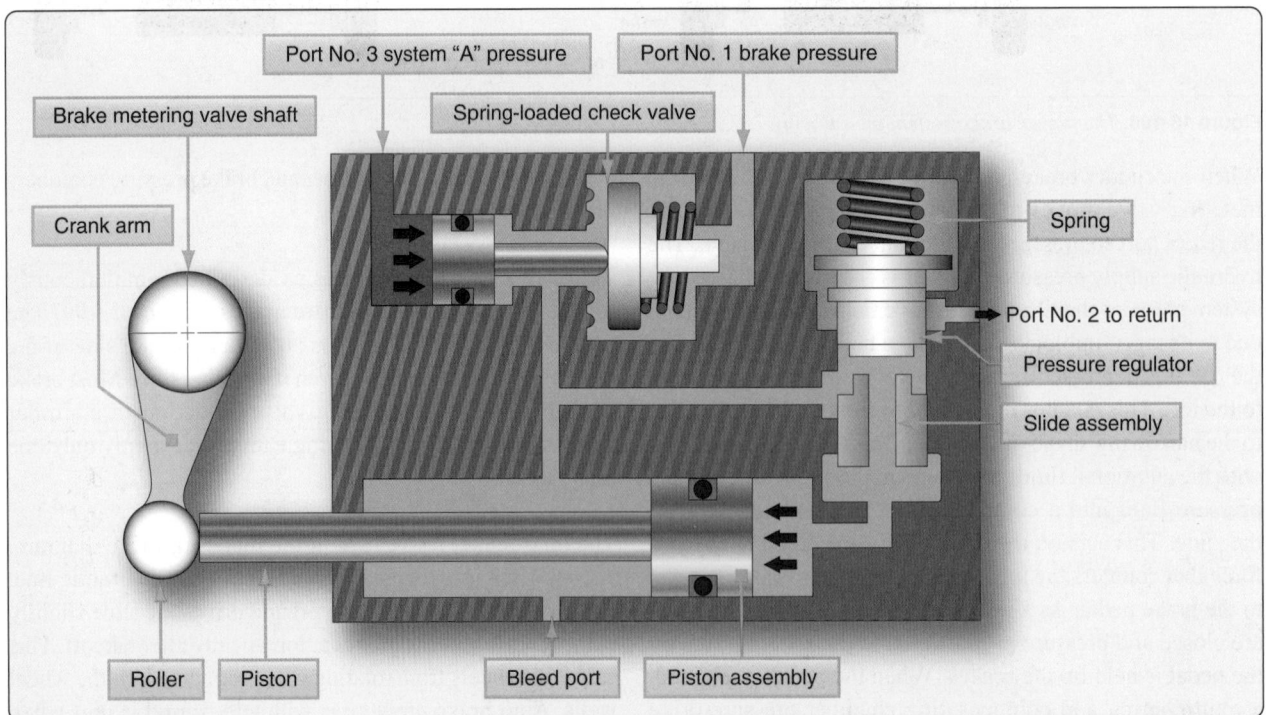

Figure 13-101. *The power brake system on a Boeing 737.*

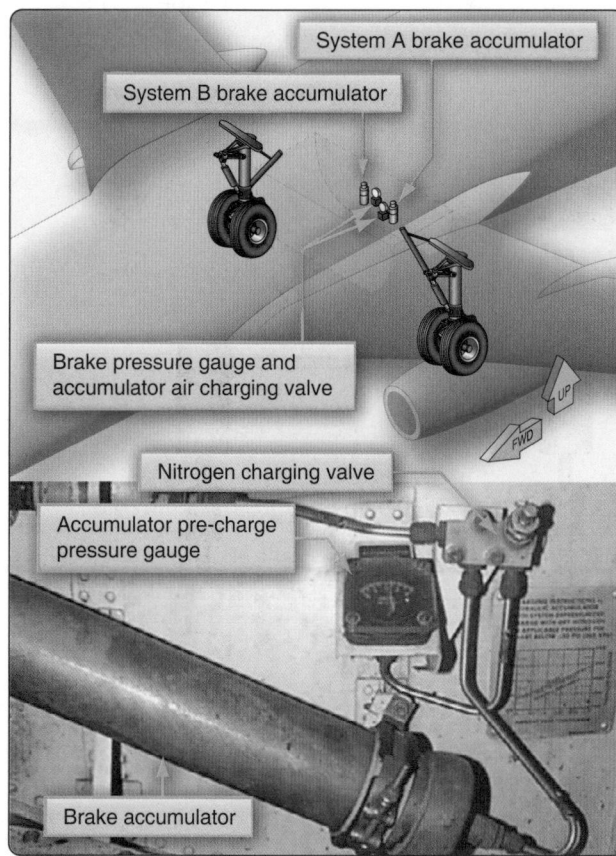

System A brake accumulator

System B brake accumulator

Brake pressure gauge and accumulator air charging valve

Nitrogen charging valve

Accumulator pre-charge pressure gauge

Brake accumulator

Figure 13-102. *Emergency brake hydraulic fluid accumulators are precharged with nitrogen to deliver brake fluid to the brakes in the event normal and alternate hydraulic sources fail.*

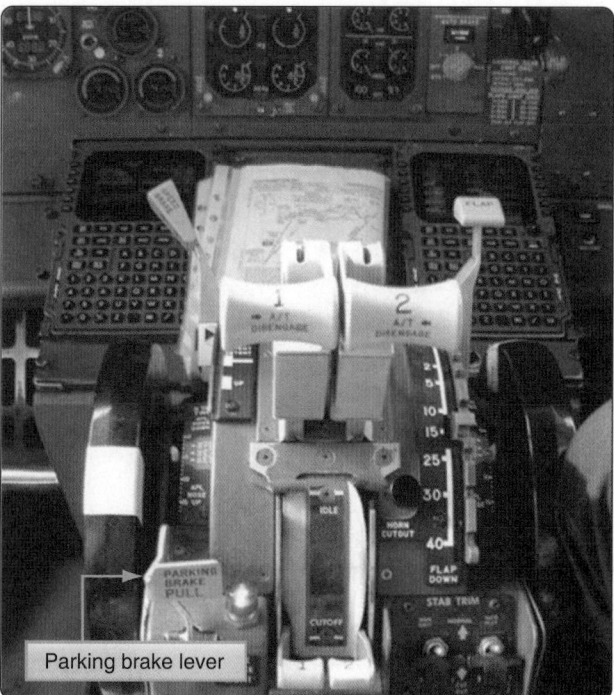

Parking brake lever

Figure 13-103. *The parking brake lever on a Boeing 737 center pedestal throttle quadrant.*

brake lever on the flight deck is pulled. *[Figure 13-103]* At the same time, a shut-off valve is closed in the common return line from the brakes to the hydraulic system. This traps the fluid in the brakes holding the rotors stationary. Depressing the pedals further releases the pedal ratchet and opens the return line valve.

Brake Deboosters

Some aircraft brake assemblies that operate on aircraft hydraulic system pressure are not designed for such high pressure. They provide effective braking through a power brake system but require less than maximum hydraulic system pressure. To supply the lower pressure, a brake debooster cylinder is installed downstream of the control valve and anti-skid valve. *[Figure 13-104]* The debooster reduces some pressure from the control valve to within the working range of the brake assembly.

Brake deboosters are simple devices that use the application of force over different sized pistons to reduce pressure. *[Figure 13-105]* Their operation can be understood through the application of the following equation:

Pressure = Force/Area

High-pressure hydraulic system input pressure acts on the small end of a piston. This develops a force proportional to the area of the piston head. The other end of the piston is larger and housed in a separate cylinder. The force from the smaller piston head is transferred to the larger area of the other end of the piston. The amount of pressure conveyed by the larger end of the piston is reduced due to the greater area over which the force is spread. The volume of output fluid increases since a larger piston and cylinder are used. The reduced pressure is delivered to the brake assembly.

The spring in the debooster aids in returning the piston to the ready position. If fluid is lost downstream of the deboost cylinder, the piston travels further down into the cylinder when the brakes are applied. The pin unseats the ball and allows fluid into the lower cylinder to replace what was lost. Once replenished, the piston rises up in the cylinder due to pressure build-up. The ball reseats as the piston travels above the pin and normal braking resumes. This function is not meant to permit leaks in the brake assemblies. Any leak discovered must be repaired by the technician.

A lockout debooster functions as a debooster and a hydraulic fuse. If fluid is not encountered as the piston moves down in the cylinder, the flow of fluid to the brakes is stopped. This prevents the loss of all system hydraulic fluid should a rupture downstream of the debooster occur. Lockout deboosters have a handle to reset the device after it closes as a fuse. If not reset, no braking action is possible.

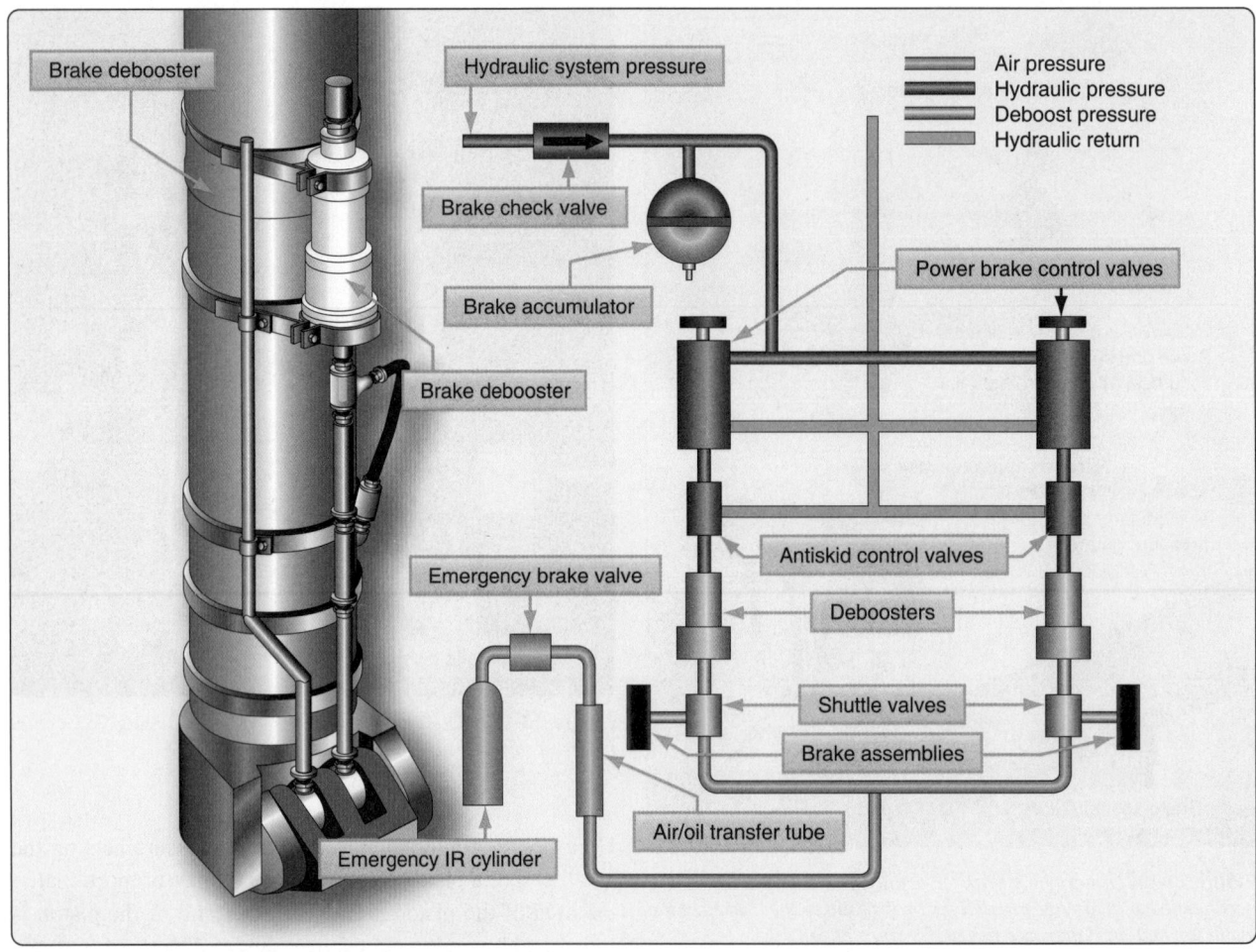

Figure 13-104. *The location of a brake debooster cylinder on a landing gear strut and the debooster's position in relation to other components of a power brake system.*

Anti-Skid

Large aircraft with power brakes require anti-skid systems. It is not possible to immediately ascertain in the flight deck when a wheel stops rotating and begins to skid, especially in aircraft with multiple-wheel main landing gear assemblies. A skid not corrected can quickly lead to a tire blowout, possible damage to the aircraft, and control of the aircraft may be lost.

System Operation

The anti-skid system not only detects wheel skid, it also detects when wheel skid is imminent. It automatically relieves pressure to the brake pistons of the wheel in question by momentarily connecting the pressurized brake fluid area to the hydraulic system return line. This allows the wheel to rotate and avoid a skid. Lower pressure is then maintained to the brake at a level that slows the wheel without causing it to skid.

Maximum braking efficiency exists when the wheels are decelerating at a maximum rate but are not skidding. If a wheel decelerates too fast, it is an indication that the brakes are about to lock and cause a skid. To ensure that this does not happen, each wheel is monitored for a deceleration rate

faster than a preset rate. When excessive deceleration is detected, hydraulic pressure is reduced to the brake on that wheel. To operate the anti-skid system, flight deck switches must be placed in the ON position. *[Figure 13-106]* After the aircraft touches down, the pilot applies and holds full pressure to the rudder brake pedals. The anti-skid system then functions automatically until the speed of the aircraft has dropped to approximately 20 mph. The system returns to manual braking mode for slow taxi and ground maneuvering.

There are various designs of anti-skid systems. Most contain three main types of components: wheel speed sensors, anti-skid control valves, and a control unit. These units work together without human interference. Some anti-skid systems provide complete automatic braking. The pilot needs only to turn on the auto brake system, and the anti-skid components slow the aircraft without pedal input. *[Figure 13-106]* Ground safety switches are wired into the circuitry for anti-skid and auto brake systems. Wheel speed sensors are located on each wheel equipped with a brake assembly. Each brake also has its own anti-skid control valve. Typically, a single control box contains the anti-skid comparative circuitry for all of the brakes on the aircraft. *[Figure 13-107]*

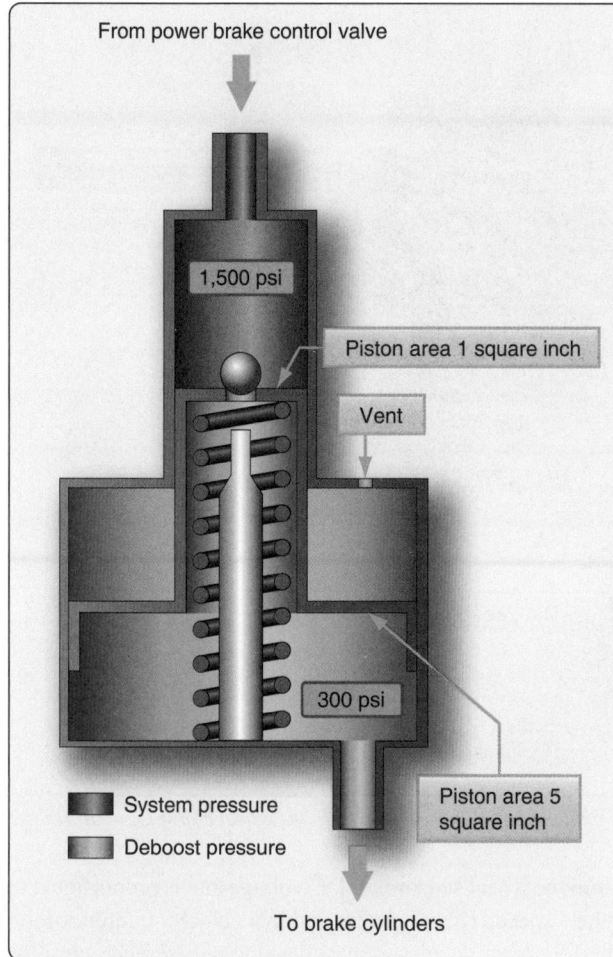

Figure 13-105. *Brake deboosters.*

Wheel Speed Sensors

Wheel speed sensors are transducers. They may be alternating current (AC) or direct current (DC). The typical AC wheel speed sensor has a stator mounted in the wheel axle. A coil around it is connected to a controlled DC source so that when energized, the stator becomes an electromagnet. A rotor that turns inside the stator is connected to the rotating wheel hub assembly through a drive coupling so that it rotates at the speed of the wheel. Lobes on the rotor and stator cause the distance between the two components to constantly change during rotation. This alters the magnetic coupling or reluctance between the rotor and stator. As the electromagnetic field changes, a variable frequency AC is induced in the stator coil. The frequency is directly proportional to the speed of rotation of the wheel. The AC signal is fed to the control unit for processing. A DC wheel speed sensor is similar, except that a DC is produced the magnitude of which is directly proportional to wheel speed. *[Figure 13-108]*

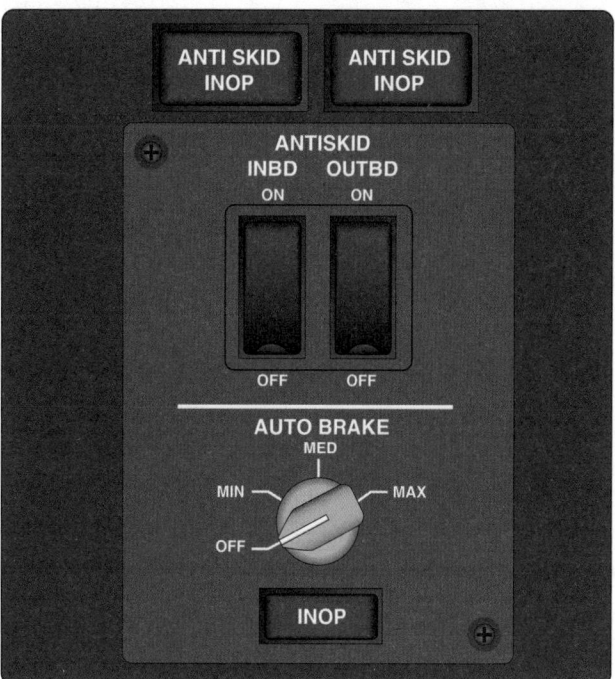

Figure 13-106. *Anti-skid switches in the flight deck.*

Control Units

The control unit can be regarded as the brain of the anti-skid system. It receives signals from each of the wheel sensors. Comparative circuits are used to determine if any of the signals indicate a skid is imminent or occurring on a particular wheel. If so, a signal is sent to the control valve of the wheel to relieve hydraulic pressure to that brake which prevents or relieves the skid. The control unit may or may not have external test switches and status indicating lights. It is common for it to be located in the avionics bay of the aircraft. *[Figure 13-109]*

The Boeing anti-skid control valve block diagram in *Figure 13-110* gives further detail on the functions of an anti-skid control unit. Other aircraft may have different logic

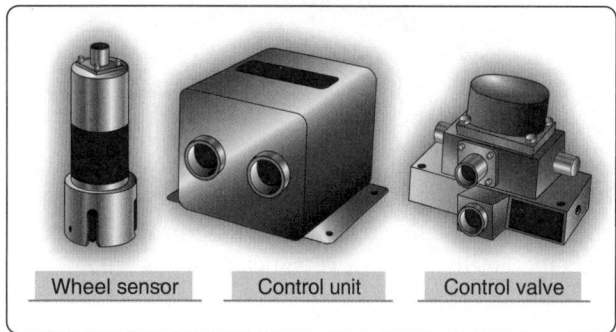

Figure 13-107. *A wheel sensor (left), a control unit (center), and a control valve (right) are components of an anti-skid system. A sensor is located on each wheel equipped with a brake assembly. An anti-skid control valve for each brake assembly is controlled from a single central control unit.*

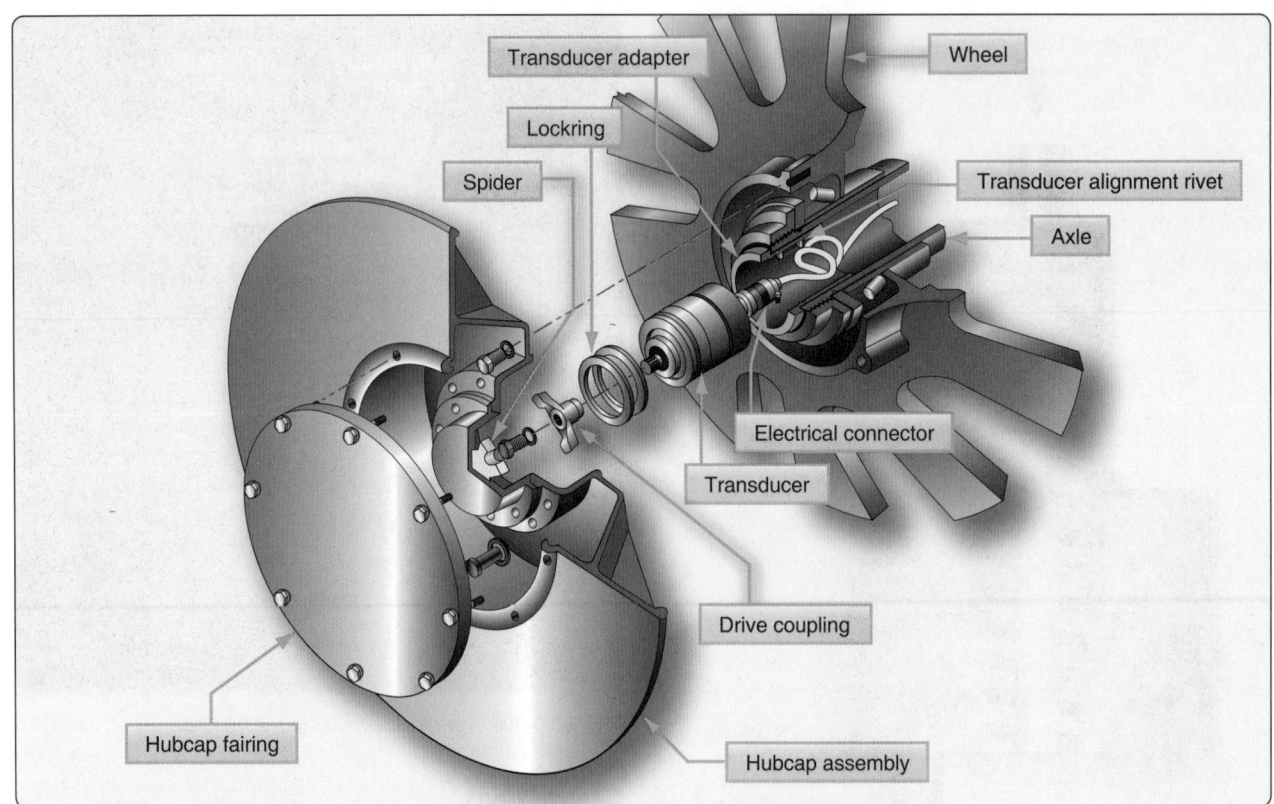

Figure 13-108. *The stator of an anti-skid wheel sensor is mounted in the axle, and the rotor is coupled to the wheel hub spider that rotates with the wheel.*

Figure 13-109. *A rack mounted anti-skid control unit from an airliner.*

to achieve similar end results. DC systems do not require an input converter since DC is received from the wheel sensors, and the control unit circuitry operates primarily with DC. Only the functions on one circuit card for one-wheel brake assembly are shown in *Figure 13-110*. Each wheel has its own identical circuitry card to facilitate simultaneous operation. All cards are housed in a single control unit that Boeing calls a control shield.

The converter shown changes the AC frequency received from the wheel sensor into DC voltage that is proportional to wheel speed. The output is used in a velocity reference loop that contains deceleration and velocity reference circuits. The converter also supplies input for the spoiler system and the locked wheel system, which is discussed at the end of this section. A velocity reference loop output voltage is produced, which represents the instantaneous velocity of the aircraft. This is compared to converter output in the velocity comparator. This comparison of voltages is essentially the comparison of the aircraft speed to wheel speed. The output from the velocity comparator is a positive or negative error voltage corresponding to whether the wheel speed is too fast or too slow for optimum braking efficiency for a given aircraft speed.

The error output voltage from the comparator feeds the pressure bias modulator circuit. This is a memory circuit that establishes a threshold where the pressure to the brakes provides optimum braking. The error voltage causes the modulator to either increase or decrease the pressure to the brakes in attempt to hold the modulator threshold. It produces a voltage output that is sent to the summing amplifier to do this. A lead output from the comparator anticipates when the tire is about to skid with a voltage that decreases the pressure to the brake. It sends this voltage to the summing amplifier as well. A transient control output from the comparator designed for rapid pressure dump when a sudden skid has occurred also

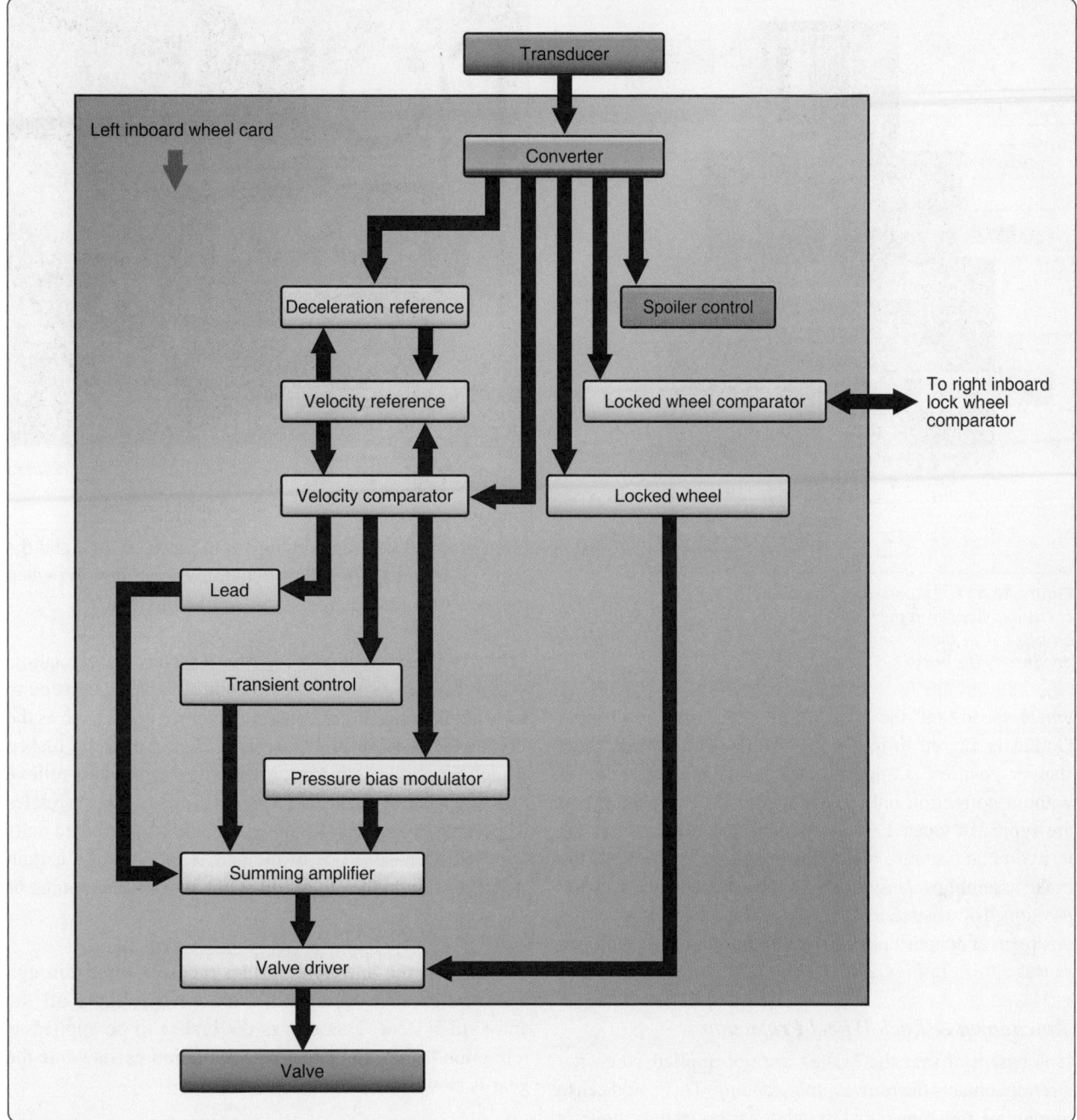

Figure 13-110. *A Boeing 737 anti-skid control unit internal block diagram.*

sends voltage to the summing amp. As the name suggests, the input voltages to the amplifier are summed, and a composite voltage is sent to the valve driver. The driver prepares the current required to be sent to the control valve to adjust the position of the valve. Brake pressure increases, decreases, or holds steady depending on this value.

Anti-Skid Control Valves

Anti-skid control valves are fast-acting, electrically controlled hydraulic valves that respond to the input from the anti-skid control unit. There is one control valve for each brake assembly. A torque motor uses the input from the valve driver to adjust the position of a flapper between two nozzles. By moving the flapper closer to one nozzle or the other, pressures are developed in the second stage of the valve. These pressures act on a spool that is positioned to build or reduce pressure to the brake by opening and blocking fluid ports. *[Figure 13-111]*

As pressure is adjusted to the brakes, deceleration slows to within the range that provides the most effective braking without skidding. The wheel sensor signal adjusts to the

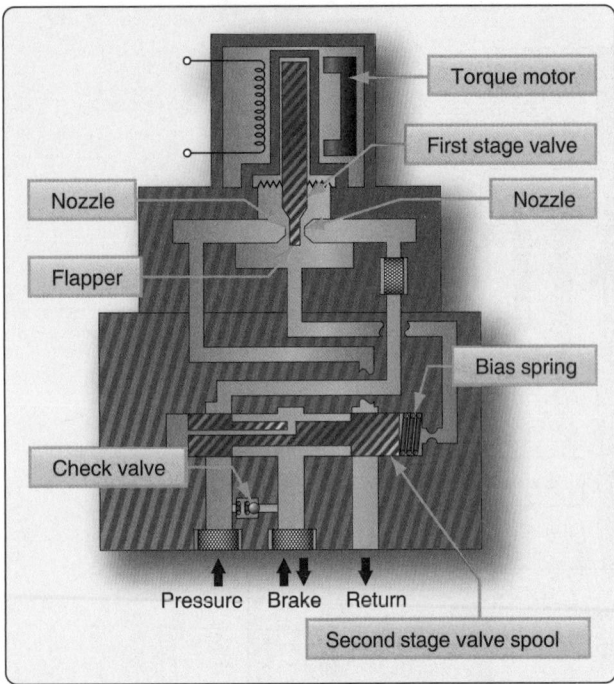

Figure 13-111. *An anti-skid control valve uses a torque motor controlled flapper in the first stage of the valve to adjust pressure on a spool in the second stage of the valve to build or relieve pressure to the brake.*

wheel speed, and the control unit processes the change. Output is altered to the control valve. The control valve flapper position is adjusted and steady braking resumes without correction until needed. Anti-skid control valves are typically located in the main wheel for close access to hydraulic pressure and return manifolds, as well as the brake assemblies. *[Figure 13-112]* Systematically, they are positioned downstream of the power brake control valves but upstream of debooster cylinders if the aircraft is so equipped as was shown in *Figure 13-104*.

Touchdown & Lock Wheel Protection

It is essential that the brakes are not applied when the aircraft contacts the runway upon landing. This could cause immediate tire blowout. A touchdown protection mode is built into most aircraft anti-skid systems to prevent this. It typically functions in conjunction with the wheel speed sensor and the air/ground safety switch on the landing gear strut (squat switch). Until the aircraft has weight on wheels, the detector circuitry signals the anti-skid control valve to open the passage between the brakes and the hydraulic system return, thus preventing pressure build-up and application of the brakes. Once the squat switch is open, the anti-skid control unit sends a signal to the control valve to close and permit brake pressure build-up. As a back-up and when the aircraft is on the ground with the strut not compressed enough to open the squat switch, a minimum wheel speed sensor signal can override and allow braking. Wheels are

Figure 13-112. *Two anti-skid control valves with associated plumbing and wiring.*

often grouped with one relying on the squat switch and the other on wheel speed sensor output to ensure braking when the aircraft is on the ground, but not before then.

Locked wheel protection recognizes if a wheel is not rotating. When this occurs, the anti-skid control valve is signaled to fully open. Some aircraft anti-skid control logic, such as the Boeing 737 shown in *Figure 13-111*, expands the locked wheel function. Comparator circuitry is used to relieve pressure when one wheel of a paired group of wheels rotates 25 percent slower than the other. Inboard and outboard pairs are used because if one of the pair is rotating at a certain speed, so should the other. If it is not, a skid is beginning or has occurred.

On takeoff, the anti-skid system receives input through a switch located on the gear selector that shuts off the anti-skid system. This allows the brakes to be applied as retraction occurs so that no wheel rotation exists while the gear is stowed.

Auto Brakes

Aircraft equipped with auto brakes typically bypass the brake control valves or brake metering valves and use a separate auto brake control valve to provide this function. In addition to the redundancy provided, auto brakes rely on the anti-skid system to adjust pressure to the brakes if required due to an impending skid. *Figure 13-113* shows a simplified diagram of the Boeing 757 brake system with the auto brake valve in relation to the main metering valve and anti-skid valves in this eight-main wheel system.

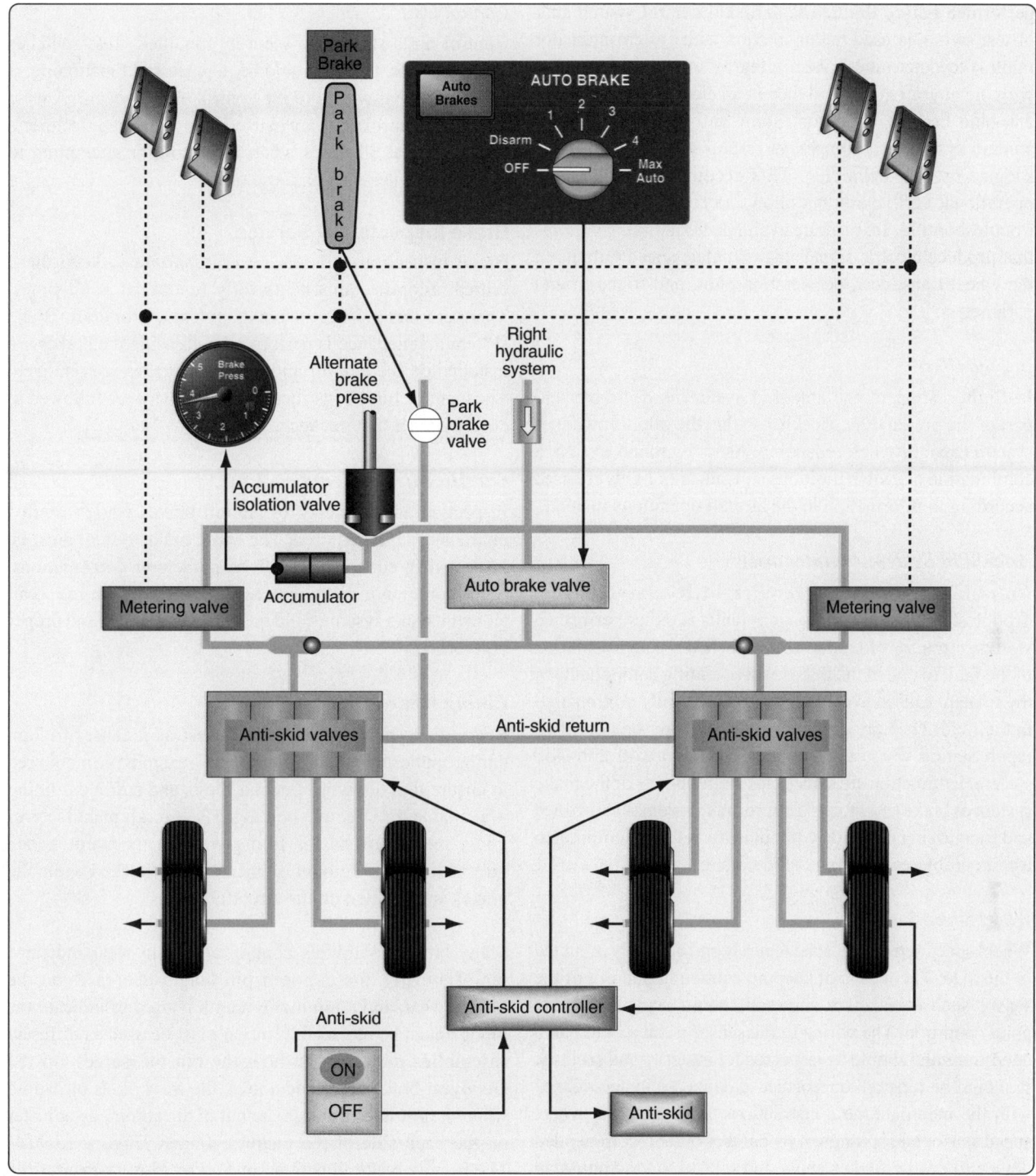

Figure 13-113. *The Boeing 757 normal brake system with auto brake and anti-skid.*

Anti-Skid System Tests

It is important to know the status of the anti-skid system prior to attempting to use it during a landing or aborted takeoff. Ground tests and in-flight tests are used. Built-in test circuits and control features allow testing of the system components and provide warnings should a particular component or part of the system become inoperative. An inoperative anti-skid system can be shut off without affecting normal brake operation.

Ground Test

Ground tests vary slightly from aircraft to aircraft. Consult the manufacturer's maintenance manual for test procedures specific to the aircraft in question.

Much of the anti-skid system testing originates from testing circuits in the anti-skid control unit. Built-in test circuits continuously monitor the anti-skid system and provide warning if a failure occurs. An operational test can be

13-65

performed before flight. The anti-skid control switch and/or test switch is used in conjunction with system indicator light(s) to determine system integrity. A test is first done with the aircraft at rest and then in an electrically simulated anti-skid braking condition. Some anti-skid control units contain system and component testing switches and lights for use by the technician. This accomplishes the same operational verification but allows an additional degree of troubleshooting. Test sets are available for anti-skid systems that produce electric signals that simulate speed outputs of the wheel transducer, deceleration rates, and flight/ground parameters.

In-Flight Test

In-flight testing of the anti-skid system is desirable and part of the pre-landing checklist so that the pilot is aware of system capability before landing. As with ground testing, a combination of switch positions and indicator lights are used according to information in the aircraft operations manual.

Anti-Skid System Maintenance

Anti-skid components require little maintenance. Troubleshooting anti-skid system faults is either performed via test circuitry or can be accomplished through isolation of the fault to one of the three main operating components of the system. Anti-skid components are normally not repaired in the field. They are sent to the manufacturer or a certified repair station when work is required. Reports of anti-skid system malfunction are sometimes malfunctions of the brake system or brake assemblies. Ensure brake assemblies are bled and functioning normally without leaks before attempting to isolate problems in the anti-skid system.

Wheel Speed Sensor

Wheel speed sensors must be securely and correctly mounted in the axle. The means of keeping contamination out of the sensor, such as sealant or a hub cap, should be in place and in good condition. The wiring to the sensor is subject to harsh conditions and should be inspected for integrity and security. It should be repaired or replaced if damaged in accordance with the manufacturer's instructions. Accessing the wheel speed sensor and spinning it by hand or other recommended device to ensure brakes apply and release via the anti-skid system is common practice.

Control Valve

Anti-skid control valve and hydraulic system filters should be cleaned or replaced at the prescribed intervals. Follow all manufacturer's instructions when performing this maintenance. Wiring to the valve must be secure, and there should be no fluid leaks.

Control Unit

Control units should be securely mounted. Test switches and indicators, if any, should be in place and functioning. It is essential that wiring to the control unit is secure. A wide variety of control units are in use. Follow the manufacturer's instructions at all times when inspecting or attempting to perform maintenance on these units.

Brake Inspection & Service

Brake inspection and service is important to keep these critical aircraft components fully functional at all times. There are many different brake systems on aircraft. Brake system maintenance is performed both while the brakes are installed on the aircraft and when the brakes are removed. The manufacturer's instructions must always be followed to ensure proper maintenance.

On Aircraft Servicing

Inspection and servicing of aircraft brakes while installed on the aircraft is required. The entire brake system must be inspected in accordance with manufacturer's instructions. Some common inspection items include: brake lining wear, air in the brake system, fluid quantity level, leaks, and proper bolt torque.

Lining Wear

Brake lining material is made to wear as it causes friction during application of the brakes. This wear must be monitored to ensure it is not worn beyond limits and sufficient lining is available for effective braking. The aircraft manufacturer gives specifications for lining wear in its maintenance information. The amount of wear can be checked while the brakes are installed on the aircraft.

Many brake assemblies contain a built-in wear indicator pin. Typically, the exposed pin length decreases as the linings wear, and a minimum length is used to indicate the linings must be replaced. Caution must be used as different assemblies may vary in how the pin measured. On the Goodyear brake described above, the wear pin is measured where it protrudes through the nut of the automatic adjuster on the back side of the piston cylinder. *[Figure 13-114]* The Boeing brake illustrated in *Figure 13-89* measures the length of the pin from the back of the pressure plate when the brakes are applied (dimension L). The manufacturer's maintenance information must be consulted to ensure brake wear pin indicators on different aircraft are read correctly.

On many other brake assemblies, lining wear is not measured via a wear pin. The distance between the disc and a portion of the brake housing when the brakes are applied is sometimes used. As the linings wear, this distance increases. The manufacturer specified at what distance the linings should be changed. *[Figure 13-115]*

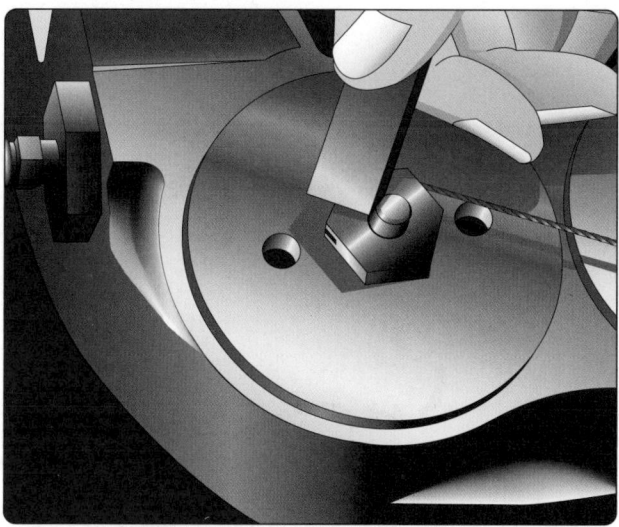

Figure 13-114. *Brake lining wear on a Goodyear brake is ascertained by measuring the wear pin of the automatic adjuster.*

On Cleveland brakes, lining wear can be measured directly, since part of the lining is usually exposed. The diameter of a number 40 twist drill is approximately equal to the minimum lining thickness allowed. *[Figure 13-116]*

Multiple disc brakes typically are checked for lining wear by applying the brakes and measuring the distance between the back of the pressure plate and the brake housing. *[Figure 13-117]* Regardless of the method particular to each brake, regular monitoring and measurement of brake wear

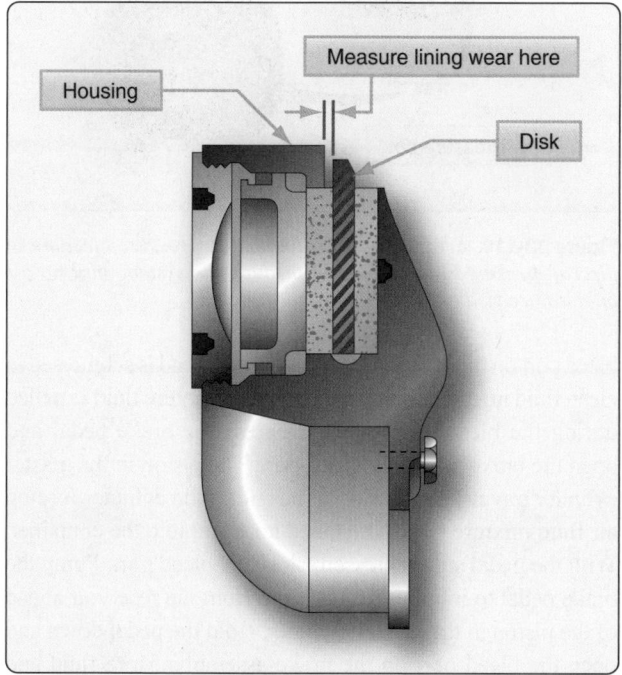

Figure 13-115. *The distance between the brake disc and the brake housing measured with the brakes applied is a means for determining brake lining wear on some brakes.*

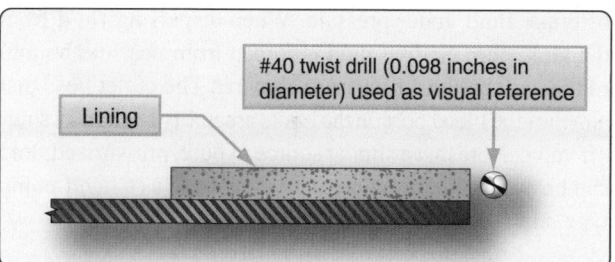

Figure 13-116. *A #40 twist drill laid next to the brake lining indicates when the lining needs to be changed on a Cleveland brake.*

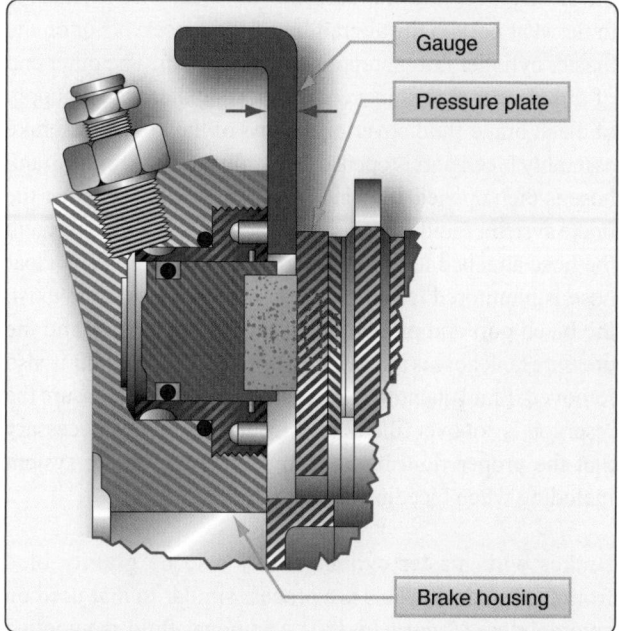

Figure 13-117. *The distance between the brake housing and the pressure plate indicates lining wear on some multiple disc brakes.*

ensures linings are replaced as they become unserviceable. Linings worn beyond limits usually require the brake assembly to be removed for replacement.

Air in the Brake System

The presence of air in the brake system fluid causes the brake pedal to feel spongy. The air can be removed by bleeding to restore firm brake pedal feel. Brake systems must be bled according to manufacturers' instructions. The method used is matched to the type of brake system. Brakes are bled by one of two methods: top down, gravity bleeding or bottom up pressure bleeding. Brakes are bled when the pedals feel spongy or whenever the brake system has been opened.

Bleeding Master Cylinder Brake Systems

Brake systems with master cylinders may be bled by gravity or pressure bleeding methods. Follow the instructions in the aircraft maintenance manual. To pressure bleed a brake system from the bottom up, a pressure pot is used. *[Figure 13-118]* This is a portable tank that contains a supply of brake fluid under pressure. When dispersing fluid from the tank, pure air-free fluid is forced from near the bottom of the tank by the air pressure above it. The outlet hose that attaches the bleed port on the brake assembly contains a shut-off valve. Note that a similar source of pure, pressurized fluid can be substituted for a pressure tank, such as a hand-pump type unit found in some hangars.

The typical pressure bleed is accomplished as illustrated in *Figure 13-119*. The hose from the pressure tank is attached to the bleed port on the brake assembly. A clear hose is attached to the vent port on the aircraft brake fluid reservoir or on the master cylinder if it incorporates the reservoir. The other end of this hose is placed in a collection container with a supply of clean brake fluid covering the end of the hose. The brake assembly bleed port is opened. The valve on the pressure tank hose is then opened allowing pure, air-free fluid to enter the brake system. Fluid containing trapped air is expelled through the hose attached to the vent port of the reservoir. The clear hose is monitored for air bubbles. When they cease to exist, the bleed port and pressure tank shutoff are closed, and the pressure tank hose is removed. The hose at the reservoir is also removed. Fluid quantity may need to be adjusted to assure the reservoir is not over filled. Note that it is absolutely necessary that the proper fluid be used to service any brake system including when bleeding air from the brake lines.

Brakes with master cylinders may also be gravity bled from the top down. This is a process similar to that used on automobiles. *[Figure 13-120]* Additional fluid is supplied to the aircraft brake reservoir so that the quantity does not exhaust while bleeding, which would cause the reintroduction of more air into the system. A clear hose is connected to the

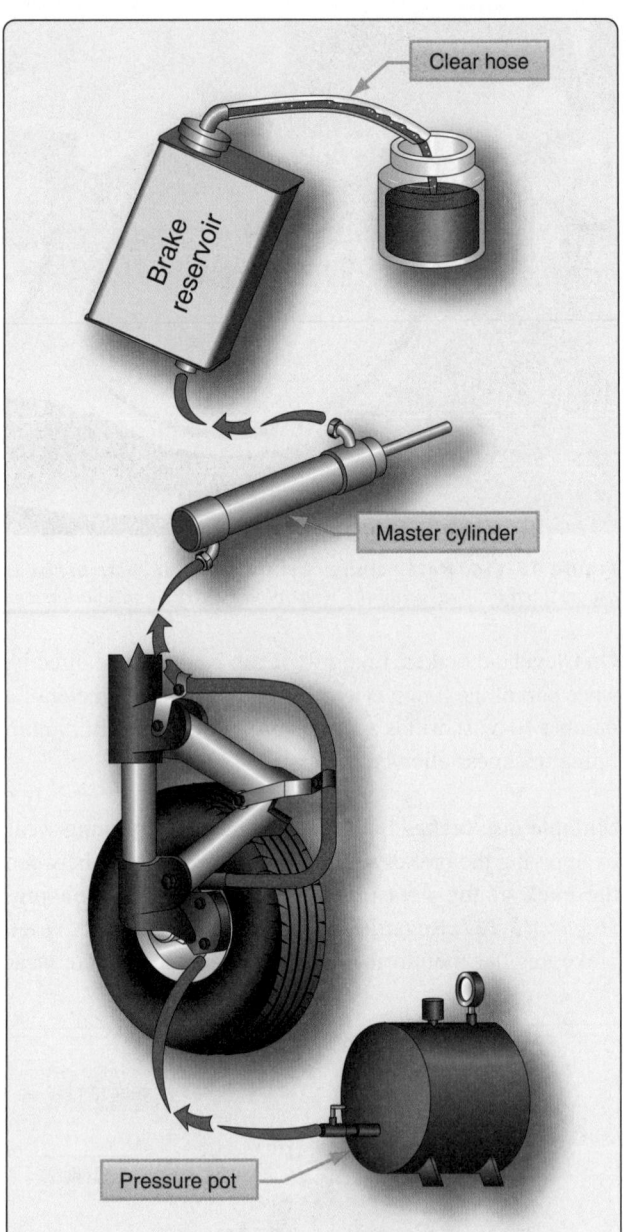

Figure 13-119. *Arrangement for bottom-up pressure bleeding of aircraft brakes. Fluid is pushed through the system until no air bubbles are visible in the hose at the top.*

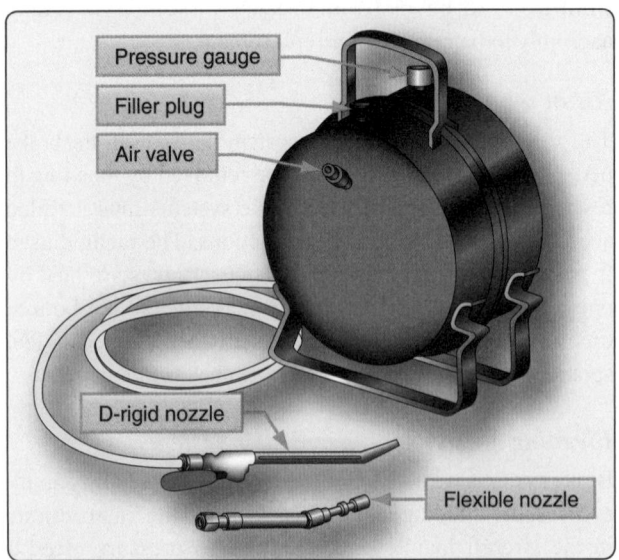

Figure 13-118. *A typical brake bleeder pot or tank contains pure brake fluid under pressure. It pushes the fluid through the brake system to displace any air that may be present.*

bleed port on the brake assembly. The other end is submersed in clean fluid in a container large enough to capture fluid expelled during the bleeding process. Depress the brake pedal and open the brake assembly bleed port. The piston in the master cylinder travels all the way to the end of the cylinder forcing air fluid mixture out of the bleed hose and into the container. With the pedal still depressed, close the bleed port. Pump the brake pedal to introduce more fluid from the reservoir ahead of the piston in the master cylinder. Hold the pedal down and open the bleed port on the brake assembly. More fluid and air is expelled through the hose into the container. Repeat this process until the fluid exiting the brake through the hose

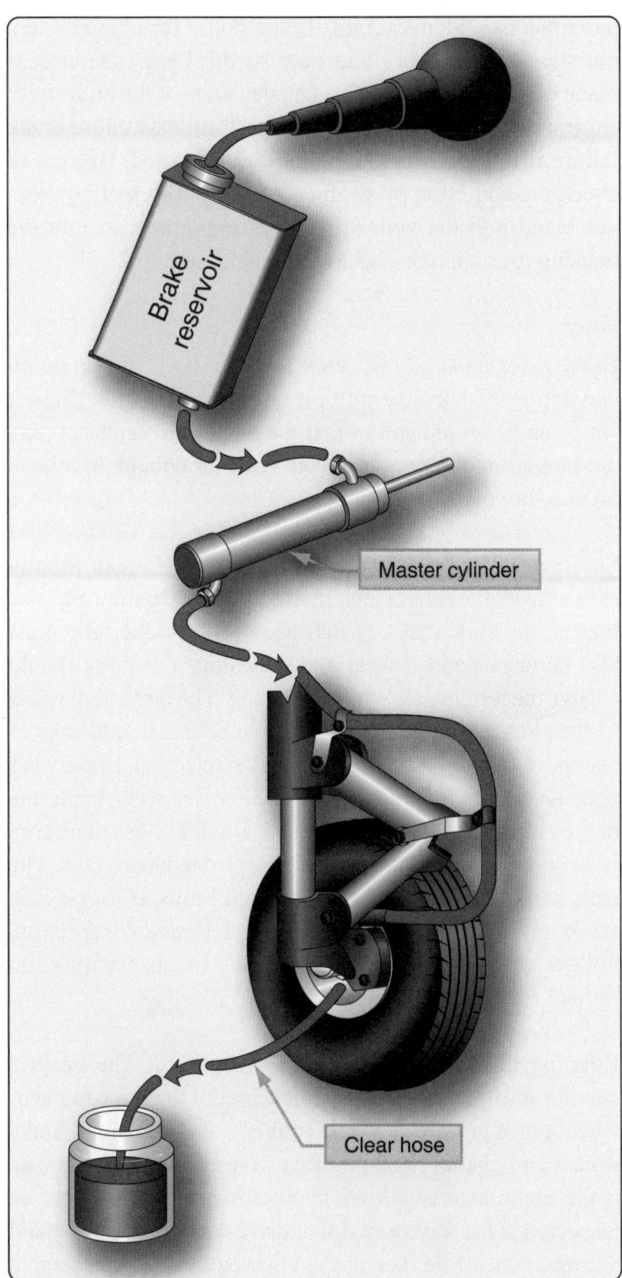

Figure 13-120. *Arrangement for top down or gravity bleeding of aircraft brakes.*

no longer contains any air. Tighten the bleed port fitting and ensure the reservoir is filled to the proper level.

Whenever bleeding the brakes, ensure that reservoirs and bleed tanks remain full during the process. Use only clean, specified fluid. Always check the brakes for proper operation, any leaks when bleeding is complete, and assure that the fluid quantity level is correct.

Bleeding Power Brake Systems

Top down brake bleeding is used in power brake systems. Power brakes are supplied with fluid from the aircraft hydraulic system. The hydraulic system should operate without air in the fluid as should the brake system. Therefore, bottom up pressure bleeding is not an option for power brakes. The trapped air in the brake system would be forced into the main hydraulic system, which is not acceptable.

Many aircraft with power brake systems accept the connection of an auxiliary hydraulic mule that can be used to establish pressure in the system for bleeding. Regardless, the aircraft system must be pressurized to bleed power brake systems. Attach a clear hose to the brake bleed port fitting on the brake assembly and immerse the other end of the hose in a container of clean hydraulic fluid. With the bleeder valve open, carefully apply the brake to allow aircraft hydraulic fluid to enter the brake system. The fluid expels the fluid contaminated with air out of the bleed hose into the container. When air is no longer visible in the hose, close the bleed valve and restore the hydraulic system to normal operation configuration.

Power brake systems on different aircraft contain many variations and a wide array of components that may affect the proper bleeding technique to be followed. Consult the manufacturer's maintenance information for the correct bleeding procedure for each aircraft. Be sure to bleed auxiliary and emergency brake systems when bleeding the normal brake system to ensure proper operation when needed.

Fluid Quantity & Type

As mentioned, it is imperative that the correct hydraulic fluid is used in each brake system. Seals in the brake system are designed for a particular hydraulic fluid. Deterioration and failure occurs when they are exposed to other fluids. Mineral-based fluid, such as MIL-H-5606 (red oil), should never be mixed with phosphate-ester based synthetic hydraulic fluid, such as Skydrol®. Contaminated brake/hydraulic systems must have all of the fluid evacuated and all seals replaced before the aircraft is released for flight.

Fluid quantity is also important. The technician is responsible for determining the method used to ascertain when the brake and hydraulic systems are fully serviced and for the maintenance of the fluid at this level. Consult the manufacturer's specifications for this information.

Inspection for Leaks

Aircraft brake systems should maintain all fluid inside lines and components and should not leak. Any evidence of a leak must be investigated for its cause. It is possible that the leak is a precursor to more significant damage that can be repaired, thus avoiding an incident or accident. *[Figure 13-121]*

Many leaks are found at brake system fittings. While

this type of leak may be fixed by tightening an obviously loose connection, the technician is cautioned against over-tightening fittings. Removal of hydraulic pressure from the brake system followed by disconnection and inspection of the connectors is recommended. Over-tightening of fitting can cause damage and make the leak worse. MS flareless fitting are particularly sensitive to over-tightening. Replace all fittings suspected of damage. Once any leak is repaired, the brake system must be re-pressurized and tested for function as well as to ensure the leak no longer exists. Occasionally, a brake housing may seep fluid through the housing body. Consult the manufacturer's maintenance manual for limits and remove any brake assembly that seeps excessively.

Proper Bolt Torque

The stress experience by the landing gear and brake system requires that all bolts are properly torqued. Bolts used to attach the brakes to the strut typically have the required torque specified in the manufacturer's maintenance manual. Check for torque specifications that may exist for any landing gear and brake bolts, and ensure they are properly tightened. Whenever applying torque to a bolt on an aircraft, use of a calibrated torque wrench is required.

Off Aircraft Brake Servicing & Maintenance

Certain servicing and maintenance of an aircraft brake assembly is performed while it has been removed from the aircraft. A close inspection of the assembly and its many parts should be performed at this time. Some of the inspection items on a typical assembly follow.

Bolt & Threaded Connections

All bolts and threaded connections are inspected. They should be in good condition without signs of wear. Self-locking nuts should still retain their locking feature. The hardware should be what is specified in the brake manufacturer's

Figure 13-121. *The cause of all aircraft brake leaks must be investigated, repaired, and tested before releasing the aircraft for flight.*

parts manual. Many aircraft brake bolts, for example, are not standard hardware and may be of closer tolerance or made of a different material. The demands of the high stress environment in which the brakes perform may cause brake failure if improper substitute hardware is used. Be sure to check the condition of all threads and O-ring seating areas machined into the housing. The fittings threaded into the housing must also be checked for condition.

Discs

Brake discs must be inspected for condition. Both rotating and stationary discs in a multiple disc brake can wear. Uneven wear can be an indication that the automatic adjusters may not be pulling the pressure plate back far enough to relieve all pressure on the disc stack.

Stationary discs are inspected for cracks. Cracks usually extend from the relief slots, if so equipped. On multiple disc brakes, the slots that key the disc to the torque tube must also be inspected for wear and widening. The discs should engage the torque tube without binding. The maximum width of the slots is given in the maintenance manual. Cracks or excessive key slot wear are grounds for rejection. Brake wear pads or linings must also be inspected for wear while the brake assembly is removed from the aircraft. Signs of uneven wear should be investigated, and the problem corrected. The pads may be replaced if worn beyond limits as long as the stationary disc upon which they mount passes inspection. Follow the manufacturer's procedures for inspections and for pad replacement.

Rotating discs must be similarly inspected. The general condition of the disc must be observed. Glazing can occur when a disc or part of a disc is overheated. It causes brake squeal and chatter. It is possible to resurface a glazed disc if the manufacturer allows it. Rotating discs must also be inspected in the drive key slot or drive tang area for wear and deformation. Little damage is allowed before replacement is required.

The pressure plate and back plate on multiple disc brakes must be inspected for freedom of movement, cracks, general condition, and warping. New linings may be riveted to the plates if the old linings are worn and the condition of the plate is good. Note that replacing brake pads and linings by riveting may require specific tools and technique as described in the maintenance manual to ensure secure attachment. Minor warping can be straightened on some brake assemblies.

Automatic Adjuster Pins

A malfunctioning automatic adjuster assembly can cause the brakes to drag on the rotating disc(s) by not fully releasing and pulling the lining away from the disc. This can lead to excessive, uneven lining wear and disc glazing. The return pin must be straight with no surface damage so it can pass through the grip without binding. Damage under the head can weaken the pin and cause failure. Magnetic inspection is sometimes used to inspect for cracks.

The components of the grip and tube assembly must be in good condition. Clean and inspect in accordance with the manufacturer's maintenance instructions. The grip must move with the force specified and must move through its full range of travel.

Torque Tube

A sound torque tube is necessary to hold the brake assembly stable on the landing gear. General visual inspection should be made for wear, burrs, and scratches. Magnetic particle inspection is used to check for cracks. The key areas should be checked for dimension and wear. All limits of damage are referenced in the manufacturer's maintenance data. The torque tube should be replaced if a limit is exceeded.

Brake Housing & Piston Condition

The brake housing must be inspected thoroughly. Scratches, gouges, corrosion, or other blemishes may be dressed out and the surface treated to prevent corrosion. Minimal material should be removed when doing so. Most important is that there are no cracks in the housing. Fluorescent dye penetrant is typically used to inspect for cracks. If a crack is found, the housing must be replaced. The cylinder area(s) of the housing must be dimensionally checked for wear. Limits are specified in the manufacturer's maintenance manual.

The brake pistons that fit into the cylinders in the housing must also be checked for corrosion, scratches, burrs, etc. Pistons are also dimensionally checked for wear limits specified in the maintenance data. Some pistons have insulators on the bottom. They should not be cracked and should be of a minimal thickness. A file can be used to smooth out minor irregularities.

Seal Condition

Brake seals are very important. Without properly functioning seals, brake operation will be compromised, or the brakes will fail. Over time, heat and pressure mold a seal into the seal groove and harden the material. Eventually, resilience is reduced and the seal leaks. New seals should be used to replace all seals in the brake assembly. Acquire seals by part number in a sealed package from a reputable supplier to avoid bogus seals and ensure the correct seals for the brake assembly in question. Check to ensure the new seals have not exceeded their shelf life, which is typically three years from the cure date.

Many brakes use back-up rings in the seal groove to support the O-ring seals and reduce the tendency of the seal to extrude into the space which it is meant to seal. These are often made of Teflon® or similar material. Back-up seals are installed on the side of the O-ring away from the fluid pressure. *[Figure 13-122]* They are often reusable.

Replacement of Brake Linings

In general aviation, replacement of brake linings is commonly done in the hangar. The general procedure used on two common brake assemblies is given. Follow the actual manufacturer's instruction when replacing brake linings on any aircraft brake assembly.

Goodyear Brakes

To replace the linings on a Goodyear single disc brake assembly, the aircraft must be jacked and supported. Detach the anti-rattle clips that help center the disc in the wheel before removing the wheel from the axle. The disc remains between the inner and outer lining when the wheel is removed. Extract the disc to provide access to the old lining pucks. These can be removed from the cavities in the housing and replaced with new pucks. Ensure the smooth braking surface of the puck contacts the disc. Reinsert the disc between the linings. Reinstall the wheel and anti-rattle clips. Tighten the axle nut in accordance with the manufacturer's instructions. Secure it with a cotter pin and lower the aircraft from the jack. *[Figure 13-123]*

Cleveland Brakes

The popular Cleveland brake uniquely features the ability to change the brake linings without jacking the aircraft or removing the wheel. On these assemblies, the torque plate is bolted to the strut while the remainder of the brake is assembled on the anchor bolts. The disc rides between the pressure plate and back plate. Linings are riveted to both plates. By unbolting the cylinder housing from the backplate,

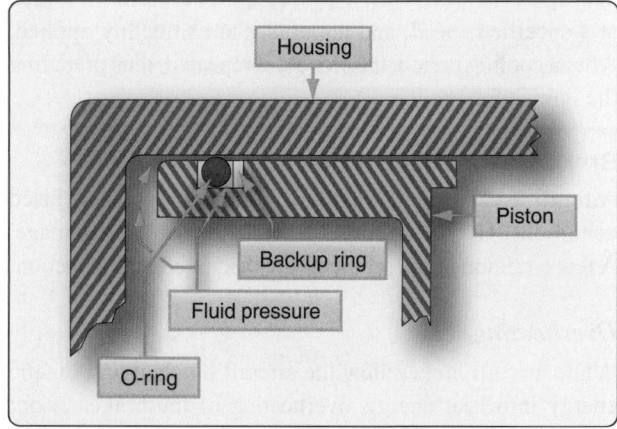

Figure 13-122. *Back-up rings are used to keep O-rings from extruding into the space between the piston and the cylinder. They are positioned on the side of the O-ring away from the fluid pressure.*

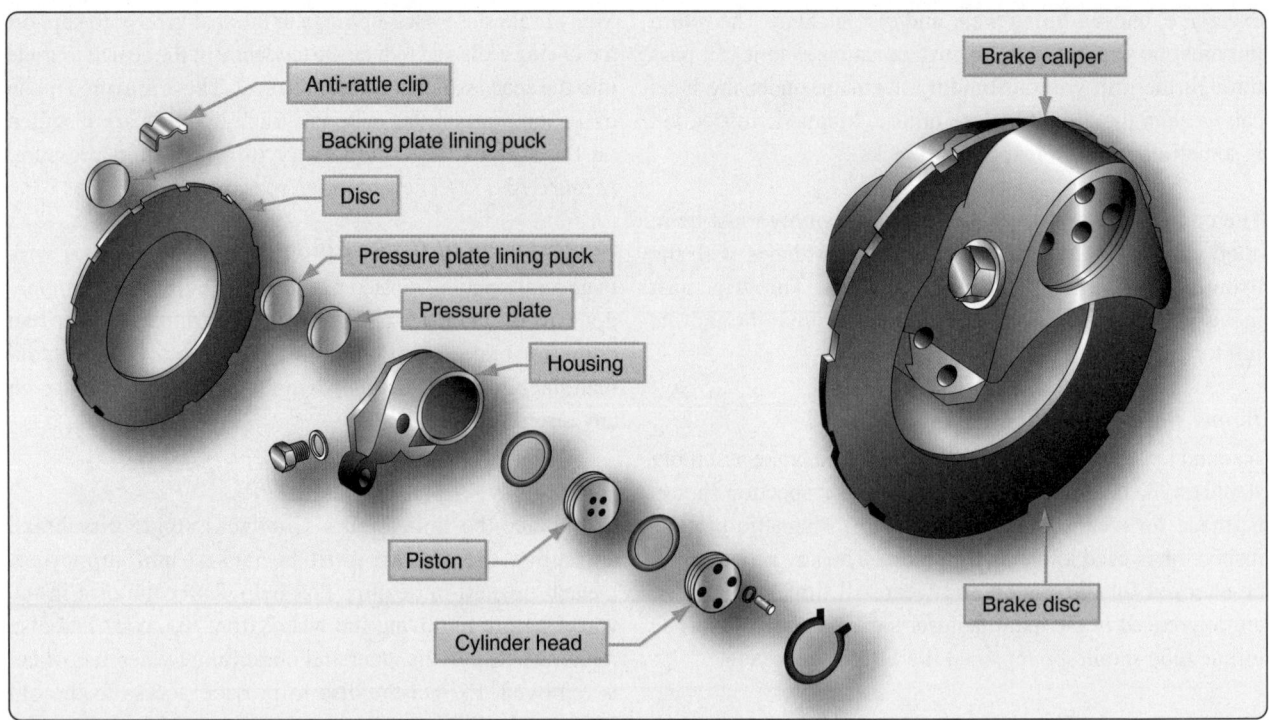

Figure 13-123. *Goodyear brake lining replacement requires that the wheel be removed from the axle to access the brake assembly. The lining pucks slip into recesses in the brake housing.*

the backplate is freed to drop away from the torque plate. The remainder of the assembly is pulled away, and the pressure plate slides off of the torque bolts. *[Figure 13-124]*

The rivets that hold the linings on the pressure plate and back plate are removed with a knockout punch. After a thorough inspection, new linings are riveted to the pressure plate and backplate using a rivet clinching tool *[Figure 13-125]* Kits are sold that supply everything needed to perform the operation. The brake is reassembled in the reverse order. Be certain to include any shims if required. The bolts holding the backplate to the cylinder assembly must be torqued according to manufacturer specifications and safetied. The manufacturer's data also provides a burn in procedure. The aircraft is taxied at a specified speed, and the brakes are smoothly applied. After a cooling period, the process is repeated, thus preparing the linings for service.

Brake Malfunctions & Damage

Aircraft brakes operate under extreme stress and varied conditions. They are susceptible to malfunction and damage. A few common brake problems are discussed in this section.

Overheating

While aircraft brakes slow the aircraft by changing kinetic energy into heat energy, overheating of the brakes is not desirable. Excessive heat can damage and distort brake parts weakening them to the point of failure. Protocol for brake usage is designed to prevent overheating. When a

brake shows signs of overheating, it must be removed from the aircraft and inspected for damage. When an aircraft is involved in an aborted takeoff, the brakes must be removed and inspected to ensure they withstood this high level of use.

The typical post-overheat brake inspection involves removal of the brake from the aircraft and disassembly of the brakes. All of the seals must be replaced. The brake housing must be checked for cracks, warping, and hardness per the maintenance manual. Any weakness or loss of heat treatment could cause the brake to fail under high-pressure braking. The brake discs must also be inspected. They must not be warped, and the surface treatment must not be damaged or transferred to an adjacent disc. Once reassembled, the brake should be bench tested for leaks and pressure tested for operation before being installed on the aircraft.

Dragging

Brake drag is a condition caused by the linings not retracting from the brake disc when the brakes are no longer being applied. It can be caused by several different factors. Brakes that drag are essentially partially on at all times. This can cause excessive lining wear and overheating leading to damage to the disc(s).

A brake may drag when the return mechanism is not functioning properly. This could be due to a weak return spring, the return pin slipping in the auto adjuster pin grip,

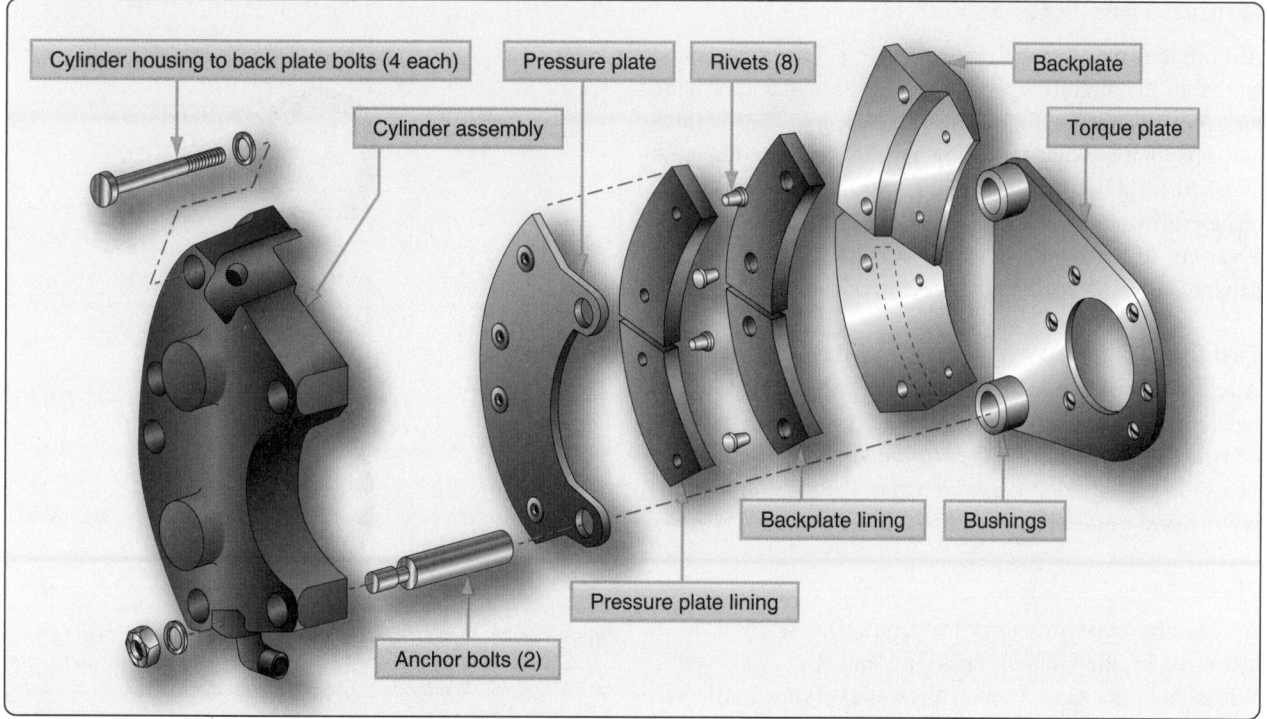

Figure 13-124. *A Cleveland brake disassembles once the four bolts holding the cylinder to the backplate are removed while the aircraft wheel remains in place. The pressure plate slides off the anchor bolts and linings can be replaced by riveting on the pressure plate and back plate.*

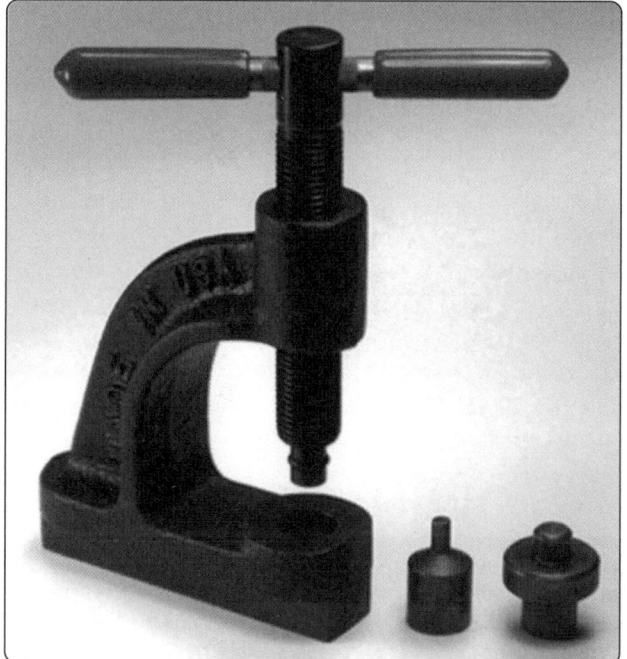

Figure 13-125. *Rivet setting tool is used to install brake linings on Cleveland brake pressure plates and back plates.*

or similar malfunction. Inspect the auto adjuster(s) and return units on the brake when dragging is reported. An overheated brake that has warped the disc also causes brake drag. Remove the brake and perform a complete inspection as discussed in the previous section. Air in the brake fluid line can also cause brake drag. Heat causes the air to expand, which pushes the brake linings against the disc prematurely. If no damage has been caused when reported, bleed the brakes to remove the air from the system to eliminate the drag.

At all times, the technician should perform inspections to ensure the proper parts are used in the brake assembly. Improper parts, especially in the retraction/adjuster assemblies, can cause the brakes to drag.

Chattering or Squealing

Brakes may chatter or squeal when the linings do not ride smoothly and evenly along the disc. A warped disc(s) in a multiple brake disc stack produces a condition wherein the brake is actually applied and removed many times per minute. This causes chattering and, at high frequency, it causes squealing. Any misalignment of the disc stack out of parallel causes the same phenomenon. Discs that have been overheated may have damage to the surface layer of the disc. Some of this mix may be transferred to the adjacent disc resulting in uneven disc surfaces that also leads to chatter or squeal. In addition to the noise produced by brake chattering and squealing, vibration is caused that may lead to further damage of the brake and the landing gear system. The technician must investigate all reports of brake chattering and squealing.

Aircraft Tires & Tubes

Aircraft tires may be tube-type or tubeless. They support the weight of the aircraft while it is on the ground and provide the necessary traction for braking and stopping. The tires also help absorb the shock of landing and cushion the roughness of takeoff, rollout, and taxi operations. Aircraft tires must be carefully maintained to perform as required. They accept a variety of static and dynamic stresses and must do so dependably in a wide range of operating conditions.

Tire Classification

Aircraft tires are classified in various ways including by: type, ply rating, whether they are tube-type or tubeless, and whether they are bias ply tires or radials. Identifying a tire by its dimensions is also used. Each of these classifications is discussed as follows.

Types

A common classification of aircraft tires is by type as classified by the United States Tire and Rim Association. While there are nine types of tires, only Types I, III, VII, and VIII, also known as a Three-Part Nomenclature tires, are still in production.

Type I tires are manufactured, but their design is no longer active. They are used on fixed gear aircraft and are designated only by their nominal overall diameter in inches. These are smooth profile tires that are obsolete for use in the modern aviation fleet. They may be found on older aircraft.

Type III tires are common general aviation tires. They are typically used on light aircraft with landing speeds of 160 miles per hour (mph) or less. Type III tires are relatively low-pressure tires that have small rim diameters when compared to the overall width of the tire. They are designed to cushion and provide flotation from a relatively large footprint. Type III tires are designated with a two-number system. The first number is the nominal section width of the tire, and the second number is the diameter of the rim the tire is designed to mount upon. *[Figure 13-126]*

Type VII tires are high performance tires found on jet aircraft. They are inflated to high-pressure and have exceptional high load carrying capability. The section width of Type VII tires is typically narrower than Type III tires. Identification of Type VII aircraft tires involves a two-number system. An X is used between the two numbers. The first number designates the nominal overall diameter of the tire. The second number designates the section width. *[Figure 13-127]*

Type VIII aircraft tires are also known as three-part nomenclature tires. *[Figure 13-128]* They are inflated to very high-pressure and are used on high-performance jet

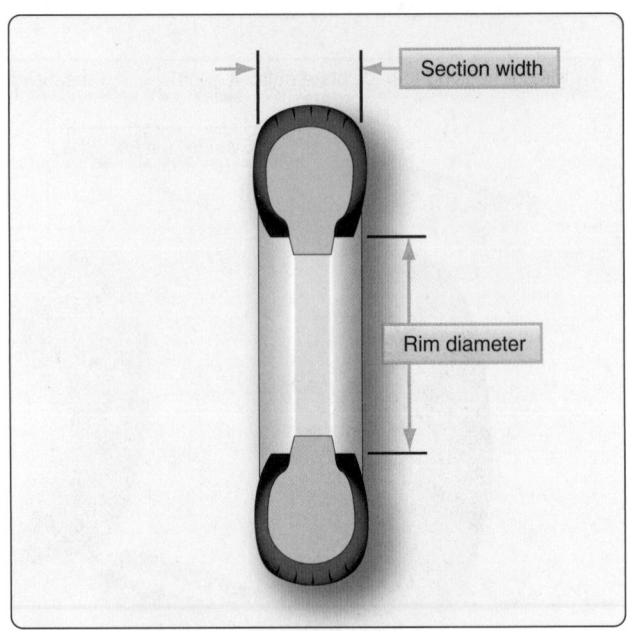

Figure 13-126. *Type III aircraft tires are identified via a two-number system with a (-) separating the numbers. The first number is the tire section width in inches. The second number is the rim diameter in inches. For example: 6.00 – 6 is a Cessna 172 tire that is 6.00 inches wide and fits on a rim that has a diameter of 6 inches.*

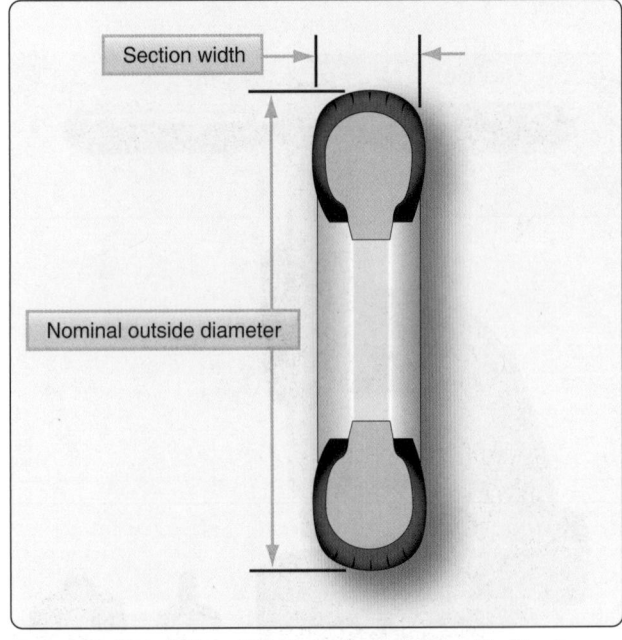

Figure 13-127. *A Type VII aircraft tire is identified by its two-number designation. The first number represents the tire's overall diameter in inches and the second number represents the section width in inches. Type VII designators separate the first and second number an "X." For example: 26 X 6.6 identifies a tire that is 26 inches in diameter with a 6.6-inch nominal width.*

aircraft. The typical Type VIII tire has relatively low profile and is capable of operating at very high speeds and very high loads. It is the most modern design of all tire types. The three-part nomenclature is a combination of Type III and Type VII nomenclature where the overall tire diameter, section

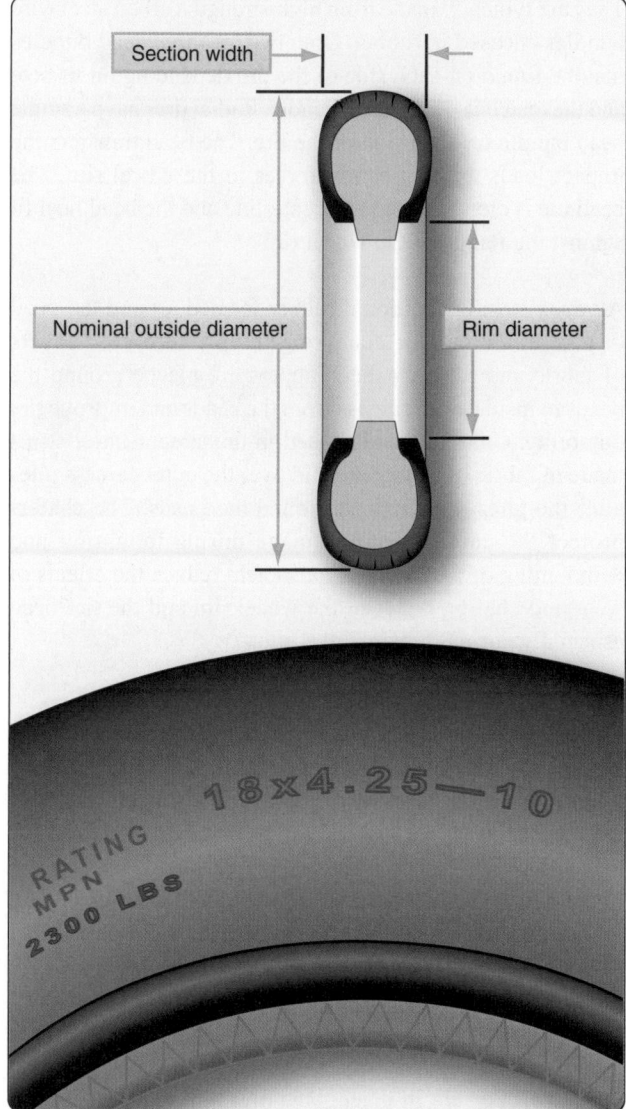

Figure 13-128. *A Type VIII or three-part nomenclature tire is identified by 3 parameters: overall diameter, section width, and rim diameter. They are arranged in that order with the first two separated by an "X" and the second two separated by a "–." For example: 18 X 4.25—10 designates a tire that is 18 inches in diameter with a 4.25-inch section width to be mounted on a 10-inch wheel rim.*

width, and rim diameter are used to identify the tire. The X and "–" symbols are used in the same respective positions in the designator.

When three-part nomenclature is used on a Type VIII tire, dimensions may be represented in inches or in millimeters. Bias tires follow the designation nomenclature and radial tires replace the "–" with the letter R. For example, 30 X 8.8 R 15 designates a Type VIII radial aircraft tire with a 30-inch tire diameter, an 8.8-inch section width to be mounted on a 15-inch wheel rim.

A few special designators may also be found for aircraft tires. When a B appears before the identifier, the tire has a wheel rim to section width ratio of 60 to 70 percent with a bead taper of 15 degrees. When an H appears before the identifier, the tire has a 60 to 70 percent wheel rim to section width ratio but a bead taper of only 5 degrees.

Ply Rating

Tire plies are reinforcing layers of fabric encased in rubber that are laid into the tire to provide strength. In early tires, the number of plies used was directly related to the load the tire could carry. Nowadays, refinements to tire construction techniques and the use of modern materials to build up aircraft tires makes the exact number of plies somewhat irrelevant when determining the strength of a tire. However, a ply rating is used to convey the relative strength of an aircraft tire. A tire with a high ply rating is a tire with high strength able to carry heavy loads regardless of the actual number of plies used in its construction.

Tube-Type or Tubeless

As stated, aircraft tires can be tube-type or tubeless. This is often used as a means of tire classification. Tires that are made to be used without a tube inserted inside have an inner liner specifically designed to hold air. Tube-type tires do not contain this inner liner since the tube holds the air from leaking out of the tire. Tires that are meant to be used without a tube have the word tubeless on the sidewall. If this designation is absent, the tire requires a tube. Consult the aircraft manufacturer's maintenance information for any allowable tire damage and the use of a tube in a tubeless tire.

Bias Ply or Radial

Another means of classifying an aircraft tire is by the direction of the plies used in construction of the tire, either bias or radial. Traditional aircraft tires are bias ply tires. The plies are wrapped to form the tire and give it strength. The angle of the plies in relation to the direction of rotation of the tire varies between 30° and 60°. In this manner, the plies have the bias of the fabric from which they are constructed facing the direction of rotation and across the tire. Hence, they are called bias tires. The result is flexibility as the sidewall can flex with the fabric plies laid on the bias. *[Figure 13-129]*

Some modern aircraft tires are radial tires. The plies in radial tires are laid at a 90° angle to the direction of rotation of the tire. This configuration puts the non-stretchable fiber of the plies perpendicular to the sidewall and direction of rotation. This creates strength in the tire allowing it to carry high loads with less deformation. *[Figure 13-130]*

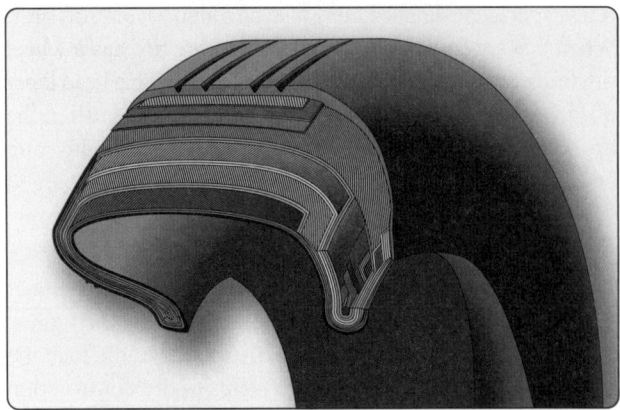

Figure 13-129. *A bias ply tire has the fabric bias oriented with and across the direction of rotation and the sidewall. Since fabric can stretch on the bias, the tire is flexible, and can absorb loads. Strength is obtained by adding plies.*

Tire Construction

An aircraft tire is constructed for the purpose it serves. Unlike an automobile or truck tire, it does not have to carry a load for a long period of continuous operation. However, an aircraft tire must absorb the high impact loads of landing and be able to operate at high speeds even if only for a short time. The deflection built into an aircraft tire is more than twice that of an automobile tire. This enables it to handle the forces during landings without being damaged. Only tires designed for an aircraft as specified by the manufacturer should be used.

It is useful to the understanding of tire construction to identify the various components of a tire and the functions contributed to the overall characteristics of a tire. Refer to *Figure 13-131* for tire nomenclature used in this discussion.

Bead

The tire bead is an important part of an aircraft tire. It anchors the tire carcass and provides a dimensioned, firm mounting surface for the tire on the wheel rim. Tire beads are strong.

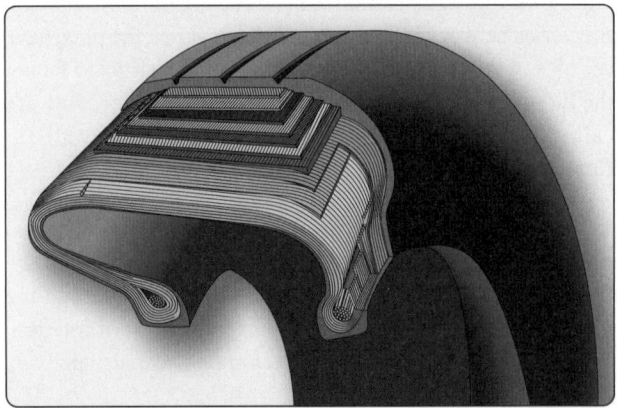

Figure 13-130. *A radial tire has the fiber strands of the ply fabric oriented with and at 90° to the direction of rotation and the tire sidewall. This restricts flexibility directionally and the flexibility of the sidewall while it strengthens the tire to carry heavy loads.*

They are typically made from high-strength carbon steel wire bundles encased in rubber. One, two, or three bead bundles may be found on each side of the tire depending on its size and the load it is designed to handle. Radial tires have a single bead bundle on each side of the tire. The bead transfers the impact loads and deflection forces to the wheel rim. The bead toe is closest to the tire centerline and the bead heel fit against the flange of the wheel rim.

An apex strip is additional rubber formed around the bead to give a contour for anchoring the ply turn-ups. Layers of fabric and rubber called flippers are placed around the beads to insulate the carcass from the beads and improve tire durability. Chafers are also used in this area. Chafer strips made of fabric or rubber are laid over the outer carcass plies after the plies are wrapped around the beads. The chafers protect the carcass from damage during mounting and demounting of the tire. They also help reduce the effects of wear and chafing between the wheel rim and the tire bead especially during dynamic operations.

Carcass Plies

Carcass plies, or casing plies as they are sometimes called, are used to form the tire. Each ply consists of fabric, usually nylon, sandwiched between two layers of rubber. The plies are applied in layers to give the tire strength and form the carcass body of the tire. The ends of each ply are anchored by wrapping them around the bead on both sides of the tire to form the ply turn-ups. As mentioned, the angle of the fiber in the ply is manipulated to create a bias tire or radial tire as desired. Typically, radial tires require fewer plies than bias tires.

Once the plies are in place, bias tires and radial tires each have their own type of protective layers on top of the plies but under the tread of the running surface of the tire. On bias tires, these single or multiple layers of nylon and rubbers are called tread reinforcing plies. On radial tires, an undertread and a protector ply do the same job. These additional plies stabilize and strengthen the crown area of the tire. They reduce tread distortion under load and increase stability of the tire at high speeds. The reinforcing plies and protector plies also help resist puncture and cutting while protecting the carcass body of the tire.

Tread

The tread is the crown area of the tire designed to come in contact with the ground. It is a rubber compound formulated to resist wear, abrasion, cutting, and cracking. It also is made to resist heat build-up. Most modern aircraft tire tread is formed with circumferential grooves that create tire ribs. The grooves provide cooling and help channel water from under the tire in wet conditions to increase adhesion to the ground surface.

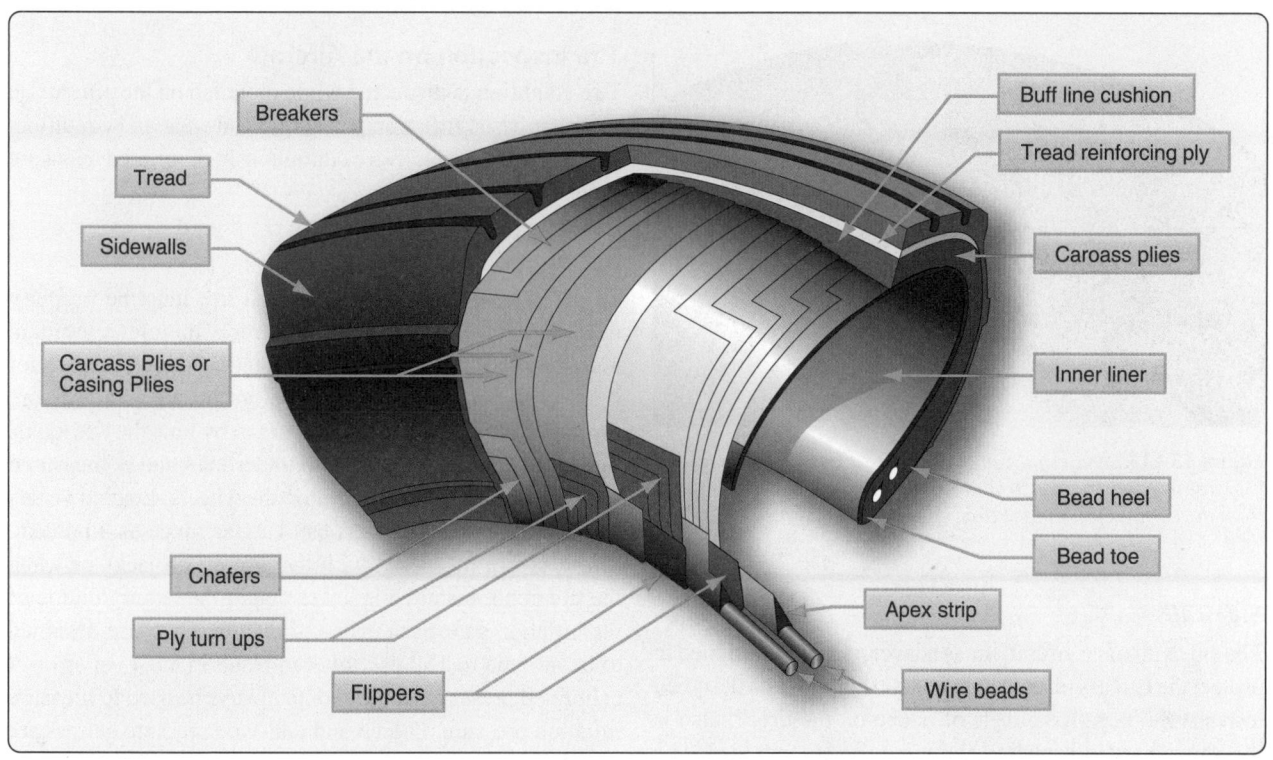

Figure 13-131. *Construction nomenclature of an aircraft tire.*

Tires designed for aircraft frequently operated from unpaved surfaces may have some type of cross-tread pattern. Older aircraft without brakes or brakes designed only to aid in taxi may not have any grooves in the tread. An all-weather tread may be found on some aircraft tires. This tread has typical circumferential ribs in the center of the tire with a diamond patterned cross tread at the edge of the tire. [*Figure 13-132*]

The tread is designed to stabilize the aircraft on the operating surface and wears with use. Many aircraft tires are designed with protective undertread layers as described above. Extra tread reinforcement is sometimes accomplished with breakers. These are layers of nylon cord fabric under the tread that strengthen the tread while protecting the carcass plies. Tires with reinforced tread are often designed to be re-treaded and used again once the tread has worn beyond limits. Consult the tire manufacturer's data for acceptable tread wear and re-tread capability for a particular tire.

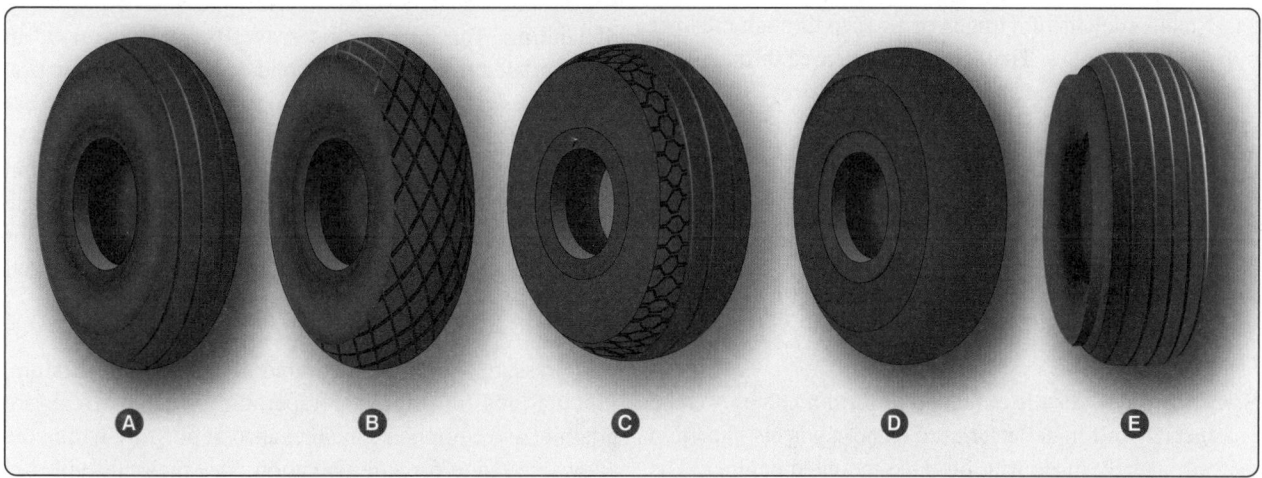

Figure 13-132. *Aircraft tire treads are designed for different uses. A is a rib tread designed for use on paved surfaces. It is the most common aircraft tire tread design. B is a diamond tread designed for unpaved runways. C is an all weather tread that combines a ribbed center tread with a diamond tread pattern of the edges. D is a smooth tread tire found on older, slow aircraft without brakes designed for stopping. E is a chine tire used on the nose gear of aircraft with fuselage mounted jet engines to deflect runway water away from the engine intake(s).*

Figure 13-133. *A sidewall vent marked by a colored dot must be kept free from obstruction to allow trapped air or nitrogen to escape from the carcass plies of the tire.*

Sidewall

The sidewall of an aircraft tire is a layer of rubber designed to protect the carcass plies. It may contain compounds designed to resist the negative effects of ozone on the tire. It also is the area where information about the tire is contained. The tire sidewall imparts little strength to the cord body. Its main function is protection.

The inner sidewall of a tire is covered by the tire inner liner. A tube-type tire has a thin rubber liner adhered to the inner surface to prevent the tube from chafing on the carcass plies. Tubeless tires are lined with a thicker, less permeable rubber. This replaces the tube and contains the nitrogen or inflation air within the tire and keeps from seeping through the carcass plies.

The inner liner does not contain 100 percent of the inflation gas. Small amounts of nitrogen or air seep through the liner into the carcass plies. This seepage is released through vent holes in the lower outer sidewall of the tires. These are typically marked with a green or white dot of paint and must be kept unobstructed. Gas trapped in the plies could expand with temperature changes and cause separation of the plies, thus weakening the tire leading to tire failure. Tube-type tires also have seepage holes in the sidewall to allow air trapped between the tube and the tire to escape. *[Figure 13-133]*

Chine

Some tire sidewalls are mounded to form a chine. A chine is a special built-in deflector used on nose wheels of certain aircraft, usually those with fuselage mounted engines. The chine diverts runway water to the side and away from the intake of the engines. *[Figure 13-132E]* Tires with a chine on both sidewalls are produced for aircraft with a single nose wheel.

Tire Inspection on the Aircraft

Tire condition is inspected while mounted on the aircraft on a regular basis. Inflation pressure, tread wear and condition, and sidewall condition are continuously monitored to ensure proper tire performance.

Inflation

To perform as designed, an aircraft tire must be properly inflated. The aircraft manufacturer's maintenance data must be used to ascertain the correct inflation pressure for a tire on a particular aircraft. Do not inflate to a pressure displayed on the sidewall of the tire or by how the tire looks. Tire pressure is checked while under load and is measured with the weight of the aircraft on the wheels. Loaded versus unloaded pressure readings can vary as much as 4 percent. Tire pressure measured with the aircraft on jacks or when the tire is not installed is lower due to the larger volume of the inflation gas space inside of the tire. On a tire designed to be inflated to 160 psi, this can result in a 6.4 psi error. A calibrated pressure gauge should always be used to measure inflation pressure. Digital and dial-type pressure gauges are more consistently accurate and preferred. *[Figure 13-134]*

Aircraft tires disperse the energy from landing, rollout, taxi, and takeoff in the form of heat. As the tire flexes, heat builds and is transferred to the atmosphere, as well as to the wheel rim through the tire bead. Heat from braking also heats the tire externally. A limited amount of heat is able to be handled by any tire beyond which structural damage occurs.

An improperly inflated aircraft tire can sustain internal damage that is not readily visible and that can lead to tire failure. Tire failure upon landing is always dangerous. An aircraft tire is designed to flex and absorb the shock of landing. Temperature rises as a result. However, an underinflated tire may flex beyond design limits of the tire. This causes excessive heat build-up that weakens the carcass construction. To ensure tire temperature is maintained within limits, tire pressure must be checked and maintained within the proper range on a daily basis or before each flight if the aircraft is only flown periodically. Important reasons for maintaining proper tire pressure are to prolong tire life and prevent tire damage.

Tire pressure should be measured at ambient temperature. Fluctuations of ambient temperature greatly affect tire pressure and complicate maintenance of pressure within the allowable range for safe operation. Tire pressure typically changes 1 percent for every 5 °F of temperature change. When aircraft are flown from one environment to another, ambient temperature differences can be vast. Maintenance personnel must ensure that tire pressure is adjusted accordingly. For

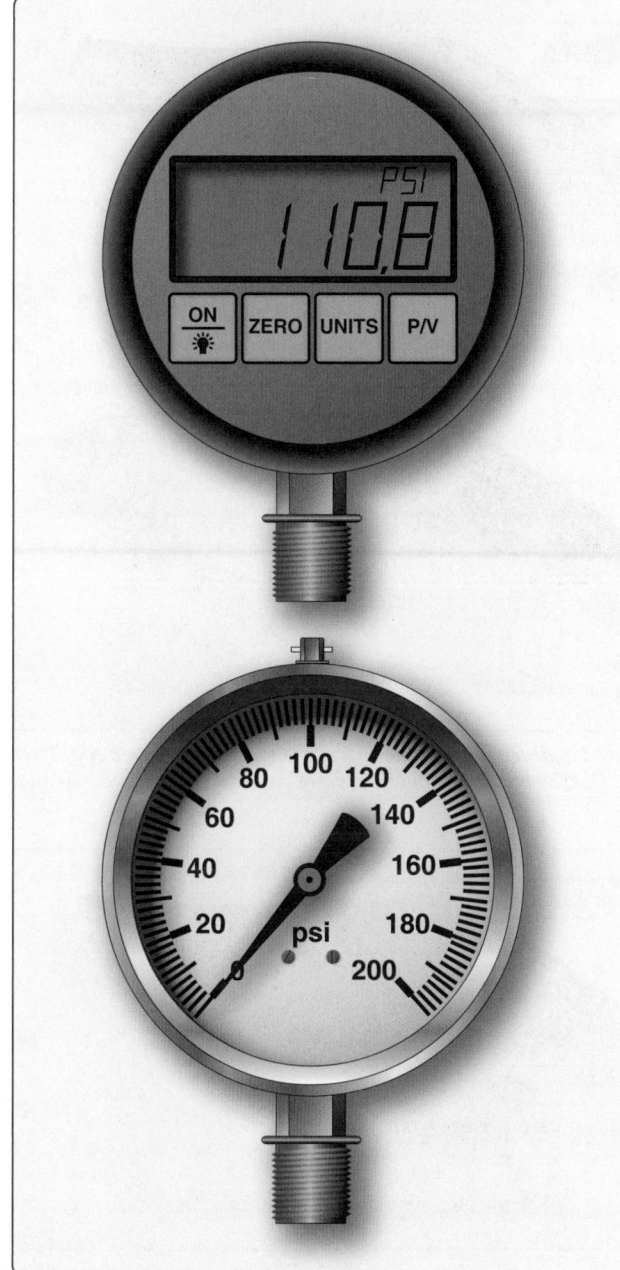

Figure 13-134. *A calibrated bourdon tube dial-type pressure gauge or a digital pressure gauge is recommended for checking tire pressure.*

When checking tire pressure, allow 3 hours to elapse after a typical landing to ensure the tire has cooled to ambient temperature. The correct tire pressure for each ambient temperature is typically provided by the manufacturer on a table or graph.

In addition to overheating, under inflated aircraft tires wear unevenly, which leads to premature tire replacement. They may also creep or slip on the wheel rim when under stress or when the brakes are applied. Severely under inflated tires can pinch the sidewall between the rim and the runway causing sidewall and rim damage. Damage to the bead and lower sidewall area are also likely. This type of abuse like any over flexing damages the integrity of the tire and it must be replaced. In dual-wheel setups, a severely underinflated tire affects both tires and both should be replaced.

Over inflation of aircraft tires is another undesirable condition. While carcass damage due to overheating does not result, adherence to the landing surface is reduced. Over a long period of time, over inflation leads to premature tread wear. Therefore, over inflation reduces the number of cycles in service before the tire must be replaced. It makes the tire more susceptible to bruises, cutting, shock damage, and blowout. *[Figure 13-135]*

Tread Condition

Condition of an aircraft tire tread is able to be determined while the tire is inflated and mounted on the aircraft. The following is a discussion of some of the tread conditions and damage that the technician may encounter while inspecting tires.

Tread Depth & Wear Pattern

Evenly worn tread is a sign of proper tire maintenance. Uneven tread wear has a cause that should be investigated and corrected. Follow all manufacturer instructions specific to the aircraft when determining the extent and serviceability of a worn tire. In the absence of this information, remove any tire that has been worn to the bottom of a tread groove along more than ⅛ of the circumference of the tire. If either the protector ply on a radial tire or the reinforcing ply on a bias tire is exposed for more than ⅛ of the tire circumference, the tire should also be removed. A properly maintained evenly worn tire usually reaches its wear limits at the centerline of the tire. *[Figure 13-136]*

Asymmetrical tread wear may be caused by the wheels being out of alignment. Follow the manufacturer's instructions while checking caster, camber, toe-in, and toe-out to correct this situation. Occasionally, asymmetrical tire wear is a result of landing gear geometry that cannot, or is not, required to be corrected. It may also be caused by regular taxiing on a single

example, an aircraft with the correct tire pressure departing Phoenix, Arizona where the ambient temperature is 100 °F arrives in Vail, Colorado where the temperature is 50 °F. The 50° difference in ambient temperature results in a 10 percent reduction in tire pressure. Therefore, the aircraft could land with underinflated tires that may be damaged due to over-temperature from flexing beyond design limits as described above. An increase in tire pressure before takeoff in Phoenix, Arizona prevents this problem as long as the tires are not inflated beyond the allowable limit provided in the maintenance data.

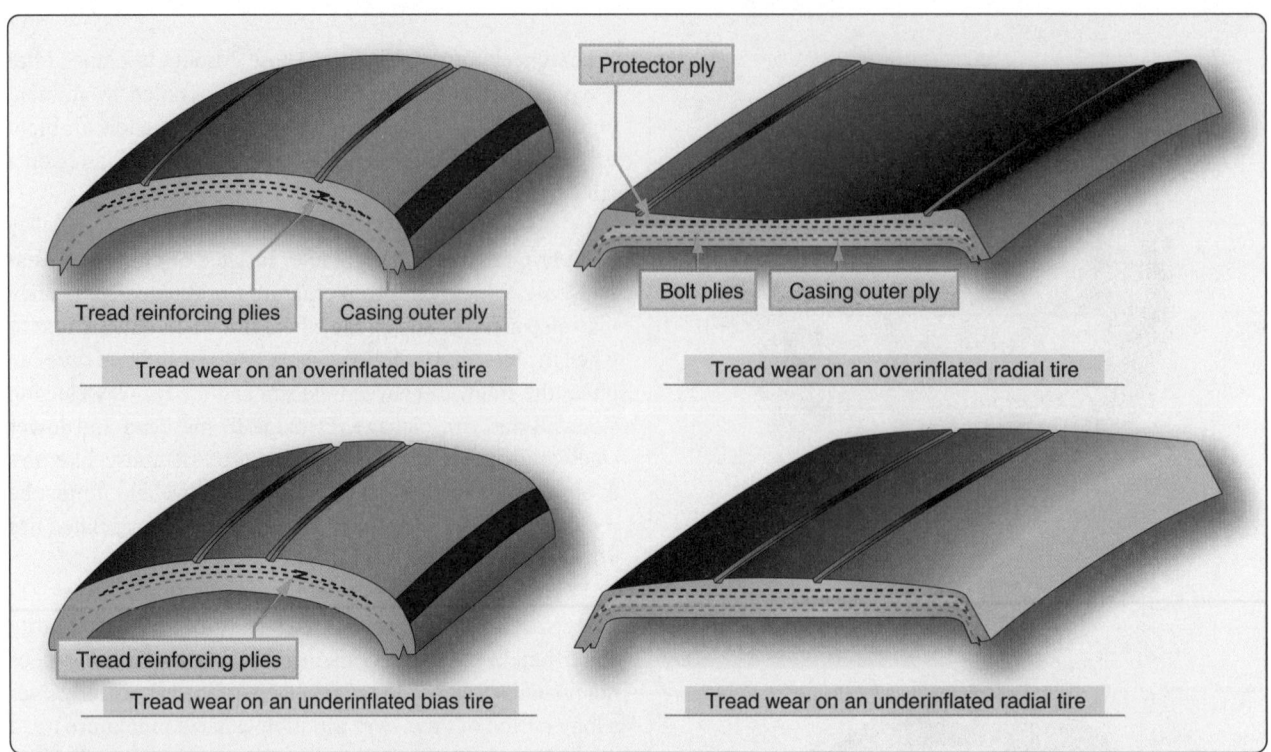

Figure 13-135. *Tires that are overinflated lack adherence to the runway and develop excess tread wear in the center of the tread. Tires that are underinflated develop excess tread wear on the tire shoulders. Overheating resulting in internal carcass damage and potential failure are possible from flexing the tire beyond design limits.*

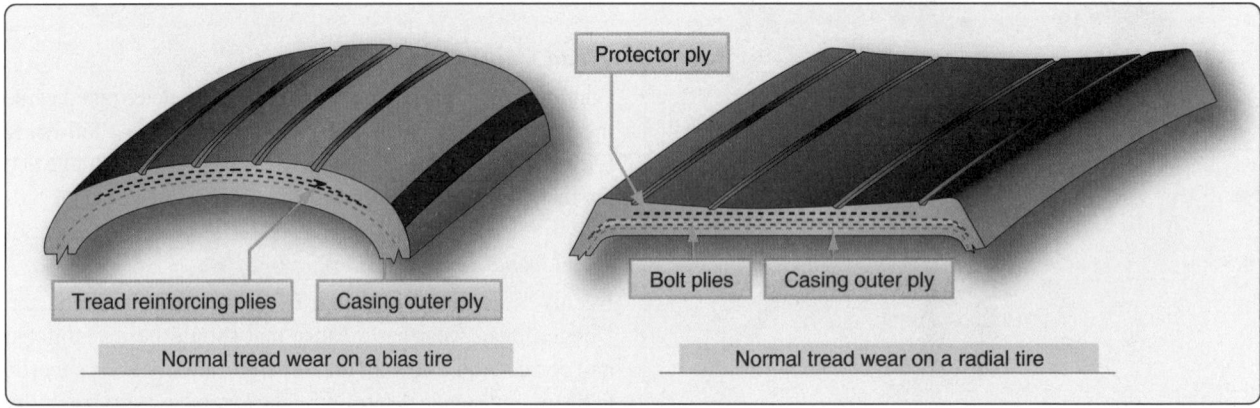

Figure 13-136. *Normal tire wear.*

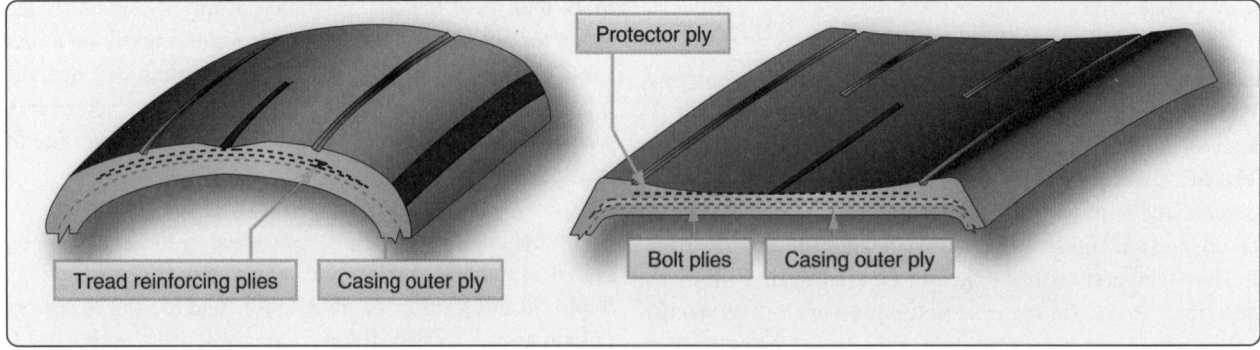

Figure 13-137. *Tread wear on a bias ply tire (left) and a radial tire (right) show wear beyond limits of serviceability but still eligible to be retreaded.*

Figure 13-138. *Marking of damaged area to enable closer inspection.*

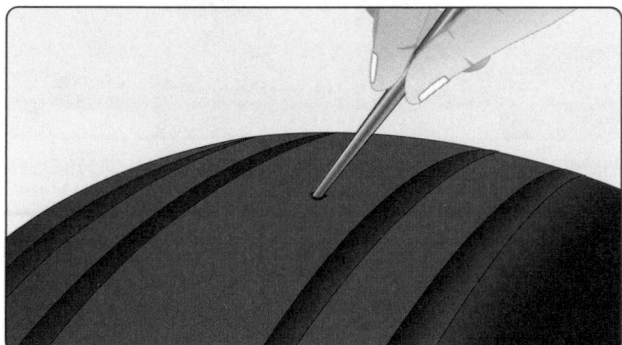

Figure 13-139. *Deflate a tire before removing or probing any area where a foreign object is lodged.*

engine or high-speed cornering while taxiing. It is acceptable to remove the tire from the wheel rim, turn it around, and remount it to even up tread wear if the tire passes all other criterion of inspection for serviceability.

Removal of a tire before it is worn beyond limits to be eligible for retreading is cost effective and good maintenance practice. Considerable traction is lost when tire tread is severely worn and must also be considered when inspecting a tire for condition. *[Figure 13-137]* Consult airframe manufacturer and tire manufacturer specifications for wear and retread limitations.

Tread Damage

In addition to tread wear, an aircraft tire should be inspected for damage. Cuts, bruises, bulges, imbedded foreign objects, chipping, and other damage must be within limits to continue the tire in service. Some acceptable methods of dealing with this type of damage are described below. All damage, suspected damage, and areas with leaks should be marked with chalk, a wax marker, paint stick, or other device before the tire is deflated or removed. Often, it is impossible to relocate these areas once the tire is deflated. Tires removed

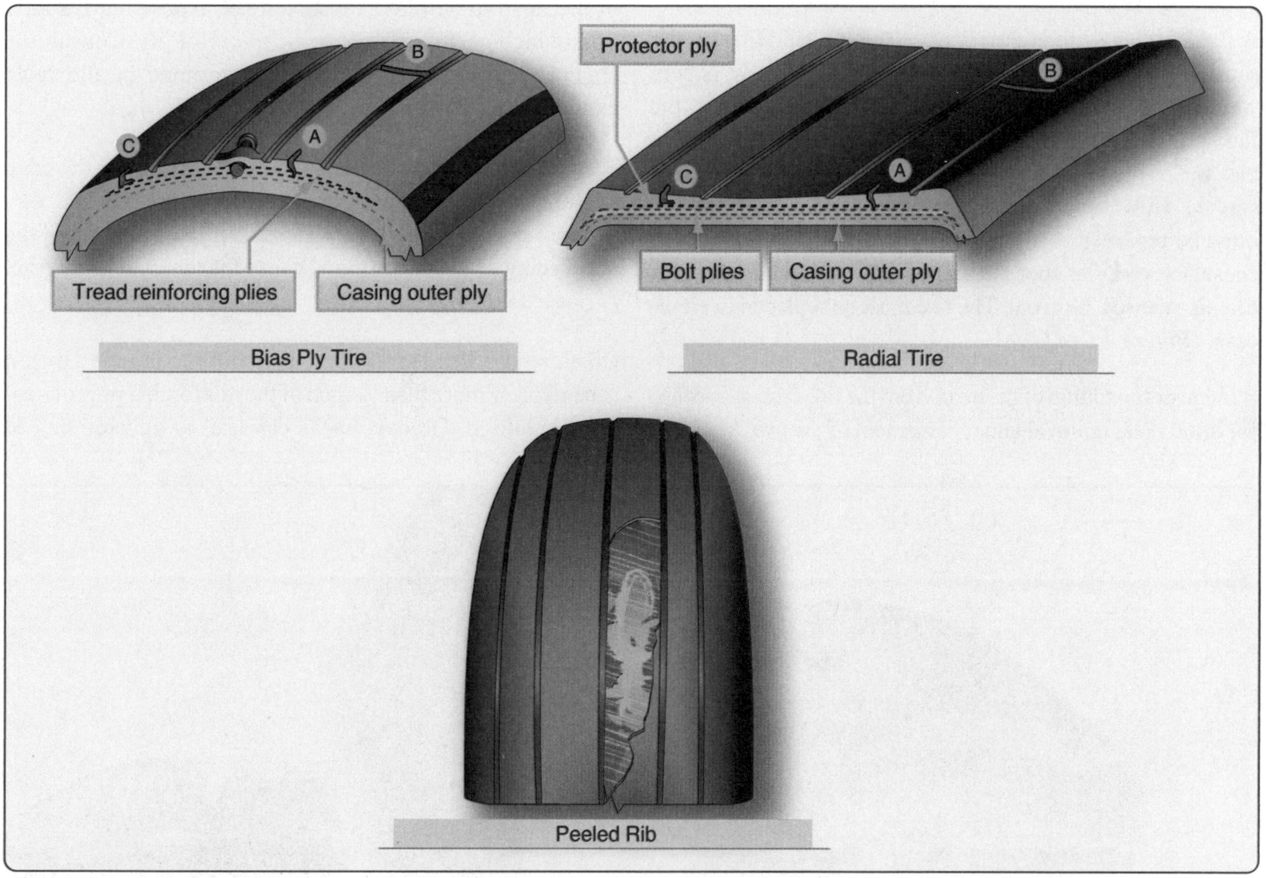

Protector ply

B

Tread reinforcing plies | Casing outer ply

Bias Ply Tire

Bolt plies | Casing outer ply

Radial Tire

Peeled Rib

Figure 13-140. *Remove an aircraft tire from service when the depth of a cut exposes the casing outer plies of a bias ply tire or the outer belt layer of a radial tire (A); a tread rib has been severed across the entire width (B); or, when undercutting occurs at the base of any cut (C). These conditions may lead to a peeled rib.*

for retread should be marked in damaged areas to enable closer inspection of the extent of the damage before the new tread is installed. *[Figure 13-138]*

Foreign objects imbedded in a tire's tread are of concern and should be removed when not imbedded beyond the tread. Objects of questionable depth should only be removed after the tire has been deflated. A blunt awl or appropriately sized screwdriver can be used to pry the object from the tread. Care must be exercised to not enlarge the damaged area with the removal tool. *[Figure 13-139]* Once removed, assess the remaining damage to determine if the tire is serviceable. A round hole caused by a foreign object is acceptable only if it is ⅜-inch or less in diameter. Embedded objects that penetrate or expose the casing cord body of a bias ply tire or the tread belt layer of a radial tire cause the tire to become unairworthy and it must be removed from service.

Cuts and tread undercutting can also render a tire unairworthy. A cut that extends across a tread rib is cause for tire removal. These can sometimes lead to a section of the rib to peel off the tire. *[Figure 13-140]* Consult the aircraft maintenance manual, airline operations manual, or other technical documents applicable to the aircraft tire in question.

A flat spot on a tire is the result of the tire skidding on the runway surface while not rotating. This typically occurs when the brakes lock on while the aircraft is moving. If the flat spot damage does not expose the reinforcing ply of a bias tire or the protector ply of a radial tire, it may remain in service. However, if the flat spot causes vibration, the tire must be removed. Landing with a brake applied can often cause a severe flat spot that exposes the tire under tread. It can also cause a blowout. The tire must be replaced in either case. *[Figure 13-141]*

A bulge or separation of the tread from the tire carcass is cause for immediate removal and replacement of the tire. Mark the

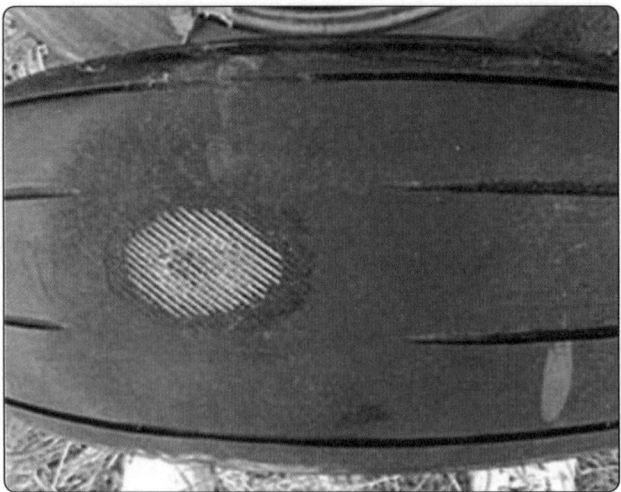

Figure 13-141. *Landing with the brake on causes a tire flat spot that exposes the under tread and requires replacement of the tire.*

area before deflation as it could easily become undetectable without air in the tire. *[Figure 13-142]*

Operation on a grooved runway can cause an aircraft tire tread to develop shallow chevron shaped cuts. These cuts are allowed for continued service, unless chunks or cuts into the fabric of the tire result. Deep chevrons that cause a chunk of the tread to be removed should not expose more than 1 square inch of the reinforcing or protector ply. Consult the applicable inspection parameters to determine the allowable extent of chevron cutting. *[Figure 13-143]*

Tread chipping and chunking sometimes occurs at the edge of the tread rib. Small amounts of rubber lost in this way are permissible. Exposure of more than 1 square inch of the reinforcing or protector ply is cause for removal of the tire. *[Figure 13-144]*

Cracking in a tread groove of an aircraft tire is generally not acceptable if more than ¼-inch of the reinforcing or protector ply is exposed. Groove cracks can lead to undercutting of

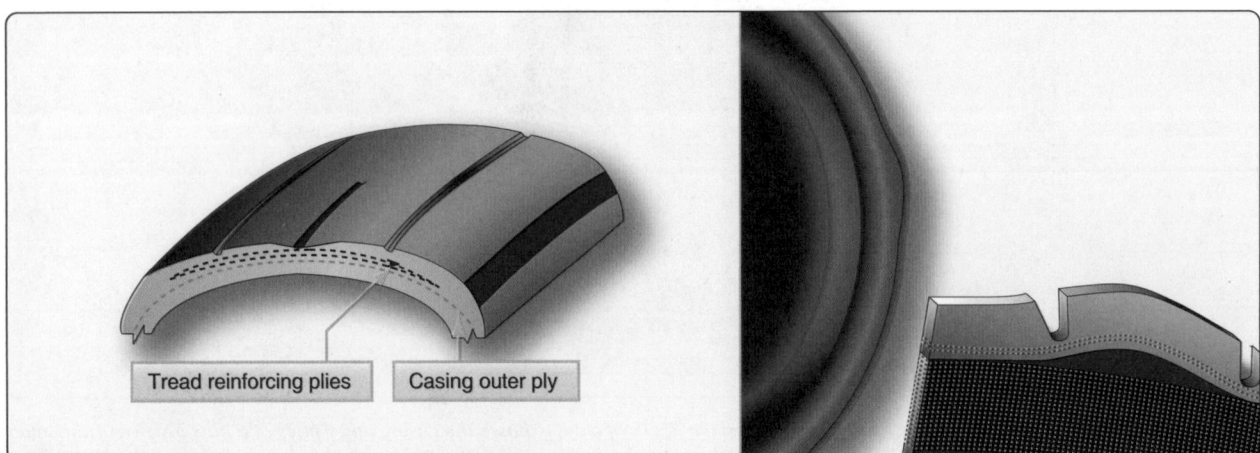

Figure 13-142. *Bulges and tread separation are cause for removal of a tire from service.*

Tread reinforcing plies   Casing outer ply

Figure 13-143. *Chevron cuts in a tire are caused by operation on grooved runway surfaces. Shallow chevron cuts are permitted on aircraft tires.*

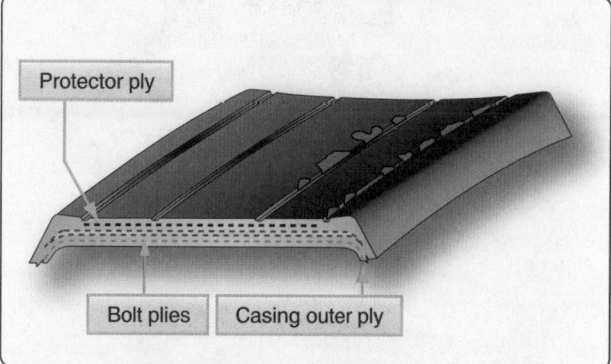

Figure 13-144. *Tread chipping and chunking of a tire requires that the tire be removed from service if more than 1 square inch of the reinforcing ply or protector ply is exposed.*

the tread, which eventually can cause the entire tread to be thrown from the tire. *[Figure 13-145]*

Oil, hydraulic fluid, solvents, and other hydrocarbon substances contaminate tire rubber, soften it, and make it spongy. A contaminated tire must be removed from service. If any volatile fluids come in contact with the tire, it is best to wash the tire or area of the tire with denatured alcohol followed by soap and water. Protect tires from contact with potentially harmful fluids by covering tires during maintenance in the landing gear area.

Tires are also subject to degradation from ozone and weather. Tires on aircraft parked outside for long periods of times can be covered for protection from the elements. *[Figure 13-146]*

Sidewall Condition

The primary function of the sidewall of an aircraft tire is protection of the tire carcass. If the sidewall cords are exposed due to a cut, gouge, snag, or other injury, the tire must be replaced. Mark the area of concern before removal of the tire. Damage to the sidewall that does not reach the cords is typically acceptable for service. Circumferential cracks or slits in the sidewall are unacceptable. A bulge in a tire sidewall indicates possible delamination of the sidewall carcass plies. The tire must immediately be removed from service.

Weather and ozone can cause cracking and checking of the sidewall. If this extends to the sidewall cords, the tire must be removed from service. Otherwise, sidewall checking as show in *Figure 13-147* does not affect the performance of the tire and it may remain in service.

Tire Removal

Removal of any tire and wheel assembly should be accomplished following all aircraft manufacturer's instructions for the procedure. Safety procedures are designed for the protection of the technician and the maintenance of aircraft parts in serviceable condition. Follow all safety

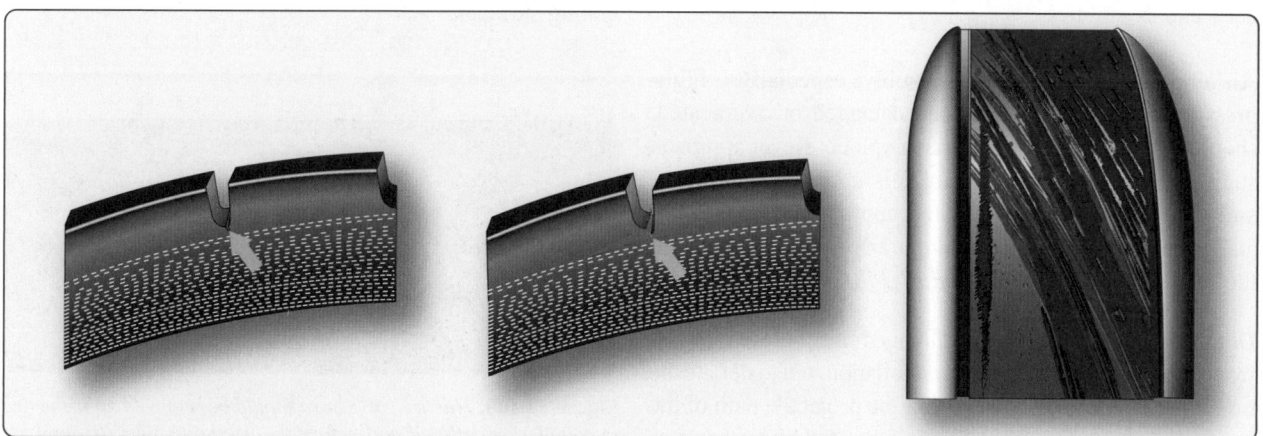

Figure 13-145. *A thrown tread can result from a groove crack or tread undercutting and must be removed from service.*

Figure 13-146. *Cover tires to protect from harmful chemicals and from the elements when parked outside for long periods of time.*

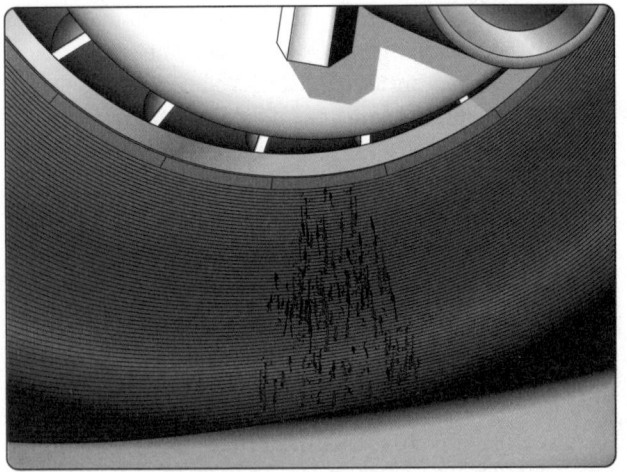

Figure 13-147. *Cracking and checking in the sidewall of a tire is acceptable for service as long as it does not extend to or expose the sidewall carcass plies.*

procedures to prevent personal injury and damage to aircraft parts and assemblies.

An aircraft tire and wheel assembly, especially a high-pressure assembly that has been damaged or overheated, should be treated as though it may explode. Never approach such a tire while its temperature is still elevated above ambient temperature. Once cooled, approach a damaged tire and wheel assembly from an oblique angle advancing toward the shoulder of the tire. [Figure 13-148]

Deflate all unserviceable and damaged tires before removal from the aircraft. Use a valve core/deflation tool to deflate the tire. Stand to the side—away from the projectile path of the valve core. A dislodged valve core propelled by internal tire

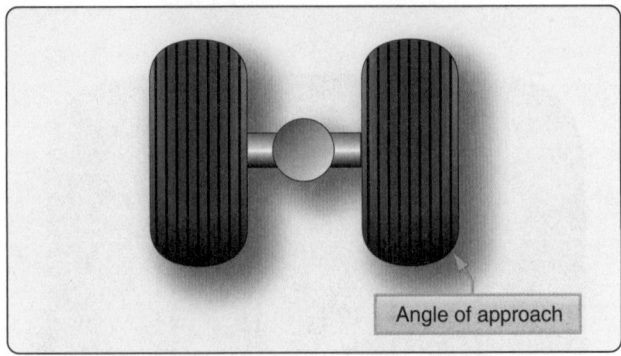

Figure 13-148. *To avoid potential injury, approach a tire/wheel assembly that has damage or has been overheated at an angle toward the tire shoulder only after it has cooled to ambient temperature.*

pressure can cause serious human injury. When completely deflated, remove the valve core. [Figure 13-149] A tire and wheel assembly in airworthy condition may be removed to access other components for maintenance without deflating the tire. This is common practice, such as when accessing the brake when the wheel assembly is immediately reinstalled.

For tracking purposes, ensure damaged areas of a tire are marked before deflation. Record all known information about an unserviceable tire and attach it to the tire for use by the retread repair station.

Once removed from the aircraft, a tire must be separated from the wheel rim upon which it is mounted. Proper equipment and technique should be followed to avoid damage to the tire and wheel. The wheel manufacturer's maintenance information is the primary source for dismounting guidelines.

The bead area of the tire sits firmly against the rim shoulder and must be broken free. Always use proper bead breaking equipment for this purpose. Never pry a tire from a wheel rim as damage to the wheel is inevitable. The wheel tie bolts must remain installed and fully tightened when the bead is broken from the rim to prevent damage to the wheel half mating surfaces.

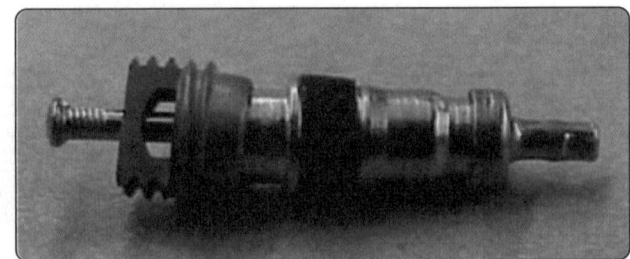

Figure 13-149. *The tire valve core should be removed after the tire is completely deflated and before the tire and wheel assembly is removed from the aircraft.*

When the bead breaking press contact surface is applied to the tire, it should be as close to the wheel as possible without touching it during the entire application of pressure. Tires and rims of different sizes require contact pads suitable for the tire. Hand presses and hydraulic presses are available. Apply the pressure and hold it to allow the bead to move on the rim. Gradually progress around the rim until the tire bead is broken free. Ring-type bead breakers apply pressure around the circumference of the entire sidewall so rotation is not required. *[Figure 13-150]* Once the bead is broken free, the wheel halves may be disassembled. *[Figure 13-151]*

Radial tires have only one bead bundle on each side of the tire. The sidewall is more flexible in this area than a bias ply tire. The proper tooling should be used, and pressure should be applied slowly to avoid heavy distortion of the sidewall. Lubrication may be applied and allowed to soak into the tire-

wheel interface. Only soapy tire solutions should be used. Never apply a hydrocarbon-based lubricant to an aircraft tire as this contaminates the rubber compound used to construct the tire. Beads on tube-type and tubeless tires are broken free in a similar manner.

Tire Inspection Off of the Aircraft

Once a tire has been removed from the wheel rim, it should be inspected for condition. It may be possible to retread the tire at an approved repair station and return it to service. A sequential inspection procedure helps ensure no parts of the tire are overlooked. Mark and record the extent of all damage. Advisory Circular (AC) 43-13-1 gives general guidelines for tire inspection and repair. Tires must only be repaired by those with the experience and equipment to do so. Most tire repairs are accomplished at a certified tire repair facility.

When inspecting a tire removed from the aircraft, pay special attention to the bead area since it must provide an air tight seal to the wheel rim and transfer forces from the tire to the rim. Inspect the bead area closely as it is where the heat is concentrated during tire operation. Surface damage to the chafer is acceptable and can be repaired when the tire is retreaded. Other damage in the bead area is usually cause for rejection. Damage to the turn-ups, ply separation at the bead, or a kinked bead are examples of bead area damage that warrant the tire be discarded. The bead area of the tire may sustain damage or have an altered appearance or texture on a tire that has been overheated. Consult a certified tire repair station or re-tread facility when in doubt about the condition observed. The wheel rim must also be inspected for damage. An effective seal without slippage, especially on tubeless tires, is dependent on the condition and integrity of the wheel in the bead seat area.

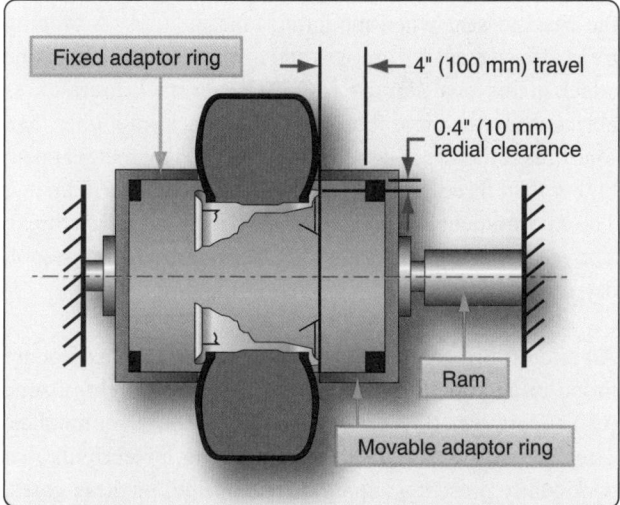

Figure 13-150. *A ring adapter applies pressure around the entire circumference of the lower sidewall of the tire to break the bead free from the wheel rim. The diameter of the adapter must be correct for the tire and the travel limited so as to not injure the tire.*

Overheating of a tire weakens it even though the damage might not be obvious. Any time a tire is involved in an aborted

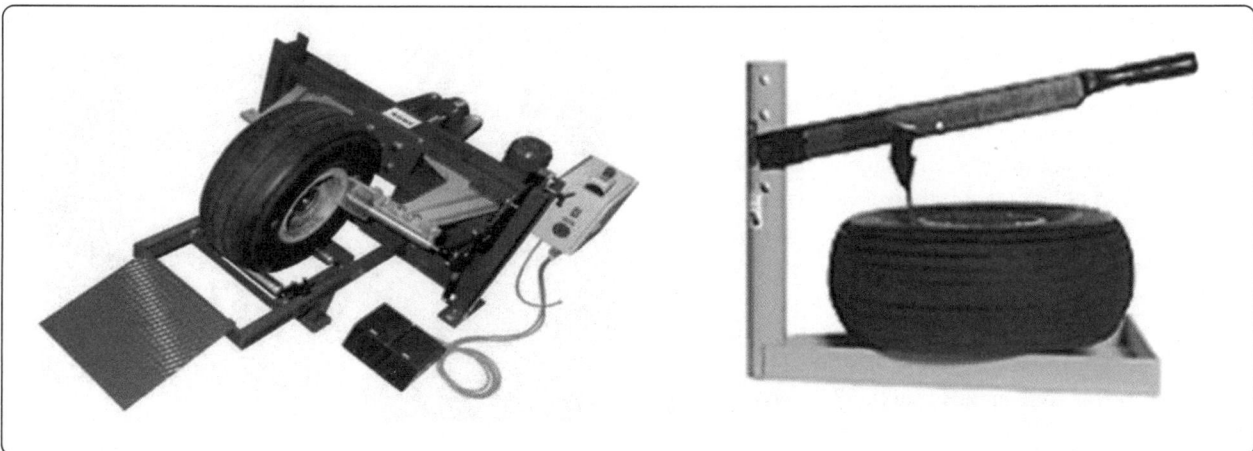

Figure 13-151. *An electrohydraulic tire bead breaker (left) used on large tires and a manual tire bead breaker (right) used on small tires.*

takeoff, severe braking, or the thermal plug in the wheel has melted to deflate the tire before explosion, the tire must be removed. On a dual installation, both tires must be removed. Even if only one tire shows obvious damage or deflates, the loads experienced by the mate are excessive. Internal damage such as ply separation, is likely. The history of having been through an overheat event is all that is required for the tire to be discarded.

Damaged or suspected damaged areas of the tire should be re-inspected while the tire is off the aircraft. Cuts can be probed to check for depth and extent of damage below the tread. In general, damage that does not exceed 40 percent of the tire plies can be repaired when the tire is retreaded. Small punctures with a diameter on the tire inner surface of less than ⅛-inch and a diameter on the outer surface of less than ¼-inch can also be repaired and retreaded. A bulge caused by ply separation is reason to discard a tire. However, a bulge caused by tread separation from the tire carcass may be repairable during retread. Exposed sidewall cord or sidewall cord damage is unacceptable, and the tire cannot be repaired or retreaded. Consult the tire manufacturer or certified retreader for clarification on damage to a tire.

Tire Repair & Retreading

The technician should follow airframe and tire manufacturer instructions to determine if a tire is repairable. Many example guidelines have also been given in this section. Nearly all tire repairs must be made at a certified tire repair facility equipped to perform the approved repair. Bead damage, ply separation, and sidewall cord exposure all require that the tire be scrapped. Inner liner condition on tubeless tires is also critical. Replacing the tube in a tube-type tire is performed by the technician as are mounting and balancing all types of aircraft tires.

Aircraft tires are very expensive. They are also extremely durable. The effective cost of a tire over its life can be reduced by having the tread replaced while the carcass is still sound, and injuries are within repairable limits. Federal Aviation Administration (FAA) certified tire retread repair stations, often the original equipment manufacturer (OEM), do this work. The technician inspects a tire to pre-qualify it for retread so that the cost of shipping it to the retread repair facility is not incurred if there is no chance to retread the tire. The tire retreader inspects and tests every tire to a level beyond the capability of the hangar or line technician. Shearography, an optical nondestructive testing method that provides detailed information about the internal integrity of the tire, is used by tire retread repair facilities to ensure a tire carcass is suitable for continued service.

Tires that are retreaded are marked as such. They are not compromised in strength and give the performance of a new tire. No limits are established for the number of times a tire can be retreaded. This is based on the structural integrity of the tire carcass. A well maintained main gear tire may be able to be retreaded a handful of times before fatigue renders the carcass un-airworthy. Some nose tires can be retread nearly a dozen times.

Tire Storage

An aircraft tire can be damaged if stored improperly. A tire should always be stored vertically so that it is resting on its treaded surface. Horizontal stacking of tires is not recommended. Storage of tires on a tire rack with a minimum 3–4-inch flat resting surface for the tread is ideal and avoids tire distortion.

If horizontal stacking of tires is necessary, it should only be done for a short time. The weight of the upper tires on the lower tires cause distortion possibly making it difficult for the bead to seat when mounting tubeless tires. A bulging tread also stresses rib grooves and opens the rubber to ozone attack in this area. [Figure 13-152] Never stack aircraft tires horizontally for more than 6 months. Stack no higher than four tires if the tire is less than 40-inches in diameter and no higher than three tires if greater than 40-inches in diameter. The environment in which an aircraft tire is stored is critical. The ideal location in which to store an aircraft tire is cool, dry, and dark, free from air currents and dirt.

An aircraft tire contains natural rubber compounds that are prone to degradation from chemicals and sunlight. Ozone (O_3) and oxygen (O_2) cause degradation of tire compounds. Tires should be stored away from strong air currents that continually present a supply of one or both of these gases. Fluorescent lights, mercury vapor lights, electric motors, battery chargers, electric welding equipment, electric generators, and similar shop equipment produce ozone and should not be operated near aircraft tires. Mounted inflated

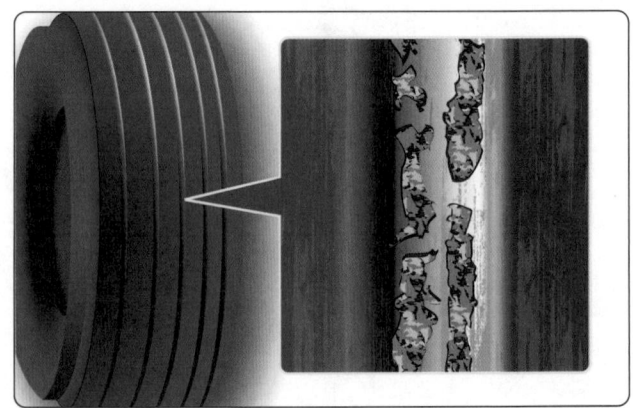

Figure 13-152. *Ozone cracking in a tire tread groove is facilitated by horizontal stacking.*

tires can be stored with up to 25 percent less pressure than operating pressure to reduce vulnerability from ozone attacks. Sodium vapor lighting is acceptable. Storage of an aircraft tire in the dark is preferred to minimize degradation from ultraviolet (UV) light. If this is not possible, wrap the tire in dark polyethylene or paper to form an ozone barrier and to minimize exposure to UV light.

Common hydrocarbon chemicals, such as fuels, oils, and solvents, should not contact a tire. Avoid rolling tires through spills on the hangar or shop floor and be sure to clean any tire immediately if contaminated. Dry the tire and store all tires in a dry place away from any moisture, which has a deteriorating effect on the rubber compounds. Moisture with foreign elements may further damage the rubber and fabric of a tire. Dirty areas must be avoided.

Tires are made to operate in a wide range of temperatures. However, storage should be at cool temperatures to minimize degradation. A general range for safe aircraft tire storage is between 32 °F and 104 °F. Temperatures below this are acceptable but higher temperatures must be avoided.

Aircraft Tubes

Many aircraft tires accept a tube inside to contain the inflation air. Tube-type tires are handled and stored in similar fashion as tubeless tires. A number of issues concerning the tubes themselves must be addressed.

Tube Construction & Selection

Aircraft tire tubes are made of a natural rubber compound. They contain the inflation air with minimal leakage. Unreinforced and special reinforced heavy-duty tubes are available. The heavy-duty tubes have nylon reinforcing fabric layered into the rubber to provide strength to resist chafing and to protect against heat such as during braking.

Tubes come in a wide range of sizes. Only the tube specified for the applicable tire size must be used. Tubes that are too small stress the tube construction.

Tube Storage & Inspection

An aircraft tire tube should be kept in the original carton until put into service to avoid deterioration through exposure to environmental elements. If the original carton is not available, the tube can be wrapped in several layers of paper to protect it. Alternately, for short time periods only, a tube may be stored in the correct size tire it is made for while inflated just enough to round out the tube. Application of talc to the inside of the tire and outside of the tube prevents sticking. Remove the tube and inspect it and the tire before permanently mounting the assembly. Regardless of storage method, always store aircraft tubes in a cool, dry, dark place away from ozone producing equipment and moving air.

When handling and storing aircraft tire tubes, creases are to be avoided. These weaken the rubber and eventually cause tube failure. Creases and wrinkles also tend to be chafe points for the tube when mounted inside the tire. Never hang a tube over a nail or peg for storage.

An aircraft tube must be inspected for leaks and damage that may eventually cause a leak or failure. To check for leaks, remove the tube from the tire. Inflate the tube just enough to have it take shape but not stretch. Immerse a small tube in a container of water and look for the source of air bubbles. A large tube may require that water be applied over the tube. Again, look for the source of bubbles. The valve core should also be wetted to inspect it for leaks.

There is no mandatory age limit for an aircraft tire tube. It should be elastic without cracks or creases in order to be consider serviceable. The valve area is prone to damage and should be inspected thoroughly. Bend the valve to ensure there are no cracks at the base where it is bonded to the tire or in the area where it passes through the hole in the wheel rim. Inspect the valve core to ensure it is tight and that it does not leak.

If an area of a tube experiences chafing to the point where the rubber is thinned, the tube should be discarded. The inside diameter of the tube should be inspected to ensure it has not been worn by contact with the toe of the tire bead. Tubes that have taken an unnatural set should be discarded. *[Figure 13-153]*

Tire Inspection

It is important to inspect the inside of a tube-type tire before installing a tube for service. Any protrusions or rough areas should be cause for concern, as these tend to abrade the tube and may cause early failure. Follow the tire, tube, and aircraft manufacturer's inspection criterion when inspecting aircraft tires and tubes.

Tire Mounting

A licensed technician is often called upon to mount an aircraft tire onto the wheel rim in preparation for service. In the case of a tube-type tire, the tube must also be mounted. The following section presents general procedures for these operations using tube-type and tubeless tires. Be sure to have the proper equipment and training to perform the work according to manufacturer's instructions.

Tubeless Tires

Aircraft tire and wheel assemblies are subject to enormous stress while in service. Proper mounting ensures tires perform to

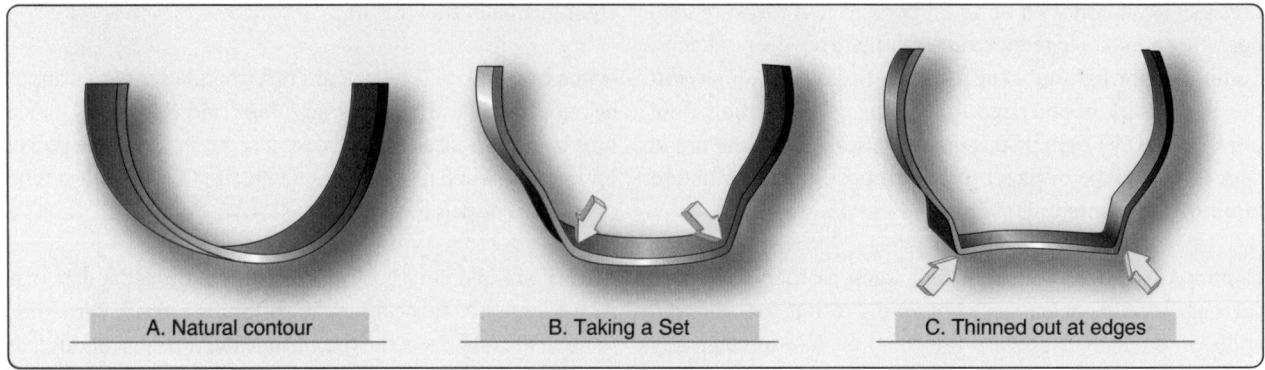

Figure 13-153. *During inspection, an aircraft tire tube should retain its natural contour. Tubes with thinned areas or that have taken a set should be discarded and replaced.*

the limits of their design. Consult and follow all manufacturer's service information including bolt torques, lubrication and balancing requirements, and inflation procedures.

As mentioned, a wheel assembly that is to have a tire mounted upon it must be thoroughly inspected to ensure it is serviceable. Pay close attention to the bead seat area, which should be smooth and free from defects. The wheel half mating surface should be in good condition. The O-ring should be lubricated and in good condition to ensure it seals the wheel for the entire life of the tire. Follow the manufacturer's instructions when inspecting wheels and the tips provided earlier in this chapter. *[Figure 13-154]*

A final inspection of the tire to be mounted should be made. Most important is to check that the tire is specified for the aircraft application. It should say tubeless on the sidewall. The part number, size, ply rating, speed rating, and technical standard order (TSO) number should also be on the sidewall and be approved for the aircraft installation. Visually check the tire for damage from shipping and handling. There should

Figure 13-154. *The wheel half O-ring for a tubeless tire wheel assembly must be in good condition and lubricated to seal for the entire life of the tire. The mating surfaces of the wheel halves must also be in good condition.*

be no permanent deformation of the tire. It should pass all inspections for cuts and other damage discussed in the previous sections of this chapter. Clean the tire bead area with a clean shop towel and soap and water or denatured alcohol. Inspect the inside of the tire for condition. There should be no debris inside the tire.

Tire beads are sometimes lubricated when mounted on aluminum wheels. Follow the manufacturer's instructions and use only the non-hydrocarbon lubricant specified. Never lubricate any tire bead with grease. Do not use lubricants with magnesium alloy wheels. Most radial tires are mounted without lubricant. The airframe manufacturer may specify lubrication for a radial tire in a few cases.

When the wheel halves and tires are ready to be mounted, thought must be given to tire orientation and the balance marks on the wheel halves and tire. Typically, the tire serial number is mounted to the outboard side of the assembly. The marks indicating the light portion of each wheel half should be opposite each other. The mark indicating the heavy spot of the wheel assembly should be mounted aligned with the light spot on the tire, which is indicated by a red mark. If the wheel lacks a mark indicating the heavy spot, align the red spot on the tire (the light point) with the valve fitting location on the wheel. A properly balanced tire and wheel assembly improves the overall performance of the tire. It promotes smooth operation free from vibration, which results in uniform tread wear and extended tire life.

When assembling the wheel halves, follow manufacturer's instructions for tie bolt tightening sequences and torque specification. Anti-seize lubricants and wet-torque values are common on wheel assemblies. Use a calibrated hand torque wrench. Never use an impact wrench on an aircraft tire assembly.

For the initial inflation of an aircraft tire and wheel assembly, the tire must be placed in a tire inflation safety cage and treated as though it may explode due to wheel or tire failure. The inflation hose should be attached to the tire valve stem, and inflation pressure should be regulated from a safe distance away. A minimum of 30 feet is recommended. Air or nitrogen should be introduced gradually as specified. Dry nitrogen keeps the introduction of water into the tire to a minimum, which helps prevent corrosion. Observe the tire seating progress on the wheel rim while it inflates. Depressurize the tire before approaching it to investigate any observed issue. *[Figure 13-155]*

Aircraft tires are typically inflated to their full specified operating pressure. Then, they are allowed to remain with no load applied for 12-hours. During this time, the tire stretches, and tire pressure decreases. A 5-10 percent reduction is normal. Upon bringing the tire up to full pressure again, less than 5 percent loss per day of pressure is allowable. More should be investigated.

Tube-Type Tires

Wheel and tire inspection should precede the mounting of any tire, including tube-type tires. The tube to be installed must also pass inspection and must be the correct size for the tire and tire must be specified for the aircraft. Tire talc is commonly used when installing tube-type tires to ensure easy mounting and free movement between the tube and tire as they inflate. *[Figure 13-156]* The technician should lightly talc the inside of the tire and the outside of the tube. Some tubes come from the factory with a light talc coating over the outside of the tube. Inflate the tube so that it just takes shape with minimal pressure. Install the tube inside the tire. Tubes are typically produced with a mark at the heavy spot

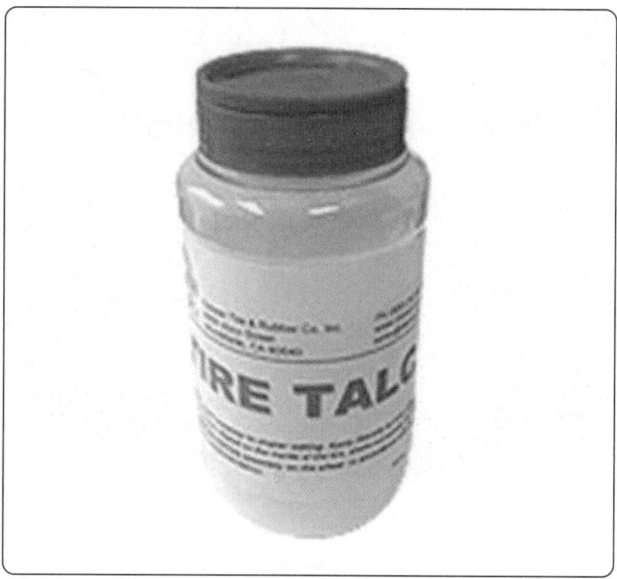

Figure 13-156. *Tire talc is used on the inside of tube-type tires and the outside of aircraft tubes. This prevents binding and allows the tube to expand without stress into place within the tire.*

of the tube. In the absence of this balance mark, it is assumed that the valve is located at the heaviest part of the tube. For proper balance, align the heavy part of the tube with the red mark on the tire (the light spot on the tire). *[Figure 13-157]*

Once wheel balance is marked and the tube balance mark and the tire balance mark are all positioned correctly, install the outboard wheel half so the valve stem of the tube passes through the valve stem opening. *[Figure 13-158]* Mate the inboard wheel half to it, being careful not to pinch the tube between the wheel rims. Install the tie bolts, tighten, and

Figure 13-155. *Modern tire inflation cages have been tested to withstand catastrophic failure of a tire and wheel assembly during inflation. All newly mounted tires should be inflated in such a cage.*

Figure 13-157. *When assembling a tube into a tire, the heavy balance mark on the tube is aligned with the light balance mark on the tire.*

Figure 13-158. *Mounting a tube type tire with the tube valve stem positioned to pass through the outboard wheel half.*

torque as specified. Inflate the assembly in a tire inflation cage.

The inflation procedure for a tube-type tire differs slightly from that of a tubeless tire. The assembly is slowly brought up.to full operating pressure. Then, it is completely deflated. Re-inflate the tire/tube assembly a second time to the specified operating pressure and allow it to remain with no load for 12-hours. This allows any wrinkles in the tube to smooth out, helps prevent the tube from being trapped under a bead, and generally evens how the tube lays within the tire to avoid any stretched areas and thinning of the tube. The holding time allows air trapped between the tube and the tire to work its way out of the assembly, typically through the tire sidewall or around the valves stem.

Tire Balancing

Once an aircraft tire is mounted, inflated, and accepted for service, it can be balanced to improve performance. Vibration is the main result of an imbalanced tire and wheel assembly. Nose wheels tend to create the greatest disturbance in the cabin when imbalanced.

Static balance is all that is required for most aircraft tires and wheels. A balance stand typically accepts the assembly on cones. The wheel is free to rotate. The heavy side moves to the bottom. *[Figure 13-159]* Temporary weights are added to eliminate the wheel from rotating and dropping the heavy side down. Once balanced, permanent weights are installed. Many aircraft wheels have provisions for securing the permanent weight to the wheel. Weights with adhesive designed to be glued to the wheel rim are also in use. Occasionally, a weight in the form of a patch glued to the inside of the tire is required. Follow all manufacturer's instructions and use only the weights specified for the wheel assembly. *[Figure 13-160]*

Some aviation facilities offer dynamic balancing of aircraft tire and wheel assemblies. While this is rarely specified by manufacturers, a well-balanced tire and wheel assembly helps provide shimmy free operation and reduces wear on brake and landing gear components, such as torque links.

Operation & Handling Tips

Aircraft tires experience longer life if operated in a manner to conserve wear and minimize damage. The most important factor impinging on tire performance and wear, as well as resistance to damage is proper inflation. Always inflate tires to the specified level before flight for maximum performance and minimal damage. An improperly inflated tire has increased potential to fail upon landing due to the high

Figure 13-159. *A typical aircraft tire and wheel balancing stand.*

Figure 13-160. *A tire balancing patch (left), adhesive wheel weights (center), and a bolted wheel weight (right) are all used to balance aircraft tire and wheel assemblies per the manufacturer's instructions.*

impact loads experienced. The following sections include other suggestions that can extend the life and the investment made in aircraft tires.

Taxiing

Needless tire damage and excessive wear can be prevented by proper handling of the aircraft during taxi. Most of the gross weight of an aircraft is on the main landing gear wheels. Aircraft tires are designed and inflated to absorb the shock of landing by deflection of the sidewalls two to three times as much as that found on an automobile tire. While this enables the tire to handle heavy loads, it also causes more working of the tread and produces scuffing action along the outer edges of the tread that results in more rapid wear. It also leaves the tire more prone to damage as the tread compound opens during this flexing.

An aircraft tire that strikes a chuck hole, a stone, or some other foreign object is more likely to sustain a cut, snag, or bruise than an automobile tire due to its more flexible nature. There is also increased risk for internal tire injury when a tire leaves the paved surface of the taxi way. These incidents should be avoided. Dual or multiple wheel main gear should be operated so that all tires remain on the paved surface so the weight of the aircraft is evenly distributed between the tires. When backing an aircraft on a ramp for parking, care should be taken to stop the aircraft before the main wheels roll off of the paved surface.

Taxiing for long distances or at high speeds increase the temperature of aircraft tires. This makes them more susceptible to wear and damage. Short taxi distances at moderate speeds are recommended. Caution should also be used to prevent riding the brakes while taxiing, which adds unnecessary heat to the tires.

Braking & Pivoting

Heavy use of aircraft brakes introduces heat into the tires. Sharp radius turns do the same and increase tread abrasion and side loads on the tire. Plan ahead to allow the aircraft to slow without heavy braking and make large radius turns to avoid these conditions. Objects under a tire are ground into the tread during a pivot. Since many aircraft are primarily maneuvered on the ground via differential braking, efforts should be made to always keep the inside wheel moving during a turn rather than pivoting the aircraft with a locked brake around a fixed main wheel tire.

Landing Field & Hangar Floor Condition

One of the main contributions made to the welfare of aircraft tires is good upkeep of airport runway and taxiway surfaces, as well as all ramp areas and hangar floors. While the technician has little input into runway and taxiway surface upkeep, known defects in the paved surfaces can be avoided and rough surfaces can be negotiated at slower than normal speeds to minimize tire damage. Ramps and hangar floors should be kept free of all foreign objects that may cause tire damage. This requires continuous diligence on the part of all aviation personnel. Do not ignore foreign object damage (FOD). When discovered, action must be taken to remove it. While FOD to engines and propellers gains significant attention, much damage to tires is avoidable if ramp areas and hangar floors are kept clean.

Takeoffs & Landings

Aircraft tires are under severe strain during takeoff and landing. Under normal conditions, with proper control and maintenance of the tires, they are able to withstand these stresses and perform as designed.

Most tire failures occur during takeoff which can be extremely dangerous. Tire damage on takeoff is often the result of running over some foreign object. Thorough preflight inspection of the tires and wheels, as well as maintenance of hangar and ramp surfaces free of foreign objects, are keys to prevention of takeoff tire failure. A flat spot caused on the way to the runway may lead to tire failure during takeoff. Heavy braking during aborted takeoffs is also a common cause of takeoff tire failure. *[Figure 13-161]*

Tire failure upon landing can have several causes. Landing with the brakes on is one. This is mitigated on aircraft with anti-skid systems but can occur on other aircraft. Other errors in judgment, such as landing too far down the runway and having to apply the brakes heavily, can cause overheating or skidding. This can lead to flat-spotting the tires or blow out.

Hydroplaning

Skidding on a wet, icy, or dry runway is accompanied by the threat of tire failure due to heat build-up and rapid tire wear damage. Hydroplaning on a wet runway may be overlooked as a damaging condition for a tire. Water building up in front of the tire provides a surface for the tire to run on and contact with the runway surface is lost. This is known as dynamic hydroplaning. Steering ability and braking action is also lost. A skid results if the brakes are applied and held.

Viscous hydroplaning occurs on runways with a thin film of water that mixes with contaminants to cause an extremely slick condition. This can also happen on a very smooth runway surface. A tire with a locked brake during viscous hydroplaning can form an area of reverted rubber or skid burn in the tread. While the tire may continue in service if the damage is not too severe, it can be cause for removal if the reinforcing tread or protector ply is penetrated. The same damage can occur while skidding on ice.

Modern runways are designed to drain water rapidly and provide good traction for tires in wet conditions. A compromise exists in that crosscut runways and textured runway surfaces cause tires to wear at a greater rate than a smooth runway. *[Figure 13-162]* A smooth landing is of great benefit to any tire. For the most part, aircraft tire handling and care is the responsibility of the pilot; however, the technician benefits from knowing the causes of tire failure and communicating this knowledge to the flight crew so that operating procedures can be modified to avoid those causes.

Figure 13-161. *Heavy braking during an aborted takeoff caused these tires to fail.*

Figure 13-162. *Crosscut runway surfaces drain water rapidly but increase tire wear.*

Chapter 14
Aircraft Fuel Systems

Basic Fuel System Requirements

All powered aircraft require fuel on board to operate the engine(s). A fuel system consisting of storage tanks, pumps, filters, valves, fuel lines, metering devices, and monitoring devices is designed and certified under strict Title 14 of the Code of Federal Regulations (14 CFR) guidelines. Each system must provide an uninterrupted flow of contaminant-free fuel regardless of the aircraft's attitude. Since fuel load can be a significant portion of the aircraft's weight, a sufficiently strong airframe must be designed. Varying fuel loads and shifts in weight during maneuvers must not negatively affect control of the aircraft in flight.

Each Federal Aviation Administration (FAA) certified aircraft is designed and constructed under regulations applicable to that type of aircraft. The certification airworthiness standards are found in 14 CFR as follows:

- 14 CFR part 23—Airworthiness Standards: Normal Category Airplanes

- 14 CFR part 25—Airworthiness Standards: Transport Category Airplanes

- 14 CFR part 27—Airworthiness Standards: Normal Category Rotorcraft

- 14 CFR part 29—Airworthiness Standards: Transport Category Rotorcraft

- 14 CFR part 31—Airworthiness Standards: Manned Free Balloons

Additional information is found in 14 CFR part 33. It addresses airworthiness standards for engines and pertains mainly to engine fuel filter and intake requirements.

Title 14 of the CFR, part 23, Airworthiness Standards: Normal Category Airplanes, section 23.2430, Fuel Systems, is summarized below. Airworthiness standards specified for air carrier and helicopter certification are similar. Although the technician is rarely involved with designing fuel systems, a review of these criteria gives insight into how an aircraft fuel system operates.

Each fuel system must be constructed and arranged to ensure fuel flow at a rate and pressure established for proper engine and auxiliary power unit (APU) functioning under each likely operating condition. This includes any maneuver for which

Figure 14-1. *Aircraft fuel systems must deliver fuel during any maneuver for which the aircraft is certified.*

certification is requested and during which the engine or APU may be in operation. *[Figure 14-1]* Each fuel system must be arranged so that no fuel pump can draw fuel from more than one tank at a time. There must also be a means to prevent the introduction of air into the system.

Each fuel system for a turbine engine powered airplane must meet applicable fuel venting requirements. 14 CFR part 34 outlines requirements that fall under the jurisdiction of the Environmental Protection Agency (EPA). A turbine engine fuel system must be capable of sustained operation throughout its flow and pressure range even though the fuel has some water in it. The standard is that the engine continues to run using fuel initially saturated with water at 80 °F having 0.75 cubic centimeters (cm) of free water per gallon added to it and then cooled to the most critical condition for icing likely to be encountered in operation.

Fuel System Independence

Each fuel system must be designed and arranged to provide independence between multiple fuel storage and supply systems so that failure of any one component in one system will not result in loss of fuel storage or supply of another system.

Fuel System Lightning Protection

The fuel system must be designed and arranged to prevent the ignition of the fuel within the system by direct lightning strikes or swept lightning strokes to areas where such occurrences are highly probable, or by corona or streamering at fuel vent outlets. A corona is a luminous discharge that occurs as a result of an electrical potential difference

Figure 14-2. *Lightning streamering at the wingtips of a jet fighter.*

isolated from personnel compartments and protected from hazards due to unintended temperature influences. The fuel storage system must provide fuel for at least one-half hour of operation at maximum continuous power or thrust and be capable of jettisoning fuel safely if required for landing. *[Figure 14-3]* Fuel jettisoning systems are also referred to as fuel dump systems. *[Figure 14-4]* A fuel dump system is a system installed in most large aircraft that allows the flight crew to jettison, or dump, fuel to lower the gross weight of the aircraft to its allowable landing weight. Boost pumps in the fuel tanks move the fuel from the tank into a fuel manifold. From the fuel manifold, it flows away from the aircraft through dump chutes in each wing tip. The fuel jettison system must be so designed and constructed that it is free from fire hazards.

Aircraft fuel tanks must be designed to prevent significant loss of stored fuel from any vent system due to fuel transfer between fuel storage or supply systems, or under likely operating conditions.

between the aircraft and the surrounding area. Streamering is a branch-like ionized path that occurs in the presence of a direct stroke or under conditions when lightning strikes are imminent. *[Figure 14-2]*

Fuel Flow

The ability of the fuel system to provide the fuel necessary to ensure each powerplant and auxiliary power unit functions properly in all likely operating conditions. It must also prevent hazardous contamination of the fuel supplied to each powerplant and auxiliary power unit.

The fuel system must provide the flightcrew with a means to determine the total useable fuel available and provide uninterrupted supply of that fuel when the system is correctly operated, accounting for likely fuel fluctuations. It should also provide a means to safely remove or isolate the fuel stored in the system from the airplane and be designed to retain fuel under all likely operating conditions and minimize hazards to the occupants during any survivable emergency landing. For level 4 airplanes, failure due to overload of the landing system must be taken into account

Fuel Storage System

Each fuel tank must be able to withstand, without failure, the loads under likely operating conditions. Each tank must be

Figure 14-3. *Fuel being jettisoned free of the airframe on a transport category aircraft.*

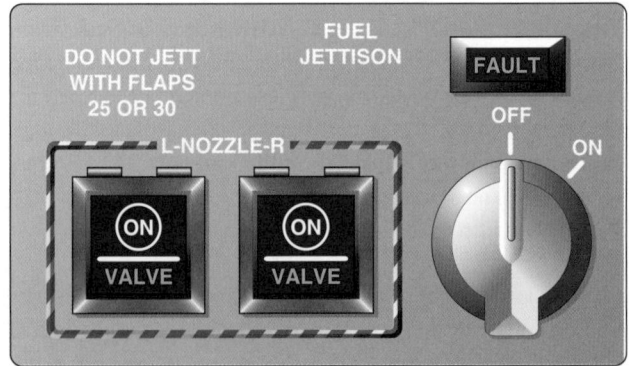
Figure 14-4. *The fuel jettison panel on a Boeing 767.*

Fuel Storage Refilling or Recharging System

Each fuel storage refilling or recharging system must be designed to prevent improper refilling or recharging; prevent contamination of the fuel stored during likely operating conditions; and prevent the occurrence of any hazard to the airplane or to persons during refilling or recharging.

Types of Aviation Fuel

Each aircraft engine is designed to burn a certain fuel. Use only the fuel specified by the manufacturer. Mixing fuels is not permitted. There are two basic types of fuel discussed in this section: reciprocating-engine fuel (also known as gasoline or AVGAS) and turbine-engine fuel (also known as jet fuel or kerosene).

Reciprocating Engine Fuel—AVGAS

Reciprocating engines burn gasoline, also known as AVGAS. It is specially formulated for use in aircraft engines. Combustion releases energy in the fuel, which is converted into the mechanical motion of the engine. AVGAS of any variety is primarily a hydrocarbon compound refined from crude oil by fractional distillation. Aviation gasoline is different from the fuel refined for use in turbine-powered aircraft. AVGAS is very volatile and extremely flammable, with a low flash point. Turbine fuel is a kerosene-type fuel with a much higher flash point, so it is less flammable.

Aircraft engines must perform throughout a wide range of demanding conditions. They must be lightweight and produce significant power in a wide range of atmospheric and engine operating temperatures. The gasoline used must support uninterrupted combustion throughout this range and must truly burn rather than explode or detonate. This ensures maximum power derivation and minimal engine wear. Over the years, AVGAS has been available in different formulas. These mostly correlate to how much energy can be produced without the fuel detonating. Larger, high-compression engines require fuel with a greater amount of potential power production without detonation than smaller low-compression engines.

Volatility

One of the most important characteristics of an aircraft fuel is its volatility. Volatility is a term used to describe how readily a substance changes from liquid into a vapor. For reciprocating engines, highly volatile fuel is desired. Liquid gasoline delivered to the engine induction system carburetor must vaporize in the carburetor to burn in the engine. Fuel with low volatility vaporizes slowly. This can cause hard engine starting, slow warm-up, and poor acceleration. It can also cause uneven fuel distribution to the cylinders and excessive dilution of the oil in the crankcase in engines equipped with oil dilution systems. However, fuel can also be too volatile,

causing detonation and vapor lock.

AVGAS is a blend of numerous hydrocarbon compounds, each with different boiling points and volatility. A straight chain of volatile compounds creates a fuel that vaporizes easily for starting, but also delivers power through the acceleration and power ranges of the engine.

Vapor Lock

Vapor lock is a condition in which AVGAS vaporizes in the fuel line or other components between the fuel tank and the carburetor. This typically occurs on warm days on aircraft with engine-driven fuel pumps that suck fuel from the tank(s). Vapor lock can be caused by excessively hot fuel, low pressure, or excessive turbulence of the fuel traveling through the fuel system. In each case, liquid fuel vaporizes prematurely and blocks the flow of liquid fuel to the carburetor.

Aircraft gasoline is refined to have a vapor pressure be between 5.5 pounds per square inch (psi) and 7.0 psi at 100 °F. At this pressure, an aircraft fuel system is designed to deliver liquid fuel to the carburetor when drawn out of the tank by an engine-driven fuel pump. But temperatures in the fuel system can exceed 100 °F under the engine cowl on a hot day. Fuel may vaporize before it reaches the carburetor, especially if it is drawn up a line under a low pressure, or if it swirls while navigating a sharp bend in the tubing. To make matters worse, when an aircraft climbs rapidly, the pressure on the fuel in the tank decreases while the fuel is still warm. This causes an increase in fuel vaporization that can also lead to vapor lock.

Various steps can be taken to prevent vapor lock. The use of boost pumps located in the fuel tank that force pressurized liquid fuel to the engine is most common.

Carburetor Icing

As fuel vaporizes, it draws energy from its surroundings to change state from a liquid to a vapor. This can be a problem if water is present. When fuel vaporizes in the carburetor, water in the fuel-air mixture can freeze and deposit inside the carburetor and fuel induction system. The fuel discharge nozzle, throttle valve, venturi, or simply the walls of the induction system all can develop ice. As the ice builds, it restricts the fuel-air flow and causes loss of engine power. In severe cases, the engine stops running. *[Figure 14-5]*

Carburetor icing is most common at ambient temperatures of 30–40 °F but can occur at much higher temperatures, especially in humid conditions. Most aircraft are equipped with carburetor heating to help eliminate this threat caused by the high volatility of the fuel and the presence of moisture. *[Figure 14-6]*

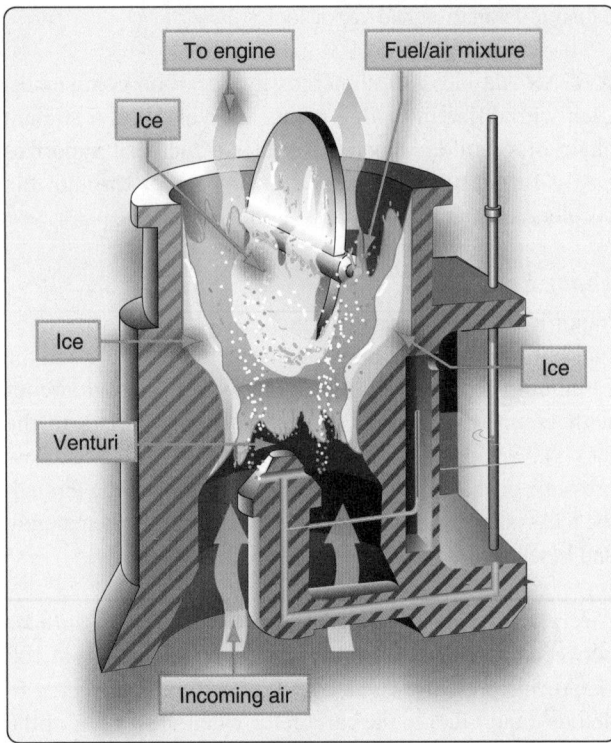

Figure 14-5. *An example of common areas where ice can form on a carburetor. The evaporation of volatile fuel takes energy from its surroundings to change state. As it does, water in the fuel-air mixture condenses and freezes.*

Aromatic Fuels

The aviation gasoline market is a relatively small part of the overall gasoline market. AVGAS producers are few. In years past, when this was less the case, considerable quantities of aromatic hydrocarbons were sometimes added to increase the rich mixture performance of AVGAS. It was used mainly in high horsepower reciprocating engines, such as military and transport category aircraft. Special hoses and seals were required for use of aromatic fuels. These additives are no longer available.

Detonation

Detonation is the rapid, uncontrolled explosion of fuel due to high pressure and temperature in the combustion chamber. The fuel-air charge ignites and explodes before the ignition system spark lights it. Occasionally, detonation occurs when the fuel is ignited via the spark plug but explodes before it is finished burning.

The engine is not designed to withstand the forces caused by detonation. It is made to turn smoothly by having the fuel-air mixture burn in the combustion chamber and propagate directionally across the top of the piston. When it does so, a smooth transfer of the force developed by the burning fuel pushes the piston down. Detonation of fuel instead sends a shock wave of force against the top of the piston, which in turn is transferred through the piston to the piston pin, to the connecting rod, and to the crankshaft. Valve operation is also affected by this shock wave. In short, the explosion of fuel detonating in the combustion chamber transfers the energy contained in the fuel harshly throughout the entire engine, causing damage.

Aviation fuels are refined and blended to avoid detonation. Each has an ignition point and burn speed at specific fuel-air mixture ratios that manufacturers rely on to design engines that can operate without detonation. An engine experiencing

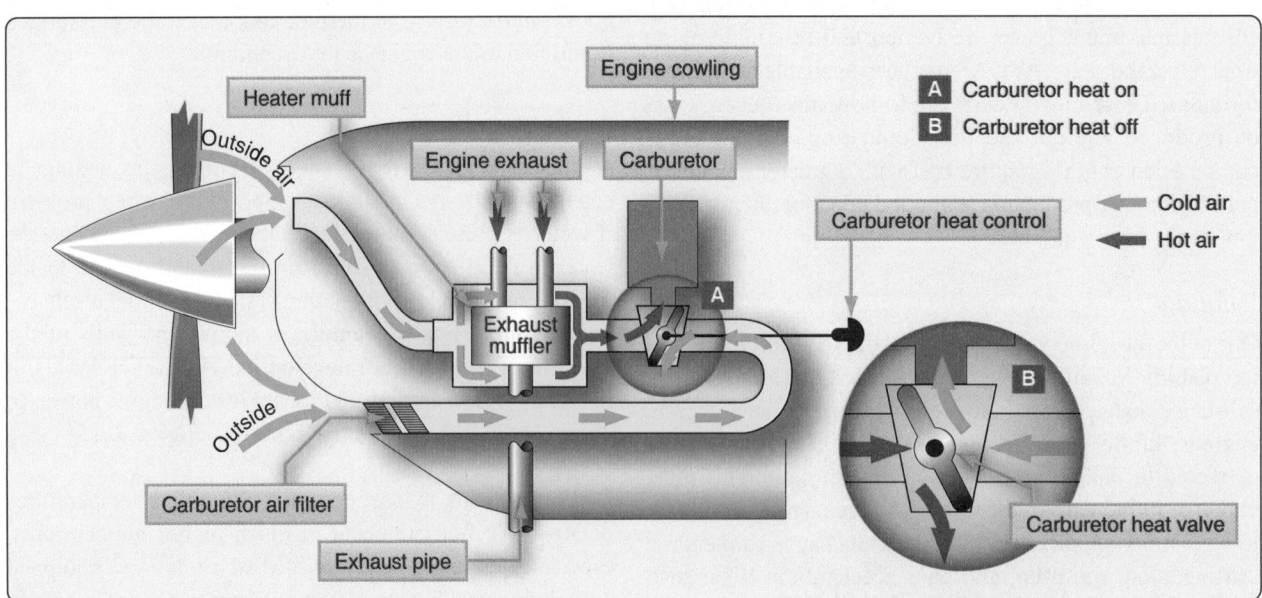

Figure 14-6. *To combat carburetor icing, air preheated by the exhaust manifold is directed into the carburetor via a push/pull control in the flight deck. The control changes the position of the air diverter butterfly in the carburetor heat valve box.*

detonation in the field should be investigated. A pinging or knocking sound is a sign of detonation. This is often more difficult to detect in an aircraft than in an automobile due to propeller tip noise. Detonation causes an increase in cylinder head temperature.

If ignored or allowed to continue, detonation can eventually lead to engine failure. Causes of detonation include incorrect fuel, already high engine temperature at high power settings, such as takeoff, preignition of the fuel, extended operations with an extremely lean mixture, and operation at high revolutions per minute (rpm) with low airspeed.

Surface Ignition & Preignition

A sharp deposit or incandescent hot spot in the combustion chamber can cause fuel to ignite before the spark plug lights it. Detonation can cause such an area to form as can a cracked spark plug insulator or a sharp valve edge. The result could be ignition of the fuel before the piston is at the proper place during its movement toward top dead center of the compression stroke. The extended burn period of the fuel can increase temperatures and pressure in the combustion chamber to the point at which the fuel detonates. The repeated incorrect flame propagation and detonation can cause serious engine damage and eventual engine failure. *[Figure 14-7]*

Maintenance personnel should ensure that the correct fuel is being used, and that the engine is being operated correctly. Spark plugs and valves should be checked for wear. Signs of deposits and detonation must also be investigated and addressed.

Octane & Performance Number Rating

Octane ratings and performance numbers are given to fuels to describe their resistance to detonation. Fuels with high

Figure 14-7. *Preignition can cause detonation and damage to the engine.*

critical pressure and high octane or performance numbers have the greatest resistance. A referencing system is used to rate the fuel. A mixture of two hydrocarbons, iso-octane (C_8H_{18}) and heptane (C_7H_{16}), is used. Various ratios of the two hydrocarbons in a mixture result in proportional antidetonation properties. The more iso-octane there is in the mixture, the higher its resistance is to detonation.

When a fuel has the same critical pressure as a reference mixture of these two hydrocarbons, it is said to have an octane rating that is the same as the percentage of the iso-octane in this reference mixture. An 80-octane fuel has the same resistance to detonation as an 80 percent iso-octane, 20 percent heptane mixture; a 90-octane fuel has the same resistance to detonation as a 90 percent iso-octane, 10 percent heptane mixture; and a 100-octane fuel has the same resistance to detonation as 100 percent pure iso-octane. So, by comparing a fuel's tendency to detonate to reference mixtures of iso-octane and heptane, octane ratings from 80 to 100 can be established. The highest-octane fuel possible with this system of measurement is 100-octane fuel.

To increase antidetonation characteristics of fuel, substances can be added. Tetraethyl lead (TEL) is the most common additive that increases the critical pressure and temperature of a fuel. However, additional additives, such as ethylene dibromide and tricresyl phosphate, must be also be added so that the TEL does not leave solid deposits in the combustion chamber.

The amount of TEL added to a fuel can be increased to raise the antidetonation characteristics from 80 to the 100-octane level and higher. References to octane characteristics above 100 percent iso-octane are made by referencing the antidetonation properties of the fuel to a mixture of pure iso-octane and specific quantities of TEL. The specific mixtures of iso-octane and TEL are assigned arbitrary octane numbers above 100. In addition to increasing the antidetonation characteristics of a fuel, TEL also lubricates the engine valves.

Performance numbers are also used to characterize the antidetonation characteristics of fuel. A performance number consists of two numbers (e.g., 80/87, 100/130, 115/145) in which higher numbers indicate a higher resistance to detonation. The first number indicates the octane rating of the fuel in a lean fuel-air mixture, and the second number indicates the octane rating of the fuel in a rich mixture.

Due to the small size of the worldwide aviation gasoline market, a single 100 octane low-lead fuel (100LL) is desired as the only AVGAS for all aircraft with reciprocating engines. This presents problems in engines originally designed to run on 80/87 fuel; the low lead 100-octane fuel still contains more lead than the 80-octane fuel. Spark plug fouling has been common and lower times between overhaul have occurred. Other engines designed for 91/96 fuel or 100/130 fuel operate satisfactorily on 100LL, which contains 2 milliliters of TEL per gallon (enough to lubricate the valves and control detonation). For environmental purposes, AVGAS with no TEL is sought for the aviation fleet of the future.

Fuel Identification

Aircraft and engine manufacturers designate approved fuels for each aircraft and engine. Consult manufacturer data and use only those fuel specified therein.

The existence of more than one fuel makes it imperative that fuel be positively identified and never introduced into a fuel system that is not designed for it. The use of dyes in fuel helps aviators monitor fuel type. 100LL AVGAS is the AVGAS most readily available and used in the United States. It is dyed blue. Some 100 octane or 100/130 fuel may still be available, but it is dyed green.

80/87 AVGAS is no longer available. It was dyed red. Many

supplemental type certificates have been issued to engine and engine/airframe combinations that permit the use of automobile gasoline in engines originally designed for red AVGAS. A relatively new AVGAS fuel, 82UL (unleaded), has been introduced for use by this group of relatively low compression engines. It is dyed purple.

115/145 AVGAS is a fuel designed for large, high performance reciprocating engines from the World War II era. It is available only by special order from refineries and is also dyed purple in color.

The color of fuel may be referred to in older maintenance manuals. All grades of jet fuel are colorless or straw colored. This distinguishes them from AVGAS of any kind that contains dye of some color. Should AVGAS fuel not be of a recognizable color, the cause should be investigated. Some color change may not affect the fuel. Other times, a color change may be a signal that fuels have been mixed or contaminated in some way. Do not release an aircraft for flight with unknown fuel onboard.

Identifying fuel and ensuring the correct fuel is delivered

Fuel Type and Grade	Color of Fuel	Equipment Control Color	Pipe Banding and Marking	Refueler Decal
AVGAS 82UL	Purple	82UL AVGAS	AVGAS 82UL	82UL / AVGAS
AVGAS 100	Green	100 AVGAS	AVGAS 100	100 / AVGAS
AVGAS 100LL	Blue	100LL AVGAS	AVGAS 100LL	100LL / AVGAS
JET A	Colorless or straw	JET A	JET A	JET A
JET A-1	Colorless or straw	JET A-1	JET A-1	JET A-1
JET B	Colorless or straw	JET B	JET B	JET B

Figure 14-8. *Color coded labeling and markings used on fueling equipment.*

into storage tanks, fuel trucks, and aircraft fuel tanks is a process aided by labeling. Decals and markings using the same colors as the AVGAS colors are used. Delivery trucks and hoses are marked as are aircraft tank fuel caps and fill areas. Jet fuel fill hose nozzles are sized too large to fit into an AVGAS tank fill opening. *Figure 14-8* shows examples of color-coded fuel labeling.

Purity

The use of filters in the various stages of transfer and storage of AVGAS removes most foreign sediment from the fuel. Once in the aircraft fuel tanks, debris should settle into the fuel tank drain sumps to be removed before flight. Filters and strainers in the aircraft fuel system can successfully capture any remaining sediment.

The purity of aviation gasoline is compromised most often by water. Water also settles into the sumps given enough time. However, water is not removed by the aircraft's filters and strainers as easily as solid particles. It can enter the fuel even when the aircraft is parked on the ramp with the fuel caps in place. Air in the tank vapor space above the liquid fuel contains water vapor. Temperature fluctuations cause the water vapor to condense on the inner surface of the tanks and settle into the liquid fuel. Eventually, this settles to the sump, but some can remain in the fuel when the aircraft is to be flown.

Proper procedure for minimizing water entering aircraft fuel is to fill the aircraft fuel tanks immediately after each flight. This minimizes the size of the vapor space above the liquid fuel and the amount of air and associated water vapor present in the tank. When excessive water is drawn into the fuel system, it passes through carburetor jets where it can interrupt the smooth operation of the engine(s).

If water is entrained or dissolved in the fuel, it cannot be removed by draining the sump(s) and filter bowls before flight. However, there may be enough water for icing to be a concern. As the aircraft climbs and fuel is drawn out of the tanks, the fuel supply cools. Entrained and dissolved water in the fuel is forced out of solution and becomes free water. If cool enough, ice crystals form rather than liquid water. These can clog filters and disrupt fuel flow to the engines. Both AVGAS and jet fuel have this type of water impurity issue leading to icing that must be monitored and treated.

Fuel anti-ice additives can be added to the bulk fuel and also directly into the aircraft fuel tank, usually during refueling. These are basically diethylene glycol solutions that work as antifreeze. They dissolve in free water as it comes out of the fuel and lower its freezing point. *[Figure 14-9]*

Turbine Engine Fuels

Aircraft with turbine engines use a type of fuel different from that of reciprocating aircraft engines. Commonly known as jet fuel, turbine engine fuel is designed for use in turbine engines and should never be mixed with aviation gasoline or introduced into the fuel system of a reciprocating aircraft engine fuel system.

The characteristics of turbine engine fuels are significantly different from those of AVGAS. Turbine engine fuels are hydrocarbon compounds of higher viscosity with much lower volatility and higher boiling points than gasoline. In the distillation process from crude oil, the kerosene cut from which jet fuel is made condenses at a higher temperature than the naphtha or gasoline cuts. The hydrocarbon molecules of turbine engine fuels are composed of more carbon than are in AVGAS. *[Figure 14-10]*

Turbine engine fuels sustain a continuous flame inside the engine. They typically have a higher sulfur content than gasoline, and various inhibitors are commonly added to them. Used to control corrosion, oxidation, ice, and microbial and bacterial growth, these additives often are already in the fuel when it arrives at the airport for use.

Turbine Fuel Volatility

The choice of turbine engine fuel reflects consideration of conflicting factors. While it is desirable to use a fuel that is low in volatility to resist vapor lock and evaporation while in the aircraft's fuel tanks, turbine engine aircraft operate in

Figure 14-9. *Fuel anti-icing products, such as Prist®, act as antifreeze for any free water in aircraft fuel. They dissolve in the water and lower its freezing point to prevent ice crystals from disrupting fuel flow.*

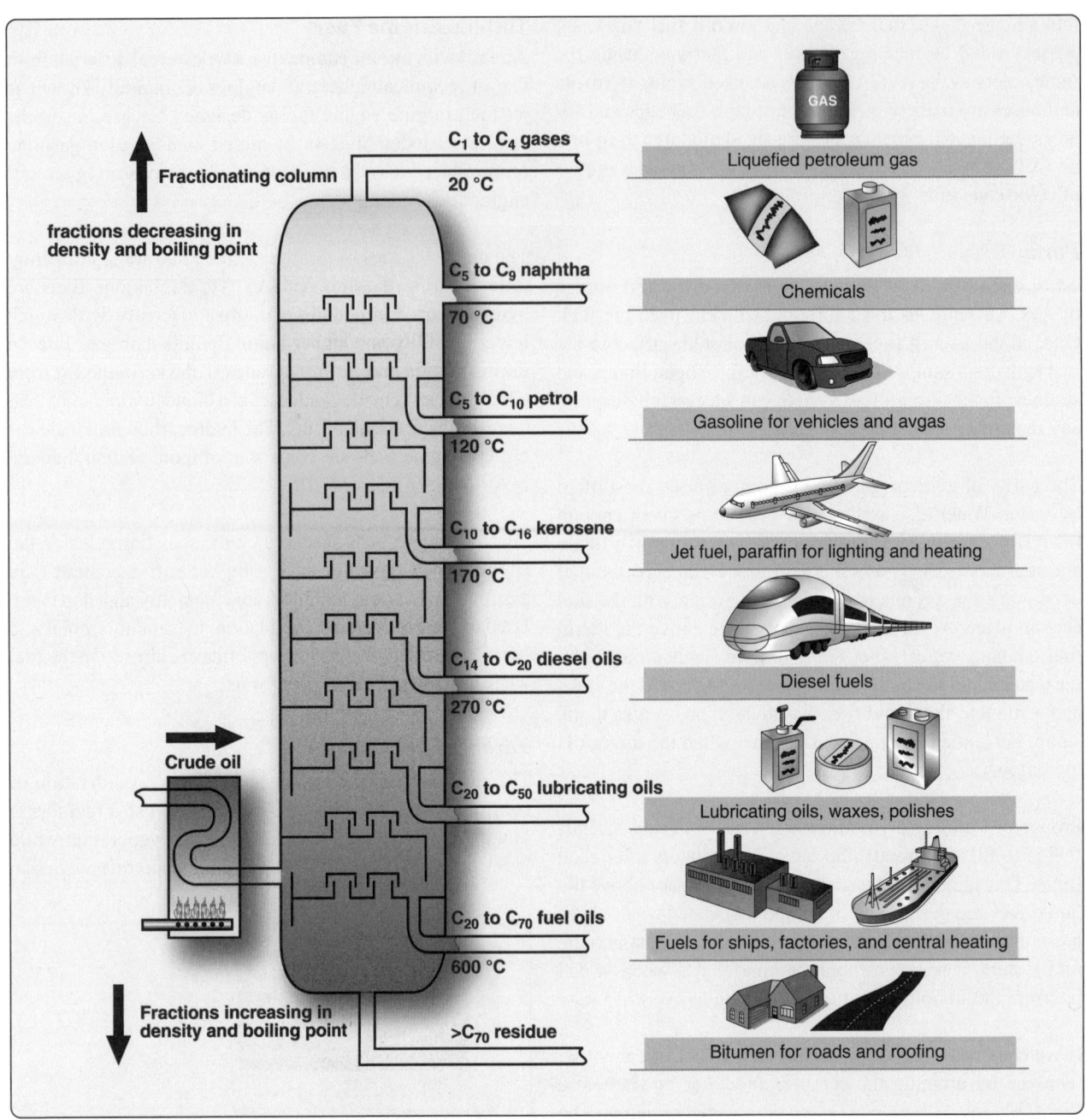

Figure 14-10. *Petroleum products are produced by distillation. Various fractions condense and are collected at different temperatures that correspond to the height of collection in the distillation tower. As can be seen, there are significant differences between turbine engine fuel and ordinary AVGAS.*

cold environments. Turbine engines must start readily and be able to restart while in flight. Fuel with high volatility makes this easier.

AVGAS has a relatively low maximum vapor pressure compared to automotive gasoline—only 7 psi. But the vapor pressure of Jet A is only 0.125 psi at standard atmospheric conditions. Jet B, a blend of Jet A and gasoline, has higher volatility with a vapor pressure between 2 and 3 psi.

Turbine Engine Fuel Types

Three basic turbine engine fuel types are available worldwide, although some countries have their own unique fuels. The first is Jet A. It is the most common turbine engine fuel available in the continental United States. Globally, Jet A-1 is the most popular. Both Jet A and Jet A-1 are fractionally distilled in the kerosene range. They have low volatility and low vapor pressure. Flashpoints range between 110 °F and 150 °F. Jet A freezes at –40 °F and Jet A-1 freezes at –52.6 °F. Most engine operations manuals permit the use of either Jet A or Jet A-1.

The third basic type of turbine engine fuel available is Jet B. It is a wide-cut fuel that is basically a blend of kerosene and gasoline. Its volatility and vapor pressure reflect this and fall between Jet A and AVGAS. Jet B is primarily available in Alaska and Canada due to its low freezing point of approximately –58 °F, and its higher volatility yields better cold weather performance.

Turbine Engine Fuel Issues

Purity issues related to turbine engine fuels are unique. While AVGAS experiences similar issues of solid particle contamination and icing concerns, the presence of water and fuel-consuming microbes is more prominent in jet fuel, which has different molecular structure and retains water in two principal ways. Some water is dissolved into the fuel. Other water also is entrained in the fuel, which is more viscous than AVGAS. The greater presence of water in jet fuel allows microbes to assemble, grow, and live on the fuel.

Since turbine engine fuels always contain water, microbial contamination is always a threat. The large tanks of many turbine engine aircraft have numerous areas where water can settle, and microbes can flourish. Areas between the fuel tank and any water that may come to rest in the bottom of the tanks is where the microbes thrive. These microorganisms form a bio-film that can clog filters, corrode tank coatings, and degrade the fuel. They can be controlled somewhat with the addition of biocides to the fuel. *[Figure 14-11]* Anti-ice additives are also known to inhibit bacterial growth.

Since the microbes are sustained by fuel and water, best practices must be followed to keep the water in fuel to a minimum. Avoid having fuel in a storage tank for a prolonged period of time on or off the aircraft. Drain sumps and monitor the fuel for settled water. Investigate all incidents of water

discovered in the fuel. In addition to water in jet fuel supporting the growth of microorganisms, it also poses a threat of icing. Follow the manufacturer's instructions for fuel handling procedures and fuel system maintenance.

Aircraft Fuel Systems

While each manufacturer designs its own fuel system, the basic fuel system requirements referenced at the beginning of this chapter yield fuel systems of similar design and function in the field. In the following sections are representative examples of various fuel systems in each class of aircraft discussed. Others are similar but not identical. Each aircraft fuel system must store and deliver clean fuel to the engine(s) at a pressure and flow rate able to sustain operations regardless of the operating conditions of the aircraft.

Small Single-Engine Aircraft Fuel Systems

Small single-engine aircraft fuel systems vary depending on factors, such as tank location and method of metering fuel to the engine. A high-wing aircraft fuel system can be designed differently from one on a low-wing aircraft. An aircraft engine with a carburetor has a different fuel system than one with fuel injection.

Gravity Feed Systems

High-wing aircraft with a fuel tank in each wing are common. With the tanks above the engine, gravity is used to deliver the fuel. A simple gravity feed fuel system is shown in *Figure 14-12*. The space above the liquid fuel is vented to maintain atmospheric pressure on the fuel as the tank empties. The two tanks are also vented to each other to ensure equal pressure when both tanks feed the

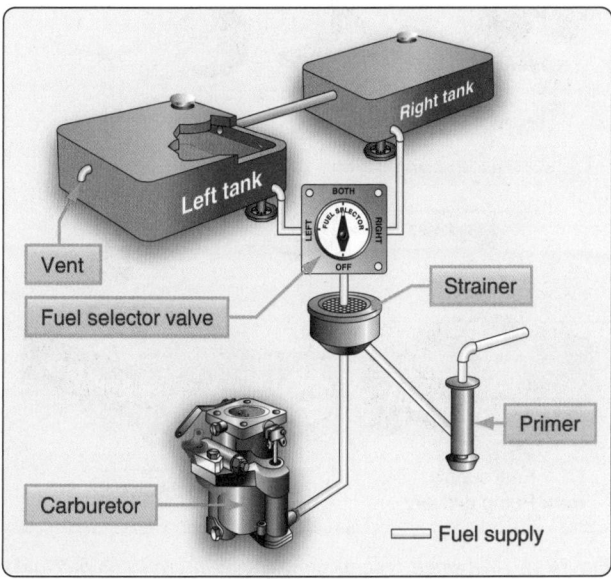

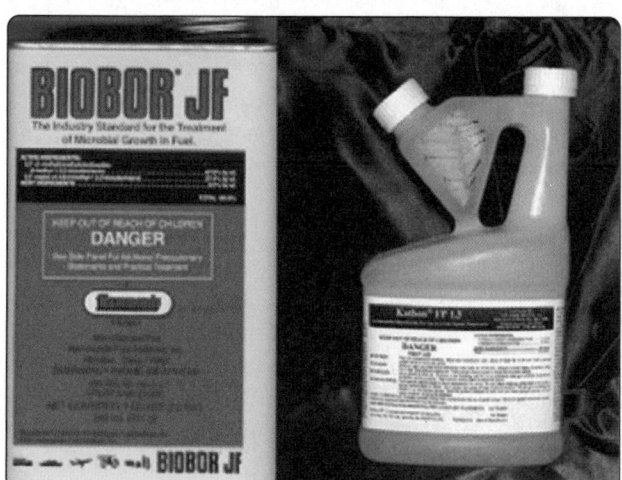

Figure 14-11. *Biocides, such as these, are often added to jet fuel to kill microbes that live on hydrocarbons.*

Figure 14-12. *The gravity-feed fuel system in a single-engine high-wing aircraft is the simplest aircraft fuel system.*

engine. A single screened outlet on each tank feeds lines that connect to either a fuel shutoff valve or multiposition selector valve. The shutoff valve has two positions: fuel ON and fuel OFF. If installed, the selector valve provides four options: fuel shutoff to the engine; fuel feed from the right-wing tank only; fuel feed from the left fuel tank only; fuel feed to the engine from both tanks simultaneously.

Downstream of the shutoff valve or selector valve, the fuel passes through a main system strainer. This often has a drain function to remove sediment and water. From there, it flows to the carburetor or to the primer pump for engine starting. Having no fuel pump, the gravity feed system is the simplest aircraft fuel system.

Pump Feed Systems

Low- and mid-wing single reciprocating engine aircraft cannot utilize gravity-feed fuel systems because the fuel tanks are not located above the engine. Instead, one or more pumps are used to move the fuel from the tanks to the engine. A common fuel system of this type is shown in *Figure 14-13*. Each tank has a line from the screened outlet to a selector valve. However, fuel cannot be drawn from both tanks simultaneously; if the fuel is depleted in one tank, the pump would draw air from that tank instead of fuel from the full tank. Since fuel is not drawn from both tanks at the same time, there is no need to connect the tank vent spaces together.

From the selector valve (LEFT, RIGHT, or OFF), fuel

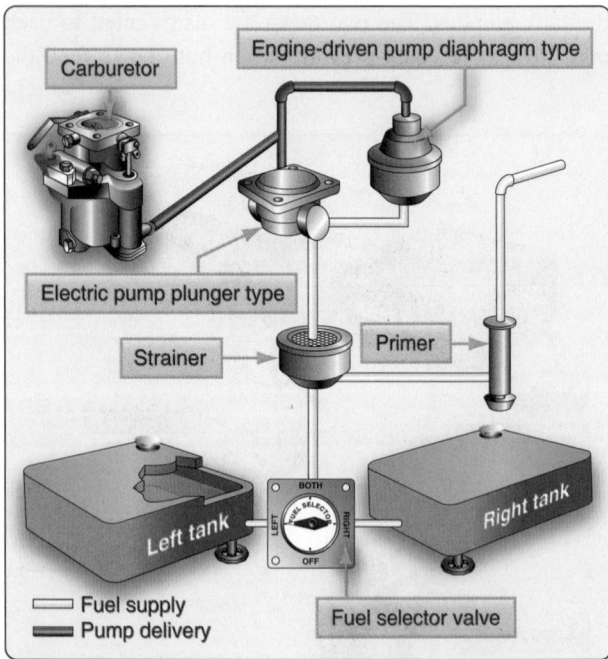

Figure 14-13. *A single reciprocating engine aircraft with fuel tanks located in wings below the engine uses pumps to draw fuel from the tanks and deliver it to the engine.*

flows through the main strainer where it can supply the engine primer. Then, it flows downstream to the fuel pumps. Typically, one electric and one engine-driven fuel pump are arranged in parallel. They draw the fuel from the tank(s) and deliver it to the carburetor. The two pumps provide redundancy. The engine-driven fuel pump acts as the primary pump. The electric pump can supply fuel should the other fail.

The electric pump also supplies fuel pressure while starting and is used to prevent vapor lock during flight at high altitude.

High-Wing Aircraft with Fuel Injection System

Some high-wing, high-performance, single-engine general aviation aircraft are equipped with a fuel system that features fuel injection rather than a carburetor. It combines gravity flow with the use of a fuel pump(s). The Teledyne-Continental system is an example. *[Figure 14-14]*

Note: Fuel injection systems spray pressurized fuel into the engine intake or directly into the cylinders. Fuel without any air mixed in is required to provide a measured, continuous spray and smooth engine operation.

Fuel pressurized by an engine-driven pump is metered as a function of engine rpm on the Teledyne-Continental system. It is first delivered from the fuel tanks by gravity to two smaller accumulator or reservoir tanks. These tanks, one for each wing tank, consolidate the liquid fuel and have a relatively small airspace. They deliver fuel through a three-way selector valve (LEFT, RIGHT, or OFF). The selector valve also acts simultaneously as a diverter of air that has been separated out of the fuel in the engine-driven fuel pump and returned to the valve. It routes the air to the vent space above the fuel in the selected reservoir tank.

An electric auxiliary fuel pump draws fuel through the selector valve. It forces the fuel through the strainer, making it available for the primer pump and the engine-driven fuel pump. This pump is typically used for starting and as a backup should the engine-driven pump fail. It is controlled by a switch in the flight deck and does not need to be operating to allow the engine-driven fuel pump access to the fuel.

The engine-driven fuel pump intakes the pressurized fuel from the electrically driven pump or from the reservoir tanks if the electric pump is not operating. It supplies a higher-than-needed volume of fuel under pressure to the fuel control. Excess fuel is returned to the pump, which pumps it through the selector valve into the appropriate reservoir tank. Fuel vapor is also returned to tanks by the pump. The fuel control unit meters the fuel according to engine rpm and mixture control inputs from the flight deck.

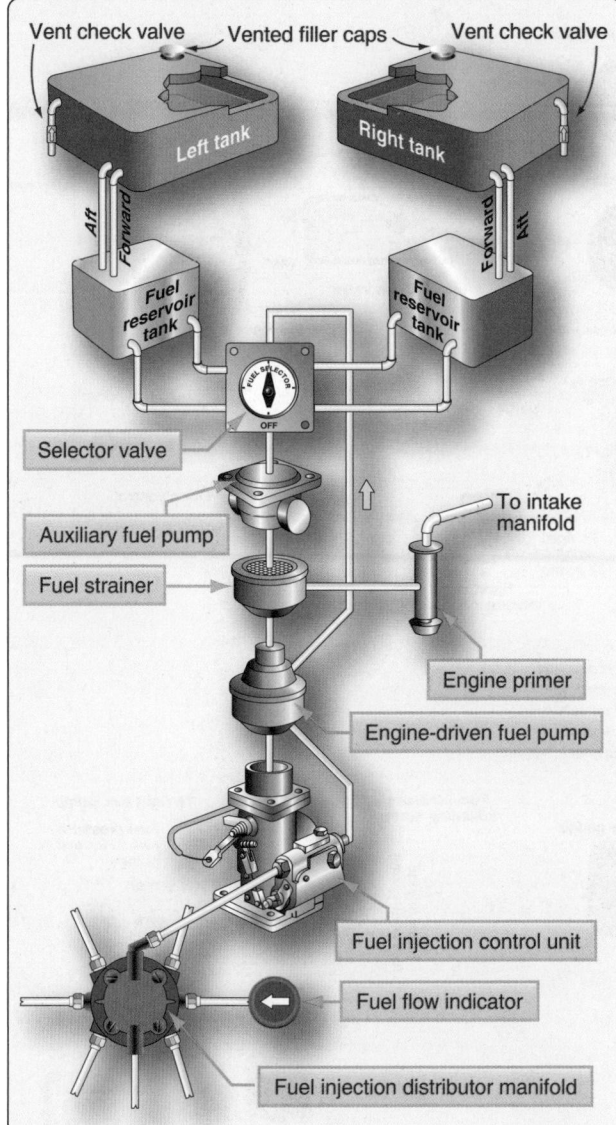

Figure 14-14. *A Teledyne-Continental fuel system featuring fuel injection used on high-wing, high-performance single-engine aircraft.*

The fuel control delivers the fuel to the distribution manifold, which divides it and provides equal, consistent fuel flow for individual fuel injector in each cylinder. *[Figure 14-15]* A fuel flow indicator tapped off of the distribution manifold provides feedback in flight deck. It senses fuel pressure but is displayed on a dial calibrated in gallons per hour.

Small Multiengine (Reciprocating) Aircraft Fuel Systems

Low-Wing Twin

The fuel system on a small, multiengine aircraft is more complicated than a single-engine aircraft but contains many of the same elements. An example system used on a low-wing aircraft is illustrated in *Figure 14-16*. It features the main

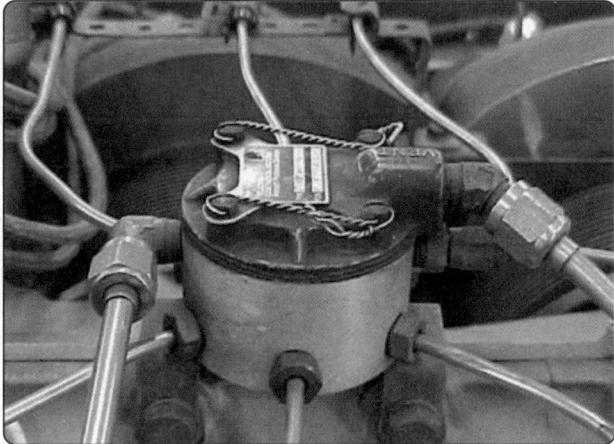

Figure 14-15. *A fuel distribution manifold for a fuel-injected engine.*

fuel tanks in the wing tips and auxiliary tanks in the wing structure. A boost pump is located at the outlet of each main tank. This pressurizes the entire fuel system from the tank to the injectors eliminating the possibility of vapor lock. An engine can operate with just its boost pump running in the event the engine-driven injection pump fails. Typically, the boost pumps are used to prime and start the engine.

Two selector valves are required on twin-engine aircraft, one for each engine. The right selector valve receives fuel from a main tank on either side of the aircraft and directs it to the right engine. The left selector valve also receives fuel from either main tank and directs it to the left engine. This allows fuel to crossfeed from one side of the aircraft to the opposite engine if desired. The selector valves can also direct fuel from the auxiliary tank to the engine on the same side. Crossfeed of fuel from auxiliary tanks is not possible. From the outlet of the selector valve, fuel flows to the strainer. On some aircraft, the strainer is built into the selector valve unit. From the strainer, fuel flows to the engine-driven fuel pump.

The engine-driven fuel pump is an assembly that also contains a vapor separator and a pressure regulating valve with an adjustment screw. The vapor separator helps eliminate air from the fuel. It returns a small amount of fuel and any vapor present back to the main fuel tank. The pump supplies pressurized fuel to the fuel control. The fuel control, one for each engine, responds to throttle and mixture control settings from the flight deck and supplies the proper amount of fuel to the fuel manifold. The manifold divides the fuel and sends it to an injector in each cylinder. A fuel pressure gauge is placed between the fuel control unit outlet and the manifold to monitor the injector-applied pressure that indicates engine power.

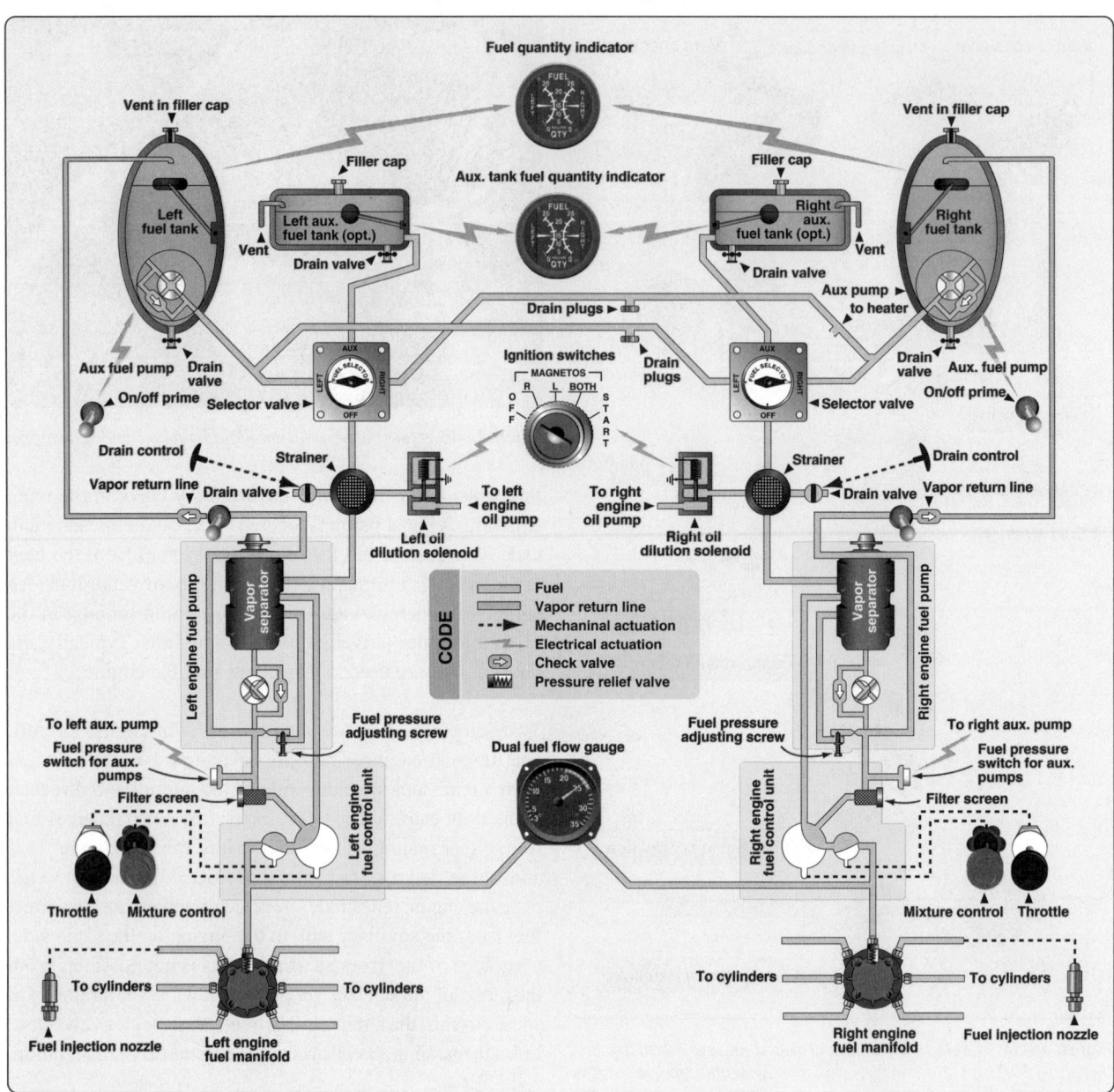

Figure 14-16. *A low-wing, twin-engine, light aircraft fuel system.*

High-Wing Twin

A simplified system on a high-wing, twin-engine aircraft that combines gravity feed with an electric fuel pump is illustrated in *Figure 14-17*. Directly downstream of the selector valves are the fuel strainers and then an electric fuel pump for each engine. This pump draws fuel from the selected tank and sends it under pressure to the inlet side of the fuel injection metering unit. The metering unit for each engine provides the proper flow of fuel to the distribution manifold which feeds the injectors.

Large Reciprocating-Engine Aircraft Fuel Systems

Large, multiengine transport aircraft powered by reciprocating radial engines are no longer produced. However, many are still in operation. They are mostly carbureted and share many features with the light aircraft systems previously discussed.

Figure 14-18 shows the fuel system of a DC-3. A selector valve for each engine allows an engine-driven pump to pull fuel from the main tank or an auxiliary tank. The fuel passes through a strainer before reaching the pump where it is delivered to the engine. The outlet of the pump can feed either engine through the use of a crossfeed line with valves controlled in the flight deck. A hand-operated wobble pump located upstream of the strainer is used to prime the system for starting. Fuel vapor lines run from the pressure carburetor to the vent space in the main and auxiliary tanks. Fuel pressure gauges are tapped off of the carburetor for power indication.

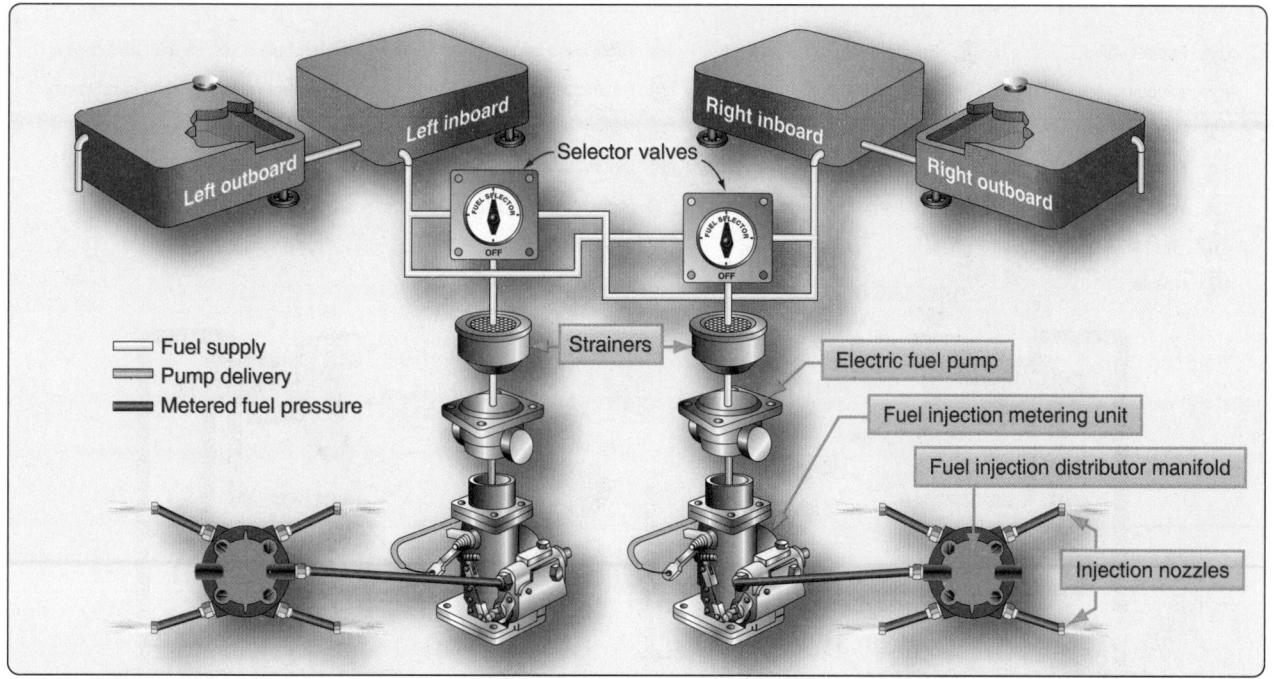

Figure 14-17. *A simple high-wing fuel injection fuel system for a light twin reciprocating-engine aircraft.*

The hand-operated wobble pumps were replaced by electric pumps on later model aircraft. A fuel pressure warning light tapped in downstream of the engine-driven fuel pump alerts the crew should fuel pressure decline.

Not all large, old aircraft have this fuel system. This is merely an example. Other aircraft share similar features and possess unique features of their own. The same is true for small reciprocating-engine aircraft. There are many systems that share features with those described above, but they also differ in some ways. Always consult the manufacturer's data when working on aircraft fuel systems and follow all instructions for service and repair. The fuel system of an aircraft provides the life blood for engine operation and must be maintained with the highest discretion.

Jet Transport Aircraft Fuel Systems

Fuel systems on large transport category jet aircraft are complex with some features and components not found in reciprocating-engine aircraft fuel systems. They typically contain more redundancy and facilitate numerous options from which the crew can choose while managing the aircraft's fuel load. Features like an onboard APU, single point pressure refueling, and fuel jettison systems, which are not needed on smaller aircraft, add to the complexity of an airliner fuel system.

Jet transport fuel systems can be regarded as a handful of fuel subsystems as follows:

1. Storage.
2. Vent.
3. Distribution.
4. Feed.
5. Indicating.

Most transport category aircraft fuel systems are very much alike. Integral fuel tanks are the norm with much of each wing's structure sealed to enable its use as a fuel tank. Center wing section or fuselage tanks are also common. These may be sealed structure or bladder type. Jet transport aircraft carry tens of thousands of pounds of fuel on board. *Figure 14-19* shows a diagram of a Boeing 777 fuel tank configuration with tank capacities.

There are optional fuel storage configurations available on the same model airliner. For example, airlines expecting to use an aircraft on transoceanic flights may order the aircraft with long-range auxiliary tanks. These additional tanks, usually located in the fuselage section of the aircraft, can alter fuel management logistics in addition to complicating the fuel system.

In addition to main and auxiliary fuel tanks, surge tanks may also be found on jet transports. These normally empty tanks located in the wing structure outboard of the main wing tanks are used for fuel overflow. A check valve allows the one-way drainage of fuel back into the main tanks. Surge tanks are also used for fuel system venting.

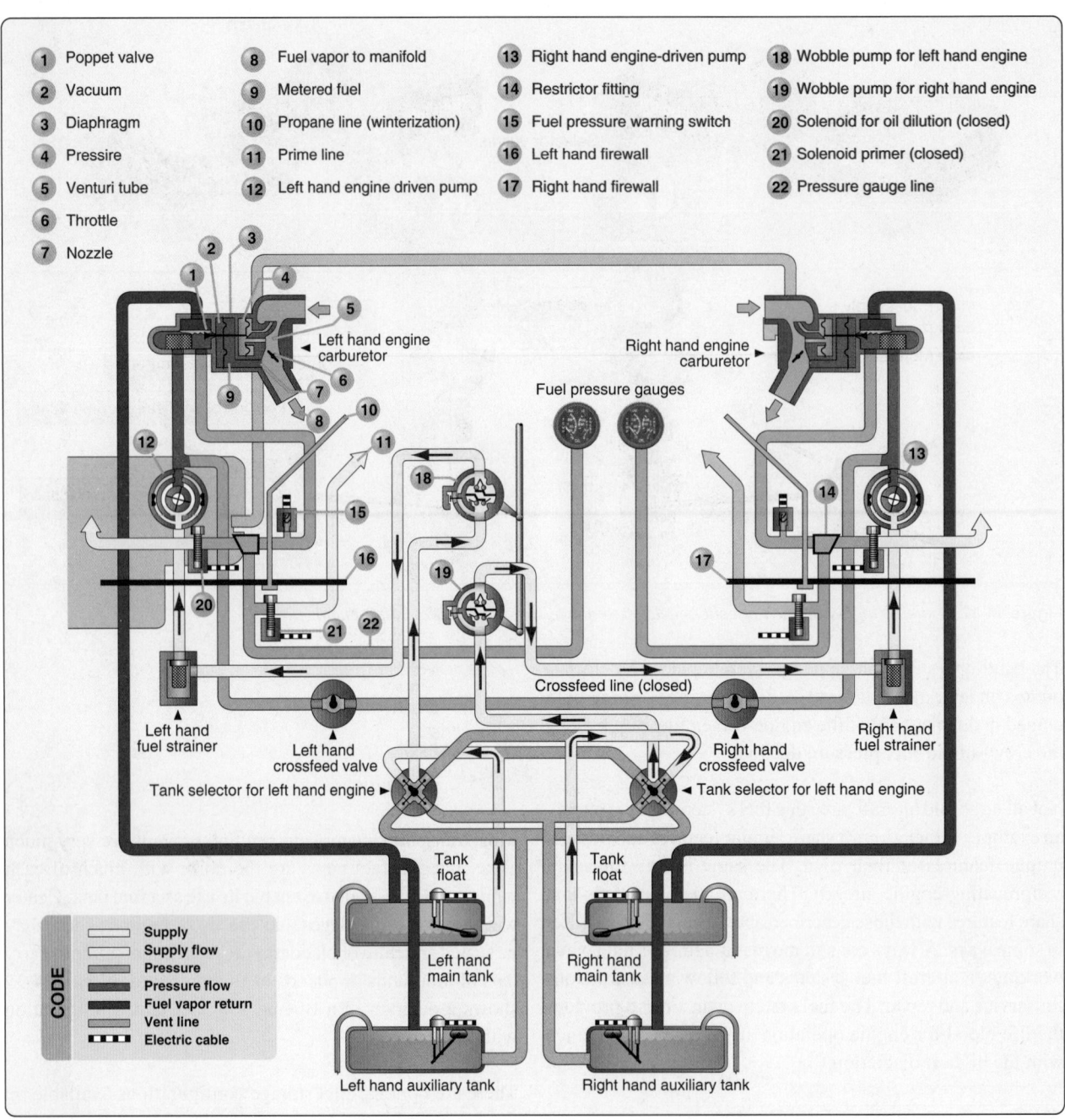

Figure 14-18. *DC-3 fuel system.*

Legend:
1. Poppet valve
2. Vacuum
3. Diaphragm
4. Pressire
5. Venturi tube
6. Throttle
7. Nozzle
8. Fuel vapor to manifold
9. Metered fuel
10. Propane line (winterization)
11. Prime line
12. Left hand engine driven pump
13. Right hand engine-driven pump
14. Restrictor fitting
15. Fuel pressure warning switch
16. Left hand firewall
17. Right hand firewall
18. Wobble pump for left hand engine
19. Wobble pump for right hand engine
20. Solenoid for oil dilution (closed)
21. Solenoid primer (closed)
22. Pressure gauge line

CODE:
- Supply
- Supply flow
- Pressure
- Pressure flow
- Fuel vapor return
- Vent line
- Electric cable

Transport category fuel systems require venting similar to reciprocating engine aircraft fuel systems. A series of vent tubing and channels exists that connects all tanks to vent space in the surge tanks (if present) or vent overboard. Venting must be configured to ensure the fuel is vented regardless of the attitude of the aircraft or the quantity of fuel on board. This sometimes requires the installation of various check valves, float valves, and multiple vent locations in the same tank. *Figure 14-20* shows the fuel vent system of a Boeing 737.

A transport category aircraft fuel distribution subsystem consists of the pressure fueling components, defueling components, transfer system, and fuel jettison or dump system. Single-point pressure fueling at a fueling station accessible by ramp refueling trucks allows all aircraft fuel tanks to be filled with one connection of the fuel hose. Leading and trailing edge wing locations are common for these stations. *Figure 14-21* shows an airliner fueling station with the fueling rig attached.

To fuel with pressure refueling, a hose nozzle is attached at the fueling station and valves to the tanks required to be filled are opened. These valves are called fueling valves

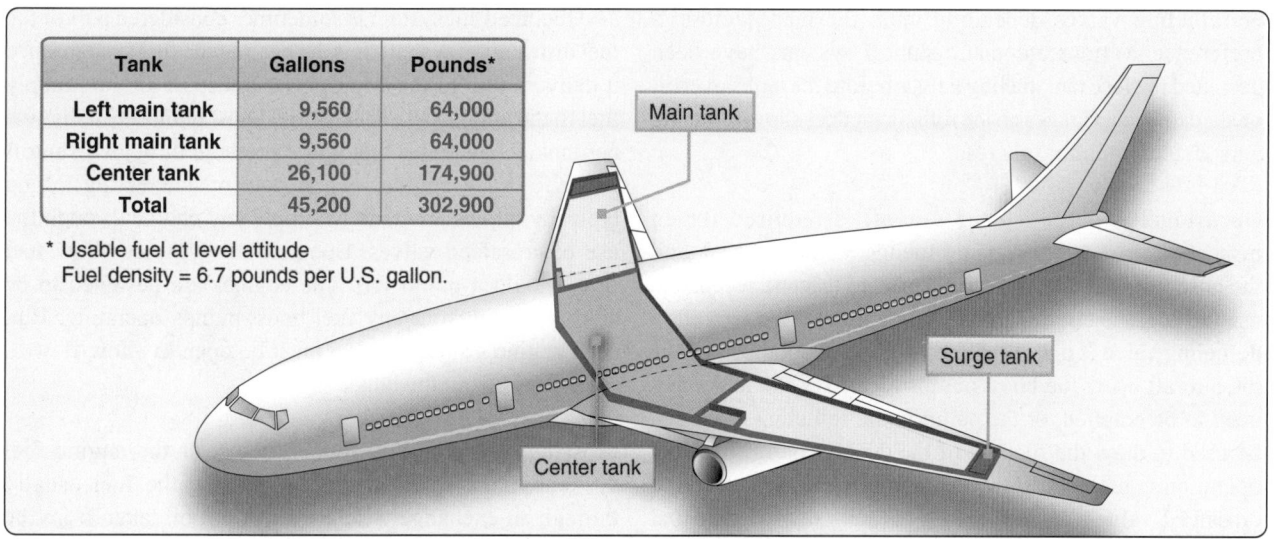

Tank	Gallons	Pounds*
Left main tank	9,560	64,000
Right main tank	9,560	64,000
Center tank	26,100	174,900
Total	45,200	302,900

* Usable fuel at level attitude
Fuel density = 6.7 pounds per U.S. gallon.

Figure 14-19. *Boeing 777 fuel tank locations and capacities.*

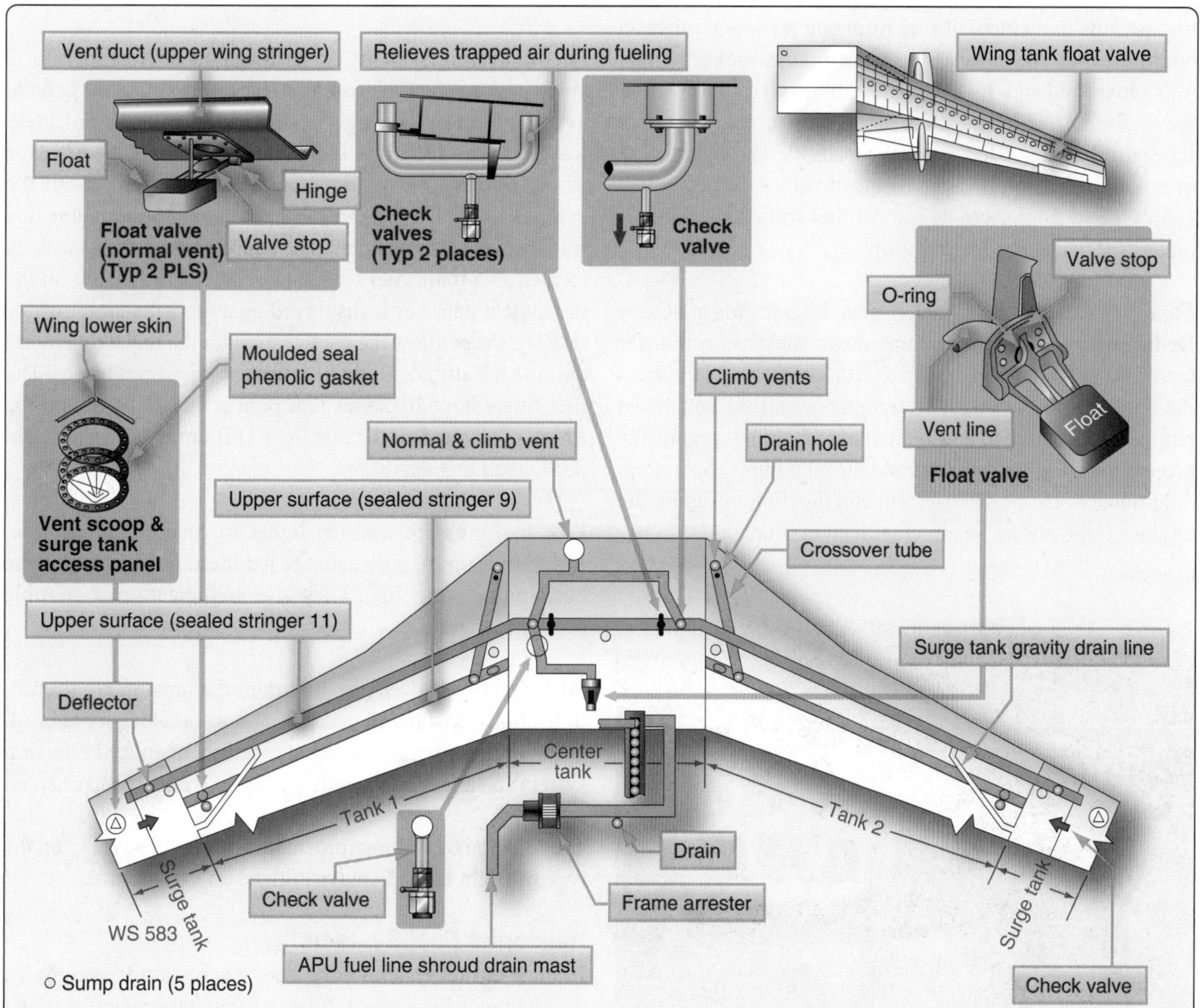

Figure 14-20. *A fuel vent system with associated float and check valves that stop fuel and keep the tanks vented regardless of the aircraft attitude.*

14-15

or refueling valves depending upon the manufacturer's preference. Various automatic shutoff systems have been designed to close tank fueling valves before the tanks overfill or are damaged. Gauges on the refueling panel allow refueling personnel to monitor progress.

Occasionally, defueling the aircraft is required for an inspection or repair. The same fueling station is used, and the hose from the fuel truck is connected to same receptacle used to fuel the aircraft. To allow fuel to exit the aircraft, a defueling valve is opened. Fuel can either be pumped out of the aircraft using the boost pumps located in the tanks that need to be emptied, or the pump in the refueling truck can be used to draw the fuel out of the tanks. Control over the operation is maintained by positioning various shutoff and crossfeed valves, as well as the defuel valve so that fuel travels from the tank to the fueling station and into the truck.

The fuel transfer system is a series of plumbing and valves that permits movement of fuel from one tank to another on board the aircraft. In-tank fuel boost pumps move the fuel into a manifold and, by opening the fuel valve (or refueling valve) for the desired tank, the fuel is transferred. Not all jet transports have such fuel transfer capability. Through the use of a fuel feed manifold and crossfeed valves, some aircraft simply allow engines to be run off fuel from any tank as a means for managing fuel location.

Figure 14-22 shows the fuel system diagram for a DC-10. Dedicated transfer boost pumps move fuel into a transfer manifold. Opening the fuel valve on one of the tanks transfers the fuel into that tank. The transfer manifold and boost pumps are also used to jettison fuel overboard by opening the proper dump valves with a transfer boost pump(s) operating. Additionally, the transfer system can function to supply the engines if the normal engine fuel feed malfunctions.

Figure 14-21. *A central pressure refueling station on a transport category aircraft allows all fuel tanks to be filled from one position.*

The fuel feed subsystem is sometimes considered part of the fuel distribution system. It is the heart of the fuel system since it delivers fuel to the engines. Jet transport aircraft supply fuel to the engines via in-tank fuel boost pumps, usually two per tank. They pump fuel under pressure through a shutoff valve for each engine. A manifold or connecting tubing typically allows any tank to supply any engine through the use of crossfeed valves. Boost pump bypasses allow fuel flow should a pump fail. The engines are designed to be able to run without any fuel boost pumps operating. But, each engine's shutoff valve must be open to allow flow to the engines from the tanks.

Most jet transport fuel feed systems, or the engine fuel systems, have some means for heating the fuel usually through an exchange with hot air or hot oil taken from the engine. *Figure 14-23* shows the fuel cooled oil cooler (FCOC) on a Rolls Royce RB211 engine, which not only heats the fuel but also cools the engine oil.

Fuel indicating systems on jet transport aircraft monitor a variety of parameters, some not normally found on general aviation aircraft. Business jet aircraft share many of these features. True fuel flow indicators for each engine are used as the primary means for monitoring fuel delivery to the engines. A fuel temperature gauge is common as are fuel filter bypass warning lights. The temperature sensor is usually located in a main fuel tank. The indicator is located on the instrument panel or is displayed on a multifunction display (MFD). These allow the crew to monitor the fuel temperature during high altitude flight in extremely frigid conditions. The fuel filters have bypasses that permit fuel flow around the filters if clogged. Indicator light(s) illuminate in the flight deck when this occurs.

Low fuel pressure warning lights are also common on jet transport aircraft. The sensors for these are located in the boost pump outlet line. They give an indication of possible boost pump failure.

Fuel quantity gauges are important features on all aircraft. Indications exist for all tanks on a transport category aircraft. Often, these use a capacitance type fuel quantity indication system and a fuel totalizer as is discussed later in this chapter.

The location of fuel instrumentation varies depending on the type of flight deck displays utilized on the aircraft.

Helicopter Fuel Systems
Helicopter fuel systems vary. They can be simple or complex depending on the aircraft. Always consult the manufacturer's manuals for fuel system description, operation, and maintenance instructions.

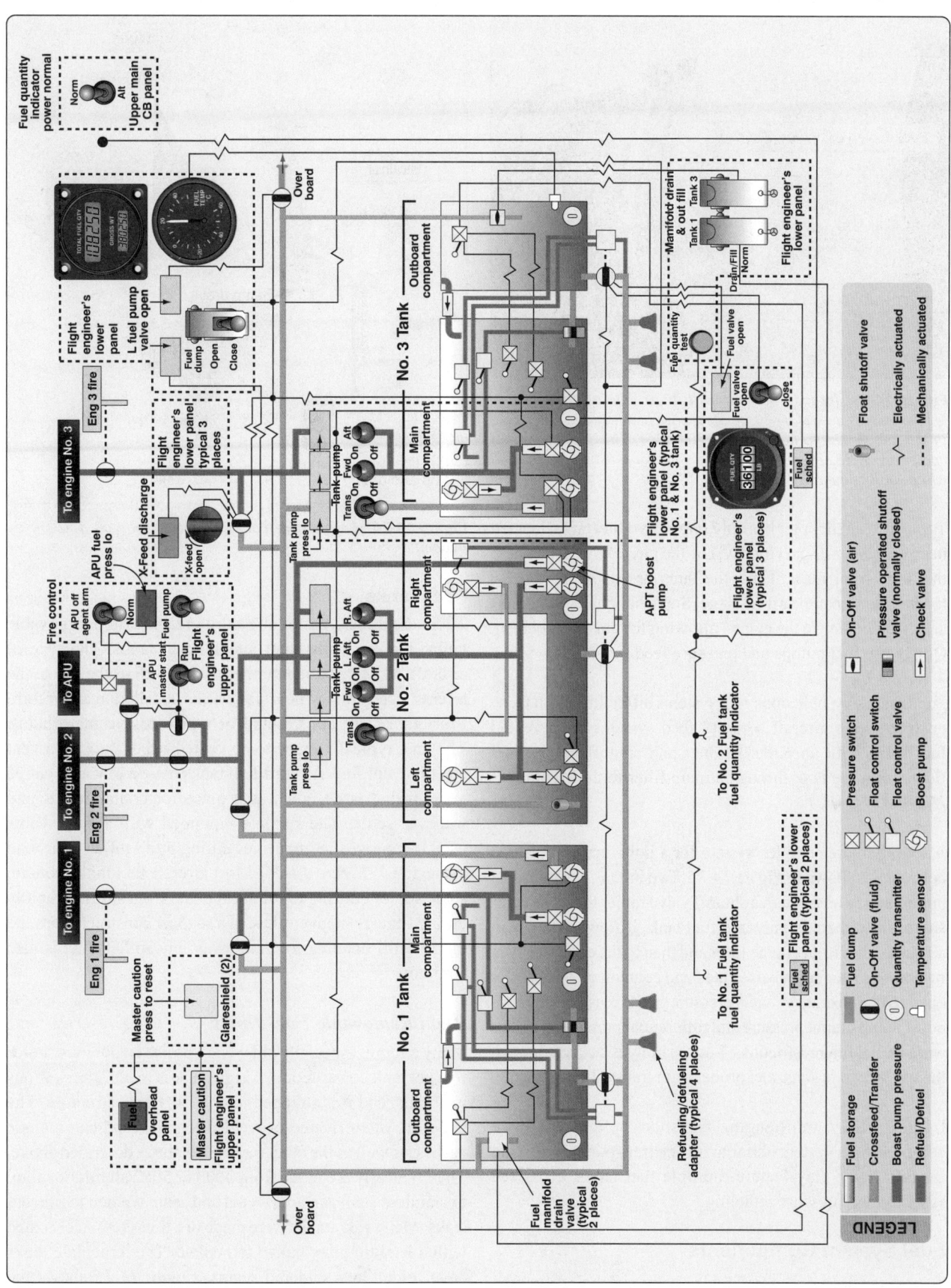

Figure 14-22. *The fuel distribution systems, components, and flight deck controls of a DC-10 airliner. Note: Fuel transfer system components and lines are used to complete the fuel dump system, the refuel/defuel system, back-up fuel delivery system, and the fuel storage system.*

Figure 14-23. *Jet transport aircraft fly at high altitudes where temperatures can reach –50 °F. Most have fuel heaters somewhere in the fuel system to help prevent fuel icing. This fuel-cooled oil cooler on an RB211 turbofan engine simultaneously heats the fuel while cooling the oil.*

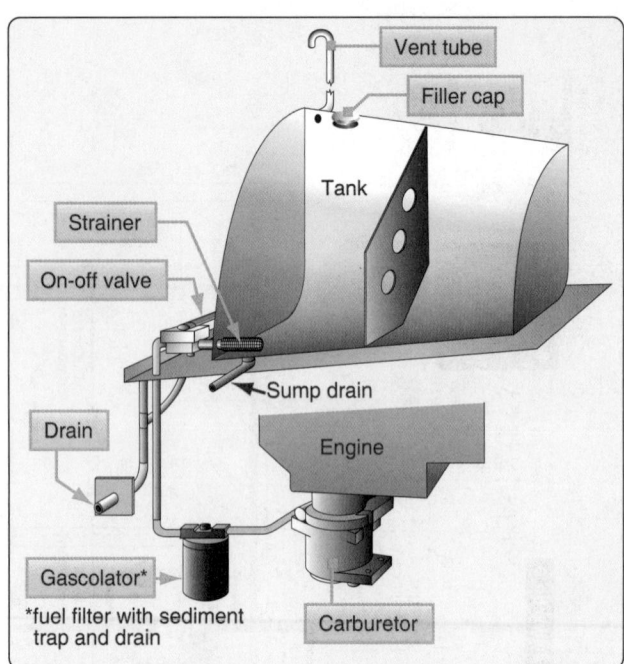

Figure 14-24. *A simple, gravity-feed fuel system on a Robinson helicopter.*

Typically, a helicopter has only one or two fuel tanks located near the center of gravity (CG) of the aircraft, which is near the main rotor mast. Thus, the tank, or tanks, are usually located in or near the aft fuselage. Some helicopter fuel tanks are mounted above the engine allowing for gravity fuel feed. Others use fuel pumps and pressure feed systems.

Fundamentally, helicopter fuel systems differ little from those on fixed-wing aircraft. Gravity-feed systems have vented fuel tanks with an outlet strainer and shutoff valve. Fuel flows from the tank through a main filter to the carburetor. *[Figure 14-24]*

A slightly more complex system for a light turbine-powered helicopter is shown in *Figure 14-25*. Two in-tank electric boost pumps send fuel through a shutoff valve rather than a selector valve, since there is only one fuel tank. It flows through an airframe filter to an engine filter and then to the engine-driven fuel pump. The fuel tank is vented and contains an electrically operated sump drain valve. A pressure gauge is used to monitor boost pump output pressure and differential pressure switches warn of fuel filter restrictions. Fuel quantity is derived through the use of two in-tank fuel probes with transmitters.

Larger, heavy, multiengine transport helicopters have complex fuel systems similar to jet transport fixed-wing aircraft. They may feature multiple fuel tanks, crossfeed systems, and pressure refueling.

Fuel System Components

To better understand aircraft fuel systems and their operation, the following discussion of various components of aircraft fuel systems is included.

Fuel Tanks

There are three types of aircraft fuel tanks: rigid removable tanks, bladder tanks, and integral fuel tanks. The type of aircraft, its design and intended use, as well as the age of the aircraft determine which fuel tank is installed in an aircraft. Most tanks are constructed of noncorrosive material(s). They are typically made to be vented either through a vent cap or a vent line. Aircraft fuel tanks have a low area called a sump that is designed as a place for contaminants and water to settle. The sump is equipped with a drain valve used to remove the impurities during preflight walk-around inspection. *[Figure 14-26]* Most aircraft fuel tanks contain some sort of baffling to subdue the fuel from shifting rapidly during flight maneuvers. Use of a scupper constructed around the fuel fill opening to drain away any spilled fuel is also common.

Rigid Removable Fuel Tanks

Many aircraft, especially older ones, utilize an obvious choice for fuel tank construction. A rigid tank is made from various materials, and it is strapped into the airframe structure. The tanks are often riveted or welded together and can include baffles, as well as the other fuel tank features described above. They typically are made from 3003 or 5052 aluminum alloy or stainless steel and are riveted and seam welded to prevent leaks. Many early tanks were made of a thin sheet steel coated with a lead/tin alloy called terneplate. The terneplate tanks have folded and soldered seams. *Figure 14-27* shows the parts of a typical rigid removable fuel tank.

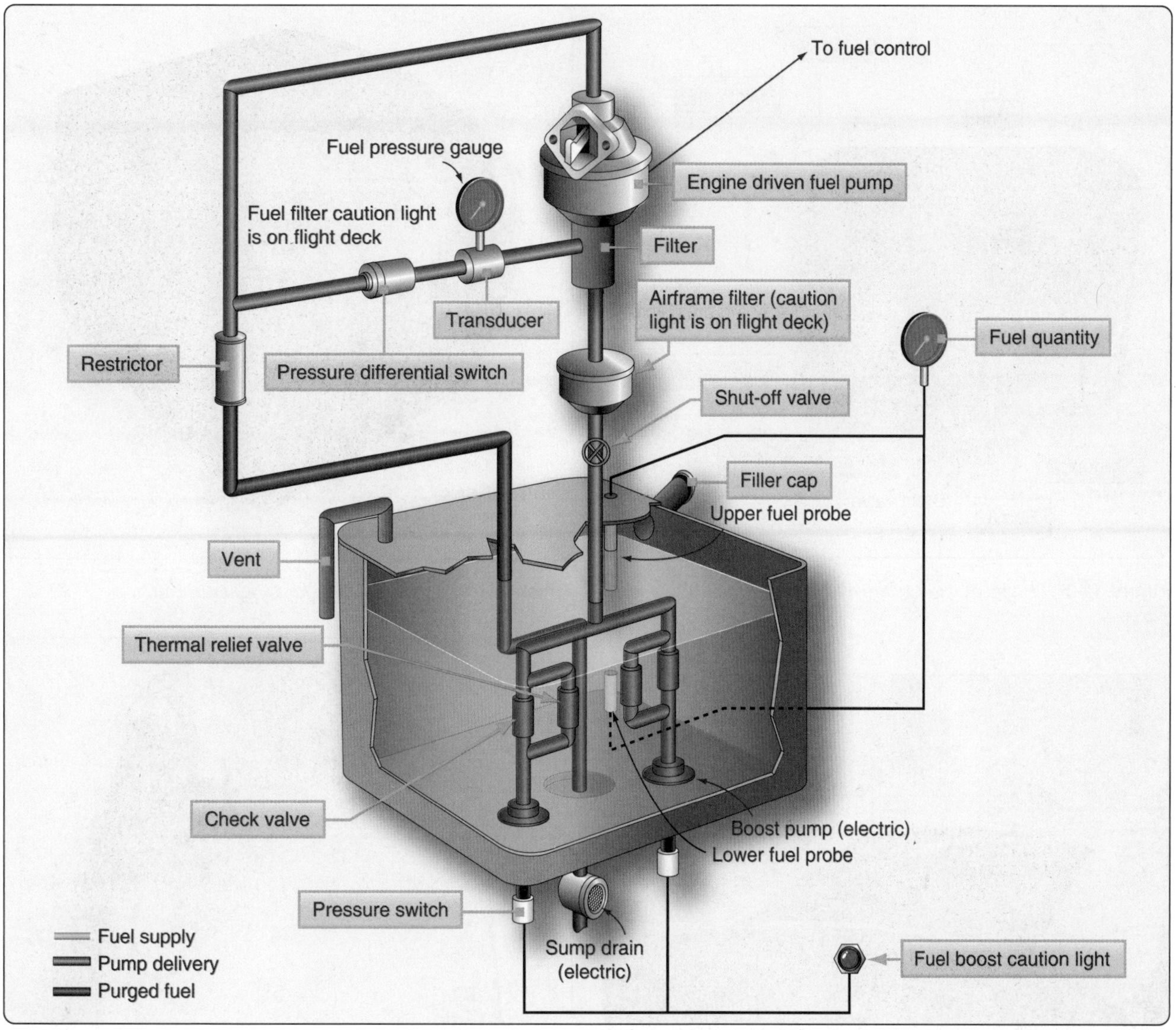

Figure 14-25. *A pressure-feed fuel system on a light turbine-powered helicopter.*

Labels in figure:
- To fuel control
- Fuel pressure gauge
- Engine driven fuel pump
- Fuel filter caution light is on flight deck
- Filter
- Transducer
- Airframe filter (caution light is on flight deck)
- Restrictor
- Pressure differential switch
- Fuel quantity
- Shut-off valve
- Filler cap
- Upper fuel probe
- Vent
- Thermal relief valve
- Check valve
- Boost pump (electric)
- Lower fuel probe
- Pressure switch
- Sump drain (electric)
- Fuel boost caution light
- Fuel supply
- Pump delivery
- Purged fuel

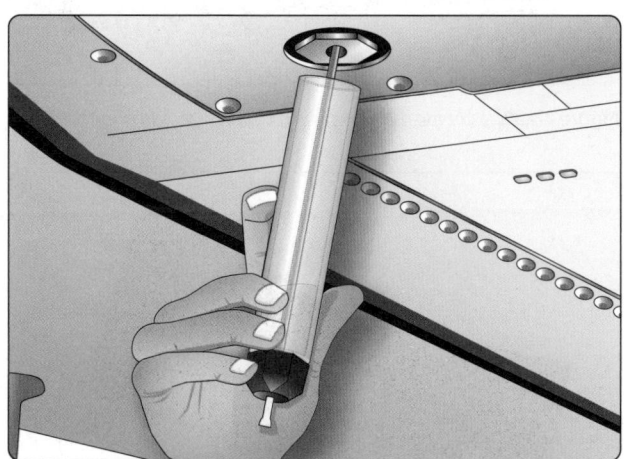

Figure 14-26. *Sumping a fuel tank with a fuel strainer that is designed to collect the sump drain material in the clear cylinder to be examined for the presence of contaminants.*

Regardless of the actual construction of removable metal tanks, they must be supported by the airframe and held in place with some sort of padded strap arrangement to resist shifting in flight. The wings are the most popular location for fuel tanks. *Figure 14-28* shows a fuel tank bay in a wing root with the tank straps. Some tanks are formed to be part of the leading edge of the wing. These are assembled using electric resistance welding and are sealed with a compound that is poured into the tank and allowed to cure. Many fuselage tanks also exist. *[Figure 14-29]* In all cases, the structural integrity of the airframe does not rely on the tank(s) being installed, so the tanks are not considered integral.

As new materials are tested and used in aircraft, fuel tanks are being constructed out of materials other than aluminum, steel, and stainless steel. *Figure 14-30* shows a rigid removable fuel tank from an ultralight category aircraft that is

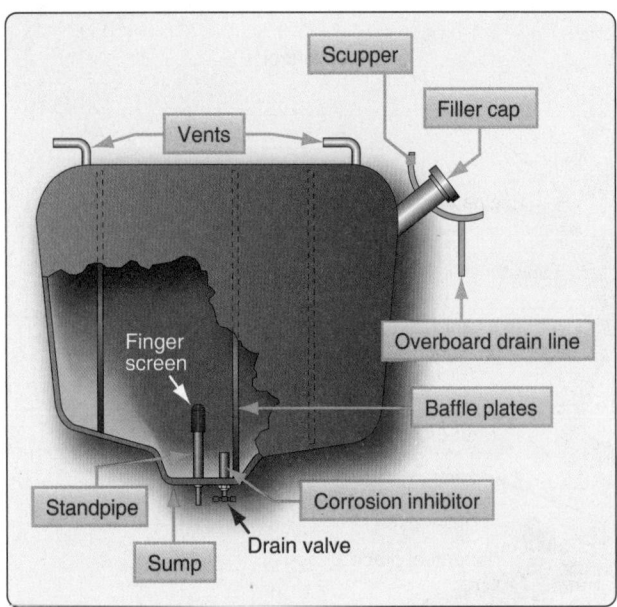

Figure 14-27. *A typical rigid removable aircraft fuel tank and its parts.*

Figure 14-28. *A fuel tank bay in the root of a light aircraft wing on a stand in a paint booth. Padded straps hold the fuel tank securely in the structure.*

Figure 14-29. *A fuselage tank for a light aircraft.*

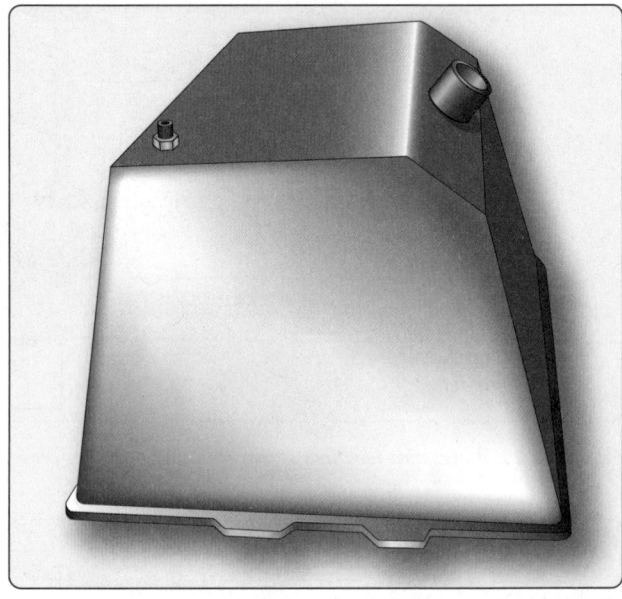

Figure 14-30. *A composite tank from a Challenger ultralight aircraft.*

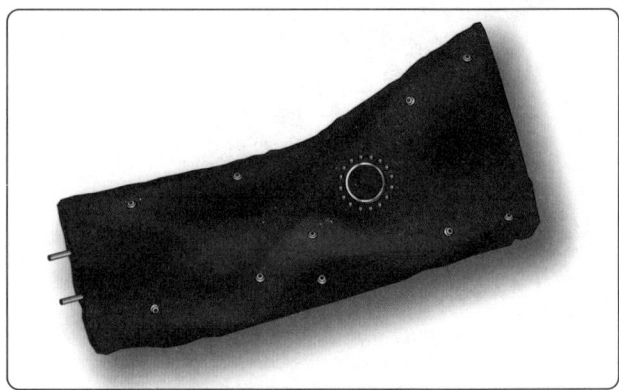

Figure 14-31. *A bladder fuel tank for a light aircraft.*

constructed from Vipel® isophthalic polyester UL 1316/UL 1746 resin and composite. Its seamless, lightweight construction may lead to the use of this type of tank in other aircraft categories in the future.

Being able to remove and repair, or replace, a fuel tank can be a great convenience if a leak or malfunction with the tank exists. Repairs to fuel tanks must be done in accordance with manufacturers' specifications. It is especially critical to follow all safety procedures when welding repairs are performed. Fuel vapors must be removed from the tank to prevent explosion. This typically involves washing out the tank with water and detergent, followed by a steam or water flush, as determined by the manufacturer. Once repaired, fuel tanks need to be pressure checked, usually while installed in the airframe, to prevent distortion while under pressure.

Bladder Fuel Tanks

A fuel tank made out of a reinforced flexible material called a bladder tank can be used instead of a rigid tank. A bladder tank contains most of the features and components of a rigid tank but does not require as large an opening in the aircraft skin to install. The tank, or fuel cell as it is sometimes called, can be rolled up and put into a specially prepared structural bay or cavity through a small opening, such as an inspection opening. Once inside, it can be unfurled to its full size. Bladder tanks must be attached to the structure with clips or other fastening devices. They should lie smooth and unwrinkled in the bay. It is especially important that no wrinkles exist on the bottom surface so that fuel contaminants are not blocked from settling into the tank sump. *[Figure 14-31]*

Bladder fuel tanks are used on aircraft of all sizes. They are strong and have a long life with seams only around installed features, such as the tank vents, sump drain, filler spout, etc. When a bladder tank develops a leak, the technician can patch it following manufacturer's instructions. The cell can also be removed and sent to a fuel tank repair station familiar with and equipped to perform such repairs.

The soft flexible nature of bladder fuel tanks requires that they remain wet. Should it become necessary to store a bladder tank without fuel in it for an extended period of time, it is common to wipe the inside of the tank with a coating of clean engine oil. Follow the manufacturer's instructions for the dry storage procedures for fuel cells.

Integral Fuel Tanks

On many aircraft, especially transport category and high-performance aircraft, part of the structure of the wings or fuselage is sealed with a fuel resistant two-part sealant to form a fuel tank. The sealed skin and structural members provide the highest volume of space available with the lowest weight. This type of tank is called an integral fuel tank since it forms a tank as a unit within the airframe structure.

Integral fuel tanks in the otherwise unused space inside the wings are most common. Aircraft with integral fuel tanks in the wings are said to have wet wings. For fuel management purposes, sometimes a wing is sealed into separate tanks and may include a surge tank or an overflow tank, which is normally empty but sealed to hold fuel when needed.

When an aircraft maneuvers, the long horizontal nature of an integral wing tank requires baffling to keep the fuel from sloshing. The wing ribs and box beam structural members

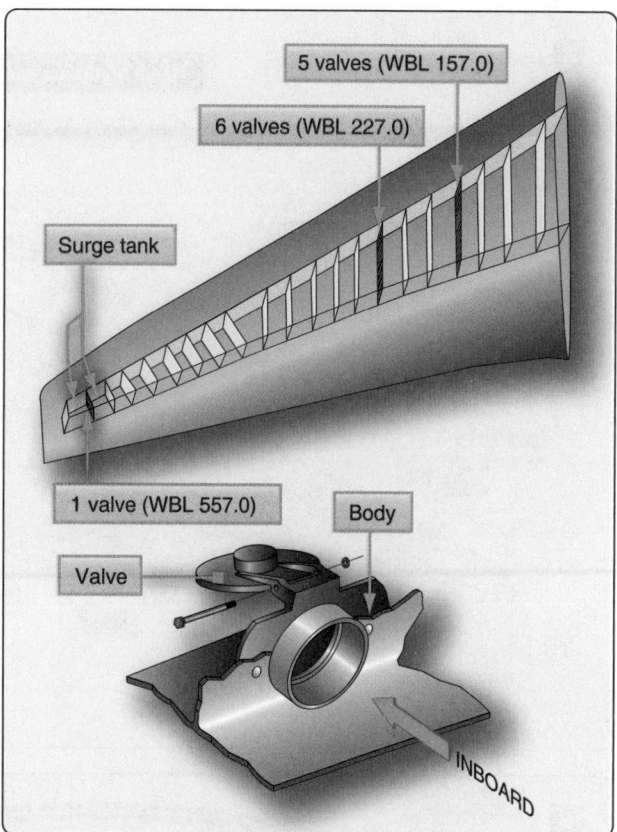

Figure 14-32. *Baffle check valves are installed in the locations shown in the integral tank rib structure of a Boeing 737 airliner. Fuel is prevented from flowing outboard during maneuvers. The tank boost pumps are located inboard of WBL 157.*

serve as baffles and others may be added specifically for that purpose. Baffle check valves are commonly used. These valves allow fuel to move to the low, inboard sections of the tank but prevent it from moving outboard. They ensure that the fuel boost pumps located in the bottom of the tanks at the lowest points above the sumps always have fuel to pump regardless of aircraft attitude. *[Figure 14-32]*

Integral fuel tanks must have access panels for inspection and repairs of the tanks and other fuel system components. On large aircraft, technicians physically enter the tank for maintenance. Transport category aircraft often have more than a dozen oval access panels or tank plates on the bottom surface of the wing for this purpose. *[Figure 14-33A]* These aluminum panels are each sealed into place with an O-ring and an aluminum gasket for electrostatic bonding. An outer clamp ring is tightened to the inner panel with screws, as shown in *Figure 14-33B*.

When entering and performing maintenance on an integral fuel tank, all fuel must be emptied from the tank and strict safety procedures must be followed. Fuel vapors must be purged from the tank and respiratory equipment must be used

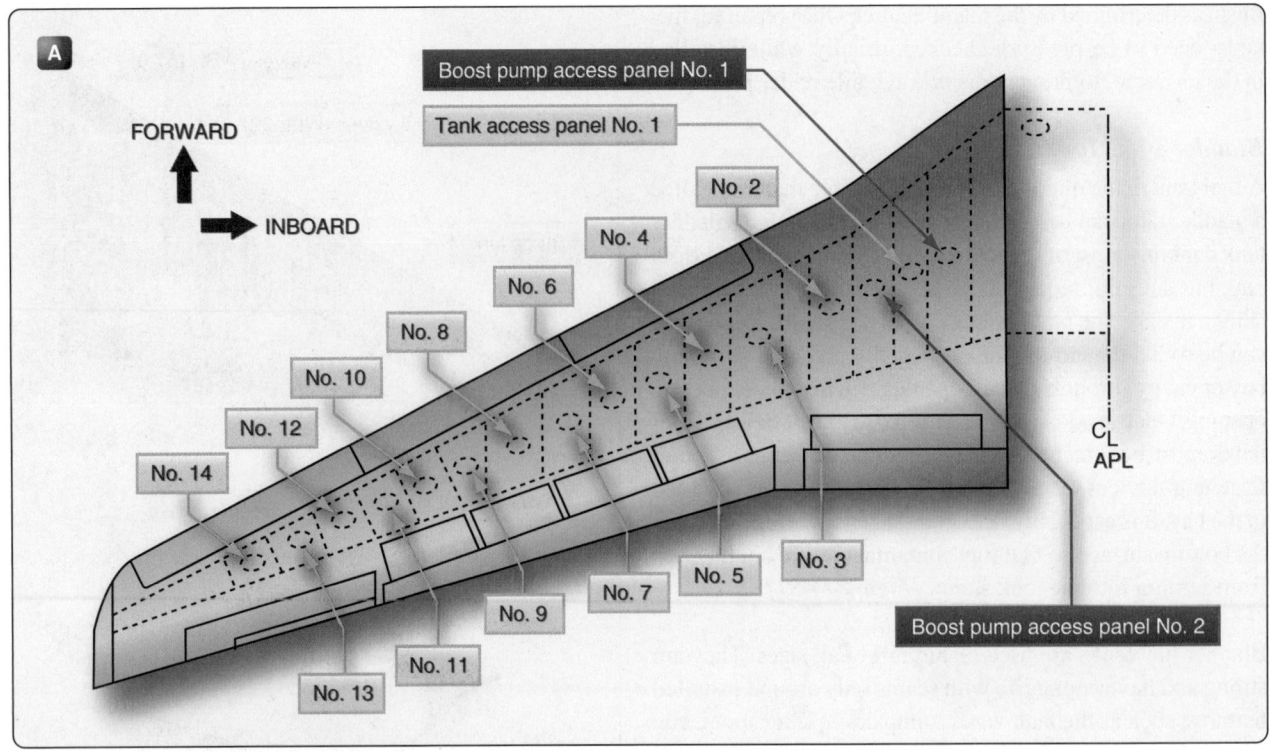

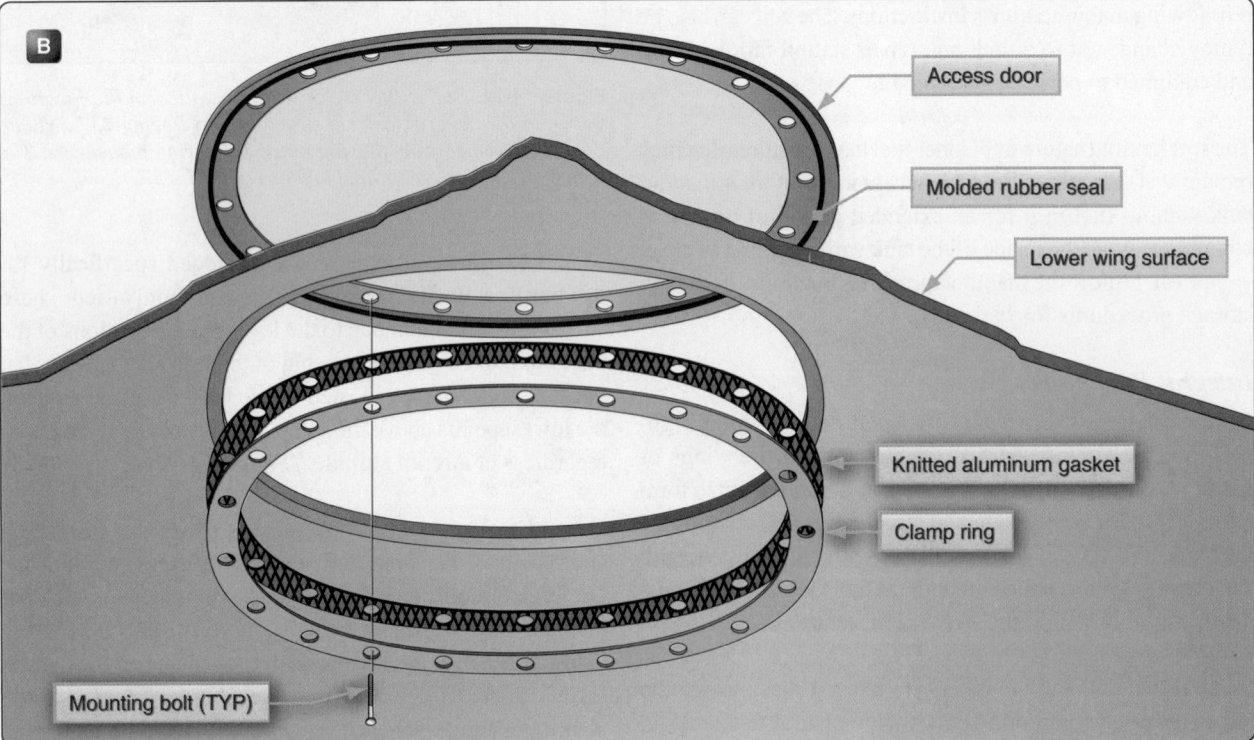

Figure 14-33. *Fuel tank access panel locations on a Boeing 737 (A), and typical fuel tank access panel seals (B).*

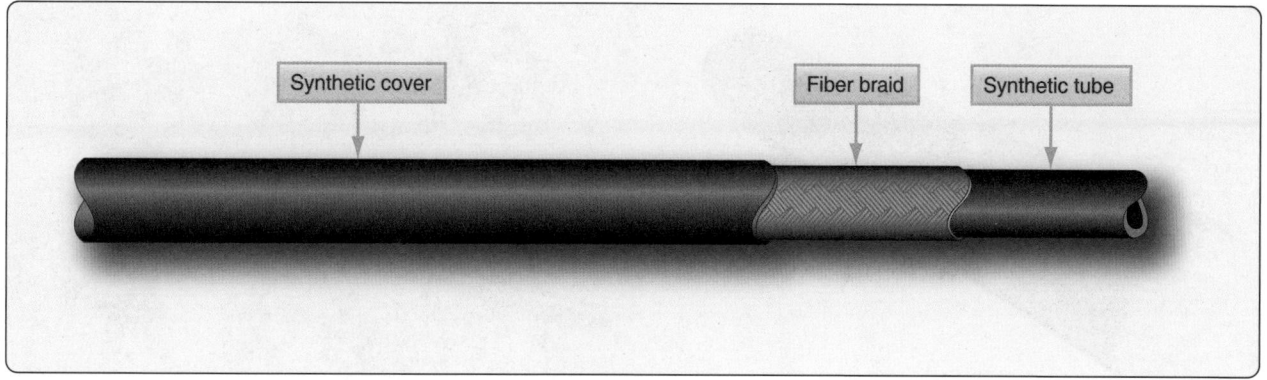

Figure 14-34. *A typical flexible aircraft fuel line with braided reinforcement.*

by the technician. A full-time spotter must be positioned just outside of the tank to assist if needed.

Aircraft using integral fuel tanks normally have sophisticated fuel systems that include in-tank boost pumps. There are usually at least two pumps in each tank that deliver fuel to the engine(s) under positive pressure. On various aircraft, these in-tank boost pumps are also used to transfer fuel to other tanks, jettison fuel, and defuel the aircraft.

Fuel Lines & Fittings

Aircraft fuel lines can be rigid or flexible depending on location and application. Rigid lines are often made of aluminum alloy and are connected with Army/Navy (AN) or military standard (MS) fittings. However, in the engine compartment, wheel wells, and other areas, subject to damage from debris, abrasion, and heat, stainless steel lines are often used.

Flexible fuel hose has a synthetic rubber interior with a reinforcing fiber braid wrap covered by a synthetic exterior. *[Figure 14-34]* The hose is approved for fuel and no other hose should be substituted. Some flexible fuel hoses have a braided stainless steel exterior. *[Figure 14-35]* The diameters of all fuel hoses and lines are determined by the fuel flow requirements of the aircraft fuel system. Flexible hoses are used in areas where vibration exists between components, such as between the engine and the aircraft structure.

Sometimes manufacturers wrap either flexible or rigid fuel

lines to provide even further protection from abrasion and especially from fire. A fire sleeve cover is held over the line with steel clamps at the end fittings. *[Figure 14-36]*

As mentioned, aircraft fuel line fittings are usually either AN or MS fittings. Both flared and flareless fittings are used. Problems with leaks at fittings can occur. Technicians are cautioned to not overtighten a leaky fitting. If the proper torque does not stop a leak, depressurize the line, disconnect the fitting and visually inspect it for a cause. The fitting or line should be replaced if needed. Replace all aircraft fuel lines and fittings with approved replacement parts from the manufacturer. If a line is manufactured in the shop, approved components must be used.

Several installation procedures for fuel hoses and rigid fuel lines exist. Hoses should be installed without twisting. The writing printed on the outside of the hose is used as a lay line to monitor fuel hose twist. Separation should be maintained between all fuel hoses and electrical wiring. Never clamp wires to a fuel line. When separation is not possible, always route the fuel line below any wiring. If a fuel leak develops, it does not drip onto the wires.

Metal fuel lines and all aircraft fuel system components need to be electrically bonded and grounded to the aircraft structure. This is important because fuel flowing through the fuel system generates static electricity that must have a place to flow to ground rather than build up. Special bonded

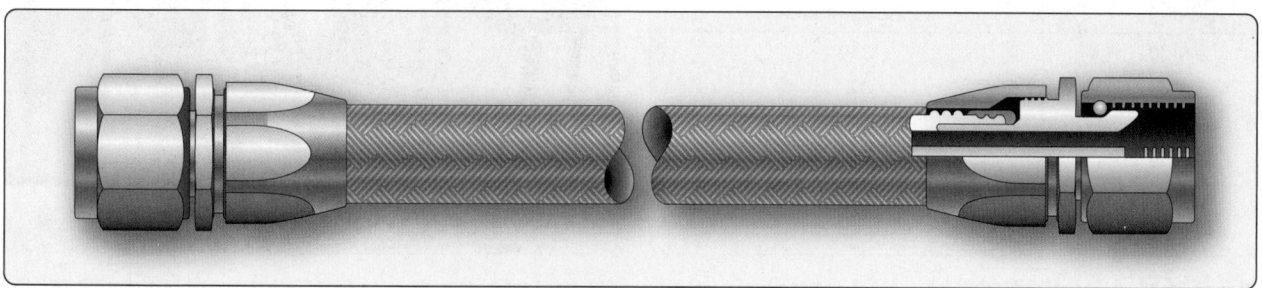

Figure 14-35. *A braided stainless steel exterior fuel line with fittings.*

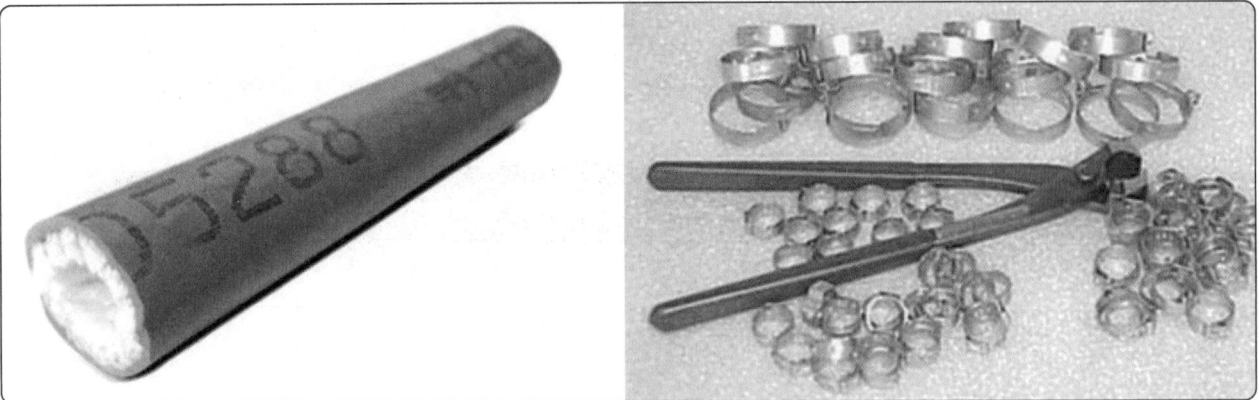

Figure 14-36. *Exterior fuel hose wrap that protects from fire, as well as abrasion, shown with the clamps and pliers used to install it.*

cushion clamps are used to secure rigid fuel lines in place. They are supported at intervals shown in *Figure 14-37*.

All fuel lines should be supported so that there is no strain on the fittings. Clamp lines so that fittings are aligned. Never draw two fittings together by threading. They should thread easily, and a wrench should be used only for tightening. Additionally, a straight length of rigid fuel line should not be made between two components or fittings rigidly mounted to the airframe. A small bend is needed to absorb any strain from vibration or expansion and contraction due to temperature changes.

Fuel Valves

There are many fuel valve uses in aircraft fuel systems. They are used to shut off fuel flow or to route the fuel to a desired location. Other than sump drain valves, light aircraft fuel systems may include only one valve, the selector valve. It incorporates the shutoff and selection features into a single valve. Large aircraft fuel systems have numerous valves. Most simply open and close and are known by different names related to their location and function in the fuel system (e.g., shutoff valve, transfer valve, crossfeed valve). Fuel valves can be manually operated, solenoid operated, or operated by electric motor.

A feature of all aircraft fuel valves is a means for positively identifying the position of the valve at all times. Hand-operated valves accomplish this through the use of detents into which a spring-loaded pin or similar protrusion locates

Support tube at least ¼ inch from edge of hole

Grommet

Bonded cushion clamp

Tubing OD (inch)	Approximate distance between supports (inches)
⅛ to ³/₁₆	9
¼ to ⁵/₁₆	12
⅜ to ½	16
⅝ to ¾	22
7 to 1¼	30
1½ to 2	40

Figure 14-37. *Rigid metallic fuel lines are clamped to the airframe with electrically bonded cushion clamps at specified intervals.*

Figure 14-38. *Detents for each position, an indicating handle, and labeling aid the pilot in knowing the position of the fuel valve.*

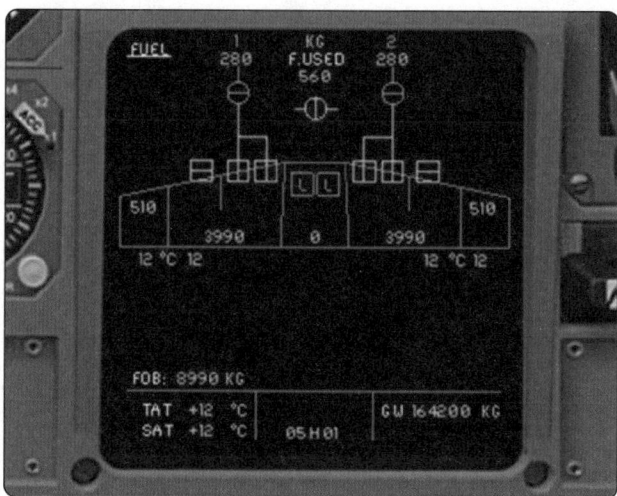

Figure 14-39. *The graphic depiction of the fuel system on this electronic centralized aircraft monitor (ECAM) fuel page includes valve position information.*

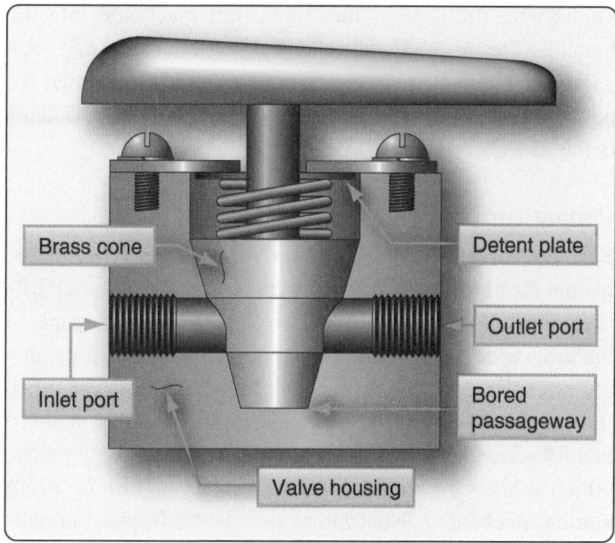

Figure 14-41. *A cone valve is open when the bored cone aligns the inlet and the outlet ports. It shuts off the flow when the un-bored portion of the cone is aligned with the inlet port(s).*

when the valve is set in each position. Combined with labels and a directional handle, this makes it easy to identify by feel and by sight that the valve is in the desired position. *[Figure 14-38]* Motor- and solenoid-operated valves use position annunciator lights to indicate valve position in addition to the switch position. Flight management system (FMS) fuel pages also display the position of the fuel valves graphically in diagrams called up on the flat screen monitors. *[Figure 14-39]* Many valves have an exterior position handle, or lever, that indicates valve position. When maintenance personnel directly observe the valve, it can be manually positioned by the technician using this same lever. *[Figure 14-40]*

Hand-Operated Valves

There are three basic types of hand-operated valves used in aircraft fuel systems. The cone-type valve and the poppet-

type valve are commonly used in light general aviation aircraft as fuel selector valves. Gate valves are used on transport category aircraft as shutoff valves. While many are motor operated, there are several applications in which gate valves are hand operated.

Cone Valves

A cone valve, also called a plug valve, consists of a machined valve housing into which a rotatable brass or nylon cone is set. The cone is manually rotated by the pilot with an attached handle. Passageways are machined through the cone so that, as it is rotated, fuel can flow from the selected source to the engine. This occurs when the passageway

Figure 14-40. *This motor-operated gate valve has a red position indicating lever that can be used by maintenance personnel to identify the position of the valve. The lever can be moved by the technician to position the valve.*

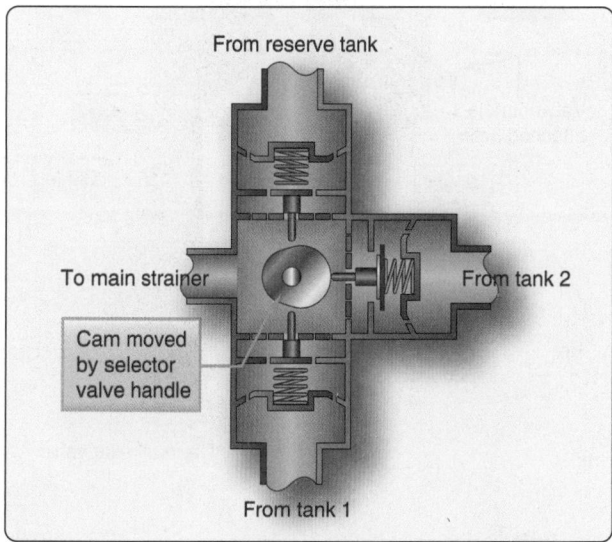

Figure 14-42. *The internal mechanism of a poppet-type fuel selector valve.*

aligns with the desired fuel input port machined into the housing. *Figure 14-41* shows a cross sectional view of a cone valve. The cone can also be rotated to a position so that the passageway(s) does not align with any fuel input port. This is the fuel OFF position of the valve.

Poppet Valves

Selector valves are also commonly the poppet type. As the handle is rotated in this valve, a cam on the attached shaft lifts the poppet off the seat of the desired port being selected. At the same time, spring-assisted poppets close off the ports that are not selected. Detents lock the valve into position when the cam pushes a poppet fully off of its seat. There is also a positive detent when the cam engages none of the poppets, which is the OFF position of the valve. *[Figure 14-42]* A similar mechanism is used in some selector valves, but balls are used instead of poppets.

Manually-Operated Gate Valves

A single selector valve is not used in complex fuel systems of transport category aircraft. Fuel flow is controlled with a series of ON/OFF, or shutoff, type valves that are plumbed between system components. Hand-operated gate valves can be used, especially as fire control valves, requiring no electrical power to shutoff fuel flow when the emergency fire handle is pulled. The valves are typically positioned in the fuel feed line to each engine. Hand-operated gate valves are also featured as ground-operated defuel valves and boost pump isolation valves, which shut off the fuel to the inlet of the boost pump, allowing it to be changed without emptying the tank.

Gate valves utilize a sealed gate or blade that slides into the path of the fuel, blocking its flow when closed. *Figure 14-43* shows a typical hand-operated gate valve.

When the handle is rotated, the actuating arm inside the valve moves the gate blade down between seals and into the fuel flow path. A thermal relief bypass valve is incorporated to relieve excess pressure buildup against the closed gate due to temperature increases.

Motor-Operated Valves

The use of electric motors to operate fuel system valves is common on large aircraft due to the remote location from the flight deck of fuel system components. The types of valves used are basically the same as the manually operated valves, but electric motors are used to actuate the units. The two most common electric motor operated fuel valves are the gate valve and the plug-type valve.

The motor-operated gate valve uses a geared, reversible electric motor to turn the actuating arm of the valve that moves the fuel gate into or out of the path of the fuel. As with the manually operated gate valve, the gate or blade is sealed. A manual override lever allows the technician to observe the

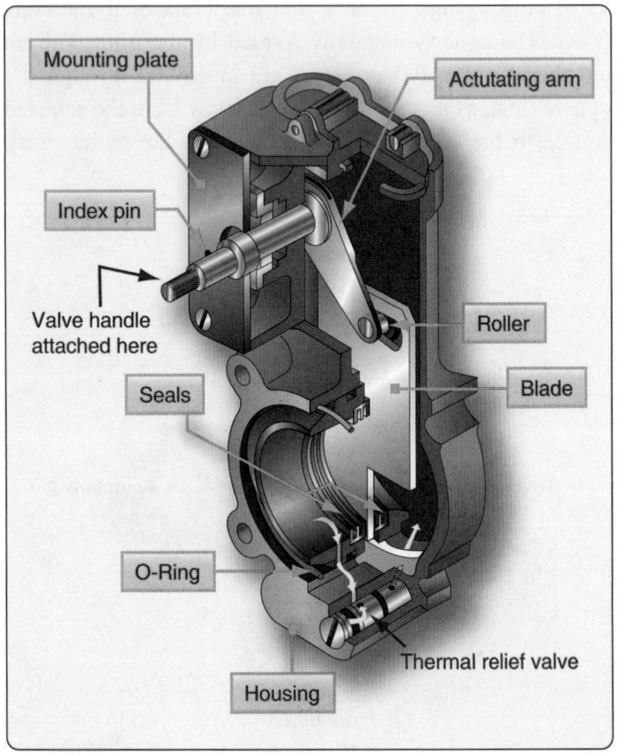

Figure 14-43. *A hand-operated gate valve used in transport category aircraft fuel systems.*

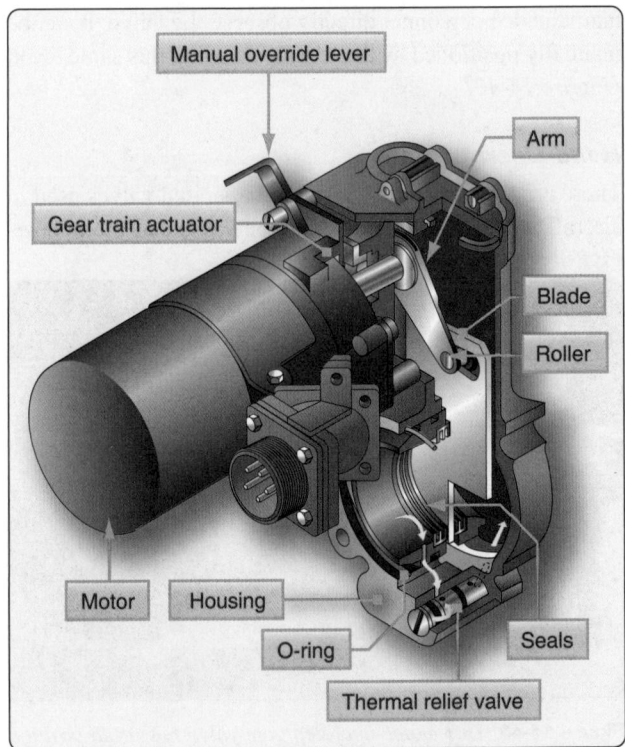

Figure 14-44. *An electric motor-driven gate valve commonly used in large aircraft fuel systems.*

position of the valve or manually position it. *[Figure 14-44]* Less common is the use of a motorized plug-type fuel valve; an electric motor is used to rotate the plug or drum rather than it being rotated manually. Regardless of the type of valve used, large aircraft fuel system valves either allow fuel to flow or shut off flow.

Solenoid-Operated Valves

An additional way to operate a remotely located fuel valve is through the use of electric solenoids. A poppet-type valve is opened via the magnetic pull developed when an opening solenoid is energized. A spring forces a locking stem into a notch in the stem of the poppet to lock the valve in the open position. Fuel then flows through the opening vacated by the poppet. To close the poppet and shut off fuel flow, a closing solenoid is energized. Its magnetic pull overcomes the force of the locking stem spring and pulls the locking stem out of the notch in the poppet stem. A spring behind the poppet forces it back onto its seat. A characteristic of solenoid-operated fuel valves is that they open and close very quickly. *[Figure 14-45]*

Fuel Pumps

Other than aircraft with gravity-feed fuel systems, all aircraft have at least one fuel pump to deliver clean fuel under pressure to the fuel metering device for each engine. Engine-driven pumps are the primary delivery device. Auxiliary pumps are used on many aircraft as well. Sometimes known as booster pumps or boost pumps, auxiliary pumps are used to provide fuel under positive pressure to the engine-driven pump and during starting when the engine-driven pump is not yet up to speed for sufficient fuel delivery. They are also used to back up the engine-driven pump during takeoff and at high altitude to guard against vapor lock. On many large aircraft, boost pumps are used to move fuel from one tank to another.

There are many different types of auxiliary fuel pumps in use. Most are electrically operated, but some hand-operated pumps are found on older aircraft. A discussion of the various pump types found in the aviation fleet follows.

Hand-Operated Fuel Pumps

Some older reciprocating engine aircraft have been equipped with hand-operated fuel pumps. They are used to back up the engine-driven pump and to transfer fuel from tank to tank. The wobble pumps, as they are known, are double-acting pumps that deliver fuel with each stroke of the pump handle. They are essentially vane-type pumps that have bored passages in the center of the pump, allowing a back-and-forth motion to pump the fuel rather than a full revolution of the vanes as is common in electrically driven or engine-driven vane-type pumps.

Figure 14-46 illustrates the mechanism found in a wobble pump. As the handle is moved down from where it is shown, the vane on the left side of the pump moves up, and the vane on the right side of the pump moves down. As the left vane moves up, it draws fuel into chamber A. Because chambers A and D are connected through the bored center, fuel is also drawn into chamber D. At the same time, the right vane forces fuel out of chamber B, through the bored passage in the center of the pump, into chamber C and out the fuel outlet through the check valve at the outlet of chamber C. When the handle is moved up again, the left vane moves down, forcing fuel

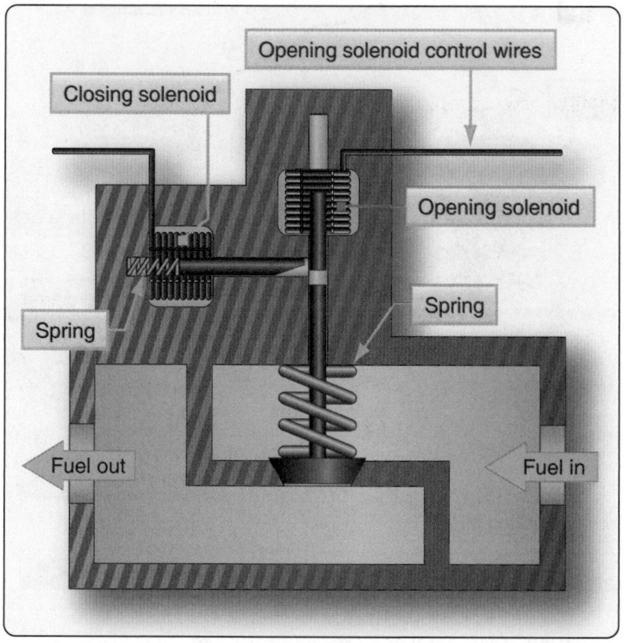

Figure 14-45. *A solenoid-operated fuel valve uses the magnetic force developed by energized solenoids to open and close a poppet.*

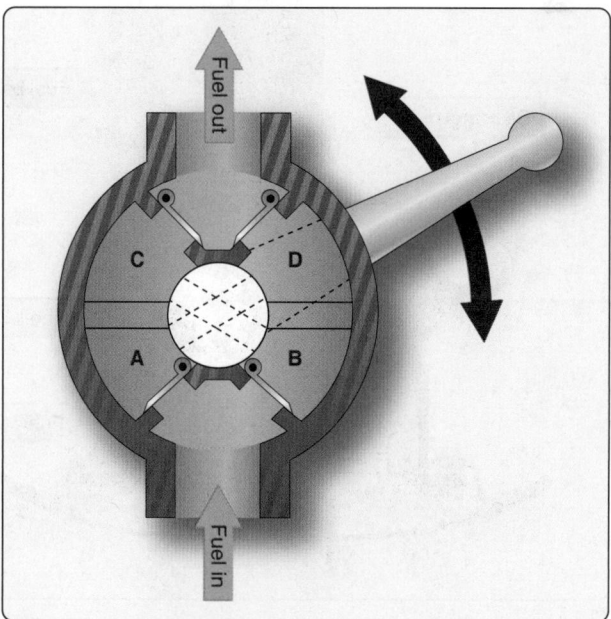

Figure 14-46. *A hand-operated wobble pump used for engine starting and fuel transfer on older transport category aircraft.*

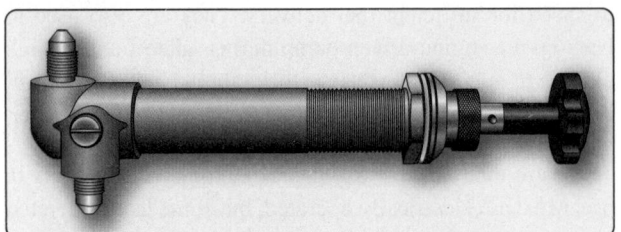

Figure 14-47. *This engine primer pump is a hand-operated piston type. It is mounted in the instrument panel and extends through the firewall where fuel intake and delivery lines are attached to the fittings on the left.*

out of chambers A and D because the check valve at the inlet of the A chamber prevents fuel from flowing back through the fuel inlet. The right vane moves up simultaneously and draws fuel into chambers B and C.

While simple with little to go wrong, a hand-operated pump requires fuel lines to be run into the flight deck to the pump, creating a potential hazard that can be avoided by the use of an electrically driven pump. Modern light reciprocating-engine aircraft usually use electric auxiliary pumps, but they often make use of a simple hand pump for priming the engine(s) during starting. These simple devices are single-acting piston pumps that pull fuel into the pump cylinder when the primer knob is pulled aft. When pushed forward, the fuel is pumped through lines to the engine cylinders. *[Figure 14-47]*

Centrifugal Boost Pumps

The most common type of auxiliary fuel pump used on aircraft, especially large and high-performance aircraft, is the centrifugal pump. It is electric motor driven and most frequently is submerged in the fuel tank or located just outside of the bottom of the tank with the inlet of the pump extending into the tank. If the pump is mounted outside the tank, a pump removal valve is typically installed so the pump can be removed without draining the fuel tank. *[Figure 14-48]*

A centrifugal boost pump is a variable displacement pump. It takes in fuel at the center of an impeller and expels it to the outside as the impeller turns. *[Figure 14-49]* An outlet check valve prevents fuel from flowing back through the pump. A fuel feed line is connected to the pump outlet. A bypass valve may be installed in the fuel feed system to allow the engine-driven pump to pull fuel from the tank if the boost pump is not operating. The centrifugal boost pump is used to supply the engine-driven fuel pump, back up the engine-driven fuel pump, and transfer fuel from tank to tank if the aircraft is so designed.

Some centrifugal fuel pumps operate at more than one speed, as selected by the pilot, depending on the phase of aircraft operation. Single-speed fuel pumps are also common. Centrifugal fuel pumps located in fuel tanks ensure positive pressure throughout the fuel system regardless of temperature, altitude, or flight attitude thus preventing vapor lock. Submerged pumps have fuel proof covers for the electric motor since the motor is in the fuel. Centrifugal

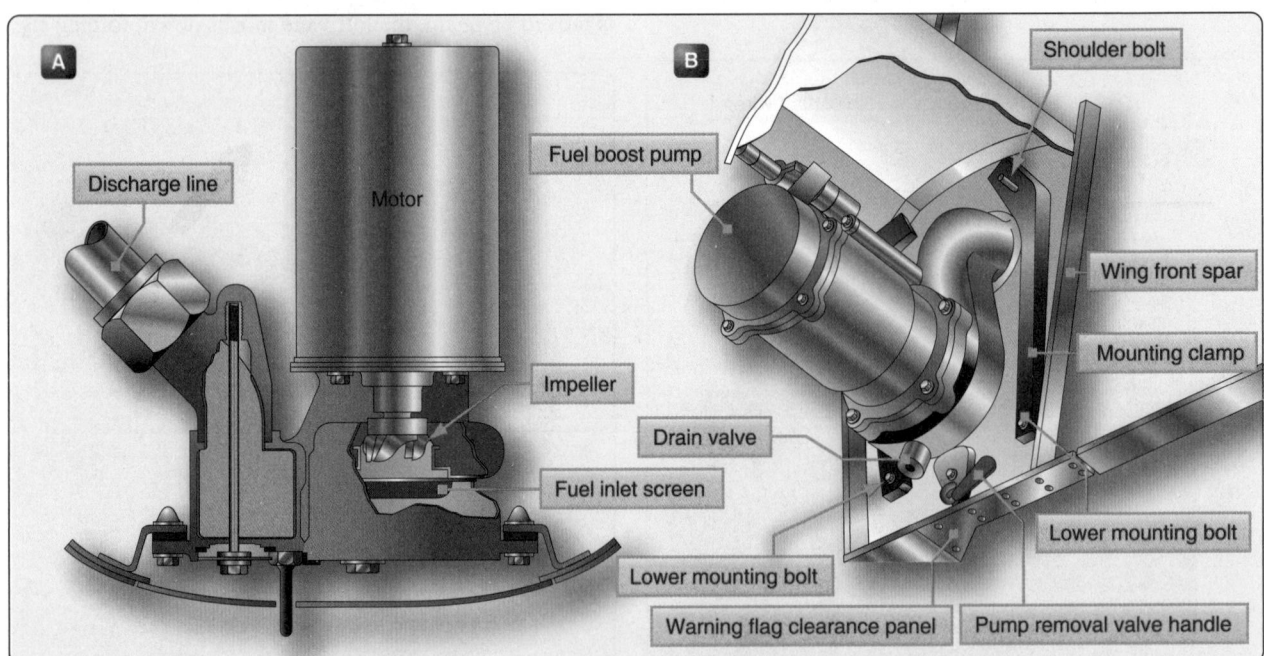

Figure 14-48. *A centrifugal fuel boost pump can be submersed in the fuel tank (A) or can be attached to the outside of the tank with inlet and outlet plumbing extending into the tank (B). The pump removal valve handle extends below the warning flag clearance panel to indicate the pump inlet is closed.*

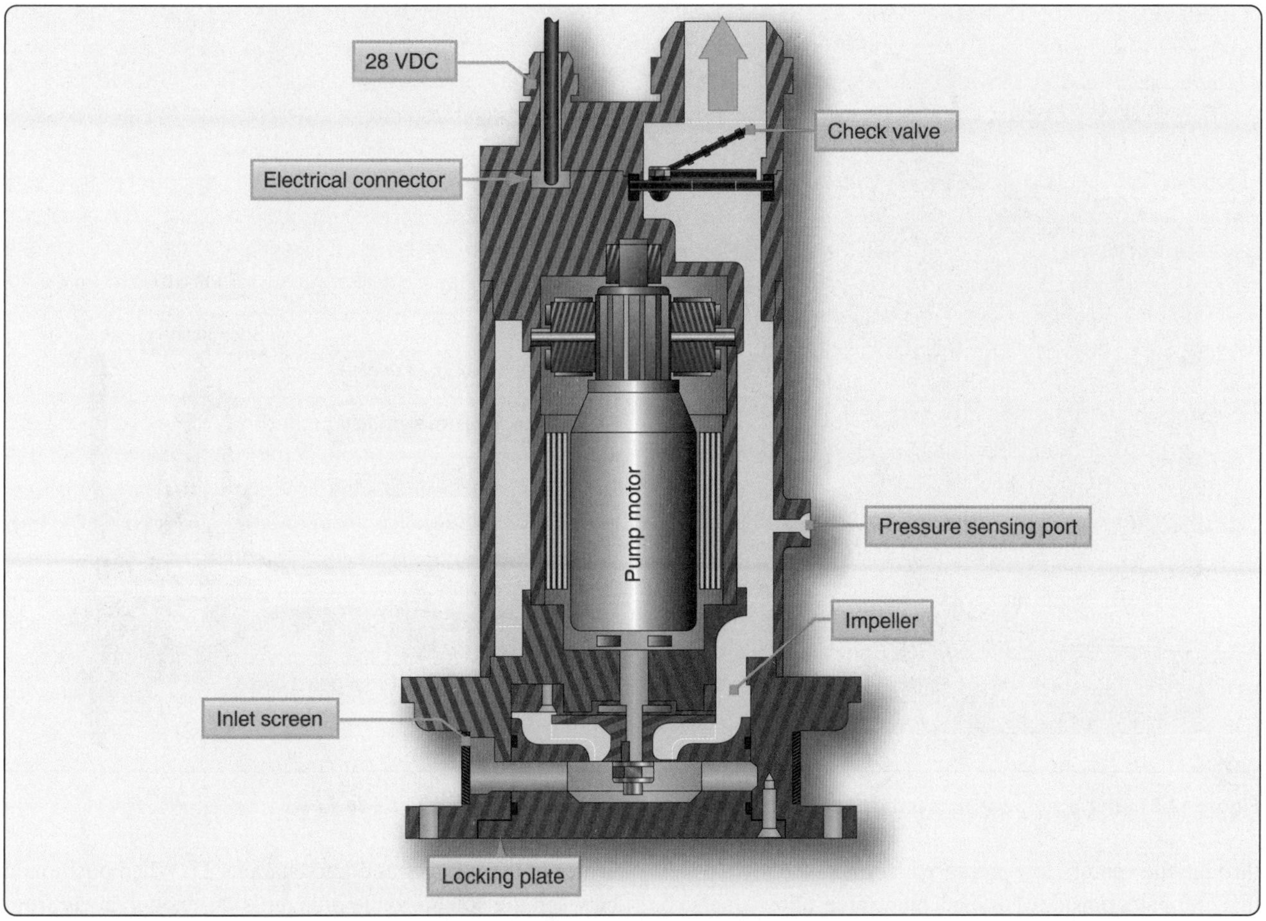

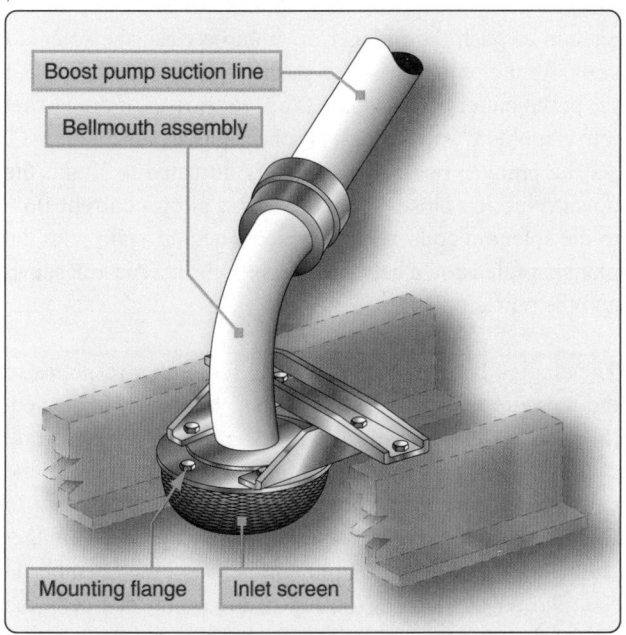

Figure 14-49. *The internal workings of a centrifugal fuel boost pump. Fuel is drawn into the center of the impeller through a screen. It is moved to the outside of the case by the impeller and out the fuel outlet tube.*

28 VDC

Check valve

Electrical connector

Pump motor

Pressure sensing port

Impeller

Inlet screen

Locking plate

Boost pump suction line

Bellmouth assembly

Mounting flange

Inlet screen

Figure 14-50. *A typical fuel boost pump inlet screen installation for a centrifugal pump mounted outside of the bottom of the tank.*

pumps mounted on the outside of the tank do not require this but have some sort of inlet that is located in the fuel. This can be a tube in which a shutoff valve is located so the pump can be changed without draining the tank. The inlet of both types of centrifugal pump is covered with a screen to prevent the ingestion of foreign matter. *[Figure 14-50]*

Ejector Pumps

Fuel tanks with in-tank fuel pumps, such as centrifugal pumps, are constructed to maintain a fuel supply to the pump inlet at all times. This ensures that the pump does not cavitate and that the pump is cooled by the fuel. The section of the fuel tank dedicated for the pump installation may be partitioned off with baffles that contain check valves, also known as flapper valves. These allow fuel to flow inboard to the pump during maneuvers but does not allow it to flow outboard.

Some aircraft use ejector pumps to help ensure that liquid fuel is always at the inlet of the pump. A relatively small diameter line circulates pump outflow back into the section of the tank where the pump is located. The fuel is directed through a venturi that is part of the ejector. As the fuel rushes

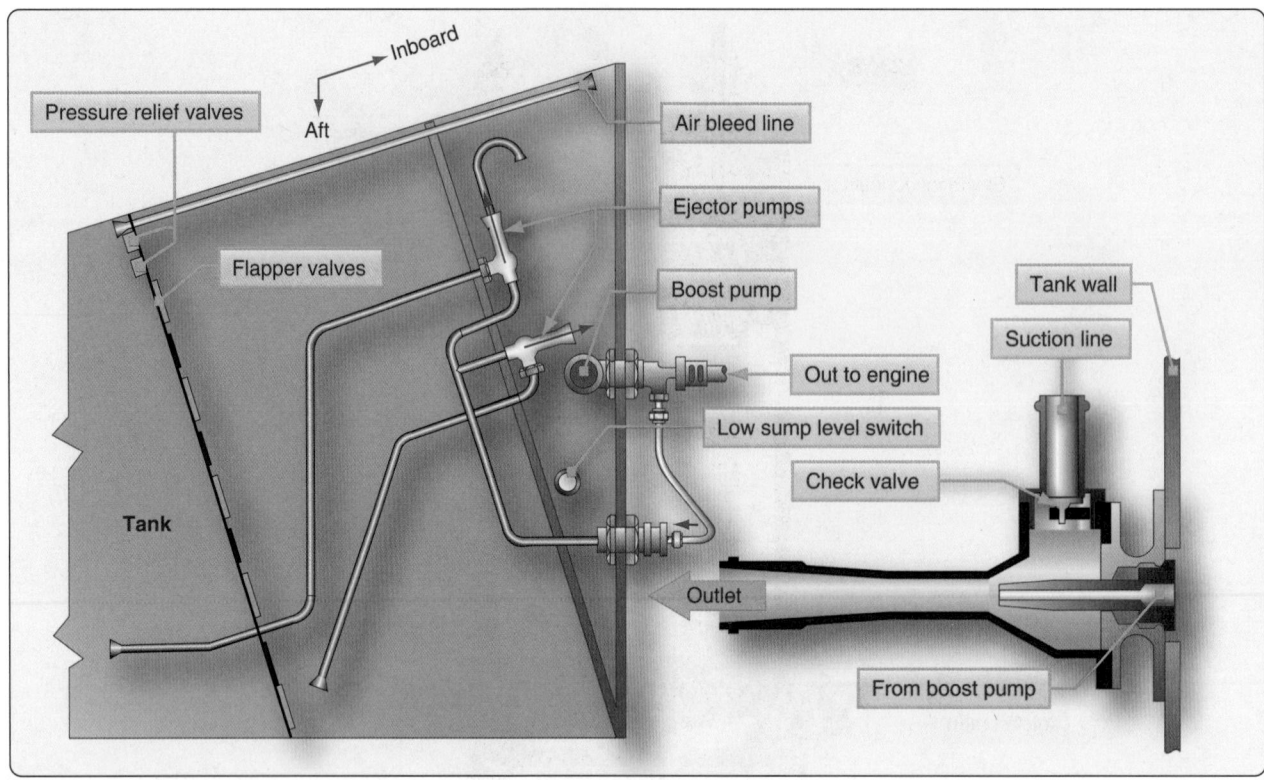

Figure 14-51. *An ejector pump uses a venturi to draw fuel into the boost pump sump area of the fuel tank.*

through the venturi, low pressure is formed. An inlet, or line that originates outside of the tank pump area, allows fuel to be drawn into the ejector assembly where it is pumped into the fuel pump tank section. Together, with baffle check valves, ejector pumps keep a positive head of fuel at the inlet of the pump. *[Figure 14-51]*

Pulsating Electric Pumps

General aviation aircraft often make use of smaller, less expensive auxiliary fuel pumps. The pulsating electric pump, or plunger-type fuel pump, is common. It is usually used in the same manner as a centrifugal fuel pump on larger aircraft, except it is located downstream of the fuel tank outlets. The pulsating electric fuel pump is plumbed in parallel with the engine-driven pump. During starting, it provides fuel before the engine-driven fuel pump is up to speed, and it can be used during takeoff as a backup. It also can be used at high altitudes to prevent vapor lock.

The pulsating electric pump uses a plunger to draw fuel in and push fuel out of the pump. It is powered by a solenoid that alternates between being energized and de-energized, which moves the plunger back and forth in a pulsating motion. *Figure 14-52* shows the internal workings of the pump. When switched ON, current travels through the solenoid coils, which pull the steel plunger down between the coils. Any fuel in chamber C is forced through the small check valve in the

center of the plunger and into chamber D. When positioned between the solenoid, the plunger is far enough away from the magnet that it no longer attracts it, and the pivot allows the contacts to open. This disrupts the current to the solenoid.

The calibrated spring shown under the plunger is then strong enough to push the plunger up from between the solenoid coils. As the plunger rises, it pushes fuel in chamber D out the pump outlet port. Also, as the plunger rises, it draws fuel into chamber C and through the check valve into chamber C. As the plunger rises, the magnet is attracted to it and the upward motion closes the points. This allows current flow to the solenoid coils, and the process begins again with the plunger pulled down between the coils, the magnet releasing, and the points opening.

The single-acting pulsating electric fuel pump responds to the pressure of the fuel at its outlet. When fuel is needed, the pump cycles rapidly with little pressure at the pump outlet. As fuel pressure builds, the pump slows because the calibrated spring meets this resistance while attempting to force the piston upwards. A spring in the center of the plunger dampens its motion. A diaphragm between the chamber D fuel and an airspace at the top of the pump dampens the output fuel pulses.

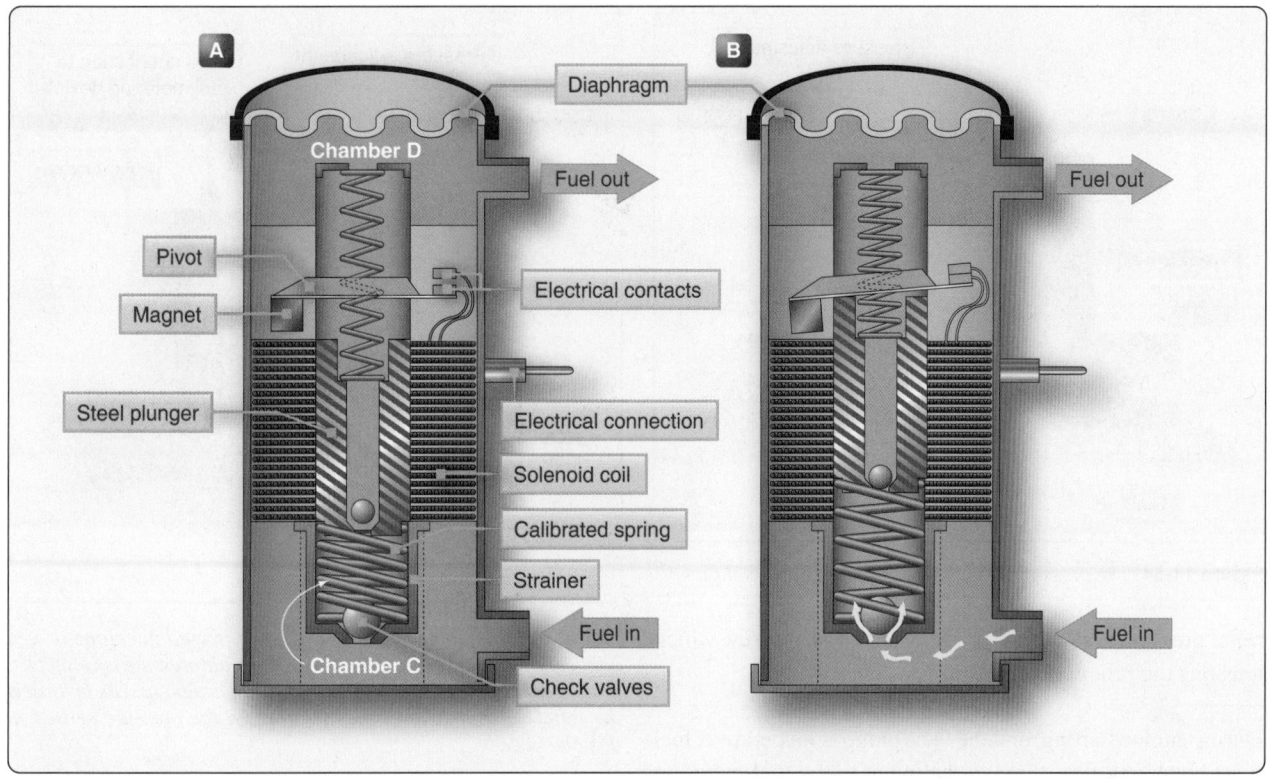

Figure 14-52. *A pulsating electric auxiliary fuel pump is used on many light reciprocating engine aircraft. In A, the pump is shown with its solenoid coil energized, which draws the plunger down between the coil. This opens the breaker points allowing the calibrated spring to push the plunger upwards, thus pumping fuel out the outlet B. This cycle repeats at a speed related to the fuel pressure buildup at the pump outlet.*

Vane-Type Fuel Pumps

Vane-type fuel pumps (engine-driven fuel pumps) are the most common types of fuel pumps found on reciprocating-engine aircraft. They are used as both engine-driven primary fuel pumps and as auxiliary or boost pumps. Regardless, the vane-type pump is a constant displacement pump that moves a constant volume of fuel with each revolution of the pump. When used as an auxiliary pump, an electric motor rotates the pump shaft. On engine-driven applications, the vane pump is typically driven by the accessory gear box.

As with all vane pumps, an eccentric rotor is driven inside a cylinder. Slots on the rotor allow vanes to slide in and out and be held against the cylinder wall by a central floating spacer pin. As the vanes rotate with the eccentric rotor, the volume space created by the cylinder wall, the rotor, and the vanes increases and then decreases. An inlet port is located where the vanes create an increasing volume space, and fuel is drawn into the pump. Further around in the rotation, the space created becomes smaller. An outlet port located there causes fuel to be forced from the cylinder. *[Figure 14-53]*

The engine-driven fuel pump delivers more fuel than the engine needs to operate. However, the constant volume of a vane pump can be excessive. To regulate flow, most vane pumps have an adjustable pressure relief feature. It uses pressure built up at the outlet of the pump to lift a valve off its seat, which returns excess fuel to the inlet side of the pump. *Figure 14-54* shows a typical vane type fuel pump with this adjustable pressure relief function. By setting the relief at a certain pressure above the engine fuel metering device air intake pressure, the correct volume of fuel is delivered. The

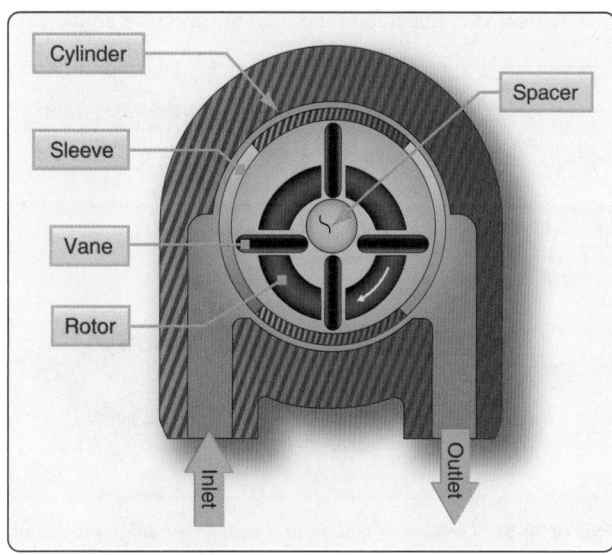

Figure 14-53. *The basic mechanism of a vane-type fuel pump.*

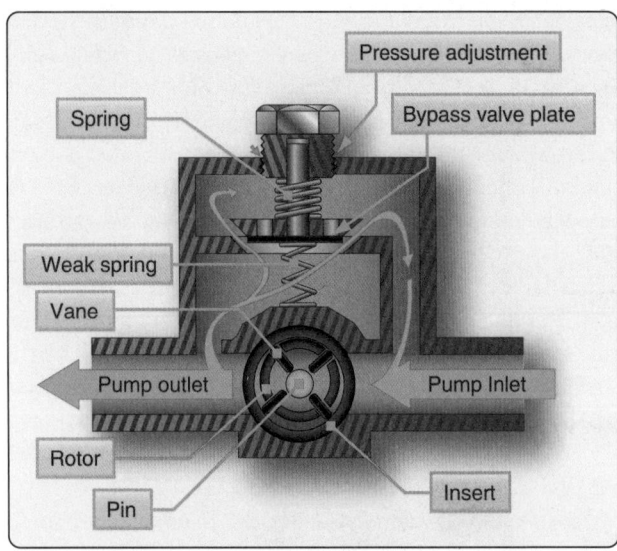

Figure 14-54. *The pressure relief valve in a vane-type fuel pump.*

relief pressure is set via the pressure adjustment screw which tensions the relief valve spring.

During engine starting, or if the vane pump is inoperative, fuel must be able to flow through the pump to the fuel metering device. This is accomplished with the use of a bypass valve inside the pump. A lightly sprung plate under the relief valve overcomes spring pressure whenever the pump's inlet fuel pressure is greater than the outlet fuel press. The plate moves down, and fuel can flow through the pump. *[Figure 14-55]*

Compensated vane-type fuel pumps are used when the vane pump is the engine-driven primary fuel pump. The relief valve setting varies automatically to provide the correct delivery of fuel as the air inlet pressure of the fuel metering device changes due to altitude or turbocharger outlet pressure. A vent chamber above a diaphragm attached to the relief mechanism is connected to the inlet air pressure source. As

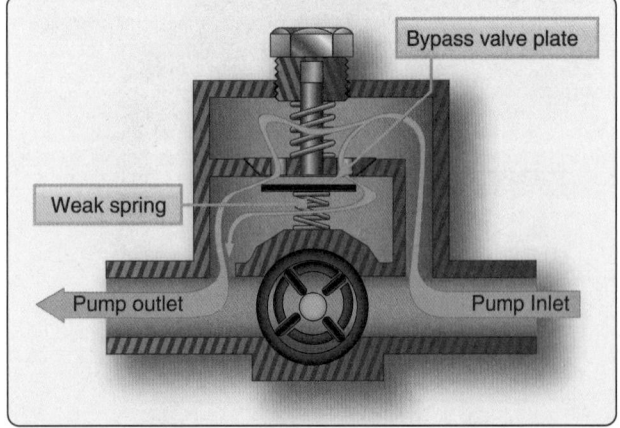

Figure 14-55. *The bypass feature in a vane-type fuel pump allows fuel to flow through the pump during starting or when the pump is inoperative.*

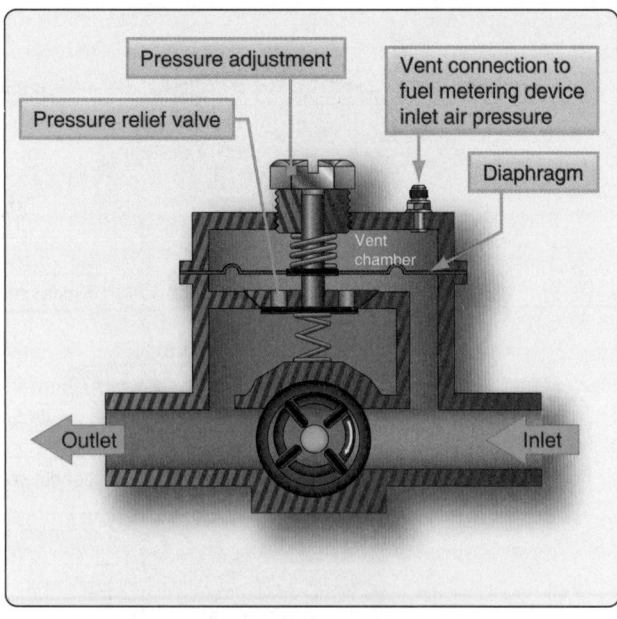

Figure 14-56. *A compensated vane pump is used in engine-driven applications. The fuel metering device inlet air pressure is connected to the vent chamber in the pump. The diaphragm assists or resists the relief valve mechanism depending on the pressure sensed in this chamber.*

air pressure varies, the diaphragm assists or resists the relief valve spring pressure, resulting in proper fuel delivery for the condition at the fuel metering device. *[Figure 14-56]*

Fuel Filters

Two main types of fuel cleaning devices are utilized on aircraft. Fuel strainers are usually constructed of relatively coarse wire mesh. They are designed to trap large pieces of debris and prevent their passage through the fuel system. Fuel strainers do not inhibit the flow of water. Fuel filters generally are usually fine mesh. In various applications, they can trap fine sediment that can be only thousands of an inch in diameter and also help trap water. The technician should

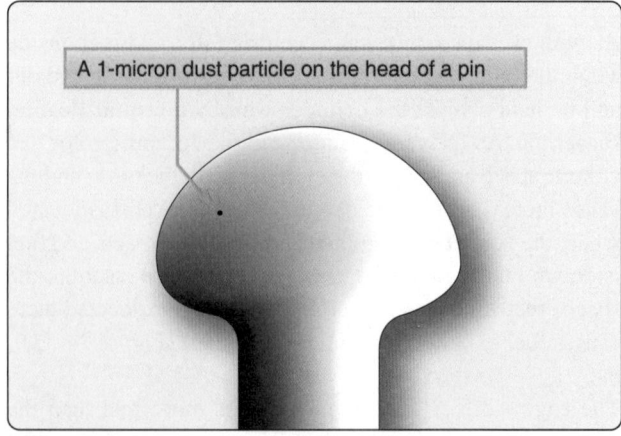

Figure 14-57. *Size comparison of 1-micron dust particle and pin head.*

be aware that the terms "strainer" and "filter" are sometimes used interchangeably. Micronic filters are commonly used on turbine-powered aircraft. This is a type of filter that captures extremely fine particles in the range of 10–25 microns. A micron is ¹⁄₁,₀₀₀ of a millimeter. *[Figure 14-57]*

All aircraft fuel systems have filters and strainers to ensure that the fuel delivered to the engine(s) is free from contaminants. The first of these is encountered at the outlet of the fuel tank. A sump is used to encourage the collection of debris in the lowest part of the tank, which can then be drained off before flight. The actual tank outlet for the fuel is positioned above this sump. Some type of screen is used to trap contaminants attempting to flow out of the tank into the fuel system. Finger screens are common on light aircraft. They effectively increase the area of the fuel tank outlet, allowing a large amount of debris to be trapped while still permitting fuel to flow. *Figure 14-58* illustrates finger screens that are screwed into a fitting welded in the tank outlet.

Fuel tank outlet screens on aircraft with more complex fuel systems are similarly designed. When in-tank boost pumps are used, the tank outlet strainer is located at the inlet to the boost pump as was shown in *Figure 14-50*. The screen's large area allows debris capture while still permitting sufficient fuel flow for operation. Regularly scheduled inspection and cleaning of these strainers are required.

An additional main strainer for the aircraft fuel system is required between the fuel tank outlet and the fuel metering device (in a carburetor or fuel-injection system). It is normally located between the fuel tank and the engine-driven fuel pump at the low point in the fuel system and is equipped with a drain for prefight sampling and draining. On light aircraft, the main strainer may be in the form of a gascolator. A gascolator is a fuel strainer, or filter, that also incorporates a sediment collection bowl. The bowl is traditionally glass to allow quick visual checks for contaminants; however, many gascolators also have opaque bowls. A gascolator has a drain, or the bowl can be removed to inspect and discard trapped debris and water. *[Figure 14-59]*

The main fuel strainer is often mounted at a low point on the engine firewall. The drain is accessible through an easy-access panel, or it simply extends through the bottom engine cowling. As with most filters or strainers, fuel is allowed to enter the unit but must travel up through the filtering element to exit. Water, which is heavier than fuel, becomes trapped and collects in the bottom of the bowl. Other debris too large to pass through the element also settles in the strainer bowl.

Higher performance light aircraft may have a main filter/strainer. *[Figure 14-60]* On twin-engine aircraft, there is a main strainer for each engine. As with single-engine aircraft, a strainer is often mounted low on the engine firewall in each nacelle.

Other larger fuel filters have double-screen construction. A cylindrical structural screen is wrapped with a fine mesh material through which inlet fuel must pass. Inside the cylinder is an additional cone-shaped screen. Fuel must pass

Figure 14-58. *Fuel tank outlet finger strainers are used in light aircraft.*

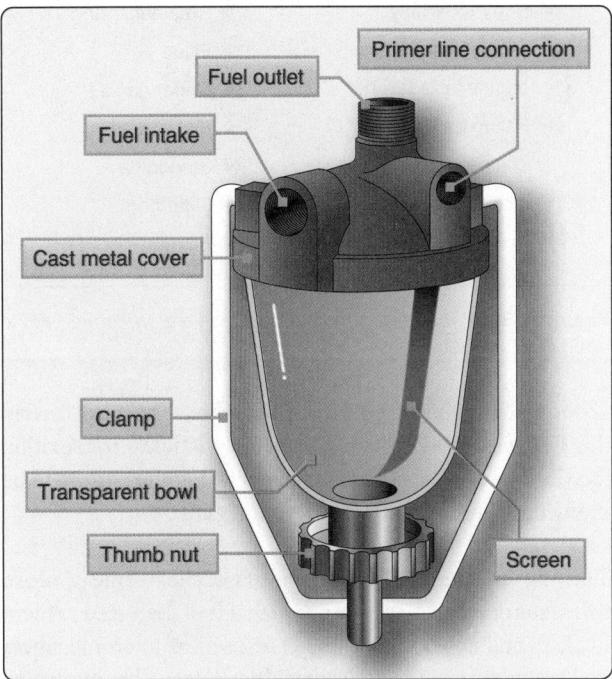

Figure 14-59. *A gascolator is the main fuel strainer between the fuel tanks and the fuel metering device on many light aircraft.*

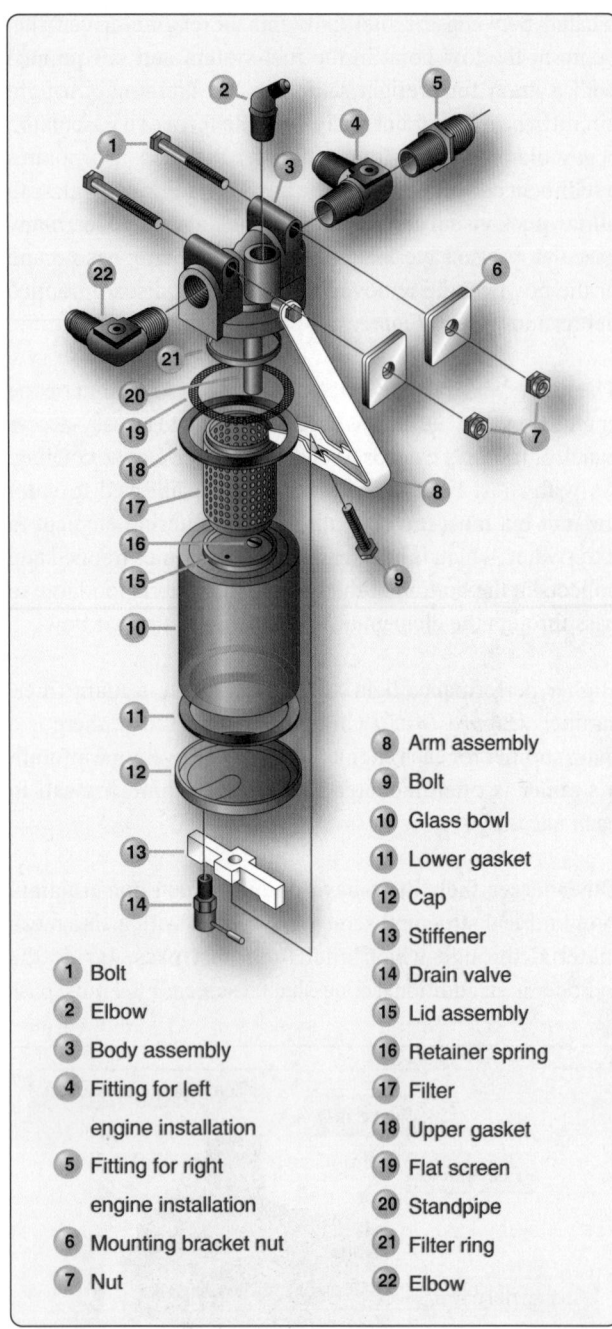

Figure 14-60. *A filter assembly on a light twin reciprocating-engine aircraft.*

① Bolt	⑧ Arm assembly
② Elbow	⑨ Bolt
③ Body assembly	⑩ Glass bowl
④ Fitting for left	⑪ Lower gasket
engine installation	⑫ Cap
⑤ Fitting for right	⑬ Stiffener
engine installation	⑭ Drain valve
⑥ Mounting bracket nut	⑮ Lid assembly
⑦ Nut	⑯ Retainer spring
	⑰ Filter
	⑱ Upper gasket
	⑲ Flat screen
	⑳ Standpipe
	㉑ Filter ring
	㉒ Elbow

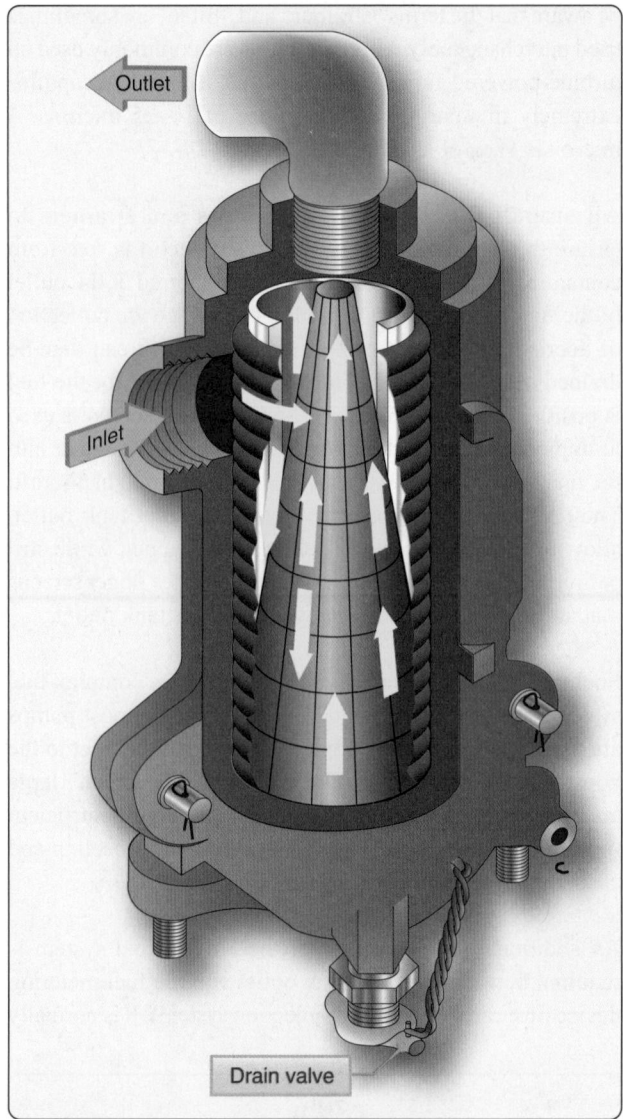

Figure 14-61. *A large-area double-screen filter passes fuel through the outer cylindrical mesh and the inner conical mesh.*

up through the cone to get to the filter outlet. The mesh used in this filter assembly prevents water and particles from exiting the filter bowl. The contaminants collect at the bottom to be drained off through a drain valve. *[Figure 14-61]*

Turbine engine fuel control units are extremely close tolerance devices. It is imperative that fuel delivered to them is clean and contaminant free. The used of micronic filters makes this possible. The changeable cellulose filter mesh type shown in *Figure 14-62* can block particles 10–200 microns

in size and absorbs water if it is present. The small size of the mesh raises the possibility of the filter being blocked by debris or water. Therefore, a relief valve is included in the filter assembly that bypasses fuel through the unit should pressure build up from blockage.

Fuel filters are often used between the engine-driven fuel pump and the fuel metering device on reciprocating, as well as turbine-engine aircraft. While these are technically part of the engine fuel system, a common type used on turbine engines is discussed here. It is also a micronic filter. It uses finely meshed discs or wafers stacked on a central core. These filters are able to withstand the higher pressure found in the engine fuel system downstream of the engine-driven pump. *[Figure 14-63]*

Indication of a filter blockage may also appear in the

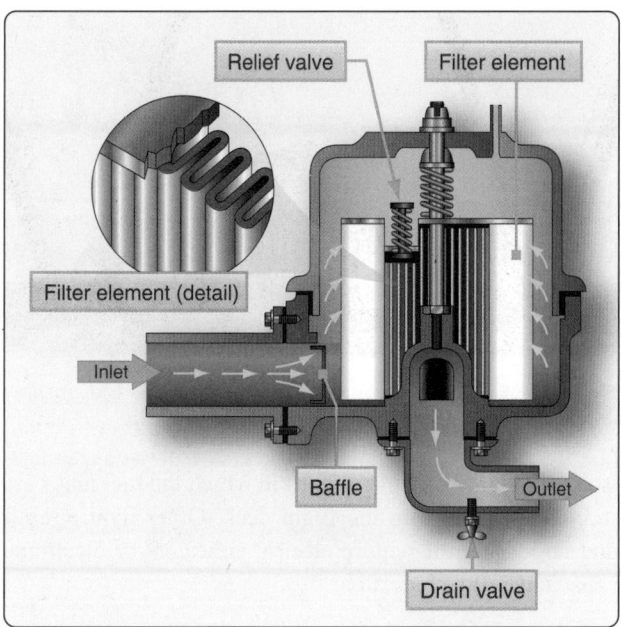

Figure 14-62. *A typical micronic fuel filter with changeable cellulose filter element.*

flight deck through the use of a bypass-activated switch or a pressure differential switch. A high fuel differential pressure indicates a blockage to the fuel filter. The bypass valve physically activates a switch that closes the circuit to the annunciator in the first type. The differential pressure type indicator compares the input pressure of the fuel filter to the output pressure. A circuit is completed when a preset difference occurs. Thus, an indicator is illuminated should a blockage cause the bypass to open or the inlet and outlet pressures to vary significantly. Fuel temperature can also be monitored for the possibility of a blockage caused by frozen water.

Fuel Heaters & Ice Prevention

Turbine powered aircraft operate at high altitude where the temperature is very low. As the fuel in the fuel tanks cools, water in the fuel condenses and freezes. It may form ice crystals in the tank or as the fuel/water solution slows and contacts the cool filter element on its way through fuel filter to the engine(s). The formation of ice on the filter element blocks the flow of fuel through the filter. A valve in the filter unit bypasses unfiltered fuel when this occurs. Fuel heaters are used to warm the fuel so that ice does not form. These heat exchanger units also heat the fuel sufficiently to melt any ice that has already formed.

The most common types of fuel heaters are air-fuel heaters and oil/fuel heaters. An air-fuel heater uses warm compressor bleed air to heat the fuel. An oil/fuel exchanger heats the fuel with hot engine oil. This latter type is often referred to as a fuel-cooled oil cooler (FCOC). *[Figure 14-23]*

Fuel heaters often operate intermittently as needed. A switch in the flight deck can direct the hot air or oil through the unit or block it. The flight crew uses the information supplied by the filter bypass indicating lights and fuel temperature gauge *[Figure 14-64]* to know when to heat the fuel. Fuel heaters can also be automatic. A built-in thermostatic device opens or closes a valve that permits the hot air or hot oil to flow into the unit to warm the fuel. *[Figure 14-65]*

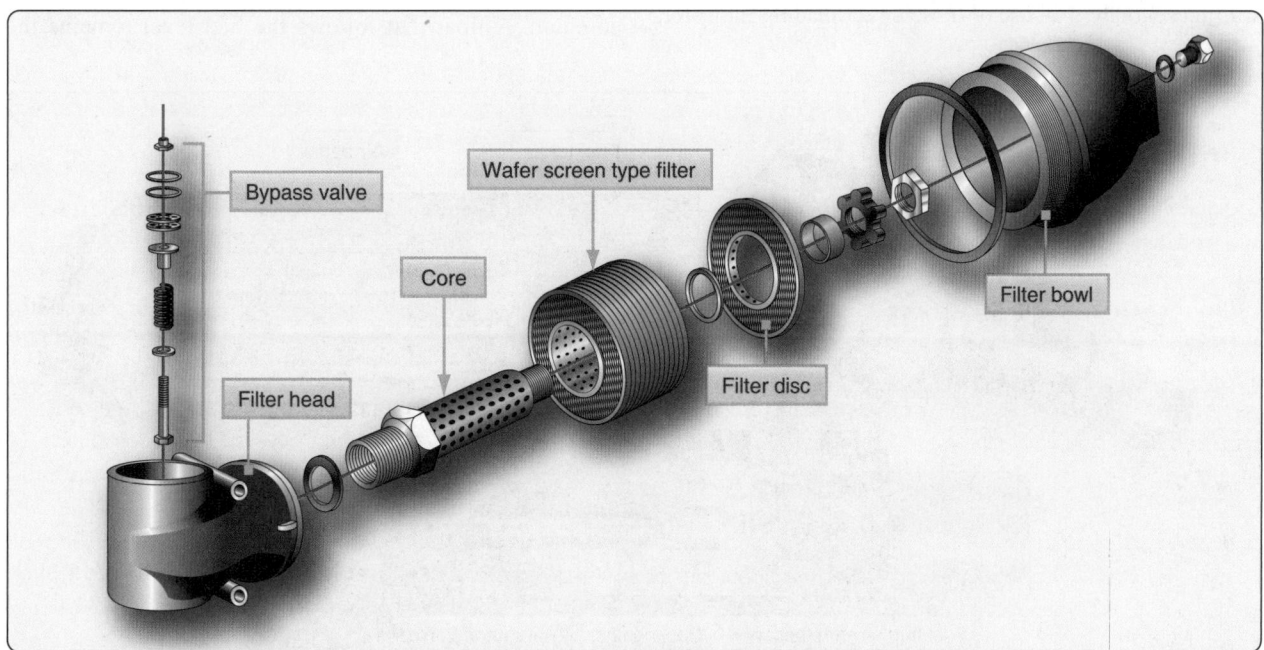

Figure 14-63. *A micronic wafer filter uses multiple screen wafers through which fuel must pass to exit the filter through the core. A spring loaded bypass valve in the filter housing unseats when the filter is clogged to continue delivery of fuel.*

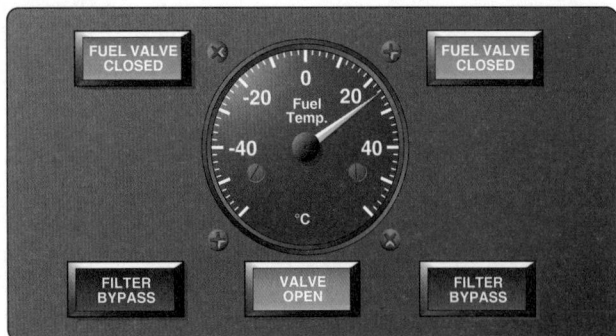

Figure 14-64. *A Boeing 737 flight deck fuel panel showing illuminated valve position indicators and fuel filter bypass lights. The fuel temperature in tank number 1 is also indicated.*

Figure 14-66. *The fuel quantity indicator on this Piper Cub is a float attached to a rod that protrudes through the fuel cap.*

Some aircraft have a hydraulic fluid cooler in one of the aircraft fuel tanks. The fluid helps warm the fuel as it cools in this type of full-time heat exchanger.

Fuel System Indicators

Aircraft fuel systems utilize various indicators. All systems are required to have some sort of fuel quantity indicator. Fuel flow, pressure, and temperature are monitored on many aircraft. Valve position indicators and various warning lights and annunciations are also used.

Fuel Quantity Indicating Systems

All aircraft fuel systems must have some form of fuel quantity indicator. These devices vary widely depending on the complexity of the fuel system and the aircraft on which they are installed. Simple indicators requiring no electrical power were the earliest type of quantity indicators and are still in use today. The use of these direct reading indicators

is possible only on light aircraft in which the fuel tanks are in close proximity to the flight deck. Other light aircraft and larger aircraft require electric indicators or electronic capacitance-type indicators.

A sight glass is a clear glass or plastic tube open to the fuel tank that fills with fuel to the same level as the fuel in the tank. It can be calibrated in gallons or fractions of a full tank that can be read by the pilot. Another type of sight gauge makes use of a float with an indicating rod attached to it. As the float moves up and down with the fuel level in the tank, the portion of the rod that extends through the fuel cap indicates the quantity of fuel in the tank. *[Figure 14-66]* These two mechanisms are combined in yet another simple fuel quantity indicator in which the float is attached to a rod that moves up or down in a calibrated cylinder. *[Figure 14-67]*

More sophisticated mechanical fuel quantity gauges are common. A float that follows the fuel level remains the

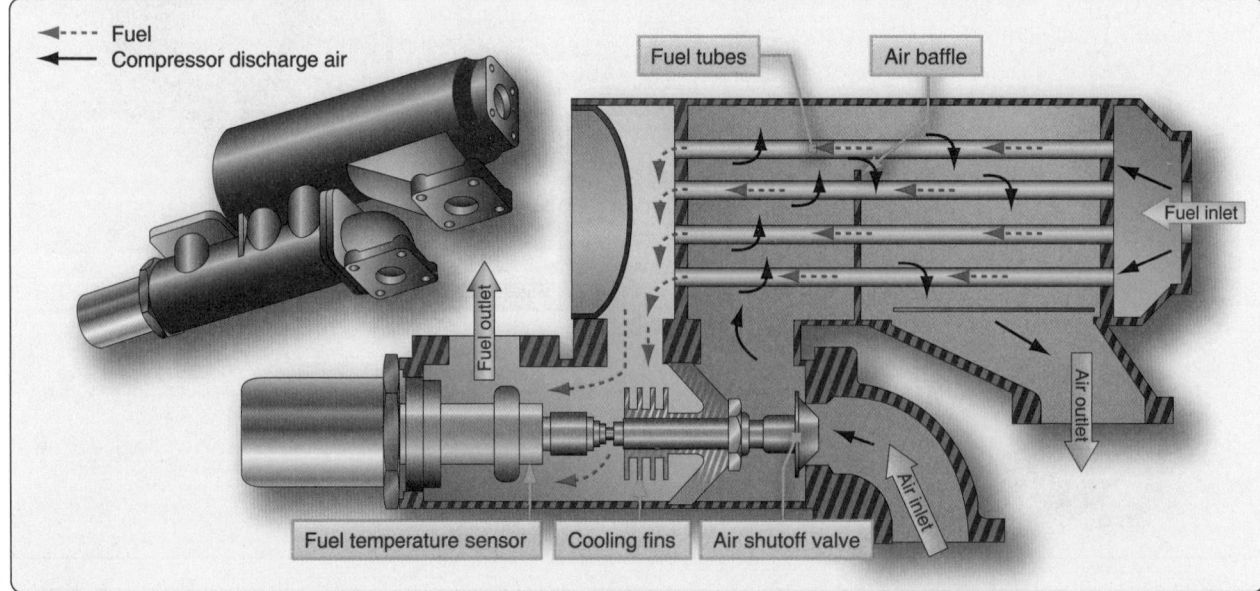

Figure 14-65. *An air-fuel heat exchanger uses engine compressor bleed air to warm the fuel on many turbine engine powered aircraft.*

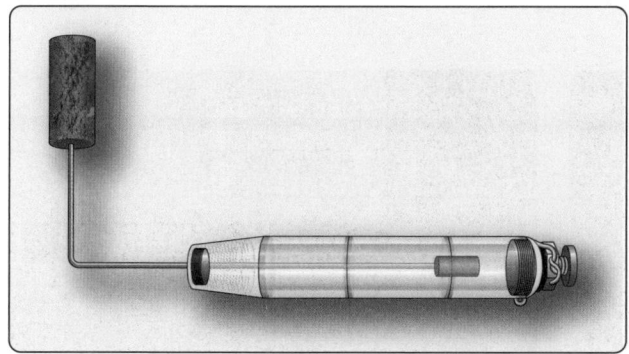

Figure 14-67. *A float-type sight gauge fuel quantity indicator.*

Electric fuel quantity indicators are more common than mechanical indicators in modern aircraft. Most of these units operate with direct current (DC) and use variable resistance in a circuit to drive a ratiometer-type indicator. The movement of a float in the tank moves a connecting arm to the wiper on a variable resistor in the tank unit. This resistor is wired in series with one of the coils of the ratiometer-type fuel gauge in the instrument panel. Changes to the current flowing through the tank unit resistor change the current flowing through one of the coils in the indicator. This alters the magnetic field in which the indicating pointer pivots. The calibrated dial indicates the corresponding fuel quantity. *[Figure 14-69]*

primary sensing element, but a mechanical linkage is connected to move a pointer across the dial face of an instrument. This can be done with a crank and pinion arrangement that drives the pointer with gears, or with a magnetic coupling, to the pointer. *[Figure 14-68]*

Digital indicators are available that work with the same variable resistance signal from the tank unit. They convert the variable resistance into a digital display in the flight deck instrument head. *[Figure 14-70]* Fully digital instrumentation systems, such as those found in a glass flight deck aircraft,

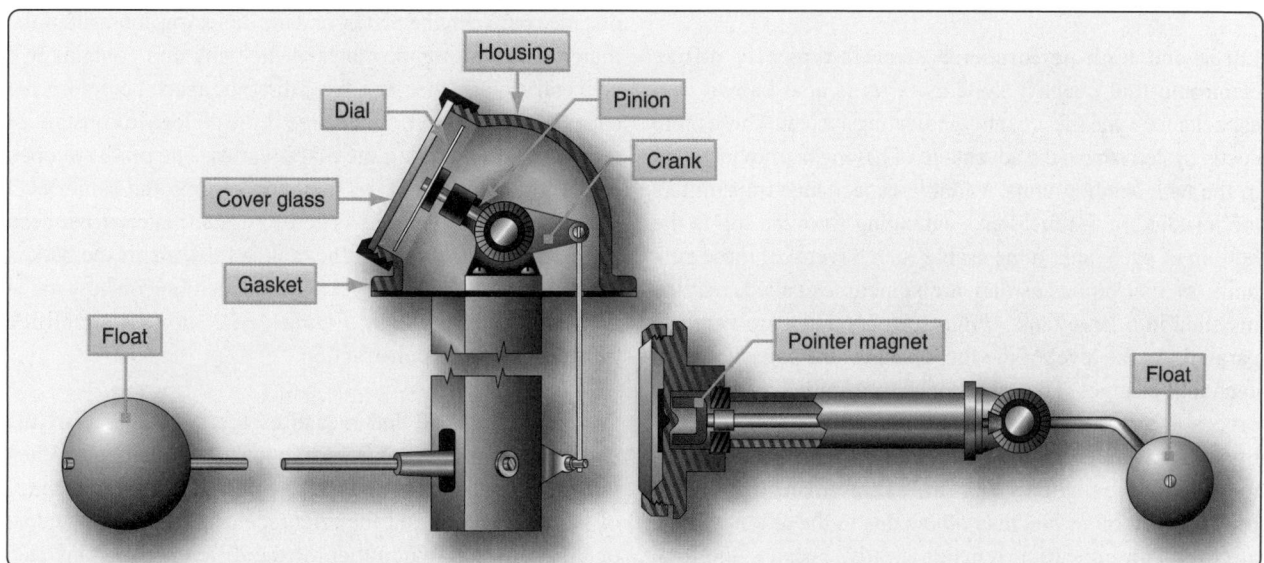

Figure 14-68. *Simple mechanical fuel indicators used on light aircraft with fuel tanks in close proximity to the pilot.*

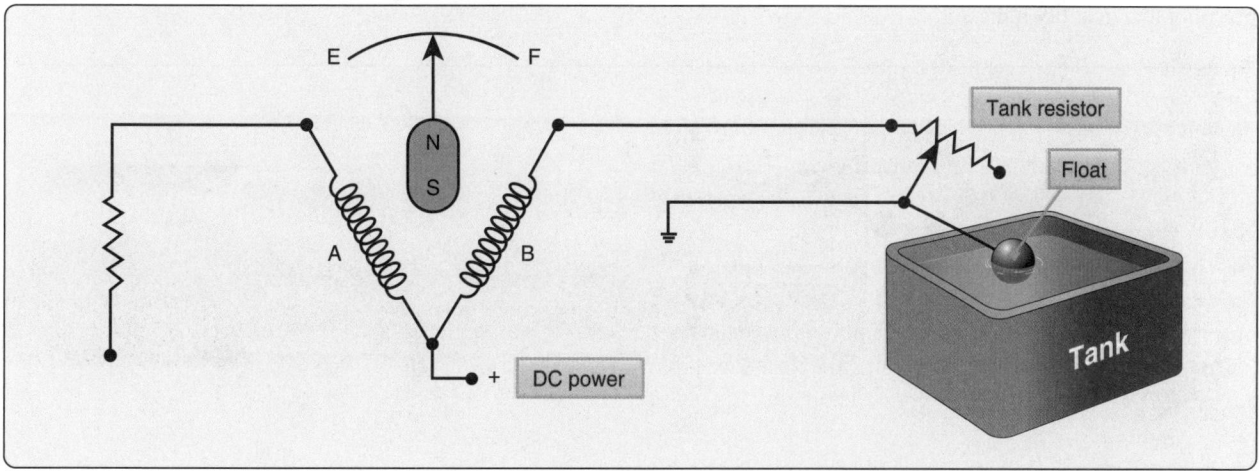

Figure 14-69. *A DC electric fuel quantity indicator uses a variable resistor in the tank unit, which is moved by a float arm.*

Figure 14-70. *Digital fuel quantity gauges that work off of variable resistance from the tank unit are shown in A and B. The fuel quantity indication of a Garmin G-1000 flat screen display is shown in C.*

convert the variable resistance into a digital signal to be processed in a computer and displayed on a flat screen panel.

Large and high-performance aircraft typically utilize electronic fuel quantity systems. This is also known as a capacitance-type fuel quantity indicating system. These more costly systems have the advantage of having no moving parts in the tank sending units. Variable capacitance transmitters are installed in the fuel tanks extending from the top to the bottom of each tank in the usable fuel. Several of these tank units, or fuel probes as they are sometimes called, may be installed in a large tank. *[Figure 14-71]* They are wired in parallel. As the level of the fuel changes, the capacitance of each unit changes. The capacitance transmitted by all of the probes in a tank is totaled and compared in a bridge circuit by a microchip computer in the tank's digital fuel quantity indicator in the flight deck. As the aircraft maneuvers, some probes are in more fuel than others due to the attitude of the aircraft. The indication remains steady, because the total capacitance transmitted by all of the probes remains the same. A trimmer is used to match the capacitance output with the precalibrated quantity indicator.

A capacitor is a device that stores electricity. The amount it can store depends on three factors: the area of its plates, the distance between the plates, and the dielectric constant of the material separating the plates. A fuel tank unit contains two concentric plates that are a fixed distance apart. Therefore, the capacitance of a unit can change if the dielectric constant of the material separating the plates varies. The units are open at the top and bottom, so they can assume the same level of fuel as is in the tanks. Therefore, the material between the plates is either fuel (if the tank is full), air (if the tank is empty), or some ratio of fuel and air depending on how much fuel remains in the tank. *Figure 14-72* shows a simplified illustration of this construction.

The bridge circuit that measures the capacitance of the tank units uses a reference capacitor for comparison. When voltage is induced into the bridge, the capacitive reactance of the tank probes and the reference capacitor can be equal or different. The magnitude of the difference is translated into an indication of the fuel quantity in the tank calibrated in pounds. Some European aircraft display this in kilograms

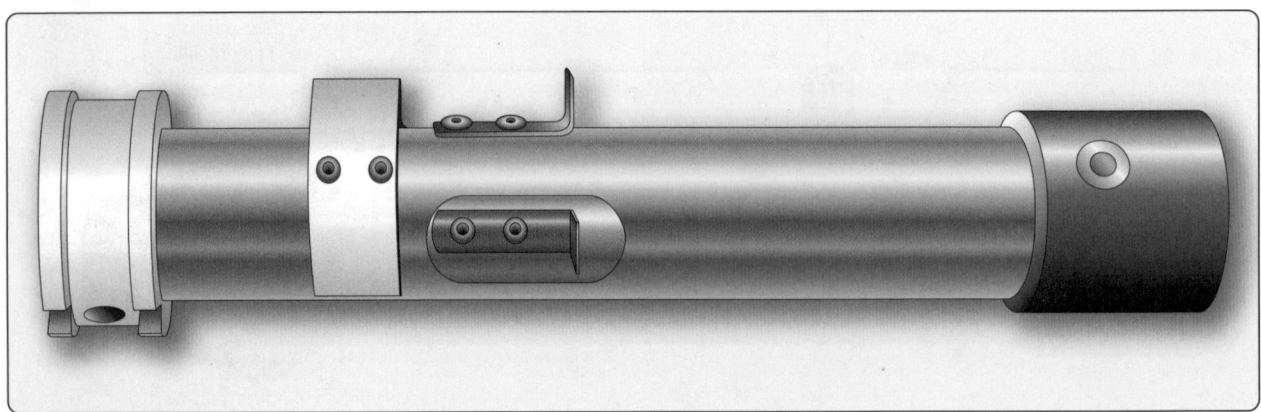

Figure 14-71. *A fuel tank transmitter for a capacitance-type fuel quantity indicating system.*

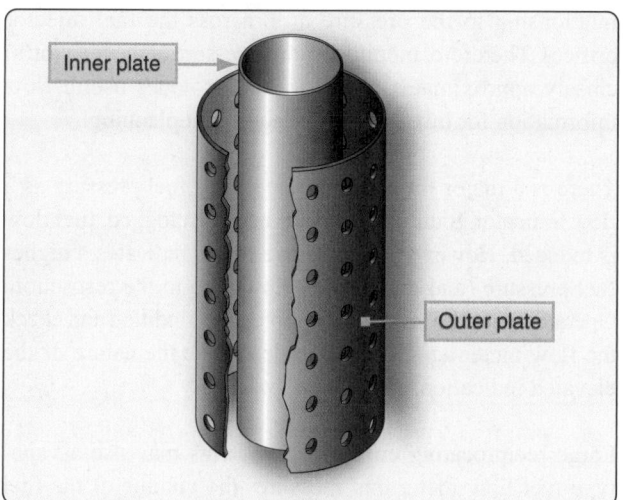

Figure 14-72. *The capacitance of tank probes varies in a capacitance-type fuel tank indicator system as the space between the inner and outer plates is filled with varying quantities of fuel and air depending on the amount of fuel in the tank.*

(kg). *Figure 14-73* represents the nature of this comparison bridge circuit.

The use of tank unit capacitors, a reference capacitor, and a microchip bridge circuit in the fuel quantity indicators is complicated by the fact that temperature affects the dielectric constant of the fuel. A compensator unit (mounted low in the tank so it is always covered with fuel) is wired into the bridge circuit. It modifies current flow to reflect temperature variations of the fuel, which affect fuel density and thus capacitance of the tank units. *[Figure 14-74]* An amplifier is also needed in older systems. The amplitude of the electric signals must be increased to move the servo motor in the analog indicator. Additionally, the dielectric constant of different turbine-engine fuels approved for a particular aircraft may also vary. Calibration is required to overcome this.

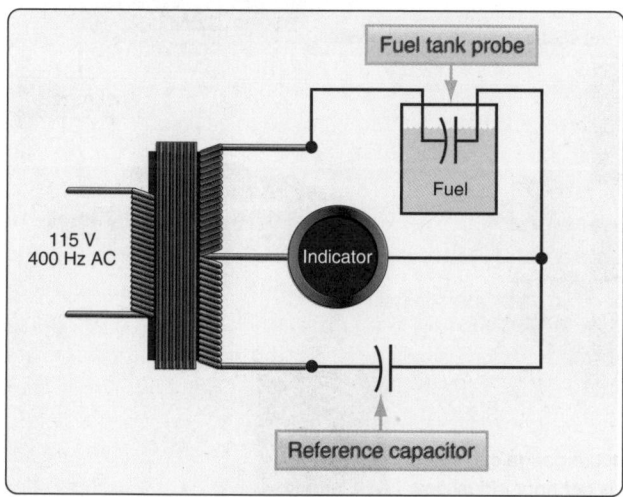

Figure 14-73. *A simplified capacitance bridge for a fuel quantity system.*

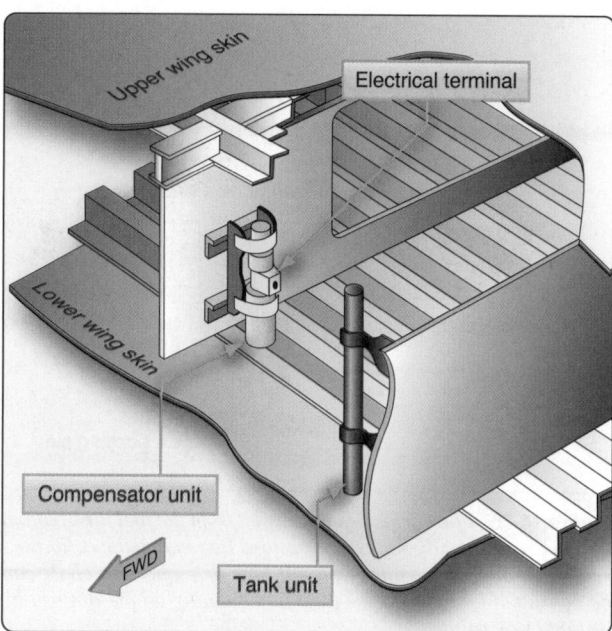

Figure 14-74. *A fuel quantity tank unit and compensator unit installed inside a wing tank.*

A fuel totalizer is part of the capacitance-type fuel quantity indication system. It is used to add the tank quantities from all indicators. This total aircraft fuel quantity can be used by the crew and by flight management computers for calculating optimum airspeed and engine performance limits for climb, cruise, descent, etc. Capacitance-type fuel quantity system test units are available for troubleshooting and ensuring proper functioning and calibration of the indicating system components.

Many aircraft with capacitance-type fuel indicating systems also use a mechanical indication system to cross-check fuel quantity indications and to ascertain the amount of fuel onboard the aircraft when electrical power is not available. A handful of fuel measuring sticks, or drip sticks, are mounted throughout each tank. When pushed and rotated, the drip stick can be lowered until fuel begins to exit the hole on the bottom of each stick. This is the point at which the top of the stick is equal to the height of the fuel. The sticks have a calibrated scale on them. By adding the indications of all of the drip sticks and converting to pounds or gallons via a chart supplied by the manufacturer, the quantity of the fuel in the tank can be ascertained. *[Figure 14-75]*

Fuel Flow Meters

A fuel flow meter indicates an engine's fuel use in real time. This can be useful to the pilot for ascertaining engine performance and for flight planning calculations. The types of fuel flow meter used on an aircraft depends primarily on the powerplant being used and the associated fuel system.

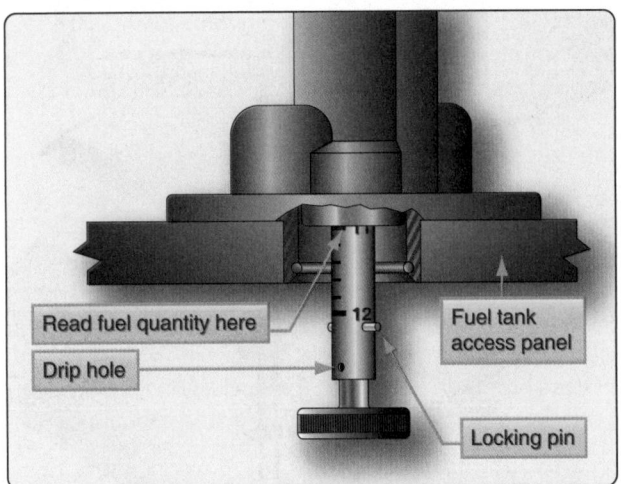

Figure 14-75. *A fuel drip stick is lowered from the fuel tank bottom until fuel drips out the hole at the bottom. By reading the calibrated scale and adding readings from all tank drip sticks, a chart can be consulted to arrive at the total fuel quantity on the aircraft by weight or by volume.*

Measuring fuel flow accurately is complicated by the fact that the fuel mass changes with temperature or with the type of fuel used in turbine engines. In light aircraft with reciprocating engines, systems have been devised to measure fuel volume. The actual mass of fuel flowing to the engine is based on an assumption of the average weight of the fuel per unit volume.

The simplest fuel flow sensing device is used in conjunction with fuel injection systems installed on horizontally opposed reciprocating engines. A pressure gauge is used but it is calibrated in gallons per hour or pounds per hour. The amount of fuel that is flowing through the fuel injectors has a direct

relationship to the pressure drop across the fuel injector orifices. Therefore, monitoring fuel pressure at the injector(s) closely approximates fuel flow and provides useful flow information for mixture control and flight planning.

There is a major limitation to the use of fuel pressure as a flow indicator. Should an injector become clogged, fuel flow is reduced. However, the pressure gauge indicates a higher fuel pressure (and greater fuel flow) due to the restriction. Operators must be aware of this potential condition and check the flow meter against EGT to determine the nature of the elevated indication. *[Figure 14-76]*

Large reciprocating engine fuel systems may use a vane-type fuel flow meter that measures the volume of the fuel consumed by the engine. The fuel flow unit is typically located between the engine-driven fuel pump and the carburetor. The entire volume of fuel delivered to the engine is made to pass through the flow meter. Inside, the fuel pushes against the vane, which counters the force of the fuel flow with a calibrated spring. The vane shaft rotates varying degrees matching the fuel flow rate through the unit. An autosyn transmitter deflects the pointer on the flight deck fuel flow gauge the same amount as the vane deflects. The dial face of the indicator is calibrated in gallons per hour or pounds per hour based on an average weight of fuel.

Since fuel fed to the engine must pass through the flow meter unit, a relief valve is incorporated to bypass the fuel around the vane should it malfunction and restrict normal fuel flow. The vane chamber is eccentric. As more fuel pushes against the vane, it rotates further around in the chamber. The volume of the chamber gradually increases to

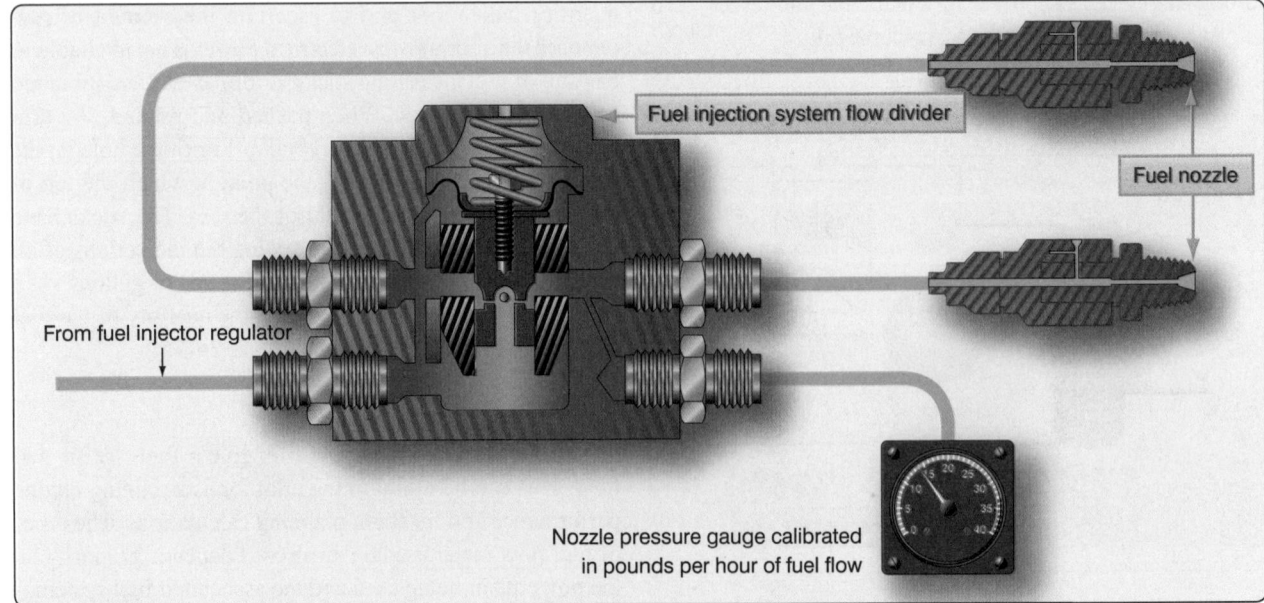

Figure 14-76. *The pressure drop across the fuel injector nozzles is used to represent fuel flow in light reciprocating-engine aircraft.*

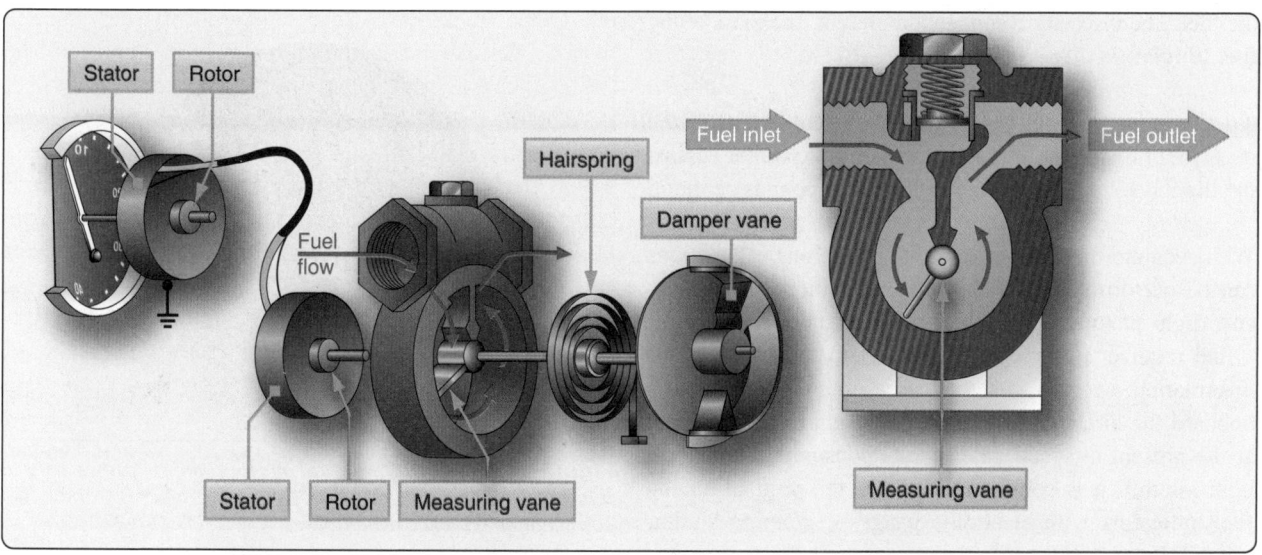

Figure 14-77. *A vane-type fuel flow meter. Greater flow volume increases deflection of the vane against a calibrated spring. An autosyn transmitter replicates the vane shaft rotation on the flight deck indicator that is calibrated in gallons or pounds of fuel flow per hour.*

permit the greater flow of fuel without restriction or pressure buildup. [Figure 14-77]

Turbine-engine aircraft experience the greatest range of fuel density from temperature variation and fuel composition. An elaborate fuel flow device is used on these aircraft. It measures fuel mass for accurate fuel flow indication in the flight deck. The mass flow indicator takes advantage of the direct relationship between fuel mass and viscosity. Fuel is swirled by a cylindrical impeller that rotates at a fixed speed. The outflow deflects a turbine just downstream of the impeller. The turbine is held with calibrated springs. Since the impeller motor swirls, the fuel at a fixed rate, any variation of the turbine deflection is caused by the volume and viscosity of

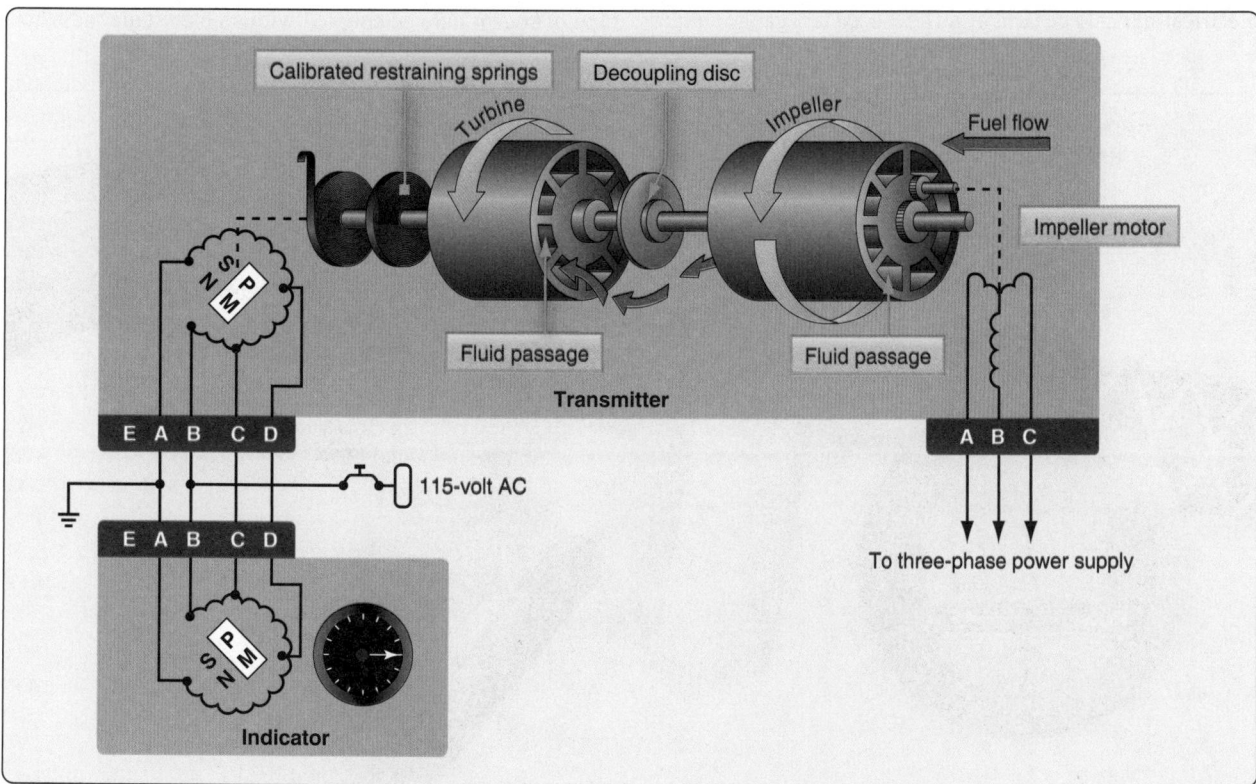

Figure 14-78. *A mass flow fuel flow indicating system used on turbine-engine aircraft uses the direct relationship between viscosity and mass to display fuel flow in pounds per hour.*

the fuel. The viscosity component represents the mass of the fuel. *[Figure 14-78]*

An alternating current (AC) synchro system is part of the mass fuel flow meter. It is used to position a pointer against the flight deck indicator scale calibrated in pounds per hour.

With accurate fuel flow knowledge, numerous calculations can be performed to aid the pilot's situational awareness and flight planning. Most high-performance aircraft have a fuel totalizer that electronically calculates and displays information, such as total fuel used, total fuel remaining onboard the aircraft, total range and flight time remaining at the present airspeed, rate of fuel consumption, etc. On light aircraft, it is common to replace the original analog fuel indicators with electronic gauges containing similar capabilities and built-in logic. Some of these fuel computers, as they are called, integrate global positioning satellite (GPS) location information. *[Figure 14-79]* Aircraft with fully digital flight decks process fuel flow data in computers and display a wide array of fuel flow related information on demand.

Relatively new types of fuel flow sensors/transmitters are available in new aircraft and for retrofit to older aircraft. One type of device found in home-built and experimental aircraft uses a turbine that rotates in the fuel flow. The higher the flow rate is, the faster the turbine rotates. A Hall effect transducer is used to convert the speed of the turbine to an electrical signal to be used by an advanced fuel gauge similar

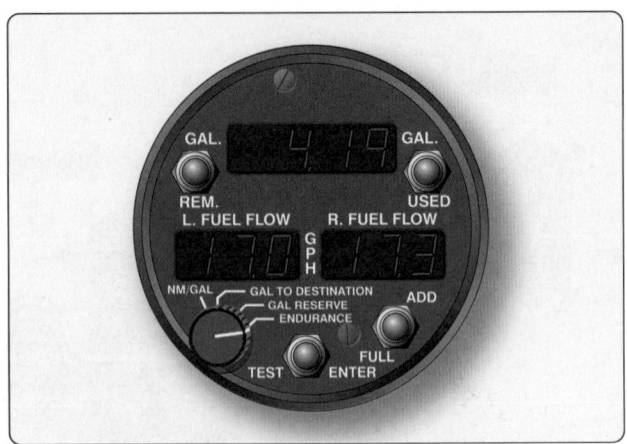

Figure 14-79. *A modern fuel management gauge uses a microprocessor to display fuel flow and numerous other fuel consumption related calculations.*

to a fuel computer to produce a variety of calculated readouts and warnings. The turbine in this unit is in line with the fuel flow but is fail-safe to allow adequate fuel flow without interruption should the unit malfunction. *[Figure 14-80]*

Another fuel flow sensor used primarily on light aircraft also detects the spinning velocity of a turbine in the fuel path. It too has a fail-safe design should the turbine malfunction. In this unit, notches in the rotor interrupt an infrared light beam between an LED and phototransistor that creates a signal proportional to the amount fuel flow. *[Figure 14-81]* This type of sensor may be coupled with an electronic indicator.

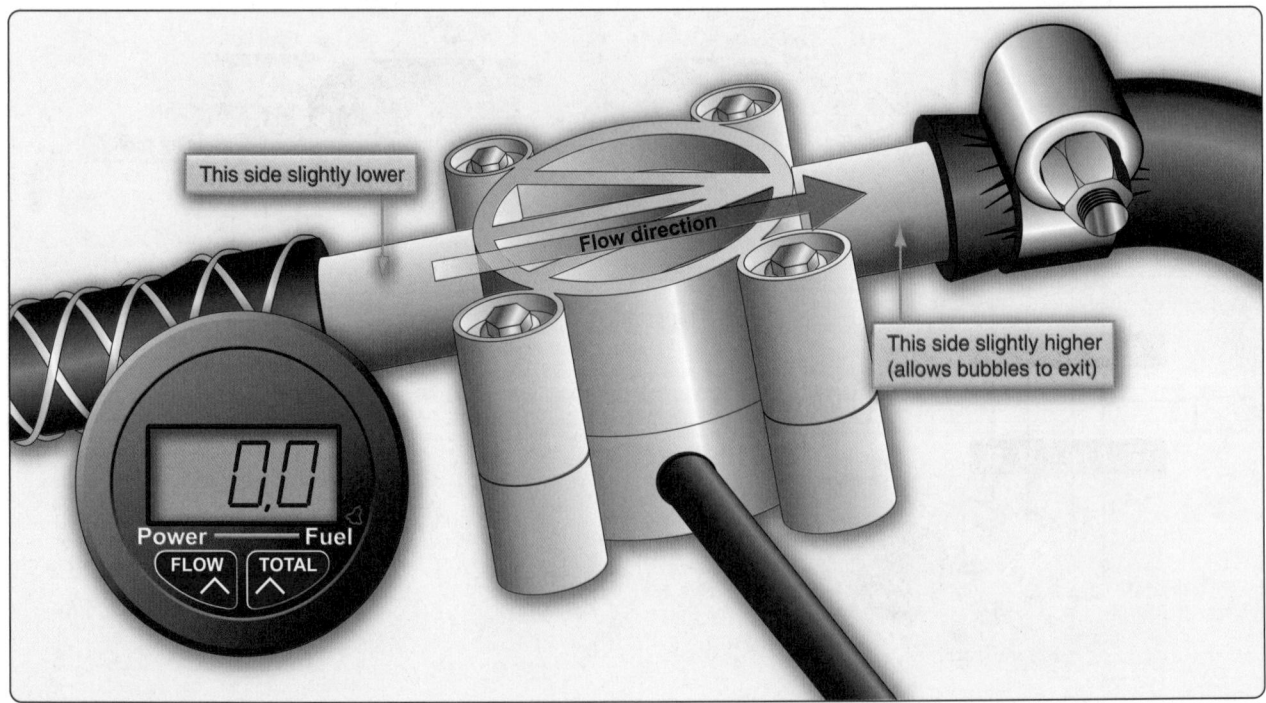

Figure 14-80. *A transducer and microprocessor for control functions are located in the base of this turbine fuel flow sensor. The gauge is menu driven with numerous display options.*

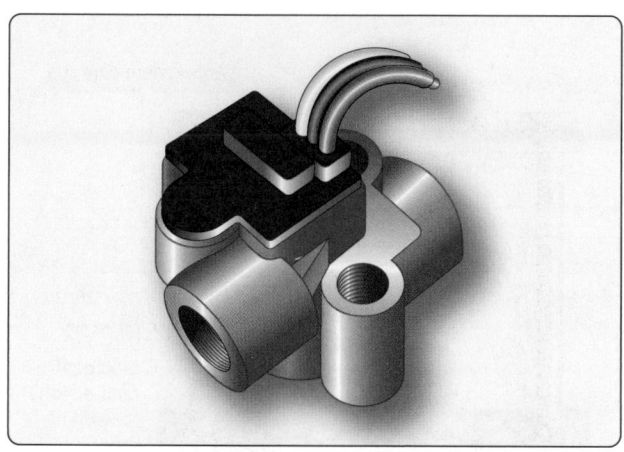

Figure 14-81. *A turbine flow transducer in this fuel flow sensor produces a current pulse signal from an opto-electronic pickup with a preamplifier.*

Increasing use of microprocessors and computers on aircraft enable the integration of fuel temperature and other compensating factors to produce highly accurate fuel flow information. Fuel flow sensing with digital output facilitates this with a high degree of reliability. Thermal dispersion technology provides flow sensing with no moving parts and digital output signals. The sensor consists of two resistance temperature detectors (RTDs). One is a reference RTD that measures the temperature of the fuel. The other is the active RTD. It is heated by an adjacent element to a temperature higher than the fuel. As the fuel flows, the active element cools proportionally to the fuel flow. The temperature difference between the two RTDs is highest at no flow.

The RTDs are connected to an electronic assembly that supplies power to the heater and uses sensing circuitry and a microprocessor to control a constant temperature difference between the heated and unheated RTDs. The electrical current to the heater is proportional to the mass flow of the fuel. As mentioned, the reference RTD is used as a temperature sensor to provide a temperature output and allow for temperature compensation of the flow measurement. *[Figure 14-82]*

Fuel Temperature Gauges

As previously mentioned, monitoring fuel temperature can inform the pilot when fuel temperature approaches that which could cause ice to form in the fuel system, especially at the fuel filter. Many large and high-performance turbine aircraft use a resistance type electric fuel temperature sender in a main fuel tank for this purpose. It can display on a traditional ratiometer gauge *[Figure 14-65]* or can be input into a computer for processing and digital display. A low fuel temperature can be corrected with the use of a fuel heater if the aircraft is so equipped. Also as mentioned, fuel temperature can be integrated into fuel flow processing calculations. Viscosity differences at varying fuel

Figure 14-82. *Fuel flow sensing units using thermal dispersion technology have no moving parts and output digital signals.*

temperatures that affect fuel flow sensing accuracy can be corrected via microprocessors and computers.

Fuel Pressure Gauges

Monitoring fuel pressure can give the pilot early warning of a fuel system related malfunction. Verification that the fuel system is delivering fuel to the fuel metering device can be critical. Simple light reciprocating-engine aircraft typically utilize a direct reading Bourdon tube pressure gauge. It is connected into the fuel inlet of the fuel metering device with a line extending to the back of the gauge in the flight deck instrument panel. A more complex aircraft may have a sensor with a transducer located at the fuel inlet to the metering device that sends electrical signals to a flight deck gauge. *[Figure 14-83]* In aircraft equipped with an auxiliary pump for starting and to back up the engine-driven pump, the fuel pressure gauge indicates the auxiliary pump pressure until the engine is started. When the auxiliary pump is switched off, the gauge indicates the pressure developed by the engine-driven pump.

More complex and larger reciprocating engine aircraft may use a differential fuel pressure gauge. It compares fuel inlet pressure to the air inlet pressure at the fuel metering device. A

Figure 14-83. *A typical fuel gauge that uses a signal from a sensing transducer to display fuel inlet pressure at the metering device.*

Figure 14-84. *A differential fuel pressure gauge used on complex and high-performance reciprocating-engine aircraft compares the fuel inlet pressure to the air inlet pressure at the fuel metering device.*

bellows type pressure gauge is normally used. *[Figure 14-84]*

Modern aircraft may use a variety of sensors including solid state types and those with digital output signals or signals that are converted to digital output. These can be processed in the instrument gauge microprocessor, if so equipped, or in a computer and sent to the display unit. *[Figure 14-85]*

Pressure Warning Signal

On aircraft of any size, visual and audible warning devices are used in conjunction with gauge indications to draw the pilot's attention to certain conditions. Fuel pressure is an important parameter that merits the use of a warning signal when it falls outside of the normal operating range. Low fuel pressure warning lights can be illuminated through the use of simple pressure sensing switches. *[Figure 14-86]* The contacts of the switch will close when fuel pressure against the diaphragm is insufficient to hold them open. This allows current to flow to the annunciator or warning light in the flight deck.

Figure 14-85. *An electronic display of fuel parameters, including fuel pressure.*

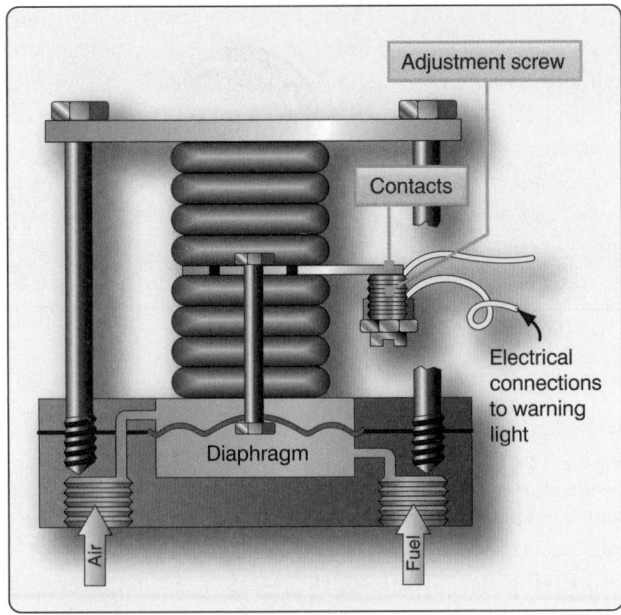

Figure 14-86. *A fuel pressure warning signal is controlled by a switch that closes when fuel pressure is low.*

Most turbine-powered aircraft utilize a low-pressure warning switch at the outlet of each fuel boost pump. The annunciator for each is typically positioned adjacent to the boost pump ON/OFF switch on the fuel panel in the flight deck. *[Figure 14-87]*

Valve-In-Transit Indicator Lights

Aircraft with multiple fuel tanks use valves and pumps to move fuel and to have it flow to desired locations, such as the engines, a certain tank, or overboard during fuel jettison. The functioning of the valves in the fuel system is critical.

Figure 14-87. *A transport category aircraft fuel panel with low pressure warning lights for each fuel boost pump.*

Some aircraft indicate to the crew when the valve is opening or closing with the use of valve-in-transit lights. Contacts in the valve control the lights that go out when the valve is fully open or when it is fully closed. Alternately, annunciator lights that show the valve position as OPEN or CLOSED are also used. Valve-in-transit and valve position indicators, or lights, are located on the fuel panel in the flight deck adjacent to the valve ON/OFF switches. *[Figure 14-88]* Sometimes the switch mechanism has the annunciator light built into it. Digital display systems graphically depict valve positions on screen.

Fuel System Repair

The integrity of an aircraft fuel system is critical and should not be compromised. Any evidence of malfunction or leak should be addressed before the aircraft is released for flight. The danger of fire, explosion, or fuel starvation in flight makes it imperative that fuel system irregularities be given top priority. Each manufacturer's maintenance and operation instructions must be used to guide the technician in maintaining the fuel system in airworthy condition. Follow the manufacturer's instructions at all times. Component

manufacturers and STC holder instructions should be used when applicable. Some general instructions for fuel system maintenance and repair are given in the following sections.

Troubleshooting the Fuel System

Knowledge of the fuel system and how it operates is essential when troubleshooting. Manufacturers produce diagrams and descriptions in their maintenance manuals to aid the technician. Study these for insight. Many manuals have troubleshooting charts or flow diagrams that can be followed. As with all troubleshooting, a logical sequence of steps to narrow the problem to a specific component or location should be followed. Defects within the system can often be located by tracing the fuel flow from the tank through the system to the engine. Each component must be functioning as designed and the cause of the defect symptom must be ruled out sequentially.

Location of Leaks & Defects

Close visual inspection is required whenever a leak or defect is suspected in a fuel system. Leaks can often be traced to the connection point of two fuel lines or a fuel line and a component. Occasionally, the component itself may have an internal leak. Fuel leaks also occur in fuel tanks and are discussed below. Leaking fuel produces a mark where it travels. It can also cause a stronger than normal odor. Gasoline may collect enough of its dye for it to be visible or an area clean of dirt may form. Jet fuel is difficult to detect at first, but it has a slow evaporation rate. Dirt and dust eventually settle into it, which makes it more visible.

When fuel leaks into an area where the vapors can collect, the leak must be repaired before flight due to the potential for fire or explosion. Repair could be deferred for external leaks that are not in danger of being ignited. However, the source of the leak should be determined and monitored to ensure it does not become worse. Follow the aircraft manufacturer's instructions on the repair of fuel leaks and the requirements that need to be met for airworthiness. Detailed visual inspection can often reveal a defect.

Fuel Leak Classification

Four basic classifications are used to describe aircraft fuel leaks: stain, seep, heavy seep, and running leak. *[Figure 14-89]* In 30 minutes, the surface area of the collected fuel from a leak is a certain size. This is used as the classification standard. When the area is less than ¾ inch in diameter, the leak is said to be a stain. From ¾ to 1½ inches in diameter, the leak is classified as a seep. Heavy seeps form an area from 1½ inches to 4 inches in diameter. Running leaks pool and actually drip from the aircraft. They may follow the contour of the aircraft for a long distance.

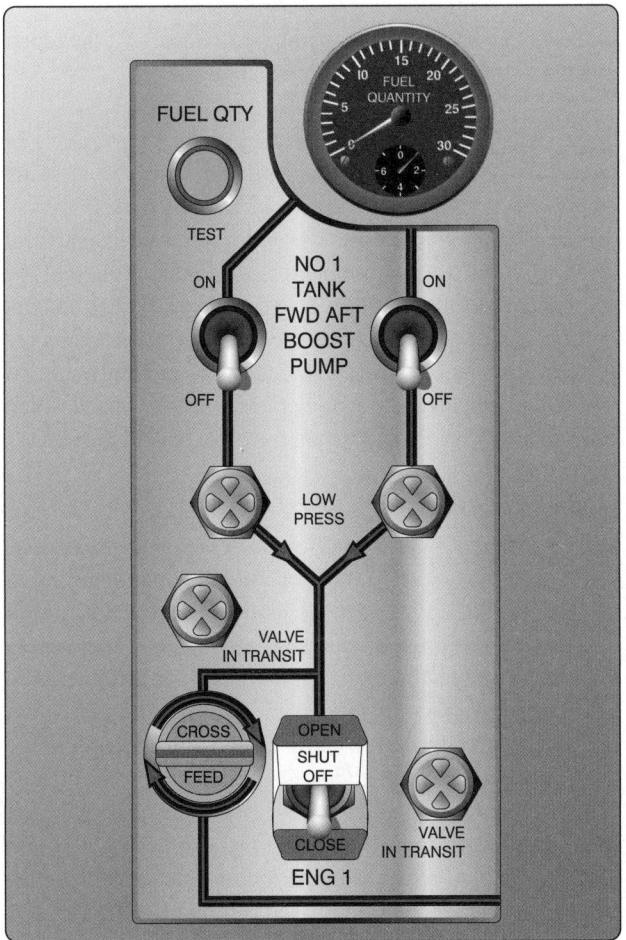

Figure 14-88. *Valve-in-transit lights are used on this section of a transport category aircraft fuel panel. Low boost pump pressure lights that look the same are also on the panel.*

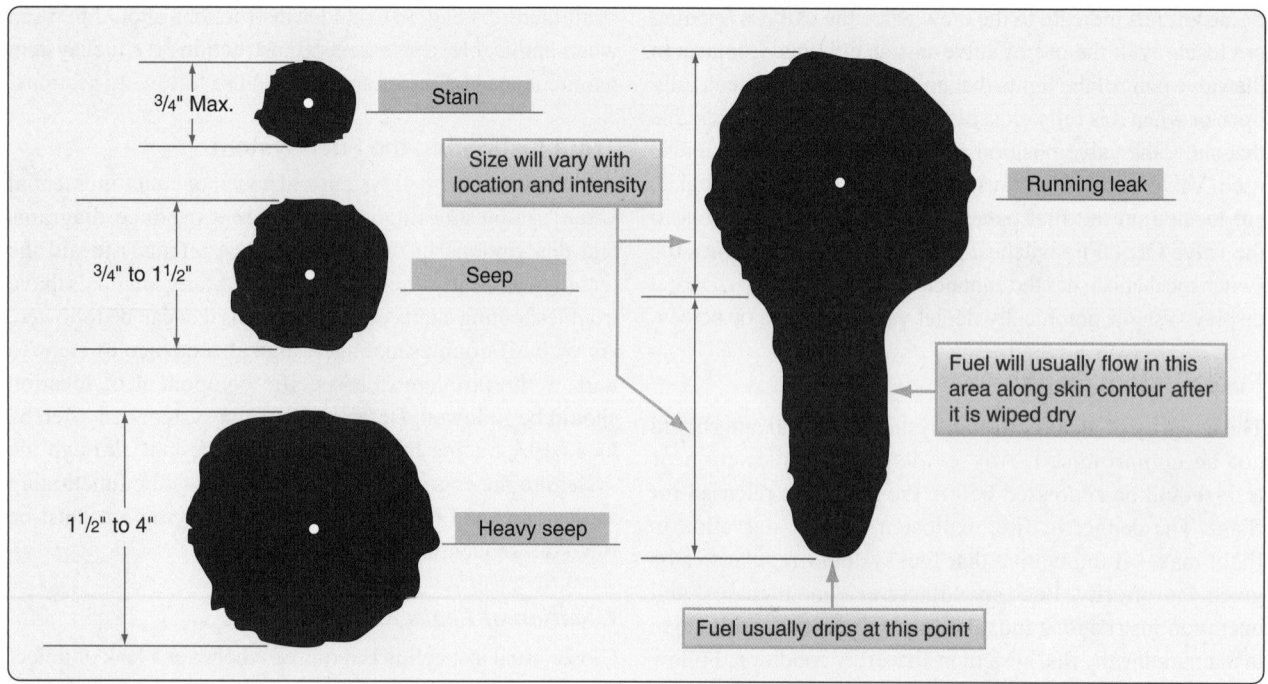

Figure 14-89. *The surface area of collected fuel from a leak is used to classify the leak into the categories shown.*

Replacement of Gaskets, Seals, & Packings

A leak can often be repaired by replacing a gasket or seal. When this occurs, or a component is replaced or reassembled after a maintenance operation, a new gasket, seal, or packing must be installed. Do not use the old one(s). Always be sure to use the correct replacement as identified by part number. Also, most gaskets, seals, and packings have a limited shelf life. They should be used only if they are within the service life stamped on the package.

Remove the entire old gasket completely and clean all mating surfaces. Clean surfaces and grooves allow a tight seal. Inspect new gaskets and seals for any flaws. Follow the manufacturer's instructions for replacement, including cleaning procedures and any sealing compound that you may need to apply during replacement. Torque assembly bolts evenly so as to provide even pressure and prevent pinching.

Fuel Tank Repair

Whether rigid removable, bladder-type, or integral, all fuel tanks have the potential to develop leaks. Repair a tank according to the manufacturer's instructions. Some general notes for repair of each tank type follow. At the time a tank is repaired, a thorough inspection should be made. Corrosion, such as that caused by water and microbes, should be identified and treated at this time, even if it is not the cause of the leak.

Rigid removable fuel tanks can be riveted, welded, or soldered together. A leak can develop at any of these types of seams or can be elsewhere on the tank. Generally, the repair must match the construction in technique.

Some metal fuel tanks experiencing minor seepage can be repaired with a sloshing procedure. An approved sloshing compound is poured into the tank, and the tank is moved so that the compound coats the entire inner surface area of the tank. Any excess compound is then poured out and the compound in the tank is allowed to cure for a specified amount of time. Minor gaps in the seams of the tank and repairs are filled in this manner. The compound is fuel resistant once dry. Check with the aircraft manufacturer to ensure that sloshing is an airworthy repair for the aircraft fuel tank in question.

Welded Tanks

Welded tank repairs are usually done by welding. These tanks can be constructed from steel or weldable aluminum, such as 3003S or 5052SO. The tank is removed from the aircraft for the repair. It must be treated to remove any fuel vapors that remain in the tank before it is welded. This is critical to avoid serious injury from explosion should the fuel vapor ignite. The manufacturer usually gives a procedure for doing this. Some common methods for purging the tank include steam cleaning, hot water purging, and inert gas purging. Most procedures involve running the steam, water, or gas through the tank for a stated period of time. Adapters may need to be fashioned or purchased for the fill port to enable proper cleaning. Follow the manufacturer's procedure for the proper time to keep the cleaning medium in the tank and for prepping the tank for welding in general.

After a seam or a damaged area is welded, you must clean the tank of any flux or debris that may have fallen into the tank. Water rinsing and acid solutions are commonly used. A leak check to ensure the repair is sound follows a welded repair. This can be done by pressurizing the tank with a specified amount of air pressure and using a soapy solution on all seams and the repaired area. Bubbles form should air escape. The amount of air pressure used for a leak check is very low. One half to 3.5 psi is common. Use an accurate regulator and pressure gauge to prevent overpressurization that could deform or otherwise damage the tank. Tanks ordinarily supported by aircraft structure when installed should be similarly supported or reinstalled in the airframe before pressurization. *Figure 14-90* shows an aircraft fuel tank being welded and the repaired tank installed in the frame of an antique aircraft.

Riveted Tanks

Riveted tanks are often repaired by riveting. The seams and rivets are coated with a fuel resistant compound when assembled to create a leak-free container. This practice is followed during a patch repair, or when repairing a seam, which may require replacing the rivets in the seam. Some minor leak repairs may only require the application of addition compound. Follow manufacturer's instructions. The compound used may be heat sensitive and require inert gas purging to prevent degradation from hot water or steam purging. Again, follow all manufacturer guidance to insure a safe airworthy repair.

Soldered Tanks

Terneplate aircraft fuel tanks that are assembled by soldering are also repaired by soldering. All patches have a minimum amount that must overlap the damaged area. Flux used in soldering must be removed from the tank after the repair with techniques similar to that used on a welded tank. Follow manufacturer's instructions.

Bladder Tanks

Bladder fuel tanks that develop leaks can also be repaired. Most commonly, they are patched using patch material, adhesive, and methods approved by the manufacturer. As with soldered tanks, the patch has a required overlap of the damaged area. Damage that penetrates completely through the bladder is repaired with an external, as well as internal, patch.

Synthetic bladder tanks have a limited service life. At some point, they seep fuel beyond acceptable limits and need to be replaced. Bladder tanks are usually required to remain wetted with fuel at all times to prevent drying and cracking of the bladder material. Storage of bladder tanks without fuel can be accomplished by coating the tanks with a substance to prevent drying, such as clean engine oil that can be flushed from the tank when ready to return to service. Follow all manufacturer's instructions for the care and repair of these common tanks. It is important to ensure that bladder tanks are correctly secured in place with the proper fasteners when reinstalling them in the aircraft after a repair.

Integral Tanks

Occasionally, an integral tank develops a leak at an access panel. This can often be repaired by transferring fuel to another tank so the panel can be removed and the seal replaced. Use of the proper sealing compound and bolt torque are required.

Other integral fuel tank leaks can be more challenging and time consuming to repair. They occur when the sealant used to seal the tank seams loses its integrity. To repair, fuel needs

Figure 14-90. *A rigid removable fuel tank with welded seams is repaired by welding.*

to be transferred or defueled out of the tank. You must enter large tanks on transport category aircraft. Preparing the tank for safe entry requires a series of steps outlined by the aircraft manufacturer. These include drying the tank and venting it of dangerous vapors. The tank is then tested with a combustible-gas indicator to be certain it can be entered safely. Clothing that does not cause static electricity and a respirator is worn.

An observer is stationed outside of the tank to assist the technician in the tank. *[Figure 14-91]* A continuous flow of ventilating air is made to flow through the tank. A checklist for fuel tank preparation for entry taken from a transport category maintenance manual is shown in *Figure 14-92*. The details of the procedures are also given in the manual.

Once the location of the leak is determined, the tank sealant is removed, and new sealant is applied. Remove old sealant with a nonmetallic scraper. Aluminum wool can be used to remove the final traces of the sealant. After cleaning the area with the recommended solvent, apply new sealant as instructed by the manufacturer. Observe cure time and leak checks as recommended before refilling the tank.

Fire Safety

Fuel vapor, air, and a source of ignition are the requirements for a fuel fire. Whenever working with fuel or a fuel system component, the technician must be vigilant to prevent these elements from coming together to cause a fire or explosion. A source of ignition is often the most controllable. In addition to removing all sources of ignition from the work area, care must be exercised to guard against static electricity. Static electricity can easily ignite fuel vapor, and its potential for igniting fuel vapor may not be as obvious as a flame or an operating electrical device. The action of fuel flowing through a fuel line can cause a static buildup as can many other situations in which one object moves past another. Always assess the work area and take steps to remove any potential static electricity ignition sources.

AVGAS is especially volatile. It vaporizes quickly due to its high vapor pressure and can be ignited very easily. Turbine engine fuel is less volatile but still possesses enormous capacity to ignite. This is especially true if atomized, such as when escaping out of a pressurized fuel hose or in a hot engine compartment on a warm day. Treat all fuels as potential fire hazards in all situations. As was discussed, empty fuel tanks have an extreme potential for ignition or explosion. Although the liquid fuel has been removed, ignitable fuel vapor can remain for a long period of time. Purging the vapor out of any empty fuel tank is an absolute necessity before any repair is initiated.

A fire extinguisher should be on hand during fuel system maintenance or whenever fuel is being handled. A fuel fire can be put out with a typical carbon dioxide (CO_2) fire extinguisher. Aim the extinguisher nozzle at the base of the flame and spray in a sweeping motion to have the agent fall over the flames to displace the oxygen and smother the fire. Dry chemical fire extinguishers rated for fuel can also be used. These leave behind a residue that requires cleanup that can be extensive and expensive. Do not use a water-type extinguisher. Fuel is lighter than water and could be spread without being extinguished. Additional precautions used to prevent fire are discussed below in the fueling–defueling section of this chapter.

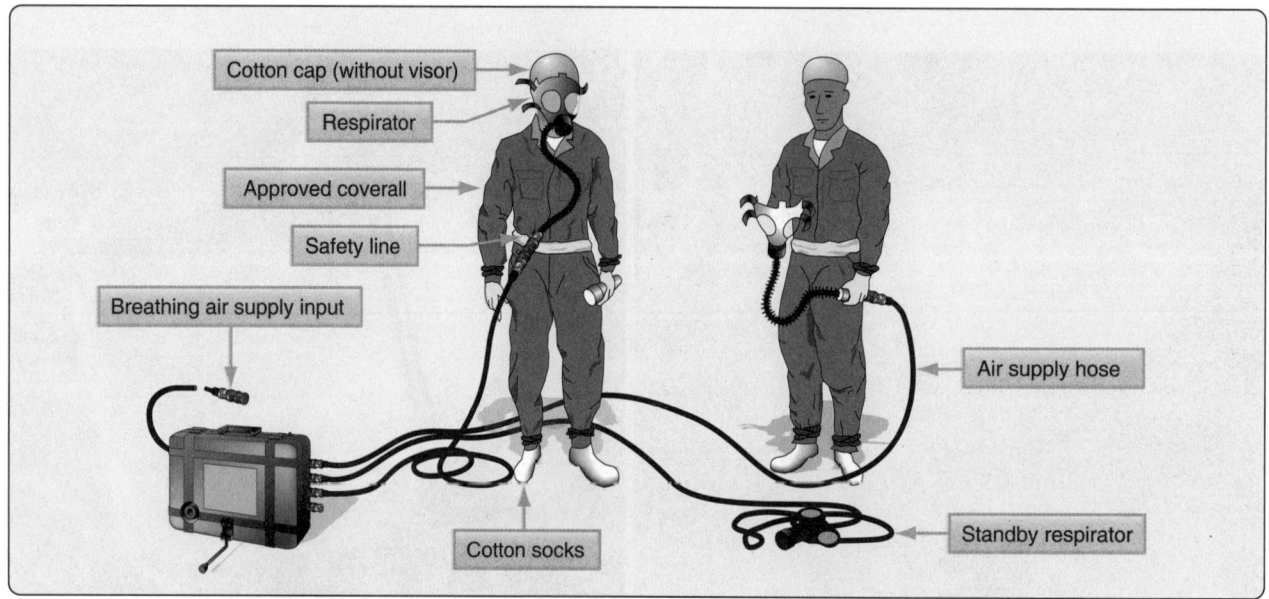

Figure 14-91. *Wear a nonstatic protective suit and respirator when entering an integral fuel tank for inspection or repair.*

This checklist must be completed prior to start of wet fuel cell entry and/or at shift change **prior** to work assignment for the continuation of tank work started by a previous shift.

Wet fuel cell entry location

Area or building: _____ Stall: _____ Airplane: _____ Tank: _____

Shift: _____ Date: _____ Supervisor: _____

1. Airplane and adjacent equipment properly grounded.
2. Area secured and warning signs positioned.
3. Boost pump switches off and circuit breakers pulled and placarded.
4. No power on airplane: battery disconnected, external power cord disconnected from airplane, and external power receptacle placarded.
5. Radio and radar equipment off (see separation distance requirements).
6. Only approved explosion-proof equipment and tools will be used for fuel cell entry (lights, blowers, pressure and test equipment, etc.).
7. Ensure requirements listed on aircraft confined space entry permit are complied with, including appropriate personal protective equipment: OSH class 110 respirator at a minimum, approved coveralls, cotton cap and foot coverings, and eye protection.
8. Trained attendant and confined space logsheet required for all wet fuel cell entries.
9. Aerators checked for cleanliness prior to use.
10. Sponges available for residual fuel mop out.
11. All plugs in use have streamers attached.
12. Mechanical ventilation (venturis or blowers) installed to ventilate all open fuel cells.

    Note: Ventilation system must remain in operation at all times while fuel cells are open. If ventilation system fails or any ill effects, such as dizziness, irriation, or excessive odors are noted, all work shall stop and fuel cells must be evacuated.

13. Shop personnel entering cells and standby observers have current "fuel cell entry" certification cards. Certification requires the following training:
    - Aircraft confined space entry safety;
    - Respirator use and maintenance; and
    - Wet fuel cell entry.
14. Fire department notified.

Meter Reading

15. Oxygen reading (%): _____ By: _____
16. Fuel vapor level reading (ppm): _____ By: _____
17. Combustible gas meter (LEL) reading: _____ By (FD): _____

I confirm that all entry requirements were met prior to any entry.

_____   _____
Signature of supervisor or designee   Date

Figure 14-92. *Fuel tank checklist entry.*

Fuel System Servicing

Maintaining aircraft fuel systems in acceptable condition to deliver clean fuel to the engine(s) is a major safety factor in aviation. Personnel handling fuel or maintaining fuel systems should be properly trained and use best practices to ensure that the fuel, or fuel system, are not the cause of an incident or accident.

Checking for Fuel System Contaminants

Continuous vigilance is required when checking aircraft fuel systems for contaminants. Daily draining of strainers and sumps is combined with periodic filter changes and inspections to ensure fuel is contaminant free. Turbine powered engines have highly refined fuel control systems through which flow hundreds of pounds of fuel per hour of operation. Sumping alone is not sufficient. Particles are suspended longer in jet fuel due to its viscosity. Engineers design a series of filters into the fuel system to trap foreign matter. Technicians must supplement these with cautious procedures and thorough visual inspections to accomplish the overall goal of delivering clean fuel to the engines.

Keeping a fuel system clean begins with an awareness of the common types of contamination. Water is the most common. Solid particles, surfactants, and microorganisms are also common. However, contamination of fuel with another fuel not intended for use on a particular aircraft is possibly the worst type of contamination.

Water

Water can be dissolved into fuel or entrained. Entrained water can be detected by a cloudy appearance to the fuel. Close examination is required. Air in the fuel tends to cause a similar cloudy condition but is near the top of the tank. The cloudiness caused by water in the fuel tends to be more towards the bottom of the tank as the water slowly settles out.

As previously discussed, water can enter a fuel system via condensation. The water vapor in the vapor space above the liquid fuel in a fuel tank condenses when the temperature changes. It normally sinks to the bottom of the fuel tank into the sump where it can be drained off before flight. *[Figure 14-93]* However, time is required for this to happen; therefore, you should wait a period of time after fueling before checking the fuel sumps so the water and sediment can settle to the drain point.

On some aircraft, a large amount of fuel needs to be drained before settled water reaches the drain valve. Awareness of this type of sump idiosyncrasy for a particular aircraft is important. The condition of the fuel and recent fueling practices need to be considered and are equally important. If the aircraft has been flown often and filled immediately after flight, there is little reason to suspect water contamination beyond what would be exposed during a routine sumping. An aircraft that has sat for a long period of time with partially full fuel tanks is a cause of concern.

It is possible that water is introduced into the aircraft fuel load during refueling with fuel that already contains water. Any suspected contamination from refueling or the general handling of the aircraft should be investigated. A change in fuel supplier may be required if water continues to be an issue despite efforts being made to keep the aircraft fuel tanks full and sumps drained on a regular basis. Fuel below freezing temperature may contain entrained water in ice form that may not settle into the sump until melted. Use of an anti-icing solution in turbine fuel tanks helps prevent filter blockage from water that condenses out of the fuel as ice during flight.

The fuel anti-ice additive level should be monitored so that

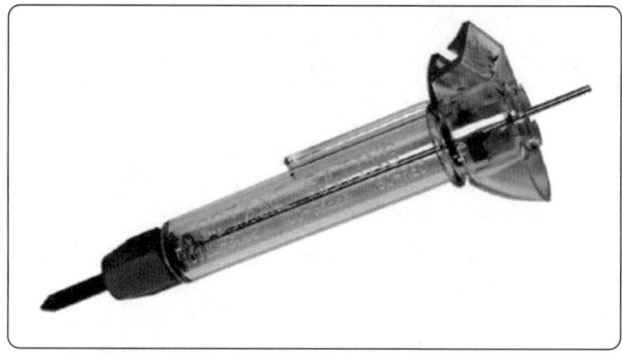

Figure 14-93. *A sump drain tool used to open and collect fuel and contaminants from the fuel system sumps. Daily sump draining is part of the procedures needed to remove water from fuel that is to be delivered to the engine(s).*

recommended quantity for the tank capacity is maintained. After repeated fueling, the level can be obscured. A field hand-held test unit can be used to check the amount of anti-ice additive already in a fuel load. *[Figure 14-94]*

Strainers and filters are designed with upward flow exits to have water collect at the bottom of the fuel bowl to be drained off. This should not be overlooked. Entrained water in small quantities that makes it to the engine usually poses no problem. Large amounts of water can disrupt engine operation. Settled water in tanks can cause corrosion. This can be magnified by microorganisms that live in the fuel/water interface. High quantities of water in the fuel can also cause discrepancies in fuel quantity probe indications.

Solid Particle Contaminants

Solid particles that do not dissolve in the fuel are common contaminants. Dirt, rust, dust, metal particles, and just about anything that can find its way into an open fuel tank is of concern. Filter elements are designed to trap these contaminants and some fall into the sump to be drained off. Pieces of debris from the inside of the fuel system may also accumulate, such as broken-off sealant, or pieces of filter elements, corrosion, etc.

Preventing solid contaminant introduction into the fuel is critical. Whenever the fuel system is open, care must be taken to keep out foreign matter. Lines should be capped immediately. Fuel tank caps should not be left open for any longer than required to refuel the tanks. Clean the area adjacent to wherever the system is opened before it is opened.

Coarse sediments are those visible to the naked eye. Should they pass beyond system filters, they can clog in fuel metering device orifices, sliding valves, and fuel nozzles. Fine sediments cannot actually be seen as individual particles. They may be detected as a haze in the fuel or they may refract light when examining the fuel. Their presence in fuel controls and metering devices is indicated by dark shellac-like marks on sliding surfaces.

The maximum amount of solid particle contamination allowable is much less in turbine engine fuel systems than in reciprocating-engine fuel systems. It is particularly important to regularly replace filter elements and investigate any unusual solid particles that collect therein. The discovery of significant metal particles in a filter could be a sign of a failing component upstream of the filter. A laboratory analysis is possible to determine the nature and possible source of solid contaminants.

Surfactants

Surfactants are liquid chemical contaminants that naturally occur in fuels. They can also be introduced during the refining or handling processes. These surface-active agents usually appear as tan to dark brown liquid when they are present in large quantities. They may even have a soapy consistency. Surfactants in small quantities are unavoidable and pose little threat to fuel system functioning. Larger quantities of surfactants do pose problems. In particular, they reduce the surface tension between water and the fuel and tend to cause water and even small particles in the fuel to remain suspended rather than settling into the sumps. Surfactants also tend to collect in filter elements making them less effective.

Surfactants are usually in the fuel when it is introduced into the aircraft. Discovery of either excessive quantities of dirt and water making their way through the system or a sudsy residue in filters and sumps may indicate their presence. The source of fuel should be investigated and avoided if found to contain a high level of these chemicals. As mentioned, slow settling rates of solids and water into sumps is a key indicator that surfactant levels are high in the fuel. Most quality fuel providers have clay filter elements on their fuel dispensing trucks and in their fixed storage and dispensing systems.

These filters, if renewed at the proper intervals, remove most surfactants through adhesion. Surfactants discovered in the aircraft systems should be traced to the fuel supply source and the use and condition of these filters. *[Figure 14-95]*

Microorganisms

The presence of microorganisms in turbine engine fuels is a critical problem. There are hundreds of varieties of these life forms that live in free water at the junction of the water and fuel in a fuel tank. They form a visible slime that is dark brown, grey, red, or black in color. This microbial growth can multiply rapidly and can cause interference with the proper functioning of filter elements and fuel quantity indicators. Moreover, the slimy water/microbe layer in contact with the fuel tank surface provides a medium for electrolytic corrosion of the tank. *[Figure 14-96]*

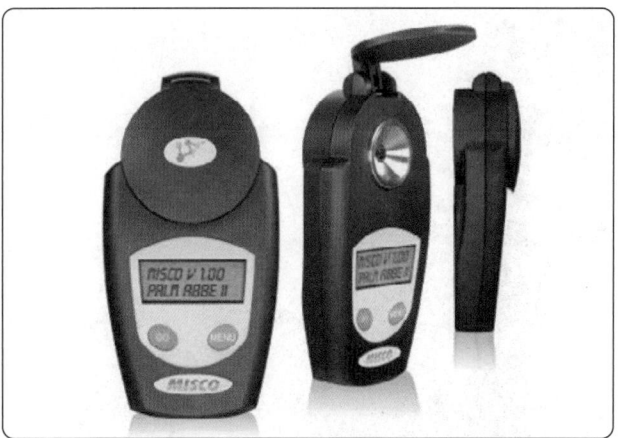

Figure 14-94. *A hand-held refractometer with digital display measures the amount of fuel anti-ice additive contained in a fuel load.*

Since the microbes live in free water and feed on fuel, the most powerful remedy for their presence is to keep water from accumulating in the fuel. Fuel 100 percent free of water is not practicable. By following best practices for sump draining and filter changes, combined with care of fuel stock tanks used to refuel aircraft, much of the potential for water to accumulate in the aircraft fuel tanks can be mitigated. The addition of biocides to the fuel when refueling also helps by killing organisms that are present.

Foreign Fuel Contamination

Aircraft engines operate effectively only with the proper fuel. Contamination of an aircraft's fuel with fuel not intended for use in that particular aircraft can have disastrous consequences. It is the responsibility of all aviators to put forth effort continuously to ensure that only the fuel designed for the operation of the aircraft's engine(s) is put into the fuel tanks. Each fuel tank receptacle or fuel cap area is clearly marked to indicate which fuel is required. *[Figure 14-97]*

If the wrong fuel is put into an aircraft, the situation must be rectified before flight. If discovered before the fuel pump is operated and an engine is started, drain all improperly filled tanks. Flush out the tanks and fuel lines with the correct fuel and then refill the tanks with the proper fuel. However, if discovered after an engine has been started or attempted to be started, the procedure is more in depth. The entire fuel system, including all fuel lines, components, metering device(s) and tanks, must be drained and flushed. If the engines have been operated, a compression test should be accomplished, and the combustion chamber and pistons should be borescope inspected. Engine oil should be drained, and all screens and filters examined for any evidence of damage. Once reassembled and the tanks have been filled with the correct fuel, a full engine run-up check should be performed before releasing the aircraft for flight.

Figure 14-95. *Clay filter elements remove surfactants. They are used in the fuel dispensing system before fuel enters the aircraft.*

Contaminated fuel caused by the introduction of small quantities of the wrong type of fuel into an aircraft may not look any different when visually inspected, making a dangerous situation more dangerous. Any person recognizing that this error has occurred must ground the aircraft. The lives of the aircraft occupants are at stake.

Detection of Contaminants

Visual inspection of fuel should always reveal a clean, bright looking liquid. Fuel should not be opaque, which could be a sign of contamination and demands further investigation. As mentioned, the technician must always be aware of the fuel's appearance, as well as when and from what sources

refueling has taken place. Any suspicion of contamination must be investigated.

In addition to the detection methods mentioned for each type of contamination above, various field and laboratory tests can be performed on aircraft fuel to expose contamination. A common field test for water contamination is performed by adding a dye that dissolves in water but not fuel to a test sample drawn from the fuel tank. The more water present in the fuel, the greater the dye disperses and colors the sample.

Another common test kit commercially available contains a grey chemical powder that changes color to pink or purple when the contents of a fuel sample contains more than 30 parts per million (ppm) of water. A 15-ppm test is available for turbine engine fuel. *[Figure 14-98]* These levels of water are considered generally unacceptable and not safe for operation of the aircraft. If levels are discovered above these amounts, time for the water to settle out of the fuel should be given or the aircraft should be defueled and refueled with acceptable fuel.

The presence and level of microorganisms in a fuel tank can also be measured with a field device. The test detects the metabolic activity of bacteria, yeast, and molds, including sulfate reducing bacteria, and other anaerobe microorganisms. This could be used to determine the amount of anti-microbial agent to be added to the fuel. The testing unit is shown in *Figure 14-99.*

Bug test kits test fuel specifically for bacteria and fungus. While other types of microorganisms may exist, this semi-quantitative test is quick and easy to perform. Treat a fuel sample with the product and match the color of the sample to the chart for an indication of the level of bacteria and fungus present. These are some of the most common types of microorganisms that grow in fuel; if growth levels of fungus and bacteria are acceptable, the fuel could be usable. *[Figure 14-100]*

Figure 14-96. *This fuel-water sample has microbial growth at the interface of the two liquids.*

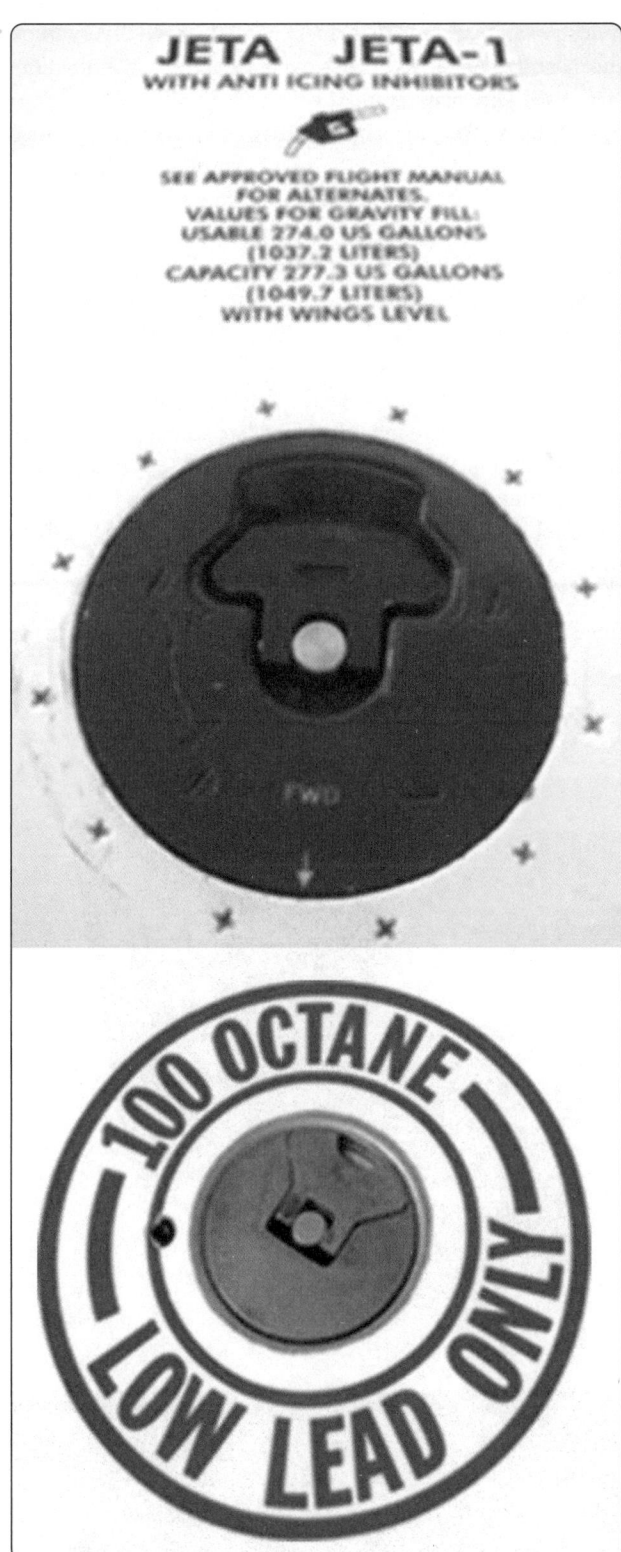

Figure 14-97. *All entry points of fuel into the aircraft are marked with the type of fuel to be used. Never introduce any other fuel into the aircraft other than that which is specified.*

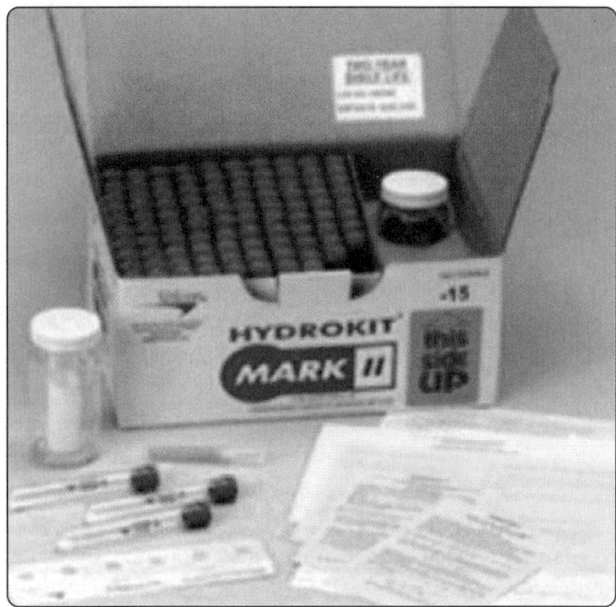

Figure 14-98. *This kit allows periodic testing for water in fuel.*

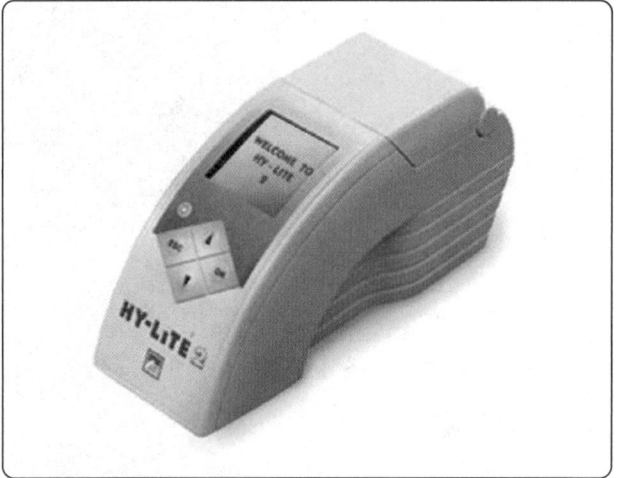

Figure 14-99. *A capture solution is put into a 1 liter sample of fuel and shaken. The solution is then put into the analyzer shown to determine the level of microorganisms in the fuel.*

Fuel trucks and fuel farms may make use of laser contaminant identification technology. All fuel exiting the storage tank going into the servicing hose is passed through the analyzer unit. Laser sensing technology determines the difference between water and solid particle contaminants. When an excessive level of either is detected, the unit automatically shuts off flow to the fueling nozzle. Thus, aircraft are fueled only with clean dry fuel. When surfactant filters are combined with contaminant identification technology and microorganism detection, chances of delivering clean fuel to the aircraft engines are good. *[Figure 14-101]*

Before various test kits were developed for use in the field by nonscientific personnel, laboratories provided complete fuel composition analysis to aviators. These services are still available. A sample is sent in a sterilized container to the lab. It can be tested for numerous factors including water, microbial growth, flash point, specific gravity, cetane index (a measure of combustibility and burning characteristics), and more. Tests for microbes involve growing cultures of whatever organisms are present in the fuel. *[Figure 14-102]*

Fuel Contamination Control

A continuous effort must be put forth by all those in the aviation industry to ensure that each aircraft is fueled only with clean fuel of the correct type. Many contaminants, both soluble and insoluble, can contaminate an aircraft's fuel supply. They can be introduced with the fuel during fueling or the contamination may occur after the fuel is onboard.

Contamination control begins long before the fuel gets pumped into an aircraft fuel tank. Many standard petroleum

industry safeguards are in place. Fuel farm and delivery truck fuel handling practices are designed to control contamination. Various filters, testing, and treatments effectively keep fuel contaminant free or remove various contaminants once discovered. However, the correct clean fuel for an aircraft should never be taken for granted. The condition of all storage tanks and fuel trucks should be monitored. All filter changes and treatments should occur regularly and on time. The fuel supplier should take pride in delivering clean, contaminant-free fuel to its customers.

Onboard aircraft fuel systems must be maintained and serviced according to manufacturer's specifications. Samples from all drains should be taken and inspected on a regular basis. Filters should be changed at the specified intervals. The fuel load should be visually inspected and tested from time to time or when there is a potential contamination issue. Particles discovered in filters should be identified and investigated if needed. Inspection of the fuel system during

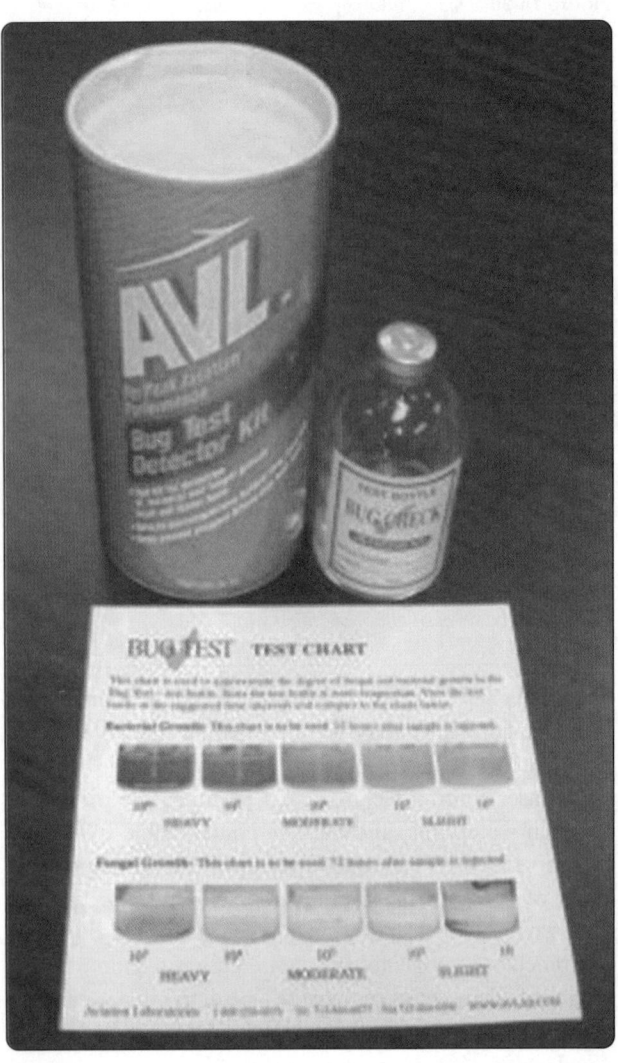

Figure 14-100. *Fuel bug test kits identify the level of bacteria and fungus present in a fuel load by comparing the color of a treated sample with a color chart.*

Figure 14-101. *This contaminant analyzer is used on fuel supply source outflow, such as that on a refueling truck. Water and solid contaminant levels are detected using laser identification technology. The valve to the fill hose is automatically closed when levels of either are elevated beyond acceptable limits.*

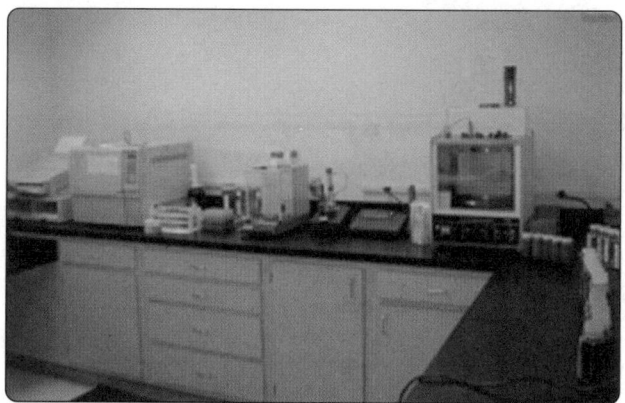

Figure 14-102. *Laboratory tests of fuel samples are available.*

periodic inspections should be treated with highest concern.

Most importantly, the choice of the correct fuel for an aircraft should never be in question. No one should ever put a fuel into an aircraft fuel tank unless absolutely certain it is the correct fuel for that aircraft and its engine(s). Personnel involved in fuel handling should be properly trained. All potential contamination situations should be investigated and remedied.

Fueling & Defueling Procedures

Maintenance technicians are often asked to fuel or defuel aircraft. Fueling procedure can vary from aircraft to aircraft. Tanks may need to be fueled in a prescribed sequence to prevent structural damage to the airframe. The proper procedure should be confirmed before fueling an unfamiliar aircraft.

Fueling

Always fuel aircraft outside, not in a hangar where fuel vapors may accumulate and increase the risk and severity of an accident. Generally, there are two types of fueling process: over-the-wing refueling and pressure refueling. Over-the-wing refueling is accomplished by opening the fuel tank cap on the upper surface of the wing or fuselage, if equipped with fuselage tanks. The fueling nozzle is carefully inserted into the fill opening and fuel is pumped into the tank. This process is similar to the process used to refuel an automobile gas tank. When finished, the cap is secured, and subsequent tanks are opened and refilled until the aircraft has the desired fuel load onboard. Pressure refueling occurs at the bottom, front, or rear of the fuel tank. A pressure refueling nozzle locks onto the fueling port at the aircraft fueling station. Fuel is pumped into the aircraft through this secured and sealed connection. Gauges are monitored to ascertain when the tanks are properly loaded. An automatic shutoff system may be part of the aircraft system. It closes the fueling valve when the tanks are full. *[Figure 14-103]*

Precautions should be used with either type of fueling. First and foremost, it is absolutely essential that the correct fuel be put in the aircraft. The type of fuel to be used is placarded near the fill port on over-the-wing systems and at the fueling station on pressure refueled aircraft. If there is any question about which fuel to use, the pilot in command, other knowledgeable personnel, or the manufacturer's maintenance/operations manual should be consulted before proceeding. An over-the-wing refueling nozzle for turbine engine fuel should be too large to fit into the fill opening on an aircraft utilizing gasoline.

Clean the area adjacent to the fill port when refueling over the wing. Ensure the fuel nozzle is also clean. Aviation fuel nozzles are equipped with static bonding wires that must be attached to the aircraft before the fuel cap is opened. *[Figure 14-104]* Open the cap only when ready to dispense the fuel. Insert the nozzle into the opening with care. The aircraft structure is much more delicate than the fuel nozzle, which could easily damage the aircraft. Do not insert the neck of the nozzle deeply enough to hit bottom. This could dent the tank, or the aircraft skin, if it is an integral tank. Exercise caution to avoid damage to the surface of the airframe by the heavy fuel hose. Lay the hose over your shoulder or use a refueling mat to protect the paint. *[Figure 14-105]*

When pressure refueling, the aircraft receptacle is part of a fueling valve assembly. When the fueling nozzle is properly connected and locked, a plunger unlocks the aircraft valve so fuel can be pumped through it. Normally, all tanks can be fueled from a single point. Valves in the aircraft fuel system are controlled at the fueling station to direct the fuel into the proper tank. *[Figure 14-106]* Ensure that the pressure developed by the refueling pump is correct for the aircraft before pumping fuel. Although similar, pressure fueling panels and their operation are different on different aircraft. Refueling personnel should be guided through the correct use of each panel. Do not guess at how the panel and associated valves operate.

When fueling from a fuel truck, precautions should be taken. If the truck is not in continuous service, all sumps should

Figure 14-103. *A float switch installed in a fuel tank can close the refueling valve when the tanks are full during pressure fueling of an aircraft. Other more sophisticated automatic shutoff systems exist.*

be drained before moving the truck, and the fuel should be visually inspected to be sure it is bright and clean. Turbine fuel should be allowed to settle for a few hours if the fuel truck tank has recently been filled or the truck has been jostled, such as when driven over a bumpy service road at the airport. Properly maneuver the fuel truck into position for refueling.

The aircraft should be approached slowly. The truck should be parked parallel to the wings and in front of the fuselage if possible. Avoid backing toward the aircraft. Set the parking brake and chock the wheels. Connect a static bonding cable from the truck to the aircraft. This cable is typically stored on a reel mounted on the truck.

There are other miscellaneous good practices that should be employed when refueling an aircraft. A ladder should be used if the refuel point is not accessible while standing on the ground. Climbing on an expensive aircraft to access the fueling ports is possible but does not give the stability of a ladder and may not be appreciated by the aircraft owner. If it is necessary to walk on the wings of the aircraft, do so only in designated areas, which are safe.

Filler nozzles should be treated as the important tools that they are. They should not be dropped or dragged across the apron. Most have attached dust caps that should be removed only for the actual fueling process and then immediately replaced. Nozzles should be clean to avoid contamination of the fuel. They should not leak and should be repaired at the earliest sign of leak or malfunction. Keep the fueling nozzle in constant contact with the filler neck spout when fueling. Never leave the nozzle in the fill spout unattended. When fueling is complete, always doublecheck the security of all fuel caps and ensure that bonding wires have been removed and stowed.

Figure 14-105. *Over-the-wing refueling a Cessna.*

Defueling

Removing the fuel contained in aircraft fuel tanks is sometimes required. This can occur for maintenance, inspection, or due to contamination. Occasionally, a change in flight plan may require defueling. Safety procedures for defueling are the same as those for fueling. Always defuel outside. Fire extinguishers should be on hand. Bonding cables should be attached to guard against static electricity buildup. Defueling should be performed by experienced personnel, and inexperienced personnel must be checked out before doing so without assistance.

Remember that there may be a sequence in defueling an aircraft's fuel tanks just as there is when fueling to avoid structural damage. Consult the manufacturer's maintenance/operations manual(s) if in doubt.

Pressure fueled aircraft normally defuel through the pressure fueling port. The aircraft's in-tank boost pumps can be used to pump the fuel out. The pump on a fuel truck can also be used to draw fuel out. These tanks can also be drained through the tank sump drains, but the large size of the tanks usually makes this impractical. Aircraft fueled over the wing are

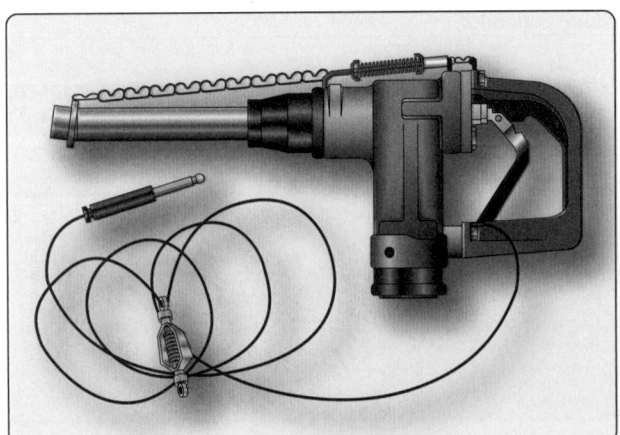

Figure 14-104. *An AVGAS fueling nozzle with static bonding grounding wire.*

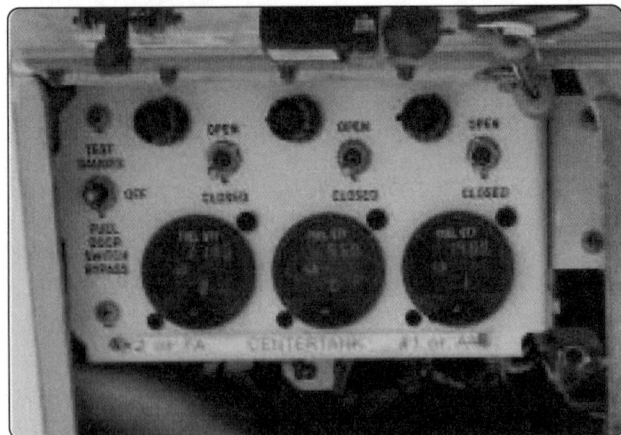

Figure 14-106. *This panel at the pressure refueling station has valve position switches and quantity gauges to be used during refueling. Valve open position lights are adjacent to the switches for each tank.*

normally drained through the tank sump drains. Follow the manufacturer's procedure for defueling the aircraft.

What to do with the fuel coming out of a tank depends on a few factors. First, if the tank is being drained due to fuel contamination or suspected contamination, it should not be mixed with any other fuel. It should be stored in a separate container from good fuel, treated if possible, or disposed of properly. Take measures to ensure that contaminated fuel is never placed onboard an aircraft or mixed with good fuel. Second, the manufacturer may have requirements for good fuel that has been defueled from an aircraft, specifying whether it can be reused and the type of storage container in which it must be stored. Above all, fuel removed from an aircraft must not be mixed with any other type of fuel.

Good fuel removed from an aircraft must be handled with all precautions used when handling any fuel. It must only be put into clean tanks and efforts must be made to keep it clean. It may be put back in the aircraft or another aircraft if the manufacturer allows. Large aircraft can often transfer fuel from a tank requiring maintenance to another tank to avoid the defueling process.

Fire Hazards When Fueling or Defueling

Due to the combustible nature of AVGAS and turbine engine fuel, the potential for fire while fueling and defueling aircraft must be addressed. Always fuel and defuel outside, not in a hangar that serves as an enclosed area for vapors to build up to a combustible level. Clothing worn by refueling personnel should not promote static electricity buildup. Synthetics, such as nylon, should be avoided. Cotton has proved to be safe for fuel handling attire.

As previously mentioned, the most controllable of the three ingredients required for fire is the source of ignition. It is absolutely necessary to prevent a source of ignition anywhere near the aircraft during fueling or refueling. Any open flame, such as a lit cigarette, must be extinguished. Operation of any electrical devices must be avoided. Radio and radar use is prohibited. It is important to note that fuel vapors proliferate well beyond the actual fuel tank opening and a simple spark, even one caused by static electricity, could be enough for ignition. Any potential for sparks must be nullified.

Spilled fuel poses an additional fire hazard. A thin layer of fuel vaporizes quickly. Small spills should be wiped up immediately. Larger spills can be flooded with water to dissipate the fuel and the potential for ignition. Do not sweep fuel that has spilled onto the ramp.

Class B fire extinguishers need to be charged and accessible nearby during the fueling and defueling processes. Fueling personnel must know exactly where they are and how to use them. In case of an emergency, the fuel truck, if used, may need to be quickly driven away from the area. For this reason alone, it should be positioned correctly on the ramp relative to the aircraft.

Chapter 15
Ice & Rain Protection

Ice Control Systems

Rain, snow, and ice are transportation's longtime enemies. Flying has added a new dimension, particularly with respect to ice. Under certain atmospheric conditions, ice can build rapidly on airfoils and air inlets. On days when there is visible moisture in the air, ice can form on aircraft leading-edge surfaces at altitudes where freezing temperatures start. Water droplets in the air can be supercooled to below freezing without actually turning into ice unless they are disturbed in some manner. This unusual occurrence is partly due to the surface tension of the water droplet not allowing the droplet to expand and freeze. However, when aircraft surfaces disturb these droplets, they immediately turn to ice on the aircraft surfaces.

The two types of ice encountered during flight are clear and rime. Clear ice forms when the remaining liquid portion of the water drop flows out over the aircraft surface, gradually freezing as a smooth sheet of solid ice. Formation occurs when droplets are large, such as in rain or in cumuliform clouds. Clear ice is hard, heavy, and tenacious. Its removal by deicing equipment is especially difficult.

Rime ice forms when water drops are small, such as those in stratified clouds or light drizzle. The liquid portion remaining after initial impact freezes rapidly before the drop has time to spread over the aircraft surface. The small frozen droplets trap air giving the ice a white appearance. Rime ice is lighter in weight than clear ice and its weight is of little significance. However, its irregular shape and rough surface decrease the effectiveness of the aerodynamic efficiency of airfoils, reducing lift and increasing drag. Rime ice is brittle and more easily removed than clear ice.

Mixed clear and rime icing can form rapidly when water drops vary in size or when liquid drops intermingle with snow or ice particles. Ice particles become imbedded in clear ice, building a very rough accumulation sometimes in a mushroom shape on leading edges. Ice may be expected to form whenever there is visible moisture in the air and temperature is near or below freezing. An exception is carburetor icing, which can occur during warm weather with no visible moisture present.

Ice or frost forming on aircraft creates two basic hazards:

1. The resulting malformation of the airfoil that could decrease the amount of lift.

2. The additional weight and unequal formation of the ice that could cause unbalancing of the aircraft, making it hard to control.

Enough ice to cause an unsafe flight condition can form in a very short period of time, thus some method of ice prevention or removal is necessary. *Figure 15-1* shows the effects of ice on a leading edge.

Icing Effects

Ice buildup increases drag and reduces lift. It causes destructive vibration and hampers true instrument readings. Control surfaces become unbalanced or frozen. Fixed slots are filled and movable slots jammed. Radio reception is hampered, and engine performance is affected. Ice, snow, and slush have a direct impact on the safety of flight. Not only because of degraded lift, reduced takeoff performance, and/or maneuverability of the aircraft, but when chunks break off, they can also cause engine failures and structural damage. Fuselage aft-mounted engines are particularly susceptible to this foreign object damage (FOD) phenomenon. Wing-mounted engines are not excluded however. Ice can be present on any part of the aircraft and, when it breaks off, there is some probability that it could go into an engine. The worst case is that ice on the wing breaks off during takeoff due to the flexing of the wing and goes directly into the engine, leading to surge, vibration, and complete thrust loss. Light snow that is loose on the wing surfaces and the fuselage can also cause engine damage leading to surge, vibration, and thrust loss.

Figure 15-1. *Formation of ice on aircraft leading edge.*

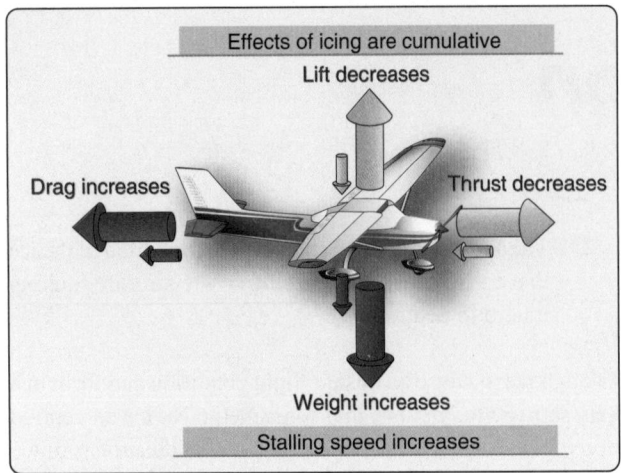

Figure 15-2. *Effects of structural icing.*

Whenever icing conditions are encountered, the performance characteristics of the airplane deteriorate. *[Figure 15-2]* Increased aerodynamic drag increases fuel consumption, reducing the airplane's range and making it more difficult to maintain speed. Decreased rate of climb must be anticipated, not only because of the decrease in wing and empennage efficiency but also because of the possible reduced efficiency of the propellers and increase in gross weight. Abrupt maneuvering and steep turns at low speeds must be avoided because the airplane stalls at higher-than-published speeds with ice accumulation. On final approach for landing, increased airspeed must be maintained to compensate for this increased stall speed. After touchdown with heavy ice accumulation, landing distances may be as much as twice the normal distance due to the increased landing speeds. In this chapter, ice prevention and ice elimination using pneumatic pressure, application of heat, and the application of fluid is discussed.

The ice and rain protection systems used on aircraft keep ice from forming on the following airplane components:

- Wing leading edges.
- Horizontal and vertical stabilizer leading edges.
- Engine cowl leading edges.
- Propellers.
- Propeller spinner.
- Air data probes.
- Flight deck windows.
- Water and waste system lines and drains.
- Antenna.

Figure 15-3 gives an overview of ice and rain protection systems installed in a large transport category aircraft. In modern aircraft, many of these systems are automatically controlled by the ice detection system and onboard computers.

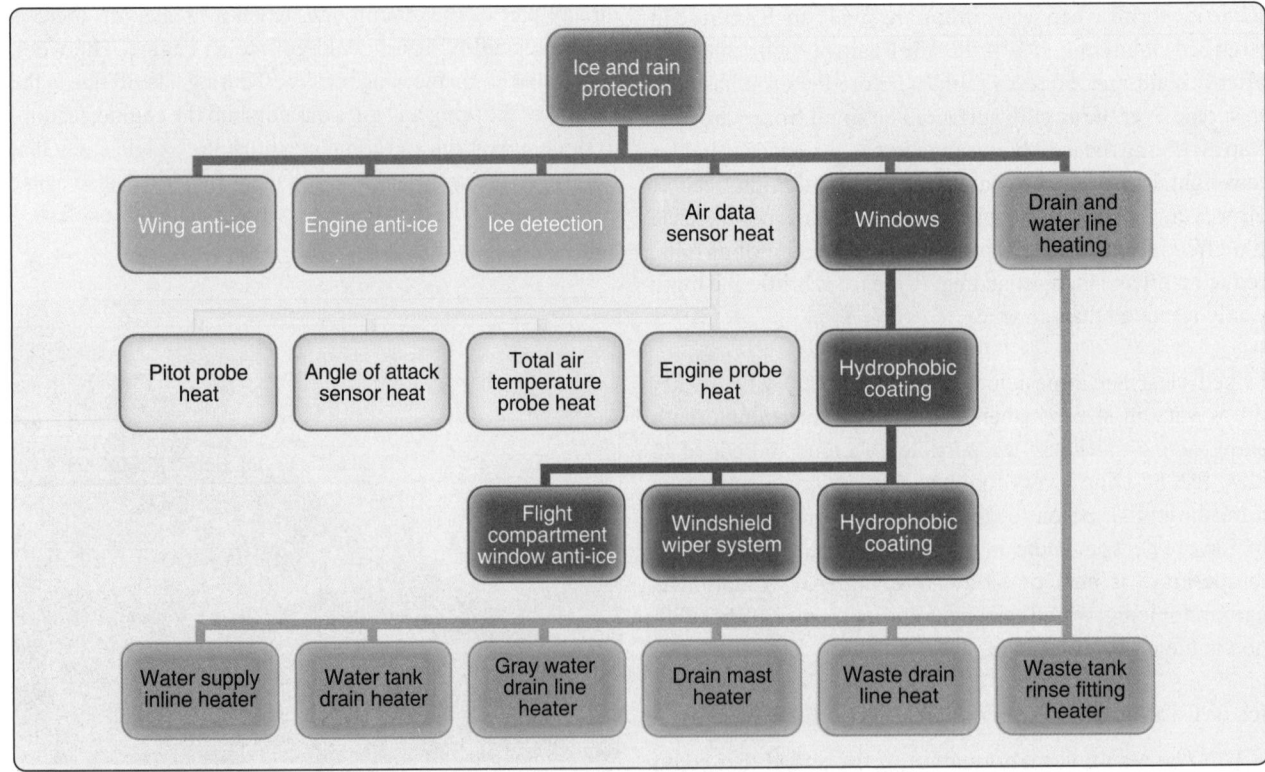

Figure 15-3. *Ice and rain protection systems.*

Figure 15-4. *An ice detector alerts the flight crew of icing conditions and, on some aircraft, automatically activates ice protection systems. One or more detectors are located on the forward fuselage.*

Ice Detector System

Ice can be detected visually, but most modern aircraft have one or more ice detector sensors that warn the flight crew of icing conditions. An annunciator light comes on to alert the flight crew. In some aircraft models, multiple ice detectors are used, and the ice detection system automatically turns on the WAI systems when icing is detected. *[Figure 15-4]*

Ice Prevention

Several means to prevent or control ice formation are used in aircraft today:

1. Heating surfaces with hot air.

2. Heating by electrical elements.

3. Breaking up ice formations, usually by inflatable boots.

4. Chemical application.

Equipment is designed for anti-icing or for deicing. Anti-icing equipment is turned on before entering icing conditions and is designed to prevent ice from forming. A surface may be anti-iced by keeping it dry, by heating to a temperature that evaporates water upon impingement, or by heating the surface just enough to prevent freezing, maintaining it running wet. Deicing equipment is designed to remove ice after it begins to accumulate typically on the wings and stabilizer leading edges. Ice may be controlled on aircraft structure by the methods described in *Figure 15-5*.

Wing & Stabilizer Anti-Icing Systems

The wing leading edges, or leading-edge slats, and horizontal and vertical stabilizer leading edges of many aircraft make and models have anti-icing systems installed to prevent the formation of ice on these components. The most common anti-icing systems used are thermal pneumatic, thermal electric, and chemical. Most general aviation (GA) aircraft equipped to fly in icing conditions use pneumatic deicing boots or a chemical anti-ice system. High-performance aircraft may have "weeping wings." Large transport-category aircraft are equipped with advanced thermal pneumatic or thermal electric anti-icing systems that are controlled automatically to prevent the formation of ice.

Location of Ice	Method of Control
Leading edge of the wing	Thermal pneumatic, thermal electric, chemical, and pneumatic (deice)
Leading edges of vertical and horizontal stabilizers	Thermal pneumatic, thermal electric, and pneumatic (deice)
Windshield, windows	Thermal pneumatic, thermal electric, and chemical
Heater and engine air inlets	Thermal pneumatic and thermal electric
Pitot and static air data sensors	Thermal electric
Propeller blade leading edge and spinner	Thermal electric and chemical
Carburetor(s)	Thermal pneumatic and chemical
Lavatory drains and potable water lines	Thermal electric

Figure 15-5. *Typical ice control methods.*

Figure 15-6. *Aircraft with thermal WAI system.*

Thermal Pneumatic Anti-icing

Thermal systems used for the purpose of preventing the formation of ice or for deicing airfoil leading edges usually use heated air ducted spanwise along the inside of the leading edge of the airfoil and distributed around its inner surface. These thermal pneumatic anti-icing systems are used for wings, leading edge slats, horizontal and vertical stabilizers, engine inlets, and more. There are several sources of heated air, including hot bleed air from the turbine compressor, engine exhaust heat exchangers, and ram air heated by a combustion heater.

Wing Anti-Ice (WAI) System

Thermal wing anti-ice (WAI or TAI) systems for business jet and large-transport category aircraft typically use hot air bled from the engine compressor. *[Figure 15-6]* Relatively large amounts of very hot air can be bled off the compressor, providing a satisfactory source of anti-icing heat. The hot air is routed through ducting, manifolds, and valves to components that need to be anti-iced. *Figure 15-7* shows a typical WAI system schematic for a business jet. The bleed air is routed to each wing leading edge by an ejector in each wing inboard area. The ejector discharges the bleed air into piccolo tubes for distribution along the leading edge. Fresh ambient air is introduced into the wing leading edge by two flush-mounted ram air scoops in each wing leading edge, one at the wing root and one near the wingtip. The ejectors entrain ambient air, reduce the temperature of the bleed air,

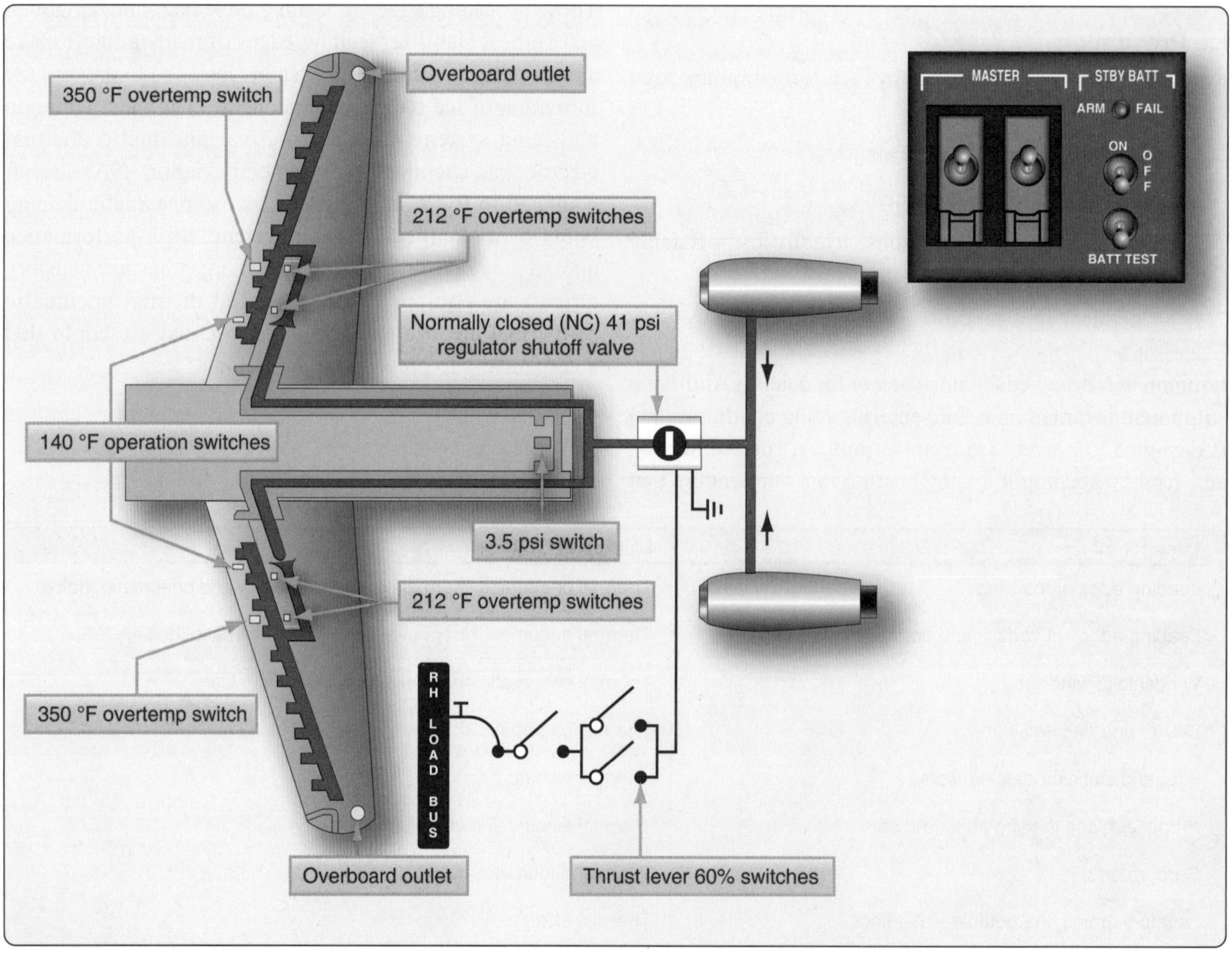

Figure 15-7 schematic labels

- 350 °F overtemp switch
- Overboard outlet
- 212 °F overtemp switches
- Normally closed (NC) 41 psi regulator shutoff valve
- 140 °F operation switches
- 3.5 psi switch
- 212 °F overtemp switches
- 350 °F overtemp switch
- Overboard outlet
- Thrust lever 60% switches
- RH LOAD BUS
- MASTER
- STBY BATT
- ARM
- FAIL
- ON
- OFF
- BATT TEST

Figure 15-7. *Thermal WAI system.*

and increase the mass airflow in the piccolo tubes. The wing leading edge is constructed of two skin layers separated by a narrow passageway. *[Figure 15-8]* The air directed against the leading edge can only escape through the passageway, after which it is vented overboard through a vent in the bottom of the wingtip.

When the WAI switch is turned on, the pressure regulator is energized, and the shutoff valve opens. When the wing leading edge temperature reaches approximately +140 °F, temperature switches turn on the operation light above the switch. If the temperature in the wing leading edge exceeds approximately +212 °F (outboard) or +350 °F (inboard), the red WING OV HT warning light on the annunciator panel illuminates.

The ducting of WAI systems usually consists of aluminum alloy, titanium, stainless steel, or molded fiberglass tubes. The tube, or duct, sections are attached to each other by bolted end flanges or by band-type V-clamps. The ducting is lagged with a fire-resistant, heat-insulating material, such as fiberglass. In some installations, thin stainless-steel expansion bellows are used. Bellows are located at strategic positions to absorb any distortion or expansion of the ducting that may occur due to temperature variations. The joined sections of ducting are hermetically sealed by sealing rings. These seals are fitted into annular recesses in the duct joint faces.

When installing a section of duct, make certain that the seal bears evenly against and is compressed by the adjacent joint's flange. When specified, the ducts should be pressure tested at the pressure recommended by the manufacturer of the aircraft concerned. Leak checks are made to detect defects in the duct that would permit the escape of heated air. The rate of leakage at a given pressure should not exceed that recommended in the aircraft maintenance manual.

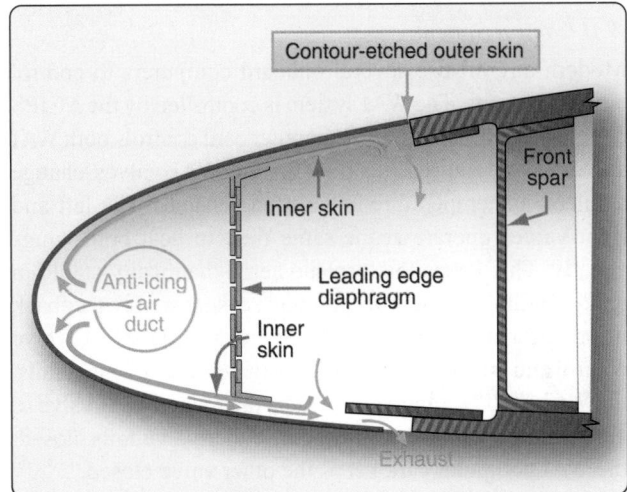

Figure 15-8. *Heated wing leading edge.*

Air leaks can often be detected audibly and are sometimes revealed by holes in the lagging or thermal insulation material. However, if difficulty arises in locating leaks, a soap-and-water solution may be used. All ducting should be inspected for security, general condition, or distortion. Lagging or insulating blankets must be checked for security and must be free of flammable fluids, such as oil or hydraulic fluid.

Leading Edge Slat Anti-Ice System

Aircraft that utilize leading edge slats often use bleed air from the engine compressor to prevent the formation of frost on these surfaces. On a modern transport category aircraft, the pneumatic system supplies bleed air for this purpose. WAI valves control the air flow from the pneumatic system to WAI ducts. The WAI ducts carry the air to the slats. Holes in the bottom of each slat let the air out.

The airfoil and cowl ice protection system (ACIPS) computer card controls the WAI valves, and pressure sensors send duct air pressure data to the computer. The aircrew can select an auto or manual mode with the WAI selector. In the auto mode, the system turns on when the ice detection system detects ice. The off and on positions are used for manual control of the WAI system. The WAI system is only used in the air, except for ground tests. The weight on wheels system and/or airspeed data disarms the system when the aircraft is on the ground. *[Figure 15-9]*

WAI Valve

The WAI valve controls the flow of bleed air from the pneumatic system to the WAI ducts. The valve is electrically controlled and pneumatically actuated. The torque motor controls operation of the valve. With no electrical power to the torque motor, air pressure on one side of the actuator holds the valve closed. Electrical current through the torque motor allows air pressure to open the valve. As the torque motor current increases, the valve opening increases. *[Figure 15-10]*

WAI Pressure Sensor

The WAI pressure sensor senses the air pressure in the WAI duct after the WAI valve. The ACIPS computer card uses the pressure information to control the WAI system.

WAI Ducts

The WAI ducts move air from the pneumatic system through the wing leading edge to the leading-edge slats. *Figure 15-9* shows that only leading edge slat sections 3, 4, and 5 on the left wing and 10, 11, and 12 on the right wing receive bleed air for WAI. Sections of the WAI ducting are perforated. The holes allow air to flow into the space inside the leading-edge slats. The air leaves the slats through holes in the bottom of each slat. Some WAI ducts have connecting "T" ducts that telescope to direct air into the slats while

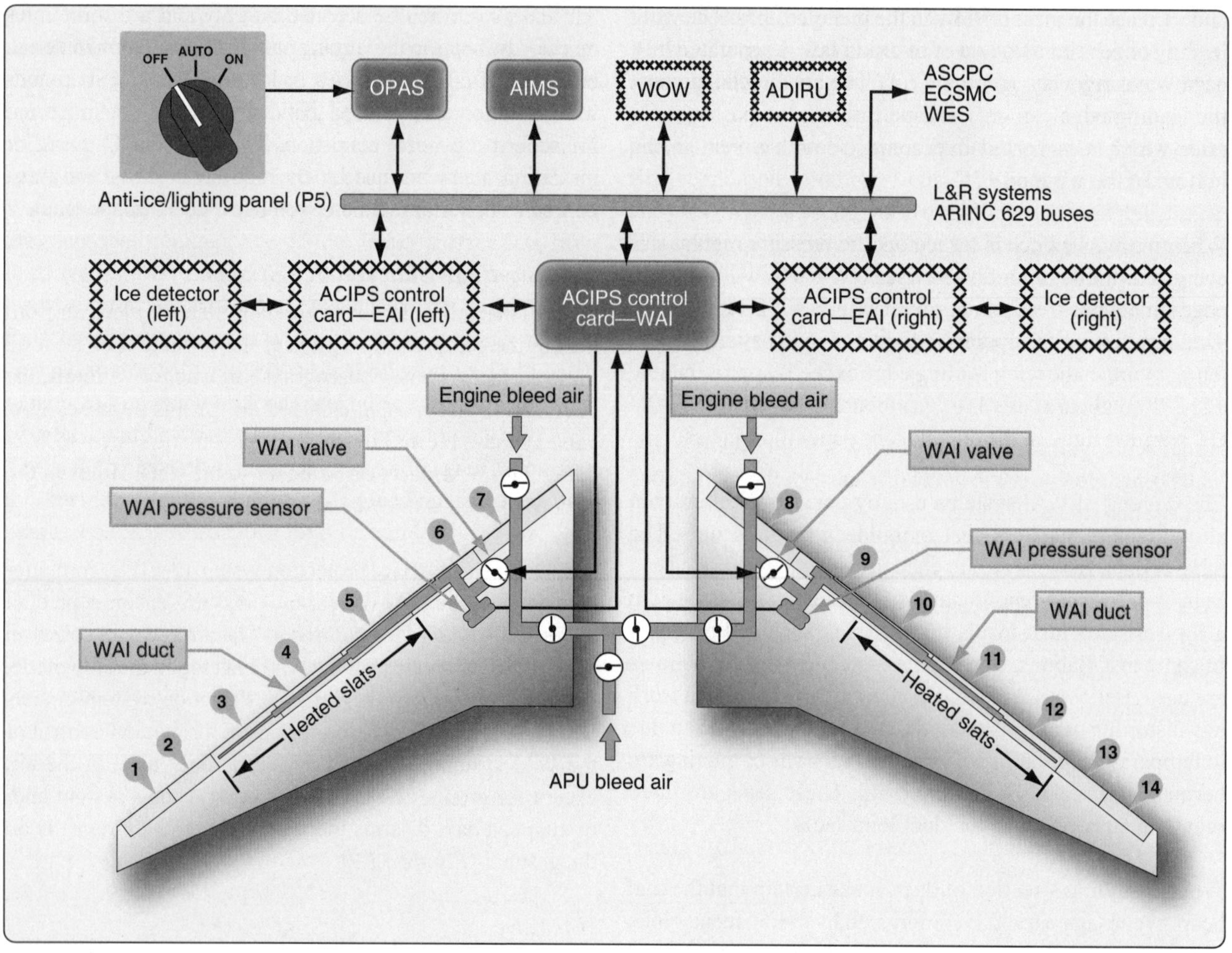

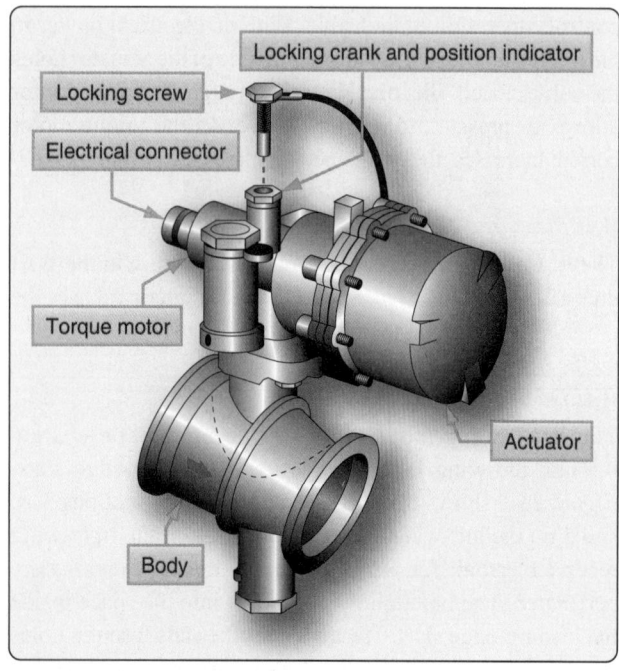

Figure 15-9. *Wing leading edge slat anti-ice system.*

Figure 15-10. *A wing anti-ice valve.*

extended. The telescoping section attached to the slat on one end, slides over the narrow diameter "T" section that is connected into the WAI duct. A seal prevents any loss of air. This arrangement allows warm air delivery to the slats while retracted, in transit, and fully deployed. *[Figure 15-11]*

WAI Control System

Modern aircraft use several onboard computers to control aircraft systems. The WAI system is controlled by the ACIPS computer card. The ACIPS computer card controls both WAI valves. The required positions of the WAI valves change as bleed air temperature and altitude change. The left and right valves operate at the same time to heat both wings equally. This keeps the airplane aerodynamically stable in icing conditions. The WAI pressure sensors supply feedback information to the WAI ACIPS computer card for WAI valve control and position indication. If either pressure sensor fails, the WAI ACIPS computer card sets the related WAI valve to either fully open or fully closed. If either valve fails closed, the WAI computer card keeps the other valve closed.

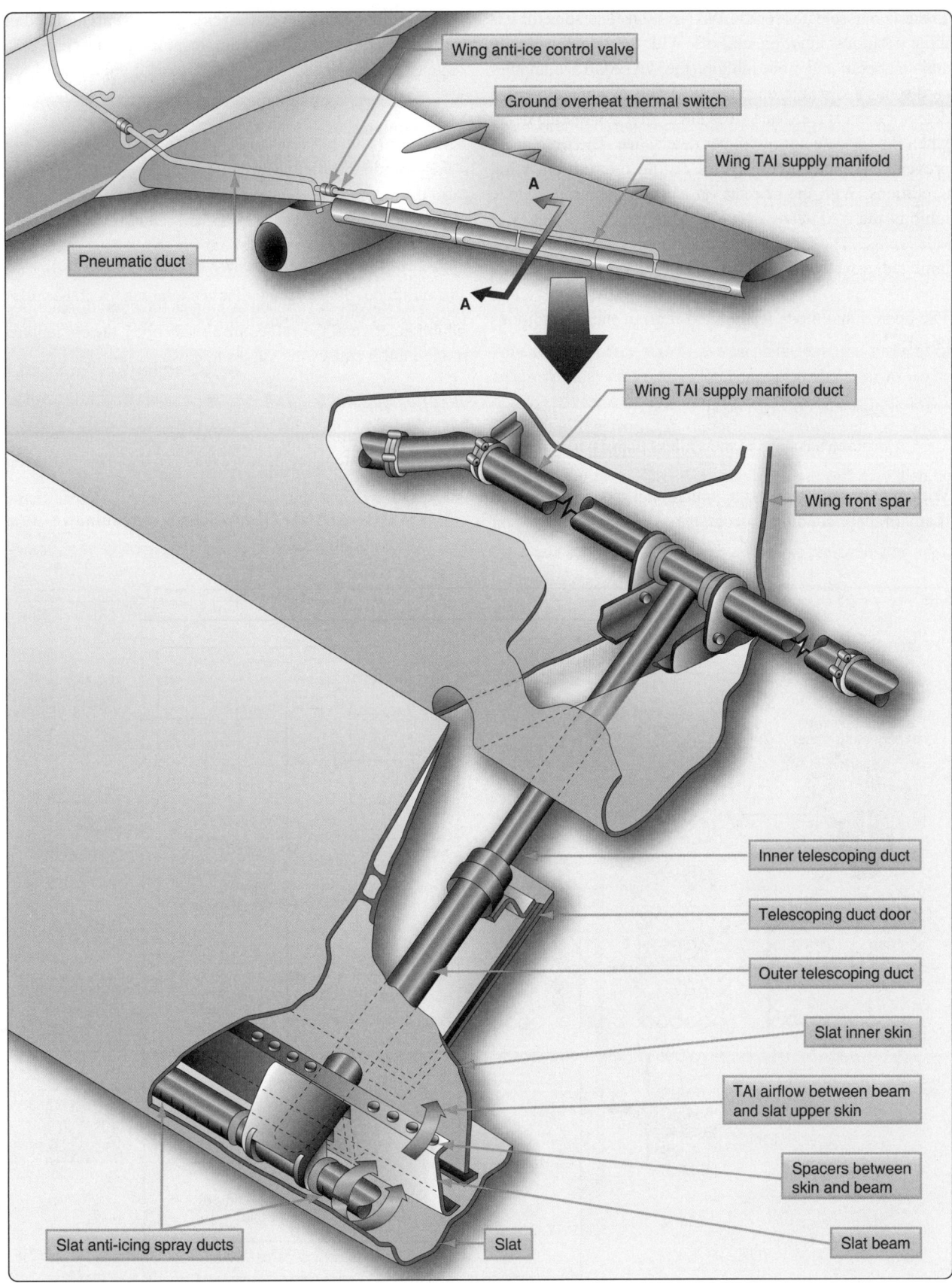

Figure 15-11. *WAI ducting.*

There is one selector for the WAI system. The selector has three positions: auto, on, and off. With the selector in auto and no operational mode inhibits, the WAI ACIPS computer card sends a signal to open the WAI valves when either ice detector detects ice. The valves close after a 3-minute delay when the ice detector no longer detects ice. The time delay prevents frequent on/off cycles during intermittent icing conditions. With the selector on and no operational mode inhibits, the WAI valves open. With the selector off, the WAI valves close. The operational mode for the WAI valves can be inhibited by many different sets of conditions. *[Figure 15-12]*

The operational mode is inhibited if all of these conditions occur:

- Auto mode is selected.
- Takeoff mode is selected.
- Airplane has been in the air less than 10 minutes.

With auto or on selected, the operational mode is inhibited if any of these conditions occur:

- Airplane on the ground (except during an initiated or periodic built-in test equipment (BITE) test).
- Total Air Temperature (TAT) is more than 50 °F (10 °C) and the time since takeoff is less than 5 minutes.
- Auto slat operation.
- Air-driven hydraulic pump operation.
- Engine start.
- Bleed air temperature less than 200 °F (93 °C).

The WAI valves stay closed as long as the operational mode inhibit is active. If the valves are already open, the operational mode inhibit causes the valves to close.

WAI Indication System

The aircrew can monitor the WAI system on the onboard computer maintenance page. *[Figure 15-13]* The following information is shown:

- WING MANIFOLD PRESS—pneumatic duct pressure in PSIG.

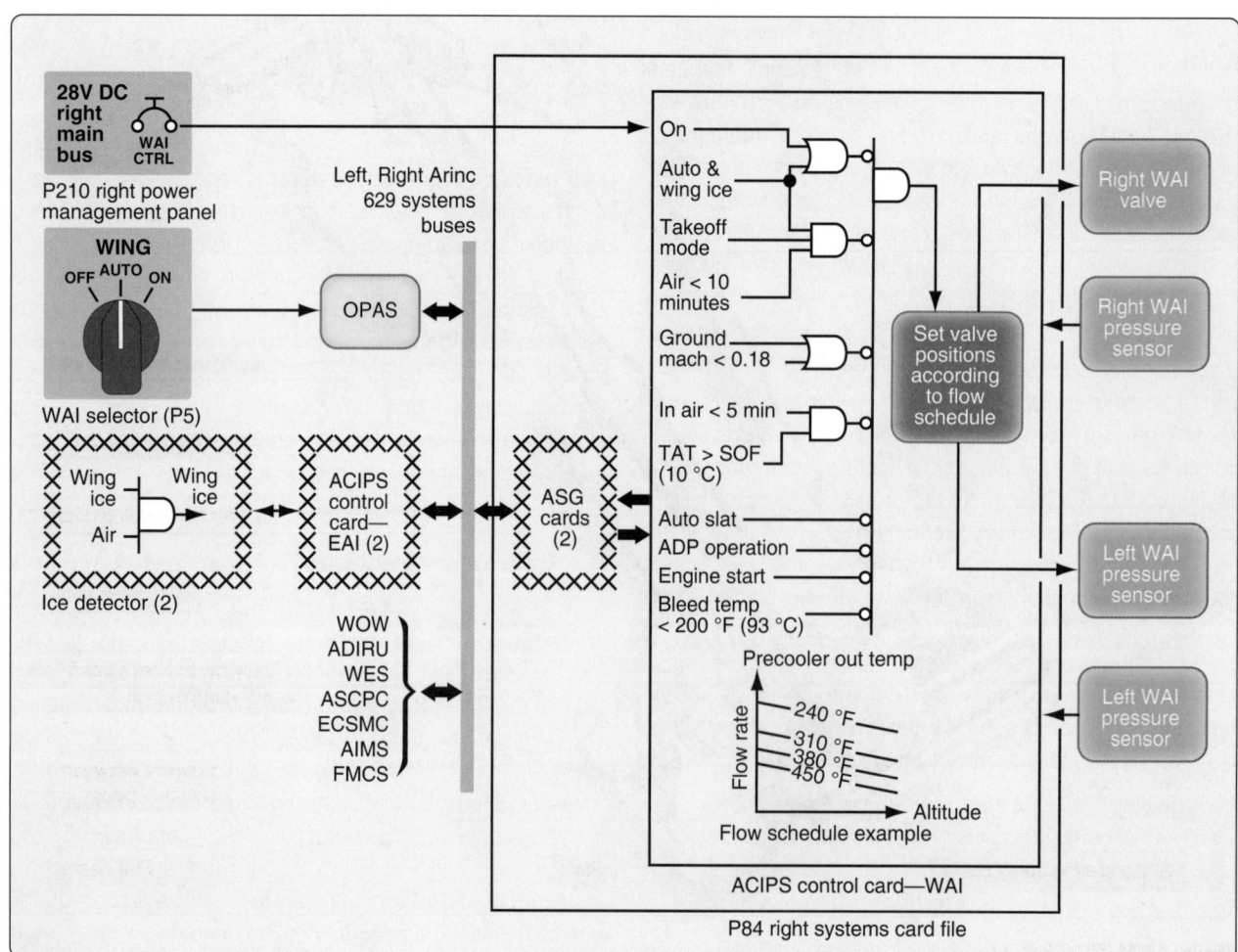

Figure 15-12. *WAI inhibit logic schematic.*

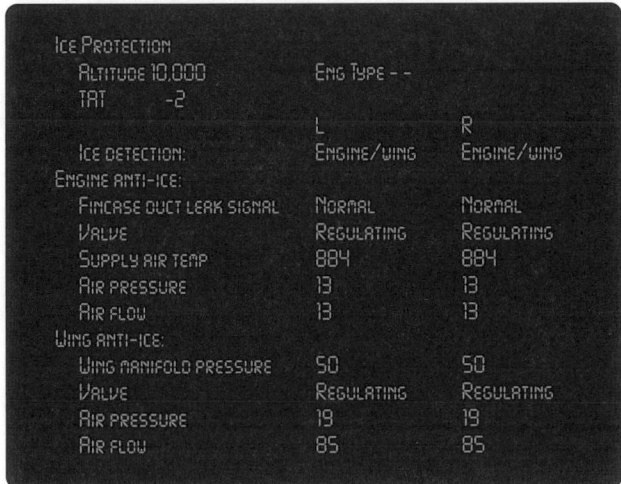

Figure 15-13. *Ice protection onboard computer maintenance page.*

- VALVE—WAI valve open, closed, or regulating.

- AIR PRESS—pressure downstream of the WAI valves in PSIG.

- AIR FLOW—air flow through the WAI valves in pounds per minute.

WAI System Built-In Test Equipment (BITE) Test

BITE circuits in the WAI ACIPS computer card continuously monitor the WAI system. Faults that affect the dispatch of the aircraft cause status messages. Other faults cause central maintenance computer system (CMCS) maintenance messages. The BITE in the WAI ACIPS computer card also performs automatic power-up and periodic tests. Faults found during these tests that affect dispatch cause status messages. Other faults cause CMCS maintenance messages. The power-up test occurs when the card gets power. BITE does a test of the card hardware and software functions and the valve and pressure sensor interfaces. The valves do not move during this test.

The periodic test occurs when all these conditions are true:

- The airplane has been on the ground between 1 and 5 minutes.

- The WAI selector is set to auto or on.

- Air-driven hydraulic pumps are not in intermittent operation.

- Bleed pressure is sufficient to open the WAI valves.

- The time since the last periodic test is more than 24 hours.

- During this test, the WAI valves cycle open and closed. This test makes sure that valve malfunctions are detected.

Thermal Electric Anti-Icing

Electricity is used to heat various components on an aircraft so that ice does not form. This type of anti-ice is typically limited to small components due to high amperage draw. Effective thermal electric anti-ice is used on most air data probes, such as pitot tubes, static air ports, TAT and AOA probes, ice detectors, and engine P2/T2 sensors. Water lines, waste water drains, and some turboprop inlet cowls are also heated with electricity to prevent ice from forming. Transport category and high-performance aircraft use thermal electric anti-icing in windshields.

In devices that use thermal electric anti-ice, current flows through an integral conductive element that produces heat. The temperature of the component is elevated above the freezing point of water, so ice cannot form. Various schemes are used, such as an internal coil wire, externally wrapped blankets or tapes, as well as conductive films and heated gaskets. A basic discussion of probe heat follows. Windshield heat and portable water heat anti-ice are discussed later in this chapter. Propeller deice boots, which also are used for anti-ice, are also thermal electric and discussed in this chapter.

Data probes that protrude into the ambient airstream are particularly susceptible to ice formation in flight. *Figure 15-14* illustrates the types and location probes that use thermal electric heat on one airliner. A pitot tube, for example, contains an internal electric element that is controlled by a switch in the flight deck. Use caution checking the function of the pitot heat when the aircraft is on the ground. The tube gets extremely hot since it must keep ice from forming at altitude in temperatures near -50 °F at speeds possibly over 500 miles per hour. An ammeter or load meter in the circuit can be used to safely determine functionality.

Simple probe heat circuits exist on GA aircraft with a switch and a circuit breaker to activate and protect the device. Advanced aircraft may have more complex circuitry in which control is by computer and flight condition of the aircraft is considered before thermal electric heaters are activated automatically. *Figure 15-15* shows such a circuit for a pitot tube. The primary flight computer (PFC) supplies signals for the air data card (ADC) to energize ground and air heat control relays to activate probe heat. Information concerning speed of the aircraft, whether it is in the air or on the ground, and if the engines are running, are factors considered by the ADC logic. Similar controls are used for other probe heaters.

Chemical Anti-Icing

Chemical anti-icing is used in some aircraft to anti-ice the leading edges of the wing, stabilizers, windshields, and propellers. The wing and stabilizer systems are often called weeping wing systems or are known by their trade name of

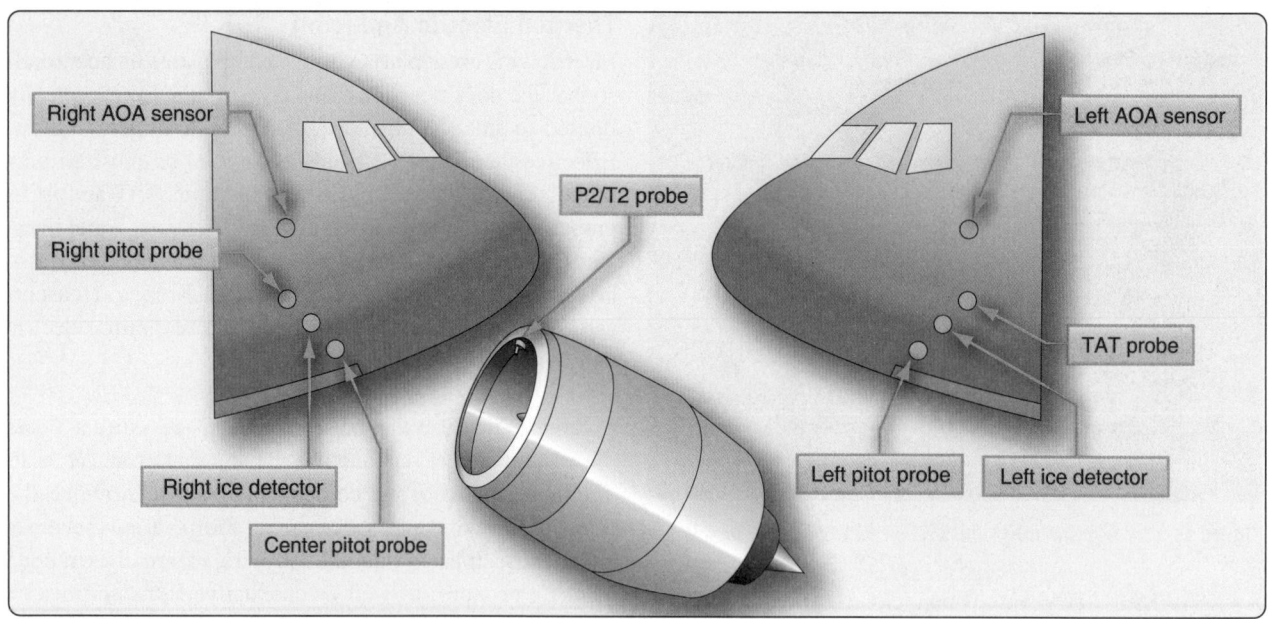

Figure 15-14. *Probes with thermal electric anti-icing on one commercial airliner.*

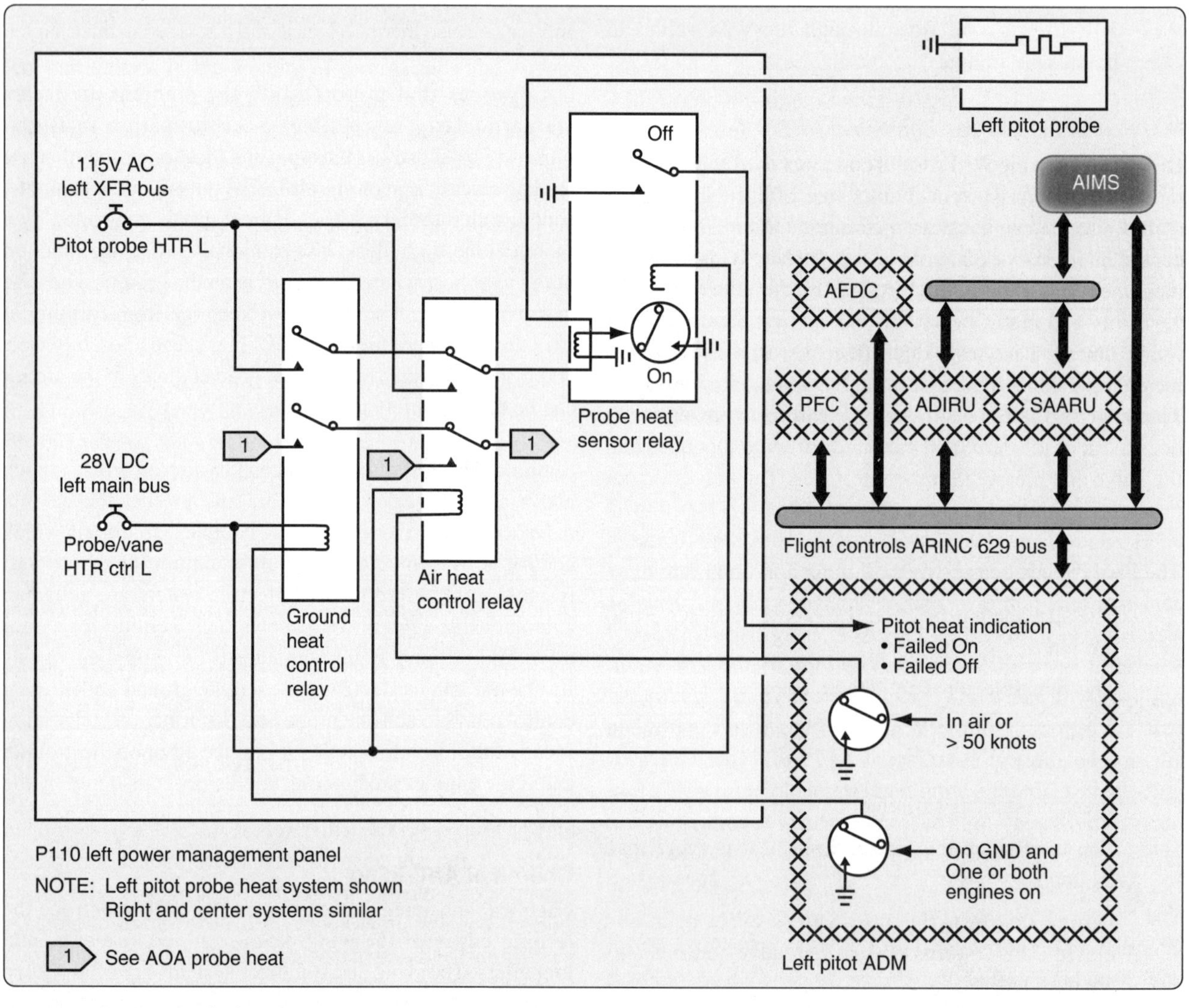

Off

115V AC
left XFR bus

Pitot probe HTR L

Left pitot probe

AIMS

AFDC

PFC   ADIRU   SAARU

28V DC
left main bus

Probe/vane
HTR ctrl L

1   1   1

On

Probe heat
sensor relay

Air heat
control relay

Ground
heat
control
relay

Flight controls ARINC 629 bus

Pitot heat indication
• Failed On
• Failed Off

In air or
> 50 knots

P110 left power management panel

NOTE: Left pitot probe heat system shown
       Right and center systems similar

1   See AOA probe heat

On GND and
One or both
engines on

Left pitot ADM

Figure 15-15. *Pitot probe heat system.*

TKS™ systems. Ice protection is based upon the freezing point depressant concept. An antifreeze solution is pumped from a reservoir through a mesh screen embedded in the leading edges of the wings and stabilizers. Activated by a switch in the flight deck, the liquid flows over the wing and tail surfaces, preventing the formation of ice as it flows. The solution mixes with the supercooled water in the cloud, depresses its freezing point, and allows the mixture to flow off of the aircraft without freezing. The system is designed to anti-ice, but it is also capable of deicing an aircraft as well. When ice has accumulated on the leading edges, the antifreeze solution chemically breaks down the bond between the ice and airframe. This allows aerodynamic forces to carry the ice away. Thus, the system clears the airframe of accumulated ice before transitioning to anti-ice protection. *Figure 15-16* shows a chemical anti-ice system.

The TKS™ weeping wing system contains formed titanium panels that are laser drilled with over 800 tiny holes (.0025-inch diameter) per square inch. These are mated with non-perforated stainless-steel rear panels and bonded to wing and stabilizer leading edges. As fluid is delivered from a central reservoir and pump, it seeps through the holes. Aerodynamic forces cause the fluid to coat the upper and lower surfaces of the airfoil. The glycol-based fluid prevents ice from adhering to the aircraft structure.

Some aircraft with weeping wing systems are certified to fly into known icing conditions. Others use it as a hedge against unexpected ice encountered in flight. The systems are basically the same. Reservoir capacity permits 1-2 hours of operation. TKS™ weeping wings are used primarily on reciprocating aircraft that lack a supply of warm bleed air for the installation of a thermal anti-ice system. However, the system is simple and effective leading to its use on some turbine powered corporate aircraft as well.

Wing & Stabilizer Deicing Systems

GA aircraft and turboprop commuter-type aircraft often use a pneumatic deicing system to break off ice after it has formed

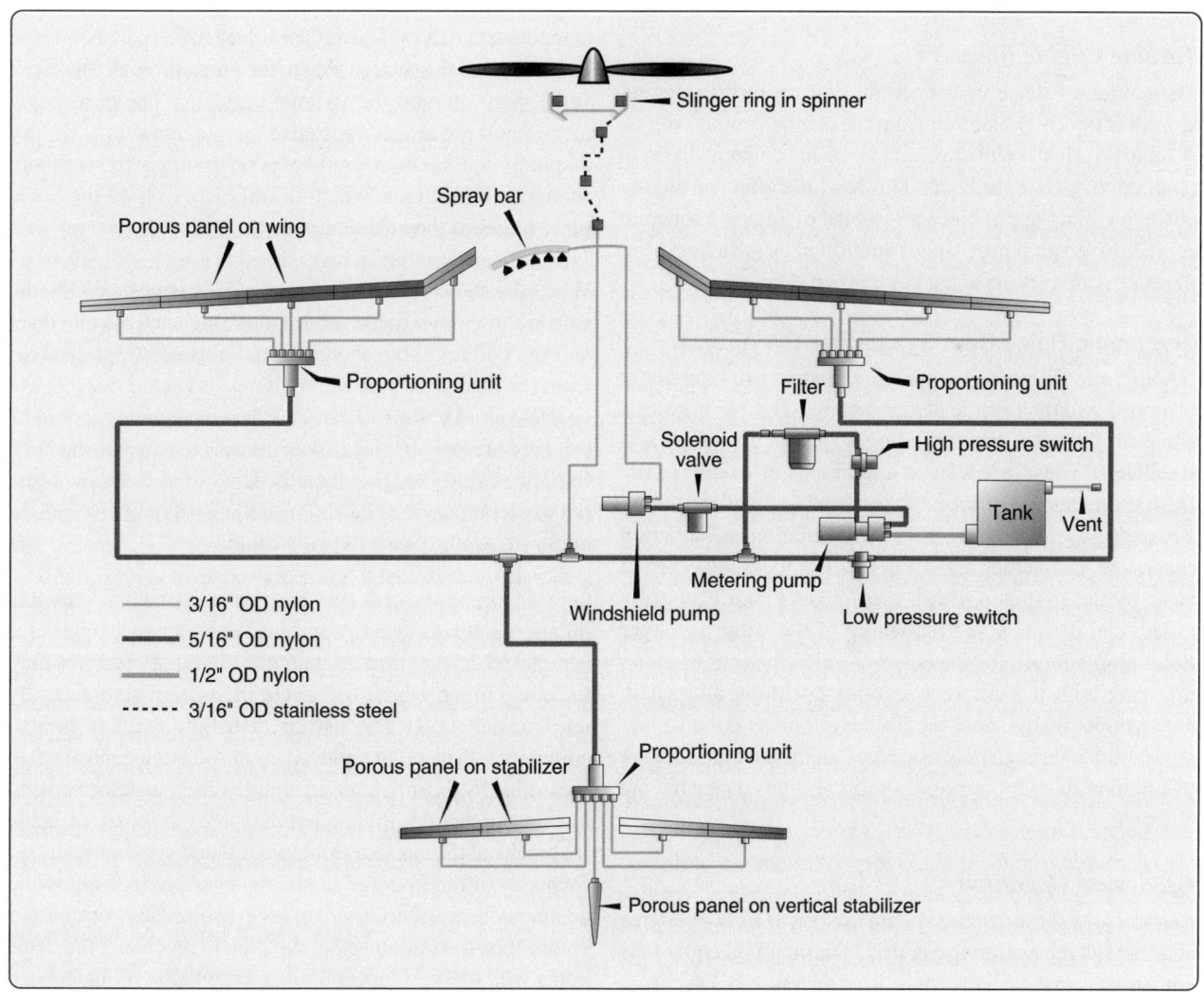

Figure 15-16. *Chemical deicing system.*

on the leading-edge surfaces. The leading edges of the wings and stabilizers have inflatable boots attached to them. The boots expand when inflated by pneumatic pressure, which breaks away ice accumulated on the boot. Most boots are inflated for 6 to 8 seconds. They are deflated by vacuum suction. The vacuum is continuously applied to hold the boots tightly against the aircraft while not in use.

Sources of Operating Air

The source of operating air for deice boot systems varies with the type of powerplant installed on the aircraft. Reciprocating engine aircraft typically use a dedicated engine-driven air pump mounted on the accessory drive gear box of the engine. The suction side of the pump is used to operate the gyroscopic instruments installed on the aircraft. It is also used to hold the deice boots tight to the aircraft when they are not inflated. The pressure side of the pump supplies air to inflate the deice boots, which breaks up ice that has formed on the wing and stabilizer leading edges. The pump operates continuously. Valves, regulators, and switches in the flight deck are used to control the flow of source air to the system.

Turbine Engine Bleed Air

The source of deice boot operating air on turbine engine aircraft is typically bleed air from the engine compressor(s). A relatively low volume of air on an intermittent basis is required to operate the boots. This has little effect on engine power enabling use of bleed air instead of adding a separate engine-driven air pump. Valves controlled by switches in the flight deck deliver air to the boots when requested.

Pneumatic Deice Boot System for GA Aircraft

GA aircraft, especially twin-engine models, are commonly equipped with pneumatic deicer systems. Rubber boots are attached with glue to the leading edges of the wings and stabilizers. These boots have a series of inflatable tubes. During operation, the tubes are inflated and deflated in an alternating cycle. *[Figure 15-17]* This inflation and deflation causes the ice to crack and break off. The ice is then carried away by the airstream. Boots used in GA aircraft typically inflate and deflate along the length of the wing. In larger turbo prop aircraft, the boots are installed in sections along the wing with the different sections operating alternately and symmetrically about the fuselage. This is done so that any disturbance to airflow caused by an inflated tube is kept to a minimum by inflating only short sections on each wing at a time.

GA System Operation

Figure 15-18 shows a deice system used on a GA twin-engine aircraft with reciprocating engines. In normal flight, all of the components in the deice system are de-energized. Discharge air from the dry air pumps is dumped overboard through the

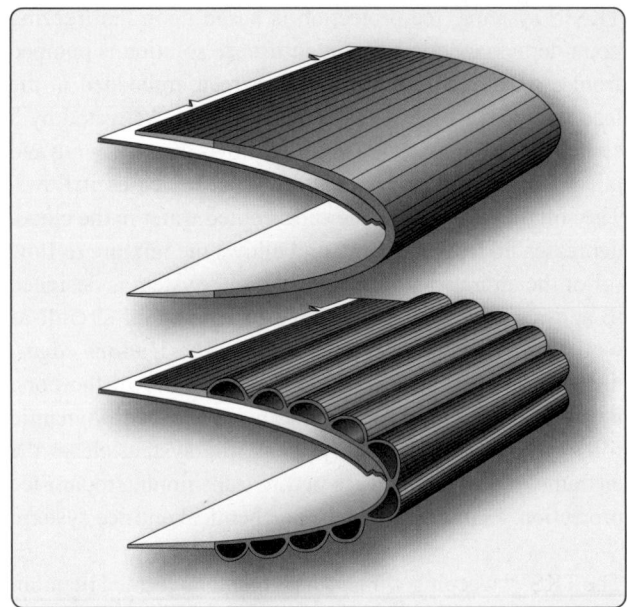

Figure 15-17. *Cross-section of a pneumatic deicing boot uninflated (top) and inflated (bottom).*

deice control valves. The deflate valve is open connecting the deice boots to the suction side of the pump through the check valve manifold and the vacuum regulator. The gyroscopic instruments are also connected to the vacuum side of the dry air pump. The vacuum regulator is set to supply the optimum suction for the gyros, which is sufficient to hold the boots tightly against the airfoil surfaces.

When the switch shown in *Figure 15-19* is pushed ON, the solenoid-operated deice control valves in each nacelle open and the deflate valve energizes and closes. Pressurized air from the discharge side of the pumps is routed through the control valves to the deice boot. When the system reaches 17 psi, pressure switches located on the deflate valve de-energize the deice control valve solenoids. The valves close and route pump air output overboard. The deflate valve opens and the boots are again connected to vacuum.

On this simple system, the pilot must manually start this inflation/deflation cycle by pushing the switch each time deice is required. Larger aircraft with more complex systems may include a timer, which will cycle the system automatically until turned OFF. The use of distributor valves is also common. A distributor valve is a multi-position control valve controlled by the timer. It routes air to different deice boots in a sequence that minimizes aerodynamic disturbances as the ice breaks of the aircraft. Boots are inflated symmetrically on each side of the fuselage to maintain control in flight while deicing occurs. Distributor valves are solenoid operated and incorporate the deflate valve function to reconnect the deice boots with the vacuum side of the pump after all have been inflated.

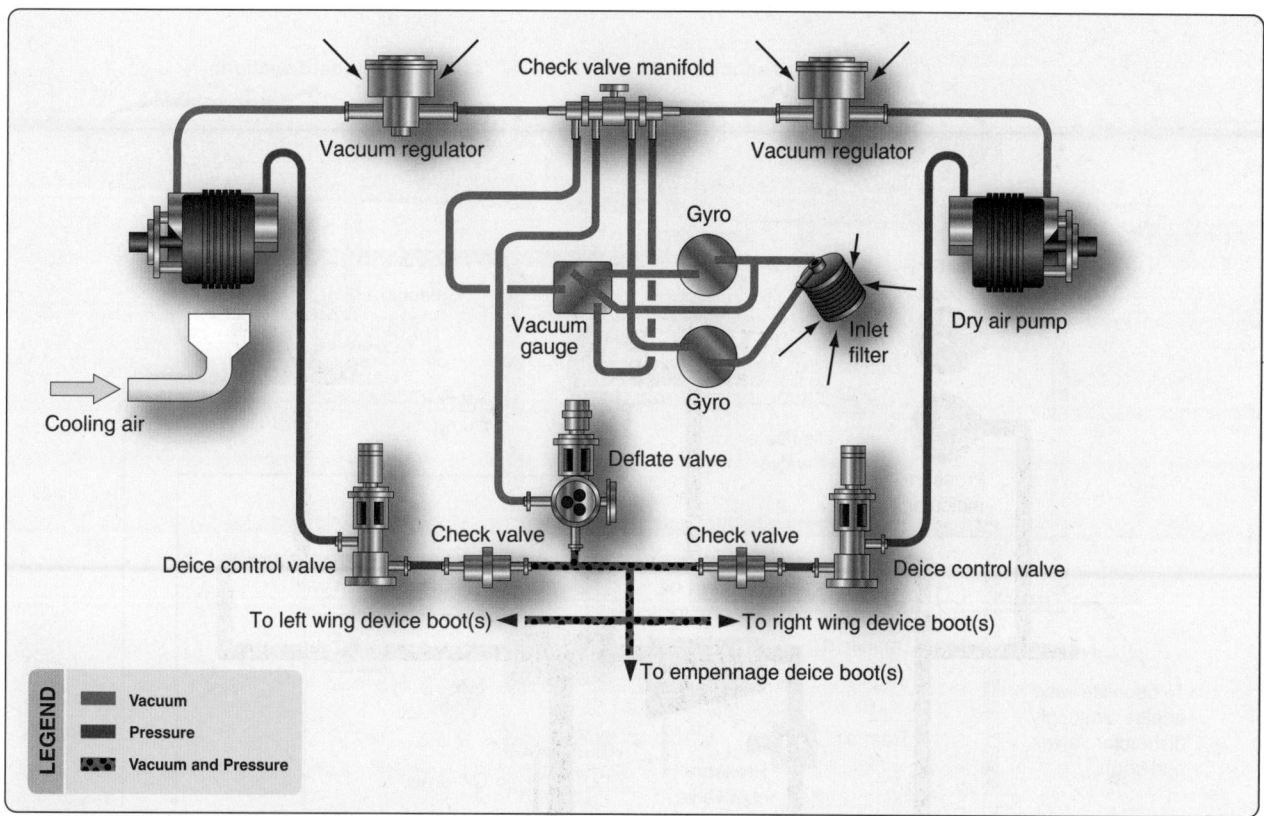

Figure 15-18. *Pneumatic deicing system for a twin engine GA aircraft with reciprocating engines.*

Figure 15-19. *Wing deice switch.*

Combining functional components of a deice system into a single unit is fairly common. *Figure 15-20* illustrates the right side of a large aircraft deice boot system. The left side is the same. In addition to the distributor valves, which combine functions of a control valve and deflate valve, the system also uses a combination unit. This unit combines the functions of a shutoff control valve for all pump supply air, as well as a pressure regulator for the system. It also contains a secondary air filter.

Deice System for Turboprop Aircraft

Figure 15-21 shows a pneumatic deice system used on a turboprop aircraft. The source of pneumatic air is engine bleed air, which is used to inflate two inboard wing boots, two

outboard boots, and horizontal stabilizer boots. Additional bleed air is routed through the brake deice valve to the brakes. A three-position switch controls the operation of the boots. This switch is spring loaded to the center OFF position. When ice has accumulated, the switch should be selected to the single-cycle (up) position and released. *[Figure 15-22]* Pressure-regulated bleed air from the engine compressors supply air through bleed air flow control units and pneumatic shutoff valves to a pneumatic control assembly that inflates the wing boots. After an inflation period of 6 seconds, an electronic timer switches the distributor in the control assembly to deflate the wing boots, and a 4-second inflation begins in the horizontal stabilizer boots. After these boots have been inflated and deflated, the cycle is complete, and all boots are again held down tightly against the wings and horizontal stabilizer by vacuum. The spring-loaded switch must be selected up again for another cycle to occur.

Each engine supplies a common bleed air manifold. To ensure the operation of the system, if one engine is inoperative, a flow control unit with check valve is incorporated in the bleed air line from each engine to prevent the loss of pressure through the compressor of the inoperative engine. If the boots fail to function sequentially, they may be operated manually by selecting the DOWN position of the same deice cycle switch. Depressing and holding it in the manual

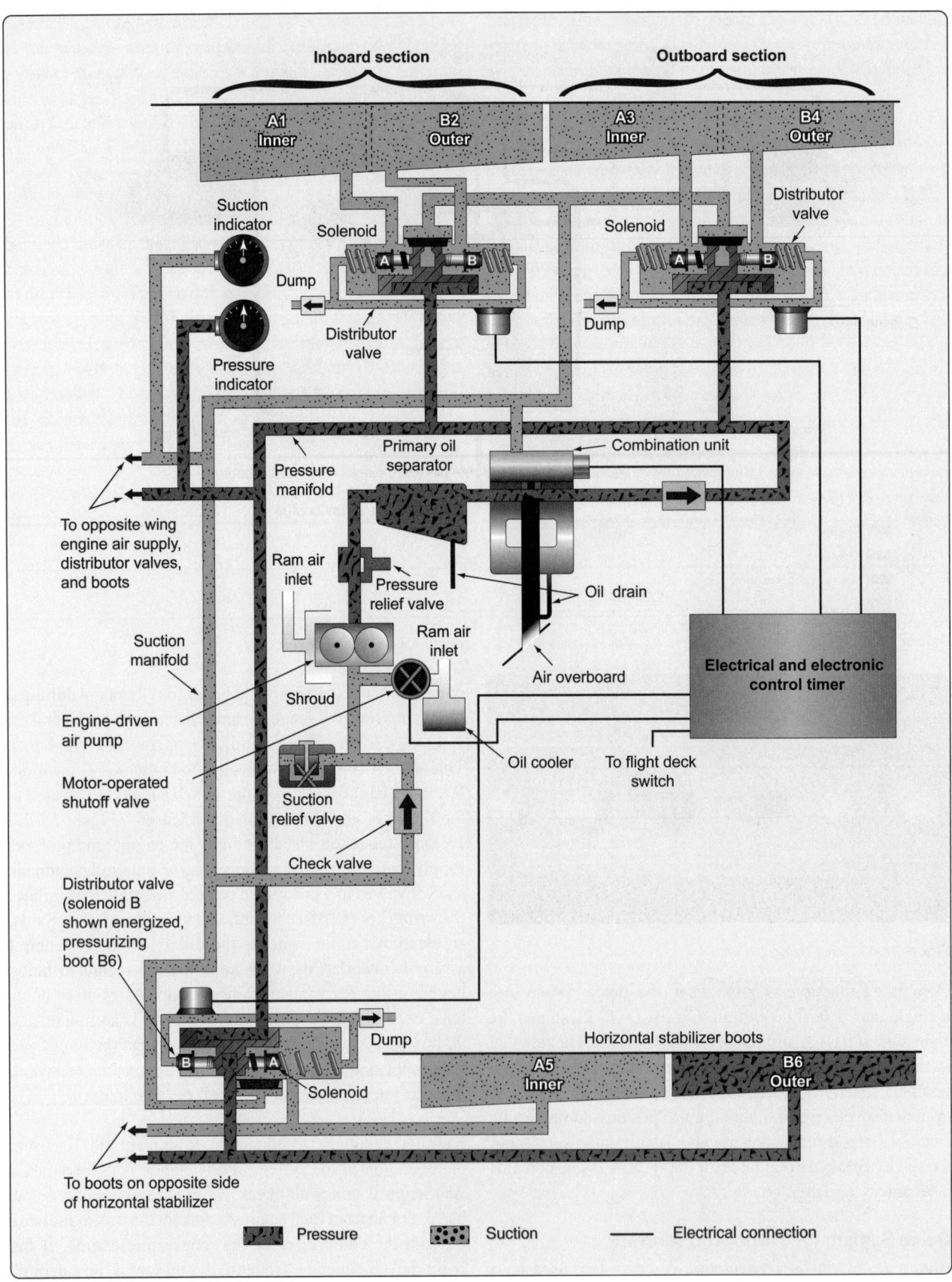

Figure 15-20. *Right-side deice boot system on a large aircraft (left side similar).*

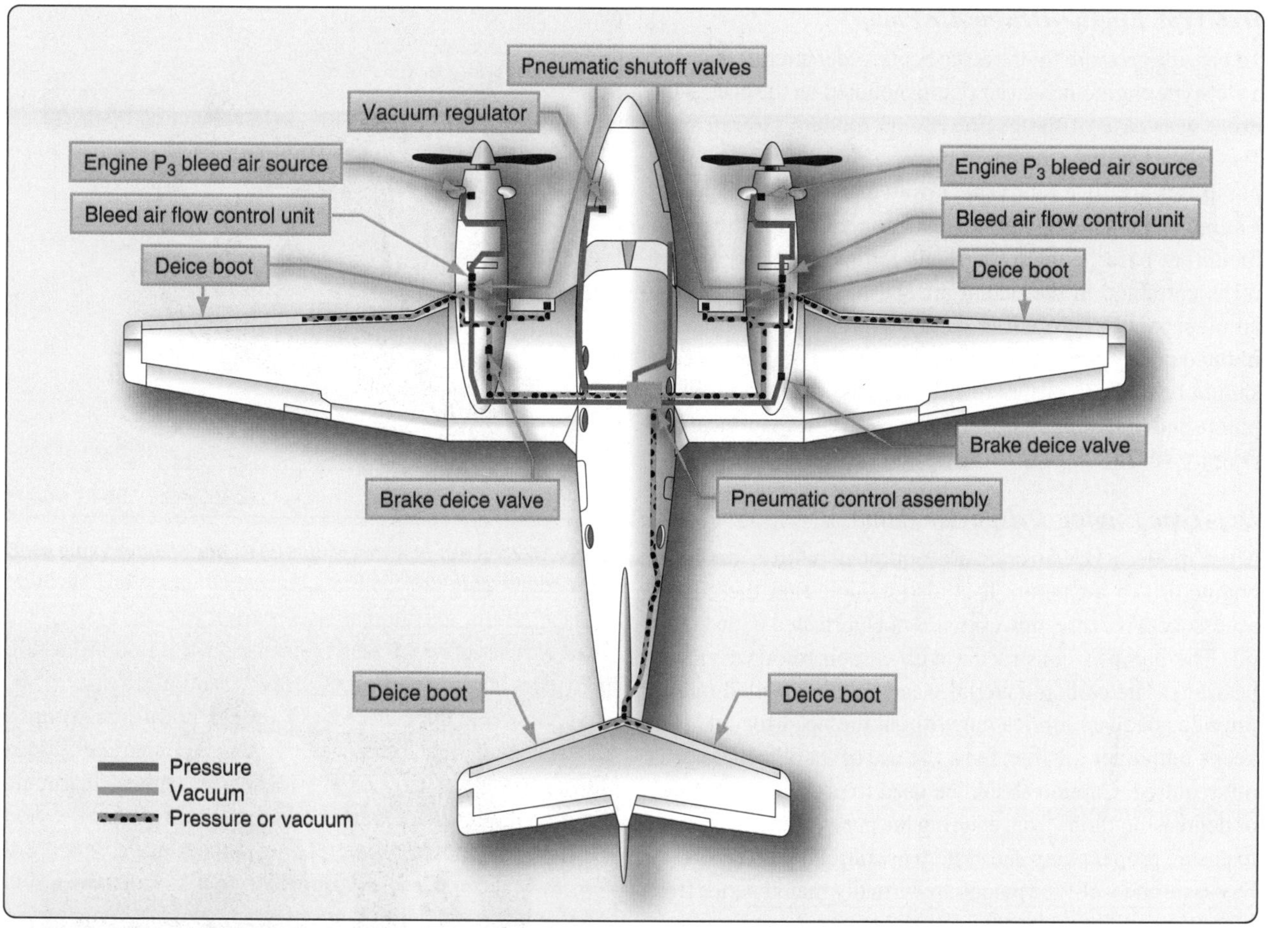

Figure 15-21. *Wing deice system for turboprop aircraft.*

Figure 15-22. *Ice protection panel on a turboprop aircraft with deice boots.*

DOWN position inflates all the boots simultaneously. When the switch is released, it returns to the (spring-loaded) off position, and each boot is deflated and held by vacuum. When operated manually, the boot should not be left inflated for more than 7 to 10 seconds, as a new layer of ice may begin to form on the expanded boots and become un-removable. If one engine is inoperative, the loss of its pneumatic pressure does not affect boot operation. Electric power to the boot system is required to inflate the boots in either single-cycle or manual operation. When electric power is lost, the vacuum holds the boots tightly against the leading edge.

Deicing System Components

Several components are used to construct all deice boot systems. The components may differ slightly in name and location within the system depending on the aircraft. Components may also combine functions to save space and weight. The basic functions of filtering, pressure regulation, distribution, and attachment to a vacuum when boots are not in use must all be present. Check valves must also be installed to prevent back flow in the system. Manifolds are common on multiengine aircraft to allow sourcing of low pressure air from both engine pumps. Note that air-pump pressure is typically expelled overboard when not needed. Bleed air is shut off by a valve when not needed for deice boot operation on turbine engine aircraft. A timer, or control unit with an automatic mode, exists on many aircraft to repeat the deice cycle periodically.

Wet-Type Engine-Driven Air Pump

To provide pressure for the deice boots, older aircraft may use a wet-type engine-driven air pump mounted on the accessory drive gear case of the engine. Some modern aircraft may also use a wet-type air pump because of its durability. The pump is typically a four vane, positive displacement pump. Engine oil passes from the accessory case through the pump mounting base flange to lubricate the pump. Some of the oil is entrained in the output air and must be removed by an oil separator before it is sent through other components in the deice system. When installing a wet-type pump, care should be taken to ensure that the oil passage in the gasket, pump, and mounting flange are aligned to ensure lubrication. *[Figure 15-23]*

Dry-Type Engine-Driven Air Pump

Most modern GA aircraft are equipped with a dry-type engine-driven air pump. It is also mounted on the engine accessory drive case; however, it is not lubricated with engine oil. The pump is constructed with carbon rotor vanes and bearings. The carbon material wears at a controlled rate to provide adequate lubrication without the need for oil. This keeps output air oil-free; thus, the use of an oil separator is not required. Caution should be used to prevent oil, grease, or degreasing fluids from entering the pump or the air system to ensure proper pump and system operation. *[Figure 15-24]* Dry-type and wet-type pumps are virtually maintenance free. Mounting bolts should be checked for security as should all hose connections. Wet-type pumps have a longer time before requiring overhaul, but dry-type pumps give the assurance that the deice system will not be contaminated with oil.

Oil Separator

An oil separator is required for each wet-type air pump. Pump output air flows through the separator where most of the oil is removed and sent back to the engine though a drain line. Some systems may include a secondary separator to ensure oil free air is delivered to the deice system. There are no moving parts in an oil separator. A convoluted (coiled or twisted) interior allows the air to pass, while the oil condenses and drains back to the engine. The only maintenance required on the separator is flushing the interior of the unit with a specified solvent. This should be done at intervals prescribed in the applicable maintenance manual. *[Figure 15-25]*

Control Valve

A control valve is a solenoid operated valve that allows air from the pump to enter the deice system. When energized by the deice switch in the flight deck, the valve opens. The control valve dumps pump air overboard when the deice system is not in use. Many control valves are built in combination with pressure relief valves that keeps the deice system safe from over pressure. *[Figure 15-26]*

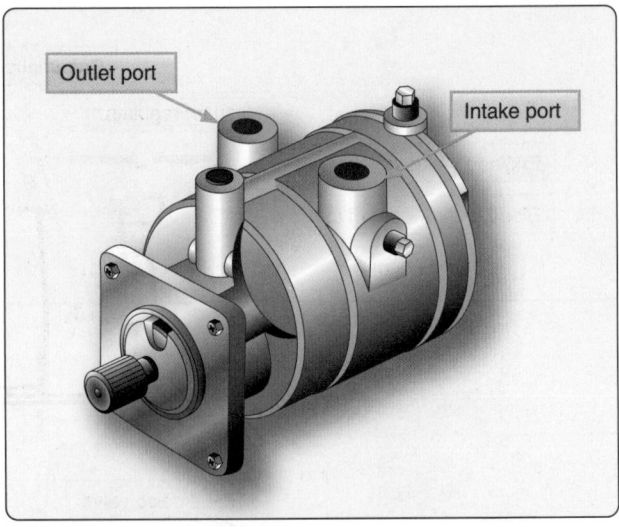

Figure 15-23. *A wet-type air pump with engine oil lubricating ports in the mounting flange.*

Figure 15-24. *Dry-type engine-driven air pump.*

Deflate Valve

All deice boot systems require a means for connecting vacuum from the air pump to the boots when the boots are not in use. This ensures the boots are held tightly deflated against the aircraft structure to provide the significant change in size and shape needed to break off accumulated ice when the boots inflate. One single deflate valve is used on simple deice boot systems. The deflate valve is solenoid operated. It is located at a point in the system where when closed, air is delivered to the boots. When open, vacuum is applied. Often, the deflate function is built into another unit, such as a distributor valve discussed next.

Distributor Valve

A distributer valve is a type of control valve used in relatively complex deice boot systems. It is an electrically-operated solenoid valve controlled by the deice boot system timer

Figure 15-25. *An oil separator used with a wet-type engine-driven air pump.*

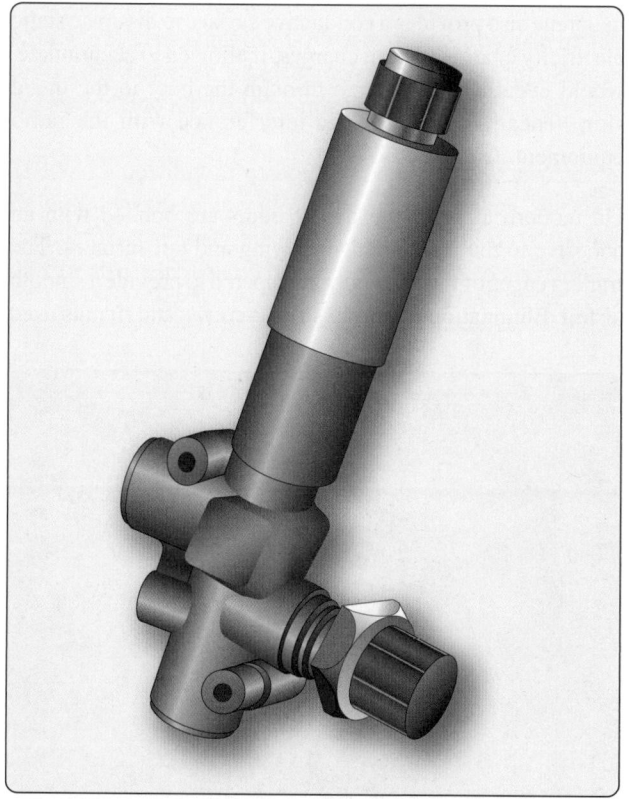

Figure 15-26. *A solenoid operated deice control valve.*

or control unit. On some systems, a distributor valve is assigned to each set of deice boots it controls. It differs from a control valve in that it has the deflate valve function built into it. Therefore, the distributor valve transfers connection of the boots from the pressure side of the air pump to the vacuum side of the pump once the proper inflation time has elapsed. The valve also dumps the unneeded air from the pump overboard.

Another type of distributor valve exists that handles the inflation and deflation of numerous sets of deice boots in a single unit. It also connects the boots to vacuum and dumps pump air when deice is not needed. A servo motor is used to position the multi-position valve. These centralized units are controlled by a timer or control unit. They inflate and deflate all of the boots on the aircraft. The timer may be built into the unit on some models.

Timer/Control Unit

All but the simplest of deice systems contain a timer or control unit. This device controls the action of the distributor valve(s) to ensure all boots are inflated in the proper sequence and for the correct duration. Six seconds of inflation is common to break off accumulated ice. The boot then must be immediately deflated so that ice does not adhere to the inflated geometry of the boot. This could cause it to fail to deflate or break off ice when the boot is re-inflated. The timer, or control unit, can also be made to cycle through the inflation and deflation of all boots periodically, thus relieving the flight crew of repetitive manual activation of the system. The function and capabilities of timers and control units vary. Consult the manufacturer's maintenance information for the performance characteristics of the timer/control unit on the aircraft in question.

The timer, or control unit, may be an independent device, or it may be built-in as part of another deice system component, such as a central distribution valve.

Note: A modern system design may use a pressure switch to signal deflation of the deice boots. When pressure builds in the boots to a preset amount, the switch signals the control valve to close and connect the boots to vacuum. However, this system retains a control unit for automatic cycling of the system at a set time interval.

Regulators & Relief Valves

Both the pressure and vacuum developed by an air pump must be regulated for use in the deice boot system. Typical boot inflation air pressure is between 15 and 20 psi. Vacuum pressure is set for the requirements of the gyroscopic instruments operated by the vacuum side of the air pump. Measured in inches of mercury, normal vacuum pressure (suction) is 4.5 to 5.5 "Hg.

Deice boot system air pressure is controlled by a pressure regulator valve located somewhere in the system downstream of the pump or oil separator, if installed. The regulator may be a stand-alone unit, or it may be combined into another deice system component. Regardless, the spring-loaded valve relieves pressure overboard when it exceeds the limit for which the system is designed.

A vacuum regulator is installed in the vacuum manifold on the suction side of the air pump to maintain the vacuum at the designed level. Also known as a suction regulating valve or similar, the spring-loaded valve contains a filter for the ambient air drawn through the valve during operation. This filter must be changed or kept clean per manufacturer's instructions. [Figure 15-27]

Manifold Assembly

In all pneumatic deice boot systems, it is necessary for check valves to be installed to prevent backflow of air in the system. The location(s) depends on system design. Sometimes, the check valve is built into another system component. On twin-engine aircraft, it is common to unite the air supplied from each engine-driven pump to provide redundancy. Check valves are required to guard against backflow should one pump fail. A manifold assemble is commonly used to join both sides of the system. [Figure 15-28] It contains the required check valves in a single assembly.

Inlet Filter

The air used in a deice boot system is ambient air drawn in upstream of the gyroscopic instruments on the suction side of the engine-driven air pump. This air must be free of contaminants for use spinning the gyros, as well as for inflation of the deice boots. To ensure clean air, an inlet filter is installed as the air intake point for the system. This filter must be regularly maintained as per manufacturer's

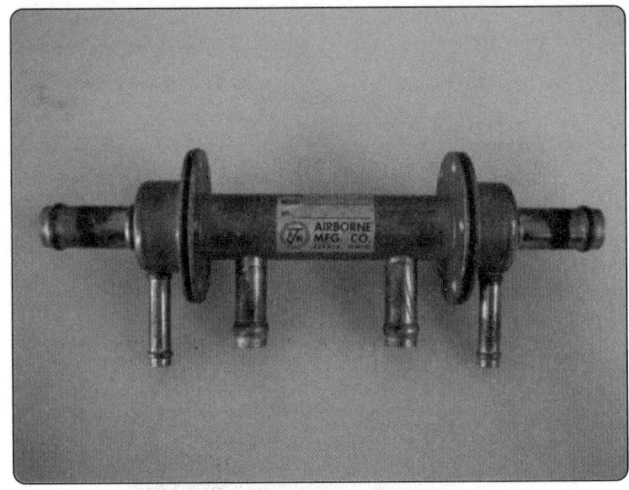

Figure 15-28. *A manifold assembly used in multiengine aircraft deice systems.*

instructions. *Figure 15-29* shows a typical inlet air filter. *Figure 15-30* shows the relationship of the vacuum regulator and inlet air filter to other system components.

Construction & Installation of Deice Boots

Deicer boots are made of soft, pliable rubber, or rubberized fabric, and contain tubular air cells. The outer ply of the deicer boot is of conductive neoprene to provide resistance to deterioration by the elements and many chemicals. The neoprene also provides a conductive surface to dissipate static electricity charges. These charges, if allowed to accumulate, would eventually discharge through the boot to the metal skin beneath, causing static interference with the radio equipment. [Figure 15-31]

On modern aircraft, the deicer boots are bonded with an adhesive to the leading edge of wing and tail surfaces. The trailing edges of this type boot are tapered to provide a smooth airfoil. Elimination of fairing strips, screws, and rivnuts used

Figure 15-27. *A vacuum regulator.*

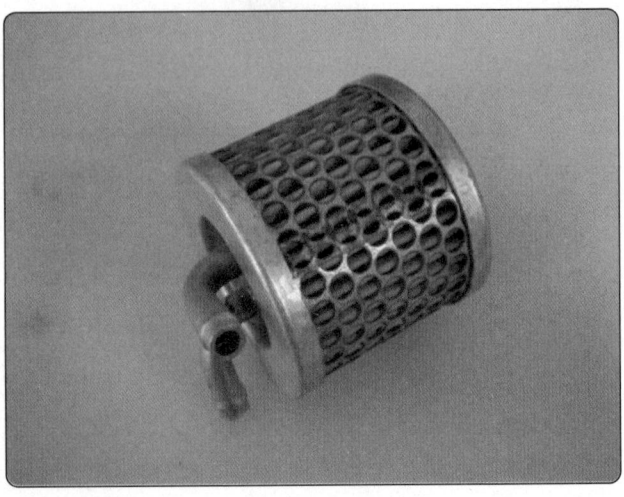

Figure 15-29. *Air filter for vacuum system.*

15-18

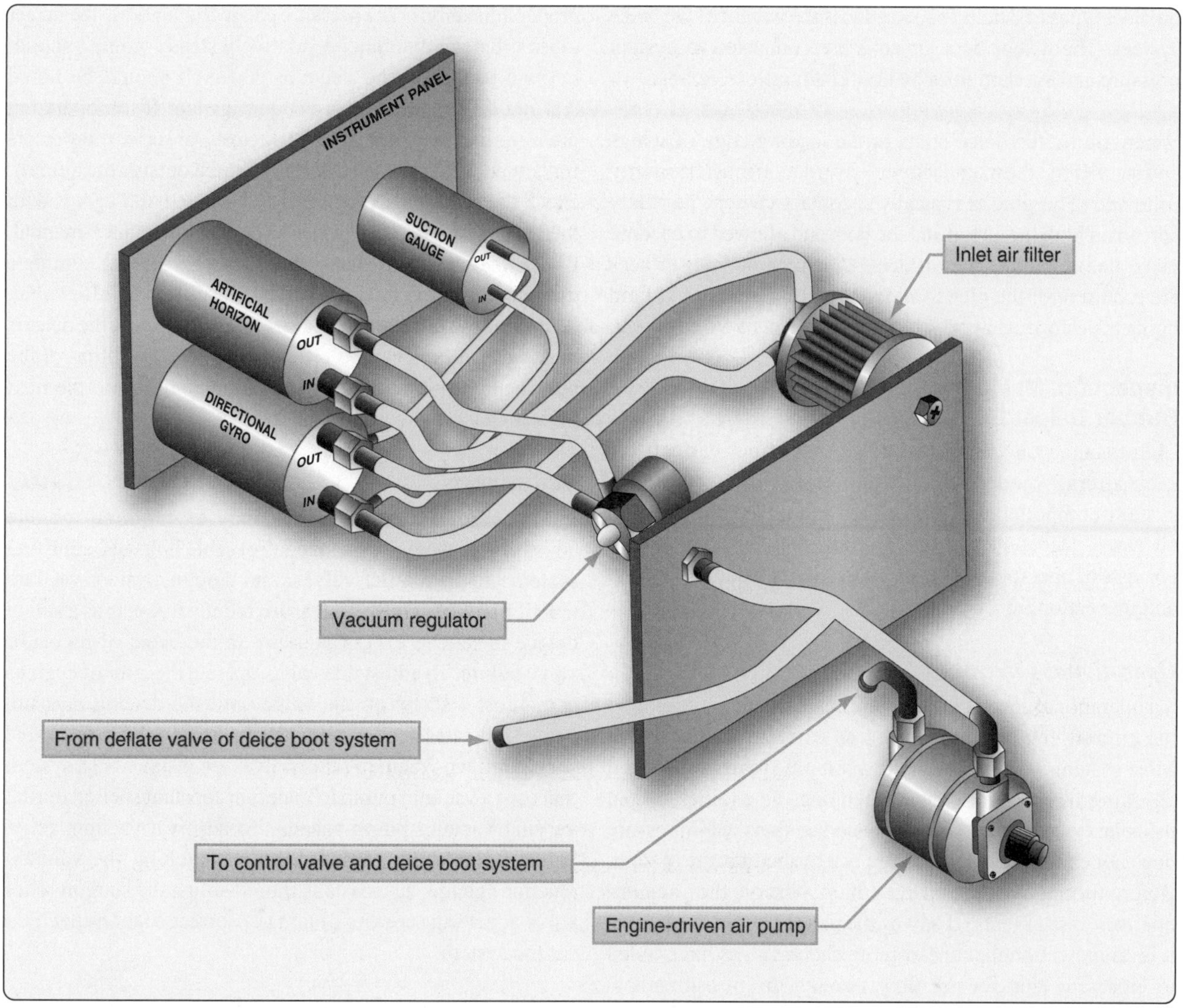

Figure 15-30. *Location of the inlet air filter in relationship to other components in a gyro instrument/pneumatic deice boot system.*

Figure 15-31. *Deicing boots inflated (left) and deflated (right).*

on older types of deicing boots reduces the weight of the deice system. The deicer boot air cells are connected to system pressure and vacuum lines by non-kinking flexible hose.

When gluing the deice boots to the leading edge of wings and stabilizers, the manufacturer's instruction must be strictly followed. The glue is typically a contact cement normally spread on both the airfoil and the boot and allowed to become tacky before mating the surfaces. Clean, paint-free surfaces are required for the glue to adhere properly. Removal of old boots is performed by re-softening the cement with solvent.

Inspection, Maintenance, & Troubleshooting of Rubber Deicer Boot Systems

Maintenance on pneumatic deicing systems varies with each aircraft model. The instructions of the airframe or system components manufacturer should be followed in all cases. Depending on the aircraft, maintenance usually consists of operational checks, adjustments, troubleshooting, and inspection.

Operational Checks

An operational check of the system can be made by operating the aircraft engines or by using an external source of air. Most systems are designed with a test plug to permit ground checking the system without operating the engines. When using an external air source, make certain that the air pressure does not exceed the test pressure established for the system. Before turning the deicing system on, observe the vacuum-operated instruments. If any of the gauges begin to operate, it is an indication that one or more check valves have failed to close and that reverse flow through the instruments is occurring. Correct the difficulty before continuing the test. If no movement of the instrument pointers occurs, turn on the deicing system. With the deicer system controls in their proper positions, check the suction and pressure gauges for proper indications. The pressure gauge fluctuates as the deicer tubes inflate and deflate. A relatively steady reading should be maintained on the vacuum gauge. It should be noted that not all systems use a vacuum gauge. If the operating pressure and vacuum are satisfactory, observe the deicers for actuation. With an observer stationed outside the aircraft, check the inflation sequence to be certain that it agrees with the sequence indicated in the aircraft maintenance manual. Check the timing of the system through several complete cycles. If the cycle time varies more than is allowable, determine the difficulty and correct it. Inflation of the deicers must be rapid to provide efficient deicing. Deflation of the boot being observed should be completed before the next inflation cycle. *[Figure 15-32]*

Adjustments

Examples of adjustments that may be required include adjusting the deicing system control cable linkages, adjusting system pressure relief valves, and deicing system vacuum (suction) relief valves. A pressure relief valve acts as a safety device to relieve excess pressure in the event of regulator valve failure. To adjust this valve, operate the aircraft engines and adjust a screw on the valve until the deicing pressure gauge indicates the specified pressure at which the valve should relieve. Vacuum relief valves are installed in a system that uses a vacuum pump to maintain constant suction during varying vacuum pump speeds. To adjust a vacuum relief valve, operate the engines. While watching the vacuum (suction) gauge, an assistant should adjust the suction relief valve adjusting screw to obtain the correct suction specified for the system.

Troubleshooting

Not all troubles that occur in a deicer system can be corrected by adjusting system components. Some troubles must be corrected by repair or replacement of system components or

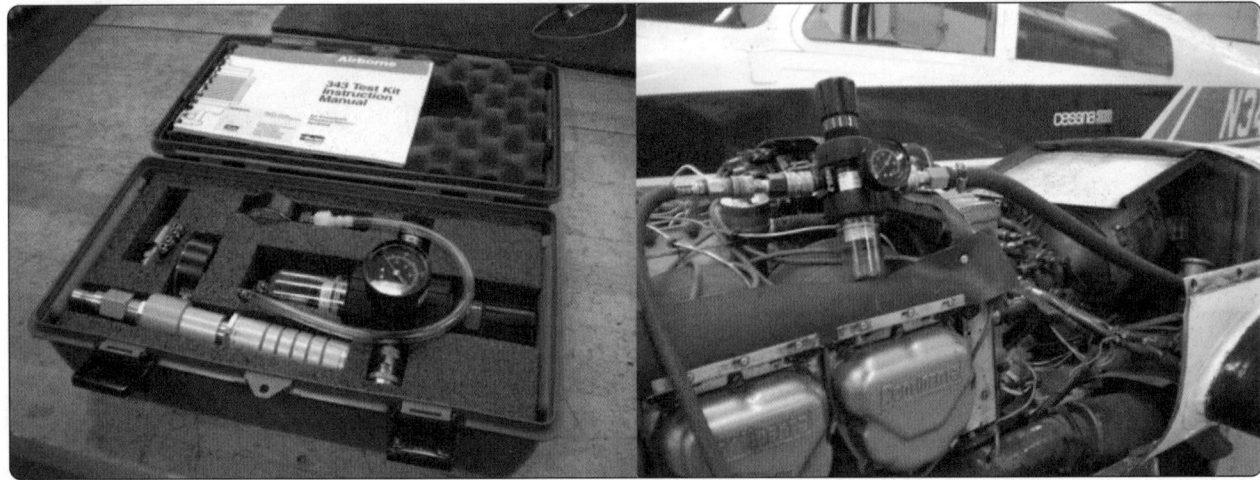

Figure 15-32. *Test equipment used to test a wing deice system (left), and test equipment installed in the aircraft for testing (right).*

by tightening loose connections. Several troubles common to pneumatic deicing systems are shown in the left-hand column of the chart in *Figure 15-33*. Note the probable causes and the remedy of each trouble listed in the chart. In addition to using troubleshooting charts, operational checks are sometimes necessary to determine the possible cause of trouble.

Inspection

During each preflight and scheduled inspection, check the deicer boots for cuts, tears, deterioration, punctures, and security; during periodic inspections, go a little further and check deicer components and lines for cracks. If weather cracking of rubber is noted, apply a coating of conductive cement. The cement, in addition to sealing the boots against weather, dissipates static electricity so that it does not puncture the boots by arcing to the metal surfaces.

Deice Boot Maintenance

The life of the deicers can be greatly extended by storing them when they are not needed and by observing these rules when they are in service:

1. Do not drag gasoline hoses over the deicers.
2. Keep deicers free of gasoline, oil, grease, dirt, and other deteriorating substances.
3. Do not lay tools on or lean maintenance equipment against the deicers.
4. Promptly repair or resurface the deicers when abrasion or deterioration is noted.
5. Wrap deice boots in paper or canvas when storing.

Problem	Causes (most of which can be identified with a 343 Test Kit)	Corrective action(s)

Boots do not inflate	• Open circuit breaker
	• Faulty deflate valve
	  Solenoid inoperable:
	    1. Improper voltage at solenoid
	    2. Blocked air vent in solenoid
	    3. Inoperative plunger
	  Diaphragm not seated
	    1. Blocked vent orifice located in rivet bottom at center of diaphragm
	    2. Dirty diaphragm seal area
	    3. Diaphragm ruptured
	• Two faulty deice control valves of faulty two-stage regulators
	• Faulty check valve
	• Relay not functioning
	• Leak in system boots
		• Reset circuit breaker
		• Check deflate valves as follows:
		  Solenoid inoperable:
		    1. Correct electrical system
		    2. Clean with alcohol or replace
		    3. Clean with alcohol or replace
		  Diaphragm not seated
		    1. Clean with .010 diameter wire and alcohol
		    2. Clean with blunt instrument and alcohol
		    3. Replace valve
		• Clean or replace valve assembly as noted above
		• Replace check valve
		• Check wiring or replace relay
		• Repair as needed

Slow boot inflation	• Lines blocked or disconnected • Low air pump capacity • One or more deice control valves not functioning properly • Deflate valve not fully closed • Ball check in deflate valve inoperative • Leaks in system or boots	• Check and replace lines • Replace air pump • Clean or replace valve assembly as noted above • Clean or replace valve assembly as noted above • Clean check valve or replace deflate valve • Repair as needed
System will not cycle	• Pressure in system not reaching specified psi to activate pressure switch • Leak in system or boots • Pressure switch on deflate valve inoperative	• Clean or replace deice control valve as noted above • Clean or replace deflate valve, as noted above • Repair as needed, tighten all hose connections • Replace switch
Slow deflation	• Low vacuum • Faulty deflate valve (indicated by temporary reduction in suction gauge reading)	• Repair as needed • Clean or replace valve assembly as noted above
No vacuum for boot hold down	• Malfunctioning deflate valve or deice valve • Leak in system or boots	• Clean or replace valve assembly as noted above • Repair as needed
Boots will not deflate during cycle	• Faulty deflate valve	• Check and replace valve
Boots appear to inflate on aircraft climb	• Vacuum source for boot holddown inoperative • Lines running through pressurized cabin loose or disconnected	• Check operation of ball check in deflate valve • Check for loose or disconnected vacuum lines and repair

Figure 15-33. *Troubleshooting guide for wing deice system.*

Thus far, preventive maintenance has been discussed. The actual work on the deicers consists of cleaning, resurfacing, and repairing. Cleaning should ordinarily be done at the same time the aircraft is washed, using a mild soap and water solution. Grease and oil can be removed with a cleaning agent, such as naptha, followed by soap and water scrubbing. Whenever the degree of wear is such that it indicates that the electrical conductivity of the deicer surface has been destroyed, it may be necessary to resurface the deicer. The resurfacing substance is a black, conductive neoprene cement. Prior to applying the resurfacing material, the deicer must be cleaned thoroughly and the surface roughened. Cold patch repairs can be made on a damaged deicer. The deicer must be relieved of its installed tension before applying the patch. The area to be patched must be clean and buffed to roughen the surface slightly. Patches are glued in place. Follow manufacturer's instructions for all repairs.

Electric Deice Boots

A few modern aircraft are equipped with electric deice boots on wing sections or on the horizontal stabilizer. These boots contain electric heating elements which are bonded to the leading edges similarly to pneumatic deice boots. When activated, the boots heat up and melt the ice off of leading edge surfaces. The elements are controlled by a sequence timer in a deice controller. Ice detector and ram air temperature probe inputs initiate operation when other flight condition parameters exist. The boot elements turn ON and OFF in paired sections to avoid aerodynamic imbalance. The system is inoperative while the aircraft is on the ground. *Figure 15-34* illustrated such a system. A benefit of electric deice boots is the conservation of engine bleed air. Current draw is limited to only those periods when de-ice is required.

Propeller Deice System

The formation of ice on the propeller leading edges, cuffs, and spinner reduces the efficiency of the powerplant system. Deice systems using electrical heating elements and systems using chemical deicing fluid are used.

Electrothermal Propeller Device System

Many propellers are deiced by an electrically heated boot on each blade. The boot, firmly cemented in place, receives current from a slip ring and brush assembly on the spinner

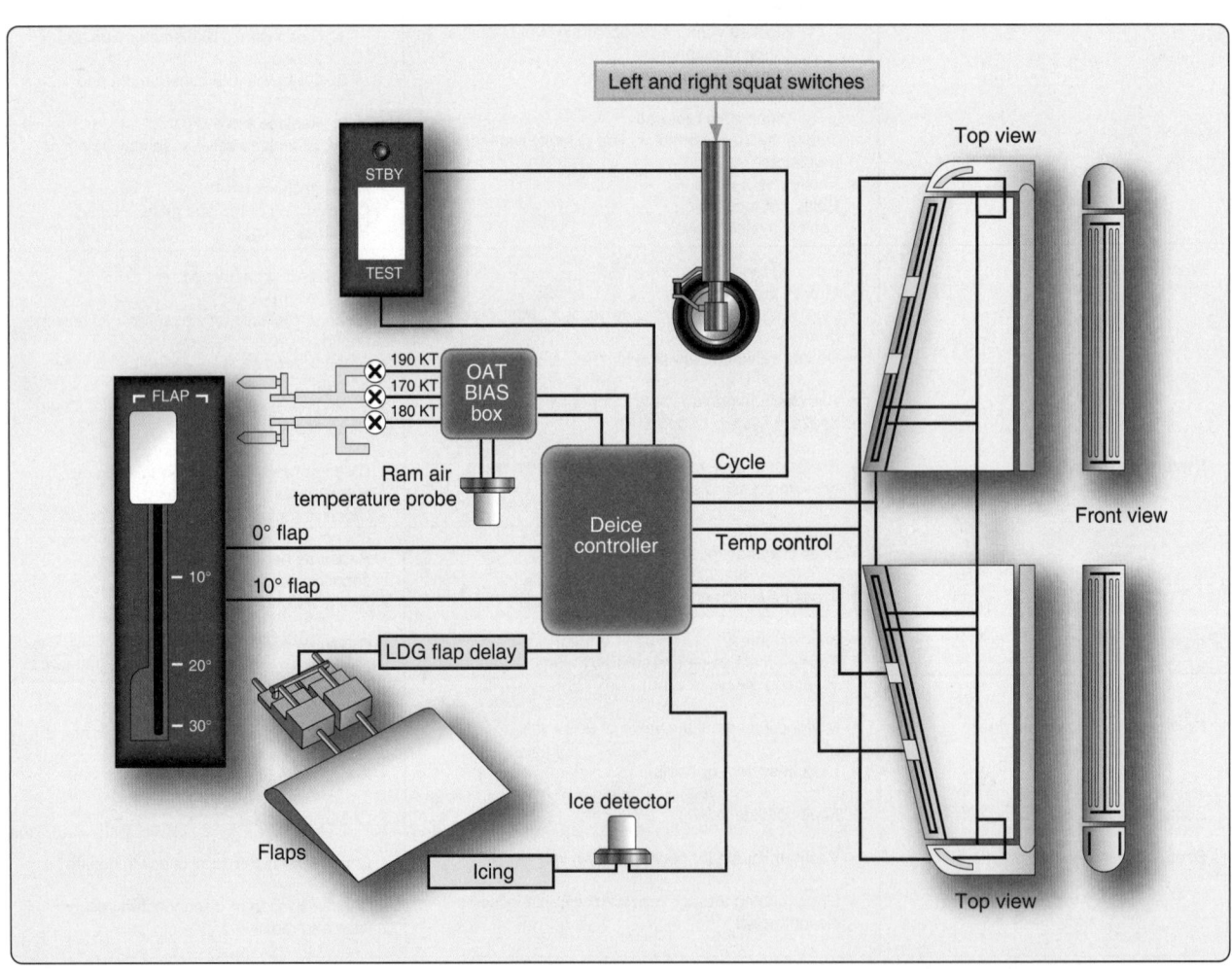

Figure 15-34. *Electric stabilizer deice system.*

bulkhead. The slip ring transmits current to the deice boot. The centrifugal force of the spinning propeller and air blast breaks the ice particles loose from the heated blades. *[Figure 15-35]*

On one aircraft model, the boots are heated in a preset sequence, which is an automatic function controlled by a timer. This sequence is as follows: 30 seconds for the right prop outer elements; 30 seconds for the right prop inner elements; 30 seconds for the left prop outer elements; and, 30 seconds for the left prop inner elements. Once the system is turned on for automatic is activated, it cycles continuously. A manual bypass of the timer is incorporated. *[Figure 15-36]*

Chemical Propeller Deice

Some aircraft models, especially single-engine GA aircraft, use a chemical deicing system for the propellers. Ice usually appears on the propeller before it forms on the wing. The glycol-based fluid is metered from a tank by a small electrically driven pump through a microfilter to the slinger rings on the prop hub. The propeller system can be a stand-alone system, or it can be part of a chemical wing and stabilizer deicing system such as the TKS™ weeping system.

Ground Deicing of Aircraft

The presence of ice on an aircraft may be the result of direct precipitation, formation of frost on integral fuel tanks after prolonged flight at high altitude, or accumulations on the landing gear following taxiing through snow or slush. In accordance with the Federal Aviation Administration (FAA) Advisory Circular (AC) 120-60, the aircraft must be free of all frozen contaminants adhering to the wings, control surfaces, propellers, engine inlets, or other critical surfaces before takeoff.

Any deposits of ice, snow, or frost on the external surfaces of an aircraft may drastically affect its performance. This may be due to reduced aerodynamic lift and increased aerodynamic drag resulting from the disturbed airflow over the airfoil surfaces, or it may be due to the weight of the deposit over the whole aircraft. The operation of an aircraft may also be seriously affected by the freezing of moisture in controls, hinges, valves, microswitches, or by the ingestion of ice into the engine. When aircraft are hangared to melt snow or frost,

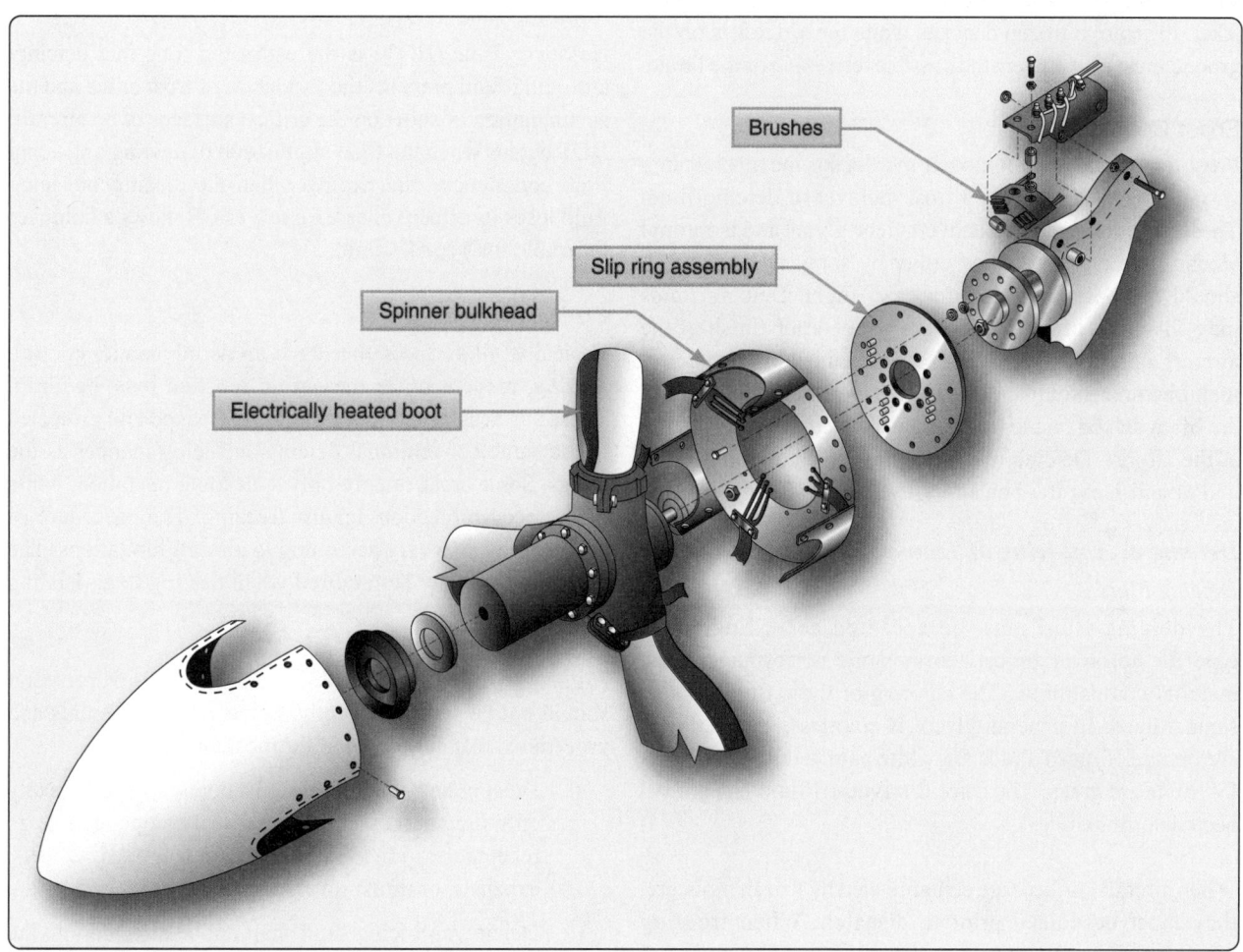

Figure 15-35. *Electro thermal propeller deice system components.*

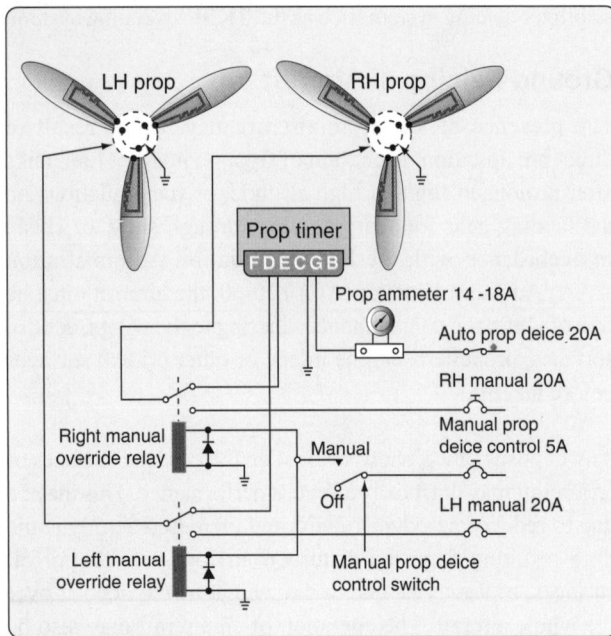

Figure 15-36. *Propeller electrical deice system schematic.*

Figure 15-37. *An American Airlines aircraft being deiced at Syracuse Hancock International Airport.*

any melted snow or ice may freeze again if the aircraft is subsequently moved into subzero temperatures. Any measures taken to remove frozen deposits while the aircraft is on the ground must also prevent the possible refreezing of the liquid.

Frost Removal

Frost deposits can be removed by placing the aircraft in a warm hangar or by using a frost remover or deicing fluid. These fluids normally contain ethylene glycol and isopropyl alcohol and can be applied either by spray or by hand. It should be applied within 2 hours of flight. Deicing fluids may adversely affect windows or the exterior finish of the aircraft, only the type of fluid recommended by the aircraft manufacturer should be used. Transport category aircraft are often deiced on the ramp or a dedicated deicing location at the airport. Deicing trucks are used to spray the deicing and/or anti-icing fluid on aircraft surfaces. *[Figure 15-37]*

Deicing & Anti-icing of Transport Type Aircraft
Deicing Fluid

The deicing fluid must be accepted according to its type for holdover times, aerodynamic performance, and material compatibility. The coloring of these fluids is also standardized. In general, glycol is colorless, Type-I fluids are orange, Type-II fluids are white/pale yellow, and Type-IV fluids are green. The color for Type-III fluid has not yet been determined.

When aircraft surfaces are contaminated by frozen moisture, they must be deiced prior to dispatch. When freezing precipitation exists, and there is a risk of contamination of

the surface at the time of dispatch, aircraft surfaces must be anti-iced. If both deicing and anti-icing are required, the procedure may be performed in one or two steps. The selection of a one- or two-step process depends upon weather conditions, available equipment, available fluids, and the holdover time to be achieved.

Holdover Time (HOT)

Holdover Time (HOT) is the estimated time that deicing/anti-icing fluid prevents the formation of frost or ice and the accumulation of snow on the critical surfaces of an aircraft. HOT begins when the final application of deicing/anti-icing fluid commences and expires when the deicing/anti-icing fluid loses its effectiveness. *Figure 15-38* shows a holdover timetable for Type IV fluid.

Critical Surfaces

Basically, all surfaces that have an aerodynamic, control, sensing, movement, or measuring function must be clean. These surfaces cannot necessarily be cleaned and protected in the same conventional deicing/anti-icing manner as the wings. Some areas require only a cleaning operation, while others need protection against freezing. The procedure of deicing may also vary according to aircraft limitations. The use of hot air may be required when deicing (e.g., landing gear or propellers).

Figure 15-39 shows critical areas on an aircraft that should not be sprayed directly. Some critical elements and procedures that are common for most aircraft are:

- Deicing/anti-icing fluids must not be sprayed directly on wiring harnesses and electrical components (e.g., receptacles, junction boxes), onto brakes, wheels, exhausts, or thrust reversers.

- Deicing/anti-icing fluid shall not be directed into the orifices of pitot heads, static ports, or directly onto

FAA Type IV Holdover Time Guidelines

Guidelines for holdover times anticipated for SAE type IV fluid mixtures as function of weather conditions and OAT.
CAUTION: This table is for use in departure planning only, and it should be used in conjunction with pretakeoff check procedures.

OAT °C	OAT °F	SAE type IV fluid concentration neat fluid water (vol. %/vol.%)	Frost*	Freezing Fog	Snow◊	Freezing drizzle***	Light free rain	Rain on cold soaked wing	Other*
above 0	above 32	100/0	18:00	1:05–2:15	0:35–1:05	0:40–1:10	0:25–0:40	0:10–0:50	
		72/25	6:00	1:05–1:45	0:30–1:05	0:35–0:50	0:15–0:30	0:05–0:35	CAUTION: no holdover time guidelines exist
		50/50	4:00	0:15–0:35	0:05–0:20	0:10–0:20	0:05–0:10		
0 through –3	32 through 27	100/0	12:00	1:05–2:15	0:30–0:55	0:40–1:10	0:15–0:40	CAUTION: clear ice may require touch for confirmation	
		75/25	5:00	1:05–2:15	0:25–0:50	0:35–0:50	0:15–0:30		
		50/50	3:00	1:15–0:35	0:05–0:15	0:10–0:20	0:05–0:15		
below –3 through –14	below 27 through 7	100/0	12:00	0:20–0:50	0:20–0:40	**0:20–0:45	**0:10–0:25		
		75/25	5:00	0:25–0:50	0:15–0:25	**0:15–0:30	**0:10–0:20		
below –14 through –25	below 7 through –13	100/0	12:00	0:15–0:40	0:15–0:30				
below –25	below –13	100/0	SAE type IV fluid may be used below –25 °C(–13 °F) if the freezing point of the fluid is at least 7 °C (13 °F) below the OAT and the aerodynamic acceptance criteria are met. Consider use of SAE type I when SAE type IV fluid cannot be used.						

°C = degrees Celsius
°F = degrees Fahrenheit
OAT= outside air temperature
VOL = volume

The responsiblity for the application of these data remains with the user.
  * During conditions that apply to aircraft protection for ACTIVE FROST.
 ** No holdover time guidelines exist for this condition below –10 °C (14 °F).
*** Use light freezing rain holdover times if positive identification of freezing drizzle is not possible.
 ‡ Snow pellets, ice pellets, heavy snow, moderate and heavy freezing rain, hail.
 ◊ Snow includes snow grains.

CAUTIONS:
- The time of protection will be shortened in heavy weather conditions: heavy precipitation rates or high moisture contents.
- High wind velocity or jet blast may reduce holdover time below the lowest time stated in the range.
- Holdover time may be reduced when aircraft skin temperature is lower than OAT.

Figure 15-38. *FAA deice holdover time guidelines.*

airstream direction detectors probes/angle of attack airflow sensors.

- All reasonable precautions shall be taken to minimize fluid entry into engines, other intakes/outlets, and control surface cavities.

- Fluids shall not be directed onto flight deck or cabin windows as this can cause crazing of acrylics or penetration of the window seals.

- Any forward area from which fluid can blow back onto windscreens during taxi or subsequent takeoff shall be free of residues prior to departure.

- If Type II, III, or IV fluids are used, all traces of the fluid on flight deck windows should be removed prior to departure, particular attention being paid to windows fitted with wipers.

- Landing gear and wheel bays shall be kept free from buildup of slush, ice, or accumulations of blown snow.

- When removing ice, snow, slush, or frost from aircraft surfaces, care shall be taken to prevent it entering and accumulating in auxiliary intakes or control surface

hinge areas (e.g., manually remove snow from wings and stabilizer surfaces forward toward the leading edge and remove from ailerons and elevators back towards the trailing edge).

Ice & Snow Removal

Probably the most difficult deposit to deal with is deep, wet snow when ambient temperatures are slightly above the freezing point. This type of deposit should be removed with a soft brush or squeegee. Use care to avoid damage to antennas, vents, stall warning devices, vortex generators, etc., that may be concealed by the snow. Light, dry snow in subzero temperatures should be blown off whenever possible; the use of hot air is not recommended, since this would melt the snow, which would then freeze and require further treatment. Moderate or heavy ice and residual snow deposits should be removed with a deicing fluid. No attempt should be made to remove ice deposits or break an ice bond by force.

After completion of deicing operations, inspect the aircraft to ensure that its condition is satisfactory for flight. All external surfaces should be examined for signs of residual snow or

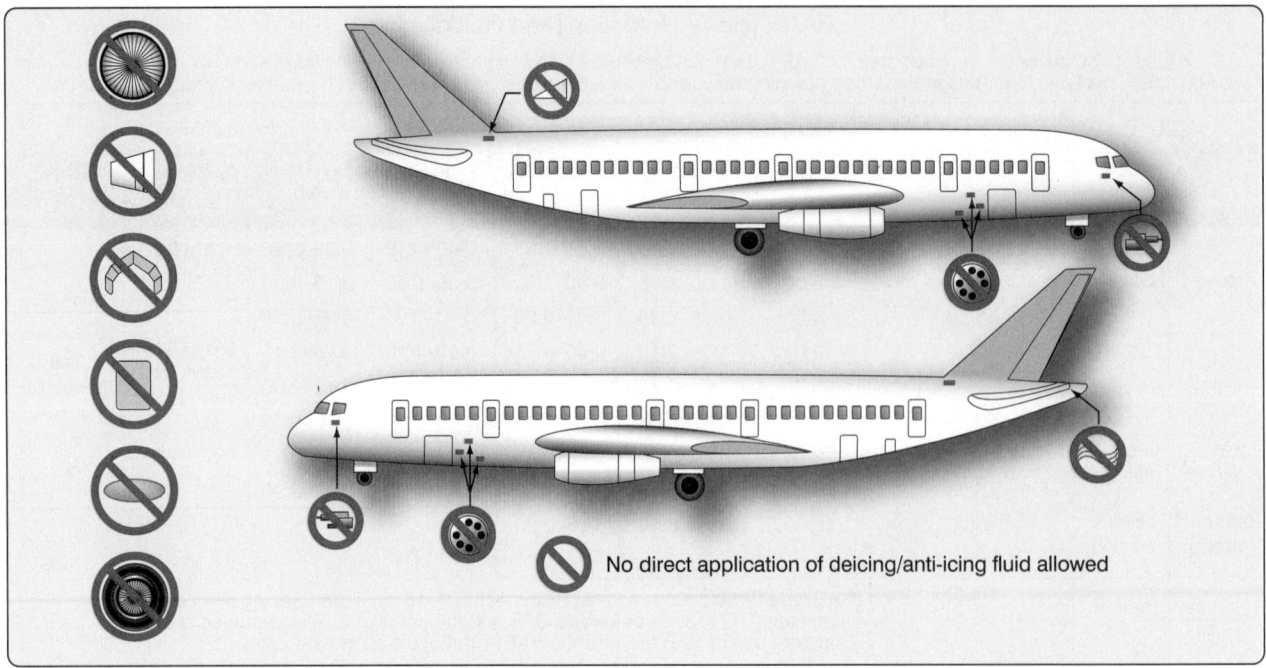

Figure 15-39. *No direct application of deicing/anti-icing fluid allowed.*

ice, particularly in the vicinity of control gaps and hinges. Check the drain and pressure sensing ports for obstructions. When it becomes necessary to physically remove a layer of snow, all protrusions and vents should be examined for signs of damage. Control surfaces should be moved to ascertain that they have full and free movement. The landing gear mechanism, doors and bay, and wheel brakes should be inspected for snow or ice deposits and the operation of up-locks and microswitches checked.

Snow or ice can enter turbine engine intakes and freeze in the compressor. If the compressor cannot be turned by hand for this reason, hot air should be blown through the engine until the rotating parts are free.

Rain Control Systems

There are several different ways to remove the rain from the windshields. Most aircraft use one or a combination of the following systems: windshield wipers, chemical rain repellent, pneumatic rain removal (jet blast), or windshields treated with a hydrophobic surface seal coating.

Windshield Wiper Systems

In an electrical windshield wiper system, the wiper blades are driven by an electric motor(s) that receive(s) power from the aircraft's electrical system. On some aircraft, the pilot's and copilot's windshield wipers are operated by separate systems to ensure that clear vision is maintained through one of the windows should one system fail. Each windshield wiper assembly consists of a wiper, wiper arm, and a wiper motor/converter. Almost all windshield wiper systems use

electrical motors. Some older aircraft might be equipped with hydraulic wiper motors. *[Figure 15-40]*

Maintenance performed on windshield wiper systems consists of operational checks, adjustments, and troubleshooting. An operational check should be performed whenever a system component is replaced or whenever the system is suspected of not working properly. During the check, make sure that the windshield area covered by the wipers is free of foreign matter and is kept wet with water. Adjustment of a windshield wiper system consists of adjusting the wiper blade tension, the angle at which the blade sweeps across the windshield, and proper parking of the wiper blades.

Chemical Rain Repellent

Water poured onto clean glass spreads out evenly. Even when the glass is held at a steep angle or subjected to air velocity, the glass remains wetted by a thin film of water. However, when glass is treated with certain chemicals, a transparent film is formed that causes the water to behave very much like mercury on glass. The water draws up into beads that cover only a portion of the glass and the area between beads is dry. The water is readily removed from the glass. This principle lends itself quite naturally to removing rain from aircraft windshields. The high-velocity slipstream continually removes the water beads, leaving a large part of the window dry.

A rain repellent system permits application of the chemical repellent by a switch or push button in the flight deck. The proper amount of repellent is applied regardless of how

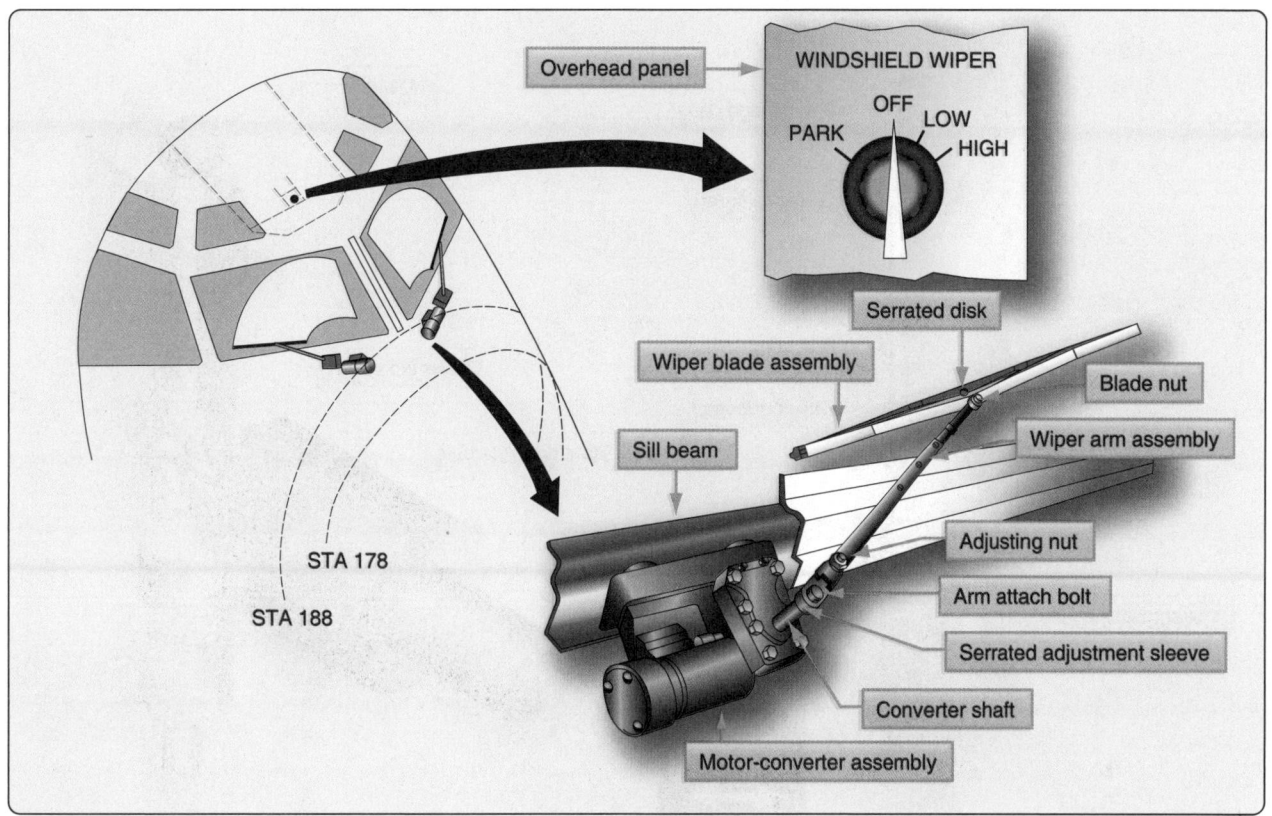

Figure 15-40. *Windshield wiper assembly/installation on a transport category aircraft. The motor-converter is mounted under the aircraft skin.*

long the switch is held. On some systems, a solenoid valve controlled by a time delay module meters the repellent to a nozzle which sprays it on the outside of the windshield. Two such units exist — one each for the forward glass of the pilot and copilot. *[Figure 15-41]*

This system should only be used in very wet conditions. The rain repellent system should not be operated on dry windows because heavy undiluted repellent restricts window visibility. Should the system be operated inadvertently, do not operate the windshield wipers or rain clearing system as this tends to increase smearing. Also, the rain repellent residues caused by application in dry weather or very light rain can cause staining or minor corrosion of the aircraft skin. To prevent this, any concentrated repellent or residue should be removed by a thorough fresh water rinse at the earliest opportunity. After application, the repellent film slowly deteriorates with continuing rain impingement. This makes periodic reapplication necessary. The length of time between applications depends upon rain intensity, the type of repellent used, and whether windshield wipers are used.

Windshield Surface Seal Coating

Some aircraft models use a surface seal coating, also called hydrophobic coating that is on the outside of the pilot's/

copilot's windshield. *[Figure 15-42]* The word hydrophobic means to repel or not absorb water. The windshield hydrophobic coating is on the external surface of the windows (windshields). The coatings cause raindrops to bead up and roll off, allowing the flight crew to see through the windshield with very little distortion. The hydrophobic windshield coating reduces the need for wipers and gives the flight crew better visibility during heavy rain.

Most new aircraft windshields are treated with surface seal coating. The manufacturer's coating process deeply penetrates the windshield surface providing hydrophobic action for quite some time. When effectiveness declines, products made to be applied in the field are used. These liquid treatments rubbed onto the surface of the windshield maintain the beading action of rain water. They must be applied periodically or as needed.

Pneumatic Rain Removal Systems

Windshield wipers characteristically have two basic problem areas. One is the tendency of the slipstream aerodynamic forces to reduce the wiper blade loading pressure on the window, causing ineffective wiping or streaking. The other is in achieving fast enough wiper oscillation to keep up with high rain impingement rates during heavy rain falls. As a result, most aircraft wiper systems fail to provide satisfactory vision in heavy rain.

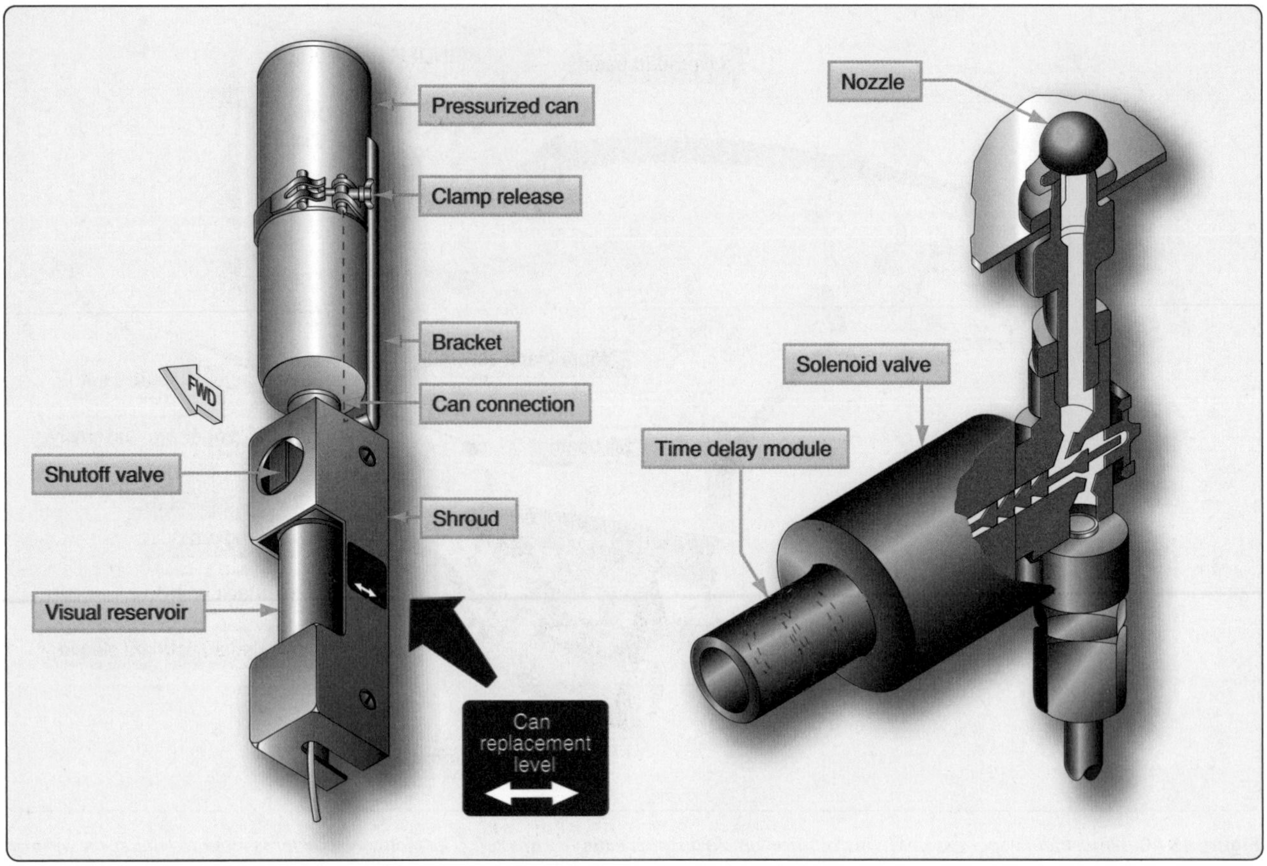

Figure 15-41. *Flight deck rain repellent canister and reservoir.*

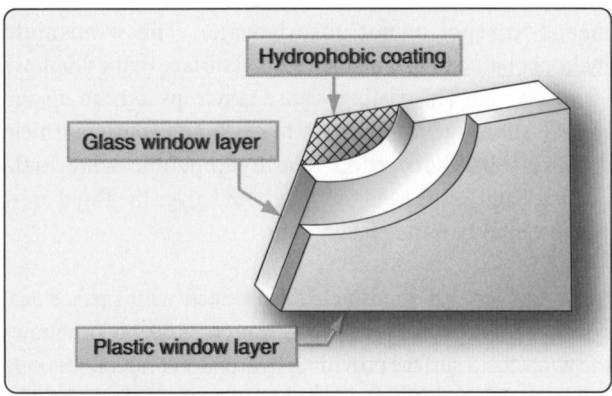

Figure 15-42. *Hydrophobic coating on windshield.*

The rain removal system shown in *Figure 15-43* controls windshield icing and removes rain by directing a flow of heated air over the windshield. This heated air serves two purposes. First, the air breaks the rain drops into small particles that are then blown away. Secondly, the air heats the windshield to prevent the moisture from freezing. The air can be supplied by an electric blower or by bleed air.

Windshield Frost, Fog, & Ice Control Systems

In order to keep windshield areas free of ice, frost, and fog, window anti-icing, deicing, and defogging systems are used. These can be electric, pneumatic, or chemical depending on the type and complexity of the aircraft. A few of these systems are discussed in this section.

Electric

High performance and transport category aircraft windshields are typically made of laminated glass, polycarbonate, or similar ply material. Typically, clear vinyl plies are also included to improve performance characteristics. The laminations create the strength and impact resistance of the windshield assembly. These are critical features for windshields as they are subject to a wide range of temperatures and pressures. They must also withstand the force of a 4-pound bird strike at cruising speed to be certified. The laminated construction facilitates the inclusion of electric heating elements into the glass layers, which are used to keep the windshield clear of ice, frost, and fog. The elements can be in the form of resistance wires or a transparent conductive material may be used as one of the window plies. To ensure enough heating is applied to the outside of the windshield, heating elements are placed on the inside of the outer glass

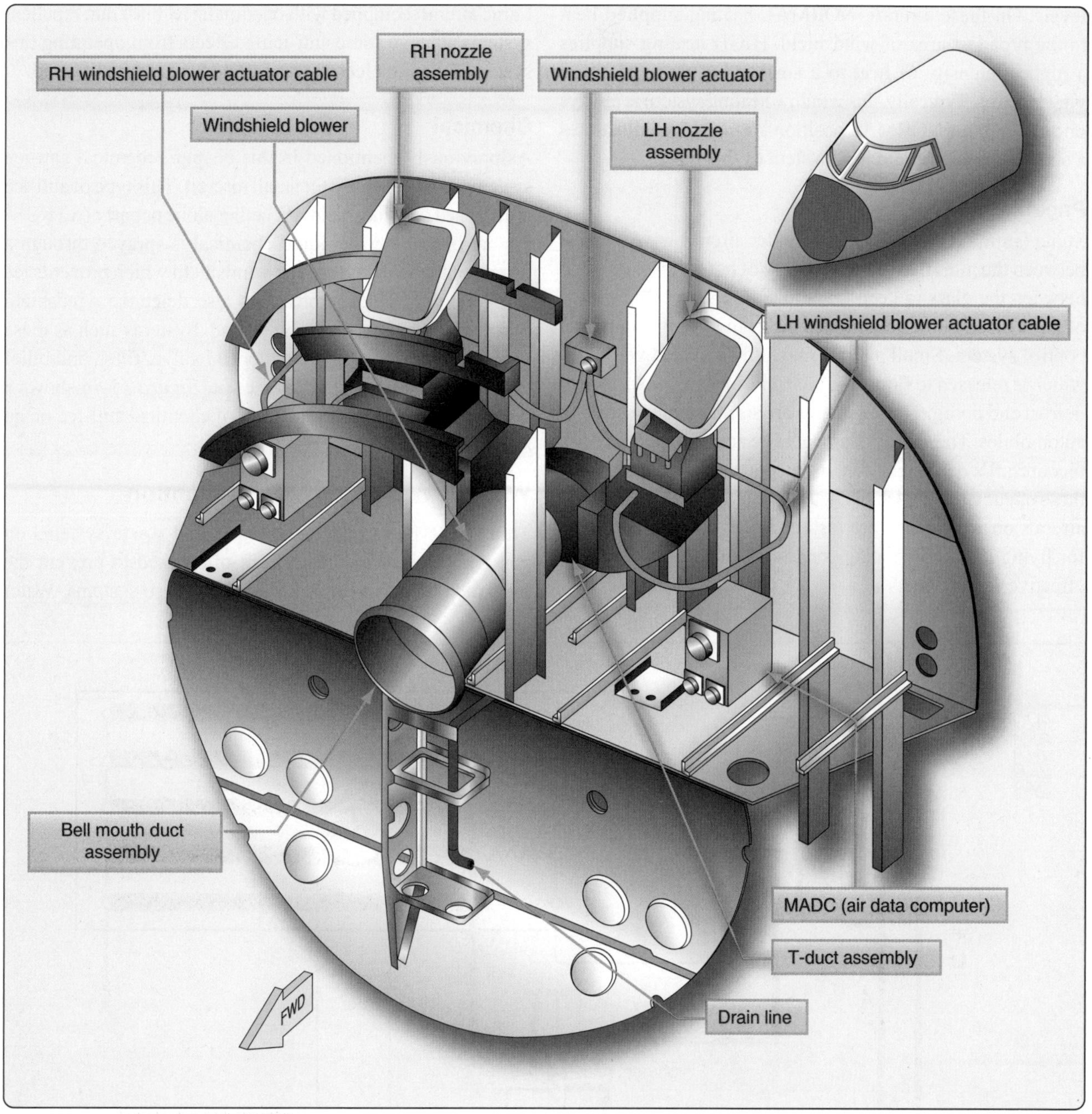

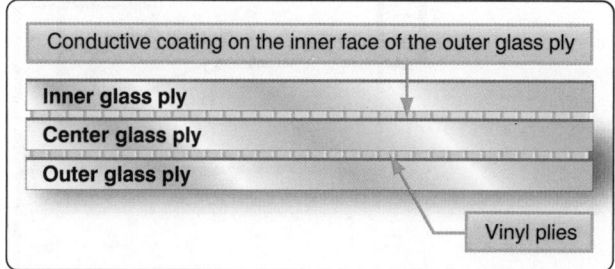

Figure 15-43. *Windshield rain and frost removal system.*

ply. Windshields are typically bonded together by the application of pressure and heat without the use of cement. *Figure 15-44* illustrates the plies in one transport category aircraft windshield.

Whether resistance wires or a laminated conductive film is used, aircraft window heat systems have transformers to supply power and feedback mechanisms, such as thermistors, to provide a window heat control unit with information used to keep operating temperature within acceptable limits. Some systems are automatic while others are controlled by flight deck switches. Separate circuits for pilot and co-pilot are common to ensure visibility in case of a malfunction. Consult

Conductive coating on the inner face of the outer glass ply
Inner glass ply
Center glass ply
Outer glass ply
Vinyl plies

Figure 15-44. *Cross-section of a transport category windshield.*

the manufacturer's maintenance information for details on the particular window heat system in question.

Some windshield heating systems can be operated at two heat

levels. On these aircraft, NORMAL heating supplied heat to the broadest area of windshield. HIGH heating supplies a higher intensity of heat to a smaller but more essential viewing area. Typically, this window heating system is always on and set in the NORMAL position. *Figure 15-45* illustrates a simplified windshield heat system of this type.

Pneumatic

Some laminated windshields on older aircraft have a space between the plies that allows the flow of hot air to be directed between the glass to keep it warm and fog free. The source of air is bleed air or conditioned air from the environmental control system. Small aircraft may utilize ducted warm air, which is released to flow over the windshield inner surface to defrost and defog. These systems are similar to those used in automobiles. The source of air could be ambient (defog only), the aircraft's heating system, or a combustion heater. While these pneumatic windshield heat systems are effective for the aircraft on which they are installed, they are not approved for flying into known icing conditions and, as such, are not effective for anti-ice.

Large aircraft equipped with pneumatic jet blast rain repellent systems achieve some anti-icing effects from operating this system although electric windshield heat is usually used.

Chemical

As previously mentioned in this chapter, chemical anti-ice systems exist generally for small aircraft. This type of anti-ice is also used on windshields. Whether alone or part of a TKS™ system or similar, the liquid chemical is sprayed through a nozzle onto the outside of the windshield which prevents ice from forming. The chemical can also deice the windshield of ice that may have already formed. Systems such as these have a fluid reservoir, pump, control valve, filter, and relief valve. Other components may exist. *Figure 15-46* shows a set of spray tubes for application of chemical anti-ice on an aircraft windshield.

Water & Waste Tank Ice Prevention

Transport type aircraft have water and waste systems on board, and electrical heaters are often used to prevent the formation of ice in the water lines of these systems. Water

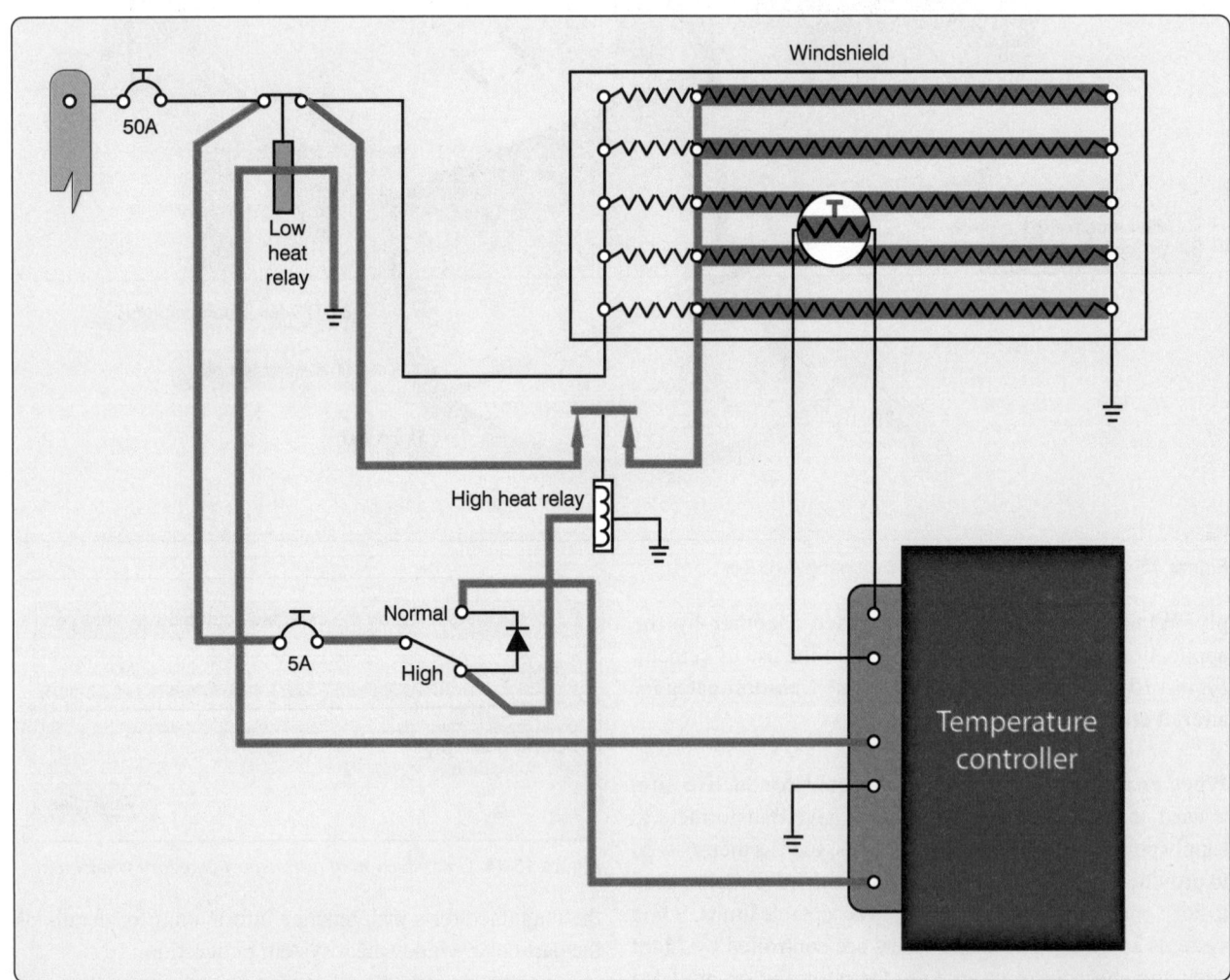

Figure 15-45. *Electric windshield heat schematic.*

Figure 15-46. *Chemical deicing spray tubes.*

lines carry water from the potable tanks to the lavatories and galleys. The waste water tanks collect the gray water from the galleys and lavatories. Heater blankets, inline heaters, or heater boots are often used to heat the water supply lines, water tank drain hoses, waste drain lines, waste tank rinse fittings, and drain masts. Thermostats in the water lines supply temperature data to the control unit that turns the electrical heaters on and off. When the temperature falls below freezing, the electrical heaters turn on and stay on until the temperature reaches a safe temperature. *Figure 15-47* is a schematic of a water supply line heater system, and *Figure 15-48* shows the location of the waste water tanks and heater blankets.

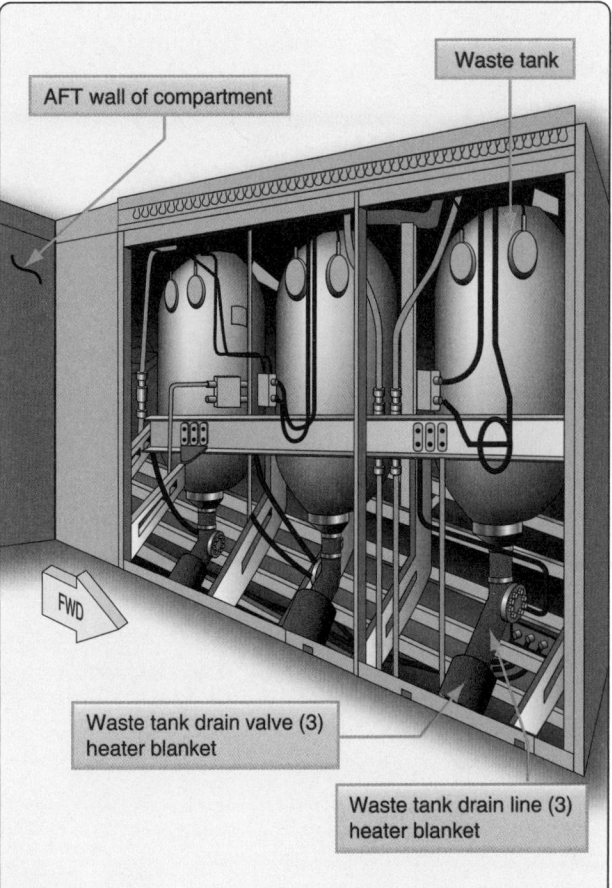

Figure 15-48. *Waste water tanks and heater blankets.*

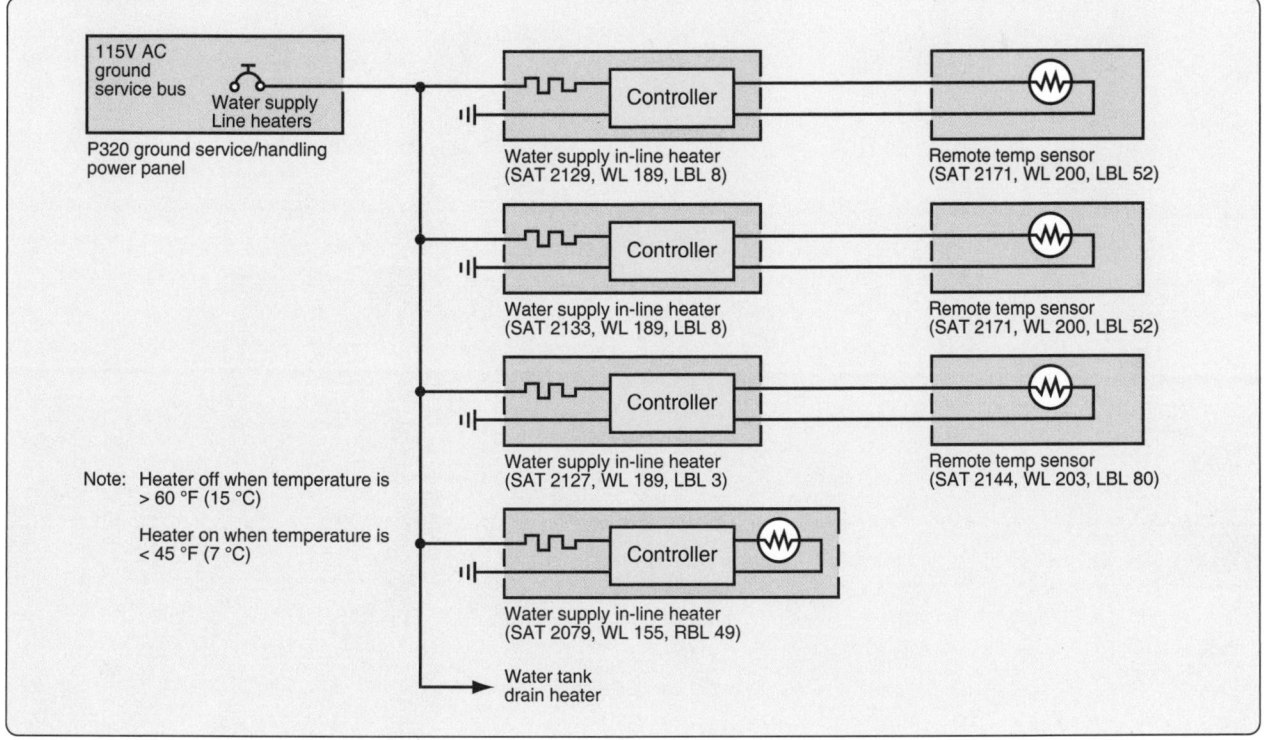

Figure 15-47. *Water supply line heater system.*

Chapter 16
Cabin Environmental Control Systems

Physiology of Flight

Composition of the Atmosphere

The mixture of gases that make up the earth's atmosphere is commonly called air. It is composed principally of 78 percent nitrogen and 21 percent oxygen. The remaining 1 percent is made up of various gases in smaller quantities. Some of these are important to human life, such as carbon dioxide, water vapor, and ozone. *Figure 16-1* indicates the respective percentage of the quantity of each gas in its relation to the total mixture.

As altitude increases, the total quantity of all the atmospheric gases reduces rapidly. However, the relative proportions of nitrogen and oxygen remain unchanged up to about 50 miles above the surface of the earth. The percentage of carbon dioxide is also fairly stable. The amounts of water vapor and ozone vary.

Nitrogen is an inert gas that is not used directly by man for life processes; however, many compounds containing nitrogen are essential to all living matter.

The small quantity of carbon dioxide in the atmosphere is utilized by plants during photosynthesis. Thus, the food supply for all animals, including man, depends on it. Carbon dioxide also helps control breathing in man and other animals.

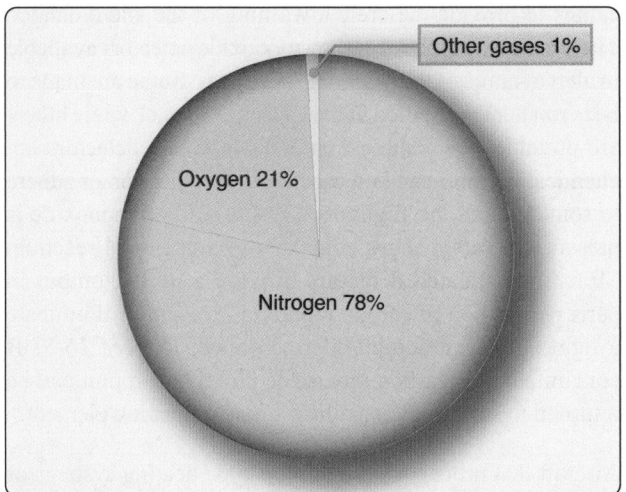

Figure 16-1. *The percentage of the various gases that comprise the atmosphere.*

The amount of water vapor in the atmosphere is variable but, even under humid conditions at sea level, it rarely exceeds 5 percent. Water also occurs in the atmosphere as ice crystals. All forms of water in the atmosphere absorb far more energy from the sun than do the other gases. Water plays an important role in the formation of weather.

Ozone is a form of oxygen. It contains three oxygen atoms per molecule, rather than the usual two. Most of the atmosphere's ozone is formed by the interaction of oxygen and the sun's rays near the top of the stratosphere in an area called the ozone layer. This is important to living organisms because ozone filters out most of the sun's harmful ultraviolet (UV) radiation. Ozone is also produced by electrical discharges, such as lightning strikes. It has a faint odor, somewhat like that of weak chlorine, that may be detected after a thunderstorm. Auroras and cosmic rays may also produce ozone. Ozone is of great consequence to living creatures on earth and to the circulation of the upper atmosphere.

Human Respiration & Circulation
Oxygen & Hypoxia

The second most prevalent substance in the atmosphere, oxygen, is essential for most living processes. Without oxygen, humans and animals die very rapidly. A reduction in the normal oxygen supply alters the human condition. It causes important changes in body functions, thought processes, and the maintainable degree of consciousness. The resultant sluggish condition of mind and body produced by insufficient oxygen is called hypoxia.

There are several scenarios that can result in hypoxia. During aircraft operations, it is brought about by a decrease in the pressure of oxygen in the lungs at high altitudes. The air contains the typical 21 percent of oxygen, but the rate at which oxygen can be absorbed into the blood depends upon the oxygen pressure. Greater pressure pushes the oxygen from the lung alveoli into the bloodstream. As the pressure is reduced, less oxygen is forced into and absorbed by the blood.

At sea level, oxygen pressure in the lungs is approximately three pounds per square inch (psi). This is sufficient to saturate the blood with oxygen and permit the mind and body to function normally. As altitude is increased, this pressure decreases. Below 7,000 feet above sea level, the available oxygen quantity and pressure remain sufficient for saturation of the blood with oxygen. Above 7,000 feet, however, the oxygen pressure becomes increasingly insufficient to saturate the blood. At 10,000 feet mean sea level (MSL), saturation of the blood with oxygen is only about 90 percent of normal. Long durations at this altitude can result in headache and fatigue, both symptoms of hypoxia. At 15,000 feet MSL, oxygen transfer to the bloodstream drops to 81 percent of saturation. This typically results in sleepiness, headache, blue lips and fingernails, and increased pulse and respiration. Worse yet, vision and judgment become impaired and safe operation of an aircraft becomes compromised. Higher in the atmosphere, decreasing pressure causes even less oxygen to enter the bloodstream; only 68 percent saturation at 22,000 feet MSL. Remaining at 25,000 feet MSL for 5 minutes, where oxygen transfer to the blood is reduced to approximately 50 percent saturation, causes unconsciousness. *[Figure 16-2]*

Hyperventilation

Another physiological phenomenon of interest to aviators is hyperventilation. Its symptoms greatly resemble hypoxia. When various cells in the body use oxygen and food delivered to them by the blood, carbon dioxide is a by-product. Blood carries this carbon dioxide to the lungs where it is exhaled.

Carbon dioxide functions in the body to regulate the depth and frequency of breathing. A high level of carbon dioxide in the blood triggers rapid, deep breathing to expel it. This also promotes the intake of a greater amount of oxygen for active cells to use. A low carbon dioxide level causes more relaxed breathing resulting in less oxygen intake. Therefore, an oxygen/carbon dioxide balance exists in the blood.

Occasionally, fear, panic, or pain triggers excessive rapid breathing in a person. With it comes a reduction of carbon

Altitude MSL (feet)	Oxygen pressure (psi)
0	3.08
5,000	2.57
10,000	2.12
15,000	1.74
20,000	1.42
25,000	1.15
30,000	0.92
35,000	0.76
40,000	0.57

Figure 16-2. *Oxygen pressure in the atmosphere at various altitudes.*

dioxide in the blood, even though the body does not need this. The lower carbon dioxide level signals the body that there is enough oxygen available and blood vessels constrict, causing hypoxia-like symptoms because insufficient oxygen is being delivered to the cells. Note that the onset of hypoxia described in the previous section occurs without the rapid breathing that accompanies hyperventilation. Hyperventilation can often be alleviated by having the person calm down and breathe normally, which restores the oxygen/carbon dioxide balance in the bloodstream.

Carbon Monoxide Poisoning

Carbon monoxide is a colorless, odorless gas produced by incomplete combustion of hydrocarbon fuels, such as those used in aviation. The human body does not require this gas to function. Its presence, however, can prevent a sufficient level of oxygen to be maintained in the body, resulting in hypoxia. This is also known as carbon monoxide poisoning. As with all forms of oxygen deprivation, extended exposure to carbon monoxide can result in unconsciousness and even death.

Hemoglobin is the substance in the blood that attaches to oxygen in the lungs and circulates it to cells in the body for use. Carbon monoxide more readily attaches itself to hemoglobin than oxygen. If carbon monoxide is present in the lungs, hemoglobin attaches to it and not oxygen. This results in cells not receiving the amount of oxygen they need. The insufficient oxygen level results in hypoxia-like symptoms.

A real danger of carbon monoxide poisoning is that long exposure to slight traces of carbon monoxide can result in oxygen deprivation just as easily as short-term exposure to a concentrated amount. The onset of its effects can be very subtle.

Carbon monoxide detectors are used in flight decks and cabins to provide the crew a warning of the silent danger. There are many types of carbon monoxide detectors available to alert aviators of the presence of this gas. Some are made to be permanently installed in the instrument panel, while others are portable. The simplest carbon monoxide detectors are chemical tabs mounted on cardboard that hang on or adhere to something in the flight deck. When carbon monoxide is present, the tab changes color due to a chemical reaction. More sophisticated detectors provide a digital output in parts per million of carbon monoxide present or illuminate a light and/or an audible alarm sounds. *[Figure 16-3]* If contaminated, a carbon monoxide portable test unit can be returned to service by installing a new indicating element.

Aircraft that utilize exhaust shroud-type heating systems or combustion heaters are more likely to have carbon monoxide introduced into the cabin from these devices. It is very

Figure 16-3. *An example of a carbon monoxide detector sold in the aviation market.*

important to discover the source of carbon monoxide if it is detected. Various leak checks and testing for cracks are performed regularly whenever a combustion source is also the source for cabin heat.

Aircraft Oxygen Systems

The negative effects of reduced atmospheric pressure at flight altitudes, forcing less oxygen into the blood, can be overcome. There are two ways this is commonly done: increase the pressure of the oxygen or increase the quantity of oxygen in the air mixture.

Large transport-category and high-performance passenger aircraft pressurize the air in the cabin. This serves to push more of the normal 21 percent oxygen found in the air into the blood for saturation. Techniques for pressurization are discussed later in this chapter. When utilized, the percentage of oxygen available for breathing remains the same; only the pressure is increased.

By increasing the quantity of oxygen available in the lungs, less pressure is required to saturate the blood. This is the basic function of an aircraft oxygen system. Increasing the level of oxygen above the 21 percent found in the atmosphere can offset the reduced pressure encountered as altitude increases. Oxygen may be regulated into the air that is breathed so as to maintain a sufficient amount for blood saturation. Normal mental and physical activity can be maintained at indicated altitudes of up to about 40,000 feet with the sole use of supplemental oxygen.

Oxygen systems that increase the quantity of oxygen in

breathing air are most commonly used as primary systems in small and medium size aircraft designed without cabin pressurization. Pressurized aircraft utilize oxygen systems as a means of redundancy should pressurization fail. Portable oxygen equipment may also be aboard for first aid purposes.

Forms of Oxygen & Characteristics
Gaseous Oxygen

Oxygen is a colorless, odorless, and tasteless gas at normal atmospheric temperatures and pressures. It transforms into a liquid at –183 °C (its boiling point). Oxygen combines readily with most elements and numerous compounds. This combining is called oxidation. Typically, oxidation produces heat. When something burns, it is actually rapidly combining with oxygen. Oxygen itself does not burn because it does not combine with itself, except to form oxygen or ozone. But, pure oxygen combines violently with petroleum products creating a significant hazard when handling these materials in close proximity to each other. Nevertheless, oxygen and various petroleum fuels combine to create the energy produced in internal combustion engines.

Pure gaseous oxygen, or nearly pure gaseous oxygen, is stored and transported in high-pressure cylinders that are typically painted green. Technicians should be cautious to keep pure oxygen away from fuel, oil, and grease to prevent unwanted combustion. Not all oxygen in containers is the same. Aviator's breathing oxygen is tested for the presence of water. This is done to avoid the possibility of it freezing in the small passage ways of valves and regulators. Ice could prevent delivery of the oxygen when needed. Aircraft often operate in subzero temperatures, increasing the possibility of icing. The water level should be a maximum of .02ml per liter of oxygen. The words "Aviator's Breathing Oxygen" should be marked clearly on any cylinders containing oxygen for this purpose. *[Figure 16-4]*

Production of gaseous oxygen for commercial or aircraft cylinders is often through a process of liquefying air. By controlling temperature and pressure, the nitrogen in the air can be allowed to boil off leaving mostly pure oxygen. Oxygen may also be produced by the electrolysis of water. Passing electric current through water separates the oxygen from the hydrogen. One further method of producing gaseous oxygen is by separating the nitrogen and oxygen in the air through the use of a molecular sieve. This membrane filters out nitrogen and some of the other gases in air, leaving nearly pure oxygen for use. Onboard oxygen sieves, or oxygen concentrators as they are sometimes called, are used on some military aircraft. Their use in civil aviation is expected.

Use of portable pulse oximeters has become more common in aviation. These devices measure the oxygen saturation

Figure 16-4. *"Aviator's breathing oxygen" is marked on all oxygen cylinders designed for this purpose.*

level of the blood. With this information, adjustments to the oxygen flow rates of onboard oxygen equipment can be made to prevent hypoxia. *Figure 16-5* shows an oximeter into

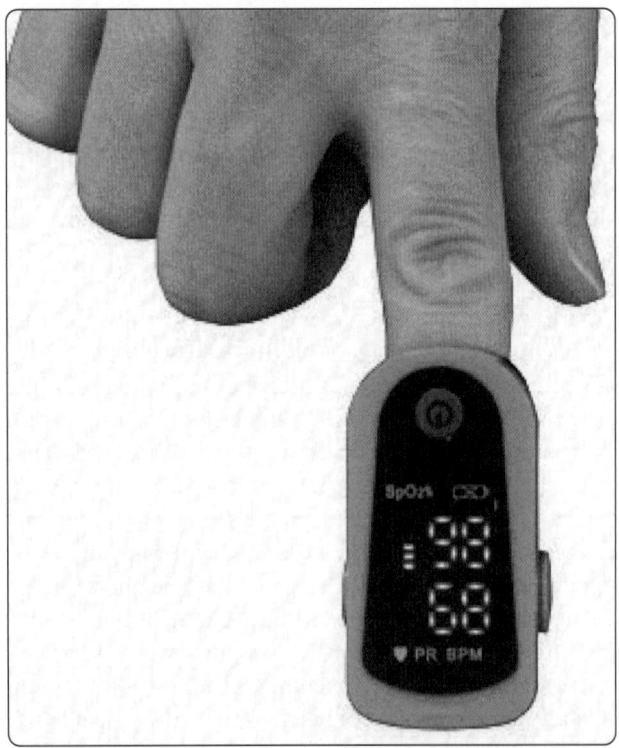

Figure 16-5. *A portable pulse-type oximeter displays percentage of oxygen saturation of the blood and heart rate. Pilots can adjust oxygen supply levels to maintain saturation and avoid hypoxia.*

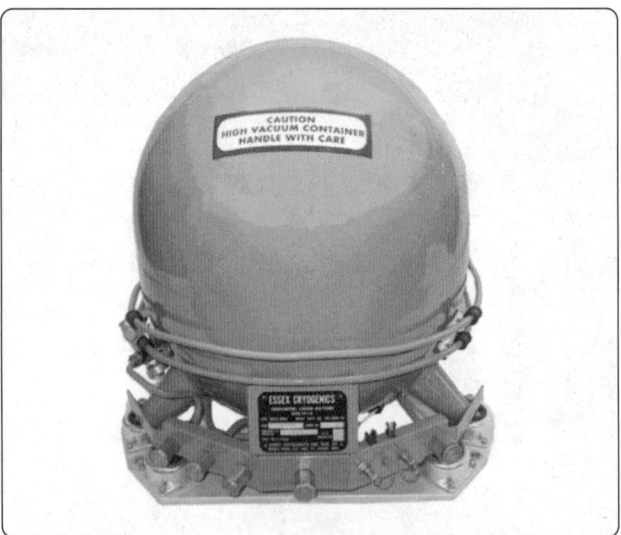

Figure 16-6. *A spherical liquid oxygen onboard container used by the military.*

which a finger is inserted to measure oxygen saturation of the blood in percentage. Heart rate is also displayed.

Liquid Oxygen

Liquid oxygen (LOX) is a pale blue, transparent liquid. Oxygen can be made liquid by lowering the temperature to below –183 °C or by placing gaseous oxygen under pressure. A combination of these is accomplished with a Dewar bottle. This special container is used to store and transport liquid oxygen. It uses an evacuated, double-walled insulation design to keep the liquid oxygen under pressure at a very low temperature. *[Figure 16-6]* A controlled amount of oxygen is allowed to vaporize and is plumbed into a gaseous oxygen delivery system downstream of a converter that is part of the container assembly.

A small quantity of LOX can be converted to an enormous amount of gaseous oxygen, resulting in the use of very little storage space compared to that needed for high-pressure gaseous oxygen cylinders. However, the difficulty of handling LOX, and the expense of doing so, has resulted in the container system used for gaseous oxygen to proliferate throughout civilian aviation. LOX is used in military aviation and some medical helicopter applications for patient oxygen.

Chemical or Solid Oxygen

Sodium chlorate has a unique characteristic. When ignited, it produces oxygen as it burns. This can be filtered and delivered through a hose to a mask that can be worn and breathed directly by the user. Solid oxygen candles, as they are called, are formed chunks of sodium chlorate wrapped inside insulated stainless-steel housings to control the heat produced when activated. The chemical oxygen supply is often ignited by a spring-loaded firing pin that when pulled, releases a hammer that smashes a cap creating a spark to light the candle. Electric ignition via a current-induced hot wire also exists. Once lit, a sodium chlorate oxygen generator cannot be extinguished. It produces a steady flow of breathable oxygen until it burns out, typically generating 10–20 minutes of oxygen. *[Figure 16-7]*

Solid oxygen generators are primarily used as backup oxygen devices on pressurized aircraft. They are one-third as heavy as gaseous oxygen systems that use heavy storage tanks for the same quantity of oxygen available. Sodium chlorate chemical oxygen generators also have a long shelf life, making them perfect as a standby form of oxygen. They are inert below 400 °F and can remain stored with little maintenance or inspection until needed, or until their expiration date is reached.

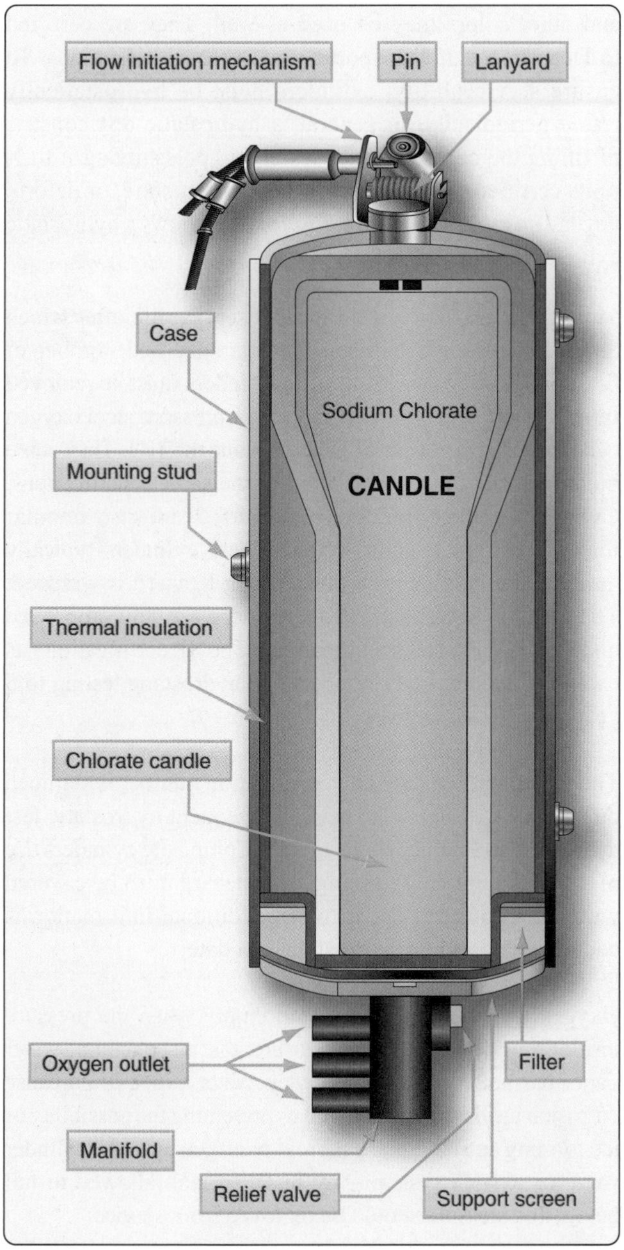

The feature of not extinguishing once lit limits the use of solid oxygen since it becomes an all-or-nothing source. The generators must be replaced if used, which can greatly increase the cost of using them as a source of oxygen for short periods of time. Moreover, chemical oxygen candles must be transported with extreme caution and as hazardous materials. They must be properly packed, and their ignition devices deactivated.

Onboard Oxygen Generating Systems (OBOGS)
The molecular sieve method of separating oxygen from the other gases in air has application in flight, as well as on the ground. The sieves are relatively light in weight and relieve the aviator of a need for ground support for the oxygen supply. Onboard oxygen generating systems on military aircraft pass bleed air from turbine engines through a sieve that separates the oxygen for breathing use. Some of the separated oxygen is also used to purge the sieve of the nitrogen and other gases that keep it fresh for use. Use of this type of oxygen production in civilian aircraft is anticipated. *[Figure 16-8]*

Oxygen Systems & Components
Built-in and portable oxygen systems are used in civilian aviation. They use gaseous or solid oxygen (oxygen generators) as suits the purpose and aircraft. LOX systems and molecular sieve oxygen systems are not discussed, as current applications on civilian aircraft are limited.

Gaseous Oxygen Systems
The use of gaseous oxygen in aviation is common; however, applications vary. On a light aircraft, it may consist of a small

Figure 16-7. *A sodium chlorate solid oxygen candle is at the core of a chemical oxygen generator.*

Flow initiation mechanism · Pin · Lanyard
Case
Sodium Chlorate
CANDLE
Mounting stud
Thermal insulation
Chlorate candle
Oxygen outlet
Filter
Manifold
Relief valve
Support screen

Figure 16-8. *This onboard oxygen generating system uses molecular sieve technology.*

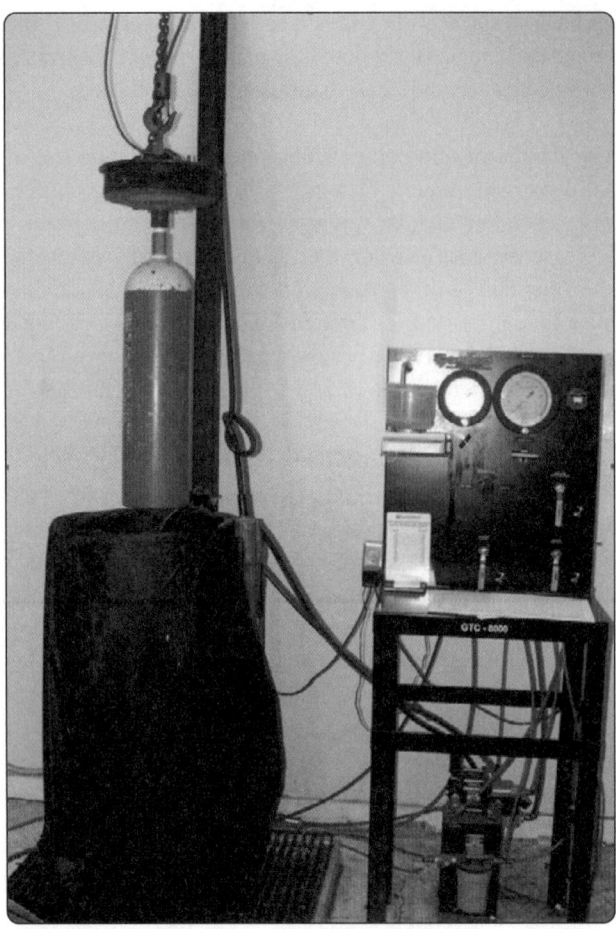

Figure 16-9. *This test stand is used for hydrostatic testing of oxygen cylinders. The water-filled cylinder is lowered into the barrel on the left where it is pressurized to the proper level as monitored via gauges mounted on the control panel. A displacement container on the top left of the control board collects water from the barrel to measure the expansion of the cylinder when pressurized to ensure it is within limits.*

carry-on portable cylinder with a single mask attached via a hose to a regulator on the bottle. Larger portable cylinders may be fitted with a regulator that divides the outlet flow for 2–4 people. Built-in oxygen systems on high performance and light twin-engine aircraft typically have a location where oxygen cylinders are installed to feed a distribution system via tubing and a regulator. The passenger compartment may have multiple breathing stations plumbed so that each passenger can individually plug in a hose and mask if oxygen is needed. A central regulator is normally controlled by the flight crew who may have their own separate regulator and oxygen cylinder. Transport category aircraft may use an elaborate built-in gaseous oxygen system as a backup system to cabin pressurization. In all of these cases, oxygen is stored as a gas at atmospheric temperature in high-pressure cylinders. It is distributed through a system with various components that are described in this section.

Oxygen Storage Cylinders

Gaseous oxygen is stored and transported in high-pressure cylinders. Traditionally, these have been heavy steel tanks rated for 1800–1850 psi of pressure and capable of maintaining pressure up to 2,400 psi. While these performed adequately, lighter weight tanks were sought. Some newer cylinders are comprised of a lightweight aluminum shell wrapped by Kevlar®. These cylinders are capable of carrying the same amount of oxygen at the same pressure as steel tanks but weigh much less. Also available are heavy-walled all-aluminum cylinders. These units are common as carry-on portable oxygen used in light aircraft.

Most oxygen storage cylinders are painted green, but yellow and other colors may be used as well. They are certified to Department of Transportation (DOT) specifications. To ensure serviceability, cylinders must be hydrostatically tested periodically. In general, a hydrostatic test consists of filling the container with water and pressurizing it to ⅝ of its certified rating. It should not leak, rupture, or deform beyond an established limit. *Figure 16-9* shows a hydrostatic cylinder testing apparatus.

Most cylinders also have a limited service life after which they can no longer be used. After a specified number of filling cycles or calendar age, the cylinders must be removed from service. The most common high-pressure steel oxygen cylinders used in aviation are the 3AA and the 3HT. They come in various sizes but are certified to the same specifications. Cylinders certified under DOT-E-8162 are also popular for their extremely light weight. These cylinders typically have an aluminum core around which Kevlar® is wrapped. The DOT-E 8162 approved cylinders are now approved under DOT-SP-8162 specifications. The SP certification has extended the required time between hydrostatic testing to 5 years (previously 3 years). [*Figure 16-10*]

The manufactured date and certification number is stamped on each cylinder near the neck. Subsequent hydrostatic test dates are also stamped there as well. Composite cylinders use placards rather than stamping. The placard must be covered with a coat of clear epoxy when additional information is added, such as a new hydrostatic test date.

Oxygen cylinders are considered empty when the pressure inside drops below 50 psi. This ensures that air containing water vapor has not entered the cylinder. Water vapor could cause corrosion inside the tank, as well as presenting the possibility of ice forming and clogging a narrow passageway in the cylinder valve or oxygen system. Any installed tank allowed to fall below this pressure should be removed from service.

Certification Type	Material	Rated pressure (psi)	Required hydrostatic test	Service life (years)	Service life (fillings)
DOT 3AA	Steel	1,800	5	Unlimited	N/A
DOT 3HT	Steel	1,850	3	24	4,380
DOT-E-8162	Composite	1,850	3	15	N/A
DOT-SP-8162	Composite	1,850	5	15	N/A
DOT 3AL	Aluminum	2,216	5	Unlimited	N/A

Figure 16-10. *Common cylinders used in aviation with some certification and testing specifications.*

Oxygen Systems & Regulators

The design of the various oxygen systems used in aircraft depends largely on the type of aircraft, its operational requirements, and whether the aircraft has a pressurization system. Systems are often characterized by the type of regulator used to dispense the oxygen: continuous-flow and demand flow. In some aircraft, a continuous-flow oxygen system is installed for both passengers and crew. The pressure demand system is widely used as a crew system, especially on the larger transport aircraft. Many aircraft have a combination of both systems that may be augmented by portable equipment.

Continuous-Flow Systems

In its simplest form, a continuous-flow oxygen system allows oxygen to exit the storage tank through a valve and passes it through a regulator/reducer attached to the top of the tank. The flow of high-pressure oxygen passes through a section of the regulator that reduces the pressure of the oxygen, which is then fed into a hose attached to a mask worn by the user. Once the valve is opened, the flow of oxygen is continuous. Even when the user is exhaling, or when the mask is not in use, a preset flow of oxygen continues until the tank valve is closed. On some systems, fine adjustment to the flow can be made with an adjustable flow indicator that is installed in the hose in line to the mask. A portable oxygen setup for a light aircraft exemplifies this type of continuous-flow system and is shown in *Figure 16-11.*

A more sophisticated continuous-flow oxygen system uses a regulator that is adjustable to provide varying amounts of oxygen flow to match increasing need as altitude increases. These regulators can be manual or automatic in design. Manual continuous-flow regulators are adjusted by the crew as altitude changes. Automatic continuous-flow regulators have a built in aneroid. As the aneroid expands with altitude, a mechanism allows more oxygen to flow though the regulator to the users. [*Figure 16-12*]

Many continuous-flow systems include a fixed location for the oxygen cylinders with permanent delivery plumbing installed to all passenger and crew stations in the cabin.

In large aircraft, separate storage cylinders for crew and passengers are typical. Fully integrated oxygen systems usually have separate, remotely mounted components to reduce pressure and regulate flow. A pressure relief valve is also typically installed in the system, as is some sort of filter and a gauge to indicate the amount of oxygen pressure

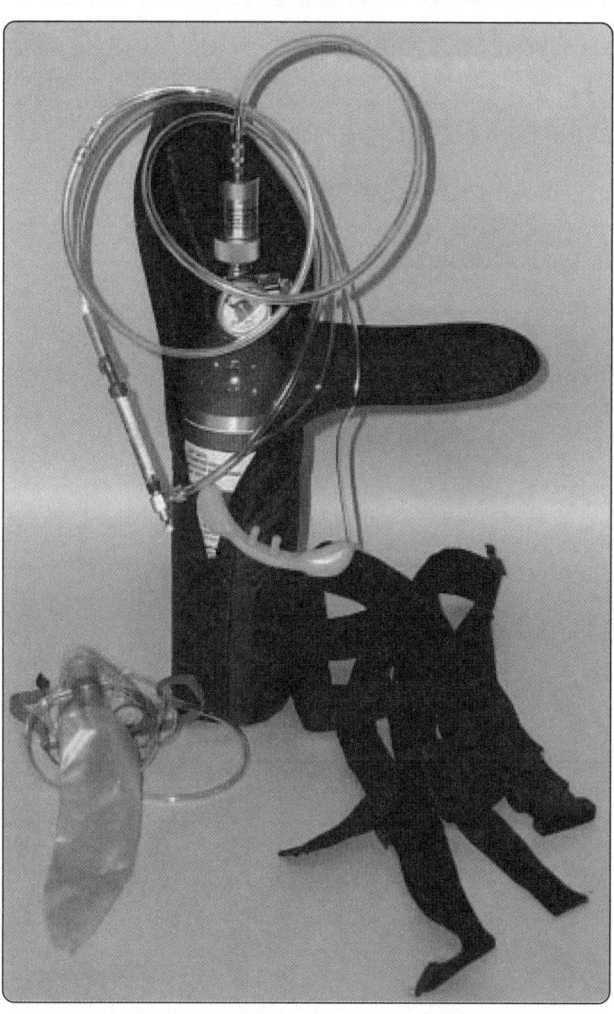

Figure 16-11. *A typical portable gaseous oxygen cylinder complete with valve, pressure gauge, regulator/reducer, hose, adjustable flow indicator, and rebreather cannula. A padded carrying case/bag can be strapped to the back of a seat in the cabin to meet certification and testing specifications.*

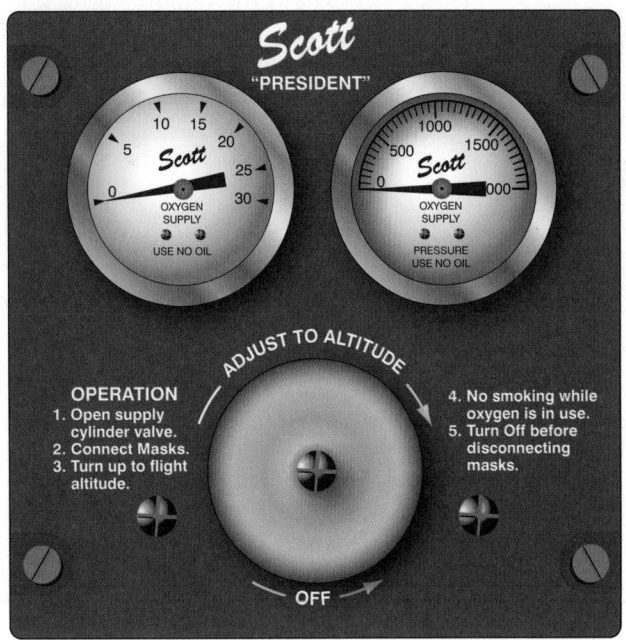

Figure 16-12. *A manual continuous flow oxygen system may have a regulator that is adjusted by the pilot as altitude varies. By turning the knob, the left gauge can be made to match the flight altitude thus increasing and decreasing flow as altitude changes.*

remaining in the storage cylinder(s). *Figure 16-13* diagrams the type of continuous-flow system that is found on small to medium sized aircraft.

Built-in continuous-flow gaseous oxygen systems accomplish a final flow rate to individual user stations through the use of a calibrated orifice in each mask. Larger diameter orifices are usually used in crew masks to provide greater flow than that for passengers. Special oxygen masks provide even greater flow via larger orifices for passengers traveling with medical conditions requiring full saturation of the blood with oxygen.

Allowing oxygen to continuously flow from the storage cylinder can be wasteful. Lowest sufficient flow rates can be accomplished through the use of rebreather apparatus. Oxygen and air that is exhaled still contains usable oxygen. By capturing this oxygen in a bag, or in a cannula with oxygen absorbing reservoirs, it can be inhaled with the next breath, reducing waste. *[Figure 16-14]*

The passenger section of a continuous-flow oxygen system may consist of a series of plug-in supply sockets fitted to the cabin walls adjacent to the passenger seats to which oxygen masks can be connected. Flow is inhibited until a passenger manually plugs in. When used as an emergency system in pressurized aircraft, depressurization automatically triggers the deployment of oxygen ready continuous-flow masks at each passenger station. A lanyard attached to the mask turns on the flow to each mask when it is pulled toward the passenger for use. The masks are normally stowed overhead in the passenger service unit (PSU). *[Figure 16-15]* Deployment of the emergency continuous-flow passenger oxygen masks may also be controlled by the crew. *[Figure 16-16]*

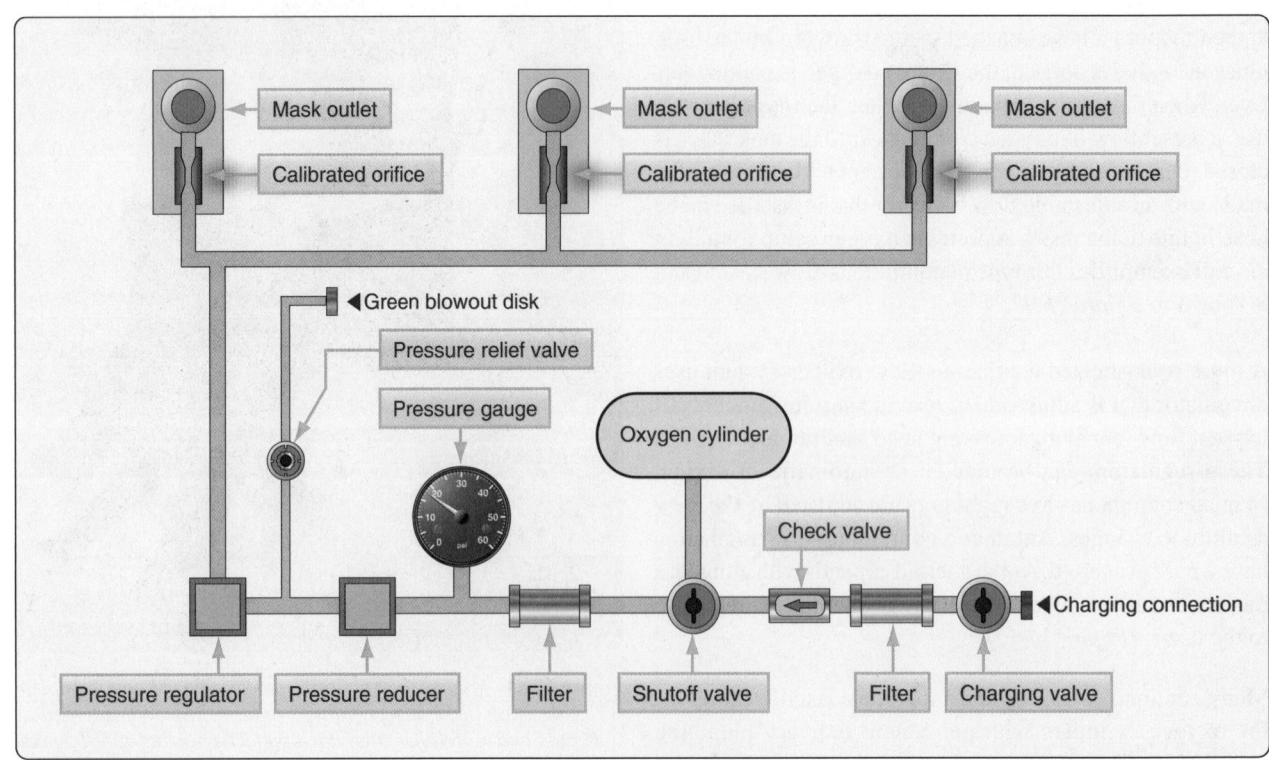

Figure 16-13. *Continuous flow oxygen system found on small to medium size aircraft.*

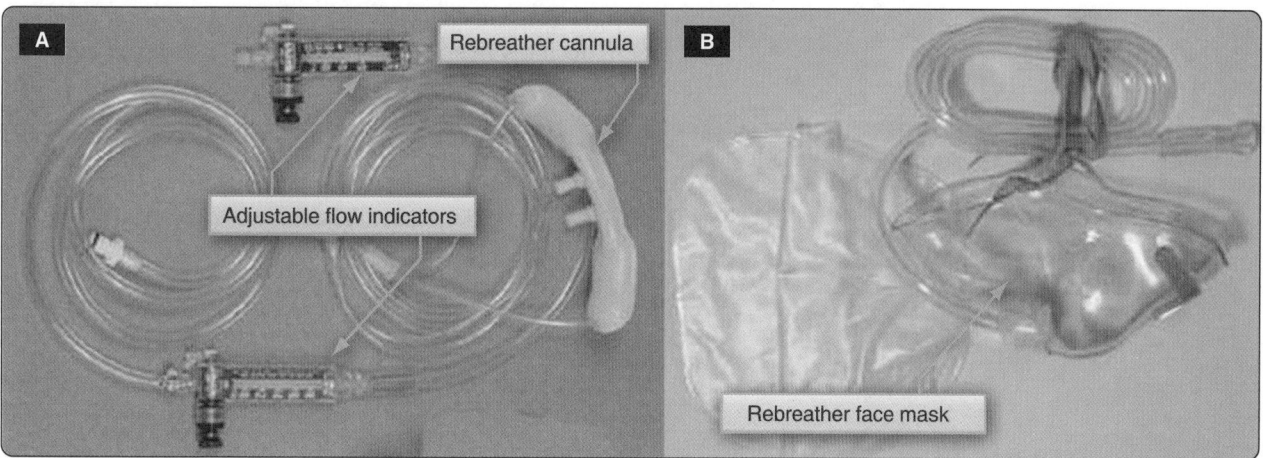

Figure 16-14. *A rebreather cannula (A) and rebreather bag (B) capture exhaled oxygen to be inhaled on the next breath. This conserves oxygen by permitting lower flow rates in continuous flow systems. The red and green devices are optional flow indicators that allow the user to monitor oxygen flow rate. The type shown also contains needle valves for final regulation of the flow rate to each user.*

Figure 16-15. *A passenger service unit (PSU) is hinged over each row of seats in an airliner. Four yellow continuous flow oxygen masks are shown deployed. They are normally stored behind a separate hinged panel that opens to allow the masks to fall from the PSU for use.*

Continuous-flow oxygen masks are simple devices made to direct flow to the nose and mouth of the wearer. They fit snugly but are not air tight. Vent holes allow cabin air to mix with the oxygen and provide escape for exhalation. In a rebreather mask, the vents allow the exhaled mixture that is not trapped in the rebreather bag to escape. This is appropriate, because this is the air-oxygen mixture that has been in the lungs the longest, so it has less recoverable oxygen to be breathed again. *[Figure 16-17]*

Demand-Flow or Pressure-Demand Systems

When oxygen is delivered only as the user inhales, or on demand, it is known as a demand-flow system or a pressure-demand system. During the hold and exhalation periods of breathing, the oxygen supply is stopped. Thus, the duration of the oxygen supply is prolonged as none is wasted. Demand-flow systems are used most frequently by

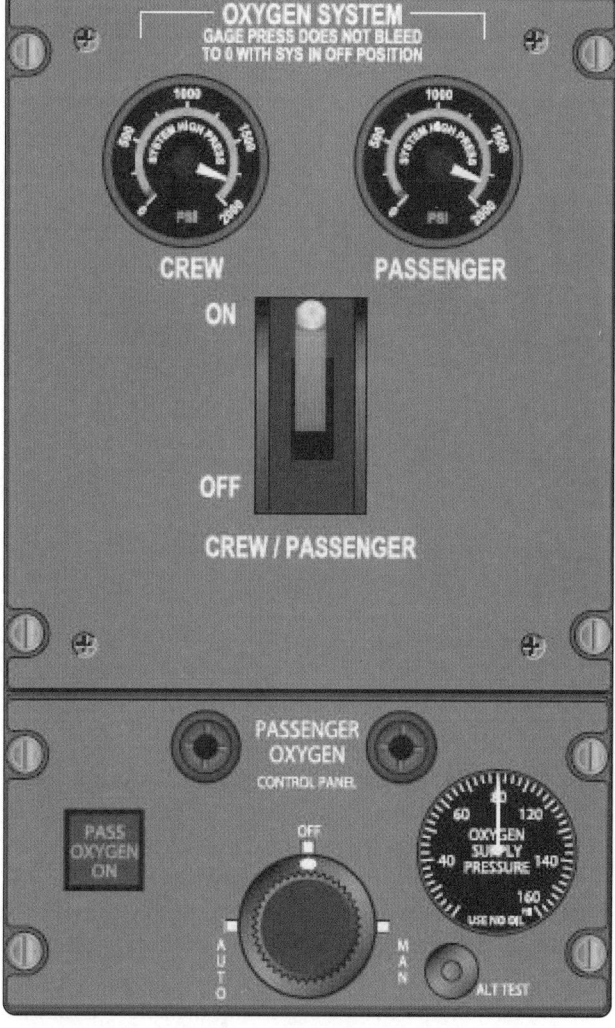

Figure 16-16. *The crew can deploy passenger emergency continuous-flow oxygen masks and supply with a switch in the flight deck.*

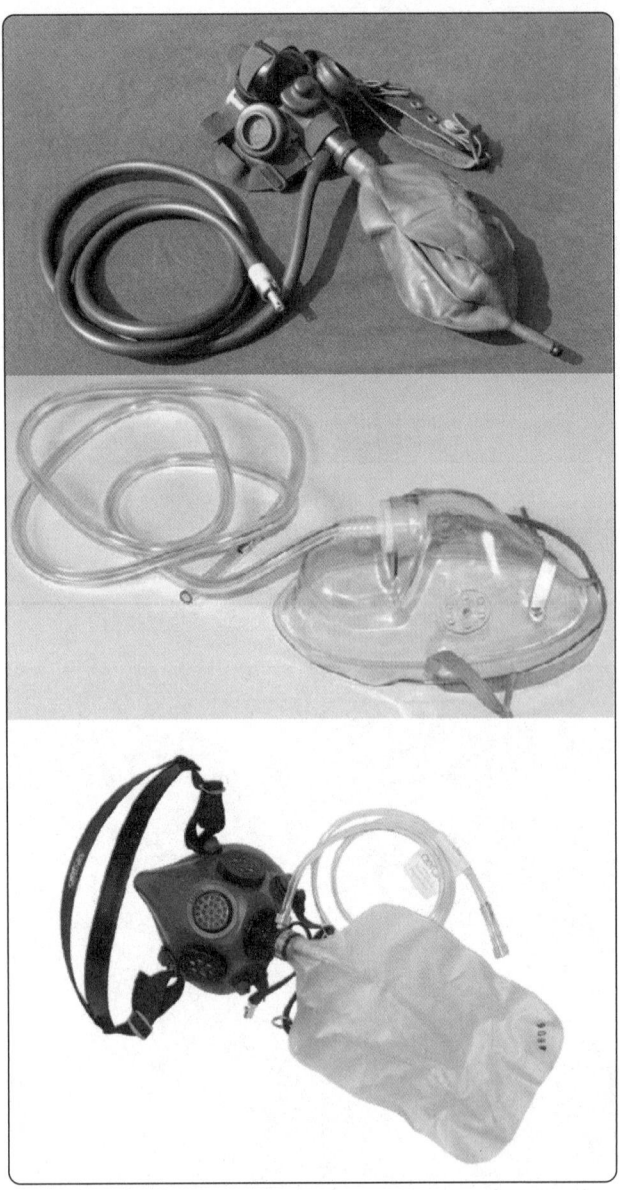

Figure 16-17. *Examples of different continuous-flow oxygen masks.*

the crew on high performance and air transport category aircraft. *[Figure 16-18]*

Demand-flow systems are similar to continuous-flow systems in that a cylinder delivers oxygen through a valve when opened. The tank pressure gauge, filter(s), pressure relief valve, and any plumbing installed to refill the cylinder while installed on the aircraft are all similar to those in a continuous-flow system. The high-pressure oxygen also passes through a pressure reducer and a regulator to adjust the pressure and flow to the user. But, demand-flow oxygen regulators differ significantly from continuous-flow oxygen regulators. They work in conjunction with close-fitting demand-type masks to control the flow of oxygen. *[Figure 16-19]*

In a demand-flow oxygen system, the system pressure-reducing valve is sometimes called a pressure regulator. This device lowers the oxygen pressure from the storage cylinder(s) to roughly 60–85 psi and delivers it to individual regulators dedicated for each user. A pressure reduction also occurs at the inlet of the individual regulator by limiting the size of the inlet orifice. There are two types of individual regulators: the diluter-demand type and the pressure-demand type. *[Figure 16-20]*

The diluter-demand type regulator holds back the flow of oxygen until the user inhales with a demand-type oxygen mask. The regulator dilutes the pure oxygen supply with cabin air each time a breath is drawn. With its control toggle switch set to normal, the amount of dilution depends on the cabin altitude. As altitude increases, an aneroid allows more oxygen and less cabin air to be delivered to the user by adjusting flows through a metering valve. At approximately 34,000 feet, the diluter-demand regulator meters 100 percent oxygen. This should not be needed unless cabin pressurization fails. Additionally, the user may select 100 percent oxygen delivery at any time by positioning the oxygen selection lever on the regulator. A built-in emergency switch also delivers 100 percent oxygen, but in a continuous flow as the demand function is bypassed. *[Figure 16-21]*

Pressure-demand oxygen systems operate similarly to diluter-demand systems, except that oxygen is delivered through the individual pressure regulator(s) under higher pressure. When the demand valve is unseated, oxygen under pressure forces its way into the lungs of the user. The demand function still operates, extending the overall supply of oxygen beyond that of a continuous-flow system. Dilution with cabin air also occurs if cabin altitude is less than 34,000 feet.

Pressure-demand regulators are used on aircraft that regularly fly at 40,000 feet and above. They are also found on many airliners and high-performance aircraft that may not typically fly that high. Forcing oxygen into the lungs under pressure ensures saturation of the blood, regardless of altitude or cabin altitude.

Both diluter-demand and pressure-demand regulators also come in mask-mounted versions. The operation is essentially the same as that of panel-mounted regulators. *[Figure 16-22]*

Flow Indicators

Flow indicators, or flow meters, are common in all oxygen systems. They usually consist of a lightweight object, or apparatus, that is moved by the oxygen stream. When flow exists, this movement signals the user in some way. *[Figure 16-23]* Many flow meters in continuous-flow oxygen systems also double as flow rate adjusters. Needle valves

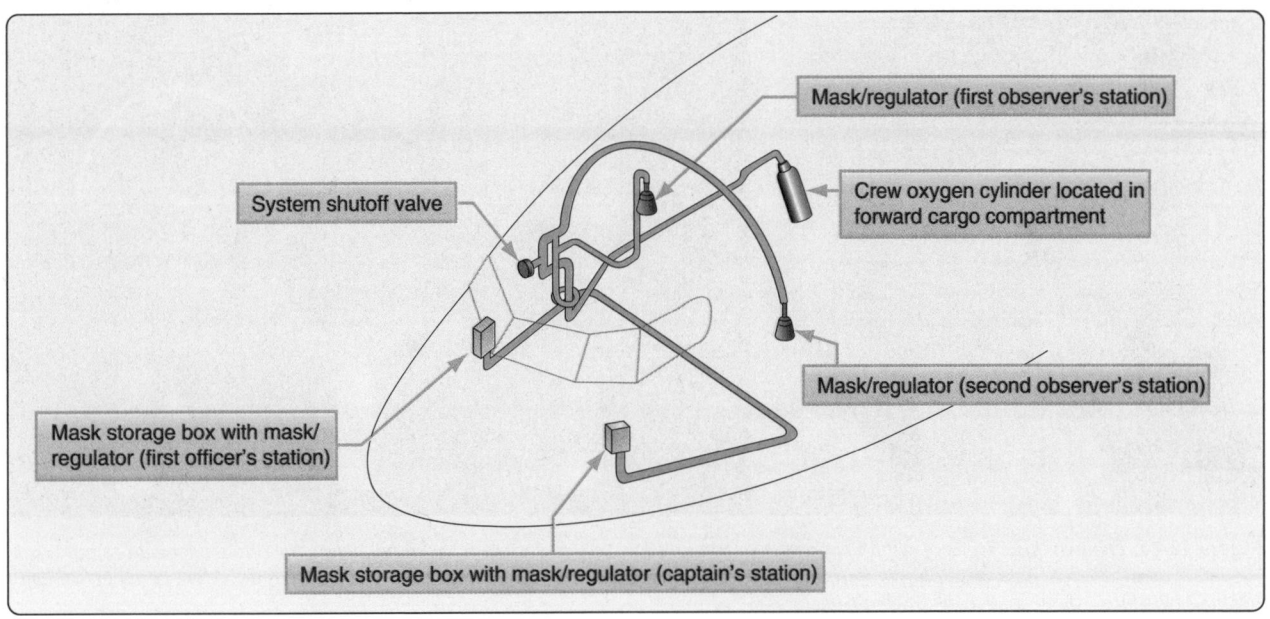

Figure 16-18. *Location of demand-flow oxygen components on a transport category aircraft.*

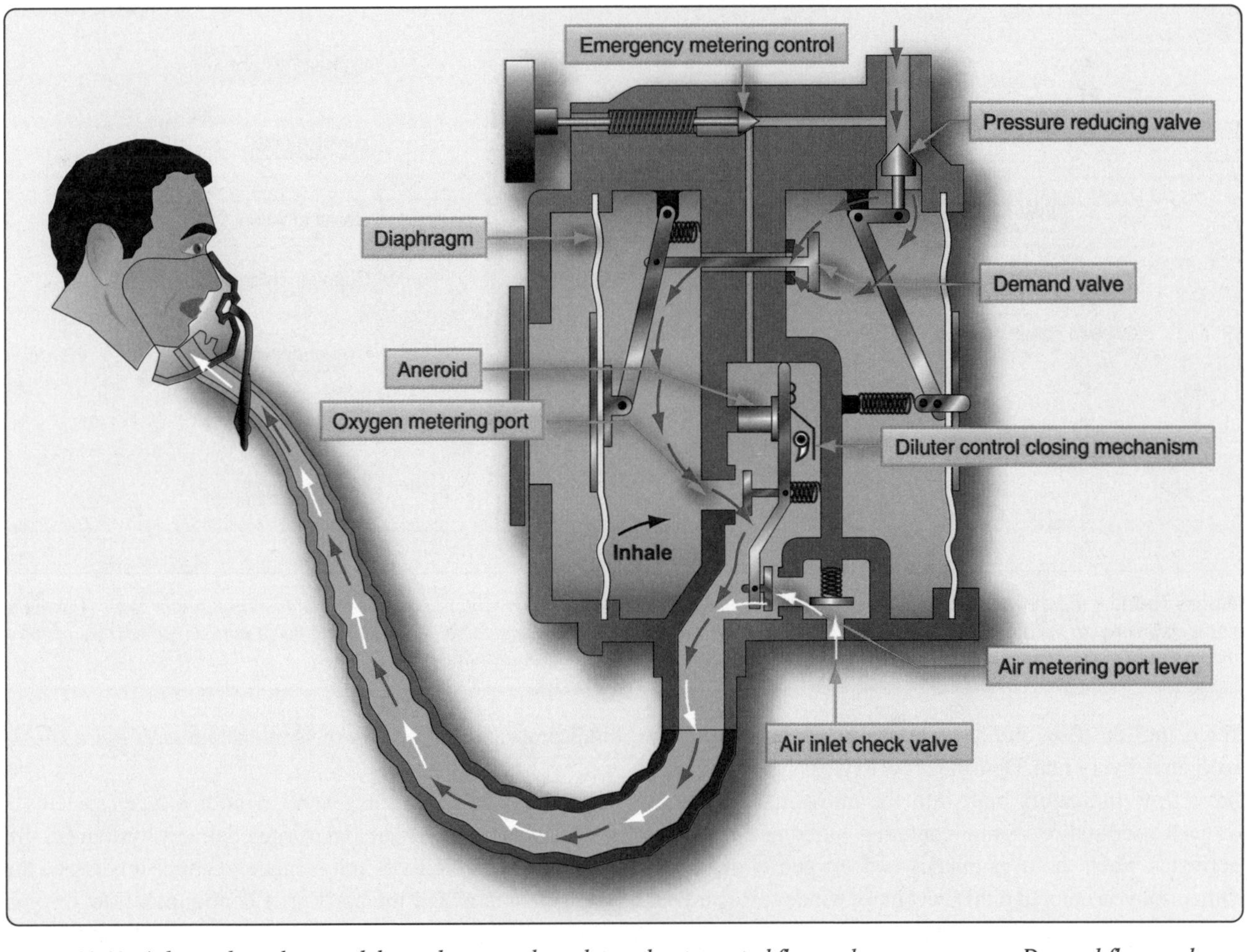

Mask/regulator (first observer's station)

System shutoff valve

Crew oxygen cylinder located in forward cargo compartment

Mask/regulator (second observer's station)

Mask storage box with mask/regulator (first officer's station)

Mask storage box with mask/regulator (captain's station)

Emergency metering control

Pressure reducing valve

Diaphragm

Demand valve

Aneroid

Oxygen metering port

Diluter control closing mechanism

Inhale

Air metering port lever

Air inlet check valve

Figure 16-19. *A demand regulator and demand-type mask work together to control flow and conserve oxygen. Demand-flow masks are close fitting so that when the user inhales, low pressure is created in the regulator, which allows oxygen to flow. Exhaled air escapes through ports in the mask, and the regulator ceases the flow of oxygen until the next inhalation.*

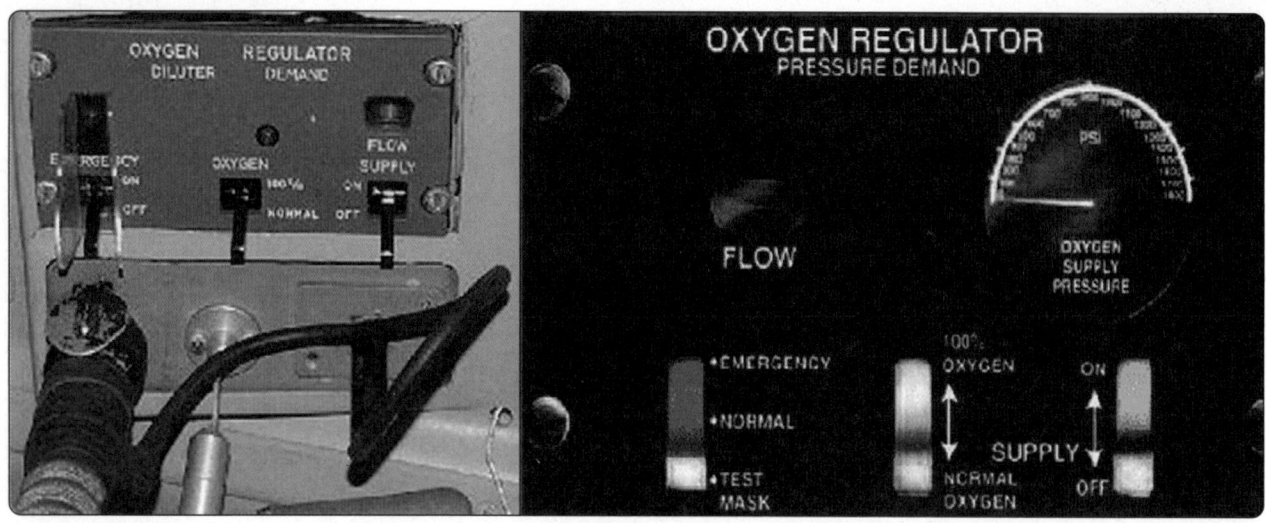

Figure 16-20. *The two basic types of regulators used in demand flow oxygen systems. The panel below the diluter demand regulator on the left is available for mask hose plug in (left), lanyard mask hanger (center), and microphone plug in (right). Most high performance demand type masks have a microphone built-in.*

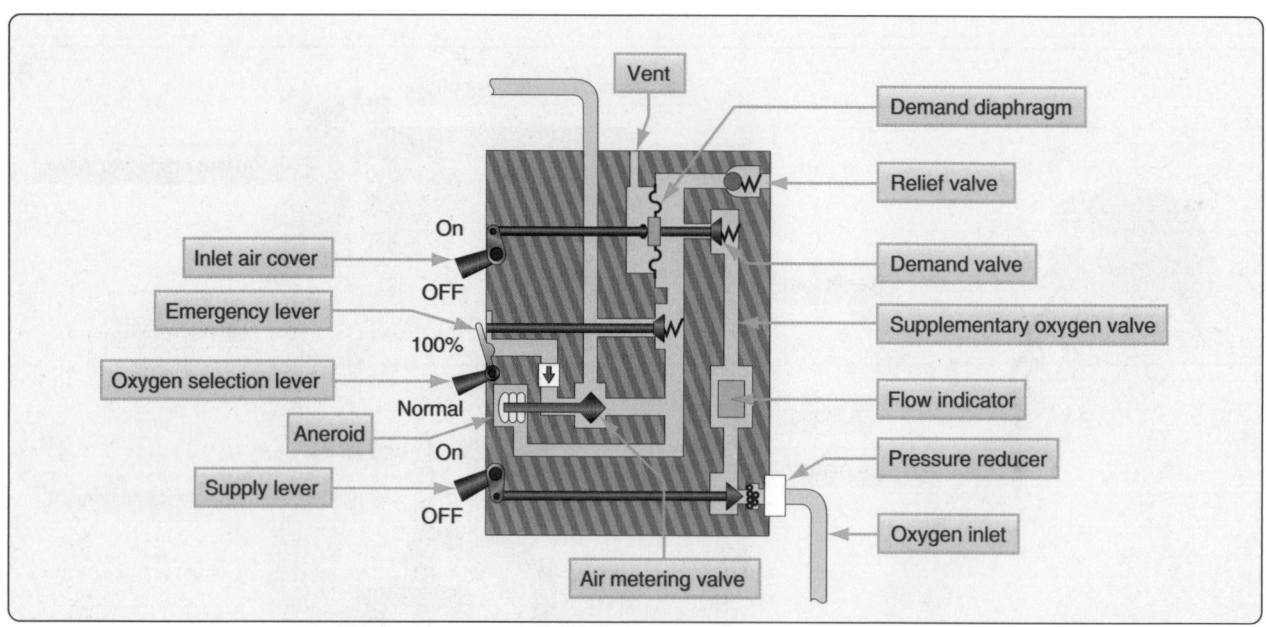

Figure 16-21. *A diluter-demand regulator operates when low pressure caused by inhalation moves the demand diaphragm. A demand valve connected to the diaphragm opens, letting oxygen flow through the metering valve. The metering valve adjusts the mixture of cabin air and pure oxygen via a connecting link to an aneroid that responds to cabin altitude.*

fitted into the flow indicator housing can fine-adjust the oxygen delivery rate. Demand-flow oxygen systems usually have flow indicators built into the individual regulators at each user station. Some contain a blinking device that activates when the user inhales and oxygen is delivered. Others move a colored pith object into a window. Regardless, flow indicators provide a quick verification that an oxygen system is functioning.

Different flow indicators are used to provide verification that the oxygen system is functioning. Other demand-flow

indicators are built into the oxygen regulators. *[Figure 16-23]*

A recent development in general aviation oxygen systems is the electronic pulse demand oxygen delivery system (EDS). A small, portable EDS unit is made to connect between the oxygen source and the mask in a continuous-flow oxygen system. It delivers timed pulses of oxygen to the wearer on demand, saving oxygen normally lost during the hold and exhale segments of the breathing cycle. Advanced pressure sensing and processing allows the unit to deliver oxygen only when an inhalation starts. It can also sense differences in

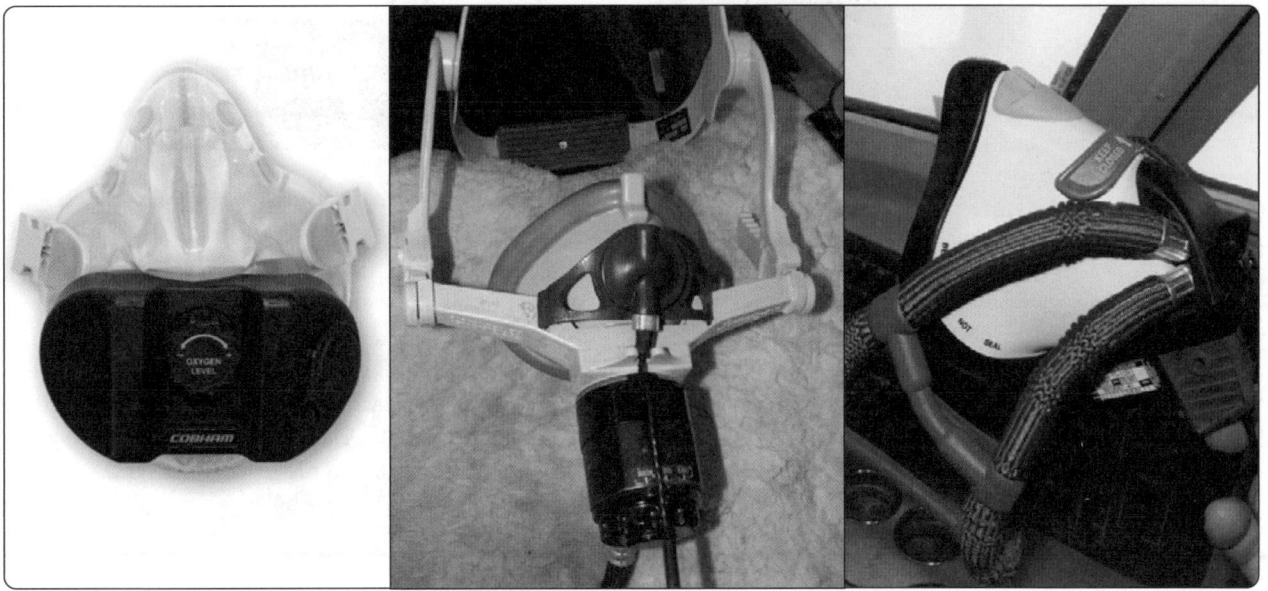

Figure 16-22. *A mask-mounted version of a miniature diluter-demand regulator designed for use in general aviation (left), a mechanical quick-donning diluter-demand mask with the regulator on the mask (center), and an inflatable quick-donning mask (right). Squeezing the red grips directs oxygen into the hollow straps.*

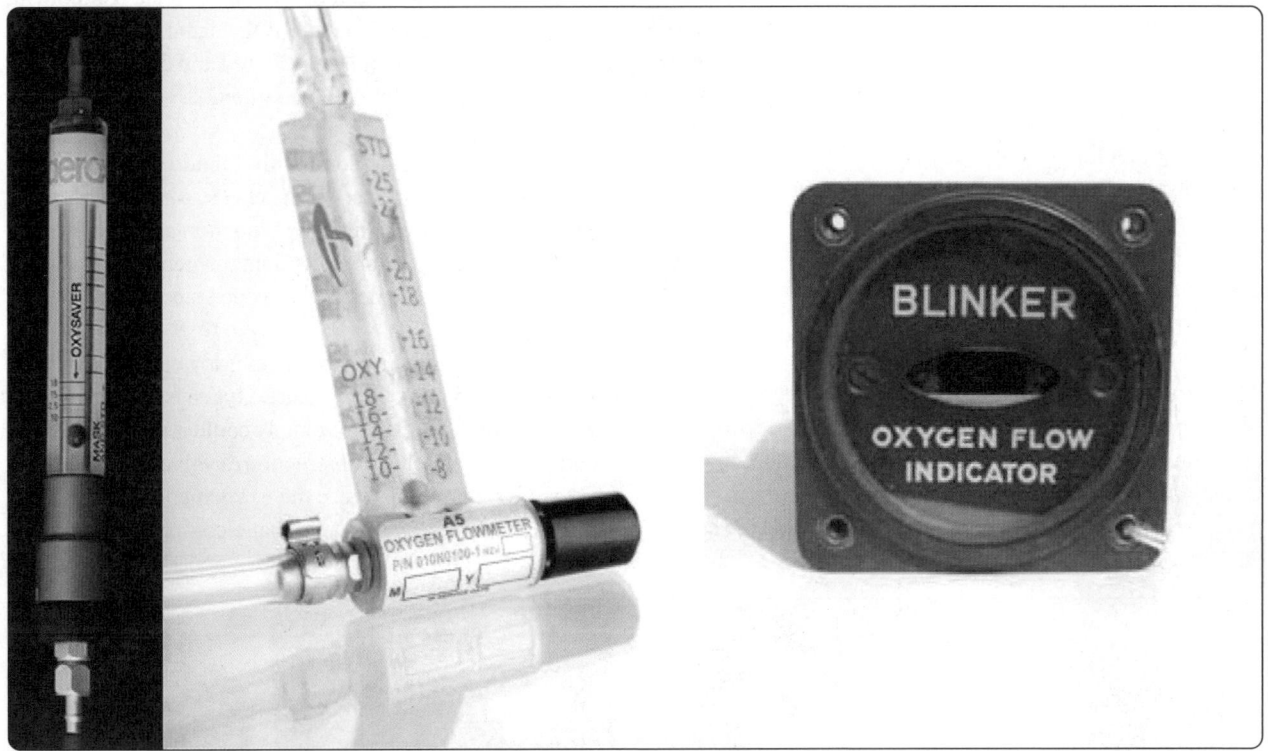

Figure 16-23. *Different flow indicators are used to provide verification that the oxygen system is functioning: continuous-flow, inline (left); continuous-flow, inline with valve adjuster (center); and old style demand flow (right).*

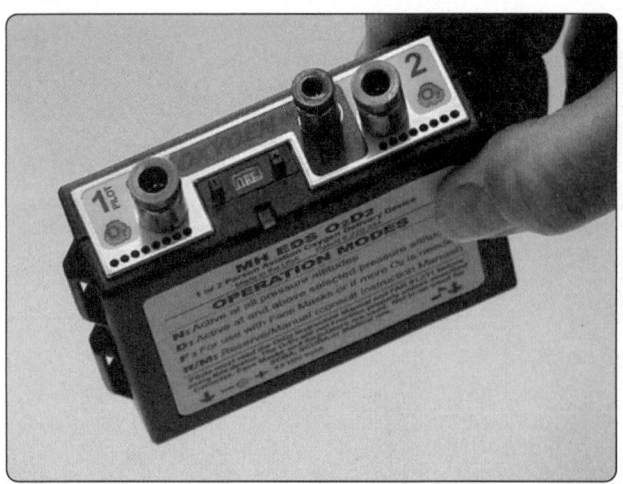

Figure 16-24. *A portable two-person electronic pulse-demand (EPD) oxygen regulating unit.*

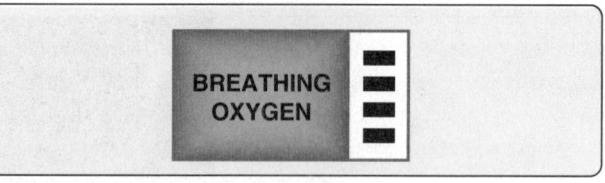

Figure 16-26. *Color-coded tape used to identify oxygen tubing.*

users' breathing cycles and physiologies and adjust the flow of oxygen accordingly. A built-in pressure-sensing device adjusts the amount of oxygen released as altitude changes. *[Figure 16-24]*

Permanently mounted EPD systems are also available. They typically integrate with an electronic valve/regulator on the oxygen cylinder and come with an emergency bypass switch to provide continuous-flow oxygen should the system malfunction. A liquid crystal display (LCD) monitor/control panel displays numerous system operating parameters and allows adjustments to the automatic settings. This type of electronic metering of oxygen has also been developed for passenger emergency oxygen use in airliners. *[Figure 16-25]*

Oxygen Plumbing & Valves
Tubing and fittings make up most of the oxygen system plumbing and connect the various components. Most lines are metal in permanent installations. High-pressure lines are usually stainless steel. Tubing in the low-pressure parts of the oxygen system is typically aluminum. Flexible plastic hosing

is used deliver oxygen to the masks; its use is increasing in permanent installations to save weight.

Installed oxygen tubing is usually identified with color-coded tape applied to each end of the tubing, and at specified intervals along its length. The tape coding consists of a green band overprinted with the words "BREATHING OXYGEN" and a black rectangular symbol overprinted on a white background border strip. *[Figure 16-26]*

Tubing-to-tubing fittings in oxygen systems are often designed with straight threads to receive flared tube connections. Tubing-to-component fittings usually have straight threads on the tubing end and external pipe threads (tapered) on the other end for attachment to the component. The fittings are typically made of the same material as the tubing (i.e., aluminum or steel). Flared and flareless fittings are both used, depending on the system.

Five types of valves are commonly found in high-pressure gaseous oxygen systems: filler, check, shutoff, pressure reducer, and pressure relief. They function as they would in any other system with one exception: oxygen system shutoff valves are specifically designed to open slowly.

The ignition point for any substances is lower in pure oxygen than it is in air. When high-pressure oxygen is allowed to rush into a low-pressure area, its velocity could reach the speed of sound. If it encounters an obstruction (a valve seat, an elbow, a piece of contaminant, etc.), the oxygen compresses. With this compression, known as adiabatic compression (since it

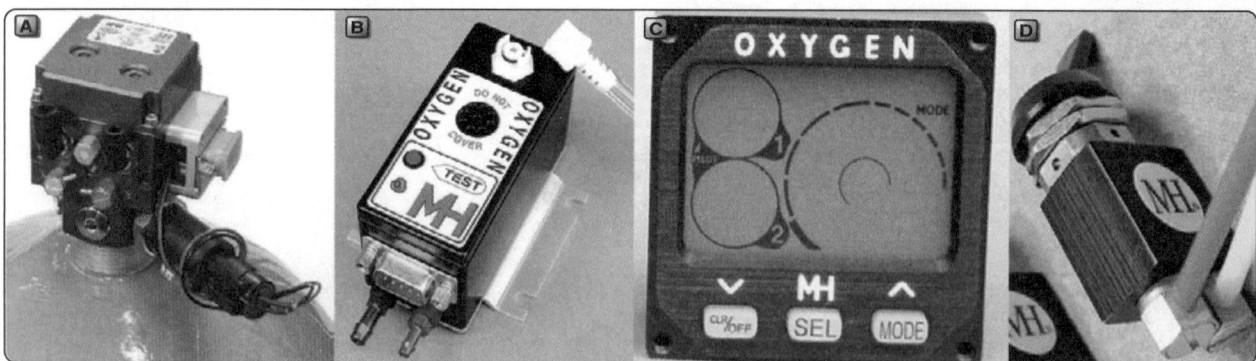

Figure 16-25. *The key components of a built-in electronic pulse demand oxygen metering system: (A) electronic regulator, (B) oxygen station distributer unit, (C) command/display unit, (D) emergency bypass switch.*

builds so quickly no heat is lost to its surroundings), comes high temperature. Under pressure, this high temperature exceeds the ignition point of the material the oxygen encounters and a fire or explosion results. A stainless-steel line, for example, would not normally burn and is used for carrying numerous fluids under high pressure. But under high pressure and temperature in the presence of 100 percent oxygen, even stainless steel can ignite.

To combat this issue, all oxygen shutoff valves are slow, opening valves designed to decrease velocity. *[Figure 16-27]*

Additionally, technicians should always open all oxygen valves slowly. Keeping oxygen from rushing into a low-pressure area should be a major concern when working with high-pressure gaseous oxygen systems.

Oxygen cylinder valves and high-pressure systems are often provided with a relief valve should the desired pressure be exceeded. Often, the valve is ported to an indicating or blowout disc. This is located in a conspicuous place, such as the fuselage skin, where it can be seen during walk-around inspection. Most blowout discs are green. The absence of the green disc indicates the relief valve has opened, and the cause should be investigated before flight. *[Figure 16-28]*

Chemical Oxygen Systems

The two primary types of chemical oxygen systems are the portable type, much like a portable carry-on gaseous oxygen cylinder, and the fully integrated supplementary oxygen

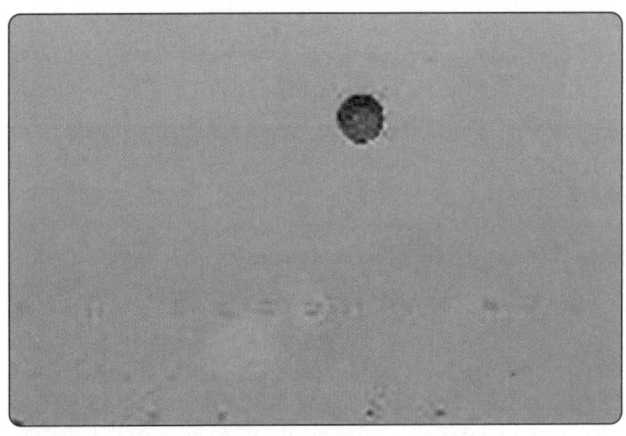

Figure 16-28. *An oxygen blowout plug on the side of the fuselage indicates when pressure relief has occurred and should be investigated.*

system used as backup on pressurized aircraft in case of pressurization failure. *[Figure 16-29]* This latter use of solid chemical oxygen generators is most common on airliners. The generators are stored in the overhead PSU attached to hoses and masks for every passenger on board the aircraft. When a depressurization occurs, or the flight crew activates a switch, a compartment door opens, and the masks and hoses fall out in front of the passengers. The action of pulling the mask down to a usable position actuates an electric current, or ignition hammer, that ignites the oxygen candle and initiates the flow of oxygen. Typically, 10 to 20 minutes of oxygen is available for each user. This is calculated to be enough time for the aircraft to descend to a safe altitude for unassisted breathing.

Chemical oxygen systems are unique in that they do not produce the oxygen until it is time to be used. This allows safer transportation of the oxygen supply with less maintenance. Chemical oxygen-generating systems also require less space

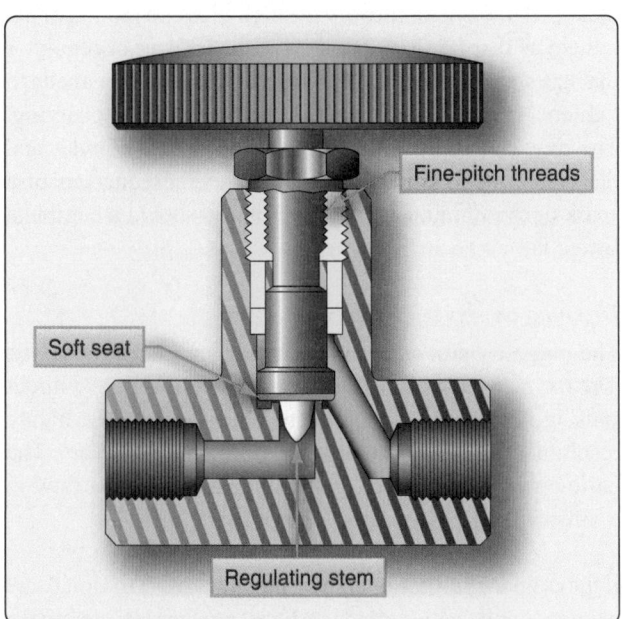

Figure 16-27. *This high-pressure oxygen system shutoff valve has fine-pitch threads and a regulating stem to slow the flow of oxygen through the valve. A soft valve seat is also included to assure the valve closes completely.*

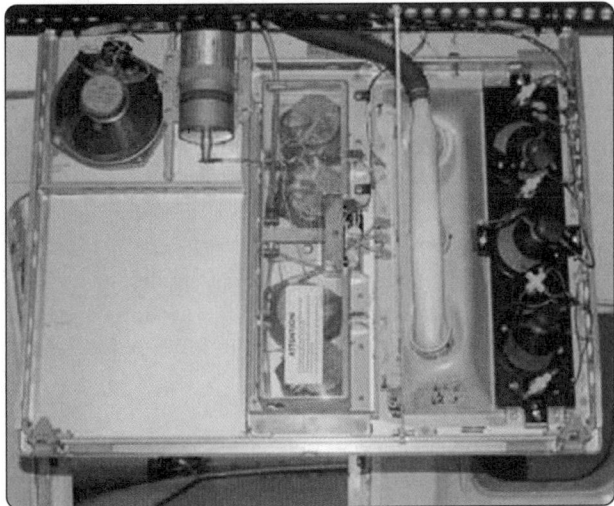

Figure 16-29. *An oxygen generator mounted in place in an overhead passenger service unit of an air transport category aircraft.*

and weigh less than gaseous oxygen systems supplying the same number of people. Long runs of tubing, fittings, regulators, and other components are avoided, as are heavy gaseous oxygen storage cylinders. Each passenger row grouping has its own fully independent chemical oxygen generator. The generators, which often weigh less than a pound, are insulated and can burn completely without getting hot. The size of the orifice opening in the hose-attach nipples regulates the continuous flow of oxygen to the users.

LOX Systems

LOX systems are rarely used in civilian aviation. They may be encountered on former military aircraft now in the civilian fleet. As mentioned, the storage of LOX requires a special container system. The plumbing arrangement to convert the liquid to a usable gas is also unique. It basically consists of a controlled heat exchange assembly of tubing and valves. Overboard pressure relief is provided for excessive temperature situations. Once gaseous, the LOX system is the same as it is in any comparable gaseous oxygen delivery system. Use of pressure-demand regulators and masks is common. Consult the manufacturer's maintenance manual for further information if a LOX system is encountered.

Oxygen System Servicing
Servicing Gaseous Oxygen

Gaseous oxygen systems are prevalent in general, corporate, and airline aviation. The use of lightweight aluminum and composite storage cylinders has improved these simple and reliable life support systems. All gaseous oxygen systems require servicing and maintenance. Various procedures and requirements to perform these functions are covered in this section.

Leak Testing Gaseous Oxygen Systems

Leaks in a continuous-flow oxygen system may be difficult to detect because the system is open at the user end. Blocking the flow of oxygen allows pressure to build and leak check procedures can be followed that are similar to those used in the high-pressure sections of the systems. Detection of leaks should be performed with oxygen-safe leak check fluid. This is a soapy liquid free from elements that might react with pure oxygen or contaminate the system. As with leak detection on an inflated tire or tube assembly, the oxygen leak detection solution is applied to the outside of fittings and mating surfaces. The formation of bubbles indicates a leak. *[Figure 16-30]*

Careful assembly of oxygen components and fittings without overtightening or undertightening is required. If a leak is found at a fitting, it should be checked for the proper torque. Tightening may not always stop the leak. If the fitting is

Figure 16-30. *Oxygen system leak check solution.*

torqued properly and a leak still exists, pressure must be released from the system and the fitting must be examined for flaws or contamination. If necessary, the fitting must be replaced. All system components, lines, and fittings must be replaced with the proper parts, which should be cleaned and inspected thoroughly before installation. Follow the manufacturer's instructions and repeat the leak check when completed.

Use caution when maintaining the high-pressure portion of a gaseous oxygen system. An open tank valve pressurizes the lines and components with up to 1,850 pounds per square inch (psi) of oxygen. Identify the high-pressure section of the system as that portion upstream of the reducer or regulator that has stainless steel tubing. No attempt should be made to tighten a leaky oxygen fitting while the system is charging. The oxygen supply should be isolated in the cylinder and the system depressurized to reduce the consequences of a spark or to minimize spillage and injury should a complete fitting failure occur.

Draining an Oxygen System

The biggest factor in draining an oxygen system is safety. The oxygen must be released into the atmosphere without causing a fire, explosion, or hazard. Draining outside is highly recommended. The exact method of draining can vary. The basic procedure involves establishing a continuous flow in a safe area until the system is empty.

If the cylinder valve is operative, close the valve to isolate the oxygen supply in the cylinder. All that remains is to empty the lines and components. This can be done without disassembling the system by letting oxygen flow from the delivery point(s). If the environment is safe to receive the oxygen, positioning a demand-flow regulator to the emergency setting delivers a continuous flow of oxygen to the mask when plugged in. Hang the mask(s) out of a window while the system drains. Plug in all mask(s) to allow oxygen to drain from a continuous-flow oxygen system. Systems without check valves can be drained by opening the refill valve.

Filling an Oxygen System

Filling procedures for oxygen systems vary. Many general aviation aircraft are set up to simply replace an empty cylinder with one that is fully charged. This is also the case with a portable oxygen system. High performance and air transport category aircraft often have built-in oxygen systems that contain plumbing designed to refill gaseous oxygen cylinders while they are in place. A general discussion of the procedure to fill this type of installation follows.

Before charging any oxygen system, consult the aircraft manufacturer's maintenance manual. The type of oxygen to be used, safety precautions, equipment to be used, and the procedures for filling and testing the system must be observed. Several general precautions should also be observed when servicing a gaseous oxygen system. Oxygen valves should be opened slowly, and filling should proceed slowly to avoid overheating. The hose from the refill source to the oxygen fill valve on the aircraft should be purged of air before it is used to transfer oxygen into the system. Pressures should also be checked frequently while refilling. Additionally, all items used for oxygen systems must be free of any grease or oil, as this will cause a fire.

Airline and fixed-base operator maintenance shops often use oxygen filler carts to service oxygen systems. These contain several large oxygen supply cylinders connected to the fill cart manifold. This manifold supplies a fill hose that attaches to the aircraft. Valves and pressure gauges allow awareness and control of the oxygen dispensing process. *[Figure 16-31]* Be sure all cylinders on the cart are aviator's breathing oxygen and that all cylinders contain at least 50 psi of oxygen pressure. Each cylinder should also be within its hydrostatic test date interval. After a cart cylinder has dispensed oxygen, the remaining pressure should be recorded. This is usually written on the outside of the cylinder with chalk or in a cylinder pressure log kept with the cart. As such, the technician can tell at a glance the status of each oxygen bottle.

No pump or mechanical device is used to transfer oxygen from the fill cart manifold to the aircraft system. Objects under pressure flow from high pressure to low pressure. Thus, by connecting the cart to the aircraft and systematically opening oxygen cylinders with increasingly higher pressure, a slow increase in oxygen flow to the aircraft can be managed.

The following is a list of steps to safely fill an aircraft oxygen system from a typical oxygen refill cart.

1. Check hydrostatic dates on all cylinders, especially those that are to be filled on the aircraft. If a cylinder is out of date, remove and replace it with a specified unit that is serviceable.

2. Check pressures on all cylinders on the cart and in the aircraft. If pressure is below 50 psi, replace the cylinder(s). On the aircraft, this may require purging the system with oxygen when completed. Best practices dictate that any low-pressure or empty cylinder(s) on the cart should also be removed and replaced when discovered.

3. Take all oxygen handling precautions to ensure a safe environment around the aircraft.

4. Ground the refill cart to the aircraft.

5. Connect the cart hose from the cart manifold to the aircraft fill port. Purge the air from the refill hose with oxygen before opening the refill valve on the aircraft. Some hoses are equipped with purge valves to do this while the hose is securely attached to the aircraft. Other hoses need to be purged while attached to the refill fitting but not fully tightened.

6. Observe the pressure on the aircraft bottle to be filled. Open it. On the refill cart, open the cylinder with the closest pressure to the aircraft cylinder that exceeds it.

7. Open the aircraft oxygen system refill valve. Oxygen will flow from cart cylinder (manifold) into the aircraft cylinder.

8. When the cylinder pressures equalize, close the cylinder on the cart, and open the cart cylinder with the next highest pressure. Allow it to flow into the

Figure 16-31. *Typical oxygen servicing cart used to fill an aircraft system.*

aircraft cylinder until the pressures equalize and flow ceases. Close the cart cylinder and proceed to the cart cylinder with the next highest pressure.

9. Continue the procedure in step 8 until the desired pressure in the aircraft cylinder is achieved.

10. Close the aircraft refill valve and close all cylinders on the cart.

11. The aircraft oxygen cylinder valve(s) should be left in the proper position for normal operations. Remotely mounted cylinders are usually left open.

12. Disconnect the refill line from the refill port on the aircraft. Cap or cover both.

13. Remove the grounding strap.

Temperature has a significant effect on the pressure of gaseous oxygen. Manufacturers typically supply a fill chart or a placard at the aircraft oxygen refill station to guide technicians in compensating for temperature/pressure variations. Technicians should consult the chart and fill cylinders to the maximum pressure listed for the prevailing ambient temperature. [Figure 16-32]

When it is hot, oxygen cylinders are filled to a higher pressure than 1,800 psi or 1,850 psi, the standard maximum pressure ratings of most high-pressure aircraft oxygen cylinders. This is allowable because at altitude the temperature and pressure of the oxygen can decrease significantly. Filling cylinders to temperature-compensated pressure values ensures a full supply of oxygen is available when needed. When filling cylinders on a cold day, compensation for temperature and pressure changes dictates that cylinders be filled to less than the maximum rated capacity to allow for pressure increases as temperature rises. Strict adherence to the temperature/pressure compensation chart values is mandatory to ensure safe storage of aircraft oxygen.

Some aircraft have temperature compensation features built into the refill valve. After setting the ambient temperature on the valve dial, the valve closes when the correct amount of oxygen pressure has been established in the aircraft cylinder. A chart can be used to ensure proper servicing.

Purging an Oxygen System
The inside of an oxygen system becomes completely saturated with oxygen during use. This is desirable to deliver clean, odor-free oxygen to the users and to prevent corrosion caused by contamination. An oxygen system needs to be purged if it has been opened or depleted for more than 2 hours, or if it is suspected that the system has been contaminated. Purging is accomplished to evacuate contaminants and to restore oxygen saturation to the inside of the system

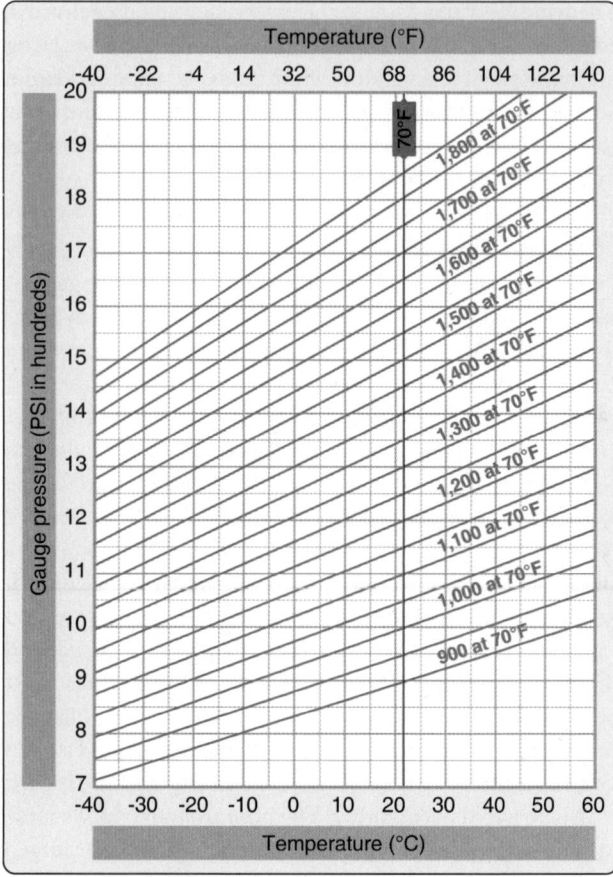

Figure 16-32. *A temperature-compensating pressure refill chart is used by the technician to ensure proper oxygen cylinder pressure in the aircraft system.*

The main cause of contamination in an oxygen system is moisture. In very cold weather, the small amount of moisture contained in the breathing oxygen can condense. With repeated charging, a significant amount of moisture may collect. Additionally, systems that are opened contain the moisture from the air that has entered. Damp charging equipment, or poor refill procedures, can also introduce water into the system. Always follow manufacturer's instructions when performing maintenance, refilling, or purging an oxygen system.

Cumulative condensation in an oxygen system cannot be entirely avoided. Purging is needed periodically. The procedure for purging may vary somewhat with each aircraft model. Generally speaking, oxygen is run through a sound oxygen system for a number of minutes at a given pressure to perform the purging. This can be as little as 10 minutes at normal delivery pressure. Other systems may require up to 30 minutes of flow at an elevated pressure. Regardless, the removal of contaminants and the resaturation of the inside of the system with oxygen is the basis for purging. It is acceptable to use nitrogen, or dry air, to blow through lines and components when performing maintenance. However, a final purging with pure oxygen is required before the system is serviceable for use.

It is important to ensure storage cylinders are refilled if they are used during the purging process. Be certain that there are no open lines and all safety caps are installed before returning the aircraft to service.

Filling LOX Systems

The use of LOX in civilian aviation is rare. The most common and safest way to fill a LOX system is to simply exchange the storage unit for one that is full. However, filling LOX on the aircraft is possible.

A portable fill cart is used, and all of the same precautions must be observed as when servicing a high-pressure gaseous oxygen system. Additionally, protection from cold burns is necessary. Due to the amount of gaseous oxygen released during the process, refilling should be accomplished outside. The servicing cart is attached to the aircraft system through a fill valve. The buildup/vent valve on the LOX container assembly is placed in the vent position. The valve on the service cart is then opened. LOX flows into the aircraft system; some vaporizes and cools the entire setup. This gaseous oxygen flows overboard through the vent valve while the system fills. When a steady stream of LOX flows from the vent valve, the system is filled. The valve is then switched to the buildup position. The aircraft refill valve and cart supply valves are closed, and the hose is removed.

Back seated valves can freeze in the open position due to the low temperature involved while LOX is being transferred. Valves should be opened completely and then closed slightly so as to not be back seated.

Inspection of Masks & Hoses

The wide varieties of oxygen masks used in aviation require periodic inspection. Mask and hose integrity ensure effective delivery of oxygen to the user when it is needed. Sometimes this is in an emergency situation. Leaks, holes, and tears are not acceptable. Most discrepancies of this type are remedied by replacement of the damaged unit.

Some continuous-flow masks are designed for disposal after use. Be sure there is a mask for each potential user on board the aircraft. Masks designed to be reused should be clean, as well as functional. This reduces the danger of infection and prolongs the life of the mask. Various mild cleaners and antiseptics that are free of petroleum products can be used. A supply of individually wrapped alcohol swabs are often kept in the flight deck.

Built-in microphones should be operational. Donning straps and fittings should be in good condition and function so that the mask is held firm to the user's face. The diameter of mask hoses in a continuous-flow system is quite a bit smaller than those used in a demand-flow system. This is because the inside diameter of the hose aids in controlling flow rate. Masks for each kind of system are made to only connect to the proper hose.

Smoke masks are required on transport aircraft and are used on some other aircraft as well. These cover the eyes, as well as the user's nose and mouth. Smoke masks are usually available within easy grasp of the crew members. They are used when the situation in the flight deck demands the increased level of protection offered. Smoke mask hoses plug into demand regulators in the same port used for regular demand type masks and operate in the same manner. Most include a built-in microphone. *[Figure 16-33]* Some portable oxygen systems are also fitted with smoke masks.

Replacing Tubing, Valves, & Fittings

The replacement of aircraft oxygen system tubing, valves, and fittings is similar to the replacement of the same components in other aircraft systems. There is, however, an added emphasis on cleanliness and compatible sealant use.

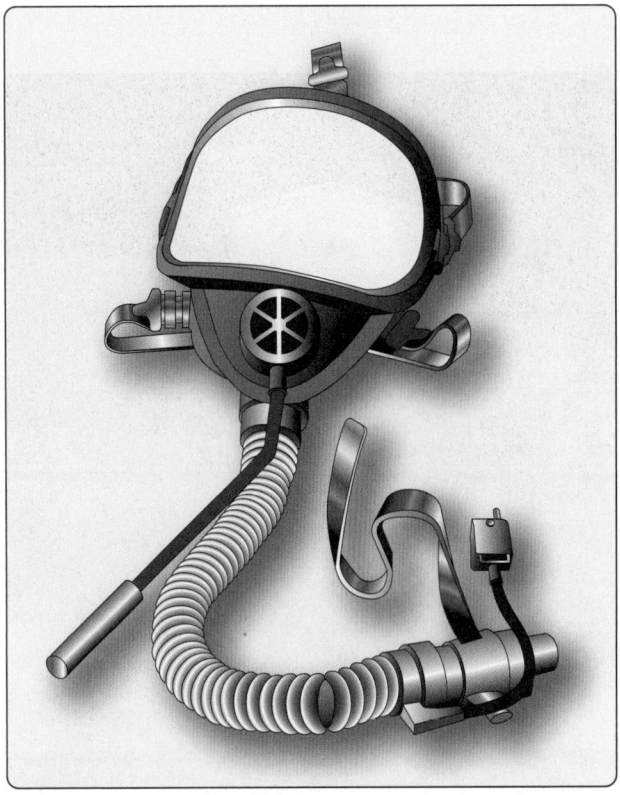

Figure 16-33. *Smoke masks cover the eyes as well as the nose and mouth of the user.*

Any oxygen system component should be cleaned thoroughly before installation. Often tubing comes with leftover residue from the bending or flaring processes. Cleaning should be accomplished with nonpetroleum-based cleansers. Trichlorethylene, acetone, and similar cleaners can be used to flush new tubing. Tubing should be blown or baked dry before installation. Follow the manufacturer's procedures for cleaning oxygen system components.

Some oxygen components make use of tapered pipe fittings. This type of connection is usually sealed with the application of thread lubricant/sealant. Typical thread sealers are petroleum based and should not be used; only oxygen compatible thread lubricant/sealers should be used. Alternatively, Teflon® tape is also used on oxygen pipe fitting connections. Be sure to begin wrapping the Teflon® tape at least two threads from the end of the fitting. This prevents any tape from coming loose and entering the oxygen system.

Prevention of Oxygen Fires or Explosions

Precautions must be observed when working with or around pure oxygen. It readily combines with other substances, some in a violent and explosive manner. As mentioned, it is extremely important to keep distance between pure oxygen and petroleum products. When allowed to combine, an explosion can result. Additionally, there are a variety of inspection and maintenance practices that should be followed to ensure safety when working with oxygen and oxygen

systems. Care should be used and, as much as possible, maintenance should be done outside.

When working on an oxygen system, it is essential that the warnings and precautions given in the aircraft maintenance manual be carefully observed. Before any work is attempted, an adequate fire extinguisher should be on hand. Cordon off the area and post "NO SMOKING" placards. Ensure that all tools and servicing equipment are clean and avoid power on checks and use of the aircraft electrical system.

Oxygen System Inspection & Maintenance

When working around oxygen and oxygen systems, cleanliness enhances safety. Clean, grease-free hands, clothes, and tools are essential. A good practice is to use only tools dedicated for work on oxygen systems. There should be absolutely no smoking or open flames within a minimum of 50 feet of the work area. Always use protective caps and plugs when working with oxygen cylinders, system components, or plumbing. Do not use any kind of adhesive tape. Oxygen cylinders should be stored in a designated, cool, ventilated area in the hangar away from petroleum products or heat sources.

Oxygen system maintenance should not be accomplished until the valve on the oxygen supply cylinder is closed and pressure is released from the system. Fittings should be unscrewed slowly to allow any residual pressure to dissipate. All oxygen lines should be marked and should have at least

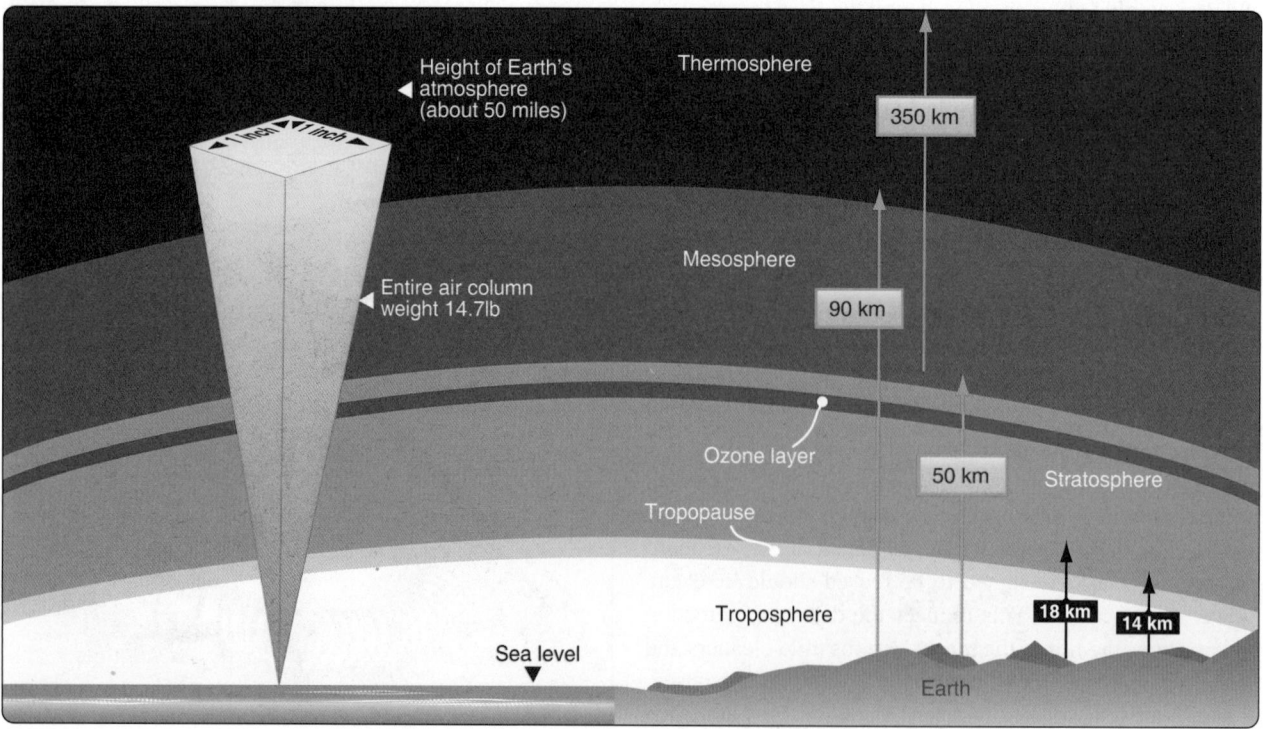

Figure 16-34. *The weight exerted by a 1 square inch column of air stretching from sea level to the top of the atmosphere is what is measured when it is said that atmospheric pressure is equal to 14.7 pounds per square inch.*

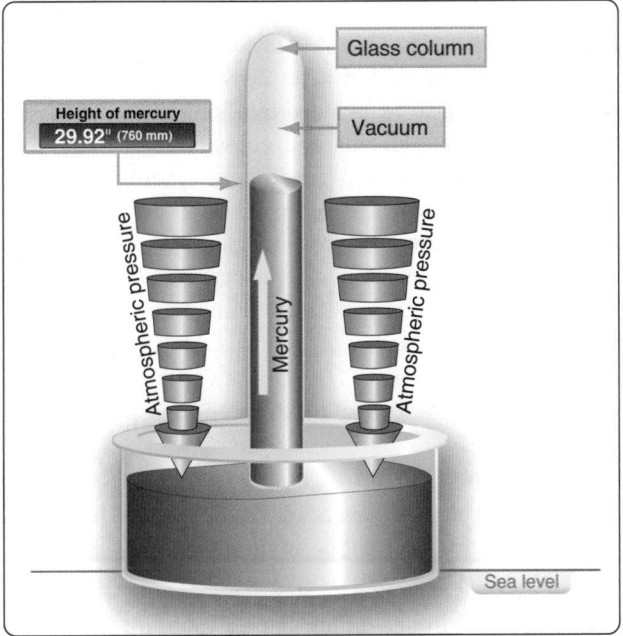

Figure 16-35. *The weight of the atmosphere pushes down on the mercury in the reservoir of a barometer, which causes mercury to rise in the column. At sea level, mercury is forced up into the column approximately 29.92 inches. Therefore, it is said that barometric pressure is 29.92 inches of mercury at sea level.*

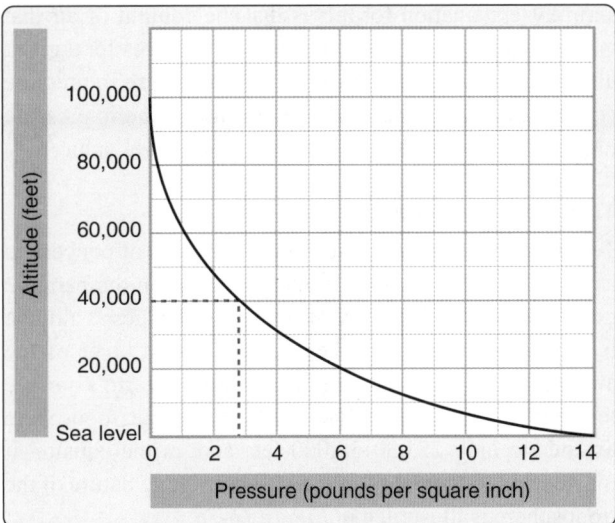

Figure 16-37. *Atmospheric pressure decreasing with altitude. At sea level the pressure is 14.7 psi, while at 40,000 feet, as the dotted lines show, the pressure is only 2.72 psi.*

2 inches of clearance from moving parts, electrical wiring, and all fluid lines. Adequate clearance must also be provided from hot ducts and other sources that might heat the oxygen. A pressure and leak check must be performed each time the system is opened for maintenance. Do not use any lubricants, sealers, cleaners, etc., unless specifically approved for oxygen system use.

Aircraft Pressurization Systems

Pressure of the Atmosphere

The gases of the atmosphere (air), although invisible, have weight. A one square inch column of air stretching from sea level into space weighs 14.7 pounds. Therefore, it can be stated that the pressure of the atmosphere, or atmospheric pressure, at sea level is 14.7 psi. *[Figure 16-34]*

Atmospheric pressure is also known as barometric pressure and is measured with a barometer. *[Figure 16-35]* Expressed in various ways, such as in inches of mercury or millimeters of mercury, these measurements come from observing the height of mercury in a column when air pressure is exerted on a reservoir of mercury into which the column is set. The column must be evacuated so air inside does not act against the mercury rising. A column of mercury 29.92 inches high weighs the same as a column of air that extends from sea level to the top of the atmosphere and has the same cross-section as the column of mercury.

Aviators often interchange references to atmospheric pressure between linear displacement (e.g., inches of mercury) and units of force (e.g., psi). Over the years, meteorology has shifted its use of linear displacement representation of atmospheric pressure to units of force. However, the unit of force nearly universally used today to represent atmospheric pressure in meteorology is the hectopascal (hPa). A hectopascal is a metric (SI) unit that expresses force in newtons per square meter. 1,013.2 hPa is equal to 14.7 psi. *[Figure 16-36]*

Atmospheric pressure decreases with increasing altitude. The

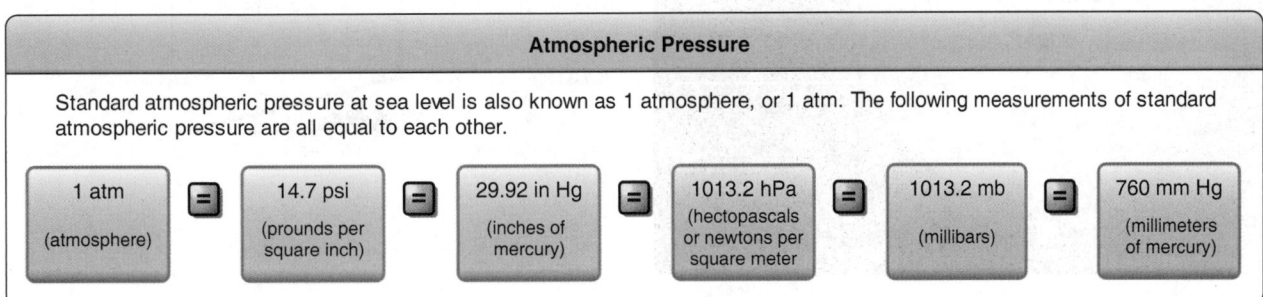

Figure 16-36. *Various equivalent representations of atmospheric pressure at sea level.*

simplest explanation for this is that the column of air that is weighed is shorter. How the pressure changes for a given altitude is shown in *Figure 16-37*. The decrease in pressure is a rapid one and, at 50,000 feet, the atmospheric pressure has dropped to almost one-tenth of the sea level value.

Temperature & Altitude

Temperature variations in the atmosphere are of concern to aviators. Weather systems produce changes in temperature near the earth's surface. Temperature also changes as altitude is increased. The troposphere is the lowest layer of the atmosphere. On average, it ranges from the earth's surface to about 38,000 feet above it. Over the poles, the troposphere extends to only 25,000–30,000 feet and, at the equator, it may extend to around 60,000 feet. This oblong nature of the troposphere is illustrated in *Figure 16-38*.

Most civilian aviation takes place in the troposphere in which temperature decreases as altitude increases. The rate of change is somewhat constant at about –2 °C or –3.5 °F for every 1,000 feet of increase in altitude. The upper boundary of the troposphere is the tropopause. It is characterized as a zone of relatively constant temperature of –57 °C or –69 °F.

Above the tropopause lies the stratosphere. Temperature increases with altitude in the stratosphere to near 0 °C before decreasing again in the mesosphere, which lies above it. The stratosphere contains the ozone layer that protects the earth's inhabitants from harmful UV rays. Some civilian flights and numerous military flights occur in the stratosphere. *Figure 16-39* diagrams the temperature variations in different layers of the atmosphere.

When an aircraft is flown at high altitude, it burns less

fuel for a given airspeed than it does for the same speed at a lower altitude. This is due to decreased drag that results from the reduction in air density. Bad weather and turbulence can also be avoided by flying in the relatively smooth air above storms and convective activity that occur in the lower troposphere. To take advantage of these efficiencies, aircraft are equipped with environmental systems to overcome extreme temperature and pressure levels. While supplemental oxygen and a means of staying warm suffice, aircraft pressurization and air conditioning systems have been developed to make high altitude flight more comfortable. *Figure 16-40* illustrates the temperatures and pressures at various altitudes in the atmosphere.

Pressurization Terms

The following terms should be understood for the discussion of pressurization and cabin environmental systems that follows:

1. Cabin altitude—given the air pressure inside the cabin, the altitude on a standard day that has the same pressure as that in the cabin. Rather than saying the pressure inside the cabin is 10.92 psi, it can be

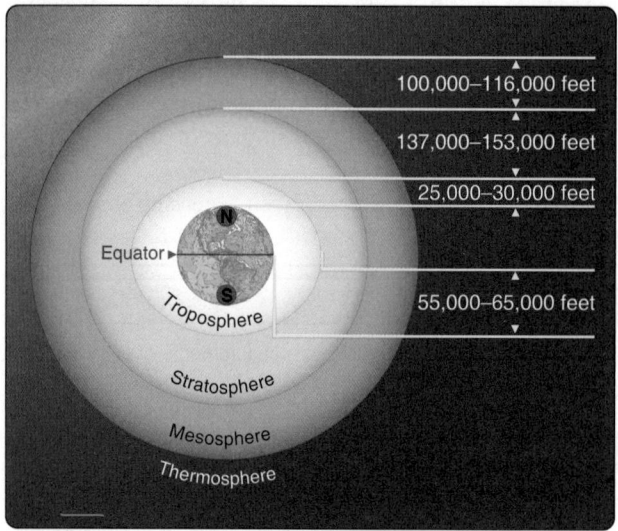

Figure 16-38. *The troposphere extends higher above the earth's surface at the equator than it does at the poles.*

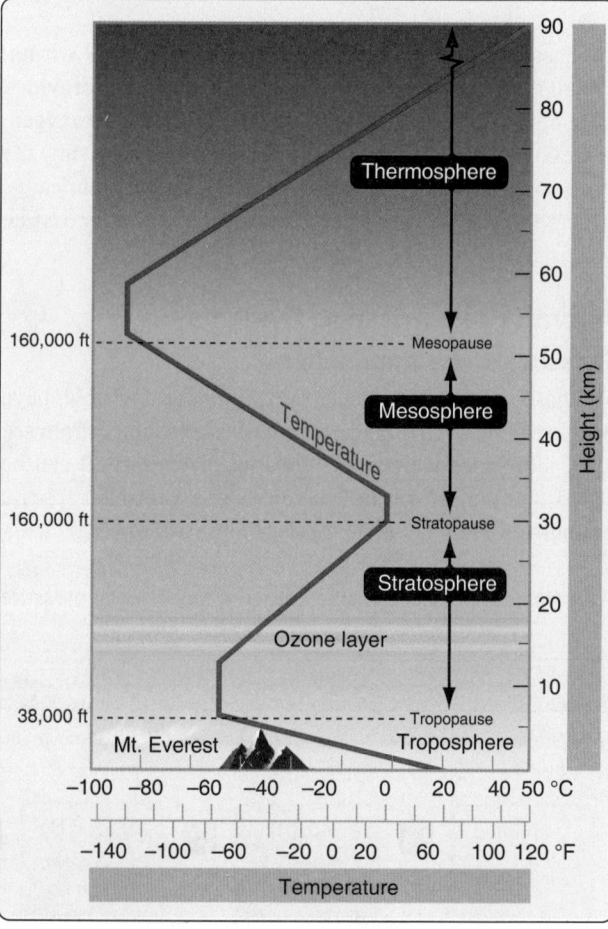

Figure 16-39. *The atmospheric layers with temperature changes depicted by the red line.*

Altitude	Pressure			Temperature	
feet	psi	hPa	in Hg	°F	°C
0	14.69	1013.2	29.92	59.0	15
1,000	14.18	977.2	28.86	55.4	13
2,000	13.66	942.1	27.82	51.9	11
3,000	13.17	908.1	26.82	48.3	9.1
4,000	12.69	875.1	25.84	44.7	7.1
5,000	12.23	843.1	24.90	41.2	5.1
6,000	11.77	812.0	23.98	37.6	3.1
7,000	11.34	781.8	23.09	34.0	1.1
8,000	10.92	752.6	22.23	30.5	−0.8
9,000	10.51	724.3	21.39	26.9	−2.8
10,000	10.10	696.8	20.58	23.3	−4.8
12,000	9.34	644.4	19.03	16.2	−8.8
14,000	8.63	595.2	17.58	9.1	−12.7
16,000	7.96	549.2	16.22	1.9	−16.7
18,000	7.34	506.0	14.94	−5.2	−29.7
20,000	6.76	465.6	13.75	−12.3	−24.6
22,000	6.21	427.9	12.64	−19.5	−28.6
24,000	5.70	392.7	11.60	−26.6	−32.5
26,000	5.22	359.9	10.63	−33.7	−36.5
28,000	4.78	329.3	9.72	−40.9	−40.5
30,000	4.37	300.9	8.89	−48.0	−44.4
32,000	3.99	274.5	8.11	−55.1	−48.4
34,000	3.63	250.0	7.38	−62.2	−52.4
36,000	3.30	227.3	6.71	−69.4	−56.3
38,000	3.00	206.5	6.10	−69.4	−56.5
40,000	2.73	187.5	5.54	−69.4	−56.5
45,000	2.14	147.5	4.35	−69.4	−56.5
50,000	1.70	116.0	3.42	−69.4	−56.5

Figure 16-40. *Cabin environmental systems establish conditions quite different from these found outside the aircraft.*

said that the cabin altitude is 8,000 feet (MSL).

2. Cabin differential pressure—the difference between the air pressure inside the cabin and the air pressure outside the cabin. Cabin pressure (psi) – ambient pressure (psi) = cabin differential pressure (psid or Δ psi).

3. Cabin rate of climb—the rate of change of air pressure inside the cabin, expressed in feet per minute (fpm) of cabin altitude change.

Pressurization Issues

Pressurizing an aircraft cabin assists in making flight possible in the hostile environment of the upper atmosphere. The degree of pressurization and the operating altitude of any aircraft are limited by critical design factors. A cabin pressurization system must accomplish several functions if it is to ensure adequate passenger comfort and safety. It must be capable of maintaining a cabin pressure altitude of approximately 8,000 feet or lower regardless of the cruising altitude of the aircraft. This is to ensure that passengers and

crew have enough oxygen present at sufficient pressure to facilitate full blood saturation. A pressurization system must also be designed to prevent rapid changes of cabin pressure, which can be uncomfortable or injurious to passengers and crew. Additionally, a pressurization system should circulate air from inside the cabin to the outside at a rate that quickly eliminates odors and to remove stale air. Cabin air must also be heated or cooled on pressurized aircraft. Typically, these functions are incorporated into the pressurization source.

To pressurize, a portion of the aircraft designed to contain air at a pressure higher than outside atmospheric pressure must be sealed. A wide variety of materials facilitate this. Compressible seals around doors combine with various other seals, grommets, and sealants to essentially establish an air tight pressure vessel. This usually includes the cabin, flight compartment, and the baggage compartments. Air is then pumped into this area at a constant rate sufficient to raise the pressure slightly above that which is needed. Control is maintained by adjusting the rate at which the air is allowed to flow out of the aircraft.

A key factor in pressurization is the ability of the fuselage to withstand the forces associated with the increase in pressure inside the structure versus the ambient pressure outside. This differential pressure can range from 3.5 psi for a single-engine reciprocating aircraft, to approximately 9 psi on high performance jet aircraft. *[Figure 16-41]* If the weight of the aircraft structure were of no concern, this would not be a problem. Making an aircraft strong for pressurization, yet also light, has been an engineering challenge met over numerous years beginning in the 1930s. The development of jet aircraft and their ability to exploit low drag flight at higher altitude made the problem even more pronounced. Today, the proliferation of composite materials in aircraft structure continues this engineering challenge.

In addition to being strong enough to withstand the pressure differential between the air inside and the air outside the cabin, metal fatigue from repeated pressurization and depressurization weakens the airframe. Some early pressurized aircraft structures failed due to this and resulted in fatal accidents. The FAA's aging aircraft program was instituted to increase inspection scrutiny of older airframes that may show signs of fatigue due to the pressurization cycle.

Aircraft of any size may be pressurized. Weight considerations when making the fuselage strong enough to endure pressurization usually limit pressurization to high performance light aircraft and larger aircraft. A few pressurized single-engine reciprocating aircraft exist, as well as many pressurized single-engine turboprop aircraft.

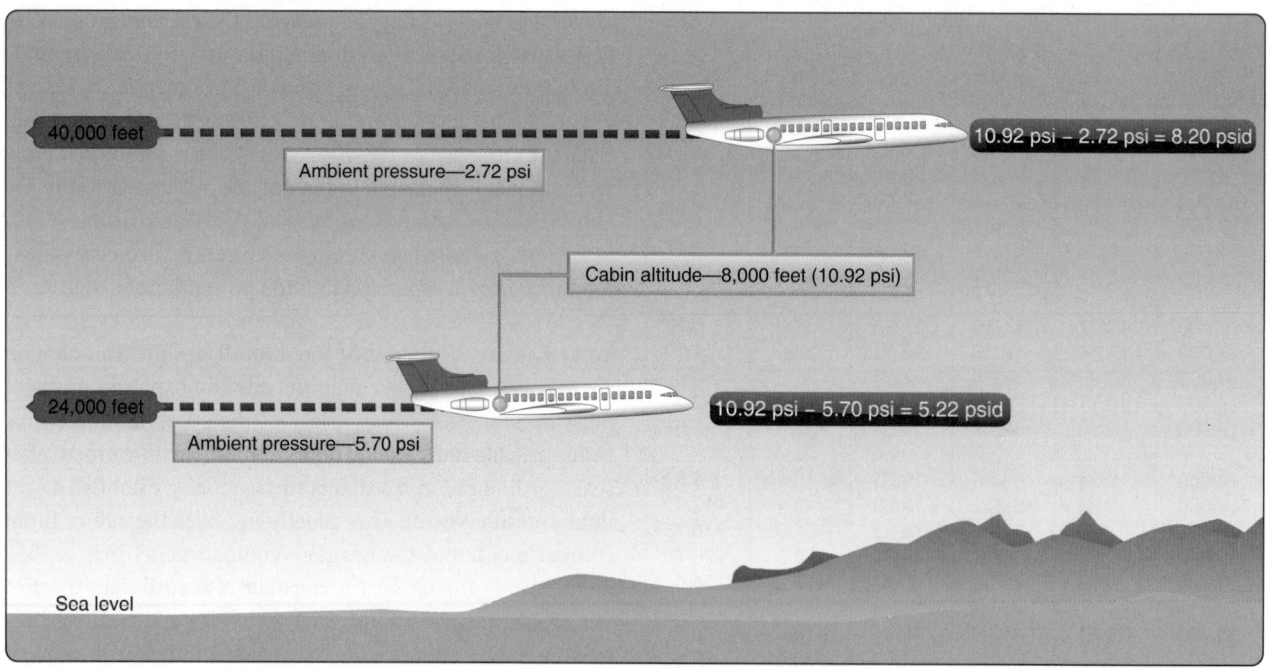

Figure 16-41. *Differential pressure (psid) is calculated by subtracting the ambient air pressure from the cabin air pressure.*

Sources of Pressurized Air

The source of air to pressurize an aircraft varies mainly with engine type. Reciprocating aircraft have pressurization sources different from those of turbine-powered aircraft. The compression of air raises its temperature. A means for keeping pressurization air cool enough is built into most pressurization systems. It may be in the form of a heat exchanger, using cold ambient air to modify the temperature of the air from the pressurization source. A full air cycle air conditioning system with expansion turbine may also be used. The latter provides the advantage of temperature control on the ground and at low altitudes where ambient air temperature may be higher than comfortable for the passengers and crew.

Reciprocating Engine Aircraft

There are three typical sources of air used to pressurize reciprocating aircraft: supercharger, turbocharger, and engine-driven compressor. Superchargers and turbochargers are installed on reciprocating engines to permit better performance at high altitude by increasing the quantity and pressure of the air in the induction system. Some of the air produced by each of these can be routed into the cabin to pressurize it.

A supercharger is mechanically driven by the engine. Despite engine performance increases due to higher induction system pressure, some of the engine output is utilized by the supercharger. Furthermore, superchargers have limited capability to increase engine performance. If supplying both the intake and the cabin with air, the engine performance ceiling is lower than if the aircraft were not pressurized.

Superchargers must be located upstream of the fuel delivery to be used for pressurization. They are found on older reciprocating engine aircraft, including those with radial engines. *[Figures 16-42 and 16-43]*

Turbochargers, sometimes known as turbosuperchargers, are driven by engine exhaust gases. They are the most common source of pressurization on modern reciprocating engine aircraft. The turbocharger impeller shaft extends through the bearing housing to support a compression impeller in a separate housing. By using some of the turbocharger

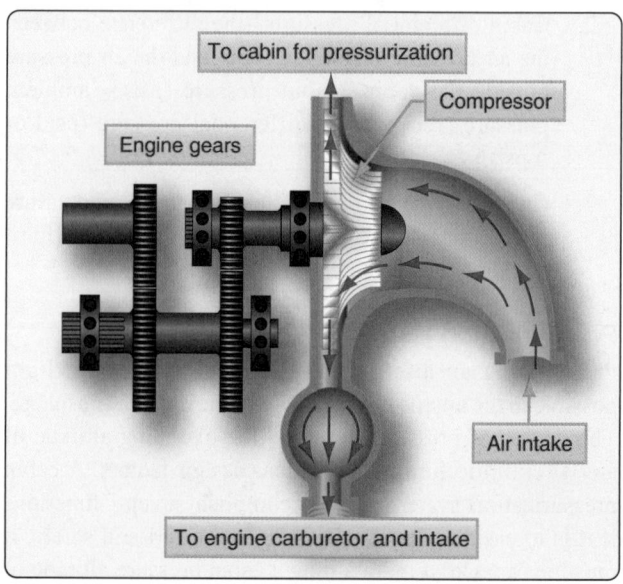

Figure 16-42. *A reciprocating engine supercharger can be used as a source of pressurization if it is upstream of carburetion.*

Figure 16-43. *The radial engine supercharger cannot be used since fuel is introduced before the supercharger impeller compresses the air.*

compressed air for cabin pressurization, less is available for the intake charge, resulting in lower overall engine performance. Nonetheless, the otherwise wasted exhaust gases are put to work in the turbocharger compressor, enabling high altitude flight with the benefits of low drag and weather avoidance in relative comfort and without the use of supplemental oxygen. *[Figures 16-44 and 16-45]*

Both superchargers and turbochargers are oil lubricated. The supercharger is part of the fuel intake system and the turbocharger is part of the exhaust system. As such, there is a risk of contamination of cabin air from oil, fuel, or exhaust fumes should a malfunction occur, a shortcoming of these pressurization sources.

A third source of air for pressurizing the cabin in reciprocating aircraft is an engine driven compressor. Either belt driven or gear driven by the accessory drive, an independent, dedicated compressor for pressurization avoids some of the potential contamination issues of superchargers and turbochargers. The compressor device does, however, add significant weight. It also consumes engine output since it is engine driven.

The roots blower is used on older, large reciprocating engine aircraft. *[Figure 16-46]* The two lobes in this compressor do not touch each other or the compressor housing. As they rotate, air enters the space between the lobes and is compressed and delivered to the cabin for pressurization. Independent engine-driven centrifugal compressors can also be found on reciprocating engine aircraft. *[Figure 16-47]* A variable ratio gear drive system is used to maintain a constant rate of airflow during changes of engine rpm.

Near maximum operating altitude, the performance of any reciprocating engine and the pressurization compressor suffer. This is due to the reduced pressure of the air at altitude that supplies the intake of each. The result is difficulty in maintaining a sufficient volume of air to the engine intake to produce power, as well as to allow enough air to the fuselage for pressurization. These are the limiting factors for determining the design ceiling of most reciprocating aircraft, which typically does not exceed 25,000 feet. Turbine engine aircraft overcome these shortcomings, permitting them to fly at much higher altitudes.

Turbine Engine Aircraft

The main principle of operation of a turbine engine involves the compression of large amounts of air to be mixed with fuel and burned. Bleed air from the compressor section of the engine is relatively free of contaminants. As such, it is a great source of air for cabin pressurization. However, the

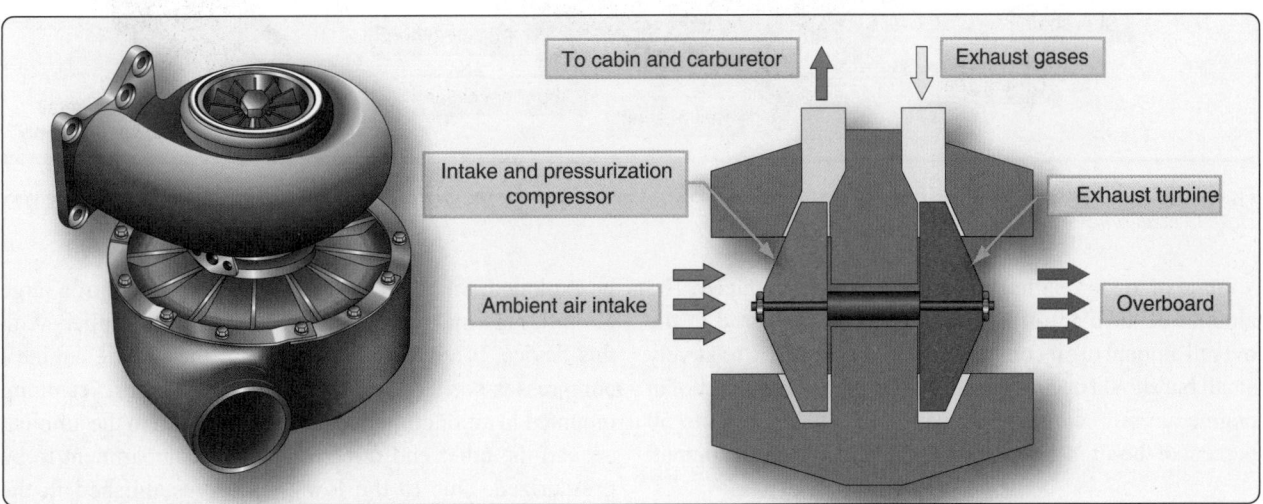

Figure 16-44. *A turbocharger used for pressurizing cabin air and engine intake air on a reciprocating engine aircraft.*

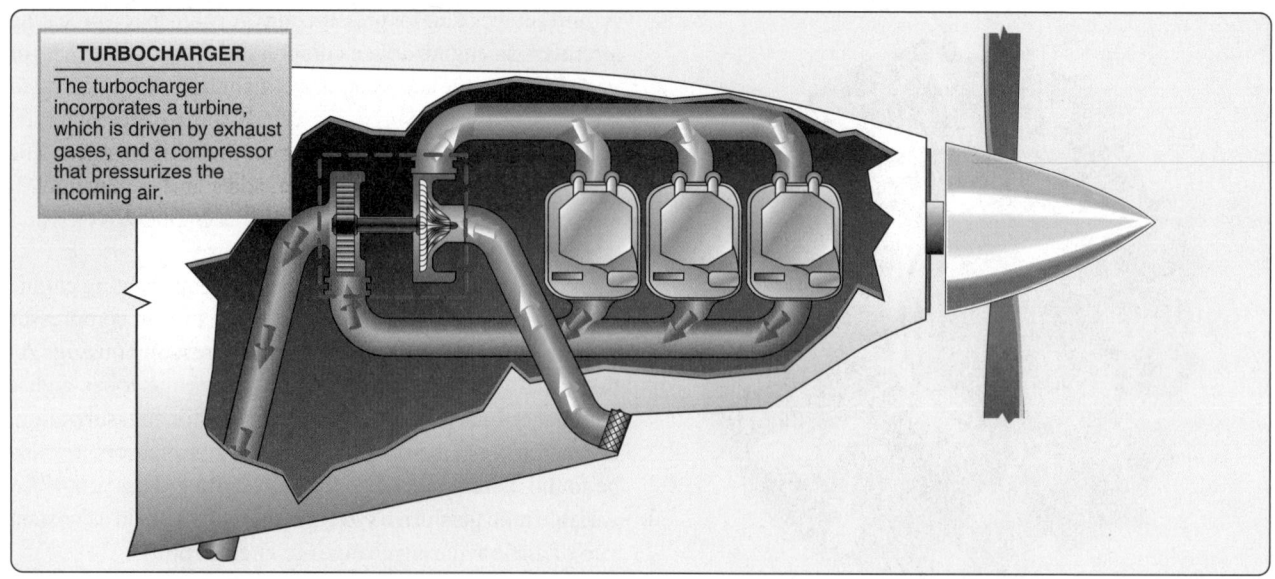

Figure 16-45. *A turbocharger installation on a reciprocating aircraft engine (top left side).*

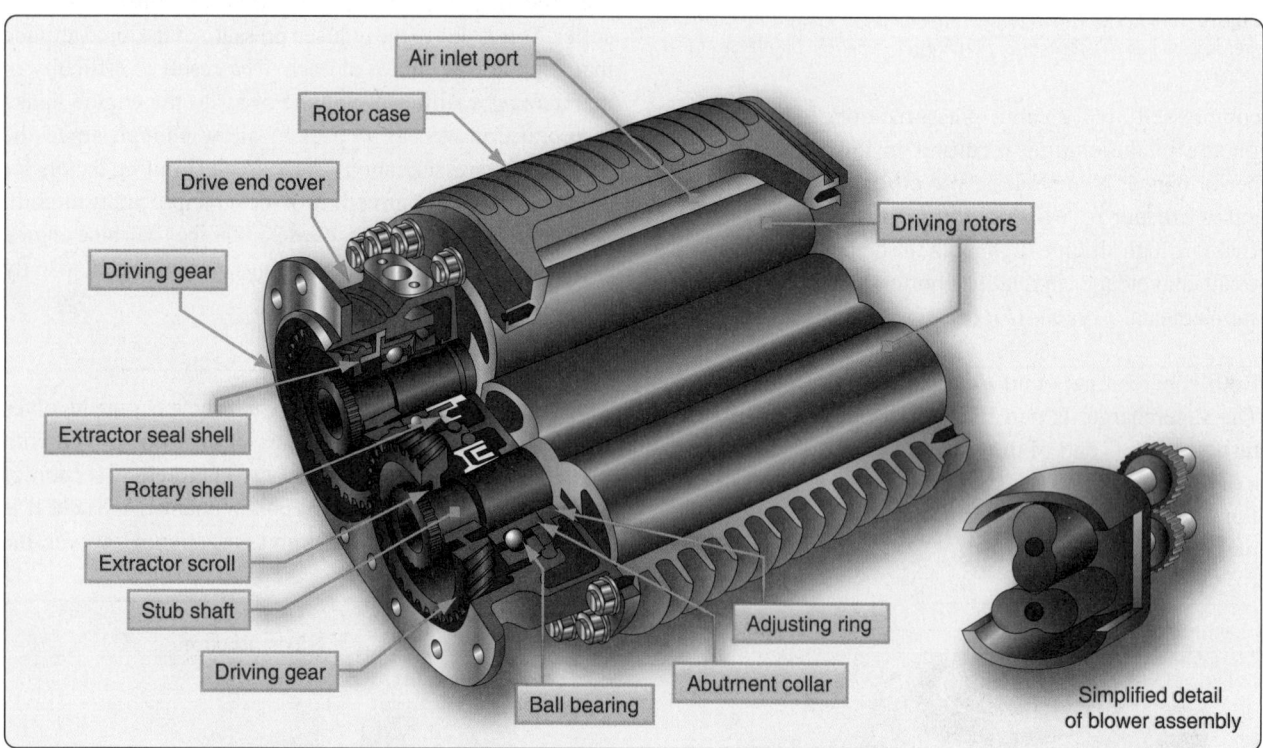

Figure 16-46. *A roots blower found on older pressurized aircraft is gear driven by the engine. It pressurizes air as the rotors rotate very close to each other without touching.*

volume of air for engine power production is reduced. The amount of air bled off for pressurization compared to the overall amount of air compressed for combustion is relatively small but should be minimized. Modern large-cabin turbofan engine aircraft contain recirculation fans to reuse up to 50 percent of the air in the cabin, maintaining high engine output.

There are different ways hot, high-pressure bleed air can

be exploited. Smaller turbine aircraft, or sections of a large aircraft, may make use of a jet pump flow multiplier. With this device, bleed air is tapped off of the turbine engine's compressor section. It is ejected into a venturi jet pump mounted in air ducting that has one end open to the ambient air and the other end directed into the compartment to be pressurized. Due to the low pressure established in the venturi by the bleed air flow, air is drawn in from outside

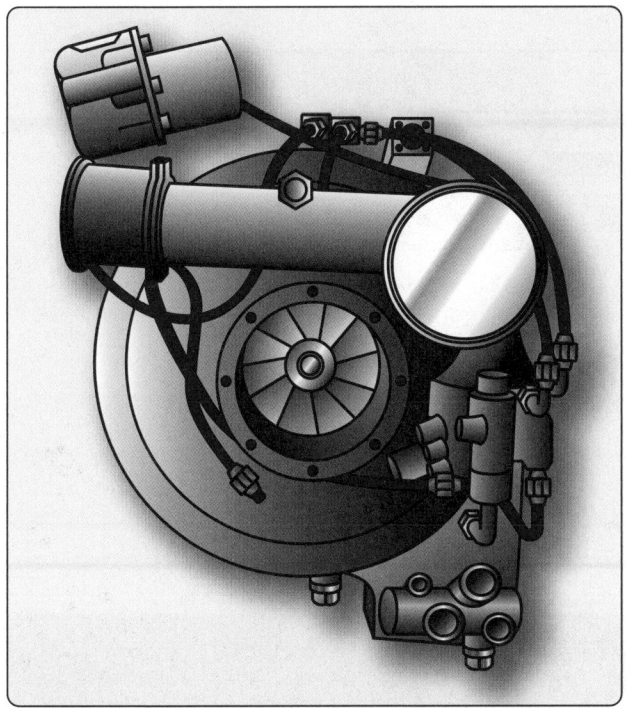

Figure 16-47. *A centrifugal cabin supercharger.*

the aircraft. It mixes with the bleed air and is delivered to the pressure vessel to pressurize it. An advantage of this type of pressurization is the lack of moving parts. *[Figure 16-48]* A disadvantage is only a relatively small volume of space can be pressurized in this manner.

Another method of pressurizing an aircraft using turbine engine compressor bleed air is to have the bleed air drive a separate compressor that has an ambient air intake. A turbine turned by bleed air rotates a compressor impellor mounted on the same shaft. Outside air is drawn in and compressed.

It is mixed with the bleed air outflow from the turbine and is sent to the pressure vessel. Turboprop aircraft often use this device, known as a turbocompressor. *[Figure 16-49]*

The most common method of pressurizing turbine-powered aircraft is with an air cycle air conditioning and pressurization system. Bleed air is used, and through an elaborate system including heat exchangers, a compressor, and an expansion turbine, cabin pressurization and the temperature of the pressurizing air are precisely controlled. This air cycle system is discussed in greater detail in the air conditioning section of this chapter. *[Figure 16-50]*

Control of Cabin Pressure
Pressurization Modes

Aircraft cabin pressurization can be controlled via two different modes of operation. The first is the isobaric mode, which works to maintain cabin altitude at a single pressure despite the changing altitude of the aircraft. For example, the flight crew may select to maintain a cabin altitude of 8,000 feet (10.92 psi). In the isobaric mode, the cabin pressure is established at the 8,000-foot level and remains at this level, even as the altitude of the aircraft fluctuates.

The second mode of pressurization control is the constant differential mode, which controls cabin pressure to maintain a constant pressure difference between the air pressure inside the cabin and the ambient air pressure, regardless of aircraft altitude changes. The constant differential mode pressure differential is lower than the maximum differential pressure for which the airframe is designed, keeping the integrity of the pressure vessel intact.

When in isobaric mode, the pressurization system maintains the cabin altitude selected by the crew. This is the condition

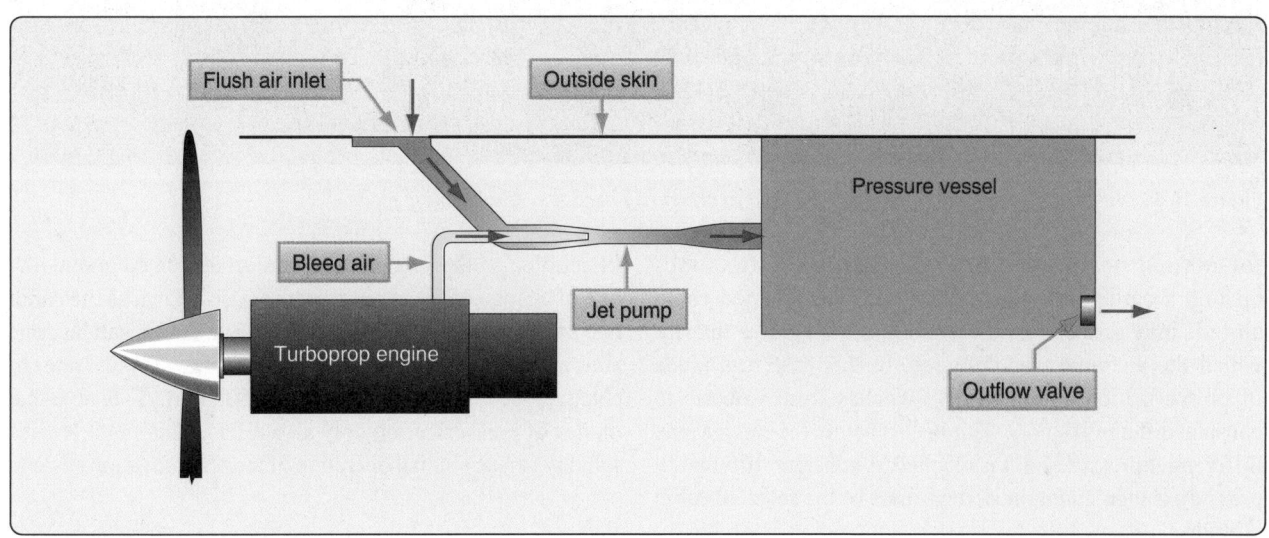

Figure 16-48. *A jet pump flow multiplier ejects bleed air into a venturi which draws air for pressurization from outside the aircraft.*

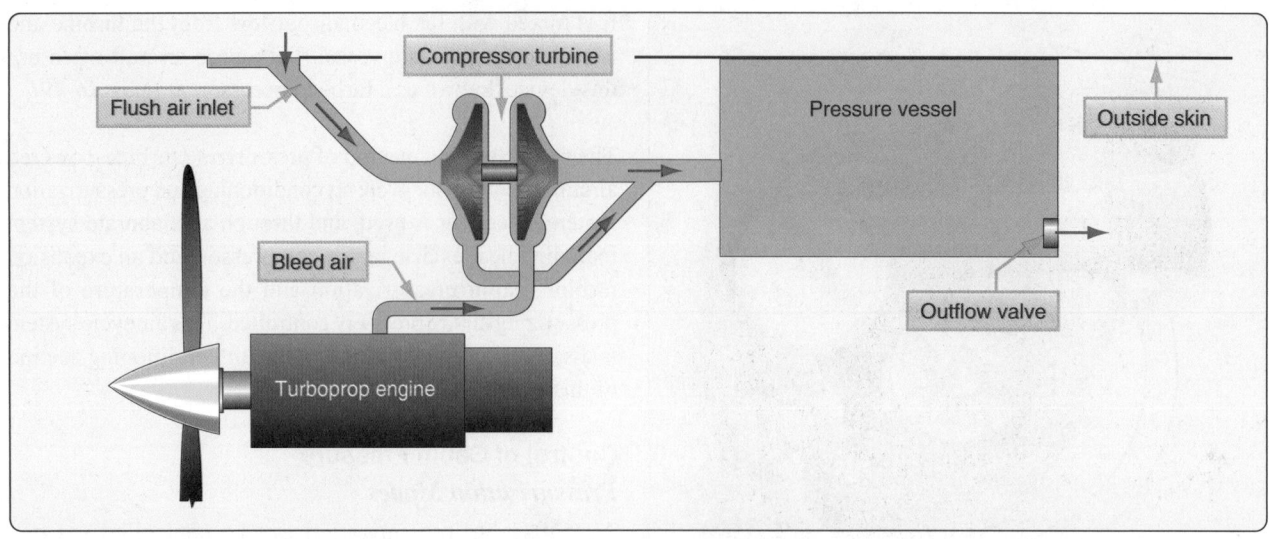

Figure 16-49. *A turbo compressor used to pressurize cabins mostly in turboprop aircraft.*

Figure 16-50. *An air cycle air conditioning system used to pressurize the cabin of a business jet.*

for normal operations. But when the aircraft climbs beyond a certain altitude, maintaining the selected cabin altitude may result in a differential pressure above that for which the airframe was designed. In this case, the mode of pressurization automatically switches from isobaric to constant differential mode. This occurs before the cabin's max differential pressure limit is reached. A constant differential pressure is then maintained, regardless of the selected cabin altitude.

In addition to the modes of operation described above, the rate of change of the cabin pressure, also known as the cabin rate of climb or descent, is also controlled. This can be done automatically or manually by the flight crew. Typical rates of change for cabin pressure are 300 to 500 fpm. Also, note that modes of pressurization may also refer to automatic versus standby versus manual operation of the pressurization system.

Cabin Pressure Controller

The cabin pressure controller is the device used to control the cabin air pressure. Older aircraft use strictly pneumatic means for controlling cabin pressure. Selections for the desired cabin altitude, rate of cabin altitude change, and barometric pressure setting are all made directly to the pressure controller from pressurization panel in the flight deck. *[Figure 16-51]*

Adjustments and settings on the pressure controller are the control input parameters for the cabin pressure regulator. The regulator controls the position of the outflow valve(s) normally located at the rear of the aircraft pressure vessel. Valve position determines the pressure level in the cabin.

Modern aircraft often combine pneumatic, electric, and electronic control of pressurization. Cabin altitude, cabin rate of change, and barometric setting are made on the cabin pressure selector of the pressurization panel in the flight deck. Electric signals are sent from the selector to the cabin pressure controller, which functions as the pressure regulator. It is remotely located out of sight near the flight deck but inside the pressurized portion of the aircraft. The signals are converted from electric to digital and are used by the controller. Cabin pressure and ambient pressure are also input to the controller, as well as other inputs. *[Figure 16-52]*

Using this information, the controller, which is essentially

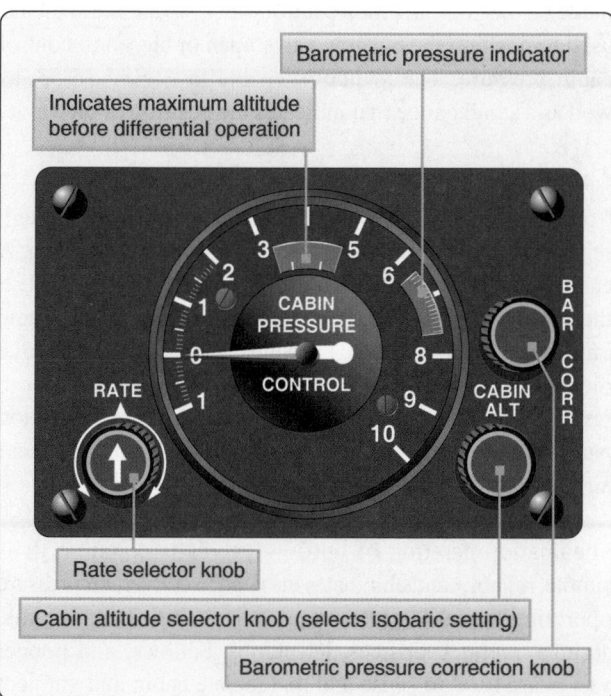

Figure 16-51. *A pressure controller for an all pneumatic cabin pressure control system.*

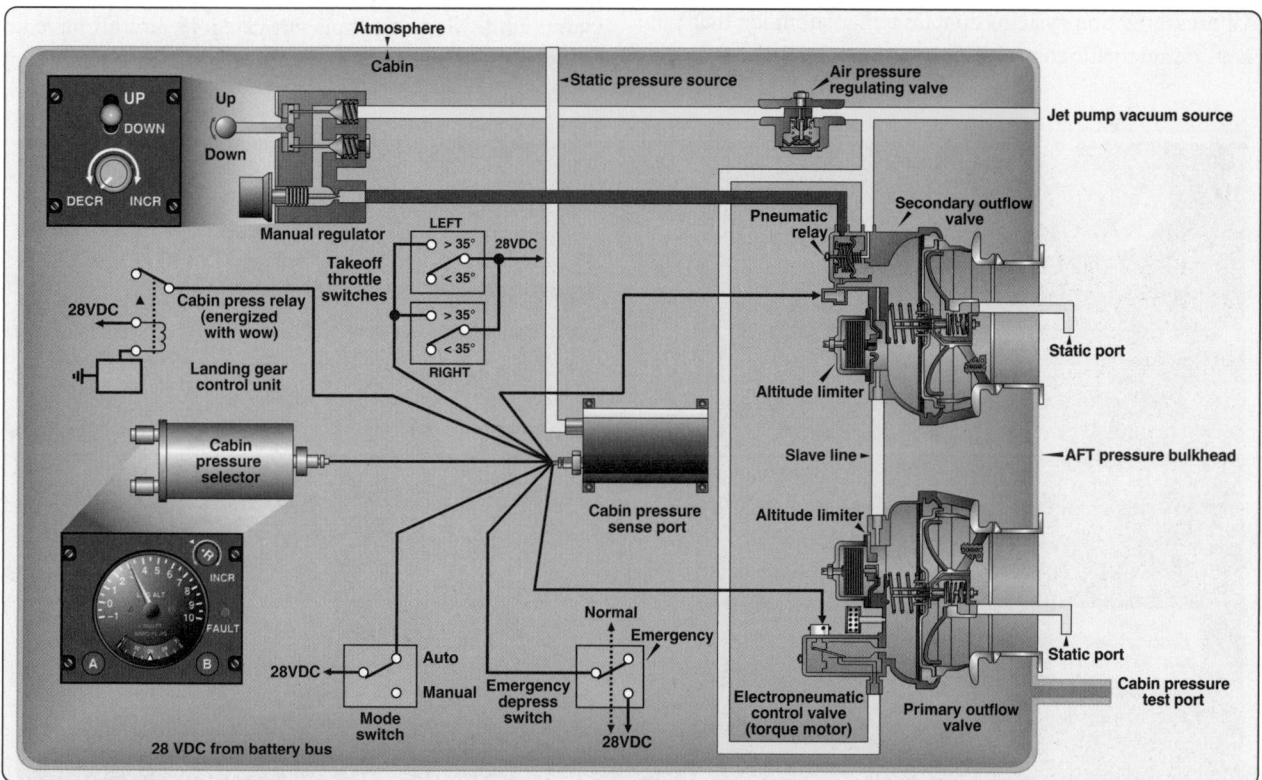

Figure 16-52. *The pressurization control system on many small transports and business jets utilizes a combination of electronic, electric, and pneumatic control elements.*

a computer, supplies pressurization logic for various stages of a flight. On many small transport and business jets, the controller's electric output signal drives a torque motor in the primary outflow valve. This modulates pneumatic airflow through the valve, which positions the valve to maintain the pressurization schedule.

On many transport category aircraft, two cabin pressure controllers, or a single controller with redundant circuitry, are used. Located in the electronics equipment bay, they receive electric input from the panel selector, as well as ambient and cabin pressure input. Flight altitude and landing field altitude information are often the crew selection choices on the pressurization control panel. Cabin altitude, rate of climb, and barometric setting are automatic through built-in logic and communication with the ADC and the flight management system (FMS). The controllers process the information and send electric signals to motors that directly position the outflow valve(s). *[Figure 16-53]*

Modern pressurization control is fully automatic once variable selections are made on the pressurization control panel if, in fact, there are any to be made. Entering or selecting a flight plan into the FMS of some aircraft automatically supplies the pressurization controller with the parameters needed to establish the pressurization schedule for the entire flight. No other input is needed from the crew.

All pressurization systems contain a manual mode that can override automatic control. This can be used in flight or on the ground during maintenance. The operator selects the manual mode on the pressurization control panel. A separate switch is used to position the outflow valve open or closed to control cabin pressure. The switch is visible in *Figure 16-53,* as well as a small gauge that indicates the position of the valve.

Cabin Air Pressure Regulator & Outflow Valve

Controlling cabin pressurization is accomplished through regulating the amount of air that flows out of the cabin. A cabin outflow valve opens, closes, or modulates to establish the amount of air pressure maintained in the cabin. Some outflow valves contain the pressure regulating and the valve mechanism in a single unit. They operate pneumatically in response to the settings on the flight deck pressurization panel that influence the balance between cabin and ambient air pressure. *[Figure 16-54]*

Pneumatic operation of outflow valves is common. It is simple, reliable, and eliminates the need to convert air pressure operating variables into some other form. Diaphragms, springs, metered orifices, jet pumps, bellows, and poppet valves are used to sense and manipulate cabin and ambient air pressures to correctly position the outflow valve without the use of electricity. Outflow valves that combine the use of electricity with pneumatic operation have all-pneumatic standby and manual modes, as shown in *Figure 16-52.*

The pressure regulating mechanism can also be found as a separate unit. Many air transport category aircraft have an outflow valve that operates electrically, using signals sent

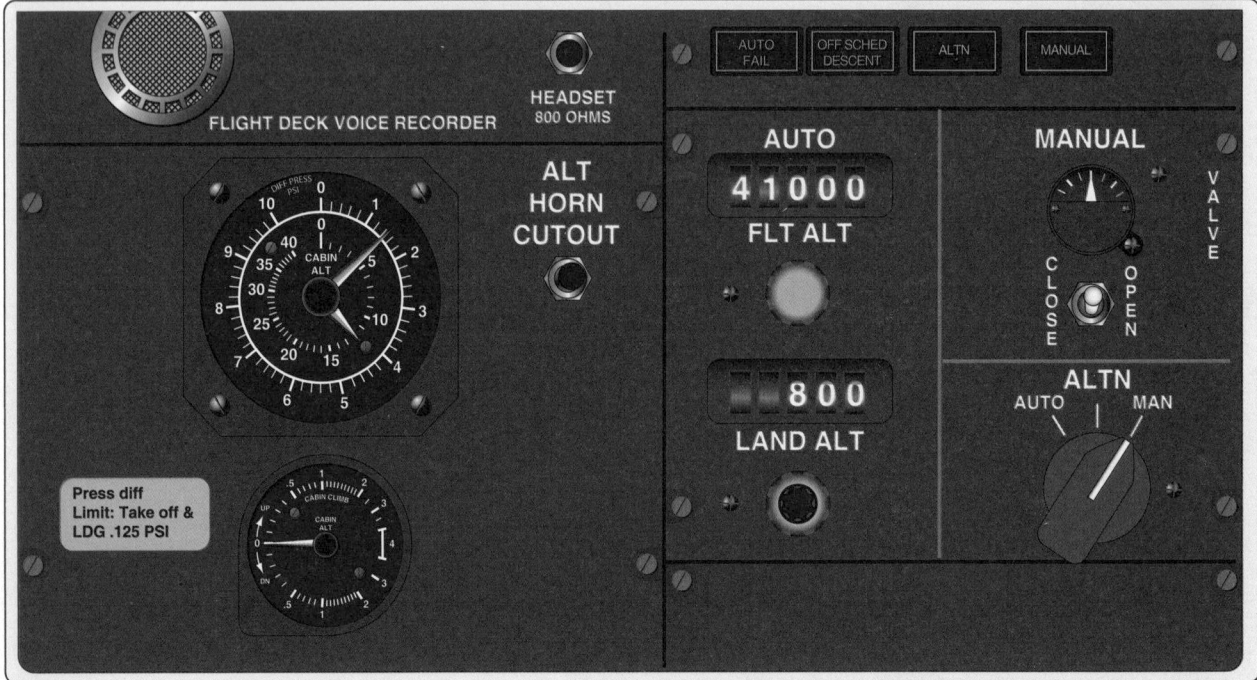

Figure 16-53. *This pressurization panel from an 800 series Boeing 737 has input selections of flight altitude and landing altitude.*

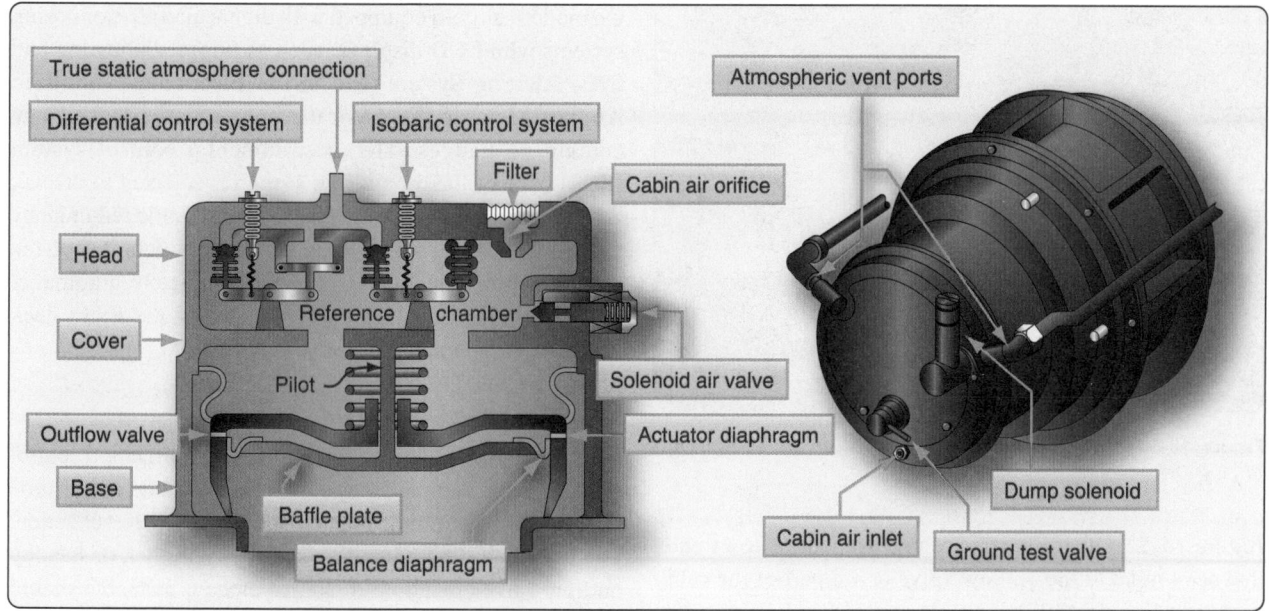

Figure 16-54. *An all-pneumatic cabin pressure regulator and outflow valve.*

from a remotely located cabin air pressure controller that acts as the pressure regulator. The controller positions the valve(s) to achieve the settings on the flight deck pressurization panel selectors according to predetermined pressurization schedules. Signals are sent to electric motors to move the valve as needed. On transports, often AC motors are used with a redundant DC motor for standby or manual operations. *[Figure 16-55]*

Cabin Air Pressure Safety Valve Operation

Aircraft pressurization systems incorporate various features to limit human and structural damage should the system malfunction or become inoperative. A means for preventing overpressurization is incorporated to ensure the structural integrity of the aircraft if control of the pressurization system is lost. A cabin air safety valve is a pressure relief valve set to open at a predetermined pressure differential. It allows air to flow from the cabin to prevent internal pressure from exceeding design limitations. *Figure 16-56* shows cabin air pressure safety valves on a large transport category aircraft. On most aircraft, safety valves are set to open between 8 and 10 psid.

Pressurization safety valves are used to prevent the over pressurization of the aircraft cabin. They open at a preset differential pressure and allow air to flow out of the cabin. Wide-body transport category aircraft cabins may have more than one cabin pressurization safety valve.

Some outflow valves incorporate the safety valve function into their design. This is common on some corporate jets when two outflow valves are used. One outflow valve operates as the primary and the other as a secondary. Both

contain a pilot valve that opens when the pressure differential increases to a preset value. This, in turn, opens the outflow valve(s) to prevent further pressurization. The outflow valves shown in *Figure 16-52* operate in this manner.

Cabin altitude limiters are also used. These close the outflow valves when the pressure in the cabin drops well below the normal cabin altitude range, preventing a further increase in cabin altitude. Some limiter functions are built into the outflow valve(s). An example of this can be seen in

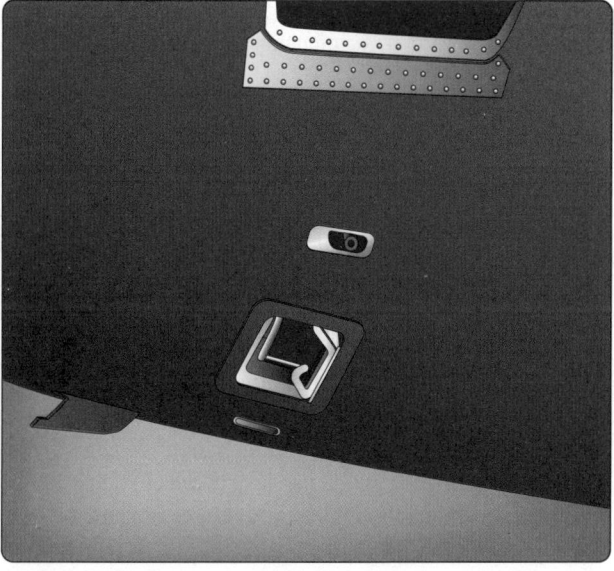

Figure 16-55. *This outflow valve on a transport category aircraft is normally operated by an AC motor controlled by a pressure controller in the electronics equipment bay. A second AC motor on the valve is used when in standby mode. A DC motor also on the valve is used for manual operation.*

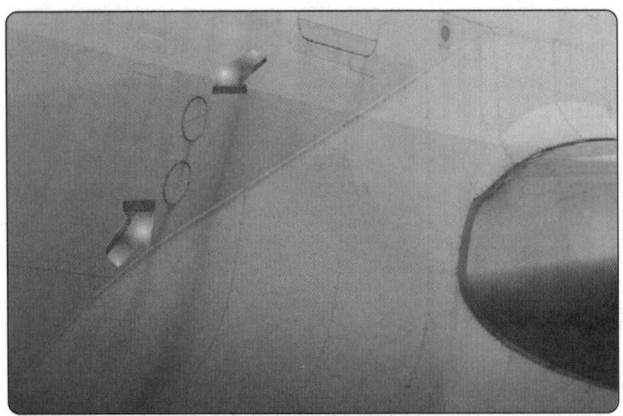

Figure 16-56. *Two pressurization safety valves are shown on a Boeing 747.*

Figure 16-52. Other limiters are independent bellows units that send input to the outflow valve or are part of the cabin pressurization controller logic.

A negative pressure relief valve is included on pressurized aircraft to ensure that air pressure outside the aircraft does not exceed cabin air pressure. The spring-loaded relief valve opens inward to allow ambient air to enter the cabin when this situation arises. Too much negative pressure can cause difficulty when opening the cabin door. If high enough, it could cause structural damage since the pressure vessel is designed for cabin pressure to be greater than ambient.

Some aircraft are equipped with pressurization dump valves. These essentially are safety valves that are operated automatically or manually by a switch in the flight deck. They are used to quickly remove air and air pressure from the cabin, usually in an abnormal, maintenance, or emergency situation.

Incorporation of an emergency pressurization mode is found on some aircraft. A valve opens when the air conditioning packs fail or emergency pressurization is selected from the flight deck. It directs a mixture of bleed air and ram air into the cabin. This combines with fully closed outflow valves to preserve some pressurization in the aircraft.

Pressurization Gauges

While all pressurization systems differ slightly, usually three flight deck indications, in concert with various warning lights and alerts, advise the crew of pressurization variables. They are the cabin altimeter, the cabin rate of climb or vertical speed indicator, and the cabin differential pressure indicator. These can be separate gauges or combined into one or two gauges. All are typically located on the pressurization panel, although sometimes they are elsewhere on the instrument panel. Outflow valve position indicator(s) are also common. *[Figure 16-57]*

On modern aircraft equipped with digital aircraft monitoring systems with LCD displays, such as Engine Indicating and Crew Alerting System (EICAS) or Electronic Centralized Aircraft Monitor (ECAM), the pressurization panel may contain no gauges. The environmental control system (ECS) page of the monitoring system is selected to display similar information. Increased use of automatic redundancy and advanced operating logic simplifies operation of the pressurization system. It is almost completely automatic. The cabin pressurization panel remains in the flight deck primarily for manual control. *[Figure 16-58]*

Pressurization Operation

The normal mode of operation for most pressurization control systems is the automatic mode. A standby mode can also be selected. This also provides automatic control of pressurization, usually with different inputs, a standby controller, or standby outflow valve operation. A manual mode is available should the automatic and standby modes fail. This allows the crew to directly position the outflow valve through pneumatic or electric control, depending on the system.

Coordination of all pressurization components during various flight segments is essential. A weight-on-wheels (WOW) switch attached to the landing gear and a throttle position switch are integral parts of many pressurization control systems. Another name for the WOW switch is "squat" switch. During ground operations and prior to

Figure 16-57. *This cabin pressurization gauge is a triple combination gauge. The long pointer operates identically to a vertical speed indicator with the same familiar scale on the left side of the gauge. It indicates the rate of change of cabin pressure. The orange PSI pointer indicates the differential pressure on the right side scale. The ALT indicator uses the same scale as the PSI pointer, but it indicates cabin altitude when ALT indicator moves against it.*

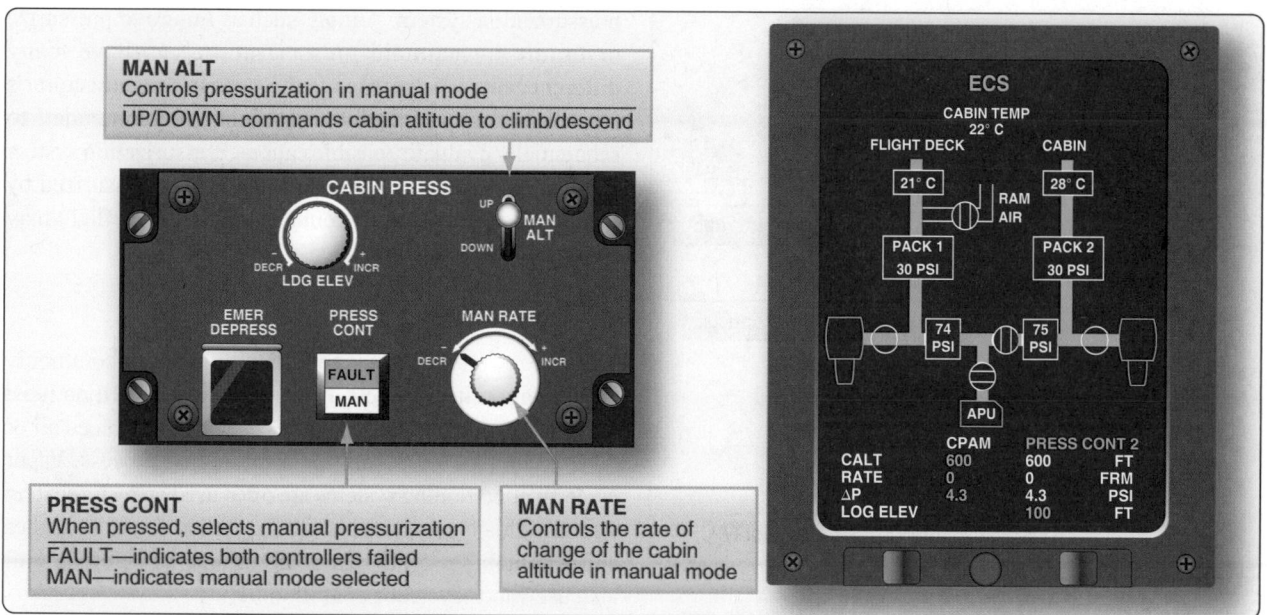

Figure 16-58. *The pressurization panel and environmental control system page on a Bombardier CRJ200 50 passenger jet have no gauges. Traditional pressurization data is presented in digital format at the bottom of the page.*

takeoff, the WOW switch typically controls the position of the pressurization safety valve, which is held in the open position until the aircraft takes off. In an advanced system, the WOW switch may give input to the pressurization controller, which in turn controls the positions and operation of all pressurization components. In other systems, the WOW switch may directly control the safety valve or a pneumatic source valve that causes the safety valve to be held open until the source is cut at takeoff when the WOW switch opens.

Throttle position switches can be used to cause a smooth transition from an unpressurized cabin to a pressurized cabin. A partial closing of the outflow valve(s) when the WOW switch is closed (on the ground) and the throttles are advanced gradually initiates pressurization during rollout. At takeoff, the rate of climb and the pressurization schedule require the outflow valve(s) to fully close. Passengers do not experience a harsh sensation from the fully closed valves because the cabin has already begun to pressurize slightly.

Once in flight, the pressurization controller automatically controls the sequence of operation of the pressurization components until the aircraft lands. When the WOW switch closes again at landing, it opens the safety valve(s) and, in some aircraft, the outflow valve(s) makes pressurizing impossible on the ground in the automatic pressurization mode. Maintenance testing of the system is done in manual mode. This allows the technician to control the position of all valves from the flight deck panel.

Air Distribution

Distribution of cabin air on pressurized aircraft is managed with a system of air ducts leading from the pressurization source into and throughout the cabin. Typically, air is ducted to and released from ceiling vents, where it circulates and flows out floor-level vents. The air then flows aft through the baggage compartments and under the floor area. It exits the pressure vessel through the outflow valve(s) mounted low, on, or near the aft pressure bulkhead. The flow of air is nearly imperceptible. Ducting is hidden below the cabin floor and behind walls and ceiling panels depending on the aircraft and system design. Valves to select pressurization air source, ventilating air, temperature trim air, as well as in line fans and jet pumps to increase flow in certain areas of the cabin, are all components of the air distribution system. Temperature sensors, overheat switches, and check valves are also common.

On turbine-powered aircraft, temperature-controlled air from the air conditioning system is the air that is used to pressurize the cabin. The final regulation of the temperature of that air is sometimes considered part of the distribution system. Mixing air-conditioned air with bleed air in a duct or a mixing chamber allows the crew to select the exact temperature desired for the cabin. The valve for mixing is controlled in the flight deck or cabin by a temperature selector. Centralized manifolds from which air can be distributed are common. *[Figure 16-59]*

Large aircraft may be divided into zones for air distribution. Each zone has its own temperature selector and associated valve to mix conditioned and bleed air so that each zone can be maintain at a temperature independent of the others.

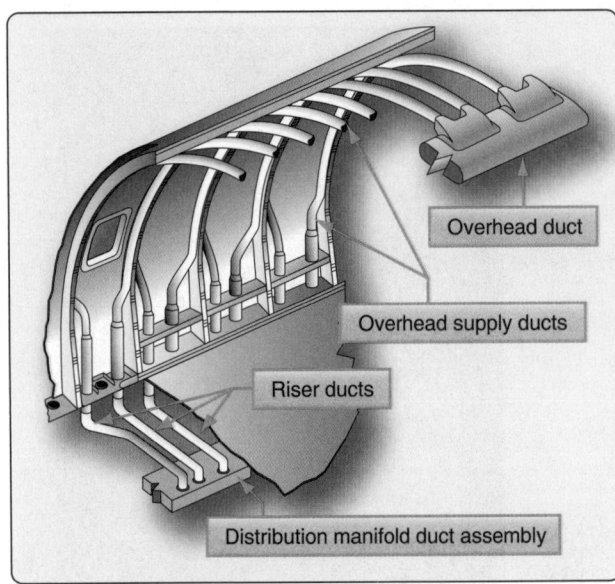

Figure 16-59. *Centralized manifolds from which air can be distributed are common.*

The air distribution system on most aircraft makes provisions for ducting and circulating cooling air to electronics equipment bays. It also contains a gasper air system. This is air ducted from the cold air manifold or duct to an overhead adjustable delivery nozzle at each passenger station. An inline fan controlled from the flight deck supplies a steady stream of gasper air that can be regulated or shut off with the delivery nozzle(s). *[Figure 16-60]*

When an aircraft is on the ground, operating the engines or the APU to provide air for air conditioning is expensive. It increases the time in service of these expensive components and expedites expensive mandatory overhauls that are performed at specified time intervals. Most high-performance, medium size, and larger turbine-powered aircraft are fitted with a receptacle in the air distribution system. To this a ground source of conditioned air can be connected via a ducting hose. The cabin can be heated or cooled through the aircraft's air distribution ducting using air from the ground source. This limits the operating time on the engines and APU. Once preflight checks and passenger boarding are completed, the ducting hose can be disconnected for taxi and flight. A check valve is used to prevent ground source air from flowing upstream into the air conditioning system. *[Figure 16-61]*

Cabin Pressurization Troubleshooting

While pressurization systems on different aircraft operate similarly with similar components, it cannot be assumed that they are the same. Even those systems constructed by a single manufacturer likely have differences when installed on different aircraft. It is important to check the aircraft manufacture's service information when troubleshooting the pressurization system. A fault, such as failure to pressurize or failure to maintain pressurization, can have many different causes. Adherence to the steps in a manufacturer's troubleshooting procedures is highly recommended to sequentially evaluate possible causes. Pressurization system test kits are available, or the aircraft can be pressurized by its normal sources during troubleshooting. A test flight may be required after maintenance.

Air Conditioning Systems

There are two types of air conditioning systems commonly used on aircraft. Air cycle air conditioning is used on most turbine-powered aircraft. It makes use of engine bleed air or APU pneumatic air during the conditioning process. Vapor cycle air conditioning systems are often used on reciprocating aircraft. This type system is similar to that found in homes and automobiles. Note that some turbine-powered aircraft also use vapor cycle air conditioning.

Air Cycle Air Conditioning

Air cycle air conditioning prepares engine bleed air to pressurize the aircraft cabin. The temperature and quantity of the air must be controlled to maintain a comfortable cabin environment at all altitudes and on the ground. The air cycle system is often called the air conditioning package or pack. It is usually located in the lower half of the fuselage or in the tail section of turbine-powered aircraft. *[Figure 16-62]*

System Operation

Even with the frigid temperatures experienced at high altitudes, bleed air is too hot to be used in the cabin without being cooled. It is let into the air cycle system and routed through a heat exchanger where ram air cools the bleed air. This cooled bleed air is directed into an air cycle machine. There, it is compressed before flowing through a secondary heat exchange that cools the air again with ram air. The bleed air then flows back into the air cycle machine where it drives an expansion turbine and cools even further. Water is then removed, and the air is mixed with bypassed bleed air for final temperature adjustment. It is sent to the cabin through the air distribution system. By examining the operation of each component in the air cycle process, a better understanding can be developed of how bleed air is conditioned for cabin use. Refer to *Figure 16-63,* which diagrams the air cycle air conditioning system of the Boeing 737.

Pneumatic System Supply

The air cycle air conditioning system is supplied with air by the aircraft pneumatic system. In turn, the pneumatic system is supplied by bleed air tap-offs on each engine compressor section or from the APU pneumatic supply. An external pneumatic air supply source may also be connected while

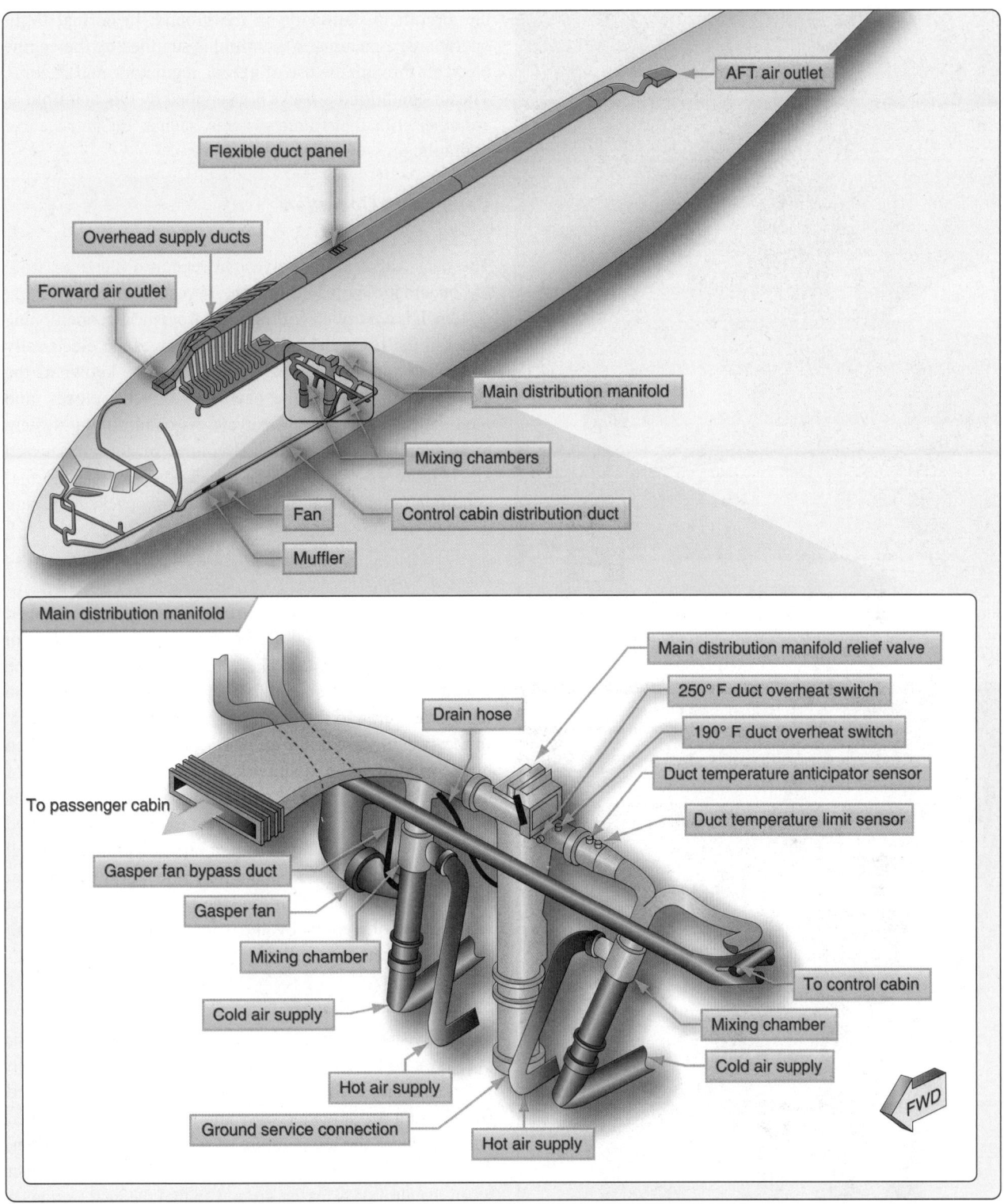

Figure 16-60. *The conditioned air distribution system on a Boeing 737. The main distribution manifold is located under the cabin floor. Riser ducts run horizontally then vertically from the manifold to supply ducts, which follow the curvature of the fuselage carrying conditioned air to be released in the cabin.*

Figure 16-61. *A duct hose installed on this airliner distributes hot or cold air from a ground-based source throughout the cabin using the aircraft's own air distribution system ducting.*

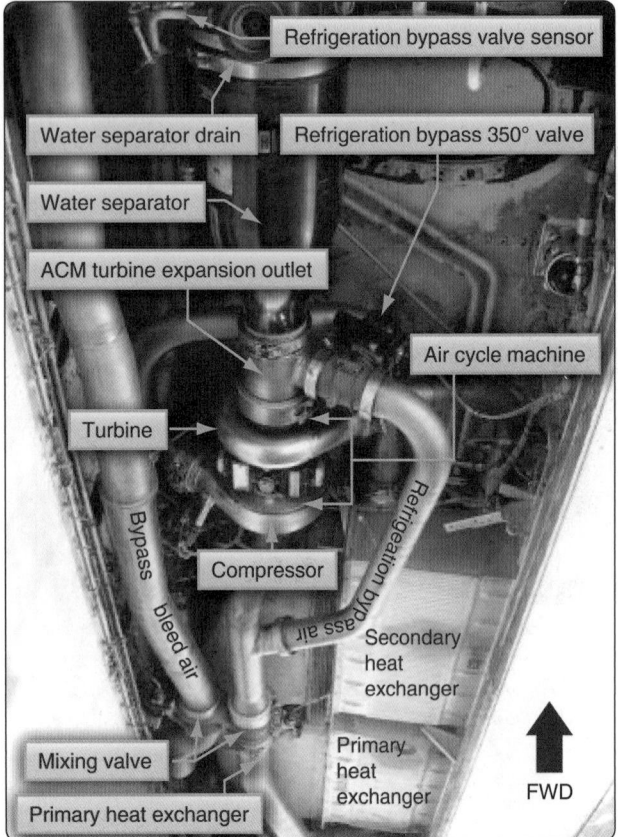

Refrigeration bypass valve sensor

Water separator drain

Refrigeration bypass 350° valve

Water separator

ACM turbine expansion outlet

Air cycle machine

Turbine

Bypass bleed air

Refrigeration bypass air

Compressor

Secondary heat exchanger

Mixing valve

Primary heat exchanger

Primary heat exchanger

FWD

Figure 16-62. *Boeing 737 air cycle system. The photo is taken looking up into the air conditioning bay located in the lower fuselage on each side of the aircraft.*

the aircraft is stationary on the ground. In normal flight operations, a pneumatic manifold is supplied by the engine bleed air through the use of valves, regulators, and ducting. The air conditioning packs are supplied by this manifold as are other critical airframe systems, such as the anti-ice and hydraulic pressurization system.

Component Operation

Pack Valve

The pack valve is the valve that regulates bleed air from the pneumatic manifold into the air cycle air conditioning system. It is controlled with a switch from the air conditioning panel in the flight deck. Many pack valves are electrically controlled and pneumatically operated. Also known as the supply shutoff valve, the pack valve opens, closes, and modulates to allow the air cycle air conditioning system to be supplied with a designed volume of hot, pressurized air. *[Figure 16-64]* When an overheat or other abnormal condition requires that the air conditioning package be shut down, a signal is sent to the pack valve to close.

Bleed Air Bypass

A means for bypassing some of the pneumatic air supplied to the air cycle air conditioning system around the system is present on all aircraft. This warm bypassed air must be mixed with the cold air produced by the air cycle system so the air delivered to the cabin is a comfortable temperature. In the system shown in *Figure 16-58,* this is accomplished by the mixing valve. It simultaneously controls the flow of bypassed air and air to be cooled to meet the requirements of the auto temperature controller. It can also be controlled manually with the cabin temperature selector in manual mode. Other air cycle systems may refer to the valve that controls the air bypassed around the air cycle cooling system as a temperature control valve, trim air pressure regulating valve, or something similar.

Primary Heat Exchanger

Generally, the warm air dedicated to pass through the air cycle system first passes through a primary heat exchanger. It acts similarly to the radiator in an automobile. A controlled flow of ram air is ducted over and through the exchanger, which reduces the temperature of the air inside the system. *[Figure 16-65]* A fan draws air through the ram air duct when the aircraft is on the ground so that the heat exchange is possible when the aircraft is stationary. In flight, ram air doors are modulated to increase or decrease ram air flow to the exchanger according to the position of the wing flaps. During slow flight, when the flaps are extended, the doors are open. At higher speeds, with the flaps retracted, the doors move toward the closed position reducing the amount of ram air to the exchanger. Similar operation is accomplished with a valve on smaller aircraft. *[Figure 16-66]*

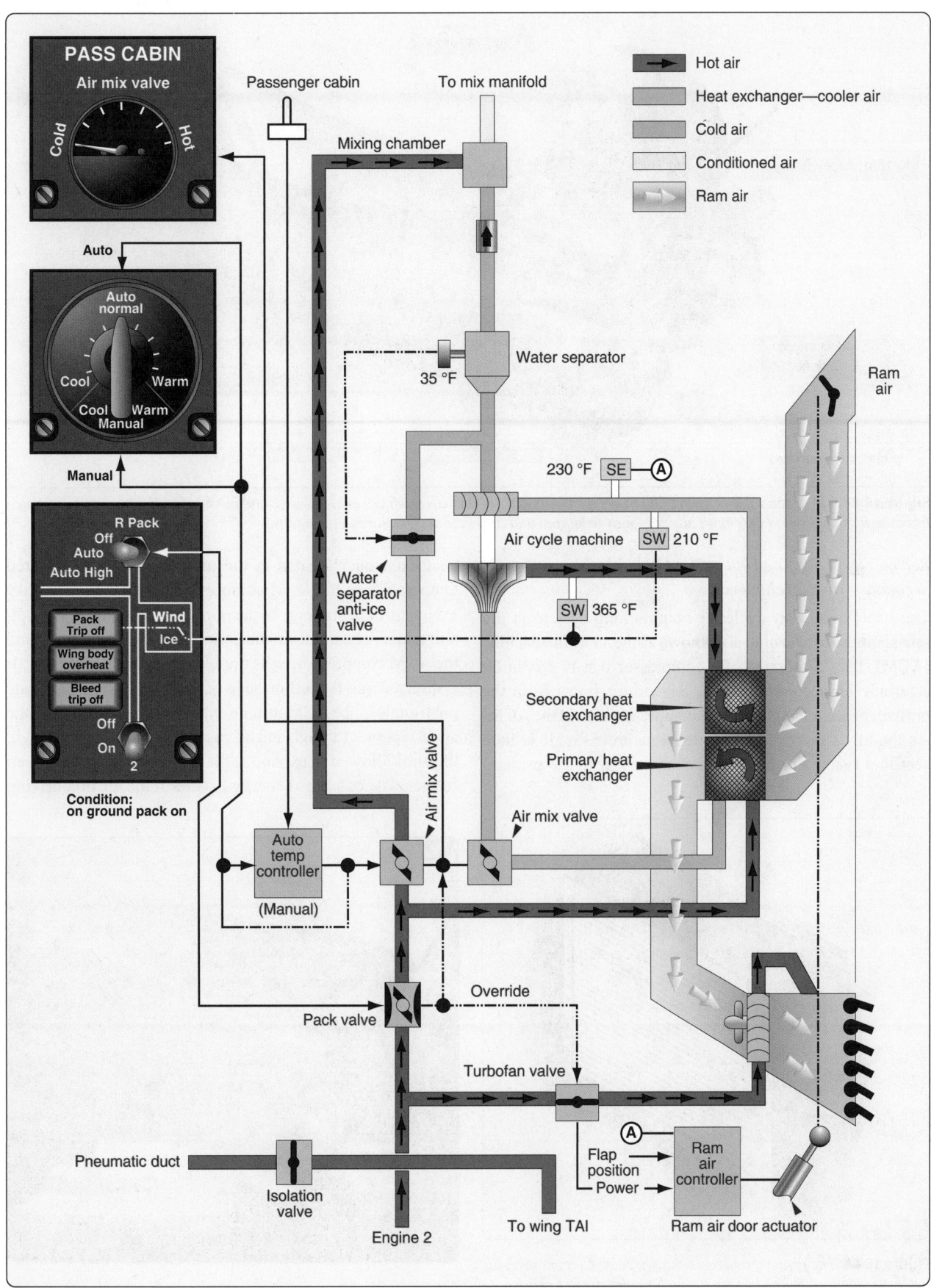

Figure 16-63. *The air cycle air conditioning system on a Boeing 737.*

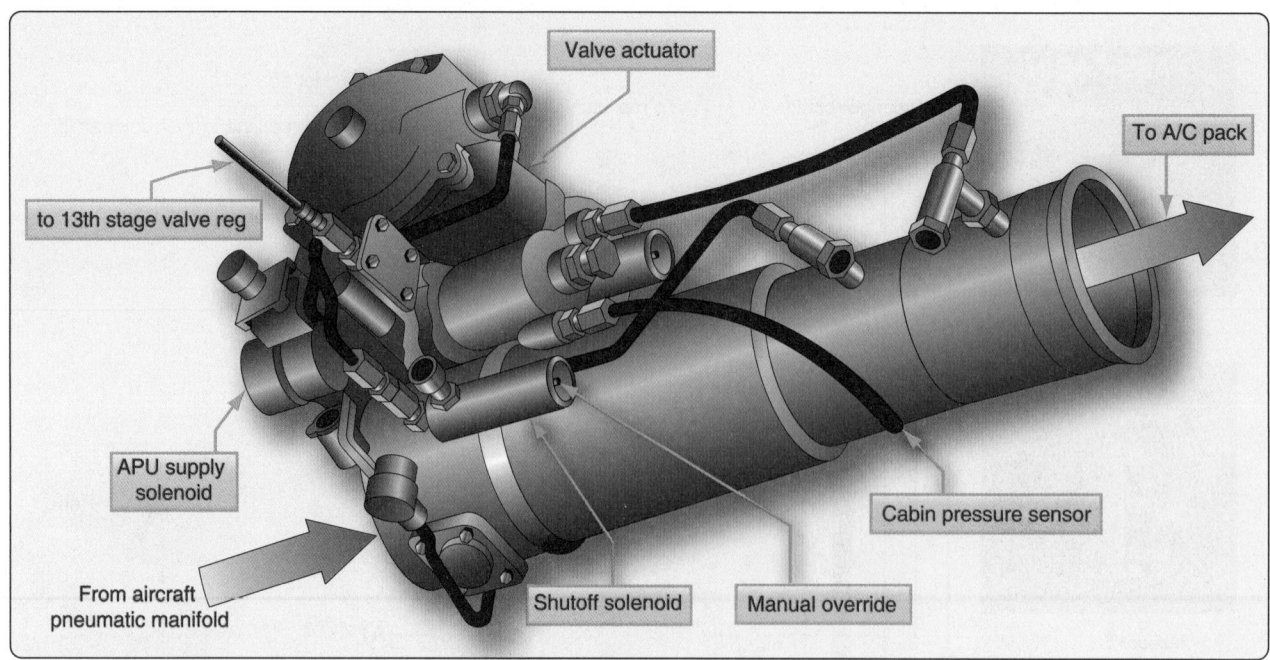

Figure 16-64. *This pack valve drawing illustrates the complexity of the valve, which opens, closes, and modulates. It is manually actuated from the flight deck and automatically responds to supply and air cycle system parameter inputs.*

Refrigeration Turbine Unit or Air Cycle Machine & Secondary Heat Exchanger

The heart of the air cycle air conditioning system is the refrigeration turbine unit, also known as the air cycle machine (ACM). It is comprised of a compressor that is driven by a turbine on a common shaft. System air flows from the primary heat exchanger into the compressor side of the ACM. As the air is compressed, its temperature rises. It is then sent to a secondary heat exchanger, similar to the primary

heat exchanger located in the ram air duct. The elevated temperature of the ACM compressed air facilitates an easy exchange of heat energy to the ram air. The cooled system air, still under pressure from the continuous system air flow and the ACM compressor, exits the secondary heat exchanger. It is directed into the turbine side of the ACM. The steep blade pitch angle of the ACM turbine extracts more energy from the air as it passes through and drives the turbine. Once through, the air is allowed to expand at the ACM outlet, cooling even further. The combined energy loss from the air first driving

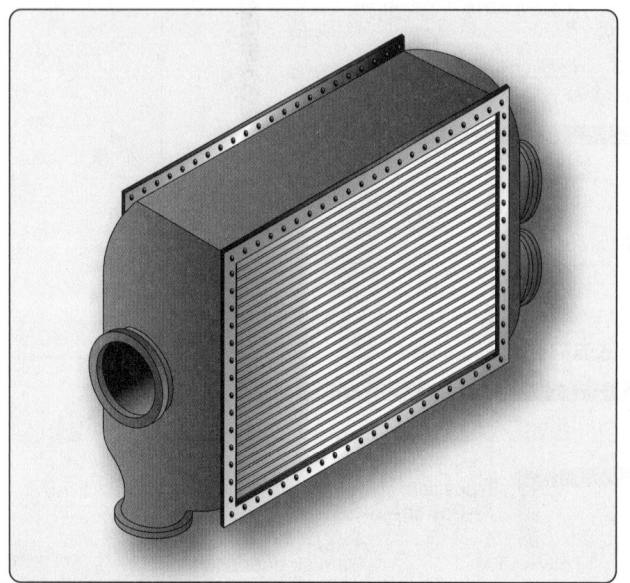

Figure 16-65. *The primary and secondary heat exchangers in an air cycle air conditioning system are of similar construction. They both cool bleed air when ram air passes over the exchanger coils and fins.*

Figure 16-66. *A ram air door controls the flow of air through the primary and secondary heat exchangers.*

the turbine and then expanding at the turbine outlet lowers the system air temperature to near freezing. *[Figure 16-67]*

Water Separator

The cool air from the air cycle machine can no longer hold the quantity of water it could when it was warm. A water separator is used to remove the water from the saturated air before it is sent to the aircraft cabin. The separator operates with no moving parts. Foggy air from the ACM enters and is forced through a fiberglass sock that condenses and coalesces the mist into larger water drops. The convoluted interior structure of the separator swirls the air and water. The water collects on the sides of the separator and drains down and out of the unit, while the dry air passes through. A bypass valve is incorporated in case of a blockage. *[Figure 16-68]*

Refrigeration Bypass Valve

As mentioned, air exiting the ACM turbine expands and cools. It becomes so cold, it could freeze the water in the water separator, thus inhibiting or blocking airflow. A temperature sensor in the separator controls a refrigeration bypass valve

designed to keep the air flowing through the water separator above freezing temperature. The valve is also identified by other names such as a temperature control valve, 35° valve, anti-ice valve, and similar. It bypasses warm air around the ACM when opened. The air is introduced into the expansion ducting, just upstream of the water separator, where it heats the air just enough to keep it from freezing. Thus, the refrigeration bypass valve regulates the temperature of the ACM discharge air so it does not freeze when passing through the water separator. This valve is visible in *Figure 16-62* and is diagrammed in the system in *Figure 16-63*.

All air cycle air conditioning systems use at least one ram air heat exchanger and an air cycle machine with expansion turbine to remove heat energy from the bleed air, but variations exist. An example of a system different from that described above is found on the McDonnell Douglas DC-10. Bleed air from the pneumatic manifold is compressed by the air cycle machine compressor before it flows to a single heat exchanger. Condensed water from the water separator is sprayed into the ram air at its entrance to the exchanger to draw additional heat

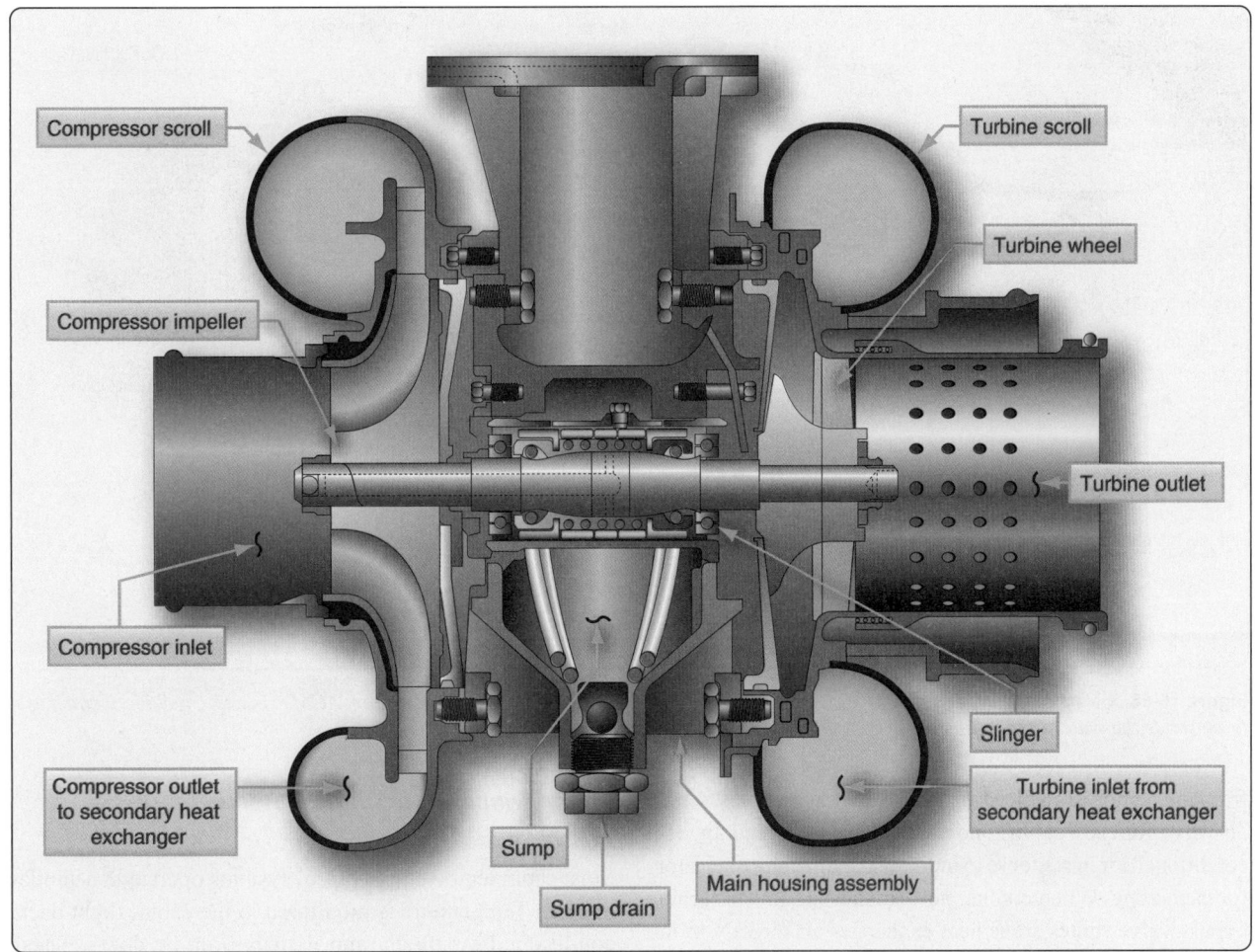

Figure 16-67. *A cutaway diagram of an air cycle machine. The main housing supports the single shaft to which the compressor and turbine are attached. Oil lubricates and cools the shaft bearings.*

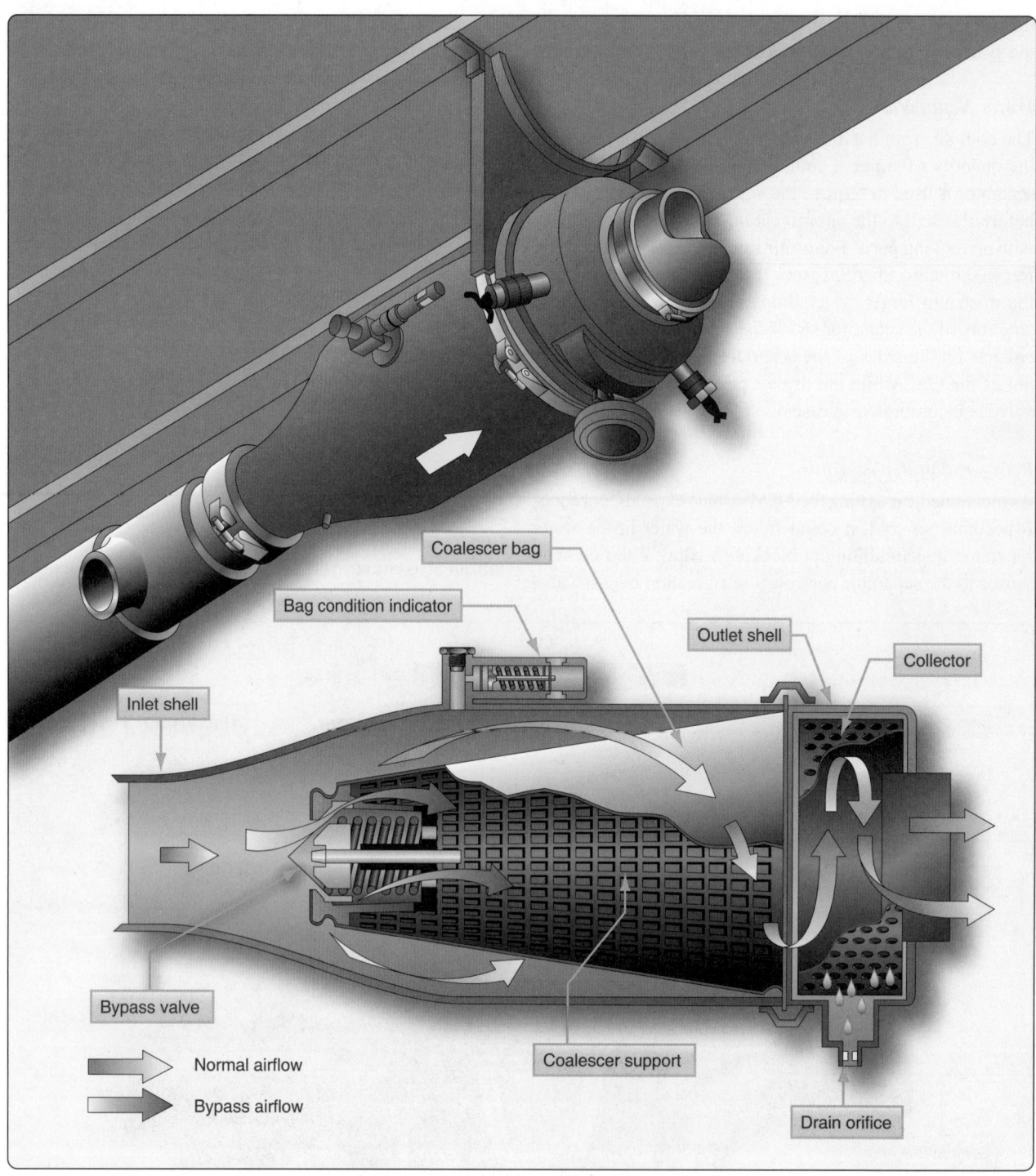

Figure 16-68. *A water separator coalesces and removes water by swirling the air/water mixture from ACM expansion turbine. Centrifugal force sends the water to the walls of the collector where it drains from the unit.*

from the compressed bleed air as the water evaporates. A trim air valve for each cabin zone mixes bypassed bleed air with conditioned air in response to individual temperature selectors for each zone. When cooling air demands are low, a turbine bypass valve routes some heat exchanger air directly to the conditioned air manifold. *[Figure 16-69]*

Cabin Temperature Control System

Typical System Operation

Most cabin temperature control systems operate in a similar manner. Temperature is monitored in the cabin, flight deck, conditioned air ducts, and distribution air ducts. These values are input into a temperature controller, or temperature control regulator, normally located in the electronics bay. A

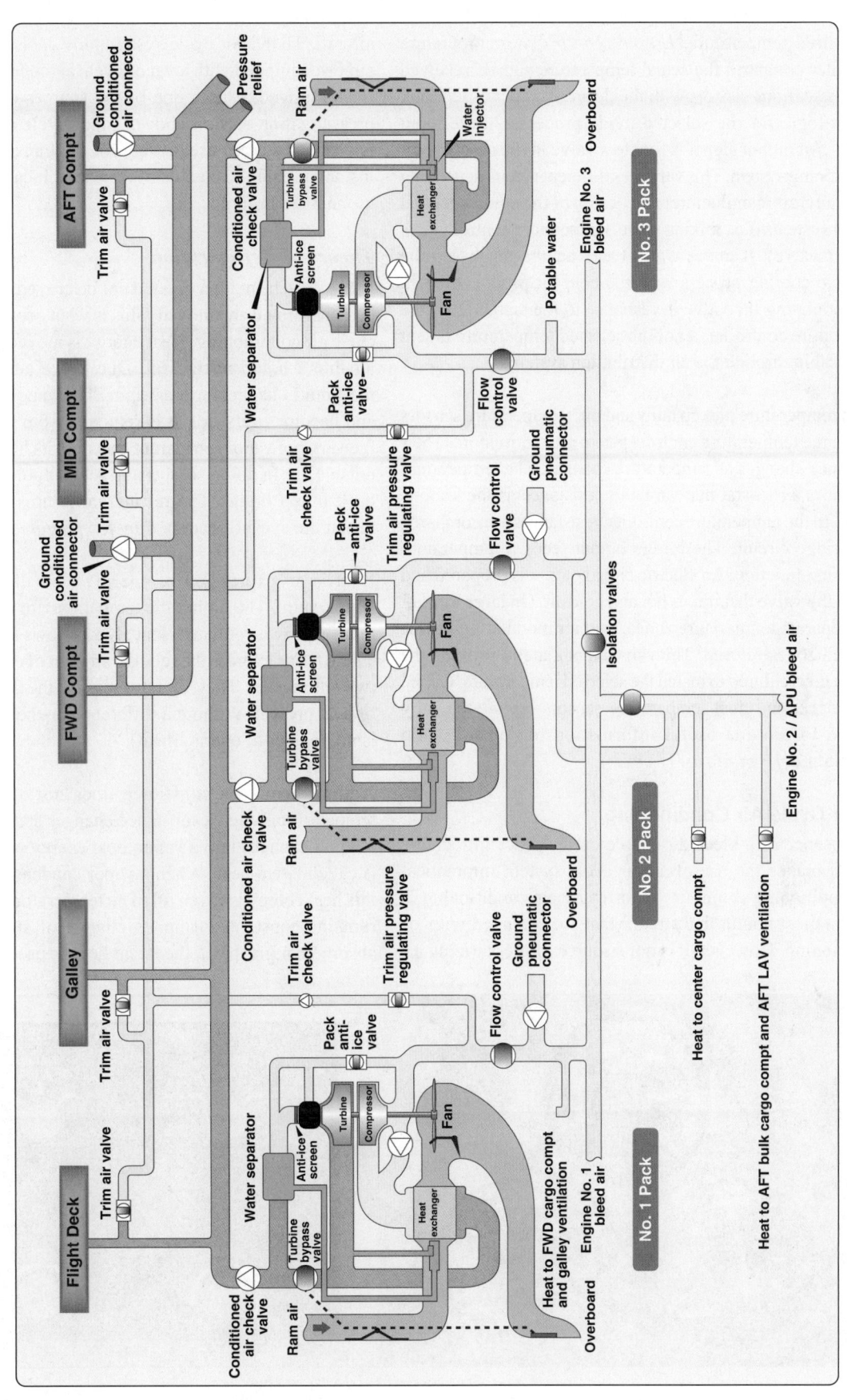

Figure 16-69. *The air cycle air conditioning system of a DC-10 transport category aircraft uses only one heat exchanger per ACM.*

temperature selector in the flight deck can be adjusted to input the desired temperature. *[Figure 16-70]* The temperature controller compares the actual temperature signals received from the various sensors with the desired temperature input. Circuit logic for the selected mode processes these input signals. An output signal is sent to a valve in the air cycle air conditioning system. This valve has different names depending on the aircraft manufacturer and design of the environmental control systems (i.e., mixing valve, temperature control valve, trim air valve). It mixes warm bleed air that bypassed the air cycle cooling process with the cold air produced by it. By modulating the valve in response to the signal from the temperature controller, air of the selected temperature is sent to the cabin through the air distribution system.

Cabin temperature pickup units and duct temperature sensors used in the temperature control system are thermistors. Their resistance changes as temperature changes. The temperature selector is a rheostat that varies its resistance as the knob is turned. In the temperature controller, resistances are compared in a bridge circuit. The bridge output feeds a temperature regulating function. An electric signal output is prepared and sent to the valve that mixes hot and cold air. On large aircraft with separate temperature zones, trim air modulating valves for each zone are used. The valves modulate to provide the correct mix required to match the selected temperature. Cabin, flight deck, and duct temperature sensors are strategically located to provide useful information to control cabin temperature. *[Figure 16-71]*

Vapor Cycle Air Conditioning

The absence of a bleed air source on reciprocating engine aircraft makes the use of an air cycle system impractical for conditioning cabin air. Vapor cycle air conditioning is used on most nonturbine aircraft that are equipped with air conditioning. However, it is not a source of pressurizing air

as the air cycle system conditioned air is on turbine powered aircraft. The vapor cycle system only cools the cabin. If an aircraft equipped with a vapor cycle air conditioning system is pressurized, it uses one of the sources discussed in the pressurization section above. Vapor cycle air conditioning is a closed system used solely for the transfer of heat from inside the cabin to outside of the cabin. It can operate on the ground and in flight.

Theory of Refrigeration

Energy can be neither created nor destroyed; however, it can be transformed and moved. This is what occurs during vapor cycle air conditioning. Heat energy is moved from the cabin air into a liquid refrigerant. Due to the additional energy, the liquid changes into a vapor. The vapor is compressed and becomes very hot. It is removed from the cabin where the very hot vapor refrigerant transfers its heat energy to the outside air. In doing so, the refrigerant cools and condenses back into a liquid. The refrigerant returns to the cabin to repeat the cycle of energy transfer. *[Figure 16-72]*

Heat is an expression of energy, typically measured by temperature. The higher the temperature of a substance, the more energy it contains. Heat always flows from hot to cold. These terms express the relative amount of energy present in two substances. They do not measure the absolute amount of heat present. Without a difference in energy levels, there is no transfer of energy (heat).

Adding heat to a substance does not always raise its temperature. When a substance changes state, such as when a liquid changes into a vapor, heat energy is absorbed. This is called latent heat. When a vapor condenses into a liquid, this heat energy is given off. The temperature of a substance remains constant during its change of state. All energy absorbed or given off, the latent heat, is used for the change

Figure 16-70. *Typical temperature selectors on a transport category aircraft temperature control panel in the flight deck (left) and a business jet (right). On large aircraft, temperature selectors may be located on control panels located in a particular cabin air distribution zone.*

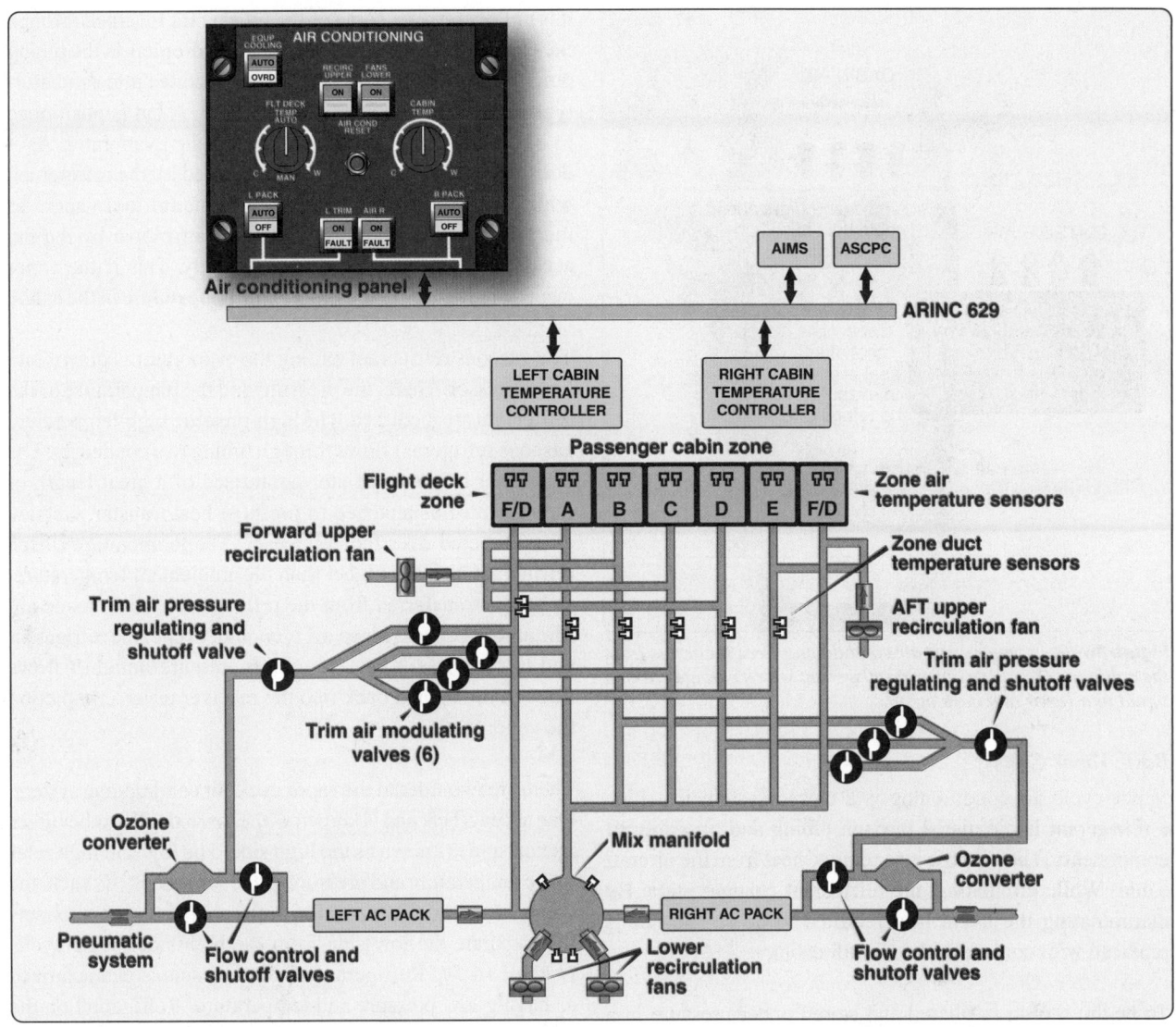

Figure 16-71. *The temperature control system of a Boeing 777 combines the use of zone and duct temperature sensors with trim air modulating valves for each zone. Redundant digital left and right cabin temperature controllers process temperature input signals from the sensors and temperature selectors on the flight deck panel and throughout the aircraft to modulate the valves.*

process. Once the change of state is complete, heat added to a substance raises the temperature of the substance. After a substance changes state into a vapor, the rise in temperature of the vapor caused by the addition of still more heat is called superheat.

The temperature at which a substance changes from a liquid into a vapor when heat is added is known as its boiling point. This is the same temperature at which a vapor condenses into a liquid when heat is removed. The boiling point of any substance varies directly with pressure. When pressure on a liquid is increased, its boiling point increases, and when pressure on a liquid is decreased, its boiling point also decreases. For example, water boils at 212 °F at normal atmospheric pressure (14.7 psi). When pressure on liquid water is increased to 20 psi, it does not boil at 212 °F. More energy is required to overcome the increase in pressure. It

boils at approximately 226.4 °F. The converse is also true. Water can also boil at a much lower temperature simply by reducing the pressure upon it. With only 10 psi of pressure upon liquid water, it boils at 194 °F. *[Figure 16-73]*

Vapor pressure is the pressure of the vapor that exists above a liquid that is in an enclosed container at any given temperature. The vapor pressure developed by various substances is unique to each substance. A substance that is said to be volatile, develops high vapor pressure at standard day temperature (59 °F). This is because the boiling point of the substance is much lower. The boiling point of tetrafluoroethane (R134a), the refrigerant used in most aircraft vapor cycle air conditioning systems, is approximately –15 °F. Its vapor pressure at 59 °F is about 71 psi. The vapor pressure of any substance varies directly with temperature.

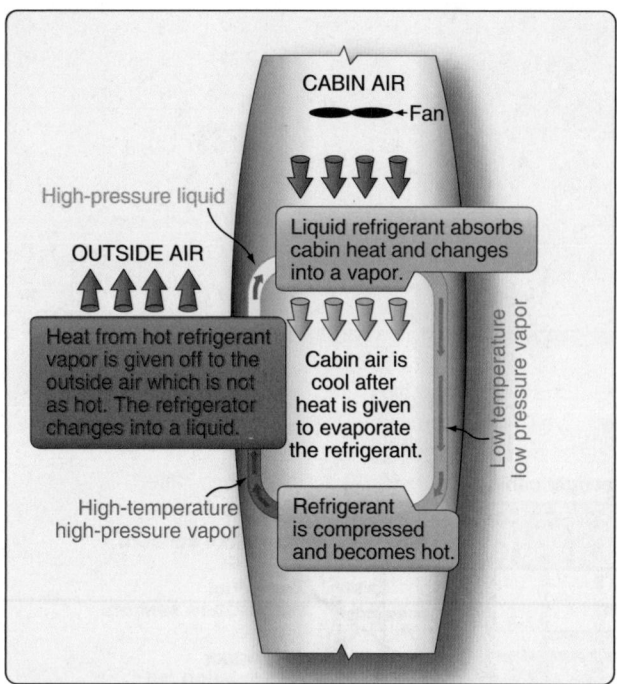

Figure 16-72. *In vapor cycle air conditioning, heat is carried from the cabin to the outside air by a refrigerant which changes from a liquid to a vapor and back again.*

Basic Vapor Cycle

Vapor cycle air conditioning is a closed system in which a refrigerant is circulated through tubing and a variety of components. The purpose is to remove heat from the aircraft cabin. While circulating, the refrigerant changes state. By manipulating the latent heat required to do so, hot air is replaced with cool air in the aircraft cabin.

To begin, R134a is filtered and stored under pressure in a reservoir known as a receiver dryer. The refrigerant is in liquid form. It flows from the receiver dryer through tubing to an expansion valve. Inside the valve, a restriction in the form of a small orifice blocks most of the refrigerant. Since

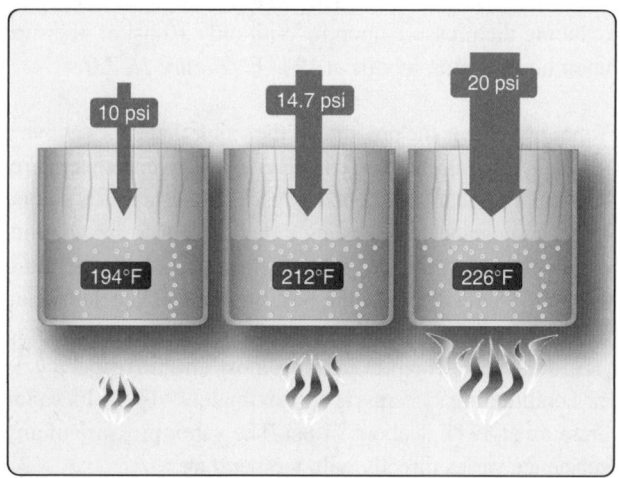

Figure 16-73. *Boiling point of water changes as pressure changes.*

it is under pressure, some of the refrigerant is forced through the orifice. It emerges as a spray of tiny droplets in the tubing downstream of the valve. The tubing is coiled into a radiator-type assembly known as an evaporator. A fan is positioned to blow cabin air over the surface of the evaporator. As it does, the heat in the cabin air is absorbed by the refrigerant, which uses it to change state from a liquid to a vapor. So much heat is absorbed that the cabin air blown by the fan across the evaporator cools significantly. This is the vapor cycle conditioned air that lowers the temperature in the cabin.

The gaseous refrigerant exiting the evaporator is drawn into a compressor. There, the pressure and the temperature of the refrigerant are increased. The high-pressure high-temperature gaseous refrigerant flows through tubing to a condenser. The condenser is like a radiator comprised of a great length of tubing with fins attached to promote heat transfer. Outside air is directed over the condenser. The temperature of the refrigerant inside is higher than the ambient air temperature, so heat is transferred from the refrigerant to the outside air. The amount of heat given off is enough to cool the refrigerant and to condense it back to a high-pressure liquid. It flows through tubing and back into the receiver dryer, completing the vapor cycle.

There are two sides to the vapor cycle air conditioning system. One accepts heat and is known as the low side. The other gives up heat and is known as the high side. The low and high refer to the temperature and pressure of the refrigerant. As such, the compressor and the expansion valve are the two components that separate the low side from the high side of the cycle. [Figure 16-74] Refrigerant on the low side is characterized as having low pressure and temperature. Refrigerant on the high side has high pressure and temperature.

Vapor Cycle Air Conditioning System Components

By examining each component in the vapor cycle air conditioning system, greater insight into its function can be gained.

Refrigerant

For many years, dichlorodifluoromethane (R12) was the standard refrigerant used in aircraft vapor cycle air conditioning systems. Some of these systems remain in use today. R12 was found to have a negative effect on the environment; in particular, it degraded the earth's protective ozone layer. In most cases, it has been replaced by tetrafluoroethane (R134a), which is safer for the environment. R12 and R134a should not be mixed, nor should one be used in a system designed for the other. Possible damage to soft components, such as hoses and seals, could result causing leaks and or malfunction. Use only the specified refrigerant when servicing vapor cycle air conditioning systems.

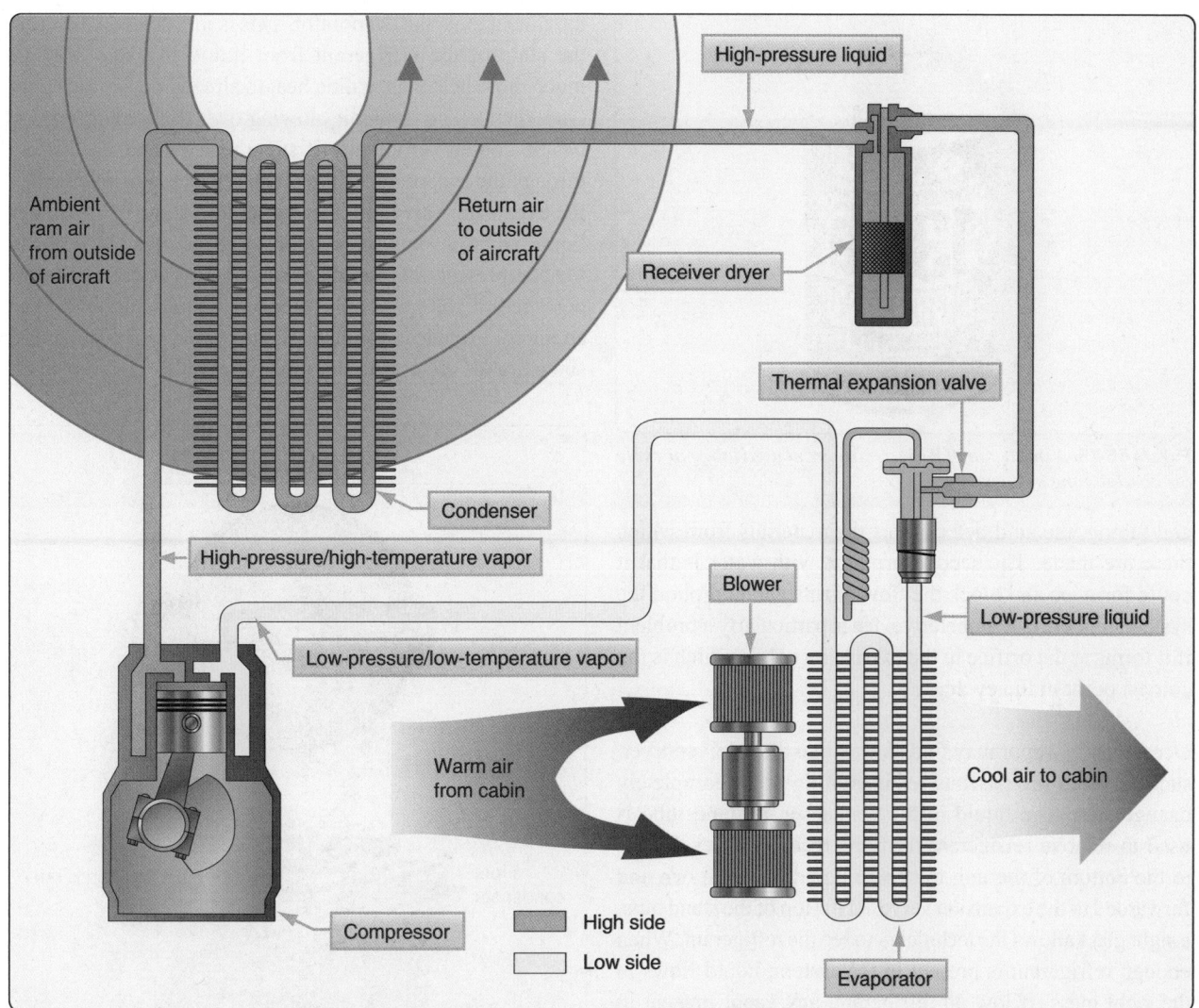

Figure 16-74. *A basic vapor cycle air conditioning system. The compressor and the expansion valve are the two components that separate the low side from the high side of the cycle. This figure illustrates this division. Refrigerant on the low side is characterized as having low pressure and temperature. Refrigerant on the high side has high pressure and temperature.*

[Figure 16-75] R12 and R134a behave so similarly that the descriptions of the R134a vapor cycle air conditioning system and components in the following paragraphs also apply to an R12 system and its components.

R134a is a halogen compound (CF3CFH2). As mentioned, it has a boiling point of approximately –15 °F. It is not poisonous to inhale in small quantities, but it does displace oxygen. Suffocation is possible if breathed in mass quantity.

Regardless of manufacturer, refrigerants are sometimes called Freon®, which is a trade name owned by the Dupont Company. Caution should be used when handling any refrigerant. Because of the low boiling points, liquid refrigerants boil violently at typical atmospheric temperatures and pressure. They rapidly absorb heat energy from all surrounding matter. If a drop lands on skin, it freezes, resulting in a burn. Similar tissue damage can result if a drop

gets in one's eye. Gloves and other skin protection, as well as safety goggles, are required when working with refrigerant.

Receiver Dryer

The receiver dryer acts as the reservoir of the vapor cycle system. It is located downstream of the condenser and upstream of the expansion valve. When it is very hot, more refrigerant is used by the system than when temperatures are moderate. Extra refrigerant is stored in the receiver dryer for this purpose.

Liquid refrigerant from the condenser flows into the receiver dryer. Inside, it passes through filters and a desiccant material. The filters remove any foreign particles that might be in the system. The desiccant captures any water in the refrigerant. Water in the refrigerant causes two major problems. First, the refrigerant and water combine to form an acid. If left in contact with the inside of the components

Figure 16-75. *A small can of R134a refrigerant used in vapor cycle air conditioning systems.*

and tubing, the acid deteriorates the materials from which these are made. The second problem with water is that it could form ice and block the flow of refrigerant around the system, rendering it inoperative. Ice is particularly a problem if it forms at the orifice in the expansion valve, which is the coldest point in the cycle.

Occasionally, vapor may find its way into the receiver dryer, such as when the gaseous refrigerant does not completely change state to a liquid in the condenser. A stand tube is used to remove refrigerant from the receiver dryer. It runs to the bottom of the unit to ensure liquid is withdrawn and forwarded to the expansion valve. At the top of the stand tube, a sight glass allows the technician to see the refrigerant. When enough refrigerant is present in the system, liquid flows in the sight glass. If low on refrigerant, any vapor present in the receiver dryer may be sucked up the stand tube causing bubbles to be visible in the sight glass. Therefore, bubbles in the sight glass indicate that the system needs to have more refrigerant added. *[Figure 16-76]*

Expansion Valve

Refrigerant exits the receiver dryer and flows to the expansion valve. The thermostatic expansion valve has an adjustable orifice through which the correct amount of refrigerant is metered to obtain optimal cooling. This is accomplished by monitoring the temperature of the gaseous refrigerant at the outlet of the next component in the cycle, the evaporator. Ideally, the expansion valve should only let the amount of refrigerant spray into the evaporator that can be completely converted to a vapor.

The temperature of the cabin air to be cooled determines the amount of refrigerant the expansion valve should spray into the evaporator. Only so much is needed to completely change the state of the refrigerant from a liquid to a vapor. Too little causes the gaseous refrigerant to be superheated by

the time it exits the evaporator. This is inefficient. Changing the state of the refrigerant from liquid to vapor absorbs much more heat than adding heat to already converted vapor (superheat). The cabin air blowing over the evaporator will not be cooled sufficiently if superheated vapor is flowing through the evaporator. If too much refrigerant is released by the expansion valve into the evaporator, some of it remains liquid when it exits the evaporator. Since it next flows to the compressor, this could be dangerous. The compressor is designed to compress only vapor. If liquid is drawn in and attempts are made to compress it, the compressor could break, since liquids are essentially incompressible.

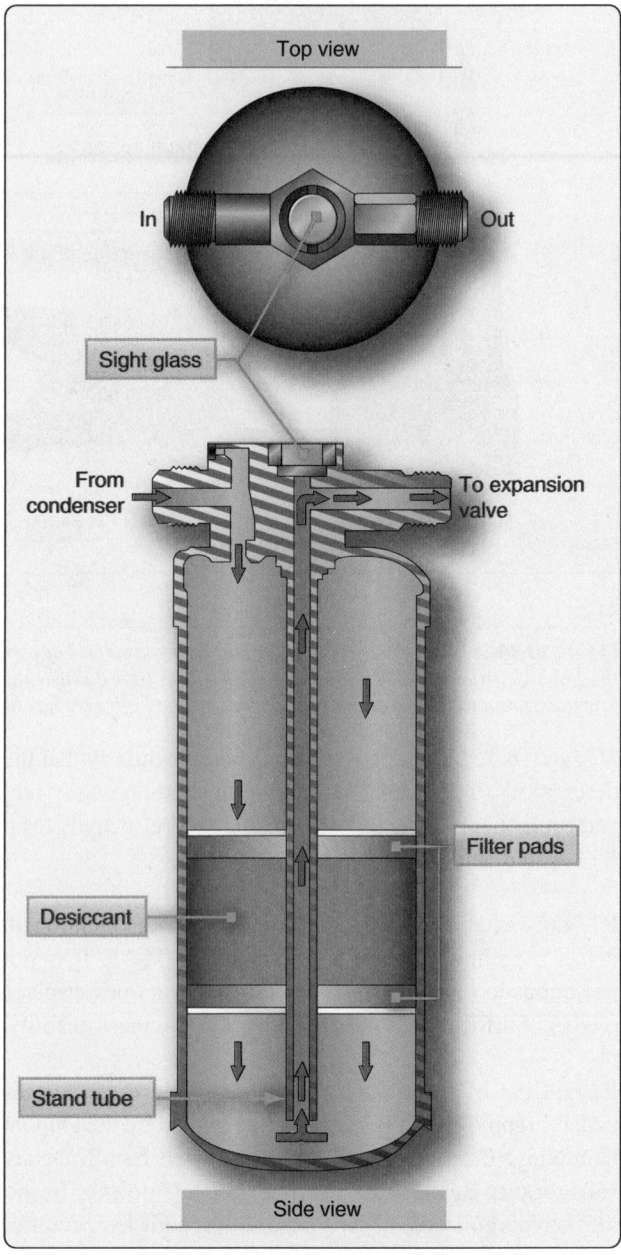

Figure 16-76. *A receiver dryer acts as reservoir and filter in a vapor cycle system. Bubbles viewed in the sight glass indicate the system is low on refrigerant and needs to be serviced.*

The temperature of superheated vapor is higher than liquid refrigerant that has not totally vaporized. A coiled capillary tube with a volatile substance inside is located at the evaporator outlet to sense this difference. Its internal pressure increases and decreases as temperature changes. The coiled end of the tube is closed and attached to the evaporator outlet. The other end terminates in the area above a pressure diaphragm in the expansion valve. When superheated refrigerant vapor reaches the coiled end of the tube, its elevated temperature increases the pressure inside the tube and in the space above the diaphragm. This increase in pressure causes the diaphragm to overcome spring tension in the valve. It positions a needle valve that increases the amount of refrigerant released by the valve. The quantity of refrigerant is increased so that the refrigerant only just evaporates, and the refrigerant vapor does not superheat.

When too much liquid refrigerant is released by the expansion valve, low-temperature liquid refrigerant arrives at the inlet of the evaporator. The result is low pressure inside the temperature bulb and above the expansion valve diaphragm. The superheat spring in the valve moves the needle valve toward the closed position, reducing the flow of refrigerant into the evaporator as the spring overcomes the lower pressure above the diaphragm. [*Figure 16-77*]

Vapor cycle air conditioning systems that have large evaporators experience significant pressure drops while refrigerant is flowing through them. Externally equalized expansion valves use a pressure tap from the outlet of the evaporator to help the superheat spring balance the diaphragm. This type of expansion valve is easily recognizable by the additional small-diameter line that comes from the evaporator into the valve (2 total). Better control of the proper amount of refrigerant allowed through the valve is attained by considering both the temperature and pressure of the evaporator refrigerant. [*Figure 16-78*]

Evaporator

Most evaporators are constructed of copper or aluminum tubing coiled into a compact unit. Fins are attached to increase surface area, facilitating rapid heat transfer between the cabin air blown over the outside of the evaporator with a fan and the refrigerant inside. The expansion valve, located at the evaporator inlet, changes the high pressure, high temperature refrigerant into low pressure, low temperature refrigerant which continues on into the evaporator. As the refrigerant absorbs heat from the cabin air, it changes into a low-pressure vapor. This is discharged from the evaporator outlet to the next component in the vapor cycle system, the compressor. The temperature and pressure pickups that regulate the expansion valve are located at the evaporator outlet.

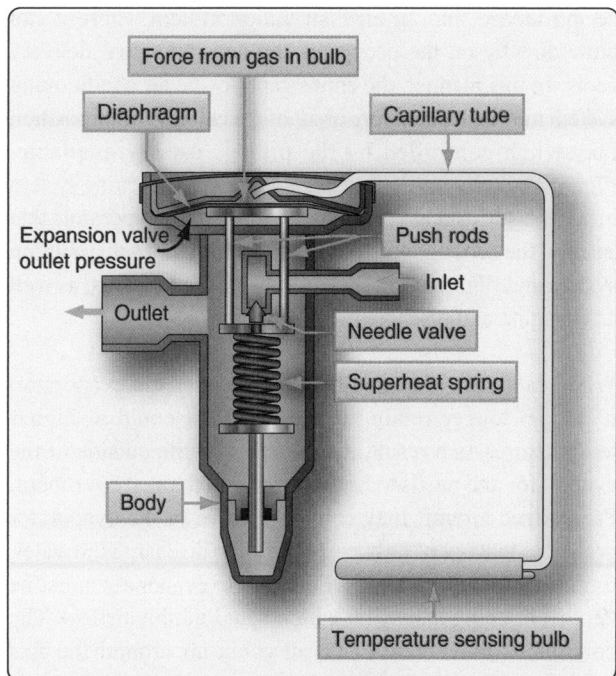

Figure 16-77. *An internally equalized expansion valve.*

The evaporator is situated in such a way that cabin air is pulled to it by a fan. The fan blows the air over the evaporator and discharges the cooled air back into the cabin. [*Figure 16-79*] This discharge can be direct when the evaporator is located in a cabin wall. A remotely located evaporator may require ducting from the cabin to the evaporator and from the evaporator back into the cabin. Sometimes the cool air produced may

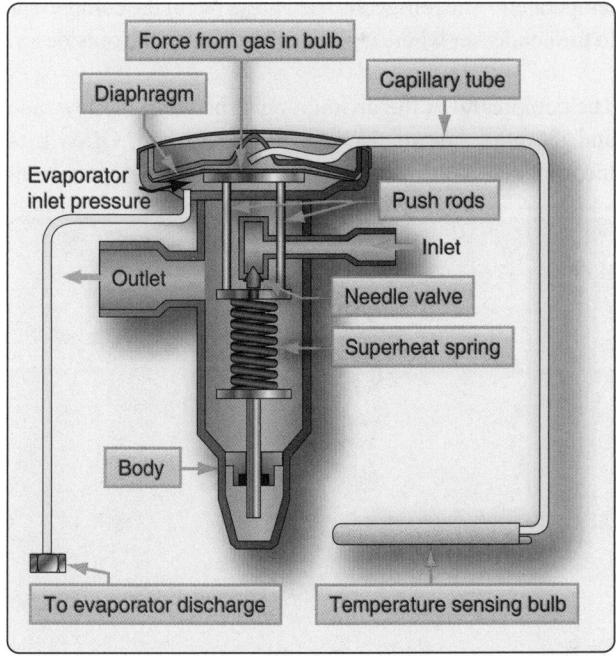

Figure 16-78. *An externally equalized expansion valve uses evaporator discharge temperature and pressure to regulate the amount of refrigerant passing through the valve and into the evaporator.*

be introduced into an air distribution system where it can blow directly on the occupants through individual delivery vents. In this manner, the entire vapor cycle air conditioning system may be located fore or aft of the cabin. A multiposition fan switch controlled by the pilot is usually available. *Figure 16-80* diagrams the vapor cycle air conditioning system in a Cessna Mustang very light jet. It has two evaporators that share in the cooling, with outlets integrated into a distribution system and flight deck mounted switches for the fans, as well as engaging and disengaging the system.

When cabin air is cooled by flowing over the evaporator, it can no longer retain the water that it could at higher temperature. As a result, it condenses on the outside of the evaporator and needs to be collected and drained overboard. Pressurized aircraft may contain a valve in the evaporator drain line that opens only periodically to discharge the water, to maintain pressurization. Fins on the evaporator must be kept from being damaged, which could inhibit airflow. The continuous movement of warm cabin air around the fins keeps condensed water from freezing. Ice on the evaporator reduces the efficiency of the heat exchange to the refrigerant.

Compressor

The compressor is the heart of the vapor cycle air conditioning system. It circulates the refrigerant around the vapor cycle system. It receives low-pressure, low-temperature refrigerant vapor from the outlet of the evaporator and compresses it. As the pressure is increased, the temperature also increases. The refrigerant temperature is raised above that of the outside air temperature. The refrigerant then flows out of the compressor to the condenser where it gives off the heat to the outside air.

The compressor is the dividing point between the low side and the high side of the vapor cycle system. Often it is incorporated with fittings or has fittings in the connecting

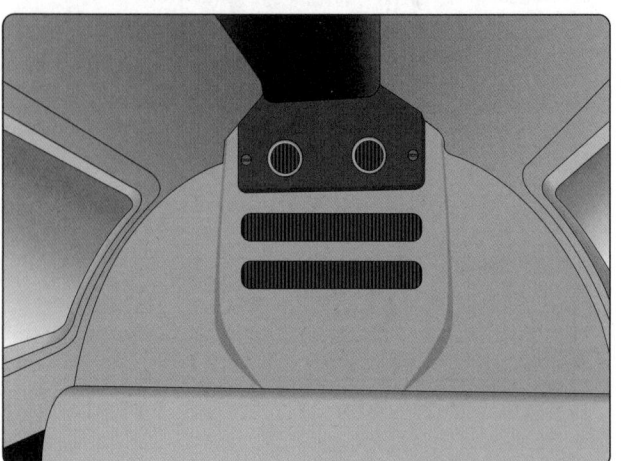

Figure 16-79. *The evaporator of this aircraft's vapor cycle conditioning system is visible in the forward cabin sidewall behind the right rudder pedal.*

lines to it that are designed to service the system with refrigerant. Access to the low and high sides of the system are required for servicing, which can be accomplished with fitting upstream and downstream of the compressor.

Modern compressors are either engine driven or driven by an electric motor. Occasionally, a hydraulically driven compressor is used. A typical engine-driven compressor, similar to that found in an automobile, is located in the engine nacelle and operated by a drive belt off of the engine crankshaft. An electromagnetic clutch engages when cooling is required, which causes the compressor to operate. When cooling is sufficient, power to the clutch is cut, and the drive pulley rotates but the compressor does not. *[Figure 16-81]*

Dedicated electric motor-driven compressors are also used on aircraft. Use of an electric motor allows the compressor to be located nearly anywhere on the aircraft, since wires can be run from the appropriate bus to the control panel and to the compressor. *[Figure 16-82]* Hydraulically-driven compressors are also able to be remotely located. Hydraulic lines from the hydraulic manifold are run through a switch-activated solenoid to the compressor. The solenoid allows fluid to the compressor or bypasses it. This controls the operation of the hydraulically driven compressor.

Regardless of how the vapor cycle air conditioning compressor is driven, it is usually a piston type pump. It requires use of a lightweight oil to lubricate and seal the unit. The oil is entrained by the refrigerant and circulates with it around the system. The crankcase of the compressor retains a supply of the oil, the level of which can be checked and adjusted by the technician. Valves exist on some compressor installations that can be closed to isolate the compressor from the remainder of the vapor cycle system while oil servicing takes place.

Condenser

The condenser is the final component in the vapor cycle. It is a radiator-like heat exchanger situated so that outside air flows over it and absorbs heat from the high-pressure, high-temperature refrigerant received from the compressor. A fan is usually included to draw the air through the condenser during ground operation. On some aircraft, outside air is ducted to the condenser. On others, the condenser is lowered into the airstream from the fuselage via a hinged panel. Often, the panel is controlled by a switch on the throttle levers. It is set to retract the compressor and streamline the fuselage when full power is required. *[Figure 16-83]*

The outside air absorbs heat from the refrigerant flowing through the condenser. The heat loss causes the refrigerant to change state back into a liquid. The high-pressure liquid refrigerant then leaves the condenser and flows to

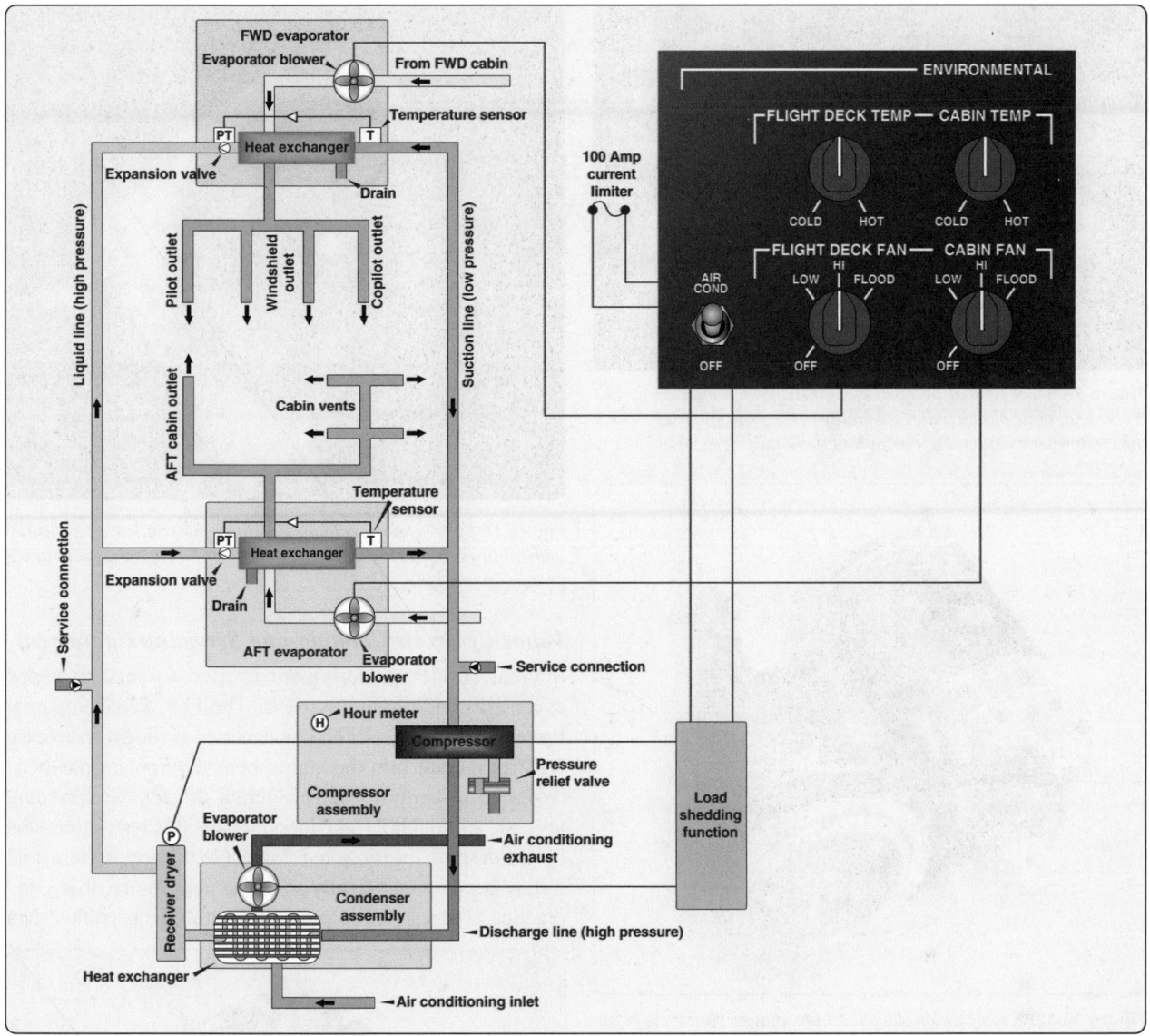

Figure 16-80. *The vapor cycle air conditioning system on a Cessna Mustang has two evaporators, one for the flight deck and one for the cabin. Each evaporator assembly contains the evaporator, a blower, a thermal expansion valve and the temperature feedback line from the outlet of the evaporator to the expansion valve.*

the receiver dryer. A properly engineered system that is functioning normally fully condenses all the refrigerant flowing through the condenser.

Service Valves

All vapor cycle air conditioning systems are closed systems; however, access is required for servicing. This is accomplished through the use of two service valves. One valve is located in the high side of the system and the other in the low side. A common type of valve used on vapor cycle systems that operate with R12 refrigerant is the Schrader valve. It is similar to the valve used to inflate tires. *[Figure 16-84]* A central valve core seats and unseats by depressing a stem attached to it. A pin in the servicing hose fitting is designed to do this when screwed onto the valve's

exterior threads. All aircraft service valves should be capped when not in use.

R134a systems use valves that are very similar to the Schrader valve in function, operation, and location. As a safety device to prevent inadvertent mixing of refrigerants, R134a valve fittings are different from Schrader valve fittings and do not attach to Schrader valve threads. The R134a valve fittings are a quick-disconnect type.

Another type of valve called a compressor isolation valve is used on some aircraft. It serves two purposes. Like the Schrader valve, it permits servicing the system with refrigerant. It also can isolate the compressor so the oil level can be checked and replenished without opening the entire

Figure 16-81. *A typical belt drive engine driven compressor. The electromagnetic clutch pulley assembly in the front starts and stops the compressor depending on cooling demand.*

Figure 16-82. *Examples of electric motor-driven vapor cycle air conditioning compressors.*

Figure 16-83. *A vapor cycle air conditioning condenser assembly with an integral fan used to pull outside air through the unit during ground operation.*

Vapor Cycle Air Conditioning Servicing Equipment

Special servicing equipment is used to service vapor cycle air conditioning systems. The U.S. Environmental Protection Agency (EPA) has declared it illegal to release R12 refrigerant into the atmosphere. Equipment has been designed to capture the refrigerant during the servicing process. Although R134a does not have this restriction, it is illegal in some locations to release it to the atmosphere, and it may become universally so in the near future. It is good practice to capture all refrigerants for future use, rather than

system and losing the refrigerant charge. These valves are usually hard mounted to the inlet and outlet of the compressor.

A compressor isolation valve has three positions. When fully open, it back seats and allows the normal flow of refrigerant in the vapor cycle. When fully closed or front seated, the valve isolates the compressor from the rest of the system and servicing with oil, or even replacement of the compressor, is possible without losing the refrigerant charge. When in an intermediate position, the valve allows access to the system for servicing. The system can be operated with the valve in this position but should be back seated for normal operation. The valve handle and service port should be capped when servicing is complete. *[Figure 16-85]*

Figure 16-84. *Cross-section of an R12 refrigerant service valve.*

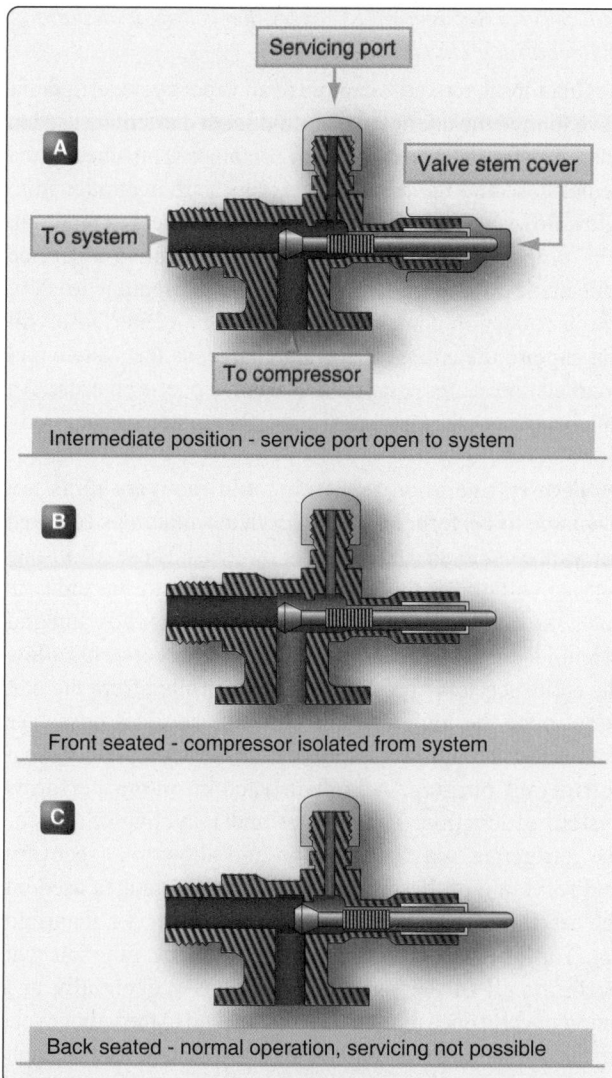

Figure 16-85. *Compressor isolation valves isolate the compressor for maintenance or replacement. They also allow normal operation and servicing of the vapor cycle air conditioning system with refrigerant.*

to waste them or to harm the environment by releasing them into the atmosphere. Capturing the refrigerant is a simple process designed into the proper servicing equipment. The technician should always be vigilant to use the approved refrigerant for the system being serviced and should follow all manufacturer's instruction.

Manifold Set, Gauges, Hoses, & Fittings
In the past, the main servicing device for vapor cycle air conditioning systems was the manifold set. It contains three hose fittings, two O-ring sealed valves, and two gauges. It is essentially a manifold into which the gauges, fittings, and valves are attached. The valves are positioned to connect or isolate the center hose with either fitting.

Hoses attach to the right and left manifold set fittings and the other ends of those hoses attach to the service valves in

the vapor cycle system. The center fitting also has a hose attached to it. The other end of this hose connects to either a refrigerant supply or a vacuum pump, depending on the servicing function to be performed. All servicing operations are performed by manipulating the valves. *[Figure 16-86]*

The gauges on the manifold set are dedicated—one for the low side of the system and the other for the high side. The low-pressure gauge is a compound gauge that indicates pressures above or below atmospheric pressure (0-gauge pressure). Below atmospheric pressure, the gauge is scaled in inches of mercury down to 30 inches. This is to indicate a vacuum. 29.92 inches of mercury equals an absolute vacuum (absolute zero air pressure). Above atmospheric pressure, gauge pressure is read in psi. The scale typically ranges from 0 to 60 psi, although some gauges extend up to 150 psi. The high-pressure gauge usually has a range from zero up to about 500 psi gauge pressure. It does not indicate vacuum (pressure lower than atmospheric). These gauges and their scales can be seen in *Figure 16-87*.

The low-pressure gauge is connected on the manifold directly to the low side fitting. The high-pressure gauge connects directly to the high side fitting. The center fitting of the manifold can be isolated from either of the gauges or the high and low service fittings by the hand valves. When these valves are turned fully clockwise, the center fitting is isolated. If the low-pressure valve is opened (turned counterclockwise), the center fitting is opened to the low-pressure gauge and the low side service line. The same is true for the high side when the high-pressure valve is opened. *[Figure 16-87]*

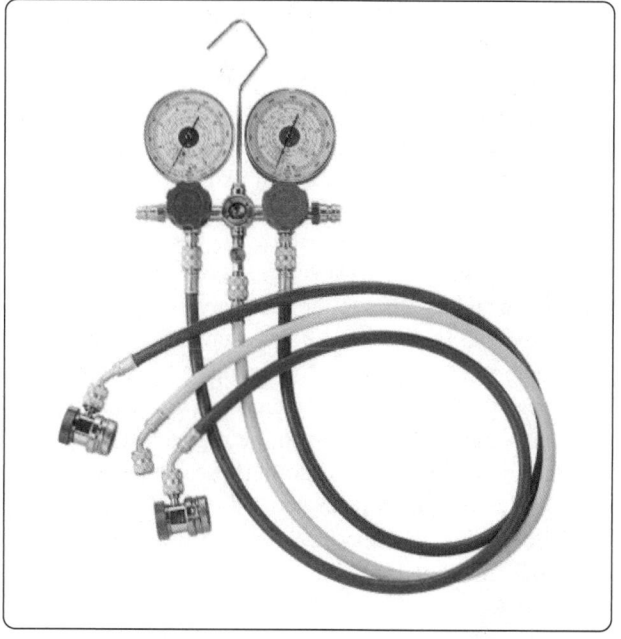

Figure 16-86. *A basic manifold set for servicing a vapor cycle air conditioning system.*

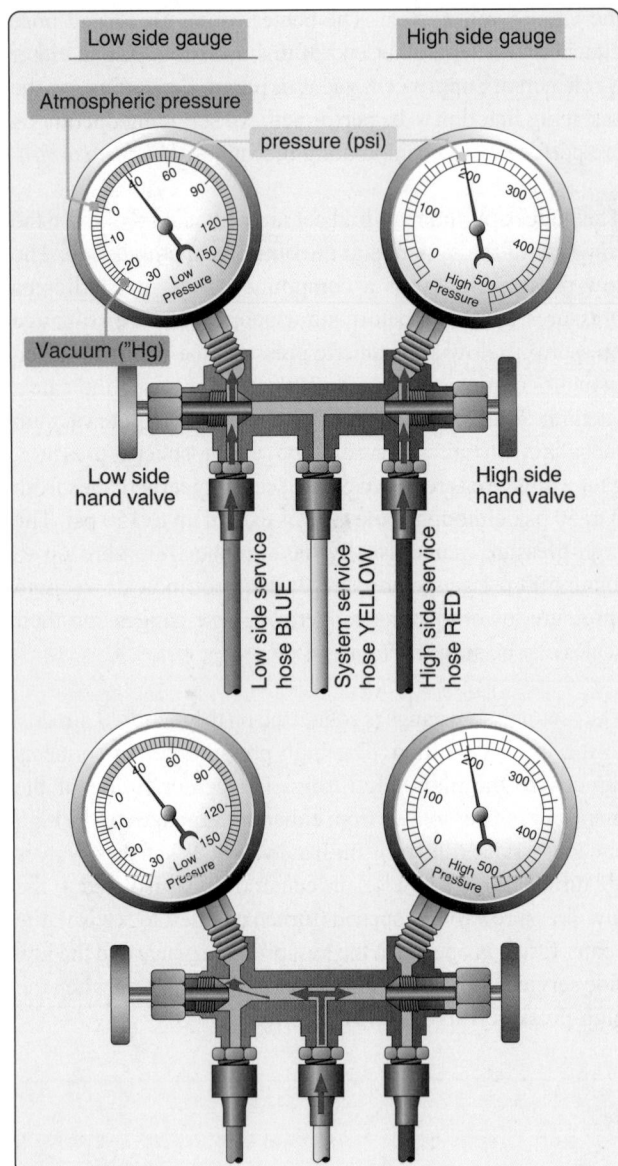

Low side gauge

High side gauge

Atmospheric pressure

pressure (psi)

Vacuum ("Hg)

Low side
hand valve

Low side service
hose BLUE

System service
hose YELLOW

High side service
hose RED

High side
hand valve

Figure 16-87. *The internal workings of a manifold set with the center fitting isolated (top). Opening a valve connects the center hose to that side of the system and the gauge (bottom).*

Special hoses are attached to the fittings of the manifold valve for servicing the system. The high-pressure charging hose is usually red and attaches to the service valve located in the high side of the system. The low-pressure hose, usually blue, attaches to the service valve that is located in the low side of the system. The center hose attaches to the vacuum pump for evacuating the system, or to the refrigerant supply for charging the system. Proper charging hoses for the refrigerant specific service valves must be used. When not using the manifold set, be sure the hoses are capped to prevent moisture from contaminating the valves.

Full Service Refrigerant Recovery, Recycling, Evacuation, & Recharging Units

Regulations that require capture of all vapor cycle refrigerant have limited the use of the manifold set. It can still be used to charge a system. The refrigerant container is attached to the center hose and the manifold set valves are manipulated to allow flow into the low or high side of the system as required. But, emptying a system of refrigerant requires a service unit made to collect it. Allowing the refrigerant to flow into a collection container attached to the center hose will not capture the entire refrigerant charge, as the system and container pressures equalize above atmospheric pressure. An independent compressor and collection system is required.

Modern refrigeration recharging and recovery units are available to perform all of the servicing functions required for vapor cycle air conditioning systems. These all-in-one service carts have the manifold set built into the unit. As such, the logic for using a manifold set still applies. Integral solenoid valves, reservoirs, filters, and smart controls allow the entire servicing procedure to be controlled from the unit panel once the high side and low side services hoses are connected. A built-in compressor enables complete system refrigerant purging. A built-in vacuum pump performs system evacuation. A container and recycling filters for the refrigerant and the lubricating oil allow total recovery and recycling of these fluids. The pressure gauges used on the service unit panel are the same as those on a manifold set. Top-of-the-line units have an automatic function that performs all of the servicing functions sequentially and automatically once the hoses are hooked up to the vapor cycle air conditioning system and the system quantity of refrigerant has been entered. *[Figure 16-88]*

Refrigerant Source

R134a comes in containers measured by the weight of the refrigerant they hold. Small 12-ounce to 2½-pound cans are common for adding refrigerant. Larger 30- and 50-pound cylinders equipped with shutoff valves are often used to charge an evacuated system, and they are used in shops that service vapor cycle systems frequently. *[Figure 16-89]* These larger cylinders are also used in the full servicing carts described above. The amount of refrigerant required for any system is measured in pounds. Check the manufacturer's service data and charge the system to the level specified using only the approved refrigerant from a known source.

Vacuum Pumps

Vacuum pumps used with a manifold set, or as part of a service cart, are connected to the vapor cycle system so that the system pressure can be reduced to a near total vacuum. The reason for doing this is to remove all of the water in the system. As mentioned, water can freeze, causing system

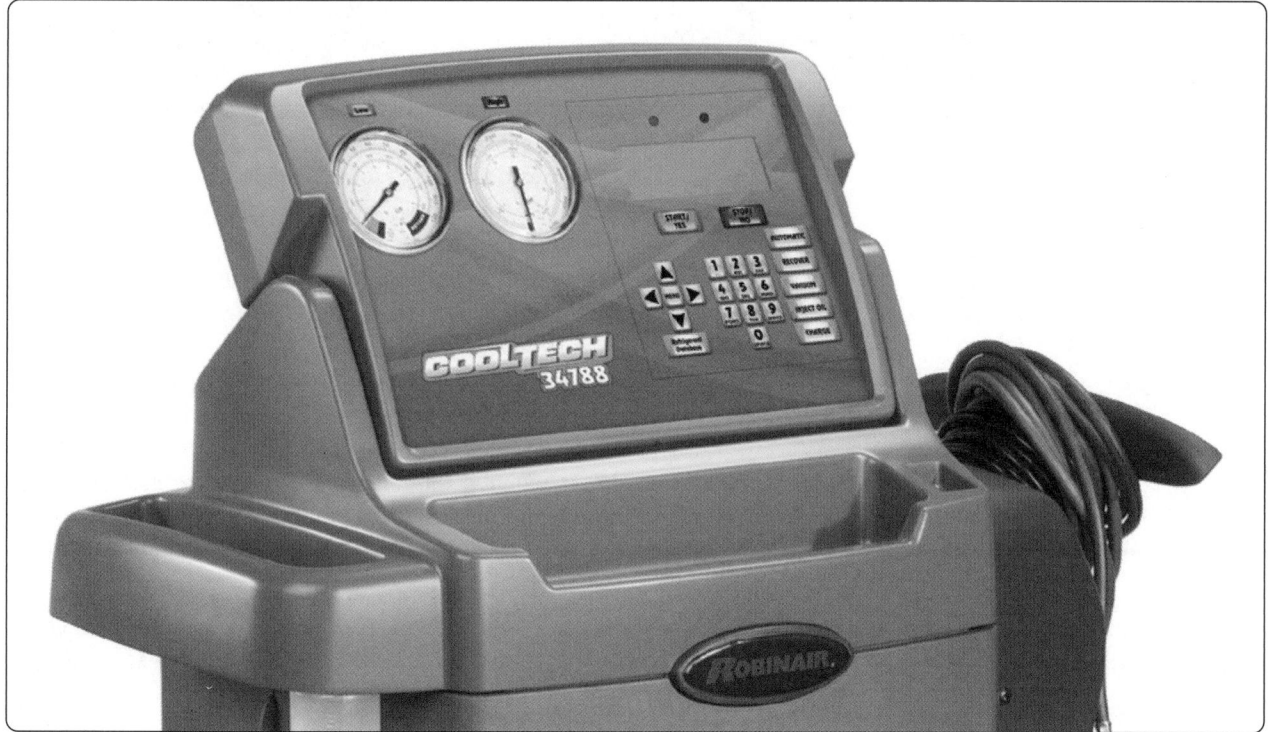

Figure 16-88. *A modern refrigerant recovery/recycle/charging service unit. Electronic control of solenoid activated valves combine with a built-in system for recovering, recycling, and recharging. A built-in vacuum pump and heated refrigerant reservoir are also included.*

Figure 16-89. *A 30 pound R134a refrigerant container with dual fittings. The fitting controlled by the blue valve wheel opens to the vapor space above the liquid refrigerant for connection to the low side of the vapor cycle system. The fitting controlled by the red valve wheel draws liquid refrigerant from the bottom of the cylinder through a stand tube. This fitting is connected to the system high side. On containers without dual fittings, the container must be inverted to deliver liquid refrigerant through a connected hose.*

malfunction and can also combine with the refrigerant to create corrosive compounds.

Once the system has been purged of its refrigerant and it is at atmospheric pressure, the vacuum pump is operated. It gradually reduces the pressure in the system. As it does, the boiling point of any water in the system is also reduced. Water boils off or is vaporized under the reduced pressure and is pulled from the system by the pump, leaving the system moisture free to be recharged with refrigerant. *[Figure 16-90]* The strength and efficiency of vacuum pumps varies as does the amount of time to hold the system at reduced pressure specified by manufacturers. Generally, the best-established vacuum is held for 15–30 minutes to ensure all water is removed from the system. Follow the manufacturer's instructions when evacuating a vapor cycle air conditioning system. *[Figure 16-91]*

Leak Detectors

Even the smallest leak in a vapor cycle air conditioning system can cause a loss of refrigerant. When operating normally, little or no refrigerant escapes. A system that requires the addition of refrigerant should be suspected of having a leak. Electronic leak detectors are safe, effective devices used to find leaks. There are many types available that are able to detect extremely small amounts of escaped refrigerant. The detector is held close to component and hose connections where most leaks occur. Audible and visual

Inches of vacuum on low side gauge (inches Hg)	Temperature at which water boils (°F)	Absolute pressure (psi)
0	212	14.696
4.92	204.98	12.279
9.23	194	10.152
15.94	176	6.866
20.72	158	4.519
24.04	140	2.888
26.28	122	1.788
27.75	104	1.066
28.67	86	0.614
28.92	80.06	0.491
29.02	75.92	0.442
29.12	71.96	0.393
29.22	69.08	0.344
29.32	64.04	0.295
29.42	59	0.246
29.52	53.06	0.196
29.62	44.96	0.147
29.74	32	0.088
29.82	21.02	0.0049
29.87	6.08	0.00245
29.91	−23.98	0.00049

Figure 16-90. *When the temperature is low, a greater amount of vacuum is needed to boil off and remove any water in the vapor cycle system.*

Figure 16-91. *A vacuum pump is used to lower the pressure in the vapor cycle air conditioning system. This reduces the boiling point of water in the system, which vaporizes and is drawn out by the pump.*

alarms signal the presence of refrigerant. A detector specified for the type of refrigerant in the system should be chosen. A good leak detector is sensitive enough to detect leaks that would result in less than ½ ounce of refrigerant to be lost per year. *[Figure 16-92]*

Other leak detection methods exist. A soapy solution can also be applied to fittings and inspected for the formation of bubbles indicating a leak. Special leak detection dyes compatible for use with refrigerant can be injected into the vapor cycle system and can be seen when they are forced

out at a leak. Many of these are made to be visible under UV light. Occasionally, a leak can be detected upon close visual inspection. Oil in the system can be forced out of a leak, leaving a visible residue that is usually on the bottom side of a leaky fitting. Old hoses may become slightly porous and leak a significant amount of refrigerant over time. Because of the length and area through which the refrigerant is lost, this type of leak may be difficult to detect, even with leak detecting methods. Visibly deteriorated hoses should be replaced.

System Servicing

Vapor cycle air conditioning systems can give many hours of reliable, maintenance-free service. Periodic visual inspections, tests, and refrigerant level and oil level checks may be all that is required for some time. Follow the manufacturer's instructions for inspection criteria and intervals.

Visual Inspection

All components of any vapor cycle system should be checked to ensure they are secure. Be vigilant for any damage, misalignment, or visual signs of leakage. The evaporator and condenser fins should be checked to ensure they are clean, unobstructed, and not folded over from an impact. Dirt and inhibited airflow through the fins can prevent effective heat exchange to and from the refrigerant. Occasionally, these units can be washed. Since the condenser often has ram air ducted to it or extends into the airstream, check for the presence of debris that may restrict airflow. Hinged units should be checked for security and wear. The mechanism to extend and retract the unit should function as specified, including the throttle position switch present on many systems. It is designed to cut power to the compressor clutch and retract the

Figure 16-92. *This electronic infrared leak detector can detect leaks that would lose less than ¼ ounce of refrigerant per year.*

condenser at full power settings. Condensers may also have a fan to pull air over them during ground operation. It should be checked to ensure it functions correctly. *[Figure 16-93]* Be sure the capillary temperature feedback sensor to the expansion valve is securely attached to the evaporator outlet. Also, check the security of the pressure sensor and thermostat sensor if the system has them. The evaporator should not have ice on the outside. This prevents proper heat exchange to the refrigerant from the warm cabin air blown over the unit. The fan blower should be checked to ensure it rotates freely. Depending on the system, it should run whenever the cooling switch is selected and should change speeds as the selector is rotated to more or less cooling. Sometimes systems low on refrigerant can cause ice on the evaporator, as can a faulty expansion valve or feedback control line. Ice formation anywhere on the outside of a vapor cycle air

conditioning system should be investigated for cause and corrected. *[Figure 16-94]*

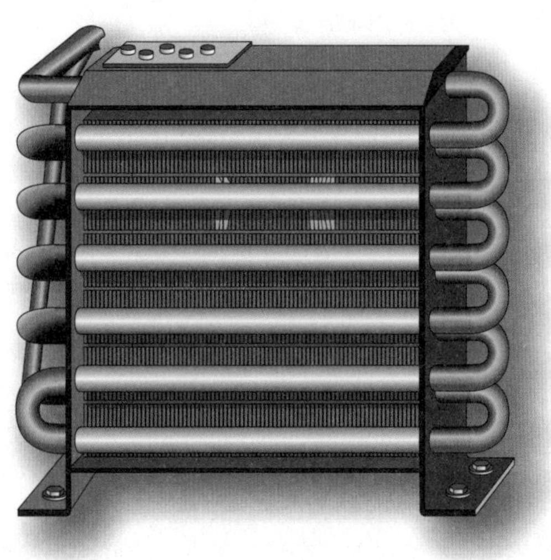

Security and alignment of the compressor is critical and should be checked during inspection. Belt-driven compressors need to have proper belt tension to function properly. Check the manufacturer's data for information on how to determine the condition and tension of the belt, as well as how to make adjustments. Oil level should be sufficient. Typically, ¼ ounce of oil is added for each pound of refrigerant added to the system. When changing a component, additional oil may need to be added to replace that which is trapped in the replaced unit. Always use the oil specified in the manufacturer's maintenance manual.

Leak Test

As mentioned under the leak detector section above, leaks in a vapor cycle air conditioning system must be discovered and repaired. The most obvious sign of a possible leak is a low refrigerant level. Bubbles present in the sight glass of the receiver dryer while the system is operating indicate more refrigerant is needed. A system check for a leak may be in order. Vapor cycle systems normally lose a small amount of refrigerant each year. No action is needed if this amount is within limits.

Occasionally, all of the refrigerant escapes from the system. No bubbles are visible in the sight glass, but the complete lack of cooling indicates the refrigerant has leaked out. To locate the leak point, the system needs to be partially charged with refrigerant so leak detection methods can be employed. About 50 psi of refrigerant in the high and low sides should be sufficient for a leak check. By introducing the refrigerant into the high side, pressure indicated on the low side gauge

Figure 16-93. *Damaged fins on a condenser.*

Figure 16-94. *Ice on the evaporator coils is cause for investigation. It prevents proper heat exchange to the refrigerant.*

verifies the orifice in the expansion valve is not clogged. When all refrigerant is lost due to a leak, the entire system should be checked. Each fitting and connection should be inspected visually and with a leak detector.

When a vapor cycle air conditioning system loses all of its refrigerant charge, air may enter the system. Water may also enter since it is in the air. This means that a full system evacuation must be performed after the leak is found and repaired. By establishing only a 50-psi charge in a depleted system, the leak(s) becomes detectable, but time and refrigerant are not wasted prior to evacuation. System evacuation is discussed below.

Performance Test

Verification of proper operation of a vapor cycle air conditioning system is often part of a performance test. This involves operating the system and checking parameters to ensure they are in the normal range. A key indication of performance is the temperature of the air that is cooled by the evaporator. This can be measured at the air outflow from the evaporator or at a nearby delivery duct outlet. An ordinary thermometer should read 40–50 °F, with the controls set to full cold after the system has been allowed to operate for a few minutes. Manufacturer's instructions include information on where to place the thermometer and the temperature range that indicates acceptable performance.

Pressures can also be observed to indicate system performance. Typically, low side pressure in a vapor cycle system operating normally is 10–50 psi, depending on ambient temperature. High side pressure is between 125 and 250 psi, again, depending on ambient temperature and the design of the system. All system performance tests are performed at a specified engine rpm (stable compressor speed) and involve a period of time to stabilize the operation of the vapor cycle. Consult the manufacturer's instructions for guidance.

Feel Test

A quick reference field test can be performed on a vapor cycle air conditioning system to gauge its health. In particular, components and lines in the high side (from the compressor to the expansion valve) should be warm or hot to the touch. The lines on both sides of the receiver dryer should be the same temperature. Low side lines and the evaporator should be cool. Ice should not be visible on the outside of the system. If any discrepancies exist, further investigation is needed. On hot, humid days, the cooling output of the vapor cycle system may be slightly compromised due to the volume of water condensing on the evaporator.

Purging the System

Purging the system means emptying it of its refrigerant charge. Since the refrigerant must be captured, a service cart with this capability should be used. By connecting the hoses to the high side and low side service valves and selecting recover, cart solenoid valves position so that a system purging compressor pumps the refrigerant out of the vapor cycle system and into a recovery tank.

Vapor cycle systems must be properly purged before opening for maintenance or component replacement. Once opened, precautions should be taken to prevent contaminants from entering the system. When suspicion exists that the system has been contaminated, such as when a component has catastrophically failed, it can be flushed clean. Special fluid flush formulated for vapor cycle air conditioning systems should be used. The receiver dryer is removed from the system for flushing and a new unit is installed, as it contains fresh filters. Follow the aircraft manufacturer's instructions.

Checking Compressor Oil

The compressor is a sealed unit in the vapor cycle system that is lubricated with oil. Any time the system is purged, it is an opportunity to check the oil quantity in the compressor crankcase. This is often done by removing a filler plug and using a dip stick. Oil quantity should be maintained within the proper range using oil recommended by the manufacturer. Be certain to replace the filler plug after checking or adding oil. *[Figure 16-95]*

Evacuating the System

Only a few drops of moisture can contaminate a vapor cycle air conditioning system. If this moisture freezes in the expansion valve, it could completely block the refrigerant flow. Water is removed from the system by evacuation. Anytime the system refrigerant charge falls below atmospheric pressure, the refrigerant is lost, or the system is opened, it must be evacuated before recharging.

Evacuating a vapor cycle air conditioning system is also known as pumping down the system. A vacuum pump is connected and pressure inside the system is reduced to vaporize any water that may exist. Continued operation of the vacuum pump draws the water vapor from the system. A typical pump used for evacuating an air conditioning system can reduce system pressure to about 29.62 "Hg (gauge pressure). At this pressure, water boils at 45 °F. Operate the vacuum pump to achieve the recommended gauge pressure. Hold this vacuum for as long as the manufacturer specifies.

As long as a vapor cycle air conditioning system retains a charge higher than atmospheric pressure, any leak forces refrigerant out of the system. The system pressure prevents air (and water vapor) from entering. Therefore, it is permissible to recharge or add refrigerant to a system that has not dropped

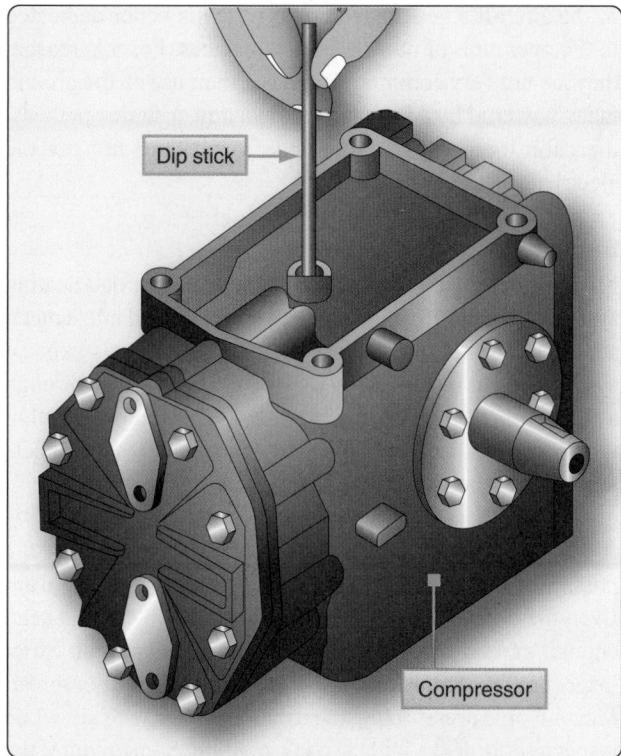

Figure 16-95. *Checking the compressor oil when the system is open.*

below atmospheric pressure without evacuating the system.

Charging the System

Charging capacity of a vapor cycle air conditioning system is measured by weight. The aircraft manufacturer's maintenance manual specifies this amount and the amount and type of oil to be put into the system when filling. Pre-weighing the refrigerant or setting the refrigerant weight into the servicing cart input ensures the system is filled to capacity.

Charging a vapor cycle air conditioning system should be undertaken immediately after evacuation of the system is completed. With the hoses still connected to the high and low side service valves, selecting charge on the service cart panel positions solenoid operated valves so that the refrigerant supply is available. First, refrigerant is released into the high side of the system. Observe the low side gauge. When the low side gauge begins to indicate pressure, it is known that refrigerant is passing through the tiny orifice in the expansion valve. As pressure builds in the high side, the flow of refrigerant into the system stops.

To complete the charge of the system, refrigerant needs to be drawn in by the compressor. A major concern is to avoid damage to the compressor by having liquid refrigerant enter the compressor inlet. After the initial release of refrigerant into the high side, the high side service valve is closed, and the remaining charge is made through the low side service

valve. The engine is started and run at a specified rpm, usually a high idle speed. Full cool is selected on the air conditioning control panel in the flight deck. As the compressor operates, it draws vapor into the low side until the correct weighed amount of refrigerant is in the system. Charging is completed with a full performance test.

Charging with a manifold set is accomplished in the same way. The manifold center hose is connected to the refrigerant source that charges the system. After opening the valve on the container (or puncturing the seal on a small can), the center hose connection on the manifold set should be loosened to allow air in the hose to escape. Once the air is bled out of the hose, the refrigerant can enter the system through whichever service valve is opened. The sequence is the same as above and all manufacturer instructions should be followed.

Oil quantity added to the system is specified by the manufacturer. Refrigerant premixed with oil is available and may be permissible for use. This eliminates the need to add oil separately. Alternately, the amount of oil to be put into the system can be selected on the servicing cart. Approximately ¼ ounce of oil for each pound of refrigerant is a standard amount; however, follow the manufacturer's specifications.

Technician Certification

The EPA requires certification of technicians that work with vapor cycle air conditioning refrigerant and equipment to ensure safe compliance with current regulations. Aircraft technicians can obtain certification or can refer vapor cycle air conditioning work to shops that specialize in this work.

Aircraft Heaters

Bleed Air Systems

Temperatures at high altitudes in which aircraft operate can be well below 0 °F. Combined with seasonally cold temperatures, this makes heating the cabin more than just a luxury. Pressurized aircraft that use air cycle air conditioning systems mix bleed air with cold air produced by the air cycle machine expansion turbine to obtain warm air for the cabin. This is discussed in the section that covers air cycle air conditioning in this chapter. Aircraft not equipped with air cycle air conditioning may be heated by one of a few possible methods.

Some turbine-powered aircraft not equipped with air cycle systems still make use of engine compressor bleed air to heat the cabin. Various arrangements exist. The bleed air is mixed with ambient air, or cabin return air, and distributed throughout the aircraft via ducting. The mixing of air can be done in a variety of ways. Mixing air valves, flow control valves, shutoff valves, and other various control valves are controlled by switches in the flight deck. One STC'd bleed air heat system uses mini-ejectors in helicopter cabins to combine bleed air with cabin air. All of these bleed air heating systems are simple and function well, as long as the valves, ducting, and controls are in operational condition.

Electric Heating Systems

Occasionally, an electric heating device is used to heat the aircraft. Electricity flowing through a heating element makes the element warm. A fan to blow air over the elements and into the cabin is used to transfer the heat. Other floor or sidewall elements simply radiate heat to warm the cabin.

Electric heating element heaters require a significant amount

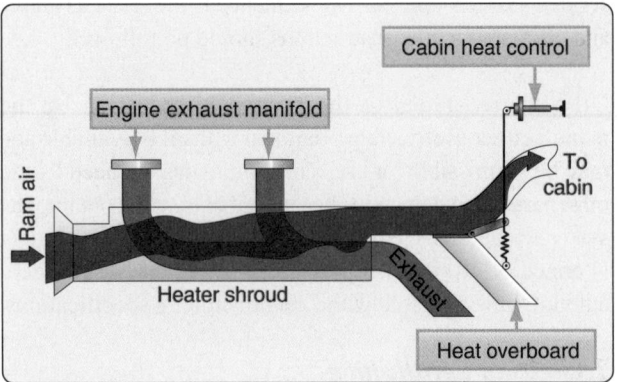

Figure 16-96. *The basic arrangement of an aircraft exhaust shroud heater.*

of the aircraft's generator output, which is better dedicated to the operation of other electrical devices. For this reason, they are not very common. However, their use on the ground when powered by a ground electrical power source preheats the cabin before passengers board and does not tax the electrical system.

Exhaust Shroud Heaters

Most single-engine light aircraft use exhaust shroud heating systems to heat the cabin. Ambient air is directed into a metal shroud, or jacket, that encases part of the engine's exhaust system. The air is warmed by the exhaust and directed through a firewall heater valve into the cabin. This simple solution requires no electrical or engine power and it makes use of heat that would otherwise be wasted. *[Figures 16-96 and 16-97]*

A major concern of exhaust shroud heat systems is the possibility that exhaust gases could contaminate the cabin air. Even the slightest crack in an exhaust manifold could send enough carbon monoxide into the cabin to be fatal. Strict inspection procedures are in place to minimize this threat. Most involve pressurizing the exhaust system with air, while inspecting for leaks with a soapy solution. Some require the exhaust to be removed and pressurized while submerged under water to detect any leaks. Frequency of exhaust heat leak detection can be every 100 hours.

Occasionally, the exhaust system is slightly modified in a

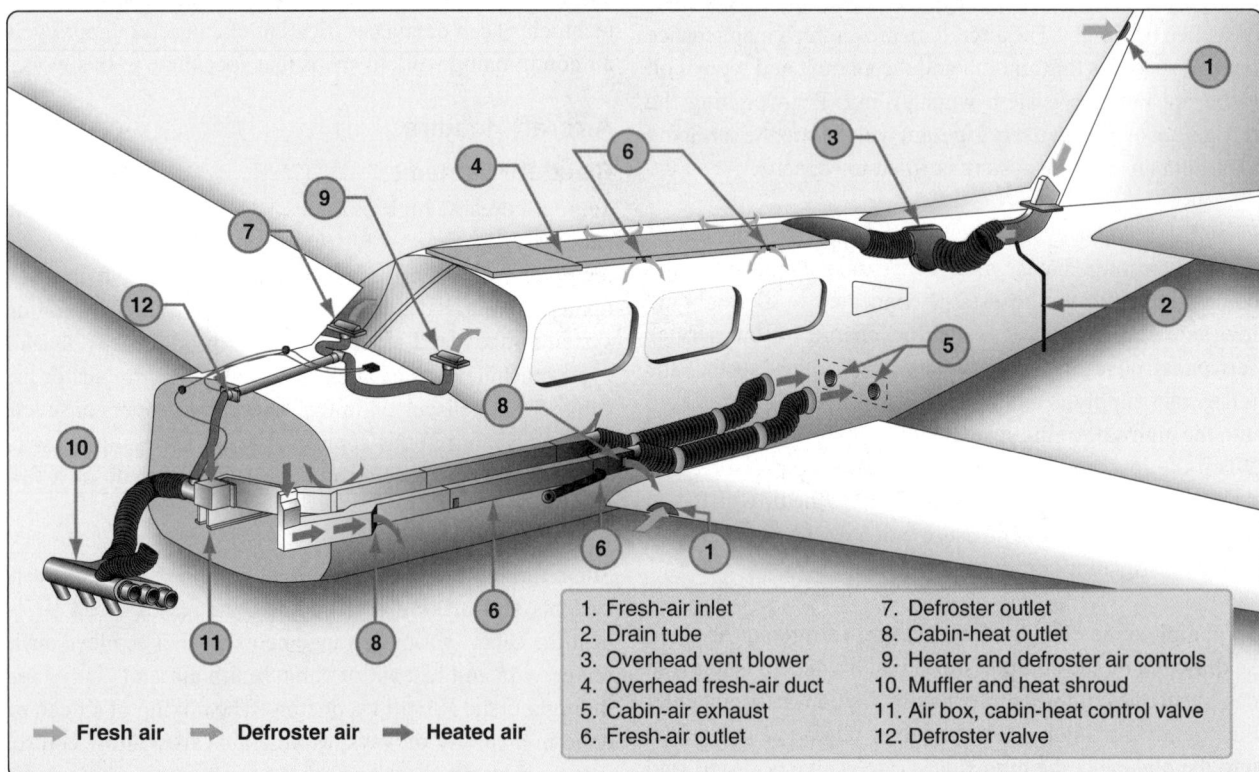

1. Fresh-air inlet
2. Drain tube
3. Overhead vent blower
4. Overhead fresh-air duct
5. Cabin-air exhaust
6. Fresh-air outlet
7. Defroster outlet
8. Cabin-heat outlet
9. Heater and defroster air controls
10. Muffler and heat shroud
11. Air box, cabin-heat control valve
12. Defroster valve

➡ Fresh air ➡ Defroster air ➡ Heated air

Figure 16-97. *The environmental system of a single-engine Piper aircraft with an exhaust shroud heating system.*

shroud heat configuration. For example, an exhaust muffler may have numerous welded studs attached, which increase heat transfer to the cabin air. Each weld point is a location for a potential leak. *[Figure 16-98]*

Regardless of age or condition, aircraft with exhaust shroud heating systems should contain a carbon monoxide detection device in the flight deck.

Combustion Heaters

An aircraft combustion heater is used on many small to medium sized aircraft. It is a heat source independent from the aircraft's engine(s), although it does use fuel from the aircraft's main fuel system. Combustion heaters are manufactured by a few different companies that supply the aviation industry. Most are similar to the description that follows. The most up to date units have electronic ignition and temperature control switches.

Combustion heaters are similar to exhaust shroud heaters in that ambient air is heated and sent to the cabin. The source of heat in this case is an independent combustion chamber located inside the cylindrical outer shroud of the heater unit. The correct amount of fuel and air are ignited in the air-tight inner chamber. The exhaust from combustion is funneled overboard. Ambient air is directed between the combustion chamber and the outer shroud. It absorbs the combustion heat by convection and is channeled into the cabin. *[Figure 16-99]* Refer to *Figure 16-100* for the following descriptions of the combustion heater subsystems and heater operation.

Combustion Air System

The air used in the combustion process is ambient air scooped from outside the aircraft, or from the compartment in which

Figure 16-99. *A modern combustion heater.*

the combustion heater is mounted. A blower ensures that the correct quantity and pressure of air are sent into the chamber.

Some units have regulators or a relief valve to ensure these parameters. The combustion air is completely separate from the air that is warmed and sent into the cabin.

Ventilating Air System

Ventilating air is the name of the air that is warmed and sent into the cabin. Typically, it comes into the combustion heater through a ram air intake. When the aircraft is on the ground, a ventilating air fan controlled by a landing gear squat switch operates to draw in the air. Once airborne, the fan ceases to operate as the ram air flow is sufficient. Ventilating air passes between the combustion chamber and the outer shroud of the combustion heater where it is warmed and sent to the cabin.

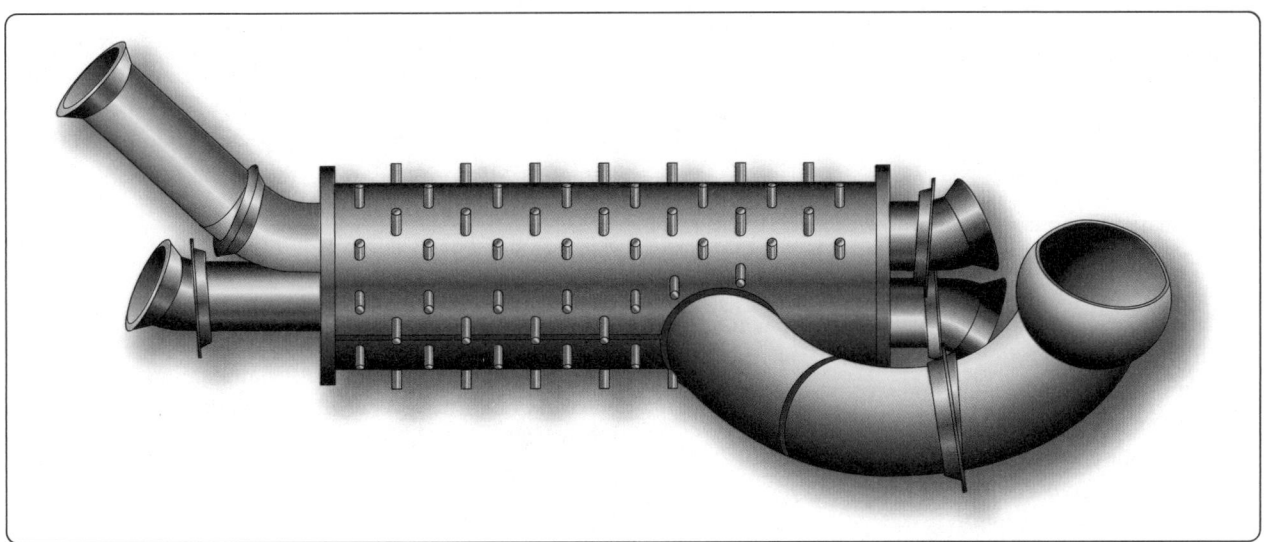

Figure 16-98. *An exhaust manifold with its shroud removed showing numerous welded studs used to increase heat transfer from the exhaust to the ambient air going to the cabin.*

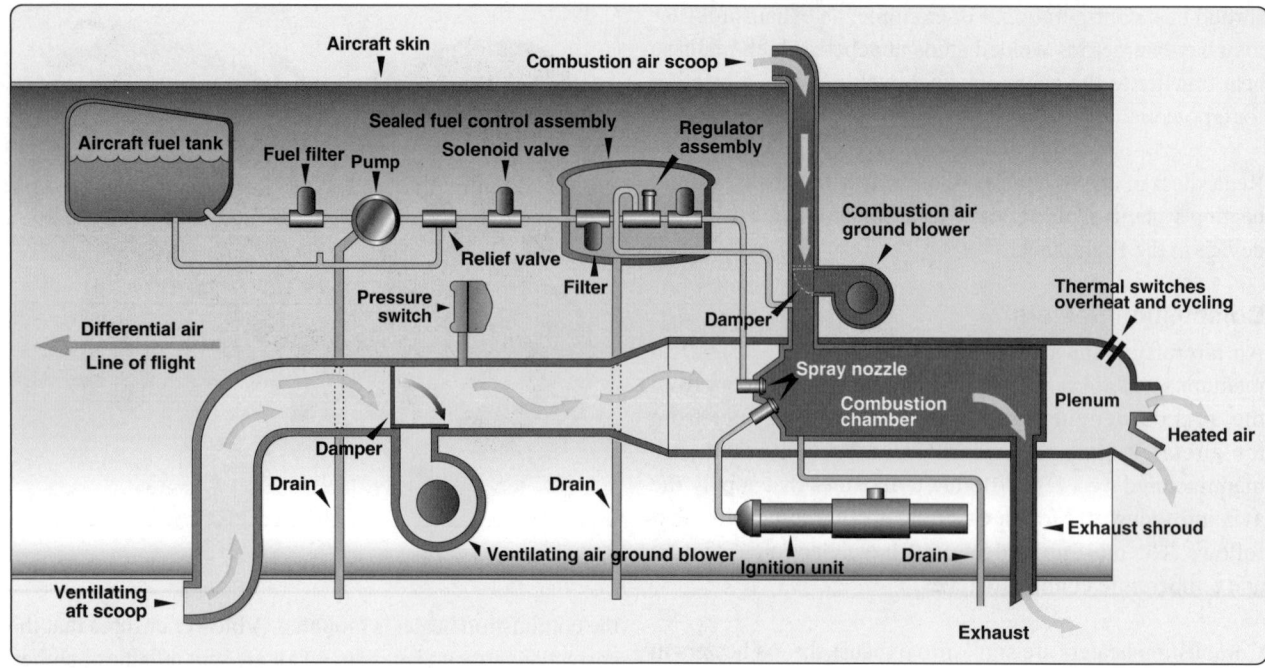

Figure 16-100. *A diagram of a typical combustion heater and its components.*

Fuel System

As mentioned, fuel for the combustion heater is drawn from an aircraft fuel tank. A constant pressure fuel pump with relief valve pulls the fuel through a filter. A main solenoid valve downstream delivers the fuel to the unit. The solenoid is controlled by the cabin heater switch in the flight deck and three safety switches located on the combustion heater. The first safety switch is a duct limit switch that keeps the valve closed should the unit not have enough ventilating airflow to keep it within the correct operating temperature range. The second is a pressure switch that must sense pressure from the combustion air fan to allow the solenoid to open. Fuel is delivered to the combustion chamber only if there is air there with which it can be mixed. Finally, an overheat switch also controls the main fuel supply solenoid. When an over temperature condition occurs, it closes the solenoid to stop the supply of fuel.

A secondary solenoid is located downstream of the main fuel supply solenoid. It is part of a fuel control unit that also houses a pressure regulator and an additional fuel filter. The valve opens and closes on command from the combustion heater thermostat. During normal operation, the heater cycles on and off by opening and closing this solenoid at the entrance to the combustion chamber. When opened, fuel flows through a nozzle that sprays it into the combustion chamber. *[Figure 16-100]*

Ignition System

Most combustion heaters have an ignition unit designed to receive aircraft voltage and step it up to fire a spark plug located in the combustion chamber. Older combustion heaters use vibrator-type ignition units. Modern units have electronic ignition. *[Figure 16-101]* The ignition is continuous when activated. This occurs when the heater switch is placed in the ON position in the flight deck, and the combustion air blower builds sufficient air pressure in the combustion chamber. Use of the proper spark plug for the combustion heater is essential. Check the manufacturer's approved data. *[Figure 16-102]*

Controls

The combustion heater controls consist of a cabin heat switch and a thermostat. The cabin heat switch starts the fuel pump, opens the main fuel supply solenoid, and turns on the combustion air fan, as well as the ventilating air fan if the aircraft is on the ground. When the combustion air fan builds pressure, it allows the ignition unit to start. The thermostat sends power to open the fuel control solenoid when heat is needed. This triggers combustion in the unit and heat is delivered to the cabin. When the preselected temperature is reached, the thermostat cuts power to the fuel control solenoid and combustion stops. Ventilating air continues to circulate and carry heat away. When the temperature level falls to that below which the thermostat is set, the combustion heater cycles on again.

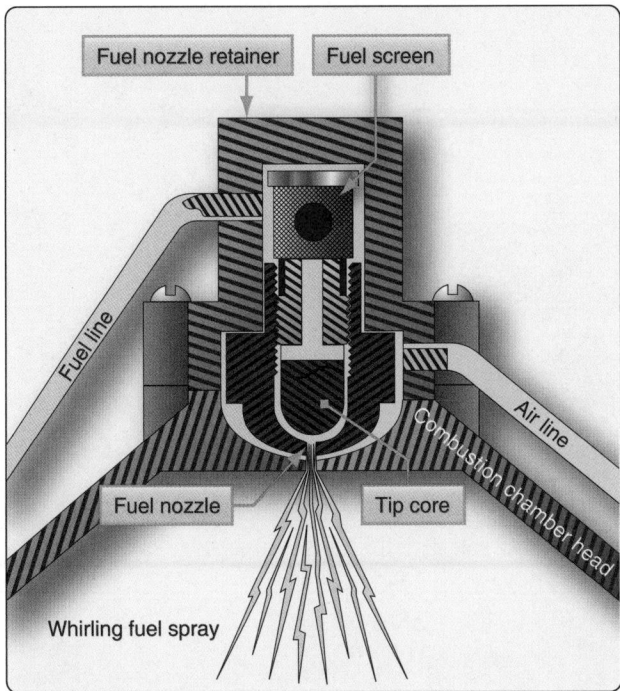

Figure 16-101. *The fuel nozzle located at the end of the combustion chamber sprays aircraft fuel, which is lit by a continuous sparking ignition system spark plug.*

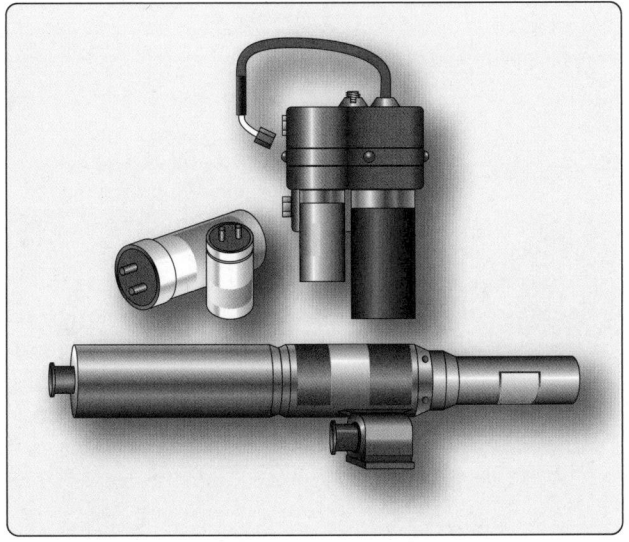

Figure 16-102. *Examples of ignition units used on combustion heaters.*

Safety Features

Various automatic combustion heater controls prevent operation of the heater when dangerous conditions exist. As stated, a duct limit switch cuts off fuel to the heater when there is not enough airflow to keep the heater duct below a preset temperature. This is usually caused by a lack of ventilating air flow. An overheat switch set at a higher temperature than the duct limit switch guards against overheat of any kind. It is designed to cut fuel to the combustion heater before an

unwanted fire occurs. When this switch activates, a light is illuminated in the flight deck and the heater cannot be restarted until maintenance determines the cause. Some heaters contain a circuit to prevent fuel from being delivered to the combustion chamber if the ignition system is not working.

Maintenance & Inspection

Maintenance of combustion heaters consists of routine items, such as cleaning filters, checking spark plug wear, and ensuring inlets are not plugged. All maintenance and inspection of combustion heaters should be accomplished in accordance with the aircraft manufacturer's instructions. Combustion heater manufacturers also produce maintenance guidelines that should be followed. Intervals between the performance of maintenance items and the time between overhauls must be followed to help ensure a properly functioning heater is available when it is needed.

Inspection of the combustion heater should be performed on schedule as provided by the manufacturer or whenever a malfunction is suspected. Inlets and outlets should be clear. All controls should be checked for freedom of operation and function. Close observation for any sign of fuel leaks or cracks in the combustion chamber and/or shroud should be made. All components should be secure. An operational check can also be made. Follow the manufacturer's inspection criteria to ensure the combustion heater is in airworthy condition.

Chapter 17
Fire Protection Systems

Introduction

Because fire is one of the most dangerous threats to an aircraft, the potential fire zones of modern multiengine aircraft are protected by a fixed fire protection system. A fire zone is an area, or region, of an aircraft designed by the manufacturer to require fire detection and/or fire extinguishing equipment and a high degree of inherent fire resistance. The term "fixed" describes a permanently installed system in contrast to any type of portable fire extinguishing equipment, such as a hand-held Halon or water fire extinguisher. A complete fire protection system on modern aircraft, and on many older aircraft, includes a fire detection system and a fire extinguishing system. Typical zones on aircraft that have a fixed fire detection and/or fire extinguisher system are:

1. Engines and auxiliary power unit (APU).

2. Cargo and baggage compartments.

3. Lavatories on transport aircraft.

4. Electronic bays.

5. Wheel wells.

6. Bleed air ducts.

To detect fires or overheat conditions, detectors are placed in the various zones to be monitored. Fires are detected in reciprocating engine and small turboprop aircraft using one or more of the following:

1. Overheat detectors.

2. Rate-of-temperature-rise detectors.

3. Flame detectors.

4. Observation by crewmembers.

In addition to these methods, other types of detectors are used in aircraft fire protection systems but are seldom used to detect engine fires. For example, smoke detectors are better suited to monitor areas where materials burn slowly or smolder, such as cargo and baggage compartments. Other types of detectors in this category include carbon monoxide detectors and chemical sampling equipment capable of detecting combustible mixtures that can lead to accumulations of explosive gases.

The complete aircraft fire protection systems of most large turbine-engine aircraft incorporate several of these different detection methods.

1. Rate-of-temperature-rise detectors.

2. Radiation sensing detectors.

3. Smoke detectors.

4. Overheat detectors.

5. Carbon monoxide detectors.

6. Combustible mixture detectors.

7. Optical detectors.

8. Observation by crew or passengers.

The types of detectors most commonly used for fast detection of fires are the rate-of-rise, optical sensor, pneumatic loop, and electric resistance systems.

Classes of Fires

The following classes of fires that are likely to occur onboard aircraft, as defined in the U.S. National Fire Protection Association (NFPA) Standard 10, Standard for Portable Fire Extinguishers, 2007 Edition, are:

1. Class A—fires involving ordinary combustible materials, such as wood, cloth, paper, rubber, and plastics.

2. Class B—fires involving flammable liquids, petroleum oils, greases, tars, oil-based paints, lacquers, solvents, alcohols, and flammable gases.

3. Class C—fires involving energized electrical equipment in which the use of an extinguishing media that is electrically nonconductive is important.

4. Class D—fires involving combustible metals, such as magnesium, titanium, zirconium, sodium, lithium, and potassium.

Requirements for Overheat & Fire Protection Systems

Fire protection systems on current-production aircraft do not rely on observation by crew members as a primary method of fire detection. An ideal fire detector system includes as many of the following features as possible:

1. No false warnings under any flight or ground condition.

2. Rapid indication of a fire and accurate location of the fire.

3. Accurate indication that a fire is out.

4. Indication that a fire has re-ignited.

5. Continuous indication for duration of a fire.

6. Means for electrically testing the detector system from the aircraft flight deck.

7. Resists damage from exposure to oil, water, vibration, extreme temperatures, or handling.

8. Light in weight and easily adaptable to any mounting position.

9. Circuitry that operates directly from the aircraft power system without inverters.

10. Minimum electrical current requirements when not indicating a fire.

11. Flight deck light that illuminates, indicating the location of the fire, and with an audible alarm system.

12. A separate detector system for each engine.

Fire Detection/Overheat Systems

A fire detection system should signal the presence of a fire. Units of the system are installed in locations where there are greater possibilities of a fire. Three detector system types in common use are the thermal switch, thermocouple, and the continuous loop.

Thermal Switch System

A number of detectors, or sensing devices, are available. Many older-model aircraft still operating have some type of thermal switch system or thermocouple system. A thermal switch system has one or more lights energized by the aircraft power system and thermal switches that control operation of the light(s). These thermal switches are heat-sensitive units that complete electrical circuits at a certain temperature. They are connected in parallel with each other but in series with the indicator lights. *[Figure 17-1]* If the temperature rises above a set value in any one section of the circuit, the thermal switch closes, completing the light circuit to indicate a fire or overheat condition. No set number of thermal switches is required; the exact number is usually determined by the aircraft manufacturer. On some installations, all the thermal detectors are connected to one light; on others, there may be one thermal switch for each indicator light.

Some warning lights are push-to-test lights. The bulb is tested by pushing it in to check an auxiliary test circuit. The circuit shown in *Figure 17-1* includes a test relay. With the relay contact in the position shown, there are two possible paths for current flow from the switches to the light. This is an additional safety feature. Energizing the test relay completes a series circuit and checks all the wiring and the light bulb. Also included in the circuit shown in *Figure 17-1* is a dimming relay. By energizing the dimming relay, the circuit is altered to

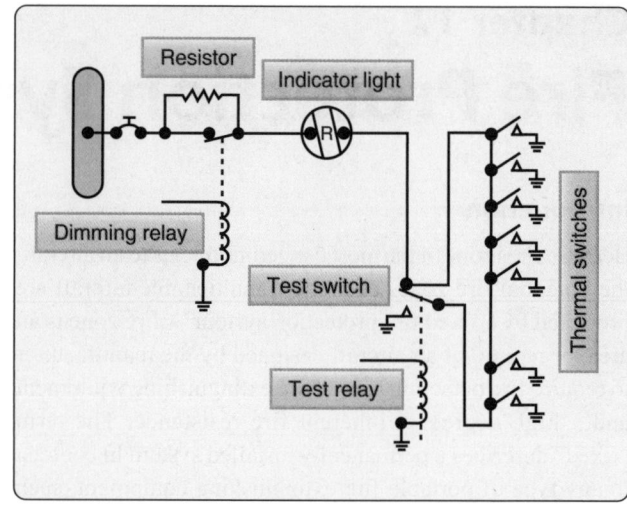

Figure 17-1. *Thermal switch fire circuit.*

include a resistor in series with the light. In some installations, several circuits are wired through the dimming relay, and all the warning lights may be dimmed at the same time.

Thermocouple System

The thermocouple fire warning system operates on an entirely different principle from the thermal switch system. A thermocouple depends on the rate of temperature rise and does not give a warning when an engine slowly overheats or a short circuit develops. The system consists of a relay box, warning lights, and thermocouples. The wiring system of these units may be divided into the following circuits:

1. Detector circuit.

2. Alarm circuit.

3. Test circuit.

These circuits are shown in *Figure 17-2*. The relay box contains two relays, the sensitive relay and the slave relay, and the thermal test unit. Such a box may contain from one to

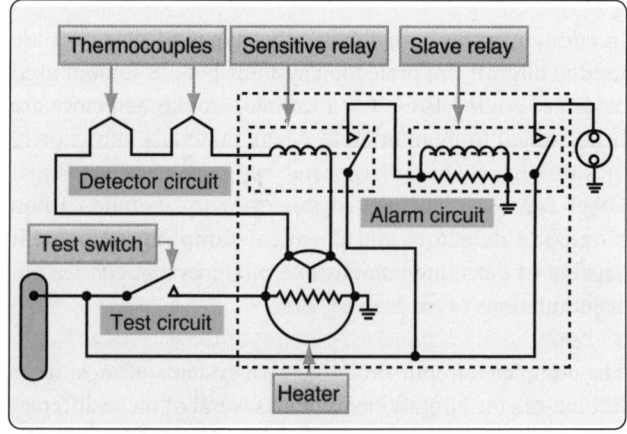

Figure 17-2. *Thermocouple fire warning circuit.*

eight identical circuits, depending on the number of potential fire zones. The relays control the warning lights. In turn, the thermocouples control the operation of the relays. The circuit consists of several thermocouples in series with each other and with the sensitive relay.

The thermocouple is constructed of two dissimilar metals, such as chromel and constantan. The point at which these metals are joined and exposed to the heat of a fire is called a hot junction. There is also a reference junction enclosed in a dead air space between two insulation blocks. A metal cage surrounds the thermocouple to give mechanical protection without hindering the free movement of air to the hot junction. If the temperature rises rapidly, the thermocouple produces a voltage because of the temperature difference between the reference junction and the hot junction. If both junctions are heated at the same rate, no voltage results. In the engine compartment, there is a normal, gradual rise in temperature from engine operation; because it is gradual, both junctions heat at the same rate and no warning signal is given. If there is a fire, however, the hot junction heats more rapidly than the reference junction. The ensuing voltage causes a current to flow within the detector circuit. Any time the current is greater than 4 milliamperes (0.004 ampere), the sensitive relay closes. This completes a circuit from the aircraft power system to the coil of the slave relay. The slave relay then closes and completes the circuit to the warning light to give a visual fire warning.

The total number of thermocouples used in individual detector circuits depends on the size of the fire zones and the total circuit resistance, which usually does not exceed 5 ohms. As shown in *Figure 17-2,* the circuit has two resistors. The resistor connected across the slave relay terminals absorbs the coil's self-induced voltage to prevent arcing across the points of the sensitive relay. The contacts of the sensitive relay are so fragile that they burn, or weld, if arcing is permitted.

When the sensitive relay opens, the circuit to the slave relay is interrupted and the magnetic field around its coil collapses. The coil then gets a voltage through self-induction but, with the resistor across the coil terminals, there is a path for any current flow as a result of this voltage, eliminating arcing at the sensitive relay contacts.

Continuous-Loop Systems

Transport aircraft almost exclusively use continuous thermal sensing elements for powerplant and wheel well protection. These systems offer superior detection performance and coverage, and they have the proven ruggedness to survive in the harsh environment of modern turbofan engines.

A continuous-loop detector or sensing system permits more complete coverage of a fire hazard area than any of the spot-type temperature detectors. Two widely used types of continuous-loop systems are the thermistor type detectors, such as the Kidde and the Fenwal systems, and the pneumatic pressure detector, such as the Lingberg system. (Lindberg system is also known as Systron-Donner and, more recently, Meggitt Safety Systems.)

Fenwal System

The Fenwal system uses a slender Inconel tube packed with thermally sensitive eutectic salt and a nickel wire center conductor. *[Figure 17-3]* Lengths of these sensing elements are connected in series to a control unit. The elements may be of equal or varying length and of the same or different temperature settings. The control unit, operating directly from the power source, impresses a small voltage on the sensing elements. When an overheat condition occurs at any point along the element length, the resistance of the eutectic salt within the sensing element drops sharply, causing current to flow between the outer sheath and the center conductor. This current flow is sensed by the control unit, which produces a signal to actuate the output relay and activate the alarms. When the fire has been extinguished or the critical temperature lowered below the set point, the Fenwal system automatically returns to standby alert, ready to detect any subsequent fire or overheat condition. The Fenwal system may be wired to employ a loop circuit. In this case, should an open circuit occur, the system still signals fire or overheat. If multiple open circuits occur, only that section between breaks becomes inoperative.

Kidde System

In the Kidde continuous-loop system, two wires are embedded in an Inconel tube filled with a thermistor core material. *[Figure 17-4]* Two electrical conductors go through the length of the core. One conductor has a ground connection to the tube, and the other conductor connects to

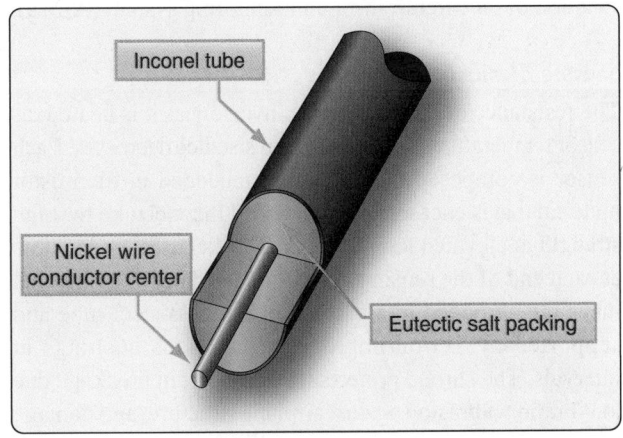

Figure 17-3. *Fenwal sensing element.*

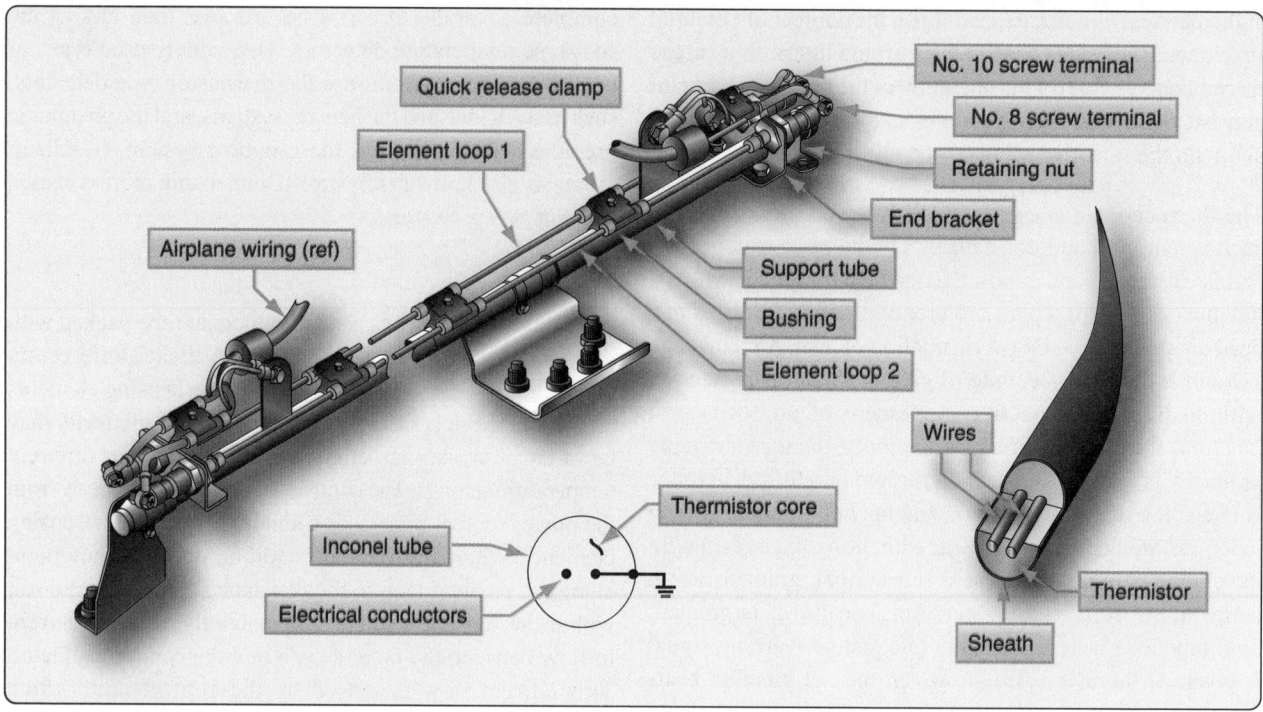

Figure 17-4. *Kidde continuous-loop system.*

the fire detection control unit. As the temperature of the core increases, electrical resistance to the ground decreases. The fire detection control unit monitors this resistance. If the resistance decreases to the overheat set point, an overheat indication occurs in the flight deck. Typically, a 10-second time delay is incorporated for the overheat indication. If the resistance decreases more to the fire set point, a fire warning occurs. When the fire or overheat condition is gone, the resistance of the core material increases to the reset point and the flight deck indications disappear. The rate of change of resistance identifies an electrical short or a fire. The resistance decreases more quickly with an electrical short than with a fire. In some aircraft, in addition to fire and overheat detection, the Kidde continuous-loop system can supply nacelle temperature data to the airplane condition monitoring function of the aircraft in-flight monitoring system (AIMS).

Sensing Element

The resistance of a sensor varies inversely as it is heated; as sensor temperature is increased, its resistance decreases. Each sensor is composed of two wires embedded in thermistor material that is encased in a heavy wall Inconel tube for high strength at elevated temperatures. The electrical connectors at each end of the sensor are ceramic insulated. The Inconel tubes are shrouded in a perforated stainless-steel tube and supported by Teflon-impregnated asbestos bushings at intervals. The shroud protects the sensor from breakage due to vibration, abrasion against airplane structure, and damage from maintenance activity.

The resistance of a sensor also varies inversely with its length, the increments of length being resistances in parallel. The heating of a short length of sensor out of a given length requires that the short length be heated above the temperature alarm point, so the total resistance of the sensor decreases to the alarm point. This characteristic permits integration of all temperatures throughout the length of the installation rather than sensing only the highest local temperature. The two wires encased within the thermistic material of each Inconel tube form a variable resistance network between themselves, between the detector wire and the Inconel tube, and between each adjacent incremental length of sensor. These variable resistance networks are monitored by the application of 28 volts direct current (DC) to the detector wire from the detector control unit.

Combination Fire & Overheat Warning

The analog signal from the thermistor-sensing element permits the control circuits to be arranged to give a two-level response from the same sensing element loop. The first is an overheat warning at a temperature level below the fire warning indicating a general engine compartment temperature rise, such as would be caused by leakage of hot bleed air or combustion gas into the engine compartment. It could also be an early warning of fire and would alert the crew to appropriate action to reduce the engine compartment temperature. The second-level response is at a level above that attainable by a leaking hot gas and is the fire warning.

Temperature Trend Indication

The analog signal produced by the sensing element loop as its temperature changes is converted to signals suitable for flight deck display to indicate engine bay temperature increases from normal. A comparison of the readings from each loop system also provides a check on the condition of the fire detection system, because the two loops should normally read alike.

System Test

The integrity of the continuous-loop fire detection system may be tested by actuating a test switch in the flight deck that switches one end of the sensing element loop from its control circuit to a test circuit built into the control unit, which simulates the sensing element resistance change due to fire. *[Figure 17-5]* If the sensing element loop is unbroken, the resistance detected by the control circuit is that of the simulated fire, and the alarm is activated. The test demonstrates, in addition to the continuity of the sensing element loop, the integrity of the alarm indicator circuit and the proper functioning of the control circuits. The thermistic properties of the sensing element remain unchanged for the life of the element (no irreversible changes take place when heated); the element functions properly as long as it is electrically connected to the control unit.

Fault Indication

Provision is made in the control unit to output a fault signal which activates a fault indicator whenever the short discriminator circuit detects a short in the sensing element loop. This is a requirement for transport category aircraft because such a short disables the fire detection system.

Dual-Loop Systems

Dual-loop systems are two complete basic fire detection systems with their output signals connected so that both must signal to result in a fire warning. This arrangement, called AND logic, results in greatly increased reliability

against false fire warnings from any cause. Should one of the two loops be found inoperative at the preflight integrity test, a flight deck selector switch disconnects that loop and allows the signal from the other loop alone to activate the fire warning. Since the single operative loop meets all fire detector requirements, the aircraft can be safely dispatched and maintenance deferred to a more convenient time. However, should one of the two loops become inoperative in flight and a fire subsequently occur, the fire signaling loop activates a flight deck fault signal that alerts the flight crew to select single-loop operation to confirm the possible occurrence of fire.

Automatic Self-Interrogation

Dual-loop systems automatically perform the loop switching and decision-making function required of the flight crew upon appearance of the fault indication in the flight deck, a function called automatic self-interrogation. Automatic self-interrogation eliminates the fault indication and assures the immediate appearance of the fire indication should fire occur while at least one loop of the dual-loop system is operative. Should the control circuit from a single-loop signal fire, the self-interrogation circuit automatically tests the functioning of the other loop. If it tests operative, the circuit suppresses the fire signal because the operative loop would have signaled if a fire existed. If, however, the other loop tests inoperative, the circuit outputs a fire signal. The interrogation and decision takes place in milliseconds, so that no delay occurs if a fire actually exists.

Support Tube Mounted Sensing Elements

For those installations where it is desired to mount the sensing elements on the engine, and in some cases, on the aircraft structure, the support tube mounted element solves the problem of providing sufficient element support points and greatly facilitates the removal and reinstallation of the sensing elements for engine or system maintenance.

Most modern installations use the support tube concept of mounting sensing elements for better maintainability, as well as increased reliability. The sensing element is attached to a prebent stainless steel tube by closely spaced clamps and bushings, where it is supported from vibration damage and protected from pinching and excessive bending. The support tube-mounted elements can be furnished with either single or dual sensing elements.

Being prebent to the designed configuration assures its installation in the aircraft precisely in its designed location, where it has the necessary clearance to be free from the possibility of the elements chafing against engine or aircraft structure. The assembly requires only a few attachment points and, should its removal for engine maintenance be

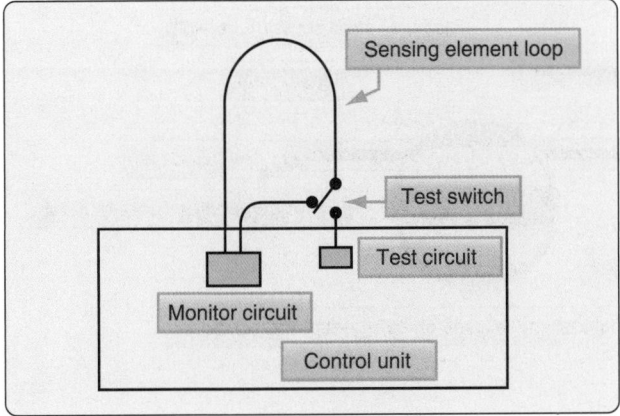

Figure 17-5. *Continuous loop fire detection system test circuit.*

necessary, it is quickly and easily accomplished. Should the assembly require repair or maintenance, it is easily replaced with another assembly, leaving the repair for the shop. Should a sensing element be damaged, it is easily replaced in the assembly.

Fire Detection Control Unit (Fire Detection Card)

The control unit for the simplest type of system typically contains the necessary electronic resistance monitoring and alarm output circuits housed in a hermetically sealed aluminum case fitted with a mounting bracket and electrical connector. For more sophisticated systems, control modules are employed that contain removable control cards with circuitry for individual hazard areas and/or unique functions. In the most advanced applications, the detection system circuitry controls all aircraft fire protection functions, including fire detection and extinguishing for engines, APUs, cargo bays, and bleed-air systems.

Pressure Type Sensor Responder Systems

Some smaller turboprop aircraft are outfitted with pneumatic single point detectors. The design of these detectors is based on the principles of gas laws. The sensing element consists

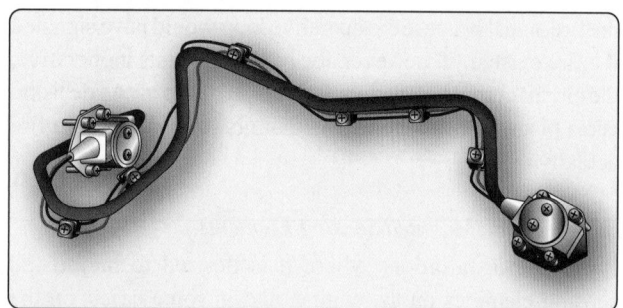

Figure 17-6. *Pneumatic dual fire/overheat detector assembly.*

of a closed, helium-filled tube connected at one end to a responder assembly. As the element is heated, the gas pressure inside the tube increases until the alarm threshold is reached. At this point, an internal switch closes and reports an alarm to the flight deck. Continuous fault monitoring is included. This type of sensor is designed as a single-sensor detection system and does not require a control unit.

Pneumatic Continuous-Loop Systems

The pneumatic continuous-loop systems are also known by their manufacturers' names Lindberg, Systron-Donner, and Meggitt Safety Systems. These systems are used for engine fire detection of transport type aircraft and have the same function as the Kidde system; however, they work on a different principle. They are typically used in a dual-loop design to increase reliability of the system.

The pneumatic detector has two sensing functions. It responds to an overall average temperature threshold and to a localized discrete temperature increase caused by impinging flame or hot gasses. Both the average and discrete temperature are factory set and are not field adjustable. *[Figure 17-6]*

Averaging Function

The fire/overheat detector serves as a fixed-volume device filled with helium gas. The helium gas pressure inside the detector increases in proportion to the absolute temperature and operates a pressure diaphragm that closes an electrical contact, actuating the alarm circuit. The pressure diaphragm within the responder assembly serves as one side of the electrical alarm contact and is the only moving part in the detector. The alarm switch is preset at an average temperature. Typical temperature ranges for average temperature settings are 200 °F (93 °C) to 850 °F (454 °C).

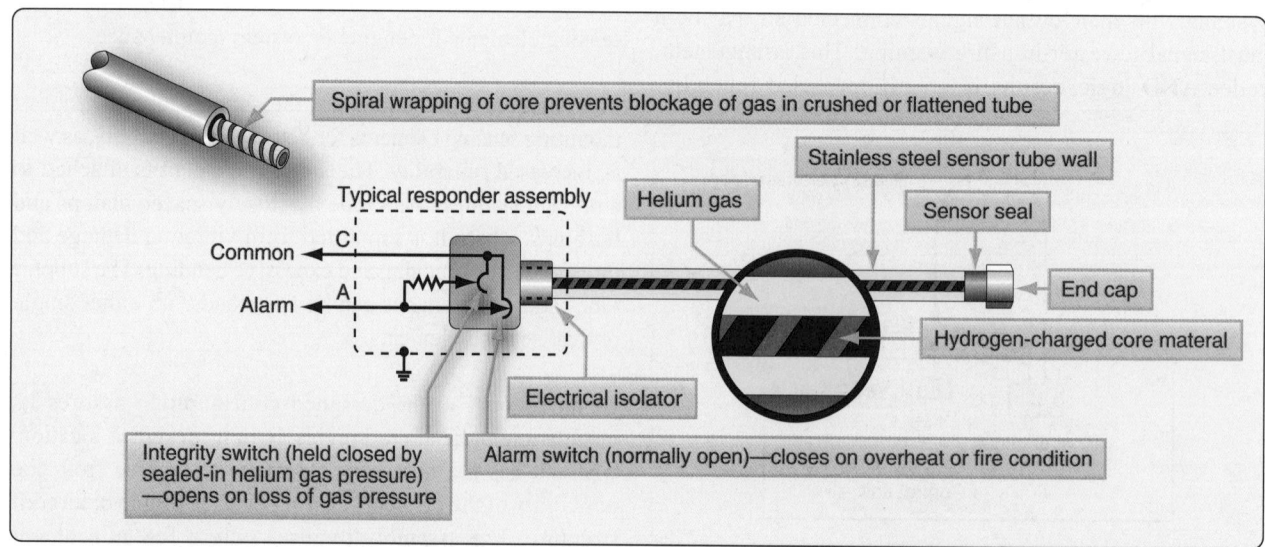

Figure 17-7. *Pneumatic pressure loop detector system.*

Discrete Function

The fire/overheat detector's sensor tube also contains a hydrogen-filled core material. *[Figure 17-7]* Large quantities of hydrogen gas are released from the detector core whenever a small section of the tube is heated to the preset discrete temperature or higher. The core outgassing increases the pressure inside the detector and actuates the alarm switch. Both the averaging and discrete functions are reversible. When the sensor tube is cooled, the average gas pressure is lowered, and the discrete hydrogen gas returns to the core material. The reduction of internal pressure allows the alarm switch to return to its normal position, opening the electrical alarm circuit.

Figure 17-8 shows a typical aircraft fire detection system in which a control module monitors two loops of up to four pneumatic detectors each, connected in parallel. The control module responds directly to an alarm condition and continuously monitors the wiring and integrity of each loop. The normally open alarm switch closes upon an overheat or fire condition, causing a short circuit between terminals A and C. During normal operation, a resistance value is maintained across the terminals by a normally closed integrity switch. Loss of sensor gas pressure opens the integrity switch, creating an open circuit across the terminals of the faulted detector. In addition to the pressure-activated alarm switch, there is a second integrity switch in the detector that is held closed by the averaging gas pressure at all temperatures down to −65 °F (−54 °C). If the detector should develop a leak, the loss of gas pressure would allow the integrity switch to open and signal a lack of detector integrity. The system then does not operate during test.

Fire Zones

Powerplant compartments are classified into zones based on the airflow through them.

1. Class A zone—area of heavy airflow past regular arrangements of similarly shaped obstructions. The power section of a reciprocating engine is usually of this type.

2. Class B zone—area of heavy airflow past aerodynamically clean obstructions. Included in this type are heat exchanger ducts, exhaust manifold shrouds, and areas where the inside of the enclosing cowling or other closure is smooth, free of pockets, and adequately drained so leaking flammables cannot puddle. Turbine engine compartments may be considered in this class if engine surfaces are aerodynamically clean and all airframe structural formers are covered by a fireproof liner to produce an aerodynamically clean enclosure surface.

3. Class C zone—area of relatively low airflow. An engine accessory compartment separated from the power section is an example of this type of zone.

4. Class D zone—area of very little or no airflow. These include wing compartments and wheel wells where little ventilation is provided.

5. Class X zone—area of heavy airflow and of unusual construction, making uniform distribution of the extinguishing agent very difficult. Areas containing deeply recessed spaces and pockets between large structural formers are of this type. Tests indicate agent requirements to be double those for Class A zones.

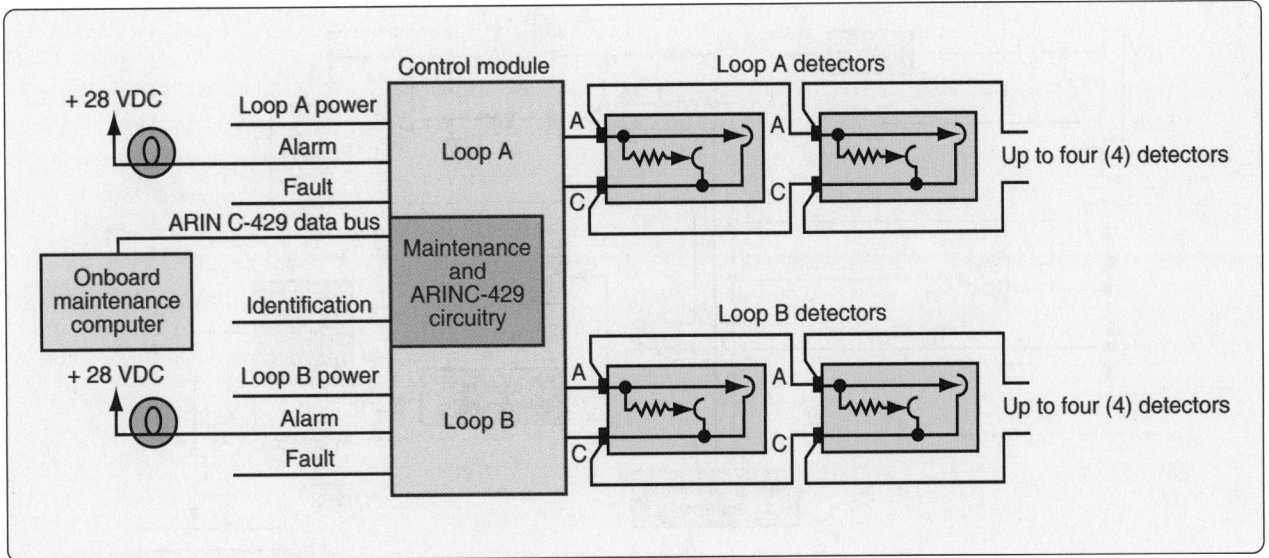

Figure 17-8. *Aircraft detection system control module.*

Smoke, Flame, & Carbon Monoxide Detection Systems

Smoke Detectors

A smoke detection system monitors the lavatories and cargo baggage compartments for the presence of smoke, which is indicative of a fire condition. Smoke detection instruments that collect air for sampling are mounted in the compartments in strategic locations. A smoke detection system is used where the type of fire anticipated is expected to generate a substantial amount of smoke before temperature changes are sufficient to actuate a heat detection system. Two common types used are light refraction and ionization.

Light Refraction Type

The light refraction type of smoke detector contains a photoelectric cell that detects light refracted by smoke particles. Smoke particles refract the light to the photoelectric cell and, when it senses enough change in the amount of light, it creates an electrical current that sets off a warning light. This type of smoke detector is referred to as a photoelectrical device.

Ionization Type

Some aircraft use an ionization type smoke detector. The system generates an alarm signal (both horn and indicator) by detecting a change in ion density due to smoke in the cabin. The system is connected to the 28-volt DC electrical power supplied from the aircraft. Alarm output and sensor sensitive checks are performed simply with the test switch on the control panel.

Flame Detectors

Optical sensors, often referred to as flame detectors, are designed to alarm when they detect the presence of prominent, specific radiation emissions from hydrocarbon flames. The two types of optical sensors available are infrared (IR) and ultraviolet (UV), based on the specific emission wavelengths that they are designed to detect. IR-based optical flame detectors are used primarily on light turboprop aircraft and helicopter engines. These sensors have proven to be very dependable and economical for these applications.

When radiation emitted by the fire crosses the airspace between the fire and the detector, it impinges on the detector front face and window. The window allows a broad spectrum of radiation to pass into the detector where it strikes the sensing device filter. The filter allows only radiation in a tight waveband centered on 4.3 micrometers in the IR band to pass on to the radiation-sensitive surface of the sensing device. The radiation striking the sensing device minutely raises its temperature causing small thermoelectric voltages to be generated. These voltages are fed to an amplifier whose output is connected to various analytical electronic processing circuits. The processing electronics are tailored exactly to the time signature of all known hydrocarbon flame sources and ignores false alarm sources, such as incandescent lights and sunlight. Alarm sensitivity level is accurately controlled by a digital circuit. *[Figure 17-9]*

Carbon Monoxide Detectors

Carbon monoxide is a colorless, odorless gas that is a byproduct of incomplete combustion. Its presence in the

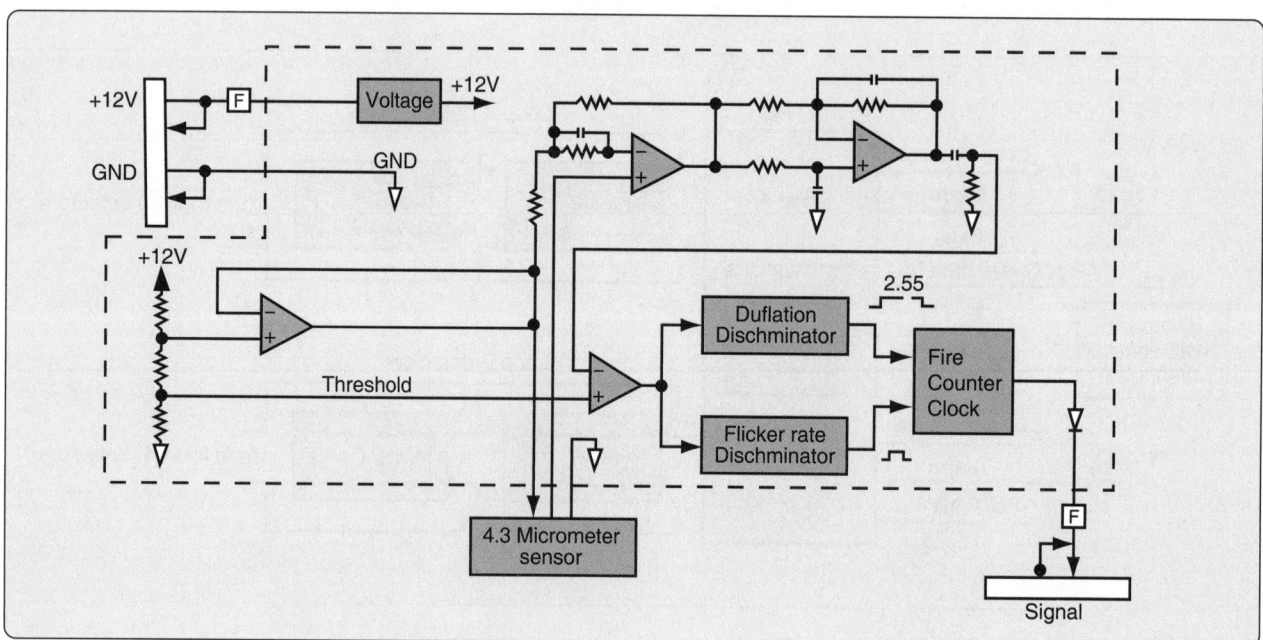

Figure 17-9. *Infrared (IR) based optical flame detector.*

breathing air of human beings can be deadly. To ensure crew and passenger safety, carbon monoxide detectors are used in aircraft cabins and flight decks. They are most often found on reciprocating engine aircraft with exhaust shroud heaters and on aircraft equipped with a combustion heater. Turbine bleed air, when used for heating the cabin, is tapped off of the engine upstream of the combustion chamber. Therefore, no threat of carbon monoxide presence is posed.

Carbon monoxide gas is found in varying degrees in all smoke and fumes of burning carbonaceous substances. Exceedingly small amounts of the gas are dangerous if inhaled. A concentration of as little as 2 parts in 10,000 may produce headache, mental dullness, and physical lethargy within a few hours. Prolonged exposure or higher concentrations may cause death.

There are several types of carbon monoxide detectors. Electronic detectors are common. Some are panel mounted and others are portable. Chemical color-change types are also common. These are mostly portable. Some are simple buttons, cards, or badges that have a chemical applied to the surface. Normally, the color of the chemical is tan. In the presence of carbon monoxide, the chemical darkens to grey or even black. The transition time required to change color is inversely related to the concentration of CO present. At 50 parts per million, the indication is apparent within 15 to 30 minutes. A concentration of 100 parts per million changes the color of the chemical in as little as 2–5 minutes. As concentration increases or duration of exposure is prolonged, the color evolves from grey to dark grey to black. If contaminated, installing a new indicating element allows a carbon monoxide portable test unit to be returned to service.

Extinguishing Agents & Portable Fire Extinguishers

There must be at least one hand held, portable fire extinguisher for use in the pilot compartment that is located within easy access of the pilot while seated. There must be at least one hand held fire extinguisher located conveniently in the passenger compartment of each airplane accommodating more than 6 and less than 30 passengers. Each extinguisher for use in a personnel compartment must be designed to minimize the hazard of toxic gas concentrations. The number of portable, hand held fire extinguishers for transport aircraft is shown in *Figure 17-10*.

Halogenated Hydrocarbons

For over 45 years, halogenated hydrocarbons (Halons) have been practically the only fire extinguishing agents used in civil transport aircraft. However, Halon is an ozone depleting and global warming chemical, and its production has been banned by international agreement. Although Halon usage has been

Passenger capacity	No. of extinguishers
7 through 30	1
31 through 60	2
61 through 200	3
201 through 300	4
301 through 400	5
401 through 500	6
501 through 600	7
601 through 700	8

Figure 17-10. *Hand held fire extinguisher requirement for transport aircraft.*

banned in some parts of the world, aviation has been granted an exemption because of its unique operational and fire safety requirements. Halon has been the fire extinguishing agent of choice in civil aviation because it is extremely effective on a per unit weight basis over a wide range of aircraft environmental conditions. It is a clean agent (no residue), electrically nonconducting, and has relatively low toxicity.

Two types of Halons are employed in aviation: Halon $1301(CBrF_3)$ a total flooding agent, and Halon 1211 $(CBrClF_2)$ a streaming agent. Class A, B, or C fires are appropriately controlled with Halons. However, do not use Halons on a class D fire. Halon agents may react vigorously with the burning metal.

Note: While Halons are still in service and are appropriate agents for these classes of fires, the production of these ozone depleting agents has been restricted. Although not required, consider replacing Halon extinguishers with Halon replacement extinguishers when discharged. Halon replacement agents found to be compliant to date include the halocarbons HCFC Blend B, HFC-227ea, and HFC-236fa.

Inert Cold Gases

Carbon dioxide (CO_2) is an effective extinguishing agent. It is most often used in fire extinguishers that are available on the ramp to fight fires on the exterior of the aircraft, such as engine or APU fires. CO_2 has been used for many years to extinguish flammable fluid fires and fires involving electrical equipment. It is noncombustible and does not react with most substances. It provides its own pressure for discharge from the storage vessel, except in extremely cold climates where a booster charge of nitrogen may be added to winterize the system. Normally, CO_2 is a gas, but it is easily liquefied by compression and cooling. After liquification, CO_2 remains in a closed container as both liquid and gas. When CO_2 is then discharged to the atmosphere, most of the liquid expands to gas. Heat absorbed by the gas during vaporization cools the remaining liquid to –110 °F, and it becomes a finely divided white solid, dry ice snow.

Carbon dioxide is about 1½ times as heavy as air, which gives it the ability to replace air above burning surfaces and maintain a smothering atmosphere. CO_2 is effective as an extinguishing agent primarily because it dilutes the air and reduces the oxygen content so that combustion is no longer supported. Under certain conditions, some cooling effect is also realized. CO_2 is considered only mildly toxic, but it can cause unconsciousness and death by suffocation if the victim is allowed to breathe CO_2 in fire extinguishing concentrations for 20 to 30 minutes. CO_2 is not effective as an extinguishing agent on fires involving chemicals containing their own oxygen supply, such as cellulose nitrate (used in some aircraft paints). Also, fires involving magnesium and titanium cannot be extinguished by CO_2.

Dry Powders

Class A, B, or C fires can be controlled by dry chemical extinguishing agents. The only all purpose (Class A, B, C rating) dry chemical powder extinguishers contain mono-ammonium phosphate. All other dry chemical powders have a Class B, C U.S – UL fire rating only. Dry powder chemical extinguishers best control class A, B, and C fire but their use is limited due to residual residue and clean up after deployment.

Water

Class A type fires are best controlled with water by cooling the material below its ignition temperature and soaking the material to prevent re-ignition.

Flight Deck & Cabin Interiors

All materials used in the flight deck and cabin must conform to strict standards to prevent fire. In case of a fire, several types of portable fire extinguishers are available to fight the fire. The most common types are Halon 1211 and water.

Extinguisher Types

Portable fire extinguishers are used to extinguish fires in the cabin or flight deck. *Figure 17-11* shows a Halon fire extinguisher used in a general aviation aircraft. The Halon extinguishers are used on electrical and flammable liquid fires. Some transport aircraft also use water fire extinguisher for use on non-electrical fires.

The following is a list of extinguishing agents and the type (class) fires for which each is appropriate.

1. Water—class A. Water cools the material below its ignition temperature and soaks it to prevent reignition.

2. Carbon dioxide—class B or C. CO_2 acts as a blanketing agent. ***Note:*** CO_2 is not recommended for hand-held extinguishers for internal aircraft use.

3. Dry chemicals—class A, B, or C. Dry chemicals are the best control agents for these types of fires.

4. Halons—only class A, B, or C.

5. Halocarbon clean agents—only class A, B, or C.

6. Specialized dry powder—class D. (Follow the recommendations of the extinguisher's manufacturer because of the possible chemical reaction between the burning metal and the extinguishing agent.)

The following hand-held extinguishers are unsuitable as cabin or flight deck equipment.

- CO_2.

- Dry chemicals (due to the potential for corrosion damage to electronic equipment, the possibility of visual obscuration if the agent were discharged into the flight deck area, and the cleanup problems from their use).

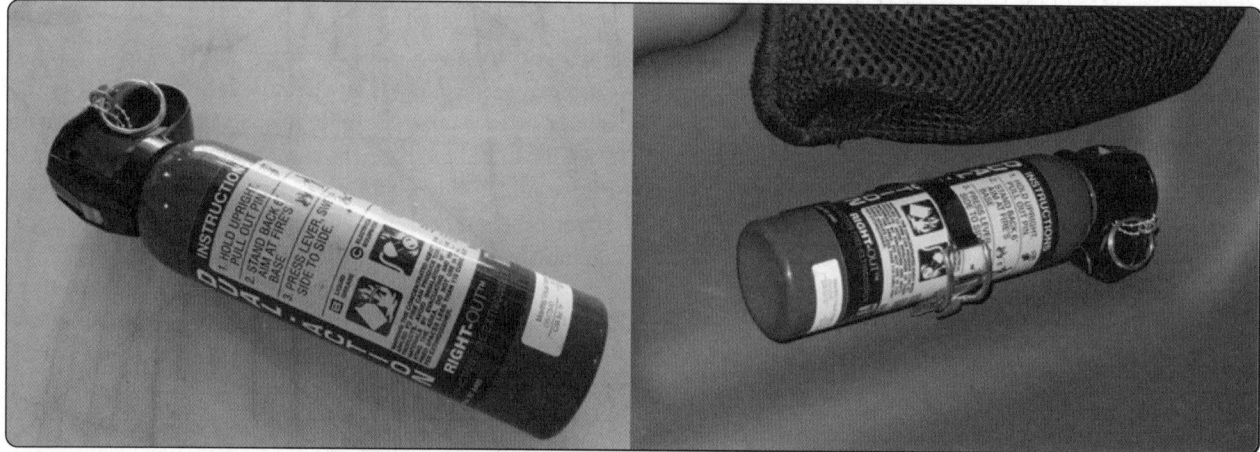

Figure 17-11. *Portable fire extinguisher.*

- Specialized dry powder (it is suitable for use in ground operations).

Installed Fire Extinguishing Systems

Transport aircraft have fixed fire extinguishing systems installed in:

1. Turbine engine compartments.
2. APU compartments.
3. Cargo and baggage compartments.
4. Lavatories.

CO$_2$ Fire Extinguishing Systems

Older aircraft with reciprocating engines used CO_2 as an extinguishing agent, but all newer aircraft designs with turbine engines use Halon or equivalent extinguishing agent, such as halocarbon clean agents.

Halogenated Hydrocarbons Fire Extinguishing Systems

The fixed fire extinguisher systems used in most engine fire and cargo compartment fire protection systems are designed to dilute the atmosphere with an inert agent that does not support combustion. Many systems use perforated tubing or discharge nozzles to distribute the extinguishing agent. High rate of discharge (HRD) systems use open-end tubes to deliver a quantity of extinguishing agent in 1 to 2 seconds. The most common extinguishing agent still used today is Halon 1301 because of its effective firefighting capability and relatively low toxicity (UL classification Group 6). Noncorrosive Halon 1301 does not affect the material it contacts and requires no cleanup when discharged. Halon 1301 is the current extinguishing agent for commercial aircraft, but a replacement is under development. Halon 1301 cannot be produced anymore because it depletes the ozone layer. Halon 1301 will be used until a suitable replacement is developed. Some military aircraft use HCL-125 and the Federal Aviation Administration (FAA) is testing HCL-125 for use in commercial aircraft.

Containers

Fire extinguisher containers (HRD bottles) store a liquid halogenated extinguishing agent and pressurized gas (typically nitrogen). They are normally manufactured from stainless steel. Depending upon design considerations, alternate materials are available, including titanium. Containers are also available in a wide range of capacities. They are produced under Department of Transportation (DOT) specifications or exemptions. Most aircraft containers are spherical in design, which provides the lightest weight possible. However, cylindrical shapes are available where space limitations are a factor. Each container incorporates a temperature/pressure sensitive safety relief diaphragm that prevents container pressure from exceeding container test pressure in the event of exposure to excessive temperatures. *[Figures 17-12 and 17-13]*

Discharge Valves

Discharge valves are installed on the containers. A cartridge (squib) and frangible disc-type valve are installed in the outlet of the discharge valve assembly. Special assemblies having solenoid-operated or manually-operated seat-type valves are also available. Two types of cartridge disc-release techniques are used. Standard release-type uses a slug driven by explosive energy to rupture a segmented closure disc. For high temperature or hermetically sealed units, a direct explosive impact-type cartridge is used that applies fragmentation impact to rupture a prestressed corrosion resistant steel diaphragm. Most containers use conventional metallic gasket seals that facilitate refurbishment following discharge. *[Figure 17-14]*

Pressure Indication

A wide range of diagnostics is utilized to verify the fire extinguisher agent charge status. A simple visual indication gauge is available, typically a helical bourdon-type indicator that is vibration resistant. *[Figure 17-13]* A combination gauge switch visually indicates actual container pressure and also provides an electrical signal if container pressure is lost, precluding the need for discharge indicators. A ground checkable diaphragm-type low-pressure switch is commonly used on hermetically sealed containers. The Kidde system has a temperature compensated pressure switch that tracks the container pressure variations with temperatures by using a hermetically sealed reference chamber.

Figure 17-12. *Built-in non-portable fire extinguisher containers (HRD bottles) on an airliner.*

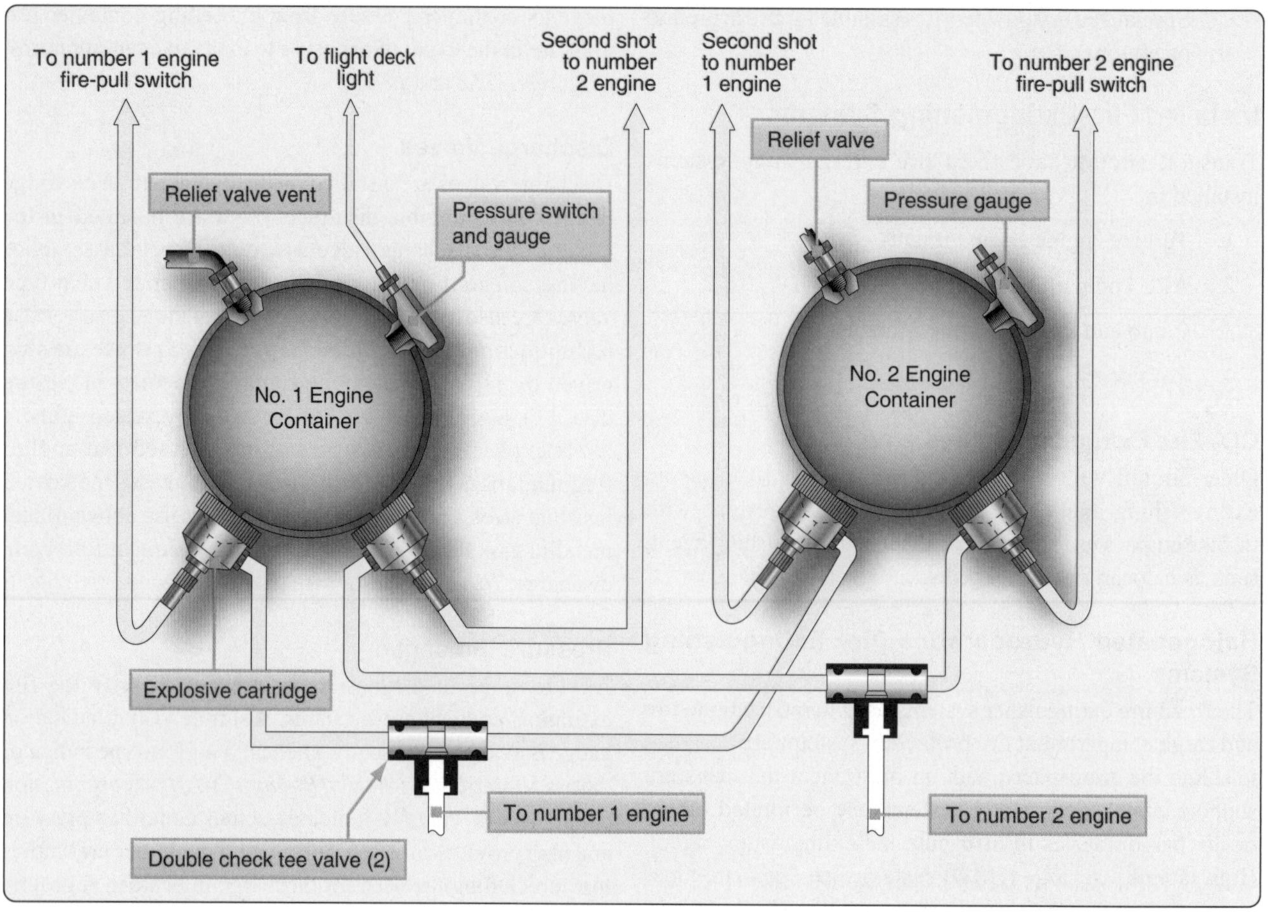

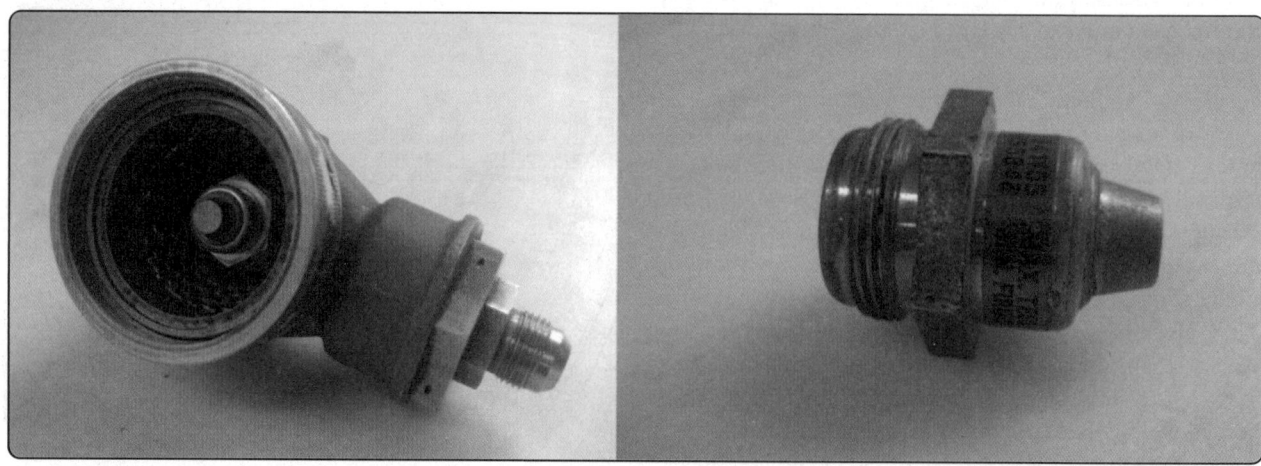

Figure 17-13. *Diagram of fire extinguisher containers (HRD bottles).*

Two-Way Check Valve

Two-way check valves are required in a two-shot system to prevent the extinguisher agent from a reserve container from backing up into the previous emptied main container. Valves are supplied with either MS-33514 or MS-33656 fitting configurations.

Discharge Indicators

Discharge indicators provide immediate visual evidence of container discharge on fire extinguishing systems. Two kinds of indicators can be furnished: thermal and discharge. Both types are designed for aircraft and skin mounting. *[Figure 17-15]*

Figure 17-14. *Discharge valve (left) and cartridge, or squib (right).*

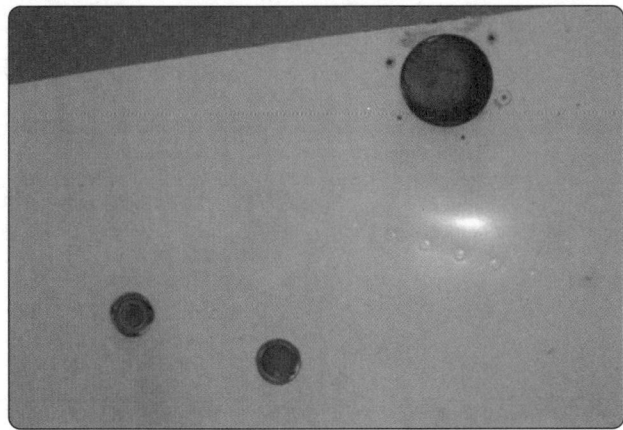

Figure 17-15. *Discharge indicators.*

Thermal Discharge Indicator (Red Disc)

The thermal discharge indicator is connected to the fire container relief fitting and ejects a red disc to show when container contents have dumped overboard due to excessive heat. The agent discharges through the opening left when the disc blows out. This gives the flight and maintenance crews an indication that the fire extinguisher container needs to be replaced before next flight.

Yellow Disc Discharge Indicator

If the flight crew activates the fire extinguisher system, a yellow disc is ejected from the skin of the aircraft fuselage. This is an indication for the maintenance crew that the fire extinguishing system was activated by the flight crew, and the fire extinguishing container needs to be replaced before next flight.

Fire Switch

The engine and APU fire switches are typically installed on the center overhead panel or center console in the flight deck. *[Figure 17-16]* When an engine fire switch is activated, the following happens: the engine stops because the fuel control shuts off, the engine is isolated from the aircraft systems, and the fire extinguishing system is activated. Some aircraft use fire switches that need to be pulled and turned to activate the system, while others use a push-type switch with a guard. To prevent accidental activation of the fire switch, a lock is installed that releases the fire switch only when a fire has been detected. This lock can be manually released by the flight crew if the fire detection system malfunctions. *[Figure 17-17]*

Cargo Fire Detection

Transport aircraft need to have the following provisions for each cargo or baggage compartment:

1. The detection system must provide a visual indication to the flight crew within 1 minute after the start of a fire.

Figure 17-16. *Engine and APU fire switches on the flight deck center overhead panel.*

2. The system must be capable of detecting a fire at a temperature significantly below that at which the structural integrity of the airplane is substantially decreased.

3. There must be means to allow the crew to check, in flight, the functioning of each fire detector circuit.

Cargo Compartment Classification
Class A

A Class A cargo or baggage compartment is one in which the presence of a fire would be easily discovered by a crewmember while at his or her station and each part of the compartment is easily accessible in flight.

Class B

A Class B cargo, or baggage compartment, is one in which there is sufficient access in flight to enable a crewmember to effectively reach any part of the compartment with the contents of a hand fire extinguisher. When the access provisions are being used, no hazardous quantity of smoke, flames, or extinguishing agent enters any compartment occupied by the crew or passengers. There is a separate approved smoke detector or fire detector system to give warning at the pilot or flight engineer station.

Class C

A Class C cargo, or baggage compartment, is one not meeting the requirements for either a Class A or B compartment but in which:

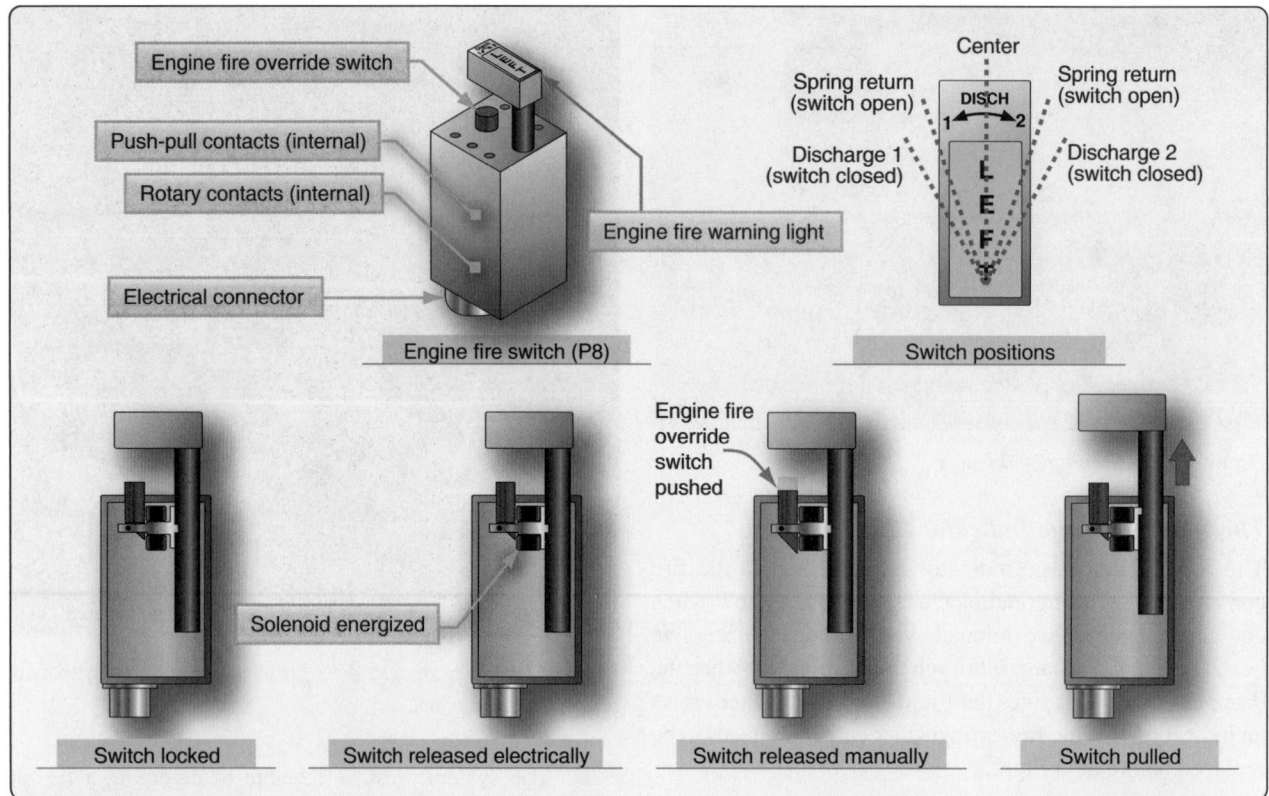

Figure 17-17. *Engine fire switch operation.*

1. There is a separate approved smoke detector or fire detector system to give warning at the pilot or flight engineer station.

2. There is an approved built-in fire extinguishing or suppression system controllable from the flight deck.

3. There are means to exclude hazardous quantities of smoke, flames, or extinguishing agent from any compartment occupied by the crew or passengers.

4. There are means to control ventilation and drafts within the compartment so that the extinguishing agent used can control any fire that may start within the compartment.

Class E

Class E cargo compartment is one on airplanes used only for the carriage of cargo and in which:

1. There is a separate approved smoke or fire detector system to give warning at the pilot or flight engineer station.

2. The controls for shutting off the ventilating airflow to, or within, the compartment are accessible to the flight crew in the crew compartment.

3. There are means to exclude hazardous quantities of smoke, flames, or noxious gases from the flight crew compartment.

4. The required crew emergency exits are accessible under any cargo loading condition.

Cargo & Baggage Compartment Fire Detection & Extinguisher System

The cargo compartment smoke detection system gives warnings in the flight deck if there is smoke in a cargo compartment. *[Figure 17-18]* Each compartment is equipped with a smoke detector. The smoke detectors monitor air in the cargo compartments for smoke. The fans bring air from the cargo compartment into the smoke detector. Before the air goes in the smoke detector, in-line water separators remove condensation and heaters increase the air temperature. *[Figure 17-19]*

Smoke Detector System

The optical smoke detector consists of source light emitting diodes (LEDs), intensity monitor photodiodes, and scatter detector photodiodes. Inside the smoke detection chamber, air flows between a source LED and a scatter detector photodiode. Usually, only a small amount of light from the LED gets to the scatter detector. If the air has smoke in it, the smoke particles reflect more light on the scatter detector. This causes an alarm signal. The intensity monitor photodiode makes sure that the source LED is on and keeps the output of the source LED constant. This configuration also finds

Figure 17-18. *Cargo fire detection warning.*

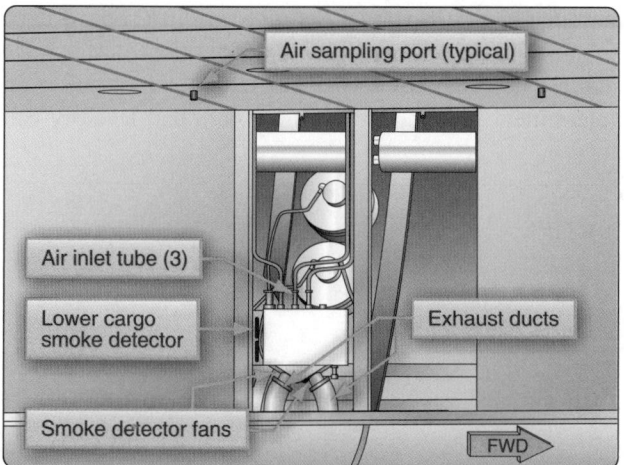

Figure 17-19. *Smoke detector installation.*

Cargo Compartment Extinguishing System

The cargo compartment extinguishing system is activated by the flight crew if the smoke detectors detect smoke in the cargo compartment. Some aircraft are outfitted with two types of fire extinguisher containers. The first system is the dump system that releases the extinguishing agent directly when the cargo fire discharge switch is activated. This action extinguishes the fire.

The second system is the metered system. After a time delay, the metered bottles discharge slowly and at a controlled rate through the filter regulator. Halon from the metered bottles replaces the extinguishing agent leakage. This keeps the correct concentration of extinguishing agent in the cargo compartment to keep the fire extinguished for 180 minutes.

The fire extinguishing bottles contain Halon 1301 or equivalent fire extinguishing agent pressurized with nitrogen. Tubing connects the bottles to discharge nozzles in the cargo compartment ceilings.

The extinguishing bottles are outfitted with squibs. The squib is an electrically operated explosive device. It is adjacent to

contamination of the LED and photodiodes. A defective diode, or contamination, causes the detector to change to the other set of diodes. The detector sends a fault message.

The smoke detector has multiple sampling ports. The fans draw air from the sampling ports through a water separator and a heater unit to the smoke detector. *[Figure 17-20]*

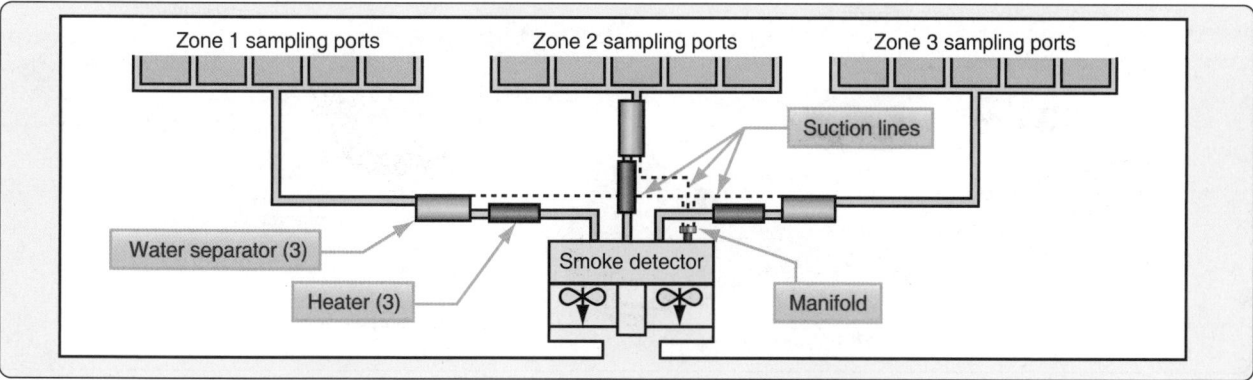

Figure 17-20. *Smoke detector system.*

a bottle diaphragm that can break. The diaphragm normally seals the pressurized bottle. When the cargo discharge switch is activated, the squib fires and the explosion breaks the diaphragm. Nitrogen pressure inside the bottle pushes the Halon through the discharge port into the cargo compartment. When the bottle discharges, a pressure switch is activated that sends an indication to the flight deck that a bottle has been discharged. Flow control valves are incorporated if the bottles can be discharged in multiple compartments. The flow control valves direct the extinguishing agent to the selected cargo compartment. *[Figure 17-21]*

The following indications occur in the flight deck if there is smoke in a cargo compartment:

- Master warning lights come on.
- Fire warning aural operates.
- A cargo fire warning message shows.
- Cargo fire warning light comes on.

The master warning lights and fire warning aural are prevented from operating during part of the takeoff operation.

Lavatory Smoke Detectors

Airplanes that have a passenger capacity of 20 or more are equipped with a smoke detector system that monitors the lavatories for smoke. Smoke indications provide a warning light in the flight deck or provide a warning light or audible warning at the lavatory and at flight attendant stations that would be readily detected by a flight attendant. Each lavatory must have a built-in fire extinguisher that discharges automatically. The smoke detector is located in the ceiling of the lavatory. *[Figure 17-22]*

Lavatory Smoke Detector System

Refer to *Figure 17-23*. The lavatory smoke detector is powered by the 28-volt DC left/right main DC bus. If there is smoke in the sensing chamber of the smoke detector, the alarm LED (red) comes on. The timing circuit makes an intermittent ground. The warning horn and lavatory call

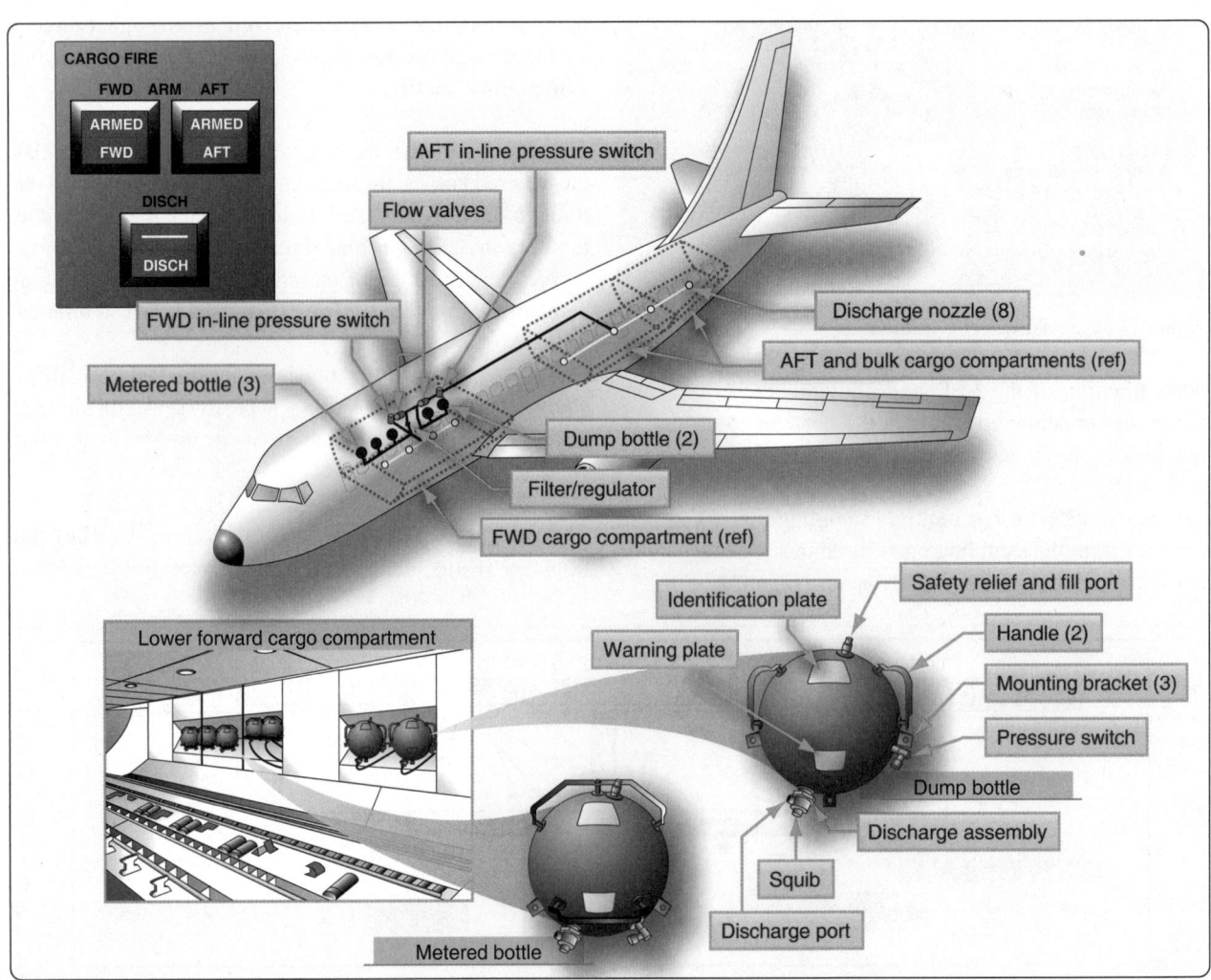

Figure 17-21. *Cargo and baggage compartment extinguishing system.*

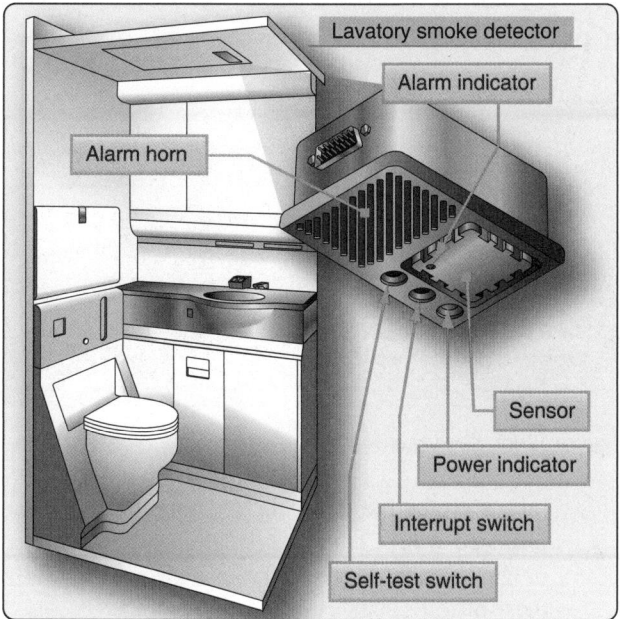

Figure 17-22. *Lavatory smoke detector diagram.*

light operate intermittently. The smoke detection circuit makes a ground for the relay. The energized relay makes a ground signal for the overhead electronics unit (OEU) in the central monitoring systems (CMS). This interface gives these indications: lavatory master call light flashes, cabin system control panel (CSCP) and cabin area control panel (CACP) pop-up window shows, and the lavatory call chime operates. Push the lavatory call reset switch or the smoke detector interrupt switch to cancel the smoke indications. If there is still

smoke in the lavatory, the alarm LED (red) stays on. All smoke indications go away automatically when the smoke is gone.

Lavatory Fire Extinguisher System

The lavatory compartment is outfitted with a fire extinguisher bottle to extinguish fires in the waste compartment. The fire extinguisher is a bottle with two nozzles. The bottle contains pressurized Halon 1301 or equivalent fire extinguishing agent. When the temperature in the waste compartment reaches approximately 170 °F, the solder that seals the nozzles melt and the Halon is discharged. Weighing the bottle is often the only way to determine if the bottle is empty or full. *[Figure 17-24]*

Fire Detection System Maintenance

Fire detector sensing elements are located in many high-activity areas around aircraft engines. Their location, together with their small size, increases the chance of damage to the sensing elements during maintenance. General maintenance of a fire detection system typically includes the inspection and servicing of damaged sections, containment of loose material that could short detector terminals, correcting connection joints and shielding, and replacement of damaged sensing elements. An inspection and maintenance program for all types of continuous-loop systems should include the following visual checks.

Note: These procedures are examples and should not be used to replace the applicable manufacturer's instructions.

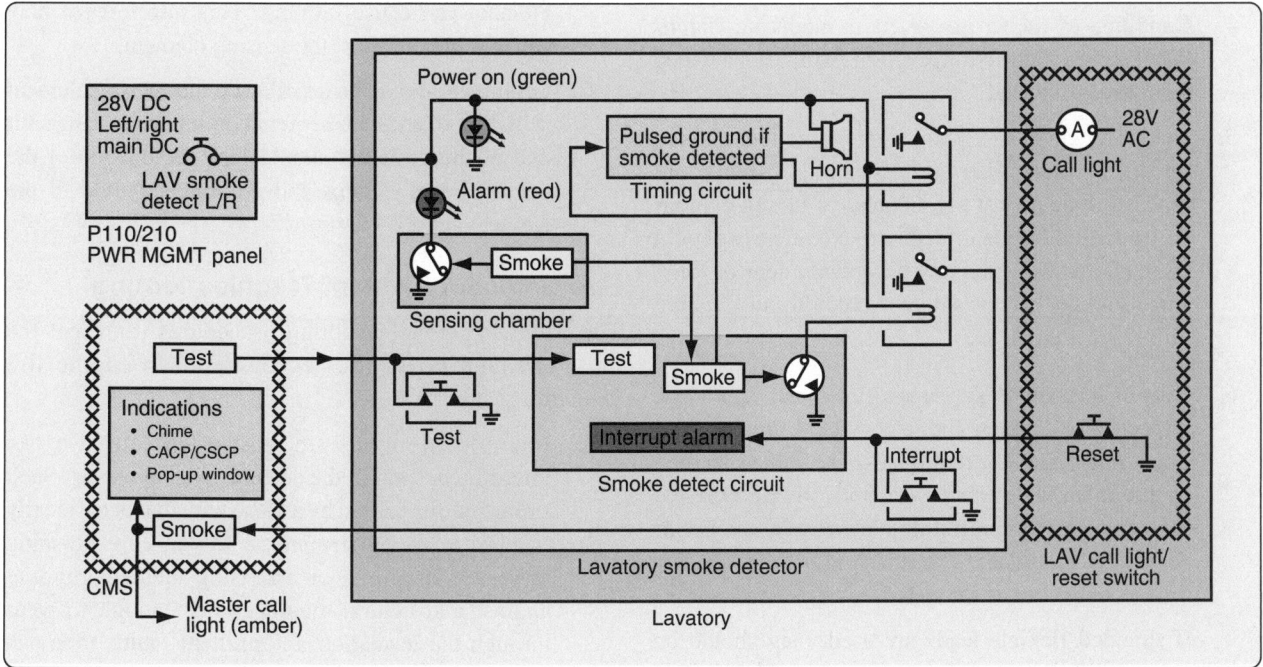

Figure 17-23. *Lavatory smoke detector diagram.*

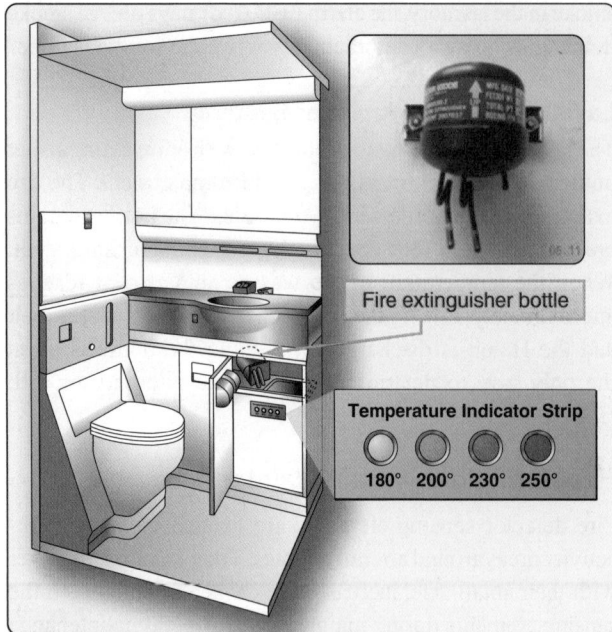

Figure 17-24. *Lavatory fire extinguishing bottle.*

Sensing elements of a continuous-loop system should be inspected for the following:

1. Cracked or broken sections caused by crushing or squeezing between inspection plates, cowl panels, or engine components.

2. Abrasion caused by rubbing of the element on cowling, accessories, or structural members.

3. Pieces of safety wire, or other metal particles, that may short the spot-detector terminals.

4. Condition of rubber grommets in mounting clamps that may be softened from exposure to oils or hardened from excessive heat.

5. Dents and kinks in sensing element sections. Limits on the element diameter, acceptable dents and kinks, and degree of smoothness of tubing contour are specified by manufacturers. No attempt should be made to straighten any acceptable dent or kink, since stresses may be set up that could cause tubing failure. *[Figure 17-25]*

6. Nuts at the end of the sensing elements should be inspected for tightness and safety wire. *[Figure 17-26]* Loose nuts should be retorqued to the value specified by the manufacturer's instructions. Some types of sensing element connection joints require the use of copper crush gaskets. These should be replaced any time a connection is separated.

7. If shielded flexible leads are used, they should be inspected for fraying of the outer braid. The braided sheath is made up of many fine metal strands woven

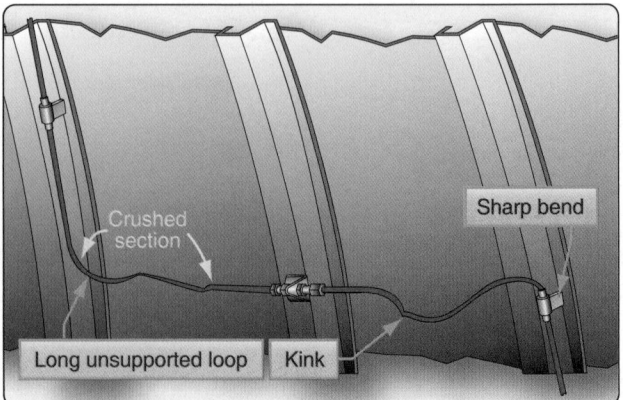

Figure 17-25. *Sensing element defects.*

into a protective covering surrounding the inner insulated wire. Continuous bending of the cable or rough treatment can break these fine wires, especially those near the connectors.

8. Sensing element routing and clamping should be inspected carefully. *[Figure 17-27]* Long, unsupported sections may permit excessive vibration that can cause breakage. The distance between clamps on straight runs, usually about 8 to 10 inches, is specified by each manufacturer. At end connectors, the first support clamp usually is located about 4 to 6 inches from the end connector fittings. In most cases, a straight run of one inch is maintained from all connectors before a bend is started, and an optimum bend radius of 3 inches is normally adhered to.

9. Interference between a cowl brace and a sensing element can cause rubbing. This interference may cause wear and short the sensing element.

10. Grommets should be installed on the sensing element so that both ends are centered on its clamp. The split end of the grommet should face the outside of the nearest bend. Clamps and grommets should fit the element snugly. *[Figure 17-28]*

Fire Detection System Troubleshooting

The following troubleshooting procedures represent the most common difficulties encountered in engine fire detection systems:

1. Intermittent alarms are most often caused by an intermittent short in the detector system wiring. Such shorts may be caused by a loose wire that occasionally touches a nearby terminal, a frayed wire brushing against a structure, or a sensing element rubbing against a structural member long enough to wear through the insulation. Intermittent faults often can be located by moving wires to recreate the short.

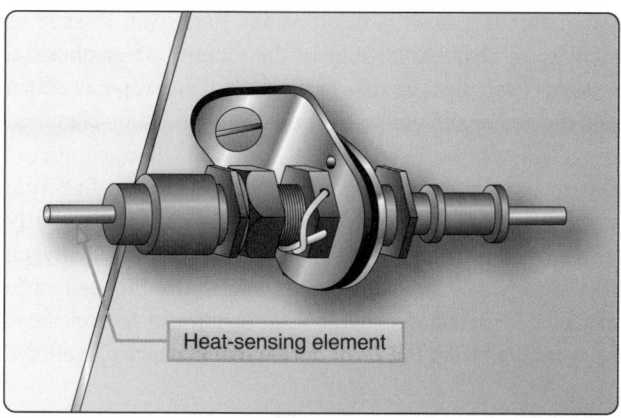

Figure 17-26. *Connector joint fitting attached to the structure.*

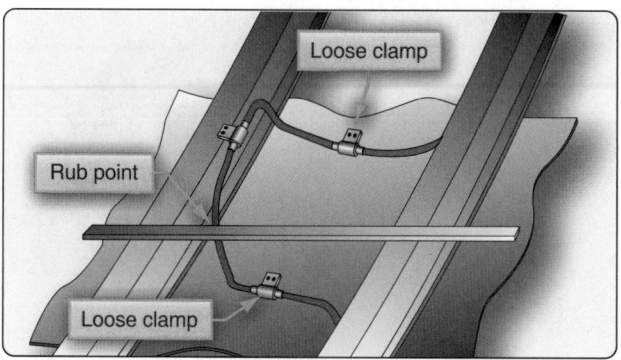

Figure 17-27. *Rubbing interference.*

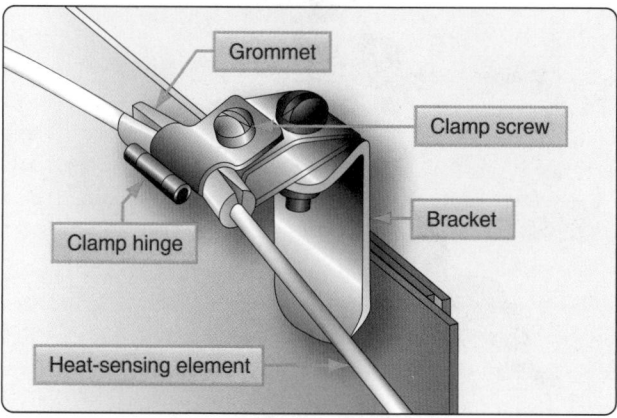

Figure 17-28. *Inspection of fire detector loop clamp.*

2. Fire alarms and warning lights can occur when no engine fire or overheat condition exists. Such false alarms can be most easily located by disconnecting the engine sensing loop connections from the control unit. If the false alarm ceases when the engine sensing loop is disconnected, the fault is in the disconnected sensing loop, which should be examined for areas that have been bent into contact with hot parts of the engine. If no bent element can be found, the shorted section can be located by isolating the connecting elements consecutively around the entire loop.

3. Kinks and sharp bends in the sensing element can cause an internal wire to short intermittently to the outer tubing. The fault can be located by checking the sensing element with an ohm meter while tapping the element in the suspected areas to produce the short.

4. Moisture in the detection system seldom causes a false fire alarm. If, however, moisture does cause an alarm, the warning persists until the contamination is removed, or boils away, and the resistance of the loop returns to its normal value.

5. Failure to obtain an alarm signal when the test switch is actuated may be caused by a defective test switch or control unit, the lack of electrical power, inoperative indicator light, or an opening in the sensing element or connecting wiring. When the test switch fails to provide an alarm, the continuity of a two-wire sensing loop can be determined by opening the loop and measuring the resistance. In a single-wire, continuous-loop system, the center conductor should be grounded.

Fire Extinguisher System Maintenance

Regular maintenance of fire extinguisher systems typically includes such items as the inspection and servicing of fire extinguisher bottles (containers), removal and reinstallation of cartridge and discharge valves, testing of discharge tubing for leakage, and electrical wiring continuity tests. The following paragraphs contain details of some of the most typical maintenance procedures.

Container Pressure Check

Fire extinguisher containers are checked periodically to determine that the pressure is between the prescribed minimum and maximum limits. Changes of pressure with ambient temperatures must also fall within prescribed limits. The graph shown in *Figure 17-29* is typical of the pressure-temperature curve graphs that provide maximum and minimum gauge readings. If the pressure does not fall within the graph limits, the extinguisher container is replaced.

Discharge Cartridges

The service life of fire extinguisher discharge cartridges is determined from the manufacturer's date stamp, which is usually placed on the face of the cartridge. The cartridge service life recommended by the manufacturer is usually in terms of years. Cartridges are available with a service life of 5 years or more. To determine the unexpired service life of a discharge cartridge, it is usually necessary to remove the electrical leads and discharge line from the plug body, which can then be removed from the extinguisher container.

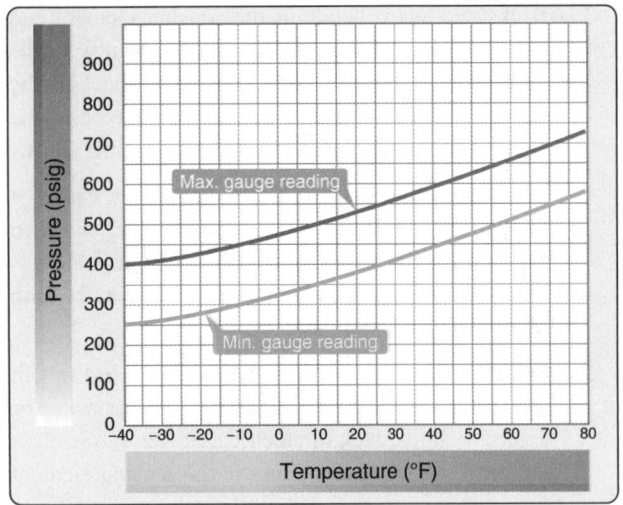

Figure 17-29. *Fire extinguisher container pressure-temperature chart.*

Agent Containers

Care must be taken in the replacement of cartridge and discharge valves. Most new extinguisher containers are supplied with their cartridge and discharge valve disassembled. Before installation on the aircraft, the cartridge must be assembled properly in the discharge valve and the valve connected to the container, usually by means of a swivel nut that tightens against a packing ring gasket. *[Figure 17-30]*

If a cartridge is removed from a discharge valve for any reason, it should not be used in another discharge valve assembly, since the distance the contact point protrudes may vary with each unit. Thus, continuity might not exist if a used plug that had been indented with a long contact point were installed in a discharge valve with a shorter contact point.

Note: The preceding material in this chapter has been largely of a general nature dealing with the principles involved and general procedures to be followed. When actually performing maintenance, always refer to the applicable maintenance manuals and other related publications pertaining to a particular aircraft.

Fire Prevention

Leaking fuel, hydraulic, deicing, or lubricating fluids can be sources of fire in an aircraft. This condition should be noted, and corrective action taken when inspecting aircraft systems. Minute pressure leaks of these fluids are particularly dangerous for they quickly produce an explosive atmospheric condition. Carefully inspect fuel tank installations for signs of external leaks. With integral fuel tanks, the external evidence may occur at some distance from where the fuel is actually escaping. Many hydraulic fluids are flammable and should not be permitted to accumulate in the structure. Sound-proofing and lagging materials may become highly

flammable if soaked with oil of any kind. Any leakage or spillage of flammable fluid in the vicinity of combustion heaters is a serious fire risk, particularly if any vapor is drawn into the heater and passes over the hot combustion chamber.

Oxygen system equipment must be kept absolutely free from traces of oil or grease, since these substances spontaneously ignite when in contact with oxygen under pressure. Oxygen servicing cylinders should be clearly marked so they cannot be mistaken for cylinders containing air or nitrogen, as explosions have resulted from this error during maintenance operations.

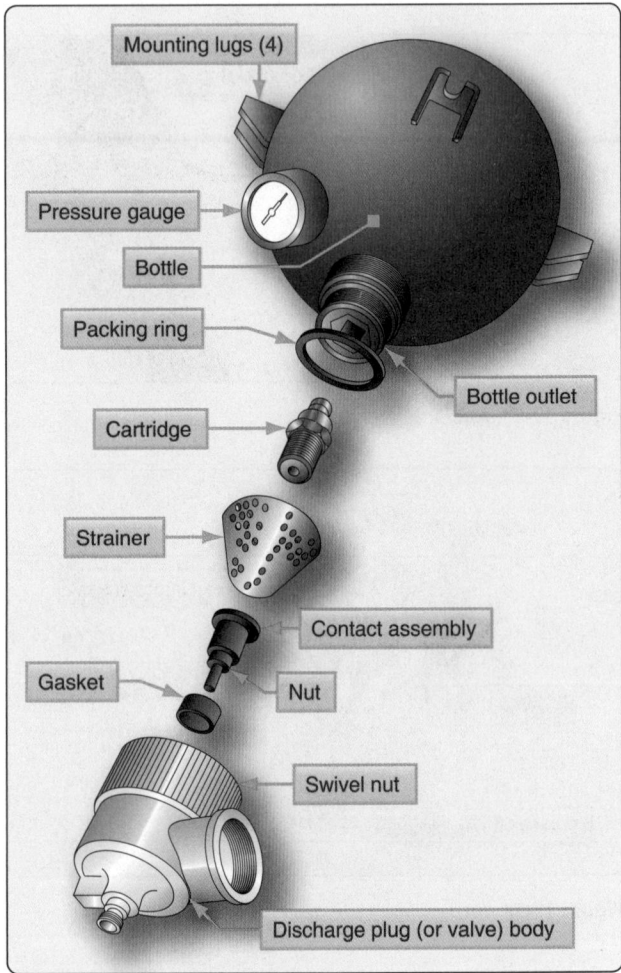

Figure 17-30. *Components of fire extinguisher container.*

Glossary

A

Aborted takeoff. A takeoff that is terminated prematurely when it is determined that some condition exists that makes takeoff or further flight dangerous.

Absolute pressure. Pressure measured from zero pressure or a vacuum.

Absolute pressure regulator. A valve used in a pneumatic system at the pump inlet to regulate the compressor inlet air pressure to prevent excessive speed variation and/or overspeeding of the compressor.

Absolute zero. The point at which all molecular motion ceases. Absolute zero is –460 °F and –273 °C.

Accumulator. A hydraulic component that consists of two compartments separated by a movable component, such as a piston, diaphragm, or bladder. One compartment is filled with compressed air or nitrogen, and the other is filled with hydraulic fluid and is connected into the system pressure manifold. An accumulator allows an incompressible fluid to be stored under pressure by the force produced by a compressible fluid. Its primary purposes are to act as a shock absorber in the system, and to provide a source of additional hydraulic power when heavy demands are placed on the system.

Actuator. A fluid power device that changes fluid pressure into mechanical motion.

ADC. Air data computer.

ADF. Automatic direction finder.

ADI. Attitude director indicator.

Advancing blade. The blade on a helicopter rotor whose tip is moving in the same direction the helicopter is moving.

Adverse yaw. A condition of flight at the beginning of a turn in which the nose of an airplane momentarily yaws in the opposite direction from the direction in which the turn is to be made.

Aerodynamic drag. The total resistance to the movement of an object through the air. Aerodynamic drag is composed of both induced drag and parasite drag. See induced drag and parasite drag.

Aerodynamic lift. The force produced by air moving over a specially shaped surface called an airfoil. Aerodynamic lift acts in a direction perpendicular to the direction the air is moving.

Aeronautical Radio Incorporated (ARINC). A corporation whose principal stockholders are the airlines. Its function is to operate certain communication links between airliners in flight and the airline ground facilities. ARINC also sets standards for communication equipment used by the airlines.

Aging. A change in the characteristics of a material with time. Certain aluminum alloys do not have their full strength when they are first removed from the quench bath after they have been heat-treated, but they gain this strength after a few days by the natural process of aging.

Agonic line. A line drawn on an aeronautical chart along which there is no angular difference between the magnetic and geographic north poles.

Air carrier. An organization or person involved in the business of transporting people or cargo by air for compensation or hire.

Air-cycle cooling system. A system for cooling the air in the cabin of a turbojet-powered aircraft. Compressor bleed air passes through two heat exchangers where it gives up some of its heat; then, it drives an expansion turbine where it loses still more of its heat energy as the turbine drives a compressor. When the air leaves the turbine, it expands and its pressure and temperature are both low.

Aircraft communication addressing and reporting system (ACARS). A two-way communication link between an airliner in flight and the airline's main ground facilities. Data is collected in the aircraft by digital sensors and is transmitted to the ground facilities. Replies from the ground may be printed out so the appropriate flight crewmember can have a hard copy of the response.

Airfoil. Any surface designed to obtain a useful reaction, or lift, from air passing over it.

Airspeed indicator. A flight instrument that measures the pressure differential between the pitot, or ram, air pressure, and the static pressure of the air surrounding the aircraft. This differential pressure is shown in units of miles per hour, knots, or kilometers per hour.

Airworthiness Directive (AD note). Airworthiness Directives (ADs) are legally enforceable rules issued by the FAA in accordance with 14 CFR part 39 to correct an unsafe condition in a product. 14 CFR part 39 defines a product as an aircraft, aircraft engine, propeller, or appliance.

Alclad. A registered trade name for clad aluminum alloy.

Alodine. The registered trade name for a popular conversion coating chemical used to produce a hard, airtight, oxide film on aluminum alloy for corrosion protection.

Alphanumeric symbols. Symbols made up of all of the letters in our alphabet, numerals, punctuation marks, and certain other special symbols.

Alternator. An electrical generator that produces alternating current. The popular DC alternator used on light aircraft produces three-phase AC in its stator windings. This AC is changed into DC by a six-diode, solid-state rectifier before it leaves the alternator.

Altimeter setting. The barometric pressure at a given location corrected to mean (average) sea level.

Altitude engine. A reciprocating engine whose rated sea-level takeoff power can be produced to an established higher altitude.

Alumel. An alloy of nickel, aluminum, manganese, and silicon that is the negative element in a thermocouple used to measure exhaust gas temperature.

Ambient pressure. The pressure of the air surrounding a person or an object.

Ambient temperature. The temperature of the air surrounding a person or an object.

American wire gauge. The system of measurement of wire size used in aircraft electrical systems.

Amphibian. An airplane with landing gear that allows it to operate from both water and land surfaces.

Amplifier. An electronic circuit in which a small change in voltage or current controls a much larger change in voltage or current.

Analog electronics. Electronics in which values change in a linear fashion. Output values vary in direct relationship to changes of input values.

Analog-type indicator. An electrical meter that indicates values by the amount a pointer moves across a graduated numerical scale.

Aneroid. The sensitive component in an altimeter or barometer that measures the absolute pressure of the air. The aneroid is a sealed, flat capsule made of thin corrugated discs of metal soldered together and evacuated by pumping all of the air out of it. Evacuating the aneroid allows it to expand or collapse as the air pressure on the outside changes.

Angle of attack. The acute angle formed between the chord line of an airfoil and the direction of the air that strikes the airfoil.

Angle of attack indicator. An instrument that measures the angle between the local airflow around the direction detector and the fuselage reference plane.

Angle of incidence. The acute angle formed between the chord line of an airfoil and the longitudinal axis of the aircraft on which it is mounted.

Annual rings. The rings that appear in the end of a log cut from a tree. The number of annual rings per inch gives an indication of the strength of the wood. The more rings there are and the closer they are together, the stronger the wood. The pattern of alternating light and dark rings is caused by the seasonal variations in the growth rate of the tree. A tree grows quickly in the spring and produces the light-colored, less dense rings. The slower growth during the summer, or latter part of the growing season, produces the dark-colored, denser rings.

Annunciator panel. A panel of warning lights in plain sight of the pilot. These lights are identified by the name of the system they represent and are usually covered with colored lenses to show the meaning of the condition they announce.

Anodizing. The electrolytic process in which a hard, airtight, oxide film is deposited on aluminum alloy for corrosion protection.

Antenna. A special device used with electronic communication and navigation systems to radiate and receive electromagnetic energy.

Anti-ice system. A system that prevents the formation of ice on an aircraft structure.

Anti-icing additive. A chemical added to the turbine-engine fuel used in some aircraft. This additive mixes with water that condenses from the fuel and lowers its freezing temperature so it will not freeze and block the fuel filters. It also acts as a biocidal agent and prevents the formation of microbial contamination in the tanks.

Antidrag wire. A structural wire inside a Pratt truss airplane wing between the spars. Antidrag wires run from the rear spar inboard, to the front spar at the next bay outboard. Antidrag wires oppose the forces that try to pull the wing forward.

Antiservo tab. A tab installed on the trailing edge of a stabilator to make it less sensitive. The tab automatically moves in the same direction as the stabilator to produce an aerodynamic force that tries to bring the surface back to a streamline position. This tab is also called an antibalance tab.

Antiskid brake system. An electrohydraulic system in an airplane's power brake system that senses the deceleration rate of every main landing gear wheel. If any wheel decelerates too rapidly, indicating an impending skid, pressure to that brake is released and the wheel stops decelerating. Pressure is then reapplied at a slightly lower value.

Antitear strip. Strips of aircraft fabric laid under the reinforcing tape before the fabric is stitched to an aircraft wing.

Arbor press. A press with either a mechanically or hydraulically operated ram used in a maintenance shop for a variety of pressing functions.

Arcing. Sparking between a commutator and brush or between switch contacts that is caused by induced current when a circuit is broken.

Area. The number of square units in a surface.

Aspect ratio. The ratio of the length, or span, of an airplane wing to its width, or chord. For a nonrectangular wing, the aspect ratio is found by dividing the square of the span of the wing by its area. Aspect Ratio = span^2 ÷ area.

Asymmetrical airfoil. An airfoil section that is not the same on both sides of the chord line.

Asymmetrical lift. A condition of uneven lift produced by the rotor when a helicopter is in forward flight. Asymmetrical lift is caused by the difference between the airspeed of the advancing blade and that of the retreating blade.

Attenuate. To weaken, or lessen the intensity of, an activity.

Attitude indicator. A gyroscopic flight instrument that gives the pilot an indication of the attitude of the aircraft relative to its pitch and roll axes. The attitude indicator in an autopilot is in the sensing system that detects deviation from a level-flight attitude.

Augmenter tube. A long, stainless steel tube around the discharge of the exhaust pipes of a reciprocating engine. Exhaust gases flow through the augmenter tube and produce a low pressure that pulls additional cooling air through the engine compartment. Heat may be taken from the augmenter tubes and directed through the leading edges of the wings for thermal anti-icing.

Autoclave. A pressure vessel inside of which air can be heated to a high temperature and pressure raised to a high value. Autoclaves are used in the composite manufacturing industry to apply heat and pressure for curing resins.

Autogyro. A heavier-than-air rotor-wing aircraft sustained in the air by rotors turned by aerodynamic forces rather than by engine power. When the name Autogyro is spelled with a capital A, it refers to a specific series of machines built by Juan de la Cierva or his successors.

Autoignition system. A system on a turbine engine that automatically energizes the igniters to provide a relight if the engine should flame out.

Automatic adjuster. A subsystem in an aircraft disc brake that compensates for disc or lining wear. Each time the brakes are applied, the automatic adjuster is reset for zero clearance, and when the brakes are released, the clearance between the discs or the disc and lining is returned to a preset value. A malfunctioning automatic adjuster in a multiple-disc brake can cause sluggish and jerky operation.

Automatic flight control system (AFCS). The full system of automatic flight control that includes the autopilot, flight director, horizontal situation indicator, air data sensors, and other avionics inputs.

Automatic pilot (autopilot). An automatic flight control device that controls an aircraft about one or more of its three axes. The primary purpose of an autopilot is to relieve the pilot of the control of the aircraft during long periods of flight.

Autosyn system. A synchro system used in remote indicating instruments. The rotors in an Autosyn system are two-pole electromagnets, and the stators are delta-connected, three-phase, distributed-pole windings in the stator housings. The rotors in the transmitters and indicators are connected in parallel and are excited with 26-volt, 400-Hz AC. The rotor in the indicator follows the movement of the rotor in the transmitter.

Auxiliary power unit (APU). A small turbine or reciprocating engine that drives a generator, hydraulic pump, and air pump. The APU is installed in the aircraft and is used to supply electrical power, compressed air, and hydraulic pressure when the main engines are not running.

Aviation snips. Compound-action hand shears used for cutting sheet metal. Aviation snips come in sets of three. One pair cuts to the left, one pair cuts to the right, and the third pair of snips cuts straight.

Aviator's oxygen. Oxygen that has had almost all of the water and water vapor removed from it.

Avionics. The branch of technology that deals with the design, production, installation, use, and servicing of electronic equipment mounted in aircraft.

Azimuth. A horizontal angular distance, measured clockwise from a fixed reference direction to an object.

B

Back course. The reciprocal of the localizer course for an ILS (Instrument Landing System). When flying a back-course approach, the aircraft approaches the instrument runway from the end on which the localizer antennas are installed.

Backhand welding. Welding in which the torch is pointed away from the direction the weld is progressing.

Backplate (brake component). A floating plate on which the wheel cylinder and the brake shoes attach on an energizing-type brake.

Backup ring. A flat leather or Teflon ring installed in the groove in which an O-ring or T-seal is placed. The backup ring is on the side of the seal away from the pressure, and it prevents the pressure extruding the seal between the piston and the cylinder wall.

Balance cable. A cable in the aileron system of an airplane that connects to one side of each aileron. When the control wheel is rotated, a cable from the flight deck pulls one aileron down and relaxes the cable going to the other aileron. The balance cable pulls the other aileron up.

Balance panel. A flat panel hinged to the leading edge of some ailerons that produces a force which assists the pilot in holding the ailerons deflected. The balance panel divides a chamber ahead of the aileron in such a way that when the aileron is deflected downward, for example, air flowing over its top surface produces a low pressure that acts on the balance panel and causes it to apply an upward force to the aileron leading edge.

Balance tab. An adjustable tab mounted on the trailing edge of a control surface to produce a force that aids the pilot in moving the surface. The tab is automatically actuated in such a way it moves in the direction opposite to the direction the control surface on which it is mounted moves.

Balanced actuator. A linear hydraulic or pneumatic actuator that has the same area on each side of the piston.

Banana oil. Nitrocellulose dissolved in amyl acetate, so named because it smells like bananas.

Bank (verb). The act of rotating an aircraft about its longitudinal axis.

Barometric scale. A small window in the dial of a sensitive altimeter in which the pilot sets the barometric pressure level from which the altitude shown on the altimeter is measured. This window is sometimes called the "Kollsman" window. base. The electrode of a bipolar transistor between the emitter and the collector. Varying a small flow of electrons moving into or out of the base controls a much larger flow of electron between the emitter and the collector.

Base. The electrode of a bipolar transistor between the emitter and the collector. Varying a small flow of electrons moving into or out of the base controls a much larger flow of electrons between the emitter and the collector.

Bead (tire component). The high-strength carbon-steel wire bundles that give an aircraft tire its strength and stiffness where it mounts on the wheel.

Bead seat area. The flat surface on the inside of the rim of an aircraft wheel on which the bead of the tire seats.

Bearing strength (sheet metal characteristic). The amount of pull needed to cause a piece of sheet metal to tear at the points at which it is held together with rivets. The bearing strength of a material is affected by both its thickness and the diameter of the rivet.

Beehive spring. A hardened-steel, coil-spring retainer used to hold a rivet set in a pneumatic rivet gun. This spring gets its name from its shape. It screws onto the end of the rivet gun and allows the set to move back and forth, but prevents it being driven from the gun.

Bend allowance. The amount of material actually used to make a bend in a piece of sheet metal. Bend allowance depends upon the thickness of the metal and the radius of the bend, and is normally found in a bend allowance chart.

Bend radius. The radius of the inside of a bend.

Bend tangent line. A line made in a sheet metal layout that indicates the point at which the bend starts.

Bernoulli's principle. The basic principle that explains the relation between kinetic energy and potential energy in fluids that are in motion. When the total energy in a column of moving fluid remains constant, any increase in the kinetic energy of the fluid (its velocity) results in a corresponding decrease in its potential energy (its pressure).

Bezel. The rim that holds the glass cover in the case of an aircraft instrument.

Bias-cut surface tape. A fabric tape in which the threads run at an angle of 45° to the length of the tape. Bias-cut tape may be stretched around a compound curve such as a wing tip bow without wrinkling.

Bilge area. A low portion in an aircraft structure in which water and contaminants collect. The area under the cabin floorboards is normally called the bilge.

Bipolar transistor. A solid-state component in which the flow of current between its emitter and collector is controlled by a much smaller flow of current into or out of its base. Bipolar transistors may be of either the NPN or PNP type.

BITE. Built-in test equipment.

Blade track. The condition of a helicopter rotor in which each blade follows the exact same path as the blade ahead of it.

Black box. A term used for any portion of an electrical or electronic system that can be removed as a unit. A black box does not have to be a physical box.

Bladder-type fuel cell. A plastic-impregnated fabric bag supported in a portion of an aircraft structure so that it forms a cell in which fuel is carried.

Bleeder. A material such as glass cloth or mat that is placed over a composite lay-up to absorb the excess resin forced out of the ply fibers when pressure is applied.

Bleeding dope. Dope whose pigments are soluble in the solvents or thinners used in the finishing system. The color will bleed up through the finished coats.

Bleeding of brakes. The maintenance procedure of removing air entrapped in hydraulic fluid in the brakes. Fluid is bled from the brake system until fluid with no bubbles flows out.

Blimp. A cigar-shaped, nonrigid lighter-than-air flying machine.

Blush. A defect in a lacquer or dope finish caused by moisture condensing on the surface before the finish dries. If the humidity of the air is high, the evaporation of the solvents cools the air enough to cause the moisture to condense. The water condensed from the air mixes with the lacquer or dope and forms a dull, porous, chalky-looking finish called blush. A blushed finish is neither attractive nor protective.

Bonding. The process of electrically connecting all isolated components to the aircraft structure. Bonding provides a path for return current from electrical components, and a low-impedance path to ground to minimize static electrical charges. Shock-mounted components have bonding braids connected across the shock mounts.

Boost pump. An electrically driven centrifugal pump mounted in the bottom of the fuel tanks in large aircraft. Boost pumps provide a positive flow of fuel under pressure to the engine for starting and serve as an emergency backup in the event an engine-driven pump should fail. They are also used to transfer fuel from one tank to another and to pump fuel overboard when it is being dumped. Boost pumps prevent vapor locks by holding pressure on the fuel in the line to the engine-driven pump. Centrifugal boost pumps have a small agitator propeller on top of the impeller to force vapors from the fuel before it leaves the tank.

Boundary layer. The layer of air that flows next to an aerodynamic surface. Because of the design of the surface and local surface roughness, the boundary layer often has a random flow pattern, sometimes even flowing in a direction opposite to the direction of flight. A turbulent boundary layer causes a great deal of aerodynamic drag.

Bourdon tube. A pressure-indicating mechanism used in most oil pressure and hydraulic pressure gages. It consists of a sealed, curved tube with an elliptical cross section. Pressure inside the tube tries to straighten it, and as it straightens, it moves a pointer across a calibrated dial. Bourdon-tube pressure gauges are used to measure temperature by measuring the vapor pressure in a sealed container of a volatile liquid, such as methyl chloride, whose vapor pressure varies directly with its temperature.

Brazing. A method of thermally joining metal parts by wetting the surface with a molten nonferrous alloy. When the molten material cools and solidifies, it holds the pieces together. Brazing materials melt at a temperature higher than 800 °F, but lower than the melting temperature of the metal on which they are used.

British thermal unit (BTU). The amount of heat energy needed to raise the temperature of one pound of pure water 1 °F.

Bucking bar. A heavy steel bar with smooth, hardened surfaces, or faces. The bucking bar is held against the end of the rivet shank when it is driven with a pneumatic rivet gun, and the shop head is formed against the bucking bar.

Buffeting. Turbulent movement of the air over an aerodynamic surface.

Bulb angle. An L-shaped metal extrusion having an enlarged, rounded edge that resembles a bulb on one of its legs.

Bulkhead. A structural partition that divides the fuselage of an aircraft into compartments, or bays.

Bungee shock cord. A cushioning material used with the nonshock absorbing landing gears installed on older aircraft. Bungee cord is made up of many small rubber bands encased in a loose-woven cotton braid.

Burnish (verb). To smooth the surface of metal that has been damaged by a deep scratch or gouge. The metal piled up at the edge of the damage is pushed back into the damage with a smooth, hard steel burnishing tool.

Burr. A sharp rough edge of a piece of metal left when the metal was sheared, punched, or drilled.

Bus. A point within an electrical system from which the individual circuits get their power.

Buttock line. A line used to locate a position to the right or left of the center line of an aircraft structure.

Butyl. Trade name for a synthetic rubber product made by the polymerization of isobutylene. Butyl withstands such potent chemicals as phosphate ester-base (Skydrol) hydraulic fluids.

C

Cage (verb). To lock the gimbals of a gyroscopic instrument so it will not be damaged by abrupt flight maneuvers or rough handling.

Calendar month. A measurement of time used by the FAA for inspection and certification purposes. One calendar month from a given day extends from that day until midnight of the last day of that month.

Calibrated airspeed (CAS). Indicated airspeed corrected for position error. See position error.

Calorie. The amount of heat energy needed to raise the temperature of one gram of pure water 1 °C.

Canted rate gyro. A rate gyro whose gimbal axis is tilted so it can sense rotation of the aircraft about its roll axis as well as its yaw axis.

Camber (wheel alignment). The amount the wheels of an aircraft are tilted, or inclined, from the vertical. If the top of the wheel tilts outward, the camber is positive. If the top of the wheel tilts inward, the camber is negative.

Canard. A horizontal control surface mounted ahead of the wing to provide longitudinal stability and control.

Cantilever wing. A wing that is supported by its internal structure and requires no external supports. The wing spars are built in such a way that they carry all the bending and torsional loads.

Cap strip. The main top and bottom members of a wing rib. The cap strips give the rib its aerodynamic shape.

Capacitance-type fuel quantity measuring system. A popular type of electronic fuel quantity indicating system that has no moving parts in the fuel tank. The tank units are cylindrical capacitors, called probes, mounted across the tank, from top to bottom. The dielectric between the plates of the probes is either fuel or the air above the fuel, and the capacitance of the probe varies with the amount of fuel in the tank. The indicator is a servo-type instrument driven by the amplified output of a capacitance bridge.

Capillary tube. A soft copper tube with a small inside diameter. The capillary tube used with vapor-pressure thermometer connects the temperature sensing bulb to the Bourdon tube. The capillary tube is protected from physical damage by enclosing it in a braided metal wire jacket.

Carbon monoxide detector. A packet of chemical crystals mounted in the aircraft flight deck or cabin where they are easily visible. The crystals change their color from yellow to green when they are exposed to carbon monoxide.

Carbon-pile voltage regulator. A type of voltage regulator used with high-output DC generators. Field current is controlled by varying the resistance of a stack of thin carbon discs. This resistance is varied by controlling the amount the stack is compressed by a spring whose force is opposed by the pull of an electromagnet. The electromagnet's strength is proportional to the generator's output voltage.

Carburizing flame. An oxyacetylene flame produced by an excess of acetylene. This flame is identified by a feather around the inner cone. A carburizing flame is also called a reducing flame.

Carcass (tire component). The layers of rubberized fabric that make up the body of an aircraft tire.

Case pressure. A low pressure that is maintained inside the case of a hydraulic pump. If a seal becomes damaged, hydraulic fluid will be forced out of the pump rather than allowing air to be drawn into the pump.

Cathode-ray tube (CRT). A display tube used for oscilloscopes and computer video displays. An electron gun emits a stream of electrons that is attracted to a positively charged inner surface of the face of the tube. Acceleration and focusing grids speed the movement of the electrons and shape the beam into a pinpoint size. Electrostatic or electromagnetic forces caused by deflection plates or coils move the beam over the face of the tube. The inside surface of the face of the tube is treated with a phosphor material that emits light when the beam of electrons strikes it.

Cavitation. A condition that exist in a hydraulic pump when there is not enough pressure in the reservoir to force fluid to the inlet of the pump. The pump picks up air instead of fluid.

CDI. Course deviation indicator.

CDU. Control display unit.

Center of gravity. The location on an aircraft about which the force of gravity is concentrated.

Center of lift. The location of the chord line of an airfoil at which all the lift forces produced by the airfoil are considered to be concentrated.

Center of pressure. The point on the chord line of an airfoil where all of the aerodynamic forces are considered to be concentrated.

Centering cam. A cam in the nose-gear shock strut that causes the piston to center when the strut fully extends. When the aircraft takes off and the strut extends, the wheel is straightened in its fore-and-aft position so it can be retracted into the wheel well.

Charging stand (air conditioning service equipment). A handy and compact arrangement of air conditioning servicing equipment. A charging stand contains a vacuum pump, a manifold gauge set, and a method of measuring and dispensing the refrigerant.

Chatter. A type of rapid vibration of a hydraulic pump caused by the pump taking in some air along with the hydraulic fluid.

Check (wood defect). Longitudinal cracks that extend across a log's annual rings.

Check valve. A hydraulic or pneumatic system component that allows full flow of fluid in one direction but blocks all flow in the opposite direction.

Chemical oxygen candle system. An oxygen system used for emergency or backup use. Solid blocks of material that release oxygen when they are burned are carried in special fireproof fixtures. When oxygen is needed, the candles are ignited with an integral igniter, and oxygen flows into the tubing leading to the masks.

Chevron seal. A form of one-way seal used in some fluid-power actuators. A chevron seal is made of a resilient material whose cross section is in the shape of the letter V. The pressure being sealed must be applied to the open side of the V.

Chromel. An alloy of nickel and chromium used as the positive element in a thermocouple for measuring exhaust gas temperature.

Circle. A closed plane figure with every point an equal distance from the center. A circle has the greatest area for its circumference of any enclosed shape.

Circuit breaker. An electrical component that automatically opens a circuit any time excessive current flows through it. A circuit breaker may be reset to restore the circuit after the fault causing the excessive current has been corrected.

Clad aluminum. A sheet of aluminum alloy that has a coating of pure aluminum rolled on one or both of its surfaces for corrosion protection.

Clamp-on ammeter. An electrical instrument used to measure current without opening the circuit through which it is flowing. The jaws of the ammeter are opened, slipped over the current-carrying wire, and then clamped shut. Current flowing through the wire produces a magnetic field which induces a voltage in the ammeter that is proportional to the amount of current.

Cleco fastener. A patented spring-type fastener used to hold metal sheets in position until they can be permanently riveted together.

Close-quarter iron. A small hand-held iron with an accurately calibrated thermostat. This iron is used for heat-shrinking polyester fabrics in areas that would be difficult to work with a large iron.

Closed angle. An angle formed in sheet metal that has been bent more than 90°.

Closed assembly time. The time elapsing between the assembly of glued joints and the application of pressure.

Closed-center hydraulic system. A hydraulic system in which the selector valves are installed in parallel with each other. When no unit is actuated, fluid circulates from the pump back to the reservoir without flowing through any of the selector valves.

Closed-center selector valve. A type of flow-control valve used to direct pressurized fluid into one side of an actuator, and at the same time, direct the return fluid from the other side of the actuator to the fluid reservoir. Closed-center selector valves are connected in parallel between the pressure manifold and the return manifold.

Coaxial. Rotating about the same axis. Coaxial rotors of a helicopter are mounted on concentric shafts in such a way that they turn in opposite directions to cancel torque.

Coaxial cable. A special type of electrical cable that consists of a central conductor held rigidly in the center of a braided outer conductor. Coaxial cable, commonly called coax, is used for attaching radio receivers and transmitters to their antenna.

Coefficient of drag. A dimensionless number used in the formula for determining induced drag as it relates to the angle of attack.

Coefficient of lift. A dimensionless number relating to the angle of attack used in the formula for determining aerodynamic lift.

Coin dimpling. A process of preparing a hole in sheet metal for flush riveting. A coining die is pressed into the rivet hole to form a sharp-edged depression into which the rivet head fits.

Collective pitch control. The helicopter control that changes the pitch of all of the rotor blades at the same time. Movement of the collective pitch control increases or decreases the lift produced by the entire rotor disc.

Collodion. Cellulose nitrate used as a film base for certain aircraft dopes.

Combustion heater. A type of cabin heater used in some aircraft. Gasoline from the aircraft fuel tanks is burned in the heater.

Compass fluid. A highly refined, water-clear petroleum product similar to kerosene. Compass fluid is used to dampen the oscillations of magnetic compasses.

Compass rose. A location on an airport where an aircraft can be taken to have its compasses "swung." Lines are painted on the rose to mark the magnetic directions in 30° increments.

Compass swinging. A maintenance procedure that minimizes deviation error in a magnetic compass. The aircraft is aligned on a compass rose, and the compensating magnets in the compass case are adjusted so the compass card indicates the direction marked on the rose. After the deviation error is minimized on all headings, a compass correction card is completed and mounted on the instrument panel next to the compass.

Compensated fuel pump. A vane-type, engine-driven fuel pump that has a diaphragm connected to the pressure regulating valve. The chamber above the diaphragm is vented to the carburetor upper deck where it senses the pressure of the air as it enters the engine. The diaphragm allows the fuel pump to compensate for altitude changes and keeps the carburetor inlet fuel pressure a constant amount higher than the carburetor inlet air pressure.

Compensator port (brake system component). A small hole between a hydraulic brake master cylinder and the reservoir. When the brakes are released, this port is uncovered and the fluid in the master cylinder is vented to the reservoir. When the brake is applied, the master-cylinder piston covers the compensator port and allows pressure in the line to the brake to build up and apply the brakes. When the brake is released, the piston uncovers the compensator port. If any fluid has been lost from the brake, the reservoir will refill the master cylinder. A restricted compensator port will cause the brakes to drag or will cause them to be slow to release.

Composite. Something made up of different materials combined in such a way that the characteristics of the resulting material are different from those of any of the components.

Compound curve. A curve formed in more than one plane. The surface of a sphere is a compound curve.

Compound gauge (air conditioning servicing equipment). A pressure gauge used to measure the pressure in the low side of an air conditioning system. A compound gauge is calibrated from zero to 30 inches of mercury vacuum, and from zero to about 150-psi positive gauge pressure.

Compressibility effect. The sudden increase in the total drag of an airfoil in transonic flight caused by formation of shock waves on the surface.

Compression failure. A type of structural failure in wood caused by the application of too great a compressive load. A compression failure shows up as a faint line running at right angles to the grain of the wood.

Compression strut. A heavy structural member, often in the form of a steel tube, used to hold the spars of a Pratt truss airplane wing apart. A compression strut opposes the compressive loads between the spars arising from the tensile loads produced by the drag and antidrag wires.

Compression wood. A defect in wood that causes it to have a high specific gravity and the appearance of an excessive growth of summerwood. In most species, there is little difference between the color of the springwood and the summerwood. Any material containing compression wood is unsuited for aircraft structural use and must be rejected.

Compressor (air conditioning system component). The component in a vapor-cycle cooling system in which the low-pressure refrigerant vapors, after they leave the evaporator, are compressed to increase both their temperature and pressure before they pass into the condenser. Some compressors are driven by electric motors, others by hydraulic motors and, in the case of most light airplanes, are belt driven from the engine.

Concave surface. A surface that is curved inward. The outer edges are higher than the center.

Condenser (air conditioning system component). The component in a vapor-cycle cooling system in which the heat taken from the aircraft cabin is given up to the ambient air outside the aircraft.

Conductor (electrical). A material that allows electrons to move freely from one atom to another within the material.

Coning angle. The angle formed between the plane of rotation of a helicopter rotor blade when it is producing lift and a line perpendicular to the rotor shaft. The degree of the coning angle is determined by the relationship between the centrifugal force acting on the blades and the aerodynamic lift produced by the blades.

Constant (mathematical). A value used in a mathematical computation that is the same every time it is used. For example, the relationship between the length of the circumference of a circle and the length of its diameter is a constant, 3.1416. This constant is called by the Greek name of Pi (π).

Constant differential mode (cabin pressurization). The mode of pressurization in which the cabin pressure is maintained a constant amount higher than the outside air pressure. The maximum differential pressure is determined by the structural strength of the aircraft cabin.

Constant-displacement pump. A fluid pump that moves a specific volume of fluid each time it rotates; the faster the pump turns, the more fluid it moves. Some form of pressure regulator or relief valve must be used with a constant-displacement pump when it is driven by an aircraft engine.

Constant-speed drive (CSD). A special drive system used to connect an alternating current generator to an aircraft engine. The drive holds the generator speed (and thus its frequency) constant as the engine speed varies.

Constantan. A copper-nickel alloy used as the negative lead of a thermocouple for measuring the cylinder head temperature of a reciprocating engine.

Contactor (electrical component). A remotely actuated, heavy-duty electrical switch. Contactors are used in an aircraft electrical system to connect the battery to the main bus.

Continuity tester. A troubleshooting tool that consists of a battery, a light bulb, and test leads. The test leads are connected to each end of the conductor under test, and if the bulb lights up, there is continuity. If it does not light up, the conductor is open.

Continuous Airworthiness Inspection Program. An inspection program that is part of a continuous airworthiness maintenance program approved for certain large airplanes (to which 14 CFR Part 125 is not applicable), turbojet multi-engine airplanes, turbopropeller-powered multi-engine airplanes, and turbine-powered rotorcraft.

Continuous-duty solenoid. A solenoid-type switch designed to be kept energized by current flowing through its coil for an indefinite period of time. The battery contactor in an aircraft electrical system is a continuous-duty solenoid. Current flows through its coil all the time the battery is connected to the electrical system.

Continuous-flow oxygen system. A type of oxygen system that allows a metered amount of oxygen to continuously flow into the mask. A rebreather-type mask is used with a continuous-flow system. The simplest form of continuous-flow oxygen system regulates the flow by a calibrated orifice in the outlet to the mask, but most systems use either a manual or automatic regulator to vary the pressure across the orifice proportional to the altitude being flown.

Continuous-loop fire-detection system. A fire-detection system that uses a continuous loop of two conductors separated with a thermistor-type insulation. Under normal temperature conditions, the thermistor material is an insulator; but if it is exposed to a fire, the thermistor changes into a conductor and completes the circuit between the two conductors, initiating a fire warning.

Control horn. The arm on a control surface to which the control cable or push-pull rod attaches to move the surface.

Control stick. The type of control device used in some airplanes. A vertical stick in the flight deck controls the ailerons by side-to-side movement and the elevators by fore-and-aft movement.

Control yoke. The movable column on which an airplane control wheel is mounted. The yoke may be moved in or out to actuate the elevators, and the control wheel may be rotated to actuate the ailerons.

Controllability. The characteristic of an aircraft that allows it to change its flight attitude in response to the pilot's movement of the flight deck controls.

Conventional current. An imaginary flow of electricity that is said to flow from the positive terminal of a power source, through the external circuit to its negative terminal. The arrowheads in semiconductor symbols point in the direction of conventional current flow.

Converging duct. A duct, or passage, whose cross-sectional area decreases in the direction of fluid flow.

Conversion coating. A chemical solution used to form an airtight oxide or phosphate film on the surface of aluminum or magnesium parts. The conversion coating prevents air from reaching the metal and keeps it from corroding.

Convex surface. A surface that is curved outward. The outer edges are lower than the center.

Coriolis effect. The change in rotor blade velocity to compensate for a change in the distance between the center of mass of the rotor blade and the axis rotation of the blade as the blades flap in flight.

Cornice brake. A large shop tool used to make straight bends across a sheet of metal. Cornice brakes are often called leaf brakes.

Corrugated metal. Sheets of metal that have been made more rigid by forming a series of parallel ridges or waves in its surface.

Cotter pin. A split metal pin used to safety a castellated or slotted nut on a bolt. The pin is passed through the hole in the shank of the bolt and the slots in the nut, and the ends of the pin are spread to prevent it backing out of the hole.

Countersinking. Preparation of a rivet hole for a flush rivet by beveling the edges of the holes with a cutter of the correct angle.

Coverite surface thermometer. A small surface-type bimetallic thermometer that calibrates the temperature of an iron used to heat-shrink polyester fabrics.

Crabbing. Pointing the nose of an aircraft into the wind to compensate for wind drift.

Crazing. A form of stress-caused damage that occurs in a transparent thermoplastic material. Crazing appears as a series of tiny, hair-like cracks just below the surface of the plastic.

Critical Mach number. The flight Mach number at which there is the first indication of supersonic airflow over any part of the aircraft structure.

Cross coat. A double coat of aircraft finishing material in which the second coat is sprayed at right angles to the first coat, before the solvents have evaporated from the first coat.

Cross-feed valve (fuel system component). A valve in a fuel system that allows any of the engines of a multi-engine aircraft to draw fuel from any fuel tank. Cross-feed systems are used to allow a multi-engine aircraft to maintain a balanced fuel condition.

Cross-flow valve. An automatic flow-control valve installed between the gear-up and gear-down lines of the landing gear of some large airplanes. When the landing gear is released from its uplocks, its weight causes it to fall faster than the hydraulic system can supply fluid to the gear-down side of the actuation cylinder. The cross-flow valve opens and directs fluid from the gear-up side into the gear-down side. This allows the gear to move down with a smooth motion.

CRT. Cathode-ray tube.

Cryogenic liquid. A liquid which boils at temperatures of less than about 110 °F (–163 °C) at normal atmospheric pressures.

Cuno filter. The registered trade name for a particular style of edge-type fluid filter. Cuno filters are made up of a stack of thin metal discs that are separated by thin scraper blades. Contaminants collect on the edge of the discs, and they are periodically scraped out and allowed to collect in the bottom of the filter case for future removal.

Current. A general term used for electrical flow. See conventional current.

Current limiter. An electrical component used to limit the amount of current a generator can produce. Some current limiters are a type of slow-blow fuse in the generator output. Other current limiters reduce the generator output voltage if the generator tries to put out more than its rated current.

Cusp. A pointed end.

Cyclic pitch control. The helicopter control that allows the pilot to change the pitch of the rotor blades individually, at a specific point in their rotation. The cyclic pitch control allows the pilot to tilt the plane of rotation of the rotor disc to change the direction of lift produced by the rotor.

D

Dacron. The registered trade name for a cloth woven from polyester fibers.

Damped oscillation. Oscillation whose amplitude decreases with time.

Database. A body of information that is available on any particular subject.

Data bus. A wire or group of wires that are used to move data within a computer system.

Debooster valve. A valve in a power brake system between the power brake control valve and the wheel cylinder. This valve lowers the pressure of the fluid going to the brake and increases its volume. A debooster valve increases the smoothness of brake application and aids in rapid release of the brakes.

Decay. The breakdown of the structure of wood fibers. Wood that shows any indication of decay must be rejected for use in aircraft structure.

Decomposition. The breakdown of the structure of wood fibers. Wood that shows any indication of decay must be rejected for use in aircraft structure.

Deciduous. A type of tree that sheds its foliage at the end of the growing season. Hardwoods come from deciduous trees.

Dedicated computer. A small digital computer, often built into an instrument or control device that contains a built-in program that causes it to perform a specific function.

Deep-vacuum pump. A vacuum pump capable of removing almost all of the air from a refrigeration system. A deep-vacuum pump can reduce the pressure inside the system to a few microns of pressure.

Deflator cap. A cap for a tire, strut, or accumulator air valve that, when screwed onto the valve, depresses the valve stem and allows the air to escape safely through a hole in the side of the cap.

Deicer system. A system that removes ice after it has formed on an aircraft.

Delamination. The separation of the layers of a laminated material.

Delivery air duct check valve. An isolation valve at the discharge side of the air turbine that prevents the loss of pressurization through a disengaged cabin air compressor.

Delta airplane. An airplane with a triangular-shaped wing. This wing has an extreme amount of sweepback on its leading edge, and a trailing edge that is almost perpendicular to the longitudinal axis of the airplane.

Delta connection (electrical connection). A method of connecting three electrical coils into a ring or, as they are drawn on a schematic diagram as a triangle, a delta (D).

Denier. A measure of the fineness of the yarns in a fabric.

Density altitude. The altitude in standard air at which the density is the same as that of the existing air.

Density ratio (σ). The ratio of the density of the air at a given altitude to the density of the air at sea level under standard conditions.

Derated (electrical specification). Reduction in the rated voltage or current of an electrical component. Derating is done to extend the life or reliability of the device.

Desiccant (air conditioning component). A drying agent used in an air conditioning system to remove water from the refrigerant. A desiccant is made of silica-gel or some similar material.

Detent. A spring-loaded pin or tab that enters a hole or groove when the device to which it is attached is in a certain position. Detents are used on a fuel valve to provide a positive means of identifying the fully on and fully off position of the valve.

Detonation. An explosion, or uncontrolled burning of the fuel-air mixture inside the cylinder of a reciprocating engine. Detonation occurs when the pressure and the temperature inside the cylinder become higher than the critical pressure and temperature of the fuel. Detonation is often confused with preignition.

Deviation error. An error in a magnetic compass caused by localized magnetic fields in the aircraft. Deviation error, which is different on each heading, is compensated by the technician "swinging" the compass. A compass must be compensated so the deviation error on any heading is no greater than 10 degrees.

Dewar bottle. A vessel designed to hold liquefied gases. It has double walls with the space between being evacuated to prevent the transfer of heat. The surfaces in the vacuum area are made heat-reflective.

Differential aileron travel. Aileron movement in which the upward-moving aileron deflects a greater distance than the one moving downward. The up aileron produces parasite drag to counteract the induced drag caused by the down aileron. Differential aileron travel is used to counteract adverse yaw.

Differential pressure. The difference between two pressures. An airspeed indicator is a differential-pressure gauge. It measures the difference between static air pressure and pitot air pressure.

Differential-voltage reverse-current cutout. A type of reverse-current cutout switch used with heavy-duty electrical systems. This switch connects the generator to the electrical bus when the generator voltage is a specific amount higher than the battery voltage.

Digital multimeter. An electrical test instrument that can be used to measure voltage, current, and resistance. The indication is in the form of a liquid crystal display in discrete numbers.

Dihedral. The positive angle formed between the lateral axis of an airplane and a line that passes through the center of the wing or horizontal stabilizer. Dihedral increases the lateral stability of an airplane.

Diluter-demand oxygen system. A popular type of oxygen system in which the oxygen is metered to the mask, where it is diluted with cabin air by an airflow-metering aneroid assembly which regulates the amount of air allowed to dilute the oxygen on the basis of cabin altitude. The mixture of oxygen and air flows only when the wearer of the mask inhales. The percentage of oxygen in the air delivered to the mask is regulated, on the basis of altitude, by the regulator. A diluter-demand regulator has an emergency position which allows 100 percent oxygen to flow to the mask, by-passing the regulating mechanism.

Dipole antenna. A half wavelength, center-fed radio antenna. The length of each of the two arms is approximately one fourth of the wavelength of the center frequency for which the antenna is designed.

Dirigible. A large, cigar-shaped, rigid, lighter-than-air flying machine. Dirigibles are made of a rigid truss structure covered with fabric. Gas bags inside the structure contain the lifting gas, which is either helium or hydrogen.

Disc area (helicopter specification). The total area swept by the blades of a helicopter main rotor.

Divergent oscillation. Oscillation whose amplitude increases with time.

Diverging duct. A duct, or passage, whose cross-sectional area increases in the direction of fluid flow.

DME. Distance measuring equipment.

Dope proofing. The treatment of a structure to be covered with fabric to keep the solvents in the dope from softening the protective coating on the structure.

Dope roping. A condition of aircraft dope brushed onto a surface in such a way that it forms a stringy, uneven surface rather than flowing out smoothly.

Double-acting actuator (hydraulic system component). A linear actuator moved in both directions by fluid power.

Double-acting hand pump (hydraulic system component). A hand-operated fluid pump that moves fluid during both strokes of the pump handle.

Doubler. A piece of sheet metal used to strengthen and stiffen a repair in a sheet metal structure.

Downtime. Any time during which an aircraft is out of commission and unable to be operated.

Downwash. Air forced down by aerodynamic action below and behind the wing of an airplane or the rotor of a helicopter. Aerodynamic lift is produced when the air is deflected downward. The upward force on the aircraft is the same as the downward force on the air.

Drag (helicopter rotor blade movement). Fore-and-aft movement of the tip of a helicopter rotor blade in its plane of rotation.

Dragging brakes. Brakes that do not fully release when the brake pedal is released. The brakes are partially applied all the time, which causes excessive lining wear and heat.

Drag wire. A structural wire inside a Pratt truss airplane wing between the spars. Drag wires run from the front spar inboard, to the rear spar at the next bay outboard. Drag wires oppose the forces that try to drag the wing backward.

Drill motor. An electric or pneumatic motor that drives a chuck that holds a twist drill. The best drill motors produce high torque, and their speed can be controlled.

Drip stick. A fuel quantity indicator used to measure the fuel level in the tank when the aircraft is on the ground. The drip stick is pulled down from the bottom of the tank until fuel drips from its opened end. This indicates that the top of the gauge inside the tank is at the level of the fuel. Note the number of inches read on the outside of the gauge at the point it contacts the bottom of the tank, and use a drip stick table to convert this measurement into gallons of fuel in the tank.

Dry air pump. An engine-driven air pump which used carbon vanes. Dry pumps do not use any lubrication, and the vanes are extremely susceptible to damage from the solid airborne particles. These pumps must be operated with filters in their inlet so they will take in only filtered air.

Dry ice. Solidified carbon dioxide. Dry ice sublimates, or changes from a solid directly into a gas, at a temperature of −110 °F (−78.5 °C).

Dry rot. Decomposition of wood fibers caused by fungi. Dry rot destroys all strength in the wood.

Ductility. The property of a material that allows it to be drawn into a thin section without breaking.

Dummy load (electrical load). A noninductive, high-power, 50-ohm resistor that can be connected to a transmission line in place of the antenna. The transmitter can be operated into the dummy load without transmitting any signal.

Duralumin. The name for the original alloy of aluminum, magnesium, manganese, and copper. Duralumin is the same as the modern 2017 aluminum alloy.

Dutch roll. An undesirable, low-amplitude coupled oscillation about both the yaw and roll axes that affects many swept wing airplanes. Dutch roll is minimized by the use of a yaw damper.

Dutchman shears. A common name for compound-action sheet metal shears.

Dynamic pressure (q). The pressure a moving fluid would have if it were stopped. Dynamic pressure is measured in pounds per square foot.

Dynamic stability. The stability that causes an aircraft to return to a condition of straight and level flight after it has been disturbed from this condition. When an aircraft is disturbed from the straight and level flight, its static stability starts it back in the correct direction; but it overshoots, and the corrective forces are applied in the opposite direction. The aircraft oscillates back and forth on both sides of the correct condition, with each oscillation smaller than the one before it. Dynamic stability is the decreasing of these restorative oscillations.

E

EADI. Electronic Attitude Director Indicator.

ECAM. Electronic Centralized Aircraft Monitor.

Eccentric brushing. A special bushing used between the rear spar of certain cantilever airplane wings and the wing attachment fitting on the fuselage. The portion of the bushing that fits through the hole in the spar is slightly offset from that which passes through the holes in the fitting. By rotating the bushing, the rear spar may be moved up or down to adjust the root incidence of the wing.

Eddy current damping (electrical instrument damping). Decreasing the amplitude of oscillations by the interaction of magnetic fields. In the case of a vertical-card magnetic compass, flux from the oscillating permanent magnet produces eddy currents in a damping disc or cup. The magnetic flux produced by the eddy currents opposes the flux from the permanent magnet and decreases the oscillations.

Edge distance. The distance between the center of a rivet hole and the edge of the sheet of metal.

EFIS. Electronic Flight Instrument System.

EHSI. Electronic Horizontal Situation Indicator.

EICAS. Engine Indicating and Crew Alerting System.

Ejector. A form of jet pump used to pick up a liquid and move it to another location. Ejectors are used to ensure that the compartment in which the boost pumps are mounted is kept full of fuel. Part of the fuel from the boost pump flowing through the ejector produces a low pressure that pulls fuel from the main tank and forces it into the boostpump sump area.

Elastic limit. The maximum amount of tensile load, in pounds per square inch, a material is able to withstand without being permanently deformed.

Electromotive force (EMF). The force that causes electrons to move from one atom to another within an electrical circuit. Electromotive force is an electrical pressure, and it is measured in volts.

Electron current. The actual flow of electrons in a circuit. Electrons flow from the negative terminal of a power source through the external circuit to its positive terminal. The arrowheads in semiconductor symbols point in the direction opposite to the flow of electron current.

ELT (emergency locator transmitter). A self-contained radio transmitter that automatically begins transmitting on the emergency frequencies any time it is triggered by a severe impact parallel to the longitudinal axis of the aircraft.

Elevator downspring. A spring in the elevator control system that produces a mechanical force that tries to lower the elevator. In normal flight, this spring force is overcome by the aerodynamic force from the elevator trim tab. But in slow flight with an aft CG position, the trim tab loses its effectiveness and the downspring lowers the nose to prevent a stall.

Elevons. Movable control surfaces on the trailing edge of a delta wing or a flying wing airplane. These surfaces operate together to serve as elevators, and differentially to act as ailerons.

EMI. Electromagnetic interference.

Empennage. The tail section of an airplane.

Enamel. A type of finishing material that flows out to form a smooth surface. Enamel is usually made of a pigment suspended in some form of resin. When the resin cures, it leaves a smooth, glossy protective surface.

Energizing brake. A brake that uses the momentum of the aircraft to increase its effectiveness by wedging the shoe against the brake drum. Energizing brakes are also called servo brakes. A single-servo brake is energizing only when moving in the forward direction, and a duo-servo brake is energizing when the aircraft is moving either forward or backward.

Epoxy. A flexible, thermosetting resin that is made by polymerization of an epoxide. Epoxy has wide application as a matrix for composite materials and as an adhesive that bonds many different types of materials. It is noted for its durability and its chemical resistance.

Equalizing resistor. A large resistor in the ground circuit of a heavy-duty aircraft generator through which all of the generator output current flows. The voltage drop across this resistor is used to produce the current in the paralleling circuit that forces the generators to share the electrical load equally.

Ethylene dibromide. A chemical compound added to aviation gasoline to convert some of the deposits left by the tetraethyl lead into lead bromides. These bromides are volatile and will pass out of the engine with the exhaust gases.

Ethylene glycol. A form of alcohol used as a coolant for liquid-cooled engines and as an anti-icing agent.

Eutectic material. An alloy or solution that has the lowest possible melting point.

Evacuation (air conditioning servicing procedure). A procedure in servicing vapor-cycle cooling systems. A vacuum pump removes all the air from the system. Evacuation removes all traces of water vapor that could condense out, freeze, and block the system.

Evaporator (air conditioning component). The component in a vapor-cycle cooling system in which heat from the aircraft cabin is absorbed into the refrigerant. As the heat is absorbed, the refrigerant evaporates, or changes from a liquid into a vapor. The function of the evaporator is to lower the cabin air temperature.

Expander-tube brake. A brake that uses hydraulic fluid inside a synthetic rubber tube around the brake hub to force rectangular blocks of brake-lining material against the rotating brake drum. Friction between the brake drum and the lining material slows the aircraft.

Expansion wave. The change in pressure and velocity of a supersonic flow of air as it passes over a surface which drops away from the flow. As the surface drops away, the air tries to follow it. In changing its direction, the air speeds up to a higher supersonic velocity and its static pressure decreases. There is no change in the total energy as the air passes through an expansion wave, and so there is no sound as there is when air passes through a shock wave.

Extruded angle. A structural angle formed by passing metal heated to its plastic state through specially shaped dies.

F

FAA Form 337. The FAA form that must be filled in and submitted to the FAA when a major repair or major alteration has been completed.

Federal Aviation Administration Flight Standards District Office (FAA FSDO). An FAA field office serving an assigned geographical area staffed with Flight Standards personnel who serve the aviation industry and the general public on matters relating to certification and operation of air carrier and general aviation aircraft.

Fading of brakes. The decrease in the amount of braking action that occurs with some types of brakes that are applied for a long period of time. True fading occurs with overheated drum-type brakes. As the drum is heated, it expands in a bell-mouthed fashion. This decreases the amount of drum in contact with the brake shoes and decreases the braking action. A condition similar to brake fading occurs when there is an internal leak in the brake master cylinder. The brakes are applied, but as the pedal is held down, fluid leaks past the piston, and the brakes slowly release.

Fairing. A part of a structure whose primary purpose is to produce a smooth surface or a smooth junction where two surfaces join.

Fairlead. A plastic or wooden guide used to prevent a steel control cable rubbing against an aircraft structure.

FCC. Federal Communications Commission.

FCC. Flight Control Computer.

Feather (helicopter rotor blade movement). Rotation of a helicopter rotor blade about its pitch-change axis.

Ferrous metal. Any metal that contains iron and has magnetic characteristics.

Fiber stop nut. A form of a self-locking nut that has a fiber insert crimped into a recess above the threads. The hole in the insert is slightly smaller than the minor diameter of the threads. When the nut is screwed down over the bolt threads, the opposition caused by the fiber insert produces a force that prevents vibration loosening the nut.

File. A hand-held cutting tool used to remove a small amount of metal with each stroke.

Fill threads. Threads in a piece of fabric that run across the width of the fabric, interweaving with the warp threads. Fill threads are often called woof, or weft, threads.

Fillet. A fairing used to give shape but not strength to an object. A fillet produces a smooth junction where two surfaces meet.

Finishing tape. Another name for surface tape. See surface tape.

Fishmouth splice. A type of splice used in a welded tubular structure in which the end of the tube whose inside diameter is the same as the outside diameter of the tube being spliced is cut in the shape of a V, or a fishmouth, and is slipped over the smaller tube welded. A fishmouth splice has more weld area than a butt splice and allows the stresses from one tube to transfer into the other tube gradually.

Fire pull handle. The handle in an aircraft flight deck that is pulled at the first indication of an engine fire. Pulling this handle removes the generator from the electrical system, shuts off the fuel and hydraulic fluid to the engine, and closes the compressor bleed air valve. The fire extinguisher agent discharge switch is uncovered, but it is not automatically closed.

Fire zone. A portion of an aircraft designated by the manufacturer to require fire-detection and/or fire-extinguishing equipment and a high degree of inherent fire resistance.

Fitting. An attachment device that is used to connect components to an aircraft structure.

Fixed fire-extinguishing system. A fire-extinguishing system installed in an aircraft.

Flameout. A condition in the operation of a gas turbine engine in which the fire in the engine unintentionally goes out.

Flap (aircraft control). A secondary control on an airplane wing that changes its camber to increase both its lift and its drag.

Flap (helicopter rotor blade movement). Up-and-down movement of the tip of a helicopter rotor blade.

Flap overload valve. A valve in the flap system of an airplane that prevents the flaps being lowered at an airspeed which could cause structural damage. If the pilot tries to extend the flaps when the airspeed is too high, the opposition caused by the air flow will open the overload valve and return the fluid to the reservoir.

Flash point. The temperature to which a material must be raised for it to ignite, but not continue to burn, when a flame is passed above it.

Flat pattern layout. The pattern for a sheet metal part that has the material used for each flat surface, and for all of the bends, marked out with bend-tangent lines drawn between the flats and bend allowances.

Flight controller. The component in an autopilot system that allows the pilot to maneuver the aircraft manually when the autopilot is engaged.

Fluid. A form of material whose molecules are able to flow past one another without destroying the material. Gases and liquids are both fluids.

Fluid power. The transmission of force by the movement of a fluid. The most familiar examples of fluid power systems are hydraulic and pneumatic systems.

Flutter. Rapid and uncontrolled oscillation of a flight control surface on an aircraft that is caused by a dynamically unbalanced condition.

Fly-by-wire. A method of control used by some modern aircraft in which control movement or pressures exerted by the pilot are directed into a digital computer where they are input into a program tailored to the flight characteristics of the aircraft. The computer output signal is sent to actuators at the control surfaces to move them the optimum amount for the desired maneuver.

Flying boat. An airplane whose fuselage is built in the form of a boat hull to allow it to land and takeoff from water. In the past, flying boats were a popular form of large airplane.

Flying wing. A type of heavier-than-air aircraft that has no fuselage or separate tail surfaces. The engines and useful load are carried inside the wing, and movable control surfaces on the trailing edge provide both pitch and roll control.

Foot-pound. A measure of work accomplished when a force of 1 pound moves an object a distance of 1 foot.

Force. Energy brought to bear on an object that tends to cause motion or to change motion.

Forehand welding. Welding in which the torch is pointed in the direction the weld is progressing.

Form drag. Parasite drag caused by the form of the object passing through the air.

Former. An aircraft structural member used to give a fuselage its shape.

FMC. Flight Management Computer.

Forward bias. A condition of operation of a semiconductor device such as a diode or transistor in which a positive voltage is connected to the P-type material and a negative voltage to the N-type material.

FPD. Freezing point depressant.

Fractional distillation. A method of separating the various components from a physical mixture of liquids. The material to be separated is put into a container and its temperature is increased. The components having the lowest boiling points boil off first and are condensed. Then, as the temperature is further raised, other components are removed. Kerosene, gasoline, and other petroleum products are obtained by fractional distillation of crude oil.

Frangible. Breakable, or easily broken.

Freon. The registered trade name for a refrigerant used in a vapor-cycle air conditioning system.

Frise aileron. An aileron with its hinge line set back from the leading edge so that when it is deflected upward, part of the leading edge projects below the wing and produces parasite drag to help overcome adverse yaw.

Full-bodied. Not thinned.

Fully articulated rotor. A helicopter rotor whose blades are attached to the hub in such a way that they are free to flap, drag, and feather. See each of these terms.

Frost. Ice crystal deposits formed by sublimation when the temperature and dew point are below freezing.

Fuel-flow transmitter. A device in the fuel line between the engine-driven fuel pump and the carburetor that measures the rate of flow of the fuel. It converts this flow rate into an electrical signal and sends it to an indicator in the instrument panel.

Fuel jettison system. A system installed in most large aircraft that allows the flight crew to jettison, or dump, fuel to lower the gross weight of the aircraft to its allowable landing weight. Boost pumps in the fuel tanks move the fuel from the tank into a fuel manifold. From the fuel manifold, it flows away from the aircraft through dump chutes is each wing tip. The fuel jettison system must be so designed and constructed that it is free from fire hazards.

Fuel totalizer. A fuel quantity indicator that gives the total amount of fuel remaining on board the aircraft on one instrument. The totalizer adds the quantities of fuel in all of the tanks.

Fungus (plural: fungi). Any of several types of plant life that include yeasts, molds, and mildew.

Fusible plugs. Plugs in the wheels of high-performance airplanes that use tubeless tires. The centers of the plugs are filled with a metal that melts at a relatively low temperature. If a takeoff is aborted and the pilot uses the brakes excessively, the heat transferred into the wheel will melt the center of the fusible plugs and allow the air to escape from the tire before it builds up enough pressure to cause an explosion.

G

Gauge (rivet). The distance between rows of rivets in a multirow seam. Gauge is also called transverse pitch.

Gauge pressure. Pressure referenced from the existing atmospheric pressure.

Galling. Fretting or pulling out chunks of a surface by sliding contact with another surface or body.

Gasket. A seal between two parts where there is no relative motion.

Gear-type pump. A constant-displacement fluid pump that contains two meshing large-tooth spur gears. Fluid is drawn into the pump as the teeth separate and is carried around the inside of the housing with teeth and is forced from the pump when the teeth come together.

Generator. A mechanical device that transforms mechanical energy into electrical energy by rotating a coil inside a magnetic field. As the conductors in the coil cut across the lines of magnetic flux, a voltage is generated that causes current to flow.

Generator series field. A set of heavy field windings in a generator connected in a series with the armature. The magnetic field produced by the series windings is used to change the characteristics of the generator.

Generator shunt field. A set of field windings in a generator connected in parallel with the armature. Varying the amount of current flowing in the shunt field windings controls the voltage output of the generator.

Gerotor pump. A form of constant-displacement gear pump. A gerotor pump uses an external-tooth spur gear that rides inside of and drives an internal-tooth rotor gear. There is one more tooth space inside the rotor than there are teeth on the drive gear. As the gears rotate, the volume of the space between two of the teeth on the inlet side of the pump increases, while the volume of the space between the two teeth on the opposite side of the pump decreases.

GHz (gigahertz). 1,000,000,000 cycles per second.

Gimbal. A support that allows a gyroscope to remain in an upright condition when its base is tilted.

Glass flight deck. An aircraft instrument system that uses a few cathode-ray-tube displays to replace a large number of mechanically actuated instruments.

Glaze ice. Ice that forms when large drops of water strike a surface whose temperature is below freezing. Glaze ice is clear and heavy.

Glide slope. The portion of an ILS (Instrument Landing System) that provides the vertical path along which an aircraft descends on an instrument landing.

Goniometer. Electronic circuitry in an ADF system that uses the output of a fixed loop antenna to sense the angle between a fixed reference, usually the nose of the aircraft, and the direction from which the radio signal is being received.

Gram. The basic unit of weight or mass in the metric system. One gram equals approximately 0.035 ounce.

Graphite. A form of carbon. Structural graphite is used in composite structure because of its strength and stiffness.

Greige (pronounced "gray"). The unshrunk condition of a polyester fabric as it is removed from the loom.

Ground effect. The increased aerodynamic lift produced when an airplane or helicopter is flown nearer than half wing span or rotor span to the ground. This additional lift is caused by an effective increase in angle of attack without the accompanying increase in induced drag, which is caused by the deflection of the downwashed air.

Ground. The voltage reference point in an aircraft electrical system. Ground has zero electrical potential. Voltage values, both positive and negative, are measured from ground. In the United Kingdom, ground is spoken of as "earth."

Ground-power unit (GPU). A service component used to supply electrical power to an aircraft when it is being operated on the ground.

Guncotton. A highly explosive material made by treating cotton fibers with nitric and sulfuric acids. Guncotton is used in making the film base of nitrate dope.

Gusset. A small plate attached to two or more members of a truss structure. A gusset strengthens the truss.

Gyro (gyroscope). The sensing device in an autopilot system. A gyroscope is a rapidly spinning wheel with its weight concentrated around its rim. Gyroscopes have two basic characteristics that make them useful in aircraft instruments: rigidity in space and precession. See rigidity in space and precession.

Gyroscopic precession. The characteristic of a gyroscope that causes it to react to an applied force as though the force were applied at a point 90° in the direction of rotation from the actual point of application. The rotor of a helicopter acts in much the same way as a gyroscope and is affected by gyroscopic precession.

H

Halon 1211. A halogenated hydrocarbon fire-extinguishing agent used in many HRD fire-extinguishing systems for powerplant protection. The technical name for Halon 1211 is bromochlorodifluoromethane.

Halon 1301. A halogenated hydrocarbon fire-extinguishing agent that is one of the best for extinguishing cabin and powerplant fires. It is highly effective and is the least toxic of the extinguishing agents available. The technical name for Halon 1301 is bromotrifluoromethane.

Hangar rash. Scrapes, bends, and dents in an aircraft structure caused by careless handling.

Hardwood. Wood from a broadleaf tree that sheds its leaves each year.

Heading indicator. A gyroscopic flight instrument that gives the pilot an indication of the heading of the aircraft.

Heat exchanger. A device used to exchange heat from one medium to another. Radiators, condensers, and evaporators are all examples of heat exchangers. Heat always moves from the object or medium having the greatest level of heat energy to a medium or object having a lower level.

Helix. A screw-like, or spiral, curve.

Hertz. One cycle per second.

Holding relay. An electrical relay that is closed by sending a pulse of current through the coil. It remains closed until the current flowing through its contacts is interrupted.

Homebuilt aircraft. Aircraft that are built by individuals as a hobby rather than by factories as commercial products. Homebuilt, or amateur-built, aircraft are not required to meet the stringent requirements imposed on the manufacture of FAA-certified aircraft.

Horsepower. A unit of mechanical power that is equal to 33,000 foot-pounds of work done in 1 minute, or 550 foot-pounds of work done in 1 second.

Hot dimpling. A process used to dimple, or indent, the hole into which a flush rivet is to be installed. Hot dimpling is done by clamping the metal between heating elements and forcing the dies through the holes in the softened metal. Hot dimpling prevents hard metal from cracking when it is dimpled.

Hot-wire cutter. A cutter used to shape blocks of Styrofoam. The wire is stretched tight between the arms of a frame and heated by electrical current. The hot wire melts its way through the foam.

HRD. High-rate-discharge.

HSI. Horizontal situation indicator.

Hydraulic actuator. The component in a hydraulic system that converts hydraulic pressure into mechanical force. The two main types of hydraulic actuators are linear actuators (cylinders and pistons) and rotary actuators (hydraulic motors).

Hydraulic fuse. A type of flow control valve that allows a normal flow of fluid in the system but, if the flow rate is excessive, or if too much fluid flows for normal operation, the fuse will shut off all further flow.

Hydraulic motor. A hydraulic actuator that converts fluid pressure into rotary motion. Hydraulic motors have an advantage in aircraft installations over electric motors, because they can operate in a stalled condition without the danger of a fire.

Hydraulic power pack. A small, self-contained hydraulic system that consists of a reservoir, pump, selector valves, and relief valves. The power pack is removable from the aircraft as a unit to facilitate maintenance and service.

Hydraulics. The system of fluid power which transmits force through an incompressible fluid.

Hydrocarbon. An organic compound that contains only carbon and hydrogen. The vast majority of fossil fuels, such as gasoline and turbine-engine fuel, are hydrocarbons.

Hydroplaning. A condition that exists when a high-speed airplane is landed on a water-covered runway. When the brakes are applied, the wheels lock up and the tires skid on the surface of the water in much the same way a water ski rides on the surface. Hydroplaning develops enough heat in a tire to ruin it.

Hydrostatic test. A pressure test used to determine the serviceability of high-pressure oxygen cylinders. The cylinders are filled with water and pressurized to 5/3 of their working pressure. Standard-weight cylinders (DOT 3AA) must by hydrostatically tested every five years, and lightweight cylinders (DOT 3HT) must be tested every three years.

Hypersonic speed. Speed of greater than Mach 5 (5 times the speed of sound).

Hyperbolic navigation. Electronic navigation systems that determine aircraft location by the time difference between reception of two signals. Signals from two stations at different locations will be received in the aircraft at different times. A line plotted between two stations along which the time difference is the same forms a hyperbola.

Hypoxia. A physiological condition in which a person is deprived of the needed oxygen. The effects of hypoxia normally disappear as soon as the person is able to breathe air containing sufficient oxygen.

ICAO. The International Civil Aeronautical Organization.

Icebox rivet. A solid rivet made of 2017 or 2024 aluminum alloy. These rivets are too hard to drive in the condition they are received from the factory, and must be heat-treated to soften them. They are heated in a furnace and then quenched in cold water. Immediately after quenching they are soft, but within a few hours at room temperature they become quite hard. The hardening can be delayed for several days by storing them in a subfreezing icebox and holding them at this low temperature until they are to be used.

IFR. Instrument flight rules.

Inch-pound. A measure of work accomplished when a force of 1 pound moves an object a distance of 1 inch.

Indicated airspeed (IAS). The airspeed as shown on an airspeed indicator with no corrections applied.

Induced current. Electrical current produced in a conductor when it is moved through or crossed by a magnetic field.

Induced drag. Aerodynamic drag produced by an airfoil when it is producing lift. Induced drag is affected by the same factors that affect induced lift.

Induction time. The time allowed an epoxy or polyurethane material between its initial mixing and its application. This time allows the materials to begin their cure.

Infrared radiation. Electromagnetic radiation whose wavelengths are longer than those of visible light.

Ingot. A large block of metal that was molded as it was poured from the furnace. Ingots are further processed into sheets, bars, tubes, or structural beams.

INS. Inertial Navigation System.

Inspection Authorization (IA). An authorization that may be issued to an experienced aviation maintenance technician who holds both an Airframe and Powerplant rating. It allows the holder to conduct annual inspections and to approve an aircraft or aircraft engine for return to service after a major repair or major alteration.

Integral fuel tank. A fuel tank which is formed by sealing off part of the aircraft structure and using it as a fuel tank. An integral wing tank is called a "wet wing." Integral tanks are used because of their large weight saving. The only way of repairing an integral fuel tank is by replacing damaged sealant and making riveted repairs, as is done with any other part of the aircraft structure.

Interference drag. Parasite drag caused by air flowing over one portion of the airframe interfering with the smooth flow of air over another portion.

Intermittent-duty solenoid. A solenoid-type switch whose coil is designed for current to flow through it for only a short period of time. The coil will overheat if current flows through it too long.

IRS. Inertial Reference System.

IRU. Inertial Reference Unit.

Iso-octane. A hydrocarbon, C_8H_{18}, which has very high critical pressure and temperature. Iso-octane is used as the high reference for measuring the antidetonation characteristics of a fuel.

Isobaric mode. The mode of pressurization in which the cabin pressure is maintained at a constant value regardless of the outside air pressure.

Isogonic line. A line drawn on an aeronautical chart along which the angular difference between the magnetic and geographic north poles is the same.

Isopropyl alcohol. A colorless liquid used in the manufacture of acetone and its derivatives and as a solvent and anti-icing agent.

J

Jackscrew. A hardened steel rod with strong threads cut into it. A jackscrew is rotated by hand or with a motor to apply a force or to lift an object.

Jet pump. A special venturi in a line carrying air from certain areas in an aircraft that need an augmented flow of air through them. High-velocity compressor bleed air is blown into the throat of a venturi where it produces a low pressure that pulls air from the area to which it is connected. Jet pumps are often used in the lines that pull air through galleys and toilet areas.

Joggle. A small offset near the edge of a piece of sheet metal. It allows one sheet of metal to overlap another sheet while maintaining a flush surface.

Jointer. A woodworking power tool used to smooth edges of a piece of wood.

K

K-factor. A factor used in sheet metal work to determine the setback for other than a 90° bend. Setback = K · (bend radius + metal thickness). For bends of less than 90°, the value of K is less than 1; for bends greater than 90°, the value of K is greater than 1.

Kevlar. A patented synthetic aramid fiber noted for its flexibility and light weight. It is to a great extent replacing fiberglass as a reinforcing fabric for composite construction.

Key (verb). To initiate an action by depressing a key or a button.

kHz (kilohertz). 1,000 cycles per second.

Kick-in pressure. The pressure at which an unloading valve causes a hydraulic pump to direct its fluid into the system manifold.

Kick-out pressure. The pressure at which an unloading valve shuts off the flow of fluid into the system pressure manifold and directs it back to the reservoir under a much reduced pressure.

Kilogram. One thousand grams.

Kinetic energy. Energy that exists because of motion.

Knot (wood defect). A hard, usually round section of a tree branch embedded in a board. The grain of the knot is perpendicular to the grain of the board. Knots decrease the strength of the board and should be avoided where strength is needed.

Knot (measure of speed). A speed measurement that is equal to one nautical mile per hour. One knot is equal to 1.15 statute mile per hour.

Kollsman window. The barometric scale window of a sensitive altimeter. See barometric scale.

Koroseal lacing. A plastic lacing material available in round or rectangular cross sections and used for holding wire bundles and tubing together. It holds tension on knots indefinitely and is impervious to petroleum products.

Kraft paper. A tough brown wrapping paper, like that used for paper bags.

L

Lacquer. A finishing material made of a film base, solvents, plasticizers, and thinners. The film base forms a tough film over the surface when it dries. The solvents dissolve the film base so it can be applied as a liquid. The plasticizers give the film base the needed resilience, and the thinners dilute the lacquer so it can be applied with a spray gun. Lacquer is sprayed on the surface as a liquid, and when the solvents and thinners evaporate, the film base remains as a tough decorative and protective coating.

Landing gear warning system. A system of lights used to indicate the condition of the landing gear. A red light illuminates when any of the gears are in an unsafe condition; a green light shows when all of the gears are down and locked, and no light is lit when the gears are all up and locked. An aural warning system is installed that sounds a horn if any of the landing gears are not down and locked when the throttles are retarded for landing.

Laminar flow. Airflow in which the air passes over the surface in smooth layers with a minimum of turbulence.

Laminated wood. A type of wood made by gluing several pieces of thin wood together. The grain of all pieces runs in the same direction.

Latent heat. Heat that is added to a material that causes a change in its state without changing its temperature.

Lateral axis. An imaginary line, passing through the center of gravity of an airplane, and extending across it from wing tip to wing tip.

Lay-up. The placement of the various layers of resin-impregnated fabric in the mold for a piece of laminated composite material.

L/D ratio. A measure of efficiency of an airfoil. It is the ratio of the lift to the total drag at a specified angle of attack.

Left-right indicator. The course-deviation indicator used with a VOR navigation system.

Lightening hole. A hole cut in a piece of structural material to get rid of weight without losing any strength. A hole several inches in diameter may be cut in a piece of metal at a point where the metal is not needed for strength, and the edges of the hole are flanged to give it rigidity. A piece of metal with properly flanged lightening holes is more rigid than the metal before the holes were cut.

Linear actuator. A fluid power actuator that uses a piston moving inside a cylinder to change pressure into linear, or straight-line, motion.

Linear change. A change in which the output is directly proportional to the input.

Loadmeter. A current meter used in some aircraft electrical systems to show the amount of current the generator or alternator is producing. Loadmeters are calibrated in percent of the generator rated output.

Localizer. The portion of an ILS (Instrument Landing System) that directs the pilot along the center line of the instrument runway.

Lodestone. A magnetized piece of natural iron oxide.

Logic flow chart. A type of graphic chart that can be made up for a specific process or procedure to help follow the process through all of its logical steps.

Longitudinal axis. An imaginary line, passing through the center of gravity of an airplane, and extending lengthwise through it from nose to tail.

Longitudinal stability. Stability of an aircraft along its longitudinal axis and about its lateral axis. Longitudinal stability is also called pitch stability.

LORAN A. Long Range Aid to Navigation. A hyperbolic navigation system that operates with frequencies of 1,950 kHz, 1,850 kHz, and 1,900 kHz.

LORAN C. The LORAN system used in aircraft. It operates on a frequency of 100 kHz.

LRU. Line replaceable unit.

Lubber line. A reference on a magnetic compass and directional gyro that represents the nose of the aircraft. The heading of the aircraft is shown on the compass card opposite the lubber line.

M

Mach number. A measurement of speed based on the ratio of the speed of the aircraft to the speed of sound under the same atmospheric conditions. An airplane flying at Mach 1 is flying at the speed of sound.

Magnetic bearing. The direction to or from a radio transmitting station measured relative to magnetic north.

Major alteration. An alteration not listed in the aircraft, aircraft engine, or propeller specifications. It is one that might appreciably affect weight, balance, structural strength performance, powerplant operation, flight characteristics, or other qualities affecting airworthiness, or that cannot be made with elementary operations.

Major repair. A repair to an aircraft structure or component that if improperly made might appreciably affect weight, balance, structural strength, performance, powerplant operation, flight characteristics, or other qualities affecting airworthiness, or that is not done according to accepted practices, or cannot be made with elementary operation.

Manifold cross-feed fuel system. A type of fuel system commonly used in large transport category aircraft. All fuel tanks feed into a common manifold, and the dump chutes and the single-point fueling valves are connected to the manifold. Fuel lines to each engine are taken from the manifold.

Manifold pressure. The absolute pressure of the air in the induction system of a reciprocating engine.

Manifold pressure gauge. A pressure gauge that measures the absolute pressure inside the induction system of a reciprocating engine. When the engine is not operating, this instrument shows the existing atmospheric pressure.

Master switch. A switch in an aircraft electrical system that can disconnect the battery from the bus and open the generator or alternator field circuit.

Matrix. The material used in composite construction to bond the fibers together and to transmit the forces into the fibers. Resins are the most widely used matrix materials.

Mean camber. A line that is drawn midway between the upper and lower camber of an airfoil section. The mean camber determines the aerodynamic characteristics of the airfoil.

MEK. Methyl-ethyl-ketone is an organic chemical solvent that is soluble in water and is used as a solvent for vinyl and nitrocellulose films. MEK is an efficient cleaner for preparing surfaces for priming or painting.

Mercerize. A treatment given to cotton thread to make it strong and lustrous. The thread is stretched while it is soaked in a solution of caustic soda.

MFD. Multi-function display.

MHz (megahertz). 1,000,000 cycles per second.

Microballoons. Tiny, hollow spheres of glass or phenolic material used to add body to a resin.

Microbial contaminants. The scum that forms inside the fuel tanks of turbine-engine-powered aircraft that is caused by micro-organisms. These micro-organisms live in water that condenses from fuel, and they feed on the fuel. The scum they form clogs fuel filters, lines, and fuel controls and holds water in contact with the aluminum alloy structure, causing corrosion.

Micro-Mesh. A patented graduated series of cloth-backed cushioned seats that contain abrasive crystals. Micro-Mesh is used for polishing and restoring transparency to acrylic plastic windows and windshields.

Micron ("micro meter"). A unit of linear measurement equal to one millionth of a meter, one thousandth of a millimeter, or 0.000039 inch. A micron is also called a micrometer.

Micronic filter. The registered trade name of a type of fluid filter whose filtering element is a specially treated cellulose paper formed into vertical convolutions, or wrinkles. Micronic filters prevent the passage of solids larger than about 10 microns, and are normally replaced with new filters rather than cleaned.

Micro-organism. An organism, normally bacteria or fungus, or microscopic size.

Microswitch. The registered trade name for a precision switch that uses a short throw of the control plunger to actuate the contacts. Microswitches are used primarily as limit switches to control electrical units automatically.

MIG welding. Metal inert gas welding is a form of electric arc welding in which the electrode is an expendable wire. MIG welding is now called GMA (gas metal arc) welding.

Mil. One thousandth of an inch (0.001 inch). Paint film thickness is usually measured in mils.

Mildew. A gray or white fungus growth that forms on organic materials. Mildew forms on cotton and linen aircraft fabric and destroys its strength.

Millivoltmeter. An electrical instrument that measures voltage in units of millivolts (thousandths of a volt).

Mist coat. A very light coat of zinc chromate primer. It is so thin that the metal is still visible, but the primer makes pencil marks easy to see.

Moisture separator. A component in a high-pressure pneumatic system that removes most of the water vapor from the compressed air. When the compressed air is used, its pressure drops, and this pressure drop causes a drop in temperature. If any moisture were allowed to remain in the air, it would freeze and block the system.

Mold line. A line used in the development of a flat pattern for a formed piece of sheet metal. The mold line is an extension of the flat side of a part beyond the radius. The mold line dimension of a part is the dimension made to the intersection of mold lines and is the dimension the part would have if its corners had no radius.

Mold point. The intersection of two mold lines of a part. Mold line dimensions are made between mold points.

Moment. A force that causes or tries to cause an object to rotate. The value of a moment is the product of the weight of an object (or the force), multiplied by the distance between the center of gravity of the object (or the point of application of the force) and the fulcrum about which the object rotates.

Monel. An alloy of nickel, copper, and aluminum or silicon.

Monocoque. A single-shell type of aircraft structure in which all of the flight loads are carried in the outside skin of the structure.

MSDS. Material Safety Data Sheets. MSDS are required by the Federal Government to be available in workplaces to inform workers of the dangers that may exist from contact with certain materials.

MSL. Mean sea level. When the letters MSL are used with an altitude, it means that the altitude is measured from mean, or average, sea level.

MTBF. Mean time between failures.

Multimeter. An electrical test instrument that consists of a single current-measuring meter and all of the needed components to allow the meter to be used to measure voltage, resistance, and current. Multimeters are available with either analog-or digital-type displays.

Multiple-disc brakes. Aircraft brakes in which one set of discs is keyed to the axle and remains stationary. Between each stationary disc there is a rotating disc that is keyed to the inside of the wheel. When the brakes are applied, the stationary discs are forced together, clamping the rotating discs between them. The friction between the discs slows the aircraft.

N

Nailing strip. A method of applying pressure to the glue in a scarf joint repair in a plywood skin. A strip of thin plywood is nailed over the glued scarf joint with the nails extending into a supporting structure beneath the skin. The strip is installed over vinyl sheeting to prevent it sticking to the skin. When the glue is thoroughly dry, the nailing strip is broken away and the nails removed.

Nap of the fabric. The ends of the fibers in a fabric. The first coat of dope on cotton or linen fabric raises the nap, and the fiber ends stick up. These ends must be carefully removed by sanding to get a smooth finish.

Naphtha. A volatile and flammable hydrocarbon liquid used chiefly as a solvent or as a cleaning fluid.

NDB. Non-directional beacons.

Negative pressure relief valve (pressurization component). A valve that opens anytime the outside air pressure is greater than the cabin pressure. It prevents the cabin altitude from ever becoming greater than the aircraft flight altitude.

Neutral axis (neutral plane). A line through a piece of material that is bent. The material in the outside of the bend is stretched and that on the inside of the bend is shrunk. The material along the neutral plane is neither shrunk nor stretched.

Neutral flame. An oxyacetylene flame produced when the ratio of oxygen and acetylene is chemically correct and there is no excess of oxygen or carbon. A neutral flame has a rounded inner cone and no feather around it.

Noise (electrical). An unwanted electrical signal within a piece of electronic equipment.

Nomex. A patented nylon material used to make the honeycomb core for certain types of sandwich materials.

Nonenergizing brake. A brake that does not use the momentum of the aircraft to increase the friction.

Nonvolatile memory. Memory in a computer that is not lost when power to the computer is lost.

Normal heptane. A hydrocarbon, C_7H_{16}, with a very low critical pressure and temperature. Normal heptane is used as the low reference in measuring the anti-detonation characteristics of a fuel.

Normal shock wave. A shock wave that forms ahead of a blunt object moving through the air at the speed of sound. The shock wave is normal (perpendicular) to the air approaching the object. Air passing through a normal shock wave is slowed to a subsonic speed and its static pressure is increased.

Normalizing. A process of strain-relieving steel that has been welded and left in a strained condition. The steel is heated to a specified temperature, usually red hot, and allowed to cool in still air to room temperature.

Nose-gear centering cam. A cam in the nose-gear shock strut that causes the piston to center when the strut fully extends. When the aircraft takes off and the strut extends, the wheel is straightened in its fore-and-aft position so it can be retracted into the wheel well.

NPN transistor. A bipolar transistor made of a thin base of P-type silicon or geranium sandwiched between a collector and an emitter, both of which are made of N-type material.

Null position. The position of an ADF loop antenna when the signal being received is canceled in the two sides of the loop and the signal strength is the weakest.

O

Oblique shock wave. A shock wave that forms on a sharp-pointed object moving through air at a speed greater than the speed of sound. Air passing through an oblique shock wave is slowed down, but not to a subsonic speed, and its static pressure is increased.

Oleo shock absorber. A shock absorber used on aircraft landing gear. The initial landing impact is absorbed by oil transferring from one compartment in the shock strut into another compartment through a metering orifice. The shocks of taxiing are taken up by a cushion of compressed air.

Octane rating. A rating of the anti-detonation characteristics of a reciprocating engine fuel. It is based on the performance of the fuel in a special test engine. When a fuel is given a dual rating such as 80/87, the first number is its anti-detonating rating with a lean fuel-air mixture, and the higher number is its rating with a rich mixture.

Open angle. An angle in which sheet metal is bent less than 90°.

Open assembly time. The period of time between the application of the glue and the assembly of the joint components.

Open-hydraulic system. A fluid power system in which the selector valves are arranged in series with each other. Fluid flows from the pump through the center of the selector valves, back into the reservoir when no unit is being actuated.

Open-center selector valve. A type of selector valve that functions as an unloading valve as well as a selector valve. Open-center selector valves are installed in series, and when no unit is actuated, fluid from the pump flows through the centers of all the valves and returns to the reservoir. When a unit is selected for actuation, the center of the selector valve is shut off and the fluid from the pump goes through the selector valve into one side of the actuator. Fluid from the other side of the actuator returns to the valve and goes back to the reservoir through the other selector valves. When the actuation is completed, the selector valve is placed in its neutral position. Its center opens, and fluid from the pump flows straight through the valve.

Open wiring. An electrical wiring installation in which the wires are tied together in bundles and clamped to the aircraft structure rather than being enclosed in conduit.

Orifice check valve. A component in a hydraulic or pneumatic system that allows unrestricted flow in one direction, and restricted flow in the opposite direction.

O-ring. A widely used type of seal made in the form of a rubber ring with a round cross section. An O-ring seals in both directions, and it can be used as a packing or a gasket.

Ornithopter. A heavier-than-air flying machine that produces lift by flapping its wings. No practical ornithopter has been built.

Oscilloscope. An electrical instrument that displays on the face of a cathode-ray tube the waveform of the electrical signal it is measuring.

Outflow valve (pressurization component). A valve in the cabin of a pressurized aircraft that controls the cabin pressure by opening to relieve all pressure above that for which the cabin pressure control is set.

Overvoltage protector. A component in an aircraft electrical system that opens the alternator field circuit any time the alternator output voltage is too high.

Oxidizing flame. An oxyacetylene flame in which there is an excess of oxygen. The inner cone is pointed and often a hissing sound is heard.

Ozone. An unstable form of oxygen produced when an electric spark passes through the air. Ozone is harmful to rubber products.

P

Packing. A seal between two parts where there is relative motion.

Paint. A covering applied to an object or structure to protect it and improve its appearance. Paint consists of a pigment suspended in a vehicle such as oil or water. When the vehicle dries by evaporation or curing, the pigment is left as a film on the surface.

Parabolic reflector. A reflector whose surface is made in the form of a parabola.

Parallel circuit. A method of connecting electrical components so that each component is in a path between the terminals of the source of electrical energy.

Paralleling circuit. A circuit in a multi-engine aircraft electrical system that controls a flow of control current which is used to keep the generators or alternators sharing the electrical load equally. The relay opens automatically to shut off the flow of paralleling current any time the output of either alternator or generator drops to zero.

Paralleling relay. A relay in multi-engine aircraft electrical system that controls a flow of control current which is used to keep the generators or alternators sharing the electrical load equally. The relay opens automatically to shut off the flow of paralleling current any time the output of either alternator or generator drops to zero.

Parasite drag. A form of aerodynamic drag caused by friction between the air and the surface over which it is flowing.

Parent metal. The metal being welded. This term is used to distinguish between the metal being welded and the welding rod.

Partial pressure. The percentage of the total pressure of a mixture of gases produced by each of the individual gases in the mixture.

Parting film. A layer of thin plastic material placed between a composite lay-up and the heating blanket. It prevents the blanket from sticking to the fabric.

Pascal's Law. A basic law of fluid power which states that the pressure in an enclosed container is transmitted equally and undiminished to all points of the container, and the force acts at right angles to the enclosing walls.

Performance number. The anti-detonation rating of a fuel that has a higher critical pressure and temperature than iso-octane (a rating of 100). Iso-octane that has been treated with varying amounts of tetraethyl lead is used as the reference fuel.

Petrolatum-zinc dust compound. A special abrasive compound used inside an aluminum wire terminal being swaged onto a piece of aluminum electrical wire. When the terminal is compressed, the zinc dust abrades the oxides from the wire, and the petrolatum prevents oxygen reaching the wire so no more oxides can form.

Petroleum fractions. The various components of a hydrocarbon fuel that are separated by boiling them off at different temperatures in the process of fractional distillation.

Phased array antenna. A complex antenna which consists of a number of elements. A beam of energy is formed by the superimposition of the signals radiating from the elements. The direction of the beam can be changed by varying the relative phase of the signals applied to each of the elements.

Phenolic plastic. A plastic material made of a thermosetting phenol-formaldehyde resin, reinforced with cloth or paper. Phenolic plastic materials are used for electrical insulators and for chemical-resistant table tops.

Pilot hole. A small hole punched or drilled in a piece of sheet metal to locate a rivet hole.

Pin knot cluster. A group of knots, all having a diameter of less than approximately $1/16$ inch.

Pinked-edge tape. Cloth tape whose edges have small V-shaped notches cut along their length. The pinked edges prevent the tape from raveling.

Pinking shears. Shears used to cut aircraft fabric with a series of small notches along the cut edge.

Pinion. A small gear that meshes with a larger gear, a sector of a gear, or a toothed rack.

Piston. A sliding plug in an actuating cylinder used to convert pressure into force and then into work.

Pitch (aircraft maneuver). Rotation of an aircraft about its lateral axis.

Pitch (rivet). The distance between the centers of adjacent rivets installed in the small row.

Pitch pocket (wood defect). Pockets of pitch that appear in the growth rings of a piece of wood.

Pitot pressure. Ram air pressure used to measure airspeed. The pitot tube faces directly into the air flowing around the aircraft. It stops the air and measures its pressure.

Plain-weave fabric. Fabric in which each warp thread passes over one fill thread and under the next. Plain-weave fabric typically has the same strength in both warp and fill directions.

Plan position indicator (PPI). A type of radar scope that shows both the direction and distance of the target from the radar antenna. Some radar antenna rotate and their PPI scopes are circular. Other antenna oscillate and their PPI scopes are fan shaped.

Planer. A woodworking power tool used to smooth the surfaces of a piece of wood.

Plasticizer. A constituent in dope or lacquer that gives its film flexibility and resilience.

Plastic media blasting (PMB). A method of removing paint from an aircraft surface by dry-blasting it with tiny plastic beads.

Plastics. The generic name for any of the organic materials produced by polymerization. Plastics can be shaped by molding or drawing.

Plenum. An enclosed chamber in which air can be held at a pressure higher than that of the surrounding air.

Ply rating. The rating of an aircraft tire that indicates its relative strength. The ply rating does not indicate the actual number of plies of fabric in the tire; it indicates the number of piles of cotton fabric needed to produce the same strength as the actual piles.

Plywood. A wood product made by gluing several pieces of thin wood veneer together. The grain of the wood in each layer runs at 90° or 45° to the grain of the layer next to it.

Pneumatics. The system of fluid power which transmits force by the use of a compressible fluid.

PNP transistor. A bipolar transistor made of a thin base of N-type silicon or germanium sandwiched between a collector and an emitter, both of which are made of P-type material.

Polyester fibers. A synthetic fiber made by the polymerization process in which tiny molecules are united to form a long chain of molecules. Polyester fibers are woven into fabrics that are known by their trade names of Dacron, Fortrel, and Kodel.

Polyester film and sheet are known as Mylar and Celenar.

Polyester resin. A thermosetting resin used as a matrix for much of the fiberglass used in composite construction.

Polyurethane enamel. A hard, chemically resistant finish used on aircraft. Polyurethane enamel is resistant to damage from all types of hydraulic fluid.

Polyvinyl chloride. A thermoplastic resin used in the manufacture of transparent tubing for electrical insulation and fluid lines which are subject to low pressures.

Position error. The error in pitot-static instruments caused by the static ports not sensing true static air pressure. Position error changes with airspeed and is usually greatest at low airspeeds.

Potential energy. Energy possessed in an object because of its position, chemical composition, shape, or configuration.

Potentiometer. A variable resistor having connections to both ends of the resistance element and to the wiper that moves across the resistance.

Pot life. The length of time a resin will remain workable after the catalyst has been added. If a catalyzed material is not used within its usable pot life, it must be discarded and a new batch mixed up.

Power. The time rate of doing work. Power is force multiplied by distance (work), divided by time.

Power brakes. Aircraft brakes that use the main hydraulic system to supply fluid for the brake actuation. Aircraft that require a large amount of fluid for their brake actuation normally use power brakes, and the volume of fluid sent to the brakes is increased by the use of deboosters.

Power control valve. A hand-operated hydraulic pump unloading valve. When the valve is open, fluid flows from the pump to the reservoir with little opposition. To actuate a unit, turn the selector valve, and manually close the power control valve. Pressurized fluid flows to the unit, and when it is completely actuated, the power control valve automatically opens.

Precession. The characteristic of a gyroscope that causes a force to be felt, not at the point of application, but at a point 90° in the direction of rotation from that point.

Preflight inspection. A required inspection to determine the condition of the aircraft for the flight to be conducted. It is conducted by the pilot-in-command.

Precipitation heat treatment. A method of increasing the strength of heat-treated aluminum alloy. After the aluminum alloy has been solution-heat-treated by heating and quenching, it is returned to the oven and heated to a temperature lower than that used for the initial heat treatment. It is held at this temperature for a specified period of time, and then removed from the oven and allowed to cool slowly.

Prepreg (preimpregnated fabric). A type of composite material in which the reinforcing fibers are encapsulated in an uncured resin. Prepreg materials must be kept refrigerated to prevent them from curing before they are used.

Press-to-test light fixture. An indicator light fixture whose lens can be pressed in to complete a circuit that tests the filament of the light bulb.

Pressure. Force per unit area. Hydraulic and pneumatic pressure are normally given in units of pounds per square inch (psi).

Pressure altitude. The altitude in standard air at which the pressure is the same as that of the existing air. Pressure altitude is read on an altimeter when the barometric scale is set to the standard sea level pressure of 29.92 inches of mercury.

Pressure-demand oxygen system. A type of oxygen system used by aircraft that fly at very high altitude. This system functions as a diluter-demand system until, at about 40,000 feet, the output to the mask is pressurized enough to force the needed oxygen into the lungs, rather than depending on the low pressure produced when the wearer of the mask inhales to pull in the oxygen. (See diluter-demand oxygen system.)

Pressure fueling. The method of fueling used by almost all transport aircraft. The fuel is put into the aircraft through a single underwing fueling port. The fuel tanks are filled to the desired quantity and in the sequence selected by the person conducting the fueling operation. Pressure fueling saves servicing time by using a single point to fuel the entire aircraft, and it reduces the chances for fuel contamination.

Pressure manifold (hydraulic system component). The portion of a fluid power system from which the selector valves receive their pressurized fluid.

Pressure plate (brake component). A strong, heavy plate used in a multiple-disc brake. The pressure plate receives the force from the brake cylinders and transmits this force to the discs.

Pressure reducing valve (oxygen system component). A valve used in an oxygen system to change high cylinder pressure to low system pressure.

Pressure relief valve (oxygen system component). A valve in an oxygen system that relieves the pressure if the pressure reducing valve should fail.

Pressure vessel. The strengthened portion of an aircraft structure that is sealed and pressurized in flight.

Primer (finishing system component). A component in a finishing system that provides a good bond between the surface and the material used for the topcoats.

Profile drag. Aerodynamic drag produced by skin friction. Profile drag is a form of parasite drag.

Progressive inspection. An inspection that may be used in place of an annual or 100-hour inspection. It has the same scope as an annual inspection, but it may be performed in increments so the aircraft will not have to be out of service for a lengthy period of time.

Pump control valve. A control valve in a hydraulic system that allows the pilot to manually direct the output of the hydraulic pump back to the reservoir when no unit is being actuated.

Pureclad. A registered trade name for clad aluminum alloy.

Purge (air conditioning system operation). To remove all of the moisture and air from a cooling system by flushing the system with a dry gaseous refrigerant.

Pusher powerplant. A powerplant whose propeller is mounted at the rear of the airplane and pushes, rather than pulls, the airplane through the air.

PVC (Polyvinylchloride). A thermoplastic resin used to make transparent tubing for insulating electrical wires.

Q

Quartersawed wood. Wood sawed from a tree in such a way that the annual rings cross the plank at an angle greater than 45°.

Quick-disconnect fitting. A hydraulic line fitting that seals the line when the fitting is disconnected. Quick-disconnect fittings are used on the lines connected to the engine-driven hydraulic pump. They allow the pump to be disconnected and an auxiliary hydraulic power system connected to perform checks requiring hydraulic power while the aircraft is in the hangar.

R

Rack-and-pinion actuator. A form of rotary actuator where the fluid acts on a piston on which a rack of gear teeth is cut. As the piston moves, it rotates a pinion gear which is mated with the teeth cut in the rack.

Radial. A directional line radiating outward from a radio facility, usually a VOR. When an aircraft is flying outbound on the 330° from the station.

Radius dimpling. A process of preparing a hole in sheet metal for flush riveting. A cone-shaped male die forces the edges of the rivet hole into the depression in a female die. Radius dimpling forms a round-edged depression into which the rivet head fits.

Range markings. Colored marks on an instrument dial that identify certain ranges of operation as specified in the aircraft maintenance or flight manual and listed in the appropriate aircraft Type Certificate Data Sheets or Aircraft Specifications. Color coding directs attention to approaching operating difficulties. Airspeed indicators and most pressure and temperature indicators are marked to show the various ranges of operation. These ranges and colors are the most generally used: Red radial line, do not exceed. Green arc, normal operating range. Yellow arc, caution range. Blue radial line, used on airspeed indicators to show best single-engine rate of climb speed. White arc, used on airspeed indicators to show flap operating range.

RDF. Radio direction finding.

Rebreather oxygen mask. A type of oxygen mask used with a continuous flow oxygen system. Oxygen continuously flows into the bottom of the loose-fitting rebreather bag on the mask. The wearer of the mask exhales into the top of the bag. The first air exhaled contains some oxygen, and this air goes into the bag first. The last air to leave the lungs contains little oxygen, and it is forced out of the bag as the bag is filled with fresh oxygen. Each time the wearer of the mask inhales, the air first exhaled, along with fresh oxygen, is taken into the lungs.

Receiver-dryer. The component in a vapor-cycle cooling system that serves as a reservoir for the liquid refrigerant. The receiver-dryer contains a desiccant that absorbs any moisture that may be in the system.

Rectangle. A plane surface with four sides whose opposite sides are parallel and whose angles are all right angles.

Rectification (arc welding condition). A condition in AC-electric arc welding in which oxides on the surface of the metal act as a rectifier and prevent electrons flowing from the metal to the electrode during the half cycle when the electrode is positive.

Reducing flame. See carburizing flame.

Reed valve. A thin, leaf-type valve mounted in the valve plate of an air conditioning compressor to control the flow of refrigerant gases into and out of the compressor cylinders.

Reinforcing tape. A narrow strip of woven fabric material placed over the fabric as it is being attached to the aircraft structure with rib lacing cord. This tape carries a large amount of the load and prevents the fabric tearing at the stitches.

Rejuvenator. A finishing material used to restore resilience to an old dope film. Rejuvenator contains strong solvents to open the dried-out film and plasticizers to restore resilience to the old dope.

Relative wind. The direction the wind strikes an airfoil.

Relay. An electrical component which uses a small amount of current flowing through a coil to produce a magnetic pull to close a set of contacts through which a large amount of current can flow. The core in a relay coil is fixed.

Relief hole. A hole drilled at the point at which two bend lines meet in a piece of sheet metal. This hole spreads the stresses caused by the bends and prevents the metal cracking.

Relief valve. A pressure-control valve that relieves any pressure over the amount for which it is set. They are damage-preventing units used in both hydraulic and pneumatic systems. In an aircraft hydraulic system, pressure relief valves prevent damaging high pressures that could be caused by a malfunctioning pressure regulator, or by thermal expansion of fluid trapped in portions of the system.

Repair. A maintenance procedure in which a damaged component is restored to its original condition, or at least to a condition that allows it to fulfill its design function.

Restrictor. A fluid power system component that controls the rate of actuator movement by restricting the flow of fluid into or out of the actuator.

Retard breaker points. A set of breaker points in certain aircraft magnetos that are used to provide a late (retarded) spark for starting the engine.

Retarder (finishing system component). Dope thinner that contains certain additives that slow its rate of evaporation enough to prevent dope blushing.

Retread. The replacement of the tread rubber on an aircraft tire.

Retreating blade. The blade on a helicopter rotor whose tip is moving in the direction opposite to that in which the helicopter is moving.

Retreating blade stall. The stall of a helicopter rotor disc that occurs near the tip of the retreating blade. A retreating blade stall occurs when the flight airspeed is high and the retreating blade airspeed is low. This results in a high angle of attack, causing the stall.

Return manifold. The portion of a fluid power system through which the fluid is returned to the reservoir.

Reverse polarity welding. DC-electric arc welding in which the electrode is positive with respect to the work.

Rib thread. A series of circumferential grooves cut into the tread of a tire. This tread pattern provides superior traction and directional stability on hard-surfaced runways.

Ribbon direction. The direction in a piece of honeycomb material that is parallel to the length of the strips of material that make up the core.

Rigid conduit. Aluminum alloy tubing used to house electrical wires in areas where they are subject to mechanical damage.

Rigidity in space. The characteristic of a gyroscope that prevents its axis of rotation tilting as the earth rotates. This characteristic is used for attitude gyro instruments.

Rime ice. A rough ice that forms on aircraft flying through visible moisture, such as a cloud, when the temperature is below freezing. Rime ice disturbs the smooth airflow as well as adding weight.

Rivet cutters. Special cutting pliers that resemble diagonal cutters except that the jaws are ground in such a way that they cut the rivet shank, or stem, off square.

Rivet set. A tool used to drive aircraft solid rivets. It is a piece of hardened steel with a recess the shape of the rivet head in one end. The other end fits into the rivet gun.

RMI. Radio magnetic indicator.

Rocking shaft. A shaft used in the mechanism of a pressure measuring instrument to change the direction of movement by 90° and to amplify the amount of movement.

Roll (aircraft maneuver). Rotation of an aircraft about its longitudinal axis.

Roots-type air compressor. A positive-displacement air pump that uses two intermeshing figure-8-shaped rotors to move the air.

Rosette weld. A method of securing one metal tube inside another by welding. Small holes are drilled in the outer tube and the inner tube is welded to it around the circumference of the holes.

Rotary actuator. A fluid power actuator whose output is rotational. A hydraulic motor is a rotary actuator.

Roving. A lightly twisted roll or strand of fibers.

RPM. Revolutions per minute.

Ruddervators. The two movable surfaces on a V-tail empennage. When these two surfaces are moved together with the in-and-out movement of the control yoke, they act as elevators, and when they are moved differentially with the rudder pedals, they act as the rudder.

S

Saddle gusset. A piece of plywood glued to an aircraft structural member. The saddle gusset has a cutout to hold a backing block or strip tightly against the skin to allow a nailing strip to be used to apply pressure to a glued joint in the skin.

Sailplane. A high-performance glider.

Sandwich material. A type of composite structural material in which a core material is bonded between face sheets of metal or resin-impregnated fabric.

Satin-weave fabric. Fabric in which the warp threads pass under one fill thread and over several others. Satin-weave fabrics are used when the lay-up must be made over complex shapes.

Scarf joint. A joint in a wood structure in which the ends to be joined are cut in a long taper, normally about 12:1, and fastened together by gluing. A glued scarf joint makes a strong splice because the joint is made along the side of the wood fibers rather than along their ends.

Schematic diagram. A diagram of an electrical system in which the system components are represented by symbols rather than drawings or pictures of the actual devices.

Schrader valve. A type of service valve used in an air conditioning system. This is a spring-loaded valve much like the valve used to put air into a tire.

Scissors. A name commonly used for torque links. See torque links.

Scrim cloth. Scrim cloth can be used in repair applications or for reinforcement of other types of materials including fiberglass, concrete and some plastics. When fully cured, the scrim cloth will add reinforcement and mimic the expansion and contraction of the surrounding substrate.

Scupper. A recess around the filler neck of an aircraft fuel tank. Any fuel spilled when the tank is being serviced collects in the scupper and drains to the ground through a drain line rather than flowing into the aircraft structure.

Sea level engine. A reciprocating engine whose rated takeoff power can be produced only at sea level.

Sector gear. A part of a gear wheel containing the hub and a portion of the rim with teeth.

Series circuit. A method of connecting electrical components in such a way that all the current flows through each of the components. There is only one path for current to flow.

Series-parallel circuit. An electrical circuit in which some of the components are connected in parallel and others are connected in series.

Selcal system. Selective calling system. Each aircraft operated by an airline is assigned a particular four-tone audio combination for identification purposes. A ground station keys the signal whenever contact with that particular aircraft is desired. The signal is decoded by the airborne selcal decoder and the crew alerted by the selcal warning system.

Selsyn system. A DC synchro system used in remote indicating instruments. The rotor in the indicator is a permanent magnet and the stator is a tapped toroidal coil. The transmitter is a circular potentiometer with DC power fed into its wiper which is moved by the object being monitored. The transmitter is connected to the indicator in such a way that rotation of the transmitter shaft varies the current in the sections of the indicator toroidal coil. The magnet in the indicator on which the pointer is mounted locks with the magnetic field produced by the coils and follows the rotation of the transmitter shaft.

Segmented-rotor brake. A heavy-duty, multiple-disc brake used on large, high-speed aircraft. Stators that are surfaced with a material that retains its friction characteristics at high temperatures are keyed to the axle. Rotors which are keyed into the wheels mesh with the stators. The rotors are made in segments to allow for cooling and for their large amounts of expansion.

Selector valve. A flow control valve used in hydraulic systems that directs pressurized fluid into one side of an actuator, and at the same time directs return fluid from the other side of the actuator back to the reservoir. There are two basic types of selector valves: open-center valves and closed-center valves. The four-port closed-center valve is the most frequently used type. See closed-center selector valve and open-center selector valve.

Selvage edge. The woven edge of fabric used to prevent the material unraveling during normal handling. The selvage edge, which runs the length of the fabric parallel to the warp threads, is usually removed from materials used in composite construction.

Semiconductor diode. A two-element electrical component that allows current to pass through it in one direction, but blocks its passage in the opposite direction. A diode acts in an electrical system in the same way a check valve acts in a hydraulic system.

Semimonocoque structure. A form of aircraft stressed skin structure. Most of the strength of a semimonocoque structure is in the skin, but the skin is supported on a substructure of formers and stringers that give the skin its shape and increase its rigidity.

Sensible heat. Heat that is added to a liquid causing a change in its temperature but not its physical state.

Sensitivity. A measure of the signal strength needed to produce a distortion-free output in a radio receiver.

Sequence valve. A valve in a hydraulic system that requires a certain action to be completed before another action can begin. Sequence valves are used to assure that the hydraulically actuated wheel-well doors are completely open before pressure is directed to the landing gear to lower it.

Servo. An electrical or hydraulic actuator connected into a flight control system. A small force on the flight deck control is amplified by the servo and provides a large force to move the control surface.

Servo amplifier. An electronic amplifier in an autopilot system that increases the signal from the autopilot enough that it can operate the servos that move the control surfaces.

Servo tab. A small movable tab built into the trailing edge of a primary control surface of an airplane. The flight deck controls move the tab in such a direction that it produces an aerodynamic force moving the surface on which it is mounted.

Setback. The distance the jaws of a brake must be set back from the mold line to form a bend. Setback for a 90° bend is equal to the inside radius of the bend plus the thickness of the metal being bent. For a bend other than 90°, a K-factor must be used. See also K-factor.

Shake (wood defect). Longitudinal cracks in a piece of wood, usually between two annual rings.

SHF. Super-high frequency.

Shear section. A necked-down section of the drive shaft of a constant-displacement engine-driven fluid pump. If the pump should seize, the shear section will break and prevent the pump from being destroyed or the engine from being damaged. Some pumps use a shear pin rather than a shear section.

Shear strength. The strength of a riveted joint in a sheet metal structure in which the rivets shear before the metal tears at the rivet holes.

Shelf life. The length of time a product is good when it remains in its original unopened container.

Shielded wire. Electrical wire enclosed in a braided metal jacket. Electromagnetic energy radiated from the wire is trapped by the braid and is carried to ground.

Shimmy. Abnormal, and often violent, vibration of the nose wheel of an airplane. Shimmying is usually caused by looseness of the nose wheel support mechanism or an unbalanced wheel.

Shimmy damper. A small hydraulic shock absorber installed between the nose wheel fork and the nose wheel cylinder attached to the aircraft structure.

Shock mounts. Resilient mounting pads used to protect electronic equipment by absorbing low-frequency, high amplitude vibrations.

Shock wave. A pressure wave formed in the air by a flight vehicle moving at a speed greater than the speed of sound. As the vehicle passes through the air, it produces sound waves that spread out in all directions. But since the vehicle is flying faster than these waves are moving, they build up and form a pressure wave at the front and rear of the vehicle. As the air passes through a shock wave it slows down, its static pressure increases, and its total energy decreases.

Shop head. The head of a rivet which is formed when the shank is upset.

Show-type finish. The type of finish put on fabric-covered aircraft intended for show. This finish is usually made up of many coats of dope, with much sanding and rubbing of the surface between coats.

Shunt winding. Field coils in an electric motor or generator that are connected in parallel with the armature.

Shuttle valve. An automatic selector valve mounted on critical components such as landing gear actuation cylinders and brake cylinders. For normal operation, system fluid flows into the actuator through the shuttle valve, but if normal system pressure is lost, emergency system pressure forces the shuttle over and emergency fluid flows into the actuator.

Sidestick controller. A flight deck flight control used on some of the fly-by-wire equipped airplanes. The stick is mounted rigidly on the side console of the flight deck, and pressures exerted on the stick by the pilot produce electrical signals that are sent to the computer that flies the airplane.

Sight glass (air conditioning system component). A small window in the high side of a vapor-cycle cooling system. Liquid refrigerant flows past the sight glass, and if the charge of refrigerant is low, bubbles will be seen. A fully charged system has no bubbles in the refrigerant.

Sight line. A line drawn on a sheet metal layout that is one bend radius from the bend-tangent line. The sight line is lined up directly below the nose of the radius bar in a cornice brake. When the metal is clamped in this position, the bend tangent line is in the correct position for the start of the bend.

Silicon controlled rectifier (SCR). A semiconductor electron control device. An SCR blocks current flow in both directions until a pulse of positive voltage is applied to its gate. It then conducts in its forward direction, while continuing to block current in its reverse direction.

Silicone rubber. An elastomeric material made from silicone elastomers. Silicone rubber is compatible with fluids that attack other natural or synthetic rubbers.

Single-acting actuator. A linear hydraulic or pneumatic actuator that uses fluid power for movement in one direction and a spring force for its return.

Single-action hand pump. A hand-operated fluid pump that moves fluid only during one stroke of the pump handle. One stroke pulls the fluid into the pump and the other forces the fluid out.

Single-disc brakes. Aircraft brakes in which a single steel disc rotates with the wheel between two brake-lining blocks. When the brake is applied, the disc is clamped tightly between the lining blocks, and the friction slows the aircraft.

Single-servo brakes. Brakes that uses the momentum of the aircraft rolling forward to help apply the brakes by wedging the brake shoe against the brake drum.

Sintered metal. A porous material made by fusing powdered metal under heat and pressure.

Skydrol hydraulic fluid. The registered trade name for a synthetic, nonflammable, phosphate ester-base hydraulic fluid used in modern high-temperature hydraulic systems.

Slat. A secondary control on an aircraft that allows it to fly at a high angle of attack without stalling. A slat is a section of leading edge of wing mounted on curved tracks that move into and out of the wing on rollers.

Slip roll former. A shop tool used to form large radius curves on sheet metal.

Slippage mark. A paint mark extending across the edge of an aircraft wheel onto a tube-type tire. When this mark is broken, it indicates the tire has slipped on the wheel, and there is a good reason to believe the tube has been damaged.

Slipstream area. For the purpose of rib stitch spacing, the slipstream area is considered to be the diameter of the propeller plus one wing rib on each side.

Slot (aerodynamic device). A fixed, nozzle-like opening near the leading edge of an airplane wing ahead of the aileron. A slot acts as a duct to force high-energy air down on the upper surface of the wing when the airplane is flying at a high angle of attack. The slot, which is located ahead of the aileron, causes the inboard portion of the wing to stall first, allowing the aileron to remain effective throughout the stall.

Slow-blow fuse. An electrical fuse that allows a large amount of current to flow for a short length of time but melts to open the circuit if more than its rated current flows for a longer period.

Smoke detector. A device that warns the flight crew of the presence of smoke in cargo and/or baggage compartments. Some smoke detectors are of the visual type, others are photoelectric or ionization devices.

Snubber. A device in a hydraulic or pneumatic component that absorbs shock and/or vibration. A snubber is installed in the line to a hydraulic pressure gauge to prevent the pointer fluctuating.

Softwood. Wood from a tree that bears cones and has needles rather than leaves.

Soldering. A method of thermally joining metal parts with a molten nonferrous alloy that melts at a temperature below 800 °F. The molten alloy is pulled up between close-fitting parts by capillary action. When the alloy cools and hardens, it forms a strong, leak-proof connection.

Solenoid. An electrical component using a small amount of current flowing through a coil to produce a magnetic force that pulls an iron core into the center of the coil. The core may be attached to a set of heavy-duty electrical contacts, or it may be used to move a valve or other mechanical device.

Solidity (helicopter rotor characteristic). The solidity of a helicopter rotor system is the ratio of the total blade area to the disc area.

Solution heat treatment. A type of heat treatment in which the metal is heated in a furnace until it has a uniform temperature throughout. It is then removed and quenched in cold water. When the metal is hot, the alloying elements enter into a solid solution with the base metal to become part of its basic structure. When the metal is quenched, these elements are locked into place.

Sonic venturi. A sonic venturi in a line between a turbine engine or turbocharger and a pressurization system. When the air flowing through the sonic venturi reaches the speed of sound, a shock wave forms across the throat of the sonic venturi and limits the flow. A sonic venturi is also called a flow limiter.

Specific heat. The number of BTUs of heat energy needed to change the temperature of one pound of a substance 1 °F.

Speed brakes. A secondary control of an airplane that produces drag without causing a change in the pitch attitude of the airplane. Speed brakes allow an airplane to make a steep descent without building up excessive forward airspeed.

Spike knot. A knot that runs through the depth of a beam perpendicular to the annual rings. Spike knots appear most frequently in quartersawed wood.

Spin. A flight maneuver in which an airplane descends in a corkscrew fashion. One wing is stalled and the other is producing lift.

Spirit level. A curved glass tube partially filled with a liquid, but with a bubble in it. When the device in which the tube is mounted is level, the bubble will be in the center of the tube.

Splayed patch (wood structure repair). A type of patch made in an aircraft plywood structure in which the edges of the patch are tapered for approximately five times the thickness of the plywood. A splayed patch is not recommended for use on plywood less than $\frac{1}{10}$ inch thick.

Split bus. A type of electrical bus that allows all of the voltage-sensitive avionic equipment to be isolated from the rest of the aircraft electrical system when the engine is being started or when the ground-power unit is connected.

Split-rocker switch. An electrical switch whose operating rocker is split so one half of the switch can be opened without affecting the other half. Split-rocker switches are used as aircraft master switches. The battery can be turned on without turning on the alternator, but the alternator cannot be turned on without also turning on the battery. The alternator can be turned off without turning off the battery, but the battery cannot be turned off without also turning off the alternator.

Split (wood defect). A longitudinal crack in a piece of wood caused by externally induced stress.

Spoilers. Flight controls that are raised up from the upper surface of a wing to destroy, or spoil, lift. Flight spoilers are used in conjunction with the ailerons to decrease lift and increase drag on the descending wing. Ground spoilers are used to produce a great amount of drag to slow the airplane on its landing roll.

Spongy brakes. Hydraulic brakes whose pedal has a spongy feel because of air trapped in the fluid.

Spontaneous combustion. Self-ignition of a material caused by heat produced in the material as it combines with oxygen from the air.

Springwood. The portion of an annual ring in a piece of wood formed principally during the first part of the growing season, the spring of the year. Springwood is softer, more porous, and lighter than the summerwood.

Square. A four-sided plane figure whose sides are all the same length, whose opposite sides are parallel, and whose angles are all right angles.

Squat switch. An electrical switch actuated by the landing gear scissors on the oleo strut. When no weight is on the landing gear, the oleo piston is extended and the switch is in one position, but when weight is on the gear, the oleo strut compresses and the switch changes its position. Squat switches are used in antiskid brake systems, landing gear safety circuits, and cabin pressurization systems.

Squib. An explosive device in the discharge valve of a high-rate-discharge container of fire-extinguishing agent. The squib drives a cutter into the seal in the container to discharge the agent.

SRM. Structural Repair Manual.

Stabilator. A flight control on the empennage of an airplane that acts as both a stabilizer and an elevator. The entire horizontal tail surface pivots and is moved as a unit.

Stability. The characteristic of an aircraft that causes it to return to its original flight condition after it has been disturbed.

Stabilons. Small wing-like horizontal surfaces mounted on the aft fuselage to improve longitudinal stability of airplanes that have an exceptionally wide center of gravity range.

Stagnation point. The point on the leading edge of a wing at which the airflow separates, with some flowing over the top of the wing and the rest below the wing.

Stall. A flight condition in which an angle of attack is reached at which the air ceases to flow smoothly over the upper surface of an airfoil. The air becomes turbulent and lift is lost.

Stall strip. A fixed device employed on the leading edge of fixed-wing aircraft to initiate flow separation at chosen locations on the wing during high-angle of attack flight, so as to improve the controllability of the aircraft when it enters stall.

Standpipe. A pipe sticking up in a tank or reservoir that allows part of the tank to be used as a reserve, or standby, source of fluid.

Starter-generator. A single-component starter and generator used on many of the smaller gas-turbine engines. It is used as a starter, and when the engine is running, its circuitry is shifted so that it acts as a generator.

Static. Still, not moving.

Static air pressure. Pressure of the ambient air surrounding the aircraft. Static pressure does not take into consideration any air movement.

Static dischargers. Devices connected to the trailing edges of control surfaces to discharge static electricity harmlessly into the air. They discharge the static charges before they can build up high enough to cause radio receiver interference.

Static stability. The characteristic of an aircraft that causes it to return to straight and level flight after it has been disturbed from that condition.

Stoddard solvent. A petroleum product, similar to naphtha, used as a solvent and a cleaning fluid.

STOL. Short takeoff and landing.

Stop drilling. A method of stopping the growth of a crack in a piece of metal or transparent plastic by drilling a small hole at the end of the crack. The stresses are spread out all around the circumference of the hole rather than concentrated at the end of the crack.

Straight polarity welding. DC-electric arc welding in which the electrode is negative with respect to the work.

Strain. A deformation or physical change in a material caused by a stress.

Stress. A force set up within an object that tries to prevent an outside force from changing its shape.

Stressed skin structure. A type of aircraft structure in which all or most of the stresses are carried in the outside skin. A stressed skin structure has a minimum of internal structure.

Stress riser. A location where the cross-sectional area of the part changes abruptly. Stresses concentrate at such a location and failure is likely. A scratch, gouge, or tool mark in the surface of a highly stressed part can change the area enough to concentrate the stresses and become a stress riser.

Stringer. A part of an aircraft structure used to give the fuselage its shape and, in some types of structure, to provide a small part of fuselage strength. Formers give the fuselage its cross-sectional shape and stringers fill in the shape between the formers.

Stroboscopic tachometer. A tachometer used to measure the speed of any rotating device without physical contact. A highly accurate variable-frequency oscillator triggers a high-intensity strobe light.

Sublimation. A process in which a solid material changes directly into a vapor without passing through the liquid stage.

Subsonic flight. Flight at an airspeed in which all air flowing over the aircraft is moving at a speed below the speed of sound.

Summerwood. The less porous, usually harder portion of an annual ring that forms in the latter part of the growing season, the summer of the year.

Sump. A low point in an aircraft fuel tank in which water and other contaminants can collect and be held until they can be drained out.

Supercooled water. Water in its liquid form at a temperature well below its natural freezing temperature. When supercooled water is disturbed, it immediately freezes.

Superheat. Heat energy that is added to a refrigerant after it changes from a liquid to a vapor.

Super heterodyne circuit. A sensitive radio receiver circuit in which a local oscillator produces a frequency that is a specific difference from the received signal frequency. The desired signal and the output from the oscillator are mixed, and they produce a single, constant intermediate frequency. This IF is amplified, demodulated, and detected to produce the audio frequency that is used to drive the speaker.

Supersonic flight. Flight at an airspeed in which all air flowing over the aircraft is moving at a speed greater than the speed of sound.

Supplemental Type Certificate (STC). An approval issued by the FAA for a modification to a type certificated airframe, engine, or component. More than one STC can be issued for the same basic alteration, but each holder must prove to the FAA that the alteration meets all the requirements of the original type certificate.

Surface tape. Strips of aircraft fabric that are doped over all seams and places where the fabric is stitched to the aircraft structure. Surface tape is also doped over the wing leading edges where abrasive wear occurs. The edges of surface tape are pink, or notched, to keep them from raveling before the dope is applied.

Surfactant. A surface active agent, or partially soluble contaminant, which is a by-product of fuel processing or of fuel additives. Surfactants adhere to other contaminants and cause them to drop out of the fuel and settle to the bottom of the fuel tank as sludge.

Surveyor's transit. An instrument consisting of a telescope mounted on a flat, graduated, circular plate on a tripod. The plate can be adjusted so it is level, and its graduations oriented to magnetic north. When an object is viewed through the telescope, its azimuth and elevation may be determined.

Swashplate. The component in a helicopter control system that consists basically of two bearing races with ball bearings between them. The lower, or nonrotating, race is tilted by the cyclic control, and the upper, or rotating, race has arms which connect to the control horns on the rotor blades. Movement of the cyclic pitch control is transmitted to the rotor blades through the swashplate. Movement of the collective pitch control raises or lowers the entire swashplate assembly to change the pitch of all the blades at the same time.

Synchro system. A remote instrument indicating system. A synchro transmitter is actuated by the device whose movement is to be measured, and it is connected electrically with wires to a synchro indicator whose pointer follows the movement of the shaft of the transmitter.

Symmetrical airfoil. An airfoil that has the same shape on both sides of its chord line, or center line.

Symmetry check. A check of an airframe to determine that the wings and tail are symmetrical about the longitudinal axis.

System-pressure regulator (hydraulic system component). A type of hydraulic system-pressure control valve. When the system pressure is low, as it is when some unit is actuated, the output of the constant-delivery pump is directed into the system. When the actuation is completed and the pressure builds up to a specified kick-out pressure, the pressure regulator shifts. A check valve seals the system off and the pressure is maintained by the accumulator. The pump is unloaded and its output is directed back into the reservoir with very little opposition. The pump output pressure drops, but the volume of flow remains the same. When the system pressure drops to the specified kick-in pressure, the regulator again shifts and directs fluid into the system. Spool-type and balanced-pressure-type system pressure regulators are completely automatic in their operation and require no attention on the part of the flight crew.

T

TACAN (Tactical Air Navigation). A radio navigation facility used by military aircraft for both direction and distance information. Civilian aircraft receive distance information from a TACAN on their DME.

Tack coat. A coat of finishing material sprayed on the surface and allowed to dry until the solvents evaporate. As soon as the solvents evaporate, a wet full-bodied coat of material is sprayed over it.

Tack rag. A clean, lintless rag, slightly damp with thinner. A tack rag is used to wipe a surface to prepare it to receive a coat of finishing material.

Tack weld. A method of holding parts together before they are permanently welded. The parts are assembled, and small spots of weld are placed at strategic locations to hold them in position.

Tacky. Slightly sticky to the touch.

Tailets. Small vertical surfaces mounted underside of the horizontal stabilizer of some airplanes to increase the directional stability.

Takeoff warning system. An aural warning system that provides audio warning signals when the thrust levers are advanced for takeoff if the stabilizer, flaps, or speed brakes are in an unsafe condition for takeoff.

Tang. A tapered shank sticking out from the blade of a knife or a file. The handle of a knife or file is mounted on the tang.

TCAS. Traffic Alert Collision Avoidance System.

Teflon. The registered trade name for a fluorocarbon resin used to make hydraulic and pneumatic seals, hoses, and backup rings.

Tempered glass. Glass that has been heat-treated to increase its strength. Tempered glass is used in bird-proof, heated windshields for high-speed aircraft.

Terminal strips. A group of threaded studs mounted in a strip of insulating plastic. Electrical wires with crimped-on terminals are placed over the studs and secured with nuts.

Terminal VOR. A low-powered VOR that is normally located on an airport.

Tetraethyl lead (TEL). A heavy, oily, poisonous liquid, $Pb(C_2H_5)^4$, that is mixed into aviation gasoline to increase its critical pressure and temperature.

Therapeutic mask adapter. A calibrated orifice in the mask adapter for a continuous-flow oxygen system that increases the flow of oxygen to a mask being used by a passenger who is known to have a heart or respiratory problem.

Thermal dimpling. See hot dimpling.

Thermal relief valve. A relief valve in a hydraulic system that relieves pressure that builds up in an isolated part of the system because of heat. Thermal relief valves are set at a higher pressure than the system pressure relief valve.

Thermistor. A special form of electrical resistor whose resistance varies with its temperature.

Thermistor material. A material with a negative temperature coefficient that causes its resistance to decrease as its temperature increases.

Thermocouple. A loop consisting of two kinds of wire, joined at the hot, or measuring, junction and at the cold junction in the instrument. The voltage difference between the two junctions is proportional to the temperature difference between the junctions. In order for the current to be meaningful, the resistance of the thermocouple is critical, and the leads are designed for a specific installation. Their length should not be altered. Thermocouples used to measure cylinder head temperature are usually made of iron and constantan, and thermocouples that measure exhaust gas temperature for turbine engines are made of chromel and alumel.

Thermocouple fire-detection system. A fire-detection system that works on the principle of the rate-of-temperature rise. Thermocouples are installed around the area to be protected, and one thermocouple is surrounded by thermal insulation that prevents its temperature changing rapidly. In the event of a fire, the temperature of all the thermocouples except the protected one will rise immediately and a fire warning will be initiated. In the case of a general overheat condition, the temperature of all the thermocouples will rise uniformly and there will be no fire warning.

Thermoplastic resin. A type of plastic material that becomes soft when heated and hardens when cooled.

Thermosetting resin. A type of plastic material that, when once hardened by heat, cannot be softened by being heated again.

Thermostatic expansion valve (TEV). The component in a vapor-cycle cooling system that meters the refrigerant into the evaporator. The amount of refrigerant metered by the TEV is determined by the temperature and pressure of the refrigerant as it leaves the evaporator coils. The TEV changes the refrigerant from a high-pressure liquid into a low-pressure liquid.

Thixotropic agents. Materials, such as microballoons, added to a resin to give it body and increase its workability.

TIG welding. Tungsten inert welding is a form of electric arc welding in which the electrode is a nonconsumable tungsten wire. TIG welding is now called GTA (gas tungsten arc) welding.

Toe-in. A condition of landing gear alignment in which the front of the tires are closer together than the rear. When the aircraft rolls forward, the wheels try to move closer together.

Toe-out. A condition of landing gear alignment in which the front of the tires are further apart than the rear. When the aircraft rolls forward, the wheels try to move farther apart.

Torque. A force that produces or tries to produce rotation.

Torque links. The hinged link between the piston and cylinder of an oleo-type landing gear shock absorber. The torque links allow the piston to move freely in and out of the landing gear cylinder, but prevent it rotating. The torque links can be adjusted to achieve and maintain the correct wheel alignment. Torque links are also called scissors and nutcrackers.

Torque tube. A tube in an aircraft control system that transmits a torsional force from the operating control to the control surface.

Torsion rod. A device in a spring tab to which the control horn is attached. For normal operation, the torsion rod acts as a fixed attachment point, but when the control surface loads are high, the torsion rod twists and allows the control horn to deflect the spring tab.

Total air pressure. The pressure a column of moving air will have if it is stopped.

TMC. Thrust management computer.

Toroidal coil. An electrical coil wound around a ring-shaped core of highly permeable material.

Total air temperature. The temperature a column of moving air will have if it is stopped.

TR unit. A transformer-rectifier unit. A TR unit reduces the voltage of AC and changes it into DC.

Tractor powerplant. An airplane powerplant in which the propeller is mounted in the front, and its thrust pulls the airplane rather than pushes it.

Trammel (verb). To square up the Pratt truss used in an airplane wing. Trammel points are set on the trammel bar so they measure the distance between the center of the front spar, at the inboard compression strut, and at the center of the rear spar at the next compression strut outboard. The drag and antidrug wires are adjusted until the distance between the center of the rear spar at the inboard compression strut and the center of the front spar at the next outboard compression strut is exactly the same as that between the first points measured.

Trammel bar. A wood or metal bar on which trammel points are mounted to compare distances.

Trammel points. A set of sharp-pointed pins that protrude from the sides of a trammel bar.

Transducer. A device that changes energy from one form to another. Commonly used transducers change mechanical movement or pressures into electrical signals.

Transformer rectifier. A component in a large aircraft electrical system used to reduce the AC voltage and change it into DC for charging the battery and for operating DC equipment in the aircraft.

Translational lift. The additional lift produced by a helicopter rotor as the helicopter changes from hovering to forward flight.

Transonic flight. Flight at an airspeed in which some air flowing over the aircraft is moving at a speed below the speed of sound, and other air is moving at a speed greater than the speed of sound.

Transverse pitch. See gauge.

Triangle. A three-sided, closed plane figure. The sum of the three angles in a triangle is always equal to 180°.

Tricresyl phosphate (TCP). A chemical compound, $(CH_3C_6H_4O)^3PO$, used in aviation gasoline to assist in scavenging the lead deposits left from the tetraethyl lead.

Trim tab. A small control tab mounted on the trailing edge of a movable control surface. The tab may be adjusted to provide an aerodynamic force to hold the surface on which it is mounted deflected in order to trim the airplane for hands-off flight at a specified airspeed.

Trimmed flight. A flight condition in which the aerodynamic forces acting on the control surfaces are balanced and the aircraft is able to fly straight and level with no control input.

Trip-free circuit breaker. A circuit breaker that opens a circuit any time an excessive amount of current flows, regardless of the position of the circuit breaker's operating handle.

Troubleshooting. A procedure used in aircraft maintenance in which the operation of a malfunctioning system is analyzed to find the reason for the malfunction and to find a method for returning the system to its condition of normal operation.

True airspeed (TAS). Airspeed shown on the airspeed indicator (indicated airspeed) corrected for position error and nonstandard air temperature and pressure.

Trunnion. Projections from the cylinder of a retractable landing gear strut about which the strut pivots retract.

Truss-type structure. A type of structure made up of longitudinal beams and cross braces. Compression loads between the main beams are carried by rigid cross braces. Tension loads are carried by stays, or wires, that go from one main beam to the other and cross between the cross braces.

Turbine. A rotary device actuated by impulse or reaction of a fluid flowing through vanes or blades that are arranges around a central shaft.

Turn and slip indicator. A rate gyroscopic flight instrument that gives the pilot an indication of the rate of rotation of the aircraft about its vertical axis. A ball in a curved glass tube shows the pilot the relationship between the centrifugal force and the force of gravity. This indicates whether or not the angle of bank is proper for the rate of turn. The turn and slip indicator shows the trim condition of the aircraft and serves as an emergency source of bank information in case the attitude gyro fails. Turn and slip indicators were formerly called needle and ball and turn and bank indicators.

Turnbuckle. A component in an aircraft control system used to adjust cable tension. A turnbuckle consists of a brass tubular barrel with right-hand threads in one end and left-hand in the other end. Control cable terminals screw into the two ends of the barrel, and turning the barrel pulls the terminals together, shortening the cable.

Twist drill. A metal cutting tool turned in a drill press or handheld drill motor. A twist drill has a straight shank and spiraled flutes. The cutting edge is ground on the end of the spiraled flutes.

Twist rope. A stripe of paint on flexible hose that runs the length of the hose. If this stripe spirals around the hose after it is installed, it indicates the hose was twisted when it was installed. Twist stripes are also called lay lines.

Two-terminal spot-type fire detection system. A fire detection system that uses individual thermoswitches installed around the inside of the area to be protected. These thermoswitches are wired in parallel between two separate circuits. A short or an open circuit can exist in either circuit without causing a fire warning.

Type Certificate Data Sheets (TCDS). The official specifications of an aircraft, engine, or propeller issued by the Federal Aviation Administration. The TCDS lists pertinent specifications for the device, and it is the responsibility of the mechanic and/or inspector to ensure, on each inspection, that the device meets these specifications.

U

UHF. Ultrahigh frequency.

Ultimate tensile strength. The tensile strength required to cause a material to break or to continue to deform under a decreasing load.

Ultraviolet-blocking dope. Dope that contains aluminum powder or some other pigment that blocks the passage of ultraviolet rays of the sun. The coat of dope protects the organic fabrics and clear dope from deterioration by these rays.

Undamped oscillation. Oscillation that continues with an unchanging amplitude once it has started.

Underslung rotor. A helicopter rotor whose center of gravity is below the point at which it is attached to the mast.

Unidirectional fabric. Fabric in which all the threads run in the same direction. These threads are often bound with a few fibers run at right angles, just enough to hold the yarns together and prevent their bunching.

Unloading valve. This is another name for system pressure regulator. See system pressure regulator.

Utility finish. The finish of an aircraft that gives the necessary tautness and fill to the fabric and the necessary protection to the metal, but does not have the glossy appearance of a show-type finish.

V

Vapor lock. A condition in which vapors form in the fuel lines and block the flow of fuel to the carburetor.

Vapor pressure. The pressure of the vapor above a liquid needed to prevent the liquid evaporating. Vapor pressure is always specified at a specific temperature.

Variable displacement pump. A fluid pump whose output is controlled by the demands of the system. These pumps normally have a built-in system pressure regulator. When the demands of the system are low, the pump moves very little fluid, but when the demands are high, the pump moves a lot of fluid. Most variable displacement pumps used in aircraft hydraulic systems are piston-type pumps.

Varnish (aircraft finishing material). A material used to produce an attractive and protective coating on wood or metal. Varnish is made of a resin dissolved in a solvent and thinned until it has the proper viscosity to spray or brush. The varnish is spread evenly over the surface to be coated, and when the solvents evaporate, a tough film is left.

Varsol. A petroleum product similar to naphtha used as a solvent and cleaning fluid.

Veneer. Thin sheets of wood "peeled" from a log. A wide-blade knife held against the surface of the log peels away the veneer as the log is rotated in the cutter. Veneer is used for making plywood. Several sheets of veneer are glued together, with the grain of each sheet placed at 45° or 90° to the grain of the sheets next to it.

Vertical axis. An imaginary line, passing vertically through the center of gravity of an airplane.

Vertical fin. The fixed vertical surface in the empennage of an airplane. The vertical fin acts as a weathervane to give the airplane directional stability.

VFR. Visual flight rules.

VHF. Very high frequency.

Vibrator-type voltage regulator. A type of voltage regulator used with a generator or alternator that intermittently places a resistance in the field circuit to control the voltage. A set of vibrating contacts puts the resistor in the circuit and takes it out several times a second.

Viscosity. The resistance of a fluid to flow. Viscosity refers to the "stiffness" of the fluid, or its internal friction.

Viscosity cup. A specially shaped cup with an accurately sized hole in its bottom. The cup is submerged in the liquid to completely fill it. It is then lifted from the liquid and the time in seconds is measured from the beginning of the flow through the hole until the first break in this flow. The viscosity of the liquid relates to this time.

Vixen file. A metal-cutting hand file that has curved teeth across its faces. Vixen files are used to remove large amounts of soft metal.

V_{NE}. Never-exceed speed. The maximum speed the aircraft is allowed to attain in any conditions of flight.

Volatile liquid. A liquid that easily changes into a vapor.

Voltmeter multiplier. A precision resistor in series with a voltmeter mechanism used to extend the range of the basic meter or to allow a single meter to measure several ranges of voltage.

VOR. Very high frequency Omni Range navigation.

VORTAC. An electronic navigation system that contains both a VOR and a TACAN facility.

Vortex (plural vortices). A whirling motion in a fluid.

Vortex generator. Small, low-aspect-ratio airfoils installed in pairs on the upper surface of a wing, on both sides of the vertical fin just ahead of the rudder, and on the underside of the vertical stabilizers of some airplanes. Their function is to pull high-energy air down to the surface to energize the boundary layer and prevent airflow separation until the surface reaches a higher angle of attack.

W

Warp clock. An alignment indicator included in a structural repair manual to show the orientation of the piles of a composite material. The ply direction is shown in relation to a reference direction.

Warp threads. Threads that run the length of the roll of fabric, parallel to the selvage edge. Warp threads are often stronger than fill threads.

Warp tracers. Threads of a different color from the warp threads that are woven into a material to identify the direction of the warp threads.

Wash in. A twist in an airplane wing that increases its angle of incidence near the tip.

Wash out. A twist in an airplane wing that decreases its angle of incidence near the tip.

Watt. The basic unit of electrical power. One watt is equal to $\frac{1}{746}$ horsepower.

Way point. A phantom location created in certain electronic navigation systems by measuring direction and distance from a VORTAC station or by latitude and longitude coordinates from Loran or GPS.

Web of a spar. The part of a spar between the caps.

Weft threads. See fill threads.

Wet-type vacuum pump. An engine-driven air pump that uses steel vanes. These pumps are lubricated by engine oil drawn in through holes in the pump base. The oil passes through the pump and is exhausted with the air. Wet-type pumps must have oil separators in their discharge line to trap the oil and return it to the engine crankcase.

Wing fences. Vertical vanes that extend chordwise across the upper surface of an airplane wing to prevent spanwise airflow.

Wing heavy. An out-of-trim flight condition in which an airplane flies hands off, with one wing low.

Wire bundle. A compact group of electrical wires held together with special wrapping devices or with waxed string. These bundles are secured to the aircraft structure with special clamps.

Woof threads. See fill threads.

Work. The product of force times distance.

Y

Yaw. Rotation of an aircraft about its vertical axis.

Yaw damper. An automatic flight control system that counteracts the rolling and yawing produced by Dutch roll. See Dutch roll. A yaw damper senses yaw with a rate gyro and moves the rudder an amount proportional to the rate of yaw, but in the opposite direction.

Yield strength. The amount of stress needed to permanently deform a material.

Z

Zener diode. A special type of solid-state diode designed to have a specific breakdown voltage and to operate with current flowing through it in its reverse direction.

Zeppelin. The name of large, rigid, lighter-than-air ships built by the Zeppelin Company in Germany prior to and during World War I.

Zero-center ammeter. An ammeter in a light aircraft

electrical system located between the battery and the main bus. This ammeter shows the current flowing into or out of the battery.

Index

S

Z